INTEGRALS AND SERIES

интегралы и ряды. дополнительные главы

INTEGRALS AND SERIES

Volume 3

More Special Functions

A.P. Prudnikov
Yu. A. Brychkov
Computing Center of the USSR Academy of Sciences, Moscow

O.I. Marichev
Byelorussian State University, Minsk

Translated from the Russian by
G.G. Gould

GORDON AND BREACH SCIENCE PUBLISHERS
New York · Philadelphia · London · Paris · Montreux · Tokyo · Melbourne

Gordon and Breach Science Publishers

Post Office Box 786
Cooper Station
New York, New York 10276
United States of America

5301 Tacony Street
Box 330
Philadelphia, Pennsylvania 19137
United States of America

Post Office Box 197
London WC2E 9PX
United Kingdom

58, rue Lhomond
75005 Paris
France

Post Office Box 161
1820 Montreux 2
Switzerland

3-14-9, Okubo
Shinjuku-ku, Tokyo 169
Japan

Private Bag 8
Camberwell, Victoria 3124
Australia

Originally published in Russian in 1986 as интегралы и ряды. дополнительные главы
© 1986 Nauka, Moscow.

Library of Congress Cataloging-in-Publication Data
(Revised for vol. 3)

Prudnikov, A. P. (Anatoliĭ Platonovich)
Integrals and series.

Translation of: Integraly i rîady.
Includes bibliographies and indexes.
Contents: v. 1. Elementary functions—v. 2. Special functions—v. 3. More special functions.
1. Integrals. 2. Series. I. Brychkov, ÎÙ. A. (ÎÙriĭ Aleksandrovich) II. Marichev, O. I. (Oleg Igorevich) III. Title.
QA308.P7813 1986 515.4′3 85-27065
ISBN 2-88124-089-5 (v. 1).
ISBN 2-88124-089-9 (v. 2).
ISBN 2-88124-097-6 (2v. set)
ISBN 2-88124-682-6 (v. 3)
ISBN 2-88124-736-9 (3v. set)

CONTENTS

PREFACE

The solution of many problems relating to various areas of science and technology reduces to the calculation of integrals and the summation of series containing elementary and special functions. As is well known, this task is considerably simplified by means of appropriate handbook literature from which we should single out the "Bateman Manuscript Project" Series "Higher transcendental functions" and "Tables of integral transforms" of world fame by A. Erdélyi *et al.* and also the handbook "Tables of integrals, sums, series and products" by I. S. Gradshtein and I. M. Ryzhik.

These handbooks have over several decades been reference manuals for theoretical and experimental physicists, research engineers, and specialists in the area of applied mathematics and cybernetics. However, they contained only formulae up to the end of the '40s: and this has led to the need for creating a more complete reference manual, in which new results are reflected. In this connection, the Nauka publications "Integrals and series. Elementary functions" and "Integrals and series. Special functions" appeared in 1981–1983. These contain results in this area of mathematical analysis that have been published in recent years.

We now call the reader's attention to a third book which contains tables of indefinite and definite integrals, finite sums and series and contain the functions of Struve, Weber, Anger, Lommel, Kelvin, Airy, Legendre, Whittaker, the hypergeometric and elliptic function, the Mathieu functions, the MacRobert function, the Meijer function, the Fox function and several others. The book also contains tables of representations of generalized hypergeometric functions, and tables of Mellin transforms of a wide class of elementary and special functions, combined with tables of special cases of the Meijer G-function. Sections are included devoted to the properties of the hypergeometric functions, the Meijer G-function and the Fox H-function. The appendix contains supplementary material which can be used in the calculation of integrals and the summation of series.

The main text is preceded by a fairly detailed list of contents, from which the required formulae can be found. The notation used is, by and large, the generally accepted notation of the mathematical literature and is listed in the indexes at the end of the book. References to formulae written in the form 2.17.9.1 denote Formula 1 of Subsection 2.17.9. Throughout, $k, l, m, n = 0, 1, 2, 3, \ldots$, if no other restrictions are indicated.

For the sake of compactness of exposition, abbreviated notation is used.

For example, the formula

$$\int_0^a \frac{x^{\alpha-1}}{(a^2-x^2)^{\mu/2}} \begin{Bmatrix} \sin bx \\ \cos bx \end{Bmatrix} P_\nu^\mu\left(\frac{x}{a}\right) dx =$$

$$= \sqrt{\pi}\left(\frac{a}{2}\right)^{\alpha-\mu+\nu} b^\delta \Gamma\left[\begin{matrix} \alpha+\delta \\ (1+\alpha+\delta-\mu-\nu)/2, \quad (2+\alpha+\delta-\mu+\nu)/2 \end{matrix}\right] \times$$

$$\times {}_2F_3\left(\frac{\alpha+1}{2}, \frac{\alpha}{2}+\delta; \delta+\frac{1}{2}, \frac{1+\alpha+\delta-\mu-\nu}{2}, \frac{2+\alpha+\delta-\mu+\nu}{2}; -\frac{a^2b^2}{4}\right)$$

$$\left[a>0;\ \operatorname{Re}\alpha>-\delta;\ \operatorname{Re}\mu<1;\ \delta=\begin{Bmatrix}1\\0\end{Bmatrix}\right]$$

is abbreviated notation for the two formulae:

$$\int_0^a \frac{x^{\alpha-1}}{(a^2-x^2)^{\mu/2}} \sin bx\, P_\nu^\mu\left(\frac{x}{a}\right) dx =$$

$$= \sqrt{\pi}\left(\frac{a}{2}\right)^{\alpha-\mu+\nu} b\Gamma\left[\begin{matrix} \alpha+1 \\ (2+\alpha-\mu-\nu)/2, \quad (3+\alpha-\mu+\nu)/2 \end{matrix}\right] \times$$

$$\times {}_2F_3\left(\frac{\alpha+1}{2}, \frac{\alpha}{2}+1; \frac{3}{2}, \frac{2+\alpha-\mu-\nu}{2}, \frac{3+\alpha-\mu+\nu}{2}; -\frac{a^2b^2}{4}\right)$$

$$\left[a>0;\ \operatorname{Re}\alpha>-1;\ \operatorname{Re}\mu<1\right]$$

(where only the upper symbols and upper expressions in the curly brackets are taken) and

$$\int_0^a \frac{x^{\alpha-1}}{(a^2-x^2)^{\mu/2}} \cos bx\, P_\nu^\mu\left(\frac{x}{a}\right) dx =$$

$$= \sqrt{\pi}\left(\frac{a}{2}\right)^{\alpha-\mu+\nu} \Gamma\left[\begin{matrix} \alpha \\ (1+\alpha-\mu-\nu)/2, \quad (2+\alpha-\mu+\nu)/2 \end{matrix}\right] \times$$

$$\times {}_2F_3\left(\frac{\alpha+1}{2}, \frac{\alpha}{2}; \frac{1}{2}, \frac{1+\alpha-\mu-\nu}{2}, \frac{2+\alpha-\mu+\nu}{2}; -\frac{a^2b^2}{4}\right)$$

$$\left[a>0;\ \operatorname{Re}\alpha>0;\ \operatorname{Re}\mu<1\right]$$

(where only the lower symbols and lower expressions in the curly brackets are taken); by definition

$$\Gamma\left[\begin{matrix} a_1, \ldots, a_m \\ b_1, \ldots, b_n \end{matrix}\right] = \prod_{k=1}^{m} \Gamma(a_k) \Big/ \prod_{l=1}^{n} \Gamma(b_l).$$

References to the main bibliographical sources are given at the end of the book. A considerable number of the results were obtained by the authors and are published for the first time.

It is our hope that the three books of the reference manual "Integrals and series" will be of help to research scientists, engineers and other specialists employing mathematical methods in their work.

The Authors

Chapter 1. **INDEFINITE INTEGRALS**

1.1. INTRODUCTION

This chapter contains indefinite integrals of special functions, also definite integrals for which one of the limits of integration is variable in those cases when the integrand is independent of it; for brevity the arbitrary constant is omitted.

Some formulae become meaningless for certain values of the parameters. If it is clear from the structure of the formula what these parameters are, no further explanation is given. Expression for the integral with these values of parameters values are given, as a rule, in the subsequent formulae.

1.2. THE GENERALIZED ZETA FUNCTION $\zeta(s, x)$, BERNOULLI POLYNOMIALS $B_n(x)$, EULER POLYNOMIALS $E_n(x)$ AND POLYLOGARITHMS $\mathrm{Li}_\nu(x)$

1.2.1. Integrals containing $\zeta(s, x)$, $B_n(x)$ and $E_n(x)$.

1. $\int \zeta(s, x)\,dx = \frac{1}{1-s}\zeta(s-1, x)$ [Re $s > 2$].

2. $\int B_n(x)\,dx = \frac{1}{n+1} B_{n+1}(x)$.

3. $\int E_n(x)\,dx = \frac{1}{n+1} E_n(x)$.

1.2.2. Integrals of the form $\int x^\alpha \mathrm{Li}_n(ax)\,dx$.

1. $$\int_0^x x^{\alpha-1} \mathrm{Li}_n(-ax)\,dx = -\frac{ax^{\alpha+1}}{\alpha+1}\,{}_{n+2}F_{n+1}\left(\begin{matrix}1, 1, \ldots, 1, \alpha+1; & -ax\\ 2, 2, \ldots, 2, \alpha+2\end{matrix}\right)$$ [Re $\alpha > -1$; $|\arg a| < \pi$].

2. $\int_0^x \frac{1}{x} \mathrm{Li}_n(-ax)\,dx = \mathrm{Li}_{n+1}(-ax)$ [$|\arg a| < \pi$].

1.2.3. Integrals of the form $\int f(x)\,\mathrm{Li}_2(x)\,dx$.

1. $$\int \frac{1}{1-x}\mathrm{Li}_2(x)\,dx = -\ln(1-x)\,\mathrm{Li}_2(x) - 2\ln(1-x)\,\mathrm{Li}_2(1-x) + 2\,\mathrm{Li}_3(1-x) - \ln x \ln^2(1-x).$$

2. $\int \frac{1}{(1-x)^2}\mathrm{Li}_2(x)\,dx = \frac{x}{1-x}\mathrm{Li}_2(x) - \frac{1}{2}\ln^2(1-x)$.

3. $\int \frac{\ln x}{x} \mathrm{Li}_2(x)\,dx = \ln x\, \mathrm{Li}_3(x) - \mathrm{Li}_4(x).$

4. $\int \frac{\ln(1-x)}{1-x} \mathrm{Li}_2(x)\,dx =$

$$= -\frac{1}{2}\ln^2(1-x)\,\mathrm{Li}_2(x) - \frac{3}{2}\ln^2(1-x)\,\mathrm{Li}_2(1-x) +$$

$$+ 3\ln(1-x)\,\mathrm{Li}_3(1-x) - 3\,\mathrm{Li}_4(1-x) - \frac{1}{2}\ln x \ln^3(1-x).$$

1.3. THE GENERALIZED FRESNEL INTEGRALS $S(x,\nu)$ AND $C(x,\nu)$

1.3.1. Integrals of the form $\int f(x) \begin{Bmatrix} S(ax,\nu) \\ C(ax,\nu) \end{Bmatrix} dx$.

1. $\int x^{\alpha-1} \begin{Bmatrix} S(ax,\nu) \\ C(ax,\nu) \end{Bmatrix} dx = \frac{x^\alpha}{\alpha} \begin{Bmatrix} S(ax,\nu) \\ C(ax,\nu) \end{Bmatrix} - \frac{a^{-\alpha}}{\alpha} \begin{Bmatrix} S(ax,\alpha+\nu) \\ C(ax,\alpha+\nu) \end{Bmatrix}.$

2. $\int x^{-\nu} \begin{Bmatrix} S(ax,\nu) \\ C(ax,\nu) \end{Bmatrix} dx = \frac{x^{1-\nu}}{1-\nu} \begin{Bmatrix} S(ax,\nu) \\ C(ax,\nu) \end{Bmatrix} \mp \frac{a^{\nu-1}}{1-\nu} \begin{Bmatrix} \cos ax \\ \sin ax \end{Bmatrix}.$

3. $\int e^{-bx} \begin{Bmatrix} S(ax,\nu) \\ C(ax,\nu) \end{Bmatrix} dx =$

$$= -\frac{e^{-bx}}{b} \begin{Bmatrix} S(ax,\nu) \\ C(ax,\nu) \end{Bmatrix} - \frac{a^\nu}{2b} \begin{Bmatrix} i \\ 1 \end{Bmatrix} [(b+ia)^{-\nu}\gamma(\nu, bx+iax) \mp (b-ia)^{-\nu}\gamma(\nu, bx-iax)].$$

4. $\int \sin bx \begin{Bmatrix} S(ax,\nu) \\ C(ax,\nu) \end{Bmatrix} dx = -\frac{1}{b}\cos bx \begin{Bmatrix} S(ax,\nu) \\ C(ax,\nu) \end{Bmatrix} +$

$$+ \frac{1}{2b}\left[\left(\frac{a}{a-b}\right)^\nu \begin{Bmatrix} S((a-b)x,\nu) \\ C((a-b)x,\nu) \end{Bmatrix} + \left(\frac{a}{a+b}\right)^\nu \begin{Bmatrix} S((a+b)x,\nu) \\ C((a+b)x,\nu) \end{Bmatrix} \right]$$

5. $\int \cos bx \begin{Bmatrix} S(ax,\nu) \\ C(ax,\nu) \end{Bmatrix} dx = \frac{1}{b}\sin bx \begin{Bmatrix} S(ax,\nu) \\ C(ax,\nu) \end{Bmatrix} \mp$

$$\mp \frac{1}{2b}\left[\left(\frac{a}{a-b}\right)^\nu \begin{Bmatrix} C((a-b)x,\nu) \\ S((a-b)x,\nu) \end{Bmatrix} - \left(\frac{a}{a+b}\right)^\nu \begin{Bmatrix} C((a+b)x,\nu) \\ S((a+b)x,\nu) \end{Bmatrix} \right]$$

1.4. THE STRUVE FUNCTIONS $\mathbf{H}_\nu(x)$ AND $\mathbf{L}_\nu(x)$

1.4.1. Integrals of the form $\int x^\lambda \mathbf{H}_\nu(ax)\,dx$.

1. $\int_{x_1}^{x_2} x^\lambda \mathbf{H}_\nu(ax)\,dx =$

$$= \pm \frac{a^{\nu+1}x^{\lambda+\nu+2}}{2^\nu \sqrt{\pi}\,(\lambda+\nu+2)\Gamma(\nu+3/2)}\, {}_2F_3\left(1, \frac{\lambda+\nu+2}{2}; \frac{3}{2}, \nu+\frac{3}{2}, \frac{\lambda+\nu+4}{2}; -\frac{a^2x^2}{4}\right) +$$

$$+ \frac{2^\lambda \pi \sec[(\lambda+\nu+1)\pi/2]}{a^{\lambda+1}\Gamma((1-\lambda+\nu)/2)\,\Gamma((1-\lambda-\nu)/2)} \begin{Bmatrix} 0 \\ 1 \end{Bmatrix}$$

$$\left[\begin{Bmatrix} x_1=0, & x_2=x; & \operatorname{Re}(\lambda+\nu) > -2 \\ x_1=x, & x_2=\infty, & a>0;\ \operatorname{Re}\lambda < 1/2;\ \operatorname{Re}(\lambda+\nu) < 0 \end{Bmatrix}\right].$$

2. $\int_{x_1}^{x_2} x^{\nu+1} \mathbf{H}_\nu(ax)\,dx = \pm \frac{x^{\nu+1}}{a} \mathbf{H}_{\nu+1}(ax)$

$$\left[\begin{Bmatrix} x_1=0, & x_2=x; & \operatorname{Re}\nu > -3/2 \\ x_1=x, & x_2=\infty; & a>0;\ \operatorname{Re}\nu < -1/2 \end{Bmatrix}\right].$$

3. $\int x^{\nu+3}\mathbf{H}_\nu(ax)\,dx = \frac{x^{\nu+3}}{a}\mathbf{H}_{\nu+1}(ax) - \frac{2x^{\nu+2}}{a^2}\mathbf{H}_{\nu+2}(ax).$

4. $\int_0^x x^{1-\nu}\mathbf{H}_\nu(ax)\,dx = \dfrac{a^{\nu-1}x}{2^{\nu-1}\sqrt{\pi}\,\Gamma(\nu+1/2)} - \dfrac{x^{1-\nu}}{a}\mathbf{H}_{\nu-1}(ax).$

5. $\int_x^\infty \dfrac{1}{x^2}\mathbf{H}_0(ax)\,dx = \dfrac{2a}{\pi}\left[\dfrac{a^2x^2}{18}\,{}_2F_3\left(\begin{matrix}1,\ 1;\ -a^2x^2/4\\ 2,\ 5/2,\ 5/2\end{matrix}\right) - \ln(2ax) - \mathbf{C} + 2\right].$

1.4.2. Integrals of the form $\int x^\lambda e^{ix}\mathbf{H}_\nu(x)\,dx$.

1. $\int_0^x x^{\pm\nu}e^{ix}\mathbf{H}_\nu(x)\,dx = \pm\dfrac{x^{1\pm\nu}e^{ix}}{2\nu\pm1}\left[\mathbf{H}_\nu(x) \mp i\mathbf{H}_{\nu\pm1}(x)\right] \mp$
$\mp \dfrac{2^{-\nu+(1\mp1)/2}}{\sqrt{\pi}\,(2\nu\pm1)\,\Gamma(\nu+1\pm1/2)}\left\{\begin{matrix}e^{i\nu\pi/2}\,\gamma(2\nu+2,\ -ix)\\ e^{ix}-1\end{matrix}\right\} \quad \left[\left\{\begin{matrix}\nu\neq-1/2;\ \operatorname{Re}\nu>-1\\ \nu\neq-1/2\end{matrix}\right\}\right].$

2. $\int_0^x x^{-1/2}e^{ix}\mathbf{H}_{-1/2}(x)\,dx = \dfrac{1}{\sqrt{2\pi}}\operatorname{Si}(2x) + \dfrac{i}{\sqrt{2\pi}}\left[\mathbf{C} + \ln 2x - \operatorname{ci}(2x)\right].$

3. $\int_0^x x^{-1/2}e^{ix}\mathbf{H}_{1/2}(x)\,dx = \dfrac{1}{\sqrt{2\pi}}\left[2\operatorname{ci}(x) - \operatorname{ci}(2x) - \mathbf{C} - \ln\dfrac{x}{2} + 2i\operatorname{Si}(x) - i\operatorname{Si}(2x)\right].$

1.4.3. Integrals of the form $\int x^\lambda \mathbf{H}_\mu(ax)\,\mathbf{H}_\nu(bx)\,dx$.

1. $\int x^\lambda \mathbf{H}_\mu(x)\,\mathbf{H}_\nu(x)\,dx = \dfrac{1}{\lambda-\mu-\nu+1}\Big\{-(\lambda+\mu+\nu-1)\int x^\lambda \mathbf{H}_{\mu-1}(x)\,\mathbf{H}_{\nu-1}(x)\,dx +$
$+ x^{\lambda+1}\left[\mathbf{H}_{\mu-1}(x)\,\mathbf{H}_{\nu-1}(x) + \mathbf{H}_\mu(x)\,\mathbf{H}_\nu(x)\right] - \dfrac{2^{-\mu+1}}{\sqrt{\pi}\,\Gamma(\mu+1/2)}\int x^{\lambda+\mu}\mathbf{H}_{\nu-1}(x)\,dx -$
$- \dfrac{2^{-\nu+1}}{\sqrt{\pi}\,\Gamma(\nu+1/2)}\int x^{\lambda+\nu}\mathbf{H}_{\mu-1}(x)\,dx\Big\}$ [see 1.4.1].

2. $\int x^{\mu+\nu+1}\mathbf{H}_\mu(x)\,\mathbf{H}_\nu(x)\,dx = \dfrac{1}{2(\mu+\nu+1)}\Big\{x^{\mu+\nu+2}\left[\mathbf{H}_\mu(x)\mathbf{H}_\nu(x) + \mathbf{H}_{\mu+1}(x)\mathbf{H}_{\nu+1}(x)\right] -$
$- \dfrac{2^{-\mu}}{\sqrt{\pi}\,\Gamma(\mu+3/2)}\int x^{2\mu+\nu+2}\mathbf{H}_\nu(x)\,dx - \dfrac{2^{-\nu}}{\sqrt{\pi}\,\Gamma(\nu+3/2)}\int x^{\mu+2\nu+2}\mathbf{H}_\mu(x)\,dx\Big\}$
[see 1.4.1].

3. $\int x^{-\mu-\nu+1}\mathbf{H}_\mu(x)\,\mathbf{H}_\nu(x)\,dx = \dfrac{1}{2(\mu+\nu-1)}\Big\{\dfrac{2^{1-\mu}}{\sqrt{\pi}\,\Gamma(\mu+1/2)}\int x^{1-\nu}\mathbf{H}_{\nu-1}(x)\,dx +$
$+ \dfrac{2^{1-\nu}}{\sqrt{\pi}\,\Gamma(\nu+1/2)}\int x^{1-\mu}\mathbf{H}_{\mu-1}(x)\,dx - x^{-\mu-\nu+2}\left[\mathbf{H}_\mu(x)\,\mathbf{H}_\nu(x) + \mathbf{H}_{\mu-1}(x)\,\mathbf{H}_{\nu-1}(x)\right]\Big\}$
[see 1.4.1].

4. $\int x\mathbf{H}_\nu(ax)\,\mathbf{H}_\nu(bx)\,dx = \dfrac{1}{a^2-b^2}\Big\{x\left[b\mathbf{H}_\nu(ax)\,\mathbf{H}_{\nu-1}(bx) - a\mathbf{H}_{\nu-1}(ax)\,\mathbf{H}_\nu(bx)\right] +$
$+ \dfrac{2^{1-\nu}}{\sqrt{\pi}\,\Gamma(\nu+1/2)}\left[a^{\nu+1}\int x^\nu \mathbf{H}_\nu(bx)\,dx - b^{\nu+1}\int x^\nu \mathbf{H}_\nu(ax)\,dx\right]\Big\}$ [see 1.4.1].

5. $\int \dfrac{1}{x}\mathbf{H}_\mu(x)\,\mathbf{H}_\nu(x)\,dx = \dfrac{1}{\mu^2-\nu^2}\Big\{x\left[\mathbf{H}_{\mu-1}(x)\,\mathbf{H}_\nu(x) - \mathbf{H}_\mu(x)\,\mathbf{H}_{\nu-1}(x)\right] +$
$+ \dfrac{2^{1-\nu}}{\sqrt{\pi}\,\Gamma(\nu+1/2)}\int x^\nu \mathbf{H}_\mu(x)\,dx - \dfrac{2^{1-\mu}}{\sqrt{\pi}\,\Gamma(\mu+1/2)}\int x^\mu \mathbf{H}_\nu(x)\,dx -$
$- (\mu-\nu)\,\mathbf{H}_\mu(x)\,\mathbf{H}_\nu(x)\Big\}$ [see 1.4.1].

6. $$\int \frac{1}{x} \mathbf{H}_\nu^2(x)\,dx = \frac{\nu-1}{\nu}\int \frac{1}{x}\mathbf{H}_{\nu-1}^2(x)\,dx - \frac{1}{2\nu}\left[\mathbf{H}_{\nu-1}^2(x)+\mathbf{H}_\nu^2(x)\right] + \frac{2^{1-\nu}}{\sqrt{\pi}\,\nu\Gamma(\nu+1/2)}\int x^{\nu-1}\mathbf{H}_{\nu-1}(x)\,dx$$ [see 1.4.1].

7. $$\int \frac{1}{x}\mathbf{H}_n^2(x)\,dx = -\frac{1}{2n}\left[\mathbf{H}_0^2(x)+\mathbf{H}_n^2(x)+2\sum_{k=1}^{n-1}\mathbf{H}_k^2(x) - \frac{2}{\sqrt{\pi}}\sum_{k=0}^{n-1}\frac{1}{2^k\Gamma(k+3/2)}\int x^k\mathbf{H}_k(x)\,dx\right]$$ [see 1.4.1].

1.4.4. Integrals of the form $\int x^\lambda J_\mu(ax)\,\mathbf{H}_\nu(bx)\,dx$.

1. $$\int x^{\mu+\nu+1}J_\mu(x)\,\mathbf{H}_\nu(x)\,dx = \frac{1}{2(\mu+\nu+1)}\left\{x^{\mu+\nu+2}\left[J_\mu(x)\,\mathbf{H}_\nu(x)+J_{\mu+1}(x)\,\mathbf{H}_{\nu+1}(x)\right] - \frac{1}{2^\nu\sqrt{\pi}\,\Gamma(\nu+3/2)}\int x^{\mu+2\nu+2}J_\mu(x)\,dx\right\}$$ [see [18], 1.8.1].

2. $$\int x^{-\mu-\nu+1}J_\mu(x)\,\mathbf{H}_\nu(x)\,dx = \frac{1}{2(\mu+\nu-1)}\left\{-x^{-\mu-\nu+2}\left[J_\mu(x)\,\mathbf{H}_\nu(x)+J_{\mu-1}(x)\,\mathbf{H}_{\nu-1}(x)\right] + \frac{1}{2^{\nu-1}\sqrt{\pi}\,\Gamma(\nu+1/2)}\int x^{-\mu+1}J_{\mu-1}(x)\,dx\right\}$$ [see [18], 1.8.1].

3. $$\int xJ_\nu(ax)\,\mathbf{H}_\nu(bx)\,dx = \frac{x}{a^2-b^2}\left[bJ_\nu(ax)\,\mathbf{H}_{\nu-1}(bx) - aJ_{\nu-1}(ax)\,\mathbf{H}_\nu(bx)\right] - \frac{b^{\nu+1}}{2^{\nu-1}\sqrt{\pi}\,\Gamma(\nu+1/2)(a^2-b^2)}\int x^\nu J_\nu(ax)\,dx$$ [see [18], 1.8.1].

4. $$= \frac{x}{a^2-b^2}\left[bJ_\nu(ax)\,\mathbf{H}_\nu'(bx) - aJ_\nu'(ax)\,\mathbf{H}_\nu(bx)\right] - \frac{ax}{a^2-b^2}\left(\frac{b}{a}\right)^{\nu+1}\left[J_\nu(ax)\,\mathbf{H}_\nu'(bx) - \mathbf{H}_\nu(bx)\,J_\nu'(ax)\right].$$

5. $$\int \frac{1}{x}J_\mu(x)\,\mathbf{H}_\nu(x)\,dx = \frac{x}{\nu^2-\mu^2}\left[J_\mu(x)\,\mathbf{H}_{\nu-1}(x) - J_{\mu-1}(x)\,\mathbf{H}_\nu(x)\right] - \frac{1}{\mu+\nu}J_\mu(x)\,\mathbf{H}_\nu(x) - \frac{1}{2^{\nu-1}\sqrt{\pi}\,\Gamma(\nu+1/2)(\nu^2-\mu^2)}\int x^\nu J_\mu(x)\,dx$$ [see [18], 1.8.1].

6. $$= \frac{x}{\nu^2-\mu^2}\left[J_\mu(x)\,\mathbf{H}_\nu'(x) - J_\nu'(x)\,\mathbf{H}_\nu(x)\right] - \frac{x}{2^{\mu-1}(\nu^2-\mu^2)\,\Gamma(\mu+1/2)}\left[(\mu+\nu-1)\,J_\mu(x)\,s_{\nu-1,\,\mu-1}(x) - J_{\mu-1}(x)\,s_{\nu,\,\mu}(x)\right].$$

7. $$\int \frac{1}{x}J_n(x)\,\mathbf{H}_n(x)\,dx = \frac{1}{2n}\left\{\frac{1}{\sqrt{\pi}}\sum_{k=0}^{n-1}\frac{1}{2^k\Gamma(k+3/2)}\int x^kJ_k(x)\,dx - \left[J_0(x)\,\mathbf{H}_0(x) - J_n(x)\,\mathbf{H}_n(x)\right] - 2\sum_{k=1}^{n-1}J_k(x)\,\mathbf{H}_k(x)\right\}$$ [see [18], 1.8.1].

1.4.5. Integrals of the form $\int x^{\lambda}\mathbf{L}_{\nu}(ax)\,dx$.

1. $$\int_0^x x^{\lambda}\mathbf{L}_{\nu}(ax)\,dx=\frac{a^{\nu+1}x^{\lambda+\nu+2}}{2^{\nu}\sqrt{\pi}\,(\lambda+\nu+2)\,\Gamma(\nu+3/2)}\times$$
$$\times\,{}_2F_3\left(1,\ \frac{\lambda+\nu}{2}+1;\ \frac{\lambda+\nu}{2}+2,\ \nu+\frac{3}{2},\ \frac{3}{2};\ \frac{a^2x^2}{4}\right)\qquad [\mathrm{Re}\,(\lambda+\nu)>-2].$$

2. $$\int_0^x x^{1\pm\nu}\mathbf{L}_{\nu}(ax)\,dx=\frac{x^{1\pm\nu}}{a}\mathbf{L}_{\nu\pm1}(ax)-\frac{2^{1-\nu}a^{\nu-1}x}{\sqrt{\pi}\,\Gamma(\nu+1/2)}\begin{Bmatrix}0\\1\end{Bmatrix}\quad\left[\begin{Bmatrix}\mathrm{Re}\,\nu>-3/2\\ \nu\text{ is arbitrary}\end{Bmatrix}\right].$$

1.4.6. Integrals of the form $\int x^{\lambda}e^{\pm x}\mathbf{L}_{\nu}(x)\,dx$.

1. $$\int_0^x x^{\nu}e^{\pm x}\mathbf{L}_{\nu}(x)\,dx=$$
$$=\frac{x^{\nu+1}e^{\pm x}}{2\nu+1}[\mathbf{L}_{\nu}(x)\mp\mathbf{L}_{\nu+1}(x)]-\frac{2^{-\nu}e^{(1\pm1)\nu\pi i}}{(2\nu+1)\,\Gamma(\nu+3/2)\sqrt{\pi}}\gamma(2\nu+2,\ \mp x)$$
$$[\mathrm{Re}\,\nu>-1;\ \nu\neq-1/2].$$

2. $$\int_0^x x^{-1/2}e^{\pm x}\mathbf{L}_{-1/2}(x)\,dx=\pm\frac{1}{\sqrt{2\pi}}[\mathrm{Ei}\,(\pm 2x)-\mathbf{C}-\ln 2x].$$

3. $$\int_0^x x^{-\nu}e^{\pm x}\mathbf{L}_{\nu}(x)\,dx=$$
$$=-\frac{x^{1-\nu}e^{\pm x}}{2\nu-1}[\mathbf{L}_{\nu}(x)\mp\mathbf{L}_{\nu-1}(x)]+\frac{2^{1-\nu}}{(2\nu-1)\,\Gamma(\nu+1/2)\sqrt{\pi}}(1-e^{\pm x})\qquad[\nu\neq1/2].$$

4. $$\int_0^x x^{-1/2}e^{\pm x}\mathbf{L}_{1/2}(x)\,dx=\frac{1}{\sqrt{2\pi}}\left[\mathrm{Ei}\,(\pm 2x)-2\mathrm{Ei}\,(\pm x)+\mathbf{C}+\ln\frac{x}{2}\right].$$

5. $$\int_{x_1}^{x_2} x^{\lambda}\left[I_{\pm\nu}(ax)-\mathbf{L}_{\nu}(ax)\right]dx=$$
$$=\frac{(-1)^{\varepsilon+1}a^{\nu+1}x^{\lambda+\nu+2}}{2^{\nu}\sqrt{\pi}\,(\lambda+\nu+2)\,\Gamma(\nu+3/2)}\,{}_2F_3\left(1,\ \frac{\lambda+\nu}{2}+1;\ \frac{3}{2},\ \nu+\frac{3}{2},\ \frac{\lambda+\nu}{2}+2;\ \frac{a^2x^2}{4}\right)+$$
$$+\frac{(-1)^{\varepsilon}a^{\pm\nu}(x/2)^{\lambda\pm\nu+1}}{(\lambda\pm\nu+1)\,\Gamma(1\pm\nu)}\,{}_1F_2\left(\frac{\lambda\pm\nu+1}{2};\ \frac{\lambda\pm\nu+3}{2},\ 1\pm\nu;\ \frac{a^2x^2}{4}\right)-$$
$$-\frac{2^{\lambda}\varepsilon\cos\nu\pi}{a^{\lambda+1}\sin[(\lambda+\nu)\pi/2]}\Gamma\begin{bmatrix}(1+\lambda\pm\nu)/2\\(1-\lambda\pm\nu)/2\end{bmatrix}$$
$$[x_1=0,\ x_2=x;\ \varepsilon=0;\ \mathrm{Re}\,(\lambda\pm\nu)>-1]\ \text{or}\ [x_1=x,\ x_2=\infty;\ \varepsilon=1;\ \mathrm{Re}\,(\lambda+\nu)<0].$$

1.5. THE ANGER FUNCTION $\mathbf{J}_\nu(x)$ AND WEBER FUNCTION $\mathbf{E}_\nu(x)$

1.5.1. Integrals of the form $\int x^\lambda \begin{Bmatrix} \mathbf{J}_\nu(ax) \\ \mathbf{E}_\nu(ax) \end{Bmatrix} dx$.

1. $\int_{x_1}^{x_2} x^\lambda \begin{Bmatrix} \mathbf{J}_\nu(ax) \\ \mathbf{E}_\nu(ax) \end{Bmatrix} dx =$

$$= \frac{2(-1)^\varepsilon x^{\lambda+1}}{(1+\lambda)\nu\pi} \sin\frac{\nu\pi}{2} \begin{Bmatrix} \cos(\nu\pi/2) \\ \sin(\nu\pi/2) \end{Bmatrix} {}_2F_3\left(1, \frac{\lambda+1}{2}; \frac{\lambda+3}{2}, 1+\frac{\nu}{2}, 1-\frac{\nu}{2}; -\frac{a^2x^2}{4}\right) \pm$$

$$\pm \frac{2(-1)^\varepsilon \ldots}{(\lambda+2)(1-\nu^2)\pi} \cos\frac{\nu\pi}{2} \begin{Bmatrix} \sin(\nu\pi/2) \\ \cos(\nu\pi/2) \end{Bmatrix} {}_2F_3\left(1, 1+\frac{\lambda}{2}; 2+\frac{\lambda}{2}, \frac{3+\nu}{2}, \frac{3-\nu}{2}; -\frac{a^2x^2}{4}\right) -$$

$$- \frac{(2/a)^{\lambda+1} \varepsilon\pi \sec\lambda\pi}{\Gamma((1-\lambda-\nu)/2)\,\Gamma((1-\lambda+\nu)/2)} \begin{Bmatrix} \sin[(\lambda-\nu)\pi/2] \\ \cos[(\lambda-\nu)\pi/2] \end{Bmatrix}$$

$[x_1=0,\ x_2=x;\ \varepsilon=0;\ \mathrm{Re}\,\lambda > -1]$ or $[x_1=x,\ x_2=\infty;\ \varepsilon=1;\ a>0;\ \mathrm{Re}\,\lambda<0]$.

1.6. THE LOMMEL FUNCTIONS $s_{\mu,\nu}(x)$ AND $S_{\mu,\nu}(x)$

1.6.1. Integrals of the form $\int x^\lambda \begin{Bmatrix} s_{\mu,\nu}(ax) \\ S_{\mu,\nu}(ax) \end{Bmatrix} dx$.

1. $\int_{x_1}^{x_2} x^\lambda s_{\mu,\nu}(ax)\,dx =$

$$= \pm \frac{a^{\mu+1}x^{\lambda+\mu+2}}{[(\mu+1)^2-\nu^2](\mu+\lambda+2)} {}_2F_3\left(1, \frac{\mu+\lambda}{2}+1; \frac{\mu+\lambda}{2}+2, \frac{\mu-\nu+3}{2}, \frac{\mu+\nu+3}{2}; -\frac{a^2x^2}{4}\right) -$$

$$- 2^{\lambda+\mu-1}\pi a^{-\lambda-1} \operatorname{cosec}\frac{\lambda+\mu}{2}\pi\,\Gamma\begin{bmatrix} (\mu-\nu+1)/2,\ (\mu+\nu+1)/2 \\ (\nu-\lambda+1)/2,\ (1-\lambda-\nu)/2 \end{bmatrix} \begin{Bmatrix} 0 \\ 1 \end{Bmatrix}$$

$\left[\begin{Bmatrix} x_1=0,\ x_2=x;\ \mathrm{Re}(\lambda+\mu)>-2;\ |\arg a|<\pi \\ x_1=x,\ x_2=\infty;\ \mathrm{Re}\,\lambda<1/2;\ \mathrm{Re}(\lambda+\mu)<0;\ a>0 \end{Bmatrix}\right]$.

2. $\int_{x_1}^{x_2} x^\lambda S_{\mu,\nu}(ax)\,dx =$

$$= \pm \frac{a^{\mu+1}x^{\lambda+\mu+2}}{[(\mu+1)^2-\nu^2](\lambda+\mu+2)} {}_2F_3\left(1, \frac{\lambda+\mu}{2}+1; \frac{\lambda+\mu}{2}+2, \frac{\mu-\nu+3}{2}, \frac{\mu+\nu+3}{2}; -\frac{a^2x^2}{4}\right) \pm$$

$$\pm \frac{2^{\mu-\nu-1}a^\nu x^{\lambda+\nu+1}}{\lambda+\nu+1}\Gamma\begin{bmatrix} -\nu,\ (\mu-\nu+1)/2 \\ (1-\mu-\nu)/2 \end{bmatrix} {}_1F_2\left(\frac{\lambda+\nu+1}{2}; \nu+1, \frac{\lambda+\nu+3}{2}; -\frac{a^2x^2}{4}\right) \pm$$

$$\pm \frac{2^{\mu+\nu-1}a^{-\nu}x^{\lambda-\nu+1}}{\lambda-\nu+1}\Gamma\begin{bmatrix} \nu,\ (\mu+\nu+1)/2 \\ (\nu-\mu+1)/2 \end{bmatrix} {}_1F_2\left(\frac{\lambda-\nu+1}{2}; 1-\nu, \frac{\lambda-\nu+3}{2}; -\frac{a^2x^2}{4}\right) -$$

$$- 2^{\lambda+\mu-1}\pi a^{-\lambda-1} \operatorname{cosec}\frac{\lambda+\mu}{2}\pi\,\Gamma\begin{bmatrix} (\lambda+\nu+1)/2,\ (\lambda-\nu+1)/2 \\ (1-\mu-\nu)/2,\ (1-\mu+\nu)/2 \end{bmatrix} \begin{Bmatrix} 0 \\ 1 \end{Bmatrix}$$

$\left[\begin{Bmatrix} x_1=0,\ x_2=x;\ \mathrm{Re}(\lambda+\mu)>-2;\ \mathrm{Re}\,\lambda>|\mathrm{Re}\,\nu|;\ |\arg a|<\pi \\ x_1=x,\ x_2=\infty;\ \mathrm{Re}(\lambda+\mu)<0,\ a>0 \end{Bmatrix},\ x>0\right]$.

3. $$\int x^\nu S_{\mu,\nu}(x)\,dx = 2^{\nu-1}\sqrt{\pi}\,\Gamma\left(\nu+\frac{1}{2}\right)[x\mathbf{H}_{\nu-1}(x)\,S_{\mu,\nu}(x) - (\mu+\nu-1)\,x\mathbf{H}_\nu(x)\,S_{\mu-1,\nu-1}(x)] + \int x^\mu \mathbf{H}_\nu(x)\,dx.$$

1.6.2. Integrals containing $J_\nu(x)$ and $s_{\mu,\nu}(x)$.

1. $$\int_0^x \left[1-a^2+\frac{\nu^2-\lambda^2}{x^2}\right] xJ_\lambda(x)\, s_{\mu,\nu}(ax)\,dx = x\left[J_\lambda(x)\, s'_{\mu,\nu}(ax) - J'_\lambda(x)\, s_{\mu,\nu}(ax)\right] - a^{\mu+1} x\left[(\lambda+\mu-1) J_\lambda(x)\, s_{\mu-1,\nu-1}(x) - J_{\lambda-1}(x)\, s_{\mu,\nu}(x)\right].$$

2. $$\int_0^x \frac{(\mu+\nu-1) J_\nu(x)\, s_{\mu-1,\nu-1}(x) - J_{\nu-1}(x)\, s_{\mu,\nu}(x)}{J_\nu^2(x)}\,dx = \frac{s_{\mu,\nu}(x)}{J_\nu(x)}.$$

3. $$\int_0^x \frac{(\mu+\nu-1) J_\nu(x)\, s_{\mu-1,\nu-1}(x) - J_{\nu-1}(x)\, s_{\mu,\nu}(x)}{s_{\mu,\nu}(x)\, J_\nu(x)}\,dx = \ln\left[-\frac{J_\nu(x)}{s_{\mu,\nu}(x)}\right].$$

4. $$\int_0^x \frac{x^{\mu-1} J_\nu(x)\,dx}{(\mu+\nu-1) J_\nu(x)\, s_{\mu-1,\nu-1}(x) - J_{\nu-1}(x)\, s_{\mu,\nu}(x)} = \ln\left[(\mu+\nu-1)\, xJ_\nu(x)\, s_{\mu-1,\nu-1}(x) - xJ_{\nu-1}(x)\, s_{\mu,\nu}(x)\right].$$

1.7. THE KELVIN FUNCTIONS $\mathrm{ber}_\nu(x)$, $\mathrm{bei}_\nu(x)$, $\mathrm{ker}_\nu(x)$ AND $\mathrm{kei}_\nu(x)$

NOTATION:

$$\begin{cases} f_\nu = \mathrm{ber}_\nu(x) \\ g_\nu = \mathrm{bei}_\nu(x) \end{cases}, \quad \begin{cases} f_\nu = \mathrm{ker}_\nu(x) \\ g_\nu = \mathrm{kei}_\nu(x) \end{cases}, \quad \begin{cases} f_\nu = \mathrm{bei}_\nu(x) \\ g_\nu = -\mathrm{ber}_\nu(x) \end{cases}, \quad \begin{cases} f_\nu = \mathrm{kei}_\nu(x) \\ g_\nu = -\mathrm{ker}_\nu(x) \end{cases}.$$

Below, f_ν, g_ν and f_ν^*, g_ν^* are any of these four pairs of functions.

1.7.1. Integrals of the form $\int x^\lambda \begin{Bmatrix} \mathrm{ber}_\nu(ax) \\ \mathrm{bei}_\nu(ax) \end{Bmatrix}$ and $\int x^\lambda \begin{Bmatrix} \mathrm{ker}_\nu(ax) \\ \mathrm{kei}_\nu(ax) \end{Bmatrix} dx$.

1. $$\int_0^x x^\lambda \begin{Bmatrix} \mathrm{ber}_\nu(ax) \\ \mathrm{bei}_\nu(ax) \end{Bmatrix} dx = \frac{a^\nu x^{\lambda+\nu+1}}{2^\nu(\lambda+\nu+1)\Gamma(\nu+1)} \begin{Bmatrix} \cos(3\nu\pi/4) \\ \sin(3\nu\pi/4) \end{Bmatrix} \times {}_1F_4\left(\frac{\lambda+\nu+1}{4};\ \frac{1}{2},\ \frac{\nu+1}{2},\ \frac{\nu}{2}+1,\ \frac{\lambda+\nu+5}{4};\ -\frac{a^4x^4}{256}\right) \mp \frac{a^{\nu+2}x^{\lambda+\nu+3}}{2^{\nu+2}(\lambda+\nu+3)\Gamma(\nu+2)} \begin{Bmatrix} \sin(3\nu\pi/4) \\ \cos(3\nu\pi/4) \end{Bmatrix} {}_1F_4\left(\frac{\lambda+\nu+3}{4};\ \frac{3}{2},\ \frac{\nu}{2}+1,\ \frac{\nu+3}{2},\ \frac{\lambda+\nu+7}{4};\ -\frac{a^4x^4}{256}\right)$$ $[\mathrm{Re}\,(\lambda+\nu) > -1]$.

2. $$\int_0^x x^\lambda \begin{Bmatrix} \mathrm{ker}_\nu(ax) \\ \mathrm{kei}_\nu(ax) \end{Bmatrix} dx = U(\nu)$$ $[\mathrm{Re}\,\lambda > |\mathrm{Re}\,\nu| - 1]$.

$$U(\nu) = \pm \frac{2^{\nu-1}x^{\lambda-\nu+1}}{a^\nu(\lambda-\nu+1)}\Gamma(\nu) \begin{Bmatrix} \cos(3\nu\pi/4) \\ \sin(3\nu\pi/4) \end{Bmatrix} \times {}_1F_4\left(\frac{\lambda-\nu+1}{4};\ \frac{\lambda-\nu+5}{4},\ \frac{1-\nu}{2},\ 1-\frac{\nu}{2},\ \frac{1}{2};\ -\frac{a^4x^4}{256}\right) + \frac{a^\nu x^{\lambda+\nu+1}}{2^{\nu+1}(\lambda+\nu+1)}\Gamma(-\nu) \begin{Bmatrix} \cos(3\nu\pi/4) \\ \sin(3\nu\pi/4) \end{Bmatrix} \times {}_1F_4\left(\frac{\lambda+\nu+1}{4};\ \frac{\lambda+\nu+5}{4},\ \frac{1+\nu}{2},\ \frac{\nu}{2}+1,\ \frac{1}{2};\ -\frac{a^4x^4}{256}\right) -$$

$$-\frac{a^{2-\nu}x^{\lambda-\nu+3}}{2^{3-\nu}(\lambda-\nu+3)}\Gamma(\nu-1)\begin{Bmatrix}\sin(3\nu\pi/4)\\ \cos(3\nu\pi/4)\end{Bmatrix}\times$$
$$\times\,{}_1F_4\left(\frac{\lambda-\nu+3}{4};\ \frac{\lambda-\nu+7}{4},\ \frac{3}{2},\ 1-\frac{\nu}{2},\ \frac{3-\nu}{2};\ -\frac{a^4x^4}{256}\right)\pm$$
$$\pm\frac{a^{\nu+2}x^{\lambda+\nu+3}}{2^{\nu+3}(\lambda+\nu+3)}\Gamma(-\nu-1)\begin{Bmatrix}\sin(3\nu\pi/4)\\ \cos(3\nu\pi/4)\end{Bmatrix}{}_1F_4\left(\frac{\lambda+\nu+3}{4};\right.$$
$$\left.\frac{\lambda+\nu+7}{4},\ \frac{3}{2},\ \frac{\nu}{2}+1,\ \frac{\nu+3}{2};\ -\frac{a^4x^4}{256}\right)\qquad [\nu\neq\ldots,-2,-1,0,1,2,\ldots],$$

$$U(n)=\pm\frac{2^{n-1}x^{\lambda-n+1}}{a^n}\sum_{k=0}^{n-1}\frac{(n-k-1)!}{k!\,(\lambda-n+2k+1)}\times$$
$$\times\begin{Bmatrix}\cos[(3n-2k)\pi/4]\\ \sin[(3n-2k)\pi/4]\end{Bmatrix}\left(-\frac{a^2x^2}{4}\right)^k\pm\frac{(-1)^n a^n x^{\lambda+n+1}}{2^{n+2}}\times$$
$$\times\sum_{k=0}^{\infty}\frac{(ax/2)^{2k}}{k!\,(n+k)!\,(\lambda+n+2k+1)}\begin{Bmatrix}\cos[(n-2k)\pi/4]\\ \sin[(n-2k)\pi/4]\end{Bmatrix}\times$$
$$\times\left[2\psi(k+1)+2\psi(n+k+1)+\frac{4}{\lambda+n+2k+1}-4\ln\frac{ax}{2}\mp\pi\operatorname{tg}^{\pm1}\frac{n-2k}{4}\pi\right]$$
$$[n=0,\ 1,\ 2,\ \ldots].$$

3. $\int x^{1+\nu}f_\nu\,dx=-\frac{x^{1+\nu}}{\sqrt{2}}(f_{\nu+1}-g_{\nu+1}).$ 4. $\int x^{1-\nu}f_\nu\,dx=\frac{x^{1-\nu}}{\sqrt{2}}(f_{\nu-1}-g_{\nu-1}).$

1.7.2. Integrals containing products of Kelvin functions.

1. $\int x\,(f_\nu g_\nu^*-g_\nu f_\nu^*)\,dx=\frac{1}{2}x\,(f_\nu' f_\nu^*-f_\nu f_\nu^{*\prime}+g_\nu' g_\nu^*-g_\nu g_\nu^{*\prime}).$

2. $\int x\,(f_\nu g_\nu^*+g_\nu f_\nu^*)\,dx=$
$$=\frac{1}{4}x^2\,(2f_\nu g_\nu^*-f_{\nu-1}g_{\nu+1}^*-f_{\nu+1}g_{\nu-1}^*+2g_\nu f_\nu^*-g_{\nu-1}f_{\nu+1}^*-g_{\nu+1}f_{\nu-1}^*).$$

3. $\int x\,(f_\nu^2+g_\nu^2)\,dx=x\,(f_\nu g_\nu'-f_\nu' g_\nu).$

4. $\int xf_\nu g_\nu\,dx=\frac{1}{4}x^2\,(2f_\nu g_\nu-f_{\nu-1}g_{\nu+1}-f_{\nu+1}g_{\nu-1}).$

5. $\int x\,(f_\nu^2-g_\nu^2)\,dx=\frac{1}{2}x^2\,(f_\nu^2-f_{\nu-1}f_{\nu+1}-g_\nu^2+g_{\nu-1}g_{\nu+1}).$

6. $\int x\left(\begin{Bmatrix}\operatorname{ber}_\nu'(x)\\ \operatorname{ker}_\nu'(x)\end{Bmatrix}^2+\begin{Bmatrix}\operatorname{bei}_\nu'(x)\\ \operatorname{kei}_\nu'(x)\end{Bmatrix}^2\right)dx=x\begin{Bmatrix}\operatorname{ber}_\nu(x)\operatorname{ber}_\nu'(x)+\operatorname{bei}_\nu(x)\operatorname{bei}_\nu'(x)\\ \operatorname{ker}_\nu(x)\operatorname{ker}_\nu'(x)+\operatorname{kei}_\nu(x)\operatorname{kei}_\nu'(x)\end{Bmatrix}.$

7. $\int x\,[\operatorname{ber}'(x)\operatorname{ker}'(x)-\operatorname{bei}'(x)\operatorname{kei}'(x)]\,dx=x\,[\operatorname{ber}_1(x)\operatorname{ker}_1(x)-\operatorname{bei}_1(x)\operatorname{kei}_1(x)].$

1.8. THE AIRY FUNCTIONS Ai(x) AND Bi(x)

1.8.1. Integrals of the form $\int f(x)\begin{Bmatrix}\operatorname{Ai}(x)\\ \operatorname{Bi}(x)\end{Bmatrix}dx$.

1. $$\int_0^x x^\lambda\begin{Bmatrix}\operatorname{Ai}(x)\\ \operatorname{Bi}(x)\end{Bmatrix}dx=x^{\lambda+1}\left[\frac{3^{-(5\pm3)/12}}{(\lambda+1)\,\Gamma(2/3)}{}_1F_2\left(\frac{\lambda+1}{3};\ \frac{\lambda+4}{3},\ \frac{2}{3};\ \frac{x^3}{9}\right)\mp\right.$$
$$\left.\mp\frac{3^{-(1\pm3)/12}}{(\lambda+2)\,\Gamma(1/3)}x\,{}_1F_2\left(\frac{\lambda+2}{3};\ \frac{\lambda+5}{3},\ \frac{4}{3};\ \frac{x^3}{9}\right)\right]\qquad[\operatorname{Re}\lambda>-1].$$

2. $$\int_0^x x^{1/2}\,\mathrm{Ai}(x)\,dx=\frac{\sqrt{\pi}\,x}{3^{2/3}}\,\Gamma\left(\frac{5}{6}\right)\left[\mathrm{Ai}(x)\,\mathbf{L}_{-2/3}\left(\frac{2}{3}x^{3/2}\right)-x^{-1/2}\,\mathrm{Ai}'(x)\,\mathbf{L}_{1/3}\left(\frac{2}{3}x^{3/2}\right)\right].$$

3. $$\int_0^x x\begin{Bmatrix}\mathrm{Ai}(x)\\ \mathrm{Bi}(x)\end{Bmatrix}dx=\begin{Bmatrix}\mathrm{Ai}'(x)\\ \mathrm{Bi}'(x)\end{Bmatrix}\pm\frac{1}{\Gamma(1/3)}\begin{Bmatrix}3^{-1/3}\\ 3^{1/6}\end{Bmatrix}.$$

4. $$\int_0^x x^2\,\mathrm{Ai}(x)\,dx=-\frac{x^2}{\pi\sqrt{3}}\,K_{4/3}\left(\frac{2}{3}x^{3/2}\right)+\frac{3^{5/6}}{2\pi}\,\Gamma\left(\frac{4}{3}\right).$$

5. $$\int_0^x x^{\lambda}e^{\pm 2x^{3/2}/3}\,\mathrm{Ai}(x)\,dx=$$
$$=\frac{3^{-2/3}}{(\lambda+1)\,\Gamma(2/3)}\,x^{\lambda+1}\,{}_2F_2\left(\frac{1}{6},\ \frac{2\lambda+2}{3};\ \frac{1}{3},\ \frac{2\lambda+5}{3};\ \pm\frac{4}{3}x^{3/2}\right)-$$
$$-\frac{3^{-1/3}}{(\lambda+2)\,\Gamma(1/3)}\,x^{\lambda+2}\,{}_2F_2\left(\frac{5}{6},\ \frac{2\lambda+4}{3};\ \frac{5}{3},\ \frac{2\lambda+7}{3};\ \pm\frac{4}{3}x^{3/2}\right)\qquad[\mathrm{Re}\,\lambda>-1].$$

6. $$\int_0^x x^{\lambda}e^{\pm 2x^{3/2}/3}\,\mathrm{Bi}(x)\,dx=$$
$$=\frac{3^{-1/6}}{(\lambda+1)\,\Gamma(2/3)}\,x^{\lambda+1}\,{}_2F_2\left(\frac{1}{6},\ \frac{2\lambda+2}{3};\ \frac{1}{3},\ \frac{2\lambda+5}{3};\ \pm\frac{4}{3}x^{3/2}\right)+$$
$$+\frac{3^{1/6}}{(\lambda+2)\,\Gamma(1/3)}\,x^{\lambda+2}\,{}_2F_2\left(\frac{5}{6},\ \frac{2\lambda+4}{3};\ \frac{5}{3},\ \frac{2\lambda+7}{3};\ \pm\frac{4}{3}x^{3/2}\right)\qquad[\mathrm{Re}\,\lambda>-1].$$

7. $$\int_0^x x^{1/2}e^{\pm 2x^{3/2}/3}\,\mathrm{Ai}(x)\,dx=\frac{2}{5}\,x^{3/2}e^{\pm 2x^{3/2}/3}\,\mathrm{Ai}(x)\mp$$
$$\mp\frac{2x^2}{15}\,e^{\pm 2x^{3/2}/3}\left[I_{-4/3}\left(\frac{2}{3}x^{3/2}\right)-I_{4/3}\left(\frac{2}{3}x^{3/2}\right)\right]\mp\frac{3^{-1/6}}{5\pi}\,\Gamma\left(\frac{1}{3}\right).$$

8. $$\int_0^x x^{1/2}e^{\pm 2x^{3/2}/3}\,\mathrm{Bi}(x)\,dx=\frac{2}{5}\,x^{3/2}e^{\pm 2x^{3/2}/3}\,\mathrm{Bi}(x)\mp$$
$$\mp\frac{2x^2}{5\sqrt{3}}\,e^{\pm 2x^{3/2}/3}\left[I_{4/3}\left(\frac{2}{3}x^{3/2}\right)+I_{-4/3}\left(\frac{2}{3}x^{3/2}\right)\right]\mp\frac{3^{1/3}}{5\pi}\,\Gamma\left(\frac{1}{3}\right).$$

9. $$\int_0^x x^{3/2}\begin{Bmatrix}J_{1/3}(ax^{3/2})\\ I_{1/3}(ax^{3/2})\end{Bmatrix}\mathrm{Ai}(x)\,dx=$$
$$=\frac{6ax}{9a^2\pm 4}\begin{Bmatrix}J_{4/3}(ax^{3/2})\\ I_{4/3}(ax^{3/2})\end{Bmatrix}\mathrm{Ai}(x)\mp\frac{3^{-1/2}\cdot 4}{\pi(9a^2+4)}\,x^{3/2}\begin{Bmatrix}J_{1/3}(ax^{3/2})\\ I_{1/3}(ax^{3/2})\end{Bmatrix}K_{4/3}\left(\frac{2}{3}x^{3/2}\right)+$$
$$+\frac{2^{2/3}\cdot 3^{5/6}a^{1/3}}{\pi(4\pm 9a^2)}.$$

2

10. $\int_0^x x^{3/2} I_{1/3}\left(\frac{2}{3}x^{3/2}\right) \operatorname{Ai}(x)\,dx = \frac{1}{12}x^{-1/2}(4x^3+1)\, I_{1/3}\left(\frac{2}{3}x^{3/2}\right)\operatorname{Ai}(x) - \frac{x^3}{3\sqrt{3}\pi} I'_{1/3}\left(\frac{2}{3}x^{3/2}\right) K'_{1/3}\left(\frac{2}{3}x^{3/2}\right) - \frac{1}{4\sqrt{3}\pi}.$

11. $\int_0^x \frac{1}{x^{3/2}} I_\nu\left(\frac{2}{3}x^{3/2}\right)\operatorname{Ai}(x)\,dx = \frac{4x}{9\nu^2-1}\left[\operatorname{Ai}(x)\, I'_\nu\left(\frac{2}{3}x^{3/2}\right) - \frac{1}{\pi}\left(\frac{x}{3}\right)^{1/2} I_\nu\left(\frac{2}{3}x^{3/2}\right) K'_{1/3}\left(\frac{2}{3}x^{3/2}\right)\right].$

12. $\int_x^\infty \frac{1}{x^{3/2}} K_\nu\left(\frac{2}{3}x^{3/2}\right)\operatorname{Ai}(x)\,dx = -\frac{4x}{9\nu^2-1}\left[\operatorname{Ai}(x)\, K'_\nu\left(\frac{2}{3}x^{3/2}\right) - \frac{1}{\pi}\left(\frac{x}{3}\right)^{1/2} K_\nu\left(\frac{2}{3}x^{3/2}\right) K'_{1/3}\left(\frac{2}{3}x^{3/2}\right)\right].$

13. $\int \frac{dx}{x^{1/2} I_{1/3}(2x^{3/2}/3)\operatorname{Ai}(x)} = \frac{2\pi}{\sqrt{3}} \ln \frac{(x/3)^{1/2} I_{1/3}(2x^{3/2}/3)}{\pi \operatorname{Ai}(x)}.$

1.8.2. Integrals containing products of Airy functions.

NOTATION: $y = a\operatorname{Ai}(x) + b\operatorname{Bi}(x)$, $y_n = a_n \operatorname{Ai}(x) + b_n \operatorname{Bi}(x)$, $n = 1, 2$; a, b, a_n, b_n are complex constants.

1. $\int x^n y_1 y_2\,dx = \frac{1}{2(2n+1)}\left[nx^{n-1}(y_1'y_2 + y_1y_2') - 2x^n y_1'y_2' + 2x^{n+1}y_1y_2 - n(n-1)\int x^{n-2}(y_1'y_2 + y_1y_2')\,dx\right].$

2. $= \frac{1}{2}\left[x^{n-1}(y_1'y_2 + y_1y_2') - (n-1)\int x^{n-2}(y_1'y_2 + y_1y_2')\,dx - 2\int x^{n-1}y_1'y_2'dx\right]$ $\quad [n \geqslant 1].$

3. $\int x^n y_1'y_2\,dx = \frac{1}{2}\left[x^n y_1 y_2 - n\int x^{n-1}y_1y_2\,dx + \frac{1}{n+1}x^{n+1}(y_1'y_2 - y_1y_2')\right].$

4. $\int x^n y_1'y_2' = \frac{1}{2(2n+3)}\left\{(n+2)\left[x^n(y_1'y_2 + y_1y_2') - n\int x^{n-1}(y_1'y_2 + y_1y_2')\,dx\right] + 2x^{n+1}(y_1'y_2' - xy_1y_2)\right\}.$

5. $\int y_1y_2\,dx = xy_1y_2 - y_1'y_2'.$

6. $\int y_1'y_2\,dx = \frac{1}{2}(y_1y_2 + xy_1'y_2 - xy_1y_2').$

7. $\int y_1'y_2'\,dx = \frac{1}{3}(y_1'y_2 + y_1y_2' + xy_1'y_2' - x^2y_1y_2).$

8. $\int xy_1y_2\,dx = \frac{1}{6}(y_1'y_2 + y_1y_2' - 2xy_1'y_2' + 2x^2y_1y_2).$

9. $\int xy_1'y_2\,dx = \frac{1}{4}(2y_1'y_2' + x^2y_1'y_2 - x^2y_1y_2').$

10. $\int xy_1'y_2'\,dx = \frac{1}{5}\left[\frac{3}{2}(xy_1'y_2 + xy_1y_2' - y_1y_2) + x^2y_1'y_2' - x^3y_1y_2\right].$

11. $\int y^2\,dx = xy^2 - y'^2$. 12. $\int y'y\,dx = \frac{1}{2}y^2$.

13. $\int y'^2\,dx = \frac{1}{3}(2y'y + xy'^2 - x^2y^2)$. 14. $\int xy^2\,dx = \frac{1}{3}(y'y - xy'^2 + x^2y^2)$.

15. $\int xy'y\,dx = \frac{1}{2}y'^2$.

16. $\int xy'^2\,dx = \frac{1}{5}\left[3\left(xy'y - \frac{1}{2}y^2\right) + x^2y'^2 - x^3y^2\right]$.

17. $\int x^2y^2\,dx = \frac{1}{5}\left[2\left(xy'y - \frac{1}{2}y^2\right) - x^2y'^2 + x^3y^2\right]$.

18. $\int x^2y'y\,dx = \frac{1}{6}(x^2y^2 - 2y'y + 2xy'^2)$.

19. $\int x^2y'^2\,dx = \frac{1}{7}(4x^2y'y - 4y'^2 + x^3y'^2 - x^4y^2)$.

20. $\int x^3y^2\,dx = \frac{1}{7}(3x^2y'y - 3y'^2 - x^3y'^2 + x^4y^2)$.

21. $\int x^3y'y\,dx = \frac{1}{5}\left(-3xy'y + \frac{3}{2}y^2 + \frac{3}{2}x^2y'^2 + x^3y^2\right)$.

22. $\int \frac{dx}{\mathrm{Ai}^2(x)} = 2\pi\left(\frac{x}{3}\right)^{1/2}\frac{1}{\mathrm{Ai}(x)}I_{1/3}\left(\frac{2}{3}x^{3/2}\right)$.

1.9. THE INTEGRAL FUNCTIONS OF BESSEL $Ji_\nu(x)$, NEUMANN $Yi_\nu(x)$ AND MACDONALD $Ki_\nu(x)$

1.9.1. Integrals of the form $\int x^\alpha Ji_\nu(x)\,dx$.

1. $\int x^{\alpha-1}Ji_\nu(x)\,dx = \frac{x^\alpha}{\alpha}Ji_\nu(x) + V_-(\nu)$,

$$V_\mp(\nu) = \frac{x^{\alpha+\nu}}{2^\nu(\alpha+\nu)\Gamma(\nu+1)}\,{}_1F_2\left(\frac{\alpha+\nu}{2};\ \frac{\alpha+\nu+2}{2};\ \nu+1;\ \mp\frac{x^2}{4}\right).$$

2. $\int x^{1\pm\nu}Ji_\nu(x)\,dx = \frac{x^{2\pm\nu}}{2\pm\nu}Ji_\nu(x) \pm \frac{x^{1\pm\nu}}{2\pm\nu}J_{\nu\pm1}(x)$.

3. $\int x^\nu Ji_\nu(x)dx =$

$$= \frac{x^{\nu+1}}{\nu+1}Ji_\nu(x) + \frac{2^{\nu-1}}{\nu+1}\sqrt{\pi}\,\Gamma\left(\nu+\frac{1}{2}\right)x\left[J_\nu(x)\mathbf{H}_{\nu-1}(x) - J_{\nu-1}(x)\mathbf{H}_\nu(x)\right].$$

4. $\int Ji_{2n}(x)\,dx = xJi_{2n}(x) + xJ_0(x) + \frac{\pi x}{2}\left[J_1(x)\mathbf{H}_0(x) - J_0(x)\mathbf{H}_1(x)\right] -$

$$-2\sum_{k=0}^{n-1}J_{2k+1}(x).$$

5. $\int Ji_{2n+1}(x)\,dx = xJi_{2n+1}(x) - J_0(x) - 2\sum_{k=1}^{n}J_{2k}(x)$.

1.9.2. Integrals of the form $\int x^\alpha \begin{Bmatrix} Yi_\nu(x) \\ Ki_\nu(x) \end{Bmatrix} dx$.

1. $\int x^{\alpha-1}\begin{Bmatrix} Yi_\nu(x) \\ Ki_\nu(x) \end{Bmatrix} dx = \frac{x^\alpha}{\alpha}\begin{Bmatrix} Yi_\nu(x) \\ Ki_\nu(x) \end{Bmatrix} \pm \frac{1}{\sin\nu\pi}\left[\begin{Bmatrix} \cos\nu\pi \\ \pi/2 \end{Bmatrix} V_\mp(\nu) -\right.$

$$\left. - \begin{Bmatrix} 1 \\ \pi/2 \end{Bmatrix} V_\mp(-\nu)\right] \quad [V_\mp \text{ see } 1.9.1.1].$$

2. $\int x^{1\pm\nu}\,Yi_\nu(x)\,dx=\frac{x^{2\pm\nu}}{2\pm\nu}\,Yi_\nu(x)\pm\frac{x^{1\pm\nu}}{2\pm\nu}\,Y_{\nu\pm1}(x).$

3. $\int x^{1\pm\nu}\,Ki_\nu(x)\,dx=\frac{x^{2\pm\nu}}{2\pm\nu}\,Ki_\nu(x)-\frac{x^{1\pm\nu}}{2\pm\nu}\,K_{\nu\pm1}(x).$

1.10. THE INCOMPLETE ELLIPTIC INTEGRALS $F(x,k)$, $E(x,k)$ AND $\Pi(x,\nu,k)$

1.10.1. Integrals with respect to the argument x.

1. $\int_0^x \sin x F(x,k)\,dx=-\cos x F(x,k)+\frac{1}{k}\arcsin(k\sin x).$

2. $\int_0^x \cos x F(x,k)\,dx=\sin x F(x,k)+\frac{1}{k}\,\mathrm{Arch}\sqrt{\frac{1-k^2\sin^2 x}{1-k^2}}-\frac{1}{k}\,\mathrm{Arch}\left(\frac{1}{\sqrt{1-k^2}}\right).$

3. $\int \sin 2xF(x,k)\,dx=\frac{1}{k^2}\left[(k^2\sin^2 x-1)F(x,k)+E(x,k)\right].$

4. $\int \frac{\sin x}{\cos^3 x}F(x,k)\,dx=\frac{1}{2}\,\mathrm{tg}^2 xF(x,k)+\frac{1}{2(1-k^2)}\left[E(x,k)-\mathrm{tg}\,x\sqrt{1-k^2\sin^2 x}\right].$

5. $\int \frac{\sin x}{\cos x}F(x,k)\,dx=\frac{1}{\cos x}F(x,k)-\frac{1}{\sqrt{1-k^2}}\ln\left(\sqrt{1-k^2}\,\mathrm{tg}\,x+\frac{\sqrt{1-k^2\sin^2 x}}{\cos x}\right).$

6. $\int \frac{\cos x}{\sin^2 x}F(x,k)\,dx=-\frac{1}{\sin x}F(x,k)-\ln\left(\mathrm{ctg}\,x+\frac{\sqrt{1-k^2\sin^2 x}}{\sin x}\right).$

7. $\int \frac{1}{\cos^2 x}F(x,k)\,dx=\mathrm{tg}\,xF(x,k)-\frac{1}{\sqrt{1-k}}\ln\left(\frac{\sqrt{1-k^2}}{\cos x}+\frac{\sqrt{1-k^2\sin^2 x}}{\cos x}\right).$

8. $\int_0^x \sqrt{1-k^2\sin^2 x}F(x,k)\,dx=F(x,k)\,E(x,k)-\frac{F^2(x,k)\,\mathbf{E}(k)}{2\mathbf{K}(k)}-\ln\frac{\theta_0[F(x,k)]}{\theta_0(0)}.$

9. $\int_0^x \frac{F(x,k)\,dx}{\sqrt{1-k^2\sin^2 x}}=\frac{1}{2}F^2(x,k)$ $\quad[0<x\leqslant\pi/2].$

10. $\int_0^x \frac{\sin 2x}{\sqrt{1-k^2\sin^2 x}}F(x,k)\,dx=\frac{2}{k^2}\left[x-F(x,k)\sqrt{1-k^2\sin^2 x}\right].$

11. $\int \frac{\sin x}{\cos^2 x\sqrt{1-k^2\sin^2 x}}F(x,k)\,dx=$

$=\frac{\sqrt{1-k^2\sin^2 x}}{(1-k^2)\cos x}F(x,k)-\ln\left(\sqrt{1-k^2}\,\mathrm{tg}\,x+\frac{\sqrt{1-k^2}}{\cos x}\right).$

12. $\int \frac{\cos x}{\sin^2 x\sqrt{1-k^2\sin^2 x}}F(x,k)\,dx=\frac{\sqrt{1-k^2\sin^2 x}}{\sin x}F(x,k)+\ln\mathrm{ctg}\frac{x}{2}.$

13. $\int \frac{\sin 2x}{(1-k^2\sin^2 x)^{3/2}}F(x,k)\,dx=-\frac{2x}{k^2}F(x,k)-\frac{2}{k^2\sqrt{1-k^2}}\,\mathrm{arctg}\left(\sqrt{1-k^2}\,\mathrm{tg}\,x\right).$

14. $$\int \frac{\cos x}{(1-k^2\sin^2 x)^{3/2}} F(x, k)\,dx = \frac{\sin x}{\sqrt{1-k^2\sin^2 x}} F(x, k) - \frac{1}{k\sqrt{1-k^2}} \operatorname{arctg} \frac{k\cos x}{\sqrt{1-k^2}}.$$

15. $$\int_0^x \sin x E(x, k)\,dx = -\cos x E(x, k) + \frac{1}{2k}\left[k\sin x\sqrt{1-k^2\sin^2 x} + \arcsin(k\sin x)\right].$$

16. $$\int_0^x \cos x E(x, k)\,dx = \sin x E(x, k) + \frac{1}{2k}\left[k\cos x\sqrt{1-k^2\sin^2 x} - (1-k^2)\operatorname{Arch}\sqrt{\frac{1-k^2\sin^2 x}{1-k^2}} - k + (1-k^2)\operatorname{Arch}\left(\frac{1}{\sqrt{1-k^2}}\right)\right].$$

17. $$\int_0^x \sqrt{1-k^2\sin^2 x}\, E(x, k)\,dx = \frac{1}{2} E^2(x, k).$$

18. $$\int_0^x \frac{E(x, k)\,dx}{\sqrt{1-k^2\sin^2 x}} = \frac{F^2(x, k)\,\mathrm{E}(k)}{2\mathrm{K}(k)} + \ln\frac{\theta_0[F(x, k)]}{\theta_0(0)}.$$

19. $$\int_0^x \frac{\sin 2x}{\sqrt{1-k^2\sin^2 x}} E(x, k)\,dx = k^{-2}\left[(2-k^2)x + k^2\sin x\cos x - 2E(x, k)\sqrt{1-k^2\sin^2 x}\right].$$

20. $$\int_0^x \sin x\,\Pi(x, \nu, k)\,dx = -\cos x\,\Pi(x, \nu, k) + \frac{1}{\sqrt{k^2-\nu}} \operatorname{arctg}\left(\sqrt{\frac{k^2-\nu}{1-k^2\sin^2 x}}\sin x\right) \qquad [\nu < k^2].$$

21. $$= -\cos x\,\Pi(x, \nu, k) + \frac{1}{\sqrt{\nu-k^2}} \operatorname{Arth}\left(\sqrt{\frac{\nu-k^2}{1-k^2\sin^2 x}}\sin x\right) \qquad [\nu > k^2].$$

22. $$\int_0^x \cos x\,\Pi(x, \nu, k)\,dx = \sin x\,\Pi(x, \nu, k) - f(x) - f(0),$$

$$f(x) = \frac{1}{2\sqrt{(1-\nu)(\nu-k^2)}} \operatorname{arctg}\left[\frac{2(1-\nu)(\nu-k^2) + (1-\nu\sin^2 x)(2k^2-\nu-\nu k^2)}{2\nu\sqrt{(1-\nu)(\nu-k^2)}\cos x\sqrt{1-k^2\sin^2 x}}\right] \qquad [(1-\nu)(\nu-k^2) > 0],$$

$$= \frac{1}{2\sqrt{(\nu-1)(\nu-k^2)}} \ln\left[\frac{2(\nu-1)(\nu-k^2) + (1-\nu\sin^2 x)(\nu+\nu k^2-2k^2)}{1-\nu\sin^2 x} + \frac{2\nu\sqrt{(\nu-1)(\nu-k^2)}\cos x\sqrt{1-k^2\sin^2 x}}{1-\nu\sin^2 x}\right] \qquad [(1-\nu)(\nu-k^2) < 0].$$

1.10.2. Integrals with respect to the modulus k.

1. $$\int kF(x, k)\,dk = E(x, k) - (1-k^2)F(x, k) + \left(\sqrt{1-k^2\sin^2 x} - 1\right)\operatorname{ctg} x.$$

2. $$\int kE(x, k)\,dk = \frac{1}{3}\left[(1+k^2)E(x, k) - (1-k^2)F(x, k) + \left(\sqrt{1-k^2\sin^2 x} - 1\right)\operatorname{ctg} x\right].$$

3. $$\int k\Pi(x, \nu, k)\,dk = (k^2-\nu)\Pi(x, \nu, k) - F(x, k) + E(x, k) + \left(\sqrt{1-k^2\sin^2 x} - 1\right)\operatorname{ctg} x.$$

1.11. THE COMPLETE ELLIPTIC INTEGRALS K(*k*), E(*k*) AND $\Pi(\pi/2, \nu, k)$

1.11.1. Integrals of the form $\int k^{\alpha}(1-k^2)^{\beta}\left\{\begin{matrix}\mathrm{K}(k)\\ \mathrm{E}(k)\end{matrix}\right\}dk$.

1. $\int k^{\alpha}\mathrm{K}(k)\,dk=J_{\alpha}$,

$$J_{\alpha}=\frac{1}{\alpha^2}\{(\alpha-1)^2 J_{\alpha-2}+k^{\alpha-1}[\mathrm{E}(k)-\alpha(1-k^2)\mathrm{K}(k)]\},$$

$$=\frac{\pi k^{\alpha+1}}{2(\alpha+1)}\,{}_3F_2\left(-\frac{1}{2},\ \frac{1}{2},\ \frac{\alpha+1}{2};\ 1,\ \frac{\alpha+3}{2};\ k^2\right),$$

$$J_1=\mathrm{E}(k)-(1-k^2)\mathrm{K}(k),$$

$$J_3=\frac{1}{9}[(4+k^2)\mathrm{E}(k)-(1-k^2)(4+3k^2)\mathrm{K}(k)],$$

$$J_5=\frac{1}{225}[(64+16k^2+9k^4)\mathrm{E}(k)-(1-k^2)(64+48k^2+45k^4)\mathrm{K}(k)].$$

$$J_{-2}=-\frac{\mathrm{E}(k)}{k}.$$

$$J_{-4}=-\frac{1}{9k^3}(1+4k^2)\mathrm{E}(k)-\frac{2(1-k^2)}{9k^3}\mathrm{K}(k).$$

2. $\int k^{\alpha}\mathrm{E}(k)\,dk=I_{\alpha}$,

$$I_{\alpha}=\frac{1}{\alpha(\alpha+2)}\{(\alpha-1)^2 I_{\alpha-2}-k^{\alpha-1}[\alpha(1-k^2)-2]\mathrm{E}(k)-k^{\alpha-1}(1-k^2)\mathrm{K}(k)\},$$

$$=\frac{\pi k^{\alpha+1}}{2(\alpha+1)}\,{}_3F_2\left(\frac{1}{2},\ \frac{1}{2},\ \frac{\alpha+1}{2};\ 1,\ \frac{\alpha+3}{2};\ k^2\right),$$

$$I_1=\frac{1}{3}[(1+k^2)\mathrm{E}(k)-(1-k^2)\mathrm{K}(k)],$$

$$I_3=\frac{1}{45}[(4+k^2+9k^4)\mathrm{E}(k)-(1-k^2)(4+3k^2)\mathrm{K}(k)],$$

$$I_5=\frac{1}{1575}[(64+16k^2+9k^4+225k^6)\mathrm{E}(k)-(1-k^2)(64+48k^2+45k^4)\mathrm{K}(k)],$$

$$I_{-2}=\frac{1}{k}[(1-k^2)\mathrm{K}(k)-2\mathrm{E}(k)],$$

$$I_{-4}=\frac{1}{9k^3}[2(k^2-2)\mathrm{E}(k)+(1-k^2)\mathrm{K}(k)].$$

3. $\int\frac{1}{1-k^2}\mathrm{E}(k)\,dk=k\mathrm{K}(k)$.

4. $\int\frac{k}{1-k^2}\mathrm{E}(k)\,dk=\mathrm{K}(k)-\mathrm{E}(k)$.

5. $\int\frac{k^3}{1-k^2}\mathrm{E}(k)\,dk=-\frac{1}{3}[(k^2-4)\mathrm{K}(k)+(k^2+4)\mathrm{E}(k)]$.

6. $\int\frac{k}{(1-k^2)^{3/2}}\mathrm{K}(k)\,dk=\frac{1}{\sqrt{1-k^2}}[\mathrm{K}(k)-\mathrm{E}(k)]$.

7. $\int\frac{k}{(1-k^2)^{3/2}}\mathrm{E}(k)\,dk=\frac{1}{\sqrt{1-k^2}}[(k^2-1)\mathrm{K}(k)+\mathrm{E}(k)]$.

8. $\int\frac{k}{(1-k^2)^{5/2}}\mathrm{E}(k)\,dk=\frac{1}{3(1-k^2)^{3/2}}[(1-k^2)\mathrm{K}(k)-(1-2k^2)\mathrm{E}(k)]$.

1.11.2. Various integrals containing $\mathbf{K}(k)$, $\mathbf{E}(k)$ and $\Pi(\pi/2, \nu, k)$

1. $\int [\mathbf{K}(k)-\mathbf{E}(k)]\frac{dk}{k} = -\mathbf{E}(k).$

2. $\int [\mathbf{E}(k)-(1-k^2)\mathbf{K}(k)]\frac{dk}{k^2(1-k^2)} = \frac{1}{k}[\mathbf{K}(k)-\mathbf{E}(k)].$

3. $\int [(1+k^2)\mathbf{E}(k)-(1-k^2)\mathbf{K}(k)]\frac{dk}{k(1-k^2)^2} = \frac{\mathbf{E}(k)}{1-k^2}.$

4. $\int \frac{k\mathbf{K}(k)\,dk}{[\mathbf{E}(k)-(1-k^2)\mathbf{K}(k)]^2} = \frac{1}{(1-k^2)\mathbf{K}(k)-\mathbf{E}(k)}.$

5. $\int k\Pi\left(\frac{\pi}{2}, \nu, k\right)dk = (k^2-\nu)\,\Pi\left(\frac{\pi}{2}, \nu, k\right)-\mathbf{K}(k)+\mathbf{E}(k).$

1.12. THE LEGENDRE FUNCTIONS $P_\nu^\mu(x)$ AND $Q_\nu^\mu(x)$

In the calculation of integrals containing the Legendre functions, the relations given in the Appendix II.18, can be used; in particular

$$P^\mu_{-\nu-1}(x) = P_\nu^\mu(x),$$

$$Q^\mu_{-\nu-1}(x) = \frac{1}{\sin(\mu-\nu)\pi}\left[\pi e^{i\mu\pi}\cos\nu\pi P_\nu^\mu(x)-\sin(\mu+\nu)\pi Q_\nu^\mu(x)\right].$$

1.12.1. Integrals of the form $\int f(x)\,P_\nu^\mu(x)\,dx$, $\int f(x)\,Q_\nu^\mu(x)\,dx$.

NOTATION: $R_\nu^\mu(x) = P_\nu^\mu(x)$ or $Q_\nu^\mu(x)$.

1. $\int \frac{(1-x)^{\lambda+\mu/2-1}}{(1+x)^{\mu/2}}P_\nu^\mu(x)\,dx = \frac{-(1-x)^\lambda}{\lambda\Gamma(1-\mu)}\,{}_3F_2\left(\begin{matrix}\lambda,\ -\nu,\ \nu+1;\ (1-x)/2\\ \lambda+1,\ 1-\mu\end{matrix}\right).$

2. $\int_0^x x^{\lambda-1}(1-x^2)^{-\mu/2}P_\nu^\mu(x)\,dx = \frac{2^\mu\sqrt{\pi}x^\lambda}{\lambda\Gamma((1-\mu-\nu)/2)\,\Gamma(1-(\mu-\nu)/2)}\times$

$$\times\,{}_3F_2\left(\begin{matrix}\lambda/2,\ (1+\mu+\nu)/2,\ (\mu-\nu)/2\\ 1+\lambda/2,\ 1/2;\ x^2\end{matrix}\right)-\frac{2^{\mu+1}\sqrt{\pi}x^{\lambda+1}}{(\lambda+1)\Gamma(-(\mu+\nu)/2)\,\Gamma(1-(\mu-\nu)/2)}\times$$

$$\times\,{}_3F_2\left(\begin{matrix}(1+\lambda)/2,\ (1+\mu-\nu)/2,\ 1+(\mu+\nu)/2\\ (3+\lambda)/2,\ 3/2;\ x^2\end{matrix}\right)\quad [0<x<1;\ \operatorname{Re}\lambda>0].$$

3. $\int \left(\frac{1\pm x}{1\mp x}\right)^{\mu/2}R_\nu^\mu(x)\,dx = \frac{1}{\nu(\nu+1)}\left(\frac{1\pm x}{1\mp x}\right)^{\mu/2}\left[(\nu x\pm\mu)R_\nu^\mu(x)-(\mu+\nu)R_{\nu-1}^\mu(x)\right].$

4. $\int (1\pm x)^{\mu/2}(1\mp x)^{\nu-\mu/2-1}R_\nu^\mu(x)\,dx = \frac{1}{2\nu(\mu-\nu)}(1\pm x)^{\mu/2}(1\mp x)^{\nu-\mu/2}\times$

$$\times\left\{(\mu-\nu-1)R_{\nu+1}^\mu(x)+[(2\nu+1)x\pm(\nu-\mu)]R_\nu^\mu(x)\right\}.$$

5. $\int (1\pm x)^{\mu/2}(1\mp x)^{-\mu/2-\nu-2}R_\nu^\mu(x)\,dx =$

$$= \frac{1}{2(\nu+1)(\mu+\nu+1)}(1\pm x)^{\mu/2}(1\mp x)^{-\mu/2-\nu-1}\times$$

$$\times\left[(\nu-\mu+1)R_{\nu+1}^\mu(x)\pm(\mu+\nu+1)R_\nu^\mu(x)\right].$$

6. $\int (1\pm x)^{-\mu/2}(1\mp x)^{\mu/2+\nu-1}R_\nu^\mu(x)\,dx =$

$$= \frac{(1\pm x)^{-\mu/2}(1\mp x)^{\mu/2+\nu}}{2\nu(\mu+\nu)}\left\{(\nu-\mu+1)R_{\nu+1}^\mu(x)-[(2\nu+1)x\pm(\mu+\nu)]R_\nu^\mu(x)\right\}.$$

7. $\int (1 \pm x)^{-\mu/2} (1 \mp x)^{\mu/2-\nu-2} R_\nu^\mu(x)\, dx =$

$$= \frac{1}{2(\nu+1)} (1 \pm x)^{-\mu/2} (1 \mp x)^{\mu/2-\nu-1} \left[R_{\nu+1}^\mu(x) \pm R_\nu^\mu(x)\right].$$

8. $\int (1-x^2)^{-\mu/2} R_\nu^\mu(x)\, dx = -(1-x^2)^{-(\mu-1)/2} R_\nu^{\mu-1}(x).$

9. $\int (1-x^2)^{\mu/2} R_\nu^\mu(x)\, dx = \frac{(1-x^2)^{(\mu+1)/2}}{(\nu-\mu)(\nu+\mu+1)} R_\nu^{\mu+1}(x).$

10. $\int (1-x^2)^{\nu/2-1} R_\nu^\mu(x)\, dx = \frac{1}{\mu^2-\nu^2}\left[(2\nu+1) x R_\nu^\mu(x) + (\mu-\nu-1) R_{\nu+1}^\mu(x)\right].$

11. $\int (1-x^2)^{-(\nu+3)/2} R_\nu^\mu(x)\, dx = \frac{(1-x^2)^{-(\nu+1)/2}}{\mu+\nu+1} R_{\nu+1}^\mu(x).$

12. $\int x(1-x^2)^{\pm\mu/2} R_\nu^\mu(x)\, dx =$

$$= \frac{(1-x^2)^{\pm\mu/2}}{(\nu\pm\mu+2)(\nu\mp\mu-1)} \{[(\nu\pm\mu+2)x^2-1)] R_\nu^\mu(x) + (\mu-\nu-1) x R_{\nu+1}^\mu(x)\}.$$

13. $\int x(1-x^2)^{(\nu-3)/2} R_\nu^\mu(x)\, dx =$

$$= \frac{(1-x^2)^{(\nu-1)/2}}{\mu^2-(\nu-1)^2} \{[(2\nu+1)x^2-1] R_\nu^\mu(x) + (\mu-\nu-1) x R_{\nu+1}^\mu(x)\}.$$

14. $\int x(1-x^2)^{-\nu/2-2} R_\nu^\mu(x)\, dx = \frac{(1-x^2)^{-\nu/2-1}}{\mu^2-(\nu+2)^2}\left[R_\nu^\mu(x) + (\mu-\nu-1) x R_{\nu+1}^\mu(x)\right].$

15. $\int x^{\nu\pm\mu-2} (1-x^2)^{\mp\mu/2} R_\nu^\mu(x)\, dx =$

$$= \frac{x^{\nu\pm\mu-1}(1-x^2)^{\mp\mu/2}}{(\nu\pm\mu)(\nu\pm\mu-1)} \{(\nu-\mu+1) x R_{\nu+1}^\mu(x) - [(2\nu+1)x^2-\nu\mp\mu] R_\nu^\mu(x)\}.$$

16. $\int x^{\pm\mu-\nu-3} (1-x^2)^{\mp\mu/2} R_\nu^\mu(x)\, dx =$

$$= \frac{x^{\pm\mu-\nu-1}(1-x^2)^{\mp\mu/2}}{(\nu\mp\mu+1)(\nu\mp\mu+2)} \left[(\nu-\mu+1) x R_{\nu+1}^\mu(x) - (\nu\mp\mu+1) R_\nu^\mu(x)\right].$$

17. $\int \ln\left|\frac{1+x}{1-x}\right| P_\nu(x)\, dx = -\frac{1-x^2}{\nu(\nu+1)}\left[\ln\left|\frac{1+x}{1-x}\right| P_\nu'(x) - \frac{2}{1-x^2} P_\nu(x)\right].$

1.12.2. Integrals containing products of Legendre functions.

NOTATION: $R_\nu^\mu(x)$, $\hat{R}_\nu^\mu(x) = P_\nu^\mu(x)$ or $Q_\nu^\mu(x)$.

1. $\int x R_\nu^\mu(x) \hat{R}_\nu^\mu(x)\, dx = \frac{1}{2\nu(\nu+1)} \{[\mu^2-(\nu+1)(\nu+x)^2] R_\nu^\mu(x) \hat{R}_\nu^\mu(x) -$

$$-(\mu-\nu-1)(\nu+1) x \left[R_\nu^\mu(x) \hat{R}_{\nu+1}^\mu(x) + R_{\nu+1}^\mu(x) \hat{R}_\nu^\mu(x)\right] -$$

$$-(\mu-\nu-1)^2 R_{\nu+1}^\mu(x) \hat{R}_{\nu+1}^\mu(x)\}.$$

2. $\int \frac{x}{(1-x^2)^{3/2}} R_\nu^\mu(x) \hat{R}_\nu^\mu(x)\, dx =$

$$= \frac{(1-x^2)^{-1/2}}{1-4\mu^2} \{[1-2\mu^2+2\nu(\nu+1)] R_\nu^\mu(x) \hat{R}_\nu^\mu(x) +$$

$$+(2\nu+1)(\mu-\nu-1) x \left[R_\nu^\mu(x) \hat{R}_{\nu+1}^\mu(x) + R_{\nu+1}^\mu(x) \hat{R}_\nu^\mu(x)\right] +$$

$$+2(\mu-\nu-1)^2 R_{\nu+1}^\mu(x) \hat{R}_{\nu+1}^\mu(x)\}.$$

3. $\int x(1-x^2)^{\pm\mu} R_\nu^\mu(x)\hat{R}_\nu^\mu(x)\,dx =$

$$=\frac{(1-x^2)^{\pm\mu}}{2(1\pm 2\mu)(\nu\pm\mu+1)(\pm\mu-\nu)}\{[\mu^2+(\nu+1)(\nu+x^2\pm 2\mu x^2)\mp$$
$$\mp\mu(1\pm 2\mu)(1-x^2)]R_\nu^\mu(x)\hat{R}_\nu^\mu(x)+(\mu-\nu-1)(\nu\pm\mu+1)x\times$$
$$\times[R_\nu^\mu(x)\hat{R}_{\nu+1}^\mu(x)+R_{\nu+1}^\mu(x)\hat{R}_\nu^\mu(x)]+(\mu-\nu-1)^2R_{\nu+1}^\mu(x)\hat{R}_{\nu+1}^\mu(x)\}.$$

4. $\int x(1-x^2)^{\nu-1}R_\nu^\mu(x)\hat{R}_\nu^\mu(x)\,dx=$

$$=\frac{(1-x^2)^\nu}{4\nu(\mu^2-\nu^2)}\{[\nu^2-\mu^2+(2\nu+1)^2x^2]R_\nu^\mu(x)\hat{R}_\nu^\mu(x)+$$
$$+(2\nu+1)(\mu-\nu-1)x[R_\nu^\mu(x)\hat{R}_{\nu+1}^\mu(x)+R_{\nu+1}^\mu(x)\hat{R}_\nu^\mu(x)]+$$
$$+(\mu-\nu-1)^2R_{\nu+1}^\mu(x)\hat{R}_{\nu+1}^\mu(x)\}.$$

5. $\int x(1-x^2)^{-\nu-2}R_\nu^\mu(x)\hat{R}_\nu^\mu(x)\,dx=$

$$=\frac{(1-x^2)^{-\nu-1}}{4(\nu+1)}\{R_\nu^\mu(x)\hat{R}_\nu^\mu(x)-[\mu^2-(\nu+1)^2]R_{\nu+1}^\mu(x)\hat{R}_{\nu+1}^\mu(x)\}.$$

1.13. THE WHITTAKER FUNCTIONS $M_{\rho,\sigma}(x)$ AND $W_{\rho,\sigma}(x)$

1.13.1. Integrals of the form $\int x^\alpha e^{\pm ax/2}M_{\rho,\sigma}(ax)\,dx$.

1. $\int_0^x x^{\alpha-1}e^{\pm ax/2}M_{\rho,\sigma}(ax)\,dx=$

$$=\frac{2a^{\sigma+1/2}x^{\alpha+\sigma+1/2}}{2\alpha+2\sigma+1}\,{}_2F_2\left(\begin{matrix}\alpha+\sigma+1/2,\ 1/2\mp\rho+\sigma;\ \pm ax\\ \alpha+\sigma+3/2,\ 2\sigma+1\end{matrix}\right)\quad [x,\ \mathrm{Re}\,(\alpha+\sigma+1/2)>0].$$

2. $\int_x^\infty x^{\alpha-1}e^{-ax/2}M_{\rho,\sigma}(ax)\,dx=$

$$=-\frac{2a^{\sigma+1/2}x^{\alpha+\sigma+1/2}}{2\alpha+2\sigma+1}\,{}_2F_2\left(\begin{matrix}\alpha+\sigma+1/2,\ \rho+\sigma+1/2;\ -ax\\ \alpha+\sigma+3/2,\ 2\sigma+1\end{matrix}\right)+$$
$$+a^{-\alpha}\Gamma\left[\begin{matrix}\rho-\alpha,\ 2\sigma+1,\ \alpha+\sigma+1/2\\ \rho+\sigma+1/2,\ \sigma-\alpha+1/2\end{matrix}\right]\quad [x,\ \mathrm{Re}\,a,\ \mathrm{Re}\,(\rho-\alpha)>0].$$

3. $\int x^{-\rho-2}e^{x/2}M_{\rho,\sigma}(x)\,dx=\frac{2}{2\sigma-2\rho-1}x^{-\rho-1}e^{x/2}M_{\rho+1,\sigma}(x).$

4. $\int x^{\sigma-1/2}e^{x/2}M_{\rho,\sigma}(x)\,dx=\frac{1}{2\sigma+1}x^\sigma e^{x/2}M_{\rho-1/2,\ \sigma+1/2}(x).$

5. $\int x^{-\sigma-1/2}e^{x/2}M_{\rho,\sigma}(x)\,dx=\frac{4\sigma}{2\sigma-2\rho+1}x^{-\sigma}e^{x/2}M_{\rho+1/2,\ \sigma-1/2}(x)\quad [\sigma\neq 0].$

6. $\int x^{\rho-2}e^{-x/2}M_{\rho,\sigma}(x)\,dx=\frac{2}{2\rho+2\sigma-1}x^{\rho-1}e^{-x/2}M_{\rho-1,\sigma}(x).$

7. $\int x^{\sigma-1/2}e^{-x/2}M_{\rho,\sigma}(x)\,dx=\frac{1}{2\sigma+1}x^\sigma e^{-x/2}M_{\rho-1/2,\ \sigma+1/2}(x).$

8. $\int x^{-\sigma-1/2}e^{-x/2}M_{\rho,\sigma}(x)\,dx=-\frac{4\sigma}{2\rho+2\sigma-1}x^{-\sigma}e^{-x/2}M_{\rho-1/2,\ \sigma-1/2}(x)\quad [\sigma\neq 0].$

1.13.2. Integrals of the form $\int x^{\alpha} e^{\pm ax/2} W_{\rho,\sigma}(ax)\,dx$.

1. $\int_0^x x^{\alpha-1} e^{\pm ax/2} W_{\rho,\sigma}(ax)\,dx =$

$$=\frac{2a^{\sigma+1/2}x^{\alpha+\sigma+1/2}}{2\alpha+2\sigma+1}\Gamma\begin{bmatrix}-2\sigma\\1/2-\rho-\sigma\end{bmatrix}{}_2F_2\left(\begin{matrix}\sigma\mp\rho+1/2,\ \alpha+\sigma+1/2;\ \pm ax\\2\sigma+1,\ \alpha+\sigma+3/2\end{matrix}\right)+$$

$$+\frac{2a^{1/2-\sigma}x^{\alpha-\sigma+1/2}}{2\alpha-2\sigma+1}\Gamma\begin{bmatrix}2\sigma\\\sigma-\rho+1/2\end{bmatrix}{}_2F_2\left(\begin{matrix}\alpha-\sigma+1/2,\ 1/2\mp\rho-\sigma;\ \pm ax\\1-2\sigma,\ \alpha-\sigma+3/2\end{matrix}\right)$$

$[\mathrm{Re}\,(\alpha+\sigma),\ \mathrm{Re}\,(\alpha-\sigma)>-1/2;\ x>0;\ |\arg a|<\pi]$.

2. $\int_x^\infty x^{\alpha-1} e^{\pm ax/2} W_{\rho,\sigma}(ax)\,dx =$

$$=-\frac{2a^{\sigma+1/2}x^{\alpha+\sigma+1/2}}{2\alpha+2\sigma+1}\Gamma\begin{bmatrix}-2\sigma\\1/2-\rho-\sigma\end{bmatrix}{}_2F_2\left(\begin{matrix}\alpha+\sigma+1/2,\ \sigma\mp\rho+1/2;\ \pm ax\\2\sigma+1,\ \alpha+\sigma+3/2\end{matrix}\right)-$$

$$-\frac{2a^{1/2-\sigma}x^{\alpha-\sigma+1/2}}{2\alpha-2\sigma+1}\Gamma\begin{bmatrix}2\sigma\\\sigma-\rho+1/2\end{bmatrix}{}_2F_2\left(\begin{matrix}\alpha-\sigma+1/2,\ 1/2\mp\rho-\sigma;\ \pm ax\\1-2\sigma,\ \alpha-\sigma+3/2\end{matrix}\right)+A_{\pm}a^{-\alpha}$$

$$\left[x>0,\ \left\{\begin{matrix}|\arg a|<3\pi/2;\ \mathrm{Re}\,(\alpha+\rho)<0\\\mathrm{Re}\,a>0\end{matrix}\right\},\ A_+=\Gamma\begin{bmatrix}\alpha+\sigma+1/2,\ \alpha-\sigma+1/2,\ -\alpha-\rho\\\sigma-\rho+1/2,\ 1/2-\rho-\sigma\end{bmatrix},\right.$$

$$\left.A_-=\Gamma\begin{bmatrix}\alpha+\sigma+1/2,\ \alpha-\sigma+1/2\\\alpha-\rho+1\end{bmatrix}\right].$$

3. $\int x^{-\rho-2} e^{x/2} W_{\rho,\sigma}(x)\,dx = \frac{4}{(2\rho+2\sigma+1)(2\rho-2\sigma+1)}x^{-\rho-1}e^{x/2}W_{\rho+1,\sigma}(x)$.

4. $\int x^{\sigma-1/2} e^{x/2} W_{\rho,\sigma}(x)\,dx = \frac{2}{2\rho+2\sigma+1}x^{\sigma}e^{x/2}W_{\rho+1/2,\ \sigma+1/2}(x)$.

5. $\int x^{-\sigma-1/2} e^{x/2} W_{\rho,\sigma}(x)\,dx = \frac{2}{2\rho-2\sigma+1}x^{-\sigma}e^{x/2}W_{\rho+1/2,\ \sigma-1/2}(x)$.

6. $\int x^{\rho-2} e^{-x/2} W_{\rho,\sigma}(x)\,dx = -x^{\rho-1}e^{-x/2}W_{\rho-1,\sigma}(x)$.

7. $\int x^{\sigma-1/2} e^{-x/2} W_{\rho,\sigma}(x)\,dx = -x^{\sigma}e^{-x/2}W_{\rho-1/2,\ \sigma+1/2}(x)$.

8. $\int x^{-\sigma-1/2} e^{-x/2} W_{\rho,\sigma}(x)\,dx = -x^{-\sigma}e^{-x/2}W_{\rho-1/2,\ \sigma-1/2}(x)$.

1.13.3. Integrals containing products of Whittaker functions.

NOTATION: $U_{\rho,\sigma}(x)$, $\hat{U}_{\rho,\sigma}(x)=M_{\rho,\sigma}(x)$ or $W_{\rho,\sigma}(x)$.

1. $\int_0^x x^{\alpha-1}\left\{\begin{matrix}M_{-\rho,\sigma}(ax)\\W_{-\rho,\sigma}(ax)\end{matrix}\right\}W_{\rho,\sigma}(ax)\,dx=\frac{ax^{\alpha+1}}{2(\alpha+1)\sigma}\Gamma^{(1\pm 1)/2}\begin{bmatrix}2\sigma+1\\1/2-\rho+\sigma\end{bmatrix}\times$

$$\times\left(-\frac{\cos\rho\pi}{\sin\sigma\pi}\right)^{(1\mp 1)/2}{}_3F_4\left(\begin{matrix}(\alpha+1)/2,\ 1/2+\rho,\ 1/2-\rho;\ a^2x^2/4\\1/2,\ (\alpha+3)/2,\ 1-\sigma,\ 1+\sigma\end{matrix}\right)-$$

$$-\frac{2\rho a^2x^{\alpha+2}}{(\alpha+2)(1-4\sigma^2)}\Gamma^{(1\pm 1)/2}\begin{bmatrix}2\sigma+1\\1/2-\rho+\sigma\end{bmatrix}\left(-\frac{\sin\rho\pi}{\cos\sigma\pi}\right)^{(1\mp 1)/2}\times$$

$$\times{}_3F_4\left(\begin{matrix}\alpha/2+1,\ 1+\rho,\ 1-\rho;\ a^2x^2/4\\3/2,\ 2-\alpha/2,\ 3/2-\sigma,\ 3/2+\sigma\end{matrix}\right)+$$

$$+\frac{a^{2\sigma+1}x^{\alpha+2\sigma+1}}{\alpha+2\sigma+1}\Gamma\begin{bmatrix}-2\sigma\\1/2-\rho-\sigma\end{bmatrix}\Gamma^{(1\mp1)/2}\begin{bmatrix}-2\sigma\\1/2+\rho-\sigma\end{bmatrix}\times$$

$$\times {}_3F_4\left(\begin{matrix}1/2+\rho+\sigma,\ 1/2-\rho+\sigma,\ \sigma+(\alpha+1)/2;\ a^2x^2/4\\ 1+\sigma,\ 1+2\sigma,\ 1/2+\sigma,\ \sigma+(\alpha+3)/2\end{matrix}\right)+$$

$$+\begin{Bmatrix}0\\1\end{Bmatrix}\frac{a^{1-2\sigma}x^{\alpha-2\sigma+1}}{1+\alpha-2\sigma}\Gamma\left[\begin{matrix}2\sigma,\quad 2\sigma\\ 1/2-\rho+\sigma,\ 1/2+\rho+\sigma\end{matrix}\right]\times$$

$$\times {}_3F_4\left(\begin{matrix}1/2+\rho-\sigma,\ 1/2-\rho-\sigma,\ (1+\alpha)/2-\sigma;\ a^2x^2/4\\ 1-\sigma,\ 1-2\sigma,\ 1/2-\sigma,\ (3+\alpha)/2-\sigma\end{matrix}\right)$$

$$\left[x>0;\ \operatorname{Re}\alpha>-1;\ \begin{Bmatrix}\operatorname{Re}(\alpha+2\sigma)>-1\\ \operatorname{Re}\alpha>2\,|\operatorname{Re}\sigma|-1\end{Bmatrix}\right].$$

2. $$\int_0^x x^{\alpha-1}\begin{Bmatrix}M_{\rho,\sigma}(-iax)\,M_{\rho,\sigma}(iax)\\ W_{\rho,\sigma}(-iax)\,W_{\rho,\sigma}(iax)\end{Bmatrix}dx=\frac{a^{2\sigma+1}x^{\alpha+2\sigma+1}}{\alpha+2\sigma+1}\Gamma^{1\mp1}\left[\begin{matrix}-2\sigma\\ 1/2-\rho-\sigma\end{matrix}\right]\times$$

$$\times {}_3F_4\left(\begin{matrix}1/2+\rho+\sigma,\ 1/2-\rho+\sigma,\ (1+\alpha)/2+\sigma;\ -a^2x^2/4\\ 1+2\sigma,\ 1/2+\sigma,\ 1+\sigma,\ \sigma+(\alpha+3)/2\end{matrix}\right)+$$

$$+\begin{Bmatrix}0\\1\end{Bmatrix}\frac{a^{1-2\sigma}x^{\alpha-2\sigma+1}}{\alpha-2\sigma+1}\Gamma^2\left[\begin{matrix}2\sigma\\ 1/2-\rho+\sigma\end{matrix}\right]\times$$

$$\times {}_3F_4\left(\begin{matrix}1/2+\rho-\sigma,\ 1/2-\rho-\sigma,\ (1+\alpha)/2-\sigma;\ -a^2x^2/4\\ 1-2\sigma,\ 1/2-\sigma,\ 1-\sigma,\ (3+\alpha)/2-\sigma\end{matrix}\right)-$$

$$-\begin{Bmatrix}0\\1\end{Bmatrix}\frac{\pi a x^{\alpha+1}}{2(\alpha+1)\sigma\sin\sigma\pi\,\Gamma(1/2-\rho+\sigma)\,\Gamma(1/2-\rho-\sigma)}\times$$

$$\times {}_3F_4\left(\begin{matrix}(1+\alpha)/2,\ 1/2+\rho,\ 1/2-\rho;\ -a^2x^2/4\\ 1/2,\ (3+\alpha)/2,\ 1-\sigma,\ 1+\sigma\end{matrix}\right)+$$

$$+\begin{Bmatrix}0\\1\end{Bmatrix}\frac{2\pi\rho a^2x^{\alpha+2}}{(\alpha+2)(1-4\sigma^2)\cos\sigma\pi\,\Gamma(1/2-\rho+\sigma)\,\Gamma(1/2-\rho-\sigma)}\times$$

$$\times {}_3F_4\left(\begin{matrix}\alpha/2+1,\ 1+\rho,\ 1-\rho;\ -a^2x^2/4\\ 3/2,\ \alpha/2+2,\ 3/2-\sigma,\ 3/2+\sigma\end{matrix}\right)\quad\left[x>0,\ \begin{Bmatrix}\operatorname{Re}(\alpha+2\sigma)>-1\\ \operatorname{Re}\alpha>2\,|\operatorname{Re}\sigma|-1\end{Bmatrix}\right].$$

3. $$\int\frac{1}{x}U_{\rho,\sigma}(x)\,\hat U_{\rho,\sigma}(x)\,dx=-U_{\rho,\sigma}(x)\frac{\partial}{\partial\rho}\hat U'_{\rho,\sigma}(x)+U'_{\rho,\sigma}(x)\frac{\partial}{\partial\rho}\hat U_{\rho,\sigma}(x).$$

4. $$\int\frac{1}{x}W_{\mu,\sigma}(x)\,W_{\rho,\sigma}(x)\,dx=\frac{1}{\mu-\rho}\left[W'_{\rho,\sigma}(x)\,W_{\mu,\sigma}(x)-W_{\rho,\sigma}(x)\,W'_{\mu,\sigma}(x)\right].$$

5. $$\int\frac{1}{x^2}U_{\rho,\sigma}(x)\hat U_{\rho,\sigma}(x)\,dx=\frac{1}{2\sigma}U_{\rho,\sigma}(x)\frac{\partial}{\partial\sigma}\hat U'_{\rho,\sigma}(x)-\frac{1}{2\sigma}U'_{\rho,\nu}(x)\frac{\partial}{\partial\sigma}\hat U_{\rho,\sigma}(x).$$

6. $$\int\frac{1}{x^2}W_{\rho,\sigma}(x)\,W_{\rho,\tau}(x)\,dx=\frac{1}{\tau^2-\sigma^2}\left[W_{\rho,\sigma}(x)\,W'_{\rho,\tau}(x)-W'_{\rho,\sigma}(x)\,W_{\sigma,\tau}(x)\right].$$

7. $$\int\left(\frac{b^2-a^2}{4}+\frac{\mu a-\rho b}{x}+\frac{\nu^2-\sigma^2}{x^2}\right)U_{\mu,\nu}(ax)\,\hat U_{\rho,\sigma}(bx)\,dx=$$
$$=U_{\mu,\nu}(ax)\,\hat U'_{\rho,\sigma}(bx)-U'_{\mu,\nu}(ax)\hat U_{\rho,\sigma}(bx).$$

8. $$\int\left(\frac{1}{2}-\frac{\rho}{x}\right)U_{\rho,\sigma}(x)\hat U_{\rho,\sigma}(x)\,dx=\left(\frac{\rho}{x}+\frac{1-4\sigma^2}{4x^2}-\frac{1}{4}\right)xU_{\rho,\sigma}(x)\,\hat U_{\rho,\sigma}(x)-$$
$$-xU'_{\rho,\sigma}(x)\,\hat U'_{\rho,\sigma}(x)+\frac{1}{2}\frac{d}{dx}\left[U_{\rho,\sigma}(x)\,\hat U_{\rho,\sigma}(x)\right].$$

9. $$\int\frac{1}{x}e^{-(a+b)x/2}M_{\rho,\sigma}(ax)\,W_{\rho+1,\sigma}(bx)\,dx=$$
$$=\frac{1}{(a-b)(\sigma-\rho-1/2)}\left[2\sigma\sqrt{\frac{a}{x}}e^{-(a+b)x/2}M_{\rho+1/2,\sigma-1/2}(ax)\,W_{\rho+1,\sigma}(bx)-\right.$$
$$\left.-\left(\rho+\sigma+\frac{1}{2}\right)\sqrt{\frac{b}{x}}e^{(a-b)x/2}M_{\rho+1,\sigma}(ax)\,W_{\rho+1/2,\sigma-1/2}(bx)\right].$$

1.14. THE CONFLUENT HYPERGEOMETRIC FUNCTIONS OF KUMMER ${}_1F_1(a; b; x)$ AND TRICOMI $\Psi(a, b; x)$

1.14.1. Integrals containing ${}_1F_1(a; b; x)$.

1. $$\int_{x_1}^{x_2} x^{\alpha-1}\,{}_1F_1(a;\ b;\ \pm\lambda x)\,dx=$$
$$=\pm\frac{x^{\alpha}}{\alpha}\,{}_2F_2(\alpha,\ a;\ \alpha+1;\ b;\ \pm\lambda x)+\lambda^{-\alpha}\Gamma\begin{bmatrix}\alpha,\ a-\alpha,\ b\\ a,\ b-\alpha\end{bmatrix}\begin{Bmatrix}0\\1\end{Bmatrix}$$
$$\left[\begin{Bmatrix}x_1=0,\ x_2=x;\ \operatorname{Re}\alpha>0\\ x_1=x,\ x_2=\infty;\ \operatorname{Re}\lambda,\ \operatorname{Re}(a-\alpha)>0\end{Bmatrix},\ x>0\right].$$

2. $$\int x^n\,{}_1F_1(a;\ b;\ x)\,dx=n!\sum_{k=1}^{n+1}\frac{(-1)^{k+1}(1-b)_k\,x^{n-k+1}}{(1-a)_k\,(n-k+1)!}\,{}_1F_1(a-k;\ b-k;\ x).$$

3. $$\int x^{a-n-2}\,{}_1F_1(a;b;\ x)\,dx=n!x^{a-n-1}\sum_{k=1}^{n+1}\frac{(-1)^k}{(1-a)_k\,(n-k+1)!}\,{}_1F_1(a-k;\ b;\ x).$$

4. $$\int x^{b+n-1}\,{}_1F_1(a;\ b;\ x)\,dx=n!\,x^{b+n}\sum_{k=1}^{n+1}\frac{(-1)^{k+1}}{(b)_k\,(n-k+1)!}\,{}_1F_1(a;\ b+k;\ x).$$

5. $$\int {}_1F_1(a;\ b;\ x)\,dx=\frac{b-1}{a-1}\,{}_1F_1(a-1;\ b-1;\ x).$$

6. $$\int x^{a-2}\,{}_1F_1(a;\ b;\ x)\,dx=\frac{1}{a-1}\,x^{a-1}\,{}_1F_1(a-1;\ b;\ x).$$

7. $$\int x^{b-1}\,{}_1F_1(a;\ b;\ x)\,dx=\frac{1}{b}\,x^b\,{}_1F_1(a;\ b+1;\ x).$$

8. $$\int_{x_1}^{x_2} x^{\alpha-1}e^{-\lambda x}\,{}_1F_1(a;\ b;\ \lambda x)\,dx=$$
$$=\pm\frac{x^{\alpha}}{\alpha}\,{}_2F_2(\alpha,\ b-a;\ \alpha+1,\ b;\ -\lambda x)+\lambda^{-\alpha}\Gamma\begin{bmatrix}\alpha,\ b,\ b-a-\alpha\\ b-a,\ b-\alpha\end{bmatrix}\begin{Bmatrix}0\\1\end{Bmatrix}$$
$$\left[\begin{Bmatrix}x_1=0,\ x_2=x;\ \operatorname{Re}\alpha>0\\ x_1=x,\ x_2=\infty;\ \operatorname{Re}\lambda,\ \operatorname{Re}(b-\alpha-a)>0\end{Bmatrix},\ x>0\right].$$

9. $$\int x^ne^{-x}\,{}_1F_1(a;\ b;\ x)\,dx=n!e^{-x}\sum_{k=1}^{n+1}\frac{(-1)^{k+1}(1-b)_kx^{n-k+1}}{(1+a-b)_k\,(n-k+1)!}\,{}_1F_1(a;\ b-k;\ x).$$

10. $$\int x^{b+n-1}e^{-x}\,{}_1F_1(a;\ b;\ x)\,dx=$$
$$=n!x^{b+n}e^{-x}\sum_{k=1}^{n+1}\frac{(-1)^{k+1}}{(b)_k\,(n-k+1)!}\,{}_1F_1(a+k;\ b+k;\ x).$$

11. $$\int x^{b-a-n-2}e^{-x}\,{}_1F_1(a;\ b;\ x)\,dx=$$
$$=n!x^{b-a-n-1}e^{-x}\sum_{k=1}^{n+1}\frac{(-1)^k}{(a-b+1)_k\,(n-k+1)!}\,{}_1F_1(a+k;\ b;\ x).$$

12. $$\int e^{-x}\,{}_1F_1(a;\ b;\ x)\,dx=\frac{b-1}{a-b+1}\,e^{-x}\,{}_1F_1(a;\ b-1;\ x).$$

13. $$\int x^{b-1}e^{-x}\,{}_1F_1(a;\ b;\ x)\,dx=\frac{1}{b}\,x^be^{-x}\,{}_1F_1(a+1;\ b+1;\ x).$$

14. $\int x^{b-a-2}e^{-x}\,{}_1F_1(a;\,b;\,x)\,dx=\frac{1}{b-a-1}\,x^{b-a-1}e^{-x}\,{}_1F_1(a+1;\,b;\,x).$

1.14.2. Integrals containing $\Psi(a,\,b;\,x)$.

1. $\int\limits_{x_1}^{x_2} x^{\alpha-1}\Psi(a,\,b;\,\lambda x)\,dx=\pm\frac{x^{\alpha}}{\alpha}\Gamma\begin{bmatrix}1-b\\ a-b+1\end{bmatrix}{}_2F_2(\alpha,\,a;\,\alpha+1,\,b;\,\lambda x)\pm$

$$\pm\frac{\lambda^{1-b}x^{\alpha-b+1}}{\alpha-b+1}\Gamma\begin{bmatrix}b-1\\ a\end{bmatrix}{}_2F_2(\alpha-b+1,\,a-b+1;\,2-b,\,\alpha-b+2;\,\lambda x)+$$

$$+\lambda^{-\alpha}\Gamma\begin{bmatrix}\alpha,\,\alpha-b+1,\,a-\alpha\\ a,\,a-b+1\end{bmatrix}\begin{Bmatrix}0\\ 1\end{Bmatrix}$$

$$\left[\begin{Bmatrix}x_1=0,\,x_2=x;\ \operatorname{Re}\alpha,\ \operatorname{Re}(\alpha-b+1)>0\\ x_1=x,\,x_2=\infty;\ \operatorname{Re}(a-\alpha)>0\end{Bmatrix},\ x>0;\ |\arg\lambda|<3\pi/2\right].$$

2. $\int x^{n}\Psi(a,\,b;\,x)\,dx=n!\sum_{k=1}^{n+1}\frac{(-1)^{k+1}x^{n-k+1}}{(1-a)_k\,(n-k+1)!}\Psi(a-k,\,b-k;\,x).$

3. $\int x^{a-n-2}\Psi(a,\,b;\,x)\,dx=n!x^{a-n-1}\sum_{k=1}^{n+1}\frac{1}{(1-a)_k\,(b-a)_k\,(n-k+1)!}\Psi(a-k,\,b;\,x).$

4. $\int x^{b+n-1}\Psi(a,\,b;\,x)\,dx=n!x^{b+n}\sum_{k=1}^{n+1}\frac{(-1)^{k+1}}{(b-a)_k\,(n-k+1)!}\Psi(a,\,b+k;\,x).$

5. $\int\Psi(a,\,b;\,x)\,dx=\frac{1}{1-a}\Psi(a-1,\,b-1;\,x).$

6. $\int x^{a-2}\Psi(a,\,b;\,x)\,dx=\frac{x^{a-1}}{(a-1)(a-b)}\Psi(a-1,\,b;\,x).$

7. $\int x^{b-1}\Psi(a,\,b;\,x)\,dx=\frac{x^{b}}{b-a}\Psi(a,\,b+1;\,x).$

8. $\int\limits_{x_1}^{x_2} x^{\alpha-1}e^{-\lambda x}\Psi(a,\,b;\,\lambda x)\,dx=$

$$=\pm\frac{x^{\alpha}}{\alpha}\Gamma\begin{bmatrix}1-b\\ 1+a-b\end{bmatrix}{}_2F_2(\alpha,\,b-a;\,\alpha+1,\,b;\,-\lambda x)\pm$$

$$\pm\frac{\lambda^{1-b}x^{\alpha-b+1}}{\alpha-b+1}\Gamma\begin{bmatrix}b-1\\ a\end{bmatrix}{}_2F_2(1-a,\,\alpha-b+1;\,2-b,\,\alpha-b+2;\,-\lambda x)+$$

$$+\lambda^{-\alpha}\Gamma\begin{bmatrix}\alpha,\,\alpha-b+1\\ \alpha+a-b+1\end{bmatrix}\begin{Bmatrix}0\\ 1\end{Bmatrix}\quad\left[\begin{Bmatrix}x_1=0,\,x_2=x,\ \operatorname{Re}\alpha,\ \operatorname{Re}(\alpha-b+1)>0;\ |\arg\lambda|<\pi\\ x_1=x,\,x_2=\infty;\ \operatorname{Re}\lambda>0\end{Bmatrix},\ x>0\right].$$

9. $\int x^{n}e^{-x}\Psi(a,\,b;\,x)\,dx=-n!e^{-x}\sum_{k=1}^{n+1}\frac{x^{n-k+1}}{(n-k+1)!}\Psi(a,\,b-k;\,x).$

10. $\int x^{b-a-n-2}e^{-x}\Psi(a,\,b;\,x)\,dx=n!x^{b-a-n-1}e^{-x}\sum_{k=1}^{n+1}\frac{1}{(n-k+1)!}\Psi(a+k,\,b;\,x).$

11. $\int x^{b+n-1}e^{-x}\Psi(a,\,b;\,x)\,dx=-n!x^{b+n}e^{-x}\sum_{k=1}^{n+1}\frac{1}{(n-k+1)!}\Psi(a+k,\,b+k;\,x).$

1.15 THE GAUSS HYPERGEOMETRIC FUNCTION ${}_2F_1(a,b;c;x)$

1.15.1. Integrals of the form $\int x^{\alpha}\,{}_2F_1(a,\ b;\ c;\ x)\,dx$.

1. $$\int_{x_1}^{x_2} x^{\alpha-1}\,{}_2F_1\left(\begin{matrix}a,\ b\\ c;\ -x\end{matrix}\right)dx=\pm\frac{x^{\alpha}}{\alpha}\,{}_3F_2\left(\begin{matrix}a,\ b,\ \alpha;\ -x\\ c,\ \alpha+1\end{matrix}\right)+\Gamma\left[\begin{matrix}\alpha,\ a-\alpha,\ b-\alpha,\ c\\ a,\ b,\ c-\alpha\end{matrix}\right]\begin{Bmatrix}0\\1\end{Bmatrix}$$

$$\left[\begin{Bmatrix}x_1=0,\ x_2=x;\ \mathrm{Re}\,\alpha>0\\ x_1=x,\ x_2=\infty;\ \mathrm{Re}\,(a-\alpha),\ \mathrm{Re}\,(b-\alpha)>0\end{Bmatrix},\ x>0\right].$$

2. $$\int x^{n}\,{}_2F_1\left(\begin{matrix}a,\ b\\ c;\ x\end{matrix}\right)dx=$$
$$=n!\sum_{k=1}^{n+1}(-1)^{k+1}\frac{(c-k)_k\,x^{n-k+1}}{(n-k+1)!\,(a-k)_k\,(b-k)_k}\,{}_2F_1\left(\begin{matrix}a-k,\ b-k\\ c-k;\ x\end{matrix}\right).$$

3. $$\int x^{a-n-2}\,{}_2F_1\left(\begin{matrix}a,\ b\\ c;\ x\end{matrix}\right)dx=n!x^{a-n-1}\sum_{k=1}^{n+1}\frac{1}{(n-k+1)!\,(a-k)_k}\,{}_2F_1\left(\begin{matrix}a-k,\ b\\ c;\ x\end{matrix}\right).$$

4. $$\int x^{c+n-1}\,{}_2F_1\left(\begin{matrix}a,\ b\\ c;\ x\end{matrix}\right)dx=n!\,x^{c+n}\sum_{k=1}^{n+1}\frac{(-1)^{k+1}}{(n-k+1)!\,(c)_k}\,{}_2F_1\left(\begin{matrix}a,\ b\\ c+k;\ x\end{matrix}\right).$$

5. $$\int {}_2F_1\left(\begin{matrix}a,\ b\\ c;\ x\end{matrix}\right)dx=\frac{c-1}{(a-1)(b-1)}\,{}_2F_1\left(\begin{matrix}a-1,\ b-1\\ c-1;\ x\end{matrix}\right).$$

6. $$\int x^{a-2}\,{}_2F_1\left(\begin{matrix}a,\ b\\ c;\ x\end{matrix}\right)dx=\frac{x^{a-1}}{a-1}\,{}_2F_1\left(\begin{matrix}a-1,\ b\\ c;\ x\end{matrix}\right).$$

7. $$\int x^{c-1}\,{}_2F_1\left(\begin{matrix}a,\ b\\ c;\ x\end{matrix}\right)dx=\frac{x^c}{c}\,{}_2F_1(a,\ b;\ c+1;\ x).$$

1.15.2. Integrals of the form $\int (1-x)^{\beta}\,{}_2F_1(a,\ b;\ c;\ x)\,dx$.

1. $$\int_x^1(1-x)^{\beta-1}{}_2F_1\left(\begin{matrix}a,\ b\\ c;\ x\end{matrix}\right)dx=\frac{(1-x)^{\beta}}{\beta}\,\Gamma\left[\begin{matrix}c,\ c-a-b\\ c-a,\ c-b\end{matrix}\right]{}_3F_2\left(\begin{matrix}\beta,\ a,\ b;\ 1-x\\ \beta+1,\ a+b-c+1\end{matrix}\right)+$$
$$+\frac{(1-x)^{\beta-a-b+c}}{\beta-a-b+c}\,\Gamma\left[\begin{matrix}c,\ a+b-c\\ a,\ b\end{matrix}\right]{}_3F_2\left(\begin{matrix}c-a,\ c-b,\ \beta-a-b+c;\ 1-x\\ \beta-a-b+c+1,\ c-a-b+1\end{matrix}\right)$$
$$[x<1;\ \mathrm{Re}\,\beta,\ \mathrm{Re}\,(\beta-a-b+c)>0].$$

2. $$\int_{-\infty}^x(1-x)^{\beta-1}{}_2F_1\left(\begin{matrix}a,\ b\\ c;\ x\end{matrix}\right)dx=\frac{(1-x)^{\beta-a}}{a-\beta}\,\Gamma\left[\begin{matrix}b-a,\ c\\ b,\ c-a\end{matrix}\right]{}_3F_2\left(\begin{matrix}a,\ a-\beta,\ c-b;\ 1/(1-x)\\ a-\beta+1,\ a-b+1\end{matrix}\right)+$$
$$+\frac{(1-x)^{\beta-b}}{b-\beta}\,\Gamma\left[\begin{matrix}a-b,\ c\\ a,\ c-b\end{matrix}\right]{}_3F_2\left(\begin{matrix}b,\ b-\beta,\ c-a;\ 1/(1-x)\\ b-\beta+1,\ b-a+1\end{matrix}\right)$$
$$[x<1;\ \mathrm{Re}\,(a-\beta),\ \mathrm{Re}\,(b-\beta)>0].$$

3. $$\int(1-x)^n\,{}_2F_1\left(\begin{matrix}a,\ b\\ c;\ x\end{matrix}\right)dx=n!\sum_{k=1}^{n+1}\frac{(c-k)_k\,(1-x)^{n-k+1}}{(n-k+1)!\,(a-k)_k\,(b-k)_k}\,{}_2F_1\left(\begin{matrix}a-k,\ b-k\\ c-k;\ x\end{matrix}\right).$$

4. $$\int(1-x)^{a+b-c+n}\,{}_2F_1\left(\begin{matrix}a,\ b\\ c;\ x\end{matrix}\right)dx=$$
$$=n!\,(1-x)^{a+b-c+n+1}\sum_{k=1}^{n+1}\frac{(c-k)_k}{(n-k+1)!\,(c-a-k)_k\,(c-b-k)_k}\,{}_2F_1\left(\begin{matrix}a,\ b\\ c-k;\ x\end{matrix}\right).$$

5. $$\int(1-x)^{a-n-2}\,{}_2F_1\left(\begin{matrix}a,\ b\\ c;\ x\end{matrix}\right)dx=$$

$$=n!\,(1-x)^{a-n-1}\sum_{k=1}^{n+1}\frac{(-1)^k (c-k)_k}{(n-k+1)!\,(a-k)_k\,(c-b-k)_k}\,{}_2F_1\left(\begin{matrix}a-k,\ b\\ c-k;\ x\end{matrix}\right).$$

6. $$\int (1-x)^{a-2}\,{}_2F_1\left(\begin{matrix}a,\ b\\ c;\ x\end{matrix}\right)dx=\frac{c-1}{(a-1)(b-c+1)}(1-x)^{a-1}\,{}_2F_1\left(\begin{matrix}a-1,\ b\\ c-1;\ x\end{matrix}\right).$$

7. $$\int (1-x)^{a+b-c}\,{}_2F_1\left(\begin{matrix}a,\ b\\ c;\ x\end{matrix}\right)dx=\frac{c-1}{(c-a-1)(c-b-1)}(1-x)^{a+b-c+1}\,{}_2F_1\left(\begin{matrix}a,\ b\\ c-1;\ x\end{matrix}\right).$$

1.15.3. Integrals of the form $\int x^{\alpha}(1-x)^{\beta}\,{}_2F_1(a,\ b;\ c;\ x)\,dx$.

1. $$\int_x^1 x^{c-1}(1-x)^{\beta-1}\,{}_2F_1\left(\begin{matrix}a,\ b\\ c;\ x\end{matrix}\right)dx=\frac{(1-x)^{\beta}}{\beta}\,\Gamma\left[\begin{matrix}c,\ c-a-b\\ c-a,\ c-b\end{matrix}\right]\times$$
$$\times\,{}_3F_2\left(\begin{matrix}\beta,\ a-c+1,\ b-c+1;\ 1-x\\ \beta+1,\ a+b-c+1\end{matrix}\right)+\frac{(1-x)^{\beta-a-b+c}}{\beta-a-b+c}\,\Gamma\left[\begin{matrix}c,\ a+b-c\\ a,\ b\end{matrix}\right]\times$$
$$\times\,{}_3F_2\left(\begin{matrix}1-a,\ 1-b,\ \beta-a-b+c;\ 1-x\\ \beta-a-b+c+1,\ c-a-b+1\end{matrix}\right)\qquad [x<1;\ \operatorname{Re}\beta,\ \operatorname{Re}(\beta-a-b+c)>0].$$

2. $$\int x^{n}(1-x)^{a-n-2}\,{}_2F_1\left(\begin{matrix}a,\ b\\ c;\ x\end{matrix}\right)dx=$$
$$=n!\,(1-x)^{a-n-1}\sum_{k=1}^{n+1}(-1)^{k+1}\frac{(c-k)_k\,x^{n-k+1}}{(n-k+1)!\,(a-k)_k\,(c-b-k)_k}\,{}_2F_1\left(\begin{matrix}a-k,\ b\\ c-k;\ x\end{matrix}\right).$$

3. $$\int x^{n}(1-x)^{a+b-c}\,{}_2F_1\left(\begin{matrix}a,\ b\\ c;\ x\end{matrix}\right)dx=$$
$$=n!\sum_{k=1}^{n+1}(-1)^{k+1}\frac{(c-k)_k\,x^{n-k+1}}{(c-a-k)_k\,(c-b-k)_k\,(n-k+1)!}(1-x)^{a+b-c+k}\,{}_2F_1\left(\begin{matrix}a,\ b\\ c+k;\ x\end{matrix}\right).$$

4. $$\int x^{a-n-2}(1-x)^{n}\,{}_2F_1\left(\begin{matrix}a,\ b\\ c;\ x\end{matrix}\right)dx=$$
$$=n!\,x^{a-n-1}\sum_{k=1}^{n+1}\frac{(1-x)^{n-k+1}}{(n-k+1)!\,(a-k)_k}\,{}_2F_1\left(\begin{matrix}a-k,\ b\\ c;\ x\end{matrix}\right).$$

5. $$\int x^{c-1}(1-x)^{n}\,{}_2F_1\left(\begin{matrix}a,\ b\\ c;\ x\end{matrix}\right)dx=$$
$$=n!\,(1-x)^{n+1}\sum_{k=1}^{n+1}\frac{x^{c+k-1}}{(n-k+1)!\,(c)_k}(1-x)^{k}\,{}_2F_1\left(\begin{matrix}a,\ b\\ c+k;\ x\end{matrix}\right).$$

6. $$\int x^{c-1}(1-x)^{a+b-c+n}\,{}_2F_1\left(\begin{matrix}a,\ b\\ c;\ x\end{matrix}\right)dx=$$
$$=n!\,(1-x)^{a+b-c+n+1}\sum_{k=1}^{n+1}\frac{x^{c+k-1}}{(n-k+1)!\,(c)_k}\,{}_2F_1\left(\begin{matrix}a+k,\ b+k\\ c+k;\ x\end{matrix}\right).$$

7. $$\int x^{c-1}(1-x)^{b-c-n-1}\,{}_2F_1\left(\begin{matrix}a,\ b\\ c;\ x\end{matrix}\right)dx=$$
$$=n!\,(1-x)^{b-c-n}\sum_{k=1}^{n+1}(-1)^{k+1}\frac{x^{c+k-1}}{(n-k+1)!\,(c)_k}\,{}_2F_1\left(\begin{matrix}a+k,\ b\\ c+k;\ x\end{matrix}\right).$$

8. $$\int x^{c-1}(1-x)^{b-c-1}\,{}_2F_1\left(\begin{matrix}a,\ b\\ c;\ x\end{matrix}\right)dx=\frac{1}{c}x^{c}(1-x)^{b-c}\,{}_2F_1\left(\begin{matrix}a+1,\ b\\ c+1;\ x\end{matrix}\right).$$

9. $$\int x^{c-1}(1-x)^{a+b-c}\,{}_2F_1\left(\begin{matrix}a,\ b\\ c;\ x\end{matrix}\right)dx=\frac{1}{c}x^{c}(1-x)^{a+b-c+1}\,{}_2F_1\left(\begin{matrix}a+1,\ b+1\\ c+1;\ x\end{matrix}\right).$$

10. $\int x^{c+n-1}(1-x)^{a+b-c}\,{}_2F_1\left(\begin{matrix}a,\ b\\ c;\ x\end{matrix}\right)dx=$

$$=n!\,x^{c+n}\sum_{k=1}^{n+1}\frac{(-1)^{k+1}}{(n-k+1)!\,(c)_k}(1-x)^{a+b-c+k}\,{}_2F_1\left(\begin{matrix}a+k,\ b+k\\ c+k;\ x\end{matrix}\right).$$

11. $\int x^{c+n-1}(1-x)^{a-c-n-1}\,{}_2F_1\left(\begin{matrix}a,\ b\\ c;\ x\end{matrix}\right)dx=$

$$=n!\,x^{c+n}\sum_{k=1}^{n+1}\frac{(-1)^{k+1}(1-x)^{a-c-k+1}}{(n-k+1)!\,(c)_k}\,{}_2F_1\left(\begin{matrix}a,\ b+k\\ c+k;\ x\end{matrix}\right).$$

12. $\int x^{c-a-n-2}(1-x)^{a+b-c}\,{}_2F_1\left(\begin{matrix}a,\ b\\ c;\ x\end{matrix}\right)dx=$

$$=n!\,x^{c-a-n-1}\sum_{k=1}^{n+1}\frac{(1-x)^{a+b-c+k}}{(n-k+1)!\,(c-a-k)_k}\,{}_2F_1\left(\begin{matrix}a+k,\ b\\ c;\ x\end{matrix}\right).$$

13. $\int x^{c-a-n-2}(1-x)^{a+b-c+n}\,{}_2F_1\left(\begin{matrix}a,\ b\\ c;\ x\end{matrix}\right)dx=$

$$=n!\,x^{c-a-n-1}\sum_{k=1}^{n+1}\frac{(1-x)^{a+b-c+k}}{(n-k+1)!\,(c-a-k)_k}\,{}_2F_1\left(\begin{matrix}a+k,\ b\\ c;\ x\end{matrix}\right).$$

14. $\int x^{c-a-2}(1-x)^{a+b-c}\,{}_2F_1\left(\begin{matrix}a,\ b\\ c;\ x\end{matrix}\right)dx=$

$$=\frac{1}{c-a-1}x^{c-a-1}(1-x)^{a+b-c+1}\,{}_2F_1\left(\begin{matrix}a+1,\ b\\ c;\ x\end{matrix}\right).$$

1.16. THE GENERALIZED HYPERGEOMETRIC FUNCTION ${}_pF_q((a_p);\ (b_q);\ x)$, THE MEIJER *G*-FUNCTION, THE MacROBERT *E*-FUNCTION AND THE FOX *H*-FUNCTION

1.16.1. Integrals containing ${}_pF_q((a_p);\ (b_q);\ x)$.

1. $\int x^{\alpha-1}\,{}_pF_q((a_p);\ (b_q);\ cx^n)\,dx=\frac{x^\alpha}{\alpha}\,{}_{p+1}F_{q+1}\left((a_p),\ \frac{\alpha}{n};\ (b_q),\ \frac{\alpha}{n}+1;\ cx^n\right).$

2. $\int {}_pF_q((a_p);\ (b_q);\ cx)\,dx=x\,{}_{p+1}F_{q+1}((a_p),\ 1;\ (b_q),\ 2;\ cx).$

3. $=\prod_{k=1}^{q}(b_k-1)\prod_{l=1}^{p}(a_l-1)^{-1}\,{}_pF_q((a_p)-1;\ (b_q)-1;\ cx).$

1.16.2. Integrals containing the Meijer *G*-function

1. $\int\limits_0^x x^{\alpha-1}G_{pq}^{mn}\left(cx\,\middle|\,\begin{matrix}(a_p)\\ (b_q)\end{matrix}\right)dx=x^\alpha G_{p+1,\ q+1}^{m,\ n+1}\left(cx\,\middle|\,\begin{matrix}1-\alpha,\ (a_p)\\ (b_q),\ -\alpha\end{matrix}\right)$

[$a_l-b_k\neq 1,\ 2,\ \ldots;\ k=1,\ \ldots,\ m;\ l=1,\ \ldots,\ n;\ \operatorname{Re}\alpha+\min\limits_{1\leqslant j\leqslant m}\operatorname{Re}b_j>0;$ 1) $c^*,\ x>0,$ $|\arg c|<c^*\pi,$ or 2) $c,\ x>0,\ c^*=0,\ p=q,\ \operatorname{Re}\mu<1,\ cx\neq 1,$ or 3) $c,\ x>0,\ c^*=0,\ p=q,$ $\operatorname{Re}\mu<2,\ cx=1,$ or 4) $c,\ x>0,\ c^*=0,\ p>q,\ \operatorname{Re}[(p-q)\alpha-\mu]>-3/2;\ c^*=m+n-(p+q)/2,$ $\mu=\sum\limits_{j=1}^{q}b_j-\sum\limits_{j=1}^{p}a_j+(p-q)/2+1$].

2. $\int x^{-a_1-1} G_{pq}^{mn}\left(cx \middle| \begin{matrix}(a_p)\\(b_q)\end{matrix}\right) dx = x^{-a_1} G_{pq}^{mn}\left(cx \middle| \begin{matrix}a_1+1,\ a_2,\ \ldots,\ a_p\\(b_q)\end{matrix}\right)$ $[n \geqslant 1]$.

3. $\int x^{-a_p-1} G_{pq}^{mn}\left(cx \middle| \begin{matrix}(a_p)\\(b_q)\end{matrix}\right) dx = -x^{-a_p} G_{pq}^{mn}\left(cx \middle| \begin{matrix}(a_{p-1}),\ a_p+1\\(b_q)\end{matrix}\right)$ $[n < p]$.

4. $\int x^{-b_1} G_{pq}^{mn}\left(cx \middle| \begin{matrix}(a_p)\\(b_q)\end{matrix}\right) dx = -x^{-b_1+1} G_{pq}^{mn}\left(cx \middle| \begin{matrix}(a_p)\\b_1-1,\ b_2,\ \ldots,\ b_q\end{matrix}\right)$ $[m \geqslant 1]$.

5. $\int x^{-b_q} G_{pq}^{mn}\left(cx \middle| \begin{matrix}(a_p)\\(b_q)\end{matrix}\right) dx = x^{-b_q+1} G_{pq}^{mn}\left(cx \middle| \begin{matrix}(a_p)\\(b_{q-1}),\ b_q-1\end{matrix}\right)$ $[m < q]$.

1.16.3. Integrals containing the MacRobert E-function.

1. $\int_0^x x^{\alpha-1} E((a_p);\ (b_q);\ cx)\, dx = -x^{\alpha} E((b_q),\ -\alpha;\ (a_p),\ 1-\alpha;\ cx)$

$\left[a_j \neq 0, -1, -2, \ldots;\ j=1, \ldots, m;\ \operatorname{Re}\left(2\alpha - \sum_{j=1}^{p} a_j + \sum_{j=1}^{p+1} b_j\right) > -1/2;\quad 1)\ p+1 > q,\right.$
$\left. x > 0,\ |\arg c| < (p-q+1)\pi/2, \text{ or } 2)\ p+1=q,\ c,\ x > 0\right]$.

2. $\int x^{-a_p} E((a_p);\ (b_q);\ x)\, dx = -x^{1-a_p} E((a_{p-1}),\ a_p-1;\ (b_q);\ x)$.

3. $\int x^{-2} E((a_p);\ (b_q);\ x)\, dx = E((a_p)-1;\ (b_q)-1;\ x)$.

1.16.4. Integrals containing the Fox H-function.

1. $\int_0^x x^{\alpha-1} H_{pq}^{mn}\left[cx \middle| \begin{matrix}[a_p,\ A_p]\\ [b_q,\ B_q]\end{matrix}\right] dx = x^{\alpha} H_{p+1,\ q+1}^{m,\ n+1}\left[cx \middle| \begin{matrix}[1-\alpha,\ 1],\ [a_p,\ A_p]\\ [b_q,\ B_q],\quad [-\alpha,\ 1]\end{matrix}\right]$

$\left[x > 0;\ |\arg c| < \left(\sum_{j=1}^{n} A_j - \sum_{j=n+1}^{p} A_j + \sum_{j=1}^{m} B_j - \sum_{j=m+1}^{q} B_j\right)\pi/2;\right.$

$\left. \operatorname{Re}\alpha + \min_{1 \leqslant j \leqslant m} \operatorname{Re}(b_j/B_j) > 0\right]$.

1.17. THE ELLIPTIC FUNCTIONS OF JACOBI AND WEIERSTRASS

1.17.1. Integrals of the form $\int f(\operatorname{sn} u)\, du$.

1. $\int \operatorname{sn}^n u\, du = \frac{1}{(n-1)k^2} \operatorname{cn} u \operatorname{dn} u \operatorname{sn}^{n-3} u +$
$+ \frac{(n-2)(1+k^2)}{(n-1)k^2} \int \operatorname{sn}^{n-2} u\, du - \frac{n-3}{(n-1)k^2} \int \operatorname{sn}^{n-4} u\, du.$

2. $\int \frac{du}{\operatorname{sn}^n u} = \frac{-1}{n-1} \frac{\operatorname{cn} u \operatorname{dn} u}{\operatorname{sn}^{n-1} u} + \frac{(n-2)(1+k^2)}{n-1} \int \frac{du}{\operatorname{sn}^{n-2} u} - \frac{(n-3)k^2}{n-1} \int \frac{du}{\operatorname{sn}^{n-4} u}.$

3. $\int \operatorname{sn} u\, du = \frac{1}{k} \ln(\operatorname{dn} u - k \operatorname{cn} u) = \frac{1}{k} \operatorname{Arsh}\left(k \frac{\operatorname{dn} u - \operatorname{cn} u}{1-k^2}\right) =$
$= -\frac{1}{k} \ln(\operatorname{dn} u + k \operatorname{cn} u).$

4. $\int \operatorname{sn}^2 u\, du = \frac{1}{k^2}[u - E(\operatorname{am} u,\ k)].$

5. $\int \operatorname{sn}^3 u\, du = \frac{1}{2k^2} \operatorname{cn} u \operatorname{dn} u + \frac{1+k^2}{2k^3} \ln(\operatorname{dn} u - k \operatorname{cn} u).$

6. $\int \frac{du}{\operatorname{sn} u} = \ln \frac{\operatorname{sn} u}{\operatorname{cn} u + \operatorname{dn} u} = \ln \frac{\operatorname{dn} u - \operatorname{cn} u}{\operatorname{sn} u}.$

7. $\int \frac{du}{\operatorname{sn}^2 u} = -\frac{\operatorname{cn} u \operatorname{dn} u}{\operatorname{sn} u} + u - E(\operatorname{am} u,\ k).$

8. $\int \frac{du}{\operatorname{sn}^3 u} = \frac{1}{2k^3} [k \operatorname{cn} u \operatorname{dn} u + (1+k^2) \ln (\operatorname{dn} u - k \operatorname{cn} u)].$

9. $$\int \frac{du}{(a+\operatorname{sn} u)^n} = \frac{-1}{(n-1)(1-a^2)(1-k^2a^2)} \frac{\operatorname{cn} u \operatorname{dn} u}{(a+\operatorname{sn} u)^{n-1}} - \frac{(2n-3)(1+k^2-2k^2a^2)a}{(n-1)(1-a^2)(1-k^2a^2)} \int \frac{du}{(a+\operatorname{sn} u)^{n-1}} + \frac{(n-2)(1+k^2-6k^2a^2)}{(n-1)(1-a^2)(1-k^2a^2)} \int \frac{du}{(a+\operatorname{sn} u)^{n-2}} + \frac{2(2n-5)k^2a}{(n-1)(1-a^2)(1-k^2a^2)} \int \frac{du}{(a+\operatorname{sn} u)^{n-3}} - \frac{(n-3)k^2}{(n-1)(1-a^2)(1-k^2a^2)} \int \frac{du}{(a+\operatorname{sn} u)^{n-4}}.$$

10. $$\int \frac{du}{(1 \pm \operatorname{sn} u)^n} = \frac{\mp 1}{(2n-1)(1-k^2)} \frac{\operatorname{cn} u \operatorname{dn} u}{(1 \pm \operatorname{sn} u)^n} + \frac{(n-1)(1-5k^2)}{(2n-1)(1-k^2)} \int \frac{du}{(1 \pm \operatorname{sn} u)^{n-1}} + \frac{2(2n-3)k^2}{(2n-1)(1-k^2)} \int \frac{du}{(1 \pm \operatorname{sn} u)^{n-2}} - \frac{(n-2)k^2}{(2n-1)(1-k^2)} \int \frac{du}{(1 \pm \operatorname{sn} u)^{n-3}}.$$

11. $$\int \frac{du}{(1 \pm k \operatorname{sn} u)^n} = \frac{\pm k}{(2n-1)(1-k^2)} \frac{\operatorname{cn} u \operatorname{dn} u}{(1 \pm k \operatorname{sn} u)^n} + \frac{(n-1)(5-k^2)}{(2n-1)(1-k^2)} \int \frac{du}{(1 \pm k \operatorname{sn} u)^{n-1}} - \frac{2(2n-3)}{(2n-1)(1-k^2)} \int \frac{du}{(1 \pm k \operatorname{sn} u)^{n-2}} + \frac{n-2}{(2n-1)(1-k^2)} \int \frac{du}{(1 \pm k \operatorname{sn} u)^{n-3}}.$$

12. $$\int \frac{du}{a+\operatorname{sn} u} = \frac{1}{a} \Pi\left(\operatorname{am} u,\ \frac{-1}{a^2},\ k\right) + \frac{1}{\sqrt{(1-a^2)(1-k^2a^2)}} \ln \frac{\sqrt{1-a^2} \operatorname{dn} u - \sqrt{1-k^2a^2} \operatorname{cn} u}{\sqrt{\operatorname{sn}^2 u - a^2}}.$$

13. $\int \frac{du}{1 \pm \operatorname{sn} u} = u - \frac{1}{1-k^2} E(\operatorname{am} u,\ k) \mp \frac{1}{1-k^2} \frac{\operatorname{cn} u \operatorname{dn} u}{1 \pm \operatorname{sn} u}.$

14. $\int \frac{du}{1 \pm k \operatorname{sn} u} = \frac{1}{1-k^2} E(\operatorname{am} u,\ k) \pm \frac{k}{1-k^2} \frac{\operatorname{cn} u \operatorname{dn} u}{1 \pm k \operatorname{sn} u}.$

1.17.2. Integrals of the form $\int f(\operatorname{cn} u)\, du$.

1. $$\int \operatorname{cn}^n u\, du = \frac{1}{(n-1)k^2} \operatorname{sn} u \operatorname{dn} u \operatorname{cn}^{n-3} u - \frac{(n-2)(1-2k^2)}{(n-1)k^2} \int \operatorname{cn}^{n-2} u\, du + \frac{(n-3)(1-k^2)}{(n-1)k^2} \int \operatorname{cn}^{n-4} u\, du.$$

2. $$\int \frac{du}{\operatorname{cn}^n u} = \frac{1}{(n-1)(1-k^2)} \frac{\operatorname{sn} u \operatorname{dn} u}{\operatorname{cn}^{n-1} u} + \frac{(n-2)(1-2k^2)}{(n-1)(1-k^2)} \int \frac{du}{\operatorname{cn}^{n-2} u} + \frac{(n-3)k^2}{(n-1)(1-k^2)} \int \frac{du}{\operatorname{cn}^{n-4} u}.$$

3. $$\int \operatorname{cn} u\, du = \frac{1}{k} \arccos (\operatorname{dn} u) = \frac{i}{k} \ln (\operatorname{dn} u - ik \operatorname{sn} u) = \frac{1}{k} \arcsin (k \operatorname{sn} u) = \frac{2}{k} \operatorname{arctg} \frac{k \operatorname{sn} u}{\operatorname{dn} u + 1}.$$

4. $\int \operatorname{cn}^2 u\, du = \frac{1}{k^2} [E(\operatorname{am} u,\ k) - (1-k^2) u].$

5. $\int \operatorname{cn}^3 u\, du = \frac{2k^2-1}{2k^3} \arcsin(k \operatorname{sn} u) + \frac{1}{2k^2} \operatorname{sn} u \operatorname{dn} u.$

6. $\int \frac{du}{\operatorname{cn} u} = \frac{1}{\sqrt{1-k^2}} \ln \frac{\sqrt{1-k^2} \operatorname{sn} u + \operatorname{dn} u}{\operatorname{cn} u} = \frac{1}{\sqrt{1-k^2}} \ln \frac{\operatorname{dn} u + \sqrt{1-k^2} \operatorname{sn} u}{\operatorname{dn} u - \sqrt{1-k^2} \operatorname{sn} u}.$

7. $\int \frac{du}{\operatorname{cn}^2 u} = \frac{1}{1-k^2} \frac{\operatorname{sn} u \operatorname{dn} u}{\operatorname{cn} u} + u - \frac{1}{1-k^2} E(\operatorname{am} u, k).$

8. $\int \frac{du}{\operatorname{cn}^3 u} = \frac{1}{2(1-k^2)} \frac{\operatorname{sn} u \operatorname{dn} u}{\operatorname{cn}^2 u} + \frac{1-2k^2}{2(1-k^2)^{3/2}} \ln \frac{\sqrt{1-k^2} \operatorname{sn} u + \operatorname{dn} u}{\operatorname{cn} u}.$

9. $\int \frac{du}{1 \pm \operatorname{cn} u} = u - E(\operatorname{am} u, k) \pm \frac{\operatorname{sn} u \operatorname{dn} u}{1 \pm \operatorname{cn} u}.$

1.17.3. Integrals of the form $\int f(\operatorname{dn} u)\, du$.

1. $\int \operatorname{dn}^n u\, du = \frac{k^2}{n-1} \operatorname{sn} u \operatorname{cn} u \operatorname{dn}^{n-3} u + \frac{(n-2)(2-k^2)}{n-1} \int \operatorname{dn}^{n-2} u\, du - \frac{(n-3)(1-k^2)}{n-1} \int \operatorname{dn}^{n-4} u\, du.$

2. $\int \operatorname{dn} u\, du = \arcsin(\operatorname{sn} u) = \operatorname{am} u.$

3. $\int \operatorname{dn}^2 u\, du = E(\operatorname{am} u, k).$

4. $\int \operatorname{dn}^3 u\, du = \frac{1}{2} [(2-k^2) \operatorname{am} u + k^2 \operatorname{sn} u \operatorname{cn} u].$

5. $\int \frac{du}{\operatorname{dn} u} = \frac{1}{\sqrt{1-k^2}} \arccos \frac{\operatorname{cn} u}{\operatorname{dn} u} = \frac{1}{\sqrt{1-k^2}} \arcsin \frac{\operatorname{sn} u}{\operatorname{dn} u}.$

6. $\int \frac{du}{\operatorname{dn}^2 u} = -\frac{k^2}{1-k^2} \frac{\operatorname{sn} u \operatorname{cn} u}{\operatorname{dn} u} + \frac{1}{1-k^2} E(\operatorname{am} u, k).$

7. $\int \frac{du}{\operatorname{dn}^3 u} = \frac{1}{2(1-k^2)^{3/2}} \left[(2-k^2) \operatorname{arctg} \frac{\sqrt{1-k^2} \operatorname{sn} u - \operatorname{cn} u}{\sqrt{1-k^2} \operatorname{sn} u + \operatorname{cn} u} - k^2 \sqrt{1-k^2} \frac{\operatorname{sn} u \operatorname{cn} u}{\operatorname{dn}^2 u} \right].$

8. $\int \frac{du}{1 \pm \operatorname{dn} u} = \frac{1}{k^2} \left[u - E(\operatorname{am} u, k) - \frac{\operatorname{cn} u (\operatorname{dn} u \mp 1)}{\operatorname{sn} u} \right].$

1.17.4. Integrals of the form $\int f(\operatorname{sn} u, \operatorname{cn} u, \operatorname{dn} u)\, du$.

1. $\int R(\operatorname{sn} u, \operatorname{cn} u, \operatorname{dn} u)\, du = \int R_1(\operatorname{sn} u)\, du + \int R_2(\operatorname{sn} u) \operatorname{cn} u\, du + \int R_3(\operatorname{sn} u) \operatorname{dn} u\, du + \int R_4(\operatorname{sn} u) \operatorname{cn} u \operatorname{dn} u\, du$

[R, R_1, R_2, R_3, R_4 are rational functions of their arguments].

2. $\int R_2(\operatorname{sn} u) \operatorname{cn} u\, du = \int R_2\left(\frac{2t}{1+k^2t^2}\right) \frac{2}{1+k^2t^2}\, dt$

$\left[\operatorname{sn} u = \frac{2t}{1+k^2t^2},\ \operatorname{cn} u\, du = \frac{2}{1+k^2t^2}\, dt,\ t = \frac{\operatorname{sn} u}{\operatorname{dn} u + 1}\right].$

3. $\int R_3(\operatorname{sn} u) \operatorname{dn} u\, du = \int R_3\left(\frac{2t}{1+t^2}\right) \frac{2}{1+t^2}\, dt$

$\left[\operatorname{sn} u = \frac{2t}{1+t^2},\ \operatorname{dn} u\, du = \frac{2}{1+t^2}\, dt,\ t = \frac{\operatorname{sn} u}{\operatorname{cn} u + 1}\right].$

4. $\int R_4(\operatorname{sn} u) \operatorname{cn} u \operatorname{dn} u\, du = \int R_4(x)\, dx$ $\quad [x = \operatorname{sn} u,\ dx = \operatorname{cn} u \operatorname{dn} u\, du].$

5. $\int \operatorname{sn} u \operatorname{cn} u \, du = -\frac{1}{k^2} \operatorname{dn} u.$

6. $\int \operatorname{sn} u \operatorname{dn} u \, du = -\operatorname{cn} u.$

7. $\int \operatorname{cn} u \operatorname{dn} u \, du = \operatorname{sn} u.$

8. $\int \operatorname{sn} u \operatorname{cn}^2 u \, du = \frac{1}{2k^3}\left[(1-k^2) \operatorname{Arch} \frac{\operatorname{dn} u}{\sqrt{1-k^2}} + k \operatorname{cn} u \operatorname{dn} u\right].$

9. $\int \operatorname{sn}^2 u \operatorname{cn} u \, du = \frac{1}{2k^3} [\operatorname{Arcsin}(k \operatorname{sn} u) - k \operatorname{sn} u \operatorname{dn} u].$

10. $\int \operatorname{dn}^2 u \operatorname{sn} u \, du = \frac{1}{2k}\left[(1-k^2) \operatorname{Arch} \frac{\operatorname{dn} u}{\sqrt{1-k^2}} - k \operatorname{cn} u \operatorname{dn} u\right].$

11. $\int \operatorname{sn}^2 u \operatorname{cn}^2 u \, du = \frac{1}{3k^4} [(2-k^2) E(\operatorname{am} u, k) - 2(1-k^2) u - k^2 \operatorname{sn} u \operatorname{cn} u \operatorname{dn} u].$

12. $\int \operatorname{sn}^2 u \operatorname{dn}^2 u \, du = \frac{1}{3k^2} [(2k^2-1) E(\operatorname{am} u, k) + (1-k^2) u - k^2 \operatorname{sn} u \operatorname{cn} u \operatorname{dn} u].$

13. $\int \operatorname{cn}^2 u \operatorname{dn}^2 u \, du = \frac{1}{3k^2} [(1+k^2) E(\operatorname{am} u, k) - (1-k^2) u + k^2 \operatorname{sn} u \operatorname{cn} u \operatorname{dn} u].$

14. $\int \frac{\operatorname{sn} u}{\operatorname{cn} u} du = \frac{1}{\sqrt{1-k^2}} \ln \frac{\operatorname{dn} u + \sqrt{1-k^2}}{\operatorname{cn} u} = \frac{1}{2\sqrt{1-k^2}} \ln \frac{\operatorname{dn} u + \sqrt{1-k^2}}{\operatorname{dn} u - \sqrt{1-k^2}}.$

15. $\int \frac{\operatorname{sn} u}{\operatorname{dn} u} du = \frac{i}{k\sqrt{1-k^2}} \ln \frac{i\sqrt{1-k^2} - k \operatorname{cn} u}{\operatorname{dn} u} =$

$$= \frac{1}{k\sqrt{1-k^2}} \operatorname{arctg} \frac{k \operatorname{cn} u}{\sqrt{1-k^2}} = \frac{1}{k\sqrt{1-k^2}} \operatorname{arctg} \frac{k\sqrt{1-k^2}(1+\operatorname{cn} u)}{k^2 \operatorname{cn} u + k^2 - 1}.$$

16. $\int \frac{\operatorname{cn} u}{\operatorname{sn} u} du = \ln \frac{1-\operatorname{dn} u}{\operatorname{sn} u} = \frac{1}{2} \ln \frac{1-\operatorname{dn} u}{1+\operatorname{dn} u} = \ln \frac{\operatorname{sn} u}{\operatorname{dn} u + 1}.$

17. $\int \frac{\operatorname{cn} u}{\operatorname{dn} u} du = -\frac{1}{k} \ln \frac{1-k \operatorname{sn} u}{\operatorname{dn} u} = \frac{1}{2k} \ln \frac{1+k \operatorname{sn} u}{1-k \operatorname{sn} u} = \frac{1}{k} \ln \frac{1+k \operatorname{sn} u}{\operatorname{dn} u}.$

18. $\int \frac{\operatorname{dn} u}{\operatorname{cn} u} du = \frac{1}{2} \ln \frac{1+\operatorname{sn} u}{1-\operatorname{sn} u} = \ln \frac{1+\operatorname{sn} u}{\operatorname{cn} u}.$

19. $\int \frac{\operatorname{dn} u}{\operatorname{sn} u} du = \frac{1}{2} \ln \frac{1-\operatorname{cn} u}{1+\operatorname{cn} u} = \ln \frac{\operatorname{sn} u}{\operatorname{cn} u + 1}.$

20. $\int \frac{\operatorname{sn}^2 u}{\operatorname{cn} u} du = \frac{1}{\sqrt{1-k^2}} \ln \frac{\sqrt{1-k^2} \operatorname{sn} u + \operatorname{dn} u}{\operatorname{cn} u} - \frac{1}{k} \arcsin(k \operatorname{sn} u).$

21. $\int \frac{\operatorname{dn}^2 u}{\operatorname{sn} u} du = \ln \frac{\operatorname{sn} u}{\operatorname{cn} u + \operatorname{dn} u} - k \operatorname{Arch} \frac{\operatorname{dn} u}{\sqrt{1-k^2}}.$

22. $\int \frac{\operatorname{cn}^2 u}{\operatorname{sn} u} du = \ln \frac{\operatorname{sn} u}{\operatorname{cn} u + \operatorname{dn} u} - \frac{1}{k} \operatorname{Arch} \frac{\operatorname{dn} u}{\sqrt{1-k^2}}.$

23. $\int \frac{\operatorname{dn}^2 u}{\operatorname{cn} u} du = k \arcsin(k \operatorname{sn} u) + \sqrt{1-k^2} \ln \frac{\operatorname{dn} u + \sqrt{1-k^2} \operatorname{sn} u}{\operatorname{cn} u}.$

24. $\int \frac{\operatorname{sn} u}{\operatorname{cn}^2 u} du = \frac{1}{1-k^2} \frac{\operatorname{dn} u}{\operatorname{cn} u}.$

25. $\int \frac{\operatorname{sn} u}{\operatorname{dn}^2 u} du = -\frac{1}{1-k^2} \frac{\operatorname{cn} u}{\operatorname{dn} u}.$

26. $\int \frac{\operatorname{cn} u}{\operatorname{sn}^2 u} du = -\frac{\operatorname{dn} u}{\operatorname{sn} u}.$

27. $\int \frac{\operatorname{cn} u}{\operatorname{dn}^2 u}\, du = \frac{\operatorname{sn} u}{\operatorname{dn} u}$.

28. $\int \frac{\operatorname{dn} u}{\operatorname{sn}^2 u}\, du = -\frac{\operatorname{cn} u}{\operatorname{sn} u}$.

29. $\int \frac{\operatorname{dn} u}{\operatorname{cn}^2 u}\, du = \frac{\operatorname{sn} u}{\operatorname{cn} u}$.

30. $\int \frac{\operatorname{cn} u \operatorname{dn} u}{\operatorname{sn} u}\, du = \ln \operatorname{sn} u$.

31. $\int \frac{\operatorname{sn} u \operatorname{dn} u}{\operatorname{cn} u}\, du = \ln \frac{1}{\operatorname{cn} u}$.

32. $\int \frac{\operatorname{sn} u \operatorname{cn} u}{\operatorname{dn} u}\, du = -\frac{1}{k^2} \ln \operatorname{dn} u$.

33. $\int \frac{\operatorname{cn} u}{\operatorname{sn} u \operatorname{dn} u}\, du = \ln \frac{\operatorname{sn} u}{\operatorname{dn} u}$.

34. $\int \frac{\operatorname{sn} u}{\operatorname{cn} u \operatorname{dn} u}\, du = \frac{1}{1-k^2} \ln \frac{\operatorname{dn} u}{\operatorname{cn} u}$.

35. $\int \frac{\operatorname{dn} u}{\operatorname{sn} u \operatorname{cn} u}\, du = \ln \frac{\operatorname{sn} u}{\operatorname{cn} u}$.

36. $\int \frac{\operatorname{sn}^2 u}{\operatorname{cn}^2 u}\, du = \frac{1}{1-k^2} \left[\frac{\operatorname{sn} u \operatorname{dn} u}{\operatorname{cn} u} - E(\operatorname{am} u, k) \right]$.

37. $\int \frac{\operatorname{cn}^2 u}{\operatorname{sn}^2 u}\, du = -\frac{\operatorname{cn} u \operatorname{dn} u}{\operatorname{sn} u} - E(\operatorname{am} u, k)$.

38. $\int \frac{\operatorname{sn}^2 u}{\operatorname{dn}^2 u}\, du = \frac{1}{k^2(1-k^2)} \left[E(\operatorname{am} u, k) - (1-k^2)\, u - k^2 \frac{\operatorname{sn} u \operatorname{cn} u}{\operatorname{dn} u} \right]$.

39. $\int \frac{\operatorname{dn}^2 u}{\operatorname{sn}^2 u}\, du = (1-k^2)\, u - \frac{\operatorname{cn} u \operatorname{dn} u}{\operatorname{sn} u} - E(\operatorname{am} u, k)$.

40. $\int \frac{\operatorname{cn}^2 u}{\operatorname{dn}^2 u}\, du = \frac{1}{k^2} \left[u - E(\operatorname{am} u, k) + k^2 \frac{\operatorname{sn} u \operatorname{cn} u}{\operatorname{dn} u} \right]$.

41. $\int \frac{\operatorname{dn}^2 u}{\operatorname{cn}^2 u}\, du = u - E(\operatorname{am} u, k) + \frac{\operatorname{sn} u \operatorname{dn} u}{\operatorname{cn} u}$.

1.17.5. Integrals containing the Weierstrass elliptic functions.

NOTATION: $\zeta(u)$ is the Weierstrass elliptic function.

1. $\int \wp^n(u)\, du =$

$$= \frac{1}{2(2n-1)} \wp^{n-2}(u)\, \wp'(u) + \frac{(2n-3)\, g_2}{4(2n-1)} \int \wp^{n-2}(u)\, du + \frac{(n-2)\, g_3}{2(2n-1)} \int \wp^{n-3}(u)\, du.$$

2. $\int \wp(u)\, du = -\zeta(u)$.

3. $\int \wp^2(u)\, du = \frac{1}{6} \wp'(u) + \frac{g_2}{12} u$.

4. $\int \wp^3(u)\, du = \frac{1}{10} \wp(u)\, \wp'(u) - \frac{3g_2}{20} \zeta(u) + \frac{g_3}{10} u$.

5. $$\int \frac{du}{[\wp(u)-\rho]^n} = \frac{-1}{(n-1)\, b} \frac{\wp'(u)}{[\wp(u)-\rho]^{n-1}} - \frac{3(2n-3)\, a}{2(n-1)\, b} \int \frac{du}{[\wp(u)-\rho]^{n-1}} -$$
$$- \frac{6(2n-4)\, a}{(n-1)\, b} \int \frac{du}{[\wp(u)-\rho]^{n-2}} - \frac{2(2n-5)}{(n-1)\, b} \int \frac{du}{[\wp(u)-\rho]^{n-1}}$$
$$\left[a = 4\rho^2 - \frac{1}{3} g_2, \quad b = 4\rho^3 - g_2\rho - g_3 \neq 0 \right].$$

6. $\int \frac{du}{\wp(u)-\wp(v)} = \frac{1}{\wp'(v)}\left[2u\zeta(v)+\ln\frac{\sigma(u-v)}{\sigma(u+v)}\right]$ $[\wp'(v) \neq 0]$.

7. $\int \frac{du}{[\wp(u)-e]^n} = \frac{-2}{3(2n-1)a}\frac{\wp(u)}{[\wp(u)-e]^n} - \frac{8(n-1)e}{(2n-1)a}\int\frac{du}{[\wp(u)-e]^{n-1}} - \frac{4(2n-3)}{3(2n-1)a}\int\frac{du}{[\wp(u)-e]^{n-2}}$

$\left[a=4e^2-\frac{1}{3}g_2,\ 4e^3-g_2e-g_3=0\right]$.

8. $\int \frac{du}{\wp(u)-e} = \frac{-2}{3a}\frac{\wp'(u)}{\wp(u)-e} - \frac{4}{3a}\zeta(u) - \frac{4e}{3a}u$ $\left[a=4e^2-\frac{1}{3}g_2,\ 4e^3-g_2e-g_3=0\right]$.

9. $\int \zeta(u)\,du = \ln\sigma(u)$.

10. $\int \frac{du}{[\wp(u)-\wp(v)]^2} = \frac{\wp''(v)}{\wp'^2(v)}\ln\frac{\sigma(u+v)}{\sigma(u-v)} - \frac{1}{\wp'^2(v)}\zeta(u+v) - \frac{1}{\wp'^2(v)}\zeta(u-v) - \left[\frac{2\wp(v)}{\wp'^2(v)}+\frac{2\wp''(v)\zeta(v)}{\wp'^3(v)}\right]$ $[\wp'(v) \neq 0]$.

CHAPTER 2. DEFINITE INTEGRALS

2.1. INTRODUCTION

This chapter contains definite integrals of specials functions not contained in [17], [18].

Integrals for which one of the limits of integration is variable in those cases when the integrand is independent of it, are located in Chapter 1. It should be noted that many of the integrals not appearing in this book can be calculated by the method set out in Chapter 8; alternatively they can be obtained as special cases of the integrals of more general form in sections 2.24–2.25. The conditions for the convergence of integrals given in the formulae ensure their existence either in the ordinary sense, or as improper integrals, or in the sense of principal value.

Some formulae become meaningless for particular values of the parameters since indeterminacies occur on the right hand sides; removal of the indeterminacies enables one to obtain formulae that hold for these values of the parameters.

At the beginning of the sections, as a rule, the integrals occur in general form where the argument of one of the integrands depends on the parameter $r>0$. For rational values of r, the right hand sides of the equalities can be expressed in terms of generalized hypergeometric functions. An example of this sort of transformation is given in [18], 2.1.1. Certain particular values of the integrals of general form occur among the formulae in the subsequent subsections.

2.2. THE GAMMA FUNCTION $\Gamma(x)$

(See also [18], 2.2)

2.2.1. Integrals along the line $(\gamma-i\infty, \gamma+i\infty)$.

(For other integrals of the form

$$\int_{\gamma-i\infty}^{\gamma+i\infty} \prod_{j,k,l,m} \frac{\Gamma(a_j+\alpha_j s)\,\Gamma(b_k-\beta_k s)}{\Gamma(c_l+\gamma_l s)\,\Gamma(d_m-\delta_m s)} z^{-s}\,ds$$

see also 8.4)

1. $$\int_{\gamma-i\infty}^{\gamma+i\infty} \Gamma[a_1+s, a_2+s, b_1-s, b_2-s]\,ds = 2\pi i\Gamma\begin{bmatrix} a_1+b_1,\ a_1+b_2,\ a_2+b_1,\ a_2+b_2 \\ a_1+a_2+b_1+b_2 \end{bmatrix}$$

[$-\operatorname{Re} a_1, -\operatorname{Re} a_2 < \gamma < \operatorname{Re} b_1, \operatorname{Re} b_2$; see also 8.4.49.20].

2. $$\int_{\gamma-i\infty}^{\gamma+i\infty} \Gamma\begin{bmatrix} a_1+s,\ a_2+s,\ b_1-s \\ 1+a_1-a_2-b_1+s \end{bmatrix} ds = \pi i\Gamma\begin{bmatrix} (a_1+b_1)/2,\ a_2+b_1 \\ 1-a_2+(a_1-b_1)/2 \end{bmatrix}$$

[$-\operatorname{Re} a_1, -\operatorname{Re} a_2 < \gamma < \operatorname{Re} b_1$].

3. $$\int_{\gamma-i\infty}^{\gamma+i\infty} \Gamma\begin{bmatrix} a+s,\ b-s \\ c+s,\ d-s \end{bmatrix} ds = 2\pi i\Gamma\begin{bmatrix} a+b,\ c+d-a-b-1 \\ c+d-1,\ c-a,\ d-b \end{bmatrix}$$

[$-\operatorname{Re} a < \gamma < \operatorname{Re} b$; $\operatorname{Re}(a+b-c-d) < -1$; see also 8.4.49.19].

4. $$\int_{\gamma-i\infty}^{\gamma+i\infty} \Gamma\begin{bmatrix} a_1+s,\ a_2+s,\ a_3+s,\ b_1-s,\ -s \\ a_1+a_2+a_3+b_1+s \end{bmatrix} ds = 2\pi i \Gamma\begin{bmatrix} a_1,\ a_2,\ a_3,\ a_1+b_1,\ a_2+b_1,\ a_3+b_1 \\ h-a_1,\ h-a_2,\ h-a_3 \end{bmatrix}$$

$[h=a_1+a_2+a_3+b_1;\ -\mathrm{Re}\,a_1,\ -\mathrm{Re}\,a_2,\ -\mathrm{Re}\,a_3 < \gamma < 0,\ \mathrm{Re}\,b_1]$.

5. $$\int_{\gamma-i\infty}^{\gamma+i\infty} \Gamma\begin{bmatrix} a+s,\ 1+a/2+s,\ b+s,\ c+s,\ d+s,\ b-a-s,\ -s \\ a/2+s,\ 1+a-c+s,\ 1+a-d+s \end{bmatrix} ds = \pi i \Gamma\begin{bmatrix} b,\ c,\ d,\ b+c-a,\ b+d-a \\ 1+a-c-d,\ b+c+d-a \end{bmatrix}$$

$[-\mathrm{Re}\,a,\ -\mathrm{Re}\,a/2-1,\ -\mathrm{Re}\,b,\ -\mathrm{Re}\,c,\ -\mathrm{Re}\,d < \gamma < \mathrm{Re}\,(b-a),\ 0]$.

6. $$\int_{\gamma-i\infty}^{\gamma+i\infty} \Gamma\begin{bmatrix} a+s,\ 1+a/2+s,\ b+s,\ c+s,\ d+s,\ e+s,\ f+s,\ b-a-s,\ -s \\ a/2+s,\ 1+a-c+s,\ 1+a-d+s,\ 1+a-e+s,\ 1+a-f+s \end{bmatrix} ds =$$

$$= \pi i \Gamma\begin{bmatrix} b,\ c,\ d,\ e,\ f,\ b+c-a,\ b+d-a,\ b+e-a,\ b+f-a \\ 1+a-d-e,\ 1+a-c-e,\ 1+a-c-d,\ 1+a-c-f,\ 1+a-d-f,\ 1+a-e-f \end{bmatrix}$$

$[2a=b+c+d+e+f-1;\ -\mathrm{Re}\,a,\ -\mathrm{Re}\,a/2-1,\ -\mathrm{Re}\,b,\ -\mathrm{Re}\,c,\ -\mathrm{Re}\,d,\ -\mathrm{Re}\,e,$ $-\mathrm{Re}\,f < \gamma < \mathrm{Re}\,(b-a),\ 0]$.

7. $$\int_{\gamma-i\infty}^{\gamma+i\infty} \Gamma\begin{bmatrix} a_1+s,\ a_2+s,\ \ldots,\ a_A+s,\ b_1-s,\ b_2-s,\ \ldots,\ b_B-s \\ c_1+s,\ c_2+s,\ \ldots,\ c_C+s,\ d_1-s,\ d_2-s,\ \ldots,\ d_D-s \end{bmatrix} z^{-s}\, ds = 2\pi i K(z)$$

$\left[\Delta=A+D-B-C,\ E=A+B-C-D,\ \nu=\sum_{j=1}^{A} a_j+\sum_{k=1}^{B} b_k-\sum_{l=1}^{C} c_l-\sum_{m=1}^{D} d_m;\right.$

$-\min_{1\leqslant j\leqslant A} \mathrm{Re}\,a_j < \mathrm{Re}\,s=\gamma < \min_{1\leqslant k\leqslant B} \mathrm{Re}\,b_k;\ A,\ B,\ C,\ D=0,\ 1,\ 2,\ \ldots;$

$0 \leqslant |\arg z| < \frac{\pi}{2}E$ (for $E>0$) or $|\arg z| = \frac{\pi}{2}E$ (for $E>0$

and $\gamma\Delta+\mathrm{Re}\,\nu-\frac{E}{2} < -1$) or $\arg z=0$ (for $E=0,\ \Delta\neq 0$ and $\gamma\Delta+\mathrm{Re}\,\nu < 1/2$)

or $\arg z=0$ (for $E=\Delta=0$ and $\mathrm{Re}\,\nu < 0,\ z\neq 1$ or $\mathrm{Re}\,\nu < -1,\ z=1$)],

$$K(z) = G_{B+C,\ A+D}^{A,\ B}\left(z \,\middle|\, \begin{matrix} 1-b_1,\ \ldots,\ 1-b_B,\ c_1,\ \ldots,\ c_C \\ a_1,\ \ldots,\ a_A,\ 1-d_1,\ \ldots,\ 1-d_D \end{matrix}\right),$$

$K(z)=\Sigma_A(z)$ $\quad[\Delta>0$ (or $\Delta=0,\ |z|<1$)],

$K(z)=\Sigma_B(1/z)$ $\quad[\Delta<0$ (or $\Delta=0,\ |z|>1$)],

$K(z)=\Sigma_A(z)=\Sigma_B(1/z)$ $\quad[z=(-1)^{A-C},\ \Delta=0,\ A\geqslant C,\ \mathrm{Re}\,\nu+C-A+1<0]$.

2.3. THE GENERALIZED ZETA FUNCTION $\zeta(s, x)$

2.3.1. Integrals of $f(x)\,\zeta(s,\ a+bx)$

1. $$\int_0^\infty x^{\alpha-1}\zeta(s,\ a+bx)\,dx = b^{-\alpha}B(\alpha,\ s-\alpha)\,\zeta(s-\alpha,\ a) \qquad [a,\ b>0;\ 0<\mathrm{Re}\,\alpha<\mathrm{Re}\,s-1].$$

2. $$\int_0^\infty x^{\alpha-1}\left[\zeta(s,\ x)-\frac{1}{x^s}\right]dx = B(\alpha,\ s-\alpha)\,\zeta(s-\alpha) \qquad [0<\mathrm{Re}\,\alpha<\mathrm{Re}\,s-1].$$

3. $$\int_0^1 \sin 2\pi x\,\zeta(s,\ x)\,dx = \frac{(2\pi)^s}{4\Gamma(s)}\operatorname{cosec}\frac{s\pi}{2} \qquad [1<\mathrm{Re}\,s<2].$$

2.4. THE POLYNOMIALS OF BERNOULLI $B_n(x)$ AND EULER $E_n(x)$

2.4.1. Integrals of $f(x) B_n(x)$.

1. $\int_0^1 B_n(x)\,dx = \delta_{n,0}$.

2. $\int_a^{a+1} B_n(x)\,dx = a^n$.

3. $\int_0^1 x^m B_n(x)\,dx = \sum_{k=0}^{n} \binom{n}{k} \frac{B_k}{m+n-k+1}$.

4. $\int_0^1 e^{ax} B_n(x)\,dx = (-1)^{n+1} \frac{e^a-1}{2a^{n+1}} \left(a \operatorname{cth} \frac{a}{2} - a - 2 \sum_{k=0}^{n} \frac{a^k}{k!} B_k \right)$ $[n \neq 0]$.

5. $\int_0^{1/2} \left\{ \begin{matrix} \sin 2m\pi x \\ \cos 2m\pi x \end{matrix} \right\} B_{2n-\delta}(x)\,dx = \pm (-1)^n \frac{(2n-\delta)!}{2(2m\pi)^{2n-\delta}}$ $\left[m, n \neq 0,\ \delta = \left\{ \begin{matrix} 1 \\ 0 \end{matrix} \right\} \right]$.

6. $\int_0^1 \left\{ \begin{matrix} \sin m\pi x \\ \cos m\pi x \end{matrix} \right\} B_n(x)\,dx = 0$ $\left[\left\{ \begin{matrix} m+n - \text{even} \\ m+n - \text{odd} \end{matrix} \right\} \right]$.

7. $\int_0^1 \left\{ \begin{matrix} \sin m\pi x \\ \cos m\pi x \end{matrix} \right\} B_{2n-\delta}(x)\,dx = \pm (-1)^n \frac{1+(-1)^m}{2} \frac{(2n-\delta)!}{(m\pi)^{2n-\delta}}$ $\left[m, n \neq 0,\ \delta = \left\{ \begin{matrix} 1 \\ 0 \end{matrix} \right\} \right]$.

8. $\int_0^1 \operatorname{ctg} \pi x B_{2n+1}(x)\,dx = (-1)^{n+1} 2 \frac{(2n+1)!}{(2\pi)^{2n+1}} \zeta(2n+1)$.

9. $\int_{-1/2}^{1/2} \left\{ \begin{matrix} \sin (2m+1) \pi x B_{2n}(x) \\ \cos (2m+1) \pi x B_{2n+1}(x) \end{matrix} \right\} dx = 0$.

2.4.2. Integrals containing products of Bernoulli polynomials.

1. $\int_0^{1/2} B_m(x) B_n(x)\,dx = (-1)^{m+1} \frac{m!\,n!}{(m+n)!\,2} B_{m+n}$ $\left[\begin{matrix} m, n \neq 0; \\ m+n - \text{even} \end{matrix} \right]$.

2. $\int_0^1 B_m(x) B_n(x)\,dx = (-1)^{m+1} \frac{m!\,n!}{(m+n)!} B_{m+n}$ $[m, n \neq 0]$.

3. $\int_0^1 B_m(x) B_n(x+a)\,dx = (-1)^{m+1} a^n m!\,n! \sum_{k=1}^{n} \frac{a^{-k}}{(m+k)!\,(n-k)!} B_{m+k}$ $[m, n \neq 0]$.

4. $$\int_0^1 B_{2m}(x)\,B_{2n}\left(x+\frac{1}{2}\right)dx=2\int_0^{1/2} B_{2m}(x)\,B_{2n}\left(x+\frac{1}{2}\right)dx=$$
$$=\frac{2^{2m+2n-1}-1}{2^{2m+2n-1}}\,\frac{(2m)!\,(2n)!}{(2m+2n)!}\,B_{2m+2n}\qquad [m,\ n\neq 0].$$

5. $$\int_0^1 B_{2m+1}(x)\,B_{2n+1}\left(x+\frac{1}{2}\right)dx=2\int_0^{1/2} B_{2m+1}(x)\,B_{2n+1}\left(x+\frac{1}{2}\right)dx=$$
$$=\frac{1-2^{2m+2n+1}}{2^{2m+2n+1}}\,\frac{(2m+1)!\,(2n+1)!}{(2m+2n+2)!}\,B_{2m+2n+2}.$$

6. $$\int_0^1 B_m(x)\,B_n(x)\,B_p(x)\,dx=$$
$$=(-1)^{p+1}\,p!\sum_{k=0}^{m+n}\left[m\binom{n}{2k}+n\binom{m}{2k}\right]\frac{(m+n-2k-1)!}{(m+n+p-2k)!}\,B_{2k}\,B_{m+n+p-2k}$$
$[m,\ n,\ p\neq 0]$.

7. $$\int_0^1 \prod_{k=1}^{m} B_{n_k}(x+a_k)\,dx=$$
$$=\sum_{k_1=1}^{n_1}\cdots\sum_{k_m=1}^{n_m}\binom{n_1}{k_1}\cdots\binom{n_m}{k_m}\frac{B_{n_1-k_1}(a_1)\ldots B_{n_m-k_m}(a_m)}{k_1+\ldots+k_m+1}$$
$[n_k\neq 0;\ k=1,\ \ldots,\ m]$.

8. $$\int_0^N \prod_{k=1}^{m} B_{n_k}\left(\frac{x}{p_k}\right)dx=\prod_{k=1}^{m} p_k^{1-n_k}\int_0^1 \prod_{k=1}^{m} B_{n_k}(x)\,dx$$
$\left[p_1,\ \ldots,\ p_m>0 \text{ are pairwise coprime integers; } N=\prod_{k=1}^{m} p_k;\ n_k\neq 0,\ k=1,\ \ldots,\ m\right]$.

2.4.3. Integrals of $f(x)\,E_n(x)$.

1. $$\int_0^1 x^m E_n(x)\,dx=4\,(-1)^n\,(2^{m+n+2}-1)\,\frac{m!\,n!}{(m+n+2)!}\,B_{m+n+2}+$$
$$+2\,(-1)^n\,m!\,n!\sum_{k=0}^{m-1}\frac{2^{n+k+2}-1}{(m-k)!\,(n+k+2)!}\,B_{n+k+2}.$$

2. $$\int_0^1 e^{ax}E_n(x)\,dx=(-1)^n\,\frac{e^a+1}{a^{n+2}}\left(a\,\mathrm{th}\,\frac{a}{2}-a-2\sum_{k=1}^{n+1}\frac{2^k-1}{k!}\,a^kB_k\right).$$

3. $$\int_0^1 \begin{Bmatrix}\sin m\pi x\\ \cos m\pi x\end{Bmatrix} E_n(x)\,dx=0\qquad \left[\begin{Bmatrix}m+n-\text{ even}\\ m+n-\text{ odd}\end{Bmatrix}\right].$$

4. $$\int_0^1 \begin{Bmatrix}\cos(2m+1)\pi x\\ \sin(2m+1)\pi x\end{Bmatrix} E_{2n+\delta}(x)\,dx=\frac{2\,(-1)^{n+\delta}\,(2n+\delta)!}{[(2m+1)\,\pi]^{2n+\delta+1}}\qquad \left[\delta=\begin{Bmatrix}1\\ 0\end{Bmatrix}\right].$$

5. $$\int_0^1 \sec\pi x\,E_{2n+1}(x)\,dx=(-1)^{n+1}\,\frac{(2n+1)!}{2^{4n+2}\,\pi^{2n+2}}\left[\zeta\left(2n+2,\ \frac{1}{4}\right)-\zeta\left(2n+2,\ \frac{3}{4}\right)\right].$$

2.4.4. Integrals of $E_m(x)\,E_n(x+a)$, $B_m(rx)\,E_n(x)$.

1. $$\int_0^1 E_m(x)\,E_n(x)\,dx = 4(-1)^n\left(2^{m+n+2}-1\right)\frac{m!\,n!}{(m+n+2)!}B_{m+n+2}.$$

2. $$\int_0^1 E_m(x)\,E_n(x+a)\,dx = 4(-1)^m a^n\,m!\,n!\sum_{k=0}^{n}\frac{2^{m+k+2}-1}{a^k(m+k+2)!\,(n-k)!}B_{m+k+2}.$$

3. $$\int_0^1 B_m(x)\,E_n(x)\,dx = 0 \qquad [m+n-\text{ odd }].$$

4. $$\int_0^1 B_{2m}\left(\frac{x}{2}\right)E_{2n+1}(x)\,dx = \frac{2^{2m+2n+2}-1}{2^{2m-1}}\,\frac{(2m)!\,(2n+1)!}{(2m+2n+2)!}B_{2m+2n+2} \qquad [m\neq 0].$$

5. $$\int_0^1 B_{2m+1}\left(\frac{x}{2}\right)E_{2n}(x)\,dx = \frac{1-2^{2m+2n+2}}{2^{2m}}\,\frac{(2m+1)!\,(2n)!}{(2m+2n+2)!}B_{2m+2n+2}.$$

2.5. THE POLYLOGARITHM $\mathrm{Li}_\nu(x)$

2.5.1. Integrals of general form

1. $$\int_0^a x^{\alpha-1}(a^r-x^r)^{\beta-1}\,\mathrm{Li}_n(-cx)\,dx = \frac{a^{\alpha+\beta r-r}}{r}\Gamma(\beta)\sum_{k=1}^{\infty}\frac{(-1)^k}{k^n}\Gamma\left[\begin{matrix}(\alpha+k)/r\\ \beta+(\alpha+k)/r\end{matrix}\right](ac)^k$$ $[r,\ \mathrm{Re}\,\beta>0;\ \mathrm{Re}\,\alpha>-1;\ |ac|<1]$.

2. $$\int_a^\infty x^{\alpha-1}(x^r-a^r)^{\beta-1}\,\mathrm{Li}_n(-cx)\,dx =$$
$$=\frac{a^{\alpha+\beta r-r}}{r}\Gamma(\beta)\sum_{k=1}^{\infty}\frac{(-1)^k}{k^n}\Gamma\left[\begin{matrix}1-\beta-(\alpha+k)/r\\ 1-(\alpha+k)/r\end{matrix}\right](ac)^k+$$
$$+\pi c^{r-\beta r-\alpha}\sum_{k=0}^{\infty}\frac{(r-\beta r-\alpha+kr)^{-n}(1-\beta)_k\,(ac)^{rk}}{k!\,\sin(\beta r+\alpha-r-kr)\pi}$$ $[r,\ \mathrm{Re}\,\beta>0;\ \mathrm{Re}(\alpha+\beta r)<r;\ |\arg c|<\pi;\ |ac|<1]$.

3. $$\int_0^\infty \frac{x^{\alpha-1}}{(x^r+z^r)^\rho}\,\mathrm{Li}_n(-cx)\,dx =$$
$$=\frac{z^{\alpha-\rho r}}{r\Gamma(\rho)}\sum_{k=1}^{\infty}\frac{(-1)^k}{k^n}\Gamma\left(\frac{\alpha+k}{r}\right)\Gamma\left(\rho-\frac{\alpha+k}{r}\right)(cz)^k-$$
$$-\pi c^{\rho r-\alpha}\sum_{k=0}^{\infty}\frac{(-1)^k(\rho)_k(\rho r-\alpha+kr)^{-n}}{k!\,\sin(\rho r-\alpha+kr)\pi}(cz)^{rk}$$ $[r>0;\ -1<\mathrm{Re}\,\alpha<r\mathrm{Re}\,\rho;\ |\arg c|<\pi;\ |cz|<1]$.

4. $$\int_0^\infty \frac{x^{\alpha-1}}{x^r-y^r}\,\mathrm{Li}_n(-cx)\,dx =$$
$$=-\frac{\pi y^{\alpha-r}}{r}\sum_{k=1}^{\infty}\frac{(-1)^k}{k^n}\operatorname{ctg}\frac{\alpha+k}{r}\pi\,(cy)^k+\pi c^{r-\alpha}\sum_{k=0}^{\infty}\frac{(r+kr-\alpha)^{-n}}{\sin(\alpha-r-kr)\pi}(cy)^{rk}$$ $[r,\ y>0;\ -1<\mathrm{Re}\,\alpha<r;\ |\arg c|<\pi;\ |cy|<1]$.

5. $$\int_0^\infty x^{\alpha-1}e^{-px^r}\mathrm{Li}_n(-cx)\,dx=\frac{(-1)^n\pi}{c^\alpha}\sum_{k=0}^\infty\frac{(-1)^k(\alpha+kr)^{-n}}{k!\sin(\alpha+kr)\pi}\left(\frac{p}{c^r}\right)^k-$$

$$-\frac{(-1)^n}{r}p^{-\alpha/r}\sum_{k=1}^\infty\frac{(-1)^k}{k^n}\Gamma\left(\frac{\alpha-k}{r}\right)\left(\frac{p}{c^r}\right)^{k/r}-$$

$$-\frac{p^{-\alpha/r}}{r^{n+1}}\sum_{k=0}^n\frac{1}{(n-k)!}\left(\ln\frac{c^r}{p}\right)^{n-k}\sum_{l=0}^{[k/2]}\frac{r^{2l}A_l}{(k-2l)!}\Gamma^{(k-2l)}\left(\frac{\alpha}{r}\right);$$

$$A_l=2\sum_{j=0}^{l-1}\frac{(-1)^j\Gamma^{(j)}(1)\,\Gamma^{(2l-j)}(1)}{j!\,(2l-j)!}+(-1)^l\left[\frac{\Gamma^{(l)}(1)}{l!}\right]^2$$

$[r,\ \mathrm{Re}\,p>0;\ \ \mathrm{Re}\,\alpha>-1;\ \ |\arg c|<\pi]$.

6. $$\int_0^\infty x^{\alpha-1}\begin{Bmatrix}\sin bx^r\\ \cos bx^r\end{Bmatrix}\mathrm{Li}_n(-cx)\,dx=$$

$$=\frac{(-1)^{n-1}\pi b^\delta}{c^{\alpha+\delta r}}\sum_{k=0}^\infty\frac{(-1)^k(2kr+\delta r+\alpha)^{-n}}{k!\sin(2kr+\delta r+\alpha)\pi}\,\frac{1}{(1/2+\delta)_k}\left(\frac{b}{2c^r}\right)^{2k}-$$

$$-\frac{2(-1)^n}{b^{\alpha/r}}\sum_{k=1}^\infty\frac{(-1)^k}{k^n}\Gamma\left(\frac{\alpha-k}{r}\right)\begin{Bmatrix}\sin[(\alpha-k)\pi/(2r)]\\ \cos[(\alpha-k)\pi/(2r)]\end{Bmatrix}\left(\frac{b}{c^r}\right)^{k/r}-$$

$$-\frac{1}{b^{\alpha/r}r^{n+1}}\left[\begin{Bmatrix}\sin[\alpha\pi/(2r)]\\ \cos[\alpha\pi/(2r)]\end{Bmatrix}\sum_{m=0}^{[n/2]}\frac{(-1)^m}{(2m)!}\left(\frac{\pi}{2}\right)^{2m}\sum_{k=0}^{n-2m}\frac{B_k}{k!}\left(\ln\frac{c^r}{b}\right)^k\pm\right.$$

$$\left.\pm\begin{Bmatrix}\cos[\alpha\pi/(2r)]\\ \sin[\alpha\pi/(2r)]\end{Bmatrix}\sum_{m=0}^{[(n-1)/2]}\frac{(-1)^m}{(2m+1)!}\left(\frac{\pi}{2}\right)^{2m+1}\sum_{k=0}^{n-2m-1}\frac{B_{k+1}}{k!}\left(\ln\frac{c^r}{b}\right)^k\right],$$

$$B_k=\sum_{l=0}^{[(n-k)/2-m]}\frac{r^{2l}A_l}{(n-k-2m-2l)!}\Gamma^{(n-k-2m-2l)}\left(\frac{\alpha}{r}\right)$$

$\left[A_l\ \text{see}\ 2.5.1.5;\ b,\ r>0;\ \ -1-\delta r<\mathrm{Re}\,\alpha<r;\ \ |\arg c|<\pi,\ \delta=\begin{Bmatrix}1\\0\end{Bmatrix}\right]$.

7. $$\int_0^\infty x^{\alpha-1}\mathrm{Li}_m(-bx^r)\,\mathrm{Li}_n(-cx)\,dx=$$

$$=\frac{(-1)^n\pi}{c^\alpha}\sum_{k=1}^\infty\frac{(-1)^k}{k^m}\,\frac{(\alpha+kr)^{-n}}{\sin(\alpha+kr)\pi}\left(\frac{b}{c^r}\right)^k+$$

$$+\frac{(-1)^{m+n}\pi}{b^{\alpha/r}r}\sum_{k=1}^\infty\frac{(-1)^k}{k^n}\left(\frac{r}{\alpha-k}\right)^m\operatorname{cosec}\frac{k-\alpha}{r}\pi\left(\frac{b}{c^r}\right)^{k/r}+$$

$$+\frac{(-1)^{m+n}r^{m-n-2}}{\alpha^{m-1}b^{\alpha/r}}\sum_{p=0}^n\frac{1}{p!}\left(\ln\frac{b}{c^r}\right)^p\sum_{l=0}^{[(n-p)/2]}A_lr^{2l}\times$$

$$\times\sum_{k=0}^{n-p-2l}\frac{(m-1)_k}{k!}\left(\frac{r}{\alpha}\right)^k\sum_{q=0}^{n-p-k-2l}\frac{(-1)^q\Gamma^{(q)}(\alpha/r)}{q!(n-p-k-q-2l)!}\Gamma^{(n-p-k-q-2l)}\left(-\frac{\alpha}{r}\right)$$

$[r>0;\ \ -r-1<\mathrm{Re}\,\alpha<0;\ \ |\arg b|,\ |\arg c|<\pi,\ |b|<|c|^r,\ A_l\ \text{see}\ 2.5.1.5]$.

2.5.2. Integrals of $A(x)\,\mathrm{Li}_n(-cx)$.

1. $$\int_0^\infty x^{\alpha-1}\mathrm{Li}_n(-cx)\,dx=\frac{(-\alpha)^{-n}\pi}{\sin\alpha\pi}c^{-\alpha}$$ $[-1<\operatorname{Re}\alpha<0;\ |\arg c|<\pi]$.

2. $$\int_0^a x^{\alpha-1}(a-x)^{\beta-1}\,\mathrm{Li}_n(-cx)\,dx=$$
$$=-a^{\alpha+\beta}c\mathrm{B}(\alpha+1,\beta)\,{}_{n+2}F_{n+1}(\alpha+1,1,1,\ldots,1;\ \alpha+\beta+1,2,\ldots,2;\ -ac)$$
$[a,\operatorname{Re}\beta>0;\ \operatorname{Re}\alpha>-1;\ |\arg c|<\pi]$.

3. $$\int_a^\infty x^{\alpha-1}(x-a)^{\beta-1}\,\mathrm{Li}_n(-cx)\,dx=$$
$$=-a^{\alpha+\beta}cB(\beta,-\alpha-\beta)\,{}_{n+2}F_{n+1}(\alpha+1,1,1,\ldots,1;\alpha+\beta+1,2,\ldots,2;-ac)-$$
$$-\frac{\pi c^{1-\alpha-\beta}}{\sin(\alpha+\beta)\pi\,(1-\alpha-\beta)^n}\,{}_{n+1}F_n(1-\beta,1-\alpha-\beta,\ldots,1-\alpha-\beta;$$
$$2-\alpha-\beta,\ldots,2-\alpha-\beta;-ac)$$ $[a,\operatorname{Re}\beta>0;\ \operatorname{Re}(\alpha+\beta)<1;\ |\arg c|<\pi]$.

4. $$\int_0^\infty\frac{x^{\alpha-1}}{(x+z)^\rho}\,\mathrm{Li}_n(-cx)\,dx=$$
$$=-cz^{\alpha-\rho+1}\mathrm{B}(\alpha+1,\rho-\alpha-1)\,{}_{n+2}F_{n+1}(\alpha+1,1,1,\ldots,1;\ 2-\rho+\alpha,2,\ldots,2;cz)+$$
$$+\frac{\pi c^{\rho-\alpha}}{(\rho-\alpha)^n\sin(\alpha-\rho)\pi}\,{}_{n+1}F_n(\rho,\rho-\alpha,\ldots,\rho-\alpha;\ 1+\rho-\alpha,\ldots,1+\rho-\alpha;\ cz)$$
$[-1<\operatorname{Re}\alpha<\operatorname{Re}\rho;\ |\arg c|,|\arg z|<\pi]$.

5. $$\int_0^\infty\frac{x^{\alpha-1}}{x-y}\,\mathrm{Li}_n(-cx)\,dx=\pi cy^\alpha\operatorname{ctg}\alpha\pi\,{}_{n+1}F_n(1,1,\ldots,1;2,\ldots,2;-cy)-$$
$$-\frac{\pi c^{1-\alpha}}{(1-\alpha)^n\sin\alpha\pi}\,{}_{n+1}F_n(1,1-\alpha,\ldots,1-\alpha;2-\alpha,\ldots,2-\alpha;-cy)$$
$[y>0;\ |\operatorname{Re}\alpha|<1;\ |\arg c|<\pi]$.

2.5.3. Integrals of $A(x)\,\mathrm{Li}_2(-cx)$.

1. $$\int_0^a x^{\alpha-1}\mathrm{Li}_2\left(\frac{x}{a}\right)dx=\frac{a^\alpha}{\alpha^2}\left[\frac{\pi^2\alpha}{6}-\psi(\alpha+1)-\mathbf{C}\right]$$ $[a>0;\ \operatorname{Re}\alpha>-1]$.

2. $$\int_0^a x^{n-1}\mathrm{Li}_2\left(\frac{x}{a}\right)dx=\frac{a^n}{n^2}\left[\frac{\pi^2 n}{6}-\sum_{k=1}^n\frac{1}{k}\right]$$ $[a>0]$.

3. $$\int_0^a \mathrm{Li}_2\left(\frac{x}{a}\right)dx=a\left(\frac{\pi^2}{6}-1\right)$$ $[a>0]$.

4. $$\int_0^a\frac{1}{(x-y)^2}\mathrm{Li}_2\left(\frac{x}{a}\right)dx=\frac{1}{y}\left[\frac{\pi^2 a}{6(y-a)}-\mathrm{Li}_2\left(\frac{a}{y}\right)-\frac{1}{2}\ln^2\left(1-\frac{a}{y}\right)\right]$$
$[0<a<y]$.

5. $$\int_0^\infty\frac{x^{-1/2}}{x+z}\mathrm{Li}_2(-cx)\,dx=\frac{4\pi}{\sqrt{z}}\mathrm{Li}_2(-\sqrt{cz})$$ $[|\arg c|,|\arg z|<\pi]$.

2.6. THE GENERALIZED FRESNEL INTEGRALS $S(x,\nu)$ AND $C(x,\nu)$

For $\nu=0$, the generalized Fresnel integrals $S(x,\nu), C(x,\nu)$ reduce to the integral sine and cosine,

$$S(x,0)=-\operatorname{si}(x),\quad C(x,0)=-\operatorname{ci}(x),$$

and for $\nu=\frac{1}{2}$, to the Fresnel integrals,

$$S\left(x,\frac{1}{2}\right)=\sqrt{2\pi}\left[\frac{1}{2}-S(x)\right],\quad C\left(x,\frac{1}{2}\right)=\sqrt{2\pi}\left[\frac{1}{2}-C(x)\right];$$

see [18] for the corresponding integrals.

2.6.1. Integrals of general form

NOTATION $\qquad \delta=\begin{Bmatrix}1\\0\end{Bmatrix}.$

1. $$\int_0^a x^{\alpha-1}(a^r-x^r)^{\beta-1}\begin{Bmatrix}S(cx,\nu)\\C(cx,\nu)\end{Bmatrix}dx=-\frac{a^{\alpha+\nu+\delta+r\beta-r}c^{\nu+\delta}}{r}\times$$
$$\times\Gamma(\beta)\sum_{k=0}^{\infty}\frac{(-1)^k}{k!(1/2+\delta)_k(\delta+\nu+2k)}\Gamma\left[\begin{matrix}(\alpha+\nu+\delta+2k)/r\\\beta+(\alpha+\nu+\delta+2k)/r\end{matrix}\right]\left(\frac{ac}{2}\right)^{2k}+$$
$$+\frac{a^{\alpha+r\beta-r}}{r}\mathrm{B}\left(\frac{\alpha}{r},\beta\right)\Gamma(\nu)\begin{Bmatrix}\sin(\nu\pi/2)\\\cos(\nu\pi/2)\end{Bmatrix}$$
$[a,\ r,\ \operatorname{Re}\alpha,\ \operatorname{Re}\beta>0;\ \operatorname{Re}(\alpha+\nu)>-\delta]$.

2. $$\int_a^{\infty} x^{\alpha-1}(x^r-a^r)^{\beta-1}\begin{Bmatrix}S(cx,\nu)\\C(cx,\nu)\end{Bmatrix}dx=-\frac{a^{\alpha+\nu+\delta+r\beta-r}c^{\nu+\delta}}{r}\Gamma(\beta)\times$$
$$\times\sum_{k=0}^{\infty}\frac{(-1)^k}{k!(1/2+\delta)_k(\delta+\nu+2k)}\Gamma\left[\begin{matrix}1-\beta-(\alpha+\nu+\delta+2k)/r\\1-(\alpha+\nu+\delta+2k)/r\end{matrix}\right]\left(\frac{ac}{2}\right)^{2k}-$$
$$-c^{r-r\beta-\alpha}\sum_{k=0}^{\infty}\frac{(ac)^{rk}(1-\beta)_k\Gamma(\alpha+\nu+r\beta-r-rk)}{k!(r-r\beta-\alpha+rk)}\begin{Bmatrix}\sin[(\alpha+\nu+r\beta-r-rk)\pi/2]\\\cos[(\alpha+\nu+r\beta-r-rk)\pi/2]\end{Bmatrix}+$$
$$+\frac{a^{\alpha+r\beta-r}}{r}\mathrm{B}\left(\beta,1-\beta-\frac{\alpha}{r}\right)\Gamma(\nu)\begin{Bmatrix}\sin(\nu\pi/2)\\\cos(\nu\pi/2)\end{Bmatrix}$$
$[a,\ c,\ r,\ \operatorname{Re}\beta>0;\ \operatorname{Re}(\alpha+r\beta+\nu)<r+2]$.

3. $$\int_0^{\infty}\frac{x^{\alpha-1}}{(x^r+z^r)^{\rho}}\begin{Bmatrix}S(cx,\nu)\\C(cx,\nu)\end{Bmatrix}dx=-\frac{c^{\nu+\delta}z^{\alpha+\nu+\delta-r\rho}}{r\Gamma(\rho)}\times$$
$$\times\sum_{k=0}^{\infty}\frac{(-1)^k}{k!(1/2+\delta)_k(\delta+\nu+2k)}\Gamma\left(\frac{\alpha+\nu+\delta+2k}{r}\right)\Gamma\left(\rho-\frac{\alpha+\nu+\delta+2k}{r}\right)\left(\frac{cz}{2}\right)^{2k}-$$
$$-c^{r\rho-\alpha}\sum_{k=0}^{\infty}\frac{(-1)^k(cz)^{rk}(\rho)_k\Gamma(\alpha+\nu-r\rho-rk)}{k!(r\rho-\alpha+rk)}\begin{Bmatrix}\sin[(\alpha+\nu-r\rho-rk)\pi/2]\\\cos[(\alpha+\nu-r\rho-rk)\pi/2]\end{Bmatrix}+$$
$$+\frac{z^{\alpha-r\rho}}{r}\mathrm{B}\left(\frac{\alpha}{r},\rho-\frac{\alpha}{r}\right)\Gamma(\nu)\begin{Bmatrix}\sin(\nu\pi/2)\\\cos(\nu\pi/2)\end{Bmatrix}$$
$[c,\ r,\ \operatorname{Re}\alpha>0;\ -\delta<\operatorname{Re}(\alpha+\nu)<r\operatorname{Re}\rho+2;\ r|\arg z|<\pi]$.

4. $$\int_0^{\infty}\frac{x^{\alpha-1}}{x^r-y^r}\begin{Bmatrix}S(cx,\nu)\\C(cx,\nu)\end{Bmatrix}dx=$$
$$=\frac{\pi c^{\nu+\delta}y^{\alpha+\nu+\delta-r}}{r}\sum_{k=0}^{\infty}\frac{(-1)^k}{k!(1/2+\delta)_k(\delta+\nu+2k)}\operatorname{ctg}\frac{\alpha+\nu+\delta+2k}{r}\pi\left(\frac{cy}{2}\right)^{2k}-$$

$$-c^{r-\alpha}\sum_{k=0}^{\infty}\frac{(cy)^{rk}\Gamma(\alpha+\nu-r-rk)}{r-\alpha+rk}\left\{\begin{matrix}\sin[(\alpha+\nu-r-rk)\pi/2]\\ \cos[(\alpha+\nu-r-rk)\pi/2]\end{matrix}\right\}+$$

$$+\frac{\pi y^{\alpha-r}}{r}\operatorname{ctg}\frac{\alpha\pi}{r}\Gamma(\nu)\left\{\begin{matrix}\sin(\nu\pi/2)\\ \cos(\nu\pi/2)\end{matrix}\right\}\qquad [c,\ r,\ y,\ \operatorname{Re}\alpha>0;\ -\delta<\operatorname{Re}(\alpha+\nu)<r+2].$$

5. $$\int_0^{\infty} x^{\alpha-1}e^{-px^r}\left\{\begin{matrix}S(cx,\ \nu)\\ C(cx,\ \nu)\end{matrix}\right\}dx=U(r)\qquad [c,\ r,\ \operatorname{Re}p,\ \operatorname{Re}\alpha>0;\ \operatorname{Re}(\alpha+\nu)>-\delta],$$

$$U(r)=-\frac{c^{\nu+\delta}}{p^{(\alpha+\nu+\delta)/r}r}\sum_{k=0}^{\infty}\frac{(-1)^k}{k!\,(\delta+1/2)_k(\nu+\delta+2k)}\Gamma\left(\frac{\alpha+\nu+\delta+2k}{r}\right)\left(\frac{c}{2p^{1/r}}\right)^{2k}+$$

$$+\frac{\Gamma(\alpha/r)\Gamma(\nu)}{rp^{\alpha/r}}\left\{\begin{matrix}\sin(\nu\pi/2)\\ \cos(\nu\pi/2)\end{matrix}\right\}\qquad [r>1],$$

$$U(r)=\frac{1}{c^{\alpha}}\sum_{k=0}^{\infty}\frac{1}{k!\,(\alpha+rk)}\Gamma(\alpha+\nu+rk)\left\{\begin{matrix}\sin[(\alpha+\nu+rk)\pi/2]\\ \cos[(\alpha+\nu+rk)\pi/2]\end{matrix}\right\}\left(-\frac{p}{c^r}\right)^k\qquad [r<1],$$

$U(1)$ see 2.6.3.1.

6. $$\int_0^{\infty} x^{\alpha-1}e^{-px^{-r}}\left\{\begin{matrix}S(cx,\ \nu)\\ C(cx,\ \nu)\end{matrix}\right\}dx=$$

$$=-\frac{c^{\nu+\delta}}{r}p^{(\alpha+\nu+\delta)/r}\sum_{k=0}^{\infty}\frac{(-1)^k}{k!\,(\delta+1/2)_k(\nu+\delta+2k)}\Gamma\left(-\frac{\alpha+\nu+\delta+2k}{r}\right)\left(\frac{cp^{1/r}}{2}\right)^{2k}+$$

$$+\frac{\Gamma(\nu)\Gamma(-\alpha/r)}{r}p^{\alpha/r}\left\{\begin{matrix}\sin(\nu\pi/2)\\ \cos(\nu\pi/2)\end{matrix}\right\}+\frac{1}{c^{\alpha}}\sum_{k=0}^{\infty}\frac{\Gamma(\alpha+\nu-rk)}{k!\,(\alpha-rk)}\left\{\begin{matrix}\sin[(\alpha+\nu-rk)\pi/2]\\ \cos[(\alpha+\nu-rk)\pi/2]\end{matrix}\right\}(-pc^r)^k$$

$[c,\ r,\ \operatorname{Re}p>0;\ \operatorname{Re}(\alpha+\nu)<2]$.

7. $$\int_0^{\infty} x^{\alpha-1}\left\{\begin{matrix}\sin bx^r\\ \cos bx^r\end{matrix}\right\}S(cx,\ \nu)\,dx=U(1)$$

$[b,\ c,\ r>0;\ \operatorname{Re}\alpha>-\delta r;\ -\delta r-1<\operatorname{Re}(\alpha+\nu)<1+\max(r,\ 1)]$,

$$U(\gamma)=-\frac{c^{\nu+\gamma}}{b^{(\alpha+\nu+\gamma)/r}r}\sum_{k=0}^{\infty}\frac{(-1)^k}{k!\,(1/2+\gamma)_k(\nu+\gamma+2k)}\times$$

$$\times\Gamma\left(\frac{\alpha+\nu+\gamma+2k}{r}\right)\left\{\begin{matrix}\sin[(\alpha+\nu+\gamma+2k)\pi/(2r)]\\ \cos[(\alpha+\nu+\gamma+2k)\pi/(2r)]\end{matrix}\right\}\left(\frac{c}{2b^{1/r}}\right)^{2k}+$$

$$+\frac{\Gamma(\nu)\Gamma(\alpha/r)}{b^{\alpha/r}r}\cos\frac{\nu-\gamma}{2}\pi\left\{\begin{matrix}\sin[\alpha\pi/(2r)]\\ \cos[\alpha\pi/(2r)]\end{matrix}\right\}\qquad [r>1],$$

$$U(\gamma)=\frac{b^{\delta}}{c^{\alpha+\delta r}}\sum_{k=0}^{\infty}\frac{(-1)^k\Gamma(\alpha+\nu+\delta r+2rk)}{k!\,(1/2+\delta)_k(\alpha+\delta r+2rk)}\cos\frac{\alpha+\nu+\delta r+2rk-\gamma}{2}\pi\left(\frac{b}{2c^r}\right)^{2k}\qquad [r<1],$$

$U(\gamma)$ for $r=1$ See also 2.6.4.1–2.

8. $$\int_0^{\infty} x^{\alpha-1}\left\{\begin{matrix}\sin bx^r\\ \cos bx^r\end{matrix}\right\}C(cx,\ \nu)\,dx=U(0)$$

$[b,\ c,\ r>0;\ \operatorname{Re}\alpha>-\delta r;\ -\delta r<\operatorname{Re}(\alpha+\nu)<1+\max(r,\ 1);\ U(\gamma)$ see 2.6.1.7].

9. $$\int_0^\infty x^{\alpha-1} \left\{ \begin{matrix} \ln(x^r+z^r) \\ \ln|x^r-z^r| \end{matrix} \right\} S(cx,\ \nu)\,dx = V(1)$$

$$\left[c,\ r,\ \operatorname{Re}\alpha > 0;\ -1 < \operatorname{Re}(\alpha+\nu) < 2,\ \left\{ \begin{matrix} r|\arg z| < \pi \\ z > 0 \end{matrix} \right\}\right],$$

$$V(\gamma) = -2^{\delta-\gamma}\pi c^{\nu+\gamma} z^{\alpha+\nu+\gamma} \sum_{k=0}^{\infty} \frac{(-1)^k}{k!\,(\delta+1/2)_k\,(\nu+\gamma+2k)\,(\alpha+\nu+\gamma+2k)} \times$$
$$\times \left\{ \begin{matrix} \operatorname{cosec}[(\alpha+\nu+\gamma+2k)\pi/r] \\ \operatorname{ctg}[(\alpha+\nu+\gamma+2k)\pi/r] \end{matrix} \right\} \left(\frac{cz}{2}\right)^{2k} -$$
$$-\frac{z^\alpha}{r}\Gamma(\nu)\cos\frac{\gamma-\nu}{2}\pi \left\{ \begin{matrix} \operatorname{cosec}(\alpha\pi/r) \\ \operatorname{ctg}(\alpha\pi/r) \end{matrix} \right\} \mp$$
$$\mp \frac{z^r}{c^{\alpha-r}} \sum_{k=0}^{\infty} \frac{(\mp 1)^k \Gamma(\alpha+\nu-r-rk)}{(k+1)(rk+r-\alpha)} \cos\frac{\gamma-\alpha-\nu+r+rk}{2}\pi\,(cz)^{rk} +$$
$$+\frac{r\Gamma(\alpha+\nu)}{2\alpha c^\alpha}\cos\frac{\gamma-\alpha-\nu}{2}\pi\left[2\psi(\alpha+\nu)+\pi\operatorname{tg}\frac{\gamma-\alpha-\nu}{2}\pi-\frac{2}{\alpha}-2\ln c\right].$$

10. $$\int_0^\infty x^{\alpha-1} \left\{ \begin{matrix} \ln(x^r+z^r) \\ \ln|x^r-z^r| \end{matrix} \right\} C(cx,\ \nu)\,dx = V(0)$$

$$\left[c,\ r,\ \operatorname{Re}\alpha > 0;\ 0 < \operatorname{Re}(\alpha+\nu) < 2,\ \left\{ \begin{matrix} r|\arg z| < \pi \\ z > 0 \end{matrix} \right\},\ V(\gamma)\ \text{see 2.6.1.9}\right].$$

11. $$\int_0^\infty x^{\alpha-1} \left\{ \begin{matrix} S(bx^r,\ \mu) \\ C(bx^r,\ \mu) \end{matrix} \right\} S(cx,\ \nu)\,dx = W(1)$$

$[b,\ c,\ r,\ \operatorname{Re}\alpha > 0;\ \operatorname{Re}(\alpha+\mu r) > -\delta r;\ \operatorname{Re}(\alpha+\nu) > -1;\ \operatorname{Re}(\alpha+\mu r+\nu) > -1-\delta r;$ $\operatorname{Re}(\alpha+\mu r+\nu) < r+1+\max(r,\ 1);$ for $r < 1$ substitute $x = t^{1/r}$].

$$W(\gamma) = -\frac{c^{\nu+\gamma}}{b^{(\alpha+\nu+\gamma)/r}} \sum_{k=0}^{\infty} \frac{(-1)^k}{k!\,(\gamma+1/2)_k\,(\alpha+\nu+\gamma+2k)\,(\nu+\gamma+2k)} \times$$
$$\times\Gamma\left(\mu+\frac{\alpha+\nu+\gamma+2k}{r}\right) \left\{ \begin{matrix} \sin[(\mu r+\alpha+\nu+\gamma+2k)\pi/(2r)] \\ \cos[(\mu r+\alpha+\nu+\gamma+2k)\pi/(2r)] \end{matrix} \right\} \times$$
$$\times\left(\frac{c}{2b^{1/r}}\right)^{2k} + \frac{\Gamma(\nu)\Gamma(\mu+\alpha/r)}{\alpha b^{\alpha/r}}\cos\frac{\nu-\gamma}{2}\pi \left\{ \begin{matrix} \sin[(\mu r+\alpha)\pi/(2r)] \\ \cos[(\mu r+\alpha)\pi/(2r)] \end{matrix} \right\} \quad [r > 1],$$

$W(\gamma)$ for $r = 1$ See also 2.6.6.1–2.

12. $$\int_0^\infty x^{\alpha-1} \left\{ \begin{matrix} S(bx^r,\ \mu) \\ C(bx^r,\ \mu) \end{matrix} \right\} C(cx,\ \nu)\,dx = W(0)$$

$[b,\ c,\ r,\ \operatorname{Re}\alpha,\ \operatorname{Re}(\alpha+\nu) > 0;\ \operatorname{Re}(\alpha+\mu r) > -\delta r;\ -\delta r <$ $< \operatorname{Re}(\alpha+\mu r+\nu) < 1+r+\max(r,\ 1);\ W(\gamma)$ see 2.6.1.11].

13. $$\int_0^\infty x^{\alpha-1}\operatorname{Ei}(-bx^r) \left\{ \begin{matrix} S(cx,\ \nu) \\ C(cx,\ \nu) \end{matrix} \right\} dx = P(r) \quad [c,\ r,\ \operatorname{Re}b,\ \operatorname{Re}\alpha > 0;\ \operatorname{Re}(\alpha+\nu) > -\delta],$$

$$P(r) = \frac{c^{\nu+\delta}}{b^{(\alpha+\nu+\delta)/r}} \sum_{k=0}^{\infty} \frac{(-1)^k}{k!\,(\delta+1/2)_k\,(\alpha+\nu+\delta+2k)\,(\nu+\delta+2k)} \times$$
$$\times\Gamma\left(\frac{\alpha+\nu+\delta+2k}{r}\right)\left(\frac{c}{2b^{1/r}}\right)^{2k} - \frac{\Gamma(\alpha/r)\,\Gamma(\nu)}{\alpha b^{\alpha/r}} \left\{ \begin{matrix} \sin(\nu\pi/2) \\ \cos(\nu\pi/2) \end{matrix} \right\} \quad [r>1],$$

$$P(r) = -\frac{b}{c^{\alpha+r}} \sum_{k=0}^{\infty} \frac{\Gamma(\alpha+\nu+r+rk)}{k!\,(k+1)^2(\alpha+r+rk)} \begin{Bmatrix} \sin[(\alpha+\nu+r+rk)\pi/2] \\ \cos[(\alpha+\nu+r+rk)\pi/2] \end{Bmatrix} \left(-\frac{b}{c^r}\right)^k +$$

$$+\frac{\Gamma(\alpha+\nu)}{\alpha c^{\alpha}} \begin{Bmatrix} \sin[(\alpha+\nu)\pi/2] \\ \cos[(\alpha+\nu)\pi/2] \end{Bmatrix} \left[\mathbf{C} - \frac{r}{\alpha} + r\psi(\alpha+\nu) \pm \frac{r\pi}{2} \operatorname{tg}^{\mp 1} \frac{\alpha+\nu}{2}\pi - \ln\frac{c^r}{b}\right]$$

$[r < 1]$,

$P(1)$ See also 2.6.7.1.

14. $\displaystyle\int_0^{\infty} x^{\alpha-1} e^{\pm bx^r} \operatorname{Ei}(\mp bx^r)\, S(cx, \nu)\, dx = Q(1)$

$\left[c,\ r,\ \operatorname{Re}\alpha > 0;\ \ -1 < \operatorname{Re}(\alpha+\nu) < r+2,\ \ \begin{Bmatrix} |\arg b| < \pi \\ b > 0 \end{Bmatrix}\right]$.

$$Q(\gamma) = \frac{\pi c^{\nu+\gamma}}{b^{(\alpha+\nu+\gamma)/r}\, r} \sum_{k=0}^{\infty} \frac{(-1)^k}{k!\,(\gamma+1/2)_k\,(\nu+\gamma+2k)} \times$$

$$\times \Gamma\left(\frac{\alpha+\nu+\gamma+2k}{r}\right) \begin{Bmatrix} \operatorname{cosec}[(\alpha+\nu+\gamma+2k)\pi/r] \\ \operatorname{ctg}[(\alpha+\nu+\gamma+2k)\pi/r] \end{Bmatrix} \left(\frac{c}{2b^{1/r}}\right)^{2k} -$$

$$-\frac{\pi\Gamma(\nu)\,\Gamma(\alpha/r)}{b^{\alpha/r} r} \cos\frac{\gamma-\nu}{2}\pi \begin{Bmatrix} \operatorname{cosec}(\alpha\pi/r) \\ \operatorname{ctg}(\alpha\pi/r) \end{Bmatrix} +$$

$$+\frac{c^{r-\alpha}}{b} \sum_{k=0}^{\infty} \frac{k!\,\Gamma(\alpha+\nu-r-rk)}{rk+r-\alpha} \cos\frac{\alpha+\nu-r-rk-\gamma}{2}\pi \left(-\frac{c^r}{b}\right)^k$$

$[r > 1]$,

$$Q(\gamma) = -\frac{1}{c^{\alpha}} \sum_{k=0}^{\infty} \frac{\Gamma(\alpha+\nu+rk)}{k!\,(\alpha+rk)} \cos\frac{\alpha+\nu+rk-\gamma}{2}\pi \times$$

$$\times\left[\psi(k+1) + \frac{r}{\alpha+rk} - r\psi(\alpha+\nu+rk) + \frac{\pi r}{2} \operatorname{tg}\frac{\alpha+\nu+rk-\gamma}{2}\pi + \ln\frac{c^r}{b}\right] \left(\pm\frac{b}{c^r}\right)^k$$

$[r < 1]$,

$Q(\gamma)$ for $r = 1$ See also 2.6.7.3–4.

15. $\displaystyle\int_0^{\infty} x^{\alpha-1} e^{\pm bx^r} \operatorname{Ei}(\mp bx^r)\, C(cx, \nu)\, dx = Q(0)$

$\left[c,\ r,\ \operatorname{Re}\alpha > 0;\ 0 < \operatorname{Re}(\alpha+\nu) < r+2,\ \begin{Bmatrix} |\arg b| < \pi \\ b > 0 \end{Bmatrix},\ Q(\gamma)\ \text{see 2.6.1.14}\right]$.

16. $\displaystyle\int_0^{\infty} x^{\alpha-1} \begin{Bmatrix} \operatorname{si}(bx^r) \\ \operatorname{ci}(bx^r) \end{Bmatrix} S(cx, \nu)\, dx = X(1)$

$[b,\ c,\ r,\ \operatorname{Re}\alpha > 0;\ -1 < \operatorname{Re}(\alpha+\nu) < r+1+\max(r,\ 1)]$,

$$X(\gamma) = \frac{c^{\nu+\gamma}}{b^{(\alpha+\nu+\gamma)/r}} \sum_{k=0}^{\infty} \frac{(-1)^k}{k!\,(\gamma+1/2)_k\,(\nu+\gamma+2k)\,(\alpha+\nu+\gamma+2k)} \times$$

$$\times \Gamma\left(\frac{\alpha+\nu+\gamma+2k}{r}\right) \begin{Bmatrix} \sin[(\alpha+\nu+\gamma+2k)\pi/(2r)] \\ \cos[(\alpha+\nu+\gamma+2k)\pi/(2r)] \end{Bmatrix} \left(\frac{c}{2b^{1/r}}\right)^{2k} -$$

$$-\frac{\Gamma(\nu)\,\Gamma(\alpha/r)}{\alpha b^{\alpha/r}} \cos\frac{\nu-\gamma}{2}\pi \begin{Bmatrix} \sin[\alpha\pi/(2r)] \\ \cos[\alpha\pi/(2r)] \end{Bmatrix}$$

$[r > 1]$,

$$X(\gamma)=-\frac{b^{\delta+2}}{2c^{\alpha+\delta r+2r}(1+2\delta)}\times$$

$$\times\sum_{k=0}^{\infty}\frac{(-1)^k\Gamma(\alpha+\nu+2r+\delta r+2rk)}{(k+1)!\,(3/2+\delta)_k(\alpha+\delta r+2r+2rk)(2+\delta+2k)}\times$$

$$\times\cos\frac{\alpha+\nu+\delta r+2r+2rk-\gamma}{2}\pi\left(\frac{b}{2c^r}\right)^{2k}+\frac{b^\delta\Gamma(\alpha+\nu+\delta r)}{(\alpha+\delta r)c^{\alpha+\delta r}}\times$$

$$\times\cos\frac{\alpha+\nu+\delta r-\gamma}{2}\pi\begin{Bmatrix}1\\R\end{Bmatrix}-\begin{Bmatrix}1\\0\end{Bmatrix}\frac{\pi\Gamma(\alpha+\nu)}{2\alpha c^\alpha}\cos\frac{\alpha+\nu-\gamma}{2}\pi,$$

$$\left[r<1;\quad R=\mathrm{C}-\frac{r}{\alpha}+r\psi(\alpha+\nu)-\frac{\pi r}{2}\operatorname{tg}\frac{\alpha+\nu-\gamma}{2}\pi-\ln\frac{c^r}{b}\right].$$

$X(\gamma)$ for $r=1$ See also 2.6.8.1–2.

17. $$\int_0^\infty x^{\alpha-1}\begin{Bmatrix}\operatorname{si}(bx^r)\\\operatorname{ci}(bx^r)\end{Bmatrix}C(cx,\nu)\,dx=X(0)$$

$[b,\ c,\ r,\ \operatorname{Re}\alpha>0;\ 0<\operatorname{Re}(\alpha+\nu)<r+1+\max(r,\ 1);\ X(\gamma)$ see 2.6.1.16].

18. $$\int_0^\infty x^{\alpha-1}\begin{Bmatrix}\operatorname{erf}(bx^r)\\\operatorname{erfc}(bx^r)\end{Bmatrix}S(cx,\nu)\,dx=Y(1)$$

$$\left[c,\ r>0,\ \begin{Bmatrix}\operatorname{Re}\alpha>-r;\ -r-1<\operatorname{Re}(\alpha+\nu)<2\\\operatorname{Re}\alpha>0;\ \operatorname{Re}(\alpha+\nu)>-1\end{Bmatrix},\ |\arg b|<\pi/4\right],$$

$$Y(\gamma)=\pm\frac{c^{\nu+\gamma}}{\sqrt{\pi}\,b^{(\alpha+\nu+\gamma)/r}}\sum_{k=0}^{\infty}\frac{(-1)^k}{k!\,(\gamma+1/2)_k(\alpha+\nu+\gamma+2k)(\nu+\gamma+2k)}\times$$

$$\times\Gamma\left(\frac{r+\alpha+\nu+\gamma+2k}{2r}\right)\left(\frac{c}{2b^{1/r}}\right)^{2k}\mp\frac{\Gamma(\nu)}{\alpha\sqrt{\pi}\,b^{\alpha/r}}\Gamma\left(\frac{r+\alpha}{2r}\right)\cos\frac{\nu-\gamma}{2}\pi+$$

$$+\begin{Bmatrix}1\\0\end{Bmatrix}\frac{\Gamma(\alpha+\nu)}{\alpha c^\alpha}\cos\frac{\alpha+\nu-\gamma}{2}\pi\qquad[r>1/2],$$

$$Y(\gamma)=\pm\frac{2b}{\sqrt{\pi}\,c^{\alpha+r}}\sum_{k=0}^{\infty}\frac{(-1)^k\Gamma(\alpha+\nu+r+2rk)}{k!\,(2k+1)(\alpha+r+2rk)}\cos\frac{\alpha+\nu+r+2rk-\gamma}{2}\pi\left(\frac{b}{c^r}\right)^{2k}+$$

$$+\begin{Bmatrix}0\\1\end{Bmatrix}\frac{\Gamma(\alpha+\nu)}{\alpha c^\alpha}\cos\frac{\alpha+\nu-\gamma}{2}\pi\qquad[r<1/2].$$

$Y(\gamma)$ for $r=1/2$ See also 2.6.9.3–4.

19. $$\int_0^\infty x^{\alpha-1}\begin{Bmatrix}\operatorname{erf}(bx^r)\\\operatorname{erfc}(bx^r)\end{Bmatrix}C(cx,\nu)\,dx=Y(0)$$

$$\left[c,\ r>0,\ \begin{Bmatrix}\operatorname{Re}\alpha>-r;\ -r<\operatorname{Re}(\alpha+\nu)<2\\\operatorname{Re}\alpha,\ \operatorname{Re}(\alpha+\nu)>0\end{Bmatrix},\ |\arg b|<\pi/4;\ Y(\gamma)\text{ see 2.6.1.18}\right].$$

20. $$\int_0^\infty x^{\alpha-1}\begin{Bmatrix}S(bx^r)\\C(bx^r)\end{Bmatrix}S(cx,\nu)\,dx=Z(1)$$

$[b,\ c,\ r>0;\ \operatorname{Re}\alpha>-r(\delta+1/2);\ -\delta r-r/2-1<\operatorname{Re}(\alpha+\nu)<2]$,

$$Z(\gamma)=\frac{c^{\nu+\gamma}}{\sqrt{2\pi}\,b^{(\alpha+\nu+\gamma)/r}}\sum_{k=0}^{\infty}\frac{(-1)^k}{k!\,(1/2+\gamma)_k(\nu+\gamma+2k)(\alpha+\nu+\gamma+2k)}\times$$

$$\times\Gamma\left(\frac{\alpha+\nu+\gamma+2k}{r}+\frac{1}{2}\right)\begin{Bmatrix}\sin\left[(2\alpha+2\nu+2\gamma+4k+r)\pi/(4r)\right]\\ \cos\left[(2\alpha+2\nu+2\gamma+4k+r)\pi/(4r)\right]\end{Bmatrix}\times$$

$$\times\left(\frac{c}{2b^{1/r}}\right)^{2k}-\frac{\Gamma(\nu)}{\alpha\sqrt{2\pi}\,b^{\alpha/r}}\Gamma\left(\frac{2\alpha+r}{2r}\right)\cos\frac{\nu-\gamma}{2}\pi\begin{Bmatrix}\sin\left[(2\alpha+r)\pi/(4r)\right]\\ \cos\left[(2\alpha+r)\pi/(4r)\right]\end{Bmatrix}+$$

$$+\frac{\Gamma(\alpha+\nu)}{4c^{\alpha}}\cos\frac{\alpha+\nu-\gamma}{2}\pi \quad [r>1],$$

$$Z(\gamma)=\frac{2^{3/2}b^{\delta+1/2}}{\sqrt{\pi}\,c^{\alpha+\delta r+r/2}}\sum_{k=0}^{\infty}\frac{(-1)^k\Gamma(\alpha+\nu+\delta r+r/2+2rk)}{k!\,(\delta+1/2)_k\,(4k+2\delta+1)\,(4rk+r+2\delta r+2\alpha)}\times$$

$$\times\cos\frac{\alpha+\nu+\delta r+r/2+2rk-\gamma}{2}\pi\left(\frac{b}{2c^r}\right)^{2k} \quad [r<1],$$

$Z(\gamma)$ for $r=1$ See also 2.6.10.1–2.

21. $\displaystyle\int_0^\infty x^{\alpha-1}\begin{Bmatrix}S(bx^r)\\ C(bx^r)\end{Bmatrix}C(cx,\ \nu)\,dx=Z(0)$

$[b,\ c,\ r>0;\ \mathrm{Re}\,\alpha>-r(\delta+1/2);\ -r(\delta+1/2)<\mathrm{Re}(\alpha+\nu)<2;\ Z(\gamma)$ see 2.6.1.20$]$.

22. $\displaystyle\int_0^\infty x^{\alpha-1}\begin{Bmatrix}\gamma(\mu,\ bx^r)\\ \Gamma(\mu,\ bx^r)\end{Bmatrix}S(cx,\ \nu)\,dx=R(1)$

$$\left[c,\ r,\ \mathrm{Re}\,b>0,\ \begin{Bmatrix}\mathrm{Re}\,\mu,\ \mathrm{Re}(\alpha+\mu r)>0;\ -1-r\,\mathrm{Re}\,\mu<\mathrm{Re}(\alpha+\nu)<2\\ \mathrm{Re}\,\alpha,\ \mathrm{Re}(\alpha+\mu r)>0;\ \mathrm{Re}(\alpha+\nu),\ \mathrm{Re}(\alpha+\mu r+\nu)>-1\end{Bmatrix}\right],$$

$$R(\gamma)=\pm\frac{c^{\nu+\gamma}}{b^{(\alpha+\nu+\gamma)/r}}\sum_{k=0}^{\infty}\frac{(-1)^k}{k!\,(\gamma+1/2)_k\,(\alpha+\nu+\gamma+2k)\,(\nu+\gamma+2k)}\times$$

$$\times\Gamma\left(\frac{\alpha+\nu+\gamma+2k}{r}+\mu\right)\left(\frac{c}{2b^{1/r}}\right)^{2k}\mp\frac{\Gamma(\nu)}{\alpha b^{\alpha/r}}\Gamma\left(\frac{\alpha}{r}+\mu\right)\cos\frac{\nu-\gamma}{2}\pi+$$

$$+\begin{Bmatrix}1\\ 0\end{Bmatrix}\frac{\Gamma(\mu)\,\Gamma(\alpha+\nu)}{\alpha c^{\alpha}}\cos\frac{\alpha+\nu-\gamma}{2}\pi \quad [r>1],$$

$$R(\gamma)=\pm\frac{b^{\mu}}{c^{\alpha+\mu r}}\sum_{k=0}^{\infty}\frac{\Gamma(\alpha+\nu+\mu r+rk)}{k!\,(\mu+k)\,(\alpha+\mu r+rk)}\cos\frac{\alpha+\nu+\mu r+rk-\gamma}{2}\pi\left(-\frac{b}{c^r}\right)^k+$$

$$+\begin{Bmatrix}0\\ 1\end{Bmatrix}\frac{\Gamma(\mu)\,\Gamma(\alpha+\nu)}{\alpha c^{\alpha}}\cos\frac{\alpha+\nu-\gamma}{2}\pi \quad [r<1],$$

$R(\gamma)$ for $r=1$ See also 2.6.11.1–2.

23. $\displaystyle\int_0^\infty x^{\alpha-1}\begin{Bmatrix}\gamma(\mu,\ bx^r)\\ \Gamma(\mu,\ bx^r)\end{Bmatrix}C(cx,\ \nu)\,dx=R(0)$

$$\left[c,\ r,\ \mathrm{Re}\,b>0,\ \begin{Bmatrix}\mathrm{Re}\,\mu,\ \mathrm{Re}(\alpha+\mu r)>0;\ -r\,\mathrm{Re}\,\mu<\mathrm{Re}(\alpha+\nu)<2\\ \mathrm{Re}\,\alpha,\ \mathrm{Re}(\alpha+\mu r),\ \mathrm{Re}(\alpha+\nu),\ \mathrm{Re}(\alpha+\mu r+\nu)>0\end{Bmatrix},\ R(\gamma)\text{ see }2.6.1.22\right].$$

2.6.2. Integrals of $A(x)\begin{Bmatrix}S(cx,\ \nu)\\ C(cx,\ \nu)\end{Bmatrix}$.

NOTATION: $\delta=\begin{Bmatrix}1\\ 0\end{Bmatrix}$.

1. $\displaystyle\int_0^\infty x^{\alpha-1}\begin{Bmatrix}S(cx,\ \nu)\\ C(cx,\ \nu)\end{Bmatrix}dx=\frac{\Gamma(\alpha+\nu)}{\alpha c^{\alpha}}\begin{Bmatrix}\sin\left[(\alpha+\nu)\pi/2\right]\\ \cos\left[(\alpha+\nu)\pi/2\right]\end{Bmatrix}\ [c,\ \mathrm{Re}\,\alpha>0;\ -\delta<\mathrm{Re}(\alpha+\nu)<2].$

2. $$\int_0^a x^{\alpha-1}(a-x)^{\beta-1}\begin{Bmatrix} S(cx,\ \nu) \\ C(cx,\ \nu) \end{Bmatrix} dx =$$
$$= -\frac{a^{\alpha+\beta+\nu+\delta-1}c^{\nu+\delta}}{\nu+\delta} B(\beta,\ \alpha+\nu+\delta)\,{}_3F_4\left(\frac{\nu+\delta}{2},\ \frac{\alpha+\nu}{2}+\delta,\ \frac{\alpha+\nu+1}{2};\right.$$
$$\left.\frac{\nu+\delta}{2}+1,\ \frac{\alpha+\beta+\nu}{2}+\delta,\ \frac{\alpha+\beta+\nu+1}{2},\ \delta+\frac{1}{2};\ -\frac{a^2c^2}{4}\right)+$$
$$+a^{\alpha+\beta-1}\mathrm{B}(\alpha,\ \beta)\,\Gamma(\nu)\begin{Bmatrix} \sin(\nu\pi/2) \\ \cos(\nu\pi/2) \end{Bmatrix}\quad [a,\ \operatorname{Re}\alpha,\ \operatorname{Re}\beta>0;\ \operatorname{Re}(\alpha+\nu)>-\delta].$$

3. $$\int_0^a x^{\alpha-1}(a^2-x^2)^{\beta-1}\begin{Bmatrix} S(cx,\ \nu) \\ C(cx,\ \nu) \end{Bmatrix} dx =$$
$$= -\frac{a^{\alpha+2\beta+\nu+\delta-2}c^{\nu+\delta}}{2(\nu+\delta)}\mathrm{B}\left(\beta,\ \frac{\alpha+\nu+\delta}{2}\right)\times$$
$$\times{}_2F_3\left(\frac{\nu+\delta}{2},\ \frac{\alpha+\nu+\delta}{2};\ \frac{\nu+\delta}{2}+1,\ \beta+\frac{\alpha+\nu+\delta}{2},\ \delta+\frac{1}{2};\ -\frac{a^2c^2}{4}\right)+$$
$$+\frac{a^{\alpha+2\beta-2}}{2}\mathrm{B}\left(\beta,\ \frac{\alpha}{2}\right)\Gamma(\nu)\begin{Bmatrix} \sin(\nu\pi/2) \\ \cos(\nu\pi/2) \end{Bmatrix}\quad [a,\ \operatorname{Re}\alpha,\ \operatorname{Re}\beta>0;\ \operatorname{Re}(\alpha+\nu)>-\delta].$$

4. $$\int_a^\infty x^{\alpha-1}(x-a)^{\beta-1}\begin{Bmatrix} S(cx,\ \nu) \\ C(cx,\ \nu) \end{Bmatrix} dx = -\frac{a^{\alpha+\beta+\nu+\delta-1}c^{\nu+\delta}}{\nu+\delta}\mathrm{B}(\beta,\ 1-\alpha-\beta-\nu-\delta)\times$$
$$\times{}_3F_4\left(\frac{\nu+\delta}{2},\ \frac{1+\alpha+\nu}{2},\ \frac{\alpha+\nu}{2}+\delta;\right.$$
$$\left.\frac{1+\alpha+\beta+\nu}{2},\ \frac{\alpha+\beta+\nu}{2}+\delta,\ 1+\frac{\nu+\delta}{2},\ \delta+\frac{1}{2};\ -\frac{a^2c^2}{4}\right)\pm$$
$$\pm\frac{c^{1-\alpha-\beta}}{1-\alpha-\beta}\Gamma(\alpha+\beta+\nu-1)\begin{Bmatrix} \cos[(\alpha+\beta+\nu)\pi/2] \\ \sin[(\alpha+\beta+\nu)\pi/2] \end{Bmatrix}\times$$
$$\times{}_3F_4\left(\frac{1-\beta}{2},1-\frac{\beta}{2},\ \frac{1-\alpha-\beta}{2};\ \frac{3-\alpha-\beta}{2},\ \frac{3-\alpha-\beta-\nu}{2},\ 1-\frac{\alpha+\beta+\nu}{2},\ \frac{1}{2};-\frac{a^2c^2}{4}\right)+$$
$$+\frac{ac^{2-\alpha-\beta}(1-\beta)}{2-\alpha-\beta}\Gamma(\alpha+\beta+\nu-2)\begin{Bmatrix} \sin[(\alpha+\beta+\nu)\pi/2] \\ \cos[(\alpha+\beta+\nu)\pi/2] \end{Bmatrix}\times$$
$$\times{}_3F_4\left(1-\frac{\beta}{2},\ \frac{3-\beta}{2},\ 1-\frac{\alpha+\beta}{2};\ 2-\frac{\alpha+\beta}{2},\ 2-\frac{\alpha+\beta+\nu}{2},\ \frac{3-\alpha-\beta-\nu}{2},\ \frac{3}{2};\ -\frac{a^2c^2}{4}\right)+$$
$$+a^{\alpha+\beta-1}\mathrm{B}(\beta,\ 1-\alpha-\beta)\,\Gamma(\nu)\begin{Bmatrix} \sin(\nu\pi/2) \\ \cos(\nu\pi/2) \end{Bmatrix}\quad [a,\ c,\ \operatorname{Re}\beta>0;\ \operatorname{Re}(\alpha+\beta+\nu)<3].$$

5. $$\int_a^\infty x^{\alpha-1}(x^2-a^2)^{\beta-1}\begin{Bmatrix} S(cx,\ \nu) \\ C(cx,\ \nu) \end{Bmatrix} dx = -\frac{a^{\alpha+2\beta+\nu+\delta-2}c^{\nu+\delta}}{2(\nu+\delta)}\mathrm{B}\left(\beta,\ 1-\beta-\frac{\alpha+\nu+\delta}{2}\right)\times$$
$$\times{}_2F_3\left(\frac{\nu+\delta}{2},\ \frac{\alpha+\nu+\delta}{2};\ \frac{\nu+\delta}{2}+1,\ \beta+\frac{\alpha+\nu+\delta}{2},\ \delta+\frac{1}{2};\ -\frac{a^2c^2}{4}\right)+$$
$$+\frac{c^{2-2\beta-\alpha}}{2-2\beta-\alpha}\Gamma(\alpha+\nu+2\beta-2)\begin{Bmatrix} \sin[(\alpha+\nu+2\beta)\pi/2] \\ \cos[(\alpha+\nu+2\beta)\pi/2] \end{Bmatrix}\times$$
$$\times{}_2F_3\left(1-\beta-\frac{\alpha}{2},\ 1-\beta;\ 2-\beta-\frac{\alpha}{2},\ 2-\beta-\frac{\alpha+\nu}{2},\ \frac{3-\alpha-\nu}{2}-\beta;\ -\frac{a^2c^2}{4}\right)+$$
$$+\frac{a^{\alpha+2\beta-2}}{2}\mathrm{B}\left(\beta,\ 1-\beta-\frac{\alpha}{2}\right)\Gamma(\nu)\begin{Bmatrix} \sin(\nu\pi/2) \\ \cos(\nu\pi/2) \end{Bmatrix}\quad [a,\ c,\ \operatorname{Re}\beta>0;\ \operatorname{Re}(\alpha+2\beta+\nu)<4].$$

6. $$\int_0^\infty \frac{x^{\alpha-1}}{(x+z)^\rho}\begin{Bmatrix}S(cx,\ \nu)\\ C(cx,\ \nu)\end{Bmatrix}dx=-\frac{c^{\nu+\delta}z^{\alpha+\nu+\delta-\rho}}{\nu+\delta}\times$$
$$\times B(\alpha+\nu+\delta,\ \rho-\alpha-\nu-\delta)\,{}_3F_4\left(\frac{\nu+\delta}{2},\ \frac{\alpha+\nu}{2}+\delta,\ \frac{\alpha+\nu+1}{2};\right.$$
$$\left.1+\frac{\alpha+\nu-\rho}{2},\ \frac{1+\alpha+\nu-\rho}{2}+\delta,\ 1+\frac{\nu+\delta}{2},\ \delta+\frac{1}{2};\ -\frac{c^2z^2}{2}\right)-$$
$$-\frac{c^{\rho-\alpha}}{\rho-\alpha}\Gamma(\alpha+\nu-\rho)\begin{Bmatrix}\sin[(\alpha+\nu-\rho)\pi/2]\\ \cos[(\alpha+\nu-\rho)\pi/2]\end{Bmatrix}\times$$
$$\times{}_3F_4\left(\frac{\rho}{2},\ \frac{1+\rho}{2},\ \frac{\rho-\alpha}{2};\ 1+\frac{\rho-\alpha}{2},\ 1-\frac{\alpha+\nu-\rho}{2},\ \frac{1-\alpha-\nu+\rho}{2},\ \frac{1}{2};\ -\frac{c^2z^2}{4}\right)\pm$$
$$\pm\frac{ac^{1+\rho-\alpha}\rho}{1+\rho-\alpha}\Gamma(\alpha+\nu-\rho-1)\begin{Bmatrix}\cos[(\alpha+\nu-\rho)\pi/2]\\ \sin[(\alpha+\nu-\rho)\pi/2]\end{Bmatrix}\times$$
$$\times{}_3F_4\left(\frac{1+\rho}{2},\ 1+\frac{\rho}{2},\ \frac{1+\rho-\alpha}{2};\ \frac{3+\rho-\alpha}{2},\ \frac{3+\rho-\alpha-\nu}{2},\ 1-\frac{\alpha+\nu-\rho}{2},\ \frac{3}{2};\ -\frac{c^2z^2}{4}\right)+$$
$$+z^{\alpha-\rho}B(\alpha,\ \rho-\alpha)\,\Gamma(\nu)\begin{Bmatrix}\sin(\nu\pi/2)\\ \cos(\nu\pi/2)\end{Bmatrix}\quad [c,\ \mathrm{Re}\,\alpha>0;\ -\delta<\mathrm{Re}\,(\alpha+\nu)<\mathrm{Re}\,\rho+2;\ |\arg z|<\pi].$$

7. $$\int_0^\infty \frac{x^{\alpha-1}}{(x^2+z^2)^\rho}\begin{Bmatrix}S(cx,\ \nu)\\ C(cx,\ \nu)\end{Bmatrix}dx=-\frac{c^{\nu+\delta}z^{\alpha+\nu+\delta-2\rho}}{2(\nu+\delta)}B\left(\frac{\alpha+\nu+\delta}{2},\ \rho-\frac{\alpha+\nu+\delta}{2}\right)\times$$
$$\times{}_2F_3\left(\frac{\nu+\delta}{2},\ \frac{\alpha+\nu+\delta}{2};\ \frac{\nu+\delta}{2}+1,\ 1-\rho+\frac{\alpha+\nu+\delta}{2},\ \delta+\frac{1}{2};\ \frac{c^2z^2}{4}\right)-$$
$$-\frac{c^{2\rho-\alpha}}{2\rho-\alpha}\Gamma(\alpha+\nu-2\rho)\begin{Bmatrix}\sin[(\alpha+\nu-2\rho)\pi/2]\\ \cos[(\alpha+\nu-2\rho)\pi/2]\end{Bmatrix}{}_2F_3\left(\rho-\frac{\alpha}{2},\ \rho;\right.$$
$$\left.\rho+1-\frac{\alpha}{2},\ 1+\rho-\frac{\alpha+\nu}{2},\ \rho+\frac{1-\alpha-\nu}{2};\ \frac{c^2z^2}{4}\right)+\frac{z^{\alpha-2\rho}}{2}B\left(\frac{\alpha}{2},\ \rho-\frac{\alpha}{2}\right)\Gamma(\nu)\begin{Bmatrix}\sin(\nu\pi/2)\\ \cos(\nu\pi/2)\end{Bmatrix}$$
$$[c,\ \mathrm{Re}\,\alpha,\ \mathrm{Re}\,z>0;\ -\delta<\mathrm{Re}\,(\alpha+\nu)<2\,\mathrm{Re}\,\rho+2].$$

8. $$\int_0^\infty \frac{x^{\alpha-1}}{x-y}\begin{Bmatrix}S(cx,\ \nu)\\ C(cx,\ \nu)\end{Bmatrix}dx=\frac{\pi c^{\nu+\delta}y^{\alpha+\nu+\delta-1}}{\nu+\delta}\operatorname{ctg}(\alpha+\nu)\pi\times$$
$$\times{}_1F_2\left(\frac{\nu+\delta}{2};\ 1+\frac{\nu+\delta}{2},\ \delta+\frac{1}{2};\ -\frac{c^2y^2}{4}\right)\pm\frac{c^{1-\alpha}}{1-\alpha}\Gamma(\alpha+\nu-1)\begin{Bmatrix}\cos[(\alpha+\nu)\pi/2]\\ \sin[(\alpha+\nu)\pi/2]\end{Bmatrix}\times$$
$$\times{}_2F_3\left(\frac{1-\alpha}{2},\ 1;\ \frac{3-\alpha}{2},\ \frac{3-\alpha-\nu}{2},\ 1-\frac{\alpha+\nu}{2};\ -\frac{c^2y^2}{4}\right)+\frac{c^{2-\alpha}y}{2-\alpha}\Gamma(\alpha+\nu-2)\times$$
$$\times\begin{Bmatrix}\sin[(\alpha+\nu)\pi/2]\\ \cos[(\alpha+\nu)\pi/2]\end{Bmatrix}{}_2F_3\left(1-\frac{\alpha}{2},\ 1;\ 2-\frac{\alpha}{2},\ 2-\frac{\alpha+\nu}{2},\ \frac{3-\alpha-\nu}{2};\ -\frac{c^2y^2}{4}\right)+$$
$$+\pi y^{\alpha-1}\operatorname{ctg}\alpha\pi\Gamma(\nu)\begin{Bmatrix}\sin(\nu\pi/2)\\ \cos(\nu\pi/2)\end{Bmatrix}\quad[c,\ y,\ \mathrm{Re}\,\alpha>0;\ -\delta<\mathrm{Re}\,(\alpha+\nu)<3].$$

9. $$\int_0^\infty \frac{x^{\alpha-1}}{x^2-y^2}\begin{Bmatrix}S(cx,\ \nu)\\ C(cx,\ \nu)\end{Bmatrix}dx=\mp\frac{\pi c^{\nu+\delta}y^{\alpha+\nu+\delta-2}}{2(\nu+\delta)}\times$$
$$\times\begin{Bmatrix}\operatorname{tg}[(\alpha+\nu)\pi/2]\\ \operatorname{ctg}[(\alpha+\nu)\pi/2]\end{Bmatrix}{}_1F_2\left(\frac{\nu+\delta}{2};\ 1+\frac{\nu+\delta}{2},\ \delta+\frac{1}{2};\ -\frac{c^2y^2}{4}\right)+$$
$$+\frac{c^{2-\alpha}}{2-\alpha}\Gamma(\alpha+\nu-2)\begin{Bmatrix}\sin[(\alpha+\nu)\pi/2]\\ \cos[(\alpha+\nu)\pi/2]\end{Bmatrix}\times$$
$$\times{}_2F_3\left(1-\frac{\alpha}{2},\ 1;\ 2-\frac{\alpha}{2},\ 2-\frac{\alpha+\nu}{2},\ \frac{3-\alpha-\nu}{2};\ -\frac{c^2y^2}{4}\right)+$$
$$+\frac{\pi y^{\alpha-2}}{2}\operatorname{ctg}\frac{\alpha\pi}{2}\Gamma(\nu)\begin{Bmatrix}\sin(\nu\pi/2)\\ \cos(\nu\pi/2)\end{Bmatrix}\quad[c,\ y,\ \mathrm{Re}\,\alpha>0;\ -\delta<\mathrm{Re}\,(\alpha+\nu)<4].$$

2.6.3. Integrals of $x^{\alpha} e^{px^{\pm r}} \begin{Bmatrix} S(cx, \nu) \\ C(cx, \nu) \end{Bmatrix}$.

NOTATION: $\delta = \begin{Bmatrix} 1 \\ 0 \end{Bmatrix}$.

1. $$\int_0^\infty x^{\alpha-1} e^{-px} \begin{Bmatrix} S(cx, \nu) \\ C(cx, \nu) \end{Bmatrix} dx =$$
$$= -\frac{c^{\nu+\delta}\Gamma(\alpha+\nu+\delta)}{p^{\alpha+\nu+\delta}(\nu+\delta)} {}_3F_2\left(\frac{\nu+\delta}{2}, \frac{\alpha+\nu+1}{2}, \frac{\alpha+\nu}{2}+\delta; \frac{\nu+\delta}{2}+1, \delta+\frac{1}{2}; -\frac{c^2}{p^2}\right)+$$
$$+\frac{\Gamma(\alpha)\Gamma(\nu)}{p^{\alpha}} \begin{Bmatrix} \sin(\nu\pi/2) \\ \cos(\nu\pi/2) \end{Bmatrix} \quad [c, \operatorname{Re} p, \operatorname{Re}\alpha > 0;\ \operatorname{Re}(\alpha+\nu) > -\delta].$$

2. $$\int_0^\infty x^{\alpha-1} e^{-px^2} \begin{Bmatrix} S(cx, \nu) \\ C(cx, \nu) \end{Bmatrix} dx =$$
$$= -\frac{c^{\nu+\delta}}{2p^{(\alpha+\nu+\delta)/2}(\nu+\delta)} \Gamma\left(\frac{\alpha+\nu+\delta}{2}\right) {}_2F_2\left(\frac{\nu+\delta}{2}, \frac{\alpha+\nu+\delta}{2}; \frac{\nu+\delta}{2}+1, \delta+\frac{1}{2}; -\frac{c^2}{4p}\right)+$$
$$+\frac{\Gamma(\nu)\Gamma(\alpha/2)}{2p^{\alpha/2}} \begin{Bmatrix} \sin(\nu\pi/2) \\ \cos(\nu\pi/2) \end{Bmatrix} \quad [c, \operatorname{Re} p, \operatorname{Re}\alpha > 0;\ \operatorname{Re}(\alpha+\nu) > -\delta].$$

3. $$\int_0^\infty x^{\alpha-1} e^{-px^{-2}} \begin{Bmatrix} S(cx, \nu) \\ C(cx, \nu) \end{Bmatrix} dx = -\frac{c^{\nu+\delta} p^{(\alpha+\nu+\delta)/2}}{2(\nu+\delta)} \Gamma\left(-\frac{\alpha+\nu+\delta}{2}\right) \times$$
$$\times {}_1F_3\left(\frac{\nu+\delta}{2}; \delta+\frac{1}{2}, \frac{\nu+\delta}{2}+1, 1+\frac{\alpha+\nu+\delta}{2}; \frac{c^2 p}{4}\right)+$$
$$+\frac{\Gamma(\alpha+\nu)}{\alpha c^{\alpha}} \begin{Bmatrix} \sin[(\alpha+\nu)\pi/2] \\ \cos[(\alpha+\nu)\pi/2] \end{Bmatrix} {}_1F_3\left(-\frac{\alpha}{2}; 1-\frac{\alpha}{2}, 1-\frac{\alpha+\nu}{2}, \frac{1-\alpha-\nu}{2}; \frac{c^2 p}{4}\right)+$$
$$+\frac{p^{\alpha/2}}{2}\Gamma(\nu)\Gamma\left(-\frac{\alpha}{2}\right) \begin{Bmatrix} \sin(\nu\pi/2) \\ \cos(\nu\pi/2) \end{Bmatrix} \quad [c, \operatorname{Re} p > 0;\ \operatorname{Re}(\alpha+\nu) < 2].$$

2.6.4. Integrals of $x^{\alpha} \begin{Bmatrix} \sin bx \\ \cos bx \end{Bmatrix} \begin{Bmatrix} S(cx, \nu) \\ C(cx, \nu) \end{Bmatrix}$.

NOTATION: $\delta = \begin{Bmatrix} 1 \\ 0 \end{Bmatrix}$.

1. $$\int_0^\infty x^{\alpha-1} \begin{Bmatrix} \sin bx \\ \cos bx \end{Bmatrix} S(cx, \nu)\, dx = U(1)$$
$$[b, c > 0;\ -\delta-1 < \operatorname{Re}(\alpha+\nu) < 2 \text{ for } b \neq c,\ -\delta-1 < \operatorname{Re}(\alpha+\nu) < 1 \text{ for } b=c],$$
$$U(\gamma) = -\frac{c^{\nu+\gamma}\Gamma(\alpha+\nu+\gamma)}{b^{\alpha+\nu+\gamma}(\nu+\gamma)} \begin{Bmatrix} \sin[(\alpha+\nu+\gamma)\pi/2] \\ \cos[(\alpha+\nu+\gamma)\pi/2] \end{Bmatrix} \times$$
$$\times {}_3F_2\left(\frac{\nu+\gamma}{2}, \frac{\alpha+\nu+1}{2}, \frac{\alpha+\nu}{2}+\gamma; \frac{\nu+\gamma}{2}+1, \gamma+\frac{1}{2}; \frac{c^2}{b^2}\right)+$$
$$+\frac{\Gamma(\nu)\Gamma(\alpha)}{b^{\alpha}} \cos\frac{\nu-\gamma}{2}\pi \begin{Bmatrix} \sin(\alpha\pi/2) \\ \cos(\alpha\pi/2) \end{Bmatrix} \quad [0 < c \leqslant b],$$
$$U(\gamma) = \frac{b^{\delta}\Gamma(\alpha+\nu+\delta)}{c^{\alpha+\delta}(\alpha+\delta)} \begin{Bmatrix} \sin[(\gamma-\alpha-\nu)\pi/2] \\ \cos[(\gamma-\alpha-\nu)\pi/2] \end{Bmatrix} \times$$
$$\times {}_3F_2\left(\frac{\alpha+\delta}{2}, \frac{\alpha+\nu+1}{2}, \frac{\alpha+\nu}{2}+\delta; \frac{\alpha+\delta}{2}+1, \delta+\frac{1}{2}; \frac{b^2}{c^2}\right) \quad [0 < b \leqslant c].$$

2. $\int\limits_0^\infty x^{\alpha-1} \begin{Bmatrix} \sin bx \\ \cos bx \end{Bmatrix} C(cx,\ \nu)\,dx = U(0)$

[$b,\ c > 0;\ -\delta < \operatorname{Re}(\alpha+\nu) < 2$ for $b \neq c$, $-\delta < \operatorname{Re}(\alpha+\nu) < 1$ for $b=c$; $U(\gamma)$ see 2.6.4.1].

2.6.5. Integrals of $x^\alpha \begin{Bmatrix} \ln(x+z) \\ \ln|x-z| \end{Bmatrix} \begin{Bmatrix} S(cx,\ \nu) \\ C(cx,\ \nu) \end{Bmatrix}$.

NOTATION: $\delta = \begin{Bmatrix} 1 \\ 0 \end{Bmatrix}$.

1. $\int\limits_0^\infty x^{\alpha-1} \begin{Bmatrix} \ln(x+z) \\ \ln|x-z| \end{Bmatrix} S(cx,\ \nu)\,dx = V(1)$

$$\left[c,\ \operatorname{Re}\alpha > 0;\ -1 < \operatorname{Re}(\alpha+\nu) < 2,\ \begin{Bmatrix} |\arg z| < \pi \\ z > 0 \end{Bmatrix}\right],$$

$$V(\gamma) = -(\mp 1)^\gamma \frac{2^{\delta-\gamma}\pi c^{\nu+\gamma} z^{\alpha+\nu+\gamma}}{(\nu+\gamma)(\alpha+\nu+\gamma)} \begin{Bmatrix} \operatorname{cosec}[(\alpha+\nu)\pi] \\ \operatorname{ctg}[(\alpha+\nu)\pi] \end{Bmatrix} \times$$
$$\times {}_2F_3\left(\frac{\nu+\gamma}{2},\ \frac{\alpha+\nu+\gamma}{2};\ \frac{\nu+\gamma}{2}+1,\ \frac{\alpha+\nu+\gamma}{2}+1,\ \delta+\frac{1}{2};\ -\frac{c^2z^2}{4}\right) \mp$$
$$\mp \frac{z\Gamma(\alpha+\nu-1)}{c^{\alpha-1}(1-\alpha)} \sin\frac{\alpha+\nu-\gamma}{2}\pi\, {}_3F_4\left(\frac{1-\alpha}{2},\ \frac{1}{2},\ 1;\right.$$
$$\left.\frac{3}{2},\ \frac{3-\alpha}{2},\ \frac{3-\alpha-\nu}{2},\ 1-\frac{\alpha+\nu}{2};\ -\frac{c^2z^2}{4}\right) - \frac{z^2\Gamma(\alpha+\nu-2)}{2c^{\alpha-2}(2-\alpha)} \times$$
$$\times \cos\frac{\alpha+\nu-\gamma}{2}\pi\, {}_3F_4\left(1-\frac{\alpha}{2},\ 1,\ 1;\ 2,\ 2-\frac{\alpha}{2},\ 2-\frac{\alpha+\nu}{2},\ \frac{3-\alpha-\nu}{2};\ -\frac{c^2z^2}{4}\right) -$$
$$- z^\alpha\Gamma(\nu)\cos\frac{\nu-\gamma}{2}\pi \begin{Bmatrix} \operatorname{cosec}\alpha\pi \\ \operatorname{ctg}\alpha\pi \end{Bmatrix} + \frac{\Gamma(\alpha+\nu)}{2\alpha c^\alpha} \cos\frac{\alpha+\nu-\gamma}{2}\pi \left[2\psi(\alpha+\nu) - \right.$$
$$\left. - \pi\operatorname{tg}\frac{\alpha+\nu-\gamma}{2}\pi - \frac{2}{\alpha} - 2\ln c\right].$$

2. $\int\limits_0^\infty x^{\alpha-1} \begin{Bmatrix} \ln(x+z) \\ \ln|x-z| \end{Bmatrix} C(cx,\ \nu)\,dx = V(0)$

$$\left[c,\ \operatorname{Re}\alpha > 0;\ 0 < \operatorname{Re}(\alpha+\nu) < 2,\ \begin{Bmatrix} |\arg z| < \pi \\ z > 0 \end{Bmatrix},\ V(\gamma) \text{ see } 2.6.5.1\right].$$

2.6.6. Integrals of $x^\alpha \begin{Bmatrix} S(bx,\ \mu) \\ C(bx,\ \mu) \end{Bmatrix} \begin{Bmatrix} S(cx,\ \nu) \\ C(cx,\ \nu) \end{Bmatrix}$.

NOTATION: $\delta = \begin{Bmatrix} 1 \\ 0 \end{Bmatrix}$.

1. $\int\limits_0^\infty x^{\alpha-1} \begin{Bmatrix} S(bx,\ \mu) \\ C(bx,\ \mu) \end{Bmatrix} S(cx,\ \nu)\,dx = W(1)$

[$b,\ c,\ \operatorname{Re}\alpha > 0;\ \operatorname{Re}(\alpha+\nu) > -1;\ \operatorname{Re}(\alpha+\mu) > -\delta;\ \operatorname{Re}(\alpha+\mu+\nu) > -\delta-1;$
$\operatorname{Re}(\alpha+\mu+\nu) < 3$ for $b \neq c$, $\operatorname{Re}(\alpha+\mu+\nu) < 2$ for $b=c$],

$$W(\gamma) = -\frac{c^{\nu+\gamma}\Gamma(\alpha+\mu+\nu+\gamma)}{b^{\alpha+\nu+\gamma}(\alpha+\nu+\gamma)(\nu+\gamma)} \begin{Bmatrix} \sin[(\alpha+\mu+\nu+\gamma)\pi/2] \\ \cos[(\alpha+\mu+\nu+\gamma)\pi/2] \end{Bmatrix} \times$$
$$\times {}_4F_3\left(\frac{\alpha+\nu+\gamma}{2},\ \frac{\nu+\gamma}{2},\ \frac{\alpha+\mu+\nu+1}{2},\ \frac{\alpha+\mu+\nu}{2}+\gamma;\right.$$
$$\left.\frac{\alpha+\nu+\gamma}{2}+1,\ \frac{\nu+\gamma}{2}+1,\ \gamma+\frac{1}{2};\ \frac{c^2}{b^2}\right) + \frac{\Gamma(\nu)\Gamma(\alpha+\mu)}{\alpha b^\alpha} \times$$
$$\times \cos\frac{\nu-\gamma}{2}\pi \begin{Bmatrix} \sin[(\alpha+\mu)\pi/2] \\ \cos[(\alpha+\mu)\pi/2] \end{Bmatrix} \qquad [0 < c \leqslant b],$$

$$W(\gamma)=\pm\frac{b^{\mu+\delta}\Gamma(\alpha+\mu+\nu+\delta)}{c^{\alpha+\mu+\delta}(\alpha+\mu+\delta)(\mu+\delta)}\begin{Bmatrix}\sin[(\alpha+\mu+\nu-\gamma)\pi/2]\\ \cos[(\alpha+\mu+\nu-\gamma)\pi/2]\end{Bmatrix}\times$$
$$\times{}_4F_3\left(\frac{\alpha+\mu+\delta}{2},\ \frac{\mu+\delta}{2},\ \frac{\alpha+\mu+\nu+1}{2},\ \frac{\alpha+\mu+\nu}{2}+\delta;\right.$$
$$\left.\frac{\alpha+\mu+\delta}{2}+1,\ \frac{\mu+\delta}{2}+1,\ \delta+\frac{1}{2};\ \frac{b^2}{c^2}\right)+\frac{\Gamma(\mu)\Gamma(\alpha+\nu)}{\alpha c^{\alpha}}\cos\frac{\alpha+\nu-\gamma}{2}\pi\begin{Bmatrix}\sin(\mu\pi/2)\\ \cos(\mu\pi/2)\end{Bmatrix}$$
$[0<b\leqslant c]$.

2. $$\int_0^\infty x^{\alpha-1}\begin{Bmatrix}S(bx,\ \mu)\\ C(bx,\ \mu)\end{Bmatrix}C(cx,\ \nu)\,dx=W(0)$$

$[b,\ c,\ \operatorname{Re}\alpha,\ \operatorname{Re}(\alpha+\nu)>0;\ \operatorname{Re}(\alpha+\mu)>-\delta;\ -\delta<\operatorname{Re}(\alpha+\mu+\nu)<3$ for $b\neq c$, $-\delta<\operatorname{Re}(\alpha+\mu+\nu)<2$ for $b=c$; $W(\gamma)$ see 2.6.6.1].

2.6.7. Integrals containing $\operatorname{Ei}(-bx^n)\begin{Bmatrix}S(cx,\ \nu)\\ C(cx,\ \nu)\end{Bmatrix}$.

NOTATION: $\delta=\begin{Bmatrix}1\\0\end{Bmatrix}$.

1. $$\int_0^\infty x^{\alpha-1}\operatorname{Ei}(-bx)\begin{Bmatrix}S(cx,\ \nu)\\ C(cx,\ \nu)\end{Bmatrix}dx=$$
$$=\frac{c^{\nu+\delta}\Gamma(\alpha+\nu+\delta)}{b^{\alpha+\nu+\delta}(\alpha+\nu+\delta)(\nu+\delta)}\,{}_4F_3\left(\frac{\alpha+\nu+\delta}{2},\ \frac{\nu+\delta}{2},\ \frac{\alpha+\nu+1}{2},\ \frac{\alpha+\nu}{2}+\delta;\right.$$
$$\left.\frac{\alpha+\nu+\delta}{2}+1,\ \frac{\nu+\delta}{2}+1,\ \delta+\frac{1}{2};\ -\frac{c^2}{b^2}\right)-\frac{\Gamma(\alpha)\Gamma(\nu)}{\alpha b^{\alpha}}\begin{Bmatrix}\sin(\nu\pi/2)\\ \cos(\nu\pi/2)\end{Bmatrix}$$
$[c, \operatorname{Re}b,\ \operatorname{Re}\alpha>0;\ \operatorname{Re}(\alpha+\nu)>-\delta]$.

2. $$\int_0^\infty x^{\alpha-1}\operatorname{Ei}(-bx^2)\begin{Bmatrix}S(cx,\ \nu)\\ C(cx,\ \nu)\end{Bmatrix}dx=\frac{c^{\nu+\delta}}{b^{(\alpha+\nu+\delta)/2}(\nu+\delta)(\alpha+\nu+\delta)}\times$$
$$\times\Gamma\left(\frac{\alpha+\nu+\delta}{2}\right){}_3F_3\left(\frac{\alpha+\nu+\delta}{2},\frac{\alpha+\nu+\delta}{2},\frac{\nu+\delta}{2};\frac{\alpha+\nu+\delta}{2}+1,\frac{\nu+\delta}{2}+1,\delta+\frac{1}{2};-\frac{c^2}{4b}\right)-$$
$$-\frac{\Gamma(\nu)\Gamma(\alpha/2)}{\alpha b^{\alpha/2}}\begin{Bmatrix}\sin(\nu\pi/2)\\ \cos(\nu\pi/2)\end{Bmatrix}$$
$[c,\ \operatorname{Re}b,\ \operatorname{Re}\alpha>0;\ \operatorname{Re}(\alpha+\nu)>-\delta]$.

3. $$\int_0^\infty x^{\alpha-1}e^{\pm bx}\operatorname{Ei}(\mp bx)\,S(cx,\ \nu)\,dx=Q(1)$$

$$\left[c,\ \operatorname{Re}\alpha>0;\ -1<\operatorname{Re}(\alpha+\nu)<3,\ \begin{Bmatrix}|\arg b|<\pi\\ b>0\end{Bmatrix}\right],$$

$$Q(\gamma)=(\mp1)^{\nu}\frac{\pi c^{\nu+\gamma}\Gamma(\alpha+\nu+\gamma)}{b^{\alpha+\nu+\gamma}(\nu+\gamma)}\begin{Bmatrix}\operatorname{cosec}[(\alpha+\nu)\pi]\\ \operatorname{ctg}[(\alpha+\nu)\pi]\end{Bmatrix}\times$$
$$\times{}_3F_2\left(\frac{\nu+\gamma}{2},\ \frac{\alpha+\nu+1}{2},\ \frac{\alpha+\nu}{2}+\gamma;\ \frac{\nu+\gamma}{2}+1,\ \gamma+\frac{1}{2};\ -\frac{c^2}{b^2}\right)+$$
$$+\frac{c^{1-\alpha}\Gamma(\alpha+\nu-1)}{b(1-\alpha)}\sin\frac{\alpha+\nu-\gamma}{2}\pi\,{}_4F_3\left(1,\ 1,\ \frac{1-\alpha}{2},\ \frac{1}{2};\right.$$
$$\left.\frac{3-\alpha}{2},\ \frac{3-\alpha-\nu}{2},\ 1-\frac{\alpha+\nu}{2};\ -\frac{c^2}{b^2}\right)+\frac{c^{2-\alpha}\Gamma(\alpha+\nu-2)}{b^2(2-\alpha)}\cos\frac{\alpha+\nu-\gamma}{2}\pi\times$$
$$\times{}_4F_3\left(1,\ 1,\ \frac{3}{2},\ 1-\frac{\alpha}{2};\ 2-\frac{\alpha}{2},\ 2-\frac{\alpha+\nu}{2},\ \frac{3-\alpha-\nu}{2};\ -\frac{c^2}{b^2}\right)-$$
$$-\frac{\pi\Gamma(\alpha)\Gamma(\nu)}{b^{\alpha}}\cos\frac{\gamma-\nu}{2}\pi\begin{Bmatrix}\operatorname{cosec}\alpha\pi\\ \operatorname{ctg}\alpha\pi\end{Bmatrix}.$$

4. $\int_0^\infty x^{\alpha-1} e^{\pm bx} \operatorname{Ei}(\mp bx)\, C(cx, \nu)\, dx = Q(0)$

$\left[c,\ \operatorname{Re}\alpha > 0;\ 0 < \operatorname{Re}(\alpha+\nu) < 3,\ \begin{Bmatrix} |\arg b| < \pi \\ b > 0 \end{Bmatrix},\ Q(\gamma) \text{ see } 2.6.7.3\right].$

2.6.8. Integrals of $x^\alpha \begin{Bmatrix} \operatorname{si}(bx) \\ \operatorname{ci}(bx) \end{Bmatrix} \begin{Bmatrix} S(cx, \nu) \\ C(cx, \nu) \end{Bmatrix}$.

NOTATION: $\delta = \begin{Bmatrix} 1 \\ 0 \end{Bmatrix}$.

1. $\int_0^\infty x^{\alpha-1} \begin{Bmatrix} \operatorname{si}(bx) \\ \operatorname{ci}(bx) \end{Bmatrix} S(cx, \nu)\, dx = X(1)$

$[b,\ c,\ \operatorname{Re}\alpha > 0;\ -1 < \operatorname{Re}(\alpha+\nu) < 3 \text{ for } b \neq c,\ -1 < \operatorname{Re}(\alpha+\nu) < 2 \text{ for } b = c],$

$$X(\gamma) = \frac{c^{\nu+\gamma}\Gamma(\alpha+\nu+\gamma)}{b^{(\alpha+\nu+\gamma)}(\nu+\gamma)(\alpha+\nu+\gamma)} \begin{Bmatrix} \sin[(\alpha+\nu+\gamma)\pi/2] \\ \cos[(\alpha+\nu+\gamma)\pi/2] \end{Bmatrix} \times$$
$$\times {}_4F_3\left(\frac{\nu+\gamma}{2}, \frac{\alpha+\nu+\gamma}{2}, \frac{\alpha+\nu+1}{2}, \frac{\alpha+\nu}{2}+\gamma;\right.$$
$$\left.\frac{\nu+\gamma}{2}+1, \frac{\alpha+\nu+\gamma}{2}+1, \gamma+\frac{1}{2}; \frac{c^2}{b^2}\right) - \frac{\Gamma(\nu)\Gamma(\alpha)}{\alpha b^\alpha} \cos\frac{\nu-\gamma}{2}\pi \begin{Bmatrix} \sin(\alpha\pi/2) \\ \cos(\alpha\pi/2) \end{Bmatrix}$$
$[0 < c \leqslant b],$

$$X(\gamma) = \frac{b^{\delta+2}\Gamma(\alpha+\nu+\delta+2)}{2c^{\alpha+\delta+2}(1+2\delta)(\alpha+\delta+2)(2+\delta)} \begin{Bmatrix} \sin[(\gamma-\alpha-\nu)\pi/2] \\ \cos[(\gamma-\alpha-\nu)\pi/2] \end{Bmatrix} \times$$
$$\times {}_5F_4\left(1, \frac{\alpha+\nu+3}{2}, \frac{\alpha+\nu}{2}+\delta+1, \frac{\alpha+\delta}{2}+1, 1+\frac{\delta}{2}; 2, \delta+\frac{3}{2}, \frac{\alpha+\delta}{2}+2, 2+\frac{\delta}{2}; \frac{b^2}{c^2}\right) +$$
$$+ \frac{b^\delta\,\Gamma(\alpha+\nu+\delta)}{c^{\alpha+\delta}(\alpha+\delta)} \begin{Bmatrix} \sin[(\gamma-\alpha-\nu)\pi/2] \\ R\cos[(\gamma-\alpha-\nu)\pi/2] \end{Bmatrix} - \begin{Bmatrix} 1 \\ 0 \end{Bmatrix} \frac{\pi\Gamma(\alpha+\nu)}{2\alpha c^\alpha} \cos\frac{\alpha+\nu-\gamma}{2}\pi,$$
$$R = \mathbf{C} - \frac{1}{\alpha} + \psi(\alpha+\nu) - \frac{\pi}{2}\operatorname{tg}\frac{\alpha+\nu-\gamma}{2}\pi - \ln\frac{c}{b} \qquad [0 < b \leqslant c].$$

2. $\int_0^\infty x^{\alpha-1} \begin{Bmatrix} \operatorname{si}(bx) \\ \operatorname{ci}(bx) \end{Bmatrix} C(cx, \nu)\, dx = X(0)$

$[b,\ c,\ \operatorname{Re}\alpha > 0;\ 0 < \operatorname{Re}(\alpha+\nu) < 3 \text{ for } b \neq c,\ 0 < \operatorname{Re}(\alpha+\nu) < 2 \text{ for } b = c;\ X(\gamma) \text{ see } 2.6.8.1].$

2.6.9. Integrals of $x^\alpha \begin{Bmatrix} \operatorname{erf}(bx^r) \\ \operatorname{erfc}(bx^r) \end{Bmatrix} \begin{Bmatrix} S(cx, \nu) \\ C(cx, \nu) \end{Bmatrix}$.

1. $\int_0^\infty x^{\alpha-1} \begin{Bmatrix} \operatorname{erf}(bx) \\ \operatorname{erfc}(bx) \end{Bmatrix} S(cx, \nu)\, dx = Y(1)$

$\left[c > 0,\ \begin{Bmatrix} \operatorname{Re}\alpha > -1;\ |\operatorname{Re}(\alpha+\nu)| < 2 \\ \operatorname{Re}\alpha > 0;\ \operatorname{Re}(\alpha+\nu) > -1 \end{Bmatrix},\ |\arg b| < \pi/4\right],$

$$Y(\gamma) = \pm \frac{c^{\nu+\gamma}}{\sqrt{\pi}\, b^{\alpha+\nu+\gamma}(\alpha+\nu+\gamma)(\nu+\gamma)} \Gamma\left(\frac{\alpha+\nu+\gamma+1}{2}\right) \times$$
$$\times {}_3F_3\left(\frac{\alpha+\nu+1}{2}, \frac{\nu+\gamma}{2}, \frac{\alpha+\nu}{2}+\gamma; \frac{\alpha+\nu+\gamma}{2}+1, \frac{\nu+\gamma}{2}+1, \gamma+\frac{1}{2}; -\frac{c^2}{4b^2}\right) \mp$$
$$\mp \frac{\Gamma(\nu)}{\alpha\sqrt{\pi}\, b^\alpha} \Gamma\left(\frac{\alpha+1}{2}\right) \cos\frac{\nu-\gamma}{2}\pi + \begin{Bmatrix} 1 \\ 0 \end{Bmatrix} \frac{\Gamma(\alpha+\nu)}{\alpha c^\alpha} \cos\frac{\alpha+\nu-\gamma}{2}\pi.$$

2. $\int\limits_0^\infty x^{\alpha-1} \left\{ \begin{matrix} \operatorname{erf}(bx) \\ \operatorname{erfc}(bx) \end{matrix} \right\} C(cx, \nu)\, dx = Y(0)$

$\left[c > 0, \ \left\{ \begin{matrix} \operatorname{Re}\alpha > -1; \ -1 < \operatorname{Re}(\alpha+\nu) < 2 \\ \operatorname{Re}\alpha, \ \operatorname{Re}(\alpha+\nu) > 0 \end{matrix} \right\}, \ |\arg b| < \pi/4, \ Y(\gamma) \text{ see } 2.6.9.1 \right].$

3. $\int\limits_0^\infty x^{\alpha-1} \left\{ \begin{matrix} \operatorname{erf}(b\sqrt{x}) \\ \operatorname{erfc}(b\sqrt{x}) \end{matrix} \right\} S(cx, \nu)\, dx = \hat{Y}(1)$

$\left[c > 0, \ \left\{ \begin{matrix} \operatorname{Re}\alpha > -1/2; \ -3/2 < \operatorname{Re}(\alpha+\nu) < 2 \\ \operatorname{Re}\alpha > 0; \ \operatorname{Re}(\alpha+\nu) > -1 \end{matrix} \right\}, \ |\arg b| < \pi/4 \right],$

$$\hat{Y}(\gamma) = \pm \frac{c^{\nu+\gamma}\Gamma(\alpha+\nu+\gamma+1/2)}{\sqrt{\pi}\, b^{2\alpha+2\nu+2\gamma}(\alpha+\nu+\gamma)(\nu+\gamma)} \times$$
$$\times {}_4F_3\left(\frac{\alpha+\nu+\gamma}{2}, \frac{\nu+\gamma}{2}, \frac{2\alpha+2\nu+3}{4}, \frac{2\alpha+2\nu+1}{4}+\gamma; \right.$$
$$\left. \frac{\alpha+\nu+\gamma}{2}+1, \frac{\nu+\gamma}{2}+1, \gamma+\frac{1}{2}; -\frac{c^2}{b^4} \right) \mp \frac{\Gamma(\nu)}{\alpha\sqrt{\pi}\, b^{2\alpha}} \times$$
$$\times \Gamma\left(\alpha+\frac{1}{2}\right) \cos\frac{\nu-\gamma}{2}\pi + \left\{ \begin{matrix} 1 \\ 0 \end{matrix} \right\} \frac{\Gamma(\alpha+\nu)}{\alpha c^\alpha} \cos\frac{\alpha+\nu-\gamma}{2}\pi.$$

4. $\int\limits_0^\infty x^{\alpha-1} \left\{ \begin{matrix} \operatorname{erf}(b\sqrt{x}) \\ \operatorname{erfc}(b\sqrt{x}) \end{matrix} \right\} C(cx, \nu)\, dx = \hat{Y}(0)$

$\left[c > 0, \ \left\{ \begin{matrix} \operatorname{Re}\alpha > -1/2; \ -1/2 < \operatorname{Re}(\alpha+\nu) < 2 \\ \operatorname{Re}\alpha, \ \operatorname{Re}(\alpha+\nu) > 0 \end{matrix} \right\}, \ |\arg b| < \pi/4; \ \hat{Y}(\gamma) \text{ see } 2.6.9.3 \right].$

2.6.10. Integrals of $x^\alpha \left\{ \begin{matrix} S(bx) \\ C(bx) \end{matrix} \right\} \left\{ \begin{matrix} S(cx, \nu) \\ C(cx, \nu) \end{matrix} \right\}$.

NOTATION: $\delta = \left\{ \begin{matrix} 1 \\ 0 \end{matrix} \right\}$.

1. $\int\limits_0^\infty x^{\alpha-1} \left\{ \begin{matrix} S(bx) \\ C(bx) \end{matrix} \right\} S(cx, \nu)\, dx = Z(1)$

$[b, c > 0; \ \operatorname{Re}\alpha > -\delta-1/2; \ -\delta-3/2 < \operatorname{Re}(\alpha+\nu) < 2$ for $b \neq c$; $-\delta-3/2 < \operatorname{Re}(\alpha+\nu) < 3/2$ for $b = c]$,

$$Z(\gamma) = \frac{c^{\nu+\gamma}\Gamma(\alpha+\nu+\gamma+1/2)}{\sqrt{2\pi}\, b^{\alpha+\nu+\gamma}(\nu+\gamma)(\alpha+\nu+\gamma)} \left\{ \begin{matrix} \sin[(2\alpha+2\nu+2\gamma+1)\pi/4] \\ \cos[(2\alpha+2\nu+2\gamma+1)\pi/4] \end{matrix} \right\} \times$$
$$\times {}_4F_3\left(\frac{\nu+\gamma}{2}, \frac{\alpha+\nu+\gamma}{2}, \frac{2\alpha+2\nu+3}{4}, \frac{2\alpha+2\nu+1}{4}+\gamma; \right.$$
$$\left. \gamma+\frac{1}{2}, \frac{\nu+\gamma}{2}+1, \frac{\alpha+\nu+\gamma}{2}+1; \frac{c^2}{b^2} \right) - \frac{\Gamma(\nu)\Gamma(\alpha+1/2)}{\alpha\sqrt{2\pi}\, b^\alpha} \cos\frac{\nu-\gamma}{2}\pi \times$$
$$\times \left\{ \begin{matrix} \sin[(2\alpha+1)\pi/4] \\ \cos[(2\alpha+1)\pi/4] \end{matrix} \right\} + \frac{\Gamma(\alpha+\nu)}{4c^\alpha} \cos\frac{\alpha+\nu-\gamma}{2}\pi \quad [0 < c \leqslant b],$$

$$Z(\gamma) = \frac{2^{3/2} b^{\delta+1/2}\Gamma(\alpha+\nu+\delta+1/2)}{3^\delta \sqrt{\pi}\, c^{\alpha+\delta+1/2}(2\alpha+2\delta+1)} \left\{ \begin{matrix} \sin[(\gamma-\alpha-\nu-1/2)\pi/2] \\ \cos[(\gamma-\alpha-\nu-1/2)\pi/2] \end{matrix} \right\} \times$$
$$\times {}_4F_3\left(\frac{1+2\delta}{4}, \frac{2\alpha+2\nu+3}{4}, \frac{2\alpha+2\nu+1}{4}+\delta, \frac{2\alpha+2\delta+1}{4}; \right.$$
$$\left. \delta+\frac{1}{2}, \frac{5+2\delta}{4}, \frac{2\alpha+2\delta+5}{4}; \frac{b^2}{c^2} \right) \quad [0 < b \leqslant c].$$

2. $\int\limits_0^\infty x^{\alpha-1}\begin{Bmatrix} S(bx) \\ C(bx) \end{Bmatrix} C(cx, \nu)\,dx = Z(0)$

[$b, c>0$; $\mathrm{Re}\,\alpha>-\delta-1/2$; $-\delta-1/2<\mathrm{Re}(\alpha+\nu)<2$ for $b\neq c$; $-\delta-1/2<\mathrm{Re}(\alpha+\nu)<3/2$ for $b=c$; $Z(\gamma)$ see 2.6.10.1].

2.6.11. Integrals of $x^\alpha \begin{Bmatrix} \gamma(\mu, bx) \\ \Gamma(\mu, bx) \end{Bmatrix} \begin{Bmatrix} S(cx, \nu) \\ C(cx, \nu) \end{Bmatrix}$.

1. $\int\limits_0^\infty x^{\alpha-1}\begin{Bmatrix} \gamma(\mu, bx) \\ \Gamma(\mu, bx) \end{Bmatrix} S(cx, \nu)\,dx = R(1)$

$\left[c,\ \mathrm{Re}\,b>0,\ \begin{Bmatrix} \mathrm{Re}\,\mu,\ \mathrm{Re}(\alpha+\mu)>0;\ -1-\mathrm{Re}\,\mu<\mathrm{Re}(\alpha+\nu)<2 \\ \mathrm{Re}\,\alpha,\ \mathrm{Re}(\alpha+\mu)>0;\ \mathrm{Re}(\alpha+\nu),\ \mathrm{Re}(\alpha+\mu+\nu)>-1 \end{Bmatrix}\right]$,

$$R(\gamma) = \pm \frac{c^{\gamma+\nu}\,\Gamma(\alpha+\gamma+\mu+\nu)}{b^{\alpha+\gamma+\nu}(\alpha+\gamma+\nu)(\gamma+\nu)}\, {}_4F_3\left(\frac{\alpha+\gamma+\nu}{2}, \frac{\gamma+\nu}{2}, \frac{\alpha+\mu+\nu+1}{2}, \frac{\alpha+\mu+\nu}{2}+\gamma; \frac{\alpha+\gamma+\nu}{2}+1, \frac{\gamma+\nu}{2}+1, \gamma+\frac{1}{2}; -\frac{c^2}{b^2}\right) \mp$$
$$\mp \frac{\Gamma(\nu)\,\Gamma(\alpha+\mu)}{\alpha b^\alpha}\cos\frac{\gamma-\nu}{2}\pi + \begin{Bmatrix} 1 \\ 0 \end{Bmatrix} \frac{\Gamma(\mu)\,\Gamma(\alpha+\nu)}{\alpha c^\alpha}\cos\frac{\alpha-\gamma+\nu}{2}\pi.$$

2. $\int\limits_0^\infty x^{\alpha-1}\begin{Bmatrix} \gamma(\mu, bx) \\ \Gamma(\mu, bx) \end{Bmatrix} C(cx, \nu)\,dx = R(0)$

$\left[c,\ \mathrm{Re}\,b>0,\ \begin{Bmatrix} \mathrm{Re}\,\mu,\ \mathrm{Re}(\alpha+\mu)>0;\ -\mathrm{Re}\,\mu<\mathrm{Re}(\alpha+\nu)<2 \\ \mathrm{Re}\,\alpha,\ \mathrm{Re}(\alpha+\nu),\ \mathrm{Re}(\alpha+\mu),\ \mathrm{Re}(\alpha+\mu+\nu)>0 \end{Bmatrix},\ R(\gamma)\ \text{see } 2.6.11.1\right]$.

2.7. THE STRUVE FUNCTIONS $\mathbf{H}_\nu(x)$ AND $\mathbf{L}_\nu(x)$

For $\nu=\pm\frac{1}{2}, \pm\frac{3}{2}, \ldots$, the functions $\mathbf{H}_\nu(x)$ and for $\nu=-\frac{1}{2}, -\frac{3}{2}, \ldots$, the functions $\mathbf{L}_\nu(x)$ reduce to the Neumann functions $Y_\nu(x)$ and the Bessel functions $J_\nu(x), I_\nu(x)$:

$$\mathbf{H}_{n+1/2}(x) = Y_{n+1/2}(x) + \frac{(x/2)^{n-1/2}}{n!\sqrt{\pi}}\sum_{k=0}^{n}(-n)_k\left(\frac{1}{2}\right)_k\left(-\frac{x^2}{4}\right)^k,$$
$$\mathbf{H}_{-n-1/2}(x) = (-1)^n J_{n+1/2}(x),$$
$$\mathbf{L}_{-n-1/2}(x) = I_{n+1/2}(x),$$

which are elementary functions for these values of the parameters. For example,

$$\mathbf{H}_{\pm 1/2}(x) = \sqrt{\frac{2}{\pi x}}\begin{Bmatrix} 1-\cos x \\ \sin x \end{Bmatrix},\quad \mathbf{L}_{-1/2}(x) = \sqrt{\frac{2}{\pi x}}\,\mathrm{sh}\,x.$$

See [17] for the corresponding integrals. For the calculation of the integrals of $\mathbf{L}_\nu(x)$, the following formula can be used:

$$\mathbf{L}_\nu(cx) = i^{-\nu-1}\mathbf{H}_\nu(icx).$$

2.7.1. Integrals of general form

1. $\int\limits_0^a x^{\alpha-1}(a^r-x^r)^{\beta-1}\begin{Bmatrix} \mathbf{H}_\nu(cx) \\ \mathbf{L}_\nu(cx) \end{Bmatrix} dx =$

$$= \frac{a^{\alpha+\nu+r\beta-r+1}c^{\nu+1}}{2^\nu\sqrt{\pi}\,r}\,\Gamma\begin{bmatrix} \beta \\ \nu+3/2 \end{bmatrix}\sum_{k=0}^{\infty}\frac{(\mp a^2c^2/4)^k}{(3/2)_k(\nu+3/2)_k}\,\Gamma\begin{bmatrix} (2k+\alpha+\nu+1)/r \\ \beta+(2k+\alpha+\nu+1)/r \end{bmatrix}$$

[$a, r, \mathrm{Re}\,\beta>0$; $\mathrm{Re}(\alpha+\nu)>-1$].

2. $$\int_a^\infty x^{\alpha-1}(x^r-a^r)^{\beta-1}\mathbf{H}_\nu(cx)\,dx=\frac{a^{\alpha+\nu+r\beta-r+1}c^{\nu+1}}{2^\nu\sqrt{\pi}\,r}\Gamma\begin{bmatrix}\beta\\ \nu+3/2\end{bmatrix}\times$$
$$\times\sum_{k=0}^\infty\frac{(-a^2c^2/4)^k}{(3/2)_k(\nu+3/2)_k}\Gamma\begin{bmatrix}1-\beta-(2k+\alpha+\nu+1)/r\\ 1-(2k+\alpha+\nu+1)/r\end{bmatrix}+$$
$$+\frac{\pi}{2}\left(\frac{c}{2}\right)^{r-r\beta-\alpha}\sum_{k=0}^\infty\frac{(1-\beta)_k}{k!}\frac{\sec[(\alpha+\nu+r\beta-r-rk)\pi/2]}{\Gamma\left(1+\frac{r+\nu-\alpha-r\beta+rk}{2}\right)\Gamma\left(1+\frac{r-\nu-\alpha-r\beta+rk}{2}\right)}\left(\frac{ac}{2}\right)^{rk}$$
$[a,\ c,\ r,\ \operatorname{Re}\beta>0;\ \operatorname{Re}(\alpha+r\beta)<r+3/2,\ r-\operatorname{Re}\nu+1]$.

3. $$\int_0^\infty\frac{x^{\alpha-1}}{(x^r+z^r)^\rho}\mathbf{H}_\nu(cx)\,dx=\frac{c^{\nu+1}z^{\alpha+\nu-r\rho+1}}{2^\nu\sqrt{\pi}\,r\Gamma(\rho)\Gamma(\nu+3/2)}\times$$
$$\times\sum_{k=0}^\infty\frac{(-c^2z^2/4)^k}{(3/2)_k(\nu+3/2)_k}\Gamma\left(\frac{2k+\alpha+\nu+1}{r}\right)\Gamma\left(\rho-\frac{2k+\alpha+\nu+1}{r}\right)+$$
$$+\frac{\pi}{2}\left(\frac{c}{2}\right)^{r\rho-\alpha}\sum_{k=0}^\infty\frac{(-1)^k(\rho)_k}{k!}\times$$
$$\times\frac{\sec[(\alpha+\nu-r\rho-rk)\pi/2]}{\Gamma\left(1+\frac{rk+r\rho+\nu-\alpha}{2}\right)\Gamma\left(1+\frac{rk+r\rho-\nu-\alpha}{2}\right)}\left(\frac{cz}{2}\right)^{rk}$$
$[c,\ r>0;\ -1<\operatorname{Re}(\alpha+\nu)<r\operatorname{Re}\rho+1;\ \operatorname{Re}(\alpha-r\rho)<3/2;\ r\,|\arg z|<\pi]$.

4. $$\int_0^\infty\frac{x^{\alpha-1}}{x^r-y^r}\mathbf{H}_\nu(cx)\,dx=$$
$$=-\frac{\sqrt{\pi}\,c^{\nu+1}y^{\alpha+\nu-r+1}}{2^\nu r\Gamma(\nu+3/2)}\sum_{k=0}^\infty\frac{(-c^2y^2/4)^k}{(3/2)_k(\nu+3/2)_k}\operatorname{ctg}\frac{2k+\alpha+\nu+1}{r}\pi+$$
$$+\frac{\pi}{2}\left(\frac{c}{2}\right)^{r-\alpha}\sum_{k=0}^\infty\frac{\sec[(\alpha+\nu-r-rk)\pi/2]}{\Gamma\left(1+\frac{rk+r+\nu-\alpha}{2}\right)\Gamma\left(1+\frac{rk+r-\nu-\alpha}{2}\right)}\left(\frac{cy}{2}\right)^{rk}$$
$[c,\ r,\ y>0;\ -1<\operatorname{Re}(\alpha+\nu)<r+1;\ \operatorname{Re}\alpha<r+3/2]$.

5. $$\int_0^\infty x^{\alpha-1}e^{-px^r}\begin{Bmatrix}\mathbf{H}_\nu(cx)\\ \mathbf{L}_\nu(cx)\end{Bmatrix}dx=U_\pm(r)$$ $[c,\ r,\ \operatorname{Re}p>0;\ \operatorname{Re}(\alpha+\nu)>-1]$,

$$U_\pm(r)=\frac{c^{\nu+1}}{2^\nu\sqrt{\pi}\,p^{(\alpha+\nu+1)/r}r\Gamma(\nu+3/2)}\sum_{k=0}^\infty\frac{(\mp1)^k}{(3/2)_k(\nu+3/2)_k}\times$$
$$\times\Gamma\left(\frac{2k+\alpha+\nu+1}{r}\right)\left(\frac{c}{2p^{1/r}}\right)^{2k}\qquad[r>1],$$

$U_+(r)=$
$$\left(\frac{c}{2}\right)^{\nu-1}\frac{p^{(1-\alpha-\nu)/r}}{\sqrt{\pi}\,r\Gamma(\nu+1/2)}\sum_{k=0}^\infty(-1)^k\left(\frac{1}{2}\right)_k\left(\frac{1}{2}-\nu\right)_k\Gamma\left(\frac{\alpha+\nu-2k-1}{r}\right)\left(\frac{2p^{1/r}}{c}\right)^{2k}+$$
$$+\frac{2^{\alpha-1}\pi}{c^\alpha}\sum_{k=0}^\infty\frac{(-1)^k}{k!}\frac{\sec[(\alpha+\nu+rk)\pi/2]}{\Gamma\left(1-\frac{\alpha-\nu+rk}{2}\right)\Gamma\left(1-\frac{\alpha+\nu+rk}{2}\right)}\left(\frac{2^rp}{c^r}\right)^k\qquad[r<1],$$

$U_\pm(1)$ see 2.7.5.2.

6. $\int\limits_0^\infty x^{\alpha-1} e^{-px^{-r}} \mathbf{H}_\nu(cx)\,dx =$

$$=\frac{c^{\nu+1}p^{(\alpha+\nu+1)/r}}{2^\nu \sqrt{\pi}\, r\Gamma(\nu+3/2)}\sum_{k=0}^{\infty}\frac{\Gamma(-(2k+\alpha+\nu+1)/r)}{(3/2)_k(\nu+3/2)_k}\left(-\frac{c^2p^{2/r}}{4}\right)^k+$$

$$+\frac{2^{\alpha-1}\pi}{c^\alpha}\sum_{k=0}^{\infty}\frac{\sec[(\alpha+\nu-rk)\pi/2]}{k!\,\Gamma(1-(\alpha-\nu-rk)/2)\,\Gamma(1-(\alpha+\nu-rk)/2)}\left(-\frac{c^r p}{2^r}\right)^k$$

$[c,\ r,\ \mathrm{Re}\,p>0;\ \mathrm{Re}\,(\alpha+\nu)<1;\ \mathrm{Re}\,\alpha<3/2]$.

7. $\int\limits_0^\infty x^{\alpha-1}\begin{Bmatrix}\sin bx^r\\ \cos bx^r\end{Bmatrix}\mathbf{H}_\nu(cx)\,dx = U(r)$

$\left[b,\ c,\ r>0;\ -1-\delta r<\mathrm{Re}\,(\alpha+\nu)<r+1;\ \mathrm{Re}\,\alpha<3/2\right.$ for $r<1$, $\mathrm{Re}\,\alpha<r+1/2$ for $r>1$; $\left.\delta=\begin{Bmatrix}1\\0\end{Bmatrix}\right]$.

$$U(r)=\frac{b^{-(\alpha+\nu+1)/r}c^{\nu+1}}{2^\nu\sqrt{\pi}\,r\Gamma(\nu+3/2)}\sum_{k=0}^{\infty}\frac{(-1)^k}{(3/2)_k(\nu+3/2)_k}\times$$

$$\times\Gamma\left(\frac{2k+\alpha+\nu+1}{r}\right)\begin{Bmatrix}\sin[(2k+\alpha+\nu+1)\pi/(2r)]\\ \cos[(2k+\alpha+\nu+1)\pi/(2r)]\end{Bmatrix}\left(\frac{c}{2b^{1/r}}\right)^{2k}\qquad [r>1],$$

$$U(r)=\left(\frac{c}{2}\right)^{\nu-1}\frac{b^{(1-\alpha-\nu)/r}}{\sqrt{\pi}\,r\Gamma(\nu+1/2)}\sum_{k=0}^{\infty}(-1)^k\left(\frac{1}{2}\right)_k\left(\frac{1}{2}-\nu\right)_k\times$$

$$\times\Gamma\left(\frac{\alpha+\nu-2k-1}{r}\right)\begin{Bmatrix}\sin[(\alpha+\nu-2k-1)\pi/(2r)]\\ \cos[(\alpha+\nu-2k-1)\pi/(2r)]\end{Bmatrix}\left(\frac{2b^{1/r}}{c}\right)^{2k}+$$

$$+\frac{2^{\alpha+\delta r-1}\pi b^\delta}{c^{\alpha+\delta r}}\sum_{k=0}^{\infty}\frac{(-1)^k}{k!(\delta+1/2)_k}\frac{\sec[(\alpha+\nu+\delta r+2rk)\pi/2]}{\Gamma\left(1-rk-\frac{\alpha-\nu+\delta r}{2}\right)\Gamma\left(1-rk-\frac{\alpha+\nu+\delta r}{2}\right)}\left(\frac{2^{r-1}b}{c^r}\right)^{2k}$$

$[r<1]$.

$U(1)$ see 2.7.7.4.

8. $\int\limits_0^\infty x^{\alpha-1}J_\mu(bx^r)\,\mathbf{H}_\nu(cx)\,dx = V(r)$

$[b,\ c,\ r>0;\ \mathrm{Re}\,(\alpha+\mu r+\nu)>-1;\ \mathrm{Re}\,(\alpha+\nu)<r+3/2;$ $\mathrm{Re}\,\alpha<2$ for $r<1$, $\mathrm{Re}\,\alpha<r+1$ for $r>1]$.

$$V(r)=\frac{1}{\sqrt{\pi}\,r\Gamma(\nu+3/2)}\left(\frac{b}{2}\right)^{-(\alpha+\nu+1)/r}\left(\frac{c}{2}\right)^{\nu+1}\times$$

$$\times\sum_{k=0}^{\infty}\frac{(-1)^k}{(3/2)_k(\nu+3/2)_k}\Gamma\begin{bmatrix}(\mu r+\alpha+\nu+2k+1)/(2r)\\ (\mu r-\alpha-\nu-2k-1)/(2r)+1\end{bmatrix}\left(\frac{c}{2^{1-1/r}b^{1/r}}\right)^{2k}\qquad [r>1].$$

$$V(r)=\frac{1}{2\sqrt{\pi}\,r\Gamma(\nu+1/2)}\left(\frac{b}{2}\right)^{(1-\alpha-\nu)/r}\left(\frac{c}{2}\right)^{\nu-1}\times$$

$$\times\sum_{k=0}^{\infty}(-1)^k\left(\frac{1}{2}\right)_k\left(\frac{1}{2}-\nu\right)_k\Gamma\begin{bmatrix}(\mu r+\alpha+\nu-2k-1)/(2r)\\ (\mu r+2k-\alpha-\nu+1)/(2r)+1\end{bmatrix}\left(\frac{2^{1-1/r}b^{1/r}}{c}\right)^{2k}+$$

$$+\frac{2^{\alpha+\mu r-\mu-1}\pi b^{\mu}}{c^{\alpha+\mu r}\Gamma(\mu+1)}\sum_{k=0}^{\infty}\frac{(-1)^k}{k!\,(\mu+1)_k}\times$$
$$\times\frac{\sec[(\alpha+\nu+\mu r+2rk)\pi/2]}{\Gamma(1-rk-(\alpha-\nu+\mu r)/2)\,\Gamma(1-rk-(\alpha+\nu+\mu r)/2)}\left(\frac{2^{r-1}b}{c^r}\right)^{2k}$$
$[r<1]$,

$V(1)$ see 2.7.14.1.

9. $\int\limits_0^a x^{\alpha-1}(a^r-x^r)^{\beta-1}[Y_\nu(cx)-\mathbf{H}_\nu(cx)]\,dx=W(1)$ $[a,\ c,\ r,\ \operatorname{Re}\beta>0;\ \operatorname{Re}\alpha>|\operatorname{Re}\nu|]$,

$$W(\gamma)=-\frac{a^{\alpha+\nu+r\beta-r+1}c^{\nu+1}}{2^\nu\sqrt{\pi}\,r}\Gamma\begin{bmatrix}\beta\\ \nu+3/2\end{bmatrix}\sum_{k=0}^{\infty}\frac{(-1)^{\gamma k}}{(3/2)_k(\nu+3/2)_k}\times$$
$$\times\Gamma\begin{bmatrix}(2k+\alpha+\nu+1)/r\\ \beta+(2k+\alpha+\nu+1)/r\end{bmatrix}\left(\frac{ac}{2}\right)^{2k}-\frac{a^{\alpha+r\beta-r}}{\pi r}\Gamma(\beta)\times$$
$$\times[\gamma\cos\nu\pi\,A(\nu)+\gamma A(-\nu)\pm(1-\gamma)\sin\nu\pi\,A(\pm\nu)],$$
$$A(\nu)=\Gamma(-\nu)\left(\frac{ac}{2}\right)^{\nu}\sum_{k=0}^{\infty}\frac{(-1)^{\gamma k}}{k!\,(\nu+1)_k}\Gamma\begin{bmatrix}(2k+\alpha+\nu)/r\\ \beta+(2k+\alpha+\nu)/r\end{bmatrix}\left(\frac{ac}{2}\right)^{2k}.$$

10. $\int\limits_0^a x^{\alpha-1}(a^r-x^r)^{\beta-1}\left[I_{\pm\nu}(cx)-\mathbf{L}_\nu(cx)\right]dx=W(0)$

$\left[a,\ c,\ r,\ \operatorname{Re}\beta>0,\ \left\{\begin{matrix}\operatorname{Re}(\alpha+\nu)>0\\ -\operatorname{Re}\alpha-1<\operatorname{Re}\nu<\operatorname{Re}\alpha\end{matrix}\right\};\ W(\gamma)\text{ see } 2.7.1.9\right]$.

11. $\int\limits_a^\infty x^{\alpha-1}(x^r-a^r)^{\beta-1}[Y_\nu(cx)-\mathbf{H}_\nu(cx)]\,dx=X(1)$

$[a,\ r,\ \operatorname{Re}\beta>0;\ \operatorname{Re}(\alpha+r\beta+\nu)<r+1;\ |\arg c|<\pi]$.

$$X(\gamma)=-\frac{a^{\alpha+\nu+r\beta-r+1}c^{\nu+1}}{2^\nu\sqrt{\pi}\,r}\Gamma\begin{bmatrix}\beta\\ \nu+3/2\end{bmatrix}\sum_{k=0}^{\infty}\frac{(-1)^{\gamma k}}{(3/2)_k(\nu+3/2)_k}\times$$
$$\times\Gamma\begin{bmatrix}1-\beta-(2k+\alpha+\nu+1)/r\\ 1-(2k+\alpha+\nu+1)/r\end{bmatrix}\left(\frac{ac}{2}\right)^{2k}-\frac{c^{r-\beta r-\alpha}(\cos\nu\pi)^{(1+\gamma\mp1\pm\gamma)/2}}{2^{r-r\beta-\alpha+1}\pi}\times$$
$$\times\sum_{k=0}^{\infty}\frac{(1-\beta)_k\,\Gamma((\alpha+\nu+\beta r-r-rk)/2)}{k!\cos[(\alpha+\nu+\beta r-r-rk)\pi/2]}\Gamma\left(\frac{\alpha-\nu-r+\beta r-rk}{2}\right)\times$$
$$\times\sin^{1-\gamma}\frac{(1-\beta+k)r-\alpha\pm\nu}{2}\pi\left(\frac{ac}{2}\right)^{rk}-\frac{a^{\alpha+r\beta-r}}{\pi r}\times$$
$$\times\Gamma(\beta)[\gamma\cos\nu\pi\,B(\nu)+\gamma B(-\nu)\pm(1-\gamma)\sin\nu\pi\,B(\pm\nu)],$$
$$B(\nu)=\Gamma(-\nu)\left(\frac{ac}{2}\right)^{\nu}\sum_{k=0}^{\infty}\frac{(-1)^{\gamma k}}{k!\,(\nu+1)_k}\Gamma\begin{bmatrix}1-\beta-(2k+\alpha+\nu)/r\\ 1-(2k+\alpha+\nu)/r\end{bmatrix}\left(\frac{ac}{2}\right)^{2k}.$$

12. $\int\limits_a^\infty x^{\alpha-1}(x^r-a^r)^{\beta-1}\left[I_{\pm\nu}(cx)-\mathbf{L}_\nu(cx)\right]dx=X(0)$

$[a,\ r,\ \operatorname{Re}\beta,\ \operatorname{Re}c>0;\ \operatorname{Re}(\alpha+r\beta+\nu)<r+1;\ X(\gamma)\text{ see } 2.7.1.11]$,

13. $$\int_0^\infty \frac{x^{\alpha-1}}{(x^r+z^r)^\rho}[Y_\nu(cx)-\mathbf{H}_\nu(cx)]\,dx=Y(1)$$

$[r>0;\ |\operatorname{Re}\nu|<\operatorname{Re}\alpha<\operatorname{Re}(r\rho-\nu)+1;\ r\,|\arg z|<\pi;\ |\arg c|<\pi]$,

$$Y(\gamma)=-\frac{z^{\alpha+\nu-r\rho+1}c^{\nu+1}}{2^\nu\sqrt{\pi}\,r\Gamma(\rho)\,\Gamma(\nu+3/2)}\sum_{k=0}^{\infty}\frac{(-1)^{\gamma k}}{(3/2)_k(\nu+3/2)_k}\Gamma\left(\frac{2k+\alpha+\nu+1}{r}\right)\times$$
$$\times\Gamma\left(\rho-\frac{2k+\alpha+\nu+1}{r}\right)\left(\frac{cz}{2}\right)^{2k}-\frac{c^{r\rho-\alpha}(\cos\nu\pi)^{(1+\gamma\mp1\pm\gamma)/2}}{2^{r\rho-\alpha+1}\pi}\times$$
$$\times\sum_{k=0}^{\infty}\frac{(-1)^k(\rho)_k}{k!}\frac{\Gamma((\alpha+\nu-r\rho-rk)/2)\,\Gamma((\alpha-\nu-r\rho-rk)/2)}{\cos[(\alpha+\nu-r\rho-rk)\pi/2]}\times$$
$$\times\sin^{1-\gamma}\frac{r\rho+rk-\alpha\pm\nu}{2}\pi\left(\frac{cz}{2}\right)^{rk}-\frac{z^{\alpha-r\rho}}{\pi r\,\Gamma(\rho)}\times$$
$$\times[\gamma\cos\nu\pi\,D(\nu)+\gamma D(-\nu)\pm(1-\gamma)\sin\nu\pi\,D(\pm\nu)],$$
$$D(\nu)=\Gamma(-\nu)\left(\frac{cz}{2}\right)^{\nu}\sum_{k=0}^{\infty}\frac{(-1)^{\gamma k}}{k!\,(\nu+1)_k}\Gamma\left(\frac{2k+\alpha+\nu}{r}\right)\Gamma\left(\rho-\frac{2k+\alpha+\nu}{r}\right)\left(\frac{cz}{2}\right)^{2k}.$$

14. $$\int_0^\infty \frac{x^{\alpha-1}}{(x^r+z^r)^\rho}\left[I_{\pm\nu}(cx)-\mathbf{L}_\nu(cx)\right]dx=Y(0)$$

$\left[r,\ \operatorname{Re}c>0;\ \operatorname{Re}(\alpha+\nu-r\rho)<1,\ \left\{\begin{matrix}\operatorname{Re}(\alpha+\nu)>0\\ -\operatorname{Re}\alpha-1<\operatorname{Re}\nu<\operatorname{Re}\alpha\end{matrix}\right\},\ r\,|\arg z|<\pi;\right.$
$\left.Y(\gamma)\text{ see } 2.7.1.13\right]$.

15. $$\int_0^\infty \frac{x^{\alpha-1}}{x^r-y^r}[Y_\nu(cx)-\mathbf{H}_\nu(cx)]\,dx=Z(1)$$

$[r,\ y>0;\ |\operatorname{Re}\nu|<\operatorname{Re}\alpha<r-\operatorname{Re}\nu+1;\ |\arg c|<\pi]$,

$$Z(\gamma)=\frac{\sqrt{\pi}\,c^{\nu+1}y^{\alpha+\nu-r+1}}{2^\nu r\Gamma(\nu+3/2)}\sum_{k=0}^{\infty}\frac{(-1)^{\gamma k}}{(3/2)_k(\nu+3/2)_k}\operatorname{ctg}\frac{2k+\alpha+\nu+1}{r}\pi\left(\frac{cy}{2}\right)^{2k}-$$
$$-\frac{c^{r-\alpha}(\cos\nu\pi)^{(1+\gamma\mp1\pm\gamma)/2}}{2^{r-\alpha+1}\pi}\sum_{k=0}^{\infty}\frac{\Gamma((\alpha+\nu-r-rk)/2)\,\Gamma((\alpha-\nu-r-rk)/2)}{\cos[(\alpha+\nu-r-rk)\pi/2]}\times$$
$$\times\sin^{1-\gamma}\frac{r+rk-\alpha\pm\nu}{2}\pi\left(\frac{cy}{2}\right)^{rk}+\frac{y^{\alpha-r}}{r}\times$$
$$\times[\gamma\cos\nu\pi\,E(\nu)+\gamma E(-\nu)\pm(1-\gamma)\sin\nu\pi\,E(\pm\nu)],$$
$$E(\nu)=\Gamma(-\nu)\left(\frac{cy}{2}\right)^{\nu}\sum_{k=0}^{\infty}\frac{(-1)^{\gamma k}}{k!\,(\nu+1)_k}\operatorname{ctg}\frac{2k+\alpha+\nu}{r}\pi\left(\frac{cy}{2}\right)^{2k}.$$

16. $$\int_0^\infty \frac{x^{\alpha-1}}{x^r-y^r}\left[I_{\pm\nu}(cx)-\mathbf{L}_\nu(cx)\right]dx=Z(0)$$

$\left[r,\ y,\ \operatorname{Re}c>0;\ \operatorname{Re}(\alpha+\nu)<r+1,\ \left\{\begin{matrix}\operatorname{Re}(\alpha+\nu)>0\\ -\operatorname{Re}\alpha-1<\operatorname{Re}\nu<\operatorname{Re}\alpha\end{matrix}\right\};\ Z(\gamma)\text{ see } 2.7.1.15\right]$.

17. $$\int_0^\infty x^{\alpha-1}e^{-px^r}[Y_\nu(cx)-\mathbf{H}_\nu(cx)]\,dx=U(1)$$

$[r,\ \operatorname{Re}p>0;\ \operatorname{Re}\alpha>|\operatorname{Re}\nu|;\ |\arg c|<\pi]$,

$$U(\gamma)=-\frac{c^{\nu+1}p^{-(\alpha+\nu+1)/r}}{2^{\nu}\sqrt{\pi}\,r\Gamma(\nu+3/2)}\sum_{k=0}^{\infty}\frac{(-1)^{\gamma k}}{(3/2)_k(\nu+3/2)_k}\Gamma\left(\frac{2k+\alpha+\nu+1}{r}\right)\left(\frac{c}{2p^{1/r}}\right)^{2k}-$$

$$-\frac{p^{-\alpha/r}}{\pi r}[\gamma\cos\nu\pi\,F(\nu)+\gamma F(-\nu)\pm(1-\gamma)\sin\nu\pi\,F(\pm\nu)],$$

$$F(\nu)=\Gamma(-\nu)\left(\frac{c}{2p^{1/r}}\right)^{\nu}\sum_{k=0}^{\infty}\frac{(-1)^{\gamma k}}{k!\,(\nu+1)_k}\Gamma\left(\frac{2k+\alpha+\nu}{r}\right)\left(\frac{c}{2p^{1/r}}\right)^{2k}\qquad [r>1],$$

$$U(\gamma)=\frac{(-1)^{\gamma}2^{\alpha-1}}{\pi c^{\alpha}}(\cos\nu\pi)^{(1+\gamma\mp1\pm\gamma)/2}\sum_{k=0}^{\infty}\frac{(-1)^k}{k!}\times$$

$$\times\frac{\Gamma((\alpha+\nu+rk)/2)\,\Gamma((\alpha-\nu+rk)/2)}{\cos[(\alpha+\nu+rk)\pi/2]}\sin^{1-\gamma}\frac{\alpha\mp\nu+rk}{2}\pi\left(\frac{2^r p}{c^r}\right)^k+$$

$$+(-1)^{\nu}\left(\frac{c}{2}\right)^{\nu-1}\frac{p^{(1-\alpha-\nu)/r}}{r\sqrt{\pi}\,\Gamma(\nu+1/2)}\sum_{k=0}^{\infty}(-1)^{\gamma k}\left(\frac{1}{2}\right)_k\left(\frac{1}{2}-\nu\right)_k\times$$

$$\times\Gamma\left(\frac{\alpha+\nu-2k-1}{r}\right)\left(\frac{2p^{1/r}}{c}\right)^{2k}\qquad [r<1].$$

18. $\int_0^{\infty}x^{\alpha-1}e^{-px^r}\left[I_{\pm}(cx)-\mathbf{L}_{\nu}(cx)\right]dx=U(0)$

$\left[r,\ \operatorname{Re}c,\ \operatorname{Re}p>0,\ \left\{\begin{matrix}\operatorname{Re}(\alpha+\nu)>0\\ -\operatorname{Re}\alpha-1<\operatorname{Re}\nu<\operatorname{Re}\alpha\end{matrix}\right\};\ U(\gamma)\ \text{see } 2.7.1.17\right].$

19. $\int_0^{\infty}x^{\alpha-1}e^{-px^{-r}}[Y_{\nu}(cx)-\mathbf{H}_{\nu}(cx)]\,dx=V(1)$

$[r,\ \operatorname{Re}p>0;\ |\arg c|<\pi;\ \operatorname{Re}(\alpha+\nu)<1],$

$$V(\gamma)=-\frac{c^{\nu+1}p^{(\alpha+\nu+1)/r}}{2^{\nu}\sqrt{\pi}\,r\Gamma(\nu+3/2)}\times$$

$$\times\sum_{k=0}^{\infty}\frac{(-1)^{\gamma k}}{(3/2)_k(\nu+3/2)_k}\Gamma\left(-\frac{2k+\alpha+\nu+1}{r}\right)\left(\frac{cp^{1/r}}{2}\right)^{2k}-\frac{p^{\alpha/r}}{r\pi}[\gamma\cos\nu\pi\,G(\nu)+$$

$$+\gamma G(-\nu)\pm(1-\gamma)\sin\nu\pi\,G(\pm\nu)]+\frac{(-1)^{\gamma}2^{\alpha-1}}{c^{\alpha}\pi}(\cos\nu\pi)^{(1+\gamma\mp1\pm\gamma)/2}\times$$

$$\times\sum_{k=0}^{\infty}\frac{(-1)^k}{k!}\frac{\Gamma((\alpha+\nu-rk)/2)\,\Gamma((\alpha-\nu-rk)/2)}{\cos[(\alpha+\nu-rk)\pi/2]}\sin^{1-\gamma}\frac{\alpha\mp\nu-rk}{2}\pi\left(\frac{c^r p}{2^r}\right)^k,$$

$$G(\nu)=\Gamma(-\nu)\left(\frac{cp^{1/r}}{2}\right)^{\nu}\sum_{k=0}^{\infty}\frac{(-1)^{\gamma k}}{k!\,(\nu+1)_k}\Gamma\left(-\frac{2k+\alpha+\nu}{r}\right)\left(\frac{cp^{1/r}}{2}\right)^{2k}.$$

20. $\int_0^{\infty}x^{\alpha-1}e^{-px^{-r}}\left[I_{\pm\nu}(cx)-\mathbf{L}_{\nu}(cx)\right]dx=V(0)$

$[r,\ \operatorname{Re}c,\ \operatorname{Re}p>0;\ \operatorname{Re}(\alpha+\nu)<1;\ V(\gamma)\ \text{see } 2.7.1.19].$

21. $\int_0^{\infty}x^{\alpha-1}\left\{\begin{matrix}\sin bx^r\\ \cos bx^r\end{matrix}\right\}[Y_{\nu}(cx)-\mathbf{H}_{\nu}(cx)]\,dx=W(1,\ \delta)$

$\left[b,\ r>0;\ |\operatorname{Re}\nu|-\delta r<\operatorname{Re}\alpha<1+r-\operatorname{Re}\nu;\ |\arg c|<\pi,\ \delta=\left\{\begin{matrix}1\\0\end{matrix}\right\}\right].$

$$W(\gamma, \delta) = -\frac{b^{-(\alpha+\nu+1)/r}c^{\nu+1}}{2^{\nu}\sqrt{\pi}\, r\Gamma(\nu+3/2)} \sum_{k=0}^{\infty} \frac{(-1)^{\gamma k}}{(3/2)_k(\nu+3/2)_k} \times$$

$$\times \Gamma\left(\frac{2k+\alpha+\nu+1}{r}\right) \cos\frac{2k+\alpha+\nu-\delta r+1}{2r}\pi\left(\frac{c}{2b^{1/r}}\right)^{2k} -$$

$$-\frac{b^{-\alpha/r}}{\pi r}\left[\gamma \cos \nu\pi\, I(\nu) + \gamma I(-\nu) \pm (1-\gamma) \sin \nu\pi\, I(\pm\nu)\right],$$

$$I(\nu) = \Gamma(-\nu)\left(\frac{c}{2b^{1/r}}\right)^{\nu} \sum_{k=0}^{\infty} \frac{(-1)^{\gamma k}}{k!\,(\nu+1)_k} \Gamma\left(\frac{2k+\alpha+\nu}{r}\right) \times$$

$$\times \cos\frac{2k+\alpha+\nu-\delta r}{2r}\pi\left(\frac{c}{2b^{1/r}}\right)^{2k} \qquad [r > 1],$$

$$W(\gamma, \delta) = (-1)^{\gamma}\left(\frac{c}{2}\right)^{\nu-1} \frac{b^{(1-\alpha-\nu)/r}}{r\sqrt{\pi}\,\Gamma(\nu+1/2)} \times$$

$$\sum_{k=0}^{\infty} (-1)^{\gamma k}\left(\frac{1}{2}\right)_k\left(\frac{1}{2}-\nu\right)_k \Gamma\left(\frac{\alpha+\nu-2k-1}{r}\right) \cos\frac{2k-\alpha-\nu+\delta r+1}{2r}\pi\left(\frac{2b^{1/r}}{c}\right)^{2k} +$$

$$+(-1)^{\gamma}\frac{2^{\alpha+\delta r-1}b^{\delta}}{c^{\alpha+\delta r}\pi}(\cos \nu\pi)^{(1+\gamma\mp 1\pm\gamma)/2} \times$$

$$\times \sum_{k=0}^{\infty} \frac{(-1)^k}{k!\,(\delta+1/2)_k} \frac{\Gamma((\alpha+\nu+\delta r)/2+rk)\,\Gamma((\alpha-\nu+\delta r)/2+rk)}{\cos[(\alpha+\nu+r\delta+2rk)\pi/2]} \times$$

$$\times \sin^{1-\gamma}\frac{\alpha\mp\nu+\delta r+2rk}{2}\pi\left(\frac{2^{r-1}b}{c^{r}}\right)^{2k} \qquad [r < 1].$$

22. $\displaystyle\int_0^{\infty} x^{\alpha-1} \sin bx^r \left[I_{\pm\nu}(cx) - \mathbf{L}_{\nu}(cx)\right] dx = W(0, 1)$

$\left[b, r, \operatorname{Re} c > 0;\ \operatorname{Re}(\alpha+\nu) < r+1,\ \begin{Bmatrix} \operatorname{Re}(\alpha+\nu) > -r \\ -1-r-\operatorname{Re}\alpha < \operatorname{Re}\nu < \operatorname{Re}\alpha+r \end{Bmatrix};\ W(\gamma, \delta) \text{ see } 2.7.1.21\right].$

23. $\displaystyle\int_0^{\infty} x^{\alpha-1} \cos bx^r \left[I_{\pm\nu}(cx) - \mathbf{L}_{\nu}(cx)\right] dx = W(0, 0)$

$\left[b, r, \operatorname{Re} c > 0;\ \operatorname{Re}(\alpha+\nu) < r+1,\ \begin{Bmatrix} \operatorname{Re}(\alpha+\nu) > 0 \\ -1-\operatorname{Re}\alpha < \operatorname{Re}\nu < \operatorname{Re}\alpha \end{Bmatrix};\ W(\gamma, \delta) \text{ see } 2.7.1.21\right].$

24. $\displaystyle\int_0^{\infty} x^{\alpha-1} J_{\mu}(bx^r)\left[Y_{\nu}(cx) - \mathbf{H}_{\nu}(cx)\right] dx = X(1)$

$[b,\ r > 0;\ |\operatorname{Re}\nu| - r\operatorname{Re}\mu < \operatorname{Re}\alpha < 3r/2+1-\operatorname{Re}\nu;\ |\arg c| < \pi],$

$$X(\gamma) = -\left(\frac{b}{2}\right)^{-(\alpha+\nu+1)/r}\left(\frac{c}{2}\right)^{\nu+1} \frac{1}{r\sqrt{\pi}\,\Gamma(\nu+3/2)} \times$$

$$\times \sum_{k=0}^{\infty} \frac{(-1)^{\gamma k}}{(3/2)_k(\nu+3/2)_k} \Gamma\left[\begin{matrix} (\mu r+\alpha+\nu+2k+1)/(2r) \\ 1+(\mu r-\alpha-\nu-2k-1)/(2r) \end{matrix}\right]\left(\frac{2^{1/r-1}c}{b^{1/r}}\right)^{2k} -$$

$$-\frac{2^{\alpha/r-1}}{b^{\alpha/r}r\pi}\left[\gamma \cos \nu\pi J(\nu) + \gamma J(-\nu) \pm (1-\gamma) \sin \nu\pi J(\pm\nu)\right],$$

$$J(\nu)=$$

$$=\Gamma(-\nu)\left(\frac{2^{1/r-1}c}{b^{1/r}}\right)^{\nu}\sum_{k=0}^{\infty}\frac{(-1)^{\nu k}}{k!\,(\nu+1)_k}\Gamma\left[\begin{matrix}(r\mu+\alpha+\nu+2k)/(2r)\\ 1+(\mu r-\alpha-\nu-2k)/(2r)\end{matrix}\right]\left(\frac{2^{1/r-1}c}{b^{1/r}}\right)^{2k}$$

$[r>1]$,

$$X(\gamma)=(-1)^{\gamma}\left(\frac{b}{2}\right)^{(1-\alpha-\nu)/r}\frac{c^{\nu-1}}{2^{\nu}\sqrt{\pi}\,r\Gamma(\nu+1/2)}\times$$

$$\sum_{k=0}^{\infty}(-1)^{\gamma k}\left(\frac{1}{2}\right)_k\left(\frac{1}{2}-\nu\right)_k\Gamma\left[\begin{matrix}(\mu r+\alpha+\nu-2k-1)/(2r)\\ 1+(\mu r-\alpha-\nu+2k+1)/(2r)\end{matrix}\right]\left(\frac{b^{1/r}}{2^{1/r-1}c}\right)^{2k}+$$

$$+(-1)^{\gamma}\frac{2^{\alpha+\mu r-\mu-1}b^{\mu}}{\pi c^{\alpha+\mu r}\Gamma(\mu+1)}(\cos\nu\pi)^{(1+\gamma\mp1\pm\gamma)/2}\times$$

$$\times\sum_{k=0}^{\infty}\frac{(-1)^k}{k!\,(\mu+1)_k}\frac{\Gamma((\alpha+\nu+\mu r)/2+rk)\Gamma((\alpha-\nu+\mu r)/2+rk)}{\cos[(\alpha+\nu+\mu r+2rk)\pi/2]}\times$$

$$\times\sin^{1-\gamma}\frac{\alpha\mp\nu+\mu r+2rk}{2}\pi\left(\frac{2^{r-1}b}{c^r}\right)^{2k}$$

$[r<1]$.

25. $\int_0^\infty x^{\alpha-1}J_\mu(bx^r)\,[I_{\pm\nu}(cx)-\mathbf{L}_\nu(cx)]\,dx=X(0)$

$\left[b,\ r,\ \operatorname{Re}c>0;\ \operatorname{Re}(\alpha+\nu)<3r/2+1,\ \left\{\begin{matrix}\operatorname{Re}(\alpha+\nu+r\mu)>0\\ -1-\operatorname{Re}(\alpha+r\mu)<\operatorname{Re}\nu<\operatorname{Re}(\alpha+r\mu)\end{matrix}\right\};\ X(\gamma)\ \text{see 2.7.1.24}\right]$.

2.7.2. Integrals of $x^\alpha\mathbf{H}_\nu(cx)$.

1. $\int_0^\infty \mathbf{H}_\nu(cx)\,dx=-\frac{1}{c}\operatorname{ctg}\frac{\nu\pi}{2}$ $\quad[c>0;\ -2<\operatorname{Re}\nu<0]$.

2. $\int_0^\infty x^{\alpha-1}\mathbf{H}_\nu(cx)\,dx=2^{\alpha-1}c^{-\alpha}\operatorname{tg}\frac{\alpha+\nu}{2}\pi\Gamma\left[\begin{matrix}(\nu+\alpha)/2\\ 1+(\nu-\alpha)/2\end{matrix}\right]$

$[c>0;\ \operatorname{Re}\alpha<3/2;\ |\operatorname{Re}(\alpha+\nu)|<1]$.

3. $\int_0^\infty x^{-\nu-1}\mathbf{H}_\nu(cx)\,dx=\frac{\pi c^\nu}{2^{\nu+1}\Gamma(\nu+1)}$ $\quad[c>0;\ \operatorname{Re}\nu>-3/2]$.

2.7.3. Integrals of $x^\alpha(z\pm x)^\beta\left\{\begin{matrix}\mathbf{H}_\nu(cx)\\ \mathbf{L}_\nu(cx)\end{matrix}\right\}$.

1. $\int_0^a x^{\alpha-1}(a-x)^{\beta-1}\left\{\begin{matrix}\mathbf{H}_\nu(cx)\\ \mathbf{L}_\nu(cx)\end{matrix}\right\}dx=$

$$=\frac{a^{\alpha+\beta+\nu}c^{\nu+1}}{2^\nu\sqrt{\pi}\,\Gamma(\nu+3/2)}\mathrm{B}(\beta,\ \alpha+\nu+1)\,{}_3F_4\left(1,\ \frac{\alpha+\nu+1}{2},\ \frac{\alpha+\nu}{2}+1;\ \frac{3}{2},\ \nu+\frac{3}{2},\ \frac{\alpha+\beta+\nu+1}{2},\ \frac{\alpha+\beta+\nu}{2}+1;\ \mp\frac{a^2c^2}{4}\right)$$

$[a,\ \operatorname{Re}\beta>0;\ \operatorname{Re}(\alpha+\nu)>-1]$.

2. $\int_a^\infty x^{\alpha-1}(x-a)^{\beta-1}\mathbf{H}_\nu(cx)\,dx=\frac{a^{\alpha+\beta+\nu}c^{\nu+1}}{2^\nu\sqrt{\pi}}\Gamma\left[\begin{matrix}\beta, & -\alpha-\beta-\nu\\ \nu+3/2, & -\alpha-\nu\end{matrix}\right]\times$

$$\times {}_3F_4\left(1, 1+\frac{\alpha+\nu}{2}, \frac{\alpha+\nu+1}{2}; 1+\frac{\alpha+\beta+\nu}{2}, \frac{\alpha+\beta+\nu+1}{2}, \frac{3}{2}, \nu+\frac{3}{2}; -\frac{a^2c^2}{4}\right)+$$

$$+\frac{\pi}{2}\left(\frac{c}{2}\right)^{1-\alpha-\beta}\frac{\operatorname{cosec}[(\alpha+\beta+\nu)\pi/2]}{\Gamma((3-\alpha-\beta+\nu)/2)\,\Gamma((3-\alpha-\beta-\nu)/2)}\times$$

$$\times {}_2F_3\left(\frac{1-\beta}{2}, 1-\frac{\beta}{2}; \frac{1}{2}, \frac{3-\alpha-\beta+\nu}{2}, \frac{3-\alpha-\beta-\nu}{2}; -\frac{a^2c^2}{4}\right)-$$

$$-\frac{\pi a}{2}\left(\frac{c}{2}\right)^{2-\alpha-\beta}\frac{(1-\beta)\sec[(\alpha+\beta+\nu)\pi/2]}{\Gamma(2-(\alpha+\beta-\nu)/2)\,\Gamma(2-(\alpha+\beta+\nu)/2)}\times$$

$$\times {}_2F_3\left(1-\frac{\beta}{2}, \frac{3-\beta}{2}; \frac{3}{2}, 2-\frac{\alpha+\beta-\nu}{2}, 2-\frac{\alpha+\beta+\nu}{2}; -\frac{a^2c^2}{4}\right)$$

[a, c, Re $\beta > 0$; Re $(\alpha+\beta) < 5/2$; Re $(\alpha+\beta+\nu) < 2$].

3. $$\int_0^\infty \frac{x^{\alpha-1}}{(x+z)^\rho}\mathbf{H}_\nu(cx)\,dx = \frac{c^{\nu+1}z^{\alpha+\nu-\rho+1}}{2^\nu\sqrt{\pi}\,\Gamma(\nu+3/2)}\,\mathrm{B}(\alpha+\nu+1, \rho-\alpha-\nu-1)\times$$

$$\times {}_3F_4\left(1, \frac{\alpha+\nu+1}{2}, \frac{\alpha+\nu}{2}+1; \frac{3+\alpha+\nu-\rho}{2}, 1+\frac{\alpha+\nu-\rho}{2}, \frac{3}{2}, \frac{3}{2}+\nu; -\frac{c^2z^2}{4}\right)+$$

$$+\frac{\pi}{2}\left(\frac{c}{2}\right)^{\rho-\alpha}\frac{\sec[(\alpha+\nu-\rho)\pi/2]}{\Gamma(1+(\rho+\nu-\alpha)/2)\,\Gamma(1+(\rho-\nu-\alpha)/2)}\times$$

$$\times {}_2F_3\left(\frac{\rho}{2}, \frac{\rho+1}{2}; 1+\frac{\rho+\nu-\alpha}{2}, 1+\frac{\rho-\nu-\alpha}{2}, \frac{1}{2}; -\frac{c^2z^2}{4}\right)-$$

$$-\frac{\pi z}{2}\left(\frac{c}{2}\right)^{1+\rho-\alpha}\frac{\rho\operatorname{cosec}[(\alpha+\nu-\rho)\pi/2]}{\Gamma((3+\rho+\nu-\alpha)/2)\,\Gamma((3+\rho-\nu-\alpha)/2)}\times$$

$$\times {}_2F_3\left(\frac{\rho+1}{2}, \frac{\rho}{2}+1; \frac{3+\rho+\nu-\alpha}{2}, \frac{3+\rho-\nu-\alpha}{2}, \frac{3}{2}; -\frac{c^2z^2}{4}\right)$$

[$c > 0$; $-1 <$ Re $(\alpha+\nu) <$ Re $\rho+1$; Re $(\alpha-\rho) < 3/2$; $|\arg z| < \pi$].

4. $$\int_0^\infty \frac{x^{\alpha-1}}{x-y}\mathbf{H}_\nu(cx)\,dx = -\pi y^{\alpha-1}\operatorname{ctg}(\alpha+\nu)\pi\,\mathbf{H}_\nu(cy)+$$

$$+\frac{2^{\alpha-2}\pi\operatorname{cosec}[(\alpha+\nu)\pi/2]}{c^{\alpha-1}\Gamma((3+\nu-\alpha)/2)\,\Gamma((3-\nu-\alpha)/2)}\,{}_1F_2\left(1; \frac{3+\nu-\alpha}{2}, \frac{3-\nu-\alpha}{2}; -\frac{c^2y^2}{4}\right)-$$

$$-\frac{\pi}{2}\left(\frac{c}{2}\right)^{2-\alpha}y\frac{\sec[(\alpha+\nu)\pi/2]}{\Gamma(2-(\alpha-\nu)/2)\,\Gamma(2-(\alpha+\nu)/2)}\,{}_1F_2\left(1; 2-\frac{\alpha-\nu}{2}, 2-\frac{\alpha+\nu}{2}; -\frac{c^2y^2}{4}\right)$$

[c, $y > 0$; $-1 <$ Re $(\alpha+\nu) < 2$; Re $\alpha < 5/2$].

2.7.4. Integrals of $x^\alpha (z^2 \pm x^2)^\beta \begin{Bmatrix}\mathbf{H}_\nu(cx)\\ \mathbf{L}_\nu(cx)\end{Bmatrix}$.

1. $$\int_0^a x^{\alpha-1}(a^2-x^2)^{\beta-1}\begin{Bmatrix}\mathbf{H}_\nu(cx)\\ \mathbf{L}_\nu(cx)\end{Bmatrix}dx=$$

$$=\frac{a^{\alpha+\nu+2\beta-1}c^{\nu+1}}{2^{\nu+1}\sqrt{\pi}\,\Gamma(\nu+3/2)}\,\mathrm{B}\left(\beta, \frac{\alpha+\nu+1}{2}\right)\times$$

$$\times {}_2F_3\left(1, \frac{\alpha+\nu+1}{2}; \frac{3}{2}, \frac{3}{2}+\nu, \frac{\alpha+\nu+1}{2}+\beta; \mp\frac{a^2c^2}{4}\right)$$ [a, Re $\beta>0$; Re $(\alpha+\nu)>-1$].

2. $$\int_0^a x^{1-\nu}(a^2-x^2)^{\beta-1}\mathbf{H}_\nu(cx)\,dx=$$

$$=2^{-\nu}a^{\beta-\nu}c^{-\beta}\Gamma\begin{bmatrix}\beta\\ \nu+1/2, \quad \beta+1/2\end{bmatrix}s_{\beta+\nu,\,\beta-\nu}(ac)$$ [a, Re $\beta > 0$].

3. $\int\limits_0^a x^{\nu+1}(a^2-x^2)^{\beta-1}\begin{Bmatrix}\mathbf{H}_\nu(cx)\\ \mathbf{L}_\nu(cx)\end{Bmatrix}dx=2^{\beta-1}a^{\beta+\nu}c^{-\beta}\Gamma(\beta)\begin{Bmatrix}\mathbf{H}_{\beta+\nu}(ac)\\ \mathbf{L}_{\beta+\nu}(ac)\end{Bmatrix}$

[a, $\operatorname{Re}\beta>0$; $\operatorname{Re}\nu>-3/2$].

4. $\int\limits_0^a x^{\nu+1}(a^2-x^2)^{-\nu-1/2}\begin{Bmatrix}\mathbf{H}_\nu(cx)\\ \mathbf{L}_\nu(cx)\end{Bmatrix}dx=\pm\frac{c^{\nu-1}}{2^\nu\sqrt{\pi}}\Gamma\left(\frac{1}{2}-\nu\right)\begin{Bmatrix}1-\cos ac\\ 1-\operatorname{ch} ac\end{Bmatrix}$

[$a>0$; $-3/2<\operatorname{Re}\nu<1/2$].

5. $\int\limits_0^a x^{\nu+1}(a^2-x^2)^{-\nu-3/2}\begin{Bmatrix}\mathbf{H}_\nu(cx)\\ \mathbf{L}_\nu(cx)\end{Bmatrix}dx=\frac{c^\nu\Gamma(-\nu-1/2)}{2^{\nu+1}\sqrt{\pi}\,a}\begin{Bmatrix}\sin ac\\ \operatorname{sh} ac\end{Bmatrix}$

[$a>0$; $-3/2<\operatorname{Re}\nu<-1/2$].

6. $$\int\limits_a^\infty x^{\alpha-1}(x^2-a^2)^{\beta-1}\mathbf{H}_\nu(cx)\,dx=\frac{a^{\alpha+\nu+2\beta-1}c^{\nu+1}}{2^{\nu+1}\sqrt{\pi}\,\Gamma(\nu+3/2)}\mathrm{B}\left(\beta,\ \frac{1-\alpha-\nu}{2}-\beta\right)\times$$
$$\times{}_2F_3\left(1,\ \frac{1+\alpha+\nu}{2};\ \frac{1+\alpha+\nu}{2}+\beta,\ \frac{3}{2},\ \frac{3}{2}+\nu;\ -\frac{a^2c^2}{4}\right)-$$
$$-\frac{\pi}{2}\left(\frac{c}{2}\right)^{2-2\beta-\alpha}\frac{\sec[(\alpha+2\beta+\nu)\pi/2]}{\Gamma(2-(\alpha-\nu)/2-\beta)\,\Gamma(2-(\alpha+\nu)/2-\beta)}\times$$
$$\times{}_1F_2\left(1-\beta;\ 2-\frac{\alpha-\nu}{2}-\beta,\ 2-\frac{\alpha+\nu}{2}-\beta;\ -\frac{a^2c^2}{4}\right)$$

[a, $\operatorname{Re}\beta>0$; $\operatorname{Re}(\alpha+2\beta)<7/2$; $\operatorname{Re}(\alpha+2\beta+\nu)<3$].

7. $$\int\limits_a^\infty x^{-\nu}(x^2-a^2)^{\beta-1}\mathbf{H}_\nu(cx)\,dx=$$
$$=\frac{2^{2\beta-\nu-2}\pi}{\sin\beta\pi}\frac{c^{\nu-2\beta+1}}{\Gamma(3/2-\beta)\,\Gamma(3/2+\nu-\beta)}\,{}_1F_2\left(1-\beta;\ \frac{3}{2}-\beta,\ \frac{3}{2}+\nu-\beta;\ -\frac{a^2c^2}{4}\right)$$

[a, $c>0$; $0<\operatorname{Re}\beta<1$; $\operatorname{Re}(\nu-2\beta)>-5/2$].

8. $$\int\limits_a^\infty x^{1-\nu}(x^2-a^2)^{\beta-1}\mathbf{H}_\nu(cx)\,dx=$$
$$=\frac{a^{\beta-\nu}c^{-\beta}}{2^\nu\pi}\Gamma(\beta)\left\{\Gamma\begin{bmatrix}1/2-\beta\\ \nu+1/2\end{bmatrix}s_{\beta+\nu,\ \beta-\nu}(ac)+2^{\nu+\beta-1}\pi\operatorname{tg}\beta\pi J_{\nu-\beta}(ac)\right\}=$$
$$=\frac{a^{\beta-\nu}c^{-\beta}}{2^\nu\pi}\Gamma(\beta)\left\{\Gamma\begin{bmatrix}1/2-\beta\\ \nu+1/2\end{bmatrix}S_{\beta+\nu,\ \beta-\nu}(ac)+2^{\nu-\beta-1}\pi Y_{\nu-\beta}(ac)\right\}$$

[a, $c>0$; $0<\operatorname{Re}\beta<1/2$; $\operatorname{Re}(2\beta-\nu)<3/2$].

9. $$\int\limits_a^\infty x^{\nu+1}(x^2-a^2)^{\beta-1}\mathbf{H}_\nu(cx)\,dx=$$
$$=\frac{2^{\beta-1}a^{\beta+\nu}\Gamma(\beta)}{c^\beta\cos(\beta+\nu)\pi}[\sin\beta\pi J_{-\nu-\beta}(ac)+\cos\nu\pi\mathbf{H}_{\nu+\beta}(ac)]$$

[a, c, $\operatorname{Re}\beta>0$; $\operatorname{Re}(\nu+\beta)<1/2$; $\operatorname{Re}(\nu+2\beta)<3/2$].

10. $\int\limits_a^\infty x^{\nu+1}(x^2-a^2)^n\mathbf{H}_\nu(cx)\,dx=(-1)^{n+1}2^n n!\,a^{\nu+n+1}c^{-n-1}\mathbf{H}_{\nu+n+1}(ac)$

[a, $c>0$; $\operatorname{Re}\nu<-2n-1/2$].

11. $$\int_a^\infty x^{-\nu}(x^2-a^2)^{-\nu-1/2}\mathbf{H}_\nu(cx)\,dx=2^{-\nu-1}\sqrt{\pi}\,a^{-2\nu}c^\nu\Gamma\left(\frac{1}{2}-\nu\right)J_\nu^2\left(\frac{ac}{2}\right)$$
$[a,\ c>0;\ |\operatorname{Re}\nu|<1/2]$.

12. $$\int_0^\infty \frac{1}{x^2+z^2}\mathbf{H}_1(cx)\,dx=\frac{\pi}{2z}[I_1(cz)-\mathbf{L}_1(cz)]$$ $[c,\ \operatorname{Re}z>0]$.

13. $$\int_0^\infty \frac{1}{x^2+z^2}\mathbf{H}_\nu(cx)\,dx=$$
$$=-\frac{\pi}{2z}\operatorname{cosec}\frac{\nu\pi}{2}\mathbf{L}_\nu(cz)+\frac{c}{1-\nu^2}\operatorname{ctg}\frac{\nu\pi}{2}\,{}_1F_2\left(1;\ \frac{3-\nu}{2},\ \frac{3+\nu}{2};\ \frac{c^2z^2}{4}\right)$$
$[c,\ \operatorname{Re}z>0;\ |\operatorname{Re}\nu|<2]$.

14. $$\int_0^\infty \frac{x^{\alpha-1}}{(x^2+z^2)^\rho}\mathbf{H}_\nu(cx)\,dx=$$
$$=\frac{c^{\nu+1}z^{\alpha+\nu-2\rho+1}}{2^{\nu+1}\sqrt{\pi}\,\Gamma(\nu+3/2)}\mathrm{B}\left(\frac{\alpha+\nu+1}{2},\ \rho-\frac{\alpha+\nu+1}{2}\right)\times$$
$$\times{}_2F_3\left(1,\ \frac{\alpha+\nu+1}{2};\ \frac{3}{2},\ \frac{3}{2}+\nu,\ \frac{3+\alpha+\nu}{2}-\rho;\ \frac{c^2z^2}{4}\right)+$$
$$+\frac{\pi}{2}\left(\frac{c}{2}\right)^{2\rho-\alpha}\frac{\sec[(\nu+\alpha-2\rho)\pi/2]}{\Gamma(1-(\alpha-\nu)/2+\rho)\,\Gamma(1-(\alpha+\nu)/2+\rho)}\times$$
$$\times{}_1F_2\left(\rho;\ 1+\rho-\frac{\alpha-\nu}{2},\ 1+\rho-\frac{\alpha+\nu}{2};\ \frac{c^2z^2}{4}\right)$$
$[c,\ \operatorname{Re}z>0;\ -1<\operatorname{Re}(\alpha+\nu)<2\operatorname{Re}\rho+1;\ \operatorname{Re}(\alpha-2\rho)<3/2]$.

15. $$\int_0^\infty \frac{x^{1-\nu}}{(x^2+z^2)^\rho}\mathbf{H}_\nu(cx)\,dx=-\frac{\pi c^{\rho-1}z^{1-\rho-\nu}}{2^\rho\Gamma(\rho)\cos\rho\pi}I_{\nu+\rho-1}(cz)+$$
$$+\frac{c^{\nu+1}z^{3-2\rho}}{2^{\nu+2}}\Gamma\left[\begin{matrix}\rho-3/2\\ \rho,\ \nu+3/2\end{matrix}\right]{}_1F_2\left(1;\ \frac{5}{2}-\rho,\ \frac{3}{2}+\nu;\ \frac{c^2z^2}{4}\right)$$
$[c,\ \operatorname{Re}z>0;\ \operatorname{Re}\rho>1/2;\ \operatorname{Re}(2\rho+\nu)>1/2]$.

16. $$\int_0^\infty \frac{x^{\nu+1}}{(x^2+z^2)^\rho}\mathbf{H}_\nu(cx)\,dx=\frac{\pi c^{\rho-1}z^{1+\nu-\rho}}{2^\rho\Gamma(\rho)\cos(\nu-\rho)\pi}[\mathbf{L}_{1+\nu-\rho}(cz)-I_{\rho-\nu-1}(cz)]$$
$[c,\ \operatorname{Re}z>0;\ \operatorname{Re}(\nu-\rho)<-1/2;\ \operatorname{Re}\nu>-3/2;\ \operatorname{Re}(\nu-2\rho)<-1/2]$.

17. $$\int_0^\infty \frac{x^{\alpha-1}}{x^2-y^2}\mathbf{H}_\nu(cx)\,dx=\frac{\pi y^{\alpha-2}}{2}\operatorname{tg}\frac{\alpha+\nu}{2}\pi\mathbf{H}_\nu(cy)+$$
$$+2^{\alpha-3}c^{2-\alpha}\operatorname{tg}\frac{\alpha+\nu}{2}\pi\Gamma\left[\begin{matrix}(\nu+\alpha)/2-1\\ 2+(\nu-\alpha)/2\end{matrix}\right]{}_1F_2\left(1;\ 2-\frac{\alpha-\nu}{2},\ 2-\frac{\alpha+\nu}{2};\ -\frac{c^2y^2}{4}\right)$$
$[c,\ y>0;\ -1<\operatorname{Re}(\alpha+\nu)<3;\ \operatorname{Re}\alpha<7/2]$.

18. $$\int_0^\infty \frac{(x+\sqrt{x^2+z^2})^{\nu+1}}{\sqrt{x^2+z^2}}\mathbf{H}_\nu(cx)\,dx=$$
$$=\frac{\sqrt{\pi}z^{\nu+1/2}}{\sqrt{c}\sin\nu\pi}\left[\operatorname{sh}\frac{cz}{2}I_{\nu+1/2}\left(\frac{cz}{2}\right)-\operatorname{ch}\frac{cz}{2}I_{-\nu-1/2}\left(\frac{cz}{2}\right)\right]$$
$[c,\ \operatorname{Re}z>0;\ -2<\operatorname{Re}\nu<0]$.

2.7.5. Integrals of $x^{\alpha}e^{-px}\begin{Bmatrix}\mathbf{H}_{\nu}(cx)\\ \mathbf{L}_{\nu}(cx)\end{Bmatrix}$.

1. $$\int_0^\infty e^{-px}\begin{Bmatrix}\mathbf{H}_n(cx)\\ \mathbf{L}_n(cx)\end{Bmatrix}dx=I_n\quad\left[A=\ln\frac{\sqrt{p^2+c^2}+c}{p},\ B=\arcsin\frac{c}{p},\ \begin{Bmatrix}\operatorname{Re}p>|\operatorname{Im}c|\\ \operatorname{Re}p>|\operatorname{Re}c|\end{Bmatrix}\right],$$

$$I_0=\frac{2}{\pi\sqrt{p^2\pm c^2}}\begin{Bmatrix}A\\ B\end{Bmatrix},\quad I_1=\pm\frac{2}{\pi p}\mp\frac{2p}{\pi c\sqrt{p^2\pm c^2}}\begin{Bmatrix}A\\ B\end{Bmatrix},$$

$$I_2=\frac{2}{\pi}\left(-\frac{2}{c}\pm\frac{c}{3p^2}+\frac{2p^2\pm c^2}{c^2\sqrt{p^2\pm c^2}}\begin{Bmatrix}A\\ B\end{Bmatrix}\right),$$

$$I_3=\frac{2}{\pi p}\left(\frac{1}{3}\pm\frac{4p^2}{c^2}\pm\frac{2c^2}{15p^2}\right)\mp\frac{2p\,(4p^2\pm3c^2)}{\pi c^3\sqrt{p^2\pm c^2}}\begin{Bmatrix}A\\ B\end{Bmatrix}.$$

2. $$\int_0^\infty x^{\alpha-1}e^{-px}\begin{Bmatrix}\mathbf{H}_{\nu}(cx)\\ \mathbf{L}_{\nu}(cx)\end{Bmatrix}dx=$$

$$=\frac{c^{\nu+1}}{2^{\nu}\sqrt{\pi}\,p^{\alpha+\nu+1}}\Gamma\begin{bmatrix}\alpha+\nu+1\\ \nu+3/2\end{bmatrix}{}_3F_2\left(1,\ \frac{\alpha+\nu+1}{2},\ \frac{\alpha+\nu}{2}+1;\ \frac{3}{2},\ \nu+\frac{3}{2};\ \mp\frac{c^2}{p^2}\right)$$

$[\operatorname{Re}(\alpha+\nu)>-1]$

3. $$\int_0^\infty\frac{1}{x}e^{-px}\begin{Bmatrix}\mathbf{H}_n(cx)\\ \mathbf{L}_n(cx)\end{Bmatrix}dx=J_n\quad\left[A=\ln\frac{\sqrt{p^2+c^2}+c}{p},\ B=\arcsin\frac{c}{p},\ \begin{Bmatrix}\operatorname{Re}p>|\operatorname{Im}c|\\ \operatorname{Re}p>|\operatorname{Re}c|\end{Bmatrix}\right]$$

$$J_1=\mp\frac{2}{\pi}\left(1-\frac{\sqrt{p^2\pm c^2}}{c}\begin{Bmatrix}A\\ B\end{Bmatrix}\right),$$

$$J_2=\frac{2}{\pi}\left(\frac{p}{c}\pm\frac{c}{3p}-\frac{\sqrt{p^2\pm c^2}}{c}\begin{Bmatrix}A\\ B\end{Bmatrix}\right),$$

$$J_3=\frac{2}{\pi}\left(\pm\frac{c^2}{15p^2}\mp\frac{4p^2}{3c^2}-\frac{7}{9}\pm\frac{4p^2\pm c^2}{3c^3}\sqrt{p^2\pm c^2}\begin{Bmatrix}A\\ B\end{Bmatrix}\right).$$

4. $$\int_0^\infty x^{\nu}e^{-px}\mathbf{L}_{\nu}(cx)\,dx=$$

$$=\frac{(2c)^{\nu}\Gamma(\nu+1/2)}{\sqrt{\pi}\,(p^2-c^2)^{\nu+1/2}}-\frac{\sqrt{2}\,c^{\nu}}{p^{\nu+1/2}}\Gamma(2\nu+1)\,(c^2-p^2)^{-(2\nu+1)/4}P_{-\nu-1/2}^{-\nu-1/2}\left(\frac{c}{p}\right)$$

$[\operatorname{Re}p>|\operatorname{Re}c|;\ \operatorname{Re}\nu>-1/2]$

5. $$\int_0^\infty e^{-px}\begin{Bmatrix}\mathbf{H}_{-n-1/2}(cx)\\ \mathbf{L}_{-n-1/2}(cx)\end{Bmatrix}dx=\frac{(\mp1)^n\left(\pm\sqrt{p^2\pm c^2}\mp p\right)^{n+1/2}}{c^{n+1/2}\sqrt{p^2\pm c^2}}\quad\left[\begin{Bmatrix}\operatorname{Re}p>|\operatorname{Im}c|\\ \operatorname{Re}p>|\operatorname{Re}c|\end{Bmatrix}\right]$$

2.7.6. Integrals of $x^{\alpha}e^{-px^{\pm2}}\begin{Bmatrix}\mathbf{H}_{\nu}(cx)\\ \mathbf{L}_{\nu}(cx)\end{Bmatrix}$.

1. $$\int_0^\infty x^{\alpha-1}e^{-px^2}\begin{Bmatrix}\mathbf{H}_{\nu}(cx)\\ \mathbf{L}_{\nu}(cx)\end{Bmatrix}dx=$$

$$=\left(\frac{c}{2}\right)^{\nu+1}\frac{p^{-(\alpha+\nu+1)/2}}{\sqrt{\pi}}\Gamma\begin{bmatrix}(\alpha+\nu+1)/2\\ \nu+3/2\end{bmatrix}{}_2F_2\left(1,\ \frac{\alpha+\nu+1}{2};\ \nu+\frac{3}{2},\ \frac{3}{2};\ \mp\frac{c^2}{4p}\right)$$

$[\operatorname{Re}p>0;\ \operatorname{Re}(\alpha+\nu)>-1]$

2. $$\int_0^\infty x^{\nu+1}e^{-px^2}\begin{Bmatrix}\mathbf{H}_{\nu}(cx)\\ \mathbf{L}_{\nu}(cx)\end{Bmatrix}dx=\begin{Bmatrix}-i\\ 1\end{Bmatrix}\frac{c^{\nu}}{(2p)^{\nu+1}}\exp\left(\mp\frac{c^2}{4p}\right)\operatorname{erf}\left(\frac{i^{(1\pm1)/2}c}{2\sqrt{p}}\right)$$

$[\operatorname{Re}p>0;\ \operatorname{Re}\nu>-3/2]$

3. $$\int_0^\infty x^{1-\nu}e^{-px^2}\begin{Bmatrix}\mathbf{H}_\nu(cx)\\ \mathbf{L}_\nu(cx)\end{Bmatrix}dx=\frac{c^{\nu+1}}{2^{\nu+2}p^{3/2}\Gamma(\nu+3/2)}\,{}_1F_1\left(1;\ \nu+\frac{3}{2};\ \mp\frac{c^2}{4p}\right)$$ [Re $p>0$].

4. $$=\begin{Bmatrix}e^{(2\nu+1)\pi i/2}\\ 1\end{Bmatrix}\frac{(2p)^{\nu-1}}{c^\nu\Gamma(1/2+\nu)}\exp\left(\mp\frac{c^2}{4p}\right)\gamma\left(\nu+\frac{1}{2},\ \mp\frac{c^2}{4p}\right)$$ [Re $p>0$].

5. $$\int_0^\infty x^{\alpha-1}e^{-p/x^2}\mathbf{H}_\nu(cx)\,dx=$$
$$=\frac{c^{\nu+1}p^{(\alpha+\nu+1)/2}}{2^{\nu+1}\sqrt{\pi}}\Gamma\begin{bmatrix}-(\alpha+\nu+1)/2\\ \nu+3/2\end{bmatrix}{}_1F_3\left(1;\ \frac{3}{2},\ \nu+\frac{3}{2},\ \frac{\alpha+\nu+3}{2};\ \frac{c^2p}{4}\right)+$$
$$+\frac{2^{\alpha-1}\pi\sec[(\alpha+\nu)\pi/2]}{c^\alpha\Gamma(1-(\alpha-\nu)/2)\,\Gamma(1-(\alpha+\nu)/2)}\,{}_0F_2\left(1-\frac{\alpha-\nu}{2},\ 1-\frac{\alpha+\nu}{2};\ \frac{c^2p}{4}\right)$$
[c, Re $p>0$; Re $\alpha<3/2$; Re $(\alpha+\nu)<1$].

2.7.7. Integrals containing the trigonometric functions and $\mathbf{H}_\nu(cx)$.

1. $$\int_0^\infty \sin bx\,\mathbf{H}_0(cx)\,dx=(c^2-b^2)_+^{-1/2}$$ [$b,\ c>0;\ b\neq c$].

2. $$\int_0^\infty \cos bx\,\mathbf{H}_0(cx)\,dx=-\frac{2}{\pi}(b^2-c^2)^{-1/2}\arcsin\frac{c}{b}$$ [$0<c<b$],
$$=\frac{2}{\pi}(c^2-b^2)^{-1/2}\ln\frac{c+\sqrt{c^2-b^2}}{b}$$ [$0<b<c$].

3. $$\int_0^\infty \frac{1}{x}\cos bx\,\mathbf{H}_0(cx)\,dx=\begin{Bmatrix}1\\0\end{Bmatrix}\left(\frac{\pi}{2}-\frac{b}{\sqrt{c^2-b^2}}\right)$$ $\left[\begin{Bmatrix}0<b<c\\ 0<c<b\end{Bmatrix}\right]$.

4. $$\int_0^\infty x^{\alpha-1}\begin{Bmatrix}\sin bx\\ \cos bx\end{Bmatrix}\mathbf{H}_\nu(cx)\,dx=U$$
$\left[-1-\delta<\mathrm{Re}(\alpha+\nu)<2;\ \mathrm{Re}\,\alpha<3/2\ \text{for}\ b\neq c,\ \mathrm{Re}\,\alpha<1/2\ \text{for}\ b=c;\ \delta=\begin{Bmatrix}1\\0\end{Bmatrix}\right]$,
$$U=\pm\frac{b^{-\alpha-\nu-1}c^{\nu+1}}{2^\nu\sqrt{\pi}}\Gamma\begin{bmatrix}\alpha+\nu+1\\ \nu+3/2\end{bmatrix}\begin{Bmatrix}\cos[(\alpha+\nu)\pi/2]\\ \sin[(\alpha+\nu)\pi/2]\end{Bmatrix}\times$$
$$\times{}_3F_2\left(\frac{\alpha+\nu+1}{2},\ \frac{\alpha+\nu}{2}+1,\ 1;\ \frac{3}{2},\ \nu+\frac{3}{2};\ \frac{c^2}{b^2}\right)$$ [$0<c<b$],
$$U=\mp\frac{b^{1-\alpha-\nu}}{\sqrt{\pi}}\left(\frac{c}{2}\right)^{\nu-1}\Gamma\begin{bmatrix}\alpha+\nu-1\\ \nu+1/2\end{bmatrix}\begin{Bmatrix}\cos[(\alpha+\nu)\pi/2]\\ \sin[(\alpha+\nu)\pi/2]\end{Bmatrix}\times$$
$$\times{}_3F_2\left(\frac{1}{2},\ \frac{1}{2}-\nu,\ 1;\ \frac{3-\alpha-\nu}{2},\ 1-\frac{\alpha+\nu}{2};\ \frac{b^2}{c^2}\right)+$$
$$+\frac{2^{\alpha+\delta-1}\pi b^\delta\sec[(\alpha+\nu+\delta)\pi/2]}{c^{\alpha+\delta}\Gamma\left(1-\frac{\alpha+\delta-\nu}{2}\right)\Gamma\left(1-\frac{\alpha+\delta+\nu}{2}\right)}\,{}_2F_1\left(\frac{\alpha+\delta-\nu}{2},\ \frac{\alpha+\delta+\nu}{2};\ \delta+\frac{1}{2};\ \frac{b^2}{c^2}\right)$$
[$0<b<c$].

5. $$\int_0^\infty x^{-\nu}\sin bx\,\mathbf{H}_\nu(cx)\,dx=\frac{\sqrt{\pi}\,(c^2-b^2)_+^{\nu-1/2}}{(2c)^\nu\Gamma(\nu+1/2)}$$ [$b,\ c>0;\ b\neq c;\ \mathrm{Re}\,\nu>-1/2$].

6. $\int_0^\infty x^{-\nu-1}\cos bx\,\mathbf{H}_\nu(cx)\,dx=\sqrt{\frac{\pi}{2c}}\,(c^2-b^2)_+^{(2\nu+1)/4}P_{\nu-1/2}^{-\nu-1/2}\left(\frac{b}{c}\right)$

$[b,\ c>0,\ b\neq c;\ \operatorname{Re}\nu>-3/2]$.

7. $\int_0^\infty x^{-1/4}\sin b\sqrt{x}\,\mathbf{H}_{1/4}(cx)\,dx=-\frac{\sqrt{\pi b}}{\sqrt{2c}}Y_{1/4}\left(\frac{b^2}{4c}\right)$ $[b,\ c>0]$.

8. $\int_0^\infty x^{-1/2}\cos b\sqrt{x}\,\mathbf{H}_0(cx)\,dx=-\frac{\pi b}{4c}\left[J_{1/4}^2\left(\frac{b^2}{8c}\right)-Y_{-1/4}^2\left(\frac{b^2}{8c}\right)\right]$ $[b,\ c>0]$.

9. $\int_0^\infty x\sin bx^2\,\mathbf{H}_0(cx)\,dx=\frac{1}{2b}\{\sin\varphi\,[C(\varphi)+S(\varphi)]+\cos\varphi\,[C(\varphi)-S(\varphi)]\}$

$[\varphi=c^2/(4b);\ b,\ c>0]$.

10. $\int_0^\infty \frac{x^{-\nu}\sin bx}{x^2+z^2}\mathbf{H}_\nu(cx)\,dx=\frac{1}{2z^{\nu+1}}e^{-bz}\mathbf{L}_\nu(cz)$ $[0<b<c;\ \operatorname{Re}\nu>-5/2]$.

2.7.8. Integrals containing the logarithmic function and $\begin{Bmatrix}\mathbf{H}_\nu(cx)\\ \mathbf{L}_\nu(cx)\end{Bmatrix}$.

1. $\int_0^a x\ln\frac{a-\sqrt{a^2-x^2}}{x}\begin{Bmatrix}\mathbf{H}_0(cx)\\ \mathbf{L}_0(cx)\end{Bmatrix}dx=\pm\frac{1}{c^2}\begin{Bmatrix}\sin ac\\ \operatorname{sh} ac\end{Bmatrix}$ $[a>0]$.

2. $\int_0^\infty \frac{x}{\sqrt{x^2+z^2}}\ln\frac{z+\sqrt{x^2+z^2}}{x}\mathbf{H}_0(cx)\,dx=-\frac{\pi}{2c}e^{-cz}$ $[c,\ \operatorname{Re}z>0]$.

2.7.9. Integrals containing the inverse trigonometric functions and $\mathbf{H}_\nu(cx)$.

1. $\int_0^a \frac{1}{\sqrt{a^2-x^2}}\cos\left[(\nu+1)\arccos\frac{x}{a}\right]\mathbf{H}_\nu(cx)\,dx=\sqrt{\frac{\pi}{ac}}\sin\frac{ac}{2}J_{\nu+1/2}\left(\frac{ac}{2}\right)$

$[a>0;\ \operatorname{Re}\nu>-2]$.

2.7.10 Integrals of $x^\alpha\mathbf{H}_\mu(bx^{\pm1})\,\mathbf{H}_\nu(cx)$.

1. $\int_0^\infty x^{-\mu-\nu}\mathbf{H}_\mu(cx)\,\mathbf{H}_\nu(cx)\,dx=\frac{c^{\mu+\nu-1}\sqrt{\pi}}{2^{\mu+\nu}}\Gamma\left[\begin{matrix}\mu+\nu\\ \mu+1/2,\ \nu+1/2,\ \mu+\nu+1/2\end{matrix}\right]$

$[c,\ \operatorname{Re}(\mu+\nu)>0;\ \operatorname{Re}\mu,\ \operatorname{Re}\nu>-3/2]$.

2. $\int_0^\infty \mathbf{H}_\nu\left(\frac{b}{x}\right)\mathbf{H}_\nu(cx)\,dx=-\frac{1}{c}J_{2\nu}(2\sqrt{bc})$ $[b,c>0;\ \operatorname{Re}\nu>-3/2]$.

3. $\int_0^\infty \frac{1}{x}\mathbf{H}_{\nu-1}\left(\frac{b}{x}\right)\mathbf{H}_\nu(cx)\,dx=-\frac{1}{\sqrt{bc}}J_{2\nu-1}(2\sqrt{bc})$ $[b,\ c>0;\ \operatorname{Re}\nu>-1/2]$.

2.7.11. Integrals containing $\operatorname{Ei}(-bx^2)\begin{Bmatrix}\mathbf{H}_\nu(cx)\\ \mathbf{L}_\nu(cx)\end{Bmatrix}$.

1. $$\int_0^\infty x^{\alpha-1}\mathrm{Ei}(-bx^2)\begin{Bmatrix}\mathbf{H}_\nu(cx)\\ \mathbf{L}_\nu(cx)\end{Bmatrix}dx=-\frac{b^{-(\alpha+\nu+1)/2}c^{\nu+1}}{2^\nu\sqrt{\pi}(\alpha+\nu+1)}\Gamma\begin{bmatrix}(\alpha+\nu+1)/2\\ \nu+3/2\end{bmatrix}\times$$
$$\times{}_3F_3\left(1,\frac{\alpha+\nu+1}{2},\frac{\alpha+\nu+1}{2};\frac{3}{2},\frac{3}{2}+\nu,\frac{\alpha+\nu+3}{2};\mp\frac{c^2}{4b}\right)\quad[\mathrm{Re}\,b>0;\ \mathrm{Re}\,(\alpha+\nu)>-1].$$

2. $$\int_0^\infty xe^{bx^2}\mathrm{Ei}(-bx^2)\mathbf{H}_0(cx)\,dx=-\frac{\pi}{2b}\exp\left(\frac{c^2}{4b}\right)\mathrm{erfc}\left(\frac{c}{2\sqrt{b}}\right)\quad[c,\ \mathrm{Re}\,b>0].$$

2.7.12. Integrals of $x^\alpha\,\mathrm{erfc}\,(bx^r)\begin{Bmatrix}\mathbf{H}_\nu(cx)\\ \mathbf{L}_\nu(cx)\end{Bmatrix}$.

1. $$\int_0^\infty x^{\alpha-1}\mathrm{erfc}\,(bx)\begin{Bmatrix}\mathbf{H}_\nu(cx)\\ \mathbf{L}_\nu(cx)\end{Bmatrix}dx=\frac{b^{-\alpha-\nu-1}c^{\nu+1}}{2^\nu\pi(\alpha+\nu+1)}\Gamma\begin{bmatrix}(\alpha+\nu)/2+1\\ \nu+3/2\end{bmatrix}\times$$
$$\times{}_3F_3\left(1,\frac{\alpha+\nu+1}{2},\frac{\alpha+\nu}{2}+1;\frac{3}{2},\nu+\frac{3}{2},\frac{\alpha+\nu+3}{2};\mp\frac{c^2}{4b^2}\right)\quad[\mathrm{Re}\,(\alpha+\nu)>-1;\ |\arg b|<\pi/4].$$

2. $$\int_0^\infty x^{\alpha-1}\mathrm{erfc}\,(b\sqrt{x})\begin{Bmatrix}\mathbf{H}_\nu(cx)\\ \mathbf{L}_\nu(cx)\end{Bmatrix}dx=\frac{c^{\nu+1}b^{-2\alpha-2\nu-2}}{2^\nu\pi(\alpha+\nu+1)}\Gamma\begin{bmatrix}\alpha+\nu+3/2\\ \nu+3/2\end{bmatrix}\times$$
$$\times{}_4F_3\left(1,\frac{\alpha+\nu+1}{2},\frac{2\alpha+2\nu+3}{4},\frac{2\alpha+2\nu+5}{4};\frac{3}{2},\nu+\frac{3}{2},\frac{\alpha+\nu+3}{2};\mp\frac{c^2}{b^4}\right)$$
$$\left[\left\{\mathrm{Re}\,b^2>\begin{Bmatrix}|\mathrm{Im}\,c|\\ |\mathrm{Re}\,c|\end{Bmatrix}\right.,\ \mathrm{Re}\,(\alpha+\nu)>-1\right].$$

2.7.13. Integrals containing $D_\mu(bx)\begin{Bmatrix}\mathbf{H}_\nu(cx)\\ \mathbf{L}_\nu(cx)\end{Bmatrix}$.

1. $$\int_0^\infty x^{\alpha-1}e^{\pm b^2x^2/4}D_\mu(bx)\,\mathbf{H}_\nu(cx)\,dx=U_\pm(1),$$

$$U_\pm(\varepsilon)=\frac{2^{(\mp\mu-\alpha-3\nu\mp1)/2-1}c^{\nu+1}}{\sqrt{\pi}\,b^{\alpha+\nu+1}\Gamma(\nu+3/2)}\begin{Bmatrix}\Gamma^{-1}(-\mu)\\ \sqrt{\pi}\end{Bmatrix}\Gamma^{\pm1}\left(\frac{-\mu\mp\nu\mp\alpha}{2}+\frac{1\mp3}{4}\right)\times$$
$$\times{}_3F_3\left(1,\frac{\alpha+\nu+1}{2},\frac{\alpha+\nu}{2}+1;\frac{3}{2},\frac{3}{2}+\nu,\frac{5\pm1}{4}+\frac{\alpha+\nu\pm\mu}{2};\pm\frac{\varepsilon c^2}{2b^2}\right)+$$
$$+\begin{Bmatrix}1\\0\end{Bmatrix}\frac{2^{\alpha+\mu-1}\pi b^\mu}{c^{\alpha+\mu}\Gamma\left(1-\frac{\alpha+\mu-\nu}{2}\right)\Gamma\left(1-\frac{\alpha+\mu+\nu}{2}\right)}\sec\frac{\alpha+\mu+\nu}{2}\pi\times$$
$$\times{}_2F_2\left(-\frac{\mu}{2},\frac{1-\mu}{2};1-\frac{\alpha+\mu-\nu}{2},1-\frac{\alpha+\nu+\mu}{2};\frac{c^2}{2b^2}\right)$$
$$\left[\mathrm{Re}\,(\alpha+\nu)>-1;\ |\arg b|<(2\pm1)\pi/4,\ \begin{Bmatrix}c>0;\ \mathrm{Re}\,(\alpha+\mu)<1-\mathrm{Re}\,\nu,\ 3/2\\ |\arg c|<\pi\end{Bmatrix}\right].$$

2. $$\int_0^\infty x^{\alpha-1}e^{-b^2x^2/4}D_\mu(bx)\,\mathbf{L}_\nu(cx)\,dx=U_-(-1)$$

$[\mathrm{Re}\,(\alpha+\nu)>-1;\ |\arg b|<\pi/4;\ |\arg c|<\pi;\ U_-(\varepsilon)$ see 2.7.13.1].

3. $$\int_0^\infty x^{-\nu}e^{-b^2x^2/4}[D_\mu(bx)-D_\mu(-bx)]\,\mathbf{H}_\nu(cx)\,dx=$$
$$=\frac{2^{1/2-\nu}c^{\mu+\nu}}{b^{\mu+1}}\sin\frac{\mu\pi}{2}\Gamma\begin{bmatrix}(\mu+1)/2\\ 1+\nu+\mu/2\end{bmatrix}{}_1F_1\left(\frac{\mu+1}{2};1+\nu+\frac{\mu}{2};-\frac{c^2}{2b^2}\right)$$
$[c,\ b>0;\ \mathrm{Re}\,\mu>-1;\ \mathrm{Re}\,(\mu+\nu)>-3/2]$.

2.7.14. Integrals containing $J_\mu(bx^{\pm r})\,\mathbf{H}_\nu(cx)$.

1. $$\int_0^\infty x^{\alpha-1} J_\mu(bx)\,\mathbf{H}_\nu(cx)\,dx = U$$

[$b, c > 0$; $\mathrm{Re}(\alpha+\mu+\nu) > -1$; $\mathrm{Re}(\alpha+\nu) < 5/2$; $\mathrm{Re}\,\alpha < 2$ for $c \neq b$, $\mathrm{Re}\,\alpha < 1$ for $c=b$],

$$U = \frac{2^\alpha c^{\nu+1}}{\sqrt{\pi}\, b^{\alpha+\nu+1}} \Gamma\left[\begin{matrix}(\alpha+\mu+\nu+1)/2 \\ \nu+3/2,\ (1+\mu-\alpha-\nu)/2\end{matrix}\right] \times$$
$$\times {}_3F_2\left(1, \frac{\alpha+\mu+\nu+1}{2}, \frac{1+\alpha+\nu-\mu}{2}; \frac{3}{2}, \frac{3}{2}+\nu; \frac{c^2}{b^2}\right) \quad [0 < c < b],$$

$$U = \frac{2^{\alpha-1} c^{\nu-1}}{\sqrt{\pi}\, b^{\alpha+\nu-1}} \Gamma\left[\begin{matrix}(\alpha+\mu+\nu-1)/2 \\ \nu+1/2,\ (3+\mu-\alpha-\nu)/2\end{matrix}\right] \times$$
$$\times {}_3F_2\left(1, \frac{1}{2}, \frac{1}{2}-\nu; \frac{3-\alpha-\mu-\nu}{2}, \frac{3+\mu-\alpha-\nu}{2}; \frac{b^2}{c^2}\right) +$$
$$+ \frac{2^{\alpha-1}\pi b^\mu}{c^{\alpha+\mu}} \frac{\sec[(\alpha+\mu+\nu)\pi/2]}{\Gamma[\mu+1,\ 1-(\alpha+\mu-\nu)/2,\ 1-(\alpha+\mu+\nu)/2]} \times$$
$$\times {}_2F_1\left(\frac{\alpha+\mu-\nu}{2}, \frac{\alpha+\mu+\nu}{2}; \mu+1; \frac{b^2}{c^2}\right) \quad [0 < b < c].$$

2. $$\int_0^\infty J_0(bx)\,\mathbf{H}_0(cx)\,dx = \left\{\begin{matrix}1\\0\end{matrix}\right\} \frac{4}{\pi(b+c)} \mathbf{K}\left(\frac{c-b}{c+b}\right) \quad \left[\left\{\begin{matrix}0<b<c\\0<c<b\end{matrix}\right\}\right].$$

3. $$\int_0^\infty x^{1/2} J_{\nu+1/2}(bx)\,\mathbf{H}_\nu(cx)\,dx = \sqrt{\frac{2}{\pi}}\, b^{-\nu-1/2} c^\nu (c^2-b^2)_+^{-1/2}$$

[$b, c > 0$, $b \neq c$; $-3/2 < \mathrm{Re}\,\nu < 1$].

4. $$\int_0^\infty x^{\mu+\nu+1} J_\mu(bx)\,\mathbf{H}_\nu(cx)\,dx = I \quad [b, c > 0;\ -3/2 < \mathrm{Re}(\mu+\nu) < 0;\ \mathrm{Re}(\mu+2\nu) < 1/2],$$

$$I = \frac{2^{\mu+\nu+1} b^\mu c^\nu \sin\mu\pi\,(c^2-b^2)^{-\mu-\nu-1}}{\cos(\mu+\nu)\pi\,\Gamma(-\mu-\nu)} +$$
$$+ \frac{2^{\mu+\nu+1} c^{\nu-1}\cos\nu\pi}{\pi^{3/2} b^{\mu+2\nu+1}} \Gamma(\mu+\nu+1/2)\,{}_2F_1\left(1, \frac{1}{2}; \frac{1}{2}-\mu-\nu; \frac{b^2}{c^2}\right) \quad [0 < b < c],$$

$$I = -\frac{2^{\mu+\nu+2} c^{\nu+1}\cos\nu\pi}{\pi^{3/2} b^{\mu+2\nu+3}} \Gamma\left(\mu+\nu+\frac{3}{2}\right) {}_2F_1\left(\mu+\nu+\frac{3}{2}, 1; \frac{3}{2}; \frac{c^2}{b^2}\right) \quad [0 < c < b].$$

5. $$\int_0^\infty x^{\mu-\nu+1} J_\mu(bx)\,\mathbf{H}_\nu(cx)\,dx = J \quad [b, c > 0;\ -3/2 < \mathrm{Re}\,\mu < \min(1/2, \mathrm{Re}\,\nu)],$$

$$J = \frac{2^{\mu-\nu+1} b^\mu \sin\mu\pi\,(c^2-b^2)^{\nu-\mu-1}}{c^\nu \cos\mu\pi\,\Gamma(\nu-\mu)} +$$
$$+ \frac{2^{\mu-\nu+1} c^{\nu-1}}{\pi b^{\mu+1}} \Gamma\left[\begin{matrix}\mu+1/2\\ \nu+1/2\end{matrix}\right] {}_2F_1\left(1, \frac{1}{2}-\nu; \frac{1}{2}-\mu; \frac{b^2}{c^2}\right) \quad [0 < b < c],$$

$$J = -\frac{2^{\mu-\nu+1} c^{\nu+1}}{\pi b^{\mu+3}} \Gamma\left[\begin{matrix}\mu+3/2\\ \nu+3/2\end{matrix}\right] {}_2F_1\left(\mu+\frac{3}{2}, 1; \frac{3}{2}+\nu; \frac{c^2}{b^2}\right) \quad [0 < c < b].$$

6. $$\int_0^\infty x^{1-\mu-\nu} J_\mu(cx)\,\mathbf{H}_\nu(cx)\,dx = \frac{(2\mu-1)\,c^{\mu+\nu-2}}{2^{\mu+\nu}(\mu+\nu-1)\,\Gamma(\mu+1/2)\,\Gamma(\nu+1/2)}$$

[$c > 0$; $-1/2 < \mathrm{Re}\,\mu < 1-\mathrm{Re}\,\nu$].

7. $$\int_0^\infty J_{2\nu}(b\sqrt{x})\,\mathbf{H}_\nu(cx)\,dx = -\frac{1}{c} Y_\nu\left(\frac{b^2}{4c}\right) \quad [b, c > 0;\ |\mathrm{Re}\,\nu| < 1].$$

8. $\int_0^\infty x^{1/2}J_{2\nu+1}(b\sqrt{x})\,\mathbf{H}_\nu(cx)\,dx=-\frac{b}{2c^2}Y_{\nu+1}\left(\frac{b^2}{4c}\right)$ $[b, c>0;\ -3/2<\operatorname{Re}\nu<1/2]$.

9. $\int_0^\infty J_{-\nu}\left(\frac{b}{x}\right)\mathbf{H}_\nu(cx)\,dx=\frac{1}{c}\left[\sin\nu\pi J_{2\nu}(2\sqrt{bc})-Y_{2\nu}(2\sqrt{bc})+\frac{2}{\pi}K_{2\nu}(2\sqrt{bc})\right]$

$[b, c>0;\ -7/2<\operatorname{Re}\nu<0]$.

10. $\int_0^\infty x^{\alpha-1}J_\mu\left(\frac{b}{x}\right)\mathbf{H}_\nu(cx)\,dx=$

$$=\frac{2^{\alpha-2\mu-1}\pi}{b^{-\mu}c^{\alpha-\mu}}\frac{\sec[(\alpha+\nu-\mu)\pi/2]}{\Gamma[\mu+1,\ 1+(\mu+\nu-\alpha)/2,\ 1+(\mu-\nu-\alpha)/2]}\times$$
$$\times{}_0F_3\left(1+\frac{\mu+\nu-\alpha}{2},\ 1+\frac{\mu-\nu-\alpha}{2},\ \mu+1;\ \frac{b^2c^2}{16}\right)+$$
$$+\frac{b^{\alpha+\nu+1}c^{\nu+1}}{2^{\alpha+2\nu+2}\sqrt{\pi}}\Gamma\left[\begin{matrix}(\mu-\alpha-\nu-1)/2\\ \nu+3/2,\ (\alpha+\mu+\nu+3)/2\end{matrix}\right]\times$$
$$\times{}_1F_4\left(1;\ \frac{3+\alpha+\nu-\mu}{2},\ \frac{3+\alpha+\mu+\nu}{2},\ \nu+\frac{3}{2},\ \frac{3}{2};\ \frac{b^2c^2}{16}\right)$$

$[b, c>0;\ -5/2<\operatorname{Re}(\alpha+\nu)<\operatorname{Re}\mu+1;\ \operatorname{Re}(\alpha-\mu)<3/2]$.

11. $\int_0^\infty xe^{-px}[J_1(cx)\,\mathbf{H}_0(cx)-J_0(cx)\,\mathbf{H}_1(cx)]\,dx=\frac{2c^2}{\pi p}(p^2+c^2)^{-3/2}$ $[\operatorname{Re}p>2\,|\operatorname{Im}c|]$.

12. $\int_0^\infty xe^{-px}[J_\nu(cx)\,\mathbf{H}'_\nu(cx)-J'_\nu(cx)\,\mathbf{H}_\nu(cx)]\,dx=\frac{2c^{2\nu}}{\pi p}(p^2+c^2)^{-\nu-1/2}$

$[\operatorname{Re}\nu>-1;\ \operatorname{Re}p>2\,|\operatorname{Im}c|]$.

13. $\int_0^\infty\left[J_{-\nu}\left(\frac{b}{x}\right)+\sin\nu\pi\,\mathbf{H}_\nu\left(\frac{b}{x}\right)\right]\mathbf{H}_\nu(cx)\,dx=\frac{1}{c}\left[\frac{2}{\pi}K_{2\nu}(2\sqrt{bc})-Y_{2\nu}(2\sqrt{bc})\right]$

$[b, c>0;\ -3/2<\operatorname{Re}\nu<0]$.

2.7.15. Integrals containing $Y_\mu(\varphi(x))\,\mathbf{H}_\nu(cx)$.

1. $\int_0^\infty Y_{\nu+1}(bx)\,\mathbf{H}_\nu(cx)\,dx=-\left\{\begin{matrix}1\\0\end{matrix}\right\}b^{-\nu-1}c^\nu$ $\left[\left\{\begin{matrix}0<b<c\\0<c<b\end{matrix}\right\},\ |\operatorname{Re}\nu|<3/2\right]$.

2. $\int_0^\infty x^{\mu-\nu+1}Y_\mu(bx)\,\mathbf{H}_\nu(cx)\,dx=\frac{2^{\mu-\nu+1}b^\mu c^{-\nu}}{\Gamma(\nu-\mu)}(c^2-b^2)_+^{\nu-\mu-1}$

$[b, c>0,\ b\neq c;\ \operatorname{Re}(\mu-\nu)<0;\ -3/2<\operatorname{Re}\mu<1/2]$.

3. $\int_0^\infty Y_{2\nu}(b\sqrt{x})\,\mathbf{H}_\nu(cx)\,dx=\frac{1}{c}J_\nu\left(\frac{b^2}{4c}\right)-\frac{2^{\nu+1}}{\pi^2c}\Gamma(\nu+1)\,S_{-\nu-1,\,\nu}\left(\frac{b^2}{4c}\right)$

$[b, c>0;\ -1<\operatorname{Re}\nu<3/4]$.

4. $\int_0^\infty[\cos bx\,J_\nu(bx)-\sin bx\,Y_\nu(bx)]\,\mathbf{H}_\nu(cx)\,dx=$

$$=\left\{\begin{matrix}1\\0\end{matrix}\right\}\frac{1}{\sqrt{bc}}P_{\nu-1/2}\left(\frac{c}{2b}\right)$$ $\left[\left\{\begin{matrix}0<2b<c\\0<c<2b\end{matrix}\right\},\ -1<\operatorname{Re}\nu<3/2\right]$.

5. $$\int_0^\infty x^{\nu+1} J_\nu(bx)\, Y_\nu(bx)\, \mathbf{H}_\nu(cx)\, dx = \frac{c^{\nu+1}}{2^{\nu+2}\pi^{3/2}b^{2\nu+3}} \Gamma\begin{bmatrix} 2\nu+3/2 \\ \nu+2 \end{bmatrix} \times$$
$$\times\, {}_2F_1\left(1, 2\nu+\frac{3}{2};\ \nu+2;\ \frac{c^2}{4b^2}\right)$$ $[0 < c < 2b;\ -3/4 < \operatorname{Re}\nu < 1/2]$.

6. $$\int_0^\infty x\,[J_\mu(bx)\, J_{\nu-\mu}(bx) - Y_\mu(bx)\, Y_{\nu-\mu}(bx)]\, \mathbf{H}_\nu(cx)\, dx = \frac{2^{1+\nu-2\mu}}{\pi b^{2\mu-\nu} c} (c^2-4b^2)_+^{-1/2} \times$$
$$\times\left[\left(c+\sqrt{c^2-4b^2}\right)^{2\mu-\nu} + \left(c-\sqrt{c^2-4b^2}\right)^{2\mu-\nu}\right]$$
$[b, c > 0,\ 2b \neq c;\ -3/2 < \operatorname{Re}\nu < 1;\ \operatorname{Re}\mu > -3/2;\ \operatorname{Re}(\mu-\nu) < 3/2]$.

7. $$\int_0^\infty x^{\mu+1} [J_\nu(bx)\, J_\mu(bx) - Y_\nu(bx)\, Y_\mu(bx)]\, \mathbf{H}_\nu(cx)\, dx =$$
$$= \frac{2^{\mu+1} b^{\mu-1/2}}{\sqrt{\pi}\, c^{\mu+1}} (c^2-4b^2)_+^{-(2\mu+1)/4} P_{\nu-1/2}^{\mu+1/2}\left(\frac{c}{2b}\right)$$
$[b, c > 0,\ 2b \neq c;\ \operatorname{Re}\nu > -3/2;\ -3/2 < \operatorname{Re}\mu < 1/2;\ -3/2 < \operatorname{Re}(\mu+\nu) < 1]$.

8. $$\int_0^\infty x^{1-\nu} [J_\nu(bx) J_{-\nu}(bx) - Y_\nu(bx)\, Y_{-\nu}(bx)]\, \mathbf{H}_\nu(cx)\, dx = \frac{2^{2-3\nu} c^{\nu-1}}{\sqrt{\pi}\, b^{2\nu}\Gamma(\nu+1/2)} (c^2-4b^2)_+^{\nu-1/2}$$
$[b, c > 0,\ 2b \neq c;\ -1/2 < \operatorname{Re}\nu < 3/2]$.

9. $$\int_0^\infty x\,[J_{\nu/2}^2(bx) - Y_{\nu/2}^2(bx)]\, \mathbf{H}_\nu(cx)\, dx = \frac{4}{\pi c}(c^2-4b^2)_+^{-1/2}$$
$[b, c > 0,\ 2b \neq c;\ -3/2 < \operatorname{Re}\nu < 1]$.

10. $$\int_0^\infty x^{\nu+1}\left[J_\nu^2(bx) - Y_\nu^2(bx)\right] \mathbf{H}_\nu(cx)\, dx = \frac{2^{3\nu+2} b^{2\nu}}{\sqrt{\pi}\, c^{\nu+1}\Gamma(1/2-\nu)} (c^2-4b^2)_+^{-\nu-1/2}$$
$[b, c > 0,\ 2b \neq c;\ -3/4 < \operatorname{Re}\nu < 1/2]$.

11. $$\int_0^\infty x\left[J_{\nu/2}^2\left(b\sqrt{x^2+z^2}-bz\right) - Y_{\nu/2}^2\left(b\sqrt{x^2+z^2}+bz\right)\right] \mathbf{H}_\nu(cx)\, dx =$$
$$= \pm\frac{4}{\pi c}\,|c^2-4b^2|^{-1/2} \begin{Bmatrix} \exp\left(-z\sqrt{c^2-4b^2}\right) \\ \sin\left(z\sqrt{4b^2-c^2}\right) \end{Bmatrix}$$ $\left[\begin{Bmatrix} 0 < 2b < c \\ 0 < c < 2b \end{Bmatrix},\ -3/2 < \operatorname{Re}\nu < 1\right]$.

2.7.16. Integrals containing products of special functions by $Y_\nu(cx) - \mathbf{H}_\nu(cx)$.

1. $$\int_0^\infty x^{\mu-\nu} J_\mu(bx)\, [Y_\nu(cx) - \mathbf{H}_\nu(cx)]\, dx =$$
$$= -\frac{2^{\mu-\nu} c^\nu}{\sqrt{\pi}\, b^{\mu+1}} \Gamma\begin{bmatrix} \mu+1/2, & 1/2+\mu-\nu \\ 1/2+\nu, & 1+\mu-\nu \end{bmatrix} {}_2F_1\left(\mu+\frac{1}{2}, \frac{1}{2};\ 1+\mu-\nu;\ 1-\frac{c^2}{b^2}\right)$$
$[b > 0;\ |\arg c| < \pi;\ -1/2 < \operatorname{Re}\mu < 3/2;\ \operatorname{Re}(\mu-\nu) > -1/2]$.

2. $$\int_0^\infty x^{1-\nu} J_\nu(bx)\, [Y_\nu(cx) - \mathbf{H}_\nu(cx)]\, dx = -\frac{2^{1-2\nu}}{b\sqrt{\pi}} \Gamma\begin{bmatrix} 1-\nu \\ \nu+1/2 \end{bmatrix} (b^2-c^2)_+^{(\nu-1)/2} P_\nu^{\nu-1}\left(\frac{b}{c}\right)$$
$[b, c > 0,\ b \neq c;\ |\operatorname{Re}\nu| < 2]$.

3. $$\int_0^\infty x J_{-\nu}(bx)\, [Y_\nu(cx) - \mathbf{H}_\nu(cx)]\, dx = -\frac{2c^\nu \cos\nu\pi}{\pi(b+c)\, b^{\nu+1}}$$ $[b > 0;\ |\arg c| < \pi;\ \operatorname{Re}\nu < 1/2]$.

4. $\int_0^\infty \frac{1}{x^{1+\sigma}} J_{-\nu-1+\sigma}\left(\frac{b}{x}\right)[Y_\nu(cx)-\mathbf{H}_\nu(cx)]\,dx = -\frac{4\cos\nu\pi}{\pi b^{(1+\sigma)/2}c^{(1-\sigma)/2}} K_{2\nu+1-\sigma}(2\sqrt{bc})$

$[\sigma=0$ or $1;\ b>0;\ |\arg c|<\pi;\ \sigma-3/2<\operatorname{Re}\nu<\sigma/2]$.

5. $\int_0^\infty x^{-2\nu-3/2} J_{\nu-1/2}\left(\frac{b}{x}\right)[Y_\nu(cx)-\mathbf{H}_\nu(cx)]\,dx =$

$= -\frac{b^{-\nu-1/2}}{2^{-5/2}\pi^{3/2}} \cos\nu\pi K_{2\nu}(\sqrt{2ibc})\, K_{2\nu}(\sqrt{-2ibc})$ $[b>0;\ |\arg c|<\pi;\ -1/2<\operatorname{Re}\nu<1/3]$.

6. $\int_0^\infty \mathbf{H}_0(b\sqrt{x})[Y_0(cx)-\mathbf{H}_0(cx)]\,dx = \frac{4b}{\pi^2 c^{3/2}}\Gamma^2\left(\frac{3}{4}\right) {}_1F_2\left(1;\ \frac{5}{4},\ \frac{5}{4};\ -\frac{b^4}{64c^2}\right) -$

$- \frac{b^3}{18\pi^2 c^{5/2}}\Gamma^2\left(\frac{1}{4}\right) {}_1F_2\left(1;\ \frac{7}{4},\ \frac{7}{4};\ -\frac{b^4}{64c^2}\right) - \frac{2}{c} J_0\left(\frac{b^2}{4c}\right)$ $[b>0;\ |\arg c|<\pi]$.

2.7.17. Integrals containing products of elementary functions by $I_{\pm\nu}(cx)-\mathbf{L}_\nu(cx)$.

1. $\int_0^\infty x^{\alpha-1}[I_{\pm\nu}(cx)-\mathbf{L}_\nu(cx)]\,dx = \frac{2^{\alpha-1}}{c^\alpha}\sec\frac{\alpha+\nu}{2}\pi \begin{Bmatrix}1\\ \cos\nu\pi\end{Bmatrix} \Gamma\begin{bmatrix}(\alpha\pm\nu)/2\\ 1-(\alpha\mp\nu)/2\end{bmatrix}$

$[\operatorname{Re} c>0;\ -1-\operatorname{Re}\nu,\ \mp\operatorname{Re}\nu<\operatorname{Re}\alpha<1-\operatorname{Re}\nu]$.

2. $\int_0^a x^{\alpha-1}(a^2-x^2)^{\beta-1}[I_{\pm\nu}(cx)-\mathbf{L}_\nu(cx)]dx = -\frac{a^{\alpha+\nu+2\beta-1}c^{\nu+1}}{2^{\nu+1}\sqrt{\pi}\,\Gamma(\nu+3/2)} \mathrm{B}\left(\frac{\alpha+\nu+1}{2},\ \beta\right)\times$

$\times {}_2F_3\left(1,\ \frac{\alpha+\nu+1}{2};\ \frac{3}{2},\ \nu+\frac{3}{2},\ \beta+\frac{\alpha+\nu+1}{2};\ \frac{a^2c^2}{4}\right) + \frac{a^{\alpha+2\beta\pm\nu-2}c^{\pm\nu}}{2^{1\pm\nu}\Gamma(1\pm\nu)} \mathrm{B}\left(\frac{\alpha\pm\nu}{2},\ \beta\right)\times$

$\times {}_1F_2\left(\frac{\alpha\pm\nu}{2};\ 1\pm\nu,\ \beta+\frac{\alpha\pm\nu}{2};\ \frac{a^2c^2}{4}\right)$ $\left[a,\ \operatorname{Re}\beta>0,\ \begin{Bmatrix}\operatorname{Re}(\alpha+\nu)>0\\ -\operatorname{Re}\alpha-1<\operatorname{Re}\nu<\operatorname{Re}\alpha\end{Bmatrix}\right]$.

3. $\int_a^\infty x^{\alpha-1}(x^2-a^2)^{\beta-1}[I_{\pm\nu}(cx)-\mathbf{L}_\nu(cx)]\,dx = -\frac{a^{\alpha+2\beta+\nu-1}c^{\nu+1}}{2^{\nu+1}\sqrt{\pi}\,\Gamma(\nu+3/2)}\times$

$\times \mathrm{B}\left(\beta,\ \frac{1-\alpha-\nu}{2}-\beta\right) {}_2F_3\left(1,\ \frac{\alpha+\nu+1}{2};\ \frac{\alpha+\nu+1}{2}+\beta,\ \frac{3}{2},\ \nu+\frac{3}{2};\ \frac{a^2c^2}{4}\right) -$

$- \frac{c^{2-2\beta-\alpha}}{2^{3-2\beta-\alpha}} \cos^{(1\mp1)/2}\nu\pi \sec\frac{\alpha+2\beta+\nu}{2}\pi\Gamma\begin{bmatrix}(\alpha\pm\nu)/2+\beta-1\\ 2-\beta-(\alpha\mp\nu)/2\end{bmatrix}\times$

$\times {}_1F_2\left(1-\beta;\ 2-\beta-\frac{\alpha+\nu}{2},\ 2-\beta-\frac{\alpha-\nu}{2};\ \frac{a^2c^2}{4}\right) +$

$+ \frac{a^{\alpha+2\beta\pm\nu-2}c^{\pm\nu}}{2^{1\pm\nu}\Gamma(1\pm\nu)} \mathrm{B}\left(\beta,\ 1-\beta-\frac{\alpha\pm\nu}{2}\right) {}_1F_2\left(\frac{\alpha\pm\nu}{2};\ \frac{\alpha\pm\nu}{2}+\beta,\ 1\pm\nu;\ \frac{a^2c^2}{4}\right)$

$[a,\ \operatorname{Re} c,\ \operatorname{Re}\beta>0;\ \operatorname{Re}(\alpha+2\beta+\nu)<3]$.

4. $\int_a^\infty x^{\nu+1}(x^2-a^2)^{\beta-1}[I_{-\nu}(cx)-\mathbf{L}_\nu(cx)]\,dx =$

$= \frac{2^{\beta-1}a^{\beta+\nu}\cos\nu\pi\Gamma(\beta)}{c^\beta\cos(\beta+\nu)\pi}[I_{-\beta-\nu}(ac)-\mathbf{L}_{\beta+\nu}(ac)]$ $[a,\ \operatorname{Re} c,\ \operatorname{Re}\beta>0;\ \operatorname{Re}(\beta+\nu)<1/2]$.

5. $$\int_0^\infty \frac{x^{\alpha-1}}{(x^2+z^2)^\rho}[I_{\pm\nu}(cx)-\mathbf{L}_\nu(cx)]\,dx=$$
$$=-\frac{c^{\nu+1}z^{\alpha+\nu-2\rho+1}}{2^{\nu+1}\sqrt{\pi}\,\Gamma(\nu+3/2)}\mathrm{B}\left(\frac{\alpha+\nu+1}{2},\rho-\frac{\alpha+\nu+1}{2}\right)\times$$
$$\times{}_2F_3\left(\frac{\alpha+\nu+1}{2},\ 1;\ \frac{3}{2},\ \nu+\frac{3}{2},\ \frac{3+\alpha+\nu}{2}-\rho;\ -\frac{c^2z^2}{4}\right)+$$
$$+\frac{c^{2\rho-\alpha}}{2^{2\rho-\alpha+1}}\cos^{(1\mp1)/2}\nu\pi\sec\frac{\alpha+\nu-2\rho}{2}\pi\Gamma\left[\begin{matrix}(\alpha\pm\nu)/2-\rho\\ 1+\rho-(\alpha\mp\nu)/2\end{matrix}\right]\times$$
$$\times{}_1F_2\left(\rho;\ 1+\rho-\frac{\alpha+\nu}{2},\ 1+\rho-\frac{\alpha-\nu}{2};\ -\frac{c^2z^2}{4}\right)+\frac{c^{\pm\nu}z^{\alpha\pm\nu-2\rho}}{2^{1\pm\nu}\Gamma(1\pm\nu)}\times$$
$$\times\mathrm{B}\left(\frac{\alpha\pm\nu}{2},\ \rho-\frac{\alpha\pm\nu}{2}\right){}_1F_2\left(\frac{\alpha\pm\nu}{2};\ 1\pm\nu,\ 1-\rho+\frac{\alpha\pm\nu}{2};\ -\frac{c^2z^2}{4}\right)$$
$$\left[\operatorname{Re}c,\ \operatorname{Re}z>0;\ \operatorname{Re}(\alpha+\nu-2\rho)<1,\ \left\{\begin{matrix}\operatorname{Re}(\alpha+\nu)>0\\ -1-\operatorname{Re}\alpha<\operatorname{Re}\nu<\operatorname{Re}\alpha\end{matrix}\right\}\right].$$

6. $$\int_0^\infty \frac{x^{\alpha-1}}{x^2-y^2}[I_{\pm\nu}(cx)-\mathbf{L}_\nu(cx)]\,dx=-\frac{\pi y^{\alpha-2}}{2}\operatorname{tg}\frac{\alpha+\nu}{2}\pi\mathbf{L}_\nu(cy)+$$
$$+\frac{c^{2-\alpha}}{2^{3-\alpha}}\cos^{(1\mp1)/2}\nu\pi\sec\frac{\alpha+\nu}{2}\pi\Gamma\left[\begin{matrix}(\alpha\pm\nu)/2-1\\ 2-(\alpha\mp\nu)/2\end{matrix}\right]\times$$
$$\times{}_1F_2\left(1;\ 2-\frac{\alpha+\nu}{2},\ 2-\frac{\alpha-\nu}{2};\ \frac{c^2y^2}{4}\right)\mp\frac{\pi y^{\alpha-2}}{2}\operatorname{ctg}\frac{\alpha\pm\nu}{2}\pi I_{\pm\nu}(cy)$$
$$\left[y,\ \operatorname{Re}c>0;\ \operatorname{Re}(\alpha+\nu)<3,\ \left\{\begin{matrix}\operatorname{Re}(\alpha+\nu)>0\\ -\operatorname{Re}\alpha-1<\operatorname{Re}\nu<\operatorname{Re}\alpha\end{matrix}\right\}\right].$$

7. $$\int_0^\infty x^{1-\nu}e^{-px^2}[I_{-\nu}(cx)-\mathbf{L}_\nu(cx)]\,dx=\frac{c^{-\nu}p^{\nu-1}}{2^{1-\nu}\Gamma(\nu+1/2)}e^{c^2/(4p)}\Gamma\left(\nu+\frac{1}{2},\ \frac{c^2}{4p}\right)$$
$$[\operatorname{Re}p>0;\ \operatorname{Re}\nu<1].$$

8. $$\int_0^\infty x^{\alpha-1}\sin bx\,[I_{\pm\nu}(cx)-\mathbf{L}_\nu(cx)]\,dx=U(1)$$
$$\left[b,\ \operatorname{Re}c>0;\ \operatorname{Re}(\alpha+\nu)<2,\ \left\{\begin{matrix}\operatorname{Re}(\alpha+\nu)>-1\\ -2-\operatorname{Re}\alpha<\operatorname{Re}\nu<\operatorname{Re}\alpha+1\end{matrix}\right\}\right],$$
$$U(\gamma)=-\frac{c^{\nu+1}}{2^\nu\sqrt{\pi}\,b^{\alpha+\nu+1}}\Gamma\left[\begin{matrix}\alpha+\nu+1\\ \nu+3/2\end{matrix}\right]\sin\frac{\gamma-\alpha-\nu}{2}\pi\times$$
$$\times{}_3F_2\left(\frac{\alpha+\nu+1}{2},\ \frac{\alpha+\nu}{2}+1,\ 1;\ \frac{3}{2},\ \nu+\frac{3}{2};\ -\frac{c^2}{b^2}\right)+$$
$$+\frac{c^{\pm\nu}}{2^{\pm\nu}b^{\alpha\pm\nu}}\Gamma\left[\begin{matrix}\alpha\pm\nu\\ 1\pm\nu\end{matrix}\right]\cos\frac{\gamma-\alpha\mp\nu}{2}\pi\ {}_2F_1\left(\frac{\alpha\pm\nu}{2},\ \frac{\alpha\pm\nu+1}{2};\ 1\pm\nu;\ -\frac{c^2}{b^2}\right).$$

9. $$\int_0^\infty x^{\alpha-1}\cos bx\,[I_{\pm\nu}(cx)-\mathbf{L}_\nu(cx)]\,dx=U(0)$$
$$\left[b,\ \operatorname{Re}c>0;\ \operatorname{Re}(\alpha+\nu)<2,\ \left\{\begin{matrix}\operatorname{Re}(\alpha+\nu)>0\\ -\operatorname{Re}\alpha-1<\operatorname{Re}\nu<\operatorname{Re}\alpha\end{matrix}\right\};\ U(\gamma)\ \text{see 2.7.17.8}\right].$$

2.7.18. Integrals containing products of special functions by $I_{\pm\nu}(cx)-\mathbf{L}_\nu(cx)$.

1. $\int_0^\infty x^{\alpha-1}J_\mu(bx)\,[I_{\pm\nu}(cx)-\mathbf{L}_\nu(cx)]\,dx=$

$$=-\frac{2^\alpha c^{\nu+1}}{\sqrt{\pi}\,b^{\alpha+\nu+1}}\Gamma\left[\begin{matrix}(\alpha+\mu+\nu+1)/2\\ \nu+3/2,\ (1+\mu-\alpha-\nu)/2\end{matrix}\right]\times$$
$$\times{}_3F_2\left(1,\ \frac{1+\alpha+\mu+\nu}{2},\ \frac{1+\alpha+\nu-\mu}{2};\ \frac{3}{2},\ \nu+\frac{3}{2};\ -\frac{c^2}{b^2}\right)+$$
$$+\frac{2^{\alpha-1}c^{\pm\nu}}{b^{\alpha\pm\nu}}\Gamma\left[\begin{matrix}(\alpha+\mu\pm\nu)/2\\ 1\pm\nu,\ 1+(\mu-\alpha\mp\nu)/2\end{matrix}\right]{}_2F_1\left(\frac{\alpha\pm\nu-\mu}{2},\ \frac{\alpha\pm\nu+\mu}{2};\ 1\pm\nu;\ -\frac{c^2}{b^2}\right)$$
$$\left[b,\ \operatorname{Re}c>0;\ \operatorname{Re}(\alpha+\nu)<5/2,\ \left\{\begin{matrix}\operatorname{Re}(\alpha+\mu+\nu)>0\\ -1-\operatorname{Re}(\alpha+\mu)<\operatorname{Re}\nu<\operatorname{Re}(\alpha+\mu)\end{matrix}\right\}\right].$$

2. $\int_0^\infty x^{\mu-\nu}J_\mu(bx)\,[I_{\pm\nu}(cx)-\mathbf{L}_\nu(cx)]\,dx=$

$$=\frac{2^{\mu-\nu}c^{\pm\nu}}{b^{1+\mu-\nu\pm\nu}}\Gamma\left[\begin{matrix}(1-\nu\pm\nu)/2+\mu\\ 1\pm\nu,\ (1+\nu\mp\nu)/2\end{matrix}\right]{}_2F_1\left(\frac{1-\nu\pm\nu}{2},\ \frac{1-\nu\pm\nu}{2}+\mu;\ 1\pm\nu;\ -\frac{c^2}{b^2}\right)$$
$$\left[b,\ \operatorname{Re}c>0,\ \left\{\begin{matrix}-1/2<\operatorname{Re}\mu<3/2\\ -1<\operatorname{Re}\mu<3/2,\ \operatorname{Re}(\mu-\nu)>-1/2\end{matrix}\right\}\right].$$

3. $\int_0^\infty x^{\mu\mp\nu+1}J_\mu(bx)\,[I_{\pm\nu}(cx)-\mathbf{L}_\nu(cx)]\,dx=$

$$=\frac{2^{\mu\mp\nu+(3\mp1)/2}c^{\nu+1}}{\pi^{(5\mp1)/4}b^{3+\mu+\nu\mp\nu}}\Gamma\left(\frac{3+\nu\mp\nu}{2}+\mu\right)\Gamma^{(\mp1-1)/2}\left(\nu+\frac{3}{2}\right)\cos^{(1\mp1)/2}\nu\pi\times$$
$$\times{}_2F_1\left(1,\ \frac{3+\nu\mp\nu}{2}+\mu;\ \frac{3+\nu\pm\nu}{2};\ -\frac{c^2}{b^2}\right)$$
$$\left[b,\ \operatorname{Re}c>0,\ \left\{\begin{matrix}-1<\operatorname{Re}\mu<1/2\\ \operatorname{Re}\mu>-1,\ -3/2-\operatorname{Re}\nu<\operatorname{Re}\mu<1/2-2\operatorname{Re}\nu\end{matrix}\right\}\right].$$

4. $\int_0^\infty x^{1\pm\nu-\mu}J_\mu(bx)\,[I_{\pm\nu}(cx)-\mathbf{L}_\nu(cx)]\,dx=$

$$=\frac{2^{1\pm\nu-\mu}c^{\nu-1}}{\pi^{(1\pm1)/4}b^{\nu\pm\nu-\mu+1}\Gamma((1-\nu\mp\nu)/2+\mu)}\Gamma^{(\pm1-1)/2}\left(\frac{1}{2}+\nu\right)\times$$
$$\times{}_2F_1\left(1,\ \frac{1\pm\nu-\nu}{2};\ \frac{1\mp\nu-\nu}{2}+\mu;\ -\frac{b^2}{c^2}\right)$$
$$[b,\ \operatorname{Re}c>0;\ \pm\operatorname{Re}\nu>-1;\ \operatorname{Re}(\mu-\nu\mp\nu)>-1/2].$$

5. $\int_0^\infty xJ_{\pm\nu}(bx)\,[I_{\pm\nu}(cx)-\mathbf{L}_\nu(cx)]\,dx=\frac{2(c/b)^{\nu+1}}{\pi(b^2+c^2)}\left\{\begin{matrix}1\\ \cos\nu\pi\end{matrix}\right\}$

$$\left[b,\ \operatorname{Re}c>0,\ \left\{\begin{matrix}-1<\operatorname{Re}\nu<1/2\\ \operatorname{Re}\nu<1/2\end{matrix}\right\}\right].$$

6. $\int_0^\infty x^{1/2}J_{\nu+1/2}(bx)\,[I_\nu(cx)-\mathbf{L}_\nu(cx)]\,dx=\sqrt{\frac{2}{\pi}}\,b^{-\nu-1/2}c^\nu(b^2+c^2)^{-1/2}$

$$[b,\ \operatorname{Re}c>0;\ |\operatorname{Re}\nu|<1].$$

7. $\int_0^\infty x^{-n}J_0(bx)\,[I_n(cx)-\mathbf{L}_n(cx)]\,dx=J_n$ $\quad[b,\ \operatorname{Re}c>0;\ z=c/\sqrt{b^2+c^2}]$,

$$J_0=\frac{2z}{\pi c}\mathbf{K}(z),\quad J_1=\frac{2}{\pi z}[\mathbf{K}(z)-\mathbf{E}(z)],$$

$$J_2=\frac{2c}{9\pi z^3}\,[(2+z^2)\,\mathbf{K}\,(z)-2\,(1+z^2)\,\mathbf{E}\,(z)],$$

$$J_3=\frac{2c^2}{225\pi z^5}\,[(8+3z^2+4z^4)\,\mathbf{K}\,(z)-(8+7z^2+8z^4)\,\mathbf{E}\,(z)].$$

8. $\int\limits_0^\infty x^{-2\nu-3/2}J_{\nu-1/2}\left(\frac{b}{x}\right)[I_\nu(cx)-\mathbf{L}_\nu(cx)]\,dx=$

$$=\frac{2^{3/2}}{\sqrt{\pi}}\,b^{-\nu-1/2}c^\nu J_{2\nu}\left(\sqrt{2bc}\right)K_{2\nu}\left(\sqrt{2bc}\right)\quad [b,\ \mathrm{Re}\,c>0;\ -1/2<\mathrm{Re}\,\nu<1].$$

9. $\int\limits_0^\infty x\mathbf{H}_0(bx)\,[I_0(cx)-\mathbf{L}_0(cx)]\,dx=\frac{4c}{\pi^2b\,(b^2+c^2)}\ln\frac{b}{c}$ $\quad[b,\ \mathrm{Re}\,c>0].$

10. $\int\limits_0^\infty xe^{-px}\,[I_1(cx)\,\mathbf{L}_0(cx)-I_0(cx)\,\mathbf{L}_1(cx)]\,dx=\frac{2c^2}{\pi p}\,(p^2-c^2)^{-3/2}$ $\quad[\mathrm{Re}\,p>2\,|\,\mathrm{Re}\,c\,|].$

11. $\int\limits_0^\infty xe^{-px}\,[I_\nu(cx)\,\mathbf{L}'_\nu(cx)-I'_\nu(cx)\,\mathbf{L}_\nu(cx)]\,dx=\frac{2c^{2\nu}}{\pi p}\,(p^2-c^2)^{-\nu-1/2}$

$[\mathrm{Re}\,p>2\,|\,\mathrm{Re}\,c\,|;\ \mathrm{Re}\,\nu>-1].$

2.7.19. Integrals containing $K_\mu(\varphi(x))\begin{Bmatrix}\mathbf{H}_\nu(cx)\\ \mathbf{L}_\nu(cx)\end{Bmatrix}$.

1. $\int\limits_0^\infty x^{\alpha-1}K_\mu(bx)\begin{Bmatrix}\mathbf{H}_\nu(cx)\\ \mathbf{L}_\nu(cx)\end{Bmatrix}dx=\frac{2^{\alpha-1}c^{\nu+1}}{\sqrt{\pi}\,b^{\alpha+\nu+1}}\Gamma\begin{bmatrix}(1+\alpha+\mu+\nu)/2,\ (1+\alpha-\mu+\nu)/2\\ \nu+3/2\end{bmatrix}\times$

$$\times{}_3F_2\left(1,\ \frac{1+\alpha+\mu+\nu}{2},\ \frac{1+\alpha-\mu+\nu}{2};\ \nu+\frac{3}{2},\ \frac{3}{2};\ \mp\frac{c^2}{b^2}\right)$$

$\left[\mathrm{Re}\,b>\begin{Bmatrix}|\,\mathrm{Im}\,c\,|\\ |\,\mathrm{Re}\,c\,|\end{Bmatrix},\ \mathrm{Re}\,(\alpha+\nu\pm\mu)>-1\right].$

2. $\int\limits_0^\infty xK_\nu(bx)\begin{Bmatrix}\mathbf{H}_\nu(cx)\\ \mathbf{L}_\nu(cx)\end{Bmatrix}dx=\left(\frac{c}{b}\right)^{\nu+1}(b^2\pm c^2)^{-1}$

$\left[\mathrm{Re}\,b>\begin{Bmatrix}|\,\mathrm{Im}\,c\,|\\ |\,\mathrm{Re}\,c\,|\end{Bmatrix},\ \mathrm{Re}\,\nu>-3/2\right].$

3. $\int\limits_0^\infty x^{(1\pm1)/2}K_{2\nu}(b\sqrt{x})\,\mathbf{H}_\nu(cx)\,dx=\frac{2^\nu b^{1\pm1}}{\pi c^{2\pm1}}\,\Gamma\left(\nu+\frac{3\pm1}{2}\right)S_{-\nu-2\mp1,\,\nu}\left(\frac{b^2}{4c}\right)$

$[c,\ \mathrm{Re}\,b>0;\ \mathrm{Re}\,\nu>-(5\pm1)/4].$

4. $\int\limits_0^\infty x^{1/2}K_{2\nu-1}(b\sqrt{x})\,\mathbf{H}_\nu(cx)\,dx=\frac{2^{\nu+1}b}{\pi c^2}\,\Gamma(\nu+1)\,S_{-\nu-2,\,\nu-1}\left(\frac{b^2}{4c}\right)$

$[c,\ \mathrm{Re}\,b>0;\ \mathrm{Re}\,\nu>-1].$

5. $\int\limits_0^\infty x^{\alpha-1}e^{\mp bx^2}K_\mu(bx^2)\,\mathbf{H}_\nu(cx)\,dx=I_\pm(1)$

$\left[|\arg b|<(2\mp1)\pi/2;\ \mathrm{Re}\,(\alpha+\nu)>2\,|\,\mathrm{Re}\,\mu\,|-1,\ \begin{Bmatrix}|\arg c|<\pi\\ c>0;\ \mathrm{Re}\,(\alpha+\nu)<2;\ \mathrm{Re}\,\alpha<5/2\end{Bmatrix}\right].$

$$I_{\mp}(\varepsilon)=\pm\left(\cos\mu\pi\operatorname{cosec}\frac{\alpha+\nu}{2}\pi\right)^{(1\mp1)/2}\frac{2^{-(\alpha+3\nu+3)/2}c^{\nu+1}}{b^{(\alpha+\nu+1)/2}}\times$$
$$\times\Gamma\left[\begin{matrix}(\alpha+\nu+1)/2+\mu,\ (\alpha+\nu+1)/2-\mu\\(\alpha+\nu)/2+1,\ \nu+3/2\end{matrix}\right]\times$$
$$\times{}_3F_3\left(1,\frac{1+\alpha+\nu}{2}+\mu,\frac{1+\alpha+\nu}{2}-\mu;\frac{3}{2},\frac{\alpha+\nu}{2}+1,\frac{3}{2}+\nu;\mp\frac{\varepsilon c^2}{8b}\right)+$$
$$+\begin{Bmatrix}0\\1\end{Bmatrix}\frac{\pi^{3/2}c^{1-\alpha}}{2^{5/2-\alpha}\sqrt{b}}\frac{\operatorname{cosec}[(\alpha+\nu)\pi/2]}{\Gamma((3-\alpha+\nu)/2)\,\Gamma((3-\alpha-\nu)/2)}\times$$
$$\times{}_2F_2\left(\frac{1}{2}+\mu,\frac{1}{2}-\mu;\frac{3-\alpha+\nu}{2},\frac{3-\alpha-\nu}{2};\frac{c^2}{8b}\right).$$

6. $$\int_0^\infty x^{\alpha-1}e^{-bx^2}K_\mu(bx^2)\,\mathbf{L}_\nu(cx)\,dx=I_-(-1)$$

[Re $b>0$; $|\arg c|<\pi$; Re $(\alpha+\nu)>2|\operatorname{Re}\mu|-1$; $I_-(\varepsilon)$ see 2.7.19.5].

7. $$\int_0^\infty xe^{bx^2}K_{\nu/2}(bx^2)\,\mathbf{H}_\nu(cx)\,dx=$$
$$=\frac{2^{-(3\nu+2)/4}\Gamma(-\nu/2)\,c^{\nu/2-1}}{\sqrt{\pi}\,b^{(\nu+2)/4}}\cos\frac{\nu\pi}{2}\exp\left(\frac{c^2}{16b}\right)W_{\nu/4,\,(\nu+2)/4}\left(\frac{c^2}{8b}\right)$$

[$c>0$; $|\arg b|<\pi$; $-3/2<\operatorname{Re}\nu<0$].

8. $$\int_0^\infty[2K_{2\nu}(b\sqrt{x})+\pi Y_{2\nu}(b\sqrt{x})]\,\mathbf{H}_\nu(cx)\,dx=\frac{\pi}{c}J_\nu\left(\frac{b^2}{4c}\right)$$

[$b,c>0$; $-1<\operatorname{Re}\nu<1/4$].

9. $$\int_0^\infty J_0(b\sqrt{x})\,K_0(b\sqrt{x})\,\mathbf{H}_0(cx)\,dx=\frac{1}{2\pi c}K_0^2\left(\frac{b^2}{4c}\right)$$ [$c>0$; $|\arg b|<\pi/4$].

10. $$\int_0^\infty x^{1-\nu}[J_{2\nu}(b\sqrt{x})-J_{-2\nu}(b\sqrt{x})]\,K_{2\nu}(b\sqrt{x})\,\mathbf{H}_\nu(cx)\,dx=$$
$$=\frac{2^\nu b^{1-2\nu}c^{2\nu-5/2}}{\sqrt{\pi}}\sin\nu\pi\,K_{\nu+1/2}\left(\frac{b^2}{2c}\right)$$ [$c>0$; $|\arg b|<\pi/4$; $|\operatorname{Re}\nu|<3/2$].

11. $$\int_0^\infty xY_\nu(b\sqrt{x})\,K_\nu(b\sqrt{x})\,\mathbf{H}_\nu(cx)\,dx=\frac{1}{2c^2}\exp\left(-\frac{b^2}{2c}\right)$$

[$c>0$; $|\arg b|<\pi/4$; Re $\nu>-3/2$].

12. $$\int_0^\infty x^\nu Y_{2\nu-1}(b\sqrt{x})\,K_{2\nu-1}(b\sqrt{x})\,\mathbf{H}_\nu(cx)\,dx=\frac{b^{2\nu-1}}{2^\nu\sqrt{\pi}\,c^{2\nu+1/2}}K_{\nu-1/2}\left(\frac{b^2}{2c}\right)$$

[$c>0$; $|\arg b|<\pi/4$; Re $\nu>-1/4$].

13. $$\int_0^\infty x^{\nu+1}Y_{2\nu}(b\sqrt{x})\,K_{2\nu}(b\sqrt{x})\,\mathbf{H}_\nu(cx)\,dx=\frac{b^{2\nu+1}}{2^{\nu+1}\sqrt{\pi}\,c^{2\nu+5/2}}K_{\nu-1/2}\left(\frac{b^2}{2c}\right)$$

[$c>0$; $|\arg b|<\pi/4$; Re $\nu>-3/4$].

14. $$\int_0^\infty\left[\cos\frac{\mu-\nu}{2}\pi J_\mu(b\sqrt{x})-\sin\frac{\mu-\nu}{2}\pi Y_\mu(b\sqrt{x})\right]K_\mu(b\sqrt{x})\,\mathbf{H}_\nu(cx)\,dx=$$
$$=\frac{1}{b^2}W_{\nu/2,\,\mu/2}\left(\frac{b^2}{2c}\right)W_{-\nu/2,\,\mu/2}\left(\frac{b^2}{2c}\right)$$ [$c>0$; $|\arg b|<\pi/4$; Re $\nu>|\operatorname{Re}\mu|-2$].

15. $$\int_0^\infty xY_\nu\left(b\sqrt{x^2+z^2}-bz\right)K_\nu\left(b\sqrt{x^2+z^2}+bz\right)\mathbf{H}_\nu(cx)\,dx=\frac{1}{2c^2}\exp\left(-cz-\frac{b}{2c}\right)$$

$[c,\ \operatorname{Re} z>0;\ |\arg b|<\pi/4;\ -1<\operatorname{Re}\nu<3]$.

16. $$\int_0^\infty x^{\alpha-1}K_\lambda(bx)K_\mu(bx)\begin{Bmatrix}\mathbf{H}_\nu(cx)\\ \mathbf{L}_\nu(cx)\end{Bmatrix}dx=\frac{2^{\alpha-2}c^{\nu+1}}{\sqrt{\pi}\,b^{\alpha+\mu+1}}\Gamma\begin{bmatrix}(\alpha+\lambda+\mu+\nu+1)/2\\ \nu+3/2,\ \alpha+\nu+1\end{bmatrix}\times$$
$$\times\Gamma\left[\frac{\alpha+\nu+\lambda-\mu+1}{2},\frac{\alpha+\nu-\lambda+\mu+1}{2},\frac{\alpha+\nu-\lambda-\mu+1}{2}\right]{}_5F_4\left(1,\frac{\alpha+\lambda+\mu+\nu+1}{2},\right.$$
$$\left.\frac{\alpha+\nu+\lambda-\mu+1}{2},\frac{\alpha+\nu-\lambda+\mu+1}{2},\frac{\alpha+\nu-\lambda-\mu+1}{2};\frac{3}{2},\nu+\frac{3}{2},\frac{\alpha+\nu+1}{2},1+\frac{\alpha+\nu}{2};\mp\frac{c^2}{4b^2}\right)$$

$\left[2\operatorname{Re} b>\begin{Bmatrix}|\operatorname{Im} c|\\ |\operatorname{Re} c|\end{Bmatrix},\ \operatorname{Re}(\alpha+\nu)>|\operatorname{Re}\lambda|+|\operatorname{Re}\mu|-1\right]$.

17. $$\int_0^\infty x^{\nu+2}K_\nu(bx)K_{\nu+1}(bx)\begin{Bmatrix}\mathbf{H}_\nu(cx)\\ \mathbf{L}_\nu(cx)\end{Bmatrix}dx=$$
$$=\frac{\sqrt{\pi}\,c^{\nu+1}}{2^{\nu+3}b^{2\nu+4}}\Gamma\begin{bmatrix}2\nu+5/2\\ \nu+2\end{bmatrix}{}_2F_1\left(1,\ 2\nu+\frac{5}{2};\ \nu+2;\ \mp\frac{c^2}{4b^2}\right)$$

$\left[2\operatorname{Re} b>\begin{Bmatrix}|\operatorname{Im} c|\\ |\operatorname{Re} c|\end{Bmatrix},\ \operatorname{Re}\nu>-5/4\right]$.

18. $$\int_0^\infty x^{-\nu}K_\nu(bx)K_{\nu+1}(bx)\begin{Bmatrix}\mathbf{H}_\nu(cx)\\ \mathbf{L}_\nu(cx)\end{Bmatrix}dx=$$
$$=\frac{\sqrt{\pi}\,c^{\nu+1}}{2^{\nu+2}b^2}\Gamma\left(\frac{1}{2}-\nu\right){}_2F_1\left(\frac{1}{2},\frac{1}{2}-\nu;\ \frac{3}{2};\ \mp\frac{c^2}{4b^2}\right)$$

$\left[2\operatorname{Re} b>\begin{Bmatrix}|\operatorname{Im} c|\\ |\operatorname{Re} c|\end{Bmatrix},\ -3/2<\operatorname{Re}\nu<1/2\right]$.

19. $$\int_0^\infty x^{-\nu}K_0(bx)K_1(bx)\begin{Bmatrix}\mathbf{H}_\nu(cx)\\ \mathbf{L}_\nu(cx)\end{Bmatrix}dx=$$
$$=\frac{\pi^{3/2}c^{\nu+1}}{2^{\nu+3}b^2\Gamma(\nu+3/2)}{}_2F_1\left(\frac{1}{2},\ \frac{1}{2};\ \nu+\frac{3}{2};\ \mp\frac{c^2}{4b^2}\right)\quad\left[2\operatorname{Re} b>\begin{Bmatrix}|\operatorname{Im} c|\\ |\operatorname{Re} c|\end{Bmatrix}\right].$$

20. $$\int_0^\infty x^{\nu+1}K_\nu^2(bx)\begin{Bmatrix}\mathbf{H}_\nu(cx)\\ \mathbf{L}_\nu(cx)\end{Bmatrix}dx=$$
$$=\frac{\sqrt{\pi}c^{\nu+1}}{2^{\nu+3}b^{2\nu+3}}\Gamma\begin{bmatrix}2\nu+3/2\\ \nu+2\end{bmatrix}{}_2F_1\left(1,\ 2\nu+\frac{3}{2};\ \nu+2;\ \mp\frac{c^2}{4b^2}\right)$$

$\left[2\operatorname{Re} b>\begin{Bmatrix}|\operatorname{Im} c|\\ |\operatorname{Re} c|\end{Bmatrix},\ \operatorname{Re}\nu>-3/4\right]$.

21. $$\int_0^\infty xK_\mu^2(bx)\mathbf{H}_0(cx)\,dx=\frac{2^{-2\mu-1}\pi}{b^{2\mu}cz}\sec\mu\pi\left[2^{2\mu}b^{2\mu-1}z-(z+c)^{2\mu}-(z-c)^{2\mu}\right]$$

$\left[z=\sqrt{4b^2+c^2};\ 2\operatorname{Re} b>\begin{Bmatrix}|\operatorname{Im} c|\\ |\operatorname{Re} c|\end{Bmatrix},\ |\operatorname{Re}\mu|<3/2\right]$.

22. $$\int_0^\infty K_1\left(be^{\pi i/4}\sqrt{x}\right)K_1\left(be^{-\pi i/4}\sqrt{x}\right)\mathbf{H}_0(cx)\,dx=$$
$$=\frac{1}{b^2}\left[\sin\varphi\operatorname{ci}(\varphi)-\cos\varphi\operatorname{si}(\varphi)\right]\qquad[\varphi=b^2/(2c);\ b,\ c>0].$$

23. $$\int_0^\infty x^\nu K_{2\nu-1}\left(be^{\pi i/4}\sqrt{x}\right)K_{2\nu-1}\left(be^{-\pi i/4}\sqrt{x}\right)\mathbf{H}_\nu(cx)\,dx=$$
$$=\frac{2^{2\nu-1/2}b^{2\nu-1}}{\sqrt{\pi}c^{2\nu+1/2}}\Gamma(\nu+1)\Gamma(2\nu+1/2)S_{-3\nu-1/2,\ \nu-1/2}\left(\frac{b^2}{2c}\right)\qquad[b,\ c>0;\ \operatorname{Re}\nu>-1/4].$$

24. $$\int_0^\infty xK_0\left(b\sqrt{x^2+z^2}+bz\right)K_0\left(b\sqrt{x^2+z^2}-bz\right)\begin{Bmatrix}\mathbf{H}_0(cx)\\ \mathbf{L}_0(cx)\end{Bmatrix}dx=$$
$$=\mp\frac{\pi}{c\sqrt{4b^2\pm c^2}}\exp\left(-z\sqrt{4b^2\pm c^2}\right)\qquad\left[2\operatorname{Re}b>\begin{Bmatrix}|\operatorname{Im}c|\\ |\operatorname{Re}c|\end{Bmatrix}\right].$$

2.8. THE FUNCTIONS OF ANGER $\mathbf{J}_\nu(x)$ AND WEBER $\mathbf{E}_\nu(x)$

2.8.1. Integrals of general form

1. $$\int_0^a x^{\alpha-1}(a^r-x^r)^{\beta-1}\begin{Bmatrix}\mathbf{J}_\nu(cx)\\ \mathbf{E}_\nu(cx)\end{Bmatrix}dx=\frac{a^{\alpha+r\beta-r}}{r}\Gamma(\beta)\times$$
$$\times\sum_{k=0}^\infty\Gamma\left[\begin{matrix}(k+\alpha)/r\\ \beta+(k+\alpha)/r,\ 1+(k+\nu)/2,\ 1+(k-\nu)/2\end{matrix}\right]\times$$
$$\times\begin{Bmatrix}\cos[(k+\nu)\pi/2]\\ \sin[(k+\nu)\pi/2]\end{Bmatrix}\left(-\frac{ac}{2}\right)^k\qquad[a,\ r,\ \operatorname{Re}\alpha,\ \operatorname{Re}\beta>0].$$

2. $$\int_a^\infty x^{\alpha-1}(x^r-a^r)^{\beta-1}\begin{Bmatrix}\mathbf{J}_\nu(cx)\\ \mathbf{E}_\nu(cx)\end{Bmatrix}dx=$$
$$=\frac{a^{\alpha+r\beta-r}}{r}\Gamma(\beta)\sum_{k=0}^\infty\Gamma\left[\begin{matrix}1-\beta-(k+\alpha)/r\\ 1-(k+\alpha)/r,\ 1+(k+\nu)/2,\ 1+(k-\nu)/2\end{matrix}\right]\times$$
$$\times\begin{Bmatrix}\cos[(k+\nu)\pi/2]\\ \sin[(k+\nu)\pi/2]\end{Bmatrix}\left(-\frac{ac}{2}\right)^k-\pi\left(\frac{c}{2}\right)^{r-\beta r-\alpha}\times$$
$$\times\sum_{k=0}^\infty\frac{(1-\beta)_k}{k!}\frac{\operatorname{cosec}(r-r\beta+rk-\alpha)\pi}{\Gamma(1+(r+\nu-\alpha-r\beta+rk)/2)\Gamma(1+(r-\nu-\alpha-r\beta+rk)/2)}\times$$
$$\times\begin{Bmatrix}\cos[(r+\nu-\alpha-r\beta+rk)\pi/2]\\ \sin[(r+\nu-\alpha-r\beta+rk)\pi/2]\end{Bmatrix}\left(\frac{ac}{2}\right)^{rk}\qquad[a,\ c,\ r,\ \operatorname{Re}\beta>0;\ \operatorname{Re}(\alpha+r\beta)<r+1].$$

3. $$\int_0^\infty\frac{x^{\alpha-1}}{(x^r+z^r)^\rho}\begin{Bmatrix}\mathbf{J}_\nu(cx)\\ \mathbf{E}_\nu(cx)\end{Bmatrix}dx=\frac{z^{\alpha-r\rho}}{r\Gamma(\rho)}\times$$
$$\times\sum_{k=0}^\infty\Gamma\left[\begin{matrix}(\alpha+k)/r,\ \rho-(\alpha+k)/r\\ 1+(k+\nu)/2,\ 1+(k-\nu)/2\end{matrix}\right]\begin{Bmatrix}\cos[(\nu+k)\pi/2]\\ \sin[(\nu+k)\pi/2]\end{Bmatrix}\left(-\frac{cz}{2}\right)^k-\pi\left(\frac{c}{2}\right)^{r\rho-\alpha}\times$$

$$\times\sum_{k=0}^{\infty}\frac{(-1)^k(\rho)_k}{k!}\frac{\operatorname{cosec}(r\rho+rk-\alpha)\pi}{\Gamma(1+(r\rho-\alpha+\nu+rk)/2)\,\Gamma(1+(r\rho-\alpha-\nu+rk)/2)}\times$$

$$\times\begin{Bmatrix}\cos[(\nu-\alpha+r\rho+rk)\pi/2]\\ \sin[(\nu-\alpha+r\rho+rk)\pi/2]\end{Bmatrix}\left(\frac{cz}{2}\right)^{rk}\qquad [c,\ r,\ \operatorname{Re}\alpha>0;\ \operatorname{Re}(\alpha-r\rho)<1;\ r\,|\arg z|<\pi].$$

4. $$\int_0^{\infty}\frac{x^{\alpha-1}}{x^r-y^r}\begin{Bmatrix}\mathbf{J}_\nu(cx)\\ \mathbf{E}_\nu(cx)\end{Bmatrix}dx=$$

$$=-\frac{\pi y^{\alpha-r}}{r}\sum_{k=0}^{\infty}\frac{1}{\Gamma(1+(k+\nu)/2)\,\Gamma(1+(k-\nu)/2)}\operatorname{ctg}\frac{k+\alpha}{r}\pi\times$$

$$\times\begin{Bmatrix}\cos[(\nu+k)\pi/2]\\ \sin[(\nu+k)\pi/2]\end{Bmatrix}\left(-\frac{cy}{2}\right)^k-\pi\left(\frac{c}{2}\right)^{r-\alpha}\times$$

$$\times\sum_{k=0}^{\infty}\frac{\operatorname{cosec}(r+rk-\alpha)\pi}{\Gamma(1+(r-\alpha+\nu+rk)/2)\,\Gamma(1+(r-\alpha-\nu+rk)/2)}\times$$

$$\times\begin{Bmatrix}\cos[(\nu-\alpha+r+rk)\pi/2]\\ \sin[(\nu-\alpha+r+rk)\pi/2]\end{Bmatrix}\left(\frac{cy}{2}\right)^{rk}\qquad [c,\ r,\ y>0;\ 0<\operatorname{Re}\alpha<1+r].$$

5. $$\int_0^{\infty}x^{\alpha-1}e^{-px^r}\begin{Bmatrix}\mathbf{J}_\nu(cx)\\ \mathbf{E}_\nu(cx)\end{Bmatrix}dx=U(r)\qquad [c,\ r,\ \operatorname{Re}p,\ \operatorname{Re}\alpha>0],$$

$$U(r)=\frac{p^{-\alpha/r}}{r}\sum_{k=0}^{\infty}\Gamma\begin{bmatrix}(\alpha+k)/r\\ 1+(k+\nu)/2,\ 1+(k-\nu)/2\end{bmatrix}\begin{Bmatrix}\cos[(\nu+k)\pi/2]\\ \sin[(\nu+k)\pi/2]\end{Bmatrix}\left(-\frac{c}{2p^{1/r}}\right)^k$$

$$[r>1],$$

$$U(r)=\pm\frac{2p^{(1-\alpha)/r}}{cr}\sum_{k=0}^{\infty}\Gamma\begin{bmatrix}(\alpha-k-1)/r\\ (1+\nu-k)/2,\ (1-\nu-k)/2\end{bmatrix}\begin{Bmatrix}\sin[(\nu-k)\pi/2]\\ \cos[(\nu-k)\pi/2]\end{Bmatrix}\left(-\frac{2p^{1/r}}{c}\right)^k+$$

$$+\pi\left(\frac{2}{c}\right)^{\alpha}\sum_{k=0}^{\infty}\frac{\operatorname{cosec}(\alpha+rk)\pi}{k!\,\Gamma(1-(\alpha-\nu+rk)/2,\ \Gamma(1-(\alpha+\nu+rk)/2)}\times$$

$$\times\begin{Bmatrix}\cos[(\nu-\alpha-rk)\pi/2]\\ \sin[(\nu-\alpha-rk)\pi/2]\end{Bmatrix}\left(-\frac{2^r p}{c^r}\right)^k\qquad [r<1],$$

$U(1)$ see 2.8.3.1.

6. $$\int_0^{\infty}x^{\alpha-1}\sin bx^r\begin{Bmatrix}\mathbf{J}_\nu(cx)\\ \mathbf{E}_\nu(cx)\end{Bmatrix}dx=V(1)$$

$[b,\ c,\ r>0;\ -r<\operatorname{Re}\alpha<r+1/2$ for $r>1,\ -r<\operatorname{Re}\alpha<\min(3/2,\ r+1)$ for $r<1]$,

$$V(\delta)=\frac{b^{-\alpha/r}}{r}\sum_{k=0}^{\infty}\Gamma\begin{bmatrix}(k+\alpha)/r\\ 1+(k+\nu)/2,\ 1+(k-\nu)/2\end{bmatrix}\times$$

$$\times\cos\frac{r\delta-\alpha-k}{2r}\pi\begin{Bmatrix}\cos[(k+\nu)\pi/2]\\ \sin[(k+\nu)\pi/2]\end{Bmatrix}\left(-\frac{c}{2b^{1/r}}\right)^k\qquad [r>1],$$

$$V(\delta)=\pm\frac{2b^{(1-\alpha)/r}}{rc}\sum_{k=0}^{\infty}\Gamma\begin{bmatrix}(\alpha-k-1)/r\\ (1+\nu-k)/2,\ (1-\nu-k)/2\end{bmatrix}\times$$

$$\times\cos\frac{r\delta+1-\alpha+k}{2r}\pi\begin{Bmatrix}\sin[(\nu-k)\pi/2]\\ \cos[(\nu-k)\pi/2]\end{Bmatrix}\left(-\frac{2b^{1/r}}{c}\right)^k+\pi b^\delta\left(\frac{2}{c}\right)^{\alpha+r\delta}\times$$

$$\times\sum_{k=0}^{\infty}\frac{(-1)^k\operatorname{cosec}(\alpha+r\delta+2rk)\pi}{k!\,(\delta+1/2)_k\,\Gamma(1-(\alpha+\nu+\delta r)/2-rk)\,\Gamma(1-(\alpha-\nu+\delta r)/2-rk)}\times$$

$$\times\begin{Bmatrix}\cos[(\nu-\alpha-r\delta-2rk)\pi/2]\\ \sin[(\nu-\alpha-r\delta-2rk)\pi/2]\end{Bmatrix}\left(\frac{2^{r-1}b}{c^r}\right)^{2k}\qquad[r<1],$$

$V(\delta)$ for $r=1$ see 2.8.4.1—2.

7. $$\int_0^\infty x^{\alpha-1}\cos bx^r\begin{Bmatrix}\mathbf{J}_\nu(cx)\\ \mathbf{E}_\nu(cx)\end{Bmatrix}dx=V(0)$$

[$b,\ c,\ r,\ \operatorname{Re}\alpha>0$; $\operatorname{Re}\alpha<r+1/2$ for $r>1$, $\operatorname{Re}\alpha<\min(3/2,\ r+1)$ for $r<1$; $V(\delta)$ see 2.8.1.6]

8. $$\int_0^\infty x^{\alpha-1}J_\mu(bx^r)\begin{Bmatrix}\mathbf{J}_\nu(cx)\\ \mathbf{E}_\nu(cx)\end{Bmatrix}dx=W(r)$$

[$b,\ c,\ r>0$; $-r\operatorname{Re}\mu<\operatorname{Re}\alpha<(3r+1)/2$ for $r>1$, $-r\operatorname{Re}\mu<\operatorname{Re}\alpha<1+\min(3r,\ 1+r)/2$ for $r<1$].

$$W(r)=\frac{2^{\alpha/r-1}}{b^{\alpha/r}r}\sum_{k=0}^{\infty}\Gamma\left[\begin{matrix}(\mu r+\alpha+k)/(2r)\\ 1+(\mu r-\alpha-k)/(2r),\ 1+(k+\nu)/2,\ 1+(k-\nu)/2\end{matrix}\right]\times$$

$$\times\begin{Bmatrix}\cos[(k+\nu)\pi/2]\\ \sin[(k+\nu)\pi/2]\end{Bmatrix}\left(-\frac{2^{1/r-1}c}{b^{1/r}}\right)^k\qquad[r>1],$$

$$W(r)=\pm\left(\frac{b}{2}\right)^{(1-\alpha)/r}\frac{1}{cr}\sum_{k=0}^{\infty}\Gamma\left[\begin{matrix}(\mu r+\alpha-k-1)/(2r)\\ 1+(\mu r-\alpha+k+1)/(2r),\ (1+\nu-k)/2,\ (1-\nu-k)/2\end{matrix}\right]\times$$

$$\times\begin{Bmatrix}\sin[(\nu-k)\pi/2]\\ \cos[(\nu-k)\pi/2]\end{Bmatrix}\left(-\frac{2^{1-1/r}b^{1/r}}{c}\right)^k+\frac{2^{\alpha+\mu r-\mu}\pi b^\mu}{c^{\alpha+\mu r}\Gamma(\mu+1)}\times$$

$$\times\sum_{k=0}^{\infty}\frac{(-1)^k\operatorname{cosec}(\alpha+\mu r+2rk)\pi}{k!\,(1+\mu)_k\,\Gamma(1-(\alpha-\nu+\mu r)/2-rk)\,\Gamma(1-(\alpha+\nu+\mu r)/2-rk)}\times$$

$$\times\begin{Bmatrix}\cos[(\nu-\alpha-\mu r-2rk)\pi/2]\\ \sin[(\nu-\alpha-\mu r-2rk)\pi/2]\end{Bmatrix}\left(\frac{2^{r-1}b}{c^r}\right)^{2k}\qquad[r<1],$$

$W(1)$ see 2.8.5.1.

2.8.2. Integrals of $A(x)\begin{Bmatrix}\mathbf{J}_\nu(cx)\\ \mathbf{E}_\nu(cx)\end{Bmatrix}$.

1. $$\int_0^\infty x^{\alpha-1}\begin{Bmatrix}\mathbf{J}_\nu(cx)\\ \mathbf{E}_\nu(cx)\end{Bmatrix}dx=\frac{2^\alpha\pi c^{-\alpha}\operatorname{cosec}\alpha\pi}{\Gamma(1-(\alpha-\nu)/2)\,\Gamma(1-(\alpha+\nu)/2)}\begin{Bmatrix}\cos[(\nu-\alpha)\pi/2]\\ \sin[(\nu-\alpha)\pi/2]\end{Bmatrix}$$

[$c>0$; $0<\operatorname{Re}\alpha<1$].

2. $$\int_0^\infty x^{\alpha-1}[\mathbf{J}_\nu(cx)\pm\mathbf{J}_{-\nu}(cx)]\,dx=$$

$$=\frac{2^\alpha\pi c^{-\alpha}}{\Gamma(1-(\alpha-\nu)/2)\,\Gamma(1-(\alpha+\nu)/2)}\begin{Bmatrix}\cos(\nu\pi/2)\operatorname{cosec}(\alpha\pi/2)\\ \sin(\nu\pi/2)\sec(\alpha\pi/2)\end{Bmatrix}$$

[$c>0$; $-(1\mp1)/2<\operatorname{Re}\alpha<(5\pm1)/4$].

3. $$\int_0^a x^{\alpha-1}(a^2-x^2)^{\beta-1}\left\{\begin{matrix}\mathbf{J}_\nu(cx)\\ \mathbf{E}_\nu(cx)\end{matrix}\right\}dx=\frac{a^{\alpha+2\beta-2}}{2\nu\pi}\mathrm{B}\left(\beta,\ \frac{\alpha}{2}\right)\left\{\begin{matrix}\sin\nu\pi\\ 1-\cos\nu\pi\end{matrix}\right\}\times$$
$$\times{}_2F_3\left(1,\ \frac{\alpha}{2};\ \beta+\frac{\alpha}{2},\ 1+\frac{\nu}{2},\ 1-\frac{\nu}{2};\ -\frac{a^2c^2}{4}\right)\pm\frac{a^{\alpha+2\beta-1}}{2(1-\nu^2)\pi}\mathrm{B}\left(\beta,\ \frac{\alpha+1}{2}\right)\times$$
$$\times\left\{\begin{matrix}\sin\nu\pi\\ 1+\cos\nu\pi\end{matrix}\right\}{}_2F_3\left(1,\ \frac{\alpha+1}{2};\ \beta+\frac{\alpha+1}{2},\ \frac{3+\nu}{2},\ \frac{3-\nu}{2};\ -\frac{a^2c^2}{4}\right)$$
$[a,\ \operatorname{Re}\alpha,\ \operatorname{Re}\beta>0]$.

4. $$\int_a^\infty x^{\alpha-1}(x^2-a^2)^{\beta-1}\left\{\begin{matrix}\mathbf{J}_\nu(cx)\\ \mathbf{E}_\nu(cx)\end{matrix}\right\}dx=\frac{a^{\alpha+2\beta-2}}{2\nu\pi}\mathrm{B}\left(\beta,\ 1-\beta-\frac{\alpha}{2}\right)\times$$
$$\times\left\{\begin{matrix}\sin\nu\pi\\ 1-\cos\nu\pi\end{matrix}\right\}{}_2F_3\left(1,\ \frac{\alpha}{2};\ \beta+\frac{\alpha}{2},\ 1+\frac{\nu}{2},\ 1-\frac{\nu}{2};\ -\frac{a^2c^2}{4}\right)\pm$$
$$\pm\frac{a^{\alpha+2\beta-1}c}{2(1-\nu^2)\pi}\mathrm{B}\left(\beta,\ \frac{1-\alpha}{2}-\beta\right)\left\{\begin{matrix}\sin\nu\pi\\ 1+\cos\nu\pi\end{matrix}\right\}\times$$
$$\times{}_2F_3\left(1,\ \frac{\alpha+1}{2};\ \beta+\frac{\alpha+1}{2},\ \frac{3+\nu}{2},\ \frac{3-\nu}{2};\ -\frac{a^2c^2}{4}\right)-$$
$$-\pi\left(\frac{c}{2}\right)^{2-2\beta-\alpha}\frac{\operatorname{cosec}(\alpha+2\beta)\pi}{\Gamma(2-\beta-(\alpha-\nu)/2)\,\Gamma(2-\beta-(\alpha+\nu)/2)}\times$$
$$\times\left\{\begin{matrix}\cos[(\nu-\alpha-2\beta)\pi/2]\\ \sin[(\nu-\alpha-2\beta)\pi/2]\end{matrix}\right\}{}_1F_2\left(1-\beta;\ 2-\beta-\frac{\alpha-\nu}{2},\ 2-\beta-\frac{\alpha+\nu}{2};\ -\frac{a^2c^2}{4}\right)$$
$[a,\ c,\ \operatorname{Re}\beta>0;\ \operatorname{Re}(\alpha+2\beta)<3]$.

5. $$\int_0^\infty\frac{x^{\alpha-1}}{(x^2+z^2)^\rho}\left\{\begin{matrix}\mathbf{J}_\nu(cx)\\ \mathbf{E}_\nu(cx)\end{matrix}\right\}dx=\frac{z^{\alpha-2\rho}}{2\nu\pi}\mathrm{B}\left(\frac{\alpha}{2},\ \rho-\frac{\alpha}{2}\right)\left\{\begin{matrix}\sin\nu\pi\\ 1-\cos\nu\pi\end{matrix}\right\}\times$$
$$\times{}_2F_3\left(1,\ \frac{\alpha}{2};\ 1-\rho+\frac{\alpha}{2},\ 1+\frac{\nu}{2},\ 1-\frac{\nu}{2};\ \frac{c^2z^2}{4}\right)\pm$$
$$\pm\frac{cz^{\alpha-2\rho+1}}{2(1-\nu^2)\pi}\mathrm{B}\left(\frac{\alpha+1}{2},\ \rho-\frac{\alpha+1}{2}\right)\left\{\begin{matrix}\sin\nu\pi\\ 1+\cos\nu\pi\end{matrix}\right\}\times$$
$$\times{}_2F_3\left(1,\ \frac{\alpha+1}{2};\ \frac{3+\alpha}{2}-\rho,\ \frac{3+\nu}{2},\ \frac{3-\nu}{2};\ \frac{c^2z^2}{4}\right)-$$
$$-\pi\left(\frac{c}{2}\right)^{2\rho-\alpha}\frac{\operatorname{cosec}(2\rho-\alpha)\pi}{\Gamma(1+\rho-(\alpha-\nu)/2)\,\Gamma(1+\rho-(\alpha+\nu)/2)}\times$$
$$\times\left\{\begin{matrix}\cos[(\nu-\alpha+2\rho)\pi/2]\\ \sin[(\nu-\alpha+2\rho)\pi/2]\end{matrix}\right\}{}_1F_2\left(\rho;\ 1+\rho-\frac{\alpha-\nu}{2},\ 1+\rho-\frac{\alpha+\nu}{2};\ \frac{c^2z^2}{4}\right)$$
$[c,\ \operatorname{Re}\alpha,\ \operatorname{Re}z>0;\ \operatorname{Re}(\alpha-2\rho)<1]$.

6. $$\int_0^\infty\frac{x^{\alpha-1}}{x^2-y^2}\left\{\begin{matrix}\mathbf{J}_\nu(cx)\\ \mathbf{E}_\nu(cx)\end{matrix}\right\}dx=-\frac{\pi y^{\alpha-2}}{2\nu\pi}\operatorname{ctg}\frac{\alpha\pi}{2}\times$$
$$\times\left\{\begin{matrix}\sin\nu\pi\\ 1-\cos\nu\pi\end{matrix}\right\}{}_1F_2\left(1;\ 1+\frac{\nu}{2},\ 1-\frac{\nu}{2};\ -\frac{c^2y^2}{4}\right)\pm$$
$$\pm\frac{\pi cy^{\alpha-1}}{2(1-\nu^2)\pi}\operatorname{tg}\frac{\alpha\pi}{2}\left\{\begin{matrix}\sin\nu\pi\\ 1+\cos\nu\pi\end{matrix}\right\}{}_1F_2\left(1;\ \frac{3+\nu}{2},\ \frac{3-\nu}{2};\ -\frac{c^2y^2}{4}\right)-\pi\left(\frac{c}{2}\right)^{2-\alpha}\times$$
$$\times\frac{\operatorname{cosec}\alpha\pi}{\Gamma(2-(\alpha-\nu)/2)\,\Gamma(2-(\alpha+\nu)/2)}\left\{\begin{matrix}\cos[(\nu-\alpha)\pi/2]\\ \sin[(\nu-\alpha)\pi/2]\end{matrix}\right\}\times$$
$$\times{}_1F_2\left(1;\ 2-\frac{\alpha-\nu}{2},\ 2-\frac{\alpha+\nu}{2};\ -\frac{c^2y^2}{4}\right)$$
$[c,\ y>0;\ 0<\operatorname{Re}\alpha<3]$.

7. $\int\limits_a^\infty x^{\nu+1}(x^2-a^2)^{\beta-1}[\mathbf{J}_\nu(cx)\pm\mathbf{J}_{-\nu}(cx)]\,dx=$

$$=\frac{a^{\nu+\beta}}{\pi c^\beta}\sin\nu\pi\Gamma(\beta)\left\{\Gamma\begin{bmatrix}(3\pm1\mp2\nu)/4-\beta\\(1\mp1-\nu\mp3\nu)/4\end{bmatrix}S_{\beta-(1\pm1)/2,\,\beta\pm\nu}(ac)-\right.$$

$$\left.-2^{\beta-1}\pi\left[\operatorname{tg}^{\mp1}\frac{\nu\pi}{2}J_{\mp\nu-1/2}(ac)+Y_{\mp\nu-1/2}(ac)\right]\right\}\quad [a,\,c,\,\operatorname{Re}\beta>0;\ \operatorname{Re}(\nu+2\beta)<(5\pm1)/4].$$

2.8.3. Integrals of $x^\alpha e^{-px^n}\begin{Bmatrix}\mathbf{J}_\nu(cx)\\\mathbf{E}_\nu(cx)\end{Bmatrix}$.

1. $\int\limits_0^\infty x^{\alpha-1}e^{-px}\begin{Bmatrix}\mathbf{J}_\nu(cx)\\\mathbf{E}_\nu(cx)\end{Bmatrix}dx=\frac{\Gamma(\alpha)}{\nu\pi}\begin{Bmatrix}\sin\nu\pi\\1-\cos\nu\pi\end{Bmatrix}\times$

$$\times p^{-\alpha}{}_3F_2\left(1,\,\frac{\alpha}{2},\,\frac{\alpha+1}{2};\,1+\frac{\nu}{2},\,1-\frac{\nu}{2};\,-\frac{c^2}{p^2}\right)\pm$$

$$\pm\frac{c\Gamma(\alpha+1)}{(1-\nu^2)\pi}\begin{Bmatrix}\sin\nu\pi\\1+\cos\nu\pi\end{Bmatrix}p^{-\alpha-1}{}_3F_2\left(1,\,\frac{\alpha+1}{2},\,\frac{\alpha}{2}+1;\,\frac{3+\nu}{2},\,\frac{3-\nu}{2};\,-\frac{c^2}{p^2}\right)$$

[$\operatorname{Re}\alpha>0;\ \operatorname{Re}p>|\operatorname{Im}c|$].

2. $\int\limits_0^\infty x^{\alpha-1}e^{-px^2}\begin{Bmatrix}\mathbf{J}_\nu(cx)\\\mathbf{E}_\nu(cx)\end{Bmatrix}dx=$

$$=\frac{\Gamma(\alpha/2)}{2\nu\pi}\begin{Bmatrix}\sin\nu\pi\\1-\cos\nu\pi\end{Bmatrix}p^{-\alpha/2}{}_2F_2\left(1,\,\frac{\alpha}{2};\,1+\frac{\nu}{2},\,1-\frac{\nu}{2};\,-\frac{c^2}{4p}\right)-$$

$$-\frac{c}{2(1-\nu^2)\pi}\Gamma\left(\frac{\alpha+1}{2}\right)\begin{Bmatrix}\sin\nu\pi\\1+\cos\nu\pi\end{Bmatrix}p^{-(\alpha+1)/2}\times$$

$$\times{}_2F_2\left(1,\,\frac{\alpha+1}{2};\,\frac{3+\nu}{2},\,\frac{3-\nu}{2};\,-\frac{c^2}{4p}\right)\quad[\operatorname{Re}p,\ \operatorname{Re}\alpha>0].$$

2.8.4. Integrals of $x^\alpha\begin{Bmatrix}\sin bx\\\cos bx\end{Bmatrix}\begin{Bmatrix}\mathbf{J}_\nu(cx)\\\mathbf{E}_\nu(cx)\end{Bmatrix}$.

1. $\int\limits_0^\infty x^{\alpha-1}\sin bx\begin{Bmatrix}\mathbf{J}_\nu(cx)\\\mathbf{E}_\nu(cx)\end{Bmatrix}dx=V(1)$

[$b,\ c>0;\ -1<\operatorname{Re}\alpha<3/2$ for $b\neq c$, $-1<\operatorname{Re}\alpha<1/2$ for $b=c$],

$$V(\delta)=\frac{\Gamma(\alpha)}{\nu\pi b^\alpha}\cos\frac{\delta-\alpha}{2}\pi\begin{Bmatrix}\sin\nu\pi\\1-\cos\nu\pi\end{Bmatrix}{}_3F_2\left(1,\,\frac{\alpha}{2},\,\frac{\alpha+1}{2};\,1+\frac{\nu}{2},\,1-\frac{\nu}{2};\,\frac{c^2}{b^2}\right)\pm$$

$$\pm\frac{c\Gamma(\alpha+1)}{(1-\nu^2)\pi b^{\alpha+1}}\sin\frac{\delta-\alpha}{2}\pi\begin{Bmatrix}\sin\nu\pi\\1+\cos\nu\pi\end{Bmatrix}{}_3F_2\left(1,\,\frac{\alpha+1}{2},\,\frac{\alpha}{2}+1;\,\frac{3+\nu}{2},\,\frac{3-\nu}{2};\,\frac{c^2}{b^2}\right)$$

[$0<c\leqslant b$],

$V(\delta)=$

$$=\pm\frac{b^{1-\alpha}\Gamma(\alpha-1)}{\nu\pi c}\sin\frac{\alpha-\delta}{2}\pi\begin{Bmatrix}\sin\nu\pi\\1+\cos\nu\pi\end{Bmatrix}{}_3F_2\left(1,\,\frac{1-\nu}{2},\,\frac{1+\nu}{2};\,\frac{3-\alpha}{2},\,1-\frac{\alpha}{2};\,\frac{b^2}{c^2}\right)+$$

$$+\frac{\nu b^{2-\alpha}\Gamma(\alpha-2)}{\pi c^2}\cos\frac{\alpha-\delta}{2}\pi\begin{Bmatrix}\sin\nu\pi\\1-\cos\nu\pi\end{Bmatrix}{}_3F_2\left(1,\,1-\frac{\nu}{2},\,1+\frac{\nu}{2};\,2-\frac{\alpha}{2},\,\frac{3-\alpha}{2};\,\frac{b^2}{c^2}\right)+$$

$$+\pi b^\delta\left(\frac{2}{c}\right)^{\alpha+\delta}\frac{\operatorname{cosec}(\alpha+\delta)\pi}{\Gamma(1-(\alpha+\nu+\delta)/2)\,\Gamma(1-(\alpha-\nu+\delta)/2)}\times$$

$$\times\begin{Bmatrix}\cos[(\nu-\alpha-\delta)\pi/2]\\\sin[(\nu-\alpha-\delta)\pi/2]\end{Bmatrix}{}_2F_1\left(\frac{\alpha+\nu+\delta}{2},\,\frac{\alpha-\nu+\delta}{2};\,\delta+\frac{1}{2};\,\frac{b^2}{c^2}\right)\quad[0<b\leqslant c].$$

2. $\int\limits_0^\infty x^{\alpha-1}\cos bx \begin{Bmatrix} \mathbf{J}_\nu(cx) \\ \mathbf{E}_\nu(cx) \end{Bmatrix} dx = V(0)$

[$b, c>0$; $0<\operatorname{Re}\alpha<3/2$ for $b\neq c$, $0<\operatorname{Re}\alpha<1/2$ for $b=c$; $V(\delta)$ see 2.8.4.1].

2.8.5. Integrals containing $J_\mu(bx)$ and $\begin{Bmatrix} \mathbf{J}_\nu(cx) \\ \mathbf{E}_\nu(cx) \end{Bmatrix}$.

1. $\int\limits_0^\infty x^{\alpha-1} J_\mu(bx) \begin{Bmatrix} \mathbf{J}_\nu(cx) \\ \mathbf{E}_\nu(cx) \end{Bmatrix} dx = U$

[$b, c>0$; $-\operatorname{Re}\mu<\operatorname{Re}\alpha<2$ for $b\neq c$, $-\operatorname{Re}\mu<\operatorname{Re}\alpha<1$ for $b=c$].

$$U=\frac{2^{\alpha-1}}{\nu\pi b^\alpha}\begin{Bmatrix}\sin\nu\pi\\1-\cos\nu\pi\end{Bmatrix}\Gamma\begin{bmatrix}(\alpha+\mu)/2\\1+(\mu-\alpha)/2\end{bmatrix}{}_3F_2\left(\frac{\alpha-\mu}{2},\frac{\alpha+\mu}{2},1;1+\frac{\nu}{2},1-\frac{\nu}{2};\frac{c^2}{b^2}\right)\pm$$
$$\pm\frac{2^\alpha c}{(1-\nu^2)\pi b^{\alpha+1}}\begin{Bmatrix}\sin\nu\pi\\1+\cos\nu\pi\end{Bmatrix}\Gamma\begin{bmatrix}(\alpha+\mu+1)/2\\(1+\mu-\alpha)/2\end{bmatrix}\times$$
$$\times{}_3F_2\left(\frac{1+\alpha-\mu}{2},\frac{1+\alpha+\mu}{2},1;\frac{3+\nu}{2},\frac{3-\nu}{2};\frac{c^2}{b^2}\right)\quad[0<c\leqslant b],$$

$$U=\pm\left(\frac{b}{2}\right)^{1-\alpha}\frac{1}{2\pi c}\begin{Bmatrix}\sin\nu\pi\\1+\cos\nu\pi\end{Bmatrix}\Gamma\begin{bmatrix}(\alpha+\mu-1)/2\\(3+\mu-\alpha)/2\end{bmatrix}\times$$
$$\times{}_3F_2\left(\frac{1-\nu}{2},\frac{1+\nu}{2},1;\frac{3-\mu-\alpha}{2},\frac{3+\mu-\alpha}{2};\frac{b^2}{c^2}\right)-$$
$$-\frac{\nu b^{2-\alpha}}{2^{3-\alpha}\pi c^2}\begin{Bmatrix}\sin\nu\pi\\1-\cos\nu\pi\end{Bmatrix}\Gamma\begin{bmatrix}(\alpha+\mu)/2-1\\2+(\mu-\alpha)/2\end{bmatrix}\times$$
$$\times{}_3F_2\left(1-\frac{\nu}{2},1+\frac{\nu}{2},1;2-\frac{\alpha+\mu}{2},2-\frac{\alpha-\mu}{2};\frac{b^2}{c^2}\right)+$$
$$+\frac{2^\alpha\pi b^\mu}{c^{\alpha+\mu}}\frac{\operatorname{cosec}(\alpha+\mu)\pi}{\Gamma[\mu+1,1-(\alpha+\mu-\nu)/2,1-(\alpha+\mu+\nu)/2]}\times$$
$$\times\begin{Bmatrix}\cos[(\nu-\alpha-\mu)\pi/2]\\\sin[(\nu-\alpha-\mu)\pi/2]\end{Bmatrix}{}_2F_1\left(\frac{\alpha+\mu-\nu}{2},\frac{\alpha+\mu+\nu}{2};\mu+1;\frac{b^2}{c^2}\right)\quad[0<b\leqslant c].$$

2. $\int\limits_0^\infty x^{\alpha-1}[J_\nu(cx)-\mathbf{J}_\nu(cx)]\,dx=-\frac{2^{\alpha-1}\sin\nu\pi}{c^\alpha\sin\alpha\pi}\Gamma\begin{bmatrix}(\alpha+\nu)/2\\1+(\nu-\alpha)/2\end{bmatrix}$

[$\operatorname{Re}(\alpha+\nu)>0$; $0<\operatorname{Re}\alpha<1$; $|\arg c|<\pi$].

3. $\int\limits_a^\infty x^{\nu+1}(x^2-a^2)^{\beta-1}[J_\nu(cx)-\mathbf{J}_\nu(cx)]\,dx=$

$$=-\frac{\sin\nu\pi a^{\nu+\beta}}{2^\nu\pi c^\beta}\Gamma(\beta)\left\{\Gamma\begin{bmatrix}(1-\nu)/2-\beta\\(1-\nu)/2\end{bmatrix}S_{\beta,\nu+\beta}(ac)+2\Gamma\begin{bmatrix}1-\beta-\nu/2\\-\nu/2\end{bmatrix}S_{\beta-1,\nu+\beta}(ac)\right\}$$

[a, $\operatorname{Re}\beta>0$; $\operatorname{Re}\nu>-1$; $\operatorname{Re}(\nu+2\beta)<1$; $|\arg c|<\pi$].

4. $\int\limits_0^\infty xJ_\nu(bx)[J_\nu(cx)-\mathbf{J}_\nu(cx)]\,dx=-\frac{\sin\nu\pi}{\pi b(b+c)}$ [$b>0$; $\operatorname{Re}\nu>-1$; $|\arg c|<\pi$].

2.9. THE LOMMEL FUNCTIONS $s_{\mu,\nu}(x)$ AND $S_{\mu,\nu}(x)$

2.9.1. Integrals of general form

1. $\int\limits_0^a x^{\alpha-1}(a^r-x^r)^{\beta-1}\begin{Bmatrix}s_{\mu,\nu}(cx)\\S_{\mu,\nu}(cx)\end{Bmatrix}dx=\frac{a^{\alpha+r\beta+\mu+1-r}c^{\mu+1}}{r[(\mu+1)^2-\nu^2]}\times$

$$\times \Gamma(\beta) \sum_{k=0}^{\infty} \frac{(-1)^k}{((\mu+\nu+3)/2)_k ((\mu-\nu+3)/2)_k} \Gamma\left[\begin{matrix}(2k+\alpha+\mu+1)/r \\ \beta+(2k+\alpha+\mu+1)/r\end{matrix}\right] \left(\frac{ac}{2}\right)^{2k} +$$

$$+\left\{\begin{matrix}0\\1\end{matrix}\right\} \frac{2^{\mu-1} a^{\alpha+r\beta-r}}{r} \Gamma(\beta) \sum_{k=0}^{\infty} \frac{(-1)^k}{k!} [A_k(\nu)+A_k(-\nu)] \left(\frac{ac}{2}\right)^{2k},$$

$$A_k(\nu) = \frac{(ac/2)^{-\nu}}{(1-\nu)_k} \Gamma\left[\begin{matrix}(1+\mu+\nu)/2,\ \nu \\ (1-\mu+\nu)/2\end{matrix}\right] \Gamma\left[\begin{matrix}(2k+\alpha-\nu)/r \\ \beta+(2k+\alpha-\nu)/r\end{matrix}\right]$$

$$\left[a,\ r,\ \operatorname{Re}\beta > 0,\ \left\{\begin{matrix}\operatorname{Re}\alpha > -\operatorname{Re}\mu-1 \\ \operatorname{Re}\alpha > -\operatorname{Re}\mu-1,\ |\operatorname{Re}\nu|\end{matrix}\right\}\right].$$

2. $$\int_a^{\infty} x^{\alpha-1} (x^r - a^r)^{\beta-1} \left\{\begin{matrix}s_{\mu,\nu}(cx) \\ S_{\mu,\nu}(cx)\end{matrix}\right\} dx = \frac{a^{\alpha+r\beta+\mu+1-r} c^{\mu+1}}{r[(\mu+1)^2-\nu^2]} \times$$

$$\times \Gamma(\beta) \sum_{k=0}^{\infty} \frac{(-1)^k}{((\mu+\nu+3)/2)_k ((\mu-\nu+3)/2)_k} \Gamma\left[\begin{matrix}1-\beta-(2k+\alpha+\mu+1)/r \\ 1-(2k+\alpha+\mu+1)/r\end{matrix}\right] \left(\frac{ac}{2}\right)^{2k} +$$

$$+ \frac{2^{\alpha+\mu+r\beta-r-3}}{\pi c^{\alpha+r\beta-r}} \Gamma\left(\frac{1+\mu-\nu}{2}\right) \Gamma\left(\frac{1+\mu+\nu}{2}\right) \times$$

$$\times \sum_{k=0}^{\infty} \frac{(1-\beta)_k}{k!} \sec\frac{\alpha+\mu-r(k-\beta+1)}{2}\pi \Gamma\left(\frac{\alpha+\nu-r(k-\beta+1)}{2}\right) \Gamma\left(\frac{\alpha-\nu-r(k-\beta+1)}{2}\right) \times$$

$$\times \left[\cos\nu\pi \mp \left\{\begin{matrix}\cos(r-r\beta-\alpha+rk)\pi \\ \cos\mu\pi\end{matrix}\right\}\right] \left(\frac{ac}{2}\right)^{rk} +$$

$$+\left\{\begin{matrix}0\\1\end{matrix}\right\} \frac{2^{\mu-1} a^{\alpha+r\beta-r}}{r} \Gamma(\beta) \sum_{k=0}^{\infty} \frac{(-1)^k}{k!} [C_k(\nu)+C_k(-\nu)] \left(\frac{ac}{2}\right)^{2k},$$

$$C_k(\nu) = \frac{(ac/2)^{-\nu}}{(1-\nu)_k} \Gamma\left[\begin{matrix}(1+\mu+\nu)/2,\ \nu \\ (1-\mu+\nu)/2\end{matrix}\right] \Gamma\left[\begin{matrix}1-(2k+\alpha-\nu)/r-\beta \\ 1-(2k+\alpha-\nu)/r\end{matrix}\right]$$

$$\left[a,\ r,\ \operatorname{Re}\beta > 0;\ \operatorname{Re}(\alpha+\mu+r\beta) < r+1,\ \left\{\begin{matrix}c > 0;\ \operatorname{Re}(\alpha+r\beta) < r+3/2 \\ |\arg c| < \pi\end{matrix}\right\}\right].$$

3. $$\int_0^{\infty} \frac{x^{\alpha-1}}{(x^r+z^r)^{\rho}} \left\{\begin{matrix}s_{\mu,\nu}(cx) \\ S_{\mu,\nu}(cx)\end{matrix}\right\} dx = \frac{c^{\mu+1} z^{\alpha-r\rho+\mu+1}}{r[(\mu+1)^2-\nu^2]\Gamma(\rho)} \times$$

$$\times \sum_{k=0}^{\infty} \frac{(-1)^k}{((\mu+\nu+3)/2)_k ((\mu-\nu+3)/2)_k} \Gamma\left(\frac{2k+\alpha+\mu+1}{r}\right) \Gamma\left(\rho-\frac{2k+\alpha+\mu+1}{r}\right) \left(\frac{cz}{2}\right)^{2k} +$$

$$+ \frac{2^{\alpha+\mu-r\rho-3}}{\pi c^{\alpha-r\rho}} \Gamma\left(\frac{1+\mu-\nu}{2}\right) \Gamma\left(\frac{1+\mu+\nu}{2}\right) \times$$

$$\times \sum_{k=0}^{\infty} \frac{(-1)^k (\rho)_k}{k!} \sec\frac{\alpha+\mu-r\rho-rk}{2}\pi \Gamma\left(\frac{\alpha+\nu-r\rho-rk}{2}\right) \Gamma\left(\frac{\alpha-\nu-r\rho-rk}{2}\right) \times$$

$$\times \left[\cos\nu\pi \mp \left\{\begin{matrix}\cos(r\rho-\alpha+rk)\pi \\ \cos\mu\pi\end{matrix}\right\}\right] \left(\frac{cz}{2}\right)^{rk} +$$

$$+\left\{\begin{matrix}0\\1\end{matrix}\right\} \frac{2^{\mu-1} z^{\alpha-r\rho}}{r\Gamma(\rho)} \sum_{k=0}^{\infty} \frac{(-1)^k}{k!} [D_k(\nu)+D_k(-\nu)] \left(\frac{cz}{2}\right)^{2k}$$

$$D_k(\nu)=\frac{(cz/2)^{-\nu}}{(1-\nu)_k}\Gamma\left[\begin{matrix}(1+\mu+\nu)/2,\ \nu\\(1-\mu+\nu)/2\end{matrix}\right]\Gamma\left(\frac{2k+\alpha-\nu}{r}\right)\Gamma\left(\rho-\frac{2k+\alpha-\nu}{r}\right)$$

$$\left[r>0;\ r\,|\arg z|<\pi;\ -1<\operatorname{Re}(\alpha+\mu)<r\operatorname{Re}\rho+1,\ \begin{Bmatrix}\operatorname{Re}(\alpha-r\rho)<3/2;\ c>0\\ \operatorname{Re}\alpha>|\operatorname{Re}\nu|;\ |\arg c|<\pi\end{Bmatrix}\right].$$

4. $$\int_0^\infty \frac{x^{\alpha-1}}{x^r-y^r}\begin{Bmatrix}s_{\mu,\nu}(cx)\\ S_{\mu,\nu}(cx)\end{Bmatrix}dx=-\frac{\pi c^{\mu+1}y^{\alpha-r+\mu+1}}{r[(\mu+1)^2-\nu^2]}\times$$

$$\times\sum_{k=0}^\infty\frac{(-1)^k}{((\mu+\nu+3)/2)_k((\mu-\nu+3)/2)_k}\operatorname{ctg}\frac{2k+\alpha+\mu+1}{r}\pi\left(\frac{cy}{2}\right)^{2k}+$$

$$+\frac{2^{\alpha+\mu-r-3}}{\pi c^{\alpha-r}}\Gamma\left(\frac{1+\mu-\nu}{2}\right)\Gamma\left(\frac{1+\mu+\nu}{2}\right)\sum_{k=0}^\infty\sec\frac{\alpha+\mu-r-rk}{2}\pi\times$$

$$\times\Gamma\left(\frac{\alpha+\nu-r-rk}{2}\right)\Gamma\left(\frac{\alpha-\nu-r-rk}{2}\right)\left[\cos\nu\pi\mp\begin{Bmatrix}\cos(r-\alpha+rk)\pi\\ \cos\mu\pi\end{Bmatrix}\right]\left(\frac{cy}{2}\right)^{rk}-$$

$$-\begin{Bmatrix}0\\1\end{Bmatrix}\frac{2^{\mu-1}\pi y^{\alpha-r}}{r}\sum_{k=0}^\infty\frac{(-1)^k}{k!}[F_k(\nu)+F_k(-\nu)]\left(\frac{cy}{2}\right)^{2k},$$

$$F_k(\nu)=\frac{(cy/2)^{-\nu}}{(1-\nu)_k}\Gamma\left[\begin{matrix}(1+\mu+\nu)/2,\ \nu\\(1-\mu+\nu)/2\end{matrix}\right]\operatorname{ctg}\frac{2k+\alpha-\nu}{r}\pi$$

$$\left[r,\ y>0;\ -1<\operatorname{Re}(\alpha+\mu)<r+1,\ \begin{Bmatrix}\operatorname{Re}\alpha<3/2+r;\ c>0\\ \operatorname{Re}\alpha>|\operatorname{Re}\nu|;\ |\arg c|<\pi\end{Bmatrix}\right].$$

5. $$\int_0^\infty x^{\alpha-1}e^{-px^r}\begin{Bmatrix}s_{\mu,\nu}(cx)\\ S_{\mu,\nu}(cx)\end{Bmatrix}dx=U(r)$$

$$\left[r,\ \operatorname{Re}p>0;\ \operatorname{Re}(\alpha+\mu)>-1,\ \begin{Bmatrix}c>0\\ \operatorname{Re}\alpha>|\operatorname{Re}\nu|;\ |\arg c|<\pi\end{Bmatrix}\right].$$

$$U(r)=\frac{c^{\mu+1}p^{-(\alpha+\mu+1)/r}}{r[(\mu+1)^2-\nu^2]}\sum_{k=0}^\infty\frac{(-1)^k}{((\mu-\nu+3)/2)_k((\mu+\nu+3)/2)_k}\times$$

$$\times\Gamma\left(\frac{2k+\alpha+\mu+1}{r}\right)\left(\frac{c}{2p^{1/r}}\right)^{2k}+\begin{Bmatrix}0\\1\end{Bmatrix}\frac{2^{\mu-1}}{rp^{\alpha/r}}\sum_{k=0}^\infty\frac{(-1)^k}{k!}[G_k(\nu)+G_k(-\nu)]\left(\frac{c}{2p^{1/r}}\right)^{2k},$$

$$G_k(\nu)=\left(\frac{2p^{1/r}}{c}\right)^{\nu}\Gamma\left[\begin{matrix}(1+\mu+\nu)/2,\ \nu\\(1-\mu+\nu)/2\end{matrix}\right]\Gamma\left(\frac{2k+\alpha-\nu}{r}\right)\frac{1}{(1-\nu)_k}\qquad[r>1],$$

$$U(r)=\frac{c^{\mu-1}}{rp^{(\alpha+\mu-1)/r}}\sum_{k=0}^\infty(-1)^k\left(\frac{1+\nu-\mu}{2}\right)_k\left(\frac{1-\nu-\mu}{2}\right)_k\times$$

$$\times\Gamma\left(\frac{\alpha+\mu-2k-1}{r}\right)\left(\frac{2p^{1/r}}{c}\right)^{2k}+\frac{2^{\alpha+\mu-3}}{c^\alpha\pi}\Gamma\left(\frac{1+\mu-\nu}{2}\right)\times$$

$$\times\Gamma\left(\frac{1+\mu+\nu}{2}\right)\sum_{k=0}^\infty\frac{(-1)^k}{k!}\sec\frac{\alpha+\mu+rk}{2}\pi\Gamma\left(\frac{\alpha+\nu+rk}{2}\right)\times$$

$$\times\Gamma\left(\frac{\alpha-\nu+rk}{2}\right)\left[\cos\nu\pi\mp\begin{Bmatrix}\cos(\alpha+rk)\pi\\ \cos\mu\pi\end{Bmatrix}\right]\left(\frac{2^rp}{c^r}\right)^k\qquad[r<1].$$

$U(1)$ see 2.9.3.1.

6. $\int_0^\infty x^{\alpha-1}\sin bx^r \begin{Bmatrix} s_{\mu,\nu}(cx) \\ S_{\mu,\nu}(cx) \end{Bmatrix} dx = V(1)$

$$\left[b,\ r>0;\ |\operatorname{Re}(\alpha+\mu)|<r+1,\ \begin{Bmatrix} c>0;\ \operatorname{Re}\alpha<\max(1,\ r)+1/2 \\ |\arg c|<\pi;\ \operatorname{Re}\alpha>|\operatorname{Re}\nu|-r \end{Bmatrix}\right],$$

$$V(\delta)=\frac{c^{\mu+1}b^{-(\alpha+\mu+1)/r}}{r[(\mu+1)^2-\nu^2]}\sum_{k=0}^{\infty}\frac{(-1)^k}{((\mu-\nu+3)/2)_k((\mu+\nu+3)/2)_k}\times$$

$$\times\Gamma\left(\frac{2k+\alpha+\mu+1}{r}\right)\cos\frac{2k+\alpha+\mu-\delta r+1}{2r}\pi\left(\frac{c}{2b^{1/r}}\right)^{2k}+$$

$$+\begin{Bmatrix}0\\1\end{Bmatrix}\frac{2^{\mu-1}}{rb^{\alpha/r}}\sum_{k=0}^{\infty}\frac{(-1)^k}{k!}[M_k(\nu)+M_k(-\nu)]\left(\frac{c}{2b^{1/r}}\right)^{2k},$$

$$M_k(\nu)=\frac{(2b^{1/r}/c)^\nu}{(1-\nu)_k}\Gamma\left[\begin{matrix}(1+\mu+\nu)/2,\ \nu\\(1-\mu+\nu)/2\end{matrix}\right]\Gamma\left(\frac{2k+\alpha-\nu}{r}\right)\cos\frac{2k+\alpha-\nu-\delta r}{2r}\pi\quad[r>1],$$

$$V(\delta)=\frac{c^{\mu-1}}{b^{(\alpha+\mu-1)/r}r}\sum_{k=0}^{\infty}(-1)^k\left(\frac{1+\nu-\mu}{2}\right)_k\left(\frac{1-\nu-\mu}{2}\right)_k\times$$

$$\times\Gamma\left(\frac{\alpha+\mu-2k-1}{r}\right)\cos\frac{2k-\alpha-\mu+\delta r+1}{2r}\pi\left(\frac{2b^{1/r}}{c}\right)^{2k}+$$

$$+\frac{2^{\alpha+\mu+\delta r-3}b^\delta}{\pi c^{\alpha+\delta r}}\Gamma\left(\frac{1+\mu-\nu}{2}\right)\Gamma\left(\frac{1+\mu+\nu}{2}\right)\times$$

$$\times\sum_{k=0}^{\infty}\frac{(-1)^k}{k!(1/2+\delta)_k}\sec\frac{\alpha+\mu+\delta r+2rk}{2}\pi\Gamma\left(\frac{\alpha+\nu+\delta r}{2}+rk\right)\Gamma\left(\frac{\alpha-\nu+\delta r}{2}+rk\right)\times$$

$$\times\left[\cos\nu\pi\mp\begin{Bmatrix}\cos(\alpha+\delta r+2rk)\pi\\\cos\mu\pi\end{Bmatrix}\right]\left(\frac{2^{r-1}b}{c^r}\right)^{2k}\quad[r<1],$$

$V(\delta)$ for $r=1$ see 2.9.4.1, 2.9.4.3.

7. $\int_0^\infty x^{\alpha-1}\cos bx^r \begin{Bmatrix} s_{\mu,\nu}(cx) \\ S_{\mu,\nu}(cx) \end{Bmatrix} dx = V(0)$

$$\left[b,\ r>0;\ -1<\operatorname{Re}(\alpha+\mu)<r+1,\ \begin{Bmatrix} c>0;\ \operatorname{Re}\alpha<\max(1,\ r)+1/2 \\ |\arg c|<\pi;\ \operatorname{Re}\alpha>|\operatorname{Re}\nu| \end{Bmatrix};\ V(\delta)\ \text{see 2.9.1.6}\right].$$

8. $\int_0^\infty x^{\alpha-1}J_\lambda(bx^r) \begin{Bmatrix} s_{\mu,\nu}(cx) \\ S_{\mu,\nu}(cx) \end{Bmatrix} dx = W(r)$

$$\left[b,\ r>0;\ -1-r\operatorname{Re}\lambda<\operatorname{Re}(\alpha+\mu)<3r/2+1,\ \begin{Bmatrix} c>0;\ \operatorname{Re}\alpha<\max(1,\ r)+(r+1)/2 \\ |\arg c|<\pi;\ \operatorname{Re}(\alpha+r\lambda)>|\operatorname{Re}\nu| \end{Bmatrix}\right],$$

$$W(r)=\left(\frac{2}{b}\right)^{(\alpha+\mu+1)/r}\frac{c^{\mu+1}}{2r[(\mu+1)^2-\nu^2]}\times$$

$$\times\sum_{k=0}^{\infty}\frac{(-1)^k}{((\mu-\nu+3)/2)_k((\mu+\nu+3)/2)_k}\Gamma\left[\begin{matrix}(\lambda r+\alpha+\mu+2k+1)/(2r)\\1+(\lambda r-\alpha-\mu-2k-1)/(2r)\end{matrix}\right]\left(\frac{2^{1/r-1}c}{b^{1/r}}\right)^{2k}+$$

$$+\begin{Bmatrix}0\\1\end{Bmatrix}\frac{2^{\mu+\alpha/r-2}}{b^{\alpha/r}r}\sum_{k=0}^{\infty}\frac{(-1)^k}{k!}[N_k(\nu)+N_k(-\nu)]\left(\frac{2^{1/r-1}c}{b^{1/r}}\right)^{2k},$$

$$N_k(\nu)=\frac{(2^{1-1/r}b^{1/r}/c)^{\nu}}{(1-\nu)_k}\Gamma\left[\begin{matrix}(1+\mu+\nu)/2,\ \nu\\(1-\mu+\nu)/2\end{matrix}\right]\Gamma\left[\begin{matrix}(\lambda r+\alpha-\nu+2k)/(2r)\\1+(\lambda r-\alpha+\nu-2k)/(2r)\end{matrix}\right]\qquad [r>1],$$

$$W(r)=\left(\frac{b}{2}\right)^{(1-\mu-\alpha)/r}\frac{c^{\mu-1}}{2r}\sum_{k=0}^{\infty}(-1)^k\left(\frac{1+\nu-\mu}{2}\right)_k\left(\frac{1-\nu-\mu}{2}\right)_k\times$$

$$\times\Gamma\left[\begin{matrix}(\lambda r+\alpha+\mu-2k-1)/(2r)\\1+(\lambda r-\alpha-\mu+2k+1)/(2r)\end{matrix}\right]\left(\frac{2^{r-1}b}{c^r}\right)^{2k/r}+\frac{2^{\alpha+\mu+\lambda r-\lambda-3}b^{\lambda}}{c^{\alpha+\lambda r}\pi}\times$$

$$\times\Gamma\left[\begin{matrix}(1+\mu-\nu)/2,\ (1+\mu+\nu)/2\\\lambda+1\end{matrix}\right]\sum_{k=0}^{\infty}\frac{(-1)^k}{k!\,(\lambda+1)_k}\sec\frac{\alpha+\mu+\lambda r+2rk}{2}\pi\times$$

$$\times\Gamma\left(\frac{\alpha+\nu+\lambda r}{2}+rk\right)\Gamma\left(\frac{\alpha-\nu+\lambda r}{2}+rk\right)\times$$

$$\times\left[\cos\nu\pi\mp\begin{Bmatrix}\cos(\alpha+\lambda r+2rk)\pi\\\cos\mu\pi\end{Bmatrix}\right]\left(\frac{2^{r-1}b}{c^r}\right)^{2k}\qquad [r<1],$$

$W(1)$ see 2.9.5.1, 2.9.5.3.

2.9.2. Integrals of $A(x)\begin{Bmatrix}s_{\mu,\nu}(cx)\\S_{\mu,\nu}(cx)\end{Bmatrix}$.

1. $$\int_0^{\infty}x^{\alpha-1}\begin{Bmatrix}s_{\mu,\nu}(cx)\\S_{\mu,\nu}(cx)\end{Bmatrix}dx=$$

$$=\frac{2^{\alpha+\mu-3}c^{-\alpha}}{\pi}\sec\frac{\alpha+\mu}{2}\pi\left(\cos\nu\pi\mp\begin{Bmatrix}\cos\alpha\pi\\\cos\mu\pi\end{Bmatrix}\right)\Gamma\left[\frac{1+\mu+\nu}{2},\frac{1+\mu-\nu}{2},\frac{\alpha+\nu}{2},\frac{\alpha-\nu}{2}\right]$$

$$\left[\operatorname{Re}(\alpha+\mu)<1,\ \begin{Bmatrix}c>0;\ \operatorname{Re}\alpha<3/2\\|\arg c|<\pi;\ \operatorname{Re}\alpha>|\operatorname{Re}\nu|\end{Bmatrix}\right].$$

2. $$\int_0^{a}x^{\alpha-1}(a^2-x^2)^{\beta-1}\begin{Bmatrix}s_{\mu,\nu}(cx)\\S_{\mu,\nu}(cx)\end{Bmatrix}dx=\frac{a^{\alpha+\mu+2\beta-1}c^{\mu+1}}{2[(\mu+1)^2-\nu^2]}\times$$

$$\times B\left(\beta,\frac{\alpha+\mu+1}{2}\right){}_2F_3\left(1,\frac{\alpha+\mu+1}{2};\frac{\alpha+\mu+1}{2}+\beta,\frac{3+\mu-\nu}{2},\frac{3+\mu+\nu}{2};-\frac{a^2c^2}{4}\right)+$$

$$+\begin{Bmatrix}0\\1\end{Bmatrix}2^{\mu-\nu-2}a^{\alpha+\nu+2\beta-2}c^{\nu}B\left(\beta,\frac{\alpha+\nu}{2}\right)\Gamma\left[\begin{matrix}-\nu,\ (1+\mu-\nu)/2\\(1-\mu-\nu)/2\end{matrix}\right]\times$$

$$\times{}_1F_2\left(\frac{\alpha+\nu}{2};\nu+1,\frac{\alpha+\nu}{2}+\beta;-\frac{a^2c^2}{4}\right)+\begin{Bmatrix}0\\1\end{Bmatrix}2^{\mu+\nu-2}a^{\alpha-\nu+2\beta-2}c^{-\nu}B\left(\beta,\frac{\alpha-\nu}{2}\right)\times$$

$$\times\Gamma\left[\begin{matrix}\nu,\ (\mu+\nu+1)/2\\(1-\mu+\nu)/2\end{matrix}\right]{}_1F_2\left(\frac{\alpha-\nu}{2};\ 1-\nu,\ \frac{\alpha-\nu}{2}+\beta;\ -\frac{a^2c^2}{4}\right)$$

$$\left[a,\ \operatorname{Re}\beta>0,\ \begin{Bmatrix}\operatorname{Re}(\alpha+\mu)>-1\\\operatorname{Re}(\alpha+\mu)>-1;\ \operatorname{Re}\alpha>|\operatorname{Re}\nu|\end{Bmatrix}\right].$$

3. $$\int_a^{\infty}x^{\alpha-1}(x^2-a^2)^{\beta-1}\begin{Bmatrix}s_{\mu,\nu}(cx)\\S_{\mu,\nu}(cx)\end{Bmatrix}dx=\frac{a^{\alpha+\mu+2\beta-1}c^{\mu+1}}{2[(\mu+1)^2-\nu^2]}B\left(\beta,\frac{1-\alpha-\mu}{2}-\beta\right)\times$$

$$\times{}_2F_3\left(1,\frac{\alpha+\mu+1}{2};\frac{\alpha+\mu+1}{2}+\beta,\frac{3+\mu-\nu}{2},\frac{3+\mu+\nu}{2};-\frac{a^2c^2}{4}\right)-$$

$$-\frac{2^{\alpha+\mu+2\beta-5}}{\pi c^{\alpha+2\beta-2}}\sec\frac{\alpha+2\beta+\mu}{2}\pi\Gamma\left[\frac{1+\mu-\nu}{2},\frac{1+\mu+\nu}{2},\frac{\alpha+\nu}{2}+\beta-1,\frac{\alpha-\nu}{2}+\beta-1\right]\times$$

$$\times\left[\cos\nu\pi\mp\begin{Bmatrix}\cos(\alpha+2\beta)\pi\\\cos\mu\pi\end{Bmatrix}\right]{}_1F_2\left(1-\beta;\ 2-\beta-\frac{\alpha+\nu}{2},\ 2-\beta-\frac{\alpha-\nu}{2};\ -\frac{a^2c^2}{4}\right)+$$

$$+\begin{Bmatrix}0\\1\end{Bmatrix} 2^{\mu+\nu-2}a^{\alpha+2\beta-\nu-2}c^{-\nu}\mathrm{B}\left(\beta,\ 1-\beta-\frac{\alpha-\nu}{2}\right)\Gamma\left[\begin{matrix}(1+\mu+\nu)/2,\ \nu\\(1-\mu+\nu)/2\end{matrix}\right]\times$$

$$\times{}_1F_2\left(\frac{\alpha-\nu}{2};\ 1-\nu,\ \beta+\frac{\alpha-\nu}{2};\ -\frac{a^2c^2}{4}\right)+\begin{Bmatrix}0\\1\end{Bmatrix} 2^{\mu-\nu-2}a^{\alpha+2\beta+\nu-2}c^{\nu}\times$$

$$\times\mathrm{B}\left(\beta,\ 1-\beta-\frac{\alpha+\nu}{2}\right)\Gamma\left[\begin{matrix}(1+\mu-\nu)/2,\ -\nu\\(1-\mu-\nu)/2\end{matrix}\right]{}_1F_2\left(\frac{\alpha+\nu}{2};1+\nu,\ \beta+\frac{\alpha+\nu}{2};-\frac{a^2c^2}{4}\right)$$

$$\left[a,\ \operatorname{Re}\beta>0;\ \operatorname{Re}(\alpha+\mu+2\beta)<3,\ \begin{Bmatrix}c>0;\ \operatorname{Re}(\alpha+2\beta)<7/2\\|\arg c|<\pi\end{Bmatrix}\right].$$

4. $$\int_a^\infty x^{1-\nu}(x^2-a^2)^{\beta-1}s_{\mu,\nu}(cx)\,dx=\frac{a^{\beta-\nu}\Gamma(\beta)}{2c^\beta}\left\{\Gamma\left[\begin{matrix}(1+\nu-\mu)/2-\beta\\(1+\nu-\mu)/2\end{matrix}\right]\times\right.$$

$$\times S_{\beta+\mu,\ \beta-\nu}(ac)-2^{\mu+\beta-1}\Gamma\left(\frac{1+\mu-\nu}{2}\right)\Gamma\left(\frac{1+\mu+\nu}{2}\right)\times$$

$$\left.\times\left[\sin\frac{\mu-\nu}{2}\pi J_{\nu-\beta}(ac)-\cos\frac{\mu-\nu}{2}\pi Y_{\nu-\beta}(ac)\right]\right\}$$

$$[c,\ \operatorname{Re}\beta>0;\ \operatorname{Re}(\nu-2\beta)>-3/2,\ \operatorname{Re}\mu-1].$$

5. $$\int_a^\infty x^{\nu+1}(x^2-a^2)^{\beta-1}\begin{Bmatrix}s_{\mu,\nu}(cx)\\S_{\mu,\nu}(cx)\end{Bmatrix}dx=\frac{a^{\beta+\nu}}{2c^\beta}\left\{\mathrm{B}\left(\beta,\ \frac{1-\mu-\nu}{2}-\beta\right)S_{\mu+\beta,\ \nu+\beta}(ac)-\right.$$

$$-\begin{Bmatrix}1\\0\end{Bmatrix}2^{\mu+\beta-1}\Gamma\left[\beta,\ \frac{1+\mu+\nu}{2},\ \frac{1+\mu-\nu}{2}\right]\times$$

$$\left.\times\left(\sin\frac{\mu+\nu}{2}\pi J_{-\nu-1/2}(ac)-\cos\frac{\mu+\nu}{2}\pi Y_{-\nu-1/2}(ac)\right)\right\}$$

$$\left[\operatorname{Re}\beta>0;\ \operatorname{Re}(2\beta+\mu+\nu)<1,\ \begin{Bmatrix}c>0;\ \operatorname{Re}(2\beta+\nu)<3/2\\|\arg c|<\pi\end{Bmatrix}\right].$$

6. $$\int_0^\infty\frac{x^{\alpha-1}}{(x^2+z^2)^\rho}\begin{Bmatrix}s_{\mu,\nu}(cx)\\S_{\mu,\nu}(cx)\end{Bmatrix}dx=\frac{c^{\mu+1}z^{\alpha-2\rho+\mu+1}}{2[(\mu+1)^2-\nu^2]}\mathrm{B}\left(\frac{\alpha+\mu+1}{2},\ \rho-\frac{\alpha+\mu+1}{2}\right)\times$$

$$\times{}_2F_3\left(1,\ \frac{\alpha+\mu+1}{2};\ \frac{3+\alpha+\mu}{2}-\rho,\ \frac{3+\mu+\nu}{2},\ \frac{3+\mu-\nu}{2};\ \frac{c^2z^2}{4}\right)+$$

$$+\frac{2^{\alpha+\mu-2\rho-3}}{\pi c^{\alpha-2\rho}}\sec\frac{\alpha+\mu-2\rho}{2}\pi\Gamma\left[\frac{1+\mu+\nu}{2},\ \frac{1+\mu-\nu}{2},\ \frac{\alpha+\nu}{2}-\rho,\ \frac{\alpha-\nu}{2}-\rho\right]\times$$

$$\times\left[\cos\nu\pi\mp\begin{Bmatrix}\cos(2\rho-\alpha)\pi\\\cos\mu\pi\end{Bmatrix}\right]{}_1F_2\left(\rho;\ 1+\rho-\frac{\alpha+\nu}{2},\ 1+\rho-\frac{\alpha-\nu}{2};\ \frac{c^2z^2}{4}\right)+$$

$$+\begin{Bmatrix}0\\1\end{Bmatrix}\frac{2^{\mu+\nu-2}z^{\alpha-2\rho-\nu}}{c^\nu}\mathrm{B}\left(\frac{\alpha-\nu}{2},\ \rho-\frac{\alpha-\nu}{2}\right)\Gamma\left[\begin{matrix}\nu,\ (1+\mu+\nu)/2\\(1-\mu+\nu)/2\end{matrix}\right]\times$$

$$\times{}_1F_2\left(\frac{\alpha-\nu}{2};\ 1-\rho+\frac{\alpha-\nu}{2},\ 1-\nu;\ \frac{c^2z^2}{4}\right)+$$

$$+\begin{Bmatrix}0\\1\end{Bmatrix}2^{\mu-\nu-2}c^\nu z^{\alpha+\nu-2\rho}\mathrm{B}\left(\frac{\alpha+\nu}{2},\ \rho-\frac{\alpha+\nu}{2}\right)\Gamma\left[\begin{matrix}-\nu,\ (1+\mu-\nu)/2\\(1-\mu-\nu)/2\end{matrix}\right]\times$$

$$\times{}_1F_2\left(\frac{\alpha+\nu}{2};\ 1-\rho+\frac{\alpha+\nu}{2},\ 1+\nu;\ \frac{c^2z^2}{4}\right)$$

$$\left[\operatorname{Re}z>0;\ -1<\operatorname{Re}(\alpha+\mu)<2\operatorname{Re}\rho+1,\ \begin{Bmatrix}\operatorname{Re}(\alpha-2\rho)<3/2;\ c>0\\\operatorname{Re}\alpha>|\operatorname{Re}\nu|;\ |\arg c|<\pi\end{Bmatrix}\right].$$

7. $\int\limits_0^\infty \frac{x^{\nu+1}}{x^2+z^2} s_{\mu,\nu}(cx)\,dx=$

$$=\frac{\pi z^\nu}{2}\sec\frac{\mu+\nu}{2}\pi\left[2^{\mu-1}\Gamma\left(\frac{\mu+\nu+1}{2}\right)\Gamma\left(\frac{\mu-\nu+1}{2}\right)I_{-\nu}(cz)+ie^{-\pi\nu i/2}s_{\mu,\nu}(icz)\right]$$

[c, $\operatorname{Re} z>0$; $-3<\operatorname{Re}(\mu+\nu)<1$; $\operatorname{Re}\nu<3/2$].

8. $\int\limits_0^\infty \frac{x^{\alpha-1}}{x^2-y^2}\begin{Bmatrix} s_{\mu,\nu}(cx)\\ S_{\mu,\nu}(cx)\end{Bmatrix}dx=\frac{\pi y^{\alpha-2}}{2}\operatorname{tg}\frac{\alpha+\mu}{2}\pi s_{\mu,\nu}(cy)-$

$$-\frac{2^{\alpha+\mu-3}y^{\alpha-2}}{\pi}\sec\frac{\alpha+\mu}{2}\pi\Gamma\left[\frac{1+\mu-\nu}{2},\ \frac{1+\mu+\nu}{2},\ \frac{\alpha+\nu}{2},\ \frac{\alpha-\nu}{2}\right]\times$$

$$\times\left(\cos\nu\pi\mp\begin{Bmatrix}\cos\alpha\pi\\ \cos\mu\pi\end{Bmatrix}\right)s_{1-\alpha,\nu}(cy)-\begin{Bmatrix}0\\1\end{Bmatrix}\frac{2^{\mu-2}\pi^2y^{\alpha-2}}{\sin\nu\pi}\times$$

$$\times\left(\operatorname{ctg}\frac{\alpha-\nu}{2}\pi\Gamma\begin{bmatrix}(1+\mu+\nu)/2\\(1-\mu+\nu)/2\end{bmatrix}J_{-\nu}(cy)-\operatorname{ctg}\frac{\alpha+\nu}{2}\pi\Gamma\begin{bmatrix}(1+\mu-\nu)/2\\(1-\mu-\nu)/2\end{bmatrix}J_\nu(cy)\right)$$

$\left[y>0;\ -1<\operatorname{Re}(\alpha+\mu)<3,\ \begin{Bmatrix}\operatorname{Re}\alpha<7/2;\ c>0\\ \operatorname{Re}\alpha>|\operatorname{Re}\nu|;\ |\arg c|<\pi\end{Bmatrix}\right]$.

9. $\int\limits_0^\infty \frac{(x+\sqrt{x^2+z^2})^{\mu+1}}{\sqrt{x^2+z^2}} s_{\mu,\nu}(cx)\,dx=\frac{2^{\mu-2}\pi z^{\mu+1}}{\sin\mu\pi}\Gamma\left(\frac{\mu+\nu+1}{2}\right)\Gamma\left(\frac{\mu-\nu+1}{2}\right)\times$

$$\times\left[I_\rho\left(\frac{cz}{2}\right)I_\sigma\left(\frac{cz}{2}\right)-I_{-\rho}\left(\frac{cz}{2}\right)I_{-\sigma}\left(\frac{cz}{2}\right)\right]$$

[$\rho=(\mu+\nu+1)/2$, $\sigma=(\mu-\nu+1)/2$; c, $\operatorname{Re} z>0$; $-1<\operatorname{Re}\mu<0$].

10. $\int\limits_a^\infty (x^2-a^2)^{-1/4}\left[(x+\sqrt{x^2-a^2})^\lambda+(x-\sqrt{x^2-a^2})^\lambda\right]S_{\mu,1/2}(cx)\,dx=$

$$=\frac{a^\lambda}{2^{\mu+1/2}\sqrt{c}}B\left(\frac{1}{4}-\frac{\lambda+\mu}{2},\ \frac{1}{4}+\frac{\lambda-\mu}{2}\right)S_{\mu+1/2,\lambda}(ac)$$

[$a>0$; $|\arg c|<\pi$; $|\operatorname{Re}\lambda|+\operatorname{Re}\mu<1/2$].

2.9.3. Integrals of $x^\alpha e^{-px^r}\begin{Bmatrix} s_{\mu,\nu}(cx)\\ S_{\mu,\nu}(cx)\end{Bmatrix}$.

1. $\int\limits_0^\infty x^{\alpha-1}e^{-px}\begin{Bmatrix} s_{\mu,\nu}(cx)\\ S_{\mu,\nu}(cx)\end{Bmatrix}dx=$

$$=\frac{c^{\mu+1}p^{-\alpha-\mu-1}}{(\mu+1)^2-\nu^2}\Gamma(\alpha+\mu+1)\,{}_3F_2\left(1,\ \frac{\alpha+\mu+1}{2},\frac{\alpha+\mu}{2}+1;\ \frac{3+\mu-\nu}{2},\frac{3+\mu+\nu}{2};-\frac{c^2}{p^2}\right)+$$

$$+\begin{Bmatrix}0\\1\end{Bmatrix}[A(\nu)+A(-\nu)],$$

$$A(\nu)=\frac{2^{\mu+\nu-1}}{c^\nu p^{\alpha-\nu}}\Gamma\begin{bmatrix}\nu,\ (1+\mu+\nu)/2,\ \alpha-\nu\\ (1-\mu+\nu)/2\end{bmatrix}{}_2F_1\left(\frac{\alpha-\nu}{2},\ \frac{\alpha-\nu+1}{2};\ 1-\nu;\ -\frac{c^2}{p^2}\right)$$

$\left[\operatorname{Re} p>0;\ \operatorname{Re}(\alpha+\mu)>-1,\ \begin{Bmatrix}c>0\\ \operatorname{Re}\alpha>|\operatorname{Re}\nu|;\ |\arg c|<\pi\end{Bmatrix}\right]$.

2. $\int\limits_0^\infty x^{\alpha-1}e^{-px^2}\begin{Bmatrix} s_{\mu,\nu}(cx)\\ S_{\mu,\nu}(cx)\end{Bmatrix}dx=$

$$=\frac{c^{\mu+1}p^{-(\alpha+\mu+1)/2}}{2[(\mu+1)^2-\nu^2]}\Gamma\left(\frac{\alpha+\mu+1}{2}\right){}_2F_2\left(1,\ \frac{\alpha+\mu+1}{2};\frac{3+\mu-\nu}{2},\frac{3+\mu+\nu}{2};-\frac{c^2}{4p}\right)+$$

$$+\begin{Bmatrix}0\\1\end{Bmatrix}[B(\nu)+B(-\nu)],$$

$$B(\nu)=\frac{2^{\mu+\nu-2}}{c^{\nu}p^{(\alpha-\nu)/2}}\Gamma\begin{bmatrix}\nu,\ (\alpha-\nu)/2,\ (1+\mu+\nu)/2\\ (1-\mu+\nu)/2\end{bmatrix}{}_1F_1\left(\frac{\alpha-\nu}{2};\ 1-\nu;\ -\frac{c^2}{4p}\right)$$

$$\left[\operatorname{Re} p>0;\ \operatorname{Re}(\alpha+\mu)>-1,\ \begin{Bmatrix}c>0\\ \operatorname{Re}\alpha>|\operatorname{Re}\nu|;\ |\arg c|<\pi\end{Bmatrix}\right].$$

3. $$\int_0^\infty x^{-1/4}e^{-p\sqrt{x}}s_{\mu,1/4}(cx)\,dx=\frac{2^{-2\mu-2}}{\sqrt{cp}}\Gamma\left(2\mu+\frac{3}{2}\right)S_{-\mu-1,1/4}\left(\frac{p^2}{4c}\right)$$

$[c,\ \operatorname{Re} p>0;\ \operatorname{Re}\mu>-7/4]$.

2.9.4. Integrals containing trigonometric or inverse trigonometric functions and $\begin{Bmatrix}s_{\mu,\nu}(cx)\\ S_{\mu,\nu}(cx)\end{Bmatrix}$.

NOTATION: $\delta=\begin{Bmatrix}1\\0\end{Bmatrix}$.

1. $$\int_0^\infty x^{\alpha-1}\begin{Bmatrix}\sin bx\\ \cos bx\end{Bmatrix}s_{\mu,\nu}(cx)\,dx=\pm\frac{b^{-\alpha-\mu-1}c^{\mu+1}}{(\mu+1)^2-\nu^2}\Gamma(\alpha+\mu+1)\times$$

$$\times\begin{Bmatrix}\cos[(\alpha+\mu)\pi/2]\\ \sin[(\alpha+\mu)\pi/2]\end{Bmatrix}{}_3F_2\left(1,\ \frac{\alpha+\mu+1}{2},\ \frac{\alpha+\mu+2}{2};\ \frac{3+\mu-\nu}{2},\ \frac{3+\mu+\nu}{2};\ \frac{c^2}{b^2}\right)$$

$[0<c<b,\ \operatorname{Re}\alpha<3/2$ or $b=c>0,\ \operatorname{Re}\alpha<1/2;\ -1-\delta<\operatorname{Re}(\alpha+\mu)<2]$,

$$=\mp b^{1-\alpha-\mu}c^{\mu-1}\Gamma(\alpha+\mu-1)\begin{Bmatrix}\cos[(\alpha+\mu)\pi/2]\\ \sin[(\alpha+\mu)\pi/2]\end{Bmatrix}\times$$

$$\times{}_3F_2\left(\frac{1+\nu-\mu}{2},\ \frac{1-\nu-\mu}{2},\ 1;\ \frac{3-\alpha-\mu}{2},\ 1-\frac{\alpha+\mu}{2};\ \frac{b^2}{c^2}\right)\mp$$

$$\mp 2^{\alpha+\mu+\delta-2}\pi b^{\delta}c^{-\alpha-\delta}\begin{Bmatrix}\operatorname{cosec}[(\alpha+\mu)\pi/2]\\ \sec[(\alpha+\mu)\pi/2]\end{Bmatrix}\times$$

$$\times\Gamma\begin{bmatrix}(\mu-\nu+1)/2,\ (\mu+\nu+1)/2\\ 1-(\nu+\alpha+\delta)/2,\ 1+(\nu-\alpha-\delta)/2\end{bmatrix}{}_2F_1\left(\frac{\alpha+\nu+\delta}{2},\ \frac{\alpha-\nu+\delta}{2};\ \delta+\frac{1}{2};\ \frac{b^2}{c^2}\right)$$

$[0<b<c,\ \operatorname{Re}\alpha<3/2$ or $c=b>0,\ \operatorname{Re}\alpha<1/2;\ -1-\delta<\operatorname{Re}(\alpha+\mu)<2]$.

2. $$\int_0^\infty x^{-\mu-1}\cos bx\, s_{\mu,\nu}(cx)\,dx=$$

$$=2^{\mu-1/2}\sqrt{\frac{\pi}{c}}\Gamma\left(\frac{1+\mu+\nu}{2}\right)\Gamma\left(\frac{1+\mu-\nu}{2}\right)(c^2-b^2)_+^{(2\mu+1)/4}P_{\nu-1/2}^{-\mu-1/2}\left(\frac{b}{c}\right)$$

$[b,\ c>0;\ \operatorname{Re}\mu>-3/2$ for $b\neq c,\ \operatorname{Re}\mu>-1/2$ for $b=c]$.

3. $$\int_0^\infty x^{\alpha-1}\begin{Bmatrix}\sin bx\\ \cos bx\end{Bmatrix}S_{\mu,\nu}(cx)\,dx=\mp b^{1-\mu-\alpha}c^{\mu-1}\Gamma(\mu+\alpha-1)\times$$

$$\times\begin{Bmatrix}\cos[(\alpha+\mu)\pi/2]\\ \sin[(\alpha+\mu)\pi/2]\end{Bmatrix}{}_3F_2\left(1,\ \frac{1+\nu-\mu}{2},\ \frac{1-\nu-\mu}{2};\ 1-\frac{\alpha+\mu}{2},\ \frac{3-\alpha-\mu}{2};\ \frac{b^2}{c^2}\right)\mp$$

$$\mp 2^{\alpha+\mu+\delta-2}\pi b^{\delta}c^{-\alpha-\delta}\begin{Bmatrix}\operatorname{cosec}[(\alpha+\mu)\pi/2]\\ \sec[(\alpha+\mu)\pi/2]\end{Bmatrix}\times$$

$$\times\Gamma\begin{bmatrix}(\alpha+\nu+\delta)/2,\ (\alpha-\nu+\delta)/2\\ (1-\mu+\nu)/2,\ (1-\mu-\nu)/2\end{bmatrix}{}_2F_1\left(\frac{\alpha+\nu+\delta}{2},\ \frac{\alpha-\nu+\delta}{2};\ \delta+\frac{1}{2};\ \frac{b^2}{c^2}\right)$$

$[b>0;\ -1-\delta<\operatorname{Re}(\alpha+\mu)<2;\ \operatorname{Re}\alpha>|\operatorname{Re}\nu|-\delta;\ |\arg c|<\pi]$.

4. $$\int_0^a \frac{1}{\sqrt{a^2-x^2}}\cos\left[(\mu+1)\arccos\frac{x}{a}\right]s_{\mu,\nu}(cx)\,dx=$$

$$=2^{\mu-2}\pi\Gamma\left(\frac{1+\mu+\nu}{2}\right)\Gamma\left(\frac{1+\mu-\nu}{2}\right)J_{(1+\mu+\nu)/2}\left(\frac{ac}{2}\right)J_{(1+\mu-\nu)/2}\left(\frac{ac}{2}\right)$$

$[a>0;\ \operatorname{Re}\mu>-2]$.

5. $\int\limits_0^a \frac{1}{\sqrt{a^2-x^2}} \cos\left[(\mu+1)\arccos\frac{x}{a}\right] S_{\mu,\nu}(cx)\,dx=$

$$=\frac{2^{\mu-2}\pi(ac)^{\mu+1}}{\sin\nu\pi\Gamma((1-\nu-\mu)/2)\Gamma((1+\nu-\mu)/2)}\left[J_\rho\left(\frac{ac}{2}\right)Y_\sigma\left(\frac{ac}{2}\right)-J_\sigma\left(\frac{ac}{2}\right)Y_\rho\left(\frac{ac}{2}\right)\right]$$

$[\rho=(\mu+\nu+1)/2,\ \sigma=(\mu-\nu+1)/2;\ a>0;\ \mathrm{Re}\,\mu>-2;\ |\mathrm{Re}\,\nu|<1]$.

6. $\int\limits_a^\infty \frac{1}{\sqrt{x^2-a^2}} \cos\left(\mu\arccos\frac{a}{x}\right) S_{\mu,\nu}(cx)\,dx=\frac{2^{\mu-1}\pi}{ac} W_{\mu/2,\,\nu/2}(iac)W_{\mu/2,\,\nu/2}(-iac)$

$[a>0;\ \mathrm{Re}\,\mu<1;\ |\arg c|<\pi]$.

2.9.5. Integrals of $x^\alpha J_\lambda(bx^{\pm1})\begin{Bmatrix}s_{\mu,\nu}(cx)\\ S_{\mu,\nu}(cx)\end{Bmatrix}$.

1. $\int\limits_0^\infty x^{\alpha-1}J_\lambda(bx)\,s_{\mu,\nu}(cx)\,dx=\frac{2^{\alpha+\mu}b^{-\alpha-\mu-1}c^{\mu+1}}{(\mu+1)^2-\nu^2}\times$

$$\times\Gamma\begin{bmatrix}(\alpha+\lambda+\mu+1)/2\\ (\lambda-\mu-\alpha+1)/2\end{bmatrix}{}_3F_2\left(1,\ \frac{1+\alpha+\mu-\lambda}{2},\ \frac{1+\alpha+\mu+\lambda}{2};\ \frac{3+\mu+\nu}{2},\ \frac{3+\mu-\nu}{2};\ \frac{c^2}{b^2}\right)$$

$[0<c<b,\ \mathrm{Re}\,\alpha<2$ or $b=c>0,\ \mathrm{Re}\,\alpha<1;\ -\mathrm{Re}\,\lambda-1<\mathrm{Re}\,(\alpha+\mu)<5/2]$,

$$=2^{\alpha+\mu-2}b^{1-\alpha-\mu}c^{\mu-1}\Gamma\begin{bmatrix}(\lambda+\mu+\alpha-1)/2\\ (\lambda-\mu-\alpha+3)/2\end{bmatrix}\times$$

$$\times\,{}_3F_2\left(1,\ \frac{1+\nu-\mu}{2},\ \frac{1-\nu-\mu}{2};\ \frac{3-\lambda-\mu-\alpha}{2},\ \frac{3+\lambda-\mu-\alpha}{2};\ \frac{b^2}{c^2}\right)+$$

$$+\frac{2^{\alpha+\mu-2}\pi b^\lambda c^{-\alpha-\lambda}}{\cos[(\alpha+\lambda+\mu)\pi/2]}\Gamma\begin{bmatrix}(\mu-\nu+1)/2,\ (\mu+\nu+1)/2\\ 1-(\alpha+\lambda+\nu)/2,\ 1-(\alpha+\lambda-\nu)/2,\ \lambda+1\end{bmatrix}\times$$

$$\times\,{}_2F_1\left(\frac{\nu+\alpha+\lambda}{2},\ \frac{\alpha+\lambda-\nu}{2};\ \lambda+1;\ \frac{b^2}{c^2}\right)$$

$[0<b<c,\ \mathrm{Re}\,\alpha<2$ or $b=c>0,\ \mathrm{Re}\,\alpha<1;\ -\mathrm{Re}\,\lambda-1<\mathrm{Re}\,(\alpha+\mu)<5/2]$.

2. $\int\limits_0^\infty x^{(1-\mu-\nu)/2}J_{(\mu-\nu+1)/2}(bx)\,s_{\mu,\nu}(cx)\,dx=$

$$=2^{(\mu-\nu-1)/2}b^{(\nu-\mu-1)/2}c^{-\nu}\Gamma\left(\frac{1+\mu-\nu}{2}\right)(c^2-b^2)_+^{(\mu+\nu-1)/2}$$

$[b,\ c>0;\ \mathrm{Re}\,(\mu+\nu)>-1$ for $b\neq c,\ \mathrm{Re}\,(\mu+\nu)>1$ for $b=c;\ -3<\mathrm{Re}\,(\mu-\nu)<2]$.

3. $\int\limits_0^\infty x^{\alpha-1}J_\lambda(bx)\,S_{\mu,\nu}(cx)\,dx=$

$$=\frac{2^{\alpha+\mu-2}\pi b^\lambda}{c^{\alpha+\lambda}}\sec\frac{\alpha+\lambda+\mu}{2}\pi\Gamma\begin{bmatrix}(\alpha+\lambda+\nu)/2,\ (\alpha+\lambda-\nu)/2\\ \lambda+1,\ (1-\mu-\nu)/2,\ (1-\mu+\nu)/2\end{bmatrix}\times$$

$${}_2F_1\left(\frac{\alpha+\lambda+\nu}{2},\ \frac{\alpha+\lambda-\nu}{2};\ \lambda+1;\ \frac{b^2}{c^2}\right)+2^{\alpha+\mu-2}b^{1-\alpha-\mu}c^{\mu-1}\times$$

$$\times\Gamma\begin{bmatrix}(\alpha+\lambda+\mu-1)/2\\ (\lambda-\mu-\alpha+3)/2\end{bmatrix}{}_3F_2\left(1,\ \frac{1+\nu-\mu}{2},\ \frac{1-\nu-\mu}{2};\ \frac{3-\lambda-\mu-\alpha}{2},\ \frac{3+\lambda-\mu-\alpha}{2};\ \frac{b^2}{c^2}\right)$$

$[b>0;\ \mathrm{Re}\,(\alpha+\lambda)>|\mathrm{Re}\,\nu|;\ -1-\mathrm{Re}\,\lambda<\mathrm{Re}\,(\alpha+\mu)<5/2;\ |\arg c|<\pi]$.

4. $\int\limits_0^\infty x^{1-(\mu+\nu)/2}J_\nu(bx)\,S_{\mu,\nu}(cx)\,dx=$

$$=\frac{\sqrt{\pi}}{2^\nu}b^{(\nu-\mu)/2-1}c^{(\mu-\nu)/2}(b^2-c^2)^{(\mu+\nu-2)/4}\Gamma\begin{bmatrix}1-(\mu+\nu)/2\\ (1-\mu+\nu)/2\end{bmatrix}P_{(\mu+\nu)/2}^{(\mu+\nu)/2-1}\left(\frac{b}{c}\right)$$

$[b>0;\ \mathrm{Re}\,(\mu-3\nu)<4;\ \mathrm{Re}\,(\mu-\nu)<1;\ -6<\mathrm{Re}\,(\mu+\nu)<4;\ |\arg c|<\pi]$.

5. $\int_0^\infty x^{(\nu-\mu)/2+1} J_{-(\mu+\nu)/2}(bx)\, S_{\mu,\nu}(cx)\, dx=$

$$=2^{\nu}\sqrt{\pi}\, b^{-(\mu+\nu)/2-1} c^{(\mu+\nu)/2}(b^2-c^2)^{(\mu-\nu-2)/4}\times$$
$$\times\Gamma\begin{bmatrix}1+(\nu-\mu)/2\\(1-\mu-\nu)/2\end{bmatrix} P^{(\mu-\nu)/2-1}_{(\mu-\nu)/2}\left(\frac{b}{c}\right)$$

[$b>0$; $\operatorname{Re}(\mu-\nu)<2$; $\operatorname{Re}(\mu+\nu)<1$; $|\arg c|<\pi$].

6. $\int_0^\infty x^{\lambda-\mu} J_\lambda(bx)\, S_{\mu,\nu}(cx)\, dx=$

$$=2^{\lambda-1} b^{\mu-\nu-\lambda-1} c^{\nu}\Gamma\begin{bmatrix}(1-\mu-\nu)/2+\lambda,\ (1-\mu+\nu)/2+\lambda\\1+\lambda-\mu\end{bmatrix}\times$$
$$\times{}_2F_1\left(\frac{1-\mu+\nu}{2}+\lambda,\ \frac{1-\mu+\nu}{2};\ 1-\mu+\lambda;\ 1-\frac{c^2}{b^2}\right)$$

[$b>0$; $-1<\operatorname{Re}\lambda<3/2$; $\operatorname{Re}(2\lambda-\mu)>|\operatorname{Re}\nu|-1$; $|\arg c|<\pi$].

7. $\int_0^\infty x^{\lambda+\nu+1} J_\lambda(bx)\, S_{\mu,\nu}(cx)\, dx=\frac{2^{\lambda+\mu+\nu}c^{\nu}}{b^{\lambda+2\nu+2}(\lambda+\nu+1)}\times$

$$\times\Gamma\begin{bmatrix}(3+\mu+\nu)/2+\lambda\\(1-\mu-\nu)/2\end{bmatrix}{}_2F_1\left(1+\lambda+\nu,\ \frac{1+\nu-\mu}{2};\ 2+\lambda+\nu;\ 1-\frac{c^2}{b^2}\right)$$

[$b>0$; $\operatorname{Re}\lambda>-1$; $-1<\operatorname{Re}(\lambda+\nu)<1/2-\operatorname{Re}\mu$; $\operatorname{Re}(\mu+\nu+2\lambda)>-3$; $|\arg c|<\pi$].

2.9.6. Integrals of $x^\alpha Y_\lambda(bx)\, s_{\mu,\nu}(cx)$.

1. $\int_0^\infty x^{1-(\mu+\nu)/2} Y_{(\mu-\nu)/2}(bx)\, s_{\mu,\nu}(cx)\, dx=$

$$=2^{(\mu-\nu)/2} b^{(\nu-\mu)/2} c^{-\nu}\Gamma\begin{bmatrix}(1+\mu-\nu)/2,\ (1+\mu+\nu)/2\\(\mu+\nu)/2\end{bmatrix}(c^2-b^2)_+^{(\mu+\nu)/2-1}$$

[$b,\ c>0$, $b\neq c$; $-3<\operatorname{Re}(\mu-\nu)<1$; $\operatorname{Re}(\mu+\nu)>0$].

2.9.7. Integrals of $x^\alpha K_\lambda(bx^r)\begin{Bmatrix}s_{\mu,\nu}(cx)\\S_{\mu,\nu}(cx)\end{Bmatrix}$.

1. $\int_0^\infty x^{\alpha-1} K_\lambda(bx)\, s_{\mu,\nu}(cx)\, dx=$

$$=\frac{2^{\alpha+\mu-1} b^{-\alpha-\mu-1} c^{\mu+1}}{(\mu+1)^2-\nu^2}\Gamma\left(\frac{\alpha+\mu+\lambda+1}{2}\right)\Gamma\left(\frac{\alpha+\mu-\lambda+1}{2}\right)\times$$
$$\times{}_3F_2\left(1,\ \frac{\alpha+\mu+\lambda+1}{2},\ \frac{\alpha+\mu-\lambda+1}{2};\ \frac{3+\mu-\nu}{2},\ \frac{3+\mu+\nu}{2};\ -\frac{c^2}{b^2}\right)$$

[$\operatorname{Re} b>|\operatorname{Im} c|$; $\operatorname{Re}(\alpha+\mu)>|\operatorname{Re}\lambda|-1$].

2. $\int_0^\infty K_{2\nu}(b\sqrt{x})\, s_{\mu,\nu}(cx)\, dx=\frac{1}{2c}\Gamma(\mu+\nu+1)\,\Gamma(\mu-\nu+1)\, S_{-\mu-1,\nu}\left(\frac{b^2}{4c}\right)$

[c, $\operatorname{Re} b>0$; $\operatorname{Re}\mu>|\operatorname{Re}\nu|-2$].

3. $\int_0^\infty x^{1/2} K_{2\nu-1}(b\sqrt{x})\, s_{\mu,\nu}(cx)\, dx=$

$$=\frac{2+\mu-\nu}{4c^2} b\Gamma(1+\mu-\nu)\,\Gamma(1+\mu+\nu)\, S_{-\mu-2,\nu-1}\left(\frac{b^2}{4c}\right)$$

[c, $\operatorname{Re} b>0$; $\operatorname{Re}(\mu+\nu)>-2$; $\operatorname{Re}(\mu-\nu)>-4$].

4. $\int_0^\infty x^{\alpha-1}K_\lambda(bx)\,S_{\mu,\nu}(cx)\,dx=$

$$=2^{\alpha+\mu-3}b^{1-\alpha-\mu}c^{\mu-1}\Gamma\left(\frac{\alpha+\mu+\lambda-1}{2}\right)\Gamma\left(\frac{\alpha+\mu-\lambda-1}{2}\right)\times$$

$$\times{}_3F_2\left(1,\ \frac{1-\mu+\nu}{2},\ \frac{1-\mu-\nu}{2};\ \frac{3-\lambda-\alpha-\mu}{2},\ \frac{3+\lambda-\alpha-\mu}{2};\ -\frac{b^2}{c^2}\right)+$$

$$+\frac{2^{\alpha+\mu-3}\pi b^\lambda}{c^{\alpha+\lambda}}\sec\frac{\alpha+\lambda+\mu}{2}\pi\Gamma\left[\begin{matrix}-\lambda,\ (\alpha+\lambda+\nu)/2,\ (\alpha+\lambda-\nu)/2\\ (1-\mu+\nu)/2,\ (1-\mu-\nu)/2\end{matrix}\right]\times$$

$$\times{}_2F_1\left(\frac{\alpha+\lambda+\nu}{2},\ \frac{\alpha+\lambda-\nu}{2};\ \lambda+1;\ -\frac{b^2}{c^2}\right)+\frac{2^{\alpha+\mu-3}\pi b^{-\lambda}}{c^{\alpha-\lambda}}\sec\frac{\alpha+\mu-\lambda}{2}\pi\times$$

$$\times\Gamma\left[\begin{matrix}\lambda,\ (\alpha+\nu-\lambda)/2,\ (\alpha-\nu-\lambda)/2\\ (1-\mu+\nu)/2,\ (1-\mu-\nu)/2\end{matrix}\right]{}_2F_1\left(\frac{\alpha+\nu-\lambda}{2},\ \frac{\alpha-\nu-\lambda}{2};\ 1-\lambda;\ -\frac{b^2}{c^2}\right)$$

[$\operatorname{Re} b>0$; $|\arg c|<\pi$; $\operatorname{Re}(\alpha+\mu)>|\operatorname{Re}\lambda|-1$; $\operatorname{Re}\alpha>|\operatorname{Re}\lambda|+|\operatorname{Re}\nu|$].

2.9.8. Integrals with respect to the index containing $S_{\mu,nix}(c)$.

1. $\int_0^\infty x\operatorname{sh}\frac{\pi x}{2}K_{ix}(b)\,S_{0,ix}(c)\,dx=\frac{\pi bc}{2(b^2+c^2)}.$

2. $\int_0^\infty x\operatorname{th}\pi x K_{ix}(b)\,S_{0,2ix}(c)\,dx=-\frac{c}{8}\sqrt{\frac{\pi}{2b}}\exp\left(\frac{c^2}{8b}-b\right)\operatorname{Ei}\left(-\frac{c^2}{8b}\right).$

3. $\int_0^\infty x\operatorname{sh}n\pi x\left|\Gamma\left(\frac{1-\mu}{2}+ix\right)\right|^2K_{inx}(b)\,S_{\mu,2ix}(c)\,dx=I_n$ [$\operatorname{Re}\mu\leqslant 1$].

$$I_1=\frac{2^{\mu-7/2}\pi^{3/2}c}{\sqrt{b}}\Gamma\left(1-\frac{\mu}{2}\right)\exp\left(\frac{c^2}{8b}-b\right)\Gamma\left(\frac{\mu}{2},\ \frac{c^2}{8b}\right),$$

$$I_2=2^{\mu-2}\pi^2\frac{b^{1-\mu}c^{\mu+1}}{b^2+c^2}.$$

2.10. THE KELVIN FUNCTIONS $\begin{Bmatrix}\operatorname{ber}_\nu(x)\\ \operatorname{bei}_\nu(x)\end{Bmatrix}$ AND $\begin{Bmatrix}\operatorname{ker}_\nu(x)\\ \operatorname{kei}_\nu(x)\end{Bmatrix}$

2.10.1. Integrals of general form.

1. $\int_0^a x^{\alpha-1}(a^r-x^r)^{\beta-1}\begin{Bmatrix}\operatorname{ber}_\nu(cx)\\ \operatorname{bei}_\nu(cx)\end{Bmatrix}dx=$

$$=\frac{a^{\alpha+\nu+r\beta-r}c^\nu}{2^\nu r}\Gamma\left[\begin{matrix}\beta\\ \nu+1\end{matrix}\right]\sum_{k=0}^\infty\frac{1}{k!\,(\nu+1)_k}\begin{Bmatrix}\cos[(3\nu-2k)\pi/4]\\ \sin[(3\nu-2k)\pi/4]\end{Bmatrix}\times$$

$$\times\Gamma\left[\begin{matrix}(2k+\alpha+\nu)/r\\ \beta+(2k+\alpha+\nu)/r\end{matrix}\right]\left(-\frac{a^2c^2}{4}\right)^k$$

[a, r, $\operatorname{Re}\beta$, $\operatorname{Re}(\alpha+\nu)>0$].

2. $\int_0^\infty x^{\alpha-1}e^{-px^r}\begin{Bmatrix}\operatorname{ber}_\nu(cx)\\ \operatorname{bei}_\nu(cx)\end{Bmatrix}dx=$

$$=\frac{c^\nu}{2^\nu p^{(\alpha+\nu)/r}r\Gamma(\nu+1)}\sum_{k=0}^\infty\frac{(-1)^k}{k!\,(\nu+1)_k}\Gamma\left(\frac{2k+\alpha+\nu}{r}\right)\begin{Bmatrix}\cos[(3\nu-2k)\pi/4]\\ \sin[(3\nu-2k)\pi/4]\end{Bmatrix}\left(\frac{c}{2p^{1/r}}\right)^{2k}$$

[$r>1$; $\operatorname{Re}p$, $\operatorname{Re}(\alpha+\nu)>0$; $r=1$ see 2.10.3.2].

3. $$\int_0^\infty x^{\alpha-1}K_\mu(bx^r)\begin{Bmatrix}\operatorname{ber}_\nu(cx)\\ \operatorname{bei}_\nu(cx)\end{Bmatrix}dx=$$

$$=\frac{2^{(\alpha+\nu)/r-\nu-2}c^\nu}{b^{(\alpha+\nu)/r}r\Gamma(\nu+1)}\sum_{k=0}^\infty\frac{(-1)^k}{k!\,(\nu+1)_k}\Gamma\left(\frac{2k+\alpha+\nu+\mu r}{2r}\right)\Gamma\left(\frac{2k+\alpha+\nu-\mu r}{2r}\right)\times$$

$$\times\begin{Bmatrix}\cos[(3\nu-2k)\pi/4]\\ \sin[(3\nu-2k)\pi/4]\end{Bmatrix}\left(\frac{2^{1/r-1}c}{b^{1/r}}\right)^{2k}$$

$[r>1;\ \operatorname{Re} b>0;\ \operatorname{Re}(\alpha+\nu)>r\,|\operatorname{Re}\mu|;\ r=1$ see 2.10.8.1].

4. $$\int_0^\infty x^{\alpha-1}\operatorname{Ei}(-bx^r)\begin{Bmatrix}\operatorname{ber}_\nu(cx)\\ \operatorname{bei}_\nu(cx)\end{Bmatrix}dx=$$

$$=-\frac{b^{-(\alpha+\nu)/r}}{\Gamma(\nu+1)}\left(\frac{c}{2}\right)^\nu\sum_{k=0}^\infty\frac{(-1)^k\Gamma((2k+\alpha+\nu)/r)}{k!\,(2k+\alpha+\nu)(\nu+1)_k}\times$$

$$\times\begin{Bmatrix}\cos[(3\nu-2k)\pi/4]\\ \sin[(3\nu-2k)\pi/4]\end{Bmatrix}\left(\frac{c}{2b^{1/r}}\right)^{2k}$$ $[r>1;\ \operatorname{Re} b,\ \operatorname{Re}(\alpha+\nu)>0;\ r=1$ see 2.10.6.1].

5. $$\int_0^a x^{\alpha-1}(a^r-x^r)^{\beta-1}\begin{Bmatrix}\operatorname{ker}_\nu(cx)\\ \operatorname{kei}_\nu(cx)\end{Bmatrix}dx=U(\nu)$$ $[a,\ r,\ \operatorname{Re}\beta>0;\ \operatorname{Re}\alpha>|\operatorname{Re}\nu|]$,

$$U(\nu)=\pm\frac{a^{\alpha+r\beta-r}\Gamma(\beta)}{2r}\sum_{k=0}^\infty\frac{1}{k!}[A_k(\nu)+A_k(-\nu)]\left(\frac{ac}{2}\right)^{2k}$$ $[\nu\neq\ldots,-2,-1,0,1,2,\ldots]$,

$$U(n)=\pm\frac{a^{\alpha+r\beta-r}\Gamma(\beta)}{2r}\sum_{k=0}^{n-1}\frac{1}{k!}A_k(n)\left(\frac{ac}{2}\right)^{2k}\pm$$

$$\pm\frac{(-1)^n a^{\alpha+r\beta+n-r}c^n\Gamma(\beta)}{2^{n+2}r}\sum_{k=0}^\infty\frac{1}{k!\,(k+n)!}\Gamma\begin{bmatrix}(2k+\alpha+n)/r\\ \beta+(2k+\alpha+n)/r\end{bmatrix}\times$$

$$\times\begin{Bmatrix}\cos[(n-2k)\pi/4]\\ \sin[(n-2k)\pi/4]\end{Bmatrix}\left[2\psi(k+1)+2\psi(k+n+1)-\frac{4}{r}\psi\left(\frac{2k+\alpha+n}{r}\right)+\right.$$

$$\left.+\frac{4}{r}\psi\left(\beta+\frac{2k+\alpha+n}{r}\right)-4\ln\frac{ac}{2}\mp\pi\operatorname{tg}^{\pm1}\frac{n-2k}{4}\pi\right]\left(\frac{ac}{2}\right)^{2k}$$ $[n=0,1,2,\ldots]$,

$$A_k(\nu)=\frac{(ac/2)^{-\nu}}{(1-\nu)_k}\Gamma\begin{bmatrix}(2k+\alpha-\nu)/r,\ \nu\\ (2k+\alpha-\nu)/r+\beta\end{bmatrix}\begin{Bmatrix}\cos[(3\nu-2k)\pi/4]\\ \sin[(3\nu-2k)\pi/4]\end{Bmatrix}.$$

6. $$\int_a^\infty x^{\alpha-1}(x^r-a^r)^{\beta-1}\begin{Bmatrix}\operatorname{ker}_\nu(cx)\\ \operatorname{kei}_\nu(cx)\end{Bmatrix}dx=V(\nu)$$ $[a,\ r,\ \operatorname{Re}\beta>0;\ |\arg c|<\pi/4]$,

$$V(\nu)=\pm\frac{a^{\alpha+r\beta-r}}{2r}\Gamma(\beta)\sum_{k=0}^\infty\frac{1}{k!}[C_k(\nu)+C_k(-\nu)]\left(\frac{ac}{2}\right)^{2k}\pm v(\nu)$$

$[\nu\neq\ldots,-2,-1,0,1,2,\ldots]$,

$$V(n)=\pm\frac{a^{\alpha+r\beta-r}}{2r}\Gamma(\beta)\sum_{k=0}^{n-1}\frac{1}{k!}C_k(n)\left(\frac{ac}{2}\right)^{2k}\pm$$

$$\pm\frac{(-1)^n a^{\alpha+r\beta+n-r}c^n}{2^{n+2}r}\Gamma(\beta)\sum_{k=0}^\infty\frac{1}{k!\,(k+n)!}\Gamma\begin{bmatrix}1-\beta-(2k+\alpha+n)/r\\ 1-(2k+\alpha+n)/r\end{bmatrix}\times$$

$$\times\begin{Bmatrix}\cos[(n-2k)\pi/4]\\ \sin[(n-2k)\pi/4]\end{Bmatrix}\left[2\psi(k+1)+2\psi(k+n+1)+\frac{4}{r}\psi\left(1-\beta-\frac{2k+\alpha+n}{r}\right)-\right.$$

$$\left.-\frac{4}{r}\psi\left(1-\frac{2k+\alpha+n}{r}\right)-4\ln\frac{ac}{2}\mp\pi\operatorname{tg}^{\pm1}\frac{n-2k}{4}\pi\right]\left(\frac{ac}{2}\right)^{2k}\pm v(n)\quad [n=0,\ 1,\ 2,\ \ldots],$$

$$C_k(\nu)=\frac{(ac/2)^{-\nu}}{(1-\nu)_k}\Gamma\begin{bmatrix}\nu,\ 1-\beta-(2k+\alpha-\nu)/r\\ 1-(2k+\alpha-\nu)/r\end{bmatrix}\begin{Bmatrix}\cos[(3\nu-2k)\pi/4]\\ \sin[(3\nu-2k)\pi/4]\end{Bmatrix},$$

$$v(\nu)=\frac{1}{4}\left(\frac{c}{2}\right)^{r-r\beta-\alpha}\sum_{k=0}^{\infty}\frac{(1-\beta)_k}{k!}\Gamma\left(\frac{\alpha+r\beta+\nu-r-rk}{2}\right)\Gamma\left(\frac{\alpha+r\beta-\nu-r-rk}{2}\right)\times$$

$$\times\begin{Bmatrix}\cos[(\alpha+r\beta+2\nu-r-rk)\pi/4]\\ \sin[(\alpha+r\beta+2\nu-r-rk)\pi/4]\end{Bmatrix}\left(\frac{ac}{2}\right)^{rk}.$$

7. $$\int_0^\infty\frac{x^{\alpha-1}}{(x^r+z^r)^\rho}\begin{Bmatrix}\operatorname{ker}_\nu(cx)\\ \operatorname{kei}_\nu(cx)\end{Bmatrix}dx=W(\nu)$$

$[r>0;\ \operatorname{Re}\alpha>|\operatorname{Re}\nu|;\ r\,|\arg z|<\pi;\ |\arg c|<\pi/4]$,

$$W(\nu)=\pm\frac{z^{\alpha-r\rho}}{2r\Gamma(\rho)}\sum_{k=0}^{\infty}\frac{1}{k!}[F_k(\nu)+F_k(-\nu)]\left(\frac{cz}{2}\right)^{2k}\pm w(\nu)$$

$[\nu\neq\ldots,-2,-1,0,1,2,\ldots]$,

$$W(n)=\pm\frac{z^{\alpha-r\rho}}{2r\Gamma(\rho)}\sum_{k=0}^{n-1}\frac{1}{k!}F_k(n)\left(\frac{cz}{2}\right)^{2k}\pm$$

$$\pm(-1)^n\frac{c^n z^{\alpha-r\rho+n}}{2^{n+2}r\Gamma(\rho)}\sum_{k=0}^{\infty}\frac{1}{k!\,(k+n)!}\Gamma\left(\frac{2k+\alpha+n}{r}\right)\times$$

$$\times\Gamma\left(\rho-\frac{2k+\alpha+n}{r}\right)\begin{Bmatrix}\cos[(n-2k)\pi/4]\\ \sin[(n-2k)\pi/4]\end{Bmatrix}\left[2\psi(k+1)+2\psi(k+n+1)-\right.$$

$$\left.-\frac{4}{r}\psi\left(\frac{2k+\alpha+n}{r}\right)+\frac{4}{r}\psi\left(\rho-\frac{2k+\alpha+n}{r}\right)-4\ln\frac{cz}{2}\mp\pi\operatorname{tg}^{\pm1}\frac{n-2k}{4}\pi\right]\times$$

$$\times\left(\frac{cz}{2}\right)^{2k}\pm w(n)\qquad [n=0,\ 1,\ 2,\ \ldots],$$

$$F_k(\nu)=\frac{(cz/2)^{-\nu}}{(1-\nu)_k}\Gamma\left[\nu,\ \frac{2k+\alpha-\nu}{r},\ \rho-\frac{2k+\alpha-\nu}{r}\right]\begin{Bmatrix}\cos[(3\nu-2k)\pi/4]\\ \sin[(3\nu-2k)\pi/4]\end{Bmatrix},$$

$$w(\nu)=\frac{1}{4}\left(\frac{c}{2}\right)^{r\rho-\alpha}\sum_{k=0}^{\infty}\frac{(-1)^k(\rho)_k}{k!}\Gamma\left(\frac{\alpha+\nu-r\rho-rk}{2}\right)\times$$

$$\times\Gamma\left(\frac{\alpha-\nu-r\rho-rk}{2}\right)\begin{Bmatrix}\cos[(\alpha+2\nu-r\rho-rk)\pi/4]\\ \sin[(\alpha+2\nu-r\rho-rk)\pi/4]\end{Bmatrix}\left(\frac{cz}{2}\right)^{rk}.$$

8. $$\int_0^\infty\frac{x^{\alpha-1}}{x^r-y^r}\begin{Bmatrix}\operatorname{ker}_\nu(cx)\\ \operatorname{kei}_\nu(cx)\end{Bmatrix}dx=X(\nu)$$ $[r,\ y>0;\ \operatorname{Re}\alpha>|\operatorname{Re}\nu|;\ |\arg c|<\pi/4]$,

$$X(\nu)=\mp\frac{\pi y^{\alpha-r}}{2r}\sum_{k=0}^{\infty}\frac{1}{k!}[G_k(\nu)+G_k(-\nu)]\left(\frac{cy}{2}\right)^{2k}\mp u(\nu)$$

$[\nu\neq\ldots,-2,-1,0,1,2,\ldots]$,

$$X(n)=\mp\frac{\pi y^{\alpha-r}}{2r}\sum_{k=0}^{n-1}\frac{1}{k!}G_k(n)\left(\frac{cy}{2}\right)^{2k}\mp$$

$$\mp(-1)^n \frac{\pi c^n y^{\alpha+n-r}}{2^{n+2}r} \sum_{k=0}^{\infty} \frac{1}{k!\,(k+n)!} \operatorname{ctg} \frac{2k+\alpha+n}{r}\pi \begin{Bmatrix} \cos[(n-2k)\pi/4] \\ \sin[(n-2k)\pi/4] \end{Bmatrix} \times$$

$$\times \Big[2\psi(k+1) + 2\psi(k+n+1) + \frac{8\pi}{r} \operatorname{cosec} \frac{2(2k+\alpha+n)}{r}\pi - 4\ln\frac{cy}{2} \mp$$

$$\mp \pi \operatorname{tg}^{\pm 1} \frac{n-2k}{4}\pi \Big] \left(\frac{cy}{2}\right)^{2k} \mp u(n) \qquad [n=0,\ 1,\ 2,\ \ldots],$$

$$G_k(\nu) = \frac{(cy/2)^{-\nu}}{(1-\nu)_k} \Gamma(\nu) \operatorname{ctg} \frac{2k+\alpha-\nu}{r}\pi \begin{Bmatrix} \cos[(3\nu-2k)\pi/4] \\ \sin[(3\nu-2k)\pi/4] \end{Bmatrix},$$

$$u(\nu) = \frac{1}{4}\left(\frac{c}{2}\right)^{r-\alpha} \sum_{k=0}^{\infty} \Gamma\left(\frac{\alpha+\nu-r-rk}{2}\right) \Gamma\left(\frac{\alpha-\nu-r-rk}{2}\right) \times$$

$$\times \begin{Bmatrix} \cos[(\alpha+2\nu-r-rk)\pi/4] \\ \sin[(\alpha+2\nu-r-rk)\pi/4] \end{Bmatrix} \left(\frac{cy}{2}\right)^{rk}.$$

9. $$\int_0^{\infty} x^{\alpha-1} e^{-px^r} \begin{Bmatrix} \operatorname{ker}_\nu(cx) \\ \operatorname{kei}_\nu(cx) \end{Bmatrix} dx = Y(\nu),$$

$[r,\ \operatorname{Re} p > 0;\ \operatorname{Re}\alpha > |\operatorname{Re}\nu|;\ |\arg c| < \pi/4;\ r=1$ see 2.10.3.9$]$,

$$Y(\nu) = \pm \frac{2^{\alpha-2}}{c^\alpha} \sum_{k=0}^{\infty} \frac{1}{k!} \Gamma\left(\frac{\alpha+\nu+rk}{2}\right) \Gamma\left(\frac{\alpha-\nu+rk}{2}\right) \times$$

$$\times \begin{Bmatrix} \cos[(\alpha+2\nu+rk)\pi/4] \\ \sin[(\alpha+2\nu+rk)\pi/4] \end{Bmatrix} \left(-\frac{2^r p}{c^r}\right)^k \qquad [r<1],$$

$$Y(\nu) = \pm \frac{p^{-\alpha/r}}{2r} \sum_{k=0}^{\infty} \frac{1}{k!} [A_k(\nu) + A_k(-\nu)] \left(\frac{c}{2p^{1/r}}\right)^{2k}$$

$[r>1;\ \nu \neq \ldots, -2, -1, 0, 1, 2, \ldots]$,

$$Y(n) = \pm \frac{p^{-\alpha/r}}{2r} \sum_{k=0}^{n-1} \frac{1}{k!} A_k(n) \left(\frac{c}{2p^{1/r}}\right)^{2k} \pm$$

$$\pm \frac{(-1)^n c^n}{2^{n+2} p^{(\alpha+n)/r}\, r} \sum_{k=0}^{\infty} \frac{1}{k!\,(k+n)!} \Gamma\left(\frac{2k+\alpha+n}{r}\right) \begin{Bmatrix} \cos[(n-2k)\pi/4] \\ \sin[(n-2k)\pi/4] \end{Bmatrix} \times$$

$$\times \Big[2\psi(k+1) + 2\psi(k+n+1) - \frac{4}{r}\psi\left(\frac{2k+\alpha+n}{r}\right) \mp \pi \operatorname{tg}^{\pm 1}\frac{n-2k}{4}\pi -$$

$$- 4\ln\frac{c}{2p^{1/r}} \Big] \left(\frac{c}{2p^{1/r}}\right)^{2k} \qquad [r>1;\ n=0,\ 1,\ 2,\ \ldots],$$

$$A_k(\nu) = \frac{(2p^{1/r}/c)^\nu}{(1-\nu)_k} \Gamma(\nu) \Gamma\left(\frac{2k+\alpha-\nu}{r}\right) \begin{Bmatrix} \cos[(3\nu-2k)\pi/4] \\ \sin[(3\nu-2k)\pi/4] \end{Bmatrix}.$$

10. $$\int_0^{\infty} x^{\alpha-1} \sin bx^r \begin{Bmatrix} \operatorname{ker}_\nu(cx) \\ \operatorname{kei}_\nu(cx) \end{Bmatrix} dx = W(\nu,\ 1)$$

$[b,\ r>0;\ \operatorname{Re}\alpha > |\operatorname{Re}\nu| - r;\ |\arg c| < \pi/4;\ r=1$ see 2.10.4.2$]$.

$$W(\nu, \delta) = \pm \frac{2^{\alpha+r\delta-2} b^\delta}{c^{\alpha+r\delta}} \sum_{k=0}^{\infty} \frac{(-1)^k}{k!\,(1/2+\delta)_k} \Gamma\left(\frac{\alpha+r\delta+\nu}{2}+rk\right) \Gamma\left(\frac{\alpha+r\delta-\nu}{2}+rk\right) \times$$

$$\times \begin{Bmatrix} \cos[(\alpha+r\delta+2\nu+2rk)\pi/4] \\ \sin[(\alpha+r\delta+2\nu+2rk)\pi/4] \end{Bmatrix} \left(\frac{2^{r-1} b}{c^r}\right)^{2k} \qquad [r<1].$$

$$W(\nu, \delta) = \pm \frac{b^{-\alpha/r}}{2r} \sum_{k=0}^{\infty} \frac{1}{k!} [C_k(\nu) + C_k(-\nu)] \left(\frac{c}{2b^{1/r}} \right)^{2k}$$

[$r > 1$; $\nu \neq \ldots, -2, -1, 0, 1, 2, \ldots$],

$$W(n, \delta) = \pm \frac{b^{-\alpha/r}}{2r} \sum_{k=0}^{n-1} \frac{1}{k!} C_k(n) \left(\frac{c}{2b^{1/r}} \right)^{2k} \pm$$

$$\pm \frac{(-1)^n c^n}{2^{n+2} b^{(\alpha+n)/r} r} \sum_{k=0}^{\infty} \frac{1}{k!\,(k+n)!} \Gamma\left(\frac{2k+\alpha+n}{r} \right) \times$$

$$\times \cos \frac{\delta r - \alpha - n - 2k}{2r} \pi \begin{Bmatrix} \cos[(n-2k)\pi/4] \\ \sin[(n-2k)\pi/4] \end{Bmatrix} \times$$

$$\times \left[2\psi(k+1) + 2\psi(k+n+1) - \frac{4}{r} \psi\left(\frac{2k+\alpha+n}{r} \right) - \right.$$

$$\left. - \frac{4}{r} \ln \frac{c^r}{2^r b} - \frac{2\pi}{r} \operatorname{tg} \frac{\delta r - \alpha - n - 2k}{2r} \pi \mp \pi \operatorname{tg}^{\pm 1} \frac{n-2k}{4} \pi \right] \left(\frac{c}{2b^{1/r}} \right)^{2k}$$

[$r > 1$; $n = 0, 1, 2, \ldots$].

$$C_k(\nu) = \frac{(2b^{1/r}/c)^{\nu}}{(1-\nu)_k} \Gamma(\nu) \Gamma\left(\frac{2k+\alpha-\nu}{r} \right) \cos \frac{\delta r + \nu - \alpha - 2k}{2r} \pi \begin{Bmatrix} \cos[(3\nu-2k)\pi/4] \\ \sin[(3\nu-2k)\pi/4] \end{Bmatrix}.$$

11. $$\int_0^{\infty} x^{\alpha-1} \cos bx^r \begin{Bmatrix} \operatorname{ker}_{\nu}(cx) \\ \operatorname{kei}_{\nu}(cx) \end{Bmatrix} dx = W(\nu, 0)$$

[$b, r > 0$; $\operatorname{Re}\alpha > |\operatorname{Re}\nu|$; $|\arg c| < \pi/4$; $r = 1$ see 2.10.4.3; $W(\nu, \delta)$ see 2.10.1.10].

12. $$\int_0^{\infty} x^{\alpha-1} J_{\mu}(bx^r) \begin{Bmatrix} \operatorname{ker}_{\nu}(cx) \\ \operatorname{kei}_{\nu}(cx) \end{Bmatrix} dx = Z(\nu)$$

[$b, r > 0$; $\operatorname{Re}(\alpha+\mu r) > |\operatorname{Re}\nu|$; $|\arg c| < \pi/4$; $r = 1$ see 2.10.7.1]

$$Z(\nu) = \pm \frac{2^{\alpha-\mu+\mu r-2} b^{\mu}}{c^{\alpha+\mu r} \Gamma(\mu+1)} \sum_{k=0}^{\infty} \frac{(-1)^k}{k!\,(\mu+1)_k} \Gamma\left(\frac{\alpha+\mu+\nu}{2} + rk \right) \Gamma\left(\frac{\alpha+\mu-\nu}{2} + rk \right) \times$$

$$\times \begin{Bmatrix} \cos[(\alpha+2\nu+r\mu+2rk)\pi/4] \\ \sin[(\alpha+2\nu+r\mu+2rk)\pi/4] \end{Bmatrix} \left(\frac{2^{r-1} b}{c^r} \right)^{2k} \quad [r < 1]$$

$$Z(\nu) = \pm \frac{2^{\alpha/r-2}}{b^{\alpha/r} r} \sum_{k=0}^{\infty} \frac{1}{k!} [D_k(\nu) + D_k(-\nu)] \left(\frac{2^{1-r} c^r}{b} \right)^{2k/r}$$

[$r > 1$; $\nu \neq \ldots, -2, -1, 0, 1, 2, \ldots$]

$$Z(n) = \pm \frac{2^{\alpha/r-2}}{b^{\alpha/r} r} \sum_{k=0}^{n-1} \frac{1}{k!} D_k(n) \left(\frac{2^{1-r} c^r}{b} \right)^{2k/r} \pm$$

$$\pm \frac{(-1)^n 2^{(\alpha+n)/r-n-2} c^n}{b^{(\alpha+n)/r} r} \sum_{k=0}^{\infty} \frac{1}{k!\,(k+n)!} \Gamma\begin{bmatrix} (\mu r+\alpha+n+2k)/(2r) \\ 1+(\mu r-\alpha-n-2k)/(2r) \end{bmatrix} \times$$

$$\times \begin{Bmatrix} \cos[(n-2k)\pi/4] \\ \sin[(n-2k)\pi/4] \end{Bmatrix} \left[2\psi(k+1) + 2\psi(k+n+1) - \frac{2}{r} \psi\left(\frac{\mu r+\alpha+n+2k}{2r} \right) - \right.$$

$$\left. - \frac{2}{r} \psi\left(1 + \frac{\mu r-\alpha-n-2k}{2r} \right) - \frac{4}{r} \ln \frac{c^r}{2^{r-1} b} \mp \pi \operatorname{tg}^{\pm 1} \frac{n-2k}{4} \pi \right] \left(\frac{2^{1-r} c^r}{b} \right)^{2k/r}$$

[$r > 1$; $n = 0, 1, 2, \ldots$]

$$D_k(\nu)=\frac{2^{\nu-\nu/r}\,b^{\nu/r}}{c^\nu(1-\nu)_k}\Gamma\begin{bmatrix}\nu,\ (\mu r+\alpha-\nu+2k)/(2r)\\ 1+(\mu r-\alpha+\nu-2k)/(2r)\end{bmatrix}\begin{Bmatrix}\cos[(3\nu-2k)\pi/4]\\ \sin[(3\nu-2k)\pi/4]\end{Bmatrix}.$$

2.10.2. Integrals containing algebraic functions and Kelvin functions.

1. $$\int_0^a x^{\alpha-1}(a^2-x^2)^{\beta-1}\begin{Bmatrix}\mathrm{ber}_\nu(cx)\\ \mathrm{bei}_\nu(cx)\end{Bmatrix}dx=$$
$$=\frac{a^{\alpha+\nu+2\beta-2}c^\nu}{2^{\nu+1}\Gamma(\nu+1)}\mathrm{B}\left(\beta,\frac{\alpha+\nu}{2}\right)\begin{Bmatrix}\cos(3\nu\pi/4)\\ \sin(3\nu\pi/4)\end{Bmatrix}\times$$
$$\times{}_2F_5\left(\frac{\alpha+\nu}{4},\frac{\alpha+\nu+2}{4};\frac{1}{2},\frac{\nu+1}{2},\frac{\nu}{2}+1,\frac{\alpha+\nu+2\beta}{4},\frac{\alpha+\nu+2\beta+2}{4};-\frac{a^4c^4}{256}\right)\mp$$
$$\mp\frac{a^{\alpha+\nu+2\beta}c^{\nu+2}}{2^{\nu+3}\Gamma(\nu+2)}\mathrm{B}\left(\beta,\frac{\alpha+\nu}{2}+1\right)\begin{Bmatrix}\sin(3\nu\pi/4)\\ \cos(3\nu\pi/4)\end{Bmatrix}{}_2F_5\left(\frac{\alpha+\nu+2}{4},\frac{\alpha+\nu}{4}+1;\right.$$
$$\left.\frac{3}{2},\frac{\nu}{2}+1,\frac{\nu+3}{2},\frac{\alpha+\nu+2\beta+2}{4},\frac{\alpha+\nu+2\beta}{4}+1;-\frac{a^4c^4}{256}\right)$$
[a, Re β, Re $(\alpha+\nu)>0$].

2. $$\int_0^a\frac{x}{\sqrt{a^2-x^2}}\mathrm{ber}(cx)\,dx=\frac{1}{c\sqrt{2}}\left(\mathrm{ch}\frac{ac}{\sqrt{2}}\sin\frac{ac}{\sqrt{2}}+\mathrm{sh}\frac{ac}{\sqrt{2}}\cos\frac{ac}{\sqrt{2}}\right)$$
[$a>0$].

3. $$\int_0^a\mathrm{ber}\left(\sqrt{x(a-x)}\right)dx=\sqrt{2}\left(\mathrm{sh}\frac{a}{2\sqrt{2}}\cos\frac{a}{2\sqrt{2}}+\mathrm{ch}\frac{a}{2\sqrt{2}}\sin\frac{a}{2\sqrt{2}}\right)$$
[$a>0$].

4. $$\int_a^\infty\frac{x}{\sqrt{x^2-a^2}}\mathrm{kei}(cx)\,dx=-\frac{\pi}{2\sqrt{2}c}e^{-ac/\sqrt{2}}\left(\cos\frac{ac}{\sqrt{2}}+\sin\frac{ac}{\sqrt{2}}\right)$$
[$a>0$; $|\arg c|<\pi/4$].

2.10.3. Integrals containing e^{-px^n} and Kelvin functions.

1. $$\int_0^\infty e^{-px}\begin{Bmatrix}\mathrm{ber}(cx)\\ \mathrm{bei}(cx)\end{Bmatrix}dx=\left[\frac{\sqrt{p^4+c^4}\pm p^2}{2(p^4+c^4)}\right]^{1/2}$$
[$\sqrt{2}\,\mathrm{Re}\,p>\mathrm{Re}\,c+|\mathrm{Im}\,c|$].

2. $$\int_0^\infty x^{\alpha-1}e^{-px}\begin{Bmatrix}\mathrm{ber}_\nu(cx)\\ \mathrm{bei}_\nu(cx)\end{Bmatrix}dx=\frac{c^\nu}{2^\nu p^{\alpha+\nu}}\Gamma\begin{bmatrix}\alpha+\nu\\ \nu+1\end{bmatrix}\begin{Bmatrix}\cos(3\nu\pi/4)\\ \sin(3\nu\pi/4)\end{Bmatrix}\times$$
$$\times{}_4F_3\left(\frac{\alpha+\nu}{4},\frac{\alpha+\nu+1}{4},\frac{\alpha+\nu+2}{4},\frac{\alpha+\nu+3}{4};\frac{1}{2},\frac{\nu+1}{2},\frac{\nu}{2}+1;-\frac{c^4}{p^4}\right)\mp$$
$$\mp\left(\frac{c}{2}\right)^{\nu+2}p^{-\alpha-\nu-2}\Gamma\begin{bmatrix}\alpha+\nu+2\\ \nu+2\end{bmatrix}\begin{Bmatrix}\sin(3\nu\pi/4)\\ \cos(3\nu\pi/4)\end{Bmatrix}\times$$
$$\times{}_4F_3\left(\frac{\alpha+\nu+2}{4},\frac{\alpha+\nu+3}{4},\frac{\alpha+\nu}{4}+1,\frac{\alpha+\nu+5}{4};\frac{3}{2},\frac{\nu+3}{2},\frac{\nu}{2}+1;-\frac{c^4}{p^4}\right)$$
[Re $(\alpha+\nu)>0$; $\sqrt{2}\,\mathrm{Re}\,p>\mathrm{Re}\,c+|\mathrm{Im}\,c|$].

3. $$\int_0^\infty e^{-px^2}\begin{Bmatrix}\mathrm{ber}_\nu(cx)\\ \mathrm{bei}_\nu(cx)\end{Bmatrix}dx=\frac{1}{2}\sqrt{\frac{\pi}{p}}J_{\nu/2}\left(\frac{c^2}{8p}\right)\begin{Bmatrix}\cos\varphi\\ \sin\varphi\end{Bmatrix}$$
[$\varphi=c^2/(8p)-3(1-\nu)\pi/4$; Re $p>0$; Re $\nu>-1$].

4. $$\int_0^\infty x^{\alpha-1}e^{-px^2}\begin{Bmatrix}\operatorname{ber}_\nu(cx)\\ \operatorname{bei}_\nu(cx)\end{Bmatrix}dx=\frac{c^\nu}{2^{\nu+1}p^{(\alpha+\nu)/2}}\Gamma\begin{bmatrix}(\alpha+\nu)/2\\ \nu+1\end{bmatrix}\begin{Bmatrix}\cos(3\nu\pi/4)\\ \sin(3\nu\pi/4)\end{Bmatrix}\times$$
$$\times{}_2F_3\left(\frac{\alpha+\nu}{4},\ \frac{\alpha+\nu+2}{4};\ \frac{1}{2},\ \frac{\nu+1}{2},\ \frac{\nu}{2}+1;\ -\frac{c^4}{64p^2}\right)\mp$$
$$\mp\frac{c^{\nu+2}}{2^{\nu+3}p^{(\alpha+\nu)/2+1}}\Gamma\begin{bmatrix}(\alpha+\nu)/2+1\\ \nu+2\end{bmatrix}\begin{Bmatrix}\sin(3\nu\pi/4)\\ \cos(3\nu\pi/4)\end{Bmatrix}\times$$
$$\times{}_2F_3\left(\frac{\alpha+\nu+2}{4},\ \frac{\alpha+\nu}{4}+1;\ \frac{3}{2},\ \frac{\nu+3}{2},\ \frac{\nu}{2}+1;\ -\frac{c^4}{64p^2}\right)$$
[Re p, Re $(\alpha+\nu)>0$].

5. $$\int_0^\infty \frac{1}{x}e^{-px^2}[1-\operatorname{ber}(cx)]\,dx=\frac{1}{2}\left[\mathrm{C}+\ln\frac{c^2}{4p}-\operatorname{ci}\left(\frac{c^2}{4p}\right)\right]$$
[Re $p>0$].

6. $$\int_0^\infty \frac{1}{x}e^{-px^2}\operatorname{bei}(cx)\,dx=\frac{1}{2}\operatorname{Si}\left(\frac{c^2}{4p}\right)$$
[Re $p>0$].

7. $$\int_0^\infty xe^{-px^2}\begin{Bmatrix}\operatorname{ber}_\nu(cx)\\ \operatorname{bei}_\nu(cx)\end{Bmatrix}dx=\frac{c}{8}\sqrt{\frac{\pi}{p^3}}\left[J_{(\nu-1)/2}(\varphi)\begin{Bmatrix}\cos\psi\\ \sin\psi\end{Bmatrix}+J_{(\nu+1)/2}(\varphi)\begin{Bmatrix}\sin\psi\\ \cos\psi\end{Bmatrix}\right]$$
[$\varphi=c^2/(8p)$, $\psi=c^2/(8p)+3\nu\pi/4$; Re $p>0$; Re $\nu>-2$].

8. $$\int_0^\infty x^{\nu+1}e^{-px^2}\begin{Bmatrix}\operatorname{ber}_\nu(cx)\\ \operatorname{bei}_\nu(cx)\end{Bmatrix}dx=\frac{c^\nu}{(2p)^{\nu+1}}\begin{Bmatrix}\cos\psi\\ \sin\psi\end{Bmatrix}$$
[$\psi=c^2/(4p)+3\nu\pi/4$; Re $p>0$; Re $\nu>-1$].

9. $$\int_0^\infty x^{\alpha-1}e^{-px}\begin{Bmatrix}\operatorname{ker}_\nu(cx)\\ \operatorname{kei}_\nu(cx)\end{Bmatrix}dx=$$
$$=\pm\frac{2^{\alpha-2}}{c^\alpha}\begin{Bmatrix}\cos[(\alpha+2\nu)\pi/4]\\ \sin[(\alpha+2\nu)\pi/4]\end{Bmatrix}\Gamma\left(\frac{\alpha+\nu}{2}\right)\Gamma\left(\frac{\alpha-\nu}{2}\right)\times$$
$$\times{}_4F_3\left(\frac{\alpha+\nu}{4},\ \frac{\alpha+\nu+2}{4},\ \frac{\alpha-\nu}{4},\ \frac{\alpha-\nu+2}{4};\ \frac{1}{4},\ \frac{1}{2},\ \frac{3}{4};\ -\frac{p^4}{c^4}\right)\mp$$
$$\mp\frac{2^{\alpha-1}p}{c^{\alpha+1}}\begin{Bmatrix}\cos[(\alpha+2\nu+1)\pi/4]\\ \sin[(\alpha+2\nu+1)\pi/4]\end{Bmatrix}\Gamma\left(\frac{\alpha+\nu+1}{2}\right)\Gamma\left(\frac{\alpha-\nu+1}{2}\right)\times$$
$$\times{}_4F_3\left(\frac{\alpha+\nu+1}{4},\ \frac{\alpha+\nu+3}{4},\ \frac{\alpha-\nu+1}{4},\ \frac{\alpha-\nu+3}{4};\ \frac{1}{2},\ \frac{3}{4},\ \frac{5}{4};\ -\frac{p^4}{c^4}\right)-$$
$$-\frac{2^{\alpha-1}p^2}{c^{\alpha+2}}\begin{Bmatrix}\sin[(\alpha+2\nu)\pi/4]\\ \cos[(\alpha+2\nu)\pi/4]\end{Bmatrix}\Gamma\left(\frac{\alpha+\nu}{2}+1\right)\Gamma\left(\frac{\alpha-\nu}{2}+1\right)\times$$
$$\times{}_4F_3\left(\frac{\alpha+\nu+2}{4},\ \frac{\alpha+\nu}{4}+1,\ \frac{\alpha-\nu+2}{4},\ \frac{\alpha-\nu}{4}+1;\ \frac{3}{4},\ \frac{5}{4},\ \frac{3}{2};\ -\frac{p^4}{c^4}\right)+$$
$$+\frac{2^\alpha p^3}{3c^{\alpha+3}}\begin{Bmatrix}\sin[(\alpha+2\nu+1)\pi/4]\\ \cos[(\alpha+2\nu+1)\pi/4]\end{Bmatrix}\Gamma\left(\frac{\alpha+\nu+3}{2}\right)\Gamma\left(\frac{\alpha-\nu+3}{2}\right)\times$$
$$\times{}_4F_3\left(\frac{\alpha+\nu+3}{4},\ \frac{\alpha+\nu+5}{4},\ \frac{\alpha-\nu+3}{4},\ \frac{\alpha-\nu+5}{4};\ \frac{5}{4},\ \frac{3}{2},\ \frac{7}{4};\ -\frac{p^4}{c^4}\right)$$
[Re $(\sqrt{2}p+c)>|\operatorname{Im} c|$; Re $\alpha>|\operatorname{Re}\nu|$].

10. $$\int_0^\infty e^{-px^2}\begin{Bmatrix}\operatorname{ker}_\nu(cx)\\ \operatorname{kei}_\nu(cx)\end{Bmatrix}dx=\pm\frac{\pi^{3/2}}{8\sqrt{p}}\sec\frac{\nu\pi}{2}\left[J_{\nu/2}(\varphi)\begin{Bmatrix}\sin\psi\\ \cos\psi\end{Bmatrix}\mp Y_{\nu/2}(\varphi)\begin{Bmatrix}\cos\psi\\ \sin\psi\end{Bmatrix}\right]$$
[$\varphi=c^2/(8p)$, $\psi=c^2/(8p)-\nu\pi/4$; Re $p>0$; $|\operatorname{Re}\nu|<1$].

11. $\int_0^\infty xe^{-px^2}\begin{Bmatrix}\operatorname{ker}(cx)\\ \operatorname{kei}(cx)\end{Bmatrix}dx=-\frac{1}{4p}\left[\operatorname{ci}(\varphi)\begin{Bmatrix}\cos\varphi\\ \sin\varphi\end{Bmatrix}\pm\operatorname{si}(\varphi)\begin{Bmatrix}\sin\varphi\\ \cos\varphi\end{Bmatrix}\right]$ $[\varphi=c^2/(4p);\ \operatorname{Re}p>0]$.

12. $\int_0^\infty x^{\alpha-1}e^{-px^2}\begin{Bmatrix}\operatorname{ker}_\nu(cx)\\ \operatorname{kei}_\nu(cx)\end{Bmatrix}dx=\pm\frac{2^{\nu-2}p^{(\nu-\alpha)/2}}{c^\nu}\begin{Bmatrix}\cos(3\nu\pi/4)\\ \sin(3\nu\pi/4)\end{Bmatrix}\Gamma(\nu)\Gamma\left(\frac{\alpha-\nu}{2}\right)\times$

$$\times{}_2F_3\left(\frac{\alpha-\nu}{4},\frac{\alpha-\nu+2}{4};\frac{1}{2},\frac{1-\nu}{2},1-\frac{\nu}{2};-\frac{c^4}{64p^2}\right)+$$

$$+\frac{c^\nu}{2^{\nu+2}p^{(\alpha+\nu)/2}}\begin{Bmatrix}\cos(3\nu\pi/4)\\ \sin(3\nu\pi/4)\end{Bmatrix}\Gamma(-\nu)\Gamma\left(\frac{\alpha+\nu}{2}\right)\times$$

$$\times{}_2F_3\left(\frac{\alpha+\nu}{4},\frac{\alpha+\nu+2}{4};\frac{1}{2},\frac{1+\nu}{2},\frac{\nu}{2}+1;-\frac{c^4}{64p^2}\right)-$$

$$-\frac{2^{\nu-4}p^{(\nu-\alpha)/2-1}}{c^{\nu-2}}\begin{Bmatrix}\sin(3\nu\pi/4)\\ \cos(3\nu\pi/4)\end{Bmatrix}\Gamma(\nu-1)\Gamma\left(\frac{\alpha-\nu}{2}+1\right)\times$$

$$\times{}_2F_3\left(\frac{\alpha-\nu+2}{4},\frac{\alpha-\nu}{4}+1;\frac{3}{2},1-\frac{\nu}{2},\frac{3-\nu}{2};-\frac{c^4}{64p^2}\right)\pm$$

$$\pm\frac{c^{\nu+2}}{2^{\nu+4}p^{(\alpha+\nu)/2+1}}\begin{Bmatrix}\sin(3\nu\pi/4)\\ \cos(3\nu\pi/4)\end{Bmatrix}\Gamma(-\nu-1)\Gamma\left(\frac{\alpha+\nu}{2}+1\right)\times$$

$$\times{}_2F_3\left(\frac{\alpha+\nu+2}{4},\frac{\alpha+\nu}{4}+1;\frac{3}{2},\frac{\nu}{2}+1,\frac{\nu+3}{2};-\frac{c^4}{64p^2}\right)\quad[\operatorname{Re}p>0;\ \operatorname{Re}\alpha>|\operatorname{Re}\nu|].$$

2.10.4. Integrals containing trigonometric or logarithmic functions and Kelvin functions.

1. $\int_0^\pi \operatorname{ber}(c\sqrt{\sin x})\,dx=\pi J_0\left(\frac{c}{\sqrt{2}}\right)I_0\left(\frac{c}{\sqrt{2}}\right)$.

2. $\int_0^\infty x^{\alpha-1}\sin bx\begin{Bmatrix}\operatorname{ker}_\nu(cx)\\ \operatorname{kei}_\nu(cx)\end{Bmatrix}dx=U(1)\quad\left[b>0;\ \operatorname{Re}\alpha>|\operatorname{Re}\nu|-1;\ |\arg c|<\pi/4,\ \delta=\begin{Bmatrix}1\\ 0\end{Bmatrix}\right]$,

$$U(\delta)=\pm\frac{2^{\alpha+\delta-2}b^\delta}{c^{\alpha+\delta}}\Gamma\left(\frac{\alpha+\delta+\nu}{2}\right)\Gamma\left(\frac{\alpha+\delta-\nu}{2}\right)\begin{Bmatrix}\cos[(\alpha+\delta+2\nu)\pi/4]\\ \sin[(\alpha+\delta+2\nu)\pi/4]\end{Bmatrix}\times$$

$$\times{}_4F_3\left(\frac{\alpha+\delta+\nu}{4},\frac{\alpha+\delta+\nu+2}{4},\frac{\alpha+\delta-\nu}{4},\frac{\alpha+\delta-\nu+2}{4};\frac{1}{2},\frac{3}{4},\frac{1}{4}+\delta;-\frac{b^4}{c^4}\right)+$$

$$+\frac{2^{\alpha+\delta-1}b^{\delta+2}}{3^\delta c^{\alpha+\delta+2}}\Gamma\left(\frac{\alpha+\delta+\nu}{2}+1\right)\Gamma\left(\frac{\alpha+\delta-\nu}{2}+1\right)\begin{Bmatrix}\sin[(\alpha+\delta+2\nu)\pi/4]\\ \cos[(\alpha+\delta+2\nu)\pi/4]\end{Bmatrix}\times$$

$$\times{}_4F_3\left(\frac{\alpha+\delta+\nu+2}{4},\frac{\alpha+\delta+\nu}{4}+1,\frac{\alpha+\delta-\nu+2}{4},\frac{\alpha+\delta-\nu}{4}+1;\frac{3}{2},\frac{5}{4},\frac{3}{4}+\delta;-\frac{b^4}{c^4}\right).$$

3. $\int_0^\infty x^{\alpha-1}\cos bx\begin{Bmatrix}\operatorname{ker}_\nu(cx)\\ \operatorname{kei}_\nu(cx)\end{Bmatrix}dx=U(0)$ $[b>0;\ \operatorname{Re}\alpha>|\operatorname{Re}\nu|;\ |\arg c|<\pi/4;\ U(\delta)$ see 2.10.4.2].

4. $\int_{-\infty}^\infty \begin{Bmatrix}\sin bx\\ \cos bx\end{Bmatrix}\operatorname{kei}\left[c\sqrt{(x-a)^2+z^2}\right]dx=$

$$=-\frac{\pi}{\sqrt{b^4+c^4}}e^{-z\zeta_+}[\zeta_-\cos\zeta_+z+\zeta_+\sin\zeta_-z]\begin{Bmatrix}\sin ab\\ \cos ab\end{Bmatrix}\quad[2\zeta_\pm^2=\sqrt{b^4+c^4}\pm b^2;\ a,b,c,\ \operatorname{Re}z>0].$$

5. $\int_0^\infty xe^{-px^2}\left[\ln\frac{cx}{2}\begin{Bmatrix}\operatorname{ber}(cx)\\ \operatorname{bei}(cx)\end{Bmatrix}-\begin{Bmatrix}\operatorname{ker}(cx)\\ \operatorname{kei}(cx)\end{Bmatrix}\right]dx=\frac{1}{2p}\left[\ln\varphi\begin{Bmatrix}\cos\varphi\\ \sin\varphi\end{Bmatrix}\mp\frac{\pi}{4}\begin{Bmatrix}\sin\varphi\\ \cos\varphi\end{Bmatrix}\right]$

$[\varphi=c^2/(4p);\ \operatorname{Re}p>0]$.

2.10.5. Integrals containing products of two Kelvin functions.

1. $$\int_0^\infty xe^{-px^2}\operatorname{ber}_\nu(cx)\operatorname{bei}_\nu(cx)\,dx=\frac{1}{4p}\sin\left(\frac{c^2}{2p}+\frac{3\nu\pi}{2}\right)J_\nu\left(\frac{c^2}{2p}\right)\qquad [\operatorname{Re}p>0;\ \operatorname{Re}\nu>-1].$$

2. $$\int_0^\infty x^{\alpha-1}e^{-px}\left[\operatorname{ber}_\nu^2(cx)+\operatorname{bei}_\nu^2(cx)\right]dx=\frac{c^{2\nu}}{2^{2\nu}p^{\alpha+2\nu}}\Gamma\begin{bmatrix}\alpha+2\nu\\ \nu+1,\ \nu+1\end{bmatrix}\times$$
$$\times{}_4F_3\left(\frac{\alpha+2\nu}{4},\frac{\alpha+2\nu+1}{4},\frac{\alpha+2\nu+2}{4},\frac{\alpha+2\nu+3}{4};\frac{\nu+1}{2},\frac{\nu}{2}+1,\ \nu+1;\ \frac{4c^4}{p^4}\right)$$
$$[\operatorname{Re}(\alpha+2\nu)>0;\ \operatorname{Re}p>\sqrt{2}\,(\operatorname{Re}c+|\operatorname{Im}c|)].$$

3. $$\int_0^\infty x^{\alpha-1}e^{-px^2}\left[\operatorname{ber}_\nu^2(cx)+\operatorname{bei}_\nu^2(cx)\right]dx=\frac{c^{2\nu}}{2^{2\nu+1}p^{\nu+\alpha/2}}\Gamma\begin{bmatrix}\nu+\alpha/2\\ \nu+1,\ \nu+1\end{bmatrix}\times$$
$$\times{}_2F_3\left(\frac{\alpha+2\nu}{4},\frac{\alpha+2\nu+2}{4};\frac{\nu+1}{2},\frac{\nu}{2}+1,\ \nu+1;\ \frac{c^4}{16p^2}\right)$$
$$[\operatorname{Re}p,\ \operatorname{Re}(\alpha+2\nu)>0].$$

4. $$\int_0^\infty xe^{-px^2}\left(\frac{d}{x\,dx}\right)^n\left[\operatorname{ber}_\nu^2(cx)+\operatorname{bei}_\nu^2(cx)\right]dx=(2p)^{n-1}I_\nu\left(\frac{c^2}{2p}\right)\qquad [\operatorname{Re}p>0;\ \operatorname{Re}\nu>n-1].$$

5. $$\int_0^\infty x^{\alpha-1}\left[\operatorname{ker}_\nu^2(cx)+\operatorname{kei}_\nu^2(cx)\right]dx=2^{\alpha-4}c^{-\alpha}\Gamma\left[\frac{\alpha}{2},\frac{\alpha+2\nu}{4},\frac{\alpha-2\nu}{4}\right]$$
$$[\operatorname{Re}\alpha>2\,|\operatorname{Re}\nu|;\ |\arg c|<\pi/4].$$

6. $$\int_0^\infty x^{\alpha-1}e^{-px}\left[\operatorname{ber}_\nu(cx)\begin{Bmatrix}\operatorname{ber}'_\nu(cx)\\ \operatorname{bei}'_\nu(cx)\end{Bmatrix}\pm\operatorname{bei}_\nu(cx)\begin{Bmatrix}\operatorname{bei}'_\nu(cx)\\ \operatorname{ber}'_\nu(cx)\end{Bmatrix}\right]dx=$$
$$=\frac{c^{2\nu\mp1}}{2^{2\nu+(1\mp1)/2}p^{\alpha+2\nu\mp1}}\Gamma\begin{bmatrix}\alpha+2\nu\mp1\\ \nu+1,\ \nu+1\mp1\end{bmatrix}\times$$
$$\times{}_4F_3\left(\frac{\alpha+2\nu+1}{4},\frac{\alpha+2\nu+2}{4},\frac{\alpha+2\nu+1\mp2}{4},\frac{\alpha+2\nu+2\mp2}{4};\right.$$
$$\left.\frac{2\nu+3\mp1}{4},\frac{2\nu+3\mp3}{4},\ \nu+1;\ \frac{4c^4}{p^4}\right)$$
$$[\operatorname{Re}(\alpha+2\nu)>\pm1;\ \operatorname{Re}p>\sqrt{2}\,(\operatorname{Re}c+|\operatorname{Im}c|)].$$

7. $$\int_0^\infty e^{-px^2}\left[\operatorname{ber}_\nu(cx)\operatorname{bei}'_\nu(cx)+\operatorname{bei}_\nu(cx)\operatorname{ber}'_\nu(cx)\right]dx=\frac{1}{2c}I_\nu\left(\frac{c^2}{2p}\right)\qquad [\operatorname{Re}p,\ \operatorname{Re}\nu>0].$$

8. $$\int_0^\infty x^{\alpha-1}e^{-px^2}\left[\operatorname{ber}_\nu(cx)\begin{Bmatrix}\operatorname{ber}'_\nu(cx)\\ \operatorname{bei}'_\nu(cx)\end{Bmatrix}\pm\operatorname{bei}_\nu(cx)\begin{Bmatrix}\operatorname{bei}'_\nu(cx)\\ \operatorname{ber}'_\nu(cx)\end{Bmatrix}\right]dx=$$
$$=\frac{c^{2\nu\mp1}}{2^{2\nu+(7\mp1)/2}p^{\nu+(\alpha\mp1)/2}}\Gamma\begin{bmatrix}\nu+(\alpha\mp1)/2\\ \nu+1,\ \nu+1\mp1\end{bmatrix}\times$$
$$\times{}_2F_3\left(\frac{\alpha+2\nu+1}{4},\frac{\alpha+2\nu+1\mp2}{4};\frac{2\nu+3\mp1}{4},\frac{2\nu+3\mp3}{4},\ \nu+1;\ \frac{c^4}{16p^2}\right)$$
$$[\operatorname{Re}p>0;\ \operatorname{Re}(\alpha+2\nu)>\pm1].$$

9. $\int\limits_0^\infty x^2 e^{-px^2} [\mathrm{ber}_\nu(cx)\,\mathrm{bei}'_\nu(cx) - \mathrm{bei}_\nu(cx)\,\mathrm{ber}'_\nu(cx)]\,dx = \frac{c}{4p^2} I_\nu\left(\frac{c^2}{2p}\right)$ [Re $p > 0$; Re $\nu > -2$].

10. $\int\limits_0^\infty x e^{-px^2} \{[\mathrm{ber}'_\nu(cx)]^2 + [\mathrm{bei}'_\nu(cx)]^2\}\,dx = \frac{8p}{c^4} I_\nu\left(\frac{c^2}{2p}\right)$ [Re p, Re $\nu > 0$].

2.10.6. Integrals of $x^\alpha \mathrm{Ei}(-bx^n) \begin{Bmatrix} \mathrm{ber}_\nu(cx) \\ \mathrm{bei}_\nu(cx) \end{Bmatrix}$.

1. $$\int\limits_0^\infty x^{\alpha-1} \mathrm{Ei}(-bx) \begin{Bmatrix} \mathrm{ber}_\nu(cx) \\ \mathrm{bei}_\nu(cx) \end{Bmatrix} dx = -\frac{b^{-\alpha-\nu} c^\nu}{2^\nu(\alpha+\nu)} \Gamma\begin{bmatrix} \alpha+\nu \\ \nu+1 \end{bmatrix} \times$$
$$\times \begin{Bmatrix} \cos(3\nu\pi/4) \\ \sin(3\nu\pi/4) \end{Bmatrix} {}_5F_4\left(\frac{\alpha+\nu}{4}, \frac{\alpha+\nu+1}{4}, \frac{\alpha+\nu+2}{4}, \frac{\alpha+\nu+3}{4}, \frac{\alpha+\nu}{4};\right.$$
$$\left.\frac{\alpha+\nu}{4}+1, \frac{1}{2}, \frac{\nu+1}{2}, \frac{\nu}{2}+1; -\frac{c^4}{b^4}\right) \pm$$
$$\pm \frac{b^{-\alpha-\nu-2} c^{\nu+2}}{2^{\nu+2}(\alpha+\nu+2)} \Gamma\begin{bmatrix} \alpha+\nu+2 \\ \nu+2 \end{bmatrix} \begin{Bmatrix} \sin(3\nu\pi/4) \\ \cos(3\nu\pi/4) \end{Bmatrix} \times$$
$$\times {}_5F_4\left(\frac{\alpha+\nu+2}{4}, \frac{\alpha+\nu+3}{4}, \frac{\alpha+\nu}{4}+1, \frac{\alpha+\nu+5}{4}, \frac{\alpha+\nu+2}{4};\right.$$
$$\left.\frac{\alpha+\nu+6}{4}, \frac{3}{2}, \frac{\nu}{2}+1, \frac{\nu+3}{2}; -\frac{c^4}{b^4}\right)$$
[Re $(\alpha+\nu) > 0$; Re $(\sqrt{2}b - c) > |\mathrm{Im}\, c|$].

2. $$\int\limits_0^\infty x^{\alpha-1} \mathrm{Ei}(-bx^2) \begin{Bmatrix} \mathrm{ber}_\nu(cx) \\ \mathrm{bei}_\nu(cx) \end{Bmatrix} dx =$$
$$= -\frac{b^{-(\alpha+\nu)/2} c^\nu}{2^\nu(\alpha+\nu)} \Gamma\begin{bmatrix} (\alpha+\nu)/2 \\ \nu+1 \end{bmatrix} \begin{Bmatrix} \cos(3\nu\pi/4) \\ \sin(3\nu\pi/4) \end{Bmatrix} \times$$
$$\times {}_3F_4\left(\frac{\alpha+\nu}{4}, \frac{\alpha+\nu}{4}, \frac{\alpha+\nu+2}{4}; \frac{\alpha+\nu}{4}+1, \frac{1}{2}, \frac{\nu+1}{2}, \frac{\nu}{2}+1; -\frac{c^4}{64b^2}\right) \pm$$
$$\pm \frac{b^{-(\alpha+\nu)/2-1} c^{\nu+2}}{2^{\nu+2}(\alpha+\nu+2)} \Gamma\begin{bmatrix} (\alpha+\nu)/2+1 \\ \nu+2 \end{bmatrix} \begin{Bmatrix} \sin(3\nu\pi/4) \\ \cos(3\nu\pi/4) \end{Bmatrix} \times$$
$$\times {}_3F_4\left(\frac{\alpha+\nu+2}{4}, \frac{\alpha+\nu}{4}+1, \frac{\alpha+\nu+2}{4}; \frac{\alpha+\nu+6}{4}, \frac{3}{2}, \frac{\nu+3}{2}, \frac{\nu}{2}+1; -\frac{c^4}{64b^2}\right)$$
[Re b, Re $(\alpha+\nu) > 0$].

3. $\int\limits_0^\infty x\, \mathrm{Ei}(-bx^2) \begin{Bmatrix} \mathrm{ber}(cx) \\ \mathrm{bei}(cx) \end{Bmatrix} dx = -\frac{2}{c^2} \begin{Bmatrix} \sin[c^2/(4b)] \\ 1 - \cos[c^2/(4b)] \end{Bmatrix}$ [Re $b > 0$].

2.10.7. Integrals of $x^\alpha J_\mu(bx) \begin{Bmatrix} \mathrm{ker}_\nu(cx) \\ \mathrm{kei}_\nu(cx) \end{Bmatrix}$.

1. $$\int\limits_0^\infty x^{\alpha-1} J_\mu(bx) \begin{Bmatrix} \mathrm{ker}_\nu(cx) \\ \mathrm{kei}_\nu(cx) \end{Bmatrix} dx = \pm \frac{2^{\alpha-2} b^\mu}{c^{\alpha+\mu}} \times$$
$$\times \begin{Bmatrix} \cos[(\alpha+\mu+2\nu)\pi/4] \\ \sin[(\alpha+\mu+2\nu)\pi/4] \end{Bmatrix} \Gamma\begin{bmatrix} (\alpha+\mu+\nu)/2,\ (\alpha+\mu-\nu)/2 \\ \mu+1 \end{bmatrix} \times$$
$$\times {}_4F_3\left(\frac{\alpha+\mu+\nu}{4}, \frac{\alpha+\mu+\nu+2}{4}, \frac{\alpha+\mu-\nu}{4}, \frac{\alpha+\mu-\nu+2}{4}; \frac{1}{2}, \frac{\mu+1}{2}, \frac{\mu}{2}+1; -\frac{b^4}{c^4}\right) +$$

$$+\frac{2^{\alpha-2}b^{\mu+2}}{c^{\alpha+\mu+2}}\begin{Bmatrix}\sin[(\alpha+\mu+2\nu)\pi/4]\\ \cos[(\alpha+\mu+2\nu)\pi/4]\end{Bmatrix}\Gamma\begin{bmatrix}(\alpha+\mu+\nu)/2+1,\ (\alpha+\mu-\nu)/2+1\\ \mu+2\end{bmatrix}\times$$
$$\times{}_4F_3\left(\frac{\alpha+\mu+\nu+2}{4},\ \frac{\alpha+\mu+\nu}{4}+1,\ \frac{\alpha+\mu-\nu+2}{4},\ \frac{\alpha+\mu-\nu}{4}+1;\ \frac{3}{2},\ \frac{\mu}{2}+1,\ \frac{\mu+3}{2};\ -\frac{b^4}{c^4}\right)$$
$[b>0;\ \operatorname{Re}(\alpha+\mu)>|\operatorname{Re}\nu|;\ |\arg c|<\pi/4]$.

2. $$\int_0^\infty J_1(bx)\begin{Bmatrix}\operatorname{ker}(cx)\\ \operatorname{kei}(cx)\end{Bmatrix}dx=\frac{1}{4b}\begin{Bmatrix}\ln(1+b^4c^{-4})\\ -2\operatorname{arctg}(b^2/c^2)\end{Bmatrix}$$ $[b>0;\ |\arg c|<\pi/4]$

2.10.8. Integrals containing $K_\mu(bx^n)$ and Kelvin functions.

1. $$\int_0^\infty x^{\alpha-1}K_\mu(bx)\begin{Bmatrix}\operatorname{ber}_\nu(cx)\\ \operatorname{bei}_\nu(cx)\end{Bmatrix}dx=$$
$$=\frac{2^{\alpha-2}c^\nu}{b^{\alpha+\nu}}\Gamma\begin{bmatrix}(\alpha+\nu+\mu)/2,\ (\alpha+\nu-\mu)/2\\ \nu+1\end{bmatrix}\begin{Bmatrix}\cos(3\nu\pi/4)\\ \sin(3\nu\pi/4)\end{Bmatrix}\times$$
$$\times{}_4F_3\left(\frac{\alpha+\nu+\mu}{4},\frac{\alpha+\nu+\mu+2}{4},\frac{\alpha+\nu-\mu}{4},\frac{\alpha+\nu-\mu+2}{4};\frac{1}{2},\frac{\nu+1}{2},\frac{\nu}{2}+1;\ -\frac{c^4}{b^4}\right)\mp$$
$$\mp\frac{2^{\alpha-2}c^{\nu+2}}{b^{\alpha+\nu+2}}\Gamma\begin{bmatrix}(\alpha+\nu+\mu)/2+1,\ (\alpha+\nu-\mu)/2+1\\ \nu+2\end{bmatrix}\begin{Bmatrix}\sin(3\nu\pi/4)\\ \cos(3\nu\pi/4)\end{Bmatrix}\times$$
$$\times{}_4F_3\left(\frac{\alpha+\nu+\mu+2}{4},\frac{\alpha+\nu+\mu}{4}+1,\ \frac{\alpha+\nu-\mu+2}{4},\frac{\alpha+\nu-\mu}{4}+1;\ \frac{3}{2},\frac{\nu}{2}+1,\ \frac{\nu+3}{2};\ -\frac{c^4}{b^4}\right)$$
$[\operatorname{Re}(\alpha+\nu)>|\operatorname{Re}\mu|;\ \operatorname{Re}(\sqrt{2}b-c)>|\operatorname{Im}c|]$.

2. $$\int_0^\infty xK_\nu(bx)\begin{Bmatrix}\operatorname{ber}_\nu(cx)\\ \operatorname{bei}_\nu(cx)\end{Bmatrix}dx=\frac{c^\nu}{b^\nu(b^4+c^4)}\left[b^2\begin{Bmatrix}\cos(3\nu\pi/4)\\ \sin(3\nu\pi/4)\end{Bmatrix}\mp c^2\begin{Bmatrix}\sin(3\nu\pi/4)\\ \cos(3\nu\pi/4)\end{Bmatrix}\right]$$
$[\operatorname{Re}\nu>-1;\ \operatorname{Re}(\sqrt{2}b-c)>|\operatorname{Im}c|]$.

3. $$\int_0^\infty x^{\alpha-1}K_\mu(bx^2)\begin{Bmatrix}\operatorname{ber}_\nu(cx)\\ \operatorname{bei}_\nu(cx)\end{Bmatrix}dx=$$
$$=\frac{2^{(\alpha-\nu)/2-3}c^\nu}{b^{(\alpha+\nu)/2}}\Gamma\begin{bmatrix}(\alpha+\nu+2\mu)/4,\ (\alpha+\nu-2\mu)/4\\ \nu+1\end{bmatrix}\begin{Bmatrix}\cos(3\nu\pi/4)\\ \sin(3\nu\pi/4)\end{Bmatrix}\times$$
$$\times{}_2F_3\left(\frac{\alpha+\nu+2\mu}{4},\ \frac{\alpha+\nu-2\mu}{4};\ \frac{1}{2},\ \frac{\nu+1}{2},\ \frac{\nu}{2}+1;\ -\frac{c^4}{64b^2}\right)+$$
$$+\frac{2^{(\alpha-\nu)/2-4}c^{\nu+2}}{b^{(\alpha+\nu)/2+1}}\Gamma\begin{bmatrix}(\alpha+\nu+2\mu+2)/4,\ (\alpha+\nu-2\mu+2)/4\\ \nu+2\end{bmatrix}\begin{Bmatrix}\sin(3\nu\pi/4)\\ \cos(3\nu\pi/4)\end{Bmatrix}\times$$
$$\times{}_2F_3\left(\frac{\alpha+\nu+2\mu+2}{4},\ \frac{\alpha+\nu-2\mu+2}{4};\ \frac{3}{2},\ \frac{\nu}{2}+1,\ \frac{\nu+3}{2};\ -\frac{c^4}{64b^2}\right)$$
$[\operatorname{Re}b>0;\ \operatorname{Re}(\alpha+\nu)>2|\operatorname{Re}\mu|]$.

4. $$\int_0^\infty x^{\alpha-1}K_\mu(bx^2)\left[\operatorname{ber}_\nu^2(cx)+\operatorname{bei}_\nu^2(cx)\right]dx=$$
$$=\frac{2^{\alpha/2-\nu-3}c^{2\nu}}{b^{\alpha/2+\nu}}\Gamma\begin{bmatrix}(\alpha+2\nu+2\mu)/4,\ (\alpha+2\nu-2\mu)/4\\ \nu+1,\ \nu+1\end{bmatrix}\times$$
$$\times{}_2F_3\left(\frac{\alpha+2\nu+2\mu}{4},\ \frac{\alpha+2\nu-2\mu}{4};\ \frac{\nu+1}{2},\ \frac{\nu}{2}+1,\ \nu+1;\ \frac{c^4}{16b^2}\right)$$
$[\operatorname{Re}b>0;\ \operatorname{Re}(\alpha+2\nu)>2|\operatorname{Re}\mu|]$.

5. $\int\limits_0^\infty x^{\nu+3-\sigma} K_{(\nu+\sigma)/2}(bx^2)\left[\operatorname{ber}_\nu^2(cx)+\operatorname{bei}_\nu^2(cx)\right]dx=$

$$=\frac{c^{\nu+1-\sigma}\sqrt{\pi}}{2^{(\nu+4-\sigma)/2}b^{\nu+5/2-\sigma}} I_{(\nu-1+\sigma)/2}\left(\frac{c^2}{2b}\right) \qquad [\sigma=0 \text{ or } 1;\ \operatorname{Re} b>0;\ \operatorname{Re}\nu>-1].$$

6. $\int\limits_0^\infty x^3 K_0(bx^2)\left[\operatorname{ber}^2(cx)+\operatorname{bei}^2(cx)\right]dx=\frac{1}{2b^2}\operatorname{ch}\frac{c^2}{2b} \qquad [\operatorname{Re} b>0].$

7. $\int\limits_0^\infty x^{\alpha-1} K_\mu(bx^2)\left[\operatorname{ber}_\nu(cx)\begin{Bmatrix}\operatorname{ber}'_\nu(cx)\\ \operatorname{bei}'_\nu(cx)\end{Bmatrix}\pm\operatorname{bei}_\nu(cx)\begin{Bmatrix}\operatorname{bei}'_\nu(cx)\\ \operatorname{ber}'_\nu(cx)\end{Bmatrix}\right]dx=$

$$=\frac{2^{(\alpha-7)/2}c^{2\nu\mp1}}{b^{\nu+(\alpha\mp1)/2}}\Gamma\left[\begin{matrix}(\alpha+2\nu+2\mu\mp1)/4,\ (\alpha+2\nu-2\mu\mp1)/4\\ \nu+1\mp1,\ \nu+1\end{matrix}\right]\times$$

$$\times\, {}_2F_3\left(\frac{\alpha+2\nu+2\mu\mp1}{4},\ \frac{\alpha+2\nu-2\mu\mp1}{4};\ \frac{\nu+1\mp1}{2},\ \frac{\nu+2\mp1}{2},\ \nu+1;\ \frac{c^4}{16b^2}\right)$$

$[\operatorname{Re} b>0;\ \operatorname{Re}(\alpha+2\nu)>2|\operatorname{Re}\mu|\pm1].$

2.11. THE AIRY FUNCTIONS $\operatorname{Ai}(x)$ AND $\operatorname{Bi}(x)$

Some integrals containing the functions $\operatorname{Ai}(x)$ and $\operatorname{Bi}(x)$ can be obtained from [18], 2.15–16 by means of the relation

$$\operatorname{Ai}(x)=\frac{1}{\pi}\sqrt{\frac{x}{3}}K_{1/3}\left(\frac{2}{3}x^{3/2}\right),$$

$$\operatorname{Bi}(x)=\sqrt{\frac{x}{3}}\left[I_{-1/3}\left(\frac{2}{3}x^{3/2}\right)+I_{1/3}\left(\frac{2}{3}x^{3/2}\right)\right].$$

2.11.1. Integrals of general form.

1. $\int\limits_0^a x^{\alpha-1}(a^r-x^r)^{\beta-1}\begin{Bmatrix}\operatorname{Ai}(cx)\\ \operatorname{Bi}(cx)\end{Bmatrix}dx=$

$$=\frac{3^{(1\mp3)/12}a^{\alpha+r\beta-r}}{2\pi r}\Gamma(\beta)\Gamma\left(\frac{1}{3}\right)\sum_{k=0}^{\infty}\frac{1}{k!\,(2/3)_k}\Gamma\left[\begin{matrix}(\alpha+3k)/r\\ \beta+(\alpha+3k)/r\end{matrix}\right]\left(\frac{a^3c^3}{9}\right)^k\mp$$

$$\mp\frac{3^{(5\mp3)/12}a^{\alpha+r\beta-r+1}c}{2\pi r}\Gamma(\beta)\Gamma\left(\frac{2}{3}\right)\times$$

$$\times\sum_{k=0}^{\infty}\frac{1}{k!\,(4/3)_k}\Gamma\left[\begin{matrix}(\alpha+3k+1)/r\\ \beta+(\alpha+3k+1)/r\end{matrix}\right]\left(\frac{a^3c^3}{9}\right)^k \qquad [a,\ r,\ \operatorname{Re}\alpha,\ \operatorname{Re}\beta>0].$$

2. $\int\limits_a^\infty x^{\alpha-1}(x^r-a^r)^{\beta-1}\operatorname{Ai}(cx)\,dx=$

$$=\frac{a^{\alpha+r\beta-r}}{2\cdot3^{1/6}\pi r}\Gamma(\beta)\Gamma\left(\frac{1}{3}\right)\sum_{k=0}^{\infty}\frac{1}{k!\,(2/3)_k}\Gamma\left[\begin{matrix}1-\beta-(\alpha+3k)/r\\ 1-(\alpha+3k)/r\end{matrix}\right]\left(\frac{a^3c^3}{9}\right)^k-$$

$$-\frac{3^{1/6}a^{\alpha+r\beta-r+1}c}{2\pi r}\Gamma(\beta)\Gamma\left(\frac{2}{3}\right)\sum_{k=0}^{\infty}\frac{1}{k!\,(4/3)_k}\Gamma\left[\begin{matrix}1-\beta-(\alpha+3k+1)/r\\ 1-(\alpha+3k+1)/r\end{matrix}\right]\left(\frac{a^3c^3}{9}\right)^k+$$

$$+\frac{c^{r-\alpha-r\beta}}{2\cdot 3^{7/6+2(r-r\beta+\alpha)/3}\pi}\sum_{k=0}^{\infty}\frac{(1-\beta)_k}{k!}\Gamma\left(\frac{\alpha+r\beta-r-rk}{3}\right)\Gamma\left(\frac{1+\alpha+r\beta-r-rk}{3}\right)\left(\frac{ac}{3^{2/3}}\right)^{rk}$$

[a, r, $\operatorname{Re}\beta>0$; $|\arg c|<\pi/3$].

3. $$\int_0^\infty \frac{x^{\alpha-1}}{(x^r+z^r)^\rho}\operatorname{Ai}(cx)\,dx=$$

$$=\frac{z^{\alpha-r\rho}}{2\cdot 3^{1/6}\pi r}\Gamma\begin{bmatrix}1/3\\ \rho\end{bmatrix}\sum_{k=0}^{\infty}\frac{1}{k!\,(2/3)_k}\Gamma\left(\frac{\alpha+3k}{r}\right)\Gamma\left(\rho-\frac{\alpha+3k}{r}\right)\left(\frac{c^3z^3}{9}\right)^k-$$

$$-\frac{3^{1/6}cz^{\alpha+1-r\rho}}{2\pi r}\Gamma\begin{bmatrix}2/3\\ \rho\end{bmatrix}\sum_{k=0}^{\infty}\frac{1}{k!\,(4/3)_k}\Gamma\left(\frac{\alpha+3k+1}{r}\right)\Gamma\left(\rho-\frac{\alpha+3k+1}{r}\right)\times$$

$$\times\left(\frac{c^3z^3}{9}\right)^k+\frac{c^{r\rho-\alpha}}{2\cdot 3^{7/6+2(r\rho-\alpha)/3}\pi}\sum_{k=0}^{\infty}\frac{(-1)^k(\rho)_k}{k!}\Gamma\left(\frac{\alpha-r\rho-rk}{3}\right)\times$$

$$\times\Gamma\left(\frac{1+\alpha-r\rho-rk}{3}\right)\left(\frac{cz}{3^{2/3}}\right)^{rk}$$

[r, $\operatorname{Re}\alpha>0$; $|\arg c|<\pi/3$; $r|\arg z|<\pi$].

4. $$\int_0^\infty \frac{x^{\alpha-1}}{x^r-y^r}\operatorname{Ai}(cx)\,dx=$$

$$=-\frac{y^{\alpha-r}}{2\cdot 3^{1/6}r}\Gamma\left(\frac{1}{3}\right)\sum_{k=0}^{\infty}\frac{1}{k!\,(2/3)_k}\operatorname{ctg}\frac{\alpha+3k}{r}\pi\left(\frac{c^3y^3}{9}\right)^k+$$

$$+\frac{3^{1/6}cy^{\alpha+1-r}}{2r}\Gamma\left(\frac{2}{3}\right)\sum_{k=0}^{\infty}\frac{1}{k!\,(4/3)_k}\operatorname{ctg}\frac{\alpha+3k+1}{r}\pi\left(\frac{c^3y^3}{9}\right)^k+$$

$$+\frac{c^{r-\alpha}}{2\cdot 3^{7/6+2(r-\alpha)/3}\pi}\sum_{k=0}^{\infty}\Gamma\left(\frac{\alpha-r-rk}{3}\right)\Gamma\left(\frac{1+\alpha-r-rk}{3}\right)\left(\frac{cy}{3^{2/3}}\right)^{rk}$$

[r, y, $\operatorname{Re}\alpha>0$; $|\arg c|<\pi/3$].

5. $$\int_0^\infty x^{\alpha-1}e^{-px^r}\begin{Bmatrix}\operatorname{Ai}(cx)\\ \operatorname{Bi}(cx)\end{Bmatrix}dx=$$

$$=\frac{3^{(1\mp 3)/12}\Gamma(1/3)}{2\pi r p^{\alpha/r}}\sum_{k=0}^{\infty}\frac{1}{k!\,(2/3)_k}\Gamma\left(\frac{\alpha+3k}{r}\right)\left(\frac{c^3}{9p^{3/r}}\right)^k\mp$$

$$\mp\frac{3^{(5\mp 3)/12}c\Gamma(2/3)}{2\pi p^{(\alpha+1)/r}r}\sum_{k=0}^{\infty}\frac{1}{k!\,(4/3)_k}\Gamma\left(\frac{\alpha+3k+1}{r}\right)\left(\frac{c^3}{9p^{3/r}}\right)^k$$

[$\operatorname{Re}\alpha$, $\operatorname{Re}p>0$; $r>3/2$; $r=3/2$ see 2.11.3.1].

6. $$\int_0^\infty x^{\alpha-1}e^{-px^r}\operatorname{Ai}(cx)\,dx=$$

$$=\frac{3^{(4\alpha-7)/6}}{2\pi c^\alpha}\sum_{k=0}^{\infty}\frac{(-1)^k}{k!}\Gamma\left(\frac{\alpha+rk}{3}\right)\Gamma\left(\frac{\alpha+rk+1}{3}\right)\left(\frac{3^{2r/3}p}{c^r}\right)^k$$

[$\operatorname{Re}\alpha$, $\operatorname{Re}p>0$; $0<r<3/2$; $|\arg c|<\pi/3$].

7. $\int_0^\infty x^{\alpha-1} \begin{Bmatrix} \sin bx^r \\ \cos bx^r \end{Bmatrix} \operatorname{Ai}(cx)\, dx = U(r)$ $\left[b,\ r>0;\ \operatorname{Re}\alpha > -r\delta,\ \delta = \begin{Bmatrix} 1 \\ 0 \end{Bmatrix};\ |\arg c| < \pi/6\right]$,

$$U(r) = \frac{\Gamma(1/3)}{2\cdot 3^{1/6}\pi b^{\alpha/r} r} \sum_{k=0}^{\infty} \frac{1}{k!\,(2/3)_k} \Gamma\left(\frac{\alpha+3k}{r}\right) \times$$

$$\times \begin{Bmatrix} \sin[(\alpha+3k)\pi/(2r)] \\ \cos[(\alpha+3k)\pi/(2r)] \end{Bmatrix} \left(\frac{c^3}{9b^{3/r}}\right)^k - \frac{3^{1/6} c\Gamma(2/3)}{2\pi r b^{(\alpha+1)/r}} \times$$

$$\times \sum_{k=0}^{\infty} \frac{1}{k!\,(4/3)_k} \Gamma\left(\frac{\alpha+3k+1}{r}\right) \begin{Bmatrix} \sin[(\alpha+3k+1)\pi/(2r)] \\ \cos[(\alpha+3k+1)\pi/(2r)] \end{Bmatrix} \left(\frac{c^3}{9b^{3/r}}\right)^k \qquad [r > 3/2],$$

$$U(r) = \frac{3^{(4\alpha+4\delta r-7)/6} b^\delta}{2\pi c^{\alpha+\delta r}} \sum_{k=0}^{\infty} \frac{(-1)^k}{k!\,(\delta+1/2)_k} \times$$

$$\times \Gamma\left(\frac{\alpha+\delta r+2rk}{3}\right) \Gamma\left(\frac{1+\alpha+\delta r+2rk}{3}\right) \left(\frac{3^{2r/3} b}{2c^r}\right)^{2k} \qquad [r < 3/2],$$

$U(3/2)$ see 2.11.4.1.

8. $\int_0^\infty x^{\alpha-1} J_\nu(bx^r) \operatorname{Ai}(cx)\, dx = V(r)$ $[b,\ r,\ \operatorname{Re}(\alpha+r\nu) > 0;\ |\arg c| < \pi/6]$.

$$V(r) = \frac{2^{\alpha/r-2}\Gamma(1/3)}{3^{1/6}\pi b^{\alpha/r} r} \sum_{k=0}^{\infty} \frac{1}{k!\,(2/3)_k} \Gamma\begin{bmatrix} (\nu r+\alpha+3k)/(2r) \\ 1+(\nu r-\alpha-3k)/(2r) \end{bmatrix} \left(\frac{2^{3/r} c^3}{9b^{3/r}}\right)^k -$$

$$- \frac{3^{1/6}\cdot 2^{(\alpha+1)/r-2} c\Gamma(2/3)}{\pi p^{(\alpha+1)/r} r} \sum_{k=0}^{\infty} \frac{1}{k!\,(4/3)_k} \Gamma\begin{bmatrix} (\nu r+\alpha+1+3k)/(2r) \\ 1+(\nu r-\alpha-1-3k)/(2r) \end{bmatrix} \left(\frac{2^{3/r} c^3}{9b^{3/r}}\right)^k$$

$$[r > 3/2],$$

$$V(r) = \frac{3^{(4\alpha+4\nu r-7)/6} b^\nu}{2^{\nu+1}\pi c^{\alpha+\nu r}\Gamma(\nu+1)} \sum_{k=0}^{\infty} \frac{(-1)^k}{k!\,(\nu+1)_k} \Gamma\left(\frac{\alpha+\nu r+2rk}{3}\right) \times$$

$$\times \Gamma\left(\frac{1+\alpha+\nu r+2rk}{3}\right) \left(\frac{3^{2r/3} b}{2c^r}\right)^{2k} \qquad [r < 3/2],$$

$V(3/2)$ see 2.11.5.1.

2.11.2. Integrals of $A(x) \begin{Bmatrix} \operatorname{Ai}(cx) \\ \operatorname{Bi}(cx) \end{Bmatrix}$.

1. $\int_0^\infty x^{\alpha-1} \operatorname{Ai}(cx)\, dx = \frac{3^{(4\alpha-1)/6-1}}{2\pi c^\alpha} \Gamma\left(\frac{\alpha}{3}\right) \Gamma\left(\frac{\alpha+1}{3}\right)$ $[\operatorname{Re}\alpha > 0;\ |\arg c| < \pi/3]$.

2. $\int_0^a x^{\alpha-1} (a^3-x^3)^{\beta-1} \begin{Bmatrix} \operatorname{Ai}(cx) \\ \operatorname{Bi}(cx) \end{Bmatrix} dx =$

$$= \frac{a^{\alpha+3\beta-3}\Gamma(1/3)}{2\cdot 3^{(11\pm 3)/12}\pi} \mathrm{B}\left(\beta, \frac{\alpha}{3}\right) {}_1F_2\left(\frac{\alpha}{3};\ \frac{2}{3}, \beta+\frac{\alpha}{3};\ \frac{a^3c^3}{9}\right) \mp$$

$$\mp \frac{a^{\alpha+3\beta-2} c\Gamma(2/3)}{2\cdot 3^{(7\pm 3)/12}\pi} \mathrm{B}\left(\beta, \frac{\alpha+1}{3}\right) {}_1F_2\left(\frac{\alpha+1}{3};\ \frac{4}{3}, \beta+\frac{\alpha+1}{3};\ \frac{a^3c^3}{9}\right)$$

$[a,\ \operatorname{Re}\alpha,\ \operatorname{Re}\beta > 0]$.

3. $\int_a^{\infty} \frac{1}{\sqrt{x-a}} \operatorname{Ai}(cx)\,dx = 2^{2/3}\sqrt{\frac{\pi}{c}}\operatorname{Ai}\left(\frac{ac}{2^{2/3}}\right)$ $[a>0;\ |\arg c|<\pi/3]$.

2.11.3. Integrals of $x^{\alpha}e^{-px^r}\left\{\begin{matrix}\operatorname{Ai}(cx)\\ \operatorname{Bi}(cx)\end{matrix}\right\}$.

1. $\int_0^{\infty} x^{\alpha-1}e^{-px^{3/2}}\left\{\begin{matrix}\operatorname{Ai}(cx)\\ \operatorname{Bi}(cx)\end{matrix}\right\}dx=$

$$=\frac{3^{(-11\mp 3)/12}\Gamma(1/3)}{\pi p^{2\alpha/3}}\Gamma\left(\frac{2\alpha}{3}\right){}_2F_1\left(\frac{\alpha}{3},\ \frac{2\alpha+3}{6};\ \frac{2}{3};\ \frac{4c^3}{9p^2}\right)\mp$$

$$\mp\frac{3^{(-7\mp 3)/12}c\Gamma(2/3)}{\pi p^{2(\alpha+1)/3}}\Gamma\left(\frac{2\alpha+2}{3}\right){}_2F_1\left(\frac{\alpha+1}{3},\ \frac{2\alpha+5}{6};\ \frac{2}{3};\ \frac{4c^3}{9p^2}\right)$$

$[\operatorname{Re}\alpha>0;\ \operatorname{Re}(3p\pm 2c^{3/2})>0;\ |\arg c|<\pi/6]$.

2.11.4. Integrals of $x^{\alpha}\left\{\begin{matrix}\sin bx^r\\ \cos bx^r\end{matrix}\right\}\operatorname{Ai}(cx)$.

NOTATION: $\delta=\left\{\begin{matrix}1\\0\end{matrix}\right\}$.

1. $\int_0^{\infty} x^{\alpha-1}\left\{\begin{matrix}\sin bx^{3/2}\\ \cos bx^{3/2}\end{matrix}\right\}\operatorname{Ai}(cx)\,dx=$

$$=\frac{3^{(4\alpha-7)/6+\delta}b^{\delta}}{2\pi c^{\alpha+3\delta/2}}\Gamma\left(\frac{2\alpha+3\delta}{6}\right)\Gamma\left(\frac{2+2\alpha+3\delta}{6}\right)\times$$

$$\times\,{}_2F_1\left(\frac{2\alpha+3\delta}{6},\frac{2+2\alpha+3\delta}{6};\ \delta+\frac{1}{2};\ -\frac{9b^2}{4c^3}\right)$$ $[b>0;\ 2\operatorname{Re}\alpha>-3\delta;\ |\arg c|<\pi/6]$.

2.11.5. Integrals of $x^{\alpha}J_{\nu}(bx^r)\operatorname{Ai}(cx)$.

1. $\int_0^{\infty} x^{\alpha-1}J_{\nu}(bx^{3/2})\operatorname{Ai}(cx)\,dx=$

$$=\frac{3^{(4\alpha-7)/6+\nu}b^{\nu}}{2^{\nu+1}\pi c^{\alpha+3\nu/2}}\Gamma\left[\begin{matrix}(2\alpha+3\nu)/6,\ (2+2\alpha+3\nu)/6\\ \nu+1\end{matrix}\right]\times$$

$$\times\,{}_2F_1\left(\frac{2\alpha+3\nu}{6},\ \frac{2+2\alpha+3\nu}{6};\ \nu+1;\ -\frac{9b^2}{4c^3}\right)$$ $[b,\ \operatorname{Re}(2\alpha+3\nu)>0;\ |\arg c|<\pi/6]$.

2.11.6. Integrals of $\operatorname{Ai}(ax+b)\operatorname{Ai}(cx+d)$.

1. $\int_{-\infty}^{\infty}\operatorname{Ai}(ax+b)\operatorname{Ai}(cx+d)\,dx=\frac{1}{\sqrt{\pi}\sqrt[3]{a^3-c^3}}\operatorname{Ai}\left(\frac{ad-bc}{\sqrt[3]{a^3-c^3}}\right)$ $[0<c<a]$.

2.12. THE INTEGRAL FUNCTIONS OF BESSEL $Ji_{\nu}(x)$, NEUMANN $Yi_{\nu}(x)$ AND MACDONALD $Ki_{\nu}(x)$

The integral functions $Ji_{\nu}(x)$, $Yi_{\nu}(x)$, $Ki_{\nu}(x)$ are defined by the formulae

$$Ji_{\nu}(x)=\int_x^{\infty}J_{\nu}(t)\frac{dt}{t},\quad Yi_{\nu}(x)=\int_x^{\infty}Y_{\nu}(t)\frac{dt}{t},\quad Ki_{\nu}(x)=\int_x^{\infty}K_{\nu}(t)\frac{dt}{t}.$$

2.12.1. Integrals of general form

1. $$\int_0^a x^{\alpha-1}(a^r-x^r)^{\beta-1} J i_\nu(cx)\,dx = -U(1) + \frac{a^{\alpha+\beta r-r}}{\nu r} \mathrm{B}\left(\frac{\alpha}{r}, \beta\right)$$

[a, r, $\operatorname{Re}\alpha$, $\operatorname{Re}\beta$, $\operatorname{Re}(\alpha+\nu)>0$; $U(\varepsilon)$ see [18], 2.12.1.1].

2. $$\int_0^a x^{\alpha-1}(a^r-x^r)^{\beta-1} \begin{Bmatrix} Y i_\nu(cx) \\ K i_\nu(cx) \end{Bmatrix} dx =$$

$$= -U_\nu(1) \pm \frac{a^{\alpha+\beta r-r}}{2\nu r} \begin{Bmatrix} 2\operatorname{ctg}(\nu\pi/2) \\ \pi\operatorname{cosec}(\nu\pi/2) \end{Bmatrix} \mathrm{B}\left(\frac{\alpha}{r}, \beta\right)$$

[a, r, $\operatorname{Re}\beta>0$; $\operatorname{Re}\alpha>|\operatorname{Re}\nu|$; $U_\nu(\varepsilon)$ see [18], 2.13.1.1].

3. $$\int_a^\infty x^{\alpha-1}(x^r-a^r)^{\beta-1} J i_\nu(cx)\,dx = -V(1) + \frac{a^{\alpha+r\beta-r}}{\nu r} \mathrm{B}\left(1-\beta-\frac{\alpha}{r}, \beta\right)$$

[a, c, r, $\operatorname{Re}\beta>0$; $\operatorname{Re}(\alpha+\beta r)<r+5/2$; $V(\varepsilon)$ see [18], 2.12.1.2].

4. $$\int_a^\infty x^{\alpha-1}(x^r-a^r)^{\beta-1} \begin{Bmatrix} Y i_\nu(cx) \\ K i_\nu(cx) \end{Bmatrix} dx =$$

$$= -V_\nu(1) \pm \frac{a^{\alpha+\beta r-r}}{2\nu r} \begin{Bmatrix} 2\operatorname{ctg}(\nu\pi/2) \\ \pi\operatorname{cosec}(\nu\pi/2) \end{Bmatrix} \mathrm{B}\left(1-\beta-\frac{\alpha}{r}, \beta\right)$$

$$\left[a,\ r,\ \operatorname{Re}\beta>0,\ \begin{Bmatrix} \operatorname{Re}(\alpha+\beta r)<r+5/2;\ c>0 \\ \operatorname{Re}c>0 \end{Bmatrix},\ V_\nu(\varepsilon) \text{ see [18], 2.13.1.2}\right].$$

5. $$\int_0^\infty \frac{x^{\alpha-1}}{(x^r+z^r)^\rho} J i_\nu(cx)\,dx = -W(1) + \frac{z^{\alpha-\rho r}}{\nu r} \mathrm{B}\left(\frac{\alpha}{r}, \rho-\frac{\alpha}{r}\right)$$

[c, r, $\operatorname{Re}\alpha>0$; $-\operatorname{Re}\nu<\operatorname{Re}\alpha<r\operatorname{Re}\rho+5/2$; $r|\arg z|<\pi$; $W(\varepsilon)$ see [18], 2.12.1.3].

6. $$\int_0^\infty \frac{x^{\alpha-1}}{(x^r+z^r)^\rho} \begin{Bmatrix} Y i_\nu(cx) \\ K i_\nu(cx) \end{Bmatrix} dx = -W_\nu(1) \pm \frac{z^{\alpha-\rho r}}{2\nu r} \begin{Bmatrix} 2\operatorname{ctg}(\nu\pi/2) \\ \pi\operatorname{cosec}(\nu\pi/2) \end{Bmatrix} \mathrm{B}\left(\frac{\alpha}{r}, \rho-\frac{\alpha}{r}\right)$$

$$\left[r>0;\ \operatorname{Re}\alpha>|\operatorname{Re}\nu|;\ r|\arg z|<\pi,\ \begin{Bmatrix} \operatorname{Re}(\alpha-r\rho)<5/2;\ c>0 \\ \operatorname{Re}c>0 \end{Bmatrix},\ W_\nu(\varepsilon) \text{ see [18], 2.13.1.3}\right].$$

7. $$\int_0^\infty \frac{x^{\alpha-1}}{x^r-y^r} J i_\nu(cx)\,dx = -X(1) - \frac{\pi y^{\alpha-r}}{\nu r} \operatorname{ctg}\frac{\alpha\pi}{r}$$

[c, r, y, $\operatorname{Re}\alpha>0$; $-\operatorname{Re}\nu<\operatorname{Re}\alpha<r+5/2$; $X(\varepsilon)$ see [18], 2.12.1.4].

8. $$\int_0^\infty \frac{x^{\alpha-1}}{x^r-y^r} \begin{Bmatrix} Y i_\nu(cx) \\ K i_\nu(cx) \end{Bmatrix} dx = -X_\nu(1) \mp \frac{\pi y^{\alpha-r}}{2\nu r} \begin{Bmatrix} 2\operatorname{ctg}(\nu\pi/2) \\ \pi\operatorname{cosec}(\nu\pi/2) \end{Bmatrix} \operatorname{ctg}\frac{\alpha\pi}{r}$$

$$\left[r,\ y>0;\ \operatorname{Re}\alpha>|\operatorname{Re}\nu|,\ \begin{Bmatrix} \operatorname{Re}\alpha<r+5/2;\ c>0 \\ \operatorname{Re}c>0 \end{Bmatrix},\ X_\nu(\varepsilon) \text{ see [18], 2.13.1.4}\right].$$

9. $\int\limits_0^\infty x^{\alpha-1}e^{-px^r} J i_\nu(cx)\,dx = U,$

$$U = -Y(1) + \frac{\Gamma(\alpha/r)}{\nu p^{\alpha/r} r} \qquad [r > 1],$$
$$U = Y(1) \qquad [r < 1],$$
U for $r=1$ see 2.12.3.5

[r, Re p, Re α, Re $(\alpha+\nu) > 0$; $c > 0$ for $r \leqslant 1$, $|\arg c| < \pi$ for $r > 1$; $Y(\varepsilon)$ see [18], 2.12.1.5].

10. $\int\limits_0^\infty x^{\alpha-1}e^{-px^r} \begin{Bmatrix} Y i_\nu(cx) \\ K i_\nu(cx) \end{Bmatrix} dx = U_1,$

$$U_1 = -Y_\nu(1) \pm \frac{p^{-\alpha/r}}{2\nu r}\Gamma\left(\frac{\alpha}{r}\right)\begin{Bmatrix} 2\,\mathrm{ctg}\,(\nu\pi/2) \\ \pi\,\mathrm{cosec}\,(\nu\pi/2) \end{Bmatrix} \qquad [r > 1],$$
$$U_1 = Y_\nu(1) \qquad [r < 1],$$
U_1 for $r=1$ see 2.12.3.6

[r, Re $p > 0$; Re $\alpha > |$Re $\nu|$, $\begin{Bmatrix} c > 0 \\ \mathrm{Re}\, c > 0 \end{Bmatrix}$, $Y_\nu(\varepsilon)$ see [18], 2.13.1.5].

11. $\int\limits_0^\infty x^{\alpha-1}e^{-px^{-r}} J i_\nu(cx)\,dx = -Z(1) + \frac{p^{\alpha/r}}{\nu r}\Gamma\left(-\frac{\alpha}{r}\right)$

[c, r, Re $p > 0$; Re $\alpha < 5/2$; $Z(\varepsilon)$ see [18], 2.12.1.6].

12. $\int\limits_0^\infty x^{\alpha-1}e^{-px^{-r}} \begin{Bmatrix} Y i_\nu(cx) \\ K i_\nu(cx) \end{Bmatrix} dx = -Z_\nu(1) \pm \frac{p^{\alpha/r}}{2\nu r}\Gamma\left(-\frac{\alpha}{r}\right)\begin{Bmatrix} 2\,\mathrm{ctg}\,(\nu\pi/2) \\ \pi\,\mathrm{cosec}\,(\nu\pi/2) \end{Bmatrix}$

$\left[r,\ \mathrm{Re}\,p > 0,\ \begin{Bmatrix} \mathrm{Re}\,\alpha < 5/2;\ c > 0 \\ \mathrm{Re}\,c > 0 \end{Bmatrix},\ Z_\nu(\varepsilon)\ \text{see [18], 2.13.1.6}\right].$

13. $\int\limits_0^\infty x^{\alpha-1} \begin{Bmatrix} \sin bx^r \\ \cos bx^r \end{Bmatrix} J i_\nu(cx)\,dx = U_2,$

$$U_2 = -P(1) + \frac{b^{-\alpha/r}}{\nu r}\Gamma\left(\frac{\alpha}{r}\right)\begin{Bmatrix} \sin[\alpha\pi/(2r)] \\ \cos[\alpha\pi/(2r)] \end{Bmatrix} \qquad [r > 1],$$
$$U_2 = P(1) \qquad [r < 1].$$
U_2 for $r=1$ see 2.12.4.1

$\left[b,\ c,\ r > 0;\ \mathrm{Re}\,\alpha > -\delta r;\ -\delta r - \mathrm{Re}\,\nu < \mathrm{Re}\,\alpha < 3/2 + \max(r, 1);\ \delta = \begin{Bmatrix} 1 \\ 0 \end{Bmatrix},\ P(\varepsilon)\ \text{see [18], 2.12.1.7}\right].$

14. $\int\limits_0^\infty x^{\alpha-1} \begin{Bmatrix} \sin bx^r \\ \cos bx^r \end{Bmatrix} Y i_\nu(cx)\,dx = V(1, 1),$

$$V(\gamma, 1) = -U_\nu(\gamma, 1) + \frac{b^{-\alpha/r}\Gamma(\alpha/r)}{\nu r \sin(\nu\pi/2)}\begin{Bmatrix} \sin[\alpha\pi/(2r)] \\ \cos[\alpha\pi/(2r)] \end{Bmatrix}\cos^\gamma\frac{\nu\pi}{2} \qquad [r > 1],$$
$$V(\gamma, 1) = U_\nu(\gamma, 1) \qquad [r < 1],$$
$V(\gamma, 1)$ for $r=1$ see 2.12.4.2—3

$\left[b,\ c,\ r > 0;\ |\mathrm{Re}\,\nu| - \delta r < \mathrm{Re}\,\alpha < 3/2 + \max(r, 1);\ \delta = \begin{Bmatrix} 1 \\ 0 \end{Bmatrix},\ U_\nu(\gamma, \varepsilon)\ \text{see [18], 2.13.1.7}\right].$

15. $\int_0^\infty x^{\alpha-1} \begin{Bmatrix} \sin bx^r \\ \cos bx^r \end{Bmatrix} Ki_\nu(cx)\,dx = -\frac{\pi}{2} V(0, 1)$

$[b, r, \operatorname{Re} c > 0;\ |\operatorname{Re}\nu| < \operatorname{Re}\alpha + \delta r;\ V(\gamma, 1)$ see 2.12.1.14].

16. $\int_0^\infty x^{\alpha-1} e^{-px^r} \begin{Bmatrix} \sin bx^r \\ \cos bx^r \end{Bmatrix} Ji_\nu(cx)\,dx = U_3,$

$$U_3 = -Q(1) + \frac{(b^2+p^2)^{-\alpha/(2r)}}{\nu r} \Gamma\left(\frac{\alpha}{r}\right) \cos\left(\frac{\delta\pi}{2} - \frac{\alpha}{r} \arccos \frac{p}{\sqrt{b^2+p^2}}\right)$$

$[r > 1]$ or $[r = 1;\ |c|^2 < |b^2+p^2|]$,

$U_3 = Q(1)$ $[r < 1]$ or $[r = 1;\ |c|^2 > |b^2+p^2|]$.

$[b, r, \operatorname{Re} p > 0;\ \operatorname{Re}\alpha, \operatorname{Re}(\alpha+\nu) > -\delta r;\ c > 0$ for $r \leqslant 1$, $|\arg c| < \pi$ for $r > 1$; $\delta = \begin{Bmatrix} 1 \\ 0 \end{Bmatrix}$,

$Q(\varepsilon)$ see [18], 2.12.1.8].

17. $\int_0^\infty x^{\alpha-1} e^{-px^r} \begin{Bmatrix} \sin bx^r \\ \cos bx^r \end{Bmatrix} Yi_\nu(cx)\,dx = W(1, 1),$

$$W(\gamma, 1) = -V_\nu(\gamma, 1) + \frac{(b^2+p^2)^{-\alpha/(2r)}}{\nu r} \Gamma\left(\frac{\alpha}{r}\right) \times$$

$$\times \cos\left(\frac{\delta\pi}{2} - \frac{\alpha}{r} \arccos \frac{p}{\sqrt{b^2+p^2}}\right) \cos^\nu \frac{\nu\pi}{2} \operatorname{cosec} \frac{\nu\pi}{2}$$

$[r > 1]$ or $[r = 1;\ |c|^2 < |b^2+p^2|]$,

$W(\gamma, 1) = V_\nu(\gamma, 1)$

$[r < 1]$ or $[r = 1;\ |c|^2 > |b^2+p^2|]$

$\left[b, c, r, \operatorname{Re} p > 0;\ \operatorname{Re}\alpha > |\operatorname{Re}\nu| - \delta r;\ \delta = \begin{Bmatrix} 1 \\ 0 \end{Bmatrix};\ V_\nu(\gamma, \varepsilon)\right.$ see [18], 2.13.1.8].

18. $\int_0^\infty x^{\alpha-1} e^{-px^r} \begin{Bmatrix} \sin bx^r \\ \cos bx^r \end{Bmatrix} Ki_\nu(cx)\,dx = -\frac{\pi}{2} W(0, 1)$

$\left[b, r, \operatorname{Re} c, \operatorname{Re} p > 0;\ \operatorname{Re}\alpha > |\operatorname{Re}\nu| - \delta r;\ \delta = \begin{Bmatrix} 1 \\ 0 \end{Bmatrix},\ W(\gamma, 1)\right.$ see 2.12.1.17.].

19. $\int_0^\infty x^{\alpha-1} \begin{Bmatrix} \ln(x^r + z^r) \\ \ln|x^r - z^r| \end{Bmatrix} Ji_\nu(cx)\,dx = R(1) + \frac{\pi z^\alpha}{\alpha\nu} \operatorname{cosec} \frac{\alpha\pi}{r}$

$\left[c, r, \operatorname{Re}\alpha, \operatorname{Re}(\alpha+\nu) > 0;\ \operatorname{Re}\alpha < 5/2,\ \begin{Bmatrix} r\ |\arg z| < \pi \\ z > 0 \end{Bmatrix},\ R(\varepsilon)\right.$ see [18], 2.12.1.9].

20. $\int_0^\infty x^{\alpha-1} \begin{Bmatrix} \ln(x^r + z^r) \\ \ln|x^r - z^r| \end{Bmatrix} Yi_\nu(cx)\,dx = W_\nu(1, 1) + \frac{\pi z^\alpha \operatorname{ctg}(\nu\pi/2)}{\alpha\nu \sin(\alpha\pi/r)}$

$\left[c, r > 0;\ |\operatorname{Re}\nu| < \operatorname{Re}\alpha < 5/2,\ \begin{Bmatrix} r\ |\arg z| < \pi \\ z > 0 \end{Bmatrix},\ W_\nu(\gamma, \varepsilon)\right.$ see [18], 2.13.1.9].

21. $\int_0^\infty x^{\alpha-1} \begin{Bmatrix} \ln(x^r + z^r) \\ \ln|x^r - z^r| \end{Bmatrix} Ki_\nu(cx)\,dx = -\frac{\pi}{2} W_\nu(0, 1) - \frac{\pi^2 z^\alpha}{2\alpha\nu} \operatorname{cosec} \frac{\alpha\pi}{r} \operatorname{cosec} \frac{\nu\pi}{2}$

$\left[r, \operatorname{Re} c > 0;\ \operatorname{Re}\alpha > |\operatorname{Re}\nu|,\ \begin{Bmatrix} r\ |\arg z| < \pi \\ z > 0 \end{Bmatrix},\ W_\nu(\gamma, \varepsilon)\right.$ see [18], 2.13.1.9].

22. $\int_0^\infty x^{\alpha-1}\operatorname{Ei}(-bx^r)\, Ji_\nu(cx)\,dx = U_4,$

$$U_4 = -X(1) - \frac{b^{-\alpha/r}}{\alpha\nu}\Gamma\left(\frac{\alpha}{r}\right) \qquad [r>1],$$

$$U_4 = X(1) \qquad [r<1],$$

U_4 for $r=1$ see 2.12.6.1

[c, r, $\operatorname{Re} b$, $\operatorname{Re}\alpha$, $\operatorname{Re}(\alpha+\nu)>0$, $X(\varepsilon)$ see [18], 2.12.1.12].

23. $\int_0^\infty x^{\alpha-1}\operatorname{Ei}(-bx^r)\begin{Bmatrix} Yi_\nu(cx) \\ Ki_\nu(cx)\end{Bmatrix} dx = U_5,$

$$U_5 = -\hat{U}_\nu(1) \mp \frac{b^{-\alpha/r}}{2\alpha\nu}\Gamma\left(\frac{\alpha}{r}\right)\begin{Bmatrix} 2\operatorname{ctg}(\nu\pi/2) \\ \pi\operatorname{cosec}(\nu\pi/2)\end{Bmatrix} \qquad [r>1],$$

$$U_5 = \hat{U}_\nu(1) \qquad [r<1],$$

U_5 for $r=1$ see 2.12.6.2

$\left[r,\ \operatorname{Re} b>0;\ \operatorname{Re}\alpha>|\operatorname{Re}\nu|,\ \begin{Bmatrix} c>0 \\ \operatorname{Re} c>0\end{Bmatrix},\ \hat{U}_\nu(\varepsilon) \text{ see [18], 2.13.1.12}\right].$

24. $\int_0^\infty x^{\alpha-1}e^{\pm bx^r}\operatorname{Ei}(\mp bx^r)\, Ji_\nu(cx)\,dx = U_6,$

$$U_6 = -Y(1) - \frac{\pi b^{-\alpha/r}}{\nu r}\Gamma\left(\frac{\alpha}{r}\right)\begin{Bmatrix} \operatorname{cosec}(\alpha\pi/r) \\ \operatorname{ctg}(\alpha\pi/r)\end{Bmatrix} \qquad [r>1],$$

$$U_6 = Y(1) \qquad [r<1].$$

U_6 for $r=1$ see 2.12.6.3

[c, r, $\operatorname{Re} b$, $\operatorname{Re}\alpha>0$; $-\operatorname{Re}\nu<\operatorname{Re}\alpha<r+5/2$; $Y(\varepsilon)$ see [18], 2.12.1.13].

25. $\int_0^\infty x^{\alpha-1}e^{\pm bx^r}\operatorname{Ei}(\mp bx^r)\, Yi_\nu(cx)\,dx = A(1, 1),$

$$A(\gamma, 1) = -\hat{U}_\nu(\gamma, 1) - \frac{\pi b^{-\alpha/r}}{\nu r}\Gamma\left(\frac{\alpha}{r}\right)\begin{Bmatrix} \operatorname{cosec}(\alpha\pi/r) \\ \operatorname{ctg}(\alpha\pi/r)\end{Bmatrix}\cos^\gamma\frac{\nu\pi}{2}\operatorname{cosec}\frac{\nu\pi}{2} \qquad [r>1],$$

$$A(\gamma, 1) = \hat{U}_\nu(\gamma, 1) \qquad [r<1],$$

$A(\gamma, 1)$ for $r=1$ see 2.12.6.4—5

[c, r, $\operatorname{Re} b>0$; $|\operatorname{Re}\nu|<\operatorname{Re}\alpha<r+5/2$; $\hat{U}_\nu(\gamma, \varepsilon)$ see [18], 2.13.1.13].

26. $\int_0^\infty x^{\alpha-1}e^{\pm bx^r}\operatorname{Ei}(\mp bx^r)\, Ki_\nu(cx)\,dx = -\frac{\pi}{2}A(0, 1)$

[r, $\operatorname{Re} b$, $\operatorname{Re} c>0$; $\operatorname{Re}\alpha>|\operatorname{Re}\nu|$; $A(\gamma, 1)$ see 2.12.1.25].

27. $\int_0^\infty x^{\alpha-1}\begin{Bmatrix} \operatorname{si}(bx^r) \\ \operatorname{ci}(bx^r)\end{Bmatrix} Ji_\nu(cx)\,dx = U_7,$

$$U_7 = -Z(1) - \frac{b^{-\alpha/r}}{\alpha\nu}\Gamma\left(\frac{\alpha}{r}\right)\begin{Bmatrix} \sin[\alpha\pi/(2r)] \\ \cos[\alpha\pi/(2r)]\end{Bmatrix} \qquad [r>1],$$

$$U_7 = Z(1) \qquad [r<1],$$

U_7 for $r=1$ see 2.12.7.1

[b, c, r, $\operatorname{Re}(\alpha+\nu)>0$; $0<\operatorname{Re}\alpha<3/2+r+\max(r, 1)$; $Z(\varepsilon)$ see [18], 2.12.1.14].

28. $\int_0^\infty x^{\alpha-1}\begin{Bmatrix} \operatorname{si}(bx^r) \\ \operatorname{ci}(bx^r)\end{Bmatrix} Yi_\nu(cx)\,dx = C(1, 1),$

$$C(\gamma,\ 1)=-\hat{V}_\nu(\gamma,\ 1)-\frac{b^{-\alpha/r}}{\alpha\nu}\Gamma\left(\frac{\alpha}{r}\right)\begin{Bmatrix}\sin[\alpha\pi/(2r)]\\ \cos[\alpha\pi/(2r)]\end{Bmatrix}\cos^\gamma\frac{\nu\pi}{2}\operatorname{cosec}\frac{\nu\pi}{2}\quad [r>1],$$

$$C(\gamma,\ 1)=\hat{V}_\nu(\gamma,\ 1)\quad [r<1],$$

$C(\gamma,\ 1)$ for $r=1$ see 2.12.7.2—3

$[b,\ c,\ r>0;\ |\operatorname{Re}\nu|<\operatorname{Re}\alpha<3/2+r+\max(r,\ 1);\ \hat{V}_\nu(\gamma,\ \varepsilon)$ see [18], 2.13 1.14].

29. $$\int_0^\infty x^{\alpha-1}\begin{Bmatrix}\operatorname{si}(bx^r)\\ \operatorname{ci}(bx^r)\end{Bmatrix}K i_\nu(cx)\,dx=-\frac{\pi}{2}C(0,\ 1)$$

$[b,\ r,\ \operatorname{Re}c>0;\ \operatorname{Re}\alpha>|\operatorname{Re}\nu|;\ C(\gamma,\ 1)$ see 2.12.1.28].

30. $$\int_0^\infty x^{\alpha-1}\begin{Bmatrix}\operatorname{erf}(bx^r)\\ \operatorname{erfc}(bx^r)\end{Bmatrix}J i_\nu(cx)\,dx=U_8,$$

$$U_8=-U(1)\mp\frac{b^{-\alpha/r}}{\alpha\nu\sqrt{\pi}}\Gamma\left(\frac{\alpha+r}{2r}\right)\quad [r>1/2],$$

$$U_8=U(1)\quad [r<1/2],$$

U_8 for $r=1/2$ see 2.12.8.4

$\left[c,\ r>0;\ \operatorname{Re}\alpha>-(1\pm 1)r/2;\ |\arg b|<\pi/4,\ \begin{Bmatrix}-r-\operatorname{Re}\nu<\operatorname{Re}\alpha<5/2\\ \operatorname{Re}(\alpha+\nu)>0\end{Bmatrix},\right.$ $U(\varepsilon)$ see [18], 2.12.1.15].

31. $$\int_0^\infty x^{\alpha-1}\begin{Bmatrix}\operatorname{erf}(bx^r)\\ \operatorname{erfc}(bx^r)\end{Bmatrix}Y i_\nu(cx)\,dx=D(1,\ 1),$$

$$D(\gamma,\ 1)=-\hat{W}_\nu(\gamma,\ 1)\mp\frac{b^{-\alpha/r}}{\alpha\nu\sqrt{\pi}}\Gamma\left(\frac{\alpha+r}{2r}\right)\cos^\gamma\frac{\nu\pi}{2}\operatorname{cosec}\frac{\nu\pi}{2}\quad [r>1/2],$$

$$D(\gamma,\ 1)=\hat{W}_\nu(\gamma,\ 1)\quad [r<1/2],$$

$D(\gamma,\ 1)$ for $r=1/2$ see 2.12.8.5—6

$\left[r>0;\ |\arg b|<\pi/4;\ \operatorname{Re}\alpha>|\operatorname{Re}\nu|-(1\pm 1)r/2,\ \begin{Bmatrix}\operatorname{Re}\alpha<5/2;\ c>0\\ c>0\end{Bmatrix},\ \hat{W}_\nu(\gamma,\ \varepsilon)\text{ see [18], 2.13.1.15}\right].$

32. $$\int_0^\infty x^{\alpha-1}\begin{Bmatrix}\operatorname{erf}(bx^r)\\ \operatorname{erfc}(bx^r)\end{Bmatrix}K i_\nu(cx)\,dx=-\frac{\pi}{2}D(0,\ 1)$$

$[r,\ \operatorname{Re}c>0;\ |\arg b|<\pi/4;\ \operatorname{Re}\alpha>|\operatorname{Re}\nu|-(1\pm 1)r/2;\ D(\gamma,\ 1)$ see 2.12.1.31].

33. $$\int_0^\infty x^{\alpha-1}e^{\mp b^2x^{2r}}\begin{Bmatrix}\operatorname{erf}(ibx^r)\\ \operatorname{erfc}(bx^r)\end{Bmatrix}J i_\nu(cx)\,dx=U_9,$$

$$U_9=-V(1)+\frac{\pi b^{-\alpha/r}}{2\nu}\begin{Bmatrix}i\sec[\alpha\pi/(2r)]\\ 2\operatorname{cosec}(\alpha\pi/r)\end{Bmatrix}\quad [r>1/2],$$

$$U_9=V(1)\quad [r<1/2],$$

U_9 for $r=1/2$ see 2.12.8.7

$[c,\ r>0;\ |\arg b|<\pi/4;\ \operatorname{Re}\alpha>-(1\pm 1)r/2;\ -\operatorname{Re}\nu-(1\pm 1)r/2<\operatorname{Re}\alpha<r+5/2;$ $V(\varepsilon)$ see [18], 2.12.1.16].

34. $$\int_0^\infty x^{\alpha-1}\begin{Bmatrix}S(bx^r)\\ C(bx^r)\end{Bmatrix}J i_\nu(cx)\,dx=U_{10},$$

$$U_{10}=-W(1)-\frac{b^{-\alpha/r}}{\alpha\nu\sqrt{2\pi}}\Gamma\left(\frac{r+2\alpha}{2r}\right)\begin{Bmatrix}\sin[(r+2\alpha)\pi/(4r)]\\ \cos[(r+2\alpha)\pi/(4r)]\end{Bmatrix}\quad [r>1],$$

$$U_{10}=W(1)$$

U_{10} for $r=1$ see 2.12.9.1 $[r<1]$.

$[b, c, r>0;\ \mathrm{Re}(\alpha+\nu)>-(2\pm1)r/2;\ -(2\pm1)r/2<\mathrm{Re}\,\alpha<5/2;\ W(\varepsilon)$ see [18], 2.12.1.17].

35. $\int_0^\infty x^{\alpha-1}\begin{Bmatrix}S(bx^r)\\ C(bx^r)\end{Bmatrix}Yi_\nu(cx)\,dx=E(1,1),$

$$E(\gamma,1)=-X_\nu(\gamma,1)-\frac{b^{-\alpha/r}}{\alpha\nu\sqrt{2\pi}}\Gamma\left(\frac{r+2\alpha}{2r}\right)\begin{Bmatrix}\sin[(r+2\alpha)\pi/(4r)]\\ \cos[(r+2\alpha)\pi/(4r)]\end{Bmatrix}\cos^\gamma\frac{\nu\pi}{2}\operatorname{cosec}\frac{\nu\pi}{2}\quad [r>1],$$

$$E(\gamma,1)=X_\nu(\gamma,1)\quad [r<1],$$

$E(\gamma,1)$ for $r=1$ see 2.12.9.2—3

$[b, c, r>0;\ |\mathrm{Re}\,\nu|-(2\pm1)r/2<\mathrm{Re}\,\alpha<5/2;\ X_\nu(\gamma,\varepsilon)$ see [18], 2.13.1.16].

36. $\int_0^\infty x^{\alpha-1}\begin{Bmatrix}S(bx^r)\\ C(bx^r)\end{Bmatrix}Ki_\nu(cx)\,dx=-\frac{\pi}{2}E(0,1)$

$[b, r, \mathrm{Re}\,c>0;\ \mathrm{Re}\,\alpha>|\mathrm{Re}\,\nu|-(2\pm1)r/2;\ E(\gamma,1)$ see 2.12.1.35].

37. $\int_0^\infty x^{\alpha-1}\begin{Bmatrix}\gamma(\mu,bx^r)\\ \Gamma(\mu,bx^r)\end{Bmatrix}Ji_\nu(cx)\,dx=U_{11},$

$$U_{11}=-X(1)\mp\frac{b^{-\alpha/r}}{\alpha\nu}\Gamma\left(\mu+\frac{\alpha}{r}\right)\quad [r>1],$$

$$U_{11}=X(1)\quad [r<1],$$

U_{11} for $r=1$ see 2.12.10.1

$\left[c, r, \mathrm{Re}\,b, \mathrm{Re}(\alpha+\mu r+\nu), \mathrm{Re}(\alpha+\mu r)>0, \begin{Bmatrix}\mathrm{Re}\,\mu>0;\ \mathrm{Re}\,\alpha<5/2\\ \mathrm{Re}(\alpha+\nu),\ \mathrm{Re}\,\alpha>0\end{Bmatrix}, X(\varepsilon)\text{ see [18], 2.12.1.18}\right].$

38. $\int_0^\infty x^{\alpha-1}\begin{Bmatrix}\gamma(\mu,bx^r)\\ \Gamma(\mu,bx^r)\end{Bmatrix}Yi_\nu(cx)\,dx=F(1,1),$

$$F(\gamma,1)=-Y_\nu(\gamma,1)\mp\frac{b^{-\alpha/r}}{\alpha\nu}\Gamma\left(\mu+\frac{\alpha}{r}\right)\cos^\nu\frac{\nu\pi}{2}\operatorname{cosec}\frac{\nu\pi}{2}\quad [r>1],$$

$$F(\gamma,1)=Y_\nu(\gamma,1)\quad [r<1],$$

$F(\gamma,1)$ for $r=1$ see 2.12.10.2—3

$\left[c, r, \mathrm{Re}\,b>0;\ \mathrm{Re}(\alpha+\mu r)>|\mathrm{Re}\,\nu|, \begin{Bmatrix}\mathrm{Re}\,\mu>0;\ \mathrm{Re}\,\alpha<5/2\\ \mathrm{Re}\,\alpha>|\mathrm{Re}\,\nu|\end{Bmatrix}, Y_\nu(\gamma,\varepsilon)\text{ see [18], 2.13.1.17}\right].$

39. $\int_0^\infty x^{\alpha-1}\begin{Bmatrix}\gamma(\mu,bx^r)\\ \Gamma(\mu,bx^r)\end{Bmatrix}Ki_\nu(cx)\,dx=-\frac{\pi}{2}F(0,1)$

$\left[r, \mathrm{Re}\,b, \mathrm{Re}\,c>0;\ \mathrm{Re}(\alpha+\mu r)>|\mathrm{Re}\,\nu|, \begin{Bmatrix}\mathrm{Re}\,\mu>0\\ \mathrm{Re}\,\alpha>|\mathrm{Re}\,\nu|\end{Bmatrix}, F(\gamma,1)\text{ see 2.12.1.38}\right]$

40. $\int_0^\infty x^{\alpha-1}e^{\pm b^2x^{2r}/4}D_\mu(bx^r)\,Ji_\nu(cx)\,dx=U_{12},$

$$U_{12}=-Y(1)+\frac{2^{-\alpha/(2r)\mp(\mu+1\pm1)/2}}{\nu b^{\alpha/r}r}\begin{Bmatrix}\Gamma^{-1}(-\mu)\\ \sqrt{\pi}\end{Bmatrix}\Gamma\left(\frac{\alpha}{r}\right)\Gamma^{\pm1}\left(\frac{1\mp1}{4}-\frac{\mu r\pm\alpha}{2r}\right)\quad [r>1/2],$$

$$U_{12}=Y(1) \qquad [r<1/2].$$

U_{12} for $r=1/2$ see 2.12.11.1

$\left[c,\ r,\ \operatorname{Re}\alpha>0;\ |\arg b|<(2\pm1)\pi/4,\ \left\{\begin{matrix}-\operatorname{Re}\nu<\operatorname{Re}\alpha<5/2-r\operatorname{Re}\mu\\ \operatorname{Re}(\alpha+\nu)>0\end{matrix}\right\},\ Y(\varepsilon)\text{ see [18], 2.12.1.19}\right].$

41. $$\int_0^\infty x^{\alpha-1}e^{\pm b^2x^{2r}/4}D_\mu(bx^r)\,Yi_\nu(cx)\,dx=G(1,\ 1),$$

$$G(\gamma,\ 1)=-Z_\nu(\gamma,\ 1)+\frac{2^{-\alpha/(2r)\mp(\mu+1\pm1)/2}}{\nu b^{\alpha/r}r}\left\{\begin{matrix}\Gamma^{-1}(-\mu)\\ \sqrt{\pi}\end{matrix}\right\}\times$$

$$\times\Gamma\left(\frac{\alpha}{r}\right)\Gamma^{\pm1}\left(\frac{1\mp1}{4}-\frac{\mu r\pm\alpha}{2r}\right)\frac{\cos^\gamma(\nu\pi/2)}{\sin(\nu\pi/2)} \qquad [r>1/2],$$

$$G(\gamma,\ 1)=Z_\nu(\gamma,\ 1) \qquad [r<1/2],$$

$G(\gamma,\ 1)$ for $r=1/2$ see 2.12.11.2—3

$\left[c,\ r>0;\ |\arg b|<(2\pm1)\pi/4,\ \left\{\begin{matrix}|\operatorname{Re}\nu|<\operatorname{Re}\alpha<5/2-r\operatorname{Re}\mu\\ |\operatorname{Re}\nu|<\operatorname{Re}\alpha\end{matrix}\right\},\ Z_\nu(\gamma,\ \varepsilon)\text{ see [18], 2.13.1.18}\right].$

42. $$\int_0^\infty x^{\alpha-1}e^{\pm b^2x^{2r}/4}D_\mu(bx^r)\,Ki_\nu(cx)\,dx=-\frac{\pi}{2}G(0,\ 1)$$

$[r,\ \operatorname{Re}c>0;\ \operatorname{Re}\alpha>|\operatorname{Re}\nu|;\ |\arg b|<(2\pm1)\pi/4;\ G(\gamma,\ 1)$ see 2.12.1.41].

43. $$\int_0^\infty x^{\alpha-1}J_\mu(bx^r)\,Ji_\nu(cx)\,dx=U_{13},\quad U_{13}=-S(1)+\frac{2^{\alpha/r-1}}{\nu b^{\alpha/r}r}\Gamma\left[\begin{matrix}(\mu r+\alpha)/(2r)\\ (\mu r-\alpha)/(2r)+1\end{matrix}\right] \qquad [r>1],$$

$$U_{13}=\frac{2^{\alpha+\mu r-\mu-1}b^\mu}{c^{\alpha+\mu r}\Gamma(\mu+1)}\times$$

$$\times\sum_{k=0}^\infty\frac{(-1)^k}{k!(\mu+1)_k(\alpha+\mu r+2rk)}\Gamma\left[\begin{matrix}(\alpha+\mu r+\nu)/2+rk\\ 1-(\alpha+\mu r-\nu)/2-rk\end{matrix}\right]\left(\frac{2^{r-1}b}{c^r}\right)^{2k} \qquad [r<1],$$

U_{13} for $r=1$ see 2.12.12.1

$[b,\ c,\ r,\ \operatorname{Re}(\alpha+\mu r+\nu),\ \operatorname{Re}(\alpha+\mu r)>0;\ \operatorname{Re}\alpha<(r+3)/2+\max(1,\ r);\ S(\varepsilon)$ see [18], 2.12.1.10].

44. $$\int_0^\infty x^{\alpha-1}J_\mu(bx^r)\left\{\begin{matrix}Yi_\nu(cx)\\ Ki_\nu(cx)\end{matrix}\right\}dx=W,$$

$$W=\mp\frac{1}{2}\left\{\begin{matrix}2\tilde U_\nu(1,\ 1)\\ \pi\tilde U_\nu(0,\ 1)\end{matrix}\right\}\pm\frac{2^{\alpha/r-2}}{\nu b^{\alpha/r}r}\Gamma\left[\begin{matrix}(\mu r+\alpha)/(2r)\\ (\mu r-\alpha)/(2r)+1\end{matrix}\right]\left\{\begin{matrix}2\operatorname{ctg}(\nu\pi/2)\\ \pi\operatorname{cosec}(\nu\pi/2)\end{matrix}\right\} \qquad [r>1],$$

$$W=\pm\frac{1}{2}\left\{\begin{matrix}2\tilde U_\nu(1,\ 1)\\ \pi\tilde U_\nu(0,\ 1)\end{matrix}\right\} \qquad [r<1],$$

W for $r=1$ see 2.12.12.2—3

$\left[b,\ r>0;\ \operatorname{Re}(\alpha+\mu r)>|\operatorname{Re}\nu|,\ \left\{\begin{matrix}\operatorname{Re}\alpha<(r+3)/2+\max(r,\ 1);\ c>0\\ \operatorname{Re}c>0\end{matrix}\right\},\ \tilde U_\nu(\gamma,\ \varepsilon)\text{ see [18], 2.13.1.10}\right].$

45. $$\int_0^\infty x^{\alpha-1}I_\mu(bx^r)\,Ki_\nu(cx)\,dx=-\frac{\pi}{2}U_\nu^+(0,\ 1)$$

$[r\leqslant1;\ |\arg b|<\pi;\ \operatorname{Re}c>0\ (\operatorname{Re}(c-b)>0$ for $r=1);\ \operatorname{Re}(\alpha+\mu r)>|\operatorname{Re}\nu|;$ $U_\nu^+(\gamma,\ \varepsilon)$ see [18], 2.13.1.10].

46. $\int_0^\infty x^{\alpha-1} J_\mu\left(\frac{b}{x^r}\right) Ji_\nu(cx)\,dx = -T(1) + \frac{b^{\alpha/r}}{2^{\alpha/r+1}\nu r}\Gamma\left[\begin{matrix}(\mu r-\alpha)/(2r)\\(\mu r+\alpha)/(2r)+1\end{matrix}\right]$

$[b, c, r>0;\ \operatorname{Re}\alpha > -3r/2;\ -3r/2-\operatorname{Re}\nu < \operatorname{Re}\alpha < 5/2+r\operatorname{Re}\mu;\ T(\varepsilon)$ see [18], 2.12.1.11].

47. $\int_0^\infty x^{\alpha-1}\begin{Bmatrix}Y_\mu(bx^r)\\K_\mu(bx^r)\end{Bmatrix} Yi_\nu(cx)\,dx = H(1, 1),$

$H(\gamma, 1) = -U_{\mu,\nu}(\gamma, 1) \mp$

$\mp\frac{2^{\alpha/r-1}\cos^\nu(\nu\pi/2)}{2\nu\pi b^{\alpha/r} r\sin(\nu\pi/2)}\Gamma\left(\frac{\alpha+\mu r}{2r}\right)\Gamma\left(\frac{\alpha-\mu r}{2r}\right)\begin{Bmatrix}2\cos[(\alpha-\mu r)\pi/(2r)]\\\pi\end{Bmatrix}$ $[r>1]$,

$H(\gamma, 1) = U_{\mu,\nu}(\gamma, 1)$ $[r<1]$,

$H(\gamma, 1)$ for $r=1$ see 2.12.12.4.

$\left[c, r>0;\ \operatorname{Re}\alpha > r|\operatorname{Re}\mu|+|\operatorname{Re}\nu|,\ \begin{Bmatrix}b>0;\ \operatorname{Re}\alpha<(r+3)/2+\max(r, 1)\\\operatorname{Re}b>0\end{Bmatrix},\right.$

$\left. U_{\mu,\nu}(\gamma, \varepsilon) \text{ see [18], 2.13.1.11}\right].$

48. $\int_0^\infty x^{\alpha-1}\begin{Bmatrix}Y_\mu(bx^r)\\K_\mu(bx^r)\end{Bmatrix} Ki_\nu(cx)\,dx = -\frac{\pi}{2}H(0, 1)$

$\left[r, \operatorname{Re}c>0;\ \operatorname{Re}\alpha > r|\operatorname{Re}\mu|+\operatorname{Re}\nu,\ \begin{Bmatrix}b>0\\\operatorname{Re}b>0\end{Bmatrix},\ H(\gamma, 1) \text{ see 2.12.1.47}\right].$

2.12.2. Integrals of $A(x)\,Ji_\nu(cx)$ and $A(x)\begin{Bmatrix}Yi_\nu(cx)\\Ki_\nu(cx)\end{Bmatrix}$.

1. $\int_0^\infty x^{\alpha-1} Ji_\nu(cx)\,dx = \frac{2^{\alpha-1}}{\alpha c^\alpha}\Gamma\left[\begin{matrix}(\alpha+\nu)/2\\1-(\alpha-\nu)/2\end{matrix}\right]$ $[c, \operatorname{Re}\alpha>0;\ -\operatorname{Re}\nu<\operatorname{Re}\alpha<5/2]$.

2. $\int_0^\infty x^{\alpha-1}\begin{Bmatrix}Yi_\nu(cx)\\Ki_\nu(cx)\end{Bmatrix}dx = \mp\frac{2^{\alpha-2}}{\alpha c^\alpha}\Gamma\left[\begin{matrix}(\alpha+\nu)/2\\1-(\alpha-\nu)/2\end{matrix}\right]\begin{Bmatrix}2\operatorname{ctg}[(\alpha-\nu)\pi/2]\\\pi\operatorname{cosec}[(\alpha-\nu)\pi/2]\end{Bmatrix}$

$\left[\operatorname{Re}\alpha > |\operatorname{Re}\nu|,\ \begin{Bmatrix}c>0;\ \operatorname{Re}\alpha<5/2\\\operatorname{Re}c>0\end{Bmatrix}\right].$

3. $\int_0^a x^{\alpha-1}(a^2-x^2)^{\beta-1} Ji_\nu(cx)\,dx = V_-(\nu) + \frac{a^{\alpha+2\beta-2}}{2\nu}\mathrm{B}\left(\frac{\alpha}{2}, \beta\right)$

$[a, \operatorname{Re}\alpha, \operatorname{Re}\beta, \operatorname{Re}(\alpha+\nu)>0]$,

$V_\mp(\nu) =$

$= -\frac{a^{\alpha+\nu+2\beta-2}c^\nu}{2^{\nu+1}\nu\Gamma(\nu+1)}\mathrm{B}\left(\frac{\alpha+\nu}{2}, \beta\right){}_2F_3\left(\frac{\alpha+\nu}{2}, \frac{\nu}{2};\ \nu+1, \frac{\nu}{2}+1, \frac{\alpha+\nu}{2}+\beta;\ \mp\frac{a^2c^2}{4}\right).$

4. $\int_0^a x^{\alpha-1}(a^2-x^2)^{\beta-1}\begin{Bmatrix}Yi_\nu(cx)\\Ki_\nu(cx)\end{Bmatrix}dx =$

$= \pm\frac{1}{\sin\nu\pi}\left[\begin{Bmatrix}\cos\nu\pi\\\pi/2\end{Bmatrix}V_\mp(\nu) - \begin{Bmatrix}1\\\pi/2\end{Bmatrix}V_\mp(-\nu)\right] \pm \frac{a^{\alpha+2\beta-2}}{4\nu}\begin{Bmatrix}2\operatorname{ctg}(\nu\pi/2)\\\pi\operatorname{cosec}(\nu\pi/2)\end{Bmatrix}\mathrm{B}\left(\frac{\alpha}{2}, \beta\right)$

$[a, \operatorname{Re}\beta>0;\ \operatorname{Re}\alpha>|\operatorname{Re}\nu|;\ V_\mp(\nu)$ see 2.12.2.3].

5. $\int_a^\infty x^{\alpha-1}(x^2-a^2)^{\beta-1} Ji_\nu(cx)\,dx = U_-(\nu) + V_- + \frac{a^{\alpha+2\beta-2}}{2\nu}B\left(1-\frac{\alpha}{2}-\beta, \beta\right)$

$[a, c, \operatorname{Re}\beta>0;\ \operatorname{Re}(\alpha+2\beta)<9/2]$,

$$U_{\mp}(\nu)=-\frac{a^{\alpha+2\beta+\nu-2}c^{\nu}}{2^{\nu+1}\nu\Gamma(\nu+1)}\mathrm{B}\left(1-\beta-\frac{\alpha+\nu}{2},\ \beta\right)\times$$

$$\times{}_2F_3((\alpha+\nu)/2,\ \nu/2;\ \beta+(\alpha+\nu)/2,\ \nu/2+1,\ \nu+1;\ \mp a^2c^2/4),$$

$$V_{\mp}=\frac{c^{2-\alpha-2\beta}}{2^{3-\alpha-2\beta}(\alpha+2\beta-2)}\Gamma\left[\begin{matrix}(\alpha+\nu)/2+\beta-1\\ 2-\beta+(\nu-\alpha)/2\end{matrix}\right]\times$$

$$\times{}_2F_3\left(1-\beta,\ 1-\beta-\frac{\alpha}{2};\ 2-\beta-\frac{\alpha}{2},\ 2-\beta-\frac{\alpha+\nu}{2},\ 2-\beta-\frac{\alpha-\nu}{2};\ \mp\frac{a^2c^2}{4}\right).$$

6. $$\int_a^\infty x^{\alpha-1}(x^2-a^2)^{\beta-1}\begin{Bmatrix}Yi_\nu(cx)\\ Ki_\nu(cx)\end{Bmatrix}dx=$$

$$=\pm\frac{1}{\sin\nu\pi}\left[\begin{Bmatrix}\cos\nu\pi\\ \pi/2\end{Bmatrix}U_{\mp}(\nu)-\begin{Bmatrix}1\\ \pi/2\end{Bmatrix}U_{\mp}(-\nu)\right]-\frac{1}{2}\begin{Bmatrix}2\operatorname{ctg}[(\alpha+2\beta-\nu)\pi/2]\\ \pi\operatorname{cosec}[(\alpha+2\beta-\nu)\pi/2]\end{Bmatrix}V_{\mp}\pm$$

$$\pm\frac{a^{\alpha+2\beta-2}}{4\nu}\begin{Bmatrix}2\operatorname{ctg}(\nu\pi/2)\\ \pi\operatorname{cosec}(\nu\pi/2)\end{Bmatrix}\mathrm{B}\left(1-\frac{\alpha}{2}-\beta,\ \beta\right)$$

$\left[a,\ \operatorname{Re}\beta>0,\ \begin{Bmatrix}c>0;\ \operatorname{Re}(\alpha+2\beta)<9/2\\ \operatorname{Re}c>0\end{Bmatrix},\ U_{\mp}(\nu),\ V_{\mp}\ \text{see } 2.12.2.5\right].$

7. $$\int_0^\infty\frac{x^{\alpha-1}}{(x^2+z^2)^\rho}Ji_\nu(cx)\,dx=W_+(\nu)+X_++\frac{z^{\alpha-2\rho}}{2\nu}\mathrm{B}\left(\frac{\alpha}{2},\ \rho-\frac{\alpha}{2}\right)$$

$[c,\ \operatorname{Re}z,\ \operatorname{Re}\alpha>0;\ -\operatorname{Re}\nu<\operatorname{Re}\alpha<2\operatorname{Re}\rho+5/2],$

$$W_\pm(\nu)=-\frac{c^\nu z^{\alpha+\nu-2\rho}}{2^{\nu+1}\nu\Gamma(\nu+1)}\mathrm{B}\left(\frac{\alpha+\nu}{2},\ \rho-\frac{\alpha+\nu}{2}\right)\times$$

$$\times{}_2F_3\left(\frac{\alpha+\nu}{2},\ \frac{\nu}{2};\ \frac{\nu}{2}+1,\ \nu+1,\ 1-\rho+\frac{\alpha+\nu}{2};\ \pm\frac{c^2z^2}{4}\right),$$

$$X_\pm=\frac{c^{2\rho-\alpha}}{2^{1+2\rho-\alpha}(\alpha-2\rho)}\Gamma\left[\begin{matrix}(\alpha+\nu)/2-\rho\\ 1+\rho+(\nu-\alpha)/2\end{matrix}\right]\times$$

$$\times{}_2F_3\left(\rho-\frac{\alpha}{2},\ \rho;\ 1+\rho-\frac{\alpha+\nu}{2},\ 1+\rho-\frac{\alpha-\nu}{2},\ 1+\rho-\frac{\alpha}{2};\ \pm\frac{c^2z^2}{4}\right).$$

8. $$\int_0^\infty\frac{x^{\alpha-1}}{(x^2+z^2)^\rho}\begin{Bmatrix}Yi_\nu(cx)\\ Ki_\nu(cx)\end{Bmatrix}dx=\pm\frac{1}{\sin\nu\pi}\left[\begin{Bmatrix}\cos\nu\pi\\ \pi/2\end{Bmatrix}W_\pm(\nu)-\begin{Bmatrix}1\\ \pi/2\end{Bmatrix}W_\pm(-\nu)\right]\pm$$

$$\pm\frac{1}{2}\begin{Bmatrix}2\operatorname{ctg}[(\nu-\alpha+2\rho)\pi/2]\\ \pi\operatorname{cosec}[(\nu-\alpha+2\rho)\pi/2]\end{Bmatrix}X_\pm\pm\frac{z^{\alpha-2\rho}}{4\nu}\begin{Bmatrix}2\operatorname{ctg}(\nu\pi/2)\\ \pi\operatorname{cosec}(\nu\pi/2)\end{Bmatrix}\mathrm{B}\left(\frac{\alpha}{2},\ \rho-\frac{\alpha}{2}\right)$$

$\left[\operatorname{Re}z>0;\ \operatorname{Re}\alpha>|\operatorname{Re}\nu|,\ \begin{Bmatrix}c>0;\ \operatorname{Re}(\alpha-2\rho)<5/2\\ \operatorname{Re}c>0\end{Bmatrix},\ W_\pm(\nu),\ X_\pm\ \text{see } 2.12.2.7\right].$

9. $$\int_0^\infty\frac{x^{\alpha-1}}{x^2-y^2}Ji_\nu(cx)\,dx=Y_-(\nu)+Z_--\frac{\pi y^{\alpha-2}}{2\nu}\operatorname{ctg}\frac{\alpha\pi}{2}$$ $[c,\ y,\ \operatorname{Re}\alpha>0;\ -\operatorname{Re}\nu<\operatorname{Re}\alpha<9/2],$

$$Y_{\mp}(\nu)=\frac{\pi c^\nu y^{\alpha+\nu-2}}{2^{\nu+1}\nu\Gamma(\nu+1)}\operatorname{ctg}\frac{\alpha+\nu}{2}\pi\,{}_1F_2\left(\frac{\nu}{2};\ \frac{\nu}{2}+1,\ \nu+1;\ \mp\frac{c^2y^2}{4}\right),$$

$$Z_{\mp}=\frac{2^{\alpha-3}}{(\alpha-2)c^{\alpha-2}}\Gamma\left[\begin{matrix}(\alpha+\nu)/2-1\\ 2+(\nu-\alpha)/2\end{matrix}\right]{}_2F_3\left(1-\frac{\alpha}{2},\ 1;\ 2-\frac{\alpha+\nu}{2},\ 2-\frac{\alpha-\nu}{2},\ 2-\frac{\alpha}{2};\ \mp\frac{c^2y^2}{4}\right).$$

10. $$\int_0^\infty\frac{x^{\alpha-1}}{x^2-y^2}\begin{Bmatrix}Yi_\nu(cx)\\ Ki_\nu(cx)\end{Bmatrix}dx=\pm\frac{1}{\sin\nu\pi}\left[\begin{Bmatrix}\cos\nu\pi\\ \pi/2\end{Bmatrix}Y_{\mp}(\nu)-\begin{Bmatrix}1\\ \pi/2\end{Bmatrix}Y_{\mp}(-\nu)\right]-$$

$$-\frac{1}{2}\begin{Bmatrix}2\operatorname{ctg}[(\alpha-\nu)\pi/2]\\ \pi\operatorname{cosec}[(\alpha-\nu)\pi/2]\end{Bmatrix}Z_{\mp}\mp\frac{\pi y^{\alpha-2}}{4\nu}\operatorname{ctg}\frac{\alpha\pi}{2}\begin{Bmatrix}2\operatorname{ctg}(\nu\pi/2)\\ \pi\operatorname{cosec}(\nu\pi/2)\end{Bmatrix}$$

$\left[y>0;\ \operatorname{Re}\alpha>|\operatorname{Re}\nu|,\ \begin{Bmatrix}c>0;\ \operatorname{Re}\alpha<9/2\\ \operatorname{Re}c>0\end{Bmatrix},\ Y_{\mp}(\nu),\ Z_{\mp}\ \text{see } 2.12.2.9\right].$

2.12.3. Integrals of $x^{\alpha}e^{-px^{\pm n}}Ji_{\nu}(cx)$ and $x^{\alpha}e^{-px^{\pm n}}\begin{Bmatrix}Yi_{\nu}(cx)\\ Ki_{\nu}(cx)\end{Bmatrix}$.

1. $$\int_0^{\infty} e^{-px}Ji_0(cx)\,dx=\frac{1}{p}\ln\frac{p+\sqrt{p^2+c^2}}{c}\qquad [c,\ \operatorname{Re} p>0].$$

2. $$\int_0^{\infty} e^{-px}Ji_{\nu}(cx)\,dx=\frac{1}{\nu p}\left[1-\left(\frac{c}{p+\sqrt{p^2+c^2}}\right)^{\nu}\right]\qquad [c,\ \operatorname{Re} p>0;\ \operatorname{Re}\nu>-1;\ \nu\neq 0].$$

3. $$\int_0^{\infty} e^{-px}\begin{Bmatrix}Yi_0(cx)\\ Ki_0(cx)\end{Bmatrix}dx=\mp\frac{1}{p\pi}\begin{Bmatrix}1\\ \pi/2\end{Bmatrix}\ln^2\frac{p+\sqrt{p^2\pm c^2}}{c}+\frac{\pi^2}{8p}\begin{Bmatrix}0\\ 1\end{Bmatrix}$$
$$\left[\operatorname{Re} p>0,\ \begin{Bmatrix}c>0\\ \operatorname{Re} c>0\end{Bmatrix}\right]$$

4. $$\int_0^{\infty} e^{-px}\begin{Bmatrix}Yi_{\nu}(cx)\\ Ki_{\nu}(cx)\end{Bmatrix}dx=\mp\frac{\operatorname{cosec}\nu\pi}{\nu p}\begin{Bmatrix}1\\ \pi/2\end{Bmatrix}\left[\left(\frac{p+\sqrt{p^2\pm c^2}}{c}\right)^{\nu}+\right.$$
$$\left.+\begin{Bmatrix}\cos\nu\pi\\ 1\end{Bmatrix}\left(\frac{p+\sqrt{p^2\pm c^2}}{c}\right)^{-\nu}-2\cos^{(3\pm 1)/2}\frac{\nu\pi}{2}\right]$$
$$\left[\operatorname{Re} p>0;\ |\operatorname{Re}\nu|<1,\ \begin{Bmatrix}c>0\\ \operatorname{Re} c>0\end{Bmatrix}\right].$$

5. $$\int_0^{\infty} x^{\alpha-1}e^{-px}Ji_{\nu}(cx)\,dx=U_-(\nu)+\frac{\Gamma(\alpha)}{\nu p^{\alpha}}\qquad [c,\ \operatorname{Re} p,\ \operatorname{Re}\alpha,\ \operatorname{Re}(\alpha+\nu)>0],$$
$$U_{\mp}(\nu)=-\frac{c^{\nu}}{2^{\nu}\nu p^{\alpha+\nu}}\Gamma\begin{bmatrix}\alpha+\nu\\ \nu+1\end{bmatrix}{}_3F_2\left(\frac{\alpha+\nu}{2},\frac{\alpha+\nu+1}{2},\frac{\nu}{2};\frac{\nu}{2}+1,\nu+1;\mp\frac{c^2}{p^2}\right).$$

6. $$\int_0^{\infty} x^{\alpha-1}e^{-px}\begin{Bmatrix}Yi_{\nu}(cx)\\ Ki_{\nu}(cx)\end{Bmatrix}dx=$$
$$=\pm\frac{1}{\sin\nu\pi}\left[\begin{Bmatrix}\cos\nu\pi\\ \pi/2\end{Bmatrix}U_{\mp}(\nu)-\begin{Bmatrix}1\\ \pi/2\end{Bmatrix}U_{\mp}(-\nu)\right]\pm\frac{\Gamma(\alpha)}{2\nu p^{\alpha}}\begin{Bmatrix}2\operatorname{ctg}(\nu\pi/2)\\ \pi\operatorname{cosec}(\nu\pi/2)\end{Bmatrix}$$
$$\left[\operatorname{Re} p>0;\ \operatorname{Re}\alpha>|\operatorname{Re}\nu|,\ \begin{Bmatrix}c>0\\ \operatorname{Re} c>0\end{Bmatrix},\ U_{\mp}(\nu)\ \text{see 2.12.3.5}\right].$$

7. $$\int_0^{\infty} x^{\alpha-1}e^{-px^2}Ji_{\nu}(cx)\,dx=V_-(\nu)+\frac{\Gamma(\alpha/2)}{2\nu p^{\alpha/2}}\qquad [\operatorname{Re} p,\ \operatorname{Re}\alpha,\ \operatorname{Re}(\alpha+\nu)>0;\ |\arg c|<\pi],$$
$$V_{\mp}(\nu)=-\frac{c^{\nu}}{2^{\nu+1}\nu p^{(\alpha+\nu)/2}}\Gamma\begin{bmatrix}(\alpha+\nu)/2\\ \nu+1\end{bmatrix}{}_2F_2\left(\frac{\alpha+\nu}{2},\frac{\nu}{2};\frac{\nu}{2}+1,\nu+1;\mp\frac{c^2}{4p}\right).$$

8. $$\int_0^{\infty} xe^{-px^2}Ji_0(cx)\,dx=\frac{1}{4p}\operatorname{Ei}\left(-\frac{c^2}{4p}\right)\qquad [\operatorname{Re} p>0;\ |\arg c|<\pi].$$

9. $$\int_0^{\infty} x^{\alpha-1}e^{-px^2}\begin{Bmatrix}Yi_{\nu}(cx)\\ Ki_{\nu}(cx)\end{Bmatrix}dx=$$
$$=\pm\frac{1}{\sin\nu\pi}\left[\begin{Bmatrix}\cos\nu\pi\\ \pi/2\end{Bmatrix}V_{\mp}(\nu)-\begin{Bmatrix}1\\ \pi/2\end{Bmatrix}V_{\mp}(-\nu)\right]\pm\frac{\Gamma(\alpha/2)}{4\nu p^{\alpha/2}}\begin{Bmatrix}2\operatorname{ctg}(\nu\pi/2)\\ \pi\operatorname{cosec}(\nu\pi/2)\end{Bmatrix}$$
$$\left[\operatorname{Re} p>0;\ \operatorname{Re}\alpha>|\operatorname{Re}\nu|,\ \begin{Bmatrix}c>0\\ \operatorname{Re} c>0\end{Bmatrix},\ V_{\mp}(\nu)\ \text{see 2.12.3.7}\right].$$

10. $\int\limits_0^\infty x^{\alpha-1}e^{-p/x^2}J i_\nu(cx)\,dx=W_+(\nu)+V_++\frac{p^{\alpha/2}}{2\nu}\Gamma\left(-\frac{\alpha}{2}\right)$ [c, $\operatorname{Re} p>0$; $\operatorname{Re}\alpha<5/2$]

$$W_\pm(\nu)=-\frac{c^\nu p^{(\alpha+\nu)/2}}{2^{\nu+1}\nu}\Gamma\begin{bmatrix}-(\alpha+\nu)/2\\ \nu+1\end{bmatrix}{}_1F_3\left(\frac{\nu}{2};\ 1+\frac{\alpha+\nu}{2},\ \nu+1,\ \frac{\nu}{2}+1;\ \pm\frac{c^2p}{4}\right),$$

$$V_\pm=\frac{2^{\alpha-1}}{\alpha c^\alpha}\Gamma\begin{bmatrix}(\alpha+\nu)/2\\ (\nu-\alpha)/2+1\end{bmatrix}{}_1F_3\left(-\frac{\alpha}{2};\ 1-\frac{\alpha}{2},\ 1-\frac{\alpha+\nu}{2},\ 1-\frac{\alpha-\nu}{2};\ \pm\frac{c^2p}{4}\right).$$

11. $\int\limits_0^\infty x^{\alpha-1}e^{-p/x^2}\begin{Bmatrix}Y i_\nu(cx)\\ K i_\nu(cx)\end{Bmatrix}dx=\pm\frac{1}{\sin\nu\pi}\left[\begin{Bmatrix}\cos\nu\pi\\ \pi/2\end{Bmatrix}W_\pm(\nu)-\begin{Bmatrix}1\\ \pi/2\end{Bmatrix}W_\pm(-\nu)\right]\pm$

$$\pm\frac{p^{\alpha/2}}{4\nu}\Gamma\left(-\frac{\alpha}{2}\right)\begin{Bmatrix}2\operatorname{ctg}(\nu\pi/2)\\ \pi\operatorname{cosec}(\nu\pi/2)\end{Bmatrix}\pm\frac{1}{2}\begin{Bmatrix}2\operatorname{ctg}[(\alpha-\nu)\pi/2]\\ \pi\operatorname{cosec}[(\alpha-\nu)\pi/2]\end{Bmatrix}V_\pm$$

$\left[\operatorname{Re} p>0,\ \begin{Bmatrix}c>0;\ \operatorname{Re}\alpha<5/2\\ \operatorname{Re} c>0\end{Bmatrix},\ W_\pm(\nu),\ V_\pm\right.$ see 2.12.3.10$\left.\right]$.

2.12.4. Integrals of $x^\alpha\begin{Bmatrix}\sin bx\\ \cos bx\end{Bmatrix}J i_\nu(cx)$ and $x^\alpha\begin{Bmatrix}\sin bx\\ \cos bx\end{Bmatrix}\begin{Bmatrix}Y i_\nu(cx)\\ K i_\nu(cx)\end{Bmatrix}$.

NOTATION: $\delta=\begin{Bmatrix}1\\ 0\end{Bmatrix}$.

1. $\int\limits_0^\infty x^{\alpha-1}\begin{Bmatrix}\sin bx\\ \cos bx\end{Bmatrix}J i_\nu(cx)\,dx=U$

[$b,\ c>0$; $\operatorname{Re}(\alpha+\nu)>-\delta$; $-\delta<\operatorname{Re}\alpha<5/2$ for $b\neq c$; $-\delta<\operatorname{Re}\alpha<3/2$ for $b=c$],

$$U=U(\nu)+\frac{\Gamma(\alpha)}{\nu b^\alpha}\begin{Bmatrix}\sin(\alpha\pi/2)\\ \cos(\alpha\pi/2)\end{Bmatrix}\qquad [0<c\leqslant b],$$

$$U=\Gamma\begin{bmatrix}(\nu+\alpha+\delta)/2\\ (\nu-\alpha-\delta)/2+1\end{bmatrix}V_0\qquad [0<b\leqslant c],$$

$$U(\nu)=-\frac{c^\nu}{2^\nu\nu b^{\alpha+\nu}}\Gamma\begin{bmatrix}\alpha+\nu\\ \nu+1\end{bmatrix}\begin{Bmatrix}\sin[(\alpha+\nu)\pi/2]\\ \cos[(\alpha+\nu)\pi/2]\end{Bmatrix}\times$$

$$\times{}_3F_2\left(\frac{\alpha+\nu}{2},\ \frac{\alpha+\nu+1}{2},\ \frac{\nu}{2};\ \frac{\nu}{2}+1,\ \nu+1;\ \frac{c^2}{b^2}\right),$$

$$V_\gamma=\frac{2^{\alpha+\delta-1}b^\delta}{(\alpha+\delta)c^{\alpha+\delta}}\,{}_3F_2\left(\frac{\alpha+\delta}{2},\ \frac{\alpha+\delta+\nu}{2},\ \frac{\alpha+\delta-\nu}{2};\ \delta+\frac{1}{2},\ \frac{\alpha+\delta}{2}+1;\ \frac{(-1)^\gamma b^2}{c^2}\right).$$

2. $\int\limits_0^\infty x^{\alpha-1}\begin{Bmatrix}\sin bx\\ \cos bx\end{Bmatrix}Y i_\nu(cx)\,dx=W$

[$b,\ c>0$; $\operatorname{Re}\alpha>|\operatorname{Re}\nu|-\delta$; $\operatorname{Re}\alpha<5/2$ for $b\neq c$; $\operatorname{Re}\alpha<3/2$ for $b=c$],

$$W=\frac{1}{\sin\nu\pi}[\cos\nu\pi\,U(\nu)-U(-\nu)]+\frac{\Gamma(\alpha)}{\nu b^\alpha}\operatorname{ctg}\frac{\nu\pi}{2}\begin{Bmatrix}\sin(\alpha\pi/2)\\ \cos(\alpha\pi/2)\end{Bmatrix}\qquad [0<c\leqslant b],$$

$$W=-\frac{1}{\pi}\Gamma\left(\frac{\alpha+\nu+\delta}{2}\right)\Gamma\left(\frac{\alpha-\nu+\delta}{2}\right)\begin{Bmatrix}\sin[(\nu-\alpha)\pi/2]\\ \cos[(\nu-\alpha)\pi/2]\end{Bmatrix}V_0$$

[$0<b\leqslant c$], [$U(\nu)$, V_γ see 2.12.4.1].

3. $\int\limits_0^\infty x^{\alpha-1}\begin{Bmatrix}\sin bx\\ \cos bx\end{Bmatrix}K i_\nu(cx)\,dx=\frac{1}{2}\Gamma\left(\frac{\alpha+\nu+\delta}{2}\right)\Gamma\left(\frac{\alpha-\nu+\delta}{2}\right)V_1$

[b, $\operatorname{Re} c>0$; $\operatorname{Re}\alpha>|\operatorname{Re}\nu|-\delta$; V_γ see 2.12.4.1].

2.12.5. Integrals of $x^\alpha\begin{Bmatrix}\ln(x^2+z^2)\\ \ln|x^2-z^2|\end{Bmatrix}J i_\nu(cx)$ and $x^\alpha\begin{Bmatrix}\ln(x^2+z^2)\\ \ln|x^2-z^2|\end{Bmatrix}\begin{Bmatrix}Y i_\nu(cx)\\ K i_\nu(cx)\end{Bmatrix}$.

1. $$\int_0^\infty x^{\alpha-1} \begin{Bmatrix} \ln(x^2+z^2) \\ \ln|x^2-z^2| \end{Bmatrix} J i_\nu(cx)\,dx=$$

$$=U_1(\nu)+V_1+\frac{2^{\alpha-1}}{\alpha c^\alpha}\Gamma\begin{bmatrix}(\alpha+\nu)/2\\(\nu-\alpha)/2+1\end{bmatrix}\left[\psi\left(\frac{\alpha+\nu}{2}\right)+\psi\left(\frac{\nu-\alpha}{2}+1\right)-\frac{2}{\alpha}-2\ln\frac{c}{2}\right]$$

$$\left[c,\ \mathrm{Re}\,\alpha,\ \mathrm{Re}(\alpha+\nu)>0;\ \mathrm{Re}\,\alpha<5/2,\ \begin{Bmatrix}\mathrm{Re}\,z>0\\z>0\end{Bmatrix}\right],$$

$$U_\gamma(\nu)=-\frac{\pi c^\nu z^{\alpha+\nu}}{2^\nu(\alpha+\nu)\nu\Gamma(\nu+1)}\begin{Bmatrix}\operatorname{cosec}[(\alpha+\nu)\pi/2]\\\operatorname{ctg}[(\alpha+\nu)\pi/2]\end{Bmatrix}\times$$

$$\times{}_2F_3\left(\frac{\alpha+\nu}{2},\ \frac{\nu}{2};\ \frac{\alpha+\nu}{2}+1,\ \frac{\nu}{2}+1,\ \nu+1;\ \mp\frac{(-1)^\gamma c^2z^2}{4}\right),$$

$$V_\gamma=\pm\frac{2^{\alpha-3}z^2}{(\alpha-2)c^{\alpha-2}}\Gamma\begin{bmatrix}(\alpha+\nu)/2-1\\-(\alpha-\nu)/2+2\end{bmatrix}\times$$

$$\times{}_3F_4\left(1-\frac{\alpha}{2},\ 1,\ 1;\ 2,\ 2-\frac{\alpha}{2},\ 2-\frac{\alpha+\nu}{2},\ 2-\frac{\alpha-\nu}{2};\ \mp\frac{(-1)^\gamma c^2z^2}{4}\right).$$

2. $$\int_0^\infty x^{\alpha-1}\begin{Bmatrix} \ln(x^2+z^2) \\ \ln|x^2-z^2| \end{Bmatrix} Y i_\nu(cx)\,dx=\frac{1}{\sin\nu\pi}[\cos\nu\pi\,U_1(\nu)-U_1(-\nu)]-$$

$$-V_1\operatorname{ctg}\frac{\alpha-\nu}{2}\pi-\frac{2^{\alpha-1}}{\alpha\pi c^\alpha}\Gamma\left(\frac{\alpha+\nu}{2}\right)\Gamma\left(\frac{\alpha-\nu}{2}\right)\cos\frac{\alpha-\nu}{2}\pi\left[\psi\left(\frac{\alpha+\nu}{2}\right)+\psi\left(\frac{\alpha-\nu}{2}\right)-\right.$$

$$\left.-\frac{2}{\alpha}-\pi\operatorname{tg}\frac{\alpha-\nu}{2}\pi-2\ln\frac{c}{2}\right]+\frac{\pi z^\alpha\operatorname{ctg}(\nu\pi/2)}{\alpha\nu\sin(\alpha\pi/2)}$$

$$\left[c>0;\ |\mathrm{Re}\,\nu|<\mathrm{Re}\,\alpha<5/2,\ \begin{Bmatrix}\mathrm{Re}\,z>0\\z>0\end{Bmatrix},\ U_\gamma(\nu),\ V_\gamma\ \text{ see } 2.12.5.1\right].$$

3. $$\int_0^\infty x^{\alpha-1}\begin{Bmatrix} \ln(x^2+z^2) \\ \ln|x^2-z^2| \end{Bmatrix} K i_\nu(cx)\,dx=-\frac{\pi}{2}\operatorname{cosec}\nu\pi\,[U_0(\nu)-U_0(-\nu)]-$$

$$-\frac{\pi}{2}V_0\operatorname{cosec}\frac{\alpha-\nu}{2}\pi+\frac{2^{\alpha-2}}{\alpha c^\alpha}\Gamma\left(\frac{\alpha+\nu}{2}\right)\Gamma\left(\frac{\alpha-\nu}{2}\right)\times$$

$$\times\left[\psi\left(\frac{\alpha+\nu}{2}\right)+\psi\left(\frac{\alpha-\nu}{2}\right)-\frac{2}{\alpha}-2\ln\frac{c}{2}\right]-\frac{\pi^2}{2\alpha\nu}\operatorname{cosec}\frac{\alpha\pi}{2}\operatorname{cosec}\frac{\nu\pi}{2}$$

$$\left[\mathrm{Re}\,c>0;\ \mathrm{Re}\,\alpha>|\mathrm{Re}\,\nu|,\ \begin{Bmatrix}\mathrm{Re}\,z>0\\z>0\end{Bmatrix},\ U_\gamma(\nu),\ V_\gamma\ \text{ see } 2.12.5.1\right].$$

2.12.6. Integrals containing $\mathrm{Ei}(-bx)\,J i_\nu(cx)$ or $\mathrm{Ei}(-bx)\begin{Bmatrix}Y i_\nu(cx)\\K i_\nu(cx)\end{Bmatrix}$.

1. $$\int_0^\infty x^{\alpha-1}\mathrm{Ei}(-bx)\,J i_\nu(cx)\,dx=U_-(\nu)-\frac{\Gamma(\alpha)}{\alpha\nu b^\alpha}\qquad [c,\ \mathrm{Re}\,b,\ \mathrm{Re}\,\alpha,\ \mathrm{Re}(\alpha+\nu)>0],$$

$$U_\mp(\nu)=\frac{c^\nu}{2^\nu\nu(\alpha+\nu)b^{\alpha+\nu}}\Gamma\begin{bmatrix}\alpha+\nu\\\nu+1\end{bmatrix}\times$$

$$\times{}_4F_3\left(\frac{\nu}{2},\ \frac{\alpha+\nu}{2},\ \frac{\alpha+\nu}{2},\ \frac{\alpha+\nu+1}{2};\ \frac{\alpha+\nu}{2}+1,\ \frac{\nu}{2}+1,\ \nu+1;\ -\frac{c^2}{b^2}\right).$$

2. $$\int_0^\infty x^{\alpha-1}\mathrm{Ei}(-bx)\begin{Bmatrix}Y i_\nu(cx)\\K i_\nu(cx)\end{Bmatrix}dx=$$

$$=\pm\frac{1}{\sin\nu\pi}\left[\begin{Bmatrix}\cos\nu\pi\\ \pi/2\end{Bmatrix}U_{\mp}(\nu)-\begin{Bmatrix}1\\ \pi/2\end{Bmatrix}U_{\mp}(-\nu)\right]\mp\frac{\Gamma(\alpha)}{2\alpha\nu b^{\alpha}}\begin{Bmatrix}2\operatorname{ctg}(\nu\pi/2)\\ \pi\operatorname{cosec}(\nu\pi/2)\end{Bmatrix}$$

$$\left[\operatorname{Re} b>0;\ \operatorname{Re}\alpha>|\operatorname{Re}\nu|,\ \begin{Bmatrix}c>0\\ \operatorname{Re} c>0\end{Bmatrix},\ U_{\mp}(\nu)\ \text{see } 2.12.6.1\right].$$

3. $$\int_0^\infty x^{\alpha-1}e^{\pm bx}\operatorname{Ei}(\mp bx)\,Ji_\nu(cx)\,dx=U_1(\nu)\pm V_1-W_1-\frac{\pi\Gamma(\alpha)}{\nu b^{\alpha}}\begin{Bmatrix}\operatorname{cosec}\alpha\pi\\ \operatorname{ctg}\alpha\pi\end{Bmatrix}$$

$$[c,\ \operatorname{Re} b,\ \operatorname{Re}\alpha>0;\ -\operatorname{Re}\nu<\operatorname{Re}\alpha<7/2],$$

$$U_\gamma(\nu)=\frac{\pi c^{\nu}}{2^{\nu}\nu b^{\alpha+\nu}}\Gamma\begin{bmatrix}\alpha+\nu\\ \nu+1\end{bmatrix}\begin{Bmatrix}\operatorname{cosec}(\alpha+\nu)\pi\\ \operatorname{ctg}(\alpha+\nu)\pi\end{Bmatrix}\times$$

$$\times{}_3F_2\left(\frac{\nu}{2},\ \frac{\alpha+\nu}{2},\ \frac{\alpha+\nu+1}{2};\ \frac{\nu}{2}+1,\ \nu+1;\ \frac{(-1)^{\gamma}c^2}{b^2}\right),$$

$$V_\gamma=\frac{2^{\alpha-2}}{(1-\alpha)\,bc^{\alpha-1}}\Gamma\begin{bmatrix}(\alpha+\nu-1)/2\\ (\nu-\alpha+3)/2\end{bmatrix}\times$$

$$\times{}_4F_3\left(\frac{1-\alpha}{2},\ \frac{1}{2},\ 1,\ 1;\ \frac{3-\alpha-\nu}{2},\ \frac{3-\alpha+\nu}{2},\ \frac{3-\alpha}{2};\ \frac{(-1)^{\gamma}c^2}{b^2}\right),$$

$$W_\gamma=\frac{2^{\alpha-3}}{(2-\alpha)\,b^2c^{\alpha-2}}\Gamma\begin{bmatrix}(\alpha+\nu)/2-1\\ (\nu-\alpha)/2+2\end{bmatrix}\times$$

$$\times{}_4F_3\left(1-\frac{\alpha}{2},\ \frac{3}{2},\ 1,\ 1;\ 2-\frac{\alpha+\nu}{2},\ 2-\frac{\alpha-\nu}{2},\ 2-\frac{\alpha}{2};\ \frac{(-1)^{\gamma}c^2}{b^2}\right).$$

4. $$\int_0^\infty x^{\alpha-1}e^{\pm bx}\operatorname{Ei}(\mp bx)\,Yi_\nu(cx)\,dx=\frac{1}{\sin\nu\pi}[\cos\nu\pi U_1(\nu)-U_1(-\nu)]\pm$$

$$\pm V_1\operatorname{tg}\frac{\alpha-\nu}{2}\pi+W_1\operatorname{ctg}\frac{\alpha-\nu}{2}\pi-\frac{\pi\Gamma(\alpha)}{\nu b^{\alpha}}\operatorname{ctg}\frac{\nu\pi}{2}\begin{Bmatrix}\operatorname{cosec}\alpha\pi\\ \operatorname{ctg}\alpha\pi\end{Bmatrix}$$

$$[c,\ \operatorname{Re} b>0;\ |\operatorname{Re}\nu|<\operatorname{Re}\alpha<7/2;\ U_\gamma(\nu),\ V_\gamma,\ W_\gamma\ \text{see } 2.12.6.3].$$

5. $$\int_0^\infty x^{\alpha-1}e^{\pm bx}\operatorname{Ei}(\mp bx)\,Ki_\nu(cx)\,dx=-\frac{\pi}{2\sin\nu\pi}[U_0(\nu)-U_0(-\nu)]\mp$$

$$\mp\frac{\pi}{2}V_0\sec\frac{\alpha-\nu}{2}\pi+\frac{\pi}{2}W_0\operatorname{cosec}\frac{\alpha-\nu}{2}\pi+\frac{\pi^2\Gamma(\alpha)}{2\nu b^{\alpha}}\operatorname{cosec}\frac{\nu\pi}{2}\begin{Bmatrix}\operatorname{cosec}\alpha\pi\\ \operatorname{ctg}\alpha\pi\end{Bmatrix}$$

$$[\operatorname{Re} b,\ \operatorname{Re} c>0;\ \operatorname{Re}\alpha>|\operatorname{Re}\nu|;\ U_\gamma(\nu),\ V_\gamma,\ W_\gamma\ \text{see } 2.12.6.3].$$

2.12.7. Integrals of $x^{\alpha}\begin{Bmatrix}\operatorname{si}(bx)\\ \operatorname{ci}(bx)\end{Bmatrix}Ji_\nu(cx)$ and $x^{\alpha}\begin{Bmatrix}\operatorname{si}(bx)\\ \operatorname{ci}(bx)\end{Bmatrix}\begin{Bmatrix}Yi_\nu(cx)\\ Ki_\nu(cx)\end{Bmatrix}$.

NOTATION: $\delta=\begin{Bmatrix}1\\ 0\end{Bmatrix}$.

1. $$\int_0^\infty x^{\alpha-1}\begin{Bmatrix}\operatorname{si}(bx)\\ \operatorname{ci}(bx)\end{Bmatrix}Ji_\nu(cx)\,dx=U$$

$$[b,\ c,\ \operatorname{Re}(\alpha+\nu)>0;\ 0<\operatorname{Re}\alpha<7/2\ \text{for } b\neq c;\ 0<\operatorname{Re}\alpha<5/2\ \text{for } b=c],$$

$$U=U(\nu)-\frac{\Gamma(\alpha)}{\alpha\nu b^{\alpha}}\begin{Bmatrix}\sin(\alpha\pi/2)\\ \cos(\alpha\pi/2)\end{Bmatrix}\qquad[0<c\leqslant b],$$

$$U=-\Gamma\begin{bmatrix}(\nu+\alpha+\delta)/2+1\\ (\nu-\alpha-\delta)/2\end{bmatrix}V_0+$$

$$+\frac{2^{\alpha+\delta-1}b^{\delta}}{(\alpha+\delta)\,c^{\alpha+\delta}}\Gamma\begin{bmatrix}(\nu+\alpha+\delta)/2\\ (\nu-\alpha-\delta)/2+1\end{bmatrix}\begin{Bmatrix}1\\ R\end{Bmatrix}-\begin{Bmatrix}1\\ 0\end{Bmatrix}\frac{2^{\alpha-2}\pi}{\alpha c^{\alpha}}\Gamma\begin{bmatrix}(\nu+\alpha)/2\\ (\nu-\alpha)/2+1\end{bmatrix},$$

$$R=\mathbf{C}+\frac{1}{2}\psi\left(\frac{\nu+\alpha}{2}\right)+\frac{1}{2}\psi\left(\frac{\nu-\alpha}{2}+1\right)-\frac{1}{\alpha}-\ln\frac{c}{2b}\qquad[0<b\leqslant c],$$

$$U(\nu)=\frac{c^{\nu}}{2^{\nu}\nu(\alpha+\nu)b^{\alpha+\nu}}\Gamma\begin{bmatrix}\alpha+\nu\\ \nu+1\end{bmatrix}\begin{Bmatrix}\sin[(\alpha+\nu)\pi/2]\\ \cos[(\alpha+\nu)\pi/2]\end{Bmatrix}\times$$

$$\times\,{}_4F_3\left(\frac{\nu}{2},\ \frac{\alpha+\nu}{2},\ \frac{\alpha+\nu}{2},\ \frac{\alpha+\nu+1}{2};\ \frac{\nu}{2}+1,\ \frac{\alpha+\nu}{2}+1,\ \nu+1;\ \frac{c^2}{b^2}\right),$$

$$V_{\gamma}=\frac{2^{\alpha+\delta}b^{\delta+2}}{(\delta+2)(2\delta+1)(\alpha+\delta+2)c^{\alpha+\delta+2}}\,{}_5F_4\left(1,\ \frac{\delta}{2}+1,\ \frac{\alpha+\delta}{2}+1,\right.$$

$$\left.\frac{\alpha+\delta+\nu}{2}+1,\ \frac{\alpha+\delta-\nu}{2}+1;\ 2,\ \frac{\delta}{2}+2,\ \frac{\alpha+\delta}{2}+2,\ \delta+\frac{3}{2};\ \frac{(-1)^{\gamma}b^2}{c^2}\right).$$

2. $$\int_0^{\infty}x^{\alpha-1}\begin{Bmatrix}\mathrm{si}\,(bx)\\ \mathrm{ci}\,(bx)\end{Bmatrix}Y i_{\nu}(cx)\,dx=W$$

[$b,\ c>0$; $\operatorname{Re}\alpha>|\operatorname{Re}\nu|$; $\operatorname{Re}\alpha<7/2$ for $b\neq c$; $\operatorname{Re}\alpha<5/2$ for $b=c$],

$$W=\frac{1}{\sin\nu\pi}[\cos\nu\pi U(\nu)-U(-\nu)]-\frac{\Gamma(\alpha)}{\alpha\nu b^{\alpha}}\operatorname{ctg}\frac{\nu\pi}{2}\begin{Bmatrix}\sin(\alpha\pi/2)\\ \cos(\alpha\pi/2)\end{Bmatrix}\qquad[0<c\leqslant b],$$

$$W=\frac{1}{\pi}\Gamma\left(\frac{\alpha+\nu+\delta}{2}+1\right)\Gamma\left(\frac{\alpha-\nu+\delta}{2}+1\right)\begin{Bmatrix}\sin[(\nu-\alpha)\pi/2]\\ \cos[(\nu-\alpha)\pi/2]\end{Bmatrix}V_0-$$

$$-\frac{2^{\alpha+\delta-1}b^{\delta}}{(\alpha+\delta)\pi c^{\alpha+\delta}}\Gamma\left(\frac{\alpha+\nu+\delta}{2}\right)\Gamma\left(\frac{\alpha-\nu+\delta}{2}\right)\begin{Bmatrix}\sin[(\nu-\alpha)\pi/2]\\ S\cos[(\nu-\alpha)\pi/2]\end{Bmatrix}+$$

$$+\begin{Bmatrix}1\\ 0\end{Bmatrix}\frac{2^{\alpha-2}}{\alpha c^{\alpha}}\Gamma\left(\frac{\alpha+\nu}{2}\right)\Gamma\left(\frac{\alpha-\nu}{2}\right)\cos\frac{\alpha-\nu}{2}\pi,$$

$$S=\mathbf{C}+\frac{1}{2}\psi\left(\frac{\alpha+\nu}{2}\right)+\frac{1}{2}\psi\left(\frac{\alpha-\nu}{2}\right)-\frac{1}{\alpha}-\frac{\pi}{2}\operatorname{tg}\frac{\alpha-\nu}{2}\pi-\ln\frac{c}{2b}\qquad[0<b\leqslant c].$$

[$U(\nu)$, V_{γ} see 2.12.7.1].

3. $$\int_0^{\infty}x^{\alpha-1}\begin{Bmatrix}\mathrm{si}\,(bx)\\ \mathrm{ci}\,(bx)\end{Bmatrix}K i_{\nu}(cx)\,dx=\frac{1}{2}\Gamma\left(\frac{\alpha+\nu+\delta}{2}+1\right)\Gamma\left(\frac{\alpha-\nu+\delta}{2}+1\right)V_1+$$

$$+\frac{2^{\alpha+\delta-2}b^{\delta}}{(\alpha+\delta)c^{\alpha+\delta}}\Gamma\left(\frac{\alpha+\nu+\delta}{2}\right)\Gamma\left(\frac{\alpha-\nu+\delta}{2}\right)\begin{Bmatrix}1\\ T\end{Bmatrix}-$$

$$-\begin{Bmatrix}1\\ 0\end{Bmatrix}\frac{2^{\alpha-3}\pi}{\alpha c^{\alpha}}\Gamma\left(\frac{\alpha+\nu}{2}\right)\Gamma\left(\frac{\alpha-\nu}{2}\right),$$

$$T=\mathbf{C}-\frac{1}{\alpha}+\frac{1}{2}\psi\left(\frac{\alpha+\nu}{2}\right)+\frac{1}{2}\psi\left(\frac{\alpha-\nu}{2}\right)-\ln\frac{c}{2b}$$

[$b,\ \operatorname{Re}c>0$; $\operatorname{Re}\alpha>|\operatorname{Re}\nu|$; V_{γ} see 2.12.7.1].

2.12.8. Integrals containing $\begin{Bmatrix}\mathrm{erf}\,(bx^r)\\ \mathrm{erfc}\,(bx^r)\end{Bmatrix}J i_{\nu}(cx)$ or $\begin{Bmatrix}\mathrm{erf}\,(bx^r)\\ \mathrm{erfc}\,(bx^r)\end{Bmatrix}\begin{Bmatrix}Y i_{\nu}(cx)\\ K i_{\nu}(cx)\end{Bmatrix}$.

1. $$\int_0^{\infty}x^{\alpha-1}\begin{Bmatrix}\mathrm{erf}\,(bx)\\ \mathrm{erfc}\,(bx)\end{Bmatrix}J i_{\nu}(cx)\,dx=$$

$$=\pm U_1(\nu)+\begin{Bmatrix}1\\ 0\end{Bmatrix}\frac{2^{\alpha-1}}{\alpha c^{\alpha}}\Gamma\begin{bmatrix}(\alpha+\nu)/2\\ (\nu-\alpha)/2+1\end{bmatrix}\mp\frac{\Gamma((\alpha+1)/2)}{\alpha\nu\sqrt{\pi}\,b^{\alpha}}$$

$$\left[c>0;\ \operatorname{Re}\alpha>-(1\pm1)/2;\ |\arg b|<\pi/4,\ \begin{Bmatrix}-1-\operatorname{Re}\nu<\operatorname{Re}\alpha<5/2\\ \operatorname{Re}(\alpha+\nu)>0\end{Bmatrix}\right],$$

$$U_{\gamma}(\nu)=\frac{c^{\nu}}{2^{\nu}\nu(\alpha+\nu)\sqrt{\pi}\,b^{\alpha+\nu}}\Gamma\begin{bmatrix}(\alpha+\nu+1)/2\\ \nu+1\end{bmatrix}\times$$

$$\times\,{}_3F_3\left(\frac{\alpha+\nu+1}{2},\ \frac{\alpha+\nu}{2},\ \frac{\nu}{2};\ \nu+1,\ \frac{\alpha+\nu}{2}+1,\ \frac{\nu}{2}+1;\ \frac{(-1)^{\gamma}c^2}{4b^2}\right).$$

2. $\int_0^\infty x^{\alpha-1}\left\{\begin{matrix}\operatorname{erf}(bx)\\ \operatorname{erfc}(bx)\end{matrix}\right\} Yi_\nu(cx)\,dx = \pm\frac{1}{\sin\nu\pi}[\cos\nu\pi U_1(\nu)-U_1(-\nu)]-$

$-\left\{\begin{matrix}1\\0\end{matrix}\right\}\frac{2^{\alpha-1}}{\alpha\pi c^\alpha}\Gamma\left(\frac{\alpha+\nu}{2}\right)\Gamma\left(\frac{\alpha-\nu}{2}\right)\cos\frac{\alpha-\nu}{2}\pi\mp\frac{\Gamma((\alpha+1)/2)}{\alpha\nu\sqrt{\pi}\,b^\alpha}\operatorname{ctg}\frac{\nu\pi}{2}$

$\left[|\arg b|<\pi/4;\ \operatorname{Re}\alpha>|\operatorname{Re}\nu|-(1\pm1)/2,\ \left\{\begin{matrix}\operatorname{Re}\alpha<5/2;\ c>0\\ c>0\end{matrix}\right\},\ U_\gamma(\nu)\ \text{see 2.12.8.1}\right].$

3. $\int_0^\infty x^{\alpha-1}\left\{\begin{matrix}\operatorname{erf}(bx)\\ \operatorname{erfc}(bx)\end{matrix}\right\} Ki_\nu(cx)\,dx = \mp\frac{\pi}{2\sin\nu\pi}[U_0(\nu)-U_0(-\nu)]+$

$+\left\{\begin{matrix}1\\0\end{matrix}\right\}\frac{2^{\alpha-2}}{\alpha c^\alpha}\Gamma\left(\frac{\alpha+\nu}{2}\right)\Gamma\left(\frac{\alpha-\nu}{2}\right)\pm\frac{\sqrt{\pi}}{2\alpha\nu b^\alpha}\operatorname{cosec}\frac{\nu\pi}{2}\Gamma\left(\frac{\alpha+1}{2}\right)$

$[\operatorname{Re}c>0;\ |\arg b|<\pi/4;\ \operatorname{Re}\alpha>|\operatorname{Re}\nu|-(1\pm1)/2;\ U_\gamma(\nu)\ \text{see 2.12.8.1}].$

4. $\int_0^\infty x^{\alpha-1}\left\{\begin{matrix}\operatorname{erf}(b\sqrt{x})\\ \operatorname{erfc}(b\sqrt{x})\end{matrix}\right\} Ji_\nu(cx)\,dx =$

$=\pm V_1(\nu)+\left\{\begin{matrix}1\\0\end{matrix}\right\}\frac{2^{\alpha-1}}{\alpha c^\alpha}\Gamma\left[\begin{matrix}(\alpha+\nu)/2\\ (\nu-\alpha)/2+1\end{matrix}\right]\mp\frac{\Gamma(\alpha+1/2)}{\alpha\nu\sqrt{\pi}\,b^{2\alpha}}$

$\left[c>0;\ \operatorname{Re}\alpha>-(1\pm1)/4;\ |\arg b|<\pi/4,\ \left\{\begin{matrix}-1/2-\operatorname{Re}\nu<\operatorname{Re}\alpha<5/2\\ \operatorname{Re}(\alpha+\nu)>0\end{matrix}\right\}\right],$

$V_\gamma(\nu)=\frac{c^\nu}{2^\nu(\alpha+\nu)\nu\sqrt{\pi}\,b^{2\alpha+2\nu}}\Gamma\left[\begin{matrix}\alpha+\nu+1/2\\ \nu+1\end{matrix}\right]\times$

$\times{}_4F_3\left(\frac{2\alpha+2\nu+1}{4},\frac{2\alpha+2\nu+3}{4},\frac{\alpha+\nu}{2},\frac{\nu}{2};\frac{\alpha+\nu}{2}+1,\frac{\nu}{2}+1,\nu+1;\frac{(-1)^\gamma c^2}{b^4}\right).$

5. $\int_0^\infty x^{\alpha-1}\left\{\begin{matrix}\operatorname{erf}(b\sqrt{x})\\ \operatorname{erfc}(b\sqrt{x})\end{matrix}\right\} Yi_\nu(cx)\,dx = \pm\frac{1}{\sin\nu\pi}[\cos\nu\pi V_1(\nu)-V_1(-\nu)]-$

$-\left\{\begin{matrix}1\\0\end{matrix}\right\}\frac{2^{\alpha-1}}{\alpha\pi c^\alpha}\Gamma\left(\frac{\alpha+\nu}{2}\right)\Gamma\left(\frac{\alpha-\nu}{2}\right)\cos\frac{\alpha-\nu}{2}\pi\mp\frac{\Gamma(\alpha+1/2)}{\alpha\nu\sqrt{\pi}\,b^{2\alpha}}\operatorname{ctg}\frac{\nu\pi}{2}$

$\left[|\arg b|<\pi/4;\ \operatorname{Re}\alpha>|\operatorname{Re}\nu|-(1\pm1)/4,\ \left\{\begin{matrix}\operatorname{Re}\alpha<5/2;\ c>0\\ c>0\end{matrix}\right\},\ V_\gamma(\nu)\ \text{see 2.12.8.4}\right].$

6. $\int_0^\infty x^{\alpha-1}\left\{\begin{matrix}\operatorname{erf}(b\sqrt{x})\\ \operatorname{erfc}(b\sqrt{x})\end{matrix}\right\} Ki_\nu(cx)\,dx = \mp\frac{\pi}{2\sin\nu\pi}[V_0(\nu)-V_0(-\nu)]+$

$+\left\{\begin{matrix}1\\0\end{matrix}\right\}\frac{2^{\alpha-2}}{\alpha c^\alpha}\Gamma\left(\frac{\alpha+\nu}{2}\right)\Gamma\left(\frac{\alpha-\nu}{2}\right)\pm\frac{\sqrt{\pi}\,\Gamma(\alpha+1/2)}{2\alpha\nu b^{2\alpha}\sin(\nu\pi/2)}$

$[\operatorname{Re}c>0;\ |\arg b|<\pi/4;\ \operatorname{Re}\alpha>|\operatorname{Re}\nu|-(1\pm1)/4;\ V_\gamma(\nu)\ \text{see 2.12.8.4}].$

7. $\int_0^\infty x^{\alpha-1}e^{\mp b^2x}\left\{\begin{matrix}\operatorname{erfi}(b\sqrt{x})\\ \operatorname{erfc}(b\sqrt{x})\end{matrix}\right\} Ji_\nu(cx)\,dx =$

$=-\frac{c^\nu}{2^\nu\nu b^{2\alpha+2\nu}}\Gamma\left[\begin{matrix}\alpha+\nu\\ \nu+1\end{matrix}\right]\left\{\begin{matrix}\operatorname{tg}(\alpha+\nu)\pi\\ \sec(\alpha+\nu)\pi\end{matrix}\right\}\times$

$\times{}_3F_2\left(\frac{\alpha+\nu}{2},\frac{\alpha+\nu+1}{2},\frac{\nu}{2};\nu+1,\frac{\nu}{2}+1;-\frac{c^2}{b^4}\right)-$

$-\frac{2^{\alpha-1/2}c^{1/2-\alpha}}{(1-2\alpha)\sqrt{\pi}\,b}\Gamma\left[\begin{matrix}(2\alpha+2\nu-1)/4\\ (2\nu-2\alpha+5)/4\end{matrix}\right]\times$

$$\times {}_4F_3\left(\frac{1-2\alpha}{4},\ \frac{1}{4},\ \frac{3}{4},\ 1;\ \frac{5-2\alpha}{4},\ \frac{5-2\alpha-2\nu}{4},\ \frac{5-2\alpha+2\nu}{4};\ -\frac{c^2}{b^4}\right)\mp$$

$$\mp\frac{2^{\alpha-5/2}c^{3/2-\alpha}}{(3-2\alpha)\sqrt{\pi}\,b^3}\Gamma\left[\begin{matrix}(2\alpha+2\nu-3)/4\\(2\nu-2\alpha+7)/4\end{matrix}\right]\times$$

$$\times {}_4F_3\left(\frac{3-2\alpha}{4},\ \frac{3}{4},\ \frac{5}{4},\ 1;\ \frac{7-2\alpha-2\nu}{4},\ \frac{7-2\alpha+2\nu}{4},\ \frac{7-2\alpha}{4};\ -\frac{c^2}{b^4}\right)+$$

$$+\frac{\pi}{2\nu b^{2\alpha}}\left\{\begin{matrix}\sec\alpha\pi\\2\operatorname{cosec}2\alpha\pi\end{matrix}\right\}$$

$[c>0;\ |\arg b|<\pi/4;\ \operatorname{Re}\alpha>-(1\pm 1)/4;\ -\operatorname{Re}\nu-(1\pm 1)/4<\operatorname{Re}\alpha<3]$.

2.12.9. Integrals of $x^{\alpha}\left\{\begin{matrix}S(bx)\\C(bx)\end{matrix}\right\}Ji_{\nu}(cx)$ and $x^{\alpha}\left\{\begin{matrix}S(bx)\\C(bx)\end{matrix}\right\}\left\{\begin{matrix}Yi_{\nu}(cx)\\Ki_{\nu}(cx)\end{matrix}\right\}$.

1. $\displaystyle\int_0^{\infty}x^{\alpha-1}\left\{\begin{matrix}S(bx)\\C(bx)\end{matrix}\right\}Ji_{\nu}(cx)\,dx=U$ $\quad[b,\ c>0;\ \operatorname{Re}\alpha,\ \operatorname{Re}(\alpha+\nu)>-(2\pm 1)/2;\ \operatorname{Re}\alpha<5/2]$,

$$U=U(\nu)+\frac{2^{\alpha-2}}{\alpha c^{\alpha}}\Gamma\left[\begin{matrix}(\alpha+\nu)/2\\(\nu-\alpha)/2+1\end{matrix}\right]-\frac{\Gamma(\alpha+1/2)}{\alpha\nu\sqrt{2\pi}\,b^{\alpha}}\left\{\begin{matrix}\sin[(2\alpha+1)\pi/4]\\\cos[(2\alpha+1)\pi/4]\end{matrix}\right\}$$

$[0<c\leqslant b]$,

$$U=\Gamma\left[\begin{matrix}(2\nu+2\alpha+2\delta+1)/4\\(2\nu-2\alpha-2\delta+3)/4\end{matrix}\right]V_0$$

$[0<b\leqslant c]$,

$$U(\nu)=\frac{c^{\nu}}{2^{\nu+1/2}\nu(\alpha+\nu)\sqrt{\pi}\,b^{\alpha+\nu}}\Gamma\left[\begin{matrix}\alpha+\nu+1/2\\\nu+1\end{matrix}\right]\left\{\begin{matrix}\cos[(1-2\alpha-2\nu)\pi/4]\\\sin[(1-2\alpha-2\nu)\pi/4]\end{matrix}\right\}\times$$

$$\times {}_4F_3\left(\frac{2\alpha+2\nu+1}{4},\ \frac{2\alpha+2\nu+3}{4},\ \frac{\nu}{2},\ \frac{\alpha+\nu}{2};\ \nu+1,\ \frac{\nu}{2}+1,\ \frac{\alpha+\nu}{2}+1;\ \frac{c^2}{b^2}\right),$$

$$V_{\gamma}=\frac{2^{\alpha+\delta}b^{\delta+1/2}}{3^{\delta}(\alpha+\delta+1/2)\sqrt{\pi}\,c^{\alpha+\delta+1/2}}\,{}_4F_3\left(\frac{2\alpha+2\delta+2\nu+1}{4},\ \frac{2\alpha+2\delta-2\nu+1}{4},\right.$$

$$\left.\frac{2\alpha+2\delta+1}{4},\ \frac{2\delta+1}{4};\ \frac{2\delta+5}{4},\ \frac{2\alpha+2\delta+5}{4},\ \delta+\frac{1}{2};\ \frac{(-1)^{\gamma}b^2}{c^2}\right).$$

2. $\displaystyle\int_0^{\infty}x^{\alpha-1}\left\{\begin{matrix}S(bx)\\C(bx)\end{matrix}\right\}Yi_{\nu}(cx)\,dx=U$ $\quad[b,\ c>0;\ |\operatorname{Re}\nu|-(2\pm 1)/2<\operatorname{Re}\alpha<5/2]$,

$$U=\frac{1}{\sin\nu\pi}[\cos\nu\pi U(\nu)-U(-\nu)]-\frac{2^{\alpha-2}}{\alpha\pi c^{\alpha}}\Gamma\left(\frac{\alpha+\nu}{2}\right)\Gamma\left(\frac{\alpha-\nu}{2}\right)\cos\frac{\alpha-\nu}{2}\pi-$$

$$-\frac{\Gamma(\alpha+1/2)}{\alpha\nu\sqrt{2\pi}b^{\alpha}}\left\{\begin{matrix}\sin[(2\alpha+1)\pi/4]\\\cos[(2\alpha+1)\pi/4]\end{matrix}\right\}\operatorname{ctg}\frac{\nu\pi}{2}$$

$[0<c\leqslant b]$,

$$U=-\frac{1}{\pi}\Gamma\left(\frac{\alpha+\nu+\delta}{2}+\frac{1}{4}\right)\Gamma\left(\frac{\alpha-\nu+\delta}{2}+\frac{1}{4}\right)\left\{\begin{matrix}\sin[(2\nu-2\alpha-1)\pi/4]\\\cos[(2\nu-2\alpha-1)\pi/4]\end{matrix}\right\}V_0$$

$[0<b\leqslant c]$,

[$U(\nu)$, V_{γ} see 2.12.9.1].

3. $\displaystyle\int_0^{\infty}x^{\alpha-1}\left\{\begin{matrix}S(bx)\\C(bx)\end{matrix}\right\}Ki_{\nu}(cx)\,dx=\frac{1}{2}\Gamma\left(\frac{\alpha+\nu+\delta}{2}+\frac{1}{4}\right)\Gamma\left(\frac{\alpha-\nu+\delta}{2}+\frac{1}{4}\right)V_1$

$[b,\ \operatorname{Re}c>0;\ \operatorname{Re}\alpha>|\operatorname{Re}\nu|-(2\pm 1)/2;\ V_{\gamma}$ see 2.12.9.1].

2.12.10. Integrals of $x^{\alpha}\left\{\begin{matrix}\gamma(\mu,\ bx)\\\Gamma(\mu,\ bx)\end{matrix}\right\}Ji_{\nu}(cx)$ and $x^{\alpha}\left\{\begin{matrix}\gamma(\mu,\ bx)\\\Gamma(\mu,\ bx)\end{matrix}\right\}\left\{\begin{matrix}Yi_{\nu}(cx)\\Ki_{\nu}(cx)\end{matrix}\right\}$.

1. $\int\limits_0^\infty x^{\alpha-1} \begin{Bmatrix} \gamma(\mu, bx) \\ \Gamma(\mu, bx) \end{Bmatrix} J i_\nu(cx)\, dx = \pm U_1(\nu) + \begin{Bmatrix} 1 \\ 0 \end{Bmatrix} \frac{2^{\alpha-1}}{\alpha c^\alpha} \times$

$\times \Gamma\left[\begin{matrix} \mu, (\alpha+\nu)/2 \\ (\nu-\alpha)/2+1 \end{matrix}\right] \mp \frac{\Gamma(\alpha+\mu)}{\alpha\nu b^\alpha}$

$\left[c, \operatorname{Re} b, \operatorname{Re}(\alpha+\mu+\nu), \operatorname{Re}(\alpha+\mu) > 0, \begin{Bmatrix} \operatorname{Re}\alpha < 5/2;\ \operatorname{Re}\mu > 0 \\ \operatorname{Re}\alpha, \operatorname{Re}(\alpha+\nu) > 0 \end{Bmatrix}\right],$

$U_\gamma(\nu) = \frac{c^\nu}{2^\nu \nu(\alpha+\nu) b^{\alpha+\nu}} \Gamma\left[\begin{matrix} \alpha+\mu+\nu \\ \nu+1 \end{matrix}\right] \times$

$\times {}_4F_3\left(\frac{\alpha+\nu}{2}, \frac{\nu}{2}, \frac{\alpha+\mu+\nu}{2}, \frac{\alpha+\mu+\nu+1}{2}; \frac{\alpha+\nu}{2}+1, \frac{\nu}{2}+1, \nu+1; \frac{(-1)^\gamma c^2}{b^2}\right).$

2. $\int\limits_0^\infty x^{\alpha-1} \begin{Bmatrix} \gamma(\mu, bx) \\ \Gamma(\mu, bx) \end{Bmatrix} Y i_\nu(cx)\, dx = \pm \frac{1}{\sin\nu\pi} [\cos\nu\pi U_1(\nu) - U_1(-\nu)] -$

$- \begin{Bmatrix} 1 \\ 0 \end{Bmatrix} \frac{2^{\alpha-1}}{\alpha\pi c^\alpha} \Gamma\left[\mu, \frac{\alpha+\nu}{2}, \frac{\alpha-\nu}{2}\right] \cos\frac{\alpha-\nu}{2}\pi \mp \frac{\Gamma(\alpha+\mu)}{\alpha\nu b^\alpha} \operatorname{ctg}\frac{\nu\pi}{2}$

$\left[c, \operatorname{Re} b > 0;\ \operatorname{Re}(\alpha+\mu) > |\operatorname{Re}\nu|, \begin{Bmatrix} \operatorname{Re}\alpha < 5/2;\ \operatorname{Re}\mu > 0 \\ \operatorname{Re}\alpha > |\operatorname{Re}\nu| \end{Bmatrix}, U_\gamma(\nu) \text{ see } 2.12.10.1\right].$

3. $\int\limits_0^\infty x^{\alpha-1} \begin{Bmatrix} \gamma(\mu, bx) \\ \Gamma(\mu, bx) \end{Bmatrix} K i_\nu(cx)\, dx = \mp \frac{\pi}{2\sin\nu\pi} [U_0(\nu) - U_0(-\nu)] +$

$+ \begin{Bmatrix} 1 \\ 0 \end{Bmatrix} \frac{2^{\alpha-2}}{\alpha c^\alpha} \Gamma\left[\mu, \frac{\alpha+\nu}{2}, \frac{\alpha-\nu}{2}\right] \pm \frac{\pi\Gamma(\alpha+\mu)}{2\alpha\nu b^\alpha \sin(\nu\pi/2)}$

$\left[\operatorname{Re} b, \operatorname{Re} c > 0;\ \operatorname{Re}(\alpha+\mu) > |\operatorname{Re}\nu|, \begin{Bmatrix} \operatorname{Re}\mu > 0 \\ \operatorname{Re}\alpha > |\operatorname{Re}\nu| \end{Bmatrix}, U_\gamma(\nu) \text{ see } 2.12.10.1\right].$

2.12.11. Integrals containing $D_\mu(b\sqrt{x})\, J i_\nu(cx)$ or $D_\mu(b\sqrt{x}) \begin{Bmatrix} Y i_\nu(cx) \\ K i_\nu(cx) \end{Bmatrix}$.

1. $\int\limits_0^\infty x^{\alpha-1} e^{\pm b^2 x/4} D_\mu(b\sqrt{x})\, J i_\nu(cx)\, dx =$

$= U_1(\nu) + \begin{Bmatrix} 1 \\ 0 \end{Bmatrix} \Gamma\left[\begin{matrix} (2\alpha+2\nu+\mu)/4 \\ (2\nu-2\alpha-\mu)/4+1 \end{matrix}\right] V_1 - \begin{Bmatrix} 1 \\ 0 \end{Bmatrix} \Gamma\left[\begin{matrix} (2\alpha+2\nu+\mu-2)/4 \\ (2\nu-2\alpha-\mu+6)/4 \end{matrix}\right] W_1$

$\left[c, \operatorname{Re}\alpha > 0;\ |\arg b| < (2\pm1)\pi/4, \begin{Bmatrix} -\operatorname{Re}\nu < \operatorname{Re}\alpha < (5-\operatorname{Re}\mu)/2 \\ \operatorname{Re}(\alpha+\nu) > 0 \end{Bmatrix}\right].$

$U_\gamma(\nu) = -\frac{2^{-\alpha-2\nu\mp(\mu+1\mp1)/2} c^\nu}{\nu b^{2\alpha+2\nu}} \begin{Bmatrix} \Gamma^{-1}(-\mu) \\ \sqrt{\pi} \end{Bmatrix} \Gamma\left[\begin{matrix} 2\alpha+2\nu \\ \nu+1 \end{matrix}\right] \times$

$\times \Gamma^{\pm1}\left(\frac{1\mp1-2\mu}{4} \mp \alpha \mp \nu\right) {}_5F_4\left(\frac{\nu}{2}, \frac{\alpha+\nu}{2}, \frac{2\alpha+2\nu+1}{4}, \frac{\alpha+\nu+1}{2}, \frac{2\alpha+2\nu+3}{4};\right.$

$\left.\frac{\nu}{2}+1, \nu+1, \frac{3\pm1}{8}+\frac{2\alpha+2\nu\pm\mu}{4}, \frac{7\pm1}{8}+\frac{2\alpha+2\nu\pm\mu}{4}; \frac{(-1)^\gamma 4c^2}{b^4}\right),$

$V_\gamma = \frac{2^{\alpha+\mu/2} b^\mu}{(2\alpha+\mu) c^{\alpha+\mu/2}} {}_5F_4\left(-\frac{\mu}{4}, \frac{1-\mu}{4}, \frac{2-\mu}{4}, \frac{3-\mu}{4}, -\frac{2\alpha+\mu}{4};\right.$

$\left.\frac{1}{2}, 1-\frac{2\alpha+\mu}{4}, 1-\frac{2\alpha+\mu+2\nu}{4}, 1-\frac{2\alpha+\mu-2\nu}{4}; \frac{(-1)^\gamma 4c^2}{b^4}\right),$

$$W_\gamma=\frac{2^{\alpha+\mu/2-2}\mu(\mu+1)b^{\mu-2}}{(2\alpha+\mu-2)c^{\alpha+\mu/2-1}}\,{}_5F_4\left(\frac{2-\mu}{4},\ \frac{3-\mu}{4},\ 1-\frac{\mu}{4},\ \frac{5-\mu}{4},\ \frac{2-2\alpha-\mu}{4};\right.$$
$$\left.\frac{3}{2},\ \frac{6-2\alpha-\mu-2\nu}{4},\ \frac{6-2\alpha-\mu+2\nu}{4},\ \frac{6-2\alpha-\mu}{4};\ \frac{(-1)^\gamma 4c^2}{b^4}\right).$$

2. $$\int_0^\infty x^{\alpha-1}e^{\pm b^2x/4}D_\mu(b\sqrt{x})\,Yi_\nu(cx)\,dx=\frac{1}{\sin\nu\pi}[\cos\nu\pi U_1(\nu)-U_1(-\nu)]-$$
$$-\left\{\begin{matrix}1\\0\end{matrix}\right\}\frac{1}{\pi}\Gamma\left(\frac{2\alpha+\mu+2\nu}{4}\right)\Gamma\left(\frac{2\alpha+\mu-2\nu}{4}\right)\cos\left(\frac{2\alpha+\mu-2\nu}{4}\pi\right)V_1+$$
$$+\left\{\begin{matrix}1\\0\end{matrix}\right\}\frac{1}{\pi}\Gamma\left(\frac{2\alpha+\mu+2\nu-2}{4}\right)\Gamma\left(\frac{2\alpha+\mu-2\nu-2}{4}\right)\sin\left(\frac{2\alpha+\mu-2\nu}{4}\pi\right)W_1+$$
$$+\frac{2^{-\alpha\mp(\mu+1\mp1)/2}}{\nu b^{2\alpha}}\left\{\begin{matrix}\Gamma^{-1}(-\mu)\\\sqrt{\pi}\end{matrix}\right\}\Gamma(2\alpha)\Gamma^{\pm1}\left(\frac{1\mp1-2\mu}{4}\mp\alpha\right)\operatorname{ctg}\frac{\nu\pi}{2}$$

$\left[c>0;\ |\arg b|<(2\pm1)\pi/4,\ \left\{\begin{matrix}|\operatorname{Re}\nu|<\operatorname{Re}\alpha<5/2-\operatorname{Re}\mu\\|\operatorname{Re}\nu|<\operatorname{Re}\alpha\end{matrix}\right\},\ U_\gamma(\nu),\ V_\gamma,\ W_\gamma\right.$ see 2.12.11.1 $\Big]$.

3. $$\int_0^\infty x^{\alpha-1}e^{\pm b^2x/4}D_\mu(b\sqrt{x})\,Ki_\nu(cx)\,dx=-\frac{\pi}{2\sin\nu\pi}[U_0(\nu)-U_0(-\nu)]+$$
$$+\frac{1}{2}\left\{\begin{matrix}1\\0\end{matrix}\right\}\Gamma\left(\frac{2\alpha+\mu+2\nu}{4}\right)\Gamma\left(\frac{2\alpha+\mu-2\nu}{4}\right)V_0-\frac{1}{2}\left\{\begin{matrix}1\\0\end{matrix}\right\}\Gamma\left(\frac{2\alpha+\mu+2\nu-2}{4}\right)\times$$
$$\times\Gamma\left(\frac{2\alpha+\mu-2\nu-2}{4}\right)W_0-\frac{2^{-\alpha\mp(\mu+1\pm1)/2}\pi}{\nu b^{2\alpha}}\left\{\begin{matrix}\Gamma^{-1}(-\mu)\\\sqrt{\pi}\end{matrix}\right\}\times$$
$$\times\frac{\Gamma(2\alpha)}{\sin(\nu\pi/2)}\Gamma^{\pm1}\left(\frac{1\mp1-2\mu}{4}\mp\alpha\right)$$

$[\operatorname{Re}c>0;\ \operatorname{Re}\alpha>|\operatorname{Re}\nu|;\ |\arg b|<(2\pm1)\pi/4;\ U_\gamma(\nu),\ V_\gamma,\ W_\gamma$ see 2.12.11.1].

2.12.12. Integrals containing products of Bessel functions by $Ji_\nu(cx)$ or $\left\{\begin{matrix}Yi_\nu(cx)\\Ki_\nu(cx)\end{matrix}\right\}$.

1. $$\int_0^\infty x^{\alpha-1}J_\mu(bx)\,Ji_\nu(cx)\,dx=U$$

$[b,\ c,\ \operatorname{Re}(\alpha+\mu+\nu),\ \operatorname{Re}(\alpha+\mu)>0;\ \operatorname{Re}\alpha<3$ for $b\ne c$; $\operatorname{Re}\alpha<2$ for $b=c]$,

$$U=U(\nu)+\frac{2^{\alpha-1}}{\nu b^\alpha}\Gamma\left[\begin{matrix}(\alpha+\mu)/2\\(\mu-\alpha)/2+1\end{matrix}\right]\qquad[0<c\leqslant b],$$

$$U=\Gamma\left[\begin{matrix}(\alpha+\mu+\nu)/2\\\mu+1,\ (\nu-\alpha-\mu)/2+1\end{matrix}\right]V_0\qquad[0<b\leqslant c],$$

$$U(\nu)=-\frac{2^{\alpha-1}c^\nu}{\nu b^{\alpha+\nu}}\Gamma\left[\begin{matrix}(\alpha+\mu+\nu)/2\\\nu+1,\ 1+(\mu-\alpha-\nu)/2\end{matrix}\right]\times$$
$$\times{}_3F_2\left(\frac{\alpha+\mu+\nu}{2},\ \frac{\alpha-\mu+\nu}{2},\ \frac{\nu}{2};\ \nu+1,\ \frac{\nu}{2}+1;\ \frac{c^2}{b^2}\right),$$

$$V_\gamma=\frac{2^{\alpha-1}b^\mu}{(\alpha+\mu)c^{\alpha+\mu}}\,{}_3F_2\left(\frac{\alpha+\mu+\nu}{2},\ \frac{\alpha+\mu-\nu}{2},\ \frac{\alpha+\mu}{2};\ \frac{\alpha+\mu}{2}+1,\ \mu+1;\ \frac{(-1)^\gamma b^2}{c^2}\right).$$

2. $$\int_0^\infty x^{\alpha-1}J_\mu(bx)\,Yi_\nu(cx)\,dx=W$$

$[b,\ c>0;\ \operatorname{Re}(\alpha+\mu)>|\operatorname{Re}\nu|;\ \operatorname{Re}\alpha<3$ for $b\ne c$; $\operatorname{Re}\alpha<2$ for $b=c]$,

$$W=\frac{1}{\sin\nu\pi}\left[\cos\nu\pi U(\nu)-U(-\nu)\right]+\frac{2^{\alpha-1}}{\nu b^{\alpha}}\Gamma\begin{bmatrix}(\alpha+\mu)/2\\(\mu-\alpha)/2+1\end{bmatrix}\operatorname{ctg}\frac{\nu\pi}{2}\quad[0<c\leqslant b],$$

$$W=-\frac{1}{\pi}\Gamma\begin{bmatrix}(\alpha+\mu+\nu)/2,\ (\alpha+\mu-\nu)/2\\ \mu+1\end{bmatrix}\cos\left(\frac{\alpha+\mu-\nu}{2}\pi\right)V_0\qquad[0<b\leqslant c],$$

$U(\nu)$, V_γ see 2.12.12.1].

3. $$\int_0^{\infty}x^{\alpha-1}\begin{Bmatrix}J_\mu(bx)\\I_\mu(bx)\end{Bmatrix}Ki_\nu(cx)\,dx=\frac{1}{2}\Gamma\begin{bmatrix}(\alpha+\mu+\nu)/2,\ (\alpha+\mu-\nu)/2\\ \mu+1\end{bmatrix}V_{(1\pm1)/2}$$

$$\left[\operatorname{Re}(\alpha+\mu)>|\operatorname{Re}\nu|,\ \begin{Bmatrix}b,\ \operatorname{Re}c>0\\ |\arg b|<\pi,\ \operatorname{Re}(c-b)>0\end{Bmatrix},\ V_\gamma\ \text{see 2.12.12.1}\right].$$

4. $$\int_0^{\infty}x^{\alpha-1}K_\mu(bx)\begin{Bmatrix}Yi_\nu(cx)\\Ki_\nu(cx)\end{Bmatrix}dx=\pm\frac{2^{\alpha-2}}{c^{\alpha}}\frac{1}{\sin\mu\pi}\left[W(\mu)-W(-\mu)\right]$$

$$\left[\operatorname{Re}b>0,\ \begin{Bmatrix}c>0\\ \operatorname{Re}c>0\end{Bmatrix},\ \operatorname{Re}\alpha>|\operatorname{Re}\mu|+|\operatorname{Re}\nu|\right],$$

$$W(\mu)=\frac{1}{\alpha+\mu}\left(\frac{b}{c}\right)^{\mu}\Gamma\begin{bmatrix}(\alpha+\mu+\nu)/2,\ (\alpha+\mu-\nu)/2\\ \mu+1\end{bmatrix}\begin{Bmatrix}\cos(\alpha+\mu-\nu)\pi\\ \pi/2\end{Bmatrix}\times$$
$$\times{}_3F_2\left(\frac{\alpha+\mu}{2},\ \frac{\alpha+\mu+\nu}{2},\ \frac{\alpha+\mu-\nu}{2};\ \frac{\alpha+\mu}{2}+1,\ \mu+1;\ \mp\frac{b^2}{c^2}\right).$$

2.13. THE LAGUERRE FUNCTION $L_\nu(x)$

Certain integrals containing the Laguerre function $L_\nu(x)$ can also be obtained from the formulae of Section 2.19 since

$$L_\nu(x)=\frac{x^{-1/2}}{\Gamma(\nu+1)}e^{x/2}M_{\nu+1/2,\,0}(x).$$

2.13.1 Integrals of general form

1. $$\int_0^{a}x^{\alpha-1}(a^r-x^r)^{\beta-1}e^{-cx}L_\nu(cx)\,dx=$$

$$=\frac{a^{\alpha+r\beta-r}}{r}\Gamma(\beta)\sum_{k=0}^{\infty}\frac{(\nu+1)_k}{(k!)^2}\Gamma\begin{bmatrix}(\alpha+k)/r\\ \beta+(\alpha+k)/r\end{bmatrix}(-ac)^k\qquad[a,\ r,\ \operatorname{Re}\alpha,\ \operatorname{Re}\beta>0].$$

2. $$\int_a^{\infty}x^{\alpha-1}(x^r-a^r)^{\beta-1}e^{-cx}L_\nu(cx)\,dx=$$

$$=\frac{a^{\alpha+r\beta-r}}{r}\Gamma(\beta)\sum_{k=0}^{\infty}\frac{(\nu+1)_k}{(k!)^2}\Gamma\begin{bmatrix}1-\beta-(\alpha+k)/r\\ 1-(\alpha+k)/r\end{bmatrix}(-ac)^k+$$

$$+\frac{c^{r-r\beta-\alpha}}{\Gamma(\nu+1)}\sum_{k=0}^{\infty}\frac{(1-\beta)_k}{k!}\Gamma\begin{bmatrix}\alpha-r(1-\beta+k),\ 1-\alpha+\nu+r(1-\beta+k)\\ 1-\alpha+r(1-\beta+k)\end{bmatrix}(ac)^{rk}$$

$$[a,\ r,\ \operatorname{Re}c,\ \operatorname{Re}\beta>0;\ \operatorname{Re}(\alpha+r\beta-\nu)<1+r\ \text{for}\ \nu\neq0,1,2,\dots].$$

3. $$\int_0^{\infty}\frac{x^{\alpha-1}}{(x^r+z^r)^{\rho}}e^{-cx}L_\nu(cx)\,dx=$$

$$=\frac{z^{\alpha-r\rho}}{r\Gamma(\rho)}\sum_{k=0}^{\infty}\frac{(\nu+1)_k}{(k!)^2}\Gamma\left(\frac{\alpha+k}{r}\right)\Gamma\left(\rho-\frac{\alpha+k}{r}\right)(-cz)^k+$$

$$+\frac{c^{r\rho-\alpha}}{\Gamma(\nu+1)}\sum_{k=0}^{\infty}\frac{(\rho)_k}{k!}\Gamma\left[\begin{matrix}\alpha-r(\rho+k),\ 1-\alpha+\nu+r(\rho+k)\\ 1-\alpha+r(\rho+k)\end{matrix}\right](-c^r z^r)^k$$

[r, $\operatorname{Re} c$, $\operatorname{Re}\alpha>0$; $r\,|\arg z|<\pi$; $\operatorname{Re}(\alpha-r\rho-\nu)<1$ for $\nu\neq 0, 1, 2, \ldots$].

4. $$\int_0^\infty \frac{x^{\alpha-1}}{x^r-y^r}e^{-cx}L_\nu(cx)\,dx=$$

$$=-\frac{\pi y^{\alpha-r}}{r}\sum_{k=0}^{\infty}\frac{(\nu+1)_k}{(k!)^2}\operatorname{ctg}\frac{\alpha+k}{r}\pi(-cy)^k+$$

$$+\frac{c^{r-\alpha}}{\Gamma(\nu+1)}\sum_{k=0}^{\infty}\Gamma\left[\begin{matrix}\alpha-r-rk,\ 1-\alpha+\nu+r+rk\\ 1-\alpha+r+rk\end{matrix}\right](cy)^{rk}$$

[r, y, $\operatorname{Re} c$, $\operatorname{Re}\alpha>0$; $\operatorname{Re}(\alpha-\nu)<1+r$ for $\nu\neq 0, 1, 2, \ldots$].

5. $$\int_0^\infty x^{\alpha-1}e^{-px^r-cx}L_\nu(cx)\,dx=U$$ [r, $\operatorname{Re} c$, $\operatorname{Re} p$, $\operatorname{Re}\alpha>0$],

$$U=\frac{p^{-\alpha/r}}{r}\sum_{k=0}^{\infty}\frac{(\nu+1)_k}{(k!)^2}\Gamma\left(\frac{\alpha+k}{r}\right)\left(-\frac{c}{p^{1/r}}\right)^k$$ [$r>1$],

$$U=\frac{p^{(\nu-\alpha+1)/r}}{c^{\nu+1}r\Gamma(-\nu)}\sum_{k=0}^{\infty}\frac{[(\nu+1)_k]^2}{k!}\Gamma\left(\frac{\alpha-\nu-k-1}{r}\right)\left(\frac{p^{1/r}}{c}\right)^k+$$

$$+\frac{c^{-\alpha}}{\Gamma(\nu+1)}\sum_{k=0}^{\infty}\frac{1}{k!}\Gamma\left[\begin{matrix}\alpha+rk,\ 1-\alpha+\nu-rk\\ 1-\alpha-rk\end{matrix}\right]\left(-\frac{p}{c^r}\right)^k$$ [$r<1$],

U for $r=1$ see 2.13.2.2.

6. $$\int_0^\infty x^{\alpha-1}e^{-px^{-r}-cx}L_\nu(cx)\,dx=\frac{p^{\alpha/r}}{r}\sum_{k=0}^{\infty}\frac{(\nu+1)_k}{(k!)^2}\Gamma\left(-\frac{\alpha+k}{r}\right)(-cp^{1/r})^k+$$

$$+\frac{c^{-\alpha}}{\Gamma(\nu+1)}\sum_{k=0}^{\infty}\frac{1}{k!}\Gamma\left[\begin{matrix}\alpha-rk,\ 1-\alpha+\nu+rk\\ 1-\alpha+rk\end{matrix}\right](-c^r p)^k$$

[r, $\operatorname{Re} c$, $\operatorname{Re} p>0$; $\operatorname{Re}(\alpha-\nu)<1$ for $\nu\neq 0, 1, 2, \ldots$].

7. $$\int_0^\infty x^{\alpha-1}e^{-cx}\left\{\begin{matrix}\sin bx^r\\ \cos bx^r\end{matrix}\right\}L_\nu(cx)\,dx=V$$

$\left[b,\ r,\ \operatorname{Re} c>0;\ \operatorname{Re}\alpha>-\delta r;\ \operatorname{Re}(\alpha-\nu)<r+1 \text{ for } \nu\neq 0, 1, 2, \ldots;\ \delta=\left\{\begin{matrix}1\\ 0\end{matrix}\right\}\right]$.

$$V=\frac{b^{-\alpha/r}}{r}\sum_{k=0}^{\infty}\frac{(\nu+1)_k}{(k!)^2}\Gamma\left(\frac{\alpha+k}{r}\right)\left\{\begin{matrix}\sin[(\alpha+k)\pi/(2r)]\\ \cos[(\alpha+k)\pi/(2r)]\end{matrix}\right\}\left(-\frac{c}{b^{1/r}}\right)^k$$ [$r>1$],

$$V=\frac{b^{(\nu-\alpha+1)/r}}{c^{\nu+1}r\Gamma(-\nu)}\sum_{k=0}^{\infty}\frac{[(\nu+1)_k]^2}{k!}\Gamma\left(\frac{\alpha-\nu-k-1}{r}\right)\left\{\begin{matrix}\sin[(\alpha-\nu-k-1)\pi/(2r)]\\ \cos[(\alpha-\nu-k-1)\pi/(2r)]\end{matrix}\right\}\left(\frac{b^{1/r}}{c}\right)^k+$$

$$+\frac{b^{\delta}}{c^{\alpha+\delta r}\Gamma(\nu+1)}\sum_{k=0}^{\infty}\frac{(-1)^k}{k!\,(\delta+1/2)_k}\Gamma\left[\begin{matrix}\alpha+\delta r+2rk,\ 1-\alpha+\nu-\delta r-2rk\\ 1-\alpha-\delta r-2rk\end{matrix}\right]\left(\frac{b}{2c^r}\right)^{2k}$$

$[r<1]$,

U for $r=1$ see 2.13.4.1.

8. $$\int_0^\infty x^{\alpha-1}e^{-cx}\left\{\begin{matrix}\ln(x^r+z^r)\\ \ln|x^r-z^r|\end{matrix}\right\}L_\nu(cx)\,dx=$$

$$=\pi z^{\alpha}\sum_{k=0}^{\infty}\frac{(\nu+1)_k}{(k!)^2(\alpha+k)}\left\{\begin{matrix}\operatorname{cosec}[(\alpha+k)\pi/r]\\ \operatorname{ctg}[(\alpha+k)\pi/r]\end{matrix}\right\}(-cz)^k\pm$$

$$\pm\frac{c^{r-\alpha}z^r}{\Gamma(\nu+1)}\sum_{k=0}^{\infty}\frac{1}{k+1}\Gamma\left[\begin{matrix}\alpha-r-rk,\ 1-\alpha+\nu+r+rk\\ 1-\alpha+r+rk\end{matrix}\right](\mp c^r z^r)^k+$$

$$+\frac{r}{c^{\alpha}}\Gamma\left[\begin{matrix}\alpha,\ 1-\alpha+\nu\\ 1-\alpha,\ \nu+1\end{matrix}\right][\psi(\alpha)+\psi(1-\alpha)-\psi(1-\alpha+\nu)-\ln c]$$

$$\left[r,\ \operatorname{Re}c,\ \operatorname{Re}\alpha>0,\ \left\{\begin{matrix}r\,|\arg z|<\pi\\ z>0\end{matrix}\right\};\ \operatorname{Re}(\alpha-\nu)<1\ \text{for}\ \nu\neq 0,\ 1,\ 2,\ \ldots\right].$$

9. $$\int_0^\infty x^{\alpha-1}e^{-cx}\operatorname{Ei}(-bx^r)L_\nu(cx)\,dx=U\qquad [r,\ \operatorname{Re}b,\ \operatorname{Re}c,\ \operatorname{Re}\alpha>0],$$

$$U=-b^{-\alpha/r}\sum_{k=0}^{\infty}\frac{(\nu+1)_k}{(k!)^2(\alpha+k)}\Gamma\left(\frac{\alpha+k}{r}\right)\left(-\frac{c}{b^{1/r}}\right)^k\qquad [r>1],$$

$$U=\frac{b^{(\nu-\alpha+1)/r}}{c^{\nu+1}\Gamma(-\nu)}\sum_{k=0}^{\infty}\frac{[(\nu+1)_k]^2}{k!\,(1-\alpha+\nu+k)}\Gamma\left(\frac{\alpha-\nu-k-1}{r}\right)\left(\frac{b^{1/r}}{c}\right)^k-$$

$$-\frac{b}{c^{\alpha+r}\Gamma(\nu+1)}\sum_{k=0}^{\infty}\frac{1}{k!\,(k+1)^2}\Gamma\left[\begin{matrix}\alpha+r+rk,\ 1-\alpha+\nu-r-rk\\ 1-\alpha-r-rk\end{matrix}\right]\left(-\frac{b}{c^r}\right)^k+$$

$$+c^{-\alpha}\Gamma\left[\begin{matrix}\alpha,\ 1-\alpha+\nu\\ 1-\alpha,\ \nu+1\end{matrix}\right]\left[\mathbf{C}+r\psi(\alpha)-r\psi(1-\alpha+\nu)+r\psi(1-\alpha)-\ln\frac{c^r}{b}\right]\qquad [r<1],$$

U for $r=1$ see 2.13.6.1.

10. $$\int_0^\infty x^{\alpha-1}e^{\pm bx^r-cx}\operatorname{Ei}(\mp bx^r)L_\nu(cx)\,dx=V$$

$$\left[r,\ \operatorname{Re}c,\ \operatorname{Re}\alpha>0,\ \left\{\begin{matrix}|\arg b|<\pi\\ b>0\end{matrix}\right\};\ \operatorname{Re}(\alpha-\nu)<r+1\ \text{for}\ \nu\neq 0,\ 1,\ 2,\ \ldots\right],$$

$$V=-\frac{\pi b^{-\alpha/r}}{r}\sum_{k=0}^{\infty}\frac{(\nu+1)_k}{(k!)^2}\Gamma\left(\frac{\alpha+k}{r}\right)\left\{\begin{matrix}\operatorname{cosec}[(\alpha+k)\pi/r]\\ \operatorname{ctg}[(\alpha+k)\pi/r]\end{matrix}\right\}\left(-\frac{c}{b^{1/r}}\right)^k\mp$$

$$\mp\frac{c^{r-\alpha}}{b\Gamma(\nu+1)}\sum_{k=0}^{\infty}k!\,\Gamma\left[\begin{matrix}\alpha-r-rk,\ 1-\alpha+\nu+r+rk\\ 1-\alpha+r+rk\end{matrix}\right]\left(\mp\frac{c^r}{b}\right)^k\qquad [r>1],$$

$$V=-\frac{c^{-\alpha}}{\Gamma(\nu+1)}\sum_{k=0}^{\infty}\frac{1}{k!}\Gamma\left[\begin{matrix}\alpha+rk,\ 1-\alpha+\nu-rk\\ 1-\alpha-rk\end{matrix}\right]\left[\psi(k+1)-r\psi(\alpha+rk)+\right.$$

$$\left.+r\psi(1-\alpha+\nu-rk)-r\psi(1-\alpha-rk)-\ln\frac{b}{c^r}\right]\left(\pm\frac{b}{c^r}\right)^k-\frac{\pi b^{(\nu-\alpha+1)/r}}{c^{\nu+1}\,r\Gamma(-\nu)}\times$$

$$\times\sum_{k=0}^{\infty}\frac{[(\nu+1)_k]^2}{k!}\Gamma\left(\frac{\alpha-\nu-k-1}{r}\right)\begin{Bmatrix}\operatorname{cosec}[(\alpha-\nu-k-1)\pi/r]\\ \operatorname{ctg}[(\alpha-\nu-k-1)\pi/r]\end{Bmatrix}\left(\frac{b^{1/r}}{c}\right)^k \qquad [r<1]$$

V for $r=1$ see 2.13.6.2.

11. $\int_0^\infty x^{\alpha-1}e^{-cx}\begin{Bmatrix}\operatorname{si}(bx^r)\\ \operatorname{ci}(bx^r)\end{Bmatrix}L_\nu(cx)\,dx=W$

$[b,\ r,\ \operatorname{Re}c,\ \operatorname{Re}\alpha>0;\ \operatorname{Re}(\alpha-\nu)<2r+1$ for $\nu\neq 0,\ 1,\ 2,\ \ldots],\ \delta=(1\pm1/2)$

$$W=-b^{-\alpha/r}\sum_{k=0}^{\infty}\frac{(\nu+1)_k}{(k!)^2(\alpha+k)}\Gamma\left(\frac{\alpha+k}{r}\right)\begin{Bmatrix}\sin[(\alpha+k)\pi/(2r)]\\ \cos[(\alpha+k)\pi/(2r)]\end{Bmatrix}\left(-\frac{c}{b^{1/r}}\right)^k \qquad [r>1]$$

$$W=\frac{b^{(\nu-\alpha+1)/r}}{c^{\nu+1}\Gamma(-\nu)}\times$$

$$\times\sum_{k=0}^{\infty}\frac{[(\nu+1)_k]^2}{k!(1-\alpha+\nu+k)}\Gamma\left(\frac{\alpha-\nu-k-1}{r}\right)\begin{Bmatrix}\sin[(\alpha-\nu-k-1)\pi/(2r)]\\ \cos[(\alpha-\nu-k-1)\pi/(2r)]\end{Bmatrix}\left(\frac{b^{1/r}}{c}\right)^k-$$

$$-\frac{b^{\delta+2}}{2\cdot3^\delta c^{\alpha+\delta r+2r}\Gamma(\nu+1)}\sum_{k=0}^{\infty}\frac{(-1)^k}{(k+1)!(\delta+3/2)_k(\delta+2k+2)}\times$$

$$\times\Gamma\begin{bmatrix}\alpha+\delta r+2r+2rk,\ 1-\alpha+\nu-\delta r-2r-2rk\\ 1-\alpha-\delta r-2r-2rk\end{bmatrix}\left(\frac{b}{2c^r}\right)^{2k}+$$

$$+\frac{b^\delta}{c^{\alpha+\delta r}}\Gamma\begin{bmatrix}\alpha+\delta r,\ 1-\alpha+\nu-\delta r\\ \nu+1,\ 1-\alpha-\delta r\end{bmatrix}\begin{Bmatrix}1\\ R\end{Bmatrix}-\begin{Bmatrix}1\\ 0\end{Bmatrix}\frac{\pi}{2c^\alpha}\Gamma\begin{bmatrix}\alpha,\ 1-\alpha+\nu\\ 1-\alpha,\ \nu+1\end{bmatrix}$$

$$R=\mathbf{C}+r\psi(\alpha)+r\psi(1-\alpha)-r\psi(1-\alpha+\nu)-\ln\frac{c^r}{b} \qquad [r<1]$$

W for $r=1$ see 2.13.7.1.

12. $\int_0^\infty x^{\alpha-1}e^{-cx}\begin{Bmatrix}\operatorname{erf}(bx^r)\\ \operatorname{erfc}(bx^r)\end{Bmatrix}L_\nu(cx)\,dx=U$

$\left[r,\ \operatorname{Re}c>0;\ |\arg b|<\pi/4,\ \begin{Bmatrix}\operatorname{Re}\alpha>-r;\ \operatorname{Re}(\alpha-\nu)<1 \text{ for } \nu\neq0,\ 1,\ 2,\ \ldots\\ \operatorname{Re}\alpha>0\end{Bmatrix}\right]$

$$U=\mp\frac{b^{-\alpha/r}}{\sqrt{\pi}}\sum_{k=0}^{\infty}\frac{(\nu+1)_k}{(k!)^2(\alpha+k)}\Gamma\left(\frac{\alpha+k+r}{2r}\right)\left(-\frac{c}{b^{1/r}}\right)^k+\begin{Bmatrix}1\\ 0\end{Bmatrix}c^{-\alpha}\Gamma\begin{bmatrix}\alpha,\ 1-\alpha+\nu\\ 1-\alpha,\ \nu+1\end{bmatrix} \qquad [r>1/2]$$

$$U=\pm\frac{b^{(\nu-\alpha+1)/r}}{\sqrt{\pi}c^{\nu+1}\Gamma(-\nu)}\sum_{k=0}^{\infty}\frac{[(\nu+1)_k]^2}{k!(1-\alpha+\nu+k)}\Gamma\left(\frac{\alpha-\nu-k+r-1}{2r}\right)\left(\frac{b^{1/r}}{c}\right)^k\pm$$

$$\pm\frac{b}{\sqrt{\pi}c^{\alpha+r}\Gamma(\nu+1)}\sum_{k=0}^{\infty}\frac{(-1)^k}{k!(k+1/2)}\Gamma\begin{bmatrix}\alpha+r+2rk,\ 1-\alpha+\nu-r-2rk\\ 1-\alpha-r-2rk\end{bmatrix}\left(\frac{b}{c^r}\right)^{2k}+$$

$$+\begin{Bmatrix}0\\ 1\end{Bmatrix}c^{-\alpha}\Gamma\begin{bmatrix}\alpha,\ 1-\alpha+\nu\\ 1-\alpha,\ \nu+1\end{bmatrix} \qquad [r<1/2]$$

U for $r=1/2$ see 2.13.8.2.

13. $\int_0^\infty x^{\alpha-1}e^{-cx}\begin{Bmatrix}S(bx^r)\\ C(bx^r)\end{Bmatrix}L_\nu(cx)\,dx=V$

$[b,\ r,\ \operatorname{Re}c>0,\ \operatorname{Re}\alpha>-r(\delta+1/2);\ \operatorname{Re}(\alpha-\nu)<1$ for $\nu\neq0,\ 1,\ 2,\ \ldots,\ \delta=(1\pm1)/2]$

$$V=-\frac{b^{-\alpha/r}}{\sqrt{2\pi}}\sum_{k=0}^{\infty}\frac{(\nu+1)_k}{(k!)^2(\alpha+k)}\Gamma\left(\frac{\alpha+k}{r}+\frac{1}{2}\right)\times$$
$$\times\begin{Bmatrix}\sin[(2\alpha+r+2k)\pi/(4r)]\\ \cos[(2\alpha+r+2k)\pi/(4r)]\end{Bmatrix}\left(-\frac{c}{b^{1/r}}\right)^k+\frac{c^{-\alpha}}{2}\Gamma\begin{bmatrix}\alpha,\ 1-\alpha+\nu\\ 1-\alpha,\ \nu+1\end{bmatrix}\quad [r>1],$$

$$V=\frac{b^{(\nu-\alpha+1)/r}}{\sqrt{2\pi}c^{\nu+1}\Gamma(-\nu)}\sum_{k=0}^{\infty}\frac{[(\nu+1)_k]^2}{k!\,(1-\alpha+\nu+k)}\Gamma\left(\frac{\alpha-\nu-k-1}{r}+\frac{1}{2}\right)\times$$
$$\times\begin{Bmatrix}\sin[(r+2\alpha-2\nu-2k-2)\pi/(4r)]\\ \cos[(r+2\alpha-2\nu-2k-2)\pi/(4r)]\end{Bmatrix}\left(\frac{b^{1/r}}{c}\right)^k+$$
$$+\frac{b^{\delta+1/2}}{\sqrt{2\pi}c^{\alpha+r\delta+r/2}\Gamma(\nu+1)}\sum_{k=0}^{\infty}\frac{(-1)^k}{k!\,(\delta+1/2)_k(\delta+1/2+2k)}\times$$
$$\times\Gamma\begin{bmatrix}\alpha+r\delta+r/2+2rk,\ 1-\alpha+\nu-r\delta-r/2-2rk\\ 1-\alpha-r\delta-r/2-2rk\end{bmatrix}\left(\frac{b}{2c^r}\right)^{2k}\quad [r<1],$$

V for $r=1$ see 2.13.9.1.

14. $$\int_0^{\infty}x^{\alpha-1}e^{-cx}\begin{Bmatrix}\gamma(\mu,bx^r)\\ \Gamma(\mu,bx^r)\end{Bmatrix}L_\nu(cx)\,dx=W$$

$$\left[r,\ \operatorname{Re}b,\ \operatorname{Re}c,\ \operatorname{Re}(\alpha+\mu r)>0,\ \begin{Bmatrix}\operatorname{Re}\mu>0,\quad \operatorname{Re}(\alpha-\nu)<1 \text{ for } \nu\neq 0,1,2,\ldots\\ \operatorname{Re}\alpha>0\end{Bmatrix}\right].$$

$$W=\mp b^{-\alpha/r}\sum_{k=0}^{\infty}\frac{(\nu+1)_k}{(k!)^2(\alpha+k)}\Gamma\left(\frac{\alpha+k}{r}+\mu\right)\left(-\frac{c}{b^{1/r}}\right)^k+\begin{Bmatrix}1\\0\end{Bmatrix}c^{-\alpha}\Gamma\begin{bmatrix}\alpha,\ \mu,\ 1-\alpha+\nu\\ 1-\alpha,\ \nu+1\end{bmatrix}$$
$$[r>1],$$

$$W=\pm\frac{b^{(\nu-\alpha+1)/r}}{c^{\nu+1}\Gamma(-\nu)}\sum_{k=0}^{\infty}\frac{[(\nu+1)_k]^2}{k!\,(1-\alpha+\nu+k)}\Gamma\left(\frac{\alpha-\nu-k-1}{r}+\mu\right)\left(\frac{b^{1/r}}{c}\right)^k\pm$$
$$\pm\frac{b^{\mu}}{c^{\alpha+\mu r}\Gamma(\nu+1)}\sum_{k=0}^{\infty}\frac{1}{k!\,(\mu+k)}\Gamma\begin{bmatrix}\alpha+\mu r+rk,\ 1-\alpha+\nu-\mu r-rk\\ 1-\alpha-\mu r-rk\end{bmatrix}\left(-\frac{b}{c^r}\right)^k+$$
$$+c^{-\alpha}\Gamma[\alpha,\ \mu,\ 1-\alpha+\nu]\,\Gamma^{-1}(1-\alpha)\,\Gamma^{-1}(\nu+1)\,(1\mp 1)/2\quad [r<1],$$

W for $r=1$ see 2.13.10.1.

15. $$\int_0^{\infty}x^{\alpha-1}e^{-cx}J_\mu(bx^r)L_\nu(cx)\,dx=U$$

$$[b,\ r,\ \operatorname{Re}c,\ \operatorname{Re}(\alpha+\mu r)>0;\ \operatorname{Re}(\alpha-\nu)<3r/2+1 \text{ for } \nu\neq 0,1,2,\ldots],$$

$$U=\frac{2^{\alpha/r-1}}{b^{\alpha/r}r}\sum_{k=0}^{\infty}\frac{(-1)^k(\nu+1)_k}{(k!)^2}\Gamma\begin{bmatrix}(\mu r+\alpha+k)/(2r)\\ (\mu r-\alpha-k)/(2r)+1\end{bmatrix}\left(\frac{2c^r}{b}\right)^{k/r}\quad [r>1],$$

$$U=\frac{1}{2rc^{\nu+1}\Gamma(-\nu)}\left(\frac{b}{2}\right)^{(\nu-\alpha+1)/r}\sum_{k=0}^{\infty}\frac{[(\nu+1)_k]^2}{k!}\times$$
$$\times\Gamma\begin{bmatrix}(\mu r+\alpha-\nu-k-1)/(2r)\\ (\mu r-\alpha+\nu+k+1)/(2r)+1\end{bmatrix}\left(\frac{b}{2c^r}\right)^{k/r}+\frac{b^{\mu}}{2^{\mu}c^{\alpha+\mu r}\Gamma(\mu+1)\Gamma(\nu+1)}\times$$
$$\times\sum_{k=0}^{\infty}\frac{(-1)^k}{k!\,(\mu+1)_k}\Gamma\begin{bmatrix}\alpha+\mu r+2rk,\ 1-\alpha+\nu-\mu r-2rk\\ 1-\alpha-\mu r-2rk\end{bmatrix}\left(\frac{b}{2c^r}\right)^{2k}\quad [r<1].$$

U for $r=1$ see 2.13.11.1.

16. $\int\limits_0^\infty x^{\alpha-1}e^{-cx}\begin{Bmatrix}Y_\mu(bx^r)\\K_\mu(bx^r)\end{Bmatrix}L_\nu(cx)\,dx=U_\mu$

$\left[r,\ \operatorname{Re}c>0;\ \operatorname{Re}\alpha>r\,|\operatorname{Re}\mu|,\ \begin{Bmatrix}b>0;\ \operatorname{Re}(\alpha-\nu)<3r/2+1\ \text{for}\ \nu\neq0,1,2,\ldots\\ \operatorname{Re}b>0\end{Bmatrix}\right].$

$$U_\mu=\mp\frac{2^{\alpha/r-2}}{b^{\alpha/r}r}\sum_{k=0}^\infty\frac{(\nu+1)_k}{(k!)^2}\Gamma\left(\frac{\mu r+\alpha+k}{2r}\right)\Gamma\left(\frac{\alpha+k-\mu r}{2r}\right)\times$$
$$\times\pi^{(-1\mp1)/2}\left(2\cos\frac{\mu r-\alpha-k}{2r}\pi\right)^{(1\pm1)/2}\left(-\frac{2^{1/r}c}{b^{1/r}}\right)^k\qquad[r>1],$$

$$U_\mu=\mp\frac{c^{-\alpha}}{2\pi\Gamma(\nu+1)}\begin{Bmatrix}2\\\pi\end{Bmatrix}\sum_{k=0}^\infty\frac{1}{k!}\left[A_k(\mu)+\begin{Bmatrix}\cos\mu\pi\\1\end{Bmatrix}A_k(-\mu)\right]\left(\mp\frac{b^2}{4c^{2r}}\right)^k\mp V_\mu,$$
$$A_k(\mu)=\left(\frac{2c^r}{b}\right)^\mu\frac{1}{(1-\mu)_k}\Gamma\left[\begin{matrix}\mu,\ \alpha-\mu r+2rk,\ 1-\alpha+\nu+\mu r-2rk\\1-\alpha+\mu r-2rk\end{matrix}\right],$$

$$V_\mu=\frac{2^{(\alpha-\nu-1)/r-2}}{b^{(\alpha-\nu-1)/r}c^{\nu+1}r\Gamma(-\nu)}\times$$
$$\times\sum_{k=0}^\infty\frac{[(\nu+1)_k]^2}{k!}\Gamma\left(\frac{\mu r+\alpha-\nu-k-1}{2r}\right)\Gamma\left(\frac{\alpha-\mu r-\nu-k-1}{2r}\right)\pi^{(-1\mp1)/2}\times$$
$$\times\left(2\cos\frac{\mu r-\alpha+\nu+k+1}{2r}\pi\right)^{(1\pm1)/2}\left(\frac{b}{2c^r}\right)^{k/r}\qquad[r<1;\ \mu\neq\ldots,-2,-1,0,1,2,\ldots],$$

$$U_m=\mp\frac{c^{-\alpha}}{2\pi\Gamma(\nu+1)}\begin{Bmatrix}2\\\pi\end{Bmatrix}\left\{\sum_{k=0}^{m-1}\frac{A_k(m)}{k!}\left(\mp\frac{b^2}{4c^{2r}}\right)^k+\right.$$
$$+\left(\pm\frac{b}{2c^r}\right)^m\sum_{k=0}^\infty\frac{1}{k!\,(m+k)!}\Gamma\left[\begin{matrix}\alpha+mr+2rk,\ 1-\alpha+\nu-mr-2rk\\1-\alpha-mr-2rk\end{matrix}\right]\times$$
$$\times\Big[\psi(k+1)+\psi(k+m+1)-2r\psi(\alpha+mr+2rk)+$$
$$+2r\psi(1-\alpha+\nu-mr-2rk)-2r\psi(1-\alpha-mr-2rk)+$$
$$\left.+2\ln\frac{2c^r}{b}\Big]\left(\mp\frac{b^2}{4c^{2r}}\right)^k\right\}+V_m\qquad[r<1;\ m=0,1,2,\ldots],$$

U_μ for $r=1$ see 2.13.12.1.

17. $\int\limits_0^\infty x^{\alpha-1}e^{-bx^r-cx}L_m^\gamma(bx^r)\,L_\nu(cx)\,dx=V\qquad[r,\ \operatorname{Re}b,\ \operatorname{Re}c,\ \operatorname{Re}\alpha>0],$

$$V=\frac{b^{-\alpha/r}}{m!\,r}\sum_{k=0}^\infty\frac{(\nu+1)_k}{(k!)^2}\left(1+\gamma-\frac{\alpha+k}{r}\right)_m\Gamma\left(\frac{\alpha+k}{r}\right)\left(-\frac{c}{b^{1/r}}\right)^k\qquad[r>1],$$

$$V=\frac{c^{-\alpha}}{m!\,\Gamma(\nu+1)}\sum_{k=0}^\infty\frac{(\gamma+k+1)_m}{k!}\Gamma\left[\begin{matrix}\alpha+rk,\ 1-\alpha+\nu-rk\\1-\alpha-rk\end{matrix}\right]\left(-\frac{b}{c^r}\right)^k+\frac{b^{(\nu-\alpha+1)/r}}{c^{\nu+1}rm!\,\Gamma(-\nu)}\times$$
$$\times\sum_{k=0}^\infty\frac{[(\nu+1)_k]^2}{k!}\left(1+\gamma+\frac{\nu-\alpha+k+1}{r}\right)_m\Gamma\left(\frac{\alpha-\nu-k-1}{r}\right)\left(\frac{b^{1/r}}{c}\right)^k\qquad[r<1],$$

V for $r=1$ see 2.13.13.1.

18. $$\int_0^\infty x^{\alpha-1}e^{-b^2x^{2r}-cx}\begin{Bmatrix}H_{2m+1}(bx^r)\\H_{2m}(bx^r)\end{Bmatrix}L_\nu(cx)\,dx=W$$

$$\left[r,\ \operatorname{Re}b>0;\ \operatorname{Re}\alpha>-\delta r;\ |\arg c|<\pi/4,\ \delta=\begin{Bmatrix}1\\0\end{Bmatrix}\right].$$

$$W=\frac{(-1)^m 2^{2m+\delta}b^\delta}{c^{\alpha+\delta r}\Gamma(\nu+1)}\times$$

$$\times\sum_{k=0}^\infty\frac{1}{k!}\left(k+\delta+\frac{1}{2}\right)_m\Gamma\begin{bmatrix}\alpha+\delta r+2kr,\ 1+\nu-\alpha-\delta r-2kr\\1-\alpha-\delta r-2kr\end{bmatrix}\left(-\frac{b^2}{c^{2r}}\right)^k+$$

$$+(-1)^m\frac{2^{2m+\delta-1}b^{(1+\nu-\alpha)/r}}{c^{\nu+1}r\Gamma(-\nu)}\sum_{k=0}^\infty\frac{[(\nu+1)_k]^2}{k!}\left(\frac{\nu+\delta r+r-\alpha+k+1}{2r}\right)_m\times$$

$$\times\Gamma\left(\frac{\alpha+\delta r-\nu-k-1}{2r}\right)\left(\frac{b^{1/r}}{c}\right)^k\qquad[r<1/2],$$

$$W=(-1)^m\frac{2^{2m+\delta-1}}{b^{\alpha/r}r}\sum_{k=0}^\infty\frac{(\nu+1)_k}{(k!)^2}\left(\frac{r+\delta r-\alpha-k}{2r}\right)_m\Gamma\left(\frac{\alpha+\delta r+k}{2r}\right)\left(-\frac{c}{b^{1/r}}\right)^k$$

$$[r>1/2],$$

W for $r=1/2$ see 2.13.14.2.

19. $$\int_0^\infty x^{\alpha-1}e^{-bx^r-cx}L_\mu(bx^r)L_\nu(cx)\,dx=U$$

$$[r,\ \operatorname{Re}b,\ \operatorname{Re}c,\ \operatorname{Re}\alpha>0;\ \operatorname{Re}(\alpha-\mu r-\nu)<1+r\ \text{for}\ \mu,\ \nu\neq0,\ 1,\ 2,\ \ldots],$$

$$U=\frac{b^{-\alpha/r}}{r\Gamma(\nu+1)\Gamma(\mu+1)}\times$$

$$\times\sum_{k=0}^\infty\frac{(\nu+1)_k}{(k!)^2}\Gamma\begin{bmatrix}(\alpha+k)/r,\ 1+\mu-(\alpha+k)/r\\1-(\alpha+k)/r\end{bmatrix}\left(-\frac{c}{b^{1/r}}\right)^k+\frac{c^{\mu r+r-\alpha}}{b^{\mu+1}\Gamma(-\mu)\Gamma(\nu+1)}\times$$

$$\times\sum_{k=0}^\infty\frac{[(1+\mu)_k]^2}{k!}\Gamma\begin{bmatrix}\alpha-r-\mu r-rk,\ 1-\alpha+\nu+r+\mu r+rk\\1-\alpha+r+\mu r+rk\end{bmatrix}\left(\frac{c^r}{b}\right)^k\qquad[r>1],$$

U for $r=1$ see 2.13.15.1.

2.13.2. Integrals of $A(x)e^{-px}L_\nu(cx)$.

1. $$\int_0^\infty e^{-px}L_\nu(cx)\,dx=\frac{(p-c)^\nu}{p^{\nu+1}}\qquad[\operatorname{Re}p,\ \operatorname{Re}(p-c)>0].$$

2. $$\int_0^\infty x^{\alpha-1}e^{-px}L_\nu(cx)\,dx=\frac{\Gamma(\alpha)}{(p-c)^\alpha}\,{}_2F_1\left(\alpha,\ \nu+1;\ 1;\ \frac{c}{c-p}\right)$$

$$[\operatorname{Re}p,\ \operatorname{Re}(p-c),\ \operatorname{Re}\alpha>0].$$

3. $$\int_0^\infty x^{\alpha-1}e^{-cx}L_\nu(cx)\,dx=c^{-\alpha}\Gamma\begin{bmatrix}\alpha,\ 1+\nu-\alpha\\1-\alpha,\ \nu+1\end{bmatrix}$$

$$[\operatorname{Re}c,\ \operatorname{Re}\alpha>0;\ \operatorname{Re}(\alpha-\nu)<1\ \text{for}\ \nu\neq0,\ 1,\ 2,\ \ldots].$$

4. $$\int_0^a x^{\alpha-1}(a-x)^{\beta-1}e^{-cx}L_\nu(cx)\,dx=$$

$$=a^{\alpha+\beta-1}\mathrm{B}(\alpha,\ \beta)\,{}_2F_2(\alpha,\ \nu+1;\ 1,\ \alpha+\beta;\ -ac)\qquad[a,\ \operatorname{Re}\alpha,\ \operatorname{Re}\beta>0].$$

5. $\int\limits_a^\infty x^{\alpha-1}(x-a)^{\beta-1}e^{-cx}L_\nu(cx)\,dx=$

$$=a^{\alpha+\beta-1}\mathrm{B}(1-\alpha-\beta,\ \beta)\,{}_2F_2(\alpha,\ \nu+1;\ 1,\ \alpha+\beta;\ -ac)+$$
$$+c^{1-\alpha-\beta}\Gamma\begin{bmatrix}\alpha+\beta-1,\ 2+\nu-\alpha-\beta\\ 2-\alpha-\beta,\ \nu+1\end{bmatrix}\times$$
$$\times{}_2F_2(1-\beta,\ 2+\nu-\alpha-\beta;\ 2-\alpha-\beta,\ 2-\alpha-\beta;\ -ac)$$

[a, $\operatorname{Re}c$, $\operatorname{Re}\beta>0$; $\operatorname{Re}(\alpha+\beta-\nu)<2$ for $\nu\neq0, 1, 2, \ldots$].

6. $\int\limits_0^\infty \frac{x^{\alpha-1}}{(x+z)^\rho}e^{-cx}L_\nu(cx)\,dx=z^{\alpha-\rho}\mathrm{B}(\alpha,\ \rho-\alpha)\,{}_2F_2(\alpha,\ \nu+1;\ 1,\ 1+\alpha-\rho;\ cz)+$

$$+c^{\rho-\alpha}\Gamma\begin{bmatrix}\alpha-\rho,\ 1+\rho+\nu-\alpha\\ \nu+1,\ 1-\alpha+\rho\end{bmatrix}{}_2F_2(\rho,\ 1+\rho+\nu-\alpha;\ 1-\alpha+\rho,\ 1-\alpha+\rho;\ cz)$$

[$\operatorname{Re}c$, $\operatorname{Re}\alpha>0$; $|\arg z|<\pi$; $\operatorname{Re}(\alpha-\rho-\nu)<1$ for $\nu\neq0, 1, 2, \ldots$]

7. $\int\limits_0^\infty \frac{x^{\alpha-1}}{x-y}e^{-cx}L_\nu(cx)\,dx=-\pi y^{\alpha-1}\operatorname{ctg}\alpha\pi\,{}_1F_1(\nu+1;\ 1;\ -cy)+$

$$+c^{1-\alpha}\Gamma\begin{bmatrix}\alpha-1,\ 2-\alpha+\nu\\ 2-\alpha,\ \nu+1\end{bmatrix}{}_2F_2(1,\ 2-\alpha+\nu;\ 2-\alpha,\ 2-\alpha;\ -cy)$$

[y, $\operatorname{Re}c$, $\operatorname{Re}\alpha>0$; $\operatorname{Re}(\alpha-\nu)<2$ for $\nu\neq0, 1, 2, \ldots$]

8. $\int\limits_0^a x^{\alpha-1}(a^2-x^2)^{\beta-1}e^{-cx}L_\nu(cx)\,dx=$

$$=\frac{a^{\alpha+2\beta-2}}{2}\mathrm{B}\left(\frac{\alpha}{2},\ \beta\right){}_3F_4\left(\frac{\alpha}{2},\ \frac{1+\nu}{2},\ 1+\frac{\nu}{2};\ \frac{1}{2},\ \frac{1}{2},\ 1,\ \frac{\alpha}{2}+\beta;\ \frac{a^2c^2}{4}\right)-$$
$$-\frac{a^{\alpha+2\beta-1}c(\nu+1)}{2}\mathrm{B}\left(\frac{\alpha+1}{2},\ \beta\right)\times$$
$$\times{}_3F_4\left(\frac{\alpha+1}{2},\ 1+\frac{\nu}{2},\ \frac{3+\nu}{2};\ 1,\ \frac{3}{2},\ \frac{3}{2},\ \beta+\frac{\alpha+1}{2};\ \frac{a^2c^2}{4}\right)$$

[a, $\operatorname{Re}\alpha$, $\operatorname{Re}\beta>0$]

9. $\int\limits_a^\infty x^{\alpha-1}(x^2-a^2)^{\beta-1}e^{-cx}L_\nu(cx)\,dx=$

$$=\frac{a^{\alpha+2\beta-2}}{2}\mathrm{B}\left(1-\frac{\alpha}{2}-\beta,\ \beta\right){}_3F_4\left(\frac{\alpha}{2},\ \frac{1+\nu}{2},\ 1+\frac{\nu}{2};\ \frac{1}{2},\ \frac{1}{2},\ 1,\ \frac{\alpha}{2}+\beta;\ \frac{a^2c^2}{4}\right)-$$
$$-\frac{a^{\alpha+2\beta-1}c(\nu+1)}{2}\mathrm{B}\left(\frac{1-\alpha}{2}-\beta,\ \beta\right)\times$$
$$\times{}_3F_4\left(\frac{1+\alpha}{2},\ 1+\frac{\nu}{2},\ \frac{3+\nu}{2};\ 1,\ \frac{3}{2},\ \frac{3}{2},\ \frac{1+\alpha}{2}+\beta;\ \frac{a^2c^2}{4}\right)+$$
$$+c^{2-\alpha-2\beta}\Gamma\begin{bmatrix}\alpha+2\beta-2,\ 3-\alpha+\nu-2\beta\\ 3-\alpha-2\beta,\ \nu+1\end{bmatrix}\times$$
$$\times{}_3F_4\left(1-\beta,\ \frac{3-\alpha+\nu}{2}-\beta,\ 2-\beta+\frac{\nu-\alpha}{2};\ 2-\frac{\alpha}{2}-\beta,\ 2-\frac{\alpha}{2}-\beta,\right.$$
$$\left.\frac{3-\alpha}{2}-\beta,\ \frac{3-\alpha}{2}-\beta;\ \frac{a^2c^2}{4}\right)$$

[a, $\operatorname{Re}c$, $\operatorname{Re}\beta>0$; $\operatorname{Re}(\alpha+2\beta-\nu)<3$ for $\nu\neq0, 1, 2, \ldots$]

10. $\int_0^\infty \frac{x^{\alpha-1}}{(x^2+z^2)^\rho} e^{-cx} L_\nu(cx)\,dx=$

$$=\frac{z^{\alpha-2\rho}}{2} \mathrm{B}\left(\frac{\alpha}{2},\, \rho-\frac{\alpha}{2}\right) {}_3F_4\left(\frac{\alpha}{2}, \frac{1+\nu}{2}, 1+\frac{\nu}{2}; \frac{1}{2}, \frac{1}{2}, 1, 1+\frac{\alpha}{2}-\rho;\, -\frac{c^2z^2}{4}\right)-$$

$$-\frac{cz^{\alpha-2\rho+1}(\nu+1)}{2} \mathrm{B}\left(\frac{\alpha+1}{2},\, \rho-\frac{\alpha+1}{2}\right)\times$$

$$\times {}_3F_4\left(\frac{\alpha+1}{2}, 1+\frac{\nu}{2}, \frac{3+\nu}{2}; 1, \frac{3}{2}, \frac{3}{2}, \frac{3+\alpha}{2}-\rho;\, -\frac{c^2z^2}{4}\right)+$$

$$+c^{2\rho-\alpha}\Gamma\begin{bmatrix}\alpha-2\rho, & 1-\alpha+\nu+2\rho \\ \nu+1, & 1-\alpha+2\rho\end{bmatrix} {}_3F_4\left(\rho,\, \frac{1-\alpha+\nu}{2}+\rho,\, 1+\rho+\frac{\nu-\alpha}{2};\right.$$

$$\left.\frac{1-\alpha}{2}+\rho,\, \frac{1-\alpha}{2}+\rho,\, 1-\frac{\alpha}{2}+\rho,\, 1-\frac{\alpha}{2}+\rho;\, -\frac{c^2z^2}{4}\right)$$

[$\mathrm{Re}\, c$, $\mathrm{Re}\, z$, $\mathrm{Re}\,\alpha>0$; $\mathrm{Re}\,(\alpha-2\rho-\nu)<1$ for $\nu\neq 0, 1, 2, \ldots$].

11. $\int_0^\infty \frac{x^{\alpha-1}}{x^2-y^2} e^{-cx} L_\nu(cx)\,dx=$

$$=-\frac{\pi y^{\alpha-2}}{2}\,\mathrm{ctg}\,\frac{\alpha\pi}{2}\, {}_2F_3\left(\frac{\nu+1}{2}, \frac{\nu}{2}+1; \frac{1}{2}, \frac{1}{2}, 1; \frac{c^2y^2}{4}\right)-$$

$$-\frac{\pi c y^{\alpha-1}(\nu+1)}{2}\,\mathrm{tg}\,\frac{\alpha\pi}{2}\, {}_2F_3\left(\frac{\nu}{2}+1,\, \frac{\nu+3}{2}; 1, \frac{3}{2}, \frac{3}{2}; \frac{c^2y^2}{4}\right)+$$

$$+c^{2-\alpha}\Gamma\begin{bmatrix}\alpha-2, & 3-\alpha+\nu \\ 3-\alpha, & \nu+1\end{bmatrix} {}_3F_4\left(1,\, \frac{3-\alpha+\nu}{2}, 2+\frac{\nu-\alpha}{2};\right.$$

$$\left.\frac{3-\alpha}{2}, \frac{3-\alpha}{2}, 2-\frac{\alpha}{2}, 2-\frac{\alpha}{2};\, \frac{c^2y^2}{4}\right)$$

[y, $\mathrm{Re}\, c$, $\mathrm{Re}\,\alpha>0$; $\mathrm{Re}\,(\alpha-\nu)<3$ for $\nu\neq 0, 1, 2, \ldots$].

12. $\int_0^a x^{\alpha-1}(\sqrt{a}-\sqrt{x})^{\beta-1} e^{-cx} L_\nu(cx)\,dx=$

$$=2a^{\alpha+(\beta-1)/2}\mathrm{B}(2\alpha, \beta)\, {}_3F_3\left(\alpha,\, \alpha+\frac{1}{2},\, \nu+1;\, 1,\, \alpha+\frac{\beta}{2},\, \alpha+\frac{\beta+1}{2};\, -ac\right)$$

[a, $\mathrm{Re}\,\alpha$, $\mathrm{Re}\,\beta>0$].

13. $\int_a^\infty x^{\alpha-1}(\sqrt{x}-\sqrt{a})^{\beta-1} e^{-cx} L_\nu(cx)\,dx=$

$$=2a^{\alpha+(\beta-1)/2}\mathrm{B}(1-\beta-2\alpha, \beta)\, {}_3F_3\left(\alpha,\, \frac{1}{2}+\alpha,\, 1+\nu;\, \alpha+\frac{\beta+1}{2},\, \alpha+\frac{\beta}{2},\, 1;\, -ac\right)+$$

$$+c^{(1-\beta)/2-\alpha}\Gamma\begin{bmatrix}\alpha+(\beta-1)/2, & (3-\beta)/2-\alpha+\nu \\ (3-\beta)/2-\alpha, & \nu+1\end{bmatrix}\times$$

$$\times {}_3F_3\left(\frac{1-\beta}{2},\, 1-\frac{\beta}{2},\, \frac{3-\beta}{2}-\alpha+\nu;\, \frac{1}{2},\, \frac{3-\beta}{2}-\alpha,\, \frac{3-\beta}{2}-\alpha;\, -ac\right)+$$

$$+\sqrt{a}\, c^{1-\alpha-\beta/2}(1-\beta)\,\Gamma\begin{bmatrix}\alpha+\beta/2-1, & 2-\alpha-\beta/2+\nu \\ 2-\alpha-\beta/2, & \nu+1\end{bmatrix}\times$$

$$\times {}_3F_3\left(1-\frac{\beta}{2},\, \frac{3-\beta}{2},\, 2-\alpha-\frac{\beta}{2}+\nu;\, \frac{3}{2},\, 2-\alpha-\frac{\beta}{2},\, 2-\alpha-\frac{\beta}{2};\, -ac\right)$$

[a, $\mathrm{Re}\, c$, $\mathrm{Re}\,\beta>0$; $\mathrm{Re}\,(2\alpha+\beta-2\nu)<3$ for $\nu\neq 0, 1, 2 \ldots$].

14. $\int\limits_0^\infty \frac{x^{\alpha-1}}{(\sqrt{x}+\sqrt{z})^\rho} e^{-cx} L_\nu(cx)\,dx=$

$$=2z^{\alpha-\rho/2}\,\mathrm{B}(2\alpha,\ \rho-2\alpha)\,{}_3F_3\left(\alpha,\alpha+\frac{1}{2},\ \nu+1;\ 1,\ 1-\frac{\rho}{2}+\alpha,\frac{1-\rho}{2}+\alpha;-cz\right)+$$

$$+c^{\rho/2-\alpha}\Gamma\begin{bmatrix}\alpha-\rho/2, & 1-\alpha+\nu+\rho/2\\ \nu+1, & 1-\alpha+\rho/2\end{bmatrix}\times$$

$$\times{}_3F_3\left(\frac{\rho}{2},\ \frac{\rho+1}{2},\ 1-\alpha+\nu+\frac{\rho}{2};\ \frac{1}{2},1-\alpha+\frac{\rho}{2},\ 1-\alpha+\frac{\rho}{2};\ -cz\right)-$$

$$-\sqrt{z}\,c^{(\rho+1)/2-\alpha}\,\rho\Gamma\begin{bmatrix}\alpha-(\rho+1)/2, & (3+\rho)/2-\alpha+\nu\\ \nu+1, & (3+\rho)/2-\alpha\end{bmatrix}\times$$

$$\times{}_3F_3\left(\frac{\rho+1}{2},\ \frac{\rho}{2}+1,\ \frac{3+\rho}{2}-\alpha+\nu;\ \frac{3}{2},\frac{3+\rho}{2}-\alpha,\ \frac{3+\rho}{2}-\alpha;\ -cz\right)$$

[$\operatorname{Re} c$, $\operatorname{Re}\alpha>0$; $|\arg z|<2\pi$; $\operatorname{Re}(2\alpha-\rho-2\nu)<2$ for $\nu\neq 0, 1, 2, \ldots$].

15. $\int\limits_0^\infty \frac{x^{\alpha-1}}{\sqrt{x}-\sqrt{y}} e^{-cx} L_\nu(cx)\,dx=$

$$=-2\pi y^{\alpha-1/2}\operatorname{ctg}2\alpha\pi\,{}_1F_1(\nu+1;\ 1;\ -cy)+c^{1/2-\alpha}\Gamma\begin{bmatrix}\alpha-1/2, & 3/2-\alpha+\nu\\ 3/2-\alpha, & \nu+1\end{bmatrix}\times$$

$$\times{}_2F_2\left(\frac{3}{2}-\alpha+\nu,\ 1;\ \frac{3}{2}-\alpha,\ \frac{3}{2}-\alpha;\ -cy\right)+$$

$$+c^{1-\alpha}\sqrt{y}\,\Gamma\begin{bmatrix}\alpha-1, & 2-\alpha+\nu\\ 2-\alpha, & \nu+1\end{bmatrix}{}_2F_2(2-\alpha+\nu,\ 1;\ 2-\alpha,\ 2-\alpha;\ -cy)$$

[y, $\operatorname{Re} c$, $\operatorname{Re}\alpha>0$; $\operatorname{Re}(\alpha-\nu)<3/2$ for $\nu\neq 0, 1, 2, \ldots$].

2.13.3. Integrals of $x^\alpha e^{\varphi(x)} L_\nu(cx)$.

1. $\int\limits_0^\infty x^{\alpha-1} e^{-px^2-cx} L_\nu(cx)\,dx=$

$$=\frac{p^{-\alpha/2}}{2}\Gamma\left(\frac{\alpha}{2}\right){}_3F_3\left(\frac{\alpha}{2},\ \frac{1+\nu}{2},1+\frac{\nu}{2};\ \frac{1}{2},\ \frac{1}{2},1;\ \frac{c^2}{4p}\right)-$$

$$-\frac{c(\nu+1)}{2p^{(\alpha+1)/2}}\Gamma\left(\frac{\alpha+1}{2}\right){}_3F_3\left(\frac{\alpha+1}{2},1+\frac{\nu}{2},\ \frac{3+\nu}{2};\ 1,\frac{3}{2},\ \frac{3}{2};\ \frac{c^2}{4p}\right)$$

[$\operatorname{Re} p$, $\operatorname{Re}\alpha>0$].

2. $\int\limits_0^\infty x^{\alpha-1} e^{-p\sqrt{x}-cx} L_\nu(cx)\,dx=\frac{2p^{2\nu-2\alpha+2}}{c^{\nu+1}}\Gamma\begin{bmatrix}2\alpha-2\nu-2\\ -\nu\end{bmatrix}\times$

$$\times{}_2F_2\left(\nu+1,\ \nu+1;\ 2-\alpha+\nu;\ \frac{3}{2}-\alpha+\nu;\ \frac{p^2}{4c}\right)+$$

$$+c^{-\alpha}\Gamma\begin{bmatrix}\alpha, & 1-\alpha+\nu\\ 1-\alpha, & \nu+1\end{bmatrix}{}_2F_2\left(\alpha,\ \alpha;\ \frac{1}{2},\ \alpha-\nu;\ \frac{p^2}{4c}\right)-$$

$$-\frac{p}{c^{\alpha+1/2}}\Gamma\begin{bmatrix}\alpha+1/2, & 1/2-\alpha+\nu\\ 1/2-\alpha, & \nu+1\end{bmatrix}{}_2F_2\left(\alpha+\frac{1}{2},\ \alpha+\frac{1}{2};\ \frac{3}{2},\ \frac{1}{2}+\alpha-\nu;\ \frac{p^2}{4c}\right)$$

[$\operatorname{Re} c$, $\operatorname{Re} p$, $\operatorname{Re}\alpha>0$].

3. $$\int_0^\infty x^{\alpha-1} e^{-p/x-cx} L_\nu(cx)\,dx = p^\alpha \Gamma(-\alpha)\,{}_1F_2(\nu+1;\ \alpha+1,\ 1;\ cp) + c^{-\alpha}\Gamma\begin{bmatrix}\alpha, & 1-\alpha+\nu\\ 1-\alpha, & \nu+1\end{bmatrix} {}_1F_2(1-\alpha+\nu;\ 1-\alpha,\ 1-\alpha;\ cp)$$
[$\operatorname{Re} c,\ \operatorname{Re} p > 0$; $\operatorname{Re}(\alpha-\nu) < 1$ for $\nu \neq 0, 1, 2, \ldots$].

4. $$\int_0^\infty x^{\alpha-1} e^{-p/x^2-cx} L_\nu(cx)\,dx = \frac{p^{\alpha/2}}{2}\Gamma\left(-\frac{\alpha}{2}\right) {}_2F_4\left(\frac{1+\nu}{2},\ 1+\frac{\nu}{2};\ \frac{1}{2},\ \frac{1}{2},\ 1,\ 1+\frac{\alpha}{2};\ -\frac{c^2p}{4}\right) - \frac{cp^{(\alpha+1)/2}}{2}\Gamma\left(-\frac{\alpha+1}{2}\right)(\nu+1)\times$$
$$\times {}_2F_4\left(1+\frac{\nu}{2},\ \frac{3+\nu}{2};\ 1,\ \frac{3}{2},\ \frac{3}{2},\ \frac{3+\alpha}{2};\ -\frac{c^2p}{4}\right) + c^{-\alpha}\Gamma\begin{bmatrix}\alpha, & 1-\alpha+\nu\\ 1-\alpha, & \nu+1\end{bmatrix} {}_2F_4\left(\frac{1-\alpha+\nu}{2},\ 1+\frac{\nu-\alpha}{2};\ \frac{1-\alpha}{2},\ \frac{1-\alpha}{2},\ 1-\frac{\alpha}{2},\ 1-\frac{\alpha}{2};\ -\frac{c^2p}{4}\right)$$
[$\operatorname{Re} c,\ \operatorname{Re} p > 0$; $\operatorname{Re}(\alpha-\nu) < 1$ for $\nu \neq 0, 1, 2, \ldots$].

5. $$\int_0^\infty x^{\alpha-1} e^{-p/\sqrt{x}-cx} L_\nu(cx)\,dx = 2p^{2\alpha}\Gamma(-2\alpha)\,{}_1F_3\left(1+\nu;\ 1,\ 1+\alpha,\ \frac{1}{2}+\alpha;\ -\frac{cp^2}{4}\right) + c^{-\alpha}\Gamma\begin{bmatrix}\alpha, & 1-\alpha+\nu\\ 1-\alpha, & 1+\nu\end{bmatrix} {}_1F_3\left(1-\alpha+\nu;\ \frac{1}{2},\ 1-\alpha,\ 1-\alpha;\ -\frac{cp^2}{4}\right) - \frac{p}{c^{\alpha-1/2}}\Gamma\begin{bmatrix}\alpha-1/2, & 3/2-\alpha+\nu\\ 3/2-\alpha, & \nu+1\end{bmatrix}\times$$
$$\times {}_1F_3\left(\frac{3}{2}-\alpha+\nu;\ \frac{3}{2},\ \frac{3}{2}-\alpha,\ \frac{3}{2}-\alpha;\ -\frac{cp^2}{4}\right)$$
[$\operatorname{Re} c,\ \operatorname{Re} p > 0$; $\operatorname{Re}(\alpha-\nu) < 1$ for $\nu \neq 0, 1, 2, \ldots$].

2.13.4. Integrals of $x^\alpha e^{-cx}\begin{Bmatrix}\sin bx^r\\ \cos bx^r\end{Bmatrix} L_\nu(cx)$.

1. $$\int_0^\infty x^{\alpha-1} e^{-cx}\begin{Bmatrix}\sin bx\\ \cos bx\end{Bmatrix} L_\nu(cx)\,dx = \frac{\Gamma(\alpha)}{b^\alpha}\begin{Bmatrix}\sin(\alpha\pi/2)\\ \cos(\alpha\pi/2)\end{Bmatrix} {}_4F_3\left(\frac{\nu+1}{2},\ 1+\frac{\nu}{2},\ \frac{\alpha}{2},\ \frac{\alpha+1}{2};\ 1,\ \frac{1}{2},\ \frac{1}{2};\ -\frac{c^2}{b^2}\right) \mp \frac{c\Gamma(\alpha+1)(\nu+1)}{b^{\alpha+1}}\begin{Bmatrix}\cos(\alpha\pi/2)\\ \sin(\alpha\pi/2)\end{Bmatrix} {}_4F_3\left(\frac{\alpha+1}{2},\ \frac{\alpha}{2}+1,\ 1+\frac{\nu}{2},\ \frac{\nu+3}{2};\ 1,\ \frac{3}{2},\ \frac{3}{2};\ -\frac{c^2}{b^2}\right)$$
[$b,\ \operatorname{Re} c > 0$; $\operatorname{Re}\alpha > -(1 \pm 1)/2$; $\operatorname{Re}(\alpha-\nu) < 2$ for $\nu \neq 0, 1, 2, \ldots$].

2. $$\int_0^\infty x^{\alpha-1} e^{-cx}\begin{Bmatrix}\sin b\sqrt{x}\\ \cos b\sqrt{x}\end{Bmatrix} L_\nu(cx)\,dx = -\frac{2b^{2\nu-2\alpha+2}}{c^{\nu+1}}\Gamma\begin{bmatrix}2\alpha-2\nu-2\\ -\nu\end{bmatrix}\begin{Bmatrix}\sin(\alpha-\nu)\pi\\ \cos(\alpha-\nu)\pi\end{Bmatrix}\times$$

$$\times {}_2F_2\left(\nu+1,\ \nu+1;\ 2-\alpha+\nu,\ \frac{3}{2}-\alpha+\nu;\ -\frac{b^2}{4c}\right)+$$

$$+\frac{b^\delta}{c^{\alpha+\delta/2}}\Gamma\begin{bmatrix}\alpha+\delta/2,\ 1-\alpha-\delta/2+\nu\\ \nu+1,\ 1-\alpha-\delta/2\end{bmatrix}\times$$

$$\times {}_2F_2\left(\alpha+\frac{\delta}{2},\ \alpha+\frac{\delta}{2};\ \delta+\frac{1}{2},\ \alpha+\frac{\delta}{2}-\nu;\ -\frac{b^2}{4c}\right)$$

$\left[b,\ \operatorname{Re} c>0;\ \operatorname{Re}\alpha>-\delta/2;\ \operatorname{Re}(\alpha-\nu)<3/2 \text{ for } \nu\neq 0,\ 1,\ 2,\ \ldots;\ \delta=\begin{Bmatrix}1\\0\end{Bmatrix}\right].$

2.13.5. Integrals of $x^\alpha e^{-cx}\begin{Bmatrix}\ln(x^n+z^n)\\ \ln|x^n-z^n|\end{Bmatrix}L_\nu(cx)$.

1. $$\int_0^\infty x^{\alpha-1}e^{-cx}\begin{Bmatrix}\ln(x+z)\\ \ln|x-z|\end{Bmatrix}L_\nu(cx)\,dx=$$

$$=\frac{\pi z^\alpha}{\alpha}\begin{Bmatrix}\operatorname{cosec}\alpha\pi\\ \operatorname{ctg}\alpha\pi\end{Bmatrix}{}_2F_2(\alpha,\ \nu+1;\ 1,\ \alpha+1;\ \pm cz)\pm$$

$$\pm c^{1-\alpha}z\Gamma\begin{bmatrix}\alpha-1,\ 2-\alpha+\nu\\ 2-\alpha,\ \nu+1\end{bmatrix}{}_3F_3(1,\ 1,\ 2-\alpha+\nu;\ 2-\alpha,\ 2-\alpha,\ 2;\ \pm cz)+$$

$$+c^{-\alpha}\Gamma\begin{bmatrix}\alpha,\ 1-\alpha+\nu\\ 1-\alpha,\ \nu+1\end{bmatrix}[\psi(\alpha)+\psi(1-\alpha)-\psi(1-\alpha+\nu)-\ln c]$$

$\left[\operatorname{Re} c,\ \operatorname{Re}\alpha>0,\ \begin{Bmatrix}|\arg z|<\pi\\ z>0\end{Bmatrix};\ \operatorname{Re}(\alpha-\nu)<1 \text{ for } \nu\neq 0,\ 1,\ 2,\ \ldots\right].$

2. $$\int_0^\infty x^{\alpha-1}e^{-cx}\begin{Bmatrix}\ln(x^2+z^2)\\ \ln|x^2-z^2|\end{Bmatrix}L_\nu(cx)\,dx=$$

$$=\frac{\pi z^\alpha}{\alpha}\begin{Bmatrix}\operatorname{cosec}(\alpha\pi/2)\\ \operatorname{ctg}(\alpha\pi/2)\end{Bmatrix}{}_3F_4\left(\frac{\alpha}{2},\ \frac{\nu+1}{2},\ \frac{\nu}{2}+1;\ \frac{1}{2},\ \frac{1}{2},\ 1,\ \frac{\alpha}{2}+1;\ \mp\frac{c^2z^2}{4}\right)\mp$$

$$\mp\frac{\pi c z^{\alpha+1}(\nu+1)}{\alpha+1}\begin{Bmatrix}\sec(\alpha\pi/2)\\ \operatorname{tg}(\alpha\pi/2)\end{Bmatrix}\times$$

$$\times {}_3F_4\left(\frac{\alpha+1}{2},\ \frac{\nu}{2}+1,\ \frac{\nu+3}{2};\ 1,\ \frac{3}{2},\ \frac{3}{2},\ \frac{\alpha+3}{2};\ \mp\frac{c^2z^2}{4}\right)\pm$$

$$\pm c^{2-\alpha}z^2\Gamma\begin{bmatrix}\alpha-2,\ 3+\nu-\alpha\\ \nu+1,\ 3-\alpha\end{bmatrix}{}_4F_5\left(1,\ 1,\ \frac{3+\nu-\alpha}{2},\ 2+\frac{\nu-\alpha}{2};\right.$$

$$\left.2,\ \frac{3-\alpha}{2},\ \frac{3-\alpha}{2},\ 2-\frac{\alpha}{2},\ 2-\frac{\alpha}{2};\ \mp\frac{c^2z^2}{4}\right)+$$

$$+\frac{2}{c^\alpha}\Gamma\begin{bmatrix}\alpha,\ 1-\alpha+\nu\\ 1-\alpha,\ \nu+1\end{bmatrix}[\psi(\alpha)+\psi(1-\alpha)-\psi(1-\alpha+\nu)-\ln c]$$

$\left[\operatorname{Re} c,\ \operatorname{Re}\alpha>0,\ \begin{Bmatrix}\operatorname{Re} z>0\\ z>0\end{Bmatrix};\ \operatorname{Re}(\alpha-\nu)<1 \text{ for } \nu\neq 0,\ 1,\ 2,\ \ldots\right].$

2.13.6. Integrals of $x^\alpha e^{\varphi(x)}\operatorname{Ei}(bx^k)L_\nu(cx)$.

1. $$\int_0^\infty x^{\alpha-1}e^{-cx}\operatorname{Ei}(-bx)L_\nu(cx)\,dx=$$

$$=-\frac{\Gamma(\alpha)}{\alpha b^\alpha}{}_3F_2\left(\alpha,\ \alpha,\ \nu+1;\ 1,\ \alpha+1;\ -\frac{c}{b}\right)\quad[\operatorname{Re} b,\ \operatorname{Re}(b+c),\ \operatorname{Re}\alpha>0].$$

2. $\int\limits_0^\infty x^{\alpha-1} e^{(\pm b-c)x} \operatorname{Ei}(\mp bx) L_\nu(cx)\,dx =$

$$= -\frac{\pi}{b^\alpha}\Gamma(\alpha)\begin{Bmatrix}\operatorname{cosec}\alpha\pi\\ \operatorname{ctg}\alpha\pi\end{Bmatrix} {}_2F_1\left(\alpha,\ \nu+1;\ 1;\ \pm\frac{c}{b}\right) \mp$$

$$\mp \frac{c^{1-\alpha}}{b}\Gamma\begin{bmatrix}\alpha-1,\ 2-\alpha+\nu\\ 2-\alpha,\ \nu+1\end{bmatrix} {}_3F_2\left(1,\ 1,\ 2-\alpha+\nu;\ 2-\alpha,\ 2-\alpha;\ \pm\frac{c}{b}\right)$$

$\left[\operatorname{Re} c,\ \operatorname{Re}\alpha > 0,\ \begin{Bmatrix}|\arg b| < \pi\\ b > 0\end{Bmatrix};\ \operatorname{Re}(\alpha-\nu) < 2 \text{ for } \nu \neq 0,\ 1,\ 2,\ \ldots\right].$

3. $\int\limits_0^\infty x^{\alpha-1} e^{-cx} \operatorname{Ei}(-bx^2) L_\nu(cx)\,dx =$

$$= -\frac{\Gamma(\alpha/2)}{\alpha b^{\alpha/2}}\, {}_4F_4\left(\frac{\alpha}{2},\ \frac{\alpha}{2},\ \frac{\nu+1}{2},\ \frac{\nu}{2}+1;\ \frac{1}{2},\ \frac{1}{2},\ 1, \frac{\alpha}{2}+1;\ \frac{c^2}{4b}\right) +$$

$$+ \frac{c(\nu+1)}{(\alpha+1)\, b^{(\alpha+1)/2}}\Gamma\left(\frac{\alpha+1}{2}\right) {}_4F_4\left(\frac{\alpha+1}{2},\ \frac{\alpha+1}{2},\ 1+\frac{\nu}{2},\ \frac{\nu+3}{2};\right.$$

$$\left. 1,\ \frac{3}{2},\ \frac{3}{2},\ \frac{\alpha+3}{2};\ \frac{c^2}{4b}\right) \qquad [\operatorname{Re} b,\ \operatorname{Re}\alpha > 0].$$

4. $\int\limits_0^\infty x^{\alpha-1} e^{\pm bx^2-cx} \operatorname{Ei}(\mp bx^2) L_\nu(cx)\,dx =$

$$= -\frac{\pi}{2b^{\alpha/2}}\Gamma\left(\frac{\alpha}{2}\right)\begin{Bmatrix}\operatorname{cosec}(\alpha\pi/2)\\ \operatorname{ctg}(\alpha\pi/2)\end{Bmatrix} {}_3F_3\left(\frac{\alpha}{2},\ \frac{1+\nu}{2},\ 1+\frac{\nu}{2};\ \frac{1}{2},\ \frac{1}{2},\ 1;\ \mp\frac{c^2}{4b}\right) \pm$$

$$\pm \frac{\pi c(\nu+1)}{2b^{(\alpha+1)/2}}\Gamma\left(\frac{\alpha+1}{2}\right)\begin{Bmatrix}\sec(\alpha\pi/2)\\ \operatorname{tg}(\alpha\pi/2)\end{Bmatrix} \times$$

$$\times {}_3F_3\left(\frac{\alpha+1}{2},\ 1+\frac{\nu}{2},\ \frac{\nu+3}{2};\ 1,\ \frac{3}{2},\ \frac{3}{2};\ \mp\frac{c^2}{4b}\right) \mp$$

$$\mp \frac{c^{2-\alpha}}{b}\Gamma\begin{bmatrix}\alpha-2,\ 3-\alpha+\nu\\ 3-\alpha,\ \nu+1\end{bmatrix} {}_4F_4\left(1,\ 1,\ \frac{3-\alpha+\nu}{2},\ 2+\frac{\nu-\alpha}{2};\right.$$

$$\left. 2-\frac{\alpha}{2},\ 2-\frac{\alpha}{2},\ \frac{3-\alpha}{2},\ \frac{3-\alpha}{2};\ \mp\frac{c^2}{4b}\right)$$

$\left[\operatorname{Re} c,\ \operatorname{Re}\alpha > 0,\ \begin{Bmatrix}|\arg b| < \pi\\ b > 0\end{Bmatrix};\ \operatorname{Re}(\alpha-\nu) < 3 \text{ for } \nu \neq\ ,\ 1,\ 2,\ \ldots\right].$

2.13.7. Integrals of $x^\alpha e^{-cx}\begin{Bmatrix}\operatorname{si}(bx^r)\\ \operatorname{ci}(bx^r)\end{Bmatrix} L_\nu(cx)$.

1. $\int\limits_0^\infty x^{\alpha-1} e^{-cx}\begin{Bmatrix}\operatorname{si}(bx)\\ \operatorname{ci}(bx)\end{Bmatrix} L_\nu(cx)\,dx = -\frac{\Gamma(\alpha)}{\alpha b^\alpha}\begin{Bmatrix}\sin(\alpha\pi/2)\\ \cos(\alpha\pi/2)\end{Bmatrix}\times$

$$\times {}_5F_4\left(\frac{\alpha}{2},\ \frac{\alpha}{2},\ \frac{\alpha+1}{2},\ \frac{\nu+1}{2},\ \frac{\nu}{2}+1;\ \frac{1}{2},\ \frac{1}{2},\ 1,\ \frac{\alpha}{2}+1;\ -\frac{c^2}{b^2}\right) \pm$$

$$\pm \frac{c\Gamma(\alpha+1)(\nu+1)}{(\alpha+1)\, b^{\alpha+1}}\begin{Bmatrix}\cos(\alpha\pi/2)\\ \sin(\alpha\pi/2)\end{Bmatrix} {}_5F_4\left(\frac{\alpha+1}{2},\ \frac{\alpha+1}{2},\ \frac{\alpha}{2}+1,\ \frac{\nu}{2}+1,\ \frac{\nu+3}{2};\right.$$

$$\left. 1,\ \frac{3}{2},\ \frac{3}{2},\ \frac{\alpha+3}{2};\ -\frac{c^2}{b^2}\right) \qquad [b,\ \operatorname{Re} c,\ \operatorname{Re}\alpha > 0;\ \operatorname{Re}(\alpha-\nu) < 3 \text{ for } \nu\neq 0,\ 1,\ 2,\ \ldots].$$

2. $\int\limits_0^\infty x^{\alpha-1}e^{-cx}\begin{Bmatrix}\operatorname{si}(b\sqrt{x})\\ \operatorname{ci}(b\sqrt{x})\end{Bmatrix}L_\nu(cx)\,dx=$

$$=-\frac{b^{2\nu-2\alpha+2}}{c^{\nu+1}(1-\alpha+\nu)}\Gamma\begin{bmatrix}2\alpha-2\nu-2\\ -\nu\end{bmatrix}\begin{Bmatrix}\sin(\alpha-\nu)\pi\\ \cos(\alpha-\nu)\pi\end{Bmatrix}\times$$

$$\times{}_3F_3\left(\nu+1,\ \nu+1,\ 1-\alpha+\nu;\ \frac{3}{2}-\alpha+\nu,\ 2-\alpha+\nu,\ 2-\alpha+\nu;\ -\frac{b^2}{4c}\right)-$$

$$-\frac{b^{\delta+2}}{2\cdot 3^\delta c^{\alpha+\delta/2+1}(2+\delta)}\Gamma\begin{bmatrix}\alpha+\delta/2+1,\ \nu-\alpha-\delta/2\\ -\alpha-\delta/2,\ \nu+1\end{bmatrix}\times$$

$$\times{}_4F_4\left(1,\ 1+\frac{\delta}{2},\ 1+\alpha+\frac{\delta}{2},\ 1+\alpha+\frac{\delta}{2};\ 2,\ 2+\frac{\delta}{2},\ 1+\alpha-\nu+\frac{\delta}{2},\ \delta+\frac{3}{2};-\frac{b^2}{4c}\right)+$$

$$+\frac{b^\delta}{c^{\alpha+\delta/2}}\Gamma\begin{bmatrix}\alpha+\delta/2,\ 1-\alpha+\nu-\delta/2\\ 1-\alpha-\delta/2,\ \nu+1\end{bmatrix}\begin{Bmatrix}1\\ R\end{Bmatrix}-\begin{Bmatrix}1\\ 0\end{Bmatrix}\frac{\pi}{2c^\alpha}\Gamma\begin{bmatrix}\alpha,\ 1+\nu-\alpha\\ 1-\alpha,\ \nu+1\end{bmatrix},$$

$$R=\mathbf{C}+\frac{1}{2}[\psi(\alpha)+\psi(1-\alpha)-\psi(1-\alpha+\nu)]-\ln\frac{\sqrt{c}}{b}$$

$\left[b,\ \operatorname{Re}c,\ \operatorname{Re}\alpha>0;\ \operatorname{Re}(\alpha-\nu)<2 \text{ for } \nu\neq 0, 1, 2, \ldots;\ \delta=\begin{Bmatrix}1\\0\end{Bmatrix}\right]$.

2.13.8. Integrals of $x^\alpha e^{-cx}\begin{Bmatrix}\operatorname{erf}(bx^r)\\ \operatorname{erfc}(bx^r)\end{Bmatrix}L_\nu(cx)$.

1. $\int\limits_0^\infty x^{\alpha-1}e^{-cx}\begin{Bmatrix}\operatorname{erf}(bx)\\ \operatorname{erfc}(bx)\end{Bmatrix}L_\nu(cx)\,dx=$

$$=\mp\frac{1}{\alpha\sqrt{\pi}\,b^\alpha}\Gamma\left(\frac{\alpha+1}{2}\right){}_4F_4\left(\frac{\alpha}{2},\frac{\alpha+1}{2},\frac{\nu+1}{2},\frac{\nu}{2}+1;\ 1,\frac{1}{2},\frac{1}{2},\frac{\alpha}{2}+1;\ \frac{c^2}{4b^2}\right)\pm$$

$$\pm\frac{c(\nu+1)}{(\alpha+1)\sqrt{\pi}b^{\alpha+1}}\Gamma\left(\frac{\alpha}{2}+1\right){}_4F_4\left(\frac{\alpha}{2}+1,\ \frac{\alpha+1}{2},\ 1+\frac{\nu}{2},\ \frac{\nu+3}{2};\ 1,\ \frac{3}{2},\ \frac{3}{2},\right.$$

$$\left.\frac{\alpha+3}{2};\ \frac{c^2}{4b^2}\right)+\begin{Bmatrix}1\\ 0\end{Bmatrix}c^{-\alpha}\Gamma\begin{bmatrix}\alpha,\ 1-\alpha+\nu\\ 1-\alpha,\ 1+\nu\end{bmatrix}$$

$\left[\operatorname{Re}c>0;\ |\arg b|<\pi/4,\ \begin{Bmatrix}\operatorname{Re}\alpha>-1;\ \operatorname{Re}(\alpha-\nu)<1 \text{ for } \nu\neq 0, 1, 2, \ldots\\ \operatorname{Re}\alpha>0\end{Bmatrix}\right]$

2. $\int\limits_0^\infty x^{\alpha-1}e^{-cx}\begin{Bmatrix}\operatorname{erf}(b\sqrt{x})\\ \operatorname{erfc}(b\sqrt{x})\end{Bmatrix}L_\nu(cx)\,dx=$

$$=\mp\frac{\Gamma(\alpha+1/2)}{\alpha\sqrt{\pi}b^{2\alpha}}\,{}_3F_2\left(\alpha,\ \alpha+\frac{1}{2},\ \nu+1;\ 1,\ \alpha+1;\ -\frac{c}{b^2}\right)+\begin{Bmatrix}1\\ 0\end{Bmatrix}c^{-\alpha}\Gamma\begin{bmatrix}\alpha,\ 1-\alpha+\nu\\ 1-\alpha,\ 1+\nu\end{bmatrix}$$

$\left[\operatorname{Re}c>0;\ |\arg b|<\pi/4,\ \begin{Bmatrix}\operatorname{Re}\alpha>-1/2;\ \operatorname{Re}(\alpha-\nu)<1 \text{ for } \nu\neq 0, 1, 2, \ldots\\ \operatorname{Re}\alpha>0\end{Bmatrix}\right]$

2.13.9. Integrals of $x^\alpha e^{-cx}\begin{Bmatrix}S(bx^r)\\ C(bx^r)\end{Bmatrix}L_\nu(cx)$.

1. $\int\limits_0^\infty x^{\alpha-1}e^{-cx}\begin{Bmatrix}S(bx)\\ C(bx)\end{Bmatrix}L_\nu(cx)\,dx=$

$$=-\frac{\Gamma(\alpha+1/2)}{\alpha\sqrt{2\pi}\,b^\alpha}\begin{Bmatrix}\sin[(2\alpha+1)\pi/4]\\ \cos[(2\alpha+1)\pi/4]\end{Bmatrix}\times$$

$$\times{}_5F_4\left(\frac{\alpha}{2},\frac{2\alpha+1}{4},\frac{2\alpha+3}{4},\frac{\nu+1}{2},\frac{\nu}{2}+1;\ 1,\frac{1}{2},\frac{1}{2},\frac{\alpha}{2}+1;\ -\frac{c^2}{b^2}\right)\pm$$

$$\pm\frac{c\Gamma(\alpha+3/2)(\nu+1)}{(\alpha+1)\sqrt{2\pi}b^{\alpha+1}}\begin{Bmatrix}\sin[(1-2\alpha)\pi/4]\\ \cos[(1-2\alpha)\pi/4]\end{Bmatrix}\times$$

$$\times{}_5F_4\left(\frac{\alpha+1}{2},\ \frac{2\alpha+3}{4},\ \frac{2\alpha+5}{4},\ \frac{\nu}{2}+1,\ \frac{\nu+3}{2};\ 1,\ \frac{3}{2},\ \frac{3}{2},\ \frac{\alpha+3}{2};\ -\frac{c^2}{b^2}\right)+$$

$$+\frac{c^{-\alpha}}{2}\Gamma\begin{bmatrix}\alpha,\ 1-\alpha+\nu\\ 1-\alpha,\ \nu+1\end{bmatrix}$$

$[b,\ \mathrm{Re}\,c>0;\ \mathrm{Re}\,\alpha>-1\mp 1/2;\ \mathrm{Re}\,(\alpha-\nu)<1$ for $\nu\neq 0,1,2,\ldots]$.

2. $$\int_0^\infty x^{\alpha-1}e^{-cx}\begin{Bmatrix}S(b\sqrt{x})\\ C(b\sqrt{x})\end{Bmatrix}L_\nu(cx)\,dx=$$

$$=-\frac{b^{2\nu-2\alpha+2}}{\sqrt{2\pi}c^{\nu+1}(1-\alpha+\nu)}\Gamma\begin{bmatrix}2\alpha-2\nu-3/2\\ -\nu\end{bmatrix}\begin{Bmatrix}\sin[(1+4\alpha-4\nu)\pi/4]\\ \cos[(1+4\alpha-4\nu)\pi/4]\end{Bmatrix}\times$$

$$\times{}_3F_3\left(\nu+1,\ \nu+1,\ 1-\alpha+\nu;\ 2-\alpha+\nu,\ \frac{7}{4}-\alpha+\nu,\ \frac{5}{4}-\alpha+\nu;\ -\frac{b^2}{4c}\right)+$$

$$+\frac{\sqrt{2}\,b^{\delta+1/2}}{3^\delta\sqrt{\pi}c^{\alpha+(2\delta+1)/4}}\Gamma\begin{bmatrix}\alpha+(2\delta+1)/4,\ \nu-\alpha+(3-2\delta)/4\\ (3-2\delta)/4-\alpha\end{bmatrix}\times$$

$$\times{}_3F_3\left(\frac{2\delta+1}{4},\ \frac{2\delta+1}{4}+\alpha,\ \frac{2\delta+1}{4}+\alpha;\ \frac{2\delta+5}{4},\ \frac{2\delta+1}{4}+\alpha-\nu,\ \delta+\frac{1}{2};\ -\frac{b^2}{4c}\right)$$

$\left[b,\ \mathrm{Re}\,c>0;\ \mathrm{Re}\,\alpha>-(2\delta+1)/4;\ \mathrm{Re}\,(\alpha-\nu)<1\right.$ for $\left.\nu\neq 0,1,2,\ldots;\ \delta=\begin{Bmatrix}1\\0\end{Bmatrix}\right]$.

2.13.10. Integrals of $x^\alpha e^{-cx}\begin{Bmatrix}\gamma(\mu,\ bx^r)\\ \Gamma(\mu,\ bx^r)\end{Bmatrix}L_\nu(cx)$.

1. $$\int_0^\infty x^{\alpha-1}e^{-cx}\begin{Bmatrix}\gamma(\mu,\ bx)\\ \Gamma(\mu,\ bx)\end{Bmatrix}L_\nu(cx)\,dx=$$

$$=\mp\frac{\Gamma(\alpha+\mu)}{\alpha b^\alpha}{}_3F_2\left(\alpha,\ \alpha+\mu,\ \nu+1;\ 1,\ \alpha+1;\ -\frac{c}{b}\right)+\begin{Bmatrix}1\\0\end{Bmatrix}c^{-\alpha}\Gamma\begin{bmatrix}\alpha,\ \mu,\ 1-\alpha+\nu\\ 1-\alpha,\ \nu+1\end{bmatrix}$$

$\left[\mathrm{Re}\,b,\ \mathrm{Re}\,c,\ \mathrm{Re}\,(\alpha+\mu)>0;\ \begin{Bmatrix}\mathrm{Re}\,\mu>0;\ \mathrm{Re}\,(\alpha-\nu)<1 \text{ for } \nu\neq 0,1,2,\ldots\\ \mathrm{Re}\,\alpha>0\end{Bmatrix}\right]$.

2. $$\int_0^\infty x^{\alpha-1}e^{-cx}\begin{Bmatrix}\gamma(\mu,\ bx^2)\\ \Gamma(\mu,\ bx^2)\end{Bmatrix}L_\nu(cx)\,dx=$$

$$=\mp\frac{\Gamma(\mu+\alpha/2)}{\alpha b^{\alpha/2}}{}_4F_4\left(\frac{\alpha}{2},\ \frac{\alpha}{2}+\mu,\ \frac{1+\nu}{2},\ 1+\frac{\nu}{2};\ 1,\ \frac{1}{2},\ \frac{1}{2},\ \frac{\alpha}{2}+1;\ \frac{c^2}{4b}\right)\pm$$

$$\pm\frac{(\nu+1)c}{(\alpha+1)b^{(\alpha+1)/2}}\Gamma\left(\frac{\alpha+1}{2}+\mu\right){}_4F_4\left(\frac{\alpha+1}{2},\ \frac{\alpha+1}{2}+\mu,\ 1+\frac{\nu}{2},\ \frac{\nu+3}{2};\right.$$

$$\left.1,\ \frac{3}{2},\ \frac{3}{2},\ \frac{\alpha+3}{2};\ \frac{c^2}{4b}\right)+\begin{Bmatrix}1\\0\end{Bmatrix}c^{-\alpha}\Gamma\begin{bmatrix}\alpha,\ \mu,\ 1-\alpha+\nu\\ 1-\alpha,\ \nu+1\end{bmatrix}$$

$\left[\mathrm{Re}\,b,\ \mathrm{Re}\,c,\ \mathrm{Re}\,(\alpha+2\mu)>0,\ \begin{Bmatrix}\mathrm{Re}\,\mu>0;\ \mathrm{Re}\,(\alpha-\nu)<1 \text{ for } \nu\neq 0,1,2,\ldots\\ \mathrm{Re}\,\alpha>0\end{Bmatrix}\right]$.

3. $$\int_0^\infty x^{\alpha-1}e^{-cx}\begin{Bmatrix}\gamma(\mu,\ b\sqrt{x})\\ \Gamma(\mu,\ b\sqrt{x})\end{Bmatrix}L_\nu(cx)\,dx=\pm\frac{b^{2\nu-2\alpha+2}}{c^{\nu+1}(1-\alpha+\nu)}\Gamma\begin{bmatrix}2\alpha+\mu-2\nu-2\\ -\nu\end{bmatrix}\times$$

$$\times{}_3F_3\left(\nu+1,\ \nu+1,\ 1+\nu-\alpha;\ 2+\nu-\alpha-\frac{\mu}{2},\ \frac{3-\mu}{2}+\nu-\alpha,\ 2+\nu-\alpha;\ \frac{b^2}{4c}\right)\pm$$

$$\pm\frac{b^\mu}{\mu c^{\alpha+\mu/2}}\Gamma\begin{bmatrix}\alpha+\mu/2,\ 1-\alpha-\mu/2+\nu\\ 1-\alpha-\mu/2,\ \nu+1\end{bmatrix}\times$$

6

$$\times {}_3F_3\left(\frac{\mu}{2},\ \alpha+\frac{\mu}{2},\ \alpha+\frac{\mu}{2};\ \frac{1}{2},\ \frac{\mu}{2}+1,\ \alpha-\nu+\frac{\mu}{2};\ \frac{b^2}{4c}\right)\mp$$

$$\mp\frac{b^{\mu+1}}{(\mu+1)\,c^{\alpha+(\mu+1)/2}}\Gamma\left[\begin{matrix}\alpha+(\mu+1)/2,\ \nu+(1-\mu)/2-\alpha\\(1-\mu)/2-\alpha,\ \nu+1\end{matrix}\right]\times$$

$$\times {}_3F_3\left(\frac{\mu+1}{2},\ \alpha+\frac{\mu+1}{2},\ \alpha+\frac{\mu+1}{2};\ \frac{3}{2},\ \frac{\mu+3}{2},\ \alpha+\frac{\mu+1}{2}-\nu;\ \frac{b^2}{4c}\right)+$$

$$+\left\{\begin{matrix}0\\1\end{matrix}\right\}c^{-\alpha}\Gamma\left[\begin{matrix}\alpha,\ \mu,\ 1-\alpha+\nu\\1-\alpha,\ \nu+1\end{matrix}\right]$$

$\left[\operatorname{Re} b,\ \operatorname{Re} c,\ \operatorname{Re}(2\alpha+\mu)>0,\ \left\{\begin{matrix}\operatorname{Re}\mu>0;\ \operatorname{Re}(\alpha-\nu)<1 \text{ for } \nu\neq 0,1,2,\ldots\\ \operatorname{Re}\alpha>0\end{matrix}\right\}\right].$

2.13.11. Integrals of $x^{\alpha}e^{-cx}J_{\mu}(bx^r)L_{\nu}(cx)$.

1. $$\int_0^\infty x^{\alpha-1}e^{-cx}J_\mu(bx)L_\nu(cx)\,dx=\frac{2^{\alpha-1}}{b^\alpha}\Gamma\left[\begin{matrix}(\alpha+\mu)/2\\(\mu-\alpha)/2+1\end{matrix}\right]\times$$

$$\times {}_4F_3\left(\frac{\alpha+\mu}{2},\ \frac{\alpha-\mu}{2},\ \frac{\nu+1}{2},\ \frac{\nu}{2}+1;\ 1,\ \frac{1}{2},\ \frac{1}{2};\ -\frac{c^2}{b^2}\right)-$$

$$-\frac{2^\alpha c(\nu+1)}{b^{\alpha+1}}\Gamma\left[\begin{matrix}(1+\alpha+\mu)/2\\(1-\alpha+\mu)/2\end{matrix}\right]\times$$

$$\times {}_4F_3\left(1+\frac{\nu}{2},\ \frac{\nu+3}{2},\ \frac{\alpha+\mu+1}{2},\ \frac{\alpha-\mu+1}{2};\ 1,\ \frac{3}{2},\ \frac{3}{2};\ -\frac{c^2}{b^2}\right)$$

$[b,\ \operatorname{Re} c,\ \operatorname{Re}(\alpha+\mu)>0;\ \operatorname{Re}(\alpha-\nu)<5/2 \text{ for } \nu\neq 0,1,2,\ldots].$

2. $$\int_0^\infty e^{-cx}J_0(b\sqrt{x})L_\nu(cx)\,dx=\left(\frac{b}{2}\right)^{2\nu}\frac{c^{-\nu-1}}{\Gamma(\nu+1)}\exp\left(-\frac{b^2}{4c}\right)$$

$[b,\ \operatorname{Re} c>0;\ \operatorname{Re}\nu>-3/2 \text{ for } \nu\neq 0,1,2,\ldots].$

3. $$\int_0^\infty x^{\alpha-1}e^{-cx}J_\mu(b\sqrt{x})L_\nu(cx)\,dx=$$

$$=\left(\frac{b}{2}\right)^{2\nu-2\alpha+2}c^{-\nu-1}\Gamma\left[\begin{matrix}\alpha+\mu/2-\nu-1\\-\nu,\ 2-\alpha+\mu/2+\nu\end{matrix}\right]\times$$

$$\times {}_2F_2\left(\nu+1,\ \nu+1;\ 2-\alpha-\frac{\mu}{2}+\nu,\ 2-\alpha+\frac{\mu}{2}+\nu;\ -\frac{b^2}{4c}\right)+$$

$$+\left(\frac{b}{2}\right)^{\mu}c^{-\alpha-\mu/2}\Gamma\left[\begin{matrix}\alpha+\mu/2,\ 1-\alpha-\mu/2+\nu\\\mu+1,\ 1-\alpha-\mu/2,\ \nu+1\end{matrix}\right]\times$$

$$\times {}_2F_2\left(\alpha+\frac{\mu}{2},\ \alpha+\frac{\mu}{2};\ \mu+1,\ \alpha+\frac{\mu}{2}-\nu;\ -\frac{b^2}{4c}\right)$$

$[b,\ \operatorname{Re} c,\ \operatorname{Re}(2\alpha+\mu)>0;\ \operatorname{Re}(\alpha-\nu)<7/4 \text{ for } \nu\neq 0,1,2,\ldots].$

4. $$\int_0^\infty x^{-\mu/2}e^{-cx}J_\mu(b\sqrt{x})L_\nu(cx)\,dx=$$

$$=\left(\frac{b}{2}\right)^{\nu-1}\frac{c^{(\mu-\nu-1)/2}}{\Gamma(\mu+\nu+1)}\exp\left(-\frac{b^2}{8c}\right)M_{(\nu-\mu+1)/2,\,(\mu+\nu)/2}\left(\frac{b^2}{4c}\right)$$

$[b,\ \operatorname{Re} c>0;\ \operatorname{Re}(\mu+2\nu)>-3/2 \text{ for } \nu\neq 0,1,2,\ldots].$

5. $$\int_0^\infty x^{-(\nu+1)/2}e^{-cx}J_{\nu+1}(b\sqrt{x})L_\nu(cx)\,dx=\left(\frac{b}{2}\right)^{\nu}\frac{\sqrt{\pi/c}}{\Gamma(\nu+1)}\exp\left(-\frac{b^2}{8c}\right)I_{\nu+1/2}\left(\frac{b^2}{8c}\right)$$

$[b,\ \operatorname{Re} c>0;\ \operatorname{Re}\nu>-5/6 \text{ for } \nu\neq 0,1,2,\ldots].$

6. $\int\limits_0^\infty x^{\nu/2}e^{-cx}J_{-\nu}(b\sqrt{x})\,L_\nu(cx)\,dx=\left(\frac{b}{2}\right)^\nu c^{-\nu-1}\exp\left(-\frac{b^2}{4c}\right)L_\nu\left(\frac{b^2}{4c}\right)$

[b, $\operatorname{Re} c>0$; $\operatorname{Re}\nu>-3/2$ for $\nu\neq 0, 1, 2, \ldots$].

7. $\int\limits_0^\infty x^{\mu/2}e^{-cx}J_\mu(b\sqrt{x})\,L_\nu(cx)\,dx=$

$$=\left(\frac{b}{2}\right)^{\nu-1}\frac{c^{-(\mu+\nu+1)/2}}{\Gamma(\nu+1)}\exp\left(-\frac{b^2}{8c}\right)W_{(\mu+\nu+1)/2,\,(\nu-\mu)/2}\left(\frac{b^2}{4c}\right)$$

[b, $\operatorname{Re} c>0$; $\operatorname{Re}\mu>-1$; $\operatorname{Re}(\mu-2\nu)<3/2$ for $\nu\neq 0, 1, 2, \ldots$].

8. $\int\limits_0^\infty x^{(2\nu+1)/4}e^{-cx}J_{\nu+1/2}(b\sqrt{x})\,L_\nu(cx)\,dx=$

$$=\frac{2^{-2\nu}b^{\nu-1/2}c^{-\nu-1}}{\Gamma(\nu+1)}\exp\left(-\frac{b^2}{8c}\right)D_{2\nu+1}\left(\frac{b}{\sqrt{2c}}\right)$$

[b, $\operatorname{Re} c>0$; $\operatorname{Re}\nu>-3/2$; $\operatorname{Re}\nu>-1$ for $\nu\neq 0, 1, 2, \ldots$].

9. $\int\limits_0^\infty x^{(2\nu-1)/4}e^{-cx}J_{\nu-1/2}(b\sqrt{x})\,L_\nu(cx)\,dx=$

$$=\frac{2^{1/2-2\nu}b^{\nu-1/2}c^{-\nu-1/2}}{\Gamma(\nu+1)}\exp\left(-\frac{b^2}{8c}\right)D_{2\nu}\left(\frac{b}{\sqrt{2c}}\right)$$

[b, $\operatorname{Re} c>0$; $\operatorname{Re}\nu>-1/2$].

10. $\int\limits_0^\infty x^{-(\nu+1)/2}e^{-cx}J_{-\nu-1}(b\sqrt{x})\,L_\nu(cx)\,dx=$

$$=\frac{(b/2)^\nu}{\sqrt{c\pi}\,\Gamma(\nu+1)}\exp\left(-\frac{b^2}{8c}\right)K_{\nu+1/2}\left(\frac{b^2}{8c}\right)$$

[b, $\operatorname{Re} c>0$; $\operatorname{Re}\nu<0$; $\operatorname{Re}\nu>-5/6$ for $\nu\neq 0, 1, 2, \ldots$].

2.13.12. Integrals of $x^\alpha e^{-cx}\begin{Bmatrix}Y_\mu(bx^r)\\K_\mu(bx^r)\end{Bmatrix}L_\nu(cx)$.

1. $\int\limits_0^\infty x^{\alpha-1}e^{-cx}\begin{Bmatrix}Y_\mu(bx)\\K_\mu(bx)\end{Bmatrix}L_\nu(cx)\,dx=$

$$=\mp\frac{2^{\alpha-2}}{\pi b^\alpha}\Gamma\left(\frac{\alpha+\mu}{2}\right)\Gamma\left(\frac{\alpha-\mu}{2}\right)\pi^{(1\mp1)/2}\left(2\cos\frac{\alpha-\mu}{2}\pi\right)^{(1\pm1)/2}\times$$

$$\times{}_4F_3\left(\frac{\alpha+\mu}{2},\frac{\alpha-\mu}{2},\frac{\nu+1}{2},\frac{\nu}{2}+1;\frac{1}{2},\frac{1}{2},1;\pm\frac{c^2}{b^2}\right)\pm$$

$$\pm\frac{2^{\alpha-1}(\nu+1)c}{\pi b^{\alpha+1}}\Gamma\left(\frac{\alpha+\mu+1}{2}\right)\Gamma\left(\frac{\alpha-\mu+1}{2}\right)\pi^{(1\mp1)/2}\left(2\sin\frac{\mu-\alpha}{2}\pi\right)^{(1\pm1)/2}\times$$

$$\times{}_4F_3\left(\frac{\alpha+\mu+1}{2},\frac{\alpha-\mu+1}{2},\frac{\nu}{2}+1,\frac{\nu+3}{2};\frac{3}{2},\frac{3}{2},1;\pm\frac{c^2}{b^2}\right)$$

$\left[\operatorname{Re} c>0;\ \operatorname{Re}\alpha>|\operatorname{Re}\mu|,\ \begin{Bmatrix}b>0;\ \operatorname{Re}(\alpha-\nu)<5/2 \text{ for } \nu\neq 0, 1, 2, \ldots\\ \operatorname{Re} b>0\end{Bmatrix}\right]$.

2. $\int\limits_0^\infty x^{\alpha-1}e^{-cx}\begin{Bmatrix}Y_\mu(b\sqrt{x})\\K_\mu(b\sqrt{x})\end{Bmatrix}L_\nu(cx)\,dx=$

$$=\mp\begin{Bmatrix}2\\\pi\end{Bmatrix}\frac{2^{\mu-1}}{\pi c^{\alpha-\mu/2}b^\mu}\Gamma\begin{bmatrix}\mu,\ \alpha-\mu/2,\ 1+\nu-\alpha+\mu/2\\1-\alpha+\mu/2,\ \nu+1\end{bmatrix}\times$$

$$\times {}_2F_2\left(\alpha-\frac{\mu}{2},\ \alpha-\frac{\mu}{2};\ 1-\mu,\ \alpha-\nu-\frac{\mu}{2};\ \mp\frac{b^2}{4c}\right)\mp$$
$$\mp\begin{Bmatrix}2\cos\mu\pi\\ \pi\end{Bmatrix}\frac{b^{\mu}}{2^{\mu+1}\pi c^{\alpha+\mu/2}}\Gamma\left[\begin{matrix}-\mu,\ \alpha+\mu/2,\ 1+\nu-\alpha-\mu/2\\ 1-\alpha-\mu/2,\ \nu+1\end{matrix}\right]\times$$
$$\times {}_2F_2\left(\alpha+\frac{\mu}{2},\ \alpha+\frac{\mu}{2};\ 1+\mu,\ \alpha+\frac{\mu}{2}-\nu;\ \mp\frac{b^2}{4c}\right)+$$
$$+\frac{2^{2\alpha-2\nu-3}}{\pi b^{2\alpha-2\nu-2}c^{\nu+1}}\Gamma\left[\begin{matrix}\alpha+\mu/2-\nu-1,\ \alpha-\nu-\mu/2-1\\ -\nu\end{matrix}\right]\times$$
$$\times\pi^{(1\mp1)/2}\left(2\cos\frac{2\alpha-\mu-2\nu}{2}\pi\right)^{(1\pm1)/2}{}_2F_2\left(\nu+1,\ \nu+1;\ 2-\alpha+\nu-\frac{\mu}{2},\right.$$
$$\left.2-\alpha+\nu+\frac{\mu}{2};\ \mp\frac{b^2}{4c}\right)$$
$$\left[\operatorname{Re}c>0;\ 2\operatorname{Re}\alpha>|\operatorname{Re}\mu|,\ \begin{Bmatrix}b>0;\ \operatorname{Re}(\alpha-\nu)<7/4 \text{ for } \nu\neq0,\ 1,\ 2,\ \ldots\\ \operatorname{Re}b>0\end{Bmatrix}\right].$$

2.13.13. Integrals of $x^{\alpha}e^{\varphi(x)}L_m^{\gamma}(bx^r)L_\nu(cx)$.

1. $$\int_0^\infty x^{\alpha-1}e^{-(b+c)x}L_m^{\gamma}(bx)\,L_\nu(cx)\,dx=$$
$$=\frac{(1+\gamma-\alpha)_m\Gamma(\alpha)}{m!b^{\alpha}}{}_3F_2\left(\alpha,\ \alpha-\gamma,\ \nu+1;\ 1,\ \alpha-\gamma-m;\ -\frac{c}{b}\right)\quad[\operatorname{Re}b,\ \operatorname{Re}c,\ \operatorname{Re}\alpha>0].$$

2. $$\int_0^\infty x^{\alpha-1}e^{-b\sqrt{x}-cx}L_m^{\gamma}(b\sqrt{x})\,L_\nu(cx)\,dx=\frac{(1+\gamma)_m}{m!c^{\alpha}}\Gamma\left[\begin{matrix}\alpha,\ 1-\alpha+\nu\\ 1-\alpha,\ \nu+1\end{matrix}\right]\times$$
$$\times {}_4F_4\left(\alpha,\ \alpha,\ \frac{1+\gamma+m}{2},\ 1+\frac{\gamma+m}{2};\ \frac{1}{2},\ \alpha-\nu,\ \frac{1+\gamma}{2},\ 1+\frac{\gamma}{2};\ \frac{b^2}{4c}\right)-$$
$$-\frac{b(2+\gamma)_m}{m!c^{\alpha+1/2}}\Gamma\left[\begin{matrix}\alpha+1/2,\ 1/2-\alpha+\nu\\ 1/2-\alpha,\ \nu+1\end{matrix}\right]{}_4F_4\left(\alpha+\frac{1}{2},\ \alpha+\frac{1}{2},\ 1+\frac{\gamma+m}{2},\ \frac{3+\gamma+m}{2};\right.$$
$$\left.1+\frac{\gamma}{2},\ \frac{3+\gamma}{2},\ \frac{3}{2},\ \frac{1}{2}+\alpha-\nu;\ \frac{b^2}{4c}\right)+\frac{2b^{2\nu-2\alpha+2}}{m!c^{\nu+1}}\Gamma\left[\begin{matrix}2\alpha-2\nu-2\\ -\nu\end{matrix}\right](3-2\alpha+\gamma+2\nu)_m\times$$
$$\times {}_4F_4\left(\nu+1,\ \nu+1,\ \frac{3+\gamma+m}{2}-\alpha+\nu,\ 2-\alpha+\frac{\gamma+m}{2}+\nu;\right.$$
$$\left.2-\alpha+\nu,\ \frac{3}{2}-\alpha+\nu,\ \frac{3+\gamma}{2}-\alpha+\nu,\ 2-\alpha+\frac{\gamma}{2}+\nu;\ \frac{b^2}{4c}\right)\quad[\operatorname{Re}b,\ \operatorname{Re}c,\ \operatorname{Re}\alpha>0].$$

2.13.14. Integrals of $x^{\alpha}e^{\varphi(x)}H_m(bx^r)L_\nu(cx)$.

NOTATION: $\delta=\begin{Bmatrix}1\\ 0\end{Bmatrix}$.

1. $$\int_0^\infty x^{\alpha-1}e^{-b^2x^2-cx}\begin{Bmatrix}H_{2m+1}(bx)\\ H_{2m}(bx)\end{Bmatrix}L_\nu(cx)\,dx=$$
$$=\frac{(-1)^m2^{2m+\delta-1}}{b^{\alpha}}\Gamma\left(\frac{\alpha+\delta}{2}\right)\left(\frac{1-\alpha+\delta}{2}\right)_m\times$$
$$\times {}_4F_4\left(\frac{\alpha}{2},\ \frac{\alpha+1}{2},\ \frac{\nu+1}{2},\ \frac{\nu}{2}+1;\ \frac{1}{2},\ \frac{1}{2},\ 1,\ \frac{1+\alpha-\delta-2m}{2};\ \frac{c^2}{4b^2}\right)-$$
$$-(-1)^m\frac{2^{2m+\delta-1}c(1+\nu)}{b^{\alpha+1}}\Gamma\left(\frac{\alpha+\delta+1}{2}\right)\left(\frac{\delta-\alpha}{2}\right)_m\times$$
$$\times {}_4F_4\left(\frac{\alpha+1}{2},\ 1+\frac{\alpha}{2},\ \frac{\nu}{2}+1,\ \frac{\nu+3}{2};\ \frac{3}{2},\ \frac{3}{2},\ 1,\ 1+\frac{\alpha-\delta-2m}{2};\ \frac{c^2}{4b^2}\right)$$
$$[\operatorname{Re}c>0;\ \operatorname{Re}\alpha>-\delta;\ |\arg b|<\pi/4].$$

2. $\int_0^\infty x^{\alpha-1}e^{-(b^2+c)x}\left\{\begin{matrix} H_{2m+1}(b\sqrt{x}) \\ H_{2m}(b\sqrt{x}) \end{matrix}\right\} L_\nu(cx)\,dx=$

$$=\frac{(-1)^m 2^{2m+\delta}}{b^{2\alpha}}\Gamma\left(\alpha+\frac{\delta}{2}\right)\left(\frac{\delta+1}{2}-\alpha\right)_m {}_3F_2\left(\alpha, \frac{1}{2}+\alpha, \nu+1; 1, \frac{1-\delta}{2}+\alpha-m; -\frac{c}{b^2}\right)$$

[$\operatorname{Re} c>0$; $\operatorname{Re}\alpha>-\delta/2$; $|\arg b|<\pi/4$].

2.13.15. Integrals of $x^\alpha e^{\varphi(x)}L_\mu(bx^k)L_\nu(cx)$.

1. $\int_0^\infty x^{\alpha-1}e^{-(b+c)x}L_\mu(bx)L_\nu(cx)\,dx=$

$$=b^{-\alpha}\Gamma\left[\begin{matrix}\alpha, 1-\alpha+\mu \\ 1-\alpha, \mu+1, \nu+1\end{matrix}\right] {}_3F_2\left(\alpha, \alpha, \nu+1; 1, 1+\alpha-\mu; -\frac{c}{b}\right)+$$

$$+\frac{c^{1-\alpha+\mu}}{b^{\mu+1}}\Gamma\left[\begin{matrix}\alpha-\mu-1, 2-\alpha+\mu+\nu \\ -\mu, 2-\alpha+\mu, \nu+1\end{matrix}\right] {}_3F_2\left(\mu+1, \mu+1, 2-\alpha+\mu+\nu; 2-\alpha+\mu,\right.$$

$$\left.2-\alpha+\mu; -\frac{c}{b}\right)$$

[$\operatorname{Re} b$, $\operatorname{Re} c$, $\operatorname{Re}\alpha>0$; $\operatorname{Re}(\alpha-\mu-\nu)<2$ for $\mu, \nu\neq 0, 1, 2, \ldots$].

2. $\int_0^\infty x^{\alpha-1}e^{-bx^2-cx}L_\mu(bx^2)L_\nu(cx)\,dx=\frac{b^{-\alpha/2}}{2}\Gamma\left[\begin{matrix}\alpha/2, 1-\alpha/2+\mu \\ 1-\alpha/2, \mu+1, \nu+1\end{matrix}\right]\times$

$$\times {}_4F_4\left(\frac{\alpha}{2}, \frac{\alpha}{2}, \frac{\nu+1}{2}, \frac{\nu}{2}+1; 1, \frac{1}{2}, \frac{1}{2}, \frac{\alpha}{2}-\mu; \frac{c^2}{4b}\right)-$$

$$-\frac{c(\nu+1)}{2b^{(\alpha+1)/2}}\Gamma\left[\begin{matrix}(\alpha+1)/2, \mu+(1-\alpha)/2 \\ (1-\alpha)/2, \mu+1, \nu+1\end{matrix}\right]\times$$

$$\times {}_4F_4\left(1+\frac{\nu}{2}, \frac{\nu+3}{2}, \frac{\alpha+1}{2}, \frac{\alpha+1}{2}; \frac{\alpha+1}{2}-\mu, 1, \frac{3}{2}, \frac{3}{2}; \frac{c^2}{4b}\right)+$$

$$+\frac{c^{2\mu-\alpha+2}}{b^{\mu+1}}\Gamma\left[\begin{matrix}\alpha-2\mu-2, 3-\alpha+2\mu+\nu \\ -\mu, 3-\alpha+2\mu, \nu+1\end{matrix}\right] {}_4F_4\left(\mu+1, \mu+1, \frac{3-\alpha+\nu}{2}+\mu,\right.$$

$$\left.2+\frac{\nu-\alpha}{2}+\mu; \frac{3-\alpha}{2}+\mu, \frac{3-\alpha}{2}+\mu, 2-\frac{\alpha}{2}+\mu, 2-\frac{\alpha}{2}+\mu; \frac{c^2}{4b}\right)$$

[$\operatorname{Re} b$, $\operatorname{Re} c$, $\operatorname{Re}\alpha>0$; $\operatorname{Re}(\alpha-2\mu-\nu)<3$ for $\mu, \nu\neq 0, 1, 2, \ldots$].

2.14. THE BATEMAN FUNCTION $k_\nu(x)$

Certain functions containing $k_\nu(x)$ can also be obtained from the formulae of Section 2.19 since

$$k_\nu(x)=\frac{1}{\Gamma(1+\nu/2)}W_{\nu/2, 1/2}(2x).$$

2.14.1. Integrals of general form.

1. $\int_0^a x^{\alpha-1}(a^r-x^r)^{\beta-1}e^{\pm cx}k_\nu(cx)\,dx=$

$$=\frac{2a^{\alpha+r\beta-r+1}c\Gamma(\beta)}{\pi r}\sin\frac{\nu\pi}{2}\sum_{k=0}^{\infty}\frac{(1\mp\nu/2)_k(\pm 2ac)^k}{k!(k+1)!}\times$$

$$\times\Gamma\left[\begin{matrix}(k+\alpha+1)/r \\ \beta+(k+\alpha+1)/r\end{matrix}\right]\left[2\psi(k+1)+\frac{1}{k+1}+\frac{1}{r}\psi\left(\beta+\frac{k+\alpha+1}{r}\right)-\right.$$

$$\left.-\frac{1}{r}\psi\left(\frac{k+\alpha+1}{r}\right)-\psi\left(\frac{1\pm 1-\nu}{2}\pm k\right)-\ln(2ac)\right]+\frac{2a^{\alpha+r\beta-r}}{\pi\nu r}\sin\frac{\nu\pi}{2}\,\mathrm{B}\left(\beta, \frac{\alpha}{r}\right)$$

[a, r, $\operatorname{Re}\alpha$, $\operatorname{Re}\beta>0$].

2. $$\int_a^\infty x^{\alpha-1}(x^r-a^r)^{\beta-1}e^{\pm cx}k_\nu(cx)\,dx=$$

$$=\frac{2a^{\alpha+r\beta-r+1}c\Gamma(\beta)}{\pi r}\sin\frac{\nu\pi}{2}\sum_{k=0}^\infty\frac{(1\mp\nu/2)_k(\pm 2ac)^k}{k!(k+1)!}\Gamma\left[\begin{matrix}1-\beta-(k+\alpha+1)/r\\1-(k+\alpha+1)/r\end{matrix}\right]\times$$

$$\times\left[2\psi(k+1)+\frac{1}{k+1}+\frac{1}{r}\psi\left(1-\beta-\frac{k+\alpha+1}{r}\right)-\frac{1}{r}\psi\left(1-\frac{k+\alpha+1}{r}\right)-\right.$$

$$\left.-\psi\left(\frac{1\pm1-\nu}{2}\pm k\right)-\ln(2ac)\right]+\frac{2a^{\alpha+r\beta-r}}{\pi\nu r}\sin\frac{\nu\pi}{2}\mathrm{B}\left(\beta,\ 1-\beta-\frac{\alpha}{r}\right)+$$

$$+\frac{(2c)^{r-r\beta-\alpha}}{\Gamma(1\mp\nu/2)}\sum_{k=0}^\infty\frac{(1-\beta)_k}{k!}\Gamma\left[\begin{matrix}\alpha-r(1-\beta+k),\ 1+\alpha-r(1-\beta+k)\\1+\alpha\pm\nu/2-r(1-\beta+k)\end{matrix}\right]\times$$

$$\times\left[\sin\frac{\nu\pi}{2}\operatorname{cosec}\left(r\beta-r-rk+\alpha+\frac{\nu}{2}\right)\pi\right]^{(1\pm1)/2}(2ac)^{rk}$$

$$\left[a,\ r,\ \operatorname{Re}\beta>0,\ \left\{\begin{matrix}|\arg c|<\pi;\ \operatorname{Re}(\alpha+r\beta+\nu/2)<r\\ \operatorname{Re}c>\end{matrix}\right\}\right].$$

3. $$\int_0^\infty\frac{x^{\alpha-1}}{(x^r+z^r)^\rho}e^{\pm cx}k_\nu(cx)\,dx=\frac{2cz^{\alpha+1-r\rho}}{\pi r\Gamma(\rho)}\sin\frac{\nu\pi}{2}\sum_{k=0}^\infty\frac{(1\mp\nu/2)_k(\pm 2cz)^k}{k!(k+1)!}\times$$

$$\times\Gamma\left(\frac{k+\alpha+1}{r}\right)\Gamma\left(\rho-\frac{k+\alpha+1}{r}\right)\times$$

$$\times\left[2\psi(k+1)+\frac{1}{k+1}+\frac{1}{r}\psi\left(\rho-\frac{k+\alpha+1}{r}\right)-\frac{1}{r}\psi\left(\frac{k+\alpha+1}{r}\right)-\right.$$

$$\left.-\psi\left(\frac{1\pm1-\nu}{2}\pm k\right)-\ln(2cz)\right]+\frac{2z^{\alpha-r\rho}}{\pi\nu r}\sin\frac{\nu\pi}{2}\mathrm{B}\left(\frac{\alpha}{r},\ \rho-\frac{\alpha}{r}\right)+$$

$$+\frac{(2c)^{r\rho-\alpha}}{\Gamma(1\mp\nu/2)}\sum_{k=0}^\infty\frac{(-1)^k(\rho)_k}{k!}\Gamma\left[\begin{matrix}\alpha-r\rho-rk,\ 1+\alpha-r\rho-rk\\1+\alpha\pm\nu/2-r\rho-rk\end{matrix}\right]\times$$

$$\times\left[\sin\frac{\nu\pi}{2}\operatorname{cosec}\left(\alpha-r\rho-rk+\frac{\nu}{2}\right)\pi\right]^{(1\pm1)/2}(2cz)^{rk}$$

$$\left[r,\ \operatorname{Re}\alpha>0,\ \left\{\begin{matrix}|\arg c|<\pi;\ \operatorname{Re}(\alpha-r\rho+\nu/2)<0\\ \operatorname{Re}c>0\end{matrix}\right\},\ r|\arg z|<\pi\right].$$

4. $$\int_0^\infty\frac{x^{\alpha-1}}{x^r-y^r}e^{\pm cx}k_\nu(cx)\,dx=$$

$$=-\frac{2cy^{\alpha-r+1}}{r}\sin\frac{\nu\pi}{2}\sum_{k=0}^\infty\frac{(1\mp\nu/2)_k(\pm 2cy)^k}{k!(k+1)!}\operatorname{ctg}\frac{k+\alpha+1}{r}\pi\left[2\psi(k+1)+\right.$$

$$\left.+\frac{1}{k+1}+\frac{2\pi}{r}\operatorname{cosec}\frac{2(k+\alpha+1)}{r}\pi-\psi\left(\frac{1\pm1-\nu}{2}\pm k\right)-\ln(2cy)\right]-$$

$$-\frac{2y^{\alpha-r}}{\nu r}\sin\frac{\nu\pi}{2}\operatorname{ctg}\frac{\alpha\pi}{r}+\frac{(2c)^{r-\alpha}}{\Gamma(1\mp\nu/2)}\sum_{k=0}^\infty\Gamma\left[\begin{matrix}\alpha-r-rk,\ 1+\alpha-r-rk\\1+\alpha\pm\nu/2-r-rk\end{matrix}\right]\times$$

$$\times\left[\sin\frac{\nu\pi}{2}\operatorname{cosec}\left(\alpha-rk-r+\frac{\nu}{2}\right)\pi\right]^{(1\pm1)/2}(2cy)^{rk}$$

$$\left[r,\ y,\ \operatorname{Re}\alpha>0,\ \left\{\begin{matrix}|\arg c|<\pi;\ \operatorname{Re}(2\alpha+\nu)<2r\\ \operatorname{Re}c>0\end{matrix}\right\}\right].$$

5. $\int\limits_0^\infty x^{\alpha-1}e^{-px^r \pm cx}k_\nu(cx)\,dx = U \qquad \left[r,\ \operatorname{Re} p,\ \operatorname{Re}\alpha>0,\ \left\{\begin{matrix}|\arg c|<\pi\\ \operatorname{Re} c>0\end{matrix}\right\}\right],$

$$U=\frac{2c}{\pi p^{(\alpha+1)/r}r}\sin\frac{\nu\pi}{2}\sum_{k=0}^{\infty}\frac{(1\mp\nu/2)_k}{k!\,(k+1)!}\Gamma\left(\frac{k+\alpha+1}{r}\right)\times$$
$$\times\left[2\psi(k+1)+\frac{1}{k+1}-\psi\left(\frac{1\pm1-\nu}{2}\pm k\right)-\frac{1}{r}\psi\left(\frac{k+\alpha+1}{r}\right)-\right.$$
$$\left.-\ln\frac{2c}{p^{1/r}}\right]\left(\pm\frac{2c}{p^{1/r}}\right)^k+\frac{2\Gamma(\alpha/r)}{\pi\nu p^{\alpha/r}r}\sin\frac{\nu\pi}{2}\qquad [r>1],$$

$$U=\frac{(2c)^{-\alpha}}{\Gamma(1\mp\nu/2)}\sum_{k=0}^{\infty}\frac{1}{k!}\Gamma\left[\begin{matrix}\alpha+rk,\ \alpha+1+rk\\ \alpha+1+rk\pm\nu/2\end{matrix}\right]\times$$
$$\times\left[\sin\frac{\nu\pi}{2}\operatorname{cosec}\left(\alpha+rk+\frac{\nu}{2}\right)\pi\right]^{(1\pm1)/2}\left(-\frac{p}{2^rc^r}\right)^k+$$
$$+\left\{\begin{matrix}1\\0\end{matrix}\right\}\frac{(2c)^{\nu/2}}{p^{(\nu+2\alpha)/(2r)}\,r\Gamma(1+\nu/2)}\sum_{k=0}^{\infty}\frac{(-\nu/2)_k(1-\nu/2)_k}{k!}\times$$
$$\times\Gamma\left(\frac{\nu+2\alpha-2k}{2r}\right)\left(-\frac{p^{1/r}}{2c}\right)^k\qquad [r<1],$$

U for $r=1$ see 2.14.2.4.

6. $$\int\limits_0^\infty x^{\alpha-1}e^{-px^{-r}\pm cx}k_\nu(cx)\,dx=\frac{2cp^{(\alpha+1)/r}}{\pi r}\sin\frac{\nu\pi}{2}\times$$
$$\times\sum_{k=0}^{\infty}\frac{(1\mp\nu/2)_k}{k!\,(k+1)!}\Gamma\left(-\frac{k+\alpha+1}{r}\right)\left[2\psi(k+1)+\frac{1}{k+1}-\right.$$
$$\left.-\psi\left(\frac{1\pm1-\nu}{2}\pm k\right)+\frac{1}{r}\psi\left(-\frac{k+\alpha+1}{r}\right)-\ln\left(2cp^{1/r}\right)\right]\times$$
$$\times(\pm2cp^{1/r})^k+\frac{(2c)^{-\alpha}}{\Gamma(1\mp\nu/2)}\sum_{k=0}^{\infty}\frac{1}{k!}\Gamma\left[\begin{matrix}\alpha-rk,\ \alpha+1-rk\\ \alpha+1-rk\pm\nu/2\end{matrix}\right]\times$$
$$\times\left[\sin\frac{\nu\pi}{2}\operatorname{cosec}\left(\alpha-rk+\frac{\nu}{2}\right)\pi\right]^{(1\pm1)/2}(-2^rc^rp)^k+\frac{2\Gamma(-\alpha/r)\,p^{\alpha/r}}{\pi\nu r}\sin\frac{\nu\pi}{2}$$
$$\left[r,\ \operatorname{Re} p>0,\ \left\{\begin{matrix}|\arg c|<\pi;\ \operatorname{Re}(2\alpha+\nu)<0\\ \operatorname{Re} c>0\end{matrix}\right\}\right].$$

7. $\int\limits_0^\infty x^{\alpha-1}e^{-cx}\left\{\begin{matrix}\sin bx^r\\ \cos bx^r\end{matrix}\right\}k_\nu(cx)\,dx=U(1)$

$$\left[b,\ r,\ \operatorname{Re} c>0;\ \operatorname{Re}\alpha>-\delta r;\ \delta=\left\{\begin{matrix}1\\0\end{matrix}\right\}\right],$$

$$U(\gamma)=\frac{2c}{\pi b^{(\alpha+1)/r}r}\sin\frac{\nu\pi}{2}\sum_{k=0}^{\infty}\frac{(-1)^{\gamma k}\left(1-(-1)^\gamma\nu/2\right)_k}{k!\,(k+1)!}\Gamma\left(\frac{k+\alpha+1}{r}\right)\times$$
$$\times\left\{\begin{matrix}\sin[(k+\alpha+1)\pi/(2r)]\\ \cos[(k+\alpha+1)\pi/(2r)]\end{matrix}\right\}\left[2\psi(k+1)+\frac{1}{k+1}-\psi\left(1-\gamma-\frac{\nu}{2}+(-1)^\gamma k\right)-\right.$$

$$-\frac{1}{r}\psi\left(\frac{k+\alpha+1}{r}\right)\mp\frac{\pi}{2r}\operatorname{tg}^{\mp 1}\frac{k+\alpha+1}{2r}\pi-\ln\frac{2c}{b^{1/r}}\Bigg]\left(\frac{2c}{b^{1/r}}\right)^k+$$

$$+\frac{2}{\nu\pi b^{\alpha/r}r}\sin\frac{\nu\pi}{2}\Gamma\left(\frac{\alpha}{r}\right)\begin{Bmatrix}\sin[\alpha\pi/(2r)]\\ \cos[\alpha\pi/(2r)]\end{Bmatrix}\quad [r>1],$$

$$U(\gamma)=\frac{b^\delta\sin^{1-\nu}(\nu\pi/2)}{(2c)^{\alpha+\delta r}\Gamma(1-(-1)^\nu\nu/2)}\times$$

$$\times\sum_{k=0}^{\infty}\frac{(-1)^k}{k!\,(\delta+1/2)_k}\Gamma\begin{bmatrix}\alpha+\delta r+2rk,\ 1+\alpha+\delta r+2rk\\ 1+\alpha+\delta r+2rk+(-1)^\nu\nu/2\end{bmatrix}\times$$

$$\times\sin^{\nu-1}\left(\alpha+\delta r+2rk+\frac{\nu}{2}\right)\pi\left(\frac{b}{2^{r+1}c^r}\right)^{2k}+(1-\gamma)\frac{(2c)^{\nu/2}}{rb^{(2\alpha+\nu)/(2r)}\Gamma(1+\nu/2)}\times$$

$$\times\sum_{k=0}^{\infty}\frac{(-\nu/2)_k(1-\nu/2)_k}{k!}\Gamma\left(\frac{2\alpha+\nu-2k}{2r}\right)\begin{Bmatrix}\sin[(2\alpha+\nu-2k)\pi/(4r)]\\ \cos[(2\alpha+\nu-2k)\pi/(4r)]\end{Bmatrix}\left(-\frac{b^{1/r}}{2c}\right)^k\quad [r<1],$$

$U(\gamma)$ for $r=1$ see 2.14.3.1.

8. $$\int_0^\infty x^{\alpha-1}e^{cx}\begin{Bmatrix}\sin bx^r\\ \cos bx^r\end{Bmatrix}k_\nu(cx)\,dx=U(0)$$

$\left[b,\ r>0;\ \operatorname{Re}\alpha>-\delta r;\ \operatorname{Re}(2\alpha+\nu)<2r;\ |\arg c|<\pi;\ \delta=\begin{Bmatrix}1\\0\end{Bmatrix};\ U(\gamma)\text{ see } 2.14.1.7\right].$

9. $$\int_0^\infty x^{\alpha-1}e^{-cx}\begin{Bmatrix}\ln(x^r+z^r)\\ \ln|x^r-z^r|\end{Bmatrix}k_\nu(cx)\,dx=V(1)\quad\left[r,\ \operatorname{Re}\alpha,\ \operatorname{Re}c>0,\ \begin{Bmatrix}r\,|\arg z|<\pi\\ z>0\end{Bmatrix}\right].$$

$$V(\gamma)=2cz^{\alpha+1}\sin\frac{\nu\pi}{2}\sum_{k=0}^{\infty}\frac{(-1)^{\nu k}(1-(-1)^\nu\nu/2)_k}{k!\,(k+1)!\,(k+\alpha+1)}\times$$

$$\times\begin{Bmatrix}\operatorname{cosec}[(k+\alpha+1)\pi/r]\\ \operatorname{ctg}[(k+\alpha+1)\pi/r]\end{Bmatrix}\Bigg[2\psi(k+1)+\frac{1}{k+1}-\psi\left(1-\gamma-\frac{\nu}{2}+(-1)^\nu k\right)+$$

$$+\frac{\pi}{r}\begin{Bmatrix}\operatorname{ctg}[(k+\alpha+1)\pi/r]\\ 2\operatorname{cosec}[2(k+\alpha+1)\pi/r]\end{Bmatrix}+\frac{1}{k+\alpha+1}-\ln(2cz)\Bigg](2cz)^k+$$

$$+\frac{2z^\alpha}{\alpha\nu}\sin\frac{\nu\pi}{2}\begin{Bmatrix}\operatorname{cosec}(\alpha\pi/r)\\ \operatorname{ctg}(\alpha\pi/r)\end{Bmatrix}\pm\frac{z^r}{(2c)^{\alpha-r}}\frac{\sin^{1-\nu}(\nu\pi/2)}{\Gamma(1-(-1)^\nu\nu/2)}\times$$

$$\times\sum_{k=0}^{\infty}\frac{(\mp1)^k}{k+1}\Gamma\begin{bmatrix}\alpha-r-rk,\ 1+\alpha-r-rk\\ 1+\alpha-r-rk+(-1)^\nu\nu/2\end{bmatrix}\sin^{\nu-1}\left(\alpha-r-rk-\frac{\nu}{2}\right)\pi(2cz)^{rk}+$$

$$+\frac{r}{(2c)^\alpha}\Gamma\begin{bmatrix}\alpha,\ \alpha+1\\ 1-(-1)^\nu\nu/2,\ 1+\alpha+(-1)^\nu\nu/2\end{bmatrix}\left[\frac{\sin(\nu\pi/2)}{\sin(\nu/2+\alpha)\pi}\right]^{1-\nu}\times$$

$$\times\left[2\psi(\alpha)+\frac{1}{\alpha}-\psi\left(1+\alpha+(-1)^\nu\frac{\nu}{2}\right)-(1-\gamma)\pi\operatorname{ctg}\frac{2\alpha+\nu}{2}\pi-\ln(2c)\right].$$

10. $$\int_0^\infty x^{\alpha-1}e^{cx}\begin{Bmatrix}\ln(x^r+z^r)\\ \ln|x^r-z^r|\end{Bmatrix}k_\nu(cx)\,dx=V(0)$$

$\left[r,\ \operatorname{Re}\alpha>0;\ |\arg c|<\pi;\ \operatorname{Re}(2\alpha+\nu)<0,\ \begin{Bmatrix}r\,|\arg z|<\pi\\ z>0\end{Bmatrix},\ V(\gamma)\text{ see } 2.14.1.9\right].$

11. $$\int_0^\infty x^{\alpha-1}e^{\pm cx}\operatorname{Ei}(-bx^r)k_\nu(cx)\,dx=U\qquad\left[r,\ \operatorname{Re}b,\ \operatorname{Re}\alpha>0;\ \begin{Bmatrix}|\arg c|<\pi\\ \operatorname{Re}c>0\end{Bmatrix}\right].$$

$$U=-\frac{2c}{\pi b^{(\alpha+1)/r}}\sin\frac{\nu\pi}{2}\sum_{k=0}^{\infty}\frac{(1\mp\nu/2)_k}{k!\,(k+1)!\,(k+\alpha+1)}\Gamma\left(\frac{k+\alpha+1}{r}\right)\times$$
$$\times\left[2\psi(k+1)+\frac{1}{k+1}-\psi\left(\frac{1\pm1-\nu}{2}\pm k\right)-\frac{1}{r}\psi\left(\frac{k+\alpha+1}{r}\right)+\frac{1}{k+\alpha+1}-\right.$$
$$\left.-\ln\frac{2c}{b^{1/r}}\right]\left(\pm\frac{2c}{b^{1/r}}\right)^k-\frac{2\Gamma(\alpha/r)}{\alpha\pi\nu b^{\alpha/r}}\sin\frac{\nu\pi}{2}\qquad[r>1],$$

$$U=-\frac{(2c)^{-\alpha-r}b}{\Gamma(1\mp\nu/2)}\sum_{k=0}^{\infty}\frac{1}{k!\,(k+1)^2}\Gamma\left[\begin{matrix}\alpha+r+rk,\ \alpha+1+r+rk\\ \alpha+1+r+rk\pm\nu/2\end{matrix}\right]\times$$
$$\times\left[\sin\frac{\nu\pi}{2}\operatorname{cosec}\left(\alpha+r+rk+\frac{\nu}{2}\right)\pi\right]^{(1\pm1)/2}\left(-\frac{b}{2^rc^r}\right)^k+\begin{Bmatrix}1\\0\end{Bmatrix}\frac{(2c)^{\nu/2}}{b^{(\nu+2\alpha)/(2r)}\Gamma(1+\nu/2)}\times$$
$$\times\sum_{k=0}^{\infty}\frac{(-\nu/2)_k(1-\nu/2)_k}{k!\,(k-\alpha-\nu/2)}\Gamma\left(\frac{\nu+2\alpha-2k}{2r}\right)\left(-\frac{b^{1/r}}{2c}\right)^k+$$
$$+(2c)^{-\alpha}\Gamma\left[\begin{matrix}\alpha,\ \alpha+1\\ 1\mp\nu/2,\ 1+\alpha\pm\nu/2\end{matrix}\right]\left[\sin\frac{\nu\pi}{2}\operatorname{cosec}\left(\alpha+\frac{\nu}{2}\right)\pi\right]^{(1\pm1)/2}\times$$
$$\times\left[\mathbf{C}+2r\psi(\alpha)+\frac{r}{\alpha}-r\psi\left(1+\alpha\pm\frac{\nu}{2}\right)-\begin{Bmatrix}1\\0\end{Bmatrix}r\pi\operatorname{ctg}\frac{2\alpha+\nu}{2}\pi-\ln\frac{2^rc^r}{b}\right]\qquad[r<1],$$

U for $r=1$ see 2.14.5.1.

12. $$\int_0^\infty x^{\alpha-1}e^{\pm bx^r-cx}\operatorname{Ei}(\mp bx^r)\,k_\nu(cx)\,dx=U(1)\quad\left[r,\ \operatorname{Re}c,\ \operatorname{Re}\alpha>0,\ \begin{Bmatrix}|\arg b|<\pi\\ b>0\end{Bmatrix}\right],$$

$$U(\gamma)=-\frac{2c}{b^{(\alpha+1)/r}r}\sin\frac{\nu\pi}{2}\sum_{k=0}^{\infty}\frac{(-1)^{\gamma k}(1-(-1)^\gamma\nu/2)_k}{k!\,(k+1)!}\Gamma\left(\frac{k+\alpha+1}{r}\right)\times$$
$$\times\begin{Bmatrix}\operatorname{cosec}[(k+\alpha+1)\pi/r]\\ \operatorname{ctg}[(k+\alpha+1)\pi/r]\end{Bmatrix}\left[2\psi(k+1)+\frac{1}{k+1}-\psi\left(1-\gamma-\frac{\nu}{2}+(-1)^\gamma k\right)-\right.$$
$$\left.-\frac{1}{r}\psi\left(\frac{k+\alpha+1}{r}\right)+\frac{\pi}{r}\begin{Bmatrix}\operatorname{ctg}[(k+\alpha+1)\pi/r]\\ 2\operatorname{cosec}[2(k+\alpha+1)\pi/r]\end{Bmatrix}-\ln\frac{2c}{b^{1/r}}\right]\left(\frac{2c}{b^{1/r}}\right)^k\mp$$
$$\mp\frac{(2c)^{r-\alpha}\sin^{1-\gamma}(\nu\pi/2)}{b\Gamma(1-(-1)^\gamma\nu/2)}\sum_{k=0}^{\infty}k!\,\Gamma\left[\begin{matrix}\alpha-r-rk,\ 1+\alpha-r-rk\\ 1+\alpha-r-rk+(-1)^\gamma\nu/2\end{matrix}\right]\times$$
$$\times\sin^{\gamma-1}\left(\alpha-r-rk+\frac{\nu}{2}\right)\pi\left(\mp\frac{2^rc^r}{b}\right)^k-\frac{2\Gamma(\alpha/r)}{\nu b^{\alpha/r}r}\sin\frac{\nu\pi}{2}\begin{Bmatrix}\operatorname{cosec}(\alpha\pi/r)\\ \operatorname{ctg}(\alpha\pi/r)\end{Bmatrix}\qquad[r>1],$$

$$U(\gamma)=-\frac{(2c)^{-\alpha}\sin^{1-\gamma}(\nu\pi/2)}{\Gamma(1-(-1)^\gamma\nu/2)}\sum_{k=0}^{\infty}\frac{1}{k!}\Gamma\left[\begin{matrix}\alpha+rk,\ 1+\alpha+rk\\ 1+\alpha+rk+(-1)^\gamma\nu/2\end{matrix}\right]\times$$
$$\times\sin^{\gamma-1}\left(\alpha+rk+\frac{\nu}{2}\right)\pi\left[\psi(k+1)-2r\psi(\alpha+rk)-\frac{r}{\alpha+rk}+\right.$$
$$+r\psi\left(1+\alpha+rk+(-1)^\gamma\frac{\nu}{2}\right)+(1-\gamma)\pi r\operatorname{ctg}\frac{2\alpha+\nu+2rk}{2}\pi+$$
$$\left.+\ln\frac{2^rc^r}{b}\right]\left(\pm\frac{b}{2^rc^r}\right)^k-(1-\gamma)\frac{\pi(2c)^{\nu/2}}{b^{(\nu+2\alpha)/(2r)}r\Gamma(1+\nu/2)}\times$$
$$\times\sum_{k=0}^{\infty}\frac{(-\nu/2)_k(1-\nu/2)_k}{k!}\Gamma\left(\frac{\nu+2\alpha-2k}{2r}\right)\begin{Bmatrix}\operatorname{cosec}[(\nu+2\alpha-2k)\pi/(2r)]\\ \operatorname{ctg}[(\nu+2\alpha-2k)\pi/(2r)]\end{Bmatrix}\left(-\frac{b^{1/r}}{2c}\right)^k$$
$$[r<1],$$

$U(\gamma)$ for $r=1$ see 2.14.5. 2—3.

13. $$\int\limits_0^\infty x^{\alpha-1} e^{\pm bx^r+cx} \operatorname{Ei}(\mp bx^r)\, k_\nu(cx)\, dx = U(0)$$

$\left[r,\ \operatorname{Re}\alpha>0;\ \operatorname{Re}(2\alpha+\nu)<2r,\ |\arg c|<\pi;\ \begin{Bmatrix} |\arg b|<\pi \\ b>0 \end{Bmatrix},\ U(\nu)\right.$ see 2.14.1.12$\left.\right]$.

14. $$\int\limits_0^\infty x^{\alpha-1} e^{-cx} \begin{Bmatrix} \operatorname{si}(bx^r) \\ \operatorname{ci}(bx^r) \end{Bmatrix} k_\nu(cx)\, dx = V(1) \qquad \left[b,\ r,\ \operatorname{Re} c,\ \operatorname{Re}\alpha>0;\ \delta=\begin{Bmatrix}1\\0\end{Bmatrix}\right].$$

$$V(\gamma) = -\frac{2c}{\pi b^{(\alpha+1)/r}} \sin\frac{\nu\pi}{2} \sum_{k=0}^{\infty} \frac{(-1)^{\nu k}(1-(-1)^\nu \nu/2)_k}{k!\,(k+1)!\,(k+\alpha+1)} \times$$
$$\times \Gamma\left(\frac{k+\alpha+1}{r}\right) \begin{Bmatrix} \sin[(k+\alpha+1)\pi/(2r)] \\ \cos[(k+\alpha+1)\pi/(2r)] \end{Bmatrix} \left[2\psi(k+1) + \frac{1}{k+1} - \right.$$
$$-\psi\left(1-\gamma-\frac{\nu}{2}+(-1)^\nu k\right) + \frac{1}{k+\alpha+1} - \frac{1}{r}\psi\left(\frac{k+\alpha+1}{r}\right) \mp$$
$$\left.\mp \frac{\pi}{2r} \operatorname{tg}^{\mp 1} \frac{k+\alpha+1}{2r}\pi - \ln\frac{2c}{b^{1/r}}\right] \left(\frac{2c}{b^{1/r}}\right)^k -$$
$$-\frac{2\Gamma(\alpha/r)}{\alpha\pi\nu b^{\alpha/r}} \sin\frac{\nu\pi}{2} \begin{Bmatrix} \sin[\alpha\pi/(2r)] \\ \cos[\alpha\pi/(2r)] \end{Bmatrix} \qquad [r>1],$$

$$V(\gamma) = -\frac{b^{\delta+2}\sin^{1-\gamma}(\nu\pi/2)}{2(2c)^{\alpha+\delta+2}3^\delta\,\Gamma(1-(-1)^\nu \nu/2)} \sum_{k=0}^{\infty} \frac{(-1)^k}{(k+1)!\,(\delta+3/2)_k\,(2k+\delta+2)} \times$$
$$\times \Gamma\begin{bmatrix} \alpha+\delta r+2r+2rk,\ 1+\alpha+\delta r+2r+2rk \\ 1+\alpha+\delta r+2r+2rk+(-1)^\nu \nu/2 \end{bmatrix} \times$$
$$\times \sin^{\gamma-1}\left(\alpha+\delta r+2r+2rk+\frac{\nu}{2}\right)\pi \left(\frac{b}{2^{r+1}c^r}\right)^{2k} +$$
$$+(1-\gamma)\frac{(2c)^{\nu/2}}{b^{(2\alpha+\nu)/(2r)}\Gamma(1+\nu/2)} \sum_{k=0}^{\infty} \frac{(-\nu/2)_k(1-\nu/2)_k}{k!\,(k-\alpha-\nu/2)} \times$$
$$\times \Gamma\left(\frac{2\alpha+\nu-2k}{2r}\right) \begin{Bmatrix} \sin[(2\alpha+\nu-2k)\pi/(4r)] \\ \cos[(2\alpha+\nu-2k)\pi/(4r)] \end{Bmatrix} \left(-\frac{b^{1/r}}{2c}\right)^k +$$
$$+\frac{b^\delta}{(2c)^{\alpha+r\delta}} \Gamma\begin{bmatrix} \alpha+r\delta,\ 1+\alpha+r\delta \\ 1+\alpha+r\delta+(-1)^\nu \nu/2,\ 1-(-1)^\nu \nu/2 \end{bmatrix} \times$$
$$\times \left[\frac{\sin(\nu\pi/2)}{\sin(\alpha+r\delta+\nu/2)\pi}\right]^{1-\nu} \begin{Bmatrix} 1 \\ R \end{Bmatrix},$$

$$R = \mathbf{C} + 2r\psi(\alpha) + \frac{r}{\alpha} - r\psi\left(1+\alpha+(-1)^\nu \frac{\nu}{2}\right) -$$
$$-(1-\gamma)\pi r \operatorname{ctg}\frac{2\alpha+\nu}{2}\pi - \ln\frac{2^r c^r}{b} \qquad [r<1].$$

$V(\gamma)$ for $r=1$ see 2.14.6.1—2.

15. $$\int\limits_0^\infty x^{\alpha-1} e^{cx} \begin{Bmatrix} \operatorname{si}(bx^r) \\ \operatorname{ci}(bx^r) \end{Bmatrix} k_\nu(cx)\, dx = V(0)$$

$[b,\ r,\ \operatorname{Re}\alpha>0;\ \operatorname{Re}(2\alpha+\nu)<4r;\ |\arg c|<\pi;\ V(\gamma)$ see 2.14.1.14$]$.

16. $$\int\limits_0^\infty x^{\alpha-1} e^{-cx} \begin{Bmatrix} \operatorname{erf}(bx^r) \\ \operatorname{erfc}(bx^r) \end{Bmatrix} k_\nu(cx)\, dx = W(1)$$

$[r,\ \operatorname{Re} c>0;\ \operatorname{Re}\alpha>-(1\pm 1)r/2;\ |\arg b|<\pi/4]$.

$$W(\gamma)=\mp\frac{2c}{\pi^{3/2}b^{(\alpha+1)/r}}\sin\frac{\nu\pi}{2}\sum_{k=0}^{\infty}\frac{(-1)^{\nu k}(1-(-1)^{\nu}\nu/2)_k}{k!\,(k+1)!\,(k+\alpha+1)}\times$$
$$\times\Gamma\left(\frac{k+r+\alpha+1}{2r}\right)\left[2\psi(k+1)+\frac{1}{k+1}-\psi\left(1-\gamma-\frac{\nu}{2}+(-1)^{\nu}k\right)-\right.$$
$$\left.-\frac{1}{2r}\psi\left(\frac{k+r+\alpha+1}{2r}\right)+\frac{1}{k+\alpha+1}-\ln\frac{2c}{b^{1/r}}\right]\left(\frac{2c}{b^{1/r}}\right)^k\mp$$
$$\mp\frac{2}{\alpha\pi^{3/2}\nu b^{\alpha/r}}\Gamma\left(\frac{r+\alpha}{2r}\right)\sin\frac{\nu\pi}{2}+$$
$$+\begin{Bmatrix}1\\0\end{Bmatrix}(2c)^{-\alpha}\Gamma\begin{bmatrix}\alpha,\ \alpha+1\\1-(-1)^{\nu}\nu/2,\ 1+\alpha+(-1)^{\nu}\nu/2\end{bmatrix}\left[\frac{\sin(\nu\pi/2)}{\sin(\alpha+\nu/2)\pi}\right]^{1-\gamma}\qquad [r>1/2],$$

$$W(\gamma)=$$
$$=\pm\frac{b\sin^{1-\gamma}(\nu\pi/2)}{\sqrt{\pi}\,(2c)^{\alpha+r}\Gamma(1-(-1)^{\nu}\nu/2)}\sum_{k=0}^{\infty}\frac{(-1)^k}{k!\,(k+1/2)}\Gamma\begin{bmatrix}\alpha+r+2rk,\ 1+\alpha+r+2rk\\1+\alpha+r+2rk+(-1)^{\nu}\nu/2\end{bmatrix}\times$$
$$\times\sin^{\gamma-1}\left(\alpha+r+2rk+\frac{\nu}{2}\right)\pi\left(\frac{b}{2^rc^r}\right)^{2k}\pm(1-\gamma)\frac{(2c)^{\nu/2}}{\sqrt{\pi}\,b^{(2\alpha+\nu)/(2r)}\Gamma(1+\nu/2)}\times$$
$$\times\sum_{k=0}^{\infty}\frac{(-\nu/2)_k(1-\nu/2)_k}{k!\,(k-\alpha-\nu/2)}\Gamma\left(\frac{r+\alpha+\nu/2-k}{2r}\right)\left(-\frac{b^{1/r}}{2c}\right)^k+$$
$$+\begin{Bmatrix}0\\1\end{Bmatrix}(2c)^{-\alpha}\Gamma\begin{bmatrix}\alpha,\ \alpha+1\\1-(-1)^{\nu}\nu/2,\ 1+\alpha+(-1)^{\nu}\nu/2\end{bmatrix}\left[\frac{\sin(\nu\pi/2)}{\sin(\alpha+\nu/2)\pi}\right]^{1-\gamma}\qquad [r<1/2],$$

$W(\gamma)$ for $r=1/2$ see 2.14.7.3—4.

17. $$\int_0^{\infty}x^{\alpha-1}e^{cx}\begin{Bmatrix}\operatorname{erf}(bx^r)\\\operatorname{erfc}(bx^r)\end{Bmatrix}k_{\nu}(cx)\,dx=W(0)$$

$\left[r>0;\ \operatorname{Re}\alpha>-(1\pm1)r/2;\ |\arg c|<\pi,\ \begin{Bmatrix}\operatorname{Re}(2\alpha+\nu)<0;\ |\arg b|<\pi/4\\|\arg b|<\pi/4\end{Bmatrix},\ W(\gamma)\text{ see } 2.14.1.16\right].$

18. $$\int_0^{\infty}x^{\alpha-1}e^{-cx}\begin{Bmatrix}S(bx^r)\\C(bx^r)\end{Bmatrix}k_{\nu}(cx)\,dx=U(1)\left[b,\ r,\ \operatorname{Re}c>0;\ \operatorname{Re}\alpha>-r(\delta+1/2);\ \delta=\begin{Bmatrix}1\\0\end{Bmatrix}\right].$$

$$U(\gamma)=-\frac{c\sqrt{2}}{\pi^{3/2}b^{(\alpha+1)/r}}\sin\frac{\nu\pi}{2}\sum_{k=0}^{\infty}\frac{(-1)^{\nu k}(1-(-1)^{\nu}\nu/2)_k}{k!\,(k+1)!\,(k+\alpha+1)}\times$$
$$\times\Gamma\left(\frac{2+2\alpha+r+2k}{2r}\right)\begin{Bmatrix}\sin[(2+2\alpha+r+2k)\pi/(4r)]\\\cos[(2+2\alpha+r+2k)\pi/(4r)]\end{Bmatrix}\times$$
$$\times\left[2\psi(k+1)+\frac{1}{k+1}-\psi\left(1-\gamma-\frac{\nu}{2}+(-1)^{\nu}k\right)-\frac{1}{r}\psi\left(\frac{2+2\alpha+r+2k}{2r}\right)\mp\right.$$
$$\left.\mp\frac{\pi}{2r}\operatorname{tg}^{\mp1}\frac{2+2\alpha+r+2k}{4r}\pi-\frac{1}{k+\alpha+1}-\ln\frac{2c}{b^{1/r}}\right]\left(\frac{2c}{b^{1/r}}\right)^k-$$
$$-\frac{\sqrt{2}}{\alpha\pi^{3/2}\nu b^{\alpha/r}}\sin\frac{\nu\pi}{2}\Gamma\left(\frac{2\alpha+r}{2r}\right)\begin{Bmatrix}\sin[(2\alpha+r)\pi/(4r)]\\\cos[(2\alpha+r)\pi/(4r)]\end{Bmatrix}+$$
$$+\frac{1}{2^{\alpha+1}c^{\alpha}}\Gamma\begin{bmatrix}\alpha,\ \alpha+1\\1+\alpha+(-1)^{\nu}\nu/2,\ 1-(-1)^{\nu}\nu/2\end{bmatrix}\left[\frac{\sin(\nu\pi/2)}{\sin(\alpha+\nu/2)\pi}\right]^{1-\gamma}\qquad [r>1],$$

$$U(\gamma)=\frac{b^{\delta+1/2}\sin^{1-\nu}(\nu\pi/2)}{\sqrt{2\pi}\,(2c)^{\alpha+\delta r+r/2}\,\Gamma\,(1-(-1)^{\nu}\nu/2)}\sum_{k=0}^{\infty}\frac{(-1)^k}{k!\,(\delta+1/2)_k\,(2k+\delta+1/2)}\times$$

$$\times\Gamma\begin{bmatrix}\alpha+\delta r+r/2+2rk,\ 1+\alpha+\delta r+r/2+2rk\\ 1+\alpha+\delta r+r/2+2rk+(-1)^{\nu}\nu/2\end{bmatrix}\times$$

$$\times\sin^{\gamma-1}\left(\alpha+\delta r+2rk+\frac{\nu+r}{2}\right)\pi\left(\frac{b}{2^{r+1}c^r}\right)^{2k}+$$

$$+\frac{(1-\gamma)\,(2c)^{\nu/2}}{\sqrt{2\pi}\,b^{(2\alpha+\nu)/(2r)}\,\Gamma\,(1+\nu/2)}\sum_{k=0}^{\infty}\frac{(-\nu/2)_k\,(1-\nu/2)_k}{k!\,(k-\alpha-\nu/2)}\times$$

$$\times\Gamma\left(\frac{2\alpha+\nu+r-2k}{2r}\right)\begin{Bmatrix}\sin\,[(2\alpha+\nu+r-2k)\,\pi/(4r)]\\ \cos\,[(2\alpha+\nu+r-2k)\,\pi/(4r)]\end{Bmatrix}\left(-\frac{b^{1/r}}{2c}\right)^k\qquad [r<1],$$

$U(\gamma)$ for $r=1$ see 2.14.8.1—2.

19. $$\int_0^{\infty} x^{\alpha-1}e^{cx}\begin{Bmatrix}S\,(bx^r)\\ C\,(bx^r)\end{Bmatrix}k_{\nu}(cx)\,dx=U(0)$$

$$\left[b,\ r>0;\ -r\,(\delta+1/2)<\operatorname{Re}\alpha<-\operatorname{Re}\nu/2;\ |\arg c|<\pi;\ \delta=\begin{Bmatrix}1\\0\end{Bmatrix};\ U(\gamma)\ \text{see 2.14.1.18}\right].$$

20. $$\int_0^{\infty} x^{\alpha-1}e^{-cx}\begin{Bmatrix}\gamma\,(\mu,\ bx^r)\\ \Gamma\,(\mu,\ bx^r)\end{Bmatrix}k_{\nu}(cx)\,dx=V\,(1)$$

$$\left[r,\ \operatorname{Re}b,\ \operatorname{Re}c,\ \operatorname{Re}(\alpha+r\mu)>0,\ \begin{Bmatrix}\operatorname{Re}\mu>0\\ \operatorname{Re}\alpha>0\end{Bmatrix}\right],$$

$$V\,(\gamma)=\mp\frac{2c}{\pi b^{(\alpha+1)/r}}\sin\frac{\nu\pi}{2}\sum_{k=0}^{\infty}\frac{(-1)^{\nu k}\,(1-(-1)^{\nu}\nu/2)_k}{k!\,(k+1)!\,(k+\alpha+1)}\times$$

$$\times\Gamma\left(\frac{k+\alpha+1}{r}+\mu\right)\left[2\psi\,(k+1)+\frac{1}{k+1}-\psi\left(1-\gamma-\frac{\nu}{2}+(-1)^{\nu}k\right)-\right.$$

$$\left.-\frac{1}{r}\psi\left(\frac{k+\alpha+1}{r}+\mu\right)+\frac{1}{k+\alpha+1}-\ln\frac{2c}{b^{1/r}}\right]\left(\frac{2c}{b^{1/r}}\right)^k\mp$$

$$\mp\frac{2}{\alpha\pi\nu b^{\alpha/r}}\Gamma\left(\frac{\alpha}{r}+\mu\right)\sin\frac{\nu\pi}{2}+$$

$$+\begin{Bmatrix}1\\0\end{Bmatrix}(2c)^{-\alpha}\,\Gamma\begin{bmatrix}\alpha,\ \alpha+1,\ \mu\\ 1-(-1)^{\nu}\nu/2,\ \alpha+1+(-1)^{\nu}\nu/2\end{bmatrix}\left[\frac{\sin\,(\nu\pi/2)}{\sin\,(\alpha+\nu/2)\,\pi}\right]^{1-\gamma}\qquad [r>1],$$

$$V\,(\gamma)=$$

$$=\pm\frac{b^{\mu}\sin^{1-\nu}(\nu\pi/2)}{(2c)^{\alpha+\mu r}\,\Gamma\,(1-(-1)^{\nu}\nu/2)}\sum_{k=0}^{\infty}\frac{1}{k!\,(\mu+k)}\,\Gamma\begin{bmatrix}\alpha+\mu r+rk,\ 1+\alpha+\mu r+rk\\ 1+\alpha+\mu r+rk+(-1)^{\nu}\nu/2\end{bmatrix}\times$$

$$\times\sin^{\gamma-1}\left(\alpha+\mu r+rk+\frac{\nu}{2}\right)\pi\left(-\frac{b}{2^rc^r}\right)^k\pm(1-\gamma)\,\frac{(2c)^{\nu/2}}{b^{(\nu+2\alpha)/(2r)}\,\Gamma\,(1+\nu/2)}\times$$

$$\times\sum_{k=0}^{\infty}\frac{(-\nu/2)_k\,(1-\nu/2)_k}{k!\,(k-\alpha-\nu/2)}\,\Gamma\left(\frac{\nu+2\alpha-2k}{2r}+\mu\right)\left(-\frac{b^{1/r}}{2c}\right)^k+$$

$$+\begin{Bmatrix}0\\1\end{Bmatrix}(2c)^{-\alpha}\,\Gamma\begin{bmatrix}\alpha,\ \alpha+1,\ \mu\\ 1-(-1)^{\nu}\nu/2,\ \alpha+1+(-1)^{\nu}\nu/2\end{bmatrix}\left[\frac{\sin\,(\nu\pi/2)}{\sin\,(\alpha+\nu/2)\,\pi}\right]^{1-\gamma}$$

$[r<1]$,

$V\,(\gamma)$ for $r=1$ see 2.14.9.1—2.

21. $\int_0^\infty x^{\alpha-1}e^{cx}\begin{Bmatrix}\gamma(\mu, bx^r)\\ \Gamma(\mu, bx^r)\end{Bmatrix}k_\nu(cx)\,dx=V(0)$

$\left[r,\ \operatorname{Re} b,\ \operatorname{Re}(\alpha+r\mu)>0;\ |\arg c|<\pi,\ \begin{Bmatrix}\operatorname{Re}\mu>0;\ \operatorname{Re}(2\alpha+\nu)<0\\ \operatorname{Re}\alpha>0\end{Bmatrix},\ V(\gamma)\ \text{see 2.14.1.20}\right].$

2.14.2. Integrals of $A(x)\,e^{\varphi(x)}k_\nu(cx)$.

1. $\int_0^\infty x^{\alpha-1}k_\nu(cx)\,dx=$

$$=(2c)^{-\alpha}\Gamma\begin{bmatrix}\alpha,\ \alpha+1\\ 1+\nu/2,\ 1+\alpha-\nu/2\end{bmatrix}{}_2F_1\left(\alpha,\ \alpha+1;\ 1+\alpha-\frac{\nu}{2};\ \frac{1}{2}\right)\qquad [\operatorname{Re} c,\ \operatorname{Re}\alpha>0].$$

2. $\int_0^\infty x^{\alpha-1}e^{\pm cx}k_\nu(cx)\,dx=$

$$=(2c)^{-\alpha}\Gamma\begin{bmatrix}\alpha,\ \alpha+1\\ 1+\alpha\pm\nu/2,\ 1\mp\nu/2\end{bmatrix}\left(\sin\frac{\nu\pi}{2}\operatorname{cosec}\frac{2\alpha+\nu}{2}\pi\right)^{(1\pm1)/2}$$

$$\left[\operatorname{Re}\alpha>0,\ \begin{Bmatrix}|\arg c|<\pi;\ \operatorname{Re}(2\alpha+\nu)<0\\ \operatorname{Re} c>0\end{Bmatrix}\right].$$

3. $\int_0^\infty e^{-px}k_\nu(cx)\,dx=\dfrac{2\sin(\nu\pi/2)}{\pi\nu(2-\nu)c}\,{}_2F_1\left(1,\ 2;\ 2-\dfrac{\nu}{2};\ \dfrac{c-p}{2c}\right)\qquad [\operatorname{Re}(c+p)>0].$

4. $\int_0^\infty x^{\alpha-1}e^{-px}k_\nu(cx)\,dx=$

$$=(2c)^{-\alpha}\Gamma\begin{bmatrix}\alpha,\ \alpha+1\\ 1+\nu/2,\ 1+\alpha-\nu/2\end{bmatrix}{}_2F_1\left(\alpha,\ \alpha+1;\ 1+\alpha-\frac{\nu}{2};\ \frac{c-p}{2c}\right)$$

$[\operatorname{Re}(c+p),\ \operatorname{Re}\alpha>0].$

5. $\int_0^\infty x^{n-3/2}e^{-px}k_{2n}(cx)\,dx=\dfrac{(-1)^n(2n-2)!\sqrt{\pi c}}{2^{2n-3/2}\,n!\,(c+p)^n}P_{2n-1}^{(1)}\left(\sqrt{\dfrac{p-c}{p+c}}\right)$

$[\operatorname{Re}(c+p)>0;\ n=1,\ 2,\ 3,\ \ldots].$

6. $\int_0^\infty e^{-px^2-cx}k_{2n}(cx)\,dx=$

$$=2^{(n-1)/2}c^np^{-(n+1)/2}\sum_{k=0}^{n-1}(-1)^k\binom{n-1}{k}\left(\frac{p}{2c^2}\right)^{k/2}\exp\left(\frac{c^2}{2p}\right)D_{k-n-1}\left(\sqrt{\frac{2}{p}}\,c\right)$$

$[\operatorname{Re} p>0].$

7. $\int_0^\infty x^{\alpha-1}e^{-p\sqrt{x}\pm cx}k_\nu(cx)\,dx=$

$$=(2c)^{-\alpha}\Gamma\begin{bmatrix}\alpha,\ \alpha+1\\ 1\mp\nu/2,\ 1+\alpha\pm\nu/2\end{bmatrix}\left(\sin\frac{\nu\pi}{2}\operatorname{cosec}\frac{2\alpha+\nu}{2}\pi\right)^{(1\pm1)/2}\times$$

$$\times{}_2F_2\left(\alpha,\ \alpha+1;\ \frac{1}{2},\ 1+\alpha\pm\nu/2;\ \mp\frac{p^2}{8c}\right)-$$

$$-\frac{p}{(2c)^{\alpha+1/2}}\Gamma\begin{bmatrix}\alpha+1/2,\ \alpha+3/2\\ 1\mp\nu/2,\ \alpha+(3\pm\nu)/2\end{bmatrix}\left(\sin\frac{\nu\pi}{2}\sec\frac{2\alpha+\nu}{2}\pi\right)^{(1\pm1)/2}\times$$

$$\times {}_2F_2\left(\alpha+\frac{1}{2},\ \alpha+\frac{3}{2};\ \frac{3}{2},\ \alpha+\frac{3\pm\nu}{2};\ \mp\frac{p^2}{8c}\right)+$$

$$+\begin{Bmatrix}1\\0\end{Bmatrix}\frac{2(2c)^{\nu/2}}{p^{\nu+2\alpha}}\Gamma\begin{bmatrix}2\alpha+\nu\\ \nu/2+1\end{bmatrix}{}_2F_2\left(-\frac{\nu}{2},\ 1-\frac{\nu}{2};\ 1-\alpha-\frac{\nu}{2},\ \frac{1-\nu}{2}-\alpha;-\frac{p^2}{8c}\right)$$

$$\left[\operatorname{Re} p,\ \operatorname{Re}\alpha>0,\quad \begin{Bmatrix}|\arg c|<\pi\\ \operatorname{Re} c>0\end{Bmatrix}\right].$$

8. $$\int_0^\infty x^{-1/2}e^{-p\sqrt{x}-cx}k_{2n}(cx)\,dx=$$

$$=(-1)^{n-1}2^{(3-6n)/4}c^{(1-2n)/4}p^{n-3/2}\exp\left(\frac{p^2}{16c}\right)W_{-(3+2n)/4,(1-2n)/4}\left(\frac{p^2}{8c}\right)$$

$$[\operatorname{Re} c,\ \operatorname{Re} p>0].$$

9. $$\int_0^\infty x^{\alpha-1}e^{-p/x\pm cx}k_\nu(cx)\,dx=\frac{2c}{\pi}p^{\alpha+1}\sin\frac{\nu\pi}{2}\Gamma(-\alpha-1)\times$$

$$\times\sum_{k=0}^\infty\frac{(1\mp\nu/2)_k(\mp 2pc)^k}{(k!)^2(2+\alpha)_k(k+1)}\left[2\psi(k+1)+\frac{1}{k+1}-\psi\left(\frac{1\pm1-\nu}{2}\pm k\right)+\right.$$

$$\left.+\psi(-k-\alpha-1)-\ln(2pc)\right]+(2c)^{-\alpha}\Gamma\begin{bmatrix}\alpha,\ \alpha+1\\ 1\mp\nu/2,\ 1+\alpha\pm\nu/2\end{bmatrix}\times$$

$$\times\left(\sin\frac{\nu\pi}{2}\operatorname{cosec}\frac{2\alpha+\nu}{2}\pi\right)^{(1\pm1)/2}{}_1F_2\left(\mp\frac{\nu}{2}-\alpha;1-\alpha,-\alpha;\mp 2pc\right)+\frac{2p^\alpha}{\nu\pi}\Gamma(-\alpha)\sin\frac{\nu\pi}{2}$$

$$\left[\operatorname{Re} p>0,\quad \begin{Bmatrix}|\arg c|<\pi; & \operatorname{Re}(2\alpha+\nu)<0\\ \operatorname{Re} c>0 & \end{Bmatrix}\right].$$

10. $$\int_0^a x^{\alpha-1}(a-x)^{\beta-1}e^{\pm cx}k_\nu(cx)\,dx=\frac{2a^{\alpha+\beta}c}{\pi}\sin\frac{\nu\pi}{2}\mathrm{B}(\alpha+1,\ \beta)\times$$

$$\times\sum_{k=0}^\infty\frac{(1\mp\nu/2)_k(\alpha+1)_k}{(k!)^2(\alpha+\beta+1)_k(k+1)}\left[2\psi(k+1)+\frac{1}{k+1}+\psi(\alpha+\beta+k+1)-\right.$$

$$\left.-\psi(\alpha+k+1)-\psi\left(\frac{1\pm1-\nu}{2}\pm k\right)-\ln(2ac)\right](\pm 2ac)^k+$$

$$+\frac{2a^{\alpha+\beta-1}}{\nu\pi}\sin\frac{\nu\pi}{2}\mathrm{B}(\alpha,\ \beta)\qquad [a,\ \operatorname{Re}\alpha,\ \operatorname{Re}\beta>0].$$

11. $$\int_a^\infty x^{\alpha-1}(x-a)^{\beta-1}e^{\pm cx}k_\nu(cx)\,dx=\frac{2a^{\alpha+\beta}c}{\pi}\sin\frac{\nu\pi}{2}\mathrm{B}(\beta,\ -\alpha-\beta)\times$$

$$\times\sum_{k=0}^\infty\frac{(1\mp\nu/2)_k(1+\alpha)_k}{(k!)^2(1+\alpha+\beta)_k(k+1)}\left[2\psi(k+1)+\frac{1}{k+1}+\psi(-\alpha-\beta-k)-\right.$$

$$\left.-\psi(-\alpha-k)-\psi\left(\frac{1\pm1-\nu}{2}\pm k\right)-\ln(2ac)\right](\pm 2ac)^k+$$

$$+\frac{2a^{\alpha+\beta-1}}{\pi\nu}\sin\frac{\nu\pi}{2}\mathrm{B}(\beta,\ 1-\alpha-\beta)\mp$$

$$\mp (2c)^{1-\alpha-\beta}\Gamma\begin{bmatrix}\alpha+\beta-1,\ \alpha+\beta\\ \alpha+\beta\pm\nu/2,\ 1\mp\nu/2\end{bmatrix}\left(\sin\frac{\nu\pi}{2}\operatorname{cosec}\frac{2\alpha+2\beta+\nu}{2}\pi\right)^{(1\pm1)/2}\times$$
$$\times {}_2F_2\left(1-\beta,\ 1-\alpha-\beta\mp\frac{\nu}{2};\ 2-\alpha-\beta,\ 1-\alpha-\beta;\ \pm 2ac\right)$$
$$\left[a,\ \operatorname{Re}\beta>0,\ \begin{Bmatrix}|\arg c|<\pi;\ \operatorname{Re}(\alpha+\beta+\nu/2)<1\\ \operatorname{Re}c>0\end{Bmatrix}\right].$$

12. $$\int_0^\infty \frac{x^{\alpha-1}}{(x+z)^\rho}e^{\pm cx}k_\nu(cx)\,dx=\frac{2cz^{\alpha-\rho+1}}{\pi}\sin\frac{\nu\pi}{2}\,\mathrm{B}(\alpha+1,\ \rho-\alpha-1)\times$$
$$\times\sum_{k=0}^\infty\frac{(\alpha+1)_k(1\mp\nu/2)_k}{(k!)^2(2+\alpha-\rho)_k\,(k+1)}\left[2\psi(k+1)+\frac{1}{k+1}+\psi(\rho-\alpha-k-1)-\right.$$
$$\left.-\psi(\alpha+k+1)-\psi\left(\frac{1\pm1-\nu}{2}\pm k\right)-\ln(2cz)\right](\mp 2cz)^k+$$
$$+\frac{2z^{\alpha-\rho}}{\pi\nu}\sin\frac{\nu\pi}{2}\,\mathrm{B}(\alpha,\ \rho-\alpha)+$$
$$+(2c)^{\rho-\alpha}\,\Gamma\begin{bmatrix}\alpha-\rho,\ 1+\alpha-\rho\\ 1+\alpha-\rho\pm\nu/2,\ 1\mp\nu/2\end{bmatrix}\left(\sin\frac{\nu\pi}{2}\operatorname{cosec}\frac{2\alpha-2\rho+\nu}{2}\pi\right)^{(1\pm1)/2}\times$$
$$\times {}_2F_2\left(\rho,\ \rho-\alpha\mp\frac{\nu}{2};\ 1+\rho-\alpha,\ \rho-\alpha;\ \mp 2cz\right)$$
$$\left[\operatorname{Re}\alpha>0,\ \begin{Bmatrix}|\arg c|<\pi;\ \operatorname{Re}(\alpha-\rho+\nu/2)<0\\ \operatorname{Re}c>0\end{Bmatrix},\ |\arg z|<\pi\right].$$

13. $$\int_0^\infty \frac{x^{\alpha-1}}{x-y}e^{\pm cx}k_\nu(cx)\,dx=-2cy^\alpha\sin\frac{\nu\pi}{2}\operatorname{ctg}\alpha\pi\sum_{k=0}^\infty\frac{(1\mp\nu/2)_k}{(k!)^2(k+1)}\times$$
$$\times\left[2\psi(k+1)+\frac{1}{k+1}+2\pi\operatorname{cosec}2\alpha\pi-\psi\left(\frac{1\pm1-\nu}{2}\pm k\right)-\ln(2cy)\right]\times$$
$$\times(\pm 2cy)^k-\frac{2y^{\alpha-1}}{\nu}\sin\frac{\nu\pi}{2}\operatorname{ctg}\alpha\pi\mp(2c)^{1-\alpha}\Gamma\begin{bmatrix}\alpha-1,\ \alpha\\ \alpha\pm\nu/2,\ 1\mp\nu/2\end{bmatrix}\times$$
$$\times\left(\sin\frac{\nu\pi}{2}\operatorname{cosec}\frac{2\alpha+\nu}{2}\pi\right)^{(1\pm1)/2}{}_2F_2\left(1,\ 1-\alpha\mp\frac{\nu}{2};\ 1-\alpha,\ 2-\alpha;\ \pm 2cy\right)$$
$$\left[y,\ \operatorname{Re}\alpha>0,\ \begin{Bmatrix}|\arg c|<\pi;\ \operatorname{Re}(2\alpha+\nu)<2\\ \operatorname{Re}c>0\end{Bmatrix}\right].$$

2.14.3. Integrals of $x^\alpha e^{-px}\begin{Bmatrix}\sin bx\\ \cos bx\end{Bmatrix}k_\nu(cx)$.

NOTATION: $\delta=\begin{Bmatrix}1\\0\end{Bmatrix}$.

1. $$\int_0^\infty x^{\alpha-1}e^{-cx}\begin{Bmatrix}\sin bx\\ \cos bx\end{Bmatrix}k_\nu(cx)\,dx=U(1)\qquad [b,\ \operatorname{Re}c>0;\ \operatorname{Re}\alpha>-\delta],$$

$$U(\gamma)=$$
$$=\frac{b^\delta}{(2c)^{\alpha+\delta}}\left[\mp\frac{\sin(\nu\pi/2)}{\sin(\alpha+\nu/2)\pi}\right]^{1-\gamma}\Gamma\begin{bmatrix}\alpha+1,\ \alpha+2\delta\\ 1+\alpha+\delta+(-1)^\gamma\nu/2,\ 1-(-1)^\gamma\nu/2\end{bmatrix}\times$$
$$\times{}_4F_3\left(\frac{\alpha+1}{2},\ \frac{\alpha}{2}+\delta,\ 1+\frac{\alpha}{2},\ \frac{1+\alpha}{2}+\delta;\right.$$
$$\left.\delta+\frac{1}{2},\ 1+\frac{2\alpha+(-1)^\gamma\nu}{4},\ \frac{2+2\alpha+(-1)^\gamma\nu}{4}+\delta;\ -\frac{b^2}{4c^2}\right).$$

2. $\int\limits_0^\infty x^{\alpha-1} e^{cx} \begin{Bmatrix} \sin bx \\ \cos bx \end{Bmatrix} k_\nu(cx)\, dx=$

$$= U(0) + \frac{(2c)^{\nu/2}}{b^{\alpha+\nu/2}} \Gamma\begin{bmatrix} \alpha+\nu/2 \\ 1+\nu/2 \end{bmatrix} \begin{Bmatrix} \sin[(2\alpha+\nu)\pi/4] \\ \cos[(2\alpha+\nu)\pi/4] \end{Bmatrix} \times$$

$$\times {}_4F_3\left(-\frac{\nu}{4}, \frac{2-\nu}{4}, \frac{2-\nu}{4}, 1-\frac{\nu}{4}; \frac{1}{2}, 1-\frac{2\alpha+\nu}{4}, \frac{2-2\alpha-\nu}{4}; -\frac{b^2}{4c^2}\right) \pm$$

$$\pm \frac{(2c)^{\nu/2-1}}{b^{\alpha+\nu/2-1}} \Gamma\begin{bmatrix} \alpha+\nu/2-1 \\ \nu/2-1 \end{bmatrix} \begin{Bmatrix} \cos[(2\alpha+\nu)\pi/4] \\ \sin[(2\alpha+\nu)\pi/4] \end{Bmatrix} \times$$

$$\times {}_4F_3\left(\frac{2-\nu}{4}, 1-\frac{\nu}{4}, 1-\frac{\nu}{4}, \frac{6-\nu}{4}; \frac{3}{2}, \frac{6-2\alpha-\nu}{4}, 1-\frac{2\alpha+\nu}{4}; -\frac{b^2}{4c^2}\right)$$

$[b > 0;\ \mathrm{Re}\,\alpha > -\delta;\ \mathrm{Re}\,(2\alpha+\nu) < 2;\ |\arg c| < \pi;\ U(\gamma)$ see 2.14.3.1].

2.14.4. Integrals of $x^\alpha e^{-px} \begin{Bmatrix} \ln(x+z) \\ \ln|x-z| \end{Bmatrix} k_\nu(cx)$.

1. $\int\limits_0^\infty x^{\alpha-1} e^{-cx} \begin{Bmatrix} \ln(x+z) \\ \ln|x-z| \end{Bmatrix} k_\nu(cx)\, dx = V(1)$ $\left[\mathrm{Re}\,c,\ \mathrm{Re}\,\alpha > 0, \begin{Bmatrix} |\arg z| < \pi \\ z > 0 \end{Bmatrix}\right]$,

$$V(\gamma) = \mp 2cz^{\alpha+1} \sin\frac{\nu\pi}{2} \begin{Bmatrix} \operatorname{cosec}\alpha\pi \\ \operatorname{ctg}\alpha\pi \end{Bmatrix} \sum_{k=0}^\infty \frac{(1-(-1)^\gamma \nu/2)_k}{k!\,(k+1)!\,(\alpha+k+1)} \times$$

$$\times \left[2\psi(k+1) + \frac{1}{k+1} - \psi\left(1-\gamma-\frac{\nu}{2}+(-1)^\gamma k\right) + \pi \begin{Bmatrix} \operatorname{ctg}\alpha\pi \\ 2\operatorname{cosec}2\alpha\pi \end{Bmatrix} + \right.$$

$$\left. + \frac{1}{\alpha+k+1} - \ln(2cz)\right] \left(\mp(-1)^\gamma 2cz\right)^k + \frac{2z^\alpha}{\alpha\nu} \sin\frac{\nu\pi}{2} \begin{Bmatrix} \operatorname{cosec}\alpha\pi \\ \operatorname{ctg}\alpha\pi \end{Bmatrix} \pm$$

$$\pm \frac{z}{(2c)^{\alpha-1}} \left[\frac{\sin(\nu\pi/2)}{\sin(\nu/2-\alpha)\pi}\right]^{1-\gamma} \Gamma\begin{bmatrix} \alpha-1,\ \alpha \\ \alpha+(-1)^\gamma \nu/2,\ 1-(-1)^\gamma \nu/2 \end{bmatrix} \times$$

$$\times {}_3F_3\left(1, 1, 1-\alpha-(-1)^\gamma \frac{\nu}{2}; 2, 1-\alpha, 2-\alpha; \mp(-1)^\gamma 2cz\right) +$$

$$+ (2c)^{-\alpha} \Gamma\begin{bmatrix} \alpha,\ \alpha+1 \\ 1-(-1)^\gamma \nu/2,\ 1+\alpha+(-1)^\gamma \nu/2 \end{bmatrix} \left[\frac{\sin(\nu\pi/2)}{\sin(\nu/2+\alpha)\pi}\right]^{1-\gamma} \times$$

$$\times \left[2\psi(\alpha) + \frac{1}{\alpha} - \psi\left(1+\alpha+(-1)^\gamma \frac{\nu}{2}\right) - (1-\gamma)\pi \operatorname{ctg}\frac{2\alpha+\nu}{2}\pi - \ln(2c)\right].$$

2. $\int\limits_0^\infty x^{\alpha-1} e^{cx} \begin{Bmatrix} \ln(x+z) \\ \ln|x-z| \end{Bmatrix} k_\nu(cx)\, dx = V(0)$

$\left[\mathrm{Re}\,\alpha > 0;\ \mathrm{Re}\,(2\alpha+\nu) < 0;\ |\arg c| < \pi;\ \begin{Bmatrix} |\arg z| < \pi \\ z > 0 \end{Bmatrix},\ V(\gamma)\ \text{see 2.14.4.1}\right]$.

2.14.5. Integrals of $x^\alpha e^{-px}\,\mathrm{Ei}(\pm bx)\, k_\nu(cx)$.

1. $\int\limits_0^\infty x^{\alpha-1} e^{\pm cx}\,\mathrm{Ei}(-bx)\, k_\nu(cx)\, dx=$

$$= \pm (2c)^{-\alpha-1} \Gamma\begin{bmatrix} \alpha+1,\ \alpha+2 \\ \alpha+2\pm\nu/2,\ 1\mp\nu/2 \end{bmatrix} \left(\sin\frac{\nu\pi}{2} \operatorname{cosec}\frac{2\alpha+\nu}{2}\pi\right)^{(1\pm1)/2} \times$$

$$\times {}_4F_3\left(1, 1, \alpha+1, \alpha+2; 2, 2, \alpha+2\pm\frac{\nu}{2}; \pm\frac{b}{2c}\right) -$$

$$-\begin{Bmatrix}1\\0\end{Bmatrix}\frac{2(2c)^{\nu/2}}{b^{\alpha+\nu/2}(2\alpha+\nu)}\Gamma\begin{bmatrix}\alpha+\nu/2\\1+\nu/2\end{bmatrix}\times$$

$$\times\,{}_3F_2\left(-\frac{\nu}{2},\,1-\frac{\nu}{2},\,-\alpha-\frac{\nu}{2};\,1-\alpha-\frac{\nu}{2},\,1-\alpha-\frac{\nu}{2};\,\frac{b}{2c}\right)+$$

$$+(2c)^{-\alpha}\Gamma\begin{bmatrix}\alpha,\ \alpha+1\\1\mp\nu/2,\ 1+\alpha\pm\nu/2\end{bmatrix}\left(\sin\frac{\nu\pi}{2}\operatorname{cosec}\frac{2\alpha+\nu}{2}\pi\right)^{(1\pm1)/2}\times$$

$$\times\left[\mathbf{C}+2\psi(\alpha)+\frac{1}{\alpha}-\psi\left(1+\alpha\pm\frac{\nu}{2}\right)-\begin{Bmatrix}1\\0\end{Bmatrix}\pi\operatorname{ctg}\frac{2\alpha+\nu}{2}\pi-\ln\frac{2c}{b}\right]$$

$$\left[\operatorname{Re} b,\ \operatorname{Re}\alpha>0,\ \begin{Bmatrix}|\arg c|<\pi\\\operatorname{Re} c>0\end{Bmatrix}\right].$$

2. $\int\limits_0^\infty x^{\alpha-1}e^{(\pm b-c)x}\operatorname{Ei}(\mp bx)\,k_\nu(cx)\,dx=U(1)\qquad\left[\operatorname{Re} c,\ \operatorname{Re}\alpha>0,\ \begin{Bmatrix}|\arg b|<\pi\\b>0\end{Bmatrix}\right],$

$$U(\gamma)=-(2c)^{-\alpha}\Gamma\begin{bmatrix}\alpha,\ \alpha+1\\1+\alpha+(-1)^\gamma\nu/2,\ 1-(-1)^\gamma\nu/2\end{bmatrix}\times$$

$$\times\left[\frac{\sin(\nu\pi/2)}{\sin(\alpha+\nu/2)\pi}\right]^{1-\gamma}\sum_{k=0}^{\infty}\frac{(\alpha)_k(\alpha+1)_k}{k!\,(1+\alpha+(-1)^\gamma\nu/2)_k}\times$$

$$\times\left[\psi(k+1)-2\psi(\alpha+k)-\frac{1}{\alpha+k}+\psi\left(1+\alpha+(-1)^\gamma\frac{\nu}{2}\right)+\right.$$

$$\left.+(1-\gamma)\pi\operatorname{ctg}\frac{2\alpha+\nu}{2}\pi+\ln\frac{2c}{b}\right]\left(\mp(-1)^\gamma\frac{b}{2c}\right)^k-$$

$$-(1-\gamma)\frac{\pi(2c)^{\nu/2}}{b^{\alpha+\nu/2}}\Gamma\begin{bmatrix}\alpha+\nu/2\\1+\nu/2\end{bmatrix}\begin{Bmatrix}\operatorname{cosec}(\alpha+\nu/2)\pi\\\operatorname{ctg}(\alpha+\nu/2)\pi\end{Bmatrix}\times$$

$$\times\,{}_2F_1\left(-\frac{\nu}{2},\,1-\frac{\nu}{2};\,1-\alpha-\frac{\nu}{2};\,\mp\frac{b}{2c}\right).$$

3. $\int\limits_0^\infty x^{\alpha-1}e^{(\pm b+c)x}\operatorname{Ei}(\mp bx)\,k_\nu(cx)\,dx=U(0)$

$$\left[\operatorname{Re}\alpha>0;\ \operatorname{Re}(2\alpha+\nu)<2;\ |\arg c|<\pi,\ \begin{Bmatrix}|\arg b|<\pi\\b>0\end{Bmatrix},\ U(\gamma)\ \text{see } 2.14.5.2\right].$$

2.14.6. Integrals of $x^\alpha e^{-px}\begin{Bmatrix}\operatorname{si}(bx)\\\operatorname{ci}(bx)\end{Bmatrix}k_\nu(cx)$.

1. $\int\limits_0^\infty x^{\alpha-1}e^{-cx}\begin{Bmatrix}\operatorname{si}(bx)\\\operatorname{ci}(bx)\end{Bmatrix}k_\nu(cx)\,dx=V(1)\qquad\left[b,\ \operatorname{Re} c,\ \operatorname{Re}\alpha>0,\ \delta=\begin{Bmatrix}1\\0\end{Bmatrix}\right],$

$$V(\gamma)=-\frac{b^{\delta+2}}{2\cdot3^\delta(2c)^{\alpha+\delta+2}(\delta+2)}\left[\mp\frac{\sin(\nu\pi/2)}{\sin(\alpha+\nu/2)\pi}\right]^{1-\gamma}\times$$

$$\times\Gamma\begin{bmatrix}\alpha+3,\ \alpha+2\delta+2\\3+\alpha+\delta+(-1)^\gamma\nu/2,\ 1-(-1)^\gamma\nu/2\end{bmatrix}\times$$

$$\times\,{}_6F_5\left(1,\,\frac{\alpha+3}{2},\,\frac{\alpha}{2}+2,\,\frac{\alpha}{2}+\delta+1,\,\frac{\alpha+3}{2}+\delta,\,\frac{\delta}{2}+1;\right.$$

$$\left.2,\,\frac{\delta}{2}+2,\,\delta+\frac{3}{2},\,\frac{2\alpha+(-1)^\gamma\nu}{4}+2,\,\frac{6+2\alpha+(-1)^\gamma\nu}{4}+\delta;\,-\frac{b^2}{4c^2}\right)+$$

$$+\frac{b^\delta}{(2c)^{\alpha+\delta}}\Gamma\begin{bmatrix}\alpha+1,\ \alpha+2\delta\\1+\alpha+\delta+(-1)^\gamma\nu/2,\ 1-(-1)^\gamma\nu/2\end{bmatrix}\left[\mp\frac{\sin(\nu\pi/2)}{\sin(\alpha+\nu/2)\pi}\right]^{1-\gamma}\begin{Bmatrix}1\\R\end{Bmatrix},$$

$$R=\mathbf{C}+2\psi(\alpha)+\frac{1}{\alpha}-\psi\left(1+\alpha+(-1)^\gamma\frac{\nu}{2}\right)-(1-\gamma)\pi\operatorname{ctg}\frac{2\alpha+\nu}{2}\pi-\ln\frac{2c}{b}.$$

2. $$\int_0^\infty x^{\alpha-1}e^{cx}\begin{Bmatrix}\mathrm{si}\,(bx)\\ \mathrm{ci}\,(bx)\end{Bmatrix}k_\nu(cx)\,dx=$$

$$=V(0)-\frac{2(2c)^{\nu/2}}{b^{\alpha+\nu/2}(2\alpha+\nu)}\begin{Bmatrix}\sin[(2\alpha+\nu)\pi/4]\\ \cos[(2\alpha+\nu)\pi/4]\end{Bmatrix}\Gamma\begin{bmatrix}\alpha+\nu/2\\ 1+\nu/2\end{bmatrix}\times$$

$$\times{}_5F_4\left(-\frac{\nu}{4},\ \frac{2-\nu}{4},\ \frac{2-\nu}{4},\ 1-\frac{\nu}{4},\ -\frac{2\alpha+\nu}{4};\ \frac{1}{2},\ 1-\frac{2\alpha+\nu}{4},\ 1-\frac{2\alpha+\nu}{4},\ \frac{2-2\alpha-\nu}{4};\ -\frac{b^2}{4c^2}\right)\pm$$

$$\pm\frac{2(2c)^{\nu/2-1}}{b^{\alpha+\nu/2-1}(2-2\alpha-\nu)}\begin{Bmatrix}\cos[(2\alpha+\nu)\pi/4]\\ \sin[(2\alpha+\nu)\pi/4]\end{Bmatrix}\Gamma\begin{bmatrix}\alpha+\nu/2-1\\ \nu/2-1\end{bmatrix}\times$$

$$\times{}_5F_4\left(\frac{2-\nu}{4},\ 1-\frac{\nu}{4},\ 1-\frac{\nu}{4},\ \frac{6-\nu}{4},\ \frac{2-2\alpha-\nu}{4};\ \frac{3}{2},\ \frac{6-2\alpha-\nu}{4},\ \frac{6-2\alpha-\nu}{4},\ 1-\frac{2\alpha+\nu}{4};\ -\frac{b^2}{4c^2}\right)$$

[b, $\mathrm{Re}\,\alpha>0$; $\mathrm{Re}\,(2\alpha+\nu)<4$; $|\arg c|<\pi$; $V(\gamma)$ see 2.14.6.1].

2.14.7. Integrals of $x^\alpha e^{-px}\begin{Bmatrix}\mathrm{erf}\,(bx^r)\\ \mathrm{erfc}\,(bx^r)\end{Bmatrix}k_\nu(cx)$.

1. $$\int_0^\infty x^{\alpha-1}e^{-cx}\begin{Bmatrix}\mathrm{erf}\,(bx)\\ \mathrm{erfc}\,(bx)\end{Bmatrix}k_\nu(cx)\,dx=W(1)$$

[$\mathrm{Re}\,c>0$; $\mathrm{Re}\,\alpha>-(1\pm1)/2$; $|\arg b|<\pi/4$],

$$W(\gamma)=\mp\frac{2c}{\pi^{3/2}b^{\alpha+1}}\sin\frac{\nu\pi}{2}\sum_{k=0}^{\infty}\frac{(1-(-1)^\gamma\nu/2)_k}{k!\,(k+1)!\,(\alpha+k+1)}\times$$

$$\times\Gamma\left(1+\frac{\alpha+k}{2}\right)\left[2\psi(k+1)+\frac{1}{k+1}-\psi\left(1-\gamma-\frac{\nu}{2}+(-1)^\gamma k\right)-\right.$$

$$\left.-\frac{1}{2}\psi\left(1+\frac{\alpha+k}{2}\right)+\frac{1}{\alpha+k+1}-\ln\frac{2c}{b}\right]\left((-1)^\gamma\frac{2c}{b}\right)^k\mp\frac{2}{\alpha\nu\pi^{3/2}b^\alpha}\Gamma\left(\frac{\alpha+1}{2}\right)\sin\frac{\nu\pi}{2}+$$

$$+\begin{Bmatrix}1\\0\end{Bmatrix}(2c)^{-\alpha}\Gamma\begin{bmatrix}\alpha,\ \alpha+1\\ 1-(-1)^\gamma\nu/2,\ 1+\alpha+(-1)^\gamma\nu/2\end{bmatrix}\left[\frac{\sin(\nu\pi/2)}{\sin(\alpha+\nu/2)\pi}\right]^{1-\gamma}.$$

2. $$\int_0^\infty x^{\alpha-1}e^{cx}\begin{Bmatrix}\mathrm{erf}\,(bx)\\ \mathrm{erfc}\,(bx)\end{Bmatrix}k_\nu(cx)\,dx=W(0)$$

$\left[\mathrm{Re}\,\alpha>-(1\pm1)/2;\ |\arg c|<\pi,\ \begin{Bmatrix}\mathrm{Re}\,(2\alpha+\nu)<0;\ |\arg b|<\pi/4\\ |\arg b|<\pi/4\end{Bmatrix},\ W(\gamma)\ \text{see 2.14.7.1}\right]$.

3. $$\int_0^\infty x^{\alpha-1}e^{-cx}\begin{Bmatrix}\mathrm{erf}\,(b\sqrt{x})\\ \mathrm{erfc}\,(b\sqrt{x})\end{Bmatrix}k_\nu(cx)\,dx=U(1)$$

[$\mathrm{Re}\,c>0$; $\mathrm{Re}\,\alpha>-(1\pm1)/4$; $|\arg b|<\pi/4$],

$$U(\gamma)=\pm\frac{2b}{\sqrt{\pi}\,(2c)^{\alpha+1/2}}\left[\frac{\sin(\nu\pi/2)}{\cos(\alpha+\nu/2)\pi}\right]^{1-\gamma}\Gamma\begin{bmatrix}\alpha+1/2,\ \alpha+3/2\\ 1-(-1)^\gamma\nu/2,\ \alpha+(3+(-1)^\gamma\nu)/2\end{bmatrix}\times$$

$$\times{}_3F_2\left(\frac{1}{2},\ \alpha+\frac{1}{2},\ \alpha+\frac{3}{2};\ \frac{3}{2},\ \alpha+\frac{3+(-1)^\gamma\nu}{2};\ (-1)^\gamma\frac{b^2}{2c}\right)+$$

$$+\begin{Bmatrix}0\\1\end{Bmatrix}(2c)^{-\alpha}\Gamma\begin{bmatrix}\alpha,\ \alpha+1\\ 1-(-1)^\gamma\nu/2,\ 1+\alpha+(-1)^\gamma\nu/2\end{bmatrix}\left[\frac{\sin(\nu\pi/2)}{\sin(\alpha+\nu/2)\pi}\right]^{1-\gamma}.$$

4. $\int_0^\infty x^{\alpha-1}e^{cx}\begin{Bmatrix}\operatorname{erf}(b\sqrt{x})\\ \operatorname{erfc}(b\sqrt{x})\end{Bmatrix}k_\nu(cx)\,dx=U(0)\mp\frac{2(2c)^{\nu/2}}{\sqrt{\pi}\,b^{2\alpha+\nu}(2\alpha+\nu)}\Gamma\left[\begin{matrix}\alpha+(\nu+1)/2\\ 1+\nu/2\end{matrix}\right]\times$

$\times\,{}_3F_2\left(-\alpha-\frac{\nu}{2},\ -\frac{\nu}{2},\ 1-\frac{\nu}{2};\ 1-\alpha-\frac{\nu}{2},\ \frac{1-\nu}{2}-\alpha;\ \frac{b^2}{2c}\right)$

$\left[\operatorname{Re}\alpha>-(1\pm 1)/4;\ |\arg c|<\pi,\ \begin{Bmatrix}\operatorname{Re}(2\alpha+\nu)<0;\ |\arg b|<\pi/4\\ |\arg b|<\pi/4\end{Bmatrix},\ U(\gamma)\ \text{see } 2.14.7.3\right].$

2.14.8 Integrals of $x^\alpha e^{-px}\begin{Bmatrix}S(bx)\\ C(bx)\end{Bmatrix}k_\nu(cx)$.

1. $\int_0^\infty x^{\alpha-1}e^{-cx}\begin{Bmatrix}S(bx)\\ C(bx)\end{Bmatrix}k_\nu(cx)\,dx=V(\gamma)$ $\left[b,\ \operatorname{Re}c>0;\ \operatorname{Re}\alpha>-\delta-1/2,\ \delta=\begin{Bmatrix}1\\ 0\end{Bmatrix}\right],$

$V(\gamma)=\frac{b^{\delta+1/2}\sqrt{2}}{\sqrt{\pi}(2c)^{\alpha+\delta+1/2}\,3^\delta}\Gamma\left[\begin{matrix}\alpha+3/2,\ \alpha+2\delta+1/2\\ 1-(-1)^\gamma\nu/2,\ \alpha+\delta+(-1)^\gamma\nu/2+3/2\end{matrix}\right]\times$

$\times\left[\mp\frac{\sin(\nu\pi/2)}{\cos(\alpha+\nu/2)\pi}\right]^{1-\gamma}{}_5F_4\left(\frac{2\delta+1}{4},\ \frac{2\alpha+3}{4},\ \frac{2\alpha+5}{4},\ \frac{2\alpha+1}{4}+\delta,\right.$

$\frac{2\alpha+3}{4}+\delta;\ \frac{2\delta+5}{4},\ \delta+\frac{1}{2},\ \frac{2\alpha+2\delta+(-1)^\gamma\nu+3}{4},$

$\left.\frac{2\alpha+2\delta+(-1)^\gamma\nu+5}{4};\ -\frac{b^2}{4c^2}\right).$

2. $\int_0^\infty x^{\alpha-1}e^{cx}\begin{Bmatrix}S(bx)\\ C(bx)\end{Bmatrix}k_\nu(cx)\,dx=$

$=V(0)-\frac{(2c)^{\nu/2}\sqrt{2}}{\sqrt{\pi}(2\alpha+\nu)\,b^{\alpha+\nu/2}}\Gamma\left[\begin{matrix}\alpha+(\nu+1)/2\\ 1+\nu/2\end{matrix}\right]\begin{Bmatrix}\sin[(2\alpha+\nu+1)\pi/4]\\ \cos[(2\alpha+\nu+1)\pi/4]\end{Bmatrix}\times$

$\times\,{}_5F_4\left(-\frac{\nu}{4},\frac{2-\nu}{4},\frac{2-\nu}{4},\ 1-\frac{\nu}{4},-\frac{2\alpha+\nu}{4};\frac{1}{2},\ 1-\frac{2\alpha+\nu}{4},\frac{3-2\alpha-\nu}{4},\frac{1-2\alpha-\nu}{4};-\frac{b^2}{4c^2}\right)-$

$-\frac{(2c)^{\nu/2-1}\sqrt{2}}{\sqrt{\pi}(2-2\alpha-\nu)\,b^{\alpha+\nu/2-1}}\Gamma\left[\begin{matrix}\alpha+(\nu-1)/2\\ \nu/2-1\end{matrix}\right]\begin{Bmatrix}\sin[(2\alpha+\nu-1)\pi/4]\\ \cos[(2\alpha+\nu-1)\pi/4]\end{Bmatrix}\times$

$\times\,{}_5F_4\left(\frac{2-\nu}{4},\ 1-\frac{\nu}{4},\ 1-\frac{\nu}{4},\frac{6-\nu}{4},\frac{2-2\alpha-\nu}{4};\ \frac{3}{2},\frac{6-2\alpha-\nu}{4},\frac{5-2\alpha-\nu}{4},\frac{3-2\alpha-\nu}{4};-\frac{b^2}{4c^2}\right)$

$[b>0;\ -(2\pm 1)/2<\operatorname{Re}\alpha<-\operatorname{Re}\nu/2;\ |\arg c|<\pi;\ V(\gamma)\ \text{see } 2.14.8.1].$

2.14.9. Integrals of $x^\alpha e^{-px}\begin{Bmatrix}\gamma(\mu,\ bx)\\ \Gamma(\mu,\ bx)\end{Bmatrix}k_\nu(cx)$.

1. $\int_0^\infty x^{\alpha-1}e^{-cx}\begin{Bmatrix}\gamma(\mu,\ bx)\\ \Gamma(\mu,\ bx)\end{Bmatrix}k_\nu(cx)\,dx=W(1)$

$\left[\operatorname{Re}b,\ \operatorname{Re}c,\ \operatorname{Re}(\alpha+\mu)>0,\ \begin{Bmatrix}\operatorname{Re}\mu>0\\ \operatorname{Re}\alpha>0\end{Bmatrix}\right],$

$W(\gamma)=\pm\frac{b^\mu}{\mu(2c)^{\alpha+\mu}}\left[\frac{\sin(\nu\pi/2)}{\sin(\alpha+\mu+\nu/2)\pi}\right]^{1-\gamma}\Gamma\left[\begin{matrix}\alpha+\mu,\ \alpha+\mu+1\\ 1-(-1)^\gamma\nu/2,\ 1+\alpha+\mu+(-1)^\gamma\nu/2\end{matrix}\right]\times$

$\times\,{}_3F_2\left(\mu,\ \alpha+\mu,\ \alpha+\mu+1;\ \mu+1,\ 1+\alpha+\mu+\frac{(-1)^\gamma\nu}{2};\ (-1)^\gamma\frac{b}{2c}\right)+$

$+\begin{Bmatrix}0\\ 1\end{Bmatrix}(2c)^{-\alpha}\Gamma\left[\begin{matrix}\alpha,\ \alpha+1,\ \mu\\ 1-(-1)^\gamma\nu/2,\ 1+\alpha+(-1)^\gamma\nu/2\end{matrix}\right]\left[\frac{\sin(\nu\pi/2)}{\sin(\alpha+\nu/2)\pi}\right]^{1-\gamma}.$

2. $\int\limits_0^\infty x^{\alpha-1}e^{cx}\left\{\begin{matrix}\gamma(\mu, bx)\\ \Gamma(\mu, bx)\end{matrix}\right\} k_\nu(cx)\,dx = W(0) \mp \frac{2(2c)^{\nu/2}}{b^{\alpha+\nu/2}(2\alpha+\nu)}\Gamma\left[\begin{matrix}\alpha+\mu+\nu/2\\ 1+\nu/2\end{matrix}\right]\times$

$\times {}_3F_2\left(-\frac{\nu}{2}, 1-\frac{\nu}{2}, -\alpha-\frac{\nu}{2}; 1-\alpha-\frac{\nu}{2}, 1-\alpha-\mu-\frac{\nu}{2}; \frac{b}{2c}\right)$

$\left[\operatorname{Re} b,\ \operatorname{Re}(\alpha+\mu)>0;\ |\arg c|<\pi,\ \left\{\begin{matrix}\operatorname{Re}\mu>0,\ \operatorname{Re}(2\alpha+\nu)<0\\ \operatorname{Re}\alpha>0\end{matrix}\right\},\ W(\gamma)\ \text{see 2.14.9.1}\right]$.

2.14.10. Integrals containing $k_\mu(bx)\,k_\nu(cx)$.

1. $\int\limits_0^\infty x^{\alpha-1}k_{-\nu}(cx)\,k_\nu(cx)\,dx = \frac{\sin(\nu\pi/2)}{(2c)^\alpha\,\nu\pi}\Gamma\left[\begin{matrix}\alpha+1,\ \alpha/2,\ \alpha/2+1\\ 1+(\alpha+\nu)/2,\ 1+(\alpha-\nu)/2\end{matrix}\right]$

[Re c, Re $\alpha>0$].

2. $\int\limits_0^a x^{\alpha-1}(a^2-x^2)^{\beta-1}k_{-\nu}(cx)\,k_\nu(cx)\,dx = -\frac{a^{\alpha+2\beta-1}c}{\sqrt{\pi}}\sin\nu\pi\,\mathrm{B}\left(\frac{\alpha+1}{2}, \beta\right)\times$

$\times {}_3F_4\left(\frac{\alpha+1}{2}, \frac{1+\nu}{2}, \frac{1-\nu}{2}; \frac{1}{2}, \frac{1}{2}, \frac{3}{2}, \beta+\frac{\alpha+1}{2}; a^2c^2\right)+$

$+\frac{2a^{\alpha+2\beta-2}}{\nu^2\pi^2}\sin^2\frac{\nu\pi}{2}\,\mathrm{B}\left(\frac{\alpha}{2}, \beta\right) - \frac{a^{\alpha+2\beta}c^2}{2\pi^2}\sin^2\frac{\nu\pi}{2}\,\mathrm{B}\left(\frac{\alpha}{2}+1, \beta\right)\times$

$\times\sum_{k=0}^\infty \frac{(\alpha/2+1)_k(1-\nu/2)_k(1+\nu/2)_k}{(\alpha/2+\beta+1)_k(3/2)_k(k!)^2(k+1)!}(ac)^{2k}\left\{\frac{1}{(k+1)^2}-6\mathbf{C}-3\psi'(k+1)+\right.$

$+\psi'\left(\frac{\alpha}{2}+k+1\right)-\psi'\left(\frac{\alpha}{2}+\beta+k+1\right)-\frac{4}{\nu^2}-\frac{\pi^2}{\sin^2(\nu\pi/2)}+$

$+\left[\psi\left(\frac{\alpha}{2}+k+1\right)-\psi\left(\frac{\alpha}{2}+\beta+k+1\right)-3\psi(k+1)+\psi\left(\frac{\nu}{2}-k\right)+\right.$

$\left.+\psi\left(-\frac{\nu}{2}-k\right)-\frac{1}{k+1}\right]^2-4\ln ac\left[3\psi(k+1)+\frac{1}{k+1}+\psi\left(\frac{\alpha}{2}+k+1\right)+\right.$

$\left.\left.+\psi\left(\frac{\alpha}{2}+\beta+k+1\right)-\psi\left(\frac{\nu}{2}-k\right)-\psi\left(-\frac{\nu}{2}-k\right)\right]+4\ln^2 ac\right\}$ [a, Re α, Re $\beta>0$].

3. $\int\limits_0^\infty \frac{1}{x}e^{-px}k_{2m}(cx)\,k_{2n}(cx)\,dx = (-p)^{m+n-2}\frac{4c^2}{(2c+p)^{m+n}}\,{}_2F_1\left(1-m, 1-n; 2; \frac{4c^2}{p^2}\right)$

[Re $(2c+p)>0$; $m, n=1, 2, 3, \ldots$].

4. $\int\limits_0^\infty \frac{1}{x}e^{-px}k_{2m}(bx)\,k_{2n}(cx)\,dx=$

$=\frac{(-1)^{m+n}4(m+n-1)!\,bc\,(b-c+p)^{n-1}(c-b+p)^{m-1}}{m!\,n!\,(b+c+p)^{m+n}}\times$

$\times{}_2F_1\left(1-m, 1-n; 1-m-n; \frac{(p+b+c)(p-b-c)}{(p+b-c)(p+c-b)}\right)$

[Re $(b+c+p)>0$; $m, n=1, 2, 3, \ldots$].

5. $\int\limits_0^\infty x^{\alpha-1}e^{-px}\prod_{l=1}^n k_{2m_l}(c_lx)\,dx = (-1)^m 2^n\frac{\Gamma(n+\alpha)}{(c+p)^{n+\alpha}}\times$

$\times\left(\prod_{l=1}^n c_l\right)F_A\left(n+\alpha; -m_1, \ldots, -m_n; 2, \ldots, 2; \frac{2c_1}{c+p}, \ldots, \frac{2c_n}{c+p}\right)$

$\left[m=\sum_{l=1}^n m_l,\ c=\sum_{l=1}^n c_l;\ m_l,\ n=1, 2, \ldots;\ \operatorname{Re}p,\ \operatorname{Re}(n+\alpha)>0\right]$

2.15. THE INCOMPLETE ELLIPTIC INTEGRALS $F(x,k)$, $E(x,k)$, $\Pi(x,\nu,k)$

2.15.1. Integrals of $f(x)\begin{Bmatrix}F(x,\ k)\\E(x,\ k)\end{Bmatrix}$.

$$\int_0^{\pi/2} \operatorname{ctg} x F(x,\ k)\,dx=\frac{\pi}{4}\mathbf{K}\left(\sqrt{1-k^2}\right)+\frac{1}{2}\ln k\,\mathbf{K}(k).$$

$$\int_0^{\pi/2}\frac{\sin 2x}{1\pm k\sin^2 x}F(x,\ k)\,dx=\pm\frac{1}{2k}\ln\frac{(1\pm k)\sqrt{k}}{2}\mathbf{K}(k)\pm\frac{\pi}{8k}\mathbf{K}\left(\sqrt{1-k^2}\right).$$

$$\int_0^{\pi/2}\frac{\sin 2x}{1-k^2\sin^2 x}\begin{Bmatrix}F(x,\ k)\\E(x,\ k)\end{Bmatrix}dx=$$

$$=-\frac{1}{2k^2}\ln(1-k^2)\begin{Bmatrix}\mathbf{K}(k)\\\mathbf{E}(k)\end{Bmatrix}+\begin{Bmatrix}0\\1\end{Bmatrix}\left(\frac{2-k^2}{k^2}\mathbf{K}(k)-\frac{2}{k^2}\mathbf{E}(k)\right).$$

$$\int_0^{\pi/2}\frac{\sin 2x}{1-(1-k)\sin^2 x}F(x,\ k')\,dx=\frac{1}{2(1-k)}\ln\frac{2}{(1+k)\sqrt{k}}\mathbf{K}\left(\sqrt{1-k^2}\right)$$

$$[k'=\sqrt{1-k^2}].$$

$$\int_0^{\pi/2}\frac{1}{\sqrt{1-k^2\sin^2 x}}E(x,\ k)\,dx=\frac{1}{2}\mathbf{E}(k)\mathbf{K}(k)-\frac{1}{4}\ln(1-k^2).$$

$$\int_0^{\pi/2}\frac{\sin 2x}{(1-k^2\sin^2 a\sin^2 x)\sqrt{1-k^2\sin^2 x}}\begin{Bmatrix}F(x,\ k)\\E(x,\ k)\end{Bmatrix}dx=$$

$$=-\frac{4}{k^2\sin 2a}\left[\operatorname{arctg}\left(\sqrt{1-k^2}\operatorname{tg} a\right)\begin{Bmatrix}\mathbf{K}(k)\\\mathbf{E}(k)\end{Bmatrix}-\frac{\pi}{2}\begin{Bmatrix}F(a,\ k)\\E(a,\ k)\end{Bmatrix}+\right.$$

$$\left.+\begin{Bmatrix}0\\1\end{Bmatrix}\frac{\pi}{2}\operatorname{ctg} a\left(1-\sqrt{1-k^2\sin^2 a}\right)\right].$$

$$\int_a^b\frac{1}{\sqrt{(\sin^2 x-\sin^2 a)(\sin^2 b-\sin^2 x)}}\begin{Bmatrix}F(x,\ k)\\E(x,\ k)\end{Bmatrix}dx=$$

$$=\frac{\sec a\operatorname{cosec} b}{2}\begin{Bmatrix}\mathbf{K}(k)\\\mathbf{E}(k)\end{Bmatrix}\mathbf{K}\left(\sqrt{1-\operatorname{tg}^2 a\operatorname{ctg}^2 b}\right)+$$

$$+\begin{Bmatrix}0\\1\end{Bmatrix}\frac{k^2\sin b}{2\cos a}\mathbf{K}\left(\sqrt{1-\sin^2 2a\operatorname{cosec}^2 2b}\right)$$

$$[k=\sqrt{1-\operatorname{ctg}^2 a\operatorname{ctg}^2 b}].$$

2.15.2. Integrals with respect to the modulus k

$$\int_0^1 k\begin{Bmatrix}F(x,\ k)\\E(x,\ k)\end{Bmatrix}dk=3^{-(1\mp 1)/2}\frac{1-\cos x}{\sin x}+\frac{1}{3}\begin{Bmatrix}0\\1\end{Bmatrix}\sin x.$$

$$\int_0^1 k\Pi(x,\ \nu,\ k)\,dk=\operatorname{tg}\frac{x}{2}-\frac{\nu^{1/2}}{2}\ln\frac{1+\nu^{1/2}\sin x}{1-\nu^{1/2}\sin x}-\nu\Pi(x,\ \nu,\ 0).$$

2.16. THE COMPLETE ELLIPTIC INTEGRALS K(x), E(x)

2.16.1. Integrals of general form

NOTATION: $\delta=\begin{Bmatrix}1\\0\end{Bmatrix}$.

1. $$\int_0^a x^{\alpha-1}(a^r-x^r)^{\beta-1}\begin{Bmatrix}\mathrm{K}(cx)\\ \mathrm{E}(cx)\end{Bmatrix}dx=$$
$$=\frac{\pi a^{\alpha+r\beta-r}}{2r}\Gamma(\beta)\sum_{k=0}^{\infty}\frac{(1/2)_k(\pm 1/2)_k}{(k!)^2}\Gamma\begin{bmatrix}(\alpha+2k)/r\\(\alpha+2k)/r+\beta\end{bmatrix}(ac)^{2k}$$
[a, r, Re α, Re $\beta>0$; $|ac|<1$]

2. $$\int_0^{\min(a,\,b)} x^{\alpha-1}(b^r-x^r)^{\beta-1}\begin{Bmatrix}\mathrm{K}(\sqrt{1-x/a})\\ \mathrm{E}(\sqrt{1-x/a})\end{Bmatrix}dx=I$$
[a, b, r, Re $\alpha>0$ (and Re $\beta>0$ for $a\geqslant$

$$I=\frac{\pi}{2}a^{\alpha}b^{\beta r-r}\sum_{k=0}^{\infty}\frac{(1-\beta)_k}{k!}\Gamma\begin{bmatrix}\alpha+rk,\ 1+rk-\delta+\alpha\\1/2+\alpha+rk,\ 3/2+\alpha-\delta+rk\end{bmatrix}\left(\frac{a}{b}\right)^{rk}$$ [$a<b$].

$$I=\frac{b^{\alpha+\beta r-r}}{2^{\delta}r}\mathrm{B}\left(\frac{\alpha}{r},\beta\right)\left[\ln\frac{16a}{b}-\frac{1}{r}\psi\left(\frac{\alpha}{r}\right)+\frac{1}{r}\psi\left(\frac{\alpha}{2}+\beta\right)\right]^{\delta}\pm$$
$$\pm\frac{b^{\alpha+\beta r-r}}{2^{\delta}r}\Gamma(\beta)\sum_{k=1}^{\infty}\frac{(1/2)_k(\delta-1/2)_k}{k!\,(k+\delta-1)!}\Gamma\begin{bmatrix}(\alpha+k)/r\\(\alpha+k)/r+\beta\end{bmatrix}\times$$
$$\times\left[2\psi(k+1)-2\psi\left(\frac{1}{2}-k\right)-\frac{1}{r}\psi\left(\frac{\alpha+k}{r}\right)+\frac{1}{r}\psi\left(\frac{\alpha+k}{r}+\beta\right)+\right.$$
$$\left.+\frac{1-\delta}{k(2k-1)}+\ln\frac{a}{b}\right]\left(\frac{b}{a}\right)^{k}$$ [$a>b$].

3. $$\int_0^a\frac{x^{\alpha-1}}{(x^r+z^r)^{\rho}}\begin{Bmatrix}\mathrm{K}(\sqrt{1-x/a})\\ \mathrm{E}(\sqrt{1-x/a})\end{Bmatrix}dx=$$
$$=\frac{\pi a^{\alpha}}{2z^{\rho r}}\sum_{k=0}^{\infty}\frac{(-1)^k(\rho)_k}{k!}\Gamma\begin{bmatrix}\alpha+rk,\ 1+\alpha-\delta+rk\\1/2+\alpha+rk,\ 3/2+\alpha-\delta+rk\end{bmatrix}\left(\frac{a}{z}\right)^{r}$$
[a, r, Re $\alpha>0$; $|z|>a$, $r|\arg z|<\pi$; $\rho=1$ for $z^r=-y^r$, $y>a$]

4. $$\int_0^a\frac{x^{\alpha-1}(a^r-x^r)^{\beta-1}}{\sqrt{x+b}}\begin{Bmatrix}\mathrm{K}(\sqrt{b/(x+b)})\\ \mathrm{E}(\sqrt{b/(x+b)})\end{Bmatrix}dx=$$
$$=\frac{\pi a^{\alpha+\beta r-r-1/2}}{2r}\Gamma(\beta)\sum_{k=0}^{\infty}\frac{(1/2)_k(3/2-\delta)_k}{(k!)^2}\Gamma\begin{bmatrix}(\alpha-k-1/2)/r\\(\alpha-k-1/2)/r+\beta\end{bmatrix}\left(-\frac{b}{a}\right)^{k}+$$
$$+\frac{a^{\beta r-r}b^{\alpha-1/2}}{2^{\delta}}\sum_{k=0}^{\infty}\frac{(1-\beta)_k}{k!}\Gamma\begin{bmatrix}\alpha+rk,\ 1+\alpha-\delta+rk,\ 1/2-\alpha-rk\\1/2+\alpha+rk\end{bmatrix}\left(\frac{b}{a}\right)^{r}$$
[a, r, Re α, Re $\beta>0$; $|b|<a$; $|\arg b|<\pi$

5. $$\int_a^{\infty}\frac{x^{\alpha-1}(x^r-a^r)^{\beta-1}}{\sqrt{x+b}}\begin{Bmatrix}\mathrm{K}(\sqrt{b/(x+b)})\\ \mathrm{E}(\sqrt{b/(x+b)})\end{Bmatrix}dx=$$

$$=\frac{\pi a^{\alpha+\beta r-r-1/2}}{2r}\Gamma(\beta)\sum_{k=0}^{\infty}\frac{(1/2)_k(3/2-\delta)_k}{(k!)^2}\Gamma\begin{bmatrix}1-\beta-(\alpha-k-1/2)/r\\1-(\alpha-k-1/2)/r\end{bmatrix}\left(-\frac{b}{a}\right)^k$$

[$a,\ r,\ \mathrm{Re}\,\beta>0;\ \mathrm{Re}\,(\alpha+\beta r)<r+1/2;\ |b|<a;\ |\arg b|<\pi$].

6. $$\int_0^\infty \frac{x^{\alpha-1}}{(x^r+z^r)^\rho\sqrt{x+b}}\begin{Bmatrix}\mathbf{K}\left(\sqrt{b/(x+b)}\right)\\\mathbf{E}\left(\sqrt{b/(x+b)}\right)\end{Bmatrix}dx=$$

$$=\frac{\pi z^{\alpha-\rho r-1/2}}{2r\Gamma(\rho)}\sum_{k=0}^{\infty}\frac{(1/2)_k(3/2-\delta)_k}{(k!)^2}\Gamma\left(\frac{2\alpha-2k-1}{2r}\right)\Gamma\left(\rho-\frac{2\alpha-2k-1}{2r}\right)\left(-\frac{b}{z}\right)^k+$$

$$+\frac{b^{\alpha-1/2}}{2^\delta z^{\rho r}}\sum_{k=0}^{\infty}\frac{(-1)^k(\rho)_k}{k!}\Gamma\begin{bmatrix}\alpha+rk,\ 1+\alpha-\delta+rk,\ 1/2-\alpha-rk\\1/2+\alpha+rk\end{bmatrix}\left(\frac{b}{z}\right)^{rk}$$

[$r>0;\ 0<\mathrm{Re}\,\alpha<r\,\mathrm{Re}\,\rho+1/2;\ |b|<|z|;\ |\arg b|,\ r|\arg z|<\pi$].

7. $$\int_0^\infty \frac{x^{\alpha-1}}{(x^r-y^r)\sqrt{x+b}}\begin{Bmatrix}\mathbf{K}\left(\sqrt{b/(x+b)}\right)\\\mathbf{E}\left(\sqrt{b/(x+b)}\right)\end{Bmatrix}dx=$$

$$=-\frac{\pi^2 y^{\alpha-r-1/2}}{2r}\sum_{k=0}^{\infty}\frac{(1/2)_k(3/2-\delta)_k}{(k!)^2}\operatorname{ctg}\frac{2\alpha-2k-1}{2r}\pi\left(-\frac{b}{y}\right)^k-$$

$$-\frac{b^{\alpha-1/2}}{2^\delta y^r}\sum_{k=0}^{\infty}\Gamma\begin{bmatrix}\alpha+rk,\ 1+\alpha-\delta+rk,\ 1/2-\alpha-rk\\1/2+\alpha+rk\end{bmatrix}\left(\frac{b}{y}\right)^{rk}$$

[$r,\ y>0;\ 0<\mathrm{Re}\,\alpha<r+1/2;\ |b|<y;\ |\arg b|<\pi$].

8. $$\int_0^\infty \frac{x^{\alpha-1}}{\sqrt{x+b}}e^{-px^r}\begin{Bmatrix}\mathbf{K}\left(\sqrt{b/(x+b)}\right)\\\mathbf{E}\left(\sqrt{b/(x+b)}\right)\end{Bmatrix}dx=$$

$$=\frac{\pi}{2r}p^{(1-2\alpha)/(2r)}\sum_{k=0}^{\infty}\frac{(1/2)_k(3/2-\delta)_k}{(k!)^2}\Gamma\left(\frac{2\alpha-2k-1}{2r}\right)(-bp^{1/r})^k+$$

$$+2^{-\delta}b^{\alpha-1/2}\sum_{k=0}^{\infty}\frac{(-b^rp)^k}{k!}\Gamma\begin{bmatrix}\alpha+rk,\ 1+\alpha-\delta+rk,\ 1/2-\alpha-rk\\1/2+\alpha+rk\end{bmatrix}$$

[$r,\ \mathrm{Re}\,p,\ \mathrm{Re}\,\alpha>0;\ |\arg b|<\pi$].

9. $$\int_0^a x^{\alpha-1}e^{-px^r}\begin{Bmatrix}\mathbf{K}\left(\sqrt{1-x/a}\right)\\\mathbf{E}\left(\sqrt{1-x/a}\right)\end{Bmatrix}dx=$$

$$=\frac{\pi a^\alpha}{2}\sum_{k=0}^{\infty}\frac{1}{k!}\Gamma\begin{bmatrix}\alpha+rk,\ 1+\alpha-\delta+rk\\1/2+\alpha+rk,\ 3/2+\alpha-\delta+rk\end{bmatrix}(-a^rp)^k$$ [$a,\ r,\ \mathrm{Re}\,\alpha>0$].

10. $$\int_0^a x^{\alpha-1}\sin bx^r\begin{Bmatrix}\mathbf{K}\left(\sqrt{1-x/a}\right)\\\mathbf{E}\left(\sqrt{1-x/a}\right)\end{Bmatrix}dx=I(1),$$

$$I(\gamma)=\frac{\pi a^{\alpha+\gamma r}b^\gamma}{2}\sum_{k=0}^{\infty}\frac{1}{(1/2+\gamma)_k\,k!}\times$$

$$\times\Gamma\begin{bmatrix}\alpha+\gamma r+2rk,\ 1+\alpha-\delta+\gamma r+2rk\\1/2+\alpha+\gamma r+2rk,\ 3/2+\alpha-\delta+\gamma r+2rk\end{bmatrix}\left(-\frac{a^{2r}b^2}{4}\right)^k$$ [$a,\ r>0;\ \mathrm{Re}\,\alpha>-r$].

11. $\int_0^a x^{\alpha-1}\cos bx^r \begin{Bmatrix} \mathbf{K}(\sqrt{1-x/a}) \\ \mathbf{E}(\sqrt{1-x/a}) \end{Bmatrix} dx = I(0)$ [a, r, $\operatorname{Re}\alpha>0$; $I(\gamma)$ see 2.16.1.10].

12. $\int_0^\infty \frac{x^{\alpha-1}}{\sqrt{x+b}}\sin ax^r \begin{Bmatrix} \mathbf{K}(\sqrt{b/(x+b)}) \\ \mathbf{E}(\sqrt{b/(x+b)}) \end{Bmatrix} dx = J(1),$

$$J(\gamma)=\frac{\pi a^{(1-2\alpha)/(2r)}}{2r}\sum_{k=0}^{\infty}\frac{(1/2)_k(3/2-\delta)_k}{(k!)^2}\Gamma\left(\frac{2\alpha-2k-1}{2r}\right)\cos\frac{\gamma r+k-\alpha+1/2}{2r}\pi(-a^{1/r}b)^k+$$

$$+\frac{a^\gamma}{2^\delta}b^{\alpha+\gamma r-1/2}\sum_{k=0}^{\infty}\frac{(-a^2b^{2r}/4)^k}{(\gamma+1/2)_k\,k!}\Gamma\left[\begin{matrix}\alpha+\gamma r+2rk,1+\alpha-\delta+\gamma r+2rk,1/2-\alpha-\gamma r-2rk\\1/2+\alpha+\gamma r+2rk\end{matrix}\right]$$

[a, $r>0$; $-r<\operatorname{Re}\alpha<r+1/2$; $|\arg b|<\pi$].

13. $\int_0^\infty \frac{x^{\alpha-1}}{\sqrt{x+b}}\cos ax^r \begin{Bmatrix} \mathbf{K}(\sqrt{b/(x+b)}) \\ \mathbf{E}(\sqrt{b/(x+b)}) \end{Bmatrix} dx = J(0)$

[a, $r>0$; $0<\operatorname{Re}\alpha<r+1/2$; $|\arg b|<\pi$; $J(\gamma)$ see 2.16.1.12].

14. $\int_0^\infty \frac{x^{\alpha-1}}{\sqrt{x+b}}J_\nu(ax^r) \begin{Bmatrix} \mathbf{K}(\sqrt{b/(x+b)}) \\ \mathbf{E}(\sqrt{b/(x+b)}) \end{Bmatrix} dx =$

$$=\frac{\pi}{4r}\left(\frac{2}{a}\right)^{(2\alpha-1)/(2r)}\sum_{k=0}^{\infty}\frac{(1/2)_k(3/2-\delta)_k}{(k!)^2}\Gamma\left[\begin{matrix}(\nu r+\alpha-k-1/2)/(2r)\\(\nu r-\alpha+k+1/2)/(2r)+1\end{matrix}\right]\left(-\frac{a^{1/r}b}{2^{1/r}}\right)^k+$$

$$+\frac{a^\nu b^{\alpha+\nu r-1/2}}{2^{\nu+\delta}\Gamma(\nu+1)}\sum_{k=0}^{\infty}\frac{(-a^2b^{2r}/4)^k}{(\nu+1)_k\,k!}\times$$

$$\times\Gamma\left[\begin{matrix}\alpha+\nu r+2rk,\ 1+\alpha-\delta+\nu r+2rk,\ 1/2-\alpha-\nu r-2rk\\1/2+\alpha+\nu r+2rk\end{matrix}\right]$$

[a, r, $\operatorname{Re}(\alpha+\nu r)>0$; $\operatorname{Re}\alpha<(3r+1)/2$; $|\arg b|<\pi$].

15. $\int_0^a x^{\alpha-1}J_\nu(bx^r) \begin{Bmatrix} \mathbf{K}(\sqrt{1-x/a}) \\ \mathbf{E}(\sqrt{1-x/a}) \end{Bmatrix} dx = \frac{\pi a^{\alpha+\nu r}b^\nu}{2^{\nu+1}\Gamma(\nu+1)}\sum_{k=0}^{\infty}\frac{1}{(\nu+1)_k\,k!}\times$

$$\times\Gamma\left[\begin{matrix}\alpha+\nu r+2rk,\ 1+\alpha-\delta+\nu r+2rk\\1/2+\alpha+\nu r+2rk,\ 3/2+\alpha-\delta+\nu r+2rk\end{matrix}\right]\left(-\frac{a^{2r}b^2}{4}\right)^k$$

[a, r, $\operatorname{Re}(\alpha+\nu r)>0$].

2.16.2. Integrals of $x^\alpha \begin{Bmatrix} \mathbf{K}(cx) \\ \mathbf{E}(cx) \end{Bmatrix}$.

1. $\int_0^a \begin{Bmatrix} \mathbf{K}(x/a) \\ \mathbf{E}(x/a) \end{Bmatrix} dx = \frac{3\pm1}{2}a\mathbf{G}+\frac{a}{2}\begin{Bmatrix}0\\1\end{Bmatrix}$ [$a>0$].

2. $\int_0^a x^{\alpha-1}\begin{Bmatrix} \mathbf{K}(x/a) \\ \mathbf{E}(x/a) \end{Bmatrix} dx = \frac{\pi a^\alpha}{2\alpha}\,{}_3F_2\left(\pm\frac{1}{2},\frac{1}{2},\frac{\alpha}{2};\ 1,\ \frac{\alpha}{2}+1;\ 1\right)$ [a, $\operatorname{Re}\alpha>0$].

3. $\int_0^a x^{-1}\left[\frac{\pi}{2}-\begin{Bmatrix} \mathbf{K}(x/a) \\ \mathbf{E}(x/a) \end{Bmatrix}\right] dx = \left(\frac{\pi}{2}-1\right)\begin{Bmatrix}0\\1\end{Bmatrix}+2\mathbf{G}-\pi\ln 2.$

4. $$\int_0^\infty x^{\alpha-1}\mathrm{K}(icx)\,dx=\frac{1}{4c^\alpha}\Gamma\begin{bmatrix}\alpha/2,\ (1-\alpha)/2,\ (1-\alpha)/2\\ 1-\alpha/2\end{bmatrix}$$ [$\operatorname{Re} c>0;\ 0<\operatorname{Re}\alpha<1$].

5. $$\int_0^\infty x^{\alpha-1}\left[\begin{Bmatrix}\pi/2\\ cx\end{Bmatrix}-\mathrm{E}(icx)\right]dx=\frac{c^{-\alpha}}{8}\Gamma\begin{bmatrix}\alpha/2,\ (1-\alpha)/2,\ -(1+\alpha)/2\\ 1-\alpha/2\end{bmatrix}$$

[$\operatorname{Re} c>0;\ -(1\pm1)<\operatorname{Re}\alpha<\mp1$].

2.16.3. Integrals of $(x\pm a)^\alpha(b\pm x)^\beta\mathrm{K}(\varrho x)$.

1. $$\int_0^a\frac{1}{x+a}\mathrm{K}\left(\frac{x}{a}\right)dx=\frac{\pi^2}{8}$$ [$a>0$].

2. $$\int_0^a\frac{1}{\sqrt{a-x}}\mathrm{K}\left(\frac{x}{a}\right)dx=2\sqrt{2a}\,\mathrm{K}(\sqrt{2}-1)\,\mathrm{K}'(\sqrt{2}-1)$$ [$a>0$].

3. $$\int_{-a}^a\frac{(a-x)^{\beta-1}}{(a+x)^\beta}\mathrm{K}\left(\frac{|x|}{a}\right)dx=\frac{1}{8\pi}\Gamma\left[\frac{\beta}{2},\frac{\beta}{2},\frac{1-\beta}{2},\frac{1-\beta}{2}\right]$$ [$a>0;\ 0<\operatorname{Re}\beta<1$].

4. $$\int_0^a\frac{(a-x)^{\beta-1}}{(a+x)^\beta}\mathrm{K}\left(\frac{x}{a}\right)dx=\frac{\pi}{8}\Gamma\begin{bmatrix}\beta/2,\ \beta/2\\ (1+\beta)/2,\ (1+\beta)/2\end{bmatrix}$$ [$a,\ \operatorname{Re}\beta>0$].

5. $$\int_a^b\frac{1}{\sqrt{(x-a)(b-x)}}\mathrm{K}\left(\frac{b-a}{ab}x\right)dx=$$
$$=\frac{8\sqrt{ab}}{a+b}\left(\frac{(z+1)(z+4)}{z^2+8z+8+4\sqrt{(z+2)(z+1)}}\right)^{1/2}\mathrm{K}(k_+)\,\mathrm{K}(k_-),$$
$$z=\frac{2(b^2-ab+a^2)-2\sqrt{ab(5ab-2a^2-2b^2)}}{(4ab-b^2-a^2)+2\sqrt{ab(5ab-2a^2-2b^2)}},$$
$$k_\pm=\left(\frac{z(\sqrt{z+1}\pm\sqrt{z+4})-2+2\sqrt{z+1}}{z(\sqrt{z+1}\pm\sqrt{z+4})+2+2\sqrt{z+1}}\right)^{1/2}$$ [$b\leqslant 2a\leqslant 2b$].

2.16.4. Integrals of $(x^2\pm a^2)^\alpha(b^2\pm x^2)^\beta\begin{Bmatrix}\mathrm{K}(cx)\\ \mathrm{E}(cx)\end{Bmatrix}$.

1. $$\int_0^a\frac{\mathrm{K}(cx)}{\sqrt{a^2-x^2}}dx=\frac{\pi^2}{4}\,{}_2F_1^2\left(\frac{1}{4},\ \frac{1}{4};\ 1;\ a^2c^2\right)$$ [$a>0$].

2. $$=\mathrm{K}^2\left(\frac{\sqrt{1+ac}-\sqrt{1-ac}}{2}\right)$$ [$a>0;\ |ac|<1$].

3. $$\int_0^a\frac{1}{\sqrt{a^2-x^2}}\begin{Bmatrix}\mathrm{K}(x/a)\\ \mathrm{E}(x/a)\end{Bmatrix}dx=\frac{1}{2}\begin{Bmatrix}2\\ 1\end{Bmatrix}\mathrm{K}^2\left(\frac{\sqrt{2}}{2}\right)+\begin{Bmatrix}0\\ 1\end{Bmatrix}\frac{\pi^2}{8\mathrm{K}^2(\sqrt{2}/2)}$$ [$a>0$].

4. $$\int_0^a x^{\alpha-1}(a^2-x^2)^{\beta-1}\begin{Bmatrix}\mathrm{K}(cx)\\ \mathrm{E}(cx)\end{Bmatrix}dx=$$
$$=\frac{\pi}{4}\mathrm{B}\left(\frac{\alpha}{2},\ \beta\right)a^{\alpha+2\beta-2}\,{}_3F_2\left(\pm\frac{1}{2},\ \frac{1}{2},\ \frac{\alpha}{2};\ 1,\ \frac{\alpha}{2}+\beta;\ a^2c^2\right)$$
[$a,\ \operatorname{Re}\alpha,\ \operatorname{Re}\beta>0;\ |\arg(1-a^2c^2)|<\pi$].

5. $\int_0^a x(a^2-x^2)^{\beta-1}\mathbf{K}(cx)\,dx=\frac{\pi\Gamma(\beta)}{4}\left(\frac{a}{c}\right)^{\beta}(1-a^2c^2)^{\beta/2}P_{-1/2}^{-\beta}(1-2a^2c^2)$

[a, $\operatorname{Re}\beta>0$; $|\arg(1-a^2c^2)|<\pi$].

6. $\int_0^a x(a^2-x^2)^{\beta-1}\mathbf{K}\left(\frac{x}{a\sqrt{2}}\right)dx=\frac{\pi^{3/2}a^{2\beta}\Gamma(\beta)}{2^{\beta+2}\Gamma^2((2\beta+3)/4)}$ [a, $\operatorname{Re}\beta>0$].

7. $\int_0^a x(a^2-x^2)^{\beta-1}\begin{Bmatrix}\mathbf{K}(x/a)\\ \mathbf{E}(x/a)\end{Bmatrix}dx=\frac{\pi a^{2\beta}}{4}\Gamma\begin{bmatrix}\beta,\ \beta+(1\mp1)/2\\ \beta+1/2,\ \beta+1\mp1/2\end{bmatrix}$ [a, $\operatorname{Re}\beta>0$].

8. $\int_0^a \frac{x}{a^2-x^2}\left[1-\mathbf{E}\left(\frac{x}{a}\right)\right]dx=1-2\ln 2$ [$a>0$].

9. $\int_0^a x^{-2\beta}(a^2-x^2)^{\beta-1}\mathbf{K}(cx)\,dx=$

$=\frac{\sqrt{\pi}}{4a}\Gamma(\beta)\Gamma\left(\frac{1}{2}-\beta\right){}_2F_1\left(\frac{1}{2}-\beta,\ \frac{1}{2};\ 1;\ a^2c^2\right)$ [$0<ac<1$; $0<\operatorname{Re}\beta<1/2$].

10. $\int_0^a x^{-2\beta}(a^2-x^2)^{\beta-1}\begin{Bmatrix}\mathbf{K}(x/a)\\ \mathbf{E}(x/a)\end{Bmatrix}dx=\frac{\pi}{4a\cos\beta\pi}\begin{Bmatrix}1\\ 2\beta\end{Bmatrix}\Gamma^2\begin{bmatrix}\beta\\ \beta+1/2\end{bmatrix}$

[$a>0$; $0<\operatorname{Re}\beta<1/2$].

11. $\int_0^a x^{2\beta-1}(a^2-x^2)^{\beta-1}\mathbf{K}\left(\frac{x}{a}\right)dx=\frac{\pi^2a^{4\beta-2}}{2^{2\beta+1}}\Gamma^2\begin{bmatrix}\beta\\ 3/4,\ \beta+1/4\end{bmatrix}$ [a, $\operatorname{Re}\beta>0$].

12. $\int_0^a x^{2-4\beta}(a^2-x^2)^{\beta-1}\mathbf{K}\left(\frac{x}{a}\right)dx=\frac{\pi a^{1-2\beta}}{2^{2\beta+3/2}}\Gamma\begin{bmatrix}\beta,\ \beta,\ 3/4-\beta\\ 3/4,\ 3/4,\ 1/4+\beta\end{bmatrix}$

[$a>0$; $0<\operatorname{Re}\beta<3/4$].

13. $\int_0^a x^{1-2\beta}(a^2-x^2)^{\beta-1}\mathbf{K}\left(\frac{x}{a}\right)dx=\frac{\pi^{5/2}}{4\sin\beta\pi}\Gamma\begin{bmatrix}(1+2\beta)/4,\ (5-2\beta)/4\\ 3/4,\ 3/4,\ (1+\beta)/2,\ 1-\beta/2\end{bmatrix}$

[$a>0$; $0<\operatorname{Re}\beta<1$].

14. $\int_0^a x^{1/2-\beta}(a^2-x^2)^{\beta-1}\mathbf{K}\left(\frac{x}{a}\right)dx=2^{\beta-7/2}\pi a^{\beta-1/2}\Gamma\begin{bmatrix}\beta/2,\ \beta/2,\ (3-2\beta)/4\\ 3/4,\ 3/4,\ (1+2\beta)/4\end{bmatrix}$

[$a>0$; $0<\operatorname{Re}\beta<3/2$].

15. $\int_0^\infty \frac{x^{\alpha-1}}{x^2+a^2}\mathbf{E}\left(i\frac{x}{a}\right)dx=\frac{a^{\alpha-2}}{2}\Gamma\begin{bmatrix}\alpha/2,\ (1-\alpha)/2,\ (3-\alpha)/2\\ 1-\alpha/2\end{bmatrix}$ [$\operatorname{Re}a>0$; $0<\operatorname{Re}\alpha<1$].

16. $\int_a^b \frac{1}{\sqrt{(x^2-a^2)(b^2-x^2)}}\mathbf{K}(cx)\,dx=\frac{2}{b(1+\sqrt{1-a^2c^2})}\mathbf{K}(k_+)\mathbf{K}'(k_-)$

[$k_\pm=ab^{-1}(1\pm\sqrt{1-b^2c^2})(1+\sqrt{1-a^2c^2})^{-1}$; $0<a<b$].

17. $\int_a^b \frac{1}{\sqrt{(x^2-a^2)(b^2-x^2)}}\mathbf{K}\left(\frac{x}{b}\right)dx=\frac{2}{a+b}\mathbf{K}\left(\sqrt{\frac{b-a}{b+a}}\right)\mathbf{K}\left(\sqrt{\frac{2a}{a+b}}\right)$

[$0<a<b$].

18. $\int_0^a \frac{1}{\sqrt{(a^2-x^2)(b^2-x^2)}} \mathbf{K}\left(\frac{x}{b}\right) dx = \frac{1}{a+b} \mathbf{K}^2\left(\sqrt{\frac{2a}{a+b}}\right)$ $[0 < a < b]$.

19. $\int_0^a (a^2-x^2)^{\beta-1} (x^2+a^2)^{3/2-\beta} \mathbf{K}\left(\frac{x}{a}\right) dx = 2^{\beta-9/2} \sqrt{\pi}\, a^2 \Gamma\begin{bmatrix}\beta/2-1, & \beta/2-1\\ \beta-3/2\end{bmatrix}$

$[a > 0;\ \operatorname{Re}\beta > 2]$.

2.16.5. Integrals of $x^\alpha (x^2 \pm a^2)^\beta (b^2 \pm x^2)^\gamma \begin{Bmatrix}\mathbf{K}(cx)\\ \mathbf{E}(cx)\end{Bmatrix}$.

NOTATION: $\delta = \begin{Bmatrix}1\\0\end{Bmatrix}$.

1. $\int_0^a \frac{x(a^2-x^2)^{\beta-1}}{(z^2 \pm x^2)^\rho} \mathbf{K}\left(\frac{x}{a}\right) dx = I(1)$,

$I(\varepsilon) = \frac{\pi a^{2\beta}}{4(z^2 \pm a^2)^\rho} \Gamma\begin{bmatrix}\beta, & \beta-\varepsilon+1\\ \beta+1/2, & \beta-\varepsilon+3/2\end{bmatrix} {}_3F_2\left(\rho, \beta, \beta-\varepsilon+1; \beta+\frac{1}{2}; \beta-\varepsilon+\frac{3}{2}; \frac{a^2}{a^2 \pm z^2}\right)$

$\left[a, \operatorname{Re}\beta > 0, \begin{Bmatrix}\operatorname{Re} z > 0\\ z > a\end{Bmatrix}\right]$.

2. $\int_0^a \frac{x(a^2-x^2)^{\beta-1}}{(z^2 \pm x^2)^\rho} \mathbf{E}\left(\frac{x}{a}\right) dx = I(0)$ $\left[a, \operatorname{Re}\beta > 0, \begin{Bmatrix}\operatorname{Re} z > 0\\ z > a\end{Bmatrix}; I(\varepsilon) \text{ see } 2.16.5.1\right]$.

3. $\int_0^a \frac{x(a^2-x^2)^{\beta-1}}{(b^2-x^2)^{\beta+1/2}} \mathbf{K}\left(\frac{x}{a}\right) dx = \frac{\sqrt{\pi}}{a^\beta b^{\beta/2}} \Gamma\begin{bmatrix}\beta\\ \beta+1/2\end{bmatrix} \frac{1}{\sqrt{b^2-a^2}} Q_{\beta-1}\left(\frac{a^2+b^2}{2ab}\right)$

$[0 < a < b;\ \operatorname{Re}\beta > 0]$.

4. $\int_0^a \frac{x}{\sqrt{a^2-x^2}\,(b^2-x^2)} \begin{Bmatrix}\mathbf{K}(x/a)\\ \mathbf{E}(x/a)\end{Bmatrix} dx =$

$= \frac{\pi a}{2b\sqrt{b^2-a^2}} \left[\mathbf{K}\left(\frac{a}{b}\right) - \begin{Bmatrix}0\\1\end{Bmatrix} \mathbf{D}\left(\frac{a}{b}\right)\right]$ $[0 < a < b]$.

5. $\int_0^a \frac{x}{\sqrt{a^2-x^2}\,(x^2+z^2)} \begin{Bmatrix}\mathbf{K}(x/a)\\ \mathbf{E}(x/a)\end{Bmatrix} dx = \frac{\pi a}{2(a^2+z^2)} \begin{Bmatrix}\mathbf{K}\left(a/\sqrt{a^2+z^2}\right)\\ \mathbf{D}\left(a/\sqrt{a^2+z^2}\right)\end{Bmatrix}$ $[a, \operatorname{Re} z > 0]$.

6. $\int_0^a x \frac{(a^2-x^2)^{\beta-1}}{(a^2+x^2)^{\beta+1/2}} \mathbf{K}\left(\frac{x}{a}\right) dx = \frac{2^{\beta-9/2}\sqrt{\pi}}{a} \Gamma\begin{bmatrix}\beta/2, & \beta/2\\ \beta+1/2\end{bmatrix}$ $[a, \operatorname{Re}\beta > 0]$.

7. $\int_0^a \frac{x}{\sqrt{(a^2-x^2)(b^2-x^2)}} \mathbf{K}\left(\frac{x}{a}\right) dx = \frac{a}{a+b} \mathbf{K}^2\left(\sqrt{\frac{2a}{a+b}}\right)$ $[0 < a < b]$.

8. $\int_0^a \frac{x(a^2-x^2)^{\beta-1}}{(b^2-x^2)^{\beta+1/2}} \mathbf{K}\left(\frac{x}{b}\right) dx = \frac{\pi a^\beta b^{1-\beta}}{4(b^2-a^2)} \Gamma(\beta) P_{-1/2}^{-\beta}\left(\frac{b^2+a^2}{b^2-a^2}\right)$ $[0 < a < b;\ \operatorname{Re}\beta > 0]$.

9. $\int_a^b x(b^2-x^2)^{\beta-1} (x^2-a^2)^{-\beta-1/2} \mathbf{K}\left(\frac{x}{b}\right) dx = \frac{a^{-\beta} b^\beta}{4\sqrt{\pi(b^2-a^2)}} \Gamma\left(\frac{1}{2}-\beta\right) \times$

$\times \Gamma(\beta) Q_{\beta-1}\left(\frac{a^2+b^2}{2ab}\right)$ $[0 < a < b;\ 0 < \operatorname{Re}\beta < 1/2]$.

10. $$\int_0^a x^{2\rho-2\beta} \frac{(a^2-x^2)^{\beta-1}}{(x^2+z^2)^{\rho}} \begin{Bmatrix} \mathbf{K}(x/a) \\ \mathbf{E}(x/a) \end{Bmatrix} dx = \frac{\pi a^{2\rho-1}}{4z^{2\beta-1}} (a^2+z^2)^{\beta-\rho-1/2} \mathrm{B}\left(\beta-\frac{1}{2}, \frac{1}{2}-\beta+\rho\right) \times$$
$$\times {}_3F_2\left(\frac{1}{2}, \frac{3}{2}-\delta, \frac{1}{2}-\beta+\rho; \frac{3}{2}-\beta, 1; \frac{z^2}{z^2+a^2}\right) + \frac{a^{2\rho-1}(a^2+z^2)^{-\rho}}{2^{\delta+1}} \Gamma\left[\begin{matrix} \beta, 1+\beta-\delta, 1/2-\beta \\ 1/2+\beta \end{matrix}\right] \times$$
$$\times {}_3F_2\left(\beta, 1+\beta-\delta, \rho; \frac{1}{2}+\beta, \frac{1}{2}+\beta; \frac{z^2}{z^2+a^2}\right)$$ [a, $\mathrm{Re}\, z > 0$; $0 < \mathrm{Re}\,\beta < \mathrm{Re}\,\rho + 1/2$].

11. $$\int_0^a x^{2-2\beta} \frac{(a^2-x^2)^{\beta-1}}{x^2-y^2} \begin{Bmatrix} \mathbf{K}(x/a) \\ \mathbf{E}(x/a) \end{Bmatrix} dx = -\frac{\pi \operatorname{tg} \beta\pi}{2} y^{1-2\beta} (a^2-y^2)^{\beta-1} \begin{Bmatrix} \mathbf{K}(y/a) \\ \mathbf{E}(y/a) \end{Bmatrix} +$$
$$+ \frac{a}{2^{\delta+1}(a^2-y^2)} \Gamma\left[\begin{matrix} \beta, 1+\beta-\delta, 1/2-\beta \\ 1/2+\beta \end{matrix}\right] {}_3F_2\left(\beta, 1+\beta-\delta, 1; \frac{1}{2}+\beta, \frac{1}{2}+\beta; \frac{y^2}{y^2-a^2}\right)$$
[$0 < y < a$; $0 < \mathrm{Re}\,\beta < 3/2$].

12. $$\int_0^a \frac{x^{1-2\rho}(a^2-x^2)^{\beta-1}}{(2a^2-x^2)^{\rho}} \mathbf{K}\left(\frac{x}{a\sqrt{2}}\right) dx = \frac{\pi}{8} a^{2\beta-4\rho} \mathrm{B}\left(\frac{\beta}{2}, 1-\rho\right) {}_3F_2\left(\frac{1}{4}, \frac{1}{4}, 1-\rho; 1, 1+\frac{\beta}{2}-\rho; 1\right)$$
[a, $\mathrm{Re}\,\beta > 0$; $\mathrm{Re}\,\rho < 1$].

13. $$\int_0^a \frac{x^{-\beta-1/2}(a^2-x^2)^{\beta-1}}{(2a^2-x^2)^{(2\beta+3)/4}} \mathbf{K}\left(\frac{x}{a\sqrt{2}}\right) dx = \frac{\sqrt{\pi}}{2^{\beta+5/2}a^3} \Gamma\left[\begin{matrix} (1-2\beta)/4, \beta \\ (3+2\beta)/4 \end{matrix}\right]$$
[$a > 0$; $0 < \mathrm{Re}\,\beta < 1/2$].

14. $$\int_0^a x^{2-2\beta-2\gamma} (a^2-x^2)^{\beta-1} (b^2-x^2)^{\gamma-1} \begin{Bmatrix} \mathbf{K}(x/b) \\ \mathbf{E}(x/b) \end{Bmatrix} dx = \frac{\pi}{4} a^{1-2\gamma} b^{1-2\beta} (b^2-a^2)^{\alpha+\beta-3/2} \mathrm{B}\left(\beta, \frac{3}{2}-\beta-\gamma\right) \times$$
$$\times {}_3F_2\left(\frac{1}{2}, \frac{3}{2}-\delta, \frac{3}{2}-\beta-\gamma; 1, \frac{3}{2}-\gamma; \frac{a^2}{a^2-b^2}\right)$$
[$\mathrm{Re}\,\beta > 0$; $\mathrm{Re}\,(\beta+\gamma) < 3/2$; $0 < a < b$].

15. $$\int_a^b x^{2-2\alpha-2\beta} (x^2-a^2)^{\alpha-1} (b^2-x^2)^{\beta-1} \begin{Bmatrix} \mathbf{K}(x/b) \\ \mathbf{E}(x/b) \end{Bmatrix} dx =$$
$$= \frac{\pi}{4} a^{1-2\beta} b^{1-2\alpha} (b^2-a^2)^{\alpha+\beta-3/2} \mathrm{B}\left(\alpha, \beta-\frac{1}{2}\right) {}_3F_2\left(\frac{1}{2}, \frac{3}{2}-\delta, \frac{3}{2}-\alpha-\beta; \frac{3}{2}-\beta, 1; \frac{a^2}{a^2-b^2}\right) + \frac{(b^2-a^2)^{\alpha-1}}{2^{\delta+1}b^{2\alpha-1}} \Gamma\left[\begin{matrix} \beta, 1+\beta-\delta, 1/2-\beta \\ 1/2+\beta \end{matrix}\right] \times$$
$$\times {}_3F_2\left(\beta, 1+\beta-\delta, 1-\alpha; \frac{1}{2}+\beta, \frac{1}{2}+\beta; \frac{a^2}{a^2-b^2}\right)$$
[$0 < a < b$; $\mathrm{Re}\,\alpha$, $\mathrm{Re}\,\beta > 0$].

2.16.6. Integrals of $A(x)\,\mathbf{K}(cx)$.

1. $$\int_0^a \frac{x}{\sqrt{a^2-x^2}\,(x-y)}\mathbf{K}\left(\frac{x}{a}\right)dx=-\frac{\pi a}{2\sqrt{y(a-y)}}\mathbf{K}\left(\sqrt{\frac{a}{y}}\right)\quad [0\leqslant a<y].$$

2. $$\int_0^a \frac{x}{\sqrt{a^2-x^2}}\,\frac{a^2-b^2-x^2}{a^2+b^2-x^2}\mathbf{K}\left(\frac{x}{a}\right)dx=\frac{a^3}{2b\sqrt{a^2+b^2}}\mathbf{E}\left(\frac{a}{\sqrt{a^2+b^2}}\right)\quad [a,\ b>0].$$

3. $$\int_0^a \frac{(b^2-x^2)^{-3/4}}{\sqrt{a^2-x^2}}\left[a(b^2-x^2)+x(b^2-a^2)\right]^{1/2}\mathbf{K}\left(\frac{x}{b}\right)dx=$$
$$=\frac{(b^2-a^2)^{1/4}}{\sqrt{a}}\left\{\left[Q_{-1/2}\left(\frac{b^2}{a\sqrt{b^2-a^2}}\right)\right]^2+\left[Q_{-3/4}\left(\frac{b^2}{a\sqrt{b^2-a^2}}\right)\right]^2\right\}\quad [0<a<b].$$

4. $$\int_0^a \frac{x^{-1/2}}{\sqrt{(b^2-x^2)\left[ab\sqrt{b^2-x^2}-\sqrt{b^2-a^2x}\right]}}\mathbf{K}\left(\frac{x}{b}\right)dx=$$
$$=\frac{1}{2b^{3/2}(b^2-a^2)^{1/4}}\left[Q_{-3/4}\left(\frac{b^2}{a\sqrt{b^2-a^2}}\right)P_{-3/4}\left(\frac{b^2}{a\sqrt{b^2-a^2}}\right)-\right.$$
$$\left.-Q_{-1/4}\left(\frac{b^2}{a\sqrt{b^2-a^2}}\right)P_{-1/4}\left(\frac{b^2}{a\sqrt{b^2-a^2}}\right)\right]\quad [0<a<b].$$

5. $$\int_0^a \frac{x^{-1}\left(\sqrt{x^2+z^2}+x\sqrt{z^2+1}\right)}{\sqrt{(a^2-x^2)(x^2+z^2)}}\mathbf{K}\left(\frac{x}{a}\right)dx=$$
$$=\frac{4\sqrt{2}}{\pi a\sqrt{z}}\left(\sqrt{a^2+z^2}-a\right)^{1/2}(\xi+1)^{-1/2}\left(\xi-\sqrt{\xi^2-1}\right)^{1/2}\mathbf{K}\left(\sqrt{\frac{\xi-1}{\xi+1}}\right)\mathbf{K}\left(\xi-\sqrt{\xi^2-1}\right)$$
$$\left[\xi^2=2\sqrt{a^2+z^2}\left(\sqrt{a^2+z^2}-a\right)/z;\ a,\ \operatorname{Re} z>0\right].$$

6. $$\int_0^a (a^2-x^2)^{\beta-1}\left[\left(x+b\sqrt{a^2-x^2}\right)^{-2\beta}+\left|x-b\sqrt{a^2-x^2}\right|^{-2\beta}\right]\mathbf{K}\left(\frac{x}{a}\right)dx=$$
$$=\frac{2^{4\beta-3}\Gamma^4(\beta)}{a\cos\beta\pi\,\Gamma^2(2\beta)}\,{}_3F_2\left(\beta,\ \beta,\ \beta;\ \beta+\frac{1}{2},\ \frac{1}{2};\ -b^2\right)\quad [0<b<1;\ 0<\operatorname{Re}\beta<1/2].$$

7. $$\int_0^a (a^2-x^2)^{\beta-1}\left[\left(x+b\sqrt{a^2-x^2}\right)^{-2\beta}-\frac{\operatorname{sgn}\left(x-b\sqrt{a^2-x^2}\right)}{\left|x-b\sqrt{a^2-x^2}\right|^{2\beta}}\right]\mathbf{K}\left(\frac{x}{a}\right)dx=$$
$$=\frac{2^{2\beta}b}{\sqrt{\pi}a}\cos\beta\pi\,\Gamma^3\left(\frac{1}{2}+\beta\right)\Gamma\begin{bmatrix}1-2\beta\\1+\beta\end{bmatrix}{}_3F_2\left(\beta+\frac{1}{2},\ \beta+\frac{1}{2},\ \beta+\frac{1}{2};\ \frac{3}{2},\ \beta+1;\ -b^2\right)$$
$$[0<b<1;\ |\operatorname{Re}\beta|<1/2].$$

8. $$\int_0^a (a^2-x^2)^{-3/4}\left[\left(bx+\sqrt{a^2-x^2}\right)^{-1/2}+(-1)^{\varepsilon}\frac{\operatorname{sgn}^{\varepsilon}\left(bx-\sqrt{a^2-x^2}\right)}{\left|bx-\sqrt{a^2-x^2}\right|^{1/2}}\right]\mathbf{K}\left(\frac{x}{a}\right)dx=$$
$$=a^{-1}\left[Q_{(2\varepsilon-3)/4}\left(b+b^{-1}\right)\right]^2\quad [\varepsilon=0\ \text{or}\ 1;\ a,\ b>0].$$

9. $$\int_0^a \frac{x^{-1/2}}{\sqrt{a^2-x^2}}\left[\left(bx+\sqrt{a^2-x^2}\right)^{-1/2}+\frac{\operatorname{sgn}^{\varepsilon}\left(bx-\sqrt{a^2-x^2}\right)}{\left|bx-\sqrt{a^2-x^2}\right|^{1/2}}\right]\mathbf{K}\left(\frac{x}{a}\right)dx=$$
$$=\frac{1}{a\sqrt{b}}P_{(2\varepsilon-3)/4}\left(b+\frac{1}{b}\right)Q_{(2\varepsilon-3)/4}\left(b+\frac{1}{b}\right)\quad [\varepsilon=0\ \text{or}\ 1;\ a,\ b>0].$$

2.16.7. Integrals of $A(x)\begin{Bmatrix}\mathbf{K}(\varphi(x))\\ \mathbf{E}(\varphi(x))\end{Bmatrix}$.

1. $$\int_0^\infty \frac{x^{\alpha-1}}{x+z}\mathbf{K}\left(\frac{2\sqrt{xz}}{x+z}\right)dx=\frac{\pi}{4}z^{\alpha-1}\Gamma\begin{bmatrix}\alpha/2,\ (1-\alpha)/2\\ (1+\alpha)/2,\ 1-\alpha/2\end{bmatrix}$$ $[\operatorname{Re} z>0;\ 0<\operatorname{Re}\alpha<1]$.

2. $$\int_0^\infty \frac{x^{\alpha-1}}{\sqrt{x^2+z^2}+z}\mathbf{K}\left(\frac{(\sqrt{x^2+z^2}-z)^2}{x^2}\right)dx=\frac{z^{\alpha-1}}{8}\Gamma\begin{bmatrix}\alpha/2,\ (1-\alpha)/2,\ (1-\alpha)/2\\ 1-\alpha/2\end{bmatrix}$$ $[\operatorname{Re} z>0;\ 0<\operatorname{Re}\alpha<1]$.

3. $$\int_0^\infty \frac{x^{\alpha-1}}{(x^2+z^2)^{1/4}}\mathbf{K}\left(\sqrt{\frac{\sqrt{x^2+z^2}-x}{2\sqrt{x^2+z^2}}}\right)dx=\frac{\sqrt{\pi}z^{\alpha-1/2}}{2^{\alpha+3/2}}\Gamma\begin{bmatrix}\alpha,\ (1-2\alpha)/4\\ (2\alpha+3)/4\end{bmatrix}$$ $[\operatorname{Re} z>0;\ 0<\operatorname{Re}\alpha<1]$.

4. $$\int_0^\infty \frac{x^{\alpha-1}}{\sqrt{x^2+z^2}+x}\mathbf{K}\left(\frac{2\sqrt[4]{(x^2+z^2)x^2}}{\sqrt{x^2+z^2}+x}\right)dx=\frac{z^{\alpha-1}}{4}\Gamma\begin{bmatrix}\alpha/2,\ (1-\alpha)/2,\ (1-\alpha)/2\\ 1-\alpha/2\end{bmatrix}$$ $[\operatorname{Re} z>0;\ 0<\operatorname{Re}\alpha<1]$.

5. $$\int_0^\infty \frac{x^{\alpha-1}}{\sqrt{(a+x)^2+z^2}}\mathbf{K}\left(\frac{2\sqrt{ax}}{\sqrt{(a+x)^2+z^2}}\right)dx=$$
$$=\frac{\sqrt{\pi}}{4}\Gamma\left(\frac{\alpha}{2}\right)\Gamma\left(\frac{1-\alpha}{2}\right)(a^2+z^2)^{(\alpha-1)/2}P_{-\alpha}\left(\frac{z}{\sqrt{a^2+z^2}}\right)$$ $[a,\ \operatorname{Re} z>0;\ 0<\operatorname{Re}\alpha<1]$.

6. $$\int_0^\infty \left(\frac{\sqrt{x^2+z^2}-z}{x^2+z^2}\right)^{1/2}\mathbf{K}\left(\frac{\sqrt{x^2+a^2}-a}{\sqrt{x^2+a^2}+a}\right)\frac{dx}{\sqrt{x^2+a^2}+a}=$$
$$=\frac{1}{a}(z-\sqrt{z^2-a^2})^{1/2}\operatorname{sech}^2\zeta\,\mathbf{K}(\operatorname{sech}\zeta)\,\mathbf{K}(\operatorname{th}\zeta)$$
$[a,\ \operatorname{Re} z>0;\ \operatorname{ch}\zeta=(a+2z^2-2z\sqrt{z^2-a^2})^{1/2}/(2a)^{1/2}]$.

7. $$\int_0^\infty \frac{x^{\alpha-1}(x^2+z^2)^{(-1\pm1)/4}}{\sqrt{\sqrt{x^2+z^2}+x}}\mathbf{K}\left(\frac{\sqrt{2x(\sqrt{x^2+z^2}-x)}}{z}\right)dx=$$
$$=\frac{\pi}{4}z^{\alpha\pm1/2-1}\Gamma\begin{bmatrix}\alpha/2,\ (2\mp1-2\alpha)/4,\ (2\mp1-2\alpha)/4\\ (2\mp1)/4,\ (2\mp1)/4,\ 1-\alpha/2\end{bmatrix}$$ $[\operatorname{Re} z>0;\ 0<\operatorname{Re}\alpha<1\mp1/2]$.

8. $$\int_0^\infty \frac{x^{\alpha-1}(x^2+z^2)^{-1/2}}{\sqrt{\sqrt{x^2+z^2}-x}}\mathbf{E}\left(\frac{\sqrt{2x(\sqrt{x^2+z^2}-x)}}{z}\right)dx=$$
$$=\pi z^{\alpha-3/2}\Gamma\begin{bmatrix}\alpha/2,\ (5-2\alpha)/4,\ (1-2\alpha)/4\\ 1-\alpha/2\end{bmatrix}$$ $[\operatorname{Re} z>0;\ 0<\operatorname{Re}\alpha<1/2]$.

9. $$\int_0^a \frac{x^{\alpha-1}}{\sqrt{\sqrt{a^2-x^2}+a}}\mathbf{K}\left(\frac{\sqrt{2(x^2-a^2+a\sqrt{a^2-x^2})}}{x}\right)dx=$$
$$=\frac{\sqrt{\pi}a^{\alpha-1/2}}{2^{5/2-\alpha}}\Gamma\begin{bmatrix}\alpha/2,\ \alpha/2\\ \alpha+1/2\end{bmatrix}$$ $[a,\ \operatorname{Re}\alpha>0]$.

10. $$\int_0^a \frac{x^{\alpha-1}}{\sqrt{a-\sqrt{a^2-x^2}}}\mathbf{E}\left(\frac{\sqrt{2(a\sqrt{a^2-x^2}-a^2+x^2)}}{x}\right)dx=$$
$$=\frac{\sqrt{\pi}a^{\alpha-1/2}}{2^{5/2-\alpha}}\Gamma\begin{bmatrix}(\alpha-1)/2,\ (\alpha+1)/2\\ \alpha+1/2\end{bmatrix}$$ $[a>0;\ \operatorname{Re}\alpha>1]$.

11. $\int_0^a \frac{x^{\alpha-1}}{a-\sqrt{a^2-x^2}} \mathbf{E}\left(\frac{2\sqrt[4]{a^2(a^2-x^2)}}{a+\sqrt{a^2-x^2}}\right) dx = \frac{\pi a^{\alpha-1}}{4} \Gamma\begin{bmatrix}\alpha/2+1, & \alpha/2-1 \\ (\alpha+1)/2, & (\alpha+1)/2\end{bmatrix}$ $[a>0;\ \operatorname{Re}\alpha>2]$.

12. $\int_0^a \frac{x^{\alpha-1}}{\sqrt{a^2-x^2}+a} \mathbf{K}\left(\frac{2\sqrt[4]{a^2(a^2-x^2)}}{\sqrt{a^2-x^2}+a}\right) dx = \frac{\pi a^{\alpha-1}}{4} \Gamma^2\begin{bmatrix}\alpha/2 \\ (\alpha+1)/2\end{bmatrix}$ $[a,\ \operatorname{Re}\alpha>0]$.

2.16.8. Integrals of $A(x)\,e^{\varphi(x)}\begin{Bmatrix}\mathbf{K}(cx)\\ \mathbf{E}(cx)\end{Bmatrix}$.

NOTATION: $\delta=\begin{Bmatrix}1\\0\end{Bmatrix}$.

1. $\int_0^a x\,(a^2-x^2)^{\beta-1} e^{-px^2} \begin{Bmatrix}\mathbf{K}(x/a)\\ \mathbf{E}(x/a)\end{Bmatrix} dx =$

$$= \frac{\pi a^{2\beta}}{4} \Gamma\begin{bmatrix}\beta, & \beta-\delta+1 \\ \beta+1/2, & \beta-\delta+3/2\end{bmatrix} e^{-a^2p}\, {}_2F_2\left(\beta,\ \beta-\delta+1;\ \beta+\frac{1}{2},\ \beta-\delta+\frac{3}{2};\ a^2p\right)$$ $[a,\ \operatorname{Re}\beta>0]$.

2. $\int_0^a \left(e^{-px}+\frac{a}{x}e^{-pa^2/x}\right) \mathbf{K}\left(\frac{x}{a}\right) dx = \frac{\pi a}{2} I_0\left(\frac{ap}{2}\right) K_0\left(\frac{ap}{2}\right)$ $[a>0]$.

3. $\int_0^a x^{-2\beta}(a^2-x^2)^{\beta-1} e^{-p/x^2} \begin{Bmatrix}\mathbf{K}(x/a)\\ \mathbf{E}(x/a)\end{Bmatrix} dx =$

$$= \frac{\pi a^{2\beta-2} p^{1/2-\beta}}{4} \Gamma\left(\beta-\frac{1}{2}\right) e^{-p/a^2}\, {}_2F_2\left(\frac{1}{2},\ \frac{3}{2}-\delta;\ 1,\ \frac{3}{2}-\beta;\ \frac{p}{a^2}\right) +$$

$$+ \frac{e^{-p/a^2}}{2^{\delta+1}a} \Gamma\begin{bmatrix}\beta, & \beta-\delta+1, & 1/2-\beta \\ \beta+1/2\end{bmatrix} {}_2F_2\left(\beta,\ \beta-\delta+1;\ \beta+\frac{1}{2},\ \beta+\frac{1}{2};\ \frac{p}{a^2}\right)$$ $[a,\ \operatorname{Re}p,\ \operatorname{Re}\beta>0]$.

4. $\int_0^a \frac{x}{(a^2-x^2)^{3/2}} \exp\left(-\frac{px^2}{a^2-x^2}\right) \mathbf{K}\left(\frac{x}{a}\right) dx = \frac{1}{4a}\sqrt{\frac{\pi}{p}}\, e^{p/2} K_0\left(\frac{p}{2}\right)$ $[a,\ \operatorname{Re}p>0]$.

5. $\int_0^a x^{-2\beta}(a^2-x^2)^{\beta-1} \exp\left(-p\frac{a^2+x^2}{a^2-x^2}\right) \mathbf{K}\left(\frac{x}{a}\right) dx =$

$$= \frac{2^{\beta-3/2}\sqrt{\pi}\, p^{\beta-1/2}}{a\Gamma(\beta)} e^{-2p} G_{23}^{31}\left(2p\,\middle|\,\begin{matrix}1, & 1 \\ 1/2-\beta, & 1/2, & 1/2\end{matrix}\right)$$ $[0<\operatorname{Re}\beta<1/2]$.

6. $\int_0^a \frac{x}{(a^2-x^2)\sqrt{b^2-x^2}} \exp\left(-p\frac{b^2-x^2}{a^2-x^2}\right) \mathbf{K}\left(\frac{x}{a}\right) dx =$

$$= \frac{1}{4\sqrt{b^2-a^2}} K_0\left(\frac{b+a}{a}\sqrt{p}\right) K_0\left(\frac{b-a}{a}\sqrt{p}\right)$$ $[0<a<b;\ \operatorname{Re}p>0]$.

7. $\int_0^a \frac{x}{(a^2-x^2)^{3/2}} \exp\left(-p\frac{x^2}{(a^2-x^2)^2}\right) \mathbf{K}\left(\frac{x}{a}\right) dx =$

$$= \frac{\pi}{4p^{3/4}}\sqrt{\frac{a}{2}} \exp\left(\frac{p}{8a^2}\right) W_{-1/4,\,0}\left(\frac{p}{4a^2}\right)$$ $[a,\ \operatorname{Re}p>0]$.

8. $$\int_0^a \frac{x}{a^2-x^2}\exp\left(-\frac{p}{\sqrt{a^2-x^2}}\right)\mathbf{K}\left(\frac{x}{a}\right)dx=\frac{1}{2}K_0^2\left(\frac{p}{2a}\right)$$ [a, $\operatorname{Re} p>0$].

9. $$\int_0^a \frac{1}{a^2-x^2}\exp\left(-\frac{px}{\sqrt{a^2-x^2}}\right)\mathbf{K}\left(\frac{x}{a}\right)dx=\frac{\pi^2}{8a}\left[J_0^2\left(\frac{p}{2}\right)+Y_0^2\left(\frac{p}{2}\right)\right]$$ [a, $\operatorname{Re} p>0$].

10. $$\int_0^a \frac{x}{(a^2-x^2)\sqrt{b^2-x^2}}\exp\left(-p\sqrt{\frac{b^2-x^2}{a^2-x^2}}\right)\mathbf{K}\left(\frac{x}{a}\right)dx=$$ $$=\frac{1}{2\sqrt{b^2-a^2}}K_0\left(\frac{p}{2a}(a+b)\right)K_0\left(\frac{p}{2a}(b-a)\right)$$ [$\operatorname{Re} p>0$; $0<a<b$].

2.16.9. Integrals of $A(x)\begin{Bmatrix}\sin\varphi(x)\\ \cos\varphi(x)\end{Bmatrix}\begin{Bmatrix}\mathbf{K}(cx)\\ \mathbf{E}(cx)\end{Bmatrix}$.

1. $$\int_0^a x(a^2-x^2)^{\beta-1}\begin{Bmatrix}\sin[b(a^2-x^2)]\\ \cos[b(a^2-x^2)]\end{Bmatrix}\mathbf{K}\left(\frac{x}{a}\right)dx=$$ $$=\frac{2^{\beta-\delta-2}\pi^{3/2}a^{2\beta+2\delta}b^{\delta}}{\Gamma^2((2\beta+2\delta+3)/4)}\Gamma(\beta+\delta)\,{}_2F_3\left(\frac{\beta+1}{2},\frac{\beta}{2}+\delta;\ \delta+\frac{1}{2},\frac{2\beta+2\delta+3}{4},\frac{2\beta+2\delta+3}{4};-\frac{a^2b}{2}\right)$$ $\left[a>0;\ \operatorname{Re}\beta>-\delta,\ \delta=\begin{Bmatrix}1\\0\end{Bmatrix}\right]$.

2. $$\int_0^a x(a^2-x^2)^{-3/2}\sin\left(b\frac{a^2+x^2}{a^2-x^2}\right)\mathbf{K}\left(\frac{x}{a}\right)dx=\frac{\pi^{3/2}S_{1/2,\,0}(b)}{8a\sqrt{b}\,\Gamma^2(3/4)}$$ [a, $b>0$].

3. $$\int_0^a x\cos(b\sqrt{a-x})\,\mathbf{K}\left(\frac{x}{a}\right)dx=a^2J_0^2\left(\frac{b\sqrt{a}}{2}\right)$$ [$a>0$].

4. $$\int_0^a x^{-1/2}\frac{a+x}{a-x}\begin{Bmatrix}\sin[b\sqrt{x}/(a-x)]\\ \cos[b\sqrt{x}/(a-x)]\end{Bmatrix}\mathbf{K}\left(\frac{x}{a}\right)dx=$$ $$=\frac{\sqrt{a}}{2}K_0\left(\frac{b}{4\sqrt{a}}\right)\begin{Bmatrix}\pi I_0(b/(4\sqrt{a}))\\ K_0(b/(4\sqrt{a}))\end{Bmatrix}$$ [a, $b>0$].

5. $$\int_0^a \frac{\sqrt{x}(x+3a)}{(x^2+a^2)^{3/2}}\begin{Bmatrix}\sin[b\sqrt{a+x}/(a-x)]\\ \cos[b\sqrt{a+x}/(a-x)]\end{Bmatrix}\mathbf{K}\left(\frac{x}{a}\right)dx=$$ $$=\frac{1}{\sqrt{2a}}K_0\left(\frac{b}{2\sqrt{2a}}\right)\begin{Bmatrix}\pi I_0(b/(2\sqrt{2a}))\\ 2K_0(b/(2\sqrt{2a}))\end{Bmatrix}$$ [a, $b>0$].

6. $$\int_0^a x(a^2-x^2)^{\beta-1}\begin{Bmatrix}\sin(b\sqrt{a^2-x^2})\\ \cos(b\sqrt{a^2-x^2})\end{Bmatrix}\mathbf{K}\left(\frac{x}{a}\right)dx=I(1),$$

$$I(\varepsilon)=\frac{\pi a^{2\beta+\delta}b^{\delta}}{4}\Gamma\left[\begin{matrix}\beta+\delta/2,\ \beta+\delta/2-\varepsilon+1\\ \beta+(\delta+1)/2,\ \beta+(\delta+3)/2-\varepsilon\end{matrix}\right]\times$$ $$\times{}_2F_3\left(\beta+\frac{\delta}{2},\ \beta+\frac{\delta}{2}-\varepsilon+1;\ \beta+\frac{\delta+1}{2},\ \delta+\frac{1}{2},\ \beta+\frac{\delta+3}{2}-\varepsilon;\ -\frac{a^2b^2}{4}\right)$$ $\left[a>0;\ \operatorname{Re}\beta>-\delta/2,\ \delta=\begin{Bmatrix}1\\0\end{Bmatrix}\right]$.

7. $\int_0^a \frac{x}{\sqrt{a^2-x^2}} \cos\left(b\sqrt{a^2-x^2}\right) \mathbf{K}\left(\frac{x}{a}\right) dx = \frac{\pi^2 a}{4} J_0^2\left(\frac{ab}{2}\right)$ $[a>0]$.

8. $\int_0^a x(a^2-x^2)^{\beta-1} \begin{Bmatrix} \sin\left(b\sqrt{a^2-x^2}\right) \\ \cos\left(b\sqrt{a^2-x^2}\right) \end{Bmatrix} \mathbf{E}\left(\frac{x}{a}\right) dx = I(0)$

$[a>0;\ \operatorname{Re}\beta>-\delta/2;\ I(\varepsilon)$ see 2.16.9.6].

9. $\int_0^a x^{-2\beta}(a^2-x^2)^{\beta-1} \sin\frac{b\sqrt{a^2-x^2}}{x} \begin{Bmatrix} \mathbf{K}(x/a) \\ \mathbf{E}(x/a) \end{Bmatrix} dx = J(1)$,

$$J(\varepsilon) = \frac{\pi b^{1-2\beta}}{2a} \Gamma(2\beta-1) \sin\left(\beta-\frac{\varepsilon}{2}\right) \pi\, {}_2F_3\left(\frac{1}{2}, \frac{3}{2}-\delta;\ 1,\ 1-\beta,\ \frac{3}{2}-\beta;\ \frac{b^2}{4}\right) +$$
$$+ \frac{b^{\varepsilon}}{2^{\delta+1}} \Gamma\left[\begin{matrix} \beta+\varepsilon/2,\ 1+\beta-\delta+\varepsilon/2,\ (1-\varepsilon)/2-\beta \\ (1+\varepsilon)/2+\beta \end{matrix}\right] \times$$
$$\times {}_2F_3\left(\beta+\frac{\varepsilon}{2},\ 1+\beta-\delta+\frac{\varepsilon}{2};\ \frac{1+\varepsilon}{2}+\beta,\ \frac{1+\varepsilon}{2}+\beta,\ \varepsilon+\frac{1}{2};\ \frac{b^2}{4}\right)$$

$[a,\ b>0;\ -1/2<\operatorname{Re}\beta<1]$.

10. $\int_0^a x^{-2\beta}(a^2-x^2)^{\beta-1} \cos\frac{b\sqrt{a^2-x^2}}{x} \begin{Bmatrix} \mathbf{K}(x/a) \\ \mathbf{E}(x/a) \end{Bmatrix} dx = J(0)$

$[a,\ b>0;\ 0<\operatorname{Re}\beta<1;\ J(\varepsilon)$ see 2.16.9.9].

11. $\int_0^a \frac{x}{a^2-x^2} \begin{Bmatrix} \sin\left(b/\sqrt{a^2-x^2}\right) \\ \cos\left(b/\sqrt{a^2-x^2}\right) \end{Bmatrix} \mathbf{K}\left(\frac{x}{a}\right) dx =$

$$= -\frac{\pi^2}{8} \begin{Bmatrix} 2J_0(b/(2a))\, Y_0(b/(2a)) \\ \left[J_0^2(b/(2a)) - Y_0^2(b/(2a))\right] \end{Bmatrix}$$ $[a,\ b>0]$.

12. $\int_0^a (a^2-x^2)^{-3/2} \sin\frac{bx}{\sqrt{a^2-x^2}} \mathbf{K}\left(\frac{x}{a}\right) dx = \frac{\pi}{2a^2} I_0(b) K_0(b)$ $[a,\ b>0]$.

13. $\int_0^a \frac{1}{a^2-x^2} \cos\frac{bx}{\sqrt{a^2-x^2}} \mathbf{K}\left(\frac{x}{a}\right) dx = \frac{1}{2a} K_0^2\left(\frac{b}{2}\right)$ $[a,\ b>0]$.

14. $\int_0^a \frac{(a^2+x^2)^{-1/2}}{a^2-x^2} \cos\frac{bx}{\sqrt{a^2-x^2}} \mathbf{K}\left(\frac{x}{a}\right) dx = \frac{2^{-3/2}}{a^2} K_0^2(b)$ $[a,\ b>0]$.

15. $\int_0^a \frac{x^{-1}}{\sqrt{a^2-x^2}} \cos\frac{b}{x\sqrt{a^2-x^2}} \mathbf{K}\left(\frac{x}{a}\right) dx = \frac{\pi}{2a} I_0\left(\frac{b}{2a^2}\right) K_0\left(\frac{b}{2a^2}\right)$ $[a,\ b>0]$.

16. $\int_0^\infty \frac{\cos bx}{\operatorname{ch} x} \mathbf{K}(\operatorname{th} x)\, dx = \frac{1}{16\pi} \left|\Gamma\left(\frac{1+ib}{4}\right)\right|^4$ $[b>0]$.

2.16.10. Integrals containing $\ln A(x)\, \mathbf{K}(cx)$.

1. $\int_0^a \frac{x}{\sqrt{a^2-x^2}\,(x^2-y^2)} \ln\frac{a^2-y^2}{a^2-x^2} \mathbf{K}\left(\frac{x}{a}\right) dx = \frac{\pi^2}{8\sqrt{a^2-y^2}} \mathbf{K}\left(\frac{y}{a}\right)$ $[0<y<a]$.

7

2. $\int_0^a \frac{1}{x\sqrt{a^2-x^2}(2a^2-x^2)} \ln\left(1-\frac{x^2}{a^2}\right) \mathbf{K}\left(\frac{x}{a}\right) dx = -\frac{\pi^{3/2}}{\sqrt{2}a^3} \ln(1+\sqrt{2})$ $[a>0]$.

3. $\int_0^a \frac{x}{a^2-x^2}\left[\ln\frac{4a}{\sqrt{a^2-x^2}} - \mathbf{K}\left(\frac{x}{a}\right)\right] dx = -\frac{1}{12}(24\ln^2 2 - \pi^2)$ $[a>0]$.

$\int_0^a \sqrt{\frac{x}{a^2-x^2}} \ln\frac{\sqrt{x}+\sqrt{a}}{\sqrt{x}-\sqrt{a}} \mathbf{K}\left(\frac{x}{a}\right) dx = \frac{\pi^2\sqrt{a}}{8} \Gamma^2\begin{bmatrix}1/4\\3/4\end{bmatrix}$ $[a>0]$.

2.16.11. Integrals containing $\mathrm{Ei}(\varphi(x))$ or $\mathrm{erfc}(\varphi(x))$ and $\mathbf{K}(cx)$.

1. $\int_0^a x\left[e^{b\sqrt{a^2-x^2}}\mathrm{Ei}(-b\sqrt{a^2-x^2}) + e^{-b\sqrt{a^2-x^2}}\mathrm{Ei}(b\sqrt{a^2-x^2})\right]\mathbf{K}\left(\frac{x}{a}\right) dx =$

$$= -\pi a^2 I_0\left(\frac{ab}{2}\right) K_0\left(\frac{ab}{2}\right) \quad [a,\ \mathrm{Re}\, b>0].$$

2. $\int_0^a \frac{x}{(a^2-x^2)^{3/2}} \exp\left[\left(b\frac{a^2+x^2}{a^2-x^2}\right)^2\right] \mathrm{erfc}\left(b\frac{a^2+x^2}{a^2-x^2}\right) \mathbf{K}\left(\frac{x}{a}\right) dx =$

$$= \frac{\Gamma^2(1/4)}{16ab^{3/2}} e^{b^2/2} W_{1/4,\,0}(b^2) \quad [a>0].$$

2.16.12. Integrals containing $J_\nu(\varphi(x))$ or $Y_\nu(\varphi(x))$ and $\mathbf{K}(cx)$.

NOTATION: $\delta = \begin{Bmatrix}1\\0\end{Bmatrix}$.

1. $\int_0^a \frac{x}{\sqrt{a^2-x^2}} J_0[b(a^2-x^2)] \mathbf{K}\left(\frac{x}{a\sqrt{2}}\right) dx = \frac{\pi}{2\sqrt{2}ab} \sin\frac{a^2b}{2} J_0\left(\frac{a^2b}{2}\right)$ $[a>0]$.

2. $\int_0^a \frac{x(a^2+x^2)}{(a^2-x^2)^{5/2}} J_0\left(\frac{bx}{a^2-x^2}\right) \mathbf{K}\left(\frac{x}{a}\right) dx = \frac{\pi\sqrt{a}}{b^{3/2}\Gamma^2(1/4)} K_0\left(\frac{b}{2a}\right)$ $[a,\ b>0]$.

3. $\int_0^a x(a^2-x^2)^{\beta-1} J_\nu(b\sqrt{a^2-x^2}) \begin{Bmatrix}\mathbf{K}(x/a)\\ \mathbf{E}(x/a)\end{Bmatrix} dx =$

$$= \frac{\pi a^{2\beta+\nu} b^\nu}{2^{\nu+2}} \Gamma\begin{bmatrix}\beta+\nu/2,\ 1+\beta-\delta+\nu/2\\ \nu+1,\ \beta+(1+\nu)/2,\ \beta-\delta+(3+\nu)/2\end{bmatrix} \times$$

$$\times {}_2F_3\left(\beta+\frac{\nu}{2},\ 1+\beta-\delta+\frac{\nu}{2};\ \nu+1,\ \beta+\frac{1+\nu}{2},\ \beta-\delta+\frac{3+\nu}{2};\ -\frac{a^2b^2}{4}\right)$$

$[a,\ \mathrm{Re}(2\beta+\nu)>0]$.

4. $\int_0^a x^{-2\beta}(a^2-x^2)^{\beta-1} J_\nu\left(\frac{b\sqrt{a^2-x^2}}{x}\right) \begin{Bmatrix}\mathbf{K}(x/a)\\ \mathbf{E}(x/a)\end{Bmatrix} dx =$

$$= \frac{2^{2\beta-3}\pi}{ab^{2\beta-1}} \Gamma\begin{bmatrix}(\nu-1)/2+\beta\\ (\nu+3)/2-\beta\end{bmatrix} {}_2F_3\left(\frac{1}{2},\ \frac{3}{2}-\delta;\ 1,\ \frac{3-\nu}{2}-\beta,\ \frac{3+\nu}{2}-\beta;\ \frac{b^2}{4}\right) +$$

$$+ \frac{b^\nu}{2^{\nu+\delta+1}a} \Gamma\begin{bmatrix}\beta+\nu/2,\ 1+\beta-\delta+\nu/2,\ (1-\nu)/2-\beta\\ \nu+1,\ (1+\nu)/2+\beta\end{bmatrix} \times$$

$$\times {}_2F_3\left(\beta+\frac{\nu}{2},\ 1+\beta-\delta+\frac{\nu}{2};\ \frac{1+\nu}{2}+\beta,\ \frac{1+\nu}{2}+\beta,\ \nu+1;\ \frac{b^2}{4}\right)$$

$[a,\ b,\ \mathrm{Re}(2\beta+\nu)>0;\ \mathrm{Re}\,\beta<5/4]$.

5. $\int\limits_0^a \frac{x}{(a^2-x^2)^{3/2}} J_0\left(\frac{bx}{\sqrt{a^2-x^2}}\right) \mathbf{K}\left(\frac{x}{a}\right) dx = \frac{1}{ab} K_0(b)$ $[a,\ b>0]$.

6. $\int\limits_0^a x(a^2-x^2)^{-5/2} J_0\left(b\frac{x\sqrt{2a^2-x^2}}{a^2-x^2}\right) \mathbf{K}\left(\frac{x}{a\sqrt{2}}\right) dx = \frac{\sqrt{\pi}}{a^3 b^{3/2}} e^{-b}$ $[a,\ b>0]$.

7. $\int\limits_0^a \frac{x\sqrt{a^2+x^2}}{(a^2-x^2)^2} \begin{Bmatrix} J_0[b(a^2+x^2)/(a^2-x^2)] \\ Y_0[b(a^2+x^2)/(a^2-x^2)] \end{Bmatrix} \mathbf{K}\left(\frac{x}{a}\right) dx =$

$= -\frac{\pi}{2^{7/2}ab}\left[\begin{Bmatrix} \cos(b/2) \\ \sin(b/2) \end{Bmatrix} Y_0\left(\frac{b}{2}\right) \pm \begin{Bmatrix} \sin(b/2) \\ \cos(b/2) \end{Bmatrix} J_0\left(\frac{b}{2}\right)\right]$ $[a,\ b>0]$.

2.16.13. Integrals containing $K_\nu(\varphi(x)) \begin{Bmatrix} \mathbf{K}(cx) \\ \mathbf{E}(cx) \end{Bmatrix}$.

1. $\int\limits_0^a \frac{x^2}{(a^2-x^2)^2} K_0\left(\frac{bx}{a^2-x^2}\right) \mathbf{K}\left(\frac{x}{a}\right) dx = \frac{\pi}{8\sqrt{ab}} S_{-1/2,\,0}\left(\frac{b}{2a}\right)$ $[a,\ \mathrm{Re}\, b>0]$.

2. $\int\limits_0^a \frac{x^2}{(a^2-x^2)^{5/2}} K_0\left(\frac{bx^2}{a^2-x^2}\right) \mathbf{K}\left(\frac{x}{a}\right) dx = \frac{\pi}{4\sqrt{2}\, a^2 b^{3/2}} S_{1/2,\,0}\left(\frac{b}{a}\right)$ $[a,\ \mathrm{Re}\, b>0]$.

3. $\int\limits_0^a \frac{x\sqrt{a^2+x^2}}{(a^2-x^2)^2} K_0\left(b\frac{a^2+x^2}{a^2-x^2}\right) \mathbf{K}\left(\frac{x}{a}\right) dx = \frac{\pi}{8\sqrt{2}ab} e^{-b/2} K_0\left(\frac{b}{2}\right)$ $[a,\ \mathrm{Re}\, b>0]$.

4. $\int\limits_0^a x(a^2-x^2)^{\beta-1} K_\nu\left(b\sqrt{a^2-x^2}\right) \begin{Bmatrix} \mathbf{K}(x/a) \\ \mathbf{E}(x/a) \end{Bmatrix} dx = X(\nu) + X(-\nu),$

$X(\nu) = \frac{\pi a^{\nu+2\beta} b^\nu}{2^{\nu+3}} \Gamma\left[\begin{matrix} -\nu,\ \beta+\nu/2,\ 1+\beta-\delta+\nu/2 \\ \beta+(\nu+1)/2,\ \beta+(\nu+3)/2-\delta \end{matrix}\right] \times$

$\times {}_2F_3\left(\beta+\frac{\nu}{2},\ 1-\delta+\beta+\frac{\nu}{2};\ \nu+1,\ \beta+\frac{\nu+1}{2},\ \beta-\delta+\frac{\nu+3}{2};\ \frac{a^2b^2}{4}\right)$

$\left[a>0;\ 2\,\mathrm{Re}\,\beta>|\mathrm{Re}\,\nu|,\ \delta=\begin{Bmatrix} 1 \\ 0 \end{Bmatrix}\right]$.

5. $\int\limits_0^a \frac{x}{(a^2-x^2)^{3/2}} K_0\left(\frac{b}{\sqrt{a^2-x^2}}\right) \mathbf{K}\left(\frac{x}{a}\right) dx = \left(\frac{\pi}{2b}\right)^{3/2} \sqrt{a}\, e^{-b/a}$ $[a,\ \mathrm{Re}\, b>0]$.

6. $\int\limits_0^a \frac{x}{(a^2-x^2)^{3/2}} K_0\left(\frac{bx}{\sqrt{a^2-x^2}}\right) \mathbf{K}\left(\frac{x}{a}\right) dx = \frac{\pi^2}{4ab}[\mathbf{H}_0(b) - Y_0(b)]$ $[a,\ \mathrm{Re}\, b>0]$.

7. $\int\limits_0^a \frac{x}{(a^2-x^2)^{3/2}} \exp\left(-\frac{bx^2}{a^2-x^2}\right) K_0\left(\frac{ba^2}{a^2-x^2}\right) \mathbf{K}\left(\frac{x}{a}\right) dx = -\frac{\pi^{3/2}}{2\sqrt{2b}\,a} \mathrm{Ei}(-2b)$

$[a,\ \mathrm{Re}\, b>0]$.

2.16.14. Integrals of $A(x)\,\mathbf{H}_0(\varphi(x))\,\mathbf{K}(cx)$.

1. $\int\limits_0^a x^{-2} \mathbf{H}_0\left(b\frac{\sqrt{a^2-x^2}}{x}\right) \mathbf{K}\left(\frac{x}{a}\right) dx = -\frac{b}{a} K_0(b)$ $[a,\ \mathrm{Re}\, b>0]$.

2. $$\int_a^{\sqrt{2a}} \frac{x}{(x^2-a^2)^{5/2}} \mathbf{H}_0\left(bx\frac{\sqrt{2a^2-x^2}}{x^2-a^2}\right) \mathbf{K}\left(\frac{x}{a\sqrt{2}}\right) dx = -\frac{\sqrt{\pi}}{4a^3b^{3/2}} e^{-b} \quad [a, \operatorname{Re} b>0].$$

2.16.15. Integrals of $A(x) S_{\mu,\nu}(\varphi(x)) \mathbf{K}(cx)$.

1. $$\int_0^a \frac{x\sqrt{a^2+x^2}}{(a^2-x^2)^2} S_{\mu,1/2}\left(b\frac{a^2+x^2}{a^2-x^2}\right) \mathbf{K}\left(\frac{x}{a}\right) dx =$$
$$= \frac{\sqrt{\pi}}{2^{\mu+9/2}ab} \Gamma\left[\begin{matrix}-\mu/2, -\mu/2\\ 1/2-\mu\end{matrix}\right] S_{\mu+1,0}(b) \quad [a, \operatorname{Re} b>0].$$

2. $$\int_0^a \frac{x}{(a^2-x^2)^{3/2}} S_{\mu,0}\left(\frac{b}{\sqrt{a^2-x^2}}\right) \mathbf{K}\left(\frac{x}{a}\right) dx = \frac{\pi}{4b} \Gamma^2\left[\begin{matrix}-\mu/2\\ (1-\mu)/2\end{matrix}\right] S_{\mu+1,0}\left(\frac{b}{a}\right)$$
$[a, \operatorname{Re} b>0]$.

3. $$\int_0^a \frac{x}{(a^2-x^2)^{5/4}} S_{-1/2,1/2}\left(\frac{b}{\sqrt{a^2-x^2}}\right) \mathbf{K}\left(\frac{x}{a}\right) dx =$$
$$= \frac{\pi^3}{16\sqrt{b}}\left[J_0^2\left(\frac{b}{2a}\right)+Y_0^2\left(\frac{b}{2a}\right)\right] \quad [a, \operatorname{Re} b>0].$$

2.16.16. Integrals of $A(x)\begin{Bmatrix}\mathbf{K}(\varphi(x))\,\mathbf{K}(cx)\\ \mathbf{E}(\varphi(x))\,\mathbf{E}(cx)\end{Bmatrix}$.

1. $$\int_0^\infty x^{1-2\rho}(2c^2x^2+1)^{-\rho}(4c^2x^2+1)\mathbf{K}^2(icx)\,dx =$$
$$= 2^{\rho-4}\sqrt{\pi}\,c^{2\rho-2}\Gamma\left[\begin{matrix}1-\rho, \rho-1/2, \rho-1/2, \rho-1/2\\ \rho, \rho\end{matrix}\right] \quad [\operatorname{Re} c>0;\ 1/2<\operatorname{Re}\rho<1].$$

2. $$\int_0^a x\begin{Bmatrix}\mathbf{K}(c\sqrt{a^2-x^2})\,\mathbf{K}(icx)\\ \mathbf{E}(c\sqrt{a^2-x^2})\,\mathbf{E}(icx)\end{Bmatrix} dx = \frac{\pi^2a^2}{16}\begin{Bmatrix}0\\1\end{Bmatrix}+\frac{\pi}{8c^2}\begin{Bmatrix}2\\1\end{Bmatrix}\int_0^{ac}\begin{Bmatrix}\mathbf{K}(x)\\ \mathbf{E}(x)\end{Bmatrix} dx \quad [a, c>0].$$

3. $$\int_0^a \frac{x}{(1+c^2x^2)(1-a^2c^2+c^2x^2)} \mathbf{E}(c\sqrt{a^2-x^2})\,\mathbf{E}(icx)\,dx = \frac{\pi}{2a^2c^4}[\mathbf{K}(a^2c^2)-\mathbf{E}(a^2c^2)]$$
$[a, c>0]$.

2.17. THE LEGENDRE FUNCTIONS OF THE FIRST KIND $P_\nu(x)$, $P_\nu^\mu(x)$

2.17.1. Integrals of $(z^m \pm x^m)^\beta P_\nu^\mu(cx)$.

1. $$\int_{-a}^a P_n^m\left(\frac{x}{a}\right) dx = \frac{2ma\,[(n/2)!]^2\,(m+n)!}{n\,[(n-m)/2]!\,[(n+m)/2]!\,(n+1)!} \quad [m=0, 2, 4, \ldots, n=2, 4, \ldots].$$

2. $$= \frac{(m+n)!\,(n+1)!\,ma\pi}{2^{2n+1}n\,[((n+1)/2)!]^2\,[(n-m)/2]!\,[(n+m)/2]!} \quad [m, n=1, 3, 5, \ldots].$$

3. $$= 0 \quad [m+n=1, 3, 5, \ldots]$$

4. $$\int_0^a x^{\alpha-1}P_\nu\left(\frac{x}{a}\right) dx = 2^{-\alpha}\sqrt{\pi}\,a^\alpha\Gamma\left[\begin{matrix}\alpha\\ (1+\alpha-\nu)/2, 1+(\alpha+\nu)/2\end{matrix}\right] \quad [a, \operatorname{Re}\alpha>0].$$

5. $$\int_0^a x^{\alpha-1}P_\nu^\mu\left(\frac{x}{a}\right) dx = 2^{2\mu-1}\sqrt{\pi}\,a^\alpha\,\Gamma\left[\begin{matrix}\alpha/2\\ (1-\mu)/2, 1+(\alpha-\mu)/2\end{matrix}\right]\times$$
$$\times {}_3F_2\left(\frac{\nu-\mu+1}{2}, -\frac{\mu+\nu}{2}, 1-\frac{\mu}{2}; 1-\mu, 1+\frac{\alpha-\mu}{2}; 1\right) \quad [a, \operatorname{Re}\alpha>0; \operatorname{Re}\mu<2].$$

6. $$\int_a^b (x-a)^{\alpha-1} P_\nu\left(\frac{x}{b}\right) dx = \Gamma(\alpha)(b^2-a^2)^{\alpha/2} P_\nu^{-\alpha}\left(\frac{a}{b}\right)$$ $[-b<a<b;\ \mathrm{Re}\,\alpha>0]$.

7. $$\int_a^b (b-x)^{\alpha-1} P_\nu\left(\frac{x}{a}\right) dx = \Gamma(\alpha)(b^2-a^2)^{\alpha/2} P_\nu^{-\alpha}\left(\frac{b}{a}\right)$$ $[0<a<b;\ \mathrm{Re}\,\alpha>0]$.

8. $$\int_{-a}^a (x+a)^{\alpha-1} P_\nu\left(\frac{x}{a}\right) dx = (2a)^\alpha \Gamma\begin{bmatrix}\alpha,\ \alpha\\ \alpha+\nu+1,\ \alpha-\nu\end{bmatrix}$$ $[a,\ \mathrm{Re}\,\alpha>0]$.

9. $$\int_{-a}^a (a-x)^{m/2} P_n^m\left(\frac{x}{a}\right) dx = 0$$ $[a>0;\ n-m=1,\ 2,\ \ldots]$.

10. $$\int_a^\infty (x+z)^{-\beta} P_\nu\left(\frac{x}{a}\right) dx = (z^2-a^2)^{(1-\beta)/2}\Gamma\begin{bmatrix}\beta+\nu,\ \beta-\nu-1\\ \beta\end{bmatrix} P_\nu^{1-\beta}\left(\frac{z}{a}\right)$$
$[a,\ \mathrm{Re}(\beta+\nu)>0;\ \mathrm{Re}(\beta-\nu)>1;\ |\arg(z+a)|<\pi]$.

11. $$\int_a^\infty (x+z)^{-3/2} P_\nu\left(\frac{x}{a}\right) dx = \sqrt{\frac{8a}{a^2-z^2}}\,\frac{\sin[(\nu+1/2)\arccos(z/a)]}{\sin(\nu+1/2)\pi}$$
$[a>0;\ -3/2<\mathrm{Re}\,\nu<1/2;\ |\arg(z+a)|<\pi]$.

12. $$\int_a^\infty (x-a)^{\alpha-1} P_\nu\left(\frac{x}{a}\right) dx = -\frac{\sin\nu\pi}{\pi}\Gamma\begin{bmatrix}\alpha,\ 1-\alpha+\nu,\ -\alpha-\nu\\ 1-\alpha\end{bmatrix}(2a)^\alpha$$
$[a>0;\ 0<\mathrm{Re}\,\alpha<-\mathrm{Re}\,\nu,\ \mathrm{Re}\,\nu+1]$.

13. $$\int_a^\infty (x-a)^{\alpha-1} P_\nu\left(\frac{x}{b}\right) dx = -\frac{\sin\nu\pi}{\pi}\Gamma[\alpha,\ 1-\alpha+\nu,\ -\alpha-\nu]\,(a^2-b^2)^{\alpha/2} P_\nu^\alpha\left(\frac{a}{b}\right)$$
$[a>b>0;\ 0<\mathrm{Re}\,\alpha<-\mathrm{Re}\,\nu,\ \mathrm{Re}\,\nu+1]$.

14. $$\int_{-a}^a (a^2-x^2)^{\alpha-1} P_\nu^\mu\left(\frac{x}{a}\right) dx =$$
$$=2^\nu\sqrt{\pi}\,a^{2\alpha-1}\Gamma\begin{bmatrix}\alpha+\mu/2,\ \alpha-\mu/2,\ (1+\mu+\nu)/2\\ \nu+1,\ (1+\mu-\nu)/2,\ \alpha+(\nu+1)/2,\ \alpha-\nu/2\end{bmatrix}$$
$[a>0;\ 2\,\mathrm{Re}\,\alpha>|\mathrm{Re}\,\mu|]$.

15. $$\int_a^\infty (x^2-a^2)^{\alpha-1} P_\nu^\mu\left(\frac{x}{a}\right) dx =$$
$$=2^{\mu-1}a^{2\alpha-1}\Gamma\begin{bmatrix}\alpha-\mu/2,\ 1-\alpha+\nu/2,\ (1-\nu)/2-\alpha\\ 1+(\nu-\mu)/2,\ (1-\mu-\nu)/2,\ 1-\alpha-\mu/2\end{bmatrix}$$
$[a,\ \mathrm{Re}(2\alpha-\mu)>0;\ \mathrm{Re}(2\alpha+\nu)<1;\ \mathrm{Re}(2\alpha-\nu)<2]$.

2.17.2. Integrals of $x^\alpha(z^2\pm x^2)^\beta P_\nu^\mu(cx)$.

1. $$\int_0^a x^{\alpha-1}(a^2-x^2)^{\beta-1} P_\nu^\mu\left(\frac{x}{a}\right) dx = 2^{\mu-1}a^{\alpha+2\beta-2}\Gamma\begin{bmatrix}\alpha/2,\ \beta-\mu/2\\ 1-\mu,\ \beta+(\alpha-\mu)/2\end{bmatrix}\times$$
$$\times\,{}_3F_2\left(\frac{1-\mu+\nu}{2},\ -\frac{\mu+\nu}{2},\ \beta-\frac{\mu}{2};\ 1-\mu,\ \beta+\frac{\alpha-\mu}{2};\ 1\right)$$
$[a,\ \mathrm{Re}\,\alpha,\ \mathrm{Re}(2\beta-\mu)>0]$.

2. $$\int_0^a x^{\nu-2\beta}(a^2-x^2)^{\beta-1}P_\nu^{\mu}\left(\frac{x}{a}\right)dx=$$
$$=-\frac{1}{\pi}\sin\frac{2\beta+\mu}{2}\pi\,\Gamma\left[\beta+\frac{\mu}{2},\ \beta-\frac{\mu}{2},\ 1-2\beta+\nu\right]a^{\nu-1}$$
[a, $\operatorname{Re}(2\beta-\mu)>0$; $\operatorname{Re}(\nu-2\beta)>-1$].

3. $$\int_0^a x^{-2\beta-\nu-1}(a^2-x^2)^{\beta-1}P_\nu^{\mu}\left(\frac{x}{a}\right)dx=\frac{2^{2\beta-1}}{a^{\nu+2}}\Gamma\left[\begin{matrix}\beta-\mu/2,\ -2\beta-\nu\\ 1-\beta-\mu/2,\ -\mu-\nu\end{matrix}\right]$$
[$a>0$; $\operatorname{Re}\mu/2<\operatorname{Re}\beta<-\operatorname{Re}\nu/2$].

4. $$\int_0^a x^{\alpha-1}(a^2-x^2)^{-\mu/2}P_\nu^{\mu}\left(\frac{x}{a}\right)dx=$$
$$=2^{\mu-\alpha}\sqrt{\pi}\,a^{\alpha-\mu}\Gamma\left[\begin{matrix}\alpha\\ (1+\alpha-\mu-\nu)/2,\ 1+(\alpha-\mu+\nu)/2\end{matrix}\right]$$
[a, $\operatorname{Re}\alpha>0$; $\operatorname{Re}\mu<1$].

5. $$\int_0^a x^{\alpha-1}(a^2-x^2)^{m/2}P_\nu^{m}\left(\frac{x}{a}\right)dx=$$
$$=\sqrt{\pi}\,(-\nu)_m(1+\nu)_m\left(\frac{a}{2}\right)^{\alpha+m}\Gamma\left[\begin{matrix}\alpha\\ (\alpha-\nu+m+1)/2,\ 1+(\alpha+\nu+m)/2\end{matrix}\right]$$
[a, $\operatorname{Re}\alpha>0$; $m=1, 2, \ldots$].

6. $$\int_{-a}^a x^l(a^2-x^2)^{m/2}P_n^{m}\left(\frac{x}{a}\right)dx=\frac{2^{n+1}l!k!\,(n+m)!\,a^{l+m+1}}{(k-n)!\,(2k+1)!\,(n-m)!}$$
[$a>0$; $l, m, n=1, 2, \ldots$; $l+m+n=2k$; $k=1, 2, \ldots$; $m\leqslant n\leqslant l+m$].

7. $=0$ [for $a>0$ and remaining values of $k, l, m, n=1, 2, \ldots$].

8. $$\int_a^\infty x^{(1\pm1)/2}(x^2-a^2)^{\beta-1}P_\nu^{\mu}\left(\frac{x}{a}\right)dx=$$
$$=2^{\mu-1}a^{2\beta-(1\mp1)/2}\Gamma\left[\begin{matrix}\beta-\mu/2,\ (1\pm\nu)/2-\beta,\ (1\mp1\mp\nu)/2-\beta\\ 1-\beta-\mu/2,\ (1-\mu\pm\nu)/2,\ (1\mp1-\mu\mp\nu)/2\end{matrix}\right]$$
[$a>0$; $\operatorname{Re}\mu/2<\operatorname{Re}\beta<(1\pm\operatorname{Re}\nu)/2,\ (1\mp1\mp\operatorname{Re}\nu)/2$].

9. $$\int_a^\infty x^{\alpha-1}(x^2-a^2)^{-\mu/2}P_\nu^{\mu}\left(\frac{x}{a}\right)dx=$$
$$=\frac{2^{\mu-\alpha-1}a^{\alpha-\mu}}{\sqrt{\pi}}\Gamma\left[\begin{matrix}(1+\mu+\nu-\alpha)/2,\ (\mu-\nu-\alpha)/2\\ 1-\alpha\end{matrix}\right]$$
[$a>0$; $\operatorname{Re}\mu<1$; $\operatorname{Re}\alpha<1+\operatorname{Re}(\mu+\nu),\ \operatorname{Re}(\mu-\nu)$].

10. $$\int_0^\infty x^{\alpha-1}|x^2-a^2|^{\mu/2}P_\nu^{\mu}\left(\frac{x}{a}\right)dx=$$
$$=\frac{a^{\alpha+\mu}}{2^{\alpha+\mu+1}\sqrt{\pi}}\Gamma\left[\begin{matrix}\alpha,\ (1-\alpha-\mu+\nu)/2,\ -(\alpha+\mu+\nu)/2\\ 1-\mu+\nu,\ -\mu-\nu\end{matrix}\right]$$
[$a>0$; $0<\operatorname{Re}\alpha<1+\operatorname{Re}(\nu-\mu),\ -\operatorname{Re}(\nu+\mu)$].

2.17.3. Integrals of $(x \pm a)^{\alpha}(b \pm x)^{\beta} P_{\nu}^{\mu}(\varphi(x))$.

1. $$\int_{-a}^{a} (x+a)^{\alpha-1}(a-x)^{\beta-1} P_{\nu}^{\mu}\left(\frac{x}{a}\right) dx = (2a)^{\alpha+\beta-1} \Gamma\left[\begin{matrix}\beta+\mu/2, \beta-\mu/2, \mu+\nu+1 \\ \alpha+\beta+\mu, \mu+1, \nu+1\end{matrix}\right] \times$$

$$\times\, {}_3F_2(\beta+\mu/2,\ \mu-\nu,\ \mu+\nu+1;\ \mu+1,\ \alpha+\beta+\mu;\ 1)$$

$[a,\ \operatorname{Re}(2\beta-\mu)>0;\ 2\operatorname{Re}\alpha>|\operatorname{Re}\mu|]$.

2. $$\int_{-a}^{a} (x+a)^{-\beta-\nu-1}(a-x)^{\beta-1} P_{\nu}^{\mu}\left(\frac{x}{a}\right) dx =$$

$$= (2a)^{-\nu-1} \Gamma\left[\begin{matrix}\beta-\mu/2,\ \mu/2-\nu-\beta,\ -\mu/2-\nu-\beta \\ 1-\beta-\mu/2,\ -\nu,\ -\mu-\nu\end{matrix}\right]$$

$[a,\ \operatorname{Re}(2\beta-\mu)>0;\ |\operatorname{Re}\mu|+2\operatorname{Re}(\beta+\nu)<0]$.

3. $$\int_{-a}^{a} (x+a)^{\alpha-1}(a-x)^{-\mu/2} P_{\nu}^{\mu}\left(\frac{x}{a}\right) dx =$$

$$= (2a)^{\alpha-\mu/2} \Gamma\left[\begin{matrix}\alpha+\mu/2,\ \alpha-\mu/2 \\ \alpha-\mu/2+\nu+1,\ \alpha-\mu/2-\nu\end{matrix}\right]$$

$[a>0;\ \operatorname{Re}\mu<1;\ 2\operatorname{Re}\alpha>|\operatorname{Re}\mu|]$.

4. $$\int_{-a}^{a} (x+a)^{\mu/2+\nu-k-1}(a-x)^{-\mu/2} P_{\nu}^{\mu}\left(\frac{x}{a}\right) dx = 0$$

$[a>0;\ \operatorname{Re}\mu<1;\ k=0,\ 1,\ 2,\ \ldots;\ k<\operatorname{Re}\nu,\ \operatorname{Re}(\mu+\nu)]$.

5. $$\int_{-a}^{a} \frac{(x+a)^{\nu}}{x-y} P_{\nu}\left(\frac{x}{a}\right) dx = -(a+y)^{\nu} Q_{\nu}\left(\frac{y}{a}\right)$$ $[\operatorname{Re}\nu>-1;\ y\notin[-a,\ a]]$.

6. $$\int_{a}^{\infty} (x-a)^{\alpha-1}(x+a)^{\pm\mu/2} P_{\nu}^{\mu}\left(\frac{x}{a}\right) dx =$$

$$= (2a)^{\alpha\pm\mu/2} \Gamma\left[\begin{matrix}\alpha-\mu/2,\ \mp\mu/2-\nu-\alpha,\ 1+\nu\mp\mu/2-\alpha \\ 1+\nu-(\mu\pm\mu)/2,\ -\nu-(\mu\pm\mu)/2,\ 1-\mu/2-\alpha\end{matrix}\right]$$

$[a>0;\ \operatorname{Re}\mu/2<\operatorname{Re}\alpha<1+\operatorname{Re}(\nu\mp\mu/2),\ -\operatorname{Re}(\nu\pm\mu/2)]$.

7. $$\int_{a}^{\infty} (x-a)^{-\mu/2}(x+a)^{\alpha-1} P_{\nu}^{\mu}\left(\frac{x}{a}\right) dx = (2a)^{\alpha-\mu/2}\, \Gamma\left[\begin{matrix}\mu/2-\nu-\alpha,\ 1+\mu/2+\nu-\alpha \\ 1-\mu/2-\alpha,\ 1+\mu/2-\alpha\end{matrix}\right]$$

$[a>0;\ \operatorname{Re}\mu<1;\ \operatorname{Re}\alpha<1+\operatorname{Re}(\mu/2+\nu),\ \operatorname{Re}(\mu/2-\nu)]$

8. $$\int_{-a}^{\infty} (x+a)^{\alpha-1}|x-a|^{\mu/2} P_{\nu}^{\mu}\left(\frac{x}{a}\right) dx =$$

$$= -\frac{(2a)^{\alpha+\mu/2}\sin\nu\pi}{\pi} \Gamma\left[\begin{matrix}\alpha+\mu/2,\ \alpha-\mu/2,\ 1-\alpha-\mu/2+\nu,\ -\alpha-\mu/2-\nu \\ 1-\mu+\nu,\ -\mu-\nu\end{matrix}\right]$$

$[a>0;\ |\operatorname{Re}\mu|/2<\operatorname{Re}\alpha<1+\operatorname{Re}(\nu-\mu/2),\ -\operatorname{Re}(\nu+\mu/2)]$.

9. $$\int_{0}^{\infty} x^{\alpha-1}|a-x|^{\nu} P_{\nu}^{\mu}\left(\frac{a+x}{|a-x|}\right) dx =$$

$$= a^{\alpha+\nu} \Gamma\left[\begin{matrix}1+\nu,\ \alpha-\mu/2,\ -\alpha-\mu/2-\nu \\ -\mu-\nu,\ 1+\alpha-\mu/2+\nu,\ 1-\alpha-\mu/2\end{matrix}\right]$$

$[a>0;\ \operatorname{Re}\mu/2<\operatorname{Re}\alpha<-\operatorname{Re}(\mu/2+\nu);\ \operatorname{Re}\nu>-1]$.

10. $\int_0^\infty x^{\alpha-1} P_\nu\left(\frac{ax^2+bx+c}{x}\right) dx=$

$$=\left(\frac{2}{\pi}\right)^{3/2}\left(\frac{c}{a}\right)^{\alpha/2}\frac{\cos\alpha\pi\sin^2\nu\pi}{\sin(\nu-\alpha)\pi\sin(\nu+\alpha)\pi}\Gamma(\alpha-\nu)Q_{1/2-\alpha}^{-\nu-1/2}(\sqrt{1+d^2})Q_\nu^\alpha(\sqrt{1+4acd^2})$$

$[(1+d^2)(1+4acd^2)=b^2d^2;\ \operatorname{Re} a,\ \operatorname{Re} c>0;\ |\operatorname{Re}\alpha|<-\operatorname{Re}\nu,\ \operatorname{Re}\nu+1]$.

11. $\int_0^a x^{\alpha-1}(a-x)^{-\mu} P_\nu^\mu\left(\frac{a+x}{2\sqrt{ax}}\right) dx=\frac{a^{\alpha-\mu}}{\sqrt{\pi}}\Gamma\left[\begin{matrix}1/2-\mu,\ \alpha-\nu/2,\ \alpha+(\nu+1)/2\\ 1+\alpha-\mu+\nu/2,\ \alpha-\mu+(1-\nu)/2\end{matrix}\right]$

$[a>0;\ \operatorname{Re}\mu<1/2;\ \operatorname{Re}\alpha>\operatorname{Re}\nu/2,\ -(1+\operatorname{Re}\nu)/2]$.

12. $\int_0^\infty x^{\alpha-1}|a-x|^\mu P_\nu^\mu\left(\frac{a+x}{2\sqrt{ax}}\right) dx=$

$$=\frac{a^{\alpha+\mu}\cos\mu\pi}{\pi^{3/2}}\Gamma\left[\begin{matrix}\mu+1/2,\ \alpha+(\nu+1)/2,\ \alpha-\nu/2,\ -\alpha-\mu-\nu/2,\ (1+\nu)/2-\mu-\alpha\\ -\mu-\nu,\ 1-\mu+\nu\end{matrix}\right]$$

$[a>0;\ \operatorname{Re}\nu/2,\ -(\operatorname{Re}\nu+1)/2<\operatorname{Re}\alpha<-\operatorname{Re}(\mu+\nu/2),\ 1/2+\operatorname{Re}(\nu/2-\mu)]$.

13. $\int_0^\infty x^{\alpha-1}(x+z)^{\nu/2} P_\nu^\mu\left(\frac{x+2z}{2\sqrt{z(x+z)}}\right) dx=\frac{z^{\alpha+\nu/2}}{\sqrt{\pi}}\Gamma\left[\begin{matrix}\alpha-\mu,\ -\alpha-\nu,\ 1/2-\alpha\\ 1-\alpha-\mu,\ -\mu-\nu\end{matrix}\right]$

$[\operatorname{Re}\mu<\operatorname{Re}\alpha<1/2,\ -\operatorname{Re}\nu;\ |\arg z|<\pi]$.

2.17.4. Integrals of $(x^m \pm a^m)^\alpha (b^2 \pm x^2)^\beta P_\nu^\mu(cx)$.

1. $\int_a^b (x-a)^{\alpha-1}(b^2-x^2)^{-\mu/2} P_\nu^\mu\left(\frac{x}{b}\right) dx=\Gamma(\alpha)(b^2-a^2)^{(\alpha-\mu)/2} P_\nu^{\mu-\alpha}\left(\frac{a}{b}\right)$

$[-b<a<b;\ \operatorname{Re}\alpha>0;\ \operatorname{Re}\mu<1]$.

2. $\int_a^b (x^2-a^2)^{-\mu/2}(b-x)^{\beta-1} P_\nu^\mu\left(\frac{x}{a}\right) dx=\Gamma(\beta)(b^2-a^2)^{(\beta-\mu)/2} P_\nu^{\mu-\beta}\left(\frac{b}{a}\right)$

$[0<a<b;\ \operatorname{Re}\beta>0;\ \operatorname{Re}\mu<1]$.

3. $\int_{-a}^a \frac{(a^2-x^2)^{\mu/2}}{x-y} P_\nu^\mu\left(\frac{x}{a}\right) dx=-2e^{-\mu\pi i}(y^2-a^2)^{\mu/2} Q_\nu^\mu\left(\frac{y}{a}\right)$

$[a>0;\ y\notin[-a,\ a];\ \operatorname{Re}\mu>-1]$.

4. $\int_{-a}^a \frac{(a^2-x^2)^{m/2}}{x-y} P_n^m\left(\frac{x}{a}\right) dx=\frac{2(-1)^{m-1}}{(y^2-a^2)^{-m/2}} Q_n^m\left(\frac{y}{a}\right)$ $[y>a>0;\ n\geqslant m\geqslant 0]$.

5. $\int_{-a}^a \frac{(a^2-x^2)^{m/2}}{(x-y)^p} P_n^m\left(\frac{x}{a}\right) dx=\frac{2(-1)^{m-1}(n+m)!}{(p-1)!(n-m)!}(y^2-a^2)^{(m-p+1)/2} Q_n^{p-m-1}\left(\frac{y}{a}\right)$

$[y>a>0;\ n\geqslant m\geqslant 0;\ p=1,\ 2,\ \ldots]$.

6. $\int_{-a}^a \frac{(a^2-x^2)^{-\mu/2}}{x-y} P_{\mu+n}^\mu\left(\frac{x}{a}\right) dx=-2e^{-\mu\pi i}(y^2-a^2)^{-\mu/2} Q_{\mu+n}^\mu\left(\frac{y}{a}\right)$

$[y\notin[-a,\ a];\ -n-1<\operatorname{Re}\mu<1;\ n=0,\ 1,\ 2,\ \ldots]$.

7. $\int_a^\infty (x-a)^{\alpha-1}(x^2-b^2)^{\mu/2}P_\nu^\mu\left(\frac{x}{b}\right)dx=$

$$=(a^2-b^2)^{(\alpha+\mu)/2}\Gamma\begin{bmatrix}\alpha,\ 1-\alpha-\mu+\nu,\ -\alpha-\mu-\nu\\ 1-\mu+\nu,\ -\mu-\nu\end{bmatrix}P_\nu^{\mu+\alpha}\left(\frac{a}{b}\right)$$

$[a>b>0;\ \operatorname{Re}\alpha>0;\ \operatorname{Re}(\alpha+\mu)-1<\operatorname{Re}\nu<-\operatorname{Re}(\alpha+\mu)]$.

8. $\int_a^\infty \frac{(x^2-a^2)^{-\mu/2}}{(x+z)^\gamma}P_\nu^\mu\left(\frac{x}{a}\right)dx=$

$$=\frac{2^{\nu+1}a^\nu}{(a+z)^{\gamma+\mu+\nu}}\Gamma\begin{bmatrix}\gamma+\mu+\nu,\ \gamma+\mu-\nu-1\\ \gamma\end{bmatrix}{}_2F_1\left(\gamma+\mu+\nu,\ \nu+1;\ \gamma+\mu;\ \frac{z-a}{z+a}\right)=$$

9. $$=(z^2-a^2)^{(1-\gamma-\mu)/2}\Gamma\begin{bmatrix}\gamma+\mu-\nu-1,\ \gamma+\mu+\nu\\ \gamma\end{bmatrix}P_\nu^{1-\gamma-\mu}\left(\frac{z}{a}\right)$$

$[a>0;\ \operatorname{Re}\mu<1;\ -\operatorname{Re}(\gamma+\mu)<\operatorname{Re}\nu<\operatorname{Re}(\gamma+\mu)-1]$.

10. $\int_{-a}^a (a^2-x^2)^{\alpha-1}(b^2-x^2)^{\mu/2}P_\nu^\mu\left(\frac{x}{b}\right)dx=$

$$=\frac{2^\mu\pi b^\mu}{a^{1-2\alpha}}\Gamma\begin{bmatrix}\alpha\\ \alpha+1/2,\ (1-\mu-\nu)/2,\ 1+(\nu-\mu)/2\end{bmatrix}{}_2F_1\left(-\frac{\mu+\nu}{2},\ \frac{1-\mu+\nu}{2};\ \alpha+\frac{1}{2};\ \frac{a^2}{b^2}\right)$$

$[-b<a<b;\ \operatorname{Re}\alpha>0]$.

11. $\int_a^\infty (x^2-a^2)^{\alpha-1}(x^2-b^2)^{\mu/2}P_\nu^\mu\left(\frac{x}{b}\right)dx=$

$$=2^{\mu-1}a^{2\alpha+\mu-\nu-2}b^{\nu+1}\Gamma\begin{bmatrix}\alpha,\ 1-\alpha+(\nu-\mu)/2,\ (1-\mu-\nu)/2-\alpha\\ 1+(\nu-\mu)/2,\ (1-\mu-\nu)/2,\ 1-\alpha-\mu\end{bmatrix}\times$$

$$\times{}_2F_1\left(\frac{1+\nu-\mu}{2},\ 1-\alpha+\frac{\nu-\mu}{2};\ 1-\alpha-\mu;\ 1-\frac{b^2}{a^2}\right)$$

$[0<b<a;\ \operatorname{Re}\alpha>0;\ \operatorname{Re}(2\alpha+\mu)-2<\operatorname{Re}\nu<-\operatorname{Re}(2\alpha+\mu)+1]$.

2.17.5. Integrals of $(x\pm a)^\alpha(b\pm x)^\beta(d\pm x)^\gamma P_\nu^\mu(cx)$.

1. $\int_a^b (x-a)^{\alpha-1}(b-x)^{\beta-1}(x+a)^{-\mu/2}P_\nu^\mu\left(\frac{x}{a}\right)dx=$

$$=(b-a)^{\alpha+\beta-\mu/2-1}\Gamma\begin{bmatrix}\alpha-\mu/2,\ \beta\\ \alpha+\beta-\mu/2,\ 1-\mu\end{bmatrix}{}_3F_2\left(-\nu,\ 1+\nu,\ \alpha-\frac{\mu}{2};\ 1-\mu,\ \alpha+\beta-\frac{\mu}{2};\ \frac{a-b}{2a}\right)$$

$[0<a<b;\ \operatorname{Re}\beta,\ \operatorname{Re}(2\alpha-\mu)>0]$.

2. $\int_a^b (x-a)^{(\mu-1)/2}(b-x)^{\mu-1/2}(x+a)^{-\mu/2}P_\nu^\mu\left(\frac{x}{a}\right)dx=$

$$=\sqrt{\pi}(b-a)^\mu\Gamma\left(\mu+\frac{1}{2}\right)P_\nu^\mu\left(\sqrt{\frac{a+b}{2a}}\right)P_\nu^{-\mu}\left(\sqrt{\frac{a+b}{2a}}\right)\quad [0<a<b;\ \operatorname{Re}\mu>-1/2].$$

3. $\int_a^b (x-a)^{(\mu-1)/2}(b-x)^{\mu-3/2}(x+a)^{-\mu/2}P_\nu^\mu\left(\frac{x}{a}\right)dx=$

$$=\frac{\sqrt{\pi}(b-a)^{\mu-3/2}}{2\sqrt{a+b}}\Gamma\left(\mu-\frac{1}{2}\right)\left[P_\nu^{1-\mu}\left(\sqrt{\frac{a+b}{2a}}\right)P_\nu^\mu\left(\sqrt{\frac{a+b}{2a}}\right)+\right.$$

$$\left.+(\mu+\nu)(1-\mu+\nu)P_\nu^{-\mu}\left(\sqrt{\frac{a+b}{2a}}\right)P_\nu^\mu\left(\sqrt{\frac{a+b}{2a}}\right)\right]\quad [0<a<b;\ \operatorname{Re}\mu>1/2].$$

4. $$\int_a^b (x-a)^{-(\mu+1)/2}(b-x)^{-\mu-1/2}(x+a)^{\mu/2}P_\nu^\mu\left(\frac{x}{a}\right)dx=\sqrt{\pi}(b-a)^{-\mu}\Gamma\left(\frac{1}{2}-\mu\right)\times$$
$$\times\left[P_\nu^\mu\left(\sqrt{\frac{a+b}{2a}}\right)\right]^2 \qquad [0<a<b;\ \operatorname{Re}\mu<1/2].$$

5. $$\int_a^b (x-a)^{-(\mu+1)/2}(b-x)^{-\mu-3/2}(x+a)^{\mu/2}P_\nu^\mu\left(\frac{x}{a}\right)dx=$$
$$=\sqrt{\frac{\pi}{2a}}(b-a)^{-1/2-\mu}\Gamma\left(-\frac{1}{2}-\mu\right)P_\nu^{\mu+1}\left(\sqrt{\frac{a+b}{2a}}\right)P_\nu^\mu\left(\sqrt{\frac{a+b}{2a}}\right)$$
$$[0<a<b;\ \operatorname{Re}\mu<-1/2].$$

6. $$\int_a^b (x-a)^{\alpha-1}(b-x)^{\beta-1}(x+b)^{-\mu/2}P_\nu^\mu\left(\frac{x}{b}\right)dx=(b-a)^{\alpha+\beta-\mu/2-1}\times$$
$$\times\Gamma\left[\begin{matrix}\alpha,\ \beta-\mu/2\\ 1-\mu,\ \alpha+\beta-\mu/2\end{matrix}\right]{}_3F_2\left(-\nu,\ 1+\nu,\ \beta-\frac{\mu}{2};\ 1-\mu,\ \alpha+\beta-\frac{\mu}{2};\ \frac{b-a}{2b}\right)$$
$$[-b<a<b;\ \operatorname{Re}\alpha,\ \operatorname{Re}\ 2\beta-\mu)>0].$$

7. $$\int_{-a}^a \frac{(x+a)^{\mu/2+\nu}(a-x)^{-\mu/2}}{x-y}P_\nu^\mu\left(\frac{x}{a}\right)dx=-2(y+a)^{\mu/2+\nu}(y-a)^{-\mu/2}e^{-\mu\pi i}Q_\nu^\mu\left(\frac{y}{a}\right)$$
$$[y\notin(-\infty,\ a];\ \operatorname{Re}\nu>-1;\ -\operatorname{Re}\nu-1<\operatorname{Re}\mu<1].$$

8. $$\int_{-a}^a \frac{x^l(a^2-x^2)^{m/2}}{x-y}P_n^m\left(\frac{x}{a}\right)dx=(-1)^{m-1}2^m y^l(y^2-a^2)^{m/2}Q_n^m\left(\frac{y}{a}\right)$$
$$[y\notin[-a,\ a];\ m\leqslant n;\ l=0,\ 1,\ \dots,\ n-m].$$

9. $$\int_{-a}^a (a-x)^{-\mu/2}(x+a)^{(\mu-1)/2}(x+z)^{\mu-1\pm1/2}P_\nu^\mu\left(\frac{x}{a}\right)dx=$$
$$=\pm 2^{(1\mp1)/2}\frac{e^{-2\mu\pi i}(z-a)^{\mu-(1\mp1)/4}}{\sqrt{\pi}(z+a)^{(1\mp1)/4}}\Gamma\left[\begin{matrix}\mu\pm1/2\\ \mu\mp\nu,\ \mu\pm\nu\pm1\end{matrix}\right]\times$$
$$\times\left[Q_{-\nu-1}^{\mu-(1\mp1)/2}\left(\sqrt{\frac{a+z}{2a}}\right)Q_\nu^\mu\left(\sqrt{\frac{a+z}{2a}}\right)+\begin{Bmatrix}0\\1\end{Bmatrix}Q_\nu^{\mu-1}\left(\sqrt{\frac{a+z}{2a}}\right)\times\right.$$
$$\left.\times Q_{-\nu-1}^\mu\left(\sqrt{\frac{a+z}{2a}}\right)\right] \qquad [a>0;\ z\notin[-a,\ a];\ -1/2<\operatorname{Re}\mu<1].$$

10. $$\int_a^\infty (x-a)^{\alpha-1}(x+a)^{\mu/2}(x+z)^{-\beta}P_\nu^\mu\left(\frac{x}{a}\right)dx=$$
$$=(2a)^{\alpha-\beta+\mu/2}\Gamma\left[\begin{matrix}\alpha-\beta-\mu/2,\ \beta-\alpha-\mu/2-\nu,\ 1-\alpha+\beta-\mu/2+\nu\\ 1-\mu+\nu,\ -\mu-\nu,\ 1-\alpha+\beta-\mu/2\end{matrix}\right]\times$$
$$\times{}_3F_2\left(\beta,\ \beta-\alpha-\frac{\mu}{2}-\nu,\ 1-\alpha+\beta-\frac{\mu}{2}+\nu;\ 1-\alpha+\beta+\frac{\mu}{2},\right.$$
$$\left.1-\alpha+\beta-\frac{\mu}{2};\ \frac{z+a}{2a}\right)+\frac{(2a)^\mu}{(z+a)^{\beta-\alpha+\mu/2}}\Gamma\left[\begin{matrix}\alpha-\mu/2,\ \beta-\alpha+\mu/2\\ \beta,\ 1-\mu\end{matrix}\right]\times$$

$$\times {}_3F_2\left(\alpha-\frac{\mu}{2},\ -\mu-\nu,\ 1-\mu+\nu;\ 1+\alpha-\beta-\frac{\mu}{2},\ 1-\mu;\ \frac{z+a}{2a}\right)$$

$[a,\ \mathrm{Re}\,(2\alpha-\mu)>0;\ \mathrm{Re}\,(\alpha-\beta+\mu/2)-1<\mathrm{Re}\,\nu<\mathrm{Re}\,(\beta-\alpha-\mu/2)]$.

11. $$\int_a^\infty (x-a)^{\alpha-1}(x+a)^{-\mu/2}(x+z)^{-\beta}P_\nu^\mu\left(\frac{x}{a}\right)dx=(z+a)^{\alpha-\beta-\mu/2}\times$$

$$\times\Gamma\begin{bmatrix}\alpha-\mu/2,\ \beta-\alpha+\mu/2\\ \beta,\ 1-\mu\end{bmatrix}{}_3F_2\left(\alpha-\frac{\mu}{2},\ -\nu,\ 1+\nu;\ 1+\alpha-\beta-\frac{\mu}{2},\ 1-\mu;\ \frac{z+a}{2a}\right)-$$

$$-\frac{\sin\nu\pi}{\pi}(2a)^{\alpha-\beta-\mu/2}\Gamma\begin{bmatrix}\alpha-\beta-\mu/2,\ \beta-\alpha+\mu/2-\nu,\ 1-\alpha+\beta+\mu/2+\nu\\ 1-\alpha+\beta-\mu/2\end{bmatrix}\times$$

$$\times {}_3F_2\left(\beta,\ \beta-\alpha+\frac{\mu}{2}-\nu,\ 1-\alpha+\beta+\frac{\mu}{2}+\nu;\ 1-\alpha+\beta-\frac{\mu}{2},\ 1-\alpha+\beta+\frac{\mu}{2};\ \frac{z+a}{2a}\right)$$

$[a,\ \mathrm{Re}\,(2\alpha-\mu)>0;\ \mathrm{Re}\,(\alpha-\beta-\mu/2)-1<\mathrm{Re}\,\nu<\mathrm{Re}\,(\beta-\alpha+\mu/2)]$.

12. $$\int_a^\infty (x-a)^{-\mu/2}(x+a)^{(\mu-1)/2}(x+z)^{\mu-1\pm1/2}P_\nu^\mu\left(\frac{x}{a}\right)dx=$$

$$=\sqrt{\pi}\,\Gamma\begin{bmatrix}1-\mu\pm\nu,\ 1\mp1-\mu\mp\nu\\ 1\mp1/2-\mu\end{bmatrix}\frac{(z-a)^{\mu-(1\mp1)/4}}{(z+a)^{(1\mp1)/4}}P_\nu^{\mu-(1\mp1)/2}\left(\sqrt{\frac{z+a}{2a}}\right)P_\nu^\mu\left(\sqrt{\frac{z+a}{2a}}\right)$$

$[a>0;\ \mathrm{Re}\,\mu<1;\ \mathrm{Re}\,(\mu\mp\nu)<1;\ \mathrm{Re}\,(\mu\mp\nu)<1\mp1;\ |\arg(z+a)|<\pi]$.

13. $$\int_a^\infty x^{\alpha-1}(x-a)^{\mu/2}(x+a)^{-\mu/2}P_\nu^\mu\left(\frac{x}{a}\right)dx=$$

$$=-\frac{\sin\nu\pi}{\pi}(2a)^\alpha\Gamma\begin{bmatrix}\alpha,\ 1-\alpha+\nu,\ -\alpha-\nu\\ 1-\alpha-\mu\end{bmatrix}{}_2F_1\left(1-\alpha+\nu,\ -\alpha-\nu;\ 1-\alpha-\mu;\ \frac{1}{2}\right)$$

$[a>0;\ 0<\mathrm{Re}\,\alpha<\mathrm{Re}\,\nu+1,\ -\mathrm{Re}\,\nu]$.

14. $$\int_a^\infty (x-a)^{-\mu-1/2}(x-b)^{-\mu-1/2}(x^2-c^2)^{\mu/2}P_\nu^\mu\left(\frac{x}{c}\right)dx=$$

$$=\frac{4c\,(a-b)^{-\mu}}{\pi}\sqrt{\frac{4c^2+(a-b)^2}{16abc^2-(a-b)^4}}\,\Gamma\begin{bmatrix}1/2-\mu\\ 1-\mu+\nu\end{bmatrix}\times$$

$$\times\frac{\sin\mu\pi\cos\nu\pi}{\sin(\mu+\nu)\pi\sin(\nu-\mu)\pi}Q_{\mu-1/2}^{-\nu-1/2}\left(\frac{2c\,(a+b)}{\sqrt{16abc^2-(a-b)^4}}\right)Q_\nu^{-\mu}\left(\frac{\sqrt{4c^2+(a-b)^2}}{2c}\right)$$

$[a+b>2c;\ a>b;\ \mathrm{Re}\,\mu-1<\mathrm{Re}\,\nu<-\mathrm{Re}\,\mu]$.

2.17.6. Integrals of $A(x)\,[P_\nu^\mu(cx)+P_\nu^\mu(-cx)]$.

1. $$\int_0^a (a^2-x^2)^{\beta-1}\left[P_\nu^\mu\left(\frac{x}{a}\right)+P_\nu^\mu\left(-\frac{x}{a}\right)\right]dx=$$

$$=2^\mu\pi a^{2\beta-1}\Gamma\begin{bmatrix}\beta+\mu/2,\ \beta-\mu/2\\ 1+(\nu-\mu)/2,\ (1-\nu-\mu)/2,\ \beta+(\nu+1)/2,\ \beta-\nu/2\end{bmatrix}$$ $[a>0;\ 2\,\mathrm{Re}\,\beta>|\mathrm{Re}\,\mu|]$.

2. $$\int_0^a x^{(1\pm1)/2}(a^2-x^2)^{\beta-1}(b^2-x^2)^{\mu/2}\left[P_\nu^\mu\left(\frac{x}{b}\right)+P_\nu^\mu\left(-\frac{x}{b}\right)\right]dx=$$

$$=\mp\frac{2^{\mu+1}\pi a^{2\beta\pm1}b^{\mu-(1\pm1)/2}\Gamma(\beta)}{\Gamma[(1-\mu\pm\nu)/2,\ (1\mp1-\mu\mp\nu)/2,\ \beta+1\pm1/2]}\times$$

$$\times {}_2F_1\left(\frac{1-\mu\mp\nu}{2},\ \frac{1\pm1\pm\nu-\mu}{2};\ \beta+1\pm\frac{1}{2};\ \frac{a^2}{b^2}\right)$$ $[0<a<|b|;\ \mathrm{Re}\,\beta>0]$.

2.17.7. Integrals of $A(x)\,e^{\varphi(x)}\,P_\nu^\mu(\chi(x))$.

1. $\int_a^\infty e^{-px} P_\nu\left(\frac{x}{a}\right) dx = \sqrt{\frac{2a}{\pi p}}\, K_{\nu+1/2}(ap)$ $\quad [a,\ \operatorname{Re} p > 0].$

2. $\int_a^\infty (x-a)^{\alpha-1}(x+a)^{-\mu/2} e^{-px} P_\nu^\mu\left(\frac{x}{a}\right) dx =$

$= -\frac{\sin \nu\pi}{\pi} p^{\mu/2-\alpha} e^{-ap} G_{23}^{31}\left(2ap \,\middle|\, \begin{matrix} 1,\ 1-\mu \\ \alpha-\mu/2,\ 1+\nu,\ -\nu \end{matrix}\right)$ $\quad [a,\ \operatorname{Re} p,\ \operatorname{Re}(2\alpha-\mu) > 0].$

3. $\int_a^\infty (x-a)^{\alpha-1}(x+a)^{\mu/2} e^{-px} P_\nu^\mu\left(\frac{x}{a}\right) dx =$

$= \frac{p^{-\alpha-\mu/2} e^{-ap}}{\Gamma(-\mu-\nu)\,\Gamma(1-\mu+\nu)} G_{23}^{31}\left(2ap \,\middle|\, \begin{matrix} \mu+1,\ 1 \\ \alpha+\mu/2,\ -\nu,\ 1+\nu \end{matrix}\right)$ $\quad [a,\ \operatorname{Re} p,\ \operatorname{Re}(2\alpha-\mu) > 0].$

4. $\int_a^\infty (x-a)^{-\mu/2}(x+a)^{\mu/2} e^{-px} P_\nu^\mu\left(\frac{x}{a}\right) dx = \frac{1}{p} W_{\mu,\ \nu+1/2}(2ap)$ $\quad [a,\ \operatorname{Re} p > 0;\ \operatorname{Re}\mu < 1].$

5. $\int_a^\infty (x^2-a^2)^{-\mu/2} e^{-px} P_\nu^\mu\left(\frac{x}{a}\right) dx = \sqrt{\frac{2a}{\pi}}\, p^{\mu-1/2} K_{\nu+1/2}(ap)$ $\quad [a,\ \operatorname{Re} p > 0;\ \operatorname{Re}\mu < 1].$

6. $\int_a^\infty x^{(1\mp 1)/2}(x^2-a^2)^{-\mu/2} e^{-px^2} P_\nu^\mu\left(\frac{x}{a}\right) dx =$

$= 2^{\mu-1} a^{-1/2} p^{(2\mu\mp 1)/4-1} e^{-a^2p/2} W_{(2\mu\pm 1)/4,\ (2\nu+1)/4}(a^2 p)$ $\quad [a,\ \operatorname{Re} p > 0;\ \operatorname{Re}\mu < 1].$

7. $\int_0^a x^{\mu+\nu-2}(a^2-x^2)^{-\mu/2} e^{-px^{-2}} P_\nu^\mu\left(\frac{x}{a}\right) dx =$

$= (2p)^{(\mu+\nu-1)/2} a^{-\mu} e^{-p/(2a^2)} D_{\mu-\nu-1}\left(\frac{\sqrt{2p}}{a}\right)$ $\quad [a,\ \operatorname{Re} p > 0;\ \operatorname{Re}\mu < 1].$

8. $\int_a^\infty \frac{e^{-p\sqrt{x+z}}}{\sqrt{x+z}} P_\nu\left(\frac{x}{a}\right) dx = \frac{\sqrt{8a}}{\pi} K_{\nu+1/2}(z_+)\, K_{\nu+1/2}(z_-)$

$[a,\ \operatorname{Re} p > 0;\ 2z_\pm = p(\sqrt{z+a} \pm \sqrt{z-a})].$

9. $\int_{-a}^\infty \frac{e^{-p\sqrt{x+a}}}{\sqrt{x+a}} P_\nu\left(\frac{x}{a}\right) dx = \frac{\sqrt{8a}}{\pi} K_{\nu+1/2}\left(p\sqrt{\frac{a}{2}}\right)$ $\quad [a,\ \operatorname{Re} p > 0].$

10. $\int_a^\infty \frac{e^{-p\sqrt{x-a}}}{\sqrt{x-a}} P_\nu\left(\frac{x}{a}\right) dx = \pi\sqrt{\frac{a}{2}}\left[J_{\nu+1/2}^2\left(p\sqrt{\frac{a}{2}}\right) + Y_{\nu+1/2}^2\left(p\sqrt{\frac{a}{2}}\right)\right]$

$[a,\ \operatorname{Re} p > 0].$

11. $\int_0^a \frac{1}{x-a} e^{-p/\sqrt{a-x}} P_\nu\left(-\frac{x}{a}\right) dx = -\frac{2\sqrt{a}}{p} W_{\nu+1/2,\ 0}\left(\frac{p}{\sqrt{a}}\right) W_{-\nu-1/2,\ 0}\left(\frac{p}{\sqrt{a}}\right)$

$[a,\ \operatorname{Re} p > 0].$

12. $\int_{-a}^{a} \frac{1}{x+a} e^{-p/\sqrt{x+a}} P_\nu\left(\frac{x}{a}\right) dx = \frac{\sqrt{8a}}{p} W_{\nu+1/2,\,0}\left(\frac{p}{\sqrt{2a}}\right) W_{-\nu-1/2,\,0}\left(\frac{p}{\sqrt{2a}}\right)$

[a, $\operatorname{Re} p > 0$].

13. $\int_{-a}^{a} (x+a)^{\mu/2+\nu-1} (a-x)^{-\mu/2} e^{-p(a-x)/(a+x)} P_\nu^\mu\left(\frac{x}{a}\right) dx =$

$= (2a)^\nu p^{(\mu-1)/2+\nu} e^{p/2} W_{(\mu-1)/2-\nu,\,\mu/2}(p)$ [a, $\operatorname{Re} p > 0$].

14. $\int_0^a x^{\alpha-1} (a^2-x^2)^{-(\alpha+\nu)/2-1} e^{-px/\sqrt{a^2-x^2}} P_\nu^\mu\left(\frac{x}{a}\right) dx =$

$= \frac{(2a)^{-\nu-2}}{\pi\Gamma(-\mu-\nu)} G_{24}^{32}\left(\frac{p^2}{4}\middle| \begin{matrix} 1-\alpha/2, (1-\alpha)/2 \\ 0, 1/2, -(\alpha+\mu+\nu)/2, (\mu-\nu-\alpha)/2 \end{matrix}\right)$

[a, $\operatorname{Re} p$, $\operatorname{Re} \alpha > 0$].

15. $\int_0^\infty e^{-px} P_\nu(ax^2+2bx+1) dx = \sqrt{\frac{2}{a}}\frac{1}{\pi} e^{bp/a} K_{\nu+1/2}(z_+) K_{\nu+1/2}(z_-)$

[$\operatorname{Re} p > 0$; $2az_\pm = p(b \pm \sqrt{b^2-2a})$].

16. $\int_0^\infty x^{-\mu/2} (x+a)^{\nu/2} (x+b)^{-(\nu+1)/2} (x+a+b)^{\mu/2} e^{-px} P_\nu^\mu\left(\sqrt{\frac{ab}{(x+a)(x+b)}}\right) dx =$

$= \sqrt{\frac{2}{p}} e^{(a+b)p/2} D_{\mu+\nu}(\sqrt{2ap}) D_{\mu-\nu-1}(\sqrt{2bp})$

[a, b, $\operatorname{Re} p > 0$; $\operatorname{Re} \mu < 1$].

2.17.8. Integrals containing the hyperbolic functions and $P_\nu^\mu(\varphi(x))$.

1. $\int_0^\infty \operatorname{ch} \sigma x \, P_\nu(a+b \operatorname{ch} x) dx =$

$= \sqrt{\frac{2d}{\pi^3}} \frac{\sin^2 \nu\pi \cos \sigma\pi}{\sin(\nu+\sigma)\pi \sin(\nu-\sigma)\pi} \Gamma(-\nu-\sigma) Q_{\sigma+1/2}^{-\nu-1/2}(\sqrt{1+d^2}) Q_\nu^{-\sigma}(\sqrt{1+b^2d^2})$

[$(1+d^2)(1+b^2d^2) = a^2d^2$; $\operatorname{Re} b > 0$; $|\operatorname{Re} \sigma| - 1 < \operatorname{Re} \nu < -|\operatorname{Re} \sigma|$].

2. $\int_0^\infty e^{-px} \operatorname{sh}^{-\mu} x P_\nu^\mu(\operatorname{ch} x) dx = \frac{2^{\mu-1}}{\sqrt{\pi}} \Gamma\begin{bmatrix} 1/2-\mu, (p+\mu-\nu)/2, (p+\mu+\nu+1)/2 \\ 1+(p-\mu+\nu)/2, (1+p-\mu-\nu)/2 \end{bmatrix}$

[$\operatorname{Re} \mu < 1/2$; $\operatorname{Re}(p+\mu-\nu) > 0$; $\operatorname{Re}(p+\mu+\nu) > -1$].

3. $\int_0^\infty e^{-px} \operatorname{sh}^{-p-\nu-2} x P_\nu^\mu(\operatorname{ch} x) dx = \frac{2^{p+\nu+1}}{\sqrt{\pi}} \Gamma\begin{bmatrix} p+1, p+\nu+3/2, -p-\mu-\nu-1 \\ -\mu-\nu, p-\mu+\nu+2 \end{bmatrix}$

[-1, $-\operatorname{Re} \nu - 3/2 < \operatorname{Re} p < -\operatorname{Re}(\mu+\nu) - 1$].

4. $\int_0^\infty e^{-px} \operatorname{sh}^{1-\mu} x P_\nu^\mu(\operatorname{ch} x) dx = \frac{2^{\mu-2} p}{\sqrt{\pi}} \Gamma\begin{bmatrix} 3/2-\mu, (p+\mu-\nu-1)/2, (p+\mu+\nu)/2 \\ (3+p-\mu+\nu)/2, (p-\mu-\nu)/2+1 \end{bmatrix}$

[$\operatorname{Re} \mu < 3/2$; $\operatorname{Re}(p+\mu-\nu) > 1$; $\operatorname{Re}(p+\mu+\nu) > 0$].

2.17.9. Integrals containing the trigonometric functions and $P_\nu^\mu(\varphi(x))$.

1. $$\int_0^a \frac{x^{\alpha-1}}{(a^2-x^2)^{\mu/2}} \begin{Bmatrix} \sin bx \\ \cos bx \end{Bmatrix} P_\nu^\mu\left(\frac{x}{a}\right) dx = \sqrt{\pi}\left(\frac{a}{2}\right)^{\alpha-\mu+\nu} b^\delta \Gamma\left[\begin{matrix}\alpha+\delta \\ (1+\alpha+\delta-\mu-\nu)/2,\ 1+(\alpha+\delta-\mu+\nu)/2\end{matrix}\right] \times$$
$$\times {}_2F_3\left(\frac{\alpha+1}{2},\ \frac{\alpha}{2}+\delta;\ \delta+\frac{1}{2},\ \frac{1+\alpha+\delta-\mu-\nu}{2},\ 1+\frac{\alpha+\delta-\mu+\nu}{2};\ -\frac{a^2b^2}{4}\right)$$
$[a>0;\ \operatorname{Re}\alpha > -\delta;\ \operatorname{Re}\mu<1;\ \delta=(1\pm1)/2]$.

2. $$\int_a^\infty (x^2-a^2)^{\mu/2} \sin bx P_\nu^\mu\left(\frac{x}{a}\right) dx = \frac{2^\mu\sqrt{\pi a}}{b^{\mu+1/2}\,\Gamma((1-\mu-\nu)/2)\,\Gamma(1+(\nu-\mu)/2)} S_{\mu+1/2,\ \nu+1/2}(ab)$$
$[a,\ b>0;\ \operatorname{Re}\mu<3/2;\ \operatorname{Re}(\mu+\nu)<1]$.

3. $$\int_a^\infty \frac{1}{\sqrt{x+a}} \begin{Bmatrix} \sin(b\sqrt{x+a}) \\ \cos(b\sqrt{x+a}) \end{Bmatrix} P_\nu\left(\frac{x}{a}\right) dx = -\pi\sqrt{\frac{a}{2}} \begin{Bmatrix} \sec \nu\pi \\ 1 \end{Bmatrix} \left[J_{\nu+1/2}\left(b\sqrt{\frac{a}{2}}\right) J_{\pm\nu\pm1/2}\left(b\sqrt{\frac{a}{2}}\right) - \begin{Bmatrix} J^2_{-\nu-1/2}(b\sqrt{a/2}) \\ Y_{\nu+1/2}(b\sqrt{a/2})\, Y_{-\nu-1/2}(b\sqrt{a/2}) \end{Bmatrix}\right]$$
$[a,\ b>0;\ -1<\operatorname{Re}\nu<0]$.

4. $$\int_a^\infty \frac{1}{\sqrt{x^2-a^2}} \sin(b\sqrt{x-a})\, P_\nu^{-1}\left(\frac{x}{a}\right) dx = \frac{\sqrt{2a}}{\pi} b \sin\nu\pi K^2_{\nu+1/2}\left(b\sqrt{\frac{a}{2}}\right)$$
$[a,\ b>0;\ -3/2<\operatorname{Re}\nu<1/2]$.

5. $$\int_a^\infty \frac{1}{\sqrt{x-a}} \cos(b\sqrt{x-a})\, P_\nu\left(\frac{x}{a}\right) dx = -\frac{\sqrt{8a}}{\pi} \sin\nu\pi\, K^2_{\nu+1/2}\left(b\sqrt{\frac{a}{2}}\right)$$
$[a,\ b>0;\ -1<\operatorname{Re}\nu<0]$.

6. $$\int_a^\infty \frac{1}{x+a} \cos\frac{b}{\sqrt{x+a}}\, P_\nu\left(\frac{x}{a}\right) dx = -\frac{\pi}{\sin\nu\pi}\, {}_1F_1\left(\nu+1;\ 1;\ ib\sqrt{\frac{a}{2}}\right) {}_1F_1\left(\nu+1;\ 1;\ -ib\sqrt{\frac{a}{2}}\right)$$
$[a,\ b>0;\ -1<\operatorname{Re}\nu<0]$.

7. $$\int_0^\infty \cos bx\, P_\nu(\operatorname{ch} x)\, dx = -\frac{\sin\nu\pi}{4\pi^2} \Gamma\left[\frac{1+\nu+ib}{2},\ \frac{1+\nu-ib}{2},\ -\frac{\nu+ib}{2},\ -\frac{\nu-ib}{2}\right]$$
$[b>0;\ -1<\operatorname{Re}\nu<0]$.

2.17.10. Integrals containing $\operatorname{erf}(\varphi(x))\, P_\nu^\mu(cx)$.

1. $$\int_0^a x^{\alpha-1}(a^2-x^2)^{-\mu/2} \operatorname{erf}(bx)\, P_\nu^\mu\left(\frac{x}{a}\right) dx =$$

$$=2^{\mu-\alpha}a^{\alpha-\mu+1}b\Gamma\begin{bmatrix}\alpha+1\\1+(\alpha-\mu-\nu)/2,\ (3+\alpha-\mu+\nu)/2\end{bmatrix}\times$$
$$\times{}_3F_3\left(\frac{1}{2},\ \frac{\alpha+1}{2},\ \frac{\alpha}{2}+1;\ \frac{3}{2},\ 1+\frac{\alpha-\mu-\nu}{2},\ \frac{3+\alpha-\mu+\nu}{2};\ -a^2b^2\right)$$
$[a>0;\ \operatorname{Re}\alpha>-1;\ \operatorname{Re}\mu<1].$

2. $$\int_a^\infty x^{\alpha-1}(x^2-a^2)^{-\mu/2}\operatorname{erf}\left(\frac{b}{x}\right)P_\nu^\mu\left(\frac{x}{a}\right)dx=$$
$$=\frac{2^{\mu-\alpha+1}a^{\alpha-\mu-1}b}{\pi}\Gamma\begin{bmatrix}1+(\mu+\nu-\alpha)/2,\ (\mu-\nu-\alpha+1)/2\\2-\alpha\end{bmatrix}\times$$
$$\times{}_3F_3\left(\frac{1}{2},\ 1+\frac{\mu+\nu-\alpha}{2},\ \frac{\mu-\nu-\alpha+1}{2};\ \frac{3}{2},\ 1-\frac{\alpha}{2},\ \frac{3-\alpha}{2};\ -\frac{b^2}{a^2}\right)$$
$[a>0;\ \operatorname{Re}\mu<1;\ \operatorname{Re}(\alpha-\mu)-2<\operatorname{Re}\nu<\operatorname{Re}(\mu-\alpha)+1].$

3. $$\int_0^a x^{\alpha-1}(a^2-x^2)^{-\mu/2}e^{b^2x^2}\operatorname{erfc}(bx)P_\nu^\mu\left(\frac{x}{a}\right)dx=$$
$$=\sqrt{\pi}\left(\frac{a}{2}\right)^{\alpha-\mu}\Gamma\begin{bmatrix}\alpha\\(1+\alpha-\mu-\nu)/2,\ 1+(\alpha-\mu+\nu)/2\end{bmatrix}\times$$
$$\times{}_2F_2\left(\frac{\alpha}{2},\ \frac{\alpha+1}{2};\ \frac{1+\alpha-\mu-\nu}{2},\ 1+\frac{\alpha-\mu+\nu}{2};\ a^2b^2\right)-$$
$$-\frac{a^{\alpha-\mu+1}b}{2^{\alpha-\mu}}\Gamma\begin{bmatrix}\alpha+1\\1+(\alpha-\mu-\nu)/2,\ (3+\alpha-\mu+\nu)/2\end{bmatrix}\times$$
$$\times{}_3F_3\left(1,\ \frac{\alpha+1}{2},\ \frac{\alpha}{2}+1;\ \frac{3}{2},\ 1+\frac{\alpha-\mu-\nu}{2},\ \frac{3+\alpha-\mu+\nu}{2};\ a^2b^2\right)$$
$[a,\ \operatorname{Re}\alpha>0;\ \operatorname{Re}\mu<1;\ |\arg b|<3\pi/4].$

4. $$\int_a^\infty (x^2-a^2)^{-\mu/2}e^{b^2x^2}\operatorname{erfc}(bx)P_\nu^\mu\left(\frac{x}{a}\right)dx=$$
$$=\frac{2^{\mu-1}}{\pi\sqrt{a}}b^{\mu-3/2}e^{a^2b^2/2}\Gamma\left(\frac{\mu+\nu+1}{2}\right)\Gamma\left(\frac{\mu-\nu}{2}\right)\times$$
$$\times W_{(1-2\mu)/4,\ (1+2\nu)/4}(a^2b^2)$$
$[a>0;\ |\arg b|<3\pi/4;\ -1-\operatorname{Re}\mu<\operatorname{Re}\nu<\operatorname{Re}\mu<1].$

2.17.11. Integrals containing $e^{b^2x^2/4}D_\lambda(bx)P_\nu^\mu(cx)$.

1. $$\int_a^\infty e^{b^2x^2/4}D_\lambda(bx)P_\nu\left(\frac{x}{a}\right)dx=$$
$$=\frac{2^{-(2\lambda+5)/4}}{\sqrt{\pi ab^3}}\Gamma\begin{bmatrix}(\nu-\lambda)/2,\ -(\nu+\lambda+1)/2\\-\lambda\end{bmatrix}e^{a^2b^2/4}W_{(2\lambda+3)/4,\ (2\nu+1)/4}\left(\frac{a^2b^2}{2}\right)$$
$[a>0;\ \operatorname{Re}\lambda<\operatorname{Re}\nu<-\operatorname{Re}\lambda-1;\ |\arg b|<3\pi/4].$

2. $$\int_a^\infty (x^2-a^2)^{-\mu/2}e^{b^2x^2/4}D_\lambda(bx)P_\nu^\mu\left(\frac{x}{a}\right)dx=$$
$$=\frac{2^{(2\mu-2\lambda-5)/4}b^{\mu-3/2}}{\sqrt{a\pi}}e^{a^2b^2/4}\Gamma\begin{bmatrix}(\mu-\lambda+\nu)/2,\ (\mu-\lambda-\nu-1)/2\\-\lambda\end{bmatrix}\times$$
$$\times W_{(3+2\lambda-2\mu)/4,\ (2\nu+1)/4}\left(\frac{a^2b^2}{2}\right)$$
$[a>0;\ \operatorname{Re}\mu<1;\ \operatorname{Re}(\lambda-\mu)<\operatorname{Re}\nu<\operatorname{Re}(\mu-\lambda)-1;\ |\arg b|<3\pi/4].$

2.17.12. Integrals containing $J_\lambda(\varphi(x))\, P_\nu^\mu(\chi(x))$.

1. $$\int_0^a x^{\alpha-1}(a^2-x^2)^{-\mu/2} J_\lambda(bx)\, P_\nu^\mu\left(\frac{x}{a}\right) dx =$$
$$=\frac{\sqrt{\pi}\, a^{\alpha+\lambda-\mu} b^\lambda}{2^{\alpha+2\lambda-\mu}} \Gamma\left[\begin{matrix}\alpha+\lambda \\ \lambda+1,\ \ (1+\alpha+\lambda-\mu-\nu)/2,\ \ 1+(\alpha+\lambda-\mu+\nu)/2\end{matrix}\right]\times$$
$$\times {}_2F_3\left(\frac{\alpha+\lambda}{2}, \frac{\alpha+\lambda+1}{2};\ \lambda+1, \frac{1+\alpha+\lambda-\mu-\nu}{2}, 1+\frac{\alpha+\lambda-\mu+\nu}{2};\ -\frac{a^2b^2}{4}\right)$$
[a, $\operatorname{Re}(\alpha+\lambda) > 0$; $\operatorname{Re}\mu < 1$].

2. $$\int_0^a x^{1/2-\mu}(a^2-x^2)^{-\mu/2} J_{\nu+1/2}(bx)\, P_\nu^\mu\left(\frac{x}{a}\right) dx =$$
$$=\sqrt{\frac{\pi}{2}}\, a^{1-\mu} b^{\mu-1/2} J_{1/2-\mu}\left(\frac{ab}{2}\right) J_{\nu+1/2}\left(\frac{ab}{2}\right)$$
[$a > 0$; $\operatorname{Re}\mu < 1$; $\operatorname{Re}(\nu-\mu) > -2$].

3. $$\int_a^\infty \sqrt{x}\, J_{\nu+1/2}(bx)\, P_\nu\left(\frac{x}{a}\right) dx = -\frac{1}{b}\sqrt{\frac{a}{2}}\left[\cos\frac{ab}{2} Y_{\nu+1/2}\left(\frac{ab}{2}\right)+\right.$$
$$\left.+\sin\frac{ab}{2} J_{\nu+1/2}\left(\frac{ab}{2}\right)\right]$$
[a, $b > 0$; $-1 < \operatorname{Re}\nu < 0$].

4. $$\int_a^\infty x^{1/2-\mu}(x^2-a^2)^{-\mu/2} J_{\nu+1/2}(bx)\, P_\nu^\mu\left(\frac{x}{a}\right) dx =$$
$$=-\sqrt{\frac{\pi}{8}}\, a^{1-\mu} b^{\mu-1/2}\left[J_{\mu-1/2}\left(\frac{ab}{2}\right) Y_{\nu+1/2}\left(\frac{ab}{2}\right)+J_{\nu+1/2}\left(\frac{ab}{2}\right) Y_{\mu-1/2}\left(\frac{ab}{2}\right)\right]$$
[a, $b > 0$; $\operatorname{Re}\mu < 1$; $-2\operatorname{Re}\mu-1 < \operatorname{Re}\nu < 2\operatorname{Re}\mu$].

5. $$\int_a^\infty x^\lambda (x^2-a^2)^{(\nu-1)/2} J_\lambda(bx)\, P_\nu^{\nu-1}\left(\frac{x}{a}\right) dx =$$
$$=\frac{2^{\lambda+\nu} a^\lambda}{\sqrt{\pi}\, b^\nu} \Gamma\left[\begin{matrix}1/2+\lambda \\ 1-\nu\end{matrix}\right] S_{\nu-\lambda,\ \nu+\lambda}(ab)$$
[a, $b > 0$; $\operatorname{Re}\lambda < 5/2$; $\operatorname{Re}(\lambda+2\nu) < 3/2$].

6. $$\int_0^a x^{\mu-5/2}(a^2-x^2)^{-\mu/2} J_{\nu+1/2}\left(\frac{b}{x}\right) P_\nu^\mu\left(\frac{x}{a}\right) dx =$$
$$=\sqrt{\frac{2}{\pi}}\,\frac{b^{\mu-3/2}}{a^\mu}\cos\left(\frac{b}{\sqrt{a}}+(\mu+\nu)\frac{\pi}{2}\right)$$
[a, $b > 0$; $0 < \operatorname{Re}\mu < 1$].

7. $$\int_{-a}^a (x+a)^{m/2} J_m(b\sqrt{a-x})\, P_n^m\left(\frac{x}{a}\right) dx=$$
$$=\frac{2^{(3m+1)/2} a^{(m-1)/2}(n+m)!}{b\,(n-m)!} J_{2n+1}(b\sqrt{2a})$$
[$a > 0$].

8. $$\int_a^\infty (x+a)^{-\mu/2} J_{-\mu}(b\sqrt{x-a})\, P_\nu^\mu\left(\frac{x}{a}\right) dx = -\frac{2^{5/2-\mu}}{\pi}\sqrt{a}\, b^{\mu-1}\sin\nu\pi\, K_{2\nu+1}(b\sqrt{2a})$$
[a, $b > 0$; $\operatorname{Re}\mu < 1$; $-\operatorname{Re}\mu-3/2 < 2\operatorname{Re}\nu < \operatorname{Re}\mu-1/2$].

9. $$\int_a^\infty (x+a)^{\mu/2} J_{-\mu-1/2}(b\sqrt{x-a})\, P_\nu^\mu\left(\frac{x}{a}\right) dx =$$
$$= \frac{2^{\mu+2} a^{(2\mu+3)/4}}{\sqrt{b\pi}\,\Gamma(1-\mu+\nu)\,\Gamma(-\mu-\nu)} K^2_{\nu+1/2}\left(b\sqrt{\frac{a}{2}}\right)$$
$[a, b > 0;\ \operatorname{Re}\mu - 3/2 < 2\operatorname{Re}\nu < -\operatorname{Re}\mu - 1/2]$.

10. $$\int_a^\infty (x+a)^{\mu/2} J_{-\mu}(b\sqrt{x-a})\, P_\nu^\mu\left(\frac{x}{a}\right) dx = \frac{2^{\mu+5/2}\sqrt{a}\, b^{-\mu-1}}{\Gamma(1-\mu+\nu)\Gamma(-\mu-\nu)} K_{2\nu+1}(b\sqrt{2a})$$
$[a, b > 0;\ \operatorname{Re}\mu < 1;\ \operatorname{Re}\mu - 3/2 < 2\operatorname{Re}\nu < -\operatorname{Re}\mu - 1/2]$.

11. $$\int_a^\infty (x-a)^{-(2\mu+1)/4} (x^2-b^2)^{\mu/2} J_{-\mu-1/2}(c\sqrt{x-a})\, P_\nu^\mu\left(\frac{x}{b}\right) dx =$$
$$= \sqrt{\frac{b}{\pi}}\, \frac{2^{\mu+2} c^{-\mu-1/2}}{\Gamma(1-\mu+\nu)\,\Gamma(-\mu-\nu)} K_{\nu+1/2}(z_+)\, K_{\nu+1/2}(z_-)$$
$[2z_\pm = c(\sqrt{a+b} \pm \sqrt{a-b});\ a > b > 0;\ \operatorname{Re}\mu < 1/2;\ \operatorname{Re}\mu/2 - 1 < \operatorname{Re}\nu < -\operatorname{Re}\mu/2]$.

12. $$\int_{-a}^a (x+a)^{-1/4}(a-x)^{-\mu/2} J_{\mu-1/2}(b\sqrt{x+a})\, P_\nu^\mu\left(\frac{x}{a}\right) dx =$$
$$= \frac{\sqrt{a\pi}\, b^{\mu-1/2}}{2^{\mu-1}} J_{1/2-\nu}(b\sqrt{2a})\, J_{\nu-1/2}(b\sqrt{2a})$$ $[a > 0;\ -1/2 < \operatorname{Re}\mu < 1]$.

13. $$\int_a^\infty \frac{(x-a)^{-\mu/2}}{\sqrt[4]{x+a}} J_{1/2-\mu}(b\sqrt{x+a})\, P_\nu^\mu\left(\frac{x}{a}\right) dx =$$
$$= -\frac{\sqrt{\pi a}\, b^{\mu-1/2}}{2^\mu \cos\nu\pi}\left[J^2_{\nu+1/2}\left(b\sqrt{\frac{a}{2}}\right) - J^2_{-\nu-1/2}\left(b\sqrt{\frac{a}{2}}\right)\right]$$
$[a, b > 0;\ \operatorname{Re}\mu < 1;\ -1 - \operatorname{Re}\mu/2 < \operatorname{Re}\nu < \operatorname{Re}\mu/2]$.

14. $$\int_a^\infty \frac{(x-a)^{\mu/2}}{\sqrt[4]{x+a}} J_{-\mu-1/2}(b\sqrt{x+a})\, P_\nu^\mu\left(\frac{x}{a}\right) dx =$$
$$= \frac{2^\mu \sqrt{\pi^3 a}\, b^{-\mu-1/2}}{\Gamma(-\mu-\nu)\,\Gamma(1-\mu+\nu)}\left[J^2_{\nu+1/2}\left(b\sqrt{\frac{a}{2}}\right) + Y^2_{\nu+1/2}\left(b\sqrt{\frac{a}{2}}\right)\right]$$
$[a, b > 0;\ \operatorname{Re}\mu/2 - 1 < \operatorname{Re}\nu < -\operatorname{Re}\mu/2]$.

15. $$\int_a^\infty (x-a)^{-\mu/2}(x+a)^{\mu-5/4} J_{-\mu-1/2}\left(\frac{b}{\sqrt{x+a}}\right) P_\nu^\mu\left(\frac{x}{a}\right) dx =$$
$$= \sqrt{\frac{2}{\pi}}\, \frac{a^\mu}{b^{\mu+1/2}}\, \Gamma\begin{bmatrix} 1-\mu+\nu, & -\mu-\nu \\ 1-\mu, & 1-\mu \end{bmatrix} {}_1F_1\left(1-\mu+\nu;\ 1-2\mu;\ \frac{ib}{\sqrt{2a}}\right) \times$$
$$\times {}_1F_1\left(1-\mu+\nu;\ 1-2\mu;\ -\frac{ib}{\sqrt{2a}}\right)$$
$[a > 0;\ \operatorname{Re}\mu < 1;\ \operatorname{Re}\mu - 1 < \operatorname{Re}\nu < -\operatorname{Re}\mu]$.

16. $$\int_{-a}^a (x+a)^{\nu-1} J_0\left(b\sqrt{\frac{a-x}{a+x}}\right) P_\nu\left(\frac{x}{a}\right) dx = \frac{2^{1-\nu} a^\nu b^{2\nu}}{\Gamma^2(\nu+1)} K_0(b)$$
$[a, b > 0;\ \operatorname{Re}\nu > -3/4]$.

17. $\int\limits_{-a}^{a}(x+a)^{-\nu-2}J_{-\mu}\left(b\sqrt{\frac{a-x}{a+x}}\right)P_{\nu}^{\mu}\left(\frac{x}{a}\right)dx=\frac{2^{\nu+2}a^{-\nu-1}b^{-2\nu-2}}{\Gamma(-\mu-\nu)\,\Gamma(-\nu)}K_{\mu}(b)$

[$a, b>0$; $\operatorname{Re}\mu<1$; $2\operatorname{Re}\nu<-|\operatorname{Re}\mu|-1/2$].

18. $\int\limits_{a}^{\infty}J_{-\mu}\left(b\sqrt{x^2-a^2}\right)P_{\nu}^{\mu}\left(\frac{x}{a}\right)dx=$

$$=\frac{2^{\mu+1/2}\sqrt{a}}{\sqrt{b}\,\Gamma(1+(\nu-\mu)/2)\,\Gamma((1-\mu-\nu)/2)}K_{\nu+1/2}(ab)$$

[$a, b>0$; $\operatorname{Re}\mu<1$; $-3/2<\operatorname{Re}\nu<1/2$].

19. $\int\limits_{0}^{a}x^{\nu-2}J_{-\mu}\left(b\sqrt{\frac{a^2-x^2}{x}}\right)P_{\nu}^{\mu}\left(\frac{x}{a}\right)dx=\frac{(ab)^{\nu-1}}{\Gamma(1-\mu+\nu)}e^{-b}$

[$a, b>0$; $\operatorname{Re}\mu<1$; $\operatorname{Re}\nu>-1/2$].

20. $\int\limits_{0}^{\infty}x^{\nu/2}J_{\nu}(b\sqrt{x})\,P_{\nu}\left(\frac{x+2z}{2\sqrt{z(x+z)}}\right)dx=$

$$=\frac{2^{\nu+2}z^{(\nu+1)/2}}{b^{\nu+1}\Gamma(-\nu)}I_{-\nu-1/2}\left(\frac{b\sqrt{z}}{2}\right)K_{\nu+1/2}\left(\frac{b\sqrt{z}}{2}\right)$$

[$b>0$; $-1<\operatorname{Re}\nu<-1/4$; $|\arg z|<\pi$].

21. $\int\limits_{0}^{\infty}x^{-(\nu+1)/2}(x+z)^{-(\nu+1)/2}J_{\nu+1}(b\sqrt{x})\,P_{\nu}\left(\frac{x+2z}{2\sqrt{z(x+z)}}\right)dx=$

$=\frac{2^{1-\nu}z^{-\nu/2}b^{\nu}}{\Gamma(\nu+1)}I_{\nu+1/2}\left(\frac{b\sqrt{z}}{2}\right)K_{\nu+1/2}\left(\frac{b\sqrt{z}}{2}\right)$ [$b>0$; $-1<\operatorname{Re}\nu<0$; $|\arg z|<\pi$].

22. $\int\limits_{0}^{\infty}x^{\nu/2}(x+z)^{\nu/2}J_{\nu}(b\sqrt{x})\,P_{\nu}\left(\frac{x+2z}{2\sqrt{z(x+z)}}\right)dx=\frac{2^{\nu+2}z^{(\nu+1)/2}}{\pi\Gamma(-\nu)\,b^{\nu+1}}K_{\nu+1/2}^{2}\left(\frac{b\sqrt{z}}{2}\right)$

[$b>0$; $-1<\operatorname{Re}\nu<-1/6$; $|\arg z|<\pi$].

23. $\int\limits_{-a}^{a}e^{bx}J_{m}\left(b\sqrt{a^2-x^2}\right)P_{n}^{m}\left(\frac{x}{a}\right)dx=\frac{2a^{n+1}b^{n}}{(n-m)!\,(2n+1)}$ [$a>0$; $n\geqslant m$].

24. $\int\limits_{0}^{a}\begin{Bmatrix}\operatorname{sh}bx\\ \operatorname{ch}bx\end{Bmatrix}J_{m}\left(b\sqrt{a^2-x^2}\right)P_{n}^{m}\left(\frac{x}{a}\right)dx=\frac{a^{n+1}b^{n}}{(n-m)!\,(2n+1)}$

$\left[a>0;\ n-m=\begin{Bmatrix}1, 3, 5, \ldots\\ 0, 2, 4, \ldots\end{Bmatrix}\right]$.

25. $\int\limits_{0}^{a}x^{\alpha-1}(a^2-x^2)^{-\mu/2}J_{\varkappa}(bx)\,J_{\lambda}(bx)\,P_{\nu}^{\mu}\left(\frac{x}{a}\right)dx=$

$$=2^{\mu-\varkappa-\lambda-\alpha}\sqrt{\pi}a^{\alpha+\varkappa+\lambda-\mu}b^{\varkappa+\lambda}\Gamma\left[\begin{matrix}\alpha\\ \varkappa+1,\ \lambda+1,\ (1+\alpha-\mu-\nu)/2,\ 1+(\alpha-\mu+\nu)/2\end{matrix}\right]\times$$
$$\times{}_4F_5\left(\begin{matrix}\alpha/2,\ (\alpha+1)/2,\ (1+\varkappa+\lambda)/2,\ 1+(\varkappa+\lambda)/2;\ -a^2b^2\\ 1+\varkappa,\ 1+\lambda,\ 1+\varkappa+\lambda,\ (1+\alpha-\mu-\nu)/2,\ 1+(\alpha-\mu+\nu)/2\end{matrix}\right)$$

[a, $\operatorname{Re}(\alpha+\varkappa+\lambda)>0$; $\operatorname{Re}\mu<1$].

2.17.13. Integrals containing $Y_\lambda(\varphi(x))\,P_\nu^\mu(cx)$.

1. $$\int_a^\infty x^{1/2}Y_{\nu+1/2}(bx)\,P_\nu\left(\frac{x}{a}\right)dx=\sqrt{\frac{a}{2b^2}}\left[\cos\frac{ab}{2}J_{\nu+1/2}\left(\frac{ab}{2}\right)-\sin\frac{ab}{2}Y_{\nu+1/2}\left(\frac{ab}{2}\right)\right]$$ $[a,\ b>0;\ -1<\operatorname{Re}\nu<0]$.

2. $$\int_a^\infty x^{1/2-\mu}(x^2-a^2)^{-\mu/2}\,Y_{\nu+1/2}(bx)\,P_\nu^\mu\left(\frac{x}{a}\right)dx=$$ $$=\sqrt{\frac{\pi}{8}}\frac{b^{\mu-1/2}}{a^{\mu-1}}\left[J_{\nu+1/2}\left(\frac{ab}{2}\right)J_{\mu-1/2}\left(\frac{ab}{2}\right)-Y_{\nu+1/2}\left(\frac{ab}{2}\right)Y_{\mu-1/2}\left(\frac{ab}{2}\right)\right]$$ $[a,\ b>0;\ -1-2\operatorname{Re}\mu<\operatorname{Re}\nu<2\operatorname{Re}\mu<2]$.

3. $$\int_a^\infty (x+a)^{-1/4}(x-a)^{-\mu/2}Y_{1/2-\mu}\left(b\sqrt{x+a}\right)P_\nu^\mu\left(\frac{x}{a}\right)dx=$$ $$=\frac{\sqrt{\pi a}\,b^{\mu-1/2}}{2^\mu}\left[J_{\nu+1/2}\left(b\sqrt{\frac{a}{2}}\right)J_{-\nu-1/2}\left(b\sqrt{\frac{a}{2}}\right)-\right.$$ $$\left.-Y_{\nu+1/2}\left(b\sqrt{\frac{a}{2}}\right)Y_{-\nu-1/2}\left(b\sqrt{\frac{a}{2}}\right)\right]$$ $[a,\ b>0;\ \operatorname{Re}\mu<1;\ -\operatorname{Re}\mu/2-1<\operatorname{Re}\nu<\operatorname{Re}\mu/2]$.

2.17.14. Integrals containing $I_\lambda(\varphi(x))\,P_\nu^\mu(cx)$.

1. $$\int_a^\infty (x-a)^{-\mu/2}(x+a)^{\mu-5/4}I_{-\mu-1/2}\left(\frac{b}{\sqrt{x+a}}\right)P_\nu^\mu\left(\frac{x}{a}\right)dx=$$ $$=\sqrt{\frac{a}{\pi}}\frac{b^{\mu-3/2}}{2^{\mu-1}}\Gamma\begin{bmatrix}-\mu-\nu,\ 1-\mu+\nu\\ 1-\mu,\ 1-\mu\end{bmatrix}M_{\nu+1/2,\,-\mu}\left(\frac{b}{\sqrt{2a}}\right)M_{-\nu-1/2,\,-\mu}\left(\frac{b}{\sqrt{2a}}\right)$$ $[a>0;\ \operatorname{Re}\mu-1<\operatorname{Re}\nu<-\operatorname{Re}\mu]$.

2. $$\int_a^\infty (x+a)^{\pm\mu/2}e^{-bx}I_{-\mu/2}(b(x-a))\,P_\nu^\mu\left(\frac{x}{a}\right)dx=$$ $$=\begin{Bmatrix}1\\0\end{Bmatrix}\frac{(2b)^{-\mu/2-1}e^{ab}}{\sqrt{\pi}}\Gamma\begin{bmatrix}\nu+(1-\mu)/2,\ -\nu-(\mu+1)/2\\ \nu-\mu+1,\ -\mu-\nu\end{bmatrix}W_{(\mu+1)/2,\,\nu+1/2}(4ab)$$ $[a,\ b>0;\ (-1\pm\operatorname{Re}\mu)/2<\operatorname{Re}\nu<(-1\mp\operatorname{Re}\mu)/2]$.

3. $$\int_0^a\begin{Bmatrix}\sin bx\\ \cos bx\end{Bmatrix}I_m\left(b\sqrt{a^2-x^2}\right)P_n^m\left(\frac{x}{a}\right)dx=\frac{(-1)^{(n-m-\delta)/2}a^{n+1}b^n}{(n-m)!\,(2n+1)}$$ $$\left[a>0;\ n-m=\begin{Bmatrix}1,\ 3,\ 5,\ \dots\\ 0,\ 2,\ 4,\ \dots\end{Bmatrix}\right].$$

2.17.15. Integrals containing $K_\lambda(\varphi(x))\,P_\nu^\mu(cx)$.

1. $$\int_a^\infty x^{1/2}K_{\nu+1/2}(bx)\,P_\nu\left(\frac{x}{a}\right)dx=\frac{1}{b}\sqrt{\frac{a}{2}}e^{-ab}K_{\nu+1/2}\left(\frac{ab}{2}\right)$$ $[a,\ \operatorname{Re}b>0]$.

2. $$\int_a^\infty x^{\mu+1/2}(x^2-a^2)^{-\mu/2}K_{\nu+1/2}(bx)\,P_\nu^\mu\left(\frac{x}{a}\right)dx=\sqrt{\frac{\pi}{2b^3}}e^{-ab/2}W_{\rho,\,\nu+1/2}(ab)$$ $[a,\ \operatorname{Re}b>0;\ \operatorname{Re}\mu<1]$.

3. $\int_a^\infty x^{\mu-3/2}(x^2-a^2)^{-\mu/2}K_{\nu+1\pm1/2}(bx)P_\nu^\mu\left(\frac{x}{a}\right)dx=$

$$=\begin{Bmatrix}\sqrt{a}\\ \sqrt{b}\end{Bmatrix}\frac{\sqrt{\pi}}{ab\sqrt{2}}e^{-ab/2}W_{\mu-(3\mp1)/4,\ \nu+(3\pm1)/4}(ab)$$ [a, Re $b>0$; Re $\mu<1$].

4. $\int_a^\infty x^{1/2-\mu}(x^2-a^2)^{-\mu/2}K_{\nu+1/2}(bx)\,P_\nu^\mu\left(\frac{x}{a}\right)dx=$

$$=\frac{a^{1-\mu}b^{\mu-1/2}}{\sqrt{2\pi}}K_{\nu+1/2}\left(\frac{ab}{2}\right)K_{\nu-1/2}\left(\frac{ab}{2}\right)$$ [a, Re $b>0$; Re $\mu<1$].

5. $\int_a^\infty (x+a)^{\mu/2}K_\mu(b\sqrt{x-a})\,P_\nu^\mu\left(\frac{x}{a}\right)dx=\frac{2^{3/2-\mu}\sqrt{a}}{b^{\mu+1}}S_{2\mu,\ 2\nu+1}(b\sqrt{2a})$

[a, Re $b>0$; Re $\mu<1$].

6. $\int_a^\infty (x-a)^{-\mu/2}K_\mu(b\sqrt{x+a})\,P_\nu^\mu\left(\frac{x}{a}\right)dx=\frac{\sqrt{a}\,b^{\mu-1}}{2^{\mu-3/2}}K_{\nu+1}(b\sqrt{2a})$

[a, Re $b>0$; Re $\mu<1$].

7. $\int_a^\infty (x-a)^{-\mu/2}(x+a)^{-1/4}K_{\mu-1/2}(b\sqrt{x+a})\,P_\nu^\mu\left(\frac{x}{a}\right)dx=$

$$=\sqrt{\frac{a}{\pi}}\frac{b^{\mu-1/2}}{2^{\mu-1}}K_{\nu+1/2}^2\left(b\sqrt{\frac{a}{2}}\right)$$ [a, Re $b>0$; Re $\mu<1$].

8. $\int_{-a}^a (x+a)^{\mu-5/4}(a-x)^{-\mu/2}K_{\mu+1/2}\left(\frac{b}{\sqrt{x+a}}\right)P_\nu^\mu\left(\frac{x}{a}\right)dx=$

$$=\frac{\sqrt{\pi a}\,b^{\mu-3/2}}{2^{\mu-1}}W_{\nu+1/2,\ \mu}\left(\frac{b}{\sqrt{2a}}\right)W_{-\nu-1/2,\ \mu}\left(\frac{b}{\sqrt{2a}}\right)$$

[a, Re $b>0$; Re $\mu<1$]

9. $\int_a^\infty x^{(1\pm1)/2}K_\mu(b\sqrt{x^2-a^2})\,P_\nu^\mu\left(\frac{x}{a}\right)dx=\sqrt{a}\,b^{-1\mp1/2}S_{\mu\pm1/2,\ \nu+1/2}(ab)$

[a, Re $b>0$; Re $\mu<1$]

10. $\int_a^\infty (x-a)^{\mu/2}K_\mu(b\sqrt{x-a})\left[P_{-\nu}^\mu\left(\frac{x}{a}\right)+P_\nu^\mu\left(\frac{x}{a}\right)\right]dx=$

$$=\frac{2^{2-\mu}\nu}{(\nu-\mu)\,b^{\mu+2}}S_{2\mu+1,\ 2\nu}(b\sqrt{2a})$$ [a, Re $b>0$; $|$Re $\mu|<2$]

11. $\int_a^\infty (x+a)^{\mu/2-1}K_\mu(b\sqrt{x-a})\left[P_{-\nu}^\mu\left(\frac{x}{a}\right)+P_\nu^\mu\left(\frac{x}{a}\right)\right]dx=$

$$=\frac{b^{-\mu}}{2^{\mu-2}}S_{2\mu-1,\ 2\nu}(b\sqrt{2a})$$ [a, Re $b>0$; Re $\mu<1$]

2. $$\int_{-a}^{a}(x+a)^{(\mu-3)/2}(a-x)^{-\mu/2}e^{-p/(x+a)}K_{\nu+1/2}\left(\frac{p}{x+a}\right)P_\nu^\mu\left(\frac{x}{a}\right)dx=$$
$$=\sqrt{\frac{\pi}{2p}}\,e^{-p/(2a)}\Gamma\left(\mu,\frac{p}{2}\right)\qquad [a,\ \operatorname{Re} p>0;\ \operatorname{Re}\mu<1].$$

2.17.16. Integrals containing $\begin{Bmatrix}\mathbf{H}_\lambda(\varphi(x))\\ \mathbf{L}_\lambda(\varphi(x))\end{Bmatrix}P_\nu^\mu(cx)$.

1. $$\int_{a}^{\infty}(x+a)^{-1/4}(x-a)^{-\mu/2}\left[Y_{\mu-1/2}\left(b\sqrt{x+a}\right)-\mathbf{H}_{\mu-1/2}\left(b\sqrt{x+a}\right)\right]P_\nu^\mu\left(\frac{x}{a}\right)dx=$$
$$=\frac{\sqrt{a\pi}\,b^{\mu-1/2}}{2^\mu}\,\frac{\sin\mu\pi}{\sin\nu\pi}\left[J^2_{\mu-1/2}\left(b\sqrt{\frac{a}{2}}\right)-Y^2_{\mu-1/2}\left(b\sqrt{\frac{a}{2}}\right)\right]$$
$$[a>0;\ \operatorname{Re}\mu<1;\ -1<\operatorname{Re}\nu<0;\ |\arg b|<\pi].$$

2. $$\int_{a}^{\infty}(x+a)^{-1/4}(x-a)^{-\mu/2}\left[I_{1/2-\mu}\left(b\sqrt{x+a}\right)-\mathbf{L}_{\mu-1/2}\left(b\sqrt{x+a}\right)\right]P_\nu^\mu\left(\frac{x}{a}\right)dx=$$
$$=\frac{\sqrt{\pi a}\,b^{\mu-1/2}}{2^{\mu-1}}\,\frac{\sin\mu\pi}{\sin 2\nu\pi}\left[I^2_{\mu-1/2}\left(b\sqrt{\frac{a}{2}}\right)-I^2_{1/2-\mu}\left(b\sqrt{\frac{a}{2}}\right)\right]$$
$$[a,\ \operatorname{Re} b>0;\ \operatorname{Re}\mu<1;\ -1<\operatorname{Re}\nu<0].$$

2.17.17. Integrals containing $S_{\varkappa,\lambda}(\varphi(x))\,P_\nu^\mu(cx)$.

1. $$\int_{0}^{a}x^{1/2-\mu}(a^2-x^2)^{-\mu/2}S_{2\mu+\nu-3/2,\ \nu+1/2}(bx)\,P_\nu^\mu\left(\frac{x}{a}\right)dx=$$
$$=-\frac{2^{2\mu+\nu-3}\sqrt{\pi}\,b^{\mu-1/2}}{a^{\mu-1}}\cos\mu\pi\,\Gamma\left(\frac{1+\mu+\nu}{2}\right)\Gamma\left(\frac{\mu-\nu}{2}\right)\times$$
$$\times\left[J_{\nu+1/2}\left(\frac{ab}{2}\right)Y_{1/2-\mu}\left(\frac{ab}{2}\right)-J_{1/2-\mu}\left(\frac{ab}{2}\right)Y_{\nu+1/2}\left(\frac{ab}{2}\right)\right]$$
$$[\operatorname{Re}\mu<1;\ |\operatorname{Re}(\mu+\nu)|<1;\ \operatorname{Re}(\mu-\nu)<2;\ |\arg b|<\pi].$$

2. $$\int_{a}^{\infty}\sqrt{x}\,(x^2-a^2)^{-\mu/2}S_{\varkappa,\ 1/2}(bx)\,P_\nu^\mu\left(\frac{x}{a}\right)dx=$$
$$=\sqrt{\frac{a}{\pi}}\,\frac{b^{\mu-1}}{2^{\varkappa-\mu+3/2}}\,\Gamma\left[\begin{matrix}(\mu+\nu-\varkappa)/2+1/4,\ (\mu-\nu-\varkappa)/2-1/4\\ 1/2-\varkappa\end{matrix}\right]S_{\varkappa-\mu+1,\ \nu+1/2}(ab)$$
$$[a>0;\ \operatorname{Re}\mu<1;\ \operatorname{Re}(\varkappa-\mu)-1/2<\operatorname{Re}\nu<\operatorname{Re}(\mu-\varkappa)-1/2;\ |\arg b|<\pi].$$

3. $$\int_{a}^{\infty}(x-a)^{-\mu/2}S_{\varkappa,\ \mu}\left(b\sqrt{x+a}\right)P_\nu^\mu\left(\frac{x}{a}\right)dx=$$
$$=\sqrt{2a}\,b^{\mu-1}\Gamma\left[\begin{matrix}(1+\mu-\varkappa)/2+\nu,\ (\mu-\varkappa-1)/2-\nu\\ (1-\mu-\varkappa)/2,\ (1+\mu-\varkappa)/2\end{matrix}\right]S_{\varkappa-\mu+1,\ 2\nu+1}\left(b\sqrt{2a}\right)$$
$$[a>0;\ \operatorname{Re}\mu<1;\ \operatorname{Re}(\varkappa-\mu)-1<2\operatorname{Re}\nu<\operatorname{Re}(\mu-\varkappa)-1;\ |\arg b|<\pi].$$

4. $$\int_{-a}^{a}(x+a)^{\mu-5/4}(a-x)^{-\mu/2}S_{\mu+2\nu+1/2,\ \mu+1/2}\left(\frac{b}{\sqrt{x+a}}\right)P_\nu^\mu\left(\frac{x}{a}\right)dx=$$
$$=2^{2\nu+3/2}\sqrt{\pi a}\,b^{\mu-3/2}\Gamma\left[\begin{matrix}\mu-\nu\\ 1/2-\nu\end{matrix}\right]W_{\nu+1/2,\ \mu}\left(\frac{e^{\pi i/2}b}{\sqrt{2a}}\right)W_{\nu+1/2,\ \mu}\left(\frac{e^{-\pi i/2}b}{\sqrt{2a}}\right)$$
$$[a>0;\ \operatorname{Re}\nu<0;\ \operatorname{Re}\nu<\operatorname{Re}\mu<1].$$

2.17.18. Integrals containing $\mathbf{K}(\varphi(x))P_\nu(cx)$.

1. $$\int_{-a}^{a}\mathbf{K}\left(\sqrt{\frac{a-x}{2a}}\right)P_\nu\left(\frac{x}{a}\right)dx=$$
$$=\frac{4a}{\pi(2\nu+1)^2}[\pi\cos\nu\pi+2\sin\nu\pi(\mathbf{C}+2\ln 2+\psi(\nu+1))]\qquad [a>0;\ \operatorname{Re}\nu\neq-1/2].$$

2. $$=-\frac{a}{\pi}\psi''\left(\frac{1}{2}\right)\qquad [a>0;\ \nu=-1/2].$$

3. $$\int_{0}^{a}\mathbf{K}\left(\sqrt{\frac{a-x}{2a}}\right)P_\nu\left(\frac{x}{a}\right)dx=$$
$$=\frac{2^{3/2}a}{(2\nu+1)^2}\left\{\Gamma\begin{bmatrix}1/4,& \nu/2+1\\ 3/4,& (\nu+1)/2\end{bmatrix}\sin\frac{\nu\pi}{2}+\Gamma\begin{bmatrix}3/4,& (\nu+1)/2\\ 1/4,& \nu/2+1\end{bmatrix}\cos\frac{\nu\pi}{2}\right\}$$
$[a>0;\ \nu\neq-1/2]$.

4. $$=\frac{a}{8}\left[\psi'\left(\frac{1}{4}\right)-\psi'\left(\frac{3}{4}\right)\right]\qquad [a>0;\ \nu=-1/2]$$

5. $$\int_{-a}^{a}(x+a)^{\nu-1/2}\mathbf{K}\left(\sqrt{\frac{a-x}{2a}}\right)P_\nu\left(\frac{x}{a}\right)dx=\sqrt{\pi}\left(\frac{a}{2}\right)^{\nu+1/2}\Gamma^3\begin{bmatrix}\nu+1/2\\ \nu+1\end{bmatrix}$$
$[a>0;\ \operatorname{Re}\nu>-1/2]$

2.17.19. Integrals containing $A(x)P_\nu^\mu(bx)P_\sigma^\rho(cx)$.

CONDITION $a>0$.

1. $$\int_{0}^{a}P_\nu\left(\frac{x}{a}\right)P_\sigma\left(\frac{x}{a}\right)dx=\frac{2a}{\pi}\,\frac{A\sin(\sigma\pi/2)\cos(\nu\pi/2)-A^{-1}\sin(\nu\pi/2)\cos(\sigma\pi/2)}{(\sigma-\nu)(\sigma+\nu+1)}$$
$$\left[a>0;\ A=\Gamma\begin{bmatrix}(\nu+1)/2,& \sigma/2+1\\ \nu/2+1,& (\sigma+1)/2\end{bmatrix}\right]$$

2. $$\int_{-a}^{a}P_\nu\left(\frac{x}{a}\right)P_\sigma\left(\frac{x}{a}\right)dx=$$
$$=\frac{2\pi\sin(\sigma-\nu)\pi+4\sin\nu\pi\sin\sigma\pi[\psi(\nu+1)-\psi(\sigma+1)]}{\pi^2(\sigma-\nu)(\sigma+\nu+1)}a\qquad [\sigma-\nu,\ \sigma+\nu+1\neq 0]$$

3. $$=2\frac{\pi^2-2\sin^2\sigma\pi\psi'(\sigma+1)}{\pi^2(2\sigma+1)}a\qquad [\nu=\sigma]$$

4. $$\int_{-a}^{a}P_l^k\left(\frac{x}{a}\right)P_n^m\left(\frac{x}{a}\right)dx=aI(k,\ l,\ m,\ n),$$
$$I(m,\ l,\ m,\ n)=\delta_{l,n}\frac{2(n+m)!}{(2n+1)(n-m)!},$$
$$I(m+2,\ l,\ m,\ n)=0\qquad [n>l\ \text{or}\ l-m=1,\ 3,\ 5,\ \ldots$$
$$I(m+2,\ n,\ m,\ n)=-\frac{(n+m)!}{(n-m-2)!}\,\frac{2}{2n+1},$$
$$I(m+2,\ n+2,\ m,\ n)=\frac{4(m+1)(n+m)!}{(n-m)!},$$
$$I(k,\ n,\ m,\ n)=(-1)^{(3m+k)/2}\frac{(n+m)!}{(n-k)!}\,\frac{2}{2n+1}$$
$[m+k=2,\ 4,\ 6,\ \ldots,\ k\geqslant m$

5. $\int_0^a x^{l+n-1} P_l^m\left(\frac{x}{a}\right) P_n^m\left(\frac{x}{a}\right) dx = 2^{-l-n-4m} a^{l+n} \frac{(l+n-1)!}{(l-m)!\,(n-m)!}.$

6. $\int_{-a}^a x^{l-n} P_l^m\left(\frac{x}{a}\right) P_n^m\left(\frac{x}{a}\right) dx = 2^{n-l} a^{l-n+1} \frac{(l+m)!}{(n-m)!} \Gamma\begin{bmatrix} n+1/2 \\ l+3/2 \end{bmatrix}$ $[l > n]$.

7. $\int_{-a}^a x P_l^m\left(\frac{x}{a}\right) P_n^m\left(\frac{x}{a}\right) dx = 0$ $[l \neq n \pm 1]$.

8. $\int_{-a}^a R(x) P_l^{2k}\left(\frac{x}{a}\right) P_n^{2m}\left(\frac{x}{a}\right) dx = 0$

[$R(x)$ is any polynomial of degree j; $l+j < n$; $k \geqslant m$; $l+n = 2, 4, 6, \ldots$].

9. $\int_{-a}^a (a \pm x)^{-p} P_l^{2k}\left(\frac{x}{a}\right) P_n^{2m}\left(\frac{x}{a}\right) dx = 0$

$[l+n = 2, 4, 6, \ldots; \; l \leqslant n; \; k-p \geqslant m; \; (l-n)/2 < p \leqslant k]$.

10. $\int_{-a}^a (x+a)^{l-n} P_l^m\left(\frac{x}{a}\right) P_n^m\left(\frac{x}{a}\right) dx = 2^{n-l} a^{l-n+1} \frac{(l+m)!}{(n-m)!} \Gamma\begin{bmatrix} n+1/2 \\ l+3/2 \end{bmatrix}$

$[l, \; n \geqslant m; \; l > n; \; m = 0, 1, 2, \ldots]$.

11. $\int_{-a}^a (x+a)^{\nu+\sigma} P_\nu\left(\frac{x}{a}\right) P_\sigma\left(\frac{x}{a}\right) dx = \frac{a^{\nu+\sigma+1}\sqrt{\pi}}{2^{\nu+\sigma}} \Gamma^2\begin{bmatrix} \nu+\sigma+1 \\ \nu+1, \; \sigma+1 \end{bmatrix} \Gamma\begin{bmatrix} \nu+\sigma+1 \\ \nu+\sigma+3/2 \end{bmatrix}$

$[\operatorname{Re}(\nu+\sigma) > -1]$.

12. $\int_0^a \frac{1}{a^2-x^2} \left[P_\nu^\mu\left(\frac{x}{a}\right)\right]^2 dx = -\frac{1}{2\mu a} \Gamma\begin{bmatrix} 1+\mu+\nu \\ 1-\mu+\nu \end{bmatrix}$ $[\operatorname{Re}\mu < 0]$.

13. $\int_{-a}^a \frac{1}{a^2-x^2} P_l^m\left(\frac{x}{a}\right) P_{l+2}^m\left(\frac{x}{a}\right) dx = \frac{(l+m-2)!}{(l-m-2)!\,ma}.$

14. $\int_{-a}^a \frac{1}{a^2-x^2} P_n^k\left(\frac{x}{a}\right) P_n^m\left(\frac{x}{a}\right) dx = \frac{(n+m)!}{(n-m)!\,ma} \delta_{k,m}$ $[0 \leqslant m \leqslant n; \; 0 \leqslant k \leqslant n]$.

15. $\int_{-a}^a \frac{1}{a^2-x^2} P_l^m\left(\frac{x}{a}\right) P_n^m\left(\frac{x}{a}\right) dx = 0$ $[0 \leqslant m \leqslant l \leqslant n; \; n-l = 1, 3, \ldots]$.

16. $\int_{-a}^a \frac{x}{a^2-x^2} P_l^k\left(\frac{x}{a}\right) P_n^m\left(\frac{x}{a}\right) dx = \frac{(l+m)!}{m\,(l-m)!} \delta_{n,\,l+2k+1}$ $[n > l]$,

17. $= \frac{(l+m-2k-1)!}{m\,(l-m-2k-1)!} \delta_{n,\,l-2k-1}$ $[n < l]$,

$[k = 0, 1, 2, \ldots; \; m \neq 0]$.

18. $\int_{-a}^a (a^2-x^2)^{-p} P_l^{2k}\left(\frac{x}{a}\right) P_n^{2m}\left(\frac{x}{a}\right) dx = 0$

$[l+n = 2, 4, 6, \ldots, \; l \leqslant n, \; k-p \geqslant m, \; l-2p < n, \; p = 1, 2, \ldots]$.

19. $\int_{-a}^{a} \sqrt{a^2-x^2} P_l^{m-1}\left(\frac{x}{a}\right) P_n^m\left(\frac{x}{a}\right) dx=0$ $[n \neq l \pm 1]$.

20. $= \frac{2(l+m+1)!}{(l-m+1)!(2l+1)(2l+3)}$ $[n=l+1]$.

21. $\int_{-a}^{a} \frac{1}{\sqrt{a^2-x^2}} P_l^k\left(\frac{x}{a}\right) P_n^m\left(\frac{x}{a}\right) dx=I(k, l, m, n)$,

$$I(m-1, n+2k+1, m, n)=I(m-1, n+2k+1, m, n)=$$
$$=I(m+1, n-2l+1, m, n)=0,$$
$$I(m+1, n+2k+1, m, n)=\frac{2(n+m)!}{(n-m)!},$$
$$I(m-1, n-2k-1, m, n)=\frac{2(n+m-2k-2)!}{(n-m-2k)!} \quad [k=0, 1, 2, \ldots].$$

22. $\int_{a}^{\infty} (x^2-a^2)^{\beta-1} P_{-\nu}^{\mu}\left(\frac{x}{a}\right) P_{\nu}^{\mu}\left(\frac{x}{a}\right) dx=$

$$=\frac{a^{2\beta-1}}{2\sqrt{\pi}} \Gamma\left[\begin{matrix}1/2-\beta, \beta-\mu, 1-\beta+\nu, 1-\beta-\nu \\ 1-\beta, 1-\beta-\mu, 1-\mu+\nu, 1-\mu-\nu\end{matrix}\right]$$
$[a>0;\ \mathrm{Re}\,\mu < \mathrm{Re}\,\beta < 1-|\mathrm{Re}\,\nu|, 1/2]$.

23. $\int_{a}^{\infty} (x^2-a^2)^{\beta-1} P_{\nu}^{\mu-1}\left(\frac{x}{a}\right) P_{\nu}^{\mu}\left(\frac{x}{a}\right) dx=$

$$=\frac{a^{2\beta-1}}{2\sqrt{\pi}} \Gamma\left[\begin{matrix}1-\beta, \beta-\mu+1/2, 1/2-\beta-\nu, 3/2-\beta+\nu \\ 3/2-\beta, 3/2-\beta-\mu, 2-\mu+\nu, 1-\mu-\nu\end{matrix}\right]$$
$[a>0;\ \mathrm{Re}\,\mu-1/2<\mathrm{Re}\,\beta<1/2-\mathrm{Re}\,\nu, 3/2+\mathrm{Re}\,\nu]$.

24. $\int_{a}^{\infty} x(x^2-a^2)^{\beta-1} \left[P_{\nu}^{\mu}\left(\frac{x}{a}\right)\right]^2 dx=\frac{a^{2\beta}}{2\sqrt{\pi}} \left[\begin{matrix}1/2-\beta, \beta-\mu, 1-\beta+\nu, -\nu-\beta \\ 1-\beta, 1-\mu+\nu, -\mu-\nu, 1-\beta-\mu\end{matrix}\right]$

$[a>0;\ \mathrm{Re}\,\mu<\mathrm{Re}\,\beta<-\mathrm{Re}\,\nu, 1+\mathrm{Re}\,\nu]$

25. $\int_{a}^{\infty} x(x^2-a^2)^{\beta-1} P_{\nu}^{-\mu}\left(\frac{x}{a}\right) P_{\nu}^{\mu}\left(\frac{x}{a}\right) dx=$

$$=-\frac{a^{2\beta}\sin\nu\pi}{2\pi^{3/2}} \Gamma\left[\begin{matrix}\beta, 1/2-\beta, -\beta-\nu, 1-\beta+\nu \\ 1-\beta+\mu, 1-\beta-\mu\end{matrix}\right] \quad [a>0;\ 0<\mathrm{Re}\,\beta<-\mathrm{Re}\,\nu, 1+\mathrm{Re}\,\nu].$$

26. $\int_{a}^{\infty} (x^2-a^2)^{-\rho/2} (x^2-b^2)^{\mu/2} P_{\nu}^{\mu}\left(\frac{x}{b}\right) P_{\sigma}^{\rho}\left(\frac{x}{a}\right) dx=\frac{a^{\mu-\nu-\rho} b^{\nu+1}}{2^{2+\mu-\rho}\pi}\times$

$$\times\Gamma\left[\begin{matrix}(1-\mu+\nu+\rho+\sigma)/2, (\nu-\mu+\rho-\sigma)/2, (\rho-\mu+\sigma-\nu)/2, (\rho-\mu-\sigma-\nu-1)/2 \\ \rho-\mu, -\mu-\nu, 1-\mu+\nu\end{matrix}\right]\times$$
$$\times {}_2F_1\left(\frac{1-\mu+\nu+\rho+\sigma}{2}; \frac{\nu+\rho-\mu-\sigma}{2}; \rho-\mu; 1-\frac{b^2}{a^2}\right)$$

$[\mathrm{Re}\,\rho<1;\ \mathrm{Re}(\rho+\sigma+\nu-\mu)>-1;\ \mathrm{Re}(\rho+\nu-\sigma-\mu), \mathrm{Re}(\rho+\sigma-\mu-\nu)>0, \mathrm{Re}(\rho-\mu-\sigma-\nu)>1;\ 0<b<a]$.

27. $\int_{a}^{\infty} (x^2-a^2)^{-\rho/2} (x^2-b^2)^{\mu/2} P_{\sigma}^{\mu}\left(\frac{x}{b}\right) P_{\sigma}^{\rho}\left(\frac{x}{a}\right) dx=$

$$=\frac{2^{\rho-\mu-2}}{\sqrt{\pi}}\Gamma\left[\begin{matrix}(\rho-\mu+1)/2+\sigma,\ (\rho-\mu-1)/2-\sigma,\ (\rho-\mu)/2\\ -\mu-\sigma,\ 1-\mu+\sigma\end{matrix}\right]\times$$
$$\times(a^2-b^2)^{(1+\mu-\rho)/2}P_\sigma^{(1+\mu-\rho)/2}\left(\frac{a^2+b^2}{2ab}\right)$$

[Re $\rho<1$; Re $(\rho-\mu+2\sigma)$, Re $(\rho-\mu-2\sigma)>1$; Re $(\rho-\mu)>0$; $0<b<a$].

28. $$\int_a^\infty (x^2-a^2)^{-\rho/2}(x^2-b^2)^{\mu/2}P_{1/2+\mu-\rho}^{\mu}\left(\frac{x}{b}\right)P_\sigma^\rho\left(\frac{x}{a}\right)dx=$$
$$=\frac{2^{\rho-\mu-2}}{\sin[(1-2\sigma)\pi/4]}\left[\begin{matrix}\rho-\mu+(2\sigma-1)/4,\ \rho-\mu-(2\sigma+3)/4\\ \rho-2\mu-1/2,\ 3/2-\rho\end{matrix}\right]\times$$
$$\times\sqrt{\frac{b}{a}}(a^2-b^2)^{(1-\rho+\mu)/2}P_{(2\sigma-1)/4}^{1+\mu-\rho}\left(\frac{2b^2-a^2}{a^2}\right)$$

[$0<b<a$; Re $\rho<1$; $-3/2<$ Re $\sigma<1/2$; 2Re $(\mu-\rho)+1/2<$ Re $\sigma<2$ Re $(\rho-\mu)-3/2$].

29. $$\int_a^\infty (x^2-a^2)^{-\rho/2}(x^2-b^2)^{\mu/2}P_\nu^\mu\left(\frac{x}{b}\right)P_{1/2+\mu-\rho}^{\rho}\left(\frac{x}{a}\right)dx=$$
$$=\frac{2^{\rho-\mu-2}}{\pi}\sqrt{\frac{a}{b}}(a^2-b^2)^{(1+\mu-\rho)/2}\times$$
$$\times\Gamma\left[\begin{matrix}(2\nu+3)/4,\ (1-2\nu)/4,\ \rho-\mu+(2\nu-1)/4,\ \rho-\mu-(2\nu-3)/4\\ -\mu-\nu,\ 1-\mu+\nu\end{matrix}\right]P_{-(2\nu+3)/4}^{1+\mu-\rho}\left(\frac{2a^2-b^2}{b^2}\right)$$

[$0<b<a$; $-3/2<$ Re $\nu<1/2$; 2Re $(\mu-\rho)+1/2<$ Re $\nu<2$Re $(\rho-\mu)+3/2$].

2.17.20. Integrals of $A(x)P_\nu^\mu(\varphi(x))P_\sigma^\rho(\chi(x))$.

1. $$\int_0^\infty \frac{x^{\alpha-1}}{(x+b)^r(x+c)^s}P_\nu^\mu\left(\sqrt{1+\frac{x}{b}}\right)P_\sigma^\rho\left(\sqrt{1+\frac{x}{c}}\right)dx=2^{\mu+\rho}b^{\alpha-r-\rho/2}c^{\rho/2-s}\times$$
$$\times\Gamma\left[\begin{matrix}(1+\nu+\rho)/2+r-\alpha,\ (\rho-\nu)/2+r-\alpha,\ \alpha-(\mu+\rho)/2\\ r+(1-\mu+\nu)/2,\ r-(\mu+\nu)/2,\ 1-\rho,\ 1-\alpha+(\rho-\mu)/2\end{matrix}\right]\times$$
$$\times{}_4F_3\left(\frac{\mu-\rho}{2}+\alpha,\ \alpha-\frac{\mu+\rho}{2},\ \frac{1-\rho+\sigma}{2}+s,\ s-\frac{\rho+\sigma}{2};\right.$$
$$\left.1-\rho,\ \frac{1-\nu-\rho}{2}+\alpha-r,\ 1+\alpha+\frac{\nu-\rho}{2}-r;\ \frac{b}{c}\right)+2^{\mu+\rho}b^{-\nu/2}c^{\nu/2+\alpha-r-s}\times$$
$$\times\Gamma\left[\begin{matrix}\nu+1/2,\ \alpha+(\nu-\rho)/2,\ r+s-\alpha+(1-\nu+\sigma)/2,\ r+s-\alpha-(\nu+\sigma)/2\\ r-(\mu+\nu)/2,\ (1-\rho+\sigma)/2+s,\ s-(\rho+\sigma)/2,\ 1+(\nu-\mu)/2,\ 1+r-\alpha-(\nu+\rho)/2\end{matrix}\right]\times$$
$$\times{}_4F_3\left(\frac{\mu-\nu}{2},\ \frac{1-\mu+\nu}{2}+r,\ \frac{1-\nu+\sigma}{2}+r+s-\alpha,\ r+s-\alpha-\frac{\nu+\sigma}{2};\right.$$
$$\left.\frac{1}{2}-\nu,\ 1-\alpha+\frac{\rho-\nu}{2},\ 1+r-\alpha-\frac{\rho+\nu}{2};\ \frac{b}{c}\right)+2^{\mu+\rho}b^{(\nu+1)/2}c^{\alpha-r-s-(\nu+1)/2}\times$$
$$\times\Gamma\left[\begin{matrix}-1/2-\nu,\ \alpha-\frac{1+\nu+\rho}{2},\ 1+r+s-\alpha+\frac{\nu+\sigma}{2},\ r+s-\alpha+\frac{1+\nu-\sigma}{2}\\ r+\frac{1-\mu+\nu}{2},\ \frac{1-\rho+\sigma}{2}+s,\ s-\frac{\rho+\sigma}{2},\ \frac{1-\mu-\nu}{2},\ \frac{3+\nu-\rho}{2}+r-\alpha\end{matrix}\right]\times$$
$$\times{}_4F_3\left(\frac{1+\mu+\nu}{2},\ r-\frac{\mu+\nu}{2},\ 1+r+s-\alpha+\frac{\nu+\sigma}{2},\ r+s-\alpha+\frac{1-\sigma+\nu}{2};\right.$$
$$\left.\frac{3}{2}+\nu,\ \frac{3+\nu+\rho}{2}-\alpha,\ \frac{3-\rho+\nu}{2}+r-\alpha;\ \frac{b}{c}\right)$$

[$r=0$ or $1/2$; $s=0$ or $1/2$; Re $(\mu+\rho)/2<$ Re $\alpha<r+s-$ Re $(\nu+\sigma)/2$, $r+s+$ Re $(1+\nu-\sigma)/2$, $r+s+$ Re $(1+\sigma-\nu)/2$, $1+r+s+$ Re $(\nu+\sigma)/2$; $|\arg b|$, $|\arg c|<\pi$].

2. $\int\limits_a^\infty x(x^2-a^2)^{\beta-1} P_{\mu-1/2}^{\nu+1/2}\left(\frac{x}{\sqrt{x^2-a^2}}\right) P_\nu^\mu\left(\frac{x}{a}\right) dx =$

$$= \frac{a^{2\beta}}{\pi\sqrt{8}} \Gamma\left[\begin{matrix} 1/4+\beta, & 1/4-\beta, & 1/4+\beta-\mu, & -1/4-\beta-\nu \\ 3/4-\beta-\mu, & 1/4+\beta-\nu, & -\mu-\nu & \end{matrix}\right]$$

[$a>0$; $-1/4$, $\operatorname{Re}\mu-1/4<\operatorname{Re}\beta<1/4$, $-1/4-\operatorname{Re}\nu$].

3. $\int\limits_a^\infty (x^2-a^2)^{-\rho/2}(x^2+z^2)^{-(\nu+1)/2} P_\nu^\mu\left(\frac{x}{\sqrt{x^2+z^2}}\right) P_\sigma^\rho\left(\frac{x}{a}\right) dx =$

$$= \frac{2^{\rho+\nu-1} a^{\mu-\rho-\nu}}{\sqrt{\pi}\, z^\mu} \Gamma\left[\begin{matrix} (1-\mu+\nu+\rho+\sigma)/2, & (\nu-\mu+\rho-\sigma)/2 \\ 1-\mu, & 1+\nu-\mu \end{matrix}\right] \times$$

$$\times {}_2F_1\left(\frac{\nu-\mu+\rho+\sigma+1}{2}, \frac{\nu-\mu+\rho-\sigma}{2}; 1-\mu; -\frac{z^2}{a^2}\right)$$

[$a>0$; $\operatorname{Re}(\mu-\nu-\rho)-1<\operatorname{Re}\sigma<\operatorname{Re}(\rho+\nu-\mu)$; $|\arg z|<\pi$].

4. $\int\limits_0^\infty x(x^2+b^2)^{-(\nu+1)/2}(x^2+c^2)^{-(\sigma+1)/2} P_\nu^\mu\left(\frac{b}{\sqrt{x^2+b^2}}\right) P_\sigma^\mu\left(\frac{c}{\sqrt{x^2+c^2}}\right) dx =$

$$= \Gamma\left[\begin{matrix} \nu+\sigma \\ 1-\mu+\nu, & 1-\mu+\sigma \end{matrix}\right] (b+c)^{-\nu-\sigma}$$ [$\operatorname{Re} b$, $\operatorname{Re} c$, $\operatorname{Re}(\nu+\sigma)>0$; $\operatorname{Re}\mu<1$].

2.17.21. Integrals containing exponential or trigonometric functions and $P_\nu^\mu(\varphi(x))P_\sigma^\rho(\chi(x))$.

1. $\int\limits_a^\infty x(x^2-a^2)^{-1/4} e^{-p\sqrt{x^2-a^2}} P_\nu^{-1/4}\left(\frac{x}{a}\right) P_\nu^{1/4}\left(\frac{x}{a}\right) dx =$

$$= \sqrt{\frac{\pi}{2p}}\frac{a}{2} H_{\nu+1/2}^{(1)}\left(\frac{ap}{2}\right) H_{\nu+1/2}^{(2)}\left(\frac{ap}{2}\right)$$ [a, $\operatorname{Re} p>0$].

2. $\int\limits_a^\infty (x^2-a^2)^{-1/4} \left\{\begin{matrix} \sin(b\sqrt{x^2-a^2}) \\ \cos(b\sqrt{x^2-a^2}) \end{matrix}\right\} P_\nu^{\pm 1/4}\left(\frac{x}{a}\right) P_\nu^{\mp 3/4}\left(\frac{x}{a}\right) dx =$

$$= a\sqrt{\frac{b}{2\pi}} \frac{1}{\Gamma((6\pm 1)/4+\nu)\,\Gamma((2\pm 1)/4-\nu)} K_{\nu+1/2}^2\left(\frac{ab}{2}\right)$$

[a, $b>0$; $(\mp 1-6)/4<\operatorname{Re}\nu<(2\pm 1)/4$].

3. $\int\limits_a^\infty (x^2-a^2)^{-1/4} \left\{\begin{matrix} \sin(b\sqrt{x^2-a^2}) \\ \cos(b\sqrt{x^2-a^2}) \end{matrix}\right\} P_{\nu-1}^{\mp 1/4}\left(\frac{x}{a}\right) P_\nu^{\mp 1/4}\left(\frac{x}{a}\right) dx =$

$$= a\sqrt{\frac{b}{2\pi}} \frac{1}{\Gamma(1\pm 1/4+\nu)\,\Gamma(1\pm 1/4-\nu)} K_{\nu-1/2}\left(\frac{ab}{2}\right) K_{\nu+1/2}\left(\frac{ab}{2}\right)$$

[a, $b>0$; $|\operatorname{Re}\nu|<1\pm 1/4$].

4. $\int\limits_a^\infty x(x^2-a^2)^{-1/4} \left\{\begin{matrix} \sin(b\sqrt{x^2-a^2}) \\ \cos(b\sqrt{x^2-a^2}) \end{matrix}\right\} \left[P_\nu^{\mp 1/4}\left(\frac{x}{a}\right)\right]^2 dx =$

$$= \sqrt{\frac{2}{\pi b}} \frac{a}{\Gamma(1\pm 1/4+\nu)\,\Gamma(\pm 1/4-\nu)} K_{\nu+1/2}^2\left(\frac{ab}{2}\right)$$ [a, $b>0$; $-1\mp 1/4<\operatorname{Re}\nu<\pm 1/4$].

5. $\int\limits_0^a \sin^{-\rho}(a-x)\sin^{\rho-1} x P_\nu^\mu(\cos x) P_\nu^\rho(\cos(a-x))\, dx =$

$$= \frac{2^{-\rho}}{\sqrt{\pi}} \Gamma\left[\begin{matrix} \rho-\mu, & 1/2-\rho \\ 1-\mu-\rho \end{matrix}\right] \sin^{-\rho} a P_\nu^\mu(\cos a)$$ [$a>0$; $\operatorname{Re}\mu<\operatorname{Re}\rho<1/2$].

6. $\int_0^a \sin^{-\rho}(a-x)\sin^{-\mu}x P_\nu^\mu(\cos x)\, P_{\mu+\nu+\rho-1}^\rho(\cos(a-x))\,dx=$

$$=\frac{1}{\sqrt{2\pi}}\mathrm{B}\left(\frac{1}{2}-\mu,\ \frac{1}{2}-\rho\right)\sin^{1/2-\mu-\rho}a P_{\nu+\rho-1/2}^{\mu+\rho-1/2}(\cos a) \quad [a>0;\ \mathrm{Re}\,\mu,\ \mathrm{Re}\,\rho<1/2]$$

2.17.22. Integrals of $x^\alpha J_\lambda(\varphi(x))P_\nu^\mu(cx)\,P_\sigma^\rho(cx)$.

1. $\int_a^\infty J_{-2\mu}\left(b\sqrt{x^2-a^2}\right)P_{\nu-1}^\mu\left(\frac{x}{a}\right)P_\nu^\mu\left(\frac{x}{a}\right)dx=$

$$=\frac{a}{\pi\Gamma(1-\mu+\nu)\,\Gamma(1-\mu-\nu)}K_{\nu-1/2}\left(\frac{ab}{2}\right)K_{\nu+1/2}\left(\frac{ab}{2}\right)$$

$[a,\ b>0;\ \mathrm{Re}\,\mu<1/2;\ |\mathrm{Re}\,\nu|<3/4]$.

2. $\int_a^\infty J_{1-2\mu}\left(b\sqrt{x^2-a^2}\right)P_\nu^{\mu-1}\left(\frac{x}{a}\right)P_\nu^\mu\left(\frac{x}{a}\right)dx=$

$$=\frac{a}{\pi\Gamma(2-\mu+\nu)\,\Gamma(1-\mu-\nu)}K_{\nu+1/2}^2\left(\frac{ab}{2}\right) \quad [a,\ b>0;\ \mathrm{Re}\,\mu<1,\ -5/4<\mathrm{Re}\,\nu<1/4].$$

3. $\int_a^\infty xJ_0\left(b\sqrt{x^2-a^2}\right)P_\nu^{-\mu}\left(\frac{x}{a}\right)P_\nu^\mu\left(\frac{x}{a}\right)dx=$

$$=-\frac{2\sin\nu\pi}{\pi b^2}W_{\mu,\ \nu+1/2}(ab)\,W_{-\mu,\ \nu+1/2}(ab) \quad [a,\ b>0;\ -3/4<\mathrm{Re}\,\nu<-1/4].$$

4. $\int_a^\infty xJ_0\left(b\sqrt{x^2-a^2}\right)\left[P_\nu\left(\frac{x}{a}\right)\right]^2dx=-\frac{2a\sin\nu\pi}{\pi^2 b}K_{\nu+1/2}^2\left(\frac{ab}{2}\right)$

$[a,\ b>0;\ -3/4<\mathrm{Re}\,\nu<-1/4]$.

5. $\int_a^\infty xJ_0\left(b\sqrt{x^2-a^2}\right)\left[P_\nu^\mu\left(\frac{x}{a}\right)\right]^2dx=$

$$=-\frac{i}{\pi b^2}W_{\mu,\ \nu+1/2}(ab)\left[W_{\mu,\ \nu+1/2}(e^{\pi i}ab)-W_{\mu,\ \nu+1/2}(e^{\pi i}ab)\right]$$

$[a,\ b>0;\ \mathrm{Re}\,\mu<1;\ -3/4<\mathrm{Re}\,\nu<-1/4]$.

6. $\int_a^\infty xJ_{-2\mu}\left(b\sqrt{x^2-a^2}\right)\left[P_\nu^\mu\left(\frac{x}{a}\right)\right]^2dx=$

$$=\frac{2a}{\pi b\Gamma(1-\mu+\nu)\,\Gamma(-\mu-\nu)}K_{\nu+1/2}^2\left(\frac{ab}{2}\right) \quad [a,\ b>0;\ \mathrm{Re}\,\mu<1/2;\ -3/4<\mathrm{Re}\,\nu<-1/4].$$

2.17.23. Integrals containing three of the functions $P_\nu^\mu(\varphi(x))$.

1. $\int_{-a}^a (a^2-x^2)^{m/2}P_k^m\left(\frac{x}{a}\right)P_l^m\left(\frac{x}{a}\right)P_n^m\left(\frac{x}{a}\right)dx=$

$$=(-1)^m 2a^{m+1}\frac{(k+m)!\,(l+m)!\,(n+m)!\,(p-m)!\,(1/2)_m(1/2)_{q-k}(1/2)_{q-l}(1/2)_{q-n}}{(k-m)!\,(l-m)!\,(n-m)!\,(p-m)!\,(p-l)!\,(p-n)!\,(3/2)_p}$$

$[2p=k+l+m+n,\ 2q=k+l+n-m$ are even; $l\geqslant m;\ m\leqslant k-l-m\leqslant n\leqslant k+l+m]$.

2. $\int_a^\infty \sqrt{\frac{x}{x^2+z^2}}\, P_\nu\left(\frac{x}{a}\right) P^\mu_{-1/4}\left(\frac{\sqrt{x^2+z^2}}{x}\right) P^\mu_{1/4}\left(\frac{\sqrt{x^2+z^2}}{x}\right) dx =$

$$=\frac{2^{1-\mu} a^{3/2}}{\sqrt{\pi}\, z} \Gamma\left[\begin{matrix}(2\nu+3)/4-\mu, & (1-2\nu)/4-\mu\\ 3/2-2\mu\end{matrix}\right]\left[P^\mu_{(2\nu-1)/4}\left(\sqrt{\frac{a^2+z^2}{a}}\right)\right]^2$$

$[a,\ \operatorname{Re} z>0;\ \operatorname{Re}\mu>1/2;\ 2\operatorname{Re}\mu-3/2<\operatorname{Re}\nu<-2\operatorname{Re}\mu+1/2]$.

2.17.24. Integrals with respect to the index of products of $P_{\sigma x+\nu}(c)$ by elementary functions.

1. $\int_{-\infty}^{\infty} P_x(c)\, dx = \sqrt{\frac{2}{1-c}}$ $[|c|<1]$.

2. $\int_0^\infty P_{-x-1/2}(c)\, dx = \frac{1}{\sqrt{2(1-c)}}$ $[|c|<1]$.

3. $\int_0^\infty \frac{\operatorname{th}\pi x}{x} P_{ix-1/2}(c)\, dx = \frac{2}{\sqrt{c+\sqrt{c^2-1}}} \mathbf{K}\left(\frac{1}{c+\sqrt{c^2-1}}\right)$ $[c>1]$.

4. $\int_0^\infty \frac{x\operatorname{th}\pi x}{x^2+z^2} P_{ix-1/2}(c)\, dx = Q_{z-1/2}(c)$ $[c>1;\ \operatorname{Re} z>0]$.

5. $\int_0^\infty x\frac{\operatorname{sh} bx}{\operatorname{ch}\pi x} P_{ix-1/2}(c)\, dx = \frac{2^{-3/2}\sin b}{(c+\cos b)^{3/2}}$ $[b>0;\ -1<c\leqslant 1]$.

6. $\int_0^\infty \frac{\operatorname{ch} bx}{\operatorname{ch}^2\pi x} P_{ix-1/2}(c)\, dx = \frac{1}{\sqrt{2(c-\cos b)}} - \frac{\sqrt{2(c-\cos b)}}{\pi}\operatorname{arctg}\sqrt{\frac{1+\cos b}{c-\cos b}}$

$[c>1]$.

7. $\int_0^\infty \frac{\operatorname{th}\pi x}{\operatorname{ch}\pi x}\operatorname{sh} bx\, P_{ix-1/2}(c)\, dx = \frac{1}{\pi}\sqrt{\frac{2}{c+\cos b}}\operatorname{arctg}\sqrt{\frac{1-\cos b}{c+\cos b}}$. $[c>1]$.

8. $\int_0^\infty \operatorname{sh}\pi x\cos bx\, P_{ix-1/2}(c)\, dx = \frac{1}{\sqrt{2(c+\operatorname{ch} b)}}$ $[b>0;\ |c|<1]$.

9. $\int_0^\infty \frac{\cos bx}{\operatorname{sh}^2\pi x} P_{ix-1/2}(c)\, dx = \frac{1}{\pi}\sqrt{\frac{2}{c-\operatorname{ch} b}}\operatorname{arctg}\sqrt{\frac{c-\operatorname{ch} b}{1+\operatorname{ch} b}}$ $[c>\operatorname{ch} b]$.

10. $= \frac{1}{\pi\sqrt{\operatorname{ch} b-c}}\ln\frac{\sqrt{\operatorname{ch} b+1}+\sqrt{\operatorname{ch} b-c}}{\sqrt{\operatorname{ch} b+1}-\sqrt{\operatorname{ch} b-c}}$ $[c<\operatorname{ch} b]$.

2.17.25. Integrals with respect to the index of products of $P^\mu_{ix-1/2}(c)$ by elementary functions.

1. $\int_0^\infty P^\mu_{ix-1/2}(c)\, dx = \sqrt{\frac{\pi}{2}}\frac{(c-1)^{-(\mu+1)/2}}{\Gamma(1/2-\mu)}(c+1)^{\mu/2}$ $[\operatorname{Re}\mu<1/2]$.

2. $\int_0^\infty \cos bx\, P^\mu_{ix-1/2}(c)\, dx = \begin{Bmatrix}1\\0\end{Bmatrix}\sqrt{\frac{\pi}{2}}\frac{(c^2-1)^{\mu/2}}{\Gamma(1/2-\mu)}(c-\operatorname{ch} b)^{-\mu-1/2}$

$\left[\begin{Bmatrix}c>\operatorname{ch} b\\ c<\operatorname{ch} b\end{Bmatrix},\ \operatorname{Re}\mu<1/2\right]$.

3. $\int_0^\infty \frac{\cos bx}{\operatorname{ch}\pi x} P^{\mu}_{ix-1/2}(c)\,dx = 2^{-\mu-1/2}(c+1)^{\mu/2}(c+\operatorname{ch} b)^{-(\mu+1)/2} P^{\mu}_{\mu}\left(\sqrt{\frac{1+\operatorname{ch} b}{c+\operatorname{ch} b}}\right).$

2.17.26. Integrals with respect to the index of products of $P_{ix-1/2}(c)$ by special functions.

1. $\int_0^\infty x \begin{Bmatrix} \operatorname{sh}\pi x \\ \operatorname{th}\pi x \end{Bmatrix} \Gamma\left[\nu+\frac{ix}{2}, \nu-\frac{ix}{2}, \frac{1}{2}-\nu+\frac{ix}{2}, \frac{1}{2}-\nu-\frac{ix}{2}\right] P_{ix-1/2}(c)\,dx=$

$= \begin{Bmatrix} 1 \\ \sec 2\nu\pi \end{Bmatrix} \frac{2\pi^2}{\sqrt{c^2-1}} \left[(c+\sqrt{c^2-1})^{1/2-2\nu} \pm (c-\sqrt{c^2-1})^{1/2-2\nu}\right]$ $[0\leqslant \operatorname{Re}\nu \leqslant 1/2]$.

2. $\int_0^\infty \cos bx \left[\psi\left(\frac{1}{2}+ix\right)+\psi\left(\frac{1}{2}-ix\right)\right] P_{ix-1/2}(c)\,dx = -\frac{\pi}{\sqrt{2(\operatorname{ch} b-c)}}$ $[c<\operatorname{ch} b]$.

3. $= \sqrt{\frac{2}{c-\operatorname{ch} b}}\left[-C-\ln 4+\frac{1}{2}\ln(c^2-1)-\ln(c-\operatorname{ch} b)\right]$ $[c>\operatorname{ch} b]$.

4. $\int_0^\infty x \frac{\operatorname{sh}(\pi x/2)}{\operatorname{ch}\pi x} \begin{Bmatrix} J_{ix}(b)+J_{-ix}(b) \\ Y_{ix}(b)+Y_{-ix}(b) \end{Bmatrix} P_{ix-1/2}(c)\,dx = \sqrt{\frac{2b}{\pi}} \begin{Bmatrix} \cos(bc-3\pi/4) \\ \sin(bc-3\pi/4) \end{Bmatrix}.$

5. $\int_0^\infty x \frac{\operatorname{th}\pi x}{\operatorname{ch}(\pi x/2)} \begin{Bmatrix} J_{ix}(b)+J_{-ix}(b) \\ Y_{ix}(b)+Y_{-ix}(b) \end{Bmatrix} P_{ix-1/2}(c)\,dx = 2\sqrt{\frac{b}{\pi}}(\pm\sin bc-\cos bc).$

6. $\int_0^\infty \frac{x}{\operatorname{ch}\pi x} \begin{Bmatrix} J^2_{ix}(b)-J^2_{-ix}(b) \\ Y^2_{ix}(b)-Y^2_{-ix}(b) \end{Bmatrix} P_{ix-1/2}(c)\,dx = \mp\frac{i}{\pi}\sqrt{\frac{2}{c+1}}\cos\left(b\sqrt{2+2c}\right).$

7. $\int_0^\infty x \operatorname{th}\pi x \left[\begin{Bmatrix} J_{ix}(a) \\ Y_{ix}(a) \end{Bmatrix} J_{-ix}(b) \mp \begin{Bmatrix} Y_{ix}(a) \\ J_{ix}(a) \end{Bmatrix} Y_{-ix}(b)\right] P_{ix-1/2}(c)\,dx=$

$= -\frac{2\sqrt{ab}}{\pi z} \begin{Bmatrix} \sin z \\ \cos z \end{Bmatrix}$ $[z=\sqrt{a^2+b^2+2abc}]$.

8. $\int_0^\infty \frac{x}{\operatorname{ch}\pi x}\left[J_{ix}(b)Y_{-ix}(b)-J_{-ix}(b)Y_{ix}(b)\right] P_{ix-1/2}(c)\,dx = \frac{i}{\pi}\sqrt{\frac{2}{c-1}}e^{-b\sqrt{2c-2}}$

$[c>1]$.

9. $\int_0^\infty x \operatorname{th}\pi x \left[J^2_{ix}(b)+Y^2_{ix}(b)\right] P_{ix-1/2}(c)\,dx = \frac{1}{\pi}\sqrt{\frac{2}{c-1}}e^{-b\sqrt{2c-2}}$ $[c>1]$.

10. $\int_0^\infty xe^{-\pi x}\operatorname{th}\pi x\left[H^{(1)}_{ix}(b)\right]^2 P_{ix-1/2}(c)\,dx = -\frac{1}{\pi}\sqrt{\frac{2}{c+1}}e^{ib\sqrt{2c+2}}$ $[c>-1]$.

11. $\int_0^\infty x \operatorname{th}\pi x\, H^{(1)}_{ix}(b)\, H^{(2)}_{ix}(b) P_{ix-1/2}(c)\,dx = \frac{1}{\pi}\sqrt{\frac{2}{c-1}}e^{-b\sqrt{2c-2}}$ $[c>1]$.

12. $\int_0^\infty xe^{\pi x}\operatorname{th}\pi x H^{(2)}_{ix}(a)\, H^{(2)}_{ix}(b)\, P_{ix-1/2}(c)\,dx = -\frac{2\sqrt{ab}}{\pi z}e^{-iz}$

$[z=\sqrt{a^2+b^2+2abc};\ \operatorname{Im} a,\ \operatorname{Im} b\leqslant 0;\ |c|<1]$.

13. $\int_0^\infty x \operatorname{th} \pi x \, [I_{ix}(b)+I_{-ix}(b)] \, P_{ix-1/2}(c)\, dx = \frac{\sin (b\sqrt{2c-2})}{\sqrt{2c-2}}$ $[c>1]$.

14. $\int_0^\infty x \frac{\operatorname{sh}(\pi x/2)}{\operatorname{ch} \pi x} K_{ix/2}(b) \, P_{ix-1/2}(c)\, dx = \sqrt{\frac{\pi}{2}}\, b^{1/4} e^{b(c^2-1)} D_{-3/2}(2c\sqrt{b})$.

15. $\int_0^\infty x \operatorname{th} \pi n x K_{ilx}(b) \, P_{ix-1/2}(c)\, dx = I_{n,\, l}$ $[c>1]$,

$$I_{1,1} = \sqrt{\frac{\pi b}{2}}\, e^{-bc}, \quad I_{1,2} = \frac{b}{4} K_0\left(b\sqrt{\frac{1+c}{2}}\right), \quad I_{2,2} = \frac{1}{4}\sqrt{\frac{\pi}{2}}\, e^{-bc}.$$

16. $\int_0^\infty x \frac{\operatorname{th} \pi x}{\operatorname{ch} \rho \pi x} K_{i\sigma x}(b) \, P_{ix-1/2}(c)\, dx = I_{\rho,\, \sigma}$ $[b>0;\ -1<c\leqslant 1]$,

$$I_{1/2,\, 1/2} = \sqrt{2\pi}\, b^{1/4} e^{bc^2-b} D_{-3/2}(2c\sqrt{b}),$$

$$I_{1/2,\, 1} = 2b^{3/4} e^{-bc^2+b} D_{-3/2}(2c\sqrt{b}), \quad I_{1,1} = -\sqrt{\frac{b}{2\pi}}\, e^{bc} \operatorname{Ei}(-b-bc).$$

17. $\int_0^\infty x \left(x^2+\frac{1}{4}\right) \frac{\operatorname{th} \pi x}{\operatorname{ch} \pi x} K_{ix}(b) \, P_{ix-1/2}(c)\, dx =$

$$= \sqrt{\frac{b}{2\pi}}\, [(1+bc-b)\, e^{-b} + b(2c-b+bc^2)\, e^{bc} \operatorname{Ei}(-b-bc)] \quad [b>0;\ 1<c<1].$$

18. $\int_0^\infty \frac{1}{\operatorname{ch} \pi x} \operatorname{Re}[K_{ix+1/2}(b)\, P_{ix-1/2}(c)]\, dx = e^{(bc-b)/2} K_0\left(\frac{b+bc}{2}\right)$ $[b>0;\ -1<c\leqslant 1]$.

19. $\int_0^\infty x \frac{\operatorname{sh}(\pi x/2)}{\operatorname{ch} \pi x} \begin{Bmatrix} J_{ix}(b)+J_{-ix}(b) \\ Y_{ix}(b)+Y_{-ix}(b) \end{Bmatrix} K_{ix}(b) \, P_{ix-1/2}(c)\, dx =$

$$= \pm \frac{1}{\sqrt{2c}}\, e^{-b\sqrt{c}} \begin{Bmatrix} \sin(b\sqrt{c}) \\ \cos(b\sqrt{c}) \end{Bmatrix}.$$

20. $\int_0^\infty x \operatorname{th} \pi x \, [I_{ix}(a) K_{ix}(b) - K_{ix}(a) I_{ix}(b)] \, P_{ix-1/2}(c)\, dx =$

$$= \begin{Bmatrix} 1 \\ 0 \end{Bmatrix} \sqrt{\frac{ab}{a^2+b^2-2abc}} \operatorname{ch}(\sqrt{a^2+b^2-2abc}) \quad \left[\begin{Bmatrix} 2abc<a^2+b^2 \\ 2abc\geqslant a^2+b^2 \end{Bmatrix}\right].$$

21. $\int_0^\infty x \operatorname{sh} \pi x K_{ix}^2(b) \, P_{ix-1/2}(c)\, dx = \frac{\pi \cos(b\sqrt{2c-2})}{2\sqrt{2(c-1)}}$ $[c>1]$.

22. $\int_0^\infty x \operatorname{th} \pi x K_{ix}(a) K_{ix}(b) \, P_{ix-1/2}(c)\, dx = \frac{\pi\sqrt{ab}}{2ze^z}$ $[z=\sqrt{a^2+b^2+2abc};\ a,\ b>0;\ -1<c\leqslant 1]$.

23. $\int_0^\infty x \frac{\operatorname{th} \pi x}{\operatorname{ch} \pi x} K_{ix}(a) K_{ix}(b) \, P_{ix-1/2}(c)\, dx = \frac{\sqrt{ab}}{2z} [e^z \operatorname{Ei}(-z-a-b) - e^{-z} \operatorname{Ei}(z-a-b)]$

$[z=\sqrt{a^2+b^2+2abc}]$.

24. $\int\limits_0^\infty x \operatorname{th} \pi x \, K_{ix}(be^{\pi i/4}) K_{ix}(be^{-\pi i/4}) P_{ix-1/2}(c)\, dx = \frac{\pi}{2\sqrt{2c}} e^{-b\sqrt{2c}}.$

25. $\int\limits_0^\infty x \operatorname{th} \pi x \, \Gamma\left(\frac{1+2ix}{4}\right) \Gamma\left(\frac{1-2ix}{4}\right) S_{1/2,\, ix}(b) P_{ix-1/2}(c)\, dx =$

$= 4\sqrt{b\pi}\,[\sin(bc)\operatorname{ci}(bc) - \cos(bc)\operatorname{si}(bc)].$

26. $\int\limits_0^\infty x \operatorname{th} \pi x \, \Gamma(\nu+ix)\Gamma(\nu-ix) S_{1-2\nu,\, 2ix}(b) P_{ix-1/2}(c)\, dx =$

$= \frac{b}{2}\Gamma^2\left(\nu+\frac{1}{2}\right) S_{-2\nu,\, 0}\left(b\sqrt{\frac{c+1}{2}}\right) \qquad [\operatorname{Re}\nu \geqslant 0].$

2.17.27. Integrals with respect to the index of products of $P^{\mu}_{ix+\nu}(c)$ by special functions.

1. $\int\limits_0^\infty \operatorname{ch} \pi x \, \Gamma\left(\frac{1}{2}-\mu+ix\right) \Gamma\left(\frac{1}{2}-\mu-ix\right) P^{\mu}_{ix-1/2}(c)\, dx =$

$= \sqrt{\frac{\pi}{2}}\, \Gamma\left(\frac{1}{2}-\mu\right)(c+1)^{-\mu/2}(c-1)^{(\mu-1)/2} \qquad \left[|\operatorname{Re}\mu| \leqslant \frac{1}{2}\right].$

2. $\int\limits_0^\infty x \operatorname{sh} \pi x \, \Gamma(\nu+ix)\Gamma(\nu-ix) P^{\mu}_{ix-1/2}(c)\, dx =$

$= 2^{\nu-1/2}\pi\Gamma\begin{bmatrix} 1/2+\nu \\ 1/2-\mu-\nu \end{bmatrix} (c-1)^{-\nu-(\mu+1)/2}(c+1)^{\mu/2} \qquad \left[\operatorname{Re}\nu \geqslant 0;\ \operatorname{Re}(\mu+2\nu) < \frac{1}{2}\right].$

3. $\int\limits_0^\infty \cos bx \, \Gamma\left(\frac{1}{2}-\mu+ix\right) \Gamma\left(\frac{1}{2}-\mu-ix\right) P^{\mu}_{ix-1/2}(c)\, dx =$

$= \sqrt{\frac{\pi}{2}}\, \Gamma\left(\frac{1}{2}-\mu\right)(c^2-1)^{-\mu/2}(c+\operatorname{ch} b)^{\mu-1/2} \qquad [b>0;\ c>1;\ |\operatorname{Re}\mu| \leqslant 1/2].$

4. $\int\limits_0^\infty \cos bx \, \Gamma\left(\frac{2\pm1-2\mu+2ix}{4}\right) \Gamma\left(\frac{2\pm1-2\mu-2ix}{4}\right) P^{\mu}_{ix-1/2}(c)\, dx =$

$= 2^{\mu\mp1/2}\pi c^{(1\pm1)/2}\Gamma\left(1\pm\frac{1}{2}-\mu\right)(c^2+\operatorname{sh}^2 b)^{-(2\pm1)/4} P^{\mu}_{\pm1/2}\left(\frac{\operatorname{ch} b}{\sqrt{c^2+\operatorname{sh}^2 b}}\right).$

5. $\int\limits_0^\infty \frac{\operatorname{ch} bx}{\operatorname{ch} \pi x} \Gamma\left(\frac{1}{2}-\mu+ix\right) \Gamma\left(\frac{1}{2}-\mu-ix\right) P^{\mu}_{ix-1/2}(c)\, dx =$

$= \sqrt{\frac{\pi}{2}}\, \Gamma\left(\frac{1}{2}-\mu\right)(c^2-1)^{-\mu/2}\left\{(c-\cos b)^{\mu-1/2} + 2^{-\mu-1/2}\frac{1}{\sqrt{\pi}}\Gamma(1-\mu)\times\right.$

$\left.\times[(c+1)(c-\cos b)]^{(2\mu-1)/4}\left[P^{\mu-1/2}_{-\mu-1/2}(z) - P^{\mu-1/2}_{-\mu-1/2}(-z)\right]\right\}$

$[z=\sqrt{1+\cos b}/\sqrt{1+c};\ c>1;\ \operatorname{Re}\mu \leqslant 1/2].$

6. $\int\limits_0^\infty \frac{\cos bx}{\operatorname{ch} \pi x} \Gamma\left(\frac{1}{2}-\mu+ix\right) \Gamma\left(\frac{1}{2}-\mu-ix\right) P^{\mu}_{ix-1/2}(c)\, dx =$

$$=2^{\mu+1/2}e^{-i\mu\pi}\Gamma(1-2\mu)(c-1)^{-\mu/2}(\operatorname{ch}b-c)^{(\mu-1)/2}Q_{-\mu}^{\mu}\left(\operatorname{ch}\frac{b}{2}\sqrt{\frac{2}{\operatorname{ch}b-c}}\right)$$

$[1<c<\operatorname{ch}b;\quad \operatorname{Re}\mu<1/2]$.

7. $$=\frac{2^{\mu}\sqrt{\pi}\Gamma(1-2\mu)}{(c^2-1)^{1/4}}\left(\frac{c-\operatorname{ch}b}{c-1}\right)^{(2\mu-1)/4}P_{\mu-1/2}^{\mu-1/2}\left(\operatorname{ch}\frac{b}{2}\sqrt{\frac{2}{c+1}}\right)$$

$[c>\operatorname{ch}b;\quad \operatorname{Re}\mu<1/2]$.

8. $$\int_0^\infty \operatorname{sh}bx\operatorname{th}\pi x\Gamma\left(\frac{1}{2}-\mu+ix\right)\Gamma\left(\frac{1}{2}-\mu-ix\right)P_{ix-1/2}^{\mu}(c)\,dx=$$

$$=\frac{-2^{\mu-1}\sqrt{\pi}\Gamma(1-2\mu)(c+\cos b)^{(2\mu-1)/4}}{(c+1)^{1/4}(c-1)^{\mu/2}}\left[P_{-\mu-1/2}^{\mu-1/2}(z)-P_{-\mu-1/2}^{\mu-1/2}(-z)\right]$$

$[z=\sqrt{1-\cos b}/\sqrt{1+c};\quad \operatorname{Re}\mu\leqslant 1/2]$.

9. $$\int_0^\infty x\operatorname{sh}\pi nx\Gamma\left[\frac{1}{2}-\mu+ix,\quad \frac{1}{2}-\mu-ix,\quad \nu+irx,\quad \nu-irx\right]P_{ix-1/2}^{\mu}(c)\,dx=I_{n,\,r}$$

$[\operatorname{Re}\mu\leqslant 1/2;\quad \operatorname{Re}\nu\geqslant 0]$,

$$I_{1,\,1/2}=2^{3/2-2\nu}\pi^{3/2}\Gamma\left(2\nu-\mu+\frac{1}{2}\right)c^{\mu-2\nu-1/2}(c^2-1)^{-\mu/2},$$

$$I_{2,\,1}=2^{\nu+1/2}\pi^2\Gamma\begin{bmatrix}1/2-\mu+\nu\\1/2-\nu\end{bmatrix}(c-1)^{\mu-\nu-1/2}(c^2-1)^{-\mu/2}$$

$[\operatorname{Re}(2\nu-\mu)<1/2]$.

10. $$\int_0^\infty x\operatorname{sh}\pi x\Gamma\begin{bmatrix}1/2-\mu+ix,\ 1/2-\mu-ix,\ \nu+ix/2,\ \nu-ix/2\\3/4-(\mu+ix)/2,\ 3/4-(\mu-ix)/2\end{bmatrix}P_{ix-1/2}^{\mu}(c)\,dx=$$

$$=2^{1-\mu}\pi\Gamma\begin{bmatrix}(1-2\mu)/4+\nu\\(3-2\mu)/4-\nu\end{bmatrix}(c^2-1)^{-\nu-1/4}$$ $[\operatorname{Re}\mu\leqslant 1/2;\quad 0\leqslant\operatorname{Re}\nu<1/2]$.

11. $$\int_0^\infty x\Gamma\left(\frac{1}{2}-\mu+ix\right)\Gamma\left(\frac{1}{2}-\mu-ix\right)\left[J_{ix}^2(b)-J_{-ix}^2(b)\right]P_{ix-1/2}^{\mu}(c)\,dx=$$

$$=-2^{(2\mu+1)/4}\pi ib^{1/2-\mu}(c+1)^{-1/4}(c-1)^{-\mu/2}J_{1/2-\mu}\left(b\sqrt{2c+2}\right)$$ $[\operatorname{Re}\mu<1/2]$.

12. $$\int_0^\infty x\operatorname{sh}\pi x\Gamma\left(\frac{1}{2}-\mu+ix\right)\Gamma\left(\frac{1}{2}-\mu-ix\right)\left[J_{ix}(b)Y_{-ix}(b)+J_{-ix}(b)Y_{ix}(b)\right]\times$$

$$\times P_{ix-1/2}^{\mu}(c)\,dx=-2^{(2\mu+1)/4}\pi b^{1/2-\mu}(c+1)^{-1/4}(c-1)^{-\mu/2}J_{1/2-\mu}\left(b\sqrt{2c+2}\right)$$

$[\operatorname{Re}\mu<1/2]$.

13. $$\int_0^\infty x\operatorname{sh}\pi x\Gamma\left(\frac{1}{2}-\mu+ix\right)\Gamma\left(\frac{1}{2}-\mu-ix\right)\left[J_{ix}(b)J_{-ix}(b)-Y_{ix}(b)Y_{-ix}(b)\right]\times$$

$$\times P_{ix-1/2}^{\mu}(c)\,dx=2^{(2\mu+1)/4}\sqrt{\pi}b^{1/2-\mu}(c+1)^{-1/4}(c-1)^{-\mu/2}Y_{1/2-\mu}\left(b\sqrt{2c+2}\right)$$

$[|\operatorname{Re}\mu|\leqslant 1/2]$.

14. $$\int_0^\infty x\Gamma\left(\frac{1}{2}-\mu+ix\right)\Gamma\left(\frac{1}{2}-\mu-ix\right)\left[Y_{ix}^2(b)-Y_{-ix}^2(b)\right]P_{ix-1/2}^{\mu}(c)\,dx=$$

$$=-2^{(2\mu+1)/4}\pi ib^{1/2-\mu}(c+1)^{-1/4}(c-1)^{-\mu/2}J_{1/2-\mu}\left(b\sqrt{2c+2}\right)$$ $[\operatorname{Re}\mu<1/2]$.

15. $\int_0^\infty xe^{\pi x}\operatorname{sh}\pi x\Gamma(\nu+ix)\Gamma(\nu-ix)H_{ix}^{(2)}(a)H_{ix}^{(2)}(b)P_{ix-1/2}^{\mu}(c)\,dx=$

$$=i\sqrt{2\pi}(1-c^2)^{-\mu/2}(ab)^{1/2-\mu}z^{\mu-1/2}H_\nu^{(2)}(z)$$

$[z=\sqrt{a^2+b^2+2abc};\ a,\ b>0;\ |c|<1;\ \operatorname{Re}\mu<1/2]$.

16. $\int_0^\infty x\operatorname{sh}\pi xK_{2ix}(b)P_{ix-1/2}^{\mu}(c)\,dx=2^{-3\mu/2-3}\pi(c+1)^{\mu/2}b^{\mu+1}\begin{Bmatrix}I_{-\mu}(z_+)\\ J_{-\mu}(z_-)\end{Bmatrix}$

$\left[\begin{Bmatrix}|c|<1\\ c>1\end{Bmatrix}\quad z_\pm=b\sqrt{\pm\frac{1-c}{2}};\quad \operatorname{Re}\mu\leqslant 0\right]$.

17. $\int_0^\infty x\frac{\operatorname{th}\pi x}{\operatorname{ch}\pi x}K_{ix}(b)P_{ix-1/2}^{1}(c)\,dx=\sqrt{\frac{b(1-c^2)}{2\pi}}\left[\frac{e^{-b}}{1+c}+be^{bc}\operatorname{Ei}(-b-bc)\right]$

$[b>0;\ -1<c\leqslant 1]$.

18. $\int_0^\infty x\left(x^2+\frac{1}{4}\right)\frac{\operatorname{th}\pi x}{\operatorname{ch}\pi x}K_{ix}(b)P_{ix-1/2}^{1}(c)\,dx=$

$$=\sqrt{\frac{b^3(1-c^2)}{2\pi}}\left[\frac{b-bc^2-3c-1}{1+c}e^{-b}-(b^2c^2+4bc-b^2+2)e^{bc}\operatorname{Ei}(-b-bc)\right]$$

$[b>0;\ -1<c<1]$.

19. $\int_0^\infty \frac{x\operatorname{th}\pi x}{(x^2+1/4)\operatorname{ch}\pi x}K_{ix}(b)P_{ix-1/2}^{1}(c)\,dx=$

$$=\frac{(2\pi b)^{-1/2}}{\sqrt{1-c^2}}\left[e^{bc}\operatorname{Ei}(-b-bc)-e^{b}\operatorname{Ei}(-2b)-e^{-b}\ln\frac{1+c}{2}\right]$$

$[b>0;\ -1<c<1]$.

20. $\int_0^\infty x\operatorname{sh}\pi nx\Gamma\left(\frac{1}{2}-\mu+ix\right)\Gamma\left(\frac{1}{2}-\mu-ix\right)K_{ilx}(b)P_{ix-1/2}^{\mu}(c)\,dx=J_{n,l},$

$$J_{1,1}=2^{-1/2}\pi^{3/2}b^{1/2-\mu}(c^2-1)^{-\mu/2}e^{-ac}\qquad[\operatorname{Re}\mu\leqslant 1/2],$$

$$J_{1,2}=2^{3\mu/2-2}\pi b^{1-\mu}(1-c)^{-\mu/2}K_\mu\left(b\sqrt{\frac{c+1}{2}}\right)$$

$[|c|<1;\ \operatorname{Re}\mu<1/2]$,

$$J_{2,2}=2^{2-\mu/2}\pi^2b^{1-\mu}(c+1)^{-\mu/2}J_{-\mu}\left(\sqrt{\frac{bc-b}{2}}\right)$$

$[c>1;\ \operatorname{Re}\mu\leqslant 1/2]$.

21. $\int_0^\infty x\operatorname{sh}\pi x\Gamma\left(\frac{2\pm1-2\mu+2ix}{4}\right)\Gamma\left(\frac{2\pm1-2\mu-2ix}{4}\right)K_{ix}(b)P_{ix-1/2}^{\mu}(c)\,dx=$

$$=2^{\mu\mp1/2}\pi^2\sqrt{b}(bc)^{(1\pm1)/2}J_{-\mu}(b\sqrt{c^2-1})\qquad[\operatorname{Re}\mu\leqslant 1\pm1/2].$$

22. $\int_0^\infty x\operatorname{sh}\frac{\pi x}{2}\Gamma\left(\frac{1}{2}-\mu+ix\right)\Gamma\left(\frac{1}{2}-\mu-ix\right)K_{ix/2}(b)P_{ix-1/2}^{\mu}(c)\,dx=$

$$=2^{1/2-\mu}\pi\Gamma\left(\frac{3}{2}-\mu\right)b^{(1-2\mu)/4}(c^2-1)^{-\mu/2}e^{b(c^2-1)}D_{\mu-3/2}(2c\sqrt{b})$$

$[\operatorname{Re}\mu\leqslant 1/2]$.

23. $\int_0^\infty x \operatorname{th} \pi x \Gamma\left(\frac{1}{2}-\mu+ix\right)\Gamma\left(\frac{1}{2}-\mu-ix\right) K_{ix}(b) P_{ix-1/2}^{\mu}(c)\,dx=$

$$=\sqrt{\frac{\pi b}{2}}\Gamma(1-\mu)(c-1)^{-\mu/2}(c+1)^{\mu/2}e^{bc}\Gamma(-\mu,\ bc+b) \quad [\operatorname{Re}\mu \leqslant 1/2].$$

24. $\int_0^\infty \Gamma\left(\frac{1}{2}-\mu+ix\right)\Gamma\left(\frac{1}{2}-\mu-ix\right)\operatorname{Re}\left[K_{ix+1/2}(b) P_{ix-1/2}^{\mu}(c)\right]dx=$

$$=\frac{\pi}{2}\Gamma\left(\frac{1}{2}-\mu\right) b^{-(\mu+1)/2}(1+c)^{-1/2}(1-c)^{-\mu/2}e^{b(c-1)/2}W_{\mu/2,\ \mu/2}(bc+b)$$

$[b>0;\quad -1<c<1]$.

25. $\int_0^\infty x \operatorname{sh}\frac{\pi x}{2}\Gamma\left(\frac{1}{2}-\mu+ix\right)\Gamma\left(\frac{1}{2}-\mu-ix\right)\left[J_{ix}(b)+J_{-ix}(b)\right]K_{ix}(b)\times$

$$\times P_{ix-1/2}^{\mu}(c)\,dx=i\sqrt{\frac{\pi}{2}}\left(\frac{2c}{b^2}\right)^{(2\mu-1)/4}(c^2-1)^{-\mu/2}\left\{e^{(1-\mu)\pi i/4}K_{\mu-1/2}(b\sqrt{2ic})-\right.$$

$$\left.-e^{(\mu-1)\pi i/4}K_{\mu-1/2}(b\sqrt{-2ic})\right\} \quad [\operatorname{Re}\mu\leqslant 1/2].$$

26. $\int_0^\infty x \operatorname{sh}\frac{\pi x}{2}\Gamma\left(\frac{1}{2}-\mu+ix\right)\Gamma\left(\frac{1}{2}-\mu-ix\right)\left[Y_{ix}(b)+Y_{-ix}(b)\right]K_{ix}(b)\times$

$$\times P_{ix-1/2}^{\mu}(c)\,dx=-\sqrt{\frac{\pi}{2}}\left(\frac{2c}{b^2}\right)^{(2\mu-1)/4}(c^2-1)^{-\mu/2}\left\{e^{(1-\mu)\pi i/4}K_{\mu-1/2}(b\sqrt{2ic})+\right.$$

$$\left.+e^{\mu\pi i/4}K_{\mu-1/2}(b\sqrt{-2ic})\right\} \quad [\operatorname{Re}\mu\leqslant 1/2].$$

27. $\int_{-\infty}^\infty (ix-\mu)\Gamma\left(\frac{1}{2}-ix\right)\Gamma\left(\frac{1}{2}-2\mu+ix\right)I_{ix-\mu}(a)K_{ix-\mu}(b)P_{ix-\mu-1/2}^{\mu}(c)\,dx=$

$$=\sqrt{2\pi}\left(\frac{ab}{z}\right)^{1/2-\mu}(1-c^2)^{-\mu/2}K_{1/2-\mu}(z) \quad [z=\sqrt{a^2+b^2+2abc};\quad |c|<1].$$

28. $\int_0^\infty x \operatorname{sh}\pi x K_{ix}(a)K_{ix}(b)P_{ix-1/2}^{\mu}(c)\,dx=$

$$=\begin{Bmatrix}1\\0\end{Bmatrix}2^{-(2\mu+7)/4}\pi^{3/2}(ab)^{(2\mu+1)/4}(c^2-1)^{\mu/2}(c-z)^{-(2\mu+1)/4}\times$$

$$\times J_{-\mu-1/2}\left(\sqrt{2abc-a^2-b^2}\right) \quad \left[z=\frac{a^2+b^2}{2ab},\quad \begin{Bmatrix}c>z\\c<z\end{Bmatrix}\right].$$

29. $\int_0^\infty x \operatorname{sh}\pi x\Gamma\left(\frac{1}{2}-\mu+ix\right)\Gamma\left(\frac{1}{2}-\mu-ix\right)K_{ix}(a)K_{ix}(b)P_{ix-1/2}^{\mu}(c)\,dx=$

$$=2^{-1/2}\pi^{3/2}(ab)^{1/2-\mu}(c^2-1)^{-\mu/2}z^{\mu-1/2}K_{1/2-\mu}(z)$$

$[z=\sqrt{a^2+b^2+2abc};\quad \operatorname{Re}\mu\leqslant 1/2;\quad |\arg a|,\quad |\arg b|<\pi/2;\quad |\arg(c-1)|<\pi]$.

30. $=2^{(2\mu-3)/4}\pi^{3/2}(c+1)^{-1/4}(c-1)^{-\mu/2}K_{1/2-\mu}(b\sqrt{2c+2})$

$[a=b;\quad \operatorname{Re}\mu\leqslant 1/2]$.

31. $\int_0^\infty x \operatorname{sh}\pi x\Gamma\begin{bmatrix}1/2-\mu+ix,\ 1/2-\mu-ix\\-\mu+ix,\ -\mu-ix\end{bmatrix}K_{ix}^2(b)P_{ix-1/2}^{\mu}(c)\,dx=$

$$=2^{-(2\mu+7)/4}\pi^{3/2}b^{\mu-1/2}c^{1/2-\mu}(c^2-1)^{\mu/2}J_{-\mu-1/2}(b\sqrt{2c-2})$$

$[\operatorname{Re}\mu\leqslant -1/2]$.

32. $\int\limits_0^\infty x \operatorname{sh} \pi x \Gamma\left(\frac{1}{2}-\mu+ix\right)\Gamma\left(\frac{1}{2}-\mu-ix\right) S_{\nu,\, ilx}(b)\, P^{\mu}_{ix-1/2}(c)\,dx=$

$$=\pi b^{1\pm 1/2} c^{(1\pm 1)/2} K_\mu\left(b\sqrt{c^2-1}\right) \qquad [\nu=\mu\pm 1/2,\ l=1;\ \operatorname{Re}\mu\leqslant 1/2]$$

33. $$=2^{\mu/2-2}\pi b^{\mu+1}(c+1)^{\mu/2} K_\mu\left(b\sqrt{\frac{c-1}{2}}\right) \qquad [\nu=2\mu,\ l=2;\ \operatorname{Re}\mu\leqslant 1/2].$$

34. $\int\limits_0^\infty x \operatorname{sh} \pi x \Gamma\left[\frac{1}{2}-\mu+ix,\ \frac{1}{2}-\mu-ix,\ \nu+\frac{ix}{2},\ \nu-\frac{ix}{2}\right] S_{1-2\nu,\, ix}(b)\, P^{\mu}_{ix-1/2}(c)\,dx=$

$$=2^{3/2-2\nu}\pi^{3/2}\Gamma\left(2\nu-\mu+\frac{1}{2}\right) b^{1-\mu}\sqrt{c}\,(c^2-1)^{-\mu/2} S_{\mu-2\nu,\ 1/2}(bc)$$

[$\operatorname{Re}\mu<1/2;\ \operatorname{Re}\nu>0$].

2.17.28. Integrals with respect to the index of products of $P_{irx-1/2}(b)\,P_{ix-1/2}(c)$ by elementary functions.

1. $\int\limits_0^\infty x\frac{\operatorname{sh}(\pi x/2)}{\operatorname{ch}\pi x} P_{(ix-1)/2}(b)\, P_{ix-1/2}(c)\,dx=$

$$=\frac{z^{3/2}}{\pi\sqrt{2}}\left\{2\mathbf{E}\left(\sqrt{\frac{1-cz}{2}}\right)-\mathbf{K}\left(\sqrt{\frac{1-cz}{2}}\right)\right\} \quad \left[z=\left(c^2+\frac{b-1}{2}\right)^{-1/2}\right].$$

2. $$\int\limits_0^\infty x\frac{\operatorname{th}\pi x}{\operatorname{ch}^2\pi x} P_{ix-1/2}(b)\, P_{ix-1/2}(c)\,dx=\frac{1}{\pi^2(c-b)}\ln\frac{c+1}{b+1}.$$

3. $$\int\limits_0^\infty x\frac{\operatorname{th}\pi x}{\operatorname{ch}^n\pi x}[P_{ix-1/2}(c)]^2\,dx=I_n,$$

$$I_1=\frac{1}{2\pi c},\qquad I_2=\frac{1}{\pi^2(1+c)}.$$

4. $$\int\limits_0^\infty x\frac{\operatorname{th}\pi x}{\operatorname{ch}\pi x} P_{ix-1/2}(b)\, P_{ix-1/2}(c)\,dx=\frac{1}{\pi(b+c)} \qquad [-1<b,\ c<1].$$

5. $$\int\limits_0^\infty x\left(x^2+\frac{1}{4}\right)\frac{\operatorname{th}\pi x}{\operatorname{ch}\pi x} P_{ix-1/2}(b)\, P_{ix-1/2}(c)\,dx=\frac{2(bc+1)}{\pi(b+c)^3} \qquad [-1<b,\ c<1].$$

2.17.29. Integrals with respect to the index of products of $P_{ix-1/2}(b)\,P_{ix-1/2}(c)$ by special functions.

1. $$\int\limits_0^\infty x \operatorname{th}\pi x\,\Gamma\left(\frac{1+2ix}{4}\right)\Gamma\left(\frac{1-2ix}{4}\right)[P_{ix-1/2}(c)]^2\,dx=\frac{2\sqrt{\pi}}{\sqrt{2c^2-1}}.$$

2. $\int\limits_0^\infty x \operatorname{th}\pi x\,\Gamma(\sigma+ix)\,\Gamma(\sigma-ix)\,[P_{ix-1/2}(c)]^2\,dx=$

$$=-\frac{c^{-\sigma-1/2}}{2}\Gamma^2\left(\sigma+\frac{1}{2}\right)P_{-\sigma-1/2}\left(\frac{c^2+1}{2c}\right) \qquad [\operatorname{Re}\sigma\geqslant 0].$$

3. $$\int\limits_0^\infty x \operatorname{sh}\frac{\pi x}{2} K_{ix}(b)\,[P_{ix-1/2}(c)]^2\,dx=\frac{\pi b}{2}J_0^2\left(\frac{b}{2}\sqrt{c^2-1}\right).$$

4. $\int_0^\infty x \operatorname{th} \pi x K_{ix}(b) [P_{ix-1/2}(c)]^2 dx = \frac{1}{\sqrt{2\pi}} e^{-bc^2} I_0(bc^2 - b).$

5. $\int_0^\infty x \operatorname{th} \pi x P_{ix-1/2}(b) [P_{ix-1/2}(c)]^2 dx = \left\{ \begin{matrix} 1 \\ 0 \end{matrix} \right\} \frac{1}{\pi} \sqrt{\frac{2c^2-b-1}{b-1}}$

$\left[c>1, \; \left\{ \begin{matrix} 1<b<2c^2-1 \\ b>2c^2-1 \end{matrix} \right\} \right].$

6. $\int_0^\infty x \frac{\operatorname{th} \pi x}{\operatorname{ch} \pi x} P_{ix-1/2}(b) [P_{ix-1/2}(c)]^2 dx = \frac{1}{\pi (b+1) \sqrt{2c^2+b-1}}$ $[b>-1].$

7. $\int_0^\infty x \frac{\operatorname{th} \pi x}{\operatorname{ch}^2 \pi x} [P_{ix-1/2}(b) P_{ix-1/2}(c)]^2 dx =$

$= \frac{1}{\pi^2 \sqrt{(c^2-1)(1-b^2)}} \arcsin \frac{\sqrt{(c^2-1)(1-b^2)}}{b+c}$ $[-1<b \leqslant 1].$

8. $= \frac{1}{\pi^2 \sqrt{(c^2-1)(b^2-1)}} \ln \frac{c\sqrt{b^2-1}+b\sqrt{c^2-1}}{\sqrt{b^2-1}+\sqrt{c^2-1}}$ $[b \geqslant 1].$

9. $\int_0^\infty x \sin \pi x P_{x-1/2}(\cos a) P_{x-1/2}(\cos b) P_{x-1/2}(\cos c) dx =$

$= \frac{1}{4\pi} \left[-\cos \frac{a+b+c}{2} \cos \frac{a+b-c}{2} \cos \frac{a-b+c}{2} \cos \frac{-a+b+c}{2} \right]^{-1/2}$

$[0<a, b, c<\pi; \; \pi<a+b+c<3\pi;$ the right hand side is a real number$].$

2.17.30. Integrals with respect to the index containing $P^{\lambda}_{irx-1/2}(b) P^{\mu}_{ix-1/2}(c)$.

1. $\int_0^\infty x \frac{\operatorname{sh} \pi x}{\operatorname{ch} 2\pi x} P_{2ix-1/2}(b) P^{\mu}_{ix-1/2}(c) dx =$

$= 2^{-\mu-11/4} (c+1)^{\mu/2} (2b^2+c-1)^{-(2\mu+3)/4} P^{\mu}_{\mu+1/2} \left(\frac{b\sqrt{2}}{\sqrt{2b^2+c-1}} \right).$

2. $\int_0^\infty x \frac{\operatorname{th} \pi x}{\operatorname{ch} \pi x} P^1_{ix-1/2}(b) P^1_{ix-1/2}(c) dx = \frac{2\sqrt{(1-b^2)(1-c^2)}}{\pi (b+c)^3}$ $[-1<b, c<1].$

3. $\int_0^\infty x \left(x^2 + \frac{1}{4} \right) \frac{\operatorname{th} \pi x}{\operatorname{ch} \pi x} P^1_{ix-1/2}(b) P^1_{ix-1/2}(c) dx = \frac{4\sqrt{(1-b^2)(1-c^2)}}{\pi (b+c)^5} (6+4bc-b^2-c^2)$

$[-1<b, c<1].$

4. $\int_0^\infty \frac{x \operatorname{th} \pi x}{(x^2+1/4) \operatorname{ch} \pi x} P^1_{ix-1/2}(b) P^1_{ix-1/2}(c) dx = \frac{1}{\pi (b+c)} \sqrt{\frac{(1-b)(1-c)}{(1+b)(1+c)}}.$

5. $\int_0^\infty x \operatorname{sh} \pi r x \, \Gamma\left(\frac{1}{2} - \mu + ix \right) \Gamma\left(\frac{1}{2} - \mu - ix \right) P^{\lambda}_{irx-1/2}(b) P^{\mu}_{ix-1/2}(c) dx =$

$= \frac{\pi}{\Gamma(\mu)} (c^2-1)^{-\mu/2} (b-c)^{\mu-1}$ $[r=1; \; \lambda=0; \; 1<b<c; \; 0<\operatorname{Re} \mu \leqslant 1/2],$

6. $= 0$ $[r=1; \; \lambda=0; \; 1<c<b; \; 0<\operatorname{Re} \mu \leqslant 1/2],$

7. $$=2^{-(3\lambda+1)/2}\sqrt{\pi}\,\Gamma\left(\frac{3}{2}-\mu\right)(b+1)^{\lambda/2}(c^2-1)^{-\mu/2}\times$$
$$\times\left(c^2+\frac{b-1}{2}\right)^{(\mu-\lambda)/2-3/4}P^{\lambda}_{\lambda-\mu+1/2}\left(c\sqrt{c^2+\frac{b-1}{2}}\right)$$ [$r=1/2$; $\operatorname{Re}\mu\leqslant 1/2$]

8. $$\int\limits_0^\infty x\,\operatorname{th}\pi nx\,\Gamma\left(\frac{1}{2}-\mu+ix\right)\Gamma\left(\frac{1}{2}-\mu-ix\right)P_{inx-1/2}(b)\,P^{\mu}_{ix-1/2}(c)\,dx=$$
$$=\Gamma(1-\mu)(c^2-1)^{-\mu/2}(b+c)^{\mu-1}$$ [$n=1$; $\operatorname{Re}\mu\leqslant 1/2$].

9. $$=2^{-(\mu+3)/2}\sqrt{\pi}\,\Gamma\left(\frac{3}{2}-2\mu\right)(c+1)^{-\mu/2}\left(b^2+\frac{c-1}{2}\right)^{(2\mu-3)/4}\times$$
$$\times P^{\mu}_{1/2-\mu}\left(\frac{b\sqrt{2}}{\sqrt{2b^2+c-1}}\right)$$ [$n=2$; $\operatorname{Re}\mu\leqslant 1/2$].

10. $$\int\limits_0^\infty x\,\operatorname{th}\pi x\,\Gamma\left(\nu+\frac{ix}{2}\right)\Gamma\left(\nu-\frac{ix}{2}\right)P_{ix-1/2}(b)\,P^{\mu}_{ix-1/2}(c)\,dx=$$
$$=2^{\mu+1}\sqrt{\pi}(c^2-1)^{-\mu/2}(b^2+c^2-1)^{-1/2}\left(b+\sqrt{b^2+c^2-1}\right)^{\mu}$$
[$\operatorname{Re}\mu\leqslant 1/2$; $\nu=(1-2\mu)/4$].

11. $$=2^{\mu}\sqrt{\pi}(c^2-1)^{-\mu/2}(b^2+c^2-1)^{-3/2}c\left[b+\sqrt{b^2+c^2-1}\right]^{\mu}\times$$
$$\times\left(b-\mu\sqrt{b^2+c^2-1}\right)$$ [$\operatorname{Re}\mu\leqslant 3/4$; $\nu=(3-2\mu)/4$].

12. $$\int\limits_0^\infty x\frac{\operatorname{sh}\pi x}{\operatorname{ch}2\pi x}\Gamma\left(\frac{1}{2}-\mu+ix\right)\Gamma\left(\frac{1}{2}-\mu-ix\right)P_{2ix-1/2}(b)\,P^{\mu}_{ix-1/2}(c)\,dx=$$
$$=2^{(6\mu-11)/4}\sqrt{\pi}\,\Gamma\left(\frac{3}{2}-2\mu\right)(c-1)^{-\mu/2}(c+1)^{-1/4}\left|b^2-\frac{c+1}{2}\right|^{(\mu-1)/2}\times$$
$$\times P^{\mu-1}_{\mu-1/2}\left(\frac{b\sqrt{2}}{\sqrt{c+1}}\right)$$ [$b\neq\sqrt{(c+1)/2}$, $\operatorname{Re}\mu\leqslant 1/2$].

13. $$\int\limits_0^\infty x\,\operatorname{sh}\pi x\,\Gamma\left[\frac{1}{2}-\mu+ix,\frac{1}{2}-\mu-ix,\ \nu+irx,\ \nu-irx\right]P^{\lambda}_{ix-1/2}(b)\,P^{\mu}_{ix-1/2}(c)dx=$$
$$=\pi\Gamma(1-\lambda-\mu)(b^2-1)^{-\lambda/2}(c^2-1)^{-\mu/2}(b+c)^{\lambda+\mu-1}$$ [$\nu=1/2-\lambda$; $r=1$; $\operatorname{Re}\lambda$, $\operatorname{Re}\mu<1/2$].

14. $$=2^{\lambda+1}\pi^{3/2}\Gamma(1-\lambda-\mu)(c^2-1)^{-\mu/2}(b^2+c^2-1)^{(\mu-1)/2}P^{\lambda}_{-\mu}\left(\frac{c}{\sqrt{b^2+c^2-1}}\right)$$
[$\nu=(1-2\lambda)/4$; $r=1/2$; $\operatorname{Re}\lambda$, $\operatorname{Re}\mu<1/2$].

15. $$=2^{\lambda}\pi^{3/2}\Gamma(2-\lambda-\mu)\,b\,(c^2-1)^{-\mu/2}(b^2+c^2-1)^{\mu/2-1}P^{\lambda}_{1-\mu}\left(\frac{c}{\sqrt{b^2+c^2-1}}\right)$$
[$\nu=(3-3\lambda)/4$; $r=1/2$; $\operatorname{Re}\lambda<3/2$; $\operatorname{Re}\mu<1/2$].

16. $$\int\limits_0^\infty x\,\operatorname{sh}\pi rx\,\Gamma\left[\frac{1}{2}-\lambda+\frac{ix}{2},\frac{1}{2}-\lambda-\frac{ix}{2},\frac{1}{2}-\mu+ix,\ \frac{1}{2}-\mu-ix\right]\times$$
$$\times P^{\lambda}_{(ix-1)/2}(b)\,P^{\mu}_{ix-1/2}(c)\,dx=2^{(1-\lambda)/2}\pi^{3/2}\Gamma\left(\frac{3}{2}-2\lambda-\mu\right)(b+1)^{-\lambda/2}\times$$
$$\times(c^2-1)^{-\mu/2}\left(c^2+\frac{b-1}{2}\right)^{(\lambda+\mu)/2-3/4}P^{\lambda}_{1/2-\lambda-\mu}\left(c\sqrt{c^2+\frac{b-1}{2}}\right)$$
[$r=1$; $\operatorname{Re}\lambda$, $\operatorname{Re}\mu<1/2$].

17. $$=2^{(6\lambda+1)/4}\pi\Gamma\left(\frac{3}{2}-\mu\right)\Gamma\left(\frac{3}{2}-2\lambda-\mu\right)(b-1)^{-\lambda/2}(b+1)^{-1/4}\times$$
$$\times(c^2-1)^{-\mu/2}\left|c^2-\frac{b+1}{2}\right|^{(\lambda+\mu-1)/2}P_{\lambda-1/2}^{\lambda+\mu-1}\left(c\sqrt{\frac{2}{b+1}}\right)$$
$[r=1/2,\ c\neq\sqrt{(b+1)/2};\ \mathrm{Re}\,\lambda,\ \mathrm{Re}\,\mu<1/2]$.

18. $$\int_0^\infty x\,\mathrm{sh}\,\pi x\,K_{2ix}(b)\,P_{ix-1/2}^{-\mu}(c)\,P_{ix-1/2}^{\mu}(c)\,dx=$$
$$=\frac{\pi b}{8}J_\mu\left(\frac{b}{2}\sqrt{c^2-1}\right)J_{-\mu}\left(\frac{b}{2}\sqrt{c^2-1}\right)\qquad [c>1].$$

19. $$=\frac{\pi b}{8}I_\mu\left(\frac{b}{2}\sqrt{1-c^2}\right)I_{-\mu}\left(\frac{b}{2}\sqrt{1-c^2}\right)\qquad [-1<c<1].$$

20. $$\int_0^\infty x\Gamma\left(\frac{1}{2}-\mu+ix\right)\Gamma\left(\frac{1}{2}-\mu-ix\right)K_{2ix}(b)\left[P_{ix-1/2}^{\mu}(c)\right]^2dx=$$
$$=\frac{\pi^2 b}{4}I_{-\mu}^2\left(\frac{b}{2}\sqrt{1-c^2}\right)\qquad [-1<c<1].$$

21. $$\int_0^\infty x\,\mathrm{sh}\,\pi nx\,\Gamma\left(\frac{1}{2}-\mu+ix\right)\Gamma\left(\frac{1}{2}-\mu-ix\right)K_{nix}(b)\left[P_{ix-1/2}^{\mu}(c)\right]^2dx=I_n$$
$[c>1;\ \mathrm{Re}\,\mu\leqslant 1/2]$,
$$I_1=\sqrt{\frac{\pi}{2}}e^{-bc^2}I_{-\mu}[b(c^2-1)],\ I_2=\frac{\pi^2 b}{4}J_{-\mu}^2\left(\frac{b}{2}\sqrt{c^2-1}\right).$$

22. $$\int_0^\infty x\frac{\mathrm{sh}\,\pi x}{\mathrm{ch}\,2\pi x}P_{2ix-1/2}(b)\,P_{ix-1/2}^{-\mu}(c)\,P_{ix-1/2}^{\mu}(c)\,dx=$$
$$=\frac{\sqrt{c^2-1}}{8z\sqrt{2b}}\left[\left(\mu+\frac{1}{4}\right)P_{-1/4}^{\mu}\left(\frac{z}{b}\right)P_{1/4}^{-\mu}\left(\frac{z}{b}\right)-\left(\mu-\frac{1}{4}\right)P_{1/4}^{\mu}\left(\frac{z}{b}\right)P_{-1/4}^{-\mu}\left(\frac{z}{b}\right)\right]$$
$[z=\sqrt{b^2+c^2-1}]$.

23. $$\int_0^\infty x\,\mathrm{th}\,\pi nx\,\Gamma\left(\frac{1}{2}-\mu+ix\right)\Gamma\left(\frac{1}{2}-\mu-ix\right)P_{inx-1/2}(b)\left[P_{ix-1/2}^{\mu}(c)\right]^2dx=J_n,$$
$$J_1=(b+1)^{-1/2}(c^2-1)^{-\mu}(b+2c^2-1)^{-1/2}\left(b+c^2+\sqrt{(b+1)(b+2c^2-1)}\right)^\mu,$$
$$J_2=2^{\mu-5/2}\sqrt{\pi}\,\Gamma\left(\frac{3}{2}-2\mu\right)z^{-1}\sqrt{\frac{c^2-1}{b}}P_{1/4}^{\mu}\left(\frac{z}{b}\right)P_{-1/4}^{\mu}\left(\frac{z}{b}\right)$$
$[z=\sqrt{b^2+c^2-1}]$.

24. $$\int_0^\infty x\,\mathrm{sh}\,\pi x\left[\Gamma\left(\frac{1}{2}-\mu+ix\right)\Gamma\left(\frac{1}{2}-\mu-ix\right)P_{ix-1/2}^{\mu}(b)\right]^2P_{ix-1/2}^{\mu}(c)\,dx=$$
$$=2^{-\mu}\sqrt{\pi}\,\Gamma\left(\frac{1}{2}-\mu\right)(b^2-1)^{-\mu}(c-1)^{-\mu/2}(c+1)^{(\mu-1)/2}(2b^2+c-1)^{\mu-1/2}$$
$[\mathrm{Re}\,\mu<1/2]$.

2.18. THE LEGENDRE FUNCTIONS OF THE SECOND KIND $Q_\nu(x)$, $Q_\nu^\mu(x)$

2.18.1. Integrals of $(z^m\pm x^m)^\beta Q_\nu^\mu(cx)$.

1. $$\int_a^\infty Q_\nu\left(\frac{x}{a}\right)dx=\frac{a}{\nu(\nu+1)}\qquad [a,\ \mathrm{Re}\,\nu>0].$$

2. $\int\limits_a^\infty x^{\alpha-1}Q_\nu\left(\frac{x}{b}\right)dx=e^{\alpha\pi i}\,\Gamma(\alpha)(a^2-b^2)^{\alpha/2}\,Q_\nu^{-\alpha}\left(\frac{a}{b}\right).$

3. $\qquad=\sqrt{\frac{\pi b}{2}}\,\Gamma(\alpha)\,\Gamma(1-\alpha+\nu)\,(a^2-b^2)^{(2\alpha-1)/4}\,P_{\alpha-1/2}^{-\nu-1/2}\left(\frac{a}{\sqrt{a^2-b^2}}\right)$

$[a>0;\ 0<\operatorname{Re}\alpha<\operatorname{Re}\nu+1;\ |\arg(a-b)|<\pi]$

4. $\int\limits_a^\infty (x-a)^{\alpha-1}\,Q_\nu\left(\frac{x}{a}\right)dx=2^{\alpha-1}a^{\alpha}\Gamma\begin{bmatrix}\alpha,\ \alpha,\ 1-\alpha+\nu\\ 1+\alpha+\nu\end{bmatrix}\quad [a>0;\ 0<\operatorname{Re}\alpha<\operatorname{Re}\nu+1].$

5. $\int\limits_a^\infty (x-a)^{\alpha-1}\,Q_\nu\left(\frac{x}{b}\right)dx=e^{\alpha\pi i}\,\Gamma(\alpha)\,(a^2-b^2)^{\alpha/2}Q_\nu^{-\alpha}\left(\frac{a}{b}\right).$

6. $\qquad=\sqrt{\frac{\pi b}{2}}\,\Gamma(\alpha)\,\Gamma(1-\alpha+\nu)\,(a^2-b^2)^{(2\alpha-1)/4}\,P_{\alpha-1/2}^{-\nu-1/2}\left(\frac{a}{\sqrt{a^2-b^2}}\right)$

$[a>0;\ 0<\operatorname{Re}\alpha<\operatorname{Re}\nu+1;\ |\arg(a-b)|<\pi].$

7. $\int\limits_a^\infty (x^2-a^2)^{\beta-1}\,Q_\nu^\mu\left(\frac{x}{a}\right)dx=2^{\mu-2}a^{2\beta-1}\,e^{\mu\pi i}\,\times$

$\times\Gamma\begin{bmatrix}1-\beta+\nu/2,\ (1+\mu+\nu)/2,\ \beta+\mu/2,\ \beta-\mu/2\\ \beta+(1+\nu)/2,\ 1+(\nu-\mu)/2\end{bmatrix}\quad [|\operatorname{Re}\mu|<2\operatorname{Re}\beta<\operatorname{Re}\nu+2]$

2.18.2. Integrals of $x^\alpha\,(z^2\pm x^2)^\beta Q_\nu^\mu\,(cx)$.

1. $\int\limits_0^a x^{\nu-2\beta}\,(a^2-x^2)^{\beta-1}\,Q_\nu^\mu\left(\frac{x}{a}\right)dx=$

$=2^{2\beta-2}\,\pi a^{\nu-1}\,\Gamma\begin{bmatrix}\beta+\mu/2,\ \beta-\mu/2,\ 1-2\beta+\nu\\ 1-\mu+\nu,\ \beta+(1+\mu)/2,\ (1-\nu)/2-\beta\end{bmatrix}$

$[a>0;\ |\operatorname{Re}\mu|/2<\operatorname{Re}\beta<(1+\operatorname{Re}\nu)/2].$

2. $\int\limits_0^a x^{-2\beta-\nu-1}(a^2-x^2)^{\beta-1}\,Q_\nu^\mu\left(\frac{x}{a}\right)dx=-\frac{2^{-\nu-3}}{\sqrt{\pi}a^{\nu+2}}\,\Gamma\,[\mu+\nu+1,\beta-\mu/2,-\beta-\nu/2]\times$

$\times\left\{\Gamma\begin{bmatrix}\beta+\mu/2,\ (1-\nu)/2-\beta\\ \beta-\mu/2+\nu,\ 1-\beta+\mu/2-\nu\end{bmatrix}+\cos(\mu+\nu)\,\pi\Gamma\begin{bmatrix}(1-\nu)/2-\beta\\ 1-\mu/2-\beta\end{bmatrix}\right\}$

$[a>0;\ |\operatorname{Re}\mu|/2<\operatorname{Re}\beta<-\operatorname{Re}\nu/2].$

3. $\int\limits_a^\infty x^{(1\pm1)/2}(x^2-a^2)^{\beta-1}\,Q_\nu^\mu\left(\frac{x}{a}\right)dx=2^{\mu-2}\,e^{\mu\pi i}a^{2\beta-(1\mp1)/2}\times$

$\times\Gamma\begin{bmatrix}(3\pm1+2\mu+2\nu)/4,\ \beta+\mu/2,\ \beta-\mu/2,\ (3\mp1+2\nu)/4-\beta\\ (3\mp1-2\mu+2\nu)/4,\ (3\pm1+2\nu)/4+\beta\end{bmatrix}$

$[a>0;\ |\operatorname{Re}\mu|/2<\operatorname{Re}\beta<(3\mp1+2\operatorname{Re}\nu)/4].$

4. $\int\limits_a^\infty x^{-2\beta-\nu-1}(x^2-a^2)^{\beta-1}\,Q_\nu^\mu\left(\frac{x}{a}\right)dx=\frac{e^{\mu\pi i}2^{2\beta-2}}{a^{\nu+2}}\,\Gamma\begin{bmatrix}\mu+\nu+1,\ \beta+\mu/2,\ \beta-\mu/2\\ 2\beta+\nu+1\end{bmatrix}$

$[a>0;\ \operatorname{Re}\nu>-3/2;\ \operatorname{Re}\beta>|\operatorname{Re}\mu|/2].$

2.18.3. Integrals of $(x\pm a)^\alpha\,(b\pm x)^\beta Q_\nu^\mu\,(\varphi(x))$.

1. $\int\limits_{-a}^a (x+a)^{-\beta-\nu-1}\,(a-x)^{\beta-1}\,Q_\nu^\mu\left(\frac{x}{a}\right)dx=$

$$=\frac{\pi(2a)^{-\nu-1}}{2\sin\mu\pi}\Gamma\begin{bmatrix}\mu/2-\nu-\beta, & -\mu/2-\nu-\beta\\ -\nu, & -\mu-\nu\end{bmatrix}\left\{\frac{\sin(\mu-\nu)\pi}{\sin(\mu+\nu)\pi}\times\right.$$

$$\left.\times\Gamma\begin{bmatrix}\beta+\mu/2\\ 1-\beta+\mu/2\end{bmatrix}+\cos\mu\pi\begin{bmatrix}\beta-\mu/2\\ 1-\beta-\mu/2\end{bmatrix}\right\}\qquad [a>0;\ |\operatorname{Re}\mu|/2<\operatorname{Re}\beta<-|\operatorname{Re}\mu|/2-\operatorname{Re}\nu].$$

2. $$\int\limits_{-a}^{a}(x+a)^{\nu-\beta}(a-x)^{\beta-1}Q_\nu^\mu\left(\frac{x}{a}\right)dx=$$

$$=\frac{\pi(2a)^{\nu}}{2\sin\mu\pi}\Gamma\begin{bmatrix}1-\beta-\mu/2+\nu, & 1-\beta+\mu/2+\nu\\ 1+\nu, & 1-\mu+\nu\end{bmatrix}\left\{\cos\mu\pi\Gamma\begin{bmatrix}\beta-\mu/2\\ 1-\beta-\mu/2\end{bmatrix}-\Gamma\begin{bmatrix}\beta+\mu/2\\ 1-\beta+\mu/2\end{bmatrix}\right\}$$

$$[a>0;\ |\operatorname{Re}\mu|/2<\operatorname{Re}\beta<1+\operatorname{Re}\nu-|\operatorname{Re}\mu|/2].$$

3. $$\int\limits_{a}^{\infty}\frac{1}{\sqrt{(x-a)(x-b)}}Q_\nu\left(\frac{x}{c}\right)dx=Q_\nu(z_+)\,Q_\nu(z_-)$$

$$\left[\operatorname{Re}\nu>-1;\ 2cz_\pm=\sqrt{a(c-b)+c(c+b)}\pm\sqrt{c(c-b)-a(c+b)}\right].$$

4. $$\int\limits_{a}^{\infty}\frac{1}{\sqrt{(x-a)(x-b)}}Q_n\left(\frac{x}{a}\right)dx=\frac{\pi^2}{4}\left[P_n\left(\sqrt{\frac{a+b}{2a}}\right)\right]^2+\left[Q_n\left(\sqrt{\frac{a+b}{2a}}\right)\right]^2$$

$$[-a<b<a].$$

5. $$\int\limits_{a}^{\infty}(x-a)^{\alpha-1}(x+a)^{\pm\mu/2}Q_\nu^\mu\left(\frac{x}{a}\right)dx=$$

$$=\frac{e^{\mu\pi i}(2a)^{\alpha\pm\mu/2}}{2}\Gamma^{(1\pm1)/2}\begin{bmatrix}1+\mu+\nu\\ 1-\mu+\nu\end{bmatrix}\Gamma\begin{bmatrix}\alpha+\mu/2, & \alpha-\mu/2, & 1-\alpha\mp\mu/2+\nu\\ 1+\alpha\pm\mu/2+\nu\end{bmatrix}$$

$$[a>0;\ |\operatorname{Re}\mu|/2<\operatorname{Re}\alpha<1+\operatorname{Re}(\nu\mp\mu/2)].$$

6. $$\int\limits_{a}^{\infty}(x-a)^{\alpha-1}(x+a)^{-\alpha-\nu-1}Q_\nu^\mu\left(\frac{x}{a}\right)dx=\frac{e^{\mu\pi i}}{2(2a)^{\nu+1}}\times$$

$$\times\Gamma\begin{bmatrix}1+\mu+\nu, & 1+\nu, & \alpha+\mu/2, & \alpha-\mu/2\\ 1+\alpha+\mu/2+\nu, & 1+\alpha-\mu/2+\nu\end{bmatrix}\qquad [a>0;\ \operatorname{Re}\nu>-1;\ \operatorname{Re}\alpha>|\operatorname{Re}\mu|/2].$$

7. $$\int\limits_{0}^{\infty}x^{\alpha-1}|a-x|^{-\nu-1}Q_\nu^\mu\left(\frac{a+x}{|a-x|}\right)dx=$$

$$=\frac{e^{\mu\pi i}a^{\alpha-\nu-1}}{2}\Gamma\begin{bmatrix}\alpha+\mu/2, & \alpha-\mu/2, & 1-\alpha+\mu/2+\nu, & 1-\alpha-\mu/2+\nu\\ 1+\nu, & 1-\mu+\nu\end{bmatrix}$$

$$[a>0;\ |\operatorname{Re}\mu|/2<\operatorname{Re}\alpha<1+\operatorname{Re}(\nu\pm\mu/2)].$$

8. $$\int\limits_{0}^{\infty}x^{\alpha-1}Q_\nu\left(\frac{ax^2+bx+c}{x}\right)dx=$$

$$=\frac{\sqrt{2\pi d}\sin\nu\pi}{\sin(\nu-\alpha)\pi}\left(\frac{c}{a}\right)^{\alpha/2}\Gamma(1+\alpha+\nu)\,P^{-\nu-1/2}_{-\alpha-1/2}\left(\sqrt{1+d^2}\right)Q_\nu^\alpha\left(\sqrt{1+4acd^2}\right)$$

$$[(1+d^2)(1+4acd^2)=b^2d^2;\ \operatorname{Re}a,\ \operatorname{Re}c>0;\ |\operatorname{Re}\alpha|<\operatorname{Re}\nu+1].$$

9. $$\int\limits_{0}^{\infty}x^{\alpha-1}Q_\nu\left(\frac{a^2+b^2+x^2}{2ax}\right)dx=$$

$$=\frac{\sqrt{\pi a}}{2^{1/2-\nu}}\Gamma\left(\frac{1+\alpha+\nu}{2}\right)\Gamma\left(\frac{1-\alpha+\nu}{2}\right)(a^2+b^2)^{(2\alpha-1)/4}P^{-\nu-1/2}_{\alpha-1/2}\left(\frac{b}{\sqrt{a^2+b^2}}\right)$$

$$[\operatorname{Re}a,\ \operatorname{Re}b>0;\ |\operatorname{Re}\alpha|<\operatorname{Re}\nu+1].$$

10. $$\int_0^a x^{\alpha-1}(a-x)^{\nu/2}Q_\nu^\mu\left(\frac{2a-x}{2\sqrt{a(a-x)}}\right)dx=e^{\mu\pi i}\sqrt{\pi}a^{\alpha+\nu/2}\times$$
$$\times\Gamma\left[\begin{matrix}\mu+\nu+1,\ \alpha+\mu,\ \alpha-\mu\\ \alpha+1/2,\ \alpha+\nu+1\end{matrix}\right]\quad [a>0;\ \operatorname{Re}\nu>-3/2;\ \operatorname{Re}\alpha>|\operatorname{Re}\mu|].$$

11. $$\int_0^\infty x^{\alpha-1}|a-x|^{-\mu}Q_\nu^\mu\left(\frac{a+x}{2\sqrt{ax}}\right)dx=$$
$$=e^{\mu\pi i}\sqrt{\pi}a^{\alpha-\mu}\Gamma\left[\begin{matrix}1/2-\mu,\ \alpha+(\nu+1)/2,\ \mu+(\nu+1)/2-\alpha\\ 1+\alpha-\mu+\nu/2,\ 1-\alpha+\nu/2\end{matrix}\right]$$
$$[a>0;\ \operatorname{Re}\mu<1/2;\ -(1+\operatorname{Re}\nu)/2<\operatorname{Re}\alpha<1/2+\operatorname{Re}(\mu+\nu)/2].$$

12. $$\int_0^\infty x^{\alpha-1}(z+x)^{\nu/2}Q_\nu^\mu\left(\frac{2z+x}{2\sqrt{z(z+x)}}\right)dx=$$
$$=\frac{e^{\mu\pi i}}{\sqrt{\pi}}\cos\mu\pi z^{\alpha+\nu/2}\Gamma\left[\begin{matrix}\mu+\nu+1,\ \alpha+\mu,\ \alpha-\mu,\ 1/2-\alpha\\ 1+\alpha+\nu\end{matrix}\right]$$
$$[|\operatorname{Re}\mu|<\operatorname{Re}\alpha<1/2;\ |\arg z|<\pi].$$

13. $$\int_0^\infty x^{\alpha-1}(z+x)^{-(\nu+1)/2}Q_\nu^\mu\left(\frac{2z+x}{2\sqrt{z(z+x)}}\right)dx=$$
$$=e^{\mu\pi i}\sqrt{\pi}z^{\alpha-(\nu+1)/2}\Gamma\left[\begin{matrix}\alpha+\mu,\ \alpha-\mu,\ 1-\alpha+\nu\\ 1-\mu+\nu,\ \alpha+1/2\end{matrix}\right]\quad [|\operatorname{Re}\mu|<\operatorname{Re}\alpha<\operatorname{Re}\nu+1;\ |\arg z|<\pi].$$

2.18.4. Integrals of $x^\alpha(x^m\pm a^m)^\beta(b^2\pm x^2)^\gamma Q_\nu^\mu(cx)$.

1. $$\int_a^\infty (x-a)^{\beta-1}(x^2-b^2)^{-\mu/2}Q_\nu^\mu\left(\frac{x}{b}\right)dx=$$
$$=\sqrt{\frac{\pi b}{2}}e^{\mu\pi i}\Gamma(\beta)\Gamma(1-\beta+\mu+\nu)(a^2-b^2)^{(2\beta-2\mu-1)/4}P_{\beta-\mu-1/2}^{-\nu-1/2}\left(\frac{a}{\sqrt{a^2-b^2}}\right)=$$

2. $$=e^{\beta\pi i}\Gamma(\beta)(a^2-b^2)^{(\beta-\mu)/2}Q_\nu^{\mu-\beta}\left(\frac{a}{b}\right)$$
$$[0<b<a;\ 0<\operatorname{Re}\beta<\operatorname{Re}(\mu+\nu)+1].$$

3. $$\int_a^\infty (x^2-a^2)^{\beta-1}(x^2-b^2)^{-\mu/2}Q_\nu^\mu\left(\frac{x}{b}\right)dx=$$
$$=2^{\mu-2}e^{\mu\pi i}a^{2\beta-\mu-\nu-2}b^{\nu+1}\Gamma\left[\begin{matrix}\beta,\ (1+\mu+\nu)/2,\ 1-\beta+(\mu+\nu)/2\\ \nu+3/2\end{matrix}\right]\times$$
$$\times{}_2F_1\left(\frac{1+\mu+\nu}{2},\ 1-\beta+\frac{\mu+\nu}{2};\ \frac{3}{2}+\nu;\ \frac{b^2}{a^2}\right)\quad [0<b<a;\ 0<\operatorname{Re}\beta<\operatorname{Re}(\mu+\nu)/2+1].$$

4. $$\int_a^\infty x^{l-\nu-k-1}(x^2-a^2)^{k-1}(x^2-b^2)^{-m/2}Q_\nu^m\left(\frac{x}{b}\right)dx=$$
$$=2^{k-1}k!\,a^{k+l-\nu-2}b^{l-m}(a^2-b^2)^{-l/2}Q_{\nu+k}^l\left(\frac{a}{b}\right)\quad [0<b<a;\ \operatorname{Re}\nu>-3/2].$$

2.18.5. Integrals of $(x-a)^\alpha(b\pm x)^\beta(x^m\pm d^m)^\gamma Q_\nu^\mu(cx)$.

1. $\int_a^b (x-a)^{-(\mu+1)/2}(b-x)^{-\mu-1/2}(x+a)^{\mu/2}Q_\nu^\mu\left(\frac{x}{a}\right)dx=$

$=\sqrt{\pi}\Gamma\left(\frac{1}{2}-\mu\right)(b-a)^{-\mu}P_\nu^\mu\left(\sqrt{\frac{a+b}{2a}}\right)Q_\nu^\mu\left(\sqrt{\frac{a+b}{2a}}\right)$ [$0<a<b$; $\operatorname{Re}\mu<1/2$].

2. $\int_a^b (x-a)^{-(\mu+1)/2}(b-x)^{-\mu-3/2}(x+a)^{\mu/2}Q_\nu^\mu\left(\frac{x}{a}\right)dx=$

$=\frac{\sqrt{\pi}(b-a)^{-\mu-1/2}}{2\sqrt{a+b}}\Gamma\left(-\frac{1}{2}-\mu\right)\left[P_\nu^{\mu+1}\left(\sqrt{\frac{a+b}{2a}}\right)Q_\nu^\mu\left(\sqrt{\frac{a+b}{2a}}\right)+\right.$

$\left.+P_\nu^\mu\left(\sqrt{\frac{a+b}{2a}}\right)Q_\nu^{\mu+1}\left(\sqrt{\frac{a+b}{2a}}\right)\right]$ [$0<a<b$; $\operatorname{Re}\mu<-1/2$].

3. $\int_a^\infty (x-b)^{-(\mu+1)/2}(x-a)^{-\mu-1/2}(x+b)^{\mu/2}Q_\nu^\mu\left(\frac{x}{b}\right)dx=$

$=e^{-\mu\pi i}\frac{(a-b)^{-\mu}}{\sqrt{\pi}}\Gamma\left(\frac{1}{2}-\mu\right)\left[Q_\nu^\mu\left(\sqrt{\frac{a+b}{2b}}\right)\right]^2$ [$a>b>0$; $\operatorname{Re}\mu<1/2$; $\operatorname{Re}(\mu+\nu)>-1$].

4. $\int_a^\infty (x-b)^{-(\mu+1)/2}(x-a)^{-\mu-3/2}(x+b)^{\mu/2}Q_\nu^\mu\left(\frac{x}{b}\right)dx=$

$=-e^{-\mu\pi i}\frac{(a-b)^{-\mu-1/2}}{\sqrt{\pi(a+b)}}\Gamma\left(-\mu-\frac{1}{2}\right)Q_\nu^{\mu+1}\left(\sqrt{\frac{a+b}{2b}}\right)Q_\nu^\mu\left(\sqrt{\frac{a+b}{2b}}\right)$

[$a>b>0$; $\operatorname{Re}\mu<-1/2$; $\operatorname{Re}(\mu+\nu)>-2$].

5. $\int_a^\infty (x-a)^{\alpha-1}(x-b)^{\alpha-1}(x^2-c^2)^{-\mu/2}Q_\nu^\mu\left(\frac{x}{c}\right)dx=$

$=2^{1-2\alpha+\mu}\sqrt{\pi}(a+b)^{2\alpha-\mu-\nu-2}c^{\nu+1}\Gamma\left[\begin{matrix}2\alpha-1,\ \mu+\nu-2\alpha+2\\ \nu+3/2\end{matrix}\right]\times$

$\times F_4\left(\frac{\mu+\nu}{2}-\alpha+1,\ \frac{\mu+\nu+3}{2}-\alpha;\ \mu+1,\ \frac{3}{2}-\alpha;\ \frac{4c^2}{(a+b)^2},\ \left(\frac{a-b}{a+b}\right)^2\right)+$

$+2^{\mu-1}(a-b)^{2\alpha-1}(a+b)^{-2\alpha-\mu-\nu}c^{2\alpha+\nu}\Gamma\left[\begin{matrix}\alpha,\ 1/2-\alpha,\ 2\alpha+\mu+\nu\\ \nu+3/2\end{matrix}\right]\times$

$\times F_4\left(\alpha+\frac{\mu+\nu}{2},\ \alpha+\frac{\mu+\nu+1}{2};\ \mu+1,\ \alpha+\frac{1}{2};\ \frac{4c^2}{(a+b)^2},\ \left(\frac{a-b}{a+b}\right)^2\right)$

[$a>b>0$; $\operatorname{Re}\alpha>0$; $\operatorname{Re}(2\alpha-\mu-\nu)<2$].

6. $\int_a^\infty (x-a)^{\mu-1/2}(x-b)^{\mu-1/2}(x^2-c^2)^{-\mu/2}Q_\nu^\mu\left(\frac{x}{c}\right)dx=$

$=2^{1/2-2\mu}(a-b)^\mu(a+b)^{\mu-1/2}c^{1/2-\mu}\sqrt{d}\frac{\sin\nu\pi}{\sin(\nu-\mu)\pi}\Gamma\left(\mu+\frac{1}{2}\right)\times$

$\times\Gamma(\mu+\nu+1)P_{-\mu-1/2}^{-\nu-1/2}\left(\sqrt{1+d^2}\right)Q_\nu^\mu\left(\frac{\sqrt{4c^2+(a-b)^2d^2}}{2c}\right)$

[$(1+d^2)(4c^2+d^2(a-b)^2)=(a+b)^2d^2$; $a+b>2c>0$; $a,\ b>0$; $\operatorname{Re}\mu>-1/2$; $\operatorname{Re}\nu>-1$].

2.18.6. Integrals of $A(x)e^{\varphi(x)}Q_\nu^\mu(cx)$.

1. $\int_0^\infty x^{-3/2}e^{-p/x}Q_n\left(\frac{x}{a}\right)dx=(-1)^n n!\sqrt{\frac{\pi}{p}}D_{-n-1}\left(\sqrt{\frac{2p}{a}}\right)D_{-n-1}\left(-\sqrt{\frac{2p}{a}}\right)$

[$\operatorname{Re}a,\ \operatorname{Re}p>0$].

2. $\int\limits_a^\infty (x-a)^{\alpha-1}(x+a)^{\mu/2} e^{-px} Q_\nu^\mu\left(\frac{x}{a}\right) dx =$

$$= \frac{e^{\mu\pi i-2ap}}{2p^{\alpha+\mu/2}} \Gamma\begin{bmatrix}\nu+\mu+1\\ \nu-\mu+1\end{bmatrix} G_{23}^{22}\left(2ap\,\middle|\,\begin{matrix}1+\mu,\ 1\\ \alpha+\mu/2,\ \nu+1,\ -\nu\end{matrix}\right)$$

$[a,\ \operatorname{Re} p>0;\ 2\operatorname{Re}\alpha>|\operatorname{Re}\mu|]$.

3. $\int\limits_a^\infty (x-a)^{\alpha-1}(x+a)^{-\mu/2} e^{-px} Q_\nu^\mu\left(\frac{x}{a}\right) dx =$

$$= \frac{e^{\mu\pi i-2ap}}{2p^{\alpha-\mu/2}} G_{23}^{22}\left(2ap\,\middle|\,\begin{matrix}1-\mu,\ 1\\ \alpha-\mu/2,\ \nu+1,\ -\nu\end{matrix}\right)$$

$[a,\ \operatorname{Re} p>0;\ 2\operatorname{Re}\alpha>|\operatorname{Re}\mu|]$.

4. $\int\limits_a^\infty (x^2-a^2)^{-(\nu+3)/2} e^{-p/(x^2-a^2)} Q_\nu^\mu\left(\frac{x}{a}\right) dx =$

$$= \frac{\sqrt{\pi}\, e^{\mu\pi i+p/(2a^2)}}{2^{\nu+2} p^{\nu/2+1}} \Gamma(\mu+\nu+1)\, W_{-(\nu+1)/2,\ -\mu/2}\left(\frac{p}{a^2}\right) \qquad [a,\ \operatorname{Re} p>0;\ \operatorname{Re}\nu>-3/2].$$

5. $\int\limits_a^\infty x(x^2-a^2)^{-\nu/2-2} e^{-p/(x^2-a^2)} Q_\nu^\mu\left(\frac{x}{a}\right) dx =$

$$= \frac{\sqrt{\pi}\, a e^{\mu\pi i+p/(2a^2)}}{2^{\nu+2} p^{(\nu+3)/2}} W_{-\nu/2,\ -\mu/2}\left(\frac{p}{a^2}\right) \qquad [a,\ \operatorname{Re} p>0;\ \operatorname{Re}\nu>-3/2].$$

6. $\int\limits_a^\infty \frac{e^{-p/\sqrt{x-a}}}{x-a} Q_\nu\left(\frac{x}{a}\right) dx =$

$$= \frac{1}{p}\sqrt{\frac{a}{2}}\,\Gamma^2(\nu+1)\, W_{-\nu-1/2,\ 0}\left(\frac{ip}{\sqrt{2a}}\right) W_{-\nu-1/2,\ 0}\left(-\frac{ip}{\sqrt{2a}}\right)$$

$[a,\ \operatorname{Re} p>0;\ \operatorname{Re}\nu>-1]$.

7. $\int\limits_a^\infty \frac{e^{-p/\sqrt{x-a}}}{x-a} Q_{-1/2}\left(\frac{x}{a}\right) dx = \frac{\pi^2}{4}\left[J_0^2\left(\frac{p}{2\sqrt{2a}}\right) + Y_0^2\left(\frac{p}{2\sqrt{2a}}\right)\right]$

$[a,\ \operatorname{Re} p>0]$.

8. $\int\limits_0^a (a^2-x^2)^{\mp(2\nu+1\pm 5)/4} e^{-px/\sqrt{a^2-x^2}} \times$

$$\times\left[\cos\frac{\mu+\nu}{2}\pi P_{\nu+1\pm 1}^\mu\left(\frac{x}{a}\right) - \frac{2}{\pi}\sin\frac{\mu+\nu}{2}\pi Q_\nu^\mu\left(\frac{x}{a}\right)\right] dx =$$

$$= \frac{2^\mu}{\sqrt{\pi}\, a}(ap)^{-(1\pm 1)/2\mp\nu}\,\Gamma\begin{bmatrix}(\mu+\nu+1)/2\\ (\nu-\mu)/2+1\end{bmatrix} S_{(1\pm 1)/2\pm\nu,\ \mu}(p) \qquad [a,\ \operatorname{Re} p>0].$$

2.18.7. Integrals containing hyperbolic or trigonometric functions and $Q_\nu^\mu(\varphi(x))$.

1. $\int\limits_0^\infty \operatorname{ch}\sigma x\, Q_\nu(a+b\operatorname{ch} x)\, dx =$

$$= \sqrt{\frac{\pi d}{2}}\,\Gamma(\nu-\sigma+1)\frac{\sin\nu\pi}{\sin(\nu+\sigma)\pi} P_{\sigma-1/2}^{-\nu-1/2}\left(\sqrt{1+d^2}\right) Q_\nu^{-\sigma}\left(\sqrt{1+b^2d^2}\right)$$

$[(1+d^2)(1+b^2d^2)=a^2d^2;\ \operatorname{Re} b>0;\ \operatorname{Re}\nu>|\operatorname{Re}\sigma|-1]$.

2. $\int_a^\infty \frac{\sin b\sqrt{x-a}}{\sqrt{x^2-a^2}} Q_\nu^1\left(\frac{x}{a}\right) dx = -\pi\sqrt{\frac{a}{2}}\, b I_{\nu+1/2}\left(b\sqrt{\frac{a}{2}}\right) K_{\nu+1/2}\left(b\sqrt{\frac{a}{2}}\right)$

$[a,\ b>0;\ \operatorname{Re}\nu>-3/2]$.

3. $\int_a^\infty \frac{\cos b\sqrt{x-a}}{\sqrt{x-a}} Q_\nu\left(\frac{x}{a}\right) dx = \pi\sqrt{2a}\, I_{\nu+1/2}\left(b\sqrt{\frac{a}{2}}\right) K_{\nu+1/2}\left(b\sqrt{\frac{a}{2}}\right)$

$[a,\ b>0;\ \operatorname{Re}\nu>-1]$.

2.18.8. Integrals of $A(x) J_\lambda(\varphi(x))\, Q_\nu^\mu(\chi(x))$.

1. $\int_a^\infty (x-a)^{\alpha-1}(x+a)^{-\mu/2} J_\lambda(b\sqrt{x-a})\, Q_\nu^\mu\left(\frac{x}{a}\right) dx =$

$$= 2^{\alpha-(\lambda+\mu)/2-1} e^{\mu\pi i} a^{\alpha+(\lambda-\mu)/2} b^\lambda \times$$
$$\times \Gamma\begin{bmatrix} 1-\alpha+(\mu-\lambda)/2+\nu,\ \alpha+(\lambda-\mu)/2,\ \alpha+(\lambda+\mu)/2 \\ 1+\lambda,\ 1+\alpha+(\lambda-\mu)/2+\nu \end{bmatrix} \times$$
$$\times {}_2F_3\left(\alpha+\frac{\lambda-\mu}{2},\ \alpha+\frac{\lambda+\mu}{2};\ \alpha+\frac{\lambda-\mu}{2}-\nu,\ 1+\lambda,\ 1+\alpha+\frac{\lambda-\mu}{2}+\nu;\ \frac{ab^2}{2}\right) +$$
$$+ 2^{2\alpha-\mu-3\nu-3} e^{\mu\pi i}\sqrt{\pi}\, a^{\nu+1} b^{2\nu+\mu-2\alpha+2}\, \Gamma\begin{bmatrix} \alpha+(\lambda-\mu)/2-\nu-1,\ 1+\mu+\nu \\ \nu+3/2,\ 2-\alpha+(\lambda+\mu)/2+\nu \end{bmatrix} \times$$
$$\times {}_2F_3\left(1+\mu+\nu,\ 1+\nu;\ 2-\alpha+\frac{\mu-\lambda}{2}+\nu,\ 2\nu+2,\ 2-\alpha+\frac{\lambda+\mu}{2}+\nu;\ \frac{ab^2}{2}\right)$$

$[a,\ b>0;\ \operatorname{Re}(2\alpha+\lambda)>|\operatorname{Re}\mu|;\ \operatorname{Re}(2\alpha-\mu-2\nu)<7/2]$.

2. $\int_a^\infty (x-a)^{-1/4}(x+a)^{-\mu/2} J_{\mu-1/2}(b\sqrt{x-a})\, Q_\nu^\mu\left(\frac{x}{a}\right) dx =$

$$= 2^{1-\mu} e^{\mu\pi i}\sqrt{\pi a}\, b^{\mu-1/2} I_{\nu+1/2}\left(b\sqrt{\frac{a}{2}}\right) K_{\nu+1/2}\left(b\sqrt{\frac{a}{2}}\right)$$

$[a,\ b>0;\ \operatorname{Re}\mu>-1/2;\ \operatorname{Re}(\mu+2\nu)>-2]$.

3. $\int_a^\infty (x-a)^{(2\mu-1)/4}(x^2-c^2)^{-\mu/2} J_{\mu-1/2}(b\sqrt{x-a})\, Q_\nu^\mu\left(\frac{x}{c}\right) dx =$

$$= 2^{1-\mu} e^{\mu\pi i} b^{\mu-1/2}\sqrt{\pi c}\, I_{\nu+1/2}\left(b\frac{\sqrt{a+c}-\sqrt{a-c}}{2}\right) K_{\nu+1/2}\left(b\frac{\sqrt{a+c}+\sqrt{a-c}}{2}\right)$$

$[a\geqslant c>0;\ \operatorname{Re}\mu>-1/2;\ \operatorname{Re}(\mu+2\nu)>-2]$.

4. $\int_a^\infty (x-a)^{\alpha-1}(x+a)^{-\mu/2} J_\lambda\left(\frac{b}{\sqrt{x-a}}\right) Q_\nu^\mu\left(\frac{x}{a}\right) dx =$

$$= \frac{2^{\alpha-\mu/2-1} e^{\pi\mu i}}{a^{\mu/2-1}} G_{24}^{31}\left(\frac{b^2}{8a}\middle|\begin{matrix}\alpha-\mu/2-\nu,\ \alpha-\mu/2+\nu+1 \\ \alpha+\mu/2,\ \alpha-\mu/2,\ \lambda/2,\ -\lambda/2\end{matrix}\right)$$

$[a,\ b>0;\ \operatorname{Re}(2\alpha-\lambda-\mu-2\nu)<2;\ 2\operatorname{Re}\alpha>|\operatorname{Re}\mu|-3/2]$.

5. $\int_a^\infty (x-a)^{(\mu-3)/2}(x+a)^{-\mu/2} J_{2\nu+1}\left(\frac{b}{\sqrt{x-a}}\right) Q_\nu^\mu\left(\frac{x}{a}\right) dx =$

$$= \frac{b^{\mu-1} e^{\mu\pi i}}{2^{3\mu/2-1} a^{\mu/2}} K_\mu\left(\frac{b}{\sqrt{2a}}\right)$$

$[a,\ b>0;\ \operatorname{Re}\mu>-1/4;\ \operatorname{Re}\nu>-1]$.

6. $\int_a^\infty (x-a)^{\mu-5/4}(x+a)^{-\mu/2} J_{\pm\mu\pm 1/2}\left(\frac{b}{\sqrt{x-a}}\right) Q_{-1/2}^{\mu}\left(\frac{x}{a}\right) dx =$

$$= e^{\mu\pi i}\pi^{\pm 1/2}\left(\frac{b}{2}\right)^{\mu-1/2}\begin{Bmatrix} I_\mu(b/\sqrt{8a})\, K_\mu(b/\sqrt{8a}) \\ K_\mu^2(b/\sqrt{8a}) \end{Bmatrix}$$

$[a,\ b>0;\ \operatorname{Re}\mu > -1/3]$.

7. $\int_a^\infty (x-a)^{-\mu-5/4}(x+a)^{\mu/2} J_{\mu-1/2}\left(\frac{b}{\sqrt{x-a}}\right) Q_{\nu}^{\mu}\left(\frac{x}{a}\right) dx =$

$$= -\frac{2^{\mu+1/2}e^{\mu\pi i}}{\sqrt{\pi}\, b^{\mu+3/2}}\Gamma[1+\mu+\nu,\ 1+\mu+\nu,\ \mu-\nu]\, W_{-\nu-1/2,\ \mu}\left(\frac{b}{\sqrt{2a}}\right)\times$$

$$\times\left[\frac{\cos\nu\pi}{\Gamma(2\mu+1)} M_{\nu+1/2,\ \mu}\left(\frac{b}{\sqrt{2a}}\right) - \frac{\cos\mu\pi}{\Gamma(1+\mu+\nu)} W_{\nu+1/2,\ \mu}\left(\frac{b}{\sqrt{2a}}\right)\right]$$

$[a,\ b>0;\ -\operatorname{Re}\nu-1,\ \operatorname{Re}\nu < \operatorname{Re}\mu < 1/3]$.

8. $\int_a^\infty (x-a)^{-\mu-5/4}(x+a)^{\mu/2} J_{1/2-\mu}\left(\frac{b}{\sqrt{x-a}}\right) Q_{\nu}^{\mu}\left(\frac{x}{a}\right) dx =$

$$= 2^{3/2}e^{\mu\pi i}\sqrt{\pi}\, a^{1-\mu} b^{\mu-5/2}\Gamma\begin{bmatrix} 1+\mu+\nu \\ 1-2\mu \end{bmatrix} M_{\nu+1/2,\ -\mu}\left(\frac{b}{\sqrt{2a}}\right) W_{-\nu-1/2,\ -\mu}\left(\frac{b}{\sqrt{2a}}\right)$$

$[a,\ b>0;\ \operatorname{Re}\mu < 1/3;\ \operatorname{Re}\nu > -3/2;\ \operatorname{Re}(\mu+\nu) > -1]$.

9. $\int_a^\infty (x^2-a^2)^{-5/4} J_{\nu+1/2}\left(\frac{b}{\sqrt{x^2-a^2}}\right) Q_{\nu}^{\mu}\left(\frac{x}{a}\right) dx =$

$$= \frac{2^{\mu-1/2}e^{\mu\pi i}}{a\sqrt{b}}\Gamma\begin{bmatrix} (1+\mu+\nu)/2 \\ 1+(\nu-\mu)/2 \end{bmatrix} K_\mu\left(\frac{b}{a}\right)$$

$[a,\ b>0;\ |\operatorname{Re}\mu| < 1;\ \operatorname{Re}\nu > -3/2]$.

10. $\int_a^\infty (x^2-a^2)^{-5/4} e^{\pm ipx/\sqrt{x^2-a^2}} J_{n+1/2}\left(\frac{b}{\sqrt{x^2-a^2}}\right) Q_{n}^{m}\left(\frac{x}{a}\right) dx =$

$$= \frac{1}{a}\sqrt{\frac{\pi}{2b}}\, e^{\pm(m+n+2)\pi i/2} H_m^{\{1 \atop 2\}}\left(\frac{\sqrt{a^2p^2-b^2}}{a}\right) Q_n^m\left(\frac{ap}{b}\right) \qquad [a,\ b,\ p>0].$$

11. $\int_0^\infty \sqrt{x}\, J_{\nu+1/2}(bx)\, Q_\nu\left(\frac{z^2+x^2}{2cx}\right) dx = \frac{\pi\sqrt{c}}{b} e^{-b\sqrt{z^2-c^2}} J_{\nu+1/2}(bc)$

$[b,\ c,\ \operatorname{Re}z > 0;\ \operatorname{Re}\nu > -1]$.

2.18.9. Integrals of $A(x)\, K_\lambda(\varphi(x))\, Q_\nu^\mu(cx)$.

1. $\int_a^\infty (x-a)^{-\mu-5/4}(x+a)^{\mu/2} K_{\mu-1/2}\left(\frac{b}{\sqrt{x-a}}\right) Q_{\nu}^{\mu}\left(\frac{x}{a}\right) dx =$

$$= 2^{\mu}e^{\mu\pi i}\sqrt{\pi a}\, b^{-\mu-3/2}\Gamma^2(1+\mu+\nu)\, W_{-\nu-1/2,\ \mu}\left(\frac{ib}{\sqrt{2a}}\right) W_{-\nu-1/2,\ \mu}\left(-\frac{ib}{\sqrt{2a}}\right)$$

$[a,\ \operatorname{Re}b > 0;\ \operatorname{Re}\nu > -3/2;\ \operatorname{Re}(\mu+\nu) > -1]$.

2. $\int_a^\infty (x-a)^{-\mu-5/4}(x+a)^{\mu/2} K_{\mu-1/2}\left(\frac{b}{\sqrt{x-a}}\right) Q_{-1/2}^{\mu}\left(\frac{x}{a}\right) dx =$

$$=\frac{2^{\mu-5/2}e^{\mu\pi i}\pi^{3/2}}{b^{\mu+1/2}}\Gamma^2\left(\mu+\frac{1}{2}\right)\left[J^2_{-\mu}\left(\frac{b}{2\sqrt{2a}}\right)+Y^2_{-\mu}\left(\frac{b}{2\sqrt{2a}}\right)\right]$$

[a, Re $b>0$; Re $\mu>-1/2$].

2.18.10. Integrals of $A(x)\,\mathbf{K}(\varphi(x))\,Q_\nu(cx)$.

1. $$\int_a^\infty \frac{1}{\sqrt{x+a}}\mathbf{K}\left(\sqrt{\frac{2a}{x+a}}\right)Q_\nu\left(\frac{x}{a}\right)dx=\frac{2\sqrt{2a}}{(2\nu+1)^2}[\psi(\nu+1)+\mathbf{C}+2\ln 2]$$

[$a>0$; Re $\nu>-1/2$].

2. $$\int_a^\infty \frac{1}{(x+a)^{3/2}}\mathbf{K}\left(\sqrt{\frac{2a}{x+a}}\right)Q_\nu\left(\frac{x}{a}\right)dx=$$

$$=\frac{1}{\sqrt{2a}\,(2\nu+1)^2}[\pi\cos\nu\pi-2\mathbf{C}-4\ln 2-2\psi(\nu+1)]$$

[$a>0$; $\nu\neq-1/2, -1, -2, \ldots$].

3. $$\int_a^\infty \frac{1}{\sqrt{x+a}}\mathbf{K}\left(\sqrt{\frac{x-a}{x+a}}\right)Q_\nu\left(\frac{x}{a}\right)dx=\frac{\sqrt{2a}\,\pi}{16}(2\nu+1)^2$$ [$a>0$; Re $\nu>-1/2$].

2.18.11. Integrals of $A(x)\,Q_\lambda^\varkappa(\varphi(x))\,Q_\nu^\mu(cx)$.

1. $$\int_0^a Q_\lambda\left(\frac{x}{a}\right)Q_\nu\left(\frac{x}{a}\right)dx=\frac{a}{(\nu-\lambda)(\nu+\lambda+1)}\left\{\psi(\lambda+1)-\psi(\nu+1)-\right.$$

$$\left.-\frac{\pi}{2}\left[(A-A^{-1})\sin\frac{\lambda+\nu}{2}\pi+(A+A^{-1})\sin\frac{\lambda-\nu}{2}\pi\right]\right\},$$

$$A=\Gamma\begin{bmatrix}(\lambda+1)/2,\ \nu/2+1\\ \lambda/2+1,\ (\nu+1)/2\end{bmatrix}$$ [a, Re λ, Re $\nu>0$].

2. $$\int_0^a Q_l\left(\frac{x}{a}\right)Q_n\left(\frac{x}{a}\right)dx=-\frac{a}{(l-n)(l+n+1)}\sum_{k=n+1}^{l}\frac{1}{k}$$ [$a>0$; $l>n$; $l+n$ even].

3. $$=-\frac{a}{(l-n)(l+n+1)}\left\{\sum_{k=n+1}^{l}\frac{1}{k}+2^{n-l-1}\frac{l!}{n!}\left[\frac{((n-1)/2)!}{(l/2)!}\right]^2\right\}$$

[$a>0$; $l>n$; $l=2, 4, 6, \ldots$; $n=1, 3, 5, \ldots$].

4. $$\int_{-a}^a Q_\lambda\left(\frac{x}{a}\right)Q_\nu\left(\frac{x}{a}\right)dx=$$

$$=\frac{2[\psi(\lambda+1)-\psi(\nu+1)](1+\cos\lambda\pi\cos\nu\pi)-\pi\sin(\lambda-\nu)\pi}{2(\nu-\lambda)(\nu+\lambda+1)}a$$

[$\lambda+\nu=-1$; $\lambda\neq\nu$; λ, $\nu\neq-1, -2, -3, \ldots$; $a>0$],

5. $$=\frac{\pi^2-2\psi'(\nu+1)(1+\cos^2\nu\pi)}{2(2\nu+1)}a$$ [$\lambda=\nu$; $\nu\neq-1, -2, -3, \ldots$; $a>0$],

6. $$=\frac{2\psi'(-\nu)\sin^2\nu\pi-\pi^2\cos^2\nu\pi}{2(2\nu+1)}a$$ [$\lambda=-\nu-1$; $\nu\neq-1, -2, -3, \ldots$; $a>0$].

7. $$\int_{-a}^a Q_l\left(\frac{x}{a}\right)Q_n\left(\frac{x}{a}\right)dx=-\begin{Bmatrix}1\\0\end{Bmatrix}\frac{2a}{(l-n)(l+n+1)}\sum_{k=n+1}^{l}\frac{1}{k}$$

$$\left[a>0,\ \begin{Bmatrix}l>n;\ l+n=2, 4, 6, \ldots\\ l\neq n;\ l+n=1, 3, 5, \ldots\end{Bmatrix}\right].$$

8. $\int\limits_{-a}^{a}\left[Q_n\left(\frac{x}{a}\right)\right]^2 dx=\frac{2a}{2n+1}\left[\frac{\pi^2}{4}+\psi'(n)\right]$ $[a>0]$.

9. $\int\limits_{a}^{\infty} Q_\lambda\left(\frac{x}{a}\right)Q_\nu\left(\frac{x}{a}\right)dx=\frac{\psi(\nu+1)-\psi(\lambda+1)}{(\nu-\lambda)(\nu+\lambda+1)}a$

$[a>0;\ \mathrm{Re}\,(\lambda+\nu)>-1;\ \lambda,\ \nu\neq-1,\ -2,\ -3,\ \ldots]$.

10. $=\frac{\psi'(\nu+1)}{2\nu+1}a$ $[\lambda=\nu;\ a>0;\ \mathrm{Re}\,\nu>-1/2]$.

11. $\int\limits_{a}^{\infty}(x^2-a^2)^{\beta-1}Q_\nu^{\mu-1}\left(\frac{x}{a}\right)Q_\nu^\mu\left(\frac{x}{a}\right)dx=$

$=-\frac{\sqrt{\pi}}{4}e^{2\mu\pi i}a^{2\beta-1}\Gamma\begin{bmatrix}\beta-1/2,\ \mu+\nu,\ \beta-\mu+1/2,\ \beta+\mu-1/2,\ 3/2-\beta+\nu\\ \beta,\ 2-\mu+\nu,\ \beta+\nu+1/2\end{bmatrix}$

$[a>0;\ 1/2,\ |\mathrm{Re}\,\mu-1/2|<\mathrm{Re}\,\beta<3/2+\mathrm{Re}\,\nu]$.

12. $\int\limits_{a}^{\infty}(x^2-a^2)^{\beta-1}Q_{\nu\pm1}^{\mu}\left(\frac{x}{a}\right)Q_\nu^\mu\left(\frac{x}{a}\right)dx=$

$=\frac{\sqrt{\pi}}{4}e^{2\mu\pi i}a^{2\beta-1}\Gamma\begin{bmatrix}\beta,\ \beta+\mu,\ \beta-\mu,\ (3\pm1)/2-\beta+\nu,\ \mu+\nu+(1\pm1)/2\\ \beta+1/2,\ \beta+\nu+(1\pm1)/2,\ (3\pm1)/2-\mu+\nu\end{bmatrix}$

$[a>0;\ |\mathrm{Re}\,\mu|<\mathrm{Re}\,\beta<(3\pm1)/2+\mathrm{Re}\,\nu]$.

13. $\int\limits_{a}^{\infty}x(x^2-a^2)^{\beta-1}\left[Q_\nu^\mu\left(\frac{x}{a}\right)\right]^2dx=$

$=\frac{\sqrt{\pi}}{4}e^{2\mu\pi i}a^{2\beta}\Gamma\begin{bmatrix}\beta,\ \beta+\mu,\ \beta-\mu,\ 1-\beta+\nu,\ 1+\mu+\nu\\ \beta+1/2,\ \beta+\nu+1,\ 1-\mu+\nu\end{bmatrix}$

$[a>0;\ |\mathrm{Re}\,\mu|<\mathrm{Re}\,\beta<1+\mathrm{Re}\,\nu]$.

14. $\int\limits_{a}^{\infty}x(x^2-a^2)^{\beta-1}Q_{-\nu-1}^{\mu}\left(\frac{x}{a}\right)Q_\nu^\mu\left(\frac{x}{a}\right)dx=$

$=\frac{\cos\mu\pi}{4\sqrt{\pi}}e^{2\mu\pi i}a^{2\beta}\Gamma\begin{bmatrix}\beta,\ \beta-\mu,\ \beta+\mu,\ 1/2-\mu,\ 1+\mu+\nu,\ \mu-\nu\\ 1+\beta+\nu,\ \beta-\nu\end{bmatrix}$

$[a>0;\ |\mathrm{Re}\,\mu|<\mathrm{Re}\,\beta<1/2]$.

15. $\int\limits_{a}^{\infty}x(x^2-a^2)^{\beta-1}Q_{\mu-1/2}^{\nu+1/2}\left(\frac{x}{\sqrt{x^2-a^2}}\right)Q_\nu^\mu\left(\frac{x}{a}\right)dx=$

$=\frac{e^{(\mu+\nu+1/2)\pi i}}{4\sqrt{2}}a^{2\beta}\Gamma\begin{bmatrix}1/4+\beta,\ 1/4-\beta,\ \beta+\mu+1/4,\ 3/4-\beta+\nu,\ \mu+\nu+1\\ 3/4-\beta+\mu,\ \beta+\nu+5/4\end{bmatrix}$

$[a>0;\ -1/4,\ -\mathrm{Re}\,\mu-1/4<\mathrm{Re}\,\beta<1/4,\ 3/4+\mathrm{Re}\,\nu]$.

2.18.12. Integrals of $A(x)J_\eta(\varphi(x))Q_\lambda^\varkappa(cx)Q_\nu^\mu(cx)$.

1. $\int\limits_{a}^{\infty}(x^2-a^2)^{-1}J_{2\nu}\left(\frac{b}{\sqrt{x^2-a^2}}\right)Q_{\nu-1}^{\mu}\left(\frac{x}{a}\right)Q_\nu^\mu\left(\frac{x}{a}\right)dx=\frac{e^{2\mu\pi i}}{2a}\Gamma\begin{bmatrix}\mu+\nu\\ 1-\mu+\nu\end{bmatrix}K_\mu^2\left(\frac{b}{2a}\right)$

$[a,\ b>0;\ |\mathrm{Re}\,\mu|<3/4;\ \mathrm{Re}\,\nu>-1/2]$.

2. $\int_a^\infty (x^2-a^2)^{-1} J_{2\nu+1}\left(\frac{b}{\sqrt{x^2-a^2}}\right) Q_\nu^{\mu+1}\left(\frac{x}{a}\right) Q_\nu^\mu\left(\frac{x}{a}\right) dx =$

$$=\frac{e^{(2\mu+1)\pi i}}{2a}\Gamma\begin{bmatrix}1+\mu+\nu\\1-\mu+\nu\end{bmatrix} K_{\mu+1}\left(\frac{b}{2a}\right) K_\mu\left(\frac{b}{2a}\right)$$

$[a, b > 0;\ \mathrm{Re}\,\nu > -1,\ -5/4 < \mathrm{Re}\,\mu < 1/4]$.

3. $\int_a^\infty x(x^2-a^2)^{-3/2} J_{2\nu+1}\left(\frac{b}{\sqrt{x^2-a^2}}\right)\left[Q_\nu^\mu\left(\frac{x}{a}\right)\right]^2 dx = \frac{e^{2\mu\pi i}}{b}\Gamma\begin{bmatrix}1-\mu+\nu\\1+\mu+\nu\end{bmatrix} K_\mu^2\left(\frac{b}{2a}\right)$

$[a, b > 0;\ \mathrm{Re}\,\nu > -1;\ |\mathrm{Re}\,\mu| < 1/4]$.

2.18.13. Integrals of $A(x) P_\lambda^\varkappa(\varphi(x))\, Q_\nu^\mu(cx)$.

1. $\int_0^a P_\lambda\left(\frac{x}{a}\right) Q_\nu\left(\frac{x}{a}\right) dx =$

$$=\frac{a}{(\lambda-\nu)(\lambda+\nu+1)}\left\{1-\cos\frac{\lambda-\nu}{2}\pi\Gamma\begin{bmatrix}\lambda/2+1,\ (\nu+1)/2\\(\lambda+1)/2,\ \nu/2+1\end{bmatrix}\right\}$$

$[a, \mathrm{Re}\,\lambda, \mathrm{Re}\,\nu > 0]$.

2. $\int_{-a}^a P_\lambda\left(\frac{x}{a}\right) Q_\nu\left(\frac{x}{a}\right) dx =$

$$=\frac{\pi-\pi\cos(\lambda-\nu)\pi-2\sin\lambda\pi\cos\nu\pi[\psi(\lambda+1)-\psi(\nu+1)]}{\pi(\lambda-\nu)(\lambda+\nu+1)}a$$

$[\lambda\neq\nu;\ a, \mathrm{Re}\,\lambda, \mathrm{Re}\,\nu > 0]$,

3. $$=-\frac{\sin 2\nu\pi\psi'(\nu+1)\,a}{\pi(2\nu+1)}$$

$[\lambda=\nu;\ a, \mathrm{Re}\,\nu > 0]$.

4. $\int_{-a}^a P_l^m\left(\frac{x}{a}\right) Q_n^m\left(\frac{x}{a}\right) dx = \frac{(-1)^m[1-(-1)^{l+n}(m+n)!]\,a}{(l-n)(l+n+1)(n-m)!}$ $[l\neq n;\ a > 0]$.

5. $\int_a^\infty P_\lambda\left(\frac{x}{a}\right) Q_\nu\left(\frac{x}{a}\right) dx = \frac{a}{(\nu-\lambda)(\nu+\lambda+1)}$ $[a, \mathrm{Re}\,(\nu-\lambda) > 0;\ \mathrm{Re}\,(\lambda+\nu) > -1]$.

6. $\int_a^\infty x(x^2-a^2)^{\beta-1} P_\nu^{\pm\mu}\left(\frac{x}{a}\right) Q_\nu^\mu\left(\frac{x}{a}\right) dx =$

$$=\frac{e^{\mu\pi i}a^{2\beta}}{4\sqrt{\pi}}\Gamma^{(1\pm1)/2}\begin{bmatrix}1+\mu+\nu\\1-\mu+\nu\end{bmatrix}\Gamma\begin{bmatrix}\beta,\ 1/2-\beta,\ \beta\mp\mu,\ 1-\beta+\nu\\1-\beta\mp\mu,\ 1+\nu+\beta\end{bmatrix}$$

$[a > 0;\ 0, \pm\mathrm{Re}\,\mu < \mathrm{Re}\,\beta < 1/2,\ 1+\mathrm{Re}\,\nu]$.

7. $\int_a^\infty x^{\alpha-1}(x^2-a^2)^{-\varkappa/2}(x^2-b^2)^{-\mu/2} P_\lambda^\varkappa\left(\frac{x}{a}\right) Q_\nu^\mu\left(\frac{x}{b}\right) dx =$

$$=e^{\mu\pi i}2^{\varkappa+\mu-2}a^{\alpha-\varkappa-\mu-\nu-1}b^{\nu+1}E\left(\frac{\mu+\nu}{2}+1,\ \frac{\mu+\nu+1}{2},\ 1+\frac{\varkappa+\lambda+\mu+\nu-\alpha}{2},\right.$$
$$\left.\frac{1-\alpha+\varkappa-\lambda+\mu+\nu}{2};\ \nu+\frac{3}{2},\ \frac{\mu+\nu-\alpha}{2}+1,\ \frac{3+\mu+\nu-\alpha}{2};\ -\frac{a^2}{b^2}\right)$$

$[|b| < a;\ \mathrm{Re}\,\varkappa < 1;\ \mathrm{Re}\,(\varkappa+\lambda+\mu+\nu-\alpha) > -2;\ \mathrm{Re}\,(\varkappa-\lambda+\mu+\nu-\alpha) > -1]$.

8. $\int_a^\infty (x^2-a^2)^{-\varkappa/2}(x^2-b^2)^{-\mu/2} P_\lambda^\varkappa\left(\frac{x}{a}\right) Q_\nu^\mu\left(\frac{x}{b}\right) dx =$

$$=\frac{2^{\varkappa+\mu-2}e^{\mu\pi i}b^{\nu+1}}{a^{\varkappa+\mu+\nu}}\Gamma\begin{bmatrix}(\varkappa+\lambda+\mu+\nu+1)/2, & (\varkappa-\lambda+\mu+\nu)/2\\ 3/2+\nu\end{bmatrix}\times$$

$$\times{}_2F_1\left(\frac{\varkappa+\lambda+\mu+\nu+1}{2},\ \frac{\varkappa-\lambda+\mu+\nu}{2};\ \frac{3}{2}+\nu;\ \frac{b^2}{a^2}\right)$$

$[|b|<a;\ \mathrm{Re}\,\varkappa<1;\ \mathrm{Re}\,(\varkappa+\lambda+\mu+\nu)>-1;\ \mathrm{Re}\,(\varkappa-\lambda+\mu+\nu)>0].$

9. $$=2^{\varkappa+\mu-2}e^{\mu\pi i}\sqrt{ab}\,(a^2-b^2)^{-(\varkappa+\mu)/2}\Gamma\left(\frac{\varkappa+\mu}{2}\right)\Gamma\left(\frac{\varkappa+\mu+1}{2}+\nu\right)\times$$

$$\times P^{-\nu-1/2}_{-(\varkappa+\mu)/2}\left(\frac{b^2+a^2}{b^2-a^2}\right)$$

$[\lambda=\nu;\ |b|<a;\ \mathrm{Re}\,\varkappa<1,\ \mathrm{Re}\,(\varkappa+\mu+2\nu)>-1;\ \mathrm{Re}\,(\varkappa+\mu)>0].$

10. $$\int_a^\infty (x^2-a^2)^{-\varkappa/2}(x^2-b^2)^{-\mu/2}P^{\varkappa}_{1/2-\varkappa-\mu}\left(\frac{x}{a}\right)Q^{\mu}_{\nu}\left(\frac{x}{b}\right)dx=$$

$$=2^{\varkappa+\mu+\nu-1/2}\sqrt{\frac{a}{\pi b}}\,(a^2-b^2)^{(1-\varkappa-\mu)/2}\Gamma\left(\frac{2\nu+3}{4}\right)Q^{\varkappa+\mu-1}_{(2\nu-1)/4}\left(\frac{2a^2}{b^2}-1\right)$$

$[|b|<a;\ \mathrm{Re}\,\varkappa<1;\ \mathrm{Re}\,\nu>-3/2;\ \mathrm{Re}\,(2\varkappa+2\mu+\nu)>1/2].$

11. $$\int_a^\infty (x^2-a^2)^{\beta-1}P^{-\nu-3/2}_{\mu-1/2}\left(\frac{x}{\sqrt{x^2-a^2}}\right)Q^{\mu}_{\nu}\left(\frac{x}{a}\right)dx=$$

$$=\frac{e^{\mu\pi i}a^{2\beta-1}}{\sqrt{8}\,(\mu+\nu+1)}\Gamma\begin{bmatrix}\beta+1/4, & \beta+\mu+1/4, & \beta-\mu+1/4, & \nu-\beta+7/4\\ \beta+3/4, & \beta+\nu+5/4, & \nu-\mu+2\end{bmatrix}$$

$[a>0;\ |\mathrm{Re}\,\mu|-1/4<\mathrm{Re}\,\beta<7/4+\mathrm{Re}\,\nu].$

12. $$\int_a^\infty (x^2-a^2)^{\beta-1}P^{-\nu-1/2}_{\mu+1/2}\left(\frac{x}{\sqrt{x^2-a^2}}\right)Q^{\mu}_{\nu}\left(\frac{x}{a}\right)dx=$$

$$=\frac{e^{\mu\pi i}a^{2\beta-1}}{\sqrt{8}\,(\mu+\nu+1)}\Gamma\begin{bmatrix}\beta-1/4, & \beta+\mu+3/4, & \beta-\mu-1/4, & 5/4-\beta+\nu\\ \beta+1/4, & \beta+\nu+3/4, & \nu-\mu+1\end{bmatrix}$$

$[a>0;\ 1/4,\ -\mathrm{Re}\,\mu-3/4,\ \mathrm{Re}\,\mu+1/4<\mathrm{Re}\,\beta<5/4+\mathrm{Re}\,\nu].$

13. $$\int_a^\infty x(x^2-a^2)^{\beta-1}P^{-\nu-1/2}_{-\mu-1/2}\left(\frac{x}{\sqrt{x^2-a^2}}\right)Q^{\mu}_{\nu}\left(\frac{x}{a}\right)dx=$$

$$=\frac{e^{\mu\pi i}a^{2\beta}}{\sqrt{8}}\Gamma\begin{bmatrix}\beta+1/4, & \beta+\mu+1/4, & \beta-\mu+1/4, & \nu-\mu+3/4\\ \beta+3/4, & \beta+\nu+5/4, & \nu-\mu+1\end{bmatrix}$$

$[a>0;\ |\mathrm{Re}\,\mu|-1/4<\mathrm{Re}\,\beta<\mathrm{Re}\,\nu+3/4].$

14. $$\int_a^\infty x(x^2-a^2)^{\beta-1}P^{\nu+1/2}_{\mu-1/2}\left(\frac{x}{\sqrt{x^2-a^2}}\right)Q^{\mu}_{\nu}\left(\frac{x}{a}\right)dx=$$

$$=\frac{e^{\mu\pi i}}{\sqrt{8\pi}}\cos\mu\pi a^{2\beta}\Gamma\begin{bmatrix}\beta+1/4, 1/4-\beta, \beta+\mu+1/4, \beta-\mu+1/4, \mu+\nu+1\\ \beta+\nu+5/4, \beta-\nu+1/4\end{bmatrix}$$

$[a>0;\ |\mathrm{Re}\,\mu|-1/4<\mathrm{Re}\,\beta<1/4],$

2.18.14. Integrals of $A(x)\left\{\begin{matrix}\sin\varphi(x)\\ \cos\varphi(x)\end{matrix}\right\}P^{\varkappa}_{\lambda}(cx)\,Q^{\mu}_{\nu}(cx)$.

1. $$\int_a^\infty x(x^2-a^2)^{\beta-1}\left\{\begin{matrix}\sin(b\sqrt{x^2-a^2})\\ \cos(b\sqrt{x^2-a^2})\end{matrix}\right\}P^{\pm\mu}_{\nu}\left(\frac{x}{a}\right)Q^{\mu}_{\nu}\left(\frac{x}{a}\right)dx=$$

$$=\frac{e^{\mu\pi i}a^{2\beta}}{4}G^{32}_{35}\left(\frac{a^2b^2}{4}\left|\begin{matrix}1-\beta, & 1-\beta\pm\mu, & 1-\beta\mp\mu\\ 1/2-\beta, & 1-\beta+\nu, & (1\pm1)/4, & -\beta-\nu, & (1\mp1)/4\end{matrix}\right.\right)$$

$[a,\ b>0;\ \mathrm{Re}\,\beta,\ \mathrm{Re}\,(\beta\mp\mu)>-(1\pm1)/4;\ \mathrm{Re}\,\beta<1;\ \mathrm{Re}\,(\beta-\nu)<3/2].$

2. $\int\limits_a^\infty x\,(x^2-a^2)^{-1/4} \left\{ \begin{matrix} \sin(b\sqrt{x^2-a^2}) \\ \cos(b\sqrt{x^2-a^2}) \end{matrix} \right\} P_\nu^{\mp 1/4}\left(\dfrac{x}{a}\right) Q_\nu^{\mp 1/4}\left(\dfrac{x}{a}\right) dx =$

$= e^{\mp \pi i/4}\,a\,\sqrt{\dfrac{\pi}{2b}}\,\Gamma^{\pm 1}\left[\begin{matrix} \nu+3/4 \\ \nu+5/4 \end{matrix}\right] I_{\nu+1/2}\left(\dfrac{ab}{2}\right) K_{\nu+1/2}\left(\dfrac{ab}{2}\right)$ $\qquad [a,\,b>0;\ \mathrm{Re}\,\nu>-3/4]$.

3. $\int\limits_a^\infty x\,(x^2-a^2)^{-1/4} \left\{ \begin{matrix} \sin(b\sqrt{x^2-a^2}) \\ \cos(b\sqrt{x^2-a^2}) \end{matrix} \right\} P_\nu^{\mp 1/4}\left(\dfrac{x}{a}\right) Q_\nu^{\pm 1/4}\left(\dfrac{x}{a}\right) dx =$

$= e^{\pm \pi i/4}a\,\sqrt{\dfrac{\pi}{2b}}\,I_{\nu+1/2}\left(\dfrac{ab}{2}\right) K_{\nu+1/2}\left(\dfrac{ab}{2}\right)$ $\qquad [a,\,b>0;\ \mathrm{Re}\,\nu>-3/4]$.

2.18.15. Integrals of $A(x)\,J_\eta(\varphi(x))\,P_\lambda^\varkappa(cx)\,Q_\nu^\mu(cx)$.

1. $\int\limits_a^\infty x J_{\mp 2\mu}(b\sqrt{x^2-a^2})\,P_\nu^{\pm\mu}\left(\dfrac{x}{a}\right) Q_\nu^\mu\left(\dfrac{x}{a}\right) dx =$

$= \dfrac{a}{b}\,e^{\mu\pi i}\,\Gamma^{(1\pm 1)/2}\left[\begin{matrix} 1+\mu+\nu \\ 1-\mu+\nu \end{matrix}\right] I_{\nu+1/2}\left(\dfrac{ab}{2}\right) K_{\nu+1/2}\left(\dfrac{ab}{2}\right)$

$[a,\,b>0;\ \mp\,\mathrm{Re}\,\mu>-1/2;\ \mathrm{Re}\,\nu>-3/4]$.

2. $\int\limits_a^\infty x J_0(b\sqrt{x^2-a^2})\,P_\nu^{\pm\mu}\left(\dfrac{x}{a}\right) Q_\nu^\mu\left(\dfrac{x}{a}\right) dx =$

$= \dfrac{e^{\mu\pi i}}{b^2}\,\Gamma\left[\begin{matrix} \mu+\nu+1 \\ 2\nu+2 \end{matrix}\right] M_{\mp\mu,\,\nu+1/2}(ab)\,W_{\pm\mu,\,\nu+1/2}(ab)$

$[a,\,b>0;\ \mp\,\mathrm{Re}\,\mu>-1;\ \mathrm{Re}\,\nu>-3/4]$.

3. $\int\limits_a^\infty x\,(x^2-a^2)^{\beta-1} J_\eta(b\sqrt{x^2-a^2})\,P_\nu^{\pm\mu}\left(\dfrac{x}{a}\right) Q_\nu^\mu\left(\dfrac{x}{a}\right) dx =$

$= \dfrac{e^{\mu\pi i}a^{2\beta}}{4\sqrt{\pi}}\,\Gamma^{(1\pm 1)/2}\left[\begin{matrix} 1+\mu+\nu \\ 1-\mu+\nu \end{matrix}\right] G_{35}^{32}\left(\dfrac{a^2b^2}{4}\,\middle|\,\begin{matrix} 1-\beta,\ 1-\beta\pm\mu,\ 1-\beta\mp\mu \\ 1/2-\beta,\ 1-\beta+\nu,\ \eta/2,\ -\beta-\nu,\ -\eta/2 \end{matrix}\right)$

$[a,\,b,\ \mathrm{Re}\,(2\beta+\eta),\ \mathrm{Re}\,(2\beta\mp 2\mu+\eta)>0;\ \mathrm{Re}\,\beta<5/4;\ \mathrm{Re}\,(\beta-\nu)<7/4]$.

4. $\int\limits_a^\infty x\,(x^2-a^2)^{-3/2} J_{2\nu+1}\left(\dfrac{b}{\sqrt{x^2-a^2}}\right) P_\nu^{-\mu}\left(\dfrac{x}{a}\right) Q_\nu^\mu\left(\dfrac{x}{a}\right) dx = \dfrac{e^{\mu\pi i}}{b}\,I_\mu\left(\dfrac{b}{2a}\right) K_\mu\left(\dfrac{b}{2a}\right)$

$[a,\,b>0;\ \mathrm{Re}\,\nu>-1;\ \mathrm{Re}\,(\mu+\nu)>-1]$.

2.18.16. Integrals with respect to the index containing $Q_{\pm ix-1/2}(c)$.

1. $\int\limits_0^\infty x\,\mathrm{sh}\,\pi x K_{2ix}(b)\,[Q_{ix-1/2}(c)+Q_{-ix-1/2}(c)]\,dx = -\dfrac{\pi^2 b}{8}\,Y_0\left(b\sqrt{\dfrac{c-1}{2}}\right)$ $\qquad [c>1]$.

2. $\int\limits_0^\infty x\,\mathrm{th}\,\pi x K_{ix}(a)\,K_{ix}(b)\,[Q_{ix-1/2}(c)+Q_{-ix-1/2}(c)]\,dx =$

$= -\dfrac{\pi\sqrt{ac}}{4z}\,[e^{-z}\,\mathrm{Ei}\,(z-a-b)-e^z\,\mathrm{Ei}\,(-z-a-b)]$ $\qquad [z=\sqrt{a^2+c^2+2abc}]$.

2.19. THE WHITTAKER FUNCTIONS $M_{\rho,\sigma}(x)$ AND $W_{\rho,\sigma}(x)$

The Whittaker functions $M_{\rho,\sigma}(x)$ and $W_{\rho,\sigma}(x)$ are related to the confluent hypergeometric functions ${}_1F_1(a;b;x)$ and $\Psi(a;b;x)$ by the relations

$$M_{\rho,\sigma}(x)=x^{\sigma+1/2}e^{-x/2}\,{}_1F_1\left(\sigma-\rho+\frac{1}{2};\ 2\sigma+1;\ x\right),$$

$$W_{\rho,\sigma}(x)=x^{\sigma+1/2}e^{-x/2}\Psi\left(\sigma-\rho+\frac{1}{2},\ 2\sigma+1;\ x\right).$$

For certain values of the parameters, the functions $M_{\rho,\sigma}(x)$ and $W_{\rho,\sigma}(x)$ reduce to other special functions; for example

$$M_{0,\sigma}(x)=2^{2\sigma}\Gamma(\sigma+1)\sqrt{x}\,I_\sigma\left(\frac{x}{2}\right),\quad W_{0,\sigma}(x)=\sqrt{\frac{x}{\pi}}K_\sigma\left(\frac{x}{2}\right).$$

For the corresponding formulae, see Ch. 7; the integrals are in other sections of this chapter and in [17, 18].

2.19.1. Integrals of general form

NOTATION:

$$T_\gamma=\Gamma^{1-\gamma^2}\begin{bmatrix}2\sigma+1\\ \rho+\sigma+1/2\end{bmatrix}\left[\Gamma\left(\frac{1}{2}-\rho+\sigma\right)\Gamma\left(\frac{1}{2}-\rho-\sigma\right)\right]^{-(\gamma^2+\gamma)/2}.$$

1. $\int_0^a x^{\alpha-1}(a^r-x^r)^{\beta-1}e^{\pm cx/2}M_{\rho,\sigma}(cx)\,dx=U(0)$ $\quad[a,\ r,\ \operatorname{Re}\beta>0;\ \operatorname{Re}(\alpha+\sigma)>-1/2]$,

$$U(\gamma)=\frac{a^{\alpha+r\beta-r+1/2}\sqrt{c}}{r}\Gamma(\beta)\sum_{k=0}^{\infty}\frac{1}{k!}\left[\frac{\Gamma^{\gamma^2}(-2\sigma)}{(1+2\sigma)_k}A_k(-\sigma)+\gamma^2\frac{\Gamma(2\sigma)}{(1-2\sigma)_k}A_k(\sigma)\right](\pm ac)^k,$$

$$A_k(\sigma)=(ac)^{-\sigma}\Gamma^{-\gamma^2}\left(\frac{1}{2}-\rho+\sigma\right)\left(\frac{1}{2}\mp\rho-\sigma\right)_k\Gamma\begin{bmatrix}(1/2+\alpha-\sigma+k)/r\\ (1/2+\alpha-\sigma+k)/r+\beta\end{bmatrix}.$$

2. $\int_0^a x^{\alpha-1}(a^r-x^r)^{\beta-1}e^{\pm cx/2}W_{\rho,\sigma}(cx)\,dx=U_\sigma(\pm1)=U_{-\sigma}(\pm1)$

$[a,\ r,\ \operatorname{Re}\beta>0;\ \operatorname{Re}\alpha>|\operatorname{Re}\sigma|-1/2]$,

$U_\sigma(\gamma)=U(\gamma)$ for $\sigma\neq n/2,\ -n/2$; $U(\gamma)$ see 2.19.1.1;

$$U_{n/2}(\pm1)=\frac{a^{\alpha+r\beta-r+1/2}\sqrt{c}}{r}\Gamma(\beta)\left\{\sum_{k=0}^{n-1}\frac{(n-k-1)!}{k!}A_k\left(\frac{n}{2}\right)(\mp ac)^k+\right.$$

$$+(-1)^n\sum_{k=0}^{\infty}\frac{A_k(-n/2)}{k!\,(n+k)!}\left[\psi(k+1)+\psi(n+k+1)-\psi\left(\frac{1\pm n}{2}-\rho\pm k\right)-\right.$$

$$\left.\left.-\frac{1}{r}\psi\left(\frac{2\alpha+n+2k+1}{2r}\right)+\frac{1}{r}\psi\left(\frac{2\alpha+n+2k+1}{2r}+\beta\right)-\ln(ac)\right](\pm ac)^k\right\}$$

$[A_k(\sigma)$ see 2.19.1.1$]$.

3. $\int_a^\infty x^{\alpha-1}(x^r-a^r)^{\beta-1}e^{-cx/2}M_{\rho,\sigma}(cx)\,dx=V(0)$

$[a,\ r,\ \operatorname{Re}c,\ \operatorname{Re}\beta>0;\ \operatorname{Re}(\alpha+r\beta-\rho)<r]$,

$$V(\gamma)=\frac{a^{\alpha+r\beta-r+1/2}\sqrt{c}}{r}\times$$

$$\times\Gamma(\beta)\sum_{k=0}^{\infty}\frac{1}{k!}\left[\frac{\Gamma^{\gamma^2}(-2\sigma)}{(1+2\sigma)_k}B_k(-\sigma)+\gamma^2\frac{\Gamma(2\sigma)}{(1-2\sigma)_k}B_k(\sigma)\right]((-1)^{1-(\gamma^2+\gamma)/2}ac)^k+$$

$$+c^{r-r\beta-\alpha}T_\gamma\sum_{k=0}^{\infty}\frac{(1-\beta)_k}{k!}\Gamma\left(\frac{1}{2}+\alpha+\sigma-(1-\beta+k)r\right)\times$$

$$\times\Gamma^{2\gamma^2-1}\left(\frac{1}{2}+(1-2\gamma^2)(\sigma-\alpha+r-r\beta+rk)\right)\times$$

$$\times\Gamma^{-\gamma^2+\gamma+1}\left((1-2\gamma^2)\rho+\frac{\gamma^2-\gamma}{2}+(\gamma^2-\gamma-1)(\alpha-r+r\beta-rk)\right)(ac)^{rk},$$

$$B_k(\sigma)=(ac)^{-\sigma}\Gamma^{-\gamma^2}\left(\frac{1}{2}-\rho+\sigma\right)\left(\frac{1}{2}-\sigma-(\gamma^2+\gamma-1)\rho\right)_k\times$$

$$\times\Gamma\left[\begin{matrix}1-(1/2+\alpha-\sigma+k)/r-\beta\\1-(1/2+\alpha-\sigma+k)/r\end{matrix}\right].$$

4. $\int\limits_a^\infty x^{\alpha-1}(x^r-a^r)^{\beta-1}e^{\pm cx/2}W_{\rho,\sigma}(cx)\,dx=V_\sigma(\pm 1)=V_{-\sigma}(\pm 1)$

$$\left[a,\ r,\ \operatorname{Re}\beta>0,\ \left\{\begin{matrix}\operatorname{Re}(\alpha+\rho+r\beta)<r;\ |\arg c|<3\pi/2\\ \operatorname{Re}c>0\end{matrix}\right\}\right],$$

$V_\sigma(\gamma)=V(\gamma)$ for $\sigma\neq n/2,\ -n/2$; $V(\gamma)$ see 2.19.1.3;

$$V_{n/2}(\pm 1)=\frac{a^{\alpha+r\beta-r+1/2}\sqrt{c}}{r}\Gamma(\beta)\left\{\sum_{k=0}^{n-1}\frac{(n-k-1)!}{k!}B_k\left(\frac{n}{2}\right)(\mp ac)^k+\right.$$

$$+(-1)^n\sum_{k=0}^{\infty}\frac{B_k(-n/2)}{k!\,(n+k)!}\left[\psi(k+1)+\psi(n+k+1)-\psi\left(\frac{1\pm n}{2}-\rho\pm k\right)+\right.$$

$$\left.\left.+\frac{1}{r}\psi\left(1-\beta-\frac{2\alpha+n+2k+1}{2r}\right)-\frac{1}{r}\psi\left(1-\frac{2\alpha+n+2k+1}{2r}\right)-\ln(ac)\right](\pm ac)^k\right\}+$$

$$+c^{r-r\beta-\alpha}\left[\Gamma\left(\frac{1+n}{2}-\rho\right)\Gamma\left(\frac{1-n}{2}-\rho\right)\right]^{(\mp 1-1)/2}\times$$

$$\times\sum_{k=0}^{\infty}\frac{(1-\beta)_k}{k!}\Gamma\left(\frac{1+n}{2}+\alpha-(1-\beta+k)r\right)\Gamma\left(\frac{1-n}{2}+\alpha-(1-\beta+k)r\right)\times$$

$$\times\Gamma^{\pm 1}\left(\frac{1\mp 1}{2}-\rho\mp(\alpha-r+r\beta-rk)\right)(ac)^{rk}\quad [B_k(\sigma)\ \text{see}\ 2.19.1.3].$$

5. $\int\limits_0^\infty\frac{x^{\alpha-1}}{(x^r+z^r)^\theta}e^{-cx/2}M_{\rho,\sigma}(cx)\,dx=W(0)$

$$[r,\ \operatorname{Re}c>0;\ -1/2-\operatorname{Re}\sigma<\operatorname{Re}\alpha<\operatorname{Re}(r\theta+\rho);\ r|\arg z|<\pi],$$

$$W(\gamma)=\frac{z^{\alpha-r\theta+1/2}\sqrt{c}}{r\,\Gamma(\theta)}\times$$

$$\times\sum_{k=0}^{\infty}\frac{1}{k!}\left[\frac{\Gamma^{\gamma^2}(-2\sigma)}{(1+2\sigma)_k}C_k(-\sigma)+\gamma^2\frac{\Gamma(2\sigma)}{(1-2\sigma)_k}C_k(\sigma)\right]\left((-1)^{1-(\gamma^2+\gamma)/2}cz\right)^k+$$

$$+c^{r\theta-\alpha}T_\gamma\sum_{k=0}^{\infty}\frac{(-1)^k(\theta)_k}{k!}\Gamma\left(\frac{1}{2}+\alpha+\sigma-r\theta-rk\right)\Gamma^{2\gamma^2-1}\left(\frac{1}{2}+(1-2\gamma^2)\times\right.$$

$$\left.\times(\sigma-\alpha+r\theta+rk)\right)\Gamma^{-\gamma^2+\gamma+1}\left((1-2\gamma^2)\rho+\frac{\gamma^2-\gamma}{2}+(\gamma^2-\gamma-1)(\alpha-r\theta-rk)\right)(cz)^{rk},$$

$$C_k(\sigma)=(cz)^{-\sigma}\Gamma^{-\gamma^2}\left(\frac{1}{2}-\rho+\sigma\right)\left(\frac{1}{2}-\sigma-(\gamma^2+\gamma-1)\rho\right)_k\times$$

$$\times\Gamma\left(\frac{1/2+\alpha-\sigma+k}{r}\right)\Gamma\left(\theta-\frac{1/2+\alpha-\sigma+k}{r}\right).$$

6. $\int\limits_0^\infty \frac{x^{\alpha-1}}{(x^r+z^r)^\theta} e^{\pm cx/2} W_{\rho,\sigma}(cx)\,dx = W_\sigma(\pm 1) = W_{-\sigma}(\pm 1)$

$\left[r>0;\ \operatorname{Re}\alpha > |\operatorname{Re}\sigma| - 1/2;\ r|\arg z| < \pi,\ \begin{cases}\operatorname{Re}(\alpha+\rho-r\theta) < 0; & |\arg c| < 3\pi/2\\ \operatorname{Re} c > 0\end{cases}\right].$

$W_\sigma(\gamma) = W(\gamma)$ for $\sigma \neq n/2,\ -n/2$; $W(\gamma)$ see 2.19.1.5;

$$W_{n/2}(\pm 1) = \frac{z^{\alpha-r\theta+1/2}\sqrt{c}}{r\Gamma(\theta)}\left\{\sum_{k=0}^{n-1}\frac{(n-k-1)!}{k!}C_k\left(\frac{n}{2}\right)(\mp cz)^k + \right.$$

$$+(-1)^n\sum_{k=0}^{\infty}\frac{C_k(-n/2)}{k!\,(n+k)!}\left[\psi(k+1)+\psi(n+k+1)-\psi\left(\frac{1\pm n}{2}-\rho\pm k\right)-\right.$$

$$\left.-\frac{1}{r}\psi\left(\frac{2\alpha+n+2k+1}{2r}\right)+\frac{1}{r}\psi\left(\theta-\frac{2\alpha+n+2k+1}{2r}\right)-\ln(cz)\right]\times$$

$$\left.\times(\pm cz)^k\right\}+c^{r\theta-\alpha}\left[\Gamma\left(\frac{1+n}{2}-\rho\right)\Gamma\left(\frac{1-n}{2}-\rho\right)\right]^{(\mp 1-1)/2}\sum_{k=0}^{\infty}\frac{(-1)^k(\theta)_k}{k!}\times$$

$$\times\Gamma\left(\frac{1+n}{2}+\alpha-r\theta-rk\right)\Gamma\left(\frac{1-n}{2}+\alpha-r\theta-rk\right)\Gamma^{\pm 1}\left(\frac{1\mp 1}{2}-\rho\mp(\alpha-r\theta-rk)\right)(cz)^{rk},$$

[$C_k(\sigma)$ see 2.19.1.5].

7. $\int\limits_0^\infty \frac{x^{\alpha-1}}{x^r-y^r} e^{-cx/2} M_{\rho,\sigma}(cx)\,dx = X(0)$

[$r,\ y,\ \operatorname{Re} c > 0;\ -1/2-\operatorname{Re}\sigma < \operatorname{Re}\alpha < \operatorname{Re}\rho + r$].

$$X(\gamma) = -\frac{\pi y^{\alpha-r+1/2}\sqrt{c}}{r}\times$$

$$\times\sum_{k=0}^{\infty}\frac{1}{k!}\left[\frac{\Gamma^{\gamma^2}(-2\sigma)}{(1+2\sigma)_k}D_k(-\sigma)+\gamma^2\frac{\Gamma(2\sigma)}{(1-2\sigma)_k}D_k(\sigma)\right]\left((-1)^{1-(\gamma^2+\gamma)/2}cy\right)^k+$$

$$+c^{r-\alpha}T_\gamma\sum_{k=0}^{\infty}\Gamma\left(\frac{1}{2}+\sigma+\alpha-r-rk\right)\Gamma^{2\gamma^2-1}\left(\frac{1}{2}+(1-2\gamma^2)(\sigma-\alpha+r+rk)\right)\times$$

$$\times\Gamma^{-\gamma^2+\gamma+1}\left((1-2\gamma^2)\rho+\frac{\gamma^2-\gamma}{2}+(\gamma^2-\gamma-1)(\alpha-r-rk)\right)(cy)^{rk},$$

$$D_k(\sigma) = (cy)^{-\sigma}\Gamma^{-\gamma^2}\left(\frac{1}{2}+\sigma-\rho\right)\left(\frac{1}{2}-\sigma-(\gamma^2+\gamma-1)\rho\right)_k\operatorname{ctg}\frac{1/2+\alpha-\sigma+k}{r}\pi.$$

8. $\int\limits_0^\infty \frac{x^{\alpha-1}}{x^r-y^r} e^{\pm cx/2} W_{\rho,\sigma}(cx)\,dx = X_\sigma(\pm 1) = X_{-\sigma}(\pm 1)$

$\left[r,\ y>0;\ \operatorname{Re}\alpha > |\operatorname{Re}\sigma| - 1/2,\ \begin{cases}\operatorname{Re}(\alpha+\rho) < r; & |\arg c| < 3\pi/2\\ \operatorname{Re} c > 0\end{cases}\right],$

$X_\sigma(\gamma) = X(\gamma)$ for $\sigma \neq n/2,\ -n/2$; $X(\gamma)$ see 2.19.1.7;

$$X_{n/2}(\pm 1) = -\frac{\pi y^{\alpha-r+1/2}\sqrt{c}}{r}\left\{\sum_{k=0}^{n-1}\frac{(n-k-1)!}{k!}D_k\left(\frac{n}{2}\right)(\mp cy)^k+\right.$$

$$+(-1)^n\sum_{k=0}^{\infty}\frac{D_k(-n/2)}{k!\,(n+k)!}\left[\psi(k+1)+\psi(n+k+1)-\psi\left(\frac{1\pm n}{2}-\rho\pm k\right)+\right.$$

$$+\frac{2\pi}{r}\operatorname{cosec}\frac{2\alpha+n+2k+1}{r}\pi-\ln(cy)\Big]\,(\pm cy)^{k}\Big\}+$$

$$+c^{r-\alpha}\left[\Gamma\left(\frac{1+n}{2}-\rho\right)\Gamma\left(\frac{1-n}{2}-\rho\right)\right]^{(\mp 1-1)/2}\times$$

$$\times\sum_{k=0}^{\infty}\Gamma\left(\frac{1+n}{2}+\alpha-r-rk\right)\Gamma\left(\frac{1-n}{2}+\alpha-r-rk\right)\Gamma^{\pm 1}\left(\frac{1\mp 1}{2}-\rho\mp(\alpha-r-rk)\right)(cy)^{rk},$$

$D_k(\sigma)$ see 2.19.1.7.

9. $\int\limits_0^{\infty} x^{\alpha-1}e^{-px^r-cx/2}M_{\rho,\sigma}(cx)\,dx=Y(0)$ $[r,\ \operatorname{Re}c,\ \operatorname{Re}p>0;\ \operatorname{Re}(\alpha+\sigma)>-1/2]$.

$$Y(\gamma)=c^{-\alpha}T_{\nu}\sum_{k=0}^{\infty}\frac{1}{k!}\Gamma\left(\frac{1}{2}+\alpha+\sigma+rk\right)\Gamma^{2\gamma^2-1}\left(\frac{1}{2}+(1-2\gamma^2)(\sigma-\alpha-rk)\right)\times$$

$$\times\Gamma^{-\gamma^2+\gamma+1}\left((1-2\gamma^2)\rho+\frac{\gamma^2-\gamma}{2}+(\gamma^2-\gamma-1)(\alpha+rk)\right)\left(-\frac{p}{c^r}\right)^{k}-$$

$$-\frac{\gamma^2-\gamma-2}{2rc^{(1-2\gamma)\rho}}\,p^{((1-2\gamma)\rho-\alpha)/r}\,\Gamma^{1-\gamma}\begin{bmatrix}2\sigma+1\\ 1/2-\rho+\sigma\end{bmatrix}\times$$

$$\times\sum_{k=0}^{\infty}\frac{(1/2+\sigma+(1-2\gamma)\rho)_k\,(1/2-\sigma+(1-2\gamma)\rho)_k}{k!}\,\Gamma\left(\frac{\alpha-(1-2\gamma)\rho-k}{r}\right)\left((-1)^{\nu}\frac{p^{1/r}}{c}\right)$$

$[r<1]$

$$Y(\gamma)=\frac{\sqrt{c}}{p^{(2\alpha+1)/(2r)}r}\sum_{k=0}^{\infty}\frac{1}{k!}\left[\frac{\Gamma^{\gamma^2}(-2\sigma)}{(1+2\sigma)_k}E_k(-\sigma)+\right.$$

$$\left.+\gamma^2\frac{\Gamma(2\sigma)}{(1-2\sigma)_k}E_k(\sigma)\right]\left((-1)^{1-(\gamma^2+\gamma)/2}\frac{c}{b^{1/r}}\right)$$

$[r>1]$

$$E_k(\sigma)=\frac{p^{\sigma/r}}{c^{\sigma}}\Gamma^{-\gamma^2}\left(\frac{1}{2}-\rho+\sigma\right)\left(\frac{1}{2}-\sigma-(\gamma^2+\gamma-1)\rho\right)_k\Gamma\left(\frac{1/2+\alpha-\sigma+k}{r}\right),$$

$Y(\gamma)$ for $r=1$ see 2.19.3.1—8.

10. $\int\limits_0^{\infty} x^{\alpha-1}e^{-px^r+cx/2}M_{\rho,\sigma}(cx)\,dx=$

$$=\frac{c^{\sigma+1/2}}{p^{(\alpha+\sigma+1/2)/r}r}\sum_{k=0}^{\infty}\frac{(1/2-\rho+\sigma)_k}{(2\sigma+1)_k\,k!}\Gamma\left(\frac{1/2+\alpha+\sigma+k}{r}\right)\left(\frac{c}{p^{1/r}}\right)$$

$[r>1;\ \operatorname{Re}p>0;\ \operatorname{Re}(\alpha+\sigma)>-1/2$

11. $\int\limits_0^{\infty} x^{\alpha-1}e^{-px^r\pm cx/2}W_{\rho,\sigma}(cx)\,dx=Y_{\sigma}(\pm 1)=Y_{-\sigma}(\pm 1)$

$[r,\ \operatorname{Re}p>0;\ \operatorname{Re}\alpha>|\operatorname{Re}\sigma|-1/2;\ |\arg c|<(1\pm 1/2)\pi$

$Y_{\sigma}(\gamma)=Y(\gamma)$ for $r<1$ or for $r>1$ and $\sigma\neq n/2,\ -n/2$; $Y(\gamma)$ see 2.19.1.

$$Y_{n/2}(\pm 1)=\frac{\sqrt{c}}{p^{(2\alpha+1)/(2r)}r}\left\{\sum_{k=0}^{n-1}\frac{(n-k-1)!}{k!}E_k\left(\frac{n}{2}\right)\left(\mp\frac{c}{p^{1/r}}\right)^{k}+\right.$$

$$+(-1)^n \sum_{k=0}^{\infty} \frac{E_k(-n/2)}{k!\,(n+k)!}\left[\psi(k+1)+\psi(n+k+1)-\psi\left(\frac{1\pm n}{2}-\rho\pm k\right)-\right.$$

$$\left.-\frac{1}{r}\psi\left(\frac{2\alpha+n+2k+1}{2r}\right)-\frac{1}{r}\ln\frac{c^r}{p}\right]\left(\pm\frac{c}{p^{1/r}}\right)^k\Bigg\}$$

[$r>1$, $E_k(\sigma)$ see 2.19.1.9].

12. $\int\limits_0^\infty x^{\alpha-1}e^{-px^{-r}-cx/2}M_{\rho,\,\sigma}(cx)\,dx=Z(0)$

[r, $\operatorname{Re} c$, $\operatorname{Re} p$, $\operatorname{Re}(\rho-\alpha)>0$],

$$Z(\gamma)=c^{-\alpha}T_\gamma\sum_{k=0}^{\infty}\frac{1}{k!}\Gamma\left(\frac{1}{r}+\alpha+\sigma-rk\right)\Gamma^{2\gamma^2-1}\left(\frac{1}{2}+(1-2\gamma^2)(\sigma-\alpha+rk)\right)\times$$

$$\times\Gamma^{-\gamma^2+\gamma+1}\left((1-2\gamma^2)\rho+\frac{\gamma^2-\gamma}{2}+(\gamma^2-\gamma-1)(\alpha-rk)\right)(-c^rp)^k+\frac{\sqrt{c}}{r}p^{(2\alpha+1)/(2r)}\times$$

$$\times\sum_{k=0}^{\infty}\frac{1}{k!}\left[\frac{\Gamma^{\gamma^2}(-2\sigma)}{(1+2\sigma)_k}F_k(-\sigma)+\gamma^2\frac{\Gamma(2\sigma)}{(1-2\sigma)_k}F_k(\sigma)\right]((-1)^{1-(\gamma^2+\gamma)/2}cp^{1/r})^k,$$

$$F_k(\sigma)=c^\sigma p^{\sigma/r}\Gamma^{-\gamma^2}\left(\frac{1}{2}-\rho+\sigma\right)\left(\frac{1}{2}-\sigma-(\gamma^2+\gamma-1)\rho\right)_k\Gamma\left(-\frac{1/2+\alpha-\sigma+k}{r}\right).$$

13. $\int\limits_0^\infty x^{\alpha-1}e^{-px^{-r}\pm cx/2}W_{\rho,\,\sigma}(cx)\,dx=Z_\sigma(\pm1)=Z_{-\sigma}(\pm1)$

$\left[r,\ \operatorname{Re} p>0,\ \begin{Bmatrix}\operatorname{Re}(\alpha+\rho)<0;\ |\arg c|<3\pi/2\\ \operatorname{Re} c>0\end{Bmatrix}\right]$,

$Z_\sigma(\gamma)=Z(\gamma)$ for $\sigma\neq n/2,\ -n/2$; $Z(\gamma)$ see 2.19.1.12;

$$Z_{n/2}(\pm1)=c^{-\alpha}\left[\Gamma\left(\frac{1-n}{2}-\rho\right)\Gamma\left(\frac{1+n}{2}-\rho\right)\right]^{(\mp1-1)/2}\times$$

$$\times\sum_{k=0}^{\infty}\frac{1}{k!}\Gamma\left(\frac{1+n}{2}+\alpha-rk\right)\Gamma\left(\frac{1-n}{2}+\alpha-rk\right)\Gamma^{\pm1}\left(\frac{1\mp1}{2}-\rho\mp(\alpha-rk)\right)(-c^rp)^k+$$

$$+\frac{\sqrt{c}}{r}p^{(2\alpha+1)/(2r)}\left\{\sum_{k=0}^{n-1}\frac{(n-k-1)!}{k!}F_k\left(\frac{n}{2}\right)(\mp cp^{1/r})^k+\right.$$

$$+(-1)^n\sum_{k=0}^{\infty}\frac{F_k(-n/2)}{k!\,(n+k)!}\left[\psi(k+1)+\psi(n+k+1)-\psi\left(\frac{1\pm n}{2}-\rho\mp k\right)+\right.$$

$$\left.\left.+\frac{1}{r}\psi\left(-\frac{2\alpha+2k+n+1}{2r}\right)-\frac{1}{r}\ln(c^rp)\right](\pm cp^{1/r})^k\right\},$$

$F_k(\sigma)$ see 2.19.1.12.

14. $\int\limits_0^\infty x^{\alpha-1}e^{-cx/2}\begin{Bmatrix}\sin bx^r\\ \cos bx^r\end{Bmatrix}M_{\rho,\,\sigma}(cx)\,dx=U(0)$

[b, r, $\operatorname{Re} c>0$; $-1/2-\delta r-\operatorname{Re}\sigma<\operatorname{Re}\alpha<\operatorname{Re}\rho+r$],

$$U(\gamma)=\frac{b^\delta}{c^{\alpha+\delta r}}T_\gamma\sum_{k=0}^{\infty}\frac{(-1)^k}{(\delta+1/2)_k\,k!}\Gamma\left(\frac{1}{2}+\alpha+\sigma+\delta r+2rk\right)\times$$

$$\times\Gamma^{2\gamma^2-1}\left(\frac{1}{2}+(1-2\gamma^2)(\sigma-\alpha-\delta r-2rk)\right)\Gamma^{-\gamma^2+\gamma+1}\Big((1-2\gamma^2)\rho+$$

$$+\frac{\gamma^2-\gamma}{2}+(\gamma^2-\gamma-1)(\alpha+\delta r+2rk)\bigg)\left(\frac{b}{2c^r}\right)^{2k}-$$

$$-\frac{\gamma^2-\gamma-2}{2c^{(1-2\gamma)\rho}r}b^{((1-2\gamma)\rho-\alpha)/r}\Gamma^{1-\gamma}\begin{bmatrix}2\sigma+1\\1/2-\rho+\sigma\end{bmatrix}\times$$

$$\times\sum_{k=0}^{\infty}\frac{(1/2+\sigma+(1-2\gamma)\rho)_k(1/2-\sigma+(1-2\gamma)\rho)_k}{k!}\Gamma\left(\frac{\alpha-(1-2\gamma)\rho-k}{r}\right)\times$$

$$\times\begin{Bmatrix}\sin[(\alpha-(1-2\gamma)\rho-k)\pi/(2r)]\\\cos[(\alpha-(1-2\gamma)\rho-k)\pi/(2r)]\end{Bmatrix}\left((-1)^{\gamma}\frac{b^{1/r}}{c}\right)^k \quad [r<1],$$

$$U(\gamma)=\frac{\sqrt{c}}{b^{(2\alpha+1)/(2r)}r}\sum_{k=0}^{\infty}\frac{1}{k!}\left[\frac{\Gamma^{\gamma}(-2\sigma)}{(1+2\sigma)_k}A_k(-\sigma)+\right.$$

$$\left.+\gamma^2\frac{\Gamma(2\sigma)}{(1-2\sigma)_k}A_k(\sigma)\right]\left((-1)^{1-(\gamma^2+\gamma)/2}\frac{c}{b^{1/r}}\right)^k \quad [r>1],$$

$$A_k(\sigma)=\frac{b^{\sigma/r}}{c^{\sigma}}\Gamma^{-\gamma^2}\left(\frac{1}{2}-\rho+\sigma\right)\left(\frac{1}{2}-\sigma-(\gamma^2+\gamma-1)\rho\right)_k\times$$

$$\times\Gamma\left(\frac{\alpha-\sigma+k+1/2}{r}\right)\begin{Bmatrix}\sin[(\alpha-\sigma+k+1/2)\pi/(2r)]\\\cos[(\alpha-\sigma+k+1/2)\pi/(2r)]\end{Bmatrix},$$

$U(\gamma)$ for $r=1$ see 2.19.6.1—3.

15. $\int_0^{\infty} x^{\alpha-1}e^{cx/2}\begin{Bmatrix}\sin bx^r\\\cos bx^r\end{Bmatrix}W_{\rho,\sigma}(cx)\,dx=U_{\sigma}(1)=U_{-\sigma}(1)$

$[b,\ r>0;\ |\operatorname{Re}\sigma|-\delta r-1/2<\operatorname{Re}\alpha<r-\operatorname{Re}\rho;\ |\arg c|<3\pi/2]$.

$U_{\sigma}(\gamma)=U(\gamma)$ for $r<1$ or for $r>1$ and $\sigma\neq n/2,\ -n/2$; $U(\gamma)$ see 2.19.1.14

$$U_{n/2}(\gamma)=\frac{\sqrt{c}}{b^{(2\alpha+1)/(2r)}r}\left\{\sum_{k=0}^{n-1}\frac{(n-k-1)!}{k!}A_k\left(\frac{n}{2}\right)\left((-1)^{(\gamma+1)/2}\frac{c}{b^{1/r}}\right)^k+\right.$$

$$+(-1)^n\sum_{k=0}^{\infty}\frac{A_k(-n/2)}{k!\,(n+k)!}\left[\psi(k+1)+\psi(n+k+1)-\right.$$

$$-\psi\left(\frac{1+\gamma n}{2}-\rho+\gamma k\right)-\frac{1}{r}\psi\left(\frac{2\alpha+n+2k+1}{2r}\right)\mp$$

$$\left.\left.\mp\frac{\pi}{2r}\operatorname{tg}^{\mp1}\frac{2\alpha+n+2k+1}{4r}\pi-\frac{1}{r}\ln\frac{c^r}{b}\right]\left((-1)^{(1-\gamma)/2}\frac{c}{b^{1/r}}\right)^k\right.$$

$[\gamma=1$ or $-1;\ r>1;\ A_k(\sigma)$ see 2.19.1.14].

16. $\int_0^{\infty} x^{\alpha-1}e^{-cx/2}\begin{Bmatrix}\sin bx^r\\\cos bx^r\end{Bmatrix}W_{\rho,\sigma}(cx)\,dx=U_{\sigma}(-1)=U_{-\sigma}(-1)$

$[b,\ r,\ \operatorname{Re}c>0;\ \operatorname{Re}\alpha>|\operatorname{Re}\sigma|-\delta r-1/2;\ U_{\sigma}(\gamma)$ see 2.19.1.15].

17. $\int_0^{\infty} x^{\alpha-1}e^{-cx/2}\begin{Bmatrix}\ln(x^r+z^r)\\\ln|x^r-z^r|\end{Bmatrix}M_{\rho,\sigma}(cx)\,dx=V(0)$

$\left[r,\ \operatorname{Re}c>0;\ -1/2-\operatorname{Re}\sigma<\operatorname{Re}\alpha<\operatorname{Re}\rho,\ \begin{Bmatrix}r\ |\arg z|<\pi\\z>0\end{Bmatrix}\right]$.

$$V(\gamma)=\pi z^{\alpha+1/2}\sqrt{c}\times$$

$$\times\sum_{k=0}^{\infty}\frac{1}{k!}\left[\frac{\Gamma^{\gamma^2}(-2\sigma)}{(1+2\sigma)_k}B_k(-\sigma)+\gamma^2\frac{\Gamma(2\sigma)}{(1-2\sigma)_k}B_k(\sigma)\right]((-1)^{1-(\gamma^2+\gamma)/2}cz)^k\pm$$

$$\pm c^{r-\alpha} z^r T_\gamma \sum_{k=0}^{\infty} \frac{(\mp 1)^k}{k+1} \Gamma\left(\frac{1}{2}+\alpha+\sigma-r-rk\right) \Gamma^{2\gamma^2-1}\left(\frac{1}{2}+(1-2\gamma^2)(\sigma-\alpha+r+rk)\right) \times$$
$$\times \Gamma^{-\gamma^2+\gamma+1}\left((1-2\gamma^2)\rho+\frac{\gamma^2-\gamma}{2}+(\gamma^2-\gamma-1)(\alpha-r-rk)\right)(cz)^{rk}+$$
$$+\frac{r}{c^\alpha} T_\gamma \Gamma\left(\frac{1}{2}+\alpha+\sigma\right) \Gamma^{2\gamma^2-1}\left(\frac{1}{2}+(1-2\gamma^2)(\sigma-\alpha)\right) \times$$
$$\times \Gamma^{-\gamma^2+\gamma+1}\left((1-2\gamma^2)\rho+\frac{\gamma^2-\gamma}{2}+(\gamma^2-\gamma-1)\alpha\right)\left[\psi\left(\frac{1}{2}+\alpha+\sigma\right)+\right.$$
$$\left.+\psi\left(\frac{1}{2}+(1-2\gamma^2)(\sigma-\alpha)\right)-\psi\left((1-2\gamma^2)\rho+\frac{\gamma^2-\gamma}{2}+(\gamma^2-\gamma-1)\alpha\right)-\ln c\right],$$

$$B_k(\sigma)=(cz)^{-\sigma}\Gamma^{-\gamma^2}\left(\frac{1}{2}-\rho+\sigma\right)\left(\frac{1}{2}-\sigma-(\gamma^2+\gamma-1)\rho\right)_k \times$$
$$\times\left(\alpha-\sigma+\frac{1}{2}+k\right)^{-1}\begin{Bmatrix}\operatorname{cosec}[(\alpha-\sigma+k+1/2)\pi/2] \\ \operatorname{ctg}[(\alpha-\sigma+k+1/2)\pi/2]\end{Bmatrix}.$$

8. $$\int_0^\infty x^{\alpha-1} e^{cx/2} \begin{Bmatrix}\ln(x^r+z^r) \\ \ln|x^r-z^r|\end{Bmatrix} W_{\rho,\sigma}(cx)\,dx = V(1)$$

$\left[r>0;\ |\operatorname{Re}\sigma|-1/2<\operatorname{Re}\alpha<-\operatorname{Re}\rho;\ |\arg c|<3\pi/2,\ \begin{Bmatrix} r\,|\arg z|<\pi \\ z>0\end{Bmatrix},\ V(\gamma)\right.$ see 2.19.1.17$\Big]$.

9. $$\int_0^\infty x^{\alpha-1} e^{-cx/2} \begin{Bmatrix}\ln(x^r+z^r) \\ \ln|x^r-z^r|\end{Bmatrix} W_{\rho,\sigma}(cx)\,dx = V(-1)$$

$\left[r,\ \operatorname{Re}c>0;\ \operatorname{Re}\alpha>|\operatorname{Re}\sigma|-1/2,\ \begin{Bmatrix} r\,|\arg z|<\pi \\ z>0\end{Bmatrix},\ V(\gamma)\right.$ see 2.19.1.17$\Big]$.

10. $$\int_0^\infty x^{\alpha-1} e^{-cx/2} \operatorname{Ei}(-bx^r) M_{\rho,\sigma}(cx)\,dx = W(0)$$

$[r,\ \operatorname{Re}b,\ \operatorname{Re}c>0;\ \operatorname{Re}(\alpha+\sigma)>-1/2]$,

$$W(\gamma)=-\frac{b}{c^{\alpha+r}} T_\gamma \sum_{k=0}^{\infty}\frac{1}{k!\,(k+1)^2}\Gamma\left(\frac{1}{2}+\alpha+\sigma+r+rk\right)\times$$
$$\times\Gamma^{2\gamma^2-1}\left(\frac{1}{2}+(1-2\gamma^2)(\sigma-\alpha-r-rk)\right)\Gamma^{-\gamma^2+\gamma+1}\left((1-2\gamma^2)\rho+\frac{\gamma^2-\gamma}{2}+\right.$$
$$\left.+(\gamma^2-\gamma-1)(\alpha+r+rk)\right)\left(-\frac{b}{c^r}\right)^k-\frac{\gamma^2-\gamma-2}{2c^{(1-2\gamma)\rho}} b^{((1-2\gamma)\rho-\alpha)/r}\Gamma^{1-\gamma}\begin{bmatrix}2\sigma+1 \\ 1/2-\rho+\sigma\end{bmatrix}\times$$
$$\times\sum_{k=0}^{\infty}\frac{(1/2+\sigma+(1-2\gamma)\rho)_k\,(1/2-\sigma+(1-2\gamma)\rho)_k}{k!\,((1-2\gamma)\rho-\alpha+k)}\Gamma\left(\frac{\alpha-(1-2\gamma)\rho-k}{r}\right)\left((-1)^\gamma\frac{b^{1/r}}{c}\right)^k+$$
$$+c^{-\alpha}T_\gamma\Gamma\left(\frac{1}{2}+\alpha+\sigma\right)\Gamma^{2\gamma^2-1}\left(\frac{1}{2}+(1-2\gamma^2)(\sigma-\alpha)\right)\Gamma^{-\gamma^2+\gamma+1}\left((1-2\gamma^2)\rho+\right.$$
$$\left.+\frac{\gamma^2-\gamma}{2}+(\gamma^2-\gamma-1)\alpha\right)\left[\mathbf{C}+r\psi\left(\frac{1}{2}+\alpha+\sigma\right)+r\psi\left(\frac{1}{2}+(1-2\gamma^2)\sigma-\alpha\right)-\right.$$
$$\left.-r\psi\left((1-2\gamma^2)\rho+\frac{\gamma^2-\gamma}{2}+(\gamma^2-\gamma-1)\alpha\right)-\ln\frac{c^r}{b}\right] \qquad [r<1],$$

$$W(\gamma)=-\frac{\sqrt{c}}{b^{(2\alpha+1)/(2r)}}\sum_{k=0}^{\infty}\frac{1}{k!}\left[\frac{\Gamma^{\gamma^2}(-2\sigma)}{(1+2\sigma)_k}C_k(-\sigma)+\right.$$
$$\left.+\gamma^2\frac{\Gamma(2\sigma)}{(1-2\sigma)_k}C_k(\sigma)\right]\left((-1)^{1-(\gamma^2+\gamma)/2}\frac{c}{b^{1/r}}\right)^k \qquad [r>1],$$

$$C_k(\sigma)=\frac{b^{\sigma/r}}{c^{\sigma}}\Gamma^{-\gamma^2}\left(\frac{1}{2}-\rho+\sigma\right)\left(\frac{1}{2}-\sigma-(\gamma^2+\gamma-1)\rho\right)_k\times$$
$$\times(\alpha-\sigma+k+1/2)^{-1}\Gamma\left(\frac{\alpha-\sigma+k+1/2}{r}\right),$$

$W(\gamma)$ for $r=1$ see 2.19.8.1—2.

21. $$\int_0^{\infty} x^{\alpha-1}e^{\pm cx/2}\,\mathrm{Ei}\,(-bx^r)\,W_{\rho,\,\sigma}(cx)\,dx=W(\pm 1)$$

$[r,\ \mathrm{Re}\,b>0;\ \mathrm{Re}\,\alpha>|\mathrm{Re}\,\sigma|-1/2;\ |\arg c|<(1\pm 1/2)\pi,\ W(\gamma)$ see 2.19.1.20$]$.

22. $$\int_0^{\infty} x^{\alpha-1}e^{-cx/2}\begin{Bmatrix}\mathrm{erf}\,(bx^r)\\ \mathrm{erfc}\,(bx^r)\end{Bmatrix}M_{\rho,\,\sigma}(cx)\,dx=X(0)$$

$\left[r>0;\ \mathrm{Re}\,(\alpha+\sigma)>-(r\pm r+1)/2;\ |\arg b|<\pi/4,\ \begin{Bmatrix}\mathrm{Re}\,(\rho-\alpha),\ \mathrm{Re}\,c>0\\ \mathrm{Re}\,c>0\end{Bmatrix}\right].$

$$X(\gamma)=\pm\frac{2b}{\sqrt{\pi}\,c^{\alpha+r}}T_{\gamma}\sum_{k=0}^{\infty}\frac{\Gamma(1/2+\alpha+\sigma+r+2rk)}{k!\,(2k+1)}\times$$
$$\times\Gamma^{2\gamma^2-1}\left(\frac{1}{2}+(1-2\gamma^2)(\sigma-\alpha-r-2rk)\right)\Gamma^{-\gamma^2+\gamma+1}\left((1-2\gamma^2)\rho+\right.$$
$$\left.+\frac{\gamma^2-\gamma}{2}+(\gamma^2-\gamma-1)(\alpha+r+2rk)\right)\left(-\frac{b^2}{c^{2r}}\right)^k\mp$$
$$\mp\frac{\gamma^2-\gamma-2}{2\sqrt{\pi}\,c^{(1-2\gamma)\rho}}b^{((1-2\gamma)\rho-\alpha)/r}\Gamma^{1-\gamma}\begin{bmatrix}2\sigma+1\\ 1/2-\rho+\sigma\end{bmatrix}\times$$
$$\times\sum_{k=0}^{\infty}\frac{(1/2+\sigma+(1-2\gamma)\rho)_k\,(1/2-\sigma+(1-2\gamma)\rho)_k}{k!\,(\rho-\alpha+k)}\Gamma\left(\frac{r+\alpha-(1-2\gamma)\rho-k}{2r}\right)\times$$
$$\times\left((-1)^{\gamma}\frac{b^{1/r}}{c}\right)^k+\begin{Bmatrix}0\\ 1\end{Bmatrix}R\qquad [r<1/2]$$

$$R=c^{-\alpha}T_{\gamma}\Gamma\left(\frac{1}{2}+\alpha+\sigma\right)\Gamma^{2\gamma^2-1}\left(\frac{1}{2}+(1-2\gamma^2)(\sigma-\alpha)\right)\times$$
$$\times\Gamma^{-\gamma^2+\gamma+1}\left((1-2\gamma^2)\rho+\frac{\gamma^2-\gamma}{2}+(\gamma^2-\gamma-1)\alpha\right)$$

$$X(\gamma)=\mp\frac{\sqrt{c}}{\sqrt{\pi}\,b^{(2\alpha+1)/(2r)}}\sum_{k=0}^{\infty}\frac{1}{k!}\left[\frac{\Gamma^{\gamma^2}(-2\sigma)}{(1+2\sigma)_k}D_k(-\sigma)+\right.$$
$$\left.+\gamma^2\frac{\Gamma(2\sigma)}{(1-2\sigma)_k}D_k(\sigma)\right]\left((-1)^{1-(\gamma^2+\gamma)/2}\frac{c}{b^{1/r}}\right)^k+\begin{Bmatrix}1\\ 0\end{Bmatrix}R\qquad [r>1/2]$$

$$D_k(\sigma)=\frac{b^{\sigma/r}}{c^{\sigma}}\Gamma^{-\gamma^2}\left(\frac{1}{2}-\rho+\sigma\right)\left(\frac{1}{2}-\sigma-(\gamma^2+\gamma-1)\rho\right)_k\times$$
$$\times(\alpha-\sigma+k+1/2)^{-1}\Gamma\left(\frac{1/2+\alpha-\sigma+k+r}{2r}\right)$$

$X(\gamma)$ for $r=1/2$ see 2.19.9.2—6.

23. $$\int_0^{\infty} x^{\alpha-1}e^{cx/2}\begin{Bmatrix}\mathrm{erf}\,(bx^r)\\ \mathrm{erfc}\,(bx^r)\end{Bmatrix}W_{\rho,\,\sigma}(cx)\,dx=X(1)$$

$\left[r>0;\ \mathrm{Re}\,\alpha>|\mathrm{Re}\,\sigma|-(r\pm r+1)/2;\ |\arg b|<\pi/4,\right.$
$\left.\begin{Bmatrix}\mathrm{Re}\,(\alpha+\rho)<0,\ |\arg c|<3\pi/2\\ |\arg c|<3\pi/2\end{Bmatrix},\ X(\gamma)\ \text{see 2.19.1.22}\right]$

24. $\int\limits_0^\infty x^{\alpha-1}e^{-cx/2}\begin{Bmatrix}\operatorname{erf}(bx^r)\\ \operatorname{erfc}(bx^r)\end{Bmatrix}W_{\rho,\sigma}(cx)\,dx=X(-1)$

$[r, \operatorname{Re} c>0;\ \operatorname{Re}\alpha>|\operatorname{Re}\sigma|-(r\pm r+1)/2;\ |\arg b|<\pi/4;\ X(\gamma)$ see 2.19.1.22].

25. $\int\limits_0^\infty x^{\alpha-1}e^{-cx/2}J_\nu(bx^r)M_{\rho,\sigma}(cx)\,dx=Y(0)$

$[b,\ r,\ \operatorname{Re} c>0;\ \operatorname{Re}(\alpha+\nu r+\sigma)>-1/2;\ \operatorname{Re}(\alpha-\rho)<3r/2]$,

$$Y(\gamma)=\frac{(b/2)^\nu}{\Gamma(\nu+1)\,c^{\alpha+\nu r}}T_\gamma\sum_{k=0}^{\infty}\frac{(-1)^k}{(\nu+1)_k\,k!}\times$$
$$\times\Gamma\left(\frac{1}{2}+\alpha+\sigma+\nu r+2rk\right)\Gamma^{2\gamma^2-1}\left(\frac{1}{2}+(1-2\gamma^2)(\sigma-\alpha-\nu r-2rk)\right)\times$$
$$\times\Gamma^{-\gamma^2+\gamma+1}\left((1-2\gamma^2)\rho+\frac{\gamma^2-\gamma}{2}+(\gamma^2-\gamma-1)(\alpha+\nu r+2rk)\right)\left(\frac{b}{2c^r}\right)^{2k}-$$
$$-\frac{\gamma^2-\gamma-2}{4rc^{(1-2\gamma)\rho}}\left(\frac{b}{2}\right)^{((1-2\gamma)\rho-\alpha)/r}\Gamma^{1-\gamma}\begin{bmatrix}2\sigma+1\\ 1/2-\rho+\sigma\end{bmatrix}\times$$
$$\times\sum_{k=0}^{\infty}\frac{(1/2+\sigma+(1-2\gamma)\rho)_k\,(1/2-\sigma+(1-2\gamma)\rho)_k}{k!}\times$$
$$\times\Gamma\begin{bmatrix}(\nu r+\alpha-(1-2\gamma)\rho-k)/(2r)\\ (\nu r-\alpha+(1-2\gamma)\rho+k)/(2r)+1\end{bmatrix}\left((-1)^\gamma\frac{b^{1/r}}{2^{1/r}c}\right)^k \quad [r<1],$$

$$Y(\gamma)=\left(\frac{2}{b}\right)^{(2\alpha+1)/(2r)}\frac{\sqrt{c}}{2r}\sum_{k=0}^{\infty}\frac{1}{k!}\left[\frac{\Gamma^\gamma(-2\sigma)}{(1+2\sigma)_k}E_k(-\sigma)+\right.$$
$$\left.+\gamma^2\frac{\Gamma(2\sigma)}{(1-2\sigma)_k}E_k(\sigma)\right]\left((-1)^{1-(\gamma^2+\gamma)/2}\frac{2^{1/r}c}{b^{1/r}}\right)^k \quad [r>1].$$

$$E_k(\sigma)=\frac{1}{c^\sigma}\left(\frac{b}{2}\right)^{\sigma/r}\Gamma^{-\gamma^2}\left(\frac{1}{2}-\rho+\sigma\right)\left(\frac{1}{2}-\sigma-(\gamma^2+\gamma-1)\rho\right)_k\times$$
$$\times\Gamma\begin{bmatrix}(\nu r+\alpha-\sigma+1/2+k)/(2r)\\ (\nu r-\alpha+\sigma-1/2-k)/(2r)+1\end{bmatrix},$$

$Y(\gamma)$ for $r=1$ see 2.19.11.1—2.

26. $\int\limits_0^\infty x^{\alpha-1}e^{\pm cx/2}J_\nu(bx^r)W_{\rho,\sigma}(cx)\,dx=Y_\sigma(\pm 1)=Y_{-\sigma}(\pm 1)$

$\left[b,\ r>0;\ \operatorname{Re}(\alpha+\nu r)>|\operatorname{Re}\sigma|-1/2,\ \begin{Bmatrix}\operatorname{Re}(\alpha+\rho)<3r/2;\ |\arg c|<3\pi/2\\ \operatorname{Re} c>0\end{Bmatrix}\right]$.

$Y_\sigma(\gamma)=Y(\gamma)$ for $r<1$ or for $r>1$ and $\sigma\neq n/2,\ -n/2$; $Y(\gamma)$ see 2.19.1.25;

$$Y_{n/2}(\pm 1)=\left(\frac{2}{b}\right)^{(2\alpha+1)/(2r)}\frac{\sqrt{c}}{2r}\left\{\sum_{k=0}^{n-1}\frac{(n-k-1)!}{k!}E_k\left(\frac{n}{2}\right)\left(\mp\frac{2^{1/r}c}{b^{1/r}}\right)^k+\right.$$
$$+(-1)^n\sum_{k=0}^{\infty}\frac{E_k(-n/2)}{k!\,(n+k)!}\left[\psi(k+1)+\psi(n+k+1)-\psi\left(\frac{1\pm n}{2}-\rho\pm k\right)-\right.$$
$$\left.-\frac{1}{2r}\psi\left(\frac{\nu r+\alpha+k+(n+1)/2}{2r}\right)-\frac{1}{2r}\psi\left(\frac{\nu r-\alpha-k-(n+1)/2}{2r}+1\right)-\frac{1}{r}\ln\frac{2c^r}{b}\right]\times$$
$$\left.\times\left(\pm\frac{2^{1/r}c}{b^{1/r}}\right)^k\right\}$$

$[r>1;\ E_k(\sigma)$ see 2.19.1.25].

27. $$\int_0^\infty x^{\alpha-1}e^{-cx/2}I_\nu(bx^r)W_{\rho,\sigma}(cx)\,dx=$$

$$=\frac{(b/2)^\nu}{\Gamma(\nu+1)c^{\alpha+\nu r}}\sum_{k=0}^{\infty}\frac{1}{(\nu+1)_k k!}\Gamma\left[\begin{matrix}1/2+\alpha+\sigma+\nu r+2rk,\ 1/2+\alpha-\sigma+\nu r+2rk\\ 1+\alpha-\rho+\nu r+2rk\end{matrix}\right]\times$$
$$\times\left(\frac{b}{2c^r}\right)^{2k}$$

$[0<r<1;\ \operatorname{Re}c>0;\ \operatorname{Re}(\alpha+\nu r)>|\operatorname{Re}\sigma|-1/2;\ |\arg b|<\pi]$.

28. $$\int_0^\infty x^{\alpha-1}e^{-bx^r-cx/2}I_\nu(bx^r)M_{\rho,\sigma}(cx)\,dx=U(0,0)$$

$[r,\ \operatorname{Re}b,\ \operatorname{Re}c>0;\ -\operatorname{Re}(\nu r+\sigma)-1/2<\operatorname{Re}\alpha<\operatorname{Re}\rho+r/2]$,

$$U(\varepsilon,\gamma)=\pi^{\varepsilon^2/2}c^{-\alpha}\Gamma^{\varepsilon^2-1}(\nu+1)T_\gamma\times$$
$$\times\sum_{k=0}^{\infty}\frac{1}{k!}\left[\frac{\Gamma^{\varepsilon^2}(-2\nu)}{(1+2\nu)_k}F_k(-\nu)+\varepsilon^2\frac{\Gamma(2\nu)}{(1-2\nu)_k}F_k(\nu)\right]\left((-1)^{1-(\varepsilon^2+\varepsilon)/2}\frac{2b}{c^r}\right)^k-$$
$$-\frac{(\gamma^2-\gamma-2)\pi^{(\varepsilon^2-\varepsilon-1)/2}}{2c^{(1-2\gamma)\rho}r}(2b)^{((1-2\gamma)\rho-\alpha)/r}\Gamma^{1-\gamma}\left[\begin{matrix}2\sigma+1\\ 1/2-\rho+\sigma\end{matrix}\right]\times$$
$$\times\cos^{(\varepsilon^2+\varepsilon)/2}\nu\pi\sum_{k=0}^{\infty}\frac{(1/2+\sigma+(1-2\gamma)\rho)_k(1/2-\sigma+(1-2\gamma)\rho)_k}{k!}\times$$
$$\times\Gamma\left(\nu-\frac{(1-2\gamma)\rho-\alpha+k}{r}\right)\Gamma^{2\varepsilon^2-1}\left(1-\varepsilon^2+(1-2\varepsilon^2)\frac{\nu r+(1-2\gamma)\rho-\alpha+k}{r}\right)\times$$
$$\times\Gamma^{-\varepsilon^2+\varepsilon+1}\left(\frac{1}{2}-(\varepsilon^2-\varepsilon-1)\frac{(1-2\gamma)\rho-\alpha+k}{r}\right)\left((-1)^\gamma\frac{(2b)^{1/r}}{c}\right)^k\quad[r<1],$$

$$F_k(\nu)=\left(\frac{2c^r}{b}\right)^\nu 2^{-2\nu\varepsilon^2}\Gamma^{-\varepsilon^2}\left(\frac{1}{2}+\nu\right)\left(\frac{1}{2}-\nu\right)_k\Gamma\left(\alpha+\sigma-\nu r+rk+\frac{1}{2}\right)\times$$
$$\times\Gamma^{2\gamma^2-1}\left(\frac{1}{2}+(1-2\gamma^2)(\sigma-\alpha+\nu r-rk)\right)\Gamma^{-\gamma^2+\gamma+1}\left((1-2\gamma^2)\rho+\frac{\gamma^2-\gamma}{2}+\right.$$
$$\left.+(\gamma^2-\gamma-1)(\alpha-\nu r+rk)\right),$$

$$U(\varepsilon,\gamma)=\frac{\sqrt{c}\,\pi^{(\varepsilon^2-\varepsilon-1)/2}}{(2b)^{(2\alpha+1)/(2r)}r}\cos^{(\varepsilon^2+\varepsilon)/2}\nu\pi\times$$
$$\times\sum_{k=0}^{\infty}\frac{1}{k!}\left[\frac{\Gamma^{\gamma^2}(-2\sigma)}{(1+2\sigma)_k}G_k(-\sigma)+\gamma^2\frac{\Gamma(2\sigma)}{(1-2\sigma)_k}G_k(\sigma)\right]\left((-1)^{1-(\gamma^2+\gamma)/2}\frac{c}{(2b)^{1/r}}\right)^k-$$
$$-\frac{(\varepsilon^2-\varepsilon-2)\pi^{\varepsilon-1/2}}{2\sqrt{2b}\,c^{\alpha-r/2}}T_\gamma\sum_{k=0}^{\infty}\frac{(1/2+\nu)_k(1/2-\nu)_k}{k!}\Gamma\left(\frac{1-r}{2}+\alpha+\sigma-rk\right)\times$$
$$\times\Gamma^{2\gamma^2-1}\left(\frac{1}{2}+(1-2\gamma^2)\left(\frac{r}{2}-\alpha+\sigma+rk\right)\right)\Gamma^{-\gamma^2+\gamma+1}\left((1-2\gamma^2)\rho+\frac{\gamma^2-\gamma}{2}+\right.$$
$$\left.+(\gamma^2-\gamma-1)\left(\alpha-rk-\frac{r}{2}\right)\right)\left((-1)^\varepsilon\frac{c^r}{2b}\right)^k\quad[r>1],$$

$$G_k(\sigma)=\frac{(2b)^{\sigma/r}}{c^\sigma}\Gamma^{-\gamma^2}\left(\frac{1}{2}-\rho+\sigma\right)\left(\frac{1}{2}-\sigma-(\gamma^2+\gamma-1)\rho\right)_k\times$$

$$\times\Gamma\left(\nu+\frac{\alpha-\sigma+k+1/2}{r}\right)\Gamma^{2\varepsilon^2-1}\left(1-\varepsilon^2+(1-2\varepsilon^2)\frac{\nu r-\alpha+\sigma-k-1/2}{r}\right)\times$$
$$\times\Gamma^{-\varepsilon^2+\varepsilon+1}\left(\frac{1}{2}+(\varepsilon^2-\varepsilon-1)\frac{\alpha-\sigma+k+1/2}{r}\right),$$

$U(\varepsilon,\gamma)$ for $r=1$ see 2.19.12.3-5.

29. $$\int_0^\infty x^{\alpha-1}e^{-bx^r\pm cx/2}I_\nu(bx^r)W_{\rho,\sigma}(cx)\,dx=U(0,\pm 1)$$

$\left[r,\ \operatorname{Re}b>0;\ \operatorname{Re}(\alpha+\nu r)>|\operatorname{Re}\sigma|-1/2,\ \begin{Bmatrix}\operatorname{Re}(\alpha+\rho)<r/2;\ |\arg c|<3\pi/2\\ \operatorname{Re}c>0\end{Bmatrix},\ U(\varepsilon,\gamma)\text{ see } 2.19.1.28\right].$

30. $$\int_0^\infty x^{\alpha-1}e^{\pm bx^r-cx/2}K_\nu(bx^r)M_{\rho,\sigma}(cx)\,dx=U(\pm 1,0)$$

$\left[r,\ \operatorname{Re}c>0;\ \operatorname{Re}\alpha>r|\operatorname{Re}\nu|+|\operatorname{Re}\sigma|-1/2,\ \begin{Bmatrix}\operatorname{Re}(\alpha-\rho)<r/2;\ |\arg b|<3\pi/2\\ \operatorname{Re}b>0\end{Bmatrix},\ U(\varepsilon,\gamma)\text{ see } 2.19.1.28\right].$

31. $$\int_0^\infty x^{\alpha-1}e^{bx^r\pm cx/2}K_\nu(bx^r)W_{\rho,\sigma}(cx)\,dx=U(1,\pm 1)$$

$\left[r>0;\ \operatorname{Re}\alpha>r|\operatorname{Re}\nu|+|\operatorname{Re}\sigma|-1/2;\ |\arg b|<3\pi/2,\ \begin{Bmatrix}\operatorname{Re}(\alpha+\rho)<r/2;\ |\arg c|<3\pi/2\\ \operatorname{Re}c>0\end{Bmatrix},\ U(\varepsilon,\gamma)\text{ see } 2.19.1.28\right].$

32. $$\int_0^\infty x^{\alpha-1}e^{-bx^r\pm cx/2}K_\nu(bx^r)W_{\rho,\sigma}(cx)\,dx=U(-1,\pm 1)$$

$[r,\ \operatorname{Re}b>0;\ \operatorname{Re}\alpha>r|\operatorname{Re}\nu|+|\operatorname{Re}\sigma|-1/2;\ |\arg c|<(1\pm 1/2)\pi;\ U(\varepsilon,\gamma)\text{ see } 2.19.1.28].$

33. $$\int_0^\infty x^{\alpha-1}e^{-cx/2}\begin{Bmatrix}Y_\nu(bx^r)\\ K_\nu(bx^r)\end{Bmatrix}M_{\rho,\sigma}(cx)\,dx=U_{\nu,\sigma}(0)$$

$\left[r,\ \operatorname{Re}c>0;\ \operatorname{Re}(\alpha+\sigma)>r|\operatorname{Re}\nu|-1/2,\ \begin{Bmatrix}b>0;\ \operatorname{Re}(\alpha-\rho)<3r/2\\ \operatorname{Re}b>0\end{Bmatrix}\right].$

$$U_{\nu,\sigma}(\gamma)=V_{\nu,\sigma}(\gamma)-\frac{\gamma^2-\gamma-2}{8r\pi c^{(1-2\gamma)\rho}}\begin{Bmatrix}-2\\ \pi\end{Bmatrix}\left(\frac{b}{2}\right)^{((1-2\gamma)\rho-\alpha)/r}\times$$
$$\times\Gamma^{1-\gamma}\begin{bmatrix}2\sigma+1\\ 1/2-\rho+\sigma\end{bmatrix}\sum_{k=0}^\infty\frac{(1/2+\sigma+(1-2\gamma)\rho)_k(1/2-\sigma+(1-2\gamma)\rho)_k}{k!}\times$$
$$\times\Gamma\left(\frac{\alpha-(1-2\gamma)\rho+\nu r-k}{2r}\right)\Gamma\left(\frac{\alpha-(1-2\gamma)\rho-\nu r-k}{2r}\right)\times$$
$$\times\cos^{(1\pm1)/2}\left(\frac{\alpha-(1-2\gamma)\rho-k-\nu r}{2r}\pi\right)\left((-1)^\gamma\frac{b^{1/r}}{2^{1/r}c}\right)^k\qquad[r<1],$$

$$V_{\nu,\sigma}(\gamma)=\frac{1}{2\pi c^\alpha}\begin{Bmatrix}-2\\ \pi\end{Bmatrix}T_\gamma\times$$
$$\times\sum_{k=0}^\infty\frac{1}{k!}\left[\frac{\Gamma(\nu)A_k(\nu)}{(1-\nu)_k}+\begin{Bmatrix}\cos\nu\pi\\ 1\end{Bmatrix}\frac{\Gamma(-\nu)A_k(-\nu)}{(1+\nu)_k}\right]\left(\mp\frac{b^2}{4c^{2r}}\right)^k$$

$[\nu\neq m,\ -m;\ m=0,1,2,\dots],$

$$V_{-m,\sigma}(\gamma)=(\mp 1)^m V_{m,\sigma}(\gamma),$$

$$V_{m,\sigma}(\gamma)=\frac{1}{2\pi c^{\alpha}}\left\{\begin{matrix}-2\\ \pi\end{matrix}\right\}T_{\gamma}\left\{\sum_{k=0}^{m-1}\frac{(m-k-1)!}{k!}A_k(m)\left(\pm\frac{b^2}{4c^{2r}}\right)^k+\right.$$

$$+(\pm 1)^m\sum_{k=0}^{\infty}\frac{A_k(-m)}{k!\,(k+m)!}\left[\psi(k+1)+\psi(k+m+1)-2r\psi\left(\frac{1}{2}+\alpha+\sigma+2rk+rm\right)-\right.$$

$$-2r\psi\left(\frac{1}{2}+(1-2\gamma^2)(\sigma-\alpha-2rk-rm)\right)+2r\psi\left((1-2\gamma^2)\rho+\frac{\gamma^2-\gamma}{2}+\right.$$

$$\left.\left.+(\gamma^2-\gamma-1)(\alpha+rm+2rk)\right)+2\ln\frac{2c^r}{b}\right]\left(\mp\frac{b^2}{4c^{2r}}\right)^k\Bigg\}\quad [m=0,1,2,\ldots],$$

$$A_k(\nu)=\left(\frac{2c^r}{b}\right)^{\nu}\Gamma\left(\frac{1}{2}+\alpha+\sigma+2rk-\nu r\right)\times$$

$$\times\Gamma^{2\gamma^2-1}\left(\frac{1}{2}+(1-2\gamma^2)(\sigma-\alpha+\nu r-2rk)\right)\times$$

$$\times\Gamma^{-\gamma^2+\gamma+1}\left((1-2\gamma^2)\rho+\frac{\gamma^2-\gamma}{2}+(\gamma^2-\gamma-1)(\alpha-\nu r+2rk)\right),$$

$$U_{\nu,\sigma}(\gamma)=\left(\frac{2}{b}\right)^{(2\alpha+1)/(2r)}\left\{\begin{matrix}-2\\ \pi\end{matrix}\right\}\frac{\sqrt{c}}{4\pi r}\times$$

$$\times\sum_{k=0}^{\infty}\frac{1}{k!}\left[\frac{\Gamma^{\gamma^2}(-2\sigma)}{(1+2\sigma)_k}B_k(-\sigma)+\gamma^2\frac{\Gamma(2\sigma)}{(1-2\sigma)_k}B_k(\sigma)\right]\left((-1)^{1-(\gamma^2+\gamma)/2}\frac{2^{1/r}c}{b^{1/r}}\right)^k$$

$[r>1;\ \gamma=0$ or $\gamma^2=1$ and $\sigma\neq n/2,\ -n/2]$,

$$U_{\nu,-n/2}(\gamma)=U_{\nu,n/2}(\gamma)=$$

$$=\left(\frac{2}{b}\right)^{(2\alpha+1)/(2r)}\left\{\begin{matrix}-2\\ \pi\end{matrix}\right\}\frac{\sqrt{c}}{4\pi r}\left\{\sum_{k=0}^{n-1}\frac{(n-k-1)!}{k!}B_k\left(\frac{n}{2}\right)\left((-1)^{(1+\gamma)/2}\frac{2^{1/r}c}{b^{1/r}}\right)^k+\right.$$

$$+(-1)^n\sum_{k=0}^{\infty}\frac{B_k(-n/2)}{k!\,(k+n)!}\left[\psi(k+1)+\psi(k+n+1)-\psi\left(\frac{1+\gamma n}{2}-\rho+\gamma k\right)-\right.$$

$$-\frac{1}{2r}\psi\left(\frac{\alpha+\nu r+k+(n+1)/2}{2r}\right)-\frac{1}{2r}\psi\left(\frac{\alpha-\nu r+k+(n+1)/2}{2r}\right)-$$

$$\left.\left.-\frac{\pi}{2r}\operatorname{tg}^{(1\pm 1)/2}\left(\frac{\alpha-\nu r+k+(n+1)/2}{2r}\pi\right)-\frac{1}{r}\ln\frac{2c^r}{b}\right]\left((-1)^{(1-\gamma)/2}\frac{2^{1/r}c}{b^{1/r}}\right)^k\right\}$$

$[n=0,\ 1,\ 2,\ \ldots;\ \gamma^2=1]$.

$$B_k(\sigma)=\left(\frac{b}{2c^r}\right)^{\sigma/r}\Gamma^{-\gamma^2}\left(\frac{1}{2}-\rho+\sigma\right)\Gamma\left(\frac{\alpha+\nu r-\sigma+k+1/2}{2r}\right)\times$$

$$\times\Gamma\left(\frac{\alpha-\nu r-\sigma+k+1/2}{2r}\right)\cos^{(1\pm 1)/2}\left(\frac{\alpha-\nu r-\sigma+k+1/2}{2r}\pi\right)\left(\frac{1}{2}-\sigma-(\gamma^2+\gamma-1)\rho\right)_k.$$

34. $\displaystyle\int_0^{\infty}x^{\alpha-1}e^{cx/2}\left\{\begin{matrix}Y_{\nu}(bx^r)\\ K_{\nu}(bx^r)\end{matrix}\right\}W_{\rho,\sigma}(cx)\,dx=U_{\nu,\sigma}(1)$

$\Big[r>0;\ \operatorname{Re}\alpha>|\operatorname{Re}\sigma|+r|\operatorname{Re}\nu|-1/2;\ |\arg c|<3\pi/2,$
$\left\{\begin{matrix}b>0;\ \operatorname{Re}(\alpha+\rho)<3r/2\\ \operatorname{Re}b>0\end{matrix}\right\}$, $U_{\nu,\sigma}(\gamma)$ see 2.19.1.33$\Big]$.

35. $$\int_0^\infty x^{\alpha-1}e^{-cx/2}\begin{Bmatrix} Y_\nu(bx^r) \\ K_\nu(bx^r)\end{Bmatrix} W_{\rho,\sigma}(cx)\,dx = U_{\nu,\sigma}(-1)$$

$$\left[r,\ \operatorname{Re} c>0;\ \operatorname{Re}\alpha>|\operatorname{Re}\sigma|+r|\operatorname{Re}\nu|-1/2;\ |\arg c|<3\pi/2,\right.$$
$$\left.\begin{Bmatrix} b>0 \\ \operatorname{Re} b>0\end{Bmatrix},\ U_{\nu,\sigma}(\nu)\ \text{ see 2.19.1.33}\right].$$

36. $$\int_0^\infty x^{\alpha-1}e^{-(bx^r+cx)/2}M_{\mu,\nu}(bx^r)\,M_{\rho,\sigma}(cx)\,dx = V(0,\ 0)$$

$$[r,\ \operatorname{Re} b,\ \operatorname{Re} c>0;\ -\operatorname{Re}(\nu r+\sigma)-(r+1)/2<\operatorname{Re}\alpha<\operatorname{Re}(\mu r+\rho)].$$

$$V(\varepsilon,\gamma)=\frac{\sqrt{b}}{c^{\alpha+r/2}}T_\gamma\sum_{k=0}^\infty\frac{1}{k!}\left[\frac{\Gamma^{\varepsilon^2}(-2\nu)}{(1+2\nu)_k}E_k(-\nu)+\varepsilon^2\frac{\Gamma(2\nu)}{(1-2\nu)_k}E_k(\nu)\right]\times$$
$$\times\left(\left((-1)^{1-(\varepsilon^2+\varepsilon)/2}\frac{b}{c^r}\right)^k-\frac{\gamma^2-\gamma-2}{2c^{(1-2\gamma)\rho}{}_r}b^{((1-2\gamma)\rho-\alpha)/r}\Gamma^{1-\gamma}\begin{bmatrix}2\sigma+1\\ 1/2-\rho+\sigma\end{bmatrix}\times\right.$$
$$\times\Gamma^{1-\varepsilon^2}\begin{bmatrix}2\nu+1\\ \mu+\nu+1/2\end{bmatrix}\left[\Gamma\left(\frac{1}{2}-\mu+\nu\right)\Gamma\left(\frac{1}{2}-\mu-\nu\right)\right]^{-(\varepsilon^2+\varepsilon)/2}\times$$
$$\times\sum_{k=0}^\infty\frac{(1/2+\sigma+(1-2\gamma)\rho)_k\,(1/2-\sigma+(1-2\gamma)\rho)_k}{k!}\Gamma\left(\frac{1}{2}+\nu-\right.$$
$$\left.-\frac{(1-2\gamma)\rho-\alpha+k}{r}\right)\Gamma^{2\varepsilon^2-1}\left(\frac{1}{2}+(1-2\varepsilon^2)\frac{\nu r+(1-2\gamma)\rho-\alpha+k}{r}\right)\times$$
$$\times\Gamma^{-\varepsilon^2+\varepsilon+1}\left((1-2\varepsilon^2)\mu+\frac{\varepsilon^2-\varepsilon}{2}-(\varepsilon^2-\varepsilon-1)\frac{(1-2\gamma)\rho-\alpha+k}{r}\right)\left((-1)^\gamma\frac{b^{1/r}}{c}\right)^k$$
$$[r<1],$$

$$E_k(\nu)=\left(\frac{c^r}{b}\right)^\nu\Gamma^{-\varepsilon^2}\left(\frac{1}{2}-\mu+\nu\right)\left(\frac{1}{2}-\nu-(\varepsilon^2+\varepsilon-1)\mu\right)_k\times$$
$$\times\Gamma\left(\alpha+\sigma-r\nu+rk+\frac{r+1}{2}\right)\Gamma^{2\gamma^2-1}\left(\frac{1}{2}+(1-2\gamma^2)\left(\sigma-\alpha+r\nu-rk-\frac{r}{2}\right)\right)\times$$
$$\times\Gamma^{-\gamma^2+\gamma+1}\left((1-2\gamma^2)\rho+\frac{\gamma^2-\gamma}{2}+(\gamma^2-\gamma-1)\left(\alpha-r\nu+rk+\frac{r}{2}\right)\right),$$

$$V(\varepsilon,\gamma)=\frac{\sqrt{c}}{b^{(2\alpha+1)/(2r)}r}\Gamma^{1-\varepsilon^2}\begin{bmatrix}2\nu+1\\ \mu+\nu+1/2\end{bmatrix}\left[\Gamma\left(\frac{1}{2}-\mu+\nu\right)\Gamma\left(\frac{1}{2}-\mu-\nu\right)\right]^{-(\varepsilon^2+\varepsilon)/2}\times$$
$$\times\sum_{k=0}^\infty\frac{1}{k!}\left[\frac{\Gamma^{\gamma^2}(-2\sigma)}{(1+2\sigma)_k}F_k(-\sigma)+\gamma^2\frac{\Gamma(2\sigma)}{(1-2\sigma)_k}F_k(\sigma)\right]\left(\left((-1)^{1-(\gamma^2+\gamma)/2}\frac{c}{b^{1/r}}\right)^k-\right.$$
$$-\frac{\varepsilon^2-\varepsilon-2}{2b^{\mu(1-2\varepsilon)}}c^{\mu r(1-2\varepsilon)-\alpha}\Gamma^{1-\varepsilon}\begin{bmatrix}2\nu+1\\ 1/2-\mu+\nu\end{bmatrix}T_\gamma\times$$
$$\times\sum_{k=0}^\infty\frac{(1/2+\nu+(1-2\varepsilon)\mu)_k\,(1/2-\nu+(1-2\varepsilon)\mu)_k}{k!}\Gamma\left(\frac{1}{2}+\alpha+\sigma-rk-\mu r(1-2\varepsilon)\right)\times$$
$$\times\Gamma^{2\gamma^2-1}\left(\frac{1}{2}+(1-2\gamma^2)(\sigma-\alpha+kr+\mu r(1-2\varepsilon))\right)\Gamma^{-\gamma^2+\gamma+1}\left((1-2\gamma^2)\rho+\right.$$
$$\left.+\frac{\gamma^2-\gamma}{2}+(\gamma^2-\gamma-1)(\alpha-kr-\mu r(1-2\varepsilon))\right)\left((-1)^\varepsilon\frac{c^r}{b}\right)^k\qquad[r>1].$$

$$F_k(\sigma)=\frac{b^{\sigma/r}}{c^\sigma}\Gamma^{-\gamma^2}\left(\frac{1}{2}-\rho+\sigma\right)\left(\frac{1}{2}-\sigma-(\gamma^2+\gamma-1)\rho\right)_k\times$$

$$\times\Gamma\left(\frac{1}{2}+\nu+\frac{\alpha-\sigma+k+1/2}{r}\right)\Gamma^{2\varepsilon^2-1}\left(\frac{1}{2}+(1-2\varepsilon^2)\frac{\nu r-\alpha+\sigma-k-1/2}{r}\right)\times$$
$$\times\Gamma^{-\varepsilon^2+\varepsilon+1}\left((1-2\varepsilon^2)\mu+\frac{\varepsilon^2-\varepsilon}{2}+(\varepsilon^2-\varepsilon-1)\frac{\alpha-\sigma+k+1/2}{r}\right)$$

$V(\varepsilon, \gamma)$ for $r=1$ see 2.19.24.2-7.

37. $\int_0^\infty x^{\alpha-1}e^{-bx^r/2\pm cx/2}M_{\mu,\nu}(bx^r)\,W_{\rho,\sigma}(cx)\,dx=V_\sigma(0, \pm1)=V_{-\sigma}(0, \pm1)$

$\left[r,\ \mathrm{Re}\,b>0;\ \mathrm{Re}(\alpha+\nu r)>|\mathrm{Re}\,\sigma|-(r+1)/2,\ \left\{\begin{matrix}\mathrm{Re}(\alpha+\rho-\mu r)<0;\ |\arg c|<3\pi/2\\ \mathrm{Re}\,c>0\end{matrix}\right\}\right]$

$V_\sigma(\varepsilon, \gamma)=V(\varepsilon, \gamma)$ for $r<1$ or for $r>1$ and $\sigma\neq n/2, -n/2$; $V(\varepsilon, \gamma)$ see 2.19.1

$$V_{n/2}(\varepsilon, \pm1)=\frac{\sqrt{c}}{b^{(2\alpha+1)/(2r)}r}\Gamma^{1-\varepsilon^2}\left[\begin{matrix}2\nu+1\\ \mu+\nu+1/2\end{matrix}\right]\times$$
$$\times\left[\Gamma\left(\frac{1}{2}-\mu+\nu\right)\Gamma\left(\frac{1}{2}-\mu-\nu\right)\right]^{-(\varepsilon^2+\varepsilon)/2}\left\{\sum_{k=0}^{n-1}\frac{(n-k-1)!}{k!}F_k\left(\frac{n}{2}\right)\left(\mp\frac{c}{b^{1/r}}\right)^k+\right.$$
$$+(-1)^n\sum_{k=0}^{\infty}\frac{F_k(-n/2)}{k!\,(n+k)!}\left[\psi(k+1)+\psi(n+k+1)-\psi\left(\frac{1\pm n}{2}-\rho\pm k\right)-\right.$$
$$-\frac{1}{r}\psi\left(\frac{1}{2}+\nu+\frac{2\alpha+n+2k+1}{2r}\right)-\frac{1}{r}\psi\left(\frac{1}{2}+(1-2\varepsilon^2)\frac{2\nu r-2\alpha-n-2k-1}{2r}\right)+$$
$$+\frac{1}{r}\psi\left((1-2\varepsilon^2)\mu+\frac{\varepsilon^2-\varepsilon}{2}+(\varepsilon^2-\varepsilon-1)\frac{2\alpha+n+2k+1}{2r}\right)-$$
$$\left.\left.-\frac{1}{r}\ln\frac{c^r}{b}\right]\left(\pm\frac{c}{b^{1/r}}\right)^k\right\}-\frac{\varepsilon^2-\varepsilon-2}{2b^{\mu(1-2\varepsilon)}}c^{\mu r(1-2\varepsilon)-\alpha}\times$$
$$\times\Gamma^{1-\varepsilon}\left[\begin{matrix}2\nu+1\\ 1/2-\mu+\nu\end{matrix}\right]\left[\Gamma\left(\frac{1+n}{2}-\rho\right)\Gamma\left(\frac{1-n}{2}-\rho\right)\right]^{-(1\pm1)/2}\times$$
$$\times\sum_{k=0}^{\infty}\frac{(1/2+\nu+(1-2\varepsilon)\mu)_k\,(1/2-\nu+(1-2\varepsilon)\mu)_k}{k!}\times$$
$$\times\Gamma\left(\frac{1+n}{2}+\alpha-rk-\mu r(1-2\varepsilon)\right)\Gamma\left(\frac{1-n}{2}+\alpha-rk-\mu r(1-2\varepsilon)\right)\times$$
$$\times\Gamma^{\pm1}\left(\frac{1\mp1}{2}-\rho\mp(\alpha-rk-\mu r(1-2\varepsilon))\right)\left((-1)^\varepsilon\frac{c^r}{b}\right)$$

[$r>1$; $F_k(\sigma)$ see 2.19.1.36]

38. $\int_0^\infty x^{\alpha-1}e^{(bx^r\pm cx)/2}W_{\mu,\nu}(bx^r)\,W_{\rho,\sigma}(cx)\,dx=V_\sigma(1, \pm1)=V_{-\sigma}(1, \pm1)$

$\left[r>0;\ \mathrm{Re}\,\alpha>r|\mathrm{Re}\,\nu|+|\mathrm{Re}\,\sigma|-(r+1)/2;\ |\arg b|<3\pi/2\right.$

$\left.\left\{\begin{matrix}\mathrm{Re}(\alpha+\rho+\mu r)<0;\ |\arg c|<3\pi/2\\ \mathrm{Re}\,c>0\end{matrix}\right\},\ V_\sigma(\varepsilon, \gamma)\ \text{see 2.19.1.37}\right]$

39. $\int_0^\infty x^{\alpha-1}e^{-bx^r/2\pm cx/2}W_{\mu,\nu}(bx^r)\,W_{\rho,\sigma}(cx)\,dx=V_\sigma(-1, \pm1)=V_{-\sigma}(-1, \pm1)$

[r, $\mathrm{Re}\,b>0$; $\mathrm{Re}\,\alpha>r|\mathrm{Re}\,\nu|+|\mathrm{Re}\,\sigma|-(r+1)/2$; $|\arg c|<(1\pm1/2)\pi$; $V_\sigma(\varepsilon, \gamma)$ see 2.19.1.37]

2.19.2. Integrals of $x^\alpha W_{\rho,\sigma}(cx)$.

1. $\int_0^\infty x^{\alpha-1}W_{\rho,\sigma}(cx)\,dx=A^\alpha_{\rho,\sigma}$ [$\mathrm{Re}\,c>0$; $\mathrm{Re}\,\alpha>|\mathrm{Re}\,\sigma|-1/2$]

$$A^{\alpha}_{\rho,\sigma}=c^{-\alpha}\Gamma\begin{bmatrix}1/2+\alpha+\sigma,\ 1/2+\alpha-\sigma\\ 1+\alpha-\rho\end{bmatrix}{}_2F_1\left(\frac{1}{2}+\alpha+\sigma,\ \frac{1}{2}+\alpha-\sigma;1+\alpha-\rho;\frac{1}{2}\right),$$

$$A^{\alpha}_{0,\sigma}=\frac{2^{2\alpha-1}}{c^{\alpha}\sqrt{\pi}}\Gamma\left(\frac{2\alpha+2\sigma+1}{4}\right)\Gamma\left(\frac{2\alpha-2\sigma+1}{4}\right),$$

$$A^{0}_{\rho,\sigma}=\frac{2^{\rho}\pi^{3/2}\sec\sigma\pi}{\Gamma((3-2\rho+2\sigma)/4)\,\Gamma((3-2\rho-2\sigma)/4)}\qquad[|\operatorname{Re}\mu|<1/2],$$

$$A^{\alpha}_{-1/2,\sigma}=\frac{\sqrt{\pi}}{2\sigma c^{\alpha}}\left(\Gamma\begin{bmatrix}\alpha-\sigma+1/2\\ \alpha-\sigma+1\end{bmatrix}-\Gamma\begin{bmatrix}\alpha+\sigma+1/2\\ \alpha+\sigma+1\end{bmatrix}\right)\qquad[\sigma\neq 0],$$

$$A^{\alpha}_{-1/2,0}=\frac{\sqrt{\pi}}{c^{\alpha}}\Gamma\begin{bmatrix}\alpha+1/2\\ \alpha+1\end{bmatrix}\left(\psi(\alpha+1)-\psi\left(\alpha+\frac{1}{2}\right)\right)\qquad[\operatorname{Re}\alpha>-1/2].$$

2.19.3. Integrals of $x^{\alpha}e^{-px}\begin{Bmatrix}M_{\rho,\sigma}(cx)\\ W_{\rho,\sigma}(cx)\end{Bmatrix}$.

1. $$\int_0^{\infty}x^{\alpha-1}e^{-px}M_{\rho,\sigma}(cx)\,dx=I_{\alpha}$$

$$[\operatorname{Re}(\alpha+\sigma)>-1/2;\ 2\operatorname{Re}p>|\operatorname{Re}c|\ \text{ or }\ \operatorname{Re}(\rho-\alpha),\ 2\operatorname{Re}p=\operatorname{Re}c>0],$$

$$I_{\alpha}=\frac{2^{\alpha+\sigma+1/2}c^{\sigma+1/2}}{(2p-c)^{\alpha+\sigma+1/2}}\Gamma\left(\frac{1}{2}+\alpha+\sigma\right){}_2F_1\left(\frac{1}{2}+\rho+\sigma,\ \frac{1}{2}+\alpha+\sigma;\ 2\sigma+1;\ \frac{2c}{c-2p}\right),$$

$$I_0=2e^{-i\rho\pi}\Gamma\begin{bmatrix}2\sigma+1\\ \rho+\sigma+1/2\end{bmatrix}\left(\frac{2p-c}{2p+c}\right)^{\rho/2}Q^{\rho}_{\sigma-1/2}\left(\frac{2p}{c}\right)=$$

$$=\sqrt{2\pi c}\,\Gamma(2\sigma+1)(4p^2-c^2)^{-1/4}\left(\frac{2p-c}{2p+c}\right)^{\rho/2}P^{-\sigma}_{\rho-1/2}\left(\frac{2p}{\sqrt{4p^2-c^2}}\right)\qquad[\operatorname{Re}\sigma>-1/2],$$

$$I_{\sigma+1/2}=2^{2\sigma+1}\Gamma(2\sigma+1)\,c^{\sigma+1/2}(2p+c)^{-\rho-\sigma-1/2}(2p-c)^{\rho-\sigma-1/2}$$
$$[\operatorname{Re}\sigma>-1/2],$$

$$I_{-\rho-2\sigma}=2^{1/2-\rho-2\sigma}\Gamma(2\sigma+1)\,\Gamma\left(\frac{1}{2}-\rho-\sigma\right)\sqrt{c}\,(2p-c)^{\rho+\sigma-1/2}\times$$
$$\times(2p+c)^{\sigma}P^{-2\sigma}_{\rho+\sigma-1/2}\left(\frac{2p+3c}{2p-c}\right)\qquad[\operatorname{Re}(\rho+\sigma)<1/2],$$

$$I_{1/2-\rho}=2^{2\sigma-\rho+1}\Gamma(2\sigma+1)\,\Gamma(\sigma-\rho+1)\sqrt{c}\,(4p^2-c^2)^{-1/2}(2p-c)^{\rho}\times$$
$$\times P^{-2\sigma}_{2\rho-1}\left(\sqrt{\frac{2p+c}{2p-c}}\right)=2^{\rho+5/4}e^{(4\rho-1)\pi i/2}\Gamma\begin{bmatrix}2\sigma+1\\ 1/2+\sigma-\rho\end{bmatrix}\sqrt{c}\times$$
$$\times(2p+c)^{-1/2}(2p-c)^{\rho-1/4}Q^{1/2-2\rho}_{2\sigma-1/2}\left(\sqrt{\frac{c+2p}{2c}}\right)\qquad[\operatorname{Re}(\sigma-\rho)>-1],$$

$$I_{-\rho-1/2}=2^{2\sigma-\rho}\Gamma(\sigma-\rho)\,\Gamma(2\sigma+1)\sqrt{c}\,(2p-c)^{\rho}P^{-2\sigma}_{2\rho}\left(\sqrt{\frac{2p+c}{2p-c}}\right)=$$
$$=2^{\rho+5/4}\Gamma\begin{bmatrix}2\sigma+1\\ 1/2+\sigma-\rho\end{bmatrix}e^{(4\rho+1)\pi i/2}c^{1/4}(2p-c)^{\rho+1/2}Q^{-2\rho-1/2}_{2\sigma-1/2}\left(\sqrt{\frac{c+2p}{2c}}\right)$$
$$[\operatorname{Re}(\sigma-\rho)>0].$$

2. $$\int_0^{\infty}x^{\alpha-1}e^{-cx/2}M_{\rho,\sigma}(cx)\,dx=c^{-\alpha}\Gamma\begin{bmatrix}2\sigma+1,\ \rho-\alpha,\ 1/2+\alpha+\sigma\\ 1/2+\rho+\sigma,\ 1/2-\alpha+\sigma\end{bmatrix}$$
$$[\operatorname{Re}c>0;\ -1/2-\operatorname{Re}\sigma<\operatorname{Re}\alpha<\operatorname{Re}\rho].$$

3. $$\int_0^{\infty}x^{(\pm1-2)/4}e^{-px}M_{\rho,\pm1/4}(cx)\,dx=\sqrt{2\pi}c^{(2\pm1)/4}\times$$
$$\times(2p-c)^{\rho-(2\pm1)/4}(2p+c)^{-\rho-(2\pm1)/4}\qquad[2\operatorname{Re}p>|\operatorname{Re}c|].$$

4. $\int_0^\infty x^{-5/4}e^{-px}M_{-1/4,\,(2n+1)/4}(cx)\,dx=2\Gamma\begin{bmatrix}n+3/2\\n/2+1\end{bmatrix}c^{1/4}Q_n\left(\sqrt{\frac{2p+c}{2c}}\right)$

[2 Re $p>|$Re $c|$].

5. $\int_0^\infty x^{\alpha-1}e^{-px}W_{\rho,\,\sigma}(cx)\,dx=J_\alpha$

[Re $(2p+c)>0$; $|\arg c|=3\pi/2$; Re $\alpha>|$Re $\sigma|-1/2$],

$$J_\alpha=\frac{1}{c^\alpha}\Gamma\begin{bmatrix}1/2+\alpha+\sigma,\ 1/2+\alpha-\sigma\\1+\alpha-\rho\end{bmatrix}{}_2F_1\left(\frac{1}{2}+\alpha+\sigma,\ \frac{1}{2}+\alpha-\sigma;\ 1+\alpha-\rho;\ \frac{c-2p}{2c}\right),$$

$$J_0=\frac{\pi}{\cos\sigma\pi}\left(\frac{2p-c}{2p+c}\right)^{\rho/2}P^{\rho}_{\sigma-1/2}\left(\frac{2p}{c}\right)=\frac{\sqrt{2\pi c}}{\cos\sigma\pi}\left(\frac{2p-c}{2p+c}\right)^{\rho/2}e^{\sigma\pi i}(4p^2-c^2)^{-1/4}\times$$
$$\times Q^{-\sigma}_{-\rho-1/2}\left(\frac{2p}{\sqrt{4p^2-c^2}}\right)\quad [|\mathrm{Re}\,\sigma|<1/2],$$

$$J_{1/2-\rho}=2^{5/4-3\rho}\Gamma(1-\rho+\sigma)\Gamma(1-\rho-\sigma)c^{1/4}(2p-c)^{\rho-1/4}(2p+c)^{-1/2}\times$$
$$\times P^{2\rho-1/2}_{2\sigma-1/2}\left(\sqrt{\frac{c+2p}{2c}}\right)=2^{2\sigma-\rho+2}\Gamma\begin{bmatrix}1-\rho+\sigma\\1/2-\rho-\sigma\end{bmatrix}\sqrt{\frac{c}{2p+c}}\times$$
$$\times(2p-c)^{\rho-1/2}e^{2\sigma\pi i}Q^{-2\sigma}_{-2\rho}\left(\sqrt{\frac{2p+c}{2p-c}}\right)\quad [\mathrm{Re}\,\rho<1-|\mathrm{Re}\,\sigma|],$$

$$J_{-\rho-1/2}=2^{3\rho-3/4}\Gamma(\rho-\sigma)\Gamma(-\rho-\sigma)c^{1/4}(2p-c)^{\rho+1/4}P^{2\rho+1/2}_{2\sigma-1/2}\left(\sqrt{\frac{c+2p}{2c}}\right)=$$
$$=2^{2\sigma-\rho+1}\Gamma\begin{bmatrix}\rho-\sigma\\1/2-\rho-\sigma\end{bmatrix}e^{2\sigma\pi i}\sqrt{c}(2p-c)^{\rho}Q^{-2\sigma}_{-2\rho-1}\left(\sqrt{\frac{2p+c}{2p-c}}\right)\quad [\mathrm{Re}\,\rho<|\mathrm{Re}\,\sigma|].$$

6. $\int_0^\infty x^{\alpha-1}e^{cx/2}W_{\rho,\,\sigma}(cx)\,dx=c^{-\alpha}\Gamma\begin{bmatrix}1/2+\alpha+\sigma,\ 1/2+\alpha-\sigma,\ -\alpha-\rho\\1/2-\rho-\sigma,\ 1/2-\rho+\sigma\end{bmatrix}$

[$|$Re $\sigma|-1/2<$ Re $\alpha<-$Re ρ; $|\arg c|<3\pi/2$].

7. $\int_0^\infty x^{\alpha-1}e^{-cx/2}W_{\rho,\,\sigma}(cx)\,dx=c^{-\alpha}\Gamma\begin{bmatrix}1/2+\alpha+\sigma,\ 1/2+\alpha-\sigma\\1+\alpha-\rho\end{bmatrix}$

[Re $\alpha>|$Re $\sigma|-1/2$; Re $c>0$].

8. $\int_0^\infty x^{\rho+2\sigma-1}e^{-3cx/2}W_{\rho,\,\sigma}(cx)\,dx=$

$=\frac{c^{-\rho-2\sigma}}{2}\Gamma\begin{bmatrix}(2\rho+6\sigma+1)/4,\ 1/2+\rho+\sigma\\(3-2\rho+2\sigma)/4\end{bmatrix}$ [Re $c>0$; Re $(\rho+\sigma)$, Re $(\rho+3\sigma)>-1/2$].

2.19.4. Integrals of $x^\alpha e^{f(x)}\left\{\begin{matrix}M_{\rho,\,\sigma}(cx)\\W_{\rho,\,\sigma}(cx)\end{matrix}\right\}$.

1. $\int_0^\infty x^{\alpha-1}e^{-p\sqrt{x}-cx/2}M_{\rho,\,\sigma}(cx)\,dx=A(0)$ [Re c, Re $p>0$; Re $(\alpha+\sigma)>-1/2$],

$$A(\gamma)=c^{-\alpha}\Gamma^{1-\gamma^2}\begin{bmatrix}2\sigma+1\\\rho+\sigma+1/2\end{bmatrix}\Gamma\left(\frac{1}{2}+\alpha+\sigma\right)\Gamma^{2\gamma^2-1}\left(\frac{1}{2}+(1-2\gamma^2)(\sigma-\alpha)\right)\times$$
$$\times\left[\Gamma\left(\frac{1}{2}-\rho-\sigma\right)\Gamma\left(\frac{1}{2}-\rho+\sigma\right)\right]^{-(\gamma^2+\gamma)/2}\times$$
$$\times\Gamma^{-\gamma^2+\gamma+1}\left((1-2\gamma^2)\rho+\frac{\gamma^2-\gamma}{2}+(\gamma^2-\gamma-1)\alpha\right)\times$$

$$\times {}_2F_2\left(\frac{1}{2}+\alpha+\sigma,\ \frac{1}{2}+\alpha-\sigma;\ \frac{1}{2},\ 1+\alpha+(\gamma^2+\gamma-1)\rho;\ (-1)^{(\gamma^2+\gamma)/2}\frac{p^2}{4c}\right)-$$

$$-\frac{p}{c^{\alpha+1/2}}\Gamma^{1-\gamma^2}\begin{bmatrix}2\sigma+1\\ \rho+\sigma+1/2\end{bmatrix}\Gamma(1+\alpha+\sigma)\,\Gamma^{2\gamma^2-1}\left(\frac{1}{2}+(1-2\gamma^2)\left(\sigma-\alpha-\frac{1}{2}\right)\right)\times$$

$$\times\left[\Gamma\left(\frac{1}{2}-\rho-\sigma\right)\Gamma\left(\frac{1}{2}-\rho+\sigma\right)\right]^{-(\gamma^2+\gamma)/2}\Gamma^{-\gamma^2+\gamma+1}\left((1-2\gamma^2)\rho+\right.$$

$$\left.+\gamma^2-\gamma+(\gamma^2-\gamma-1)\alpha-\frac{1}{2}\right){}_2F_2\left(\alpha+\sigma+1,\ \alpha-\sigma+1;\right.$$

$$\left.\frac{3}{2},\ \frac{3}{2}+\alpha+(\gamma^2+\gamma-1)\rho;\ (-1)^{(\gamma^2+\gamma)/2}\frac{p^2}{4c}\right)-$$

$$-\frac{\gamma^2-\gamma-2}{c^{(1-2\gamma)\rho}}p^{2\rho(1-2\gamma)-2\alpha}\Gamma^{1-\gamma}\begin{bmatrix}2\sigma+1\\ 1/2+\sigma-\rho\end{bmatrix}\times$$

$$\times\Gamma(2\alpha-2(1-2\gamma)\rho)\,{}_2F_2\left(\frac{1}{2}+\sigma+(1-2\gamma)\rho,\ \frac{1}{2}-\sigma+(1-2\gamma)\rho;\right.$$

$$\left.1-\alpha+(1-2\gamma)\rho,\ \frac{1}{2}-\alpha+(1-2\gamma)\rho;\ \frac{(-1)^{\gamma}p^2}{4c}\right).$$

2. $$\int_0^\infty x^{\sigma-1+\varepsilon/2}e^{-p\sqrt{x}-cx/2}M_{\rho,\sigma}(cx)\,dx=$$

$$=2^{1-\rho-3\sigma-\varepsilon}c^{(1-\rho-\sigma-\varepsilon)/2}p^{\rho-\sigma-1}\Gamma(4\sigma+\varepsilon+1)\exp\left(\frac{p^2}{8c}\right)\times$$

$$\times W_{-(3\sigma+\rho+\varepsilon)/2,\ (\sigma-\rho+\varepsilon)/2}\left(\frac{p^2}{4c}\right)$$

[$\varepsilon=0$ or 1; $\operatorname{Re} c$, $\operatorname{Re} p>0$; $\operatorname{Re}\sigma>-(1+\varepsilon)/4$].

3. $$\int_0^\infty x^{\sigma-1}e^{-p\sqrt{x}-cx/2}M_{-3\sigma,\ \sigma}(cx)\,dx=\frac{\Gamma(4\sigma+1)}{\sqrt{\pi}}c^{\sigma}p^{-4\sigma}\exp\left(\frac{p^2}{8c}\right)K_{2\sigma}\left(\frac{p^2}{8c}\right)$$

[$\operatorname{Re} p$, $\operatorname{Re} c>0$; $\operatorname{Re}\sigma>-1/4$].

4. $$\int_0^\infty x^{\alpha-1}e^{-p\sqrt{x}\pm cx/2}W_{\rho,\sigma}(cx)\,dx=A(\pm 1)$$

[$\operatorname{Re} p>0$; $\operatorname{Re}\alpha>|\operatorname{Re}\sigma|-1/2$; $|\arg c|<(1\pm 1/2)\pi$; $A(\gamma)$ see 2.19.4.1].

5. $$\int_0^\infty x^{\alpha-1}e^{-p/x-cx/2}M_{\rho,\sigma}(cx)\,dx=B(0)$$

[$\operatorname{Re} c$, $\operatorname{Re} p$, $\operatorname{Re}(\rho-\sigma)>0$],

$$B(\gamma)=c^{-\alpha}\Gamma\left(\frac{1}{2}+\alpha+\sigma\right)\Gamma^{1-\gamma^2}\begin{bmatrix}2\sigma+1\\ \rho+\sigma+1/2\end{bmatrix}\left[\Gamma\left(\frac{1}{2}-\rho-\sigma\right)\Gamma\left(\frac{1}{2}-\rho+\sigma\right)\right]^{-(\gamma^2+\gamma)/2}\times$$

$$\times\Gamma^{2\gamma^2-1}\left(\frac{1}{2}+(1-2\gamma^2)(\sigma-\alpha)\right)\Gamma^{-\gamma^2+\gamma+1}\left((1-2\gamma^2)\rho+\frac{\gamma^2-\gamma}{2}+(\gamma^2-\gamma-1)\alpha\right)\times$$

$$\times {}_1F_2\left(-(\gamma^2+\gamma-1)\rho-\alpha;\ \frac{1}{2}-\alpha+\sigma,\ \frac{1}{2}-\alpha-\sigma;\ (-1)^{(\gamma^2+\gamma)/2}cp\right)+$$

$$+c^{1/2-\sigma}p^{\alpha-\sigma+1/2}\Gamma\left(-\frac{1}{2}-\alpha-\sigma\right)\Gamma^{\gamma^2}\begin{bmatrix}-2\sigma\\ 1/2-\rho-\sigma\end{bmatrix}\times$$

$$\times {}_1F_2\left(\frac{1}{2}+\sigma-(\gamma^2+\gamma-1)\rho;\ 2\sigma+1,\ \frac{3}{2}+\alpha+\sigma;\ (-1)^{(\gamma^2+\gamma)/2}cp\right)\pm$$

$$\pm\gamma c^{1/2-\sigma}p^{1/2+\alpha-\sigma}\Gamma\begin{bmatrix}\sigma-\alpha-1/2,\ 2\sigma\\ 1/2-\rho+\sigma\end{bmatrix}\times$$

$$\times {}_1F_2\left(\frac{1}{2}\mp\rho-\sigma;\ 1-2\sigma,\ \frac{3}{2}+\alpha-\sigma;\ \mp cp\right).$$

6. $\int_0^\infty x^{\alpha-1} e^{-p/x\pm cx/2} W_{\rho,\sigma}(cx)\,dx = B(\pm 1)$

$\left[\operatorname{Re} p>0,\ \begin{Bmatrix}\operatorname{Re}(\alpha+\rho)<0;\ |\arg c|<3\pi/2\\ \operatorname{Re} c>0\end{Bmatrix},\ B(\gamma)\ \text{see 2.19.4.5}\right].$

7. $\int_0^\infty x^{\mp\rho-2} e^{-p/x\pm cx/2} W_{\rho,\sigma}(cx)\,dx=$

$=2^{1-(1\pm1)\rho}\sqrt{c}\,p^{\mp\rho-1/2}\begin{Bmatrix}S_{2\rho,\,2\sigma}(2\sqrt{cp})\\ K_{2\sigma}(2\sqrt{cp})\end{Bmatrix}$ $[\operatorname{Re} p>0;\ |\arg c|<(1\pm1/2)\pi]$.

8. $\int_0^\infty x^{-\sigma-2} e^{-p/x+cx/2} W_{\sigma,\sigma}(cx)\,dx=$

$=\frac{\sqrt{\pi c}}{p^{\sigma+1/2}}\Gamma\left(2\sigma+\frac{1}{2}\right)\left[\mathbf{H}_{2\sigma}(2\sqrt{cp})-Y_{2\sigma}(2\sqrt{cp})\right]$ $[\operatorname{Re} p>0;\ |\arg c|<3\pi/2]$.

9. $\int_0^\infty x^{-3\sigma-3/2} e^{-p/x+cx/2} W_{\sigma,\sigma}(cx)\,dx=$

$=\frac{c^{\sigma+1/2}}{2p^{2\sigma}}\Gamma\left(2\sigma+\frac{1}{2}\right)H_{2\sigma}^{(1)}(\sqrt{cp})\,H_{2\sigma}^{(2)}(\sqrt{cp})$

$[\operatorname{Re} p>0;\ \operatorname{Re}\sigma>-1/4;\ |\arg c|<3\pi/2]$.

10. $\int_0^\infty x^{3\sigma-3/2} e^{-p/x-cx/2} W_{\sigma,\sigma}(cx)\,dx=\frac{2c^{1/2-\sigma}}{\sqrt{\pi}\,p^{2\sigma}}K_{2\sigma}^2(\sqrt{cp})$ $[\operatorname{Re} c,\ \operatorname{Re} p>0]$.

11. $\int_0^\infty x^{-1}e^{-p/x+cx/2} W_{-1/2,\,\sigma}(cx)\,dx=$

$=\frac{\pi^{3/2}\sqrt{cp}}{4\sigma}\left[H_{\sigma+1/2}^{(1)}(\sqrt{cp})\,H_{\sigma-1/2}^{(2)}(\sqrt{cp})+H_{\sigma-1/2}^{(1)}(\sqrt{cp})\,H_{\sigma+1/2}^{(2)}(\sqrt{cp})\right]$

$[\operatorname{Re} p>0;\ |\arg c|<3\pi/2]$.

12. $\int_0^\infty x^{-1} e^{-p/x-cx/2} W_{1/2,\,\sigma}(cx)\,dx=2\sqrt{\frac{2cp}{\pi}}K_{\sigma-1/2}(\sqrt{cp})\,K_{\sigma+1/2}(\sqrt{cp})$

$[\operatorname{Re} c,\ \operatorname{Re} p>0]$.

2.19.5. Integrals of $A(x)\,e^{\pm cx/2}\begin{Bmatrix}M_{\rho,\sigma}(cx)\\ W_{\rho,\sigma}(cx)\end{Bmatrix}$.

1. $\int_0^a x^{\alpha-1}(a-x)^{\beta-1} e^{\pm cx/2} M_{\rho,\sigma}(cx)\,dx=U(0)$ $[a,\ \operatorname{Re}\beta>0;\ \operatorname{Re}(\alpha+\sigma)>-1/2]$,

$$U(\gamma)=a^{\alpha+\beta+\sigma-1/2}c^{\sigma+1/2}\Gamma^{\gamma^2}\begin{bmatrix}-2\sigma\\ 1/2-\rho-\sigma\end{bmatrix}B\left(\beta,\ \frac{1}{2}+\alpha+\sigma\right)\times$$
$$\times{}_2F_2\left(\frac{1}{2}\mp\rho+\sigma,\ \frac{1}{2}+\alpha+\sigma;\ 2\sigma+1,\ \frac{1}{2}+\alpha+\beta+\sigma;\ \pm ac\right)+$$
$$+\gamma^2a^{\alpha+\beta-\sigma-1/2}c^{1/2-\sigma}\Gamma\begin{bmatrix}\beta,\ 2\sigma,\ 1/2+\alpha-\sigma\\ 1/2-\rho+\sigma,\ 1/2+\alpha+\beta-\sigma\end{bmatrix}\times$$
$$\times{}_2F_2\left(\frac{1}{2}\mp\rho-\sigma,\ \frac{1}{2}+\alpha-\sigma;\ 1-2\sigma,\ \frac{1}{2}+\alpha+\beta-\sigma;\ \pm ac\right).$$

2. $$\int_0^a x^{\rho-\beta-1}(a-x)^{\beta-1}e^{-cx/2}M_{\rho,\sigma}(cx)\,dx=$$
$$=\mathrm{B}\left(\beta,\ \rho+\sigma-\beta+\frac{1}{2}\right)a^{\rho-1}e^{-ac/2}M_{\rho-\beta,\,\sigma}(ac)$$
$[a,\ \operatorname{Re}\beta>0;\ \operatorname{Re}(\rho+\sigma-\beta)>-1/2]$.

3. $$\int_0^a x^{\sigma-1/2}(a-x)^{\beta-1}e^{\pm cx/2}M_{\rho,\sigma}(cx)\,dx=$$
$$=a^{\sigma+(\beta-1)/2}c^{-\beta/2}\,\mathrm{B}(\beta,\ 2\sigma+1)\,e^{\pm ac/2}M_{\rho\pm\beta/2,\,\sigma+\beta/2}(ac)$$ $[a,\ \operatorname{Re}\beta>0;\operatorname{Re}\sigma>-1/2]$.

4. $$\int_0^a x^{-1}(a-x)^{\rho-1}e^{-cx/2}M_{\rho,\sigma}(cx)\,dx=\sqrt{\pi c}\,a^{\rho-1/2}\Gamma\begin{bmatrix}\rho,\ 2\sigma+1\\ \rho+\sigma+1/2\end{bmatrix}e^{-ac/2}I_\sigma\left(\frac{ac}{2}\right)$$
$[a,\ \operatorname{Re}\rho>0;\ \operatorname{Re}\sigma>-1/2]$.

5. $$\int_0^a x^{\alpha-1}(a-x)^{\beta-1}e^{\pm cx/2}W_{\rho,\sigma}(cx)\,dx=U(\pm 1)$$
$[a,\ \operatorname{Re}\beta>0;\ \operatorname{Re}\alpha>|\operatorname{Re}\sigma|-1/2;\ U(\gamma)$ see 2.19.5.1].

6. $$\int_0^a x^{-\beta-\rho-1}(a-x)^{\beta-1}e^{cx/2}W_{\rho,\sigma}(cx)\,dx=$$
$$=a^{-\rho-1}\Gamma\begin{bmatrix}\beta,\ 1/2-\beta-\rho+\sigma,\ 1/2-\beta-\rho-\sigma\\ 1/2-\rho+\sigma,\ 1/2-\rho-\sigma\end{bmatrix}W_{\beta+\rho,\,\sigma}(ac)$$
$[a,\ \operatorname{Re}\beta>0;\ \operatorname{Re}(\beta+\rho)<1/2-|\operatorname{Re}\sigma|]$.

7. $$\int_0^a x^{\rho-\beta-1}(a-x)^{\beta-1}e^{-cx/2}W_{\rho,\sigma}(cx)\,dx=$$
$$=\frac{a^{\rho-1}\Gamma(\beta)}{\cos(\beta-\rho+\sigma)\pi}e^{-ac/2}\left\{\sin\beta\pi\Gamma\begin{bmatrix}1/2-\beta+\rho+\sigma\\ 2\sigma+1\end{bmatrix}M_{\rho-\beta,\,\sigma}(ac)+\right.$$
$$\left.+\cos(\rho-\sigma)\pi W_{\rho-\beta,\,\sigma}(ac)\right\}$$ $[a,\ \operatorname{Re}\beta>0;\ \operatorname{Re}(\beta-\rho)<1/2-|\operatorname{Re}\sigma|]$.

8. $$\int_a^\infty x^{\alpha-1}(x-a)^{\beta-1}e^{-cx/2}M_{\rho,\sigma}(cx)\,dx=V(0)$$ $[a,\ \operatorname{Re}\beta,\ \operatorname{Re}c>0;\ \operatorname{Re}(\alpha+\beta-\rho)<1]$,

$$V(\gamma)=a^{\alpha+\beta+\sigma-1/2}c^{\sigma+1/2}\Gamma^{\gamma^2}\begin{bmatrix}-2\sigma\\ 1/2-\rho-\sigma\end{bmatrix}\mathrm{B}\left(\beta,\ \frac{1}{2}-\alpha-\beta-\sigma\right)\times$$
$$\times{}_2F_2\left(\frac{1}{2}+\sigma-(\gamma^2+\gamma-1)\rho,\ \frac{1}{2}+\alpha+\sigma;\ \frac{1}{2}+\alpha+\beta+\sigma,\ 2\sigma+1;\ (-1)^{1-(\gamma^2+\gamma)/2}ac\right)+$$
$$+\gamma^2a^{\alpha+\beta-\sigma-1/2}c^{1/2-\sigma}\Gamma\begin{bmatrix}\beta,\ 2\sigma,\ 1/2-\alpha-\beta+\sigma\\ 1/2-\rho+\sigma,\ 1/2-\alpha+\sigma\end{bmatrix}\times$$
$$\times{}_2F_2\left(\frac{1}{2}\mp\rho-\sigma,\ \frac{1}{2}+\alpha-\sigma;\ \frac{1}{2}+\alpha+\beta-\sigma,\ 1-2\sigma;\ \pm ac\right)+$$
$$+c^{1-\alpha-\beta}\Gamma^{1-\gamma^2}\begin{bmatrix}2\sigma+1\\ \rho+\sigma+1/2\end{bmatrix}\left[\Gamma\left(\frac{1}{2}-\rho-\sigma\right)\Gamma\left(\frac{1}{2}-\rho+\sigma\right)\right]^{-(\gamma^2+\gamma)/2}\times$$
$$\times\Gamma\left(\alpha+\beta+\sigma-\frac{1}{2}\right)\Gamma^{2\gamma^2-1}\left(\frac{1}{2}+(1-2\gamma^2)(\sigma-\alpha-\beta+1)\right)\times$$
$$\times\Gamma^{-\gamma^2+\gamma+1}\left((1-2\gamma^2)\rho+\frac{\gamma^2-\gamma}{2}+(\gamma^2-\gamma-1)(\alpha+\beta-1)\right)\times$$
$$\times{}_2F_2\left(1-\beta,\ 1-\alpha-\beta-(\gamma^2+\gamma-1)\rho;\ \frac{3}{2}-\alpha-\beta+\sigma,\ \frac{3}{2}-\alpha-\beta-\sigma;\ (-1)^{1-(\gamma^2+\gamma)/2}ac\right).$$

9. $\int\limits_a^\infty x^{-\sigma-1/2}(x-a)^{\beta-1}e^{-cx/2}M_{\rho,\sigma}(cx)\,dx=$

$$=a^{(\beta-1)/2-\sigma}c^{-\beta/2}\Gamma\begin{bmatrix}\beta,\ 2\sigma+1,\ \rho+\sigma-\beta+1/2\\ \rho+\sigma+1/2,\ 2\sigma-\beta+1\end{bmatrix}e^{-ac/2}M_{\rho-\beta/2,\ \sigma-\beta/2}(ac)$$

[a, $\operatorname{Re}\beta$, $\operatorname{Re}c>0$; $\operatorname{Re}(\rho+\sigma-\beta)>-1/2$].

10. $\int\limits_a^\infty x^{\alpha-1}(x-a)^{\beta-1}e^{\pm cx/2}W_{\rho,\sigma}(cx)\,dx=V(\pm 1)$

$\left[a,\ \operatorname{Re}\beta>0,\ \begin{cases}\operatorname{Re}(\alpha+\beta+\rho)<1;\ |\arg c|<3\pi/2\\ \operatorname{Re}c>0\end{cases},\ V(\gamma)\ \text{see } 2.19.5.8\right]$.

11. $\int\limits_a^\infty x^{\pm\sigma-1/2}(x-a)^{\beta-1}e^{cx/2}W_{\rho,\sigma}(cx)\,dx=$

$$=a^{\pm\sigma+(\beta-1)/2}c^{-\beta/2}\mathrm{B}\left(\beta,\ \frac{1}{2}-\beta-\rho\mp\sigma\right)e^{ac/2}W_{\rho+\beta/2,\ \sigma\pm\beta/2}(ac)$$

[a, $\operatorname{Re}\beta>0$; $\operatorname{Re}(\beta+\rho\pm\sigma)<1/2$; $|\arg c|<3\pi/2$].

12. $\int\limits_a^\infty x^{\pm\sigma-1/2}(x-a)^{\beta-1}e^{-cx/2}W_{\rho,\sigma}(cx)\,dx=$

$$=\Gamma(\beta)\,a^{\pm\sigma+(\beta-1)/2}c^{-\beta/2}e^{-ac/2}W_{\rho-\beta/2,\ \sigma\pm\beta/2}(ac)$$

[a, $\operatorname{Re}c$, $\operatorname{Re}\beta>0$].

13. $\int\limits_a^\infty x^{\rho-\beta-1}(x-a)^{\beta-1}e^{-cx/2}W_{\rho,\sigma}(cx)\,dx=\Gamma(\beta)\,a^{\rho-1}e^{-ac/2}W_{\rho-\beta,\ \sigma}(ac)$

[a, $\operatorname{Re}c$, $\operatorname{Re}\beta>0$].

14. $\int\limits_0^\infty \frac{x^{\alpha-1}}{(x+z)^\theta}e^{-cx/2}M_{\rho,\sigma}(cx)\,dx=W(0)$

[$\operatorname{Re}c>0$; $-1/2-\operatorname{Re}\sigma<\operatorname{Re}\alpha<\operatorname{Re}(\theta+\rho)$; $|\arg z|<\pi$],

$$W(\gamma)=c^{\sigma+1/2}z^{\alpha-\theta+\sigma+1/2}\Gamma^{\gamma^2}\begin{bmatrix}-2\sigma\\ 1/2-\rho-\sigma\end{bmatrix}\mathrm{B}\left(\frac{1}{2}+\alpha+\sigma,\ \theta-\alpha-\sigma-\frac{1}{2}\right)\times$$

$$\times{}_2F_2\left(\frac{1}{2}+\alpha+\sigma,\ \frac{1}{2}+\sigma-(\gamma^2+\gamma-1)\rho;\ 2\sigma+1,\ \frac{3}{2}+\alpha-\theta+\sigma;\ (-1)^{(\gamma^2+\gamma)/2}cz\right)+$$

$$+\gamma^2c^{1/2-\sigma}z^{\alpha-\theta-\sigma+1/2}\Gamma\begin{bmatrix}2\sigma,\ 1/2+\alpha-\sigma,\ \theta-\alpha+\sigma-1/2\\ \theta,\ 1/2-\rho+\sigma\end{bmatrix}\times$$

$$\times{}_2F_2\left(\frac{1}{2}\mp\rho-\sigma,\ \frac{1}{2}+\alpha-\sigma;\ \frac{3}{2}+\alpha-\theta-\sigma,\ 1-2\sigma;\ \mp cz\right)+$$

$$+c^{\theta-\alpha}\Gamma^{1-\gamma^2}\begin{bmatrix}2\sigma+1\\ \rho+\sigma+1/2\end{bmatrix}\left[\Gamma\left(\frac{1}{2}-\rho-\sigma\right)\Gamma\left(\frac{1}{2}-\rho+\sigma\right)\right]^{-(\gamma^2+\gamma)/2}\times$$

$$\times\Gamma(1/2+\alpha-\theta+\sigma)\,\Gamma^{2\gamma^2-1}(1/2+(1-2\gamma^2)(\sigma-\alpha+\theta))\times$$

$$\times\Gamma^{-\gamma^2+\gamma+1}\left((1-2\gamma^2)\rho+\frac{\gamma^2-\gamma}{2}+(\gamma^2-\gamma-1)(\alpha-\theta)\right)\times$$

$$\times{}_2F_2\left(\theta,\ \theta-\alpha-(\gamma^2+\gamma-1)\rho;\ \frac{1}{2}-\alpha+\theta+\sigma,\ \frac{1}{2}-\alpha+\theta-\sigma;\ (-1)^{(\gamma^2+\gamma)/2}cz\right).$$

15. $\int\limits_0^\infty \frac{x^{\sigma-1/2}}{(x+z)^\theta}e^{-cx/2}M_{\rho,\sigma}(cx)\,dx=$

$$=c^{(\theta-1)/2}z^{\sigma-\theta/2}\Gamma\begin{bmatrix}2\sigma+1,\ \theta+\rho-\sigma-1/2\\ \theta\end{bmatrix}e^{cz/2}W_{(1-\theta)/2-\rho,\ \sigma+(1-\theta)/2}(cz)$$

[$\operatorname{Re}c>0$; $-1/2<\operatorname{Re}\sigma<\operatorname{Re}(\theta+\rho)-1/2$; $|\arg z|<\pi$].

16. $$\int_0^\infty \frac{x^{\sigma+n+1/2}}{x+z} e^{-cx/2} M_{\rho,\sigma}(cx)\,dx=$$

$$=(-1)^{n+1}\Gamma(2\sigma+1)\,\Gamma\left(\frac{1}{2}+\rho-\sigma\right) z^{\sigma+n+1/2} e^{cz/2} W_{-\rho,\sigma}(cz)$$

$[\operatorname{Re} c>0;\ -n/2-1<\operatorname{Re}\sigma<\operatorname{Re}\rho-n-1/2;\ |\arg z|<\pi]$.

17. $$\int_0^\infty \frac{x^{\alpha-1}}{(x+z)^\theta} e^{\pm cx/2} W_{\rho,\sigma}(cx)\,dx=W(\pm 1)$$

$\left[\operatorname{Re}\alpha>|\operatorname{Re}\sigma|-1/2;\ |\arg z|<\pi,\ \begin{cases}\operatorname{Re}(\alpha-\theta+\rho)<0;\ |\arg c|<3\pi/2\\ \operatorname{Re} c>0\end{cases},\ W(\gamma)\ \text{see 2.19.5.14}\right]$.

18. $$\int_0^\infty \frac{x^{\theta+\rho-2}}{(x+z)^\theta} e^{-cx/2} W_{\rho,\sigma}(cx)\,dx=$$

$$=\Gamma\left[\begin{matrix}\theta+\rho+\sigma-1/2,\ \theta+\rho-\sigma-1/2\\ \theta\end{matrix}\right] z^{\rho-1} e^{cz/2} W_{1-\theta-\rho,\sigma}(cz)$$

$[\operatorname{Re} c>0;\ \operatorname{Re}(\theta+\rho)>|\operatorname{Re}\sigma|+1/2;\ |\arg z|<\pi]$.

19. $$\int_0^\infty \frac{x^{\alpha-1}}{x-y} e^{-cx/2} M_{\rho,\sigma}(cx)\,dx=X(0)$$ $[y,\ \operatorname{Re} c>0;\ -1/2-\operatorname{Re}\sigma<\operatorname{Re}\alpha<\operatorname{Re}\rho+1]$,

$$X(\gamma)=\pi c^{\sigma+1/2} y^{\alpha+\sigma-1/2}\Gamma^{\gamma^2}\left[\begin{matrix}-2\sigma\\ 1/2-\rho-\sigma\end{matrix}\right]\operatorname{tg}(\alpha+\sigma)\pi\times$$
$$\times{}_1F_1\left(\frac{1}{2}+\sigma-(\gamma^2+\gamma-1)\rho;\ 2\sigma+1;\ (-1)^{1-(\gamma^2+\gamma)/2}cy\right)+$$
$$+\gamma^2\pi c^{1/2-\sigma} y^{\alpha-\sigma-1/2}\Gamma\left[\begin{matrix}2\sigma\\ 1/2-\rho+\sigma\end{matrix}\right]\operatorname{tg}(\alpha-\sigma)\pi\times$$
$$\times{}_1F_1\left(\frac{1}{2}\mp\rho-\sigma;\ 1-2\sigma;\ \pm cy\right)+c^{1-\alpha}\Gamma^{1-\gamma^2}\left[\begin{matrix}2\sigma+1\\ \rho+\sigma+1/2\end{matrix}\right]\times$$
$$\times\left[\Gamma\left(\frac{1}{2}-\rho-\sigma\right)\Gamma\left(\frac{1}{2}-\rho+\sigma\right)\right]^{-(\gamma^2+\gamma)/2}\Gamma\left(\alpha+\sigma-\frac{1}{2}\right)\times$$
$$\times\Gamma^{2\gamma^2-1}\left(\frac{1}{2}+(1-2\gamma^2)(\sigma-\alpha+1)\right)\Gamma^{-\gamma^2+\gamma+1}\left((1-2\gamma^2)\rho+\frac{\gamma^2-\gamma}{2}+\right.$$
$$\left.+(\gamma^2-\gamma-1)(\alpha-1)\right){}_2F_2\left(1,\ 1-\alpha-(\gamma^2+\gamma-1)\rho;\right.$$
$$\left.\frac{3}{2}-\alpha+\sigma,\ \frac{3}{2}-\alpha-\sigma;\ (-1)^{1-(\gamma^2+\gamma)/2}cy\right).$$

20. $$\int_0^\infty \frac{x^{\alpha-1}}{x-y} e^{\pm cx/2} W_{\rho,\sigma}(cx)\,dx=X(\pm 1)$$

$\left[y>0;\ \operatorname{Re}\alpha>|\operatorname{Re}\sigma|-1/2,\ \begin{cases}\operatorname{Re}(\alpha+\rho)<1;\ |\arg c|<3\pi/2\\ \operatorname{Re} c>0\end{cases},\ X(\gamma)\ \text{see 2.19.5.19}\right]$.

21. $$\int_0^\infty x^{\alpha-1}\frac{(\sqrt{x+z}\pm\sqrt{z})^\nu}{(x+z)^r} e^{-cx/2} M_{\rho,\sigma}(cx)\,dx=Y(0)$$

$[r=0$ or $1/2;\ \operatorname{Re} c>0;\ -\operatorname{Re}(2\sigma+\nu\mp\nu)-1<2\operatorname{Re}\alpha<\operatorname{Re}(2\rho-\nu)+2r]$.

$$Y(\gamma)=(\pm\nu)^{1-2r}c^{(1-\nu)/2-\alpha}z^{1/2-r}\Gamma(\alpha+\nu/2+\sigma)\times$$
$$\times\Gamma^{2\gamma^2-1}\left((2\gamma^2-1)\left(\alpha+\frac{\nu}{2}-\sigma\right)+1-\gamma^2\right)\Gamma^{-\gamma^2+\gamma+1}\left(\frac{1}{2}+(\gamma^2-\gamma-1)\times\right.$$
$$\left.\times\left(\alpha+\frac{\nu}{2}\right)-(2\gamma^2-1)\rho\right)\Gamma^{1-\gamma^2}\left[\begin{matrix}2\sigma+1\\ \rho+\sigma+1/2\end{matrix}\right]\left[\Gamma\left(\frac{1}{2}-\rho-\sigma\right)\times\right.$$
$$\left.\times\Gamma\left(\frac{1}{2}-\rho+\sigma\right)\right]^{-(\gamma^2+\gamma)/2}{}_3F_3\left(\frac{1-\nu}{2},\ \frac{1+\nu}{2},\ \frac{1-\nu}{2}-\alpha-(\gamma^2+\gamma-1)\rho;\right.$$
$$\left.\frac{3}{2}-2r,\ 1-\alpha-\frac{\nu}{2}-\sigma,\ 1-\alpha-\frac{\nu}{2}-\sigma;\ (-1)^{(\gamma^2+\gamma)/2}cz\right)+$$
$$+(\pm\nu)^{2r}c^{2r-\alpha-\nu/2}z^r\Gamma\left(\frac{1+\nu}{2}+\alpha+\sigma-2r\right)\Gamma^{2\gamma^2-1}\left((2\gamma^2-1)\times\right.$$
$$\left.\times\left(\alpha+\frac{\nu}{2}-\sigma-2r\right)+\frac{1}{2}\right)\Gamma^{-\gamma^2+\gamma+1}\left((\gamma^2-\gamma-1)\left(\alpha+\frac{\nu}{2}-2r\right)-\right.$$
$$\left.-(2\gamma^2-1)\rho+\frac{\gamma^2-\gamma}{2}\right)\Gamma^{1-\gamma^2}\left[\begin{matrix}2\sigma+1\\ \rho+\sigma+1/2\end{matrix}\right]\left[\Gamma\left(\frac{1}{2}-\rho-\sigma\right)\Gamma\left(\frac{1}{2}-\rho+\sigma\right)\right]^{-(\gamma^2+\gamma)/2}\times$$
$$\times{}_3F_3\left(2r+\frac{\nu}{2},\ 2r-\frac{\nu}{2},\ 2r-\alpha-\frac{\nu}{2}-(\gamma^2+\gamma-1)\rho;\right.$$
$$\left.2r+\frac{1}{2},\ 2r-\alpha+\frac{1-\nu}{2}-\sigma;\ 2r-\alpha+\frac{1-\nu}{2}+\sigma;\ (-1)^{(\gamma^2+\gamma)/2}cz\right)+$$
$$+2^{2\alpha+\nu+2\sigma+1}(\mp\nu)^{1-2r}c^{\sigma+1/2}z^{\alpha+(\nu+1)/2+\sigma-r}\times$$
$$\times\Gamma\left[\begin{matrix}2r-2\alpha-\nu-2\sigma-1,\ \alpha+\sigma+(\nu\mp\nu+1)/2\\ (1-\nu\mp\nu)/2-\alpha-\sigma\end{matrix}\right]\Gamma^{\gamma^2}\left[\begin{matrix}-2\sigma\\ 1/2-\rho-\sigma\end{matrix}\right]\times$$
$$\times{}_3F_3\left(\frac{1}{2}+\alpha+\sigma,\ \frac{1}{2}+\alpha+\nu+\sigma,\ \frac{1}{2}+\sigma-(\gamma^2+\gamma-1)\rho;\right.$$
$$\left.\alpha+\frac{\nu}{2}+\sigma+1,\ \frac{3+\nu}{2}+\alpha+\sigma-2r,\ 2\sigma+1;\ (-1)^{(\gamma^2+\gamma)/2}cz\right)+$$
$$+2^{2\alpha+\nu-2\sigma+1}\gamma^2(\mp\nu)^{1-2r}c^{1/2-\sigma}z^{\alpha+(\nu+1)/2-\sigma-r}\times$$
$$\times\Gamma\left[\begin{matrix}2r-\nu-2\alpha+2\sigma-1,\ \alpha-\sigma+(\nu\mp\nu+1)/2,\ 2\sigma\\ (1-\nu\mp\nu)/2-\alpha+\sigma,\ 1/2-\rho+\sigma\end{matrix}\right]\times$$
$$\times{}_3F_3\left(1/2+\alpha-\sigma,\ 1/2+\alpha+\nu-\sigma,\ 1/2-\sigma-\gamma\rho;\right.$$
$$\left.1-2\sigma,\ 1+\alpha+\frac{\nu}{2}-\sigma,\ \alpha+\frac{3+\nu}{2}-\sigma-2r;\ (-1)^{(\gamma+1)/2}cz\right).$$

22. $$\int_0^\infty x^{\alpha-1}\frac{(\sqrt{x+z}\pm\sqrt{z})^\nu}{(x+z)^r}e^{cx/2}W_{\rho,\sigma}(cx)\,dx=Y(1)$$

[$r=0$ or $1/2$; $|\arg c|<3\pi/2$; $2|\mathrm{Re}\,\sigma|-\mathrm{Re}(\nu\mp\nu)-1<2\,\mathrm{Re}\,\alpha<2r-\mathrm{Re}(\nu+2\rho)$, $Y(\gamma)$ see 2.19.5.21]

23. $$\int_0^\infty x^{\alpha-1}\frac{(\sqrt{x+z}\pm\sqrt{z})^\nu}{(x+z)^r}e^{-cx/2}W_{\rho,\sigma}(cx)\,dx=Y(-1)$$

[$r=0$ or $1/2$; $\mathrm{Re}\,c>0$; $\mathrm{Re}(2\alpha+\nu\mp\nu)>2|\mathrm{Re}\,\sigma|-1$; $Y(\gamma)$ see 2.19.5.21].

24. $$\int_0^\infty x^{\alpha-1}\frac{(\sqrt{x+z}\pm\sqrt{x})^\nu}{(x+z)^r}e^{-cx/2}M_{\rho,\sigma}(cx)\,dx=Z(0)$$

[$r=0$ or $1/2$; $\mathrm{Re}\,c>0$; $-1/2-\mathrm{Re}\,\sigma<\mathrm{Re}\,\alpha<\mathrm{Re}(\rho\mp\nu/2)+r$],

$$Z(\gamma)=2^{\pm\nu}c^{r-\alpha\mp\nu/2}z^{(\nu\mp\nu)/2}\times$$
$$\times\Gamma\left(\frac{1\pm\nu}{2}+\alpha+\sigma-r\right)\Gamma^{2\gamma^2-1}\left((2\gamma^2-1)\left(\alpha-\sigma-r\pm\frac{\nu}{2}\right)+\frac{1}{2}\right)\times$$
$$\times\Gamma^{-\gamma^2+\gamma+1}\left((\gamma^2-\gamma-1)\left(\alpha\pm\frac{\nu}{2}-r\right)-(2\gamma^2-1)\rho+\frac{\gamma^2-\gamma}{2}\right)\times$$
$$\times\Gamma^{1-\gamma^2}\begin{bmatrix}2\sigma+1\\ \rho+\sigma+1/2\end{bmatrix}\left[\Gamma\left(\frac{1}{2}-\rho-\sigma\right)\Gamma\left(\frac{1}{2}-\rho+\sigma\right)\right]^{-(\gamma^2+\gamma)/2}\times$$
$$\times{}_3F_3\left(r\mp\frac{\nu}{2},\ r+\frac{1\mp\nu}{2},\ r-\alpha\mp\frac{\nu}{2}-(\gamma^2+\gamma-1)\rho;\right.$$
$$\left.1\mp\nu,\ r+\frac{1\mp\nu}{2}-\alpha+\sigma,\ r+\frac{1\mp\nu}{2}-\alpha-\sigma;\ (-1)^{(\gamma^2+\gamma)/2}cz\right)+$$
$$+2^{2r-2\alpha-2\sigma-1}(\mp\nu)^{1-2r}c^{\sigma+1/2}z^{\alpha+\sigma+(\nu+1)/2-r}\times$$
$$\times\Gamma\begin{bmatrix}2\alpha+2\sigma+1,\ r-\alpha-\sigma-(\pm\nu+1)/2\\ (3\mp\nu)/2+\alpha+\sigma-r\end{bmatrix}\Gamma^{\gamma^2}\begin{bmatrix}-2\sigma\\ 1/2-\rho-\sigma\end{bmatrix}\times$$
$$\times{}_3F_3\left(\alpha+\sigma+\frac{1}{2},\ \alpha+\sigma+1,\ \frac{1}{2}+\sigma-(\gamma^2+\gamma-1)\rho;\ \frac{3-\nu}{2}+\alpha+\sigma-r,\right.$$
$$\left.\frac{3+\nu}{2}+\alpha+\sigma-r,\ 2\sigma+1;\ (-1)^{(\gamma^2+\gamma)/2}cz\right)+$$
$$+2^{2r-2\alpha+2\sigma-1}\gamma^2(\mp\nu)^{1-2r}c^{1/2-\sigma}z^{\alpha+(\nu+1)/2-\sigma-r}\times$$
$$\times\Gamma\begin{bmatrix}2\alpha-2\sigma+1,\ \sigma-\alpha-(1\pm\nu)/2+r,\ 2\sigma\\ (3\mp\nu)/2+\alpha-\sigma-r,\ 1/2-\rho+\sigma\end{bmatrix}\times$$
$$\times{}_3F_3\left(\frac{1}{2}+\alpha-\sigma,\ 1+\alpha-\sigma,\ \frac{1}{2}-\sigma-\gamma\rho;\ \frac{3-\nu}{2}+\alpha-\sigma-r,\right.$$
$$\left.\frac{3+\nu}{2}+\alpha-\sigma-r,\ 1-2\sigma;\ (-1)^{(\nu+1)/2}cz\right).$$

25. $$\int_0^\infty x^{\alpha-1}\frac{(\sqrt{x+z}\pm\sqrt{x})^\nu}{(x+z)^r}e^{cx/2}W_{\rho,\sigma}(cx)\,dx=Z(1)$$

[$r=0$ or $1/2$; $|\arg c|<3\pi/2$; $|\operatorname{Re}\sigma|-1/2<\operatorname{Re}\alpha<r-\operatorname{Re}(\rho\pm\nu/2)$; $Z(\gamma)$ see 2.19.5.24].

26. $$\int_0^\infty x^{\alpha-1}\frac{(\sqrt{x+z}\pm\sqrt{x})^\nu}{(x+z)^r}e^{-cx/2}W_{\rho,\sigma}(cx)\,dx=Z(-1)$$

[$r=0$ or $1/2$; $\operatorname{Re}c>0$; $\operatorname{Re}\alpha>|\operatorname{Re}\sigma|-1/2$; $Z(\gamma)$ see 2.19.5.24].

2.19.6. Integrals of $x^\alpha e^{\pm cx/2}\begin{Bmatrix}\sin bx^r\\ \cos bx^r\end{Bmatrix}\begin{Bmatrix}M_{\rho,\sigma}(cx)\\ W_{\rho,\sigma}(cx)\end{Bmatrix}$.

1. $$\int_0^\infty x^{\alpha-1}e^{-cx/2}\begin{Bmatrix}\sin bx\\ \cos bx\end{Bmatrix}M_{\rho,\sigma}(cx)\,dx=$$
$$=\frac{c^{\sigma+1/2}}{b^{\alpha+\sigma+1/2}}\Gamma\left(\alpha+\sigma+\frac{1}{2}\right)\begin{Bmatrix}\sin[(2\alpha+2\sigma+1)\pi/4]\\ \cos[(2\alpha+2\sigma+1)\pi/4]\end{Bmatrix}\times$$
$$\times{}_4F_3\left(\frac{2\rho+2\sigma+1}{4},\frac{2\rho+2\sigma+3}{4},\frac{2\alpha+2\sigma+1}{4},\frac{2\alpha+2\sigma+3}{4};\sigma+\frac{1}{2},\sigma+1,\frac{1}{2};-\frac{c^2}{b^2}\right)\mp$$
$$\mp\frac{c^{\sigma+3/2}(1/2+\rho+\sigma)}{b^{\alpha+\sigma+3/2}(2\sigma+1)}\Gamma\left(\alpha+\sigma+\frac{3}{2}\right)\begin{Bmatrix}\cos[(2\alpha+2\sigma+1)\pi/4]\\ \sin[(2\alpha+2\sigma+1)\pi/4]\end{Bmatrix}\times$$
$$\times{}_4F_3\left(\frac{2\alpha+2\sigma+3}{4},\frac{2\alpha+2\sigma+5}{4},\frac{2\rho+2\sigma+3}{4},\frac{2\rho+2\sigma+5}{4};\sigma+1,\ \sigma+\frac{3}{2},\frac{3}{2};-\frac{c^2}{b^2}\right)$$

[b, $\operatorname{Re}c>0$; $-1\mp1/2-\operatorname{Re}\sigma<\operatorname{Re}\alpha<1+\operatorname{Re}\rho$].

2. $\int_0^\infty x^{\alpha-1}e^{cx/2}\begin{Bmatrix}\sin bx\\ \cos bx\end{Bmatrix} W_{\rho,\sigma}(cx)\,dx=U(1)$

$\left[b>0;\ |\operatorname{Re}\sigma|-1\mp1/2<\operatorname{Re}\alpha<1-\operatorname{Re}\rho;\ |\arg c|<3\pi/2;\ \delta=\begin{Bmatrix}1\\0\end{Bmatrix}\right].$

$$U(\gamma)=\frac{b^\delta}{c^{\alpha+\delta}}\left[\Gamma\left(\frac{1}{2}-\rho-\sigma\right)\Gamma\left(\frac{1}{2}-\rho+\sigma\right)\right]^{-(\gamma+1)/2}\Gamma\left(\frac{1}{2}+\alpha+\sigma+\delta\right)\times$$
$$\times\Gamma\left(\frac{1}{2}+\alpha-\sigma+\delta\right)\Gamma^\gamma\left(\frac{1-\gamma}{2}-\rho-\gamma(\alpha+\delta)\right){}_4F_3\left(\frac{3+2\alpha+2\sigma}{4},\ \frac{1+2\alpha+2\sigma}{4}+\delta,\right.$$
$$\left.\frac{3+2\alpha-2\sigma}{4},\frac{1+2\alpha-2\sigma}{4}+\delta;\ \delta+\frac{1}{2},1+\frac{\alpha+\rho\gamma}{2},\frac{1+\alpha+\rho\gamma}{2}+\delta;\ -\frac{b^2}{c^2}\right)+$$
$$+\frac{(\gamma+1)c^\rho}{2b^{\alpha+\rho}}\Gamma(\alpha+\rho)\begin{Bmatrix}\sin[(\alpha+\rho)\pi/2]\\ \cos[(\alpha+\rho)\pi/2]\end{Bmatrix}\times$$
$$\times{}_4F_3\left(\frac{1-2\rho+2\sigma}{4},\ \frac{3-2\rho+2\sigma}{4},\ \frac{1-2\rho-2\sigma}{4},\ \frac{3-2\rho-2\sigma}{4};\ \frac{1}{2},1-\frac{\alpha+\rho}{2},\frac{1-\alpha-\rho}{2};\right.$$
$$\left.-\frac{b^2}{c^2}\right)\pm\frac{\gamma+1}{2}\frac{c^{\rho-1}}{b^{\alpha+\rho-1}}\left[\left(\frac{1}{2}-\rho\right)^2-\sigma^2\right]\Gamma(\alpha+\rho-1)\begin{Bmatrix}\cos[(\alpha+\rho)\pi/2]\\ \sin[(\alpha+\rho)\pi/2]\end{Bmatrix}\times$$
$$\times{}_4F_3\left(\frac{3-2\rho+2\sigma}{4},\frac{5-2\rho+2\sigma}{4},\frac{3-2\rho-2\sigma}{4},\frac{5-2\rho-2\sigma}{4};\frac{3}{2},\frac{3-\alpha-\rho}{2},\right.$$
$$\left.1-\frac{\alpha+\rho}{2};\ -\frac{b^2}{c^2}\right).$$

3. $\int_0^\infty x^{\alpha-1}e^{-cx/2}\begin{Bmatrix}\sin bx\\ \cos bx\end{Bmatrix} W_{\rho,\sigma}(cx)\,dx=U(-1)$

[b, $\operatorname{Re}c>0$; $\operatorname{Re}\alpha>|\operatorname{Re}\sigma|-1\mp1/2$; $U(\gamma)$ see 2.19.6.2].

4. $\int_0^\infty x^{\alpha-1}e^{-cx/2}\begin{Bmatrix}\sin b\sqrt{x}\\ \cos b\sqrt{x}\end{Bmatrix} M_{\rho,\sigma}(cx)\,dx=V(0)$

$\left[b,\ \operatorname{Re}c>0;\ -\operatorname{Re}\sigma-(\delta+1)/2<\operatorname{Re}\alpha<\operatorname{Re}\rho+1/2,\ \delta=\begin{Bmatrix}1\\0\end{Bmatrix}\right].$

$$V(\gamma)=\frac{b^\delta}{c^{\alpha+\delta/2}}\Gamma^{1-\gamma^2}\begin{bmatrix}2\sigma+1\\ \rho+\sigma+1/2\end{bmatrix}\left[\Gamma\left(\frac{1}{2}-\rho-\sigma\right)\Gamma\left(\frac{1}{2}-\rho+\sigma\right)\right]^{-(\gamma^2+\gamma)/2}\times$$
$$\times\Gamma\left(\frac{1+\delta}{2}+\alpha+\sigma\right)\Gamma^{2\gamma^2-1}\left(\frac{1}{2}+(1-2\gamma^2)\left(\sigma-\alpha-\frac{\delta}{2}\right)\right)\times$$
$$\times\Gamma^{-\gamma^2+\gamma+1}\left((1-2\gamma^2)\rho+\frac{\gamma^2-\gamma}{2}+(\gamma^2-\gamma-1)\left(\alpha+\frac{\delta}{2}\right)\right)\times$$
$$\times{}_2F_2\left(\frac{1+\delta}{2}+\alpha+\sigma,\frac{1+\delta}{2}+\alpha-\sigma;\ 1+\alpha+\frac{\delta}{2}+(\gamma^2+\gamma-1)\rho,\ \delta+\frac{1}{2};\right.$$
$$\left.(-1)^{1-(\gamma^2+\gamma)/2}\frac{b^2}{4c}\right)-\frac{\gamma^2-\gamma-2}{c^{(1-2\gamma)\rho}}b^{2(1-2\gamma)\rho-2\alpha}\times$$
$$\times\Gamma^{1-\gamma}\begin{bmatrix}2\sigma+1\\ 1/2-\rho+\sigma\end{bmatrix}\Gamma(2\alpha-2(1-2\gamma)\rho)\begin{Bmatrix}\sin(\alpha-(1-2\gamma)\rho)\pi\\ \cos(\alpha-(1-2\gamma)\rho)\pi\end{Bmatrix}\times$$
$$\times{}_2F_2\left(\frac{1}{2}+\sigma+(1-2\gamma)\rho,\frac{1}{2}-\sigma+(1-2\gamma)\rho;\right.$$
$$\left.1-\alpha+(1-2\gamma)\rho,\ \frac{1}{2}-\alpha+(1-2\gamma)\rho;\ (-1)^{\gamma+1}\frac{b^2}{4c}\right).$$

5. $\int_0^\infty x^{-\sigma-1/2}e^{-cx/2}\sin b\sqrt{x}\, M_{\rho,\sigma}(cx)\,dx=$

$$=2^{1-\rho-\sigma}\sqrt{\pi}\,b^{\rho+\sigma-1}c^{(\sigma-\rho)/2}\Gamma\begin{bmatrix}3-2\sigma\\ \rho+\sigma+1/2\end{bmatrix}\exp\left(-\frac{b^2}{8c}\right)W_{(\rho-3\sigma+1)/2,\,(\rho+\sigma-1)/2}\left(\frac{b^2}{4c}\right)$$

[b, Re c, Re $(\rho+\sigma)>0$].

6. $\int_0^\infty x^{\sigma-1}e^{-cx/2}\sin b\sqrt{x}\, M_{3\sigma,\sigma}(cx)\,dx=$

$$=\sqrt{\pi}\left(\frac{b}{2}\right)^{2\sigma-1}c^{1/2-2\sigma}\exp\left(-\frac{b^2}{8c}\right)M_{3\sigma,\sigma}\left(\frac{b^2}{4c}\right)$$

[b, Re $c>0$; Re $\sigma>-1/4$].

7. $\int_0^\infty x^{-\sigma-1}e^{-cx/2}\cos b\sqrt{x}\, M_{\rho,\sigma}(cx)\,dx=$

$$=\sqrt{\pi}\left(\frac{b}{2}\right)^{\rho+\sigma-1}c^{(1-\rho+\sigma)/2}\Gamma\begin{bmatrix}2\sigma+1\\ \rho+\sigma+1/2\end{bmatrix}\exp\left(-\frac{b^2}{8c}\right)W_{(\rho-3\sigma)/2,\,(\rho+\sigma)/2}\left(\frac{b^2}{4c}\right)$$

[b, Re $c>0$; Re$(\rho+\sigma)>-1/2$].

8. $\int_0^\infty x^{\alpha-1}e^{cx/2}\begin{Bmatrix}\sin b\sqrt{x}\\ \cos b\sqrt{x}\end{Bmatrix}W_{\rho,\sigma}(cx)\,dx=V(1)$

[$b>0$; $|\text{Re}\,\sigma|-1\mp1/2<\text{Re}\,\alpha<1/2-\text{Re}\,\rho$; $|\arg c|<3\pi/2$; $V(\gamma)$ see 2.19.6.4].

9. $\int_0^\infty x^{\alpha-1}e^{-cx/2}\begin{Bmatrix}\sin b\sqrt{x}\\ \cos b\sqrt{x}\end{Bmatrix}W_{\rho,\sigma}(cx)\,dx=V(-1)$

[b, Re $c>0$; Re $\alpha>|\text{Re}\,\sigma|-1\mp1/2$; $V(\gamma)$ see 2.19.6.4].

10. $\int_0^\infty x^{-\sigma+\delta/2-1}e^{cx/2}\begin{Bmatrix}\sin b\sqrt{x}\\ \cos b\sqrt{x}\end{Bmatrix}W_{3\sigma-\delta,\sigma}(cx)\,dx=$

$$=2^{2\sigma-\delta+1}\sqrt{\pi}\,b^{-2\sigma+\delta-1}c^{2\sigma-\delta+1/2}\exp\left(\frac{b^2}{8c}\right)W_{3\sigma-\delta,\,\sigma}\left(\frac{b^2}{4c}\right)$$

$\left[b>0;\ \text{Re}\,\sigma<(\delta+1)/4;\ |\arg c|<3\pi/2;\ \delta=\begin{Bmatrix}1\\0\end{Bmatrix}\right]$.

11. $\int_0^\infty x^{\alpha-1}e^{-cx/2}\begin{Bmatrix}\sin(b/\sqrt{x})\\ \cos(b/\sqrt{x})\end{Bmatrix}M_{\rho,\sigma}(cx)\,dx=W(0)$

$\left[b,\ \text{Re}\,c>0;\ -\text{Re}\,\sigma-1<\text{Re}\,\alpha<\text{Re}\,\rho+\delta/2,\ \delta=\begin{Bmatrix}1\\0\end{Bmatrix}\right]$

$$W(\gamma)=-2b^{2\alpha+2\sigma+1}c^{\sigma+1/2}\begin{Bmatrix}\cos(\alpha+\sigma)\pi\\ \sin(\alpha+\sigma)\pi\end{Bmatrix}\Gamma(-2\alpha-2\sigma-1)\,\Gamma^{\gamma^2}\begin{bmatrix}-2\sigma\\ 1/2-\rho-\sigma\end{bmatrix}\times$$

$$\times{}_1F_3\left(\frac{1}{2}+\sigma-(\gamma^2+\gamma-1)\rho;\ 2\sigma+1,\ \alpha+\sigma+1,\ \alpha+\sigma+\frac{3}{2};\ (-1)^{(\gamma^2+\gamma)/2}\frac{b^2c}{4}\right)-$$

$$-2\gamma^2b^{2\alpha-2\sigma+1}c^{1/2-\sigma}\begin{Bmatrix}\cos(\alpha-\sigma)\pi\\ \sin(\alpha-\sigma)\pi\end{Bmatrix}\Gamma\begin{bmatrix}2\sigma,\ 2\sigma-2\alpha-1\\ 1/2-\rho+\sigma\end{bmatrix}\times$$

$$\times{}_1F_3\left(\frac{1}{2}-\sigma-\gamma\rho;\ 1-2\sigma,\ \alpha-\sigma+1,\ \alpha-\sigma+\frac{3}{2};\ (-1)^{(1+\gamma)/2}\frac{b^2c}{4}\right)+$$

$$+b^{\delta}c^{\delta/2-\alpha}\Gamma^{1-\gamma^2}\begin{bmatrix}2\sigma+1\\ \rho+\sigma+1/2\end{bmatrix}\left[\Gamma\left(\frac{1}{2}-\rho-\sigma\right)\Gamma\left(\frac{1}{2}-\rho+\sigma\right)\right]^{-(\gamma^2+\gamma)/2}\times$$

$$\times\Gamma\left(\frac{1-\delta}{2}+\alpha+\sigma\right)\Gamma^{2\gamma^2-1}\left(\frac{1}{2}+(1-2\gamma^2)\left(\sigma-\alpha+\frac{\delta}{2}\right)\right)\times$$

$$\times\Gamma^{-\gamma^2+\gamma+2}\left((1-2\gamma^2)\rho+\frac{\gamma^2-\gamma}{2}+(\gamma^2-\gamma-1)\left(\alpha-\frac{\delta}{2}\right)\right)\times$$

$$\times{}_1F_3\left(\frac{\delta}{2}-\alpha-(\gamma^2+\gamma-1)\rho;\ \frac{1+\delta}{2}-\alpha+\sigma,\ \frac{1+\delta}{2}-\alpha-\sigma,\ \delta+\frac{1}{2};\ (-1)^{(\gamma^2+\gamma)/2}\frac{b^2c}{4}\right).$$

12. $\int\limits_0^\infty x^{\alpha-1}e^{cx/2}\begin{Bmatrix}\sin(b/\sqrt{x})\\ \cos(b/\sqrt{x})\end{Bmatrix}W_{\rho,\sigma}(cx)\,dx=W(1)$

[$b>0$; $|\operatorname{Re}\sigma|-1<\operatorname{Re}\alpha<\operatorname{Re}\rho+\delta/2$; $|\arg c|<3\pi/2$; $W(\gamma)$ see 2.19.6.11].

13. $\int\limits_0^\infty x^{\alpha-1}e^{-cx/2}\begin{Bmatrix}\sin(b/\sqrt{x})\\ \cos(b/\sqrt{x})\end{Bmatrix}W_{\rho,\sigma}(cx)\,dx=W(-1)$

[b, $\operatorname{Re}c>0$; $\operatorname{Re}\alpha>|\operatorname{Re}\sigma|-1$; $W(\gamma)$ see 2.19.6.11].

2.19.7. Integrals of $x^\alpha e^{\pm cx/2}\ln\varphi(x)\begin{Bmatrix}M_{\rho,\sigma}(cx)\\ W_{\rho,\sigma}(cx)\end{Bmatrix}$.

1. $\int\limits_0^\infty x^{\alpha-1}e^{-cx/2}\ln x\, M_{\rho,\sigma}(cx)\,dx=A(0)$ [$\operatorname{Re}c>0$; $-1/2-\operatorname{Re}\sigma<\operatorname{Re}\alpha<\operatorname{Re}\rho$].

$$A(\gamma)=c^{-\alpha}\Gamma^{1-\gamma^2}\begin{bmatrix}2\sigma+1\\ \rho+\sigma+1/2\end{bmatrix}\left[\Gamma\left(\frac{1}{2}-\rho-\sigma\right)\Gamma\left(\frac{1}{2}-\rho+\sigma\right)\right]^{-(\gamma^2+\gamma)/2}\times$$

$$\times\Gamma\left(\frac{1}{2}+\alpha+\sigma\right)\Gamma^{2\gamma^2-1}\left(\frac{1}{2}+(1-2\gamma^2)(\sigma-\alpha)\right)\times$$

$$\times\Gamma^{-\gamma^2+\gamma+1}\left((1-2\gamma^2)\rho+\frac{\gamma^2-\gamma}{2}+(\gamma^2-\gamma-1)\alpha\right)\left[\psi\left(\frac{1}{2}+\alpha+\sigma\right)+\right.$$

$$\left.+\psi\left(\frac{1}{2}+(1-2\gamma^2)(\sigma-\alpha)\right)-\psi\left((1-2\gamma^2)\rho+\frac{\gamma^2-\gamma}{2}+(\gamma^2-\gamma-1)\alpha\right)-\ln c\right].$$

2. $\int\limits_0^\infty x^{\alpha-1}e^{\pm cx/2}\ln x\, W_{\rho,\sigma}(cx)\,dx=A(\pm 1)$

$\left[\operatorname{Re}\alpha>|\operatorname{Re}\sigma|-1/2,\ \begin{Bmatrix}\operatorname{Re}(\alpha+\rho)<0;\ |\arg c|<3\pi/2\\ \operatorname{Re}c>0\end{Bmatrix},\ A(\gamma)\text{ see 2.19.7.1}\right].$

3. $\int\limits_0^\infty x^{\alpha-1}e^{-cx/2}\begin{Bmatrix}\ln(x+z)\\ \ln|x-z|\end{Bmatrix}M_{\rho,\sigma}(cx)\,dx=V(0)$

$\left[\operatorname{Re}c>0;\ -1/2-\operatorname{Re}\sigma<\operatorname{Re}\alpha<\operatorname{Re}\rho,\ \begin{Bmatrix}|\arg z|<\pi\\ z>0\end{Bmatrix}\right].$

$$V(\gamma)=\pm\frac{\pi c^{\sigma+1/2}z^{\alpha+\sigma+1/2}}{\alpha+\sigma+1/2}\Gamma^{\gamma^2}\begin{bmatrix}-2\sigma\\ 1/2-\rho-\sigma\end{bmatrix}\begin{Bmatrix}\sec(\alpha+\sigma)\pi\\ \operatorname{tg}(\alpha+\sigma)\pi\end{Bmatrix}\times$$

$$\times{}_2F_2\left(\frac{1}{2}+\sigma-(\gamma^2+\gamma-1)\rho,\ \alpha+\sigma+\frac{1}{2};\ 2\sigma+1,\ \alpha+\sigma+\frac{3}{2};\ \pm(-1)^{(\gamma^2+\gamma)/2}cz\right)\pm$$

$$\pm\frac{\gamma^2\pi c^{1/2-\sigma}}{\alpha-\sigma+1/2}z^{\alpha-\sigma+1/2}\Gamma\begin{bmatrix}2\sigma\\ 1/2-\rho+\sigma\end{bmatrix}\begin{Bmatrix}\sec(\alpha-\sigma)\pi\\ \operatorname{tg}(\alpha-\sigma)\pi\end{Bmatrix}\times$$

$$\times{}_2F_2\left(\frac{1}{2}-\sigma-\gamma\rho,\ \alpha-\sigma+\frac{1}{2};\ \alpha-\sigma+\frac{3}{2},\ 1-2\sigma;\ \pm(-1)^{(\gamma+1)/2}cz\right)\pm$$

$$\pm c^{1-\alpha}z\Gamma^{1-\gamma^2}\begin{bmatrix}2\sigma+1\\ \rho+\sigma+1/2\end{bmatrix}\left[\Gamma\left(\frac{1}{2}-\rho-\sigma\right)\Gamma\left(\frac{1}{2}-\rho+\sigma\right)\right]^{-(\gamma^2+\gamma)/2}\times$$

$$\times\Gamma\left(\alpha+\sigma-\frac{1}{2}\right)\Gamma^{2\gamma^2-1}\left(\frac{1}{2}+(1-2\gamma^2)(\sigma-\alpha+1)\right)\times$$

$$\times\Gamma^{-\gamma^2+\gamma+1}\left((1-2\gamma^2)\rho+\frac{\gamma^2-\gamma}{2}+(\gamma^2-\gamma-1)(\alpha-1)\right)\times$$

$$\times {}_3F_3\left(1,\ 1,\ 1-\alpha-(\gamma^2+\gamma-1)\rho;\ 2,\ \frac{3}{2}-\alpha+\sigma,\ \frac{3}{2}-\alpha-\sigma;\ \mp(-1)^{(\gamma^2+\gamma)/2}cz\right)+$$
$$+c^{-\alpha}\Gamma^{1-\gamma^2}\left[\begin{matrix}2\sigma+1\\ \rho+\sigma+1/2\end{matrix}\right]\left[\Gamma\left(\frac{1}{2}-\rho-\sigma\right)\Gamma\left(\frac{1}{2}-\rho+\sigma\right)\right]^{-(\gamma^2+\gamma)/2}\times$$
$$\times\Gamma\left(\frac{1}{2}+\alpha+\sigma\right)\Gamma^{2\gamma^2-1}\left(\frac{1}{2}+(1-2\gamma^2)(\sigma-\alpha)\right)\times$$
$$\times\Gamma^{-\gamma^2+\gamma+1}\left((1-2\gamma^2)\rho+\frac{\gamma^2-\gamma}{2}+(\gamma^2-\gamma-1)\alpha\right)\times$$
$$\times\left[\psi\left(\frac{1}{2}+\alpha+\sigma\right)+\psi\left(\frac{1}{2}+(1-2\gamma^2)(\sigma-\alpha)\right)-\right.$$
$$\left.-\psi\left((1-2\gamma^2)\rho+\frac{\gamma^2-\gamma}{2}+(\gamma^2-\gamma-1)\alpha\right)-\ln c\right].$$

4. $$\int_0^\infty x^{\alpha-1}e^{cx/2}\left\{\begin{matrix}\ln(x+z)\\ \ln|x-z|\end{matrix}\right\}W_{\rho,\sigma}(cx)\,dx=V(1)$$

$\left[|\operatorname{Re}\sigma|-1/2<\operatorname{Re}\alpha<-\operatorname{Re}\rho;\ |\arg c|<3\pi/2,\ \left\{\begin{matrix}|\arg z|<\pi\\ z>0\end{matrix}\right\},\ V(\gamma)\ \text{see 2.19.7.3}\right].$

5. $$\int_0^\infty x^{\alpha-1}e^{-cx/2}\left\{\begin{matrix}\ln(x+z)\\ \ln|x-z|\end{matrix}\right\}W_{\rho,\sigma}(cx)\,dx=V(-1)$$

$\left[\operatorname{Re}c>0;\ \operatorname{Re}\alpha>|\operatorname{Re}\sigma|-1/2,\ \left\{\begin{matrix}|\arg z|<\pi\\ z>0\end{matrix}\right\},\ V(\gamma)\ \text{see 2.19.7.3}\right].$

2.19.8. Integrals of $x^\alpha e^{\pm cx/2}\operatorname{Ei}(-bx)\left\{\begin{matrix}M_{\rho,\sigma}(cx)\\ W_{\rho,\sigma}(cx)\end{matrix}\right\}$.

1. $$\int_0^\infty x^{\alpha-1}e^{-cx/2}\operatorname{Ei}(-bx)M_{\rho,\sigma}(cx)\,dx=W(0)$$ $[\operatorname{Re}b,\ \operatorname{Re}c>0;\ \operatorname{Re}(\alpha+\sigma)>-1/2],$

$$W(\gamma)=-\frac{c^{\sigma+1/2}}{b^{\alpha+\sigma+1/2}(\alpha+\sigma+1/2)}\Gamma^{\gamma^2}\left[\begin{matrix}-2\sigma\\ 1/2-\rho-\sigma\end{matrix}\right]\Gamma\left(\alpha+\sigma+\frac{1}{2}\right)\times$$
$$\times{}_3F_2\left(\frac{1}{2}+\sigma-(\gamma^2+\gamma-1)\rho,\ \alpha+\sigma+\frac{1}{2},\ \alpha+\sigma+\frac{1}{2};\ \alpha+\sigma+\frac{3}{2},\ 2\sigma+1;\right.$$
$$\left.(-1)^{1-(\gamma^2+\gamma)/2}\frac{c}{b}\right)-\frac{\gamma^2c^{1/2-\sigma}}{b^{\alpha-\sigma+1/2}(\alpha-\sigma+1/2)}\Gamma\left[\begin{matrix}2\sigma,\ 1/2+\alpha-\sigma\\ 1/2-\rho+\sigma\end{matrix}\right]\times$$
$$\times{}_3F_2\left(\frac{1}{2}\mp\rho-\sigma,\ \alpha-\sigma+\frac{1}{2},\ \alpha-\sigma+\frac{1}{2};\ 1-2\sigma,\ \alpha-\sigma+\frac{3}{2};\ \pm\frac{c}{b}\right).$$

2. $$\int_0^\infty x^{\alpha-1}e^{\pm cx/2}\operatorname{Ei}(-bx)W_{\rho,\sigma}(cx)\,dx=W(\pm1)$$

$[\operatorname{Re}b>0;\ \operatorname{Re}\alpha>|\operatorname{Re}\sigma|-1/2;\ |\arg c|<(1\pm1/2)\pi;\ W(\gamma)\ \text{see 2.19.8.1}].$

3. $$\int_0^\infty x^{\sigma-1/2}e^{(b-c/2)x}\operatorname{Ei}(-bx)M_{\rho,\sigma}(cx)\,dx=-c^{-\sigma-1/2}\Gamma(2\sigma+1)\times$$
$$\times\mathrm{B}\left(2\sigma+1,\ \frac{1}{2}+\rho-\sigma\right){}_2F_1\left(1,\ 2\sigma+1;\ \frac{3}{2}+\rho+\sigma;\ 1-\frac{c}{b}\right)$$
$[\operatorname{Re}c>0;\ -1/2<\operatorname{Re}\sigma<\operatorname{Re}\rho+1/2;\ |\arg b|<3\pi/2].$

2.19.9. Integrals of $x^\alpha e^{px}\left\{\begin{matrix}\operatorname{erf}(b\sqrt{a\pm x})\\ \operatorname{erfc}(b\sqrt{a\pm x})\end{matrix}\right\}\left\{\begin{matrix}M_{\rho,\sigma}(cx)\\ W_{\rho,\sigma}(cx)\end{matrix}\right\}$.

1. $$\int_0^\infty x^{-3/2}e^{-px}\operatorname{erf}(b\sqrt{x})\,M_{\rho,\,1/4}(cx)\,dx=$$
$$=\frac{b\,(2c)^{3/2}\,(2p+2b^2+c)^{1/4-\rho}}{(2p+c)\,(2p+2b^2-c)^{3/4-\rho}}\,{}_2F_1\left(1,\ \frac{3}{4}-\rho;\ \frac{3}{2};\ \frac{4b^2c}{(2p+c)\,(2p+2b^2-c)}\right)$$
$[2\operatorname{Re}p>|\operatorname{Re}c|;\ |\arg b|<\pi/4]$.

2. $$\int_0^\infty x^{\alpha-1}e^{-cx/2}\operatorname{erf}(b\sqrt{x})\,M_{\rho,\,\sigma}(cx)\,dx=-\frac{2c^{\sigma+1/2}\Gamma(\alpha+\sigma+1)}{\sqrt{\pi}\,b^{2\alpha+2\sigma+1}\,(2\alpha+2\sigma+1)}\times$$
$$\times\,{}_3F_2\left(\rho+\sigma+\frac{1}{2},\ \alpha+\sigma+\frac{1}{2},\ \alpha+\sigma+1;\ \alpha+\sigma+\frac{3}{2},\ 2\sigma+1;\ -\frac{c}{b^2}\right)+$$
$$+c^{-\alpha}\Gamma\begin{bmatrix}2\sigma+1,\ 1/2+\alpha+\sigma,\ \rho-\alpha\\ 1/2+\rho+\sigma,\ 1/2-\alpha+\sigma\end{bmatrix}$$
$[\operatorname{Re}c>0;\ -1-\operatorname{Re}\sigma<\operatorname{Re}\alpha<\operatorname{Re}\rho;\ |\arg b|<\pi/4]$.

3. $$\int_0^\infty x^{\sigma-1}e^{-cx/2}\operatorname{erf}(b\sqrt{x})\,M_{-\sigma,\,\sigma}(cx)\,dx=$$
$$=\frac{b^{4\sigma}\,\Gamma(2\sigma)}{\sqrt{\pi}c^{\sigma}\,(b^2+c)^{2\sigma}}\,{}_2F_1\left(\frac{1}{2}-2\sigma,\ 2\sigma;\ 2\sigma+1;\ \frac{b^2}{b^2+c}\right)$$
$[\operatorname{Re}c,\ \operatorname{Re}\sigma>0;\ |\arg b|<\pi/4]$.

4. $$\int_0^\infty x^{\alpha-1}e^{\pm cx/2}\operatorname{erfc}(b\sqrt{x})\,M_{\rho,\,\sigma}(cx)\,dx=\frac{2c^{\sigma+1/2}\Gamma(\alpha+\sigma+1)}{\sqrt{\pi}\,b^{2\alpha+2\sigma+1}\,(2\alpha+2\sigma+1)}\times$$
$$\times\,{}_3F_2\left(\frac{1}{2}\mp\rho+\sigma,\ \alpha+\sigma+\frac{1}{2},\ \alpha+\sigma+1;\ \alpha+\sigma+\frac{3}{2},\ 2\sigma+1;\ \pm\frac{c}{b^2}\right)$$
$\left[\operatorname{Re}(\alpha+\sigma)>1/2;\ |\arg b|<\pi/4,\ \begin{Bmatrix}\operatorname{Re}(b^2-c)>0 \text{ or } b=\sqrt{c},\ \operatorname{Re}(\alpha-\rho)<1/2\\ \operatorname{Re}c>0\end{Bmatrix}\right]$.

5. $$\int_0^\infty x^{\alpha-1}e^{cx/2}\begin{Bmatrix}\operatorname{erf}(b\sqrt{x})\\ \operatorname{erfc}(b\sqrt{x})\end{Bmatrix}W_{\rho,\,\sigma}(cx)\,dx=X(1)$$
$\left[\operatorname{Re}\alpha>|\operatorname{Re}\sigma|+(\mp 3-1)/4;\ |\arg b|<\pi/4;\ \begin{Bmatrix}\operatorname{Re}(\alpha+\rho)<0;\ |\arg c|<3\pi/2\\ |\arg c|<3\pi/2\end{Bmatrix}\right]$,

$$X(\gamma)=\pm\frac{2b}{\sqrt{\pi}\,c^{\alpha+1/2}}\left[\Gamma\left(\frac{1}{2}-\rho-\sigma\right)\Gamma\left(\frac{1}{2}-\rho+\sigma\right)\right]^{-(\gamma+1)/2}\times$$
$$\times\,\Gamma(1+\alpha+\sigma)\,\Gamma(1+\alpha-\sigma)\,\Gamma^{\gamma}\left(\frac{1-\gamma}{2}-\rho-\gamma\left(\alpha+\frac{1}{2}\right)\right)\times$$
$$\times\,{}_3F_2\left(1+\alpha+\sigma,\ 1+\alpha-\sigma,\ \frac{1}{2};\ \frac{3}{2},\ \frac{3}{2}+\alpha+\gamma\rho;\ (-1)^{(1-\gamma)/2}\frac{b^2}{c}\right)\pm$$
$$\pm\frac{(\gamma+1)\,c^{\rho}}{2\sqrt{\pi}\,b^{2\alpha+2\rho}\,(\rho-\alpha)}\,\Gamma\left(\frac{1}{2}+\alpha+\rho\right)\times$$
$$\times\,{}_3F_2\left(\frac{1}{2}-\rho+\sigma,\ \frac{1}{2}-\rho-\sigma,\ \rho-\alpha;\ 1-\alpha+\rho,\ \frac{1}{2}-\alpha-\rho;\ \frac{b^2}{c}\right)+$$
$$+\begin{Bmatrix}0\\1\end{Bmatrix}c^{-\alpha}\left[\Gamma\left(\frac{1}{2}-\rho-\sigma\right)\Gamma\left(\frac{1}{2}-\rho+\sigma\right)\right]^{-(\gamma+1)/2}\times$$
$$\times\,\Gamma\left(\frac{1}{2}+\alpha+\sigma\right)\Gamma\left(\frac{1}{2}+\alpha-\sigma\right)\Gamma^{\gamma}\cdot\left(\frac{1-\gamma}{2}-\rho-\gamma\alpha\right).$$

6. $\int_0^\infty x^{\alpha-1}e^{-cx/2}\begin{Bmatrix}\operatorname{erf}(b\sqrt{x})\\ \operatorname{erfc}(b\sqrt{x})\end{Bmatrix}W_{\rho,\sigma}(cx)\,dx=X(-1)$

[$\operatorname{Re} c>0$; $\operatorname{Re}\alpha>|\operatorname{Re}\sigma|+(\mp 3-1)/4$; $|\arg b|<\pi/4$; $X(\gamma)$ see 2.19.9.5].

7. $\int_0^\infty x^{\alpha-1}e^{(\mp b^2-c/2)x}\begin{Bmatrix}\operatorname{erfi}(b\sqrt{x})\\ \operatorname{erfc}(b\sqrt{x})\end{Bmatrix}M_{\rho,\sigma}(cx)\,dx=B(0)$

[$\operatorname{Re} c>0$; $-\operatorname{Re}\sigma-(3\pm 1)/4<\operatorname{Re}\alpha<\operatorname{Re}\rho+1/2$; $|\arg b|<(2\mp 1)\pi/4$].

$$B(\gamma)=\pm\frac{2b}{\sqrt{\pi}\,c^{\alpha+1/2}}\Gamma^{1-\gamma^2}\left[\begin{matrix}2\sigma+1\\ \rho+\sigma+1/2\end{matrix}\right]\left[\Gamma\left(\frac{1}{2}-\rho-\sigma\right)\Gamma\left(\frac{1}{2}-\rho+\sigma\right)\right]^{-(\gamma^2+\gamma)/2}\times$$
$$\times\Gamma(\alpha+\sigma+1)\,\Gamma^{2\gamma^2-1}(\gamma^2+(1-2\gamma^2)(\sigma-\alpha))\times$$
$$\times\Gamma^{-\gamma^2+\gamma+1}\left((1-2\gamma^2)\rho+\gamma^2-\gamma-\frac{1}{2}+(\gamma^2-\gamma-1)\alpha\right)\times$$
$$\times{}_3F_2\left(1,\alpha+\sigma+1,\alpha-\sigma+1;\frac{3}{2},\frac{3}{2}+\alpha+(\gamma^2+\gamma-1)\rho;\ \mp(-1)^{(\gamma^2+\gamma)/2}\frac{b^2}{c}\right)-$$
$$-\frac{\gamma^2-\gamma-2}{2c^{(1-2\gamma)\rho}}b^{2(1-2\gamma)\rho-2\alpha}\Gamma^{1-\gamma}\left[\begin{matrix}2\sigma+1\\ 1/2-\rho+\sigma\end{matrix}\right]\times$$
$$\times\Gamma(\alpha-(1-2\gamma)\rho)\begin{Bmatrix}\operatorname{tg}(\alpha-(1-2\gamma)\rho)\pi\\ \sec(\alpha-(1-2\gamma)\rho)\pi\end{Bmatrix}\times$$
$$\times{}_3F_2\left(\frac{1}{2}+\sigma+(1-2\gamma)\rho,\ \frac{1}{2}-\sigma+(1-2\gamma)\rho,\ \frac{1}{2}-\alpha+(1-2\gamma)\rho;\right.$$
$$\left.1-\alpha+(1-2\gamma)\rho,\ \frac{1}{2}-\alpha+(1-2\gamma)\rho;\ \mp(-1)^{\gamma}\frac{b^2}{c}\right)+$$
$$+\begin{Bmatrix}0\\ 1\end{Bmatrix}c^{-\alpha}\Gamma^{1-\gamma^2}\left[\begin{matrix}2\sigma+1\\ \rho+\sigma+1/2\end{matrix}\right]\left[\Gamma\left(\frac{1}{2}-\rho-\sigma\right)\Gamma\left(\frac{1}{2}-\rho+\sigma\right)\right]^{-(\gamma^2+\gamma)/2}\times$$
$$\times\Gamma\left(\frac{1}{2}+\alpha+\sigma\right)\Gamma^{2\gamma^2-1}\left(\frac{1}{2}+(1-2\gamma^2)(\sigma-\alpha)\right)\times$$
$$\times\Gamma^{-\gamma^2+\gamma+1}\left((1-2\gamma^2)\rho+\frac{\gamma^2-\gamma}{2}+(\gamma^2-\gamma-1)\alpha\right)\times$$
$$\times{}_2F_1\left(\alpha+\sigma+\frac{1}{2},\ \alpha-\sigma+\frac{1}{2};\ 1+\alpha+(\gamma^2+\gamma-1)\rho;\ (-1)^{(\gamma^2+\gamma)/2}\frac{b^2}{c}\right).$$

8. $\int_0^\infty x^{\sigma-1}e^{(b^2-c/2)x}\operatorname{erfc}(b\sqrt{x})\,M_{\rho,\sigma}(cx)\,dx=$

$$=\frac{c^{\sigma+1/2}}{2^{4\sigma}b^{4\sigma+1}}\Gamma\left[\begin{matrix}4\sigma+1,\ \rho-\sigma+1/2\\ \rho+\sigma+1\end{matrix}\right]{}_2F_1\left(\rho+\sigma+\frac{1}{2},\ 2\sigma+\frac{1}{2};\ \rho+\sigma+1;\ 1-\frac{c}{b^2}\right)$$

[$\operatorname{Re} c>0$; $-1/4<\operatorname{Re}\sigma<\operatorname{Re}\rho+1/2$; $|\arg b|<3\pi/4$].

9. $\int_0^\infty x^{\sigma-1/2}e^{(b^2-c/2)x}\operatorname{erfc}(b\sqrt{x})\,M_{\rho,\sigma}(cx)\,dx=$

$$=\frac{c^{\sigma+1/2}}{2^{4\sigma+1}\sqrt{\pi}\,b^{4\sigma+2}}\Gamma\left[\begin{matrix}4\sigma+2,\ \rho-\sigma\\ \rho+\sigma+3/2\end{matrix}\right]{}_2F_1\left(\rho+\sigma+\frac{1}{2},\ 2\sigma+\frac{3}{2};\ \rho+\sigma+\frac{3}{2};\ 1-\frac{c}{b^2}\right)$$

[$\operatorname{Re} c>0$; $-1/2<\operatorname{Re}\sigma<\operatorname{Re}\rho$; $|\arg b|<3\pi/4$].

10. $\int_0^\infty x^{\alpha-1}e^{(\mp b^2+c/2)x}\begin{Bmatrix}\operatorname{erfi}(b\sqrt{x})\\ \operatorname{erfc}(b\sqrt{x})\end{Bmatrix}W_{\rho,\sigma}(cx)\,dx=B(1)$

[$|\operatorname{Re}\sigma|-(3\pm 1)/4<\operatorname{Re}\alpha<1/2-\operatorname{Re}\rho$; $|\arg b|<(2\mp 1)\pi/4$; $|\arg c|<3\pi/2$; $B(\gamma)$ see 2.19.9.7].

11. $\int_0^\infty x^{\alpha-1} e^{(\mp b^2 - c/2)x} \begin{Bmatrix} \operatorname{erfi}(b\sqrt{x}) \\ \operatorname{erfc}(b\sqrt{x}) \end{Bmatrix} W_{\rho,\sigma}(cx)\,dx = B(-1)$

$[\operatorname{Re} c > 0;\ \operatorname{Re}\alpha > |\operatorname{Re}\sigma| - (3 \pm 1)/4;\ |\arg b| < (2 \mp 1)\pi/4;\ B(\gamma)$ see 2.19.9.7].

12. $\int_0^a x^{\sigma-1/2} e^{-x/(2a)} \operatorname{erf}\left(\sqrt{\frac{a-x}{a}}\right) M_{\rho,\sigma}\left(\frac{x}{a}\right) dx =$

$= \frac{a^{\sigma+1/2}}{e} \Gamma\begin{bmatrix} 2\sigma+1 \\ 2\sigma+5/2 \end{bmatrix} {}_1F_1\left(\frac{3}{2}-\rho+\sigma;\ 2\sigma+\frac{5}{2};\ 1\right)$ $[a > 0;\ \operatorname{Re}\sigma > -1/2]$.

13. $\int_0^a x^{\sigma-1/2} e^{(c/2-b^2)x} \operatorname{erf}(b\sqrt{a-x})\, M_{-\sigma-1,\sigma}(cx)\,dx =$

$= a^{2\sigma+3/2} bc^{\sigma+1/2} \Gamma\begin{bmatrix} 2\sigma+1 \\ 2\sigma+5/2 \end{bmatrix} {}_1F_1(1;\ 2\sigma+5/2;\ ac-ab^2)$ $[a > 0;\ \operatorname{Re}\sigma > -1/2]$.

2.19.10. Integrals of $x^\alpha e^{px} D_\nu(b(x+z)^{\pm 1/2}) \begin{Bmatrix} M_{\rho,\sigma}(cx) \\ W_{\rho,\sigma}(cx) \end{Bmatrix}$.

1. $\int_0^\infty x^{\alpha-1} e^{(\pm b^2 - 2c)x/4} D_\nu(b\sqrt{x})\, M_{\rho,\sigma}(cx)\,dx = U(0)$

$\left[\operatorname{Re}(\alpha+\sigma) > -1/2;\ |\arg b| < (2 \pm 1)\pi/4,\ \begin{Bmatrix} \operatorname{Re} c,\ \operatorname{Re}(2\rho-2\alpha-\nu) > 0 \\ \operatorname{Re} c > 0 \end{Bmatrix}\right]$.

$$U(\gamma) = \frac{2^{\mp(\nu+1)/2-\alpha-\sigma}\pi^{(1\mp 1)/4} c^{\sigma+1/2}}{b^{2\alpha+2\sigma+1}} \Gamma^{(\mp 1-1)/2}(-\nu)\,\Gamma(2\alpha+2\sigma+1)\times$$
$$\times \Gamma^{\gamma^2}\begin{bmatrix} -2\sigma \\ 1/2-\rho-\sigma \end{bmatrix} \Gamma^{\pm 1}\left(\frac{1\mp 3-2\nu}{4} \mp \alpha \mp \sigma\right)\times$$
$$\times {}_3F_2\left(\frac{1}{2}+\sigma-(\gamma^2+\gamma-1)\rho,\ \frac{1}{2}+\alpha+\sigma,\ 1+\alpha+\sigma;\ 2\sigma+1,\ \frac{5\pm 1\pm 2\nu}{4}+\alpha+\sigma;\right.$$
$$\left.\pm(-1)^{(\gamma^2+\gamma)/2}\frac{2c}{b^2}\right) + \gamma^2 \frac{2^{\mp(\nu+1)/2-\alpha+\sigma}\pi^{(1\mp 1)/4} c^{1/2-\sigma}}{b^{2\alpha-2\sigma+1}} \Gamma^{(\mp 1-1)/2}(-\nu)\times$$
$$\times \Gamma\begin{bmatrix} 2\alpha-2\sigma+1,\ 2\sigma \\ 1/2-\rho+\sigma \end{bmatrix} \Gamma^{\pm 1}\left(\frac{1\mp 3-2\nu}{4} \mp \alpha \pm \sigma\right)\times$$
$$\times {}_3F_2\left(\frac{1}{2}-\sigma-\gamma\rho,\ \frac{1}{2}+\alpha-\sigma,\ 1+\alpha-\sigma;\ 1-2\sigma,\ \frac{5\pm 1\pm 2\nu}{4}+\alpha-\sigma;\ \pm(-1)^{(\gamma+1)/2}\frac{2c}{b^2}\right)+$$
$$+ \begin{Bmatrix} 1 \\ 0 \end{Bmatrix} \frac{b^\nu}{c^{\alpha+\nu/2}} \Gamma^{1-\gamma^2}\begin{bmatrix} 2\sigma+1 \\ 1/2+\rho+\sigma \end{bmatrix} \left[\Gamma\left(\frac{1}{2}-\rho+\sigma\right)\Gamma\left(\frac{1}{2}-\rho-\sigma\right)\right]^{-(\gamma^2+\gamma)/2}\times$$
$$\times \Gamma\left(\frac{1+\nu}{2}+\alpha+\sigma\right) \Gamma^{2\gamma^2-1}\left(\frac{1}{2}+(1-2\gamma^2)\left(\sigma-\alpha-\frac{\nu}{2}\right)\right)\times$$
$$\times \Gamma^{-\gamma^2+\gamma+1}\left((1-2\gamma^2)\rho+\frac{\gamma^2-\gamma}{2}+(\gamma^2-\gamma-1)\left(\alpha+\frac{\nu}{2}\right)\right)\times$$
$$\times {}_3F_2\left(-\frac{\nu}{2},\ \frac{1-\nu}{2},\ -\alpha-\frac{\nu}{2}-(\gamma^2+\gamma-1)\rho;\right.$$
$$\left.\frac{1-\nu}{2}-\alpha-\sigma,\ \frac{1-\nu}{2}-\alpha+\sigma;\ (-1)^{(\gamma^2+\gamma)/2}\frac{2c}{b^2}\right).$$

2. $\int_0^\infty x^{\sigma-1/2} e^{(b^2-2c)x/4} D_\nu(b\sqrt{x})\, M_{\rho,\sigma}(cx)\,dx =$

$$=\frac{2^{\nu/2-2\sigma}\sqrt{\pi}\,c^{\sigma+1/2}}{b^{4\sigma+2}}\Gamma\begin{bmatrix}4\sigma+2,\ \rho-\sigma-(\nu+1)/2\\ -\nu/2,\ 1+\rho+\sigma-\nu/2\end{bmatrix}\times$$

$$\times{}_2F_1\left(\frac{1}{2}+\rho+\sigma,\ 2\sigma+\frac{3}{2};\ 1+\rho+\sigma-\frac{\nu}{2};\ 1-\frac{2c}{b^2}\right)$$

$[\operatorname{Re} c>0;\ \operatorname{Re}\sigma>-1/2;\ \operatorname{Re}(2\rho-2\sigma-\nu)>1;\ |\arg b|<3\pi/4]$.

3. $\int\limits_0^\infty x^{\sigma-1}e^{(b^2-2c)x/4}D_\nu(b\sqrt{x})\,M_{\rho,\sigma}(cx)\,dx=$

$$=\frac{2^{(\nu+1)/2-2\sigma}\sqrt{\pi}\,c^{\sigma+1/2}}{b^{4\sigma+1}}\Gamma\begin{bmatrix}4\sigma+1,\ \rho-\sigma-\nu/2\\ (1-\nu)/2,\ \rho+\sigma+(1-\nu)/2\end{bmatrix}\times$$

$$\times{}_2F_1\left(\frac{1}{2}+\rho+\sigma,\ 2\sigma+\frac{1}{2};\ \rho+\sigma+\frac{1-\nu}{2};\ 1-\frac{2c}{b^2}\right)$$

$[\operatorname{Re} c,\ \operatorname{Re}(2\rho-2\sigma-\nu)>0;\ \operatorname{Re}\sigma>-1/4;\ |\arg b|<3\pi/4]$.

4. $\int\limits_0^\infty x^{\alpha-1}e^{(\pm b^2+2c)x/4}D_\nu(b\sqrt{x})\,W_{\rho,\sigma}(cx)\,dx=U(1)$

$\left[\operatorname{Re}\alpha>|\operatorname{Re}\sigma|-1/2;\ |\arg b|<(2\pm1)\pi/4,\ \begin{Bmatrix}\operatorname{Re}(2\alpha+2\rho+\nu)<0;\ |\arg c|<3\pi/2\\ |\arg c|<3\pi/2\end{Bmatrix},\right.$

$\left. U(\gamma)\quad \text{see } 2.19.10.1\right]$.

5. $\int\limits_0^\infty x^{\alpha-1}e^{(\pm b^2-2c)x/4}D_\nu(b\sqrt{x})\,W_{\rho,\sigma}(cx)\,dx=U(-1)$

$[\operatorname{Re} c>0;\ \operatorname{Re}\alpha>|\operatorname{Re}\sigma|-1/2;\ |\arg b|<(2\pm1)\pi/4;\ U(\gamma)\quad \text{see } 2.19.10.1]$.

6. $\int\limits_0^\infty x^{\rho-3/2}e^{b^2/(4x)-cx/2}D_{2\rho-2\sigma-2}\left(\frac{b}{\sqrt{x}}\right)M_{\rho,\sigma}(cx)\,dx=$

$$=2^{(\sigma-3\rho)/2}\sqrt{\pi}\,b^{\rho+\sigma}c^{(\sigma-\rho+1)/2}\,\Gamma\begin{bmatrix}2\sigma+1,\ 1/2-\rho-\sigma\\ 2\sigma-2\rho+2\end{bmatrix}\times$$

$$\times\left[\mathbf{H}_{-\rho-\sigma}(b\sqrt{2c})-Y_{-\rho-\sigma}(b\sqrt{2c})\right]$$

$[\operatorname{Re} c,\ \operatorname{Re}(2\rho+2\sigma-\nu)>0;\ |\arg b|<3\pi/4]$.

7. $\int\limits_0^\infty \frac{x^{\sigma-1/2}}{\sqrt{x+z}}D_\nu\left(\sqrt{2c(x+z)}\right)M_{\sigma-1/2,\sigma}(cx)\,dx=$

$$=\frac{2^{\nu/2}z^{\sigma-1/4}}{c^{1/4}}\Gamma\begin{bmatrix}-(\nu+1)/2,\ 2\sigma+1\\ (1-\nu)/2\end{bmatrix}W_{(3+2\nu)/4-\sigma,\ \sigma+1/4}(cz)$$

$[\operatorname{Re} c>0;\ \operatorname{Re}\nu<-1;\ \operatorname{Re}\sigma>-1/2;\ |\arg z|<\pi]$.

2.19.11. Integrals containing $e^{\pm cx/2}J_\nu(bx^{\pm r})\begin{Bmatrix}M_{\rho,\sigma}(cx)\\ W_{\rho,\sigma}(cx)\end{Bmatrix}$.

1. $\int\limits_0^\infty x^{\alpha-1}e^{-cx/2}J_\nu(bx)\,M_{\rho,\sigma}(cx)\,dx=$

$$=\frac{2^{\alpha+\sigma-1/2}c^{\sigma+1/2}}{b^{\alpha+\sigma+1/2}}\Gamma\begin{bmatrix}(\nu+\alpha+\sigma+1/2)/2\\ (\nu-\alpha-\sigma+3/2)/2\end{bmatrix}{}_4F_3\left(\frac{2\rho+2\sigma+1}{4},\ \frac{2\rho+2\sigma+3}{4},\right.$$

$$\left.\frac{2\alpha+2\nu+2\sigma+1}{4},\ \frac{2\alpha-2\nu+2\sigma+1}{4};\ \frac{1}{2},\ \sigma+\frac{1}{2},\ \sigma+1;\ -\frac{c^2}{b^2}\right)-$$

$$-\frac{2^{\alpha+\sigma-1/2}c^{\sigma+3/2}(2\rho+2\sigma+1)}{b^{\alpha+\sigma+3/2}(2\sigma+1)}\Gamma\begin{bmatrix}(\nu+\alpha+\sigma+3/2)/2\\ (\nu-\alpha-\sigma+1/2)/2\end{bmatrix}\times$$

$$\times {}_4F_3\left(\frac{2\rho+2\sigma+3}{4},\ \frac{2\rho+2\sigma+5}{4};\ \frac{2\alpha+2\nu+2\sigma+3}{4},\ \frac{2\alpha-2\nu+2\sigma+3}{4};\ \frac{3}{2},\ \sigma+1,\ \sigma+\frac{3}{2};\ -\frac{c^2}{b^2}\right)\quad [b,\ \operatorname{Re} c>0;\ -1/2-\operatorname{Re}(\nu+\sigma)<\operatorname{Re}\alpha<\operatorname{Re}\rho+3/2].$$

2. $$\int_0^\infty x^{\alpha-1}e^{\pm cx/2}J_\nu(bx)\,W_{\rho,\sigma}(cx)\,dx=$$

$$=\frac{(b/2)^\nu}{c^{\alpha+\nu}}\left[\Gamma\left(\frac{1}{2}-\rho-\sigma\right)\Gamma\left(\frac{1}{2}-\rho+\sigma\right)\right]^{-(1\pm1)/2}\Gamma\left[\begin{matrix}1/2+\alpha+\nu+\sigma,\ 1/2+\alpha+\nu-\sigma\\ \nu+1\end{matrix}\right]\times$$
$$\times\Gamma^{\pm1}\left(\frac{1\mp1}{2}-\rho\mp\alpha\mp\nu\right){}_4F_3\left(\frac{2\alpha+2\nu+2\sigma+1}{4},\ \frac{2\alpha+2\nu+2\sigma+3}{4},\ \frac{2\alpha+2\nu-2\sigma+1}{4},\right.$$
$$\left.\frac{2\alpha+2\nu-2\sigma+3}{4};\ \nu+1,\ \frac{1\pm\rho+\alpha+\nu}{2},\ 1+\frac{\alpha\pm\rho+\nu}{2};\ -\frac{b^2}{c^2}\right)+$$
$$+\frac{(\gamma+1)\,c^\rho}{2^{2-\alpha-\rho}\,b^{\alpha+\rho}}\Gamma\left[\begin{matrix}(\nu+\alpha+\rho)/2\\ (\nu-\alpha-\rho)/2+1\end{matrix}\right]{}_4F_3\left(\frac{1-2\rho+2\sigma}{4},\ \frac{3-2\rho+2\sigma}{4},\right.$$
$$\left.\frac{1-2\rho-2\sigma}{4},\ \frac{3-2\rho-2\sigma}{4};\ 1-\frac{\alpha+\nu+\rho}{2},\ 1-\frac{\alpha-\nu+\rho}{2},\ \frac{1}{2};\ -\frac{b^2}{c^2}\right)-$$
$$-\frac{(\gamma+1)\,c^{\rho-1}\,[(1-2\rho)^2-4\sigma^2]}{2^{5-\alpha-\rho}\,b^{\alpha+\rho-1}}\Gamma\left[\begin{matrix}(\nu+\alpha+\rho-1)/2\\ (\nu-\alpha-\rho+3)/2\end{matrix}\right]{}_4F_3\left(\frac{3-2\rho+2\sigma}{4},\ \frac{5-2\rho+2\sigma}{4},\right.$$
$$\left.\frac{3-2\rho-2\sigma}{4},\ \frac{5-2\rho-2\sigma}{4};\ \frac{3-\alpha-\nu-\rho}{2},\ \frac{3-\alpha+\nu-\rho}{2},\ \frac{3}{2};\ -\frac{b^2}{c^2}\right)$$
$$\left[b>0;\ \operatorname{Re}(\alpha+\nu)>|\operatorname{Re}\sigma|-1/2,\ \left\{\begin{matrix}\operatorname{Re}(\alpha+\rho)<3/2;\ |\arg c|<3\pi/2\\ \operatorname{Re} c>0\end{matrix}\right\}\right].$$

3. $$\int_0^\infty x^{\alpha-1}e^{-cx/2}J_\nu(b\sqrt{x})\,M_{\rho,\sigma}(cx)\,dx=Y(0)$$

$$[b,\ \operatorname{Re} c>0;\ -\operatorname{Re}(\nu/2+\sigma)-1/2<\operatorname{Re}\alpha<\operatorname{Re}\rho+3/4];$$

$$Y(\gamma)=\frac{(b/2)^\nu}{c^{\alpha+\nu/2}}\Gamma^{1-\gamma^2}\left[\begin{matrix}2\sigma+1\\ \rho+\sigma+1/2\end{matrix}\right]\left[\Gamma\left(\frac{1}{2}-\rho-\sigma\right)\Gamma\left(\frac{1}{2}-\rho+\sigma\right)\right]^{-(\gamma^2+\gamma)/2}\times$$
$$\times\Gamma\left[\begin{matrix}(\nu+1)/2+\alpha+\sigma\\ \nu+1\end{matrix}\right]\Gamma^{2\gamma^2-1}\left(\frac{1}{2}+(1-2\gamma^2)\left(\sigma-\alpha-\frac{\nu}{2}\right)\right)\times$$
$$\times\Gamma^{-\gamma^2+\gamma+1}\left((1-2\gamma^2)\,\rho+\frac{\gamma^2-\gamma}{2}+(\gamma^2-\gamma-1)\left(\alpha+\frac{\nu}{2}\right)\right)\times$$
$$\times{}_2F_2\left(\frac{1+\nu}{2}+\alpha+\sigma,\ \frac{1+\nu}{2}+\alpha-\sigma;\ 1+\alpha+\frac{\nu}{2}+(\gamma^2+\gamma-1)\rho,\ \nu+1;\right.$$
$$\left.(-1)^{1-(\gamma^2+\gamma)/2}\frac{b^2}{4c}\right)-\frac{\gamma^2-\gamma-2}{2c^{(1-2\gamma)\rho}}\left(\frac{b}{2}\right)^{2(1-2\gamma)\rho-2\alpha}\times$$
$$\times\Gamma^{1-\gamma}\left[\begin{matrix}2\sigma+1\\ 1/2-\rho+\sigma\end{matrix}\right]\Gamma\left[\begin{matrix}\nu/2+\alpha-(1-2\gamma)\rho\\ \nu/2+(1-2\gamma)\rho-\alpha+1\end{matrix}\right]\times$$
$$\times{}_2F_2\left(\frac{1}{2}+\sigma+(1-2\gamma)\rho,\ \frac{1}{2}-\sigma+(1-2\gamma)\rho;\right.$$
$$\left.1+(1-2\gamma)\rho-\frac{\nu}{2}-\alpha,\ 1+(1-2\gamma)\rho+\frac{\nu}{2}-\alpha;\ (-1)^{\gamma+1}\frac{b^2}{4c}\right).$$

4. $$\int_0^\infty x^{\sigma-(\nu+1)/2}e^{-cx/2}J_\nu(b\sqrt{x})\,M_{\rho,\sigma}(cx)\,dx=$$

$$=\left(\frac{b}{2}\right)^{\rho-\sigma-3/2} c^{(2\nu-2\rho-2\sigma+1)/4}\,\Gamma\begin{bmatrix}2\sigma+1\\ \nu+\rho-\sigma+1/2\end{bmatrix}\times$$

$$\times\exp\left(-\frac{b^2}{8c}\right) M_{(\rho+3\sigma-\nu+1/2)/2,\ (\rho-\sigma+\nu-1/2)/2}\left(\frac{b^2}{4c}\right)$$

[b, $\operatorname{Re} c>0$; $-1/2<\operatorname{Re}\sigma<\operatorname{Re}(\rho+\nu/2)+1/4$].

5. $$\int_0^\infty x^{(\nu-1)/2-\sigma} e^{-cx/2} J_\nu(b\sqrt{x})\, M_{\rho,\sigma}(cx)\,dx=$$

$$=\left(\frac{b}{2}\right)^{\rho+\sigma-3/2} c^{(2\sigma-2\rho-2\nu+1)/4}\,\Gamma\begin{bmatrix}2\sigma+1\\ \rho+\sigma+1/2\end{bmatrix}\times$$

$$\times\exp\left(-\frac{b^2}{8c}\right) W_{(\rho+\nu-3\sigma+1/2)/2,\ (\rho-\nu+\sigma-1/2)/2}\left(\frac{b^2}{4c}\right)$$

[b, $\operatorname{Re} c>0$; $-1<\operatorname{Re}\nu<2\operatorname{Re}(\rho+\sigma)+1/2$].

6. $$\int_0^\infty x^{-1/2} e^{-cx/2} J_{2\sigma}(b\sqrt{x})\, M_{\rho,\sigma}(cx)\,dx=$$

$$=\frac{(b/2)^{2\rho-1}}{c^\rho}\,\Gamma\begin{bmatrix}2\sigma+1\\ \rho+\sigma+1/2\end{bmatrix}\exp\left(-\frac{b^2}{4c}\right)$$

[b, $\operatorname{Re} c>0$; $\operatorname{Re}\rho>1/4$; $\operatorname{Re}\sigma>-1/2$].

7. $$\int_0^\infty x^{(\rho-\sigma)/2-1} e^{-cx/2} J_{\rho+\sigma-1}(b\sqrt{x})\, M_{\rho,\sigma}(cx)\,dx=$$

$$=\frac{2^{3/2-2\rho}\, b^{\rho+\sigma-1}}{c^{\rho-1/2}}\,\Gamma\begin{bmatrix}2\sigma+1\\ \rho+\sigma+1/2\end{bmatrix}\exp\left(-\frac{b^2}{8c}\right) D_{2\rho-2\sigma-1}\left(\frac{b}{\sqrt{2c}}\right)$$

[b, $\operatorname{Re} c$, $\operatorname{Re}(\rho+\sigma)>0$].

8. $$\int_0^\infty x^{(\rho-\sigma-1)/2} e^{-cx/2} J_{\rho+\sigma}(b\sqrt{x})\, M_{\rho,\sigma}(cx)\,dx=$$

$$=\frac{2^{1-2\rho}\, b^{\rho+\sigma-1}}{c^{\rho}}\,\Gamma\begin{bmatrix}2\sigma+1\\ \rho+\sigma+1/2\end{bmatrix}\exp\left(-\frac{b^2}{8c}\right) D_{2\rho-2\sigma}\left(\frac{b}{\sqrt{2c}}\right)$$

[b, $\operatorname{Re} c>0$; $\operatorname{Re}(\rho+\sigma)>-1/2$].

9. $$\int_0^\infty x^{(2\sigma-2\rho-3)/4} e^{-cx/2} J_{3\sigma-\rho-1/2}(b\sqrt{x})\, M_{\rho,\sigma}(cx)\,dx=$$

$$=\frac{2^{1/2-\rho-\sigma}}{\sqrt{\pi}\, c^\sigma}\, b^{\rho+\sigma-1/2}\,\Gamma\begin{bmatrix}2\sigma+1\\ \rho+\sigma+1/2\end{bmatrix}\exp\left(-\frac{b^2}{8c}\right) K_{\rho-\sigma}\left(\frac{b^2}{8c}\right)$$

[b, $\operatorname{Re} c>0$; $\operatorname{Re}(3\sigma-\rho)>-1/2$; $\operatorname{Re}(\sigma-3\rho)<1$].

10. $$\int_0^\infty x^{-(2\rho+2\sigma+3)/4} e^{-cx/2} J_{\rho+3\sigma+1/2}(b\sqrt{x})\, M_{\rho,\sigma}(cx)\,dx=$$

$$=\sqrt{\pi}\left(\frac{b}{2}\right)^{\rho-\sigma-1/2} c^\sigma\,\Gamma\begin{bmatrix}2\sigma+1\\ \rho+\sigma+1/2\end{bmatrix}\exp\left(-\frac{b^2}{8c}\right) I_{\rho+\sigma}\left(\frac{b^2}{8c}\right)$$

[b, $\operatorname{Re} c>0$; $\operatorname{Re}\sigma>-1/2$; $\operatorname{Re}(3\rho+\sigma)>-1$].

11. $$\int_0^\infty x^{-1} e^{-cx/2} J_{2\sigma-1}(b\sqrt{x})\, M_{\sigma,\sigma}(cx)\,dx=$$

$$=\sqrt{\pi}\left(\frac{b^2}{4c}\right)^{\sigma-1/2}\Gamma\begin{bmatrix}2\sigma+1\\ 2\sigma+1/2\end{bmatrix}\operatorname{erfc}\left(\frac{b}{2\sqrt{c}}\right)$$

[b, $\operatorname{Re} c$, $\operatorname{Re}\sigma>0$].

12. $\int_0^\infty x^{-3/4} e^{-cx/2} J_{2\sigma-1/2}(b\sqrt{x})\, M_{\sigma-1/2,\,\sigma}(cx)\,dx=$

$$=2^{5/2-2\sigma}\sigma b^{2\sigma-3/2}\,c^{1/2-\sigma}\operatorname{erfc}\left(\frac{b}{2\sqrt{c}}\right) \qquad [b,\ \operatorname{Re} c,\ \operatorname{Re}\sigma>0].$$

13. $\int_0^\infty x^{\alpha-1} e^{\pm cx/2} J_\nu(b\sqrt{x})\, W_{\rho,\,\sigma}(cx)\,dx=Y(\pm 1)$

$\left[b>0;\ \operatorname{Re}(\alpha+\nu/2)>|\operatorname{Re}\sigma|-1/2,\ \begin{Bmatrix}\operatorname{Re}(\alpha+\rho)<3/4;\ |\arg c|<3\pi/2\\ \operatorname{Re} c>0\end{Bmatrix},\ Y(\gamma)\ \text{see } 2.19.11.3\right].$

14. $\int_0^\infty x^{\pm\sigma+(\nu-1)/2} e^{cx/2} J_\nu(b\sqrt{x})\, W_{\rho,\,\sigma}(cx)\,dx=$

$$=2^{3/2+\rho\pm\sigma}\,b^{-\rho\mp\sigma-3/2}\,c^{(1-2\nu+2\rho\mp2\sigma)/4}\,\Gamma\begin{bmatrix}1+\nu\pm2\sigma\\ 1/2-\rho\mp\sigma\end{bmatrix}\times$$

$$\times\exp\left(\frac{b^2}{8c}\right) W_{(\rho\mp3\sigma-\nu-1/2)/2,\ (\rho\pm\sigma+\nu+1/2)/2}\left(\frac{b^2}{4c}\right)$$

$[b>0;\ |\arg c|<3\pi/2;\ \operatorname{Re}\nu>-1;\ -1<\operatorname{Re}(\nu\pm2\sigma)<1/2-2\operatorname{Re}\rho].$

15. $\int_0^\infty x^{(2\rho\pm2\sigma-3)/4} e^{cx/2} J_{\rho\pm3\sigma-1/2}(b\sqrt{x})\, W_{\rho,\,\sigma}(cx)\,dx=$

$$=\frac{2^{\rho\mp\sigma+1/2}\,b^{-\rho\pm\sigma-1/2}}{\sqrt{\pi}\,c^{\pm\sigma}}\,\Gamma\begin{bmatrix}\rho\pm\sigma+1/2\\ -\rho\pm\sigma+1/2\end{bmatrix}\exp\left(\frac{b^2}{8c}\right) K_{\rho\pm\sigma}\left(\frac{b^2}{8c}\right)$$

$[b>0;\ \operatorname{Re}(\rho\pm3\sigma),\ \operatorname{Re}(\rho\pm\sigma)>-1/2;\ \operatorname{Re}(3\rho\pm\sigma)<1;\ |\arg c|<3\pi/2].$

16. $\int_0^\infty x^{\pm\sigma+(\nu-1)/2} e^{-cx/2} J_\nu(b\sqrt{x})\, W_{\rho,\,\sigma}(cx)\,dx=$

$$=\left(\frac{b}{2}\right)^{\rho\mp\sigma-3/2} c^{(1-2\nu-2\rho\mp2\sigma)/4}\,\Gamma\begin{bmatrix}\nu\pm2\sigma+1\\ \nu-\rho\pm\sigma+3/2\end{bmatrix}\times$$

$$\times\exp\left(-\frac{b^2}{8c}\right) M_{(\nu+\rho\pm3\sigma+1/2)/2,\ (\nu-\rho\pm\sigma+1/2)/2}\left(\frac{b^2}{4c}\right)$$

$[b,\ \operatorname{Re} c>0;\ \operatorname{Re}\nu,\ \operatorname{Re}(\nu\pm2\sigma)>-1].$

17. $\int_0^\infty x^{(-2\rho\pm2\sigma-3)/4} e^{-cx/2} J_{-\rho\pm3\sigma-1/2}(b\sqrt{x})\, W_{\rho,\,\sigma}(cx)\,dx=$

$$=2^{1/2-\rho\mp\sigma}\sqrt{\pi}\,b^{\rho\pm\sigma-1/2}\,c^{\mp\sigma}\exp\left(-\frac{b^2}{8c}\right) I_{-\rho\pm\sigma}\left(\frac{b^2}{8c}\right)$$

$[b,\ \operatorname{Re} c>0;\ \operatorname{Re}(-\rho\pm3\sigma),\ \operatorname{Re}(-\rho\pm\sigma)>-1/2].$

18. $\int_0^\infty x^{\alpha-1} e^{-cx/2} J_\nu\left(\frac{b}{\sqrt{x}}\right) M_{\rho,\,\sigma}(cx)\,dx=V(0)$

$[b,\ \operatorname{Re} c>0;\ -5/4-\operatorname{Re}\sigma<\operatorname{Re}\alpha<\operatorname{Re}(\rho+\nu/2)].$

$$V(\gamma)=\left(\frac{b}{2}\right)^{2\alpha+2\sigma+1} c^{\sigma+1/2}\,\Gamma\begin{bmatrix}(\nu-1)/2-\alpha-\sigma\\ \alpha+\sigma+(\nu+3)/2\end{bmatrix}\Gamma^{\gamma^2}\begin{bmatrix}-2\sigma\\ 1/2-\rho-\sigma\end{bmatrix}\times$$

$$\times{}_1F_3\left(\frac{1}{2}+\sigma-(\gamma^2+\gamma-1)\rho;\ 2\sigma+1,\ \alpha+\sigma+\frac{\nu+3}{2},\ \frac{3-\nu}{2}+\alpha+\sigma;\right.$$

$$(-1)^{(\gamma^2+\gamma)/2}\frac{b^2c}{4}\Big)+\gamma^2\left(\frac{b}{2}\right)^{2\alpha-2\sigma+1}c^{1/2-\sigma}\Gamma\begin{bmatrix}(\nu-1)/2-\alpha+\sigma,\ 2\sigma\\ \alpha-\sigma+(\nu+3)/2,\ 1/2-\rho+\sigma\end{bmatrix}\times$$
$$\times{}_1F_3\left(\frac{1}{2}-\sigma-\gamma\rho;\ 1-2\sigma,\ \alpha-\sigma+\frac{\nu+3}{2},\ \alpha-\sigma-\frac{\nu-3}{2};\ (-1)^{(1+\gamma)/2}\frac{b^2c}{4}\right)+$$
$$+\left(\frac{b}{2}\right)^{\nu}c^{\nu/2-\alpha}\Gamma^{1-\gamma^2}\begin{bmatrix}2\sigma+1\\ \rho+\sigma+1/2\end{bmatrix}\left[\Gamma\left(\frac{1}{2}-\rho-\sigma\right)\Gamma\left(\frac{1}{2}-\rho+\sigma\right)\right]^{-(\gamma^2+\gamma)/2}\times$$
$$\times\Gamma\begin{bmatrix}(1-\nu)/2+\alpha+\sigma\\ \nu+1\end{bmatrix}\Gamma^{2\gamma^2-1}\left(\frac{1}{2}+(1-2\gamma^2)\left(\sigma+\frac{\nu}{2}-\alpha\right)\right)\times$$
$$\times\Gamma^{-\gamma^2+\gamma+1}\left((1-2\gamma^2)\rho+\frac{\gamma^2-\gamma}{2}+(\gamma^2-\gamma-1)\left(\alpha-\frac{\nu}{2}\right)\right)\times$$
$$\times{}_1F_3\left(\frac{\nu}{2}-\alpha-(\gamma^2+\gamma-1)\rho;\ \frac{1+\nu}{2}-\alpha+\sigma,\ \frac{1+\nu}{2}-\alpha-\sigma,\ \nu+1;\ (-1)^{(\gamma^2+\gamma)/2}\frac{b^2c}{4}\right).$$

19. $$\int_0^\infty x^{\alpha-1}e^{\pm cx/2}J_\nu\left(\frac{b}{\sqrt{x}}\right)W_{\rho,\sigma}(cx)\,dx=V(\pm 1)$$

$\left[b>0;\ \mathrm{Re}\,\alpha>|\mathrm{Re}\,\sigma|-5/4,\ \begin{Bmatrix}\mathrm{Re}\,(\alpha+\rho-\nu/2)<0;\ |\arg c|<3\pi/2\\ \mathrm{Re}\,c>0\end{Bmatrix},\ V(\gamma)\ \text{see 2.19.11.18}\right].$

20. $$\int_0^\infty x^{\alpha-1}e^{-cx/2}J_\mu(b\sqrt{x})\,J_\nu(b\sqrt{x})\,M_{\rho,\sigma}(cx)\,dx=W(0)$$

$[b,\ \mathrm{Re}\,c>0;\ -1-\mathrm{Re}\,(\mu+\nu+2\sigma)<2\,\mathrm{Re}\,\alpha<2\,\mathrm{Re}\,\rho+1],$

$$W(\gamma)=\left(\frac{b}{2}\right)^{\mu+\nu}\frac{1}{c^{\alpha+(\mu+\nu)/2}}\Gamma^{1-\gamma^2}\begin{bmatrix}2\sigma+1\\ \rho+\sigma+1/2\end{bmatrix}\left[\Gamma\left(\frac{1}{2}-\rho-\sigma\right)\times\right.$$
$$\left.\times\Gamma\left(\frac{1}{2}-\rho+\sigma\right)\right]^{-(\gamma^2+\gamma)/2}\Gamma\begin{bmatrix}\alpha+\sigma+(\mu+\nu+1)/2\\ \mu+1,\ \nu+1\end{bmatrix}\Gamma^{2\gamma^2-1}\left(\frac{1}{2}+(1-2\gamma^2)\times\right.$$
$$\left.\times\left(\sigma-\alpha-\frac{\mu+\nu}{2}\right)\right)\Gamma^{-\gamma^2+\gamma+1}\left((1-2\gamma^2)\rho+\frac{\gamma^2-\gamma}{2}+(\gamma^2-\gamma-1)\left(\alpha+\frac{\mu+\nu}{2}\right)\right)\times$$
$$\times{}_4F_4\left(\frac{\mu+\nu+1}{2},\ \frac{\mu+\nu}{2}+1,\ \alpha+\sigma+\frac{\mu+\nu+1}{2},\ \alpha-\sigma+\frac{\mu+\nu+1}{2};\right.$$
$$\mu+\nu+1,\ \nu+1,\ \mu+1,\ \alpha+\frac{\mu+\nu}{2}+1+(\gamma^2+\gamma-1)\rho;$$
$$\left.(-1)^{1-(\gamma^2+\gamma)/2}\frac{b^2}{c}\right)-\frac{\gamma^2-\gamma-2}{2}\left(\frac{b}{2}\right)^{2(1-2\gamma)\rho-2\alpha}c^{(2\gamma-1)\rho}\Gamma^{1-\gamma}\begin{bmatrix}2\sigma+1\\ 1/2-\rho+\sigma\end{bmatrix}\times$$
$$\times\Gamma\begin{bmatrix}1+2\rho(1-2\gamma)-2\alpha,\ (\mu+\nu)/2+\alpha-\rho(1-2\gamma)\\ 1+\frac{\mu+\nu}{2}-\alpha+\rho(1-2\gamma),\ 1+(\nu-\mu)/2-\alpha+\rho(1-2\gamma),\ 1+\frac{\mu-\nu}{2}-\alpha+\rho(1-2\gamma)\end{bmatrix}\times$$
$$\times{}_4F_4\left(\frac{1}{2}+\rho(1-2\gamma)-\alpha,\ 1+\rho(1-2\gamma)-\alpha,\ \frac{1}{2}+\sigma+\rho(1-2\gamma),\ \frac{1}{2}-\sigma+\rho(1-2\gamma);\right.$$
$$1+\frac{\mu+\nu}{2}+\rho(1-2\gamma)-\alpha,\ 1+\frac{\mu-\nu}{2}+\rho(1-2\gamma)-\alpha,\ 1+\frac{\nu-\mu}{2}+\rho(1-2\gamma)-\alpha,$$
$$\left.1-\frac{\mu+\nu}{2}+\rho(1-2\gamma)-\alpha;\ (-1)^{\gamma+1}\frac{b^2}{c}\right).$$

1. $$\int_0^\infty x^{\alpha-1}e^{\pm cx/2}J_\mu(b\sqrt{x})\,J_\nu(b\sqrt{x})\,W_{\rho,\sigma}(cx)\,dx=W(\pm 1)$$

$\left[b>0;\ \mathrm{Re}\,(2\alpha+\mu+\nu)>2\,|\mathrm{Re}\,\sigma|-1,\ \begin{Bmatrix}\mathrm{Re}\,(\alpha+\rho)<1/2;\ |\arg c|<3\pi/2\\ \mathrm{Re}\,c>0\end{Bmatrix},\ W(\gamma)\ \text{see 2.19.11.20}\right].$

22. $\int\limits_0^\infty x^{-1/2} e^{cx/2} \operatorname{Ei}(-cx) J_{2\sigma}(b\sqrt{x}) M_{\rho,\sigma}(cx)\,dx = -\left(\frac{2}{b}\right)^{2\rho+1} c^\rho \times$

$$\times\Gamma\begin{bmatrix}2\sigma+1\\1/2+\rho+\sigma\end{bmatrix}\exp\left(\frac{b^2}{4c}\right)\Gamma\left(\frac{1}{2}+\rho-\sigma,\ \frac{b^2}{4c}\right)\gamma\left(\frac{1}{2}+\rho+\sigma,\ \frac{b^2}{4c}\right)$$

[b, $\operatorname{Re} c>0$; $\operatorname{Re}\rho>-5/4$; $\operatorname{Re}\sigma>-1/2$].

2.19.12. Integrals containing $I_\nu(bx^r)\begin{Bmatrix}M_{\rho,\sigma}(cx)\\W_{\rho,\sigma}(cx)\end{Bmatrix}$.

1. $\int\limits_0^\infty x^{\alpha-1} I_\nu\left(\frac{cx}{2}\right) W_{\rho,\sigma}(cx)\,dx = C(0)$ [$\operatorname{Re} c>0$; $|\operatorname{Re}\sigma|-\operatorname{Re}\nu-1/2<\operatorname{Re}\alpha<1/2-\operatorname{Re}\rho$],

$$C(\gamma)=\frac{c^{-\alpha}}{2^{2\nu+1}}\left(-\frac{2\sin\nu\pi}{\pi}\right)^{1-\gamma}\Gamma\begin{bmatrix}-\nu,\ 1/2+\alpha+\nu+\sigma,\ 1/2+\alpha+\nu-\sigma,\ -\alpha-\nu-\rho\\1/2-\rho+\sigma,\ 1/2-\rho-\sigma\end{bmatrix}\times$$

$$\times{}_3F_2\left(\frac{1}{2}+\nu,\ \frac{1}{2}+\alpha+\nu+\sigma,\ \frac{1}{2}+\alpha+\nu-\sigma;\ 1+2\nu,\ 1+\alpha+\nu+\rho;\ 1\right)+$$

$$+\gamma\frac{c^{-\alpha}}{2^{1-2\nu}}\Gamma\begin{bmatrix}\nu,\ 1/2+\alpha-\nu+\sigma,\ 1/2+\alpha-\nu-\sigma,\ \nu-\alpha-\rho\\1/2-\rho+\sigma,\ 1/2-\rho-\sigma\end{bmatrix}\times$$

$$\times{}_3F_2\left(\frac{1}{2}-\nu,\ \frac{1}{2}+\alpha-\nu+\sigma,\ \frac{1}{2}+\alpha-\nu-\sigma;\ 1-2\nu,\ 1+\alpha-\nu+\rho;\ 1\right)+$$

$$+c^{-\alpha}\pi^{\gamma-1/2}\Gamma(\alpha+\nu+\rho)\,\Gamma^{2\gamma-1}(1-\gamma+(1-2\gamma)(\nu-\alpha-\rho))\times$$

$$\times\Gamma^{1-2\gamma}\left(\frac{1}{2}-(1-2\gamma)(\alpha+\rho)\right){}_3F_2\left(\frac{1}{2}-\rho+\sigma,\ \frac{1}{2}-\rho-\sigma,\right.$$

$$\left.\frac{1}{2}-\alpha-\rho;\ 1-\alpha+\nu-\rho,\ 1-\alpha-\nu-\rho;\ 1\right).$$

2. $\int\limits_0^\infty x^{\alpha-1}e^{-cx/2} I_\nu(bx)\,W_{\rho,\sigma}(cx)\,dx=$

$$=\frac{(b/2)^\nu}{c^{\alpha+\nu}}\Gamma\begin{bmatrix}1/2+\alpha+\nu+\sigma,\ 1/2+\alpha+\nu-\sigma\\\nu+1,\ 1+\alpha+\nu-\rho\end{bmatrix}{}_4F_3\left(\frac{1+2\alpha+2\nu+2\sigma}{4},\ \frac{3+2\alpha+2\nu+2\sigma}{4},\right.$$

$$\left.\frac{1+2\alpha+2\nu-2\sigma}{4},\ \frac{3+2\alpha+2\nu-2\sigma}{4};\ \frac{1+\alpha+\nu-\rho}{2},\ \frac{\alpha+\nu-\rho}{2}+1,\ \nu+1;\ \frac{b^2}{c^2}\right)$$

[$\operatorname{Re} c>|\operatorname{Re} b|$; $\operatorname{Re}(\alpha+\nu)>|\operatorname{Re}\sigma|-1/2$]

3. $\int\limits_0^\infty x^{\alpha-1}e^{-(b+c/2)x} I_\nu(bx)\,M_{\rho,\sigma}(cx)\,dx=U(0,\ 0)$

[$\operatorname{Re} b$, $\operatorname{Re} c>0$; $-\operatorname{Re}(\nu+\sigma)-1/2<\operatorname{Re}\alpha<\operatorname{Re}\rho+1/2$]

$$U(\varepsilon,\ \gamma)=\frac{b^\nu}{2^{\nu+1}c^{\alpha+\nu}}\left(-\frac{2}{\pi}\sin\nu\pi\right)^{1-\varepsilon^2}\Gamma^{1-\gamma^2}\begin{bmatrix}2\sigma+1\\\rho+\sigma+1/2\end{bmatrix}\times$$

$$\times\left[\Gamma\left(\frac{1}{2}-\rho+\sigma\right)\Gamma\left(\frac{1}{2}-\rho-\sigma\right)\right]^{-(\gamma^2+\gamma)/2}\Gamma(-\nu)\,\Gamma\left(\alpha+\nu+\sigma+\frac{1}{2}\right)\times$$

$$\times\Gamma^{2\gamma^2-1}\left(\frac{1}{2}+(1-2\gamma^2)(\sigma-\alpha-\nu)\right)\Gamma^{-\gamma^2+\gamma+1}\left((1-2\gamma^2)\rho+\frac{\gamma^2-\gamma}{2}+\right.$$

$$\left.+(\gamma^2-\gamma-1)(\alpha+\nu)\right){}_3F_2\left(\frac{1}{2}+\nu,\ \frac{1}{2}+\alpha+\nu+\sigma,\ \frac{1}{2}+\alpha+\nu-\sigma;\ 1+2\nu,\right.$$

$$\left.1+\alpha+\nu+(\gamma^2+\gamma-1)\rho;\ (-1)^{1-(\varepsilon^2+\varepsilon+\gamma^2+\gamma)/2}\frac{2b}{c}\right)+$$

$$+\varepsilon^2 \frac{2^{\nu-1}}{b^\nu c^{\alpha-\nu}} \Gamma^{1-\gamma^2}\left[\begin{matrix}2\sigma+1\\ \rho+\sigma+1/2\end{matrix}\right]\left[\Gamma\left(\frac{1}{2}-\rho+\sigma\right)\Gamma\left(\frac{1}{2}-\rho-\sigma\right)\right]^{-(\gamma^2+\gamma)/2}\times$$

$$\times\Gamma(\nu)\,\Gamma\left(\alpha-\nu+\sigma+\frac{1}{2}\right)\Gamma^{2\gamma^2-1}\left(\frac{1}{2}+(1-2\gamma^2)(\sigma+\nu-\alpha)\right)\times$$

$$\times\Gamma^{-\gamma^2+\gamma+1}\left((1-2\gamma^2)\rho+\frac{\gamma^2-\gamma}{2}+(\gamma^2-\gamma-1)(\alpha-\nu)\right)\times$$

$$\times{}_3F_2\left(\frac{1}{2}-\nu,\ \frac{1}{2}+\alpha-\nu+\sigma,\ \frac{1}{2}+\alpha-\nu-\sigma;\ 1-2\nu,\ 1+\alpha-\nu+(\gamma^2+\gamma-1)\rho;\right.$$

$$\left.(-1)^{(1-\varepsilon-\gamma^2-\gamma)/2}\frac{2b}{c}\right)-\frac{(\gamma^2-\gamma-2)\,\pi^{(\varepsilon^2-\varepsilon-1)/2}}{2c^{(1-2\gamma)\rho}}(2b)^{(1-2\gamma)\rho-\alpha}\times$$

$$\times\Gamma^{1-\gamma}\left[\begin{matrix}2\sigma+1\\ 1/2-\rho+\sigma\end{matrix}\right]\Gamma(\nu-(1-2\gamma)\rho+\alpha)\cos^{(\varepsilon^2+\varepsilon)/2}\nu\pi\times$$

$$\times\Gamma^{2\varepsilon^2-1}\left(1-\varepsilon^2+(1-2\varepsilon^2)(\nu+(1-2\gamma)\rho)\right)\Gamma^{-\varepsilon^2+\varepsilon+1}\left(\frac{1}{2}-(\varepsilon^2-\varepsilon-1)((1-2\gamma)\rho-\alpha)\right)\times$$

$$\times{}_3F_2\left(\frac{1}{2}+\sigma+(1-2\gamma)\rho,\ \frac{1}{2}-\sigma+(1-2\gamma)\rho,\ (1-2\gamma)\rho-\alpha+\frac{1}{2};\right.$$

$$\left.1+\nu+(1-2\gamma)\rho-\alpha,\ 1-\nu+(1-2\gamma)\rho-\alpha;\ (-1)^{1-\gamma-(\varepsilon^2+\varepsilon)/2}\frac{2b}{c}\right).$$

4. $$\int_0^\infty x^{\alpha-1}e^{(-b\pm c/2)x}I_\nu(bx)W_{\rho,\sigma}(cx)\,dx=U(0,\ \pm 1)$$

$$\left[\operatorname{Re} b>0;\ \operatorname{Re}(\alpha+\nu)>|\operatorname{Re}\sigma|-1/2,\ \left\{\begin{matrix}\operatorname{Re}(\alpha+\rho)<1/2;\ |\arg c|<3\pi/2\\ \operatorname{Re} c>0\end{matrix}\right\}\right.$$

$\left.U(\varepsilon,\gamma)\ \text{see 2.19.12.3}\right]$.

5. $$\int_0^\infty x^{\alpha-1}e^{(b-c/2)x}I_\nu(bx)\,W_{\rho,\sigma}(cx)\,dx=$$

$$=\frac{b^\nu}{2^\nu c^{\alpha+\nu+1/2}}\Gamma\left[\begin{matrix}\alpha+\nu+\sigma+1/2,\ \alpha+\nu-\sigma+1/2\\ \nu+1,\ \alpha+\nu-\rho+1\end{matrix}\right]\times$$

$$\times{}_3F_2\left(\nu+\frac{1}{2},\ \alpha+\nu+\sigma+\frac{1}{2},\ \alpha+\nu-\sigma+\frac{1}{2};\ 2\nu+1,\ \alpha+\nu-\rho+1;\ -\frac{2b}{c}\right)$$

$[\operatorname{Re} c,\ \operatorname{Re}(c-b)>0;\ \operatorname{Re}(\alpha+\nu)>|\operatorname{Re}\sigma|-1/2]$.

6. $$\int_0^\infty x^{\alpha-1}e^{-cx/2}I_\nu(b\sqrt{x})\,W_{\rho,\sigma}(cx)\,dx=$$

$$=\frac{(b/2)^\nu}{c^{\alpha+\nu/2}}\Gamma\left[\begin{matrix}(1+\nu)/2+\alpha+\sigma,\ \ (1+\nu)/2+\alpha-\sigma\\ \nu+1,\ \alpha-\rho+\nu/2+1\end{matrix}\right]\times$$

$$\times{}_2F_2\left(\frac{1+\nu}{2}+\alpha+\sigma,\ \frac{1+\nu}{2}+\alpha-\sigma;\ \nu+1,\ \alpha-\rho+\frac{\nu}{2}+1;\ \frac{b^2}{4c}\right)$$

$[\operatorname{Re} c>0;\ \operatorname{Re}(\alpha+\nu/2)>|\operatorname{Re}\sigma|-1/2;\ |\arg b|<\pi]$.

7. $$\int_0^\infty x^{\alpha-1}e^{-b\sqrt{x}-cx/2}I_\nu(b\sqrt{x})\,M_{\rho,\sigma}(cx)\,dx=D(0)$$

$[\operatorname{Re} b,\ \operatorname{Re} c>0;\ -1/2-\operatorname{Re}(\sigma+\nu/2)<\operatorname{Re}\alpha<\operatorname{Re}\rho+1/4]$,

$$D(\gamma)=\frac{(b/2)^{\nu}}{c^{\alpha+\nu/2}}\Gamma\begin{bmatrix}(1+\nu)/2+\alpha+\sigma\\ \nu+1\end{bmatrix}\Gamma^{1-\gamma^2}\begin{bmatrix}2\sigma+1\\ \rho+\sigma+1/2\end{bmatrix}\times$$

$$\times\left[\Gamma\left(\frac{1}{2}-\rho-\sigma\right)\Gamma\left(\frac{1}{2}-\rho+\sigma\right)\right]^{-(\gamma^2+\gamma)/2}\Gamma^{2\gamma^2-1}\left(\frac{1}{2}+(1-2\gamma^2)\left(\sigma-\alpha-\frac{\nu}{2}\right)\right)\times$$

$$\times\Gamma^{-\gamma^2+\gamma+1}\left((1-2\gamma^2)\rho+\frac{\gamma^2-\gamma}{2}+(\gamma^2-\gamma-1)\left(\alpha+\frac{\nu}{2}\right)\right)\times$$

$$\times{}_4F_4\left(\frac{2\nu+1}{4},\ \frac{2\nu+3}{4},\ \frac{\nu+1}{2}+\alpha+\sigma,\ \frac{\nu+1}{2}+\alpha-\sigma;\right.$$

$$\left.1/2,\ \nu+1/2,\ \nu+1,\ 1+\alpha+\nu/2+(\gamma^2+\gamma-1)\rho;\ (-1)^{(\gamma^2+\gamma)/2}b^2/c\right)-$$

$$-\frac{b^{\nu+1}}{2^{\nu}c^{\alpha+(\nu+1)/2}}\Gamma\begin{bmatrix}1+\nu/2+\alpha+\sigma\\ \nu+1\end{bmatrix}\Gamma^{1-\gamma^2}\begin{bmatrix}2\sigma+1\\ \rho+\sigma+1/2\end{bmatrix}\times$$

$$\times\left[\Gamma\left(\frac{1}{2}-\rho-\sigma\right)\Gamma\left(\frac{1}{2}-\rho+\sigma\right)\right]^{-(\gamma^2+\gamma)/2}\Gamma^{2\gamma^2-1}\left(\frac{1}{2}+(1-2\gamma^2)\left(\sigma-\alpha-\frac{\nu+1}{2}\right)\right)\times$$

$$\times\Gamma^{-\gamma^2+\gamma+1}\left((1-2\gamma^2)\rho+\frac{\gamma^2-\gamma}{2}+(\gamma^2-\gamma-1)\left(\alpha+\frac{\nu+1}{2}\right)\right)\times$$

$$\times{}_4F_4\left(\frac{2\nu+3}{4},\ \frac{2\nu+5}{4},\ 1+\frac{\nu}{2}+\alpha+\sigma,\ 1+\frac{\nu}{2}+\alpha-\sigma;\right.$$

$$\left.3/2,\ \nu+1,\ \nu+3/2,\ (3+\nu)/2+\alpha+(\gamma^2+\gamma-1)\rho;\ (-1)^{(\gamma^2+\gamma)/2}b^2/c\right)-$$

$$-\frac{(\gamma^2-\gamma-2)(2b)^{2(1-2\gamma)\rho-2\alpha}}{\sqrt{\pi}\,c^{(1-2\gamma)\rho}}\Gamma^{1-\gamma}\begin{bmatrix}2\sigma+1\\ 1/2-\rho+\sigma\end{bmatrix}\times$$

$$\times\Gamma\begin{bmatrix}1/2+2(1-2\gamma)\rho-2\alpha,\ \nu+2\alpha-2(1-2\gamma)\rho\\ 1+\nu-2\alpha+2(1-2\gamma)\rho\end{bmatrix}{}_4F_4\left(\frac{1}{2}+\sigma+(1-2\gamma)\rho,\right.$$

$$\frac{1}{2}-\sigma+(1-2\gamma)\rho,\ \frac{1}{4}+(1-2\gamma)\rho-\alpha,\ \frac{3}{4}+(1-2\gamma)\rho-\alpha;\ \frac{1+\nu}{2}-\alpha+(1-2\gamma)\rho,$$

$$\left.1+\frac{\nu}{2}-\alpha+(1-2\gamma)\rho,\ \frac{1-\nu}{2}-\alpha+(1-2\gamma)\rho,\ 1-\frac{\nu}{2}-\alpha+(1-2\gamma)\rho;\ (-1)^{\nu}\frac{b^2}{c}\right).$$

8. $\int\limits_0^\infty x^{\alpha-1}e^{-b\sqrt{x}\pm cx/2}I_\nu(b\sqrt{x})\,W_{\rho,\sigma}(cx)\,dx=D(\pm 1)$

$\left[\operatorname{Re} b>0;\ \operatorname{Re}(2\alpha+\nu)>2\,|\operatorname{Re}\sigma|-1,\ \begin{Bmatrix}\operatorname{Re}(\alpha+\rho)<1/4;\ |\arg c|<3\pi/2\\ \operatorname{Re} c>0\end{Bmatrix},\right.$

$\left. D(\gamma)\ \text{ see } 2.19.12.7\right].$

2.19.13. Integrals containing $\begin{Bmatrix}Y_\nu(bx^r)\\ K_\nu(bx^r)\end{Bmatrix}\begin{Bmatrix}M_{\rho,\sigma}(cx)\\ W_{\rho,\sigma}(cx)\end{Bmatrix}$.

1. $\int\limits_0^\infty x^{\alpha-1}K_\nu\left(\frac{cx}{2}\right)W_{\rho,\sigma}(cx)\,dx=C(1)$

[$\operatorname{Re} c>0;\ \operatorname{Re}\alpha>|\operatorname{Re}\nu|+|\operatorname{Re}\sigma|-1/2;\ C(\gamma)$ see 2.19.12.1].

2. $\int\limits_0^\infty x^{\alpha-1}e^{-cx/2}\begin{Bmatrix}Y_\nu(bx)\\ K_\nu(bx)\end{Bmatrix}M_{\rho,\sigma}(cx)\,dx=\begin{Bmatrix}-2\\ \pi\end{Bmatrix}\frac{2^{\alpha+\sigma-3/2}c^{\sigma+1/2}}{\pi b^{\alpha+\sigma+1/2}}\times$

$$\times\Gamma\left(\frac{2\alpha+2\nu+2\sigma+1}{4}\right)\Gamma\left(\frac{2\alpha-2\nu+2\sigma+1}{4}\right)\cos^{(1\pm1)/2}\left(\frac{2\alpha-2\nu+2\sigma+1}{4}\pi\right)\times$$

$$\times{}_4F_3\left(\frac{2\alpha+2\nu+2\sigma+1}{4},\ \frac{2\alpha-2\nu+2\sigma+1}{4},\ \frac{2\rho+2\sigma+1}{4},\ \frac{2\rho+2\sigma+3}{4};\right.$$

$$\left.\frac{1}{2},\ \sigma+\frac{1}{2},\ \sigma+1;\ \mp\frac{c^2}{b^2}\right)-\begin{Bmatrix}2\\ \pi\end{Bmatrix}\frac{2^{\alpha-\sigma-3/2}c^{\sigma+3/2}(2\rho+2\sigma+1)}{\pi b^{\alpha+\sigma+3/2}(2\sigma+1)}\times$$

$$\times\Gamma\left(\frac{2\alpha+2\nu+2\sigma+3}{4}\right)\Gamma\left(\frac{2\alpha-2\nu+2\sigma+3}{4}\right)\sin^{(1\pm1)/2}\left(\frac{2\alpha-2\nu+2\sigma+1}{4}\pi\right)\times$$
$$\times{}_4F_3\left(\frac{2\alpha+2\nu+2\sigma+3}{4},\ \frac{2\alpha-2\nu+2\sigma+3}{4},\ \frac{2\rho+2\sigma+3}{4},\ \frac{2\rho+2\sigma+5}{4};\ \frac{3}{2},\ \sigma+1,\ \sigma+\frac{3}{2};\ \mp\frac{c^2}{b^2}\right)$$

$$\left[\operatorname{Re} c>0;\ \operatorname{Re}(\alpha+\sigma)>|\operatorname{Re}\nu|-1/2,\ \begin{Bmatrix} b>0;\ \operatorname{Re}(\alpha-\rho)<3/2 \\ \operatorname{Re} b>0\end{Bmatrix}\right].$$

3. $\displaystyle\int_0^\infty x^{\alpha-1}e^{cx/2}\begin{Bmatrix}Y_\nu(bx)\\K_\nu(bx)\end{Bmatrix}W_{\rho,\sigma}(cx)\,dx=E(1)$

$$\left[\operatorname{Re}\alpha>|\operatorname{Re}\nu|+|\operatorname{Re}\sigma|-1/2;\ |\arg c|<3\pi/2,\ \begin{Bmatrix} b>0;\ \operatorname{Re}(\alpha+\rho)<3/2 \\ \operatorname{Re} b>0\end{Bmatrix}\right],$$

$$E(\gamma)=\begin{Bmatrix}-2\\ \pi\end{Bmatrix}\frac{2^{\nu-1}}{\pi b^\nu c^{\alpha-\nu}}\Gamma\left[\nu,\ \frac{1}{2}+\alpha-\nu+\sigma,\ \frac{1}{2}+\alpha-\nu-\sigma\right]\times$$
$$\times\Gamma^\gamma\left(\frac{1-\gamma}{2}-\rho-\gamma(\alpha-\nu)\right)\left[\Gamma\left(\frac{1}{2}-\rho+\sigma\right)\Gamma\left(\frac{1}{2}-\rho-\sigma\right)\right]^{-(\gamma+1)/2}\times$$
$$\times{}_4F_3\left(\frac{1+2\alpha-2\nu+2\sigma}{4},\ \frac{3+2\alpha-2\nu+2\sigma}{4},\ \frac{1+2\alpha-2\nu-2\sigma}{4},\ \frac{3+2\alpha-2\nu-2\sigma}{4};\ 1-\nu,\ \frac{1+\alpha-\nu+\gamma\rho}{2},\ 1+\frac{\alpha-\nu+\gamma\rho}{2};\ \mp\frac{b^2}{c^2}\right)+$$
$$+\begin{Bmatrix}-2\cos\nu\pi\\ \pi\end{Bmatrix}\frac{b^\nu}{2^{\nu+1}\pi c^{\alpha+\nu}}\Gamma\left[-\nu,\ \frac{1}{2}+\alpha+\nu+\sigma,\ \frac{1}{2}+\alpha+\nu-\sigma\right]\times$$
$$\times\Gamma^\gamma\left(\frac{1-\gamma}{2}-\rho-\gamma(\alpha+\nu)\right)\left[\Gamma\left(\frac{1}{2}-\rho+\sigma\right)\Gamma\left(\frac{1}{2}-\rho-\sigma\right)\right]^{-(\gamma+1)/2}\times$$
$$\times{}_4F_3\left(\frac{1+2\alpha+2\nu+2\sigma}{4},\ \frac{3+2\alpha+2\nu+2\sigma}{4},\ \frac{1+2\alpha+2\nu-2\sigma}{4},\ \frac{3+2\alpha+2\nu-2\sigma}{4};\ 1+\nu,\ \frac{1+\alpha+\nu+\gamma\rho}{2},\ 1+\frac{\alpha+\nu+\gamma\rho}{2};\ \mp\frac{b^2}{c^2}\right)+$$
$$+\frac{(\gamma+1)c^\rho}{4\pi}\begin{Bmatrix}-2\\ \pi\end{Bmatrix}\left(\frac{2}{b}\right)^{\alpha+\rho}\Gamma\left(\frac{\alpha+\nu+\rho}{2}\right)\Gamma\left(\frac{\alpha-\nu+\rho}{2}\right)\cos^{(1\pm1)/2}\left(\frac{\alpha-\nu+\rho}{2}\pi\right)\times$$
$$\times{}_4F_3\left(\frac{1-2\rho+2\sigma}{4},\ \frac{3-2\rho+2\sigma}{4},\ \frac{1-2\rho-2\sigma}{4},\ \frac{3-2\rho-2\sigma}{4};\ \frac{1}{2},\ 1-\frac{\alpha+\nu+\rho}{2},\ 1-\frac{\alpha-\nu+\rho}{2};\ \mp\frac{b^2}{c^2}\right)-\frac{(\gamma+1)c^{\rho-1}}{2^4\pi}\begin{Bmatrix}-2\\ \pi\end{Bmatrix}\left(\frac{2}{b}\right)^{\alpha+\rho-1}\times$$
$$\times\Gamma\left(\frac{\alpha+\nu+\rho-1}{2}\right)\Gamma\left(\frac{\alpha-\nu+\rho-1}{2}\right)\sin^{(1\pm1)/2}\left(\frac{\alpha-\nu+\rho}{2}\pi\right)[(1-2\rho)^2-4\sigma^2]\times$$
$$\times{}_4F_3\left(\frac{3-2\rho+2\sigma}{4},\ \frac{5-2\rho+2\sigma}{4},\ \frac{3-2\rho-2\sigma}{4},\ \frac{5-2\rho-2\sigma}{4};\ \frac{3}{2},\ \frac{3-\alpha-\nu-\rho}{2},\ \frac{3-\alpha+\nu-\rho}{2};\ \mp\frac{b^2}{c^2}\right).$$

4. $\displaystyle\int_0^\infty x^{\alpha-1}e^{-cx/2}\begin{Bmatrix}Y_\nu(bx)\\K_\nu(bx)\end{Bmatrix}W_{\rho,\sigma}(cx)\,dx=E(-1)$

$$\left[\operatorname{Re} c>0;\ \operatorname{Re}\alpha>|\operatorname{Re}\nu|+|\operatorname{Re}\sigma|-1/2;\ |\arg c|<3\pi/2,\ \begin{Bmatrix} b>0 \\ \operatorname{Re} b>0\end{Bmatrix},\ E(\gamma)\ \text{see 2.19.13.3}\right].$$

5. $\int\limits_0^\infty x^{\alpha-1} e^{(\pm b-c/2)x} K_\nu(bx) M_{\rho,\sigma}(cx)\,dx = U(\pm 1, 0)$

$$\left[\operatorname{Re} c>0;\ \operatorname{Re}(\alpha+\sigma)>|\operatorname{Re}\nu|-1/2,\ \begin{Bmatrix}\operatorname{Re}(\alpha-\rho)<1/2; & |\arg b|<3\pi/2\\ \operatorname{Re} b>0\end{Bmatrix}.\ U(\varepsilon,\gamma)\ \text{see 2.19.12.3}\right].$$

6. $\int\limits_0^\infty x^{\rho-3/2} e^{-(b+c/2)x} K_\sigma(bx) M_{\rho,\sigma}(cx)\,dx =$

$$= \frac{c^{\sigma+1/2}\sqrt{\pi}}{(2b)^{\rho+\sigma}}\Gamma\begin{bmatrix}\rho,\ \rho+2\sigma\\ \rho+\sigma+1/2\end{bmatrix} {}_2F_1\left(\rho,\ \rho+2\sigma;\ 2\sigma+1;\ -\frac{c}{2b}\right)$$

$[\operatorname{Re} b,\ \operatorname{Re} c,\ \operatorname{Re}\rho,\ \operatorname{Re}(\rho+2\sigma)>0]$.

7. $\int\limits_0^\infty x^{\alpha-1} e^{(b\pm c/2)x} K_\nu(bx) W_{\rho,\sigma}(cx)\,dx = U(1, \pm 1)$

$$\left[\operatorname{Re}\alpha>|\operatorname{Re}\nu|+|\operatorname{Re}\sigma|-1/2;\ |\arg b|<3\pi/2,\ \begin{Bmatrix}\operatorname{Re}(\alpha+\rho)<1/2; & |\arg c|<3\pi/2\\ \operatorname{Re} c>0\end{Bmatrix}.\ U(\varepsilon,\gamma)\ \text{see 2.19.12.3}\right].$$

8. $\int\limits_0^\infty x^{\alpha-1} e^{(-b\pm c/2)x} K_\nu(bx) W_{\rho,\sigma}(cx)\,dx = U(-1, \pm 1)$

$[\operatorname{Re} b>0;\ \operatorname{Re}\alpha>|\operatorname{Re}\nu|+|\operatorname{Re}\sigma|-1/2;\ |\arg c|<(1\pm 1/2)\pi;\ U(\varepsilon,\gamma)$ see 2.19.12.3].

9. $\int\limits_0^\infty x^{-\rho-3/2} e^{(c/2-b)x} K_\sigma(bx) W_{\rho,\sigma}(cx)\,dx =$

$$= 2^{3\rho-\sigma+1}\pi b^{\rho-\sigma} c^{\sigma+1/2}\Gamma\begin{bmatrix}-\rho,\ 2\sigma-\rho,\ -2\sigma-\rho\\ 1/2-\rho,\ 1/2+\sigma-\rho,\ 1/2-\sigma-\rho\end{bmatrix}\times$$

$$\times {}_2F_1\left(-\rho,\ 2\sigma-\rho;\ -2\rho;\ 1-\frac{c}{2b}\right)\quad [\operatorname{Re} b>0;\ \operatorname{Re}\rho<-2|\operatorname{Re}\sigma|;\ |\arg c|<3\pi/2].$$

10. $\int\limits_0^\infty x^{\alpha-1} e^{-cx/2} \begin{Bmatrix} Y_\nu(b\sqrt{x})\\ K_\nu(b\sqrt{x})\end{Bmatrix} M_{\rho,\sigma}(cx)\,dx = F(0)$

$$\left[\operatorname{Re} c>0;\ 2\operatorname{Re}(\alpha+\sigma)>|\operatorname{Re}\nu|-1,\ \begin{Bmatrix} b>0; & \operatorname{Re}(\alpha-\rho)<3/4\\ \operatorname{Re} b>0\end{Bmatrix}\right],$$

$$F(\gamma) = \begin{Bmatrix}-2\\ \pi\end{Bmatrix}\frac{1}{2\pi c^\alpha}\Gamma^{1-\gamma^2}\begin{bmatrix}2\sigma+1\\ \rho+\sigma+1/2\end{bmatrix}\left[\Gamma\left(\frac{1}{2}-\rho+\sigma\right)\times\right.$$

$$\left.\times\Gamma\left(\frac{1}{2}-\rho-\sigma\right)\right]^{-(\gamma^2+\gamma)/2}\times$$

$$\times\left\{\left(\frac{2\sqrt{c}}{b}\right)^\nu \Gamma(\nu)\,\Gamma\left(\frac{1-\nu}{2}+\alpha+\sigma\right)\Gamma^{2\gamma^2-1}\left(\frac{1}{2}+(1-2\gamma^2)\left(\sigma-\alpha+\frac{\nu}{2}\right)\right)\times\right.$$

$$\times\Gamma^{-\gamma^2+\gamma+1}\left((1-2\gamma^2)\rho+\frac{\gamma^2-\gamma}{2}+(\gamma^2-\gamma-1)\left(\alpha-\frac{\nu}{2}\right)\right)\times$$

$$\times {}_2F_2\left(\frac{1-\nu}{2}+\alpha+\sigma,\ \frac{1-\nu}{2}+\alpha-\sigma;\ 1-\nu,\ 1+\alpha-\frac{\nu}{2}+(\gamma^2+\gamma-1)\rho;\right.$$

$$\left.\mp(-1)^{(\gamma^2+\gamma)/2}\frac{b^2}{4c}\right)+$$

$$+\left(\frac{b}{2\sqrt{c}}\right)^{\nu}\Gamma(-\nu)\begin{Bmatrix}\cos\nu\pi\\ 1\end{Bmatrix}\Gamma\left(\frac{1+\nu}{2}+\alpha+\sigma\right)\Gamma^{2\gamma^2-1}\left(\frac{1}{2}+(1-2\gamma^2)\left(\sigma-\alpha-\frac{\nu}{2}\right)\right)\times$$

$$\times\Gamma^{-\gamma^2+\gamma+1}\left((1-2\gamma^2)\rho+\frac{\gamma^2-\gamma}{2}+(\gamma^2-\gamma-1)\left(\alpha+\frac{\nu}{2}\right)\right)\times$$

$$\times{}_2F_2\left(\frac{1+\nu}{2}+\alpha+\sigma,\ \frac{1+\nu}{2}+\alpha-\sigma;\ 1+\nu,\ 1+\alpha+\frac{\nu}{2}+(\gamma^2+\gamma-1)\rho;\right.$$

$$\left.\left.\mp(-1)^{(\gamma^2+\gamma)/2}\frac{b^2}{4c}\right)\right\}-\frac{\gamma^2-\gamma-2}{4\pi c^{(1-2\gamma)\rho}}\begin{Bmatrix}-2\cos(\alpha-(1-2\gamma)\rho-\nu/2)\pi\\ \pi\end{Bmatrix}\times$$

$$\times\left(\frac{b}{2}\right)^{2(1-2\gamma)\rho-2\alpha}\Gamma^{1-\gamma}\begin{bmatrix}2\sigma+1\\ 1/2-\rho+\sigma\end{bmatrix}\Gamma\left(\alpha-(1-2\gamma)\rho+\frac{\nu}{2}\right)\times$$

$$\times\Gamma\left(\alpha-(1-2\gamma)\rho-\frac{\nu}{2}\right){}_2F_2\left(\frac{1}{2}+\sigma+(1-2\gamma)\rho,\ \frac{1}{2}-\sigma+(1-2\gamma)\rho;\right.$$

$$\left.1-\alpha+(1-2\gamma)\rho-\frac{\nu}{2},\ 1-\alpha+(1-2\gamma)\rho+\frac{\nu}{2};\ \mp(-1)^{\nu}\frac{b^2}{4c}\right).$$

11. $$\int_0^\infty x^{\sigma+(\pm\nu-1)/2}e^{-cx/2}Y_\nu\left(b\sqrt{x}\right)M_{\rho,\sigma}(cx)\,dx=$$

$$=\frac{1}{\pi}\left(\frac{b}{2}\right)^{\rho-\sigma-3/2}c^{(1/2\mp\nu-\rho-\sigma)/2}\left[\Gamma(2\sigma+1)\Gamma\left(\frac{1}{2}-\rho-\sigma\right)\right]^{-(1\mp1)/2}\times$$

$$\times\exp\left(-\frac{b^2}{8c}\right)\left\{\Gamma\begin{bmatrix}2\sigma\pm\nu+1\\ 3/2\pm\nu-\rho+\sigma\end{bmatrix}\cos\left(2\sigma+\frac{\nu\mp\nu}{2}\right)\pi\times\right.$$

$$\left.\times M_{\varphi,\psi}\left(\frac{b^2}{4c}\right)+\sin\left(\sigma-\rho-\frac{\nu\mp\nu}{2}\right)\pi W_{\varphi,\psi}\left(\frac{b^2}{4c}\right)\right\}$$

$[2\varphi=3\sigma\pm\nu+\rho+1/2,\ 2\psi=\sigma\pm\nu-\rho+1/2;\ b,\ \mathrm{Re}\,c>0;\ \mathrm{Re}\,(2\sigma\pm\nu)>-1;\ -1/2<\mathrm{Re}\,\sigma<1/4+\mathrm{Re}\,(\rho\mp\nu/2)]$.

12. $$\int_0^\infty x^{-(2n+1)/4}e^{-cx/2}Y_{2\sigma+n-1/2}\left(b\sqrt{x}\right)M_{\rho,\sigma}(cx)\,dx=$$

$$=(-1)^n2^{\sigma-\rho+3/2}c^{(\sigma-\rho+n)/2}b^{\rho-\sigma-1/2}\times$$

$$\times\Gamma\begin{bmatrix}2\sigma+1\\ \rho+\sigma+1/2\end{bmatrix}\exp\left(-\frac{b^2}{8c}\right)W_{(\rho+\sigma-n+1)/2,\ (\rho+\sigma+n-1)/2}\left(\frac{b^2}{4c}\right)$$

$[b,\ \mathrm{Re}\,c>0;\ n\geqslant-1;\ \mathrm{Re}\,\sigma>-1/2;\ \mathrm{Re}\,\rho>-n/2]$.

13. $$\int_0^\infty x^{(2\nu-2n-1)/4}e^{-cx/2}Y_\nu\left(b\sqrt{x}\right)M_{\rho,\ (1-2n)/4}(cx)\,dx=$$

$$=(-1)^n2^{(7-2n)/4-\rho}b^{\rho+(2n-7)/4}c^{(2n+1)/8-(\nu+\rho)/2}\times$$

$$\times\Gamma\begin{bmatrix}3/2-n\\ (3-2n)/4+\rho\end{bmatrix}\exp\left(-\frac{b^2}{8c}\right)W_{\varphi,\psi}\left(\frac{b^2}{4c}\right)$$

$[2\varphi=\rho+\nu-(3n-5)/4,\ 2\psi=\rho-\nu+(2n-3)/4;\ b,\ \mathrm{Re}\,c>0;\ \mathrm{Re}\,\nu>(6n-7)/4;\ \mathrm{Re}\,(\nu-2\rho)<n]$.

14. $$\int_0^\infty x^{\sigma+(\nu-1)/2}e^{-cx/2}K_\nu\left(b\sqrt{x}\right)M_{\rho,\sigma}(cx)\,dx=$$

$$=2^{\sigma-\rho+1/2}b^{\rho-\sigma-3/2}c^{(1/2-\nu-\rho-\sigma)/2}\Gamma(2\sigma+1)\times$$

$$\times\Gamma(\nu+2\sigma+1)\exp\left(\frac{b^2}{8c}\right)W_{\varphi,\psi}\left(\frac{b^2}{4c}\right)$$

$[2\varphi=1/2-\nu-\rho-3\sigma,\ 2\psi=1/2+\nu-\rho+\sigma;\ \mathrm{Re}\,b,\ \mathrm{Re}\,c>0;\ \mathrm{Re}\,\sigma>-1/2;\ \mathrm{Re}\,(\nu+2\sigma)>-1]$.

15. $\int\limits_0^\infty x^{\alpha-1} e^{cx/2} \left\{ \begin{matrix} Y_\nu(b\sqrt{x}) \\ K_\nu(b\sqrt{x}) \end{matrix} \right\} W_{\rho,\sigma}(cx)\, dx = F(1)$

$\left[2\operatorname{Re}\alpha > |\operatorname{Re}\nu| + 2|\operatorname{Re}\sigma| - 1;\ |\arg c| < 3\pi/2;\ \left\{ \begin{matrix} b>0;\ \operatorname{Re}(\alpha+\rho)<3/4 \\ \operatorname{Re} b>0 \end{matrix} \right\},\ F(\gamma)\ \text{see 2.19.13.10}\right].$

16. $\int\limits_0^\infty x^{\alpha-1} e^{-cx/2} \left\{ \begin{matrix} Y_\nu(b\sqrt{x}) \\ K_\nu(b\sqrt{x}) \end{matrix} \right\} W_{\rho,\sigma}(cx)\, dx = F(-1)$

$[\operatorname{Re} b,\ \operatorname{Re} c>0;\ 2\operatorname{Re}\alpha > |\operatorname{Re}\nu| + 2|\operatorname{Re}\sigma| - 1;\ F(\gamma)\ \text{see 2.19.13.10}].$

17. $\int\limits_0^\infty x^{-1/2} J_{2\nu+2\sigma}(b\sqrt{x})\, K_\nu\left(\frac{cx}{2}\right) M_{\rho,\sigma}(cx)\, dx =$

$$= \frac{1}{b}\Gamma\left[\begin{matrix} 2\sigma+1 \\ \rho+\nu+\sigma+1 \end{matrix}\right] W_{(\rho-\sigma)/2,\ (\rho-2\nu-2\sigma)/2}\left(\frac{b^2}{2c}\right) \times$$

$$\times M_{(\rho+\sigma)/2,\ (\rho+2\nu+2\sigma)/2}\left(\frac{b^2}{2c}\right) \qquad [b,\ \operatorname{Re} c>0;\ \operatorname{Re}\rho>-1/4;\ \operatorname{Re}\sigma,\ \operatorname{Re}(\nu+\sigma)>-1/2].$$

18. $\int\limits_0^\infty x^{\alpha-1} e^{-cx/2} I_\mu(b\sqrt{x})\, K_\nu(b\sqrt{x})\, M_{\rho,\sigma}(cx)\, dx = G(0)$

$[\operatorname{Re} c>0;\ (|\operatorname{Re}\nu| - \operatorname{Re}(\mu+2\sigma)-1)/2 < \operatorname{Re}\alpha < \operatorname{Re}\rho + 1/2],$

$$G(\gamma) = \frac{2^{-\mu-1} b^\mu}{c^{\alpha+\mu/2}\Gamma(\mu+1)} \Gamma^{1-\gamma^2}\left[\begin{matrix} 2\sigma+1 \\ \rho+\sigma+1/2 \end{matrix}\right] \times$$

$$\times\left[\Gamma\left(\frac{1}{2}-\rho+\sigma\right)\Gamma\left(\frac{1}{2}-\rho-\sigma\right)\right]^{-(\gamma^2+\gamma)/2} \times$$

$$\times[A(\nu)+A(-\nu)] - \frac{\gamma^2-\gamma-2}{4c^{(1-2\gamma)\rho}}\left(\frac{b}{2}\right)^{2(1-2\gamma)\rho-2\alpha} \Gamma^{1-\gamma}\left[\begin{matrix} 2\sigma+1 \\ 1/2-\rho+\sigma \end{matrix}\right] \times$$

$$\times\Gamma\left[\begin{matrix} (\mu+\nu)/2+\alpha-(1-2\gamma)\rho,\ (\mu-\nu)/2+\alpha-(1-2\gamma)\rho,\ 2(1-2\gamma)\rho-2\alpha+1 \\ 1+(\mu-\nu)/2-\alpha+(1-2\gamma)\rho,\ 1+(\mu+\nu)/2-\alpha+(1-2\gamma)\rho \end{matrix}\right] \times$$

$$\times {}_4F_4\left(\frac{1}{2}+(1-2\gamma)\rho-\alpha,\ 1+(1-2\gamma)\rho-\alpha,\ \frac{1}{2}+\sigma+(1-2\gamma)\rho,\ \frac{1}{2}-\sigma+(1-2\gamma)\rho;\right.$$

$$1+(1-2\gamma)\rho-\alpha-\frac{\mu+\nu}{2},\ 1+(1-2\gamma)\rho-\alpha-\frac{\mu-\nu}{2},\ 1+(1-2\gamma)\rho-\alpha+$$

$$\left.+\frac{\mu-\nu}{2},\ 1+(1-2\gamma)\rho-\alpha+\frac{\mu+\nu}{2};\ (-1)^{(\gamma^2+\gamma)/2}\frac{b^2}{c}\right),$$

$$A(\nu) = \left(\frac{b}{2\sqrt{c}}\right)^\nu \Gamma(-\nu)\,\Gamma\left(\alpha+\sigma+\frac{\mu+\nu+1}{2}\right) \times$$

$$\times\Gamma^{2\gamma^2-1}\left(\frac{1}{2}+(1-2\gamma^2)\left(\sigma-\alpha-\frac{\mu+\nu}{2}\right)\right) \times$$

$$\times\Gamma^{-\gamma^2+\gamma+1}\left((1-2\gamma^2)\rho+\frac{\gamma^2-\gamma}{2}+(\gamma^2-\gamma-1)\left(\alpha+\frac{\mu+\nu}{2}\right)\right) \times$$

$$\times {}_4F_4\left(\frac{\mu+\nu}{2}+1,\ \frac{\mu+\nu+1}{2},\ \frac{\mu+\nu+1}{2}+\alpha+\sigma,\ \frac{\mu+\nu+1}{2}+\alpha-\sigma;\right.$$

$$\left.\mu+1,\ \mu+\nu+1,\ \nu+1,\ 1+\alpha+\frac{\mu+\nu}{2}+(\gamma^2+\gamma-1)\rho,\ (-1)^{(\gamma^2+\gamma)/2}\frac{b^2}{c}\right).$$

19. $\int\limits_0^\infty x^{\alpha-1} e^{\pm cx/2} I_\mu(b\sqrt{x}) K_\nu(b\sqrt{x}) W_{\rho,\sigma}(cx)\,dx = G(\pm 1)$

$\left[\operatorname{Re}(2\alpha+\mu)>|\operatorname{Re}\nu|+2|\operatorname{Re}\sigma|-1;\ |\arg b|<\pi,\ \begin{Bmatrix}\operatorname{Re}(\alpha+\rho)<1/2;\ |\arg c|<3\pi/2\\ \operatorname{Re}c>0\end{Bmatrix},\right.$

$\left. G(\gamma)\ \text{see } 2.19.13.18\right].$

2.19.14. Integrals of $x^\alpha e^{\pm cx/2}\mathbf{H}_\nu(b\sqrt{x})\begin{Bmatrix}M_{\rho,\sigma}(cx)\\ W_{\rho,\sigma}(cx)\end{Bmatrix}$.

1. $\int\limits_0^\infty x^{\alpha-1} e^{-cx/2}\mathbf{H}_\nu(b\sqrt{x}) M_{\rho,\sigma}(cx)\,dx = H(0)$

$[b,\ \operatorname{Re}c>0;\ -\operatorname{Re}(\sigma+\nu/2)-1<\operatorname{Re}\alpha<\operatorname{Re}(\rho-\nu/2)+1/2,\ \operatorname{Re}\rho+3/4],$

$$H(\gamma)=\frac{b^{\nu+1}}{2^\nu\sqrt{\pi}c^{\alpha+(\nu+1)/2}}\Gamma\begin{bmatrix}\alpha+\sigma+\nu/2+1\\ \nu+3/2\end{bmatrix}\times$$

$$\times\Gamma^{1-\gamma^2}\begin{bmatrix}2\sigma+1\\ \rho+\sigma+1/2\end{bmatrix}\left[\Gamma\left(\frac{1}{2}-\rho+\sigma\right)\Gamma\left(\frac{1}{2}-\rho-\sigma\right)\right]^{-(\gamma^2+\gamma)/2}\times$$

$$\times\Gamma^{2\gamma^2-1}\left(\gamma^2+(1-2\gamma^2)\left(\sigma-\alpha-\frac{\nu}{2}\right)\right)\Gamma^{-\gamma^2+\gamma+1}\left((1-2\gamma^2)\rho+\frac{1}{2}+(\gamma^2-\gamma-1)\times\right.$$

$$\left.\times\left(\alpha+\frac{\nu}{2}+1\right)\right){}_3F_3\left(1,\ 1+\alpha+\sigma+\frac{\nu}{2},\ 1+\alpha-\sigma+\frac{\nu}{2};\ \frac{3}{2},\ \frac{3}{2}+\nu,\right.$$

$$\left.\frac{3+\nu}{2}+\alpha+(\gamma^2+\gamma-1)\rho;\ (-1)^{1-(\gamma^2+\gamma)/2}\frac{b^2}{4c}\right)-\frac{(\gamma^2-\gamma-2)\pi}{2c^{(1-2\gamma)\rho}}\left(\frac{b}{2}\right)^{2(1-2\gamma)\rho-2\alpha}\times$$

$$\times\sec\left(\frac{2\alpha+\nu-2(1-2\gamma)\rho}{2}\pi\right)\Gamma^{1-\gamma}\begin{bmatrix}2\sigma+1\\ 1/2-\rho+\sigma\end{bmatrix}\times$$

$$\times\left[\Gamma\left(1+\frac{\nu}{2}-\alpha+(1-2\gamma)\rho\right)\Gamma\left(1-\frac{\nu}{2}-\alpha+(1-2\gamma)\rho\right)\right]^{-1}\times$$

$$\times{}_2F_2\left(\frac{1}{2}+\sigma+(1-2\gamma)\rho,\ \frac{1}{2}-\sigma+(1-2\gamma)\rho;\right.$$

$$\left.1+\frac{\nu}{2}+(1-2\gamma)\rho-\alpha,\ 1-\frac{\nu}{2}+(1-2\gamma)\rho-\alpha;\ (-1)^{1-\gamma}\frac{b^2}{4c}\right).$$

2. $\int\limits_0^\infty x^{\alpha-1} e^{\pm cx/2}\mathbf{H}_\nu(b\sqrt{x}) W_{\rho,\sigma}(cx)\,dx = H(\pm 1)$

$\left[b>0;\ \operatorname{Re}(2\alpha+\nu)>2|\operatorname{Re}\sigma|-2,\ \begin{Bmatrix}\operatorname{Re}(\alpha+\rho)<(1-\operatorname{Re}\nu)/2,\ 3/4;\ |\arg c|<3\pi/2\\ \operatorname{Re}c>0\end{Bmatrix},\right.$

$\left. H(\gamma)\ \text{see } 2.19.14.1\right].$

3. $\int\limits_0^\infty x^{-1/2} e^{cx/2}\mathbf{H}_{2\sigma}(b\sqrt{x}) W_{\rho,\sigma}(cx)\,dx =$

$$=2^{\rho-\sigma+1}b^{\sigma-\rho-1}c^{(\rho-\sigma)/2}\Gamma\begin{bmatrix}-\rho-\sigma\\ 1/2-\rho+\sigma,\ 1/2-\rho-\sigma\end{bmatrix}\times$$

$$\times\exp\left(\frac{b^2}{8c}\right)W_{(\rho+\sigma)/2,\ (\rho+\sigma+1)/2}\left(\frac{b^2}{4c}\right)$$

$[b>0;\ -3/4<\operatorname{Re}\sigma<-\operatorname{Re}\rho;\ \operatorname{Re}\rho<1/4;\ |\arg c|<3\pi/2].$

4. $\int\limits_0^\infty x^{-1/2} e^{cx/2} \mathbf{H}_{2\sigma}(b\sqrt{x})\, W_{-\sigma-1/2,\,\sigma}(cx)\,dx =$

$$= \frac{\pi (b/2)^{2\sigma}}{c^{\sigma+1/2}\Gamma(2\sigma+1)} \exp\left(\frac{b^2}{4c}\right) \operatorname{erfc}\left(\frac{b}{2\sqrt{c}}\right) \quad [b>0;\ \operatorname{Re}\sigma>-3/4;\ |\arg c|<3\pi/2].$$

2.19.15. Integrals of $A(x)\, e^{\pm cx/2} P_n(ax^{\pm r}-b) \begin{Bmatrix} M_{\rho,\sigma}(cx) \\ W_{\rho,\sigma}(cx) \end{Bmatrix}$.

1. $\int\limits_0^a x^{\alpha-1} e^{-cx/2} P_n\left(\frac{2x}{a}-1\right) M_{\rho,\sigma}(cx)\,dx = U(0) \quad [a>0;\ \operatorname{Re}(\alpha+\sigma)>-1/2],$

$$U(\gamma) = (-1)^n a^{\alpha+\sigma+1/2} c^{\sigma+1/2} \Gamma^{\gamma^2}\begin{bmatrix} -2\sigma \\ 1/2-\rho-\sigma \end{bmatrix} \frac{(1/2-\alpha-\sigma)_n}{(\alpha+\sigma+1/2)_{n+1}} \times$$
$$\times {}_3F_3\left(\frac{1}{2}+\sigma-(\gamma^2+\gamma-1)\rho,\ \frac{1}{2}+\alpha+\sigma,\ \frac{1}{2}+\alpha+\sigma;\right.$$
$$\left. 2\sigma+1,\ \frac{3}{2}+\alpha+\sigma+n,\ \frac{1}{2}+\alpha+\sigma-n;\ (-1)^{1-(\gamma^2+\gamma)/2} ac\right) +$$
$$+\gamma^2 (-1)^n a^{\alpha-\sigma+1/2} c^{1/2-\sigma} \Gamma\begin{bmatrix} 2\sigma \\ 1/2-\rho+\sigma \end{bmatrix} \frac{(1/2-\alpha+\sigma)_n}{(1/2+\alpha-\sigma)_{n+1}} \times$$
$$\times {}_3F_3\left(\frac{1}{2}-\sigma-(\gamma^2+\gamma-1)\rho,\ \frac{1}{2}+\alpha-\sigma,\ \frac{1}{2}+\alpha-\sigma;\right.$$
$$\left. 1-2\sigma,\ \frac{3}{2}+\alpha-\sigma+n,\ \frac{1}{2}+\alpha-\sigma-n;\ (-1)^{(1-\gamma)/2} ac\right).$$

2. $\int\limits_0^a x^{\alpha-1} e^{\pm cx/2} P_n\left(\frac{2x}{a}-1\right) W_{\rho,\sigma}(cx)\,dx = U(\pm 1)$

$[a>0;\ \operatorname{Re}\alpha > |\operatorname{Re}\sigma|-1/2;\ U(\gamma)$ see 2.19.15.1].

3. $\int\limits_0^a x^{\alpha-1} e^{-cx/2} P_n\left(\frac{2a}{x}-1\right) M_{\rho,\sigma}(cx)\,dx = V(0)$

$[a>0;\ \operatorname{Re}(\alpha+\sigma)>n-1/2],$

$$V(\gamma) = (-1)^{n+1} a^{\alpha+\sigma+1/2} c^{\sigma+1/2} \Gamma^{\gamma^2}\begin{bmatrix} -2\sigma \\ 1/2-\rho-\sigma \end{bmatrix} \frac{(3/2+\alpha+\sigma)_n}{(-1/2-\alpha-\sigma)_{n+1}} \times$$
$$\times {}_3F_3\left(\frac{1}{2}+\sigma-(\gamma^2+\gamma-1)\rho,\ \frac{1}{2}+\alpha+\sigma-n,\ \frac{3}{2}+\alpha+\sigma+n;\right.$$
$$\left. \frac{3}{2}+\alpha+\sigma,\ \frac{3}{2}+\alpha+\sigma,\ 2\sigma+1;\ (-1)^{1-(\gamma^2+\gamma)/2} ac\right) +$$
$$+\gamma^2 \frac{(-1)^{n+1} a^{\alpha-\sigma+1/2} c^{1/2-\sigma}}{n!} \Gamma\begin{bmatrix} 2\sigma \\ 1/2-\rho+\sigma \end{bmatrix} \frac{(3/2+\alpha-\sigma)_n}{(\sigma-\alpha-1/2)_{n+1}} \times$$
$$\times {}_3F_3\left(\frac{1}{2}-\sigma-(\gamma^2+\gamma-1)\rho,\ \frac{1}{2}+\alpha-\sigma-n,\ \frac{3}{2}+\alpha-\sigma+n;\right.$$
$$\left. \frac{3}{2}+\alpha-\sigma,\ \frac{3}{2}+\alpha-\sigma,\ 1-2\sigma;\ (-1)^{(1-\gamma)/2} ac\right).$$

4. $\int\limits_0^a x^{\alpha-1} e^{\pm cx/2} P_n\left(\frac{2a}{x}-1\right) W_{\rho,\sigma}(cx)\,dx = V(\pm 1)$

$[a>0;\ \operatorname{Re}\alpha > |\operatorname{Re}\sigma|+n-1/2;\ V(\gamma)$ see 2.19.15.3].

5. $\int\limits_a^\infty x^{\alpha-1}e^{-cx/2}P_n\left(\frac{2x}{a}-1\right)M_{\rho,\,\sigma}(cx)\,dx=W(0)$ [a, $\operatorname{Re} c>0$; $\operatorname{Re}(\alpha-\rho)<-n$],

$$W(\gamma)=(-1)^{n-1}a^{\alpha+\sigma+1/2}c^{\sigma+1/2}\Gamma^{\gamma^2}\begin{bmatrix}-2\sigma\\1/2-\rho-\sigma\end{bmatrix}\frac{(1/2-\alpha-\sigma)_n}{(1/2+\alpha+\sigma)_{n+1}}\times$$
$$\times\,{}_3F_3\left(\frac{1}{2}+\alpha+\sigma,\ \frac{1}{2}+\alpha+\sigma,\ \frac{1}{2}+\sigma-(\gamma^2+\gamma-1)\rho;\right.$$
$$\left.1+2\sigma,\ \frac{1}{2}+\alpha+\sigma-n,\ \frac{3}{2}+\alpha+\sigma+n;\ (-1)^{1-(\gamma^2+\gamma)/2}\,ac\right)+$$
$$+\gamma^2(-1)^{n-1}a^{1/2+\alpha-\sigma}c^{1/2-\sigma}\Gamma\begin{bmatrix}2\sigma\\1/2-\rho+\sigma\end{bmatrix}\frac{(1/2-\alpha+\sigma)_n}{(1/2+\alpha-\sigma)_{n+1}}\times$$
$$\times\,{}_3F_3\left(\frac{1}{2}+\alpha-\sigma,\ \frac{1}{2}+\alpha-\sigma,\ \frac{1}{2}-\sigma-(\gamma^2+\gamma-1)\rho;\right.$$
$$\left.1-2\sigma,\ \frac{1}{2}+\alpha-\sigma-n,\ \frac{3}{2}+\alpha-\sigma+n;\ (-1)^{(1-\gamma)/2}\,ac\right)+$$
$$+\frac{c^{-\alpha-n}(2n)!}{a^n(n!)^2}\Gamma\left(\frac{1}{2}+\alpha+\sigma+n\right)\Gamma^{1-\gamma^2}\begin{bmatrix}2\sigma+1\\1/2+\rho+\sigma\end{bmatrix}\times$$
$$\times\left[\Gamma\left(\frac{1}{2}-\rho+\sigma\right)\Gamma\left(\frac{1}{2}-\rho-\sigma\right)\right]^{-(\gamma^2+\gamma)/2}\Gamma^{2\gamma^2-1}\left(\frac{1}{2}+(1-2\gamma^2)(\sigma-\alpha-n)\right)\times$$
$$\times\,\Gamma^{-\gamma^2+\gamma+1}\left((1-2\gamma^2)\rho+\frac{\gamma^2-\gamma}{2}+(\gamma^2-\gamma-1)(\alpha+n)\right)\times$$
$$\times\,{}_3F_3\left(-n,\ -n,\ -\alpha-n-(\gamma^2+\gamma-1)\rho;\right.$$
$$\left.-2n,\ \frac{1}{2}-\alpha-\sigma-n,\ \frac{1}{2}-\alpha+\sigma-n;\ (-1)^{1-(\gamma^2+\gamma)/2}\,ac\right).$$

6. $\int\limits_a^\infty x^{\alpha-1}e^{\pm cx/2}P_n\left(\frac{2x}{a}-1\right)W_{\rho,\,\sigma}(cx)\,dx=W(\pm 1)$

$\left[a>0,\ \begin{Bmatrix}\operatorname{Re}(\alpha+\rho)<-n;\ |\arg c|<3\pi/2\\ \operatorname{Re} c>0\end{Bmatrix},\ W(\gamma)\text{ see } 2.19.15.5\right]$.

7. $\int\limits_a^\infty x^{\alpha-1}e^{-cx/2}P_n\left(\frac{2a}{x}-1\right)M_{\rho,\,\sigma}(cx)\,dx=X(0)$ [a, $\operatorname{Re} c$, $\operatorname{Re}(\rho-\alpha)>0$],

$$X(\gamma)=(-1)^n a^{\alpha+\sigma+1/2}c^{\sigma+1/2}\Gamma^{\gamma^2}\begin{bmatrix}-2\sigma\\1/2-\rho-\sigma\end{bmatrix}\frac{(3/2+\alpha+\sigma)_n}{(-1/2-\alpha-\sigma)_{n+1}}\times$$
$$\times\,{}_3F_3\left(\frac{1}{2}+\alpha+\sigma-n,\ \frac{3}{2}+\alpha+\sigma+n,\ \frac{1}{2}+\sigma-(\gamma^2+\gamma-1)\rho;\right.$$
$$\left.\frac{3}{2}+\alpha+\sigma,\ \frac{3}{2}+\alpha+\sigma,\ 1+2\sigma;\ (-1)^{1-(\gamma^2+\gamma)/2}\,ac\right)+$$
$$+(-1)^n\gamma^2a^{\alpha-\sigma+1/2}c^{1/2-\sigma}\Gamma\begin{bmatrix}2\sigma\\1/2-\rho+\sigma\end{bmatrix}\frac{(3/2+\alpha-\sigma)_n}{(\sigma-\alpha-1/2)_{n+1}}\times$$
$$\times\,{}_3F_3\left(\frac{1}{2}+\alpha-\sigma-n,\ \frac{3}{2}+\alpha-\sigma+n,\ \frac{1}{2}-\sigma-(\gamma^2+\gamma-1)\rho;\right.$$
$$\left.\frac{3}{2}+\alpha-\sigma,\ \frac{3}{2}+\alpha-\sigma,\ 1-2\sigma;\ (-1)^{(1-\gamma)/2}\,ac\right)+$$
$$+\frac{(-1)^n}{c^\alpha}\Gamma\left(\frac{1}{2}+\alpha+\sigma\right)\Gamma^{1-\gamma^2}\begin{bmatrix}2\sigma+1\\\rho+\sigma+1/2\end{bmatrix}\times$$

$$\times\left[\Gamma\left(\frac{1}{2}-\rho+\sigma\right)\Gamma\left(\frac{1}{2}-\rho-\sigma\right)\right]^{-(\gamma^2+\gamma)/2}\Gamma^{2\gamma^2-1}\left(\frac{1}{2}+(1-2\gamma^2)(\sigma-\alpha)\right)\times$$

$$\times\,\Gamma^{-\gamma^2+\gamma+1}\left((1-2\gamma^2)\rho+\frac{\gamma^2-\gamma}{2}+(\gamma^2-\gamma-1)\alpha\right)\times$$

$$\times\,{}_3F_3\left(-n,\ n+1,\ -\alpha-(\gamma^2+\gamma-1)\rho;\ 1,\ \frac{1}{2}-\alpha-\sigma,\ \frac{1}{2}-\alpha+\sigma;\ (-1)^{1-(\gamma^2+\gamma)/2}ac\right)$$

8. $$\int_a^\infty x^{\alpha-1}e^{\pm cx/2}P_n\left(\frac{2a}{x}-1\right)W_{\rho,\sigma}(cx)\,dx=X(\pm 1)$$

$$\left[a>0,\ \begin{cases}\operatorname{Re}(\alpha+\rho)<0;\ |\arg c|<3\pi/2\\ \operatorname{Re}c>0\end{cases},\ X(\gamma)\ \text{see 2.19.15.7}\right]$$

9. $$\int_0^a x^{\alpha-1}e^{-cx/2}P_{2n+\varepsilon}\left(\sqrt{\frac{x}{a}}\right)M_{\rho,\sigma}(cx)\,dx=Y(0)$$

$$[\varepsilon=0\ \text{or}\ 1;\ a>0;\ \operatorname{Re}(\alpha+\sigma)>-(\varepsilon+1)/2]$$

$$Y(\gamma)=(-1)^n a^{\alpha+\sigma+1/2}c^{\sigma+1/2}\Gamma^{\gamma^2}\left[\begin{matrix}-2\sigma\\ 1/2-\rho-\sigma\end{matrix}\right]\frac{(\varepsilon/2-\alpha-\sigma)_n}{(\alpha+\sigma+(\varepsilon+1)/2)_{n+1}}\times$$

$$\times\,{}_3F_3\left(\frac{1}{2}+\sigma-(\gamma^2+\gamma-1)\rho,\ \frac{1}{2}+\alpha+\sigma,\ 1+\alpha+\sigma;\right.$$

$$\left.1+\alpha+\sigma-\frac{\varepsilon}{2}-n,\ 2\sigma+1,\ \alpha+\sigma+n+\frac{\varepsilon+3}{2};\ (-1)^{1-(\gamma^2+\gamma)/2}ac\right)+$$

$$+\gamma^2(-1)^n a^{\alpha-\sigma+1/2}c^{1/2-\sigma}\Gamma\left[\begin{matrix}2\sigma\\ 1/2-\rho+\sigma\end{matrix}\right]\frac{(\varepsilon/2-\alpha+\sigma)_n}{(\alpha-\sigma+(\varepsilon+1)/2)_{n+1}}\times$$

$$\times\,{}_3F_3\left(\frac{1}{2}-\sigma-\gamma\rho,\ \frac{1}{2}+\alpha-\sigma,\ 1+\alpha-\sigma;\ 1+\alpha-\sigma-\frac{\varepsilon}{2}-n,\right.$$

$$\left.1-2\sigma,\ \alpha-\sigma+\frac{\varepsilon+3}{2}+n;\ (-1)^{(1-\gamma)/2}ac\right)$$

10. $$\int_0^a x^{\alpha-1}e^{\pm cx/2}P_{2n+\varepsilon}\left(\sqrt{\frac{x}{a}}\right)W_{\rho,\sigma}(cx)\,dx=Y(\pm 1)$$

$$[\varepsilon=0\ \text{or}\ 1;\ a>0;\ \operatorname{Re}\alpha>|\operatorname{Re}\sigma|-(\varepsilon+1)/2;\ Y(\gamma)\ \text{see 2.19.15.9}]$$

11. $$\int_0^a x^{\alpha-1}e^{-cx/2}P_{2n+\varepsilon}\left(\sqrt{\frac{a}{x}}\right)M_{\rho,\sigma}(cx)\,dx=Z(0)$$

$$[\varepsilon=0\ \text{or}\ 1;\ a>0;\ \operatorname{Re}(\alpha+\sigma)>n+(\varepsilon-1)/2]$$

$$Z(\gamma)=(-1)^{n+1}a^{\alpha+\sigma+1/2}c^{\sigma+1/2}\Gamma^{\gamma^2}\left[\begin{matrix}-2\sigma\\ 1/2-\rho-\sigma\end{matrix}\right]\frac{(\alpha+\sigma+\varepsilon/2+1)_n}{((\varepsilon-1)/2-\alpha-\sigma)_{n+1}}\times$$

$$\times\,{}_3F_3\left(\frac{1}{2}+\sigma-(\gamma^2+\gamma-1)\rho,\ \alpha+\sigma-n+\frac{1-\varepsilon}{2},\ \alpha+\sigma+n+1+\frac{\varepsilon}{2};\right.$$

$$\left.\alpha+\sigma+1,\ \alpha+\sigma+\frac{3}{2},\ 2\sigma+1;\ (-1)^{1-(\gamma^2+\gamma)/2}ac\right)+$$

$$+\gamma^2(-1)^{n+1}a^{\alpha-\sigma+1/2}c^{1/2-\sigma}\Gamma\left[\begin{matrix}2\sigma\\ 1/2-\rho+\sigma\end{matrix}\right]\frac{(\alpha-\sigma+\varepsilon/2+1)_n}{((\varepsilon-1)/2-\alpha+\sigma)_{n+1}}\times$$

$$\times\,{}_3F_3\left(\frac{1}{2}-\sigma-\gamma\rho,\ \alpha-\sigma-n+\frac{1-\varepsilon}{2},\ \alpha-\sigma+n+1+\frac{\varepsilon}{2};\right.$$

$$\left.1+\alpha-\sigma,\ \frac{3}{2}+\alpha-\sigma,\ 1-2\sigma;\ (-1)^{(1-\gamma)/2}ac\right)$$

12. $$\int_0^a x^{\alpha-1}e^{\pm cx/2}P_{2n+\varepsilon}\left(\sqrt{\frac{a}{x}}\right)W_{\rho,\sigma}(cx)\,dx=Z(\pm 1)$$

[$\varepsilon=0$ or 1; $a>0$; $\operatorname{Re}\alpha>|\operatorname{Re}\sigma|+n+(\varepsilon-1)/2$; $Z(\gamma)$ see 2.19.15.11].

13. $$\int_a^\infty x^{\alpha-1}e^{-cx/2}P_{2n+\varepsilon}\left(\sqrt{\frac{x}{a}}\right)M_{\rho,\sigma}(cx)\,dx=U(0)$$

[$\varepsilon=0$ or 1; a, $\operatorname{Re}c>0$; $\operatorname{Re}(\alpha-\rho)<-\varepsilon/2-n$],

$$U(\gamma)=(-1)^{n+1}a^{\alpha+\sigma+1/2}c^{\sigma+1/2}\Gamma^{\gamma^2}\begin{bmatrix}-2\sigma\\1/2-\rho-\sigma\end{bmatrix}\frac{(\varepsilon/2-\alpha-\sigma)_n}{((1+\varepsilon)/2+\alpha+\sigma)_{n+1}}\times$$
$$\times\,{}_3F_3\left(\frac{1}{2}+\sigma-(\gamma^2+\gamma-1)\rho,\ \frac{1}{2}+\alpha+\sigma,\ 1+\alpha+\sigma;\ 1-\frac{\varepsilon}{2}+\alpha+\sigma-n,\right.$$
$$\left.\frac{3+\varepsilon}{2}+\alpha+\sigma+n,\ 2\sigma+1;\ (-1)^{1-(\gamma^2+\gamma)/2}ac\right)+$$
$$+\gamma^2(-1)^{n+1}a^{\alpha-\sigma+1/2}c^{1/2-\sigma}\Gamma\begin{bmatrix}2\sigma\\1/2-\rho+\sigma\end{bmatrix}\frac{(\varepsilon/2-\alpha+\sigma)_n}{((1+\varepsilon)/2+\alpha-\sigma)_{n+1}}\times$$
$$\times\,{}_3F_3\left(\frac{1}{2}-\sigma-\gamma\rho,\ \frac{1}{2}+\alpha-\sigma,\ 1+\alpha-\sigma;\right.$$
$$\left.1-\frac{\varepsilon}{2}+\alpha-\sigma-n,\ \frac{3+\varepsilon}{2}+\alpha-\sigma+n,\ 1-2\sigma;\ (-1)^{(1-\gamma)/2}ac\right)+$$
$$+\frac{(n+\varepsilon+1/2)_n}{n!\,a^{n+\varepsilon/2}}c^{-\varepsilon/2-\alpha-n}\Gamma^{1-\gamma^2}\begin{bmatrix}2\sigma+1\\\rho+\sigma+1/2\end{bmatrix}\times$$
$$\times\left[\Gamma\left(\frac{1}{2}-\rho-\sigma\right)\Gamma\left(\frac{1}{2}-\rho+\sigma\right)\right]^{-(\gamma^2+\gamma)/2}\times$$
$$\times\,\Gamma\left(\frac{\varepsilon+1}{2}+\alpha+\sigma+n\right)\Gamma^{2\gamma^2-1}\left(\frac{1}{2}+(1-2\gamma^2)\left(\sigma-\frac{\varepsilon}{2}-\alpha-n\right)\right)\times$$
$$\times\,\Gamma^{-\gamma^2+\gamma+1}\left((1-2\gamma^2)\rho+\frac{\gamma^2-\gamma}{2}+(\gamma^2-\gamma-1)\left(\alpha+n+\frac{\varepsilon}{2}\right)\right)\times$$
$$\times\,{}_3F_3\left(-n,\ -\frac{\varepsilon}{2}-\alpha-n-(\gamma^2+\gamma-1)\rho,\ \frac{1}{2}-\varepsilon-n;\right.$$
$$\left.\frac{1}{2}-\varepsilon-2n,\ \frac{1-\varepsilon}{2}-\alpha-\sigma-n,\ \frac{1-\varepsilon}{2}-\alpha+\sigma-n;\ (-1)^{1-(\gamma^2+\gamma)/2}ac\right).$$

14. $$\int_a^\infty x^{\alpha-1}e^{\pm cx/2}P_{2n+\varepsilon}\left(\sqrt{\frac{x}{a}}\right)W_{\rho,\sigma}(cx)\,dx=U(\pm 1)$$

$\left[\varepsilon=0 \text{ or } 1;\ a>0,\ \begin{cases}\operatorname{Re}(\alpha+\rho)<-\varepsilon/2-n; & |\arg c|<3\pi/2\\ \operatorname{Re}c>0\end{cases},\ U(\gamma) \text{ see } 2.19.15.13\right].$

15. $$\int_a^\infty x^{\alpha-1}e^{-cx/2}P_{2n+\varepsilon}\left(\sqrt{\frac{a}{x}}\right)M_{\rho,\sigma}(cx)\,dx=V(0)$$

[$\varepsilon=0$ or 1; a, $\operatorname{Re}c>0$; $\operatorname{Re}(\alpha-\rho)<\varepsilon/2$],

$$V(\gamma)=(-1)^n a^{\alpha+\sigma+1/2}c^{\sigma+1/2}\Gamma^{\gamma^2}\begin{bmatrix}-2\sigma\\1/2-\rho-\sigma\end{bmatrix}\frac{(\varepsilon/2+\alpha+\sigma+1)_n}{((\varepsilon-1)/2-\alpha-\sigma)_{n+1}}\times$$
$$\times\,{}_3F_3\left(\frac{1-\varepsilon}{2}+\alpha+\sigma-n,\ \frac{1}{2}+\sigma-(\gamma^2+\gamma-1)\rho,\ \frac{\varepsilon}{2}+\alpha+\sigma+n+1;\right.$$
$$\left.\alpha+\sigma+1,\ \alpha+\sigma+\frac{3}{2},\ 2\sigma+1;\ (-1)^{1-(\gamma^2+\gamma)/2}ac\right)+$$
$$+\gamma^2(-1)^n a^{\alpha-\sigma+1/2}c^{1/2-\sigma}\Gamma\begin{bmatrix}2\sigma\\1/2-\rho+\sigma\end{bmatrix}\frac{(\varepsilon/2+\alpha-\sigma+1)_n}{((\varepsilon-1)/2-\alpha+\sigma)_{n+1}}\times$$

$$\times {}_3F_3\left(\frac{1-\varepsilon}{2}+\alpha-\sigma-n,\ \frac{1}{2}-\sigma-\gamma\rho,\ \frac{\varepsilon}{2}+\alpha-\sigma+n+1;\right.$$

$$\left.\alpha-\sigma+1,\ \alpha-\sigma+\frac{3}{2},\ 1-2\sigma;\ (-1)^{(1-\gamma)/2}ac\right)+$$

$$+\frac{(-1)^n 2^{\varepsilon}(1/2)_{n+\varepsilon}}{n!}a^{\varepsilon/2}c^{\varepsilon/2-\alpha}\Gamma^{1-\gamma^2}\left[\begin{matrix}2\sigma+1\\ \rho+\sigma+1/2\end{matrix}\right]\times$$

$$\times\left[\Gamma\left(\frac{1}{2}-\rho-\sigma\right)\Gamma\left(\frac{1}{2}-\rho+\sigma\right)\right]^{-(\gamma^2+\gamma)/2}\Gamma\left(\alpha+\sigma+\frac{1-\varepsilon}{2}\right)\times$$

$$\times\Gamma^{2\gamma^2-1}\left(\frac{1}{2}+(1-2\gamma^2)\left(\sigma-\alpha+\frac{\varepsilon}{2}\right)\right)\times$$

$$\times\Gamma^{-\gamma^2+\gamma+1}\left((1-2\gamma^2)\rho+\frac{\gamma^2-\gamma}{2}+(\gamma^2-\gamma-1)\left(\alpha-\frac{\varepsilon}{2}\right)\right)\times$$

$$\times {}_3F_3\left(-n,\ \frac{\varepsilon}{2}-\alpha-(\gamma^2+\gamma-1)\rho,\ \frac{1}{2}+\varepsilon+n;\right.$$

$$\left.\frac{1}{2}+\varepsilon,\ \frac{1+\varepsilon}{2}-\alpha-\sigma,\ \frac{1+\varepsilon}{2}-\alpha+\sigma;\ (-1)^{1-(\gamma^2+\gamma)/2}ac\right).$$

16. $\displaystyle\int_a^\infty x^{\alpha-1}e^{\pm cx/2}P_{2n+\varepsilon}\left(\sqrt{\frac{a}{x}}\right)W_{\rho,\sigma}(cx)\,dx=V(\pm1)$

$\left[\varepsilon=0 \text{ or } 1;\ a>0,\ \left\{\begin{matrix}\operatorname{Re}(\alpha+\rho)<\varepsilon/2;\ |\arg c|<3\pi/2\\ \operatorname{Re}c>0\end{matrix}\right\},\ V(\gamma) \text{ see } 2.19.15.15\right].$

2.19.16. Integrals of $A(x)e^{px}L_n^{\lambda}(a+bx)\left\{\begin{matrix}M_{\rho,\sigma}(cx)\\ W_{\rho,\sigma}(cx)\end{matrix}\right\}$.

1. $\displaystyle\int_0^a x^{\sigma-1/2}(a-x)^{\lambda}e^{cx/2}L_n^{\lambda}(ba-bx)M_{\rho,\sigma}(cx)\,dx=$

$$=\frac{a^{\lambda/2+\sigma}}{n!c^{(\lambda+1)/2}}\Gamma\left[\begin{matrix}2\sigma+1,\ \lambda+n+1\\ \lambda+2\sigma+2\end{matrix}\right]e^{ac/2}M_{\varphi,\psi}(ac)$$

$[\varphi=\rho+n+(\lambda+1)/2,\ \psi=\sigma+(\lambda+1)/2;\ a>0;\ \operatorname{Re}\lambda,\ 2\operatorname{Re}\sigma>-1].$

2. $\displaystyle\int_0^\infty x^{\alpha-1}e^{-(b+c/2)x}L_n^{\lambda}(bx)M_{\rho,\sigma}(cx)\,dx=W(0)$ $\quad[\operatorname{Re}b,\ \operatorname{Re}c>0;\ \operatorname{Re}(\alpha+\sigma)>-1/2],$

$$W(\gamma)=\frac{c^{\sigma+1/2}}{b^{\alpha+\sigma+1/2}}\Gamma^{\gamma^2}\left[\begin{matrix}-2\sigma\\ 1/2-\rho-\sigma\end{matrix}\right]\Gamma\left(\alpha+\sigma+\frac{1}{2}\right)\left(\frac{1}{2}-\alpha+\lambda-\sigma\right)_n\times$$

$$\times {}_3F_2\left(\frac{1}{2}+\alpha-\lambda+\sigma,\ \frac{1}{2}+\alpha+\sigma,\ \frac{1}{2}+\sigma-(\gamma^2+\gamma-1)\rho;\right.$$

$$\left.2\sigma+1,\ \frac{1}{2}+\alpha-\lambda+\sigma-n;\ (-1)^{1-(\gamma^2+\gamma)/2}\frac{c}{b}\right)+$$

$$+\gamma^2\frac{c^{1/2-\sigma}}{b^{\alpha-\sigma+1/2}}\Gamma\left[\begin{matrix}2\sigma,\ 1/2+\alpha-\sigma\\ 1/2-\rho+\sigma\end{matrix}\right]\left(\frac{1}{2}-\alpha+\lambda+\sigma\right)_n\times$$

$$\times {}_3F_2\left(\frac{1}{2}+\alpha-\lambda-\sigma,\ \frac{1}{2}+\alpha-\sigma,\ \frac{1}{2}-\sigma-\gamma\rho;\right.$$

$$\left.1-2\sigma,\ \frac{1}{2}+\alpha-\lambda-\sigma-n;\ (-1)^{1-\gamma}\frac{c}{b}\right).$$

3. $\displaystyle\int_0^\infty x^{\alpha-1}e^{(-b\pm c/2)/x}L_n^{\lambda}(bx)W_{\rho,\sigma}(cx)\,dx=W(\pm1)$

$[\operatorname{Re}b>0;\ \operatorname{Re}\alpha>|\operatorname{Re}\sigma|-1/2;\ |\arg c|<3\pi/2,\ W(\gamma) \text{ see } 2.19.16.2].$

2.19.17. Integrals of $x^{\alpha}e^{-px}H_n(b\sqrt{x})\begin{Bmatrix}M_{\rho,\sigma}(cx)\\W_{\rho,\sigma}(cx)\end{Bmatrix}$.

1. $\int\limits_0^\infty x^{\alpha-1}e^{-(b^2+c/2)x}H_{2n+\varepsilon}(b\sqrt{x})\,M_{\rho,\sigma}(cx)\,dx=X(\gamma)$

[$\varepsilon=0$ or 1; $\operatorname{Re}c>0$; $\operatorname{Re}(\alpha+\sigma)>-(\varepsilon+1)/2$; $|\arg b|<\pi/4$];

$$X(\gamma)=\frac{(-1)^n2^{n+\varepsilon}b^{\varepsilon}}{c^{\alpha+\varepsilon/2}}(2n+2\varepsilon-1)!!\,\Gamma^{1-\gamma^2}\begin{bmatrix}2\sigma+1\\\rho+\sigma+1/2\end{bmatrix}\times$$
$$\times\left[\Gamma\left(\frac{1}{2}-\rho+\sigma\right)\Gamma\left(\frac{1}{2}-\rho-\sigma\right)\right]^{-(\gamma^2+\gamma)/2}\Gamma\left(\alpha+\sigma+\frac{\varepsilon+1}{2}\right)\Gamma^{2\gamma^2-1}\left(\frac{1}{2}+\right.$$
$$\left.+(1-2\gamma^2)\left(\sigma-\alpha-\frac{\varepsilon}{2}\right)\right)\Gamma^{-\gamma^2+\gamma+1}\left((1-2\gamma^2)\rho+\frac{\gamma^2-\gamma}{2}+(\gamma^2-\gamma-1)\left(\alpha+\frac{\varepsilon}{2}\right)\right)\times$$
$$\times{}_3F_2\left(\frac{1}{2}+\varepsilon+n,\ \frac{1+\varepsilon}{2}+\alpha+\sigma,\ \frac{1+\varepsilon}{2}+\alpha-\sigma;\right.$$
$$\left.\frac{1}{2}+\varepsilon,\ 1+\alpha+\frac{\varepsilon}{2}+(\gamma^2-\gamma-1)\rho;\ (-1)^{1-(\gamma^2+\gamma)/2}\frac{b^2}{c}\right)-$$
$$-(-1)^n\frac{2^{2n+\varepsilon-1}(\gamma^2-\gamma-2)}{c^{(1-2\gamma)\rho}}b^{2(1-2\gamma)\rho-2\alpha}\times$$
$$\times\Gamma^{1-\gamma}\begin{bmatrix}2\sigma+1\\1/2-\rho+\sigma\end{bmatrix}\Gamma\left(\frac{\varepsilon}{2}+\alpha-(1-2\gamma)\rho\right)\left(\frac{1+\varepsilon}{2}-\alpha+(1-2\gamma)\rho\right)_n\times$$
$$\times{}_3F_2\left(\frac{1}{2}+\sigma+(1-2\gamma)\rho,\ \frac{1}{2}-\sigma+(1-2\gamma)\rho,\ \frac{1+\varepsilon}{2}-\alpha+(1-2\gamma)\rho+n;\right.$$
$$\left.\frac{1+\varepsilon}{2}-\alpha+(1-2\gamma)\rho,\ 1-\alpha-\frac{\varepsilon}{2}+(1-2\gamma)\rho;\ (-1)^{1-\gamma}\frac{b^2}{c}\right).$$

2. $\int\limits_0^\infty x^{\alpha-1}e^{(-b^2\pm c/2)x}H_{2n+\varepsilon}(b\sqrt{x})\,W_{\rho,\sigma}(cx)\,dx=X(\pm1)$

[$\varepsilon=0$ or 1; $\operatorname{Re}\alpha>|\operatorname{Re}\sigma|-(\varepsilon+1)/2$; $|\arg b|<\pi/4$; $|\arg c|<(1\pm1/2)\pi$, $X(\gamma)$ see 2.19.17.1].

2.19.18. Integrals of $A(x)e^{\pm cx/2}C_n^{\lambda}(ax^{\pm r}-b)\begin{Bmatrix}M_{\rho,\sigma}(cx)\\W_{\rho,\sigma}(cx)\end{Bmatrix}$.

1. $\int\limits_0^a x^{\alpha-1}(a-x)^{\lambda-1/2}e^{-cx/2}C_n^{\lambda}\left(\frac{2x}{a}-1\right)M_{\rho,\sigma}(cx)\,dx=U(0)$

[$a>0$; $\operatorname{Re}\lambda$, $\operatorname{Re}(\alpha+\sigma)>-1/2$],

$$U(\gamma)=\frac{(-1)^na^{\alpha+\lambda+\sigma}c^{\sigma+1/2}}{n!}\Gamma^{\gamma^2}\begin{bmatrix}-2\sigma\\1/2-\rho-\sigma\end{bmatrix}(2\lambda)_n\Gamma\begin{bmatrix}\lambda+1/2,\ \alpha+\sigma+1/2\\\alpha+\lambda+\sigma+n+1\end{bmatrix}\times$$
$$\times(\lambda-\alpha-\sigma)_n\,{}_3F_3\left(\frac{1}{2}+\sigma-(\gamma^2+\gamma-1)\rho,\ \frac{1}{2}+\alpha+\sigma,\ 1+\alpha-\lambda+\sigma;\right.$$
$$\left.2\sigma+1,\ 1+\alpha+\lambda+\sigma+n,\ 1+\alpha-\lambda+\sigma-n;\ (-1)^{1-(\gamma^2+\gamma)/2}ac\right)+$$
$$+\gamma^2\frac{(-1)^na^{\alpha+\lambda-\sigma}c^{1/2-\sigma}}{n!}(2\lambda)_n\Gamma\begin{bmatrix}2\sigma,\ \lambda+1/2,\ \alpha-\sigma+1/2\\1/2-\rho+\sigma,\ \alpha+\lambda-\sigma+n+1\end{bmatrix}(\lambda+\sigma-\alpha)_n\times$$
$$\times{}_3F_3\left(\frac{1}{2}-\sigma-(\gamma^2+\gamma-1)\rho,\ \frac{1}{2}+\alpha-\sigma,\ 1+\alpha-\lambda-\sigma;\right.$$
$$\left.1-2\sigma,\ 1+\alpha+\lambda-\sigma+n,\ 1+\alpha-\lambda-\sigma-n;\ (-1)^{1-(\gamma^2+\gamma)/2}ac\right).$$

2. $\int\limits_0^a x^{\alpha-1}(a-x)^{\lambda-1/2}e^{\pm cx/2}C_n^\lambda\left(\frac{2x}{a}-1\right)W_{\rho,\,\sigma}(cx)\,dx=U(\pm 1)$

$[a>0;\ \operatorname{Re}\lambda>-1/2;\ \operatorname{Re}\alpha>|\operatorname{Re}\sigma|-1/2;\ U(\gamma)$ see 2.19.18.1].

3. $\int\limits_0^a x^{\alpha-1}(a-x)^{\lambda-1/2}e^{-cx/2}C_n^\lambda\left(\frac{2a}{x}-1\right)M_{\rho,\,\sigma}(cx)\,dx=V(0)$

$[a>0;\ \operatorname{Re}\lambda>-1/2;\ \operatorname{Re}(\alpha+\sigma)>n-1/2]$,

$$V(\gamma)=\frac{a^{\alpha+\lambda+\sigma}c^{\sigma+1/2}}{n!}\Gamma^{\gamma^2}\begin{bmatrix}-2\sigma\\ 1/2-\rho-\sigma\end{bmatrix}\times$$
$$\times\Gamma\begin{bmatrix}\lambda+1/2,\ \alpha+\sigma-n+1/2\\ \alpha+\lambda+\sigma+1\end{bmatrix}\left(\alpha+2\lambda+\sigma+\frac{1}{2}\right)_n(2\lambda)_n\times$$
$$\times{}_3F_3\left(\frac{1}{2}+\sigma-(\gamma^2+\gamma-1)\rho,\ \frac{1}{2}+\alpha+\sigma-n,\ \frac{1}{2}+\alpha+2\lambda+\sigma+n;\right.$$
$$\left.\frac{1}{2}+\alpha+2\lambda+\sigma,\ 1+\alpha+\lambda+\sigma,\ 2\sigma+1;\ (-1)^{1-(\gamma^2+\gamma)/2}ac\right)+$$
$$+\gamma^2\frac{a^{\alpha+\lambda-\sigma}c^{1/2-\sigma}}{n!}\Gamma\begin{bmatrix}2\sigma,\ \lambda+1/2,\ \alpha-\sigma-n+1/2\\ 1/2-\rho+\sigma,\ \alpha+\lambda-\sigma+1\end{bmatrix}(2\lambda)_n\left(\alpha+2\lambda-\sigma+\frac{1}{2}\right)_n\times$$
$$\times{}_3F_3\left(\frac{1}{2}-\sigma-\gamma\rho,\ \frac{1}{2}+\alpha-\sigma-n,\ \frac{1}{2}+\alpha+2\lambda-\sigma+n;\right.$$
$$\left.\frac{1}{2}+\alpha+2\lambda-\sigma,\ 1+\alpha+\lambda-\sigma,\ 1-2\sigma;\ (-1)^{(1-\gamma)/2}ac\right).$$

4. $\int\limits_0^a x^{\alpha-1}(a-x)^{\lambda-1/2}e^{\pm cx/2}C_n^\lambda\left(\frac{2a}{x}-1\right)W_{\rho,\,\sigma}(cx)\,dx=V(\pm 1)$

$[a>0;\ \operatorname{Re}\lambda>-1/2;\ \operatorname{Re}\alpha>|\operatorname{Re}\sigma|+n-1/2;\ V(\gamma)$ see 2.19.18.3].

5. $\int\limits_a^\infty x^{\alpha-1}(x-a)^{\lambda-1/2}e^{-cx/2}C_n^\lambda\left(\frac{2x}{a}-1\right)M_{\rho,\,\sigma}(cx)\,dx=W(0)$

$[a,\ \operatorname{Re}c>0;\ \operatorname{Re}\lambda>-1/2;\ \operatorname{Re}(\alpha+\lambda-\rho)<1/2-n]$.

$$W(\gamma)=\frac{a^{\alpha+\lambda+\sigma}c^{\sigma+1/2}}{n!}\Gamma^{\gamma^2}\begin{bmatrix}-2\sigma\\ 1/2-\rho-\sigma\end{bmatrix}\Gamma\begin{bmatrix}\lambda+1/2,\ -\alpha-\lambda-\sigma-n\\ 1/2-\alpha-\sigma\end{bmatrix}(2\lambda)_n(\lambda-\alpha-\sigma)_n\times$$
$$\times{}_3F_3\left(1+\alpha-\lambda+\sigma,\ \frac{1}{2}+\alpha+\sigma,\ \frac{1}{2}+\sigma-(\gamma^2+\gamma-1)\rho;\right.$$
$$\left.2\sigma+1,\ 1+\alpha-\lambda+\sigma-n,\ 1+\alpha+\lambda+\sigma+n;\ (-1)^{1-(\gamma^2+\gamma)/2}ac\right)+$$
$$+\gamma^2\frac{a^{\alpha+\lambda-\sigma}c^{1/2-\sigma}}{n!}\Gamma\begin{bmatrix}2\sigma,\ \lambda+1/2,\ \sigma-\alpha-\lambda-n\\ 1/2-\rho+\sigma,\ 1/2-\alpha+\sigma\end{bmatrix}(2\lambda)_n(\lambda-\alpha+\sigma)_n\times$$
$$\times{}_3F_3\left(1+\alpha-\lambda-\sigma,\ \frac{1}{2}+\alpha-\sigma,\ \frac{1}{2}-\sigma-(\gamma^2+\gamma-1)\rho;\right.$$
$$\left.1-2\sigma,\ 1+\alpha-\lambda-\sigma-n,\ 1+\alpha+\lambda-\sigma+n;\ (-1)^{(1-\gamma)/2}ac\right)+$$
$$+\frac{c^{1/2-\alpha-\lambda-n}(2\lambda)_{2n}}{n!a^n(\lambda+1/2)_n}\Gamma(\alpha+\lambda+\sigma+n)\times$$
$$\times\Gamma^{1-\gamma^2}\begin{bmatrix}2\sigma+1\\ 1/2+\rho+\sigma\end{bmatrix}\left[\Gamma\left(\frac{1}{2}-\rho+\sigma\right)\Gamma\left(\frac{1}{2}-\rho-\sigma\right)\right]^{-(\gamma^2+\gamma)/2}\times$$
$$\times\Gamma^{2\gamma^2-1}(1-\gamma^2+(1-2\gamma^2)(\sigma-\alpha-\lambda-n))\times$$

$$\times\Gamma^{-\gamma^2+\gamma+1}\left((1-2\gamma^2)\rho+\frac{1}{2}+(\gamma^2-\gamma-1)(\alpha+\lambda+n)\right)\times$$
$$\times{}_3F_3\left(1-2\lambda-n,\ \frac{1}{2}-\lambda-n,\ \frac{1}{2}-\alpha-\lambda-n-(\gamma^2+\gamma-1)\rho;\right.$$
$$\left.1-2\lambda-2n,\ 1-\alpha-\lambda-\sigma-n,\ 1-\alpha-\lambda+\sigma-n;\ (-1)^{1-(\gamma^2+\gamma)/2}ac\right).$$

6. $$\int_a^\infty x^{\alpha-1}(x-a)^{\lambda-1/2}e^{\pm cx/2}C_n^\lambda\left(\frac{2x}{a}-1\right)W_{\rho,\sigma}(cx)\,dx=W(\pm 1)$$

$\left[a>0;\ \operatorname{Re}\lambda>-1/2,\ \begin{cases}\operatorname{Re}(\alpha+\lambda+\rho)<1/2-n;\ |\arg c|<3\pi/2\\ \operatorname{Re}c>0\end{cases},\ W(\gamma)\ \text{see } 2.19.18.5\right].$

7. $$\int_a^\infty x^{\alpha-1}(x-a)^{\lambda-1/2}e^{-cx/2}C_n^\lambda\left(\frac{2a}{x}-1\right)M_{\rho,\sigma}(cx)\,dx=X(0)$$

$[a,\ \operatorname{Re}c>0;\ \operatorname{Re}\lambda>-1/2;\ \operatorname{Re}(\alpha+\lambda-\rho)<1/2],$

$$X(\gamma)=\frac{(-1)^n a^{\alpha+\lambda+\sigma}c^{\sigma+1/2}}{n!}\Gamma^{\gamma^2}\begin{bmatrix}-2\sigma\\ 1/2-\rho-\sigma\end{bmatrix}\times$$
$$\times\Gamma\begin{bmatrix}\lambda+1/2 & -\alpha-\lambda-\sigma\\ 1/2 & -\alpha-\sigma+n\end{bmatrix}(2\lambda)_n\left(\frac{1}{2}+\alpha+2\lambda+\sigma\right)_n\times$$
$$\times{}_3F_3\left(\frac{1}{2}+\alpha+\sigma-n,\ \frac{1}{2}+\alpha+2\lambda+\sigma+n,\ \frac{1}{2}+\sigma-(\gamma^2+\gamma-1)\rho;\right.$$
$$\left.1+\alpha+\lambda+\sigma,\ \frac{1}{2}+\alpha+2\lambda+\sigma,\ 1+2\sigma;\ (-1)^{1-(\gamma^2+\gamma)/2}ac\right)+$$
$$+\gamma^2\frac{(-1)^n a^{\alpha+\lambda-\sigma}c^{1/2-\sigma}}{n!}\Gamma\begin{bmatrix}2\sigma,\ \lambda+1/2,\ \sigma-\alpha-\lambda\\ 1/2-\rho+\sigma,\ 1/2-\alpha+\sigma+n\end{bmatrix}(2\lambda)_n\left(\frac{1}{2}+\alpha+2\lambda-\sigma\right)_n\times$$
$$\times{}_3F_3\left(\frac{1}{2}+\alpha-\sigma-n,\ \frac{1}{2}+\alpha+2\lambda-\sigma+n,\ \frac{1}{2}-\sigma-\gamma\rho;\right.$$
$$\left.1+\alpha+\lambda-\sigma,\ \frac{1}{2}+\alpha+2\lambda-\sigma,\ 1-2\sigma;\ (-1)^{(1-\gamma)/2}ac\right)+$$
$$+\frac{(-1)^n c^{1/2-\alpha-\lambda}}{n!}(2\lambda)_n\Gamma(\alpha+\lambda+\sigma)\,\Gamma^{1-\gamma^2}\begin{bmatrix}2\sigma+1\\ \rho+\sigma+1/2\end{bmatrix}\times$$
$$\times\left[\Gamma\left(\frac{1}{2}-\rho+\sigma\right)\Gamma\left(\frac{1}{2}-\rho-\sigma\right)\right]^{-(\gamma^2+\gamma)/2}\Gamma^{2\gamma^2-1}(1-\gamma^2+(1-2\gamma^2)(\sigma-\alpha-\lambda))\times$$
$$\times\Gamma^{-\gamma^2+\gamma+1}\left((1-2\gamma^2)\rho+\frac{1}{2}+(\gamma^2-\gamma-1)(\alpha+\lambda)\right)\times$$
$$\times{}_3F_3\left(\frac{1}{2}-\lambda-n,\ \frac{1}{2}+\lambda+n,\ \frac{1}{2}-\alpha-\lambda-(\gamma^2+\gamma-1)\rho;\right.$$
$$\left.\lambda+\frac{1}{2},\ 1-\alpha-\lambda-\sigma,\ 1-\alpha-\lambda+\sigma;\ (-1)^{1-(\gamma^2+\gamma)/2}ac\right).$$

8. $$\int_a^\infty x^{\alpha-1}(x-a)^{\lambda-1/2}e^{\pm cx/2}C_n^\lambda\left(\frac{2a}{x}-1\right)W_{\rho,\sigma}(cx)\,dx=X(\pm 1)$$

$\left[a>0;\ \operatorname{Re}\lambda>-1/2,\ \begin{cases}\operatorname{Re}(\alpha+\lambda+\rho)<1/2;\ |\arg c|<3\pi/2\\ \operatorname{Re}c>0\end{cases},\ X(\gamma)\ \text{see } 2.19.18.7\right].$

9. $$\int_0^a x^{\alpha-1}(a-x)^{\lambda-1/2}e^{-cx/2}C_{2n+\varepsilon}^\lambda\left(\sqrt{\frac{x}{a}}\right)M_{\rho,\sigma}(cx)\,dx=Y(0)$$

$[\varepsilon=0$ or $1;\ a>0;\ \operatorname{Re}\lambda>-1/2;\ \operatorname{Re}(\alpha+\sigma)>-(\varepsilon+1)/2],$

$$Y(\gamma)=\frac{(-1)^n a^{\alpha+\lambda+\sigma}c^{\sigma+1/2}}{(2n+\varepsilon)!}(2\lambda)_{2n+\varepsilon}\times$$
$$\times\Gamma\begin{bmatrix}\lambda+1/2,\ \alpha+\sigma+(\varepsilon+1)/2\\ \alpha+\lambda+\sigma+n+\varepsilon/2+1\end{bmatrix}\Gamma^{\gamma^2}\begin{bmatrix}-2\sigma\\ 1/2-\rho-\sigma\end{bmatrix}\left(\frac{\varepsilon}{2}-\alpha-\sigma\right)_n\times$$
$$\times{}_3F_3\left(\frac{1}{2}+\sigma-(\gamma^2+\gamma-1)\rho,\ \alpha+\sigma+\frac{1}{2},\ 1+\alpha+\sigma;\ 1+\alpha+\sigma-\frac{\varepsilon}{2}-n,\right.$$
$$\left.2\sigma+1,\ \alpha+\lambda+\sigma+n+\frac{\varepsilon}{2}+1;\ (-1)^{1-(\gamma^2+\gamma)/2}ac\right)+$$
$$+\gamma^2\frac{(-1)^n a^{\alpha+\lambda-\sigma}c^{1/2-\sigma}}{(2n+\varepsilon)!}(2\lambda)_{2n+\varepsilon}\times$$
$$\times\Gamma\begin{bmatrix}2\sigma,\ \lambda+1/2,\ \alpha-\sigma+(\varepsilon+1)/2\\ 1/2-\rho+\sigma,\ \alpha+\lambda-\sigma+\varepsilon/2+n+1\end{bmatrix}\left(\frac{\varepsilon}{2}-\alpha+\sigma\right)_n\times$$
$$\times{}_3F_3\left(\frac{1}{2}-\sigma-\gamma\rho,\ \alpha-\sigma+\frac{1}{2},\ 1+\alpha-\sigma;\right.$$
$$\left.1+\alpha-\sigma-\frac{\varepsilon}{2}-n,\ 1-2\sigma,\ \alpha+\lambda-\sigma+\frac{\varepsilon}{2}+n+1;\ (-1)^{(1-\gamma)/2}ac\right).$$

10. $$\int_0^a x^{\alpha-1}(a-x)^{\lambda-1/2}e^{\pm cx/2}C_{2n+\varepsilon}^{\lambda}\left(\sqrt{\frac{x}{a}}\right)W_{\rho,\sigma}(cx)\,dx=Y(\pm 1)$$

[$\varepsilon=0$ or 1; $a>0$; $\operatorname{Re}\lambda>-1/2$; $\operatorname{Re}\alpha>|\operatorname{Re}\sigma|-(\varepsilon+1)/2$; $Y(\gamma)$ see 2.19.18.9]

11. $$\int_0^a x^{\alpha-1}(a-x)^{\lambda-1/2}e^{-cx/2}C_{2n+\varepsilon}^{\lambda}\left(\sqrt{\frac{a}{x}}\right)M_{\rho,\sigma}(cx)\,dx=Z(0)$$

[$\varepsilon=0$ or 1; $a>0$; $\operatorname{Re}\lambda>-1/2$; $\operatorname{Re}(\alpha+\sigma)>n+(\varepsilon-1)/2$],

$$Z(\gamma)=\frac{a^{\alpha+\lambda+\sigma}c^{\sigma+1/2}}{(2n+\varepsilon)!}(2\lambda)_{2n+\varepsilon}\Gamma\begin{bmatrix}\lambda+1/2,\alpha+\sigma-n+(1-\varepsilon)/2\\ \alpha+\lambda+\sigma-\varepsilon/2+1\end{bmatrix}\times$$
$$\times\Gamma^{\gamma^2}\begin{bmatrix}-2\sigma\\ 1/2-\rho-\sigma\end{bmatrix}\left(\alpha+\lambda+\sigma+\frac{\varepsilon+1}{2}\right)_n\times$$
$$\times{}_3F_3\left(\frac{1}{2}+\sigma-(\gamma^2+\gamma-1)\rho,\ \alpha+\sigma-n+\frac{1-\varepsilon}{2},\ \alpha+\lambda+\sigma+n+\frac{\varepsilon+1}{2};\right.$$
$$\left.\alpha+\lambda+\sigma+\frac{1}{2},\ 2\sigma+1,\ \alpha+\lambda+\sigma+1;\ (-1)^{1-(\gamma^2+\gamma)/2}ac\right)+$$
$$+\gamma^2\frac{a^{\alpha+\lambda-\sigma}c^{1/2-\sigma}}{(2n+\varepsilon)!}(2\lambda)_{2n+\varepsilon}\Gamma\begin{bmatrix}2\sigma,\lambda+1/2,\alpha-\sigma-n+(1-\varepsilon)/2\\ 1/2-\rho+\sigma,1+\alpha+\lambda-\sigma-\varepsilon/2\end{bmatrix}\left(\alpha+\lambda-\sigma+\frac{\varepsilon+1}{2}\right)_n\times$$
$$\times{}_3F_3\left(\frac{1}{2}-\sigma-\gamma\rho,\ \alpha-\sigma-n+\frac{1-\varepsilon}{2},\ \alpha+\lambda-\sigma+n+\frac{\varepsilon+1}{2};\right.$$
$$\left.\alpha+\lambda-\sigma+\frac{1}{2},\ \alpha+\lambda-\sigma+1,\ 1-2\sigma;\ (-1)^{(1-\gamma)/2}ac\right).$$

12. $$\int_0^a x^{\alpha-1}(a-x)^{\lambda-1/2}e^{\pm cx/2}C_{2n+\varepsilon}^{\lambda}\left(\sqrt{\frac{a}{x}}\right)W_{\rho,\sigma}(cx)\,dx=Z(\pm 1)$$

[$\varepsilon=0$ or 1; $a>0$; $\operatorname{Re}\lambda>-1/2$; $\operatorname{Re}\alpha>|\operatorname{Re}\sigma|+n+(\varepsilon-1)/2$; $Z(\gamma)$ see 2.19.18.11].

13. $$\int_a^\infty x^{\alpha-1}(x-a)^{\lambda-1/2}e^{-cx/2}C_{2n+\varepsilon}^{\lambda}\left(\sqrt{\frac{x}{a}}\right)M_{\rho,\sigma}(cx)\,dx=U(0)$$

[$\varepsilon=0$ or 1; a, $\operatorname{Re}c>0$; $\operatorname{Re}\lambda>-1/2$; $\operatorname{Re}(\alpha+\lambda-\rho)<(1-\varepsilon)/2-n$],

$$U(\gamma)=\frac{a^{\alpha+\lambda+\sigma}c^{\sigma+1/2}}{(2n+\varepsilon)!}(2\lambda)_{2n+\varepsilon}\Gamma\begin{bmatrix}\lambda+1/2, & -\varepsilon/2-\alpha-\lambda-\sigma-n\\ (1-\varepsilon)/2-\alpha-\sigma\end{bmatrix}\times$$
$$\times\Gamma^{\gamma^2}\begin{bmatrix}-2\sigma\\ 1/2-\rho-\sigma\end{bmatrix}\left(\frac{\varepsilon}{2}-\alpha-\sigma\right)_n {}_3F_3\left(\frac{1}{2}+\sigma-(\gamma^2+\gamma-1)\rho, \frac{1}{2}+\alpha+\sigma, 1+\alpha+\sigma;\right.$$
$$\left.1-\frac{\varepsilon}{2}+\alpha+\sigma-n,\ 2\sigma+1,\ 1+\frac{\varepsilon}{2}+\alpha+\lambda+\sigma+n;\ (-1)^{1-(\gamma^2+\gamma)/2}ac\right)+$$
$$+\gamma^2\frac{a^{\alpha+\lambda-\sigma}c^{1/2-\sigma}}{(2n+\varepsilon)!}(2\lambda)_{2n+\varepsilon}\Gamma\begin{bmatrix}\lambda+1/2, & \sigma-\alpha-\lambda-\varepsilon/2-n, & 2\sigma\\ (1-\varepsilon)/2-\alpha+\sigma, & 1/2-\rho+\sigma\end{bmatrix}\left(\frac{\varepsilon}{2}-\alpha+\sigma\right)_n\times$$
$$\times{}_3F_3\left(\frac{1}{2}-\sigma-\gamma\rho,\ \frac{1}{2}+\alpha-\sigma,\ 1+\alpha-\sigma;\ 1-\frac{\varepsilon}{2}+\alpha-\sigma-n,\ 1-2\sigma,\ 1+\frac{\varepsilon}{2}+\right.$$
$$\left.+\alpha+\lambda-\sigma+n;\ (-1)^{(1-\gamma)/2}ac\right)+\frac{2^{2n+\varepsilon}c^{(1-\varepsilon)/2-\alpha-\lambda-n}}{(2n+\varepsilon)!a^{n+\varepsilon/2}}(\lambda)_{2n+\varepsilon}\Gamma^{1-\gamma^2}\begin{bmatrix}2\sigma+1\\ \rho+\sigma+1/2\end{bmatrix}\times$$
$$\times\left[\Gamma\left(\frac{1}{2}-\rho-\sigma\right)\Gamma\left(\frac{1}{2}-\rho+\sigma\right)\right]^{-(\gamma^2+\gamma)/2}\Gamma\left(\frac{\varepsilon}{2}+\alpha+\lambda+\sigma+n\right)\times$$
$$\times\Gamma^{2\gamma^2-1}\left(1-\gamma^2+(1-2\gamma^2)\left(\sigma-\frac{\varepsilon}{2}-\alpha-\lambda-n\right)\right)\times$$
$$\times\Gamma^{-\gamma^2+\gamma+1}\left((1-2\gamma^2)\rho+\frac{1}{2}+(\gamma^2-\gamma-1)\left(\alpha+\lambda+\frac{\varepsilon}{2}+n\right)\right)\times$$
$$\times{}_3F_3\left(\frac{1}{2}-\lambda-n,\ \frac{1-\varepsilon}{2}-\alpha-\lambda-n-(\gamma^2+\gamma-1)\rho,\ 1-\lambda-\varepsilon-n;\right.$$
$$\left.1-\lambda-\varepsilon-2n,\ 1-\alpha-\lambda-\sigma-\frac{\varepsilon}{2}-n,\ 1-\alpha-\lambda+\sigma-\frac{\varepsilon}{2}-n;\ (-1)^{1-(\gamma^2+\gamma)/2}ac\right).$$

14. $\displaystyle\int_a^\infty x^{\alpha-1}(x-a)^{\lambda-1/2}e^{\pm cx/2}C_{2n+\varepsilon}^{\lambda}\left(\sqrt{\frac{x}{a}}\right)W_{\rho,\sigma}(cx)\,dx=U(\pm 1)$

$\left[\varepsilon=0 \text{ or } 1;\ a>0;\ \operatorname{Re}\lambda>-1/2,\ \begin{Bmatrix}\operatorname{Re}(\alpha+\lambda+\rho)<(1-\varepsilon)/2-n;\ |\arg c|<3\pi/2\\ \operatorname{Re} c>0\end{Bmatrix},U(\gamma)\text{ see } 2.19.18.13\right].$

15. $\displaystyle\int_a^\infty x^{\alpha-1}(x-a)^{\lambda-1/2}e^{-cx/2}C_{2n+\varepsilon}^{\lambda}\left(\sqrt{\frac{a}{x}}\right)M_{\rho,\sigma}(cx)\,dx=V(0)$

$[\varepsilon=0 \text{ or } 1;\ a,\ \operatorname{Re} c>0;\ \operatorname{Re}\lambda>-1/2;\ \operatorname{Re}(\alpha+\lambda-\rho)<(\varepsilon+1)/2],$

$$V(\gamma)=\frac{(-1)^n a^{\alpha+\lambda+\sigma}c^{\sigma+1/2}}{(2n+\varepsilon)!}(2\lambda)_{2n+\varepsilon}\Gamma\begin{bmatrix}\lambda+1/2, & \varepsilon/2-\alpha-\lambda-\sigma\\ (1+\varepsilon)/2-\alpha-\sigma+n\end{bmatrix}\times$$
$$\times\Gamma^{\gamma^2}\begin{bmatrix}-2\sigma\\ 1/2-\rho-\sigma\end{bmatrix}\left(\frac{\varepsilon+1}{2}+\alpha+\lambda+\sigma\right)_n {}_3F_3\left(\frac{1-\varepsilon}{2}+\alpha+\sigma-n,\ \frac{1}{2}+\sigma-(\gamma^2+\gamma-1)\rho,\right.$$
$$\left.\frac{\varepsilon+1}{2}+\alpha+\lambda+\sigma+n;\ \frac{1}{2}+\alpha+\lambda+\sigma,\ 1+\alpha+\lambda+\sigma,\ 1+2\sigma;\ (-1)^{1-(\gamma^2+\gamma)/2}ac\right)+$$
$$+\gamma^2\frac{(-1)^n a^{\alpha+\lambda-\sigma}c^{1/2-\sigma}}{(2n+\varepsilon)!}(2\lambda)_{2n+\varepsilon}\Gamma\begin{bmatrix}\lambda+1/2, & \varepsilon/2-\alpha-\lambda+\sigma, & 2\sigma\\ (1+\varepsilon)/2-\alpha+\sigma+n, & 1/2-\rho+\sigma\end{bmatrix}\times$$
$$\times\left(\frac{1+\varepsilon}{2}+\alpha+\lambda-\sigma\right)_n {}_3F_3\left(\frac{1-\varepsilon}{2}+\alpha-\sigma-n,\ \frac{1}{2}-\sigma-(\gamma^2+\gamma-1)\rho,\right.$$
$$\left.\frac{\varepsilon+1}{2}+\alpha+\lambda-\sigma+n;\ \frac{1}{2}+\alpha+\lambda-\sigma,\ 1+\alpha+\lambda-\sigma,\ 1-2\sigma;\ (-1)^{(1-\gamma)/2}ac\right)+$$
$$+\frac{(-1)^n 2^{\varepsilon}a^{\varepsilon/2}c^{(1+\varepsilon)/2-\alpha-\lambda}}{n!}(\lambda)_{n+\varepsilon}\Gamma^{1-\gamma^2}\begin{bmatrix}2\sigma+1\\ 1/2+\rho+\sigma\end{bmatrix}\times$$
$$\times\left[\Gamma\left(\frac{1}{2}-\rho-\sigma\right)\Gamma\left(\frac{1}{2}-\rho+\sigma\right)\right]^{-(\gamma^2+\gamma)/2}\Gamma\left(\alpha+\lambda+\sigma-\frac{\varepsilon}{2}\right)\times$$
$$\times\Gamma^{2\gamma^2-1}\left(1-\gamma^2+(1-2\gamma^2)\left(\sigma-\alpha-\lambda+\frac{\varepsilon}{2}\right)\right)\times$$

$$\times\Gamma^{-\gamma^2+\gamma+1}\left((1-2\gamma^2)\rho+\frac{1}{2}+(\gamma^2-\gamma-1)\left(\alpha+\lambda-\frac{\varepsilon}{2}\right)\right)\times$$
$$\times{}_3F_3\left(\frac{1}{2}-\lambda-n,\frac{1+\varepsilon}{2}-\alpha-\lambda-(\gamma^2+\gamma-1)\rho,\frac{1}{2}+\varepsilon+n;\right.$$
$$\left.\frac{1}{2}+\varepsilon,1+\frac{\varepsilon}{2}-\alpha-\lambda-\sigma,1+\frac{\varepsilon}{2}-\sigma-\lambda+\sigma;(-1)^{1-(\gamma^2+\gamma)/2}ac\right).$$

16. $$\int_a^\infty x^{\alpha-1}(x-a)^{\lambda-1/2}e^{\pm cx/2}C_{2n+\varepsilon}^{\lambda}\left(\sqrt{\frac{a}{x}}\right)W_{\rho,\sigma}(cx)\,dx=V(\pm1)$$

$\left[\varepsilon=0\text{ or }1;\ a>0;\ \operatorname{Re}\lambda>-1/2,\ \begin{Bmatrix}\operatorname{Re}(\alpha+\lambda+\rho)<(\varepsilon+1)/2;\ |\arg c|<3\pi/2\\ \operatorname{Re}c>0\end{Bmatrix},\ V(\gamma)\text{ see 2.19.18.15}\right].$

2.19.19. Integrals of $A(x)e^{\pm cx/2}P_n^{(\mu,\nu)}(ax^{\pm1}-1)\begin{Bmatrix}M_{\rho,\sigma}(cx)\\ W_{\rho,\sigma}(cx)\end{Bmatrix}$.

1. $$\int_0^a x^{\alpha-1}(a-x)^{\mu}e^{-cx/2}P_n^{(\mu,\nu)}\left(\frac{2x}{a}-1\right)M_{\rho,\sigma}(cx)\,dx=W(0)$$

$[a>0;\ \operatorname{Re}\mu>-1;\ \operatorname{Re}(\alpha+\sigma)>-1/2]$,

$$W(\gamma)=$$
$$=\frac{(-1)^n a^{\alpha+\mu+\sigma+1/2}c^{\sigma+1/2}}{n!}\Gamma^{\gamma^2}\begin{bmatrix}-2\sigma\\ 1/2-\rho-\sigma\end{bmatrix}\mathrm{B}\left(\mu+n+1,\alpha+\sigma+\frac{1}{2}\right)\left(\frac{1}{2}-\alpha+\nu-\sigma\right)_n\times$$
$$\times{}_3F_3\left(\frac{1}{2}+\sigma-(\gamma^2+\gamma-1)\rho,\ \frac{1}{2}+\alpha+\sigma,\frac{1}{2}+\alpha-\nu+\sigma;\right.$$
$$\left.2\sigma+1,\frac{3}{2}+\alpha+\mu+\sigma+n,\frac{1}{2}+\alpha-\nu+\sigma-n;(-1)^{1-(\gamma^2+\gamma)/2}ac\right)+$$
$$+\gamma^2\frac{(-1)^n a^{\alpha+\mu-\sigma+1/2}c^{1/2-\sigma}}{n!}\Gamma\begin{bmatrix}2\sigma,\ \mu+n+1,\ \alpha-\sigma+1/2\\ 1/2-\rho+\sigma,\ \alpha+\mu-\sigma+n+3/2\end{bmatrix}\times$$
$$\times\left(\frac{1}{2}-\alpha+\nu+\sigma\right)_n{}_3F_3\left(\frac{1}{2}-\sigma-\rho\gamma,\frac{1}{2}+\alpha-\sigma,\frac{1}{2}+\alpha-\nu-\sigma;\right.$$
$$\left.1-2\sigma,\frac{3}{2}+\alpha+\mu-\sigma+n,\ \frac{1}{2}+\alpha-\nu-\sigma-n;\ (-1)^{(1-\gamma)/2}ac\right).$$

2. $$\int_0^a x^{\alpha-1}(a-x)^{\mu}e^{\pm cx/2}P_n^{(\mu,\nu)}\left(\frac{2x}{a}-1\right)W_{\rho,\sigma}(cx)\,dx=W(\pm1)$$

$[a>0;\ \operatorname{Re}\mu>-1;\ \operatorname{Re}\alpha>|\operatorname{Re}\sigma|-1/2;\ W(\gamma)$ see 2.19.19.1].

3. $$\int_0^a x^{\alpha-1}(a-x)^{\mu}e^{-cx/2}P_n^{(\mu,\nu)}\left(\frac{2a}{x}-1\right)M_{\rho,\sigma}(cx)\,dx=X(0)$$

$[a>0;\ \operatorname{Re}\mu>-1;\ \operatorname{Re}(\alpha+\sigma)>n-1/2]$,

$$X(\gamma)=\frac{a^{\alpha+\mu+\sigma+1/2}c^{\sigma+1/2}}{n!}\times$$
$$\times\Gamma^{\gamma^2}\begin{bmatrix}-2\sigma\\ 1/2-\rho-\sigma\end{bmatrix}\left(\alpha+\mu+\nu+\sigma+\frac{3}{2}\right)_n\mathrm{B}\left(\mu+n+1,\ \frac{1}{2}+\alpha+\sigma-n\right)\times$$
$$\times{}_3F_3\left(\frac{1}{2}+\sigma-(\gamma^2+\gamma-1)\rho,\frac{1}{2}+\alpha+\sigma-n,\frac{3}{2}+\alpha+\mu+\nu+\sigma+n;\right.$$
$$\left.\frac{3}{2}+\alpha+\mu+\nu+\sigma,\frac{3}{2}+\alpha+\mu+\sigma,\ 1+2\sigma;\ (-1)^{1-(\gamma^2+\gamma)/2}ac\right)+$$
$$+\gamma^2\frac{a^{\alpha+\mu-\sigma+1/2}c^{1/2-\sigma}}{n!}\Gamma\begin{bmatrix}2\sigma,\ \mu+n+1,\ 1/2+\alpha-\sigma-n\\ 1/2-\rho+\sigma,\ 3/2+\alpha+\mu-\sigma\end{bmatrix}\left(\frac{3}{2}+\alpha+\mu+\nu-\sigma\right)_n\times$$

$$\times{}_3F_3\left(\frac{1}{2}-\sigma-\gamma\rho,\ \frac{1}{2}+\alpha-\sigma-n,\ \frac{3}{2}+\alpha+\mu+\nu-\sigma+n;\right.$$
$$\left.\frac{3}{2}+\alpha+\mu+\nu-\sigma,\ \frac{3}{2}+\alpha+\mu-\sigma,\ 1-2\sigma;\ (-1)^{(1-\gamma)/2}ac\right).$$

4. $$\int_0^a x^{\alpha-1}(a-x)^{\mu}e^{\pm cx/2}P_n^{(\mu,\nu)}\left(\frac{2a}{x}-1\right)W_{\rho,\sigma}(cx)\,dx=X(\pm 1)$$

[$a>0$; $\operatorname{Re}\mu>-1$; $\operatorname{Re}\alpha>|\operatorname{Re}\sigma|+n-1/2$; $X(\gamma)$ see 2.19.19.3].

5. $$\int_a^\infty x^{\alpha-1}(x-a)^{\mu}e^{-cx/2}P_n^{(\mu,\nu)}\left(\frac{2x}{a}-1\right)M_{\rho,\sigma}(cx)\,dx=Y(0)$$

[a, $\operatorname{Re}c>0$; $\operatorname{Re}\mu>-1$; $\operatorname{Re}(\alpha+\mu-\rho)<-n$],

$$Y(\gamma)=\frac{a^{\alpha+\mu+\sigma+1/2}c^{\sigma+1/2}}{n!}\times$$
$$\times\Gamma^{\gamma^2}\begin{bmatrix}-2\sigma\\ 1/2-\rho-\sigma\end{bmatrix}\left(\frac{1}{2}-\alpha+\nu-\sigma\right)_n\mathrm{B}\left(\mu+n+1,\ -n-\alpha-\mu-\sigma-\frac{1}{2}\right)\times$$
$$\times{}_3F_3\left(\frac{1}{2}+\alpha-\nu+\sigma,\ \frac{1}{2}+\alpha+\sigma,\ \frac{1}{2}+\sigma-(\gamma^2+\gamma-1)\rho;\right.$$
$$\left.2\sigma+1,\ \frac{1}{2}+\alpha-\nu+\sigma-n,\ \frac{3}{2}+\alpha+\mu+\sigma+n;\ (-1)^{1-(\gamma^2+\gamma)/2}ac\right)+$$
$$+\gamma^2\frac{a^{\alpha+\mu-\sigma+1/2}c^{1/2-\sigma}}{n!}\Gamma\begin{bmatrix}2\sigma,\ \mu+n+1,\ \sigma-\alpha-\mu-n-1/2\\ 1/2-\rho+\sigma,\ 1/2-\alpha+\sigma\end{bmatrix}\left(\frac{1}{2}-\alpha+\nu+\sigma\right)_n\times$$
$$\times{}_3F_3\left(\frac{1}{2}+\alpha-\nu-\sigma,\ \frac{1}{2}+\alpha-\sigma,\ \frac{1}{2}-\sigma-\gamma\rho;\ 1-2\sigma,\ \frac{1}{2}+\alpha-\nu-\sigma-n,\right.$$
$$\left.\frac{3}{2}+\alpha+\mu-\sigma+n;\ (-1)^{(1-\gamma)/2}ac\right)+\frac{a^{-n}c^{-\alpha-\mu-n}}{n!}(1+\mu+\nu+n)_n\times$$
$$\times\Gamma\left(\frac{1}{2}+\alpha+\mu+\sigma+n\right)\Gamma^{1-\gamma^2}\begin{bmatrix}2\sigma+1\\ \rho+\sigma+1/2\end{bmatrix}\left[\Gamma\left(\frac{1}{2}-\rho-\sigma\right)\times\right.$$
$$\left.\times\Gamma\left(\frac{1}{2}-\rho+\sigma\right)\right]^{-(\gamma^2+\gamma)/2}\Gamma^{2\gamma^2-1}\left(\frac{1}{2}+(1-2\gamma^2)(\sigma-\alpha-\mu-n)\right)\times$$
$$\times\Gamma^{-\gamma^2+\gamma+1}\left((1-2\gamma^2)\rho+\frac{\gamma^2-\gamma}{2}+(\gamma^2-\gamma-1)(\alpha+\mu+n)\right)\times$$
$$\times{}_3F_3\left(-\mu-\nu-n,\ -\mu-n,\ -\alpha-\mu-n-(\gamma^2+\gamma-1)\rho;\ -\mu-\nu-2n,\right.$$
$$\left.\frac{1}{2}-\alpha-\mu-\sigma-n,\ \frac{1}{2}-\alpha-\mu+\sigma-n;\ (-1)^{1-(\gamma^2+\gamma)/2}ac\right).$$

6. $$\int_a^\infty x^{\alpha-1}(x-a)^{\mu}e^{\pm cx/2}P_n^{(\mu,\nu)}\left(\frac{2x}{a}-1\right)W_{\rho,\sigma}(cx)\,dx=Y(\pm 1)$$

$\left[a>0;\ \operatorname{Re}\mu>-1,\ \begin{cases}\operatorname{Re}(\alpha+\mu+\rho)<-n;\ |\arg c|<3\pi/2\\ \operatorname{Re}c>0\end{cases},\ Y(\gamma)\ \text{see 2.19.19.5}\right].$

7. $$\int_a^\infty x^{\alpha-1}(x-a)^{\mu}e^{-cx/2}P_n^{(\mu,\nu)}\left(\frac{2a}{x}-1\right)M_{\rho,\sigma}(cx)\,dx=Z(0)$$

[a, $\operatorname{Re}c>0$; $\operatorname{Re}\mu>-1$; $\operatorname{Re}(\alpha+\mu-\rho)<0$],

$$Z(\gamma)=\frac{(-1)^n a^{\alpha+\mu+\sigma+1/2}c^{\sigma+1/2}}{n!}\times$$
$$\times\Gamma^{\gamma^2}\begin{bmatrix}-2\sigma\\ 1/2-\rho-\sigma\end{bmatrix}\left(\frac{3}{2}+\alpha+\mu+\nu+\sigma\right)_n\mathrm{B}\left(\mu+n+1,\ -\frac{1}{2}-\alpha-\mu-\sigma\right)\times$$

$$\times {}_3F_3\left(\frac{1}{2}+\alpha+\sigma-n,\ \frac{3}{2}+\alpha+\mu+\nu+\sigma+n,\ \frac{1}{2}+\sigma-(\gamma^2+\gamma-1)\rho;\right.$$
$$\left.\frac{3}{2}+\alpha+\mu+\sigma,\ \frac{3}{2}+\alpha+\mu+\nu+\sigma,\ 1+2\sigma;\ (-1)^{1-(\gamma^2+\gamma)/2}ac\right)+$$
$$+\gamma^2\,\frac{(-1)^n a^{\alpha+\mu-\sigma+1/2}\,c^{1/2-\sigma}}{n!}\,\Gamma\begin{bmatrix}2\sigma,\ \mu+n+1,\ \sigma-\alpha-\mu-1/2\\ 1/2-\rho+\sigma,\ 1/2+n-\alpha+\sigma\end{bmatrix}\times$$
$$\times\left(\frac{3}{2}+\alpha+\mu+\nu-\sigma\right)_n {}_3F_3\left(\frac{1}{2}+\alpha-\sigma-n,\ \frac{3}{2}+\alpha+\mu+\nu-\sigma+n,\ \frac{1}{2}-\sigma-\gamma\rho;\right.$$
$$\left.\frac{3}{2}+\alpha+\mu-\sigma,\ \frac{3}{2}+\alpha+\mu+\nu-\sigma,\ 1-2\sigma;\ (-1)^{(1-\gamma)/2}ac\right)+$$
$$+\frac{(-1)^n c^{-\alpha-\mu}}{n!}(1+\nu)_n\,\Gamma\left(\frac{1}{2}+\alpha+\mu+\sigma\right)\Gamma^{1-\gamma^2}\begin{bmatrix}2\sigma+1\\ \rho+\sigma+1/2\end{bmatrix}\times$$
$$\times\left[\Gamma\left(\frac{1}{2}-\rho+\sigma\right)\Gamma\left(\frac{1}{2}-\rho-\sigma\right)\right]^{-(\gamma^2+\gamma)/2}\Gamma^{2\gamma^2-1}\left(\frac{1}{2}+(1-2\gamma^2)(\sigma-\alpha-\mu)\right)\times$$
$$\times\Gamma^{-\gamma^2+\gamma+1}\left((1-2\gamma^2)\rho+\frac{\gamma^2-\gamma}{2}+(\gamma^2-\gamma-1)(\alpha+\mu)\right)\times$$
$$\times {}_3F_3\left(-\mu-n,\ 1+\nu+n,\ -\alpha-\mu-(\gamma^2+\gamma-1)\rho;\right.$$
$$\left.1+\nu,\ \frac{1}{2}-\alpha-\mu-\sigma,\ \frac{1}{2}-\alpha-\mu+\sigma;\ (-1)^{1-(\gamma^2+\gamma)/2}ac\right).$$

8. $$\int_a^\infty x^{\alpha-1}(x-a)^\mu e^{\pm cx/2} P_n^{(\mu,\,\nu)}\left(\frac{2a}{x}-1\right) W_{\rho,\,\sigma}(cx)\,dx = Z(\pm 1)$$

$$\left[a,\ \operatorname{Re}\mu>-1,\ \begin{Bmatrix}\operatorname{Re}(\alpha+\mu+\rho)<0;\ |\arg c|<3\pi/2\\ \operatorname{Re}c>0\end{Bmatrix},\ Z(\gamma)\ \text{see 2.19.19.7}\right].$$

2.19.20. Integrals of $A(x)\,e^{\pm cx/2}\begin{Bmatrix}\mathbf{K}(\varphi(x))\\ \mathbf{E}(\varphi(x))\end{Bmatrix}\begin{Bmatrix}M_{\rho,\,\sigma}(cx)\\ W_{\rho,\,\sigma}(cx)\end{Bmatrix}$.

1. $$\int_0^a x^{\alpha-1}e^{-cx/2}\begin{Bmatrix}\mathbf{K}\left(\sqrt{1-x/a}\right)\\ \mathbf{E}\left(\sqrt{1-x/a}\right)\end{Bmatrix} M_{\rho,\,\sigma}(cx)\,dx = U(0) \qquad [a>0,\ \operatorname{Re}(\alpha+\sigma)>-1/2],$$

$$U(\gamma)=\frac{\pi}{2}\,a^{\alpha+\sigma+1/2}\,c^{\sigma+1/2}\,\Gamma\begin{bmatrix}\alpha+\sigma+1/2,\ \alpha+\sigma+1\mp 1/2\\ \alpha+\sigma+1,\ \alpha+\sigma+(3\mp 1)/2\end{bmatrix}\Gamma^{\gamma^2}\begin{bmatrix}-2\sigma\\ 1/2-\rho-\sigma\end{bmatrix}\times$$
$$\times {}_3F_3\left(\frac{1}{2}+\sigma-(\gamma^2+\gamma-1)\rho,\ \alpha+\sigma+\frac{1}{2},\ \alpha+\sigma+1\mp\frac{1}{2};\right.$$
$$\left.\alpha+\sigma+1,\ \alpha+\sigma+\frac{3\mp 1}{2},\ 2\sigma+1;\ (-1)^{1-(\gamma^2+\gamma)/2}ac\right)+$$
$$+\gamma^2\,\frac{\pi}{2}\,a^{\alpha-\sigma+1/2}\,c^{1/2-\sigma}\,\Gamma\begin{bmatrix}\alpha-\sigma+1/2,\ \alpha-\sigma+1\mp 1/2,\ 2\sigma\\ \alpha-\sigma+1,\ \alpha-\sigma+(3\mp 1)/2,\ 1/2-\rho+\sigma\end{bmatrix}\times$$
$$\times {}_3F_3\left(\frac{1}{2}-\sigma-\gamma\rho,\ \alpha-\sigma+\frac{1}{2},\ \alpha-\sigma+1\mp\frac{1}{2};\right.$$
$$\left.\alpha-\sigma+1,\ \alpha-\sigma+\frac{3\mp 1}{2},\ 1-2\sigma;\ (-1)^{(1-\gamma)/2}ac\right).$$

2. $$\int_0^a x^{\alpha-1}e^{cx/2}\begin{Bmatrix}\mathbf{K}\left(\sqrt{1-x/a}\right)\\ \mathbf{E}\left(\sqrt{1-x/a}\right)\end{Bmatrix} W_{\rho,\,\sigma}(cx)\,dx = U(1)$$

$$[a>0;\ \operatorname{Re}\alpha>|\operatorname{Re}\sigma|-1/2;\ U(\gamma)\ \text{see 2.19.20.1}].$$

3. $\int_0^a x^{\alpha-1}e^{-cx/2}\begin{Bmatrix}\mathbf{K}(\sqrt{1-x/a})\\ \mathbf{E}(\sqrt{1-x/a})\end{Bmatrix} W_{\rho,\,\sigma}(cx)\,dx=U(-1)$

[$a>0$; $\operatorname{Re}\alpha>|\operatorname{Re}\sigma|-1/2$; $U(\gamma)$ see 2.19.20.1].

4. $\int_0^\infty \frac{x^{\alpha-1}}{\sqrt{x+z}}e^{-cx/2}\begin{Bmatrix}\mathbf{K}(\sqrt{z/(x+z)})\\ \mathbf{E}(\sqrt{z/(x+z)})\end{Bmatrix} M_{\rho,\,\sigma}(cx)\,dx=V(0)$

[$\operatorname{Re}c>0$; $-1/2-\operatorname{Re}\sigma<\operatorname{Re}\alpha<\operatorname{Re}\rho+1/2$; $|\arg z|<\pi$],

$$V(\gamma)=2^{(\mp1-1)/2}c^{\sigma+1/2}z^{\alpha+\sigma}\Gamma\begin{bmatrix}-\alpha-\sigma,\ \alpha+\sigma+1/2,\ \alpha+\sigma+1\mp1/2\\ \alpha+\sigma+1\end{bmatrix}\times$$
$$\times\Gamma^{\gamma^2}\begin{bmatrix}-2\sigma\\ 1/2-\rho-\sigma\end{bmatrix}{}_3F_3\left(\frac{1}{2}+\sigma-(\gamma^2+\gamma-1)\rho,\ \alpha+\sigma+\frac{1}{2},\ \alpha+\sigma+1\mp\frac{1}{2};\right.$$
$$\left.2\sigma+1,\ \alpha+\sigma+1,\ \alpha+\sigma+1;\ (-1)^{(\gamma^2+\gamma)/2}cz\right)+$$
$$+2^{(\mp1-1)/2}\gamma^2z^{\alpha-\sigma}c^{1/2-\sigma}\Gamma\begin{bmatrix}2\sigma,\ \sigma-\alpha,\ \alpha-\sigma+1/2,\ \alpha-\sigma+1\mp1/2\\ \alpha-\sigma+1,\ 1/2-\rho+\sigma\end{bmatrix}\times$$
$$\times{}_3F_3\left(\frac{1}{2}-\sigma-\gamma\rho,\ \alpha-\sigma+\frac{1}{2},\ \alpha-\sigma+1\mp\frac{1}{2};\ 1-2\sigma,\ \alpha-\sigma+1,\ \alpha-\sigma+1;\right.$$
$$\left.(-1)^{(1+\gamma)/2}cz\right)+\frac{\pi}{2}c^{1/2-\alpha}\Gamma^{1-\gamma^2}\begin{bmatrix}2\sigma+1\\ \rho+\sigma+1/2\end{bmatrix}\times$$
$$\times\left[\Gamma\left(\frac{1}{2}-\rho-\sigma\right)\Gamma\left(\frac{1}{2}-\rho+\sigma\right)\right]^{-(\gamma^2+\gamma)/2}\Gamma(\alpha+\sigma)\times$$
$$\times\Gamma^{2\gamma^2-1}(1-\gamma^2+(1-2\gamma^2)(\sigma-\alpha))\,\Gamma^{-\gamma^2+\gamma+1}\left((1-2\gamma^2)\rho+\frac{1}{2}+(\gamma^2-\gamma-1)\alpha\right)\times$$
$$\times{}_3F_3\left(\frac{1}{2},\ 1\mp\frac{1}{2},\ \frac{1}{2}-(\gamma^2+\gamma-1)\rho-\alpha;\ 1,\ 1-\alpha+\sigma,\ 1-\alpha-\sigma;\ (-1)^{(\gamma^2+\gamma)/2}cz\right).$$

5. $\int_0^\infty \frac{x^{\alpha-1}}{\sqrt{x+z}}e^{cx/2}\begin{Bmatrix}\mathbf{K}(\sqrt{z/(x+z)})\\ \mathbf{E}(\sqrt{z/(x+z)})\end{Bmatrix} W_{\rho,\,\sigma}(cx)\,dx=V(1)$

[$|\operatorname{Re}\sigma|-1/2<\operatorname{Re}\alpha<1/2-\operatorname{Re}\rho$; $|\arg c|<3\pi/2$; $|\arg z|<\pi$; $V(\gamma)$ see 2.19.20.4].

6. $\int_0^\infty \frac{x^{\alpha-1}}{\sqrt{x+z}}e^{-cx/2}\begin{Bmatrix}\mathbf{K}(\sqrt{z/(x+z)})\\ \mathbf{E}(\sqrt{z/(x+z)})\end{Bmatrix} W_{\rho,\,\sigma}(cx)\,dx=V(-1)$

[$\operatorname{Re}c>0$; $\operatorname{Re}\alpha>|\operatorname{Re}\sigma|-1/2$; $|\arg z|<\pi$; $V(\gamma)$ see 2.19.20.4].

2.19.21. Integrals of $A(x)\,e^{\pm cx/2}P_\nu^\mu(\varphi(x))\begin{Bmatrix}M_{\rho,\,\sigma}(cx)\\ W_{\rho,\,\sigma}(cx)\end{Bmatrix}$.

1. $\int_0^\infty x^{\alpha-1}(x+2z)^{\pm\mu/2}e^{-cx/2}P_\nu^\mu\left(\frac{x}{z}+1\right)M_{\rho,\,\sigma}(cx)\,dx=W(0)$

[$\operatorname{Re}c>0$; $\operatorname{Re}(\mu/2-\sigma)-1/2<\operatorname{Re}\alpha<\operatorname{Re}(\rho-\nu\mp\mu/2)$, $\operatorname{Re}(\rho+\nu\mp\mu/2)+1$; $|\arg z|<\pi/2$],

$$W(\gamma)=c^{\sigma+1/2}(2z)^{\alpha+\sigma+(1\pm\mu)/2}[\Gamma(1-\mu+\nu)\Gamma(-\mu-\nu)]^{-(1\pm1)/2}\times$$
$$\times\left(-\frac{\sin\nu\pi}{\pi}\right)^{(1\mp1)/2}\Gamma\begin{bmatrix}(\mp\mu-1)/2-\alpha-\nu-\sigma,\ (1\mp\mu)/2-\alpha+\nu-\sigma,\ (1-\mu)/2+\alpha+\sigma\\ (1-\mu)/2-\alpha-\sigma\end{bmatrix}\times$$
$$\times\Gamma^{\gamma^2}\begin{bmatrix}-2\sigma\\ 1/2-\rho-\sigma\end{bmatrix}{}_3F_3\left(\frac{1-\mu}{2}+\alpha+\sigma,\ \frac{1}{2}+\sigma-(\gamma^2+\gamma-1)\rho,\ \frac{1+\mu}{2}+\alpha+\sigma;\right.$$
$$\left.2\sigma+1,\ \frac{3\pm\mu}{2}+\alpha+\nu+\sigma,\ \frac{1\pm\mu}{2}+\alpha-\nu+\sigma;\ (-1)^{(\gamma^2+\gamma)/2}2cz\right)+$$

$$+\gamma^2 c^{1/2-\sigma}(2z)^{\alpha-\sigma+(1\pm\mu)/2}\left[\Gamma(1-\mu+\nu)\Gamma(-\mu-\nu)\right]^{-(1\pm1)/2}\left(-\frac{\sin\nu\pi}{\pi}\right)^{(1\mp1)/2}\times$$

$$\times\Gamma\left[\begin{matrix}2\sigma,\ (\mp\mu-1)/2-\alpha-\nu+\sigma,\ (1\mp\mu)/2-\alpha+\nu+\sigma,\ (1-\mu)/2+\alpha-\sigma\\ 1/2-\rho+\sigma,\ (1-\mu)/2-\alpha+\sigma\end{matrix}\right]\times$$

$$\times{}_3F_3\left(\frac{1-\mu}{2}+\alpha-\sigma,\ \frac{1}{2}-\sigma-\gamma\rho,\ \frac{1+\mu}{2}+\alpha-\sigma;\right.$$

$$\left.1-2\sigma,\ \frac{3\pm\mu}{2}+\alpha+\nu-\sigma,\ \frac{1\pm\mu}{2}-\alpha-\nu-\sigma;\ (-1)^{(\gamma+1)/2}2cz\right)+$$

$$+\frac{2^\nu}{\sqrt{\pi}}c^{-\alpha-\nu\mp\mu/2}z^{-\nu}\Gamma\left[\begin{matrix}\nu+1/2,\ \alpha+\nu+\sigma+(1\pm\mu)/2\\ 1-\mu+\nu\end{matrix}\right]\times$$

$$\times\Gamma^{1-\gamma^2}\left[\begin{matrix}2\sigma+1\\ 1/2+\rho+\sigma\end{matrix}\right]\left[\Gamma\left(\frac{1}{2}-\rho+\sigma\right)\Gamma\left(\frac{1}{2}-\rho-\sigma\right)\right]^{-(\gamma^2+\gamma)/2}\times$$

$$\times\Gamma^{2\gamma^2-1}\left(\frac{1}{2}+(1-2\gamma^2)\left(\sigma-\alpha-\nu\mp\frac{\mu}{2}\right)\right)\Gamma^{-\gamma^2+\gamma+1}\left((1-2\gamma^2)\rho+\frac{\gamma^2-\gamma}{2}+\right.$$

$$\left.+(\gamma^2-\gamma-1)\left(\alpha+\nu\pm\frac{\mu}{2}\right)\right){}_3F_3\left(-\nu,\ \mp\frac{\mu}{2}-\alpha-\nu-(\gamma^2+\gamma-1)\rho,\ \mp\mu-\nu;\right.$$

$$\left.-2\nu,\ \frac{1\mp\mu}{2}-\alpha-\nu-\sigma,\ \frac{1\mp\mu}{2}-\alpha-\nu+\sigma;\ (-1)^{(\gamma^2+\gamma)/2}2cz\right)+$$

$$+\frac{2^{-\nu-1}}{\sqrt{\pi}}c^{1-\alpha+\nu\mp\mu/2}z^{\nu+1}\Gamma\left[\begin{matrix}-\nu-1/2,\ \alpha-\nu+\sigma-(1\mp\mu)/2\\ -\mu-\nu\end{matrix}\right]\times$$

$$\times\Gamma^{1-\gamma^2}\left[\begin{matrix}2\sigma+1\\ 1/2+\rho+\sigma\end{matrix}\right]\left[\Gamma\left(\frac{1}{2}-\rho+\sigma\right)\Gamma\left(\frac{1}{2}-\rho-\sigma\right)\right]^{-(\gamma^2+\gamma)/2}\times$$

$$\times\Gamma^{2\gamma^2-1}\left(\frac{3}{2}+(1-2\gamma^2)\left(\sigma-\alpha+\nu\mp\frac{\mu}{2}\right)-2\gamma^2\right)\times$$

$$\times\Gamma^{-\gamma^2+\gamma+1}\left((1-2\gamma^2)\rho+1+\frac{\gamma-\gamma^2}{2}+(\gamma^2-\gamma-1)\left(\alpha-\nu\pm\frac{\mu}{2}\right)\right)\times$$

$$\times{}_3F_3\left(\nu+1,\ 1\mp\frac{\mu}{2}-\alpha+\nu-(\gamma^2+\gamma-1)\rho,\ 1\mp\mu+\nu;\right.$$

$$\left.2\nu+2,\ \frac{3\mp\mu}{2}-\alpha+\nu-\sigma,\ \frac{3\mp\mu}{2}-\alpha+\nu+\sigma;\ (-1)^{(\gamma^2+\gamma)/2}2cz\right).$$

2. $$\int_0^\infty x^{-1/2}(x+2z)^{\pm\sigma}e^{-cx/2}P_\nu^{-2\sigma}\left(\frac{x}{z}+1\right)M_{\rho,\sigma}(cx)\,dx=$$

$$=\mp c^{\mp\sigma-1/2}\left(\frac{\sin\nu\pi}{\pi}\right)^{(1\pm1)/2}\Gamma\left[\begin{matrix}2\sigma+1,\ 1/2+\nu+\rho\mp\sigma,\ \rho-\nu\mp\sigma-1/2\\ \rho\end{matrix}\right]\times$$

$$\times\Gamma^{(1\mp1)/2}\left[\begin{matrix}\rho\\ 2\sigma-\nu,\ 2\sigma+\nu+1,\ \rho+\sigma+1/2\end{matrix}\right]e^{cz}W_{1/2-\rho\pm\sigma,\ 1/2+\nu}(2cz)$$

[$\operatorname{Re} c>0$; $\operatorname{Re}\sigma>-1/2$; $\operatorname{Re}(\rho\mp\mu)>|\operatorname{Re}\nu+1/2|$; $|\arg z|<\pi/2$].

3. $$\int_0^\infty x^{\alpha-1}(x+2z)^{\pm\mu/2}e^{cx/2}P_\nu^\mu\left(\frac{x}{z}+1\right)W_{\rho,\sigma}(cx)\,dx=W(1)$$

[$\operatorname{Re}\mu/2+|\operatorname{Re}\sigma|-1/2<\operatorname{Re}\alpha<-\operatorname{Re}(\rho+\nu\pm\mu/2)$, $\operatorname{Re}(\nu-\rho\mp\mu/2)+1$; $|\arg c|<3\pi/2$, $|\arg z|<\pi$; $W(\gamma)$ see 2.19.21.1].

4. $$\int_0^\infty x^{\alpha-1}(x+2z)^{\pm\mu/2}e^{-cx/2}P_\nu^\mu\left(\frac{x}{z}+1\right)W_{\rho,\sigma}(cx)\,dx=W(-1)$$

[$\operatorname{Re} c>0$; $\operatorname{Re}(\alpha-\mu/2)>|\operatorname{Re}\sigma|-1/2$; $|\arg z|<\pi$; $W(\gamma)$ see 2.19.21.1].

5. $$\int_0^\infty x^{-\sigma-(\mu+1)/2}(x+2z)^{\pm\mu/2}e^{-cx/2}P^{\mu}_{\rho+\sigma+(\mu\mp\mu-3)/2}\left(\frac{x}{z}+1\right)W_{\rho,\sigma}(cx)\,dx=$$
$$=c^{(2\rho+2\sigma+2\mu\mp2\mu-3)/4}(2z)^{(2\rho-2\sigma-1)/4}\Gamma\begin{bmatrix}1-\mu-2\sigma\\3/2-\mu-\rho-\sigma\end{bmatrix}e^{cz}W_{\varphi,\psi}(2cz)$$
$$\left[2\varphi=\frac{1}{2}-\rho+\sigma+\mu\pm\mu,\ 2\psi=\rho+3\sigma-\frac{3}{2}+\mu\mp\mu;\ \operatorname{Re}c>0;\ \operatorname{Re}\mu,\ \operatorname{Re}(\mu+2\sigma)<1;\ |\arg z|<\pi\right].$$

6. $$\int_0^\infty x^{\alpha-1}(x+z)^{-r}e^{-cx/2}P^{\mu}_{\nu}\left(\sqrt{\frac{x+z}{z}}\right)M_{\rho,\sigma}(cx)\,dx=X(0)$$

$[r=0$ or $1/2;\ \operatorname{Re}c>0;\ \operatorname{Re}(\mu/2-\sigma)-1/2<\operatorname{Re}\alpha<\operatorname{Re}(\rho-\nu/2)+r,\ \operatorname{Re}(\rho+\nu/2)+r+1/2;\ |\arg z|<\pi]$,

$$X(\gamma)=2^{\mu}c^{\sigma+1/2}z^{\alpha+\sigma-r+1/2}\times$$
$$\times\Gamma\begin{bmatrix}r-\alpha-\sigma-(\nu+1)/2,\ r+\nu/2-\alpha-\sigma,\ \alpha+\sigma+(1-\mu)/2\\r-(\mu+\nu)/2,\ r+(1-\mu+\nu)/2,\ (1-\mu)/2-\alpha-\sigma\end{bmatrix}\Gamma^{\gamma^2}\begin{bmatrix}-2\sigma\\1/2-\rho-\sigma\end{bmatrix}\times$$
$$\times{}_3F_3\left(\frac{1-\mu}{2}+\alpha+\sigma,\ \frac{1+\mu}{2}+\alpha+\sigma,\ \frac{1}{2}+\sigma-(\gamma^2+\gamma-1)\rho;\right.$$
$$\left.2\sigma+1,\ \frac{3+\nu}{2}+\alpha+\sigma-r,\ 1+\alpha+\sigma-\frac{\nu}{2}-r;\ (-1)^{(\gamma^2+\gamma)/2}cz\right)+$$
$$+\gamma^2c^{1/2-\sigma}z^{\alpha-\sigma+1/2-r}\,\Gamma\begin{bmatrix}2\sigma,\ r-\alpha+\sigma-\frac{\nu+1}{2},\ r-\alpha+\sigma+\frac{\nu}{2},\ \alpha-\sigma+\frac{1-\mu}{2}\\\frac{1}{2}-\rho+\sigma,\ r-\frac{\mu+\nu}{2},\ r+\frac{1-\mu+\nu}{2},\ \frac{1-\mu}{2}-\alpha+\sigma\end{bmatrix}\times$$
$$\times{}_3F_3\left(\frac{1-\mu}{2}+\alpha-\sigma,\ \frac{1+\mu}{2}+\alpha-\sigma,\ \frac{1}{2}-\sigma-\gamma\rho;\right.$$
$$\left.1-2\sigma,\ \frac{3+\nu}{2}+\alpha-\sigma-r,\ 1+\alpha-\sigma-\frac{\nu}{2}-r;\ (-1)^{(\gamma+1)/2}cz\right)+$$
$$+\frac{2^{\nu}c^{r-\alpha-\nu/2}z^{-\nu/2}}{\sqrt{\pi}}\Gamma^{1-\gamma^2}\begin{bmatrix}2\sigma+1\\1/2+\rho+\sigma\end{bmatrix}\times$$
$$\times\left[\Gamma\left(\frac{1}{2}-\rho+\sigma\right)\Gamma\left(\frac{1}{2}-\rho-\sigma\right)\right]^{-(\gamma^2+\gamma)/2}\Gamma\begin{bmatrix}\nu+1/2,\ (1+\nu)/2+\alpha+\sigma-r\\1-\mu+\nu\end{bmatrix}\times$$
$$\times\Gamma^{2\gamma^2-1}\left(\frac{1}{2}+(1-2\gamma^2)\left(\sigma-\alpha-\frac{\nu}{2}+r\right)\right)\times$$
$$\times\Gamma^{-\gamma^2+\gamma+1}\left((1-2\gamma^2)\rho+\frac{\gamma^2-\gamma}{2}+(\gamma^2-\gamma-1)\left(\alpha+\frac{\nu}{2}-r\right)\right)\times$$
$$\times{}_3F_3\left(r-\frac{\mu+\nu}{2},\ r-\alpha-\frac{\nu}{2}-(\gamma^2+\gamma-1)\rho,\ \frac{\mu-\nu}{2}+r;\right.$$
$$\left.\frac{1}{2}-\nu,\ \frac{1-\nu}{2}+r-\alpha-\sigma,\ \frac{1-\nu}{2}+r-\alpha+\sigma;\ (-1)^{(\gamma^2+\gamma)/2}cz\right)+$$
$$+\frac{2^{-\nu-1}c^{r-\alpha+(\nu+1)/2}z^{(\nu+1)/2}}{\sqrt{\pi}}\Gamma^{1-\gamma^2}\begin{bmatrix}2\sigma+1\\1/2+\rho+\sigma\end{bmatrix}\times$$
$$\times\left[\Gamma\left(\frac{1}{2}-\rho+\sigma\right)\Gamma\left(\frac{1}{2}-\rho-\sigma\right)\right]^{-(\gamma^2+\gamma)/2}\Gamma\begin{bmatrix}-\nu-1/2,\ \alpha+\sigma-r-\nu/2\\-\mu-\nu\end{bmatrix}\times$$
$$\times\Gamma^{2\gamma^2-1}\left(1-\gamma^2+(1-2\gamma^2)\left(\sigma+\frac{\nu}{2}+r-\alpha\right)\right)\times$$
$$\times\Gamma^{-\gamma^2+\gamma+1}\left((1-2\gamma^2)\rho+\frac{1}{2}+(\gamma^2-\gamma-1)\left(\alpha-\frac{\nu}{2}-r\right)\right)\times$$

$$\times {}_3F_3\left(r+\frac{1-\mu+\nu}{2},\ r-\alpha+\frac{\nu+1}{2}-(\gamma^2+\gamma-1)\rho,\ \frac{1+\mu+\nu}{2}+r;\right.$$
$$\left.\frac{3}{2}+\nu,\ 1+\frac{\nu}{2}+r-\sigma-\alpha,\ 1+\frac{\nu}{2}+r+\sigma-\alpha;\ (-1)^{(\gamma^2+\gamma)/2}cz\right).$$

7. $$\int_0^\infty x^{-1/2}e^{-cx/2}P_\nu^{-2\sigma}\left(\sqrt{\frac{x+z}{z}}\right)M_{\rho,\sigma}(cx)\,dx=$$
$$=2^{-2\sigma}c^{-3/4}z^{-1/4}\Gamma\begin{bmatrix}2\sigma+1,\ \rho+\nu/2,\ \rho-(\nu+1)/2\\ \rho+\sigma+1/2,\ \sigma+(\nu+1)/2,\ \sigma-\nu/2\end{bmatrix}e^{cz/2}W_{3/4-\rho,\,(2\nu+1)/4}(cz)$$
$[\operatorname{Re} c,\ \operatorname{Re}(2\rho+\nu)>0;\ \operatorname{Re}(2\rho-\nu)>-1;\ |\arg z|<\pi]$.

8. $$\int_0^\infty x^{\alpha-1}(x+z)^{-r}e^{\pm cx/2}P_\nu^\mu\left(\sqrt{\frac{x+z}{z}}\right)W_{\rho,\sigma}(cx)\,dx=X(\pm 1)$$
$\left[r=0 \text{ or } \frac{1}{2};\ \operatorname{Re}(\alpha-\mu/2)>|\operatorname{Re}\sigma|-1/2;\ |\arg z|<\pi,\right.$
$\left.\begin{cases}\operatorname{Re}(\alpha+\rho)<r-\operatorname{Re}\nu/2,\ r+(1+\operatorname{Re}\nu)/2;\ |\arg c|<3\pi/2\\ \operatorname{Re} c>0\end{cases},\ X(\gamma)\ \text{see 2.19.21.6}\right].$

9. $$\int_0^\infty x^{-\sigma-(\mu+1)/2}(x+z)^{-(1\mp 1)/4}e^{-cx/2}P^\mu_{\mu+2\rho+2\sigma-3}\left(\sqrt{\frac{x+z}{z}}\right)W_{\rho,\sigma}(cx)\,dx=$$
$$=2^\mu c^{(\mu+\rho+\sigma)/2-(7\pm 1)/8}z^{(\rho-\sigma)/2-(5\mp 1)/8}\Gamma\begin{bmatrix}1-\mu-2\sigma\\ (7\mp 1)/4-\mu-\rho-\sigma\end{bmatrix}e^{cz/2}W_{\varphi,\psi}(cz)$$
$\left[2\varphi=\mu-\rho+\sigma+(3\pm 1)/4,\ 2\psi=\mu+\rho+3\sigma-(7\pm 1)/4;\ \operatorname{Re} c>0,\ \begin{cases}\operatorname{Re}\mu,\ \operatorname{Re}(\mu+2\sigma)<1\\ \operatorname{Re}\mu,\ \operatorname{Re}\sigma>0\end{cases}\right].$

2.19.22. Integrals of $A(x)e^{\pm cx/2}Q_\nu^\mu(\varphi(x))\begin{Bmatrix}M_{\rho,\sigma}(cx)\\ W_{\rho,\sigma}(cx)\end{Bmatrix}$.

1. $$\int_0^\infty x^{\alpha-1}(x+2z)^{\pm\mu/2}e^{-cx/2}Q_\nu^\mu\left(\frac{x}{z}+1\right)M_{\rho,\sigma}(cx)\,dx=Y(0)$$
$[\operatorname{Re} c>0;\ |\operatorname{Re}\mu|/2-\operatorname{Re}\sigma-1/2<\operatorname{Re}\alpha<\operatorname{Re}(\rho\mp\mu/2+\nu)+1;\ |\arg z|<\pi]$,

$$Y(\gamma)=\frac{e^{\mu\pi i}}{2}c^{\sigma+1/2}(2z)^{\alpha+\sigma+(1\pm\mu)/2}\Gamma^{(1\pm 1)/2}\begin{bmatrix}1+\mu+\nu\\ 1-\mu+\nu\end{bmatrix}\times$$
$$\times\Gamma\begin{bmatrix}\nu-\alpha-\sigma+(1\mp\mu)/2,\ \alpha+\sigma+(1+\mu)/2,\ \alpha+\sigma+(1-\mu)/2\\ \nu+\alpha+\sigma+(3\pm\mu)/2\end{bmatrix}\Gamma^{\gamma^2}\begin{bmatrix}-2\sigma\\ 1/2-\rho-\sigma\end{bmatrix}\times$$
$$\times {}_3F_3\left(\frac{1}{2}+\sigma-(\gamma^2+\gamma-1)\rho,\ \alpha+\sigma+\frac{1+\mu}{2},\ \alpha+\sigma+\frac{1-\mu}{2};\right.$$
$$\left.2\sigma+1,\ \alpha+\nu+\sigma+\frac{3\pm\mu}{2},\ \alpha-\nu+\sigma+\frac{1\pm\mu}{2};\ (-1)^{(\gamma^2+\gamma)/2}2cz\right)+$$
$$+\frac{\gamma^2e^{\mu\pi i}}{2}c^{1/2-\sigma}(2z)^{\alpha-\sigma+(1\pm\mu)/2}\Gamma^{(1\pm 1)/2}\begin{bmatrix}1+\mu+\nu\\ 1-\mu+\nu\end{bmatrix}\times$$
$$\times\Gamma\begin{bmatrix}2\sigma,\ \nu+\sigma-\alpha+(1\mp\mu)/2,\ \alpha-\sigma+(1+\mu)/2,\ \alpha-\sigma+(1-\mu)/2\\ 1/2-\rho+\sigma,\ \alpha+\nu-\sigma+(3\pm\mu)/2\end{bmatrix}\times$$
$$\times {}_3F_3\left(\frac{1}{2}-\sigma-\gamma\rho,\ \alpha-\sigma+\frac{1+\mu}{2},\ \alpha-\sigma+\frac{1-\mu}{2};\right.$$
$$\left.1-2\sigma,\ \alpha+\nu-\sigma+\frac{3\pm\mu}{2},\ \alpha-\nu-\sigma+\frac{1\pm\mu}{2};\ (-1)^{(\gamma+1)/2}2cz\right)+$$
$$+\frac{e^{\mu\pi i}\sqrt{\pi}}{2^{\nu+1}}c^{1-\alpha+\nu\mp\mu/2}z^{\nu+1}\Gamma\begin{bmatrix}\mu+\nu+1,\ \alpha-\nu+\sigma+(\pm\mu-1)/2\\ \nu+3/2\end{bmatrix}\times$$

$$\times\Gamma^{1-\gamma^2}\begin{bmatrix}2\sigma+1\\1/2+\rho+\sigma\end{bmatrix}\left[\Gamma\left(\frac{1}{2}-\rho+\sigma\right)\Gamma\left(\frac{1}{2}-\rho-\sigma\right)\right]^{-(\gamma^2+\gamma)/2}\times$$
$$\times\Gamma^{2\gamma^2-1}\left(\frac{1}{2}+(1-2\gamma^2)\left(1-\alpha+\nu+\sigma\mp\frac{\mu}{2}\right)\right)\times$$
$$\times\Gamma^{-\gamma^2+\gamma+1}\left((1-2\gamma^2)\rho+\frac{\gamma^2-\gamma}{2}+(\gamma^2-\gamma-1)\left(\alpha-\nu-\rho-1\pm\frac{\mu}{2}\right)\right)\times$$
$$\times{}_3F_3\left(\nu+1,\ 1\mp\mu+\nu,\ 1-\alpha+\nu\mp\frac{\mu}{2}-(\gamma^2+\gamma-1)\rho;\right.$$
$$\left.2\nu+2,\ \frac{3\mp\mu}{2}-\alpha+\nu+\sigma,\ \frac{3\mp\mu}{2}-\alpha+\nu-\sigma;\ (-1)^{(\gamma^2+\gamma)/2}2cz\right).$$

2. $$\int_0^\infty x^{\sigma+(\mu-1)/2}(x+2z)^{\mu/2}e^{-cx/2}Q^{\mu}_{\rho-\sigma-\mu-3/2}\left(\frac{x}{z}+1\right)M_{\rho,\sigma}(cx)\,dx=$$
$$=\frac{e^{\mu\pi i}}{2}c^{(2\rho-2\mu-2\sigma-3)/4}(2z)^{(2\sigma+2\mu+2\rho-1)/4}\times$$
$$\times\Gamma\begin{bmatrix}1+\mu+2\sigma,\ 2\sigma+1,\ \rho-\sigma-1/2,\ \rho-\sigma-1/2\\\rho+\sigma+1/2,\ \rho-\sigma-2\mu-1/2\end{bmatrix}e^{cz}W_{\varphi,\psi}(2cz)$$

[$2\varphi=1/2-\mu-\rho-\sigma$, $2\psi=\rho-\mu-3\sigma-3/2$; $\operatorname{Re}c>0$; $\operatorname{Re}\sigma>-1/2$; $\operatorname{Re}(\rho-\sigma)>1/2$; $\operatorname{Re}(2\sigma+\mu)>-1$; $|\arg z|<\pi$].

3. $$\int_0^\infty x^{\alpha-1}(x+2z)^{\pm\mu/2}e^{cx/2}Q^{\mu}_{\nu}\left(\frac{x}{z}+1\right)W_{\rho,\sigma}(cx)\,dx=Y(1)$$

[$|\operatorname{Re}\mu|/2+|\operatorname{Re}\sigma|-1/2<\operatorname{Re}\alpha<\operatorname{Re}(\nu\mp\mu/2-\rho)+1$; $|\arg c|<3\pi/2$; $|\arg z|<\pi$; $Y(\gamma)$ see 2.19.22.1].

4. $$\int_0^\infty x^{\alpha-1}(x+2z)^{\pm\mu/2}e^{-cx/2}Q^{\mu}_{\nu}\left(\frac{x}{z}+1\right)W_{\rho,\sigma}(cx)\,dx=Y(-1)$$

[$\operatorname{Re}c>0$; $\operatorname{Re}\alpha>|\operatorname{Re}\mu|/2+|\operatorname{Re}\sigma|-1/2$; $|\arg z|<\pi$; $Y(\gamma)$ see 2.19.22.1].

5. $$\int_0^\infty x^{\alpha-1}(x+z)^{-r}e^{-cx/2}Q^{\mu}_{\nu}\left(\sqrt{\frac{x+z}{z}}\right)M_{\rho,\sigma}(cx)\,dx=Z(0)$$

[$r=0$ or $1/2$; $\operatorname{Re}c>0$; $|\operatorname{Re}\mu|/2-\operatorname{Re}\sigma-1/2<\operatorname{Re}\alpha<\operatorname{Re}(\rho+\nu/2)+r+1/2$; $|\arg z|<\pi$],
$$Z(\gamma)=2^{\mu-1}e^{\mu\pi i}c^{\sigma+1/2}z^{\alpha+\sigma+1/2-r}\times$$
$$\times\Gamma\begin{bmatrix}(\mu+\nu)/2-r+1,\ \nu/2-\alpha-\sigma+r,\ (1+\mu)/2+\alpha+\sigma,\ (1-\mu)/2+\alpha+\sigma\\(1-\mu+\nu)/2+r,\ (3+\nu)/2+\alpha+\sigma-r\end{bmatrix}\times$$
$$\times\Gamma^{\gamma^2}\begin{bmatrix}-2\sigma\\1/2-\rho-\sigma\end{bmatrix}{}_3F_3\left(\frac{1+\mu}{2}+\alpha+\sigma,\ \frac{1-\mu}{2}+\alpha+\sigma,\ \frac{1}{2}+\sigma-(\gamma^2+\gamma-1)\rho;\ \frac{3+\nu}{2}+\alpha+\sigma-r,\right.$$
$$\left.1+2\sigma,\ 1+\alpha+\sigma-r-\frac{\nu}{2};\ (-1)^{(\gamma^2+\gamma)/2}cz\right)+2^{\mu-1}\gamma^2e^{\mu\pi i}c^{1/2-\sigma}z^{1/2+\alpha-\sigma-r}\times$$
$$\times\Gamma\begin{bmatrix}2\sigma,\ (\mu+\nu)/2-r+1,\ \nu/2-\alpha+\sigma+r,\ (1+\mu)/2+\alpha-\sigma,\ (1-\mu)/2+\alpha-\sigma\\1/2-\rho+\sigma,\ (1-\mu+\nu)/2+r,\ (3+\nu)/2+\alpha-\sigma-r\end{bmatrix}\times$$
$$\times{}_3F_3\left(\frac{1+\mu}{2}+\alpha-\sigma,\ \frac{1-\mu}{2}+\alpha-\sigma,\ \frac{1}{2}-\sigma-\gamma\rho;\ \frac{3+\nu}{2}+\alpha-\sigma-r,\ 1-2\sigma,\ 1+\alpha-\sigma-r-\frac{\nu}{2};\right.$$
$$\left.(-1)^{(1+\gamma)/2}cz\right)+\frac{e^{\mu\pi i}\sqrt{\pi}}{2^{\nu+1}}c^{r+(\nu+1)/2-\alpha}z^{(\nu+1)/2}\,\Gamma\begin{bmatrix}\mu+\nu+1,\ \alpha+\sigma-r-\nu/2\\3/2+\nu\end{bmatrix}\times$$
$$\times\Gamma^{1-\gamma^2}\begin{bmatrix}2\sigma+1\\1/2+\rho+\sigma\end{bmatrix}\left[\Gamma\left(\frac{1}{2}-\rho+\sigma\right)\Gamma\left(\frac{1}{2}-\rho-\sigma\right)\right]^{-(\gamma^2+\gamma)/2}\times$$
$$\times\Gamma^{2\gamma^2-1}\left(1-\gamma^2+(1-2\gamma^2)\left(\sigma+r+\frac{\nu}{2}-\alpha\right)\right)\Gamma^{-\gamma^2+\gamma+1}\left((1-2\gamma^2)\rho+\frac{1}{2}+(\gamma^2-\gamma-1)\times\right.$$

$$\times\left(\alpha-r-\frac{\nu}{2}\right)\right) {}_3F_3\left(\frac{1+\mu+\nu}{2}+r,\ \frac{1-\mu+\nu}{2}+r,\ \frac{1+\nu}{2}+r-\alpha-(\gamma^2+\gamma-1)\rho;\right.$$
$$\left.\frac{3}{2}+\nu,\ 1-\alpha+\sigma+r+\frac{\nu}{2},\ 1-\alpha-\sigma+r+\frac{\nu}{2};\ (-1)^{(\gamma^2+\gamma)/2}cz\right).$$

6. $$\int_0^\infty x^{\alpha-1}(x+z)^{-r}e^{-cx/2}Q_{2\rho-2\alpha+2r-2}^{2\sigma-2\alpha+1}\left(\sqrt{\frac{x+z}{z}}\right)M_{\rho,\sigma}(cx)\,dx=$$
$$=-2^{2\sigma-2\alpha}e^{(2\sigma-2\alpha)\pi i}c^{(\rho+\sigma-1)/2+r-\alpha}z^{(\rho+\sigma-1)/2}\times$$
$$\times\Gamma\begin{bmatrix}2\alpha,\ 2\sigma+1,\ \rho+\sigma-2\alpha+1/2\\ \rho+\sigma+1/2\end{bmatrix}e^{cz}W_{\varphi,\psi}(cz)$$

[$r=0$ or $1/2$; $2\varphi=\sigma-\rho-2r-2\alpha+2$, $2\psi=\rho-\sigma+2r-2\alpha-1$; $\operatorname{Re} c>0$; $0<\operatorname{Re}\alpha<\operatorname{Re}\rho+r-1/4$; $\operatorname{Re}\sigma>-1/2$; $|\arg z|<\pi$].

7. $$\int_0^\infty x^{\alpha-1}(x+z)^{-r}e^{\pm cx/2}\,Q_\nu^\mu\left(\sqrt{\frac{x+z}{z}}\right)W_{\rho,\sigma}(cx)\,dx=Z(\pm 1)$$

[$r=0$ or $1/2$; $\operatorname{Re}\alpha>|\operatorname{Re}\mu|/2+|\operatorname{Re}\sigma|-1/2$; $|\arg z|<\pi$, $\left\{\begin{matrix}\operatorname{Re}(\alpha+\rho)<\operatorname{Re}\nu/2+r+1/2; |\arg c|<3\pi/2\\ \operatorname{Re} c>0\end{matrix}\right\}$, $Z(\gamma)$ see 2.19.22.5].

2.19.23. Integrals of $A(x)\begin{Bmatrix}M_{\mu,\nu}(cx+z)\\W_{\mu,\nu}(cx+z)\end{Bmatrix}\begin{Bmatrix}M_{\rho,\sigma}(cx)\\M_{\rho,\sigma}(cx)\end{Bmatrix}$.

1. $$\int_0^\infty x^{\alpha-1}\begin{Bmatrix}M_{\mu,\nu}(cx)\\W_{\mu,\nu}(cx)\end{Bmatrix}W_{\rho,\sigma}(cx)\,dx=$$
$$=c^{-\alpha}\Gamma^{(1\mp1)/2}\begin{bmatrix}-2\nu\\1/2-\mu-\nu\end{bmatrix}\Gamma\begin{bmatrix}1+\alpha+\nu+\sigma,\ 1+\alpha+\nu-\sigma,\ -1/2-\alpha-\nu-\rho\\1/2-\rho+\sigma,\ 1/2-\rho-\sigma\end{bmatrix}\times$$
$$\times{}_3F_2\left(\frac{1}{2}+\mu+\nu,\ 1+\alpha+\nu+\sigma,\ 1+\alpha+\nu-\sigma;\ 1+2\nu,\ \frac{3}{2}+\alpha+\nu+\rho;\ 1\right)+$$
$$+\begin{Bmatrix}0\\1\end{Bmatrix}c^{-\alpha}\Gamma\begin{bmatrix}2\nu,\ 1+\alpha-\nu+\sigma,\ 1+\alpha-\nu-\sigma,\ \nu-\alpha-\rho-1/2\\1/2-\rho+\sigma,\ 1/2-\rho-\sigma,\ 1/2-\mu+\nu\end{bmatrix}\times$$
$$\times{}_3F_2\left(\frac{1}{2}+\mu-\nu,\ 1+\alpha-\nu+\sigma,\ 1+\alpha-\nu-\sigma;\ 1-2\nu,\ \frac{3}{2}+\alpha-\nu+\rho;\ 1\right)+$$
$$+c^{-\alpha}\Gamma^{(1\pm1)/2}\begin{bmatrix}2\nu+1\\\mu+\nu+1/2\end{bmatrix}\Gamma\left(\frac{1}{2}+\alpha+\nu+\rho\right)\Gamma^{\mp1}\left(\frac{1}{2}\pm(\nu-\alpha-\rho)\right)\times$$
$$\times\Gamma^{\pm1}\left(\frac{1\mp1}{2}\pm(\mu-\alpha-\rho)\right){}_3F_2\left(\frac{1}{2}-\rho+\sigma,\ \frac{1}{2}-\rho-\sigma,\ \mu-\alpha-\rho;\right.$$
$$\left.\frac{1}{2}-\alpha+\nu-\rho,\ \frac{1}{2}-\alpha-\nu-\rho;\ 1\right)$$

[$\operatorname{Re} c>0$, $\left\{\begin{matrix}|\operatorname{Re}\sigma|-\operatorname{Re}\nu-1<\operatorname{Re}\alpha<\operatorname{Re}(\mu-\rho)\\ \operatorname{Re}\alpha>|\operatorname{Re}\nu|+|\operatorname{Re}\sigma|-1\end{matrix}\right\}$].

2. $$\int_0^\infty x^{\alpha-1}\begin{Bmatrix}M_{-\rho,\sigma}(cx)\\W_{-\rho,\sigma}(cx)\end{Bmatrix}W_{\rho,\sigma}(cx)\,dx=$$
$$=\frac{1}{2c^\alpha}\Gamma\begin{bmatrix}\alpha+1,\ (\alpha+1)/2+\sigma,\ (\alpha+1)/2-\sigma\\1+\alpha/2-\rho,\ 1+\alpha/2+\rho\end{bmatrix}\left(-\Gamma\begin{bmatrix}2\sigma+1\\1/2-\rho+\sigma\end{bmatrix}\frac{\cos(\alpha/2-\sigma)\pi}{\sin(\alpha/2+\rho)\pi}\right)^{(1\pm1)/2}$$

[$\operatorname{Re} c>0$, $\left\{\begin{matrix}-1,\ -1-2\operatorname{Re}\sigma<\operatorname{Re}\alpha<-2\operatorname{Re}\rho\\ \operatorname{Re}\alpha>2|\operatorname{Re}\sigma|-1\end{matrix}\right\}$].

3. $$\int_0^\infty x^{-1} W_{\mu,\sigma}(cx)\, W_{\rho,\sigma}(cx)\, dx =$$
$$= \frac{\pi}{(\rho-\mu)\sin 2\sigma\pi}\left\{\left[\Gamma\left(\frac{1}{2}-\mu-\sigma\right)\Gamma\left(\frac{1}{2}-\rho+\sigma\right)\right]^{-1} - \right.$$
$$\left. - \left[\Gamma\left(\frac{1}{2}-\mu+\sigma\right)\Gamma\left(\frac{1}{2}-\rho-\sigma\right)\right]^{-1}\right\}$$
$[\operatorname{Re} c > 0; |\operatorname{Re}\sigma| < 1/2; \mu \neq \rho; \sigma \neq 0]$

4. $$\int_0^\infty x^{-1} W^2_{\rho,\sigma}(cx)\, dx = \frac{\pi[\psi(1/2-\rho+\sigma)-\psi(1/2-\rho-\sigma)]}{\sin 2\sigma\pi\,\Gamma(1/2-\rho+\sigma)\,\Gamma(1/2-\rho-\sigma)}$$
$[\operatorname{Re} c > 0; |\operatorname{Re}\sigma| < 1/2; \sigma \neq 0]$.

5. $$\int_0^\infty x^{-1} W^2_{\rho,0}(cx)\, dx = \frac{\psi'(1/2-\rho)}{\Gamma^2(1/2-\rho)}$$ $[\operatorname{Re} c > 0]$.

6. $$\int_0^\infty x^{\alpha-1} M_{\nu+m+1/2,\,\nu}(cx)\, M_{\sigma+n+1/2,\,\sigma}(cx)\, dx =$$
$$= \frac{(-1)^{m+n}\, m!\,\Gamma(\alpha+\nu-\sigma+1)}{c^\alpha (2\nu+1)_m (2\sigma+1)_n} \sum_{k=0}^{n} \binom{n}{k}\binom{\alpha-\nu+\sigma}{m-k}\Gamma\begin{bmatrix}\alpha+\nu+\sigma+k+1\\ \alpha+\nu-\sigma-n+k+1\end{bmatrix}$$
$[\operatorname{Re} c > 0, \operatorname{Re}(\alpha+\nu+\sigma) > -1]$.

7. $$\int_0^\infty x^{\alpha-1} M_{\rho,\sigma}(-icx)\, M_{\rho,\sigma}(icx)\, dx =$$
$$= \frac{1}{2c^\alpha}\Gamma\begin{bmatrix}2\sigma+1,\ 2\sigma+1,\ (1+\alpha)/2+\sigma,\ \rho-\alpha/2,\ -\rho-\alpha/2\\ 1/2-\rho+\sigma,\ 1/2+\rho+\sigma,\ (1-\alpha)/2+\sigma,\ -\alpha\end{bmatrix}$$
$[c > 0; -1-2\operatorname{Re}\sigma < \operatorname{Re}\alpha < -2|\operatorname{Re}\rho|]$.

8. $$\int_0^\infty x^{\alpha-1} W_{\rho,\sigma}(-icx)\, W_{\rho,\sigma}(icx)\, dx =$$
$$= \frac{1}{2c^\alpha}\Gamma\begin{bmatrix}\alpha+1,\ (\alpha+1)/2+\sigma,\ (\alpha+1)/2-\sigma,\ -\rho-\alpha/2\\ 1+\alpha/2-\rho,\ 1/2-\rho+\sigma,\ 1/2-\rho-\sigma\end{bmatrix}$$
$[2|\operatorname{Re}\sigma|-1 < \operatorname{Re}\alpha < -2\operatorname{Re}\rho; |\arg c| < \pi]$.

9. $$\int_0^a x^{\alpha-1}(a^2-x^2)^{\beta-1}\begin{Bmatrix}M_{-\rho,\sigma}(cx)\\ W_{-\rho,\sigma}(cx)\end{Bmatrix} W_{\rho,\sigma}(cx)\, dx =$$
$$= \frac{a^{\alpha+2\beta-1}c}{4\sigma} B\left(\beta, \frac{\alpha+1}{2}\right)\Gamma^{(1\pm1)/2}\begin{bmatrix}2\sigma+1\\ 1/2-\rho+\sigma\end{bmatrix}\left(-\frac{\cos\rho\pi}{\sin\sigma\pi}\right)^{(1\mp1)/2}\times$$
$$\times {}_3F_4\left(\frac{1}{2}+\rho, \frac{1}{2}-\rho, \frac{\alpha+1}{2}; \frac{1}{2}, 1-\sigma, 1+\sigma, \beta+\frac{\alpha+1}{2}; \frac{a^2c^2}{4}\right) -$$
$$- \frac{a^{\alpha+2\beta}c^2\rho}{1-4\sigma^2} B\left(\beta, \frac{\alpha}{2}+1\right)\Gamma^{(1\pm1)/2}\begin{bmatrix}2\sigma+1\\ 1/2-\rho+\sigma\end{bmatrix}\left(-\frac{\sin\rho\pi}{\cos\sigma\pi}\right)^{(1\mp1)/2}\times$$
$$\times {}_3F_4\left(1+\rho, 1-\rho, \frac{\alpha}{2}+1; \frac{3}{2}, \frac{3}{2}-\sigma, \frac{3}{2}+\sigma, 1+\beta-\frac{\alpha}{2}; \frac{a^2c^2}{4}\right) +$$
$$+ \frac{a^{\alpha+2\beta+2\sigma-1}c^{2\sigma+1}}{2} B\left(\beta, \sigma+\frac{\alpha+1}{2}\right)\Gamma\begin{bmatrix}-2\sigma\\ 1/2-\rho-\sigma\end{bmatrix}\Gamma^{(1\mp1)/2}\begin{bmatrix}-2\sigma\\ 1/2+\rho-\sigma\end{bmatrix}\times$$
$$\times {}_3F_4\left(\frac{1}{2}+\rho+\sigma, \frac{1}{2}-\rho+\sigma, \sigma+\frac{\alpha+1}{2}; 1+\sigma, \frac{1}{2}+\sigma, 1+2\sigma, \beta+\sigma+\frac{\alpha+1}{2}; \frac{a^2c^2}{4}\right) +$$

$$+\begin{Bmatrix}0\\1\end{Bmatrix}\frac{a^{\alpha+2\beta-2\sigma-1}c^{1-2\sigma}}{2}\mathrm{B}\left(\beta,\frac{\alpha+1}{2}-\sigma\right)\Gamma\begin{bmatrix}2\sigma,\ 2\sigma\\1/2-\rho+\sigma,\ 1/2+\rho+\sigma\end{bmatrix}\times$$
$$\times{}_3F_4\left(\frac{1}{2}+\rho-\sigma,\ \frac{1}{2}-\rho-\sigma,\ \frac{1+\alpha}{2}-\sigma;\ 1-\sigma,\ 1-2\sigma,\ \frac{1}{2}-\sigma,\ \frac{1+\alpha}{2}+\beta-\sigma;\ \frac{a^2c^2}{4}\right)$$
$$\left[a,\ \mathrm{Re}\,\beta>0;\ \mathrm{Re}\,\alpha>-1,\ \begin{Bmatrix}\mathrm{Re}\,(\alpha+2\sigma)>-1\\\mathrm{Re}\,\alpha>2\,|\,\mathrm{Re}\,\sigma\,|-1\end{Bmatrix}\right].$$

10. $$\int_0^a x^{\alpha-1}(a^2-x^2)^{\beta-1}\begin{Bmatrix}M_{\rho,\sigma}(-icx)\,M_{\rho,\sigma}(icx)\\W_{\rho,\sigma}(-icx)\,W_{\rho,\sigma}(icx)\end{Bmatrix}dx=$$
$$=\frac{a^{\alpha+2\beta+2\sigma-1}c^{2\sigma+1}}{2}\mathrm{B}\left(\beta,\sigma+\frac{\alpha+1}{2}\right)\Gamma^{1\mp1}\begin{bmatrix}-2\sigma\\1/2-\rho-\sigma\end{bmatrix}\times$$
$$\times{}_3F_4\left(\frac{1}{2}+\rho+\sigma,\ \frac{1}{2}-\rho+\sigma,\ \frac{1+\alpha}{2}+\sigma;\ 2\sigma+1,\ \frac{1}{2}+\sigma,\ 1+\sigma,\ \frac{1+\alpha}{2}+\beta+\sigma;\ -\frac{a^2c^2}{4}\right)+$$
$$+\begin{Bmatrix}0\\1\end{Bmatrix}\frac{a^{\alpha+2\beta-2\sigma-1}c^{1-2\sigma}}{2}\mathrm{B}\left(\beta,\frac{1+\alpha}{2}-\sigma\right)\Gamma^2\begin{bmatrix}2\sigma\\1/2-\rho+\sigma\end{bmatrix}\times$$
$$\times{}_3F_4\left(\frac{1}{2}+\rho-\sigma,\ \frac{1}{2}-\rho-\sigma,\ \frac{1+\alpha}{2}-\sigma;\ 1-2\sigma,\ \frac{1}{2}-\sigma,\ 1-\sigma,\right.$$
$$\left.\frac{1+\alpha}{2}+\beta-\sigma;\ -\frac{a^2c^2}{4}\right)-\begin{Bmatrix}0\\1\end{Bmatrix}\frac{\pi a^{\alpha+2\beta-1}c}{4\sigma\sin\sigma\pi}\Gamma\begin{bmatrix}\beta,\ (\alpha+1)/2\\\beta+(\alpha+1)/2,\ 1/2-\rho+\sigma,\ 1/2-\rho-\sigma\end{bmatrix}\times$$
$$\times{}_3F_4\left(\frac{1}{2}+\rho,\ \frac{1}{2}-\rho,\ \frac{1+\alpha}{2};\ \frac{1}{2},\ 1-\sigma,\ 1+\sigma,\ \beta+\frac{1+\alpha}{2};\ -\frac{a^2c^2}{4}\right)+$$
$$+\begin{Bmatrix}0\\1\end{Bmatrix}\frac{\pi a^{\alpha+2\beta}c^2\rho}{(1-4\sigma^2)\cos\sigma\pi}\Gamma\begin{bmatrix}\beta,\ \alpha/2+1\\\beta+\alpha/2+1,\ 1/2-\rho+\sigma,\ 1/2-\rho-\sigma\end{bmatrix}\times$$
$$\times{}_3F_4\left(1+\rho,\ 1-\rho,\ \frac{\alpha}{2}+1;\ \frac{3}{2},\ \frac{3}{2}-\sigma,\ \frac{3}{2}+\sigma,\ \frac{\alpha}{2}+\beta+1;\ -\frac{a^2c^2}{4}\right)$$
$$\left[a,\ \mathrm{Re}\,\beta>0,\ \begin{Bmatrix}\mathrm{Re}\,(\alpha+2\sigma)>-1\\\mathrm{Re}\,\alpha>2\,|\,\mathrm{Re}\,\sigma\,|-1\end{Bmatrix}\right].$$

11. $$\int_0^a x^{\sigma-1/2}(a-x)^{\nu-1/2}M_{\mu,\nu}(ca-cx)\,M_{\rho,\sigma}(cx)\,dx=$$
$$=a^{\nu+\sigma}\mathrm{B}\,(2\nu+1,\ 2\sigma+1)\,M_{\mu+\rho,\ \nu+\sigma+1/2}(ac)\qquad[a>0;\ \mathrm{Re}\,\nu,\ \mathrm{Re}\,\sigma>-1/2].$$

12. $$\int_0^a x^{1/2-\sigma}(a-x)^{1/2-\nu}M_{\mu,\nu}(ca-cx)\,M_{\rho,\sigma}(cx)\,dx=$$
$$=\frac{a^{-\nu-\sigma}}{c^2}\Gamma\begin{bmatrix}2\nu+2\sigma+1\\2\nu+1,\ 2\sigma+1\end{bmatrix}M_{\mu+\rho,\ \nu+\sigma+1/2}(ac)\qquad[a>0].$$

13. $$\int_a^\infty x^{-1}(x-a)^{\beta-1}\,W_{\rho,\sigma}(-icx)\,W_{\rho,\sigma}(icx)\,dx=$$
$$=\frac{a^{\beta+2\rho-1}c^{2\rho}}{2^{\beta+2\rho}\sqrt{\pi}}\Gamma\begin{bmatrix}\beta\\1/2-\rho+\sigma,\ 1/2-\rho-\sigma\end{bmatrix}\times$$
$$\times E\left(1-\frac{\beta}{2}-\rho,\ \frac{1-\beta}{2}-\rho,\ \frac{1}{2}-\rho+\sigma,\ \frac{1}{2}-\rho-\sigma:\ 1-2\rho:\ \frac{a^2c^2}{4}\right)$$
$$[a,\ \mathrm{Re}\,\beta>0;\ \mathrm{Re}\,(\beta+2\rho)<1;\ |\arg c\,|<\pi].$$

14. $$\int_0^\infty x^{\nu-3/2}(cx+z)^{-\nu-1/2}\,W_{\rho,\nu}(cx+z)\,W_{\rho,\sigma}(cx)\,dx=$$
$$=z^{3/2-\nu}c^{1/2-\nu}\,\Gamma\begin{bmatrix}\nu-\sigma,\ \nu+\sigma\\1/2+\nu-\rho\end{bmatrix}W_{\rho,\sigma}(z)\qquad[\mathrm{Re}\,c>0;\ \mathrm{Re}\,\nu>|\,\mathrm{Re}\,\sigma\,|;\ |\arg z\,|<\pi].$$

15. $$\int_0^\infty \frac{1}{x+z} |W_{\rho,\sigma}(e^{\pi i} cx)|^{-2} dx =$$

$$= \Gamma\left(\frac{1}{2}-\rho \mp \sigma\right) \Gamma\left(\frac{3}{2}-\rho \pm \sigma\right) \frac{W_{\rho-1/2,\,\sigma\pm 1/2}(cz)}{\sqrt{cz}\, W_{\rho,\sigma}(cz)}$$

$$\left[c > 0;\ |\arg z| < \pi\ \ \begin{cases} \operatorname{Re}\rho - 1/2 < \operatorname{Re}\sigma < 0 \\ \operatorname{Re}\rho - 1/2 < \operatorname{Re}\sigma < 1/2 - \operatorname{Re}\rho;\ \ \operatorname{Re}\sigma > 0 \end{cases}\right].$$

2.19.24. Integrals of $x^{\alpha} e^{f(x)} \prod_{j,\,k} M_{\mu_j,\,\nu_j}(b_j x^{\pm 1})\, W_{\rho_k,\,\sigma_k}(c_k x^{\mp 1})$.

1. $$\int_0^\infty x^{-1} e^{-px} M_{\mu,\sigma}(bx)\, M_{\rho,\sigma}(cx)\, dx =$$

$$= 2^{2\sigma+1} (bc)^{\sigma+1/2}\, \Gamma(2\sigma+1)\, (2p+b+c)^{-\mu-\rho}\, (2p+c-b)^{\mu-\sigma-1/2}\, (2p+b-c)^{\rho-\sigma-1/2} \times$$

$$\times {}_2F_1\left(\frac{1}{2}-\mu+\sigma,\ \frac{1}{2}-\rho+\sigma;\ 2\sigma+1;\ \frac{4bc}{(2p+c-b)(2p+b-c)}\right)$$

$$[\operatorname{Re}\sigma > -1/2;\ 2\operatorname{Re}p > |\operatorname{Re}b| + |\operatorname{Re}c|].$$

2. $$\int_0^\infty x^{\alpha-1} e^{-(b+c)x/2} M_{\mu,\nu}(bx)\, M_{\rho,\sigma}(cx)\, dx = V(0,\ 0)$$

$$[\operatorname{Re}b,\ \operatorname{Re}c > 0;\ -\operatorname{Re}(\nu+\sigma)-1 < \operatorname{Re}\alpha < \operatorname{Re}(\mu+\rho)],$$

$$V(\varepsilon,\gamma) = \frac{b^{\nu+1/2}}{c^{\alpha+\nu+1/2}} \Gamma^{1-\gamma^2}\left[\begin{matrix}2\sigma+1\\ \rho+\sigma+1/2\end{matrix}\right] \left[\Gamma\left(\frac{1}{2}-\rho+\sigma\right)\Gamma\left(\frac{1}{2}-\rho-\sigma\right)\right]^{-(\gamma^2+\gamma)/2} \times$$

$$\times \Gamma^{\varepsilon^2}\left[\begin{matrix}-2\nu\\ 1/2-\mu-\nu\end{matrix}\right] \Gamma(\alpha+\nu+\sigma+1)\, \Gamma^{2\gamma^2-1}(\gamma^2+(1-2\gamma^2)(\sigma-\alpha-\nu)) \times$$

$$\times \Gamma^{-\gamma^2+\gamma+1}\left((1-2\gamma^2)\rho + \gamma^2-\gamma-\frac{1}{2}+(\gamma^2-\gamma-1)(\alpha+\nu)\right) \times$$

$$\times {}_3F_2\left(\frac{1}{2}+\nu-(\varepsilon^2+\varepsilon-1)\mu,\ 1+\alpha+\nu+\sigma,\ 1+\alpha+\nu-\sigma;\right.$$

$$\left. 1+2\nu,\ \frac{3}{2}+\alpha+\nu+(\gamma^2+\gamma-1)\rho;\ (-1)^{1-(\varepsilon^2+\varepsilon+\gamma^2+\gamma)/2}\frac{b}{c}\right) +$$

$$+\varepsilon^2 \frac{b^{1/2-\nu}}{c^{\alpha-\nu+1/2}} \Gamma^{1-\gamma^2}\left[\begin{matrix}2\sigma+1\\ \rho+\sigma+1/2\end{matrix}\right] \left[\Gamma\left(\frac{1}{2}-\rho+\sigma\right)\Gamma\left(\frac{1}{2}-\rho-\sigma\right)\right]^{-(\gamma^2+\gamma)/2} \times$$

$$\times \Gamma\left[\begin{matrix}2\nu,\ \alpha-\nu+\sigma+1\\ 1/2-\mu+\nu\end{matrix}\right] \Gamma^{2\gamma^2-1}(\gamma^2+(1-2\gamma^2)(\sigma-\alpha+\nu)) \times$$

$$\times \Gamma^{-\gamma^2+\gamma+1}\left((1-2\gamma^2)\rho+\gamma^2-\gamma-\frac{1}{2}+(\gamma^2-\gamma-1)(\alpha-\nu)\right) \times$$

$$\times {}_3F_2\left(\frac{1}{2}-\nu-\varepsilon\mu,\ 1+\alpha-\nu+\sigma,\ 1+\alpha-\nu-\sigma;\ 1-2\nu,\right.$$

$$\left.\frac{3}{2}+\alpha-\nu+(\gamma^2+\gamma-1)\rho;\ (-1)^{(1-\varepsilon-\gamma^2-\gamma)/2}\frac{b}{c}\right) -$$

$$-\frac{\gamma^2-\gamma-2}{2c^{(1-2\gamma)\rho}} b^{(1-2\gamma)\rho-\alpha}\, \Gamma^{1-\gamma}\left[\begin{matrix}2\sigma+1\\ 1/2-\rho+\sigma\end{matrix}\right] \Gamma^{1-\varepsilon^2}\left[\begin{matrix}2\nu+1\\ \mu+\nu+1/2\end{matrix}\right] \times$$

$$\times \left[\Gamma\left(\frac{1}{2}-\mu+\nu\right)\Gamma\left(\frac{1}{2}-\mu-\nu\right)\right]^{-(\varepsilon^2+\varepsilon)/2} \Gamma\left(\frac{1}{2}+\nu-(1-2\gamma)\rho+\alpha\right) \times$$

$$\times \Gamma^{2\varepsilon^2-1}\left(\frac{1}{2}+(1-2\varepsilon^2)(\nu+(1-2\gamma)\rho-\alpha)\right) \times$$

$$\times\Gamma^{-\varepsilon^2+\varepsilon+1}\left((1-2\varepsilon^2)\mu+\frac{\varepsilon^2-\varepsilon}{2}-(\varepsilon^2-\varepsilon-1)((1-2\gamma)\rho-\alpha)\right)\times$$

$$\times{}_3F_2\left(\frac{1}{2}+\sigma+(1-2\gamma)\rho,\ \frac{1}{2}-\sigma+(1-2\gamma)\rho,\ (1-2\gamma)\rho-\alpha-(\varepsilon^2+\varepsilon-1)\mu;\right.$$
$$\left.\frac{1}{2}+\nu+(1-2\gamma)\rho-\alpha,\ \frac{1}{2}-\nu+(1-2\gamma)\rho-\alpha;\ (-1)^{1-\gamma-(\varepsilon^2+\varepsilon)/2}\frac{b}{c}\right).$$

3. $$\int_0^\infty x^{\alpha-1}e^{(-b\pm c)x/2}M_{\mu,\nu}(bx)\,W_{\rho,\sigma}(cx)\,dx=V(0,\ \pm1)$$

$\left[\operatorname{Re} b>0;\ \operatorname{Re}(\alpha+\nu)>|\operatorname{Re}\sigma|-1,\ \begin{Bmatrix}\operatorname{Re}(\alpha-\mu+\rho)<0;\ |\arg c|<3\pi/2\\ \operatorname{Re} c>0\end{Bmatrix},\ V(\varepsilon,\gamma)\right.$ see 2.19.24.2].

4. $$\int_0^\infty x^{\alpha-1}e^{(b-c)x/2}M_{\mu,\nu}(bx)W_{\rho,\sigma}(cx)dx=\frac{b^{\nu+1/2}}{c^{\alpha+\nu+1/2}}\Gamma\begin{bmatrix}\alpha+\nu+\sigma+1,\ \alpha+\nu-\sigma+1\\ 3/2+\alpha+\nu-\rho\end{bmatrix}\times$$
$$\times{}_3F_2\left(\frac{1}{2}-\mu+\nu,\ \alpha+\nu+\sigma+1,\ \alpha+\nu-\sigma+1;\ 2\nu+1,\ \frac{3}{2}+\alpha+\nu-\rho;\ \frac{b}{c}\right)$$

[$\operatorname{Re} c,\ \operatorname{Re}(c-b)>0;\ \operatorname{Re}(\alpha+\nu)>|\operatorname{Re}\sigma|-1$].

5. $$\int_0^\infty x^{\nu\pm\sigma-1}e^{(c-b)x/2}M_{\mu,\nu}(bx)\,W_{\rho,\sigma}(cx)\,dx=$$
$$=\frac{b^{\nu+1/2}}{c^{2\nu\pm\sigma+1/2}}\Gamma\begin{bmatrix}2\nu+1,\ 1+2\nu\pm2\sigma,\ \mu-\nu-\rho\mp\sigma\\ 1/2-\rho\mp\sigma,\ 1+\mu+\nu-\rho\pm\sigma\end{bmatrix}\times$$
$$\times{}_2F_1\left(\frac{1}{2}+\mu+\nu,\ 1+2\nu\pm2\sigma;\ 1+\mu+\nu-\rho\pm\sigma;\ 1-\frac{b}{c}\right)$$

[$\operatorname{Re} b>0;\ \operatorname{Re}\nu>-1/2;\ -1/2<\operatorname{Re}(\nu\pm\sigma)<\operatorname{Re}(\mu-\rho);\ |\arg c|<3\pi/2$].

6. $$\int_0^\infty x^{\alpha-1}e^{(b\pm c)x/2}W_{\mu,\nu}(bx)\,W_{\rho,\sigma}(cx)\,dx=V(1,\ \pm1)$$

$\left[\operatorname{Re}\alpha>|\operatorname{Re}\nu|+|\operatorname{Re}\sigma|-1;\ |\arg b|<3\pi/2,\ \begin{Bmatrix}\operatorname{Re}(\alpha+\mu+\rho)<0;\ |\arg c|<3\pi/2\\ \operatorname{Re} c>0\end{Bmatrix},\right.$

$V(\varepsilon,\gamma)$ see 2.19.24.2].

7. $$\int_0^\infty x^{\alpha-1}e^{(-b\pm c)x/2}W_{\mu,\nu}(bx)\,W_{\rho,\sigma}(cx)\,dx=V(-1,\ \pm1)$$

[$\operatorname{Re} b>0;\ \operatorname{Re}\alpha>|\operatorname{Re}\nu|+|\operatorname{Re}\sigma|-1;\ |\arg c|<(1\pm1/2)\pi,\ V(\varepsilon,\gamma)$ see 2.19.24.2].

8. $$\int_0^\infty x^{\alpha-1}e^{-b/(2x)-cx/2}M_{\mu,\nu}\left(\frac{b}{x}\right)M_{\rho,\sigma}(cx)\,dx=U(0,\ 0)$$

[$\operatorname{Re} b,\ \operatorname{Re} c>0;\ -1/2-\operatorname{Re}(\mu+\sigma)<\operatorname{Re}\alpha<\operatorname{Re}(\nu+\rho)+1/2$],

$$U(\varepsilon,\gamma)=b^{\nu+1/2}c^{\nu-\alpha+1/2}\Gamma^{1-\gamma^2}\begin{bmatrix}2\sigma+1\\ \rho+\sigma+1/2\end{bmatrix}\left[\Gamma\left(\frac{1}{2}-\rho+\sigma\right)\times\right.$$
$$\left.\times\Gamma\left(\frac{1}{2}-\rho-\sigma\right)\right]^{-(\gamma^2+\gamma)/2}\Gamma^{\varepsilon^2}\begin{bmatrix}-2\nu\\ 1/2-\mu-\nu\end{bmatrix}\Gamma(\alpha-\nu+\sigma)\times$$
$$\times\Gamma^{2\gamma^2-1}(1-\gamma^2+(1-2\gamma^2)(\sigma-\alpha+\nu))\times$$
$$\times\Gamma^{-\gamma^2+\gamma+1}\left((1-2\gamma^2)\rho+\frac{1}{2}+(\gamma^2-\gamma-1)(\alpha-\nu)\right){}_2F_3\left(\frac{1}{2}+\nu-(\varepsilon^2+\varepsilon-1)\mu,\right.$$
$$\left.\frac{1}{2}-\alpha+\nu-(\gamma^2+\gamma-1)\rho;\ 1+2\nu,\ 1-\alpha+\nu-\sigma,\ 1-\alpha+\nu+\sigma;\ (-1)^{(\varepsilon^2+\varepsilon+\gamma^2+\gamma)/2}bc\right)+$$

$$+\varepsilon^2 b^{1/2-\nu} c^{1/2-\alpha-\nu} \Gamma^{1-\gamma^2} \begin{bmatrix} 2\sigma+1 \\ \rho+\sigma+1/2 \end{bmatrix} \left[\Gamma\left(\frac{1}{2}-\rho+\sigma\right) \times\right.$$

$$\left.\times \Gamma\left(\frac{1}{2}-\rho-\sigma\right)\right]^{-(\gamma^2+\gamma)/2} \Gamma\begin{bmatrix} 2\nu,\ \alpha+\nu+\sigma \\ 1/2-\mu+\nu \end{bmatrix} \Gamma^{2\gamma^2-1}(1-\gamma^2+(1-2\gamma^2)\times$$

$$\times(\sigma-\alpha-\nu))\,\Gamma^{-\gamma^2+\gamma+1}\left((1-2\gamma^2)\rho+\frac{1}{2}+(\gamma^2-\gamma-1)(\alpha+\nu)\right)\times$$

$$\times {}_2F_3\left(\frac{1}{2}-\nu-\mu\varepsilon,\ \frac{1}{2}-\alpha-\nu-(\gamma^2+\gamma-1)\rho;\ 1-2\nu,\ 1-\alpha-\nu-\sigma,\ 1-\alpha-\nu+\sigma;\right.$$

$$\left.(-1)^{(1+\varepsilon+\gamma^2+\gamma)/2} bc\right)+b^{\alpha+\sigma+1/2} c^{\sigma+1/2} \Gamma^{1-\varepsilon^2} \begin{bmatrix} 2\nu+1 \\ \mu+\nu+1/2 \end{bmatrix} \times$$

$$\times\left[\Gamma\left(\frac{1}{2}-\mu+\nu\right)\Gamma\left(\frac{1}{2}-\mu-\nu\right)\right]^{-(\varepsilon^2+\varepsilon)/2} \Gamma^{\gamma^2} \begin{bmatrix} -2\sigma \\ 1/2-\rho-\sigma \end{bmatrix} \Gamma(\nu-\alpha-\sigma)\times$$

$$\times\Gamma^{2\varepsilon^2-1}(1-\varepsilon^2+(1-2\varepsilon^2)(\alpha+\nu+\sigma))\,\Gamma^{-\varepsilon^2+\varepsilon+1}\left((1-2\varepsilon^2)\mu+\frac{1}{2}-(\varepsilon^2-\varepsilon-1)(\alpha+\sigma)\right)\times$$

$$\times {}_2F_3\left(\frac{1}{2}+\sigma-(\gamma^2+\gamma-1)\rho,\ \frac{1}{2}+\alpha+\sigma-(\varepsilon^2+\varepsilon-1)\mu;\right.$$

$$\left.2\sigma+1,\ 1+\alpha-\nu+\sigma,\ 1+\alpha+\nu+\sigma;\ (-1)^{(\varepsilon^2+\varepsilon+\gamma^2+\gamma)/2} bc\right)+$$

$$+\gamma^2 b^{\alpha-\sigma+1/2} c^{1/2-\sigma} \Gamma^{1-\varepsilon^2} \begin{bmatrix} 2\nu+1 \\ \mu+\nu+1/2 \end{bmatrix} \left[\Gamma\left(\frac{1}{2}-\mu+\nu\right)\Gamma\left(\frac{1}{2}-\mu-\nu\right)\right]^{-(\varepsilon^2+\varepsilon)/2}\times$$

$$\times\Gamma\begin{bmatrix} 2\sigma,\ \nu-\alpha+\sigma \\ 1/2-\rho+\sigma \end{bmatrix} \Gamma^{2\varepsilon^2-1}(1-\varepsilon^2+(1-2\varepsilon^2)(\alpha-\sigma+1))\times$$

$$\times\Gamma^{-\varepsilon^2+\varepsilon+1}\left((1-2\varepsilon^2)\mu+\frac{1}{2}-(\varepsilon^2-\varepsilon-1)(\alpha-\sigma)\right) {}_2F_3\left(\frac{1}{2}-\sigma-\gamma\rho,\right.$$

$$\left.\frac{1}{2}+\alpha-\sigma-(\varepsilon^2+\varepsilon-1)\mu;\ 1-2\sigma,\ 1+\alpha-\nu-\sigma,\ 1+\alpha+\nu-\sigma;\ (-1)^{(\varepsilon^2+\varepsilon+\gamma+1)/2} bc\right).$$

9. $$\int_0^\infty x^{\alpha-1} e^{-b/(2x)\pm cx/2} M_{\mu,\nu}\left(\frac{b}{x}\right) W_{\rho,\sigma}(cx)\,dx = U(0,\ \pm 1)$$

$\left[\operatorname{Re} b>0;\ \operatorname{Re}(\alpha+\mu)>|\operatorname{Re}\sigma|-1/2,\ \begin{cases}\operatorname{Re}(\alpha-\nu+\rho)<1/2;\ |\arg c|<3\pi/2 \\ \operatorname{Re} c>0\end{cases},\ U(\varepsilon,\gamma)\ \text{see } 2.19.24.8\right].$

10. $$\int_0^\infty x^{\alpha-1} e^{b/(2x)\pm cx/2} W_{\mu,\nu}\left(\frac{b}{x}\right) W_{\rho,\sigma}(cx)\,dx = U(1,\ \pm 1)$$

$\left[\operatorname{Re}(\alpha-\mu)>|\operatorname{Re}\sigma|-1/2;\ |\arg b|<3\pi/2,\ \begin{cases}\operatorname{Re}(\alpha+\rho)<1/2-|\operatorname{Re}\nu|;\ |\arg c|<3\pi/2 \\ \operatorname{Re} c>0\end{cases},\ U(\varepsilon,\gamma)\ \text{see } 2.19.24.8\right].$

11. $$\int_0^\infty x^{\alpha-1} e^{-b/(2x)\pm cx/2} W_{\mu,\nu}\left(\frac{b}{x}\right) W_{\rho,\sigma}(cx)\,dx = U(-1,\ \pm 1)$$

$\left[\operatorname{Re} b>0,\ \begin{cases}\operatorname{Re}(\alpha+\rho)<1/2-|\operatorname{Re}\nu|;\ |\arg c|<3\pi/2 \\ \operatorname{Re} c>0\end{cases},\ U(\varepsilon,\gamma)\ \text{see } 2.19.24.8\right].$

12. $$\int_0^a x^{\sigma-1/2}(a-x)^{\nu-1/2} e^{(c-b)x/2} M_{\mu,\nu}(ba-bx)\,M_{\rho,\sigma}(cx)\,dx=$$

$$=a^{\nu+\sigma} b^{\nu+1/2} c^{\sigma+1/2}(b-c)^{\mu+\rho}\,\mathrm{B}(2\nu+1,\ 2\sigma+1)\,e^{ac/2} M_{\rho+\nu+1/2,\ \nu+\sigma+1/2}(ab-ac)$$

$[\mu+\nu+\rho+\sigma=-1;\ a>0;\ \operatorname{Re}\nu,\ \operatorname{Re}\sigma>-1/2].$

13. $\int\limits_0^\infty x^{\alpha-1} e^{-px} \prod\limits_{k=1}^{n} M_{\rho_k,\ \sigma_k}(c_k x)\, dx =$

$$= \prod_{k=1}^{n} c_k^{\sigma_k+1/2} (p+Q)^{-\alpha-R}\, \Gamma(\alpha+R)\, F_A^{(n)}\left(\alpha+R;\ \frac{1}{2}-\rho_1+\sigma_1,\ \frac{1}{2}-\rho_2+\sigma_2,\ \ldots\right.$$

$$\left.\ldots,\ \frac{1}{2}-\rho_n+\sigma_n;\ 2\sigma_1+1,\ 2\sigma_2+1,\ \ldots,\ 2\sigma_n+1;\ \frac{c_1}{p+Q},\ \ldots,\ \frac{c_n}{p+Q}\right)$$

$\left[2Q=\sum\limits_{k=1}^{n} c_k,\ R=\sum\limits_{k=1}^{n} \sigma_k+\frac{n}{2};\ \operatorname{Re}(\alpha+R)>0;\ 2\operatorname{Re} p>\sum\limits_{k=1}^{n} |\operatorname{Re} c_k|\right]$

2.19.25. Integrals of $x^\alpha \cos ax\, M_{\mu,\ \nu}(bx)\, M_{\rho,\ \sigma}(cx)$.

1. $\int\limits_0^\infty x^{-1} \cos bx\, M_{\rho,\ 0}(-icx)\, M_{\rho,\ 0}(icx)\, dx = \begin{Bmatrix} 1 \\ 0 \end{Bmatrix} \frac{c}{b} \cos \rho\pi\, P_{\rho-1/2}\left(\frac{2c^2}{b^2}-1\right)$

$\left[\begin{Bmatrix} 0<b<c \\ 0<c<b \end{Bmatrix},\ |\operatorname{Re}\rho|<1/2\right]$

2.19.26. Integrals of $x^\alpha J_\nu(bx^{\pm r}) \prod\limits_{j,\ k} M_{\rho_j,\ \sigma_j}(b_j x)\, W_{\rho_k,\ \sigma_k}(c_k x)$.

1. $\int\limits_0^\infty x^{\alpha-1} J_\nu(bx)\, M_{\rho,\ \sigma}(-icx)\, M_{\rho,\ \sigma}(icx)\, dx = I$

$[b,\ c>0;\ -1-\operatorname{Re}(\nu+2\sigma)<\operatorname{Re}\alpha<3/2-2|\operatorname{Re}\rho|:$ also $\operatorname{Re}\alpha<1/2$ for $b=c)]$,

$$I = \frac{2^{\alpha+2\sigma} c^{2\sigma+1}}{b^{\alpha+2\sigma+1}}\, \Gamma\begin{bmatrix} (1+\alpha+\nu)/2+\sigma \\ (1-\alpha+\nu)/2-\sigma \end{bmatrix} {}_4F_3\left(\frac{1+\alpha-\nu}{2}+\sigma,\ \frac{1+\alpha+\nu}{2}+\sigma,\ \frac{1}{2}-\rho+\sigma,\right.$$

$$\left.\frac{1}{2}+\rho+\sigma;\ 2\sigma+1,\ \sigma+\frac{1}{2},\ \sigma+1;\ \frac{c^2}{b^2}\right) \qquad [c<b]$$

$$I = \frac{b^\nu}{2^{\nu+1} c^{\alpha+\nu}}\, \Gamma\begin{bmatrix} 2\sigma+1,\ 2\sigma+1,\ \rho-(\alpha+\nu)/2,\ -\rho-(\alpha+\nu)/2,\ (1+\alpha+\nu)/2+\sigma \\ 1/2-\rho+\sigma,\ 1/2+\rho+\sigma,\ -\alpha-\nu,\ (1-\alpha-\nu)/2+\sigma,\ \nu+1 \end{bmatrix} \times$$

$$\times {}_4F_3\left(\frac{1+\alpha+\nu}{2}-\sigma,\ \frac{1+\alpha+\nu}{2}+\sigma,\ 1+\frac{\alpha+\nu}{2},\ \frac{1+\alpha+\nu}{2};\right.$$

$$\left.1+\rho+\frac{\alpha+\nu}{2},\ 1-\rho+\frac{\alpha+\nu}{2},\ \nu+1;\ \frac{b^2}{c^2}\right)+$$

$$+\frac{2^{\alpha+2\rho-1} c^{2\rho}}{b^{\alpha+2\rho}}\, \Gamma^2\begin{bmatrix} 2\sigma+1 \\ 1/2+\rho+\sigma \end{bmatrix} \Gamma\begin{bmatrix} (\alpha+\nu)/2+\rho \\ 1-(\alpha-\nu)/2-\rho \end{bmatrix} \times$$

$$\times {}_4F_3\left(\frac{1}{2}-\rho-\sigma,\ \frac{1}{2}-\rho+\sigma,\ 1-\rho,\ \frac{1}{2}-\rho;\ 1-2\rho,\ 1-\rho-\frac{\alpha+\nu}{2},\ 1-\frac{\alpha-\nu}{2}-\rho;\ \frac{b^2}{c^2}\right)+$$

$$+\frac{2^{\alpha-2\rho-1}}{b^{\alpha-2\rho} c^{2\rho}}\, \Gamma^2\begin{bmatrix} 1+2\sigma \\ 1/2-\rho+\sigma \end{bmatrix} \Gamma\begin{bmatrix} (\alpha+\nu)/2-\rho \\ 1-(\alpha-\nu)/2+\rho \end{bmatrix} \times$$

$$\times {}_4F_3\left(\frac{1}{2}+\rho-\sigma,\ \frac{1}{2}+\rho+\sigma,\ 1+\rho,\ \frac{1}{2}+\rho;\ 1+2\rho,\ 1-\frac{\alpha+\nu}{2}+\rho,\ 1-\frac{\alpha-\nu}{2}+\rho;\ \frac{b^2}{c^2}\right)$$

$[b<c]$

2. $\int\limits_0^\infty J_{2\sigma}(bx)\, M_{\rho,\ \sigma}(-icx)\, M_{\rho,\ \sigma}(icx)\, dx =$

$$= \frac{c\,(c^2-b^2)_+^{-1/2}}{b^{2\rho+1}}\, \Gamma\begin{bmatrix} 2\sigma+1,\ 2\sigma+1 \\ 1/2-\rho+\sigma,\ 1/2+\rho+\sigma \end{bmatrix} \left[\left(c+\sqrt{c^2-b^2}\right)^{2\rho}+\left(c-\sqrt{c^2-b^2}\right)^{2\rho}\right]$$

$[b\neq c;\ b,\ c>0;\ |\operatorname{Re}\rho|<1/4;\ \operatorname{Re}\sigma>-1/2]$

3. $$\int_0^\infty x^{-\sigma-1/2} J_{\sigma-1/2}(bx)\, M_{\rho,\sigma}(-icx)\, M_{\rho,\sigma}(icx)\, dx =$$
$$= \frac{2^{1/2-\sigma}\sqrt{\pi}\, c^{1-\sigma}}{b^{3/2-\sigma}} (c^2-b^2)_+^{\sigma/2} \Gamma\begin{bmatrix} 2\sigma+1,\ 2\sigma+1 \\ 1/2-\rho+\sigma,\ 1/2+\rho+\sigma \end{bmatrix} P^{-\sigma}_{\rho-1/2}\left(\frac{2c^2}{b^2}-1\right)$$
$[b \neq c;\ b,\ c > 0;\ \operatorname{Re}\sigma > -1/2,\ 2\,|\operatorname{Re}\rho\,| - 1]$.

4. $$\int_0^\infty J_{2\sigma}(bx)\, M_{-\rho,\sigma}(cx)\, W_{\rho,\sigma}(cx)\, dx =$$
$$= \frac{c}{b^{2\rho+1}\sqrt{b^2+c^2}} \Gamma\begin{bmatrix} 2\sigma+1 \\ 1/2-\rho+\sigma \end{bmatrix} (\sqrt{b^2+c^2}+c)^{2\rho}$$
$[b,\ \operatorname{Re} c > 0;\ \operatorname{Re}\rho < 1/4;\ \operatorname{Re}\sigma > -1/2]$.

5. $$\int_0^\infty x^{\alpha-1} J_\nu(bx) \begin{Bmatrix} M_{-\rho,\sigma}(cx) \\ W_{-\rho,\sigma}(cx) \end{Bmatrix} W_{\rho,\sigma}(cx)\, dx =$$
$$= \frac{b^\nu}{2^{\nu+1} c^{\alpha+\nu}} \Gamma^{(1\pm1)/2}\begin{bmatrix} 2\sigma+1 \\ 1/2-\rho+\sigma \end{bmatrix} \Gamma\begin{bmatrix} -(\alpha+\nu)/2-\rho,\ 1+\alpha+\nu,\ (1+\alpha+\nu)/2+\sigma \\ (1-\alpha-\nu)/2+\sigma,\ 1+(\alpha+\nu)/2-\rho,\ \nu+1 \end{bmatrix} \times$$
$$\times \left[-\frac{\sin(\rho+(\alpha+\nu)/2)\pi}{\cos(\sigma-(\alpha+\nu)/2)\pi}\right]^{(1\mp1)/2} {}_4F_3\left(\frac{1+\alpha+\nu}{2}-\sigma,\ \frac{1+\alpha+\nu}{2}+\sigma,\ \frac{1+\alpha+\nu}{2},\right.$$
$$\left. 1+\frac{\alpha+\nu}{2};\ 1+\rho+\frac{\alpha+\nu}{2},\ 1-\rho+\frac{\alpha+\nu}{2},\ \nu+1;\ -\frac{b^2}{c^2}\right) +$$
$$+ \begin{Bmatrix} 1 \\ 0 \end{Bmatrix} \frac{2^{\alpha+2\rho-1} c^{2\rho}}{b^{\alpha+2\rho}} \Gamma\begin{bmatrix} 2\sigma+1,\ \rho+(\alpha+\nu)/2 \\ 1/2+\rho+\sigma,\ 1+(\nu-\alpha)/2-\rho \end{bmatrix} \times$$
$$\times {}_4F_3\left(\frac{1}{2}-\rho-\sigma,\ \frac{1}{2}-\rho+\sigma,\ \frac{1}{2}-\rho,\ 1-\rho;\ 1-2\rho,\ 1-\rho-\frac{\alpha+\nu}{2},\ 1-\rho+\frac{\nu-\alpha}{2};\ -\frac{b^2}{c^2}\right)$$
$\left[b,\ \operatorname{Re} c > 0,\ \begin{Bmatrix} \operatorname{Re}(\alpha+\nu),\ \operatorname{Re}(\alpha+\nu+2\sigma) > -1;\ \operatorname{Re}(\alpha+2\rho) < 3/2 \\ \operatorname{Re}(\alpha+\nu) > 2\,|\operatorname{Re}\sigma\,| - 1 \end{Bmatrix}\right]$.

6. $$\int_0^\infty x^{2\sigma-\nu} J_\nu(bx)\, M_{-\rho,\sigma}(cx)\, W_{\rho,\sigma}(cx)\, dx =$$
$$= \frac{2^{2\rho+2\sigma-\nu} c^{2\rho}}{b^{2\rho+2\sigma-\nu+1}} \Gamma\begin{bmatrix} 2\sigma+1 \\ 1/2+\nu-\rho-\sigma \end{bmatrix} {}_3F_2\left(\frac{1}{2}-\rho+\sigma,\ \frac{1}{2}-\rho,\ 1-\rho;\ 1-2\rho,\right.$$
$$\left.\frac{1}{2}+\nu-\rho-\sigma;\ -\frac{b^2}{c^2}\right)$$
$[b,\ \operatorname{Re} c > 0;\ \operatorname{Re}\sigma > -1/2;\ \operatorname{Re}(2\rho+2\sigma-\nu) < 1/2]$.

7. $$\int_0^\infty x^{-1} J_0(bx)\, M_{-\rho,\sigma}(cx)\, W_{\rho,\sigma}(cx)\, dx =$$
$$= e^{-\rho\pi i} \Gamma\begin{bmatrix} 2\sigma+1 \\ \rho+\sigma+1/2 \end{bmatrix} P^{\rho}_{\sigma-1/2}\left(\frac{\sqrt{b^2+c^2}}{c}\right) Q^{\rho}_{\sigma-1/2}\left(\frac{\sqrt{b^2+c^2}}{c}\right) =$$
$$= \frac{e^{-\sigma\pi i} c}{b} \Gamma\begin{bmatrix} 2\sigma+1 \\ \sigma-\rho+1/2 \end{bmatrix} P^{-\sigma}_{\rho-1/2}\left(\frac{\sqrt{b^2+c^2}}{b}\right) Q^{\sigma}_{-\rho-1/2}\left(\frac{\sqrt{b^2+c^2}}{b}\right)$$
$[b,\ \operatorname{Re} c > 0;\ \operatorname{Re}\rho < 3/4;\ \operatorname{Re}\sigma > -1/2]$.

8. $$\int_0^\infty x^{-1} J_0(bx)\, W_{-\rho,\sigma}(cx)\, W_{\rho,\sigma}(cx)\, dx =$$
$$= \frac{\pi}{2\cos\sigma\pi} P^{\rho}_{\sigma-1/2}\left(\frac{\sqrt{b^2+c^2}}{c}\right) P^{-\rho}_{\sigma-1/2}\left(\frac{\sqrt{b^2+c^2}}{c}\right) =$$
$$= \frac{e^{2\sigma\pi i} c}{b\cos\sigma\pi\,\Gamma(1/2+\rho-\sigma)\,\Gamma(1/2-\rho-\sigma)} Q^{-\sigma}_{-\rho-1/2}\left(\frac{\sqrt{b^2+c^2}}{b}\right) Q^{-\sigma}_{\rho-1/2}\left(\frac{\sqrt{b^2+c^2}}{b}\right)$$
$[b,\ \operatorname{Re} c > 0;\ |\operatorname{Re}\sigma\,| < 1/2]$.

9. $$\int_0^\infty x^{\alpha-1}J_\nu(bx)\,W_{\rho,\,\sigma}(-icx)\,W_{\rho,\,\sigma}(icx)\,dx=\frac{2^{\alpha+2\rho-1}c^{2\rho}}{b^{\alpha+2\rho}}\Gamma\left[\begin{matrix}(\alpha+\nu)/2+\rho\\ 1+(\nu-\alpha)/2-\rho\end{matrix}\right]\times$$
$$\times{}_4F_3\left(\frac{1}{2}-\rho,\,1-\rho,\,\frac{1}{2}-\rho+\sigma,\,\frac{1}{2}-\rho-\sigma;\;1-\rho-\frac{\alpha+\nu}{2},\,1-\rho+\frac{\nu-\alpha}{2},\;1-2\rho;\;\frac{b^2}{c^2}\right)+$$
$$+\frac{b^\nu}{2^{\nu+1}c^{\alpha+\nu}}\Gamma\left[\begin{matrix}-\rho-(\alpha+\nu)/2,\;\alpha+\nu+1,\;(\alpha+\nu+1)/2+\sigma,\;(\alpha+\nu+1)/2-\sigma\\ \nu+1,\;1+(\alpha+\nu)/2-\rho,\;1/2-\rho+\sigma,\;1/2-\rho-\sigma\end{matrix}\right]\times$$
$$\times{}_4F_3\left(\frac{\alpha+\nu+1}{2},\,\frac{\alpha+\nu}{2}+1,\,\frac{\alpha+\nu+1}{2}+\sigma,\,\frac{\alpha+\nu+1}{2}-\sigma;\right.$$
$$\left.\nu+1,\,1+\rho+\frac{\alpha+\nu}{2},\;1-\rho+\frac{\alpha+\nu}{2};\;\frac{b^2}{c^2}\right)$$
$[b>0;\ 2\,|\mathrm{Re}\,\sigma|-\mathrm{Re}\,\nu-1<\mathrm{Re}\,\alpha<-2\,\mathrm{Re}\,\rho;\ |\arg c|<\pi]$.

10. $$\int_0^\infty x^{-1}J_{4\sigma}(b\sqrt{x})\,M_{-\rho,\,\sigma}(cx)\,W_{\rho,\,\sigma}(cx)\,dx=\Gamma\left[\begin{matrix}2\sigma+1\\ \sigma-\rho+1/2\end{matrix}\right]I_{\sigma-\rho}\left(\frac{b^2}{8c}\right)\times$$
$$\times K_{\sigma+\rho}\left(\frac{b^2}{8c}\right)$$
$[b,\ \mathrm{Re}\,c>0;\ \mathrm{Re}\,\rho<3/8;\ \mathrm{Re}\,\sigma>-1/4]$.

11. $$\int_0^\infty x^{-1}J_{2\nu+2\sigma}(b\sqrt{x})\,M_{\mu,\,\nu}(cx)\,W_{\rho,\,\sigma}(cx)\,dx=$$
$$=\frac{4c}{b^2}\Gamma\left[\begin{matrix}2\nu+1\\ \mu+\nu-\rho+\sigma+1\end{matrix}\right]M_{(\mu+\nu+\rho-\sigma)/2,\;(\mu+\nu-\rho+\sigma)/2}\left(\frac{b^2}{4c}\right)\times$$
$$\times W_{(\mu-\nu+\rho+\sigma)/2,\;(\nu-\mu+\rho+\sigma)/2}\left(\frac{b^2}{4c}\right)$$
$[b,\ \mathrm{Re}\,c>0;\ \mathrm{Re}\,\nu,\ \mathrm{Re}\,(\nu+\sigma)>-1/2;\ \mathrm{Re}(\rho-\mu)<3/4]$.

12. $$\int_0^\infty x^{-3}J_{-2\rho}\left(\frac{b}{x}\right)M_{-\rho,\,\sigma}(cx)\,W_{\rho,\,\sigma}(cx)\,dx=$$
$$=\frac{2c}{b}\Gamma\left[\begin{matrix}2\sigma+1\\ 1/2-\rho+\sigma\end{matrix}\right]J_{2\sigma}(\sqrt{2bc})\,K_{2\sigma}(\sqrt{2bc})$$
$[b,\ \mathrm{Re}\,c>0;\ \mathrm{Re}\,\rho<1/2;\ \mathrm{Re}\,\sigma>-1/4]$.

13. $$\int_0^\infty x^{-3}J_{2\rho}\left(\frac{b}{x}\right)W_{-\rho,\,\sigma}(cx)\,W_{\rho,\,\sigma}(cx)\,dx=$$
$$=-\frac{2c}{b}\{\sin(\sigma-\rho)\pi J_{2\sigma}(\sqrt{2bc})+\cos(\sigma-\rho)\pi Y_{2\sigma}(\sqrt{2bc})\}\,K_{2\sigma}(\sqrt{2bc})$$
$[b,\ \mathrm{Re}\,c>0;\ |\mathrm{Re}\,\sigma|<1/4]$.

14. $$\int_0^\infty x^{-3}J_{-2\rho}\left(\frac{b}{x}\right)W_{\rho,\,\sigma}(-icx)\,W_{\rho,\,\sigma}(icx)\,dx=$$
$$=\frac{4c}{b\Gamma(1/2-\rho+\sigma)\,\Gamma(1/2-\rho-\sigma)}K_{2\sigma}(\sqrt{2ibc})\,K_{2\sigma}(\sqrt{-2ibc})$$
$[b>0;\ |\arg c|<\pi;\ \mathrm{Re}\,\rho<1/2;\ |\mathrm{Re}\,\sigma|<1/4]$.

15. $$\int_0^\infty J_{2\sigma}(bx)\,M_{-\rho,\,\sigma}\left[c\left(\sqrt{x^2+z^2}-z\right)\right]W_{\rho,\,\sigma}\left[c\left(\sqrt{x^2+z^2}+z\right)\right]dx=$$
$$=\frac{b^{-2\rho-1}c}{\sqrt{b^2+c^2}}\Gamma\left[\begin{matrix}2\sigma+1\\ 1/2-\rho+\sigma\end{matrix}\right](\sqrt{b^2+c^2}+c)^{2\rho}\exp\left(-z\sqrt{b^2+c^2}\right)$$
$[b,\ \mathrm{Re}\,c,\ \mathrm{Re}\,z>0;\ \mathrm{Re}\,\rho<1/4;\ \mathrm{Re}\,\sigma>-1/2]$.

2.19.27. Integrals of $x^{\alpha}\begin{Bmatrix} Y_\nu(bx^{\pm r}) \\ K_\nu(bx^{\pm r}) \end{Bmatrix}\begin{Bmatrix} M_{\rho,\sigma}(-icx)\,M_{\rho,\sigma}(icx) \\ W_{-\rho,\sigma}(cx)\,W_{\rho,\sigma}(cx) \end{Bmatrix}$.

1. $$\int_0^\infty x^{-1}K_0(b\sqrt{x})\,M_{\rho,\sigma}(-icx)\,M_{\rho,\sigma}(icx)\,dx=\frac{2c}{b^2}\Gamma^2(2\sigma+1)\,W_{-\sigma,\rho}\left(-\frac{ib^2}{4c}\right)\times$$ $$\times W_{-\sigma,\rho}\left(\frac{ib^2}{4c}\right) \quad [c,\ \operatorname{Re} b>0;\ \operatorname{Re}\sigma>-1/2].$$

2. $$\int_0^\infty x^{-1}K_0(b\sqrt{x})\,M_{\rho,0}(-icx)\,M_{\rho,0}(icx)\,dx=\frac{\pi}{8}\left[J_\rho^2\left(\frac{b^2}{8c}\right)+Y_\rho^2\left(\frac{b^2}{8c}\right)\right]$$ $$[c,\ \operatorname{Re} b>0].$$

3. $$\int_0^\infty x^{-3}Y_{2\rho}\left(\frac{b}{x}\right)W_{-\rho,\sigma}(cx)\,W_{\rho,\sigma}(cx)\,dx=$$ $$=\frac{2c}{b}\left[\cos(\rho-\sigma)\pi J_{2\sigma}(\sqrt{2bc})+\sin(\rho-\sigma)\pi Y_{2\sigma}(\sqrt{2bc})\right]K_{2\sigma}(\sqrt{2bc})$$ $$[b,\ \operatorname{Re} c>0;\ |\operatorname{Re}\sigma|<1/4].$$

4. $$\int_0^\infty x^{-3}K_{2\rho}\left(\frac{b}{x}\right)W_{-\rho,\sigma}(cx)\,W_{\rho,\sigma}(cx)\,dx=\frac{2c}{b}K_{2\sigma}(\sqrt{2ibc})\,K_{2\sigma}(\sqrt{-2ibc})$$ $$[\operatorname{Re} b,\ \operatorname{Re} c>0].$$

2.19.28. Integrals with respect to the index containing $M_{\rho,\sigma}(c)$ or $W_{\rho,\sigma}(c)$.

Condition: $b,\ c>0$.

1. $$\int_{-\infty}^\infty e^{ixy}\Gamma\left(\frac{1}{2}+\sigma+ix\right)\Gamma\left(\frac{1}{2}+\sigma-ix\right)M_{ix,\sigma}(c)\,dx=$$ $$=2^{-2\sigma}\pi c^{\sigma+1/2}\,\Gamma(2\sigma+1)\operatorname{ch}^{-2\sigma-1}\frac{y}{2}\exp\left(\frac{c}{2}\operatorname{th}\frac{y}{2}\right) \quad [\operatorname{Re}\sigma>-1/2;\ |\operatorname{Im} y|<\pi].$$

2. $$\int_{-\infty}^\infty e^{ixy}\Gamma\left(\frac{1}{2}+\sigma+ix\right)\Gamma\left(\frac{1}{2}+\sigma-ix\right)M_{ix,\sigma}(b)\,M_{ix,\sigma}(c)\,dx=$$ $$=\frac{\pi\sqrt{bc}\,\Gamma^2(2\sigma+1)}{\operatorname{ch}(y/2)}\exp\left[\frac{1}{2}(b+c)\operatorname{th}\frac{y}{2}\right]_{2\sigma}\left(\frac{\sqrt{bc}}{\operatorname{ch}(y/2)}\right) \quad [\operatorname{Re}\sigma>-1/2;\ |\operatorname{Im} y|<\pi].$$

3. $$\int_{-i\infty}^{i\infty}\Gamma\left[\frac{1}{2}+\mu+\sigma+x,\ \frac{1}{2}+\mu+\sigma-x,\ \frac{1}{2}-\mu+\sigma+x,\ \frac{1}{2}-\mu+\sigma-x\right]\times$$ $$\times M_{\mu+ix,\sigma}(b)\,M_{\mu-ix,\sigma}(c)\,dx=$$ $$=\frac{\pi^{3/2}(bc)^{\sigma+1/2}}{2^{4\sigma}(b+c)^{2\sigma+1}}\Gamma\begin{bmatrix}2\sigma+1,\ 2\sigma+2\mu+1,\ 2\sigma-2\mu+1\\ 2\sigma+3/2\end{bmatrix}M_{2\mu,\,2\sigma+1/2}(b+c)$$ $$[\operatorname{Re}\sigma>|\operatorname{Re}\mu|-1/2].$$

4. $$\int_{-\infty}^\infty\frac{1}{\Gamma(1+2\sigma+2x)\,\Gamma(1+2\nu-2x)}M_{-1/2-\nu-x,\,\nu-x}(c)\,M_{-1/2-\sigma+x,\,\sigma+x}(c)\,dx=$$ $$=\frac{2^{\nu+\sigma-3/2}\sqrt{c}\,e^{2c}}{\Gamma(2\nu+2\sigma+1)}M_{-\nu-\sigma-3/2,\,\nu+\sigma}(2c) \quad [\operatorname{Re}(\nu+\sigma)>-1/2].$$

5. $$\int_{-\infty}^{\infty} \frac{\sin[(2k+1)2\pi x]}{\sin 2\pi x \Gamma(1+2\sigma+2x)\Gamma(1+2\nu-2x)} M_{\mu-x,\,\nu-x}(c)\, M_{\rho+x,\,\sigma+x}(c)\, dx =$$
$$= \frac{2^{\nu+\sigma-3/2}\sqrt{c}\, e^{2c}}{\Gamma(2\nu+2\sigma+1)} M_{\mu+\rho-1/2,\,\nu+\sigma}(2c) \quad [\operatorname{Re}(\nu+\sigma)>-1/2].$$

6. $$\int_{-\infty}^{\infty} \frac{c^{-2x}}{\Gamma[1+2\nu+2x,\ 1+2\sigma+2x,\ 1/2-\mu-\nu-2x,\ 1/2-\rho-\sigma-2x]} \times$$
$$\times M_{\mu+x,\,\nu+x}(c)\, M_{\rho+x,\,\sigma+x}(c)\, dx =$$
$$= \frac{c^{\nu+\sigma+1} e^{c}}{2} \Gamma\left[\begin{matrix} \nu-\mu-\rho+\sigma \\ 1/2-\mu+\nu,\quad 1/2-\rho+\sigma,\quad 1/2+2\nu-\rho-\sigma,\quad 1/2-\mu-\nu+2\sigma \end{matrix}\right] \times$$
$$\times {}_2F_2\left(\frac{\nu-\mu-\rho+\sigma}{2}, \frac{1-\mu+\nu-\rho+\sigma}{2}; \frac{1}{2}-\mu-\nu+2\sigma, \frac{1}{2}+2\nu-\rho-\sigma;\ 4c\right)$$
$$[\operatorname{Re}(\nu-\mu-\rho+\sigma)>0].$$

7. $$\int_{0}^{\infty} x \operatorname{sh} n\pi x\, \Gamma\left(\frac{1}{2}-\rho+ix\right) \Gamma\left(\frac{1}{2}-\rho-ix\right) K_{inx}(b)\, W_{\rho,\,ix}(c)\, dx = I_n$$
$$[\operatorname{Re}\rho \leqslant 1/2],$$
$$I_1 = \left(\frac{\pi}{2}\right)^{3/2} \Gamma(1-\rho)\, b^{1/2-\rho} c\, (b+c)^{\rho-1} e^{-b-c},$$
$$I_2 = 2^{2\rho-2} \pi^2 b^{1-2\rho} c^{\rho} \exp\left(-\frac{2c^2+b^2}{4c}\right).$$

8. $$\int_{0}^{\infty} x \operatorname{sh} 2\pi x\, \Gamma(-\rho+2ix)\, \Gamma(-\rho-2ix)\, S_{2\rho+1,\,2ix}(b)\, W_{\rho,\,ix}(c)\, dx =$$
$$= \frac{\pi c^{\rho}}{8 b^{2\rho-1}} \Gamma\left(\frac{1}{2}-2\rho\right) \exp\left(\frac{b^2-2c^2}{4c}\right) \Gamma\left(2\rho+\frac{1}{2}, \frac{b^2}{4c}\right) \quad [\operatorname{Re}\rho<0].$$

9. $$\int_{0}^{\infty} x \operatorname{th} 2\pi x\, \Gamma\left(\frac{1}{2}-\rho+ix\right) \Gamma\left(\frac{1}{2}-\rho-ix\right) P_{-1/2+2ix}(b)\, W_{\rho,\,ix}(c)\, dx =$$
$$= 2^{\rho-3/4} \sqrt{\pi} c^{3/4} \Gamma\left(\frac{3}{2}-2\rho\right) e^{(b^2-1)c/2} D_{2\rho-3/2}\left(b\sqrt{2c}\right) \quad [\operatorname{Re} c>0;\ \operatorname{Re}\rho<1/2].$$

10. $$\int_{0}^{\infty} x \operatorname{sh} \pi x\, \Gamma\left(\frac{1}{2}-\rho+ix\right) \Gamma\left(\frac{1}{2}-\rho-ix\right) P^{\rho}_{ix-1/2}(b)\, W_{\rho,\,ix}(c)\, dx =$$
$$= \frac{\pi c}{2} \left(\frac{b+1}{b-1}\right)^{\rho/2} e^{-bc/2} \quad [\operatorname{Re}\rho \leqslant 1/2].$$

11. $$\int_{0}^{\infty} x \operatorname{sh} 2\pi x\, \Gamma\left(\frac{1\pm1}{2}-2\rho+2ix\right) \Gamma\left(\frac{1\pm1}{2}-2\rho-2ix\right) P^{2\rho\mp1/2}_{2ix-1/2}(b)\, W_{\rho,\,ix}(c)\, dx =$$
$$= 2^{-2\rho-1\pm1/2} \pi b^{(1\pm1)/2} c^{1\pm(1\mp2\rho)/4} (b^2-1)^{\pm1/4-\rho} e^{c/2-b^2c} \quad [\operatorname{Re}\rho \leqslant (1\pm1)/4].$$

12. $$\int_{0}^{\infty} x \operatorname{sh} \pi x\, \Gamma\left[\frac{1}{2}-\mu+ix, \frac{1}{2}-\mu-ix, \frac{1}{2}-\rho+ix, \frac{1}{2}-\rho-ix\right] P^{\mu}_{ix-1/2}(b) \times$$
$$\times W_{\rho,\,ix}(c)\, dx = \left(\frac{2}{c}\right)^{(\mu-1)/2} \pi \Gamma(1-\rho) \Gamma(1-\mu-\rho)\, (b+1)^{-1/2} (b-1)^{-\mu/2} e^{(b-1)c/4} \times$$
$$\times W_{\rho-(\mu+1)/2,\,-\mu/2}\left(\frac{b+1}{2} c\right) \quad [\operatorname{Re}\mu,\ \operatorname{Re}\rho<1/2].$$

13. $$\int_0^\infty x \operatorname{sh} 2\pi x \Gamma\left[\frac{1}{2}-\mu+2ix, \frac{1}{2}-\mu-2ix, \frac{1}{2}-\rho+ix, \frac{1}{2}-\rho-ix\right]\times$$
$$\times P^{\mu}_{2ix-1/2}(b)\, W_{\rho,\, ix}(c)\, dx = 2^{\rho-(2\mu+3)/4}\pi^{3/2}c^{(3-2\mu)/4}\Gamma\left(\frac{3}{2}-\mu-2\rho\right)\times$$
$$\times(b^2-1)^{-\mu/2}\exp\left(\frac{b^2-1}{2}c\right)D_{\mu+2\rho-3/2}\left(b\sqrt{2c}\right) \quad [\operatorname{Re}\mu,\ \operatorname{Re}\rho<1/2].$$

14. $$\int_0^\infty x \operatorname{sh} 2\pi x \Gamma\left[2\rho+2ix,\ 2\rho-2ix,\ \frac{1}{2}-\rho+ix, \frac{1}{2}-\rho-ix\right]P^{1/2-2\rho}_{2ix-1/2}(b)\times$$
$$\times W_{\rho,\, ix}(c)\, dx = 2^{2\rho-3/2}\pi^2 c^{\rho+1/2}(b^2-1)^{\rho-1/4}\exp\left(\frac{2b^2-1}{2}c\right)\operatorname{erfc}\left(b\sqrt{c}\right) \quad [\operatorname{Re}\rho>0].$$

15. $$\int_0^\infty x \operatorname{sh} 2\pi x \Gamma\left[\frac{1}{2}-\rho+2ix, \frac{1}{2}-\rho-2ix, \frac{1+2\rho}{4}+ix, \frac{1+2\rho}{4}-ix\right]P^{\rho}_{2ix-1/2}(b)\times$$
$$\times W_{\rho,\, ix}(c)\, dx = 2^{-\rho-1}\pi^2 c^{3/2-\rho}(b^2-1)^{-\rho/2}\exp\left(\frac{2b^2-1}{2}c^2\right)\operatorname{erfc}(bc) \quad [|\operatorname{Re}\rho|<1/2].$$

16. $$\int_{-\infty}^\infty \frac{1}{\operatorname{ch}\pi x}W_{ix,\,0}(b)\,W_{-ix,\,0}(c)\,dx = \frac{2\sqrt{bc}}{b+c}\exp\left(-\frac{b+c}{2}\right).$$

17. $$\int_{-\infty}^\infty \Gamma(ix)\,\Gamma(1+2\sigma+ix)\,W_{1/2+\sigma+ix,\,\sigma}(b)\,W_{-1/2-\sigma-ix,\,\sigma}(c)\,dx=$$
$$=2\sqrt{\pi}\,\Gamma(2\sigma+1)\,(bc)^{\sigma+1/2}(b+c)^{-2\sigma-1/2}K_{2\sigma+1/2}\left(\frac{b+c}{2}\right).$$

18. $$\int_{-\infty}^\infty \Gamma\left(\frac{1}{2}+\sigma-x\right)\Gamma\left(\frac{1}{2}+\sigma+x\right)W_{-x,\,\sigma}(b)\,W_{x,\,\sigma}(c)\,dx=$$
$$=2\pi i\,\Gamma(2\sigma+1)\frac{(bc)^{\sigma+1/2}}{(b+c)^{2\sigma+1}}W_{0,\,2\sigma+1/2}(b+c).$$

19. $$\int_0^\infty x \operatorname{sh} 2\pi x\,\Gamma(-\rho+2ix)\,\Gamma(-\rho-2ix)\,W_{\rho,\,ix}(b)\,W_{\rho+1/2,\,ix}(c)\,dx=$$
$$=\frac{\pi b^{1/2-\rho}c^{1-\rho}}{2^{4\rho+2}}(b+c)^{2\rho-1/2}\Gamma\left(\frac{1}{2}-2\rho\right)e^{-(b+c)/2} \quad [\operatorname{Re}\rho\leqslant 0].$$

20. $$\int_0^\infty x \operatorname{sh} 2\pi x\,K_{2ix}(b)\,W_{\rho,\,ix}(c)\,W_{-\rho,\,ix}(c)\,dx=$$
$$=\frac{\pi b^{1-2\rho}c}{8\sqrt{b^2-c^2}}\left[\left(c+i\sqrt{b^2-c^2}\right)^{2\rho}+\left(c-i\sqrt{b^2-c^2}\right)^{2\rho}\right]=\frac{\pi bc}{4\sqrt{b^2-c^2}}\cos\left(2\rho\arccos\frac{c}{b}\right) \quad [b>c],$$
$$=0 \quad [b<c].$$

21. $$\int_0^\infty x \operatorname{sh} 2\pi x\,\Gamma\left(\frac{1}{2}-\rho+ix\right)\Gamma\left(\frac{1}{2}-\rho-ix\right)K_{2ix}(b)\,W_{\rho,\,ix}(ic)\,W_{\rho,\,ix}(-ic)\,dx=$$
$$=\frac{\pi^2 b^{1-2\rho}c}{4\sqrt{b^2+c^2}}\left(c+\sqrt{b^2+c^2}\right)^{2\rho} \quad [\operatorname{Re}\rho<1/2].$$

2.20. THE CONFLUENT HYPERGEOMETRIC FUNCTIONS OF KUMMER ${}_1F_1(a;b;x)$ AND TRICOMI $\Psi(a,b;x)$

The confluent hypergeometric functions ${}_1F_1(a;b;x)$ and $\Psi(a,b;x)$ are related to the Whittaker functions $M_{\rho,\sigma}(x)$ and $W_{\rho,\sigma}(x)$ by the relations

$$ {}_1F_1(a;\ b;\ \pm x)=x^{-b/2}e^{\pm x/2}M_{b/2-a,\ (b-1)/2}(x), $$
$$ \Psi(a,\ b;\ x)=x^{-b/2}e^{x/2}W_{b/2-a,\ (b-1)/2}(x). $$

For the calculation of integrals containing the functions ${}_1F_1(a;b;x)$ and $\Psi(a,b;x)$, one has to go over to the functions $M_{\rho,\sigma}(x)$ and $W_{\rho,\sigma}(x)$ and use the formulae of section 2.19; see also the formulae of section 2.22 "The hypergeometric function ${}_pF_q((a_p);(b_q);x)$" for $p=q=1$.

2.21. THE GAUSS HYPERGEOMETRIC FUNCTION ${}_2F_1(a,b;c;x)$

2.21.1. Integrals of $A(x){}_2F_1(a,\ b;\ c;\ \varphi(x))$.

1. $$\int_0^\infty x^{\alpha-1}{}_2F_1(a,\ b;\ c;\ -\omega x)\,dx=\omega^{-\alpha}\Gamma\begin{bmatrix} c,\ \alpha,\ a-\alpha,\ b-\alpha \\ a,\ b,\ c-\alpha \end{bmatrix}$$

$[0<\operatorname{Re}\alpha<\operatorname{Re}a,\ \operatorname{Re}b;\ |\arg\omega|<\pi]$.

2. $$\int_0^\infty x^{\alpha-1}{}_2F_1(a,\ b;\ c;\ 1-\omega x)\,dx=\omega^{-\alpha}\Gamma\begin{bmatrix} c,\ \alpha,\ a-\alpha,\ b-\alpha,\ c-a-b+\alpha \\ a,\ b,\ c-a,\ c-b \end{bmatrix}$$

$[0,\ \operatorname{Re}(a+b-c)<\operatorname{Re}\alpha<\operatorname{Re}a,\ \operatorname{Re}b;\ |\arg\omega|<\pi]$.

3. $$\int_0^y x^{\alpha-1}{}_2F_1\left(a,\ b;\ c;\ \frac{x}{y}\right)dx+$$
$$+y^a\Gamma\begin{bmatrix} 1-b,\ c \\ c-a,\ a-b+1 \end{bmatrix}\int_y^\infty x^{\alpha-a-1}{}_2F_1\left(a,\ a-c+1;\ a-b+1;\ \frac{y}{x}\right)dx=$$
$$=y^\alpha\Gamma\begin{bmatrix} 1-b,\ c,\ \alpha,\ a-\alpha \\ a,\ 1-b+\alpha,\ c-\alpha \end{bmatrix}$$

$[y>0;\ \operatorname{Re}(c-a-b)>-1;\ 0<\operatorname{Re}\alpha<\operatorname{Re}a;\ b\neq 1,\ 2,\ \ldots]$.

4. $$\int_0^y x^{\alpha-1}(y-x)^{\beta-1}{}_2F_1(a,\ b;\ c;\ -\omega x)\,dx=\mathrm{B}(\alpha,\ \beta)\,y^{\alpha+\beta-1}{}_3F_2(a,\ b,\ \alpha;\ c,\ \alpha+\beta;\ -\omega y)$$

$[y,\ \operatorname{Re}\alpha,\ \operatorname{Re}\beta>0;\ |\arg(1+\omega y)|<\pi]$.

5. $$\int_0^y x^{\alpha-1}(y-x)^{\beta-1}{}_2F_1\left(a,b;\ c;\ \frac{x}{y}\right)dx=\mathrm{B}(\alpha,\ \beta)\,y^{\alpha+\beta-1}{}_3F_2(a,\ b,\ \alpha;\ c,\ \alpha+\beta;\ 1)$$

$[y,\ \operatorname{Re}\alpha,\ \operatorname{Re}\beta,\ \operatorname{Re}(c-a-b+\beta)>0]$.

6. $$=y^{c+\beta-1}\Gamma\begin{bmatrix} c,\ \beta,\ c-a-b+\beta \\ c-a+\beta,\ c-b+\beta \end{bmatrix}$$ $[\alpha=c;\ \operatorname{Re}\alpha,\ \operatorname{Re}\beta,\ \operatorname{Re}(\alpha+\beta-a-b)>0]$.

7. $$=y^{\alpha+\beta-1}\Gamma\begin{bmatrix} c,\ \alpha,\ \beta,\ c-a-\alpha \\ c-a,\ c-\alpha,\ \alpha+\beta \end{bmatrix}$$ $[b=\alpha+\beta;\ \operatorname{Re}\alpha,\ \operatorname{Re}\beta,\ \operatorname{Re}(c-a-\alpha)>0]$.

8. $$=y^{\alpha+b-c-n}\frac{(c-\alpha)_n}{(c)_n}\Gamma\begin{bmatrix} b-c+1,\ \alpha \\ c-\alpha,\ b-c+\alpha+1 \end{bmatrix}$$

$[a=-n,\ n=0,\ 1,\ 2,\ \ldots;\ \beta=b-c-n+1;\ \operatorname{Re}\alpha,\ \operatorname{Re}\beta>0]$.

9. $$=\frac{y^{a-b}\sqrt{\pi}}{2^a}\Gamma\begin{bmatrix} c,\ a-c+1,\ c-a/2-b \\ (a+1)/2,\ a/2-b+1,\ c-a/2 \end{bmatrix}$$

$[\alpha=a-c+1,\ \beta=c-b;\ \operatorname{Re}b<\operatorname{Re}c<\operatorname{Re}a+1;\ \operatorname{Re}(2c-2b-a)>0]$.

10. $\int_0^y x^{\alpha-1}(y-x)^{\beta-1}{}_2F_1(a, b; c; 1-\omega x)\,dx=$

$$=y^{\alpha+\beta-1}\Gamma\begin{bmatrix} c, c-a-b, \alpha, \beta \\ c-a, c-b, \alpha+\beta \end{bmatrix}{}_3F_2(a, b, \alpha; a+b-c+1, \alpha+\beta; \omega y)+$$
$$+\omega^{c-a-b}y^{c-a-b+\alpha+\beta-1}\Gamma\begin{bmatrix} c, a+b-c, \beta, c-a-b+\alpha \\ a, b, c-a-b+\alpha+\beta \end{bmatrix}\times$$
$$\times{}_3F_2(c-a, c-b, c-a-b+\alpha; c-a-b+1, c-a-b+\alpha+\beta; \omega y)$$

[y, $\mathrm{Re}\,\alpha$, $\mathrm{Re}\,\beta$, $\mathrm{Re}\,(c-a-b+\alpha)>0$; $|\arg\omega|<\pi$].

11. $\int_0^y x^{\alpha-1}(y-x)^{c-1}{}_2F_1\left(a, b; c; 1-\frac{x}{y}\right)dx=y^{c+\alpha-1}\Gamma\begin{bmatrix} c, \alpha, c-a-b+\alpha \\ c-a+\alpha, c-b+\alpha \end{bmatrix}$

[y, $\mathrm{Re}\,c$, $\mathrm{Re}\,\alpha$, $\mathrm{Re}\,(c-a-b+\alpha)>0$].

12. $\int_0^y x^{-1/2}(\sqrt{y}+\sqrt{x})^{2a}\,{}_2F_1\left(a, a+1; 2; \frac{x}{y}\right)dx=\frac{y^{a+1/2}}{(1-a)\,2^{2a+1}}$ $\quad y>0$; $\mathrm{Re}\,a<1$].

13. $\int_y^\infty x^{\alpha-1}(x-y)^{\beta-1}\,{}_2F_1(a, b; c; -\omega x)\,dx=$

$$=y^{\alpha+\beta-1}\mathrm{B}(\beta, 1-\alpha-\beta)\,{}_3F_2(a, b, \alpha; c, \alpha+\beta; -\omega y)+$$
$$+\omega^{1-\alpha-\beta}\Gamma\begin{bmatrix} c, a-\alpha-\beta+1, b-\alpha-\beta+1, \alpha+\beta-1 \\ a, b, c-\alpha-\beta+1 \end{bmatrix}\times$$
$$\times{}_3F_2(1-\beta, a-\alpha-\beta+1, b-\alpha-\beta+1; 2-\alpha-\beta, c-\alpha-\beta+1; -\omega y)$$

[y, $\mathrm{Re}\,\beta>0$; $\mathrm{Re}\,(\alpha+\beta-a)$, $\mathrm{Re}\,(\alpha+\beta-b)<1$; $|\arg\omega|<\pi$].

14. $\int_y^\infty x^{\alpha-1}(x-y)^{\beta-1}\,{}_2F_1(a, b; c; 1-\omega x)\,dx=\omega^{-a}y^{\alpha+\beta-a-1}\times$

$$\times\Gamma\begin{bmatrix} c, b-a, \beta, a-\alpha-\beta+1 \\ b, c-a, a-\alpha+1 \end{bmatrix}{}_3F_2\left(a, c-b, a-\alpha-\beta+1; a-\alpha+1, a-b+1; \frac{1}{\omega y}\right)+$$
$$+\omega^{-b}y^{\alpha+\beta-b-1}\,\Gamma\begin{bmatrix} c, a-b, \beta, b-\alpha-\beta+1 \\ a, c-b, b-\alpha+1 \end{bmatrix}\times$$
$$\times{}_3F_2\left(b, c-a, b-\alpha-\beta+1; b-a+1, b-\alpha+1; \frac{1}{\omega y}\right)$$

[y, $\mathrm{Re}\,\beta>0$; $\mathrm{Re}\,(\alpha+\beta-a)$, $\mathrm{Re}\,(\alpha+\beta-b)<1$; $|\arg\omega|<\pi$].

15. $\int_0^\infty \frac{x^{\alpha-1}}{(x+z)^\rho}\,{}_2F_1(a, b; c; -\omega x)\,dx=z^{\alpha-\rho}\mathrm{B}(\alpha, \rho-\alpha)\,{}_3F_2(a, b, \alpha; c, \alpha-\rho+1; \omega z)+$

$$+\omega^{\rho-\alpha}\,\Gamma\begin{bmatrix} c, a-\alpha+\rho, b-\alpha+\rho, \alpha-\rho \\ a, b, c-\alpha+\rho \end{bmatrix}{}_3F_2(a-\alpha+\rho, b-\alpha+\rho, \rho; c-\alpha+\rho,$$
$$\rho-\alpha+1; \omega z)$$

[$\mathrm{Re}\,\alpha$, $\mathrm{Re}\,(a-\alpha+\rho)$, $\mathrm{Re}\,(b-\alpha+\rho)>0$; $|\arg\omega|$, $|\arg z|<\pi$].

16. $\int_0^\infty \frac{x^{c-1}}{(x+z)^\rho}\,{}_2F_1(a, b; c; -\omega x)\,dx=$

$$=\omega^{\rho-c}\Gamma\begin{bmatrix} c, a-c+\rho, b-c+\rho \\ \rho, a+b-c+\rho \end{bmatrix}{}_2F_1(a-c+\rho, b-c+\rho; a+b-c+\rho; 1-\omega z)$$

[$\mathrm{Re}\,(a+\rho)$, $\mathrm{Re}\,(b+\rho)>\mathrm{Re}\,c>0$; $|\arg\omega|$, $|\arg z|<\pi$].

17. $$\int\limits_0^\infty \frac{x^{\alpha-1}}{(x+z)^\rho}\,{}_2F_1(a, b; c; 1-\omega x)\,dx = \omega^{-\alpha} z^{-\rho}\Gamma\begin{bmatrix} c, \alpha, a-\alpha, b-\alpha, c-a-b+\alpha \\ a, b, c-a, c-b \end{bmatrix}\times$$
$$\times{}_3F_2\left(\alpha, \rho, c-a-b+\alpha; \alpha-a+1, \alpha-b+1; -\frac{1}{\omega z}\right)+$$
$$+\omega^{-a} z^{\alpha-\rho-a}\Gamma\begin{bmatrix} b-a, c, \alpha-a, a-\alpha+\rho \\ b, c-a, \rho \end{bmatrix}{}_3F_2\Big(a, c-b, a-\alpha+\rho; a-\alpha+1,$$
$$a-b+1; -\frac{1}{\omega z}\Big)+\omega^{-b} z^{\alpha-\rho-b}\,\Gamma\begin{bmatrix} a-b, c, \alpha-b, b-\alpha+\rho \\ a, c-b, \rho \end{bmatrix}\times$$
$$\times{}_3F_2\left(b, c-a, b-\alpha+\rho; b-a+1, b-\alpha+1; -\frac{1}{\omega z}\right)$$
[$\operatorname{Re}\alpha$, $\operatorname{Re}(c-a-b+\alpha)$, $\operatorname{Re}(a-\alpha+\rho)$, $\operatorname{Re}(b-\alpha+\rho)>0$; $|\arg\omega|$, $|\arg z|<\pi$].

18. $$\int\limits_0^\infty \frac{x^{\alpha-1}}{x-y}\,{}_2F_1(a, b; c; -\omega x)\,dx = \omega^{1-\alpha}\Gamma\begin{bmatrix} c, a-\alpha+1, b-\alpha+1, \alpha-1 \\ a, b, c-\alpha+1 \end{bmatrix}\times$$
$$\times{}_3F_2(1, a-\alpha+1, b-\alpha+1; 2-\alpha, c-\alpha+1; -\omega y)-$$
$$-\pi y^{\alpha-1}\operatorname{ctg}\alpha\pi\,{}_2F_1(a, b; c; -\omega y)$$ [y, $\operatorname{Re}\alpha>0$; $\operatorname{Re}(\alpha-a)$, $\operatorname{Re}(\alpha-b)<1$; $|\arg\omega|<\pi$].

19. $$\int\limits_0^\infty \frac{x^{\alpha-1}}{x-y}\,{}_2F_1(a, b; c; 1-\omega x)\,dx =$$
$$=\pi\omega^{-a}y^{\alpha-a-1}\operatorname{ctg}(a-\alpha)\pi\,\Gamma\begin{bmatrix} c, b-a \\ b, c-a \end{bmatrix}{}_2F_1\left(a, c-b; a-b+1; \frac{1}{\omega y}\right)+$$
$$+\pi\omega^{-b}y^{\alpha-b-1}\operatorname{ctg}(b-\alpha)\pi\Gamma\begin{bmatrix} c, a-b \\ a, c-b \end{bmatrix}{}_2F_1\left(b, c-a; b-a+1; \frac{1}{\omega y}\right)-$$
$$-\frac{\omega^{-\alpha}}{y}\Gamma\begin{bmatrix} c, \alpha, a-\alpha, b-\alpha, c-a-b+\alpha \\ a, b, c-a, c-b \end{bmatrix}\times$$
$$\times{}_3F_2\left(1, \alpha, c-a-b+\alpha; \alpha-a+1, \alpha-b+1; \frac{1}{\omega y}\right)$$
[y, $\operatorname{Re}\alpha$, $\operatorname{Re}(c-a-b+\alpha)>0$; $\operatorname{Re}(\alpha-a)$, $\operatorname{Re}(\alpha-b)<1$; $|\arg\omega|<\pi$].

20. $$\int\limits_0^y \frac{x^{c-1}(y-x)^{\beta-1}}{(1-xz)^\rho}\,{}_2F_1(a, b; c; \omega x)\,dx =$$
$$=\mathrm{B}(c, \beta)\frac{y^{c+\beta-1}}{(1-yz)^\rho}F_3\left(\rho, a, \beta, b, c+\beta; \frac{yz}{yz-1}, \omega y\right)$$
[y, $\operatorname{Re}c$, $\operatorname{Re}\beta>0$; $|\arg(1-\omega y)|$, $|\arg(1-z)|<\pi$].

21. $$\int\limits_0^y x^{c-1}(y-x)^{\beta-1}(1-\omega x)^{a-c-\beta}\,{}_2F_1(a, b; c; \omega x)\,dx =$$
$$=y^{c+\beta-1}(1+\omega y)^a(1-\omega y)^{-c}\,\mathrm{B}(c, \beta)\,{}_2F_1(a, b+\beta; c+\beta; \omega y)$$
[y, $\operatorname{Re}c$, $\operatorname{Re}\beta>0$; $|\arg(1-\omega y)|<\pi$].

22. $$\int\limits_0^y \frac{x^{c-1}(y-x)^{\beta-1}}{(1-zx)^\rho}\,{}_2F_1\left(a, b; c; \frac{x}{y}\right)dx =$$
$$=\frac{y^{c+\beta-1}}{(1-yz)^\rho}\Gamma\begin{bmatrix} c, \beta, c-a-b+\beta \\ c-a+\beta, c-b+\beta \end{bmatrix}{}_3F_2\left(\beta, \rho, c-a-b+\beta; c-a+\beta, c-b+\beta; \frac{yz}{yz-1}\right)$$
[y, $\operatorname{Re}c$, $\operatorname{Re}\beta$, $\operatorname{Re}(c-a-b+\beta)>0$; $|\arg(1-yz)|<\pi$].

23. $$\int_0^{\min(y,\,z)} x^{\alpha-1}(z-x)^{\beta-1}(y-x)^{c-1}\,{}_2F_1\left(a,\,b;\,c;\,1-\frac{x}{y}\right)dx=$$
$$=y^{c-1}z^{\alpha+\beta-1}\Gamma\begin{bmatrix}c,\ \alpha,\ \beta,\ c-a-b\\ c-a,\ c-b,\ \alpha+\beta\end{bmatrix}{}_3F_2\left(a-c+1,\,b-c+1,\,\alpha;\,a+b-c+1,\,\alpha+\beta;\,\frac{z}{y}\right)+$$
$$+y^{a+b-1}z^{c-a-b+\alpha+\beta-1}\,\Gamma\begin{bmatrix}c,\ a+b-c,\ \beta,\ c-a-b+\alpha\\ a,\ b,\ c-a-b+\alpha+\beta\end{bmatrix}\times$$
$$\times\,{}_3F_2\left(1-a,\,1-b,\,c-a-b+\alpha;\,c-a-b+1,\,c-a-b+\alpha+\beta;\,\frac{z}{y}\right)$$
$[0<z<y;\ \operatorname{Re}\alpha,\ \operatorname{Re}\beta,\ \operatorname{Re}(c-a-b+\alpha)>0].$

24. $$=y^{c+\alpha-1}z^{\beta-1}\Gamma\begin{bmatrix}c,\ \alpha,\ c-a-b+\alpha\\ c-a+\alpha,\ c-b+\alpha\end{bmatrix}\times$$
$$\times\,{}_3F_2\left(\alpha,\,1-\beta,\,c-a-b+\alpha;\,c-a+\alpha,\,c-b+\alpha;\,\frac{y}{z}\right)$$
$[0<y<z;\ \operatorname{Re}c,\ \operatorname{Re}\alpha,\ \operatorname{Re}(c-a-b+\alpha)>0].$

25. $$\int_z^y x^{\alpha-1}(x-z)^{\beta-1}(y-x)^{c-1}\,{}_2F_1\left(a,\,b;\,c;\,1-\frac{x}{y}\right)dx=$$
$$=y^{c-1}z^{\alpha+\beta-1}\Gamma\begin{bmatrix}c,\ c-a-b,\ \beta,\ 1-\alpha-\beta\\ c-a,\ c-b,\ 1-\alpha\end{bmatrix}\times$$
$$\times\,{}_3F_2\left(a-c+1,\,b-c+1,\,\alpha;\,a+b-c+1,\,\alpha+\beta;\,\frac{z}{y}\right)+$$
$$+y^{a+b-1}z^{c-a-b+\alpha+\beta-1}\Gamma\begin{bmatrix}c,\ a+b-c,\ \beta,\ a+b-c-\alpha-\beta+1\\ a,\ b,\ a+b-c-\alpha+1\end{bmatrix}\times$$
$$\times\,{}_3F_2\left(1-a,\,1-b,\,c-a-b+\alpha;\,c-a-b+1,\,c-a-b+\alpha+\beta;\,\frac{z}{y}\right)+$$
$$+y^{c+\alpha+\beta-2}\Gamma\begin{bmatrix}c,\ \alpha+\beta-1,\ c-a-b+\alpha+\beta-1\\ c-a+\alpha+\beta-1,\ c-b+\alpha+\beta-1\end{bmatrix}\times$$
$$\times\,{}_3F_2\left(1-\beta,\,a-c-\alpha-\beta+2,\,b-c-\alpha-\beta+2;\right.$$
$$\left.2-\alpha-\beta,\,a+b-c-\alpha-\beta+2;\,\frac{z}{y}\right)$$
$[0<z<y;\ \operatorname{Re}c,\ \operatorname{Re}\beta>0].$

26. $$\int_0^y \frac{x^{\alpha-1}(y-x)^{c-1}}{(x+z)^{\rho}}\,{}_2F_1\left(a,\,b;\,c;\,1-\frac{x}{y}\right)dx=$$
$$=y^{c+\alpha-1}z^{-\rho}\Gamma\begin{bmatrix}c,\ \alpha,\ c-a-b+\alpha\\ c-a+\alpha,\ c-b+\alpha\end{bmatrix}{}_3F_2\left(\alpha,\,\rho,\,c-a-b+\alpha;\,c-a+\alpha,\,c-b+\alpha;\,-\frac{y}{z}\right)$$
$[y,\ \operatorname{Re}c,\ \operatorname{Re}\alpha,\ \operatorname{Re}(c-a-b+\alpha)>0;\ |\arg z|<\pi].$

27. $$\int_0^y \frac{x^{\alpha-1}(y-x)^{c-1}}{x-z}\,{}_2F_1\left(a,\,b;\,c;\,1-\frac{x}{y}\right)dx=$$
$$=-\pi y^{c-1}z^{\alpha-1}\operatorname{ctg}\alpha\pi\,\Gamma\begin{bmatrix}c,\ c-a-b\\ c-a,\ c-b\end{bmatrix}{}_2F_1\left(a-c+1,\,b-c+1;\,a+b-c+1;\,\frac{z}{y}\right)+$$
$$+\pi y^{a+b-1}z^{c-a-b+\alpha-1}\operatorname{ctg}(a+b-c-\alpha)\pi\times$$
$$\times\,\Gamma\begin{bmatrix}c,\ a+b-c\\ a,\ b\end{bmatrix}{}_2F_1\left(1-a,\,1-b;\,c-a-b+1;\,\frac{z}{y}\right)+$$
$$+y^{c+\alpha-2}\Gamma\begin{bmatrix}c,\ \alpha-1,\ c-a-b+\alpha-1\\ c-a+\alpha-1,\ c-b+\alpha-1\end{bmatrix}\times$$
$$\times\,{}_3F_2\left(1,\,a-c-\alpha+2,\,b-c-\alpha+2;\,2-\alpha,\,a+b-c-\alpha+2;\,\frac{z}{y}\right)$$
$[0<z<y;\ \operatorname{Re}c,\ \operatorname{Re}\alpha,\ \operatorname{Re}(c-a-b+\alpha)>0].$

28. $$= -\frac{y^{c+\alpha-1}}{z}\Gamma\left[\begin{matrix}c, \alpha, c-a-b+\alpha\\ c-a+\alpha, c-b+\alpha\end{matrix}\right]\times$$
$$\times\, {}_3F_2\left(1, \alpha, c-a-b+\alpha;\; c-a+\alpha, c-b+\alpha;\; \frac{y}{z}\right)$$
$[0<y<z;\ \operatorname{Re} c,\ \operatorname{Re}\alpha,\ \operatorname{Re}(c-a-b+\alpha)>0]$.

29. $$\int_0^y x^{\alpha-1}(y-x)^{\beta-1}\,{}_2F_1(a, b; c; \omega x(y-x))\,dx=$$
$$= y^{\alpha+\beta-1}B(\alpha, \beta)\,{}_4F_3\left(a, b, \alpha, \beta;\; c, \frac{\alpha+\beta}{2}, \frac{\alpha+\beta+1}{2};\; \frac{\omega y^2}{4}\right)$$
$[y,\ \operatorname{Re}\alpha,\ \operatorname{Re}\beta>0]$.

30. $$\int_0^y x^{\alpha-1}(y-x)^{c-\alpha-1}(1-wx-zx)^{-2a}\,{}_2F_1\left(a, a+\frac{1}{2};\; c;\; \frac{4wzx^2}{(1-wx-zx)^2}\right)dx=$$
$$= y^{c-1}B(\alpha, c-\alpha)\,F_4(2a, \alpha;\; c, c;\; yw, yz)$$
$[y>0;\ \operatorname{Re} c>\operatorname{Re}\alpha>0;\ \sqrt{y|w|}+\sqrt{y|z|}<1]$.

2.21.2. Integrals of $A(x)e^{-px\pm r}{}_2F_1(a, b; c; \varphi(x))$.

1. $$\int_0^\infty x^{\alpha-1}e^{-px}\,{}_2F_1(a, b; c; -\omega x)\,dx=$$
$$= \omega^{-\alpha}\Gamma\left[\begin{matrix}c, \alpha, a-\alpha, b-\alpha\\ a, b, c-\alpha\end{matrix}\right]{}_2F_2\left(\alpha, \alpha-c+1;\; \alpha-a+1, \alpha-b+1;\; \frac{p}{\omega}\right)+$$
$$+\frac{p^{a-\alpha}}{\omega^a}\Gamma\left[\begin{matrix}c, b-a, \alpha-a\\ b, c-a\end{matrix}\right]{}_2F_2\left(a, a-c+1;\; a-\alpha+1, a-b+1;\; \frac{p}{\omega}\right)+$$
$$+\frac{p^{b-\alpha}}{\omega^b}\Gamma\left[\begin{matrix}c, a-b, \alpha-b\\ a, c-b\end{matrix}\right]{}_2F_2\left(b, b-c+1;\; b-a+1, b-\alpha+1;\; \frac{p}{\omega}\right)$$
$[\operatorname{Re} p,\ \operatorname{Re}\alpha>0;\ |\arg\omega|<\pi]$

2. $$\int_0^\infty x^{c-1}e^{-px}\,{}_2F_1(a, b; c; -\omega x)\,dx=$$
$$= \frac{\Gamma(c)}{p^c}\left(\frac{p}{\omega}\right)^{(a+b-1)/2}e^{p/(2\omega)}W_{(1-a-b)/2,\ (a-b)/2}\left(\frac{p}{\omega}\right)$$
$[\operatorname{Re} c,\ \operatorname{Re} p>0;\ |\arg\omega|<\pi]$

3. $$\int_0^\infty x^{c-1}e^{-px}\,{}_2F_1(a, 1-a; c; -\omega x)\,dx=\sqrt{\frac{p}{\pi\omega}}\,\frac{\Gamma(c)}{(2p)^c}K_{a-1/2}\left(\frac{p}{2\omega}\right)$$
$[\operatorname{Re} c,\ \operatorname{Re} p>0;\ |\arg\omega|<\pi]$

4. $$\int_0^\infty x^{c-1}e^{-px}\,{}_2F_1\left(\frac{1}{2}, 1; c; -\omega x\right)dx=\sqrt{\frac{\pi}{\omega}}\,\frac{\Gamma(c)}{p^{c-1/2}}\,e^{p/\omega}\operatorname{erfc}\left(\sqrt{\frac{p}{\omega}}\right)$$
$[\operatorname{Re} c,\ \operatorname{Re} p>0;\ |\arg\omega|<\pi]$

5. $$\int_0^\infty x^{\alpha-1}e^{-px}\,{}_2F_1(a, b; c; 1-\omega x)\,dx=$$
$$= \omega^{-\alpha}\Gamma\left[\begin{matrix}c, \alpha, a-\alpha, b-\alpha, c-a-b-\alpha\\ a, b, c-a, c-b\end{matrix}\right]\times$$
$$\times\, {}_2F_2\left(\alpha, c-a-b-\alpha;\; \alpha-a+1, \alpha-b+1;\; -\frac{p}{\omega}\right)+$$

$$+\frac{p^{a-\alpha}}{\omega^a}\Gamma\begin{bmatrix}c,\ b-a,\ \alpha-a\\ b,\ c-a\end{bmatrix}{}_2F_2\left(a,\ c-b;\ a-b+1,\ a-\alpha+1;\ -\frac{p}{\omega}\right)+$$

$$+\frac{p^{b-\alpha}}{\omega^b}\Gamma\begin{bmatrix}c,\ a-b,\ \alpha-b\\ a,\ c-b\end{bmatrix}{}_2F_2\left(b,\ c-a;\ b-a+1,\ b-\alpha+1;\ -\frac{p}{\omega}\right)$$

[$\operatorname{Re} p,\ \operatorname{Re}\alpha,\ \operatorname{Re}(c-a-b+\alpha)>0;\ |\arg\omega|<\pi$].

6. $$\int_0^\infty x^{\alpha-1}e^{-p/x}\,{}_2F_1(a,\ b;\ c;\ -\omega x)\,dx=$$

$$=\omega^{-\alpha}\Gamma\begin{bmatrix}c,\ \alpha,\ a-\alpha,\ b-\alpha\\ a,\ b,\ c-\alpha\end{bmatrix}{}_2F_2(a-\alpha,\ b-\alpha;\ 1-\alpha,\ c-\alpha;\ \omega p)+$$

$$+p^{\alpha}\Gamma(-\alpha)\,{}_2F_2(a,\ b;\ c,\ \alpha+1;\ \omega p)$$ [$\operatorname{Re} p,\ \operatorname{Re}(a-\alpha),\ \operatorname{Re}(b-\alpha)>0;\ |\arg\omega|<\pi$].

7. $$\int_0^\infty e^{-p\sqrt{x}}\,{}_2F_1\left(a,\ b;\ \frac{3}{2};\ -\omega x\right)dx=\frac{2p^{a+b-2}}{\omega^{(a+b)/2}}S_{1-a-b,\ a-b}\left(\frac{p}{\sqrt{\omega}}\right)$$

[$\operatorname{Re} p>0;\ |\arg\omega|<\pi$].

8. $$\int_0^\infty x^{\alpha-1}e^{-p\sqrt{x}}\,{}_2F_1(a,\ b;\ c;\ -\omega x)\,dx=$$

$$=\frac{2p^{2a-2\alpha}}{\omega^a}\Gamma\begin{bmatrix}c,\ b-a,\ 2\alpha-2a\\ b,\ c-a\end{bmatrix}{}_2F_3\left(a,\ a-c+1;\ a-b+1,\ a-\alpha+1,\ a-\alpha+\frac{1}{2};\ -\frac{p^2}{4\omega}\right)+$$

$$+\frac{2p^{2b-2\alpha}}{\omega^b}\Gamma\begin{bmatrix}c,\ a-b,\ 2\alpha-2b\\ a,\ c-b\end{bmatrix}\times$$

$$\times{}_2F_3\left(b,\ b-c+1;\ b-a+1,\ b-\alpha+1,\ b-\alpha+\frac{1}{2};\ -\frac{p^2}{4\omega}\right)+$$

$$+\omega^{-\alpha}\Gamma\begin{bmatrix}c,\ \alpha,\ a-\alpha,\ b-\alpha\\ a,\ b,\ c-\alpha\end{bmatrix}{}_2F_3\left(\alpha,\ \alpha-c+1;\ \alpha-a+1,\ \alpha-b+1,\ \frac{1}{2};\ -\frac{p^2}{4\omega}\right)-$$

$$-\frac{p}{\omega^{\alpha+1/2}}\Gamma\begin{bmatrix}c,\ \alpha+1/2,\ a-\alpha-1/2,\ b-\alpha-1/2\\ a,\ b,\ c-\alpha-1/2\end{bmatrix}\times$$

$$\times{}_2F_3\left(\alpha-c+\frac{3}{2},\ \alpha+\frac{1}{2};\ \frac{3}{2},\ \alpha-a+\frac{3}{2},\ \alpha-b+\frac{3}{2};\ -\frac{p^2}{4\omega}\right)$$

[$\operatorname{Re} p,\ \operatorname{Re}\alpha>0;\ |\arg\omega|<\pi$].

9. $$\int_0^\infty x^{-1/2}e^{-p\sqrt{x}}\,{}_2F_1\left(a,\ b;\ \frac{1}{2};\ -\omega x\right)dx=2\omega^{-(a+b)/2}p^{a+b-1}S_{1-a-b,\ a-b}\left(\frac{p}{\sqrt{\omega}}\right)$$

[$\operatorname{Re} p>0;\ |\arg\omega|<\pi$].

10. $$\int_0^\infty x^{\alpha-1}e^{-p\sqrt{x}}\,{}_2F_1(a,\ b;\ c;\ 1-\omega x)\,dx=$$

$$=\frac{2p^{2a-2\alpha}}{\omega^a}\Gamma\begin{bmatrix}c,\ b-a,\ 2\alpha-2a\\ b,\ c-a\end{bmatrix}{}_2F_3\left(a,\ c-b;\ a-b+1,\ a-\alpha+1,\ a-\alpha+\frac{1}{2};\ \frac{p^2}{4\omega}\right)+$$

$$+\frac{2p^{2b-2\alpha}}{\omega^b}\Gamma\begin{bmatrix}c,\ a-b,\ 2\alpha-2b\\ a,\ c-b\end{bmatrix}{}_2F_3\left(b,\ c-a;\ b-a+1,\ b-\alpha+1,\ b-\alpha+\frac{1}{2};\ \frac{p^2}{4\omega}\right)+$$

$$+\omega^{-\alpha}\Gamma\begin{bmatrix}c,\ \alpha,\ a-\alpha,\ b-\alpha,\ c-a-b+\alpha\\ a,\ b,\ c-a,\ c-b\end{bmatrix}\times$$

$$\times{}_2F_3\left(\alpha,\ c-a-b+\alpha;\ \frac{1}{2},\ \alpha-a+1,\ \alpha-b+1;\ \frac{p^2}{4\omega}\right)-$$

$$-\frac{p}{\omega^{\alpha+1/2}}\Gamma\begin{bmatrix}c,\ \alpha+1/2,\ a-\alpha-1/2,\ b-\alpha-1/2,\ c-a-b+\alpha+1/2\\ a,\ b,\ c-a,\ c-b\end{bmatrix}\times$$

$$\times\,{}_2F_3\left(\alpha+\frac{1}{2},\ c-a-b+\alpha+\frac{1}{2};\ \frac{3}{2},\ \alpha-a+\frac{3}{2},\ \alpha-b+\frac{3}{2};\ \frac{p^2}{4\omega}\right)$$

[Re p, Re α, Re $(c-a-b+\alpha)>0$; $|\arg\omega|>\pi$].

11. $$\int_0^\infty x^{\alpha-1}e^{-p/\sqrt{x}}\,{}_2F_1(a,\ b;\ c;\ -\omega x)\,dx=$$

$$=\omega^{-\alpha}\Gamma\begin{bmatrix}c,\ \alpha,\ a-\alpha,\ b-\alpha\\ a,\ b,\ c-\alpha\end{bmatrix}{}_2F_3\left(a-\alpha,\ b-\alpha;\ \frac{1}{2},\ 1-\alpha,\ c-\alpha;\ -\frac{p^2\omega}{4}\right)+$$

$$+2p^{2\alpha}\Gamma(-2\alpha)\,{}_2F_3\left(a,\ b;\ c,\ \alpha+1,\ \alpha+\frac{1}{2};\ -\frac{p^2\omega}{4}\right)-$$

$$-\omega^{1/2-\alpha}p\Gamma\begin{bmatrix}c,\ \alpha-1/2,\ a-\alpha+1/2,\ b-\alpha+1/2\\ a,\ b,\ c-\alpha+1/2\end{bmatrix}\times$$

$$\times{}_2F_3\left(a-\alpha+\frac{1}{2},\ b-\alpha+\frac{1}{2};\ \frac{3}{2},\ \frac{3}{2}-\alpha,\ c-\alpha+\frac{1}{2};\ -\frac{p^2\omega}{4}\right)$$

[Re p, Re $(a-\alpha)$, Re $(b-\alpha)>0$; $|\arg\omega|<\pi$].

12. $$\int_0^y x^{\alpha-1}(y-x)^{c-1}e^{-px}\,{}_2F_1\left(a,\ b;\ c;\ 1-\frac{x}{y}\right)dx=$$

$$=y^{c+\alpha-1}\Gamma\begin{bmatrix}c,\ \alpha,\ c-a-b+\alpha\\ c-a+\alpha,\ c-b+\alpha\end{bmatrix}{}_2F_2(\alpha,\ c-a-b+\alpha;\ c-a+\alpha,\ c-b+\alpha;\ -py)$$

[y, Re c, Re α, Re $(c-a-b+\alpha)>0$].

13. $$\int_0^y x^{\alpha-1}(y-x)^{c-1}e^{-p\sqrt{x}}\,{}_2F_1\left(a,\ b;\ c;\ 1-\frac{x}{y}\right)dx=$$

$$=y^{c+\alpha-1}\Gamma\begin{bmatrix}c,\ \alpha,\ c-a-b+\alpha\\ c-a+\alpha,\ c-b+\alpha\end{bmatrix}{}_2F_3\left(\alpha,\ c-a-b+\alpha;\ c-a+\alpha,\ c-b+\alpha,\ \frac{1}{2};\ \frac{p^2y}{4}\right)-$$

$$-py^{c+\alpha-1/2}\Gamma\begin{bmatrix}c,\ \alpha+1/2,\ c-a-b+\alpha+1/2\\ c-a+\alpha+1/2,\ c-b+\alpha+1/2\end{bmatrix}\times$$

$$\times{}_2F_3\left(\alpha+\frac{1}{2},\ c-a-b+\alpha+\frac{1}{2};\ c-a+\alpha+\frac{1}{2},\ c-b+\alpha+\frac{1}{2},\ \frac{3}{2};\ \frac{p^2y}{4}\right)$$

[y, Re c, Re α, Re $(c-a-b+\alpha)>0$].

14. $$\int_y^\infty x^{\alpha-1}(x-y)^{c-1}e^{-px}\,{}_2F_1\left(a,\ b;\ c;\ 1-\frac{x}{y}\right)dx=$$

$$=y^{c+\alpha-1}\Gamma\begin{bmatrix}c,\ a-c-\alpha+1,\ b-c-\alpha+1\\ 1-\alpha,\ a+b-c-\alpha+1\end{bmatrix}{}_2F_2(\alpha,\ c-a-b+\alpha;\ c-a+\alpha,\ c-b+\alpha;\ -py)+$$

$$+p^{a-c-\alpha+1}y^a\Gamma\begin{bmatrix}c,\ b-a,\ c-a+\alpha-1\\ b,\ c-a\end{bmatrix}{}_2F_2(1-b,\ a-c+1;\ a-b+1,\ a-c-\alpha+2;\ -py)+$$

$$+p^{b-c-\alpha+1}y^b\Gamma\begin{bmatrix}c,\ a-b,\ c-b+\alpha-1\\ a,\ c-b\end{bmatrix}{}_2F_2(1-a,\ b-c+1;\ b-a+1,\ b-c-\alpha+2;\ -py)$$

[y, Re c, Re $p>0$].

15. $$\int_y^\infty x^{\alpha-1}(x-y)^{c-1}e^{-p\sqrt{x}}\,{}_2F_1\left(a,\ b;\ c;\ 1-\frac{x}{y}\right)dx=$$

$$=y^{c+\alpha-1}\Gamma\begin{bmatrix}c,\ a-c-\alpha+1,\ b-c-\alpha+1\\ 1-\alpha,\ a+b-c-\alpha+1\end{bmatrix}{}_2F_3\left(\alpha,\ c-a-b+\alpha;\ c-a+\alpha,\ c-b+\alpha,\ \frac{1}{2};\right.$$

$$\left.\frac{p^2y}{4}\right)+2p^{2(a-c-\alpha+1)}y^a\Gamma\left[\begin{matrix}c,\ b-a,\ 2(c-a+\alpha-1)\\ c-a,\ b\end{matrix}\right]\times$$

$$\times{}_2F_3\left(1-b,\ a-c+1;\ a-b+1,\ a-c-\alpha+\frac{3}{2},\ a-c-\alpha+2;\ \frac{p^2y}{4}\right)+$$

$$+2p^{2(b-c-\alpha+1)}y^b\Gamma\left[\begin{matrix}c,\ a-b,\ 2(c-b+\alpha-1)\\ c-b,\ a\end{matrix}\right]\times$$

$$\times{}_2F_3\left(1-a,\ b-c+1;\ b-a+1,\ b-c-\alpha+\frac{3}{2},\ b-c-\alpha+2;\ \frac{p^2y}{4}\right)-$$

$$-py^{c+\alpha-1/2}\Gamma\left[\begin{matrix}c,\ a-c-\alpha+1/2,\ b-c-\alpha+1/2\\ 1/2-\alpha,\ a+b-c-\alpha+1/2\end{matrix}\right]\times$$

$$\times{}_2F_3\left(\alpha+\frac{1}{2},\ c-a-b+\alpha+\frac{1}{2};\ c-a+\alpha+\frac{1}{2},\ c-b+\alpha+\frac{1}{2};\ \frac{3}{2};\ \frac{p^2y}{4}\right)$$

[y, $\operatorname{Re} c$, $\operatorname{Re} p>0$].

16. $$\int_y^\infty x^{-1/2}(x-y)^{c-1}e^{-p\sqrt{x}}\,{}_2F_1\left(a,\ c-a-\frac{1}{2};\ c;\ 1-\frac{x}{y}\right)dx=$$

$$=\frac{2^{c+1/2}\Gamma(c)}{\sqrt{\pi}}p^{1/2-c}y^{(2c-1)/4}K_{c-2a-1/2}(p\sqrt{y})$$

[$\operatorname{Re} c$, $\operatorname{Re} p>0$].

17. $$\int_0^\infty \frac{x^{c-1}e^{-px}}{(x+w)^a(x+z)^b}\,{}_2F_1\left(a,\ b;\ c;\ \frac{x(x+w+z)}{(x+w)(x+z)}\right)dx=$$

$$=\frac{\Gamma(c)}{p^{c-a-b}}\Psi(a,\ a+b-c+1;\ pw)\,\Psi(b,\ a+b-c+1;\ pz)$$

[$\operatorname{Re} c$, $\operatorname{Re} p>0$; $|\arg w|$, $|\arg z|<\pi$].

18. $$\int_0^\infty \frac{e^{-px}}{(x+w)^a(x+z)^a}\,{}_2F_1\left(a,\ a;\ 1;\ \frac{x(x+w+z)}{(x+w)(x+z)}\right)dx=$$

$$=\frac{(wz)^{1/2-a}}{\pi}e^{p(w+z)/2}K_{a-1/2}\left(\frac{pw}{2}\right)K_{a-1/2}\left(\frac{pz}{2}\right)$$

[$\operatorname{Re} p>0$; $|\arg w|$, $|\arg z|<\pi$].

19. $$\int_0^\infty \frac{x^{c-1}e^{-px}}{(x+w)^a(x+z)^{c-a-1/2}}\,{}_2F_1\left(a,\ c-a-\frac{1}{2};\ c;\ \frac{x(x+w+z)}{(x+w)(x+z)}\right)dx=$$

$$=\frac{2^{c-1/2}\Gamma(c)}{\sqrt{p}}e^{p(w+z)/2}D_{-2a}(\sqrt{2pw})\,D_{2a-2c+1}(\sqrt{2pz})$$

[$\operatorname{Re} c$, $\operatorname{Re} p>0$; $|\arg w|$, $|\arg z|<\pi$].

20. $$\int_{\gamma-i\infty}^{\gamma+i\infty} x^{-b}e^{px}\,{}_2F_1\left(a,\ b;\ c;\ 1-\frac{1}{x}\right)dx=2\pi i\Gamma\left[\begin{matrix}c\\ b,\ c-a\end{matrix}\right]p^{b-1}\Psi(a,\ a+b-c+1;\ p)$$

[$\operatorname{Re} b$, $\operatorname{Re}(c-a)>0$; $\gamma>1/2$].

2.21.3. Integrals of $A(x)\begin{Bmatrix}\sin\sigma x^{\pm1/2}\\ \cos\sigma x^{\pm1/2}\end{Bmatrix}{}_2F_1(a,\ b;\ c;\ \varphi(x))$.

NOTATION $\delta=\begin{Bmatrix}1\\0\end{Bmatrix}$.

1. $$\int_0^\infty \sin\sigma\sqrt{x}\,{}_2F_1\left(a,\ b;\ \frac{3}{2};\ -\omega x\right)dx=\frac{\pi(\sigma/2)^{a+b-2}\omega^{-(a+b)/2}}{\Gamma(a)\Gamma(b)}K_{a-b}\left(\frac{\sigma}{\sqrt{\omega}}\right)$$

[$\sigma>0$; $\operatorname{Re} a$, $\operatorname{Re} b>1/2$; $|\arg\omega|<\pi$].

2. $$\int_0^\infty x^{\alpha-1} \begin{Bmatrix} \sin \sigma \sqrt{x} \\ \cos \sigma \sqrt{x} \end{Bmatrix} {}_2F_1(a, b; c; -\omega x)\, dx =$$
$$= \frac{2\sigma^{2a-2\alpha}}{\omega^a} \Gamma \begin{bmatrix} b-a, c, 2\alpha-2a \\ b, c-a \end{bmatrix} \begin{Bmatrix} \sin(\alpha-a)\pi \\ \cos(\alpha-a)\pi \end{Bmatrix} \times$$
$$\times {}_2F_3\left(a, a-c+1; a-b+1, a-\alpha+\frac{1}{2}, a-\alpha+1; \frac{\sigma^2}{4\omega}\right) +$$
$$+ \frac{2\sigma^{2b-2\alpha}}{\omega^b} \Gamma \begin{bmatrix} a-b, c, 2\alpha-2b \\ a, c-b \end{bmatrix} \begin{Bmatrix} \sin(\alpha-b)\pi \\ \cos(\alpha-b)\pi \end{Bmatrix} \times$$
$$\times {}_2F_3\left(b, b-c+1; b-a+1, b-\alpha+\frac{1}{2}, b-\alpha+1; \frac{\sigma^2}{4\omega}\right) +$$
$$+ \frac{\sigma^\delta}{\omega^{\alpha+\delta/2}} \Gamma \begin{bmatrix} c, \alpha+\delta/2, a-\alpha-\delta/2, b-\alpha-\delta/2 \\ a, b, c-\alpha-\delta/2 \end{bmatrix} \times$$
$$\times {}_2F_3\left(\alpha+\frac{\delta}{2}, \alpha+\frac{\delta}{2}-c+1; \alpha+\frac{\delta}{2}-a+1, \alpha+\frac{\delta}{2}-b+1, \delta+\frac{1}{2}; \frac{\sigma^2}{4\omega}\right)$$

[$\sigma > 0$; $\operatorname{Re} \alpha > -\delta/2$; $\operatorname{Re}(\alpha-a)$, $\operatorname{Re}(\alpha-b) < 1/2$; $|\arg \omega| < \pi$].

3. $$\int_0^\infty x^{-1/2} \cos \sigma \sqrt{x}\, {}_2F_1\left(a, b; \frac{1}{2}; -\omega x\right) dx = \frac{2^{2-a-b} \pi \sigma^{a+b-1}}{\Gamma(a)\Gamma(b)\omega^{(a+b)/2}} K_{a-b}\left(\frac{\sigma}{\sqrt{\omega}}\right)$$

[σ, $\operatorname{Re} a$, $\operatorname{Re} b > 0$; $|\arg \omega| < \pi$].

4. $$\int_0^\infty x^{\alpha-1} \begin{Bmatrix} \sin(\sigma/\sqrt{x}) \\ \cos(\sigma/\sqrt{x}) \end{Bmatrix} {}_2F_1(a, b; c; -\omega x)\, dx =$$
$$= \frac{\sigma^\delta}{\omega^{\alpha-\delta/2}} \Gamma \begin{bmatrix} c, \alpha-\delta/2, a-\alpha+\delta/2, b-\alpha+\delta/2 \\ a, b, c-\alpha+\delta/2 \end{bmatrix} \times$$
$$\times {}_2F_3\left(a-\alpha+\frac{\delta}{2}, b-\alpha+\frac{\delta}{2}; c-\alpha+\frac{\delta}{2}, 1-\alpha+\frac{\delta}{2}, \delta+\frac{1}{2}; \frac{\sigma^2\omega}{4}\right) \mp$$
$$\mp 2\sigma^{2\alpha} \Gamma(-2\alpha) \begin{Bmatrix} \sin \alpha\pi \\ \cos \alpha\pi \end{Bmatrix} {}_2F_3\left(a, b; c, \alpha+\frac{1}{2}, \alpha+1; \frac{\sigma^2\omega}{4}\right)$$

[$\sigma>0$; $\operatorname{Re} \alpha > -1/2$; $\operatorname{Re}(a-\alpha)$, $\operatorname{Re}(b-\alpha) > -\delta/2$; $|\arg \omega| < \pi$].

5. $$\int_0^\infty x^{\alpha-1} \begin{Bmatrix} \sin \sigma \sqrt{x} \\ \cos \sigma \sqrt{x} \end{Bmatrix} {}_2F_1(a, b; c; 1-\omega x)\, dx =$$
$$= \frac{\sigma^\delta}{\omega^{\alpha+\delta/2}} \Gamma \begin{bmatrix} c, \alpha+\delta/2, a-\alpha-\delta/2, b-\alpha-\delta/2, c-a-b+\alpha+\delta/2 \\ a, b, c-a, c-b \end{bmatrix} \times$$
$$\times {}_2F_3\left(\alpha+\frac{\delta}{2}, c-a-b+\alpha+\frac{\delta}{2}; \alpha+\frac{\delta}{2}-a+1, \alpha+\frac{\delta}{2}-b+1, \delta+\frac{1}{2}; -\frac{\sigma^2}{4\omega}\right) +$$
$$+ \frac{2\sigma^{2a-2\alpha}}{\omega^a} \begin{Bmatrix} \sin(\alpha-a)\pi \\ \cos(\alpha-a)\pi \end{Bmatrix} \Gamma \begin{bmatrix} c, b-a \\ b, c-a \end{bmatrix} {}_2F_3\left(a, c-b; a-b+1;\right.$$
$$\left. a-\alpha+\frac{1}{2}, a-\alpha+1; -\frac{\sigma^2}{4\omega}\right) + \frac{2\sigma^{2b-2\alpha}}{\omega^b} \begin{Bmatrix} \sin(\alpha-b)\pi \\ \cos(\alpha-b)\pi \end{Bmatrix} \times$$
$$\times \Gamma \begin{bmatrix} a-b, c \\ a, c-b \end{bmatrix} {}_2F_3\left(b, c-a; b-a+1, b-\alpha+\frac{1}{2}, b-\alpha+1; -\frac{\sigma^2}{4\omega}\right)$$

[$\sigma>0$; $\operatorname{Re} \alpha$, $\operatorname{Re}(c-a-b+\alpha) > -\delta/2$; $\operatorname{Re}(\alpha-a)$, $\operatorname{Re}(\alpha-b) < 1/2$; $|\arg \omega| < \pi$].

6. $$\int_0^y x^{\alpha-1}(y-x)^{c-1}\begin{Bmatrix}\sin\sigma\sqrt{x}\\ \cos\sigma\sqrt{x}\end{Bmatrix}{}_2F_1\left(a,\ b;\ c;\ 1-\frac{x}{y}\right)dx=$$
$$=\sigma^{\delta}y^{c+\alpha+\delta/2-1}\Gamma\begin{bmatrix}c,\ \alpha+\delta/2,\ c-a-b+\alpha+\delta/2\\ c-a+\alpha+\delta/2,\ c-b+\alpha+\delta/2\end{bmatrix}\times$$
$$\times{}_2F_3\left(\alpha+\frac{\delta}{2},\ c-a-b+\alpha+\frac{\delta}{2};\ c-a+\alpha+\frac{\delta}{2},\ c-b+\alpha+\frac{\delta}{2},\ \delta+\frac{1}{2};\ -\frac{\sigma^2y}{4}\right)$$
[y, $\operatorname{Re} c>0$; $\operatorname{Re}\alpha$, $\operatorname{Re}(c-a-b+\alpha)>-\delta/2$].

7. $$\int_y^\infty x^{\alpha-1}(x-y)^{c-1}\begin{Bmatrix}\sin\sigma\sqrt{x}\\ \cos\sigma\sqrt{x}\end{Bmatrix}{}_2F_1\left(a,\ b;\ c;\ 1-\frac{x}{y}\right)dx=$$
$$=\sigma^{\delta}y^{c+\alpha+\delta/2-1}\Gamma\begin{bmatrix}c,\ a-c-\alpha-\delta/2+1,\ b-c-\alpha-\delta/2+1\\ 1-\alpha-\delta/2,\ a+b-c-\alpha-\delta/2+1\end{bmatrix}\times$$
$$\times{}_2F_3\left(\alpha+\frac{\delta}{2},\ c-a-b+\alpha+\frac{\delta}{2};\ c-a+\alpha+\frac{\delta}{2},\ c-b+\alpha+\frac{\delta}{2},\ \delta+\frac{1}{2};\ -\frac{\sigma^2y}{4}\right)-$$
$$-2\sigma^{2(a-c-\alpha+1)}y^a\Gamma\begin{bmatrix}b-a,\ c\\ b,\ c-a\end{bmatrix}\begin{Bmatrix}\sin(c-a+\alpha)\pi\\ \cos(c-a+\alpha)\pi\end{Bmatrix}\times$$
$$\times{}_2F_3\left(1-b,\ a-c+1;\ a-b+1,\ a-c-\alpha+\frac{3}{2},\ a-c-\alpha+2;\ -\frac{\sigma^2y}{4}\right)-$$
$$-2\sigma^{2(b-c-\alpha+1)}y^b\Gamma\begin{bmatrix}a-b,\ c\\ a,\ c-b\end{bmatrix}\begin{Bmatrix}\sin(c-b+\alpha)\pi\\ \cos(c-b+\alpha)\pi\end{Bmatrix}\times$$
$$\times{}_2F_3\left(1-a,\ b-c+1;\ b-a+1,\ b-c-\alpha+\frac{3}{2},\ b-c-\alpha+2;\ -\frac{\sigma^2y}{4}\right)$$
[y, σ, $\operatorname{Re} c>0$; $\operatorname{Re}(c-a+\alpha)$, $\operatorname{Re}(c-b+\alpha)<3/2$].

2.21.4. Integrals of $A(x)J_\nu(\sigma x^{\pm r})\,{}_2F_1(a,\ b;\ c;\ \varphi(x))$.

1. $$\int_0^\infty x^{\alpha-1}J_\nu(\sigma\sqrt{x})\,{}_2F_1(a,\ b;\ c;\ -\omega x)\,dx=$$
$$=\frac{(\sigma/2)^\nu}{\omega^{\alpha+\nu/2}}\Gamma\begin{bmatrix}c,\ \alpha+\nu/2,\ a-\alpha-\nu/2,\ b-\alpha-\nu/2\\ a,\ b,\ \nu+1,\ c-\alpha-\nu/2\end{bmatrix}\times$$
$$\times{}_2F_3\left(\alpha+\nu/2,\ \alpha+\nu/2-c+1;\ \nu+1,\ \alpha+\nu/2-a+1,\ \alpha+\nu/2-b+1;\ \frac{\sigma^2}{4\omega}\right)+$$
$$+\frac{(\sigma/2)^{2a-2\alpha}}{\omega^a}\Gamma\begin{bmatrix}c,\ b-a,\ \alpha+\nu/2-a\\ b,\ c-a,\ a-\alpha+\nu/2+1\end{bmatrix}\times$$
$$\times{}_2F_3\left(a,\ a-c+1;\ a-b+1,\ a-\alpha-\frac{\nu}{2}+1,\ a-\alpha+\frac{\nu}{2}+1;\ \frac{\sigma^2}{4\omega}\right)+$$
$$+\frac{(\sigma/2)^{2b-2\alpha}}{\omega^b}\Gamma\begin{bmatrix}c,\ a-b,\ \alpha+\nu/2-b\\ a,\ c-b,\ b-\alpha+\nu/2+1\end{bmatrix}\times$$
$$\times{}_2F_3\left(b,\ b-c+1;\ b-a+1,\ b-\alpha-\frac{\nu}{2}+1,\ b-\alpha+\frac{\nu}{2}+1;\ \frac{\sigma^2}{4\omega}\right)$$
[σ, $\operatorname{Re}(2\alpha+\nu)>0$; $\operatorname{Re}(\alpha-a)$, $\operatorname{Re}(\alpha-b)<3/4$; $|\arg\omega|<\pi$].

2. $$\int_0^\infty x^{(c-1)/2}J_{c-1}(\sigma\sqrt{x})\,{}_2F_1(a,\ b;\ c;\ -\omega x)\,dx=$$
$$=\frac{2^{c-a-b+2}\sigma^{a+b-c-1}}{\omega^{(a+b)/2}}\Gamma\begin{bmatrix}c\\ a,\ b\end{bmatrix}K_{a-b}\left[\frac{\sigma}{\sqrt{\omega}}\right]$$
[$0<\operatorname{Re} c<2\operatorname{Re} a+1/2$, $2\operatorname{Re} b+1/2$; $|\arg\omega|<\pi$].

3. $$\int_0^\infty x^{c-a/2-1}J_{2c-a-2}(\sigma\sqrt{x})\,{}_2F_1(a,\ b;\ c;\ -\omega x)\,dx=$$
$$=\frac{2^{2-a}\omega^{1/2-c}}{\sqrt{\pi\sigma}}\Gamma\begin{bmatrix}c\\ a,\ b\end{bmatrix}K^2_{c-a-1/2}\left(\frac{\sigma}{2\sqrt{\omega}}\right)$$
[$1<\operatorname{Re}(2c-a)<2\operatorname{Re}a+1/2,\ \ 2\operatorname{Re}b+1/2;\ \ |\arg\omega|<\pi$].

4. $$\int_0^\infty x^{(3c-2a-3)/4}J_{(c-1)/2-a}(\sigma\sqrt{x})\,{}_2F_1\left(a,\ \frac{c}{2};\ c;\ -\omega x\right)dx=$$
$$=\frac{2^{(3c-1)/2-a}}{\pi}i\sigma^{a-(c+3)/2}\omega^{(1-c)/2}\Gamma\left(\frac{c+1}{2}\right)\Gamma(c-a)\times$$
$$\times W_{(1-c)/2,\ a-c/2}\left(\frac{\sigma}{\sqrt{\omega}}\right)\left[W_{(1-c)/2,\ a-c/2}\left(e^{-\pi i}\frac{\sigma}{\sqrt{\omega}}\right)-W_{(1-c)/2,\ a-c/2}\left(e^{\pi i}\frac{\sigma}{\sqrt{\omega}}\right)\right]$$
[$\operatorname{Re}a<\operatorname{Re}c<2\operatorname{Re}a+2/3;\ \ |\arg\omega|<\pi$].

5. $$\int_0^\infty x^{(3c-2a-3)/4}J_{a+(1-c)/2}(\sigma\sqrt{x})\,{}_2F_1\left(a,\ \frac{c}{2};\ c;\ -\omega x\right)dx=$$
$$=\frac{2^{(3c+1)/2-a}\sigma^{a-(c+3)/2}}{\omega^{(c-1)/2}}\Gamma\begin{bmatrix}(c+1)/2\\ 2a-c+1\end{bmatrix}M_{(c-1)/2,\ a-c/2}\left(\frac{\sigma}{\sqrt{\omega}}\right)W_{(1-c)/2,\ a-c/2}\left(\frac{\sigma}{\sqrt{\omega}}\right)$$
[$\sigma>0;\ \ -1<\operatorname{Re}c<2\operatorname{Re}a+2/3;\ \ |\arg\omega|<\pi$].

6. $$\int_0^\infty x^{\alpha-1}J_\nu\left(\frac{\sigma}{\sqrt{x}}\right){}_2F_1(a,\ b;\ c;\ -\omega x)\,dx=$$
$$=\left(\frac{\sigma}{2}\right)^{2\alpha}\Gamma\begin{bmatrix}\nu/2-\alpha\\ \nu/2+\alpha+1\end{bmatrix}{}_2F_3\left(a,\ b;\ c,\ \alpha-\frac{\nu}{2}+1,\ \alpha+\frac{\nu}{2}+1;\ \frac{\sigma^2\omega}{4}\right)+$$
$$+\left(\frac{\sigma}{2}\right)^{\nu}\omega^{\nu/2-\alpha}\Gamma\begin{bmatrix}c,\ \alpha-\nu/2,\ a-\alpha+\nu/2,\ b-\alpha+\nu/2\\ a,\ b,\ \nu+1,\ c-\alpha+\nu/2\end{bmatrix}\times$$
$$\times{}_2F_3\left(a-\alpha+\frac{\nu}{2},\ b-\alpha+\frac{\nu}{2};\ \nu+1,\ \frac{\nu}{2}-\alpha+1,\ \frac{\nu}{2}-\alpha+c;\ \frac{\sigma^2\omega}{4}\right)$$
[$\sigma>0;\ \ -3/4<\operatorname{Re}\alpha<\operatorname{Re}(a+\nu/2),\ \ \operatorname{Re}(b+\nu/2);\ \ |\arg\omega|<\pi$].

7. $$\int_0^\infty x^{(c-3)/2}J_\nu\left(\frac{\sigma}{\sqrt{x}}\right){}_2F_1\left(\frac{c}{2},\ \frac{c+1}{2};\ c;\ -\omega x\right)dx=$$
$$=2^c\omega^{(1-c)/2}I_{(c+\nu-1)/2}\left(\frac{\sigma\sqrt{\omega}}{2}\right)K_{(c-\nu-1)/2}\left(\frac{\sigma\sqrt{\omega}}{2}\right)$$
[$\operatorname{Re}c>-1/2;\ \ \operatorname{Re}\nu>-1;\ \ |\arg\omega|<\pi$].

8. $$\int_0^\infty x^{2a-(c+3)/2}J_{c-1}\left(\frac{\sigma}{\sqrt{x}}\right){}_2F_1\left(a,\ a+\frac{1}{2};\ c;\ -\omega x\right)dx=$$
$$=2^c\sigma^{2a-c}\omega^{1/2-a}\Gamma\begin{bmatrix}c-1\\ 2a\end{bmatrix}I_{c-1}\left(\frac{\sigma\sqrt{\omega}}{2}\right)K_{2a-c}\left(\frac{\sigma\sqrt{\omega}}{2}\right)$$
[$\operatorname{Re}a<\operatorname{Re}c<4\operatorname{Re}a+1/2;\ \ |\arg\omega|<\pi$].

9. $$\int_0^\infty x^{\alpha-1}J_\nu(\sigma\sqrt{x})\,{}_2F_1(a,\ b;\ c;\ 1-\omega x)\,dx=$$
$$=\frac{(\sigma/2)^{2a-2\alpha}}{\omega^a}\Gamma\begin{bmatrix}b-a,\ c,\ \alpha+\nu/2-a\\ b,\ c-a,\ a-\alpha+\nu/2+1\end{bmatrix}\times$$

$$\times {}_2F_3\left(a,\ c-b;\ a-b+1,\ a-\alpha+\frac{\nu}{2}+1,\ a-\alpha-\frac{\nu}{2}+1;\ -\frac{\sigma^2}{4\omega}\right)+$$

$$+\frac{(\sigma/2)^{2b-2\alpha}}{\omega^b}\Gamma\left[\begin{matrix}c,\ a-b,\ \alpha+\nu/2-b\\ a,\ c-b,\ b-\alpha+\nu/2+1\end{matrix}\right]\times$$

$$\times {}_2F_3\left(b,\ c-a;\ b-a+1,\ b-\alpha+\frac{\nu}{2}+1,\ b-\alpha-\frac{\nu}{2}+1;\ -\frac{\sigma^2}{4\omega}\right)+$$

$$+\frac{(\sigma/2)^{\nu}}{\omega^{\alpha+\nu/2}}\Gamma\left[\begin{matrix}c,\ \alpha+\nu/2,\ a-\alpha-\nu/2,\ b-\alpha-\nu/2,\ c-a-b+\alpha+\nu/2\\ a,\ b,\ c-a,\ c-b,\ \nu+1\end{matrix}\right]\times$$

$$\times {}_2F_3\left(\alpha+\frac{\nu}{2},\ c-a-b+\alpha+\frac{\nu}{2};\ \nu+1,\ \alpha+\frac{\nu}{2}-a+1,\ \alpha+\frac{\nu}{2}-b+1;\ -\frac{\sigma^2}{4\omega}\right)$$

[σ, $\operatorname{Re}(2\alpha+\nu)$, $\operatorname{Re}(c-a-b+\alpha+\nu/2)>0$; $\operatorname{Re}(\alpha-a)$, $\operatorname{Re}(\alpha-b)<3/4$; $|\arg\omega|<\pi$].

10. $$\int_0^y x^{\alpha-1}(y-x)^{c-1}J_\nu(\sigma\sqrt{x})\,{}_2F_1\left(a,\ b;\ c;\ 1-\frac{x}{y}\right)dx=$$

$$=\left(\frac{\sigma}{2}\right)^{\nu}y^{c+\alpha+\nu/2-1}\Gamma\left[\begin{matrix}c,\ \alpha+\nu/2,\ c-a-b+\alpha+\nu/2\\ \nu+1,\ c-a+\alpha+\nu/2,\ c-b+\alpha+\nu/2\end{matrix}\right]\times$$

$$\times {}_2F_3\left(\alpha+\frac{\nu}{2},\ c-a-b+\alpha+\frac{\nu}{2};\ \nu+1,\ c-a+\alpha+\frac{\nu}{2},\ c-b+\alpha+\frac{\nu}{2};\ -\frac{\sigma^2 y}{4}\right)$$

[y, $\operatorname{Re}c$, $\operatorname{Re}(2\alpha+\nu)$, $\operatorname{Re}(c-a-b+\alpha+\nu/2)>0$].

11. $$\int_y^\infty x^{\alpha-1}(x-y)^{c-1}J_\nu(\sigma\sqrt{x})\,{}_2F_1\left(a,\ b;\ c;\ 1-\frac{x}{y}\right)dx=$$

$$=\left(\frac{\sigma}{2}\right)^{2(a-c-\alpha+1)}y^{a}\Gamma\left[\begin{matrix}c,\ b-a,\ c-a+\alpha+\nu/2-1\\ c-a,\ b,\ a-c-\alpha+\nu/2+2\end{matrix}\right]\times$$

$$\times {}_2F_3\left(1-b,\ a-c+1;\ a-b+1,\ a-c-\alpha-\frac{\nu}{2}+2,\ a-c-\alpha+\frac{\nu}{2}+2;\ -\frac{\sigma^2 y}{4}\right)+$$

$$+\left(\frac{\sigma}{2}\right)^{2(b-c-\alpha+1)}y^{b}\Gamma\left[\begin{matrix}c,\ a-b,\ c-b+\alpha+\nu/2-1\\ a,\ c-b,\ b-c-\alpha+\nu/2+2\end{matrix}\right]\times$$

$$\times {}_2F_3\left(1-a,\ b-c+1;\ b-a+1,\ b-c-\alpha-\frac{\nu}{2}+2,\ b-c-\alpha+\frac{\nu}{2}+2;\ -\frac{\sigma^2 y}{4}\right)+$$

$$+\left(\frac{\sigma}{2}\right)^{\nu}y^{c+\alpha+\nu/2-1}\Gamma\left[\begin{matrix}c,\ a-c-\alpha-\nu/2+1,\ b-c-\alpha-\nu/2+1\\ \nu+1,\ 1-\alpha-\nu/2,\ a+b-c-\alpha-\nu/2+1\end{matrix}\right]\times$$

$$\times {}_2F_3\left(\alpha+\frac{\nu}{2},\ c-a-b+\alpha+\frac{\nu}{2};\ \nu+1,\ c-a+\alpha+\frac{\nu}{2},\ c-b+\alpha+\frac{\nu}{2};\ -\frac{\sigma^2 y}{4}\right)$$

[y, σ, $\operatorname{Re}c>0$; $\operatorname{Re}(c-a+\alpha)$, $\operatorname{Re}(c-b+\alpha)<7/4$].

12. $$\int_0^\infty \frac{x^{\alpha-1}}{(x+z)^{2a}}J_\nu(\sigma x)\,{}_2F_1\left(a,\ \frac{c}{2};\ c;\ \frac{4xz}{(x+z)^2}\right)dx=\frac{(b/2)^{2a-\alpha}}{2}\Gamma\left[\begin{matrix}(\alpha+\nu)/2-a\\ a+(\nu-\alpha)/2+1\end{matrix}\right]\times$$

$$\times {}_2F_3\left(a,\ a+\frac{1-c}{2};\ \frac{c+1}{2},\ a-\frac{\alpha+\nu}{2}+1,\ a+\frac{\nu-\alpha}{2}+1;\ -\frac{\sigma^2z^2}{4}\right)+$$

$$+\frac{\sigma^{\nu}z^{\alpha+\nu-2a}}{2^{\nu+1}}\Gamma\left[\begin{matrix}(c+1)/2,\ (c+1)/2-a,\ (\alpha+\nu)/2,\ a-(\alpha+\nu)/2\\ a,\ \nu+1,\ (c-\alpha-\nu+1)/2,\ (c+\alpha+\nu+1)/2-a\end{matrix}\right]\times$$

$$\times {}_2F_3\left(\frac{\alpha+\nu-c+1}{2},\ \frac{\alpha+\nu}{2};\ \nu+1,\ \frac{\alpha+\nu}{2}-a+1,\ \frac{\alpha+\nu+c+1}{2}-a;\ -\frac{\sigma^2z^2}{4}\right)$$

[σ, $\operatorname{Re}(\alpha+\nu)>0$; $\operatorname{Re}\alpha-3/2<2\operatorname{Re}a<\operatorname{Re}c+2$ $|\arg z|<\pi$].

13. $\int\limits_0^\infty \frac{x^{(c+1)/2}}{(x+z)^{2a}} J_{(c-1)/2}(\sigma x)\, {}_2F_1\left(a, \frac{c}{2}; c; \frac{4xz}{(x+z)^2}\right) dx =$

$$= 2^{c-2a}\sigma^{2a-c-1} z^{(1-c)/2}\Gamma\begin{bmatrix}(c+1)/2, & (c+1)/2-a\\ a\end{bmatrix} J_{(c-1)/2}(\sigma z)$$

[$2\operatorname{Re} a-2$, $-1<\operatorname{Re} c<4\operatorname{Re} a$; $|\arg z|<\pi$].

2.21.5. Integrals of $A(x)\begin{Bmatrix}Y_\nu(\sigma\sqrt{x})\\ I_\nu(\sigma\sqrt{x})\end{Bmatrix}{}_2F_1(a, b; c; \varphi(x))$.

1. $\int\limits_0^\infty x^{\alpha-1} Y_\nu(\sigma\sqrt{x})\, {}_2F_1(a, b; c; -\omega x)\, dx =$

$$= -\frac{(\sigma/2)^{2a-2\alpha}}{\pi\omega^a}\cos\left(a-\alpha+\frac{\nu}{2}\right)\pi\Gamma\begin{bmatrix}b-a, c, \alpha+\nu/2-a, a-\nu/2-b\\ b, c-a\end{bmatrix}\times$$

$$\times {}_2F_3\left(a, a-c+1; a-b+1, a-\alpha-\frac{\nu}{2}+1, b-\alpha+\frac{\nu}{2}+1; \frac{\sigma^2}{4\omega}\right)-$$

$$-\frac{(\sigma/2)^{2b-2\alpha}}{\pi\omega^b}\cos\left(b-\alpha+\frac{\nu}{2}\right)\pi\Gamma\begin{bmatrix}a-b, c, \alpha-\nu/2-a, \alpha+\nu/2-b\\ a, c-b\end{bmatrix}\times$$

$$\times {}_2F_3\left(b, b-c+1; b-a+1, a-\alpha+\frac{\nu}{2}+1, b-\alpha-\frac{\nu}{2}+1; \frac{\sigma^2}{4\omega}\right)-$$

$$-\frac{(\sigma/2)^\nu}{\pi\omega^{\nu/2}}\cos\nu\pi\Gamma\begin{bmatrix}c, -\nu, \alpha+\nu/2, a-\alpha-\nu/2, b-\alpha-\nu/2\\ a, b, c-\alpha-\nu/2\end{bmatrix}\times$$

$$\times {}_2F_3\left(\alpha+\frac{\nu}{2}, \alpha+\frac{\nu}{2}-c+1; \nu+1, \alpha+\frac{\nu}{2}-a+1, \alpha+\frac{\nu}{2}-b+1; \frac{\sigma^2}{4\omega}\right)-$$

$$-\frac{\omega^{\nu/2}}{\pi(\sigma/2)^\nu}\Gamma\begin{bmatrix}c, \nu, \alpha-\nu/2, a-\alpha+\nu/2, b-\alpha+\nu/2\\ a, b, c-\alpha+\nu/2\end{bmatrix}\times$$

$$\times {}_2F_3\left(\alpha-\frac{\nu}{2}, \alpha-\frac{\nu}{2}-c+1; 1-\nu, \alpha-\frac{\nu}{2}-a+1, \alpha-\frac{\nu}{2}-b+1; \frac{\sigma^2}{4\omega}\right)$$

[$\sigma>0$; $\operatorname{Re}(\alpha-a)$, $\operatorname{Re}(\alpha-b)<3/4$; $2\operatorname{Re}\alpha>|\operatorname{Re}\nu|$; $|\arg\omega|<\pi$].

2. $\int\limits_0^\infty x^{(3-2a)/4} Y_{1/2-a}(\sigma\sqrt{x})\, {}_2F_1\left(\frac{1}{2}, a; \frac{3}{2}; -\omega x\right) dx =$

$$= \frac{(\sigma/2)^{b-3/2}}{\sqrt{\pi}\,\omega\Gamma(a)} K_{a-1/2}\left(\frac{\sigma}{2\sqrt{\omega}}\right) K_{a-3/2}\left(\frac{\sigma}{2\sqrt{\omega}}\right)\quad [\sigma>0;\ 1<\operatorname{Re} a<2;\ |\arg\omega|<\pi].$$

3. $\int\limits_0^\infty x^{(c-1)/2} Y_{c-2}(\sigma\sqrt{x})\, {}_2F_1\left(1, 2c-\frac{5}{2}; c; -\omega x\right) dx =$

$$= \frac{2(\sigma/2)^{c-2}}{\sqrt{\pi}\,\omega^{1/2-c}}\Gamma\begin{bmatrix}c\\ 2c-5/2\end{bmatrix} K_{c-2}^2\left(\frac{\sigma}{2\sqrt{\omega}}\right)\quad [\sigma>0;\ 3/2<\operatorname{Re} c<5/2;\ |\arg\omega|<\pi].$$

4. $\int\limits_0^\infty x^{(\nu+1)/2} Y_\nu(\sigma\sqrt{x})\, {}_2F_1\left(1, a; \frac{3}{2}; -\omega x\right) dx =$

$$= \frac{\sqrt{\pi}(\sigma/2)^{a-3/2}}{\omega^{(2a+2\nu+3)/4}\Gamma(a)} K_{a-\nu-3/2}\left(\frac{\sigma}{\sqrt{\omega}}\right)$$

[$\sigma>0$; $-3/2<\operatorname{Re}\nu<1/2$; $\operatorname{Re}(2a-\nu)>3/2$; $|\arg\omega|<\pi$].

5. $\int\limits_0^y x^{\alpha-1}(y-x)^{c-1}I_\nu(\sigma\sqrt{x})\,{}_2F_1\left(a,\ b;\ c;\ 1-\frac{x}{y}\right)dx=$

$$=\left(\frac{\sigma}{2}\right)^\nu y^{c+\alpha+\nu/2-1}\Gamma\begin{bmatrix}c,\ \alpha+\nu/2,\ c-a-b+\alpha+\nu/2\\ \nu+1,\ c-a+\alpha+\nu/2,\ c-b+\alpha+\nu/2\end{bmatrix}\times$$

$$\times\,{}_2F_3\left(\alpha+\frac{\nu}{2},\ c-a-b+\alpha+\frac{\nu}{2};\ \nu+1,\ c-a+\alpha+\frac{\nu}{2},\ c-b+\alpha+\frac{\nu}{2};\ \frac{\sigma^2 y}{4}\right)$$

$[y,\ \mathrm{Re}\, c,\ \mathrm{Re}\,(2\alpha+\nu),\ \mathrm{Re}\,(c-a-b+\alpha+\nu/2)>0].$

2.21.6. Integrals of $A(x)\,K_\nu(\varphi(x))\,{}_2F_1(a,\ b;\ c;\ \chi(x))$.

1. $\int\limits_0^\infty x^{\alpha-1}K_\nu(\sigma\sqrt{x})\,{}_2F_1(a,\ b;\ c;\ -\omega x)\,dx=$

$$=\frac{(\sigma/2)^{2a-2\alpha}}{2\omega^a}\Gamma\begin{bmatrix}b-a,\ c,\ \alpha+\nu/2-a,\ \alpha-\nu/2-a\\ b,\ c-a\end{bmatrix}\times$$

$$\times\,{}_2F_3\left(a,\ a-c+1;\ a-b+1,\ a-\alpha+\frac{\nu}{2}+1,\ a-\alpha-\frac{\nu}{2}+1;\ -\frac{\sigma^2}{4\omega}\right)+$$

$$+\frac{(\sigma/2)^{2b-2\alpha}}{2\omega^b}\Gamma\begin{bmatrix}a-b,\ c,\ \alpha+\nu/2-b,\ \alpha-\nu/2-b\\ a,\ c-b\end{bmatrix}\times$$

$$\times\,{}_2F_3\left(b,\ b-c+1;\ b-a+1,\ b-\alpha+\frac{\nu}{2}+1,\ b-\alpha-\frac{\nu}{2}+1;\ -\frac{\sigma^2}{4\omega}\right)+$$

$$+\frac{(\sigma/2)^\nu}{2\omega^{\alpha+\nu/2}}\Gamma\begin{bmatrix}c,\ -\nu,\ \alpha+\nu/2,\ a-\alpha-\nu/2,\ b-\alpha-\nu/2\\ a,\ b,\ c-\alpha-\nu/2\end{bmatrix}\times$$

$$\times\,{}_2F_3\left(\alpha+\frac{\nu}{2},\ \alpha+\frac{\nu}{2}-c+1;\ \nu+1,\ \alpha+\frac{\nu}{2}-a+1,\ \alpha+\frac{\nu}{2}-b+1;\ -\frac{\sigma^2}{4\omega}\right)+$$

$$+\frac{(\sigma/2)^{-\nu}}{2\omega^{\alpha-\nu/2}}\Gamma\begin{bmatrix}c,\ \nu,\ \alpha-\nu/2,\ a-\alpha+\nu/2,\ b-\alpha+\nu/2\\ a,\ b,\ c-\alpha+\nu/2\end{bmatrix}\times$$

$$\times\,{}_2F_3\left(\alpha-\frac{\nu}{2},\ \alpha-\frac{\nu}{2}-c+1;\ 1-\nu,\ \alpha-\frac{\nu}{2}-a+1,\ \alpha-\frac{\nu}{2}-b+1;\ -\frac{\sigma^2}{4\omega}\right)$$

$[\mathrm{Re}\,\sigma>0;\ 2\mathrm{Re}\,\alpha>|\mathrm{Re}\,\nu|;\ |\arg\omega|<\pi].$

2. $\int\limits_0^\infty x^{(c-1)/2}K_{c-1}(\sigma\sqrt{x})\,{}_2F_1(a,\ b;\ c;\ -\omega x)\,dx=$

$$=\frac{2^{c+1}\sigma^{a+b-c-1/2}}{\omega^{(a+b)/2}}\Gamma(c)\,S_{1-a-b,\ a-b}\left(\frac{\sigma}{\sqrt{\omega}}\right)\quad[\mathrm{Re}\, c,\ \mathrm{Re}\,\sigma>0;\ |\arg\omega|<\pi].$$

3. $\int\limits_0^\infty x^{\alpha-1}K_\nu\left(\frac{\sigma}{\sqrt{x}}\right)\,{}_2F_1(a,\ b;\ c;\ -\omega x)\,dx=$

$$=\frac{(\sigma/2)^\nu}{2\omega^{\alpha-\nu/2}}\Gamma\begin{bmatrix}c,\ -\nu,\ \alpha-\nu/2,\ a-\alpha+\nu/2,\ b-\alpha+\nu/2\\ a,\ b,\ c-\alpha+\nu/2\end{bmatrix}\times$$

$$\times\,{}_2F_3\left(a-\alpha+\frac{\nu}{2},\ b-\alpha+\frac{\nu}{2};\ \nu+1,\ 1-\alpha+\frac{\nu}{2},\ c-\alpha+\frac{\nu}{2};\ -\frac{\sigma^2\omega}{4}\right)+$$

$$+\frac{(\sigma/2)^{-\nu}}{2\omega^{\alpha+\nu/2}}\Gamma\begin{bmatrix}c,\ \nu,\ \alpha+\nu/2,\ a-\alpha-\nu/2,\ b-\alpha-\nu/2\\ a,\ b,\ c-\alpha-\nu/2\end{bmatrix}\times$$

$$\times\,{}_2F_3\left(a-\alpha-\frac{\nu}{2},\ b-\alpha-\frac{\nu}{2};\ 1-\nu,\ 1-\alpha-\frac{\nu}{2},\ c-\alpha-\frac{\nu}{2};\ -\frac{\sigma^2\omega}{4}\right)+$$

$$+\frac{(\sigma/2)^{2\alpha}}{2}\Gamma\left(\frac{\nu}{2}-\alpha\right)\Gamma\left(-\frac{\nu}{2}-\alpha\right){}_2F_3\left(a,\ b;\ c,\alpha-\frac{\nu}{2}+1,\ \alpha+\frac{\nu}{2}+1;\ -\frac{\sigma^2\omega}{4}\right)$$

$[\mathrm{Re}\,\sigma>0;\ 2\mathrm{Re}\,(a-\alpha),\ 2\mathrm{Re}\,(b-\alpha)>|\mathrm{Re}\,\nu|;\ |\arg\omega|<\pi].$

4. $\int\limits_0^\infty x^{\alpha-1}K_\nu(\sigma\sqrt{x})\,{}_2F_1(a,\,b;\,c;\,1-\omega x)\,dx=$

$$=\frac{(\sigma/2)^{2a-2\alpha}}{2\omega^a}\Gamma\begin{bmatrix}b-a,\ c,\ \alpha+\nu/2-a,\ \alpha-\nu/2-a\\ b,\ c-a\end{bmatrix}\times$$

$$\times{}_2F_3\left(a,\ c-b;\ a-b+1,\ a-\alpha-\frac{\nu}{2}+1,\ a-\alpha+\frac{\nu}{2}+1;\ \frac{\sigma^2}{4\omega}\right)+$$

$$+\frac{(\sigma/2)^{2b-2\alpha}}{2\omega^b}\Gamma\begin{bmatrix}a-b,\ c,\ \alpha+\nu/2-b,\ \alpha-\nu/2-b\\ a,\ c-b\end{bmatrix}\times$$

$$\times{}_2F_3\left(b,\ c-a;\ b-a+1,\ b-\alpha-\frac{\nu}{2}+1,\ b-\alpha+\frac{\nu}{2}+1;\ \frac{\sigma^2}{4\omega}\right)+$$

$$+\frac{(\sigma/2)^{\nu}}{2\omega^{\alpha+\nu/2}}\Gamma\begin{bmatrix}c,\ -\nu,\ \alpha+\nu/2,\ a-\alpha-\nu/2,\ b-\alpha-\nu/2,\ c-a-b+\alpha+\nu/2\\ a,\ b,\ c-a,\ c-b\end{bmatrix}\times$$

$$\times{}_2F_3\left(\alpha+\frac{\nu}{2},\ c-a-b+\alpha+\frac{\nu}{2};\ \nu+1,\ \alpha+\frac{\nu}{2}-a+1,\ \alpha+\frac{\nu}{2}-b+1;\ \frac{\sigma^2}{4\omega}\right)+$$

$$+\frac{(\sigma/2)^{-\nu}}{2\omega^{\alpha-\nu/2}}\Gamma\begin{bmatrix}c,\ \nu,\ \alpha-\nu/2,\ a-\alpha+\nu/2,\ b-\alpha+\nu/2,\ c-a-b+\alpha-\nu/2\\ a,\ b,\ c-a,\ c-b\end{bmatrix}\times$$

$$\times{}_2F_3\left(\alpha-\frac{\nu}{2},c-a-b+\alpha-\frac{\nu}{2};\ 1-\nu,\ \alpha-\frac{\nu}{2}-a+1,\ \alpha-\frac{\nu}{2}-b+1;\ \frac{\sigma^2}{4\omega}\right)$$

[$\operatorname{Re}\sigma>0$; $2\operatorname{Re}\alpha,\ 2\operatorname{Re}(c-a-b+\alpha)>|\operatorname{Re}\nu|$; $|\arg\omega|<\pi$].

5. $\int\limits_0^y x^{\alpha-1}(y-x)^{c-1}K_\nu(\sigma\sqrt{x})\,{}_2F_1\left(a,\ b;\ c;\ 1-\frac{x}{y}\right)dx=$

$$=2^{\nu-1}\sigma^{-\nu}y^{c+\alpha-\nu/2-1}\Gamma\begin{bmatrix}c,\ \nu,\ \alpha-\nu/2,\ c-a-b+\alpha-\nu/2\\ c-a+\alpha-\nu/2,\ c-b+\alpha-\nu/2\end{bmatrix}\times$$

$$\times{}_2F_3\left(\alpha-\frac{\nu}{2},\ c-a-b+\alpha-\frac{\nu}{2};\ 1-\nu,\ c-a+\alpha-\frac{\nu}{2},\ c-b+\alpha-\frac{\nu}{2};\frac{\sigma^2y}{4}\right)+$$

$$+2^{-\nu-1}\sigma^{\nu}y^{c+\alpha+\nu/2-1}\Gamma\begin{bmatrix}c,\ -\nu,\ \alpha+\nu/2,\ c-a-b+\alpha+\nu/2\\ c-a+\alpha+\nu/2,\ c-b+\alpha+\nu/2\end{bmatrix}\times$$

$$\times{}_2F_3\left(\alpha+\frac{\nu}{2},\ c-a-b+\alpha+\frac{\nu}{2};\ \nu+1,\ c-a+\alpha+\frac{\nu}{2},\ c-b+\alpha+\frac{\nu}{2};\ \frac{\sigma^2y}{4}\right)$$

[$y,\ \operatorname{Re}c>0$; $2\operatorname{Re}\alpha,\ 2\operatorname{Re}(c-a-b+\alpha)>|\operatorname{Re}\nu|$].

6. $\int\limits_y^\infty x^{\alpha-1}(x-y)^{c-1}K_\nu(\sigma\sqrt{x})\,{}_2F_1\left(a,\ b;\ c;\ 1-\frac{x}{y}\right)dx=$

$$=\frac{(\sigma/2)^{2(a-c-\alpha+1)}y^a}{2}\Gamma\begin{bmatrix}b-a,\ c,\ c-a+\alpha-\nu/2-1,\ c-a+\alpha+\nu/2-1\\ b,\ c-a\end{bmatrix}\times$$

$$\times{}_2F_3\left(1-b,\ a-c+1;\ a-b+1,\ a-c-\alpha+\frac{\nu}{2}+2,\ a-c-\alpha-\frac{\nu}{2}+2;\ \frac{\sigma^2y}{4}\right)+$$

$$+\frac{(\sigma/2)^{2(b-c-\alpha+1)}y^b}{2}\Gamma\begin{bmatrix}a-b,\ c,\ c-b+\alpha-\nu/2-1,\ c-b+\alpha+\nu/2-1\\ a,\ c-b\end{bmatrix}\times$$

$$\times{}_2F_3\left(1-a,\ b-c+1;\ b-a+1,\ b-c-\alpha+\frac{\nu}{2}+2,\ b-c-\alpha-\frac{\nu}{2}+2;\ \frac{\sigma^2y}{4}\right)+$$

$$+2^{\nu-1}\sigma^{-\nu}y^{c+\alpha-\nu/2-1}\Gamma\begin{bmatrix}c,\ \nu,\ a-c-\alpha+\nu/2+1,\ b-c-\alpha+\nu/2+1\\ 1-\alpha+\nu/2,\ a+b-c-\alpha+\nu/2+1\end{bmatrix}\times$$

$$\times{}_2F_3\left(\alpha-\frac{\nu}{2},\ c-a-b+\alpha-\frac{\nu}{2};\ 1-\nu,\ c-a+\alpha-\frac{\nu}{2},\ c-b+\alpha-\frac{\nu}{2};\ \frac{\sigma^2y}{4}\right)+$$

$$+2^{-\nu-1}\sigma^{\nu}y^{c+\alpha+\nu/2-1}\Gamma\begin{bmatrix}c, -\nu, a-c-\alpha-\nu/2+1, b-c-\alpha-\nu/2+1\\1-\alpha-\nu/2, a+b-c-\alpha-\nu/2+1\end{bmatrix}\times$$
$$\times {}_2F_3\left(\alpha+\frac{\nu}{2}, c-a-b+\alpha+\frac{\nu}{2}; \nu+1, c-a+\alpha+\frac{\nu}{2}, c-b+\alpha+\frac{\nu}{2}; \frac{\sigma^2 y}{4}\right)$$

[Re c, Re $\sigma > 0$].

7. $$\int_0^\infty \frac{x^{c-1}e^{zx}}{(x+y)^{(a+b-c)/2}}K_{(a+b-c)/2}(xz+yz)\,{}_2F_1\left(a, b; c; -\frac{x}{y}\right)dx=$$
$$=\frac{y^{c-(a+b+1)/2}}{2^{(c+1)/2}\sqrt{\pi}}\cos\frac{a+b-c}{2}\pi\,\Gamma\left[c, \frac{a-b-c+1}{2}, \frac{b-a-c+1}{2}\right]W_{c/2, (a-b)/2}(2yz)$$

[y, Re $c > 0$; Re $(a-b-c)$, Re $(b-a-c) > -1$; $|\arg z| < 3\pi/2$].

2.21.7. Integrals of $A(x)\,\mathrm{K}(\sigma x)\,{}_2F_1(a, b; c; \varphi(x))$.

1. $$\int_0^y \frac{x^{-1}}{\sqrt{y^2-x^2}}\mathrm{K}\left(\frac{x}{y}\right){}_2F_1\left(a, \frac{1}{2}-a; \frac{1}{2}; 1-\frac{y^2}{x^2}\right)dx=\frac{\pi}{2y}\Gamma\begin{bmatrix}a, 1/2-a\\1/2+a, 1-a\end{bmatrix}$$

[$y > 0$; $0 < \mathrm{Re}\, a < 1/2$].

2. $$\int_0^y \frac{x(y^2-x^2)^{c-1}}{(z^2-x^2)^a}\mathrm{K}\left(\frac{x}{z}\right){}_2F_1\left(a, -\frac{1}{2}; c; \frac{y^2-x^2}{z^2-x^2}\right)dx=$$
$$=\frac{\pi y^{2c}z^{-2a}}{4c}\,{}_2F_1\left(a+\frac{1}{2}, \frac{1}{2}; c+1; \frac{y^2}{z^2}\right)$$

[$z > y > 0$; Re $c > 0$].

3. $$\int_0^y x(y^2-x^2)^{c-1}\mathrm{K}(\sigma x)\,{}_2F_1\left(c+\frac{1}{2}, c+\frac{1}{2}; c; \omega(y^2-x^2)\right)dx=$$
$$=\frac{\pi y^{2c}}{4c}(1-\omega y^2)^{-c}{}_2F_1\left(\frac{1}{2}, \frac{1}{2}; c+1; y^2(\sigma^2+\omega-\sigma^2\omega y^2)\right)$$

[y, Re $c > 0$; $|\arg(1-\omega y^2)| < \pi$].

2.21.8. Integrals containing Whittaker functions and ${}_2F_1(a,b;c;\omega x)$.

1. $$\int_0^\infty x^{\alpha-1}e^{-\sigma/(2x)}M_{c-\alpha, a-\alpha+1/2}\left(\frac{\sigma}{x}\right){}_2F_1(a, b; c; -\omega x)\,dx=$$
$$=\frac{b^{(a-1)/2}}{\omega^{\alpha-(a-1)/2}}\Gamma\begin{bmatrix}2a-2\alpha, a+b-2\alpha, c\\a, a+c-2\alpha\end{bmatrix}e^{\sigma\omega/2}W_{\varphi,\psi}(\sigma\omega)$$

[$2\varphi=2\alpha-a-2b+1$, $2\psi=a-2\alpha$; Re σ, Re c, Re $(a-\alpha)$, Re $(a+b-2\alpha) > 0$; $|\arg\omega| < \pi$].

2. $$\int_y^\infty x^{(a+b-c-1)/2}(x-y)^{c-1}e^{\sigma x/2}W_{\rho, (a+b-c)/2}(\sigma x)\,{}_2F_1\left(a, b; c; 1-\frac{x}{y}\right)dx=$$
$$=\frac{y^{(a+b-1)/2}}{\sigma^{c/2}}\Gamma\begin{bmatrix}c, (a-b-c+1)/2-\rho, (b-a-c+1)/2-\rho\\(a+b-c+1)/2-\rho, (c-a-b+1)/2-\rho\end{bmatrix}e^{\sigma y}W_{\rho+c/2, (a-b)/2}(\sigma y)$$

[y, Re $c > 0$; Re $(c+2\rho) < 1-|\mathrm{Re}\,(a-b)|$; $|\arg\sigma| < 3\pi/2$].

2.21.9. Integrals containing products of two functions ${}_2F_1(a,b;c;\varphi(x))$.

1. $$\int_0^y {}_2F_1\left(1-a, 1-b; 2-c; \frac{x}{y}\right){}_2F_1\left(a, b; c; \frac{x}{y}\right)dx=$$
$$=\frac{(c-1)y\{[\psi(a)-\psi(b)]\sin a\pi\sin b\pi+[\psi(c-a)-\psi(c-b)]\sin(c-a)\pi\sin(c-b)\pi\}}{(b-a)\sin c\pi\sin(c-a-b)\pi}$$

[$c\neq 1$; $a\neq b$; $|\mathrm{Re}\,(c-a-b)| < 1$].

2. $$=\frac{\sin a\pi \sin b\pi}{\pi (a-b) \sin (a+b) \pi} [\psi (a)+\psi (1-a)-\psi (b)-\psi (1-b)]$$ $[c=1;\ a\neq b;\ 0 < \operatorname{Re} (a+b) < 2]$.

3. $$=\frac{\operatorname{tg} a\pi}{2\pi} [\zeta (2,\ a)-\zeta (2,\ 1-a)]$$ $[c=1,\ b=a;\ 0 < \operatorname{Re} a < 1]$.

4. $$\int_0^y {}_2F_1\left(1-a,\ 2-b;\ 3-c;\ \frac{x}{y}\right) {}_2F_1\left(a,\ b;\ c;\ \frac{x}{y}\right) dx=$$ $$=\frac{a (c-1) (c-2) \sin (c-a) \pi \sin (c-b) \pi}{(b-1) (c-a-1) \sin c\pi \sin (c-a-b) \pi} [\psi (a)-\psi (c-b)]$$ $[|\operatorname{Re} (c-a-b)| < 1]$.

5. $$\int_0^y x^{c-1} (y-x)^{a+b-c} \left[{}_2F_1\left(a,\ b;\ c;\ \frac{x}{y}\right)\right]^2 dx=$$ $$=\frac{\pi y^{a+b}}{(a-b) \sin (c-a-b) \pi} \Gamma \begin{bmatrix} c,\ c \\ a,\ b,\ c-a,\ c-b \end{bmatrix} [\psi (a)+\psi (c-a)-\psi (b)-\psi (c-b)]$$ $[a\neq b;\ |\operatorname{Re} (c-a-b)| < 1]$.

6. $$=\frac{\pi y^{2a}}{\sin (c-2a) \pi} \Gamma^2 \begin{bmatrix} c \\ a,\ c-a \end{bmatrix} [\psi' (a)-\psi' (c-a)]$$ $[b=a;\ |\operatorname{Re} (c-2a)| < 1]$.

7. $$\int_0^\infty x^{c-1} {}_2F_1 (a,\ b;\ c;\ -\sigma x)\, {}_2F_1 (a',\ b';\ c,\ -\omega x)\, dx=$$ $$=\sigma^{-a}\omega^{a-c}\Gamma \begin{bmatrix} c,\ c,\ a+a'-c,\ a+b'-c,\ a'+b-c,\ b+b'-c \\ a,\ b,\ a',\ b',\ a+a'+b+b'-2c \end{bmatrix}\times$$ $$\times {}_2F_1 \left(a+a'-c,\ a+b-c;\ a+a'+b+b'-2c;\ 1-\frac{\omega}{\sigma}\right)$$ $[\operatorname{Re} c,\ \operatorname{Re} (a+a'-c),\ \operatorname{Re} (a+b'-c),\ \operatorname{Re} (a'+b-c),\ \operatorname{Re} (b+b'-c) > 0;\ |\arg \sigma|,\ |\arg \omega| < \pi]$.

8. $$\int_0^\infty x^{-1/2}\, {}_2F_1 (a,\ b;\ c;\ -\sigma x)\, {}_2F_1 \left(a,\ b;\ c;\ -\frac{\omega}{x}\right) dx=$$ $$=\sqrt{\frac{\pi}{\sigma}} \Gamma \begin{bmatrix} a-1/2,\ b-1/2,\ c \\ a,\ b,\ c-1/2 \end{bmatrix} {}_2F_1 (2a-1,\ 2b-1;\ 2c-1;\ -\sqrt{\sigma\omega})$$ $[\operatorname{Re} a,\ \operatorname{Re} b > 1/2;\ |\arg \sigma|,\ |\arg \omega| < \pi]$.

9. $$\int_0^y x^{c-1} (y-x)^{c'-1} {}_2F_1 (a,\ b;\ c;\ -\omega x)\, {}_2F_1 (a,\ b;\ c';\ \omega (y-x))\, dx=$$ $$=y^{c+c'-1} \mathrm{B} (c,\ c')\, {}_4F_3 \left(\begin{matrix} a,\ b,\ (a+b)/2,\ (a+b+1)/2;\ \omega^2 y^2 \\ a+b,\ (c+c')/2,\ (c+c'+1)/2 \end{matrix}\right)$$ $[b\neq -a;\ y,\ \operatorname{Re} c,\ \operatorname{Re} c' > 0;\ |\arg (1-\omega y)|,\ |\arg (1+\omega y)| < \pi]$.

10. $$=\frac{y^{c+c'-1}}{2} \mathrm{B} (c,\ c') \left[1+{}_3F_2 \left(\begin{matrix} 1/2,\ a,\ -a;\ \omega^2 y^2 \\ (c+c')/2,\ (c+c'+1)/2 \end{matrix}\right)\right]$$ $[b=-a,\ y,\ \operatorname{Re} c,\ \operatorname{Re} c' > 0;\ |\arg (1-\omega y)|,\ |\arg (1+\omega y)| < \pi]$.

11. $$\int_0^y x^{c-1} (y-x)^{c'-1} {}_2F_1 (a,\ b;\ c;\ \sigma x)\, {}_2F_1 (c+c'-a,\ c+c'-b;\ c';\ \omega (y-x))\, dx=$$ $$=y^{c+c'-1} \mathrm{B} (c,\ c') (1-\omega y)^{2a-c-c'}\, {}_2F_1 (a,\ b;\ c+c';\ (\sigma+\omega-\sigma\omega) y)$$ $[y,\ \operatorname{Re} c,\ \operatorname{Re} c' > 0;\ |\arg (1-\sigma y)|,\ |\arg (1-\omega y)| < \pi]$.

12. $\int\limits_0^y x^{\alpha-1}(y-x)^{c-1}\,{}_2F_1\left(a,\ b;\ c;\ 1-\frac{x}{y}\right){}_2F_1(a'\ \ b';\ c';\ 1-\omega x)\,dx=$

$$=y^{\alpha+c-1}\Gamma\left[\begin{matrix}c,\ c',\ c'-a'-b',\ \alpha,\ c-a-b+\alpha\\ c-a+\alpha,\ c-b+\alpha,\ c'-a',\ c'-b'\end{matrix}\right]\times$$

$$\times{}_4F_3\left(\begin{matrix}a',\ b',\ \alpha,\ c-a-b+\alpha;\ \omega y\\ c-a+\alpha,\ c-b+\alpha,\ a'+b'-c'+1\end{matrix}\right)+\omega^{c'-a'-b'}y^{c+c'-a'-b'+\alpha-1}\times$$

$$\times\Gamma\left[\begin{matrix}c,\ c',\ a'+b'-c,\ c'-a'-b'+\alpha,\ c+c'-a-a'-b-b'+\alpha\\ a',\ b',\ c+c'-a-a'-b'+\alpha,\ c+c'-a'-b-b'+\alpha\end{matrix}\right]\times$$

$$\times{}_4F_3\left(\begin{matrix}c'-a',\ c'-b',\ c'-a'-b'+\alpha,\ c+c'-a-a'-b-b'+\alpha;\ \omega y\\ c'-a'-b'+1,\ c+c'-a-a'-b'+\alpha,\ c+c'-a'-b-b'+\alpha\end{matrix}\right)$$

[y, $\operatorname{Re} c$, $\operatorname{Re}\alpha$, $\operatorname{Re}(c-a-b+\alpha)$, $\operatorname{Re}(c'-a'-b'+\alpha)$, $\operatorname{Re}(c+c-a-a'-b-b'+\alpha)>0$; $|\arg(1-\omega y)|<\pi$].

13. $\int\limits_\sigma^\omega x^\alpha(x-\sigma)^{c-1}(\omega-x)^{c'-1}\,{}_2F_1\left(a,\ b;\ c;\ 1-\frac{x}{\sigma}\right){}_2F_1\left(a',\ b';\ c';\ 1-\frac{x}{\omega}\right)dx=$

$$=\mathrm{B}(c,\ c')(\omega-\sigma)^{c+c'-1}\sigma^A\omega^B\,{}_2F_1\left(C,\ D;\ c+c';\ 1-\frac{\sigma}{\omega}\right)$$

[$0<\sigma<\omega$; $\operatorname{Re} c$, $\operatorname{Re} c'>0$],

α	a	b	a'	b'	A	B	C	D
0	a	$c'-a'$	a'	$c-a$	a	$-a$	$a+a'$	c
0	a	$c'-a$	a'	$c-a'$	a	$-a$	$a+a'$	$a+c-a'$
0	a	$c+c'-a'$	a'	b'	$c+c'-$ $-a'-b'$	$a'+b'-$ $-c-c'$	$c+c'-a'$	$c-a+$ $+c'-b'$
$a'+b'-c'$	$c-b+a'+$ $+b'-c'$	b	a'	b'	$c-b+a'+$ $+b'-c'$	$b-c$	$c-b+a'$	$c-b+b'$
$a'+b'-c'$	a	$c+b'$	a'	b'	$a'+b'$	$-c$	$c+b'$	$c-a+a'$
$b-c+$ $+a'-c'$	$a'-c'$	b	a'	b'	$b-b'$	$a'+b'-$ $-c-c'$	$b+c'-a'$	$b-b'+c'$
$b-c+$ $+a'-c'$	$b+a'-b'$	b	a'	b'	$b-b'$	$a'+b'-$ $-c-c'$	$c-b+$ $+c'-a'$	$b-b'+c'$
$b-c+$ $+a'-c'$	$-b'$	b	a'	b'	$b+a'-c'$	$-c$	$c+b'$	$b+a'$
$b-c+$ $+a'-c'$	a'	b	a'	b	$b+a'-c'$	$-c$	c	$b+a'$

14. $\int\limits_0^y x^{c-1}(y-x)^{c'-1}(1-\omega x)^{-a'}\,{}_2F_1(a,\ b;\ c;\ \omega x)\,{}_2F_1\left(a',\ b-c;\ c';\ \frac{\omega(y-x)}{1-\omega x}\right)dx=$

$=\mathrm{B}(c,\ c')\,y^{c+c'-1}\,{}_2F_1(a+a',\ b;\ c+c';\ \omega y)$ [y, $\operatorname{Re} c$, $\operatorname{Re} c'>0$; $|\arg(1-\omega y)|<\pi$].

2.21.10. Integrals with respect to the integrals containing ${}_2F_1(a,b;c;x)$.

1. $\int\limits_{\gamma-i\infty}^{\gamma+i\infty}\Gamma\left[\begin{matrix}s\\ s+d\end{matrix}\right]{}_2F_1(a,\ s;\ c;\ -x)\,ds=0$ [γ, $\operatorname{Re} d>0$; $0<\operatorname{Re} a<\operatorname{Re} c$; $|\arg x|<\pi$].

2. $\int\limits_{\gamma-i\infty}^{\gamma+i\infty}\Gamma\left[\begin{matrix}b-s\\ 1-s\end{matrix}\right]{}_2F_1(a,\ s;\ c;\ -x)\,ds=2\pi i x^{-b}\Gamma\left[\begin{matrix}c,\ a-b\\ a,\ 1-b,\ c-b\end{matrix}\right]$

[$\gamma<1$; $0<\gamma<\operatorname{Re} b<\operatorname{Re} a<\operatorname{Re} c$; $|\arg x|<\pi$].

3. $$\int_{\gamma-i\infty}^{\gamma+i\infty} \Gamma\begin{bmatrix} a-s \\ s+d \end{bmatrix} {}_2F_1(a,\ s;\ c;\ -x)\,ds=0 \quad [\gamma,\ \operatorname{Re} d>0;\ 0<\operatorname{Re} a<\operatorname{Re} c;\ |\arg x|<\pi].$$

4. $$\int_{\gamma-i\infty}^{\gamma+i\infty} \Gamma\begin{bmatrix} b+s \\ c+s \end{bmatrix} {}_2F_1(a,\ s;\ c+s;\ -x)\,ds=0$$

$$[\operatorname{Re} a>0;\ \gamma>-\operatorname{Re} b;\ \operatorname{Re}(c-a-b)>1;\ |\arg(x+1)|<\pi].$$

5. $$\int_{\gamma-i\infty}^{\gamma+i\infty} \Gamma\begin{bmatrix} b+s,\ d-s \\ c+s,\ 1-s \end{bmatrix} {}_2F_1(a,\ s;\ c+s;\ -x)\,ds=2\pi i\,(1+x)^{-d}\,\Gamma\begin{bmatrix} b+d,\ a-d \\ c,\ 1-d,\ a \end{bmatrix}$$

$$[\operatorname{Re} a,\ \operatorname{Re}(a-d)>0;\ -\operatorname{Re} b<\gamma<\operatorname{Re} d<1;\ |\arg(x+1)|<\pi].$$

6. $$\int_{\gamma-i\infty}^{\gamma+i\infty} z^s\Gamma(2c-1+s)\,\Gamma(-s)\,{}_2F_1(2c-1+s,\ -s;\ c;\ -x)\,ds=$$

$$=2\pi i\,\Gamma(2c-1)\,(1+2z+4xz+z^2)^{1/2-c} \quad [1-2\operatorname{Re} c<\gamma<0;\ |\arg(x+1)|<\pi].$$

7. $$\int_{\gamma-i\infty}^{\gamma+i\infty} (-z)^s\Gamma\begin{bmatrix} a+s,\ b'+s,\ -s \\ c+s \end{bmatrix} {}_2F_1(a+s,\ b;\ c+s;\ w)\,ds=$$

$$=2\pi i\Gamma\begin{bmatrix} a,\ b' \\ c \end{bmatrix} F_1(a,\ b,\ b';\ c;\ w,\ z) \quad [-\operatorname{Re} a,\ -\operatorname{Re} b'<\gamma<0].$$

8. $$\int_{\gamma-i\infty}^{\gamma+i\infty} (-z)^s\Gamma\begin{bmatrix} a+s,\ b'+s,\ -s \\ c'+s \end{bmatrix} {}_2F_1(a+s,\ b;\ c;\ w)\,ds=$$

$$=2\pi i\Gamma\begin{bmatrix} a,\ b' \\ c' \end{bmatrix} F_2(a,\ b,\ b';\ c,\ c';\ w,\ z) \quad [-\operatorname{Re} a,\ -\operatorname{Re} b'<\gamma<0].$$

9. $$\int_{\gamma-i\infty}^{\gamma+i\infty} (-z)^s\Gamma\begin{bmatrix} a'+s,\ b'+s,\ -s \\ c+s \end{bmatrix} {}_2F_1(a,\ b;\ c+s;\ w)\,ds=$$

$$=2\pi i\Gamma\begin{bmatrix} a',\ b' \\ c \end{bmatrix} F_3(a,\ a',\ b,\ b';\ c;\ w,\ z) \quad [-\operatorname{Re} a',\ -\operatorname{Re} b'<\gamma<0].$$

10. $$\int_{\gamma-i\infty}^{\gamma+i\infty} (-z)^s\Gamma\begin{bmatrix} a+s,\ b+s,\ -s \\ c'+s \end{bmatrix} {}_2F_1(a+s,\ b+s;\ c;\ w)\,ds=$$

$$=2\pi i\Gamma\begin{bmatrix} a,\ b \\ c' \end{bmatrix} F_4(a,\ b;\ c,\ c';\ w,\ z) \quad [-\operatorname{Re} a,\ -\operatorname{Re} b<\gamma<0].$$

11. $$\int_0^\infty x\operatorname{sh}2\pi x\,\Gamma(a+ix)\,\Gamma(a-ix)\,K_{2ix}(b)\,{}_2F_1(a+ix,\ a-ix;\ c;\ -z)\,dx=$$

$$=2^{c-2a-2}\pi^2b^{2a-c+1}z^{(1-c)/2}\Gamma(c)\,J_{c-1}\left(b\sqrt{z}\right) \quad [b>0;\ \operatorname{Re} c>2\operatorname{Re} a>0;\ |\arg z|<\pi]$$

12. $$\int_{-\infty}^{\infty} (c-1+ix)\,\Gamma\left(\frac{1}{2}-ix\right)\Gamma\left(2c-\frac{3}{2}+ix\right) I_{c-1+ix}(a)\,K_{c-1+ix}(b)\times$$

$$\times{}_2F_1\left(2c-\frac{3}{2}+ix,\ \frac{1}{2}-ix;\ c;\ -z\right)dx=2^{c-1/2}\sqrt{\pi}\,\Gamma(c)\left(\frac{ab}{w}\right)^{c-1/2}K_{c-1/2}(w)$$

$$[w=\sqrt{a^2+b^2+2ab\,(1+2z)};\ |\arg(z+1)|<\pi].$$

13. $\int\limits_{-\infty}^{\infty} \frac{\sin (2n+1)\pi x}{\sin \pi x \Gamma(c+x)\Gamma(d-x)} {}_2F_1(a, b; c+x; z) \times$

$$\times {}_2F_1(c+d-a-1, c+d-b-1; d-x; z)\,dx = \frac{2^{c+d-2}(1-2z)^{a+b-c-d+1}}{\Gamma(c+d-1)} {}_2F_1(a, b; c+d-1; 4z(z-1))$$

[$\operatorname{Re}(c+d) > 1$; $|z| < 1$].

2.22. THE GENERALIZED HYPERGEOMETRIC FUNCTION ${}_pF_q((a_p); (b_q); x)$ AND HYPERGEOMETRIC FUNCTIONS OF TWO VARIABLES

(see also 2.23—24 and 2.19—21)

2.22.1. Integrals of general form

NOTATION:

$k, l, m, n, p, q = 0, 1, 2, \ldots$; $k, l \neq 0$; $m \leqslant n+1$; $p \leqslant q+1$; $r = l/k$ [k, l are coprime]

$$\mu = \sum_{j=1}^{p} a_j - \sum_{j=1}^{q} b_j + \frac{q-p+1}{2}, \quad \rho = \sum_{j=1}^{m} c_j - \sum_{j=1}^{n} d_j + \frac{n-m+1}{2}$$

[a_j, b_j, c_j, d_j are complex parameters, $a_j, c_j \neq 0, -1, -2, \ldots$],

$\varphi = q-p-r(n-m+1)+1$, $g = 2^{-1}[(1-l)(m-n+1)+(1-k)(p-q+1)]$,

σ, ω — are complex variables

1°. $\operatorname{Re}(\alpha - c_j - ra_i) < 0$ [$j = 1, 2, \ldots, m$; $i = 1, 2, \ldots, p$];

2°. $(m-n-1)\operatorname{Re}(\alpha - ra_i) - \operatorname{Re}\rho > -3/2$ [$i = 1, 2, \ldots, p$];

3°. $(p-q-1)\operatorname{Re}(\alpha - c_j) - r\operatorname{Re}\mu > -3r/2$ [$j = 1, 2, \ldots, p$];

4°. $|\varphi| + 2\operatorname{Re}[(q-p+1)(n-m+1)\alpha + r(n-m+1)(\mu-1)+(q-p+1)(\rho-1)] > 0$;

5°. $|\varphi| - 2\operatorname{Re}[(q-p+1)(n-m+1)\alpha + r(n-m+1)(\mu-1)+(q-p+1)(\rho-1)] > 0$;

6°. If $k = n-m+1$, $l = q-p+1$, then $|\arg(1 - z_0\sigma^{-l}\omega^k)| <$ where $z_0 = \left(\frac{p-q+1}{n-m+1}\right)^{(n-m+1)(q-p+1)} e^{[(m-n)(p-q)-1]\pi i}$, where for $\operatorname{Re}[\mu+\rho+(n-m+1)\alpha] < 1$ the following value can hold $\sigma^{q-p+1}\omega^{m-n-1} = z_0$.

7°. $\operatorname{Re}(\alpha + c_j) > 0$ [$j = 1, 2, \ldots, m$];

8°. $\operatorname{Re}(\alpha - ra_j) < 0$ [$j = 1, 2, \ldots, p$];

9°. $\operatorname{Re}[(p-q-1) - r\mu] > -3r/2$;

10°. $\operatorname{Re}[(n-m+1)\alpha - \rho] > -3/2$.

11°. One of the following three conditions holds;

$\lambda_c > 0$ or $\lambda_c = 0$, $\lambda_s \neq 0$, $\operatorname{Re}(\alpha+\mu) < 3/2$ or $\lambda_c = \lambda_s = 0$, $\operatorname{Re}(\alpha+\mu) < 1/2$,

where $\lambda_c = (q-p+1)|\omega|^{1/(q-p+1)} \cos \frac{|\arg\omega| + (q-p)\pi}{q-p+1} + |\sigma|\cos(\arg\sigma)$,

$$\lambda_s = (q-p+1)|\omega|^{1/(q-p+1)} \operatorname{sgn}(\arg\omega) \sin \frac{|\arg\omega| + (q-p)\pi}{q-p+1} + \sin(\arg\sigma).$$

[$\arg\sigma \cdot \arg\omega \neq 0$].

1. $\int\limits_0^{\infty} x^{\alpha-1} {}_mF_n((c_m); (d_n); -\sigma x)\, {}_pF_q((a_p); (b_q); -\omega x^{l/k})\,dx =$

$$= (2\pi)^g k^{\mu} l^{\rho+\alpha(n-m+1)-1} \sigma^{-\alpha} \Gamma\left[\begin{matrix}(b_q), (d_n)\\(a_p), (c_m)\end{matrix}\right] \times$$

$$\times G^{k+lm,\ kp+l}_{kp+ln+l,\ kq+k+lm}\left(\frac{k^{k(p-q-1)}\omega^k}{l^{l(m-n-1)}\sigma^l}\,\middle|\,\begin{matrix}\Delta(l, 1-\alpha), \Delta(k, 1-(a_p)), \Delta(l, (d_n)-\alpha)\\ \Delta(k, 0), \Delta(l, (c_m)-\alpha), \Delta(k, 1-(b_q))\end{matrix}\right),$$

if any of the following four groups of conditions holds:

1) $mp \neq 0$; $m=n$ or $n+1$; $p=q$ or $q+1$; $|\arg\sigma| < (m-n+1)\pi/2$; $|\arg\omega| < (p-q+1)\pi/2$; $\operatorname{Re}\alpha > 0$; 1°;
2) $m > 0$; $m=n$ or $n+1$; $p=q-1$ or q; $|\arg\sigma| < (m-n+1)\pi/2$; $|\arg\omega| = (p-q+1)\pi/2$; $\operatorname{Re}\alpha > 0$; 1°, 3°;
3) $m=n-1$ or n; $p > 0$; $p=q$ or $q+1$; $|\arg\sigma| = (m-n+1)\pi/2$; $|\arg\omega| < (p-q+1)\pi/2$; $\operatorname{Re}\alpha > 0$; 1°, 2°;
4) $m=n-1$ or n; $p=q-1$ or q; $|\arg\sigma| = (m-n+1)\pi/2$; $|\arg\omega| = (p-q+1)\pi/2$; $\operatorname{Re}\alpha > 0$; 1°—3°, 5°, 6° [when $m=n=0$ or $p=q=0$ see 2.22.1.3].

2. $$\int_0^\infty x^{\alpha-1}{}_mF_n((c_m);\ (d_n);\ -\sigma x^{-1})\,{}_pF_q((a_p);\ (b_q);\ -\omega x^{l/k})\,dx =$$
$$= (2\pi)^g k^\mu l^{\rho+\alpha(m-n-1)-1}\sigma^\alpha \Gamma\left[\begin{matrix}(b_q),\ (d_n)\\(a_p),\ (c_m)\end{matrix}\right] G^{k+l,\ kp+lm}_{kp+lm,\ kq+k+ln+l}\left(\sigma^l\omega^k \times \frac{k^{k(p-q-1)}}{l^{l(n-m+1)}}\left|\begin{matrix}\Delta(k,\ (a_p)),\ \Delta(l,\ 1-\alpha-(c_m))\\ \Delta(l,\ -\alpha),\ \Delta(k,\ 0),\ \Delta(k,\ 1-(b_q)),\ \Delta(l,\ 1-\alpha-(d_n))\end{matrix}\right.\right),$$
if any of the following four groups of conditions holds;
1) $mp \neq 0$; $m=n$ or $n+1$; $p=q$ or $q+1$; $|\arg\sigma| < (m-n+1)\pi/2$; $|\arg\omega| < (p-q+1)\pi/2$; 7°, 8°;
2) $m=n$ or $n+1$; $p=q-1$ or q; $|\arg\sigma| < (m-n+1)\pi/2$; $|\arg\omega| = (p-q+1)\pi/2$; 7°—9°;
3) $m=n-1$ or n; $p=q$ or $q+1$; $|\arg\sigma| = (m-n+1)\pi/2$; $|\arg\omega| < (p-q+1)\pi/2$; 7°, 8°, 10°;
4) $m=n-1$ or n; $p=q-1$ or q; $|\arg\sigma| = (m-n+1)\pi/2$; $|\arg\omega| = (p-q+1)\pi/2$; 7°—10° [when $m=n=0$ or $p=q=0$ see 2.22.1.4].

3. $$\int_0^\infty x^{\alpha-1}e^{-\sigma x}\,{}_pF_q((a_p);\ (b_q);\ -\omega x^{l/k})\,dx = \frac{k^\mu l^{\alpha-1/2}\sigma^{-\alpha}}{(\sqrt{2\pi})^{l+(k-1)(p-q+1)-1}}\Gamma\left[\begin{matrix}(b_q)\\(a_p)\end{matrix}\right] \times$$
$$\times G^{k,\ kp+l}_{kp+l,\ kq+k}\left(\frac{k^{k(p-q-1)}\omega^k l^l}{\sigma^l}\left|\begin{matrix}\Delta(l,\ 1-\alpha),\ \Delta(k,\ 1-(a_p))\\ \Delta(k,\ 0),\ \Delta(k,\ 1-(b_q))\end{matrix}\right.\right),$$
if any of the following five groups of conditions holds;
1) $q-p-r+1 > 0$; $\operatorname{Re}\alpha$, $\operatorname{Re}\sigma > 0$; $\arg\omega$ — is arbitrary
2) $p=q$ or $q+1$; $\operatorname{Re}\alpha$, $\operatorname{Re}\sigma > 0$; $|\arg\omega| < (p-q+1)\pi/2$;
3) $k=1$; $l=q-p+1$; $p < q-1$; $\operatorname{Re}\alpha$, $\operatorname{Re}\sigma > 0$; $|\arg\omega| < 2\pi$; 6° for $m=n=0$ and 11°;
4) $k=1$; $l=r=2$; $p=q-1$; $\operatorname{Re}\alpha$, $\operatorname{Re}\sigma > 0$; $0 < |\arg\omega| < 2\pi$; 6° for $m=n=0$ and 11°;
5) $k=l=r=1$; $p=q$; $\operatorname{Re}\alpha$, $\operatorname{Re}\sigma > 0$; $\pi/2 < |\arg\omega| < 3\pi/2$; 6° for $m=n=0$ and 11°.

4. $$\int_0^\infty x^{\alpha-1}e^{-\sigma/x}\,{}_pF_q((a_p);\ (b_q);\ -\omega x^{l/k})\,dx = \frac{k^\mu\sigma^\alpha}{(\sqrt{2\pi})^{l+(k-1)(p-q+1)-1}}\Gamma\left[\begin{matrix}(b_q)\\(a_p)\end{matrix}\right] \times$$
$$\times G^{k+l,\ kp}_{kp,\ kq+k+l}\left(\frac{k^{k(p-q-1)}\sigma^l\omega^k}{l^l}\left|\begin{matrix}\Delta(k,\ (a_p))\\ \Delta(l,\ -\alpha),\ \Delta(k,\ 0),\ \Delta(k,\ 1-(b_q))\end{matrix}\right.\right)$$
[$p=q$ or $q+1$; $\operatorname{Re}\sigma > 0$; $|\arg\omega| < (p-q+1)\pi/2$; $0 < \operatorname{Re}\alpha < r \min\limits_{1 \leqslant j \leqslant p} \operatorname{Re} a_j$].

2.22.2. Integrals of $A(x)\,{}_pF_q((a_p);\ (b_q);\ \varphi(x))$ (see also 2.23.1).

1. $$\int_0^a x^{\alpha-1}(a-x)^{\beta-1}\,{}_pF_q((a_p);\ (b_q);\ \omega x)\,dx = a^{\alpha+\beta-1}B(\alpha,\ \beta)\,{}_{p+1}F_{q+1}\left(\begin{matrix}(a_p),\ \alpha;\ a\omega\\(b_q);\ \alpha+\beta\end{matrix}\right)$$
[a, $\operatorname{Re}\alpha$, $\operatorname{Re}\beta > 0$; $p \leqslant q+1$].

2. $$\int_0^\infty \frac{x^{\alpha-1}}{(x+z)^\rho}\, {}_pF_q\left(\begin{matrix}-n,\ (a_{p-1});\ \omega x/(x+z)\\ (b_q)\end{matrix}\right) dx=$$

$$= z^{\alpha-\rho} B(\alpha,\ \rho-\alpha)\ {}_{p+1}F_{q+1}\left(\begin{matrix}-n,\ \alpha,\ (a_{p-1});\ \omega\\ \rho,\ (b_q)\end{matrix}\right)$$

[$\operatorname{Re}\rho > \operatorname{Re}\alpha > 0$; $|\arg z| < \pi$ for $p \leqslant q+1$; $|\arg z|$, $|\arg(1-\omega)| < \pi$ for $p=q+1$].

3. $$\int_0^a x^{(b+c)/2-1}(a-x)^{(b+c)/2-2}\ {}_1F_2\left(\frac{b+c-1}{2};\ b,\ c;\ \omega x\right) dx=$$

$$=\sqrt{\pi}\left(\frac{a}{y}\right)^{(b+c)/2-1} \Gamma\left[\begin{matrix}b,\ c,\ (b+c)/2-1\\ (b+c-1)/2\end{matrix}\right] I_{b-1}(\sqrt{a\omega})\, I_{c-1}(\sqrt{a\omega})$$

[$\operatorname{Re} b$, $\operatorname{Re} c > 0$].

2.22.3. Integrals of $e^{f(x)}\,{}_pF_q((a_p);\ (b_q);\ cx)$ (see also 2.23.2).

1. $$\int_0^\infty x^{\alpha-1} e^{-\sigma x^{1/l}}\ {}_pF_q((a_p);\ (b_q);\ -\omega x)\, dx=$$

$$= l\sigma^{-l\alpha}\,\Gamma(l\alpha)\ {}_{l+p}F_q\left((a_p),\ \Delta(l,\ l\alpha);\ (b_q);\ -\frac{l^l\omega}{\sigma^l}\right),$$

if any of the following five groups of conditions holds;

1) $l < q-p+1$; $\operatorname{Re}\alpha$, $\operatorname{Re}\sigma > 0$; $\arg\omega$ is arbitrary

2) $p=q$ or $q+1$; $\operatorname{Re}\alpha$, $\operatorname{Re}\sigma > 0$; $|\arg\omega| < (p-q+1)\pi/2$;

3) $l=q-p+1$, $p < q-1$; $\operatorname{Re}\alpha$, $\operatorname{Re}\sigma > 0$; $|\arg\omega| < 2\pi$ and 6° (for $m=n=0$), 11° with α replaced by $l\alpha$:

4) $l=2$; $p=q+1$; $\operatorname{Re}\alpha$, $\operatorname{Re}\sigma > 0$; $0 < |\arg\omega| < 2\pi$ and 6° (for $m=n=0$), 11° with α replaced by $l\alpha$:

5) $l=1$; $p=q$; $\operatorname{Re}\alpha$, $\operatorname{Re}\sigma > 0$; $\pi/2 < |\arg\omega| < 3\pi/2$ and 6° (for $m=n=0$), 11° with α replaced by $l\alpha$:

[conditions 6° and 11° see 2.22.1].

2. $$\int_0^\infty x^{c-1} e^{-\sigma\sqrt{x}}\ {}_3F_2\left(\begin{matrix}1,\ a,\ b;\ -\omega x\\ c,\ c+1/2\end{matrix}\right) dx=$$

$$=2\Gamma(2c)\,\sigma^{a+b-2c}\,\omega^{-(a+b)/2}\, S_{1-a-b,\ a-b}\left(\frac{\sigma}{\sqrt{\omega}}\right)$$

[$\operatorname{Re} c$, $\operatorname{Re}\sigma > 0$; $|\arg\omega| < \pi$].

3. $$\int_0^\infty x^{c-1} e^{-\sigma\sqrt{x}}\ {}_4F_3\left(\begin{matrix}a,\ b,\ (a+b)/2,\ (a+b+1)/2;\ -\omega x\\ a+b,\ c,\ c+1/2\end{matrix}\right) dx=$$

$$=2\Gamma(2c)\sigma^{a+b-2c-1}\,\omega^{(1-a-b)/2}\, W_{(1-a-b)/2,\ (a-b)/2}\left(\frac{i\sigma}{\sqrt{\omega}}\right) W_{(1-a-b)/2,\ (a-b)/2}\left(-\frac{i\sigma}{\sqrt{\omega}}\right)$$

[$\operatorname{Re} c$, $\operatorname{Re}\sigma > 0$; $|\arg\omega| < \pi$].

4. $$\int_0^\infty x^{c-1} e^{-\sigma\sqrt{x}}\ {}_4F_3\left(\begin{matrix}a,\ b,\ n-a,\ n-b;\ -\omega x\\ n-1/2,\ c,\ c+1/2\end{matrix}\right) dx=$$

$$=\frac{\pi\Gamma(2c)}{4\sigma^{2c-1}\sqrt{\omega}}\left[\frac{\sigma}{16i\sqrt{\omega}(a-1)(1-b)}\right]^{n-1}\left\{e^{(n/2-b)\pi i} H^{(1)}_{a-b}\left(\frac{\sigma}{2\sqrt{\omega}}\right)\times\right.$$

$$\left.\times H^{(2)}_{a+b-n}\left(\frac{\sigma}{2\sqrt{\omega}}\right)-(-1)^n e^{(b-n/2)\pi i} H^{(1)}_{a+b-n}\left(\frac{\sigma}{2\sqrt{\omega}}\right) H^{(2)}_{a-b}\left(\frac{\sigma}{2\sqrt{\omega}}\right)\right\}$$

[$n=1$ or 2; $\operatorname{Re} c$, $\operatorname{Re}\sigma > 0$; $|\arg\omega| < \pi$].

5. $$\int_0^\infty x^{c-1} e^{-\sigma x} {}_2F_2\left(\begin{matrix} a, b; \omega x \\ c, d \end{matrix}\right) dx = \frac{\Gamma(c)}{\sigma^c} {}_2F_1\left(a, b; d; \frac{\omega}{\sigma}\right)$$

[Re $c > 0$; Re $\sigma >$ Re ω, 0].

6. $$\int_0^\infty x^{c-1} e^{-\sigma x} {}_2F_2\left(\begin{matrix} -n, n+2d-1 \\ c, d; \omega x \end{matrix}\right) dx = n\Gamma(c) B(n, 2d-1) \sigma^{-c} C_n^{d-1/2}\left(1-\frac{2\omega}{\sigma}\right)$$

[Re c, Re $\sigma > 0$].

7. $$\int_0^\infty x^{c-1} e^{-\sigma x} {}_2F_2\left(\begin{matrix} (2d-1)/4, (2d+1)/4; \omega x \\ c, d \end{matrix}\right) dx =$$

$$= \frac{2^{d-1/2} \sigma^{(2d-1)/4-c}}{\sqrt{\pi}\, \omega^{(2d-1)/4}} \Gamma\left[\begin{matrix} c, d \\ d-1/2 \end{matrix}\right] Q_{d-3/2}\left(\sqrt{\frac{\sigma}{\omega}}\right)$$

[Re $c > 0$; Re $\sigma >$ Re ω, 0].

8. $$\int_0^\infty x^{c-1} e^{-\sigma x} {}_2F_2\left(\begin{matrix} -n, n+(1\pm 1)/2; \omega x \\ 1\pm 1/2, c \end{matrix}\right) dx = \frac{\Gamma(c)}{\sigma^c} \left\{\begin{matrix} [\sin(2n+1)\varphi]/[(2n+1)\varphi] \\ \cos 2n\varphi \end{matrix}\right\}$$

[$\varphi = \arcsin\sqrt{\omega/\sigma}$; Re c, Re $\sigma > 0$].

9. $$\int_0^\infty x^{c-1} e^{-\sigma x} {}_2F_2\left(\begin{matrix} -n, n+1; \omega x \\ 1, c \end{matrix}\right) dx = \sigma^{-c}\Gamma(c) P_n\left(1-\frac{2\omega}{\sigma}\right)$$

[Re c, Re $\sigma > 0$].

10. $$\int_0^\infty x^{c-1} e^{-2\omega x - \sigma\sqrt{x}} {}_2F_2\left(\begin{matrix} a, 2c-a; \omega x \\ c, c+1/2 \end{matrix}\right) dx =$$

$$= 2^{1-2c} \omega^{-c} \Gamma(2c) e^{\sigma^2/(16\omega)} D_{-a}\left(\frac{\sigma}{2\sqrt{\omega}}\right) D_{a-c}\left(\frac{\sigma}{2\sqrt{\omega}}\right)$$

[Re c, Re $\omega > 0$].

11. $$\int_0^\infty x^{c-1} e^{-\sigma x} {}_0F_2(c, d; -\omega x) dx = \Gamma(c)\Gamma(d) \sigma^{(d-1)/2-c} \omega^{(1-d)/2} J_{d-1}\left(2\sqrt{\frac{\omega}{\sigma}}\right)$$

[Re c, Re $\sigma > 0$].

12. $$\int_0^\infty x^{c-1} e^{-\sigma x} {}_0F_2\left(1\pm\frac{1}{2}, c; -\omega x\right) dx = \frac{\Gamma(c)}{\sigma^c} \left\{\begin{matrix} [\sigma/(4\omega)]^{1/2} \sin\left(2\sqrt{\sigma/\omega}\right) \\ \cos\left(2\sqrt{\sigma/\omega}\right) \end{matrix}\right\}$$

[Re c, Re $\sigma > 0$].

13. $$\int_0^\infty x^{c-3/2} e^{-\sigma\sqrt{x}} {}_0F_2(c, 2c-1; -\omega x) dx =$$

$$= \frac{2\omega^{3/2-c}}{\sigma^2} \Gamma(c)\Gamma(2c-1) e^{-2\omega/\sigma^2} I_{\nu-1}\left(\frac{2\omega}{\sigma^2}\right)$$

[Re $c > 1/2$; Re $\sigma > 0$].

14. $\int_0^\infty x^{c-1} e^{-\sigma\sqrt{x}} {}_0F_3\left(c, c+\frac{1}{2}, d; -\omega x\right) dx =$

$= 2^{2-d}\,\Gamma(d)\,\Gamma(2c)\,\sigma^{d-2c-1}\omega^{(1-d)/2}\,J_{d-1}\left(\frac{4\sqrt{\omega}}{\sigma}\right)$ [Re c, Re $\sigma > 0$].

15. $\int_0^\infty x^{c-1}e^{-\sigma x}\,{}_1F_2\left(\frac{d}{2}; c, d; \omega x\right) dx =$

$= 2^{d-1}\,\Gamma(c)\,\Gamma\left(\frac{d+1}{2}\right)\sigma^{(d-1)/2-c}\,\omega^{(1-d)/2}\,e^{\omega/(2\sigma)}\,I_{d-1/2}\left(\frac{\omega}{2\sigma}\right)$

[Re c, Re $\sigma > 0$].

16. $\int_0^\infty x^{c-1}e^{-\sigma x}\,{}_1F_2(1; c, d; \omega x)\,dx = (d-1)\,\Gamma(c)\,\sigma^{d-c-1}\,\omega^{1-d}\,e^{\omega/\sigma}\gamma\left(d-1, \frac{\omega}{\sigma}\right)$

[Re c, Re $\sigma > 0$; Re $d > 1$].

17. $\int_0^\infty x^{c-1}e^{-\sigma x}\,{}_1F_2(-n; c, d; \omega x)\,dx = \frac{n!\,\Gamma(c)}{(d)_n\,\sigma^c}\,L_n^{d-1}\left(\frac{\omega}{\sigma}\right)$ [Re c, Re $\sigma > 0$].

18. $\int_0^\infty x^{c-1}e^{-\sigma\sqrt{x}}\,{}_1F_2(a; c, c+1/2; -\omega x)\,dx = 2\Gamma(2c)\,\sigma^{2a-2c}(\sigma^2+4\omega)^{-a}$

[Re c, Re $(\sigma \pm 2\sqrt{\omega}) > 0$].

19. $\int_0^\infty x^{c-1}e^{-\sigma\sqrt[3]{x}}\,{}_1F_3\left(\begin{matrix}a; & & -\omega x\\ c, & c+1/3, & c+2/3\end{matrix}\right) dx = 3\Gamma(3c)\,\omega^{a-c}(\sigma^3+27\omega)^{-a}$

[Re c, Re $(2\sigma - 3\omega^{1/3}) > 0$].

20. $\int_0^\infty x^{c-1}e^{-\sigma\sqrt{x}}\,{}_1F_4\left(\begin{matrix}1; \pm\omega x\\ 3/2, c, c+1/2, d\end{matrix}\right) dx =$

$= 2^{1/2-a}\sqrt{\pi}\,\sigma^{a-2c-1/2}\,\omega^{(1-2a)/4}\,\Gamma(a)\,\Gamma(2c)\begin{Bmatrix}\mathbf{L}_{a-3/2}(4\sqrt{\omega}/\sigma)\\ \mathbf{H}_{a-3/2}(4\sqrt{\omega}/\sigma)\end{Bmatrix}$

[Re c, Re $\sigma > 0$].

2.22.4. Integrals containing Bessel functions and ${}_pF_q((a_p); (b_q); cx)$ (see also 2.23.3).

1. $\int_0^\infty x^{\alpha-1}J_\nu(\sigma\sqrt{x})\,{}_pF_q((a_p); (b_q); -\omega x)\,dx =$

$= \frac{2^\nu\omega^{\nu/2-\alpha}}{\sigma^\nu}\,\Gamma\left[\begin{matrix}\alpha-\nu/2, (a_p)+\nu/2-\alpha, (b_q)\\ \nu+1, (b_q)+\nu/2-\alpha, (a_p)\end{matrix}\right]\times$

$\times\,{}_pF_{q+2}\left(\begin{matrix}(a_p)+\nu/2-\alpha; & 4\sigma^{-2}\omega\\ 1+\nu/2-\alpha, \nu+1, (b_q)+\nu/2-\alpha\end{matrix}\right)+$

$+\left(\frac{2}{\sigma}\right)^{2\alpha}\Gamma\left[\begin{matrix}\nu/2-\alpha\\ \alpha+\nu/2+1\end{matrix}\right]{}_pF_{q+2}\left(\begin{matrix}(a_p); & 4\sigma^{-2}\omega\\ 1+\alpha-\nu/2, 1+\alpha+\nu/2, (b_q)\end{matrix}\right)$

[$p=q+1$; σ, Re $(2\alpha+\nu) > 0$, Re $\alpha < 3/4 + \min_{1\leqslant j\leqslant p}$ Re a_j; $|\arg\omega| < \pi$], or [$p=q$; Re $\omega > 0$], or [$p=q-1$; Re $\alpha <$

$< 3/4 + \min_{1\leqslant j\leqslant p}$ Re a_j; $\omega > 0$; Re$\left(2\alpha + \sum_{j=1}^p a_j - \sum_{k=1}^q b_k\right) < 1$].

2. $\int_0^\infty x^{(c+d)/2-1} K_{c-d}(\sigma\sqrt{x})\, {}_3F_2\left(\begin{matrix}1,\ a,\ b;\ -\omega x\\ c,\ d\end{matrix}\right) dx =$

$$=\frac{2^{c+d-1}\sigma^{a+b-c-d}}{\omega^{(a+b)/2}}\Gamma(c)\,\Gamma(d)\, S_{1-a-b,\ a-b}\left(\frac{\sigma}{\sqrt{\omega}}\right)$$

[Re c, Re d, Re $\sigma > 0$; $|\arg\omega| < \pi$].

3. $\int_0^\infty x^{c-\nu/2-1} J_\nu(\sigma\sqrt{x})\, {}_2F_2(a,\ b;\ c,\ d;\ -\omega x)\, dx =$

$$=\frac{2^\nu \omega^{\nu-c}\,\Gamma(c)\,\Gamma(d)}{\sigma^\nu}[A(a,\ b)+A(b,\ a)],$$

$$A(a,\ b)=\Gamma\left[\begin{matrix}b-a,\\ d-a,\ b,\ a-c+\nu+1\end{matrix}\right] {}_2F_2\left(\begin{matrix}a,\ a-d+1;\ -\sigma^2/(4\omega)\\ a-b+1,\ a-c+\nu+1\end{matrix}\right)$$

[σ, Re c, Re $\omega > 0$; Re $(2c-2a-\nu)$, Re $(2c-2b-\nu) < 3/2$].

4. $\int_0^\infty x^{(c-1)/2} J_{c-1}(\sigma\sqrt{x})\, {}_2F_2(a,\ b;\ c,\ d;\ -\omega x)\, dx =$

$$=\frac{(\sigma/2)^{a+b-c-4}}{\omega^{(a+b-3)/2}}\Gamma\left[\begin{matrix}c,\ d\\ a,\ b\end{matrix}\right] e^{-\sigma^2/(8\omega)} W_{(a+b+3)/2-c,\ (a-b)/2}\left(\frac{\sigma^2}{4\omega}\right)$$

[σ, Re c, Re $\omega > 0$; Re $(c-2a)$, Re $(c-2b) < 1/2$].

5. $\int_0^\infty x^{(c+d)/2-1} J_{c-d}(\sigma\sqrt{x})\, {}_2F_2(a,\ b;\ c,\ d;\ -\omega x)\, dx =$

$$=\frac{(\sigma/2)^{a+b-c-d-1}}{\pi\omega^{(a+b-1)/2}}\Gamma\left[\begin{matrix}c,\ d\\ a,\ b\end{matrix}\right] e^{-\sigma^2/(8\omega)}\Big\{\Gamma(a)\,\Gamma(b-a)\sin(c-a)\pi\times$$

$$\times M_{(a+b-1)/2,\ (a-b)/2}\left(\frac{\sigma^2}{4\omega}\right)+\Gamma(b)\,\Gamma(a-b)\sin(c-b)\pi\, M_{(a+b-1)/2,\ (b-a)/2}\left(\frac{\sigma^2}{4\omega}\right)\Big\}$$

[Re c, σ, Re $\omega > 0$; Re $(c+d-2a)$, Re $(c+d-2b) < 3/2$].

6. $\int_0^\infty x^{(c-1)/2} K_{c-1}(\sigma\sqrt{x})\, {}_0F_2(1,\ c;\ \omega x)\, dx = \frac{2^c\Gamma(c)}{\sigma^{c+1}} e^{4\omega/\sigma^2}$ [Re c, Re $\sigma > 0$].

7. $\int_0^\infty x K_0(\sigma\sqrt{x})\, {}_0F_2(2,\ 2;\ \omega x)\, dx = \frac{2}{\sigma^2\omega}\left[\mathrm{Ei}\left(\frac{4\omega}{\sigma^2}\right)-\ln\frac{4\omega}{\sigma^2}-\mathbf{C}\right]$ [Re $c > 0$].

8. $\int_0^\infty x^{c-1} J_{c-1}(\sigma\sqrt{x})\, K_{c-1}(\tau\sqrt{x})\, {}_0F_2(c,\ 2c-1;\ -\omega x)\, dx =$

$$=\frac{2^{2-c}\Gamma(c)\,\Gamma(2c-1)}{(\sigma^2+\tau^2)\,\omega^{c-1}} e^{4\omega(\sigma^2-\tau^2)/(\sigma^2+\tau^2)^2} J_{c-1}\left(\frac{8\sigma\tau\omega}{(\sigma^2+\tau^2)^2}\right)$$ [Re $c > 1/2$; Re $\tau > |\mathrm{Im}\,\sigma|$].

9. $\int_0^\infty x^{d-1} I_\nu(\sigma\sqrt{x})\, K_\nu(\tau\sqrt{x})\, {}_1F_2\left(\frac{c}{2};\ c,\ d;\ -\omega x\right) dx =$

$$=\frac{2^{2d-1}(\sigma\tau)^\nu}{(\sigma^2+\tau^2+2\omega)^{d+\nu}}\Gamma\left[\begin{matrix}d;\ d+\nu\\ \nu+1\end{matrix}\right]\times$$

$$\times F_4\left(\frac{d+\nu}{2},\ \frac{d+\nu+1}{2};\ \nu+1,\ \frac{c+1}{2};\ \frac{4\sigma^2\tau^2}{(\sigma^2+\tau^2+2\omega)^2},\ \frac{4\omega}{(\sigma^2+\tau^2+2\omega)^2}\right)$$

[Re d, Re $(d+\nu) > 0$; Re $\tau > |\mathrm{Re}\,\sigma| + |\mathrm{Im}\sqrt{\omega}|$].

10. $\int\limits_0^\infty x^{\alpha-1}K_\nu(\sigma\sqrt{x})\,K_\nu(\tau\sqrt{x})\,{}_1F_2\left(\frac{c}{2};c,d;-\omega x\right)dx=2^{2d-2}\Gamma(d)\,[A_\nu+A_{-\nu}]$,

$$A_\nu=\frac{(\sigma\tau)^\nu\Gamma(-\nu)\,\Gamma(d+\nu)}{(\sigma^2+\tau^2+2\omega)^{d+\nu}}\times$$
$$\times F_4\left(\frac{d+\nu}{2},\ \frac{d+\nu+1}{2};\ \nu+1,\ \frac{c+1}{2};\ \frac{4\sigma^2\tau^2}{(\sigma^2+\tau^2+2\omega)^2},\ \frac{4\omega}{(\sigma^2+\tau^2+2\omega)^2}\right)$$

[Re $d>|\mathrm{Re}\,\nu|$; Re $(\tau+\sigma)>|\mathrm{Im}\,\sqrt{\omega}|$].

2.22.5. Integrals containing various special functions and ${}_pF_q((a_p);(b_q);cx)$ (see also 2.23.4–6).

1. $\int\limits_0^a x^{\alpha-1}(a-x)^\beta P_n^{(\gamma,\beta)}\left(1-\frac{2x}{a}\right){}_pF_q((a_p);(b_q);\omega x)\,dx=$

$$=\frac{a^{\alpha+\beta}(1-\alpha+\gamma)_n}{n!}B(\alpha,\beta+n+1)\,{}_{p+2}F_{q+2}\left(\begin{matrix}(a_p),\ \alpha,\ \alpha-\gamma;\ a\omega\\(b_q),\ \alpha+\beta+n+1,\ \alpha-\gamma-n\end{matrix}\right)$$

[a, Re $\alpha>0$; Re $\beta>-1$].

2. $\int\limits_0^a x^{1/4}P_{-1/2}\left(\sqrt{\frac{a}{x}}\right){}_3F_2\left(1,\frac{7}{4},\frac{9}{4};\frac{3}{2},\frac{3}{2};-\omega x\right)dx=\frac{8\sqrt{2}\,a^{5/4}}{15\,(1+\omega a)}$ [$a>0$].

3. $\int\limits_0^a x^{b_1+b_2-\alpha-1}(a-x)^{\alpha+a_1-b_1-b_2-1}\,{}_2F_1\left(\alpha-b_1,\alpha-b_2;\alpha+a_1-b_1-b_2;1-\frac{a}{x}\right)\times$

$$\times{}_3F_2((a_3);(b_2);\omega x)\,dx=a^{a_1-1}\,\Gamma\left[\begin{matrix}b_1,\ b_2,\ \alpha+a_1-b_1-b_2\\\alpha,\ a_1\end{matrix}\right]{}_2F_1(a_2,\ a_3;\ \alpha;\ a\omega)$$

[Re b_1, Re b_2, Re $(\alpha+a_1-b_1-b_2)>0$; $|\arg(1-a\omega)|<\pi$].

4. $\int\limits_0^a x^{b_1+b_2-\alpha-1}(a-x)^{\alpha+a_1-b_1-b_2-1}\,{}_2F_1\left(\alpha-b_1,\ \alpha-b_2;\ \alpha+a_2-b_1-b_2;1-\frac{a}{x}\right)\times$

$$\times{}_2F_2(a_1,\ a_2;\ b_1,b_2;\omega x)\,dx=a^{a_1-1}\,\Gamma\left[\begin{matrix}b_1,\ b_2,\ \alpha+a_2-b_1-b_2\\\alpha,\ a_1\end{matrix}\right]{}_1F_1(a_2;\ \alpha;\ a\omega)$$

[Re b_1, Re b_2, Re $(\alpha+a_2-b_1-b_2)>0$].

5. $\int\limits_0^a x^{\beta+n}(a-x)^\beta P_n^{(\alpha,\beta)}\left(1-\frac{2x}{a}\right){}_1F_2\left(\beta-\alpha+1;n+\beta-\alpha+1,\ n+\beta+\frac{3}{2};-\omega x\right)dx=$

$$=\frac{(-1)^n\sqrt{\pi}}{n!}(\beta-\alpha+1)_n\,\Gamma\left(n+\beta+\frac{3}{2}\right)a^{\beta+1/2}\omega^{-n-\beta-1/2}J^2_{n+\beta+1}(a\omega)$$

[$a>0$; Re $\beta>-1$].

6. $\int\limits_0^\infty x^{b/2-1}e^{-\sigma x/2}W_{1+b/2-a,\,(1-b)/2}(\sigma x)\,{}_1F_2(a;\ 1,\ b;\ \omega x)\,dx=\Gamma\left[\begin{matrix}b\\a\end{matrix}\right]\sigma^{-b/2}e^{\omega/\sigma}$

[ω, Re $\sigma>0$].

7. $\int\limits_0^\infty x^{2b-2}\,{}_1F_2^2(c-b;\ b,\ c;\ -\omega x)\,dx=\frac{2^{2c-4b}\omega^{1-2b}}{\sqrt{\pi}}\,\Gamma\left[\begin{matrix}2b-1,\ c,\ c,\ 1/2-2b+c\\c-2b+1,\ 2c-2b\end{matrix}\right]$

[Re $b>1/2$; Re $(c-2b)>-1/2$].

8. $\int\limits_0^\infty x^{b+b'-2}\,{}_1F_2(a;\ b,\ 2a;\ -\omega x)\,{}_1F_2(a';\ b',\ 2a';\ -\omega x)\,dx=0$

[Re $(b+b')>1$; $b-b'=\pm1,\ \pm3,\ \pm5,\ \ldots$; $\omega>0$].

9. $$=\frac{\omega^{1-b-d}}{2^{2b-1}\sqrt{\pi}}\Gamma\left[\begin{matrix}b,\ 2b-1,\ a+a'-2b+1,\ a+1/2,\ a'+1/2\\ a-b+1,\ a'-b+1,\ a+a'-b+1/2\end{matrix}\right]$$

[$b'=b$; $\omega>0$; $\operatorname{Re}(a+a'-2b)>-1$; $\operatorname{Re} b>1/2$].

2.22.6. Integrals with respect to the parameters containing ${}_pF_q((a_p);(b_q);x)$.

1. $$\int_{-\infty}^{\infty}\frac{1}{\Gamma[a+x,\ a-x]}\,{}_2F_2\left(\begin{matrix}a/2,\ (a+1)/2;\ -\omega\\ (a+x)/2,\ (a+x+1)/2\end{matrix}\right){}_2F_2\left(\begin{matrix}a/2,\ (a+1)/2;\ \omega\\ (a-x)/2,\ (a-x+1)/2\end{matrix}\right)dx=$$
$$=\frac{2^{2a-2}}{\Gamma(2a-1)}\,{}_1F_2\left(\begin{matrix}a/2+3/4;\ 4\omega^2\\ a/2-1/4,\ a+1/2\end{matrix}\right)$$ [$\operatorname{Re} a>1/2$].

2. $$\int_{-\infty}^{\infty}\frac{1}{\Gamma[a+x,\ a-x,\ b+x,\ b-x]}\,{}_2F_2\left(\begin{matrix}a-1/2,\ a;\ -\omega\\ (a+x)/2,\ (a+x+1)/2\end{matrix}\right)\times$$
$$\times{}_2F_2\left(\begin{matrix}b-1/2,\ b;\ \omega\\ (b+x)/2,\ (b+x+1)/2\end{matrix}\right)dx=\frac{2^{2a+2b-4}}{\sqrt{\pi}}\Gamma\left[\begin{matrix}a+b-3/2\\ 2a-1,\ 2b-1,\ a+b-1\end{matrix}\right]\times$$
$$\times{}_5F_6\left(\begin{matrix}\frac{2a+2b-3}{6},\ \frac{2a+2b-1}{6},\ \frac{2a+2b+1}{6},\ \frac{2a+2b-3}{4},\ \frac{2a+2b-1}{4},\ -\frac{27\omega^2}{4}\\ \frac{a+b-1}{4},\ \frac{a+b}{4},\ \frac{a+b+1}{4},\ \frac{a+b+2}{4},\ a+b-\frac{3}{2},\ a+b-1\end{matrix}\right)$$

[$\operatorname{Re}(a+b)>3/2$].

3. $$\int_{-\infty}^{\infty}\frac{1}{\Gamma[a+x,\ b-x,\ c+x,\ d-x]}\begin{Bmatrix}\sin x\pi\\ \cos x\pi\end{Bmatrix}{}_2F_2\left(\begin{matrix}(a+b)/2,\ a';\ \omega\\ a+x,\ b-x\end{matrix}\right)\times$$
$$\times{}_2F_2\left(\begin{matrix}(c+d)/2,\ c';\ \omega\\ c+x,\ d-x\end{matrix}\right)dx=\frac{1}{2\Gamma[(a+b)/2,\ (c+d)/2,\ a+d-1]}\times$$
$$\times\Gamma\begin{Bmatrix}\sin[(b-a)\pi/2]\\ \cos[(b-a)\pi/2]\end{Bmatrix}{}_1F_1\left(\begin{matrix}a'+c';\ \omega\\ a+d-1\end{matrix}\right)$$ [$a+d=b+c$; $\operatorname{Re}(a+b+c+d)>2$].

2.22.7. Hypergeometric functions of two variables.

1. $$\int_0^{\infty}x^{\alpha-1}e^{-px}\Phi_1(a,\ b;\ c;\ \omega,\ zx)\,dx=\Gamma(\alpha)\,p^{-\alpha}F_1\left(a,\ b,\ \alpha;\ c;\ \omega,\ \frac{z}{p}\right)$$

[$\operatorname{Re} p$, $\operatorname{Re}(p-z)$, $\operatorname{Re}\alpha>0$].

2. $$\int_0^{\infty}x^{\alpha-1}e^{-px}\Phi_2(b,\ b';\ c;\ \omega x,\ z)\,dx=\Gamma(\alpha)\,p^{-\alpha}\Xi_1\left(b,\ b',\ \alpha;\ c;\ \frac{\omega}{p},\ z\right)$$

[$\operatorname{Re}\alpha$, $\operatorname{Re} p$, $\operatorname{Re}(p-\omega)>0$].

3. $$\int_0^{\infty}x^{\alpha-1}e^{-px}\Phi_2(b,\ b';\ c;\ \omega x,\ zx)\,dx=\Gamma(\alpha)\,p^{-\alpha}F_1\left(\alpha,\ b,\ b';\ c;\ \frac{\omega}{p},\ \frac{z}{p}\right)$$

[$\operatorname{Re}\alpha$, $\operatorname{Re} p$, $\operatorname{Re}(p-\omega)$, $\operatorname{Re}(p-z)>0$].

4. $$\int_0^{\infty}x^{c-1}e^{-px}\Phi_2(b,\ b';\ c;\ \omega x,\ zx)\,dx=\Gamma(c)\,p^{b+b'-c}(p-\omega)^{-b}(p-z)^{-b'}$$

[$\operatorname{Re} c$, $\operatorname{Re} p$, $\operatorname{Re}(p-\omega)$, $\operatorname{Re}(p-z)>0$].

5. $$\int_0^{\infty}x^{\alpha-1}e^{-px}\Phi_3(b;\ c;\ \omega,\ zx)\,dx=\Gamma(\alpha)\,p^{-\alpha}\Phi_2\left(b,\ \alpha;\ c;\ \omega,\ \frac{z}{p}\right)$$

[$\operatorname{Re}\alpha$, $\operatorname{Re} p$, $\operatorname{Re}(p-z)>0$].

6. $\int_0^\infty x^{\alpha-1}e^{-px}\Phi_3(b;\ c;\ \omega x,\ z)\,dx=\Gamma(\alpha)\,p^{-\alpha}\,\Xi_2\left(\alpha,\ b;\ c;\ \frac{\omega}{p},\ z\right)$

[Re α, Re p, Re $(p-\omega)>0$].

7. $\int_0^\infty x^{\alpha-1}e^{-px}\Phi_3(b;\ c;\ \omega x,\ zx)\,dx=\Gamma(\alpha)\,p^{-\alpha}\Phi_1\left(\alpha,\ b;\ c;\ \frac{\omega}{p},\ \frac{z}{p}\right)$

[Re α, Re p, Re $(p-\omega)>0$].

8. $\int_0^\infty x^{c-1}e^{-px}\Phi_3(b;\ c;\ \omega x,\ zx)\,dx=\Gamma(c)\,p^{b-c}(p-\omega)^{-b}e^{z/p}$

[Re c, Re p, Re $(p-\omega)>0$].

9. $\int_0^\infty x^{\alpha-1}e^{-p\sqrt{x}}\Phi_3(b;\ c;\ \omega,\ zx)\,dx=2\Gamma(2\alpha)\,p^{-2\alpha}\Xi_1\left(\alpha,\ b,\ \alpha+\frac{1}{2};\ c;\ \frac{4z}{p^2},\ \omega\right)$

[Re $\alpha>3/4$; Re $p>2\,|\,\mathrm{Re}\sqrt{z}\,|$].

10. $\int_0^\infty x^{\alpha-1}e^{-px}\Psi_1(a,\ b;\ c,\ c';\ \omega,\ zx)\,dx=\Gamma(\alpha)\,p^{-\alpha}F_2\left(a,\ b,\ \alpha;\ c,\ c';\ \omega,\ \frac{z}{p}\right)$

[Re α, Re p, Re $(p-z)>0$].

11. $\int_0^\infty x^{\alpha-1}e^{-px}\Psi_2(a;\ c,\ c';\ \omega x,\ z)\,dx=\Gamma(\alpha)\,p^{-\alpha}\Psi_1\left(a,\ \alpha;\ c,\ c';\ \frac{\omega}{p},\ z\right)$

[Re α, Re p, Re $(p-\omega)>0$].

12. $\int_0^\infty x^{\alpha-1}e^{-px}\Psi_2(a;\ c,\ c';\ \omega x,\ zx)\,dx=\Gamma(\alpha)\,p^{-\alpha}F_4\left(\alpha,\ a;\ c,\ c';\ \frac{\omega}{p},\ \frac{z}{p}\right)$

[Reα, Rep, Re $(p-\omega)$, Re $(p-z)>0$].

13. $\int_0^\infty x^{\alpha-1}e^{-px}\Xi_1(a,\ a',\ b;\ c;\ \omega,\ zx)\,dx=\Gamma(\alpha)\,p^{-\alpha}F_3\left(a,\ a',\ b,\ \alpha;\ c;\ \omega,\ \frac{z}{p}\right)$

[Reα, Rep, Re $(p-z)>0$].

14. $\int_0^\infty x^{\alpha-1}e^{-px}\Xi_2(a,\ b;\ c;\ \omega,\ zx)\,dx=\Gamma(\alpha)\,p^{-\alpha}\Xi_1\left(a,\ \alpha,\ b;\ c;\ \omega,\ \frac{z}{p}\right)$

[Reα, Rep, Re $(p-z)>0$].

15. $\int_0^\infty x^{\alpha-1}e^{-p\sqrt{x}}\Xi_2(a,\ b;\ c;\ \omega,\ zx)\,dx=2\Gamma(2\alpha)\,p^{-2\alpha}F_3\left(a,\ \alpha,\ b,\ \alpha+\frac{1}{2};\ c;\ \omega,\ \frac{4z}{p^2}\right)$

[Re$\alpha>3/4$; Re$p>2\,|\,\mathrm{Re}\sqrt{z}\,|$].

2.23 THE MacROBERT E-FUNCTION $E(p;a_r:q;b_s:x)$

The relation

$$E(p;\ a_r:q;\ b_s:cx)=\Gamma\begin{bmatrix}(a_p)\\(b_q)\end{bmatrix}{}_pF_q\left((a_p);\ (b_q);\ -\frac{1}{cx}\right)\qquad [q\geqslant p-1]$$

enables one to obtain integrals containing the E-function from the formulae of section 2.22 as well. But here, the condition $q\geqslant p-1$ for the generalized hypergeometric function is replaced by the conditions $q\leqslant p+1$,

$|\arg c| \leqslant (p-q+1)\pi/2$ for the E-function. Integrals containing the E-function can also be obtained from section 2.24, since it is a particular case of the Meijer G-function. The following notation is used for the E-function:

$$E(p;\ a_r:q;\ b_s:x) \equiv E((a_p);\ (b_q);\ x) \equiv E\left(\begin{matrix}(a_p);\ x\\ (b_q)\end{matrix}\right) \equiv E\left(\begin{matrix}(a_p)\\ (b_q);\ x\end{matrix}\right).$$

2.23.1. Integrals of $A(x)\,E((a_p);\ (b_q);\ \varphi(x))$.

1. $$\int_0^\infty x^{\alpha-1} E((a_p);\ (b_q);\ cx)\,dx = \Gamma\left[\begin{matrix}\alpha+(a_p),\ -\alpha\\ \alpha+(b_q)\end{matrix}\right] c^{-\alpha}$$

$\left[0 \leqslant q < p+1;\ -\min\limits_{1\leqslant k\leqslant p} \operatorname{Re} a_k < \operatorname{Re}\alpha < 0\right]$ or

$$\left[q=p+1;\ -1/4+\operatorname{Re}\left(\sum_{k=1}^{p} a_k - \sum_{l=1}^{p+1} b_l\right)\Big/2,\ -\min_{1\leqslant k\leqslant p}\operatorname{Re} a_k < \operatorname{Re}\alpha < 0\right].$$

2. $$\int_a^\infty x^{\alpha-1}(x^{1/m}-a^{1/m})^{\beta-1} E((a_p);\ (b_q);\ cx)\,dx =$$
$$= \Gamma(\beta)\, m^{1-\beta} a^{\alpha+(\beta-1)/m} E\left(\begin{matrix}(a_p),\ \Delta(m,\ 1-m\alpha-\beta);\ ac\\ (b_q),\ \Delta(m,\ 1-m\alpha)\end{matrix}\right)$$

$[a,\ \operatorname{Re}\beta > 0;\ \operatorname{Re}(m\alpha+\beta) < 1;\ m=1,\ 2,\ \ldots]$.

3. $$\int_0^a x^{\alpha-1}(a-x)^{\beta-1} E((a_p);\ (b_q);\ cx^{-k}(a-x)^{-l})\,dx =$$
$$= \frac{\sqrt{2\pi}\, k^{\alpha-1/2}\, l^{\beta-1/2}}{(k+l)^{\alpha+\beta-1/2}}\, a^{\alpha+\beta-1}\, E\left(\begin{matrix}(a_p),\ \Delta(k,\ \alpha),\ \Delta(l,\ \beta);\ k^{-k} l^{-l}(k+l)^{k+l} a^{-k-l} c\\ (b_q),\ \Delta(k+l,\ \alpha+\beta)\end{matrix}\right)$$

$[a,\ \operatorname{Re}\alpha,\ \operatorname{Re}\beta > 0;\ k,\ l=1,\ 2,\ \ldots]$.

4. $$\int_0^\infty x^{-\rho-1/2}(a+bx+cx^2)^\rho E\left((a_p);\ (b_q);\ \left(\frac{a+bx+cx^2}{x}\right)^n\right) dx =$$
$$= \sqrt{\frac{\pi}{nc}}\,(b+2\sqrt{ac})^{\rho+1/2} E\left(\begin{matrix}(a_p),\ \Delta(n,\ -\rho-1/2);\ (b+2\sqrt{ac})^n\\ (b_q),\ \Delta(n,\ -\rho)\end{matrix}\right)$$

$[a,\ c > 0;\ \operatorname{Re}\rho < -1/2]$.

5. $$\int_0^\infty x^{1/2-\rho}(a+bx+cx^2)^\rho E\left((a_p);\ (b_q);\ \left(\frac{a+bx+cx^2}{x}\right)^n\right) dx =$$
$$= \frac{\sqrt{\pi}}{2(nc)^{3/2}}(b+2\sqrt{ac})^{\rho+3/2} E\left(\begin{matrix}(a_p),\ \Delta(n,\ -\rho-3/2);\ (b+2\sqrt{ac})^n\\ (b_q),\ \Delta(n,\ -\rho)\end{matrix}\right) +$$
$$+ \frac{1}{c}\sqrt{\frac{\pi a}{n}}\,(b+2\sqrt{ac})^{\rho+1/2} E\left(\begin{matrix}(a_p),\ \Delta(n,\ -\rho-1/2);\ (b+2\sqrt{ac})^n\\ (b_q),\ \Delta(n,\ -\rho)\end{matrix}\right)$$

$[q < p+1;\ \operatorname{Re} a \geqslant 0;\ \operatorname{Re} c > 0;\ \operatorname{Re}\rho < -3/2]$.

2.23.2. Integrals containing exponential or hyperbolic functions and $E((a_p);\ (b_q);\ cx)$.

1. $$\int_0^\infty x^{\alpha-1} e^{-sx} E((a_p);\ (b_q);\ cx)\,dx =$$
$$= \pi \operatorname{cosec}\alpha\pi \left[s^{-\alpha} E\left(\begin{matrix}(a_p);\ e^{\pm\pi i}c/s\\ (b_q),\ 1-\alpha\end{matrix}\right) - c^{-\alpha} E\left(\begin{matrix}(a_p)+\alpha;\ e^{\pm\pi i}c/s\\ (b_q)+\alpha,\ 1+\alpha\end{matrix}\right)\right]$$

$\left[q < p+1;\ \operatorname{Re}s > 0;\ |\arg c| < (p-q+1)\pi/2;\ \operatorname{Re}\alpha > -\min\limits_{1\leqslant k\leqslant p}\operatorname{Re} a_k\right]$

or $\left[q=p+1;\ c,\ \operatorname{Re} s>0;\ \operatorname{Re}\alpha>-\min_{1\leqslant k\leqslant p}\operatorname{Re}a_k,\ -\frac{1}{4}+\operatorname{Re}\left(\sum_{k=1}^{p}a_k-\sum_{l=1}^{p+1}b_l\right)\Big/2\right]$.

2. $$\int_0^\infty x^{\alpha-1}e^{-sx^{-1/m}}E((a_p);\ (b_q);\ cx)\,dx=\frac{m^{-m\alpha+1/2}s^{m\alpha}}{(2\pi)^{(m-1)/2}}E\left(\begin{matrix}(a_p),\ \Delta(m,\ -m\alpha)\\(b_q);\ c(s/m)^m\end{matrix}\right)$$

$[q\leqslant p+1;\ \operatorname{Re} s>0;\ |\arg c|\leqslant(p-q+1)\pi/2;\ \operatorname{Re}\alpha<0]$.

3. $$\int_0^\infty \operatorname{ch}\frac{3x}{2}(a+b\operatorname{ch}x)^\theta\,E((a_p);\ (b_q);\ (a+b\operatorname{ch}x)^n)\,dx=$$

$$=\sqrt{\frac{\pi}{2n^3b^3}}(a+b)^{\theta+3/2}E\left(\begin{matrix}(a_p),\ \Delta(n,\ -\theta-3/2);\ (a+b)^n\\(b_q),\ \Delta(n,\ -\theta)\end{matrix}\right)+$$

$$+\sqrt{\frac{\pi}{2}}\frac{1}{nb}(a+b)^{\theta+1/2}E\left(\begin{matrix}(a_p),\ \Delta(n,\ -\theta-1/2);\ (a+b)^n\\(b_q),\ \Delta(n,\ -\theta)\end{matrix}\right)$$

$[q<p+1;\ \operatorname{Re} b>0;\ \operatorname{Re}\theta<-3/2]$.

2.23.3. Integrals containing Bessel functions and $E((a_p);(b_q);cx)$.

1. $$\int_0^\infty x^{\alpha-1}J_\nu(bx^{-1/(2m)})\,E((a_p);\ (b_q);\ cx)\,dx=$$

$$=(2\pi)^{-m}\left(\frac{b}{2m}\right)^{2m\alpha}\left[\exp\left(-\frac{2m\alpha+\nu+1}{2}\pi i\right)\times\right.$$

$$\times E\left(\begin{matrix}(a_p),\ \Delta(m,\ \nu/2-m\alpha),\ \Delta(m,\ -\nu/2-m\alpha);\ e^{-m\pi i}(2m)^{-2m}b^{2m}c\\(b_q)\end{matrix}\right)+$$

$$\left.+\exp\left(\frac{2m\alpha+\nu+1}{2}\pi i\right)E\left(\begin{matrix}(a_p),\Delta(m,\nu/2-m\alpha),\Delta(m,-\nu/2-m\alpha);\ e^{m\pi i}(2m)^{-2m}b^{2m}c\\(b_q)\end{matrix}\right)\right]$$

$[q<p+1;\ b>0;\ |\arg c|<(p-q+1)\pi/2;\ -3/(4m)-\min_{1\leqslant k\leqslant p}\operatorname{Re}a_k<\operatorname{Re}\alpha<\operatorname{Re}\nu/(2m)]$ or

$\left[q=p+1;\ b,\ c>0;\ -3/(4m)-\min_{1\leqslant k\leqslant p}\operatorname{Re}a_k,\ \operatorname{Re}\left(\sum_{k=1}^{p}a_k-\sum_{l=1}^{p+1}b_l\right)\Big/2-\right.$

$\left.-(m+1)/(4m)<\operatorname{Re}\alpha<\operatorname{Re}\nu/(2m)\right]$.

2. $$\int_0^\infty x^{\alpha-1}K_\nu(bx^{-1/(2m)})\,E((a_p);\ (b_q);\ cx)\,dx=$$

$$=\frac{\pi^{1-m}}{2m(2\alpha+1)}\left(\frac{b}{m}\right)^{2m\alpha}E\left(\begin{matrix}(a_p),\ \Delta(m,\ \nu/2-m\alpha),\ \Delta(m,\ -\nu/2-m\alpha)\\(b_q);\qquad (2m)^{-2m}b^{2m}c\end{matrix}\right)$$

$[q\leqslant p+1;\ \operatorname{Re} b>0;\ |\arg c|\leqslant(p-q+1)\pi/2;\ \operatorname{Re}\alpha<-|\operatorname{Re}\nu|/(2m)]$.

3. $$\int_0^\infty x^{\alpha-1}e^{-bx^{-1/m}}K_\nu(bx^{-1/m})\,E((a_p);\ (b_q);\ cx)dx=$$

$$=2^{m\alpha+(1-m)/2}\pi^{1-m/2}\left(\frac{b}{m}\right)^{m\alpha}E\left(\begin{matrix}(a_p),\ \Delta(m,\ \nu-m\alpha),\ \Delta(m,\ -\nu-m\alpha)\\(b_q),\ \Delta(m,\ 1/2-m\alpha);\ (2b/m)^m c\end{matrix}\right)$$

$[q\leqslant p+1;\ \operatorname{Re} b>0;\ |\arg c|\leqslant(p-q+1)\pi/2;\ \operatorname{Re}\alpha<-|\operatorname{Re}\nu|/m]$.

4. $\displaystyle\int_0^\infty x^{\alpha-1} K_\mu(bx^{-(2m)^{-1}}) K_\nu(bx^{-(2m)^{-1}}) E((a_p);\ (b_q);\ cx)\,dx = \frac{\pi^{3/2-m} b^{2m\alpha}}{2^m m^{2m\alpha+1/2}} \times$

$$\times E\left(\begin{matrix}(a_p),\ \Delta(m,\ (\nu+\mu)/2-m\alpha),\ \Delta(m,\ (\nu-\mu)/2-m\alpha),\ \Delta(m,\ (\mu-\nu)/2-m\alpha),\ \Delta(m,\ -(\mu+\nu)/2-m\alpha)\\ (b_q),\ \Delta(2m,\ -2m\alpha);\qquad (b/m)^{2m}c\end{matrix}\right)$$

$[q\leqslant p+1;\ \mathrm{Re}\,b>0;\ |\arg c|\leqslant(p-q+1)\pi/2;\ \mathrm{Re}\,\alpha<-(|\mathrm{Re}\,\mu|+|\mathrm{Re}\,\nu|)/(2m)]$.

2.23.4. Integrals containing Legendre functions and $E((a_p);(b_q);cx)$.

$$\int_a^\infty x^{\alpha-1}(x^{1/m}-a^{1/m})^{-\mu/2} P_\nu^\mu\left(\left(\frac{x}{a}\right)^{1/(2m)}\right) E((a_p);\ (b_q);\ cx)\,dx =$$

$$=(2m)^\mu a^{\alpha-\mu/(2m)} E\left(\begin{matrix}(a_p),\ \Delta(m,\ (\mu+\nu+1)/2),\ \Delta(m,\ (\mu-\nu)/2)\\ (b_q),\ \Delta(2m,\ 1);\qquad ac\end{matrix}\right)$$

$[q\leqslant p+1;\ a>0;\ |\arg c|\leqslant(p-q+1)\pi/2;\ \mathrm{Re}\,\mu,\ \mathrm{Re}\,(2m\alpha-\mu-\nu)<1;\ \mathrm{Re}\,(2m\alpha-\mu+\nu)<0]$.

2. $\displaystyle\int_a^\infty x^{\alpha-1}(x^{1/m}-a^{1/m})^{-\mu/2} P_\nu^\mu\left(2\left(\frac{a}{x}\right)^{1/m}-1\right) E((a_p);\ (b_q);\ cx)\,dx =$

$$=m^\mu a^{\alpha-\mu/(2m)} E\left(\begin{matrix}(a_p),\ \Delta(m,\ \mu-m\alpha),\ \Delta(m,\ -m\alpha);\ ac\\ (b_q),\ \Delta(m,\ 1+\nu-m\alpha),\ \Delta(m,\ -\nu-m\alpha)\end{matrix}\right)$$

$[q\leqslant p+1;\ a>0;\ |\arg c|\leqslant(p-q+1)\pi/2;\ \mathrm{Re}\,\alpha,\ \mathrm{Re}\,(m\alpha+\mu)<0;\ \mathrm{Re}\,\mu<1]$.

2.23.5. Integrals containing Whittaker functions and $E((a_p);(b_q);cx)$.

1. $\displaystyle\int_0^\infty x^{\alpha-1} e^{-sx^{-1/m}/2} W_{\rho,\sigma}(sx^{-1/m}) E((a_p);\ (b_q);\ cx)\,dx =$

$$=\frac{m^{\rho-m\alpha+1/2} s^{m\alpha}}{(2\pi)^{(m-1)/2}} E\left(\begin{matrix}(a_p),\ \Delta(m,\ 1/2+\sigma-m\alpha),\ \Delta(m,\ 1/2-\sigma-m\alpha)\\ (b_q),\ \Delta\ \ ,\ 1-\rho-m\alpha);\qquad (s/m)^m c\end{matrix}\right)$$

$[q\leqslant p+1;\ \mathrm{Re}\,s>0;\ |\arg c|\leqslant(p-q+1)\pi/2;\ m\mathrm{Re}\,\alpha+|\mathrm{Re}\,\sigma|<1/2]$.

2. $\displaystyle\int_0^\infty x^{\alpha-1} W_{\rho,\sigma}(bx^{-1/(2m)}) W_{-\rho,\sigma}(bx^{-1/(2m)}) E((a_p);\ (b_q);\ cx)\,dx =$

$$=\frac{b^{2m\alpha} m^{1/2-2m\alpha}}{2^{m(2\alpha+1)-1}\pi^{m-1/2}} E\left(\begin{matrix}(a_p),\ \Delta(2m,\ 1-2m\alpha),\ \Delta(m,\ 1/2+\sigma-m\alpha),\ \Delta(m,\ 1/2-\sigma-m\alpha)\\ (b_q),\ \Delta(m,\ 1+\rho-m\alpha),\ \Delta(m,\ 1-\rho-m\alpha);\ (2m)^{-2m} b^{2m} c\end{matrix}\right)$$

$[q\leqslant p+1;\ \mathrm{Re}\,b>0;\ |\arg c|\leqslant(p-q+1)\pi/2;\ m\mathrm{Re}\,\alpha+|\mathrm{Re}\,\sigma|<1/2]$.

2.23.6. Integrals containing products of two E-functions.

1. $\displaystyle\int_0^\infty x^{\alpha-1} E((c_m);\ (d_n);\ \sigma x) E((a_p);\ (b_q);\ \omega x^{l/k})\,dx =$

$$=(2\pi)^\lambda k^\mu l^{\nu+(m-n-1)\alpha-1}\sigma^{-\alpha}\times$$

$$\times G^{kp+l,\ k+lm}_{k+kq+lm,\ kp+l+ln}\left(\frac{k^{k(q-p+1)}\omega^k}{l^{l(n-m+1)}\sigma^l}\left|\begin{matrix}\Delta(k,\ 1),\ \Delta(l,\ 1-\alpha-(c_m)),\ \Delta(k,\ (b_q))\\ \Delta(l,\ -\alpha),\ \Delta(k,\ (a_p)),\ \Delta(l,\ 1-\alpha-(d_n))\end{matrix}\right.\right)$$

$[\lambda=(m-n-1)(1-l)/2+(p-q-1)(1-k)/2,$

$$\mu=\sum_{j=1}^p a_j-\sum_{j=1}^q b_j+\frac{q-p+1}{2},\quad \nu=\sum_{j=1}^m c_j-\sum_{j=1}^n d_j+\frac{n-m+1}{2};$$

conditions see 2.24.1.1 for $s=v=m,\ t=n=1,\ u=n+1,\ m=q=p,\ p=q+1;\ (c_u)=1,\ (d_n)$; $(d_v)=(c_m);\ (a_p)=1,\ (b_q);\ (b_q)=(a_p)]$.

2. $$\int_0^\infty x^{\alpha-1} E((c_m);\ (d_n);\ \tau x^{-1})\, E((a_p);\ (b_q);\ \omega x^{l/k})\, dx = (2\pi)^{\lambda} k^{\mu} l^{\nu+(n-m+1)\alpha-1} \tau^{\alpha} \times$$

$$\times G^{kp+lm,\ k+l}_{kq+k+ln+l,\ kp+lm}\left(\frac{k^{k(q-p+1)} \tau^{l} \omega^{k}}{l^{l(m-n-1)}} \middle| \begin{matrix} \Delta(k,\ 1),\ \Delta(l,\ 1-\alpha), \Delta(k,\ (b_q)),\ \Delta(l,\ (d_n)-\alpha) \\ \Delta(k,\ (a_p)),\qquad \Delta(l,\ (c_m)-\alpha) \end{matrix} \right)$$

[λ, μ, ν see 2.23.6.1 conditions see 2.24.1.1 for $s=n=1$, $t=u=m$, $v=n+1$, $m=q=p$, $p=q+1$; $(c_u)=1-(c_m)$; $(d_v)=0,\ 1-(d_n)$; $(a_p)=1,\ (b_q)$; $(b_q)=(a_p)$, $\sigma=\tau^{-1}$].

2.24. THE MEIJER G- FUNCTION $G_{pq}^{mn}\left(x \middle| \begin{matrix}(a_p)\\(b_q)\end{matrix}\right)$

In section 2.24.1, an integral of general form is given that contains as particular cases for various values of the parameters many integrals with elementary and special functions; some of these integrals are included in this handbook.

2.24.1. Integrals of general form

NOTATION:

$$m,\ n,\ p,\ q,\ s,\ t,\ u,\ v = 0,\ 1,\ 2,\ \ldots;\ k,\ l = 1,\ 2,\ 3,\ \ldots$$

$$0 \leqslant m \leqslant q,\quad 0 \leqslant n \leqslant p,\quad 0 \leqslant s \leqslant v,\quad 0 \leqslant t \leqslant u,\quad r = l/k$$

[k, l are coprime]

$$b^* = s + t - \frac{u+v}{2},\quad c^* = m + n - \frac{p+q}{2},$$

$$\mu = \sum_{j=1}^{q} b_j - \sum_{j=1}^{p} a_j + \frac{p-q}{2} + 1,\quad \rho = \sum_{j=1}^{v} d_j - \sum_{j=1}^{u} c_j + \frac{u-v}{2} + 1$$

[a_j, b_j, c_j, d_j are complex parameters

$$\varphi = q - p - r(v-u),\quad \eta = 1 - \alpha(v-u) - \mu - \rho$$

[α is a complex parameter

σ, ω are complex variables; $\sigma,\ \omega \neq 0$.

1°. $a_i - b_j$, $c_g - d_h \neq 1,\ 2,\ \ldots$ [$i=1, 2, \ldots, n$; $j=1, 2, \ldots, m$; $g=1, 2, \ldots, t$; $h=1, 2, \ldots, s$];

2°. $\mathrm{Re}(\alpha + d_h + rb_j) > 0$ [$j=1, 2, \ldots, m$; $h=1, 2, \ldots, s$];

3°. $\mathrm{Re}(\alpha + c_g + ra_i) < r + 1$ [$i=1, 2, \ldots, n$; $g=1, 2, \ldots, t$];

4°. $(p-q)\,\mathrm{Re}(\alpha + c_g - 1) - r\,\mathrm{Re}\,\mu > -3r/2$ [$g=1, 2, \ldots, t$];

5°. $(p-q)\,\mathrm{Re}(\alpha + d_h) - r\,\mathrm{Re}\,\mu > -3r/2$ [$h=1, 2, \ldots, s$];

6°. $(u-v)\,\mathrm{Re}(\alpha + ra_i - r) - \mathrm{Re}\,\rho > -3/2$ [$i=1, 2, \ldots, n$];

7°. $(u-v)\,\mathrm{Re}(\alpha + rb_j) - \mathrm{Re}\,\rho > -3/2$ [$j=1, 2, \ldots, m$];

8°. $|\varphi| + 2\,\mathrm{Re}\,[(q-p)(v-u)\alpha + r(v-u)(\mu-1) + (q-p)(\rho-1)] > 0$;

9°. $|\varphi| - 2\,\mathrm{Re}\,[(q-p)(v-u)\alpha + r(v-u)(\mu-1) + (q-p)(\rho-1)] > 0$;

10°. $|\arg\sigma| < b^*\pi$;

11°. $|\arg\sigma| = b^*\pi$;

12°. $|\arg\omega| < c^*\pi$;

13°. $|\arg\omega| = c^*\pi$;

14°. If $\varphi = 0$ and $c^* + r(b^*-1) \leqslant 0$, then $|\arg(1 - z_0\sigma^{-l}\omega^{k})| < \pi$, where $z_0 = r^{l(v-u)} \exp[-(lb^* + kc^*)\pi i]$, and for $\mathrm{Re}\,[\mu + \rho + \alpha(v-u)] < 1$ the following value is taken $\sigma^{l}\omega^{-k} = z_0$.

15°. One of the following three conditions holds:

$$\lambda_c > 0 \text{ or } \lambda_c = 0,\ \lambda_s \neq 0,\ \mathrm{Re}\,\eta > -1 \text{ or } \lambda_c = \lambda_s = 0,\ \mathrm{Re}\,\eta > 0,$$

where

$$\lambda_c\,(q-p)\,|\,\omega\,|^{1/(q-p)}\cos\psi+(v-u)\,|\,\sigma\,|^{1/(v-u)}\cos\theta,$$

$$\psi=\frac{1}{q-p}\,[|\arg\omega|+(q-m-n)\pi],\quad \theta=\frac{1}{v-u}\,[|\arg\sigma|+(v-s-t)\,\pi];$$

$$\lambda_s=(q-p)\,|\,\omega\,|^{1/(q-p)}\operatorname{sgn}(\arg\omega)\sin\psi+(v-u)\,|\,\sigma\,|^{1/(v-u)}\operatorname{sgn}(\arg\sigma)\sin\theta$$

under the condition $\arg\sigma\cdot\arg\omega\neq 0$;

if $\arg\sigma=0$, $\arg\omega\neq 0$, then $\lambda_s=\lambda_s^{+}\lambda_s^{-}$,

if $\arg\sigma\neq 0$, $\arg\omega=0$, then $\lambda_s=\tilde{\lambda}_s^{+}\tilde{\lambda}_s^{-}$,

if $\arg\sigma=\arg\omega=0$, then $\lambda_s=\lambda_s^{++}\lambda_s^{+-}$,

where $\lambda_s^{\pm}=\lim\limits_{\arg\sigma\to\pm 0}\lambda_s$, $\tilde{\lambda}_s^{\pm}=\lim\limits_{\arg\omega\to\pm 0}\lambda_s$, $\lambda_s^{+\pm}=\lim\limits_{\substack{\arg\omega\to+0\\ \arg\sigma\to\pm 0}}\lambda_s$.

1. $$\int_0^\infty x^{\alpha-1}\,G_{uv}^{st}\left(\sigma x\left|\begin{matrix}(c_u)\\(d_v)\end{matrix}\right.\right)G_{pq}^{mn}\left(\omega x^{l/k}\left|\begin{matrix}(a_p)\\(b_q)\end{matrix}\right.\right)dx=$$

$$=\frac{k^{\mu}l^{\rho+\alpha(v-u)-1}\sigma^{-\alpha}}{(2\pi)^{b^*(l-1)+c^*(k-1)}}\,G_{kp+lv,\;kq+lu}^{km+lt,\;kn+ls}\left(\frac{\omega^k k^{k(p-q)}}{\sigma^l l^{l(u-v)}}\right)\times$$

$$\times\left|\begin{matrix}\Delta(k,a_1),\ldots,\Delta(k,a_n),\Delta(l,1-\alpha-d_1),\ldots,\Delta(l,1-\alpha-d_v),\Delta(k,a_{n+1}),\ldots,\Delta(k,a_p)\\ \Delta(k,b_1),\ldots,\Delta(k,b_m),\Delta(l,1-\alpha-c_1),\ldots,\Delta(l,1-\alpha-c_u),\Delta(k,b_{m+1}),\ldots,\Delta(k,b_q)\end{matrix}\right),$$

If one of the following conditions holds (if $ms=0$ or $nt=0$, then 2° or 3° respectively are ommitted):

1) $mnst\neq 0$; b^*, $c^*>0$; 1°—3°, 10°, 12°;
2) $u=v$; $b^*=0$; c^*, $\sigma>0$; $\operatorname{Re}\rho<1$; 1°—3°, 12°;
3) $p=q$; $c^*=0$; b^*, $\omega>0$; $\operatorname{Re}\mu<1$; 1°—3°, 10°;
4) $p=q$; $u=v$; $b^*=c^*=0$; σ, $\omega>0$; $\operatorname{Re}\mu$, $\operatorname{Re}\rho<1$; $\sigma^l\neq\omega^k$; 1°—3°;
5) $p=q$; $u=v$; $b^*=c^*=0$; σ, $\omega>0$; $\operatorname{Re}(\mu+\rho)<1$; $\sigma^l=\omega^k$; 1°—3°;
6) $p>q$; s, $b^*>0$; $c^*\geqslant 0$; 1°—3°, 5°, 10°, 13°;
7) $p<q$; t, $b^*>0$; $c^*\geqslant 0$; 1°—4°, 10°, 13°;
8) $u>v$; m, $c^*>0$; $b^*\geqslant 0$; 1°—3°, 7°, 11°, 12°;
9) $u<v$; n, $c^*>0$; $b^*\geqslant 0$; 1°—3°, 6°, 11°, 12°;
10) $p>q$; $u=v$; $b^*=0$; $c^*\geqslant 0$; $\sigma>0$; $\operatorname{Re}\rho<1$; 1°—3°, 5°, 13°;
11) $p<q$; $u=v$; $b^*=0$; $c^*\geqslant 0$; $\sigma>0$; $\operatorname{Re}\rho<1$; 1°—4°, 13°;
12) $p=q$; $u>v$; $b^*\geqslant 0$; $c^*=0$; $\omega>0$; $\operatorname{Re}\mu<1$; 1°—3°, 7°, 11°;
13) $p=q$, $u<v$; $b^*\geqslant 0$; $c^*=0$; $\omega>0$; $\operatorname{Re}\mu<1$; 1°—3°, 6°, 11°;
14) $p<q$; $u>v$; b^*, $c^*\geqslant 0$; 1°—4°, 7°, 11°, 13°;
15) $p>q$; $u<v$; b^*, $c^*\geqslant 0$; 1°—3°, 5°, 6°, 11°, 13°;
16) $p>q$; $u>v$; b^*, $c^*\geqslant 0$; 1°—3°, 5°, 7°, 8°, 11°, 13°, 14°;
17) $p<q$; $u<v$; b^*, $c^*\geqslant 0$; 1°—4°, 6°, 9°, 11°, 13°, 14°;
18) $t=0$; s, b^*, $\varphi>0$; 1°, 2°, 10°;
19) $s=0$; t, $b^*>0$; $\varphi<0$; 1°, 3°, 10°;
20) $n=0$; m, $c^*>0$; $\varphi<0$; 1°, 2°, 12°;
21) $m=0$; n, c^*, $\varphi>0$; 1°, 3°, 12°;
22) $st=0$; b^*, $c^*>0$; 1°—3°, 10°, 12°;
23) $mn=0$; b^*, $c^*>0$; 1°—3°, 10°, 12°;
24) $m+n>p$; $t=\varphi=0$; s, $b^*>0$; $c^*<0$; $|\arg\omega|<(m+n-p+1)\pi$; 1°, 2°, 10°, 14°, 15°;
25) $m+n>q$; $s=\varphi=0$; t, $b^*>0$; $c^*<0$; $|\arg\omega|<(m+n-q+1)\pi$; 1°, 3°, 10°, 14°, 15°;

26) $p=q-1$; $t=\varphi=0$; s, $b^*>0$; $c^*\geqslant 0$; $c^*\pi<|\arg\omega|<(c^*+1)\pi$; 1°, 2°, 10°, 14°, 15°;

27) $p=q+1$; $s=\varphi=0$; t, $b^*>0$; $c^*\geqslant 0$; $c^*\pi<|\arg\omega|<(c^*+1)\pi$; 1°, 3°, 10°, 14°, 15°;

28) $p<q-1$; $t=\varphi=0$; s, $b^*>0$; $c^*\geqslant 0$; $c^*\pi<|\arg\omega|<(m+n-p+1)\pi$; 1°, 2°, 10°, 14°, 15°;

29) $p>q+1$; $s=\varphi=0$; t, $b^*>0$; $c^*\geqslant 0$; $c^*\pi<|\arg\omega|<(m+n-q+1)\pi$; 1°, 3°, 10°, 14°, 15°;

30) $n=\varphi=0$; $s+t>u$; m, $c^*>0$; $b^*<0$; $|\arg\sigma|<(s+t-u+1)\pi$; 1°, 2°, 12°, 14°, 15°;

31) $m=\varphi=0$; $s+t>v$; n, $c^*>0$; $b^*<0$; $|\arg\sigma|<(s+t-v+1)\pi$; 1°, 3°, 12°, 14°, 15°;

32) $n=\varphi=0$; $u=v-1$; m, $c^*>0$; $b^*\geqslant 0$; $b^*\pi<|\arg\sigma|<(b^*+1)\pi$; 1°, 2°, 12°, 14°, 15°;

33) $m=\varphi=0$; $u=v+1$; n, $c^*>0$; $b^*\geqslant 0$; $b^*\pi<|\arg\sigma|<(b^*+1)\pi$; 1°, 3°, 12°, 14°, 15°;

34) $n=\varphi=0$; $u<v-1$; m, $c^*>0$; $b^*\geqslant 0$; $b^*\pi<|\arg\sigma|<(s+t-u+1)\pi$; 1°, 2°, 12°, 14°, 15°;

35) $m=\varphi=0$; $u>v+1$; n, $c^*>0$; $b^*\geqslant 0$; $b^*\pi<|\arg\sigma|<(s+t-v+1)\pi$; 1°, 3°, 12°, 14°, 15°.

2. $$\int_0^\infty G_{uv}^{st}\left(\sigma x\left|\begin{matrix}(c_u)\\(d_v)\end{matrix}\right.\right)G_{pq}^{mn}\left(\omega x\left|\begin{matrix}(a_p)\\(b_q)\end{matrix}\right.\right)dx=\sigma^{-1}G_{p+v,\,q+u}^{m+t,\,n+s}\left(\frac{\omega}{\sigma}\left|\begin{matrix}(a_n),\,-(d_v),\,a_{n+1},\,\dots,\,a_p\\(b_m),\,-(c_u),\,b_{m+1},\,\dots,\,b_q\end{matrix}\right.\right)$$

[see 2.24.1.1 for $k=l=\alpha=1$].

3. $$\int_0^\infty x^{\alpha-1}G_{uv}^{st}\left(\sigma+x\left|\begin{matrix}(c_u)\\(d_v)\end{matrix}\right.\right)G_{pq}^{mn}\left(\omega x\left|\begin{matrix}(a_p)\\(b_q)\end{matrix}\right.\right)dx=$$

$$=\sum_{k=0}^\infty\frac{(-\sigma)^k}{k!}G_{p+v+1,\,q+u+1}^{m+t,\,n+s+1}\left(\omega\left|\begin{matrix}1-\alpha,\,(a_n),\,1+k-\alpha-(d_v),\,a_{n+1},\,\dots,\,a_p\\(b_m),\,1+k-\alpha-(c_u),\,1+k-\alpha,\,b_{m+1},\,\dots,\,b_q\end{matrix}\right.\right)$$

$\left[b^*,\,c^*>0;\ |\arg\sigma|<\pi;\ |\arg\omega|<c^*\pi;\ -\min\limits_{1\leqslant j\leqslant m}b_j<\operatorname{Re}\alpha<2-\max\limits_{1\leqslant i\leqslant n}\operatorname{Re}a_i-\max\limits_{1\leqslant k\leqslant t}\operatorname{Re}c_k\right]$.

4. $$\int_{-\infty}^\infty G_{pq}^{mn}\left(\omega\left|\begin{matrix}(a_{p-2}),\,a_{p-1}+x,\,a_p+x\\(b_{q-2}),\,b_{q-1}+x,\,b_q+x\end{matrix}\right.\right)dx=$$

$$=\Gamma\left[\begin{matrix}a_{p-1}+a_p-b_{q-1}-b_q-1\\a_{p-1}-b_{q-1},\,a_{p-1}-b_q,\,a_p-b_{q-1},\,a_p-b_q\end{matrix}\right]G_{p-2,\,q-2}^{m,\,n}\left(\omega\left|\begin{matrix}(a_{p-2})\\(b_{q-2})\end{matrix}\right.\right)$$

$\left[c^*>0;\ |\arg\omega|<c^*\pi;\ \operatorname{Re}\left(\sum\limits_{j=1}^p a_j-\sum\limits_{j=1}^q b_j\right)>1\right]$.

2.24.2. Integrals of $A(x)\,G_{pq}^{mn}\left(\varphi(x)\left|\begin{matrix}(a_p)\\(b_q)\end{matrix}\right.\right)$.

NOTATION: see 2.24.1.

1. $$\int_0^\infty x^{\alpha-1}G_{pq}^{mn}\left(\omega x\left|\begin{matrix}(a_p)\\(b_q)\end{matrix}\right.\right)dx=$$

$$=\omega^{-\alpha}\Gamma\left[\begin{matrix}(b_m)+\alpha,\,1-(a_n)-\alpha\\a_{n+1}+\alpha,\,\dots,\,a_p+\alpha,\,1-b_{m+1}-\alpha,\,\dots,\,1-b_q-\alpha\end{matrix}\right]$$

$\left[m^2+n^2\neq 0;\ c^*>0;\ |\arg\omega|<c^*\pi;\ \omega\neq 0;\ -\min\limits_{1\leqslant j\leqslant m}\operatorname{Re}b_j<\operatorname{Re}\alpha<1-\max\limits_{1\leqslant j\leqslant n}\operatorname{Re}a_j\right]$ or

$[p\neq q;\ c^*\geqslant 0;\ |\arg\omega|=c^*\pi;\ -\min\limits_{1\leqslant j\leqslant m}\operatorname{Re} b_j<\operatorname{Re}\alpha<1-\max\limits_{1\leqslant j\leqslant n}\operatorname{Re} a_j,\ \operatorname{Re}[\mu+(q-p)\alpha]<3/2]$

or $\left[p=q;\ c^*=0;\ \omega>0;\ -\min\limits_{1\leqslant j\leqslant m}\operatorname{Re} b_j<\operatorname{Re}\alpha<1-\max\limits_{1\leqslant j\leqslant n}\operatorname{Re} a_j,\ \operatorname{Re}\sum\limits_{j=1}^{q}(b_j-a_j)<0\right].$

2. $$\int_0^a x^{\alpha-1}(a-x)^{\beta-1}G_{pq}^{mn}\left(\omega x^{l/k}\middle|\begin{matrix}(a_p)\\(b_q)\end{matrix}\right)dx=$$
$$=\frac{k^{\mu}l^{-\beta}\Gamma(\beta)}{(2\pi)^{c^*(k-1)}a^{1-\alpha-\beta}}G_{kp+l,\ kq+l}^{km,\ kn+l}\left(\frac{\omega^k a^l}{k^{k(q-p)}}\middle|\begin{matrix}\Delta(l,\ 1-\alpha),\ \Delta(k,\ (a_p))\\\Delta(k,\ (b_q)),\ \Delta(l,\ 1-\alpha-\beta)\end{matrix}\right)$$
[conditions see 2.24.1.1 for $\sigma=a^{-1}$, $s=u=v=1$, $t=d_1=0$, $c_1=\beta$].

3. $$\int_a^\infty x^{\alpha-1}(x-a)^{\beta-1}G_{pq}^{mn}\left(\omega x^{l/k}\middle|\begin{matrix}(a_p)\\(b_q)\end{matrix}\right)dx=$$
$$=\frac{k^{\mu}l^{-\beta}\Gamma(\beta)}{(2\pi)^{c^*(k-1)}a^{1-\alpha-\beta}}G_{kp+l,\ kq+l}^{km+l,\ kn}\left(\frac{\omega^k a^l}{k^{k(q-p)}}\middle|\begin{matrix}\Delta(k,\ (a_p)),\ \Delta(l,\ 1-\alpha)\\\Delta(l,\ 1-\alpha-\beta),\ \Delta(k,\ (b_q))\end{matrix}\right)$$
[conditions see 2.24.1.1. for $\sigma=a^{-1}$, $s=d_1=0$, $t=u=v=1$, $c_1=\beta$].

4. $$\int_0^\infty \frac{x^{\alpha-1}}{(x+z)^{\lambda}}G_{pq}^{mn}\left(\omega x^{l/k}\middle|\begin{matrix}(a_p)\\(b_q)\end{matrix}\right)dx=$$
$$=\frac{k^{\mu}l^{\lambda-1}z^{\alpha-\lambda}}{(2\pi)^{c^*(k-1)+l-1}\Gamma(\lambda)}G_{kp+l,\ kq+l}^{km+l,\ kn+l}\left(\frac{\omega^k z^l}{k^{k(q-p)}}\middle|\begin{matrix}\Delta(l,\ 1-\alpha),\ \Delta(k,\ (a_p))\\\Delta(l,\ \lambda-\alpha),\ \Delta(k,\ (b_q))\end{matrix}\right)$$
[conditions see 2.24.1.1 for $\sigma=z^{-1}$, $s=t=u=v=1$, $c_1=1-\lambda$, $d_1=0$].

5. $$\int_0^\infty \frac{x^{\alpha-1}}{x-y}G_{pq}^{mn}\left(\omega x^{l/k}\middle|\begin{matrix}(a_p)\\(b_q)\end{matrix}\right)dx=$$
$$=-\frac{\pi k^{\mu}y^{\alpha-1}}{(2\pi)^{c^*(k-1)}}G_{kp+2l,\ kq+2l}^{km+l,\ kn+l}\left(\frac{\omega^k y^l}{k^{k(q-p)}}\middle|\begin{matrix}\Delta(l,\ 1-\alpha),\ \Delta(k,\ (a_p)),\ \Delta(l,\ 1/2-\alpha)\\\Delta(l,\ 1-\alpha),\ \Delta(k,\ (b_q)),\ \Delta(l,\ 1/2-\alpha)\end{matrix}\right)$$
[conditions see 2.24.1.1 for $\sigma=y^{-1}$, $s=t=1$, $u=v=2$, $c_1=d_1=0$, $c_2=d_2=1/2$].

6. $$\int_0^\infty \frac{x^{\alpha-1}}{(x+z)^{\beta}}G_{pq}^{mn}\left(\omega x^{-k}(x+z)^l\middle|\begin{matrix}(a_p)\\(b_q)\end{matrix}\right)dx=I,$$
$$I=\frac{\sqrt{2\pi}\,k^{\alpha-1/2}l^{1/2-\beta}z^{\alpha-\beta}}{(l-k)^{\alpha-\beta+1/2}}G_{p+l,\ q+l}^{m+l,\ n}\left(\frac{\omega l^l}{k^k}\left(\frac{z}{l-k}\right)^{l-k}\middle|\begin{matrix}(a_p),\ \Delta(l,\ \beta)\\\Delta(k,\ \alpha),\ \Delta(l-k,\ \beta-\alpha),\ (b_q)\end{matrix}\right)$$
$\left[0<k<l;\ c^*>0;\ |\arg(\omega z^{l-k})|<c^*\pi;\ (l-k)\left(1-\max\limits_{1\leqslant j\leqslant n}\operatorname{Re} a_j\right)+\operatorname{Re}(\beta-\alpha),\ k-k\max\limits_{1\leqslant j\leqslant n}\operatorname{Re} a_j+\operatorname{Re}\alpha>0\right],$
$$I=\frac{\sqrt{2\pi}\,k^{\alpha-1/2}l^{1/2-\beta}z^{\alpha-\beta}}{(k-l)^{1/2+\alpha-\beta}}\times$$
$$\times G_{p+k,\ q+k}^{m+k,\ n+k-l}\left(\frac{\omega l^l}{k^k}\left(\frac{z}{k-l}\right)^{l-k}\middle|\begin{matrix}\Delta(k-l,\ \alpha-\beta+1),\ (a_p),\ \Delta(l,\ \beta)\\\Delta(k,\ \alpha),\ (b_q)\end{matrix}\right)$$
$\left[0<l<k;\ c^*>0;\ |\arg(\omega z^{l-k})|<c^*\pi;\ (k-l)\min\limits_{1\leqslant j\leqslant m}\operatorname{Re} b_j+\operatorname{Re}(\beta-\alpha),\ k-k\max\limits_{1\leqslant j\leqslant n}\operatorname{Re} a_j+\operatorname{Re}\alpha>0\right],$

$$I=\sqrt{2\pi}\,(kz)^{\alpha-\beta}\,\Gamma(\beta-\alpha)\,G^{m+k,\,n}_{p+k,\,q+k}\left(\omega\,\middle|\,\begin{matrix}(a_p),\ \Delta(k,\ \beta)\\ \Delta(k,\ \alpha),\ (b_q)\end{matrix}\right)$$

$\left[l=k>0;\ c^*>0;\ |\arg\omega|<c^*\pi;\ \operatorname{Re}(\beta-\alpha),\ k-k\max\limits_{1\leqslant j\leqslant n}\operatorname{Re}a_j+\operatorname{Re}\alpha>0\right].$

7. $$\int_0^1\frac{x^{\alpha-1}(1-x)^{\beta-1}}{[1+ax+b(1-x)]^{\alpha+\beta}}\,G^{mn}_{pq}\left(\frac{\omega x^l(1-x)^k}{[1+ax+b(1-x)]^{l+k}}\,\middle|\,\begin{matrix}(a_p)\\ (b_q)\end{matrix}\right)dx=$$

$$=\frac{\sqrt{2\pi}\,k^{\alpha-1/2}\,l^{\beta-1/2}}{(1+a)^{\alpha}(1+b)^{\beta}(l+k)^{\alpha+\beta-1/2}}\times$$

$$\times G^{m,\,n+k+l}_{p+k+l,\,q+k+l}\left(\frac{\omega k^k l^l}{(1+a)^l(1+b)^k(k+l)^{k+l}}\,\middle|\,\begin{matrix}\Delta(l,\ 1-\alpha),\ \Delta(k,\ 1-\beta),\ (a_p)\\ (b_q),\ \Delta(k+l,\ 1-\alpha-\beta)\end{matrix}\right)$$

$\left[m>0;\ p<q;\ \operatorname{Re}\alpha+l\min\limits_{1\leqslant j\leqslant m}\operatorname{Re}b_j,\ \operatorname{Re}\beta+k\min\limits_{1\leqslant j\leqslant m}\operatorname{Re}b_j>0\right]$ or $\left[n>0;\ p>q;\ c^*\geqslant 0;\ |\arg\omega|=c^*\pi;\ \operatorname{Re}\alpha+l\min\limits_{1\leqslant j\leqslant m}\operatorname{Re}b_j,\ \operatorname{Re}\beta+k\min\limits_{1\leqslant j\leqslant m}\operatorname{Re}b_j>0;\ \operatorname{Re}((q-p)\alpha+l\mu)<3l/2;\ \operatorname{Re}((q-p)\beta+k\mu)<3k/2\right].$

8. $$\int_0^\infty\frac{x^{\alpha-1}}{\sqrt{x+z}}(\sqrt{x}+\sqrt{x+z})^{2\beta}\,G^{mn}_{pq}\left(\omega x^k(\sqrt{x}+\sqrt{x+z})^{2l}\,\middle|\,\begin{matrix}(a_p)\\ (b_q)\end{matrix}\right)dx=J,$$

$$J=\frac{(2\pi)^{1-k-l}\,k^{2\alpha-1/2}\,z^{\alpha+\beta-1/2}}{\sqrt{\pi}\,(k+l)^{\alpha+\beta}(k-l)^{\alpha-\beta}}\,G^{m+k+l,\,n+2k}_{p+2k,\,q+2k}\left(\frac{\omega k^{2k}}{(k-l)^{k-l}}\times\right.$$

$$\left.\times\left(\frac{z}{k+l}\right)^{k+l}\,\middle|\,\begin{matrix}\Delta(2k,\ 1-2\alpha),\quad (a_p)\\ \Delta(k+l,\ 1/2-\alpha-\beta),\ (b_q),\ \Delta(k-l,\ 1/2+\beta-\alpha)\end{matrix}\right)\qquad [l<k],$$

$$J=\frac{(2\pi)^{1-2k}\,k^{2\alpha-1/2}z^{\alpha+\beta-1/2}}{\sqrt{\pi}\,(l+k)^{\alpha+\beta}(l-k)^{\alpha-\beta}}\times$$

$$\times G^{m+k+l,\,n+2k}_{p+k+l,\,q+k+l}\left(\frac{\omega k^{2k}}{(l-k)^{k-l}}\left(\frac{z}{k+l}\right)^{k+l}\,\middle|\,\begin{matrix}\Delta(2k,\ 1-2\alpha),\ (a_p),\ \Delta(l-k,\ 1/2+\alpha-\beta)\\ \Delta(k+l,\ 1/2-\alpha-\beta),\ (b_q)\end{matrix}\right)$$

$[l>k],$

$$J=\frac{\pi^{1-2k}k^{\alpha-\beta-1/2}z^{\alpha+\beta-1/2}}{2^{\alpha+\beta+2k-3/2}\,\Gamma(1/2+\alpha-\beta)}\times$$

$$\times G^{m+2k,n+2k}_{p+2k,\,q+2k}\left(\omega\left(\frac{z}{2}\right)^{2k}\,\middle|\,\begin{matrix}\Delta(2k,\ 1-2\alpha),\quad (a_p)\\ \Delta(2k,\ 1/2-\alpha-\beta),\ (b_q)\end{matrix}\right)\qquad [l=k]$$

$[mn\neq 0;\ c^*>0;\ |\arg(\omega z^{k+l})|<c^*\pi;\ |\arg z|<\pi;\ \operatorname{Re}(\alpha+kb_j)>0,\ j=1,\ 2,\ \ldots,\ m;\ \operatorname{Re}[\alpha+\beta+(k+l)(a_i-1)]<1/2,\ i=1,\ 2,\ \ldots,\ n]$ or $[m>0;\ q>p;\ c^*\geqslant 0;\ |\arg(\omega z^{k+l})|=c^*\pi;\ |\arg z|<\pi;\ \operatorname{Re}(\alpha+kb_j)>0,\ j=1,\ 2,\ \ldots,\ m;\ \operatorname{Re}[\alpha+\beta+(k+l)(a_i-1)]<1/2,\ i=1,\ 2,\ \ldots,\ n;\ \operatorname{Re}[(q-p)(\alpha+\beta-1/2)+(k+l)\mu]<3(k+l)/2]$ or $[n>0;\ q<p;\ c^*\geqslant 0;\ |\arg(\omega z^{k+l})|=c^*\pi;\ |\arg z|<\pi;\ \operatorname{Re}(\alpha+kb_j)>0,\ j=1,\ 2,\ \ldots,\ m;\ \operatorname{Re}[\alpha+\beta+(k+l)(a_i-1)]<1/2,\ i=1,\ 2,\ \ldots,\ n;\ \operatorname{Re}[(q-p)\alpha+k\mu]<3k/2].$

9. $$\int_0^\infty x^{\alpha-1}(ax^2+bx+c)^{3/2-\alpha}\,G^{mn}_{pq}\left(\left(\frac{ax^2+bx+c}{x}\right)^l\,\middle|\,\begin{matrix}(a_p)\\ (b_q)\end{matrix}\right)dx=$$

$$=\frac{\sqrt{\pi}}{2\,(al)^{3/2}}\,G^{m+l,\,n}_{p+l,\,q+l}\left((b+2\sqrt{ac})^l\,\middle|\,\begin{matrix}(a_p)+(3-\alpha)/l,\ \Delta(l,\ 3/2)\\ \Delta(l,\ 0),\ (b_q)+(3-\alpha)/l\end{matrix}\right)+$$

$$+\frac{1}{a}\sqrt{\frac{\pi c}{l}}\,G^{m+l,\,n}_{p+l,\,q+l}\left((b+2\sqrt{ac})^l\,\middle|\,\begin{matrix}(a_p)+(2-\alpha)/l,\ \Delta(l,\ 1/2)\\ \Delta(l,\ 0),\ (b_q)+(2-\alpha)/l\end{matrix}\right)$$

$[p\neq q;\ c^*,\ \operatorname{Re}a>0;\ \operatorname{Re}c\geqslant 0;\ \operatorname{Re}[l(1-a_j)+\alpha]>3;\ j=1,\ 2,\ \ldots,\ n;\ l=1,\ 2,\ \ldots].$

10. $$\int_0^\infty x^{\alpha-1}(ax^2+bx+c)^{1/2-\alpha} G_{pq}^{mn}\left(\left(\frac{ax^2+bx+c}{x}\right)^{-l}\middle|\begin{matrix}(a_p)\\(b_q)\end{matrix}\right)dx=$$

$$=\sqrt{\frac{\pi}{al}}(b+2\sqrt{ac})^{1-\alpha}G_{p+l,\ q+l}^{m,\ n+l}\left((b+2\sqrt{ac})^{-l}\middle|\begin{matrix}\Delta(l,\ 2-\alpha),\ (a_p)\\(b_q),\ \Delta(l,\ 3/2-\alpha)\end{matrix}\right)$$

$[p \neq q;\ c^*,\ \operatorname{Re} a,\ \operatorname{Re} c > 0;\ \operatorname{Re}(lb_j+\alpha) > 1;\ j=1,\ 2,\ \ldots,\ m;\ l=1,\ 2,\ \ldots]$.

2.24.3. Integrals containing exponential, hyperbolic or trigonometric functions and $G_{pq}^{mn}\left(\omega x^{l/k}\middle|\begin{matrix}(a_p)\\(b_q)\end{matrix}\right)$.

NOTATION: see 2.24.1.

1. $$\int_0^\infty x^{\alpha-1}e^{-\sigma x}G_{pq}^{mn}\left(\omega x^{l/k}\middle|\begin{matrix}(a_p)\\(b_q)\end{matrix}\right)dx=$$

$$=\frac{k^{\mu}l^{\alpha-1/2}\sigma^{-\alpha}}{(2\pi)^{(l-1)/2+c^*(k-1)}}G_{kp+l,\ kq}^{km,\ kn+l}\left(\frac{\omega^k l^l}{\sigma^l k^{k(q-p)}}\middle|\begin{matrix}\Delta(l,\ 1-\alpha),\ \Delta(k,\ (a_p))\\\Delta(k,\ (b_q))\end{matrix}\right)$$

[conditions see 2.24.1.1 for $s=v=1,\ t=u=d_1=0$].

2. $$\int_0^\infty \operatorname{ch}\frac{3x}{2}(a+b\operatorname{ch}x)^\theta G_{pq}^{mn}\left((a+b\operatorname{ch}x)^l\middle|\begin{matrix}(a_p)\\(b_q)\end{matrix}\right)dx=$$

$$=\sqrt{\frac{\pi}{2b^3l^3}}G_{p+l,\ q+l}^{m+l,\ n}\left((a+b)^l\middle|\begin{matrix}(a_p)+(2\theta+3)/(2l),\ \Delta(l,\ 3/2)\\\Delta(l,\ 0),\ (b_q)+(2\theta+3)/(2l)\end{matrix}\right)+$$

$$+\sqrt{\frac{\pi}{2}}\frac{1}{bl}G_{p+l,\ q+l}^{m+l,\ n}\left((a+b)^l\middle|\begin{matrix}(a_p)+(2\theta+1)/(2l),\ \Delta(n,\ 1/2)\\\Delta(l,\ 0),\ (b_q)+(2\theta+1)/(2l)\end{matrix}\right)$$

$[p \neq q;\ c^* > 0;\ \operatorname{Re} b > 0;\ \operatorname{Re}[l(1-a_j)-\theta] > 3/2;\ j=1,\ 2,\ \ldots,\ n;\ l=1,\ 2,\ \ldots]$.

3. $$\int_0^\infty x^{\tilde{\alpha}-1}\begin{Bmatrix}\sin bx\\\cos bx\end{Bmatrix}G_{pq}^{mn}\left(\omega x^{2l/k}\middle|\begin{matrix}(a_p)\\(b_q)\end{matrix}\right)dx=\frac{k^{\mu}(2l)^{\tilde{\alpha}-1/2}b^{-\tilde{\alpha}}}{2(2\pi)^{c^*(k-1)-1/2}}\times$$

$$\times G_{kp+2l,\ kq}^{km,\ kn+l}\left(\frac{\omega^k(2l)^{2l}}{b^{2l}k^{k(q-p)}}\middle|\begin{matrix}\Delta(l,\ (3\mp 1-2\tilde{\alpha})/4),\ \Delta(k,\ (a_p)),\ \Delta(l,\ (3\pm 1-2\tilde{\alpha})/4)\\\Delta(k,\ (b_q))\end{matrix}\right)$$

[conditions see 2.24.1.1 for $\alpha=\tilde{\alpha}/2,\ \sigma=b^2/4,\ s=1,\ t=u=0,\ v=2,\ d_1=1/2,\ d_2=(1\mp 1)/4$].

4. $$\int_0^{\pi/2}\sin^{\alpha-1}x\cos^{\beta-1}x e^{i(\alpha+\beta)x}G_{pq}^{mn}\left(\omega e^{ilx}\sin^l x\middle|\begin{matrix}(a_p)\\(b_q)\end{matrix}\right)dx=$$

$$=e^{i\pi\alpha/2}l^{-\beta}\,\Gamma(\beta)\,G_{p+l,\ q+l}^{m,\ n+l}\left(\omega e^{i\pi l/2}\middle|\begin{matrix}\Delta(l,\ 1-\alpha),\ (a_p)\\(b_q),\ \Delta(l,\ 1-\alpha-\beta)\end{matrix}\right)$$

$[p < q;\ m > 0;\ \operatorname{Re}\beta,\ \operatorname{Re}(lb_j+\alpha) > 0;\ j=1,\ 2,\ \ldots,\ m]$ or $[q < p;\ n > 0;\ c^* \geqslant 0;\ |\arg\omega| = c^*\pi;$ $\operatorname{Re}\beta,\ \operatorname{Re}(lb_j+\alpha) > 0,\ j=1,\ 2,\ \ldots,\ m;\ \operatorname{Re}[(p-q)\alpha-l\mu] > -3l/2]$.

5. $$\int_0^{\pi/2}\sin^{\alpha-1}x\cos^{\beta-1}x e^{i(\alpha+\beta)x}G_{pq}^{mn}\left(\omega e^{i(k+l)x}\sin^l x\cos^k x\middle|\begin{matrix}(a_p)\\(b_q)\end{matrix}\right)dx=$$

$$=e^{i\pi\alpha/2}\sqrt{2\pi}\frac{l^{\alpha-1/2}k^{\beta-1/2}}{(k+l)^{\alpha+\beta-1/2}}\times$$

$$\times G_{p+k+l,\ q+k+l}^{m,\ n+k+l}\left(\frac{\omega k^k l^l e^{i\pi l/2}}{(k+l)^{k+l}}\middle|\begin{matrix}\Delta(l,\ 1-\alpha),\ \Delta(k,\ 1-\beta),\ (a_p)\\(b_q),\ \Delta(k+l,\ 1-\alpha-\beta)\end{matrix}\right)$$

$[p < q;\ m > 0;\ \operatorname{Re}(\alpha+lb_j),\ \operatorname{Re}(\beta+kb_j) > 0;\ j=1,\ 2,\ \ldots,\ m]$ or $[n > 0;\ q < p;\ c^* \geqslant 0;$ $|\arg\omega| = c^*\pi;\ \operatorname{Re}(\alpha+lb_j),\ \operatorname{Re}(\beta+kb_j) > 0;\ j=1,\ 2,\ \ldots,\ m;\ \operatorname{Re}[l\mu+(q-p)\alpha] < 3l/2;$ $\operatorname{Re}[k\mu+(q-p)\beta] < 3k/2]$

2.24.4. Integrals containing Bessel functions and $G_{pq}^{mn}\left(\omega x^{l/k}\,\middle|\,\begin{matrix}(a_p)\\(b_q)\end{matrix}\right)$.

NOTATION: see 2.24.1.

1. $$\int_0^\infty x^{\tilde{\alpha}-1} J_\nu(bx)\, G_{pq}^{mn}\left(\omega x^{2l/k}\,\middle|\,\begin{matrix}(a_p)\\(b_q)\end{matrix}\right) dx =$$
$$= \frac{k^\mu (2l)^{\tilde{\alpha}-1}}{(2\pi)^{c^*(k-1)}\, b^{\tilde{\alpha}}}\, G_{kp+2l,\,kq}^{km,\,kn+l}\left(\frac{\omega^k (2l)^{2l}}{b^{2l} k^{k(q-p)}}\,\middle|\,\begin{matrix}\Delta\left(l,\,1-\frac{\tilde{\alpha}+\nu}{2}\right),\ \Delta(k,(a_p)),\ \Delta\left(l,\,1-\frac{\tilde{\alpha}-\nu}{2}\right)\\ \Delta(k,\,(b_q))\end{matrix}\right)$$
[conditions see 2.24.1.1 for $\alpha=\tilde{\alpha}/2$, $\sigma=b^2/4$, $s=1$, $t=u=0$, $v=2$, $d_1=\nu/2$, $d_2=-\nu/2$].

2. $$\int_0^\infty x^{\tilde{\alpha}-1} Y_\nu(bx)\, G_{pq}^{mn}\left(\omega x^{2l/k}\,\middle|\,\begin{matrix}(a_p)\\(b_q)\end{matrix}\right) dx = \frac{k^\mu (2l)^{\tilde{\alpha}-1}}{(2\pi)^{c^*(k-1)}\, b^{\tilde{\alpha}}}\, G_{kp+3l,\,kq+l}^{km,\,kn+2l}\left(\omega^k \times\right.$$
$$\left.\times \frac{(2l)^{2l}}{b^{2l} k^{k(q-p)}}\,\middle|\,\begin{matrix}\Delta\left(l,\,1-\frac{\tilde{\alpha}+\nu}{2}\right),\ \Delta\left(l,\,1-\frac{\tilde{\alpha}-\nu}{2}\right),\ \Delta(k,(a_p)),\ \Delta\left(l,\,\frac{3-\tilde{\alpha}+\nu}{2}\right)\\ \Delta(k,\,(b_q)),\ \Delta\left(l,\,\frac{3-\tilde{\alpha}+\nu}{2}\right)\end{matrix}\right)$$
[conditions see 2.24.1.1 for $\alpha=\tilde{\alpha}/2$, $\sigma=b^2/4$, $s=2$, $t=0$, $u=1$, $v=3$, $c_1=d_3=(1-\nu)/2$, $d_1=-\nu/2$, $d_2=\nu/2$].

3. $$\int_0^\infty x^{\tilde{\alpha}-1} K_\nu(bx)\, G_{pq}^{mn}\left(\omega x^{2l/k}\,\middle|\,\begin{matrix}(a_p)\\(b_q)\end{matrix}\right) dx = \frac{\pi k^\mu (2l)^{\tilde{\alpha}-1}}{(2\pi)^{c^*(k-1)+l}\, b^{\tilde{\alpha}}} \times$$
$$\times G_{kp+2l,\,kq}^{km,\,kn+2l}\left(\frac{\omega^k (2l)^{2l}}{b^{2l} k^{k(q-p)}}\,\middle|\,\begin{matrix}\Delta(l,\,1-(\tilde{\alpha}+\nu)/2),\ \Delta(l,\,1-(\tilde{\alpha}-\nu)/2),\ \Delta(k,\,(a_p))\\ \Delta(k,\,(b_q))\end{matrix}\right)$$
[conditions see 2.24.1.1 for $\alpha=\tilde{\alpha}/2$, $\sigma=b^2/4$, $s=v=2$, $t=u=0$, $d_1=-\nu/2$, $d_2=\nu/2$].

2.24.5. Integrals containing orthogonal polynomials and $G_{pq}^{mn}\left(\omega x^{l/k}\,\middle|\,\begin{matrix}(a_p)\\(b_q)\end{matrix}\right)$.

NOTATION: see 2.24.1.

1. $$\int_0^a x^{\tilde{\alpha}-1} (a^2-x^2)^{\lambda-1/2} C_r^\lambda\left(\frac{x}{a}\right) G_{pq}^{mn}\left(\omega x^{2l/k}\,\middle|\,\begin{matrix}(a_p)\\(b_q)\end{matrix}\right) dx = \frac{(2\lambda)_r \Gamma(\lambda+1/2)\, k^\mu l^{-\lambda-1/2}}{2(2\pi)^{c^*(k-1)} r!\, a^{1-\tilde{\alpha}-2\lambda}} \times$$
$$\times G_{kp+2l,\,kq+2l}^{km,\,kn+2l}\left(\frac{\omega^k k^{k(p-q)}}{\sigma^{2l}}\,\middle|\,\begin{matrix}\Delta(2l,\,1-\tilde{\alpha}),\ \Delta(k,\,(a_p))\\ \Delta(k,(b_q)),\ \Delta(l,\,(1-\tilde{\alpha}-r)/2-\lambda),\ \Delta(l,\,(1-\tilde{\alpha}+r)/2)\end{matrix}\right)$$
[conditions see 2.24.1.1 for $\alpha=\tilde{\alpha}/2$, $\sigma=a^{-2}$, $s=u=v=2$, $t=0$, $c_1=\lambda+(1+r)/2$, $c_2=(1-r)/2$, $d_1=0$, $d_2=1/2$; $r=0,\ 1,\ 2,\ \ldots$].

2. $$\int_0^a x^{\alpha-1} (a-x)^{\beta-1} P_r^{(\lambda,\,\nu)}(1-bx)\, G_{pq}^{mn}\left(\omega x^l\,\middle|\,\begin{matrix}(a_p)\\(b_q)\end{matrix}\right) dx = \frac{4(\lambda+1)_r \Gamma(\beta)\, a^{\alpha+\beta-1}}{l\beta\, r!} \times$$
$$\times \sum_{j=0}^r \frac{(-r)_j (1+\lambda+\nu+r)_j}{(\lambda+1)_j\, j!}\left(\frac{ab}{2}\right)^j G_{p+l,\,q+l}^{m,\,n+l}\left(a^l\omega\,\middle|\,\begin{matrix}\Delta(l,\,1-\alpha-j),\ (a_p)\\ (b_q),\ \Delta(l,\,1-\alpha-\beta-j)\end{matrix}\right)$$
[$p\leqslant q$; a, c^*, $\operatorname{Re}\beta>0$; $|\arg\omega|<c^*\pi$; $\operatorname{Re}\alpha+l\min\limits_{1\leqslant j\leqslant m}\operatorname{Re} b_j > 0$].

3. $$\int_0^a x^{\alpha-1}(a-x)^{\lambda} P_r^{(\lambda,\nu)}\left(\frac{2x}{a}-1\right) G_{pq}^{mn}\left(\omega x^{l/k}\left|\begin{matrix}(a_p)\\(b_q)\end{matrix}\right.\right)dx=\frac{\Gamma(\lambda+r+1)\,k^{\mu}a^{\alpha+\lambda}}{(2\pi)^{c^*(k-1)}l^{\lambda+1}r!}\times$$
$$\times G_{kp+2l,\ kq+2l}^{km,\ kn+2l}\left(\frac{\omega^k a^l}{k^{k(q-p)}}\left|\begin{matrix}\Delta(l,1-\alpha),\ \Delta(l,1-\alpha+\nu),\ \Delta(k,(a_p))\\ \Delta(k,(b_q)),\ \Delta(l,1-\alpha+\nu+r),\ \Delta(l,-\alpha-\lambda-r)\end{matrix}\right.\right)$$
[conditions see 2.24.1.1 for $\sigma=a^{-1}$, $s=u=v=2$, $t=0$, $c_1=\lambda+r+1$, $c_2=-\nu-r$, $d_1=0$, $d_2=-\nu$; $r=0, 1, 2, \ldots$].

2.24.6. Integrals containing Legendre functions and $G_{pq}^{mn}\left(\omega x^{l/k}\left|\begin{matrix}(a_p)\\(b_q)\end{matrix}\right.\right)$.

NOTATION: see 2.24.1.

1. $$\int_0^a x^{\tilde{\alpha}-1}(a^2-x^2)^{-\lambda/2}P_{\nu}^{\lambda}\left(\frac{x}{a}\right)G_{pq}^{mn}\left(\omega x^{2l/k}\left|\begin{matrix}(a_p)\\(b_q)\end{matrix}\right.\right)dx=\frac{k^{\mu}(2l)^{\lambda-1}a^{\tilde{\alpha}-\lambda}}{(2\pi)^{c^*(k-1)}}\times$$
$$\times G_{kp+2l,\ kq+2l}^{km,\ kn+2l}\left(\frac{a^{2l}\omega^k}{k^{k(q-p)}}\left|\begin{matrix}\Delta(2l,1-\tilde{\alpha}),\ \Delta(k,(a_p))\\ \Delta(k,(b_q)),\ \Delta(l,(\lambda-\tilde{\alpha}-\nu)/2),\ \Delta(l,(1+\lambda-\tilde{\alpha}+\nu)/2)\end{matrix}\right.\right)$$
[conditions see 2.24.1.1 for $\alpha=\tilde{\alpha}/2$, $\sigma=a^{-2}$, $s=u=v=2$, $t=0$, $c_1=(1-\lambda-\nu)/2$, $c_2=1+(\nu-\lambda)/2$, $d_1=0$, $d_2=1/2$].

2. $$\int_a^{\infty} x^{\tilde{\alpha}-1}(x^2-a^2)^{-\lambda/2}P_{\nu}^{\lambda}\left(\frac{x}{a}\right)G_{pq}^{mn}\left(\omega x^{2l/k}\left|\begin{matrix}(a_p)\\(b_q)\end{matrix}\right.\right)dx=\frac{k^{\mu}(2l)^{\lambda-1}a^{\tilde{\alpha}-\lambda}}{(2\pi)^{c^*(k-1)}}\times$$
$$\times G_{kp+2l,\ kq+2l}^{km+2l,\ kn}\left(\frac{a^{2l}\omega^k}{k^{k(q-p)}}\left|\begin{matrix}\Delta(k,(a_p)),\ \Delta(2l,1-\tilde{\alpha})\\ \Delta(l,(\lambda-\tilde{\alpha}-\nu)/2),\ \Delta(l,(1+\lambda-\tilde{\alpha}+\nu)/2),\ \Delta(k,(b_q))\end{matrix}\right.\right)$$
[conditions see 2.24.1.1 for $\alpha=\tilde{\alpha}/2$, $\sigma=a^{-2}$, $t=u=v=2$, $s=0$, $c_1=(1-\lambda-\nu)/2$, $c_2=1+(\nu-\lambda)/2$, $d_1=0$, $d_2=1/2$].

2.24.7. Integrals containing Whittaker functions and $G_{pq}^{mn}\left(\omega x^{l/k}\left|\begin{matrix}(a_p)\\(b_q)\end{matrix}\right.\right)$.

NOTATION: see 2.24.1.

1. $$\int_0^{\infty} x^{\alpha-1}e^{-\sigma x/2}M_{\lambda,\nu}(\sigma x)\,G_{pq}^{mn}\left(\omega x^{l/k}\left|\begin{matrix}(a_p)\\(b_q)\end{matrix}\right.\right)dx=\frac{k\ \ l^{\alpha+\lambda-1/2}\sigma^{-\alpha}}{(2\pi)^{(l-1)/2+c^*(k-1)}}\Gamma\left[\begin{matrix}2\nu+1\\ \lambda+\nu+1/2\end{matrix}\right]\times$$
$$\times G_{kp+2l,\ kq+l}^{km+l,\ kn+l}\left(\frac{\omega^k l^l}{\sigma^l k^{k(q-p)}}\left|\begin{matrix}\Delta(l,1/2-\alpha-\nu),\ \Delta(k,(a_p)),\ \Delta(l,1/2-\alpha+\nu)\\ \Delta(l,\lambda-\alpha),\ \Delta(k,(b_q))\end{matrix}\right.\right)$$
[conditions see 2.24.1.1 for $s=t=u=1$, $v=2$, $c_1=1-\lambda$, $d_1=1/2+\nu$, $d_2=1/2-\nu$].

2. $$\int_0^{\infty} x^{\alpha-1}e^{-\sigma x/2}W_{\lambda,\nu}(\sigma x)\,G_{pq}^{mn}\left(\omega x^{l/k}\left|\begin{matrix}(a_p)\\(b_q)\end{matrix}\right.\right)dx=\frac{k^{\mu}l^{\alpha+\lambda-1/2}\sigma^{-\alpha}}{(2\pi)^{(l-1)/2+c^*(k-1)}}\times$$
$$\times G_{kp+2l,\ kq+l}^{km,\ kn+2l}\left(\frac{\omega^k l^l}{\sigma^l k^{k(q-p)}}\left|\begin{matrix}\Delta(l,1/2-\alpha-\nu),\ \Delta(l,1/2-\alpha+\nu),\ \Delta(k,(a_p))\\ \Delta(k,(b_q)),\ \Delta(l,\lambda-\alpha)\end{matrix}\right.\right)$$
[conditions see 2.24.1.1 for $s=v=2$, $t=0$, $u=1$, $c_1=1-\lambda$, $d_1=1/2+\nu$, $d_2=1/2-\nu$].

3. $$\int_0^{\infty} x^{\alpha-1}e^{\sigma x/2}W_{\lambda,\nu}(\sigma x)\,G_{pq}^{mn}\left(\omega x^{l/k}\left|\begin{matrix}(a_p)\\(b_q)\end{matrix}\right.\right)dx=$$
$$=\frac{(2\pi)^{3(1-l)/2+c^*(1-k)}k^{\mu}l^{\alpha-\lambda-1/2}\sigma^{-\alpha}}{\Gamma(1/2-\lambda-\nu)\,\Gamma(1/2-\lambda+\nu)}\times$$
$$\times G_{kp+2l,\ kq+l}^{km+l,\ kn+2l}\left(\frac{\omega^k l^l}{\sigma^l k^{k(q-p)}}\left|\begin{matrix}\Delta(l,1/2-\alpha-\nu),\ \Delta(l,1/2-\alpha+\nu),\ \Delta(k,(a_p))\\ \Delta(l,-\alpha-\lambda),\ \Delta(k,(b_q))\end{matrix}\right.\right)$$
[conditions see 2.24.1.1 for $s=v=2$, $t=u=1$, $c_1=\lambda+1$, $d_1=1/2+\nu$, $d_2=1/2-\nu$].

2.24.8. Integrals containing the Gauss function ${}_2F_1(a, b; c; x)$ and $G_{pq}^{mn}\left(\varphi(x)\left|\begin{matrix}(a_p)\\(b_q)\end{matrix}\right.\right)$.

NOTATION: see 2.24.1.

1. $$\int_0^\infty x^{\alpha-1}{}_2F_1(a, b; c; 1-\sigma x)\, G_{pq}^{mn}\left(\omega x^{l/k}\left|\begin{matrix}(a_p)\\(b_q)\end{matrix}\right.\right) dx=$$
$$=\frac{k^{\mu} l^{c-2}\sigma^{-\rho}}{(2\pi)^{2(l-1)+c^*(k-1)}}\Gamma\begin{bmatrix}c\\a, b, c-a, c-b\end{bmatrix}\times$$
$$\times G_{kp+2l,\ kq+2l}^{km+2l,\ kn+2l}\left(\frac{\omega^k}{\sigma^l k^{k(q-p)}}\left|\begin{matrix}\Delta(l, 1-\alpha),\ \Delta(l, 1+a+b-c-\alpha),\ \Delta(k, (a_p))\\ \Delta(l, a-\alpha),\ \Delta(l, b-\alpha),\ \Delta(k, (b_q))\end{matrix}\right.\right)$$
[conditions see 2.24.1.1 for $s=t=u=v=2,\ c_1=1-a,\ c_2=1-b,\ d_1=0,\ d_2=c-a-b$].

2. $$\int_0^d x^{\alpha-1}(d-x)^{c-1}{}_2F_1\left(a, b; c; 1-\frac{x}{d}\right) G_{pq}^{mn}\left(\omega x^{l/k}\left|\begin{matrix}(a_p)\\(b_q)\end{matrix}\right.\right) dx=\frac{k^{\mu} l^{-c}\Gamma(c)}{(2\pi)^{c^*(k-1)}d^{1-c-\alpha}}\times$$
$$\times G_{kp+2l,\ kq+2l}^{km,\ kn+2l}\left(\frac{\omega^k d^l}{k^{k(q-p)}}\left|\begin{matrix}\Delta(l, 1-\alpha),\ \Delta(l, 1+a+b-c-\alpha),\ \Delta(k, (a_p))\\ \Delta(k, (b_q)),\ \Delta(l, 1+a-c-\alpha),\ \Delta(l, 1+b-c-\alpha)\end{matrix}\right.\right)$$
[conditions see 2.24.1.1 for $\sigma=d^{-1},\ s=u=v=2,\ t=0,\ c_1=c-a,\ c_2=c-b,\ d_1=0,\ d_2=c-a-b$].

3. $$\int_d^\infty x^{\alpha-1}(x-d)^{c-1}{}_2F_1\left(a, b; c; 1-\frac{x}{d}\right) G_{pq}^{mn}\left(\omega x^{l/k}\left|\begin{matrix}(a_p)\\(b_q)\end{matrix}\right.\right) dx=\frac{k^{\mu} l^{-c}\Gamma(c)}{(2\pi)^{c^*(k-1)}d^{1-c-\alpha}}\times$$
$$\times G_{kp+2l,\ kq+2l}^{km+2l,\ kn}\left(\frac{\omega^k d^l}{k^{k(q-p)}}\left|\begin{matrix}\Delta(k, (a_p)),\ \Delta(l, 1-\alpha),\ \Delta(l, 1+a+b-c-\alpha)\\ \Delta(l, 1+a-c-\alpha),\ \Delta(l, 1+b-c-\alpha),\ \Delta(k, (b_q))\end{matrix}\right.\right)$$
[conditions see 2.24.1.1 for $\sigma=d^{-1},\ t=u=v=2,\ s=0,\ c_1=c-a,\ c_2=c-b,\ d_1=0,\ d_2=c-a-b$].

4. $$\int_0^\infty x^{\alpha-1}{}_2F_1(a, b; c; -\sigma x)\, G_{pq}^{mn}\left(\omega x^{l/k}\left|\begin{matrix}(a_p)\\(b_q)\end{matrix}\right.\right) dx=\frac{k^{\mu} l^{a+b-c-1}\sigma^{-\alpha}}{(2\pi)^{l-1+c^*(k-1)}}\Gamma\begin{bmatrix}c\\a, b\end{bmatrix}\times$$
$$\times G_{kp+2l,\ kq+2l}^{km+2l,\ kn+l}\left(\frac{\sigma^{-l}\omega^k}{k^{k(q-p)}}\left|\begin{matrix}\Delta(l, 1-\alpha),\ \Delta(k, (a_p)),\ \Delta(l, c-\alpha)\\ \Delta(l, a-\alpha),\ \Delta(l, b-\alpha),\ \Delta(k, (b_q))\end{matrix}\right.\right)$$
[conditions see 2.24.1.1 for $s=1,\ t=u=v=2,\ c_1=1-a,\ c_2=1-b,\ d_1=0,\ d_2=1-c$].

5. $$\int_0^a x^{\alpha-1}(a-x)^{\alpha-c}{}_2F_1\left(b, 1-b; c; \frac{x}{a}\right) G_{pq}^{mn}\left(\omega(ax-x^2)^{l/k}\left|\begin{matrix}(a_p)\\(b_q)\end{matrix}\right.\right) dx=$$
$$=\frac{(2\pi)^{c^*(1-k)+1}k^{\mu}}{2^{2\alpha}\sqrt{l}\,a^{c-2\alpha}}\Gamma\begin{bmatrix}c\\(b+c)/2,\ (1-b+c)/2\end{bmatrix}\times$$
$$\times G_{kp+2l,\ kq+2l}^{km,\ kn+2l}\left(\frac{\omega^k a^{2l}}{2^{2l}k^{k(q-p)}}\left|\begin{matrix}\Delta(l, 1-\alpha),\ \Delta(l, c-\alpha),\ \Delta(k, (a_p))\\ \Delta(k, (b_q)),\ \Delta(l, (1-b+c)/2-\alpha),\ \Delta(l, (b+c)/2-\alpha)\end{matrix}\right.\right)$$
[$p<q;\ a, m>0;\ \mathrm{Re}\,(k\alpha+lb_j),\ \mathrm{Re}\,(k+k\alpha-kc+lb_j)>0;\ j=1, 2, \ldots, m$] or [$p>q;\ a, n>0;\ c^*\geqslant 0;$ $|\arg\omega|=c^*\pi;\ \mathrm{Re}\,(k\alpha+lb_j),\ \mathrm{Re}\,(k+k\alpha-kc+lb_j)>0;\ j=1, 2, \ldots, m;\ (p-q)k\,\mathrm{Re}\,\alpha-l\,\mathrm{Re}\,\mu>$ $>-3l/2;\ (p-q)k(1+\mathrm{Re}\,\alpha-\mathrm{Re}\,c)-l\,\mathrm{Re}\,\mu>-3l/2$].

6. $$\int_0^a x^{\alpha-1}(a-x)^{\alpha-1}{}_2F_1\left(a, b; \frac{a+b+1}{2}; \frac{x}{a}\right) G_{pq}^{mn}\left(\omega(ax-x^2)^{l/k}\left|\begin{matrix}(a_p)\\(b_q)\end{matrix}\right.\right) dx=$$
$$=\frac{(2\pi)^{c^*(1-k)+1}k^{\mu}a^{2\alpha-1}}{2^{2\alpha}\sqrt{l}}\Gamma\begin{bmatrix}(a+b+1)/2\\(a+1)/2,\ (b+1)/2\end{bmatrix}\times$$
$$\times G_{kp+2l,\ kq+2l}^{km,\ kn+2l}\left(\frac{\omega^k a^{2l}}{2^{2l}k^{k(q-p)}}\left|\begin{matrix}\Delta(l, 1-\alpha),\ \Delta(l, (a+b+1)/2-\alpha),\ \Delta(k, (a_p))\\ \Delta(k, (b_q)),\ \Delta(l, (a+1)/2-\alpha),\ \Delta(l, (b+1)/2-\alpha)\end{matrix}\right.\right)$$
[$a, c^*>0;\ |\arg\omega|<c^*\pi;\ \mathrm{Re}\,(k\alpha+lb_j)>0;\ \mathrm{Re}\,(2lb_j+2k\alpha-ka-kb)>-k;\ j=1, 2, \ldots, m$].

7. $\int_0^{\pi/2} \sin^{\alpha-1} x \cos^{c-1} x\, e^{i(\alpha+c)x} {}_2F_1(a, b; c; e^{ix}\cos x)\, G_{pq}^{mn}\left(\omega e^{il(x-\pi/2)}\sin^l x \middle| \begin{matrix}(a_p)\\(b_q)\end{matrix}\right) dx =$

$$= \frac{e^{\alpha\pi i/2}\Gamma(c)}{l^c} G_{p+2k,\, q+2k}^{m,\, n+2k}\left(\omega \middle| \begin{matrix}\Delta(l, 1-\alpha),\ \Delta(l, 1+a+b-c-\alpha),\ (a_p)\\ (b_q),\ \Delta(l, 1+a-c-\alpha),\ \Delta(l, 1+b-c-\alpha)\end{matrix}\right)$$

[$p<q$; $\operatorname{Re} c$, $\operatorname{Re}(c-a-b)$, $\operatorname{Re}(\alpha+lb_j)>0$; $j=1, 2, \ldots, m$] or [$q<p$; $n>0$; $c^*\geqslant 0$; $|\arg\omega|=c^*\pi$; $\operatorname{Re} c$, $\operatorname{Re}(c-a-b)$, $\operatorname{Re}(\alpha+lb_j)>0$; $j=1, 2, \ldots, m$; $\operatorname{Re}[(p-q)\alpha-l\mu]>-3l/2$].

2.24.9. Integrals containing ${}_sF_t((c_s); (d_t); -\sigma x)\, G_{pq}^{mn}\left(\omega x^{l/k} \middle| \begin{matrix}(a_p)\\(b_q)\end{matrix}\right)$.

NOTATION: see 2.24.1.

1. $\int_0^\infty x^{\alpha-1} {}_sF_t\left(\begin{matrix}(c_s); -\sigma x\\(d_t)\end{matrix}\right) G_{pq}^{mn}\left(\omega x^{l/k} \middle| \begin{matrix}(a_p)\\(b_q)\end{matrix}\right) dx = \frac{k^\mu l^\eta \sigma^{-\alpha}}{(2\pi)^{c^*(k-1)+(1+s-t)(l-1)/2}} \Gamma\left[\begin{matrix}(d_t)\\(c_s)\end{matrix}\right] \times$

$$\times G_{kp+tl+l,\ kq+sl}^{km+sl,\ kn+l}\left(\frac{\omega^k k^{k(p-q)}}{\sigma^l l^{l(s-t-1)}} \middle| \begin{matrix}\Delta(l, 1-\alpha),\ \Delta(k, (a_p)),\ \Delta(l, (d_t)-\alpha)\\ \Delta(l, (c_s)-\alpha),\ \Delta(k, (b_q))\end{matrix}\right)$$

[conditions see 2.24.1.1 for $s=1$, $t=u=s$, $v=t+1$, $(c_u)=1-(c_s)$, $(d_v)=0, 1-(d_t)$].

2.25. THE FOX H-FUNCTION $H_{pq}^{mn}\left[x\begin{matrix}[a_p, A_p]\\ [b_q, B_q]\end{matrix}\right]$

The conditions for convergence of integrals given in this section are only sufficient ones; other convergence conditions can also be given.

2.25.1. Integrals of general form

NOTATION:

$$a^* = \sum_{j=1}^n A_j - \sum_{j=n+1}^p A_j + \sum_{j=1}^m B_j - \sum_{j=m+1}^q B_j,$$

$$b^* = \sum_{j=1}^t C_j - \sum_{j=t+1}^u C_j + \sum_{j=1}^s D_j - \sum_{j=s+1}^v D_j.$$

1. $\int_0^\infty x^{\alpha-1} H_{uv}^{st}\left[\sigma x \middle| \begin{matrix}[c_u, C_u]\\ [d_v, D_v]\end{matrix}\right] H_{pq}^{mn}\left[\omega x^r \middle| \begin{matrix}[a_p, A_p]\\ [b_q, B_q]\end{matrix}\right] dx =$

$$= \sigma^{-\alpha} H_{p+v,\ q+u}^{m+t,\ n+s}\left[\frac{\omega}{\sigma^r} \middle| \begin{matrix}[a_n, A_n],\ [1-d_v-\alpha D_v, rD_v],\ (a_{n+1}, A_{n+1}), \ldots, (a_p, A_p)\\ [b_m, B_m],\ [1-c_u-\alpha C_u, rC_u],\ (b_{m+1}, B_{m+1}), \ldots, (b_q, B_q)\end{matrix}\right]$$

[$mnst \neq 0$; a^*, b^*, $r>0$; $|\arg\sigma|<b^*\pi/2$; $|\arg\omega|<a^*\pi/2$; $\operatorname{Re}\alpha + r\min_{1\leqslant j\leqslant m}\operatorname{Re}(b_j/B_j) + \min_{1\leqslant h\leqslant s}\operatorname{Re}(d_h/D_h) > 0$; $\operatorname{Re}\alpha + r\max_{1\leqslant j\leqslant n}\operatorname{Re}((a_j-1)/A_j) + \max_{1\leqslant h\leqslant t}\operatorname{Re}((c_h-1)/C_h) < 0$; $A_j(b_h+k) \neq B_h(a_j-l-1)$; $k, l=0, 1, \ldots$; $j=1, 2, \ldots, n$; $h=1, 2, \ldots, m$; $C_j(d_h+k) \neq D_h(c_j-l-1)$; $k, l=0, 1, \ldots$; $j=1, 2, \ldots, t$; $h=1, 2, \ldots, s$].

2.25.2. Integrals containing elementary functions and the H-function.

NOTATION:

$$a^* = \sum_{j=1}^n A_j - \sum_{j=n+1}^p A_j + \sum_{j=1}^m B_j - \sum_{j=m+1}^q B_j.$$

1. $$\int_0^\infty x^{\alpha-1} H_{pq}^{mn}\left[\omega x \middle| \begin{matrix} [a_p, & A_p] \\ [b_q, & B_q] \end{matrix}\right] dx =$$

$$= \omega^{-\alpha}\Gamma\begin{bmatrix} b_1+B_1\alpha, \ldots, b_m+B_m\alpha, 1-a_1-A_1\alpha, \ldots, 1-a_n-A_n\alpha \\ a_{n+1}+A_{n+1}\alpha, \ldots, a_p+A_p\alpha, 1-b_{m+1}-B_{m+1}\alpha, \ldots, 1-b_q-B_q\alpha \end{bmatrix}$$

$$\left[mn \neq 0;\ a^* > 0;\ |\arg\omega| < a^*\pi/2;\ -\min_{1\leqslant j\leqslant m} \operatorname{Re}(b_j/B_j) < \operatorname{Re}\alpha < \min_{1\leqslant j\leqslant n} \operatorname{Re}((1-a_j)/A_j)\right].$$

2. $$\int_0^a x^{\alpha-1}(a-x)^{\beta-1} H_{pq}^{mn}\left[\omega x^r (a-x)^\rho \middle| \begin{matrix} [a_p, & A_p] \\ [b_q, & B_q] \end{matrix}\right] dx =$$

$$= a^{\alpha+\beta-1} H_{p+2,\, q+1}^{m,\, n+2}\left[\omega a^{r+\rho} \middle| \begin{matrix} (1-\alpha, r), (1-\beta, \rho), [a_p, A_p] \\ [b_q, B_q], (1-\alpha-\beta, r+\rho) \end{matrix}\right]$$

$$\left[mn \neq 0;\ r, \rho \geqslant 0;\ a, a^* > 0;\ |\arg\omega| < a^*\pi/2;\ \operatorname{Re}\alpha + r \min_{1\leqslant j\leqslant m} \operatorname{Re}(b_j/B_j) > 0;\right.$$
$$\left.\operatorname{Re}\alpha + \rho \min_{1\leqslant j\leqslant m} \operatorname{Re}(b_j/B_j) > 0\right].$$

3. $$\int_0^\infty x^{\alpha-1} e^{-\sigma x} H_{pq}^{mn}\left[\omega x^r \middle| \begin{matrix} [a_p, & A_p] \\ [b_q, & B_q] \end{matrix}\right] dx = \sigma^{-\alpha} H_{p+1,\, q}^{m,\, n+1}\left[\frac{\omega}{\sigma^r} \middle| \begin{matrix} (1-\alpha, r), [a_p, A_p] \\ [b_q, B_q] \end{matrix}\right]$$

$$\left[mn \neq 0;\ a^*, r, \operatorname{Re}\sigma > 0;\ |\arg\omega| < a^*\pi/2;\ \operatorname{Re}\alpha + r \min_{1\leqslant j\leqslant m} \operatorname{Re}(b_j/B_j) > 0\right].$$

4. $$\int_0^\infty x^{\alpha-1} \begin{Bmatrix} \sin\sigma x \\ \cos\sigma x \end{Bmatrix} H_{pq}^{mn}\left[\omega x^r \middle| \begin{matrix} [a_p, & A_p] \\ [b_q, & B_q] \end{matrix}\right] dx =$$

$$= \frac{2^{\alpha-1}\sqrt{\pi}}{\sigma^\alpha} H_{p+2,\, q}^{m,\, n+1}\left[\omega\left(\frac{2}{\sigma}\right)^r \middle| \begin{matrix} ((3\mp 1-2\alpha)/4, r/2), [a_p, A_p], ((3\pm 1-2\alpha)/4, r/2) \\ [b_q, B_q] \end{matrix}\right]$$

$$\left[a^*, r, \sigma > 0;\ |\arg\omega| < a^*\pi/2;\ \operatorname{Re}\alpha + r \min_{1\leqslant j\leqslant m} \operatorname{Re}(b_j/B_j) > (-1\mp 1)/2;\right.$$
$$\left.\operatorname{Re}\alpha + r \max_{1\leqslant j\leqslant n} \operatorname{Re}((a_j-1)/A_j) < 1\right].$$

2.25.3. Integrals containing special functions and the H-function.

1. $$\int_0^\infty \Gamma\begin{bmatrix} (a+ix)/2, (a-ix)/2 \\ ix, -ix \end{bmatrix} H_{p,\, q+2}^{m+2,\, n}\left[\omega \middle| \begin{matrix} [a_p, A_p] \\ (b+ix, r), (b-ix, r), [b_q, B_q] \end{matrix}\right] dx =$$

$$= 2^{b-a+1}\sqrt{\pi}\, H_{p,\, q+2}^{m+2,\, n}\left[\frac{\omega}{2^r} \middle| \begin{matrix} [a_p, A_p] \\ (a+b, r), (b+1/2, r), [b_q, B_q] \end{matrix}\right]$$

$$\left[a^*, r, \operatorname{Re} a > 0;\ |\arg\omega| < a^*\pi/2;\ \operatorname{Re} b - r \max_{1\leqslant j\leqslant n} ((a_j-1)/A_j) > 0\right].$$

2. $$\int_0^\infty x^{\alpha-1} J_\nu(\sigma x) H_{pq}^{mn}\left[\omega x^r \middle| \begin{matrix} [a_p, & A_p] \\ [b_q, & B_q] \end{matrix}\right] dx =$$

$$= \frac{2^{\alpha-1}}{\sigma^\alpha} H_{p+2,\, q}^{m,\, n+1}\left[\omega\left(\frac{2}{\sigma}\right)^r \middle| \begin{matrix} (1-(\alpha+\nu)/2, r/2), [a_p, A_p], (1-(\alpha-\nu)/2, r/2) \\ [b_q, B_q] \end{matrix}\right]$$

$$\left[a^*, r, \sigma > 0;\ |\arg\omega| < a^*\pi/2;\ \operatorname{Re}(\alpha+\nu) + r \min_{1\leqslant j\leqslant m} \operatorname{Re}(b_j/B_j) > 0,\right.$$
$$\left.\operatorname{Re}\alpha + r \max_{1\leqslant j\leqslant n} \operatorname{Re}((a_j-1)/A_j) < 3/2\right].$$

3. $\int_0^{\infty} x^{\alpha-1} J_\mu(\sigma x) J_\nu(\sigma x) H_{pq}^{mn}\left[\omega x^r \middle| \begin{matrix}[a_p, A_p]\\ [b_q, B_q]\end{matrix}\right] dx =$

$$= \frac{2^{\alpha-1}}{\sigma^\alpha} H_{p+4,\, q+1}^{m+1,\, n+1}\left[\omega\left(\frac{2}{\sigma}\right)^r \middle| \begin{matrix}(1-(\alpha+\mu+\nu)/2,\ r/2),\ [a_p, A_p],\\ (1-\alpha,\ r),\ [b_q, B_q]\end{matrix}\right.$$
$$\left.(1-(\alpha-\mu-\nu)/2,\ r/2), (1-(\alpha+\mu-\nu)/2,\ r/2),\ (1-(\alpha-\mu+\nu)/2,\ r/2)\right]$$

$\left[a^*,\ r,\ \sigma>0;\ |\arg\omega|<a^*\pi/2;\ \operatorname{Re}(\alpha+\mu+\nu)+r\min_{1\leqslant j\leqslant m}\operatorname{Re}(b_j/B_j)>0.\right.$
$\left.\operatorname{Re}\alpha+r\max_{1\leqslant j\leqslant n}\operatorname{Re}((a_j-1)/A_j)<1\right].$

4. $\int_0^{\infty} x^{\alpha-1} K_\nu(\sigma x) H_{pq}^{mn}\left[\omega x^r \middle| \begin{matrix}[a_p, A_p]\\ [b_q, B_q]\end{matrix}\right] dx =$

$$= \frac{2^{\alpha-2}}{\sigma^\alpha} H_{p+2,\, q}^{m,\, n+2}\left[\omega\left(\frac{2}{\sigma}\right)^r \middle| \begin{matrix}(1-(\alpha-\nu)/2,\ r/2),\ (1-(\alpha+\nu)/2,\ r/2),\ [a_p, A_p]\\ [b_q, B_q]\end{matrix}\right]$$

$\left[a^*,\ r,\ \operatorname{Re}\sigma>0;\ |\arg\omega|<a^*\pi/2;\ \operatorname{Re}\alpha-|\operatorname{Re}\nu|+r\min_{1\leqslant j\leqslant m}\operatorname{Re}(b_j/B_j)>0\right].$

5. $\int_0^{\infty} x^{\alpha-1} {}_sF_t((c_s);\ (d_t);\ -\sigma x) H_{pq}^{mn}\left[\omega x^r \middle| \begin{matrix}[a_p, A_p]\\ [b_q, B_q]\end{matrix}\right] dx =$

$$= \Gamma\left[\begin{matrix}(d_t)\\ (c_s)\end{matrix}\right] \sigma^{-\alpha} H_{p+t+1,\, q+s}^{m+s,\, n+1}\left[\frac{\omega}{\sigma^r} \middle| \begin{matrix}(1-\alpha,\ r),\ [a_p, A_p],\ [d_t-\alpha,\ r]\\ [c_s-\alpha,\ r],\ [b_q, B_q]\end{matrix}\right]$$

$\left[a^*,\ r>0;\ s+1>t;\ |\arg\sigma|<(s-t+1)\pi/2;\ |\arg\omega|<a^*\pi/2;\ \operatorname{Re}\alpha+r\min_{1\leqslant j\leqslant m}\operatorname{Re}(b_j/B_j)>0;\right.$
$\left.\operatorname{Re}\alpha-\min_{1\leqslant j\leqslant s}\operatorname{Re}c_j+r\max_{1\leqslant j\leqslant n}\operatorname{Re}((a_j-1)/A_j)<0\right].$

2.26. THE THETA-FUNCTIONS $\theta_j(x, q)$

2.26.1. Integrals of $f(x)\,\theta_j(x, q)$.

1. $\int_0^{\pi} \theta_j(nx, q)\, dx = 0$ $\quad [j=1,\ 2;\ n=1,\ 2,\ \ldots]$.

2. $\int_0^{\pi} \left\{\begin{matrix}\sin 2mx\, \theta_1((2n+1)x, q)\\ \cos 2mx\, \theta_2((2n+1)x, q)\end{matrix}\right\} dx = 0.$

3. $\int_0^{\pi} P_{2n+1}(\cos x)\, \theta_1(x, q)\, dx = 0.$

4. $\int_0^{\pi} P_{2n}(\cos x)\, \theta_2(x, q)\, dx = 0.$

5. $\int_0^{\pi} \theta_j(nx, q)\, dx = \pi$ $\quad [j=3, 4;\ n=1, 2, \ldots]$

6. $\int_0^{\pi/2} \cos 2mx \left\{\begin{matrix}\theta_3(nx, q)\\ \theta_4(nx, q)\end{matrix}\right\} dx = (\pm 1)^{m/n} \frac{\pi}{2} q^{m^2/n^2}$ $\quad [m/n=0,\ 1,\ 2,\ \ldots]$,

7. $\qquad = 0$ $\quad$ in the remaining cases

8. $\int_0^{\pi} \sin 2mx\theta_j(nx, q)\,dx = 0$ $[j=3, 4]$.

9. $\int_0^{\pi} \cos 2mx \begin{Bmatrix} \theta_3(nx, q) \\ \theta_4(nx, q) \end{Bmatrix} dx = (\pm 1)^{m/n}\pi q^{m^2/n^2}$ $[m/n = 0, 1, 2, \ldots]$,

10. $= 0$ in the remaining cases

11. $\int_0^{\pi} \cos(2m+1)x\theta_j(nx, q) = 0$ $[j=3, 4]$.

12. $\int_0^{\pi/2} \frac{q\cos 2x - q^2}{1-2q\cos 2x + q^2}\theta_3(x, q)\,dx = \frac{\pi}{4}q^{-1/4}\theta_2(0, q)$.

13. $\int_0^{\pi} P_{2n+1}(\cos x) \begin{Bmatrix} \theta_3(x, q) \\ \theta_4(x, q) \end{Bmatrix} dx = 0$.

2.26.2. Integrals containing products of $\theta_j(ax, q)$.

1. $\int_0^{\pi} \theta_1(mx, p)\,\theta_2(nx, q)\,dx = 0$ $[m, n = 1, 2, \ldots]$.

2. $\int_0^{\pi} \theta_j((2m+1)x, p)\,\theta_j((2m+1)(2n+1)x, q)\,dx = (-1)^{nj}\pi\theta_2(0, p^{4n^2+4n+1}q)$ $[j=1, 2]$.

3. $\int_0^{\pi} \theta_2((2m+1)x, p)\,\theta_j(nx, q)dx = 0$ $[j=3, 4; n=1, 2, \ldots]$.

4. $\int_0^{\pi} \theta_j((2m+1)x, p)\,\theta_j((2m+1)x, q)\,dx = \pi\theta_3(0, pq)$ $[j=3, 4]$.

5. $\int_0^{\pi} \theta_j(2mx, p)\,\theta_4(nx, q)\,dx = 0$ $[j=1, 2; m=1, 2, \ldots]$.

6. $\int_0^{\pi} \theta_2(2mx, p)\,\theta_j(nx, q)\,dx = (-1)^{m(j-1)/n}\pi\theta_2\left(0, pq^{4m^2/n^2}\right)$ $[j=3, 4; m/n = 1, 2, \ldots]$.

7. $\int_0^{\pi} \theta_3(mx, p)\,\theta_4(nx, q)\,dx = \pi\theta_4\left(0, pq^{m^2/n^2}\right)$ $[m/n = 1, 2, \ldots]$.

2.26.3. Integrals with respect to q containing $\theta_j(x, aq)$.

1. $\int_0^{a} q^{\alpha-1} \begin{Bmatrix} \theta_1(x, q/a) \\ \theta_2(x, q/a) \end{Bmatrix} dq = \frac{\pi a^{\alpha}\alpha^{-1/2}}{\operatorname{ch}\sqrt{\alpha}\,\pi} \begin{Bmatrix} \operatorname{sh}(2\sqrt{\alpha}\,x) \\ \operatorname{sh}\left[2\sqrt{\alpha}(\pi - 2x)\right] \end{Bmatrix}$ $[\operatorname{Re}\alpha > 0;\ -(\pi \pm \pi)/4 \leqslant x \leqslant (3\pi \mp \pi)/4]$.

2. $\int_0^{a} q^{\alpha-1} \begin{Bmatrix} \theta_3(x, q/a) \\ \theta_4(x, q/a) \end{Bmatrix} dq = \frac{\pi a^{\alpha}\alpha^{-1/2}}{\operatorname{sh}\sqrt{\alpha}\,\pi} \begin{Bmatrix} \operatorname{ch}\left[\sqrt{\alpha}(\pi - 2x)\right] \\ \operatorname{ch}(2\sqrt{\alpha}\,x) \end{Bmatrix}$ $[\operatorname{Re}\alpha > 0;\ -(\pi \mp \pi)/4 \leqslant x \leqslant (3\pi \pm \pi)/4]$.

3. $\int_0^a q^{\alpha-1}\theta_3\left(\frac{q}{a},\frac{q^\beta}{a^\beta}\right)dq=\frac{\pi^2 a^\alpha}{2\sqrt{\alpha\beta\pi^2+1}}\left[\operatorname{th}\left(\sqrt{\alpha\beta\pi^2+1}+\frac{1}{\beta}\right)+\operatorname{th}\left(\sqrt{\alpha\beta\pi^2+1}-\frac{1}{\beta}\right)\right]$

[$\operatorname{Re}\alpha>-1/(\beta\pi^2)$; $\beta>0$].

4. $\int_0^a \frac{1}{q}\ln^{\alpha-1}\frac{a}{q}\left\{\begin{matrix}\theta_1(x,\ q/a)\\ \theta_2(x,\ q/a)\end{matrix}\right\}dq=2^{2\alpha+1}\Gamma(\alpha)\sum_{k=0}^{\infty}\frac{(\mp 1)^k}{(2k+1)^{2\alpha}}\left\{\begin{matrix}\sin(2k+1)x\\ \cos(2k+1)x\end{matrix}\right\}$

[$\operatorname{Re}a>0$; $\operatorname{Re}\alpha>1/2$; $-(1\pm 1)\pi/4<x<(3\mp 1)\pi/4$].

5. $=2^{2\alpha-1}\pi^{2\alpha-1/2}\Gamma\left(\frac{1}{2}-\alpha\right)\left[\zeta\left(1-2\alpha,\ \frac{3\pm 3}{8}+\frac{x}{2\pi}\right)+\right.$
$\left.+\zeta\left(1-2\alpha,\ \frac{5\mp 3}{8}-\frac{x}{2\pi}\right)-\zeta\left(1-2\alpha,\ \frac{3\mp 1}{8}+\frac{x}{2\pi}\right)-\zeta\left(1-2\alpha,\ \frac{5\pm 1}{8}-\frac{x}{2\pi}\right)\right]$

[$\operatorname{Re}a$, $\operatorname{Re}\alpha>0$; $-(1\pm 1)\pi/4<x<(3\mp 1)\pi/4$].

6. $\int_0^a \frac{1}{q}\ln^{\alpha-1}\frac{a}{q}\left\{\begin{matrix}\theta_3(x,\ q/a)-1\\ \theta_4(x,\ q/a)-1\end{matrix}\right\}dq=2\Gamma(\alpha)\sum_{k=1}^{\infty}\frac{(\pm 1)^k}{k^{2\alpha}}\cos 2kx$

[$\operatorname{Re}a$, $\operatorname{Re}\alpha>0$; $-(1\mp 1)\pi/4<x<(3\pm 1)\pi/4$],

7. $=\pi^{2\alpha-1/2}\Gamma\left(\frac{1}{2}-\alpha\right)\left[\zeta\left(1-2\alpha,\ \frac{1\mp 1}{4}+\frac{x}{\pi}\right)+\zeta\left(1-2\alpha,\ \frac{3\pm 1}{4}-\frac{x}{\pi}\right)\right]$

[$\operatorname{Re}a>0$; $\operatorname{Re}\alpha>1/2$; $-(1\mp 1)\pi/4<x<(3\pm 1)\pi/4$].

8. $\int_0^a \frac{1}{q}\ln^{\alpha-1}\frac{a}{q}\left\{\begin{matrix}\theta_1(\pi/2,\ q/a)\\ \theta_2(0,\ q/a)\end{matrix}\right\}dq=2(2^\alpha-1)\Gamma(\alpha)\zeta(2\alpha)$ [$\operatorname{Re}a>0$; $\operatorname{Re}\alpha>1/2$].

9. $\int_0^a \frac{1}{q}\ln^{\alpha-1}\frac{a}{q}\left\{\begin{matrix}\theta_3(0,\ q/a)-1\\ \theta_4(\pi/2,\ q/a)-1\end{matrix}\right\}dq=2\Gamma(\alpha)\zeta(2\alpha)$ [$\operatorname{Re}a>0$; $\operatorname{Re}\alpha>1/2$].

2.26.4. Integrals with respect to q, containing $\hat{\theta}_j(x, aq)$.

1. $\int_0^\infty q^{\alpha-1}\left\{\begin{matrix}\hat{\theta}_1(x,\ aq)\\ \hat{\theta}_2(x,\ aq)\end{matrix}\right\}dq=\frac{2^{2\alpha-1}a^{-\alpha}}{\sqrt{\pi}}\Gamma\left(\frac{1}{2}-\alpha\right)\left[\zeta\left(1-2\alpha,\ \frac{x}{2}+\frac{1\pm 1}{8}\right)-\right.$
$\left.-\zeta\left(1-2\alpha,\ \frac{x}{2}+\frac{5\pm 1}{8}\right)+\zeta\left(1-2\alpha,\ \frac{3\mp 1}{8}-\frac{x}{2}\right)-\zeta\left(1-2\alpha,\ \frac{7\mp 1}{8}-\frac{x}{2}\right)\right]$

[$\operatorname{Re}a>0$; $\operatorname{Re}\alpha<0$; $-(1\pm 1)/4<x<(3\mp 1)/4$].

2. $\int_0^\infty q^{\alpha-1}\left\{\begin{matrix}\hat{\theta}_3(x,\ aq)\\ \hat{\theta}_4(x,\ aq)\end{matrix}\right\}dq=\frac{\Gamma(1/2-\alpha)}{\sqrt{\pi}\,a^\alpha}\left[\zeta\left(1-2\alpha,\ \frac{1\mp 1}{4}+x\right)-\zeta\left(1-2\alpha,\ \frac{3\pm 1}{4}-x\right)\right]$

[$\operatorname{Re}a>0$; $\operatorname{Re}\alpha<0$; $-(1\mp 1)/4<x<(3\pm 1)/4$].

3. $\int_0^\infty e^{-pq}\left\{\begin{matrix}\hat{\theta}_1(x,\ aq)\\ \hat{\theta}_2(x,\ aq)\end{matrix}\right\}dq=\frac{\mp 1}{\sqrt{ap}}\operatorname{sech}\sqrt{\frac{p}{a}}\left\{\begin{matrix}\operatorname{ch}(2x\sqrt{p/a})\\ \operatorname{ch}[(1-2x)\sqrt{p/a}]\end{matrix}\right\}$

[$\operatorname{Re}a$, $\operatorname{Re}p>0$; $-(1\pm 1)/4\leqslant x\leqslant(3\mp 1)/4$].

4. $\int_0^\infty e^{-pq}\left\{\begin{matrix}\hat{\theta}_3(x,\ aq)\\ \hat{\theta}_4(x,\ aq)\end{matrix}\right\}dq=\frac{\pm 1}{\sqrt{ap}}\operatorname{cosech}\sqrt{\frac{p}{a}}\left\{\begin{matrix}\operatorname{sh}[(1-2x)\sqrt{p/a}]\\ \operatorname{sh}(2x\sqrt{p/a})\end{matrix}\right\}$

[$\operatorname{Re}a$, $\operatorname{Re}p>0$; $-(1\mp 1)/4\leqslant x\leqslant(3\pm 1)/4$].

5. $\int_0^\infty e^{-(p-b^2/a)q}\,\hat{\theta}_3(bq,\ aq)\,dq=\frac{1}{2\sqrt{ap}}\left(\operatorname{th}\frac{\sqrt{ap}+b}{a}-\operatorname{th}\frac{\sqrt{ap}-b}{a}+2\right)$

[$a,\ b>0$; $\operatorname{Re}p>b^2/a^2$].

2.27. THE MATHIEU FUNCTIONS

This section contains integrals in which there occur the Mathieu functions of the first kind $\mathrm{ce}_n(x,q)$, $\mathrm{se}_n(x,q)$ and the second kind $\mathrm{fe}_n(x,q)$, $\mathrm{ge}_n(x,q)$, and the modified Mathieu functions of the first kind $\mathrm{Ce}_n(x,q)$, $\mathrm{Se}_n(x,q)$ and the second kind $\mathrm{Fe}_n(x,q)$, $\mathrm{Ge}_n(x,q)$. In some formulae, the parameter q in the notation for the Mathieu functions is omitted for brevity; here $k=2\sqrt{q}$.

For definitions and notations see Appendix II.24.

2.27.1. Integrals containing elementary functions and $\mathrm{ce}_n(x,q)$ or $\mathrm{se}_n(x,q)$.

1. $$\int_0^{\pi/2}\left\{\begin{matrix}\mathrm{ch}\,(k\sin a\sin x)\\ \cos\,(k\cos a\cos x)\end{matrix}\right\}\mathrm{ce}_{2n}(x,q)\,dx=\frac{\pi A_0^{(2n)}}{2}\left\{\begin{matrix}\mathrm{ce}_{2n}^{-1}\,(0)\\ \mathrm{ce}_{2n}^{-1}\,(\pi/2)\end{matrix}\right\}\mathrm{ce}_{2n}\,(a).$$

2. $$\int_0^{\pi/2}\left\{\begin{matrix}\cos x\,\mathrm{ch}\,(k\sin a\sin x)\\ \sin\,(k\cos a\cos x)\end{matrix}\right\}\mathrm{ce}_{2n+1}\,(x,q)\,dx=$$
$$=\pm\frac{\pi}{4}A_1^{(2n+1)}\left\{\begin{matrix}\sec a/\mathrm{ce}_{2n+1}\,(0)\\ k/\mathrm{ce}'_{2n+1}\,(\pi/2)\end{matrix}\right\}\mathrm{ce}_{2n+1}\,(a).$$

3. $$\int_0^{\pi/2}\left\{\begin{matrix}\mathrm{sh}\,(k\sin a\sin x)\\ \sin x\cos\,(k\cos a\cos x)\end{matrix}\right\}\mathrm{se}_{2n+1}\,(x,q)\,dx=$$
$$=\frac{\pi}{4}B_1^{(2n+1)}\left\{\begin{matrix}k/\mathrm{se}'_{2n+1}\,(0)\\ \mathrm{cosec}\,a/\mathrm{se}_{2n+1}\,(\pi/2)\end{matrix}\right\}\mathrm{se}_{2n+1}\,(a).$$

4. $$\int_0^{\pi/2}\left\{\begin{matrix}\cos x\,\mathrm{sh}\,(k\sin a\sin x)\\ \sin x\sin\,(k\cos a\cos x)\end{matrix}\right\}\mathrm{se}_{2n+2}\,(x,q)\,dx=$$
$$=\pm\frac{\pi k}{8}B_2^{(2n+2)}\left\{\begin{matrix}\sec a/\mathrm{se}'_{2n+2}\,(0)\\ \mathrm{cosec}\,a/\mathrm{se}'_{2n+2}\,(\pi/2)\end{matrix}\right\}\mathrm{se}_{2n+2}\,(a).$$

5. $$\int_0^{\pi}\left\{\begin{matrix}\mathrm{ch}\,(k\,\mathrm{ch}\,a\sin x)\\ \mathrm{ch}\,(k\,\mathrm{sh}\,a\cos x)\end{matrix}\right\}\mathrm{ce}_{2n}\,(x,q)\,dx=(-1)^n\,\pi A_0^{(2n)}\left\{\begin{matrix}\mathrm{ce}_{2n}^{-1}\,(0,q)\\ \mathrm{ce}_{2n}^{-1}\,(\pi/2,q)\end{matrix}\right\}\mathrm{Ce}_{2n}\,(\pi,-q).$$

6. $$\int_0^{\pi}\left\{\begin{matrix}\cos\,(k\,\mathrm{ch}\,a\cos x)\\ \cos\,(k\cos a\cos x)\end{matrix}\right\}\mathrm{ce}_{2n}\,(x,q)\,dx=\frac{\pi A_0^{(2n)}}{\mathrm{ce}_{2n}\,(\pi/2,q)}\left\{\begin{matrix}\mathrm{Ce}_{2n}\,(a,q)\\ \mathrm{ce}_{2n}\,(a,q)\end{matrix}\right\}.$$

7. $$\int_0^{\pi}\mathrm{sh}\,(k\,\mathrm{ch}\,a\cos x)\,\mathrm{ce}_{2n+1}\,(x,q)\,dx=\frac{(-1)^n\,\pi kA_1^{(2n+1)}}{2\,\mathrm{ce}'_{2n+1}\,(\pi/2,q)}\mathrm{Se}_{2n+1}\,(a,-q).$$

8. $$\int_0^{\pi}\left\{\begin{matrix}\sin\,(k\,\mathrm{ch}\,a\cos x)\\ \sin\,(k\cos a\cos x)\end{matrix}\right\}\mathrm{ce}_{2n+1}\,(x,q)\,dx=-\frac{\pi kA_1^{(2n+1)}}{2\,\mathrm{ce}'_{2n+1}\,(\pi/2,q)}\left\{\begin{matrix}\mathrm{Ce}_{2n+1}\,(a,q)\\ \mathrm{ce}_{2n+1}\,(a,q)\end{matrix}\right\}.$$

9. $$\int_0^{\pi}\left\{\begin{matrix}\mathrm{sh}\,(k\,\mathrm{ch}\,a\sin x)\\ \mathrm{sh}\,(k\sin a\sin x)\end{matrix}\right\}\mathrm{se}_{2n+1}\,(x,q)\,dx=\frac{\pi kB_1^{(2n+1)}}{2\mathrm{se}'_{2n+1}\,(0,q)}\left\{\begin{matrix}(-1)^n\,\mathrm{Ce}_{2n+1}\,(a,-q)\\ \mathrm{se}_{2n+1}\,(a,q)\end{matrix}\right\}.$$

10. $$\int_0^{\pi}\left\{\begin{matrix}\cos\,(k\,\mathrm{sh}\,a\sin x)\,\mathrm{ce}_{2n}\,(x,q)\\ \sin\,(k\,\mathrm{sh}\,a\sin x)\,\mathrm{se}_{2n+1}\,(x,q)\end{matrix}\right\}dx=\frac{\pi}{2}\left\{\begin{matrix}2A_0^{(2n)}\,\mathrm{Ce}_{2n}\,(a,q)/\mathrm{ce}_{2n}\,(0,q)\\ kB_1^{(2n+1)}\,\mathrm{Se}_{2n+1}\,(a,q)/\mathrm{se}'_{2n+1}\,(0,q)\end{matrix}\right\}.$$

11. $$\int_0^{\pi} \cos x \begin{Bmatrix} \cos(k \operatorname{sh} a \sin x) \\ \operatorname{ch}(k \operatorname{ch} a \sin x) \end{Bmatrix} \operatorname{ce}_{2n+1}(x, q)\, dx = \frac{\pi A_1^{(2n+1)}}{2 \operatorname{ce}_{2n+1}(0, q)} \begin{Bmatrix} \operatorname{sech} a \operatorname{Ce}_{2n+1}(a, q) \\ (-1)^n \operatorname{cosech} a \operatorname{Se}_{2n+1}(a, -q) \end{Bmatrix}.$$

12. $$\int_0^{\pi} \sin x \begin{Bmatrix} \cos(k \operatorname{ch} a \cos x) \\ \operatorname{ch}(k \operatorname{sh} a \cos x) \end{Bmatrix} \operatorname{se}_{2n+1}(x, q)\, dx = \frac{\pi B_1^{(2n+1)}}{2 \operatorname{se}_{2n+1}(\pi/2, q)} \begin{Bmatrix} \operatorname{cosech} a \operatorname{Se}_{2n+1}(a, q) \\ (-1)^n \operatorname{sech} a \operatorname{Ce}_{2n+1}(a, -q) \end{Bmatrix}.$$

13. $$\int_0^{\pi} \cos x \begin{Bmatrix} \sin(k \operatorname{sh} a \sin x) \\ \operatorname{sh}(k \operatorname{ch} a \sin x) \end{Bmatrix} \operatorname{se}_{2n+2}(x, q)\, dx = \frac{\pi k B_2^{(2n+2)}}{4 \operatorname{se}'_{2n+2}(0, q)} \begin{Bmatrix} \operatorname{sech} a \operatorname{Se}_{2n+2}(a, q) \\ (-1)^n \operatorname{cosech} a \operatorname{Se}_{2n+2}(a, -q) \end{Bmatrix}.$$

14. $$\int_0^{\pi} \sin x \begin{Bmatrix} \sin(k \operatorname{ch} a \cos x) \\ \operatorname{sh}(k \operatorname{sh} a \cos x) \end{Bmatrix} \operatorname{se}_{2n+2}(x, q)\, dx = -\frac{\pi k B_2^{(2n+2)}}{4 \operatorname{se}'_{2n+2}(\pi/2, q)} \begin{Bmatrix} \operatorname{cosech} a \operatorname{Se}_{2n+2}(a, q) \\ (-1)^n \operatorname{sech} a \operatorname{Se}_{2n+2}(a, -q) \end{Bmatrix}.$$

15. $$\int_0^{2\pi} \begin{Bmatrix} \sin[z \cos(x-w)] \\ \cos[z \cos(x-w)] \end{Bmatrix} \operatorname{ce}_{2n}(x)\, dx = \begin{Bmatrix} 0 \\ 1 \end{Bmatrix} 2\pi A_0^{(2n)} \frac{\operatorname{ce}_{2n}(u) \operatorname{Ce}_{2n}(v)}{\operatorname{ce}_{2n}(0) \operatorname{ce}_{2n}(\pi/2)}$$

$[z = k\sqrt{\operatorname{ch}^2 v - \sin^2 u},\ \operatorname{tg} w = \operatorname{tg} u \operatorname{th} v]$.

16. $$\int_0^{2\pi} \begin{Bmatrix} \sin[z \cos(x-w)] \\ \cos[z \cos(x-w)] \end{Bmatrix} \operatorname{ce}_{2n+1}(x)\, dx = \begin{Bmatrix} 1 \\ 0 \end{Bmatrix} \pi k A_1^{(2n+1)} \frac{\operatorname{ce}_{2n+1}(u) \operatorname{Ce}_{2n+1}(v)}{\operatorname{ce}_{2n+1}(0) \operatorname{ce}'_{2n+1}(\pi/2)}$$

$[w,\ z$ see 2.27.1.15$]$.

17. $$\int_0^{2\pi} \begin{Bmatrix} \sin[z \cos(x-w)] \\ \cos[z \cos(x-w)] \end{Bmatrix} \operatorname{se}_{2n+1}(x)\, dx = \begin{Bmatrix} 1 \\ 0 \end{Bmatrix} \pi k B_1^{(2n+1)} \frac{\operatorname{se}_{2n+1}(u) \operatorname{Se}_{2n+1}(v)}{\operatorname{se}_{2n+1}(0) \operatorname{se}_{2n+1}(\pi/2)}$$

$[w,\ z$ see 2.27.1.15$]$.

18. $$\int_0^{2\pi} \begin{Bmatrix} \sin[z \cos(x-w)] \\ \cos[z \cos(x-w)] \end{Bmatrix} \operatorname{se}_{2n+2}(x)\, dx = -\begin{Bmatrix} 0 \\ 1 \end{Bmatrix} \frac{\pi k^2 B_2^{(2n+2)} \operatorname{se}_{2n+2}(u) \operatorname{Se}_{2n+2}(v)}{\operatorname{se}'_{2n+2}(0) \operatorname{se}'_{2n+2}(\pi/2)}$$

$[w,\ z$ see 2.27.1.15$]$.

19. $$\int_0^{2\pi} \exp\left(k \begin{Bmatrix} \sin a \sin x \\ i \cos a \cos x \end{Bmatrix}\right) \operatorname{ce}_{2n}(x)\, dx = 2\pi A_0^{(2n)} \begin{Bmatrix} \operatorname{ce}_{2n}^{-1}(0) \\ \operatorname{ce}_{2n}^{-1}(\pi/2) \end{Bmatrix} \operatorname{ce}_{2n}(a).$$

20. $$\int_0^{2\pi} \begin{Bmatrix} \cos x \exp(k \sin a \sin x) \\ \exp(ik \cos a \cos x) \end{Bmatrix} \operatorname{ce}_{2n+1}(x)\, dx = \pi A_1^{(2n+1)} \begin{Bmatrix} \sec a / \operatorname{ce}_{2n+1}(0) \\ ik / \operatorname{ce}'_{2n+1}(\pi/2) \end{Bmatrix} \operatorname{ce}_{2n+1}(a).$$

21. $$\int_0^{2\pi} \begin{Bmatrix} \exp(k \sin a \sin x) \\ \sin x \exp(ik \cos a \cos x) \end{Bmatrix} \operatorname{se}_{2n+1}(x)\, dx = \pi B_1^{(2n+1)} \begin{Bmatrix} k / \operatorname{se}'_{2n+1}(0) \\ \operatorname{cosec} a / \operatorname{se}_{2n+1}(\pi/2) \end{Bmatrix} \operatorname{se}_{2n+1}(a).$$

22. $\int_0^{\pi} \begin{Bmatrix} \cos x \exp(k \sin a \sin x) \\ \sin x \exp(ik \cos a \cos x) \end{Bmatrix} \mathrm{se}_{2n+2}(x)\,dx =$

$$= \frac{\pi k}{2} B_2^{(2n+2)} \begin{Bmatrix} \sec a/\mathrm{se}'_{2n+2}(0) \\ i \operatorname{cosec} a/\mathrm{se}'_{2n+2}(\pi/2) \end{Bmatrix} \mathrm{se}_{2n+2}(a).$$

23. $\int_0^{2\pi} e^{ikw} \begin{Bmatrix} 1 \\ \partial w/\partial u \end{Bmatrix} \mathrm{ce}_{2n}(x)\,dx = \frac{2\pi A_0^{(2n)}}{k\,\mathrm{ce}_{2n}(0)\,\mathrm{ce}_{2n}(\pi/2)} \begin{Bmatrix} k\,\mathrm{ce}_{2n}(u) \\ i\,\mathrm{ce}'_{2n}(u) \end{Bmatrix} \mathrm{Ce}_{2n}(v)$

$$\left[w = \frac{1}{8} e^{v} \cos(x-u) + \frac{1}{8} \cos(x+u) = \frac{1}{4} (\operatorname{ch} v \cos x \cos u + \operatorname{sh} v \sin x \sin u)\right].$$

24. $\int_0^{2\pi} e^{ikw} \begin{Bmatrix} 1 \\ \partial w/\partial u \end{Bmatrix} \mathrm{ce}_{2n+1}(x)\,dx =$

$$= -\frac{4\pi A_1^{(2n+1)}}{k^3\,\mathrm{ce}_{2n+1}(0)\,\mathrm{ce}'_{2n+1}(\pi/2)} \begin{Bmatrix} ik^2\,\mathrm{ce}_{2n+1}(u) \\ 2\,\mathrm{ce}'_{2n+1}(u) \end{Bmatrix} \mathrm{Ce}_{2n+1}(v) \qquad [w \text{ see } 2.27.1.23].$$

25. $\int_0^{2\pi} e^{ikw} \begin{Bmatrix} 1 \\ \partial w/\partial u \end{Bmatrix} \mathrm{se}_{2n+1}(x)\,dx =$

$$= \pm \frac{\pi B_1^{(2n+1)}}{\mathrm{se}'_{n+1}(0)\,\mathrm{se}_{2n+1}(\pi/2)} \begin{Bmatrix} k\,\mathrm{se}_{2n+1}(u) \\ \mathrm{se}'_{2n+1}(u) \end{Bmatrix} \mathrm{Se}_{2n+1}(v) \qquad [w \text{ see } 2.27.1.23].$$

26. $\int_0^{2\pi} e^{ikw} \begin{Bmatrix} 1 \\ \partial w/\partial u \end{Bmatrix} \mathrm{se}_{2n+2}(x)\,dx =$

$$= -\frac{\pi B_2^{(2n+2)}}{k\,\mathrm{se}_{2n+2}(0)\,\mathrm{se}'_{2n+2}(\pi/2)} \begin{Bmatrix} ik^2\,\mathrm{se}_{2n+2}(u) \\ 2\,\mathrm{se}'_{2n+2}(u) \end{Bmatrix} \mathrm{Se}_{2n+2}(v) \qquad [w \text{ see } 2.27.1.23].$$

2.27.2. Integrals containing special functions and $\mathrm{ce}_n(x, q)$ or $\mathrm{se}_n(x, q)$.

1. $\int_0^{a} J_0(k(\cos a - \cos x)) \begin{Bmatrix} \mathrm{ce}_{2n}(x) \\ \mathrm{ce}_{2n+1}(x) \end{Bmatrix} dx = \frac{\pi}{8} \begin{Bmatrix} 4\left[A_0^{(2n)}/\mathrm{ce}_{2n}(\pi/2)\right]^2 \mathrm{fe}_{2n}(x) \\ k^2\left[A_1^{(2n+1)}/\mathrm{ce}'_{2n+1}(\pi/2)\right]^2 \mathrm{fe}_{2n+1}(x) \end{Bmatrix}$

$$\left[\mathrm{fe}_{2n}(\pi/2) = \mathrm{ce}_{2n}(\pi/2),\ \mathrm{fe}'_{2n+1}(\pi/2) = \mathrm{ce}'_{2n+1}(\pi/2)\right].$$

2. $\int_0^{a} \frac{\sin x}{\cos a - \cos x} J_1(k(\cos a - \cos x)) \begin{Bmatrix} \mathrm{se}_{2n+1}(x) \\ \mathrm{se}_{2n+2}(x) \end{Bmatrix} dx =$

$$= \frac{\operatorname{cosec} a}{32k} \begin{Bmatrix} 4\pi k^2 \left[B_1^{(2n+1)}/\mathrm{se}_{2n+1}(\pi/2)\right]^2 \mathrm{ge}_{2n+1}(a) - 32\,\mathrm{se}'_{2n+1}(a) \\ \pi k^4 \left[B_2^{(2n+2)}/\mathrm{se}'_{2n+2}(\pi/2)\right]^2 \mathrm{ge}_{2n+2}(a) - 32\,\mathrm{se}'_{2n+2}(a) \end{Bmatrix}$$

$$\left[\mathrm{ge}_{2n+1}(\pi/2) = \mathrm{se}_{2n+1}(\pi/2),\ \mathrm{ge}'_{2n+2}(\pi/2) = \mathrm{se}'_{2n+2}(\pi/2)\right].$$

3. $\int_0^{\pi} J_0\left(k \sqrt{\frac{\cos 2z + \cos 2x}{2}}\right) \mathrm{ce}_{2n}(x)\,dx = \frac{\pi \left(A_0^{(2n)}\right)^2}{\mathrm{ce}_{2n}(0)\,\mathrm{ce}_{2n}(\pi/2)} \mathrm{ce}_{2n}(z).$

4. $$\int_0^{\pi} \frac{\cos x}{\sqrt{\cos 2z+\cos 2x}} J_1\left(k\sqrt{\frac{\cos 2z+\cos 2x}{2}}\right) \mathrm{ce}_{2n+1}(x)\,dx = -\frac{\pi k\left(A_1^{(2n+1)}\right)^2}{4\sqrt{2}\cos z\,\mathrm{ce}_{2n+1}(0)\,\mathrm{ce}'_{2n+1}(\pi/2)}\mathrm{ce}_{2n+1}(z).$$

5. $$\int_0^{\pi} \frac{\sin x}{\sqrt{\cos 2z+\cos 2x}} J_1\left(k\sqrt{\frac{\cos 2z+\cos 2x}{2}}\right) \mathrm{se}_{2n+1}(x)\,dx = \frac{\pi k\left(B_1^{(2n+1)}\right)^2}{4\sqrt{2}\sin z\,\mathrm{se}'_{2n+1}(0)\,\mathrm{se}_{2n+1}(\pi/2)}\mathrm{se}_{2n+1}(z).$$

6. $$\int_0^{\pi} \frac{\sin 2x}{\cos 2z+\cos 2x} J_2\left(k\sqrt{\frac{\cos 2z+\cos 2x}{2}}\right) \mathrm{se}_{2n+2}(x)\,dx = -\frac{\pi k^2\left(B_2^{(2n+2)}\right)^2}{8\sin 2z\,\mathrm{se}'_{2n+2}(0)\,\mathrm{se}'_{2n+2}(\pi/2)}\mathrm{se}_{2n+2}(z).$$

7. $$\int_0^{\pi} \frac{\sin x}{\sqrt{\mathrm{ch}\, 2z+\cos 2x}} Y_1\left(k\sqrt{\frac{\mathrm{ch}\, 2z+\cos 2x}{2}}\right) \mathrm{se}_{2n+1}(x)\,dx = \frac{\pi k\left(B_1^{(2n+1)}\right)^2}{4\sqrt{2}\,\mathrm{sh}\, z\,\mathrm{se}'_{2n+1}(0)\,\mathrm{se}_{2n+1}(\pi/2)}\mathrm{Gey}_{2n+1}(z).$$

8. $$\int_0^{\pi} \frac{\sin 2x}{\mathrm{ch}\, 2z+\cos 2x} Y_2\left(k\sqrt{\frac{\mathrm{ch}\, 2z+\cos 2x}{2}}\right) \mathrm{se}_{2n+2}(x)\,dx = -\frac{\pi k^3\left(B_2^{(2n+2)}\right)^2}{8\,\mathrm{sh}\, 2z\,\mathrm{se}'_{2n+2}(0)\,\mathrm{se}'_{2n+2}(\pi/2)}\mathrm{Gey}_{2n+2}(z).$$

9. $$\int_0^{2\pi} J_0(k(\cos z-\cos x)) \begin{Bmatrix}\mathrm{ce}_{2n}(x)\\ \mathrm{ce}_{2n+1}(x)\end{Bmatrix} dx = 2\pi \begin{Bmatrix}\left[A_0^{(2n)}/\mathrm{ce}_{2n}(\pi/2)\right]^2 \mathrm{ce}_{2n}(z)\\ q\left[A_1^{(2n+1)}/\mathrm{ce}'_{2n+1}(\pi/2)\right]^2 \mathrm{ce}_{2n+1}(z)\end{Bmatrix}.$$

10. $$\int_0^{2\pi} \frac{\sin x}{\cos z-\cos x} J_1(k(\cos z-\cos x)) \begin{Bmatrix}\mathrm{se}_{2n+1}(x)\\ \mathrm{se}_{2n+2}(x)\end{Bmatrix} dx = \frac{\pi k}{8\sin z}\begin{Bmatrix}4\left[B_1^{(2n+1)}/\mathrm{se}_{2n+1}(\pi/2)\right]^2 \mathrm{se}_{2n+1}(z)\\ k^2\left[B_2^{(2n+2)}/\mathrm{se}'_{2n+2}(\pi/2)\right]^2 \mathrm{se}_{2n+2}(z)\end{Bmatrix}.$$

11. $$\int_0^{2\pi} \begin{Bmatrix}J_{2m}(k\cos x)\,\mathrm{ce}_{2n}(x)\\ J_{2m+1}(k\cos x)\,\mathrm{ce}_{2n+1}(x)\end{Bmatrix} dx = \pm\frac{(-1)^m\pi}{2}\begin{Bmatrix}2A_0^{(2n)}A_{2m}^{(2n)}/\mathrm{ce}_{2n}(\pi/2)\\ kA_1^{(2n+1)}A_{2m+1}^{(2n+1)}/\mathrm{ce}'_{2n+1}(\pi/2)\end{Bmatrix}.$$

12. $$\int_0^{2\pi} \mathrm{tg}\, x J_{2m}(k\cos x)\begin{Bmatrix}\mathrm{se}_{2n+1}(x)\\ \mathrm{se}_{2n+2}(x)\end{Bmatrix} dx = \mp\frac{(-1)^m\pi k}{8m(2m+1)}\begin{Bmatrix}4mB_1^{(2n+1)}B_{2m+1}^{(2n+1)}/\mathrm{se}_{2n+1}(\pi/2)\\ 2k(2m+1)B_2^{(2n+2)}B_{2m}^{(2n+2)}/\mathrm{se}'_{2n+2}(\pi/2)\end{Bmatrix}.$$

13. $$\int_0^{2\pi} Y_0\left(k\sqrt{\frac{\cos 2z+\cos 2x}{2}}\right)\mathrm{ce}_{2n}(x)\,dx = \frac{2\pi\left(A_0^{(2n)}\right)^2}{\mathrm{ce}_{2n}(0)\,\mathrm{ce}_{2n}(\pi/2)}\mathrm{Fey}_{2n}(z).$$

14. $$\int_0^{2\pi} H_0^{(j)}(k(\operatorname{ch} z-\cos x))\,\mathrm{ce}_n(x)\,dx=\frac{(-1)^{j+1}4i\,\mathrm{Me}_n^{(j)}(z)}{\mathrm{Me}_n^{(j)}(0)} \qquad [j=1,\ 2].$$

15. $$\int_0^{2\pi} H_0^{(1)}(k(\operatorname{sh} z-i\sin x))\,\mathrm{ce}_{2n}(x)\,dx=4i\,\frac{\mathrm{ce}_{2n}^2(\pi/2)\,\mathrm{Fek}_{2n}(z)}{\mathrm{ce}_{2n}(0)\,\mathrm{Fek}_{2n}(0)}.$$

16. $$\int_0^{2\pi} H_0^{(1)}(k(\operatorname{sh} z-i\sin x))\,\mathrm{se}_{2n+1}(x)\,dx=4\frac{\mathrm{se}_{2n+1}^2(\pi/2)\,\mathrm{Gek}_{2n+1}(z)}{\mathrm{se}'_{2n+1}(0)\,\mathrm{Gek}_{2n+1}(0)}.$$

17. $$\int_0^{2\pi} \frac{\cos x}{\operatorname{sh} z-i\sin x} H_1^{(1)}(k(\operatorname{sh} z-i\sin x))\,\mathrm{ce}_{2n+1}(x)\,dx=\frac{4i(\mathrm{ce}'_{2n+1}(\pi/2))^2\,\mathrm{Fek}_{2n+1}(z)}{k\,\mathrm{ce}_{2n+1}(0)\,\mathrm{Fek}'_{2n+1}(0)}.$$

18. $$\int_0^{2\pi} \frac{\cos x}{\operatorname{sh} z-i\sin x} H_1^{(1)}(k(\operatorname{sh} z-i\sin x))\,\mathrm{se}_{2n+2}(x)\,dx=\frac{4(\mathrm{se}'_{2n+2}(\pi/2))^2\,\mathrm{Gek}_{2n+2}(z)}{k\cos z\,\mathrm{se}'_{2n+2}(0)\,\mathrm{Gek}_{2n+2}(0)}.$$

19. $$\int_0^{2\pi} \frac{\cos x}{\sqrt{\operatorname{sh}^2 z+\cos^2 x}} H_1^{(1)}\left(k\sqrt{\operatorname{sh}^2 z+\cos^2 x}\right)\mathrm{ce}_{2n+1}(x)\,dx=$$
$$=\frac{4i\,\mathrm{ce}'_{2n+1}(\pi/2)\,\mathrm{Fek}_{2n+1}(z)}{k\operatorname{ch} z\,\mathrm{Fek}'_{2n+1}(0)}.$$

20. $$\int_0^{2\pi} H_0^{(j)}(k(\operatorname{ch} z-\cos x))\,\mathrm{ce}_{2n}(x)\,dx=\frac{\pi}{2}\left(\frac{A_0^{(2n)}}{\mathrm{ce}_{2n}(\pi/2)}\right)^2\mathrm{Me}_{2n}^{(j)}(z).$$

21. $$\int_0^{2\pi} H_0^{(j)}(k(\operatorname{ch} z-\cos x))\,\mathrm{ce}_{2n+1}(x)\,dx=\frac{\pi k^2}{2}\left(\frac{A_1^{(2n+1)}}{\mathrm{ce}'_{2n+1}(\pi/2)}\right)^2\mathrm{Me}_{2n+1}^{(j)}(z).$$

22. $$\int_0^{2\pi} \frac{\sin x}{\operatorname{ch} z-\cos x} H_1^{(j)}(k(\operatorname{ch} z-\cos x))\,\mathrm{se}_{2n+1}(x)\,dx=$$
$$=\frac{\pi q^{1/2}}{\operatorname{sh} z}\left(\frac{B_1^{(2n+1)}}{\mathrm{se}_{2n+1}(\pi/2)}\right)^2\mathrm{Ne}_{2n+1}^{(j)}(z).$$

23. $$\int_0^{2\pi} \frac{\sin x}{\operatorname{ch} z-\cos x} H_1^{(j)}(k(\operatorname{ch} z-\cos x))\,\mathrm{se}_{2n+2}(x)\,dx=$$
$$=\frac{\pi k^3}{8\operatorname{sh} z}\left(\frac{B_2^{(2n+2)}}{\mathrm{se}'_{2n+2}(\pi/2)}\right)^2\mathrm{Ne}_{2n+2}^{(j)}(z).$$

24. $$\int_0^{2\pi} \frac{\sin 2x}{\operatorname{sh}^2 z+\cos^2 x} H_2^{(1)}\left(k\sqrt{\operatorname{sh}^2 z+\cos^2 x}\right)\mathrm{se}_{2n}(x)\,dx=$$
$$=\frac{16i}{k^2}\operatorname{cosech} 2z\,\mathrm{se}'_{2n}\left(\frac{\pi}{2}\right)\frac{\mathrm{Ne}_{2n}^{(1)}(z)}{\mathrm{Ne}_{2n}^{(1)}(0)}.$$

25. $$\int_0^{2\pi} I_0(k(\sin z+\sin x))\,\mathrm{ce}_{2n}(x)\,dx=2\pi\left(\frac{A_0^{(2n)}}{\mathrm{ce}_{2n}(0)}\right)^2\mathrm{ce}_{2n}(z).$$

26. $\int_0^{2\pi} I_0(k(\sin z+\sin x))\,\mathrm{se}_{2n+1}(x)\,dx=\frac{\pi k^2}{8}\left(\frac{B_1^{(2n+1)}}{\mathrm{se}'_{2n+1}(0)}\right)^2\mathrm{se}_{2n+1}(z).$

27. $\int_0^{2\pi}\frac{\cos x}{\sin z+\sin x}I_1(k(\sin z+\sin x))\,\mathrm{se}_{2n}(x)\,dx=\frac{\pi k^3}{8\cos z}\left(\frac{B_2^{(2n)}}{\mathrm{se}_{2n}(0)}\right)^2\mathrm{se}_{2n}(z).$

28. $\int_0^{2\pi}\frac{\cos x}{\sin z+\sin x}I_1(k(\sin z+\sin x))\,\mathrm{ce}_{2n+1}(x)\,dx=\frac{\pi k}{2\cos z}\left(\frac{A_1^{(2n+1)}}{\mathrm{ce}_{2n+1}(0)}\right)^2\mathrm{ce}_{2n+1}(z).$

2.27.3. Integrals containing $\mathrm{Ce}_n(x, q)$ or $\mathrm{Se}_n(x, q)$.

1. $\int_0^{\infty} e^{ik\,\mathrm{ch}\,a\,\mathrm{ch}\,x}\begin{Bmatrix}\mathrm{Ce}_{2n}(x)\\ \mathrm{Ce}_{2n+1}(x)\end{Bmatrix}dx=\pm\frac{\pi}{2}\begin{Bmatrix}2A_0^{(2n)}\,\mathrm{Fek}_{2n}(a)/\mathrm{ce}_{2n}(\pi/2)\\ kA_1^{(2n+1)}\,\mathrm{Fek}_{2n+1}(a)/\mathrm{ce}'_{2n+1}(\pi/2)\end{Bmatrix}.$

2. $\int_0^{\infty}\mathrm{sh}\,x\,e^{ik\,\mathrm{ch}\,a\,\mathrm{ch}\,x}\begin{Bmatrix}\mathrm{Se}_{2n+1}(x)\\ \mathrm{Se}_{2n+2}(x)\end{Bmatrix}dx=\frac{\pi i}{4\,\mathrm{sh}\,a}\begin{Bmatrix}2B_1^{(2n+1)}\,\mathrm{Gek}_{2n+1}(a)/\mathrm{se}_{2n+1}(\pi/2)\\ kB_2^{(2n+2)}\,\mathrm{Gek}_{2n+2}(a)/\mathrm{se}'_{2n+2}(\pi/2)\end{Bmatrix}.$

3. $\int_0^{\infty}\mathrm{sh}\,x\begin{Bmatrix}\sin(k\,\mathrm{ch}\,a\,\mathrm{ch}\,x)\\ \cos(k\,\mathrm{ch}\,a\,\mathrm{ch}\,x)\end{Bmatrix}\mathrm{Se}_{2n+1}(x)\,dx=-\frac{\pi B_1^{(2n+1)}\,\mathrm{cosech}\,a}{4\,\mathrm{se}_{2n+1}(\pi/2)}\begin{Bmatrix}\mathrm{Se}_{2n+1}(a)\\ \mathrm{Gey}_{2n+1}(a)\end{Bmatrix}.$

4. $\int_0^{\infty}\mathrm{sh}\,x\begin{Bmatrix}\sin(k\,\mathrm{ch}\,a\,\mathrm{ch}\,x)\\ \cos(k\,\mathrm{ch}\,a\,\mathrm{ch}\,x)\end{Bmatrix}\mathrm{Se}_{2n+2}(x)\,dx=-\frac{\pi kB_2^{(2n+2)}\,\mathrm{cosech}\,a}{8\,\mathrm{se}'_{2n+2}(\pi/2)}\begin{Bmatrix}\mathrm{Gey}_{2n+2}(a)\\ \mathrm{Se}_{2n+2}(a)\end{Bmatrix}.$

5. $\int_0^{\infty}\begin{Bmatrix}\sin(k\,\mathrm{ch}\,a\,\mathrm{ch}\,x)\\ \cos(k\,\mathrm{ch}\,a\,\mathrm{ch}\,x)\end{Bmatrix}\mathrm{Ce}_{2n}(x)\,dx=\pm\frac{\pi A_0^{(2n)}}{2\,\mathrm{ce}_{2n}(\pi/2)}\begin{Bmatrix}\mathrm{Ce}_{2n}(a)\\ \mathrm{Fey}_{2n}(a)\end{Bmatrix}.$

6. $\int_0^{\infty}\begin{Bmatrix}\sin(k\,\mathrm{ch}\,a\,\mathrm{ch}\,x)\\ \cos(k\,\mathrm{ch}\,a\,\mathrm{ch}\,x)\end{Bmatrix}\mathrm{Ce}_{2n+1}(x)\,dx=\frac{\pi kA_1^{(2n+1)}}{4\,\mathrm{ce}'_{2n+1}(\pi/2)}\begin{Bmatrix}\mathrm{Fey}_{2n+1}(a)\\ \mathrm{Ce}_{2n+1}(a)\end{Bmatrix}.$

7. $\int_0^{\infty}\mathrm{sh}\,x\begin{Bmatrix}\cos(k\cos a\,\mathrm{ch}\,x)\,\mathrm{Ce}_{2n}(x)\\ \sin(k\cos a\,\mathrm{ch}\,x)\,\mathrm{Ce}_{2n+1}(x)\end{Bmatrix}dx=$

$$=\frac{1}{k^2\sin a}\begin{Bmatrix}k\,\mathrm{ce}_{2n}(\pi/2)\,\mathrm{ce}_{2n}(a)/A_0^{(2n)}\\ 2\mathrm{ce}'_{2n+1}(\pi/2)\,\mathrm{ce}_{2n+1}(a)/A_1^{(2n+1)}\end{Bmatrix}-\int_0^{\pi/2}\sin x\begin{Bmatrix}\cos(k\cos a\cos x)\,\mathrm{ce}_{2n}(x)\\ \sin(k\cos a\cos x)\,\mathrm{ce}_{2n+1}(x)\end{Bmatrix}dx.$$

8. $\int_0^{\infty}\mathrm{sh}\,x\begin{Bmatrix}\cos(k\,\mathrm{ch}\,a\,\mathrm{ch}\,x)\,\mathrm{Ce}_{2n}(x)\\ \sin(k\,\mathrm{ch}\,a\,\mathrm{ch}\,x)\,\mathrm{Ce}_{2n+1}(x)\end{Bmatrix}dx=-\int_0^{\pi}\sin x\begin{Bmatrix}\cos(k\,\mathrm{ch}\,a\cos x)\,\mathrm{ce}_{2n}(x)\\ \sin(k\,\mathrm{ch}\,a\cos x)\,\mathrm{ce}_{2n+1}(x)\end{Bmatrix}dx.$

9. $\int_0^{\infty}\begin{Bmatrix}\cos(k\cos a\,\mathrm{ch}\,x)\,\mathrm{Se}_{2n+1}(x)\\ \sin(k\cos a\,\mathrm{ch}\,x)\,\mathrm{Se}_{2n+2}(x)\end{Bmatrix}dx=$

$$=\frac{4}{k^2}\begin{Bmatrix}k\,\mathrm{se}_{2n+1}(\pi/2)\,\mathrm{se}_{2n+1}(a)/B_1^{(2n+1)}\\ 2\,\mathrm{se}'_{2n+2}(\pi/2)\,\mathrm{se}_{2n+2}(a)/B_2^{(2n+2)}\end{Bmatrix}-\int_0^{\pi/2}\begin{Bmatrix}\cos(k\cos a\cos x)\,\mathrm{se}_{2n+1}(x)\\ \sin(k\cos a\cos x)\,\mathrm{se}_{2n+2}(x)\end{Bmatrix}dx.$$

10. $\int_0^{\infty}\begin{Bmatrix}\cos(k\,\mathrm{ch}\,a\,\mathrm{ch}\,x)\,\mathrm{Se}_{2n+1}(x)\\ \sin(k\,\mathrm{ch}\,a\,\mathrm{ch}\,x)\,\mathrm{Se}_{2n+2}(x)\end{Bmatrix}dx=-\int_0^{\pi/2}\begin{Bmatrix}\cos(k\,\mathrm{ch}\,a\cos x)\,\mathrm{se}_{2n+1}(x)\\ \sin(k\,\mathrm{ch}\,a\cos x)\,\mathrm{se}_{2n+2}(x)\end{Bmatrix}dx.$

11. $$\int_0^\infty \begin{Bmatrix} \sin(k\cos a \operatorname{ch} x)\operatorname{Ce}_{2n}(x) \\ \cos(k\cos a \operatorname{ch} x)\operatorname{Ce}_{2n+1}(x) \end{Bmatrix} dx =$$

$$= \frac{\pi}{4}\begin{Bmatrix} 2A_0^{(2n)}[\operatorname{ce}_{2n}(a)-\operatorname{fe}_{2n}(a)]/\operatorname{ce}_{2n}(\pi/2) \\ kA_1^{(2n+1)}[\operatorname{ce}_{2n+1}(a)-\operatorname{fe}_{2n+1}(a)]/\operatorname{ce}'_{2n+1}(\pi/2) \end{Bmatrix}$$

$[\operatorname{fe}_{2n}(\pi/2)=\operatorname{ce}_{2n}(\pi/2),\quad \operatorname{fe}'_{2n+1}(\pi/2)=\operatorname{ce}'_{2n+1}(\pi/2)]$.

12. $$\int_0^\infty \operatorname{sh} x \begin{Bmatrix} \sin(k\cos a \operatorname{ch} x)\operatorname{Se}_{2n+1}(x) \\ \cos(k\cos a \operatorname{ch} x)\operatorname{Se}_{2n+2}(x) \end{Bmatrix} dx =$$

$$= \mp\frac{\pi}{8}\begin{Bmatrix} 2B_2^{(2n+1)}[\operatorname{se}_{2n+1}(a)-\operatorname{ge}_{2n+1}(a)]/\operatorname{se}_{2n+1}(\pi/2) \\ kB_2^{(2n+2)}[\operatorname{se}_{2n+2}(a)-\operatorname{ge}_{2n+2}(a)]/\operatorname{se}'_{2n+2}(\pi/2) \end{Bmatrix}.$$

$[\operatorname{ge}_{2n+1}(\pi/2)=\operatorname{se}_{2n+1}(\pi/2),\quad \operatorname{ge}'_{2n+2}(\pi/2)=\operatorname{se}'_{2n+2}(\pi/2)]$.

2.27.4. Integrals containing $fe_n(x, q)$ or $ge_n(x, q)$.

1. $$\int_0^{\pi/2} \sin x \begin{Bmatrix} \sin(k\cos a\cos x) & [\operatorname{fe}_{2n}(x)-\operatorname{ce}_{2n}(x)] \\ \cos(k\cos a\cos x) & [\operatorname{fe}_{2n+1}(x)-\operatorname{ce}_{2n+1}(x)] \end{Bmatrix} dx =$$

$$= \int_0^\infty \operatorname{sh} x \begin{Bmatrix} \sin(k\cos a \operatorname{ch} x)\operatorname{Ce}_{2n}(x) \\ \cos(k\cos a \operatorname{ch} x)\operatorname{Ce}_{2n+1}(x) \end{Bmatrix} dx$$

$[\operatorname{fe}_{2n}(\pi/2)=\operatorname{ce}_{2n}(\pi/2),\quad \operatorname{fe}'_{2n+1}(\pi/2)=\operatorname{ce}'_{2n+1}(\pi/2)]$.

2. $$\int_0^{\pi/2} \sin x \begin{Bmatrix} \sin(k\operatorname{ch} a\cos x)[\operatorname{fe}_{2n}(x)-\operatorname{ce}_{2n}(x)] \\ \cos(k\operatorname{ch} a\cos x)[\operatorname{fe}_{2n+1}(x)-\operatorname{ce}_{2n+1}(x)] \end{Bmatrix} dx =$$

$$= \int_0^\infty \operatorname{sh} x \begin{Bmatrix} \sin(k\operatorname{ch} a \operatorname{ch} x)\operatorname{Ce}_{2n}(x) \\ \cos(k\operatorname{ch} a \operatorname{ch} x)\operatorname{Ce}_{2n+1}(x) \end{Bmatrix} dx - \frac{\operatorname{cosech} a}{k}\begin{Bmatrix} \operatorname{ce}_{2n}(\pi/2)\operatorname{Ce}_{2n}(a)/A_0^{(2n)} \\ \operatorname{ce}'_{2n+1}(\pi/2)\operatorname{Ce}_{2n+1}(a)/A_1^{(2n+1)} \end{Bmatrix}$$

$[\operatorname{fe}_{2n}(\pi/2)=\operatorname{ce}_{2n}(\pi/2),\quad \operatorname{fe}'_{2n+1}(\pi/2)=\operatorname{ce}'_{2n+1}(\pi/2)]$.

3. $$\int_0^{\pi/2} \begin{Bmatrix} \sin(k\cos a\cos x)[\operatorname{ge}_{2n+1}(x)-\operatorname{se}_{2n+1}(x)] \\ \cos(k\cos a\cos x)[\operatorname{ge}_{2n+2}(x)-\operatorname{se}_{2n+2}(x)] \end{Bmatrix} dx =$$

$$= \int_0^\infty \begin{Bmatrix} \sin(k\cos a \operatorname{ch} x)\operatorname{Se}_{2n+1}(x) \\ \cos(k\cos a \operatorname{ch} x)\operatorname{Se}_{2n+2}(x) \end{Bmatrix} dx$$

$[\operatorname{ge}_{2n+1}(\pi/2)=\operatorname{se}_{2n+1}(\pi/2),\quad \operatorname{ge}'_{2n+2}(\pi/2)=\operatorname{se}'_{2n+2}(\pi/2)]$.

4. $$\int_0^{\pi/2} \begin{Bmatrix} \sin(k\operatorname{ch} a\cos x)[\operatorname{ge}_{2n+1}(x)-\operatorname{se}_{2n+1}(x)] \\ \cos(k\operatorname{ch} a\cos x)[\operatorname{ge}_{2n+2}(x)-\operatorname{se}_{2n+2}(x)] \end{Bmatrix} dx =$$

$$= \int_0^\infty \begin{Bmatrix} \sin(k\operatorname{ch} a \operatorname{ch} x)\operatorname{Se}_{2n+1}(x) \\ \cos(k\operatorname{ch} a \operatorname{ch} x)\operatorname{Se}_{2n+2}(x) \end{Bmatrix} dx - 2\begin{Bmatrix} \operatorname{se}_{2n+1}(\pi/2)\operatorname{Se}_{2n+1}(a)/B_1^{(2n+1)} \\ \operatorname{se}'_{2n+2}(\pi/2)\operatorname{Se}_{2n+2}(a)/B_2^{(2n+2)} \end{Bmatrix}$$

$[\operatorname{ge}_{2n+1}(\pi/2)=\operatorname{se}_{2n+1}(\pi/2),\quad \operatorname{ge}'_{2n+2}(\pi/2)=\operatorname{se}_{2n+2}(\pi/2)]$.

2.27.5. Integrals containing $\mathrm{Fe}_n(x, q)$ or $\mathrm{Ge}_n(x, q)$.

1. $$\int_0^\infty \operatorname{sh} x \begin{Bmatrix} \sin(k \operatorname{ch} a \operatorname{ch} x) \\ \cos(k \operatorname{ch} a \operatorname{ch} x) \end{Bmatrix} \mathrm{Fe}_{2n}(x)\,dx = -\frac{\pi}{2k \operatorname{sh} a}\left(\frac{A_0^{(2n)}}{\mathrm{ce}_{2n}(\pi/2)}\right)^{\pm 1} \begin{Bmatrix} \mathrm{fe}'_{2n}(0)\,\mathrm{Fey}_{2n}(a)/\mathrm{ce}_{2n}(0) \\ C_{2n}(q)\,\mathrm{Ce}_{2n}(a) \end{Bmatrix}.$$

2. $$\int_0^\infty \operatorname{sh} x \begin{Bmatrix} \sin(k \operatorname{ch} a \operatorname{ch} x) \\ \cos(k \operatorname{ch} a \operatorname{ch} x) \end{Bmatrix} \mathrm{Fe}_{2n+1}(x)\,dx = \pm\frac{\pi}{2k^2 \operatorname{sh} a}\left(\frac{\mathrm{ce}'_{2n+1}(\pi/2)}{A_1^{(2n+1)}}\right)^{\pm 1} \begin{Bmatrix} 2C_{2n+1}(q)\,\mathrm{Ce}_{2n+1}(a) \\ k\,\mathrm{fe}'_{2n+1}(0)\,\mathrm{Fey}_{2n+1}(a)/\mathrm{ce}_{2n+1}(0) \end{Bmatrix}.$$

3. $$\int_0^\infty \begin{Bmatrix} \sin(k \operatorname{ch} a \operatorname{ch} x) \\ \cos(k \operatorname{ch} a \operatorname{ch} x) \end{Bmatrix} \mathrm{Ge}_{2n+1}(x)\,dx = \pm\frac{\pi}{4k}\left(\frac{B_1^{(2n+1)}}{\mathrm{se}_{2n+1}(\pi/2)}\right)^{\pm 1} \begin{Bmatrix} k^2\,\mathrm{ge}_{2n+1}(0)\,\mathrm{Gey}_{2n+1}(a)/\mathrm{se}'_{2n+1}(0) \\ 4S_{2n+1}(q)\,\mathrm{Se}_{2n+1}(a) \end{Bmatrix}.$$

4. $$\int_0^\infty \begin{Bmatrix} \sin(k \operatorname{ch} a \operatorname{ch} x) \\ \cos(k \operatorname{ch} a \operatorname{ch} x) \end{Bmatrix} \mathrm{Ge}_{2n+2}(x)\,dx = \pm\frac{\pi}{4k^2}\left(\frac{\mathrm{se}'_{2n+2}(\pi/2)}{B_2^{(2n+2)}}\right)^{\pm 1} \begin{Bmatrix} 2S_{2n+2}(q)\,\mathrm{Se}_{2n+2}(a) \\ k\,\mathrm{ge}_{2n+2}(0)\,\mathrm{Gey}_{2n+2}(a)/\mathrm{se}'_{2n+2}(0) \end{Bmatrix}.$$

5. $$\int_0^\infty \operatorname{sh} x \sin(k \cos b \operatorname{ch} x)\,\mathrm{Fe}_{2n}(x)\,dx = \frac{\mathrm{ce}_{2n}(\pi/2)}{k \sin b\, A_0^{(2n)}}\left[\frac{\pi C_{2n}(q)}{2}\mathrm{ce}_{2n}(b) - \mathrm{fe}_{2n}(b)\right] \qquad [0<b<\pi].$$

6. $$\int_0^\infty \operatorname{sh} x \cos(k \cos b \operatorname{ch} x)\,\mathrm{Fe}_{2n}(x)\,dx = -\frac{\pi C_{2n}(q)\,\mathrm{ce}_{2n}(\pi/2)}{2k \sin b A_0^{(2n)}\,\mathrm{ce}_{2n}(0)}\mathrm{Fey}_{2n}(0)\,\mathrm{ce}_{2n}(b) \qquad [0<b<\pi].$$

7. $$\int_0^\infty \operatorname{sh} x \sin(k \cos b \operatorname{ch} x)\,\mathrm{Fe}_{2n+1}(x)\,dx = \frac{\pi C_{2n+1}(q)\,\mathrm{c}'_{2n+1}(\pi/2)}{k^2 \sin b A_1^{(2n+1)}\,\mathrm{ce}_{2n+1}(0)}\mathrm{Fey}_{2n+1}(0)\,\mathrm{ce}_{2n+1}(b) \qquad [0<b<\pi].$$

8. $$\int_0^\infty \operatorname{sh} x \cos(k \cos b \operatorname{ch} x)\,\mathrm{Fe}_{2n+1}(x)\,dx = \frac{2\,\mathrm{ce}'_{2n+1}(\pi/2)}{k^2 \sin b A_1^{(2n+1)}}\left[\frac{\pi C_{2n+1}(q)}{2}\mathrm{ce}_{2n+1}(b) - \mathrm{fe}_{2n+1}(b)\right] \qquad [0<b<\pi].$$

9. $$\int_0^\infty \sin(k \cos b \operatorname{ch} x)\,\mathrm{Ge}_{2n+1}(x)\,dx = -\frac{2\,\mathrm{se}_{2n+1}(\pi/2)}{kB_1^{(2n+1)}}\left[\frac{\pi S_{2n+1}(q)}{2}\mathrm{se}_{2n+1}(b) - \mathrm{ge}_{2n+1}(b)\right] \qquad [0<b<\pi].$$

10. $\int_0^\infty \cos(k \cos b \operatorname{ch} x) \operatorname{Ge}_{2n+1}(x)\, dx =$

$$= -\frac{\pi S_{2n+1}(q) \operatorname{se}_{2n+1}(\pi/2)}{k B_1^{(2n+1)} \operatorname{se}_{2n+1}(0)} \operatorname{Gey}_{2n+1}(0) \operatorname{se}_{2n+1}(b) \qquad [0<b<\pi],$$

11. $\int_0^\infty \sin(k \cos b \operatorname{ch} x) \operatorname{Ge}_{2n+2}(x)\, dx =$

$$= -\frac{8 S_{2n+2}(q) \operatorname{se}'_{2n+2}(\pi/2)}{k^3 B_2^{(2n+2)} \operatorname{se}'_{2n+2}(0)} \operatorname{Ge}_{2n+2}(0) \operatorname{se}_{2n+2}(b) \qquad [0<b<\pi].$$

12. $\int_0^\infty \cos(k \cos b \operatorname{ch} x) \operatorname{Ge}_{2n+2}(x)\, dx =$

$$= -\frac{4 \operatorname{se}'_{2n+2}(\pi/2)}{k^2 B_2^{(2n+2)}} \left[\frac{\pi S_{2n+2}(q)}{2} \operatorname{se}_{2n+2}(b) - \operatorname{ge}_{2n+2}(b)\right] \qquad [0<b<\pi].$$

2.27.6. Integrals containing products of Mathieu functions.

1. $\int_0^{2\pi} \operatorname{ce}_m(x) \operatorname{ce}_n(x)\, dx = \pi \delta_{m,n}.$

2. $\int_0^{2\pi} \operatorname{se}_m(x) \operatorname{se}_n(x)\, dx = \pi \delta_{m,n} \qquad [m, n \neq 0].$

3. $\int_0^{2\pi} \operatorname{ce}_m(x) \operatorname{se}_n(x)\, dx = 0 \qquad [m, n \neq 0].$

4. $\int_0^{2\pi} \begin{Bmatrix} \operatorname{ce}_m(x) \operatorname{ce}'_n(x) \\ \operatorname{se}_m(x) \operatorname{se}'_n(x) \end{Bmatrix} dx = 0.$

5. $\int_0^{2\pi} \begin{Bmatrix} \operatorname{ce}_{2m}(x) \operatorname{se}'_{2n}(x) \\ \operatorname{ce}'_{2m}(x) \operatorname{se}_{2n}(x) \end{Bmatrix} dx = \pm 2\pi \sum_{l=0}^{\infty} l A_{2l}^{(2m)} B_{2l}^{(2n)}.$

6. $\int_0^{2\pi} \begin{Bmatrix} \operatorname{ce}_{2m+1}(x) \operatorname{se}'_{2n+1}(x) \\ \operatorname{ce}'_{2m+1}(x) \operatorname{se}_{2n+1}(x) \end{Bmatrix} dx = \pm \pi \sum_{l=0}^{\infty} (2l+1) A_{2l+1}^{(2m+1)} B_{2l+1}^{(2n+1)}.$

7. $\int_0^{2\pi} \sin 2x \begin{Bmatrix} \operatorname{ce}_n(x) \\ \operatorname{se}_n(x) \end{Bmatrix}^2 dx = 0.$

8. $\int_0^{2\pi} \cos 2x \begin{Bmatrix} \operatorname{ce}_n(x) \operatorname{ce}'_n(x) \\ \operatorname{se}_n(x) \operatorname{se}'_n(x) \end{Bmatrix} dx = 0.$

9. $\int_0^{2\pi} \cos 2x \operatorname{ce}_{2n}^2(x)\, dx = \pi A_0^{(2n)} A_2^{(2n)} + \pi \sum_{l=0}^{\infty} A_{2l}^{(2n)} A_{2l+2}^{(2n)}.$

10. $\int_0^{2\pi} \cos 2x \operatorname{ce}_{2n+1}^2(x)\, dx = \frac{\pi}{2} (A_1^{(2n+1)})^2 + \pi \sum_{l=0}^{\infty} A_{2l+1}^{(2n+1)} A_{2l+3}^{(2n+1)}.$

11. $$\int_0^{2\pi} \cos 2x \operatorname{se}^2_{2n+1}(x)\,dx = -\frac{\pi}{2}\left(B_1^{(2n+1)}\right)^2 + \pi \sum_{l=0}^{\infty} B_{2l+1}^{(2n+1)} B_{2l+3}^{(2n+1)}.$$

12. $$\int_0^{2\pi} \cos 2x \operatorname{se}^2_{2n+2}(x)\,dx = \pi \sum_{l=0}^{\infty} B_{2l+2}^{(2n+2)} B_{2l+4}^{(2n+2)}.$$

2.28. THE FUNCTIONS (x), $\nu(x, \rho)$, $\mu(x, \lambda)$, $\mu(x, m, n)$, $\lambda(x, a)$

2.28.1. Integrals containing $\nu(cx)$, $\nu(cx, \rho)$.

1. $$\int_0^a x^{\alpha-1} \nu\left(\frac{x}{a}\right) dx = a^{\alpha} \int_0^{\infty} \frac{dx}{(\alpha+x)\,\Gamma(x+1)}$$ [a, $\operatorname{Re}\alpha>0$].

2. $$\int_0^a (a-x)^{\beta-1} \nu\left(\frac{x}{a}\right) dx = a^{\beta}\,\Gamma(\beta)\,\nu(1, \beta)$$ [a, $\operatorname{Re}\beta>0$].

3. $$\int_0^{\infty} e^{-px} \nu(cx)\,dx = \left(p \ln \frac{p}{c}\right)^{-1}$$ [$\operatorname{Re} p>c$].

4. $$\int_0^{\infty} x^n e^{-px} \nu(cx)\,dx = \frac{1}{p^{n+1}} \sum_{k=0}^{n} a_k k! \left(\ln \frac{p}{c}\right)^{-k-1}$$ [$\operatorname{Re} p>c$; a_k are the coefficients of the expansion $(s+1)_n = \sum_{k=0}^{n} a_k s^k$].

5. $$\int_0^{\infty} e^{-px^2} \nu(cx)\,dx = \sqrt{\frac{\pi}{p}}\,\nu\left(\frac{c^2}{4p}\right)$$ [$\operatorname{Re} p>0$].

6. $$\int_0^{\infty} \frac{e^{-px}}{e^{cx}-1}\,\nu(cx)\,dx = \frac{1}{c}\int_0^{\infty} \zeta\left(x+1,\ 1-\frac{p}{c}\right) dx$$ [$\operatorname{Re} p>0$].

7. $$\int_0^{\infty} e^{-px} \nu(cx, \rho)\,dx = \frac{c^{\rho}}{p^{\rho+1}} \left(\ln \frac{p}{c}\right)^{-1}$$ [$\operatorname{Re} p>c$; $\operatorname{Re}\rho>-1$].

8. $$\int_0^{\infty} x^n e^{-px} \nu(cx, \rho)\,dx = \frac{c^{\rho}}{p^{n+\rho+1}} \sum_{k=0}^{n} b_k k! \left(\ln \frac{p}{c}\right)^{-k-1}$$ [$\operatorname{Re} p>c$; $\operatorname{Re}\rho>-1$; b_k are the coefficients of the expansion $(s+\rho+1)_n = \sum_{k=0}^{n} b_k s^k$].

9. $$\int_0^{\infty} e^{-px^2} \nu(cx, \rho)\,dx = \sqrt{\frac{\pi}{p}}\,\nu\left(\frac{c^2}{4p},\ \frac{\rho}{2}\right)$$ [$\operatorname{Re} p>0$; $\operatorname{Re}\rho>-1$].

10. $$\int_0^{\infty} x e^{-px^2} \nu(cx, \rho)\,dx = \frac{\sqrt{\pi}}{8cp^{3/2}}\,\nu\left(\frac{c^2}{4p},\ \frac{\rho-1}{2}\right)$$ [$\operatorname{Re} p>0$; $\operatorname{Re}\rho>-1$].

2.28.2. Integrals containing $\mu(cx, \lambda)$, $\mu(cx, m, n)$.

1. $$\int_0^\infty e^{-px}\mu(cx, \lambda)\,dx = \frac{\Gamma(\lambda+1)}{p}\left(\ln\frac{p}{c}\right)^{-\lambda-1}$$ [Re $p>c$; Re $\lambda>-1$].

2. $$\int_0^\infty x^n e^{-px}\mu(cx, \lambda)\,dx = \frac{1}{p^{n+1}}\sum_{k=0}^{n} a_k\Gamma(k+\lambda+1)\left(\ln\frac{p}{c}\right)^{-k-\lambda-1}$$ $\left[\text{Re } p>c;\ \text{Re }\lambda>-1;\ a_k \text{ are the coefficients of the expansion } (s+1)_n=\sum_{k=0}^{n} a_k s^k\right]$.

3. $$\int_0^\infty e^{-px^2}\mu(cx, \lambda)\,dx = 2^\lambda\sqrt{\frac{\pi}{p}}\,\mu\left(\frac{c^2}{4p}, \lambda\right)$$ [Re $p>0$; Re $\lambda>-1$].

4. $$\int_0^a \frac{1}{x}\ln^\beta\frac{a}{x}\,\mu(cx, \lambda)\,dx = \Gamma(\beta+1)\,\mu(ac, \lambda-\beta-1)$$ [$a>0$; Re $\lambda>-1$].

5. $$\int_0^\infty e^{-px^2}\mu(cx, m, 2n)\,dx = 2^{m-1}\sqrt{\frac{\pi}{p}}\,\mu\left(\frac{c^2}{4p}, m, n\right)$$ [Re $p>0$].

2.28.3. Integrals containing $\lambda(cx, a)$.

1. $$\int_0^\infty x^{-3/2}e^{-p/x}\lambda(cx, a)\,dx = \frac{1}{2}\sqrt{\frac{\pi}{p}}\,\lambda(2\sqrt{cp}, 2a)$$ [a, Re $p>0$].

2. $$\int_0^\infty x^{-5/2}e^{-p/x}\lambda(cx, a)\,dx = \frac{\sqrt{\pi c}}{2p}\left[\lambda(2\sqrt{cp}, 2a+1) - \lambda(2\sqrt{cp}, 1)\right]$$ [a, Re $p>0$].

Chapter 3. DEFINITE INTEGRALS OF PIECEWISE-CONTINUOUS FUNCTIONS

3.1. INTRODUCTION

This chapter contains the Laplace transform

$$F(p)=\int_0^\infty e^{-px} f(x)\,dx$$

of certain piecewise-continuous functions defined by different formulae over separate intervals of the positive real semi-axis.

3.2. PIECEWISE-CONSTANT FUNCTIONS

3.2.1. Bounded functions.

	$f(x)$		$F(p)$
1	μ_1	$[na<x<na+b_1]$	$\frac{1}{p(1-e^{-ap})}\left[\mu_1-\mu_m e^{-ap}+\sum_{k=1}^{m-1}(\mu_{k+1}-\mu_k)e^{-b_{k+1}p}\right]$
	μ_2	$[na+b_1<x<na+b_2]$	
	. . . .		
	μ_m	$[na+b_m<x<(n+1)a]$	
		$[b_1<b_2<\ldots<b_m<a]$	
2	1	$[2na<x<(2n+1)a]$	$\frac{1}{p(1+e^{-ap})}$
	0	$[(2n+1)a<x<(2n+2)a]$	
3	0	$[2na<x<(2n+1)a]$	$\frac{1}{p(1+e^{ap})}$
	1	$[(2n+1)a<x<(2n+2)a]$	
4	1	$[(2n)^2a<x<(2n+1)^2a]$	$\frac{1}{p}\theta_0(0,e^{-ap})$
	-1	$[(2n+1)^2a<x<(2n+2)^2a]$	
5	μ_0	$[0<x<b_1]$	$\frac{1}{p(1-e^{-ap})}\Big[\mu_0-(\mu_0+\mu_m)e^{-ap}+$
	μ_1	$[na+b_1<x<na+b_2]$	$+\mu_0 e^{-(a+b_1)p}+$

	$f(x)$		$F(p)$
5	μ_2 . . . . μ_m	$[na+b_2 < x < na+b_3]$ . . . . $[na+b_m < x < (n+1)a]$	$+\sum_{k=0}^{m-1}(\mu_{k+1}-\mu_k)\,e^{-b_{k+1}p}\Big]$
6	0 1 -1	$[0 < x < a]$ $[(2n+1)a < x < (2n+2)a]$ $[(2n+2)a < x < (2n+3)a]$	$\frac{1-e^{-ap}}{p(1+e^{ap})}$
7	0 1 0	$[0 < x < a]$ $[(4n+1)a < x < (4n+3)a]$ $[(4n+3)a < x < (4n+5)a]$	$\frac{\operatorname{sech} ap}{2p}$
8	1 -1 1	$[0 < x < a]$ $[(4n+1)a < x < (4n+3)a]$ $[(4n+3)a < x < (4n+5)a]$	$\frac{1-\operatorname{sech} ap}{p}$
9	0 1 0 -1	$[0 < x < a$ и $4na < x < (4n+1)a]$ $[(4n+1)a < x < (4n+2)a]$ $[(4n+2)a < x < (4n+3)a]$ $[(4n+3)a < x < (4n+4)a]$	$\frac{1-e^{-ap}}{p(e^{ap}+e^{-ap})}$
10	0 1 0 -1 0	$[0 < x < a$ и $8na < x < (8n+1)a]$ $[(8n+1)a < x < (8n+3)a]$ $[(8n+3)a < x < (8n+5)a]$ $[(8n+5)a < x < (8n+7)a]$ $[(8n+7)a < x < (8n+8)a]$	$\frac{\operatorname{sh} ap}{p \operatorname{ch} 2ap}$

3.2.2. Unbounded functions.

	$f(x)$		$F(p)$
1	n	$[na < x < (n+1)a]$	$\frac{1}{p(e^{ap}-1)}$
2	$n+1$	$[na < x < (n+1)a]$	$\frac{1}{p(1-e^{-ap})}$
3	$2n+1$	$[na < x < (n+1)a]$	$\frac{1}{p}\operatorname{cth}\frac{p}{2}$
4	$n(n-1)$	$[na < x < (n+1)a]$	$\frac{2}{p^2(e^{ap}-1)}$
5	n^2	$[na < x < (n+1)a]$	$\frac{e^{ap}+1}{p(e^{ap}-1)^2}$
6	n^3	$[na < x < (n+1)a]$	$\frac{e^{2ap}+4e^{ap}+1}{p(e^{p}-1)^3}$
7	n^m	$[na < x < (n+1)a]$	$(-1)^m\frac{1-e^{-ap}}{p}\frac{d^m}{ap^m}\frac{1}{1-e^{-ap}}$

	$f(x)$	$F(p)$
8	$n\mu_1$ $[na<x<na+b_1]$ $n\mu_2$ $[na+b_1<x<na+b_2]$ . . . . $n\mu_m$ $[na+b_{m-1}<x<(n+1)a]$ $[b_1<b_2<\ldots<b_{m-1}<a]$	$\frac{e^{-ap}}{p(1-e^{-ap})^2}[\mu_1(1-e^{-b_1p})+$ $+\mu_m(e^{-b_{m-1}p}-e^{-ap})+$ $+\sum_{k=1}^{m-2}\mu_{k+1}(e^{-b_kp}-e^{-b_{k+1}p})$
9	λ $[0<x<b]$ $n\mu+\nu$ $[na+b<x<(n+1)a+b]$	$\frac{\lambda}{p}(1-e^{-bp})+$ $+\frac{\nu e^{-bp}-(\nu-\mu)e^{-(2a+b)p}}{p(1-e^{-ap})}$
10	0 $[0<x<a]$ $n+1$ $[(2n+1)a<x<(2n+3)a]$	$\frac{\operatorname{cosech} ap}{2p}$
11	μ_0 $[0<x<b]$ $n\mu_1+\nu_1$ $[na+b<x<na+c]$ $n\mu_2+\nu_2$ $[na+c<x<(n+1)a+b]$	$\frac{1}{p}\Big\{\mu_0(1-e^{-bp})+\frac{1}{1-e^{-ap}}\times$ $\times[\nu_1e^{-bp}+(\nu_2-\nu_1)e^{-cp}-$ $-\nu_2e^{-(a+b)p}]+\frac{e^{-ap}}{(1-e^{-ap})^2}\times$ $\times[\mu_1e^{-bp}+(\mu_2-\mu_1)e^{-cp}-$ $-\mu_2e^{-(a+b)p}]\Big\}$
12	0 $[0<x<a]$ $2n+1$ $[(4n+1)a<x<(4n+3)a]$ $-2n-2$ $[(4n+3)a<x<(4n+5)a]$	$\frac{\operatorname{sh} ap}{2p\operatorname{ch}^2 ap}$
13	n $[n^2a<x<(n+1)^2a]$	$\frac{1}{2p}[\theta_3(0,\ e^{-ap})-1]$
14	$2n+1$ $[n^2a<x<(n+1)^2a]$	$\frac{1}{p}\theta_3(0,\ e^{-ap})$
15	0 $[0<x<a/4]$ $n+1$ $[(n+1/2)^2a<x<(n+3/2)^2a]$	$\frac{1}{2p}\theta_2(0,\ e^{-ap})$
16	$n\mu+\nu$ $[a\ln(nb+c)<x<a\ln(nb+b+c)]$	$\frac{1}{p}\left[\frac{\mu}{b^{ap}}\zeta\left(ap,\ \frac{c}{b}\right)-\right.$ $\left.-\frac{\mu}{c^{ap}}+\frac{\nu}{(b+c)^{ap}}\right]$
17	$n\mu+\nu$ $[a\ln n<x<a\ln(n+1)]$	$\frac{\mu}{p}\zeta(ap)+\frac{\nu}{p}$
18	$n+1$ $[a\ln(2n+1)<x<a\ln(2n+3)]$	$\frac{1}{p}(1-2^{2-ap})\zeta(ap-1)$
19	$\mu+b^n\nu$ $[na<x<(n+1)a]$	$\frac{(1-e^{-ap})}{p}\left(\mu+\nu\frac{1-e^{-ap}}{1-be^{-ap}}\right)$
20	$\binom{n}{m}$ $[na<x<(n+1)a]$	$\frac{1}{p(e^{ap}-1)^m}$
21	$\sum_{1\leqslant n\leqslant e^x}\frac{1}{n^\nu}$	$\frac{1}{p}\zeta(p+\nu)$ $[\operatorname{Re} p>-\operatorname{Re}\nu+1]$

3.3. SOME PIECEWISE-CONTINUOUS FUNCTIONS

3.3.1. Power functions.

	$f(x)$		$F(p)$
1	$\lambda_1 x$ $\lambda_2 x$ $\cdots\cdots$ $\lambda_m x$	$[na < x < na+b_1]$ $[na+b_1 < x < na+b_2]$ $\cdots\cdots$ $[na+b_{m-1} < x < (n+1)a]$	$\frac{1}{p^2(1-e^{-ap})^2}[\lambda_1+(ap\lambda_1-ap\lambda_m-\lambda_m-\lambda_1)e^{-ap}+\lambda_m e^{-2ap}+\sum_{k=1}^{m-1}(\lambda_{k+1}-\lambda_k)(1+pb_k)e^{-b_k p}+\sum_{k=1}^{m-1}(\lambda_{k+1}-\lambda_k)\times(ap-b_k p-1)e^{-(a+b_k)p}]$
2	x 0	$[na < x < na+b]$ $[na+b < x < (n+1)a]$	$\frac{1-(bp+1)e^{-bp}}{p^2(1-e^{-ap})}+\frac{a}{p}e^{-ap}\frac{1-e^{-bp}}{(1-e^{-ap})^2}$
3	$\lambda x+n\mu+\nu$	$[na < x < (n+1)a]$	$\frac{\lambda}{p^2}+\frac{\nu}{p}+\frac{\mu e^{-ap}}{p(1-e^{-ap})}$
4	$\lambda_1 x+n\mu_1+\nu_1$ $\lambda_2 x+n\mu_2+\nu_2$	$[na < x < na+b]$ $[na+b < x < (n+1)a]$	$\frac{\lambda_2}{p^2}+(\lambda_1-\lambda_2)\frac{1-(bp+1)e^{-bp}}{p^2(1-e^{-ap})}+(\lambda_1-\lambda_2)\frac{a}{p}e^{-ap}\frac{1-e^{-bp}}{(1-e^{-ap})^2}+\frac{1}{p(1-e^{-ap})^2}\times[\nu_1+(\mu_1-\nu_1-\nu_2)e^{-ap}+(\nu_2-\mu_2)e^{-2ap}+(\nu_2-\nu_1)e^{-bp}+(\mu_2-\mu_1-\nu_2+\nu_1)e^{-(a+b)p}]$
5	$x-na$ b 0	$[na < x < na+b]$ $[na+b < x < na+c]$ $[na+c < x < (n+1)a]$	$\frac{1-e^{-bp}-bpe^{-cp}}{p^2(1-e^{-ap})}$
6	$\frac{x-na}{b}$ $\frac{x-na-c}{b-c}$ 0	$[na < x < na+b]$ $[na+b < x < na+c]$ $[na+c < x < (n+1)a]$	$\frac{c(1-e^{-bp})-b(1-e^{-cp})}{b(c-b)p^2(1-e^{-ap})}$
7	$\frac{x-2na}{b}$	$[2na < x < 2na+b]$	$\frac{a(1-e^{-bp})-b(1-e^{-ap})}{b(a-b)p^2(1-e^{-2ap})}$

	$f(x)$	$F(p)$
	$\dfrac{x-(2n+1)a}{b-a}$ $[2na+b<x<(2n+1)a]$ 0 $[(2n+1)a<x<(2n+2)a]$	
8	$x-4na$ $[4na<x<(4n+1)a]$ $-x+4na+2a$ $[(4n+1)a<x<(4n+2)a]$ 0 $[(4n+\ )a<x<(4n+4)a]$	$\dfrac{1}{p^2(1+e^{-2ap})}\operatorname{th}\dfrac{ap}{2}$
9	$x-na$ $[na<x<na+b]$ $-x+na+b$ $[na+b<x<na+2b]$ 0 $[na+2b<x<(n+1)a]$	$\dfrac{(1-e^{-bp})^2}{p^2(1-e^{-ap})}$
10	$-b$ $[na<x<(2n+1)a-b]$ $x-(2n+1)a$ $[(2n+1)a-b<x<(2n+\ )a+b]$ b $[(2n+1)a+b<x<(n+1)a]$	$2\dfrac{\operatorname{sh}bp-bp(1+e^{-2ap})}{p^2(1-e^{-2ap})}$
11	$x-2na$ $[2na<x<2na+b]$ b $[2na+b<x<(2n+2)a-b]$ $-x+(2n+2)a$ $[(2n+2)a-b<x<(2n+2)a]$	$\dfrac{(1-e^{-bp})[1-e^{-(2a-b)p}]}{p^2(1-e^{-2ap})}$
12	$\lambda_0 x$ $[0<x<b_1]$ $\lambda_1 x$ $[na+b_1<x<na+b_2]$ . . . . $\lambda_m x$ $[na+b_m<x<(n+1)a]$ $[b_1<b_2<\ldots<b_m<a]$	$\dfrac{1}{p^2(1-e^{-ap})^2}\times$ $\times[\lambda_0+(\lambda_0-\lambda_0 a-\lambda_m a)p-$ $-2\lambda_0-\lambda_m]e^{-ap}+$ $+(\lambda_0+\lambda_m)(ap-p+1)e^{-ap}+$ $+\lambda_0(ap+b_1p+1)e^{-b_1p}-$ $-\lambda_0 e^{-(a+b_1)p}+$ $+\sum_{k=0}^{m-1}(\lambda_{k+1}-\lambda_k)\times$ $\times(b_{k+1}p+1)e^{-b_{k+1}p}+$ $+\sum_{k=0}^{m-1}(\lambda_{k+1}-\lambda_k)\times$ $\times(p-b_{k+1}p-1)e^{-(a+b_{k+1})p}$
13	0 $[0<x<a$ и $(4n+3)a<x<(4n+5)a]$ x $[(4n+1)a<x<(4n+3)a]$	$\dfrac{1+ap\operatorname{th}ap}{2p^2\operatorname{ch}ap}$
14	0 $[0<x<a]$ $x+(-1)^n(x-2na-2a)$ $[(2n+1)a<x<(2n+3)a]$	$\dfrac{\operatorname{sech}ap}{p^2}$

	$f(x)$	$F(p)$
15	$a-(-1)^n(2na+a-x)$ $[2a<x<(2n+2)a]$	$\dfrac{\operatorname{th} ap}{p^2}$
16	0 $[0<x<a]$ $(n+1)(x-na-a)$ $[(2n+1)a<x<(2n+3)a]$	$\dfrac{\operatorname{cosech} ap}{2p^2}$
17	$nx-n(n+1)a$ $[2na<x<(2n+2)a]$	$\dfrac{1}{p^2(e^{2ap}-1)}$
18	$(2n+1)x-n(n+1)a$ $[na<x<(n+1)a]$	$\dfrac{1}{p^2}\operatorname{cth}\dfrac{ap}{2}$
19	$\dfrac{1}{4}[1-(-1)^n](2x-a)+\dfrac{(-1)^n}{2}na$ $[na<x<(n+1)a]$	$\dfrac{1}{p^2(e^{ap}+1)}$
20	$(x-na)^2$ $[na<x<(n+1)a]$	$\dfrac{2}{p^3}-\dfrac{a(2p+a)}{p^2(e^{ap}-1)}$

3.3.2. Various functions.

	$f(x)$	$F(p)$
1	$[a(x-na)-(x-na)^2]^\nu$ $[na<x<(n+1)a]$ $[\operatorname{Re}\nu>-1]$	$\dfrac{\pi\Gamma(2\nu+1)}{\Gamma(\nu+1/2)}\operatorname{cosech}\dfrac{ap}{2}\left(\dfrac{a}{4p}\right)^{\nu+1/2}\times$ $\times I_{\nu+1/2}\left(\dfrac{ap}{2}\right)$
2	$\pm[2a(x-2na)-(x-2na)^2]^\nu$ $\left[\begin{Bmatrix}2na<x<(2n+1)a\\(2n+1)a<x<(2n+2)a\end{Bmatrix}, \operatorname{Re}\nu>-1\right]$	$\pi\dfrac{\Gamma(2\nu+1)}{\Gamma(\nu+1/2)}\left(\dfrac{a}{2p}\right)^{\nu+1/2}\times$ $\times\operatorname{cosech} ap\,\mathbf{L}_{\nu+1/2}(ap)$
3	$\sum\limits_{1\leqslant n\leqslant e^x}(x-\ln n)^\nu$ $[\operatorname{Re}\nu>-1]$	$\dfrac{\Gamma(\nu+1)}{p^{\nu+1}}\zeta(p)$
4	b^n $[na<x<(n+1)a]$	$\dfrac{e^{ap}-1}{p(e^{ap}-b)}$ $[\operatorname{Re}p>\operatorname{Re}\ln b]$
5	nb^{n-1} $[na<x<(n+1)a]$	$\dfrac{e^{ap}-1}{p(e^{ap}-b)^2}$ $[\operatorname{Re}p>\operatorname{Re}\ln b]$
6	n^2b^{n-1} $[na<x<(n+1)a]$	$\dfrac{(e^{ap}-1)(e^{ap}+b)}{p(e^{ap}-b)^3}$ $[\operatorname{Re}p>\operatorname{Re}\ln b]$

	$f(x)$	$F(p)$
7	$\frac{1-b^n}{1-b}x-$ $-a\frac{1-(n+1)b^n+nb^{n+1}}{(1-b)^2}$ $[na<x<(n+1)a]$	$\frac{1}{p^2(e^{ap}-b)}$
8	$c^n \sin bn$ $[na<x<(n+1)a]$	$\frac{c\sin b\,(e^{ap}-1)}{p(e^{2ap}-2ce^{ap}\cos b+c^2)}$ $[\operatorname{Re} p>\operatorname{Re}\ln c+\lvert\operatorname{Im} b\rvert]$
9	$c^n \cos bn$ $[na<x<(n+1)a]$	$\frac{(e^{ap}-1)(e^{ap}-c\cos b)}{p(e^{2ap}-2ce^{ap}\cos b+c^2)}$ $[\operatorname{Re} p>\operatorname{Re}\ln c+\lvert\operatorname{Im} b\rvert]$

Chapter 4. MULTIPLE INTEGRALS

4.1. INTRODUCTION

This chapter contains double and multiple integrals involving special functions. Multiple integrals involving other special functions, as well as a number of formulae of a general character by means of which multiple integrals can be reduced to single ones, are given in [17, 18].

4.2. DOUBLE INTEGRALS

4.2.1. Integrals containing the functions $\mathbf{H}_\nu(x)$, $\mathbf{J}_\nu(x)$, $\operatorname{ber}_\nu(x)$, $\operatorname{bei}_\nu(x)$, $s_{\mu,\nu}(x)$.

1. $$\int_0^\infty\int_0^\infty x^{\alpha-1}y^{\alpha+1}e^{-ax^4-by^4}\,\mathbf{H}_\nu(cxy)\,dx\,dy=$$
$$=\frac{2^{-\nu-4}}{\sqrt{\pi}}\Gamma\begin{bmatrix}(\alpha+\nu+1)/4,\ (\alpha+\nu+3)/4\\ \nu+3/2\end{bmatrix}a^{-(\alpha+\nu+1)/4}b^{-(\alpha+\nu+3)/4}\times$$
$$\times c^{\nu+1}\,{}_2F_2\left(\begin{matrix}1,\ (\alpha+\nu+2)/2;\ -c^2/(8\sqrt{ab})\\ 3/2,\ \nu+3/2\end{matrix}\right)\qquad[\operatorname{Re}(\alpha+\nu)>-1].$$

2. $$\int_0^\infty\int_0^\infty x^{\alpha-1}y^{\beta-1}e^{-ax^2-by^2}\,J_\nu(cxy)\,dx\,dy=$$
$$=\frac{1}{4}\cos\frac{\nu\pi}{2}\,\Gamma\begin{bmatrix}\alpha/2,\ \beta/2\\ 1+\nu/2,\ 1-\nu/2\end{bmatrix}a^{-\alpha/2}b^{-\beta/2}\,{}_3F_2\left(\begin{matrix}1,\ \alpha/2,\ \beta/2;\ -c^2/(4ab)\\ 1+\nu/2,\ 1-\nu/2\end{matrix}\right)+$$
$$+\frac{1}{8}\sin\frac{\nu\pi}{2}\,\Gamma\begin{bmatrix}(\alpha+1)/2,\ (\beta+1)/2\\ (3+\nu)/2,\ (3-\nu)/2\end{bmatrix}a^{-(\alpha+1)/2}b^{-(\beta+1)/2}\times$$
$$\times\,{}_3F_2\left(\begin{matrix}1,\ (\alpha+1)/2,\ (\beta+1)/2;\ -c^2/(4ab)\\ (3+\nu)/2,\ (3-\nu)/2\end{matrix}\right)\qquad[\operatorname{Re}\alpha,\ \operatorname{Re}\beta>0].$$

3. $$\int_0^\infty\int_0^\infty x^{\alpha-1}y^{\beta-1}e^{-ax^2-by^2}\operatorname{ber}(cxy)\,dx\,dy=$$
$$=\frac{\Gamma(\alpha/2)\,\Gamma(\beta/2)}{4a^{\alpha/2}b^{\beta/2}}\,{}_4F_3\left(\begin{matrix}\alpha/4,\ (\alpha+2)/4,\ \beta/4,\ (\beta+2)/4\\ 1/2,\ 1/2,\ 1;\ -c^4/(16a^2b^2)\end{matrix}\right)\qquad[\operatorname{Re}\alpha,\ \operatorname{Re}\beta>0].$$

4. $$\int_0^\infty\int_0^\infty x^{\alpha-1}y^{\beta-1}e^{-ax^2-by^2}\operatorname{bei}(cxy)\,dx\,dy=$$
$$=\frac{c^2\Gamma(\alpha/2+1)\,\Gamma(\beta/2+1)}{16a^{\alpha/2+1}b^{\beta/2+1}}\,{}_4F_3\left(\begin{matrix}(\alpha+2)/4,\ (\alpha+4)/4,\ (\beta+2)/4,\ (\beta+4)/4\\ 1,\ 3/2,\ 3/2;\ -c^4/(16a^2b^2)\end{matrix}\right)$$
$$[\operatorname{Re}\alpha,\ \operatorname{Re}\beta>-1].$$

5. $$\int_0^\infty\int_0^\infty x^{\alpha-1}y^{\beta-1}e^{-ax^2-by^2}\left[\operatorname{ber}_\nu^2(cxy)+\operatorname{bei}_\nu^2(cxy)\right]dx\,dy=$$
$$=\frac{2^{-2\nu-2}c^{2\nu}}{a^{\alpha/2+\nu}b^{\beta/2+\nu}}\Gamma\left[\begin{matrix}\alpha/2+\nu,\ \beta/2+\nu\\ \nu+1,\ \nu+1\end{matrix}\right]\times$$
$$\times{}_4F_3\left(\begin{matrix}(\alpha+2\nu)/4,\ (\alpha+2\nu+2)/4,\ (\beta+2\nu)/4,\ (\beta+2\nu+2)/4\\ (\nu+1)/2,\ \nu/2+1,\ \nu+1;\ c^4/(4a^2b^2)\end{matrix}\right)$$
$[\operatorname{Re}(\alpha+2\nu),\ \operatorname{Re}(\beta+2\nu)>0]$.

6. $$\int_0^\infty\int_0^\infty x^{\alpha}y^{\beta}e^{-ax^2-by^2}\left[\operatorname{ber}_\nu(cxy)\operatorname{ber}_\nu'(cxy)+\operatorname{bei}_\nu(cxy)\operatorname{bei}_\nu'(cxy)\right]dx\,dy=$$
$$=\frac{2^{-2\nu-2}c^{2\nu-1}}{a^{\alpha/2+\nu}b^{\beta/2+\nu}}\Gamma\left[\begin{matrix}\alpha/2+\nu,\ \beta/2+\nu\\ \nu,\ \nu+1\end{matrix}\right]\times$$
$$\times{}_4F_3\left(\begin{matrix}(\alpha+2\nu)/4,\ (\alpha+2\nu+2)/4,\ (\beta+2\nu)/4,\ (\beta+2\nu+2)/4\\ \nu/2,\ (\nu+1)/2,\ \nu+1;\ c^4/(4a^2b^2)\end{matrix}\right)$$
$[\operatorname{Re}\nu,\ \operatorname{Re}(\alpha/2+\nu),\ \operatorname{Re}(\beta/2+\nu)>0]$.

7. $$\int_0^\infty\int_0^\infty x^{\alpha-1}y^{\beta-1}e^{-ax^2-by^2}s_{\mu,\nu}(cxy)\,dx\,dy=$$
$$=\frac{a^{-(\alpha+\mu+1)/2}b^{-(\beta+\mu+1)/2}c^{\mu+1}}{4[(\mu+1)^2-\nu^2]}\Gamma\left(\frac{\alpha+\mu+1}{2}\right)\times$$
$$\times\Gamma\left(\frac{\beta+\mu+1}{2}\right){}_3F_2\left(\begin{matrix}1,\ (\alpha+\mu+1)/2,\ (\beta+\mu+1)/2;\ -c^2/(4ab)\\ (\mu-\nu+3)/2,\ (\mu+\nu+3)/2\end{matrix}\right)$$
$[\operatorname{Re}(\alpha+\mu),\ \operatorname{Re}(\beta+\mu)>-1]$.

4.2.2. Integrals containing ${}_1F_1(a;\ b;\ x)$.

1. $$\iint\limits_{\substack{x,\,y\geqslant 0\\ x+y\leqslant 1}} x^{\alpha-1}y^{\beta-1}(1-x-y)^{\gamma-1}{}_1F_1(a;\ b;\ cxy)\,{}_1F_1(a;\ b;\ -cxy)\,dx\,dy=$$
$$=\Gamma\left[\begin{matrix}\alpha,\ \beta,\ \gamma\\ \alpha+\beta+\gamma\end{matrix}\right]{}_6F_7\left(\begin{matrix}a,\ b-a,\ \Delta(2,\ \alpha),\ \Delta(2,\ \beta);\ c^2/64\\ b,\ b/2,\ (b+1)/2,\ \Delta(4,\ \alpha+\beta+\gamma)\end{matrix}\right)$$
$[\operatorname{Re}\alpha,\ \operatorname{Re}\beta,\ \operatorname{Re}\gamma>0]$.

2. $$\int_0^1\int_0^1 x^{b}(1-x)^{b'-1}(1-y)^{b-1}f(xy)\,{}_1F_1(a;\ b;\ x(1-y))\,{}_1F_1(a';\ b';\ 1-x)\,dx\,dy=$$
$$=\mathrm{B}(b,\ b')\int_0^1(1-t)^{b+b'-1}f(t)\,{}_1F_1(a+a';\ b+b';\ 1-t)\,dt.$$

3. $$\int_0^\infty\int_0^\infty x^{b-1}y^{b'-1}f(x+y)\,{}_1F_1(a;\ b;\ x)\,{}_1F_1(a';\ b';\ y)\,dx\,dy=$$
$$=\mathrm{B}(b,\ b')\int_0^\infty t^{b+b'-1}f(t)\,{}_1F_1(a+a';\ b+b';\ t)\,dt.$$

4. $$\int_0^\infty\int_0^\infty x^{b-1}y^{b}e^{-px^2-qy^2}{}_1F_1\left(\begin{matrix}a;\ cxy\\ b\end{matrix}\right){}_1F_1\left(\begin{matrix}a;\ -cxy\\ b\end{matrix}\right)dx\,dy=$$
$$=\frac{\Gamma(b/2)\,\Gamma((b+1)/2)}{4p^{b/2}q^{(b+1)/2}}\,{}_2F_1\left(\begin{matrix}a,\ b-a\\ b;\ c^2/(4pq)\end{matrix}\right)$$
$[\operatorname{Re}b>0]$.

5. $$\int_0^\infty\int_0^\infty x^{2b-1}y^{b/2}e^{-px^2-qy^2}\,{}_1F_1\left(\begin{matrix}a;\ cxy\\ b\end{matrix}\right)\,{}_1F_1\left(\begin{matrix}a;\ -cxy\\ b\end{matrix}\right)dx\,dy=$$
$$=\frac{\Gamma(b)\,\Gamma((b+1)/2)}{4p^bq^{(b+1)/2}}\,{}_2F_1\left(\begin{matrix}a,\ b-a\\ b/2;\ c^2/(4pq)\end{matrix}\right)\qquad [\operatorname{Re} b>0].$$

6. $$\int_0^\infty\int_0^\infty x^{2b-1}y^{(b-1)/2}e^{-px^2-qy^2}\,{}_1F_1\left(\begin{matrix}a;\ cxy\\ b\end{matrix}\right)\,{}_1F_1\left(\begin{matrix}a;\ -cxy\\ b\end{matrix}\right)dx\,dy=$$
$$=\frac{\Gamma(b)\,\Gamma(b/2)}{4p^bq^{b/2}}\,{}_2F_1\left(\begin{matrix}a,\ b-a\\ (b+1)/2;\ c^2/(4pq)\end{matrix}\right)\qquad [\operatorname{Re} b>0].$$

4.2.3. Integrals containing ${}_2F_1(a,\ b;\ c;\ x)$.

1. $$\int_0^1\int_0^1 (xy)^{c-1}(1-x)^{d-c-1}(1-y)^{a-b-d}(1-ux)^{-a}(1-vy)^{-a}\times$$
$$\times\,{}_2F_1\left(a,\ b;\ c;\ \frac{uv\,xy}{(1-ux)(1-vy)}\right)dx\,dy=$$
$$=\Gamma\left[\begin{matrix}c,\ c,\ d-c,\ a-b-d+1\\ d,\ a-b+c-d+1\end{matrix}\right]F_4(a,\ c;\ d,\ a-b+c-d+1;\ u-uv,\ v-uv)$$
$$[\operatorname{Re} c,\ \operatorname{Re}(d-c),\ \operatorname{Re}(a-b-d+1)>0;\ |u|+|v|<1].$$

2. $$\int_0^1\int_0^1 (xy)^{b-1}(1-x)^{c-1}(1-y)^{d-1}(1-ux-vy)^{-2a}\,{}_2F_1\left(a,\,a+1;\,b;\,\frac{4uv\,xy}{(1-ux-vy)^2}\right)dx\,dy=$$
$$=\Gamma\left[\begin{matrix}b,\ b,\ c,\ d\\ c+b,\ d+b\end{matrix}\right]F_4(2a,\ b;\ c+b,\ d+b;\ u,\ v)$$
$$[\operatorname{Re} b,\ \operatorname{Re} c,\ \operatorname{Re} d>0;\ \sqrt{|u|}+\sqrt{|v|}<1].$$

3. $$\int_0^1\int_0^1 (xy)^{c-1}(1-x)^{d-1}(1-y)^{a+b-2c-d}(1-ux-vy)^{-a}\,{}_2F_1\left(a,b;c;-\frac{uv\,xy}{1-ux-vy}\right)dx\,dy=$$
$$=\Gamma\left[\begin{matrix}c,\ c,\ d,\ a+b-2c-d+1\\ c+d,\ a+b+c-d+1\end{matrix}\right]F_4(a,\ c;\ c+d,\ a+b-c-d+1;\ u-uv,\ v-uv)$$
$$[\operatorname{Re} c,\ \operatorname{Re} d,\ \operatorname{Re}(a+b-2c-d+1)>0;\ |u|+|v|+|uv|<1].$$

4.2.4. Integrals containing ${}_pF_q((a_p);\ (b_q);\ x)$.

1. $$\int_0^1\int_0^1 f(xy)(1-x)^{\alpha-1}y^{\alpha}(1-y)^{\beta-1}\,{}_pF_q\left(\begin{matrix}(a_p);\ c(1-x)^m y^m(1-y)^n\\ (b_q)\end{matrix}\right)dx\,dy=$$
$$=\mathrm{B}(\alpha,\beta)\int_0^1 f(t)(1-t)^{\alpha+\beta-1}\,{}_{p+m+n}F_{q+m+n}\left(\begin{matrix}(a_p),\ \Delta(m,\alpha),\ \Delta(n,\beta);\ cu(1-t)^{m+n}\\ (b_q),\ \Delta(m+n,\ \alpha+\beta)\end{matrix}\right)dt$$
$$[u=m^mn^n(m+n)^{-m-n}].$$

2. $$\int_0^1\int_0^1 x^{\alpha}(1-x)^{\beta-1}y^{\alpha+\beta}(1-y)^{\gamma-1}(1-xy)^{\delta}\,{}_pF_q\left(\begin{matrix}(a_p);c(1-x)^m y^m(1-y)^n\\ (b_q)\end{matrix}\right)dx\,dy=$$
$$=\Gamma\left[\begin{matrix}\alpha+1,\ \beta,\ \gamma,\ \beta+\gamma+\delta\\ \beta+\gamma,\ \alpha+\beta+\gamma+\delta+1\end{matrix}\right]\times$$
$$\times\,{}_{p+2m+2n}F_{q+2m+2n}\left(\begin{matrix}(a_p),\ \Delta(m,\beta),\ \Delta(n,\gamma),\ \Delta(m+n,\ \beta+\gamma+\delta);\ cu\\ (b_q),\ \Delta(m+n,\ \beta+\gamma),\ \Delta(m+n,\ \alpha+\beta+\gamma+\delta+1)\end{matrix}\right)$$
$$[\operatorname{Re}\alpha>-1;\ \operatorname{Re}\beta,\ \operatorname{Re}\gamma,\ \operatorname{Re}(\beta+\gamma+\delta)>0,\ u=m^mn^n(m+n)^{-m-n}].$$

3. $$\iint\limits_{\substack{x,\,y\geqslant 0\\ x+y\leqslant 1}} x^{\alpha-1}y^{\beta-1}f(x+y)\,{}_pF_q\left(\begin{matrix}(a_p);\ cx^my^n\\(b_q)\end{matrix}\right)dx\,dy=$$
$$=\mathrm{B}(\alpha,\beta)\int\limits_0^\infty t^{\alpha+\beta-1}f(t)\,{}_{p+m+n}F_{q+m+n}\left(\begin{matrix}(a_p),\ \Delta(m,\alpha),\ \Delta(n,\beta);\ cut^{m+n}\\(b_q),\ \Delta(m+n,\ \alpha+\beta)\end{matrix}\right)dt$$
$[\mathrm{Re}\,\alpha,\ \mathrm{Re}\,\beta>0;\ u=m^mn^n(m+n)^{-m-n}]$.

4. $$\iint\limits_{\substack{x,\,y\geqslant 0\\ x+y\leqslant 1}} x^{\alpha-1}y^{\beta-1}(1-x-y)^{\gamma-1}\,{}_pF_q\left(\begin{matrix}(a_p);\ cx^my^n\\(b_q)\end{matrix}\right)dx\,dy=$$
$$=\Gamma\left[\begin{matrix}\alpha,\ \beta,\ \gamma\\ \alpha+\beta+\gamma\end{matrix}\right]{}_{p+m+n}F_{q+m+n}\left(\begin{matrix}(a_p),\ \Delta(m,\alpha),\ \Delta(n,\beta);\ cu\\(b_q),\ \Delta(m+n,\ \alpha+\beta)\end{matrix}\right)$$
$[\mathrm{Re}\,\alpha,\ \mathrm{Re}\,\beta,\ \mathrm{Re}\,\gamma>0;\ u=m^nn^n(m+n)^{-m-n}]$.

5. $$\int\limits_0^\infty\int\limits_0^\infty x^{\alpha-1}y^{\beta-1}e^{-ax-by}\,{}_pF_q\left(\begin{matrix}(a_p);\ cx^my^n\\(b_q)\end{matrix}\right)dx\,dy=$$
$$=\frac{\Gamma(\alpha)\Gamma(\beta)}{a^\alpha b^\beta}\,{}_{p+m+n}F_q\left(\begin{matrix}(a_p),\ \Delta(m,\alpha),\ \Delta(n,\beta)\\(b_q);\ c(m/a)^m\ (n/b)^n\end{matrix}\right)$$
$[\mathrm{Re}\,a,\ \mathrm{Re}\,b,\ \mathrm{Re}\,\alpha,\ \mathrm{Re}\,\beta>0;\ q>p+m+n-1$ or $|\arg(-c)|<(p-q+1)\pi/2$ for $p=q,\ q+1]$.

6. $$\int\limits_0^\infty\int\limits_0^\infty x^{\alpha-1}y^{\beta-1}e^{-a\sqrt{x}-b\sqrt{y}}\,{}_pF_q\left(\begin{matrix}(a_p);\ cxy\\(b_q)\end{matrix}\right)dx\,dy=$$
$$=4\frac{\Gamma(2\alpha)\Gamma(2\beta)}{a^{2\alpha}b^{2\beta}}\,{}_{p+4}F_q\left(\begin{matrix}(a_p),\ \alpha,\ \alpha+1/2,\ \beta,\ \beta+1/2\\(b_q);\qquad 16c^2/(a^2b^2)\end{matrix}\right)$$
$[q\geqslant p+3;\ \mathrm{Re}\,a,\ \mathrm{Re}\,b,\ \mathrm{Re}\,\alpha,\ \mathrm{Re}\,\beta>0]$.

7. $$\int\limits_0^\infty\int\limits_0^\infty x^{\alpha-1}y^{\alpha}e^{-ax^2-by^2}\,{}_pF_q\left(\begin{matrix}(a_p);\ cxy\\(b_q)\end{matrix}\right)dx\,dy=$$
$$=\frac{\sqrt{\pi}\,\Gamma(\alpha)}{2^{\alpha+1}a^{\alpha/2}b^{(\alpha+1)/2}}\,{}_{p+1}F_q\left(\begin{matrix}(a_p),\ \alpha\\(b_q);\ c/(2\sqrt{ab})\end{matrix}\right)\qquad [p\leqslant q;\ \mathrm{Re}\,a,\ \mathrm{Re}\,b,\ \mathrm{Re}\,\alpha>0].$$

8. $$\int\limits_0^\infty\int\limits_0^\infty x^{\alpha-1}y^{\beta-1}(x+y)^{\gamma}e^{-x-y}\,{}_pF_q\left(\begin{matrix}(a_p);\ cx^my^n\\(b_q)\end{matrix}\right)dx\,dy=$$
$$=\Gamma\left[\begin{matrix}\alpha,\ \beta,\ \alpha+\beta+\gamma\\ \alpha+\beta\end{matrix}\right]{}_{p+2m+2n}F_{q+m+n}\left(\begin{matrix}(a_p),\ \Delta(m,\alpha),\ \Delta(n,\beta),\ \Delta(m+n,\ \alpha+\beta+\gamma)\\(b_q),\ \Delta(m+n,\ \alpha+\beta);\qquad cm^mn^n\end{matrix}\right)$$
$[\mathrm{Re}\,\alpha,\ \mathrm{Re}\,\beta,\ \mathrm{Re}(\alpha+\beta+\gamma)>0;\ q>p+m+n-1$ or $|\arg(-c)|<(p-q+1)\pi/2$ for $p=q,\ q+1]$.

4.3. MULTIPLE INTEGRALS

4.3.1. Integrals containing ${}_pF_q((a_p);\ (b_q);\ x)$.

NOTATION: $d\boldsymbol{x}=dx_1dx_2\ \ldots\ dx_n$.

1. $$\int\limits_0^1\ldots\int\limits_0^1\left[\prod_{k=1}^n x_k^{\alpha_k-1}(1-x_k)^{\beta_k-1}\right]{}_pF_q\left(\begin{matrix}(a_p);\ cx_1\ldots x_n\\(b_q)\end{matrix}\right)d\boldsymbol{x}=$$
$$=\left(\prod_{k=1}^n\mathrm{B}(\alpha_k,\ \beta_k)\right){}_{p+n}F_{q+n}\left(\begin{matrix}(a_p),\ (\alpha_n);\ c\\(b_q),\ (\alpha_n+\beta_n)\end{matrix}\right)\qquad [\mathrm{Re}\,\alpha_k,\ \mathrm{Re}\,\beta_k>0,\ k=1,\ \ldots,\ n].$$

2. $$\int_0^1 \cdots \int_0^1 \left(\prod_{k=1}^{n} \frac{x_k^{\alpha_k-1}}{(1+x_k)^{2\alpha_k}}\right) {}_pF_q\left(\begin{matrix}(a_p);\ cx_1 \ldots x_n\\ (b_q)\end{matrix}\right) d\boldsymbol{x} =$$
$$=\left(\prod_{k=1}^{n} \frac{\Gamma^2(\alpha_k)}{\Gamma(2\alpha_k)}\right) {}_{p+n}F_{q+n}\left(\begin{matrix}(a_p),\ (\alpha_n);\ c/4\\ (b_q),\ (\alpha_n)+1\end{matrix}\right) \quad [c,\ \operatorname{Re}\alpha_k>0;\ k=1,\ 2,\ \ldots, n].$$

3. $$\int \cdots \int_{\substack{x_1\geqslant 0, \ldots, x_n\geqslant 0\\ x_1+\ldots+x_n\leqslant 1}} \left(\prod_{k=1}^{n} x_k^{\alpha_k-1}\right) f(x_1+\ldots+x_n)\, {}_pF_q\left(\begin{matrix}(a_p);\ cx_1^{m_1}\ldots x_n^{m_n}\\ (b_q)\end{matrix}\right) d\boldsymbol{x} =$$
$$=\Gamma\left[\begin{matrix}\alpha_1, \ldots, \alpha_n\\ \alpha_1+\ldots+\alpha_n\end{matrix}\right] \int_0^1 t^{\alpha-1} f(t)\, {}_{p+m}F_{q+m}\left(\begin{matrix}(a_p),\ \Delta(m_1, \alpha_1), \ldots, \Delta(m_n, \alpha_n)\\ (b_q),\ \Delta(m, \alpha);\qquad cut^m\end{matrix}\right) dt$$
$$\left[\alpha=\alpha_1+\ldots+\alpha_n,\ m=m_1+\ldots+m_n,\ u=m^{-m}\prod_{k=1}^{n} m_k^{m_k}\right].$$

4. $$\int_0^\infty \cdots \int_0^\infty \left(\prod_{k=1}^{n} x_k^{\alpha_k-1}\right) e^{-a_1x_1-\ldots-a_nx_n}\, {}_pF_q\left(\begin{matrix}(a_p);\ cx_1^{m_1}\ldots x_n^{m_n}\\ (b_q)\end{matrix}\right) d\boldsymbol{x} =$$
$$=\left(\prod_{k=1}^{n} \frac{\Gamma(\alpha_k)}{a_k^{\alpha_k}}\right) {}_{p+m}F_q\left(\begin{matrix}(a_p),\ \Delta(m_1, \alpha_1), \ldots, \Delta(m_n, \alpha_n)\\ (b_q);\ c(m_1/a_1)^{m_1}\ldots(m_n/a_n)^{m_n}\end{matrix}\right)$$
$$[m=m_1+\ldots+m_n;\ \operatorname{Re} a_k,\ \operatorname{Re}\alpha_k>0,\ k=1, 2, \ldots, n].$$

5. $$\int_0^\infty \cdots \int_0^\infty {}_2F_1\left(\begin{matrix}a,\ b\\ c;\ -\lambda/(x_1\ldots x_n)\end{matrix}\right) \left[\prod_{k=1}^{n} x^{k/(n+1)-1}\, {}_2F_1\left(\begin{matrix}a,\ b\\ c;\ -x_k\end{matrix}\right)\right] d\boldsymbol{x} =$$
$$=(2\pi)^n (n+1)^{(n+1)(1-a-b+c)-1}\, \Gamma^{n+1}\left[\begin{matrix}c\\ a,\ b\end{matrix}\right] \Gamma\left[\begin{matrix}(n+1)a-n,\ (n+1)b-n\\ (n+1)c-n\end{matrix}\right] \times$$
$$\times {}_2F_1\left(\begin{matrix}na+a-n,\ nb+b-n\\ nc+c-n;\ -\lambda^{1/(n+1)}\end{matrix}\right) \quad [\operatorname{Re} a,\ \operatorname{Re} b>n/(n+1);\ \lambda>0;\ n=1, 2, \ldots].$$

4.3.2. Integrals over a sphere.

NOTATION: $\boldsymbol{x}=(x_1,\ldots,x_n)$, $\boldsymbol{a}=(a_1,\ldots,a_n)$, $|\boldsymbol{x}|=\sqrt{x_1^2+\ldots+x_n^2}$, $\boldsymbol{ax}=a_1x_1+\ldots+a_nx_n$; $\boldsymbol{\sigma}=(\sigma_1,\ldots,\sigma_n)$, $|\boldsymbol{\sigma}|=1$, $d\sigma$ is an element of area on the sphere $|\boldsymbol{\sigma}|=1$; $R_m(\boldsymbol{\sigma})$ is the restriction of the homogeneous harmonic polynomial (i.e. satisfying the Laplace equation $\left(\frac{\partial^2}{\partial x_1^2}+\ldots+\frac{\partial^2}{\partial x_n^2}\right)R_m(\boldsymbol{x})=0$) of order m in n variables $x_1,\ldots,x_n$ on the sphere $|\boldsymbol{\sigma}|=1$.

1. $$\int_{|\sigma|=1} f(\boldsymbol{x\sigma})\, R_m(\sigma)\, d\sigma = \frac{2\pi^{(n-1)/2}}{\Gamma((n-1)/2)} R_m\left(\frac{\boldsymbol{x}}{|\boldsymbol{x}|}\right) \int_{-1}^{1} f(|\boldsymbol{x}|t)(1-t^2)^{(n-3)/2} g_m(t)\, dt$$
$$\left[g_m(t)=T_m(t) \text{ for } n=2,\ g_m(t)=\binom{m+n-3}{m}^{-1} C_m^{n/2-1}(t) \text{ for } n\geqslant 3\right].$$

2. $$\int_{|\sigma|=1} f(x\sigma)\, d\sigma = \frac{2\pi^{(n-1)/2}}{\Gamma((n-1)/2)} \int_{-1}^{1} f(|x|t)(1-t^2)^{(n-3)/2} dt.$$

3. $$\int_{|\sigma|=1} (-i\boldsymbol{x\sigma})^\alpha R_m(\sigma)\, d\sigma = -\frac{i^m \pi^{n/2-1} |\boldsymbol{x}|^{\alpha-m}}{2^\alpha} \sin\alpha\pi\, \Gamma\left[\begin{matrix}\alpha+1,\ (m-\alpha)/2\\ (m+n+\alpha)/2\end{matrix}\right] R_m(\boldsymbol{x})$$
$$[\operatorname{Re}\alpha>-1;\ (-it)^\alpha=|t|^\alpha e^{-i\alpha\pi(\operatorname{sgn} t)/2}].$$

4. $$\int\limits_{|\sigma|=1} |x\sigma|^{\alpha} R_m(\sigma)\,d\sigma = \begin{Bmatrix}1\\0\end{Bmatrix} \frac{\pi^{n/2}}{2^{\alpha-1}} |x|^{\alpha-m}\, \Gamma\begin{bmatrix}\alpha+1\\ (\alpha+m+n)/2,\ 1+(\alpha-m)/2\end{bmatrix} R_m(x)$$

$\left[\begin{Bmatrix} m=2,\ 4,\ 6,\ \ldots;\ m\neq\alpha+2,\ \alpha+4,\ \ldots;\ \operatorname{Re}\alpha>-1 \\ m=1,\ 3,\ 5,\ \ldots\ \text{or}\ m=\alpha+2,\ \alpha+4,\ \ldots;\ \operatorname{Re}\alpha>-1\end{Bmatrix}\right].$

5. $$\int\limits_{|\sigma|=1} e^{ix\sigma} R_m(\sigma)\,d\sigma = (2\pi)^{n/2} i^m |x|^{1-n/2-m} J_{m+n/2-1}(|x|)\, R_m(x).$$

6. $$\int\limits_{|\sigma|=1} \sigma_1^{j_1}\ldots\sigma_n^{j_n}\,d\sigma = \frac{2}{\Gamma((n+j_1+\ldots+j_n)/2)} \prod_{k=1}^{n} \frac{1+(-1)^{j_k}}{2}\, \Gamma\left(\frac{1+j_k}{2}\right).$$

7. $$\int\limits_{|\sigma|=1} \frac{d\sigma}{Q^{n/2}} = \frac{2\pi^{n/2}}{\Gamma(n/2)\sqrt{\det Q}}$$ [Q is a positive definite quadratic form].

8. $$\int\limits_{|\sigma|=1} \frac{d\sigma}{(a_1\sigma_1^2+\ldots+a_n\sigma_n^2)^{n/2-m}} = \frac{2\pi^{n/2}}{\Gamma(m)\Gamma(n/2-m)} \int\limits_0^\infty \frac{t^{m-1}\,dt}{\sqrt{(t+a_1)\ldots(t+a_n)}}$$

$[a_k>0,\ k=1,\ 2,\ \ldots,\ n;\ 0<m<n/2]$.

4.3.3. Various integrals.

NOTATION: see 4.3.2.

1. $$\int\limits_{-\infty}^{\infty}\ldots\int\limits_{-\infty}^{\infty} |x|^{-\alpha-n}(1-e^{iax})^l\,dx = \frac{\pi^{n/2+1}}{2^\alpha\Gamma(\alpha/2+1)\Gamma((\alpha+n)/2)} A$$

$$\left[A=\operatorname{cosec}\frac{\alpha\pi}{2}\sum_{k=0}^{l}(-1)^{k-1}\binom{l}{k}k^\alpha \text{ for } \alpha\neq 2,\ 4,\ \ldots;\right.$$

$$\left. A=\frac{2(-1)^{\alpha/2}}{\pi}\sum_{k=0}^{l}(-1)^{k-1}\binom{l}{k}k^\alpha \ln k \text{ for } \alpha=2,\ 4,\ldots;\ 0<\alpha<l\right].$$

2. $$\int\limits_{|x|\leqslant r} \frac{dx}{|x-a|^{n-\alpha}(r^2-|x|^2)^{\alpha/2}} = \frac{\pi^{n/2+1}}{\Gamma(n/2)}\operatorname{cosec}\frac{\alpha\pi}{2}$$ $[|a|<r;\ 0<\alpha<2]$.

3. $$\int\limits_{\substack{x_1,\ldots,x_n\geqslant 0\\ x_1+\ldots+x_n\leqslant 1}}\!\!\ldots\!\!\int x_1^{\alpha_1-1}\ldots x_n^{\alpha_n-1}(1-x_1-\ldots-x_n)^{\gamma-\alpha_1-\ldots-\alpha_n-1}(1-x_1z_1)^{\beta_1}\ldots$$

$$\ldots(1-x_nz_n)^{\beta_n}\,dx = \Gamma\begin{bmatrix}\alpha_1,\ \ldots,\ \alpha_n,\ \gamma-\alpha_1-\ldots-\alpha_n\\ \gamma\end{bmatrix} F_B(\alpha_1,\ \ldots,\ \alpha_n,\ -\beta_1,\ \ldots$$

$\ldots,\ -\beta_n,\ \gamma;\ z_1,\ \ldots,\ z_n)$ $[\operatorname{Re}\alpha_k,\ \operatorname{Re}(\gamma-\alpha_1-\ldots-\alpha_n)>0;\ |\arg(1-z_k)|<\pi;\ k=1,\ 2,\ \ldots,\ n]$.

4. $$\int\limits_{\substack{x_1,\ldots,x_n\geqslant 0\\ x_1+\ldots+x_n\leqslant 1}}\!\!\ldots\!\!\int x_1^{\alpha_1-1}\ldots x_n^{\alpha_n-1}(1-x_1-\ldots-x_n)^{\gamma-\alpha_1-\ldots-\alpha_n-1}(1-x_1z_1-\ldots$$

$$\ldots-x_nz_n)^\beta\,dx = \Gamma\begin{bmatrix}\alpha_1,\ \ldots,\ \alpha_n,\ \gamma-\alpha_1-\ldots-\alpha_n\\ \gamma\end{bmatrix} F_D(-\beta,\alpha_1,\ \ldots,\ \alpha_n,\ \gamma;\ z_1,\ \ldots,\ z_n)$$

$[\operatorname{Re}\alpha_k,\ \operatorname{Re}(\gamma-\alpha_1-\ldots-\alpha_n)>0;\ |\arg(1-z_k)|<\pi;\ k=1,\ 2,\ \ldots,\ n]$.

5. $$\int\limits_0^1\ldots\int\limits_0^1 x_1^{\alpha_1-1}\ldots x_n^{\alpha_n-1}(1-x_1)^{\gamma_1-\alpha_1-1}\ldots(1-x_n)^{\gamma_n-\alpha_n-1}(1-x_1z_1-\ldots-x_nz_n)^\beta\,dx =$$

$$=\Gamma\begin{bmatrix}\alpha_1,\ \ldots,\ \alpha_n,\ \gamma_1-\alpha_1,\ \ldots,\ \gamma_n-\alpha_n\\ \gamma_1,\ \ldots,\ \gamma_n\end{bmatrix} F_A(-\beta,\ \alpha_1,\ \ldots,\ \alpha_n,\ \gamma_1,\ \ldots,\ \gamma_n;\ z_1,\ \ldots,\ z_n)$$

$[\operatorname{Re}\alpha_k,\ \operatorname{Re}(\gamma_k-\alpha_k)>0;\ |\arg(1-z_k)|<\pi;\ k=1,\ 2,\ \ldots,\ n]$.

Chapter 5. FINITE SUMS

5.1. THE NUMBERS AND POLYNOMIALS OF BERNOULLI $B_n, B_n(x)$ AND EULER $E_n, E_n(x)$

5.1.1. Sums containing B_n.

1. $$\sum_{k=0}^{n}\binom{m}{k}t^kB_k=t^m\left[B_m\left(\frac{1}{t}\right)+(-1)^n\binom{m}{n}\int_0^1 B_n(\tau)\,B_{m-n}\left(\tau+\frac{1}{t}\right)d\tau\right].$$

2. $$\sum_{k=0}^{n}(-1)^k\frac{t^{2k}}{(2k)!}B_{2k}=\frac{t}{2}\operatorname{ctg}\frac{t}{2}+\frac{(-1)^n t^{2n+1}}{(2n)!\,2}\operatorname{cosec}\frac{t}{2}\int_0^1\cos t\left(\tau-\frac{1}{2}\right)B_{2n}(\tau)\,d\tau.$$

3. $$\sum_{k=0}^{n}\frac{(a)_{2k}}{(2k)!}B_{2k}=a\left[\zeta(a+1)-\frac{1}{2}+(a+1)\frac{(a+2)_{2n}}{(2n+1)!}\int_1^\infty\tau^{-a-2n-2}B_{2n+1}(\tau-[\tau])d\tau\right]$$

$[n\geqslant 1;\ a\neq 0;\ \operatorname{Re} a>-2n-1]$.

4. $$\sum_{k=0}^{2n+1}\binom{2n+1}{k}\frac{(n+k)!}{(2n+k+1)!}B_{2n-k+1}=0.$$

5. $$\sum_{k=0}^{2n+1}\binom{2n+1}{k}\frac{(n+k)!\,(1-2^{2n-k+2})}{(2n+k+1)!\,(2n-k+2)}B_{2n-k+2}=0.$$

6. $$\sum_{k=0}^{n}\frac{(2n+2k+1)!\,B_{2m+2k+2\sigma+2}}{(2k+\sigma)!\,(2n-2k+1)!\,(2m+2k+2\sigma+2)!}=(-1)^\sigma\frac{(2n+2)!}{(4n+2\sigma+3)!\,2}\delta_{m,n}$$

$[0\leqslant m\leqslant n;\ \sigma=0$ or $1]$.

7. $$\sum_{k=0}^{n}\frac{(2n+2k+2\sigma)!\,2^{2m+\sigma+2}\,(1-2^{2m+2k+2\sigma+2})\,B_{2m+2k+2\sigma+2}}{(2k)!\,(2n-2k)!\,(2m+2k+2\sigma+1)!\,(2m+2k+2\sigma+2)}=$$

$$=(-1)^{\sigma+1}\frac{2^{2n+\sigma}\,(2n+\sigma)!}{(4n+2\sigma+1)!}\delta_{m,n}\qquad [0\leqslant m\leqslant n;\ \sigma=0 \text{ or } 1].$$

8. $$\sum_{k=1}^{[n/2]}\binom{2n}{2k}B_{2k}B_{2n-2k}=\frac{1+(-1)^n}{4}\binom{2n}{n}B_n^2-\left(n+\frac{1}{2}\right)B_{2n}.$$

9. $$\sum_{k=0}^{n}\sum_{l=0}^{n}\frac{(2^{2k}-1)\,(2^{2l}-1)}{(2k)!\,(2l)!\,(4n-2k-2l+2)!}\,|B_{2k}B_{2l}B_{4n-2k-2l+2}|=\frac{2^{4n}-1}{(4n)!\,4}\,|B_{4n}|.$$

10. $$\sum_{k=1}^{n}\sum_{l=1}^{n}\frac{(2^{2k}-1)\,(2^{2l}-1)}{(2k)!\,(2l)!\,(4n-2k-2l+4)!}\,|B_{2k}B_{2l}B_{4n-2k-2l+4}|=$$

$$=\frac{2^{4n+2}-1}{(4n+2)!\,4}B_{4n+2}-\frac{(2^{2n+2}-1)^2}{[(2n+2)!]^2}B_{2n+2}^2.$$

5.1.2. Sums containing $B_n(x)$.

1. $\sum_{k=1}^{n} B_m(k)=nB_m(n+1)+\frac{m}{m+1}[(-1)^{m+1}B_{m+1}-B_{m+1}(n+1)]$.

2. $\sum_{k=0}^{n-1} B_m\left(x\pm\frac{k}{n}\right)=n^{1-m}B_m\left(nx-\frac{1\mp 1}{2}n\right)$.

3. $\sum_{k=0}^{2n} (-1)^k B_m\left(x\pm\frac{k}{2n}\right)=-2^{-m}n^{1-m}mE_{m-1}(2nx-n\pm n)+B_m\left(x+\frac{1\pm 1}{2}\right)$.

4. $\sum_{k=0}^{n-1} B_p\left(\frac{x+mk}{n}\right)=\left(\frac{m}{n}\right)^{p-1}\sum_{k=0}^{m-1} B_p\left(\frac{x+nk}{m}\right)$.

5. $\sum_{k=0}^{2n-1} (-1)^k B_{p+1}\left(\frac{x+mk}{2n}\right)=-\frac{p+1}{2}\left(\frac{m}{2n}\right)^p\sum_{k=0}^{m-1} E_p\left(\frac{x+2nk}{m}\right)$.

6. $\sum_{k=1}^{n-1} kB_{2m}\left(\frac{kl}{n}\right)=\frac{n}{2}(n^{1-2m}-1)B_{2m}$ [l, n are coprime].

7. $\sum_{k=0}^{n}\binom{n}{k}t^kB_{k+m}(x)=(-1)^mt^{n-m}\sum_{k=0}^{m}(\mp t)^k\binom{m}{n}B_{k+n}\left(x\pm\frac{1}{t}\right)$.

8. $\sum_{k=0}^{n}\frac{(-x)^{-k}}{(n-k)!}B_k(x)=\frac{(-x)^{-n}}{n!}\int_{-x}^{\infty}\frac{\tau^n}{e^\tau-1}d\tau$ [$x\leqslant 0$].

9. $\sum_{k=0}^{[n/2]}\binom{n}{2k}t^{2k}B_{2k+n}(x)=$

$$=\frac{(-1)^n}{2}t^{n-m}\sum_{k=0}^{m}\binom{m}{k}t^k\left[B_{k+n}\left(x-\frac{1}{t}\right)+(-1)^{n-m+k}B_{k+n}\left(x+\frac{1}{t}\right)\right].$$

10. $\sum_{k=0}^{[n/2]}\binom{n}{2k}\frac{t^{2k}}{(2k+1)!}B_{2k+m}(x)=$

$$=\frac{(-1)^{m-1}}{2n+2}t^{n-m+1}\sum_{k=0}^{m-1}\frac{(1-m)_k}{k!}t^k\left[B_{k+n+1}\left(x+\frac{1}{t}\right)+\right.$$

$$\left.+(-1)^{n-m+k+1}B_{k+n+1}\left(x-\frac{1}{t}\right)\right].$$

11. $\sum_{k=0}^{\max(m,n)}\left[m\binom{n}{2k}+n\binom{m}{2k}\right]\frac{B_{2k}B_{m+n-2k}(x)}{m+n-2k}=$

$$=B_m(x)B_n(x)+(-1)^m\frac{m!n!}{(m+n)!}B_{m+n} \quad [m+n\geqslant 2].$$

12. $\sum_{k=0}^{\max(m,n)}\left[m\binom{n}{2k}+n\binom{m}{2k}\right]B_{2k}B_{m+n-2k-1}(x)=mB_{m-1}(x)B_n(x)+nB_m(x)B_{n-1}(x)$.

13. $$\sum_{k=0}^{n}(\pm 1)^k\binom{n}{k}B_k(x)B_{n-k}(y)=(\pm 1)^n\left[n\left(x\pm y-\frac{1\pm 1}{2}\right)\times\right.$$
$$\left.\times B_{n-1}(x\pm y)-(n-1)B_n(x\pm y)\right].$$

14. $$\sum_{k=0}^{n}\binom{n}{k}t^kB_k(x)\,B_{n-k}(y)=\sum_{k=0}^{n}\binom{n}{k}t^kB_k\left(\frac{y-a}{t}+x\right)B_{n-k}(a)$$ [a is arbitrary;].

5.1.3. Sums containing E_n.

1. $$\sum_{k=0}^{n}(\pm 1)^k\binom{n}{k}E_k=\frac{(-2)^{n+1}}{n+1}(2^{n+1}-1)\,B_{n+1}.$$

2. $$\sum_{k=0}^{[n/2]}\binom{n}{2k}E_{2k}=\frac{2^{n+1}}{n+1}(2^{n+1}-1)\,B_{n+1}+2\delta_{0,\,n}.$$

3. $$\sum_{k=0}^{n}\frac{(2n+2k+2\sigma)!E_{2k+2m+2}}{(2k+\sigma)!\,(2n-2k)!\,2^{2k+\sigma}(2k+2m+2)!}=\frac{(-1)^{m+n+1}(2n-2m+2\sigma-2)!}{2^{2n+\sigma}(n+m+1)!(n-m+\sigma-1)!}$$
[$0\leqslant m\leqslant n$; $\sigma=0$ or 1].

4. $$\sum_{k=0}^{n}\binom{2n}{2k}B_{2k}E_{2n-2k}=(1+2^{2n-1}-2^{4n-1})\,B_{2n}.$$

5. $$\sum_{k=0}^{n}\binom{2n+1}{2k+1}\frac{2^{2k+1}(1-2^{2k+2})}{k+1}B_{2k+2}E_{2n-2k}=E_{2n+2}.$$

5.1.4. Sums containing $E_n(x)$.

1. $$\sum_{k=0}^{n}\binom{n}{k}t^kE_{k+m}(x)=(-1)^mt^{n-m}\sum_{k=0}^{m}(-1)^k\binom{m}{k}t^kE_{k+n}\left(x+\frac{1}{t}\right).$$

2. $$\sum_{k=0}^{2n+1}(-1)^kE_m\left(x\pm\frac{k}{2n+1}\right)=\pm(2n+1)^{-m}E_m\left((2n+1)\,x-(2n+1)\frac{1\mp 1}{2}\right)\mp$$
$$\mp E_m\left(x+\frac{1\pm 1}{2}\right).$$

3. $$\sum_{k=0}^{n-1}(-1)^kE_p\left(\frac{x+mk}{n}\right)=\left(\frac{m}{n}\right)^p\sum_{k=0}^{m-1}(-1)^kE_p\left(\frac{x+nk}{m}\right)$$ [m, n are odd].

4. $$\sum_{k=0}^{n-1}E_p\left(\frac{x+2mk}{n}\right)=-\frac{2}{p+1}\left(\frac{2m}{n}\right)^p\sum_{k=0}^{2m-1}(-1)^kB_{p+1}\left(\frac{x+nk}{2m}\right).$$

5. $$\sum_{k=0}^{n}(\pm 1)^k\binom{n}{k}E_k(x)E_{n-k}(y)=2(\pm 1)^n\left[\left(\frac{1\pm 1}{2}\mp x-y\right)E_n(x\pm y)\pm E_{n+1}(x\pm y)\right].$$

6. $$\sum_{k=0}^{n}\binom{n}{k}t^kE_k(x)\,E_{n-k}(y)=\sum_{k=0}^{n}\binom{n}{k}t^kE_k\left(x+\frac{y-a}{t}\right)E_{n-k}(a)$$ [a is arbitrary;].

7. $\sum_{k=0}^{m+n}\left[\binom{m}{k}+\binom{n}{k}\right]2^{-k}E_kE_{m+n-k}(x)=2^{m+n}\left[E_m\left(\frac{x}{2}\right)E_n\left(\frac{x+1}{2}\right)+\right.$
$\left.+E_m\left(\frac{x+1}{2}\right)E_n\left(\frac{x}{2}\right)\right].$

8. $\sum_{k=0}^{n}(\pm 1)^k\binom{n}{k}B_k(x)E_{n-k}(y)=(\pm 1)^{n-1}2^nB_n\left(\frac{x\pm y}{2}\right)+(-1)^nB_n(x-y)\left\{\begin{matrix}0\\2\end{matrix}\right\}.$

9. $\sum_{k=0}^{n}\binom{n}{k}t^kB_k(x)E_{n-k}(y)=\sum_{k=0}^{n}\binom{n}{k}t^kB_k(x+a)E_{n-k}(y+at)$ [a – любое].

5.2. THE LEGENDRE FUNCTIONS $P_\nu^\mu(x)$ AND $Q_\nu^\mu(x)$

(For $\mu=0$, $\nu=n$ see also [18], 4.3)

5.2.1. Sums of the form $\sum a_kP_{\nu\pm mk}^{\mu+lk}(x)$, $\sum a_kQ_{\nu\pm mk}^{\mu+lk}(x)$.

1. $\sum_{k=1}^{n}(\pm 1)^k(2k+2\nu+1)R_{k+\nu}^\mu(z)=\frac{1}{1\mp z}[(\pm\nu\mp\mu+1)R_{\nu+1}^\mu(z)-(\mu+\nu+1)R_\nu^\mu(z)+$
$+(\pm\mu\mp\nu\mp n\mp 1)R_{n+\nu+1}^\mu(z)+(-1)^n(\mu+\nu+n+1)R_{\nu+n}^\mu(z)]$
$[R_\nu^\mu(z)=P_\nu^\mu(z)$ or $Q_\nu^\mu(z)]$.

2. $\sum_{k=0}^{n-1}(\pm 1)^k(2k+1)Q_k^n(x)=(-1)^nn!(x\mp 1)^{-n/2-1}[(x\pm 1)^{n/2}-(\pm 2)^n(x\pm 1)^{-n/2}]$
$[x>1]$.

3. $\sum_{k=0}^{[(n-1)/2]}(2n-4k-1)Q_{n-2k-1}(x)=\frac{1}{2}\left[\frac{1}{x-1}-\frac{(-1)^n}{x+1}\right]+Q_n'(x)$ $[x>1]$.

4. $\sum_{k=0}^{[(n-1)/2]}(2n-2k+1)Q_{n-2k}(x)=\frac{(-1)^n-1}{2}+\frac{x}{2}\left[\frac{1}{x-1}-\frac{(-1)^n}{x+1}+2Q_n'(x)\right]-nQ_n(x)$
$[x>1;\ n\geqslant 1]$.

5. $\sum_{k=0}^{n-1}(4n-4k-1)Q_{2n-2k-1}^m(x)=(x^2-1)^{-1/2}[Q_{2n}^{m+1}(x)-Q_0^{m-1}(x)]$ $[x>1]$.

6. $\sum_{k=1}^{n}(4n-4k+1)Q_{2n-2k}^m(x)=\frac{x}{\sqrt{x^2-1}}[Q_{2n}^{m+1}(x)-Q_0^{m+1}(x)]+m[Q_{2n}^m(x)-Q_0^m(x)]-$
$-2nQ_{2n}^m(x)$ $[x>1]$.

7. $\sum_{k=0}^{2n}\frac{(\pm 1)^k}{k!}P_n^{k-n}(x)=\frac{(\pm 1)^n}{n!}x^n$ $[-1\leqslant x\leqslant 1]$.

8. $\sum_{k=0}^{[(n-1)/2]}\frac{(-1)^k}{(2k+n+1)!}P_n^{2k+1}(x)=-\frac{1}{n!2}\sin(n\arccos x)$ $[-1<x<1]$.

9. $\sum_{k=0}^{[n/2]}\frac{(-1)^k}{(2k+n)!}P_n^{2k}(x)=\frac{1}{n!2}[\cos(n\arccos x)+P_n(x)]$ $[-1<x<1]$.

10. $\sum_{k=0}^{n} t^k \binom{n}{k} P_{k+\nu}^{-\nu}(z) = t^{-\nu}(1-2tz+t^2)^{(n+\nu)/2} P_{n+\nu}^{-\nu}\left(\frac{1+tz}{\sqrt{1+2tz+t^2}}\right)$.

11. $\sum_{k=0}^{n} \frac{t^k}{(2m+k)!} \binom{n}{k} P_{k+m}^{m}(z) = \frac{n!}{(2m+n)!} t^{-m}(1+2tz+t^2)^{(m+n)/2} \times$

$\times P_{m+n}^{m}\left(\frac{1+tz}{\sqrt{1+2tz+t^2}}\right)$.

12. $\sum_{k=0}^{[(n-m)/2]} (2n-4k+1) \frac{2^{-2k}(n-k)!}{k!(2n-2k+1)!} P_{n-2k}^{m}(x) = \frac{x^{n-m}}{(n-m)!}(1-x^2)^{m/2}$ $[-1 \leqslant x \leqslant 1]$.

5.2.2. Sums containing products of Legendre functions.

1. $\sum_{k=1}^{n} \frac{(n-k)!}{(n+k)!} P_n^k(x) P_n^k(y) = \frac{1}{2} P_n\left(xy+\sqrt{(1-x^2)(1-y^2)}\right) - \frac{1}{2} P_n(x) P_n(y)$

$[-1 \leqslant x,\ y \leqslant 1]$.

2. $\sum_{k=m}^{m+n} (2k+1) P_k^{m-1}(x) P_k^m(x) = \frac{1}{\sqrt{1-x^2}} P_{m+n+1}^m(x) P_{m+n}^m(x)$ $[-1 < x < 1;\ m \geqslant 1]$.

3. $\sum_{k=0}^{n} (2k+1)[Q_k(x)]^2 = (n+1)[Q_n(x) Q_{n+1}'(x) - Q_n'(x) Q_{n+1}(x)] - Q_0'(x)$ $[x > 1]$.

4. $\sum_{k=0}^{n-1} (2k+1)[Q_k^n(x)]^2 = \frac{(n!)^2}{2(2n+1)} (x^2-1)^{-n-1} [(x+1)^{2n+1} - (x-1)^{2n+1} - 2^{2n+1}]$

$[x > 1;\ n \geqslant 1]$.

5. $\sum_{k=1}^{n} \frac{(n-k)!}{(n+k)!} \cos ka P_n^k(x) P_n^k(y) = \frac{1}{2} P_n\left(xy+\sqrt{(1-x^2)(1-y^2)} \cos a\right) - \frac{1}{2} P_n(x) P_n(y)$

$[-1 < x,\ y < 1]$.

6. $\sum_{k=0}^{2n} \begin{Bmatrix} \sin ka \\ \cos ka \end{Bmatrix} P_n^{(k-n)}(x) P_n^{(n-k)}(y) = \begin{Bmatrix} \sin na \\ \cos na \end{Bmatrix} P_n\left(xy-\sqrt{(1-x^2)(1-y^2)} \cos a\right)$

$[-1 < x,\ y < 1]$.

7. $\sum_{k=0}^{2n} (-1)^k \frac{(2n-k)!}{k!} \begin{Bmatrix} \sin ka \\ \cos ka \end{Bmatrix} P_n^{(k-n)}(x) P_n^{(k-n)}(y) =$

$= (-1)^n \begin{Bmatrix} \sin na \\ \cos na \end{Bmatrix} P_n\left(xy-\sqrt{(1-x^2)(1-y^2)} \cos a\right)$ $[-1 < x,\ y < 1]$.

5.3. THE GENERALIZED HYPERGEOMETRIC FUNCTION ${}_pF_q((a_p); (b_q); x)$ AND THE MEIJER G-FUNCTION

5.3.1. Sums of the form $\sum \alpha_k\, {}_{p+1}F_q(-k-m, (a_p); (b_q); x)$.

1. $\sum_{k=0}^{n} {}_{p+1}F_q\left(\begin{matrix} -k, (a_p) \\ (b_q);\ x \end{matrix}\right) = \left(\prod_{i=1}^{q}(b_i-1)\right)\left(\prod_{j=1}^{p}(a_j-1)\right)^{-1} \times$

$\times x^{-1}\left[1 - {}_{p+1}F_q\left(\begin{matrix} -n-1, (a_p)-1 \\ (b_q)-1;\ x \end{matrix}\right)\right]$.

2. $$\sum_{k=1}^{n} k\,{}_{p+1}F_q\left(\begin{matrix}-k,\,(a_p)\\(b_q);\,x\end{matrix}\right)=$$
$$=\left(\prod_{i=1}^{q}(b_i-1)(b_i-2)\right)\left(\prod_{j=1}^{p}(a_j-1)(a_j-2)\right)^{-1}x^{-2}\left[1-\left(\prod_{i=1}^{p}(a_i-2)\right)\left(\prod_{j=1}^{q}(b_j-2)\right)^{-1}x+\right.$$
$$\left.+n\,{}_{p+1}F_q\left(\begin{matrix}-n-2,\,(a_p)-2\\(b_q)-2;\,x\end{matrix}\right)-(n+1)\,{}_{p+1}F_q\left(\begin{matrix}-n-1,\,(a_p)-2\\(b_q)-2;\,x\end{matrix}\right)\right].$$

3. $$\sum_{k=0}^{n}\binom{n}{k}t^k\,{}_{p+1}F_q\left(\begin{matrix}-k,\,(a_p);\,x\\(b_q)\end{matrix}\right)=(t+1)^n\,{}_{p+1}F_q\left(\begin{matrix}-n,\,(a_p);\,tx/(1+t)\\(b_q)\end{matrix}\right).$$

4. $$\sum_{k=0}^{n}(-1)^k\binom{n}{k}\,{}_{p+1}F_q\left(\begin{matrix}-k,\,(a_p)\\(b_q);\,x\end{matrix}\right)=\frac{\Pi\,(a_p)_n}{\Pi\,(b_q)_n}x^n.$$

5. $$\sum_{k=0}^{n}(-1)^k\binom{n}{k}\,{}_{p+1}F_q\left(\begin{matrix}-2k,\,(a_p)\\(b_q);\,x\end{matrix}\right)=\frac{\Pi\,(a_p)_n}{\Pi\,(b_q)_n}(2x)^n\,{}_{p+1}F_q\left(\begin{matrix}-n,\,(a_p)+n\\(b_q)+n;\,x/2\end{matrix}\right).$$

6. $$\sum_{k=0}^{n}\frac{(\alpha)_k}{k!}\,{}_{p+1}F_q\left(\begin{matrix}-k,\,(a_p)\\(b_q);\,x\end{matrix}\right)=\frac{(\alpha+1)_n}{n!}\,{}_{p+2}F_{q+1}\left(\begin{matrix}-n,\,\alpha,\,(a_p)\\\alpha+1,\,(b_q);\,x\end{matrix}\right).$$

7. $$\sum_{k=0}^{n}\binom{n}{k}(\alpha)_{n-k}(\beta)_k\,{}_{p+1}F_q\left(\begin{matrix}-k,\,(a_p);\,x\\(b_q)\end{matrix}\right)=$$
$$=(\alpha+\beta)_n\,{}_{p+2}F_{q+1}\left(\begin{matrix}-n,\,\beta,\,(a_p);\,x\\\alpha+\beta,\,(b_q)\end{matrix}\right).$$

8. $$\sum_{k=0}^{n}(-1)^k\frac{(\alpha)_k}{(\alpha-n-1/2)_k}\,{}_{p+1}F_q\left(\begin{matrix}-k,\,(a_p);\,x\\(b_q)\end{matrix}\right)=$$
$$=\frac{(2n+1)!}{n!2^{2n}(3/2-\alpha)_n}\,{}_{p+2}F_{q+1}\left(\begin{matrix}-n,\,\alpha,\,(a_p);\,x\\3/2,\,(b_q)\end{matrix}\right).$$

9. $$\sum_{k=0}^{n}(-1)^k\binom{n}{k}\frac{(\alpha)_k}{(\beta)_k}\,{}_{p+1}F_q\left(\begin{matrix}-k,\,(a_p);\,x\\(b_q)\end{matrix}\right)=$$
$$=(-1)^n\frac{(\beta-\alpha)_n}{(\beta)_n}\,{}_{p+2}F_{q+1}\left(\begin{matrix}-n,\,\alpha,\,(a_p);\,x\\\alpha-\beta-n+1,\,(b_q)\end{matrix}\right).$$

10. $$\sum_{k=0}^{n}(-1)^k\binom{n}{k}^2\binom{2n}{k}^{-1}\,{}_{p+1}F_q\left(\begin{matrix}-k,\,(a_p)\\(b_q);\,x\end{matrix}\right)=\binom{2n}{n}^{-1}\,{}_{p+2}F_{q+1}\left(\begin{matrix}-n,\,-n,\,(a_p)\\1,\,(b_q);\,x\end{matrix}\right).$$

11. $$\sum_{k=0}^{n}(-1)^k\binom{n}{k}\,{}_{p+1}F_q\left(\begin{matrix}-m-k,\,(a_p)\\(b_q);\,x\end{matrix}\right)=\frac{\Pi\,(a_p)_n}{\Pi\,(b_q)_n}x^n\,{}_{p+1}F_q\left(\begin{matrix}-m,\,(a_p)+n\\(b_q)+n;\,x\end{matrix}\right).$$

12. $$\sum_{k=0}^{n}(-1)^k\binom{n}{k}\frac{(\alpha)_{m+k}}{(m+k)!}\,{}_{p+1}F_q\left(\begin{matrix}-m-k,\,(a_p);\,x\\(b_q)\end{matrix}\right)=$$
$$=(-1)^n\frac{(\alpha-n)_{m+n}}{(m+n)!}\,{}_{p+2}F_{q+1}\left(\begin{matrix}-m-n,\,\alpha,\,(a_p);\,x\\\alpha-n,\,(b_q)\end{matrix}\right).$$

13. $\sum_{k=0}^{n}(-1)^k\binom{n}{k}(k\alpha+1)^{n-1}\,{}_{p+1}F_q\left(\begin{matrix}-k,\ (a_p);\ x/(k\alpha+1)\\ (b_q)\end{matrix}\right)=$

$$=\frac{\Pi\,(a_p)_n}{\Pi\,(b_q)_n}\,\frac{x^n}{n\alpha+1}.$$

5.3.2. Sums of the form $\sum\alpha_k\,{}_{p+2}F_q(-k,\ \nu+k,\ (a_p);\ (b_q);\ x)$.

1. $\sum_{k=0}^{n}(2k+1)\,{}_{p+2}F_{q+1}\left(\begin{matrix}-k,\ k+1,\ (a_p)\\ 1,\ (b_q);\ x\end{matrix}\right)=-\frac{n+1}{2x}\prod_{i=1}^{p}(a_i-1)^{-1}\prod_{j=1}^{q}(b_j-1)\times$

$$\times\left[{}_{p+2}F_{q+1}\left(\begin{matrix}-n-1,\ n+2,\ (a_p)-1\\ (b_q)-1,\ 1;\ x\end{matrix}\right)-{}_{p+2}F_{q+1}\left(\begin{matrix}-n,\ n+1,\ (a_p)-1\\ (b_q)-1,\ 1;\ x\end{matrix}\right)\right].$$

2. $\sum_{k=0}^{n}(-1)^k\binom{n}{k}(2k+\nu)\frac{(\nu)_k}{(n+\nu+1)_k}\,{}_{p+2}F_q\left(\begin{matrix}-k,\ \nu+k,\ (a_p)\\ (b_q);\ x\end{matrix}\right)=$

$$=(\nu)_{n+1}\frac{\Pi\,(a_p)_n}{\Pi\,(b_q)_n}x^n.$$

5.3.3. Sums of the form $\sum\alpha_k\,{}_{p+1}F_q(-k,\ (a_p)+k;\ (b_q)+k;\ x)$.

1. $\sum_{k=0}^{n}\binom{n}{k}\frac{(2\alpha+2n-1)_{2k}}{(\alpha+n)_k}(-x)^k\frac{\Pi\,(a_p)_k}{\Pi\,(b_q)_k}\,{}_{p+2}F_q\left(\begin{matrix}-k,\ \alpha+k,\ (a_p)+k\\ (b_q)+k;\ x\end{matrix}\right)=$

$$={}_{p+2}F_q\left(\begin{matrix}-2n,\ 2\alpha+2n-1,\ (a_p)\\ (b_q);\ x\end{matrix}\right).$$

2. $\sum_{k=0}^{n}(-1)^k\binom{n}{k}\frac{\Pi\,(a_p)_k}{\Pi\,(b_q)_k}(4x)^k\,{}_{p+1}F_q\left(\begin{matrix}-k,\ (a_p)+k\\ (b_q)+k;\ x\end{matrix}\right)={}_{p+1}F_q\left(\begin{matrix}-2n,\ (a_p)\\ (b_q);\ 2x\end{matrix}\right).$

3. $\sum_{k=0}^{n}(-1)^k\binom{n}{k}\frac{\Pi\,(a_p)_k}{\Pi\,(b_q)_k}(k+\alpha-n)_{n+m}x^k\,{}_{p+1}F_{q+1}\left(\begin{matrix}-k,\ (a_p)+k;\ x\\ \alpha,\ (b_q)+k\end{matrix}\right)=$

$$=(\alpha)_m\,{}_{p+1}F_{q+1}\left(\begin{matrix}-m-n,\ (a_p);\ x\\ \alpha-n,\ (b_q)\end{matrix}\right).$$

5.3.4. Sums of the form $\sum\alpha_k\,{}_{p+1}F_q(-k,\ (a_p)-mk;\ (b_q)-nk;\ x)$.

1. $\sum_{k=0}^{n}(-1)^k\binom{n}{k}\frac{(1-\alpha)_k}{(\beta)_k}\,{}_{p+1}F_{q+1}\left(\begin{matrix}-k,\ (a_p);\ x\\ \alpha-k,\ (b_q)\end{matrix}\right)=$

$$=\frac{(\alpha+\beta-1)_n}{(\beta)_n}\,{}_{p+1}F_{q+1}\left(\begin{matrix}-n,\ (a_p);\ x\\ \alpha+\beta-1,\ (b_q)\end{matrix}\right).$$

2. $\sum_{k=0}^{n}\binom{n}{k}t^k\,{}_{p+2}F_q\left(\begin{matrix}-k/2,\ (1-k)/2,\ (a_p)\\ (b_q);\ x\end{matrix}\right)=$

$$=(t+1)^n\,{}_{p+2}F_q\left(\begin{matrix}-n/2,\ (1-n)/2,\ (a_p)\\ (b_q);\ xt^2/(t+1)^2\end{matrix}\right).$$

3. $\sum_{k=0}^{2n}(-1)^k(ka+1)^{2n-1}\binom{2n}{k}\,{}_{p+2}F_q\left(\begin{matrix}-k/2,\ (1-k)/2,\ (a_p)\\ (b_q);\ x/(ka+1)^2\end{matrix}\right)=\frac{(1/2)_n}{2na+1}x^n\frac{\Pi\,(a_p)_n}{\Pi\,(b_q)_n}.$

4. $\sum_{k=0}^{n}\binom{n}{k}x^{-k}\frac{\Pi\,(1-b_q)_k}{\Pi(1-a_p)_k}\,{}_{p+1}F_q\left(\begin{matrix}-k,\ (a_p)-k\\ (b_q)-k;\ x\end{matrix}\right)=\frac{\Pi\,(1-b_q)_n}{\Pi\,(1-a_p)_n}x^{-n}.$

5. $$\sum_{k=0}^{n} \frac{t^k}{(n-k)!\,(2k+\sigma)!} \frac{\Pi(1-b_q)_k}{\Pi(1-a_p)_k}\, {}_{p+2}F_q\left(\begin{matrix}-k-\sigma/2,\ -k+(1-\sigma)/2,\ (a_p)-k \\ (b_q)-k;\ x\end{matrix}\right)= \frac{t^n}{(2n+\sigma)} \frac{\Pi(1-b_q)_n}{\Pi(1-a_p)_n}\, {}_{p+2}F_q\left(\begin{matrix}-n-\sigma/2,\ -n+(1-\sigma)/2,\ (a_p)-n \\ (b_q)-n;\ x+4/t\end{matrix}\right)$$
$[\sigma=0$ or $1]$.

6. $$\sum_{k=0}^{n} \binom{n}{k} \frac{(-x)^{-k}}{(1-\alpha)_k} (ka+1)^{n-k-1} \frac{\Pi(1-b_q)_k}{\Pi(1-a_p)_k}\, {}_{p+2}F_q\left(\begin{matrix}-k,\ \alpha-k,\ (a_p)-k \\ (b_q)-k;\ (ka+1)x\end{matrix}\right)= \frac{(-x)^n}{(1-\alpha)_n (na+1)} \frac{\Pi(1-b_q)_n}{\Pi(1-a_p)_n}.$$

7. $$\sum_{k=0}^{n} \binom{n}{k} t^k \frac{\Pi(1-b_q)_k}{\Pi(1-a_p)_k}\, {}_{p+1}F_q\left(\begin{matrix}-k,\ (a_p)-k \\ (b_q)-k;\ x\end{matrix}\right)= t^n \frac{\Pi(1-b_q)_n}{\Pi(1-a_p)_n}\, {}_{p+1}F_q\left(\begin{matrix}-n,\ (a_p)-n \\ (b_q)-n;\ x-1/t\end{matrix}\right).$$

8. $$\sum_{k=0}^{n} \binom{n}{k} \frac{(\alpha)_k}{(\beta)_k} x^{-k} \frac{\Pi(1-b_q)_k}{\Pi(1-a_p)_k}\, {}_{p+1}F_q\left(\begin{matrix}-k,\ (a_p)-k \\ (b_q)-k;\ x\end{matrix}\right)= \frac{(\alpha)_n}{(\beta)_n} x^{-n} \frac{\Pi(1-b_q)_n}{\Pi(1-a_p)_n}\, {}_{p+2}F_{q+1}\left(\begin{matrix}-n,\ (a_p)-n,\ \beta-\alpha \\ (b_q)-n,\ 1-\alpha-n;\ x\end{matrix}\right).$$

9. $$\sum_{k=0}^{n} \binom{n}{k} \frac{x^{-k}}{(\beta)_k} \frac{\Pi(1-b_q)_k}{\Pi(1-a_p)_k}\, {}_{p+2}F_q\left(\begin{matrix}-k,\ (a_p)-k,\ \alpha \\ (b_q)-k;\ x\end{matrix}\right)= \frac{x^{-n}}{(\beta)_n} \frac{\Pi(1-b_q)_n}{\Pi(1-a_p)_n}\, {}_{p+2}F_q\left(\begin{matrix}-n,\ (a_p)-n,\ \alpha-\beta-n+1 \\ (b_q)-n;\ x\end{matrix}\right).$$

10. $$\sum_{k=0}^{n} (-1)^k \binom{n}{k} \frac{(1-\beta)_k}{(k+m)!}\, {}_{p+1}F_{q+1}\left(\begin{matrix}-k-m,\ (a_p);\ x \\ \beta-k,\ (b_q)\end{matrix}\right)= \frac{(\beta)_{m+n}}{(m+n)!\,(\beta)_m}\, {}_{p+1}F_{q+1}\left(\begin{matrix}-m-n,\ (a_p);\ x \\ \beta,\ (b_q)\end{matrix}\right).$$

5.3.5. Sums of the form $\sum \alpha_k\, {}_pF_q((a_p)+k(c_p);\ (b_q)+k(d_q);\ x)$.

1. $$\sum_{k=0}^{n} {}_{p+1}F_q\left(\begin{matrix}\alpha+k,\ (a_p) \\ (b_q);\ x\end{matrix}\right)= \left(\prod_{i=1}^{q}(b_i-1)\right)\left(\prod_{j=1}^{p}(a_j-1)\right)^{-1} x^{-1}\left[{}_{p+1}F_q\left(\begin{matrix}\alpha+n,\ (a_p)-1 \\ (b_q)-1;\ x\end{matrix}\right)- {}_{p+1}F_q\left(\begin{matrix}\alpha-1,\ (a_p)-1 \\ (b_q)-1;\ x\end{matrix}\right)\right].$$

2. $$\sum_{k=0}^{n} (-1)^k \binom{n}{k} \frac{(\alpha)_k}{(\beta)_k}\, {}_{p+m}F_q\left(\begin{matrix}(a_p),\ \Delta(m,\ \alpha+k) \\ (b_q);\ x\end{matrix}\right)= \frac{(\beta-\alpha)_n}{(\beta)_n}\, {}_{p+2m}F_{q+m}\left(\begin{matrix}(a_p),\ \Delta(m,\ \alpha),\ \Delta(m,\ \alpha-\beta+1) \\ (b_q),\ \Delta(m,\ \alpha-\beta-n-1);\ x\end{matrix}\right).$$

3. $$\sum_{k=0}^{n}(-1)^k\binom{n}{k}\frac{(\alpha)_k}{(\beta)_k}\,{}_pF_{q+m}\left(\begin{matrix}(a_p);\ x\\(b_q),\ \Delta(m,\ \beta+k)\end{matrix}\right)=$$
$$=\frac{(\beta-\alpha)_n}{(\beta)_n}\,{}_{p+m}F_{q+2m}\left(\begin{matrix}(a_p),\ \ \Delta(m,\ \beta-\alpha+n);\ x\\(b_q),\ \ \Delta(m,\ \beta-\alpha),\ \ \Delta(m,\ \beta+n)\end{matrix}\right).$$

4. $$\sum_{k=0}^{n}(-1)^k\binom{n}{k}\frac{(\alpha)_k}{(\beta)_k}\,{}_{p+m}F_{q+m}\left(\begin{matrix}(a_p),\ \ \Delta(m,\ \alpha+k);\ x\\(b_q),\ \Delta(m,\ \beta+k)\end{matrix}\right)=$$
$$=\frac{(\beta-\alpha)_n}{(\beta)_n}\,{}_{p+m}F_{q+m}\left(\begin{matrix}(a_p),\ \ \Delta(m,\ \alpha);\ x\\(b_q),\ \ \Delta(m,\ \beta+n)\end{matrix}\right).$$

5. $$\sum_{k=0}^{n}\binom{n}{k}\frac{x^k}{(\beta)_k}\frac{\Pi(a_p)_k}{\Pi(b_q)_k}\,{}_pF_q\left(\begin{matrix}(a_p)+k;\ x\\(b_q)+k\end{matrix}\right)={}_{p+1}F_{q+1}\left(\begin{matrix}(a_p),\ \beta+n;\ x\\(b_q),\ \beta\end{matrix}\right).$$

6. $$\sum_{k=0}^{n}\frac{(4x)^k}{(n-k)!\,(2k+\sigma)!}\,\frac{\Pi(a_p)_k}{\Pi(b_q)_k}\,F\left(\begin{matrix}(a_p)+k;\ x\\(b_q)+k\end{matrix}\right)=$$
$$=\frac{1}{n!}\,{}_{p+1}F_{q+1}\left(\begin{matrix}n+\sigma+1/2,\ (a_p);\ x\\\sigma+1/2,\ (b_q)\end{matrix}\right)\qquad [\sigma=0\ \text{ or }\ 1].$$

7. $$\sum_{k=0}^{n}\binom{n}{k}(-x)^k\frac{(\beta-\alpha)_k}{(\beta-n)_k}\,\frac{\Pi(a_p)_k}{\Pi(b_q)_k}\,{}_{p+1}F_q\left(\begin{matrix}(a_p)+k,\ \alpha\\(b_q)+k;\ x\end{matrix}\right)={}_{p+2}F_{q+1}\left(\begin{matrix}(a_p),\ \alpha-n,\ \beta\\(b_q),\ \beta-n;\ x\end{matrix}\right).$$

8. $$\sum_{k=1}^{n}\frac{x^k}{(\beta)_{2k}}\,\frac{\Pi(a_p)_k}{\Pi(b_q)_k}\,{}_pF_{q+1}\left(\begin{matrix}(a_p)+k;\ x\\(b_q)+k,\ \ 2k+\beta+1\end{matrix}\right)=\frac{x}{\beta(\beta+1)}\prod_{i=1}^{p}a_i\prod_{j=1}^{q}b_j^{-1}\times$$
$$\times\,{}_pF_{q+1}\left(\begin{matrix}(a_p)+1;\ x\\(b_q)+1,\ \beta+2\end{matrix}\right)-\frac{x^{n+1}}{(\beta)_{2n+2}}\,\frac{\Pi(a_p)_{n+1}}{\Pi(b_q)_{n+1}}\,{}_pF_{q+1}\left(\begin{matrix}(a_p)+n+1;\ x\\(b_q)+n+1,\ \ 2n+\beta+2\end{matrix}\right).$$

9. $$\sum_{k=0}^{n}\frac{x^k}{(2k)!}\,\frac{\Pi(a_p)_k}{\Pi(b_q)_k}\,{}_pF_{q+1}\left(\begin{matrix}(a_p)+k;\ x\\(b_q)+k,\ \ 2k+2\end{matrix}\right)=$$
$$={}_pF_{q+1}\left(\begin{matrix}(a_p);\ x\\(b_q),\ \ 1\end{matrix}\right)-\frac{x^{n+1}}{(2n+2)!}\,\frac{\Pi(a_p)_{n+1}}{\Pi(b_q)_{n+1}}\,{}_pF_{q+1}\left(\begin{matrix}(a_p)+n+1;\ x\\(b_q)+n+1,\ \ 2n+3\end{matrix}\right).$$

10. $$\sum_{k=0}^{n}\binom{2n}{n-k}\frac{(-x)^k}{(2k)!}\,\frac{\Pi(a_p)_k}{\Pi(b_q)_k}\,{}_pF_{q+1}\left(\begin{matrix}(a_p)+k;\ x\\(b_q)+k,\ \ 2k+1\end{matrix}\right)=$$
$$=\frac{1}{2}\binom{2n}{n}\left[{}_pF_{q+1}\left(\begin{matrix}(a_p);\ x\\(b_q),\ \ n+1\end{matrix}\right)-{}_pF_{q+1}\left(\begin{matrix}(a_p);\ x\\(b_q),\ \ 1\end{matrix}\right)\right].$$

11. $$\sum_{k=0}^{[n/2]}\frac{(n\alpha+2k\alpha+1)^{k-1}}{k!\,(n-2k)!}\left(-\frac{x}{4}\right)^k\frac{\Pi(a_p)_k}{\Pi(b_q)_k}\,{}_{p+2}F_q\left(\begin{matrix}-n/2+k,\ (1-n)/2+k,\ (a_p)+k\\(b_q)+k;\ \ (n\alpha-2k\alpha+1)\,x\end{matrix}\right)=$$
$$=\frac{1}{n!\,(n\alpha+1)}.$$

12. $$\sum_{k=0}^{n}(-1)^k\binom{2n}{n-k}\frac{(x/4)^k}{(k!)^2}\,\frac{\Pi(a_p)_k}{\Pi(b_q)_k}\,{}_{p+1}F_{q+2}\left(\begin{matrix}(a_p)+k,\ \ k+1/2;\ x\\(b_q)+k,\ \ k+1,\ \ 2k+1\end{matrix}\right)=$$
$$=\frac{1}{2}\binom{2n}{n}\left[{}_{p+1}F_{q+2}\left(\begin{matrix}(a_p),\ \ 1/2;\ x\\(b_q),\ \ 1,\ \ n+1\end{matrix}\right)+{}_{p+1}F_{q+2}\left(\begin{matrix}(a_p),\ \ 1/2;\ \ x\\(b_q),\ \ 1,\ \ 1\end{matrix}\right)\right].$$

13. $\sum_{k=0}^{n}\binom{n}{k}\frac{(\beta-1)_k(-x)^k}{(\beta+n)_k(\beta-1)_{2k}}\frac{\Pi(a_p)_k}{\Pi(b_q)_k}\,{}_pF_{q+1}\left(\begin{matrix}(a_p)+k;\ x\\(b_q)+k,\ 2k+\beta\end{matrix}\right)={}_pF_{q+1}\left(\begin{matrix}(a_p);\ x\\(b_q),\ n+\beta\end{matrix}\right).$

14. $\sum_{k=0}^{n}\frac{(-x)^k}{(\beta)_{2k}}\frac{\Pi(a_p)_k}{\Pi(b_q)_k}\,{}_pF_{q+1}\left(\begin{matrix}(a_p)+k;\ x\\(b_q)+k,\ 2k+\beta\end{matrix}\right)={}_{p+1}F_{q+2}\left(\begin{matrix}(a_p),\ (\beta-1)/2;\ x\\(b_q),\ \beta-1,\ (\beta+1)/2\end{matrix}\right)-$

$-\frac{(-x)^{n+1}}{(\beta)_{2n+2}}\frac{\Pi(a_p)_{n+1}}{\Pi(b_q)_{n+1}}\,{}_{p+1}F_{q+2}\left(\begin{matrix}(a_p)+n+1,\ n+(\beta+1)/2;\ x\\(b_q)+n+1,\ n+(\beta+3)/2,\ 2n+\beta+1\end{matrix}\right).$

5.3.6. Sums of the form $\sum \alpha_k\, {}_pF_q((a_p)-k(c_p);(b_q)-k(d_q);x)$.

1. $\sum_{k=0}^{n}{}_{p+1}F_q\left(\begin{matrix}(c_p),\ \alpha-k\\(b_q);\ x\end{matrix}\right)=$

$=\left(\prod_{i=1}^{p}(a_i-1)\right)^{-1}\left(\prod_{j=1}^{q}(b_j-1)\right)x^{-1}\left[{}_{p+1}F_q\left(\begin{matrix}(a_p)-1,\ \alpha\\(b_q)-1;\ x\end{matrix}\right)-\right.$

$\left.-{}_{p+1}F_q\left(\begin{matrix}(a_p)-1,\ \alpha-n-1\\(b_q)-1;\ x\end{matrix}\right)\right].$

2. $\sum_{k=1}^{n}k\,{}_{p+1}F_q\left(\begin{matrix}(a_p),\ \alpha-k\\(b_q);\ x\end{matrix}\right)=\left(\prod_{i=1}^{p}(a_i-1)(a_i-2)\right)^{-1}\left(\prod_{j=1}^{q}(b_j-1)(b_j-2)\right)\times$

$\times x^{-2}\left[{}_{p+1}F_q\left(\begin{matrix}(a_p)-2,\ \alpha-1\\(b_q)-2;\ x\end{matrix}\right)+n\,{}_{p+1}F_q\left(\begin{matrix}(a_p)-2,\ \alpha-n-2\\(b_q)-2;\ x\end{matrix}\right)-\right.$

$\left.-(n+1)\,{}_{p+1}F_q\left(\begin{matrix}(a_p)-2,\ \alpha-n-1\\(b_q)-2;\ x\end{matrix}\right)\right].$

3. $\sum_{k=0}^{n}(-1)^k\binom{n}{k}\frac{(1-\beta)_k}{(1-\alpha)_k}\,{}_{p+m}F_q\left(\begin{matrix}(a_p),\ \Delta(m,\ \alpha-k)\\(b_q);\ x\end{matrix}\right)=$

$=\frac{(\beta-\alpha)_n}{(1-\alpha)_n}\,{}_{p+2m}F_{q+m}\left(\begin{matrix}(a_p),\ \Delta(m,\ \alpha-\beta+1),\ \Delta(m,\ \alpha-n)\\(b_q),\ \Delta(m,\ \alpha-\beta-n+1);\ x\end{matrix}\right).$

4. $\sum_{k=0}^{n}(-1)^k\binom{n}{k}\frac{(1-\beta)_k}{(1-\alpha)_k}\,{}_pF_{q+m}\left(\begin{matrix}(a_p);\ x\\(b_q),\ \Delta(m,\ \beta-k)\end{matrix}\right)=$

$=\frac{(\beta-\alpha)_n}{(1-\alpha)_n}\,{}_{p+m}F_{q+2m}\left(\begin{matrix}(a_p),\ \Delta(m,\ \beta-\alpha+n);\ x\\(b_q),\ \Delta(m,\ \beta),\ \Delta(m,\ \beta-\alpha)\end{matrix}\right).$

5. $\sum_{k=0}^{n}(-1)^k\binom{n}{k}\frac{(1-\beta)_k}{(2-\beta-n)_k}\,{}_pF_{q+1}\left(\begin{matrix}(a_p);\ x\\(b_q),\ \beta-k\end{matrix}\right)=$

$=\frac{(-x)^n}{(\beta)_n(\beta-1)_n}\frac{\Pi(a_p)_n}{\Pi(b_q)_n}\,{}_pF_{q+1}\left(\begin{matrix}(a_p)+n;\ x\\(b_q)+n,\ \beta+n\end{matrix}\right).$

6. $\sum_{k=0}^{n}(-1)^k\binom{n}{k}\frac{(1-\beta)_k}{(1-\alpha)_k}\,{}_{p+m}F_{q+m}\left(\begin{matrix}(a_p),\ \Delta(m,\ \alpha-k);\ x\\(b_q),\ \Delta(m,\ \beta-k)\end{matrix}\right)=$

$=\frac{(\beta-\alpha)_n}{(1-\alpha)_n}\,{}_{p+m}F_{q+m}\left(\begin{matrix}(a_p),\ \Delta(m,\ \alpha-n);\ x\\(b_q),\ \Delta(m,\ \beta)\end{matrix}\right).$

7. $$\sum_{k=0}^{n}\binom{n}{k}(\alpha)_k x^{-k}\frac{\Pi(1-b_q)_k}{\Pi(1-a_p)_k}\,{}_pF_q\left(\begin{matrix}(a_p)-k;\ x\\(b_q)-k\end{matrix}\right)=$$
$$=(\alpha)_n x^{-n}\frac{\Pi(1-b_q)_n}{\Pi(1-a_p)_n}\,{}_{p+1}F_{q+1}\left(\begin{matrix}(a_p)-n,\ 1-\alpha;\ x\\(b_q)-n,\ 1-\alpha-n\end{matrix}\right).$$

8. $$\sum_{k=0}^{n}(-1)^k\binom{n}{k}\frac{(1-\beta)_k}{(2-\beta-n)_k}\,{}_{p+1}F_{q+1}\left(\begin{matrix}(a_p),\ \alpha-k;\ x\\(b_q),\ \beta-k\end{matrix}\right)=$$
$$=\frac{(\beta-\alpha)_n x^n}{(\beta)_n(\beta-1)_n}\frac{\Pi(a_p)_n}{\Pi(b_q)_n}\,{}_{p+1}F_{q+1}\left(\begin{matrix}(a_p)+n,\ \alpha;\ x\\(b_q)+n,\ \beta+n\end{matrix}\right).$$

9. $$\sum_{k=0}^{n}\binom{n}{k}(-x)^{-k}\frac{\Pi(1-b_q)_k}{\Pi(1-a_p)_k}\,{}_{p+1}F_q\left(\begin{matrix}(a_p)-k,\ \alpha\\(b_q)-k;\ x\end{matrix}\right)=$$
$$=\frac{\Pi(1-b_q)_n}{\Pi(1-a_p)_n}(-x)^{-n}\,{}_{p+1}F_q\left(\begin{matrix}(a_p)-n,\ \alpha-n\\(b_q)-n;\ x\end{matrix}\right).$$

5.3.7. Various sums containing ${}_pF_q((a_p);\ (b_q);\ x)$.

1. $$\sum_{k=1}^{n}\frac{\prod\limits_{\substack{i=1\\i\neq k}}^{n}\Gamma(b_i-b_k)\prod\limits_{i=1}^{m}\Gamma(b_k-a_i+1)}{\prod\limits_{i=n+1}^{q}\Gamma(b_k-b_i+1)\prod\limits_{i=m+1}^{p}\Gamma(a_i-b_k)}x^{b_k}\times$$
$$\times{}_pF_{q-1}\left(\begin{matrix}b_k-(a_p)+1;\ (-1)^{p-m-n}x\\b_k-b_1+1,\ \ldots,\ b_k-b_{k-1}+1,\ b_k-b_{k+1}+1,\ \ldots,\ b_k-b_q+1\end{matrix}\right)=$$
$$=G_{pq}^{nm}\left(x\,\middle|\,\begin{matrix}(a_p)\\(b_q)\end{matrix}\right)$$
$[p<q]$ or $[p=q;\ |x|<1]$.

2. $$\sum_{k=1}^{n}\frac{\prod\limits_{\substack{i=1\\i\neq k}}^{n}\Gamma(a_k-a_i)\prod\limits_{i=1}^{m}\Gamma(b_i-a_k+1)}{\prod\limits_{i=n+1}^{p}\Gamma(a_i-a_k+1)\prod\limits_{i=m+1}^{q}\Gamma(a_k-b_i)}x^{a_k}\times$$
$$\times{}_qF_{p-1}\left(\begin{matrix}1-a_k+(b_q);\ (-1)^{q-m-n}/x\\a_1-a_k+1,\ \ldots,\ a_{k-1}-a_k+1,\ a_{k+1}-a_k+1,\ \ldots,\ a_p-a_k+1\end{matrix}\right)=$$
$$=xG_{pq}^{mn}\left(x\,\middle|\,\begin{matrix}(a_p)\\(b_q)\end{matrix}\right)$$
$[q<p]$ or $[q=p;\ |x|>1]$.

5.3.8. Sums containing the G-function

NOTATION: $k,\ m,\ n,\ p,\ q,\ r,\ s=0,\ 1,\ 2,\ \ldots$

1. $$\sum_{k=0}^{s}(-1)^k\binom{s}{k}G_{p+r,\ q+r}^{m,\ n+r}\left(x\,\middle|\,\begin{matrix}\Delta(r,\ \alpha-k),\ (a_p)\\(b_q),\ \Delta(r,\ \beta-k)\end{matrix}\right)=$$
$$=\frac{(\alpha-\beta)_s}{r^s}G_{p+r,\ q+r}^{m,\ n+r}\left(x\,\middle|\,\begin{matrix}\Delta(r,\ \alpha),\ (a_p)\\(b_q),\ \Delta(r,\ \beta-1)\end{matrix}\right)$$
$[\mathrm{Re}\,(\alpha-\beta+s)>0;\ m+n>(p+q)/2]$.

2. $$\sum_{k=0}^{s} \frac{(-1)^k}{r^k} \binom{s}{k} (\alpha-\beta+s)_k G_{p,\, q+2r}^{m+r,\, n}\left(x \middle| \begin{matrix}(a_p)\\ \Delta(r,\alpha),\ (b_q),\ \Delta(r,\beta-k)\end{matrix}\right)=$$

$$=(-1)^s G_{p,\, q+2r}^{m+r,\, n}\left(x \middle| \begin{matrix}(a_p)\\ \Delta(r,\alpha+s),\ (b_q),\ \Delta(r,\beta-s)\end{matrix}\right)$$ [Re $\alpha<1$; $m+n>(p+q)/2$].

3. $$\sum_{k=0}^{s} \frac{(-r)^k}{(\beta)_k} \binom{s}{k} G_{p+r,\, q}^{m,\, n+r}\left(x \middle| \begin{matrix}\Delta(r,\alpha-k),\ (a_p)\\ (b_q)\end{matrix}\right)=$$

$$=\frac{(-r)^s}{(\beta)_s} G_{p+2r,\, q+r}^{m,\, n+2r}\left(x \middle| \begin{matrix}\Delta(r,\alpha),\ \Delta(r,\alpha+\beta-1),\ (a_p)\\ (b_q),\ \Delta(r,\alpha+\beta+s-1)\end{matrix}\right)$$

[Re $(\alpha+\beta+s)>1$; $m+n>(p+q-r)/2$].

4. $$\sum_{k=0}^{s} r^k \binom{s}{k} G_{p+r,\, q+r}^{m,\, n}\left(x \middle| \begin{matrix}(a_p),\ \Delta(r,\alpha+k)\\ (b_q),\ \Delta(r,\beta+k)\end{matrix}\right)=$$

$$=(\alpha-\beta)_s G_{p+r,\, q+r}^{m,\, n}\left(x \middle| \begin{matrix}(a_p),\ \Delta(r,\alpha+s)\\ (b_q),\ \Delta(r,\beta)\end{matrix}\right)$$

[Re $(\alpha-\beta+s)>0$; $m+n>(p+q)/2+r$].

5. $$\sum_{k=0}^{s} \frac{(-r)^k}{(\beta-\alpha+1)_k} \binom{s}{k} G_{p,\, q+r}^{m+r,\, n}\left(x \middle| \begin{matrix}(a_p)\\ \Delta(r,\beta+k),\ (b_q)\end{matrix}\right)=$$

$$=\frac{(-r)^s}{(\beta-\alpha+1)_s} G_{p+r,\, q+2r}^{m+2r,\, n}\left(x \middle| \begin{matrix}(a_p),\ \Delta(r,\alpha-s)\\ \Delta(r,\alpha),\ \Delta(r,\beta),\ (b_q)\end{matrix}\right)$$ [Re $\alpha<s+1$; $m+n>(p+q-r)/2$].

6. $$\sum_{k=0}^{s} \frac{(-1)^k}{r^k} \binom{s}{k} (\beta-\alpha-s)_k G_{p+2r,\, q}^{m,\, n+r}\left(x \middle| \begin{matrix}\Delta(r,\alpha),\ (a_p),\ \Delta(r,\beta+k)\\ (b_q)\end{matrix}\right)=$$

$$=(-1)^s G_{p+2r,\, q}^{m,\, n+r}\left(x \middle| \begin{matrix}\Delta(r,\alpha-s),\ (a_p),\ \Delta(r,\beta+s)\\ (b_q)\end{matrix}\right)$$ [Re $\alpha>0$; $m+n>(p+q)/2$].

7. $$\sum_{k=0}^{s} \frac{(-r^2)^k (\gamma)_{s-k}}{(1+\gamma)_k} \binom{s}{k} G_{p+r,\, q+r}^{m,\, n+r}\left(x \middle| \begin{matrix}\Delta(r,\alpha-k),\ (a_p)\\ (b_q),\ \Delta(r,\alpha+k)\end{matrix}\right)=$$

$$=\frac{(-r^2)^s}{(1+\gamma)_s} G_{p+2r,\, q+r}^{m,\, n+2r}\left(x \middle| \begin{matrix}\Delta(r,\alpha+\gamma),\ \Delta(r,\alpha-\gamma-s),\ (a_p)\\ (b_q),\ \Delta(r,\alpha-\gamma),\ \Delta(r,\alpha+\gamma+s)\end{matrix}\right)$$ [$m+n>(p+q)/2$].

8. $$\sum_{k=0}^{s} \frac{(-r^2)^k (\gamma)_{s-k}}{(1+\gamma)_k} \binom{s}{k} G_{p+r,\, q+r}^{m+r,\, n}\left(x \middle| \begin{matrix}(a_p),\ \Delta(r,\alpha-k)\\ \Delta(r,\alpha+k),\ (b_q)\end{matrix}\right)=$$

$$=\frac{(-r^2)^s}{(1+\gamma)_s} G_{p+2r,\, q+2r}^{m+2r,\, n}\left(x \middle| \begin{matrix}(a_p),\ \Delta(r,\alpha-\gamma-s),\ \Delta(r,\alpha+\gamma)\\ \Delta(r,\alpha-\gamma),\ \Delta(r,\alpha+\gamma+s),\ (b_q)\end{matrix}\right)$$ [$m+n>(p+q)/2$].

9. $$\sum_{k=0}^{s} \frac{(-4r^2)^k (2\beta)_{2s-k}}{(2\beta)_k} \binom{s}{k} G_{p+2r,\, q+2r}^{m+2r,\, n}\left(x \middle| \begin{matrix}(a_p),\ \Delta(2r,2\alpha-k)\\ \Delta(2r,2\alpha+k),\ (b_q)\end{matrix}\right)=$$

$$=(-4r^2)^s G_{p+2r,\, q+2r}^{m+2r,\, n}\left(x \middle| \begin{matrix}(a_p),\ \Delta(r,\alpha+\beta),\ \Delta(r,\alpha-\beta-s+1/2)\\ \Delta(r,\alpha+\beta+s),\ \Delta(r,\alpha-\beta+1/2),\ (b_q)\end{matrix}\right)$$

[$m+n>(p+q)/2$].

10. $$\sum_{k=0}^{s} \frac{(-r)^k (\beta+s)_k}{(\beta+1)_{2k}} \binom{s}{k} G_{p+r,\, q+r}^{m+r,\, n}\left(x \middle| \begin{matrix}(a_p),\ \Delta(r,\alpha+k)\\ \Delta(r,\alpha+2k),\ (b_q)\end{matrix}\right)=$$

$$=\frac{(-r)^s \Gamma(\beta+1)}{(\beta+2s)\Gamma(\beta+s)} G_{p+r,\, q+r}^{m+r,\, n}\left(x \middle| \begin{matrix}(a_p),\ \Delta(r,\alpha-\beta-s)\\ \Delta(r,\alpha-\beta),\ (b_q)\end{matrix}\right)$$ [$m+n>(p+q)/2$].

11. $$\sum_{k=0}^{n} (-1)^k \binom{n}{k} G_{pq}^{mn}\left(x \left| \begin{matrix} (a_{p-1}),\ a_p+k \\ b_1+k,\ b_2,\ \ldots,\ b_q \end{matrix}\right.\right) = (-1)^n (a_p - b_1)_n G_{pq}^{mn}\left(x \left| \begin{matrix} (a_{p-1}),\ a_p+n \\ (b_q) \end{matrix}\right.\right)$$ $[n < p;\ q \geqslant m \geqslant 1;\ \operatorname{Re}(b_1 - a_p) < n]$.

5.3.9. Sums containing the Neumann polynomials $O_n(x)$.

1. $$\sum_{k=0}^{n} (-1)^k \binom{n}{k} O_{2k+m}(x) = 2^n \frac{d^n O_{m+n}(x)}{dx^n}.$$

2. $$\sum_{k=0}^{n} O_k'(x) = \frac{1}{2}\left[\frac{1}{x} + \frac{1}{x^2} - O_n(x) - O_{n+1}(x)\right].$$

3. $$\sum_{k=0}^{n} O_{2k+\sigma}'(x) = \frac{1}{2x^{2-\sigma}} - \frac{1}{2} O_{2n+\sigma+1}(x)$$ $[\sigma = 0$ or $1]$.

5.3.10. Various sums

1. $$\sum_{k=0}^{n} \zeta\left(s,\ v \pm \frac{k}{n}\right) = n^s \zeta\left(s,\ nv - \frac{n \mp n}{2}\right) + \zeta\left(s,\ v + \frac{1 \pm 1}{2}\right).$$

2. $$\sum_{k=0}^{n} \zeta\left(s,\ \frac{k}{n}\right) = n^s \zeta(s).$$

3. $$\sum_{k=0}^{n} (-1)^k \binom{n}{k} (\mu)_k \left(\frac{z}{2}\right)^{-k} \mathbf{H}_{k+\nu}(z) = \frac{z^{\nu+1}}{2^\nu \sqrt{\pi}} \frac{(\nu-\mu+3/2)_n}{\Gamma(n+\nu+3/2)}\, {}_2F_3\left(\begin{matrix} 1,\ \nu-\mu+n+3/2;\ -z^2/4 \\ 3/2,\ \nu-\mu+3/2,\ \nu+n+3/2 \end{matrix}\right).$$

4. $$\sum_{k=0}^{n} (-1)^k \binom{n}{k} x^{k/2} W_{\mu+k/2,\ \nu+k/2}(x) = (-1)^n (\nu-\mu+1/2)_n W_{\mu-n/2,\ \nu+n/2}(x).$$

5. $$\sum_{k=0}^{n} \binom{n}{k} \frac{1}{(\mu-\nu+1/2)_k} W_{\mu+k,\ \nu}(x) = \frac{1}{(\mu-\nu+1/2)_n} W_{\mu+n/2,\ \nu-n/2}(x).$$

6. $$\sum_{k=0}^{n} \frac{(-1)^k}{(n-k)!} \binom{n}{k} \left(\frac{1}{2}-\mu-\nu\right)_k \left(\frac{1}{2}-\mu+\nu\right)_k W_{\mu-k,\ \nu}(x) = (-1)^n \left(\frac{1}{2}-\mu-\nu\right)_n \left(\frac{1}{2}-\mu+\nu\right)_n W_{\mu-n,\ \nu}(x)$$

7. $$\sum_{k=0}^{n} (-1)^k \frac{(1-b)_k}{k!} L_{n-k}^{1-b-n+k}(-x)\, {}_1F_1\left(\begin{matrix} a-k;\ x \\ b-k \end{matrix}\right) = \frac{(a)_n}{(b)_n} \frac{x^n}{n!}\, {}_1F_1\left(\begin{matrix} a+n;\ x \\ b+n \end{matrix}\right).$$

8. $$\sum_{k=0}^{n} \frac{x^k}{k!} \frac{(a)_k}{(b)_k} L_{n-k}^{b-n+k-1}(x)\, {}_1F_1\left(\begin{matrix} a+k;\ x \\ b+k \end{matrix}\right) = \frac{(-1)^n (1-b)_n}{n!}\, {}_1F_1\left(\begin{matrix} a-n;\ x \\ b-n \end{matrix}\right).$$

9. $$\sum_{k=0}^{2n} \binom{2n}{k} \frac{(a)_k (2a-2n-1)_k}{(a+1/2)_k (2a)_k (2a-2n-1)_{2k}} (-x)^k\, {}_1F_2\left(\begin{matrix} a+k;\ x \\ a+k+1/2,\ 2a-2n+2k \end{matrix}\right) = \Gamma\left[\begin{matrix} a+1/2,\ 2a \\ a \end{matrix}\right] x^{1-2a} I_{a-1/2}^2(\sqrt{x}).$$

Chapter 6. SERIES

6.1. INTRODUCTION

In this chapter sums of series are given that contain certain special functions. Under a transformation of these series, in particular, by reducing them to a form adopted in this book, one can make use of the formulae in Appendix II. The formulae of section 6.8 enable one to find sums of hypergeometric functions and, by means of the formulae of section 7.3, sums of series with various elementary and special functions.

6.2. THE GENERALIZED ZETA-FUNCTION $\zeta(s, v)$

6.2.1. Series of the form $\sum a_k t^k \zeta(s \pm k, v)$.

1. $\sum_{k=2}^{\infty} t^k \zeta(k, v) = t[\psi(v) - \psi(v-t)]$ $[|t| < v \leqslant 1]$.

2. $\sum_{k=2}^{\infty} \zeta(k, v) = \frac{1}{v-1}$.

3. $\sum_{k=1}^{\infty} t^{2k+\sigma} \zeta(2k+\sigma, v) = \frac{t}{2}[2\sigma\psi(v) + (-1)^{\sigma}\psi(v+t) - \psi(v-t)]$

$[|t| < v \leqslant 1;\ \sigma = 0$ or $1]$.

4. $\sum_{k=2}^{\infty} \frac{t^k}{k} \zeta(k, v) = t\psi(v) + \ln \frac{\Gamma(v-t)}{\Gamma(v)}$ $[|t| < v \leqslant 1]$.

5. $\sum_{k=2}^{\infty} \frac{k-1}{k(k+1)} \zeta(k, v) = 2\ln\Gamma(v) + (1-2v)\ln(v-1) - \ln(2\pi) + 2v - 2$

$[\arg|v-1| < \pi;\ v \neq 0, -1, -2, \ldots]$.

6. $\sum_{k=0}^{\infty} \frac{(s)_k}{k!} t^k \zeta(s+k, v) = \zeta(s, v-t)$ $[|t| < |v|]$.

7. $\sum_{k=0}^{\infty} \frac{t^k}{k!} \zeta(s-k, v) = e^{vt}\Phi(e^t, s, v) - \frac{\Gamma(1-s)}{(-t)^{1-s}}$ $[-2\pi < t < 0;\ s,\ 1-v \neq 1, 2, \ldots]$.

8. $$\sum_{k=1}^{\infty} \frac{2^{-2k}}{k(2k+1)} \zeta(2k, v) = (2v-1)\ln\left(v-\frac{1}{2}\right) - 2\ln\Gamma(v) + \ln(2\pi) - 2v + 1$$ $[\operatorname{Re} v > -1/2]$.

9. $$\sum_{k=0}^{\infty} \frac{(s)_{2k}}{(2k+1)!\,2^{2k}} \zeta(s+2k, v) = \frac{1}{s-1}\left(v-\frac{1}{2}\right)^{1-s}$$ $\left[v \neq \frac{1}{2};\ s \neq 1;\ \left|\arg\left(v-\frac{1}{2}\right) < \pi\right|\right]$.

6.3. THE NUMBERS AND POLYNOMIALS OF BERNOULLI B_n, $B_n(x)$ AND EULER E_n, $E_n(x)$

6.3.1. Series of the form $\sum a_k B_k$.

1. $$\sum_{k=0}^{\infty} \frac{t^k}{k!} B_k = \frac{t}{2}\left(\operatorname{cth}\frac{t}{2} - 1\right)$$ $[|t| < 2\pi]$.

2. $$\sum_{k=0}^{\infty} (\pm 1)^k \frac{t^{2k}}{(2k)!} B_{2k} = \frac{t}{2} \begin{Bmatrix} \operatorname{cth}(t/2) \\ \operatorname{ctg}(t/2) \end{Bmatrix}$$ $[|t| < 2\pi]$.

3. $$\sum_{k=1}^{\infty} (\pm 1)^k \frac{t^{2k}}{(2k)!\,k} B_{2k} = 2\ln\left[\frac{2}{t} \begin{Bmatrix} \operatorname{sh}(t/2) \\ \sin(t/2) \end{Bmatrix}\right]$$ $[|t| < 2\pi]$.

4. $$\sum_{k=0}^{\infty} (-1)^k \frac{t^{2k}}{(2k)!\,(k+1)} B_{2k+2} = \frac{1}{2}\operatorname{cosec}^2\frac{t}{2} - \frac{2}{t^2}.$$

5. $$\sum_{k=0}^{\infty} \frac{t^{2k}}{(2k)!\,(2k+n)} B_{2k} = \frac{t}{2n+2} + t^{-n}\int_0^t \frac{\tau^n}{e^\tau - 1}\,d\tau$$ $[|t| < 2\pi;\ n = 1, 2, \ldots]$.

6. $$\sum_{k=0}^{\infty} (-1)^k \frac{t^{2k}}{(2k)!\,(2k-1)} B_{2k} = \frac{t}{2}\int_0^t \frac{1}{\tau}\operatorname{ctg}\frac{\tau}{2}\,d\tau.$$

7. $$\sum_{k=0}^{\infty} (\pm 1)^k \frac{t^{2k+1}}{(2k+1)!} B_{2k} = \frac{1}{2}\int_0^t \tau \begin{Bmatrix} \operatorname{cth}(\tau/2) \\ \operatorname{ctg}(\tau/2) \end{Bmatrix} d\tau$$ $[|t| < 2\pi]$.

8. $$\sum_{k=1}^{\infty} \frac{t^{2k}}{(2k+1)!\,k} B_{2k} = 2 + \frac{t}{2} - \frac{\pi^2}{3t} - 2\ln t + \frac{2}{t}\operatorname{Li}_2(e^{-t}).$$

9. $$\sum_{k=1}^{\infty} (-1)^k \frac{t^{2k}}{(2k+1)!\,k} B_{2k} = \frac{2}{t}\operatorname{Cl}_2(t) + 2\ln|t| - 2$$ $[|t| < 2\pi]$.

10. $$\sum_{k=1}^{\infty} (-1)^k \frac{t^{2k}}{(4k)!\,k} B_{2k} = 2\ln\frac{1}{2\left(\operatorname{ch}\sqrt{t/2} - \cos\sqrt{t/2}\right)}.$$

6.3.2. Series of the form $\sum a_k B_k(x+ky)$.

1. $$\sum_{k=0}^{\infty} \frac{t^k}{k!} B_k(x+ky) = \frac{ue^{xu}}{(1-yu)(e^u - 1)}$$ $[t = ue^{-yu}]$.

2. $$\sum_{k=0}^{\infty} \frac{t^k}{k!\,(k+n)} B_{k+n}(x) = (n-1)!\,(-t)^{-n} - e^{tx}\Phi(e^t, 1-n, x)$$ $[|t| < 2\pi;\ n = 2, 3, \ldots]$.

3. $\sum_{k=0}^{\infty}(\pm 1)^k \frac{t^{2k}}{(2k)!} B_{2k}(x) = \frac{t}{2}\begin{Bmatrix}\operatorname{cosech}(t/2)\operatorname{ch} t(x-1/2)\\ \operatorname{cosec}(t/2)\cos t(x-1/2)\end{Bmatrix}$ $[|t| < 2\pi]$.

4. $\sum_{k=0}^{\infty}(\pm 1)^k \frac{t^{2k+1}}{(2k+1)!} B_{2k+1}(x) = \frac{t}{2}\begin{Bmatrix}\operatorname{cosech}(t/2)\operatorname{sh} t(x-1/2)\\ \operatorname{cosec}(t/2)\sin t(x-1/2)\end{Bmatrix}$ $[|t| < 2\pi]$.

5. $\sum_{k=0}^{\infty}(-1)^k (2n)^k B_k(x)\cos\left(2nx+\frac{k\pi}{2}\right) = n\operatorname{ctg} n$ $[n=1, 2, 3;\ 0 < x < 1]$.

6. $\sum_{k=0}^{\infty}(-1)^k (2n+1)^k B_k(x)\cos\left[(2n+1)x+\frac{k\pi}{2}\right] = \frac{2n+1}{2}\operatorname{ctg}\frac{2n+1}{2}$

$[n=0, 1, 2;\ 0 < x < 1]$.

6.3.3. Series of the form $\sum a_k E_k$.

1. $\sum_{k=0}^{\infty}(\pm 1)^k \frac{t^k}{k!} E_k = \begin{Bmatrix}\operatorname{sech} t\\ \operatorname{sech} t\end{Bmatrix}$ $[|t| < \pi/2]$.

2. $\sum_{k=0}^{\infty}(\pm 1)^k \frac{t^{2k}}{(2k)!} E_{2k} = \begin{Bmatrix}\operatorname{sech} t\\ \sec t\end{Bmatrix}$ $[|t| < \pi/2]$.

3. $\sum_{k=0}^{\infty}\frac{t^{2k}}{(2k)!\,(2k+n)} E_{2k} = t^{-n}\int_0^t \frac{\tau^{n-1}}{\operatorname{ch}\tau}\,d\tau$ $[0 < t < \pi/2;\ n=1, 2, \ldots]$.

4. $\sum_{k=0}^{\infty}(\pm 1)^k \frac{t^{2k+1}}{(2k+1)!} E_{2k} = \begin{Bmatrix}\operatorname{arctg}\operatorname{sh} t\\ \ln|\operatorname{tg}[(2t+\pi)/4]|\end{Bmatrix}$ $[|t| < \pi/2]$.

6.3.4. Series of the form $\sum a_k E_k(x+ky)$.

1. $\sum_{k=0}^{\infty}\frac{t^k}{k!} E_k(x+ky) = \frac{2e^{xu}}{(1-yu)(e^u+1)}$ $[t=ue^{-yu}]$.

2. $\sum_{k=0}^{\infty}(\pm 1)^k \frac{t^{2k}}{(2k)!} E_{2k}(x) = \begin{Bmatrix}\operatorname{sech}(t/2)\operatorname{ch} t(x-1/2)\\ \sec(t/2)\cos t(x-1/2)\end{Bmatrix}$ $[|t| < \pi]$.

3. $\sum_{k=0}^{\infty}(\pm 1)^k \frac{t^{2k+1}}{(2k+1)!} E_{2k+1}(x) = \begin{Bmatrix}\operatorname{sech}(t/2)\operatorname{sh} t(x-1/2)\\ \sec(t/2)\sin t(x-1/2)\end{Bmatrix}$ $[|t| < \pi]$.

6.4. THE FUNCTIONS OF STRUVE $\mathbf{H}_\nu(x)$, WEBER $\mathbf{E}_\nu(x)$ AND ANGER $\mathbf{J}_\nu(x)$

6.4.1. Series containing $\mathbf{H}_\nu(x)$.

1. $\sum_{k=1}^{\infty}\frac{1}{k^{\nu+2n+1}}\mathbf{H}_\nu(kx) =$

$$=\frac{(-1)^{n+1}}{(2n+1)!\,2\sqrt{\pi}}(2\pi)^{2n+1}\left(\frac{x}{2}\right)^{\nu}\sum_{k=0}^{2n+1}\binom{2n+1}{k}B_{2n-k+1}\left(\frac{x}{2\pi}\right)^k\Gamma\begin{bmatrix}(k+1)/2\\(k+\nu)/2+1\end{bmatrix}$$

$[0 \leqslant x \leqslant 2\pi;\ \operatorname{Re}\nu > -1/2]$.

2. $$\sum_{k=1}^{\infty} \frac{(-1)^k}{k^{\nu+2n+1}} \mathbf{H}_\nu(kx) =$$

$$= \frac{(-1)^{n+1}}{(2n+1)!\,2\sqrt{\pi}} (2\pi)^{2n+1} \left(\frac{x}{2}\right)^{\nu} \sum_{k=0}^{2n+1} \binom{2n+1}{k} \frac{B_{2n-k+1}}{2^k} \sum_{l=0}^{k} \binom{k}{l} \left(\frac{x}{\pi}\right)^{l} \Gamma\left[\begin{matrix}(l+1)/2\\ l/2+\nu+1\end{matrix}\right]$$

$[-\pi < x \leqslant \pi;\ \operatorname{Re}\nu > -1/2]$.

3. $$\sum_{k=1}^{\infty} \frac{(\pm 1)^k}{k^{\nu+1}} \mathbf{H}_\nu(kx) = -\frac{(x/2)^{\nu+1}}{\sqrt{\pi}\,\Gamma(\nu+3/2)} + \frac{\pi x^\nu}{2^{\nu+1}\Gamma(\nu+1)} \left\{\begin{matrix}1\\0\end{matrix}\right\}$$

$[-(1\mp 1)\pi/2 < x < (3\pm 1)\pi/2;\ \operatorname{Re}\nu > -3/2]$.

4. $$\sum_{k=1}^{\infty} \frac{(\pm 1)^k}{k^{\nu-1}(k^2-a^2)} \mathbf{H}_\nu(kx) = -\frac{\pi \operatorname{cosec} a\pi}{2a^\nu} \left\{\begin{matrix}\cos a\pi\\1\end{matrix}\right\} \mathbf{H}_\nu(ax) + \frac{\pi}{2a^\nu} \left\{\begin{matrix}1\\0\end{matrix}\right\} J_\nu(ax)$$

$[-(\pi\mp\pi)/2 < x < (3\pm 1)\pi/2;\ \operatorname{Re}\nu > -3/2]$.

5. $$\sum_{k=1}^{\infty} \frac{(-1)^k}{k^{\nu+1}(k^2-a^2)} \mathbf{H}_\nu(kx) = \frac{(x/2)^{\nu+1}}{a^2\sqrt{\pi}\,\Gamma(\nu+3/2)} - \frac{\pi \operatorname{cosec} a\pi}{2a^{\nu+2}} \mathbf{H}_\nu(ax)$$

$[0 < x < \pi;\ \operatorname{Re}\nu > -7/2]$.

6. $$\sum_{\substack{k=1\\k\neq n}}^{\infty} \frac{k}{k^2-n^2} \mathbf{H}_0(kx) = \frac{\pi}{2} J_0(nx) - \frac{x}{4}\left[\mathbf{E}_1(nx) - \mathbf{E}_{-1}(nx)\right] + \frac{1}{4n}\mathbf{E}_0(nx) \quad [0 < x < 2\pi].$$

7. $$\sum_{k=2}^{\infty} \frac{(-1)^k k}{k^2-1} \mathbf{H}_0(kx) = -\frac{1}{4}\mathbf{E}_0(x) + \frac{x}{2}\mathbf{E}_1(x) \quad [-\pi < x < \pi].$$

8. $$\sum_{k=0}^{\infty} \frac{(-1)^k}{(2k+1)^{\nu+2n}} \mathbf{H}_\nu((2k+1)x) =$$

$$= \frac{(-1)^{n-1}}{(2n-1)!\,2\sqrt{\pi}} \left(\frac{\pi}{2}\right)^{2n} \left(\frac{x}{2}\right)^{\nu} \sum_{k=0}^{2n-1} \binom{2n-1}{k} \left(\frac{2x}{\pi}\right)^{k} E_{2n-k-1}\, \Gamma\left[\begin{matrix}(k+1)/2\\ k/2+\nu+1\end{matrix}\right]$$

$[-\pi/2 < x < \pi/2;\ \operatorname{Re}\nu > -1/2;\ n = 1, 2, \ldots]$.

9. $$\sum_{k=0}^{\infty} \frac{1}{(2k+1)^{\nu+2n+1}} \mathbf{H}_\nu((2k+1)x) =$$

$$= \frac{(-1)^n}{(2n-1)!\,2\sqrt{\pi}} \left(\frac{\pi}{2}\right)^{2n+1} \left(\frac{x}{2}\right)^{\nu} \sum_{k=0}^{2n} \binom{2n}{k} E_{2n-k} \sum_{l=0}^{k} (-1)^l \binom{k}{l} \left(\frac{2x}{\pi}\right)^{l} \Gamma\left[\begin{matrix}(l+1)/2\\ l/2+\nu+1\end{matrix}\right]$$

$[0 < x < \pi;\ \operatorname{Re}\nu > -1/2]$.

10. $$\sum_{k=0}^{\infty} \frac{(-1)^k}{(2k+1)^{\nu+2}} \mathbf{H}_\nu((2k+1)x) = \frac{\sqrt{\pi}\,(x/2)^{\nu+1}}{2\Gamma(\nu+3/2)} \quad [-\pi/2 \leqslant x \leqslant \pi/2;\ \operatorname{Re}\nu > -1/2].$$

11. $$\sum_{k=0}^{\infty} \frac{(-1)^k}{(2k+1)^\nu[(2k+1)^2-a^2]} \mathbf{H}_\nu((2k+1)x) = \frac{\pi}{4a^{\nu+1}} \sec\frac{a\pi}{2}\, \mathbf{H}_\nu(ax)$$

$[-\pi/2 < x < \pi/2;\ \operatorname{Re}\nu > -1/2]$.

12. $$\sum_{k=1}^{\infty} \frac{\sin ka}{k^\nu} \mathbf{H}_\nu(kx) = \frac{\sqrt{\pi}}{\Gamma(\nu+1/2)} (2x)^{-\nu} (x^2-a^2)^{\nu-1/2} \left\{\begin{matrix}0\\1\end{matrix}\right\} \quad \left[\left\{\begin{matrix}0 < x < a\\ a < x \leqslant \pi\end{matrix}\right\}\right].$$

13. $\sum_{k=1}^{\infty} \frac{\cos ka}{k^{\nu+1}} \mathbf{H}_\nu(kx) = -\frac{(x/2)^{\nu+1}}{\sqrt{\pi}\,\Gamma(\nu+3/2)}$ $[0 < x < a \leqslant \pi;\ \operatorname{Re}\nu > -1]$.

14. $= -\frac{(x/2)^{\nu+1}}{\sqrt{\pi}\,\Gamma(\nu+3/2)} \left[1 - \frac{\pi}{x}\left(1 - \frac{a^2}{x^2}\right)^{\nu+1/2}\right] {}_2F_1\left(\begin{matrix}1/2,\ \nu+1/2;\ 1-a^2/x^2\\ \nu+3/2\end{matrix}\right)$
$[0 < a < x < \pi;\ \operatorname{Re}\nu > -1]$.

6.4.2. Series containing $\mathbf{E}_\nu(x)$.

1. $\sum_{k=1}^{\infty} \frac{(\pm 1)^k}{k} \mathbf{E}_{2n}(kx) = -\left\{\begin{matrix}1\\0\end{matrix}\right\} \frac{\pi}{2} \delta_{0,n} - \frac{x}{(4n^2-1)\pi}$ $[-(\pi\mp\pi)/2 < x < (3\pi\pm\pi)/2]$.

2. $\sum_{k=1}^{\infty} \frac{(\pm 1)^k}{k^2} \mathbf{E}_{2n+1}(kx) = \frac{(2n+1)x^2}{4(2n-1)(2n+3)\pi} - \left\{\begin{matrix}1\\0\end{matrix}\right\} \frac{\pi x}{4} \delta_{0,n} \pm \frac{1}{6}\left\{\begin{matrix}2\\1\end{matrix}\right\} \frac{\pi}{2n+1}$
$[-(\pi\mp\pi)/2 < x < (3\pi\pm\pi)/2]$.

3. $\sum_{k=1}^{\infty} \frac{1}{k^2-a^2} \mathbf{E}_{2n+1}(kx) = -\frac{\pi}{2a}[\operatorname{ctg} a\pi\, \mathbf{E}_{2n+1}(ax) + J_{2n+1}(ax)] + \frac{1}{(2n+1)a^2\pi}$
$[0 \leqslant x \leqslant 2\pi]$.

4. $\sum_{k=1}^{\infty} \frac{(-1)^k}{k^2-a^2} \mathbf{E}_{2n+1}(kx) = \frac{1}{(2n+1)a^2\pi} - \frac{\pi}{2a} \operatorname{cosec} a\pi\, \mathbf{E}_{2n+1}(ax)$ $[-\pi \leqslant x \leqslant \pi]$.

5. $\sum_{\substack{k=1\\k\neq m}}^{\infty} \frac{1}{k^2-m^2} \mathbf{E}_{2n+1}(kx) = \frac{1}{m^2(2n+1)\pi} + \frac{1}{4m^2} \mathbf{E}_{2n+1}(mx) -$
$- \frac{\pi}{2m} J_{2n}(mx) - \frac{x}{2m}[\mathbf{E}_{2n}(mx) - \mathbf{E}_{2n+2}(mx)]$ $[0 < x < 2\pi]$.

6. $\sum_{k=2}^{\infty} \frac{(-1)^k}{k^2-1} \mathbf{E}_{2n+1}(kx) = \frac{1}{(2n+1)\pi} - \frac{1}{4} \mathbf{E}_{2n+1}(x) + \frac{x}{4}[\mathbf{E}_{2n}(x) - \mathbf{E}_{2n+2}(x)]$
$[-\pi \leqslant x \leqslant \pi]$.

7. $\sum_{k=1}^{\infty} \frac{k}{k^2-a^2} \mathbf{E}_{2n}(kx) = \frac{\pi}{2}[\operatorname{ctg} a\pi\, \mathbf{E}_{2n}(ax) - J_{2n}(ax)]$ $[0 < x < 2\pi]$.

8. $\sum_{\substack{k=1\\k\neq m}}^{\infty} \frac{k}{k^2-m^2} \mathbf{E}_{2n}(kx) = \frac{x}{4}[\mathbf{E}_{2n+1}(mx) - \mathbf{E}_{2n-1}(mx)] - \frac{1}{4m} \mathbf{E}_{2n}(mx) - \frac{\pi}{2} J_{2n}(mx)$
$[0 < x < 2\pi]$.

9. $\sum_{k=0}^{\infty} \frac{1}{2k+1} \mathbf{E}_{2n}((2k+1)x) = -\frac{\pi}{4} \delta_{0,n}$ $[0 < x < \pi]$.

10. $\sum_{k=0}^{\infty} \frac{(-1)^k}{2k+1} \mathbf{E}_{2n+1}((2k+1)x) = \frac{1}{2(2n+1)}$ $[-\pi/2 \leqslant x \leqslant \pi/2]$.

11. $\sum_{k=0}^{\infty} \frac{(-1)^k}{(2k+1)^2} \mathbf{E}_{2n+1}((2k+1)x) = \frac{\pi x}{8} \delta_{0,n}$ $[-\pi/2 \leqslant x \leqslant \pi/2]$.

12. $$\sum_{k=0}^{\infty} \frac{(-1)^k}{(2k+1)^3} E_{2n+1}((2k+1)x) = \frac{\pi^2}{16(2n+1)} - \frac{(n-2)x^2}{4(4n^2-1)(2n+3)} \quad [-\pi/2 \leqslant x \leqslant \pi/2].$$

13. $$\sum_{k=0}^{\infty} \frac{1}{(2k+1)^2} E_{2n+1}((2k+1)x) = \frac{\pi}{4(2n+1)} - \frac{x}{8} \quad [0 \leqslant x < \pi].$$

14. $$\sum_{k=1}^{\infty} \frac{(\pm 1)^k}{k^2} [\mathbf{E}_\nu(kx) - \mathbf{E}_{-\nu}(kx)] = \frac{2}{\nu\pi} \sin^2 \frac{\nu\pi}{2} \left(\frac{x^2}{4-\nu^2} \pm \frac{\pi}{6} \begin{Bmatrix} 2 \\ 1 \end{Bmatrix} \right) - \begin{Bmatrix} 1 \\ 0 \end{Bmatrix} \frac{x}{1-\nu^2} \sin \nu\pi \quad [0 < x < (3 \pm 1)\pi/2].$$

15. $$\sum_{k=1}^{\infty} \frac{(\pm 1)^k}{k^2 - a^2} [\mathbf{E}_\nu(kx) - \mathbf{E}_{-\nu}(kx)] = \frac{2}{a^2 \nu\pi} \sin^2 \frac{\nu\pi}{2} - \frac{\pi}{2a} \begin{Bmatrix} \operatorname{ctg} a\pi \\ \operatorname{cosec} a\pi \end{Bmatrix} [\mathbf{E}_\nu(ax) - \mathbf{E}_{-\nu}(ax)] + \frac{\pi}{2a} \begin{Bmatrix} 1 \\ 0 \end{Bmatrix} \operatorname{tg} \frac{\nu\pi}{2} [\mathbf{E}_\nu(ax) + \mathbf{E}_{-\nu}(ax)] \quad [0 < x < \pi].$$

6.4.3. Series containing $\mathbf{J}_\nu(x)$.

1. $$\sum_{k=1}^{\infty} \frac{(\pm 1)^k}{k} [\mathbf{J}_\nu(kx) - \mathbf{J}_{-\nu}(kx)] = \mp \frac{x}{\pi(1-\nu^2)} \sin \nu\pi + \begin{Bmatrix} 1 \\ 0 \end{Bmatrix} \frac{2}{\nu} \sin^2 \frac{\nu\pi}{2} \quad [0 < x < (3 \pm 1)\pi/2].$$

2. $$\sum_{k=1}^{\infty} \frac{(\pm 1)^k}{k} [\mathbf{J}_{k\nu}(z) - \mathbf{J}_{-k\nu}(z)] = -\frac{\nu\pi}{2} \mathbf{H}_0(z) + \begin{Bmatrix} \nu\pi \\ 0 \end{Bmatrix} \quad [0 < \nu < (3 \pm 1)/2].$$

Other similar series containing $\mathbf{J}_\nu(z)$ can be obtained from the corresponding series for the Weber functions $\mathbf{E}_\nu(z)$ by means of the relations

$$\mathbf{E}_\nu(z) - \mathbf{E}_{-\nu}(z) = \operatorname{tg} \frac{\nu\pi}{2} [\mathbf{J}_\nu(z) + \mathbf{J}_{-\nu}(z)],$$

$$\mathbf{E}_\nu(z) + \mathbf{E}_{-\nu}(z) = -\operatorname{ctg} \frac{\nu\pi}{2} [\mathbf{J}_\nu(z) - \mathbf{J}_{-\nu}(z)].$$

6.5. THE LEGENDRE FUNCTIONS $P_\nu^\mu(x)$ AND $Q_\nu^\mu(x)$

The following relations may prove to be useful for the summation of series with the Legendre functions:

$$P_\nu^k(z) = (\pm 1)^k \frac{\Gamma(\nu+k+1)}{\Gamma(\nu-k+1)} P_\nu^{-k}(z) \quad \left[\begin{Bmatrix} |\arg(z+1)|,\ |\arg(z-1)| < \pi \\ z = x,\ -1 < x < 1 \end{Bmatrix} \right].$$

$$Q_\nu^k(z) = (\pm 1)^k \frac{\Gamma(\nu+k+1)}{\Gamma(\nu-k+1)} Q_\nu^{-k}(z) \quad \left[\begin{Bmatrix} |\arg z|,\ |\arg(z+1)|,\ |\arg(z-1)| < \pi \\ z = x,\ -1 < x < 1 \end{Bmatrix} \right].$$

6.5.1. Series of the form $\sum \alpha_k P_{\nu+k\sigma}^{\mu+k\rho}(x)$.

1. $$\sum_{k=0}^{\infty} P_k^\mu(x) = \frac{\sqrt{\pi/2}}{\Gamma(1/2-\mu)} (1-x)^{-\frac{\mu}{2}-\frac{1}{2}} (1+x)^{\mu/2} \quad [-1 < x < 1;\ \operatorname{Re}\mu < 1/2].$$

2. $$\sum_{k=1}^{\infty} P_{k-1/2}^\mu(x) = \frac{\sqrt{\pi/2}}{\Gamma(1/2-\mu)} (1-x)^{-\frac{\mu}{2}-\frac{1}{2}} (1+x)^{\mu/2} - \frac{1}{2} P_{-1/2}^\mu(x) \quad [-1 < x < 1;\ \operatorname{Re}\mu < 1/2].$$

3. $$\sum_{k=1}^{\infty} (-1)^k P^{\mu}_{k\nu-1/2}(\cos x) = -\frac{1}{2} P^{\mu}_{-1/2}(\cos x) \qquad [0<x,\ \nu x<\pi;\ \mu<1/2].$$

4. $$\sum_{k=1}^{\infty} (-1)^k P_{k\nu-1/2}(\cos x) = \frac{1-\delta_{0,m}}{\nu\sqrt{2}} \times$$
$$\times \sum_{k=1}^{m} \left(\cos\frac{2k-1}{\nu}\pi - \cos x\right)^{-1/2} - \frac{1}{2} P_{-1/2}(\cos x) \qquad [(2m-1)\pi<\nu x<(2m+1)\pi].$$

5. $$\sum_{k=0}^{\infty} t^k P^n_{k+n}(x) = (2n-1)!!\,\frac{(1-x^2)^{n/2}}{(1-2tx+t^2)^{n+1/2}} \qquad [-1<x<1;\ |t|<1].$$

6. $$\sum_{k=0}^{\infty} (-1)^k (2k+1) P_{(k+1/2)\nu-1/2}(\cos x) =$$
$$= \frac{1}{\nu^2\sqrt{2}} \sum_{k=0}^{m-1} (-1)^{k+1} \sin\frac{2k+1}{\nu}\pi \left(\cos\frac{2k+1}{\nu}\pi - \cos x\right)^{-3/2} \qquad [(2m-1)\pi<\nu x<(2m+1)\pi].$$

7. $$\sum_{k=0}^{\infty} \frac{(-1)^k}{2k+1} P_{(k+1/2)\nu-1/2}(\cos x) = \frac{\pi}{4} P_{-1/2}(\cos x) \qquad [0<x<\pi;\ 0\leqslant \nu x<\pi].$$

8. $$\sum_{k=1}^{\infty} \frac{(-1)^k}{k^2-a^2} P^{\mu}_{k\nu-1/2}(x) = \frac{1}{2a^2} P^{\mu}_{-1/2}(x) - \frac{\pi}{2a}\operatorname{cosec} a\pi\, P^{\mu}_{a\nu-1/2}(x)$$
$$[-1<x<1;\ \mu<5/2;\ \nu\geqslant 0].$$

9. $$\sum_{k=0}^{\infty} (-1)^k \frac{2k+1}{(k-a)(k+a+1)} P^{\mu}_{k}(x) = -\frac{\pi}{\sin a\pi} P^{\mu}_{a}(x) \qquad [-1<x<1;\ \mu\leqslant 0].$$

10. $$\sum_{k=0}^{\infty} \frac{t^k}{(k+n)!} P^n_k(x) = (-1)^n e^{tx} J_n\left(t\sqrt{1-x^2}\right).$$

11. $$\sum_{k=0}^{\infty} \frac{t^k}{(k+n)!} P^n_{k+n}(x) =$$
$$= \frac{2^n\Gamma^2(n+1/2)}{n!\,\pi}(1-x^2)^{n/2} e^{tx}\,{}_2F_3\left(\begin{matrix} n+1/2,\ n+1/2;\ -t^2(1-x^2)/4 \\ n/2+1,\ (n+1)/2,\ 1/2 \end{matrix}\right).$$

In the formulae 6.5.1.12–14 for $-1\leqslant x\leqslant 1$, one must make the substitutions $z\to x, \sqrt{z^2-1}\to\sqrt{1-x^2}$.

12. $$\sum_{k=0}^{\infty} \frac{t^k}{(k+2n)!} P^n_{k+n}(z) = t^{-n} e^{tz} I_n\left(t\sqrt{z^2-1}\right).$$

13. $$\sum_{k=0}^{\infty} \frac{(\pm 1)^k t^{2k}}{(2k+2n)!} P^n_{2k+n}(z) = t^{-n} \left\{\begin{matrix} \operatorname{ch} tz\, I_n\left(t\sqrt{z^2-1}\right) \\ \cos tz\, J_n\left(t\sqrt{z^2-1}\right) \end{matrix}\right\}.$$

14. $$\sum_{k=0}^{\infty} \frac{(\pm 1)^k t^{2k+1}}{(2k+2n+1)!} P^n_{2k+n+1}(z) = t^{-n} \left\{\begin{matrix} \operatorname{sh} tz\, I_n\left(t\sqrt{z^2-1}\right) \\ \sin tz\, J_n\left(t\sqrt{z^2-1}\right) \end{matrix}\right\}.$$

15. $$\sum_{k=0}^{\infty} \frac{k!}{(k+n)!} t^k P_k^{n}(\cos x) = \frac{1}{\rho^{1/2}} \left(\frac{1-t\cos x+\rho^{1/2}}{t\sin x} \right)^n$$

$[\rho = 1-2t\cos x+t^2,\ 0<x<\pi]$.

16. $$\sum_{k=0}^{\infty} \frac{(\nu-\mu+1)_k}{k!} t^k P_{k+\nu}^{\mu}(z) = (1-2tz+t^2)^{-(\nu+1)/2} P_{\nu}^{\mu}\left(\frac{z-t}{\sqrt{1-2tz+t^2}} \right)$$

[z is arbitrary;].

17. $$\sum_{k=0}^{\infty} \frac{(2\nu+1)_k}{k!} P_{k+\nu}^{-\nu}(z) = \frac{2^{-\nu}}{\Gamma(\nu+1)} \frac{(\pm z^2 \mp 1)^{\nu/2}}{(1-2tz+t^2)^{\nu+1/2}}$$

$\left[\begin{Bmatrix} |\arg(z+1)|,\ |\arg(z-1)|<\pi \\ z=x,\ -1<x<1 \end{Bmatrix} \right]$.

18. $$\sum_{k=0}^{\infty} \frac{(1-\mu+\nu)_k (\nu+1)_k}{k!\,(\lambda)_k} t^k P_{k+\nu}^{\mu}(x) =$$
$$= \frac{1}{\Gamma(1-\mu)} \left(\frac{x+1}{2} \right)^{\mu/2-\nu-1} \left(\frac{x-1}{2} \right)^{-\mu/2} F_4\left(1-\mu+\nu,\ \nu+1;\ \lambda,\ 1-\mu;\ \frac{2t}{x+1},\ \frac{x-1}{x+1} \right).$$

19. $$\sum_{k=0}^{\infty} \frac{t^k}{k!} P_{\nu}^{k+\mu}(z) = \left(1+2t\frac{z}{\sqrt{\pm z^2 \mp 1}} \pm t^2 \right)^{-\mu/2} P_{\nu}^{\mu}\left(z \pm t\sqrt{\pm z^2 \mp 1} \right)$$

$\left[\begin{Bmatrix} |\arg(z+1)|,\ |\arg(z-1)|<\pi \\ z=x,\ -1<x<1 \end{Bmatrix} \right]$.

20. $$\sum_{k=1}^{\infty} \frac{(\pm 1)^k}{(\nu+1)_k} P_{\nu}^{k}(z) = \frac{1}{2}\left(z \pm \sqrt{z^2-1} \right)^{\nu} - \frac{1}{2} P_{\nu}(z)$$ [$\operatorname{Re} z>0$].

21. $$\sum_{k=1}^{\infty} \frac{(\pm 1)^k}{(-\nu)_k} P_{\nu}^{k}(z) = \frac{1}{2}\left(z \pm \sqrt{z^2-1} \right)^{-\nu-1} - \frac{1}{2} P_{\nu}(z)$$ [$\operatorname{Re} z>0$].

22. $$\sum_{k=0}^{\infty} \frac{(1-\mu+\nu)_k (-\mu-\nu)_k}{k!} t^k P_{\nu}^{\mu-k}(x) =$$
$$= \frac{1}{\Gamma(1-\mu)} \left(\frac{4}{x^2-1} \right)^{\mu/2} {}_2F_1\left[1-\mu+\nu,\ -\mu-\nu;\ 1-\mu;\ \frac{1}{2}\left(1-x+t\sqrt{x^2-1}\right) \right].$$

23. $$\sum_{k=0}^{\infty} \frac{t^k}{k!} P_{k+\nu}^{k}(x) = [1-2t(1-x^2)]^{-(\nu+1)/2} P_{\nu}\left(\frac{x}{\sqrt{1-2t(1-x^2)}} \right).$$

.5.2. Series of the form $\sum \alpha_k Q_{\nu+k\sigma}^{\mu+k\rho}(x)$.

1. $$\sum_{k=0}^{\infty} t^k Q_k(z) = (1-2tz+t^2)^{-1/2} \ln \frac{z-t+(1-2tz+t^2)^{1/2}}{(z^2-1)^{1/2}}$$ $[|t|<e^{|x|};\ z=\operatorname{ch}(x+iy)]$.

2. $$\sum_{k=0}^{\infty} t^k Q_k(x) = (1-2tx+t^2)^{-1/2} \ln \frac{x-t+(1-2tx+t^2)^{1/2}}{(1-x^2)^{1/2}}$$ $[-1<x<1]$.

3. $$\sum_{k=1}^{\infty} Q_{k-1/2}^{\mu}(z) = \frac{1}{2}\sqrt{\frac{\pi}{2}}\,\Gamma\left(\mu+\frac{1}{2}\right) \frac{(z+1)^{\mu/2}}{(z-1)^{(\mu+1)/2}} - \frac{1}{2} Q_{-1/2}^{\mu}(z)$$

$[\operatorname{Re}\mu>-1/2;\ |\arg(z+1)|,\ |\arg(z-1)|<\pi]$.

4. $\sum_{k=n}^{\infty} (\pm 1)^k (2k+1) Q_k^n (x) = (\mp 2)^n n! (x \pm 1)^{-n/2} (x \mp 1)^{-n/2-1}$ $[x > 1]$.

5. $\sum_{k=0}^{\infty} (2k+1) Q_k (z) = \frac{1}{z-1}$ $[|z| > 1]$.

6. $\sum_{k=n}^{\infty} (\pm 1)^k (2k+1) \frac{(k+n)!}{(k-n)!} Q_k^m (x) = (\pm 1)^n (-2)^m m! \frac{(x \pm 1)^m}{(x \mp 1)^{n+m/2+1}}$

$[m = -n, -n+1, \ldots; x > 1]$.

7. $\sum_{k=0}^{\infty} \frac{(-2\sqrt{z^2-1})^{-k}}{k!} Q_{k+n}^{k+m} (z) = \frac{\sqrt{\pi}}{2^{n+1}} \Gamma \begin{bmatrix} m+n+1 \\ n+3/2 \end{bmatrix} z^{-m-n-1} (z^2-1)^{m/2}$

$[|z|, |z^2-1| > 1]$.

8. $\sum_{k=0}^{\infty} \frac{(a)_k}{k! (b)_k} (-2\sqrt{z^2-1})^{-k} Q_{k+n}^{k+m} (z) = \frac{\sqrt{\pi}}{2^{n+1}} \Gamma \begin{bmatrix} m+n+1 \\ m+3/2 \end{bmatrix} \times$

$\times z^{-m-n-1} (z^2-1)^{m/2} {}_3F_2 \left(\begin{matrix} b-a, (m+n+1)/2, (m+n+2)/2 \\ b, \quad n+3/2; \quad 1/z^2 \end{matrix} \right)$ $[|z|, |z^2-1| > 1]$.

6.5.3. Series of the form $\sum \alpha_k \cos(ka+b) P_{\nu+k\sigma}^{\mu+k\rho}(x)$.

1. $\sum_{k=1}^{\infty} \cos ka P_{k\nu-1/2}^{\mu} (\cos x) =$

$$= \frac{\sqrt{\pi/2}}{\nu\Gamma(1/2-\mu)} \sin^{\mu} x \sum_{k=k_-}^{k_+} \alpha_k \left(\cos \frac{2k\pi+a}{\nu} - \cos x \right)^{-\mu-1/2} - \frac{1}{2} P_{-1/2}^{\mu} (\cos x)$$

$[0 < x < \pi;\ \operatorname{Re} \mu < 1/2;\ \operatorname{Re} \mu < -1/2$ for $x = \pm a/\nu;\ k_{\pm} = \pm [(\nu x \mp a)/(2\pi)],\ \alpha_k = 1$ for $k \neq k_{\pm}$, $\alpha_{k_{\pm}} = \frac{1}{2}$ for $(\nu\pi \mp a)/(2\pi) = \ldots, -2, -1, 0, 1, 2, \ldots,\ \alpha_{k_{\pm}} = 1$ for $(\nu\pi \mp a)/(2\pi) \neq \ldots$

$\ldots, -2, -1, 0, 1, 2, \ldots]$.

2. $\sum_{k=1}^{\infty} (-1)^k \cos ka P_{k-1/2}^{\mu} (-\cos x) = \begin{Bmatrix} 1 \\ 0 \end{Bmatrix} \frac{\sqrt{\pi/2}}{\Gamma(1/2-\mu)} \sin^{\mu} x (\cos a - \cos x)^{-\mu-1/2} -$

$- \frac{1}{2} P_{-1/2}^{\mu} (-\cos x)$ $\left[\begin{Bmatrix} 0 < x < a < \pi \\ 0 < a < x < \pi \end{Bmatrix}, \operatorname{Re} \mu < \frac{1}{2} \right]$.

3. $\sum_{k=0}^{\infty} \cos (2k+1) a P_k^{\mu} (\cos x) = \begin{Bmatrix} 1 \\ 0 \end{Bmatrix} \frac{\sqrt{\pi/2}}{\Gamma(1/2-\mu)} \sin^{\mu} x (\cos 2a - \cos x)^{-\mu-1/2}$

$\left[\begin{Bmatrix} 0 < 2a < x < \pi \\ 0 < x < 2a < \pi \end{Bmatrix}, \operatorname{Re} \mu < \frac{1}{2} \right]$.

4. $\sum_{k=1}^{\infty} \frac{(\pm 1)^k}{(\nu+1)_k} \cos ka P_{\nu}^k (z) = \frac{1}{2} (z \pm \sqrt{z^2-1} \cos a)^{\nu} - \frac{1}{2} P_{\nu} (z)$ $[\operatorname{Re} z > 0]$

5. $\sum_{k=1}^{\infty} \frac{(\pm 1)^k}{(-\nu)_k} \cos ka P_{\nu}^k (z) = \frac{1}{2} (z \pm \sqrt{z^2-1} \cos a)^{-\nu-1} - \frac{1}{2} P_{\nu} (z)$ $[\operatorname{Re} z > 0]$

6.5.4. Series of the form $\sum \alpha_k \begin{Bmatrix} \sin(ka+b) \\ \cos(ka+b) \end{Bmatrix} Q_{\nu+k\sigma}^{\mu+k\rho}(x)$.

1. $$\sum_{k=0}^{\infty} (\mp 1)^k \begin{Bmatrix} \sin(2k+1)a \\ \cos(2k+1)a \end{Bmatrix} Q_k(x) = (2x \pm 2\cos 2a)^{-1/2} \operatorname{arctg} \sqrt{\frac{1 \mp \cos 2a}{x \pm \cos 2a}}$$ $[x > 1]$.

2. $$\sum_{k=1}^{\infty} \cos ka\, Q_{k-1/2}^{\mu}(z) = \frac{\sqrt{\pi}}{2^{3/2}} e^{i\mu\pi} \Gamma\left(\mu+\frac{1}{2}\right) \frac{(z^2-1)^{\mu/2}}{(z-\cos a)^{\mu+1/2}} - \frac{1}{2} Q_{-1/2}^{\mu}(z)$$ $[|\arg(z+1)|, |\arg(z-1)| < \pi;\ \operatorname{Re}\mu > -1/2]$.

3. $$\sum_{k=1}^{\infty} \cos ka Q_{k-1/2}(\cos x) = \frac{\pi}{2^{3/2}} (\cos a - \cos x)^{-1/2} - \frac{1}{2} Q_{-1/2}(\cos x)$$ $[a < x]$.

4. $$\sum_{k=1}^{\infty} (-1)^k \cos ka Q_{k-1/2}(x) = \frac{\pi}{2} (x + \cos a)^{-1/2} - \frac{1}{2} Q_{-1/2}(x)$$ $[x > 1]$.

5. $$\sum_{k=1}^{\infty} (\pm 1)^k \cos ka Q_{(k-1)/2}(x) = (2x - 2\cos 2a)^{-1/2} \left[\frac{\pi}{2} \pm \operatorname{arctg} \frac{\sqrt{2}\cos a}{(x-\cos 2a)^{1/2}}\right] - \frac{1}{2} Q_{-1/2}(x)$$ $[x > 1]$.

6.5.5. Series of the form $\sum \alpha_k P_{\nu_1+k\sigma_1}^{\mu_1+k\rho_1}(x) P_{\nu_2+k\sigma_2}^{\mu_2+k\rho_2}(y)$.

1. $$\sum_{k=n}^{\infty} (2k+1) P_k^m(x) P_k^n(y) = (-1)^n (m+n)! \frac{(1-x^2)^{m/2} (y^2-1)^{n/2}}{(y-x)^{m+n+1}}.$$

2. $$\sum_{k=0}^{\infty} (-1)^k \frac{2k+1}{(k-a)(k+a+1)} P_k^{\mu}(\cos x) P_k^{\nu}(\cos y) = -\frac{\pi}{\sin a\pi} P_a^{\mu}(\cos x) P_a^{\nu}(\cos y)$$ $[-\pi < x \pm y < \pi;\ \mu, \nu \leqslant 0]$.

3. $$\sum_{k=0}^{\infty} (-1)^k \frac{2k+2n+1}{(k+n-a)(k+n+a+1)} P_{k+n}^n(\cos x) P_{k+n}^{-n}(\cos y) = (-1)^{n-1} \frac{\pi}{\sin a\pi} P_a^n(\cos x) P_a^{-n}(\cos y)$$ $[0 < x, y, x+y < \pi]$.

4. $$\sum_{k=1}^{\infty} (\pm 1)^k P_{\nu}^k(w) P_{\nu}^{-k}(z) = \frac{1}{2} P_{\nu}\left(wz \pm \sqrt{(w^2-1)(z^2-1)}\right) - \frac{1}{2} P_{\nu}(w) P_{\nu}(z)$$ $[\operatorname{Re} w, \operatorname{Re} z > 0;\ |\arg(w-1)|, |\arg(z-1)| < \pi]$.

5. $$\sum_{k=1}^{\infty} (\pm 1)^k P_{\nu}^k(\cos x) P_{\nu}^{-k}(\cos y) = \frac{1}{2} P_{\nu}(\cos(x \mp y)) - \frac{1}{2} P_{\nu}(\cos x) P_{\nu}(\cos y)$$ $[-\pi \leqslant x, y, x+y \leqslant \pi]$.

6. $$\sum_{k=1}^{\infty} (-1)^k P_{\nu}^{2k}(w) P_{\nu}^{-2k}(z) = \frac{1}{2} P_{\nu}(wz) - \frac{1}{2} P_{\nu}(w) P_{\nu}(z)$$ $[\operatorname{Re} w, \operatorname{Re} z > 0;\ |\arg(w-1)|, |\arg(z-1)| < \pi]$ or $[w = \cos\varphi,\ z = \cos\theta;\ 0 \leqslant \varphi, \theta, \varphi+\theta < \pi]$.

7. $\sum_{k=1}^{\infty} \frac{(-1)^k}{k^2-a^2} P_\nu^{k\mu}(x) P_\nu^{-k\mu}(x) = \frac{1}{2a^2}[P_\nu(x)]^2 - \frac{\pi}{2a \sin a\pi} P_\nu^{a\mu}(x) P_\nu^{-a\mu}(x)$ $[0 < \mu < 1]$.

6.5.6. Series of the form $\sum \alpha_k Q_{\nu_1+k\sigma_1}^{\mu_1+k\rho_1}(x)\, Q_{\nu_2+k\sigma_2}^{\mu_2+k\rho_2}(y)$.

1. $\sum_{k=0}^{\infty} (2k+1) Q_k(w) Q_k(z) = \frac{1}{2(w-z)} \ln \frac{(w-1)(z+1)}{(w+1)(z-1)}$.

2. $\sum_{k=0}^{\infty} (2k+1) [Q_k(x)]^2 = \frac{1}{x^2-1}$ $[x > 1]$.

3. $\sum_{k=n}^{\infty} (2k+1) Q_k(x) Q_k^n(x) = (-1)^{n+1} \frac{2^n n!}{n+1} (x^2-1)^{-n/2} Q_n'(x)$ $[x > 1]$.

4. $\sum_{k=0}^{\infty} (2k+1) Q_k(x) Q_k^n(x) = \frac{(-1)^n n!}{2(n+1)} (x^2-1)^{n/2} [(x-1)^{-n-1} - (x+1)^{-n-1}]$ $[x > 1]$.

5. $\sum_{k=n}^{\infty} (2k+1) Q_k^n(x) Q_k^m(x) = (-1)^{n+1} 2^n \frac{n!}{(n+m+1)!} (x^2-1)^{-(n+1)/2} Q_n^{m+1}(x)$

$[x > 1;\ m = -n,\ -n+1,\ \ldots]$.

6. $\sum_{k=0}^{\infty} (2k+1) [Q_k^n(x)]^2 = \frac{(n!)^2 (x^2-1)^n}{2(2n+1)} [(x-1)^{-2n-1} - (x+1)^{-2n-1}]$ $[x > 1]$.

7. $\sum_{k=n}^{\infty} (2k+1) \frac{(k+n)!}{(k-n)!} Q_k(x) Q_k^m(x) =$

$= (-1)^{n+1} 2^n \frac{(n!)^2 (n+m)!}{(2n+m+1)!} (x^2-1)^{-(n+1)/2} Q_n^{n+m+1}(x)$ $[x > 1;\ m = -n,\ -n+1,\ \ldots]$.

8. $\sum_{k=n}^{\infty} (2k+1) \frac{(k+n)!}{(k-n)!} Q_k^n(x) Q_k^m(x) = (-1)^{n+1} 2^n \frac{n!\,(2n)!\,(n+m)!}{(3n+m+1)!} \times$

$\times (x^2-1)^{-(n+1)/2} Q_n^{2n+m+1}(x)$ $[x > 1;\ m = -n,\ -n+1,\ \ldots]$.

9. $\sum_{k=n}^{\infty} (2k+1) \frac{(k+n)!}{(k-n)!} [Q_k^m(x)]^2 =$

$= \frac{(-1)^{n+1} 2^n n!\, [(n+m)!]^2}{(2n+2m+1)!} (x^2-1)^{-(n+1)/2} Q_n^{n+2m+1}(x)$ $[x > 1;\ m = -n,\ -n+1,\ \ldots]$.

10. $\sum_{k=n}^{\infty} (2k+1) \frac{(k+n)!}{(k-n)!} [Q_k(x)]^2 = \frac{2^{2n} (n!)^4}{(2n+1)!} (x^2-1)^{-n-1}$ $[x > 1]$.

11. $\sum_{k=0}^{\infty} (-1)^k (2k-2\nu+1) \frac{\Gamma(k-2\nu+1)}{k!} Q_{k-\nu}^{-\mu}(w) Q_{\nu-k-2}^{\mu}(z) =$

$= \frac{\pi\Gamma(1-\nu)}{2 \sin \nu\pi\, \Gamma(\mu+\nu)} \left(\frac{w+z}{2}\right)^{\nu-1} Q_{\nu-1}^{\mu}\left(\frac{wz+1}{w+z}\right)$

$\left[\operatorname{Re} w,\ \operatorname{Re} z,\ \operatorname{Re} \frac{wz}{wz+1} \geqslant 1;\ |w-\sqrt{w^2-1}| < |z-\sqrt{z^2-1}|\right]$.

6.5.7. Series of the form $\sum \alpha_k P_{\nu_1+k\sigma_1}^{\mu_1+k\rho_1}(x)\, Q_{\nu_2+k\sigma_2}^{\mu_2+k\rho_2}(y)$.

1. $$\sum_{k=0}^{\infty} P_k(x)\, Q_k(y) = \frac{1}{\sqrt{(x+1)(y-1)}}\,[\mathbf{K}(k) - F(\varphi, k)]$$
$$[1<x<y;\ k=\sqrt{(x-1)(y+1)}/\sqrt{(x+1)(y-1)},\quad \sin\varphi=\sqrt{y-1}/\sqrt{y+1}].$$

2. $$\sum_{k=0}^{\infty} (2k+1)\, P_k(x)\, Q_k(y) = \frac{1}{y-x} \qquad [-1\leqslant x\leqslant 1;\ y>1].$$

3. $$\sum_{k=n}^{\infty} (2k+1)\, P_k^n(x)\, Q_k(y) = n!\, \frac{(1-x^2)^{n/2}}{(y-x)^{n+1}} \qquad [-1\leqslant x\leqslant 1;\ y>1].$$

4. $$\sum_{k=0}^{\infty} (2k+1)\, P_k(x)\, Q_k^n(y) = (-1)^n\, n!\, \frac{(y^2-1)^{n/2}}{(y-x)^{n+1}} \qquad [-1\leqslant x\leqslant 1;\ y>1].$$

5. $$\sum_{k=n}^{\infty} (2k+1)\, P_k^n(x)\, Q_k^m(y) = (-1)^m (m+n)!\, \frac{(1-x^2)^{n/2}(y^2-1)^{m/2}}{(y-x)^{m+n+1}}$$
$$[-1\leqslant x\leqslant 1;\ y>1;\ m=-n,\ -n+1, \ldots].$$

6. $$\sum_{k=n}^{\infty} (2k+1)\, \frac{(k-n)!}{(k+n)!}\, P_k^n(x)\, Q_k^n(y) = (-1)^n\, \frac{(1-x^2)^{n/2}(y^2-1)^{-n/2}}{y-x}$$
$$[-1\leqslant x\leqslant 1;\ y>1].$$

7. $$\sum_{k=1}^{\infty} (\pm 1)^k P_\nu^k(x)\, Q_\nu^k(y) = \frac{1}{2} Q_\nu\left(xy \mp \sqrt{(1-x^2)(1-y^2)}\right) - \frac{1}{2} P_\nu(x)\, Q_\nu(y)$$
$$[0<x<1;\ -1<y<1].$$

8. $$\sum_{k=1}^{\infty} (\pm 1)^k P_\nu^k(x)\, Q_\nu^k(y) = \frac{1}{2} Q_\nu\left(xy \pm \sqrt{(x^2-1)(y^2-1)}\right) - \frac{1}{2} P_\nu(x) Q_\nu(y) \qquad [1<x<y].$$

9. $$\sum_{k=1}^{\infty} (-1)^k P_\nu^{2k}(x)\, Q_\nu^{-2k}(y) = \frac{1}{2} Q_\nu(xy) - \frac{1}{2} P_\nu(x)\, Q_\nu(y)$$
$$[0<x<1;\ -1<y<1] \text{ or } [1<y<x].$$

10. $$\sum_{k=1}^{\infty} (\pm 1)^k P_\nu^{-k}(x)\, Q_\nu^k(y) = \frac{1}{2} Q_\nu\left[xy \mp \sqrt{(x^2-1)(y^2-1)}\right] - \frac{1}{2} P_\nu(x)\, Q_\nu(y)$$
$$[1<y<x;\ \nu\neq -1,\ -2,\ -3, \ldots].$$

11. $$\sum_{k=0}^{\infty} (-1)^{k+1} (2k+2\nu+1)\, \frac{\Gamma(k+2\nu+1)}{k!}\, P_{k+\nu}^{\mu}(w)\, Q_{k+\nu}^{-\mu}(z) =$$
$$= \frac{\Gamma(\nu+1)}{2\Gamma(\mu-\nu)\sin\nu\pi} \left(\frac{w+z}{2}\right)^{-\nu-1} P_{-\nu-1}^{\mu}\left(\frac{wz+1}{w+z}\right)$$
$$\left[\operatorname{Re} w,\ \operatorname{Re} z,\ \operatorname{Re}\frac{wz+1}{wz}\geqslant 1;\ |z-\sqrt{z^2-1}|<|w-\sqrt{w^2-1}|\right].$$

12. $$\sum_{k=0}^{\infty} (-1)^{k+1} (2k+2\nu+1)\, \frac{\Gamma(k+2\nu+1)}{k!}\, P_{k+\nu}^{-\nu}(w)\, Q_{k+\nu}^{\nu}(z) =$$
$$= \frac{\pi \operatorname{cosec} \nu\pi}{\Gamma(-2\nu-1)}\, \frac{(w^2-1)^{\nu/2}(z^2-1)^{\nu/2}}{(w+z)^{2\nu+1}} \qquad [|z-\sqrt{z^2-1}|<|w-\sqrt{w^2-1}|].$$

6.5.8. Series of the form $\sum \alpha_k \cos(ka+b) P_{\nu_1+k\sigma_1}^{\mu_1+k\rho_1}(x) P_{\nu_2+k\sigma_2}^{\mu_2+k\rho_2}(y)$.

1. $$\sum_{k=0}^{\infty} (-1)^k \frac{\Gamma(k-\mu+1)}{\Gamma(k+\mu+1)} \cos(2k+1)a \left[P_k^{\mu}(x)\right]^2 =$$
$$= \frac{1}{2}(1-x^2)^{-1/2} P_{-\mu-1/2}\left(1-\frac{2\cos^2 a}{1-x^2}\right) \quad [\cos a<(1-x^2)^{1/2};\ \operatorname{Re}\mu<1/2].$$
$$= \frac{\cos\mu\pi}{\pi}(1-x^2)^{-1/2} Q_{-\mu-1/2}\left(\frac{2\cos^2 a}{1-x^2}-1\right) \quad [\cos a>(1-x^2)^{1/2};\ \operatorname{Re}\mu<1/2].$$

2. $$\sum_{k=0}^{\infty} (-1)^k \cos 2ka P_{k-1/2}^{\mu}(x) P_{k-1/2}^{-\mu}(x) =$$
$$= \frac{1}{2}(1-x^2)^{-1/2} P_{\mu-1/2}\left(\frac{2\cos^2 a}{1-x^2}-1\right) + \frac{1}{2} P_{-1/2}^{\mu}(x) P_{-1/2}^{-\mu}(x)$$
$$[0<\cos a<(1-x^2)^{1/2}].$$

3. $$\sum_{k=1}^{\infty} (\pm 1)^k \cos ka P_{\nu}^{k}(\cos x) P_{\nu}^{-k}(\cos y) = \frac{1}{2} P_{\nu}(\cos x\cos y \pm \sin x\sin y\cos a) -$$
$$-\frac{1}{2} P_{\nu}(\cos x) P_{\nu}(\cos y) \quad [0\leqslant x,\ y<\pi;\ x+y<\pi].$$

4. $$\sum_{k=1}^{\infty} (\pm 1)^k \cos ka P_{\nu}^{k}(w) P_{\nu}^{-k}(z) = \frac{1}{2} P_{\nu}\left(w \pm \sqrt{(z^2-1)(w^2-1)}\cos a\right) - \frac{1}{2} P_{\nu}(w) P_{\nu}(z)$$
$$[\operatorname{Re} w,\ \operatorname{Re} z>0;\ |\arg(w-1)|,\ |\arg(z-1)|<\pi].$$

5. $$\sum_{k=0}^{\infty} \cos(2k+1) a P_{\nu}^{k+1/2}(x) P_{\nu}^{-k-1/2}(x) = \frac{1}{2} P_{\nu}(x^2-(1-x^2)\cos 2a) \quad [0<x<1].$$

6. $$\sum_{k=1}^{\infty} (-1)^k \frac{\Gamma(\nu-k+1)}{\Gamma(\nu+k+1)} \cos ka P_{\nu}^{k}(w) P_{\nu}^{k}(z) =$$
$$= \frac{1}{2} P_{\nu}\left(wz-\sqrt{(w^2-1)(z^2-1)}\cos a\right) - \frac{1}{2} P_{\nu}(w) P_{\nu}(z)$$
$$[\operatorname{Re} w,\ \operatorname{Re} z>0;\ |\arg(w-1)|,\ |\arg(z-1)|<\pi].$$

6.5.9. Series of the form $\sum \alpha_k \cos ka Q_{\nu_1+k\sigma_1}^{\mu_1+k\rho_1}(x) Q_{\nu_2+k\sigma_2}^{\mu_2+k\rho_2}(y)$.

1. $$\sum_{k=1}^{\infty} (-1)^k \cos 2ka \left[Q_{k-1/2}(z)\right]^2 = \frac{\pi}{2(z^2-\sin^2 a)^{1/2}} \mathbf{K}\left(\frac{\cos a}{\sqrt{z^2-\sin^2 a}}\right) -$$
$$-\frac{1}{2} Q_{-1/2}^2(z) \quad [z>1].$$

2. $$\sum_{k=1}^{\infty} (-1)^k \frac{\Gamma(k-\mu+1/2)}{\Gamma(k+\mu+1/2)} \cos 2ka \left[Q_{k-1/2}^{\mu}(z)\right]^2 =$$
$$= \frac{\pi^2}{4}\sec\mu\pi e^{2\mu\pi i}(z^2-1)^{-1/2} P_{-\mu-1/2}\left(1+\frac{2\cos^2 a}{z^2-1}\right) - \frac{1}{2}\frac{\Gamma(1/2-\mu)}{\Gamma(1/2+\mu)}\left[Q_{-1/2}^{\mu}(z)\right]^2$$
$$[z>1].$$

3. $\sum_{k=1}^{\infty} (-1)^k \cos 2kaQ_{k-1/2}(z)\, Q_{-k-1/2}(z) =$

$$= \frac{\pi}{2}(z^2 - \sin^2 a)^{-1/2}\, \mathbf{K}\left(\frac{\cos a}{\sqrt{z^2 - \sin^2 a}}\right) - \frac{1}{2}[Q_{-1/2}(z)]^2 \quad [z>1].$$

6.5.10. Series of the form $\sum \alpha_k \cos(ka+b)\, P^{\mu_1 + k\rho_1}_{\nu_1 + k\sigma_1}(x)\, Q^{\mu_2 + k\rho_2}_{\nu_2 + k\sigma_2}(y)$.

1. $\sum_{k=0}^{\infty} (-1)^k \cos(2k+1)\, aP_k(x)\, Q_k(x) = \frac{(x^2 - \sin^2 a)^{-1/2}}{2}\, \mathbf{K}\left(\frac{\cos a}{\sqrt{x^2 - \sin^2 a}}\right) \quad [x>1].$

2. $\sum_{k=1}^{\infty} (-1)^k \cos 2ka \frac{\Gamma(k-\mu+1/2)}{\Gamma(k+\mu+1/2)} P^{\mu}_{k-1/2}(z)\, Q^{\mu}_{k-1/2}(z) =$

$$= \frac{e^{\mu\pi i}}{2}(z^2-1)^{-1/2} Q_{-\mu-1/2}\left(\frac{2\cos^2 a}{z^2-1}+1\right) - \frac{\Gamma(1/2-\mu)}{\Gamma(1/2+\mu)} P^{\mu}_{-1/2}(z)\, Q^{\mu}_{-1/2}(z) \quad [z>1].$$

3. $\sum_{k=1}^{\infty} (\pm 1)^k \cos kaP^{-k}_{\nu}(x)\, Q^{k}_{\nu}(y) = \frac{1}{2} Q_{\nu}\left(xy \pm \sqrt{(x^2-1)(y^2-1)} \cos a\right) - \frac{1}{2} P_{\nu}(x)\, Q_{\nu}(y)$

$[1<x<y;\ \operatorname{Im} a = 0;\ \nu \neq -1,\ -2,\ \ldots].$

4. $\sum_{k=1}^{\infty} (\pm 1)^k \cos kaP^{-k}_{\nu}(x)\, Q^{k}_{\nu}(y) = \frac{1}{2} Q_{\nu}\left(xy \pm \sqrt{(1-x^2)(1-y^2)} \cos a\right) -$

$$-\frac{1}{2} P_{\nu}(x)\, Q_{\nu}(y) \quad [-(1\pm 1)/2 < x < 1;\ -(1\mp 1)/2 < y < 1;\ \operatorname{Im} a = 0].$$

6.6. THE KUMMER CONFLUENT HYPERGEOMETRIC FUNCTION ${}_1F_1(a; b; x)$

To reduce series to a form occurring in this handbook, one can use the relation

$${}_1F_1(a;\ b;\ x) = e^x {}_1F_1(b-a;\ b;\ -x).$$

If the series contains only hypergeometric functions of the form ${}_1F_1(-k; b; x)$, then for the corresponding formulae after the substitution

$${}_1F_1(-k;\ b;\ x) = \frac{k!}{(b)_k} L^{b-1}_k(x)$$

see Ch. 5 of [18]. Furthermore, one can use the series for the generalized hypergeometric function ${}_pF_q((a_p); (b_q); x)$ for $p=q=1$ (see 6.8).

6.6.1. Series of the form $\sum \alpha_k\, {}_1F_1(a_k;\ b_k;\ x)$.

1. $\sum_{k=0}^{\infty} \frac{(c)_k}{k!\,(b)_k} t^k {}_1F_1\left(\begin{matrix} a; & x \\ b+k & \end{matrix}\right) = \Gamma\left[\begin{matrix} b \\ c,\ b-c \end{matrix}\right] x^{1-b} \int_0^x \tau^{c-1} (x-\tau)^{b-c-1} e^{t\tau/x} \times$

$$\times {}_1F_1\left(\begin{matrix} a; & x-\tau \\ b-c & \end{matrix}\right) d\tau \quad [0 < \operatorname{Re} c < \operatorname{Re}(b-1)].$$

2. $\sum_{k=0}^{\infty} \frac{(a)_k}{k!\,(b)_k} t^k {}_1F_1\left(\begin{matrix} a; & x \\ b+k & \end{matrix}\right) = \Phi_2(a,\ a;\ b;\ x,\ t).$

3. $\sum_{k=0}^{\infty} \frac{(b-a)_k}{k!\,(b)_k} t^k {}_1F_1\left(\begin{matrix} a; & x \\ b+k & \end{matrix}\right) = e^t {}_1F_1\left(\begin{matrix} a; & x-t \\ b & \end{matrix}\right).$

4. $$\sum_{k=0}^{\infty} \frac{(1-b)_k}{k!} t^k {}_1F_1\left(\begin{matrix} a; & x \\ b-k & \end{matrix}\right) = (1-t)^{b-1} {}_1F_1\left(\begin{matrix} a; & x-tx \\ b & \end{matrix}\right).$$

5. $$\sum_{k=0}^{\infty} \frac{(b-a)_k}{k!} t^k {}_1F_1\left(\begin{matrix} a-k \\ b;\ x \end{matrix}\right) = (1-t)^{a-b} e^x {}_1F_1\left(\begin{matrix} b-a \\ b;\ x(t-1)^{-1} \end{matrix}\right) \quad [\operatorname{Re}(1-t)^{-1} > 1/2].$$

6. $$\sum_{k=0}^{\infty} \frac{(c)_k}{k!} t^k {}_1F_1\left(\begin{matrix} a-k \\ b;\ x \end{matrix}\right) = \Gamma\left[\begin{matrix} b \\ c,\ b-c \end{matrix}\right] (1-t)^{-c} x^{1-b} \int_0^x \tau^{b-c-1} (x-\tau)^{c-1} \times$$
$$\times e^{(\tau-x)t/(1-t)} {}_1F_1\left(\begin{matrix} a; & \tau \\ b-c & \end{matrix}\right) d\tau$$

$[\operatorname{Re} c,\ \operatorname{Re}(b-c) > 0;\ |t| < 1;\ |\arg x| < 3\pi/4]$.

7. $$\sum_{k=0}^{\infty} \frac{(c)_k}{k!} {}_1F_1\left(\begin{matrix} a-k \\ b;\ x \end{matrix}\right) = \Gamma\left[\begin{matrix} b \\ b-c \end{matrix}\right] x^{-c} {}_1F_1\left(\begin{matrix} a;\ x \\ b-c \end{matrix}\right) \quad [\operatorname{Re}(b-c) > 0].$$

8. $$\sum_{k=0}^{\infty} \frac{x^{2k}}{k!\,(b)_k} {}_1F_1\left(\begin{matrix} a+b; & x \\ b+k & \end{matrix}\right) = e^{-x} {}_2F_2\left(\begin{matrix} (a+b)/2,\ (a+b-1)/2; & 4x \\ b,\ a+b-1 & \end{matrix}\right).$$

9. $$\sum_{k=0}^{\infty} (-1)^k \frac{x^{2k}}{k!\,(2-a)_k} {}_1F_1\left(\begin{matrix} 2-2a;\ x \\ 2-a+k \end{matrix}\right) = {}_1F_2\left(\begin{matrix} a/2; & -x^2 \\ 1/2,\ (3-a)/2 & \end{matrix}\right) +$$
$$+\frac{1-a}{2-a} x\, {}_1F_2\left(\begin{matrix} (a+1)/2; & -x^2 \\ 3/2; & 2-a/2 \end{matrix}\right).$$

10. $$\sum_{k=0}^{\infty} (-1)^k \frac{(a)_k (b-a)_k}{k!\,(b)_k (b)_{2k}} x^k y^k {}_1F_1\left(\begin{matrix} a+k; & x+y \\ b+2k & \end{matrix}\right) = {}_1F_1\left(\begin{matrix} a; & x \\ b & \end{matrix}\right) {}_1F_1\left(\begin{matrix} a; & y \\ b & \end{matrix}\right).$$

11. $$\sum_{k=0}^{\infty} (-1)^k \frac{(a)_k}{k!\,(b)_{2k}} x^k y^k {}_1F_1\left(\begin{matrix} a+k; & x+y \\ b+2k & \end{matrix}\right) = \Phi_2(a, a; b; x, y).$$

12. $$\sum_{k=0}^{\infty} \frac{(a)_{2k}}{k!\,(b)_k (b)_{2k}} x^k y^k {}_1F_1\left(\begin{matrix} a+2k; & x+y \\ b+2k & \end{matrix}\right) = \Psi_2(a; b, b; x, y).$$

13. $$\sum_{k=0}^{\infty} \frac{(a-c)_k (c)_k}{k!\,(b)_k (b)_{2k}} x^{2k} {}_1F_1\left(\begin{matrix} a+2k; & x \\ b+2k & \end{matrix}\right) = {}_1F_1\left(\begin{matrix} c; & x \\ b & \end{matrix}\right) {}_1F_1\left(\begin{matrix} a-c; & x \\ b & \end{matrix}\right).$$

14. $$\sum_{k=0}^{\infty} \frac{(-1)^k}{k!\,(a)_k (c+2k-1)} t^{2k} {}_1F_1\left(\begin{matrix} 1;\ x \\ c+2k \end{matrix}\right) =$$
$$= \Gamma(a)\, t^{1-a} x^{a-c} e^x \int_0^x \tau^{c-a-1} e^{-\tau} J_{a-1}\left(\frac{2t\tau}{x}\right) d\tau.$$

15. $$\sum_{k=0}^{\infty} \frac{(a)_k}{k!\,(2a)_{2k}} x^k y^k {}_1F_1\left(\begin{matrix} a+k; & x-y \\ 2a+2k & \end{matrix}\right) = e^{-y} {}_1F_1\left(\begin{matrix} a; & x+y \\ 2a & \end{matrix}\right).$$

16. $$\sum_{k=0}^{\infty}\frac{(2a-1)_k(b)_k(c)_k}{k!\,(a-1/2)_k(2a-b)_k(2a-c)_k}\left(-\frac{x}{4}\right)^k {}_1F_1\left(\begin{matrix}a+k;\ x\\ 2a+2k\end{matrix}\right)={}_2F_2\left(\begin{matrix}a,\ 2a-b-c;\ x\\ 2a-b,\ 2a-c\end{matrix}\right).$$

17. $$\sum_{k=0}^{\infty}\frac{(a)_k(b-a)_k(b-1/2)_k(1/2)_k}{k!\,(b)_k(2b-1)_{4k}}(-4x^2)^k {}_1F_1\left(\begin{matrix}2a+2k;\ x\\ 2b+4k\end{matrix}\right)={}_1F_1^2\left(\begin{matrix}a;\ x/2\\ b\end{matrix}\right).$$

6.6.2. Series of the form $\sum \alpha_k\, {}_1F_1(a_k;\ b_k;\ x)\, {}_1F_1(a_k';\ b_k';\ y)$.

1. $$\sum_{k=0}^{\infty}\frac{(a)_k(a')_k}{k!\,(b)_k(b')_k}(-x)^k {}_1F_1\left(\begin{matrix}a+k;\ x\\ b+k\end{matrix}\right){}_1F_1\left(\begin{matrix}a'+k;\ x\\ b'+k\end{matrix}\right)=e^x\,{}_2F_2\left(\begin{matrix}b-a,\ b'-a'\\ b,\ b';\ -x\end{matrix}\right).$$

2. $$\sum_{k=0}^{\infty}\frac{(a)_k}{k!\,(b)_k(b')_k}x^ky^k {}_1F_1\left(\begin{matrix}a+k;\ x\\ b+k\end{matrix}\right){}_1F_1\left(\begin{matrix}a+k;\ y\\ b'+k\end{matrix}\right)=\Psi_2(a;\ b,\ b';\ x,\ y).$$

3. $$\sum_{k=0}^{\infty}(-1)^k\frac{(a)_k}{k!\,(b)_k^2}x^{2k} {}_1F_1\left(\begin{matrix}a+k;\ x\\ b+k\end{matrix}\right){}_1F_1\left(\begin{matrix}a+k;\ -x\\ b+k\end{matrix}\right)={}_2F_3\left(\begin{matrix}a/2,\ (a+1)/2;\ -x^2\\ b/2,\ (b+1)/2,\ b\end{matrix}\right).$$

4. $$\sum_{k=0}^{\infty}\frac{(a)_k}{k!\,(b)_k(b')_k}x^{2k} {}_1F_1\left(\begin{matrix}a+k;\ x\\ b+k\end{matrix}\right){}_1F_1\left(\begin{matrix}a+k;\ x\\ b'+k\end{matrix}\right)=$$
$$={}_3F_3\left(\begin{matrix}a,\ (b+b')/2,\ (b+b'-1)/2;\ 4x\\ b,\ b',\ b+b'-1\end{matrix}\right).$$

5. $$\sum_{k=0}^{\infty}(-1)^k\frac{(a)_k}{k!\,(b)_k(2-b)_k}x^{2k} {}_1F_1\left(\begin{matrix}a+k;\ x\\ b+k\end{matrix}\right){}_1F_1\left(\begin{matrix}a+k;\ -x\\ 2-b+k\end{matrix}\right)=$$
$$={}_2F_3\left(\begin{matrix}a/2,\ (a+1)/2;\ -x^2\\ 1/2,\ (3-b)/2,\ (1+b)/2\end{matrix}\right)+\frac{a(1-b)x}{b(2-b)}\,{}_2F_3\left(\begin{matrix}1+a/2,\ (1+a)/2;\ -x^2\\ 3/2,\ 2-b/2,\ 1+b/2\end{matrix}\right).$$

6. $$\sum_{k=0}^{\infty}\frac{(2a)_k}{k!\,(a+1/2)_k^2}x^{2k} {}_1F_1^2\left(\begin{matrix}2a+k;\ x\\ a+k+1/2\end{matrix}\right)={}_1F_1\left(\begin{matrix}a;\ 4x\\ a+1/2\end{matrix}\right).$$

7. $$\sum_{k=0}^{\infty}\frac{(2a)_k}{k!\,(a)_k(a+1)_k}x^{2k} {}_1F_1\left(\begin{matrix}2a+k;\ x\\ a+k+1\end{matrix}\right){}_1F_1\left(\begin{matrix}2a+k;\ x\\ a+k\end{matrix}\right)={}_1F_1\left(\begin{matrix}a+1/2;\ 4x\\ a+1\end{matrix}\right).$$

8. $$\sum_{k=0}^{\infty}\frac{(a)_k(b-1)_k(b-a)_k}{k!\,(b-1)_{2k}(b)_{2k}}x^ky^k {}_1F_1\left(\begin{matrix}a+k;\ x\\ b+2k\end{matrix}\right){}_1F_1\left(\begin{matrix}a+k;\ y\\ b+2k\end{matrix}\right)={}_1F_1\left(\begin{matrix}a;\ x+y\\ b\end{matrix}\right).$$

9. $$\sum_{k=0}^{\infty}(-1)^k\frac{(a)_k(a')_k(b-1)_k}{k!\,(b-1)_{2k}(b)_{2k}}x^ky^k {}_1F_1\left(\begin{matrix}a+k;\ x\\ b+2k\end{matrix}\right){}_1F_1\left(\begin{matrix}a'+k;\ y\\ b+2k\end{matrix}\right)=$$
$$=\Phi_2(a,\ a';\ b;\ x,\ y).$$

10. $$\sum_{k=0}^{\infty}(-1)^k\frac{(a)_k(2a)_k(4a-1)_k}{k!\,(4a)_{2k}(4a-1)_{2k}}x^{2k} {}_1F_1\left(\begin{matrix}a+k;\ x\\ 4a+2k\end{matrix}\right){}_1F_1\left(\begin{matrix}2a+k;\ -x/2\\ 4a+2k\end{matrix}\right)=$$
$$={}_1F_2\left(\begin{matrix}a;\ x^2/16\\ 2a,\ 2a+1/2\end{matrix}\right).$$

6.6.3. Various series containing ${}_1F_1(a;b;x)$.

1. $$\sum_{k=0}^{\infty}\frac{(1-a)_k}{(2k+\delta)!}t^k\left\{\begin{matrix}\sin(k+1)\varphi\\ \cos k\varphi\end{matrix}\right\}{}_1F_1\left(\begin{matrix}a+2\delta+k;\ x\\ 2k+2\delta+1\end{matrix}\right)=$$
$$=2^{a-1-3(1-\delta)/2}\frac{\Gamma(a)}{\sqrt{\pi}}x^{-\delta/2}\cos^{\delta}\frac{\varphi}{2}\exp\left[\frac{x}{2}\left(1+\cos^2\frac{\varphi}{2}\right)\right]\times$$
$$\times\left[D_{1-2a-\delta}\left(-\sqrt{2x}\sin\frac{\varphi}{2}\right)+(-1)^{\delta}D_{1-2a-\delta}\left(\sqrt{2x}\sin\frac{\varphi}{2}\right)\right]\qquad\left[\delta=\left\{\begin{matrix}1\\0\end{matrix}\right\}\right].$$

2. $$\sum_{k=0}^{\infty}\frac{(a)_k(c-1)_k}{k!\,(c-1)_{2k}}(-y)^kI_{c+2k-1}(x)\,{}_1F_1\left(\begin{matrix}a+k;\ y\\ c+2k\end{matrix}\right)=\frac{(x/2)^{c-1}}{\Gamma(c)}\Phi_3\left(a;\ c;\ y,\ \frac{x^2}{4}\right).$$

3. $$\sum_{k=0}^{\infty}\frac{y^k}{(b)_k}L_k^{b-a-1}(x)\,{}_1F_1\left(\begin{matrix}a;\ y\\ b+k\end{matrix}\right)=\Gamma(b)(xy)^{(1-b)/2}e^yJ_{b-1}(2\sqrt{xy}).$$

4. $$\sum_{k=0}^{\infty}\frac{(b-a)_k}{(b-1)_{2k}}y^kC_k^{(b-1)/2}(x)\,{}_1F_1\left(\begin{matrix}a+k;\ y\\ b+2k\end{matrix}\right)=e^{(1+x)y/2}\,{}_1F_1\left(\begin{matrix}a-b/2;\ (1-x)y/2\\ b/2\end{matrix}\right)$$
$$[-1\leqslant x\leqslant 1].$$

5. $$\sum_{k=0}^{\infty}\frac{(a)_k(2b-a)_k(1/2)_k}{(2b)_k(4b-1)_{4k}}(2y)^{2k}C_{2k}^{2b-1/2}(x)\,{}_1F_1\left(\begin{matrix}2a+2k;\ y\\ 4b+4k\end{matrix}\right)=$$
$$={}_1F_1\left(\begin{matrix}a;\ (1+x)y/2\\ 2b\end{matrix}\right){}_1F_1\left(\begin{matrix}a;\ (1-x)y/2\\ 2b\end{matrix}\right).$$

6. $$\sum_{k=0}^{\infty}\frac{(a)_k(b-1)_k\,k!}{(c)_k(b-c)_k(b-1)_{2k}}t^kP_k^{(c-1,\,b-c-1)}(x)\,P_k^{(c-1,\,b-c-1)}(y)\,{}_1F_1\left(\begin{matrix}a+k;\ t\\ b+k\end{matrix}\right)=$$
$$=\Psi_2\left(a;\ c,\ b-c;\ \frac{t}{4}(1-x)(1-y),\ \frac{t}{4}(1+x)(1+y)\right).$$

6.7. THE GAUSS HYPERGEOMETRIC FUNCTION ${}_2F_1(a,b;c;x)$

To reduce series to a form occurring in this handbook, one can use the relations

$$\begin{aligned}{}_2F_1(a,\ b;\ c;\ x)&=(1-x)^{c-a-b}{}_2F_1(c-a,\ c-b;\ c;\ x)\\ &=(1-x)^{-a}{}_2F_1\left(a,\ c-b;\ c;\ \frac{x}{x-1}\right)\\ &=(1-x)^{-b}{}_2F_1\left(c-a,\ b;\ c;\ \frac{x}{x-1}\right).\end{aligned}$$

If the series contains only hypergeometric functions of the form ${}_2F_1(-k,b;c;x)$, the corresponding formulae then go over to classical polynomials according to the formulae of 7.3.1 see Ch. 5 of [18]. Furthermore, one can use the series for the generalized hypergeometric function ${}_pF_q((a_p);(b_q);x)$ for $p=2$, $q=1$ (see 6.8).

6.7.1. Series of the form $\sum\alpha_k\,{}_2F_1(a_k,\ b_k;\ c_k;\ x)$.

1. $$\sum_{k=0}^{\infty}\frac{(a)_k}{k!\,(c)_k}t^k{}_2F_1\left(\begin{matrix}a,\ b;\ x\\ c+k\end{matrix}\right)=\Xi_1(a,\ a,\ b;\ c;\ x,\ t).$$

2. $$\sum_{k=0}^{\infty} \frac{(a')_k (b')_k}{k!\,(c)_k} t^k {}_2F_1\begin{pmatrix} a,\ b;\ x \\ c+k \end{pmatrix} = F_3(a,\ a',\ b,\ b';\ x,\ t) \qquad [|t|,\ |x|<1].$$

3. $$\sum_{k=0}^{\infty} \frac{(c-a)_k}{k!} t^k {}_2F_1\begin{pmatrix} a-k,\ b \\ c; \quad x \end{pmatrix} = \frac{(1-t)^{a-c}}{(1-x)^b} {}_2F_1\begin{pmatrix} c-a,\ b \\ c;\ (t-1)^{-1}(1-x)^{-1}x \end{pmatrix}$$
$[|x|<1;\ \mathrm{Re}\,(1-t)^{-1}>1/2]$.

4. $$\sum_{k=0}^{\infty} \frac{(a)_k (b')_k}{k!\,(c')_k} t^k {}_2F_1\begin{pmatrix} a+k,\ b \\ c; \quad x \end{pmatrix} = F_2(a,\ b,\ b';\ c,\ c';\ x,\ t) \qquad [|t|+|x|<1].$$

5. $$\sum_{k=0}^{\infty} \frac{(a)_k (b)_k}{k!\,(c')_k} t^k {}_2F_1\begin{pmatrix} a+k,\ b+k \\ c; \quad x \end{pmatrix} = F_4(a,\ b;\ c,\ c';\ x,\ t) \qquad [\sqrt{|t|}+\sqrt{|x|}<1].$$

6. $$\sum_{k=0}^{\infty} \frac{(a)_k (b')_k}{k!\,(c)_k} t^k {}_2F_1\begin{pmatrix} a+k,\ b \\ c+k;\ x \end{pmatrix} = F_1(a,\ b,\ b';\ c;\ x,\ t) \qquad [|t|,\ |x|<1].$$

7. $$\sum_{k=0}^{\infty} \frac{(a)_k (b')_k}{k!\,(c)_k} x^k {}_2F_1\begin{pmatrix} a+k,\ b \\ c+k;\ x \end{pmatrix} = {}_2F_1\begin{pmatrix} a,\ b+b' \\ c; \quad x \end{pmatrix}.$$

8. $$\sum_{k=0}^{\infty} \frac{(a)_k (c-b)_k}{k!\,(c)_k} t^k {}_2F_1\begin{pmatrix} a+k,\ b \\ c+k;\ x \end{pmatrix} = (1-t)^{-a} {}_2F_1\begin{pmatrix} a,\ b \\ c;\ (1-t)^{-1}(x-t) \end{pmatrix}$$
$[|t|,\ |x|<1]$.

9. $$\sum_{k=0}^{\infty} \frac{(c-a)_k}{k!\,(c)_k} t^k {}_2F_1\begin{pmatrix} a+k,\ b \\ c+k;\ x \end{pmatrix} = (1-x)^{-b}\, \Xi_1\left(a,\ a,\ b;\ c;\ \frac{x}{x-1},\ t\right).$$

10. $$\sum_{k=0}^{\infty} \frac{(a)_k (b)_k}{k!\,(c)_k} t^k {}_2F_1\begin{pmatrix} a+k,\ b+k \\ c+k; \quad x \end{pmatrix} = {}_2F_1\begin{pmatrix} a,\ b \\ c;\ t+x \end{pmatrix} \qquad [|t|,\ |x|,\ |t+x|<1].$$

11. $$\sum_{k=0}^{\infty} \frac{(b)_k (b')_k}{k!\,(c)_k} t^k {}_2F_1\begin{pmatrix} a+k,\ b+k \\ c+k; \quad x \end{pmatrix} = F_1(b,\ b',\ a-b';\ c;\ x,\ x-t)$$
$[|t|,\ |x|,\ |t-x|<1]$.

12. $$\sum_{k=0}^{\infty} (-1)^k \frac{(a)_k (b)_k}{k!\,(a+b+1/2)_k} \left(\frac{x}{2}\right)^{2k} {}_2F_1\begin{pmatrix} a+k,\ b+k \\ a+b+k+1/2;\ x \end{pmatrix} = {}_2F_1\begin{pmatrix} 2a,\ 2b \\ a+b+1/2;\ x/4 \end{pmatrix}.$$

13. $$\sum_{k=0}^{\infty} \frac{(a)_k (b)_k (c-a)_k}{k!\,(c)_{2k}} x^k y^k {}_2F_1\begin{pmatrix} a+k,\ b+k \\ c+2k;\ x+y-xy \end{pmatrix} = F_1(a,\ b,\ b;\ c;\ x,\ y)$$
$[|x|,\ |y|,\ |x+y-xy|<1]$.

14. $$\sum_{k=0}^{\infty} \frac{(a)_k (b)_k (c-a-b)_k}{k!\,(c)_{2k}} x^k y^k {}_2F_1\begin{pmatrix} a+k,\ b+k \\ c+2k;\ x+y-xy \end{pmatrix} = F_3(a,\ a,\ b,\ b;\ c;\ x,\ y)$$
$[|x|,\ |y|,\ |x+y-xy|<1]$.

15. $$\sum_{k=0}^{\infty} \frac{(a)_k (b)_k (c-a)_k (c-b)_k}{k!\,(c+1/2)_k (c)_{2k}} x^{2k} {}_2F_1\begin{pmatrix} a+k,\ b+k \\ c+2k;\ 2x-x^2 \end{pmatrix} = {}_2F_1\begin{pmatrix} 2a,\ 2b \\ 2c; \quad x \end{pmatrix}.$$

16. $$\sum_{k=0}^{\infty} \frac{(a)_k (b)_k (c-a)_k (c-b)_k}{k!\,(c)_k (c)_{2k}} x^k y^k {}_2F_1\begin{pmatrix} a+k,\ b+k \\ c+2k;\ x+y-xy \end{pmatrix} =$$
$$= {}_2F_1\begin{pmatrix} a,\ b \\ c;\ x \end{pmatrix} {}_2F_1\begin{pmatrix} a,\ b \\ c;\ y \end{pmatrix} \qquad [|x|,\ |y|<1;\ |x+y| \text{ or } |x+y-xy|<1].$$

17. $\sum_{k=0}^{\infty}(-1)^k\frac{(a)_k(b)_k(c-a)_k}{k!\,(c)_{2k}}x^{2k}{}_2F_1\left(\begin{matrix}a+k,\ 2b+2k\\c+2k;\ x\end{matrix}\right)={}_2F_1\left(\begin{matrix}a,\ b\\c;\ 2x-x^2\end{matrix}\right).$

18. $\sum_{k=0}^{\infty}\frac{(a)_k(b)_{2k}}{k!\,(c)_{2k}}x^ky^k{}_2F_1\left(\begin{matrix}a+k,\ b+2k\\c+2k;\ x-y\end{matrix}\right)=F_1(b,\ a,\ a;\ c;\ x,\ -y)$

$[|x|+|y|+|xy|<1].$

19. $\sum_{k=0}^{\infty}\frac{(a)_k(c-a)_k(b)_{2k}}{k!\,(c)_k(c)_{2k}}x^ky^k{}_2F_1\left(\begin{matrix}a+k,\ b+2k\\c+2k;\ x-y\end{matrix}\right)=F_2(b,\ a,\ a;\ c,\ c;\ x,\ -y)$

$[|x|+|y|+|xy|<1].$

20. $\sum_{k=0}^{\infty}(-1)^k\frac{(a)_k(b-b')_k(b')_k(c-a)_k}{k!\,(c)_k(c)_{2k}}x^{2k}{}_2F_1\left(\begin{matrix}a+k,\ b+2k\\c+2k;\ x\end{matrix}\right)=$

$={}_2F_1\left(\begin{matrix}a,\ b-b'\\c;\ x\end{matrix}\right){}_2F_1\left(\begin{matrix}a,\ b'\\c;\ x\end{matrix}\right).$

21. $\sum_{k=0}^{\infty}\frac{(a)_{2k}(b)_{2k}}{k!\,(c)_k(c)_{2k}}x^ky^k{}_2F_1\left(\begin{matrix}a+2k,\ b+2k\\c+2k;\ x+y\end{matrix}\right)=F_4(a,\ b;\ c,\ c;\ x,\ y)$

$[\sqrt{|x|}+\sqrt{|y|}<1].$

22. $\sum_{k=0}^{\infty}\frac{(a)_k(b)_k(a+b-c+1/2)_k}{k!\,(c)_{2k}}(2x)^{2k}{}_2F_1\left(\begin{matrix}2a+2k,\ 2b+2k\\c+2k;\ x\end{matrix}\right)={}_2F_1\left(\begin{matrix}a,\ b\\c;\ 4x-4x^2\end{matrix}\right).$

23. $\sum_{k=0}^{\infty}(-1)^k\frac{(a)_k(b_k)(c-a)_k(c-b)_k}{k!\,(c+k-1/2)_k(c)_{2k}}x^{2k}{}_2F_1\left(\begin{matrix}2a+2k,\ 2b+2k\\2c+4k;\ x\end{matrix}\right)={}_2F_1\left(\begin{matrix}a,\ b\\c;\ 2x-x^2\end{matrix}\right).$

24. $\sum_{k=0}^{\infty}\frac{(a)_k(b)_k(c-a)_k(c-b)_k(1/2)_k}{k!\,(c)_k(c+k-1/2)_k(c)_{2k}}x^{2k}{}_2F_1\left(\begin{matrix}2a+2k,\ 2b+2k\\2c+4k;\ x\end{matrix}\right)={}_2F_1^2\left(\begin{matrix}a,\ b\\c;\ x\end{matrix}\right).$

25. $\sum_{k=0}^{\infty}\frac{(a)_k(b+1)_k(c-a)_k(c-b)_k(1/2)_k}{k!\,(c)_k(c+k-1/2)_k(c)_{2k}}x^{2k}{}_2F_1\left(\begin{matrix}2a+2k,\ 2b+2k+1\\2c+4k;\ x\end{matrix}\right)=$

$={}_2F_1\left(\begin{matrix}a,\ b\\c;\ x\end{matrix}\right){}_2F_1\left(\begin{matrix}a,\ b+1\\c;\ x\end{matrix}\right).$

6.7.2. Series of the form $\sum\alpha_k\,{}_2F_1\left(\begin{matrix}a_k,\ b_k\\c_k;\ x\end{matrix}\right){}_2F_1\left(\begin{matrix}a_k',\ b_k'\\c_k';\ y\end{matrix}\right)$.

1. $\sum_{k=0}^{\infty}\frac{(a)_k(b)_k(b')_k}{k!\,(c)_k(c')_k}x^ky^k{}_2F_1\left(\begin{matrix}a+k,\ b+k\\c+k;\ x\end{matrix}\right){}_2F_1\left(\begin{matrix}a+k,\ b+k\\c+k;\ y\end{matrix}\right)=$

$=F_2(a,b,b';\ c,\ c';\ x,\ y)\qquad[|x|+|y|<1].$

2. $\sum_{k=0}^{\infty}\frac{(a)_k(b)_k(a+b-c-c'+1)_k}{k!\,(c)_k(c')_k}x^ky^k{}_2F_1\left(\begin{matrix}a+k,\ b+k\\c+k;\ x\end{matrix}\right){}_2F_1\left(\begin{matrix}a+k,\ b+k\\c'+k;\ y\end{matrix}\right)=$

$=F_4(a,\ b;\ c,\ c';\ x-xy,\ y-xy)\qquad[\sqrt{|x|(1+|y|)}+\sqrt{|y|(1+|x|)}<1].$

3. $\sum_{k=0}^{\infty}\frac{(a)_k(b)_k(b')_k(c-a)_k}{k!\,(c+k-1)_k(c)_{2k}}x^{2k}{}_2F_1\left(\begin{matrix}a+k,\ b+k\\c+2k;\ x\end{matrix}\right){}_2F_1\left(\begin{matrix}a+k,\ b'+k\\c+2k;\ x\end{matrix}\right)=$

$={}_2F_1\left(\begin{matrix}a,\ b+b'\\c;\ x\end{matrix}\right).$

4. $$\sum_{k=0}^{\infty} \frac{(a)_k (b)_k (b')_k (c-a)_k}{k!\,(c+k-1)_k (c)_{2k}} x^k y^k\, {}_2F_1\left(\begin{matrix} a+k,\ b+k \\ c+2k;\ x \end{matrix}\right) {}_2F_1\left(\begin{matrix} a+k,\ b'+k \\ c+2k;\ y \end{matrix}\right) = F_1(a,\ b,\ b';\ c;\ x,\ y) \qquad [|x|,\ |y|<1].$$

5. $$\sum_{k=0}^{\infty} (-1)^k \frac{(a)_k (b)_k (c-a)_k (c-b)_k}{k!\,(c+k-1)_k (c)_{2k}} x^k y^k\, {}_2F_1\left(\begin{matrix} a+k,\ b+k \\ c+2k;\ x \end{matrix}\right) {}_2F_1\left(\begin{matrix} a+k,\ b+k \\ c+2k;\ y \end{matrix}\right) = {}_2F_1\left(\begin{matrix} a,\ b \\ c;\ x+y-xy \end{matrix}\right) \qquad [|x|,\ |y|,\ |x+y-xy|<1].$$

6. $$\sum_{k=0}^{\infty} (-1)^k \frac{(a)_k (b)_k (c)_k}{k!\,(2a)_k (2b)_k} x^{2k}\, {}_2F_1\left(\begin{matrix} a+k,\ c+k \\ 2a+k;\ x \end{matrix}\right) {}_2F_1\left(\begin{matrix} b+k,\ c+k \\ 2b+k;\ -x \end{matrix}\right) = {}_4F_3\left(\begin{matrix} (a+b)/2,\ (a+b+1)/2,\ c/2,\ (c+1)/2 \\ a+1/2,\ b+1/2,\ a+b;\ x^2 \end{matrix}\right).$$

7. $$\sum_{k=0}^{\infty} (-1)^k \frac{(a)_k (b)_k^2}{k!\,(c)_k^2} x^{2k}\, {}_2F_1\left(\begin{matrix} a+k,\ b+k \\ c+k;\ x \end{matrix}\right) {}_2F_1\left(\begin{matrix} a+k,\ b+k \\ c+k;\ -x \end{matrix}\right) = {}_4F_3\left(\begin{matrix} a/2,\ (a+1)/2,\ b,\ c-b \\ c,\ c/2,\ (c+1)/2;\ x^2 \end{matrix}\right).$$

8. $$\sum_{k=0}^{\infty} (-1)^k \frac{(a)_k (a')_k (b)_k (b')_k}{k!\,(c+k-1)_k (c)_{2k}} x^k y^k\, {}_2F_1\left(\begin{matrix} a+k,\ b+k \\ c+2k;\ x \end{matrix}\right) {}_2F_1\left(\begin{matrix} a'+k,\ b'+k \\ c+2k;\ y \end{matrix}\right) = F_3(a,\ a',\ b,\ b';\ c;\ x,\ y) \qquad [|x|,\ |y|<1].$$

9. $$\sum_{k=0}^{\infty} \frac{(a)_k (b)_k (c-a)_k (c-b)_k (-1/2)_k}{k!\,(c+1/2)_k (c+k-1)_k (c)_{2k}} x^{2k}\, {}_2F_1^2\left(\begin{matrix} a+k,\ b+k \\ c+2k;\ x \end{matrix}\right) = {}_2F_1\left(\begin{matrix} 2a,\ 2b \\ 2c;\ x \end{matrix}\right).$$

10. $$\sum_{k=0}^{\infty} \frac{(a)_k (b+1)_k (c-a)_k (c-b)_k (-1/2)_k}{k!\,(c+1/2)_k (c+k-1)_k (c)_{2k}} x^{2k}\, {}_2F_1\left(\begin{matrix} a+k,\ b+k \\ c+2k;\ x \end{matrix}\right) \times {}_2F_1\left(\begin{matrix} a+k,\ b+k+1 \\ c+2k;\ x \end{matrix}\right) = {}_2F_1\left(\begin{matrix} 2a,\ 2b+1 \\ 2c;\ x \end{matrix}\right).$$

11. $$\sum_{k=0}^{\infty} \frac{(a)_k^2 (b)_k (b-1)_k (c-a-b+1/2)_k}{k!\,(a+b-1/2)_k (c+k-1)_k (c)_{2k}} x^{2k}\, {}_2F_1\left(\begin{matrix} a+k,\ b+k \\ c+2k;\ x \end{matrix}\right) \times {}_2F_1\left(\begin{matrix} a+k,\ b+k-1 \\ c+2k;\ x \end{matrix}\right) = {}_3F_2\left(\begin{matrix} 2a,\ 2b-1,\ a+b-1 \\ 2a+2b-2,\ c;\ x \end{matrix}\right).$$

12. $$\sum_{k=0}^{\infty} \frac{(a)_k^2 (b)_k (b-1)_k (-1/2)_k}{k!\,(a+b-1/2)_k (a+b+k-2)_k (a+b-1)_{2k}} x^{2k}\, {}_2F_1\left(\begin{matrix} a+k,\ b+k \\ a+b+2k-1;\ x \end{matrix}\right) \times {}_2F_1\left(\begin{matrix} a+k,\ b+k-1 \\ a+b+2k-1;\ x \end{matrix}\right) = {}_2F_1\left(\begin{matrix} 2a,\ 2b-1 \\ 2a+2b-2;\ x \end{matrix}\right).$$

13. $$\sum_{k=0}^{\infty} \frac{(a)_k^2 (b)_k (b-1)_k (a-b+1/2)_k}{k!\,(a+b-1/2)_k (2a+k-1)_k (2a)_{2k}} x^{2k}\, {}_2F_1\left(\begin{matrix} a+k,\ b+k \\ 2a+2k;\ x \end{matrix}\right) \times {}_2F_1\left(\begin{matrix} a+k,\ b+k-1 \\ 2a+2k;\ x \end{matrix}\right) = {}_2F_1\left(\begin{matrix} 2b-1,\ a+b-1 \\ 2a+2b-2;\ x \end{matrix}\right).$$

14. $$\sum_{k=0}^{\infty} \frac{(a)_k^2 (b)_k (b-1)_k (2a+2b-3)_k}{k!\,(2a+2b-2)_{2k}^2} x^{2k}\, {}_2F_1\left(\begin{matrix} a+k,\ b+k;\ x \\ 2a+2b+2k-2 \end{matrix}\right) \times {}_2F_1\left(\begin{matrix} a+k,\ b+k-1;\ x \\ 2a+2b+2k-2 \end{matrix}\right) = {}_3F_2\left(\begin{matrix} 2a,\ 2b-1,\ a+b-1;\ x \\ 2a+2b-2,\ 2a+2b-2 \end{matrix}\right).$$

15. $$\sum_{k=0}^{\infty} \frac{(k+1)\,(a)_k^2\,(b-1)_k}{(2a+2b+k-2)_k\,(2a+2b-1)_{2k}}\, x^{2k}\, {}_2F_1\left(\begin{matrix}a+k,\ b+k;\ x\\ 2a+2b+2k-1\end{matrix}\right)\times$$
$$\times\, {}_2F_1\left(\begin{matrix}a+k,\ b+k-1;\ x\\ 2a+2b+2k-1\end{matrix}\right) = {}_3F_2\left(\begin{matrix}2a,\ 2b-1,\ a+b-1;\ x\\ 2a+2b-1,\ 2a+2b-2\end{matrix}\right).$$

16. $$\sum_{k=0}^{\infty} \frac{(a)_k^2\,(b)_k^2\,(2a+2b-1)_k}{k!\,(2a+2b)_{2k}^2}\, x^{2k}\, {}_2F_1^2\left(\begin{matrix}a+k,\ b+k;\ x\\ 2a+2b+2k\end{matrix}\right) = {}_3F_2\left(\begin{matrix}2a,\ 2b,\ a+b;\ x\\ 2a+2b,\ 2a+2b\end{matrix}\right).$$

17. $$\sum_{k=0}^{\infty} \frac{(a)_k^2\,(b)_k^2}{k!\,(2a+2b+k)_k\,(2a+2b+1)_{2k}}\, x^{2k}\, {}_2F_1^2\left(\begin{matrix}a+k,\ b+k;\ x\\ 2a+2b+2k+1\end{matrix}\right) =$$
$$= {}_3F_2\left(\begin{matrix}2a,\ 2b,\ a+b;\ x\\ 2a+2b,\ 2a+2b+1\end{matrix}\right).$$

6.7.3. Various series containing ${}_2F_1(a, b; c; x)$.

1. $$\sum_{k=1}^{\infty} \frac{(a)_k}{k!}\, z^{k/2} \cos k\varphi\, {}_2F_1\left(\begin{matrix}a+k,\ a\\ 1+k;\ z\end{matrix}\right) = \frac{1}{2}\left(1-2z^{1/2}\cos\varphi+z\right)^{-a} - \frac{1}{2}\, {}_2F_1\left(\begin{matrix}a,\ a\\ 1;\ z\end{matrix}\right).$$

2. $$\sum_{k=0}^{\infty} (-1)^k \frac{(a)_k\,(b_k)\,(c-1)_k}{k!\,(c-1)_{2k}}\, y^k I_{c+2k-1}(x)\, {}_2F_1\left(\begin{matrix}a+k,\ b+k\\ c+2k;\ y\end{matrix}\right) =$$
$$= \frac{(x/2)^{c-1}}{\Gamma(c)}\, \Xi_2\left(a,\ b;\ c;\ y,\ \frac{x^2}{4}\right).$$

3. $$\sum_{k=0}^{\infty} \frac{(a)_k}{(b-1)_k}\, y^{k/2} C_k^{b-1}(x)\, {}_2F_1\left(\begin{matrix}a+k,\ a-b+1\\ b+k;\ y\end{matrix}\right) = \left(1-2x\sqrt{y}+y\right)^{-a}$$
$[\operatorname{Re} a > -1]$.

4. $$\sum_{k=0}^{\infty} (-1)^k \frac{(a)_k}{(\mu+1)_k}\, P_k^{(\mu,\ \nu-k)}(x)\, {}_2F_1\left(\begin{matrix}a+k,\ b\\ c;\ y\end{matrix}\right) =$$
$$= 2^{-a}\, F_2\left(a,\ b,\ \mu+\nu+1;\ c,\ \mu;\ \frac{y}{2},\ \frac{1-x}{4}\right).$$

5. $$\sum_{k=0}^{\infty} \frac{(a)_k\,(b)_k}{(\mu+1)_k\,(\nu+1)_k}\, t^k P_k^{(\mu,\ \nu)}(x)\, {}_2F_1\left(\begin{matrix}a+k,\ b+k\\ c;\ z\end{matrix}\right) =$$
$$= F_C^3\left(a,\ b;\ c,\ \mu+1,\ \nu+1;\ z,\ \frac{x-1}{2}t,\ \frac{x+1}{2}t\right)$$
$$\left[|z|^{1/2} + \left|\frac{x-1}{2}t\right|^{1/2} + \left|\frac{x+1}{2}t\right|^{1/2} < 1\right].$$

6. $$\sum_{k=0}^{\infty} (-1)^k \frac{(a)_k}{(c)_k}\, y^k P_k^{(b-1,\ \nu-k)}(x)\, {}_2F_1\left(\begin{matrix}a+k,\ b+k\\ c+k;\ y\end{matrix}\right) = {}_2F_1\left(\begin{matrix}a,\ b+\gamma\\ c;\ (1-x)\,y/2\end{matrix}\right).$$

7. $$\sum_{k=0}^{\infty} \frac{(a)_k}{(c+k-1)_k}\, y^k P_k^{(d,\ b-1)}(x)\, {}_2F_1\left(\begin{matrix}a+k,\ b+k\\ c+2k;\ y\end{matrix}\right) = \left(1-\frac{1+x}{2}y\right)^{-a}$$
$[-1 \leqslant x \leqslant 1;\ y \neq 1;\ |\arg(1-y)| < \pi]$.

8. $$\sum_{k=0}^{\infty} \frac{(a)_k (b)_k}{k!\,(c)_k (c')_k} x^k y^k \, {}_1F_1\left(\begin{matrix}a+k\\ c'+k;\ x\end{matrix}\right) {}_2F_1\left(\begin{matrix}a+k,\ b+k\\ c+k;\ y\end{matrix}\right) = \Psi_1(a,\ b,\ c,\ c';\ y,\ x).$$

9. $$\sum_{k=0}^{\infty} (-1)^k \frac{(a)_k (a')_k (b)_k}{k!\,(c+k-1)_k (c)_{2k}} x^k y^k \, {}_1F_1\left(\begin{matrix}a'+k\\ c+2k;\ x\end{matrix}\right) {}_2F_1\left(\begin{matrix}a+k,\ b+k\\ c+2k;\ y\end{matrix}\right) =$$
$$= \Xi_1(a,\ a',\ b;\ c;\ y,\ x).$$

10. $$\sum_{k=0}^{\infty} \frac{(a)_k (b)_k (c-a)_k}{k!\,(c+k-1)_k (c)_{2k}} x^k y^k \, {}_1F_1\left(\begin{matrix}a+k\\ c+2k;\ x\end{matrix}\right) {}_2F_1\left(\begin{matrix}a+k,\ b+k\\ c+2k;\ y\end{matrix}\right) = \Phi_1(a, b;\ c;\ y,\ x).$$

6.8. THE GENERALIZED HYPERGEOMETRIC FUNCTION ${}_pF_q((a_p); (b_q); x)$

6.8.1. Series of the form $\sum \alpha_k t^k \, {}_pF_q((a_p) \pm k\,(c_p);\ (b_q) \pm k\,(d_q);\ x)$.

1. $$\sum_{k=1}^{\infty} \frac{t^k}{k} \, {}_{p+1}F_q\left(\begin{matrix}-k,\ (a_p)\\ (b_q);\ x\end{matrix}\right) =$$
$$= \frac{tx}{t-1}\left(\prod_{i=1}^{p} a_i\right)\left(\prod_{j=1}^{q} b_j\right)^{-1} {}_{p+2}F_{q+1}\left(\begin{matrix}1,\ 1,\ (a_p)+1\\ 2,\ (b_q)+1;\ (t-1)^{-1}tx\end{matrix}\right) - \ln(1-t)$$
$[|t| < 1]$.

2. $$\sum_{k=0}^{\infty} \frac{t^k}{k!} \, {}_{p+1}F_q\left(\begin{matrix}-k,\ (a_p)\\ (b_q);\ x\end{matrix}\right) = e^t \, {}_pF_q\left(\begin{matrix}(a_p);\ -tx\\ (b_q)\end{matrix}\right).$$

3. $$\sum_{k=0}^{\infty} \frac{(\alpha)_k}{k!} t^k \, {}_{p+1}F_q\left(\begin{matrix}-k,\ (a_p)\\ (b_q);\ x\end{matrix}\right) = (1-t)^{-\alpha} \, {}_{p+1}F_q\left(\begin{matrix}(a_p),\ \alpha\\ (b_q);\ t(t-1)^{-1}x\end{matrix}\right) \qquad [|t| < 1].$$

4. $$\sum_{k=0}^{\infty} \frac{(k\alpha+1)^k}{k!} t^k \, {}_{p+1}F_q\left(\begin{matrix}-k,\ (a_p)\\ (b_q);\ (k\alpha+1)^{-1}x\end{matrix}\right) = \frac{e^u}{1-\alpha u} \, {}_pF_q\left(\begin{matrix}(a_p);\ -ux\\ (b_q)\end{matrix}\right)$$
$[t = ue^{-\alpha u};\ |\alpha u e^{1-\alpha u}| < 1]$.

5. $$\sum_{k=0}^{\infty} \frac{(k\alpha+1)^{k-1}}{k!} t^k \, {}_{p+1}F_q\left(\begin{matrix}-k,\ (a_p)\\ (b_q);\ (k\alpha+1)^{-1}x\end{matrix}\right) = e^u \, {}_{p+1}F_{q+1}\left(\begin{matrix}(a_p),\ 1/\alpha;\ -ux\\ (b_q),\ 1+1/\alpha\end{matrix}\right)$$
$[t = ue^{-\alpha u};\ |\alpha u e^{1-\alpha u}| < 1]$.

6. $$\sum_{k=0}^{\infty} \frac{(\alpha)_k}{k!} t^k \, {}_{p+2}F_q\left(\begin{matrix}-k,\ \alpha+k,\ (a_p)\\ (b_q);\ x\end{matrix}\right) = (1-t)^{-\alpha} \, {}_{p+2}F_q\left(\begin{matrix}(a_p),\ \alpha/2,\ (\alpha+1)/2\\ (b_q);\ -4t(1-t)^{-2}x\end{matrix}\right)$$
$[|t| < 1]$.

7. $$\sum_{k=0}^{\infty} \frac{(\beta)_{2k}}{k!\,(\beta)_k} t^k \, {}_{p+1}F_{q+1}\left(\begin{matrix}-k,\ (a_p);\ x\\ \beta+k,\ (b_q)\end{matrix}\right) =$$
$$= \frac{1}{\sqrt{1-4t}}\left(\frac{2}{1+\sqrt{1-4t}}\right)^{\beta-1} {}_pF_q\left(\begin{matrix}(a_p);\ -4t(1+\sqrt{1-4t})^{-2}x\\ (b_q)\end{matrix}\right) \qquad [|t| < 1/4].$$

8. $$\sum_{k=0}^{\infty} \frac{(k\alpha+\beta)_k}{k!} t^k \, {}_{p+1}F_{q+1}\left(\begin{matrix}-k,\ (a_p);\ x\\ k\alpha+\beta,\ (b_q)\end{matrix}\right) = \frac{(1+v)^{\beta}}{1-\alpha v} \, {}_pF_q\left(\begin{matrix}(a_p);\ -xv\\ (b_q)\end{matrix}\right)$$
$[t = v(1+v)^{-\alpha-1},\ v(0) = 0;\ |t| < 1]$.

9. $$\sum_{k=1}^{\infty} \frac{t^k}{k} {}_{p+2}F_q\left(\begin{matrix}-k/2,\ (1-k)/2,\ (a_p)\\ (b_q);\ x\end{matrix}\right)=$$
$$=\left(\prod_{i=1}^{p} a_i\right)\left(\prod_{j=1}^{q} b_j\right)^{-1} \frac{t^2x}{4(1-t)^2} {}_{p+3}F_{q+1}\left(\begin{matrix}1,\ 1,\ 3/2,\ (a_p)+1\\ 2,\ (b_q)+1;\ t^2(1-t)^{-2}x\end{matrix}\right) - \ln(1-t)$$
$[|t|<1]$.

10. $$\sum_{k=0}^{\infty} \frac{t^k}{k!} {}_{p+2}F_q\left(\begin{matrix}-k/2,\ (1-k)/2,\ (a_p)\\ (b_q);\ x\end{matrix}\right)=e^t {}_pF_q\left(\begin{matrix}(a_p);\ t^2x/4\\ (b_q)\end{matrix}\right).$$

11. $$\sum_{k=0}^{\infty} \frac{(\alpha)_k}{k!} t^k {}_{p+2}F_q\left(\begin{matrix}-k/2,\ (1-k)/2,\ (a_p)\\ (b_q);\ x\end{matrix}\right)=(1-t)^{-\alpha} {}_{p+2}F_q\left(\begin{matrix}(a_p),\ \alpha/2,\ (\alpha+1)/2\\ (b_q);\ t^2(1-t)^{-2}x\end{matrix}\right)$$
$[|t|<1]$.

12. $$\sum_{k=0}^{\infty} \frac{t^k}{k!} \frac{\Pi(b_q)_k}{\Pi(a_p)_k} {}_{p+r+1}F_{q+s}\left(\begin{matrix}-k,\ 1-k-(a_p),\ (c_r)\\ 1-k-(b_q),\ (d_s);\ x\end{matrix}\right)=$$
$$={}_qF_p\left(\begin{matrix}(b_q);\ t\\ (a_p)\end{matrix}\right) {}_rF_s\left(\begin{matrix}(c_r);\ (-1)^{p+q+1}tx\\ (d_s)\end{matrix}\right)$$ $[|t|<1]$.

13. $$\sum_{k=0}^{\infty} \frac{(\alpha)_k t^k}{k!} \frac{\Pi(b_q)_k}{\Pi(a_p)_k} {}_{p+1}F_q\left(\begin{matrix}-k,\ 1-k-(a_p)\\ 1-k-(b_q);\ x\end{matrix}\right)=u^{\alpha} {}_{q+1}F_p\left(\begin{matrix}(b_q),\ \alpha\\ (a_p);\ t/u\end{matrix}\right)$$
$[u=1+(-1)^{p+q}tx]$.

14. $$\sum_{k=0}^{\infty} \frac{t^k}{k!} \frac{\Pi(b_q)_k}{\Pi(a_p)_k} {}_{p+2}F_q\left(\begin{matrix}-k,\ \alpha+k,\ 1-k-(a_p)\\ 1-k-(b_q);\ x\end{matrix}\right)=$$
$$=\frac{1}{v}\left(\frac{1+v}{2}\right)^{1-\alpha} {}_qF_p\left(\begin{matrix}(b_q);\ x^{-1}(1-v)/2\\ (a_p)\end{matrix}\right)$$ $[v=(1+4(-1)^{p+q}tx)^{1/2}]$.

15. $$\sum_{k=0}^{\infty} \frac{t^k}{k!} \frac{\Pi(1-b_q)_k}{\Pi(1-a_p)_k} {}_{p+2}F_q\left(\begin{matrix}-k,\ \alpha+nk,\ (a_p)-k\\ (b_q)-k;\ x\end{matrix}\right)=$$
$$=\frac{(1+u)^{1-\alpha}}{1+(n+1)u} {}_qF_p\left(\begin{matrix}1-(b_q);\ t(1+u)^{-n}\\ 1-(a_p)\end{matrix}\right)$$ $[u(1+u)^n=(-1)^{p+q}tx;\ n\geqslant 1]$.

16. $$\sum_{k=0}^{\infty} \frac{(\alpha)_k}{k!} t^k {}_{p+r+1}F_{q+r+1}\left(\begin{matrix}-k,\ \Delta(r,\ k+\alpha),\ (a_p)\\ \Delta(r+1,\ \alpha),\ (b_q);\ r^rx\end{matrix}\right)=$$
$$=(1-t)^{-\alpha} {}_pF_q\left(\begin{matrix}(a_p);\ -(r+1)^{r+1}t(1-t)^{-r-1}x\\ (b_q)\end{matrix}\right)$$ $[|t|<1]$.

17. $$\sum_{k=0}^{\infty} \frac{(\alpha)_k}{k!} t^k {}_{p+1}F_q\left(\begin{matrix}(a_p),\ \alpha+k\\ (b_q);\ x\end{matrix}\right)=(1-t)^{-\alpha} {}_{p+1}F_q\left(\begin{matrix}(a_p),\ \alpha\\ (b_q);\ (1-t)^{-1}x\end{matrix}\right)$$
$[|t|<1;\ |x|<1$ for $p\leqslant q]$.

18. $$\sum_{k=0}^{\infty} \frac{(\beta)_k}{k!} {}_{p+1}F_q\left(\begin{matrix}(a_p),\ \alpha-k\\ (b_q);\ x\end{matrix}\right)=$$
$$=\left(\prod_{i=1}^{p}\Gamma(a_i-\beta)\right)\left(\prod_{j=1}^{q}\Gamma(b_j-\beta)\right)^{-1} x^{-\beta} {}_{p+1}F_q\left(\begin{matrix}(a_p)-\beta,\ \alpha\\ (b_q)-\beta;\ x\end{matrix}\right)$$
$[\operatorname{Re}\beta<0;\ \operatorname{Re}b_j>\operatorname{Re}\beta;\ j=1,\ \ldots,\ q]$.

19. $$\sum_{k=0}^{\infty} \frac{t^k}{k!} \frac{\Pi(a_p)_k}{\Pi(b_q)_k} {}_pF_q\left(\begin{matrix}(a_p)+k;\ x\\(b_q)+k\end{matrix}\right) = {}_pF_q\left(\begin{matrix}(a_p);\ t+x\\(b_q)\end{matrix}\right).$$

20. $$\sum_{k=0}^{\infty} \frac{x^k}{k!\,(\alpha)_k} \frac{\Pi(a_p)_k}{\Pi(b_q)_k} {}_pF_{q+1}\left(\begin{matrix}(a_p)+k;\ x\\(b_q)+k,\ \beta\end{matrix}\right) = {}_{p+2}F_{q+3}\left(\begin{matrix}(a_p),\ (\alpha+\beta-1)/2,\ (\alpha+\beta)/2;\ 4x\\(b_q),\ \alpha,\ \beta,\ \alpha+\beta-1\end{matrix}\right).$$

21. $$\sum_{k=0}^{\infty} \frac{(\alpha)_k(-x)^k}{k!\,(\beta)_k} \frac{\Pi(a_p)_k}{\Pi(b_q)_k} {}_pF_q\left(\begin{matrix}(a_p)+k;\ x\\(b_q)+k\end{matrix}\right) = {}_{p+1}F_{q+1}\left(\begin{matrix}(a_p),\ \beta-\alpha;\ x\\(b_q),\ \beta\end{matrix}\right)$$

$[p \leqslant q]$ or $[p=q+1;\ |x|<1]$.

22. $$\sum_{k=0}^{\infty} \frac{(\beta)_k}{k!} x^k \frac{\Pi(a_p)_k}{\Pi(b_q)_k} {}_{p+1}F_q\left(\begin{matrix}(a_p)+k,\ \alpha\\(b_q)+k;\ x\end{matrix}\right) = {}_{p+1}F_q\left(\begin{matrix}(a_p),\ \alpha+\beta\\(b_q);\ x\end{matrix}\right).$$

23. $$\sum_{k=0}^{\infty} \frac{(\beta-\alpha)_k}{k!\,(\beta)_k} (-x)^k {}_pF_q\left(\begin{matrix}(a_p)+k;\ x\\(b_q)+k\end{matrix}\right) = {}_{p+1}F_{q+1}\left(\begin{matrix}(a_p),\ \alpha;\ x\\(b_q),\ \beta\end{matrix}\right).$$

24. $$\sum_{k=1}^{\infty} \frac{(\pm x)^k}{(2k)!} \frac{\Pi(a_p)_k}{\Pi(b_q)_k} {}_pF_{q+1}\left(\begin{matrix}(a_p)+k;\ x\\(b_q)+k,\ 2k+1\end{matrix}\right) = \frac{1}{2}\,{}_pF_{q+1}\left(\begin{matrix}(a_p);\ (x\pm x)/2\\(b_q),\ 1/2\end{matrix}\right) - \frac{1}{2}\,{}_pF_{q+1}\left(\begin{matrix}(a_p);\ x\\(b_q),\ 1\end{matrix}\right).$$

25. $$\sum_{k=0}^{\infty} \frac{x^k}{(2k+1)!} \frac{\Pi(a_p)_k}{\Pi(b_q)_k} {}_pF_{q+1}\left(\begin{matrix}(a_p)+k;\ x\\(b_q)+k,\ 2k+2+\sigma\end{matrix}\right) = {}_pF_{q+1}\left(\begin{matrix}(a_p);\ x\\(b_q),\ (3+\sigma)/2\end{matrix}\right)$$

$[\sigma=0$ or $1]$.

26. $$\sum_{k=0}^{\infty} \frac{(\pm x)^k}{(2k)!} \frac{\Pi(a_p)_k}{\Pi(b_q)_k} {}_pF_{q+1}\left(\begin{matrix}(a_p)+k;\ x\\(b_q)+k,\ 2k+2\end{matrix}\right) = {}_pF_{q+1}\left(\begin{matrix}(a_p);\ (x\pm x)/2\\(b_q),\ 1\end{matrix}\right).$$

27. $$\sum_{k=0}^{\infty} \frac{(1-\alpha)_k}{(\alpha)_k(2k)!} x^k \frac{\Pi(a_p)_k}{\Pi(b_q)_k} {}_pF_{q+1}\left(\begin{matrix}(a_p)+k;\ x\\(b_q)+k,\ 2k+1\end{matrix}\right) = \frac{1}{2}\,{}_pF_{q+1}\left(\begin{matrix}(a_p);\ x\\(b_q),\ \alpha\end{matrix}\right) + \frac{1}{2}\,{}_pF_{q+1}\left(\begin{matrix}(a_p);\ x\\(b_q),\ 1\end{matrix}\right).$$

28. $$\sum_{k=0}^{\infty} \frac{(1-\alpha)_k}{(\alpha)_k(2k)!} (-x)^k \frac{\Pi(a_p)_k}{\Pi(b_q)_k} {}_pF_{q+1}\left(\begin{matrix}(a_p)+k;\ x\\(b_q)+k,\ 2k+1\end{matrix}\right) = \frac{1}{2}\,{}_{p+1}F_{q+2}\left(\begin{matrix}(a_p),\ \alpha-1/2;\ x\\(b_q),\ \alpha,\ 1/2\end{matrix}\right) + \frac{1}{2}\,{}_pF_{q+1}\left(\begin{matrix}(a_p);\ x\\(b_q),\ 1\end{matrix}\right).$$

29. $$\sum_{k=0}^{\infty} \frac{x^k}{(1/2+\sigma)_{2k}} \frac{\Pi(a_p)_k}{\Pi(b_q)_k} {}_pF_{q+1}\left(\begin{matrix}(a_p)+k;\ x\\(b_q)+k,\ 2k+3/2+\sigma\end{matrix}\right) = {}_pF_{q+1}\left(\begin{matrix}(a_p);\ x\\(b_q),\ 1/2+\sigma\end{matrix}\right)$$

$[\sigma=0$ or $1]$.

30. $$\sum_{k=0}^{\infty} \frac{(-x)^k}{(2k+1)!\,(2k+1)} \frac{\Pi(a_p)_k}{\Pi(b_q)_k} {}_pF_{q+1}\left(\begin{matrix}(a_p)+k;\ x\\(b_q)+k,\ 2k+2\end{matrix}\right) = {}_{p+1}F_{q+1}\left(\begin{matrix}(a_p),\ 1;\ x\\(b_q),\ 3/2,\ 3/2\end{matrix}\right).$$

31. $$\sum_{k=0}^{\infty} \frac{(\beta-1)_k}{k!\,(\beta-1)_{2k}} (-x)^k \frac{\Pi(a_p)_k}{\Pi(b_q)_k} {}_pF_{q+1}\left(\begin{matrix}(a_p)+k;\ x\\(b_q)+k,\ 2k+\beta\end{matrix}\right) = 1.$$

32. $\sum_{k=0}^{\infty} \frac{x^k}{(\beta-1)_{2k}} \frac{\Pi(a_p)_k}{\Pi(b_q)_k} {}_pF_{q+1}\left(\begin{matrix}(a_p)+k; & x\\ (b_q)+k, & 2k+\beta\end{matrix}\right) = {}_pF_{q+1}\left(\begin{matrix}(a_p); & x\\ (b_q), & \beta-1\end{matrix}\right).$

33. $\sum_{k=0}^{\infty} \frac{(-x)^k}{(\beta)_{2k}} \frac{\Pi(a_p)_k}{\Pi(b_q)_k} {}_pF_{q+1}\left(\begin{matrix}(a_p)+k; & x\\ (b_q)+k, & 2k+\beta\end{matrix}\right) = {}_{p+1}F_{q+2}\left(\begin{matrix}(a_p), (\beta-1)/2; & x\\ (b_q), (\beta+1)/2, \beta-1\end{matrix}\right).$

34. $\sum_{k=1}^{\infty} \frac{x^k}{k(\beta+k-1)(\beta)_{2k-1}} \frac{\Pi(a_p)_k}{\Pi(b_q)_k} {}_pF_{q+1}\left(\begin{matrix}(a_p)+k; & x\\ (b_q)+k, & 2k+\beta\end{matrix}\right) =$

$$= -\frac{\partial}{\partial\beta} {}_pF_{q+1}\left(\begin{matrix}(a_p); & x\\ (b_q), & \beta\end{matrix}\right).$$

35. $\sum_{k=0}^{\infty} \frac{(2\beta-2-\sigma)_{2k}}{(2k+\sigma)!\,(\beta-1)_{2k}} x^k \frac{\Pi(a_p)_k}{\Pi(b_q)_k} {}_pF_{q+1}\left(\begin{matrix}(a_p)+k; & x\\ (b_q)+k, & 2k+\beta\end{matrix}\right) = {}_pF_{q+1}\left(\begin{matrix}(a_p); & x\\ (b_q), & \sigma+1/2\end{matrix}\right)$

[Re $\beta > 3/2$; $\sigma = 0$ or 1].

36. $\sum_{k=0}^{\infty} \frac{(-x)^k}{(\beta-1)_{2k}} \frac{\Pi(a_p)_k}{\Pi(b_q)_k} {}_pF_{q+1}\left(\begin{matrix}(a_p)+k; & x\\ (b_q)+k, & 2k+\beta\end{matrix}\right) =$

$$= {}_pF_{q+1}\left(\begin{matrix}(a_p); & x\\ (b_q), & \beta-1\end{matrix}\right) - \frac{2x}{\beta(\beta-1)}\left(\prod_{i=1}^{p} a_i\right)\left(\prod_{j=1}^{q} b_j\right)^{-1} {}_{p+1}F_{q+2}\left(\begin{matrix}(a_p)+1, \beta/2; & x\\ (b_q)+1, \beta, 1+\beta/2\end{matrix}\right).$$

37. $\sum_{k=0}^{\infty} \frac{(\alpha)_k(\alpha-1)_k}{k!\,(\beta-\alpha)_k(\alpha-1)_{2k}} x^k \frac{\Pi(a_p)_k}{\Pi(b_q)_k} {}_pF_{q+1}\left(\begin{matrix}(a_p)+k; & x\\ (b_q)+k, 2k+\beta\end{matrix}\right) = {}_pF_{q+1}\left(\begin{matrix}(a_p) & x\\ (b_q), & \beta-\alpha\end{matrix}\right).$

38. $\sum_{k=0}^{\infty} \frac{(\alpha)_k}{k!\,(2\alpha)_{2k}} x^{2k} \frac{\Pi(a_p)_{2k}}{\Pi(b_q)_{2k}} {}_pF_q\left(\begin{matrix}(a_p)+2k; & x\\ (b_q)+2k\end{matrix}\right) = {}_{p+1}F_{q+1}\left(\begin{matrix}(a_p), \alpha; & 2x\\ (b_q), 2\alpha\end{matrix}\right)$

[$|x| < 1$].

39. $\sum_{k=0}^{\infty} \frac{(-1)^k}{(2k)!}\left(\frac{x}{4}\right)^{2k} \frac{\Pi(a_p)_{2k}}{\Pi(b_q)_{2k}} {}_{p+1}F_{q+1}\left(\begin{matrix}(a_p)+2k, 2k+3/2; & x\\ (b_q)+2k, 4k+3\end{matrix}\right) = {}_pF_q\left(\begin{matrix}(a_p); & x/2\\ (b_q)\end{matrix}\right).$

40. $\sum_{k=0}^{\infty} \frac{(\alpha+\beta k)_k}{k!} t^k {}_{p+1}F_{q+1}\left(\begin{matrix}(a_p), \alpha+(\beta+1)k\\ (b_q), \alpha+\beta k; x\end{matrix}\right) =$

$$= \frac{(1+v)^{\alpha}}{1-\beta v} {}_pF_q\left(\begin{matrix}(a_p); & x(v+1)\\ (b_q)\end{matrix}\right)$$ [$v = t(1+v)^{\beta+1}$, $v(0) = 0$].

41. $\sum_{k=0}^{\infty} \frac{(a)_k(b)_{kp}}{k!\,(a+b)_{kp+k}} (-x)^k {}_pF_p\left(\begin{matrix}\Delta(p, kp+b); & x\\ \Delta(p, kp+a+b)\end{matrix}\right) = {}_{p+1}F_{p+1}\left(\begin{matrix}\Delta(p+1, b); & x\\ \Delta(p+1, a+b)\end{matrix}\right)$

[Re $b > 0$].

6.8.2. Series containing trigonometric functions and ${}_pF_q((a_p); (b_q); x)$.

1. $\sum_{k=0}^{\infty} \frac{(\pm y)^k}{(2k)!} \cos kx \frac{\Pi(a_p)_k}{\Pi(b_q)_k} {}_pF_{q+1}\left(\begin{matrix}(a_p)+k; & y\\ (b_q)+k, & 2k+1\end{matrix}\right) =$

$$= \frac{1}{2} {}_pF_{q+1}\left(\begin{matrix}(a_p); \\ (b_q), & 1/2; \end{matrix}\; y\begin{Bmatrix}\cos^2(x/2)\\ \sin^2(x/2)\end{Bmatrix}\right) + \frac{1}{2} {}_pF_{q+1}\left(\begin{matrix}(a_p); & y\\ (b_q), & 1\end{matrix}\right).$$

2. $$\sum_{k=0}^{\infty} \frac{(\mp y)^k}{(2k+1)!} \begin{Bmatrix} \sin(2k+1)x \\ \cos(2k+1)x \end{Bmatrix} \frac{\Pi(a_p)_k}{\Pi(b_q)_k} {}_pF_{q+1}\left(\begin{matrix}(a_p)+k; & y \\ (b_q)+k, & 2k+2\end{matrix}\right) = \begin{Bmatrix} \sin x \\ \cos x \end{Bmatrix} {}_pF_{q+1}\left(\begin{matrix}(a_p); \\ (b_q), 3/2;\end{matrix}\; y \begin{Bmatrix} \sin x \\ \cos x \end{Bmatrix}\right).$$

3. $$\sum_{k=0}^{\infty} \frac{(\pm y)^k}{(2k+1)!} \sin(k+1)x \frac{\Pi(a_p)_k}{\Pi(b_q)_k} {}_pF_{q+1}\left(\begin{matrix}(a_p)+k; & y \\ (b_q)+k, & 2k+3\end{matrix}\right) = \sin x\, {}_pF_{q+1}\left(\begin{matrix}(a_p); \\ (b_q), 3/2;\end{matrix}\; y \begin{Bmatrix} \cos^2(x/2) \\ \sin^2(x/2) \end{Bmatrix}\right).$$

4. $$\sum_{k=0}^{\infty} \frac{(\pm y)^k}{(2k)!} \begin{Bmatrix} \sin(2k+1)x \\ \cos(2k+1)x \end{Bmatrix} \frac{\Pi(a_p)_k}{\Pi(b_q)_k} {}_pF_{q+1}\left(\begin{matrix}(a_p)+k; & y \\ (b_q)+k, & 2k+2\end{matrix}\right) = \begin{Bmatrix} \sin x \\ \cos x \end{Bmatrix} {}_pF_{q+1}\left(\begin{matrix}(a_p); \\ (b_q), 1/2;\end{matrix}\; y \begin{Bmatrix} \cos^2 x \\ \sin^2 x \end{Bmatrix}\right).$$

6.8.3. Series containing special functions and ${}_pF_q((a_p);(b_q);x)$.

1. $$\sum_{k=0}^{\infty} \frac{1}{k!} \gamma(\beta+k, x)\, {}_{p+1}F_{q+1}\left(\begin{matrix}-k, (a_p); & y \\ \beta, (b_q) & \end{matrix}\right) = \frac{x^\beta}{\beta} {}_pF_{q+1}\left(\begin{matrix}(a_p); & xy \\ (b_q), & \beta+1\end{matrix}\right)$$ [Re $\beta > 0$].

2. $$\sum_{k=0}^{\infty} (2k+\mu+\nu) \frac{(\mu+\nu)_k}{k!} J_{k+\mu}(x) J_{k+\nu}(x)\, {}_{p+2}F_q\left(\begin{matrix}-k, k+\mu+\nu, (a_p) \\ (b_q); \quad y\end{matrix}\right) = \frac{\mu+\nu}{\Gamma(\mu+1)\Gamma(\nu+1)} \left(\frac{x}{2}\right)^{\mu+\nu} {}_{p+2}F_{q+2}\left(\begin{matrix}(\mu+\nu+1)/2, (\mu+\nu+2)/2, (a_p) \\ \mu+1, \nu+1, (b_q); -x^2y\end{matrix}\right).$$

3. $$\sum_{k=0}^{\infty} (2k+\nu) \frac{(\nu)_k}{k!} J_{2k+\nu}(x)\, {}_{p+2}F_q\left(\begin{matrix}-k, k+\nu, (a_p) \\ (b_q); \quad y\end{matrix}\right) = \frac{(x/2)^\nu}{\Gamma(\nu)} {}_pF_q\left(\begin{matrix}(a_p); & -x^2y/4 \\ (b_q) & \end{matrix}\right).$$

4. $$\sum_{k=0}^{\infty} \frac{(x/2)^k}{k!} J_{k+\nu}(x)\, {}_{p+1}F_q\left(\begin{matrix}-k, (a_p) \\ (b_q); y\end{matrix}\right) = \frac{(x/2)^\nu}{\Gamma(\nu+1)} {}_pF_{q+1}\left(\begin{matrix}(a_p); & -x^2y/4 \\ (b_q), & \nu+1\end{matrix}\right).$$

5. $$\sum_{k=0}^{\infty} (-1)^k \left(k+\frac{\nu}{2}\right) \frac{(\nu)_k}{k!} I_{k+\nu/2}(x)\, {}_{p+2}F_q\left(\begin{matrix}-k, k+\nu, (a_p) \\ (b_q); y\end{matrix}\right) = \frac{(x/2)^{\nu/2}}{\Gamma(\nu/2)} e^{-x}\, {}_{p+1}F_q\left(\begin{matrix}(a_p), (\nu+1)/2 \\ (b_q); 2xy\end{matrix}\right).$$

6. $$\sum_{k=0}^{\infty} \frac{(-x)^k}{(2k)!} \frac{\Pi(a_p)_k}{\Pi(b_q)_k} P_k(x)\, {}_pF_{q+1}\left(\begin{matrix}(a_p)+k; & y \\ (b_q)+k, & 2k+2\end{matrix}\right) = {}_pF_{q+1}\left(\begin{matrix}(a_p); & (1-x)y/2 \\ (b_q), & 1\end{matrix}\right)$$ [$|x| \leqslant 1$].

7. $$\sum_{k=0}^{\infty} (2k+1) Q_k(x)\, {}_{p+2}F_q\left(\begin{matrix}-k, k+1, (a_p) \\ (b_q); y\end{matrix}\right) = -\frac{1}{1-x} {}_{p+2}F_q\left(\begin{matrix}(a_p), 1, 1 \\ (b_q); 2(1-x)^{-1}y\end{matrix}\right).$$

8. $$\sum_{k=0}^{\infty} \frac{(1-\beta)_k}{(\alpha-\beta+1)_k} L_k^{\alpha-\beta}(x)\, {}_{p+2}F_{q+1}\left(\begin{matrix}-n,\ \alpha,\ (a_p)\\ \beta-k,\ (b_q);\ y\end{matrix}\right) = \Gamma\left[\begin{matrix}\alpha-\beta+1\\ \alpha+1\end{matrix}\right] x^{\beta-1}\, {}_{p+1}F_{q+1}\left(\begin{matrix}-n,\ (a_p)\\ \beta,\ (b_q);\ xy\end{matrix}\right).$$

9. $$\sum_{k=0}^{\infty} \frac{(\beta)_k}{k!\,(\beta)_{2k}} \left(\frac{y}{x}\right)^k L_k^{-2k-\beta}(x)\, {}_pF_{q+1}\left(\begin{matrix}(a_p)+k;\ y\\ (b_q)+k,\ 2k+\beta+1\end{matrix}\right) = {}_pF_q\left(\begin{matrix}(a_p);\ -y/x\\ (b_q)\end{matrix}\right).$$

10. $$\sum_{k=0}^{\infty} \frac{x^k}{k!} L_n^{k+\alpha}(x)\, {}_{p+1}F_q\left(\begin{matrix}-k,\ (a_p)\\ (b_q);\ y\end{matrix}\right) = \frac{(\alpha+1)_n}{n!} e^x\, {}_{p+1}F_{q+1}\left(\begin{matrix}(a_p),\ \alpha+n+1\\ (b_q),\ \alpha+1;\ -xy\end{matrix}\right).$$

11. $$\sum_{k=0}^{\infty} \frac{(y/4)^k}{k!\,(1/2)_k} \frac{\Pi\,(a_p)_k}{\Pi\,(b_q)_k} T_k(x)\, {}_pF_{q+1}\left(\begin{matrix}(a_p)+k;\ y\\ (b_q)+k,\ 2k+1\end{matrix}\right) = \frac{1}{2}\, {}_pF_{q+1}\left(\begin{matrix}(a_p);\ (x+1)\,y/2\\ (b_q),\ 1/2\end{matrix}\right) - \frac{1}{2}\, {}_pF_{q+1}\left(\begin{matrix}(a_p);\ y/2\\ (b_q),\ 1\end{matrix}\right) \qquad [-1 \leqslant x \leqslant 1].$$

12. $$\sum_{k=0}^{\infty} \frac{y^k}{(\beta-1)_{2k}} \frac{\Pi\,(a_p)_k}{\Pi\,(b_q)_k} C_k^{(\beta-1)/2}(x)\, {}_pF_{q+1}\left(\begin{matrix}(a_p)+k;\ y\\ (b_q)+k,\ 2k+\beta\end{matrix}\right) = {}_pF_{q+1}\left(\begin{matrix}(a_p);\ (1+x)\,y/2\\ (b_q),\ \beta/2\end{matrix}\right) \qquad [-1 \leqslant x \leqslant 1].$$

13. $$\sum_{k=0}^{\infty} \frac{(\beta-1)_k}{(\beta-1)_{2k}\,(\gamma-1)_k} y^k \frac{\Pi\,(a_p)_k}{\Pi\,(b_q)_k} P_k^{(\beta-\gamma,\ \gamma-2)}(x)\, {}_pF_{q+1}\left(\begin{matrix}(a_p)+k;\ y\\ (b_q)+k,\ 2k+\beta\end{matrix}\right) = {}_pF_{q+1}\left(\begin{matrix}(a_p);\ (1+x)\,y/2\\ (b_q),\ \gamma-1\end{matrix}\right).$$

6.8.4. Series containing products of ${}_pF_q((a_p);(b_q);x)$.

1. $$\sum_{k=0}^{\infty} \frac{(\beta)_k}{k!\,(\beta)_{2k}} (-y)^k\, {}_{p+2}F_q\left(\begin{matrix}-k,\ k+\beta,\ (a_p)\\ (b_q);\ x\end{matrix}\right) \frac{\Pi\,(c_r)_k}{\Pi\,(d_s)_k} \times {}_rF_{s+1}\left(\begin{matrix}(c_r)+k;\ y\\ (d_s)+k,\ 2k+\beta+1\end{matrix}\right) = {}_{p+r}F_{q+s}\left(\begin{matrix}(a_p),\ (c_r);\ xy\\ (b_q),\ (d_s)\end{matrix}\right).$$

2. $$\sum_{k=0}^{\infty} \frac{(2\beta)_k}{k!\,(\beta)_k\,(\beta+1/2)_k} \left(-\frac{y}{4}\right)^k {}_{p+2}F_q\left(\begin{matrix}-k,\ 2\beta+k,\ (a_p)\\ (b_q);\ x\end{matrix}\right) \frac{\Pi\,(c_r)_k}{\Pi\,(d_s)_k} \times {}_rF_{s+1}\left(\begin{matrix}(c_r)+k;\ y\\ (d_s)+k,\ 2k+2\beta+1\end{matrix}\right) = {}_{p+r}F_{q+s}\left(\begin{matrix}(a_p),\ (c_r);\ xy\\ (b_q),\ (d_s)\end{matrix}\right).$$

6.9. VARIOUS HYPERGEOMETRIC FUNCTIONS

6.9.1. Series containing ${}_2F_2(a,\ b;\ c,\ d;\ x)$.

1. $$\sum_{k=0}^{\infty} (-1)^k \frac{(a)_k^2\,(b)_k^2\,(c-1)_k}{k!\,(a+b+1/2)_k^2\,(c-1)_{2k}\,(c)_{2k}} x^{2k}\, {}_2F_2^2\left(\begin{matrix}a+k,\ b+k;\ x\\ a+b+k+1/2,\ c+2k\end{matrix}\right) = {}_3F_3\left(\begin{matrix}2a,\ 2b,\ a+b;\ x\\ 2a+2b,\ a+b+1/2,\ c\end{matrix}\right).$$

2. $$\sum_{k=0}^{\infty}(-1)^k\frac{(a)_k(b)_k(b')_k(c-1)_k(c-a)_k}{k!\,(2b)_k(2b')_k(c-1)_{2k}(c)_{2k}}x^{2k}{}_2F_2\left(\begin{matrix}a+k,\ b+k;\ x\\ 2b+k,\ c+2k\end{matrix}\right)\times$$
$$\times{}_2F_2\left(\begin{matrix}a+k,\ b'+k;\ -x\\ 2b'+k,\ c+2k\end{matrix}\right)={}_4F_5\left(\begin{matrix}a/2,\ (a+1)/2,\ (b+b')/2,\ (b+b'+1)/2;\ x^2/4\\ b+1/2,\ b'+1/2,\ b+b',\ c/2,\ (c+1)/2\end{matrix}\right).$$

3. $$\sum_{k=0}^{\infty}(-1)^k\frac{(a)_k(b)_k^2(d-1)_k(d-a)_k}{k!\,(c)_k^2(d-1)_{2k}(d)_{2k}}x^{2k}{}_2F_2\left(\begin{matrix}a+k,\ b+k;\ x\\ c+k,\ d+2k\end{matrix}\right){}_2F_2\left(\begin{matrix}a+k,\ b+k;\ -x\\ c+k,\ d+2k\end{matrix}\right)=$$
$$={}_4F_5\left(\begin{matrix}a/2,\ (a+1)/2,\ b,\ c-b;\ x^2/4\\ c,\ c/2,\ (c+1)/2,\ d/2,\ (d+1)/2\end{matrix}\right).$$

4. $$\sum_{k=0}^{\infty}(-1)^k\frac{(a)_k^2(b)_k^2(c-1)_k}{k!\,(a+b-1/2)_k(a+b+1/2)_k(c-1)_{2k}(c)_{2k}}\times$$
$$\times x^{2k}{}_2F_2\left(\begin{matrix}a+k,\ b+k;\ x\\ a+b+k-1/2,\ c+2k\end{matrix}\right){}_2F_2\left(\begin{matrix}a+k,\ b+k;\ x\\ a+b+k+1/2,\ c+2k\end{matrix}\right)=$$
$$={}_3F_3\left(\begin{matrix}2a,\ 2b,\ a+b;\ x\\ a+b+1/2,\ 2a+2b-1,\ c\end{matrix}\right).$$

5. $$\sum_{k=0}^{\infty}(-1)^k\frac{(a)_k(b)_k(b')_k(c-1)_k(c-a)_k}{k!\,(2b)_k(2b')_k(c-1)_{2k}(c)_{2k}}x^{2k}{}_2F_2\left(\begin{matrix}a+k,\ b+k;\ x\\ 2b+k,\ c+2k\end{matrix}\right)\times$$
$$\times{}_2F_2\left(\begin{matrix}a+k,\ b'+k;\ -x\\ 2b'+k,\ c+2k\end{matrix}\right)={}_4F_5\left(\begin{matrix}a/2,\ (a+1)/2,\ (b+b')/2,\ (b+b'+1)/2;\ x^2/4\\ b+1/2,\ b'+1/2,\ b+b',\ c/2,\ (c+1)/2\end{matrix}\right).$$

6. $$\sum_{k=0}^{\infty}(-1)^k\frac{(a)_k(b)_k(b-c+1)_k(d-1)_k(d-a)_k}{k!\,(c)_k(2-c)_k(d-1)_{2k}(d)_{2k}}x^{2k}{}_2F_2\left(\begin{matrix}a+k,\ b+k;\ x\\ c+k,\ d+2k\end{matrix}\right)\times$$
$$\times{}_2F_2\left(\begin{matrix}a+k,\ b-c+k+1;\ -x\\ 2-c+k,\ d+2k\end{matrix}\right)={}_4F_5\left(\begin{matrix}a/2,\ (a+1)/2,\ b-c/2+1/2,\ c/2-b+1/2;\ x^2/4\\ (c+1)/2,\ (3-c)/2,\ d/2,\ (d+1)/2,\ 1/2\end{matrix}\right)+$$
$$+\frac{a(2b-c)(1-c)}{cd(2-c)}x\,{}_4F_5\left(\begin{matrix}a/2+1,\ (a+1)/2,\ 1+b-c/2,\ 1-b+c/2;\ x^2/4\\ c/2+1,\ 2-c/2,\ 1+d/2,\ (d+1)/2,\ 3/2\end{matrix}\right).$$

7. $$\sum_{k=0}^{\infty}(-1)^k\frac{(a)_k(b)_k(c-1)_k(d-a-b)_k}{k!\,(c-1)_{2k}(c)_{2k}(d)_k}x^{2k}{}_1F_1\left(\begin{matrix}d-a-b+k\\ c+2k;\ x\end{matrix}\right)\times$$
$$\times{}_2F_2\left(\begin{matrix}a+k,\ b+k\\ c+2k,\ d+k;\ x\end{matrix}\right)={}_2F_2\left(\begin{matrix}d-a,\ d-b\\ c,\ d;\ x\end{matrix}\right).$$

8. $$\sum_{k=0}^{\infty}(-1)^k\frac{(a)_k(b)_k(c-1)_k(c-b)_k}{k!\,(2a)_k(c-1)_{2k}(c)_{2k}}x^{2k}{}_1F_1\left(\begin{matrix}b+k;\ -x/2\\ c+2k\end{matrix}\right)\times$$
$$\times{}_2F_2\left(\begin{matrix}a+k,\ b+k;\ x\\ 2a+k,\ c+2k\end{matrix}\right)={}_2F_3\left(\begin{matrix}b/2,\ (b+1)/2;\ x^2/16\\ a+1/2,\ c/2,\ (c+1)/2\end{matrix}\right).$$

6.9.2. Series containing ${}_3F_2(a_1,\ a_2,\ a_3;\ b_1,\ b_2;\ x)$.

1. $$\sum_{k=0}^{\infty}\frac{(1-b)_k}{k!\,(1-a)_k}t^k{}_3F_2\left(\begin{matrix}-k,\ a-k,\ c\\ b-k,\ d;\ x\end{matrix}\right)={}_1F_1(1-b;\ 1-a;\ t)\,{}_1F_1(c;\ d;\ -tx).$$

2. $$\sum_{k=0}^{\infty} \frac{(1-c)_k (f)_k}{k!\,(1-a)_k} t^k {}_3F_2\left(\begin{matrix}-k,\ a-k,\ b\\ c-k,\ d;\ x\end{matrix}\right) = F_2(f,\ 1-c,\ b;\ 1-a,\ d;\ t,\ -tx).$$

3. $$\sum_{k=0}^{\infty} \frac{(1-c)_k (1-d)_k}{k!\,(f)_k} t^k {}_3F_2\left(\begin{matrix}-k,\ a,\ b;\ x\\ c-k,\ d-k\end{matrix}\right) = F_3(1-c,\ a,\ 1-d,\ b;\ f;\ t,\ -tx).$$

4. $$\sum_{k=0}^{\infty} \frac{(a)_k (b)_k (c-a)_k (c-b)_k (c+1/2)_k (1/2)_k}{k!\,(c)_k (c)_{2k} (c-1/2)_{2k}} x^{2k} {}_3F_2\left(\begin{matrix}2a+2k,\ 2b+2k,\ 2c+2k\\ 2c+4k,\ 2c+2k+1;\ x\end{matrix}\right) =$$
$$= {}_2F_1\left(\begin{matrix}a,\ b\\ c;\ x\end{matrix}\right) {}_3F_2\left(\begin{matrix}a,\ b,\ c+1/2\\ c,\ c-1/2;\ x\end{matrix}\right).$$

5. $$\sum_{k=0}^{\infty} t^k C_k^{b/2}(x)\, {}_3F_2\left(\begin{matrix}-k,\ a,\ b\\ c,\ (c+1)/2;\ y\end{matrix}\right) =$$
$$= \rho^{-c} F_4\left(a,\ b;\ \frac{c+1}{2},\ \frac{c+1}{2};\ \frac{1}{2} ty\rho^{-2}(t-x+\rho),\ \frac{1}{2} ty\rho^{-2}(t-x-\rho)\right)$$
$[\rho=(1-2tx+t^2)^{1/2}]$.

6. $$\sum_{k=0}^{\infty} \frac{t^k}{k!} C_k^{(a+b)/2}(x)\, {}_3F_2\left(\begin{matrix}-k,\ a,\ b;\ y\\ a+b,\ (a+b+1)/2\end{matrix}\right) =$$
$$= \rho^{-a-b} \Gamma^2\left[\begin{matrix}1-a,\ a+b\\ b\end{matrix}\right] C_{-a}^{(a+b)/2}\left(\frac{\omega-yt}{\rho}\right) C_{-a}^{(a+b)/2}\left(\frac{\omega+yt}{\rho}\right)$$
$[\rho=(1-2tx+t^2)^{1/2},\ \omega=(1-2tx(1-y)+t^2(1-y)^2)^{1/2}]$.

7. $$\sum_{k=0}^{\infty} \frac{(a)_k (b)_k}{(c-1/2)_k (c+1/2)_k} \left(-\frac{x}{4}\right)^k C_k^{c-1/2}(x)\, {}_3F_2\left(\begin{matrix}a+k,\ b+k,\ c+k;\ y\\ c+k+1/2,\ 2c+2k\end{matrix}\right) =$$
$$= {}_2F_1\left(\begin{matrix}a,\ b;\ (1-x)y/2\\ c+1/2\end{matrix}\right)$$
$[\operatorname{Re} a,\ \operatorname{Re} b > 0;\ -1 \leqslant x \leqslant 1]$.

8. $$\sum_{k=0}^{\infty} \frac{(a)_k (b)_k (c)_k (d-1)_k (d-a)_k (e-b-c)_k}{k!\,(e)_k (d-1)_{2k} (d)_{2k}} x^{2k} {}_2F_1\left(\begin{matrix}a+k,\ e-b-c\\ d+2k;\ x\end{matrix}\right) \times$$
$$\times {}_3F_2\left(\begin{matrix}a+k,\ b+k,\ c+k\\ d+2k,\ e+k;\ x\end{matrix}\right) = {}_3F_2\left(\begin{matrix}a,\ e-b,\ e-c\\ d,\ e;\ x\end{matrix}\right).$$

9. $$\sum_{k=0}^{\infty} \frac{(a)_k (b)_k (c-1)_k (c-a)_k (c-b)_k (-1/2)_k}{k!\,(c-1/2)_k (c-1)_{2k} (c)_{2k}} x^{2k} {}_2F_1\left(\begin{matrix}a+k,\ b+k\\ c+2k;\ x\end{matrix}\right) \times$$
$$\times {}_3F_2\left(\begin{matrix}a+k,\ b+k,\ c+k+1/2\\ c+k-1/2,\ c+2k;\ x\end{matrix}\right) = {}_2F_1\left(\begin{matrix}2a,\ 2b\\ 2c-1;\ x\end{matrix}\right).$$

10. $$\sum_{k=0}^{\infty} \frac{(a)_k^2 (b)_k^2 (c)_k (d-c)_k}{k!\,(a+b-1/2)_k (a+b+1/2)_k (d+k-1)_k (d)_{2k}} x^{2k} \times$$
$$\times {}_3F_2\left(\begin{matrix}a+k,\ b+k,\ c+k;\ x\\ a+b+k-1/2,\ d+2k\end{matrix}\right) {}_3F_2\left(\begin{matrix}a+k,\ b+k,\ c+k;\ x\\ a+b+k+1/2,\ d+2k\end{matrix}\right) =$$
$$= {}_4F_3\left(\begin{matrix}2a,\ 2b,\ a+b,\ c;\ x\\ 2a+2b-1,\ a+b+1/2,\ d\end{matrix}\right).$$

11. $$\sum_{k=0}^{\infty} \frac{(a)_k^2\,(b)_k^2\,(c)_k\,(d-c)_k}{k!\,(a+b+1/2)_k^2\,(d+k-1)_k\,(d)_{2k}}\, x^{2k}\, {}_3F_2\left(\begin{matrix} a+k,\ b+k,\ c+k;\ x \\ a+b+k+1/2,\ d+2k \end{matrix}\right) = {}_4F_3\left(\begin{matrix} 2a,\ 2b,\ a+b,\ c;\ x \\ 2a+2b,\ a+b+1/2,\ d \end{matrix}\right)$$

6.9.3. Series containing various hypergeometric functions.

1. $$\sum_{k=0}^{\infty} \frac{x^{2k}}{k!\,(a)_k^2\,(b+k-1)_k\,(b)_{2k}}\, {}_0F_2\left(\begin{matrix} x \\ a+k,\ b+2k \end{matrix}\right) {}_2F_2\left(\begin{matrix} -x \\ a+k,\ b+2k \end{matrix}\right) = {}_0F_5\left(\begin{matrix} -x^2/16 \\ a,\ a/2,\ (a+1)/2,\ b/2,\ (b+1)/2 \end{matrix}\right).$$

2. $$\sum_{k=0}^{\infty} \frac{x^{2k}}{k!\,(a)_k\,(2-a)_k\,(b+k-1)_k\,(b)_{2k}}\, {}_0F_2\left(\begin{matrix} x \\ a+k,\ b+2k \end{matrix}\right) {}_0F_2\left(\begin{matrix} -x \\ 2-a+k,\ c+2k \end{matrix}\right) = {}_0F_5\left(\begin{matrix} -x^2/16 \\ 1/2,\ (1+a)/2,\ (3-a)/2,\ b/2,\ (b+1)/2 \end{matrix}\right) + \frac{2\,(1-a)\,x}{a\,(2-a)\,b}\, {}_0F_5\left(\begin{matrix} -x^2/16 \\ 3/2,\ 1+a/2,\ 2-a/2,\ 1+b/2,\ (b+1)/2 \end{matrix}\right).$$

3. $$\sum_{k=0}^{\infty} \frac{(a)_k^2}{k!\,(b)_k^2\,(c+k-1)_k\,(c)_{2k}}\, x^{2k}\, {}_1F_2\left(\begin{matrix} a+k;\ x \\ b+k,\ c+2k \end{matrix}\right) {}_1F_2\left(\begin{matrix} a+k;\ -x \\ b+k,\ c+2k \end{matrix}\right) = {}_2F_5\left(\begin{matrix} a,\ b-a;\ x^2/16 \\ b,\ b/2,\ (b+1)/2,\ c/2,\ (c+1)/2 \end{matrix}\right).$$

4. $$\sum_{k=0}^{\infty} \frac{(a)_k\,(b)_k}{k!\,(2a)_k\,(2b)_k\,(c+k-1)_k\,(c)_{2k}}\, x^{2k}\, {}_1F_2\left(\begin{matrix} a+k;\ x \\ 2a+k,\ c+2k \end{matrix}\right) {}_1F_2\left(\begin{matrix} b+k;\ -x \\ 2b+k,\ c+2k \end{matrix}\right) = {}_2F_5\left(\begin{matrix} (a+b)/2,\ (a+b+1)/2;\ x^2/16 \\ a+1/2,\ b+1/2,\ a+b,\ c/2,\ (c+1)/2 \end{matrix}\right).$$

5. $$\sum_{k=0}^{\infty} (-1)^k \frac{(a)_k\,(c-a)_k}{k!\,(b)_k^2\,(c+k-1)_k\,(c)_{2k}}\, x^{2k}\, {}_1F_2\left(\begin{matrix} a+k;\ x \\ b+k,\ c+2k \end{matrix}\right) {}_1F_2\left(\begin{matrix} a+k;\ -x \\ b+k,\ c+2k \end{matrix}\right) = {}_2F_5\left(\begin{matrix} a/2,\ (a+1)/2;\ -x^2/4 \\ b,\ b/2,\ (b+1)/2,\ c/2,\ (c+1)/2 \end{matrix}\right).$$

6. $$\sum_{k=0}^{\infty} t^k L_k^c(x)\, {}_2F_3\left(\begin{matrix} -k/2,\ (1-k)/2;\ y \\ 1/2,\ 1+c/2,\ (1+c)/2 \end{matrix}\right) = \frac{\Gamma(c+1)}{2}\,(1-t)^{-c-1}\left(\frac{2txy^{1/2}}{1-t}\right)^{-c}\left[\exp\left(\frac{2ty^{1/2}}{1-t}\right) J_c\left(\frac{2\sqrt{2txy^{1/2}}}{1-t}\right) + \exp\left(-\frac{2ty^{1/2}}{1-t}\right) I_c\left(\frac{2\sqrt{2txy^{1/2}}}{1-t}\right)\right].$$

7. $$\sum_{k=0}^{\infty} \frac{t^k}{k!} H_k(x)\, {}_3F_1\left(\begin{matrix} -k/2,\ (1-k)/2,\ a \\ 1/2;\ y \end{matrix}\right) = \frac{e^{2tx-t^2}}{(1+t^2y)^a} \times {}_1F_1\left(a;\ \frac{1}{2};\ \frac{t^2\,(t-x)^2\,y}{1+t^2y}\right).$$

8. $$\sum_{k=1}^{\infty} \frac{1}{k^{2a}}\, {}_4F_3\left(\begin{matrix} a,\ a+1/2,\ b,\ b+1/2 \\ 2b,\ 2b-1/2,\ 3/2;\ 1/k^2 \end{matrix}\right) = \frac{2^{4b-3}\,\Gamma(2b-1/2)}{\sqrt{\pi}\,(2a-1)\,\Gamma(4b-2a-1)}$$
$[2b+1/2>2a>1]$.

9. $$\sum_{k=0}^{\infty} t^k L_k^c(x)\, {}_4F_3\left(\begin{matrix} -k/2,\ (1-k)/2,\ a/2,\ (a+1)/2 \\ 1/2,\ b/2+1,\ (b+1)/2;\ y \end{matrix}\right)=$$
$$=\frac{(1-t)^{a-b-1}}{2}\exp\left(-\frac{tx}{1-t}\right)\left[(1-t-ty^{1/2})^{-a}\, {}_1F_1\left(a;\, b+1;\, \frac{-t(1-t)^{-1}xy^{1/2}}{1-t-ty^{1/2}}\right)+\right.$$
$$\left.+(1-t+ty^{1/2})^{-a}\, {}_1F_1\left(a;\ b+1;\ \frac{t(1-t)^{-1}xy^{1/2}}{1-t-ty^{1/2}}\right)\right].$$

10. $$\sum_{k=0}^{\infty} \frac{(a)_k (b)_k}{(\mu+1)_k (\nu+1)_k} t^k P_k^{(\mu,\, \nu)}(x)\, {}_4F_3\left(\begin{matrix} a+k,\ b+k,\ (c+d-1)/2,\ (c+d)/2 \\ c,\ d,\ c+d-1;\ y \end{matrix}\right)=$$
$$=F_C^{(4)}\left(a,\ b;\ c,\ d,\ \mu+1,\ \nu+1;\ y,\ y;\ \frac{x-1}{2}t,\ \frac{x+1}{2}t\right)$$
$$\left[2|y|^{1/2}+\left|\frac{x-1}{2}t\right|^{1/2}+\left|\frac{x+1}{2}t\right|^{1/2}<1\right].$$

6.10. THE MacROBERT *E*-FUNCTION $E(p; a_r: q; b_s: z)$

6.10.1. Series of the form $\sum \alpha_k E((a_p) \pm mk;\ (b_q) \pm nk;\ z)$.

1. $$\sum_{k=0}^{\infty} \frac{t^k}{k!} E\left(\begin{matrix} (a_p)+k:\ z \\ (b_q)+k \end{matrix}\right)=E\left(\begin{matrix} (a_p):\ z/(1-tz) \\ (b_q) \end{matrix}\right) \quad [|tz|<1;\ |\arg z/(1-tz)|<\pi].$$

2. $$\sum_{k=0}^{\infty} \frac{t^k}{k!} E\left(\begin{matrix} (a_p),\ a+k \\ (b_q):\ z \end{matrix}\right)=(1-t)^{-a} E\left(\begin{matrix} (a_p),\ a \\ (b_q):\ z-tz \end{matrix}\right) \quad [|t|<1;\ |\arg z|,\ |\arg(1-t)z|<\pi].$$

3. $$\sum_{k=0}^{\infty} \frac{(b-c)_k}{k!\,(b)_k} z^{-k} E\left(\begin{matrix} (a_p)+k:\ z \\ (b_q)+k,\ c+k \end{matrix}\right)=\Gamma\left[\begin{matrix} b \\ c \end{matrix}\right] E\left(\begin{matrix} (a_p):\ z \\ (b_q),\ b \end{matrix}\right) \quad [\mathrm{Re}\, c>0;\ |\arg z|<\pi].$$

4. $$\sum_{k=0}^{\infty} \frac{(2a-2)_k}{k!} (-z)^{-k} E\left(\begin{matrix} (a_p)+k,\ a+k:\ z \\ (b_q)+k,\ 2a+2k-1 \end{matrix}\right)=\Gamma\left[\begin{matrix} a \\ 2a-1 \end{matrix}\right] E\left(\begin{matrix} (a_p):\ z \\ (b_q) \end{matrix}\right)$$
$$[|\arg z|<\pi].$$

5. $$\sum_{k=0}^{\infty} \frac{t^k}{k!} E\left(\begin{matrix} (a_p),\ \Delta(m,\ a+k) \\ (b_q):\ z \end{matrix}\right)=\left(1-\frac{t}{m}\right)^{-a} E\left(\begin{matrix} (a_p),\ \Delta(m,\ a) \\ (b_q):\ (1-t/m)^m z \end{matrix}\right)$$
$$[|t|<m;\ |\arg z|,\ |\arg((1-t/m)^m z)|<\pi].$$

6. $$\sum_{k=0}^{\infty} \frac{(c)_k}{k!} E\left(\begin{matrix} (a_p),\ \Delta(m,\ a+k):\ z \\ (b_q),\ \Delta(m,\ b+k) \end{matrix}\right)=\Gamma\left[\begin{matrix} b-a-c \\ b-a \end{matrix}\right] m^c E\left(\begin{matrix} (a_p),\ \Delta(m,\ a):\ z \\ (b_q),\ \Delta(m,\ b-c) \end{matrix}\right)$$
$$[\mathrm{Re}\,(b-a-c)>0;\ |\arg z|<\pi].$$

7. $$\sum_{k=0}^{\infty} \frac{t^k}{k!} E\left(\begin{matrix} (a_p):\ z \\ (b_q),\ \Delta(m,\ b-k) \end{matrix}\right)=\left(1+\frac{t}{m}\right)^{b-1} E\left(\begin{matrix} (a_p):\ (1+t/m)^{-m} z \\ (b_q),\ \Delta(m,\ b) \end{matrix}\right)$$
$$[|t|<m;\ |\arg z|,\ |\arg((1+t/m)^{-m} z)|<\pi].$$

8. $$\sum_{k=1}^{\infty} z^{-k} \cos 2ka\, E\left(\begin{matrix} 1/2+k,\ (a_p)+k:\ z \\ 1+2k,\ (b_q)+k \end{matrix}\right)=$$
$$=\frac{\sqrt{\pi}}{2} E\left(\begin{matrix} (a_p):\ z/\sin^2 a \\ (b_q) \end{matrix}\right)-\frac{1}{2} E\left(\begin{matrix} 1/2,\ (a_p):\ z \\ 1,\ (b_q) \end{matrix}\right) \quad [0 \leqslant a \leqslant \pi/2;\ |\arg z|<\pi].$$

9. $\sum_{k=0}^{\infty} z^{-k} \sin(2k+1) aE\left(\begin{matrix}3/2+k, & (a_p)+k: z\\ 2+2k, & (b_q)+k\end{matrix}\right)=\frac{\sqrt{\pi}}{2}\sin aE\left(\begin{matrix}(a_p): z/\sin^2 a\\ (b_q)\end{matrix}\right)$

[$0 \leqslant a \leqslant \pi/2$; $|\arg z| < \pi$].

6.10.2. Series containing products of E-functions.

1. $\sum_{k=0}^{\infty} \frac{z^{-2k}}{k!\,(c)_k} E^2\left(\begin{matrix}a+k, \ b+k, \ c+k\\ a+b+1/2+k: z\end{matrix}\right)=$

$$=2\sqrt{\pi}\,\Gamma\left[\begin{matrix}a, \ b, \ c\\ a+1/2, \ b+1/2\end{matrix}\right] E\left(\begin{matrix}2a, \ 2b, \ a+b, \ c\\ 2a+2b, \ a+b+1/2: z\end{matrix}\right)$$

[Re a, Re b, Re $c > 0$; $|\arg z| < 3\pi/2$].

2. $\sum_{k=0}^{\infty} \frac{z^{-2k}}{k!\,(c)_k} E\left(\begin{matrix}a+k, \ b+k, \ c+k\\ a+b-1/2+k: z\end{matrix}\right) E\left(\begin{matrix}a+k, \ b+k, \ c+k\\ a+b+1/2+k: z\end{matrix}\right)=$

$$=\sqrt{\pi}\,\Gamma\left[\begin{matrix}a, \ b, \ c\\ a+1/2, \ b+1/2\end{matrix}\right] E\left(\begin{matrix}2a, \ 2b, \ a+b, \ c\\ a+b+1/2, \ 2a+2b-1: z\end{matrix}\right)$$

[Re a, Re b, Re $c > 0$; Re $(2a+2b)>1$; $|\arg z| < 3\pi/2$].

3. $\sum_{k=0}^{\infty} \frac{z^{-2k}}{k!\,(c)_k} E\left(\begin{matrix}a+k, \ b+k, \ c+k\\ a+b-1/2+k: z\end{matrix}\right) E\left(\begin{matrix}a-1+k, \ b+k, \ c+k\\ a+b-1/2+k: z\end{matrix}\right)=$

$$=\sqrt{\pi}\,\Gamma\left[\begin{matrix}a-1, \ b, \ c\\ a-1/2, \ b+1/2\end{matrix}\right] E\left(\begin{matrix}2a-1, \ 2b, \ a+b-1, \ c\\ a+b-1/2, \ 2a+2b-2: z\end{matrix}\right)$$

[Re b, Re $c > 0$; Re $a > 1$; $|\arg z| < 3\pi/2$].

4. $\sum_{k=0}^{\infty} \frac{z^{-2k}}{k!\,(c)_k} E\left(\begin{matrix}a+k, \ b+k, \ c+k\\ a+b+1/2+k: \ z\end{matrix}\right) E\left(\begin{matrix}1/2-a+k, \ 1/2-b+k, \ c+k\\ 3/2-a-b+k: \qquad z\end{matrix}\right)=$

$$=\frac{\cos(a-b)\pi}{\pi^{3/2}}\Gamma\left[a, \ \frac{1}{2}-a, \ b, \ \frac{1}{2}-b, \ c\right] E\left(\begin{matrix}1/2+a-b, \ 1/2-a+b, \ c, \ 1/2\\ 1/2+a+b, \ 3/2-a-b: \qquad z\end{matrix}\right)$$

[$0 <$ Re a, Re $b < 1/2$; Re $c > 0$; $|\arg z| < 3\pi/2$].

5. $\sum_{k=0}^{\infty} \frac{z^{-2k}}{k!\,(a)_k} E\left(\begin{matrix}a+k, \ d-b-c+k\\ z\end{matrix}\right) E\left(\begin{matrix}a+k, \ b+k, \ c+k\\ d+k: \qquad z\end{matrix}\right)=$

$$=\Gamma\left[\begin{matrix}a, \ b, \ c, \ d-b-c\\ d-b, \ d-c\end{matrix}\right] E\left(\begin{matrix}a, \ d-b, \ d-c\\ d: \qquad z\end{matrix}\right)$$

[Re a, Re $(d-b-c) > 0$; $|\arg z| < 3\pi/2$].

6. $\sum_{k=0}^{\infty} \frac{z^{-2k}}{k!\,(a)_k} E\left(\begin{matrix}a+k, \ b+k\\ z\end{matrix}\right) E\left(\begin{matrix}a+k, \ c+k\\ z\end{matrix}\right)=B(b, c)\, E\left(\begin{matrix}a, \ b+c\\ z\end{matrix}\right)$

[Re $a > 0$; $|\arg z| < 3\pi/2$].

6.11. THE MEIJER G-FUNCTION: $G_{pq}^{mn}\left(z\,\middle|\,\begin{matrix}(a_p)\\ (b_q)\end{matrix}\right)$

NOTATION: $c^* = m+n-\frac{p+q}{2}$.

6.11.1. Series of the form $\sum \alpha_k t^k G_{pq}^{mn}\left(z\,\middle|\,\begin{matrix}(a_p) \pm k\,(c_p)\\ (b_q) \pm k\,(d_q)\end{matrix}\right)$.

1. $\sum_{k=0}^{\infty} \frac{t^k}{k!} G_{pq}^{mn}\left(z\,\middle|\,\begin{matrix}(a_p)\\ b+k, \ (b_{q-1})\end{matrix}\right)=(1-t)^{-b} G_{pq}^{mn}\left((1-t)\,z\,\middle|\,\begin{matrix}(a_p)\\ b, \ (b_{q-1})\end{matrix}\right)$

[$|t| < 1$; $m \geqslant 1$; $|\arg z| < c^*\pi$].

2. $\sum_{k=0}^{\infty} \frac{t^k}{k!} G_{pq}^{mn}\left(z \middle| \begin{matrix}(a_p)\\(b_{q-1}),\ b+k\end{matrix}\right)=(1+t)^{-b} G_{pq}^{mn}\left((1+t)z \middle| \begin{matrix}(a_p)\\(b_{q-1}),\ b\end{matrix}\right)$

[$|t|<1;\ m<q;\ |\arg z|<c^*\pi$].

3. $\sum_{k=0}^{\infty} \frac{t^k}{k!} G_{pq}^{mn}\left(z \middle| \begin{matrix}a-k,\ (a_{p-1})\\(b_q)\end{matrix}\right)=(1-t)^{a-1} G_{pq}^{mn}\left(\frac{z}{1-t} \middle| \begin{matrix}a,\ (a_{p-1})\\(b_q)\end{matrix}\right)$

[$|t|<1;\ n\geqslant 1;\ |\arg z|<c^*\pi$].

4. $\sum_{k=0}^{\infty} \frac{t^k}{k!} G_{pq}^{mn}\left(z \middle| \begin{matrix}(a_{p-1}),\ a-k\\(b_q)\end{matrix}\right)=(1+t)^{a-1} G_{pq}^{mn}\left(\frac{z}{1+t} \middle| \begin{matrix}(a_{p-1}),\ a\\(b_q)\end{matrix}\right)$

[$|t|<1;\quad |\arg z|<c^*\pi$].

5. $\sum_{k=0}^{\infty} \frac{1}{k!} G_{p+1,\ q}^{m,\ n+1}\left(z \middle| \begin{matrix}a-k,\ (a_p)\\(b_q)\end{matrix}\right)=z^{a-1}\frac{\prod_{j=1}^{m}\Gamma(b_j-a+1)\prod_{j=1}^{n}\Gamma(a-a_j)}{\prod_{j=m+1}^{q}\Gamma(a-b_j)\prod_{j=n+1}^{p}\Gamma(a_j-a+1)}$

[$0\leqslant m\leqslant q;\ 0\leqslant n\leqslant p;\ |\arg z|<(c^*+1/2)\pi$].

6. $\sum_{k=0}^{\infty} (-1)^k (k+\nu) G_{pq}^{mn}\left(z \middle| \begin{matrix}a-\nu-k,\ (a_{p-2}),\ a+\nu+k\\(b_q)\end{matrix}\right)=$

$=\frac{1}{2} G_{pq}^{mn}\left(z \middle| \begin{matrix}a-\nu,\ (a_{p-2}),\ a+\nu-1\\(b_q)\end{matrix}\right)$

[$1\leqslant n\leqslant p-1;\ |\arg z|<c^*\pi;\ \nu\neq 0,\ \pm 1,\ \pm 2,\ \ldots$].

7. $\sum_{k=0}^{\infty} \frac{t^k}{k!} G_{pq}^{mn}\left(z \middle| \begin{matrix}a,\ (a_{p-1})+k\\(b_q)+k\end{matrix}\right)=$

$=(1-t)^{a-1} G_{pq}^{mn}\left(\frac{z}{1-t} \middle| \begin{matrix}a,\ (a_{p-1})\\(b_q)\end{matrix}\right)$

[$|\arg z|<c^*\pi;\ |t|<1;\ |t\sqrt{z}-1/\sqrt{z}|<c^*\pi/2$].

6.11.2. Series containing trigonometric functions and the G-function.

1. $\sum_{k=0}^{\infty} \sin(2k+1)\varphi\, G_{p+2,\ q+2}^{m+1,\ n+1}\left(z \middle| \begin{matrix}1-k,\ (a_p),\ 2+k\\3/2,\ (b_q),\ 1\end{matrix}\right)=$

$=\frac{\sqrt{a}}{2}\sin\varphi\, G_{pq}^{mn}\left(\frac{z}{\sin^2\varphi} \middle| \begin{matrix}(a_p)\\(b_q)\end{matrix}\right)$

[$|\arg z|<c^*\pi;\ 0\leqslant\varphi\leqslant\pi$].

2. $\sum_{k=0}^{\infty} z^{-k}\sin(2k+1)\varphi\, G_{p+1,\ q+1}^{m+1,\ n}\left(z \middle| \begin{matrix}a,\ (a_{p-1})+k,\ a+2k+1\\a+k+1/2,\ (b_q)+k\end{matrix}\right)=$

$=\frac{\sqrt{\pi}}{2}\sin^{2a-1}\varphi\, G_{pq}^{mn}\left(\frac{z}{\sin^2\varphi} \middle| \begin{matrix}(a_p)\\(b_q)\end{matrix}\right)$ [$|\arg z|<c^*\pi;\quad 0\leqslant\varphi\leqslant\pi$].

3. $\sum_{k=1}^{\infty} z^{-k}\cos k\varphi\, G_{p+1,\ q+1}^{m+1,\ n}\left(z\left|\begin{matrix}a,\ (a_{p-1})+k,\ a+2k\\ a+k-1/2,\ (b_q)+k\end{matrix}\right.\right)=$

$=\frac{\sqrt{\pi}}{2}\sin^{2a-2}\frac{\varphi}{2}\, G_{pq}^{mn}\left(z\operatorname{cosec}^2\frac{\varphi}{2}\left|\begin{matrix}a,\ (a_{p-1})\\ (b_q)\end{matrix}\right.\right)-\frac{1}{2}\, G_{p+1,\ q+1}^{m+1,\ n}\left(z\left|\begin{matrix}a,\ (a_{p-1}),\ a\\ a-1/2,\ (b_q)\end{matrix}\right.\right)$

$[|\arg z|<c^*\pi;\quad 0\leqslant\varphi\leqslant\pi]$.

6.12. VARIOUS SERIES

6.12.1. Series containing the Neumann polynomials $O_n(x)$

1. $\sum_{k=0}^{\infty} J_k(x)\, O_k(y)=\frac{1}{2y}J_0(x)+\frac{1}{2(y-x)}$ $[|x|<|y|]$.

2. $\sum_{k=1}^{\infty} J_k(x)\,[O_{k+1}(y)-O_{k-1}(y)]=\frac{1}{(x-y)^2}-\frac{1}{y^2}J_0(x)$.

3. $\sum_{k=1}^{\infty} k^2 J_k(x)\, O(y)=\frac{x(y+x)}{(y-x)^3}+\frac{x^2}{y-x}$.

4. $\sum_{k=1}^{\infty} k\, J_k(kx)\, O_k(ky)=\frac{1}{2(1-y^2)}\left(\frac{1}{y-x}-\frac{1}{y}-x\frac{1+xy}{1-x^2}\right)$.

Chapter 7. THE HYPERGEOMETRIC FUNCTIONS: PROPERTIES, REPRESENTATIONS, PARTICULAR VALUES

7.1. INTRODUCTION

In this chapter we set out the main properties, representations and particular values of the hypergeometric functions of one and several variables. In view of their importance for applications, the cases $p=q+1$ and $p=q$ of the generalized hypergeometric functions ${}_pF_q((a_p);\ (b_q);z)$ have been given the greatest attention; in this connection, a sizeable section has been set aside for tables of expressions of these functions in terms of various elementary and special functions corresponding to various relations between the parameters (a_p), (b_q) with arbitrary argument z. Those cases where not all the parameters are fixed are called representations; if, on the other hand, all the parameters take numerical values, then these expressions are called particular values of ${}_pF_q((a_p);(b_q);z)$.

It should be noted that the formulae for the representations of ${}_pF_q((a_p);(b_q);-z)$ can be obtained from the corresponding formulae for ${}_pF_q((a_p);(b_q);z)$ by replacing z with $-z$; it has been appropriate in a number of cases to give the expressions for the functions ${}_pF_q((a_p);(b_q);-z)$ as well.

A general formula for calculating the values of ${}_{q+1}F_q((a_{q+1});(b_q);z_0)$ at the point $z_0=1$ for arbitrary parameters (a_{q+1}) and (b_q) is known only for the case $q=1$ (see 7.3.5). The analogous formulae expressing the values of ${}_{q+1}F_q((a_{q+1});(b_q);z)$ at the other points z_0 are available only under certain relations between the parameters and for particular values of z_0.

Tables 7.3–18 can be substantially extended by using the various properties of ${}_pF_q((a_p);(b_q);z)$ of section 7.2.

The results set out in this chapter can be used for calculating the values of sums and series after reducing them to the form $\sum_{k!} \frac{(a_p)^k}{k!(b_q)_k} z^k$ by means of the relations of Appendix II, also for calculating definite and indefinite integrals based on the formulae of Chapters 1 and 2.

7.2. THE MAIN PROPERTIES OF THE HYPERGEOMETRIC FUNCTIONS

7.2.1. The Gauss hypergeometric function ${}_2F_1(a,b;c;z)$.

The Gauss hypergeometric series is defined by the formula

1. $${}_2F_1(a,\ b;\ c;\ z) \equiv F(a,\ b;\ c;\ z) \equiv {}_2F_1\left(\begin{matrix} a,\ b;\ z \\ c \end{matrix}\right) \equiv {}_2F_1\left(\begin{matrix} a,\ b \\ c;\ z \end{matrix}\right) = \sum_{k=0}^{\infty} \frac{(a)_k (b)_k}{(c)_k\, k!} z^k,$$

where $c \neq 0, -1, -2, \ldots$ It converges if one of the following conditions holds

1) $|z|<1$;
2) $|z|=1,\quad \mathrm{Re}\,(c-a-b)>0$;
3) $|z|=1,\quad z\neq 1,\quad -1<\mathrm{Re}\,(c-a-b)\leqslant 0$;

and diverges in the remaining cases.

The Gauss hypergeometric function is defined for $|z|<1$ as the sum of the Gauss series, while for $|z|\geqslant 1$, it is defined as its analytic continuation. In order to select the principal branch of the analytic continuation (also denoted by the symbol ${}_2F_1(a,b;c;z)$) satisfying the condition $|\arg(1-z)|<\pi$, the cut $[1,\infty)$ is drawn in the complex z-plane. In particular, the analytic continuation can be obtained by means of the Euler integral representation

2. $${}_2F_1(a, b; c; z)=\Gamma\begin{bmatrix} c \\ b, c-b \end{bmatrix}\int_0^1 t^{b-1}(1-t)^{c-b-1}(1-tz)^{-a}\,dt$$

[$\operatorname{Re} c > \operatorname{Re} b > 0$; $|\arg(1-z)|<\pi$],

or the Mellin–Barnes representation

3. $${}_2F_1(a, b; c; z)=\Gamma\begin{bmatrix} c \\ a, b \end{bmatrix}\frac{1}{2\pi i}\int_{\gamma-i\infty}^{\gamma+i\infty}\Gamma\begin{bmatrix} s, a-s, b-s \\ c-s \end{bmatrix}(-z)^{-s}\,ds$$

[$0<\operatorname{Re} s=\gamma<\operatorname{Re} a$, $\operatorname{Re} b$; $|\arg(-z)|<\pi$].

4. $$=\Gamma\begin{bmatrix} c \\ a, b, c-a, c-b \end{bmatrix}\frac{1}{2\pi i}\int_{\gamma-i\infty}^{\gamma+i\infty}\Gamma[s, s+c-a-b, a-s, b-s](1-z)^{-s}ds$$

[0, $\operatorname{Re}(a+b-c)<\operatorname{Re} s<\operatorname{Re} a$, $\operatorname{Re} b$; $|\arg(1-z)|<\pi$],

and also by means of the Gauss series in neighbourhoods of the singular points $z=1$ and $z=\infty$.

5. $${}_2F_1(a, b; c; z)=\Gamma\begin{bmatrix} c, c-a-b \\ c-a, c-b \end{bmatrix}{}_2F_1(a, b; 1+a+b-c; 1-z)+$$
$$+\Gamma\begin{bmatrix} c, a+b-c \\ a, b \end{bmatrix}(1-z)^{c-a-b}{}_2F_1(c-a, c-b; 1+c-a-b; 1-z)$$

[$c-a-b\neq\pm n$; $|\arg(1-z)|<\pi$].

6. $${}_2F_1(a, b; c; z)=\Gamma\begin{bmatrix} c, b-a \\ b, c-a \end{bmatrix}(-z)^{-a}{}_2F_1\left(a, 1+a-c; 1+a-b; \frac{1}{z}\right)+$$
$$+\Gamma\begin{bmatrix} c, a-b \\ a, c-b \end{bmatrix}(-z)^{-b}{}_2F_1\left(b, 1+b-c; 1+b-a; \frac{1}{z}\right)$$

[$a-b\neq\pm n$; $|\arg(-z)|<\pi$]

(for the cases $c-a-b=\pm n$, $a-b=\pm$ see 7.2.3.14—23, 78—83 and 7.3.1).

Formulae for change of parameters and argument:

7. $${}_2F_1(a, b; c; z)={}_2F_1(b, a; c; z)=(1-z)^{-a}{}_2F_1\left(a, c-b; c; \frac{z}{z-1}\right)=$$
$$=(1-z)^{c-a-b}{}_2F_1(c-a, c-b; c; z).$$

For other transformation formulae for the functions ${}_2F_1(a,b;c;z)$ see 7.3.1.

8. ${}_2F_1(0, b; c; z)={}_2F_1(a, b; c; 0)=1.$

9. $$\lim_{c\to -m}\frac{1}{\Gamma(c)}\,{}_2F_1(a, b; c; z)=\frac{(a)_{m+1}(b)_{m+1}}{(m+1)!}z^{m+1}\times$$
$$\times{}_2F_1(a+m+1, b+m+1; m+2; z).$$

Differentiation formulae:

10. $$\frac{d^n}{dz^n}{}_2F_1(a, b; c; z)=\frac{(a)_n(b)_n}{(c)_n}{}_2F_1(a+n; b+n; c+n; z).$$

11. $$\frac{d^n}{dz^n}[z^{a+n-1}{}_2F_1(a, b; c; z)]=(a)_n z^{a-1}{}_2F_1(a+n, b; c; z).$$

12. $$\frac{d^n}{dz^n}[z^{c-1}{}_2F_1(a, b; c; z)]=(-1)^n(1-c)_n z^{c-n-1}{}_2F_1(a, b; c-n; z).$$

13. $$\frac{d^n}{dz^n}[(1-z)^{a+n-1}{}_2F_1(a, b; c; z)]=(-1)^n\frac{(a)_n(c-b)_n}{(c)_n}(1-z)^{a-1}\times$$
$$\times{}_2F_1(a+n, b; c+n; z).$$

14. $\frac{d^n}{dz^n}[(1-z)^{a+b-c}\,{}_2F_1(a,\ b;\ c;\ z)]=\frac{(c-a)_n(c-b)_n}{(c)_n}(1-z)^{a+b-c-n}\times$
$\times\,{}_2F_1(a,\ b;\ c+n;\ z).$

15. $\frac{d^n}{dz^n}[z^{c-1}(1-z)^{b-c+n}\,{}_2F_1(a,\ b;\ c;\ z)]=$
$=(-1)^n(1-c)_n z^{c-n-1}(1-z)^{b-c}\,{}_2F_1(a-n,\ b;\ c-n;\ z).$

16. $\frac{d^n}{dz^n}[z^{c-1}(1-z)^{a+b-c}\,{}_2F_1(a,\ b;\ c;\ z)]=$
$=(-1)^n(1-c)_n z^{c-n-1}(1-z)^{a+b-c-n}\,{}_2F_1(a-n,\ b-n;\ c-n;\ z).$

17. $\frac{d^n}{dz^n}[z^{c-a+n-1}(1-z)^{a+b-c}\,{}_2F_1(a,\ b;\ c;\ z)]=$
$=(c-a)_n z^{c-a-1}(1-z)^{a+b-c-n}\,{}_2F_1(a-n,\ b;\ c;\ z).$

18. $\frac{d^n}{dz^n}\left[z^{-a}\,{}_2F_1\left(a,\ b;\ c;\ \frac{1}{z}\right)\right]=(-1)^n(a)_n z^{-a-n}\,{}_2F_1\left(a+n,\ b;\ c;\ \frac{1}{z}\right).$

19. $\frac{d^n}{dz^n}\left[z^{-a}(z-1)^{a+b-c}\,{}_2F_1\left(a,\ b;\ c;\ \frac{1}{z}\right)\right]=$
$=(-1)^n(c-b)_n z^{-a}(z-1)^{a+b-c-n}\,{}_2F_1\left(a,\ b-n;\ c;\ \frac{1}{z}\right).$

20. $\frac{d^n}{dz^n}\left[z^{-a}(z-1)^{a+n-1}\,{}_2F_1\left(a,\ b;\ c;\ \frac{1}{z}\right)\right]=$
$=\frac{(a)_n(c-b)_n}{(c)_n}z^{-a-n}(z-1)^{a-1}\,{}_2F_1\left(a+n,\ b;\ c+n;\ \frac{1}{z}\right).$

21. $\frac{d^n}{dz^n}\left[z^{-a}(z-1)^{a-c+n}\,{}_2F_1\left(a,\ b;\ c;\ \frac{1}{z}\right)\right]=(1-c)_n z^{-a}(z-1)^{a-c}\times$
$\times\,{}_2F_1\left(a,\ b-n;\ c-n;\ \frac{1}{z}\right).$

22. $\frac{d^{2n+\sigma}}{dz^{2n+\sigma}}\,{}_2F_1\left(-n+\frac{1}{2},\ b;\ c;\ z^2\right)=$
$=(-4)^n\left(\frac{1}{2}\right)_n\left(\sigma+\frac{1}{2}\right)_n\frac{(b)_{n+\sigma}}{(c)_{n+\sigma}}z^\sigma\,{}_2F_1\left(n+\sigma+\frac{1}{2},\ \ b+n+\sigma;\ c+n+\sigma;\ z^2\right)$
$[\sigma=0$ or $1]$.

23. $\frac{d^{2n+\sigma}}{dz^{2n+\sigma}}\left[z\,{}_2F_1\left(-n-\sigma+\frac{3}{2},\ b;\ c;\ z^2\right)\right]=$
$=(-4)^n\left(\sigma-\frac{1}{2}\right)_n\left(\frac{3}{2}\right)_n\frac{(b)_n}{(c)_n}z^{1-\sigma}\,{}_2F_1\left(n+\frac{3}{2},\ b+n;\ c+n;\ z^2\right)$ $[\sigma=0$ or $1]$.

24. $\frac{d^{2n+\sigma}}{dz^{2n+\sigma}}\left[z^{2c-1}\,{}_2F_1\left(c-n-\sigma+\frac{1}{2},\ b;\ c;\ z^2\right)\right]=$
$=(-1)^\sigma(1-2c)_{2n+\sigma}z^{2c-2n-\sigma-1}\,{}_2F_1\left(c+\frac{1}{2},\ b;\ c-n;\ z^2\right)$ $[\sigma=0$ or $1]$.

25. $\frac{d^{2n+\sigma}}{dz^{2n+\sigma}}\left[z^{2c-2}\,{}_2F_1\left(c-n-\frac{1}{2},\ b;\ c;\ z^2\right)\right]=$
$=(-1)^\sigma(2-2c)_{2n+\sigma}z^{2c-2n-\sigma-2}\,{}_2F_1\left(c-\frac{1}{2},\ b;\ c-n-\sigma;\ z^2\right)$ $[\sigma=0$ or $1]$.

26. $\frac{d^{2n+\sigma}}{dz^{2n+\sigma}}\left[(1-z^2)^{b+n-1/2}\,{}_2F_1\left(c+n-\frac{1}{2},\ b;\ c;\ z^2\right)\right]=$
$=(-4)^n\left(\frac{1}{2}\right)_n\left(\sigma+\frac{1}{2}\right)_n\frac{(c-b)_{n+\sigma}}{(c)_{n+\sigma}}z^\sigma(1-z^2)^{b-n-\sigma-1/2}\,{}_2F_1\left(c-\frac{1}{2},\ b;\ c+n+\sigma;\ z^2\right)$
$[\sigma=0$ or $1]$.

27. $$\frac{d^{2n+\sigma}}{dz^{2n+\sigma}}\left[z(1-z^2)^{b+n+\sigma-3/2}{}_2F_1\left(c+n+\sigma-\frac{3}{2},\ b;\ c;\ z^2\right)\right]=$$
$$=(-4)^n\left(\sigma-\frac{1}{2}\right)_n\left(\frac{3}{2}\right)_n\frac{(c-b)_n}{(c)_n}z^{1-\sigma}(1-z^2)^{b-n-3/2}{}_2F_1\left(c-\frac{3}{2},\ b;\ c+n;\ z^2\right)$$
[$\sigma=0$ or 1].

28. $$\frac{d^{2n+\sigma}}{dz^{2n+\sigma}}\left[z^{2c-1}(1-z^2)^{b-c+n+\sigma-1/2}{}_2F_1\left(n+\sigma-\frac{1}{2},\ b;\ c;\ z^2\right)\right]=$$
$$=(-1)^\sigma(1-2c)_{2n+\sigma}z^{2c-2n-\sigma-1}(1-z^2)^{b-c-n-1/2}\times$$
$$\times{}_2F_1\left(-n-\frac{1}{2},\ b-n;\ c-n;\ z^2\right)$$
[$\sigma=0$ or 1].

29. $$\frac{d^{2n+\sigma}}{dz^{2n+\sigma}}\left[z^{2c-2}(1-z^2)^{b-c+n+1/2}{}_2F_1\left(n+\frac{1}{2},\ b;\ c;\ z^2\right)\right]=$$
$$=(-1)^\sigma(2-2c)_{2n+\sigma}z^{2c-2n-\sigma-2}(1-z^2)^{b-c-n-\sigma+1/2}\times$$
$$\times{}_2F_1\left(-n-\sigma+\frac{1}{2},\ b-n-\sigma;\ c-n-\sigma;\ z^2\right)$$
[$\sigma=0$ or 1].

30. $$\frac{d^{2n+\sigma}}{dz^{2n+\sigma}}{}_2F_1\left(a,\ b;\ \frac{1}{2};\ z^2\right)=$$
$$=2^{2n+2\sigma}(a)_{n+\sigma}(b)_{n+\sigma}z^\sigma{}_2F_1\left(a+n+\sigma,\ b+n+\sigma;\ \frac{1}{2}+\sigma;\ z^2\right)$$ [$\sigma=0$ or 1].

31. $$\frac{d^{2n+\sigma}}{dz^{2n+\sigma}}\left[z\,{}_2F_1\left(a,\ b;\ \frac{3}{2};\ z^2\right)\right]=$$
$$=2^{2n}(a)_n(b)_n z^{1-\sigma}{}_2F_1\left(a+n,\ b+n;\ \frac{3}{2}-\sigma;\ z^2\right)$$ [$\sigma=0$ or 1].

32. $$\frac{d^{2n+\sigma}}{dz^{2n+\sigma}}\left[(1-z^2)^{a+b-1/2}{}_2F_1\left(a,\ b;\ \frac{1}{2};z^2\right)\right]=$$
$$=2^{2n+2\sigma}\left(\frac{1}{2}-a\right)_{n+\sigma}\left(\frac{1}{2}-b\right)_{n+\sigma}z^\sigma(1-z^2)^{a+b-2n-\sigma-1/2}\times$$
$$\times{}_2F_1\left(a-n,\ b-n;\ \sigma+\frac{1}{2};\ z^2\right)$$
[$\sigma=0$ or 1].

33. $$\frac{d^{2n+\sigma}}{dz^{2n+\sigma}}\left[z(1-z^2)^{a+b-3/2}{}_2F_1\left(a,\ b;\ \frac{3}{2};\ z^2\right)\right]=$$
$$=2^{2n}\left(\frac{3}{2}-a\right)_n\left(\frac{3}{2}-b\right)_n z^{1-\sigma}(1-z^2)^{a+b-2n-\sigma-3/2}\times$$
$$\times{}_2F_1\left(a-n-\sigma,\ b-n-\sigma;\ \frac{3}{2}-\sigma;\ z^2\right)$$
[$\sigma=0$ or 1].

34. $$\frac{d^{2n+\sigma}}{dz^{2n+\sigma}}\left[(1-z^2)^{a+n+\sigma-1}\,{}_2F_1\left(a,\ a+\sigma-\frac{1}{2};\ \frac{1}{2};\ z^2\right)\right]=$$

$$=(-4)^{n+\sigma}(a)_{n+\sigma}(1-\sigma-a)_{n+\sigma}z^{\sigma}(1-z^2)^{a-n-1}{}_2F_1\left(a+\sigma,\ a+\sigma-\frac{1}{2};\ \sigma+\frac{1}{2};\ z^2\right)$$
[$\sigma=0$ or 1].

35. $$\frac{d^{2n+\sigma}}{dz^{2n+\sigma}}\left[z(1-z^2)^{a+n-1}\,{}_2F_1\left(a,\ a-\sigma+\frac{1}{2};\ \frac{3}{2};\ z^2\right)\right]=$$

$$=(-4)^{n}(a)_{n}(1+\sigma-a)_{n}z^{1-\sigma}(1-z^2)^{a-n-\sigma-1}{}_2F_1\left(a-\sigma,\ a-\sigma+\frac{1}{2};\ \frac{3}{2}-\sigma;\ z^2\right)$$
[$\sigma=0$ or 1].

36. $$\frac{d^{2n+\sigma}}{dz^{2n+\sigma}}\left[(1-z^2)^{n+\sigma}\,{}_2F_1\left(1,\ a;\ \frac{1}{2};\ z^2\right)\right]=$$

$$=(-4)^{n+\sigma}(n+\sigma)!\left(\frac{1}{2}-a\right)_{n+\sigma}z^{\sigma}{}_2F_1\left(n+\sigma+1,\ a;\ \sigma+\frac{1}{2};\ z^2\right)$$ [$\sigma=0$ or 1].

For derivatives with respect to the parameters, see 7.2.357–62.

The Gauss differential equation

37. $z(1-z)u''+[c-(a+b+1)z]u'-abu=0$

has the general solution

$$u(z)=C_1u_1(z)+C_2u_2(z),$$

where C_1, C_2 are arbitrary constants; for $c\neq 0, \pm 1, \pm 2, \ldots$ one can choose as u_1, u_2 the functions ${}_2F_1(a,b;c;z)$, $z^{1-c}{}_2F_1(1+a-c,1+b-c;2-c;z)$ or any of the four terms on the right hand side of formulae 7.2.1.5–6 or any of the 12 functions obtained from the 16 named by means of formulae 7.2.1.7.

7.2.2. The confluent hypergeometric functions of Kummer ${}_1F_1(a;b;z)$, Tricomi $\Psi(a,b;z)$ and Whittaker $M_{\rho,\sigma}(z)$, $W_{\rho,\sigma}(z)$.

The confluent hypergeometric functions of Kummer ${}_1F_1(a:b;z)$ and Tricomi $\Psi(a,b;z)$ are defined by the formulae

1. $${}_1F_1(a;\ b;\ z)=\sum_{k=0}^{\infty}\frac{(a)_k}{(b)_k}\frac{z^k}{k!}$$ [$|z|<\infty;\ b\neq 0,\ -1,\ -2,\ \ldots$].

2. $$\Psi(a,\ b;\ z)=\Gamma\begin{bmatrix}1-b\\ a-b+1\end{bmatrix}{}_1F_1(a;\ b;\ z)+\Gamma\begin{bmatrix}b-1\\ a\end{bmatrix}z^{1-b}{}_1F_1(a-b+1;\ 2-b;z)$$
[$|z|<\infty;\ b\neq 0,\ \pm 1,\ \pm 2,\ \ldots$].

3. $$\Psi(a,\ n+1;\ z)=\lim_{b\to n+1}\Psi(a,\ b;\ z)=$$

$$=\frac{(-1)^{n-1}}{n!\,\Gamma(a-n)}\left\{{}_1F_1(a;\ n+1;\ z)\ln z+\sum_{k=0}^{\infty}\frac{(a)_k}{(n+1)_k}[\psi(a+k)-\psi(k+1)-\psi(n+k+1)]\frac{z^k}{k!}\right\}+$$

$$+\frac{(n-1)!}{\Gamma(a)}\sum_{k=0}^{n-1}\frac{(a-n)_k}{(1-n)_k}\frac{z^{k-n}}{k!},$$

4. $\Psi(a,\ 1-n;\ z)=z^n\Psi(a+n,\ n+1;\ z)$ [$n=0,\ 1,\ 2,\ \ldots$].

The Whittaker functions $M_{\rho,\sigma}(z)$, $W_{\rho,\sigma}(z)$ are defined by the formulae

5. $M_{\rho,\sigma}(z)=z^{b/2}e^{-z/2}{}_1F_1(a;\ b;\ z)$, $W_{\rho,\sigma}(z)=z^{b/2}e^{-z/2}\Psi(a,\ b;\ z)$
[$a=1/2-\rho+\sigma,\ b=2\sigma+1;\ |\arg z|<\pi$].

Integral representations:

6. $${}_1F_1(a;\ b;\ z) = \Gamma\begin{bmatrix} b \\ a,\ b-a \end{bmatrix} \int_0^1 t^{a-1}(1-t)^{b-a-1} e^{zt}\,dt \qquad [\operatorname{Re} b > \operatorname{Re} a > 0].$$

7. $$\Psi(a,\ b;\ z) = \frac{1}{\Gamma(a)} \int_0^\infty t^{a-1}(1+t)^{b-a-1} e^{-zt}\,dt \qquad [\operatorname{Re} a,\ \operatorname{Re} z > 0].$$

Transformation formulae:

8. $${}_1F_1(a;\ b;\ z) = e^{z}{}_1F_1(b-a;\ b;\ -z).$$

9. $${}_1F_1(a; b; x) = \Gamma\begin{bmatrix} b \\ b-a \end{bmatrix} e^{\varepsilon a\pi i}\,\Psi(a,\ b;\ x) + \Gamma\begin{bmatrix} b \\ a \end{bmatrix} e^{\varepsilon(a-b)\pi i} e^{x}\,\Psi(b-a,\ b; -x) \qquad [\varepsilon = \operatorname{sgn} x].$$

10. $$\lim_{b \to -n} \frac{1}{\Gamma(b)}\,{}_1F_1(a;\ b;\ z) = \frac{(a)_{n+1}}{(n+1)!} z^{n+1}\,{}_1F_1(a+n+1;\ n+2;\ z).$$

11. $$\Psi(a,\ b;\ z) = z^{1-b}\Psi(a-b+1,\ 2-b;\ z).$$

12. $$\Psi(a,\ b;\ e^{2m\pi i} z) = e^{-2mb\pi i}\Psi(a,\ b;\ z) + (1-e^{-2mb\pi i})\Gamma\begin{bmatrix} 1-b \\ a-b+1 \end{bmatrix} {}_1F_1(a;\ b;\ z) \qquad [m = 0,\ \pm 1,\ \pm 2,\ \ldots].$$

13. $${}_1F_1(a;\ b;\ z) = \lim_{\alpha\to\infty} {}_2F_1\left(a,\ \alpha;\ b;\ \frac{z}{\alpha}\right).$$

14. $$\Psi(a,\ b;\ z) = z^{-a} \lim_{\alpha\to\infty} {}_2F_1\left(a,\ a-b+1;\ \alpha;\ 1-\frac{\alpha}{z}\right).$$

Evenness and symmetry formulae:

15. $$M_{\rho,\,\sigma}(z) = e^{\pm i\pi(\sigma+1/2)} M_{-\rho,\,\sigma}(-z) \qquad \left[\begin{Bmatrix} \operatorname{Im} z > 0 \\ \operatorname{Im} z < 0 \end{Bmatrix}\right].$$

16. $$W_{\rho,\,\sigma}(z) = W_{\rho,\,-\sigma}(z) = \Gamma\begin{bmatrix} -2\sigma \\ 1/2-\rho-\sigma \end{bmatrix} M_{\rho,\,\sigma}(z) + \Gamma\begin{bmatrix} 2\sigma \\ 1/2-\rho+\sigma \end{bmatrix} M_{\rho,\,-\sigma}(z).$$

Differentiation formulae:

17. $$\frac{d^n}{dz^n}\,{}_1F_1(a;\ b;\ z) = \frac{(a)_n}{(b)_n}\,{}_1F_1(a+n;\ b+n;\ z).$$

18. $$\frac{d^n}{dz^n}\left[z^{a+n-1}{}_1F_1(a;\ b;\ z)\right] = (a)_n z^{a-1}\,{}_1F_1(a+n;\ b;\ z).$$

19. $$\frac{d^n}{dz^n}\left[z^{b-1}{}_1F_1(a;\ b;\ z)\right] = (-1)^n (1-b)_n z^{b-n-1}\,{}_1F_1(a;\ b-n;\ z).$$

20. $$\frac{d^n}{dz^n}\left[e^{-z}{}_1F_1(a;\ b;\ z)\right] = (-1)^n \frac{(b-a)_n}{(b)_n} e^{-z}\,{}_1F_1(a;\ b+n;\ z).$$

21. $$\frac{d^n}{dz^n}\left[z^{b-1}e^{-z}{}_1F_1(a;\ b;\ z)\right] = (-1)^n (1-b)_n z^{b-n-1}e^{-z}{}_1F_1(a-n;\ b-n;\ z).$$

22. $$\frac{d^n}{dz^n}\left[z^{b-a+n-1}e^{-z}{}_1F_1(a;\ b;\ z)\right] = (b-a)_n z^{b-a-1}e^{-z}{}_1F_1(a-n;\ b;\ z).$$

23. $$\frac{d^n}{dz^n}\left[z^{-a}{}_1F_1\left(a;\ b;\ \frac{1}{z}\right)\right] = (-1)^n (a)_n z^{-a-n}{}_1F_1\left(a+n;\ b;\ \frac{1}{z}\right).$$

24. $$\frac{d^n}{dz^n}\left[z^{a-b}e^{-1/z}\,{}_1F_1\left(a;\ b;\ \frac{1}{z}\right)\right] = (-1)^n (b-a)_n z^{a-b-n} e^{-1/z}\,{}_1F_1\left(a-n;\ b;\ \frac{1}{z}\right).$$

25. $$\frac{d^{2n+\sigma}}{dz^{2n+\sigma}}\,{}_1F_1\left(-n+\frac{1}{2};\ c;\ z^2\right) = (-4)^n \frac{(1/2)_n(\sigma+1/2)_n}{(c)_{n+\sigma}} z^{\sigma} \times {}_1F_1\left(n+\sigma+\frac{1}{2};\ c+n+\sigma;\ z^2\right) \qquad [\sigma = 0 \text{ or } 1].$$

26. $$\frac{d^{2n+\sigma}}{dz^{2n+\sigma}}\left[z\,{}_1F_1\left(-n-\sigma+\frac{3}{2};\ c;\ z^2\right)\right]=$$
$$=(-4)^m\frac{(\sigma-1/2)_n\,(3/2)_n}{(c)_n}z^{1-\sigma}{}_1F_1\left(n+\frac{3}{2};\ c+n;\ z^2\right)\qquad[\sigma=0\ \text{or}\ 1]$$

27. $$\frac{d^{2n+\sigma}}{dz^{2n+\sigma}}\left[z^{2c-1}{}_1F_1\left(c-n-\sigma+\frac{1}{2};\ c;\ z^2\right)\right]=$$
$$=(-1)^\sigma(1-2c)_{2n+\sigma}z^{2c-2n-\sigma-1}{}_1F_1\left(c+\frac{1}{2};\ c-n;\ z^2\right)\qquad[\sigma=0\ \text{or}\ 1]$$

28. $$\frac{d^{2n+\sigma}}{dz^{2n+\sigma}}\left[z^{2c-2}{}_1F_1\left(c-n-\frac{1}{2};\ c;\ z^2\right)\right]=$$
$$=(-1)^\sigma(2-2c)_{2n+\sigma}z^{2c-2n-\sigma-2}\ {}_1F_1\left(c-\frac{1}{2};\ c-n-\sigma;\ z^2\right)\qquad[\sigma=0\ \text{or}\ 1]$$

29. $$\frac{d^{2n+\sigma}}{dz^{2n+\sigma}}{}_1F_1\left(a;\ \frac{1}{2};\ z^2\right)=2^{2n+2\sigma}(a)_{n+\sigma}z^\sigma\,{}_1F_1\left(a+n+\sigma;\ \sigma+\frac{1}{2};\ z^2\right)$$
$[\sigma=0$ or $1]$

30. $$\frac{d^{2n+\sigma}}{dz^{2n+\sigma}}\left[z\,{}_1F_1\left(a;\ \frac{3}{2};\ z^2\right)\right]=2^{2n}(a)_nz^{1-\sigma}{}_1F_1\left(a+n;\ \frac{3}{2}-\sigma;\ z^2\right)\qquad[\sigma=0\ \text{or}\ 1]$$

31. $$\frac{d^{2n+\sigma}}{dz^{2n+\sigma}}\left[e^{-z^2}{}_1F_1\left(c+n-\frac{1}{2};\ c;\ z^2\right)\right]=$$
$$=(-1)^\sigma 2^{2n}\frac{(1/2)_n\,(\sigma+1/2)_n}{(c)_{n+\sigma}}z^\sigma e^{-z^2}{}_1F_1\left(c-\frac{1}{2};\ c+n+\sigma;\ z^2\right)\qquad[\sigma=0\ \text{or}\ 1]$$

32. $$\frac{d^{2n+\sigma}}{dz^{2n+\sigma}}\left[ze^{-z^2}{}_1F_1\left(c+n+\sigma-\frac{3}{2};\ c;\ z^2\right)\right]=2^{2n}\frac{(\sigma-1/2)_n(3/2)_n}{(c)_n}z^{1-\sigma}e^{-z^2}\times$$
$$\times{}_1F_1\left(c-\frac{3}{2};\ c+n;\ z^2\right)\qquad[\sigma=0\ \text{or}\ 1]$$

33. $$\frac{d^{2n+\sigma}}{dz^{2n+\sigma}}\left[z^{2c-1}e^{-z^2}{}_1F_1\left(n+\sigma-\frac{1}{2};\ c;\ z^2\right)\right]=$$
$$=(-1)^\sigma(1-2c)_{2n+\sigma}z^{2c-2n-\sigma-1}e^{-z^2}{}_1F_1\left(-n-\frac{1}{2};\ c-n;\ z^2\right)\qquad[\sigma=0\ \text{or}\ 1]$$

34. $$\frac{d^{2n+\sigma}}{dz^{2n+\sigma}}\left[z^{2c-2}e^{-z^2}{}_1F_1\left(n+\frac{1}{2};\ c;\ z^2\right)\right]=$$
$$=(-1)^\sigma(2-2c)_{2n+\sigma}z^{2c-2n-\sigma-2}e^{-z^2}{}_1F_1\left(-n-\sigma+\frac{1}{2};\ c-n-\sigma;\ z^2\right)\qquad[\sigma=0\ \text{or}\ 1]$$

35. $$\frac{d^{2n+\sigma}}{dz^{2n+\sigma}}\left[e^{-z^2}{}_1F_1\left(a;\ \frac{1}{2};\ z^2\right)\right]=(-4)^{n+\sigma}\left(\frac{1}{2}-a\right)_{n+\sigma}z^\sigma e^{-z^2}{}_1F_1\left(a-n;\ \sigma+\frac{1}{2};\ z^2\right)$$
$[\sigma=0$ or $1]$

36. $$\frac{d^{2n+\sigma}}{dz^{2n+\sigma}}\left[ze^{-z^2}{}_1F_1\left(a;\ \frac{3}{2};\ z^2\right)\right]=(-4)^n\left(\frac{3}{2}-a\right)_nz^{1-\sigma}e^{-z^2}\times$$
$$\times{}_1F_1\left(a-n-\sigma;\ \frac{3}{2}-\sigma;\ z^2\right)\qquad[\sigma=0\ \text{or}\ 1].$$

37. $$\frac{d^n}{dz^n}\Psi(a,\ b;\ z)=(-1)^n(a)_n\Psi(a+n,b+n;\ z).$$

38. $$\frac{d^n}{dz^n}[z^{a+n-1}\Psi(a,\ b;\ z)]=(a)_n(a-b+1)_nz^{a-1}\Psi(a+n,\ b;\ z).$$

39. $$\frac{d^n}{dz^n}[z^{b-1}\Psi(a,\ b;\ z)]=(-1)^n(a-b+1)_nz^{b-n-1}\Psi(a,\ b-n;\ z).$$

40. $$\frac{d^n}{dz^n}[e^{-z}\Psi(a,\ b;\ z)]=(-1)^ne^{-z}\Psi(a,\ b+n;\ z).$$

41. $\frac{d^n}{dz^n}[z^{b-a+n-1}e^{-z}\Psi(a,\ b;\ z)]=(-1)^n z^{b-a-1}e^{-z}\Psi(a-n,\ b;\ z).$

42. $\frac{d^n}{dz^n}\left[z^{\pm\sigma-1/2}e^{-z/2}W_{\rho,\sigma}(z)\right]=(-1)^n z^{\pm\sigma-(n+1)/2}e^{-z/2}W_{\rho+n/2,\ \sigma\mp n/2}(z).$

For derivatives with respect to the parameters, see 7.2.3.57–61.

The Kummer differential equation

43. $zu''+(b-z)u'-au=0$

has the general solution

$$u(z)=C_1\,{}_1F_1(a;\ b;\ z)+C_2\Psi(a,\ b;\ z),$$

where C_1, C_2 are arbitrary constants.

7.2.3. The generalized hypergeometric function ${}_pF_q((a_p);(b_q);z)$.

The generalized hypergeometric series of order (p,q), $p,q=0,1,2,\ldots,$ is defined by the formula

1. $${}_pF_q(a_1,\ \ldots,\ a_p;\ b_1,\ \ldots,\ b_q;\ z)\equiv{}_pF_q((a_p);\ (b_q);\ z)\equiv{}_pF_q\left(\begin{matrix}(a_p);\ z\\(b_q)\end{matrix}\right)\equiv$$
$$\equiv{}_pF_q\left(\begin{matrix}(a_p)\\(b_q);\ z\end{matrix}\right)=\sum_{k=0}^{\infty}\frac{(a_1)_k(a_2)_k\ldots(a_p)_k}{(b_1)_k\ldots(b_q)_k}\frac{z^k}{k!},$$

where $b_j\neq 0,-1,-2,\ldots\ j=1,2,\ldots,q$; the quantities a_i and b_j are called the upper and lower parameters respectively. This series converges if one of the following conditions holds:

1) $p\leqslant q,\ |z|<\infty;$

2) $p=q+1,\ |z|<1;$

3) $p=q+1,\ |z|=1,\ \operatorname{Re}\psi_q=\sum_{k=1}^{q}b_k-\sum_{j=1}^{q+1}a_j>0;$

4) $p=q+1,\ |z|=1,\ z\neq 1,\ -1<\operatorname{Re}\psi_q\leqslant 0;$

and diverges in the remaining cases. If one of the upper parameters is equal to zero or a negative integer, then the series terminates, turning into a hypergeometric polynomial (see 7.2.3.4–5).

The generalized hypergeometric function of order (p,q) is defined as the sum of a generalized hypergeometric series in the region of convergence, while for $p=q+1, |z|\geqslant 1$, it is defined as an analytic continuation of this series. To select the principal branch of this continuation (also denoted by the symbol ${}_{q+1}F_q((a_{q+1});\ (b_q);z))$ satisfying the condition $|\arg(1-z)|<\pi$, the cut $[1,\infty)$ is drawn in the complex z-plane. In particular, the analytic continuation can be obtained by means of the integral representation 7.2.3.10, the Mellin–Barnes integral 7.2.3.13 or by means of the series expansions 7.2.3.77–79 and 7.2.3.80–82 in neighbourhoods of the singular points $z=\infty$ and $z=1$ respectively, where in a neighbourhood of $z=1$, these series for the function ${}_{q+1}F_q((a_{q+1});(b_q);z)$ are not hypergeometric for $q>1$.

2. The function ${}_pF_q((a_p);(b_q);z)$ is symmetric with respect to the upper and lower parameters, that is, it does not depend on the order of the arrangement $a_1,a_2,\ldots,a_p$ in (a_p) or $b_1,b_2,\ldots,b_q$ in (b_q).

3. ${}_pF_q((a_p);\ (b_q);\ 0)={}_pF_q(0,\ (a_{p-1});\ (b_q);\ z)=1.$

If some upper parameter is equal to $-n,\ n=0,1,2,\ldots,$ then the function ${}_pF_q((a_p);(b_q);z)$ turns into a polynomial of degree n, representable in the form

4. $${}_pF_q\left(\begin{matrix}-n,\ (a_{p-1});\ z\\b,\ (b_{q-1})\end{matrix}\right)=\frac{(-z)^n}{(b)_n}\frac{\Pi(a_{p-1})_n}{\Pi(b_{q-1})_n}\times$$
$$\times{}_{q+1}F_{p-1}\left(\begin{matrix}-n,\ 1-n-b,\ 1-n-(b_{q-1})\\1-n-(a_{p-1});\ (-1)^{p+q-1}z^{-1}\end{matrix}\right)\quad[b\neq 0,\ -1,\ -2,\ldots].$$

5. $${}_pF_q\left(\begin{matrix}-n,\ (a_{p-1});\ z\\ -m,\ (b_{q-1})\end{matrix}\right)=\frac{(m-n)!}{m!}z^n\frac{\Pi(a_{p-1})_n}{\Pi(b_{q-1})_n}\times$$
$$\times{}_{q+1}F_{p-1}\left(\begin{matrix}-n,\ 1+m-n,\ 1-n-(b_{q-1})\\ 1-n-(a_{p-1});\ (-1)^{p+q-1}z^{-1}\end{matrix}\right)+(-1)^{m+n}\frac{(m-n)!\,n!}{m!\,(m+1)!}\times$$
$$\times z^{m+1}\frac{\Pi(a_{p-1})_{m+1}}{\Pi(b_{q-1})_{m+1}}{}_pF_q\left(\begin{matrix}1+m-n,\ 1+m+(a_{p-1})\\ m+2,\ 1+m+(b_{q-1});\ z\end{matrix}\right)\quad [m>n$$

If some lower parameter is equal to $-m$, $m=0,1,2,\ldots$, and the parameter $-n$, $n=0,1,2,\ldots$, is not among the upper parameters, or there is such a parameter but $n>m$, then the function ${}_pF_q((a_p);(b_q);z)$ is undefined; however, the following limit exists:

6. $$\lim_{b\to-m}\frac{1}{\Gamma(b)}{}_pF_q\left(\begin{matrix}(a_p);\ z\\ (b_{q-1}),\ b\end{matrix}\right)=\frac{z^{m+1}}{(m+1)!}\frac{\Pi(a_p)_{m+1}}{\Pi(b_{q-1})_{m+1}}{}_pF_q\left(\begin{matrix}(a_p)+m+1;\ z\\ (b_{q-1})+m+1,\ m+2\end{matrix}\right)$$

If for r values of the upper parameters there are r values of the lower parameters equal to them, then the order of the function ${}_pF_q((a_p);(b_q);z)$ is reduced to $(p-r,q-r)$:

7. $${}_pF_q\left(\begin{matrix}(a_{p-r}),\ (c_r);\ z\\ (b_{q-r}),\ (c_r)\end{matrix}\right)={}_{p-r}F_{q-r}\left(\begin{matrix}(a_{p-r});\ z\\ (b_{q-r})\end{matrix}\right).$$

8. For fixed values of the argument z, the function $\Gamma^{-1}[(b_q)]\,{}_pF_q((a_p);(b_q);z)$ is an entire analytic function of the parameters.

Integral representations:

9. $${}_pF_q((a_p);\ (b_q);\ z)=$$
$$=\Gamma\left[\begin{matrix}b_q\\ a_p,\ b_q-a_p\end{matrix}\right]\int_0^1 t^{a_p-1}(1-t)^{b_q-a_p-1}\,{}_{p-1}F_{q-1}((a_{p-1});\ (b_{q-1});\ tz)\,dt$$
$[p\leqslant q+1;\ \mathrm{Re}\,b_q>\mathrm{Re}\,a_p>0;$ otherwise $|\arg(1-z)|<\pi$ for $p=q+1]$

10. $${}_{q+1}F_q\left(\begin{matrix}(a_q),\ \sigma\\ (b_q);\ z\end{matrix}\right)=$$
$$=\Gamma\left[\begin{matrix}(b_q)\\ (a_q),\ (b_q-a_q)\end{matrix}\right]\int_0^1\cdots\int_0^1\prod_{k=1}^{q}t_k^{a_k-1}(1-t_k)^{b_k-a_k-1}(1-t_1\ldots t_qz)^{-\sigma}dt_1\ldots dt_q$$
$[\mathrm{Re}\,b_k>\mathrm{Re}\,a_k>0;\ k=1,\ 2,\ \ldots,\ q;\ |\arg(1-z)|<\pi]$

11. $${}_pF_{p+r}\left(\begin{matrix}(a_p);\ z\\ (b_p),\ (c_r)\end{matrix}\right)=$$
$$=\Gamma\left[\begin{matrix}(b_p)\\ (a_p),\ (b_p-a_p)\end{matrix}\right]\int_0^1\cdots\int_0^1\prod_{k=1}^{p}t_k^{a_k-1}(1-t_k)^{b_k-a_k-1}\,{}_0F_r((c_r);\ t_1\ldots t_pz)\,dt_1\ldots dt$$
$[\mathrm{Re}\,b_k>\mathrm{Re}\,a_k>0;\ k=1,\ 2,\ \ldots,\ p$

The Mellin–Barnes integrals:

12. $${}_pF_q((a_p);\ (b_q);\ z)=\Gamma\left[\begin{matrix}(b_q)\\ (a_p)\end{matrix}\right]\frac{1}{2\pi i}\int_{L_{-\infty}}\Gamma\left[\begin{matrix}s,\ (a_p)-s\\ (b_q)-s\end{matrix}\right](-z)^{-s}ds$$
$[a_j\neq0,\ -1,\ -2,\ \ldots;\ j=1,\ 2,\ \ldots,\ p;\ s=-k\in D^+,\ s=a_j+k\in D^-,\ k=0,\ 1,\ 2,\ \ldots;\ L_-$ see 8.4.51.1]

13. $${}_{q+1}F_q((a_{q+1});(b_q);\ z)=\Gamma\left[\begin{matrix}(b_q)\\ (a_{q+1})\end{matrix}\right]\frac{1}{2\pi i}\int_{\gamma-i\infty}^{\gamma+i\infty}\Gamma\left[\begin{matrix}s,\ (a_{q+1})-s\\ (b_q)-s\end{matrix}\right](-z)^{-s}ds$$
$\left[0<\mathrm{Re}\,s=\gamma<\min_{1\leqslant j\leqslant q+1}\mathrm{Re}\,a_j;\ |\arg(-z)|<\pi\right]$

For representations in terms of the Meijer G-function, see 8.4.51.1.

Certain cases where the order of ${}_pF_q((a_p);(b_q);z)$ is lowered:

14. $${}_pF_q\left(\begin{matrix}(a_{p-1}),\ \sigma+1;\ z\\(b_{q-1}),\ \sigma\end{matrix}\right)= {}_{p-1}F_{q-1}\left(\begin{matrix}(a_{p-1});\ z\\(b_{q-1})\end{matrix}\right)+\frac{z}{\sigma}\prod_{j=1}^{p-1}a_j\prod_{k=1}^{q-1}b_k^{-1}\,{}_{p-1}F_{q-1}\left(\begin{matrix}(a_{p-1})+1;\ z\\(b_{q-1})+1\end{matrix}\right).$$

15. $${}_pF_q\left(\begin{matrix}(a_{p-1}),\ \sigma+m;\ z\\(b_{q-1}),\ \sigma\end{matrix}\right)=\sum_{k=0}^{m}\binom{m}{k}\frac{1}{(\sigma)_k}\frac{\Pi\,(a_{p-1})_k}{\Pi\,(b_{q-1})_k}\,{}_{p-1}F_{q-1}\left(\begin{matrix}(a_{p-1})+k;\ z\\(b_{q-1})+k\end{matrix}\right).$$

16. $${}_pF_q\left(\begin{matrix}(a_{p-2}),\ \rho+m,\ \sigma+n;\ z\\(b_{q-2}),\ \rho,\ \sigma\end{matrix}\right)=\sum_{j=0}^{m}\sum_{k=0}^{n}(-1)^{j+k}\frac{(-m)_j\,(-n)_k\,(\sigma+n)_j}{j!\,k!\,(\rho)_j\,(\sigma)_{j+k}}\times$$
$$\times\frac{\Pi\,(a_{p-2})_{j+k}}{\Pi\,(b_{q-2})_{j+k}}\,{}_{p-2}F_{q-2}\left(\begin{matrix}(a_{p-2})+j+k;\ z\\(b_{q-2})+j+k\end{matrix}\right).$$

17. $${}_pF_q\left(\begin{matrix}(a_{p-1}),\ 1;\ z\\(b_{q-1}),\ 2\end{matrix}\right)=z^{-1}\prod_{j=1}^{q-1}(b_j-1)\prod_{k=1}^{p-1}(a_k-1)^{-1}\left[{}_{p-1}F_{q-1}\left(\begin{matrix}(a_{p-1})-1;\ z\\(b_{q-1})-1\end{matrix}\right)-1\right]$$
$[a_1,\ \ldots,\ a_{p-1}\neq 1]$.

18. $${}_pF_q\left(\begin{matrix}(a_{p-1}),\ 1;\ z\\(b_{q-1}),\ 3\end{matrix}\right)=2z^{-2}\prod_{j=1}^{q-1}(b_j-2)\,(b_j-1)\prod_{k=1}^{p-1}(a_k-2)^{-1}\,(a_k-1)^{-1}\times$$
$$\times\left[{}_{p-1}F_{q-1}\left(\begin{matrix}(a_{p-1})-2;\ z\\(b_{q-1})-2\end{matrix}\right)-z\prod_{k=1}^{p-1}(a_k-2)\prod_{j=1}^{q-1}(b_j-2)^{-1}-1\right]$$
$[a_1,\ \ldots,\ a_{p-1}\neq 1,\ 2;\ b_1,\ \ldots,\ b_{q-1}\neq 2]$.

19. $${}_pF_q\left(\begin{matrix}(a_{p-1}),\ 1;\ z\\(b_{q-1}),\ n+1\end{matrix}\right)=(-1)^{p-q}\frac{\Pi\,(1-b_{q-1})_n}{\Pi(1-a_{p-1})_n}\times$$
$$\times\left[{}_{p-1}F_{q-1}\left(\begin{matrix}(a_{p-1})-n;\ z\\(b_{q-1})-n\end{matrix}\right)-\sum_{l=0}^{n-1}\frac{\Pi\,(a_{p-1}-n)_l}{\Pi\,(b_{q-1}-n)_l}\frac{z^l}{l!}\right].$$

20. $${}_pF_q\left(\begin{matrix}(a_{p-2}),\ \rho,\ \sigma;\ z\\(b_{q-2}),\ \rho+1,\ \sigma+1\end{matrix}\right)=\frac{1}{\sigma-\rho}\left[\sigma\,{}_{p-1}F_{q-1}\left(\begin{matrix}(a_{p-2}),\ \rho;\ z\\(b_{q-2}),\ \rho+1\end{matrix}\right)-\right.$$
$$\left.-\rho\,{}_{p-1}F_{q-1}\left(\begin{matrix}(a_{p-2}),\ \sigma;\ z\\(b_{q-2}),\ \sigma+1\end{matrix}\right)\right]$$ $[\rho\neq\sigma]$.

21. $${}_pF_q\left(\begin{matrix}(a_{p-2}),\ \rho,\ \sigma;\ z\\(b_{q-2}),\ \rho+n,\ \sigma+1\end{matrix}\right)=\frac{(\rho)_n}{(\rho-\sigma)_n}\,{}_{p-1}F_{q-1}\left(\begin{matrix}(a_{p-2}),\ \sigma;\ z\\(b_{q-2}),\ \sigma+1\end{matrix}\right)-$$
$$-\frac{(\rho)_n\sigma}{(\rho-\sigma)_n}\sum_{k=1}^{n}\frac{(\rho-\sigma-1)_k}{(\rho)_k}\,{}_{p-1}F_{q-1}\left(\begin{matrix}(a_{p-2}),\ \rho;\ z\\(b_{q-2}),\ \rho+k\end{matrix}\right)$$ $[\rho\neq\sigma]$.

22. $${}_pF_q\left(\begin{matrix}(a_{p-n}),\ (\sigma_n);\ z\\(b_{q-n}),\ (\sigma_n)+1\end{matrix}\right)=\sum_{k=1}^{n}{}_{p-n+1}F_{q-n+1}\left(\begin{matrix}(a_{p-n}),\ (\sigma_k);\ z\\(b_{q-n}),\ (\sigma_k)+1\end{matrix}\right)\prod_{\substack{j=1\\j\neq k}}^{n}\frac{\sigma_j}{\sigma_j-\sigma_k}.$$

23. $${}_pF_q\left(\begin{matrix}(a_{p-n}),\ (\sigma_n);\ z\\(b_{q-n}),\ (\sigma_n+m_n)\end{matrix}\right)=\prod_{j=1}^{n}\frac{(\sigma_j)^{m_j}}{(m_j-1)!}\sum_{k=1}^{n}\sum_{j_1=0}^{m_1-1}\cdots$$
$$\cdots\sum_{j_n=0}^{m_n-1}\frac{1}{\sigma_k+j_k}\prod_{l=1}^{n}\frac{(1-m_l)_{j_l}}{j_l!}\prod_{\substack{i=1\\i\neq k}}^{n}\frac{1}{\sigma_i+j_i-\sigma_k-j_k}\,{}_{p-n+1}F_{q-n+1}\left(\begin{matrix}(a_{p-n}),\ \sigma_k+j_k;\ z\\(b_{q-n}),\ \sigma_k+j_k+1\end{matrix}\right)$$

$[m_n=1,\ 2,\ldots$ and all the σ_i are distinct; if $\sigma_i-\sigma_k=N=1,\ 2,\ \ldots,$ then $m_k<N]$.

24. $${}_{np+1}F_{nq+n}\left(\begin{matrix}1,\ \Delta(n,(a_p)+m);\ z\\ \Delta(n,m+1),\ \Delta(n,(b_q)+m)\end{matrix}\right)=m!\,\frac{\Pi(b_q)_m}{\Pi(a_p)_m}\,n^{(p-q-1)m-1}\times$$
$$\times z^{-m/n}\sum_{k=0}^{n-1}e^{-2\pi imk/n}\,{}_pF_q\left((a_p);(b_q);\ n^{q-p+1}e^{2k\pi i/n}z^{1/n}\right).$$

Functions ${}_pF_q((a_p);(b_q);z)$, the upper parameters of which differ from the lower ones by integers, are connected by recurrence relations of the following type:

25. $$\sigma\,{}_pF_q\left(\begin{matrix}(a_{p-2}),\ \rho,\ \sigma+1\\ (b_q);\ z\end{matrix}\right)-\rho\,{}_pF_q\left(\begin{matrix}(a_{p-2}),\ \rho+1,\ \sigma\\ (b_q);\ z\end{matrix}\right)=(\sigma-\rho)\,{}_pF_q\left(\begin{matrix}(a_{p-2}),\ \rho,\ \sigma\\ (b_q);\ z\end{matrix}\right).$$

26. $$\sigma\,{}_pF_q\left(\begin{matrix}(a_{p-1}),\ \rho;\ z\\ (b_{q-1}),\ \sigma\end{matrix}\right)-\rho\,{}_pF_q\left(\begin{matrix}(a_{p-1}),\ \rho+1;\ z\\ (b_{q-1}),\ \sigma+1\end{matrix}\right)=(\sigma-\rho)\,{}_pF_q\left(\begin{matrix}(a_{p-1}),\ \rho;\ z\\ (b_{q-1}),\ \sigma+1\end{matrix}\right).$$

27. $$\sigma\,{}_pF_q\left(\begin{matrix}(a_p);\ z\\ \rho+1,\sigma,(b_{q-2})\end{matrix}\right)-\rho\,{}_pF_q\left(\begin{matrix}(a_p);\ z\\ \rho,\sigma+1,(b_{q-2})\end{matrix}\right)=(\sigma-\rho)\,{}_pF_q\left(\begin{matrix}(a_p);\ z\\ \rho+1,\sigma+1,(b_{q-2})\end{matrix}\right).$$

28. $$\tau(\rho-\sigma)\,{}_pF_q\left(\begin{matrix}(a_{p-2}),\ \rho,\ \sigma\\ (b_{q-1}),\ \tau;\ z\end{matrix}\right)-\rho(\tau-\sigma)\,{}_pF_q\left(\begin{matrix}(a_{p-2}),\ \rho+1,\ \sigma\\ (b_{q-1}),\ \tau+1;\ z\end{matrix}\right)+$$
$$+\sigma(\tau-\rho)\,{}_pF_q\left(\begin{matrix}(a_{p-2}),\ \rho,\ \sigma+1\\ (b_{q-1}),\ \tau+1;\ z\end{matrix}\right)=0.$$

29. $$\sigma(\tau-\rho)\,{}_pF_q\left(\begin{matrix}(a_{p-1}),\ \rho;\ z\\ \sigma,\ \tau+1,\ (b_{q-2})\end{matrix}\right)-\tau(\sigma-\rho)\,{}_pF_q\left(\begin{matrix}(a_{p-1}),\ \rho;\ z\\ \sigma+1,\ \tau,\ (b_{q-2})\end{matrix}\right)+$$
$$+\rho(\sigma-\tau)\,{}_pF_q\left(\begin{matrix}(a_{p-1}),\ \rho+1;\ z\\ \sigma+1,\ \tau+1,\ (b_{q-2})\end{matrix}\right)=0.$$

30. $$\prod_{j=1}^{q}b_j\left[{}_pF_q\left(\begin{matrix}(a_{p-1}),\sigma\\ (b_q);\ z\end{matrix}\right)-{}_pF_q\left(\begin{matrix}(a_{p-1}),\sigma+1\\ (b_q);\ z\end{matrix}\right)\right]+z\prod_{j=1}^{p-1}a_j\,{}_pF_q\left(\begin{matrix}(a_{p-1})+1,\sigma+1\\ (b_q)+1;\ z\end{matrix}\right)=0.$$

31. $$\sigma(\sigma+1)\prod_{j=1}^{q-1}b_j\left[{}_pF_q\left(\begin{matrix}(a_p);z\\ (b_{q-1}),\ \sigma\end{matrix}\right)-{}_pF_q\left(\begin{matrix}(a_p);\ \ z\\ (b_{q-1}),\ \sigma+1\end{matrix}\right)\right]=z\prod_{j=1}^{p}a_j\times$$
$$\times\,{}_pF_q\left(\begin{matrix}(a_p)+1;\ \ z\\ (b_{q-1})+1,\ \sigma+2\end{matrix}\right).$$

32. $$\prod_{j=1}^{q}b_j\left[{}_pF_q\left(\begin{matrix}(a_{p-2}),\ \rho,\ \sigma+1\\ (b_q);\ \ z\end{matrix}\right)-{}_pF_q\left(\begin{matrix}(a_{p-2}),\ \rho+1,\ \sigma\\ (b_q);\ \ z\end{matrix}\right)\right]+$$
$$+(\sigma-\rho)\,z\prod_{j=1}^{p-2}a_j\,{}_pF_q\left(\begin{matrix}(a_{p-2})+1,\ \rho+1,\ \sigma+1\\ (b_q)+1;\ \ z\end{matrix}\right)=0.$$

33. $$\sigma(\sigma+1)\prod_{j=1}^{q-1}b_j\left[{}_pF_q\left(\begin{matrix}(a_{p-1}),\ \rho;\ z\\ (b_{q-1}),\ \sigma\end{matrix}\right)-{}_pF_q\left(\begin{matrix}(a_{p-1}),\ \rho+1;z\\ (b_{q-1}),\ \sigma+1\end{matrix}\right)\right]+$$
$$+(\sigma-\rho)\,z\prod_{j=1}^{p}a_j\,{}_pF_q\left(\begin{matrix}(a_{p-1})+1,\ \rho+1;\ z\\ (b_{q-1})+1,\ \sigma+2\end{matrix}\right)=0.$$

34. $$\prod_{j=1}^{q-1}b_j\left[\tau\,{}_pF_q\left(\begin{matrix}(a_{p-2}),\ \rho,\ \sigma\\ (b_{q-1}),\ \tau;\ z\end{matrix}\right)-\rho\,{}_pF_q\left(\begin{matrix}(a_{p-2}),\ \rho+1,\ \sigma+1\\ (b_{q-1}),\ \tau+1;\ z\end{matrix}\right)-\right.$$
$$\left.-(\tau-\rho)\,{}_pF_q\left(\begin{matrix}(a_{p-2}),\ \rho,\ \sigma+1\\ (b_{q-1}),\ \tau+1;\ z\end{matrix}\right)\right]+z\rho\prod_{j=1}^{p-2}a_j\,{}_pF_q\left(\begin{matrix}(a_{p-2})+1,\ \rho+1,\ \sigma+1\\ (b_{q-1})+1,\ \tau+1;\ z\end{matrix}\right)=0.$$

35. $${}_pF_q\left(\begin{matrix}(a_{p-2}),\ \mu+1,\ \nu+1;\ z\\ \rho+1,\sigma+1,\tau+1,(b_{q-3})\end{matrix}\right)=\frac{\rho\sigma(\mu-\tau)(\nu-\tau)}{\mu\nu(\rho-\tau)(\sigma-\tau)}\,{}_pF_q\left(\begin{matrix}(a_{p-2}),\ \mu,\ \nu;\ z\\ \rho,\sigma,\tau+1,(b_{q-3})\end{matrix}\right)+$$
$$+\frac{\rho\tau(\mu-\sigma)(\nu-\sigma)}{\mu\nu(\rho-\sigma)(\tau-\sigma)}\,{}_pF_q\left(\begin{matrix}(a_{p-2}),\ \mu,\ \nu;\ z\\ \rho,\sigma+1,\tau,(b_{q-3})\end{matrix}\right)+\frac{\sigma\tau(\mu-\rho)(\nu-\rho)}{\mu\nu(\sigma-\rho)(\tau-\rho)}\,{}_pF_q\left(\begin{matrix}(a_{p-2}),\mu,\nu;z\\ \rho+1,\sigma,\tau,(b_{q-3})\end{matrix}\right).$$

36. $${}_pF_q\left(\begin{matrix}(a_{p-3}),\ \mu,\ \nu,\ \rho;\ z\\ \sigma,\ \tau,\ (b_{q-2})\end{matrix}\right)=\frac{\mu\nu(\sigma-\rho)(\tau-\rho)}{\sigma\tau(\mu-\rho)(\nu-\rho)}\,{}_pF_q\left(\begin{matrix}(a_{p-3}),\ \mu+1,\ \nu+1,\ \rho;\ z\\ \sigma+1,\ \tau+1,\ (b_{q-2})\end{matrix}\right)+$$
$$+\frac{\mu\rho(\sigma-\nu)(\tau-\nu)}{\sigma\tau(\mu-\nu)(\rho-\nu)}\,{}_pF_q\left(\begin{matrix}(a_{p-3}),\ \mu+1,\ \nu,\ \rho+1;\ z\\ \sigma+1,\ \tau+1,\ (b_{q-2})\end{matrix}\right)+\frac{\nu\rho(\sigma-\mu)(\tau-\mu)}{\sigma\tau(\nu-\mu)(\rho-\mu)}\times$$
$$\times\,{}_pF_q\left(\begin{matrix}(a_{p-3}),\ \mu,\ \nu+1,\ \rho+1;\ z\\ \sigma+1,\ \tau+1,\ (b_{q-2})\end{matrix}\right).$$

37. $$\left[\sigma+z\sum_{j=1}^{q}(a_j-b_j)\right]{}_{q+1}F_q\left(\begin{matrix}(a_q),\ \sigma;\ z\\ (b_q)\end{matrix}\right)+$$
$$+z\sum_{j=1}^{q}\frac{(b_j-\sigma)\prod\limits_{k=1}^{q}(b_j-a_k)}{b_j\prod\limits_{\substack{k=1\\k\neq j}}^{q}(b_j-b_k)}\,{}_{q+1}F_q\left(\begin{matrix}(a_q),\ \sigma;\ z\\ b_1,\ \ldots,\ b_{k-1},\ b_k+1,\ b_{k+1},\ldots,b_q\end{matrix}\right)=$$
$$=\sigma(1-z)\,{}_{q+1}F_q\left(\begin{matrix}(a_q),\ \sigma+1;\ z\\ (b_q)\end{matrix}\right).$$

Relations of special type:

38. $${}_pF_q\left(\begin{matrix}(a_p);\ \ z\\ (b_{q-2}),\ \sigma,\ 1-\sigma\end{matrix}\right)+{}_pF_q\left(\begin{matrix}(a_p);\ \ z\\ (b_{q-2}),\ -\sigma,\ 1+\sigma\end{matrix}\right)=2\,{}_pF_q\left(\begin{matrix}(a_p);\ \ z\\ (b_{q-2}),\ 1-\sigma,\ 1+\sigma\end{matrix}\right).$$

39. $${}_pF_q\left(\begin{matrix}(a_{p-1}),\ \sigma;\ \ z\\ (b_{q-2}),\ -\sigma,\ 1+\sigma\end{matrix}\right)-2\,{}_pF_q\left(\begin{matrix}(a_{p-1}),\ \sigma;\ \ z\\ (b_{q-2}),\ 1-\sigma,\ 1+\sigma\end{matrix}\right)=-{}_{p-1}F_{q-1}\left(\begin{matrix}(a_{p-1});\ z\\ (b_{q-2}),\ 1-\sigma\end{matrix}\right).$$

40. $${}_pF_q\left(\begin{matrix}(a_{p-1}),\ \sigma;\ z\\ (b_{q-1}),\ 1+\sigma\end{matrix}\right)+{}_pF_q\left(\begin{matrix}(a_{p-1}),\ -\sigma;\ z\\ (b_{q-1}),\ 1-\sigma\end{matrix}\right)=2\,{}_{p+1}F_{q+1}\left(\begin{matrix}(a_{p-1}),\ -\sigma,\ \sigma;\ z\\ (b_{q-1}),\ 1-\sigma,\ 1+\sigma\end{matrix}\right).$$

41. $${}_pF_q\left(\begin{matrix}(a_{p-2}),\ \sigma,\ 1-\sigma\\ (b_q);\ z\end{matrix}\right)+{}_pF_q\left(\begin{matrix}(a_{p-2}),\ -\sigma,\ 1+\sigma\\ (b_q);\ z\end{matrix}\right)=2\,{}_pF_q\left(\begin{matrix}(a_{p-2}),\ -\sigma,\ \sigma\\ (b_q);\ z\end{matrix}\right).$$

Decomposition of ${}_pF_q$ into even and odd parts and its generalization:

42. $${}_pF_q((a_p);\ (b_q);\ z)=A^+(z)+A^-(z),$$
$$A^{\pm}(z)=\frac{1}{2}\left[{}_pF_q((a_p);\ (b_q);\ z)\pm{}_pF_q((a_p);\ (b_q);\ -z)\right],$$
$$A^+(z)={}_{2p}F_{2q+1}\left(\begin{matrix}(a_p)/2,\ ((a_p)+1)/2;\ 4^{p-q-1}z^2\\ (b_q)/2,\ ((b_q)+1)/2,\ 1/2\end{matrix}\right).$$
$$A^-(z)=z\prod_{j=1}^{p}a_j\prod_{k=1}^{q}b_k^{-1}\,{}_{2p}F_{2q+1}\left(\begin{matrix}(a_p)/2+1,\ ((a_p)+1)/2;\ 4^{p-q-1}z^2\\ (b_q)/2+1,\ ((b_q)+1)/2,\ 3/2\end{matrix}\right).$$

43. $${}_pF_q((a_p);\ (b_q);\ z)=$$
$$=\sum_{k=0}^{n-1}\frac{z^k}{k!}\frac{\Pi(a_p)_k}{\Pi(b_q)_k}\,{}_{np+1}F_{nq+n}\left(\begin{matrix}1,\ \Delta(n,a_1+k),\ \ldots,\ \Delta(n,a_p+k);\ z^n/n^{n(1+q-p)}\\ \Delta(n,\ k+1),\ \Delta(n,\ b_1+k),\ \ldots,\ \Delta(n,\ b_q+k)\end{matrix}\right).$$

Product of functions ${}_pF_q((a_p);(b_q);z)$:

44. $${}_pF_q((a_p);\ (b_q);\ cz)\,{}_rF_s((\alpha_r);\ (\beta_s);\ dz)=\sum_{k=0}^{\infty}c_kz^k,$$
$$c_k=\frac{d^k}{k!}\frac{\Pi(\alpha_r)_k}{\Pi(\beta_s)_k}\,{}_{p+s+1}F_{q+r}\left(\begin{matrix}-k,\ 1-(\beta_s)-k,\ (a_p);\ (-1)^{r+s+1}c/d\\ 1-(\alpha_r)-k,\ (b_q)\end{matrix}\right)$$

или

$$c_k=\frac{c^k}{k!}\frac{\Pi(a_p)_k}{\Pi(b_q)_k}\,{}_{q+r+1}F_{p+s}\left(\begin{matrix}-k,\ 1-(b_q)-k,\ (\alpha_r);\ (-1)^{p+q+1}d/c\\ 1-(a_p)-k,\ (\beta_s)\end{matrix}\right),$$

when ${}_pF_q((a_p);(b_q);z)$ and ${}_rF_s((\alpha_r);(\beta_s);z)$ are not polynomials. If ${}_pF_q((a_p);(b_q);z)$ is a polynomial of degree m (for example when $a_p=-m$), then for $k=0,1,\ldots,m$, the values of c_k are defined by the same formulae and

$$c_k=\frac{d^k}{k!}\frac{\Pi(\alpha_r)_k}{\Pi(\beta_s)_k}\,{}_{p+s+1}F_{q+r}\left(\begin{matrix}-m,\ 1-(\beta_s)-k,\ (a_{p-1}),\ -k;\ (-1)^{r+s+1}c/d\\ (b_q),\ 1-(\alpha_r)-k\end{matrix}\right)$$

If ${}_pF_q((a_p);(b_q);z)$ and ${}_rF_s((\alpha_r);(\beta_s);z)$ are polynomials (for example, when $a_p=-m$, $\alpha_r=-n$, $m\leqslant n$), then for $k=0,1,\ldots,m$ and $k=m+1,m+2,\ldots$, the corresponding formulae for c_k given above remain valid and

$$c_{n+k}=\frac{(-m)_k c^k(-d)^n}{k!}\frac{\Pi(a_{p-1})_k\,\Pi(\alpha_{r-1})_n}{\Pi(b_q)_k\,\Pi(\beta_s)_n}\times$$
$$\times{}_{p+s+1}F_{q+r}\left(\begin{matrix}k-m,\ (a_{p-1})+k,\ -n,\ 1-(\beta_s)-n\\ (b_q)+k,\ k+1,\ 1-(\alpha_{r-1})-n;\ (-1)^{r+s+1}c/d\end{matrix}\right)\qquad [k=1,\ 2,\ \ldots,\ m].$$

Confluence formulae:

45. $$\lim_{|\sigma|\to\infty}{}_pF_q\left((a_{p-1}),\ \sigma;\ (b_q);\ \frac{z}{\sigma}\right)={}_{p-1}F_q((a_{p-1});\ (b_q);\ z)\qquad [q\geqslant p-1].$$

46. $$\lim_{|\sigma|\to\infty}{}_pF_q((a_p);\ (b_{q-1}),\ \sigma;\ \sigma z)={}_pF_{q-1}((a_p);\ (b_{q-1});\ z)$$
$$[q\geqslant p+1;\ |z|<\infty]\ \text{or}\ [q=p;\ |z|<1;\ \operatorname{Re}\sigma\geqslant 0].$$

Differentiation with respect to the argument:

47. $$\frac{d^n}{dz^n}\,{}_pF_q((a_p);\ (b_q);\ z)=\frac{\Pi(a_p)_n}{\Pi(b_q)_n}\,{}_pF_q((a_p)+n;\ (b_q)+n;\ z).$$

48. $$\frac{d^n}{dz^n}[z^\sigma\,{}_pF_q((a_p);(b_q);z)]=(-1)^n(-\sigma)_n z^{\sigma-n}\,{}_{p+1}F_{q+1}(\sigma+1,(a_p);\sigma-n+1,(b_q);z)$$
$$[\sigma-n+1\neq 0,\ -1,\ -2,\ \ldots],$$

49. $$=\frac{n!\,\Pi(a_p)_{n-\sigma}}{(n-\sigma)!\,\Pi(b_q)_{n-\sigma}}\,{}_{p+1}F_{q+1}(n+1,\ (a_p)+n-\sigma;\ n-\sigma+1,\ (b_q)+n-\sigma;\ z)$$
$$[\sigma-n+1=0,\ -1,\ -2,\ \ldots].$$

50. $$\frac{d^n}{dz^n}[z^{\sigma+n-1}\,{}_pF_q((a_{p-1}),\ \sigma;\ (b_q);\ z)]=(\sigma)_n z^{\sigma-1}\,{}_pF_q((a_{p-1}),\ \sigma+n;\ (b_q);\ z).$$

51. $$\frac{d^n}{dz^n}[z^{\sigma-1}\,{}_pF_q((a_p);\ (b_{q-1}),\ \sigma;\ z)]=(\sigma-n)_n z^{\sigma-n-1}\,{}_pF_q((a_p);\ (b_{q-1}),\ \sigma-n;\ z).$$

52. $$\frac{d^n}{dz^n}\left[z^n\,{}_pF_q\left(-n,\ (a_{p-1});\ \frac{1}{2},\ (b_{q-1});\ z\right)\right]=n!\ {}_{p+1}F_{q+1}\left(-n,\ n+1,\ (a_{p-1})\ \ \frac{1}{2},\ 1,\ (b_{q-1});\ z\right).$$

53. $$\frac{d^n}{dz^n}[z^n\,{}_pF_q(-n,\ 1,\ (a_{p-2});\ n+1,\ (b_{q-1});\ z)]=n!\ {}_{p-1}F_{q-1}(-n,(a_{p-2});\ (b_{q-1});\ z).$$

54. $$\frac{d^n}{dz^n}[z^\sigma\,{}_pF_q(-n,\ (a_{p-1});\ (b_q);\ z)]=(-1)^n(-\sigma)_n z^{\sigma-n}\,{}_{p+1}F_{q+1}(-n,\sigma+1,\ (a_{p-1});\ \sigma-n+1,\ (b_q);\ z).$$

55. $$\frac{d^n}{dz^n}\left[z^\sigma\,{}_pF_q\left(\begin{matrix}\Delta(r,\ -n),\ (a_{p-r})\\ (b_q);\quad z^m\end{matrix}\right)\right]=(-1)^n(-\sigma)_n z^{\sigma-n}\times$$
$$\times{}_{p+m}F_{q+m}\left(\begin{matrix}\Delta(r,\ -n),\ \Delta(m,\ \sigma+1),\ (a_{p-r})\\ \Delta(m,\ \sigma-n+1),\ (b_q);\ z^m\end{matrix}\right).$$

56. $$\frac{d^n}{dz^n}[e^{-z}\,{}_pF_q(-n,\ (a_{p-1});\ (b_q);\ z)]=$$
$$=(-1)^n e^{-z}\sum_{k=0}^{n}\frac{(-n)_k z^k}{k!\,(b_1)_k\cdots(b_q)_k}\,{}_{p+1}F_q(-n,\ k-n,\ (a_{p-1})+k;\ (b_q)+k;\ z).$$

Differentiation of ${}_pF_q((a_p);(b_q);z)$ with respect to the parameters does not, in general, lead to hypergeometric functions:

57. $$\frac{\partial}{\partial\alpha}\,{}_pF_q\left(\frac{\alpha+\nu}{\beta},\ (a_{p-1});\ (b_q);\ z\right)=$$
$$=\frac{\alpha+\nu}{\beta^2}\prod_{j=1}^{p-1}a_j\prod_{k=1}^{q}b_k^{-1}\sum_{l=0}^{\infty}\frac{(a_1+1)_l\ldots(a_{p-1}+1)_l}{(b_1+1)_l\ldots(b_q+1)_l}\left(\frac{\alpha+\nu}{\beta}+1\right)_l\frac{z^{l+1}}{(l+1)!}\times$$
$$\times\left[\psi\left(\frac{\alpha+\nu}{\beta}+l+1\right)-\psi\left(\frac{\alpha+\nu}{\beta}\right)\right].$$

58. $$\frac{\partial}{\partial\alpha}\,{}_pF_q\left((a_p);\ (b_{q-1}),\ \frac{\alpha+\nu}{\beta};\ z\right)=-\frac{z}{\alpha+\nu}\prod_{j=1}^{p}a_j\prod_{k=1}^{q-1}b_k^{-1}\times$$
$$\times\sum_{l=0}^{\infty}\frac{(a_1+1)_l\ldots(a_p+1)_l}{(b_1+1)_l\ldots(b_{q-1}+1)_l\,((\alpha+\nu)/\beta+1)_l}\frac{z^l}{(l+1)!}\left[\psi\left(\frac{\alpha+\nu}{\beta}+l+1\right)-\psi\left(\frac{\alpha+\nu}{\beta}\right)\right].$$

Certain particular cases are an exception:

59. $$\frac{\partial}{\partial\alpha}\,{}_pF_q\left(\begin{matrix}\alpha+\nu,\ (a_{p-1});\ z\\ \alpha+\nu+1,\ (b_{q-1})\end{matrix}\right)=$$
$$=\frac{z}{(\alpha+\nu+1)^2}\prod_{j=1}^{p-1}a_j\prod_{k=1}^{q-1}b_k^{-1}\,{}_{p+1}F_{q+1}\left(\begin{matrix}\alpha+\nu+1,\ \alpha+\nu+1,\ (a_{p-1})+1;\ z\\ \alpha+\nu+2,\ \alpha+\nu+2,\ (b_{q-1})+1\end{matrix}\right).$$

60. $$\frac{\partial}{\partial\alpha}\,{}_pF_q\left(\begin{matrix}\alpha+\nu+1,\ (a_{p-1});\ z\\ \alpha+\nu,\ (b_{q-1})\end{matrix}\right)=$$
$$=-\frac{z}{(\alpha+\nu)^2}\prod_{j=1}^{p-1}a_j\prod_{k=1}^{q-1}b_k^{-1}\,{}_{p+1}F_{q+1}\left(\begin{matrix}\alpha+\nu,\ \alpha+\nu,\ (a_{p-1})+1;\ z\\ \alpha+\nu+1,\ \alpha+\nu+1,\ (b_{q-1})+1\end{matrix}\right).$$

61. $$\frac{\partial}{\partial\alpha}\,{}_1F_1(1;\ \alpha+\nu;\ z)=-\frac{ze^z}{(\alpha+\nu)^2}\,{}_2F_2(\alpha+\nu,\ \alpha+\nu;\ \alpha+\nu+1,\ \alpha+\nu+1;\ -z).$$

62. $$\frac{\partial}{\partial\alpha}\,{}_2F_1(1,\ \alpha;\ \alpha+\nu;\ z)=\frac{\nu z(1-z)^{\nu-1}}{(\alpha+\nu)^2}\,{}_3F_2(\alpha+\nu,\ \alpha+\nu,\ \nu+1;\ \alpha+\nu+1,\ \alpha+\nu+1;\ z).$$

63. $$\frac{\partial}{\partial\alpha}\,{}_1F_2\left(1;\ \alpha+1,\ \alpha+\frac{3}{2};\ -\frac{z^2}{4}\right)\Bigg|_{\alpha=0}=\frac{2}{z}[\sin z\,\mathrm{ci}(z)-\cos z\,\mathrm{Si}(z)-$$
$$-\sin z\,(\mathbf{C}+\ln z-1)].$$

64. $$\frac{\partial}{\partial\alpha}\,{}_1F_2\left(1;\ \alpha+1,\ \alpha+\frac{1}{2};\ -\frac{z^2}{4}\right)\Bigg|_{\alpha=0}=$$
$$=2[\cos z\,\mathrm{ci}(z)+\sin z\,\mathrm{Si}(z)-\cos z\,(\mathbf{C}+\ln z)].$$

65. $$\frac{\partial}{\partial\alpha}\,{}_1F_2\left(1;\ \alpha+1,\ \alpha+n+1;\ -\frac{z^2}{4}\right)\Bigg|_{\alpha=0}=$$
$$=n!\left(\frac{2}{z}\right)^n\left[\pi Y_n(z)+\left(\psi(n+1)-\mathbf{C}-2\ln\frac{z}{2}\right)J_n(z)+\sum_{k=0}^{n-1}\frac{(n-k-1)!}{k!}\left(\frac{z}{2}\right)^{2k-n}\right].$$

The differential equation satisfied by ${}_pF_q((a_p);(b_q);z)$ has the form

66. $$\left[\prod_{k=0}^{q}\left(z\frac{d}{dz}+b_k-1\right)-z\sum_{j=1}^{p}\left(z\frac{d}{dz}+a_j\right)\right]u=0,\ b_0=1.$$

This equation of order $\max(p,q+1)$ has two ($z=0,\infty$ for $p\leqslant q$) or three ($z=0,1,\infty$ for $p=q+1$) singular points. If $p\leqslant q$, then the singular point $z=0$ is regular, while $z=\infty$ is a non-regular point (essential singularity); if $p=q+1$, then all three singular points $z=0,1,\infty$ are regular.

We introduce the notation:

$(b^*)=b_0, b_1, \ldots, b_q$ is a $q+1$-dimensional vector for which $b_0=1$,

67. $u=u_k^0(z)=z^{1-b_k}\,{}_pF_q\left(1+(a_p)-b_k;\ 1+(b_q^*)'-b_k;\ z\right)$ $[k=0, 1, 2, \ldots, q;\ b_k-b_l\neq 0, \pm 1, \pm 2, \ldots, k\neq l]$,

68. $u=u_j^\infty(z)=z^{-a_j}\,{}_{q+1}F_{p-1}\left(1+a_j-(b_q^*);\ 1+a_j-(a_p)';\ (-1)^{q-p+1}z^{-1}\right)$ $[j=1, 2, \ldots, p;\ a_j-a_k\neq 0, \pm 1, \pm 2, \ldots, k\neq j]$,

where the prime $'$ in $1+(b_q^*)'-b_k$ and $1+a_j-(a_p)'$ means that the term $1+b_h-b_k$ for $h=k$ and $1+a_j-a_h$ for $h=j$ are missing;

69. $u=R_{jq}(z)=G_{q+2,\,q+2}^{2,\,q+2}\left(z\,\middle|\,\begin{matrix}1-(a_{q+1}),\ 1-b_j\\ 0,\ 1-b_j,\ 1-(b_q)\end{matrix}\right)$ $[j=1, 2, \ldots, q]$,

70. $u=\xi_q(z, \psi_q)$ $[\psi_q\neq 0, \pm 1, \pm 2, \ldots]$,

71. $u=\varphi_q(z)$ $[\psi_q=0, \pm 1, \pm 2, \ldots]$,

where the function $\xi_q(z, \psi_q)$ can be expressed in terms of the Meijer G-function,

72. $\xi_q(z, \psi_q)=G_{q+1,\,q+1}^{q+1,\,0}\left(z\,\middle|\,\begin{matrix}1-(a_{q+1})\\ 0,\ 1-(b_q)\end{matrix}\right)+e^{\mp\pi i\psi_q}G_{q+1,\,q+1}^{0,\,q+1}\left(z\,\middle|\,\begin{matrix}1-(a_{q+1})\\ 0,\ 1-(b_q)\end{matrix}\right)$ $\left[\left\{\begin{matrix}-\pi<\arg(1-z)\leqslant 0\\ 0\leqslant\arg(1-z)<\pi\end{matrix}\right\}\right]$,

or in the form of the series

73. $\xi_q(z, \psi_q)=(1-z)^{\psi_q}\sum_{k=0}^{\infty}\frac{c_{kq}}{\Gamma(\psi_q+k+1)}(1-z)^k$ $[|1-z|<1]$,

$$\psi_q=\sum_{k=1}^{q}b_k-\sum_{j=1}^{q+1}a_j.$$

Here $c_{0q}=1, c_{k0}=0, k=1,2,\ldots$, the remaining coefficients c_{kq} being defined by the formulae

$c_{11}=R(\psi_1),\ 2c_{21}=R(\psi_1+1)\,c_{11},\ \ldots,$

$kc_{k1}=R(\psi_1+k-1)\,c_{k-1,\,1};$

$c_{12}=\Delta R(\psi_2-1)-Q(\psi_2),$

$2c_{22}=[\Delta R(\psi_2)-Q(\psi_2+1)]\,c_{12}-R(\psi_2),$

. . . .

$kc_{k2}=[\Delta R(\psi_2+k-2)-Q(\psi_2+k-1)]\,c_{k-1,\,2}-R(\psi_2+k-2)\,c_{k-2};$

$$c_{1q}=\frac{\Delta^{q-1}R(\psi_q-q+1)}{(q-1)!}-\frac{\Delta^{q-2}Q(\psi_q-q+2)}{(q-2)!},$$

$$2c_{2q}=\left[\frac{\Delta^{q-1}R(\psi_q-q+2)}{(q-1)!}-\frac{\Delta^{q-2}Q(\psi_q-q+3)}{(q-2)!}\right]c_{1q}-\left[\frac{\Delta^{q-2}R(\psi_q-q+2)}{(q-2)!}-\frac{\Delta^{q-3}Q(\psi_q-q+3)}{(q-3)!}\right],$$

. . . .

$$kc_{kq}=\sum_{j=1}^{q-1}(-1)^{q-j-1}\left[\frac{\Delta^{j}R(\psi_q+k-q)}{j!}-\frac{\Delta^{j-1}Q(\psi_q+k-q+1)}{(j-1)!}\right]c_{k+j-q,\,q}+(-1)^{q-1}R(\psi_q+k-q)\,c_{k-q,\,q},$$

where Δ is the difference operator of the form

74. $\Delta^0 f(x)=f(x),\quad \Delta f(x)=f(x+1)-f(x),$

75. $\Delta^n f(x)=\Delta(\Delta^{n-1}f(x))=\sum_{k=0}^{n}(-1)^{n-k}\binom{n}{k}f(x+k),$

76. $R(x)=\prod_{j=1}^{q+1}(x+a_j),\quad Q(x)=x\prod_{j=1}^{q}(x+b_j-1),$

where the following relations hold:

$$c_{kq}=\xi_q(1,\,-k),$$

$$\xi_q(z,\,-m)=\sum_{k=0}^{\infty}\frac{c_{k+m,\,q}}{k!}(1-z)^k\qquad [|1-z|<1;\ m=0,\,1,\,2,\,\ldots].$$

The function $\varphi_q(z)$ is the same as the right hand sides of any of the corresponding formulae 7.2.3.81–82 given below.

The functions $\xi_q(z,\psi_q)$ and $\varphi_q(z)$ have, in general, power (for $\mathrm{Re}\,\psi_q<0$, $\psi_q\neq-1,-2,\ldots$, and logarithmic-power singularities respectively at the point $z=1$, while all the functions $R_{jq}(z)$ are continuous at this point.

The functions $u_k^0(z)$, $k=0,1,2,\ldots,q$; $u_j^\infty(z)$, $j=1,2,\ldots,p$; $R_{jq}(z)$, $j=1,2,\ldots,q$, and $\xi_k(z,\psi_q)$ (for $\psi_q\neq0,\pm1,\pm2,\ldots$) or $\varphi_q(z)$ (for $\psi_q=0,\pm1,\pm2,\ldots$) then form three fundamental systems of solutions of the equation 7.2.3.66 respectively in neighbourhoods of the singular points $z=0$ for $p\leqslant q+1$, $z=\infty$ for $p\geqslant q+1$ and $z=1$ for $p=q+1$.

In the case $p=q+1$, this equation is invariant under a simultaneous exchange of the positions of z and z^{-1}, b_k and $1-a_{k+1}$, while $q+2$ of its solutions $u_0^0(z)={}_{q+1}F_q((a_{q+1});(b_q);z)$ and $u_j^\infty(z)$, or $R_{jq}(z)$, $\xi_q(z,\psi_q)$ are related respectively by the formulae 7.2.3.77–79 and 7.2.3.80–83.

The representations of the function ${}_{q+1}F_q((a_{q+1});(b_q);z)$ in neighbourhoods of the singular points $z=\infty$ and $z=1$ have the following form.

If $a_j-a_k\neq0,\,\pm1,\,\pm2,\,\ldots;\ j,\ k=1,2,\,\ldots,q+1;\ j\neq k;\ |\arg(-z)|<\pi$, then

77. $${}_{q+1}F_q((a_{q+1});\ (b_q);\ z)=\Gamma\begin{bmatrix}(b_q)\\(a_{q+1})\end{bmatrix}\sum_{k=1}^{q+1}\Gamma\begin{bmatrix}a_k,\ (a_{q+1})'-a_k\\(b_q)-a_k\end{bmatrix}\times$$
$$\times(e^{\pi i}z^{-1})^{a_k}{}_{q+1}F_q(1+a_k-(b_q),\ a_k;\ 1+a_k-(a_{q+1})';\ z^{-1});$$

the prime $'$ means that the term $1+a_k-a_j$ is missing for $k=j$.

If $a_{n+1}-a_n=m_n$, where $m_n=0,\,1,\,2,\,\ldots,\ n=1,\,2,\,\ldots,\ r-1,\ r=2,\,3,\,\ldots$ $\ldots,\ q+1,\ m_r=\infty$ and $a_j-a_k\neq0,\ \pm1,\ \pm2,\ \ldots,\ j,\ k=r,\ r+1,\ \ldots,\ q+1,\ j\neq k$, $|\arg(-z)|<\pi$, then

78. $${}_{q+1}F_q((a_{q+1});\ (b_q);\ z)=$$
$$=\Gamma\begin{bmatrix}(b_q)\\(a_{q+1})\end{bmatrix}\left\{\sum_{n=r+1}^{q+1}\Gamma\begin{bmatrix}a_n,\ (a_{q+1})'-a_n\\(b_q)-a_n\end{bmatrix}\left(\frac{e^{\pi i}}{z}\right)^{a_n}{}_{q+1}F_q\left(\begin{matrix}a_n,\ 1+a_n-(b_q);\ 1/z\\1+a_n-(a_{q+1})'\end{matrix}\right)-\right.$$
$$\left.-\sum_{n=1}^{r}\sum_{l=0}^{m_n-1}\frac{(-1)^{H_{nl}}\Gamma(a_n+l)\prod\limits_{k=n+1}^{q+1}\Gamma(a_k-a_n-l)}{\prod\limits_{k=1}^{n}(l+a_n-a_k)!\prod\limits_{k=1}^{q}\Gamma(b_k-a_n-l)}\left(\frac{e^{\pi i}}{z}\right)^{a_n+l}X_{n-1}\right\},$$

$$H_{nl}=n(1+l)+\sum_{k=1}^{[n/2]}m_{2k-1},$$

where

$$X_{n-1}=\sum_{m=0}^{n-1}\frac{(-\ln(e^{-\pi i}z))^m}{m!}F_{n-m-1};\quad F_k=\sum_{m=0}^{k}(-1)^m M_{k-m}\Phi_m,\quad M_0=\Phi_0=1,$$

$$M_k=\sum_{k_1+\ldots+k_n=k}\prod_{j=1}^{n}P_{j,k_j},\quad P_{j,\nu}=2\sum_{m=0}^{[\nu/2]}(-1)^{m+1}\frac{2^{2m-1}-1}{(2m)!}\pi^{2m}B_{2m}N_{j,\nu-2m}$$

[B_{2m} is the Bernoulli number]

$$P_{j,0}=1,\ N_{j,k}=-\sum_{m=1}^{k}N_{j,k-m}\frac{\Gamma^{(m)}(1+l+a_n-a_j)}{m!\Gamma(1+l+a_n-a_j)},\quad N_{j,0}=1,$$

$$\Phi_k=\sum_{m=0}^{k}T_mA_{k-m},\quad T_k=-\sum_{m=1}^{k}T_{k-m}D_m,\quad T_0=1,$$

$$A_k=\sum_{k_1+\ldots+k_{q-n+1}=k}\ \prod_{j=1}^{q-n+1}\frac{\Gamma^{(k_j)}(a_{j+n}-a_n-l)}{k_j!\Gamma(a_{j+n}-a_n-l)},\quad A_0=1,$$

$$D_k=\sum_{k_1+\ldots+k_q=k}\ \prod_{j=1}^{q}\frac{\Gamma^{(k_j)}(b_j-a_n-l)}{k_j!\Gamma(b_j-a_n-l)},\quad D_0=1.$$

In particular, for $r=2, m_1=n$

79. $${}_{q+1}F_q(a_1+n,(a_q);(b_q);z)=$$
$$=\Gamma\left[\begin{matrix}(b_q)\\a_1+n,\ (a_q)\end{matrix}\right]\left\{\sum_{k=2}^{q}\Gamma\left[\begin{matrix}a_k,\ (a_q)'-a_k,\ a_1-a_k+n\\(b_q)-a_k\end{matrix}\right]\left(\frac{e^{\pi i}}{z}\right)^{a_k}\times\right.$$
$$\times{}_{q+1}F_q\left(\begin{matrix}a_k,\ 1-(b_q)+a_k;\ 1/z\\1-(a_q)'+a_k,\ 1-a_1-n+a_k\end{matrix}\right)+(n-1)!\,\Gamma\left[\begin{matrix}a_1,\ (a_q)'-a_1\\(b_q)-a_1\end{matrix}\right]\left(\frac{e^{\pi i}}{z}\right)^{a_1}\times$$
$$\times\sum_{k=0}^{n-1}\frac{(a_1)_k}{(1-n)_k k!}\frac{\Pi(1+a_1-(b_q))_k}{\Pi(1+a_1-(a_q)')_k}z^{-k}+\Gamma\left[\begin{matrix}a_1,\ (a_q)'-a_1\\(b_q)-a_1\end{matrix}\right]\left(\frac{e^{\pi i}}{z}\right)^{a_1+n}\times$$
$$\times\sum_{k=0}^{\infty}\frac{(a_1)_{n+k}}{k!\,(n+k)!}\frac{\Pi(1+a_1-(b_q))_{n+k}}{\Pi(1+a_1-(a_q)')_{n+k}}z^{-k}\left[\ln(e^{-\pi i}z)+\psi(k+1)+\psi(n+k+1)+\right.$$
$$\left.\left.+\sum_{j=2}^{q}\psi(a_j-a_1-n-k)-\sum_{j=1}^{q}\psi(b_j-a_1-n-k)-\psi(a_1+n+k)\right]\right\}$$

The following relations hold in a neighbourhood of the point $z=1$:

80. $${}_{q+1}F_q((a_{q+1});(b_q);z)=-\frac{\pi}{\sin\psi_q\pi}\Gamma\left[\begin{matrix}(b_q)\\(a_{q+1})\end{matrix}\right][\xi_q(z,\ \psi_q)+\zeta_q(z)]$$

[$\psi_q\neq 0, \pm 1, \pm 2, \ldots;\ |\arg(1-z)|<\pi$]

81. $${}_{q+1}F_q((a_{q+1});(b_q);z)=\sum_{k=0}^{m-1}\frac{\Pi(a_{q+1})_k}{\Pi(b_q)_k\,k!}\,{}_{q+1}F_q((a_{q+1})+k;\ (b_q)+k;\ 1)(z-1)^k+$$
$$+(-1)^{m+1}\Gamma\left[\begin{matrix}(b_q)\\(a_{q+1})\end{matrix}\right]\left\{\xi_q(z,\ m)\ln(1-z)-(1-z)^m\sum_{k=0}^{\infty}d_k^+(1-z)^k\right\}$$

[$\psi_q=m;\ m=0, 1, 2, \ldots;\ |1-z|<1;\ |\arg(1-z)|<\pi;\ \mathrm{Re}\,a_j>-m;\ a_j\neq 0, -1, -2, \ldots;\ j=1, 2, \ldots, q+1$]

82. $${}_{q+1}F_q((a_{q+1});(b_q);z)=\Gamma\left[\begin{matrix}(b_q)\\(a_{q+1})\end{matrix}\right]\times$$
$$\times\left\{(m-1)!(1-z)^{-m}\sum_{k=0}^{m-1}\frac{c_{kq}}{(1-m)_k}(1-z)^k+(-1)^{m+1}\xi_q(z,-m)\times\right.$$
$$\left.\times\ln(1-z)+(-1)^m\sum_{k=0}^{\infty}d_k^-(1-z)^k\right\}$$
$[\psi_q=-m;\ m=1,2,\ldots;\ |1-z|<1;\ |\arg(1-z)|<\pi;\ a_j\neq 0,-1,-2,\ldots;\ j=1,2,\ldots,q+1]$,

where the function $\zeta_q(z)$ is representable in one of the following forms:

$$\zeta_q(z)=\sum_{k=0}^{\infty}b_{kq}(z-1)^k \qquad [|1-z|<1],$$

$$b_{kq}=\frac{1}{2\pi i}\int_{\gamma-i\infty}^{\gamma+i\infty}\frac{\xi_q(t,\psi_q)}{(t-1)^{k+1}}dt \qquad [0<\gamma<1],$$

$$\zeta_q(z)=-G_{q+2,q+2}^{1,q+1}\left(z\left|\begin{matrix}1-(a_{q+1}),\ \psi_q\\0,\ 1-(b_q),\ \psi_q\end{matrix}\right.\right)-G_{q+1,q+1}^{q+1,0}\left(z\left|\begin{matrix}1-(a_{q+1})\\0,\ 1-(b_q)\end{matrix}\right.\right) \qquad [|z|<1],$$

$$\zeta_q(z)=\frac{1}{2\pi i}\int_{\gamma-i\infty}^{\gamma+i\infty}\frac{\xi_q(t,\psi_q)}{t-z}dt$$
$[0<\gamma<1;\ z$ is to the right of $(\gamma-i\infty,\gamma+i\infty)]$,

83. $$\zeta_q(z)=-\frac{1}{\pi^2}\sum_{j=1}^{q}\frac{\sin((a_{q+1})-b_j)\pi}{\sin((b_q)'-b_j)\pi}R_{jq}(z);$$

the functions $R_{jq}(z)$ are defined by the equation 7.2.3.69,

$$\sin((b_q)'-b_j)\pi=\prod_{\substack{k=1\\k\neq j}}^{q}\sin(b_k-b_j)\pi.$$

The coefficients $d_k^{\pm}$ are calculated from the formula

$$d_k^{\pm}=\frac{1}{2\pi i}\int_{\gamma-i\infty}^{\gamma+i\infty}\frac{\xi(t,\pm m)\ln(1-t)}{(1-t)^{(m\pm m)/2+k+1}}dt \qquad [0<\gamma<1].$$

If

$$\prod_{k=1}^{m-q}R(1-b_j+k)=0 \quad \left(\prod_{k=1}^{m}R(1-b_j-k)=0\right) \qquad [j=0,1,2,\ldots,q],$$

where $R(x)$ has the form 7.2.3.76, then the coefficient of $\ln(1-z)$ in the formula 7.2.3.81 (7.2.3.82) vanishes and this formula contains no logarithmic term.

In particular, the following relation holds for $q=0$ and $q=1$

$$\xi_0(z,\psi_0)=\frac{(1-z)^{-a_1}}{\Gamma(1-a_1)} \qquad [\psi_0=-a_1],$$

$$\xi_1(z,\psi_1)=\frac{(1-z)^{\psi_1}}{\Gamma(1+\psi_1)}{}_2F_1(b_1-a_1,b_1-a_2;1+\psi_1;1-z) \qquad [\psi_1=b_1-a_1-a_2],$$

$$c_{k,1}=\frac{(b_1-a_1)_k(b_1-a_2)_k}{k!},$$

$$b_{0q}=\Gamma\left[\begin{matrix}(a_{q+1})\\(b_q)\end{matrix}\right]{}_{q+1}F_q((a_{q+1});\ (b_q);\ 1) \qquad [\operatorname{Re}\psi_q>0],$$

$$b_{k-1,\,0}=0,$$

$$b_{k,\,1}=(-1)^{k+1}\Gamma\left[\begin{matrix}a_1,\ a_2\\b_1-a_1,\ b_1-a_2,\ 1-\psi_1\end{matrix}\right]\frac{(a_1)_k\,(a_2)_k}{(1-\psi_1)_k\,k!}$$

and the equations 7.2.3.80–82 are transformed into the formulae 7.2.1.5 and 7.3.1.29–31, 79–80. If $q>1$, then the functions $\xi_q(z,\psi_q)$ and $\zeta_q(z)$ are not expressed in terms of hypergeometric functions of the argument $1-z$, that is, in a neighbourhood of the point $z=1$, the function ${}_{q+1}F_q((a_{q+1});(b_q);z)$ has a non-hypergeometric type behaviour when $q\geqslant 2$.

In conclusion, we give a representation of the generalized hypergeometric function in the form of a series in hypergeometric polynomials, generalizing formula 7.2.1.7:

84. $${}_{q+1}F_q((a_q),\ \sigma;\ (b_q);\ wz)=(1-w)^{-\sigma}\sum_{k=0}^{\infty}\frac{(\sigma)_k}{k!}\left(\frac{w}{w-1}\right)^k{}_{q+1}F_q(-k,\ (a_q);\ (b_q);\ z).$$

See [1, 10, 11, 20, 21, 32, 33, 46, 49, 59] with regard to other properties and applications of the generalized hypergeometric function.

7.2.4. Hypergeometric functions of several variables.

NOTATION:

1. $$F_1(a,b,b';c;w,z)=\sum_{k,\,l=0}^{\infty}\frac{(a)_{k+l}\,(b)_k\,(b')_l}{(c)_{k+l}}\frac{w^kz^l}{k!\,l!} \qquad [|w|,\ |z|<1],$$

2. $$F_2(a,b,b';c,c';w,z)=\sum_{k,\,l=0}^{\infty}\frac{(a)_{k+l}\,(b)_k\,(b')_l}{(c)_k\,(c')_l}\frac{w^kz^l}{k!\,l!} \qquad [|w|+|z|<1],$$

3. $$F_3(a,a',b,b';c;w,z)=\sum_{k,\,l=0}^{\infty}\frac{(a)_k\,(a')_l\,(b)_k\,(b')_l}{(c)_{k+l}}\frac{w^kz^l}{k!\,l!} \qquad [|w|,\ |z|<1],$$

4. $$F_4(a,b;c,c';w,z)=\sum_{k,\,l=0}^{\infty}\frac{(a)_{k+l}\,(b)_{k+l}}{(c)_k\,(c')_l}\frac{w^kz^l}{k!\,l!}$$

$[\sqrt{|w|}+\sqrt{|z|}<1]$ – are the Appell hypergeometric functions of two variables.

5. $$\Phi_1(a,b;c;w,z)=\sum_{k,\,l=0}^{\infty}\frac{(a)_{k+l}\,(b)_k}{(c)_{k+l}}\frac{w^kz^l}{k!\,l!} \qquad [|w|<1],$$

6. $$\Phi_2(b,b';c;w,z)=\sum_{k,\,l=0}^{\infty}\frac{(b)_k\,(b')_l}{(c)_{k+l}}\frac{w^kz^l}{k!\,l!},$$

7. $$\Phi_3(b;c;w,z)=\sum_{k,\,l=0}^{\infty}\frac{(b)_k}{(c)_{k+l}}\frac{w^kz^l}{k!\,l!},$$

8. $$\Psi_1(a,b;c,c';w,z)=\sum_{k,\,l=0}^{\infty}\frac{(a)_{k+l}\,(b)_k}{(c)_k\,(c')_l}\frac{w^kz^l}{k!\,l!} \qquad [|w|<1],$$

9. $$\Psi_2(a;c,c';w,z)=\sum_{k,\,l=0}^{\infty}\frac{(a)_{k+l}}{(c)_k\,(c')_l}\frac{w^kz^l}{k!\,l!},$$

10. $$\Xi_1(a,a',b;c;w,z)=\sum_{k,\,l=0}^{\infty}\frac{(a)_k\,(a')_l\,(b)_k}{(c)_{k+l}}\frac{w^kz^l}{k!\,l!} \qquad [|w|<1],$$

11. $\Xi_2(a, b; c; w, z) = \sum_{k, l=0}^{\infty} \frac{(a)_k (b)_k}{(c)_{k+l}} \frac{w^k z^l}{k! l!}$ $[|w|<1]$ are confluent hypergeometric functions of two variables.

12. $F_A^{(n)}(a, b_1, \ldots, b_n; c_1, \ldots, c_n; z_1, \ldots, z_n) =$

$$= \sum_{k_1, \ldots, k_n=0}^{\infty} \frac{(a)_{k_1+\ldots+k_n} (b_1)_{k_1} \ldots (b_n)_{k_n}}{(c_1)_{k_1} \ldots (c_n)_{k_n}} \frac{z_1^{k_1} \ldots z_n^{k_n}}{k_1! \ldots k_n!} \qquad \left[\sum_{j=1}^{n} |z_j|<1\right],$$

13. $F_B^{(n)}(a_1, \ldots, a_n, b_1, \ldots, b_n; c; z_1, \ldots, z_n) =$

$$= \sum_{k_1, \ldots, k_n=0}^{\infty} \frac{(a_1)_{k_1} \ldots (a_n)_{k_n} (b_1)_{k_1} \ldots (b_n)_{k_n}}{(c)_{k_1+\ldots+k_n}} \frac{z_1^{k_1} \ldots z_n^{k_n}}{k_1! \ldots k_n!} \qquad [|z_j|<1, j=1, 2, \ldots, n],$$

14. $F_C^{(n)}(a, b; c_1, \ldots, c_n; z_1, \ldots, z_n) =$

$$= \sum_{k_1, \ldots, k_n=0}^{\infty} \frac{(a)_{k_1+\cdots+k_n} (b)_{k_1+\cdots+k_n}}{(c_1)_{k_1} \ldots (c_n)_{k_n}} \frac{z_1^{k_1} \ldots z_n^{k_n}}{k_1! \ldots k_n!} \qquad \left[\sum_{j=1}^{n} \sqrt{|z_j|}<1\right],$$

15. $F_D^{(n)}(a, b_1, \ldots, b_n; c; z_1, \ldots, z_n) = \sum_{k_1, \ldots, k_n=0}^{\infty} \frac{(a)_{k_1+\cdots+k_n} (b_1)_{k_1} \ldots (b_n)_{k_n}}{(c)_{k_1+\cdots+k_n}} \frac{z_1^{k_1} \ldots z_n^{k_n}}{k_1! \ldots k_n!}$

$[|z_j|<1; j=1, 2, \ldots, n]$ are the Lauricella hypergeometric functions.

Concerning the properties of other hypergeometric functions of several variables, see, for example, [31, 40, 41].

Symmetry formulae:

16. $F_1(a, b, b'; c; z, w) = F_1(a, b', b; c; w, z)$,
17. $F_2(a, b, b'; c, c'; z, w) = F_2(a, b', b; c', c; w, z)$,
18. $F_3(a, a', b, b'; c; z, w) = F_3(a', a, b', b; c; w, z)$,
19. $F_4(a, b; c, c'; z, w) = F_4(b, a; c, c'; z, w) = F_4(a, b; c', c; w, z)$,
20. $\Phi_2(b, b'; c; z, w) = \Phi_2(b', b; c; w, z)$,
21. $\Psi_1(a, b; c, c'; z, w) = \Psi_1(a, b; c', c; w, z)$,
22. $\Psi_2(a; c, c'; z, w) = \Psi_2(a; c', c; w, z)$,
23. $\Xi_1(a, a', b; c; z, w) = \Xi_1(a', a, b; c; w, z)$,
24. $\Xi_2(a, b; c; z, w) = \Xi_2(b, a; c; z, w)$.

Interdependence formulae:

25. $F_1(a, b, b'; c; w, z) = (1-w)^{-b}(1-z)^{-b'} F_1\left(c-a, b, b'; c; \frac{w}{w-1}, \frac{z}{z-1}\right) =$

26. $\qquad = (1-w)^{-a} F_1\left(a, c-b-b', b'; c; \frac{w}{w-1}, \frac{z-w}{1-w}\right) =$

27. $\qquad = (1-w)^{c-a-b}(1-z)^{-b'} F_1\left(c-a, c-b-b', b'; c; w, \frac{w-z}{1-z}\right)$,

28. $\lim_{\varepsilon \to 0} F_1\left(a, b, \frac{1}{\varepsilon}; c; w, \varepsilon z\right) = \Phi_1(a, b; c; w, z)$,

29. $\lim_{\varepsilon \to 0} F_1\left(\frac{1}{\varepsilon}, b, b'; c; \varepsilon w, \varepsilon z\right) = \Phi_2(b, b'; c; w, z)$,

30. $\lim_{\varepsilon \to 0} F_1\left(\frac{1}{\varepsilon}, b, \frac{1}{\varepsilon}; c; \varepsilon w, \varepsilon^2 z\right) = \Phi_3(b; c; w, z)$,

31. $F_2(a, b, b'; c, c'; w, z) = (1-w)^{-a} F_2\left(a, c-b, b'; c, c'; \frac{w}{w-1}, \frac{z}{1-w}\right) =$

32. $\qquad = (1-w-z)^{-a} F_2\left(a, c-b, c'-b'; c, c'; \frac{w}{w+z-1}, \frac{z}{w+z-1}\right)$,

33. $\lim\limits_{\varepsilon\to 0} F_2\left(a, b, \frac{1}{\varepsilon}; c, c'; w, \varepsilon z\right) = \Psi_1(a, b; c, c'; w, z)$,

34. $\lim\limits_{\varepsilon\to 0} F_2\left(a, \frac{1}{\varepsilon}, \frac{1}{\varepsilon}; c, c'; \varepsilon w, \varepsilon z\right) = \Psi_2(a; c, c'; w, z)$,

35. $F_3(a, a', b, b'; c; w, z) =$

$$= \sum \Gamma\begin{bmatrix} c, \rho-\mu, \sigma-\nu \\ \rho, \sigma, c-\mu-\nu \end{bmatrix} (-w)^{-\mu}(-z)^{-\nu} F_2\left(1+\mu+\nu-c, \mu, \nu; 1+\mu-\rho, 1+\nu-\sigma; \frac{1}{w}, \frac{1}{z}\right)$$

[here the sum consists of four terms in which μ, ν, ρ, σ are respectively equal to a, a', b, b'; a, b', b, a'; b, a', a, b' and b, b', a, a'].

36. $F_3(a, a', b, b'; a+a'; w, z) = (1-z)^{-b'} F_1\left(a, b, b'; a+a'; w, \frac{z}{z-1}\right)$,

37. $\lim\limits_{\varepsilon\to 0} F_3\left(a, a', b, \frac{1}{\varepsilon}; c; w, \varepsilon z\right) = \Xi_1(a, a', b; c; w, z)$,

38. $\lim\limits_{\varepsilon\to 0} F_3\left(a, \frac{1}{\varepsilon}, b, \frac{1}{\varepsilon}; c; w, \varepsilon^2 z\right) = \Xi_2(a, b; c; w, z)$,

39. $F_4(a, b; c, c'; w, z) = \Gamma\begin{bmatrix} c', b-a \\ c'-a, b \end{bmatrix} (-z)^{-a} F_4\left(a, 1+a-c'; c, 1+a-b; \frac{w}{z}, \frac{1}{z}\right) +$

$$+ \Gamma\begin{bmatrix} c', a-b \\ c'-b, a \end{bmatrix} (-z)^{-b} F_4\left(1+b-c', b; c, 1+b-a; \frac{w}{z}, \frac{1}{z}\right),$$

40. $F_A^{(2)}(a, b_1, b_2; c_1, c_2; z_1, z_2) = F_2(a, b_1, b_2; c_1, c_2; z_1, z_2)$,

41. $F_C^{(2)}(a, b; c_1, c_2; z_1, z_2) = F_4(a, b; c_1, c_2; z_1, z_2)$.

Integral representations:

42. $F_1(a, b, b'; c; w, z) = \Gamma\begin{bmatrix} c \\ a, c-a \end{bmatrix} \int\limits_0^1 \frac{u^{a-1}(1-u)^{c-a-1}}{(1-uw)^b(1-uz)^{b'}} du$ [Re a, Re $(c-a)>0$],

43. $F_1(a, b, b'; c; w, z) = \Gamma\begin{bmatrix} c \\ b, b', c-b-b' \end{bmatrix} \times$

$$\times \iint\limits_{\substack{u, v\geqslant 0 \\ u+v\leqslant 1}} u^{b-1}v^{b'-1}(1-u-v)^{c-b-b'-1}(1-uw-vz)^{-a} du\, dv$$ [Re b, Re b', Re$(c-b-b')>0$],

44. $F_2(a, b, b'; c, c'; w, z) = \Gamma\begin{bmatrix} c, c' \\ b, b', c-b, c'-b' \end{bmatrix} \times$

$$\times \int\limits_0^1\int\limits_0^1 u^{b-1}v^{b'-1}(1-u)^{c-b-1}(1-v)^{c'-b'-1}(1-uw-vz)^{-a} du\, dv$$

[Re b, Re b', Re $(c-b)$, Re $(c'-b')>0$],

45. $F_3(a, a', b, b'; c; w, z) = \Gamma\begin{bmatrix} c \\ b, b', c-b-b' \end{bmatrix} \times$

$$\times \iint\limits_{\substack{u, v\geqslant 0 \\ u+v\leqslant 1}} u^{b-1}v^{b'-1}(1-u-v)^{c-b-b'-1}(1-uw)^{-a}(1-vz)^{-a'} du\, dv$$

[Re b, Re b', Re $(c-b-b')>0$],

46. $F_3(a, a', b, b'; a+a'; w, z) = \Gamma\begin{bmatrix} a+a' \\ a, a' \end{bmatrix} \times$

$$\times \int\limits_0^1 u^{a-1}(1-u)^{a'-1}(1-uw)^{-b}(1-(1-u)z)^{-b'} du$$ [Re a, Re $a'>0$],

47. $F_4(a, b; c, c'; w(1-z), z(1-w)) = \Gamma\begin{bmatrix} c, c' \\ a, b, c-a, c'-b \end{bmatrix} \times$
$$\times \int_0^1\int_0^1 \frac{u^{a-1}v^{b-1}(1-u)^{c-a-1}(1-v)^{c'-b-1}\,du\,dv}{(1-uw)^{c+c'-a-1}(1-vz)^{c+c'-b-1}(1-uw-vz)^{a+b-c-c'+1}}$$
[Re a, Re b, Re $(c-a)$, Re $(c'-b)>0$],

48. $\Phi_1(a, b; c; w, z) = \Gamma\begin{bmatrix} c \\ a, c-a \end{bmatrix} \int_0^1 u^{a-1}(1-u)^{c-a-1}(1-uw)^{-b}e^{uz}\,du$

[Re a, Re $(c-a)>0$],

49. $\Phi_2(b, b'; c; w, z) = \Gamma\begin{bmatrix} c \\ b, b', c-b-b' \end{bmatrix} \times$
$$\times \iint\limits_{\substack{u, v\geqslant 0\\ u+v\leqslant 1}} u^{b-1}v^{b'-1}(1-u-v)^{c-b-b'-1}e^{uw+vz}\,du\,dv$$
[Re b, Re b', Re $(c-b-b')>0$],

50. $\Phi_3(b; c; w, z) = \Gamma\begin{bmatrix} c \\ b, 1/2, c-b-1/2 \end{bmatrix} \times$
$$\times \int_0^1\int_0^1 u^{c-3/2}v^{b-1}(1-u)^{-1/2}(1-v)^{c-b-3/2}e^{uvw}\operatorname{ch}\left(2\sqrt{(1-u)z}\right)du\,dv$$
[Re $b>0$; Re c, Re $(c-b)>1/2$],

51. $\Psi_1(a, b; c, c'; w, z) = \Gamma\begin{bmatrix} c, c' \\ b, c-b, a, c'-a \end{bmatrix} \times$
$$\times \int_0^1\int_0^1 u^{b-1}(1-u)^{c-b-1}v^{a-1}(1-v)^{c'-a-1}(1-uw)^{-a}e^{\frac{vz}{1-uw}}\,du\,dv$$
[Re b, Re $(c-b)$, Re a, Re $(c'-a)>0$],

52. $\Xi_1(a, a', b; c; w, z) = \Gamma\begin{bmatrix} c \\ a, a', c-a-a' \end{bmatrix} \times$
$$\times \iint\limits_{\substack{u, v\geqslant 0\\ u+v\leqslant 1}} u^{a-1}v^{a'-1}(1-u-v)^{c-a-a'-1}(1-uw)^{-b}e^{vz}\,du\,dv$$
[Re a, Re a', Re $(c-a-a')>0$],

53. $\Xi_2(a, b; c; w, z) = \Gamma\begin{bmatrix} c \\ a, c-a-1/2, 1/2 \end{bmatrix} \times$
$$\times \int_0^1\int_0^1 u^{c-3/2}v^{a-1}(1-u)^{-1/2}(1-v)^{c-a-3/2}(1-uvw)^{-b}\operatorname{ch}\left(2\sqrt{(1-u)z}\right)du\,dv$$
[Re $a>0$; Re c, Re $(c-a)>1/2$],

54. $F_A^{(n)}(a, b_1, \ldots, b_n; c_1, \ldots, c_n; z_1, \ldots, z_n) = \Gamma\begin{bmatrix} (c_n) \\ (b_n), (c_n-b_n) \end{bmatrix} \times$
$$\times \int_0^1 \cdots \int_0^1 u_1^{b_1-1}\ldots u_n^{b_n-1}(1-u_1)^{c_1-b_1-1}\ldots(1-u_n)^{c_n-b_n-1}\times$$
$$\times(1-u_1z_1-\ldots-u_nz_n)^{-a}\,du_1\ldots du_n$$
[Re b_1, ..., Re b_n, Re (c_1-b_1), ..., Re $(c_n-b_n)>0$],

55. $F_B^{(n)}(a_1, \ldots, a_n, b_1, \ldots, b_n; c; z_1, \ldots, z_n) = \Gamma\begin{bmatrix} c \\ a_1, \ldots, a_n, c-a_1-\ldots-a_n \end{bmatrix} \times$
$$\times \int\limits_{\substack{u_1, \ldots, u_n\geqslant 0\\ 1-u_1-\ldots-u_n\geqslant 0}} \cdots \int u_1^{a_1-1}\ldots u_n^{a_n-1}(1-u_1-\ldots-u_n)^{c-a_1-\ldots-a_n-1}\times$$
$$\times(1-u_1z_1)^{-b_1}\ldots(1-u_nz_n)^{-b_n}\,du_1\ldots du_n$$
[Re a_1, ..., Re a_n, Re $(c-a_1-\ldots-a_n)>0$; $|\arg(1-z_1)|, \ldots, |\arg(1-z_n)|<\pi$],

56. $F_D^{(n)}(a, b_1, \ldots, b_n; c; z_1, \ldots, z_n)=\Gamma\begin{bmatrix} c \\ b_1, \ldots, b_n, c-b_1-\ldots-b_n\end{bmatrix}\times$

$$\times\int\limits_{\substack{u_1, \ldots, u_n\geqslant 0\\ 1-u_1-\ldots-u_n\geqslant 0}}\cdots\int u_1^{b_1-1}\ldots u_n^{b_n-1}(1-u_1-\ldots-u_n)^{c-b_1-\ldots-b_n-1}\times$$

$$\times(1-u_1z_1-\ldots-u_nz_n)^{-a}\,du_1\ldots du_n \quad [\operatorname{Re} b_1, \ldots, \operatorname{Re} b_n, \operatorname{Re}(c-b_1-\ldots-b_n)>0],$$

57. $F_D^{(n)}(a, b_1, \ldots, b_n; c; z_1, \ldots, z_n)=\Gamma\begin{bmatrix} c \\ a, c-a\end{bmatrix}\times$

$$\times\int\limits_0^1 u^{a-1}(1-u)^{c-a-1}(1-uz_1)^{-b_1}\ldots(1-uz_n)^{-b_n}\,du$$

$[\operatorname{Re} c>\operatorname{Re} a>0;\ |\arg(1-z_1)|, \ldots, |\arg(1-z_n)|<\pi]$.

Representations in terms of the functions ${}_pF_q((a_p);(b_q);z)$:

58. $F_1(a, b, b'; c; w, 0)=F_1(a, b, 0; c; w, z)={}_2F_1(a, b; c; w)$,

59. $F_1(a, b, b'; c; 0, z)=F_1(a, 0, b'; c; w, z)={}_2F_1(a, b'; c; z)$,

60. $F_1(a; b, b'; c; w, 1)=\Gamma\begin{bmatrix} c, c-a-b' \\ c-a, c-b'\end{bmatrix}{}_2F_1(a, b; c-b'; w)$,

61. $F_1(a, b, b'; c; z, z)={}_2F_1(a, b+b'; c; z)$,

62. $F_1(a, b, b; c; z, -z)={}_3F_2\left(\begin{matrix} a/2, (a+1)/2, b \\ c/2, (c+1)/2; z^2\end{matrix}\right)$,

63. $F_1(a, b, b'; b+b'; w, z)=(1-z)^{-a}\,{}_2F_1\left(a, b; b+b'; \dfrac{w-z}{1-z}\right)$,

64. $F_2(a, b, b'; c, c'; w, 0)=F_2(a, b, 0; c, c'; w, z)={}_2F_1(a, b; c; w)$,

65. $F_2(a, b, b'; c, c'; 0, z)=F_2(a, 0, b'; c, c'; w, z)={}_2F_1(a, b'; c'; z)$,

66. $F_2(a, b, b; c, c; z, -z)={}_4F_3\left(\begin{matrix} a/2, (a+1)/2, b, c-b \\ c/2, (c+1)/2, c; z^2\end{matrix}\right)$,

67. $F_2(a, b, b'; 2b, 2b'; z, -z)={}_4F_3\left(\begin{matrix} a/2, (a+1)/2, (b+b')/2, (b+b'+1)/2 \\ b+1/2, b'+1/2, b+b'; \quad z^2\end{matrix}\right)$,

68. $F_2(a, b, c-b; c, c; z, z)=(1-z)^{-a}\,{}_4F_3\left(\begin{matrix} a/2, (a+1)/2, b, c-b; z^2(1-z)^{-2} \\ c/2, (c+1)/2, c\end{matrix}\right)$,

69. $F_2(a, b, b'; 2b, 2b'; z, z)=(1-z)^{-a}\,{}_4F_3\left(\begin{matrix} a/2, (a+1)/2, (b+b')/2, (b+b'+1)/2 \\ b+1/2, b'+1/2, b+b'; z^2(1-z)^{-2}\end{matrix}\right)$,

70. $F_2(a, b, b'; b, c; w, z)=(1-w)^{-a}\,{}_2F_1\left(a, b'; c; \dfrac{z}{1-w}\right)$,

71. $F_2(a, b, b'; a, a; w, z)=(1-w)^{-b}(1-z)^{-b'}\,{}_2F_1\left(b, b'; a; \dfrac{wz}{(1-w)(1-z)}\right)$,

72. $F_3(a, a', b, b'; c; w, 0)=F_3(a, 0, b, b'; c; w, z)=F_3(a, a', b, 0; c; w, z)=$
$={}_2F_1(a, b; c; w)$,

73. $F_3(a, a', b, b'; c; 0, z)=F_3(0, a', b, b'; c; w, z)=F_3(a, a', 0, b'; c; w, z)=$
$={}_2F_1(a', b'; c; z)$,

74. $F_3\left(a, a', b, b'; c; z, \dfrac{z}{z-1}\right)=$

$$=z^{1-c}(1-z)^{a'}\Gamma\begin{bmatrix} c, b'-a', c-a-b-a' \\ b', c-a-a', c-b-a'\end{bmatrix}{}_3F_2\left(\begin{matrix} 1-b', 1+a'+a-c, 1+a'+b-c; 1-z \\ 1+a'-b', 1+a'+a+b-c\end{matrix}\right)+$$

$$+z^{1-c}(1-z)^{b'}\Gamma\begin{bmatrix} c, a'-b', c-a-b-b' \\ a', c-a-b', c-b-b'\end{bmatrix}{}_3F_2\left(\begin{matrix} 1-a', 1+b'+a-c, 1+b'+b-c; 1-z \\ 1+b'-a', 1+b'+b+a-c\end{matrix}\right)+$$

$$+z^{1-c}(1-z)^{c-a-b}\times$$

$$\times\Gamma\begin{bmatrix} c, a'+a+b-c, b'+a+b-c \\ a'+b'+a+b-c, a, b\end{bmatrix}{}_3F_2\left(\begin{matrix} 1+c-a-b-a'-b', 1-a, 1-b; 1-z \\ 1+c-a-b-a', 1+c-a-b-b'\end{matrix}\right),$$

75. $F_3(a, b, b, b; c; z, -z) = {}_4F_3\left(\begin{matrix} a, b, (a+b)/2, (a+b+1)/2 \\ a+b, c/2, (c+1)/2; z^2 \end{matrix}\right)$,

76. $F_3(a, c-a, b, c-b; c; w, z) = (1-z)^{a+b-c}\, {}_2F_1(a, b; c; w+z-wz)$,

77. $F_4(a, b; c, c'; w, 0) = {}_2F_1(a, b; c; w)$,

78. $F_4(a, b; c, c'; 0, z) = {}_2F_1(a, b; c'; z)$.

79. $F_4(a, b; b, b; w, z) = (1-w-z)^{-a}\, {}_2F_1\left(\frac{a}{2}, \frac{a+1}{2}; b; \frac{4wz}{(1-w-z)^2}\right)$,

80. $F_4\left(a, a+\frac{1}{2}; c, \frac{1}{2}; w, z\right) = \frac{1}{2}(1+\sqrt{z})^{-2a}\, {}_2F_1\left(a, a+\frac{1}{2}; c; w(1+\sqrt{z})^{-2}\right) +$
$+\frac{1}{2}(1-\sqrt{z})^{-2a}\, {}_2F_1\left(a, a+\frac{1}{2}; c; w(1-\sqrt{z})^{-2}\right)$,

81. $F_4(a, b; c, c'; z, z) = {}_4F_3\left(\begin{matrix} a, b, (c+c'-1)/2, (c+c')/2 \\ c, c', c+c'-1; 4z \end{matrix}\right)$,

82. $F_4(a, b; a, b; z, z) = \frac{1}{\sqrt{1-4z}}\left(\frac{1+\sqrt{1-4z}}{2}\right)^{2-a-b}$,

83. $F_4(a, b; c, c; z, -z) = {}_4F_3\left(\begin{matrix} a/2, (a+1)/2, b/2, (b+1)/2 \\ c/2, (c+1)/2, c; -4z^2 \end{matrix}\right)$,

84. $F_4(-n, b; c, c; (1-z)^2, z^2) = \frac{(c-b)_n}{(c)_n}\, {}_3F_2\left(\begin{matrix} -n, b, c-1/2; 4z(1-z) \\ 2c-1, b-c-n+1 \end{matrix}\right)$,

85. $F_4(a, b; c, a+b-c+1; w(1-z), z(1-w)) = {}_2F_1(a, b; c; w)\times$
$\times {}_2F_1(a, b; a+b-c+1; z)$,

86. $F_4(a, c+c'-a-1; c, c'; w(1-z), z(1-w)) = {}_2F_1(a, c+c'-a-1; c; w)\times$
$\times {}_2F_1(a, c+c'-a-1; c'; z)$,

87. $F_4\left(a, b; a, b; \frac{-w}{(1-w)(1-z)}, \frac{-z}{(1-w)(1-z)}\right) = \frac{(1-w)^b(1-z)^a}{1-wz}$,

88. $F_4\left(a, b; b, b; \frac{-w}{(1-w)(1-z)}, \frac{-z}{(1-w)(1-z)}\right) = (1-w)^a(1-z)^a\, {}_2F_1\left(\begin{matrix} a, 1+a-b \\ b; wz \end{matrix}\right)$

89. $F_4\left(a, b; 1+a-b, b; \frac{-w}{(1-w)(1-z)}, \frac{-z}{(1-w)(1-z)}\right) =$
$= (1-z)^a\, {}_2F_1\left(a, b; 1+a-b; -\frac{w(1-z)}{1-w}\right)$,

90. $\Psi_2(b, b'; b+b'; w, z) = e^z\, {}_1F_1(b; b+b'; w-z)$,

91. $\Psi_2(a; a, a; w, z) = \Gamma(a)(wz)^{(1-a)/2} e^{w+z} I_{a-1}(2\sqrt{wz})$,

92. $\Psi_2(a; c, c'; z, z) = {}_3F_3\left(\begin{matrix} a, (c+c')/2, (c+c'-1)/2 \\ c, c', c+c'-1; 4z \end{matrix}\right)$.

7.3. THE FUNCTIONS ${}_1F_0(a; z)$ AND ${}_2F_1(a, b; c; z)$.

7.3.1. Representations of ${}_1F_0(a; z)$ and ${}_2F_1(a, b; c; z)$.

1. ${}_1F_0(a; z) = (1-z)^{-a}$.

2. ${}_2F_1(a, b; c; z)\, {}_2F_1(-a, -b; -c; z) + \frac{ab(a-c)(b-c)}{c^2(1-c^2)} z^2 \times$
$\times {}_2F_1(1-a, 1-b; 2-c; z)\, {}_2F_1(1+a, 1+b; 2+c; z) = 1$.

	$a \; b$	c	${}_2F_1(a, b; c; z)$
3	$a \; b$	c	$(1-z)^{-a}{}_2F_1\left(a, c-b; c; \frac{z}{z-1}\right)$
4	$a \; b$	c	$(1-z)^{c-a-b}{}_2F_1(c-a, c-b; c; z)$
5	$a \; b$	c	$\Gamma\begin{bmatrix} c, c-a-b \\ c-a, c-b \end{bmatrix} {}_2F_1(a, b; a+b-c+1; 1-z)+$ $+\Gamma\begin{bmatrix} c, a+b-c \\ a, b \end{bmatrix} (1-z)^{c-a-b}{}_2F_1(c-a, c-b; c-a-$ $-b+1; 1-z)$ $[\lvert\arg(1-z)\rvert<\pi]$
6	$a \; b$	c	$\Gamma\begin{bmatrix} c, b-a \\ b, c-a \end{bmatrix} (-z)^{-a}{}_2F_1\left(a, 1-c+a; 1-b+a; \frac{1}{z}\right)+$ $+\Gamma\begin{bmatrix} c, a-b \\ a, c-b \end{bmatrix} (-z)^{-b}{}_2F_1\left(b, 1-c+b; 1-a+b; \frac{1}{z}\right)$ $[\lvert\arg(-z)\rvert<\pi]$
7	$a \; b$	c	$\Gamma\begin{bmatrix} c, b-a \\ b, c-a \end{bmatrix} (1-z)^{-a}{}_2F_1\left(a, c-b; a-b+1; \frac{1}{1-z}\right)+$ $+\Gamma\begin{bmatrix} c, a-b \\ a, c-b \end{bmatrix} (1-z)^{-b}{}_2F_1\left(b, c-a; b-a+1; \frac{1}{1-z}\right)$ $[\lvert\arg(1-z)\rvert<\pi]$
8	$a \; b$	c	$\Gamma\begin{bmatrix} c, c-a-b \\ c-a, c-b \end{bmatrix} z^{-a}{}_2F_1\left(a, 1+a-c; a+b-c+1; 1-\frac{1}{z}\right)+$ $+\Gamma\begin{bmatrix} c, a+b-c \\ a, b \end{bmatrix} z^{a-c}(1-z)^{c-a-b}\times$ $\times{}_2F_1\left(c-a, 1-a; 1+c-a-b; 1-\frac{1}{z}\right)$ $[\lvert\arg(1-z)\rvert, \; \lvert\arg z\rvert<\pi]$
9	$a \; b$	c	$\frac{(1-c)_m}{(1+a-c)_m} z^{-m} \sum_{k=0}^{m} \binom{m}{k} (z-1)^{m-k}{}_2F_1(a, b-k; c-m; z)$ $[\mathrm{Re}\,(c-a)>m]$
10	$a \; b$	c	$\frac{(1-c)_m}{(1-a)_m} z^{-m} \sum_{k=0}^{m} (-1)^k \binom{m}{k} {}_2F_1(a-m, b-k; c-m; z)$ $[\mathrm{Re}\,a>m]$
11	$a \; b$	c	$[c-2a-(b-a)z]^{-1}[a(z-1){}_2F_1(a+1, b; c; z)+$ $+(c-a){}_2F_1(a-1, b; c; z)]$
12	$a \; b$	c	$(a-b)^{-1}[a{}_2F_1(a+1, b; c; z)-b{}_2F_1(a, b+1; c; z)]$
13	$a \; b$	c	$(c-a-b)^{-1}[a(z-1){}_2F_1(a+1, b; c; z)+$ $+(c-b){}_2F_1(a, b-1; c; z)]$
14	$a \; b$	c	$[a-(c-b)z]^{-1}[a(1-z){}_2F_1(a+1, b; c; z)-$ $-zc^{-1}(c-a)(c-b){}_2F_1(a, b; c+1; z)]$
15	$a \; b$	c	$(1+a-c)^{-1}[a{}_2F_1(a+1, b; c; z)-(c-1){}_2F_1(a, b; c-1; z)]$
16	$a \; b$	c	$[(b-a)(1-z)]^{-1}[(c-a){}_2F_1(a-1, b; c; z)+$ $+(b-c){}_2F_1(a, b-1; c; z)]$
17	$a \; b$	c	$[c(1-z)]^{-1}[c{}_2F_1(a-1, b; c; z)+(b-c)z{}_2F_1(a, b; c+1; z)]$
18	$a \; b$	c	$[a-1-(c-b-1)z]^{-1}[(a-c){}_2F_1(a-1, b; c; z)+$ $+(c-1)(1-z){}_2F_1(a, b; c-1; z)]$
19	$a \; b$	c	$[c-1-(2c-a-b-1)z]^{-1}[(c-1)(1-z){}_2F_1(a, b; c-1; z)-$ $-c^{-1}(c-a)(c-b)z{}_2F_1(a, b; c+1; z)]$
20	$a \; b$	$m+2$	$\frac{(m+1)!z^{-m-1}}{(1-a)_{m+1}(1-b)_{m+1}} \lim_{c\to -m} [\Gamma^{-1}(c){}_2F_1(a-m-1, b-m-$ $-1; c; z)]$
21	$a \; b$	$b-n$	$(1-z)^{-a}{}_2F_1\left(-n, a; b-n; \frac{z}{z-1}\right)=$

	a b	c	${}_2F_1(a,\ b;\ c;\ z)$
22			$=(-1)^n \frac{(a)_n}{(1-b)_n}(1-z)^{-a-n}{}_2F_1(-n, b-a-n; 1-a-n; 1-z)=$
23			$=\frac{(1+a-b)_n}{(1-b)_n}(-z)^n (1-z)^{-a-n}{}_2F_1\left(-n, 1-b; 1+a-b; \frac{1}{z}\right)=$
24			$=\frac{(1+a-b)_n}{(1-b)_n}(1-z)^{-a}{}_2F_1\left(-n,\ a;\ 1+a-b;\ \frac{1}{1-z}\right)=$
25			$=\frac{(a)_n}{(1-b)_n}(-z)^n(1-z)^{-a-n}{}_2F_1\left(-n, 1-b; 1-a-n; 1-\frac{1}{z}\right)=$
26			$=(1-z)^{-a-n}{}_2F_1(-n,\ b-a-n;\ b-n;\ z)$
27	a b	b	$(1-z)^{-a}$
28	a b	$b+1$	$bz^{-b}B_z(b,\ 1-a)$
29	a b	$a+b-m$	$\Gamma\begin{bmatrix} m,\ a+b-m \\ a,\ b \end{bmatrix}(1-z)^{-m}\sum_{k=0}^{m-1}\frac{(a-m)_k(b-m)_k}{k!(1-m)_k}(1-z)^k-$ $-(-1)^m\Gamma\begin{bmatrix} a+b-m \\ a-m,\ b-m \end{bmatrix}\sum_{k=0}^{\infty}\frac{(a)_k(b)_k}{k!\,(k+m)!}(1-z)^k\times$ $\times[\ln(1-z)-\psi(k+1)-\psi(k+m+1)+\psi(a+k)+\psi(b+k)]$ $[\lvert\arg(1-z)\rvert<\pi;\ \lvert 1-z\rvert<1]$
30	a b	$a+b$	$\Gamma\begin{bmatrix} a+b \\ a,\ b \end{bmatrix}\sum_{k=0}^{\infty}\frac{(a)_k(b)_k}{(k!)^2}[2\psi(k+1)-\psi(a+k)-$ $-\psi(b+k)-\ln(1-z)](1-z)^k$ $[\lvert\arg(1-z)\rvert<\pi;\ \lvert 1-z\rvert<1]$
31	a b	$a+b+m$	$\Gamma\begin{bmatrix} m,\ a+b+m \\ a+m,\ b+m \end{bmatrix}\sum_{k=0}^{m-1}\frac{(a)_k(b)_k}{k!\,(1-m)_k}(1-z)^k-$ $-\Gamma\begin{bmatrix} a+b+m \\ a,\ b \end{bmatrix}(z-1)^m\sum_{k=0}^{\infty}\frac{(a+m)_k(b+m)_k}{k!(m+k)!}(1-z)^k\times$ $\times[\ln(1-z)-\psi(k+1)-\psi(k+m+1)+\psi(a+k+m)+$ $+\psi(b+k+m)]$ $[\lvert\arg(1-z)\rvert<\pi;\ \lvert 1-z\rvert<1]$
32	a b	$a+b-\frac{1}{2}$	$(1-z)^{-1/2}{}_2F_1\left(2a-1,\ 2b-1;\ a+b-\frac{1}{2};\ \frac{1-\sqrt{1-z}}{2}\right)=$
33			$=(1-z)^{-1/2}\left(\frac{1+\sqrt{1-z}}{2}\right)^{1-2a}{}_2F_1\left(2a-1,\ a-b+\frac{1}{2};\right.$ $\left.a+b-\frac{1}{2};\ \frac{\sqrt{1-z}-1}{\sqrt{1-z}+1}\right)=$
34			$=(1-z)^{-1/2}(\sqrt{1-z}+\sqrt{-z})^{1-2a}\times$ $\times{}_2F_1(2a-1,\ a+b-1;\ 2a+2b-2;\ 2\sqrt{z^2-z}+2z)$
35	a b	$a+b-\frac{1}{2}$	$2^{a+b-3/2}\Gamma\left(a+b-\frac{1}{2}\right)(-z)^{(3-2a-2b)/4}(1-z)^{-1/2}\times$ $\times P_{b-a-1/2}^{3/2-a-b}(\sqrt{1-z})$ $[\lvert\arg(-z)\rvert,\ \lvert\arg(1-z)\rvert<\pi;\ \operatorname{Re}\sqrt{1-z}>0]$
36	a b	$a+b-\frac{1}{2}$	$2^{a+b-3/2}\Gamma\left(a+b-\frac{1}{2}\right)x^{(3-2a-2b)/4}(1-x)^{-1/2}\times$ $\times P_{b-a-1/2}^{3/2-a-b}(\sqrt{1-x})$ $[z=x,\ 0<x<1]$

	a b	c	${}_2F_1(a,\ b;\ c;\ z)$
37	a b	$a+b+\frac{1}{2}$	${}_2F_1\left(2a,\ 2b;\ a+b+\frac{1}{2};\ \frac{1-\sqrt{1-z}}{2}\right)=$
38			$=\left(\frac{1+\sqrt{1-z}}{2}\right)^{-2a}{}_2F_1\Big(2a,\ a-b+\frac{1}{2};$
39			$a+b+\frac{1}{2};\ \frac{\sqrt{1-z}-1}{\sqrt{1-z}+1}\Big)=(\sqrt{1-z}+\sqrt{-z})^{-2a}\times$ $\times{}_2F_1(2a,\ a+b;\ 2a+2b;\ 2\sqrt{z^2-z}+2z)$
40	a b	$a+b+\frac{1}{2}$	$2^{a+b-1/2}\Gamma\left(a+b+\frac{1}{2}\right)(-z)^{(1-2a-2b)/4}P^{1/2-a-b}_{a-b-1/2}(\sqrt{1-z})$ $[\lvert\arg(-z)\rvert,\ \lvert\arg(1-z)\rvert<\pi]$
41	a b	$a+b+\frac{1}{2}$	$2^{a+b-1/2}\Gamma\left(a+b+\frac{1}{2}\right)x^{(1-2a-2b)/4}P^{1/2-a-b}_{a-b-1/2}(\sqrt{1-x})$ $[z=x,\ 0<x<1]$
42	a b	$a+b+\frac{3}{2}$	$\frac{\Gamma(a+b+3/2)}{2a+1}\left(-\frac{z}{4}\right)^{-(2a+2b+1)/4}\Big\{(-z)^{1/2}\times$ $\times P^{1/2-a-b}_{b-a-1/2}(\sqrt{1-z})-2bP^{-1/2-a-b}_{b-a+1/2}(\sqrt{1-z})\Big\}$ $[\lvert\arg(-z)\rvert,\ \lvert\arg(1-z)\rvert<\pi]$
43	a b	$a+b+\frac{3}{2}$	$\frac{\Gamma(a+b+3/2)}{2a+1}\left(\frac{x}{4}\right)^{-(2a+2b+1)/4}\Big\{\sqrt{x}P^{1/2-a-b}_{b-a-1/2}(\sqrt{1-x})-$ $-2bP^{-1/2-a-b}_{b-a+1/2}(\sqrt{1-x})\Big\}$ $[z=x,\ 0<x<1]$
44	a b	$\frac{a+b}{2}$	$-\frac{z}{2}\Gamma\left(\frac{a+b}{2}\right)(z^2-z)^{-(a+b+2)/4}\Big[P^{1-(a+b)/2}_{(a-b)/2}(1-2z)+$ $+P^{1-(a+b)/2}_{(a-b)/2-1}(1-2z)\Big]$ $[\lvert\arg(-z)\rvert,\ \lvert\arg(1-z)\rvert<\pi]$
45	a b	$\frac{a+b}{2}$	$\frac{x}{2}\Gamma\left(\frac{a+b}{2}\right)(x-x^2)^{-(a+b+2)/4}\Big[P^{1-(a+b)/2}_{(a-b)/2}(1-2x)+$ $+P^{1-(a+b)/2}_{(a-b)/2-1}(1-2x)\Big]$ $[z=x,\ 0<x<1]$
46	a b	$\frac{a+b+1}{2}$	${}_2F_1\left(\frac{a}{2},\ \frac{b}{2};\ \frac{a+b+1}{2};\ 4z(1-z)\right)=$
47			$=(1-2z)\,{}_2F_1\left(\frac{a+1}{2},\ \frac{b+1}{2};\ \frac{a+b+1}{2};\ 4z(1-z)\right)=$
48			$=(1-2z)^{-a}{}_2F_1\left(\frac{a}{2},\ \frac{a+1}{2};\ \frac{a+b+1}{2};\ \frac{4z(z-1)}{(2z-1)^2}\right)=$
49			$=(\sqrt{1-z}+\sqrt{-z})^{-2a}{}_2F_1\times$ $\times\left(a,\ \frac{a+b}{2};\ a+b;\ \frac{4\sqrt{z^2-z}}{(\sqrt{1-z}+\sqrt{-z})^2}\right)$
50	a b	$\frac{a+b+1}{2}$	$\Gamma\left(\frac{a+b+1}{2}\right)(z^2-z)^{(1-a-b)/4}P^{(1-a-b)/2}_{(a-b-1)/2}(1-2z)$ $[\lvert\arg z\rvert,\ \lvert\arg(z-1)\rvert<\pi]$
51	a b	$\frac{a+b+1}{2}$	$\Gamma\left(\frac{a+b+1}{2}\right)(x-x^2)^{(1-a-b)/4}P^{(1-a-b)/2}_{(a-b-1)/2}(1-2x)$ $[z=x,\ 0<x<1]$
52	a b	$\frac{a+b}{2}+1$	$\frac{1}{z(b-a)}\Gamma\left(\frac{a+b}{2}+1\right)(z^2-z)^{(2-a-b)/4}\Big[P^{1-(a+b)/2}_{(a-b)/2}(1-2z)-$ $-P^{1-(a+b)/2}_{(a-b)/2-1}(1-2z)\Big]$ $[\lvert\arg(-z)\rvert,\ \lvert\arg(1-z)\rvert<\pi]$
53	a b	$\frac{a+b}{2}+1$	$\frac{1}{x(b-a)}\Gamma\left(\frac{a+b}{2}+1\right)(x-x^2)^{(2-a-b)/4}\Big[P^{1-(a+b)/2}_{(a-b)/2}(1-2x)-$ $-P^{1-(a+b)/2}_{(a-b)/2-1}(1-2x)\Big]$ $[z=x,\ 0<x<1]$

	a b	c	${}_2F_1(a, b; c; z)$
54	a b	$a-b+1$	$(1-z)^{-a}{}_2F_1\left(\frac{a}{2}, \frac{a+1}{2}-b; a-b+1; -\frac{4z}{(1-z)^2}\right)=$
55			$=\frac{1+z}{(1-z)^{a+1}}{}_2F_1\left(\frac{a+1}{2}, \frac{a}{2}+1-b; a-b+1; -\frac{4z}{(1-z)^2}\right)=$
56			$=(1+z)^{-a}{}_2F_1\left(\frac{a}{2}, \frac{a+1}{2}; a-b+1; \frac{4z}{(1+z)^2}\right)=$
57			$=(1-z)^{1-2b}(1+z)^{2b-a-1}{}_2F_1\left(\frac{a+1}{2}-b, \frac{a}{2}+1-b; a-b+1;\right.$
58			$\left.\frac{4z}{(1+z)^2}\right)=(1\pm\sqrt{z})^{-2a}{}_2F_1\left(a, a-b+\frac{1}{2}; 2a-2b+1;\right.$ $\left.\pm\frac{4\sqrt{z}}{(1\pm\sqrt{z})^2}\right)$
59	a b	$a-b+1$	$\Gamma(a-b+1)z^{(b-a)/2}(1-z)^{-b}P_{-b}^{b-a}\left(\frac{1+z}{1-z}\right)$ $[\lvert\arg(1-z)\rvert<\pi;\quad z\notin(-\infty, 0)]$
60	a b	$a-b+1$	$\Gamma(a-b+1)(-x)^{(b-a)/2}(1-x)^{-b}P_{-b}^{b-a}\left(\frac{1+x}{1-x}\right)$ $[z=x,\quad -\infty<x<0]$
61	a b	$a-b+1$	$e^{\pi i(1-2b)/2}\Gamma\begin{bmatrix}a-b+1\\a\end{bmatrix}\frac{z^{(2b-2a-1)/4}}{\sqrt{\pi}}(1-z)^{1/2-b}\times$ $\times Q_{a-b-1/2}^{b-1/2}\left(\frac{z+1}{2\sqrt{z}}\right)\quad\left[\left\lvert\arg\frac{(1-\sqrt{z})^2}{2\sqrt{z}}\right\rvert<\pi\right]$
62	a b	$a-b+2$	$\frac{\Gamma(a-b+2)}{b-1}z^{(b-a-1)/2}(1-z)^{-b}\left[aP_{-b}^{b-a-1}\left(\frac{1+z}{1-z}\right)-\right.$ $\left.-\sqrt{z}P_{-b}^{b-a}\left(\frac{1+z}{1-z}\right)\right]\quad[\lvert\arg z\rvert<\pi]$
63	a b	$a-b+2$	$\frac{\Gamma(a-b+2)}{b-1}(-x)^{(b-a-1)/2}(1-x)^{-b}\left[aP_{-b}^{b-a-1}\left(\frac{1+x}{1-x}\right)-\right.$ $\left.-\sqrt{-x}P_{-b}^{b-a}\left(\frac{1+x}{1-x}\right)\right]\quad[z=x,\quad -\infty<x<0]$
64	a b	$2b$	$(1-z)^{-a/2}{}_2F_1\left(\frac{a}{2}, b-\frac{a}{2}; b+\frac{1}{2}; \frac{z^2}{4(z-1)}\right)=$
65			$=\left(1-\frac{z}{2}\right)(1-z)^{-(a+1)/2}{}_2F_1\left(b+\frac{1-a}{2}, \frac{a+1}{2}; b+\frac{1}{2};\right.$
66			$\left.\frac{z^2}{4(z-1)}\right)=\left(1-\frac{z}{2}\right)^{-a}{}_2F_1\left(\frac{a}{2}, \frac{a+1}{2}; b+\frac{1}{2}; \frac{z^2}{(2-z)^2}\right)=$
67			$=(1-z)^{b-a}\left(1-\frac{z}{2}\right)^{a-2b}{}_2F_1\left(b-\frac{a}{2}, b+\frac{1-a}{2}; b+\frac{1}{2};\right.$ $\left.\frac{z^2}{(2-z)^2}\right)$
68	a b	$2b$	$(1-z)^{-a/2}{}_2F_1\left(a, 2b-a; b+\frac{1}{2}; -\frac{(1-\sqrt{1-z})^2}{4\sqrt{1-z}}\right)=$
69			$=\left(\frac{1+\sqrt{1-z}}{2}\right)^{-2a}{}_2F_1\left[a, a-b+\frac{1}{2}; b+\frac{1}{2};\right.$ $\left.\left(\frac{1-\sqrt{1-z}}{1+\sqrt{1-z}}\right)^2\right]$

	a	b	c	${}_2F_1(a, b; c; z)$
70	a	b	$2b$	$2^{2b-1}\Gamma\left(\frac{1}{2}+b\right)z^{1/2-b}(1-z)^{(2b-2a-1)/4}\times P_{a-b-1/2}^{1/2-b}\left(\frac{2-z}{2\sqrt{1-z}}\right)$ $\left[\operatorname{Re}\left(\frac{2-z}{\sqrt{1-z}}\right)>0\right]$
71	a	b	$2b$	$\frac{2^{2b}}{\sqrt{\pi}}\Gamma\begin{bmatrix}b+1/2\\2b-a\end{bmatrix}z^{-b}(1-z)^{(b-a)/2}e^{i\pi(a-b)}\times Q_{b-1}^{b-a}\left(\frac{2}{z}-1\right)$
72	a	b	$2b$	$\frac{2^{2b}}{\sqrt{\pi}}\Gamma\begin{bmatrix}b+1/2\\a\end{bmatrix}(-z)^{-b}(1-z)^{(b-a)/2}e^{i\pi(b-a)}\times Q_{b-1}^{a-b}\left(1-\frac{2}{z}\right)$ $[\lvert\arg(-z)\rvert,\ \lvert\arg(1\mp z)\rvert<\pi]$
73	a	b	$\frac{1}{2}$	$\frac{2^{a+b-3/2}}{\sqrt{\pi}}\Gamma\left(a+\frac{1}{2}\right)\Gamma\left(b+\frac{1}{2}\right)(z-1)^{(1-2a-2b)/4}\times\left[P_{a-b-1/2}^{1/2-a-b}(\sqrt{z})+P_{a-b-1/2}^{1/2-a-b}(-\sqrt{z})\right]$ $[\lvert\arg z\rvert,\ \lvert\arg(z-1)\rvert<\pi]$
74	a	b	$\frac{1}{2}$	$\frac{2^{a+b-3/2}}{\sqrt{\pi}}\Gamma\left(a+\frac{1}{2}\right)\Gamma\left(b+\frac{1}{2}\right)(1-x)^{(1-2a-2b)/4}\times\left[P_{a-b-1/2}^{1/2-a-b}(\sqrt{x})+P_{a-b-1/2}^{1/2-a-b}(-\sqrt{x})\right]$ $[z=x,\ 0<x<1]$
75	a	b	$\frac{1}{2}$	$\frac{2^{a-b-1}}{\sqrt{\pi}}e^{\pm(b-a)\pi i/2}\Gamma\left(a+\frac{1}{2}\right)\Gamma(1-b)\times(1-z)^{-(a+b)/2}\left[P_{a+b-1}^{b-a}\left(\sqrt{\frac{z}{z-1}}\right)+P_{a+b-1}^{b-a}\left(-\sqrt{\frac{z}{z-1}}\right)\right]$ $\left[\lvert\arg z\rvert<\pi,\ \begin{Bmatrix}\operatorname{Im} z>0\\\operatorname{Im} z<0\end{Bmatrix}\right]$
76	a	b	$\frac{1}{2}$	$\frac{2^{a-b-1}}{\sqrt{\pi}}\Gamma\left(a+\frac{1}{2}\right)\Gamma(1-b)(1-x)^{-(a+b)/2}\times\left[P_{a+b-1}^{b-a}\left(\sqrt{\frac{x}{x-1}}\right)+P_{a+b-1}^{b-a}\left(-\sqrt{\frac{x}{x-1}}\right)\right]$ $[z=x,\ -\infty<x<0]$
77	a	b	$\frac{3}{2}$	$\frac{2^{a+b-5/2}}{\sqrt{\pi z}}\Gamma\left(a-\frac{1}{2}\right)\Gamma\left(b-\frac{1}{2}\right)(z-1)^{(3-2a-2b)/4}\times\left[P_{a-b-1/2}^{3/2-a-b}(-\sqrt{z})-P_{a-b-1/2}^{3/2-a-b}(\sqrt{z})\right]$ $[z\notin[0,1]]$
78	a	b	$\frac{3}{2}$	$\frac{2^{a+b-5/2}}{\sqrt{\pi x}}\Gamma\left(a-\frac{1}{2}\right)\Gamma\left(b-\frac{1}{2}\right)(1-x)^{(3-2a-2b)/4}\times\left[P_{a-b-1/2}^{3/2-a-b}(-\sqrt{x})+P_{a-b-1/2}^{3/2-a-b}(\sqrt{x})\right]$ $[z=x,\ 0<x<1]$

	a	b	c	${}_2F_1(a, b; c; z)$
79	a	a	c	$\Gamma\begin{bmatrix} c \\ a,\ c-a \end{bmatrix}(-z)^{-a}\sum_{k=0}^{\infty}\frac{(a)_k(1+a-c)_k}{(k!)^2}\times[\ln(-z)+2\psi(k+1)-\psi(a+k)-\psi(c-a-k)]z^{-k}$ $[\lvert\arg(-z)\rvert<\pi;\ \lvert z\rvert>1;\ c-a\neq 0,\ \pm1,\ \pm2,\ \ldots]$
80	a	$a+m$	c	$\Gamma\begin{bmatrix} c \\ a+m,\ c-a \end{bmatrix}(-z)^{-a-m}\times\sum_{k=0}^{\infty}\frac{(a)_{k+m}(1+a-c)_{k+m}}{k!(k+m)!}z^{-k}[\ln(-z)+\psi(k+m+1)+\psi(k+1)-\psi(a+k+m)-\psi(c-a-m-k)]+(-z)^{-a}\Gamma\begin{bmatrix} c \\ a+m \end{bmatrix}\sum_{k=0}^{m-1}\frac{(a)_k}{k!}\Gamma\begin{bmatrix} m-k \\ c-a-k \end{bmatrix}z^{-k}$ $[\lvert\arg(-z)\rvert<\pi;\ \lvert z\rvert>1;\ c-a\neq 0,\ \pm1,\ \pm2,\ \ldots]$
81	a	$a+m$	$a+m+l+1$	$(a+m)_{l+1}(-z)^{-a}\Big\{(-1)^{l+1}z^{-m}\times\sum_{k=l+1}^{\infty}\frac{(a)_{k+m}(n-l-1)!}{k!(k+m)!}z^{-k}+\sum_{k=0}^{m-1}\frac{(m-k-1)!(a)_k}{(m+l-k)!k!}z^{-k}+\frac{(-z)^{-m}}{(l+m)!}\times\sum_{k=0}^{l}\frac{(a)_{k+m}}{k!}\frac{(-m-l)_{k+m}}{(k+m)!}z^{-k}[\ln(-z)+\psi(1+k)+\psi(1+k+m)-\psi(a+k+m)-\psi(l+1-k)]\Big\}$ $[\lvert\arg(-z)\rvert<\pi;\ \lvert z\rvert>1]$
82	$\frac{m}{2}+k-n$	$\frac{m}{2}+k$	k	$(-1)^n\frac{(k-1)!}{4}m^2\left[\left(\frac{m}{2}\right)_k\right]^{-2}\left(\frac{m}{2}+1\right)_n\times\left(1-\frac{m}{2}-k\right)_n\left(\frac{d}{dz}\right)^{k-n-1}\left[z^{-n/2}(1-z)^{-m/2-1}\times P_{-m/2-1}^{-n}\left(\frac{1+z}{1-z}\right)\right]$
83	a	$-a$	$\frac{1}{2}$	$\cos(2a\arcsin\sqrt{z})$
84	a	$1-a$	c	$(1-z)^{c-1}{}_2F_1\left(\frac{c-a}{2},\frac{c+a-1}{2};c;4z(1-z)\right)=$
85				$=(1-z)^{c-1}(1-2z)\,{}_2F_1\left(\frac{a+c}{2},\frac{1+c-a}{2};c;4z(1-z)\right)=$
86				$=(1-z)^{c-1}(1-2z)^{a-c}{}_2F_1\left(\frac{c-a}{2},\frac{1+c-a}{2};c;\frac{4z(z-1)}{(1-2z)^2}\right)=$
87				$=(1-z)^{c-1}(\sqrt{1-z}+\sqrt{-z})^{2-2a-2c}\times{}_2F_1\left(a+c-1,c-\frac{1}{2};2c-1;\frac{4\sqrt{z^2-z}}{(\sqrt{1-z}+\sqrt{-z})^2}\right)$

	a	b	c	${}_2F_1(a, b; c; z)$
88	a	$1-a$	c	$\Gamma(c)(-z)^{(1-c)/2}(1-z)^{(c-1)/2}P_{-a}^{1-c}(1-2z)$ [$\lvert\arg(-z)\rvert, \lvert\arg(1-z)\rvert<\pi$]
89	a	$1-a$	c	$\Gamma(c)x^{(1-c)/2}(1-x)^{(c-1)/2}P_{-a}^{1-c}(1-2x)$ [$z=x,\ 0<x<1$]
90	a	$1-a$	$\frac{1}{2}$	$\frac{1}{\sqrt{1-z}}\cos\left[(2a-1)\arcsin\sqrt{z}\right]$
91	a	$1-a$	$\frac{3}{2}$	$\frac{1}{(2a-1)\sqrt{z}}\sin\left[(2a-1)\arcsin\sqrt{z}\right]$
92	a	$2-a$	c	$\frac{\Gamma(c)}{1-a}\left(\frac{z}{z-1}\right)^{(1-c)/2}\left[\left(\frac{z}{z-1}\right)^{1/2}P_{-a}^{2-c}(1-2z)-\right.$ $\left.-(a+c-2)P_{-a}^{1-c}(1-2z)\right]$ [$\lvert\arg(-z)\rvert, \lvert\arg(1-z)\rvert<\pi$]
93	a	$2-a$	c	$\frac{\Gamma(c)}{1-a}\left(\frac{x}{1-x}\right)^{(1-c)/2}\left[\left(\frac{x}{1-x}\right)^{1/2}P_{-a}^{2-c}(1-2x)-\right.$ $\left.-(a+c-2)P_{-a}^{1-c}(1-2x)\right]$ [$z=x,\ 0<x<1$]
94	a	$2-a$	$\frac{3}{2}$	$\frac{1}{2(a-1)\sqrt{z-z^2}}\sin\left[2(a-1)\arcsin\sqrt{z}\right]$
95	a	$3-a$	c	$\frac{\Gamma(c)z^{-2}}{2(a-1)(a-2)}\left(1-\frac{1}{z}\right)^{c/2-2}\{(a+c-3)\times$ $\times P_{-a}^{2-c}(1-2z)-[a+c-3+2z(1-a)]\times$ $\times P_{1-a}^{2-c}(1-2z)\}$ [$\lvert\arg(-z)\rvert, \lvert\arg(1-z)\rvert<\pi$]
96	a	$3-a$	c	$\frac{\Gamma(c)x^{-2}}{2(a-1)(a-2)}\left(\frac{1}{x}-1\right)^{c/2-2}\{[a+c-3+$ $+2x(1-a)]P_{1-a}^{2-c}(1-2x)-(a+c-3)P_{a}^{2-c}(1-2x)\}$ [$z=x,\ 0<x<1$]
97	a	$a+\frac{1}{2}$	c	$(1-z)^{-a}{}_2F_1\left(2a,\ 2c-2a-1;\ c;\ \frac{\sqrt{1-z}-1}{2\sqrt{1-z}}\right)=$
98				$=\left(\frac{1+\sqrt{1-z}}{2}\right)^{-2a}{}_2F_1\left(2a,\ 2a-c+1;\ c;\ \frac{1-\sqrt{1-z}}{1+\sqrt{1-z}}\right)=(1\pm\sqrt{z})^{-2a}{}_2F_1\left(2a,\ c-\frac{1}{2};\right.$
99				$\left.2c-1;\ \pm\frac{2\sqrt{z}}{1\pm\sqrt{z}}\right)$
100	a	$a+\frac{1}{2}$	c	$2^{c-1}\Gamma(c)z^{(1-c)/2}(1-z)^{(c-1)/2-a}P_{2a-c}^{1-c}[(1-z)^{-1/2}]$ [$\lvert\arg z\rvert, \lvert\arg(1-z)\rvert<\pi$]
101	a	$a+\frac{1}{2}$	c	$2^{c-1}\Gamma(c)(-x)^{(1-c)/2}(1-x)^{(c-1)/2-a}\times$ $\times P_{2a-c}^{1-c}[(1-x)^{-1/2}]$ [$z=x,\ -\infty<x<0$]
102	a	$a+\frac{1}{2}$	c	$e^{(c-2a-1/2)\pi i}\Gamma\begin{bmatrix}c\\2a\end{bmatrix}\frac{2^{c-1/2}z^{(1-2c)/4}}{\sqrt{\pi}}\times$ $\times(1-z)^{(2c-1)/4-a}Q_{c-3/2}^{2a-c+1/2}\left(\frac{1}{\sqrt{z}}\right)$ [$\lvert z\rvert<1$]
103	a	$a+\frac{1}{2}$	$2a-1$	$\frac{1}{2a-1}[y+2(1-a)(y-2)]y^{2a}(2-y)^{-3}$ [$zy^2=4(y-1)$]

	a	b	c	${}_2F_1(a, b; c; z)$
104	a	$a+\frac{1}{2}$	$2a$	$\frac{1}{\sqrt{1-z}}\left(\frac{2}{1+\sqrt{1-z}}\right)^{2a-1}$
105	a	$a+\frac{1}{2}$	$2a+1$	$\left(\frac{2}{1+\sqrt{1-z}}\right)^{2a}$
106	a	$a+\frac{1}{2}$	$\frac{1}{2}$	$\frac{1}{2}\left[(1+\sqrt{z})^{-2a}+(1-\sqrt{z})^{-2a}\right]$
107	a	$a+\frac{1}{2}$	$\frac{3}{2}$	$\frac{1}{2(2a-1)\sqrt{z}}\left[(1-\sqrt{z})^{1-2a}-(1+\sqrt{z})^{1-2a}\right]$
108	a	$a+\frac{1}{2}$	$\frac{4a+5}{6}$	$(1-9z)^{-2a/3}\,{}_2F_1\left(\frac{a}{3}, \frac{a}{3}+\frac{1}{2}; \frac{4a+5}{6}; -\frac{27z(1-z)^2}{(1-9z)^2}\right)$
109	a	$a+\frac{1}{2}$	$\frac{4a+5}{6}$	$[1+(-z)^{3/2}]^{-2a}\,{}_2F_1\left(\frac{2a}{3}, \frac{2a+1}{3}; \frac{4a+2}{3}; \frac{2(-z)^{3/2}(3-z^3)}{(1+(-z)^{3/2})^3}\right)$
110	a	$a+\frac{1}{2}$	$\frac{4a+2}{3}$	$\left(1-\frac{9z}{8}\right)^{-2a/3}{}_2F_1\left(\frac{a}{3}, \frac{a}{3}+\frac{1}{2}; \frac{a}{3}+\frac{5}{6}; -\frac{27z^2(1-z)}{(9z-8)^2}\right)$
111	a	$3a-\frac{3}{2}$	$\frac{3}{2}$	$\left(1-\frac{z}{9}\right)\left(1+\frac{z}{3}\right)^{1/2-3a}{}_2F_1\left(a-\frac{1}{6}, a+\frac{1}{6}; \frac{3}{2}; \frac{z(z-9)^2}{(z+3)^3}\right)$
112	a	$3a-\frac{1}{2}$	$4a$	$\left(1-\frac{z}{4}\right)^{1/2-3a}{}_2F_1\left(a-\frac{1}{6}, a+\frac{1}{6}; 2a+\frac{1}{2}; -\frac{27z^2}{(z-4)^3}\right)$
113	a	$3a-\frac{1}{2}$	$\frac{1}{2}$	$(1-z)^{1/3-2a}\,{}_2F_1\left(a-\frac{1}{6}, \frac{1}{3}-a; \frac{1}{2}; -\frac{z(z-9)^2}{27(z-1)^2}\right)$
114	a	$1-3a$	$\frac{3}{2}-2a$	$(1-4z)^{3a-1}\,{}_2F_1\left(\frac{1}{3}-a, \frac{2}{3}-a; \frac{3}{2}-2a; \frac{27z}{(4z-1)^3}\right)$
115	a	$1-3a$	$\frac{1}{2}$	$(1-z)^{a-1/3}{}_2F_1\left(\frac{1}{3}-a, a-\frac{1}{6}; \frac{1}{2}; \frac{z(9-8z)^2}{27(1-z)}\right)$
116	a	$3-3a$	$\frac{3}{2}$	$\left(1-\frac{8z}{9}\right)\left(1-\frac{4z}{3}\right)^{3a-4}{}_2F_1\left(\frac{4}{3}-a, \frac{5}{3}-a; \frac{3}{2}; \frac{z(9-8z)^2}{(4z-3)^3}\right)$
117	a	$\frac{a}{2}+1$	$\frac{a}{2}$	$(1+z)(1-z)^{-a-1}$
118	n	b	c	$\frac{2(1-z)^{-b}}{B(b, c-b)}\left(1-\frac{1}{z}\right)^{c-n}Q_{n-1}^{(b-n,\, c-b-n)}\left(1-\frac{2}{z}\right)$
119	1	b	c	$z^{1-c}(1-z)^{c-b-1}(c-1)B_z(c-1, b-c+1)$
120	1	b	c	$\frac{c-1}{b-1}\left[\sum_{k=0}^{m-1}\frac{(c-b)_k}{(2-b)_k}(1-z)^{-k-1}-\frac{(c-b)_m}{(1-b)_m}\times(1-z)^{-m}{}_2F_1(1, b-m; c; z)\right]=$

	a	b	c	${}_2F_1(a, b; c; z)$
				$=z^{-1}\sum_{k=1}^{m}\frac{(1-c)_k}{(1+b-c)_k}\left(\frac{z-1}{z}\right)^{k-1}+\frac{(1-c)_m}{(1+b-c)_m}\times\left(\frac{z-1}{z}\right)^m {}_2F_1(1, b; c-m; z)$
121	1	b	$b-m$	$\frac{b-m-1}{b-1}\left[\sum_{k=0}^{m-1}\frac{(-m)_k}{(2-b)_k}(1-z)^{-k-1}-\frac{m!}{(1-b)_m}\times(z-1)^{-m-1}\right]$
122	1	b	$b+1$	$b\Phi(z, 1, b)$
123	1	b	m	$(m-1)!(-z)^{1-m}\frac{1}{(1-b)_{m-1}}\left[(1-z)^{m-b-1}-\sum_{k=0}^{m-2}\frac{(b-m+1)_k}{k!}z^k\right]$ $[m=1, 2, 3, \ldots; m-b\neq 1, 2, 3, \ldots]$
124	1	b	m	$\frac{1}{(-b)_m}\left(\frac{z-1}{z}\right)^{m-1}\left[\Gamma(1-b)(1-z)^{-b}+z^{-1}\times\sum_{k=1}^{m-1}(-b)_k\left(\frac{z-1}{z}\right)^{-k}\right]$ $[b\neq 1, 2, \ldots, m-1]$
125	1	b	2	$\frac{1}{(b-1)z}[(1-z)^{1-b}-1]$ $[b\neq 1]$
126	1	b	$\frac{3}{2}$	$\frac{\Gamma(1-b)\,2^{1/2-b}}{\sqrt{\pi z}}(1-z)^{(1-2b)/4}Q_{1/2-b}^{b-1/2}(\sqrt{z})$
127	1	$\frac{1}{2}-m$	$\frac{3}{2}$	$\frac{(2m-1)!!}{(2m)!!}\frac{(1-z)^m}{2\sqrt{z}}\ln\frac{1+\sqrt{z}}{1-\sqrt{z}}+\sum_{k=1}^{m}a_k z^{k-1}(1-z)^{m-k},\ a_k=\frac{(2m-1)!!\,(k-1)!}{2^{m-k+1}m!}\times\sum_{j=k}^{m}(-1)^{k-j}\binom{m}{j}\frac{(2j-2k-1)!!}{(2j-1)!!}$
128	1	n	m	$\frac{(m-1)!}{(m-n-1)!z}\left\{\sum_{k=1}^{m-n-1}\frac{(m-n-k-1)!}{(m-k-1)!}\left(\frac{z-1}{z}\right)^{k-1}-\frac{z}{(n-1)!}\left(\frac{z-1}{z}\right)^{m-n-1}\left[\sum_{k=1}^{n-1}\frac{z^{-k}}{n-k}+z^{-n}\ln(1-z)\right]\right\}$ $[m>n]$

	a	b	c	${}_2F_1(a, b; c; z)$
129	1	n	m	$(m-1)(1-z)^{m-n-1}\sum_{k=0}^{n-m}\frac{(m-n)_k}{k!(m+k-1)}z^k$ [$m\leqslant n$]
130	1	n	m	$\frac{(1-m)_n}{(n-1)!}\left[-(-z)^{1-m}(1-z)^{m-n-1}\ln(1-z)+(-z)^{1-n}\sum_{k=0}^{m-n-1}\frac{(n-m+1)_k}{k!}\sum_{j=1}^{n+k-1}\frac{1}{j}z^{j-k-1}\right]$ [$1<n<m$]
131	1	$\frac{m}{n}$	$\frac{m}{n}+1$	$-\frac{m}{n}z^{-m/n}\sum_{k=0}^{n-1}\exp\left(-\frac{2\pi ikm}{n}\right)\ln\left[1-z^{1/n}\times\exp\left(\frac{2\pi ik}{n}\right)\right]=-\frac{m}{n}z^{-m/n}\left\{\ln(1-z^{1/n})+\frac{(-1)^m}{2}[1+(-1)^n]\ln(1+z^{1/n})+\sum_{k=1}^{[(n-1)/2]}\left[\cos\frac{2\pi km}{n}\ln\left(1-2z^{1/n}\cos\frac{2\pi k}{n}+z^{2/n}\right)-2\sin\frac{2\pi km}{n}\operatorname{arctg}\frac{z^{1/n}\sin(2\pi k/n)}{1-z^{1/n}\cos(2\pi k/n)}\right]\right\}$ [$m, n=1, 2, 3, \ldots;\ m\leqslant n$]
132	1	$m+n$	$n+\frac{1}{2}$	$\frac{n-1/2}{m+n-1}\left[\sum_{k=0}^{m-1}\frac{(1/2-m)_k}{(2-m-n)_k}(1-z)^{-k-1}+\frac{(1/2)_m}{(n)_m}\times(1-z)^{-m}{}_2F_1\left(1, n; n+\frac{1}{2}; z\right)\right]$
133	1	n	$n+\frac{1}{2}$	$\frac{(1/2)_n}{(n-1)!z^n}\left[2\sqrt{\frac{z}{1-z}}\arcsin\sqrt{z}-\sum_{k=1}^{n-1}\frac{(k-1)!z^k}{(1/2)_k}\right]$
134	1	$n-\frac{1}{2}$	$n+\frac{1}{2}$	$\frac{2n-1}{z^n}\left[\sqrt{z}\operatorname{Arth}\sqrt{z}-\sum_{k=1}^{n-1}\frac{z^k}{2k-1}\right]$
135	1	n	$n+1$	$-\frac{n}{z^n}\left[\ln(1-z)+\sum_{k=1}^{n-1}\frac{z^k}{k}\right]$ [$n=1, 2, 3, \ldots$]
136	1	1	m	$\frac{m-1}{(z-1)^2}z\left[\sum_{k=2}^{m-1}\frac{1}{m-k}\left(\frac{z-1}{z}\right)^k-\left(\frac{z-1}{z}\right)^m\ln(1-z)\right]$ [$m=2, 3, 4, \ldots$]
137	1	$\frac{1}{2}$	$m+\frac{1}{2}$	$\frac{(1/2)_m}{(m-1)!}\frac{(z-1)^{m-1}}{z^m}\left[2\sqrt{z}\operatorname{Arth}\sqrt{z}+\sum_{k=1}^{m-1}\frac{(k-1)!}{(1/2)_k}\left(\frac{z}{z-1}\right)^k\right]$

	a	b	c	${}_2F_1(a, b; c; z)$
138	1	1	$m+\frac{1}{2}$	$\frac{2m-1}{z}\left(\frac{z-1}{z}\right)^{m-1}\left[\sqrt{\frac{z}{1-z}}\arcsin\sqrt{z}+\sum_{k=1}^{m-1}\frac{1}{2k-1}\left(\frac{z}{z-1}\right)^{k}\right]$
139	1	1	$m+\frac{1}{2}$	$\frac{2m-1}{z}\left(\frac{1-z}{z}\right)^{m-3/2}\left[(-1)^{m-1}\arcsin\sqrt{z}-\sqrt{\frac{z}{1-z}}\sum_{k=0}^{m-2}\frac{1}{2k+1}\left(\frac{z}{z-1}\right)^{k}\right]$ [$m=1, 2, 3, \ldots$]
140	$-n$	b	c	$\sum_{k=0}^{n}\frac{(-n)_k\,(b)_k}{(c_k)\,k!}z^k=\frac{z^{1-c}}{(c)_n}(1-z)^{n+c-b}\times\left(\frac{d}{dz}\right)^n\left[z^{n+c-1}(1-z)^{b-c}\right]$
141	$-n$	b	c	$\frac{n!}{(c)_n}P_n^{(c-1,\,\beta)}(1-2z)$ [$\beta=b-c-n$]
142	$-n$	b	c	$\frac{n!z^n}{(c)_n}P_n^{(\alpha,\,\beta)}\left(1-\frac{2}{z}\right)$ [$\alpha=-n-b,\ \beta=b-c-n$]
143	$-n$	b	c	$\frac{n!(1-z)^n}{(c)_n}P_n^{(c-1,\,\beta)}\left(\frac{1+z}{1-z}\right)$ [$\beta=-n-b$]
144	$-n$	b	$-2n-2$	$\frac{n!\Gamma(1-b)}{2(2n+2)!}(-z)^{n+2}(1-z)^{-(b+n+3)/2}\times\left\{-(b+2n+2)P_n^{b+n+1}\left(1-\frac{2}{z}\right)+\left[b+2n+2-\frac{2(n+1)}{z}\right]P_{n+1}^{b+n+1}\left(1-\frac{2}{z}\right)\right\}$ [$\lvert\arg(1-z)\rvert<\pi$]
145	$-n$	b	$-2n-2$	$\frac{n!\Gamma(1-b)}{2(2n+2)!}(-x)^{n+2}(x-1)^{-(b+n+3)/2}\times\left\{(b+2n+2)P_n^{b+n+1}\left(1-\frac{2}{x}\right)-\left[b+2n+2-\frac{2(n+1)}{x}\right]P_{n+1}^{b+n+1}\left(1-\frac{2}{x}\right)\right\}$ [$z=x,\ 1<x<\infty$]
146	$-n$	b	$-2n-1$	$\frac{n!\,\Gamma(1-b)}{(2n+1)!}(-z)^n(1-z)^{-(b+n)/2}\left[(1-z)^{-1/2}\times P_n^{b+n+1}\left(1-\frac{2}{z}\right)+(b+2n+1)P_n^{b+n}\left(1-\frac{2}{z}\right)\right]$ [$\lvert\arg(1-z)\rvert<\pi$]
147	$-n$	b	$-2n-1$	$\frac{n!\Gamma(1-b)}{(2n+1)!}(-x)^n(x-1)^{-(b+n)/2}\left[(x-1)^{-1/2}\times P_n^{b+n+1}\left(1-\frac{2}{x}\right)+(b+2n+1)P_n^{b+n}\left(1-\frac{2}{x}\right)\right]$ [$z=x,\ 1<x<\infty$]
148	$-n$	b	$-2n$	$\frac{n!}{(2n)!}\Gamma(1-b)(-z)^n(1-z)^{-(n+b)/2}P_n^{n+b}\left(1-\frac{2}{z}\right)$ [$\lvert\arg(1-z)\rvert<\pi$]
149	$-n$	b	$-2n$	$\frac{n!}{(2n)!}\Gamma(1-b)(-x)^n(x-1)^{-(n+b)/2}P_n^{n+b}\left(1-\frac{2}{x}\right)$ [$z=x,\ 1<x<\infty$]

	a	b	c	${}_2F_1(a, b; c; z)$
150	$-n$	b	$2b-1$	$-\frac{\Gamma(1-b)}{2(2b-1)_n} z^{1-b}(1-z)^{(b+n)/2-1}\left[P_{b-1}^{b+n}\left(1-\frac{2}{z}\right)+ P_{b-2}^{b+n}\left(1-\frac{2}{z}\right)\right]$ $[\lvert\arg z\rvert,\ \lvert\arg(1-z)\rvert<\pi]$
151	$-n$	b	$2b-1$	$\frac{\Gamma(1-b)}{2(2b-1)_n} x^{1-b}(x-1)^{(b+n)/2-1}\left[P_{b-1}^{b+n}\left(1-\frac{2}{x}\right)+ P_{b-2}^{b+n}\left(1-\frac{2}{x}\right)\right]$ $[z=x,\ 1<x<\infty]$
152	$-n$	b	$2b$	$\frac{n!\ 2^{-2n}z^n}{(b+1/2)_n} C_n^{\lambda}\left(1-\frac{2}{z}\right)$ $\left[\lambda=\frac{1}{2}-b-n\right]$
153	$-n$	b	$-b-n$	$\frac{n!}{(b+1)_n} z^{n/2}\left[C_n^{b+1}\left(\frac{z+1}{2\sqrt{z}}\right)-\sqrt{z}\, C_{n-1}^{b+1}\left(\frac{z+1}{2\sqrt{z}}\right)\right]$ $[n\geqslant 1]$
154	$-n$	b	$1-b-n$	$\frac{n!z^{n/2}}{(b)_n} C_n^b\left(\frac{z+1}{2\sqrt{z}}\right)$
155	$-n$	b	$1-b-n$	$\frac{n!\,(1-z)^n}{(1-2b-2n)_n} C_n^{\lambda}\left(\frac{1+z}{1-z}\right)$ $\left[\lambda=\frac{1}{2}-b-n\right]$
156	$-n$	b	$2-b-n$	$\frac{n!}{(b-1)_n} z^{(n-1)/2}\left[\sqrt{z}\, C_n^b\left(\frac{z+1}{2\sqrt{z}}\right)-C_{n-1}^b\left(\frac{z+1}{2\sqrt{z}}\right)\right]$ $[n\geqslant 1]$
157	$-n$	b	$b-n-\frac{1}{2}$	$\frac{(2n+1)!\,(-z)^{n+1/2}}{2^{2n+1}(1/2-b-n)_{2n+1}\sqrt{1-z}} C_{2n+1}^{1/2-n-b}\left(\sqrt{\frac{z-1}{z}}\right)$
158	$-n$	b	$b-n-\frac{1}{2}$	$-\frac{(2n+1)!}{(2-2b)_{2n+1}\sqrt{1-z}} C_{2n+1}^{b-n-1}(\sqrt{1-z})$
159	$-n$	b	$b-n+\frac{1}{2}$	$\frac{(2n)!}{(1-2b)_{2n}} C_{2n}^{b-n}(\sqrt{1-z})$
160	$-n$	b	$b-n+\frac{1}{2}$	$\frac{(2n)!}{(b-n+1/2)_{2n}}\left(-\frac{z}{4}\right)^n C_{2n}^{1/2-n-b}\left(\sqrt{\frac{z-1}{z}}\right)$
161	$-n$	b	$\frac{b-n+1}{2}$	$\frac{(-1)^n\, n!}{(1-b)_n} C_n^{(b-n)/2}(1-2z)$
162	$-n$	b	$\frac{1}{2}$	$\frac{n!}{(1-b)_n} C_{2n}^{b-n}(\sqrt{z})$
163	$-n$	b	$\frac{3}{2}$	$-\frac{n!}{2(1-b)_{n+1}\sqrt{z}} C_{2n+1}^{b-n-1}(\sqrt{z})$
164	$-n$	m	$m+1$	$I_{m,n},\ I_{2,n}=\frac{2z^{-2}}{(n+1)(n+2)}[1-(1+z+nz)(1-z)^{n+1}]$, $I_{3,n}=\frac{3z^{-3}}{(n+1)(n+2)(n+3)}\{2-[2(1-z)^2+ z(n+3)(nz+2)](1-z)^{n+1}\}$, $(m+1)(n+1)I_{m,n}=(m+1)I_{m,n+1}+z(n+1)mI_{m+1,n}$, $I_{m,0}=1,\quad I_{1,n}=\frac{n!}{n+1} z^{-1}[1-(1-z)^{n+1}]$

	a	b	c	${}_2F_1(a,\ b;\ c;\ z)$
165	$-n$	$-\frac{1}{2}-n$	$-1-2n$	$(-1)^n 2^{-2n} z^n U_n\left(1-\frac{2}{z}\right)$
166	$-n$	$-\frac{1}{2}-n$	$\frac{3}{2}$	$\frac{(1-z)^n}{n+1} U_n\left(\frac{1+z}{1-z}\right)$
167	$-n$	$-n$	$-\frac{1}{2}-2n$	$(-1)^n \frac{(2n+1)!}{\Gamma(2n+3/2)} \sqrt{\pi} \left(\frac{z}{4}\right)^{n+1/2} (z-1)^{-1/2} \times P_{2n+1}\left(\sqrt{\frac{z-1}{z}}\right)$
168	$-n$	$-n$	$-2n$	$\frac{(-1)^n n!\, z^n}{2^{2n} (1/2)_n} P_n\left(1-\frac{2}{z}\right)$
169	$-n$	$-n$	$\frac{1}{2}-2n$	$(-1)^n \frac{[(2n)!]^2}{(4n)!} (4z)^n P_{2n}\left(\sqrt{\frac{z-1}{z}}\right)$
170	$-n$	$-n$	$\frac{1}{2}$	$\frac{n!}{(1/2)_n} (z-1)^n P_{2n}\left(\sqrt{\frac{z}{z-1}}\right)$
171	$-n$	$-n$	1	$(1-z)^n P_n\left(\frac{1+z}{1-z}\right)$
172	$-n$	$-n$	$\frac{3}{2}$	$\frac{n!\,(z-1)^{n+1/2}}{(3/2)_n \sqrt{z}} P_{2n+1}\left(\sqrt{\frac{z}{z-1}}\right)$
173	$-n$	$\frac{1}{2}-n$	$1-2n$	$(-1)^n 2^{1-2n} z^n\, T_n\left(1-\frac{2}{z}\right)$ $[n \geqslant 1]$
174	$-n$	$\frac{1}{2}-n$	$\frac{1}{2}$	$(1-z)^n T_n\left(\frac{1+z}{1-z}\right)$
175	$-n$	$\frac{1}{2}$	$\frac{1}{2}-n$	$2^{2n} z^{n/2} \frac{(n!)^2}{(2n)!} P_n\left(\frac{z+1}{2\sqrt{z}}\right)$
176	$-n$	$\frac{1}{2}$	1	$(1-z)^{n/2} P_n\left(\frac{2-z}{2\sqrt{1-z}}\right)$
177	$-n$	1	c	$(1-c)\, z^{1-c} (z-1)^{n+c-1} B_{1-1/z}(1-c-n,\ n+1)$
178	$-n$	1	$-m$	$\frac{m+1}{n+1} \sum_{k=0}^{m-n} \frac{(n-m)_k}{(n+2)_k} (1-z)^{-k-1} + \frac{(-1)^n n!}{(-m)_n} z^{m+1} \times (z-1)^{n-m-1}$
179	$-n$	1	m	$-\frac{n!\,(z-1)^{m-2}}{(m)_n z^{m-1}} \left[(1-z)^{n+1} - \sum_{k=0}^{m-2} \frac{(n+1)_k}{k!} \left(\frac{z}{z-1}\right)^k\right] =$
180				$= \frac{n!}{(m)_n} (-z)^{1-m} \left[(1-z)^{m+n-1} - \sum_{k=0}^{m-2} \frac{(1-m-n)_k}{k!} z^k\right]$
181	$-n$	n	$\frac{1}{2}$	$T_n(1-2z) = \frac{(-1)^n}{2} [(\sqrt{z}+\sqrt{z-1})^{2n} + (\sqrt{z}-\sqrt{z-1})^{2n}]$
182	$-n$	n	$\frac{1}{2}$	$(-1)^n T_{2n}(\sqrt{z})$

	a	b	c	${}_2F_1(a,\ b;\ c;\ z)$
183	$-n$	n	$\frac{1}{2}$	$T_{2n}(\sqrt{1-z})$
184	$-n$	$n+\frac{1}{2}$	$\frac{1}{2}$	$(-1)^n \frac{n!}{(1/2)_n} P_{2n}(\sqrt{z})$
185	$-n$	$n+\frac{1}{2}$	1	$P_{2n}(\sqrt{1-z})$
186	$-n$	$n+1$	$\frac{1}{2}$	$(-1)^n U_{2n}(\sqrt{z})$
187	$-n$	$n+1$	$\frac{1}{2}$	$\frac{1}{\sqrt{1-z}} T_{2n+1}(\sqrt{1-z})$
188	$-n$	$n+1$	1	$P_n(1-2z)$
189	$-n$	$n+1$	$\frac{3}{2}$	$\frac{1}{2n+1} U_{2n}(\sqrt{1-z})$
190	$-n$	$n+1$	$\frac{3}{2}$	$\frac{(-1)^n}{(2n+1)\sqrt{z}} T_{2n+1}(\sqrt{z})$
191	$-n$	$n+\frac{3}{2}$	$\frac{1}{2}$	$\frac{(-1)^n\, n!}{(3/2)_n} P'_{2n+1}(\sqrt{z})$
192	$-n$	$n+\frac{3}{2}$	1	$\frac{1}{\sqrt{1-z}} P_{2n+1}(\sqrt{1-z})$
193	$-n$	$n+\frac{3}{2}$	$\frac{3}{2}$	$\frac{(-1)^n\, n!}{(3/2)_n \sqrt{z}} P_{2n+1}(\sqrt{z})$
194	$-n$	$n+2$	$\frac{3}{2}$	$\frac{1}{n+1} U_n(1-2z)$
195	$-n$	$n+2$	$\frac{3}{2}$	$\frac{(-1)^n}{2(n+1)\sqrt{z}} U_{2n+1}(\sqrt{z})$
196	$-n$	$n+2$	$\frac{3}{2}$	$\frac{1}{2(n+1)\sqrt{1-z}} U_{2n+1}(\sqrt{1-z})$
197	$-n$	$n+\frac{5}{2}$	$\frac{3}{2}$	$\frac{(-1)^n\, n!}{3(5/2)_n \sqrt{z}} P'_{2n+2}(\sqrt{z})$
198	$-\frac{n}{2}$	b	$b+\frac{1-n}{2}$	$\frac{(-1)^n\, n!}{(1-2b)_n} C_n^{b-n/2}(\sqrt{1-z})$
199	$-\frac{n}{2}$	$\frac{n}{2}$	$\frac{1}{2}$	$T_n(\sqrt{1-z})$
200	$-\frac{n}{2}$	$\frac{n+1}{2}$	1	$P_n(\sqrt{1-z})$
201	$-\frac{n}{2}$	$\frac{n}{2}+1$	$\frac{3}{2}$	$\frac{1}{n+1} U_n(\sqrt{1-z})$
202	$-\frac{n}{2}$	$\frac{1-n}{2}$	c	$\frac{n!}{(2c-1)_n}(1-z)^{n/2} C_n^{c-1/2}[(1-z)^{-1/2}]$
203	$-\frac{n}{2}$	$\frac{1-n}{2}$	c	$(-1)^n \frac{n!}{(c)_n}\left(\frac{z}{4}\right)^{n/2} C_n^{1-c-n}\left(\frac{1}{\sqrt{z}}\right)$

	a	b	c	${}_2F_1\,(a,\ b;\ c;\ z)$
204	$-\frac{n}{2}$	$\frac{1-n}{2}$	$-\frac{1}{2}-n$	$\frac{(n!)^2}{(2n+1)!}\,2^n z^{n/2} P'_{n+1}\left(\frac{1}{\sqrt{z}}\right)$
205	$-\frac{n}{2}$	$\frac{1-n}{2}$	$-n$	$2^{-n} z^{n/2} U_n\left(\frac{1}{\sqrt{z}}\right)=2^{-n-1}(1-z)^{-1/2}\times$ $\times\left[\left(1+\sqrt{1-z}\right)^{n+1}-\left(1-\sqrt{1-z}\right)^{n+1}\right]$
206	$-\frac{n}{2}$	$\frac{1-n}{2}$	$\frac{1}{2}-n$	$\frac{n!\,z^{n/2}}{2^n\,(1/2)_n}\,P_n\left(\frac{1}{\sqrt{z}}\right)$
207	$-\frac{n}{2}$	$\frac{1-n}{2}$	$1-n$	$2^{1-n} z^{n/2} T_n\left(\frac{1}{\sqrt{z}}\right)=$
208				$=2^{-n-1}\left[\left(1+\sqrt{1-z}\right)^n+\left(1-\sqrt{1-z}\right)^n\right]\quad [n\geqslant 1]$
209	$-\frac{n}{2}$	$\frac{1-n}{2}$	$\frac{1}{2}$	$(1-z)^{n/2}\,T_n\left(\frac{1}{\sqrt{1-z}}\right)$
210	$-\frac{n}{2}$	$\frac{1-n}{2}$	1	$(1-z)^{n/2}\,P_n\left(\frac{1}{\sqrt{1-z}}\right)$
211	$-\frac{n}{2}$	$\frac{1-n}{2}$	$\frac{3}{2}$	$\frac{(1-z)^{n/2}}{n+1}\,U_n\left(\frac{1}{\sqrt{1-z}}\right)$

7.3.2. Special values of ${}_2F_1\,(a,b;c;z)$.

(See also 7.3.3 with z replaced by $-z$).

	a	b	c	${}_2F_1\,(a,\ b;\ c;\ z)$
1	$-\frac{1}{2}$	$-\frac{1}{2}$	$\frac{1}{2}$	$\sqrt{1-z}+\sqrt{z}\arcsin\sqrt{z}$
2	$-\frac{1}{2}$	$\frac{1}{2}$	1	$\frac{2}{\pi}\,\mathbf{E}\,(\sqrt{z})$
3	$-\frac{1}{2}$	$\frac{1}{2}$	$\frac{3}{2}$	$\frac{1}{2}\left(\frac{\arcsin\sqrt{z}}{\sqrt{z}}+\sqrt{1-z}\right)$
4	$-\frac{1}{2}$	$\frac{1}{2}$	2	$\frac{4}{3\pi}\left[2\mathbf{K}\,(\sqrt{z})-(1+z)\,\mathbf{D}\,(\sqrt{z})\right]$
5	$-\frac{1}{2}$	$\frac{1}{2}$	$\frac{5}{2}$	$\frac{3}{16z}\left[(1+2z)\sqrt{1-z}-(1-4z)\frac{\arcsin\sqrt{z}}{\sqrt{z}}\right]$
6	$-\frac{1}{2}$	$\frac{1}{2}$	3	$\frac{16}{45\pi z}\left[(2-7z-3z^2)\,\mathbf{D}\,(\sqrt{z})-(1-9z)\,\mathbf{K}\,(\sqrt{z})\right]$
7	$-\frac{1}{2}$	$\frac{1}{2}$	$\frac{7}{2}$	$\frac{5}{128z^2}\left[3(1-4z+8z^2)\frac{\arcsin\sqrt{z}}{\sqrt{z}}-(3-10z-8z^2)\sqrt{1-z}\right]$
8	$-\frac{1}{2}$	$\frac{1}{2}$	4	$\frac{32}{525\pi z^2}\left[4\,(1-4z+15z^2)\,\mathbf{K}(\sqrt{z})-(8-33z+58z^2+\right.$ $\left.+15z^3)\,\mathbf{D}\,(\sqrt{z})\right]$

	a	b	c	${}_2F_1(a, b; c; z)$
9	$-\frac{1}{2}$	1	$\frac{1}{2}$	$1-\sqrt{z}\,\mathrm{Arth}\sqrt{z}$
10	$-\frac{1}{2}$	1	$\frac{3}{2}$	$\frac{1}{2}\left[1+(1-z)\frac{\mathrm{Arth}\sqrt{z}}{\sqrt{z}}\right]$
11	$-\frac{1}{2}$	1	2	$\frac{2}{3z}[1-(1-z)^{3/2}]$
12	$-\frac{1}{2}$	1	$\frac{5}{2}$	$\frac{3}{8z}\left[1+z-(1-z)^2\frac{\mathrm{Arth}\sqrt{z}}{\sqrt{z}}\right]$
13	$-\frac{1}{2}$	1	3	$\frac{8}{15z^2}\left[(1-z)^{5/2}-1+\frac{5}{2}z\right]$
14	$-\frac{1}{2}$	1	$\frac{7}{2}$	$\frac{5}{48z^2}\left[3(1-z)^3\frac{\mathrm{Arth}\sqrt{z}}{\sqrt{z}}-3+8z+3z^2\right]$
15	$-\frac{1}{2}$	1	4	$\frac{2}{35z^3}[8-28z+35z^2-8(1-z)^{7/2}]$
16	$-\frac{1}{2}$	$\frac{3}{2}$	$\frac{1}{2}$	$(1-2z)(1-z)^{-1/2}$
17	$-\frac{1}{2}$	$\frac{3}{2}$	1	$\frac{2}{\pi}[\mathbf{K}(\sqrt{z})-2z\mathbf{D}(\sqrt{z})]$
18	$-\frac{1}{2}$	$\frac{3}{2}$	2	$\frac{4}{3\pi}[\mathbf{K}(\sqrt{z})+(1-2z)\mathbf{D}(\sqrt{z})]$
19	$-\frac{1}{2}$	$\frac{3}{2}$	$\frac{5}{2}$	$\frac{3}{8z}\left[\frac{\arcsin\sqrt{z}}{\sqrt{z}}-(1-2z)\sqrt{1-z}\right]$
20	$-\frac{1}{2}$	$\frac{3}{2}$	3	$\frac{16}{15\pi z}[(1+z)\mathbf{K}(\sqrt{z})-2(1-z+z^2)\mathbf{D}(\sqrt{z})]$
21	$-\frac{1}{2}$	$\frac{3}{2}$	$\frac{7}{2}$	$\frac{5}{32z^2}\left[(3-4z+4z^2)\sqrt{1-z}-3(1-2z)\frac{\arcsin\sqrt{z}}{\sqrt{z}}\right]$
22	$-\frac{1}{2}$	$\frac{3}{2}$	4	$\frac{32}{105\pi z^2}[(8-19z+9z^2-6z^3)\mathbf{D}(\sqrt{z})-$ $-(4-9z-3z^2)\mathbf{K}(\sqrt{z})]$
23	$-\frac{1}{2}$	2	$\frac{1}{2}$	$\frac{(1-z)^{-1}}{2}[2-3z-3\sqrt{z}(1-z)\mathrm{Arth}\sqrt{z}]$
24	$-\frac{1}{2}$	2	1	$\frac{1}{2}(2-3z)(1-z)^{-1/2}$
25	$-\frac{1}{2}$	2	$\frac{3}{2}$	$\frac{1}{4}\left[3+(1-3z)\frac{\mathrm{Arth}\sqrt{z}}{\sqrt{z}}\right]$
26	$-\frac{1}{2}$	2	$\frac{5}{2}$	$\frac{3}{16z}\left[(1+3z)(1-z)\frac{\mathrm{Arth}\sqrt{z}}{\sqrt{z}}-1+3z\right]$
27	$-\frac{1}{2}$	2	3	$\frac{4}{15z^2}[2-(2+3z)(1-z)^{3/2}]$
28	$-\frac{1}{2}$	2	$\frac{7}{2}$	$\frac{5}{32z^2}\left[3-2z+3z^2-3(1-z)^2(1+z)\frac{\mathrm{Arth}\sqrt{z}}{\sqrt{z}}\right]$

	a	b	c	${}_2F_1(a, b; c; z)$
29	$-\frac{1}{2}$	2	4	$\frac{8}{35z^3}[(4+3z)(1-z)^{5/2}-4+7z]$
30	$-\frac{1}{2}$	$\frac{5}{2}$	$\frac{1}{2}$	$\frac{1}{3}(3-12z+8z^2)(1-z)^{-3/2}$
31	$-\frac{1}{2}$	$\frac{5}{2}$	1	$\frac{2}{3\pi(1-z)}[(3-4z)\mathbf{K}(\sqrt{z})-z(7-8z)\mathbf{D}(\sqrt{z})]$
32	$-\frac{1}{2}$	$\frac{5}{2}$	$\frac{3}{2}$	$\frac{1}{3}(3-4z)(1-z)^{-1/2}$
33	$-\frac{1}{2}$	$\frac{5}{2}$	2	$\frac{4}{9\pi}[4\mathbf{K}(\sqrt{z})+(1-8z)\mathbf{D}(\sqrt{z})]$
34	$-\frac{1}{2}$	$\frac{5}{2}$	3	$\frac{16}{45\pi z}[(2+3z-8z^2)\mathbf{D}(\sqrt{z})-(1-4z)\mathbf{K}(\sqrt{z})]$
35	$-\frac{1}{2}$	$\frac{5}{2}$	$\frac{7}{2}$	$\frac{5}{48z^2}\left[\frac{3\arcsin\sqrt{z}}{\sqrt{z}}-\frac{3-z-10z^2+8z^3}{\sqrt{1-z}}\right]$
36	$-\frac{1}{2}$	$\frac{5}{2}$	4	$\frac{32}{105\pi z^2}[2(2-z+2z^2)\mathbf{K}(\sqrt{z})-$ $-(8-5z-5z^2+8z^3)\mathbf{D}(\sqrt{z})]$
37	$-\frac{1}{2}$	3	$\frac{1}{2}$	$\frac{(1-z)^{-2}}{8}[8-25z+15z^2-15\sqrt{z}(1-z)^2\operatorname{Arth}\sqrt{z}]$
38	$-\frac{1}{2}$	3	1	$\frac{1}{8}(8-24z+15z^2)(1-z)^{-3/2}$
39	$-\frac{1}{2}$	3	$\frac{3}{2}$	$\frac{(1-z)^{-1}}{16}\left[13-15z+3(1-5z)(1-z)\frac{\operatorname{Arth}\sqrt{z}}{\sqrt{z}}\right]$
40	$-\frac{1}{2}$	3	2	$\frac{1}{4}(4-5z)(1-z)^{-1/2}$
41	$-\frac{1}{2}$	3	$\frac{5}{2}$	$\frac{3}{64z}\left[(1+6z-15z^2)\frac{\operatorname{Arth}\sqrt{z}}{\sqrt{z}}-1+15z\right]$
42	$-\frac{1}{2}$	3	$\frac{7}{2}$	$\frac{5}{128z^2}\left[3(1+2z+5z^2)(1-z)\frac{\operatorname{Arth}\sqrt{z}}{\sqrt{z}}-\right.$ $\left.-3-4z+15z^2\right]$
43	$-\frac{1}{2}$	3	4	$\frac{2}{35z^3}[8-(8+12z+15z^2)(1-z)^{3/2}]$
44	$-\frac{1}{2}$	$\frac{7}{2}$	$\frac{1}{2}$	$\frac{1}{5}(5-30z+40z^2-16z^3)(1-z)^{-5/2}$
45	$-\frac{1}{2}$	$\frac{7}{2}$	1	$\frac{2}{15\pi}(1-z)^{-2}[(15-41z+24z^2)\mathbf{K}(\sqrt{z})-$ $-2z(19-44z+24z^2)\mathbf{D}(\sqrt{z})]$
46	$-\frac{1}{2}$	$\frac{7}{2}$	$\frac{3}{2}$	$\frac{1}{15}(15-40z+24z^2)(1-z)^{-3/2}$
47	$-\frac{1}{2}$	$\frac{7}{2}$	2	$\frac{4}{15\pi(1-z)}[(7-8z)\mathbf{K}(\sqrt{z})+(1-16z+16z^2)\mathbf{D}(\sqrt{z})]$

	a	b	c	${}_2F_1(a, b; c; z)$
48	$-\frac{1}{2}$	$\frac{7}{2}$	$\frac{5}{2}$	$\frac{1}{5}(5-6z)(1-z)^{-1/2}$
49	$-\frac{1}{2}$	$\frac{7}{2}$	3	$\frac{16}{225\pi z}[2(1+4z-24z^2)\mathbf{D}(\sqrt{z})-(1-24z)\mathbf{K}(\sqrt{z})]$
50	$-\frac{1}{2}$	$\frac{7}{2}$	4	$\frac{32}{525\pi z^2}[(8+9z+16z^2-48z^3)\mathbf{D}(\sqrt{z})-$ $-(4+5z-24z^2)\mathbf{K}(\sqrt{z})]$
51	$-\frac{1}{2}$	4	$\frac{1}{2}$	$\frac{(1-z)^{-3}}{48}(48-231z+280z^2-105z^3)-\frac{35}{16}\sqrt{z}\,\mathrm{Arth}\sqrt{z}$
52	$-\frac{1}{2}$	4	1	$\frac{1}{16}(16-72z+90z^2-35z^3)(1-z)^{-5/2}$
53	$-\frac{1}{2}$	4	$\frac{3}{2}$	$\frac{(1-z)^{-2}}{96}(81-190z+105z^2)+\frac{5}{32}(1-7z)\frac{\mathrm{Arth}\sqrt{z}}{\sqrt{z}}$
54	$-\frac{1}{2}$	4	2	$\frac{1}{24}(24-60z+35z^2)(1-z)^{-3/2}$
55	$-\frac{1}{2}$	4	$\frac{5}{2}$	$\frac{(1-z)^{-1}}{128z}\left[3(1+10z-35z^2)(1-z)\frac{\mathrm{Arth}\sqrt{z}}{\sqrt{z}}-\right.$ $\left.-3-100z-105z^2\right]$
56	$-\frac{1}{2}$	4	3	$\frac{1}{6}(6-7z)(1-z)^{-1/2}$
57	$-\frac{1}{2}$	4	$\frac{7}{2}$	$\frac{5}{768z^2}\left[3(1+3z+15z^2-35z^3)\frac{\mathrm{Arth}\sqrt{z}}{\sqrt{z}}-\right.$ $\left.-3-10z+105z^2\right]$
58	$\frac{1}{8}$	1	$\frac{9}{8}$	$\frac{1}{8x}\left[\ln\frac{1+x}{1-x}+2\,\mathrm{arctg}\,x+\sqrt{2}\,\mathrm{arctg}\frac{2^{1/2}x}{1-x^2}-\right.$ $\left.-2^{-1/2}\ln\frac{1-2^{1/2}x+x^2}{1+2^{1/2}x+x^2}\right]$ $[x=z^{1/8}]$
59	$\frac{1}{6}$	1	$\frac{7}{6}$	$\frac{1}{12x}\left[2\ln\frac{1+x}{1-x}-\ln\frac{1-x+x^2}{1+x+x^2}+2\sqrt{3}\,\mathrm{arctg}\frac{3^{1/2}x}{1-x^2}\right]$ $[x=z^{1/6}]$
60	$\frac{1}{6}$	1	$\frac{13}{6}$	$\frac{7}{6x^6}-\frac{7}{36x^7}(1-x^6)\left[\ln\frac{1+x}{1-x}-\frac{1}{2}\ln\frac{1-x+x^2}{1+x+x^2}+\right.$ $\left.+\sqrt{3}\,\mathrm{arctg}\frac{3^{1/2}x}{1-x^2}\right]$ $[x=z^{1/6}]$
61	$\frac{1}{5}$	1	$\frac{6}{5}$	$\frac{1}{20x}\left\{(\sqrt{5}+1)\ln\left[1+(\sqrt{5}+1)\frac{x}{2}+x^2\right]-\right.$ $-(\sqrt{5}-1)\ln\left[1-(\sqrt{5}-1)\frac{x}{2}+x^2\right]+$ $+2(10-2\sqrt{5})^{1/2}\,\mathrm{arctg}\frac{(10-2\sqrt{5})^{1/2}x}{4+(\sqrt{5}+1)x}+$ $\left.+2(10+2\sqrt{5})^{1/2}\,\mathrm{arctg}\frac{(10+2\sqrt{5})^{1/2}x}{4-(\sqrt{5}-1)x}-4\ln(1-x)\right\}$ $[x=z^{1/5}]$

	a	b	c	${}_2F_1(a, b; c; z)$
62	$\frac{1}{5}$	1	$\frac{11}{5}$	$\frac{6}{5x^5}-\frac{3}{50x^6}(1-x^5)\Big\{(1+\sqrt{5})\ln\left[1+(\sqrt{5}+1)\frac{x}{2}+x^2\right]-(\sqrt{5}-1)\ln\left[1-(\sqrt{5}-1)\frac{x}{2}+x^2\right]+2(10-2\sqrt{5})^{1/2}\operatorname{arctg}\frac{(10-2\sqrt{5})^{1/2}x}{4+(\sqrt{5}+1)x}+2(10+2\sqrt{5})^{1/2}\operatorname{arctg}\frac{(10+2\sqrt{5})^{1/2}x}{4-(\sqrt{5}-1)x}-4\ln(1-x)\Big\}$ $[x=z^{1/5}]$
63	$\frac{1}{4}$	$\frac{3}{4}$	$\frac{1}{2}$	$\frac{1}{2}\left[(1-\sqrt{z})^{-1/2}+(1+\sqrt{z})^{-1/2}\right]=(2-2z)^{-1/2}(1+\sqrt{1-z})^{1/2}$
64	$\frac{1}{4}$	$\frac{3}{4}$	1	$\frac{2}{\pi}(1-z)^{-1/4}\,\mathbf{K}\left(\sqrt{\frac{1-(1-z)^{-1/2}}{2}}\right)$
65	$\frac{1}{4}$	$\frac{3}{4}$	$\frac{3}{2}$	$2^{1/2}\left[1+\sqrt{1-z}\right]^{-1/2}$
66	$\frac{1}{4}$	1	$\frac{5}{4}$	$\frac{1}{4x}\left[\ln\frac{1+x}{1-x}+2\operatorname{arctg}x\right]$ $[x=z^{1/4}]$
67	$\frac{1}{4}$	1	$\frac{9}{4}$	$\frac{5}{16x^5}\left\{4x-(1-x^4)\left[\ln\frac{1+x}{1-x}+2\operatorname{arctg}x\right]\right\}$ $[x=z^{1/4}]$
68	$\frac{1}{3}$	$\frac{2}{3}$	$\frac{3}{2}$	$\frac{3}{2}\sqrt{\frac{3}{z}}\,r$ [r is the middle real root of the equation $r^3-r+\frac{2}{3}\sqrt{\frac{z}{3}}=0$]
69	$\frac{1}{3}$	1	$\frac{4}{3}$	$\frac{1}{6x}\left[\ln(1+x+x^2)-2\ln(1-x)+2\sqrt{3}\operatorname{arctg}\frac{3^{1/2}x}{2+x}\right]$ $[x=z^{1/3}]$
70	$\frac{1}{3}$	1	$\frac{7}{3}$	$\frac{4}{9x^4}\left\{3x-(1-x^3)\left[\frac{1}{2}\ln(1+x+x^2)-\ln(1-x)+\sqrt{3}\operatorname{arctg}\frac{3^{1/2}x}{2+x}\right]\right\}$ $[x=z^{1/3}]$
71	$\frac{3}{8}$	1	$\frac{11}{8}$	$\frac{3}{8x^3}\left[\ln\frac{1+x}{1-x}-2\operatorname{arctg}x-2^{-1/2}\ln\frac{1+2^{1/2}x+x^2}{1-2^{1/2}x+x^2}+\sqrt{2}\operatorname{arctg}\frac{2^{1/2}x}{1-x^2}\right]$ $[x=z^{1/8}]$
72	$\frac{2}{5}$	1	$\frac{7}{5}$	$\frac{1}{10x^2}\Big[\ln(1-x^5)-5\ln(1-x)-\sqrt{5}\ln\frac{1+2^{-1}(\sqrt{5}+1)x+x^2}{1-2^{-1}(\sqrt{5}-1)x+x^2}+2(10-2\sqrt{5})^{1/2}\operatorname{arctg}\frac{(10+2\sqrt{5})^{1/2}x}{4-(\sqrt{5}-1)x}-2(10+2\sqrt{5})^{1/2}\operatorname{arctg}\frac{(10-2\sqrt{5})^{1/2}x}{4+(\sqrt{5}+1)x}\Big]$ $[x=z^{1/5}]$
73	$\frac{2}{5}$	1	$\frac{12}{5}$	$\frac{7}{5x^5}-\frac{7}{50x^7}(1-x^5)\Big[\ln(1-x^5)-5\ln(1-x)-\sqrt{5}\ln\frac{1+2^{-1}(\sqrt{5}+1)x+x^2}{1-2^{-1}(\sqrt{5}-1)x+x^2}+2(10-2\sqrt{5})^{1/2}\operatorname{arctg}\frac{(10+2\sqrt{5})^{1/2}x}{4-(\sqrt{5}-1)x}-$

	a	b	c	${}_2F_1(a, b; c; z)$
				$-2(10+2\sqrt{5})^{1/2}\operatorname{arctg}\dfrac{(10-2\sqrt{5})^{1/2}x}{4+(\sqrt{5}+1)x}\Big]$ $[x=z^{1/5}]$
74	$\frac{1}{2}$	$\frac{1}{2}$	$-\frac{1}{2}$	$(1-2z)(1-z)^{-3/2}$
75	$\frac{1}{2}$	$\frac{1}{2}$	1	$\frac{2}{\pi}\mathbf{K}(\sqrt{z})$
76	$\frac{1}{2}$	$\frac{1}{2}$	$\frac{3}{2}$	$\dfrac{\arcsin\sqrt{z}}{\sqrt{z}}$
77	$\frac{1}{2}$	$\frac{1}{2}$	2	$\frac{4}{\pi}[\mathbf{K}(\sqrt{z})-\mathbf{D}(\sqrt{z})]$
78	$\frac{1}{2}$	$\frac{1}{2}$	$\frac{5}{2}$	$\frac{3}{4z}\left[\sqrt{1-z}-(1-2z)\dfrac{\arcsin\sqrt{z}}{\sqrt{z}}\right]$
79	$\frac{1}{2}$	$\frac{1}{2}$	3	$\frac{16}{9\pi z}[2(1-2z)\mathbf{D}(\sqrt{z})-(1-3z)\mathbf{K}(\sqrt{z})]$
80	$\frac{1}{2}$	$\frac{1}{2}$	$\frac{7}{2}$	$\frac{15}{64z^2}\left[(3-8z+8z^2)\dfrac{\arcsin\sqrt{z}}{\sqrt{z}}-3(1-2z)\sqrt{1-z}\right]$
81	$\frac{1}{2}$	$\frac{1}{2}$	4	$\frac{32}{75\pi z^2}\left[(4-11z+15z^2)\mathbf{K}(\sqrt{z})-(8-23z+23z^2)\mathbf{D}(\sqrt{z})\right]$
82	$\frac{1}{2}$	1	$-\frac{1}{2}$	$(1-3z)(1-z)^{-2}$
83	$\frac{1}{2}$	1	$\frac{3}{2}$	$\dfrac{\operatorname{Arth}\sqrt{z}}{\sqrt{z}}=\dfrac{1}{2\sqrt{z}}\ln\dfrac{1+\sqrt{z}}{1-\sqrt{z}}$
84	$\frac{1}{2}$	1	2	$\dfrac{2}{1+\sqrt{1-z}}$
85	$\frac{1}{2}$	1	$\frac{5}{2}$	$\frac{3}{2z}\left[1-(1-z)\dfrac{\operatorname{Arth}\sqrt{z}}{\sqrt{z}}\right]$
86	$\frac{1}{2}$	1	3	$\frac{4}{3z^2}[2(1-z)^{3/2}-2+3z]$
87	$\frac{1}{2}$	1	$\frac{7}{2}$	$\frac{5}{8z^2}\left[3(1-z)^2\dfrac{\operatorname{Arth}\sqrt{z}}{\sqrt{z}}-3+5z\right]$
88	$\frac{1}{2}$	1	4	$\frac{2}{5z^3}[8-20z+15z^2-8(1-z)^{5/2}]$
89	$\frac{1}{2}$	$\frac{3}{2}$	$-\frac{1}{2}$	$(1-4z)(1-z)^{-5/2}$
90	$\frac{1}{2}$	$\frac{3}{2}$	1	$\dfrac{2}{\pi(1-z)}\mathbf{E}(\sqrt{z})$
91	$\frac{1}{2}$	$\frac{3}{2}$	2	$\frac{4}{\pi}\mathbf{D}(\sqrt{z})$
92	$\frac{1}{2}$	$\frac{3}{2}$	$\frac{5}{2}$	$\frac{3}{2z}\left[\dfrac{\arcsin\sqrt{z}}{\sqrt{z}}-\sqrt{1-z}\right]$
93	$\frac{1}{2}$	$\frac{3}{2}$	3	$\frac{16}{3\pi z}[\mathbf{K}(\sqrt{z})-(2-z)\mathbf{D}(\sqrt{z})]$
94	$\frac{1}{2}$	$\frac{3}{2}$	$\frac{7}{2}$	$\frac{15}{16z^2}\left[(3-2z)\sqrt{1-z}-(3-4z)\dfrac{\arcsin\sqrt{z}}{\sqrt{z}}\right]$
95	$\frac{1}{2}$	$\frac{3}{2}$	4	$\frac{32}{15\pi z^2}[(8-13z+3z^2)\mathbf{D}(\sqrt{z})-2(2-3z)\mathbf{K}(\sqrt{z})]$

	a	b	c	${}_2F_1(a, b; c; z)$
96	$\frac{1}{2}$	2	$-\frac{1}{2}$	$(1-5z)(1-z)^{-3}$
97	$\frac{1}{2}$	2	1	$\frac{1}{2}(2-z)(1-z)^{-3/2}$
98	$\frac{1}{2}$	2	$\frac{3}{2}$	$\frac{1}{2}\left[\frac{1}{1-z}+\frac{\operatorname{Arth}\sqrt{z}}{\sqrt{z}}\right]$
99	$\frac{1}{2}$	2	$\frac{5}{2}$	$\frac{3}{4z}\left[(1+z)\frac{\operatorname{Arth}\sqrt{z}}{\sqrt{z}}-1\right]$
100	$\frac{1}{2}$	2	3	$\frac{4}{3z^2}[2-(2+z)\sqrt{1-z}]$
101	$\frac{1}{2}$	2	$\frac{7}{2}$	$\frac{5}{16z^2}\left[9-3z-3(1-z)(3+z)\frac{\operatorname{Arth}\sqrt{z}}{\sqrt{z}}\right]$
102	$\frac{1}{2}$	2	4	$\frac{8}{5z^3}[(4+z)(1-z)^{3/2}-4+5z]$
103	$\frac{1}{2}$	$\frac{5}{2}$	$-\frac{1}{2}$	$(1-6z)(1-z)^{-7/2}$
104	$\frac{1}{2}$	$\frac{5}{2}$	1	$\frac{2}{3\pi}(1-z)^{-2}[(3-z)\mathbf{K}(\sqrt{z})-2z(2-z)\mathbf{D}(\sqrt{z})]$
105	$\frac{1}{2}$	$\frac{5}{2}$	$\frac{3}{2}$	$\frac{1}{3}(3-2z)(1-z)^{-3/2}$
106	$\frac{1}{2}$	$\frac{5}{2}$	2	$\frac{4}{3\pi(1-z)}[\mathbf{K}(\sqrt{z})+(1-2z)\mathbf{D}(\sqrt{z})]$
107	$\frac{1}{2}$	$\frac{5}{2}$	3	$\frac{16}{9\pi z}[2(1+z)\mathbf{D}(\sqrt{z})-\mathbf{K}(\sqrt{z})]$
108	$\frac{1}{2}$	$\frac{5}{2}$	$\frac{7}{2}$	$\frac{5}{8z^2}\left[\frac{3\arcsin\sqrt{z}}{\sqrt{z}}-(3+2z)\sqrt{1-z}\right]$
109	$\frac{1}{2}$	$\frac{5}{2}$	4	$\frac{32}{15\pi z^2}[(4-z)\mathbf{K}(\sqrt{z})-(8-3z-2z^2)\mathbf{D}(\sqrt{z})]$
110	$\frac{1}{2}$	3	$-\frac{1}{2}$	$(1-7z)(1-z)^{-4}$
111	$\frac{1}{2}$	3	1	$\frac{1}{8}(8-8z+3z^2)(1-z)^{-5/2}$
112	$\frac{1}{2}$	3	$\frac{3}{2}$	$\frac{1}{8}\left[\frac{5-3z}{(1-z)^2}+3\frac{\operatorname{Arth}\sqrt{z}}{\sqrt{z}}\right]$
113	$\frac{1}{2}$	3	2	$\frac{1}{4}(4-3z)(1-z)^{-3/2}$
114	$\frac{1}{2}$	3	$\frac{5}{2}$	$\frac{3}{16z}\left[(1+3z)\frac{\operatorname{Arth}\sqrt{z}}{\sqrt{z}}-\frac{1-3z}{1-z}\right]$
115	$\frac{1}{2}$	3	$\frac{7}{2}$	$\frac{15}{64z^2}\left[(3+2z+3z^2)\frac{\operatorname{Arth}\sqrt{z}}{\sqrt{z}}-3(1+z)\right]$
116	$\frac{1}{2}$	3	4	$\frac{2}{5z^3}[8-(8+4z+3z^2)\sqrt{1-z}]$
117	$\frac{1}{2}$	$\frac{7}{2}$	$-\frac{1}{2}$	$(1-8z)(1-z)^{-9/2}$
118	$\frac{1}{2}$	$\frac{7}{2}$	1	$\frac{2}{15\pi}(1-z)^{-3}[(15-11z+4z^2)\mathbf{K}(\sqrt{z})-z(23-23z+8z^2)\mathbf{D}(\sqrt{z})]$

	a	b	c	${}_2F_1(a, b; c; z)$
119	$\frac{1}{2}$	$\frac{7}{2}$	$\frac{3}{2}$	$\frac{1}{15}(15-20z+8z^2)(1-z)^{-5/2}$
120	$\frac{1}{2}$	$\frac{7}{2}$	2	$\frac{4}{15\pi}(1-z)^{-2}\left[2(3-2z)\,\mathbf{K}(\sqrt{z})+(3-13z+8z^2)\,\mathbf{D}(\sqrt{z})\right]$
121	$\frac{1}{2}$	$\frac{7}{2}$	$\frac{5}{2}$	$\frac{1}{5}(5-4z)(1-z)^{-3/2}$
122	$\frac{1}{2}$	$\frac{7}{2}$	3	$\frac{16}{45\pi z}(1-z)^{-1}\left[(2+3z-8z^2)\,\mathbf{D}(\sqrt{z})-(1-4z)\,\mathbf{K}(\sqrt{z})\right]$
123	$\frac{1}{2}$	$\frac{7}{2}$	4	$\frac{32}{75\pi z^2}\left[(8+7z+8z^2)\,\mathbf{D}(\sqrt{z})-4(1+z)\,\mathbf{K}(\sqrt{z})\right]$
124	$\frac{1}{2}$	4	$-\frac{1}{2}$	$(1-9z)(1-z)^{-\frac{5}{2}}$
125	$\frac{1}{2}$	4	1	$\frac{1}{16}(16-24z+18z^2-5z^3)(1-z)^{-7/2}$
126	$\frac{1}{2}$	4	$\frac{3}{2}$	$\frac{1}{48}\left[\frac{33-40z+15z^2}{(1-z)^3}+15\frac{\operatorname{Arth}\sqrt{z}}{\sqrt{z}}\right]$
127	$\frac{1}{2}$	4	2	$\frac{1}{8}(8-12z+5z^2)(1-z)^{-5/2}$
128	$\frac{1}{2}$	4	$\frac{5}{2}$	$\frac{1}{32z}\left[3(1+5z)\frac{\operatorname{Arth}\sqrt{z}}{\sqrt{z}}-\frac{3-22z+15z^2}{(1-z)^2}\right]$
129	$\frac{1}{2}$	4	3	$\frac{1}{6}(6-5z)(1-z)^{-3/2}$
130	$\frac{1}{2}$	4	$\frac{7}{2}$	$\frac{5}{128z^2}\left[3(1+2z+5z^2)\frac{\operatorname{Arth}\sqrt{z}}{\sqrt{z}}-\frac{3+4z-15z^2}{1-z}\right]$
131	$\frac{3}{5}$	1	$\frac{8}{5}$	$\frac{3}{20x^3}\Big\{-4\ln(1-x)-(\sqrt{5}-1)\ln\left[1+(\sqrt{5}+1)\frac{x}{2}+x^2\right]+$ $+(\sqrt{5}+1)\ln\left[1-(\sqrt{5}-1)\frac{x}{2}+x^2\right]-$ $-2(10-2\sqrt{5})^{1/2}\operatorname{arctg}\frac{(10+2\sqrt{5})^{1/2}x}{4-(\sqrt{5}-1)x}+$ $+2(10+2\sqrt{5})^{1/2}\operatorname{arctg}\frac{(10-2\sqrt{5})^{1/2}x}{4+(\sqrt{5}+1)x}\Big\}$ $[x=z^{1/5}]$
132	$\frac{3}{5}$	1	$\frac{13}{5}$	$\frac{8}{5x^5}-\frac{6}{25x^8}(1-x^5)\Big\{-4\ln(1-x)-(\sqrt{5}-1)\ln\Big[1+$ $+(\sqrt{5}+1)\frac{x}{2}+x^2\Big]-(\sqrt{5}+1)\ln\left[1-(\sqrt{5}-1)\frac{x}{2}+x^2\right]+$ $+2(10+2\sqrt{5})^{1/2}\operatorname{arctg}\frac{(10-2\sqrt{5})^{1/2}x}{4+(\sqrt{5}+1)x}-$ $-2(10-2\sqrt{5})^{1/2}\operatorname{arctg}\frac{(10+2\sqrt{5})^{1/2}x}{4-(\sqrt{5}-1)x}\Big\}$ $[x=z^{1/5}]$
133	$\frac{5}{8}$	1	$\frac{13}{8}$	$\frac{5}{8x^5}\Big[\ln\frac{1+x}{1-x}+2\operatorname{arctg}x-2^{-1/2}\ln\frac{1+2^{1/2}x+x^2}{1-2^{1/2}x+x^2}-$ $-\sqrt{2}\operatorname{arctg}\frac{2^{1/2}x}{1-x^2}\Big]$ $[x=z^{1/8}]$
134	$\frac{3}{4}$	1	$\frac{7}{4}$	$\frac{3}{4x^3}\left[\ln\frac{1+x}{1-x}-2\operatorname{arctg}x\right]$ $[x=z^{1/4}]$

	a	b	c	${}_2F_1(a, b; c; z)$	
135	$\frac{3}{4}$	1	$\frac{11}{4}$	$\frac{7}{16x^7}\left\{4x^3-3(1-x^4)\left[\ln\frac{1+x}{1-x}-2\operatorname{arctg} x\right]\right\}$	$[x=z^{1/4}]$
136	$\frac{3}{4}$	$\frac{5}{4}$	$\frac{1}{2}$	$\frac{1}{2}\left[(1-\sqrt{z})^{-3/2}+(1+\sqrt{z})^{-3/2}\right]$	
137	$\frac{3}{4}$	$\frac{5}{4}$	$\frac{3}{2}$	$\frac{1}{\sqrt{z}}\left[(1-\sqrt{z})^{-1/2}-(1+\sqrt{z})^{-1/2}\right]$	
138	$\frac{2}{3}$	1	$\frac{5}{3}$	$\frac{1}{3x^2}\left[\ln(1-x^3)-3\ln(1-x)-2\sqrt{3}\operatorname{arctg}\frac{3^{1/2}x}{2+x}\right]$	$[x=z^{1/3}]$
139	$\frac{2}{3}$	1	$\frac{8}{3}$	$\frac{5}{3x^3}-\frac{5}{9x^5}(1-x^3)\left[\ln(1-x^3)-3\ln(1-x)-2\sqrt{3}\operatorname{arctg}\frac{3^{1/2}x}{2+x}\right]$	$[x=z^{1/3}]$
140	$\frac{4}{5}$	1	$\frac{9}{5}$	$\frac{1}{5x^4}\left[\ln(1-x^5)-5\ln(1-x)-\sqrt{5}\ln\frac{1-2^{-1}(\sqrt{5}-1)x+x^2}{1+2^{-1}(\sqrt{5}+1)x+x^2}-2(10+2\sqrt{5})^{1/2}\operatorname{arctg}\frac{(10+2\sqrt{5})^{1/2}x}{4-(\sqrt{5}-1)x}-2(10-2\sqrt{5})^{1/2}\operatorname{arctg}\frac{(10-2\sqrt{5})^{1/2}x}{4+(\sqrt{5}+1)x}\right]$	$[x=z^{1/5}]$
141	$\frac{4}{5}$	1	$\frac{14}{5}$	$\frac{9}{5x^5}-\frac{9}{25x^9}(1-x^5)\left\{\ln(1-x^5)-5\ln(1-x)-\sqrt{5}\ln\frac{1-2^{-1}(\sqrt{5}-1)x+x^2}{1+2^{-1}(\sqrt{5}-1)x+x^2}-2(10+2\sqrt{5})^{1/2}\operatorname{arctg}\frac{(10+2\sqrt{5})^{1/2}x}{4-(\sqrt{5}-1)x}-2(10-2\sqrt{5})^{1/2}\operatorname{arctg}\frac{(10-2\sqrt{5})^{1/2}x}{4+(\sqrt{5}+1)x}\right\}$	$[x=z^{1/5}]$
142	$\frac{5}{6}$	1	$\frac{11}{6}$	$\frac{5}{12x^5}\left[2\ln\frac{1+x}{1-x}+\ln\frac{1+x+x^2}{1-x+x^2}-2\sqrt{3}\operatorname{arctg}\frac{3^{1/2}x}{1-x^2}\right]$	$[x=z^{1/6}]$
143	$\frac{5}{6}$	1	$\frac{17}{6}$	$\frac{11}{6x^6}+\frac{55}{36x^{11}}(1-x^6)\left[\ln\frac{1-x}{1+x}+\ln\frac{1-x+x^2}{1+x+x^2}+\sqrt{3}\operatorname{arctg}\frac{3^{1/2}x}{1-x^2}\right]$	$[x=z^{1/6}]$
144	$\frac{7}{8}$	1	$\frac{15}{8}$	$\frac{7}{8x^7}\left[\ln\frac{1+x}{1-x}+2^{-1/2}\ln\frac{1+2^{1/2}x+x^2}{1-2^{1/2}x+x^2}-2\operatorname{arctg} x-\sqrt{2}\operatorname{arctg}\frac{2^{1/2}x}{1-x^2}\right]$	$[x=z^{1/8}]$
145	1	1	$-\frac{1}{2}$	$(1-z)^{-2}\left[1-4z-\frac{3z^{3/2}\arcsin\sqrt{z}}{\sqrt{1-z}}\right]$	
146	1	1	$\frac{1}{2}$	$(1-z)^{-1}\left[1+\frac{\sqrt{z}\arcsin\sqrt{z}}{\sqrt{1-z}}\right]$	
147	1	1	$\frac{3}{2}$	$\frac{\arcsin\sqrt{z}}{\sqrt{z(1-z)}}$	
148	1	1	2	$-\frac{1}{z}\ln(1-z)$	
149	1	1	$\frac{5}{2}$	$\frac{3}{z}\left[1-\sqrt{\frac{1-z}{z}}\arcsin\sqrt{z}\right]$	

	a	b	c	${}_2F_1(a, b; c; z)$
150	1	1	3	$\frac{2}{z^2}[z+(1-z)\ln(1-z)]$
151	1	1	$\frac{7}{2}$	$\frac{5}{3z^2}\left[3(1-z)^{3/2}\frac{\arcsin\sqrt{z}}{\sqrt{z}}-3+4z\right]$
152	1	1	4	$\frac{3}{2z^3}[z(3z-2)-2(1-z)^2\ln(1-z)]$
153	1	1	5	$\frac{2}{3z^4}[6(1-z)^3\ln(1-z)+6z-15z^2+11z^3]$
154	1	$\frac{3}{2}$	$-\frac{1}{2}$	$(1-6z-3z^2)(1-z)^{-3}$
155	1	$\frac{3}{2}$	$\frac{1}{2}$	$(1+z)(1-z)^{-2}$
156	1	$\frac{3}{2}$	2	$\frac{2}{z}\left(\frac{1}{\sqrt{1-z}}-1\right)$
157	1	$\frac{3}{2}$	$\frac{5}{2}$	$\frac{3}{z}\left[\frac{\operatorname{Arth}\sqrt{z}}{\sqrt{z}}-1\right]$
158	1	$\frac{3}{2}$	3	$4(1+\sqrt{1-z})^{-2}$
159	1	$\frac{3}{2}$	$\frac{7}{2}$	$\frac{5}{2z^2}\left[3-2z-3(1-z)\frac{\operatorname{Arth}\sqrt{z}}{\sqrt{z}}\right]$
160	1	$\frac{3}{2}$	4	$\frac{2}{z^3}[8(1-z)^{3/2}-8+12z-3z^2]$
161	1	2	$-\frac{1}{2}$	$\frac{(1-z)^{-3}}{2}\left[2-14z-3z^2-\frac{15z^{3/2}\arcsin\sqrt{z}}{\sqrt{1-z}}\right]$
162	1	2	$\frac{1}{2}$	$\frac{(1-z)^{-2}}{2}\left[2+z+3\sqrt{\frac{z}{1-z}}\arcsin\sqrt{z}\right]$
163	1	2	$\frac{3}{2}$	$\frac{(1-z)^{-1}}{2}\left[1+\frac{\arcsin\sqrt{z}}{\sqrt{z(1-z)}}\right]$
164	1	2	$\frac{5}{2}$	$\frac{3}{2z}\left[\frac{\arcsin\sqrt{z}}{\sqrt{z(1-z)}}-1\right]$
165	1	2	3	$-\frac{2}{z^2}[z+\ln(1-z)]$
166	1	2	$\frac{7}{2}$	$\frac{5}{2z^2}\left[3-z-3\sqrt{\frac{1-z}{z}}\arcsin\sqrt{z}\right]$
167	1	2	4	$\frac{3}{z^3}[z(2-z)+2(1-z)\ln(1-z)]$
168	1	$\frac{5}{2}$	$-\frac{1}{2}$	$(1-9z-9z^2+z^3)(1-z)^{-4}$
169	1	$\frac{5}{2}$	$\frac{1}{2}$	$\frac{1}{3}(3+6z-z^2)(1-z)^{-3}$
170	1	$\frac{5}{2}$	$\frac{3}{2}$	$\frac{1}{3}(3-z)(1-z)^{-2}$

	a	b	c	${}_2F_1(a, b; c; z)$
171	1	$\frac{5}{2}$	2	$\frac{2}{3z}[(1-z)^{-3/2}-1]$
172	1	$\frac{5}{2}$	3	$\frac{4}{3z^2}\left[\frac{2}{\sqrt{1-z}}-2-z\right]$
173	1	$\frac{5}{2}$	$\frac{7}{2}$	$\frac{5}{z^2}\left[\frac{\operatorname{Arth}\sqrt{z}}{\sqrt{z}}-1-\frac{z}{3}\right]$
174	1	$\frac{5}{2}$	4	$\frac{2}{z^3}[8-4z-z^2-8\sqrt{1-z}]$
175	1	3	$-\frac{1}{2}$	$\frac{(1-z)^{-4}}{8}\left[8-80z-39z^2+6z^3-\frac{105z^{3/2}\arcsin\sqrt{z}}{\sqrt{1-z}}\right]$
176	1	3	$\frac{1}{2}$	$\frac{(1-z)^{-3}}{8}\left[8+9z-2z^2+15\sqrt{\frac{z}{1-z}}\arcsin\sqrt{z}\right]$
177	1	3	$\frac{3}{2}$	$\frac{(1-z)^{-2}}{8}\left[5-2z+\frac{3\arcsin\sqrt{z}}{\sqrt{z(1-z)}}\right]$
178	1	3	2	$\frac{1}{2}(2-z)(1-z)^{-2}$
179	1	3	$\frac{5}{2}$	$\frac{3}{8z}(1-z)^{-1}\left[\frac{\arcsin\sqrt{z}}{\sqrt{z(1-z)}}-1+2z\right]$
180	1	3	$\frac{7}{2}$	$\frac{5}{8z^2}\left[\frac{3\arcsin\sqrt{z}}{\sqrt{z(1-z)}}-3-2z\right]$
181	1	3	4	$-\frac{3}{2z^3}[z(z+2)+2\ln(1-z)]$
182	1	$\frac{7}{2}$	$-\frac{1}{2}$	$\frac{1}{5}(5-60z-90z^2+20z^3-3z^4)(1-z)^{-5}$
183	1	$\frac{7}{2}$	$\frac{1}{2}$	$\frac{1}{5}(5+15z-5z^2+z^3)(1-z)^{-4}$
184	1	$\frac{7}{2}$	$\frac{3}{2}$	$\frac{1}{15}(15-10z+3z^2)(1-z)^{-3}$
185	1	$\frac{7}{2}$	2	$\frac{2}{5z}\left[(1-z)^{-5/2}-1\right]$
186	1	$\frac{7}{2}$	$\frac{5}{2}$	$\frac{1}{5}(5-3z)(1-z)^{-2}$
187	1	$\frac{7}{2}$	3	$\frac{4}{15z^2}[2(1-z)^{-3/2}-2-3z]$
188	1	$\frac{7}{2}$	4	$\frac{2}{5z^3}\left[\frac{8}{\sqrt{1-z}}-8-4z-3z^2\right]$
189	1	$\frac{7}{2}$	$\frac{9}{2}$	$-\frac{7}{15z^3}\left[15+5z+3z^2-\frac{\operatorname{Arth}\sqrt{z}}{\sqrt{z}}\right]$
190	1	4	$-\frac{1}{2}$	$\frac{(1-z)^{-5}}{16}\left[16-208z-165z^2+50z^3-8z^4-315\frac{z^{3/2}}{\sqrt{1-z}}\arcsin\sqrt{z}\right]$
191	1	4	$\frac{1}{2}$	$\frac{(1-z)^{-4}}{48}\left[48+87z-38z^2+8z^3+105\sqrt{\frac{z}{1-z}}\arcsin\sqrt{z}\right]$
192	1	4	$\frac{3}{2}$	$\frac{(1-z)^{-3}}{48}\left[33-26z+8z^2+\frac{15\arcsin\sqrt{z}}{\sqrt{z(1-z)}}\right]$

	a	b	c	${}_2F_1(a, b; c; z)$
193	1	4	2	$\frac{1}{3z}\left[\frac{1}{(1-z)^3}-1\right]$
194	1	4	$\frac{5}{2}$	$\frac{(1-z)^{-2}}{16z}\left[\frac{3\arcsin\sqrt{z}}{\sqrt{z(1-z)}}-3+14z-8z^2\right]$
195	1	4	3	$\frac{1}{3}(3-2z)(1-z)^{-2}$
196	1	4	$\frac{7}{2}$	$\frac{5}{48z^2}(1-z)^{-1}\left[\frac{3\arcsin\sqrt{z}}{\sqrt{z(1-z)}}-3-2z+8z^2\right]$
197	1	4	$\frac{9}{2}$	$\frac{7}{48z^4}\left[15\sqrt{\frac{z}{1-z}}\arcsin\sqrt{z}-15z-10z^2-8z^3\right]$
198	$\frac{3}{2}$	$\frac{3}{2}$	$-\frac{1}{2}$	$(1-8z-8z^2)(1-z)^{-7/2}$
199	$\frac{3}{2}$	$\frac{3}{2}$	$\frac{1}{2}$	$(1+2z)(1-z)^{-5/2}$
200	$\frac{3}{2}$	$\frac{3}{2}$	1	$\frac{2}{\pi}(1-z)^{-2}\left[(1+z)\mathbf{K}(\sqrt{z})-2z\mathbf{D}(\sqrt{z})\right]=$
201				$=\frac{2}{\pi}\left[(1-\sqrt{z})(1-z)\right]^{-1}\mathbf{E}\left(\frac{2\sqrt[4]{z}}{1+z}\right)$
202	$\frac{3}{2}$	$\frac{3}{2}$	2	$\frac{4}{\pi}(1-z)^{-1}\left[\mathbf{K}(\sqrt{z})-\mathbf{D}(\sqrt{z})\right]$
203	$\frac{3}{2}$	$\frac{3}{2}$	$\frac{5}{2}$	$\frac{3}{z}\left[(1-z)^{-1/2}-\frac{\arcsin\sqrt{z}}{\sqrt{z}}\right]$
204	$\frac{3}{2}$	$\frac{3}{2}$	3	$\frac{16}{\pi z}\left[2\mathbf{D}(\sqrt{z})-\mathbf{K}(\sqrt{z})\right]$
205	$\frac{3}{2}$	$\frac{3}{2}$	$\frac{7}{2}$	$\frac{15}{4z^2}\left[(3-2z)\frac{\arcsin\sqrt{z}}{\sqrt{z}}-3\sqrt{1-z}\right]$
206	$\frac{3}{2}$	$\frac{3}{2}$	4	$\frac{32}{3\pi z^2}\left[(4-3z)\mathbf{K}(\sqrt{z})-(8-7z)\mathbf{D}(\sqrt{z})\right]$
207	$\frac{3}{2}$	2	$-\frac{1}{2}$	$(1-10z-15z^2)(1-z)^{-4}$
208	$\frac{3}{2}$	2	$\frac{1}{2}$	$(1+3z)(1-z)^{-3}$
209	$\frac{3}{2}$	2	1	$\frac{1}{2}(2+z)(1-z)^{-5/2}$
210	$\frac{3}{2}$	2	$\frac{5}{2}$	$\frac{3}{2z}\left(\frac{1}{1-z}-\frac{\operatorname{Arth}\sqrt{z}}{\sqrt{z}}\right)$
211	$\frac{3}{2}$	2	3	$\frac{4}{z^2}\left[\frac{2-z}{\sqrt{1-z}}-2\right]=\frac{4}{\sqrt{1-z}}(1+\sqrt{1-z})^{-2}$
212	$\frac{3}{2}$	2	$\frac{7}{2}$	$\frac{15}{4z^2}\left[(3-z)\frac{\operatorname{Arth}\sqrt{z}}{\sqrt{z}}-3\right]$
213	$\frac{3}{2}$	2	4	$\frac{8}{z^3}\left[4-3z-(4-z)\sqrt{1-z}\right]=8(1+\sqrt{1-z})^{-3}$
214	$\frac{3}{2}$	$\frac{5}{2}$	$-\frac{1}{2}$	$(1-12z-24z^2)(1-z)^{-9/2}$
215	$\frac{3}{2}$	$\frac{5}{2}$	$\frac{1}{2}$	$(1+4z)(1-z)^{-7/2}$

	a	b	c	${}_2F_1(a, b; c; z)$
216	$\frac{3}{2}$	$\frac{5}{2}$	1	$\frac{2}{3\pi}(1-z)^{-3}[(3+5z)\,\mathbf{K}(\sqrt{z})-z(7+z)\,\mathbf{D}(\sqrt{z})]$
217	$\frac{3}{2}$	$\frac{5}{2}$	2	$\frac{4}{3\pi}(1-z)^{-2}[2\mathbf{K}(\sqrt{z})-(1+z)\,\mathbf{D}(\sqrt{z})]$
218	$\frac{3}{2}$	$\frac{5}{2}$	3	$\frac{16}{3\pi z}(1-z)^{-1}[\mathbf{K}(\sqrt{z})-(2-z)\,\mathbf{D}(\sqrt{z})]$
219	$\frac{3}{2}$	$\frac{5}{2}$	$\frac{7}{2}$	$\frac{5}{2z^2}\left(\frac{3-z}{\sqrt{1-z}}-3\frac{\arcsin\sqrt{z}}{\sqrt{z}}\right)$
220	$\frac{3}{2}$	$\frac{5}{2}$	4	$\frac{32}{3\pi z^2}[(8-z)\,\mathbf{D}(\sqrt{z})-4\mathbf{K}(\sqrt{z})]$
221	$\frac{3}{2}$	3	$-\frac{1}{2}$	$(1-14z-35z^2)(1-z)^{-5}$
222	$\frac{3}{2}$	3	$\frac{1}{2}$	$(1+5z)(1-z)^{-4}$
223	$\frac{3}{2}$	3	1	$\frac{1}{8}(8+8z-z^2)(1-z)^{-7/2}$
224	$\frac{3}{2}$	3	2	$\frac{1}{4}(4-z)(1-z)^{-5/2}$
225	$\frac{3}{2}$	3	$\frac{5}{2}$	$\frac{3}{8z}\left[\frac{1+z}{(1-z)^2}-\frac{\operatorname{Arth}\sqrt{z}}{\sqrt{z}}\right]$
226	$\frac{3}{2}$	3	$\frac{7}{2}$	$\frac{15}{16z^2}\left[\frac{3-z}{1-z}-(3+z)\frac{\operatorname{Arth}\sqrt{z}}{\sqrt{z}}\right]$
227	$\frac{3}{2}$	3	4	$\frac{2}{z^3}[(8-4z-z^2)(1-z)^{-1/2}-8]$
228	$\frac{3}{2}$	$\frac{7}{2}$	$-\frac{1}{2}$	$(1-16z-48z^2)(1-z)^{-11/2}$
229	$\frac{3}{2}$	$\frac{7}{2}$	$\frac{1}{2}$	$(1+6z)(1-z)^{-9/2}$
230	$\frac{3}{2}$	$\frac{7}{2}$	1	$\frac{2}{15\pi}(1-z)^{-4}[(15+34z-z^2)\,\mathbf{K}(\sqrt{z})-2z(19+6z-z^2)\,\mathbf{D}(\sqrt{z})]$
231	$\frac{3}{2}$	$\frac{7}{2}$	2	$\frac{4}{15\pi}(1-z)^{-3}[(9-z)\,\mathbf{K}(\sqrt{z})-(3+7z-2z^2)\,\mathbf{D}(\sqrt{z})]$
232	$\frac{3}{2}$	$\frac{7}{2}$	$\frac{5}{2}$	$\frac{1}{5}(5-2z)(1-z)^{-5/2}$
233	$\frac{3}{2}$	$\frac{7}{2}$	3	$\frac{16}{15\pi z}(1-z)^{-2}[(1+z)\,\mathbf{K}(\sqrt{z})-2(1-z+z^2)\,\mathbf{D}(\sqrt{z})]$
234	$\frac{3}{2}$	$\frac{7}{2}$	4	$\frac{32}{15\pi z^2}(1-z)^{-1}[(4-z)\,\mathbf{K}(\sqrt{z})-(8-3z-2z^2)\,\mathbf{D}(\sqrt{z})]$
235	$\frac{3}{2}$	4	$-\frac{1}{2}$	$(1-18z-63z^2)(1-z)^{-6}$
236	$\frac{3}{2}$	4	$\frac{1}{2}$	$(1+7z)(1-z)^{-5}$
237	$\frac{3}{2}$	4	1	$\frac{1}{16}(16+24z-6z^2+z^3)(1-z)^{-9/2}$
238	$\frac{3}{2}$	4	2	$\frac{1}{8}(8-4z+z^2)(1-z)^{-7/2}$
239	$\frac{3}{2}$	4	$\frac{5}{2}$	$\frac{1}{16z}\left[(3+8z-3z^2)(1-z)^{-3}-3\frac{\operatorname{Arth}\sqrt{z}}{\sqrt{z}}\right]$

	a	b	c	${}_2F_1(a, b; c; z)$
240	$\frac{3}{2}$	4	3	$\frac{1}{2}(2-z)(1-z)^{-5/2}$
241	$\frac{3}{2}$	4	$\frac{7}{2}$	$\frac{5}{32z^2}\left[\frac{3-2z+3z^2}{(1-z)^2}-3(1+z)\frac{\operatorname{Arth}\sqrt{z}}{\sqrt{z}}\right]$
242	2	2	$-\frac{1}{2}$	$\frac{(1-z)^{-4}}{4}\left[4-48z-61z^2-15(5+2z)\frac{z^{3/2}\arcsin\sqrt{z}}{\sqrt{1-z}}\right]$
243	2	2	$\frac{1}{2}$	$\frac{(1-z)^{-3}}{4}\left[4+11z+3(3+2z)\sqrt{\frac{z}{1-z}}\arcsin\sqrt{z}\right]$
244	2	2	1	$(1+z)(1-z)^{-3}$
245	2	2	$\frac{3}{2}$	$\frac{(1-z)^{-2}}{4}\left[3+\frac{1+2z}{\sqrt{z(1-z)}}\arcsin\sqrt{z}\right]$
246	2	2	$\frac{5}{2}$	$\frac{3}{4z}(1-z)^{-1}\left[1-\frac{1-2z}{\sqrt{z(1-z)}}\arcsin\sqrt{z}\right]$
247	2	2	3	$\frac{2}{z^2}\left[\frac{z}{1-z}+\ln(1-z)\right]$
248	2	2	$\frac{7}{2}$	$\frac{15}{4z^2}\left[\frac{3-2z}{\sqrt{z(1-z)}}\arcsin\sqrt{z}-3\right]$
249	2	2	4	$-\frac{6}{z^3}[2z+(2-z)\ln(1-z)]$
250	2	$\frac{5}{2}$	$-\frac{1}{2}$	$(1-15z-45z^2-5z^3)(1-z)^{-5}$
251	2	$\frac{5}{2}$	$\frac{1}{2}$	$(1+6z+z^2)(1-z)^{-4}$
252	2	$\frac{5}{2}$	1	$\frac{1}{2}(2+3z)(1-z)^{-7/2}$
253	2	$\frac{5}{2}$	$\frac{3}{2}$	$\frac{1}{3}(3+z)(1-z)^{-3}$
254	2	$\frac{5}{2}$	3	$\frac{4}{3z^2}[2-(2-3z)(1-z)^{-3/2}]=$ $=\frac{4}{3}(1-z)^{-3/2}(1+\sqrt{1-z})^{-2}(1+2\sqrt{1-z})$
255	2	$\frac{5}{2}$	$\frac{7}{2}$	$\frac{5}{2z^2}\left(\frac{3-2z}{1-z}-3\frac{\operatorname{Arth}\sqrt{z}}{\sqrt{z}}\right)$
256	2	$\frac{5}{2}$	4	$\frac{8}{z^3}\left(\frac{4-3z}{\sqrt{1-z}}-4+z\right)=\frac{8}{\sqrt{1-z}}(1+\sqrt{1-z})^{-3}$
257	2	$\frac{5}{2}$	5	$\frac{16}{z^4}(1-\sqrt{1-z})^4$
258	2	$\frac{7}{2}$	$-\frac{1}{2}$	$(1-20z-90z^2-20z^3+z^4)(1-z)^{-6}$
259	2	$\frac{7}{2}$	$\frac{1}{2}$	$\frac{1}{5}(5+45z+15z^2-z^3)(1-z)^{-5}$
260	2	$\frac{7}{2}$	1	$\frac{1}{2}(2+5z)(1-z)^{-9/2}$
261	2	$\frac{7}{2}$	$\frac{3}{2}$	$\frac{1}{15}(15+10z-z^2)(1-z)^{-4}$
262	2	$\frac{7}{2}$	$\frac{5}{2}$	$\frac{1}{5}(5-z)(1-z)^{-3}$

	a	b	c	${}_2F_1(a, b; c; z)$
263	2	$\frac{7}{2}$	3	$\frac{4}{15z^2}[2-(2-5z)(1-z)^{-5/2}]$
264	2	$\frac{7}{2}$	4	$\frac{8}{5z^3}\left[4+z-\frac{4-5z}{(1-z)^{3/2}}\right]$
265	2	3	$-\frac{1}{2}$	$\frac{(1-z)^{-5}}{16}\left[16-272z-659z^2-30z^3-105(5+4z)\times\frac{z^{3/2}\arcsin\sqrt{z}}{\sqrt{1-z}}\right]$
266	2	3	$\frac{1}{2}$	$\frac{(1-z)^{-4}}{16}\left[16+83z+6z^2+15(3+4z)\sqrt{\frac{z}{1-z}}\arcsin\sqrt{z}\right]$
267	2	3	1	$(1+2z)(1-z)^{-4}$
268	2	3	$\frac{3}{2}$	$\frac{(1-z)^{-3}}{16}\left[13+2z+\frac{3(1+4z)}{\sqrt{z(1-z)}}\arcsin\sqrt{z}\right]$
269	2	3	$\frac{5}{2}$	$\frac{3}{16z}(1-z)^{-2}\left[1+2z-\frac{1-4z}{\sqrt{z(1-z)}}\arcsin\sqrt{z}\right]$
270	2	3	$\frac{7}{2}$	$\frac{15}{16z^2}(1-z)^{-1}\left[3-2z-\frac{3-4z}{\sqrt{z(1-z)}}\arcsin\sqrt{z}\right]$
271	2	3	4	$\frac{3}{z^3}\left[z+\frac{z}{1-z}+2\ln(1-z)\right]$
272	2	4	$-\frac{1}{2}$	$\frac{(1-z)^{-6}}{32}\left[32-704z-2553z^2-260z^3+20z^4-315(5+6z)\frac{z^{3/2}\arcsin\sqrt{z}}{\sqrt{1-z}}\right]$
273	2	4	$\frac{1}{2}$	$\frac{(1-z)^{-5}}{32}\left[32+247z+40z^2-4z^3+105(1+2z)\sqrt{\frac{z}{1-z}}\arcsin\sqrt{z}\right]$
274	2	4	1	$(1+3z)(1-z)^{-5}$
275	2	4	$\frac{3}{2}$	$\frac{(1-z)^{-4}}{96}\left[81+28z-4z^2+15(1+6z)\frac{\arcsin\sqrt{z}}{\sqrt{z(1-z)}}\right]$
276	2	4	$\frac{5}{2}$	$\frac{(1-z)^{-3}}{32z}\left[3+16z-4z^2-3(1-6z)\frac{\arcsin\sqrt{z}}{\sqrt{z(1-z)}}\right]$
277	2	4	3	$\frac{1}{3}(3-z)(1-z)^{-3}$
278	2	4	$\frac{7}{2}$	$\frac{5}{32z^2}(1-z)^{-2}\left[3-4z+4z^2-3(1-2z)\frac{\arcsin\sqrt{z}}{\sqrt{z(1-z)}}\right]$
279	$\frac{5}{2}$	$\frac{5}{2}$	$-\frac{1}{2}$	$(1-18z-72z^2-16z^3)(1-z)^{-11/2}$
280	$\frac{5}{2}$	$\frac{5}{2}$	$\frac{1}{2}$	$\frac{1}{3}(3+24z+8z^2)(1-z)^{-9/2}$
281	$\frac{5}{2}$	$\frac{5}{2}$	1	$\frac{2}{3\pi}(1-z)^{-4}[(3+10z+3z^2)\mathbf{K}(\sqrt{z})-8z(1+z)\mathbf{D}(\sqrt{z})]$
282	$\frac{5}{2}$	$\frac{5}{2}$	$\frac{3}{2}$	$\frac{1}{3}(3+2z)(1-z)^{-7/2}$
283	$\frac{5}{2}$	$\frac{5}{2}$	2	$\frac{4}{9\pi}(1-z)^{-3}[(5+3z)\mathbf{K}(\sqrt{z})-(1+7z)\mathbf{D}(\sqrt{z})]$
284	$\frac{5}{2}$	$\frac{5}{2}$	3	$\frac{16}{9\pi z}(1-z)^{-2}[2(1-2z)\mathbf{D}(\sqrt{z})-(1-3z)\mathbf{K}(\sqrt{z})]$

	a	b	c	${}_2F_1(a, b; c; z)$
285	$\frac{5}{2}$	$\frac{5}{2}$	$\frac{7}{2}$	$\frac{5}{3z^2}\left[\frac{3\arcsin\sqrt{z}}{\sqrt{z}}-(3-4z)(1-z)^{-3/2}\right]$
286	$\frac{5}{2}$	$\frac{5}{2}$	4	$\frac{32}{3\pi z^2}(1-z)^{-1}\left[(4-3z)\,\mathrm{K}(\sqrt{z})-(8-7z)\,\mathrm{D}(\sqrt{z})\right]$
287	$\frac{5}{2}$	3	$-\frac{1}{2}$	$(1-21z-105z^2-35z^3)(1-z)^{-6}$
288	$\frac{5}{2}$	3	$\frac{1}{2}$	$(1+10z+5z^2)(1-z)^{-5}$
289	$\frac{5}{2}$	3	1	$\frac{1}{8}(8+24z+3z^2)(1-z)^{-9/2}$
290	$\frac{5}{2}$	3	$\frac{3}{2}$	$(1+z)(1-z)^{-4}$
291	$\frac{5}{2}$	3	2	$\frac{1}{4}(4+z)(1-z)^{-7/2}$
292	$\frac{5}{2}$	3	$\frac{7}{2}$	$\frac{5}{8z^2}\left[3\frac{\operatorname{Arth}\sqrt{z}}{\sqrt{z}}-\frac{3-5z}{(1-z)^2}\right]$
293	$\frac{5}{2}$	3	4	$\frac{2}{z^3}\left[8-(8-12z+3z^2)(1-z)^{-3/2}\right]$
294	$\frac{5}{2}$	$\frac{7}{2}$	$-\frac{1}{2}$	$(1-24z-144z^2-64z^3)(1-z)^{-13/2}$
295	$\frac{5}{2}$	$\frac{7}{2}$	$\frac{1}{2}$	$(1+12z+8z^2)(1-z)^{-11/2}$
296	$\frac{5}{2}$	$\frac{7}{2}$	1	$\frac{2}{15\pi}(1-z)^{-5}\left[(15+74z+39z^2)\,\mathrm{K}(\sqrt{z})-\right.$ $\left.-z(43+82z+3z^2)\,\mathrm{D}(\sqrt{z})\right]$
297	$\frac{5}{2}$	$\frac{7}{2}$	$\frac{3}{2}$	$\frac{1}{3}(3+4z)(1-z)^{-9/2}$
298	$\frac{5}{2}$	$\frac{7}{2}$	2	$\frac{4}{15\pi}(1-z)^{-4}\left[8(1+z)\mathrm{K}(\sqrt{z})-(1+14z+z^2)\,\mathrm{D}(\sqrt{z})\right]$
299	$\frac{5}{2}$	$\frac{7}{2}$	3	$\frac{16}{45\pi z}(1-z)^{-3}\left[(2-7z-3z^2)\,\mathrm{D}(\sqrt{z})-(1-9z)\,\mathrm{K}(\sqrt{z})\right]$
300	$\frac{5}{2}$	$\frac{7}{2}$	4	$\frac{32}{15\pi z^2}(1-z)^{-2}\left[(8-13z+3z^2)\,\mathrm{D}(\sqrt{z})-2(2-3z)\,\mathrm{K}(\sqrt{z})\right]$
301	$\frac{5}{2}$	4	$-\frac{1}{2}$	$(1-27z-189z^2-105z^3)(1-z)^{-7}$
302	$\frac{5}{2}$	4	$\frac{1}{2}$	$\frac{1}{3}(3+42z+35z^2)(1-z)^{-6}$
303	$\frac{5}{2}$	4	1	$\frac{1}{16}(16+72z+18z^2-z^4)(1-z)^{-11/2}$
304	$\frac{5}{2}$	4	$\frac{3}{2}$	$\frac{1}{3}(3+5z)(1-z)^{-5}$
305	$\frac{5}{2}$	4	2	$\frac{1}{24}(24+12z-z^2)(1-z)^{-9/2}$
306	$\frac{5}{2}$	4	3	$\frac{1}{6}(6-z)(1-z)^{-7/2}$
307	$\frac{5}{2}$	4	$\frac{7}{2}$	$\frac{5}{48z^2}\left[3\frac{\operatorname{Arth}\sqrt{z}}{\sqrt{z}}-\frac{3-8z-3z^2}{(1-z)^3}\right]$

	a	b	c	${}_2F_1(a, b; c; z)$
308	3	3	$-\frac{1}{2}$	$\frac{(1-z)^{-6}}{64}\Big[64-1536z-6885z^2-2038z^3-$ $-105(35+56z+8z^2)\frac{z^{3/2}\arcsin\sqrt{z}}{\sqrt{1-z}}\Big]$
309	3	3	$\frac{1}{2}$	$\frac{(1-z)^{-5}}{64}\Big[64+607z+274z^2+15(15+40z+8z^2)\times$ $\times\sqrt{\frac{z}{1-z}}\arcsin\sqrt{z}\Big]$
310	3	3	1	$(1+4z+z^2)(1-z)^{-5}$
311	3	3	$\frac{3}{2}$	$\frac{(1-z)^{-4}}{64}\Big[5(11+10z)+3(3+24z+8z^2)\frac{\arcsin\sqrt{z}}{\sqrt{z(1-z)}}\Big]$
312	3	3	2	$\frac{1}{2}(2+z)(1-z)^{-4}$
313	3	3	$\frac{5}{2}$	$\frac{3}{64z}(1-z)^{-3}\Big[1+14z-(1-8z-8z^2)\frac{\arcsin\sqrt{z}}{\sqrt{z(1-z)}}\Big]$
314	3	3	$\frac{7}{2}$	$\frac{15}{64z^2}(1-z)^{-2}\Big[(3-8z+8z^2)\frac{\arcsin\sqrt{z}}{\sqrt{z(1-z)}}-3(1-2z)\Big]$
315	3	3	4	$-\frac{3}{2z^3}[z(2-3z)(1-z)^{-2}+2\ln(1-z)]$
316	3	$\frac{7}{2}$	$-\frac{1}{2}$	$(1-28z-210z^2-140z^3-7z^4)(1-z)^{-7}$
317	3	$\frac{7}{2}$	$\frac{1}{2}$	$(1+15z+15z^2+z^3)(1-z)^{-6}$
318	3	$\frac{7}{2}$	1	$\frac{1}{8}(8+40z+15z^2)(1-z)^{-11/2}$
319	3	$\frac{7}{2}$	$\frac{3}{2}$	$\frac{1}{5}(5+10z+z^2)(1-z)^{-5}$
320	3	$\frac{7}{2}$	2	$\frac{1}{4}(4+3z)(1-z)^{-9/2}$
321	3	$\frac{7}{2}$	$\frac{5}{2}$	$\frac{1}{5}(5+z)(1-z)^{-4}$
322	3	$\frac{7}{2}$	4	$\frac{2}{5z^3}[(8-20z+15z^2)(1-z)^{-5/2}-8]$
323	3	4	$-\frac{1}{2}$	$\frac{(1-z)^{-7}}{128}\Big[128-3968z-26223z^2-14702z^3-$ $-280z^4-315(35+84z+24z^2)z^{3/2}\frac{\arcsin\sqrt{z}}{\sqrt{1-z}}\Big]$
324	3	4	$\frac{1}{2}$	$\frac{(1-z)^{-6}}{128}\Big[128+1779z+1518z^2+40z^3+$ $+105(5+20z+8z^2)\sqrt{\frac{z}{1-z}}\arcsin\sqrt{z}\Big]$
325	3	4	1	$(1+6z+3z^2)(1-z)^{-6}$
326	3	4	$\frac{3}{2}$	$\frac{(1-z)^{-5}}{128}\Big[113+194z+8z^2+15(1+12z+8z^2)\frac{\arcsin\sqrt{z}}{\sqrt{z(1-z)}}\Big]$
327	3	4	2	$(1+z)(1-z)^{-5}$

	a	b	c	${}_2F_1(a, b; c; z)$
328	3	4	$\frac{5}{2}$	$\frac{(1-z)^{-4}}{128z}\left[3+94z+8z^2-3(1-12z-24z^2)\frac{\arcsin\sqrt{z}}{\sqrt{z(1-z)}}\right]$
329	3	4	$\frac{7}{2}$	$\frac{5}{128z^2}(1-z)^{-3}\left[3(1-4z+8z^2)\frac{\arcsin\sqrt{z}}{\sqrt{z(1-z)}}-3+10z+8z^2\right]$
330	$\frac{7}{2}$	$\frac{7}{2}$	$-\frac{1}{2}$	$\frac{1}{5}(5-160z-1440z^2-1280z^3-128z^4)(1-z)^{-15/2}$
331	$\frac{7}{2}$	$\frac{7}{2}$	$\frac{1}{2}$	$\frac{1}{5}(5+90z+120z^2+16z^3)(1-z)^{-13/2}$
332	$\frac{7}{2}$	$\frac{7}{2}$	1	$\frac{2}{15\pi}(1-z)^{-6}\left[(15+113z+113z^2+15z^3)\mathbf{K}(\sqrt{z})-2z(23+82z+23z^2)\mathbf{D}(\sqrt{z})\right]$
333	$\frac{7}{2}$	$\frac{7}{2}$	$\frac{3}{2}$	$\frac{1}{15}(15+40z+8z^2)(1-z)^{-11/2}$
334	$\frac{7}{2}$	$\frac{7}{2}$	2	$\frac{4}{75\pi}(1-z)^{-5}\left[(39+74z+15z^2)\mathbf{K}(\sqrt{z})-(3+82z+43z^2)\mathbf{D}(\sqrt{z})\right]$
335	$\frac{7}{2}$	$\frac{7}{2}$	$\frac{5}{2}$	$\frac{1}{5}(5+2z)(1-z)^{-9/2}$
336	$\frac{7}{2}$	$\frac{7}{2}$	3	$\frac{16}{225\pi z}(1-z)^{-4}\left[2(1-6z-19z^2)\mathbf{D}(\sqrt{z})-(1-34z-15z^2)\mathbf{K}(\sqrt{z})\right]$
337	$\frac{7}{2}$	$\frac{7}{2}$	4	$\frac{32}{75\pi z^2}(1-z)^{-3}\left[(4-11z+15z^2)\mathbf{K}(\sqrt{z})-(8-23z+23z^2)\mathbf{D}(\sqrt{z})\right]$
338	$\frac{7}{2}$	4	$-\frac{1}{2}$	$(1-36z-378z^2-420z^3-63z^4)(1-z)^{-8}$
339	$\frac{7}{2}$	4	$\frac{1}{2}$	$(1+21z+35z^2+7z^3)(1-z)^{-7}$
340	$\frac{7}{2}$	4	1	$\frac{1}{16}(16+120z+90z^2+5z^3)(1-z)^{-13/2}$
341	$\frac{7}{2}$	4	$\frac{3}{2}$	$\frac{1}{3}(3+10z+3z^2)(1-z)^{-6}$
342	$\frac{7}{2}$	4	2	$\frac{1}{8}(8+12z+z^2)(1-z)^{-11/2}$
343	$\frac{7}{2}$	4	$\frac{5}{2}$	$\frac{1}{5}(5+3z)(1-z)^{-5}$
344	$\frac{7}{2}$	4	3	$\frac{1}{6}(6+z)(1-z)^{-9/2}$
345	4	4	$-\frac{1}{2}$	$\frac{(1-z)^{-8}}{256}\left[256-10240z-99021z^2-102592z^3-13628z^4-315(105+378z+216z^2+16z^3)\frac{z^{3/2}\arcsin\sqrt{z}}{\sqrt{1-z}}\right]$
346	4	4	$\frac{1}{2}$	$\frac{(1-z)^{-7}}{256}\left[256+5175z+8132z^2+1452z^3+35(35+210z+168z^2+16z^3)\sqrt{\frac{z}{1-z}}\arcsin\sqrt{z}\right]$
347	4	4	1	$(1+9z+9z^2+z^3)(1-z)^{-7}$

	a	b	c	${}_2F_1(a, b; c; z)$
348	4	4	$\frac{3}{2}$	$\frac{(1-z)^{-6}}{256}\left[7(33+104z+28z^2)+ +5(5+90z+120z^2+16z^3)\frac{\arcsin\sqrt{z}}{\sqrt{z(1-z)}}\right]$
349	4	4	2	$\frac{1}{3}(3+6z+z^2)(1-z)^{-6}$
350	4	4	$\frac{5}{2}$	$\frac{3}{256z}(1-z)^{-5}\left[1+68z+36z^2- -(1-18z-72z^2-16z^3)\frac{\arcsin\sqrt{z}}{\sqrt{z(1-z)}}\right]$
351	4	4	3	$\frac{1}{3}(3+z)(1-z)^{-5}$
352	4	4	$\frac{7}{2}$	$\frac{5}{768z^2}(1-z)^{-4}\left[3(1-6z+24z^2+16z^3)\times \times\frac{\arcsin\sqrt{z}}{\sqrt{z(1-z)}}-3+16z+92z^2\right]$

7.3.3. Representations of ${}_2F_1(a,b;c;-z)$.
(See also 7.3.1. with z replaced by $-z$).

	a	b	c	${}_2F_1(a, b; c; -z)$
1	a	$a+\frac{1}{2}$	$\frac{1}{2}$	$(1+z)^{-a}\cos(2a\,\mathrm{arctg}\sqrt{z})$
2	a	$a+\frac{1}{2}$	$\frac{3}{2}$	$\frac{(1+z)^{1/2-a}}{(2a-1)\sqrt{z}}\sin[(2a-1)\,\mathrm{arctg}\sqrt{z}]$
3	a	$-a$	$\frac{1}{2}$	$\frac{1}{2}[(\sqrt{1+z}+\sqrt{z})^{2a}+(\sqrt{1+z}-\sqrt{z})^{2a}]$
4	a	$1-a$	$\frac{1}{2}$	$\frac{1}{2\sqrt{1+z}}[(\sqrt{1+z}+\sqrt{z})^{2a-1}+(\sqrt{1+z}-\sqrt{z})^{2a-1}]$
5	a	$1-a$	$\frac{3}{2}$	$\frac{1}{2(2a-1)\sqrt{z}}[(\sqrt{1+z}+\sqrt{z})^{2a-1}-(\sqrt{1+z}-\sqrt{z})^{2a-1}]$
6	a	$2-a$	$\frac{3}{2}$	$\frac{1}{4(a-1)\sqrt{z(1+z)}}[(\sqrt{1+z}+\sqrt{z})^{2a-2}- -(\sqrt{1+z}-\sqrt{z})^{2a-2}]$
7	$\frac{m}{n}$	1	$\frac{m}{n}+1$	$-\frac{m}{n}z^{-m/n}\sum_{k=0}^{n-1}\exp\left[-(2k+1)\frac{\pi i m}{n}\right]\times \times\ln\left\{1-z^{1/n}\exp\left[(2k+1)\frac{\pi i}{n}\right]\right\}=$
8				$=-\frac{m}{n}z^{-m/n}\left\{\frac{(-1)^m}{2}[1-(-1)^n]\ln(1+z^{1/n})+ +\sum_{k=0}^{[n/2]-1}\left[\cos\left[(2k+1)\frac{\pi m}{n}\right]\ln\left\{1-2z^{1/n}\cos\left[(2k+1)\frac{\pi}{n}\right]+\right.\right.$

	a	b	c	${}_2F_1(a, b; c; -z)$
				$+z^{2/n}\Big\} - 2\sin\left[(2k+1)\frac{\pi m}{n}\right]\times$ $\times\operatorname{arctg}\frac{z^{1/n}\sin[(2k+1)\pi/n]}{1-z^{1/n}\cos[(2k+1)\pi/n]}\Big]\Big\}+$ $+m\sum_{k=1}^{[(m-1)/n]}\frac{(-1)^{k-1}z^{-k}}{m-kn}$ $[m, n=1, 2, 3, \ldots; m\leqslant n]$
9	$\frac{m}{2}$	1	$\frac{m}{2}+1$	$m\sum_{k=1}^{[(m-1)/2]}\frac{(-1)^{k-1}z^{-k}}{m-2k}+mz^{-m/2}\left[\sin\frac{m\pi}{2}\operatorname{arctg}\sqrt{z}-\right.$ $\left.-\frac{1}{2}\cos\frac{m\pi}{2}\ln(1-z)\right]$
10	$\frac{m}{3}$	1	$\frac{m}{3}+1$	$m\sum_{k=1}^{[(m-1)/3]}\frac{(-1)^{k-1}z^{-k}}{m-3k}+\frac{m}{3}z^{-m/3}\times$ $\times\left[2\sin\frac{m\pi}{3}\operatorname{arctg}\frac{z^{1/3}\sqrt{3}}{2-z^{1/3}}-\cos\frac{m\pi}{3}\ln\left(1-z^{1/3}+z^{2/3}\right)+\right.$ $\left.+(-1)^{m-1}\ln\left(1+z^{1/3}\right)\right]$
11	$\frac{m}{4}$	1	$\frac{m}{4}+1$	$m\sum_{k=1}^{[(m-1)/4]}\frac{(-1)^{k-1}z^{-k}}{m-4k}+\frac{m}{2}z^{-m/4}\times$ $\times\left[\sin\frac{m\pi}{4}\operatorname{arctg}\frac{z^{1/4}}{\sqrt{2}-z^{1/4}}+\right.$ $+\sin\frac{3m\pi}{4}\operatorname{arctg}\frac{z^{1/4}}{\sqrt{2}+z^{1/4}}-\frac{1}{2}\cos\frac{m\pi}{4}\ln(\sqrt{z}-\sqrt[4]{4z}+1)+$ $\left.+\frac{1}{2}\cos\frac{3m\pi}{4}\ln\left(\sqrt{z}+\sqrt[4]{4z}+1\right)\right]$

7.3.4. Particular values of ${}_2F_1(a,b;c;-z)$.
(See also 7.3.2. with z replaced by $-z$).

	a	b	c	${}_2F_1(a, b; c; -z)$
1	$\frac{1}{8}$	1	$\frac{9}{8}$	$\frac{1}{16x}\left\{a\left[2\operatorname{arctg}\frac{ax}{1-x^2}-\ln\frac{1-ax+x^2}{1+ax+x^2}\right]+\right.$ $\left.+b\left[2\operatorname{arctg}\frac{bx}{1-x^2}-\ln\frac{1-bx+x^2}{1+bx+x^2}\right]\right\}$ $[x=z^{1/8},\ a=(2-\sqrt{2})^{1/2},\ b=(2+\sqrt{2})^{1/2}]$
2	$\frac{1}{6}$	1	$\frac{7}{6}$	$\frac{1}{12x}\left[4\operatorname{arctg}x+2\operatorname{arctg}\frac{x}{1-x^2}-3^{-1/2}\ln\frac{1-3^{1/2}x+x^2}{1+3^{1/2}x+x^2}\right]$ $[x=z^{1/6}]$

	a b	c	$_2F_1(a, b; c; -z)$
3	$\frac{1}{6}$ 1	$\frac{13}{6}$	$\frac{7}{72x^7}\left\{-12x+(1+x^6)\times\left[4\,\mathrm{arctg}\,x+2\,\mathrm{arctg}\,\frac{x}{1-x^2}-3^{-1/2}\ln\frac{1-3^{1/2}x+x^2}{1+3^{1/2}x+x^2}\right]\right\}$ $[x=z^{1/6}]$
4	$\frac{1}{4}$ 1	$\frac{5}{4}$	$\frac{\sqrt{2}}{8x}\left[\ln\frac{1+2^{1/2}x+x^2}{1-2^{1/2}x+x^2}+2\,\mathrm{arctg}\,\frac{2^{1/2}x}{1-x^2}\right]$ $[x=z^{1/4}]$
5	$\frac{1}{4}$ 1	$\frac{9}{4}$	$\frac{5}{32x^5}\left\{-8x+\sqrt{2}\,(1+x^4)\left[\ln\frac{1+2^{1/2}x+x^2}{1-2^{1/2}x+x^2}+2\,\mathrm{arctg}\,\frac{2^{1/2}x}{1-x^2}\right]\right\}$ $[x=z^{1/4}]$
6	$\frac{3}{8}$ 1	$\frac{11}{8}$	$\frac{3}{16x^3}\left\{a\left[\ln\frac{1+bx+x^2}{1-bx+x^2}-2\,\mathrm{arctg}\,\frac{bx}{1-x^2}\right]+b\left[2\,\mathrm{arctg}\,\frac{ax}{1-x^2}-\ln\frac{1+ax+x^2}{1-ax+x^2}\right]\right\}$ $[x=z^{1/8},\ a=(2-\sqrt{2})^{1/2},\ b=(2+\sqrt{2})^{1/2}]$
7	$\frac{1}{2}$ $\frac{1}{2}$	$\frac{3}{2}$	$\frac{1}{\sqrt{z}}\ln(\sqrt{1+z}+\sqrt{z})=\frac{\mathrm{Arsh}\sqrt{z}}{\sqrt{z}}$
8	$\frac{1}{2}$ 1	$\frac{3}{2}$	$\frac{1}{\sqrt{z}}\,\mathrm{arctg}\,\sqrt{z}$
9	$\frac{3}{4}$ $\frac{5}{4}$	$\frac{1}{2}$	$2^{-1/2}(1+z)^{-3/2}\left[(\sqrt{1+z}+1)^{1/2}-\sqrt{z}\,(\sqrt{1+z}-1)^{1/2}\right]$
10	$\frac{3}{4}$ $\frac{5}{4}$	$\frac{3}{2}$	$\frac{\sqrt{2}}{\sqrt{z}}(1+z)^{-1/2}(\sqrt{1+z}-1)^{1/2}$
11	$\frac{3}{4}$ 1	$\frac{7}{4}$	$\frac{3\sqrt{2}}{8x^3}\left[2\,\mathrm{arctg}\,\frac{2^{1/2}x}{1-x^2}-\ln\frac{1+2^{1/2}x+x^2}{1-2^{1/2}x+x^2}\right]$ $[x=z^{1/4}]$
12	$\frac{3}{4}$ 1	$\frac{11}{4}$	$\frac{21\sqrt{2}}{32x^7}(1-x^4)\left[\ln\frac{1+2^{1/2}x+x^2}{1-2^{1/2}x+x^2}-2\,\mathrm{arctg}\,\frac{2^{1/2}x}{1-x^2}\right]$ $[x=z^{1/4}]$
13	$\frac{5}{6}$ 1	$\frac{11}{6}$	$\frac{5}{12x^5}\left[3^{-1/2}\ln\frac{1-3^{1/2}x+x^2}{1+3^{1/2}x+x^2}+2\,\mathrm{arctg}\,\frac{x}{1-x^2}+4\,\mathrm{arctg}\,x\right]$ $[x=z^{1/6}]$
14	$\frac{5}{6}$ 1	$\frac{17}{6}$	$-\frac{11}{6x^6}+\frac{55}{36x^{11}}(x^6+1)\times\left[\mathrm{arctg}\,\frac{x}{1-x^2}+2\,\mathrm{arctg}\,x+\frac{\sqrt{3}}{2}\ln\frac{1-3^{1/2}x+x^2}{1+3^{1/2}x+x^2}\right]$ $[x=z^{1/6}]$
15	$\frac{5}{8}$ 1	$\frac{13}{8}$	$\frac{5}{16x^5}\left\{b\left[2\,\mathrm{arctg}\,\frac{ax}{1-x^2}+\ln\frac{1+ax+x^2}{1-ax+x^2}\right]-a\left[2\,\mathrm{arctg}\,\frac{bx}{1-x^2}+\ln\frac{1+bx+x^2}{1-bx+x^2}\right]\right\}$ $[x=z^{1/8},\ a=(2-\sqrt{2})^{1/2},\ b=(2+\sqrt{2})^{1/2}]$
16	$\frac{7}{8}$ 1	$\frac{15}{8}$	$\frac{7}{16x^7}\left\{a\left[2\,\mathrm{arctg}\,\frac{ax}{1-x^2}+\ln\frac{1-ax+x^2}{1+ax+x^2}\right]+b\left[2\,\mathrm{arctg}\,\frac{bx}{1-x^2}+\ln\frac{1-bx+x^2}{1+bx+x^2}\right]\right\}$ $[x=z^{1/8},\ a=(2-\sqrt{2})^{1/2},\ b=(2+\sqrt{2})^{1/2}]$
17	1 1	$\frac{3}{2}$	$\frac{\ln(\sqrt{z}+\sqrt{1+z})}{\sqrt{z(1+z)}}=\frac{\mathrm{Arsh}\sqrt{z}}{\sqrt{z(1+z)}}$

7.3.5. Values of ${}_2F_1(a, b; c; 1)$.

1. ${}_2F_1(a, c; c; 1) = {}_1F_0(a; 1) = 0$ [Re $a<0$].

2. ${}_2F_1(a, b; c; 1) = \Gamma\begin{bmatrix} c, & c-a-b \\ c-a, & c-b \end{bmatrix}$ [Re $(c-a-b)>0$].

3. ${}_2F_1(l, m; n; 1) = \frac{(l-n)_l}{(l+m-n)_l}$ [$n > l + m$].

4. ${}_2F_1(-n, b; c; 1) = \frac{(c-b)_n}{(c)_n}$ [Re $(c-b) > -n$].

7.3.6. Values of ${}_2F_1(a, b; c; -1)$.

1. ${}_2F_1(a, b; a-b; -1) =$
$$= 2^{-a}\sqrt{\pi}\,\Gamma(a-b)\left\{\frac{1}{\Gamma(a/2-b)\,\Gamma[(a+1)/2]} + \frac{a}{2\Gamma[(a+1)/2-b]\,\Gamma(a/2+1)}\right\}.$$

2. ${}_2F_1(a, b; 1+a-b; -1) = 2^{-a}\sqrt{\pi}\,\Gamma\begin{bmatrix} 1+a-b \\ (1+a)/2, \; 1+a/2-b \end{bmatrix}.$

3. ${}_2F_1(a, b; 2+a-b; -1) =$
$$= \frac{2^{-a}\sqrt{\pi}}{b-1}\left\{\Gamma\begin{bmatrix} 2+a-b \\ a/2, \; (3+a)/2-b \end{bmatrix} - \Gamma\begin{bmatrix} 2+a-b \\ (1+a)/2, \; 1+a/2-b \end{bmatrix}\right\}.$$

4. ${}_2F_1(a, b; 3+a-b; -1) = \frac{\sqrt{\pi}}{2^a(1-b)(2-b)}\left\{\Gamma\begin{bmatrix} 3+a-b \\ (1+a)/2, \; 1+a/2-b \end{bmatrix} -\right.$
$$\left. - a\Gamma\begin{bmatrix} 3+a-b \\ 1+a/2, \; (3+a)/2-b \end{bmatrix} + \frac{a(a+1)}{4}\Gamma\begin{bmatrix} 3+a-b \\ (a+3)/2, \; 2+a/2-b \end{bmatrix}\right\}.$$

5. $b\,{}_2F_1(a, a+b; a+1; -1) + a\,{}_2F_1(b, a+b; b+1; -1) = \Gamma\begin{bmatrix} a+1, & b+1 \\ a+b \end{bmatrix}$
[$a+1, \; b+1 \neq 0, -1, -2, \ldots$].

6. $(1-a)\,{}_2F_1(1, a; 2-b; -1) + (1-b)\,{}_2F_1(1, b; 2-a; -1) = 2^{1-a-b}\,\Gamma\begin{bmatrix} 2-a, & 2-b \\ 2-a-b \end{bmatrix}$
[$2-a, \; 2-b \neq 0, -1, -2, \ldots$].

7. ${}_2F_1(1, a; -a-n; -1) = 2^{-n-2a-2}\,\Gamma\begin{bmatrix} 1-a, & -n-a \\ -n-2a \end{bmatrix} + \frac{1}{2}\sum_{k=0}^{n+1}(-1)^k\frac{(a)_k}{(-a-n)_k}$
[Re $a < -n$; $n = -1, 0, 1, 2, \ldots$].

8. ${}_2F_1(1, a; -a+n; -1) = 2^{n-2a-2}\,\Gamma\begin{bmatrix} 1-a, & n-a \\ n-2a \end{bmatrix} - \frac{1}{2}\sum_{k=1}^{n-2}(-1)^k\frac{(1+a-n)_k}{(1-a)_k}$
[Re $a < 0$; $n = 2, 3, 4, \ldots$].

9. ${}_2F_1(1, a; a+1; -1) = a\beta(a)$.

10. ${}_2F_1\left(1, \frac{m}{n}+l; \frac{m}{n}+l+1; -1\right) = (-1)^l\,\frac{m+nl}{n}\left\{\frac{\pi}{2}\operatorname{cosec}\frac{m\pi}{n} -\right.$
$$\left. -2\sum_{k=0}^{[n/2]-1}\cos\left[(2k+1)\frac{m\pi}{n}\right]\ln\left[\sin(2k+1)\frac{\pi}{2n}\right] - \sum_{k=0}^{l-1}\frac{(-1)^k}{k+m/n}\right\}$$
[$n = 2, 3, 4, \ldots$; $m = 1, 2, \ldots, n-1$; $l = 0, 1, 2, \ldots$].

11. ${}_2F_1(1, m; m+1; -1) = (-1)^{m-1}m\left[\ln 2 + \sum_{k=1}^{m-1}\frac{(-1)^k}{k}\right].$

12. ${}_2F_1(1, a; a+2; -1) = 2a(a+1)\beta(a) - a - 1$.

13. ${}_2F_1(1, a; a+3; -1) = 2a(a+1)(a+2)\beta(a) - a^2 - 7a/2 - 3$.

14. ${}_2F_1(1,\ a;\ a+4;\ -1)=(a)_4\left[\frac{4}{3}\beta(a)-\frac{7}{6a}+\frac{2}{3(a+1)}-\frac{1}{6(a+2)}\right].$

15. ${}_2F_1(2,a;b;-1)-\frac{b-1}{2(a-1)}(a+b-3)\,{}_2F_1(1,a-1;b-1;-1)=\frac{(b-2)(b-1)}{2(1-a)}$ $[\operatorname{Re}(b-a)>1]$.

16. ${}_2F_1(2,\ a;\ a+1;\ -1)=a(a-1)\beta(a-1)-a/2.$

17. ${}_2F_1(-n,\ a;\ b;\ -1)=\frac{n!}{(b)_n}P_n^{(b-1,\ a-b-n)}(3)$ $[0<\operatorname{Re}b<\operatorname{Re}a-n+1]$.

18. $=\frac{n!}{(b)_n}(-2)^n P_n^{(-a-n,\ b-1)}(0)$ $[\operatorname{Re}b>0;\ \operatorname{Re}a<1-n]$.

19. ${}_2F_1(-n,-m;1-m;-1)=(-1)^{m-n}\frac{n!}{(m-1)!}(m-n-1)!\left[1-2^m\sum_{k=0}^{m-n-1}\frac{(-m)_k}{k!}2^{-k}\right]$ $[m>n]$.

20. ${}_2F_1(-n,m;1+m;-1)=\frac{(-1)^m m!n!}{(m+n)!}-2^n\sum_{k=1}^{m}\frac{(-m)_k 2^k}{(n+1)_k}$ $[m=1,2,3,\ldots]$.

21. ${}_2F_1(-2n,\ 2m-2n+1;\ -m;\ -1)=\frac{2^{2n-\varepsilon}(m-n)!\,\Gamma(n+1/2)}{m!\sqrt{\pi}}$ $[\varepsilon=0$ for $2n\leqslant m;\ \varepsilon=1$ for $2n=m+1,\ m+2,\ldots,\ 2m+1]$.

22. ${}_2F_1(-n,\ 2;\ 1;\ -1)=2^{n-1}(n+2).$

	a	b	c	${}_2F_1(a,b;c;-1)$
23	$\frac{1}{6}$	1	$\frac{13}{6}$	$\frac{7\sqrt{3}}{54}[\pi\sqrt{3}-6\sqrt{3}\ln(2+\sqrt{3})-3\sqrt{3}]$
24	$\frac{1}{4}$	1	$\frac{5}{4}$	$\frac{\sqrt{2}}{8}[\pi+2\ln(1+\sqrt{2})]$
25	$\frac{1}{4}$	1	$\frac{9}{4}$	$\frac{5\sqrt{2}}{16}[\pi+2\ln(1+\sqrt{2})-2\sqrt{2}]$
26	$\frac{1}{3}$	1	$\frac{4}{3}$	$\frac{\sqrt{3}}{9}(\pi+\sqrt{3}\ln 2)$
27	$\frac{1}{3}$	1	$\frac{7}{3}$	$\frac{4}{9}(3-2\ln 2-3\sqrt{3}\,\pi)$
28	$\frac{1}{3}$	1	$\frac{10}{3}$	$\frac{7\sqrt{3}}{162}(16\pi+16\sqrt{3}\ln 2-33\sqrt{3})$
29	$\frac{1}{2}$	1	$\frac{3}{2}$	$\frac{\pi}{4}$
30	$\frac{1}{2}$	1	$\frac{5}{2}$	$\frac{3}{4}(\pi-2)$
31	$\frac{1}{2}$	1	$\frac{7}{2}$	$\frac{5}{8}(3\pi-8)$
32	$\frac{2}{3}$	1	$\frac{5}{3}$	$\frac{2\sqrt{3}}{9}(\pi-\sqrt{3}\ln 2)$
33	$\frac{2}{3}$	1	$\frac{8}{3}$	$\frac{5\sqrt{3}}{27}(4\pi-4\ln 2-3\sqrt{3})$
34	$\frac{2}{3}$	1	$\frac{11}{3}$	$\frac{4\sqrt{3}}{81}(40\pi-40\sqrt{3}\ln 2-39\sqrt{3})$

	a	b	c	${}_2F_1(a, b; c; -1)$
35	$\frac{3}{4}$	1	$\frac{7}{4}$	$\frac{3\sqrt{2}}{8}[\pi-2\ln(1+\sqrt{2})]$
36	$\frac{3}{4}$	1	$\frac{11}{4}$	$\frac{7\sqrt{2}}{16}(3\pi-6\ln(1+\sqrt{2})-2\sqrt{2})$
37	$\frac{5}{6}$	1	$\frac{17}{6}$	$\frac{11}{18}[5\pi-5\sqrt{3}\ln(2+\sqrt{3})-3]$
38	1	1	2	$\ln 2$
39	1	1	3	$4\ln 2-2$
40	1	1	4	$\frac{3}{2}(8\ln 2-5)$
41	1	1	5	$\frac{32}{3}(3\ln 2-2)$
42	1	2	3	$2(1-\ln 2)$

7.3.7. Values of ${}_2F_1\left(a, b; c; \frac{1}{2}\right)$.

1. $${}_2F_1\left(a, b; c; \frac{1}{2}\right)=2^a\,{}_2F_1(a, c-b; c; -1) \quad [\operatorname{Re}(b-a)>-1].$$

2. $${}_2F_1\left(a, b; \frac{a+b+1-m}{2}; \frac{1}{2}\right)=$$
$$=2^{b-1}\Gamma\left[\begin{matrix}(a+b+1-m)/2\\ b\end{matrix}\right]\sum_{k=0}^{m}\binom{m}{k}\Gamma\left[\begin{matrix}(b+k)/2\\ (1+a+k-m)/2\end{matrix}\right].$$

3. $${}_2F_1\left(a, b; \frac{a+b-1}{2}; \frac{1}{2}\right)=\sqrt{\pi}\left\{\Gamma\left[\begin{matrix}(a+b+1)/2\\ (a+1)/2,\ (b+1)/2\end{matrix}\right]+2\Gamma\left[\begin{matrix}(a+b-1)/2\\ a,\ b\end{matrix}\right]\right\}.$$

4. $${}_2F_1\left(a, b; \frac{a+b}{2}; \frac{1}{2}\right)=\sqrt{\pi}\left\{\Gamma\left[\begin{matrix}(a+b)/2\\ b/2,\ (a+1)/2\end{matrix}\right]+\Gamma\left[\begin{matrix}(a+b)/2\\ a/2,\ (b+1)/2\end{matrix}\right]\right\}.$$

5. $${}_2F_1\left(a, b; \frac{a+b+1}{2}; \frac{1}{2}\right)=\sqrt{\pi}\Gamma\left[\begin{matrix}(a+b+1)/2\\ (a+1)/2,\ (b+1)/2\end{matrix}\right].$$

6. $${}_2F_1\left(a, b; \frac{a+b}{2}+1; \frac{1}{2}\right)=\frac{2\sqrt{\pi}}{a-b}\left\{\Gamma\left[\begin{matrix}(a+b)/2+1\\ a/2,\ (b+1)/2\end{matrix}\right]-\Gamma\left[\begin{matrix}(a+b)/2+1\\ b/2,\ (a+1)/2\end{matrix}\right]\right\}.$$

7. $${}_2F_1\left(a, -a; b; \frac{1}{2}\right)=\frac{\sqrt{\pi}}{2^b}\left\{\Gamma\left[\begin{matrix}b\\ (a+b)/2,\ (b-a+1)/2\end{matrix}\right]+\right.$$
$$\left.+\Gamma\left[\begin{matrix}b\\ (a+b+1)/2,\ (b-a)/2\end{matrix}\right]\right\}.$$

8. $${}_2F_1\left(a, 1-a; b; \frac{1}{2}\right)=2^{1-b}\sqrt{\pi}\,\Gamma\left[\begin{matrix}b\\ (a+b)/2,\ (1+b-a)/2\end{matrix}\right].$$

9. $${}_2F_1\left(a, 2-a; b; \frac{1}{2}\right)=$$
$$=\frac{\sqrt{\pi}}{2^{b-2}(a-1)}\left\{\Gamma\left[\begin{matrix}b\\ (a+b)/2-1,\ (1+b-a)/2\end{matrix}\right]-\Gamma\left[\begin{matrix}b\\ (a+b-1)/2,\ (b-a)/2\end{matrix}\right]\right\}.$$

10. $${}_2F_1\left(a, 3-a; b; \frac{1}{2}\right)=$$
$$=\frac{\sqrt{\pi}\,\Gamma(b)}{2^{b-3}(a-1)(a-2)}\left\{\frac{b-2}{\Gamma[(b-a+1)/2]\,\Gamma[(a+b)/2-1]}-\frac{2}{\Gamma[(b-a)/2]\,\Gamma[(a+b-3)/2]}\right\}.$$

11. $${}_2F_1\left(a,\ 4-a;\ b;\ \frac{1}{2}\right)=\frac{\sqrt{\pi}\,\Gamma(b)}{2^{b-4}(a-1)(a-2)(a-3)}\times$$
$$\times\left\{\frac{2b-a-3}{\Gamma[(a+b)/2-2]\,\Gamma[(b-a+1)/2]}-\frac{2b+a-7}{\Gamma[(a+b-3)/2]\,\Gamma[(b-a)/2]}\right\}.$$

12. $${}_2F_1\left(a,\ 5-a;\ b;\ \frac{1}{2}\right)=\frac{\sqrt{\pi}\,\Gamma(b)}{2^{b-5}(a-1)(a-2)(a-3)(a-4)}\times$$
$$\times\left\{\frac{2(b-3)^2-(a-2)(a-3)}{\Gamma[(a+b)/2-2]\,\Gamma[(b-a+1)/2]}-\frac{4(b-3)}{\Gamma[(a+b-5)/2]\,\Gamma[(b-a)/2]}\right\}.$$

13. $${}_2F_1\left(a,\ 6-a;\ b;\ \frac{1}{2}\right)=\frac{\sqrt{\pi}\,\Gamma(b)}{2^{b-6}(a-1)(a-2)(a-3)(a-4)(a-5)}\times$$
$$\times\left\{\frac{4b^2-2ab-a^2+13a-22b+20}{\Gamma[(a+b)/2-3]\,\Gamma[(b-a+1)/2]}-\frac{4b^2+2ab-a^2-a-34b+62}{\Gamma[(a+b-5)/2]\,\Gamma[(b-a)/2]}\right\}.$$

14. $${}_2F_1\left(1,\ a;\ \frac{a-m}{2};\ \frac{1}{2}\right)=\frac{(-2)^{m-1}((a-m)/2+1)_{m-1}}{(1-a)_{m-1}}\times$$
$$\times\left\{\sqrt{\pi}\,\Gamma\left[\begin{matrix}1+(a-m)/2\\(a-m+1)/2\end{matrix}\right]+\sum_{k=0}^{m-1}\frac{(a-m)_k}{(1+(a-m)/2)_k}2^{-k}\right\}\qquad [m=1,\ 2,\ 3,\ \ldots].$$

15. $${}_2F_1\left(1,\ a;\ \frac{a+m}{2};\ \frac{1}{2}\right)=\frac{(-1)^m\,2^{-m}(a)_m}{(1-(a+m)/2)_m}\times$$
$$\times\left\{\sqrt{\pi}\,\Gamma\left[\begin{matrix}1+(a+m)/2\\(a+m+1)/2\end{matrix}\right]-\sum_{k=1}^{m}\frac{(-(a+m)/2)_k}{(1-a-m)_k}2^k\right\}\qquad [m=0,\ 1,\ 2,\ldots].$$

16. $${}_2F_1\left(a,\ a;\ a+1;\ \frac{1}{2}\right)=2^a a\beta(a).$$

17. $${}_2F_1\left(1,\ 1;\ b;\ \frac{1}{2}\right)=2(b-1)\beta(b-1).$$

18. $${}_2F_1\left(1,\ 2;\ b;\ \frac{1}{2}\right)=2(b-1)[1-2(b-2)\beta(b-1)].$$

19. $${}_2F_1\left(1,\ 3;\ b;\ \frac{1}{2}\right)=(b-1)[7-2b+4(b-2)(b-3)\beta(b-1)].$$

20. $${}_2F_1\left(1,\ 4;\ b;\ \frac{1}{2}\right)=\frac{2}{3}(b-1)[2b^2-15b+29-4(b-2)(b-3)(b-4)\beta(b-1)].$$

21. $${}_2F_1\left(1,\ 5;\ b;\ \frac{1}{2}\right)=\frac{b-1}{6}[293-208b+50b^2-4b^3+$$
$$+8(b-2)(b-3)(b-4)(b-5)\beta(b-1)].$$

22. $${}_2F_1\left(2,\ 2;\ b;\ \frac{1}{2}\right)=4(b-1)[3-b+(b-2)(2b-5)\beta(b-1)].$$

23. $${}_2F_1\left(2,\ 3;\ b;\ \frac{1}{2}\right)=2(b-1)(b-2)[7-2b+4(b-3)^2\beta(b-2)].$$

24. $${}_2F_1\left(2,\ 4;\ b;\ \frac{1}{2}\right)=\frac{2}{3}(b-1)(b-2)[4b^2-32b+65-$$
$$-4(b-3)(b-4)(2b-7)\beta(b-2)].$$

25. $${}_2F_1\left(3,\ 3;\ b;\ \frac{1}{2}\right)=2(b-1)(b-2)(b-3)[7-2b+2(2b^2-14b+25)\beta(b-3)].$$

	a	b	c	${}_2F_1(a, b; c; 1/2)$
26	$\frac{1}{3}$	$\frac{2}{3}$	1	$\frac{3\sqrt[3]{4}}{8\pi^2}\Gamma^3\left(\frac{1}{3}\right)$
27	1	$\frac{5}{4}$	2	$8\sqrt[4]{2}$
28	1	3	2	3
29	1	4	2	14/3
30	1	4	3	8/3
31	1	5	2	15/2
32	1	5	3	11/3
33	1	5	4	5/2
34	2	2	1	12
35	2	3	1	32
36	2	4	1	80
37	2	4	3	20/3
38	3	3	1	104
39	3	3	2	20

7.3.8. Values of ${}_2F_1(-n, b; c; 2)$.

1. $${}_2F_1(-n, a; 2a-1; 2) = \frac{1}{\sqrt{\pi}}\Gamma\left(a-\frac{1}{2}\right)\left\{\frac{1+(-1)^n}{2}\Gamma\begin{bmatrix}(n+1)/2\\ a+(n-1)/2\end{bmatrix} - \frac{1-(-1)^n}{2}\Gamma\begin{bmatrix}n/2+1\\ a+n/2\end{bmatrix}\right\}.$$

2. $${}_2F_1(-n, a; 2a; 2) = \frac{n!2^{-n-1}}{(n/2)!}[1+(-1)^n]\Gamma\begin{bmatrix}a+1/2\\ a+(n+1)/2\end{bmatrix}.$$

3. $${}_2F_1(-n, a; 2a+1; 2) = \frac{1}{\sqrt{\pi}}\Gamma\left(a+\frac{1}{2}\right)\left\{\frac{1+(-1)^n}{2}\Gamma\begin{bmatrix}(n+1)/2\\ a+(n+1)/2\end{bmatrix} + \frac{1-(-1)^n}{2}\Gamma\begin{bmatrix}n/2+1\\ a+n/2+1\end{bmatrix}\right\}.$$

4. $${}_2F_1(-n, a; -2n-2; 2) = 2^{2n+2}\frac{n!}{(2n+2)!}\left[(a+n+1)\left(\frac{a+1}{2}\right)_{n+1} - 2\left(\frac{a}{2}\right)_{n+2}\right].$$

5. $${}_2F_1(-n, a; -2n-1; 2) = 2^{2n+1}\frac{n!}{(2n+1)!}\left[\left(\frac{a+1}{2}\right)_{n+1} - \left(\frac{a}{2}\right)_{n+1}\right].$$

6. $${}_2F_1(-n, a; -2n; 2) = 2^{2n}\frac{n!}{(2n)!}\left(\frac{a+1}{2}\right)_n.$$

7. $${}_2F_1(-n, a; 1; 2) = (-1)^n\frac{(a)_n}{n!}\,{}_2F_1(-n, 1-a; 1-a-n; -1).$$

8. $${}_2F_1(-n, 1; m; 2) = 2^{1-m}\left[(-1)^n\frac{n!}{(m)_n} - \sum_{k=0}^{m-2}(-1)^k\frac{(1-m)_{k+1}}{k!\,(n+k+1)}\right].$$

9. $${}_2F_1(-n, 1; -2n-1; 2) = \frac{n!\,(n+1)!}{(2n+1)!}2^{2n+1} - 1.$$

10. $${}_2F_1(-n, 1; -2n; 2) = \frac{(n!)^2}{(2n)!}2^{2n}.$$

11. $${}_2F_1(-n, 2; -2n-2; 2) = \frac{(2n+3)(n+3)}{n+1} - \frac{n!\,(n+2)!}{(2n+2)!}2^{2n+3}.$$

12. $${}_2F_1(-n, 2; -2n-1; 2) = 2n+3 - \frac{[(n+1)!]^2}{(2n+2)!}2^{2n+2}.$$

7.3.9. Values of ${}_2F_1(a, b; c; z_0)$ for $z_0 \neq \pm 1, 2^{\pm 1}$.

1. $${}_2F_1\left(a, \frac{a+1}{3}; 2\frac{a+1}{3}; e^{\pm\pi i/3}\right) = \Gamma\begin{bmatrix}2(a+1)/3, a/3\\ 2/3, a\end{bmatrix}3^{a/2-1}e^{\pm\pi ai/6} \quad [\mathrm{Re}\, a < 2].$$

	a	b	c	${}_2F_1(a, b; c; -8)$
2	$-n$	$\frac{8}{3}-2n$	$-\frac{1}{3}$	$2^{2n}(4-3n)_n \frac{(-5/6)_n+(n/3-3/4)(-1/2)_{n-1}}{(-1/3)_n(-2/3)_n}$ $[n=2, 3, 4, \ldots]$
3	$-n$	$\frac{4}{3}-2n$	$\frac{1}{3}$	$2^{2n-1}(2-3n)_n \frac{2(-1/6)_n+(-1/2)_n}{(-1/3)_n(1/3)_n}$ $[n=1, 2, 3, \ldots]$
4	$-n$	$-\frac{1}{3}-2n$	$\frac{2}{3}$	$(-27)^n$
5	$-n$	$\frac{2}{3}-2n$	$\frac{2}{3}$	$2^{2n-1}(1-3n)_n \frac{2(1/6)_n+(1/2)_n}{(1/3)_n(2/3)_n}$ $[n=1, 2, 3, \ldots]$
6	$-n$	$-\frac{2}{3}-2n$	$\frac{4}{3}$	$(-27)^n \frac{(5/6)_n}{(3/2)_n}$
7	$-n$	$\frac{1}{3}-\frac{n}{2}$	$\frac{1}{3}$	$(-1)^{[(n+1)/2]}3^{2n-[n/2]-1}\left[1+2\frac{(1/2)_{n/2}}{(1/6)_{n/2}}\right]^{1+2[n/2]-n}$
8	$-n$	$\frac{1}{6}-\frac{n}{2}$	$\frac{2}{3}$	$2\cdot 3^{(3n-1)/2}\cos\frac{3n+1}{6}\pi$
9	$-2n$	$-\frac{1}{6}-n$	$\frac{4}{3}$	$(-1)^n \frac{3^{3n}}{2n+1}$
10	$-2n-1$	$-\frac{5}{6}-n$	$\frac{5}{3}$	$(-1)^{n+1}\frac{3^{3n+2}}{2n+3}$
11	$-2n-1$	$-\frac{7}{6}-n$	$\frac{7}{3}$	$(-1)^{n+1}\frac{5\cdot 3^{3n+2}}{(2n+3)(2n+5)}$

12. ${}_2F_1\left(-\frac{n}{2}, \frac{1-n}{2}; -m; -2\right)=\frac{n!}{m!}(m-n)!\, C_n^{(m-n+1)}$ [$C_n^{(r)}$ are the Fibonacci numbers].

13. ${}_2F_1\left(2, a; \frac{5-a}{2}; -\frac{1}{2}\right)=1-\frac{a}{3}.$

14. ${}_2F_1\left(a, 1-\frac{a}{2}; a+\frac{1}{2}; -\frac{1}{3}\right)=\frac{\sqrt{\pi}}{3^{a/2}}\Gamma\left[\begin{matrix}a+1/2\\(a+1)/2, (a+1)/2\end{matrix}\right].$

15. ${}_2F_1\left(a, a+\frac{1}{2}; \frac{3}{2}-2a; -\frac{1}{3}\right)=\frac{2}{\sqrt{\pi}}\left(\frac{9}{8}\right)^{2a}\Gamma\left[\begin{matrix}4/3, 3/2-2a\\4/3-2a\end{matrix}\right].$

16. ${}_2F_1\left(\frac{1}{2}, a; \frac{2a+3}{4}; \frac{1-\sqrt{2}}{2}\right)=2^{-a/2}\sqrt{\pi}\,\Gamma\left[\begin{matrix}(2a+3)/4\\(a+2)/4, (a+3)/4\end{matrix}\right].$

17. ${}_2F_1\left(\frac{3}{2}, a; \frac{2a+5}{4}; \frac{1-\sqrt{2}}{2}\right)=$

$$=2^{2-a/2}\sqrt{\pi}\left\{\Gamma\left[\begin{matrix}(2a+5)/4\\(a+1)/4, (a+2)/4\end{matrix}\right]-\Gamma\left[\begin{matrix}(2a+5)/4\\a/4, (a+3)/4\end{matrix}\right]\right\}.$$

18. ${}_2F_1\left(a, a+\frac{1}{3}; \frac{4}{3}-a; -\frac{1}{8}\right)=\left(\frac{2}{3}\right)^{3a}\Gamma\left[\begin{matrix}2/3-a, 4/3-a\\2/3, 4/3-2a\end{matrix}\right].$

19. ${}_2F_1\left(a, \frac{1-a}{3}; \frac{4a+5}{6}; -\frac{1}{8}\right)=2^{-a}\sqrt{\pi}\,\Gamma\left[\begin{matrix}(4a+2)/3\\a+1/2, (a+2)/3\end{matrix}\right].$

20. ${}_2F_1\left(a, \frac{2-a}{3}; \frac{4a+7}{6}; -\frac{1}{8}\right)=\frac{2^{-a}3\sqrt{\pi}}{2a-1}\left\{\Gamma\left[\begin{matrix}(4a+4)/3\\a, (2a+5)/6\end{matrix}\right]-\Gamma\left[\begin{matrix}(4a+4)/3\\a+1/2, (a+1)/3\end{matrix}\right]\right\}$ $[a\neq 1/2]$.

21. ${}_2F_1\left(\frac{1}{2}, \frac{1}{2}; \frac{3}{2}; -\frac{1}{8}\right)=\sqrt{2}\ln 2.$

22. $${}_2F_1\left(a, \frac{2-a}{3}; \frac{2a+5}{6}; \frac{4-3\sqrt{2}}{8}\right)=\left(\frac{2}{3}\right)^{a/2}\sqrt{\pi}\,\Gamma\begin{bmatrix}(2a+5)/6\\(a+3)/6,\ (a+5)/6\end{bmatrix}.$$

23. $${}_2F_1\left(a, \frac{4-a}{3}; \frac{2a+7}{6}; \frac{4-3\sqrt{2}}{8}\right)=$$
$$=\frac{2^{-a/2}3\sqrt{\pi}}{a-1}\left\{\Gamma\begin{bmatrix}(2a+4)/3\\a/2,\ (a+5)/6\end{bmatrix}-\Gamma\begin{bmatrix}(2a+4)/3\\(a+1)/2,\ (a+2)/6\end{bmatrix}\right\}\qquad [a\neq 1].$$

24. $${}_2F_1\left(1, 1; \frac{3}{4}; \frac{4-3\sqrt{2}}{8}\right)=\sqrt{2}\ln 2.$$

25. $${}_2F_1\left(a, 2-3a; \frac{3}{2}-a; \frac{2-\sqrt{3}}{4}\right)=\frac{3^{3a/2}}{2^{2a-1}\sqrt{\pi}}\Gamma\begin{bmatrix}4/3,\ 3/2-a\\4/3-a\end{bmatrix}.$$

26. $${}_2F_1\left(a, a+\frac{1}{2}; a+\frac{5}{6}; \frac{1}{9}\right)=\left(\frac{3}{4}\right)^{a}\sqrt{\pi}\,\Gamma\begin{bmatrix}5/6+2a/3\\1/2+a/3,\ 5/6+a/3\end{bmatrix}$$
$$[5/6+2a/3\neq 0, -1, -2, \ldots].$$

27. $${}_2F_1\left(a, 1-2a; \frac{4}{3}-a; \frac{1}{9}\right)=3^{-a}\Gamma\begin{bmatrix}2/3-a,\ 4/3-a\\2/3,\ 4/3-2a\end{bmatrix}.$$

28. $${}_2F_1\left(-n, \frac{1}{4}-n; 2n+\frac{7\mp 2}{4}; \frac{1}{9}\right)=\frac{2^{6n}}{3^{5n}}\frac{((7\mp 2)/4)_{2n}}{(1\mp 1/3)_n\,((15\mp 2)/12)_n}.$$

29. $${}_2F_1\left(a, 4-a; \frac{5}{2}; \frac{2-\sqrt{2}}{4}\right)=$$
$$=\frac{3\pi}{2^{3/2}(a-2)}\left\{\left[\Gamma\left(\frac{a+1}{4}\right)\Gamma\left(\frac{7-a}{4}\right)\right]^{-1}-\left[\Gamma\left(\frac{a+3}{4}\right)\Gamma\left(\frac{5-a}{4}\right)\right]^{-1}\right\}\qquad [a\neq 2].$$

30. $${}_2F_1\left(2, 2; \frac{5}{2}; \frac{2-\sqrt{2}}{4}\right)=\frac{3}{2}(4-\pi).$$

31. $${}_2F_1\left(\frac{1}{2}, a; \frac{5}{2}-2a; \frac{1}{4}\right)=\frac{2^{2a}}{3}\sqrt{\pi}\,\Gamma\begin{bmatrix}5/2-2a\\3/2-a,\ 3/2-a\end{bmatrix}.$$

32. $${}_2F_1\left(-2n, -3n-1; -2n-\frac{1}{2}; \frac{1}{4}\right)=\frac{(3n+1)!}{n!\,(3/2)_{2n}}2^{-4n}.$$

33. $${}_2F_1\left(-n, \frac{1}{2}; 2n+\frac{3}{2}; \frac{1}{4}\right)=\left(\frac{9}{2}\right)^{2n}\frac{(1/2)_n}{(2n+3/2)_n}.$$

34. $${}_2F_1\left(a, 1-2a; a+2; \frac{1}{3}\right)=\left(\frac{2}{3}\right)^{2a}(a+1).$$

35. $${}_2F_1\left(\frac{1}{2}, a; 3a; \frac{3}{4}\right)=2\left(\frac{4}{27}\right)^{a}\sqrt{\pi}\,\Gamma\begin{bmatrix}3a\\2a,\ a+1/2\end{bmatrix}.$$

36. $${}_2F_1\left(a, \frac{2a+1}{4}; a+\frac{1}{2}; 2\sqrt{2}-2\right)=\left(\frac{2+\sqrt{2}}{2}\right)^{a}\sqrt{\pi}\,\Gamma\begin{bmatrix}(2a+3)/4\\(a+2)/4,\ (a+3)/4\end{bmatrix}.$$

37. $${}_2F_1\left(a, \frac{2a+3}{4}; a+\frac{3}{2}; 2\sqrt{2}-2\right)=$$
$$=\frac{(2+\sqrt{2})^{a}}{2^{a-2}}\sqrt{\pi}\left\{\Gamma\begin{bmatrix}(2a+5)/4\\(a+1)/4,\ (a+2)/4\end{bmatrix}-\Gamma\begin{bmatrix}(2a+5)/4\\a/4,\ (a+3)/4\end{bmatrix}\right\}.$$

38. $${}_2F_1\left(a, a+\frac{1}{2}; \frac{4a+2}{3}; \frac{8}{9}\right)=\left(\frac{3}{2}\right)^{2a}\sqrt{\pi}\,\Gamma\begin{bmatrix}(4a+2)/3\\(a+2)/3,\ a+1/2\end{bmatrix}.$$

39. $${}_2F_1\left(a, a+\frac{1}{2}; \frac{4a}{3}+1; \frac{8}{9}\right)=2^{1-2a}3^{2a}\sqrt{\pi}\left\{\Gamma\left[\begin{matrix}4a/3\\ a, (2a+3)/6\end{matrix}\right]-\Gamma\left[\begin{matrix}4a/3\\ a+1/2, a/3\end{matrix}\right]\right\}.$$

40. $${}_2F_1\left(a, a+\frac{1}{2}; 4a-1; \frac{8}{9}\right)=$$
$$=2^{2-6a}3^{2a}\sqrt{\pi}\left\{3\Gamma\left[\begin{matrix}4a-1\\ a, 3a-1/2\end{matrix}\right]-\Gamma\left[\begin{matrix}4a-1\\ a+1/2, 3a-1\end{matrix}\right]\right\}.$$

41. $${}_2F_1\left(a, a+\frac{1}{2}; 4a; \frac{8}{9}\right)=2^{1-6a}3^{2a}\sqrt{\pi}\,\Gamma\left[\begin{matrix}4a\\ a+1/2, 3a\end{matrix}\right].$$

42. $${}_2F_1\left(-n, \frac{n}{2}+1; \frac{4}{3}; \frac{8}{9}\right)=\frac{1+(-1)^n}{2}(-3)^{-n/2}\frac{(1/2)_{n/2}}{(7/6)_{n/2}}.$$

43. $${}_2F_1\left(a, \frac{4a+1}{6}; \frac{4a+1}{3}; 12\sqrt{2}-16\right)=\left(\frac{2+\sqrt{2}}{2}\right)^{2a}\sqrt{\pi}\,\Gamma\left[\begin{matrix}(2a+2)/3\\ (a+4)/6, (a+1)/2\end{matrix}\right].$$

44. $${}_2F_1\left(a, \frac{4a+3}{6}; \frac{4a}{3}+1; 12\sqrt{2}-16\right)=$$
$$=\left(\frac{2+\sqrt{2}}{2}\right)^{2a}\frac{\sqrt{\pi}}{2}\left\{3\Gamma\left[\begin{matrix}2a/3+1\\ a/2+1, (a+3)/6\end{matrix}\right]-\Gamma\left[\begin{matrix}2a/3+1\\ (a+1)/2, a/6+1\end{matrix}\right]\right\}.$$

45. $${}_2F_1\left(a, 2a-\frac{3}{2}; 4a-3; 12\sqrt{2}-16\right)=$$
$$=\frac{(2+\sqrt{2})^{2a}}{2^{4a-2}}\sqrt{\pi}\left\{3\Gamma\left[\begin{matrix}2a-1\\ a/2, (3a-1)/2\end{matrix}\right]-\Gamma\left[\begin{matrix}2a-1\\ (a+1)/2, 3a/2-1\end{matrix}\right]\right\}.$$

46. $${}_2F_1\left(a, 2a-\frac{1}{2}; 4a-1; 12\sqrt{2}-16\right)=\frac{(2+\sqrt{2})^{2a}}{2^{4a-1}}\sqrt{\pi}\,\Gamma\left[\begin{matrix}2a\\ (a+1)/2, 3a/2\end{matrix}\right].$$

47. $${}_2F_1\left(-2n, n+\frac{1}{2}; n+1; \frac{4}{3}\right)=3^{-2n}.$$

48. $${}_2F_1\left(-n, \frac{1}{2}; m; 4\right)=S(n, m),\ mS(n, m)=(n+m)S(n, m+1)-nS(n-1, m+1),$$
$$S\left(2n, n+\frac{3}{2}\right)=\frac{(1/2)_n(3/2)_n}{(5/6)_n(7/6)_n},\quad S(2n+1, n+2)=0,\quad S(2n, n)=3,$$
$$S(2n+1, n)=-\frac{5n+2}{n},\quad S(2n, n+1)=1,\ S(2n+1, n+1)=-1,$$
$$S(2n, n+2)=\frac{n+1}{2n+1},\quad S(2n+1, n+2)=0,\quad S(2n, n+3)=\frac{3(n+1)(n+2)}{2(2n+1)(2n+3)},$$
$$S(2n+1, n+3)=\frac{n+2}{2(2n+3)},\quad S(2n, n+4)=\frac{(n+2)(n+3)(11n+10)}{4(2n+1)(2n+3)(2n+5)},$$
$$S(2n+1, n+4)=\frac{5(n+2)(n+3)}{4(2n+3)(2n+5)},$$
$$S(2n, n+5)=\frac{(n+2)(n+3)(n+4)(43n+35)}{8(2n+1)(2n+3)(2n+5)(2n+7)},$$
$$S(2n+1, n+5)=\frac{21(n+2)(n+3)(n+4)}{8(2n+3)(2n+5)(2n+7)}.$$

49. $${}_2F_1\left(-2n, \frac{3}{2}; n+2; 4\right)=n+1.$$

7.4. THE FUNCTION $_3F_2(a_1, a_2, a_3; b_1, b_2; z)$

7.4.1. Representations of $_3F_2(a_1, a_2, a_3; b_1, b_2; z)$.

	a_1	a_2	a_3	b_1	b_2	$_3F_2(a_1, a_2, a_3; b_1, b_2; z)$
1	a	b	c	d	e	$\frac{1}{1-z}\left\{\left[1+\frac{d+e-b-c}{1-a}z\right]\times\right.$ $\times{}_3F_2(a-1, b, c; d, e; z)-\frac{z}{e(1-a)}\times$ $\times[(d+e-b-c)(1+e-a)+(d-b)\times$ $\times(d-c)]\,{}_3F_2(a-1, b, c; d, e+1; z)+$ $+\frac{(d-b)(d-c)(1+d-a)}{de(1-a)}\,{}_3F_2(a-1, b, c;$ $\left.d+1, e+1; z)\right\}$
2	a	b	c	$a-n$	d	$\frac{1}{(1-a)_n}\sum_{k=0}^{n}(-1)^k\binom{n}{k}(1-a)_{n-k}\times$ $\times\frac{(b)_k(c)_k}{(d)_k}z^k{}_2F_1(b+k, c+k; d+k; z)$
3	a	b	c	$a-1$	d	$_2F_1(b, c; d; z)+\frac{bcz}{(a-1)d}\times$ $\times{}_2F_1(b+1, c+1; d+1; z)$
4	a	b	c	$a+1$	d	$\frac{a}{1+a-d}\,{}_2F_1(b, c; d; z)-\frac{d-1}{1+a-d}\times$ $\times{}_3F_2(a, b, c; a+1, d-1; z)$
5	a	b	c	$a+1$	$b+1$	$\frac{1}{b-a}[b\,{}_2F_1(a, c; a+1; z)-$ $-a\,{}_2F_1(b, c; b+1; z)]$
6	a	b	c	$1+a-b$	$1+a-c$	$(1-z)^{-a}\,{}_3F_2\left(\frac{a}{2}, \frac{a+1}{2}, 1+a-b-c;\right.$ $\left.1+a-b, 1+a-c; -\frac{4z}{(1-z)^2}\right)$
7	1	a	b	2	c	$\frac{c-1}{(a-1)(b-1)z}[{}_2F_1(a-1, b-1;$ $c-1; z)-1]$
8	1	a	b	3	c	$\frac{2(c-2)_2}{(a-2)_2(b-2)_2z^2}\times$ $\times[{}_2F_1(a-2, b-2; c-2; z)-1]-$ $-\frac{2(c-1)}{(a-1)(b-1)z}$
9	a	$1-a$	b	$\frac{a+b+1}{2}$	$\frac{b-a}{2}+1$	$(1-4z)^{-b}\,{}_3F_2\left(\frac{b}{3}, \frac{b+1}{3}, \frac{b+2}{3};\right.$ $\left.\frac{a+b+1}{2}, \frac{b-a}{2}+1; -\frac{27z}{(1-4z)^3}\right)$
10	a	$1-a$	$\frac{1}{2}$	b	$2-b$	$_2F_1\left(a, 1-a; 2-b; \frac{1-\sqrt{1-z}}{2}\right)\times$ $\times{}_2F_1\left(a, 1-a; b; \frac{1-\sqrt{1-z}}{2}\right)$
11	a	$a+\frac{1}{2}$	b	c	$2a+b-c+1$	$\left[\frac{2}{z}(1-\sqrt{1-z})\right]^{2a}{}_3F_2\left(2a, c-b,\right.$

	a_1	a_2	a_3	b_1	b_2	${}_3F_2(a_1, a_2, a_3; b_1, b_2; z)$
						$2a-c+1;\ c,\ 2a+b-c+1;\ 1-\frac{2}{z}(1-\sqrt{1-z})\Big)$
12	a	$a+\frac{1}{2}$	b	$2a$	$b+1$	$\left[\frac{2}{z}(1-\sqrt{1-z})\right]^{2b}\times$ $\times{}_2F_1\Big(b,\ 1+2b-2a;\ b+1;\ 1-\frac{2}{z}(1-\sqrt{1-z})\Big)$
13	a	$a+\frac{1}{2}$	b	$2a$	$b+1$	$\left(\frac{1+\sqrt{1-z}}{2}\right)^{-b}\times$ $\times{}_2F_1\Big(b,\ 2a-b;\ b+1;\ 1-\frac{1-\sqrt{1-z}}{2}\Big)$
14	a	$\frac{a-1}{2}$	b	$a-1$	$\frac{a+b+1}{2}$	$(1-z)^{-b/2}\,{}_2F_1\Big(\frac{b}{2},\ \frac{a-b-1}{2};\ \frac{a+b+1}{2};\ z\Big)$
15	a	$\frac{a}{2}+1$	$b-\frac{1}{2}$	$\frac{a}{2}$	$a-b+\frac{3}{2}$	$(1-z)\,{}_2F_1\Big(a+1,\ b+\frac{1}{2};\ a-b+\frac{3}{2};\ z\Big)$
16	a	$\frac{a}{2}+1$	b	$\frac{a}{2}$	$a-b+1$	$(1-z)^{-2b-1}\,{}_3F_2\Big(a-2b-1,\ \frac{a+1}{2}-b,\ -b-1;\ \frac{a-1}{2}-b,\ a-b+1;\ z\Big)$
17	a	b	$a+b-\frac{1}{2}$	$2a$	$2b$	$(1-z/4)^{1/2-a-b}\,{}_3F_2\Big(\frac{2a+2b-1}{6},\ \frac{2a+2b+1}{6},\ \frac{2a+2b+3}{6};\ a+\frac{1}{2},\ b+\frac{1}{2};\ \frac{27z^2}{(4-z)^3}\Big)$
18	a	b	$\frac{a+b-1}{2}$	$a+b-1$	$\frac{a+b}{2}$	${}_2F_1\Big(\frac{a}{2},\ \frac{b-1}{2};\ \frac{a+b}{2};\ z\Big)\times$ $\times{}_2F_1\Big(\frac{a}{2},\ \frac{b+1}{2};\ \frac{a+b}{2};\ z\Big)$
19	a	b	$\frac{a+b}{2}$	$a+b-1$	$\frac{a+b+1}{2}$	${}_2F_1\Big(\frac{a}{2},\ \frac{b}{2};\ \frac{a+b-1}{2};\ z\Big)\times$ $\times{}_2F_1\Big(\frac{a}{2},\ \frac{b}{2};\ \frac{a+b+1}{2};\ z\Big)=$
20						$={}_2F_1\Big(\frac{a-1}{2},\ \frac{b-1}{2};\ \frac{a+b-1}{2};\ z\Big)\times$ $\times{}_2F_1\Big(\frac{a+1}{2},\ \frac{b+1}{2};\ \frac{a+b+1}{2};\ z\Big)$
21	a	b	$\frac{a+b}{2}$	$a+b$	$\frac{a+b-1}{2}$	$2\,{}_2F_1\Big(\frac{a}{2};\ \frac{b}{2};\ \frac{a+b-1}{2};\ z\Big)\times$ $\times{}_2F_1\Big(\frac{a}{2},\ \frac{b}{2};\ \frac{a+b+1}{2};\ z\Big)-$ $-{}_2F_1^2\Big(\frac{a}{2},\ \frac{b}{2};\ \frac{a+b+1}{2};\ z\Big)$
22	a	b	$\frac{a+b}{2}$	$a+b$	$\frac{a+b+1}{2}$	${}_2F_1^2\Big(\frac{a}{2},\ \frac{b}{2};\ \frac{a+b+1}{2};\ z\Big)=$

	a_1	a_2	a_3	b_1	b_2	${}_3F_2(a_1, a_2, a_3; b_1, b_2; z)$
3						$= {}_2F_1^2\left(a, b; \frac{a+b+1}{2}; \frac{1-\sqrt{1-z}}{2}\right)=$
4						$=\sqrt{1-z}\,{}_2F_1\left(\frac{a}{2}, \frac{b}{2}; \frac{a+b+1}{2}; z\right)\times {}_2F_1\left(\frac{a+1}{2}, \frac{b+1}{2}; \frac{a+b+1}{2}; z\right)$
5	a	b	$\frac{a+b+1}{2}$	$a+b$	$\frac{a+b}{2}+1$	$\frac{a}{a-b}\,{}_2F_1\left(\frac{a+1}{2}, \frac{b}{2}; \frac{a+b}{2}; z\right)\times {}_2F_1\left(\frac{a+1}{2}, \frac{b}{2}; \frac{a+b}{2}+1; z\right)-\frac{b}{a-b}\,{}_2F_1\left(\frac{a}{2}, \frac{b+1}{2}; \frac{a+b}{2}; z\right)\times {}_2F_1\left(\frac{a}{2}, \frac{b+1}{2}; \frac{a+b}{2}+1; z\right)$
6	a	b	$\frac{a+b+1}{2}$	$a+b+1$	$\frac{a+b}{2}+1$	$\frac{a}{a-b}\,{}_2F_1^2\left(\frac{a+1}{2}, \frac{b}{2}; \frac{a+b}{2}+1; z\right)-\frac{b}{a-b}\,{}_2F_1^2\left(\frac{a}{2}, \frac{b+1}{2}; \frac{a+b}{2}+1; z\right)$
7	a	b	$\frac{a+b}{2}+1$	$a+b$	$\frac{a+b+1}{2}$	$2\frac{a+b-1}{a+b}\,{}_2F_1\left(\frac{a}{2}, \frac{b}{2}; \frac{a+b-1}{2}; z\right)\times {}_2F_1\left(\frac{a}{2}, \frac{b}{2}; \frac{a+b+1}{2}; z\right)-\frac{a+b-2}{a+b}\,{}_2F_1^2\left(\frac{a}{2}, \frac{b}{2}; \frac{a+b+1}{2}; z\right)$
8	a	$a+\frac{1}{3}$	$a+\frac{2}{3}$	$\frac{3a}{2}$	$\frac{3a+1}{2}$	$\frac{(1-x)^{3a}}{2x+1}$ $[4(1-x)^3 z=-27x]$
9	a	$a+\frac{1}{3}$	$a+\frac{2}{3}$	$\frac{3a+1}{2}$	$\frac{3a}{2}+1$	$(1-x)^{3a}$ $[4(1-x)^3 z=-27x]$
0	a	$a+\frac{1}{3}$	$a+\frac{2}{3}$	$\frac{1}{3}$	$\frac{2}{3}$	$\frac{1}{3}\left[(1-x)^{-3a}+\left(1-xe^{2\pi i/3}\right)^{-3a}+\left(1-xe^{4\pi i/3}\right)^{-3a}\right]$ $[x=z^{1/3}]$
1	a	$a+\frac{1}{3}$	$a+\frac{2}{3}$	$\frac{2}{3}$	$\frac{4}{3}$	$\frac{1}{3x(3a-1)}\left[(1-x)^{1-3a}+e^{-2\pi i/3}\times\left(1-xe^{2\pi i/3}\right)^{1-3a}+e^{-4\pi i/3}\left(1-xe^{4\pi i/3}\right)^{1-3a}\right]$ $[x=z^{1/3}]$
2	a	$a+\frac{1}{3}$	$a+\frac{2}{3}$	$\frac{4}{3}$	$\frac{5}{3}$	$\frac{2}{3x^2(3a-1)(3a-2)}\left[(1-x)^{2-3a}+e^{-4\pi i/3}\left(1-xe^{2\pi i/3}\right)^{2-3a}+e^{-2\pi i/3}\left(1-xe^{4\pi i/3}\right)^{2-3a}\right]$ $[x=z^{1/3}]$
3	$\frac{m}{j}$	$\frac{n}{l}$	1	$\frac{m}{j}+1$	$\frac{n}{l}+1$	$\frac{mn}{nj-lm}\left\{z^{-n/l}\sum_{k=0}^{l-1}\exp\left(-\frac{2\pi ikn}{l}\right)\times\ln\left[1-z^{1/l}\exp\left(\frac{2\pi ik}{l}\right)\right]-z^{-m/j}\sum_{k=0}^{j-1}\exp\left(-\frac{2\pi ikm}{j}\right)\times\right.$

	a_1	a_2	a_3	b_1	b_2	${}_3F_2(a_1, a_2, a_3; b_1, b_2; z)$
						$\times \ln\left[1-z^{1/j}\exp\left(\frac{2\pi ik}{j}\right)\right]+ + l\sum_{k=1}^{p}\frac{1}{n-kl}z^{-k}\Big\}$ $[m\leqslant j,\ nj\neq lm,\ p=0, 1, 2, \ldots, pl<n\leqslant \leqslant (p+1)l]$
34	1	m	n	$m+1$	$n+1$	$\frac{mn}{(n-m)z^n}\left[(1-z^{n-m})\ln(1-z)+ +\sum_{k=1}^{n-1}\frac{z^k}{k}-z^{n-m}\sum_{k=1}^{m-1}\frac{z^k}{k}\right]$
35	1	1	$2n+2$	$n+2$	$n+2$	$\frac{[(n+1)!]^2}{(2n+1)!(-z)^{n+1}}\Big\{\ln(1-z)P_n(1-2z)+ +\sum_{k=1}^{n}\frac{(n+k)!(-z)^k}{(n-k)!(k!)^2}[\psi(n+k)+ +\psi(n-k)-2\psi(n)]\Big\}$
36	1	$\frac{3+i}{2}$	$\frac{3-i}{2}$	$\frac{3}{2}$	2	$\frac{1}{z}\left(\frac{\operatorname{ch}\arcsin\sqrt{z}}{\cos\arcsin\sqrt{z}}-1\right)$
37	$-n$	a	$2a+n$	$a+\frac{1}{2}$	$2a$	$\left[\frac{n!}{(2a)_n}C_n^a(\sqrt{1-z})\right]^2$
38	$-n$	$\frac{1}{2}$	$n+1$	1	1	$[P_n(\sqrt{1-z})]^2$

7.4.2. Particular values of ${}_3F_2(a_1, a_2, a_3; b_1, b_2; z)$.

NOTATION:

$$\psi_1(z)=\frac{4}{\pi^2}[\mathbf{K}(x)]^2,$$

$$\psi_2(z)=\frac{4}{\pi^2}\left\{[\mathbf{K}(x)]^2-\frac{x^2}{1-x^2}[\mathbf{K}(x)-\mathbf{D}(x)]^2\right\},$$

$$\psi_3(z)=\frac{2}{\pi^2}\Big\{3[\mathbf{K}(x)]^2-4\frac{1-x^2+x^4}{(1-x^2)^2}[\mathbf{K}(x)-\mathbf{D}(x)]^2+\frac{1}{(1-x^2)^2}\times$$
$$\times[2(1-2x^2)\mathbf{D}(x)-(1-3x^2)\mathbf{K}(x)]^2\Big\},$$

$$x=\sqrt{\frac{1-\sqrt{1-z}}{2}}.$$

	a_1	a_2	a_3	b_1	b_2	${}_3F_2(a_1, a_2, a_3; b_1, b_2; z)$
1	$-\frac{1}{2}$	$\frac{1}{2}$	$\frac{1}{2}$	1	1	$(1-z)\psi_1(z)+z\psi_2(z)-\frac{z}{4}\psi_3(z)$
2	$-\frac{1}{2}$	$\frac{1}{2}$	$\frac{1}{2}$	1	2	$\frac{1}{6}[4(1-z)\psi_1(z)+2(1+2z)\psi_2(z)-z\psi_3(z)]$

	a_1	a_2	a_3	b_1	b_2	${}_3F_2(a_1, a_2, a_3; b_1, b_2; z)$
3	$-\frac{1}{2}$	$\frac{1}{2}$	$\frac{1}{2}$	1	3	$\frac{2}{135z}[4(2+7z-9z^2)\psi_1(z)-2(4-19z-18z^2)\psi_2(z)+z(1-9z)\psi_3(z)]$
4	$-\frac{1}{2}$	$\frac{1}{2}$	$\frac{1}{2}$	2	2	$\frac{1}{9}[4(1-z)\psi_1(z)+2(1+2z)\psi_2(z)+(3-z)\psi_3(z)]$
5	$-\frac{1}{2}$	$\frac{1}{2}$	$\frac{1}{2}$	2	3	$\frac{4}{135z}[4(1-z+3z^2)\psi_2(z)-4(1-4z+3z^2)\psi_1(z)+z(22-3z)\psi_3(z)]$
6	$-\frac{1}{2}$	$\frac{1}{2}$	$\frac{1}{2}$	3	3	$\frac{16}{675z}[2(7-22z+6z^2)\psi_2(z)-12(2-3z+z^2)\psi_1(z)+(10+52z-3z^2)\psi_3(z)]$
7	$-\frac{1}{2}$	$\frac{1}{2}$	1	$\frac{3}{2}$	$\frac{3}{2}$	$\frac{1}{4\sqrt{z}}[\sqrt{z}+(1-z)\operatorname{Arth}\sqrt{z}+\operatorname{Li}_2(\sqrt{z})-\operatorname{Li}_2(-\sqrt{z})]$
8	$-\frac{1}{2}$	$\frac{1}{2}$	1	$\frac{3}{2}$	2	$\frac{1}{3z}[(2+z)\sqrt{1-z}+3\sqrt{z}\arcsin\sqrt{z}-2]$
9	$-\frac{1}{2}$	$\frac{1}{2}$	1	$\frac{3}{2}$	$\frac{5}{2}$	$\frac{3}{16z}\left\{(1-z^2)\frac{\operatorname{Arth}\sqrt{z}}{\sqrt{z}}-1+z+2\sqrt{z}[\operatorname{Li}_2(\sqrt{z})-\operatorname{Li}_2(-\sqrt{z})]\right\}$
10	$-\frac{1}{2}$	$\frac{1}{2}$	1	$\frac{3}{2}$	3	$\frac{4}{45z^2}[2-15z-(2-14z-3z^2)\sqrt{1-z}+15z^{3/2}\arcsin\sqrt{z}]$
11	$-\frac{1}{2}$	$\frac{1}{2}$	1	2	2	$\frac{4}{9\pi z}[2(3+5z)\mathbf{K}(\sqrt{z})-3\pi-2z(7+z)\mathbf{D}(\sqrt{z})]$
12	$-\frac{1}{2}$	$\frac{1}{2}$	1	2	$\frac{5}{2}$	$\frac{1}{8z}\left[(13+2z)\sqrt{1-z}+3(1+4z)\frac{\arcsin\sqrt{z}}{\sqrt{z}}-16\right]$
13	$-\frac{1}{2}$	$\frac{1}{2}$	1	2	3	$\frac{32}{45\pi z}\left[8(1+z)\mathbf{K}(\sqrt{z})-(1+14z+z^2)\mathbf{D}(\sqrt{z})-\frac{15\pi}{4}\right]$
14	$-\frac{1}{2}$	$\frac{1}{2}$	1	$\frac{5}{2}$	$\frac{5}{2}$	$\frac{9}{64z^{3/2}}\{(5-4z-z^2)\operatorname{Arth}\sqrt{z}+2(1+2z)[\operatorname{Li}_2(\sqrt{z})-\operatorname{Li}_2(-\sqrt{z})]-9\sqrt{z}+z^{3/2}\}$
15	$-\frac{1}{2}$	$\frac{1}{2}$	1	$\frac{5}{2}$	3	$\frac{1}{30z^2}[(16+83z+6z^2)\sqrt{1-z}+15\sqrt{z}(3+4z)\times\arcsin\sqrt{z}-16-120z]$
16	$-\frac{1}{2}$	$\frac{1}{2}$	1	3	3	$\frac{128}{675\pi z^2}[(15+74z+39z^2)\mathbf{K}(\sqrt{z})-z(43+82z+3z^2)\mathbf{D}(\sqrt{z})]-\frac{16}{45z^2}(4+15z)$
17	$-\frac{1}{2}$	$\frac{1}{2}$	$\frac{3}{2}$	1	1	$(1-2z)\psi_1(z)+2z\psi_2(z)-\frac{z}{2}\psi_3(z)$
18	$-\frac{1}{2}$	$\frac{1}{2}$	$\frac{3}{2}$	1	2	$\frac{1}{3}[4(1-z)\psi_1(z)-(1-4z)\psi_2(z)-z\psi_3(z)]$
19	$-\frac{1}{2}$	$\frac{1}{2}$	$\frac{3}{2}$	1	3	$\frac{2}{45z}[8(1-z+3z^2)\psi_2(z)-8(1-4z+3z^2)\psi_1(z)-z(1+6z)\psi_3(z)]$
20	$-\frac{1}{2}$	$\frac{1}{2}$	$\frac{3}{2}$	2	2	$\frac{1}{9}[8(1-z)\psi_1(z)+4(1+2z)\psi_2(z)-(3+2z)\psi_3(z)]$
21	$-\frac{1}{2}$	$\frac{1}{2}$	$\frac{3}{2}$	2	3	$\frac{4}{45z}[4(1+z-2z^2)\psi_1(z)-2(2-7z-4z^2)\psi_2(z)-z(7+2z)\psi_3(z)]$

	a_1	a_2	a_3	b_1	b_2	${}_3F_2(a_1, a_2, a_3; b_1, b_2; z)$
22	$-\frac{1}{2}$	$\frac{1}{2}$	$\frac{3}{2}$	3	3	$\frac{32}{675z}[2(13-7z-6z^2)\psi_1(z)-(11-56z-12z^2)\psi_2(z)-(15+23z+3z^2)\psi_3(z)]$
23	$-\frac{1}{2}$	$\frac{1}{2}$	2	1	1	$\frac{1}{\pi}[2\mathbf{K}(\sqrt{z})-3z\mathbf{D}(\sqrt{z})]$
24	$-\frac{1}{2}$	$\frac{1}{2}$	2	1	$\frac{3}{2}$	$\frac{1}{4}\left(\frac{\arcsin\sqrt{z}}{\sqrt{z}}+3\sqrt{1-z}\right)$
25	$-\frac{1}{2}$	$\frac{1}{2}$	2	1	$\frac{5}{2}$	$\frac{3}{32z}\left[(1+4z)\frac{\arcsin\sqrt{z}}{\sqrt{z}}-(1-6z)\sqrt{1-z}\right]$
26	$-\frac{1}{2}$	$\frac{1}{2}$	2	1	3	$\frac{8}{45\pi z}[2(1+6z)\mathbf{K}(\sqrt{z})-(4+z+9z^2)\mathbf{D}(\sqrt{z})]$
27	$-\frac{1}{2}$	$\frac{1}{2}$	2	$\frac{3}{2}$	$\frac{3}{2}$	$\frac{1}{8\sqrt{z}}[3\sqrt{z}+3(1-z)\operatorname{Arth}\sqrt{z}+\mathrm{Li}_2(\sqrt{z})-\mathrm{Li}_2(-\sqrt{z})]$
28	$-\frac{1}{2}$	$\frac{1}{2}$	2	$\frac{3}{2}$	$\frac{5}{2}$	$\frac{3}{32z^{3/2}}\{\sqrt{z}+3z^{3/2}-(1-4z+3z^2)\operatorname{Arth}\sqrt{z}+2z[\mathrm{Li}_2(\sqrt{z})-\mathrm{Li}_2(-\sqrt{z})]\}$
29	$-\frac{1}{2}$	$\frac{1}{2}$	2	$\frac{3}{2}$	3	$\frac{2}{45z^2}[(4+2z+9z^2)\sqrt{1-z}-4+15z^{3/2}\arcsin\sqrt{z}]$
30	$-\frac{1}{2}$	$\frac{1}{2}$	2	$\frac{5}{2}$	$\frac{5}{2}$	$\frac{9}{128z^{3/2}}\{5\sqrt{z}+3z^{3/2}-(1-4z+3z^2)\operatorname{Arth}\sqrt{z}-2(1-2z)[\mathrm{Li}_2(\sqrt{z})-\mathrm{Li}_2(-\sqrt{z})$
31	$-\frac{1}{2}$	$\frac{1}{2}$	2	$\frac{5}{2}$	3	$\frac{1}{60z^2}[32-(32-29z-18z^2)\sqrt{1-z}-15\sqrt{z}(3-4z)\arcsin\sqrt{z}$
32	$-\frac{1}{2}$	$\frac{1}{2}$	2	3	3	$\frac{64}{675\pi z^2}[z(71-46z-9z^2)\mathbf{D}(\sqrt{z})-2(15+14z-21z^2)\mathbf{K}(\sqrt{z})+15\pi$
33	$-\frac{1}{2}$	$\frac{1}{2}$	$\frac{5}{2}$	1	1	$\frac{(1-z)^{-1}}{12}[4(3-10z+8z^2)\psi_1(z)+2z(13-16z)\psi_2(z)-z(7-8z)\psi_3(z)]$
34	$-\frac{1}{2}$	$\frac{1}{2}$	$\frac{5}{2}$	1	$\frac{3}{2}$	$\frac{2}{\pi}\mathbf{K}(\sqrt{z})-\frac{8}{3\pi}z\mathbf{D}(\sqrt{z})$
35	$-\frac{1}{2}$	$\frac{1}{2}$	$\frac{5}{2}$	1	2	$\frac{1}{9}[2(5-8z)\psi_1(z)-(1-16z)\psi_2(z)-4z\psi_3(z)]$
36	$-\frac{1}{2}$	$\frac{1}{2}$	$\frac{5}{2}$	1	3	$\frac{2}{135z}[8(1+11z-12z^2)\psi_1(z)-2(4+11z-48z^2)\psi_2(z)+z(1-24z)\psi_3(z)]$
37	$-\frac{1}{2}$	$\frac{1}{2}$	$\frac{5}{2}$	$\frac{3}{2}$	$\frac{3}{2}$	$\frac{1}{3}\left[\frac{\arcsin\sqrt{z}}{\sqrt{z}}+2\sqrt{1-z}\right]$
38	$-\frac{1}{2}$	$\frac{1}{2}$	$\frac{5}{2}$	$\frac{3}{2}$	2	$\frac{4}{9\pi}[5\mathbf{K}(\sqrt{z})-(1+4z)\mathbf{D}(\sqrt{z})]$
39	$-\frac{1}{2}$	$\frac{1}{2}$	$\frac{5}{2}$	$\frac{3}{2}$	3	$\frac{16}{135\pi z}[(1+21z)\mathbf{K}(\sqrt{z})-2(1+4z+6z^2)\mathbf{D}(\sqrt{z})]$
40	$-\frac{1}{2}$	$\frac{1}{2}$	$\frac{5}{2}$	2	2	$\frac{1}{27}[32(1-z)\psi_1(z)-2(1-16z)\psi_2(z)-(3+8z)\psi_3(z)]$
41	$-\frac{1}{2}$	$\frac{1}{2}$	$\frac{5}{2}$	2	3	$\frac{8}{135z}[(2+3z+16z^2)\psi_2(z)-2(1-9z+8z^2)\psi_1(z)-4z(1+z)\psi_3(z)]$

	a_1	a_2	a_3	b_1	b_2	$_3F_2(a_1, a_2, a_3; b_1, b_2; z)$
42	$-\frac{1}{2}$	$\frac{1}{2}$	$\frac{5}{2}$	3	3	$\frac{16}{2025z}[8(1+11z-12z^2)\psi_1(z)-2(19-49z-48z^2)\psi_2(z)+(30-59z-24z^2)\psi_3(z)]$
43	$-\frac{1}{2}$	$\frac{1}{2}$	3	1	1	$\frac{(1-z)^{-1}}{4\pi}[(8-9z)\mathbf{K}(\sqrt{z})-z(14-15z)\mathbf{D}(\sqrt{z})]$
44	$-\frac{1}{2}$	$\frac{1}{2}$	3	1	$\frac{3}{2}$	$\frac{1}{16}\left[\frac{13-15z}{\sqrt{1-z}}+3\frac{\arcsin\sqrt{z}}{\sqrt{z}}\right]$
45	$-\frac{1}{2}$	$\frac{1}{2}$	3	1	2	$\frac{1}{2\pi}[4\mathbf{K}(\sqrt{z})-5z\mathbf{D}(\sqrt{z})]$
46	$-\frac{1}{2}$	$\frac{1}{2}$	3	1	$\frac{5}{2}$	$\frac{3}{128z}\left[(1+12z)\frac{\arcsin\sqrt{z}}{\sqrt{z}}-(1-30z)\sqrt{1-z}\right]$
47	$-\frac{1}{2}$	$\frac{1}{2}$	3	$\frac{3}{2}$	$\frac{3}{2}$	$\frac{1}{32\sqrt{z}}\{15\sqrt{z}+(11-15z)\operatorname{Arth}\sqrt{z}+3[\operatorname{Li}_2(\sqrt{z})-\operatorname{Li}_2(-\sqrt{z})]\}$
48	$-\frac{1}{2}$	$\frac{1}{2}$	3	$\frac{3}{2}$	2	$\frac{1}{8}\left[3\frac{\arcsin\sqrt{z}}{\sqrt{z}}+5\sqrt{1-z}\right]$
49	$-\frac{1}{2}$	$\frac{1}{2}$	3	$\frac{3}{2}$	$\frac{5}{2}$	$\frac{3}{128z^{3/2}}\{\sqrt{z}+15z^{3/2}-(1-16z+15z^2)\operatorname{Arth}\sqrt{z}+6z[\operatorname{Li}_2(\sqrt{z})-\operatorname{Li}_2(-\sqrt{z})]\}$
50	$-\frac{1}{2}$	$\frac{1}{2}$	3	2	2	$\frac{1}{3\pi}[7\mathbf{K}(\sqrt{z})-(2+5z)\mathbf{D}(\sqrt{z})]$
51	$-\frac{1}{2}$	$\frac{1}{2}$	3	2	$\frac{5}{2}$	$\frac{3}{64z}\left[(1+10z)\sqrt{1-z}-(1-12z)\frac{\arcsin\sqrt{z}}{\sqrt{z}}\right]$
52	$-\frac{1}{2}$	$\frac{1}{2}$	3	$\frac{5}{2}$	$\frac{5}{2}$	$\frac{9}{512z^{3/2}}\{9\sqrt{z}+15z^{3/2}-5(1-4z+3z^2)\operatorname{Arth}\sqrt{z}-2(1-6z)[\operatorname{Li}_2(\sqrt{z})-\operatorname{Li}_2(-\sqrt{z})]\}$
53	$-\frac{1}{2}$	1	1	$\frac{1}{2}$	2	$\frac{1}{3}\left[2(1-\sqrt{z}\operatorname{Arth}\sqrt{z})-\frac{1}{z}\ln(1-z)\right]$
54	$-\frac{1}{2}$	1	1	$\frac{1}{2}$	3	$\frac{2}{15z}\left[3+\frac{3-5z}{z}\ln(1-z)+4z(1-\sqrt{z}\operatorname{Arth}\sqrt{z})\right]$
55	$-\frac{1}{2}$	1	1	$\frac{3}{2}$	2	$\frac{1}{3}\left[1+\frac{1}{z}\ln(1-z)+(3-z)\frac{\operatorname{Arth}\sqrt{z}}{\sqrt{z}}\right]$
56	$-\frac{1}{2}$	1	1	$\frac{3}{2}$	3	$\frac{2}{15z}\left[2(5-z)\sqrt{z}\operatorname{Arth}\sqrt{z}-1+2z+\frac{5z-1}{z}\ln(1-z)\right]$
57	$-\frac{1}{2}$	1	1	2	2	$\frac{4}{9z}\left[4-(4-z)\sqrt{1-z}+3\ln\frac{1+\sqrt{1-z}}{2}\right]$
58	$-\frac{1}{2}$	1	1	2	$\frac{5}{2}$	$\frac{1}{4z}\left[(3+6z-z^2)\frac{\operatorname{Arth}\sqrt{z}}{\sqrt{z}}+4\ln(1-z)-3+z\right]$
59	$-\frac{1}{2}$	1	1	2	3	$\frac{4}{45z^2}\left[6+25z-2(3+14z-2z^2)\sqrt{1-z}+30z\ln\frac{1+\sqrt{1-z}}{2}\right]$
60	$-\frac{1}{2}$	1	1	$\frac{5}{2}$	3	$\frac{1}{5z}\left[(15+10z-z^2)\frac{\operatorname{Arth}\sqrt{z}}{\sqrt{z}}-13+z+\frac{2(1+5z)}{z}\ln(1-z)\right]$

	a_1	a_2	a_3	b_1	b_2	${}_3F_2\,(a_1,\ a_2,\ a_3;\ b_1,\ b_2;\ z)$
61	$-\frac{1}{2}$	1	1	3	3	$\frac{16}{225z^2}\left[(61+25z)-(61+48z-4z^2)\sqrt{1-z}+15(2+5z)\ln\frac{1+\sqrt{1-z}}{2}\right].$
62	$-\frac{1}{2}$	1	$\frac{3}{2}$	$\frac{1}{2}$	$\frac{1}{2}$	$\frac{1-2z}{1-z}-2\sqrt{z}\,\mathrm{Arth}\,\sqrt{z}$
63	$-\frac{1}{2}$	1	$\frac{3}{2}$	$\frac{1}{2}$	2	$\frac{2}{3z}\left[(1+2z)\sqrt{1-z}-1\right]$
64	$-\frac{1}{2}$	1	$\frac{3}{2}$	$\frac{1}{2}$	$\frac{5}{2}$	$\frac{3}{4z}\left[(1-z^2)\frac{\mathrm{Arth}\,\sqrt{z}}{\sqrt{z}}-1+z\right]$
65	$-\frac{1}{2}$	1	$\frac{3}{2}$	$\frac{1}{2}$	3	$\frac{4}{15z^2}\left[6-5z-2(3+2z)(1-z)^{3/2}\right]$
66	$-\frac{1}{2}$	1	$\frac{3}{2}$	2	2	$\frac{4}{9\pi z}\left[3\pi-2(3-z)\,\mathbf{K}(\sqrt{z})+4z(2-z)\,\mathbf{D}(\sqrt{z})\right]$
67	$-\frac{1}{2}$	1	$\frac{3}{2}$	2	$\frac{5}{2}$	$\frac{1}{4z}\left[8-(5-2z)\sqrt{1-z}-3\frac{\arcsin\sqrt{z}}{\sqrt{z}}\right]$
68	$-\frac{1}{2}$	1	$\frac{3}{2}$	2	3	$\frac{8}{45\pi z}\left[15\pi-4(9-z)\,\mathbf{K}(\sqrt{z})+4(3+7z-2z^2)\,\mathbf{D}(\sqrt{z})\right]$
69	$-\frac{1}{2}$	1	$\frac{3}{2}$	$\frac{5}{2}$	$\frac{5}{2}$	$\frac{9}{32z^{3/2}}\left\{7\sqrt{z}+z^{3/2}-2\left[\mathrm{Li}_2(\sqrt{z})-\mathrm{Li}_2(-\sqrt{z})\right]-(3-4z+z^2)\,\mathrm{Arth}\,\sqrt{z}\right\}$
70	$-\frac{1}{2}$	1	$\frac{3}{2}$	$\frac{5}{2}$	3	$\frac{1}{5z^2}\left[8+20z-(8+9z-2z^2)\sqrt{1-z}-15\sqrt{z}\arcsin\sqrt{z}\right]$
71	$-\frac{1}{2}$	1	$\frac{3}{2}$	3	3	$\frac{64}{15z^2}+\frac{16}{3z}-\frac{128}{225\pi z^2}\left[(15+34z-z^2)\,\mathbf{K}(\sqrt{z})-2z(19+6z-z^2)\,\mathbf{D}(\sqrt{z})\right]$
72	$-\frac{1}{2}$	1	2	$\frac{1}{2}$	3	$\frac{2}{5z}\left[2z(1-\sqrt{z}\,\mathrm{Arth}\,\sqrt{z})-1-\frac{1}{z}\ln(1-z)\right]$
73	$-\frac{1}{2}$	1	2	$\frac{3}{2}$	3	$\frac{2}{15z}\left[1+\frac{1}{z}\ln(1-z)+3z+(5-3z)\sqrt{z}\,\mathrm{Arth}\,\sqrt{z}\right]$
74	$-\frac{1}{2}$	1	2	$\frac{5}{2}$	3	$\frac{1}{10z}\left[11+3z-\frac{4}{z}\ln(1-z)-(15-10z+3z^2)\frac{\mathrm{Arth}\,\sqrt{z}}{\sqrt{z}}\right]$
75	$-\frac{1}{2}$	1	2	3	3	$\frac{8}{225z^2}\left[4(23-11z+3z^2)\sqrt{1-z}-92+75z-60\ln\frac{1+\sqrt{1-z}}{2}\right]$
76	$-\frac{1}{2}$	1	$\frac{5}{2}$	$\frac{1}{2}$	$\frac{1}{2}$	$\frac{1}{3}\left[\frac{3-13z+8z^2}{(1-z)^2}-8\sqrt{z}\,\mathrm{Arth}\,\sqrt{z}\right]$
77	$-\frac{1}{2}$	1	$\frac{5}{2}$	$\frac{1}{2}$	$\frac{3}{2}$	$\frac{1}{3}\left[\frac{3-4z}{1-z}-4\sqrt{z}\,\mathrm{Arth}\,\sqrt{z}\right]$
78	$-\frac{1}{2}$	1	$\frac{5}{2}$	$\frac{1}{2}$	2	$\frac{2}{9z}\left[\frac{1+4z-8z^2}{\sqrt{1-z}}-1\right]$
79	$-\frac{1}{2}$	1	$\frac{5}{2}$	$\frac{1}{2}$	3	$\frac{4}{45z^2}\left[2(3+4z+8z^2)\sqrt{1-z}-6-5z\right]$
80	$-\frac{1}{2}$	1	$\frac{5}{2}$	$\frac{3}{2}$	$\frac{3}{2}$	$\frac{1}{3}\left[2+(1-2z)\frac{\mathrm{Arth}\,\sqrt{z}}{\sqrt{z}}\right]$
81	$-\frac{1}{2}$	1	$\frac{5}{2}$	$\frac{3}{2}$	2	$\frac{2}{9z}\left[1-(1-4z)\sqrt{1-z}\right]$

	a_1	a_2	a_3	b_1	b_2	${}_3F_2(a_1, a_2, a_3; b_1, b_2; z)$
82	$-\frac{1}{2}$	1	$\frac{5}{2}$	$\frac{3}{2}$	3	$\frac{4}{45z^2}[2+5z-2(1+4z)(1-z)^{3/2}]$
83	$-\frac{1}{2}$	1	$\frac{5}{2}$	2	2	$\frac{4}{27\pi z}[3\pi-2(3-4z)\mathbf{K}(\sqrt{z})+2z(7-8z)\mathbf{D}(\sqrt{z})]$
84	$-\frac{1}{2}$	1	$\frac{5}{2}$	2	3	$\frac{8}{135\pi z}[15\pi-8(3-2z)\mathbf{K}(\sqrt{z})-4(3-13z+8z^2)\mathbf{D}(\sqrt{z})]$
85	$-\frac{1}{2}$	1	$\frac{5}{2}$	3	3	$\frac{16}{675\pi z^2}[8(15-11z+4z^2)\mathbf{K}(\sqrt{z})-8z(23-23z+8z^2)\mathbf{D}(\sqrt{z})-60\pi+75\pi z]$
86	$-\frac{1}{2}$	1	3	$\frac{1}{2}$	2	$\frac{1}{4}[(4-5z)(1-z)^{-1}-5\sqrt{z}\,\operatorname{Arth}\sqrt{z}]$
87	$-\frac{1}{2}$	1	3	$\frac{3}{2}$	2	$\frac{1}{8}\left[5+(3-5z)\frac{\operatorname{Arth}\sqrt{z}}{\sqrt{z}}\right]$
88	$-\frac{1}{2}$	1	3	2	2	$\frac{1}{6z}[2-(2-5z)\sqrt{1-z}]$
89	$-\frac{1}{2}$	1	3	2	$\frac{5}{2}$	$\frac{3}{32z}\left[1+5z-(1-6z+5z^2)\frac{\operatorname{Arth}\sqrt{z}}{\sqrt{z}}\right]$
90	$-\frac{1}{2}$	$\frac{3}{2}$	$\frac{3}{2}$	$\frac{1}{2}$	$\frac{1}{2}$	$(1-6z+4z^2)(1-z)^{3/2}$
91	$-\frac{1}{2}$	$\frac{3}{2}$	$\frac{3}{2}$	$\frac{1}{2}$	1	$\frac{2}{\pi}(1-z)^{-1}[(1-2z)\mathbf{K}(\sqrt{z})-z(3-4z)\mathbf{D}(\sqrt{z})]$
92	$-\frac{1}{2}$	$\frac{3}{2}$	$\frac{3}{2}$	$\frac{1}{2}$	2	$\frac{4}{3\pi}[2\mathbf{K}(\sqrt{z})-(1+4z)\mathbf{D}(\sqrt{z})]$
93	$-\frac{1}{2}$	$\frac{3}{2}$	$\frac{3}{2}$	$\frac{1}{2}$	$\frac{5}{2}$	$\frac{3}{4z}\left[\frac{1+z-2z^2}{\sqrt{1-z}}-\frac{\arcsin\sqrt{z}}{\sqrt{z}}\right]$
94	$-\frac{1}{2}$	$\frac{3}{2}$	$\frac{3}{2}$	$\frac{1}{2}$	3	$\frac{16}{15\pi z}[(6-z-4z^2)\mathbf{D}(\sqrt{z})-(3-2z)\mathbf{K}(\sqrt{z})]$
95	$-\frac{1}{2}$	$\frac{3}{2}$	$\frac{3}{2}$	1	1	$\frac{(1-z)^{-1}}{4}[4(1-4z+4z^2)\psi_1(z)+2(5-8z)z\psi_2(z)-z(3-4z)\psi_3(z)]$
96	$-\frac{1}{2}$	$\frac{3}{2}$	$\frac{3}{2}$	1	2	$\frac{1}{3}[2(1-4z)\psi_1(z)+(1+8z)\psi_2(z)-2z\psi_3(z)]$
97	$-\frac{1}{2}$	$\frac{3}{2}$	$\frac{3}{2}$	1	3	$\frac{2}{15z}[8(1+z-2z^2)\psi_1(z)-2(4+z-8z^2)\psi_2(z)+z(1-4z)\psi_3(z)]$
98	$-\frac{1}{2}$	$\frac{3}{2}$	$\frac{3}{2}$	2	2	$\frac{1}{9}[16(1-z)\psi_1(z)-2(5-8z)\psi_2(z)+(3-4z)\psi_3(z)]$
99	$-\frac{1}{2}$	$\frac{3}{2}$	$\frac{3}{2}$	2	3	$\frac{8}{45z}[(6-11z+8z^2)\psi_2(z)-2(3-7z+4z^2)\psi_1(z)+z(3-2z)\psi_3(z)]$
100	$-\frac{1}{2}$	$\frac{3}{2}$	$\frac{3}{2}$	3	3	$\frac{16}{225z}[2(1-21z+8z^2)\psi_2(z)-16(2-3z+z^2)\psi_1(z)+(30+11z-4z^2)\psi_3(z)]$
101	$-\frac{1}{2}$	$\frac{3}{2}$	2	$\frac{1}{2}$	$\frac{1}{2}$	$(1-5z+3z^2)(1-z)^{-2}-3\sqrt{z}\,\operatorname{Arth}\sqrt{z}$
102	$-\frac{1}{2}$	$\frac{3}{2}$	2	$\frac{1}{2}$	1	$\frac{1}{2}(2-9z+6z^2)(1-z)^{-3/2}$
103	$-\frac{1}{2}$	$\frac{3}{2}$	2	$\frac{1}{2}$	$\frac{5}{2}$	$\frac{3}{8z}\left[1+3z-(1+3z^2)\frac{\operatorname{Arth}\sqrt{z}}{\sqrt{z}}\right]$

	a_1	a_2	a_3	b_1	b_2	${}_3F_2(a_1, a_2, a_3; b_1, b_2; z)$
104	$-\frac{1}{2}$	$\frac{3}{2}$	2	$\frac{1}{2}$	3	$\frac{4}{5z^2}\left[(2+z+2z^2)\sqrt{1-z}-2\right]$
105	$-\frac{1}{2}$	$\frac{3}{2}$	2	1	1	$\frac{(1-z)^{-1}}{\pi}\left[(2-3z)\,\mathbf{K}(\sqrt{z})-z(5-6z)\,\mathbf{D}(\sqrt{z})\right]$
106	$-\frac{1}{2}$	$\frac{3}{2}$	2	1	$\frac{5}{2}$	$\frac{3}{16z}\left[(1+6z)\sqrt{1-z}-\frac{\arcsin\sqrt{z}}{\sqrt{z}}\right]$
107	$-\frac{1}{2}$	$\frac{3}{2}$	2	1	3	$\frac{8}{15\pi z}\left[(4+z-6z^2)\,\mathbf{D}(\sqrt{z})-(2-3z)\,\mathbf{K}(\sqrt{z})\right]$
108	$-\frac{1}{2}$	$\frac{3}{2}$	2	$\frac{5}{2}$	$\frac{5}{2}$	$\frac{9}{64z^{3/2}}\left\{2\left[\mathrm{Li}_2(\sqrt{z})-\mathrm{Li}_2(-\sqrt{z})\right]-3\sqrt{z}(1-z)-(1-4z+3z^2)\,\mathrm{Arth}\sqrt{z}\right\}$
109	$-\frac{1}{2}$	$\frac{3}{2}$	2	$\frac{5}{2}$	3	$\frac{1}{10z^2}\left[(16-7z+6z^2)\sqrt{1-z}+15\sqrt{z}\arcsin\sqrt{z}-16\right]$
110	$-\frac{1}{2}$	$\frac{3}{2}$	2	3	3	$\frac{64}{225\pi z^2}\left[(30+23z+3z^2)\,\mathbf{K}(\sqrt{z})-15\pi-z(61-11z+6z^2)\,\mathbf{D}(\sqrt{z})\right]$
111	$-\frac{1}{2}$	$\frac{3}{2}$	$\frac{5}{2}$	$\frac{1}{2}$	$\frac{1}{2}$	$\frac{1}{3}(3-30z+40z^2-16z^3)(1-z)^{-5/2}$
112	$-\frac{1}{2}$	$\frac{3}{2}$	$\frac{5}{2}$	$\frac{1}{2}$	1	$\frac{2}{3\pi}(1-z)^{-2}\left[(3-13z+8z^2)\,\mathbf{K}(\sqrt{z})-2z(5-14z+8z^2)\,\mathbf{D}(\sqrt{z})\right]$
113	$-\frac{1}{2}$	$\frac{3}{2}$	$\frac{5}{2}$	$\frac{1}{2}$	2	$\frac{4}{9\pi}(1-z)^{-1}\left[(5-8z)\,\mathbf{K}(\sqrt{z})-(1+12z-16z^2)\,\mathbf{D}(\sqrt{z})\right]$
114	$-\frac{1}{2}$	$\frac{3}{2}$	$\frac{5}{2}$	$\frac{1}{2}$	3	$\frac{16}{45\pi z}\left[(3+8z)\,\mathbf{K}(\sqrt{z})-2(3+2z+8z^2)\,\mathbf{D}(\sqrt{z})\right]$
115	$-\frac{1}{2}$	$\frac{3}{2}$	$\frac{5}{2}$	1	1	$\frac{(1-z)^{-2}}{12}\left[4(3-17z+31z^2-16z^3)\,\psi_1(z)+2z(15-50z+32z^2)\,\psi_2(z)-z(10-27z+16z^2)\,\psi_3(z)\right]$
116	$-\frac{1}{2}$	$\frac{3}{2}$	$\frac{5}{2}$	1	2	$\frac{(1-z)^{-1}}{18}\left[4(4-17z+16z^2)\,\psi_1(z)+2(1+22z-32z^2)\,\psi_2(z)-z(13-16z)\,\psi_3(z)\right]$
117	$-\frac{1}{2}$	$\frac{3}{2}$	$\frac{5}{2}$	1	3	$\frac{2}{45z}\left[4(2+3z+16z^2)\,\psi_2(z)-4(2-3z+16z^2)\,\psi_1(z)-z(1+16z)\,\psi_3(z)\right]$
118	$-\frac{1}{2}$	$\frac{3}{2}$	$\frac{5}{2}$	2	2	$\frac{1}{27}\left[4(7-16z)\,\psi_1(z)-4(1-16z)\,\psi_2(z)+(3-16z)\,\psi_3(z)\right]$
119	$-\frac{1}{2}$	$\frac{3}{2}$	$\frac{5}{2}$	2	3	$\frac{4}{135z}\left[4(3+13z-16z^2)\,\psi_1(z)-4(3+7z-16z^2)\,\psi_2(z)+z(9-16z)\,\psi_3(z)\right]$
120	$-\frac{1}{2}$	$\frac{3}{2}$	$\frac{5}{2}$	3	3	$\frac{16}{675z}\left[2(29-34z+32z^2)\,\psi_2(z)-4(7-23z+16z^2)\,\psi_1(z)-(30-19z+16z^2)\,\psi_3(z)\right]$
121	$-\frac{1}{2}$	$\frac{3}{2}$	3	$\frac{1}{2}$	$\frac{1}{2}$	$\frac{1}{4}\left[\frac{4-33z+40z^2-15z^3}{(1-z)^3}-15\sqrt{z}\,\mathrm{Arth}\sqrt{z}\right]$
122	$-\frac{1}{2}$	$\frac{3}{2}$	3	$\frac{1}{2}$	1	$\frac{1}{8}(8-56z+75z^2-30z^3)(1-z)^{-5/2}$
123	$-\frac{1}{2}$	$\frac{3}{2}$	3	$\frac{1}{2}$	2	$\frac{1}{4}(4-15z+10z^2)(1-z)^{-3/2}$

	a_1	a_2	a_3	b_1	b_2	${}_3F_2(a_1, a_2, a_3; b_1, b_2; z)$
124	$-\frac{1}{2}$	$\frac{3}{2}$	3	$\frac{1}{2}$	$\frac{5}{2}$	$\frac{3}{32z}\left[\frac{1+10z-15z^2}{1-z}-(1+15z^2)\frac{\operatorname{Arth}\sqrt{z}}{\sqrt{z}}\right]$
125	$-\frac{1}{2}$	$\frac{3}{2}$	3	1	1	$\frac{(1-z)^{-2}}{4\pi}[(8-25z+15z^2)\mathbf{K}(\sqrt{z})-2z(11-27z+15z^2)\mathbf{D}(\sqrt{z})]$
126	$-\frac{1}{2}$	$\frac{3}{2}$	3	1	2	$\frac{(1-z)^{-1}}{2\pi}[(4-5z)\mathbf{K}(\sqrt{z})-z(9-10z)\mathbf{D}(\sqrt{z})]$
127	$-\frac{1}{2}$	$\frac{3}{2}$	3	1	$\frac{5}{2}$	$\frac{3}{64z}\left[\frac{1+21z-30z^2}{\sqrt{1-z}}-\frac{\arcsin\sqrt{z}}{\sqrt{z}}\right]$
128	$-\frac{1}{2}$	$\frac{3}{2}$	3	2	2	$\frac{1}{3\pi}[5\mathbf{K}(\sqrt{z})+2(1-5z)\mathbf{D}(\sqrt{z})]$
129	$-\frac{1}{2}$	$\frac{3}{2}$	3	2	$\frac{5}{2}$	$\frac{3}{32z}\left[\frac{\arcsin\sqrt{z}}{\sqrt{z}}-(1-10z)\sqrt{1-z}\right]$
130	$-\frac{1}{2}$	$\frac{3}{2}$	3	$\frac{5}{2}$	$\frac{5}{2}$	$\frac{9}{256z^{3/2}}\{2[\operatorname{Li}_2(\sqrt{z})-\operatorname{Li}_2(-\sqrt{z})]+ +3(1+4z-5z^2)\operatorname{Arth}\sqrt{z}-7\sqrt{z}+15z^{3/2}\}$
131	$-\frac{1}{2}$	2	2	$\frac{1}{2}$	1	$\frac{1}{4}[(4-15z+9z^2)(1-z)^{-2}-9\sqrt{z}\operatorname{Arth}\sqrt{z}]$
132	$-\frac{1}{2}$	2	2	$\frac{1}{2}$	3	$\frac{2}{5z}\left[1+3z+\frac{1}{z}\ln(1-z)-3z^{3/2}\operatorname{Arth}\sqrt{z}\right]$
133	$-\frac{1}{2}$	2	2	1	1	$\frac{1}{4}(4-14z+9z^2)(1-z)^{-3/2}$
134	$-\frac{1}{2}$	2	2	1	$\frac{3}{2}$	$\frac{1}{8}\left[\frac{7-9z}{1-z}+(1-9z)\frac{\operatorname{Arth}\sqrt{z}}{\sqrt{z}}\right]$
135	$-\frac{1}{2}$	2	2	1	$\frac{5}{2}$	$\frac{3}{32z}\left[1+9z-(1-2z+9z^2)\frac{\operatorname{Arth}\sqrt{z}}{\sqrt{z}}\right]$
136	$-\frac{1}{2}$	2	2	1	3	$\frac{2}{15z^2}[(4+2z+9z^2)\sqrt{1-z}-4]$
137	$-\frac{1}{2}$	2	2	$\frac{3}{2}$	3	$\frac{1}{15z}\left[-2+9z-\frac{2}{z}\ln(1-z)+(5-9z)\sqrt{z}\operatorname{Arth}\sqrt{z}\right]$
138	$-\frac{1}{2}$	2	2	3	3	$\frac{16}{225z^2}\left[31-(31+8z-9z^2)\sqrt{1-z}+30\ln\frac{1+\sqrt{1-z}}{2}\right]$
139	$-\frac{1}{2}$	2	$\frac{5}{2}$	$\frac{1}{2}$	$\frac{1}{2}$	$\frac{1}{3}(3-27z+32z^2-12z^3)(1-z)^{-3}-4\sqrt{z}\operatorname{Arth}\sqrt{z}$
140	$-\frac{1}{2}$	2	$\frac{5}{2}$	$\frac{1}{2}$	1	$\frac{1}{2}(2-15z+20z^2-8z^3)(1-z)^{-5/2}$
141	$-\frac{1}{2}$	2	$\frac{5}{2}$	$\frac{1}{2}$	$\frac{3}{2}$	$\frac{1}{3}\left[\frac{3-10z+6z^2}{(1-z)^2}-6\sqrt{z}\operatorname{Arth}\sqrt{z}\right]$
142	$-\frac{1}{2}$	2	$\frac{5}{2}$	$\frac{1}{2}$	3	$\frac{4}{15z^2}\left[2-\frac{2-z-4z^2+8z^3}{\sqrt{1-z}}\right]$
143	$-\frac{1}{2}$	2	$\frac{5}{2}$	1	1	$\frac{(1-z)^{-2}}{3\pi}[2(3-10z+6z^2)\mathbf{K}(\sqrt{z})-z(17-43z+24z^2)\mathbf{D}(\sqrt{z})]$
144	$-\frac{1}{2}$	2	$\frac{5}{2}$	1	$\frac{3}{2}$	$\frac{1}{6}(6-19z+12z^2)(1-z)^{-3/2}$
145	$-\frac{1}{2}$	2	$\frac{5}{2}$	1	3	$\frac{8}{45\pi z}[(44-9z-26z^2)\mathbf{D}(\sqrt{z})-(22-13z)\mathbf{K}(\sqrt{z})]$

	a_1	a_2	a_3	b_1	b_2	${}_3F_2(a_1, a_2, a_3; b_1, b_2; z)$
146	$-\frac{1}{2}$	2	$\frac{5}{2}$	$\frac{3}{2}$	$\frac{3}{2}$	$\frac{1}{6}\left[(5-6z)(1-z)^{-1}+(1-6z)\frac{\operatorname{Arth}\sqrt{z}}{\sqrt{z}}\right]$
147	$-\frac{1}{2}$	2	$\frac{5}{2}$	$\frac{3}{2}$	3	$\frac{4}{45z^2}\left[(2+z+12z^2)\sqrt{1-z}-2\right]$
148	$-\frac{1}{2}$	2	2	$\frac{5}{2}$	3	$\frac{1}{20z}\left[(15+10z-9z^2)\frac{\operatorname{Arth}\sqrt{z}}{\sqrt{z}}-7+9z+\frac{8}{z}\ln(1-z)\right]$
149	$-\frac{1}{2}$	2	$\frac{5}{2}$	3	3	$\frac{64}{675\pi z^2}\left[15\pi-2(15+4z-6z^2)\mathbf{K}(\sqrt{z})+\right.$ $\left.+z(31+19z-24z^2)\mathbf{D}(\sqrt{z})\right]$
150	$-\frac{1}{2}$	2	3	$\frac{1}{2}$	1	$\frac{1}{16}\left[(16-99z+120z^2-45z^3)(1-z)^{-3}-45\sqrt{z}\operatorname{Arth}\sqrt{z}\right]$
151	$-\frac{1}{2}$	2	3	1	1	$\frac{1}{16}(16-88z+144z^2-45z^3)(1-z)^{-5/2}$
152	$-\frac{1}{2}$	2	3	1	$\frac{3}{2}$	$\frac{1}{32}\left[(29-78z+45z^2)(1-z)^{-2}+3(1-15z)\frac{\operatorname{Arth}\sqrt{z}}{\sqrt{z}}\right]$
153	$-\frac{1}{2}$	2	3	1	$\frac{5}{2}$	$\frac{3}{128z}\left[\frac{1+36z-45z^2}{1-z}-(1-6z+45z^2)\frac{\operatorname{Arth}\sqrt{z}}{\sqrt{z}}\right]$
154	$-\frac{1}{2}$	$\frac{5}{2}$	$\frac{5}{2}$	$\frac{1}{2}$	$\frac{1}{2}$	$\frac{1}{9}(9-144z+280z^2-224z^3+64z^4)(1-z)^{-7/2}$
155	$-\frac{1}{2}$	$\frac{5}{2}$	$\frac{5}{2}$	$\frac{1}{2}$	1	$\frac{2}{9\pi}(1-z)^{-3}\left[(9-69z+84z^2-32z^3)\mathbf{K}(\sqrt{z})-\right.$ $\left.-z(33-153z+176z^2-64z^3)\mathbf{D}(\sqrt{z})\right]$
156	$-\frac{1}{2}$	$\frac{5}{2}$	$\frac{5}{2}$	$\frac{1}{2}$	$\frac{3}{2}$	$\frac{1}{9}(9-60z+80z^2-32z^3)(1-z)^{-5/2}$
157	$-\frac{1}{2}$	$\frac{5}{2}$	$\frac{5}{2}$	$\frac{1}{2}$	2	$\frac{4}{27\pi}(1-z)^{-2}\left[2(7-26z+16z^2)\mathbf{K}(\sqrt{z})-\right.$ $\left.-(1+41z-112z^2+64z^3)\mathbf{D}(\sqrt{z})\right]$
158	$-\frac{1}{2}$	$\frac{5}{2}$	$\frac{5}{2}$	$\frac{1}{2}$	3	$\frac{16(1-z)^{-1}}{135\pi z}\left[(6-7z-48z^2+64z^3)\mathbf{D}(\sqrt{z})-\right.$ $\left.-(3-20z+32z^2)\mathbf{K}(\sqrt{z})\right]$
159	$-\frac{1}{2}$	$\frac{5}{2}$	$\frac{5}{2}$	1	1	$\frac{(1-z)^{-3}}{36}\left[4(9-70z+186z^2-186z^3+64z^4)\psi_1(z)+\right.$ $+2(44-249z+324z^2-128z^3)z\psi_2(z)-$ $\left.-(33-142z+170z^2-64z^3)z\psi_3(z)\right]$
160	$-\frac{1}{2}$	$\frac{5}{2}$	$\frac{5}{2}$	1	2	$\frac{(1-z)^{-2}}{54}\left[4(13-72z+126z^2-64z^3)\psi_1(z)+\right.$ $\left.+2(1+66z-204z^2+128z^3)\psi_2(z)-(43-110z+64z^2)z\psi_3(z)\right]$
161	$-\frac{1}{2}$	$\frac{5}{2}$	$\frac{5}{2}$	1	$\frac{3}{2}$	$\frac{2}{9\pi}(1-z)^{-2}\left[(9-27z+16z^2)\mathbf{K}(\sqrt{z})-\right.$ $\left.-2z(12-29z+16z^2)\mathbf{D}(\sqrt{z})\right]$
162	$-\frac{1}{2}$	$\frac{5}{2}$	$\frac{5}{2}$	1	3	$\frac{2(1-z)^{-1}}{135z}\left[4(2+15z-66z^2+64z^3)\psi_1(z)-\right.$ $\left.-2(4-3z-84z^2+128z^3)\psi_2(z)+z(1-50z+64z^2)\psi_3(z)\right]$
163	$-\frac{1}{2}$	$\frac{5}{2}$	$\frac{5}{2}$	$\frac{3}{2}$	$\frac{3}{2}$	$\frac{1}{9}(9-26z+16z^2)(1-z)^{-3/2}$
164	$-\frac{1}{2}$	$\frac{5}{2}$	$\frac{5}{2}$	$\frac{3}{2}$	2	$\frac{4}{27\pi}(1-z)^{-1}\left[(13-16z)\mathbf{K}(\sqrt{z})+(1-30z+32z^2)\mathbf{D}(\sqrt{z})\right]$

	a_1	a_2	a_3	b_1	b_2	${}_3F_2(a_1, a_2, a_3; b_1, b_2; z)$
165	$-\frac{1}{2}$	$\frac{5}{2}$	$\frac{5}{2}$	$\frac{3}{2}$	3	$\frac{16}{135\pi z}\left[(1+16z)\,\mathrm{K}(\sqrt{z})-2(1-z+16z^2)\,\mathbf{D}(\sqrt{z})\right]$
166	$-\frac{1}{2}$	$\frac{5}{2}$	$\frac{5}{2}$	2	2	$\frac{(1-z)^{-1}}{81}[4(19-74z+64z^2)\,\psi_1(z)+$ $+2(1+100z-128z^2)\,\psi_2(z)+(3-58z+64z^2)\,\psi_3(z)]$
167	$-\frac{1}{2}$	$\frac{5}{2}$	$\frac{5}{2}$	2	3	$\frac{8}{405z}[2(3+2z+64z^2)\,\psi_2(z)-2(3-22z+64z^2)\,\psi_1(z)+$ $+z(3-32z)\,\psi_3(z)]$
168	$-\frac{1}{2}$	$\frac{5}{2}$	$\frac{5}{2}$	3	3	$\frac{32}{2025z}[4(11+21z-32z^2)\,\psi_1(z)-(59+36z-128z^2)\,\psi_2(z)+$ $+(15+13z-32z^2)\,\psi_3(z)]$
169	$-\frac{1}{2}$	$\frac{5}{2}$	3	$\frac{1}{2}$	$\frac{1}{2}$	$\frac{1}{3}(3-42z+73z^2-55z^3+15z^4)(1-z)^{-4}-5\sqrt{z}\,\mathrm{Arth}\sqrt{z}$
170	$-\frac{1}{2}$	$\frac{5}{2}$	3	$\frac{1}{2}$	1	$\frac{1}{8}(8-88z+175z^2-140z^3+40z^4)(1-z)^{-7/2}$
171	$-\frac{1}{2}$	$\frac{5}{2}$	3	$\frac{1}{2}$	$\frac{3}{2}$	$\frac{1}{6}\left[\frac{6-33z+40z^2-15z^3}{(1-z)^3}-15\sqrt{z}\,\mathrm{Arth}\sqrt{z}\right]$
172	$-\frac{1}{2}$	$\frac{5}{5}$	3	$\frac{1}{2}$	2	$\frac{1}{12}(12-75z+100z^2-40z^3)(1-z)^{-5/2}$
173	$-\frac{1}{2}$	$\frac{5}{2}$	3	1	1	$\frac{(1-z)^{-3}}{12\pi}\left[(24-131z+159z^2-60z^3)\,\mathrm{K}(\sqrt{z})-\right.$ $\left.-z(74-295z+333z^2-120z^3)\,\mathbf{D}(\sqrt{z})\right]$
174	$-\frac{1}{2}$	$\frac{5}{2}$	3	1	$\frac{3}{2}$	$\frac{1}{8}(8-40z+51z^2-20z^3)(1-z)^{-5/2}$
175	$-\frac{1}{2}$	$\frac{5}{2}$	3	1	2	$\frac{(1-z)^{-2}}{6\pi}\left[2(6-17z+10z^2)\mathrm{K}(\sqrt{z})-z(31-73z+40z^2)\mathbf{D}(\sqrt{z})\right]$
176	$-\frac{1}{2}$	$\frac{5}{2}$	3	$\frac{3}{2}$	$\frac{3}{2}$	$\frac{1}{24}\left[\frac{21-53z+30z^2}{(1-z)^2}+3(1-10z)\,\frac{\mathrm{Arth}\sqrt{z}}{\sqrt{z}}\right]$
177	$-\frac{1}{2}$	$\frac{5}{2}$	3	$\frac{3}{2}$	2	$\frac{1}{12}(12-33z+20z^2)(1-z)^{-3/2}$
178	$-\frac{1}{2}$	$\frac{5}{2}$	3	2	2	$\frac{(1-z)^{-1}}{9\pi}\left[(17-20z)\,\mathrm{K}(\sqrt{z})+(2-39z+40z^2)\,\mathbf{D}(\sqrt{z})\right]$
179	$-\frac{1}{2}$	3	3	$\frac{1}{2}$	1	$\frac{1}{64}\left[(64-607z+1095z^2-825z^3+225z^4)(1-z)^{-4}-\right.$ $\left.-225\sqrt{z}\,\mathrm{Arth}\sqrt{z}\right]$
180	$-\frac{1}{2}$	3	3	$\frac{1}{2}$	2	$\frac{1}{32}\left[(32-165z+200z^2-75z^3)(1-z)^{-3}-75\sqrt{z}\,\mathrm{Arth}\sqrt{z}\right]$
181	$-\frac{1}{2}$	3	3	1	1	$\frac{1}{64}(64-512z+1000z^2-792z^3+225z^4)(1-z)^{-7/2}$
182	$-\frac{1}{2}$	3	3	1	$\frac{3}{2}$	$\frac{1}{128}\left[(119-519z+609z^2-225z^3)(1-z)^{-3}+\right.$ $\left.+9(1-25z)\,\frac{\mathrm{Arth}\sqrt{z}}{\sqrt{z}}\right]$
183	$-\frac{1}{2}$	3	3	1	2	$\frac{1}{32}(32-152z+192z^2-75z^3)(1-z)^{-5/2}$
184	$-\frac{1}{2}$	3	3	1	$\frac{5}{2}$	$\frac{3}{512z}\left[(1+151z-393z^2+225z^3)(1-z)^{-2}-\right.$ $\left.-(1-18z+225z^2)\,\frac{\mathrm{Arth}\sqrt{z}}{\sqrt{z}}\right]$

	a_1	a_2	a_3	b_1	b_2	${}_3F_2(a_1, a_2, a_3; b_1, b_2; z)$
185	$-\frac{1}{2}$	3	3	$\frac{3}{2}$	2	$\frac{1}{64}\left[(55-134z+75z^2)(1-z)^{-2}+3(3-25z)\frac{\operatorname{Arth}\sqrt{z}}{\sqrt{z}}\right]$
186	$-\frac{1}{2}$	3	3	2	2	$\frac{1}{16}(16-42z+25z^2)(1-z)^{-3/2}$
187	$-\frac{1}{2}$	3	3	2	$\frac{5}{2}$	$\frac{3}{256z}\left[(1+18z-75z^2)\frac{\operatorname{Arth}\sqrt{z}}{\sqrt{z}}-\frac{1-68z+75z^2}{1-z}\right]$
188	$\frac{1}{8}$	$\frac{3}{8}$	1	$\frac{9}{8}$	$\frac{11}{8}$	$\frac{3}{16x^3}\left\{(1+x^2)\left[2\operatorname{arctg}x+2^{-1/2}\ln\frac{1+2^{1/2}x+x^2}{1-2^{1/2}x+x^2}\right]-(1-x^2)\left[\ln\frac{1+x}{1-x}+\sqrt{2}\operatorname{arctg}\frac{2^{1/2}x}{1-x^2}\right]\right\}$ $[x=z^{1/8}]$
189	$\frac{1}{8}$	$\frac{5}{8}$	1	$\frac{9}{8}$	$\frac{13}{8}$	$\frac{5}{32x^5}\left\{2^{-1/2}(1+x^4)\left[\ln\frac{1+2^{1/2}x+x^2}{1-2^{1/2}x+x^2}+2\operatorname{arctg}\frac{2^{1/2}x}{1-x^2}\right]-(1-x^4)\left[\ln\frac{1+x}{1-x}+2\operatorname{arctg}x\right]\right\}$ $[x=z^{1/8}]$
190	$\frac{1}{8}$	$\frac{7}{8}$	1	$\frac{9}{8}$	$\frac{15}{8}$	$\frac{7}{48x^7}\left\{(1+x)^6\left[2\operatorname{arctg}x+\sqrt{2}\operatorname{arctg}\frac{2^{1/2}x}{1-x^2}\right]-(1-x^6)\left[\ln\frac{1+x}{1-x}+2^{-1/2}\ln\frac{1+2^{1/2}x+x^2}{1-2^{1/2}x+x^2}\right]\right\}$ $[x=z^{1/8}]$
191	$\frac{1}{6}$	$\frac{1}{3}$	1	$\frac{7}{6}$	$\frac{4}{3}$	$\frac{1}{x}\left\{\frac{1}{3}\ln(1-x)-\frac{1}{6}\ln(1+x+x^2)-3^{-1/2}\operatorname{arctg}\frac{3^{1/2}x}{2+x}+\frac{\sqrt{x}}{3}\left[\ln\frac{1+\sqrt{x}}{1-\sqrt{x}}-\frac{1}{2}\ln\frac{1-\sqrt{x}+x}{1+\sqrt{x}+x}+3^{1/2}\operatorname{arctg}\frac{\sqrt{3x}}{1-x}\right]\right\}$ $[x=z^{1/3}]$
192	$\frac{1}{6}$	$\frac{2}{3}$	1	$\frac{7}{6}$	$\frac{5}{3}$	$\frac{2}{x^2}\left\{3^{-3/2}\operatorname{arctg}\frac{3^{1/2}x}{x+2}+\left(\frac{x}{3}\right)^{3/2}\operatorname{arctg}\frac{\sqrt{3x}}{1-x}+\frac{1}{18}\left[2(1+x^{3/2})\ln(1+\sqrt{x})+2(1-x^{3/2})\ln(1-\sqrt{x})-\ln(1+x+x^2)-x^{3/2}\ln\frac{1-\sqrt{x}+x}{1+\sqrt{x}+x}\right]\right\}$ $[x=z^{1/3}]$
193	$\frac{1}{6}$	$\frac{2}{3}$	1	$\frac{5}{3}$	$\frac{13}{6}$	$\frac{7}{54x^{7/2}}\left\{(1-x^{3/2})^2\ln[(1+\sqrt{x}+x)(1-\sqrt{x})^{-2}]+(1+x^{3/2})^2\ln\frac{1+2\sqrt{x}+x}{1-\sqrt{x}+x}\right\}+\frac{7\cdot 3^{1/2}}{27}x^{-7/2}(1+x^{3/2})^2\times\operatorname{arctg}\frac{\sqrt{3x}}{1-x}-\frac{28\sqrt{3}}{27}x^{-2}\operatorname{arctg}\frac{\sqrt{3x}}{2+\sqrt{x}}-\frac{14}{9x^3}$ $[x=z^{1/3}]$
194	$\frac{1}{4}$	$\frac{3}{4}$	1	$\frac{5}{4}$	$\frac{7}{4}$	$\frac{3}{8x^3}\left[(1-x^2)\ln\frac{1-x}{1+x}+2(1+x^2)\operatorname{arctg}x\right]$ $[x=z^{1/4}]$
195	$\frac{1}{4}$	$\frac{3}{4}$	1	$\frac{7}{4}$	$\frac{9}{4}$	$\frac{15}{32x^5}\left[(1-x^2)^2\ln\frac{1+x}{1-x}+2(1+x^2)^2\operatorname{arctg}x-4x\right]$ $[x=z^{1/4}]$
196	$\frac{1}{4}$	1	1	$\frac{5}{4}$	2	$\frac{1}{3z}\left[\ln(1-z)+z^{3/4}\ln\frac{1+z^{1/4}}{1-z^{1/4}}+2\operatorname{arctg}z\right]$
197	$\frac{1}{3}$	$\frac{2}{3}$	1	$\frac{4}{3}$	$\frac{5}{3}$	$\frac{x-1}{3x^2}[\ln(1-x^3)-3\ln(1-x)]+\frac{2(x+1)}{x^2\sqrt{3}}\operatorname{arctg}\frac{3^{1/2}x}{2+x}$ $[x=z^{1/3}]$

	a_2 a_2 a_3	b_1 b_2	${}_3F_2\,(a_1,\ a_2,\ a_3;\ b_1,\ b_2;\ z)$
198	$\frac{1}{3}$ $\frac{5}{6}$ 1	$\frac{4}{3}$ $\frac{11}{6}$	$\frac{5}{9x}\left[\frac{1}{2}\ln(1+x+x^2)-\ln(1-x)+3^{1/2}\operatorname{arctg}\frac{3^{1/2}x}{2+x}\right]-$ $-\frac{5}{9x^{5/2}}\left[\ln\frac{1+\sqrt{x}}{1-\sqrt{x}}+\frac{1}{2}\ln\frac{1+\sqrt{x}+x}{1-\sqrt{x}+x}-3^{1/2}\operatorname{arctg}\frac{\sqrt{3x}}{1-x}\right]$ $[x=z^{1/3}]$
199	$\frac{1}{3}$ 1 1	$\frac{4}{3}$ 2	$\frac{1}{4z}\left[2\cdot 3^{1/2}z^{2/3}\operatorname{arctg}\frac{3^{1/2}z^{1/3}}{2+z^{1/3}}+(2+z^{2/3})\ln(1-z)-\right.$ $\left.-3z^{2/3}\ln\left(1-z^{1/3}\right)\right]$
200	$\frac{3}{8}$ $\frac{5}{8}$ 1	$\frac{11}{8}$ $\frac{13}{8}$	$\frac{15}{16x^5}\left\{2(1+x^2)\left[2^{-1/2}\operatorname{arctg}\frac{2^{1/2}x}{1-x^2}-\operatorname{arctg}x\right]-\right.$ $\left.-(1-x^2)\left[\ln\frac{1+x}{1-x}-2^{-1/2}\ln\frac{1+2^{1/2}x+x^2}{1-2^{1/2}x+x^2}\right]\right\}$ $[x=z^{1/8}]$
201	$\frac{3}{8}$ $\frac{7}{8}$ 1	$\frac{11}{8}$ $\frac{15}{8}$	$\frac{21}{32x^7}\left\{2^{-1/2}(1+x^4)\left[2\operatorname{arctg}\frac{2^{1/2}x}{1-x^2}-\ln\frac{1+2^{1/2}x+x^2}{1-2^{1/2}x+x^2}\right]-\right.$ $\left.-(1-x^4)\left[\ln\frac{1+x}{1-x}-2\operatorname{arctg}x\right]\right\}$ $[x=z^{1/8}]$
202	$\frac{1}{2}$ $\frac{1}{2}$ $\frac{1}{2}$	1 1	$\psi_1(z)$
203	$\frac{1}{2}$ $\frac{1}{2}$ $\frac{1}{2}$	1 2	$\psi_2(z)$
204	$\frac{1}{2}$ $\frac{1}{2}$ $\frac{1}{2}$	1 3	$\frac{2}{27z}[8(1-z)\psi_1(z)-4(2-5z)\psi_2(z)+z\psi_3(z)]$
205	$\frac{1}{2}$ $\frac{1}{2}$ $\frac{1}{2}$	2 2	$\psi_3(z)$
206	$\frac{1}{2}$ $\frac{1}{2}$ $\frac{1}{2}$	2 3	$\frac{4}{27z}[2(2-5z)\psi_2(z)-4(1-z)\psi_1(z)+13z\psi_3(z)]$
207	$\frac{1}{2}$ $\frac{1}{2}$ $\frac{1}{2}$	3 3	$\frac{32}{27z}[(1-4z)\psi_2(z)-2(1-z)\psi_1(z)+(1+3z)\psi_3(z)]$
208	$\frac{1}{2}$ $\frac{1}{2}$ 1	$\frac{3}{2}$ $\frac{3}{2}$	$\frac{1}{2\sqrt{z}}[\operatorname{Li}_2(\sqrt{z})-\operatorname{Li}_2(-\sqrt{z})]$
209	$\frac{1}{2}$ $\frac{1}{2}$ 1	$\frac{3}{2}$ 2	$\frac{2}{z}[\sqrt{1-z}+\sqrt{z}\arcsin\sqrt{z}-1]$
210	$\frac{1}{2}$ $\frac{1}{2}$ 1	$\frac{3}{2}$ $\frac{5}{2}$	$\frac{3}{4z}\left\{(1-z)\frac{\operatorname{Arth}\sqrt{z}}{\sqrt{z}}-1+\sqrt{z}[\operatorname{Li}_2(\sqrt{z})-\operatorname{Li}_2(-\sqrt{z})]\right\}$
211	$\frac{1}{2}$ $\frac{1}{2}$ 1	$\frac{3}{2}$ 3	$\frac{4}{9z^2}[2-9z+6z^{3/2}\arcsin\sqrt{z}-2(1-4z)\sqrt{1-z}]$
212	$\frac{1}{2}$ $\frac{1}{2}$ 1	2 2	$\frac{4}{\pi z}[2(1+z)\mathbf{K}(\sqrt{z})-4z\mathbf{D}(\sqrt{z})-\pi]=$
213			$=\frac{4}{z}\left[\frac{2}{\pi}(1+\sqrt{z})\mathbf{E}\left(\frac{2\sqrt[4]{z}}{1+\sqrt{z}}\right)-1\right]$
214	$\frac{1}{2}$ $\frac{1}{2}$ 1	2 $\frac{5}{2}$	$\frac{3}{2z}\left[3\sqrt{1-z}+(1+2z)\frac{\arcsin\sqrt{z}}{\sqrt{z}}-4\right]$
215	$\frac{1}{2}$ $\frac{1}{2}$ 1	2 3	$\frac{8}{9\pi z}[4(5+3z)\mathbf{K}(\sqrt{z})-4(1+7z)\mathbf{D}(\sqrt{z})-9\pi]$

	a_1	a_2	a_3	b_1	b_2	${}_3F_2(a_1, a_2, a_3; b_1, b_2; z)$
216	$\frac{1}{2}$	$\frac{1}{2}$	1	$\frac{5}{2}$	$\frac{5}{2}$	$\frac{9}{4z}\left\{(1-z)\frac{\operatorname{Arth}\sqrt{z}}{\sqrt{z}}-2+\frac{1+z}{2\sqrt{z}}[\operatorname{Li}_2(\sqrt{z})-\operatorname{Li}_2(-\sqrt{z})]\right\}$
217	$\frac{1}{2}$	$\frac{1}{2}$	1	$\frac{5}{2}$	3	$\frac{2}{3z^2}\left[(4+11z)\sqrt{1-z}+3(3+2z)\sqrt{z}\arcsin\sqrt{z}-4-18z\right]$
218	$\frac{1}{2}$	$\frac{1}{2}$	1	3	3	$\frac{16}{27\pi z^2}\left[8(3+10z+3z^2)\mathbf{K}(\sqrt{z})-64z(1+z)\mathbf{D}(\sqrt{z})-12\pi-27\pi z\right]$
219	$\frac{1}{2}$	$\frac{1}{2}$	$\frac{3}{2}$	1	1	$\frac{(1-z)^{-1}}{4}[4(1-2z)\psi_1(z)+6z\psi_2(z)-z\psi_3(z)]$
220	$\frac{1}{2}$	$\frac{1}{2}$	$\frac{3}{2}$	1	2	$2\psi_1(z)-\psi_2(z)$
221	$\frac{1}{2}$	$\frac{1}{2}$	$\frac{3}{2}$	1	3	$\frac{2}{9z}[2(4-z)\psi_2(z)-8(1-z)\psi_1(z)-z\psi_3(z)]$
222	$\frac{1}{2}$	$\frac{1}{2}$	$\frac{3}{2}$	2	2	$2\psi_2(z)-\psi_3(z)$
223	$\frac{1}{2}$	$\frac{1}{2}$	$\frac{3}{2}$	2	3	$\frac{8}{9z}[2(1-z)\psi_1(z)-(2-5z)\psi_2(z)-2z\psi_3(z)]$
224	$\frac{1}{2}$	$\frac{1}{2}$	$\frac{3}{2}$	3	3	$\frac{16}{27z}[8(1-z)\psi_1(z)-2(1-7z)\psi_2(z)-(6+5z)\psi_3(z)]$
225	$\frac{1}{2}$	$\frac{1}{2}$	2	1	1	$\frac{(1-z)^{-1}}{\pi}\left[(2-z)\mathbf{K}(\sqrt{z})-z\mathbf{D}(\sqrt{z})\right]$
226	$\frac{1}{2}$	$\frac{1}{2}$	2	1	$\frac{3}{2}$	$\frac{1}{2}\left(\frac{\arcsin\sqrt{z}}{\sqrt{z}}+\frac{1}{\sqrt{1-z}}\right)$
227	$\frac{1}{2}$	$\frac{1}{2}$	2	1	$\frac{5}{2}$	$\frac{3}{8z}\left[(1+2z)\frac{\arcsin\sqrt{z}}{\sqrt{z}}-\sqrt{1-z}\right]$
228	$\frac{1}{2}$	$\frac{1}{2}$	2	1	3	$\frac{8}{9\pi z}\left[(2+3z)\mathbf{K}(\sqrt{z})-(4+z)\mathbf{D}(\sqrt{z})\right]$
229	$\frac{1}{2}$	$\frac{1}{2}$	2	$\frac{3}{2}$	$\frac{3}{2}$	$\frac{1}{4\sqrt{z}}\left[\operatorname{Li}_2(\sqrt{z})-\operatorname{Li}_2(-\sqrt{z})+2\operatorname{Arth}\sqrt{z}\right]$
230	$\frac{1}{2}$	$\frac{1}{2}$	2	$\frac{3}{2}$	$\frac{5}{2}$	$\frac{3}{8z}\left\{1-(1-z)\frac{\operatorname{Arth}\sqrt{z}}{\sqrt{z}}+\sqrt{z}[\operatorname{Li}_2(\sqrt{z})-\operatorname{Li}_2(-\sqrt{z})]\right\}$
231	$\frac{1}{2}$	$\frac{1}{2}$	2	$\frac{3}{2}$	3	$\frac{4}{9z^2}\left[(2+z)\sqrt{1-z}-2+3z^{3/2}\arcsin\sqrt{z}\right]$
232	$\frac{1}{2}$	$\frac{1}{2}$	2	$\frac{5}{2}$	$\frac{5}{2}$	$\frac{9}{16z^{3/2}}\left\{2\sqrt{z}-(1-z)\left[\operatorname{Li}_2(\sqrt{z})-\operatorname{Li}_2(-\sqrt{z})\right]\right\}$
233	$\frac{1}{2}$	$\frac{1}{2}$	2	$\frac{5}{2}$	3	$\frac{1}{3z^2}\left[8-(8-5z)\sqrt{1-z}-3(3-2z)\sqrt{z}\arcsin\sqrt{z}\right]$
234	$\frac{1}{2}$	$\frac{1}{2}$	2	3	3	$\frac{64}{27\pi z^2}\left[3\pi-(6+5z-3z^2)\mathbf{K}(\sqrt{z})+z(13-5z)\mathbf{D}(\sqrt{z})\right]$
235	$\frac{1}{2}$	$\frac{1}{2}$	$\frac{5}{2}$	1	1	$\frac{(1-z)^{-2}}{12}[4(3-9z+5z^2)\psi_1(z)+2z(11-8z)\psi_2(z)-$ $-z(4-3z)\psi_3(z)]$
236	$\frac{1}{2}$	$\frac{1}{2}$	$\frac{5}{2}$	1	$\frac{3}{2}$	$\frac{2}{3\pi}(1-z)^{-1}\left[(3-2z)\mathbf{K}(\sqrt{z})-z\mathbf{D}(\sqrt{z})\right]$
237	$\frac{1}{2}$	$\frac{1}{2}$	$\frac{5}{2}$	1	2	$\frac{(1-z)^{-1}}{6}[4(2-3z)\psi_1(z)-2(1-4z)\psi_2(z)-z\psi_3(z)]$
238	$\frac{1}{2}$	$\frac{1}{2}$	$\frac{5}{2}$	1	3	$\frac{2}{27z}[4(2+7z)\psi_1(z)-8(1+2z)\psi_2(z)+z\psi_3(z)]$

	a_1	a_2	a_3	b_1	b_2	${}_3F_2(a_1, a_2, a_3; b_1, b_2; z)$
239	$\frac{1}{2}$	$\frac{1}{2}$	$\frac{5}{2}$	$\frac{3}{2}$	$\frac{3}{2}$	$\frac{1}{3}\left[2\frac{\arcsin\sqrt{z}}{\sqrt{z}}+\frac{1}{\sqrt{1-z}}\right]$
240	$\frac{1}{2}$	$\frac{1}{2}$	$\frac{5}{2}$	$\frac{3}{2}$	2	$\frac{4}{3\pi}[2\mathbf{K}(\sqrt{z})-\mathbf{D}(\sqrt{z})]$
241	$\frac{1}{2}$	$\frac{1}{2}$	$\frac{5}{2}$	$\frac{3}{2}$	3	$\frac{16}{27\pi z}[(1+6z)\mathbf{K}(\sqrt{z})-(2+5z)\mathbf{D}(\sqrt{z})]$
242	$\frac{1}{2}$	$\frac{1}{2}$	$\frac{5}{2}$	2	2	$\frac{1}{3}[4\psi_1(z)-\psi_3(z)]$
243	$\frac{1}{2}$	$\frac{1}{2}$	$\frac{5}{2}$	2	3	$\frac{4}{27z}[4(1+2z)\psi_2(z)-4(1-z)\psi_1(z)-5z\psi_3(z)]$
244	$\frac{1}{2}$	$\frac{1}{2}$	$\frac{5}{2}$	3	3	$\frac{16}{81z}[4(1-z)\psi_1(z)-2(5-8z)\psi_2(z)+(6-7z)\psi_3(z)]$
245	$\frac{1}{2}$	$\frac{1}{2}$	3	1	1	$\frac{(1-z)^{-2}}{4\pi}[(8-9z+3z^2)\mathbf{K}(\sqrt{z})-2z(3-2z)\mathbf{D}(\sqrt{z})]$
246	$\frac{1}{2}$	$\frac{1}{2}$	3	1	$\frac{3}{2}$	$\frac{1}{8}\left[3\frac{\arcsin\sqrt{z}}{\sqrt{z}}+\frac{5-4z}{(1-z)^{3/2}}\right]$
247	$\frac{1}{2}$	$\frac{1}{2}$	3	1	2	$\frac{(1-z)^{-1}}{2\pi}[(4-3z)\mathbf{K}(\sqrt{z})-z\mathbf{D}(\sqrt{z})]$
248	$\frac{1}{2}$	$\frac{1}{2}$	3	1	$\frac{5}{2}$	$\frac{3}{32z}\left[(1+6z)\frac{\arcsin\sqrt{z}}{\sqrt{z}}-\frac{1-5z}{\sqrt{1-z}}\right]$
249	$\frac{1}{2}$	$\frac{1}{2}$	3	$\frac{3}{2}$	$\frac{3}{2}$	$\frac{1}{8}\left\{\frac{1}{1-z}+4\frac{\operatorname{Arth}\sqrt{z}}{\sqrt{z}}+\frac{3}{2\sqrt{z}}[\operatorname{Li}_2(\sqrt{z})-\operatorname{Li}_2(-\sqrt{z})]\right\}$
250	$\frac{1}{2}$	$\frac{1}{2}$	3	$\frac{3}{2}$	2	$\frac{1}{4}\left[3\frac{\arcsin\sqrt{z}}{\sqrt{z}}+\frac{1}{\sqrt{1-z}}\right]$
251	$\frac{1}{2}$	$\frac{1}{2}$	3	$\frac{3}{2}$	$\frac{5}{2}$	$\frac{3}{32z^{3/2}}\{\sqrt{z}-(1-5z)\operatorname{Arth}\sqrt{z}+3z[\operatorname{Li}_2(\sqrt{z})-\operatorname{Li}_2(-\sqrt{z})]\}$
252	$\frac{1}{2}$	$\frac{1}{2}$	3	2	2	$\frac{1}{\pi}[3\mathbf{K}(\sqrt{z})-2\mathbf{D}(\sqrt{z})]$
253	$\frac{1}{2}$	$\frac{1}{2}$	3	2	$\frac{5}{2}$	$\frac{3}{16z}\left[\sqrt{1-z}-(1-6z)\frac{\arcsin\sqrt{z}}{\sqrt{z}}\right]$
254	$\frac{1}{2}$	$\frac{1}{2}$	3	$\frac{5}{2}$	$\frac{5}{2}$	$\frac{9}{64z^{3/2}}\{4\sqrt{z}-2(1-z)\operatorname{Arth}\sqrt{z}-(1-3z)[\operatorname{Li}_2(\sqrt{z})-$ $-\operatorname{Li}_2(-\sqrt{z})]\}$
255	$\frac{1}{2}$	1	1	$\frac{1}{4}$	$\frac{3}{4}$	$\frac{1}{1-z}+\frac{z^{1/4}}{2}[(1-\sqrt{z})^{-3/2}\arcsin\sqrt[4]{z}-$ $-(1+\sqrt{z})^{-3/2}\ln(z^{1/4}+\sqrt{z^{1/2}+1})]$
256	$\frac{1}{2}$	1	1	$\frac{3}{2}$	2	$2\frac{\operatorname{Arth}\sqrt{z}}{\sqrt{z}}+\frac{1}{z}\ln(1-z)$
257	$\frac{1}{2}$	1	1	$\frac{3}{2}$	3	$\frac{2}{3z}\left[4\sqrt{z}\operatorname{Arth}\sqrt{z}-1-\frac{1}{z}(1-3z)\ln(1-z)\right]$
258	$\frac{1}{2}$	1	1	2	2	$\frac{4}{z}\left[1-\sqrt{1-z}+\ln\frac{1+\sqrt{1-z}}{2}\right]$
259	$\frac{1}{2}$	1	1	2	$\frac{5}{2}$	$\frac{3}{z}\left[(1+z)\frac{\operatorname{Arth}\sqrt{z}}{\sqrt{z}}-1+\ln(1-z)\right]$

	a_1 a_2 a_3	b_1 b_2	${}_3F_2(a_1, a_2, a_3; b_1, b_2; z)$
260	$\frac{1}{2}$ 1 1	2 3	$\frac{4}{3z^2}\left[2+3z-2(1+2z)\sqrt{1-z}+6z\ln\frac{1+\sqrt{1-z}}{2}\right]$
261	$\frac{1}{2}$ 1 1	$\frac{5}{2}$ 3	$\frac{2}{z}\left[\frac{1+3z}{z}\ln(1-z)+2(3+z)\frac{\operatorname{Arth}\sqrt{z}}{\sqrt{z}}-5\right]$
262	$\frac{1}{2}$ 1 1	3 3	$\frac{16}{9z^2}\left[11-(11+4z)\sqrt{1-z}+3(2+3z)\ln\frac{1+\sqrt{1-z}}{2}\right]$
263	$\frac{1}{2}$ 1 $\frac{3}{2}$	2 2	$\frac{8}{\pi z}\left[\frac{\pi}{2}-\mathbf{E}(\sqrt{z})\right]$
264	$\frac{1}{2}$ 1 $\frac{3}{2}$	2 $\frac{5}{2}$	$\frac{3}{z}\left[2-\frac{\arcsin\sqrt{z}}{\sqrt{z}}-\sqrt{1-z}\right]$
265	$\frac{1}{2}$ 1 $\frac{3}{2}$	2 3	$\frac{8}{3\pi z}\left[3\pi-8\mathbf{K}(\sqrt{z})+4(1+z)\mathbf{D}(\sqrt{z})\right]$
266	$\frac{1}{2}$ 1 $\frac{3}{2}$	$\frac{5}{2}$ $\frac{5}{2}$	$\frac{9}{4z^{3/2}}\left[3\sqrt{z}-(1-z)\operatorname{Arth}\sqrt{z}-\operatorname{Li}_2(\sqrt{z})+\operatorname{Li}_2(-\sqrt{z})\right]$
267	$\frac{1}{2}$ 1 $\frac{3}{2}$	$\frac{5}{2}$ 3	$\frac{4}{z^2}\left[2+3z-(2+z)\sqrt{1-z}-3\sqrt{z}\arcsin\sqrt{z}\right]$
268	$\frac{1}{2}$ 1 $\frac{3}{2}$	3 3	$\frac{16}{9\pi z^2}\left[12\pi+9\pi z-8(3+5z)\mathbf{K}(\sqrt{z})+8z(7+z)\mathbf{D}(\sqrt{z})\right]$
269	$\frac{1}{2}$ 1 2	$\frac{3}{2}$ 3	$\frac{2}{3z}\left[1+\frac{1}{z}\ln(1-z)+2\sqrt{z}\operatorname{Arth}\sqrt{z}\right]$
270	$\frac{1}{2}$ 1 2	$\frac{5}{2}$ 3	$\frac{2}{z}\left[2-\frac{1}{z}\ln(1-z)-(3-z)\frac{\operatorname{Arth}\sqrt{z}}{\sqrt{z}}\right]$
271	$\frac{1}{2}$ 1 2	3 3	$\frac{8}{9z^2}\left[4(4-z)\sqrt{1-z}-16+9z-12\ln\frac{1+\sqrt{1-z}}{2}\right]$
272	$\frac{1}{2}$ 1 $\frac{5}{2}$	$\frac{3}{2}$ $\frac{3}{2}$	$\frac{1}{3}\left[\frac{1}{1-z}+2\frac{\operatorname{Arth}\sqrt{z}}{\sqrt{z}}\right]$
273	$\frac{1}{2}$ 1 $\frac{5}{2}$	$\frac{3}{2}$ 2	$\frac{2}{3z}\left[1-\frac{1-2z}{\sqrt{1-z}}\right]$
274	$\frac{1}{2}$ 1 $\frac{5}{2}$	$\frac{3}{2}$ 3	$\frac{4}{9z^2}\left[2+3z-2(1+2z)\sqrt{1-z}\right]$
275	$\frac{1}{2}$ 1 $\frac{5}{2}$	2 2	$\frac{4}{3\pi z}\left[\pi-2\mathbf{K}(\sqrt{z})+4z\mathbf{D}(\sqrt{z})\right]$
276	$\frac{1}{2}$ 1 $\frac{5}{2}$	2 3	$\frac{8}{9\pi z}\left[3\pi-4\mathbf{K}(\sqrt{z})-4(1-2z)\mathbf{D}(\sqrt{z})\right]$
277	$\frac{1}{2}$ 1 $\frac{5}{2}$	3 3	$\frac{16}{27\pi z^2}\left[8(3-z)\mathbf{K}(\sqrt{z})-16z(2-z)\mathbf{D}(\sqrt{z})-12\pi+9\pi z\right]$
278	$\frac{1}{2}$ 1 3	$\frac{3}{2}$ 2	$\frac{1}{4}\left[\frac{1}{1-z}+3\frac{\operatorname{Arth}\sqrt{z}}{\sqrt{z}}\right]$
279	$\frac{1}{2}$ 1 3	2 2	$\frac{1}{2z}\left[2-\frac{2-3z}{\sqrt{1-z}}\right]$
280	$\frac{1}{2}$ 1 3	2 $\frac{5}{2}$	$\frac{3}{8z}\left[1-(1-3z)\frac{\operatorname{Arth}\sqrt{z}}{\sqrt{z}}\right]$
281	$\frac{1}{2}$ $\frac{3}{2}$ $\frac{3}{2}$	1 1	$\frac{(1-z)^{-2}}{4}\left[4(1-3z+z^2)\psi_1(z)+2z(5-2z)\psi_2(z)-\right.$ $\left.-z(2-z)\psi_3(z)\right]$

	a_1	a_2	a_3	b_1	b_2	${}_3F_2(a_1, a_2, a_3; b_1, b_2; z)$
282	$\frac{1}{2}$	$\frac{3}{2}$	$\frac{3}{2}$	1	2	$\frac{(1-z)^{-1}}{2}[2(1+2z)\psi_2(z)-4z\psi_1(z)-z\psi_3(z)]$
283	$\frac{1}{2}$	$\frac{3}{2}$	$\frac{3}{2}$	1	3	$\frac{2}{3z}[4(2+z)\psi_1(z)-4(2+z)\psi_2(z)+z\psi_3(z)]$
284	$\frac{1}{2}$	$\frac{3}{2}$	$\frac{3}{2}$	2	2	$4\psi_1(z)-4\psi_2(z)+\psi_3(z)$
285	$\frac{1}{2}$	$\frac{3}{2}$	$\frac{3}{2}$	2	3	$\frac{4}{3z}[4(1-z)\psi_2(z)-4(1-z)\psi_1(z)+z\psi_3(z)]$
286	$\frac{1}{2}$	$\frac{3}{2}$	$\frac{3}{2}$	3	3	$\frac{16}{9z}[(6+z)\psi_3(z)-4(1-z)\psi_1(z)-2(1+2z)\psi_2(z)]$
287	$\frac{1}{2}$	$\frac{3}{2}$	2	1	1	$\frac{(1-z)^{-2}}{\pi}[2\mathbf{K}(\sqrt{z})-z(3-z)\mathbf{D}(\sqrt{z})]$
288	$\frac{1}{2}$	$\frac{3}{2}$	2	1	$\frac{5}{2}$	$\frac{3}{4z}\left(\frac{1+z}{\sqrt{1-z}}-\frac{\arcsin\sqrt{z}}{\sqrt{z}}\right)$
289	$\frac{1}{2}$	$\frac{3}{2}$	2	1	3	$\frac{8}{3\pi z}[(4+z)\mathbf{D}(\sqrt{z})-2\mathbf{K}(\sqrt{z})]$
290	$\frac{1}{2}$	$\frac{3}{2}$	2	$\frac{5}{2}$	$\frac{5}{2}$	$\frac{9}{8z^{3/2}}[\mathrm{Li}_2(\sqrt{z})-\mathrm{Li}_2(-\sqrt{z})-\sqrt{z}-(1-z)\,\mathrm{Arth}\sqrt{z}]$
291	$\frac{1}{2}$	$\frac{3}{2}$	2	$\frac{5}{2}$	3	$\frac{2}{z^2}[(4-z)\sqrt{1-z}+3\sqrt{z}\arcsin\sqrt{z}-4]$
292	$\frac{1}{2}$	$\frac{3}{2}$	2	3	3	$\frac{64}{9\pi z^2}[2(3+2z)\mathbf{K}(\sqrt{z})-z(11-z)\mathbf{D}(\sqrt{z})-3\pi]$
293	$\frac{1}{2}$	$\frac{3}{2}$	$\frac{5}{2}$	1	1	$\frac{(1-z)^{-3}}{12}[4(3-10z+6z^2-2z^3)\psi_1(z)+$ $+2z(16-11z+4z^2)\psi_2(z)-z(7-6z+2z^2)\psi_3(z)]$
294	$\frac{1}{2}$	$\frac{3}{2}$	$\frac{5}{2}$	1	2	$\frac{(1-z)^{-2}}{6}[2(1+6z-4z^2)\psi_2(z)+4(1-4z+2z^2)\psi_1(z)-$ $-z(3-2z)\psi_3(z)]$
295	$\frac{1}{2}$	$\frac{3}{2}$	$\frac{5}{2}$	1	3	$\frac{(1-z)^{-1}}{9z}[4(4+z+4z^2)\psi_2(z)-8(2-z+2z^2)\psi_1(z)-$ $-2z(1+2z)\psi_3(z)]$
296	$\frac{1}{2}$	$\frac{3}{2}$	$\frac{5}{2}$	2	2	$\frac{(1-z)^{-1}}{3}[4(1-2z)\psi_1(z)-2(1-4z)\psi_2(z)+(1-2z)\psi_3(z)]$
297	$\frac{1}{2}$	$\frac{3}{2}$	$\frac{5}{2}$	2	3	$\frac{8}{9z}[2(1+2z)\psi_1(z)-2(1+2z)\psi_2(z)+z\psi_3(z)]$
298	$\frac{1}{2}$	$\frac{3}{2}$	$\frac{5}{2}$	3	3	$\frac{32}{27z}[(7-4z)\psi_2(z)-4(1-z)\psi_1(z)-(3-z)\psi_3(z)]$
299	$\frac{1}{2}$	$\frac{3}{2}$	3	1	1	$\frac{(1-z)^{-3}}{4\pi}[(8-z+z^2)\mathbf{K}(\sqrt{z})-z(14-9z+3z^2)\mathbf{D}(\sqrt{z})]$
300	$\frac{1}{2}$	$\frac{3}{2}$	3	1	2	$\frac{(1-z)^{-2}}{2\pi}[2(2-z)\mathbf{K}(\sqrt{z})-z(5-3z)\mathbf{D}(\sqrt{z})]$
301	$\frac{1}{2}$	$\frac{3}{2}$	3	1	$\frac{5}{2}$	$\frac{3}{16z}\left[\frac{1+4z-3z^2}{(1-z)^{3/2}}-\frac{\arcsin\sqrt{z}}{\sqrt{z}}\right]$
302	$\frac{1}{2}$	$\frac{3}{2}$	3	2	2	$\frac{(1-z)^{-1}}{\pi}[(2-3z)\mathbf{D}(\sqrt{z})+\mathbf{K}(\sqrt{z})]$
303	$\frac{1}{2}$	$\frac{3}{2}$	3	2	$\frac{5}{2}$	$\frac{3}{8z}\left[\frac{\arcsin\sqrt{z}}{\sqrt{z}}-\frac{1-3z}{\sqrt{1-z}}\right]$

	a_1	a_2	a_3	b_1	b_2	${}_3F_2(a_1, a_2, a_3; b_1, b_2; z)$
304	$\frac{1}{2}$	$\frac{3}{2}$	3	$\frac{5}{2}$	$\frac{5}{2}$	$\frac{9}{32z^{3/2}}\left[\mathrm{Li}_2(\sqrt{z})-\mathrm{Li}_2(-\sqrt{z})+(1+3z)\,\mathrm{Arth}\sqrt{z}-3\sqrt{z}\right]$
305	$\frac{1}{2}$	2	2	1	1	$\frac{1}{4}(4-2z+z^2)(1-z)^{-5/2}$
306	$\frac{1}{2}$	2	2	1	$\frac{3}{2}$	$\frac{1}{4}\left[(3-z)(1-z)^{-2}+\frac{\mathrm{Arth}\sqrt{z}}{\sqrt{z}}\right]$
307	$\frac{1}{2}$	2	2	1	$\frac{5}{2}$	$\frac{3}{8z}\left[\frac{1+z}{1-z}-(1-z)\frac{\mathrm{Arth}\sqrt{z}}{\sqrt{z}}\right]$
308	$\frac{1}{2}$	2	2	1	3	$\frac{2}{3z^2}\left[\frac{4-2z+z^2}{\sqrt{1-z}}-4\right]$
309	$\frac{1}{2}$	2	2	$\frac{3}{2}$	3	$\frac{2}{3z}\left[\sqrt{z}\,\mathrm{Arth}\sqrt{z}-1-\frac{1}{z}\ln(1-z)\right]$
310	$\frac{1}{2}$	2	2	$\frac{5}{2}$	3	$\frac{1}{z}\left[(3+z)\frac{\mathrm{Arth}\sqrt{z}}{\sqrt{z}}-1+\frac{2}{z}\ln(1-z)\right]$
311	$\frac{1}{2}$	2	2	3	3	$\frac{16}{9z^2}\left[5-(5+z)\sqrt{1-z}+6\ln\frac{1+\sqrt{1-z}}{2}\right]$
312	$\frac{1}{2}$	2	$\frac{5}{2}$	1	1	$\frac{(1-z)^{-3}}{3\pi}\left[(6+z+z^2)\mathbf{K}(\sqrt{z})-z(11-5z+2z^2)\mathbf{D}(\sqrt{z})\right]$
313	$\frac{1}{2}$	2	$\frac{5}{2}$	1	$\frac{3}{2}$	$\frac{1}{6}(6-5z+2z^2)(1-z)^{-5/2}$
314	$\frac{1}{2}$	2	$\frac{5}{2}$	1	3	$\frac{8(1-z)^{-1}}{9\pi z}\left[(2+z)\mathbf{K}(\sqrt{z})-(4-3z+2z^2)\mathbf{D}(\sqrt{z})\right]$
315	$\frac{1}{2}$	2	$\frac{5}{2}$	$\frac{3}{2}$	$\frac{3}{2}$	$\frac{1}{3}\left[\frac{2-z}{(1-z)^2}+\frac{\mathrm{Arth}\sqrt{z}}{\sqrt{z}}\right]$
316	$\frac{1}{2}$	2	$\frac{5}{2}$	$\frac{3}{2}$	3	$\frac{4}{9z^2}\left[\frac{2-z+2z^2}{\sqrt{1-z}}-2\right]$
317	$\frac{1}{2}$	2	$\frac{5}{2}$	3	3	$\frac{64}{27\pi z^2}\left[3\pi-(6+z)\mathbf{K}(\sqrt{z})+z(5+2z)\mathbf{D}(\sqrt{z})\right]$
318	$\frac{1}{2}$	2	3	1	1	$\frac{1}{16}(16-8z+10z^2-3z^3)(1-z)^{-7/2}$
319	$\frac{1}{2}$	2	3	1	$\frac{3}{2}$	$\frac{1}{16}\left[3\frac{\mathrm{Arth}\sqrt{z}}{\sqrt{z}}+(13-8z+3z^2)(1-z)^{-3}\right]$
320	$\frac{1}{2}$	2	3	1	$\frac{5}{2}$	$\frac{3}{32z}\left[(1+6z-3z^2)(1-z)^{-2}-(1-3z)\frac{\mathrm{Arth}\sqrt{z}}{\sqrt{z}}\right]$
321	$\frac{1}{2}$	$\frac{5}{2}$	$\frac{5}{2}$	1	1	$\frac{(1-z)^{-4}}{36}[4(9-28z+15z^2-15z^3+4z^4)\psi_1(z)+$ $+2z(50-24z+27z^2-8z^3)\psi_2(z)-z(24-19z+14z^2-4z^3)\psi_3(z)]$
322	$\frac{1}{2}$	$\frac{5}{2}$	$\frac{5}{2}$	1	$\frac{3}{2}$	$\frac{2}{9\pi}(1-z)^{-3}[(9-3z+2z^2)\mathbf{K}(\sqrt{z})-$ $-z(15-11z+4z^2)\mathbf{D}(\sqrt{z})]$
323	$\frac{1}{2}$	$\frac{5}{2}$	$\frac{5}{2}$	1	2	$\frac{(1-z)^{-3}}{18}[4(4-15z+12z^2-4z^3)\psi_1(z)+$ $+2(1+21z-21z^2+8z^3)\psi_2(z)-z(10-11z+4z^2)\psi_3(z)]$
324	$\frac{1}{2}$	$\frac{5}{2}$	$\frac{5}{2}$	1	3	$\frac{(1-z)^{-2}}{27z}[8(2-9z^2+4z^3)\psi_1(z)-$ $-4(4-6z-15z^2+8z^3)\psi_2(z)+2z(1-8z+4z^2)\psi_3(z)]$

	a_1	a_2	a_3	b_1	b_2	${}_3F_2(a_1, a_2, a_3; b_1, b_2; z)$
325	$\frac{1}{2}$	$\frac{5}{2}$	$\frac{5}{2}$	$\frac{3}{2}$	$\frac{3}{2}$	$\frac{1}{9}(9-10z+4z^2)(1-z)^{-5/2}$
326	$\frac{1}{2}$	$\frac{5}{2}$	$\frac{5}{2}$	$\frac{3}{2}$	2	$\frac{4}{9\pi}(1-z)^{-2}\left[2(2-z)\mathbf{K}(\sqrt{z})+(1-7z+4z^2)\mathbf{D}(\sqrt{z})\right]$
327	$\frac{1}{2}$	$\frac{5}{2}$	$\frac{5}{2}$	$\frac{3}{2}$	3	$\frac{16(1-z)^{-1}}{27\pi z}\left[(1+2z)\mathbf{K}(\sqrt{z})-(2-3z+4z^2)\mathbf{D}(\sqrt{z})\right]$
328	$\frac{1}{2}$	$\frac{5}{2}$	$\frac{5}{2}$	2	2	$\frac{(1-z)^{-2}}{9}[4(2-7z+4z^2)\psi_1(z)+2z(11-8z)\psi_2(z)+(1-6z+4z^2)\psi_3(z)]$
329	$\frac{1}{2}$	$\frac{5}{2}$	$\frac{5}{2}$	2	3	$\frac{(1-z)^{-1}}{27z}[8(2-z+8z^2)\psi_2(z)-16(1-2z+4z^2)\psi_1(z)+4z(1-4z)\psi_3(z)]$
330	$\frac{1}{2}$	$\frac{5}{2}$	$\frac{5}{2}$	3	3	$\frac{32}{81z}[2(5+4z)\psi_1(z)-(13+8z)\psi_2(z)+(3+2z)\psi_3(z)]$
331	$\frac{1}{2}$	$\frac{5}{2}$	3	1	1	$\frac{(1-z)^{-4}}{12\pi}\left[(24+13z+14z^2-3z^3)\mathbf{K}(\sqrt{z})-2z(25-9z+11z^2-3z^3)\mathbf{D}(\sqrt{z})\right]$
332	$\frac{1}{2}$	$\frac{5}{2}$	3	1	$\frac{3}{2}$	$\frac{1}{8}(8-8z+7z^2-2z^3)(1-z)^{-7/2}$
333	$\frac{1}{2}$	$\frac{5}{2}$	3	1	2	$\frac{(1-z)^{-3}}{6\pi}\left[(12-7z+3z^2)\mathbf{K}(\sqrt{z})-z(19-17z+6z^2)\mathbf{D}(\sqrt{z})\right]$
334	$\frac{1}{2}$	$\frac{5}{2}$	3	$\frac{3}{2}$	$\frac{3}{2}$	$\frac{1}{12}\left[(9-8z+3z^2)(1-z)^{-3}+3\frac{\operatorname{Arth}\sqrt{z}}{\sqrt{z}}\right]$
335	$\frac{1}{2}$	$\frac{5}{2}$	3	$\frac{3}{2}$	2	$\frac{1}{4}(4-5z+2z^2)(1-z)^{-5/2}$
336	$\frac{1}{2}$	$\frac{5}{2}$	3	2	2	$\frac{(1-z)^{-2}}{3\pi}\left[(5-3z)\mathbf{K}(\sqrt{z})+2(1-5z+3z^2)\mathbf{D}(\sqrt{z})\right]$
337	$\frac{1}{2}$	3	3	1	1	$\frac{1}{64}(64+72z^2-40z^3+9z^4)(1-z)^{-9/2}$
338	$\frac{1}{2}$	3	3	1	$\frac{3}{2}$	$\frac{1}{64}\left[(55-31z+33z^2-9z^3)(1-z)^{-4}+9\frac{\operatorname{Arth}\sqrt{z}}{\sqrt{z}}\right]$
339	$\frac{1}{2}$	3	3	1	2	$\frac{1}{32}(32-40z+32z^2-9z^3)(1-z)^{-7/2}$
340	$\frac{1}{2}$	3	3	1	$\frac{5}{2}$	$\frac{3}{128z}\left[(1+31z-25z^2+9z^3)(1-z)^{-3}-(1-9z)\frac{\operatorname{Arth}\sqrt{z}}{\sqrt{z}}\right]$
341	$\frac{1}{2}$	3	3	$\frac{3}{2}$	2	$\frac{1}{32}\left[9\frac{\operatorname{Arth}\sqrt{z}}{\sqrt{z}}+(23-24z+9z^2)(1-z)^{-3}\right]$
342	$\frac{1}{2}$	3	3	2	2	$\frac{1}{16}(16-22z+9z^2)(1-z)^{-5/2}$
343	$\frac{1}{2}$	3	3	2	$\frac{5}{2}$	$\frac{3}{64z}\left[(1+9z)\frac{\operatorname{Arth}\sqrt{z}}{\sqrt{z}}-\frac{1-14z+9z^2}{(1-z)^2}\right]$
344	$\frac{2}{3}$	1	1	$\frac{5}{3}$	2	$\frac{2}{z}\ln(1-z)+\frac{1}{z^{2/3}}\left[\ln(1-z)-3\ln(1-z^{1/3})-2\sqrt{3}\operatorname{arctg}\frac{3^{1/2}z^{1/3}}{2+z^{1/3}}\right]$

	a_1	a_2	a_3	b_1	b_2	${}_3F_2(a_1, a_2, a_3; b_1, b_2; z)$
345	$\frac{2}{3}$	1	$\frac{4}{3}$	$\frac{5}{3}$	$\frac{7}{3}$	$\frac{4}{x^4}\left\{x+\frac{1-x^2}{6}[3\ln(1-x)-\ln(1-x^3)]-\frac{1+x^2}{\sqrt{3}}\operatorname{arctg}\frac{3^{1/2}x}{2+x}\right\}$ $[x=z^{1/3}]$
346	$\frac{3}{4}$	1	1	$\frac{7}{4}$	2	$\frac{3}{z}\left\{\ln(1-z)+z^{1/4}\left[\ln\frac{1+z^{1/4}}{1-z^{1/4}}-2\operatorname{arctg} z^{1/4}\right]\right\}$
347	$\frac{3}{4}$	1	$\frac{5}{4}$	$\frac{7}{4}$	$\frac{9}{4}$	$\frac{15}{8x^5}\left[4x-2(1+x^2)\operatorname{arctg}x-(1-x^2)\ln\frac{1+x}{1-x}\right]$ $[x=z^{1/4}]$
348	$\frac{4}{5}$	1	$\frac{6}{5}$	$\frac{9}{5}$	$\frac{11}{5}$	$\frac{12}{x^5}+\frac{12}{5x^6}(1-x^2)\left\{\ln(1-x)+\frac{6}{x^4}(\sqrt{5}-1)\ln\left[1-\frac{x}{2}(\sqrt{5}-1)+x^2\right]-\frac{6}{x^4}(\sqrt{5}+1)\ln\left[1+\frac{x}{2}(\sqrt{5}+1)+x^2\right]\right\}-\frac{24}{5x^6}(1+x^2)\left[\frac{1}{4}(10+2\sqrt{5})^{1/2}\operatorname{arctg}\frac{(10+2\sqrt{5})^{1/2}x}{4-(\sqrt{5}-1)x}+\frac{1}{4}(10-2\sqrt{5})^{1/2}\operatorname{arctg}\frac{(10-2\sqrt{5})^{1/2}x}{4+(\sqrt{5}+1)x}\right]$ $[x=z^{1/5}]$
349	$\frac{5}{6}$	1	$\frac{7}{6}$	$\frac{11}{6}$	$\frac{13}{6}$	$\frac{35}{24x^7}\left\{12x-(1-x^2)\left[\ln\frac{1+x+x^2}{1-x+x^2}+2\ln\frac{1+x}{1-x}\right]-2\sqrt{3}(1+x^2)\operatorname{arctg}\frac{3^{1/2}x}{1-x^2}\right\}$ $[x=z^{1/6}]$
350	$\frac{7}{8}$	1	$\frac{9}{8}$	$\frac{15}{8}$	$\frac{17}{8}$	$\frac{63}{16x^9}\left\{8x+(1-x^2)\left[\ln\frac{1-x}{1+x}+2^{-1/2}\ln\frac{1-2^{1/2}x+x^2}{1+2^{1/2}x+x^2}\right]-2(1+x^2)\left[\operatorname{arctg}x+2^{-1/2}\operatorname{arctg}\frac{2^{1/2}x}{x^2-1}\right]\right\}$ $[x=z^{1/8}]$
351	1	1	1	$\frac{1}{2}$	2	$\frac{\arcsin\sqrt{z}}{\sqrt{z}}\left(\frac{2}{\sqrt{1-z}}-\frac{\arcsin\sqrt{z}}{\sqrt{z}}\right)$
352	1	1	1	$\frac{1}{2}$	3	$\frac{1}{z}\left[3+(3-2z)\frac{\arcsin^2\sqrt{z}}{z}-6\sqrt{\frac{1-z}{z}}\arcsin\sqrt{z}\right]$
353	1	1	1	$\frac{3}{2}$	2	$\frac{\arcsin^2\sqrt{z}}{z}$
354	1	1	1	$\frac{3}{2}$	3	$\frac{1}{z}\left(2\sqrt{\frac{1-z}{z}}\arcsin\sqrt{z}-\frac{1-2z}{z}\arcsin^2\sqrt{z}-1\right)$
355	1	1	1	2	2	$z^{-1}\operatorname{Li}_2(z)$
356	1	1	1	2	$\frac{5}{2}$	$\frac{3}{z}\left(\arcsin^2\sqrt{z}-2+2\sqrt{\frac{1-z}{z}}\arcsin\sqrt{z}\right)$
357	1	1	1	2	3	$\frac{2}{z}\left[\operatorname{Li}_2(z)-1-\frac{1-z}{z}\ln(1-z)\right]$
358	1	1	1	$\frac{5}{2}$	3	$\frac{3}{z}\left(6\sqrt{\frac{1-z}{z}}\arcsin\sqrt{z}+\frac{1+2z}{z}\arcsin^2\sqrt{z}-7\right)$
359	1	1	1	3	3	$\frac{4}{z^2}[(1+z)\operatorname{Li}_2(z)-3z-2(1-z)\ln(1-z)]$
360	1	1	1	4	4	$\frac{9}{z^2}\left[(1+2z)^2\operatorname{Li}_2(z)-4z-\frac{23}{4}z^2+3(1-z^2)\ln(1-z)\right]$

	a_1	a_2	a_3	b_1	b_2	${}_3F_2(a_1, a_2, a_3; b_1, b_2; z)$
361	1	1	$\frac{3}{2}$	$\frac{1}{2}$	$\frac{1}{2}$	$\frac{1+2z}{(1-z)^2}+(2+z)(1-z)^{-5/2}\sqrt{z}\arcsin\sqrt{z}$
362	1	1	$\frac{3}{2}$	$\frac{1}{2}$	2	$\frac{2}{1-z}+\frac{1}{z}\ln(1-z)$
363	1	1	$\frac{3}{2}$	$\frac{1}{2}$	$\frac{5}{2}$	$\frac{3}{z}\left[\frac{2-z}{\sqrt{1-z}}\frac{\arcsin\sqrt{z}}{\sqrt{z}}-2\right]$
364	1	1	$\frac{3}{2}$	$\frac{1}{2}$	3	$-\frac{2}{z}\left[3+\frac{3-z}{z}\ln(1-z)\right]$
365	1	1	$\frac{3}{2}$	2	2	$-\frac{4}{z}\ln\frac{1+\sqrt{1-z}}{2}=\frac{4}{z}\ln\frac{2(1-\sqrt{1-z})}{z}$
366	1	1	$\frac{3}{2}$	2	$\frac{5}{2}$	$\frac{3}{z}\left[2-2\frac{\operatorname{Arth}\sqrt{z}}{\sqrt{z}}-\ln(1-z)\right]$
367	1	1	$\frac{3}{2}$	2	3	$\frac{4}{z^2}\left[2\sqrt{1-z}-2+z-2z\ln\frac{1+\sqrt{1-z}}{2}\right]$
368	1	1	$\frac{3}{2}$	$\frac{5}{2}$	3	$\frac{6}{z}\left[3-4\frac{\operatorname{Arth}\sqrt{z}}{\sqrt{z}}-\frac{1+z}{z}\ln(1-z)\right]$
369	1	1	$\frac{3}{2}$	3	3	$\frac{16}{z^2}\left[3\sqrt{1-z}-3+z-(2+z)\ln\frac{1+\sqrt{1-z}}{2}\right]$
370	1	1	2	$\frac{1}{2}$	3	$\frac{1}{z}\left[2\frac{3-z}{\sqrt{z(1-z)}}\arcsin\sqrt{z}-3-\frac{3}{z}\arcsin^2\sqrt{z}\right]$
371	1	1	2	$\frac{3}{2}$	3	$\frac{1}{z}\left(1+\frac{1}{z}\arcsin^2\sqrt{z}-2\sqrt{\frac{1-z}{z}}\arcsin\sqrt{z}\right)$
372	1	1	2	$\frac{5}{2}$	3	$\frac{3}{z}\left(3-2\sqrt{\frac{1-z}{z}}\arcsin\sqrt{z}-\frac{1}{z}\arcsin^2\sqrt{z}\right)$
373	1	1	2	3	3	$\frac{4}{z^2}[2z-\mathrm{Li}_2(z)+(1-z)\ln(1-z)]$
374	1	1	$\frac{5}{2}$	$\frac{1}{2}$	$\frac{1}{2}$	$\frac{(1-z)^{-3}}{3}\left[3+13z-z^2+(8+8z-z^2)\sqrt{\frac{z}{1-z}}\arcsin\sqrt{z}\right]$
375	1	1	$\frac{5}{2}$	$\frac{1}{2}$	$\frac{3}{2}$	$\frac{(1-z)^{-2}}{3}\left[3+(4-z)\sqrt{\frac{z}{1-z}}\arcsin\sqrt{z}\right]$
376	1	1	$\frac{5}{2}$	$\frac{1}{2}$	2	$\frac{1}{3}\left[4(1-z)^{-2}+\frac{1}{z}\ln(1-z)\right]$
377	1	1	$\frac{5}{2}$	$\frac{1}{2}$	3	$\frac{2(3+z)}{3z}\left[\frac{1}{1-z}+\frac{1}{z}\ln(1-z)\right]$
378	1	1	$\frac{5}{2}$	$\frac{3}{2}$	$\frac{3}{2}$	$\frac{(1-z)^{-1}}{3}\left[1+\frac{2-z}{\sqrt{z(1-z)}}\arcsin\sqrt{z}\right]$
379	1	1	$\frac{5}{2}$	$\frac{3}{2}$	2	$\frac{1}{3}\left[\frac{2}{1-z}-\frac{1}{z}\ln(1-z)\right]$
380	1	1	$\frac{5}{2}$	$\frac{3}{2}$	3	$-\frac{2}{3z}\left[1+\frac{1+z}{z}\ln(1-z)\right]$
318	1	1	$\frac{5}{2}$	2	2	$\frac{4}{3z}\left[\frac{1}{\sqrt{1-z}}-1-\ln\frac{1+\sqrt{1-z}}{2}\right]$
382	1	1	$\frac{5}{2}$	2	3	$\frac{4}{3z^2}\left[2-z-2\sqrt{1-z}-2z\ln\frac{1+\sqrt{1-z}}{2}\right]$

	a_1	a_2	a_3	b_1	b_2	${}_3F_2(a_1, a_2, a_3; b_1, b_2; z)$
383	1	1	$\frac{5}{2}$	2	$\frac{7}{2}$	$\frac{5}{9z^2}\left[6+2z-3z\ln(1-z)-6\frac{\operatorname{Arth}\sqrt{z}}{\sqrt{z}}\right]$
384	1	1	$\frac{5}{2}$	3	3	$\frac{16}{3z^2}\left[1-\sqrt{1-z}+(2-z)\ln\frac{1+\sqrt{1-z}}{2}\right]$
385	1	1	3	$\frac{1}{2}$	2	$\frac{(1-z)^{-2}}{4}\left[4-z+(5-2z)\sqrt{\frac{z}{1-z}}\arcsin\sqrt{z}\right]$
386	1	1	3	$\frac{3}{2}$	2	$\frac{(1-z)^{-1}}{4}\left[1+(3-2z)\frac{\arcsin\sqrt{z}}{\sqrt{z(1-z)}}\right]$
387	1	1	3	2	2	$\frac{1}{2}\left[\frac{1}{1-z}-\frac{1}{z}\ln(1-z)\right]$
388	1	1	3	2	$\frac{5}{2}$	$\frac{3}{4z}\left[1-(1-2z)\frac{\arcsin\sqrt{z}}{\sqrt{z(1-z)}}\right]$
389	1	1	3	2	4	$\frac{3}{4z^3}[2z+z^2+2(1-z^2)\ln(1-z)]$
390	1	1	4	2	5	$\frac{2}{9z^4}[6z+3z^2+2z^3+6(1-z^3)\ln(1-z)]$
391	1	$\frac{5}{4}$	$\frac{7}{4}$	2	$\frac{5}{2}$	$\frac{8}{z}[\sqrt{2}(1+\sqrt{1-z})^{-1/2}-1]$
392	1	$\frac{3}{2}$	$\frac{3}{2}$	$\frac{1}{2}$	$\frac{1}{2}$	$(1+6z+z^2)(1-z)^{-3}$
393	1	$\frac{3}{2}$	$\frac{3}{2}$	$\frac{1}{2}$	2	$\frac{2}{z}\left[1-\frac{1-2z}{(1-z)^{3/2}}\right]$
394	1	$\frac{3}{2}$	$\frac{3}{2}$	$\frac{1}{2}$	$\frac{5}{2}$	$\frac{3}{z}\left[\frac{2-z}{1-z}-2\frac{\operatorname{Arth}\sqrt{z}}{\sqrt{z}}\right]$
395	1	$\frac{3}{2}$	$\frac{3}{2}$	$\frac{1}{2}$	3	$\frac{4}{z^2}\left[\frac{2(3-2z)}{\sqrt{1-z}}-6+z\right]$
396	1	$\frac{3}{2}$	$\frac{3}{2}$	2	2	$\frac{8}{\pi z}\left[\mathbf{K}(\sqrt{z})-\frac{\pi}{2}\right]$
397	1	$\frac{3}{2}$	$\frac{3}{2}$	2	$\frac{5}{2}$	$\frac{6}{z}\left[\frac{\arcsin\sqrt{z}}{\sqrt{z}}-1\right]$
398	1	$\frac{3}{2}$	$\frac{3}{2}$	2	3	$\frac{8}{\pi z}[4\mathbf{K}(\sqrt{z})-4\mathbf{D}(\sqrt{z})-\pi]$
399	1	$\frac{3}{2}$	$\frac{3}{2}$	$\frac{5}{2}$	$\frac{5}{2}$	$\frac{9}{2z^{3/2}}[\mathrm{Li}_2(\sqrt{z})-\mathrm{Li}_2(-\sqrt{z})-2\sqrt{z})]$
400	1	$\frac{3}{2}$	$\frac{3}{2}$	$\frac{5}{2}$	3	$\frac{12}{z^2}[2\sqrt{1-z}+2\sqrt{z}\arcsin\sqrt{z}-2-z]$
401	1	$\frac{3}{2}$	$\frac{3}{2}$	3	3	$\frac{16}{\pi z^2}[8(1+z)\mathbf{K}(\sqrt{z})-16z\mathbf{D}(\sqrt{z})-4\pi-\pi z]=$
402						$=\frac{16}{\pi z^2}\left[-4\pi-\pi z+8(1+\sqrt{z})\mathbf{E}\left(\frac{2\sqrt[4]{z}}{1+\sqrt{z}}\right)\right]$
403	1	$\frac{3}{2}$	2	$\frac{1}{2}$	$\frac{1}{2}$	$\frac{(1-z)^{-3}}{2}\left[2+12z+z^2+3(2+3z)\sqrt{\frac{z}{1-z}}\arcsin\sqrt{z}\right]$
404	1	$\frac{3}{2}$	2	$\frac{1}{2}$	$\frac{5}{2}$	$\frac{3(1-z)^{-1}}{2z}\left[2-z-\frac{2-3z}{\sqrt{z(1-z)}}\arcsin\sqrt{z}\right]$

	a_1	a_2	a_3	b_1	b_2	$_3F_2(a_1, a_2, a_3; b_1, b_2; z)$
405	1	$\frac{3}{2}$	2	$\frac{1}{2}$	3	$\frac{2}{z}\left[\frac{3-z}{1-z}+\frac{3}{z}\ln(1-z)\right]$
406	1	$\frac{3}{2}$	2	$\frac{5}{2}$	3	$\frac{6}{z}\left[2\frac{\operatorname{Arth}\sqrt{z}}{\sqrt{z}}+\frac{1}{z}\ln(1-z)-1\right]$
407	1	$\frac{3}{2}$	2	3	3	$\frac{8}{z^2}\left[4-z-4\sqrt{1-z}+4\ln\frac{1+\sqrt{1-z}}{2}\right]$
408	1	$\frac{3}{2}$	$\frac{5}{2}$	$\frac{1}{2}$	$\frac{1}{2}$	$\frac{1}{3}(3+33z+13z^2-z^3)(1-z)^{-4}$
409	1	$\frac{3}{2}$	$\frac{5}{2}$	$\frac{1}{2}$	2	$\frac{2}{3z}\left[1-\frac{1-4z}{(1-z)^{5/2}}\right]$
410	1	$\frac{3}{2}$	$\frac{5}{2}$	$\frac{1}{2}$	3	$\frac{4}{3z^2}\left[6+z-2\frac{3-4z}{(1-z)^{3/2}}\right]$
411	1	$\frac{3}{2}$	$\frac{5}{2}$	2	2	$\frac{4(1-z)^{-1}}{3\pi z}[2\mathbf{K}(\sqrt{z})-2z\mathbf{D}(\sqrt{z})-\pi+\pi z]$
412	1	$\frac{3}{2}$	$\frac{5}{2}$	2	3	$\frac{8}{3\pi z}[4\mathbf{D}(\sqrt{z})-\pi]$
413	1	$\frac{3}{2}$	$\frac{5}{2}$	3	3	$\frac{16}{3\pi z^2}[4\pi-\pi z-8\mathbf{K}(\sqrt{z})+8z\mathbf{D}(\sqrt{z})]$
414	1	$\frac{3}{2}$	$\frac{5}{2}$	3	$\frac{7}{2}$	$\frac{20}{3z^2}\left(6-z-3\sqrt{1-z}+3\frac{\arcsin\sqrt{z}}{\sqrt{z}}\right)$
415	1	$\frac{3}{2}$	3	$\frac{1}{2}$	$\frac{1}{2}$	$\frac{(1-z)^{-4}}{8}\left[8+82z+17z^2-2z^3+15(2+5z)\sqrt{\frac{z}{1-z}}\arcsin\sqrt{z}\right]$
416	1	$\frac{3}{2}$	3	$\frac{1}{2}$	2	$\frac{1}{2}(2+3z-z^2)(1-z)^{-3}$
417	1	$\frac{3}{2}$	3	$\frac{1}{2}$	$\frac{5}{2}$	$\frac{3(1-z)^{-2}}{8z}\left[2-z+2z^2-\frac{2-5z}{\sqrt{z(1-z)}}\arcsin\sqrt{z}\right]$
418	1	$\frac{3}{2}$	3	2	2	$\frac{1}{2z}\left[\frac{2-z}{(1-z)^{3/2}}-2\right]$
419	1	$\frac{3}{2}$	3	2	$\frac{5}{2}$	$\frac{3}{4z}\left[\frac{\operatorname{Arth}\sqrt{z}}{\sqrt{z}}-\frac{1-2z}{1-z}\right]$
420	1	2	2	$\frac{1}{2}$	3	$\frac{1}{z}\left[\frac{3}{z}\arcsin^2\sqrt{z}+\frac{3-z}{1-z}-2(3-4z)(1-z)^{-3/2}\frac{\arcsin\sqrt{z}}{\sqrt{z}}\right]$
421	1	2	2	$\frac{3}{2}$	3	$\frac{1}{z}\left[2\frac{\arcsin\sqrt{z}}{\sqrt{z(1-z)}}-1-\frac{1}{z}\arcsin^2\sqrt{z}\right]$
422	1	2	2	$\frac{5}{2}$	3	$\frac{3}{z}\left(\frac{1}{z}\arcsin^2\sqrt{z}-1\right)$
423	1	2	2	3	3	$\frac{4}{z^2}[\operatorname{Li}_2(z)-z]$
424	1	2	$\frac{5}{2}$	$\frac{1}{2}$	$\frac{1}{2}$	$\frac{(1-z)^{-4}}{6}\left[6+72z+28z^2-z^3+3(8+24z+3z^2)\times\sqrt{\frac{z}{1-z}}\arcsin\sqrt{z}\right]$
425	1	2	$\frac{5}{2}$	$\frac{1}{2}$	$\frac{3}{2}$	$\frac{(1-z)^{-3}}{6}\left[6+10z-z^2+3(4+z)\sqrt{\frac{z}{1-z}}\arcsin\sqrt{z}\right]$

	a_1	a_2	a_3	b_1	b_2	${}_3F_2(a_1, a_2, a_3; b_1, b_2; z)$
426	1	2	$\frac{5}{2}$	$\frac{1}{2}$	3	$-\frac{2}{z}\left[\frac{1}{z}\ln(1-z)+\frac{1}{3}(3-6z-z^2)(1-z)^{-2}\right]$
427	1	2	$\frac{5}{2}$	$\frac{3}{2}$	$\frac{3}{2}$	$\frac{(1-z)^{-2}}{6}\left[4-z+\frac{2+z}{\sqrt{z(1-z)}}\arcsin\sqrt{z}\right]$
428	1	2	$\frac{5}{2}$	$\frac{3}{2}$	3	$\frac{2}{3z}\left[\frac{1+z}{1-z}+\frac{1}{z}\ln(1-z)\right]$
429	1	2	$\frac{5}{2}$	3	3	$-\frac{8}{3z^2}\left[4\ln\frac{1+\sqrt{1-z}}{2}+z\right]$
430	1	$\frac{5}{2}$	$\frac{5}{2}$	$\frac{1}{2}$	$\frac{1}{2}$	$\frac{1}{9}(9+180z+190z^2+4z^3+z^4)(1-z)^{-5}$
431	1	$\frac{5}{2}$	$\frac{5}{2}$	$\frac{1}{2}$	$\frac{3}{2}$	$\frac{1}{9}(9+39z-z^2+z^3)(1-z)^{-4}$
432	1	$\frac{5}{2}$	$\frac{5}{2}$	$\frac{1}{2}$	2	$\frac{2}{9z}\left[1-\frac{1-8z-8z^2}{(1-z)^{7/2}}\right]$
433	1	$\frac{5}{2}$	$\frac{5}{2}$	$\frac{1}{2}$	3	$\frac{4}{9z^2}\left[2\frac{3-8z+8z^2}{(1-z)^{5/2}}-6+z\right]$
434	1	$\frac{5}{2}$	$\frac{5}{2}$	$\frac{3}{2}$	$\frac{3}{2}$	$\frac{1}{9}(9-2z+z^2)(1-z)^{-3}$
435	1	$\frac{5}{2}$	$\frac{5}{2}$	$\frac{3}{2}$	2	$\frac{2}{9z}\left[\frac{1+2z}{(1-z)^{5/2}}-1\right]$
436	1	$\frac{5}{2}$	$\frac{5}{2}$	$\frac{3}{2}$	3	$\frac{4}{9z^2}\left[2-z-2\frac{1-2z}{(1-z)^{3/2}}\right]$
437	1	$\frac{5}{2}$	$\frac{5}{2}$	2	2	$\frac{4(1-z)^{-2}}{9\pi z}\left[2(1+z)\mathbf{K}(\sqrt{z})-4z\mathbf{D}(\sqrt{z})-\pi(1-z)^2\right]$
438	1	$\frac{5}{2}$	$\frac{5}{2}$	2	3	$\frac{8(1-z)^{-1}}{9\pi z}\left[4\mathbf{K}(\sqrt{z})-4\mathbf{D}(\sqrt{z})-\pi+\pi z\right]$
439	1	$\frac{5}{2}$	$\frac{5}{2}$	3	3	$\frac{16}{9\pi z^2}\left[8\mathbf{K}(\sqrt{z})-4\pi-\pi z\right]$
440	1	$\frac{5}{2}$	3	$\frac{1}{2}$	$\frac{1}{2}$	$\frac{(1-z)^{-5}}{24}\left[24+480z+440z^2-z^3+2z^4+15(8+40z+15z^2)\sqrt{\frac{z}{1-z}}\arcsin\sqrt{z}\right]$
441	1	$\frac{5}{2}$	3	$\frac{1}{2}$	$\frac{3}{2}$	$\frac{(1-z)^{-4}}{24}\left[24+84z-5z^2+2z^3+15(4+3z)\sqrt{\frac{z}{1-z}}\arcsin\sqrt{z}\right]$
442	1	$\frac{5}{2}$	3	$\frac{1}{2}$	2	$\frac{1}{6}(6+21z-4z^2+z^3)(1-z)^{-4}$
443	1	$\frac{5}{2}$	3	$\frac{3}{2}$	$\frac{3}{2}$	$\frac{(1-z)^{-3}}{24}\left[18-5z+2z^2+3(2+3z)\frac{\arcsin\sqrt{z}}{\sqrt{z(1-z)}}\right]$
444	1	$\frac{5}{2}$	3	$\frac{3}{2}$	2	$\frac{1}{6}(6-3z+z^2)(1-z)^{-3}$
445	1	$\frac{5}{2}$	3	2	2	$\frac{1}{6z}\left[\frac{2+z}{(1-z)^{5/2}}-2\right]$
446	1	3	3	$\frac{1}{2}$	2	$\frac{(1-z)^{-4}}{32}\left[32+85z-16z^2+4z^3-15(5+2z)\sqrt{\frac{z}{1-z}}\arcsin\sqrt{z}\right]$
447	1	3	3	$\frac{3}{2}$	2	$\frac{(1-z)^{-3}}{32}\left[23-12z+4z^2+3(3+2z)\frac{\arcsin\sqrt{z}}{\sqrt{z(1-z)}}\right]$

	a_1	a_2	a_3	b_1	b_2	${}_3F_2(a_1, a_2, a_3; b_1, b_2; z)$
448	1	3	3	2	2	$\frac{1}{4}(1-z)^{-3}(4-3z+z^2)$
449	1	3	3	2	$\frac{5}{2}$	$\frac{3(1-z)^{-2}}{32z}\left[(1+2z)\frac{\arcsin\sqrt{z}}{\sqrt{z(1-z)}}-1+8z-4z^2\right]$
450	$\frac{3}{2}$	$\frac{3}{2}$	$\frac{3}{2}$	$\frac{1}{2}$	$\frac{1}{2}$	$(1+10z+4z^2)(1-z)^{-7/2}$
451	$\frac{3}{2}$	$\frac{3}{2}$	$\frac{3}{2}$	$\frac{1}{2}$	1	$\frac{2}{\pi}(1-z)^{-3}\left[(1+5z+2z^2)\,\mathbf{K}(\sqrt{z})-z(3+5z)\,\mathbf{D}(\sqrt{z})\right]$
452	$\frac{3}{2}$	$\frac{3}{2}$	$\frac{3}{2}$	$\frac{1}{2}$	2	$\frac{4}{\pi}(1-z)^{-2}\left[(1-3z)\,\mathbf{D}(\sqrt{z})+2z\mathbf{K}(\sqrt{z})\right]$
453	$\frac{3}{2}$	$\frac{3}{2}$	$\frac{3}{2}$	$\frac{1}{2}$	$\frac{5}{2}$	$\frac{3}{z}\left[2\frac{\arcsin\sqrt{z}}{\sqrt{z}}-(2-3z)(1-z)^{-3/2}\right]$
454	$\frac{3}{2}$	$\frac{3}{2}$	$\frac{3}{2}$	$\frac{1}{2}$	3	$\frac{16(1-z)^{-1}}{\pi z}\left[(3-2z)\,\mathbf{K}(\sqrt{z})-(6-5z)\,\mathbf{D}(\sqrt{z})\right]$
455	$\frac{3}{2}$	$\frac{3}{2}$	$\frac{3}{2}$	1	1	$\frac{(1-z)^{-3}}{4}\left[4(1-2z-2z^2)\,\psi_1(z)+6z(2+z)\,\psi_2(z)-3z\psi_3(z)\right]$
456	$\frac{3}{2}$	$\frac{3}{2}$	$\frac{3}{2}$	1	2	$\frac{(1-z)^{-2}}{2}\left[4(1-2z)\,\psi_1(z)-2(1-4z)\,\psi_2(z)-z\psi_3(z)\right]$
457	$\frac{3}{2}$	$\frac{3}{2}$	$\frac{3}{2}$	1	3	$\frac{2(1-z)^{-1}}{z}\left[2(4-z)\,\psi_2(z)-4(2-z)\,\psi_1(z)-z\psi_3(z)\right]$
458	$\frac{3}{2}$	$\frac{3}{2}$	$\frac{3}{2}$	2	2	$(1-z)^{-1}\left[6\psi_2(z)-4\psi_1(z)-\psi_3(z)\right]$
459	$\frac{3}{2}$	$\frac{3}{2}$	$\frac{3}{2}$	2	3	$\frac{16}{z}\left[\psi_1(z)-\psi_2(z)\right]$
460	$\frac{3}{2}$	$\frac{3}{2}$	$\frac{3}{2}$	3	3	$\frac{32}{z}\left[\psi_2(z)-\psi_3(z)\right]$
461	$\frac{3}{2}$	$\frac{3}{2}$	2	$\frac{1}{2}$	$\frac{1}{2}$	$(1+14z+9z^2)(1-z)^{-4}$
462	$\frac{3}{2}$	$\frac{3}{2}$	2	$\frac{1}{2}$	1	$\frac{1}{2}(2+11z+2z^2)(1-z)^{-7/2}$
463	$\frac{3}{2}$	$\frac{3}{2}$	2	$\frac{1}{2}$	$\frac{5}{2}$	$\frac{3}{z}\left[\frac{\operatorname{Arth}\sqrt{z}}{\sqrt{z}}-\frac{1-2z}{(1-z)^2}\right]$
464	$\frac{3}{2}$	$\frac{3}{2}$	2	$\frac{1}{2}$	3	$\frac{4}{z^2}\left[6-\frac{6-9z+2z^2}{(1-z)^{3/2}}\right]$
465	$\frac{3}{2}$	$\frac{3}{2}$	2	1	1	$\frac{(1-z)^{-3}}{\pi}\left[(2+5z+z^2)\,\mathbf{K}(\sqrt{z})-z(5+3z)\,\mathbf{D}(\sqrt{z})\right]$
466	$\frac{3}{2}$	$\frac{3}{2}$	2	1	$\frac{5}{2}$	$\frac{3}{2z}\left[\frac{\arcsin\sqrt{z}}{\sqrt{z}}-\frac{1-2z}{(1-z)^{3/2}}\right]$
467	$\frac{3}{2}$	$\frac{3}{2}$	2	1	3	$\frac{8(1-z)^{-1}}{\pi z}\left[(2-z)\,\mathbf{K}(\sqrt{z})-(4-3z)\,\mathbf{D}(\sqrt{z})\right]$
468	$\frac{3}{2}$	$\frac{3}{2}$	2	$\frac{5}{2}$	$\frac{5}{2}$	$\frac{9}{4z^{3/2}}\left[2\operatorname{Arth}\sqrt{z}-\operatorname{Li}_2(\sqrt{z})+\operatorname{Li}_2(-\sqrt{z})\right]$
469	$\frac{3}{2}$	$\frac{3}{2}$	2	$\frac{5}{2}$	3	$\frac{12}{z^2}\left[2-2\sqrt{1-z}-\sqrt{z}\arcsin\sqrt{z}\right]$
470	$\frac{3}{2}$	$\frac{3}{2}$	2	3	3	$\frac{64}{\pi z^2}\left[\pi-(2+z)\,\mathbf{K}(\sqrt{z})+3z\mathbf{D}(\sqrt{z})\right]$

	a_1	a_2	a_3	b_1	b_2	${}_3F_2(a_1, a_2, a_3; b_1, b_2; z)$
471	$\frac{3}{2}$	$\frac{3}{2}$	$\frac{5}{2}$	$\frac{1}{2}$	$\frac{1}{2}$	$(1+18z+16z^2)(1-z)^{-9/2}$
472	$\frac{3}{2}$	$\frac{3}{2}$	$\frac{5}{2}$	$\frac{1}{2}$	1	$\frac{2}{3\pi}(1-z)^{-4}\left[(3+26z+19z^2)\,\mathrm{K}(\sqrt{z})-2z(5+18z+z^2)\,\mathrm{D}(\sqrt{z})\right]$
473	$\frac{3}{2}$	$\frac{3}{2}$	$\frac{5}{2}$	$\frac{1}{2}$	2	$\frac{4}{3\pi}(1-z)^{-3}\left[(1+7z)\,\mathrm{K}(\sqrt{z})+(1-7z-2z^2)\,\mathrm{D}(\sqrt{z})\right]$
474	$\frac{3}{2}$	$\frac{3}{2}$	$\frac{5}{2}$	$\frac{1}{2}$	3	$\frac{16(1-z)^{-2}}{3\pi z}\left[2(3-5z+z^2)\,\mathrm{D}(\sqrt{z})-(3-5z)\,\mathrm{K}(\sqrt{z})\right]$
475	$\frac{3}{2}$	$\frac{3}{2}$	$\frac{5}{2}$	1	1	$\frac{(1-z)^{-4}}{12}[4(3-2z-17z^2+z^3)\psi_1(z)+$ $+6z(6+10z-z^2)\psi_2(z)-z(10+7z-2z^2)\psi_3(z)]$
476	$\frac{3}{2}$	$\frac{3}{2}$	$\frac{5}{2}$	1	2	$\frac{(1-z)^{-3}}{6}[4(2-5z)\psi_1(z)-2(1-11z+z^2)\psi_2(z)-z(4-z)\psi_3(z)]$
477	$\frac{3}{2}$	$\frac{3}{2}$	$\frac{5}{2}$	1	3	$\frac{2(1-z)^{-2}}{3z}[4(2-2z-z^2)\psi_1(z)-2(4-4z-3z^2)\psi_2(z)+$ $+z(1-2z)\psi_3(z)]$
478	$\frac{3}{2}$	$\frac{3}{2}$	$\frac{5}{2}$	2	2	$\frac{(1-z)^{-2}}{3}[2(2+z)\psi_2(z)-4z\psi_1(z)-\psi_3(z)]$
479	$\frac{3}{2}$	$\frac{3}{2}$	$\frac{5}{2}$	2	3	$\frac{4(1-z)^{-1}}{3z}[2(2+z)\psi_2(z)-4\psi_1(z)-z\psi_3(z)]$
480	$\frac{3}{2}$	$\frac{3}{2}$	$\frac{5}{2}$	3	3	$\frac{32}{3z}[2\psi_1(z)-3\psi_2(z)+\psi_3(z)]$
481	$\frac{3}{2}$	$\frac{3}{2}$	3	$\frac{1}{2}$	$\frac{1}{2}$	$(1+22z+25z^2)(1-z)^{-5}$
482	$\frac{3}{2}$	$\frac{3}{2}$	3	$\frac{1}{2}$	1	$\frac{1}{8}(8+72z+27z^2-2z^3)(1-z)^{-9/2}$
483	$\frac{3}{2}$	$\frac{3}{2}$	3	$\frac{1}{2}$	2	$\frac{1}{4}(4+13z-2z^2)(1-z)^{-7/2}$
484	$\frac{3}{2}$	$\frac{3}{2}$	3	$\frac{1}{2}$	$\frac{5}{2}$	$\frac{3}{4z}\left[\frac{\mathrm{Arth}\sqrt{z}}{\sqrt{z}}-\frac{1-4z-z^2}{(1-z)^3}\right]$
485	$\frac{3}{2}$	$\frac{3}{2}$	3	1	1	$\frac{(1-z)^{-4}}{4\pi}\left[(8+31z+10z^2-z^3)\,\mathrm{K}(\sqrt{z})-\right.$ $\left.-2z(11+14z-z^2)\,\mathrm{D}(\sqrt{z})\right]$
486	$\frac{3}{2}$	$\frac{3}{2}$	3	1	2	$\frac{(1-z)^{-3}}{2\pi}\left[(4+5z-z^2)\,\mathrm{K}(\sqrt{z})-z(9-z)\,\mathrm{D}(\sqrt{z})\right]$
487	$\frac{3}{2}$	$\frac{3}{2}$	3	1	$\frac{5}{2}$	$\frac{3}{8z}\left[\frac{\arcsin\sqrt{z}}{\sqrt{z}}-\frac{1-5z+z^2}{(1-z)^{5/2}}\right]$
488	$\frac{3}{2}$	$\frac{3}{2}$	3	2	2	$\frac{(1-z)^{-2}}{\pi}\left[(3-z)\,\mathrm{K}(\sqrt{z})-2\mathrm{D}(\sqrt{z})\right]$
489	$\frac{3}{2}$	$\frac{3}{2}$	3	2	$\frac{5}{2}$	$\frac{3}{4z}\left[(1-z)^{-3/2}-\frac{\arcsin\sqrt{z}}{\sqrt{z}}\right]$
490	$\frac{3}{2}$	$\frac{3}{2}$	3	$\frac{5}{2}$	$\frac{5}{2}$	$\frac{9}{16z^{3/2}}\left[\frac{2\sqrt{z}}{1-z}-\mathrm{Li}_2(\sqrt{z})+\mathrm{Li}_2(-\sqrt{z})\right]$
491	$\frac{3}{2}$	2	2	$\frac{1}{2}$	$\frac{1}{2}$	$\frac{(1-z)^{-4}}{4}\left[4+62z+39z^2+3(6+23z+6z^2)\sqrt{\frac{z}{1-z}}\arcsin\sqrt{z}\right]$
492	$\frac{3}{2}$	2	2	$\frac{1}{2}$	1	$(1+8z+3z^2)(1-z)^{-4}$

	a_1	a_2	a_3	b_1	b_2	${}_3F_2\,(a_1,\ a_2,\ a_3;\ b_1,\ b_2;\ z)$
493	$\frac{3}{2}$	2	2	$\frac{1}{2}$	$\frac{5}{2}$	$\frac{3(1-z)^{-2}}{4z}\left[\frac{2-5z+6z^2}{\sqrt{z(1-z)}}\arcsin\sqrt{z}-2+5z\right]$
494	$\frac{3}{2}$	2	2	$\frac{1}{2}$	3	$-\frac{2}{z}\left[\frac{3-5z}{(1-z)^2}+\frac{3}{z}\ln(1-z)\right]$
495	$\frac{3}{2}$	2	2	1	1	$\frac{1}{4}(4+10z+z^2)(1-z)^{-7/2}$
496	$\frac{3}{2}$	2	2	1	$\frac{5}{2}$	$\frac{3}{4z}\left[\frac{\operatorname{Arth}\sqrt{z}}{\sqrt{z}}-\frac{1-3z}{(1-z)^2}\right]$
497	$\frac{3}{2}$	2	2	1	3	$\frac{2}{z^2}\left[4-\frac{4-6z+z^2}{(1-z)^{3/2}}\right]$
498	$\frac{3}{2}$	2	2	$\frac{5}{2}$	3	$-\frac{6}{z}\left[\frac{\operatorname{Arth}\sqrt{z}}{\sqrt{z}}+\frac{1}{z}\ln(1-z)\right]$
499	$\frac{3}{2}$	2	2	3	3	$\frac{16}{z^2}\left[\sqrt{1-z}-1-2\ln\frac{1+\sqrt{1-z}}{2}\right]$
500	$\frac{3}{2}$	2	$\frac{5}{2}$	$\frac{1}{2}$	$\frac{1}{2}$	$(1+25z+35z^2+3z^3)(1-z)^{-5}$
501	$\frac{3}{2}$	2	$\frac{5}{2}$	$\frac{1}{2}$	1	$\frac{1}{2}(2+21z+12z^2)(1-z)^{-9/2}$
502	$\frac{3}{2}$	2	$\frac{5}{2}$	$\frac{1}{2}$	3	$\frac{4}{z^2}\left[\frac{2-5z+4z^2}{(1-z)^{5/2}}-2\right]$
503	$\frac{3}{2}$	2	$\frac{5}{2}$	1	1	$\frac{(1-z)^{-4}}{3\pi}\left[2(3+14z+7z^2)\,\mathbf{K}(\sqrt{z})-z(17+30z+z^2)\,\mathbf{D}(\sqrt{z})\right]$
504	$\frac{3}{2}$	2	$\frac{5}{2}$	1	3	$\frac{8(1-z)^{-2}}{3\pi z}\left[(4-7z+z^2)\,\mathbf{D}(\sqrt{z})-2(1-2z)\,\mathbf{K}(\sqrt{z})\right]$
505	$\frac{3}{2}$	2	$\frac{5}{2}$	3	3	$\frac{64}{3\pi z^2}\left[2\mathbf{K}(\sqrt{z})-z\mathbf{D}(\sqrt{z})-\pi\right]$
506	$\frac{3}{2}$	2	3	$\frac{1}{2}$	$\frac{1}{2}$	$\frac{(1-z)^{-5}}{16}\left[16+406z+505z^2+18z^3+15(6+37z+20z^2)\sqrt{\frac{z}{1-z}}\arcsin\sqrt{z}\right]$
507	$\frac{3}{2}$	2	3	$\frac{1}{2}$	1	$(1+13z+10z^2)(1-z)^{-5}$
508	$\frac{3}{2}$	2	3	$\frac{1}{2}$	$\frac{5}{2}$	$\frac{3(1-z)^{-3}}{16z}\left[(2-7z+20z^2)\frac{\arcsin\sqrt{z}}{\sqrt{z(1-z)}}-2+11z+6z^2\right]$
509	$\frac{3}{2}$	2	3	1	1	$\frac{1}{16}(16+72z+18z^2-z^3)(1-z)^{-9/2}$
510	$\frac{3}{2}$	2	3	1	$\frac{5}{2}$	$\frac{3}{16z}\left[\frac{\operatorname{Arth}\sqrt{z}}{\sqrt{z}}-(1-8z-z^2)(1-z)^{-3}\right]$
511	$\frac{3}{2}$	$\frac{5}{2}$	$\frac{5}{2}$	$\frac{1}{2}$	$\frac{1}{2}$	$\frac{1}{3}(3+96z+184z^2+32z^3)(1-z)^{-11/2}$
512	$\frac{3}{2}$	$\frac{5}{2}$	$\frac{5}{2}$	$\frac{1}{2}$	1	$\frac{2}{3\pi}(1-z)^{-5}\left[(3+46z+67z^2+12z^3)\,\mathbf{K}(\sqrt{z})-z(11+82z+35z^2)\,\mathbf{D}(\sqrt{z})\right]$
513	$\frac{3}{2}$	$\frac{5}{2}$	$\frac{5}{2}$	$\frac{1}{2}$	2	$\frac{4}{9\pi}(1-z)^{-4}\left[4(1+8z+3z^2)\mathbf{K}(\sqrt{z})+(1-18z-31z^2)\mathbf{D}(\sqrt{z})\right]$

	a_1	a_2	a_3	b_1	b_2	${}_3F_2(a_1, a_2, a_3; b_1, b_2; z)$
514	$\frac{3}{2}$	$\frac{5}{2}$	$\frac{5}{2}$	$\frac{1}{2}$	3	$\frac{16(1-z)^{-3}}{9\pi z}\left[(3-7z+12z^2)\,\mathbf{K}(\sqrt{z})-(6-17z+19z^2)\,\mathbf{D}(\sqrt{z})\right]$
515	$\frac{3}{2}$	$\frac{5}{2}$	$\frac{5}{2}$	1	1	$\frac{(1-z)^{-5}}{36}[4(9+20z-111z^2-24z^3+z^4)\,\psi_1(z)+$ $+2z(53+237z+27z^2-2z^3)\,\psi_2(z)-z(33+83z-10z^2-z^3)\,\psi_3(z)]$
516	$\frac{3}{2}$	$\frac{5}{2}$	$\frac{5}{2}$	1	2	$\frac{(1-z)^{-4}}{18}[4(5-9z-12z^2+z^3)\,\psi_1(z)-$ $-2(1-30z-18z^2+2z^3)\,\psi_2(z)-z(14+2z-z^2)\,\psi_3(z)]$
517	$\frac{3}{2}$	$\frac{5}{2}$	$\frac{5}{2}$	1	3	$\frac{(1-z)^{-3}}{9z}[4(4-9z+12z^2+2z^3)\,\psi_2(z)-$ $-8(2-6z+6z^2+z^3)\,\psi_1(z)-2z(1+z+z^2)\,\psi_3(z)]$
518	$\frac{3}{2}$	$\frac{5}{2}$	$\frac{5}{2}$	2	2	$\frac{(1-z)^{-3}}{9}[4(2-6z+z^2)\,\psi_1(z)+$ $+2(1+10z-2z^2)\,\psi_2(z)-(1+3z-z^2)\,\psi_3(z)]$
519	$\frac{3}{2}$	$\frac{5}{2}$	$\frac{5}{2}$	2	3	$\frac{(1-z)^{-2}}{9z}[16(1-z-z^2)\,\psi_1(z)-8(2-3z-2z^2)\,\psi_2(z)-4z^2\psi_3(z)]$
520	$\frac{3}{2}$	$\frac{5}{2}$	$\frac{5}{2}$	3	3	$\frac{16(1-z)^{-1}}{9z}[2(5-2z)\,\psi_2(z)-4(2-z)\,\psi_1(z)-(2-z)\,\psi_3(z)]$
521	$\frac{3}{2}$	$\frac{5}{2}$	3	$\frac{1}{2}$	$\frac{1}{2}$	$(1+39z+95z^2+25z^3)(1-z)^{-6}$
522	$\frac{3}{2}$	$\frac{5}{2}$	3	$\frac{1}{2}$	1	$\frac{1}{8}(8+136z+159z^2+12z^3)(1-z)^{-11/2}$
523	$\frac{3}{2}$	$\frac{5}{2}$	3	$\frac{1}{2}$	2	$\frac{1}{4}(4+27z+4z^2)(1-z)^{-9/2}$
524	$\frac{3}{2}$	$\frac{5}{2}$	3	1	1	$\frac{(1-z)^{-5}}{12\pi}\left[(24+181z+166z^2+13z^3)\,\mathbf{K}(\sqrt{z})-\right.$ $\left.-z(74+259z+52z^2-z^3)\,\mathbf{D}(\sqrt{z})\right]$
525	$\frac{3}{2}$	$\frac{5}{2}$	3	1	2	$\frac{(1-z)^{-4}}{6\pi}\left[4(3+8z+z^2)\,\mathbf{K}(\sqrt{z})-z(31+18z-z^2)\,\mathbf{D}(\sqrt{z})\right]$
526	$\frac{3}{2}$	$\frac{5}{2}$	3	2	2	$\frac{(1-z)^{-3}}{3\pi}\left[(7+z)\,\mathbf{K}(\sqrt{z})-(2+7z-z^2)\,\mathbf{D}(\sqrt{z})\right]$
527	$\frac{3}{2}$	3	3	$\frac{1}{2}$	$\frac{1}{2}$	$\frac{(1-z)^{-6}}{64}\left[64+2622z+6219z^2+1490z^3+\right.$ $\left.+15(30+295z+328z^2+40z^3)\sqrt{\frac{z}{1-z}}\arcsin\sqrt{z}\right]$
528	$\frac{3}{2}$	3	3	$\frac{1}{2}$	1	$(1+21z+33z^2+5z^3)(1-z)^{-6}$
529	$\frac{3}{2}$	3	3	$\frac{1}{2}$	2	$\frac{1}{2}(2+17z+5z^2)(1-z)^{-5}$
530	$\frac{3}{2}$	3	3	$\frac{1}{2}$	$\frac{5}{2}$	$\frac{3(1-z)^{-4}}{64z}\left[(2-9z+72z^2+40z^3)\frac{\arcsin\sqrt{z}}{\sqrt{z(1-z)}}-2+29z+78z^2\right]$
531	$\frac{3}{2}$	3	3	1	1	$\frac{1}{64}(64+512z+360z^2+8z^3+z^4)(1-z)^{-11/2}$
532	$\frac{3}{2}$	3	3	1	2	$\frac{1}{32}(32+72z+z^3)(1-z)^{-9/2}$
533	$\frac{3}{2}$	3	3	1	$\frac{5}{2}$	$\frac{3}{64z}\left[\frac{\operatorname{Arth}\sqrt{z}}{\sqrt{z}}-\frac{1-25z-25z^2+z^3}{(1-z)^4}\right]$

	a_1	a_2	a_3	b_1	b_2	${}_3F_2\,(a_1,\ a_2,\ a_3;\ b_1,\ b_2;\ z)$
534	$\frac{3}{2}$	3	3	2	2	$\frac{1}{16}(16-2z+z^2)(1-z)^{-7/2}$
535	$\frac{3}{2}$	3	3	2	$\frac{5}{2}$	$\frac{3}{32z}\left[(1+8z-z^2)(1-z)^{-3}-\frac{\operatorname{Arth}\sqrt{z}}{\sqrt{z}}\right]$
536	2	2	2	$\frac{1}{2}$	1	$\frac{(1-z)^{-4}}{8}\left[8+69z+28z^2+3(9+22z+4z^2)\sqrt{\frac{z}{1-z}}\arcsin\sqrt{z}\right]$
537	2	2	2	$\frac{1}{2}$	3	$\frac{1}{z}\left[(6-14z+11z^2)(1-z)^{-5/2}\frac{\arcsin\sqrt{z}}{\sqrt{z}}-\right.$ $\left.-\frac{3}{z}\arcsin^2\sqrt{z}-\frac{3-6z}{(1-z)^2}\right]$
538	2	2	2	1	1	$(1+4z+z^2)(1-z)^{-4}$
539	2	2	2	1	$\frac{3}{2}$	$\frac{(1-z)^{-3}}{8}\left[7+8z+(1+10z+4z^2)\frac{\arcsin\sqrt{z}}{\sqrt{z(1-z)}}\right]$
540	2	2	2	1	$\frac{5}{2}$	$\frac{3(1-z)^{-2}}{8z}\left[(1-2z+4z^2)\frac{\arcsin\sqrt{z}}{\sqrt{z(1-z)}}-1+4z\right]$
541	2	2	2	1	3	$-\frac{2}{z}\left[\frac{1-2z}{(1-z)^2}+\frac{1}{z}\ln(1-z)\right]$
542	2	2	2	$\frac{3}{2}$	3	$\frac{1}{z}\left[\frac{1}{1-z}+\frac{1}{z}\arcsin^2\sqrt{z}-(2-3z)(1-z)^{-3/2}\frac{\arcsin\sqrt{z}}{\sqrt{z}}\right]$
543	2	2	2	$\frac{5}{2}$	3	$\frac{3}{z}\left[\frac{\arcsin\sqrt{z}}{\sqrt{z(1-z)}}-\frac{1}{z}\arcsin^2\sqrt{z}\right]$
544	2	2	2	3	3	$-\frac{4}{z^2}[\operatorname{Li}_2(z)+\ln(1-z)]$
545	2	2	$\frac{5}{2}$	$\frac{1}{2}$	$\frac{1}{2}$	$\frac{(1-z)^{-5}}{4}\left[4+116z+176z^2+19z^3+\right.$ $\left.+3(8+56z+39z^2+2z^3)\sqrt{\frac{z}{1-z}}\arcsin\sqrt{z}\right]$
546	2	2	$\frac{5}{2}$	$\frac{1}{2}$	1	$(1+15z+15z^2+z^3)(1-z)^{-5}$
547	2	2	$\frac{5}{2}$	$\frac{1}{2}$	$\frac{3}{2}$	$\frac{(1-z)^{-4}}{12}\left[12+76z+17z^2+\right.$ $\left.+3(12+21z+2z^2)\sqrt{\frac{z}{1-z}}\arcsin\sqrt{z}\right]$
548	2	2	$\frac{5}{2}$	$\frac{1}{2}$	3	$\frac{2}{3z^2}[3\ln(1-z)+z(3-6z+11z^2)(1-z)^{-3}]$
549	2	2	$\frac{5}{2}$	1	1	$\frac{1}{4}(4+22z+9z^2)(1-z)^{-9/2}$
550	2	2	$\frac{5}{2}$	1	$\frac{3}{2}$	$\frac{1}{3}(3+8z+z^2)(1-z)^{-4}$
551	2	2	$\frac{5}{2}$	1	3	$\frac{2}{3z^2}\left[\frac{4-10z+9z^2}{(1-z)^{5/2}}-4\right]$
552	2	2	$\frac{5}{2}$	$\frac{3}{2}$	$\frac{3}{2}$	$\frac{(1-z)^{-3}}{12}\left[5(2+z)+(2+11z+2z^2)\frac{\arcsin\sqrt{z}}{\sqrt{z(1-z)}}\right]$
553	2	2	$\frac{5}{2}$	$\frac{3}{2}$	3	$-\frac{2}{3z}\left[\frac{1-3z}{(1-z)^2}+\frac{1}{z}\ln(1-z)\right]$

	a_1	a_2	a_3	b_1	b_2	${}_3F_2(a_1, a_2, a_3; b_1, b_2; z)$
554	2	2	$\frac{5}{2}$	3	3	$\frac{16}{3z^2}\left[\frac{1}{\sqrt{1-z}}-1+2\ln\frac{1+\sqrt{1-z}}{2}\right]$
555	2	2	3	$\frac{1}{2}$	1	$\frac{(1-z)^{-5}}{32}\left[32+473z+428z^2+12z^3+\right.$ $\left.+15(9+38z+16z^2)\sqrt{\frac{z}{1-z}}\arcsin\sqrt{z}\right]$
556	2	2	3	1	1	$(1+7z+4z^2)(1-z)^{-5}$
557	2	2	3	1	$\frac{3}{2}$	$\frac{(1-z)^{-4}}{32}\left[29+72z+4z^2+3(1+18z+16z^2)\frac{\arcsin\sqrt{z}}{\sqrt{z(1-z)}}\right]$
558	2	2	3	1	$\frac{5}{2}$	$\frac{3(1-z)^{-3}}{32z}\left[(1-2z+16z^2)\frac{\arcsin\sqrt{z}}{\sqrt{z(1-z)}}-1+12z+4z^2\right]$
559	2	$\frac{5}{2}$	$\frac{5}{2}$	$\frac{1}{2}$	$\frac{1}{2}$	$\frac{1}{3}(3+132z+370z^2+132z^3+3z^4)(1-z)^{-6}$
560	2	$\frac{5}{2}$	$\frac{5}{2}$	$\frac{1}{2}$	1	$\frac{1}{2}(2+39z+56z^2+8z^3)(1-z)^{-11/2}$
561	2	$\frac{5}{2}$	$\frac{5}{2}$	$\frac{1}{2}$	$\frac{3}{2}$	$\frac{1}{3}(3+35z+25z^2+z^3)(1-z)^{-5}$
562	2	$\frac{5}{2}$	$\frac{5}{2}$	$\frac{1}{2}$	3	$\frac{4}{3z^2}\left[2-\frac{2-7z+8z^2-8z^3}{(1-z)^{7/2}}\right]$
563	2	$\frac{5}{2}$	$\frac{5}{2}$	1	1	$\frac{(1-z)^{-5}}{3\pi}\left[(6+53z+60z^2+9z^3)\mathbf{K}(\sqrt{z})-\right.$ $\left.-z(19+82z+27z^2)\mathbf{D}(\sqrt{z})\right]$
564	2	$\frac{5}{2}$	$\frac{5}{2}$	1	$\frac{3}{2}$	$\frac{1}{6}(6+23z+6z^2)(1-z)^{-9/2}$
565	2	$\frac{5}{2}$	$\frac{5}{2}$	1	3	$\frac{8(1-z)^{-3}}{9\pi z}\left[(2-3z+9z^2)\mathbf{K}(\sqrt{z})-(4-11z+15z^2)\mathbf{D}(\sqrt{z})\right]$
566	2	$\frac{5}{2}$	$\frac{5}{2}$	$\frac{3}{2}$	$\frac{3}{2}$	$\frac{1}{9}(9+14z+z^2)(1-z)^{-4}$
567	2	$\frac{5}{2}$	$\frac{5}{2}$	$\frac{3}{2}$	3	$\frac{4}{9z^2}\left[\frac{2-5z+6z^2}{(1-z)^{5/2}}-2\right]$
568	2	$\frac{5}{2}$	$\frac{5}{2}$	3	3	$\frac{64(1-z)^{-1}}{9\pi z^2}\left[\pi-\pi z-(2-3z)\mathbf{K}(\sqrt{z})-z\mathbf{D}(\sqrt{z})\right]$
569	2	$\frac{5}{2}$	3	$\frac{1}{2}$	$\frac{1}{2}$	$\frac{(1-z)^{-6}}{16}\left[16+744z+2040z^2+659z^3+6z^4+\right.$ $\left.+15(8+88z+115z^2+20z^3)\sqrt{\frac{z}{1-z}}\arcsin\sqrt{z}\right]$
570	2	$\frac{5}{2}$	3	$\frac{1}{2}$	1	$(1+24z+45z^2+10z^3)(1-z)^{-6}$
571	2	$\frac{5}{2}$	3	$\frac{1}{2}$	$\frac{3}{2}$	$\frac{(1-z)^{-5}}{16}\left[16+180z+117z^2+2z^3+\right.$ $\left.+15(4+13z+4z^2)\sqrt{\frac{z}{1-z}}\arcsin\sqrt{z}\right]$
572	2	$\frac{5}{2}$	3	1	1	$\frac{1}{16}(16+152z+138z^2+9z^3)(1-z)^{-11/2}$
573	2	$\frac{5}{2}$	3	1	$\frac{3}{2}$	$(1+5z+2z^2)(1-z)^{-5}$

	a_1	a_2	a_3	b_1	b_2	${}_3F_2(a_1, a_2, a_3; b_1, b_2; z)$
574	2	$\frac{5}{2}$	3	$\frac{3}{2}$	$\frac{3}{2}$	$\frac{(1-z)^{-4}}{48}\left[42+61z+2z^2+3(2+21z+12z^2)\frac{\arcsin\sqrt{z}}{\sqrt{z(1-z)}}\right]$
575	2	3	3	$\frac{1}{2}$	1	$\frac{(1-z)^{-6}}{128}\left[128+3165z+5886z^2+1216z^3+15(45+320z+296z^2+32z^3)\sqrt{\frac{z}{1-z}}\arcsin\sqrt{z}\right]$
576	2	3	3	1	1	$(1+12z+15z^2+2z^3)(1-z)^{-6}$
577	2	3	3	1	$\frac{3}{2}$	$\frac{(1-z)^{-5}}{128}\left[119+602z+224z^2+3(3+96z+184z^2+32z^3)\frac{\arcsin\sqrt{z}}{\sqrt{z(1-z)}}\right]$
578	2	3	3	1	$\frac{5}{2}$	$\frac{3(1-z)^{-4}}{128z}\left[(1+72z^2+32z^3)\frac{\arcsin\sqrt{z}}{\sqrt{z(1-z)}}-1+42z+64z^2\right]$
579	$\frac{5}{2}$	$\frac{5}{2}$	$\frac{5}{2}$	$\frac{1}{2}$	$\frac{1}{2}$	$\frac{1}{9}(9+504z+1864z^2+1024z^3+64z^4)(1-z)^{-13/2}$
580	$\frac{5}{2}$	$\frac{5}{2}$	$\frac{5}{2}$	$\frac{1}{2}$	1	$\frac{2(1-z)^{-6}}{9\pi}\left[(9+243z+663z^2+341z^3+24z^4)\mathbf{K}(\sqrt{z})-2z(18+261z+320z^2+41z^3)\mathbf{D}(\sqrt{z})\right]$
581	$\frac{5}{2}$	$\frac{5}{2}$	$\frac{5}{2}$	$\frac{1}{2}$	$\frac{3}{2}$	$\frac{1}{9}(9+138z+152z^2+16z^3)(1-z)^{-11/2}$
582	$\frac{5}{2}$	$\frac{5}{2}$	$\frac{5}{2}$	$\frac{1}{2}$	2	$\frac{4(1-z)^{-5}}{27\pi}\left[(13+166z+181z^2+24z^3)\mathbf{K}(\sqrt{z})+(1-52z-259z^2-74z^3)\mathbf{D}(\sqrt{z})\right]$
583	$\frac{5}{2}$	$\frac{5}{2}$	$\frac{5}{2}$	$\frac{1}{2}$	3	$\frac{16(1-z)^{-4}}{27\pi z}\left[2(3-11z+9z^2-25z^3)\mathbf{D}(\sqrt{z})-(3-14z-13z^2-24z^3)\mathbf{K}(\sqrt{z})\right]$
584	$\frac{5}{2}$	$\frac{5}{2}$	$\frac{5}{2}$	1	1	$\frac{(1-z)^{-6}}{12}\left[4(3+23z-57z^2-72z^3-2z^4)\psi_1(z)+2z(17+174z+126z^2-2z^3)\psi_2(z)-z(12+74z+26z^2-7z^3)\psi_3(z)\right]$
585	$\frac{5}{2}$	$\frac{5}{2}$	$\frac{5}{2}$	1	$\frac{3}{2}$	$\frac{2(1-z)^{-5}}{9\pi}\left[(9+60z+53z^2+6z^3)\mathbf{K}(\sqrt{z})-z(27+82z+19z^2)\mathbf{D}(\sqrt{z})\right]$
586	$\frac{5}{2}$	$\frac{5}{2}$	$\frac{5}{2}$	1	2	$\frac{(1-z)^{-5}}{54}\left[4(14+6z-114z^2-11z^3)\psi_1(z)-2(1-84z-225z^2-7z^3)\psi_2(z)-z(47+71z-13z^2)\psi_3(z)\right]$
587	$\frac{5}{2}$	$\frac{5}{2}$	$\frac{5}{2}$	1	3	$\frac{(1-z)^{-4}}{27z}\left[8(2-3z+3z^2-17z^3)\psi_1(z)-4(4-12z-9z^2-28z^3)\psi_2(z)+2z(1-14z-2z^2)\psi_3(z)\right]$
588	$\frac{5}{2}$	$\frac{5}{2}$	$\frac{5}{2}$	$\frac{3}{2}$	$\frac{3}{2}$	$\frac{1}{9}(9+22z+4z^2)(1-z)^{-9/2}$
589	$\frac{5}{2}$	$\frac{5}{2}$	$\frac{5}{2}$	$\frac{3}{2}$	2	$\frac{4(1-z)^{-4}}{27\pi}\left[2(7+14z+3z^2)\mathbf{K}(\sqrt{z})-(1+30z+17z^2)\mathbf{D}(\sqrt{z})\right]$
590	$\frac{5}{2}$	$\frac{5}{2}$	$\frac{5}{2}$	$\frac{3}{2}$	3	$\frac{16(1-z)^{-3}}{27\pi z}\left[(1+z+6z^2)\mathbf{K}(\sqrt{z})-(2-5z+11z^2)\mathbf{D}(\sqrt{z})\right]$

	a_1	a_2	a_3	b_1	b_2	${}_3F_2\,(a_1,\ a_2,\ a_3;\ b_1\ b_2;\ z)$
591	$\frac{5}{2}$	$\frac{5}{2}$	$\frac{5}{2}$	2	2	$\frac{(1-z)^{-4}}{27}[4(7-17z-5z^2)\psi_1(z)+6z(13+2z)\psi_2(z)-(1+16z-2z^2)\psi_3(z)]$
592	$\frac{5}{2}$	$\frac{5}{2}$	$\frac{5}{2}$	2	3	$\frac{4(1-z)^{-3}}{27z}[2(2-4z+11z^2)\psi_2(z)-4(1-4z+6z^2)\psi_1(z)-z(1+2z)\psi_3(z)]$
593	$\frac{5}{2}$	$\frac{5}{2}$	$\frac{5}{2}$	3	3	$\frac{16(1-z)^{-2}}{27z}[4(3-4z)\psi_1(z)-2(7-10z)\psi_2(z)+(2-3z)\psi_3(z)]$
594	$\frac{5}{2}$	$\frac{5}{2}$	$\frac{5}{2}$	3	5	$\frac{256(1-x^2)^{-4}}{81\pi^2x^4}[2(1-2x^2)\mathbf{D}(x)-(1-3x^2)\mathbf{K}(x)]^2$ $[x=2^{-1/2}(1-\sqrt{1-z})^{1/2}]$
595	$\frac{5}{2}$	$\frac{5}{2}$	3	$\frac{1}{2}$	$\frac{1}{2}$	$\frac{1}{3}(3+204z+938z^2+700z^3+75z^4)(1-z)^{-7}$
596	$\frac{5}{2}$	$\frac{5}{2}$	3	$\frac{1}{2}$	1	$\frac{1}{8}(8+248z+643z^2+248z^3+8z^4)(1-z)^{-13/2}$
597	$\frac{5}{2}$	$\frac{5}{2}$	3	$\frac{1}{2}$	$\frac{3}{2}$	$\frac{1}{3}(3+57z+85z^2+15z^3)(1-z)^{-6}$
598	$\frac{5}{2}$	$\frac{5}{2}$	3	$\frac{1}{2}$	2	$\frac{1}{12}(12+159z+136z^2+8z^3)(1-z)^{-11/2}$
599	$\frac{5}{2}$	$\frac{5}{2}$	3	1	1	$\frac{(1-z)^{-6}}{12\pi}[(24+341z+663z^2+243z^3+9z^4)\mathbf{K}(\sqrt{z})-2z(41+320z+261z^2+18z^3)\mathbf{D}(\sqrt{z})]$
600	$\frac{5}{2}$	$\frac{5}{2}$	3	1	$\frac{3}{2}$	$\frac{1}{8}(8+56z+39z^2+2z^3)(1-z)^{-11/2}$
601	$\frac{5}{2}$	$\frac{5}{2}$	3	1	2	$\frac{(1-z)^{-5}}{6\pi}[(12+67z+46z^2+3z^3)\mathbf{K}(\sqrt{z})-z(35+82z+11z^2)\mathbf{D}(\sqrt{z})]$
602	$\frac{5}{2}$	$\frac{5}{2}$	3	$\frac{3}{2}$	$\frac{3}{2}$	$\frac{1}{3}(3+10z+3z^2)(1-z)^{-5}$
603	$\frac{5}{2}$	$\frac{5}{2}$	3	$\frac{3}{2}$	2	$\frac{1}{12}(12+21z+2z^2)(1-z)^{-9/2}$
604	$\frac{5}{2}$	$\frac{5}{2}$	3	2	2	$\frac{(1-z)^{-4}}{9\pi}[(19+26z+3z^2)\mathbf{K}(\sqrt{z})-2(1+18z+5z^2)\mathbf{D}(\sqrt{z})]$
605	$\frac{5}{2}$	3	3	$\frac{1}{2}$	$\frac{1}{2}$	$\frac{(1-z)^{-7}}{64}\left[64+4712z+22104z^2+16475z^3+1690z^4+15(40+680z+1563z^2+680z^3+40z^4)\sqrt{\frac{z}{1-z}}\arcsin\sqrt{z}\right]$
606	$\frac{5}{2}$	3	3	$\frac{1}{2}$	1	$(1+38z+126z^2+70z^3+5z^4)(1-z)^{-7}$
607	$\frac{5}{2}$	3	3	$\frac{1}{2}$	$\frac{3}{2}$	$\frac{(1-z)^{-6}}{64}\left[64+1236z+1851z^2+314z^3+15(20+115z+88z^2+8z^3)\sqrt{\frac{z}{1-z}}\arcsin\sqrt{z}\right]$
608	$\frac{5}{2}$	3	3	$\frac{1}{2}$	2	$\frac{1}{2}(2+33z+40z^2+5z^3)(1-z)^{-6}$
609	$\frac{5}{2}$	3	3	1	1	$\frac{1}{64}(64+1024z+1864z^2+504z^3+9z^4)(1-z)^{-13/2}$
610	$\frac{5}{2}$	3	3	1	$\frac{3}{2}$	$(1+9z+9z^2+z^3)(1-z)^{-6}$

	a_1	a_2	a_3	b_1	b_2	${}_3F_2(a_1, a_2, a_3; b_1, b_2: z)$
611	$\frac{5}{2}$	3	3	1	2	$\frac{1}{32}(32+184z+96z^2+3z^3)(1-z)^{-11/2}$
612	$\frac{5}{2}$	3	3	$\frac{3}{2}$	$\frac{3}{2}$	$\frac{(1-z)^{-5}}{64}\left[58+199z+58z^2+3(2+39z+56z^2+8z^3)\frac{\arcsin\sqrt{z}}{\sqrt{z(1-z)}}\right]$
613	$\frac{5}{2}$	3	3	$\frac{3}{2}$	2	$\frac{1}{2}(2+5z+z^2)(1-z)^{-5}$
614	$\frac{5}{2}$	3	3	2	2	$\frac{1}{16}(16+18z+z^2)(1-z)^{-9/2}$
615	3	3	3	$\frac{1}{2}$	1	$\frac{(1-z)^{-7}}{512}\left[512+20\,689z+70\,878z^2+40\,144z^3+2912z^4+\right.$ $+15(225+2600z+4584z^2+1536z^3+64z^4)\times$ $\left.\times\sqrt{\frac{z}{1-z}}\arcsin\sqrt{z}\right]$
616	3	3	3	$\frac{1}{2}$	2	$\frac{(1-z)^{-6}}{256}\left[256+4251z+5220z^2+668z^3+\right.$ $\left.+15(75+370z+232z^2+16z^3)\sqrt{\frac{z}{1-z}}\arcsin\sqrt{z}\right]$
617	3	3	3	1	1	$(1+20z+48z^2+20z^3+z^4)(1-z)^{-7}$
618	3	3	3	1	$\frac{3}{2}$	$\frac{(1-z)^{-6}}{512}\left[485+4614z+4752z^2+544z^3+\right.$ $\left.+3(9+504z+1864z^2+1024z^3+64z^4)\frac{\arcsin\sqrt{z}}{\sqrt{z(1-z)}}\right]$
619	3	3	3	1	2	$\frac{1}{2}(2+15z+12z^2+z^3)(1-z)^{-6}$
620	3	3	3	1	$\frac{5}{2}$	$\frac{3(1-z)^{-5}}{512z}\left[(1+8z+360z^2+512z^3+64z^4)\frac{\arcsin\sqrt{z}}{\sqrt{z(1-z)}}-\right.$ $\left.-1+162z+624z^2+160z^3\right]$
621	3	3	3	$\frac{3}{2}$	2	$\frac{(1-z)^{-5}}{256}\left[229+592z+124z^2+3(9+138z+152z^2+16z^3)\times\right.$ $\left.\times\frac{\arcsin\sqrt{z}}{\sqrt{z(1-z)}}\right]$
622	3	3	3	2	2	$\frac{1}{4}(4+7z+z^2)(1-z)^{-5}$
623	3	3	3	2	$\frac{5}{2}$	$\frac{3(1-z)^{-4}}{256z}\left[1+68z+36z^2-(1-18z-72z^2-16z^3)\times\right.$ $\left.\times\frac{\arcsin\sqrt{z}}{\sqrt{z(1-z)}}\right]$

7.4.3. Particular values of ${}_3F_2(a_1,\ a_2,\ a_3;\ b_1,\ b_2;\ -z)$. (See also 7.4.2).

	a_1	a_2	a_3	b_1	b_2	${}_3F_2(a_1,\ a_2,\ a_3;\ b_1,\ b_2;\ -z)$
1	$\frac{1}{8}$	$\frac{3}{8}$	1	$\frac{9}{8}$	$\frac{11}{8}$	$\frac{3}{32x^3}\left[2(ax^2-b)\operatorname{arctg}\frac{ax}{1-x^2}+2(bx^2+a)\operatorname{arctg}\frac{bx}{1-x^2}+(bx^2-a)\ln\frac{1+bx+x^2}{1-bx+x^2}+(ax^2+b)\ln\frac{1+ax+x^2}{1-ax+x^2}\right]$ $\left[a=(2-\sqrt{2})^{1/2},\ b=(2+\sqrt{2})^{1/2},\ x=z^{1/8}\right]$
2	$\frac{1}{8}$	$\frac{5}{8}$	1	$\frac{9}{8}$	$\frac{13}{8}$	$\frac{5}{64x^5}\left[2(ax^4-b)\operatorname{arctg}\frac{ax}{1-x^2}+2(bx^4+a)\operatorname{arctg}\frac{bx}{1-x^2}-(bx^4-a)\ln\frac{1-bx+x^2}{1+bx+x^2}-(ax^4+b)\ln\frac{1-ax+x^2}{1+ax+x^2}\right]$ $\left[a=(2-\sqrt{2})^{1/2},\ b=(2+\sqrt{2})^{1/2},\ x=z^{1/8}\right]$
3	$\frac{1}{8}$	$\frac{7}{8}$	1	$\frac{9}{8}$	$\frac{15}{8}$	$\frac{7}{96x^7}\left\{a\left[(1+x^6)\ln\frac{1+ax+x^2}{1-ax+x^2}-2(1-x^6)\operatorname{arctg}\frac{ax}{1-x^2}\right]+b\left[(1+x^6)\ln\frac{1+bx+x^2}{1-bx+x^2}-2(1-x^6)\operatorname{arctg}\frac{bx}{1-x^2}\right]\right\}$ $\left[a=(2-\sqrt{2})^{1/2},\ b=(2+\sqrt{2})^{1/2},\ x=z^{1/8}\right]$
4	$\frac{1}{6}$	$\frac{1}{3}$	1	$\frac{7}{6}$	$\frac{4}{3}$	$\frac{1}{6x}\left\{\ln(1-x+x^2)-2\ln(1+x)-2\sqrt{3}\operatorname{arctg}\frac{3^{1/2}x}{2-x}+\sqrt{x}\left[4\operatorname{arctg}\sqrt{x}+2\operatorname{arctg}\frac{\sqrt{x}}{1-x}-3^{-1/2}\ln\frac{1-\sqrt{3x}+x}{1+\sqrt{3x}+x}\right]\right\}$ $\left[x=z^{1/3}\right]$
5	$\frac{1}{6}$	$\frac{2}{3}$	1	$\frac{7}{6}$	$\frac{5}{3}$	$\frac{1}{9x^2}\left\{x^{3/2}\left[4\operatorname{arctg}\sqrt{x}-3^{-1/2}\ln\frac{1-\sqrt{3x}+x}{1+\sqrt{3x}+x}+2\operatorname{arctg}\frac{\sqrt{x}}{1-x}\right]+\ln\frac{(1+x)^2}{1-x+x^2}-2\sqrt{3}\operatorname{arctg}\frac{3^{1/2}x}{2-x}\right\}$ $\left[x=z^{1/3}\right]$
6	$\frac{1}{4}$	$\frac{3}{4}$	1	$\frac{5}{4}$	$\frac{7}{4}$	$\frac{3}{2^{5/2}x^3}\left[(1+x^2)\ln\frac{1+2^{1/2}x+x^2}{1-2^{1/2}x+x^2}-2(1-x^2)\operatorname{arctg}\frac{2^{1/2}x}{1-x^2}\right]$ $\left[x=z^{1/4}\right]$
7	$\frac{1}{4}$	$\frac{3}{4}$	1	$\frac{7}{4}$	$\frac{9}{4}$	$\frac{15}{64x^5}\left[\sqrt{2}(x^4+2x^2-1)\ln\frac{1+2^{1/2}x+x^2}{1-2^{1/2}x+x^2}+2\sqrt{2}(x^4-2x^2-1)\operatorname{arctg}\frac{2^{1/2}x}{1-x^8}+8x\right]$ $\left[x=z^{1/4}\right]$
8	$\frac{1}{4}$	1	1	$\frac{5}{4}$	2	$-\frac{1}{3z}\left\{\ln(1+z)+\frac{z^{3/4}}{\sqrt{2}}\left[\ln\frac{1-\sqrt[4]{4z}+\sqrt{z}}{1+\sqrt[4]{4z}+\sqrt{z}}-2\operatorname{arctg}\frac{\sqrt[4]{4z}}{1-\sqrt{z}}\right]\right\}$
9	$\frac{1}{3}$	$\frac{5}{6}$	1	$\frac{4}{3}$	$\frac{11}{6}$	$\frac{5}{9x}\left[\ln(1+x)-\frac{1}{2}\ln(1-x+x^2)+\sqrt{3}\operatorname{arctg}\frac{3^{1/2}x}{2-x}\right]-\frac{5}{18x^{5/2}}\left[3^{-1/2}\ln\frac{1-\sqrt{3x}+x}{1+\sqrt{3x}+x}+2\operatorname{arctg}\frac{\sqrt{x}}{1-x}+4\operatorname{arctg}\sqrt{x}\right]$ $\left[x=z^{1/3}\right]$

	a_1	a_2	a_3	b_1	b_2	${}_3F_2(a_1, a_2, a_3; b_1, b_2; -z)$
10	$\frac{3}{8}$	$\frac{5}{8}$	1	$\frac{11}{8}$	$\frac{13}{8}$	$\frac{15}{32x^5}\left\{a\left[(1+x^2)\ln\frac{1+bx+x^2}{1-bx+x^2}+2(1-x^2)\operatorname{arctg}\frac{bx}{1-x^2}\right]-b\left[(1+x^2)\ln\frac{1+ax+x^2}{1-ax+x^2}+2(1-x^2)\operatorname{arctg}\frac{ax}{1-x^2}\right]\right\}$ $\left[a=(2-\sqrt{2})^{1/2},\ b=(2+\sqrt{2})^{1/2},\ x=z^{1/8}\right]$
11	$\frac{3}{8}$	$\frac{7}{8}$	1	$\frac{11}{8}$	$\frac{15}{8}$	$\frac{21}{64x^7}\left\{(ax^4+b)\left[\ln\frac{1+bx+x^2}{1-bx+x^2}-2\operatorname{arctg}\frac{bx}{1-x^2}\right]-(bx^4-a)\left[\ln\frac{1+ax+x^2}{1-ax+x^2}-2\operatorname{arctg}\frac{ax}{1-x^2}\right]\right\}$ $\left[a=(2-\sqrt{2})^{1/2},\ b=(2+\sqrt{2})^{1/2},\ x=z^{1/8}\right]$
12	$\frac{1}{2}$	$\frac{2}{3}$	$\frac{4}{3}$	$\frac{3}{2}$	$\frac{3}{2}$	$\frac{3\sqrt{3}}{2\sqrt{z}}\operatorname{arctg}x$ $\quad[4z=27x^2(1+x)^2]$
13	$\frac{1}{2}$	1	1	$\frac{3}{2}$	$\frac{3}{2}$	$\frac{1-x^2}{2x}[\operatorname{Li}_2(x)-\operatorname{Li}_2(-x)]$ $\quad[z(1-x^2)^2=4x^2]$
14	$\frac{3}{4}$	1	1	$\frac{7}{4}$	2	$-\frac{3}{z}\left\{\ln(1+z)+\frac{z^{1/4}}{\sqrt{2}}\left[\ln\frac{1+\sqrt[4]{4z}+\sqrt{z}}{1-\sqrt[4]{4z}+\sqrt{z}}-2\operatorname{arctg}\frac{\sqrt[4]{4z}}{1-\sqrt{z}}\right]\right\}$
15	$\frac{3}{4}$	1	$\frac{5}{4}$	$\frac{7}{4}$	$\frac{9}{4}$	$\frac{15}{2x^4}-\frac{15\sqrt{2}}{16x^5}\left[2(1-x^2)\operatorname{arctg}\frac{2^{1/2}x}{1-x^2}+(1+x^2)\times\ln\frac{1+2^{1/2}x+x^2}{1-2^{1/2}x+x^2}\right]$ $\quad[x=z^{1/4}]$
16	$\frac{5}{6}$	1	$\frac{7}{6}$	$\frac{11}{6}$	$\frac{13}{6}$	$-\frac{35}{2x^6}-\frac{35}{8\cdot 3^{3/2}x^7}(1-x^2)\ln\frac{1-3^{1/2}x+x^2}{1+3^{1/2}x+x^2}+\frac{35}{12x^7}(1+x^2)\left[\operatorname{arctg}\frac{x}{1-x^2}+2\operatorname{arctg}x\right]$ $\quad[x=z^{1/6}]$
17	$\frac{7}{8}$	1	$\frac{9}{8}$	$\frac{15}{8}$	$\frac{17}{8}$	$\frac{63}{32x^9}\left\{-16x+(1-x^2)\left[a\ln\frac{1+ax+x^2}{1-ax+x^2}+b\ln\frac{1+bx+x^2}{1-bx+x^2}\right]+2(1+x^2)\left[a\operatorname{arctg}\frac{ax}{1-x^2}+b\operatorname{arctg}\frac{bx}{1-x^2}\right]\right\}$ $\left[a=(2-\sqrt{2})^{1/2},\ b=(2+\sqrt{2})^{1/2},\ x=z^{1/8}\right]$

7.4.4. Values of ${}_3F_2(a_1, a_2, a_3; b_1, b_2; 1)$.

1. $${}_3F_2\left(\begin{matrix}a,\ b,\ c;\ 1\\ d,\ e\end{matrix}\right)=\Gamma\left[\begin{matrix}d,\ d+e-a-b-c\\ d+e-a-b,\ d-c\end{matrix}\right]{}_3F_2\left(\begin{matrix}e-a,\ e-b,\ c;\ 1\\ d+e-a-b,\ e\end{matrix}\right)$$
$[\operatorname{Re}(d+e-a-b-c),\ \operatorname{Re}(d-c)>0]$.

2. $${}_3F_2\left(\begin{matrix}a,\ b,\ c;\ 1\\ d,\ e\end{matrix}\right)=\Gamma\left[\begin{matrix}d,\ e,\ s\\ a,\ b+s,\ c+s\end{matrix}\right]{}_3F_2\left(\begin{matrix}d-a,\ e-a,\ s;\ 1\\ b+s,\ c+s\end{matrix}\right)$$
$[s=d+e-a-b-c;\ \operatorname{Re}a,\ \operatorname{Re}s>0]$.

3. $${}_3F_2\left(\begin{matrix}a,\ b,\ c;\ 1\\ d,\ e\end{matrix}\right)=\Gamma\left[\begin{matrix}d,\ e,\ 1-a,\ c-b\\ d-b,\ e-b,\ 1+b-a,\ c\end{matrix}\right]{}_3F_2\left(\begin{matrix}b,\ 1+b-d,\ 1+b-e;\ 1\\ 1+b-a,\ 1+b-c\end{matrix}\right)+\Gamma\left[\begin{matrix}d,\ e,\ 1-a,\ b-c\\ d-c,\ e-c,\ 1+c-a,\ b\end{matrix}\right]{}_3F_2\left(\begin{matrix}c,\ 1+c-d,\ 1+c-e;\ 1\\ 1+c-a,\ 1+c-b\end{matrix}\right)$$
$[\operatorname{Re}(d+e-a-b-c)>0]$.

4. $${}_3F_2\left(\begin{matrix}a,\ b,\ c;\ 1\\ d,\ e\end{matrix}\right)=\Gamma\left[\begin{matrix}e-a-b,\ e\\ e-a,\ e-b\end{matrix}\right]{}_3F_2\left(\begin{matrix}a,\ b,\ d-c;\ 1\\ d,\ 1+a+b-e\end{matrix}\right)-$$
$$-\Gamma\left[\begin{matrix}a+b-e,\ d,\ e,\ d+e-a-b-c\\ a,\ b,\ d-c,\ d+e-a-b\end{matrix}\right]{}_3F_2\left(\begin{matrix}e-a,\ e-b,\ d+e-a-b-c;\ 1\\ 1+e-a-b,\ d+e-a-b\end{matrix}\right)$$
$[\operatorname{Re}(d+e-a-b-c),\ \operatorname{Re}(1+c-e)>0]$.

5. $${}_3F_2\left(\begin{matrix}a,\ b,\ c;\ 1\\ d,\ e\end{matrix}\right)=\Gamma\left[\begin{matrix}1+a-d,\ 1+b-d,\ 1+c-d,\ d,\ e\\ a,\ b,\ c,\ 1+e-d,\ 2-d\end{matrix}\right]\times$$
$$\times{}_3F_2\left(\begin{matrix}1+a-d,\ 1+b-d,\ 1+c-d;\ 1\\ 1+e-d,\ 2-d\end{matrix}\right)+$$
$$+\Gamma\left[\begin{matrix}1+a-d,\ 1+c-d\\ 1-d,\ 1+a+c-d\end{matrix}\right]{}_3F_2\left(\begin{matrix}a,\ c,\ e-b;\ 1\\ 1+a+c-d,\ e\end{matrix}\right)$$
$[\operatorname{Re}(d+e-a-b-c),\ \operatorname{Re}(1+b-d)>0]$.

6. $${}_3F_2\left(\begin{matrix}a,\ b,\ c;\ 1\\ d,\ e\end{matrix}\right)=$$
$$=\Gamma\left[\begin{matrix}1+a-d,\ 1+b-d,\ 1+c-d,\ d,\ e\\ a,\ b,\ c,\ 1+e-d,\ 2-d\end{matrix}\right]{}_3F_2\left(\begin{matrix}1+a-d,\ 1+b-d,\ 1+c-d;\ 1\\ 1+e-d,\ 2-d\end{matrix}\right)+$$
$$+\Gamma\left[\begin{matrix}1+a-d,\ 1+b-d,\ 1+c-d,\ e\\ 1-d,\ 1+a+b-d,\ 1+a+c-d,\ e-a\end{matrix}\right]{}_3F_2\left(\begin{matrix}a,\ 1+a-d,\ 1+a+b+c-d-e;\ 1\\ 1+a+b-d,\ 1+a+c-d\end{matrix}\right)$$
$[\operatorname{Re}(d+e-a-b-c),\ \operatorname{Re}(e-a)>0]$.

7. $$\frac{b}{\Gamma(1+a-d)\,\Gamma(1-a-c)}\,{}_3F_2\left(\begin{matrix}a,\ a+b,\ a+c;\ 1\\ a+1,\ 1+a-d\end{matrix}\right)+$$
$$+\frac{a}{\Gamma(1+b-c)\,\Gamma(1-b-d)}\,{}_3F_2\left(\begin{matrix}b,\ b+a,\ b+d;\ 1\\ b+1,\ 1+b-c\end{matrix}\right)=\Gamma\left[\begin{matrix}a+1,\ b+1\\ a+b,\ 1-c,\ 1-d\end{matrix}\right]$$
$[\operatorname{Re}(a+b+c+d)<2]$.

8. $$cd\Gamma\left[\begin{matrix}a+b\\ a-d,\ b-c\end{matrix}\right]\left\{\frac{b}{\Gamma(b+d)\,\Gamma(c-b)}\,{}_3F_2\left(\begin{matrix}a+b,\ a+c,\ a;\ 1\\ 1+a-d,\ 1+a\end{matrix}\right)+\right.$$
$$\left.+\frac{a}{\Gamma(a+c)\,\Gamma(d-a)}\,{}_3F_2\left(\begin{matrix}a+b,\ b+d,\ b;\ 1\\ 1+b-c,\ 1+b\end{matrix}\right)\right\}+$$
$$+ab\Gamma\left[\begin{matrix}c+d\\ d-a,\ c-b\end{matrix}\right]\left\{\frac{c}{\Gamma(a+c)\,\Gamma(b-c)}\,{}_3F_2\left(\begin{matrix}b+d,\ c+d,\ d;\ 1\\ 1+d-a,\ 1+d\end{matrix}\right)+\right.$$
$$\left.+\frac{d}{\Gamma(c+d)\,\Gamma(a-d)}\,{}_3F_2\left(\begin{matrix}c+d,\ a+c,\ c;\ 1\\ 1+c-b,\ 1+c\end{matrix}\right)\right\}=$$
$$=\Gamma\left[\begin{matrix}a+1,\ b+1,\ c+1,\ d+1\\ a+c,\ b+d,\ a-d,\ d-a,\ b-c,\ c-b\end{matrix}\right]$$
$[\operatorname{Re}(a+b+c+d)<2]$.

9. $${}_3F_2\left(\begin{matrix}a,\ b,\ c;\ 1\\ d,\ a-n\end{matrix}\right)=\Gamma\left[\begin{matrix}d-b-c,\ d\\ d-b,\ d-c\end{matrix}\right]{}_3F_2\left(\begin{matrix}-n,\ b,\ c;\ 1\\ 1+b+c-d,\ a-n\end{matrix}\right)$$
$[a\neq 0,\ -1,\ -2,\dots;\ \operatorname{Re}(d-b-c)>n]$.

10. $${}_3F_2\left(\begin{matrix}a,\ b,\ c;\ 1\\ d,\ a-1\end{matrix}\right)=\Gamma\left[\begin{matrix}d,\ d-b-c\\ d-b,\ d-c\end{matrix}\right]\left[1-\frac{bc}{(a-1)(1+b+c-d)}\right]$$
$[a\neq 0,\ -1,\ -2,\dots;\ \operatorname{Re}(d-b-c)>1]$.

11. $${}_3F_2\left(\begin{matrix}a,\ b,\ c;\ 1\\ a+1,\ d\end{matrix}\right)=\Gamma(1-c)\,\Gamma(d)\left\{\Gamma\left[\begin{matrix}a+1,\ b-a\\ b,\ d-a,\ a-c+1\end{matrix}\right]-\right.$$
$$\left.-\frac{a}{\Gamma(b-c+1)\,\Gamma(d-b)\,(b-a)}\,{}_3F_2\left(\begin{matrix}b,\ b-d+1,\ b-a;\ 1\\ b-a+1,\ b-c+1\end{matrix}\right)\right\}$$
$[\operatorname{Re}(b+c-d)<1]$.

12. $${}_3F_2\left(\begin{matrix}a,\ b,\ c;\ 1\\ d,\ 1+a+b+c-d\end{matrix}\right)=\Gamma\left[\begin{matrix}1+a+b+c-d,\ 1+c-d,\ d,\ d-a-b\\ 1+a+c-d,\ 1+b+c-d,\ d-a,\ d-b\end{matrix}\right]+$$
$$+\frac{1}{d-c-1}\Gamma\left[\begin{matrix}d,\ a+b+c-d+1\\ a,\ b,\ c+1\end{matrix}\right]{}_3F_2\left(\begin{matrix}a+c-d+1,\ b+c-d+1,\ 1;\ 1\\ c-d+2,\ c+1\end{matrix}\right)$$
$[\operatorname{Re}(d-a-b)>0]$.

3. $${}_3F_2\left(\begin{matrix}a,\ b,\ c;\ 1\\ a-n,\ a+1\end{matrix}\right)=n!\,\Gamma\left[\begin{matrix}a+1,\ a-b-c-n+1\\ a-b+1,\ a-c+1\end{matrix}\right]\sum_{k=0}^{n}\frac{(a-b-n)_k\,(a-c-n)_k}{k!\,(a-n)_k}$$
$[\operatorname{Re} a>n+1;\ \operatorname{Re}(a-b-c)>n-1]$.

14. $${}_3F_2\left(\begin{matrix}a,\ b,\ c;\ 1\\ a-n,\ b+m\end{matrix}\right)=$$
$$=\Gamma\left[\begin{matrix}1-c,\ b+m\\ 1+b-c\end{matrix}\right]\frac{(1+b-a)_n}{(m-1)!\,(1-a)_n}\,{}_3F_2\left(\begin{matrix}1-m,\ b,\ 1+b-a+n;\ 1\\ 1+b-a,\ 1+b-c\end{matrix}\right)$$
$[\operatorname{Re} c<m-n]$.

15. $${}_3F_2\left(\begin{matrix}a,\ b,\ c;\ 1\\ a-n,\ b+1\end{matrix}\right)=\frac{(1+b-a)_n}{(1-a)_n}\Gamma\left[\begin{matrix}1-c,\ b+1\\ 1+b-c\end{matrix}\right]$$ $[\operatorname{Re} c<1-n]$.

16. $${}_3F_2\left(\begin{matrix}a,\ b,\ c;\ 1\\ a+1,\ b+1\end{matrix}\right)=\frac{ab}{a-b}\Gamma(1-c)\left\{\Gamma\left[\begin{matrix}b\\ 1+b-c\end{matrix}\right]-\Gamma\left[\begin{matrix}a\\ 1+a-c\end{matrix}\right]\right\}$$
$[a\neq b,\ c\neq 1,\ \operatorname{Re} c<2]$.

17. $${}_3F_2\left(\begin{matrix}a,\ b,\ c;\ 1\\ a+2,\ b+2\end{matrix}\right)=\frac{1}{a-b}\left[\frac{a(a+1)}{a-b-1}\Gamma\left[\begin{matrix}2-c,\ 2+b\\ 2+b-c\end{matrix}\right]+\frac{b(b+1)}{a-b+1}\times\right.$$
$$\left.\times\Gamma\left[\begin{matrix}2-c,\ 2+a\\ 2+a-c\end{matrix}\right]+\frac{2(a+1)(b+1)\Gamma(1-c)}{(a-b-1)(a-b+1)}\left\{b\Gamma\left[\begin{matrix}a+1\\ a-c+1\end{matrix}\right]-a\Gamma\left[\begin{matrix}b+1\\ b-c+1\end{matrix}\right]\right\}\right]$$
$[a\neq b-1,\ b,\ b+1,\ c\neq 1,\ 2,\ 3;\ \operatorname{Re} c<4]$.

18. $${}_3F_2\left(\begin{matrix}a,\ b,\ c;\ 1\\ (a+b+1)/2,\ 2c\end{matrix}\right)=\sqrt{\pi}\,\Gamma\left[\begin{matrix}c+1/2,\ (a+b+1)/2,\ (1-a-b)/2+c\\ (a+1)/2,\ (b+1)/2,\ (1-a)/2+c,\ (1-b)/2+c\end{matrix}\right]$$
$[\operatorname{Re}(2c-a-b)>-1]$.

19. $${}_3F_2\left(\begin{matrix}a,\ b,\ c;\ 1\\ a+b+1/2,\ c+1/2\end{matrix}\right)=\sqrt{\pi}\,\Gamma\left[\begin{matrix}a+b+1/2,\ c+1/2,\ 1/2+c-a-b\\ a+1/2,\ b+1/2,\ 1/2+c-a,\ 1/2+c-b\end{matrix}\right]$$
$[\operatorname{Re}(c-a-b)>-1/2]$.

20. $${}_3F_2\left(\begin{matrix}a,\ b,\ c;\ 1\\ 1+a-b,\ a-c\end{matrix}\right)=\frac{\sqrt{\pi}}{2^a}\Gamma\left[\begin{matrix}a-c,\ 1+a-b\\ 1+a-b-c\end{matrix}\right]\times$$
$$\times\left\{\Gamma\left[\begin{matrix}a/2+1-b-c\\ a/2-c,\ (a+1)/2,\ a/2+1-b\end{matrix}\right]+\Gamma\left[\begin{matrix}(a+1)/2-b-c\\ (a+1)/2-c,\ a/2,\ (a+1)/2-b\end{matrix}\right]\right\}$$
$[\operatorname{Re}(a-2b-2c)>-1]$.

21. $${}_3F_2\left(\begin{matrix}a,\ b,\ c;\ 1\\ 1+a-b,\ 1+a-c\end{matrix}\right)=$$
$$=\frac{\sqrt{\pi}}{2^a}\Gamma\left[\begin{matrix}1+a-b,\ 1+a-c,\ 1+a/2-b-c\\ (1+a)/2,\ 1+a/2-b,\ 1+a/2-c,\ 1+a-b-c\end{matrix}\right]$$ $[\operatorname{Re}(a-2b-2c)>-2]$.

22. $${}_3F_2\left(\begin{matrix}a,\ b,\ c;\ 1\\ 2+a-b,\ 2+a-c\end{matrix}\right)=$$
$$=\frac{1}{2(b-1)(c-1)}\Gamma\left[\begin{matrix}2+a-b,\ 2+a-c\\ a,\ 2+a-b-c\end{matrix}\right]\left\{\Gamma\left[\begin{matrix}a/2,\ 2+a/2-b-c\\ 1+a/2-b,\ 1+a/2-c\end{matrix}\right]-\right.$$
$$\left.-\Gamma\left[\begin{matrix}(a+1)/2,\ (5+a)/2-b-c\\ (3+a)/2-b,\ (3+a)/2-c\end{matrix}\right]\right\}$$ $[\operatorname{Re}(a-2b-2c)>-4]$.

23. $${}_3F_2\left(\begin{matrix}a,\ a-n,\ b;\ 1\\ a-n+1,\ c\end{matrix}\right)=\frac{(a-n)(b-a)_n(n-1)!}{(a-b)(1-a)_n(c-a)_n}\Gamma\left[\begin{matrix}c,\ c-a-b+1\\ c-a,\ c-b\end{matrix}\right]\times$$
$$\times\sum_{k=0}^{n-1}\frac{(1-a)_k\,(c-a)_k}{k!\,(b-a+1)_k}$$ $[\operatorname{Re}(a+b-c)<1]$.

24. $${}_3F_2\left(\begin{matrix}a,\ 1-a,\ b;\ 1\\ c,\ 1+2b-c\end{matrix}\right)=2^{1-2b}\pi\,\Gamma\left[\begin{matrix}c,\ 1+2b-c\\ (a+c)/2,\ (1+a-c)/2+b,\ (1+c-a)/2,\ d\end{matrix}\right]$$
$[d=1+b-(a+c)/2,\ \operatorname{Re} b>0]$.

25. $${}_3F_2\left(\begin{matrix}a,\ a+1/2,\ b;\ 1\\ c,\ c+1/2\end{matrix}\right)=$$
$$=4^{a+c-b}\Gamma\left[\begin{matrix}2c,\ 2c-2a-b\\ b,\ 4c-2a-2b\end{matrix}\right]{}_3F_2\left(\begin{matrix}c-b,\ c-b+1/2,\ 2c-2a-b;\ 1\\ 2c-a-b,\ 2c-a-b+1/2\end{matrix}\right)=$$

26. $$=\Gamma\left[\begin{matrix}2c,\ 2c-2a-b\\ 2c-2a,\ 2c-b\end{matrix}\right]{}_2F_1(2a,\ b;\ 2c-b;\ -1)$$
$[0<\operatorname{Re} b<2\operatorname{Re}(c-a)]$.

27. $${}_3F_2\left(\begin{matrix}1,\ a,\ b;\ 1\\ c,\ d\end{matrix}\right)+\frac{ab(2+a+b-c-d)}{cd[ab-(c-1)(d-1)]}\,{}_3F_2\left(\begin{matrix}2,\ a+1,\ b+1;\ 1\\ c+1,\ d+1\end{matrix}\right)=$$
$$=\frac{(1-c)(d-1)}{ab-(c-1)(d-1)}$$
$[\operatorname{Re}(a+b-c-d)<-2]$.

28. $${}_3F_2\left(\begin{matrix}1,\ a,\ b;\ 1\\ c,\ 2+a+b-c\end{matrix}\right)=\frac{1+a+b-c}{(1+a-c)(1+b-c)}\left\{1-c+\Gamma\left[\begin{matrix}c,\ 1+a+b-c\\ a,\ b\end{matrix}\right]\right\}.$$

29. $${}_3F_2\left(\begin{matrix}1,\ a,\ b;\ 1\\ 2,\ c\end{matrix}\right)=\frac{c-1}{(a-1)(b-1)}\left\{\Gamma\left[\begin{matrix}c-1,\ 1+c-a-b\\ c-a,\ c-b\end{matrix}\right]-1\right\}$$
$[a,\ b,\ c\neq 1;\ \operatorname{Re}(c-a-b)>-1]$.

30. $${}_3F_2\left(\begin{matrix}1,\ a,\ b;\ 1\\ 3,\ c\end{matrix}\right)=\frac{2(c-2)_2}{(a-2)_2(b-2)_2}\left\{\Gamma\left[\begin{matrix}c-2,\ 2+c-a-b\\ c-a,\ c-b\end{matrix}\right]-1\right\}-\frac{2(c-1)}{(a-1)(b-1)}$$
$[a,\ b,\ c\neq 1,2;\ \operatorname{Re}(c-a-b)>-2]$.

31. $${}_3F_2\left(\begin{matrix}1,\ a,\ b;\ 1\\ -m-a,\ -m-b\end{matrix}\right)=\frac{1}{2}\sum_{k=0}^{m+1}\frac{(a)_k(b)_k}{(-a-m)_k(-b-m)_k}+$$
$$+\frac{2^{2m-1}\sqrt{\pi}}{(1+m+2a)_m(1+m+2b)_m}\Gamma\left[\begin{matrix}1-a,\ 1-b,\ -a-b-m-1/2\\ -a-b-m,\ 1/2-a-m,\ 1/2-b-m\end{matrix}\right]$$
$[\operatorname{Re}(a+b)<-m-1/2;\ m=-2,\ -1,\ 0,\ 1,\ldots]$.

32. $${}_3F_2\left(\begin{matrix}1,\ a,\ b;\ 1\\ m-a,\ m-b\end{matrix}\right)=-\frac{1}{2}\sum_{k=1}^{m-2}\frac{(1+a-m)_k(1+b-m)_k}{(1-a)_k(1-b)_k}+$$
$$+\frac{\sqrt{\pi}(m-2a)_m(m-2b)_m(m-a-b)_m}{2^{2m+1}(m-a-b-1/2)_m}\Gamma\left[\begin{matrix}1-a,\ 1-b,\ 2m-a-b-1/2\\ 1/2+m-a,\ 1/2+m-b,\ 2m-a-b\end{matrix}\right]$$
$[\operatorname{Re}(a+b)<m-1/2,\ m=2,\ 3,\ 4,\ldots]$.

33. $${}_3F_2\left(\begin{matrix}1,\ a,\ b;\ 1\\ 1+a,\ 1+b\end{matrix}\right)=\frac{ab}{b-a}[\psi(b)-\psi(a)]$$ $[b\neq a]$.

34. $$=a^2\psi'(a)$$ $[b=a]$.

35. $${}_3F_2\left(\begin{matrix}1,\ a,\ b;\ 1\\ 2+a,\ 2+b\end{matrix}\right)=\frac{(a+1)(b+1)}{(a-b-1)_3}\{(a-b)(a+b+1)-2ab[\psi(a)-\psi(b)]\}$$
$[b\neq a-1,\ a,\ a+1]$.

36. $${}_3F_2\left(\begin{matrix}2,\ a,\ b;\ 1\\ 2+a,\ 2+b\end{matrix}\right)=\frac{(a)_2(b)_2(a+b-1)}{(a-b-1)_3}\left[\psi(a)-\psi(b)-\frac{2(a-b)}{a+b-1}\right]$$
$[b\neq a-1,\ a,\ a+1]$.

37. $${}_3F_2\left(\begin{matrix}3,\ a,\ b;\ 1\\ 2+a,\ 2+b\end{matrix}\right)=\frac{(a-1)_3(b-1)_3}{(a-b-1)_3}\left[\psi(b)-\psi(a)+\frac{(a-b)(a+b-3)}{2(a-1)(b-1)}\right]$$
$[b\neq a-1,\ a,\ a+1]$.

38. $${}_3F_2\left(\begin{matrix}1,\ a,\ b;\ 1\\ 1+a,\ 2+a-b\end{matrix}\right)=\frac{1+a-b}{b-1}\left[\psi(a)-\psi(2+a-2b)-\psi\left(\frac{a+1}{2}\right)+\right.$$
$$\left.+\psi\left(\frac{3+a}{2}-b\right)\right]$$ $[\operatorname{Re}(a-2b)>-2]$.

39. $${}_3F_2\left(\begin{matrix}1,\ a,\ b;\ 1\\ 2,\ 3\end{matrix}\right)=\frac{2}{(a-1)(b-1)}\left\{\Gamma\left[\begin{matrix}4-a-b\\ 3-a,\ 3-b\end{matrix}\right]-1\right\}$$ $[a,\ b\neq 1;\ \operatorname{Re}(a+b)<4]$.

40. $${}_3F_2\left(\begin{matrix}1,\ 1,\ a;\ 1\\ 2,\ b\end{matrix}\right)=\frac{b-1}{a-1}[\psi(b-1)-\psi(b-a)]\qquad [a\neq 1;\ \operatorname{Re}(b-a)>0],$$

41. $$=(b-1)\,\psi'(b-1)\qquad [a=1;\ \operatorname{Re} b>1].$$

42. $${}_3F_2\left(\begin{matrix}1,\ 1,\ a;\ 1\\ 3,\ b\end{matrix}\right)=\frac{2(b-1)(b-a)}{(a-2)_2}[\psi(b-a+1)-\psi(b-1)]+\frac{2(b-1)}{a-1}$$
$$[a\neq 1,\ 2;\ \operatorname{Re}(b-a)>-1],$$

43. $$=2(b-2)+2(b-1)^2\psi'(b)\qquad [a=1;\ \operatorname{Re} b>0],$$

44. $$=2(b-1)[1-(b-2)\psi'(b-1)]\qquad [a=2;\ \operatorname{Re} b>1].$$

45. $${}_3F_2\left(\begin{matrix}1,\ 2,\ a;\ 1\\ 3,\ b\end{matrix}\right)=\frac{2(b-2)_2}{(a-2)_2}[\psi(b-2)-\psi(b-a)]-\frac{2(b-1)}{a-1}$$
$$[a\neq 1,\ 2;\ \operatorname{Re}(b-a)>0].$$

46. $$=2(b-2)_2\,\psi'(b-2)-2(b-1)\qquad [a=2;\ \operatorname{Re} b>2].$$

47. $${}_3F_2\left(\begin{matrix}a,\ b,\ (a+b)/2;\ 1\\ a+b,\ (a+b+1)/2\end{matrix}\right)=\pi\left\{\Gamma\left[\begin{matrix}(a+b+1)/2\\ (a+1)/2,\quad (b+1)/2\end{matrix}\right]\right\}^2.$$

48. $${}_3F_2\left(\begin{matrix}a,\ b,\ 2a-b+1;\ 1\\ a+1,\ 2a+1\end{matrix}\right)=\Gamma\left[\begin{matrix}2a+1\\ b,\ 2a-b+1\end{matrix}\right][\beta(b)+\beta(2a-b+1)].$$

49. $${}_3F_2\left(\begin{matrix}a,\ a,\ b;\ 1\\ a+1,\ a+1\end{matrix}\right)=a\Gamma\left[\begin{matrix}a+1,\ 1-b\\ 1+a-b\end{matrix}\right][\psi(1+a-b)-\psi(a)]\qquad [\operatorname{Re} b<2].$$

50. $${}_3F_2\left(\begin{matrix}a,\ a,\ b;\ 1\\ a+2,\ a+2\end{matrix}\right)=\Gamma\left[\begin{matrix}2+a,\ 1-b\\ 2+a-b\end{matrix}\right]\{a(a+1)(b-2a-1)[\psi(2+a)-\psi(2+a-b)]+$$
$$+2a^2+(1-b)(1-2a^2)\}\qquad [b\neq 1,\ 2,\ 3;\ \operatorname{Re} b<4].$$

51. $${}_3F_2\left(\begin{matrix}a,\ a+1,\ b;\ 1\\ a+2,\ a+3\end{matrix}\right)=\frac{a+1}{2}\Gamma\left[\begin{matrix}a+3,\ 1-b\\ 2+a-b\end{matrix}\right]\left\{2a[\psi(a+1)-\psi(2+a-b)]+\right.$$
$$\left.+\frac{(1-b)(2+2a-b)}{2+a-b}\right\}\qquad [b\neq 1,\ 2,\ 3;\ \operatorname{Re} b<4],$$

52. $${}_3F_2\left(\begin{matrix}a,\ a/2,\ b,\ 1\\ a/2+1,\ a-b+1\end{matrix}\right)=\frac{\sqrt{\pi}}{2^a}\Gamma\left[\begin{matrix}a-b+1,\ 1-b,\ a/2+1\\ (a+1)/2,\ a/2-b+1,\ a/2-b+1\end{matrix}\right]\qquad [\operatorname{Re} b<1].$$

53. $${}_3F_2\left(\begin{matrix}a,\ a/2+1,\ b;\ 1\\ a/2,\ 1+a-b\end{matrix}\right)=0\qquad [a\neq 0;\ \operatorname{Re} b<0].$$

54. $${}_3F_2\left(\begin{matrix}a,\ a+1/2,\ b;\ 1\\ a+3/2,\ a+2\end{matrix}\right)=(2a+1)(a+1)\Gamma(1-b)\left\{\Gamma\left[\begin{matrix}a+1\\ a-b+1\end{matrix}\right]-\right.$$
$$\left.-2a\Gamma\left[\begin{matrix}a+1/2\\ a-b+3/2\end{matrix}\right]+a\Gamma\left[\begin{matrix}a+1\\ a-b+2\end{matrix}\right]\right\}\qquad [b\neq 1,\ 2;\ \operatorname{Re} b<3].$$

55. $${}_3F_2\left(\begin{matrix}a,\ a-m,\ a-n;\ 1\\ a-m+1,\ a-n+1\end{matrix}\right)=\frac{n!\,(n-a)}{(1-a)_{n-1}\sin a\pi}\qquad [m=n,\ a\neq 1,\ 0,\ -1,\ -2,\ \ldots;\ \operatorname{Re} a<2].$$

56. $$=0\qquad [m\neq n,\ a\neq 1,\ 0,\ -1,\ -2,\ \ldots;\ \operatorname{Re} a<2;\ m,\ n=1,\ 2,\ 3,\ \ldots].$$

57. $${}_3F_2\left(\begin{matrix}a,\ a,\ a;\ 1\\ 1,\ 1\end{matrix}\right)=\Gamma\left(1-\frac{3a}{2}\right)\Gamma^{-3}\left(1-\frac{a}{2}\right)\cos\frac{a\pi}{2}\qquad [\operatorname{Re} a<2/3].$$

58. $${}_3F_2\left(\begin{matrix}a,\ a,\ a;\ 1\\ 2,\ 2\end{matrix}\right)=(1-a)^{-3}\left\{\Gamma\left(2-\frac{3a}{2}\right)\Gamma^{-3}\left(1-\frac{a}{2}\right)\cos\frac{a\pi}{2}-\right.$$
$$\left.-\Gamma\left[\begin{matrix}(5-3a)/2\\ (1-a)/2\end{matrix}\right]\Gamma^{-2}\left(\frac{3-a}{2}\right)\sin\frac{a\pi}{2}\right\}\qquad [a\neq 1;\ \operatorname{Re} a<4/3].$$

59. $${}_3F_2\left(\begin{matrix}a,\ a+1/3,\ a+2/3;\ 1\\ 1/3,\ 2/3\end{matrix}\right)=2\cdot 3^{-3a/2-1}\cos\frac{a\pi}{2}\qquad [\operatorname{Re} a<0].$$

60. $${}_3F_2\left(\begin{matrix}a,\ a+1/3,\ a+2/3;\ 1\\ 2/3,\ 4/3\end{matrix}\right)=\frac{2}{1-3a}3^{-(3a+1)/2}\cos(3a+1)\frac{\pi}{6}\qquad [\operatorname{Re} a<1/3].$$

61. $${}_3F_2\left(\begin{matrix}a,\ a+1/3,\ a+2/3;\ 1\\ 4/3,\ 5/3\end{matrix}\right)=\frac{4}{(1-3a)(2-3a)}3^{-3a/2}\cos(3a+2)\frac{\pi}{6}$$
$$[a\neq 1/3;\ \operatorname{Re} a<2/3]$$

62. $_3F_2\left(\begin{matrix}1,\ a,\ -a;\ 1\\ 1+a,\ 1-a\end{matrix}\right)=\frac{1}{2}+\frac{\pi a}{2}\operatorname{ctg} a\pi.$

63. $_3F_2\left(\begin{matrix}1,\ a,\ 1-a;\ 1\\ 1+a,\ 2-a\end{matrix}\right)=\frac{\pi a(1-a)}{1-2a}\operatorname{ctg} a\pi.$

64. $_3F_2\left(\begin{matrix}1,\ a,\ a;\ 1\\ 2+a,\ 2+a\end{matrix}\right)=(a+1)^2\,[2a^2\psi'(a)-2a-1].$

65. $_3F_2\left(\begin{matrix}1,\ a,\ a+1/2;\ 1\\ a+3/2,\ a+2\end{matrix}\right)=(2a+1)(a+1)\left[2a\psi\left(a+\frac{1}{2}\right)-2a\psi(a)-1\right].$

66. $_3F_2\left(\begin{matrix}1,\ a,\ a+1;\ 1\\ a+2,\ a+3\end{matrix}\right)=\frac{(a+1)(a+2)}{2}\,[2a+1-2a(a+1)\psi'(a+1)].$

67. $_3F_2\left(\begin{matrix}2,\ a,\ a;\ 1\\ a+2,\ a+2\end{matrix}\right)=-a^2(a+1)^2\,[(2a-1)\psi'(a)+2].$

68. $_3F_2\left(\begin{matrix}2,\ a,\ a+1/2;\ 1\\ a+3/2,\ a+2\end{matrix}\right)=a(2a+1)(a+1)\left[(2a-1)\psi(a)-(2a-1)\psi\left(a-\frac{1}{2}\right)-1\right].$

69. $_3F_2\left(\begin{matrix}2,\ a,\ a+1;\ 1\\ a+2,\ a+3\end{matrix}\right)=\frac{(a+1)^2(a+2)}{2}\,[2a^2\psi'(a+1)-2a+1].$

70. $_3F_2\left(\begin{matrix}3,\ a,\ a;\ 1\\ a+2,\ a+2\end{matrix}\right)=\frac{a^2(a+1)^2}{2}\,[2(a-1)^2\psi'(a)-2a+3].$

71. $_3F_2\left(\begin{matrix}3,\ a,\ a+1;\ 1\\ a+2,\ a+3\end{matrix}\right)=\frac{(a+1)^2(a+2)}{4}\,[2a^2-3a+2-2a(a-1)\psi'(a+1)].$

72. $_3F_2\left(\begin{matrix}1/2,\ 1,\ a;\ 1\\ 3/2,\ 2-a\end{matrix}\right)=\frac{\pi}{4}\Gamma\left[\begin{matrix}1-a,\ 2-a\\ 3/2-a,\ 3/2-a\end{matrix}\right]$ [Re $a<1$].

73. $_3F_2\left(\begin{matrix}1/2,\ 1/2,\ a;\ 1\\ 1,\ 1\end{matrix}\right)=\sqrt{\pi}\,\Gamma\left[\begin{matrix}(3+2a)/4,\ (3-2a)/4\\ 3/4,\ 3/4,\ (1+a)/2,\ 1-a/2\end{matrix}\right]$ [Re $a<1$].

74. $_3F_2\left(\begin{matrix}1,\ m,\ n;\ 1\\ m+1,\ n+1\end{matrix}\right)=\frac{mn}{m-n}\sum_{k=1}^{m-n}\frac{1}{k+n-1}$ [$m>n$].

75. $_3F_2\left(\begin{matrix}1,\ m/n,\ m/n+l;\ 1\\ m/n+1,\ m/n+l+1\end{matrix}\right)=\frac{m(m+nl)}{nl}\sum_{k=0}^{l-1}\frac{1}{kn+m}.$

76. $_3F_2\left(\begin{matrix}1,\ 1,\ m;\ 1\\ 2,\ m+1\end{matrix}\right)=\frac{m}{m-1}\sum_{k=1}^{m-1}\frac{1}{k}.$

77. $_3F_2\left(\begin{matrix}1,\ 1,\ m/n+l;\ 1\\ 2,\ m/n+l+1\end{matrix}\right)=\frac{m+nl}{m+nl-n}\left[n\sum_{k=1}^{l-1}\frac{1}{kn+m}+\psi\left(\frac{m}{n}+1\right)+C\right]$

[$l=0,\ 1,\ 2,\ \ldots;\ m=1,\ 2,\ \ldots,\ n-1;\ n=2,\ 3,\ 4,\ \ldots$].

78. $_3F_2\left(\begin{matrix}1/2,\ 1,\ 1;\ 1\\ n/2,\ (n+1)/2\end{matrix}\right)=2^{n-3}(n-1)\left[\ln 2+\sum_{k=1}^{n-3}\frac{(3-n)_k}{k\cdot k!}(1-2^{-k})\right]$

[$n=3,\ 4,\ 5,\ \ldots$].

79. $_3F_2\left(\begin{matrix}1,\ 1,\ 1;\ 1\\ n,\ n\end{matrix}\right)=\frac{(2n-4)!}{6}\left[\frac{n-1}{(n-2)!}\right]^2\left[\pi^2-18\sum_{k=0}^{n-3}\frac{(k!)^2}{(2k+2)!}\right]$ [$n=2,\ 3,\ 4,\ \ldots$].

80. $_3F_2\left(\begin{matrix}1,\ 1,\ 3/2;\ 1\\ n/2,\ (n+1)/2\end{matrix}\right)=$

$$=(n-1)(n-2)\left[\frac{1+(-1)^n(2^{n-3}-1)}{2(n-3)}-2^{n-4}\ln 2-2^{n-4}\sum_{k=1}^{n-4}\frac{(3-n)_k}{k\cdot k!}(1-2^{-k})\right]$$

[$n=4,\ 5,\ 6,\ \ldots$].

81. $${}_3F_2\left(\begin{matrix}-n, a, b; 1\\ c, d\end{matrix}\right)=\frac{(c-a)_n(d-a)_n}{(c)_n(d)_n}\,{}_3F_2\left(\begin{matrix}-n, a, a+b-c-d-n+1; 1\\ a-c-n+1, a-d-n+1\end{matrix}\right)=$$

82. $$=\frac{(a)_n(c+d-a-b)_n}{(c)_n(d)_n}\,{}_3F_2\left(\begin{matrix}-n, c-a, d-a; 1\\ 1-a-n, c+d-a-b\end{matrix}\right)=$$

83. $$=\frac{(c+d-a-b)_n}{(c)_n}\,{}_3F_2\left(\begin{matrix}-n, d-a, d-b; 1\\ d, c+d-a-b\end{matrix}\right)=$$

84. $$=(-1)^n\frac{(a)_n(b)_n}{(c)_n(d)_n}\,{}_3F_2\left(\begin{matrix}-n, 1-c-n, 1-d-n; 1\\ 1-a-n, 1-b-n\end{matrix}\right)=$$

85. $$=(-1)^n\frac{(d-a)_n(d-b)_n}{(c)_n(d)_n}\,{}_3F_2\left(\begin{matrix}-n, 1-d-n, a+b-c-d-n+1; 1\\ a-d-n+1, b-d-n+1\end{matrix}\right)=$$

86. $$=\frac{(c-a)_n}{(c)_n}\,{}_3F_2\left(\begin{matrix}-n, a, d-b; 1\\ d, a-c-n+1\end{matrix}\right)=$$

87. $$=\frac{(c-a)_n(b)_n}{(c)_n(d)_n}\,{}_3F_2\left(\begin{matrix}-n, d-b, 1-c-n; 1\\ 1-b-n, a-c-n+1\end{matrix}\right).$$

88. $${}_3F_2\left(\begin{matrix}-n, a, b; 1\\ c, 1+a+b-c-n\end{matrix}\right)=\frac{(c-a)_n(c-b)_n}{(c)_n(c-a-b)_n}.$$

89. $${}_3F_2\left(\begin{matrix}-n, a, b; 1\\ c, 2+a+b-c-n\end{matrix}\right)=\frac{(c-a-1)_n(c-b)_n}{(c)_n(c-a-b-1)_n}\left[1+\frac{nb}{(c-a-1)(b-c-n+1)}\right].$$

90. $${}_3F_2\left(\begin{matrix}-n, a, b; 1\\ a-l, b-m\end{matrix}\right)=0 \qquad [l+m=1, 2, 3, \dots, n-1],$$

91. $$=R \qquad [l+m=n],$$

92. $$=R\,[(a-l+n)(l+m)+m(b-a-m)] \qquad [l+m=n+1],$$

93. $$=\frac{R}{2}\,[m(m-1)(b-m+n)(b-m+n+1)+l(l-1)(a-l-m+n)\times$$
$$\times(a-l-m+n+1)+2lm(b-m+n)(a-l-m+n+1)] \qquad [l+m=n+2],$$
$$R=\frac{(-1)^n\, n!}{(1-a)_l\,(1-b)_m}.$$

94. $${}_3F_2\left(\begin{matrix}-n, a, b; 1\\ a+l, b+m\end{matrix}\right)=$$
$$=n!\,(a)_l\,(b)_m\left[\frac{1}{(l-1)!\,(a)_{n+1}\,(b-a)_m}\,{}_3F_2\left(\begin{matrix}1-l, a, 1+a-b-m; 1\\ 1+a+n, 1+a-b\end{matrix}\right)+\right.$$
$$\left.+\frac{1}{(m-1)!\,(b)_{n+1}\,(a-b)_l}\,{}_3F_2\left(\begin{matrix}1-m, b, 1+b-a-l; 1\\ 1+b+n, 1+b-a\end{matrix}\right)\right]$$
$$[a\neq 0, -1, \dots, -n-l-1;\ b\neq 0, -1, \dots, -n-m-1;\ a+b\neq 1-l, 2-l, \dots, m-1].$$

95. $${}_3F_2\left(\begin{matrix}-n, a, b; 1\\ 1+a, b-m\end{matrix}\right)=\frac{n!\,(1+a-b)_m}{(1+a)_n\,(1-b)_m} \qquad [m=0, 1, 2, \dots, n],$$

96. $$=\frac{n!\,a}{(1-b)_{n+1}}\left[\frac{(1+a-b)_{n+1}}{(a)_{n+1}}-1\right] \qquad [m=n+1].$$

97. $${}_3F_2\left(\begin{matrix}-n, a, b; 1\\ 1-a-n, 1-b-n\end{matrix}\right)=\left\{\begin{matrix}1\\0\end{matrix}\right\}\frac{(-4)^m\,(1-a-b-n)_m\,(1/2)_m}{(1-a-n)_m\,(1-b-n)_m}$$
$$\left[m=\frac{n}{2},\ \left\{\begin{matrix}n=0, 2, 4, \dots\\ n=1, 3, 5, \dots\end{matrix}\right\}\right].$$

98. $${}_3F_2\left(\begin{matrix}-n, a, b; 1\\ 1+2a-n, 1-a+b\end{matrix}\right)=\frac{(2a-b-2n)\,(-a)_n\,(b-2a)_n}{(2a-b)\,(-2a)_n\,(1+b-a)_n}.$$

99. $${}_3F_2\left(\begin{matrix}-n, a, b; 1\\ 1+a+n, 1+a-b\end{matrix}\right)=\frac{(1+a)_n\,(1+a/2-b)_n}{(1+a/2)_n\,(1+a-b)_n}.$$

100. $${}_3F_2\left(\begin{matrix}-n, a, b; 1\\ 2a, (1+b-n)/2\end{matrix}\right)=\left\{\begin{matrix}1\\0\end{matrix}\right\}\frac{(1/2)_m\,((1-b)/2+a)_m}{((1-b)/2)_m\,(a+1/2)_m}$$
$$\left[a\neq 0,\ m=\frac{n}{2},\ \left\{\begin{matrix}n=0, 2, 4, \dots\\ n=1, 3, 5, \dots\end{matrix}\right\}\right].$$

101. $${}_3F_2\left(\begin{matrix}-n,\ a,\ b;\ 1\\ -2n,\ (a+b+1)/2\end{matrix}\right)=\frac{((a+1)/2)_n\,((b+1)/2)_n}{(1/2)_n\,((a+b+1)/2)_n}.$$

102. $${}_3F_2\left(\begin{matrix}-n,\ a,\ b;\ 1\\ (a-n)/2,\ (1+a-n)/2\end{matrix}\right)={}_2F_1(-n,\ 2b;\ a-n;\ 2)=$$

103. $$=\frac{(2b-a+1)_n}{(1-a)_n}\,{}_2F_1(-n,\ 2b;\ 2b-a+1;\ -1)\qquad [a\neq 0,\ -1,\ -2,\ \ldots].$$

104. $${}_3F_2\left(\begin{matrix}-n,\ a,\ 1-a;\ 1\\ b,\ 1-b-2n\end{matrix}\right)=\frac{2^{2n+1}}{(b)_{2n}}\left(\frac{a+b}{2}\right)_n\left(\frac{b-a+1}{2}\right)_n.$$

105. $${}_3F_2\left(\begin{matrix}-n,\ 1,\ a;\ 1\\ b,\ a-m\end{matrix}\right)=\frac{(b-1)\,(a-m-1)}{(b+n-1)\,(a-1)}\,{}_3F_2\left(\begin{matrix}-m,\ 1,\ 2-b;\ 1\\ 2-b-n,\ 2-a\end{matrix}\right).$$

106. $${}_3F_2\left(\begin{matrix}-n,\ a,\ a/2+1;\ 1\\ a/2,\ b\end{matrix}\right)=(b-a-n-1)\,\frac{(b-a)_{n-1}}{(b)_n}.$$

107. $${}_3F_2\left(\begin{matrix}-n,\ a,\ a/2+1;\ 1\\ a/2,\ 1+a+n\end{matrix}\right)=0.$$

108. $${}_3F_2\left(\begin{matrix}-n,\ a,\ a+1/2;\ 1\\ (1+2a-n)/3,\ (2+2a-n)/3\end{matrix}\right)=$$
$$=(-1)^n\,\frac{((2a-n)/3)_n}{(1-2a)_n}\,{}_2F_1\left(-n,\ \frac{2}{3}(2a-n);\ 1-\frac{2}{3}(a+n);\ -8\right).$$

109. $${}_3F_2\left(\begin{matrix}-n,\ 1,\ 1;\ 1\\ l,\ m\end{matrix}\right)=\frac{l-1}{n+l-1}\,{}_3F_2\left(\begin{matrix}-n,\ m-1,\ 1;\ 1\\ m,\ 2-n-l\end{matrix}\right)$$
$$[m=1,\ 2,\ 3,\ \ldots;\ l=1,\ 2,\ 3\ \ldots\ \text{or}\ l=-n,\ -n-1,\ -n-2,\ \ldots].$$

110. $${}_3F_2\left(\begin{matrix}-n,\ 2,\ 2;\ 1\\ 1,\ a\end{matrix}\right)=\frac{(1-a)\,(2-a)\,(a-n-3)}{(n+a-1)\,(n+a-2)\,(n+a-3)}.$$

111. $${}_3F_2\left(\begin{matrix}-n,\ -n,\ m;\ 1\\ 1,\ m+1\end{matrix}\right)=\frac{(2n)!}{(n!)^2}\,{}_3F_2\left(\begin{matrix}-n,\ -n,\ m+1;\ 1\\ m+2,\ -2n\end{matrix}\right).$$

112. $${}_3F_2\left(\begin{matrix}-n,\ -n,\ 1;\ 1\\ -2n,\ m\end{matrix}\right)=\frac{(n!)^2}{(2n)!}\,{}_3F_2\left(\begin{matrix}-n,\ -n,\ m-1;\ 1\\ 1,\ m\end{matrix}\right).$$

113. $${}_3F_2\left(\begin{matrix}-n/2,\ (1-n)/2,\ a;\ 1\\ b,\ b+1/2\end{matrix}\right)=\frac{(2b-a)_n}{(2b)_n}\,{}_2F_1(-n,\ a;\ 2b-a;\ -1).$$

114. $${}_3F_2\left(\begin{matrix}-n/2,\ (1-n)/2,\ a;\ 1\\ b,\ 3/2+a-b-n\end{matrix}\right)=4^{-n}\,\frac{(2b-2a-1)_n\,(2b+n-1)_n}{(b)_n\,(b-a-1/2)_n}.$$

115. $${}_3F_2\left(\begin{matrix}-n/2,\ (1-n)/2,\ a;\ 1\\ a+1,\ 1/2-m\end{matrix}\right)=\frac{n!\,(1/2+a)_m}{(1/2)_m\,(1+2a)_n}\qquad [n\leqslant 2m\leqslant 2n],$$

116. $$=\frac{n!}{(1/2)_{n+1}}\left[\frac{(1/2+a)_{n+1}}{(1+2a)_n}-\frac{a}{2^n}\right]\qquad [m=n+1].$$

117. $${}_3F_2\left(\begin{matrix}-n/2,\ (1-n)/2,\ a;\ 1\\ (1+a-n)/2,\ 1+(a-n)/2\end{matrix}\right)=(-1)^n\,\frac{(a)_n}{(-a)_n}.$$

118. $${}_3F_2\left(\begin{matrix}-n/2,\ (1-n)/2,\ a;\ 1\\ (1+a-n)/3,\ (2+a-n)/3\end{matrix}\right)=$$
$$=2^{-n}\,\frac{((2a-2n)/3)_n}{(1-a)_n}\,{}_2F_1\left(-n,\ \frac{a-n}{3};\ 1-\frac{2a+n}{3};\ -8\right).$$

119. $${}_3F_2\left(\begin{matrix}-n/2,\ (1-n)/2,\ -m;\ 1\\ 1,\ -m-n\end{matrix}\right)=2^{-n}\,{}_3F_2\left(\begin{matrix}-n,\ -n,\ m+1;\ -1\\ 1,\ -n-m\end{matrix}\right)=$$

120. $$=2^n\,\frac{(1/2)_n}{n!}\,{}_3F_2\left(\begin{matrix}-n,\ -n/2,\ (1-n)/2;\ 1\\ 1/2-n,\ -n-m\end{matrix}\right).$$

121. $${}_3F_2\left(\begin{matrix}-n/3,\ (1-n)/3,\ (2-n)/3;\ 1\\ a,\ a+1/3\end{matrix}\right)=$$
$$=\frac{(a-1/3)_n}{(3a-1)_n}\,{}_2F_1\left(-\frac{n}{2},\ \frac{1-n}{2};\ \frac{4}{3}-a-n;\ \frac{4}{3}\right).$$

122. $${}_3F_2\left(\begin{matrix}-n/3,\ (1-n)/3,\ (2-n)/3;\ 1\\ a,\ 2-n-2a\end{matrix}\right)={}_2F_1\left(-n,\ a-\frac{1}{2};\ 2-n-2a;\ 4\right).$$

123. $${}_3F_2\left(\begin{matrix}-2n/3,\ (1-2n)/3,\ (2-2n)/3;\ 1\\ a,\ 1/2-n\end{matrix}\right)=9^{-n}{}_2F_1(-n,\ 2-2n-2a;\ a;\ 4).$$

	a_1	a_2	a_3	b_1	b_2	${}_3F_2(a_1,\ a_2,\ a_3;\ b_1,\ b_2;\ 1)$
124	$\frac{1}{8}$	$\frac{1}{4}$	1	$\frac{5}{4}$	$\frac{9}{4}$	$\frac{\sqrt{2}}{8}[\pi+\sqrt{2}\ln 2+2\ln(1+\sqrt{2})]$
125	$\frac{1}{8}$	$\frac{3}{8}$	1	$\frac{9}{8}$	$\frac{11}{8}$	$\frac{3}{16}[\pi+2\sqrt{2}\ln(1+\sqrt{2})]$
126	$\frac{1}{8}$	$\frac{1}{2}$	1	$\frac{9}{8}$	$\frac{3}{2}$	$\frac{1}{12}[(1+\sqrt{2})\pi+4\ln 2+2\sqrt{2}\ln(1+\sqrt{2})]$
127	$\frac{1}{8}$	$\frac{5}{8}$	1	$\frac{9}{8}$	$\frac{13}{8}$	$\frac{5\sqrt{2}}{32}[\pi+2\ln(1+\sqrt{2})]$
128	$\frac{1}{8}$	$\frac{3}{4}$	1	$\frac{9}{8}$	$\frac{7}{4}$	$\frac{3\sqrt{2}}{40}[(1+\sqrt{2})\pi+\sqrt{2}\ln 2+2\ln(1+\sqrt{2})]$
129	$\frac{1}{8}$	$\frac{7}{8}$	1	$\frac{9}{8}$	$\frac{15}{8}$	$\frac{7}{48}(1+\sqrt{2})\pi$
130	$\frac{1}{8}$	1	1	$\frac{9}{8}$	2	$\frac{1}{14}[(1+\sqrt{2})\pi+8\ln 2+2\sqrt{2}\ln(1+\sqrt{2})]$
131	$\frac{1}{6}$	$\frac{1}{3}$	1	$\frac{7}{6}$	$\frac{4}{3}$	$3^{-3/2}(\pi+2\sqrt{3}\ln 2)$
132	$\frac{1}{6}$	$\frac{2}{3}$	1	$\frac{7}{6}$	$\frac{5}{3}$	$\frac{4\sqrt{3}}{27}(\pi+\sqrt{3}\ln 2)$
133	$\frac{1}{6}$	$\frac{5}{6}$	1	$\frac{7}{6}$	$\frac{11}{6}$	$\frac{5\sqrt{3}}{24}\pi$
134	$\frac{1}{4}$	$\frac{1}{4}$	$\frac{1}{2}$	$\frac{5}{4}$	$\frac{5}{4}$	$\frac{\sqrt{\pi}}{16\sqrt{2}}\Gamma^2\left(\frac{1}{4}\right)$
135	$\frac{1}{4}$	$\frac{1}{4}$	1	$\frac{5}{4}$	$\frac{5}{4}$	$\frac{1}{16}\zeta\left(2,\ \frac{1}{4}\right)=1{,}07483307\ldots$
136	$\frac{1}{4}$	$\frac{3}{8}$	1	$\frac{5}{4}$	$\frac{11}{8}$	$\frac{3\sqrt{2}}{8}[(\sqrt{2}-1)\pi-\sqrt{2}\ln 2+2\ln(1+\sqrt{2})]$
137	$\frac{1}{4}$	$\frac{1}{2}$	1	$\frac{5}{4}$	$\frac{3}{2}$	$\frac{1}{4}(\pi+2\ln 2)$
138	$\frac{1}{4}$	$\frac{1}{2}$	1	$\frac{5}{4}$	$\frac{5}{2}$	$\frac{3}{10}(\pi+2\ln 2-1)$
139	$\frac{1}{4}$	$\frac{1}{2}$	1	$\frac{3}{2}$	$\frac{9}{4}$	$\frac{5}{12}(\pi+2\ln 2-2)$
140	$\frac{1}{4}$	$\frac{1}{2}$	1	$\frac{9}{4}$	$\frac{5}{2}$	$\frac{1}{2}(2\pi+4\ln 2-7)$
141	$\frac{1}{4}$	$\frac{5}{8}$	1	$\frac{5}{4}$	$\frac{13}{8}$	$\frac{5\sqrt{2}}{24}[\pi-\sqrt{2}\ln 2+2\ln(1+\sqrt{2})]$
142	$\frac{1}{4}$	$\frac{3}{4}$	1	$\frac{5}{4}$	$\frac{7}{4}$	$\frac{3\pi}{8}$

	a_1	a_2	a_3	b_1	b_2	$_3F_2(a_1, a_2, a_3; b_1, b_2; 1)$
143	$\frac{1}{4}$	$\frac{3}{4}$	1	$\frac{5}{4}$	$\frac{11}{4}$	$\frac{7}{48}(3\pi-2)$
144	$\frac{1}{4}$	$\frac{3}{4}$	1	$\frac{7}{4}$	$\frac{9}{4}$	$\frac{15}{16}(\pi-2)$
145	$\frac{1}{4}$	$\frac{7}{8}$	1	$\frac{5}{4}$	$\frac{15}{8}$	$\frac{7\sqrt{2}}{40}[(1+\sqrt{2})\pi-\sqrt{2}\ln 2-2\ln(1+\sqrt{2})]$
146	$\frac{1}{4}$	1	1	$\frac{5}{4}$	2	$\frac{\pi}{6}+\ln 2$
147	$\frac{1}{4}$	1	1	2	$\frac{9}{4}$	$\frac{5}{6}(\pi+6\ln 2-6)$
148	$\frac{1}{4}$	1	$\frac{3}{2}$	$\frac{5}{4}$	$\frac{5}{2}$	$\frac{3}{20}(4+\pi+2\ln 2)$
149	$\frac{1}{4}$	1	$\frac{3}{2}$	$\frac{9}{4}$	$\frac{5}{2}$	$\frac{3}{4}(6-\pi-2\ln 2)$
150	$\frac{1}{4}$	1	$\frac{7}{4}$	$\frac{5}{4}$	$\frac{11}{4}$	$\frac{7}{72}(3\pi+4)$
151	$\frac{1}{4}$	1	$\frac{7}{4}$	$\frac{9}{4}$	$\frac{11}{4}$	$\frac{35}{144}(14-3\pi)$
152	$\frac{1}{3}$	$\frac{1}{2}$	1	$\frac{4}{3}$	$\frac{3}{2}$	$\frac{\sqrt{3}}{6}(\pi-4\sqrt{3}\ln 2+3\sqrt{3}\ln 3)$
153	$\frac{1}{3}$	$\frac{2}{3}$	1	$\frac{4}{3}$	$\frac{5}{3}$	$\frac{2\pi}{3\sqrt{3}}$
154	$\frac{1}{3}$	$\frac{2}{3}$	1	$\frac{4}{3}$	$\frac{8}{3}$	$\frac{5\sqrt{3}}{72}(2\pi-\sqrt{3})$
155	$\frac{1}{3}$	$\frac{2}{3}$	1	$\frac{5}{3}$	$\frac{7}{3}$	$\frac{4\sqrt{3}}{9}(\pi-\sqrt{3})$
156	$\frac{1}{3}$	$\frac{5}{6}$	1	$\frac{4}{3}$	$\frac{11}{6}$	$\frac{10\sqrt{3}}{27}(\pi-\sqrt{3}\ln 2)$
157	$\frac{1}{3}$	1	1	$\frac{4}{3}$	2	$\frac{\sqrt{3}}{12}(\pi+3\sqrt{3}\ln 3)$
158	$\frac{1}{3}$	1	$\frac{5}{3}$	$\frac{4}{3}$	$\frac{8}{3}$	$\frac{5\sqrt{3}}{72}(3\sqrt{3}+2\pi)$
159	$\frac{3}{8}$	$\frac{1}{2}$	1	$\frac{11}{8}$	$\frac{3}{2}$	$\frac{3}{4}[(\sqrt{2}-1)\pi+4\ln 2-2\sqrt{2}\ln(1+\sqrt{2})]$
160	$\frac{3}{8}$	$\frac{5}{8}$	1	$\frac{11}{8}$	$\frac{13}{8}$	$\frac{15}{16}(\sqrt{2}-1)\pi$
161	$\frac{3}{8}$	$\frac{3}{4}$	1	$\frac{11}{8}$	$\frac{7}{4}$	$\frac{3\sqrt{2}}{8}[\pi+\sqrt{2}\ln 2-2\ln(1+\sqrt{2})]$
162	$\frac{3}{8}$	$\frac{7}{8}$	1	$\frac{11}{8}$	$\frac{15}{8}$	$\frac{21\sqrt{2}}{32}[\pi-2\ln(1+\sqrt{2})]$
163	$\frac{3}{8}$	1	1	$\frac{11}{8}$	2	$\frac{3}{10}[(\sqrt{2}-1)\pi+8\ln 2-2\sqrt{2}\ln(1+\sqrt{2})]$
164	$\frac{1}{2}$	$\frac{1}{2}$	$\frac{1}{2}$	1	1	$\frac{1}{4\pi^3}\Gamma^4\left(\frac{1}{4}\right)=1,39320393\ldots$
165	$\frac{1}{2}$	$\frac{1}{2}$	$\frac{1}{2}$	1	$\frac{3}{2}$	$\frac{4}{\pi}\mathbf{G}$
166	$\frac{1}{2}$	$\frac{1}{2}$	$\frac{1}{2}$	$\frac{3}{2}$	$\frac{3}{2}$	$\frac{\pi}{2}\ln 2$

	a_1	a_2	a_3	b_1	b_2	${}_3F_2(a_1, a_2, a_3; b_1, b_2; 1)$
167	$\frac{1}{2}$	$\frac{1}{2}$	1	$\frac{3}{2}$	$\frac{3}{2}$	$\frac{\pi^2}{8}$
168	$\frac{1}{2}$	$\frac{1}{2}$	1	$\frac{3}{2}$	$\frac{5}{2}$	$\frac{3}{16}(\pi^2-8)$
169	$\frac{1}{2}$	$\frac{1}{2}$	1	2	2	$\frac{4}{\pi}(4-\pi)$
170	$\frac{1}{2}$	$\frac{1}{2}$	1	$\frac{5}{2}$	$\frac{5}{2}$	$\frac{9}{16}(\pi^2-8)$
171	$\frac{1}{2}$	$\frac{1}{2}$	1	$\frac{7}{2}$	$\frac{7}{2}$	$\frac{225}{256}\left(3\pi^2-\frac{256}{9}\right)$
172	$\frac{1}{2}$	$\frac{1}{2}$	$\frac{3}{2}$	1	$\frac{5}{2}$	$\frac{3}{2\pi}(2\mathbf{G}+1)$
173	$\frac{1}{2}$	$\frac{1}{2}$	2	1	3	$\frac{40}{9\pi}$
174	$\frac{1}{2}$	$\frac{1}{2}$	2	$\frac{5}{2}$	$\frac{5}{2}$	$\frac{9}{8}$
175	$\frac{1}{2}$	$\frac{1}{2}$	$\frac{5}{2}$	1	$\frac{7}{2}$	$\frac{5}{32\pi}(18\mathbf{G}+13)$
176	$\frac{1}{2}$	$\frac{1}{2}$	3	1	4	$\frac{356}{75\pi}$
177	$\frac{1}{2}$	$\frac{1}{2}$	$\frac{7}{2}$	1	$\frac{9}{2}$	$\frac{7}{128\pi}(50\mathbf{G}+43)$
178	$\frac{1}{2}$	$\frac{5}{8}$	1	$\frac{3}{2}$	$\frac{13}{8}$	$\frac{5}{4}[(\sqrt{2}-1)\pi-4\ln 2+2\sqrt{2}\ln(1+\sqrt{2})]$
179	$\frac{1}{2}$	$\frac{2}{3}$	1	$\frac{3}{2}$	$\frac{5}{3}$	$\frac{\sqrt{3}}{3}(\pi+4\sqrt{3}\ln 2-3\sqrt{3}\ln 3)$
180	$\frac{1}{2}$	$\frac{3}{4}$	1	$\frac{3}{2}$	$\frac{7}{4}$	$\frac{3}{4}(\pi-2\ln 2)$
181	$\frac{1}{2}$	$\frac{3}{4}$	1	$\frac{3}{2}$	$\frac{11}{4}$	$\frac{7}{20}(3\pi-6\ln 2-2)$
182	$\frac{1}{2}$	$\frac{3}{4}$	1	$\frac{7}{4}$	$\frac{5}{2}$	$\frac{3}{2}(\pi-2\ln 2-1)$
183	$\frac{1}{2}$	$\frac{7}{8}$	1	$\frac{3}{2}$	$\frac{15}{8}$	$\frac{7}{12}[(1+\sqrt{2})\pi-4\ln 2-2\sqrt{2}\ln(1+\sqrt{2})]$
184	$\frac{1}{2}$	1	1	$\frac{3}{2}$	2	$2\ln 2$
185	$\frac{1}{2}$	1	1	2	$\frac{5}{2}$	$3(2\ln 2-1)$
186	$\frac{1}{2}$	1	1	$\frac{5}{2}$	3	$16\ln 2-10$
187	$\frac{1}{2}$	1	$\frac{5}{4}$	$\frac{3}{2}$	$\frac{9}{4}$	$\frac{5}{12}(8-\pi-2\ln 2)$
188	$\frac{1}{2}$	1	$\frac{5}{4}$	$\frac{9}{4}$	$\frac{5}{2}$	$\frac{5}{2}(5-\pi-2\ln 2)$
189	$\frac{1}{2}$	1	$\frac{3}{2}$	2	2	$\frac{4}{\pi}(\pi-2)$
190	$\frac{1}{2}$	1	$\frac{7}{4}$	$\frac{3}{2}$	$\frac{11}{4}$	$\frac{7}{60}(8+3\pi-6\ln 2)$

	a_1	a_2	a_3	b_1	b_2	${}_3F_2(a_1, a_2, a_3; b_1, b_2; 1)$
191	$\frac{1}{2}$	1	$\frac{7}{4}$	$\frac{5}{2}$	$\frac{11}{4}$	$\frac{7}{10}(7-3\pi+6\ln 2)$
192	$\frac{1}{2}$	1	2	$\frac{3}{2}$	3	$\frac{2}{3}(1+2\ln 2)$
193	$\frac{1}{2}$	1	$\frac{7}{2}$	$\frac{5}{2}$	$\frac{9}{2}$	$\frac{77}{60}$
194	$\frac{5}{8}$	$\frac{7}{8}$	1	$\frac{13}{8}$	$\frac{15}{8}$	$\frac{35}{16}[\pi-2\sqrt{2}\ln(1+\sqrt{2})]$
195	$\frac{5}{8}$	1	1	$\frac{13}{8}$	2	$\frac{5}{6}[8\ln 2-(\sqrt{2}-1)\pi-2\sqrt{2}\ln(1+\sqrt{2})]$
196	$\frac{2}{3}$	1	1	$\frac{5}{3}$	2	$3\ln 3-\frac{\pi}{\sqrt{3}}$
197	$\frac{2}{3}$	1	$\frac{4}{3}$	$\frac{5}{3}$	$\frac{7}{3}$	$\frac{4\sqrt{3}}{9}(3\sqrt{3}-\pi)$
198	$\frac{3}{4}$	$\frac{3}{4}$	1	$\frac{11}{4}$	$\frac{11}{4}$	$\frac{441}{32}(\pi^2+2\pi-16)$
199	$\frac{3}{4}$	$\frac{7}{8}$	1	$\frac{7}{4}$	$\frac{15}{8}$	$\frac{21\sqrt{2}}{8}[\pi-\sqrt{2}\ln 2-2\ln(1+\sqrt{2})]$
200	$\frac{3}{4}$	1	1	$\frac{7}{4}$	2	$\frac{3}{2}(6\ln 2-\pi)$
201	$\frac{3}{4}$	1	1	2	$\frac{11}{4}$	$\frac{7}{6}(18\ln 2-3\pi-2)$
202	$\frac{3}{4}$	1	$\frac{5}{4}$	$\frac{7}{4}$	$\frac{9}{4}$	$\frac{15}{8}(4-\pi)$
203	$\frac{3}{4}$	1	$\frac{5}{4}$	$\frac{9}{4}$	$\frac{11}{4}$	$\frac{35}{16}(10-3\pi)$
204	$\frac{3}{4}$	1	$\frac{3}{2}$	$\frac{7}{4}$	$\frac{5}{2}$	$\frac{3}{4}(4-\pi+2\ln 2)$
205	$\frac{3}{4}$	1	$\frac{3}{2}$	$\frac{5}{2}$	$\frac{11}{4}$	$\frac{21}{4}(2-\pi+2\ln 2)$
206	$\frac{4}{5}$	1	$\frac{6}{5}$	$\frac{9}{5}$	$\frac{11}{5}$	$\frac{12}{5}\left(5-\pi\sqrt{2+\sqrt{5}}\right)$
207	$\frac{5}{6}$	1	$\frac{7}{6}$	$\frac{11}{6}$	$\frac{13}{6}$	$\frac{35}{12}(6-\pi\sqrt{3})$
208	$\frac{7}{8}$	1	1	$\frac{15}{8}$	2	$\frac{7}{2}[8\ln 2-(1+\sqrt{2})\pi+2\sqrt{2}\ln(1+\sqrt{2})]$
209	$\frac{7}{8}$	1	$\frac{9}{8}$	$\frac{15}{8}$	$\frac{17}{8}$	$\frac{63}{16}[8-(1+\sqrt{2})\pi]$
210	1	1	1	$\frac{3}{2}$	2	$\frac{\pi^2}{4}$
211	1	1	1	$\frac{3}{2}$	$\frac{5}{2}$	$\frac{3}{2}(\pi-2)$
212	1	1	1	2	2	$\frac{\pi^2}{6}$
213	1	1	1	2	3	$\frac{\pi^2}{3}-2$
214	1	1	1	$\frac{5}{2}$	$\frac{5}{2}$	$9(\pi-3)$

	a_1	a_2	a_3	b_1	b_2	${}_3F_2(a_1, a_2, a_3; b_1, b_2; 1)$
215	1	1	1	3	3	$\frac{4}{3}(\pi^2-9)$
216	1	1	1	$\frac{7}{2}$	$\frac{7}{2}$	$\frac{25}{6}(\pi-5)$
217	1	1	1	4	4	$\frac{9}{4}(4\pi^2-39)$
218	1	1	1	5	5	$\frac{8}{3}(20\pi^2-197)$
219	1	1	$\frac{9}{8}$	2	$\frac{17}{8}$	$\frac{9}{2}[16-(1+\sqrt{2})\pi-8\ln 2-2\sqrt{2}\ln(1+\sqrt{2})]$
220	1	1	$\frac{5}{4}$	2	$\frac{9}{4}$	$\frac{5}{2}(8-\pi-6\ln 2)$
221	1	1	$\frac{5}{4}$	$\frac{9}{4}$	3	$\frac{10}{3}(15-2\pi-12\ln 2)$
222	1	1	$\frac{4}{3}$	$\frac{7}{3}$	3	$32-2\pi\sqrt{3}-18\ln 3$
223	1	1	$\frac{11}{8}$	2	$\frac{19}{8}$	$\frac{11}{18}[16-3(\sqrt{2}-1)\pi-24\ln 2+6\sqrt{2}\ln(1+\sqrt{2})]$
224	1	1	$\frac{3}{2}$	2	2	$4\ln 2$
225	1	1	$\frac{3}{2}$	2	$\frac{5}{2}$	$6(1-\ln 2)$
226	1	1	$\frac{3}{2}$	2	$\frac{7}{2}$	$\frac{5}{6}(5-6\ln 2)$
227	1	1	$\frac{3}{2}$	$\frac{5}{2}$	3	$6-12\ln 2$
228	1	1	$\frac{3}{2}$	3	$\frac{7}{2}$	$\frac{10}{3}(17-24\ln 2)$
229	1	1	$\frac{13}{8}$	2	$\frac{21}{8}$	$\frac{13}{50}[16-5(\sqrt{2}-1)\pi-40\ln 2+10\sqrt{2}\ln(1+\sqrt{2})]$
230	1	1	$\frac{5}{3}$	2	$\frac{8}{3}$	$\frac{5\sqrt{3}}{12}[3\sqrt{3}(1-\ln 3)-\pi]$
231	1	1	$\frac{5}{3}$	$\frac{8}{3}$	3	$\frac{5}{2}(4+\pi\sqrt{3}-9\ln 3)$
232	1	1	$\frac{7}{4}$	2	$\frac{11}{4}$	$\frac{7}{18}(8+3\pi-18\ln 2)$
233	1	1	$\frac{7}{4}$	$\frac{11}{4}$	3	$\frac{11}{9}(7+6\pi-36\ln 2)$
234	1	1	$\frac{15}{8}$	2	$\frac{23}{8}$	$\frac{15}{196}[32+7(1+\sqrt{2})\pi-112\ln 2-28\sqrt{2}\ln(1+\sqrt{2})]$
235	1	1	2	3	3	$8-\frac{2}{3}\pi^2$
236	1	1	$\frac{5}{2}$	2	$\frac{7}{2}$	$\frac{10}{9}(4-3\ln 2)$
237	1	1	$\frac{5}{2}$	3	$\frac{7}{2}$	$\frac{10}{9}(12\ln 2-7)$
238	1	1	3	2	4	$\frac{9}{4}$

	a_1	a_2	a_3	b_1	b_2	${}_3F_2(a_1, a_2, a_3; b_1, b_2; 1)$
239	1	1	3	2	5	$\frac{5}{3}$
240	1	1	4	2	5	$\frac{22}{9}$
241	1	1	4	3	5	$\frac{14}{9}$
242	1	1	5	2	6	$\frac{125}{48}$
243	1	$\frac{3}{2}$	$\frac{3}{2}$	3	3	$\frac{16}{\pi}(16-5\pi)$
244	1	2	2	3	3	$\frac{2}{3}\pi^2-4$

7.4.5. Values of ${}_3F_2(a_1, a_2, a_3; b_1, b_2; -1)$.

1. $$\frac{bc}{\Gamma(1+a-d)}\,{}_3F_2\left(\begin{matrix}a, a+b, a+c; -1\\ 1+a-d, 1+a\end{matrix}\right)+ac\Gamma\left[\begin{matrix}c-b\\ a+c, 1-b-d\end{matrix}\right]\times$$ $$\times{}_3F_2\left(\begin{matrix}b, a+b, b+d; -1\\ 1+b-c, 1+b\end{matrix}\right)+$$ $$+ab\Gamma\left[\begin{matrix}b-c\\ a+b, 1-c-d\end{matrix}\right]{}_3F_2\left(\begin{matrix}c, a+c, c+d; -1\\ 1+c-b, 1+c\end{matrix}\right)=\Gamma\left[\begin{matrix}a+1, b+1, c+1\\ a+b, a+c, 1-d\end{matrix}\right].$$

2. $${}_3F_2\left(\begin{matrix}a, b, c; -1\\ 1+a-b, 1+a-c\end{matrix}\right)=\Gamma\left[\begin{matrix}1+a-b, 1+a-c\\ 1+a, 1+a-b-c\end{matrix}\right]{}_3F_2\left(\begin{matrix}1/2, b, c; 1\\ a/2+1, (a+1)/2\end{matrix}\right).$$

3. $${}_3F_2\left(\begin{matrix}1, a, b; -1\\ 1-2n-a, 1-2n-b\end{matrix}\right)=\frac{(-1)^n}{2(a+n)_n(b+n)_n}\Gamma\left[\begin{matrix}1-a, 1-b\\ 1-2n-a-b\end{matrix}\right]-$$ $$-\frac{(-1)^n(a)_n(b)_n}{2(a+n)_n(b+n)_n}+\sum_{k=0}^{n}(-1)^k\frac{(a)_k(b)_k}{(1-2n-a)_k(1-2n-b)_k}$$ $[\operatorname{Re}(a+b)\leqslant 1/2-2n]$.

4. $${}_3F_2\left(\begin{matrix}1, a, b; -1\\ 1+2n-a, 1+2n-b\end{matrix}\right)=$$ $$=(-1)^n\frac{(1+n-a)_n(1+n-b)_n}{2}\left\{(1+2n-a-b)_n\Gamma\left[\begin{matrix}1-a, 1-b\\ 1+4n-a-b\end{matrix}\right]+\right.$$ $$\left.+\frac{1}{(1-a)_n(1-b)_n}\right\}-\sum_{k=1}^{n-1}(-1)^k\frac{(a-2n)_k(b-2n)_k}{(1-a)_k(1-b)_k}$$ $[n=1, 2, 3, \ldots;\ \operatorname{Re}(a+b)\leqslant 1/2+2n]$.

5. $${}_3F_2\left(\begin{matrix}1, a, b; -1\\ a+1, b+1\end{matrix}\right)=\frac{ab}{a-b}[\beta(b)-\beta(a)]$$ $[b\neq a]$, $$=-a^2\beta'(a)$$ $[b=a]$.

6. $${}_3F_2\left(\begin{matrix}1, m, n; -1\\ m+1, n+1\end{matrix}\right)=$$ $$=\frac{mn}{m-n}[(-1)^m-(-1)^n]\ln 2+\frac{(-1)^m mn}{m-n}\sum_{k=1}^{m-n}\frac{(-1)^k}{k}-(-1)^n mn\sum_{k=1}^{n-1}\frac{(-1)^k}{k(k+m-n)}$$ $[n<m]$.

7. $${}_3F_2\left(\begin{matrix}a,\ a/2+1,\ b;\ -1\\ a/2,\ 1+a-b\end{matrix}\right)=\frac{\sqrt{\pi}}{2^a}\Gamma\left[\begin{matrix}1+a-b\\ 1+a/2,\ (a+1)/2-b\end{matrix}\right].$$

8. $${}_3F_2\left(\begin{matrix}a,\ a+1/3,\ a+2/3;\ -1\\ 1/3,\ 2/3\end{matrix}\right)=\frac{2}{3}(2^{-1-3a}+\cos a\pi) \qquad [\operatorname{Re} a<1/3].$$

9. $${}_3F_2\left(\begin{matrix}a,\ a+1/3,a+2/3;\ -1\\ 2/3,\ 4/3\end{matrix}\right)=\frac{2}{3(1-3a)}\left(2^{-3a}+\cos\frac{1+3a}{3}\pi\right) \qquad [\operatorname{Re} a<0].$$

10. $${}_3F_2\left(\begin{matrix}a,\ a+1/3,\ a+2/3;\ -1\\ 4/3,\ 5/3\end{matrix}\right)=\frac{4}{3(1-3a)(2-3a)}\left[2^{1-3a}+\cos(2+3a)\frac{\pi}{3}\right] \qquad [\operatorname{Re} a<-1/3].$$

11. $${}_3F_2\left(\begin{matrix}1,\ a,\ a;\ -1\\ 1-a,\ 1-a\end{matrix}\right)=\frac{1}{2}+2^{2a-1}\sqrt{\pi}\,\Gamma\left[\begin{matrix}1-a\\ 1/2-a\end{matrix}\right].$$

12. $${}_3F_2\left(\begin{matrix}1,\ a,\ -a;\ -1\\ 1+a,\ 1-a\end{matrix}\right)=\frac{1}{2}+\frac{\pi a}{2\sin a\pi}.$$

13. $${}_3F_2\left(\begin{matrix}1,\ a,\ 1-a;\ -1\\ 1+a,\ 2-a\end{matrix}\right)=\frac{a(1-a)}{2a-1}\left[\frac{\pi}{\sin a\pi}-2\beta(a)\right] \qquad [a\neq 1/2].$$

14. $${}_3F_2\left(\begin{matrix}2,\ a,\ 2-a;\ -1\\ 1+a,\ 3-a\end{matrix}\right)=\frac{a(a-2)}{2}[\beta(1-a)+\beta(a-1)].$$

15. $${}_3F_2\left(\begin{matrix}1,\ 1,\ a;\ -1\\ 2,\ 3-a\end{matrix}\right)=\frac{2-a}{2(1-a)}[\psi(2-a)+\mathbf{C}] \qquad [a\neq 1;\ \operatorname{Re} a<3/2].$$

16. $${}_3F_2\left(\begin{matrix}1,\ 1,\ a;\ -1\\ 2,\ a+1\end{matrix}\right)=\frac{a}{1-a}(\beta(a)-\ln 2).$$

17. $${}_3F_2\left(\begin{matrix}-n,\ a,\ a/2+1;\ -1\\ a/2,\ 1+a+n\end{matrix}\right)=\frac{(a+1)_n}{((a+1)/2)_n}.$$

18. $${}_3F_2\left(\begin{matrix}-n,\ a,\ -a-n;\ -1\\ a+1,\ 1-a-n\end{matrix}\right)=\frac{2(a+n)}{2a+n}\,{}_2F_1(-n,\ a;\ a+1;\ -1).$$

19. $${}_3F_2\left(\begin{matrix}-n,\ -n,\ m;\ -1\\ 1,\ 1-m-n\end{matrix}\right)=2^n\,{}_3F_2\left(\begin{matrix}-n/2,\ (1-n)/2,\ -m;\ 1\\ 1,\ -m-n\end{matrix}\right)=$$

20. $$=2^{2n}\frac{(1/2)_n}{n!}\,{}_3F_2\left(\begin{matrix}-n,\ -n/2,\ (1-n)/2;\ 1\\ 1/2-n,\ -m-n\end{matrix}\right).$$

21. $${}_3F_2\left(\begin{matrix}-n,\ -n,\ -n;\ -1\\ 1,\ 1\end{matrix}\right)=2^n\,{}_3F_2\left(\begin{matrix}-n/2,\ (1-n)/2,\ n+1;\ 1\\ 1,\ 1\end{matrix}\right).$$

22. $${}_3F_2\left(\begin{matrix}-n,\ -n-1/2,\ 1/2;\ -1\\ 1/2-n,\ 3/2\end{matrix}\right)=\frac{2n+1}{n+1}\,{}_2F_1\left(-n,\ \frac{1}{2};\ \frac{3}{2};\ -1\right).$$

23. $${}_3F_2\left(\begin{matrix}-n,\ 2,\ 2;\ -1\\ 1,\ 1\end{matrix}\right)=2^{n-2}(n+1)(n+4).$$

24. $${}_3F_2\left(\begin{matrix}1,\ (1+i)/2,\ (1-i)/2;\ -1\\ (3+i)/2,\ (3-i)/2\end{matrix}\right)=0{,}851682\ldots$$

25. $${}_3F_2\left(\begin{matrix}1,\ (2+i)/4,\ (2-i)/4;\ -1\\ (6+i)/4,\ (6-i)/4\end{matrix}\right)=0{,}89831195\ldots$$

26. $${}_3F_2\left(\begin{matrix}2,\ 1+i,\ 1-i;\ -1\\ 2+i,\ 2-i\end{matrix}\right)=0{,}539222\ldots$$

27. $${}_3F_2\left(\begin{matrix}2,\ 1+i/2,\ 1-i/2;\ -1\\ 2+i/2,\ 2-i/2\end{matrix}\right)=0{,}645160\ldots$$

	a_1	a_2	a_3	b_1	b_2	${}_3F_2(a_1, a_2, a_3; b_1, b_2; -1)$
28	$\frac{1}{6}$	$\frac{1}{3}$	1	$\frac{7}{6}$	$\frac{4}{3}$	$\frac{1}{9}[(3-\sqrt{3})\pi-3\ln 2+\sqrt{3}\ln(2+\sqrt{3})]$
29	$\frac{1}{6}$	$\frac{2}{3}$	1	$\frac{7}{6}$	$\frac{5}{3}$	$\frac{2}{27}[(3-\sqrt{3})\pi+3\ln 2+\sqrt{3}\ln(2+\sqrt{3})]$
30	$\frac{1}{6}$	$\frac{5}{6}$	1	$\frac{7}{6}$	$\frac{11}{6}$	$\frac{5\sqrt{3}}{36}\ln(2+\sqrt{3})$
31	$\frac{1}{4}$	$\frac{1}{2}$	1	$\frac{5}{4}$	$\frac{3}{2}$	$\frac{1}{4}[(\sqrt{2}-1)\pi-2\sqrt{2}\ln(1+\sqrt{2})]$
32	$\frac{1}{4}$	$\frac{1}{2}$	1	$\frac{5}{4}$	$\frac{5}{2}$	$\frac{\sqrt{2}}{20}[(10-3\sqrt{2})\pi-3\sqrt{2}+20\ln(1+\sqrt{2})]$
33	$\frac{1}{4}$	$\frac{1}{2}$	1	$\frac{3}{2}$	$\frac{9}{4}$	$\frac{5\sqrt{2}}{24}[2\sqrt{2}+(1-\sqrt{2})\pi+2\ln(1+\sqrt{2})]$
34	$\frac{1}{4}$	$\frac{3}{4}$	1	$\frac{5}{4}$	$\frac{7}{4}$	$\frac{3}{2\sqrt{2}}\ln(1+\sqrt{2})$
35	$\frac{1}{4}$	$\frac{3}{4}$	1	$\frac{5}{4}$	$\frac{11}{4}$	$\frac{7\sqrt{2}}{96}[2\sqrt{2}-3\pi+18\ln(1+\sqrt{2})]$
36	$\frac{1}{4}$	$\frac{3}{4}$	1	$\frac{7}{4}$	$\frac{9}{4}$	$\frac{15\sqrt{2}}{32}[2\sqrt{2}-\pi+2\ln(1+\sqrt{2})]$
37	$\frac{1}{4}$	1	1	$\frac{5}{4}$	2	$\frac{\sqrt{2}}{6}[2\ln(1+\sqrt{2})+\pi-\sqrt{2}\ln 2]$
38	$\frac{1}{4}$	1	1	2	$\frac{9}{4}$	$\frac{5\sqrt{2}}{6}[\pi-\sqrt{2}\ln 2+2\ln(1+\sqrt{2})]$
39	$\frac{1}{4}$	1	$\frac{3}{2}$	$\frac{5}{4}$	$\frac{5}{2}$	$\frac{3\sqrt{2}}{40}[(2+\sqrt{2})\pi-4\sqrt{2}+4\ln(1+\sqrt{2})]$
40	$\frac{1}{4}$	1	$\frac{3}{2}$	$\frac{9}{4}$	$\frac{5}{2}$	$\frac{3\sqrt{2}}{8}[(3-\sqrt{2})\pi-6\sqrt{2}+6\ln(1+\sqrt{2})]$
41	$\frac{1}{4}$	1	$\frac{7}{4}$	$\frac{5}{4}$	$\frac{11}{4}$	$\frac{7\sqrt{2}}{72}(3\pi-2\sqrt{2})$
42	$\frac{1}{4}$	1	$\frac{7}{4}$	$\frac{9}{4}$	$\frac{11}{4}$	$\frac{35\sqrt{2}}{288}[3\pi-14\sqrt{2}+18\ln(1+\sqrt{2})]$
43	$\frac{1}{3}$	$\frac{1}{2}$	1	$\frac{4}{3}$	$\frac{3}{2}$	$\left(\frac{\sqrt{3}}{3}-\frac{1}{2}\right)\pi+\ln 2$
44	$\frac{1}{3}$	$\frac{2}{3}$	1	$\frac{4}{3}$	$\frac{5}{3}$	$\frac{4}{3}\ln 2$
45	$\frac{1}{3}$	$\frac{2}{3}$	1	$\frac{4}{3}$	$\frac{8}{3}$	$\frac{5\sqrt{3}}{54}(9\sqrt{3}-7\pi+13\sqrt{3}\ln 2)$
46	$\frac{1}{3}$	$\frac{2}{3}$	1	$\frac{5}{3}$	$\frac{7}{3}$	$\frac{4\sqrt{3}}{27}(3\sqrt{3}-2\pi+4\sqrt{3}\ln 2)$
47	$\frac{1}{3}$	$\frac{5}{6}$	1	$\frac{4}{3}$	$\frac{11}{6}$	$\frac{5}{27}[(\sqrt{3}-3)\pi+\sqrt{3}\ln(2+\sqrt{3})]$
48	$\frac{1}{3}$	1	1	$\frac{4}{3}$	2	$\frac{\pi}{2\sqrt{3}}$
49	$\frac{1}{3}$	1	$\frac{5}{3}$	$\frac{4}{3}$	$\frac{8}{3}$	$\frac{5\sqrt{3}}{72}(4\pi-3\sqrt{3})$

	a_1	a_2	a_3	b_1	b_2	${}_3F_2(a_1, a_2, a_3; b_1, b_2; -1)$
50	$\frac{1}{2}$	$\frac{1}{2}$	1	$\frac{3}{2}$	$\frac{3}{2}$	G
51	$\frac{1}{2}$	$\frac{1}{2}$	1	$\frac{5}{2}$	$\frac{5}{2}$	$\frac{9}{8}(4-\pi)$
52	$\frac{1}{2}$	$\frac{1}{2}$	1	$\frac{9}{2}$	$\frac{9}{2}$	$\frac{1}{576}(3675\pi-10976)$
53	$\frac{1}{2}$	$\frac{2}{3}$	1	$\frac{3}{2}$	$\frac{5}{3}$	$\frac{1}{3}[(3-2\sqrt{3})\pi+6\ln 2]$
54	$\frac{1}{2}$	$\frac{3}{4}$	1	$\frac{3}{2}$	$\frac{7}{4}$	$\frac{3}{4}[(1-\sqrt{2})\pi-2\sqrt{2}\ln(1+\sqrt{2})]$
55	$\frac{1}{2}$	$\frac{3}{4}$	1	$\frac{7}{4}$	$\frac{5}{2}$	$\frac{3}{4}[2+(1-2\sqrt{2})\pi+4\sqrt{2}\ln(1+\sqrt{2})]$
56	$\frac{1}{2}$	1	1	$\frac{3}{2}$	$\frac{3}{2}$	$\frac{1}{8}[\pi^2-4\ln^2(1+\sqrt{2})]$
57	$\frac{1}{2}$	1	1	$\frac{3}{2}$	2	$\frac{\pi}{2}-\ln 2$
58	$\frac{1}{2}$	1	1	2	$\frac{5}{2}$	$3(1-\ln 2)$
59	$\frac{1}{2}$	1	1	$\frac{5}{2}$	3	$10-2\pi-4\ln 2$
60	$\frac{1}{2}$	1	$\frac{5}{4}$	$\frac{3}{2}$	$\frac{9}{4}$	$\frac{\sqrt{2}}{24}[12(\sqrt{2}-1)+3(2+\sqrt{2})\pi-8\sqrt{2}]$
61	$\frac{1}{2}$	1	$\frac{5}{4}$	$\frac{9}{4}$	$\frac{5}{2}$	$\frac{5\sqrt{2}}{16}[(8-3\sqrt{2})\pi-20\sqrt{2}+4\sqrt{2}\ln 2+16\ln(1+\sqrt{2})]$
62	$\frac{1}{2}$	1	2	$\frac{3}{2}$	3	$\frac{1}{3}(\pi-2+2\ln 2)$
63	$\frac{1}{2}$	1	$\frac{7}{2}$	$\frac{5}{2}$	$\frac{9}{2}$	$\frac{7}{120}(15\pi-32)$
64	$\frac{2}{3}$	1	1	$\frac{5}{3}$	2	$\frac{2\sqrt{3}}{3}(\pi-2\sqrt{3}\ln 2)$
65	$\frac{2}{3}$	1	$\frac{4}{3}$	$\frac{5}{3}$	$\frac{7}{3}$	$\frac{4\sqrt{3}}{9}(2\pi-3\sqrt{3})$
66	$\frac{3}{4}$	1	1	$\frac{7}{4}$	2	$\frac{3\sqrt{2}}{2}[\pi-2\ln(1+\sqrt{2})-\sqrt{2}\ln 2]$
67	$\frac{3}{4}$	1	1	2	$\frac{11}{4}$	$\frac{7\sqrt{2}}{12}[2\sqrt{2}+3\pi-6\sqrt{2}\ln 2-6\ln(1+\sqrt{2})]$
68	$\frac{3}{4}$	1	$\frac{5}{4}$	$\frac{7}{4}$	$\frac{9}{4}$	$\frac{15\sqrt{2}}{8}(\pi-2\sqrt{2})$
69	$\frac{3}{4}$	1	$\frac{5}{4}$	$\frac{9}{4}$	$\frac{11}{4}$	$\frac{35\sqrt{2}}{32}[3\pi-10\sqrt{2}+6\ln(1+\sqrt{2})]$
70	$\frac{3}{4}$	1	$\frac{3}{2}$	$\frac{7}{4}$	$\frac{5}{2}$	$\frac{3\sqrt{2}}{8}[(2+\sqrt{2})\pi-4\sqrt{2}-4\ln(1+\sqrt{2})]$
71	$\frac{3}{4}$	1	$\frac{3}{2}$	$\frac{5}{2}$	$\frac{11}{4}$	$\frac{21}{4}[(1-\sqrt{2})\pi-2+2\sqrt{2}\ln(1+\sqrt{2})]$
72	$\frac{4}{5}$	1	$\frac{6}{5}$	$\frac{9}{5}$	$\frac{11}{5}$	$\frac{12}{25}[\pi(50+10\sqrt{5})^{1/2}-25]$
73	$\frac{5}{6}$	1	$\frac{7}{6}$	$\frac{11}{6}$	$\frac{13}{6}$	$\frac{35}{6}(\pi-3)$

	a_1	a_2	a_3	b_1	b_2	${}_3F_2(a_1, a_2, a_3; b_1, b_2; -1)$
74	$\frac{7}{8}$	1	$\frac{9}{8}$	$\frac{15}{8}$	$\frac{17}{8}$	$\frac{63}{8}[\pi(1+\sqrt{2})^{1/2}-4]$
75	1	1	1	2	2	$\frac{\pi^2}{12}$
76	1	1	1	2	3	$\frac{1}{6}(\pi^2+12-24\ln 2)$
77	1	1	1	3	3	$12-16\ln 2$
78	1	1	1	4	4	$\frac{1}{4}(63-6\pi^2)$
79	1	1	$\frac{5}{4}$	2	$\frac{9}{4}$	$\frac{5\sqrt{2}}{2}[\pi-4\sqrt{2}+\sqrt{2}\ln 2+2\ln(1+\sqrt{2})]$
80	1	1	$\frac{5}{4}$	$\frac{9}{4}$	3	$\frac{10}{3}[2\sqrt{2}\pi+2\ln 2+\sqrt{2}\ln(1+\sqrt{2})-15]$
81	1	1	$\frac{4}{3}$	2	$\frac{7}{3}$	$\frac{4\sqrt{3}}{3}(\pi-3\sqrt{3}+2\sqrt{3}\ln 2)$
82	1	1	$\frac{4}{3}$	$\frac{7}{3}$	3	$4\sqrt{3}\pi+16\ln 2-32$
83	1	1	$\frac{3}{2}$	2	$\frac{5}{2}$	$\frac{3}{2}(2\ln 2-4+\pi)$
84	1	1	$\frac{3}{2}$	2	$\frac{7}{2}$	$5\pi-15\ln 2-5$
85	1	1	$\frac{3}{2}$	$\frac{5}{2}$	3	$6\pi-18$
86	1	1	$\frac{3}{2}$	3	$\frac{7}{2}$	$\frac{10}{3}(3\pi-6\ln 2-5)$
87	1	1	$\frac{5}{3}$	2	$\frac{8}{3}$	$\frac{5\sqrt{3}}{12}(2\pi-3\sqrt{3})$
88	1	1	$\frac{5}{3}$	$\frac{8}{3}$	3	$\frac{5}{2}(2\sqrt{3}\pi-8\ln 2-5)$
89	1	1	$\frac{7}{4}$	2	$\frac{11}{4}$	$\frac{7\sqrt{2}}{18}[3\pi-4\sqrt{2}+3\sqrt{2}\ln 2-6\ln(1+\sqrt{2})]$
90	1	1	$\frac{7}{4}$	$\frac{11}{4}$	3	$-\frac{14}{3}[13+2\ln 2+\sqrt{2}\ln(1+\sqrt{2})]$
91	1	1	2	3	3	$\frac{1}{3}(\pi^2+24\ln 2-24)$
92	1	1	$\frac{5}{2}$	2	$\frac{7}{2}$	$\frac{5}{18}(6\ln 2+8-3\pi)$
93	1	1	$\frac{5}{2}$	3	$\frac{7}{2}$	$\frac{10}{9}(3\pi+12\ln 2-17)$
94	1	1	3	2	4	$\frac{3}{4}$
95	1	1	4	2	5	$\frac{2}{9}(12\ln 2-5)$
96	1	1	5	2	6	$\frac{35}{48}$
97	$\frac{5}{4}$	$\frac{7}{4}$	2	$\frac{9}{4}$	$\frac{11}{4}$	$\frac{35\sqrt{2}}{32}[4\ln(1+\sqrt{2})-\pi]$
98	2	2	2	3	3	$\frac{\pi^2}{3}-4\ln 2$

7.4.6. Values of ${}_3F_2(a_1, a_2, a_3; b_1, b_2; z_0)$ for $z_0 \neq \pm 1$.

1. $${}_3F_2\left(\begin{matrix}1/2, 1/2, 1/2; -1/4\\ 3/2, 3/2\end{matrix}\right)=\frac{\pi^2}{10}.$$

2. $${}_3F_2\left(\begin{matrix}1/2, 1, 1; -1/4\\ 3/2, 3/2\end{matrix}\right)=\frac{\pi^2}{6}-3\ln^2\left(\frac{\sqrt{5}-1}{2}\right).$$

3. $${}_3F_2\left(\begin{matrix}a, 1-a, 3a-1; -1/8\\ 2a, a+1/2\end{matrix}\right)=\frac{2^{3a-3}}{\pi}\left\{\Gamma\left[\begin{matrix}a/2, a+1/2\\ 3a/2\end{matrix}\right]\right\}^2.$$

4. $${}_3F_2\left(\begin{matrix}-n, a, 2-a; 3/4\\ 3/2, -3n-1\end{matrix}\right)=\frac{((4-a)/3)_n\,((2+a)/3)_n}{(2/3)_n\,(4/3)_n}.$$

5. $${}_3F_2\left(\begin{matrix}-n, a, 2-3a; 3/4\\ 1/2, -3n\end{matrix}\right)=\frac{(a)_n\,(1-a)_n}{(1/3)_n\,(2/3)_n}.$$

6. $${}_3F_2\left(\begin{matrix}-n, a, 3a+n-1; 3/4\\ (3a-1)/2, 3a/2\end{matrix}\right)=\frac{n!}{(3a-1)_n}A,$$

$$A=\begin{cases}(a)_m/m! & [n=3m],\\ -(a)_m/m! & [n=3m+1],\\ 0 & [n=3m+2].\end{cases}$$

7. $${}_3F_2\left(\begin{matrix}-n, a, 3a+n; 3/4\\ 3a/2, (3a+1)/2\end{matrix}\right)=\left\{\begin{matrix}1\\0\end{matrix}\right\}\frac{n!\,(a+1)_{n/3}}{(n/3)!\,(3a+1)_n}\qquad\left[\left\{\begin{matrix}n=3, 6, 9, \ldots\\ n\neq 3, 6, 9, \ldots\end{matrix}\right\}\right].$$

8. $${}_3F_2\left(\begin{matrix}-n, a, a+1/2; 4/3\\ 2a, (2+2a-n)/3\end{matrix}\right)=\left\{\begin{matrix}1\\0\end{matrix}\right\}\frac{(1/3)_m\,(2/3)_m}{((2+2a)/3)_m\,((1-2a)/3)_m}$$

$$\left[\left\{\begin{matrix}n=3m, m=1, 2, 3, \ldots\\ n\neq 3, 6, 9, \ldots\end{matrix}\right\}\right].$$

9. $${}_3F_2\left(\begin{matrix}-n, a, 3a+n; 3/2\\ 3a/2, (3a+1)/2\end{matrix}\right)=\frac{(-1)^{[n/3]}-(-1)^{[(n-1)/3]}}{2\Gamma(n/3+1)\,\Gamma(1-a-n/3)\,(3a)_n}\,n!\,\Gamma(1-a).$$

10. $${}_3F_2\left(\frac{1}{2}, \frac{1}{2}, 1; \frac{3}{2}, \frac{3}{2}; z\right)=F(z),\quad F(-1)=\mathbf{G},$$

$$F(9-4\sqrt{5})=\frac{\sqrt{5}+2}{24}\left[\pi^2-2\ln^2(\sqrt{5}-2)\right],$$

$$F(3-2\sqrt{2})=\frac{\sqrt{2}+1}{16}\left[\pi^2-4\ln^2(\sqrt{2}-1)\right],$$

$$F\left(\frac{3-\sqrt{5}}{2}\right)=\frac{\sqrt{5}+1}{24}\left[\pi^2-9\ln^2\left(\frac{\sqrt{5}-1}{2}\right)\right],\quad F(1)=\frac{\pi^2}{8}.$$

11. $${}_3F_2(1, 1, 1; 2, 2; z)=\frac{1}{z}\operatorname{Li}_2(z)=\Phi(z);\quad \Phi\left(-\frac{1+\sqrt{5}}{2}\right)=$$

$$=\frac{\sqrt{5}-1}{4}\left[\frac{\pi^2}{5}-\ln^2\left(\frac{\sqrt{5}+1}{2}\right)\right],$$

$$\Phi(-1)=\frac{\pi^2}{12},\quad \Phi\left(\frac{1-\sqrt{5}}{2}\right)=\frac{\sqrt{5}+1}{2}\left[\frac{\pi^2}{15}-\frac{1}{2}\ln^2\left(\frac{\sqrt{5}-1}{2}\right)\right],$$

$$\Phi\left(\frac{3-\sqrt{5}}{2}\right)=\frac{\sqrt{5}+3}{2}\left[\frac{\pi^2}{15}-\ln^2\left(\frac{\sqrt{5}-1}{2}\right)\right],\quad \Phi\left(\frac{1}{2}\right)=\frac{\pi^2}{6}-\ln^2 2,$$

$$\Phi\left(\frac{\sqrt{5}-1}{2}\right)=\frac{\sqrt{5}+1}{2}\left[\frac{\pi^2}{10}-\ln^2\left(\frac{\sqrt{5}-1}{2}\right)\right],\quad \Phi(1)=\frac{\pi^2}{6}.$$

7.5. THE FUNCTION ${}_4F_3(a_1, a_2, a_3, a_4; b_1, b_2, b_3; z)$

7.5.1. Representations of ${}_4F_3(a_1, a_2, a_3, a_4; b_1, b_2, b_3; z)$.

	a_1	a_2	a_3	a_4	b_1	b_2	b_3	${}_4F_3(a_1, a_2, a_3, a_4; b_1, b_2, b_3; z)$
1	a	b	c	d	$a+1$	$b+1$	e	$\frac{b(a-e)}{e(a-b)}\,{}_3F_2(a, c, d; a+1, e+1; z)-\frac{a(b-e)}{e(a-b)}\,{}_3F_2(b, c, d; b+1, e+1; z)$
2	a	b	c	d	$a+1$	$b+1$	$c+1$	$\frac{ab}{(a-c)(b-c)}\,{}_2F_1(c, d; c+1; z)+\frac{bc}{(b-a)(c-a)}\,{}_2F_1(a, d; a+1; z)+\frac{ac}{(a-b)(c-b)}\times {}_2F_1(b, d; b+1; z)$
3	a	b	c	$a+b-c$	$\frac{a+b}{2}$	$\frac{a+b+1}{2}$	$a+b$	$(1-x)^a\,{}_2F_1(a, c\ a+b; x)\times {}_2F_1(a, a+b-c; a+b; x)$ $[x^2=4z(x-1)]$
4	a	$\frac{a}{2}+1$	b	c	$\frac{a}{2}$	$1+a-b$	$1+a-c$	$(1+z)(1-z)^{-a-1}\times {}_3F_2\left(\frac{a}{2}+1, \frac{a+1}{2}, 1+a-b-c; 1+a-b, 1+a-c; -\frac{4z}{(1-z)^2}\right)$
5	a	b	$\frac{a+b}{2}$	$\frac{a+b+1}{2}$	$a+b$	c	$a+b-c+1$	${}_2F_1\left(a, b; c; \frac{1-\sqrt{1-z}}{2}\right)\times {}_2F_1\left(a, b; a+b-c+1; \frac{1-\sqrt{1-z}}{2}\right)$
6	a	$a+\frac{1}{2}$	b	$b+\frac{1}{2}$	$\frac{1}{2}$	c	$c+\frac{1}{2}$	$\frac{1}{2}\left[{}_2F_1(2a, 2b; 2c; \sqrt{z})+{}_2F_1(2a, 2b; 2c; -\sqrt{z})\right]$
7	a	$a+\frac{1}{2}$	b	$b+\frac{1}{2}$	$\frac{3}{2}$	c	$c+\frac{1}{2}$	$\frac{2c-1}{2(2a-1)(2b-1)\sqrt{z}}\times\left[{}_2F_1(2a-1, 2b-1; 2c-1; \sqrt{z})-{}_2F_1(2a-1, 2b-1; 2c-1; -\sqrt{z})\right]$
8	a	$\frac{2a}{3}+1$	b	$a-b+\frac{1}{2}$	$\frac{2a}{3}$	$2b$	$2a-2b+1$	$2^{2a-1}\frac{(8+z)}{(4-z)^{a+1}}\,{}_3F_2\left(\frac{a+1}{3}, \frac{a+2}{3}, \frac{a}{3}+1; b+\frac{1}{2}, a-b+1; \frac{27z^2}{(4-z)^3}\right)$

	a_1	a_2	a_3	a_4	b_1	b_2	b_3	${}_4F_3(a_1, a_2, a_3, a_4; b_1, b_2, b_3; z)$
9	a	$a+\frac{1}{4}$	$a+\frac{1}{2}$	$a+\frac{3}{4}$	b	$2a+\frac{1}{2}$	$2a-b+\frac{3}{2}$	${}_2F_1\left(a, a+\frac{1}{2}; b; \frac{1-\sqrt{1-z}}{2}\right){}_2F_1\left(a, a+\frac{1}{2}; 2a-b+\frac{3}{2}; \frac{1-\sqrt{1-z}}{2}\right)$
10	a	$\frac{1}{2}+a$	$\frac{1}{2}-a$	$1-a$	$\frac{1}{2}$	1	1	$P_{2a-1}(\sqrt{1-z}-\sqrt{-z})\times P_{2a-1}(\sqrt{1-z}+\sqrt{-z})$
11	$-n$	a	$\frac{1}{2}+a$	$2a+n$	$2a$	b	$2a-b+1$	$\frac{(-1)^n(n!)^2}{(b)_n(2a-b+1)_n}\times P_n^{(b-1,\, 2a-b)}(\sqrt{1-z})\times P_n^{(b-1,\, 2a-b)}(-\sqrt{1-z})$
12	$-n$	$-n$	$-n$	$\frac{1}{2}-n$	$\frac{1}{2}$	$-2n$	$\frac{1}{2}-2n$	$\frac{(2n)!}{(4n)!}(n!)^2 2^{2n} z^n\times P_{2n}\left(\frac{\sqrt{1-z}+1}{\sqrt{-z}}\right)\times P_{2n}\left(\frac{\sqrt{1-z}-1}{\sqrt{-z}}\right)$
13	$-n$	$-n$	$-n$	$\frac{1}{2}-n$	$\frac{3}{2}$	$-2n$	$-\frac{1}{2}-2n$	$\frac{(2n)!(n!)^2 2^{2n}}{(4n+1)!}z^n\times P_{2n+1}\left(\frac{\sqrt{1-z}+1}{\sqrt{-z}}\right)\times P_{2n+1}\left(\frac{\sqrt{1-z}-1}{\sqrt{-z}}\right)$

7.5.2. Particular values of ${}_4F_3(a_1, a_2, a_3, a_4; b_1, b_2, b_3; z)$.

	a_1	a_2	a_3	a_4	b_1	b_2	b_3	${}_4F_3(a_1, a_2, a_3, a_4; b_1, b_2, b_3; z)$
1	$\frac{1}{4}$	$\frac{1}{2}$	$\frac{3}{4}$	1	$\frac{5}{4}$	$\frac{3}{2}$	$\frac{7}{4}$	$\frac{3}{4}z^{-3/4}[(1+z^{1/4})^2\ln(1+z^{1/4})-(1-z^{1/4})^2\ln(1-z^{1/4})]-\frac{3}{2\sqrt{z}}\ln(1+\sqrt{z})+\frac{3}{2z^{3/4}}(\sqrt{z}-1)\operatorname{arctg} z^{1/4}$
2	$\frac{3}{4}$	1	1	$\frac{5}{4}$	$\frac{7}{4}$	2	$\frac{9}{4}$	$\frac{15}{2z^{5/4}}\left[(1+z^{1/2})\ln\frac{1+z^{1/4}}{1-z^{1/4}}+2z^{1/4}\ln(1-z)+2(1-z^{1/2})\operatorname{arctg} z^{1/4}-4z^{1/4}\right]$
3	1	1	3	3	2	$\frac{7}{2}$	4	$\frac{5}{16z}\left[13+\frac{15}{z}-\frac{6}{z^{3/2}}(1+2z)\sqrt{1-z}\times\arcsin\sqrt{z}-\frac{9}{z^2}\arcsin^2\sqrt{z}\right]$

	a_1 a_2 a_3 a_4	b_1 b_2 b_3	${}_4F_3(a_1, a_2, a_3, a_4; b_1, b_2, b_3; z)$
4	$\frac{4}{3}$ $\frac{5}{3}$ 2 2	$\frac{7}{3}$ $\frac{8}{3}$ 3	$\frac{10}{3z^{8/3}}\left[3z(z^{1/3}-4)\ln(1-z^{1/3})+ +z^{2/3}(6+4z^{1/3}-z^{2/3})\ln(1-z)- -2\sqrt{3}\,z(z^{1/3}+4)\operatorname{arctg}\frac{\sqrt{3}\,z^{1/3}}{2+z^{1/3}}\right]$
5	2 2 2 2	1 1 1	$(1-z)^{-5}(z^3+11z^2+11z+1)$

7.5.3. Values of ${}_4F_3(a_1, a_2, a_3, a_4; b_1, b_2, b_3; 1)$.

1. $${}_4F_3\left(\begin{matrix}a, b, c, d; 1\\ a+1, b+1, c-1\end{matrix}\right)=$$
$$=\frac{ab\Gamma(1-d)}{(c-1)(b-a)}\left\{(c-a-1)\Gamma\left[\begin{matrix}a\\ 1+a-d\end{matrix}\right]-(c-b-1)\Gamma\left[\begin{matrix}b\\ 1+b-d\end{matrix}\right]\right\}\quad[\operatorname{Re} d<1].$$

2. $${}_4F_3\left(\begin{matrix}a, a/2+1, b, c; 1\\ a/2, 1+a-b, 1+a-c\end{matrix}\right)=$$
$$=\Gamma\left[\begin{matrix}(1+a)/2, 1+a-b, 1+a-c, (1+a)/2-b-c\\ 1+a, (1+a)/2-b, (1+a)/2-c, 1+a-b-c\end{matrix}\right]\quad[\operatorname{Re}(a-2b-2c)>-1].$$

3. $${}_4F_3\left(\begin{matrix}1, a, b, c; 1\\ a+1, b+1, c+1\end{matrix}\right)=-abc\left[\frac{\psi(a)}{(b-a)(c-a)}+\frac{\psi(b)}{(a-b)(c-b)}+\frac{\psi(c)}{(a-c)(b-c)}\right]$$
$[a\neq b,\ b\neq c,\ a\neq c]$.

4. $$=\frac{a^2b}{(a-b)^2}[\psi(a)-\psi(b)]-\frac{a^2b}{a-b}\psi'(a)\quad[c=a,\ a\neq b].$$

5. $$=-\frac{a^3}{2}\psi''(a)\quad[c=b=a].$$

6. $${}_4F_3\left(\begin{matrix}1, a, b, c; 1\\ 3-a, 3-b, 3-c\end{matrix}\right)=\frac{1}{2(a-1)(b-1)(c-1)}\times$$
$$\times\left\{\Gamma\left[\begin{matrix}3-a, 3-b, 3-c, 4-a-b-c\\ 3-a-b, 3-a-c, 3-b-c\end{matrix}\right]-(2-a)(2-b)(2-c)\right\}\quad[\operatorname{Re}(a+b+c)<4].$$

7. $${}_4F_3\left(\begin{matrix}1, a, b, c; 1\\ 2, 2+a-b, 2+a-c\end{matrix}\right)=$$
$$=\frac{(1+a-b)(1+a-c)}{(a-1)(b-1)(c-1)}\left\{\Gamma\left[\begin{matrix}(a+1)/2, 1+a-b, 1+a-c, (a+5)/2-b-c\\ a, (a+3)/2-b, (a+3)/2-c, 2+a-b-c\end{matrix}\right]-1\right\}$$
$[\operatorname{Re}(a-2b-2c)>-5]$.

8. $${}_4F_3\left(\begin{matrix}1, 1, a, b; 1\\ 2, 2, a-n\end{matrix}\right)=\frac{a-n-1}{(a-1)(b-1)}[\psi(a-n-1)-\psi(a-1)-\psi(2-b)-\mathbf{C}]$$
$[\operatorname{Re} b<2-n]$.

9. $${}_4F_3\left(\begin{matrix}a, a, a, b; 1\\ a+1, a+1, a+1\end{matrix}\right)=$$
$$=\frac{a^3}{2}\mathrm{B}(a, 1-b)\{\psi'(a)-\psi'(1+a-b)+[\psi(a)-\psi(1+a-b)]^2\}\quad[\operatorname{Re} b<3].$$

10. $${}_4F_3\left(\begin{matrix}a, 2a+1, a+3/2, b; 1\\ a+1/2, a+2, 2+2a-b\end{matrix}\right)=$$
$$=\frac{\sqrt{\pi}}{2^{2a+1}}\Gamma\left[\begin{matrix}2+a, 1-b, 2+2a-b\\ 3/2+a, 1+a-b, 2+a-b\end{matrix}\right]\quad[\operatorname{Re} b<0].$$

11. $${}_4F_3\left(\begin{matrix}a,\ a+1/2,\ b,\ b+1/2;\ 1\\ 1/2,\ a-b+1/2,\ a-b+1\end{matrix}\right)=$$
$$=2^{-2a-1}\Gamma\left[\begin{matrix}2a-2b+1\\ a-2b+1\end{matrix}\right]\left\{\Gamma\left[\begin{matrix}1/2-2b\\ 1/2+a-2b\end{matrix}\right]+\frac{\sqrt{\pi}}{\Gamma(a+1/2)}\right\}\qquad [\operatorname{Re} b<1/4].$$

12. $${}_4F_3\left(\begin{matrix}a,\ a+1/2,\ b,\ b+1/2;\ 1\\ 3/2,\ a-b+3/2,\ a-b+1\end{matrix}\right)=$$
$$=\frac{2^{-2a}}{(2a-1)(2b-1)}\Gamma\left[\begin{matrix}2a-2b+2\\ a-2b+3/2\end{matrix}\right]\left\{\Gamma\left[\begin{matrix}3/2-2b\\ a-2b+1\end{matrix}\right]-\frac{\sqrt{\pi}}{\Gamma(a)}\right\}\qquad [\operatorname{Re} b<3/4].$$

13. $${}_4F_3\left(\begin{matrix}1,\ 1,\ a,\ b;\ 1\\ 2,\ 1+a,\ 2+a-b\end{matrix}\right)=$$
$$=\frac{a(1+a-b)}{(a-1)(b-1)}\left[\psi\left(\frac{a+1}{2}\right)-\psi(a)+\psi(1+a-b)-\psi\left(\frac{a+3}{2}-b\right)\right]\qquad [\operatorname{Re}(a-2b)>-3].$$

14. $${}_4F_3\left(\begin{matrix}a,\ a+1/4,\ a+1/2,\ a+3/4;\ 1\\ b_1,\ b_2,\ b_3\end{matrix}\right)=$$
$$=n!\prod_{k=1}^{n}(k-4a)^{-1}\left(2^{n-4a-2}+2^{n/2-2a-1}\cos\frac{n+4a}{4}\pi\right)$$

[the vector b_1, b_2, b_3 consists of the three components of the vector $(1+n)/4$, $(2+n)/4$, $(3+n)/4$, $(4+n)/4$, that are not equal to 1; $n=0, 1, 2, 3$; $\operatorname{Re} a<n/4$].

15. $${}_4F_3\left(\begin{matrix}a,\ a,\ a,\ a/2+1;\ 1\\ 1,\ 1,\ a/2\end{matrix}\right)=\frac{2^{1-a}\sqrt{\pi}}{a}\Gamma\left[\begin{matrix}(1-3a)/2\\ a/2,\ 1-a,\ (1-a)/2,\ (1-a)/2\end{matrix}\right]\qquad [\operatorname{Re} a<1/3].$$

16. $${}_4F_3\left(\begin{matrix}1/2,\ 1,\ a,\ 1-a;\ 1\\ 3/2,\ 1+a,\ 2-a\end{matrix}\right)=\frac{\pi a(1-a)}{(1-2a)^2}\left\{\operatorname{ctg} a\pi-\frac{2}{\pi}[\mathbf{C}+2\ln 2+\psi(1-a)]\right\}\qquad [a\neq 1/2].$$

17. $$=7\zeta(3)/8.\qquad [a=1/2].$$

18. $${}_4F_3\left(\begin{matrix}-n,\ a,\ b,\ c;\ 1\\ a-k,\ b-l,\ c-m\end{matrix}\right)=0\qquad [k+l+m=0,\ 1,\ 2,\ \ldots,\ n-1].$$

19. $$=\frac{n!}{(1-a)_k(1-b)_l(1-c)_m}\qquad [k+l+m=n].$$

20. $${}_4F_3\left(\begin{matrix}-n,\ a,\ b,\ c;\ 1\\ a+1,\ b-l,\ c-m\end{matrix}\right)=n!\frac{(1+a-b)_l(1+a-c)_m}{(1+a)_n(1-b)_l(1-c)_m}\qquad [l+m=0,\ 1,\ 2,\ \ldots,\ n].$$

21. $${}_4F_3\left(\begin{matrix}-n,\ a,\ b,\ c;\ 1\\ a-n+1,\ b-n+1,\ c+n-1\end{matrix}\right)=$$
$$=\frac{n(n-1)(n+c-a-1)(n+c-b-1)}{(a-n+1)(b-n+1)(c+n-1)(c+n)}\,{}_4F_3\left(\begin{matrix}2-n,\ a+1,\ b+1,\ c;\ 1\\ a-n+2,\ b-n+2,\ c+n+1\end{matrix}\right).$$

22. $${}_4F_3\left(\begin{matrix}-n,\ a,\ b,\ c;\ 1\\ -a-n,\ a+b+1/2,\ a+c+1/2\end{matrix}\right)=$$
$$=(-1)^n\frac{(a+1/2)_n(a+b+c-n+1/2)_{2n}}{(a+b+1/2)_n(a+c+1/2)_n(1/2-a-b-c)_n}\,{}_3F_2\left(\begin{matrix}-n,\ 2b,\ 2c;\ 1\\ -2a-2n,\ a+b+c+1/2\end{matrix}\right).$$

23. $${}_4F_3\left(\begin{matrix}-n,\ a,\ b,\ c;\ 1\\ (a-n)/2,\ (1+a-n)/2,\ b+c+1/2\end{matrix}\right)=$$
$$=\frac{(1+2c-a)_n}{(1-a)_n}\,{}_3F_2\left(\begin{matrix}-n,\ 2c,\ c-b+1/2;\ 1\\ b+c+1/2,\ 2c-a+1\end{matrix}\right)=$$

24. $$={}_3F_2\left(\begin{matrix}-n,\ 2b,\ 2c;\ 1\\ a-n,\ b+c+1/2\end{matrix}\right)=$$

25. $$=\frac{(1/2+b+c-a)_n}{(1-a)_n}\,{}_3F_2\left(\begin{matrix}-n,\ b-c+1/2,\ c-b+1/2;\ 1\\ b+c+1/2,\ 1/2+a-b-c-n\end{matrix}\right).$$

26. $${}_4F_3\left(\begin{matrix}-n,\ a,\ a/2+1,\ b;\ 1\\ a/2,\ 1+a-b,\ c\end{matrix}\right)=$$
$$=\frac{(c-2b-1)_n}{(c)_n}\,{}_4F_3\left(\begin{matrix}-n,\ a-2b-1,\ (a+1)/2-b,\ -b-1;\ 1\\ (a-1)/2-b,\ 1+a-b,\ c-2b-1\end{matrix}\right)=$$

27. $$=\frac{c+n}{c}\,{}_3F_2\left(\begin{matrix}-n,\ a+1,\ b+1;\ 1\\ c+1,\ 1+a-b\end{matrix}\right).$$

28. $${}_4F_3\left(\begin{matrix}-n,\ a,\ 1-a,\ b;\ 1\\ 1-b-n,\ c,\ 1+2b-c\end{matrix}\right)=\frac{((a+c-1)/2)_n\,((c-a)/2)_n\,(2b)_n}{(b)_n\,(b+1/2)_n\,(c)_n}\times$$
$$\times{}_4F_3\left(\begin{matrix}-n,\ 1+b-(a+c)/2,\ b+(1+a-c)/2,\ 1-c-n;\ 1\\ (3-a-c)/2-n,\ 1+(a-c)/2-n,\ 1+2b-c\end{matrix}\right).$$

29. $${}_4F_3\left(\begin{matrix}-n,\ a,\ a/2+1,\ b;\ 1\\ a/2,\ 1+a+n,\ 1+a-b\end{matrix}\right)=\frac{(1+a)_n\,((1+a)/2-b)_n}{((1+a)/2)_n\,(1+a-b)_n}.$$

30. $${}_4F_3\left(\begin{matrix}-n,\ a,\ a/2+1,\ b;\ 1\\ a/2,\ 1+a-b,\ 2+2b-n\end{matrix}\right)=$$
$$=\frac{(a-2b-1)_n}{(-2b-1)_n}\,{}_3F_2\left(\begin{matrix}-n,\ (a+1)/2,\ a-2b+n-1;\ 1\\ 1+a-b,\ (a-1)/2-b\end{matrix}\right)=$$

31. $$=\frac{(a-2b-1)_n\,(-b-1)_n\,(a-2b+2n-1)}{(1+a-b)_n\,(-2b-1)_n\,(a-2b-1)}.$$

32. $${}_4F_3\left(\begin{matrix}-n,\ a,\ a+1/2,\ b;\ 1\\ 2a,\ (b-n+1)/2,\ (b-n)/2+1\end{matrix}\right)=\frac{(2a-b)_n\,(b-n)}{(1-b)_n\,(b+n)}.$$

33. $${}_4F_3\left(\begin{matrix}-n,\ a,\ a+1/2,\ b;\ 1\\ 2a+1,\ (b-n)/2,\ (b-n+1)/2\end{matrix}\right)=\frac{(1+2a-b)_n}{(1-b)_n}.$$

34. $${}_4F_3\left(\begin{matrix}-n,\ a,\ a+1/2,\ b;\ 1\\ 2a+1,\ (b-n+1)/2,\ (b-n)/2+1\end{matrix}\right)=\frac{(1+2a-b)_n\,(2a-b-n)\,(b-n)}{(1-b)_n\,(2a-b+n)\,(b+n)}.$$

35. $${}_4F_3\left(\begin{matrix}-n,\ a,\ b,\ -1/2-a-b-n;\ 1\\ -a-n,\ -b-n,\ a+b+1/2\end{matrix}\right)=\frac{(2a+1)_n\,(2b+1)_n\,(a+b+1)_n}{(a+1)_n\,(b+1)_n\,(2a+2b+1)_n}.$$

36. $${}_4F_3\left(\begin{matrix}-n,\ a,\ b,\ 1/2-a-b-n;\ 1\\ -a-n,\ 1-b-n,\ a+b+1/2\end{matrix}\right)=\frac{(2a+1)_n\,(2b)_n\,(a+b)_n}{(a+1)_n\,(b)_n\,(2a+2b)_n}.$$

37. $${}_4F_3\left(\begin{matrix}-n,\ a,\ b,\ 1/2-a-b-n;\ 1\\ 1-a-n,\ 1-b-n,\ a+b\pm1/2\end{matrix}\right)=\frac{(2a)_n\,(2b)_n\,(a+b)_n}{(a)_n\,(b)_n\,(2a+2b-(1\mp1)/2)_n}.$$

38. $${}_4F_3\left(\begin{matrix}-n,\ a,\ b,\ 3/2-a-b-n;\ 1\\ 1-a-n,\ 1-b-n,\ a+b+1/2\end{matrix}\right)=\frac{(2a)_n\,(2b)_n\,(a+b)_n\,(2a+2b-1)}{(a)_n\,(b)_n\,(2a+2b-1)_n\,(2a+2b+2n-1)}.$$

39. $${}_4F_3\left(\begin{matrix}-n,\ a,\ b,\ 3/2-a-b-n;\ 1\\ 1-a-n,\ 2-b-n,\ a+b-1/2\end{matrix}\right)=\frac{(2a)_n\,(2b-1)_n\,(a+b-1)_n}{(a)_n\,(b-1)_n\,(2a+2b-2)_n}.$$

40. $${}_4F_3\left(\begin{matrix}-n,\ a,\ b,\ 5/2-a-b-n;\ 1\\ 2-a-n,\ 2-b-n,\ a+b-1/2\end{matrix}\right)=$$
$$=\frac{(2a-1)_n\,(2b-1)_n\,(a+b-1)_n\,(2a+2b-3)}{(a-1)_n\,(b-1)_n\,(2a+2b-3)_n\,(2a+2b+2n-3)}.$$

41. $${}_4F_3\left(\begin{matrix}-n,\ 1+n,\ a,\ a+1/2;\ 1\\ 1/2,\ b,\ 2a-b+2\end{matrix}\right)=\frac{1}{2(a-b+1)}\left[\frac{(1-b)_{n+1}}{(2a-b+2)_n}-\frac{(b-2a-1)_{n+1}}{(b)_n}\right].$$

42. $${}_4F_3\left(\begin{matrix}-n,\ 2+n,\ a,\ a+1/2;\ 1\\ 3/2,\ b,\ 2a-b+2\end{matrix}\right)=$$
$$=\frac{1}{2(n+1)(a-b+1)(1-2a)}\left[\frac{(1-b)_{n+2}}{(2a-b+2)_n}-\frac{(b-2a-1)_{n+2}}{(b)_n}\right].$$

43. $${}_4F_3\left(\begin{matrix}-n,\ 1,\ 1,\ a;\ 1\\ 2,\ b,\ 1+a-b-n\end{matrix}\right)=$$
$$=\frac{(b-1)(a-b-n)}{(n+1)(a-1)}\,[\psi(n+b)+\psi(1+a-b)-\psi(b-1)-\psi(a-b-n)].$$

44. $${}_4F_3\left(\begin{matrix}-n,\ 1,\ a,\ 2-a;\ 1\\ n+3,\ a+1,\ 3-a\end{matrix}\right)=\frac{n+2}{2(n+1)(1-a)^2}\left\{\frac{[(n+1)!]^2}{(3-a)_n(1+a)_n}-a(2-a)\right\}.$$

45. $${}_4F_3\left(\begin{matrix}-n,\ 1,\ 1,\ n+3;\ 1\\ 2,\ 2,\ m\end{matrix}\right)=\frac{m-1}{(n+1)(n+2)}[\mathbf{C}-\psi(m-1)+2\psi(n+2)]$$
$[m=2,\ 3,\ 4,\ \ldots,\ n+3]$.

46. $${}_4F_3\left(\begin{matrix}-n,\ 1,\ 1,\ 2m+n-1;\ 1\\ 2,\ m,\ m\end{matrix}\right)=\frac{2(m-1)^2}{(n+1)(2m+n-2)}[\psi(m+n)-\psi(m-1)].$$

47. $${}_4F_3\left(\begin{matrix}-n,\ -n,\ -n-1/2,\ -n-1/2;\ 1\\ 1,\ 3/2,\ 3/2\end{matrix}\right)=\frac{1}{8(n+1)}\{2(4n+3)!\,[(2n+2)!]^{-2}+$$
$$+(-1)^n(2n+1)!\,[(n+1)!]^{-2}+[1+(-1)^n](2n+1)!\,(2n+2)!\,[(n+1)!]^{-4}\}.$$

48. $${}_4F_3\left(\begin{matrix}-n,\ -n,\ -n,\ -n;\ 1\\ 2,\ 2,\ 2\end{matrix}\right)=$$
$$=\frac{(n+1)^{-3}}{2}\,{}_4F_3\left(\begin{matrix}-n-1,\ -n-1,\ -n-1,\ -n-1;\ 1\\ 1,\ 1,\ 1\end{matrix}\right).$$

49. $${}_4F_3\left(\begin{matrix}-n,\ -n,\ 1/2-n,\ 1/2-n;\ 1\\ 1,\ 1/2,\ 1/2\end{matrix}\right)=$$
$$=\frac{1}{4}\{(4n)!\,[(2n)!]^{-2}+(-1)^n(2n)!\,(n!)^{-2}+[1+(-1)^n]\,[(2n)!]^2\,(n!)^{-4}\}.$$

50. $${}_4F_3\left(\begin{matrix}-n,\ -3/4,\ -1/4,\ 5/8;\ 1\\ -3/8,\ 1/2,\ 1/2-n\end{matrix}\right)=-\frac{n!\,(1/2+n)_{n-1}}{2(2n)!}.$$

51. $${}_4F_3\left(\begin{matrix}-n,\ -1/4,\ 1/4,\ 9/8;\ 1\\ 1/8,\ 3/2,\ 1/2-n\end{matrix}\right)=\frac{n!\,(1/2+n)_n}{(2n+1)!}.$$

52. $${}_4F_3\left(\begin{matrix}-n,\ 1/2,\ 1,\ 1;\ 1\\ 1/2,\ 3/2,\ 3/2-n\end{matrix}\right)=\frac{1-2n}{1+2n}.$$

53. $${}_4F_3\left(\begin{matrix}-n/2,\ (1-n)/2,\ a,\ b;\ 1\\ c,\ c+1/2,\ a+b-2c-n+1\end{matrix}\right)=$$
$$=\frac{(a)_n}{(2c)_n}\,{}_3F_2\left(\begin{matrix}-n,\ a-2c-n+1,\ 2c-a-b;\ 1\\ 1-a-n,\ a+b-2c-n+1\end{matrix}\right).$$

54. $${}_4F_3\left(\begin{matrix}-n/2,\ (1-n)/2,\ a,\ b;\ 1\\ c,\ 1-c-n,\ a+b+1/2\end{matrix}\right)=\frac{(c-a)_n}{(c)_n}\,{}_3F_2\left(\begin{matrix}-n,\ 2a,\ a+b;\ 1\\ 2a+2b,\ a-c-n+1\end{matrix}\right).$$

55. $${}_4F_3\left(\begin{matrix}-n/2,\ (1-n)/2,\ a,\ a+1/2;\ 1\\ b,\ b+1/2,\ c\end{matrix}\right)=\frac{(2b-2a)_n}{(2b)_n}\,{}_3F_2\left(\begin{matrix}-n,\ 2a,\ c-1/2;\ 2\\ 2a-2b-n+1,\ 2c-1\end{matrix}\right).$$

56. $${}_4F_3\left(\begin{matrix}-n/2,\ (1-n)/2,\ a,\ b;\ 1\\ 1/2-m,\ a+1,\ b+1\end{matrix}\right)=\frac{n!}{(a-b)(1/2)_m}\left[\frac{a(1/2+b)_m}{(1+2b)_n}-\frac{b(1/2+a)_m}{(1+2a)_n}\right]$$
$[n/2\leqslant m\leqslant n]$.

57. $${}_4F_3\left(\begin{matrix}-n/2,\ (1-n)/2,\ a,\ b;\ 1\\ -n,\ (a+b+1)/2,\ (a+b)/2+1\end{matrix}\right)=\frac{(a)_{n+1}-(b)_{n+1}}{(a-b)(a+b+1)_n}.$$

58. $${}_4F_3\left(\begin{matrix}-n/2,\ (1-n)/2,\ a,\ -a;\ 1\\ 1/2,\ b,\ 1-b-n\end{matrix}\right)=\frac{(b+a)_n+(b-a)_n}{2(b)_n}.$$

59. $${}_4F_3\left(\begin{matrix}-n/2,\ (1-n)/2,\ (1-n)/2,\ (1-n)/2;\ 1\\ 1/2,\ -(n+1)/2,\ -(n+1)/2\end{matrix}\right)=\frac{2^{n-3}(n+4)}{n+1}\qquad [n=3,\ 4,\ 5,\ \ldots].$$

60. $${}_4F_3\left(\begin{matrix}-n/3,\ (1-n)/3,\ (2-n)/3,\ a;\ 1\\ b,\ b+1/3,\ b+2/3\end{matrix}\right)=$$
$$=\frac{(3b-3a)_n}{(3b)_n}\,{}_3F_2\left(\begin{matrix}-n,\ a,\ 3a-3b+1;\ 3/4\\ (3a-3b-n+1)/2,\ (3a-3b-n)/2+1\end{matrix}\right).$$

61. $_4F_3\left(\begin{matrix}1/2,\ 1,\ (1+i)/2,\ (1-i)/2;\ 1\\ 3/2,\ (3+i)/2,\ (3-i)/2\end{matrix}\right)=1,095402\ldots$

62. $_4F_3\left(\begin{matrix}1/2,\ 1,\ (2+i)/4,\ (2-i)/4;\ 1\\ 3/2,\ (6+i)/4,\ (6-i)/4\end{matrix}\right)=1,06337193\ldots$

63. $_4F_3\left(\begin{matrix}1,\ 1,\ 1+i,\ 1-i;\ 1\\ 2,\ 2+i,\ 2-i\end{matrix}\right)=1,343732\ldots$

64. $_4F_3\left(\begin{matrix}1,\ 1,\ 1+i/2,\ 1-i/2;\ 1\\ 2,\ 2+i/2,\ 2-i/2\end{matrix}\right)=1,241646\ldots$

	a_1	a_2	a_3	a_4	b_1	b_2	b_3	$_4F_3(a_1,\ a_2,\ a_3,\ a_4;\ b_1,\ b_2,\ b_3;\ 1)$
65	$\frac{1}{4}$	$\frac{1}{4}$	$\frac{1}{4}$	$\frac{1}{2}$	$\frac{5}{4}$	$\frac{5}{4}$	$\frac{5}{4}$	$\frac{(2\pi)^{-1/2}}{128}\Gamma^2\left(\frac{1}{4}\right)\left[\pi^2+\psi'\left(\frac{1}{4}\right)-\psi'\left(\frac{3}{4}\right)\right]$
66	$\frac{1}{4}$	$\frac{1}{2}$	$\frac{3}{4}$	1	$\frac{5}{4}$	$\frac{3}{2}$	$\frac{7}{4}$	$\frac{3}{2}\ln 2$
67	$\frac{1}{4}$	$\frac{1}{2}$	$\frac{3}{4}$	1	$\frac{5}{4}$	$\frac{7}{4}$	$\frac{5}{2}$	$\frac{3}{10}(1+8\ln 2-\pi)$
68	$\frac{1}{4}$	$\frac{1}{2}$	$\frac{3}{4}$	1	$\frac{3}{2}$	$\frac{7}{4}$	$\frac{9}{4}$	$\frac{5}{8}(4\ln 2-\pi+2)$
69	$\frac{1}{4}$	$\frac{1}{2}$	$\frac{3}{4}$	1	$\frac{3}{2}$	$\frac{9}{4}$	$\frac{11}{4}$	$\frac{7}{12}(7+6\ln 2-3\pi)$
70	$\frac{1}{4}$	$\frac{1}{2}$	$\frac{3}{4}$	1	$\frac{7}{4}$	$\frac{9}{4}$	$\frac{5}{2}$	$\frac{3}{4}(4\ln 2-3\pi+8)$
71	$\frac{1}{4}$	$\frac{1}{2}$	$\frac{3}{4}$	1	$\frac{9}{4}$	$\frac{5}{2}$	$\frac{11}{4}$	$\frac{7}{4}(10-3\pi)$
72	$\frac{1}{4}$	$\frac{1}{2}$	1	1	$\frac{5}{4}$	$\frac{3}{2}$	2	$\frac{\pi}{3}$
73	$\frac{1}{4}$	$\frac{1}{2}$	1	1	$\frac{5}{4}$	2	$\frac{5}{2}$	$\frac{1}{5}(3+2\pi-6\ln 2)$
74	$\frac{1}{4}$	$\frac{1}{2}$	1	1	$\frac{3}{2}$	2	$\frac{9}{4}$	$\frac{10}{3}(1-\ln 2)$
75	$\frac{1}{4}$	$\frac{1}{2}$	1	1	2	$\frac{9}{4}$	$\frac{5}{2}$	$17-2\pi-14\ln 2$
76	$\frac{1}{4}$	$\frac{1}{2}$	1	$\frac{7}{4}$	$\frac{5}{4}$	$\frac{3}{2}$	$\frac{11}{4}$	$\frac{7}{90}(3\pi+9\ln 2-2)$
77	$\frac{1}{4}$	$\frac{1}{2}$	1	$\frac{7}{4}$	$\frac{9}{4}$	$\frac{5}{2}$	$\frac{11}{4}$	$\frac{7}{12}(3\pi+12\ln 2-16)$
78	$\frac{1}{4}$	$\frac{3}{4}$	1	1	$\frac{5}{4}$	$\frac{7}{4}$	2	$\pi-3\ln 2$
79	$\frac{1}{4}$	$\frac{3}{4}$	1	1	$\frac{7}{4}$	2	$\frac{9}{4}$	$\frac{5}{4}(\pi-12\ln 2+6)$
80	$\frac{1}{4}$	$\frac{3}{4}$	1	1	2	$\frac{9}{4}$	$\frac{11}{4}$	$\frac{35}{18}(13-18\ln 2)$
81	$\frac{1}{4}$	$\frac{3}{4}$	1	$\frac{3}{2}$	$\frac{5}{4}$	$\frac{7}{4}$	$\frac{5}{2}$	$\frac{3}{10}(2\pi-\ln 2-2)$
82	$\frac{1}{4}$	$\frac{3}{4}$	1	$\frac{3}{2}$	$\frac{7}{4}$	$\frac{9}{4}$	$\frac{5}{2}$	$\frac{3}{8}(7\pi+4\ln 2-22)$
83	$\frac{1}{4}$	$\frac{3}{4}$	1	$\frac{3}{2}$	$\frac{9}{4}$	$\frac{5}{2}$	$\frac{11}{4}$	$\frac{7}{4}(3\pi+6\ln 2-13)$

	a_1	a_2	a_3	a_4	b_1	b_2	b_3	${}_4F_3(a_1, a_2, a_3, a_4; b_1, b_2, b_3; 1)$
84	$\frac{1}{4}$	1	1	$\frac{3}{2}$	$\frac{5}{4}$	2	$\frac{5}{2}$	$\frac{1}{5}(\pi+12\ln 2-6)$
85	$\frac{1}{4}$	1	1	$\frac{3}{2}$	2	$\frac{9}{4}$	$\frac{5}{2}$	$4\pi+18\ln 2-24$
86	$\frac{1}{4}$	1	1	$\frac{7}{4}$	$\frac{5}{4}$	2	$\frac{11}{4}$	$\frac{7}{27}(9\ln 2-2)$
87	$\frac{1}{4}$	1	$\frac{3}{2}$	$\frac{7}{4}$	$\frac{5}{4}$	$\frac{5}{2}$	$\frac{11}{4}$	$\frac{7}{30}(8+9\ln 2-3\pi)$
88	$\frac{1}{3}$	$\frac{1}{2}$	1	1	$\frac{4}{3}$	$\frac{3}{2}$	2	$\frac{1}{12}(2\sqrt{3}\,\pi-48\ln 2+27\ln 3)$
89	$\frac{1}{3}$	$\frac{2}{3}$	1	1	$\frac{4}{3}$	$\frac{5}{3}$	2	$\frac{1}{2}(\pi\sqrt{3}-3\ln 3)$
90	$\frac{1}{3}$	$\frac{2}{3}$	1	1	$\frac{5}{3}$	2	$\frac{7}{3}$	$\frac{2\sqrt{3}}{3}(2\sqrt{3}+\pi-3\sqrt{3}\ln 3)$
91	$\frac{1}{3}$	$\frac{2}{3}$	1	$\frac{7}{3}$	$\frac{4}{3}$	$\frac{5}{3}$	$\frac{10}{3}$	$\frac{7\sqrt{3}}{90}(20\pi-\sqrt{3})$
92	$\frac{1}{2}$	$\frac{1}{2}$	$\frac{1}{2}$	$\frac{1}{2}$	$\frac{3}{2}$	$\frac{3}{2}$	$\frac{3}{2}$	$\frac{\pi}{4}\left(\frac{\pi^2}{12}+\ln^2 2\right)$
93	$\frac{1}{2}$	$\frac{1}{2}$	$\frac{1}{2}$	1	$\frac{3}{2}$	$\frac{3}{2}$	$\frac{3}{2}$	$\frac{7}{8}\zeta(3)$
94	$\frac{1}{2}$	$\frac{1}{2}$	$\frac{1}{2}$	1	$\frac{5}{2}$	$\frac{5}{2}$	$\frac{5}{2}$	$\frac{27}{64}(32-3\pi^2)$
95	$\frac{1}{2}$	$\frac{1}{2}$	1	1	2	$\frac{5}{2}$	$\frac{5}{2}$	$\frac{9}{2}(3-4\ln 2)$
96	$\frac{1}{2}$	$\frac{2}{3}$	1	1	$\frac{3}{2}$	$\frac{5}{3}$	2	$\pi\sqrt{3}+8\ln 2-9\ln 3$
97	$\frac{1}{2}$	$\frac{3}{4}$	1	1	$\frac{3}{2}$	$\frac{7}{4}$	2	$3(\pi-4\ln 2)$
98	$\frac{1}{2}$	$\frac{3}{4}$	1	1	$\frac{7}{4}$	2	$\frac{5}{2}$	$6\pi-30\ln 2+3$
99	$\frac{1}{2}$	$\frac{3}{4}$	1	1	2	$\frac{5}{2}$	$\frac{11}{4}$	$\frac{7}{5}(6\pi-42\ln 2-11)$
100	$\frac{1}{2}$	$\frac{3}{4}$	1	$\frac{5}{4}$	$\frac{3}{2}$	$\frac{7}{4}$	$\frac{9}{4}$	$\frac{5}{2}(\pi-\ln 2-2)$
101	$\frac{1}{2}$	$\frac{3}{4}$	1	$\frac{5}{4}$	$\frac{7}{4}$	$\frac{9}{4}$	$\frac{5}{2}$	$\frac{15}{2}(\pi-3)$
102	$\frac{1}{2}$	$\frac{3}{4}$	1	$\frac{5}{4}$	$\frac{9}{4}$	$\frac{5}{2}$	$\frac{11}{4}$	$\frac{21}{4}(3\pi+4\ln 2-12)$
103	$\frac{1}{2}$	1	1	1	2	2	$\frac{5}{2}$	$\frac{1}{8}(12-\pi^2)$
104	$\frac{1}{2}$	1	1	$\frac{5}{4}$	$\frac{3}{2}$	2	$\frac{9}{4}$	$\frac{5}{3}(\pi+8\ln 2-8)$
105	$\frac{1}{2}$	1	1	$\frac{5}{4}$	2	$\frac{9}{4}$	$\frac{5}{2}$	$10\pi+50\ln 2-65$
106	$\frac{1}{2}$	1	1	$\frac{7}{4}$	$\frac{3}{2}$	2	$\frac{11}{4}$	$\frac{7}{45}(36\ln 2-3\pi-8)$
107	$\frac{1}{2}$	1	1	$\frac{5}{2}$	$\frac{3}{2}$	3	$\frac{7}{2}$	$\frac{10}{9}$

	a_1	a_2	a_3	a_4	b_1	b_2	b_3	${}_4F_3(a_1, a_2, a_3, a_4; b_1, b_2, b_3; 1)$
108	$\frac{1}{2}$	1	$\frac{5}{4}$	$\frac{7}{4}$	$\frac{3}{2}$	$\frac{9}{4}$	$\frac{11}{4}$	$\frac{7}{6}(8-2\pi-\ln 2)$
109	$\frac{1}{2}$	1	$\frac{3}{2}$	$\frac{9}{4}$	$\frac{5}{4}$	2	3	$\frac{8}{5}$
110	$\frac{2}{3}$	1	1	$\frac{4}{3}$	$\frac{5}{3}$	2	$\frac{7}{3}$	$12(\ln 3-1)$
111	$\frac{3}{4}$	$\frac{3}{4}$	$\frac{3}{4}$	1	$\frac{11}{4}$	$\frac{11}{4}$	$\frac{11}{4}$	$\frac{3^3\cdot 7^3}{256}(128-12\pi-6\pi^2+\pi^3)$
112	$\frac{3}{4}$	1	1	$\frac{5}{4}$	$\frac{7}{4}$	2	$\frac{9}{4}$	$15(3\ln 2-2)$
113	$\frac{3}{4}$	1	1	$\frac{3}{2}$	$\frac{7}{4}$	2	$\frac{5}{2}$	$24\ln 2-3\pi-6$
114	$\frac{3}{4}$	1	$\frac{5}{4}$	$\frac{3}{2}$	$\frac{7}{4}$	$\frac{9}{4}$	$\frac{5}{2}$	$\frac{15}{2}(4-\pi-\ln 2)$
115	$\frac{5}{6}$	1	1	$\frac{7}{6}$	$\frac{11}{6}$	2	$\frac{13}{6}$	$\frac{35}{2}(3\ln 3+4\ln 2-6)$
116	1	1	$\frac{5}{4}$	$\frac{3}{2}$	2	$\frac{9}{4}$	$\frac{5}{2}$	$90-15\pi-60\ln 2$
117	1	1	$\frac{5}{4}$	$\frac{7}{4}$	2	$\frac{9}{4}$	$\frac{11}{4}$	$\frac{35}{9}(4-3\pi-9\ln 2)$
118	1	1	$\frac{4}{3}$	$\frac{3}{2}$	2	$\frac{7}{3}$	$\frac{5}{2}$	$24-6\sqrt{3}\pi+48\ln 2-54\ln 3$
119	1	1	$\frac{4}{3}$	$\frac{5}{3}$	$\frac{7}{3}$	$\frac{8}{3}$	3	$\frac{20}{9}(1-\sqrt{3}\pi)$
120	1	1	$\frac{3}{2}$	$\frac{3}{2}$	2	2	2	$16\left(\ln 2-\frac{2}{\pi}\mathbf{G}\right)$
121	1	1	$\frac{3}{2}$	$\frac{5}{3}$	2	$\frac{5}{2}$	$\frac{8}{3}$	$\frac{15}{4}(7+\sqrt{3}\pi-16\ln 2+9\ln 3)$
122	1	1	$\frac{3}{2}$	$\frac{7}{4}$	2	$\frac{5}{2}$	$\frac{11}{4}$	$\frac{7}{3}(10-3\pi)$
123	1	1	1	1	3	3	3	$8(10-\pi^2)$
124	1	1	1	1	5	5	5	$\frac{32}{27}(630\pi^2-1)$
125	1	1	$\frac{5}{2}$	$\frac{5}{2}$	2	3	3	$\frac{32}{9}\left[2\ln 2+1-\frac{2}{\pi}(2\mathbf{G}+1)\right]$
126	1	1	$\frac{7}{2}$	$\frac{7}{2}$	2	4	4	$\frac{4}{25}\left[36\ln 2+27-\frac{4}{\pi}(18\mathbf{G}+13)\right]$
127	1	1	$\frac{9}{2}$	$\frac{9}{2}$	2	5	5	$\frac{64}{1225}\left[100\ln 2+\frac{275}{3}-\frac{8}{\pi}(25\mathbf{G}+19)\right]$

7.5.4. Values of ${}_4F_3(a_1, a_2, a_3, a_4; b_1, b_2, b_3; -1)$.

1. $${}_4F_3\left(\begin{matrix}a, b, c, d; & -1\\ 1+a-b, 1+a-c, 1+a-d\end{matrix}\right)= \Gamma\left[\begin{matrix}1+a-b, 1+a-c\\ 1+a, 1+a-b-c\end{matrix}\right] {}_3F_2\left(\begin{matrix}1+a/2-d, b, c; 1\\ 1+a/2, 1+a-d\end{matrix}\right).$$

2. $${}_4F_3\left(\begin{matrix}a, a/2+1, b, c; & -1\\ a/2, 1+a-b, 1+a-c\end{matrix}\right)=\Gamma\left[\begin{matrix}1+a-b, 1+a-c\\ 1+a, 1+a-b-c\end{matrix}\right]$$

[Re $(a-2b-2c)>-1$].

3. $${}_4F_3\left(\begin{matrix}1,\ a,\ b,\ c; & -1\\ a+1,\ b+1,\ c+1\end{matrix}\right)=abc\left[\frac{\beta(a)}{(b-a)(c-a)}+\frac{\beta(b)}{(a-b)(c-b)}+\frac{\beta(c)}{(a-c)(b-c)}\right]$$ $[a\neq b,\ b\neq c,\ a\neq c]$,

4. $$=\frac{a^2b}{(a-b)^2}[\beta(b)-\beta(a)]+\frac{a^2b}{a-b}\beta'(a)$$ $[c=a,\ a\neq b]$,

5. $$=\frac{a^3}{2}\beta''(a)$$ $[c=b=a]$.

6. $${}_4F_3\left(\begin{matrix}a,\ 2a+1,\ a+3/2,\ b; & -1\\ a+1/2,\ a+2,\ 2+2a-b\end{matrix}\right)=\Gamma\left[\begin{matrix}2+2a-b,\ 2+a\\ 2+a-b,\ 2+2a\end{matrix}\right]$$ $[\operatorname{Re} b<1/2]$.

7. $${}_4F_3\left(\begin{matrix}1,\ 3/2,\ a,\ b; & -1\\ 1/2,\ 2-a,\ 2-b\end{matrix}\right)=\Gamma\left[\begin{matrix}2-a,\ 2-b\\ 2-a-b\end{matrix}\right].$$

8. $${}_4F_3\left(\begin{matrix}1,\ 3/2,\ a,\ 1-a; & -1\\ 1/2,\ 1+a,\ 2-a\end{matrix}\right)=\frac{\pi a(1-a)}{\sin a\pi}.$$

9. $${}_4F_3\left(\begin{matrix}1/2,\ 1,\ a,\ 1-a; & -1\\ 3/2,\ 1+a,\ 2-a\end{matrix}\right)=\frac{\pi a(1-a)}{(1-2a)^2}[\operatorname{cosec} a\pi-1].$$

10. $${}_4F_3\left(\begin{matrix}-n,\ a,\ a/2+1,\ b; & -1\\ a/2,\ 1+a-b,\ 1+a+n\end{matrix}\right)=\frac{(1+a)_n}{(1+a-b)_n}.$$

11. $${}_4F_3\left(\begin{matrix}-n,\ 1,\ a,\ 1-a; & -1\\ n+2,\ a+1,\ 2-a\end{matrix}\right)=\frac{2^{2n+2}a(1-a)\,n!\,(n+1)!}{(2n+1)!\,[(2n+1)^2-(1-2a)^2]}.$$

12. $${}_4F_3\left(\begin{matrix}-n,\ 1,\ a,\ 2-a; & -1\\ n+3,\ a+1,\ 3-a\end{matrix}\right)=$$
$$=\frac{(n+2)\,a\,(2-a)}{2(n+1)(1-a)^2}\left[1-\frac{[(n+1)!]^2}{(a)_{n+1}(2-a)_{n+1}}\sum_{k=0}^{n+1}\frac{(1-a)_k(a-1)_k}{(2k)!}2^{2k}\right].$$

13. $${}_4F_3\left(\begin{matrix}-n,\ 1-n/2,\ a,\ -a-n; & -1\\ -n/2,\ 1+a,\ 1-a-n\end{matrix}\right)=0.$$

14. $${}_4F_3\left(\begin{matrix}-n,\ 2,\ 2,\ 2; & -1\\ 1,\ 1,\ 1\end{matrix}\right)=2^{n-3}(n+2)(n^2+7n+4).$$

15. $${}_4F_3\left(\begin{matrix}1/2,\ 1,\ (1+i)/2,\ (1-i)/2; & -1\\ 3/2,\ (3+i)/2,\ (3-i)/2\end{matrix}\right)=0{,}94477614\ldots$$

16. $${}_4F_3\left(\begin{matrix}1,\ 1,\ 1+i/2,\ 1-i/2; & -1\\ 2,\ 2+i/2,\ 2-i/2\end{matrix}\right)=0{,}885097\ldots$$

17. $${}_4F_3\left(\begin{matrix}1,\ 1,\ 1+i,\ 1-i; & -1\\ 2,\ 2+i,\ 2-i\end{matrix}\right)=0{,}847134\ldots$$

	a_1	a_2	a_3	a_4	b_1	b_2	b_3	${}_4F_3(a_1, a_2, a_3, a_4;\ b_1, b_2, b_3;\ -1)$
18	$\frac{1}{4}$	$\frac{1}{2}$	$\frac{3}{4}$	1	$\frac{5}{4}$	$\frac{3}{2}$	$\frac{7}{4}$	$\frac{3}{4}(\sqrt{2}-1)\pi$
19	$\frac{1}{4}$	$\frac{1}{2}$	$\frac{3}{4}$	1	$\frac{5}{4}$	$\frac{7}{4}$	$\frac{5}{2}$	$\frac{3}{20}[(8\sqrt{2}-7)\pi-2-30\sqrt{2}\ln(1+\sqrt{2})]$

	a_1	a_2	a_3	a_4	b_1	b_2	b_3	${}_4F_3(a_1, a_2, a_3, a_4; b_1, b_2, b_3; -1)$
20	$\frac{1}{4}$	$\frac{1}{2}$	1	1	$\frac{5}{4}$	$\frac{3}{2}$	2	$\frac{1}{6}[(2\sqrt{2}-3)\pi+2\ln 2+4\sqrt{2}\ln(1+\sqrt{2})]$
21	$\frac{1}{4}$	$\frac{3}{4}$	1	1	$\frac{5}{4}$	$\frac{7}{4}$	2	$\frac{\sqrt{2}}{4}[2\sqrt{2}\ln 2-\pi+6\ln(1+\sqrt{2})]$
22	$\frac{1}{4}$	$\frac{1}{2}$	1	1	$\frac{5}{4}$	2	$\frac{5}{2}$	$\frac{1}{5}[5\ln 2+(2\sqrt{2}-3)\pi+4\sqrt{2}\ln(1+\sqrt{2})-3]$
23	$\frac{1}{4}$	$\frac{3}{4}$	1	$\frac{3}{2}$	$\frac{5}{4}$	$\frac{7}{4}$	$\frac{5}{2}$	$\frac{3}{20}[4-(1+\sqrt{2})\pi+16\ln(1+\sqrt{2})]$
24	$\frac{1}{4}$	1	1	$\frac{3}{2}$	$\frac{5}{4}$	2	$\frac{5}{2}$	$\frac{1}{10}[12-(3-2\sqrt{2})\pi-10\ln 2+4\sqrt{2}\ln(1+\sqrt{2})]$
25	$\frac{1}{4}$	1	$\frac{3}{2}$	$\frac{7}{4}$	$\frac{5}{4}$	$\frac{5}{2}$	$\frac{11}{4}$	$\frac{7}{60}[3(3-2\sqrt{2})\pi-16+18\sqrt{2}\ln(1+\sqrt{2})]$
26	$\frac{1}{3}$	$\frac{1}{2}$	1	1	$\frac{4}{3}$	$\frac{3}{2}$	2	$2\ln 2-\left(1-\frac{\sqrt{3}}{2}\right)\pi$
27	$\frac{1}{3}$	$\frac{2}{3}$	1	1	$\frac{4}{3}$	$\frac{5}{3}$	2	$4\ln 2-\frac{\pi}{\sqrt{3}}$
28	$\frac{1}{3}$	$\frac{2}{3}$	1	1	$\frac{5}{3}$	2	$\frac{7}{3}$	$\frac{4}{3}(3-\sqrt{3}\pi+2\ln 2)$
29	$\frac{1}{3}$	$\frac{2}{3}$	1	$\frac{7}{3}$	$\frac{4}{3}$	$\frac{5}{3}$	$\frac{10}{3}$	$\frac{7}{60}(16\ln 2-3)$
30	$\frac{1}{2}$	$\frac{1}{2}$	$\frac{1}{2}$	1	$\frac{5}{2}$	$\frac{5}{2}$	$\frac{5}{2}$	$\frac{27}{128}(12\pi-\pi^3-64)$
31	$\frac{1}{2}$	$\frac{2}{3}$	1	1	$\frac{3}{2}$	$\frac{5}{3}$	2	$8\ln 2+2(1-\sqrt{3})\pi$
32	$\frac{1}{2}$	$\frac{3}{4}$	1	1	$\frac{3}{2}$	$\frac{7}{4}$	2	$\frac{3}{2}[(1-2\sqrt{2})\pi+2\ln 2+4\sqrt{2}\ln(1+\sqrt{2})]$
33	$\frac{1}{2}$	$\frac{3}{4}$	1	1	$\frac{7}{4}$	2	$\frac{5}{2}$	$3(1-2\sqrt{2})\pi+12\sqrt{2}\ln(1+\sqrt{2})-3$
34	$\frac{1}{2}$	$\frac{3}{4}$	1	$\frac{5}{4}$	$\frac{3}{2}$	$\frac{7}{4}$	$\frac{9}{4}$	$\frac{5}{4}[4-(1-2\sqrt{2})\pi+2\sqrt{2}\ln(1+\sqrt{2})]$
35	$\frac{1}{2}$	1	1	1	2	2	$\frac{5}{2}$	$\frac{1}{8}(6\pi-\pi^2-12)$
36	$\frac{1}{2}$	1	1	$\frac{5}{4}$	$\frac{3}{2}$	2	$\frac{9}{4}$	$\frac{5}{6}[16+(1-6\sqrt{2})\pi-6\ln 2-12\sqrt{2}\ln(1+\sqrt{2})]$
37	$\frac{1}{2}$	1	1	$\frac{7}{4}$	$\frac{3}{2}$	2	$\frac{11}{4}$	$\frac{7}{45}[8+3(3-\sqrt{2})\pi-15\ln 2+6\sqrt{2}\ln(1+\sqrt{2})]$
38	$\frac{1}{2}$	1	1	$\frac{5}{2}$	$\frac{3}{2}$	3	$\frac{7}{2}$	$\frac{10}{9}(5-6\ln 2)$

	a_1	a_2	a_3	a_4	b_1	b_2	b_3	${}_4F_3(a_1, a_2, a_3, a_4; b_1, b_2, b_3; -1)$
39	$\frac{2}{3}$	1	1	$\frac{4}{3}$	$\frac{5}{3}$	2	$\frac{7}{3}$	$12-16\ln 2$
40	$\frac{3}{4}$	1	1	$\frac{5}{4}$	$\frac{7}{4}$	2	$\frac{9}{4}$	$15[2-\ln 2-\sqrt{2}\ln(1+\sqrt{2})]$
41	$\frac{3}{4}$	1	1	$\frac{3}{2}$	$\frac{7}{4}$	2	$\frac{5}{2}$	$6+3(\sqrt{2}-1)\pi-9\ln 2+6\sqrt{2}\ln(1+\sqrt{2})$
42	$\frac{3}{4}$	1	$\frac{5}{4}$	$\frac{3}{2}$	$\frac{7}{4}$	$\frac{9}{4}$	$\frac{5}{2}$	$\frac{15}{4}[(2\sqrt{2}-1)\pi-8+4\sqrt{2}\ln(1+\sqrt{2})]$
43	$\frac{5}{6}$	1	1	$\frac{7}{6}$	$\frac{11}{6}$	2	$\frac{13}{6}$	$\frac{35\sqrt{3}}{3}[3\sqrt{3}-\sqrt{3}\ln 2-\ln(2+\sqrt{3})]$
44	1	1	1	1	3	3	3	$4(3\zeta(3)+24\ln 2-20)$
45	1	1	$\frac{5}{4}$	$\frac{3}{2}$	2	$\frac{9}{4}$	$\frac{5}{2}$	$\frac{15}{2}[(2\sqrt{2}-1)\pi+4\sqrt{2}\ln(1+\sqrt{2})-12]$
46	1	1	$\frac{5}{4}$	$\frac{7}{4}$	2	$\frac{9}{4}$	$\frac{11}{4}$	$\frac{35\sqrt{2}}{18}[3\pi-16\sqrt{2}-6\ln(1+\sqrt{2})]$
47	1	1	$\frac{4}{3}$	$\frac{3}{2}$	2	$\frac{7}{3}$	$\frac{5}{2}$	$12[4\ln 2-(1-\sqrt{3})\pi-5]$
48	1	1	$\frac{4}{3}$	$\frac{5}{3}$	$\frac{7}{3}$	$\frac{8}{3}$	3	$160\ln 2-110$
49	1	1	$\frac{3}{2}$	$\frac{5}{3}$	2	$\frac{5}{2}$	$\frac{8}{3}$	$\frac{15}{4}[2(2-\sqrt{3})\pi+8\ln 2-7]$
50	1	1	$\frac{3}{2}$	$\frac{7}{4}$	2	$\frac{5}{2}$	$\frac{11}{4}$	$\frac{7}{2}[2\ln 2-3(\sqrt{2}-1)\pi-4+6\sqrt{2}\ln(1+\sqrt{2})]$

7.5.5. Values of ${}_4F_3(a_1, a_2, a_3, a_4; b_1, b_2, b_3; z_0)$ for $z_0 \neq \pm 1$.

$$ {}_4F_3\left(\begin{matrix}-n/2,\ (1-n)/2,\ 1/3-n,\ (22-9n)/21;\ -27\\ 5/6,\ 4/3,\ (1-9n)/21\end{matrix}\right)=\frac{(-8)^n}{1-9n}. $$

$$ {}_4F_3\left(\begin{matrix}1,\ 1,\ 1,\ 1;\ 1/2\\ 2,\ 2,\ 2\end{matrix}\right)=\frac{1}{12}(21\zeta(3)-2\pi^2\ln 2+4\ln^3 2). $$

$$ {}_4F_3\left(\begin{matrix}1,1,1,1;\ (3-\sqrt{5})/2\\ 2,\ 2,\ 2\end{matrix}\right)=\frac{2}{3-\sqrt{5}}\left(\frac{4}{5}\zeta(3)+\frac{\pi^2}{15}\ln\frac{3-\sqrt{5}}{2}-\frac{1}{12}\ln^3\frac{3-\sqrt{5}}{2}\right). $$

7.6 THE FUNCTION ${}_5F_4(a_1,\ldots,a_5;b_1,\ldots,b_4;z)$

7.6.1. Particular values of ${}_5F_4(a_1,\ldots,a_5;b_1,\ldots,b_4;\pm z)$.

	a_1	a_2	a_3	a_4	a_5	b_1	b_2	b_3	b_4	${}_5F_4(a_1,\ldots,a_5;b_1,\ldots,b_4;z)$
1	$\frac{1}{4}$	$\frac{1}{2}$	$\frac{3}{4}$	1	1	$\frac{3}{2}$	$\frac{7}{4}$	2	$\frac{9}{4}$	$\frac{5}{4z^{5/4}}\left[2(z-6\sqrt{z}+1)\operatorname{arctg} z^{1/4}+(z+6\sqrt{z}+1)\ln\frac{1+z^{1/4}}{1-z^{1/4}}\right]+\frac{5}{z}[(1-\sqrt{z})\ln(1+\sqrt{z})+(1+\sqrt{z})\ln(1-\sqrt{z})-1$
2	1	1	1	3	3	2	2	4	4	$\frac{9}{4z^3}\left[(1+z^2)\operatorname{Li}_2(z)-(1-z^2)\ln(1-z)-2z-\frac{3}{4}z^2\right.$

3. $${}_5F_4\left(\begin{matrix}1/4,\ 1/2,\ 3/4,\ 1,\ 1;\ -z\\ 3/2,\ 7/4,\ 2,\ 9/4\end{matrix}\right)=\frac{5\sqrt{2}}{8}z^{-5/4}\times$$
$$\times\left[2(z+6\sqrt{z}-1)\operatorname{arctg}\frac{\sqrt[4]{4z}}{1-\sqrt{z}}+(z-6\sqrt{z}-1)\ln\frac{\sqrt{z}+\sqrt[4]{4z}+1}{\sqrt{z}-\sqrt[4]{4z}+1}\right]-$$
$$-10\operatorname{arctg}\sqrt{z}+5z^{-1}[1-\ln(1+z)]$$

7.6.2. Values of ${}_5F_4(a_1,\ldots,a_5;b_1,\ldots,b_4;1)$.

1. $${}_5F_4\left(\begin{matrix}a,\ b,\ c,\ d,\ e;\ 1\\ 1+a-b,\ 1+a-c,\ 1+a-d,\ 1+a-e\end{matrix}\right)=$$
$$=\Gamma\left[\begin{matrix}1+a-b,\ 1+a-c,\ 1+a-d,\ 1+a-b-c-d\\ 1+a,\ 1+a-b-c,\ 1+a-b-d,\ 1+a-c-d\end{matrix}\right]\times$$
$$\times{}_4F_3\left(\begin{matrix}1+a/2-e,\ b,\ c,\ d;\ 1\\ 1+a/2,\ b+c+d-a,\ 1+a-e\end{matrix}\right)$$

2. $${}_5F_4\left(\begin{matrix}a,\ a+1/2,\ b,\ c,\ d;\ 1\\ 2a,\ e,\ e+1/2,\ b+c+d-2e+1\end{matrix}\right)=$$
$$=\Gamma\left[\begin{matrix}2e,\ 2e-b-c,\ 2e-b-d,\ 2e-c-d\\ 2e-b,\ 2e-c,\ 2e-d,\ 2e-b-c-d\end{matrix}\right]{}_4F_3\left(\begin{matrix}2e-2a,\ b,\ c,\ d;\ 1\\ 2e-b,\ 2e-c,\ 2e-d\end{matrix}\right)$$

3. $${}_5F_4\left(\begin{matrix}a,\ a/2+1,\ b,\ c,\ d;\ 1\\ a/2,\ 1+a-b,\ 1+a-c,\ 1+a-d\end{matrix}\right)=$$
$$=\Gamma\left[\begin{matrix}1+a-b,\ 1+a-c,\ 1+a-d,\ 1+a-b-c-d\\ 1+a,\ 1+a-b-c,\ 1+a-b-d,\ 1+a-c-d\end{matrix}\right]\quad[\operatorname{Re}(a-b-c-d)>-1]$$

4. $${}_5F_4\left(\begin{matrix}1,\ a,\ b,\ c,\ d;\ 1\\ a+1,\ b+1,\ c+1,\ d+1\end{matrix}\right)=I_\psi,$$
$$I_\psi=-abcd\left[\frac{\psi(a)}{(b-a)(c-a)(d-a)}+\frac{\psi(b)}{(a-b)(c-b)(d-b)}+\frac{\psi(c)}{(a-c)(b-c)(d-c)}+\right.$$
$$\left.+\frac{\psi(d)}{(a-d)(b-d)(c-d)}\right]\quad[a,\ b,\ c,\ d\ \text{are distinct}],$$
$$I_\psi=a^2bc\left[\frac{b+c-2a}{(a-b)^2(a-c)^2}\psi(a)+\frac{\psi'(a)}{(a-b)(a-c)}+\frac{\psi(b)}{(a-b)^2(b-c)}+\frac{\psi(c)}{(a-c)^2(c-b)}\right]$$
$$[a=d,\ a\neq b,\ b\neq c,\ c\neq a$$
$$I_\psi=\frac{a^2b^2}{(b-a)^3}\{2[\psi(a)-\psi(b)]+(b-a)[\psi'(a)+\psi'(b)]\}\quad[a=d,\ b=c,\ a\neq b$$

$$I_\psi = \frac{a^3 b}{(a-b)^3}\left[\psi(a) - \psi(b) - (a-b)\psi'(a) + \frac{(a-b)^2}{2}\psi''(a)\right] \quad [a=c=d,\ a\neq b],$$

$$I_\psi = \frac{a^4}{6}\psi'''(a) \quad [a=b=c=d].$$

5. $${}_5F_4\left(\begin{matrix}1,\ a,\ a+1/2,\ b,\ b+1/2; & 1\\ 1/2-a-n,\ 1-a-n,\ 1/2-b-n,\ 1-b-n\end{matrix}\right) = \sum_{k=0}^{[n/2]} \frac{(2a)_{2k}(2b)_{2k}}{(1-2a-2n)_k(1-2b-2n)_k} - \frac{[1+(-1)^n]}{4}(2a)_n(2b)_n + $$
$$+\frac{1}{4}\Gamma\left[\begin{matrix}1-2a,\ 1-2b\\ 1-2a-2b-2n\end{matrix}\right]\left\{\frac{\sqrt{\pi}\,2^{4n}}{(4a+2n)_{2n}(4b+2n)_{2n}}\Gamma\left[\begin{matrix}1/2-2a-2b-2n\\ 1/2-2a-2n,\ 1/2-2b-2n\end{matrix}\right] + \right.$$
$$\left. + \frac{(-1)^n}{(2a+n)_n(2b+n)_n}\right\} \quad [\mathrm{Re}\,(a+b) < 1/4-n].$$

6. $${}_5F_4\left(\begin{matrix}1,\ a,\ a+1/2,\ b,\ b+1/2; & 1\\ 1/2+n-a,\ 1+n-a,\ 1/2+n-b,\ 1+n-b\end{matrix}\right) =$$
$$= -\sum_{k=1}^{[(n-1)/2]} \frac{(2a-2n)_{2k}(2b-2n)_{2k}}{(1-2a)_{2k}(1-2b)_{2k}} - [1+(-1)^n]\frac{(2a-2n)_n(2b-2n)_n}{4(1-2a)_n(1-2b)_n} +$$
$$+\frac{(1+2n-2a-2b)_{2n}}{4}\Gamma\left[\begin{matrix}1-2a,\ 1-2b\\ 1+4n-2a-2b\end{matrix}\right]\left\{\frac{\sqrt{\pi}(1+2n-4a)_{2n}(1+2n-4b)_{2n}}{2^{4n}(1/2+2n-2a-2b)_{2n}}\times\right.$$
$$\left.\times\Gamma\left[\begin{matrix}1/2+4n-2a-2b\\ 1/2+2n-2a,\ 1/2+2n-2b\end{matrix}\right] + (-1)^n(1+n-2a)_n(1+n-2b)_n\right\}$$
$$[\mathrm{Re}(a+b) < n+1/4,\ n=1,\ 2,\ 3,\ \ldots].$$

7. $${}_5F_4\left(\begin{matrix}1,\ a,\ -a,\ b,\ -b; & 1\\ 1+a,\ 1-a,\ 1+b,\ 1-b\end{matrix}\right) = \frac{\pi ab}{2(a^2-b^2)}[a\operatorname{ctg} b\pi - b\operatorname{ctg} a\pi] + \frac{1}{2} \quad [a\neq b],$$

8. $$= \frac{\pi a}{4}\operatorname{ctg} a\pi + \frac{\pi^2 a^2}{4\sin^2 a\pi} + \frac{1}{2} \quad [b=a].$$

9. $${}_5F_4\left(\begin{matrix}1,\ a,\ a,\ 1-a,\ 1-a; & 1\\ 1+a,\ 1+a,\ 2-a,\ 2-a\end{matrix}\right) = \frac{\pi a^2(1-a)^2}{(1-2a)^3}\left[\frac{\pi(1-2a)}{\sin^2 a\pi} - 2\operatorname{ctg} a\pi\right] \quad [a\neq 1/2].$$

10. $$= \frac{\pi^4}{96} \quad [a=1/2].$$

11. $${}_5F_4\left(\begin{matrix}2,\ a,\ a,\ 2-a,\ 2-a; & 1\\ 1+a,\ 1+a,\ 3-a,\ 3-a\end{matrix}\right) = \frac{a^2(2-a)^2}{4(1-a)}[\psi'(a-1) - \psi'(1-a)] \quad [a\neq 1].$$

12. $${}_5F_4\left(\begin{matrix}1/2,\ 1/2,\ 1,\ a,\ 1-a; & 1\\ 3/2,\ 3/2,\ 1+a,\ 2-a\end{matrix}\right) = \frac{\pi a(1-a)}{(1-2a)^3}\left[\operatorname{ctg} a\pi - \frac{1-2a}{2}\pi\right] \quad [a\neq 1/2].$$

13. $${}_5F_4\left(\begin{matrix}1,\ 1,\ 1,\ a,\ 2-a; & 1\\ 2,\ 2,\ 1+a,\ 3-a\end{matrix}\right) = \frac{a(2-a)}{6(1-a)^4}[3 - \pi^2(1-a)^2 + 3\pi(1-a)\operatorname{ctg} a\pi].$$

14. $${}_5F_4\left(\begin{matrix}-n,\ a,\ b,\ c,\ d; & 1\\ 1+a-b,\ 1+a-c,\ 1+a-d,\ 1+a+n\end{matrix}\right) = \frac{(1+a)_n(1+a/2-d)_n}{(1+a/2)_n(1+a-d)_n}\,{}_4F_3\left(\begin{matrix}-n,\ a/2,\ 1+a-b-c,\ d; & 1\\ 1+a-b,\ 1+a-c,\ d-a/2-n\end{matrix}\right).$$

15. $${}_5F_4\left(\begin{matrix}-n,\ a,\ a+1/2,\ b,\ c; & 1\\ (b-n)/2,\ (1+b-n)/2,\ d,\ 1+2a+c-d\end{matrix}\right) = \frac{(1+2a-b)_n}{(1-b)_n}\,{}_4F_3\left(\begin{matrix}-n,\ 2a,\ d-c,\ 1+2a-d; & 1\\ 1+2a-b,\ d,\ 1+2a+c-d\end{matrix}\right).$$

16. $${}_5F_4\left(\begin{matrix}-n,\ a,\ a+1/2,\ b,\ c; & 1\\ 2a,\ (b+c)/2,\ (b+c+1)/2,\ 1+2a-b-c-n\end{matrix}\right)=$$
$$=\frac{(c)_n\,(c+2b-2a)_n}{(b+c)_n\,(b+c-2a)_n}\,{}_5F_4\left(\begin{matrix}-n,\ b/2,\ (b+1)/2,\ b+c-2a,\ n+c+2b-2a; & 1\\ b+c+n,\ 1-c-n,\ b+c/2-a,\ b+(c+1)/2-a\end{matrix}\right)$$

17. $${}_5F_4\left(\begin{matrix}-n,\ a,\ a/2+1,\ b,\ c; & 1\\ a/2,\ 1+a-b,\ 1+a-c,\ 1+a+n\end{matrix}\right)=\frac{(1+a)_n\,(1+a-b-c)_n}{(1+a-b)_n\,(1+a-c)_n}.$$

18. $${}_5F_4\left(\begin{matrix}-n,\ 1/2-n,\ -1/4,\ 1/4,\ 9/8; & 1\\ 1/4-n,\ 3/4-n,\ 1/8,\ 3/2\end{matrix}\right)=0 \qquad [n=1,\ 2,3,\ \ldots]$$

19. $${}_5F_4\left(\begin{matrix}2,\ 1+i/2,\ 1+i/2,\ 1-i/2,\ 1-i/2; & 1\\ 2+i/2,\ 2+i/2,\ 2-i/2,\ 2-i/2\end{matrix}\right)=1{,}289128\ldots$$

20. $${}_5F_4\left(\begin{matrix}2,\ 1+i,\ 1+i,\ 1-i,\ 1-i; & 1\\ 2+i,\ 2+i,\ 2-i,\ 2-i\end{matrix}\right)=1{,}588468\ldots$$

	a_1	a_2	a_3	a_4	a_5	b_1	b_2	b_3	b_4	${}_5F_4(a_1, \ldots, a_5;\ b_1, \ldots, b_4;\ 1)$
21	$\frac{1}{8}$	$\frac{3}{8}$	$\frac{5}{8}$	$\frac{7}{8}$	1	$\frac{9}{8}$	$\frac{11}{8}$	$\frac{13}{8}$	$\frac{15}{8}$	$\frac{35\pi}{32\,(2+\sqrt{2})}$
22	$\frac{1}{4}$	$\frac{1}{4}$	$\frac{1}{4}$	$\frac{1}{4}$	$\frac{1}{2}$	$\frac{5}{4}$	$\frac{5}{4}$	$\frac{5}{4}$	$\frac{5}{4}$	$\frac{\sqrt{\pi}}{512\sqrt{2}}\,\Gamma^2\left(\frac{1}{4}\right)\left[3\pi^2+\psi'\left(\frac{1}{4}\right)-\psi'\left(\frac{3}{4}\right)\right]$
23	$\frac{1}{4}$	$\frac{1}{4}$	$\frac{3}{4}$	$\frac{3}{4}$	1	$\frac{5}{4}$	$\frac{5}{4}$	$\frac{7}{4}$	$\frac{7}{4}$	$\frac{9\pi}{32}\,(\pi-2)$
24	$\frac{1}{4}$	$\frac{1}{2}$	$\frac{1}{2}$	$\frac{3}{4}$	1	$\frac{5}{4}$	$\frac{3}{2}$	$\frac{3}{2}$	$\frac{7}{4}$	$\frac{3\pi}{8}\,(4-\pi)$
25	$\frac{1}{4}$	$\frac{1}{2}$	$\frac{3}{4}$	$\frac{3}{4}$	1	$\frac{5}{4}$	$\frac{3}{2}$	$\frac{7}{4}$	$\frac{7}{4}$	1,014678...
26	$\frac{1}{4}$	$\frac{1}{2}$	$\frac{3}{4}$	1	1	$\frac{5}{4}$	$\frac{3}{2}$	$\frac{7}{4}$	2	$6\ln 2-\pi$
27	$\frac{1}{4}$	$\frac{1}{2}$	$\frac{3}{4}$	1	1	$\frac{5}{4}$	$\frac{3}{2}$	2	$\frac{11}{4}$	$\frac{7}{90}\,(144\ln 2-27\pi-2)$
28	$\frac{1}{4}$	$\frac{1}{2}$	$\frac{3}{4}$	1	1	$\frac{5}{4}$	$\frac{7}{4}$	2	$\frac{5}{2}$	$\frac{3}{5}\,(22\ln 2-4\pi-1)$
29	$\frac{1}{4}$	$\frac{1}{2}$	$\frac{3}{4}$	1	1	$\frac{5}{4}$	2	$\frac{5}{2}$	$\frac{11}{4}$	$\frac{7}{15}\,(54\ln 2-9\pi-7)$
30	$\frac{1}{4}$	$\frac{1}{2}$	$\frac{3}{4}$	1	1	$\frac{3}{2}$	$\frac{7}{4}$	2	$\frac{9}{4}$	$\frac{5}{2}\,(8\ln 2-\pi-2)$
31	$\frac{1}{4}$	$\frac{1}{2}$	$\frac{3}{4}$	1	1	$\frac{3}{2}$	2	$\frac{9}{4}$	$\frac{11}{4}$	$\frac{7}{18}\,(108\ln 2-9\pi-44)$
32	$\frac{1}{4}$	$\frac{1}{2}$	$\frac{3}{4}$	1	1	$\frac{7}{4}$	2	$\frac{9}{4}$	$\frac{5}{2}$	$54\ln 2-3\pi-27$
33	$\frac{1}{4}$	$\frac{1}{2}$	$\frac{3}{4}$	1	1	2	$\frac{9}{4}$	$\frac{5}{2}$	$\frac{11}{4}$	$\frac{7}{3}\,(54\ln 2-37)$
34	$\frac{1}{4}$	$\frac{1}{2}$	$\frac{3}{4}$	1	$\frac{7}{4}$	$\frac{5}{4}$	$\frac{7}{4}$	$\frac{5}{2}$	$\frac{11}{4}$	$\frac{7}{20}\,(4+12\ln 2-3\pi)$
35	$\frac{1}{4}$	$\frac{1}{2}$	1	1	$\frac{7}{4}$	$\frac{5}{4}$	2	$\frac{5}{2}$	$\frac{11}{4}$	$\frac{7}{45}\,(19-18\ln 2)$
36	$\frac{1}{4}$	$\frac{1}{2}$	1	1	$\frac{7}{4}$	$\frac{3}{2}$	2	$\frac{9}{4}$	$\frac{11}{4}$	$\frac{7}{54}\,(86-72\ln 2-9\pi)$

	a_1	a_2	a_3	a_4	a_5	b_1	b_2	b_3	b_4	${}_5F_4(a_1, \ldots, a_5; b_1, \ldots, b_4; 1)$
37	$\frac{1}{4}$	$\frac{1}{2}$	1	1	$\frac{7}{4}$	2	$\frac{9}{4}$	$\frac{5}{2}$	$\frac{11}{4}$	$\frac{7}{9}(67-9\pi-54\ln 2)$
38	$\frac{1}{4}$	$\frac{3}{4}$	1	1	$\frac{3}{2}$	$\frac{5}{4}$	$\frac{7}{4}$	2	$\frac{5}{2}$	$\frac{3}{5}(3\pi-14\ln 2+2)$
39	$\frac{1}{4}$	$\frac{3}{4}$	1	1	$\frac{3}{2}$	$\frac{5}{4}$	2	$\frac{5}{2}$	$\frac{11}{4}$	$\frac{7}{30}(9\pi-72\ln 2+26)$
40	$\frac{1}{4}$	$\frac{3}{4}$	1	1	$\frac{3}{2}$	$\frac{7}{4}$	2	$\frac{9}{4}$	$\frac{5}{2}$	$\frac{3}{2}(26-\pi-32\ln 2)$
41	$\frac{1}{4}$	$\frac{3}{4}$	1	1	$\frac{3}{2}$	2	$\frac{9}{4}$	$\frac{5}{2}$	$\frac{11}{4}$	$\frac{7}{6}(104-9\pi-108\ln 2)$
42	$\frac{1}{4}$	1	1	$\frac{3}{2}$	$\frac{7}{4}$	2	$\frac{9}{4}$	$\frac{5}{2}$	$\frac{11}{4}$	$\frac{7}{18}(27\pi+144\ln 2-182)$
43	$\frac{1}{2}$	$\frac{1}{2}$	$\frac{1}{2}$	$\frac{1}{2}$	$\frac{1}{2}$	$\frac{3}{2}$	$\frac{3}{2}$	$\frac{3}{2}$	$\frac{3}{2}$	$\frac{\pi}{48}[\pi^2\ln 2+4\ln^3 2+6\zeta(3)]$
44	$\frac{1}{2}$	$\frac{1}{2}$	$\frac{1}{2}$	$\frac{1}{2}$	1	$\frac{5}{2}$	$\frac{5}{2}$	$\frac{5}{2}$	$\frac{5}{2}$	$\frac{27}{256}(\pi^4+30\pi^2-384)$
45	$\frac{1}{2}$	$\frac{1}{2}$	1	$\frac{5}{2}$	$\frac{5}{2}$	$\frac{3}{2}$	$\frac{3}{2}$	$\frac{7}{2}$	$\frac{7}{2}$	$\frac{25}{576}(9\pi^2-64)$
46	$\frac{1}{2}$	$\frac{3}{4}$	1	1	$\frac{5}{4}$	$\frac{3}{2}$	$\frac{7}{4}$	2	$\frac{9}{4}$	$5\pi-50\ln 2+20$
47	$\frac{1}{2}$	$\frac{3}{4}$	1	1	$\frac{5}{4}$	$\frac{3}{2}$	2	$\frac{9}{4}$	$\frac{11}{4}$	$\frac{7}{6}(3\pi-96\ln 2+58)$
48	$\frac{1}{2}$	$\frac{3}{4}$	1	1	$\frac{5}{4}$	$\frac{7}{4}$	2	$\frac{9}{4}$	$\frac{5}{2}$	$135-150\ln 2$
49	$\frac{1}{2}$	$\frac{3}{4}$	1	1	$\frac{5}{4}$	2	$\frac{9}{4}$	$\frac{5}{2}$	$\frac{11}{4}$	$7(47-3\pi-54\ln 2)$
50	$\frac{1}{2}$	1	1	1	$\frac{3}{2}$	2	2	$\frac{5}{2}$	$\frac{5}{2}$	$\frac{3}{4}(84-5\pi^2-48\ln 2)$
51	$\frac{1}{2}$	1	1	$\frac{5}{4}$	$\frac{7}{4}$	2	$\frac{9}{4}$	$\frac{5}{2}$	$\frac{11}{4}$	$\frac{7}{3}(12\pi+66\ln 2-83)$
52	$\frac{2}{3}$	1	1	1	$\frac{4}{3}$	$\frac{5}{3}$	2	2	$\frac{7}{3}$	$\frac{4}{27}(27-3\sqrt{3}\pi-\pi^2)$
53	$\frac{3}{4}$	1	1	$\frac{5}{4}$	$\frac{3}{2}$	$\frac{7}{4}$	2	$\frac{9}{4}$	$\frac{5}{2}$	$15\pi+150\ln 2-150$
54	$\frac{3}{4}$	1	1	$\frac{5}{4}$	$\frac{3}{2}$	2	$\frac{9}{4}$	$\frac{5}{2}$	$\frac{11}{4}$	$\frac{35}{2}(3\pi-24\ln 2-26)$
55	1	1	1	1	1	3	3	3	3	$\frac{16}{45}(\pi^4+150\pi^2-525)$
56	1	1	1	$\frac{3}{2}$	$\frac{3}{2}$	$\frac{5}{2}$	$\frac{5}{2}$	3	3	$72\pi^4+12\pi^2-828$
57	1	1	1	3	3	2	2	4	4	$\frac{3}{16}(4\pi^2-33)$

7.6.3. Values of ${}_5F_4(a_1, \ldots, a_5; b_1, \ldots, b_4; -1)$.

1. ${}_5F_4\left(\begin{matrix}1, a, b, c, d; -1\\ a+1, b+1, c+1, d+1\end{matrix}\right)=-I_\beta$

[I_β see 7.6.2.4, where $\psi(x)$ should be replaced by $\beta(x)$].

2. $${}_5F_4\left(\begin{matrix}1,\ a,\ a,\ -a,\ -a;\ -1\\ 1+a,\ 1+a,\ 1-a,\ 1-a\end{matrix}\right)=\frac{\pi a}{4\sin a\pi}+\frac{\pi^2a^2\cos a\pi}{4\sin^2 a\pi}+\frac{1}{2}.$$

3. $${}_5F_4\left(\begin{matrix}1/2,\ 1/2,\ 1,\ a,\ 1-a;\ -1\\ 3/2,\ 3/2,\ 1+a,\ 2-a\end{matrix}\right)=-\frac{\pi a\,(1-a)}{(1-2a)^3}\left[\operatorname{cosec} a\pi-\frac{2}{\pi}\,\beta(a)-\frac{1-2a}{\pi}\,\beta'\left(\frac{1}{2}\right)\right]$$ $[a\neq 1/2]$.

4. $${}_5F_4\left(\begin{matrix}1,\ 1,\ 1,\ a,\ 2-a;\ -1\\ 2,\ 2,\ 1+a,\ 3-a\end{matrix}\right)=\frac{a\,(a-2)}{12\,(a-1)^4}\left[6+\pi^2\,(1-a)^2-\frac{6\pi\,(1-a)}{\sin a\pi}\right].$$

5. $${}_5F_4\left(\begin{matrix}1/2,\ 1/2,\ 1,\ (1+i)/2,\ (1-i)/2;\ -1\\ 3/2,\ 3/2,\ (3+i)/2,\ (3-i)/2\end{matrix}\right)=0{,}980248\ldots$$

6. $${}_5F_4\left(\begin{matrix}1/2,\ 1/2,\ 1,\ (2+i)/4,\ (2-i)/4;\ -1\\ 3/2,\ 3/2,\ (6+i)/4,\ (6-i)/4\end{matrix}\right)=0{,}986580\ldots$$

	a_1	a_2	a_3	a_4	a_5	b_1	b_2	b_3	b_4	${}_5F_4(a_1,\ \ldots,\ a_5;\ b_1,\ \ldots,\ b_4;\ -1)$
7	$\frac{1}{4}$	$\frac{1}{4}$	$\frac{3}{4}$	$\frac{3}{4}$	1	$\frac{5}{4}$	$\frac{5}{4}$	$\frac{7}{4}$	$\frac{7}{4}$	$0{,}99339507\ldots$
8	$\frac{1}{4}$	$\frac{1}{2}$	$\frac{1}{2}$	$\frac{3}{4}$	1	$\frac{5}{4}$	$\frac{3}{2}$	$\frac{3}{2}$	$\frac{7}{4}$	$\frac{3}{4}[\sqrt{2}\ln(1+\sqrt{2})-\mathbf{G}]$
9	$\frac{1}{4}$	$\frac{1}{2}$	$\frac{3}{4}$	1	1	$\frac{5}{4}$	$\frac{3}{2}$	$\frac{7}{4}$	2	$\frac{1}{2}[(4\sqrt{2}-3)\,\pi-2\ln 2-4\sqrt{2}\ln(1+\sqrt{2})]$
10	$\frac{1}{4}$	$\frac{1}{2}$	$\frac{3}{4}$	1	1	$\frac{3}{2}$	$\frac{7}{4}$	2	$\frac{9}{4}$	$\frac{5\sqrt{2}}{48}[35\pi-72\ln(1+\sqrt{2})+2-2\ln 2]$
11	$\frac{1}{2}$	$\frac{1}{2}$	1	$\frac{5}{2}$	$\frac{5}{2}$	$\frac{3}{2}$	$\frac{3}{2}$	$\frac{7}{2}$	$\frac{7}{2}$	$\frac{25}{144}(18\mathbf{G}-11)$
12	$\frac{1}{2}$	1	$\frac{3}{2}$	$\frac{3}{2}$	$\frac{9}{4}$	$\frac{5}{4}$	2	2	3	$\frac{16}{5}\left(\frac{4}{\pi}-1\right)$
13	$\frac{2}{3}$	1	1	1	$\frac{4}{3}$	$\frac{5}{3}$	2	2	$\frac{7}{3}$	$\frac{2}{27}(12\sqrt{3}\,\pi-54-\pi^2)$
14	1	1	1	$\frac{3}{2}$	$\frac{3}{2}$	$\frac{5}{2}$	$\frac{5}{2}$	3	3	$828-432\ln 2-576\mathbf{G}$
15	1	1	1	3	3	2	2	4	4	$\frac{3}{16}(2\pi^2-15)$

7.6.4. Values of ${}_5F_4(a_1,\ \ldots,\ a_5;\ b_1,\ \ldots,\ b_4;\ z_0)$ for $z_0\neq\pm 1$.

1. $${}_5F_4\left(\begin{matrix}-n,\ a,\ a/3+1,\ b,\ 1-b & 1/4\\ a/3,\ a+2n+1,\ (a-b)/2+1,\ (a+b+1)/2\end{matrix}\right)=\frac{((a+1)/2)_n\,(a/2+1)_n}{((a-b)/2+1)_n\,((a+b+1)/2)_n}.$$

2. $${}_5F_4\left(\begin{matrix}-n,\ a,\ 2a/3+1,\ b,\ a-b+1/2;\ 4\\ 2a/3,\ a+n/2+1,\ 2b,\ 2a-2b+1\end{matrix}\right)=$$
$$=\frac{1+(-1)^n}{2}\,\frac{2^{-n}n!\,(a+1)_{n/2}}{(n/2)!\,(b+1/2)_{n/2}\,(a-b+1)_{n/2}}$$ $[a,\ b\neq 0]$.

7.7. THE FUNCTION ${}_6F_5(a_1,\ \ldots,\ a_6;\ b_1,\ \ldots,\ b_5;\ z)$

7.7.1. Representations of ${}_6F_5(a_1,\ \ldots,\ a_6;\ b_1,\ \ldots,\ b_5;\ z)$.

1. $${}_6F_5\left(\begin{matrix}a,\ a+1/3,\ a+2/3,\ b,\ b+1/3,\ b+2/3;\ z\\ 1/3,\ 2/3,\ c,\ c+1/3,\ c+2/3\end{matrix}\right)=\frac{1}{3}\sum_{k=0}^{2}{}_2F_1\,(3a,\ 3b;\ 3c;\ (e^{2\pi ik}z)^{1/3}).$$

7.7.2. Values of ${}_6F_5(a_1, \ldots, a_6; b_1, \ldots, b_5; 1)$.

1. $${}_6F_5\left(\begin{matrix}1,\ 3/2,\ a,\ a,\ 1-a,\ 1-a;\ 1\\ 1/2,\ 1+a,\ 1+a,\ 2-a,\ 2-a\end{matrix}\right)=\frac{a^2(1-a)^2}{1-2a}[\psi'(a)-\psi'(1-a)].$$

2. $${}_6F_5\left(\begin{matrix}2,\ 2,\ a,\ a,\ 2-a,\ 2-a;\ 1\\ 1,\ 1+a,\ 1+a,\ 3-a,\ 3-a\end{matrix}\right)=\frac{\pi a^2(2-a)^2}{4(1-a)}\left[\operatorname{ctg} a\pi+\frac{\pi(1-a)}{\sin^2 a\pi}\right].$$

3. $${}_6F_5\left(\begin{matrix}-n,\ a,\ a+1/3,\ a+2/3,\ b,\ b+1/2;\ 1\\ c,\ d,\ (2b-n)/3,\ (1+2b-n)/3,\ (2+2b-n)/3\end{matrix}\right)=$$ $$=\frac{(3a-2b+1)_n}{(1-2b)_n}\,{}_4F_3\left(\begin{matrix}-n,\ 3a,\ c-d+1/2,\ d-c+1/2;\ 1/4\\ c,\ d,\ 3a-2b+1\end{matrix}\right)\quad [c+d=3a+3/2].$$

4. $${}_6F_5\left(\begin{matrix}-n,\ a,\ a+1/3,\ a+2/3,\ b,\ b+1/2;\ 1\\ (3a+1)/2,\ 3a/2+1,\ (2b-n)/3,\ (1+2b-n)/3,\ (2+2b-n)/3\end{matrix}\right)=\frac{(1+3a-2b)_n}{(1-2b)_n}.$$

5. $${}_6F_5\left(\begin{matrix}-n,\ 1,\ a,\ 2-a,\ b,\ 2-b;\ 1\\ n+2,\ a+1,\ 3-a,\ b+1,\ 3-b\end{matrix}\right)=\frac{(n+2)!\,n!}{2(b-a)(a+b-2)}\left[\frac{a(2-a)}{(b-1)^2(b+1)_n(3-b)_n}-\right.$$ $$\left.-\frac{b(2-b)}{(a-1)^2(a+1)_n(3-a)_n}\right]+\frac{(n+2)(2-a)(2-b)ab}{2(n+1)(a-1)^2(b-1)^2}.$$

6. $${}_6F_5\left(\begin{matrix}-n,\ 1-n/3,\ a,\ b,\ 1-2a-2b-n,\ 1/2-a-b-n;\ 1\\ -n/3,\ 1-2a-n,\ 1-2b-n,\ 1-a-b-n,\ a+b+1/2\end{matrix}\right)=0\quad [a,\ b\neq 0, -1, -2 \ldots].$$

7. $${}_6F_5\left(\begin{matrix}-n/2,\ (1-n)/2,\ a,\ a+1/3,\ a+2/3,\ b;\ 1\\ (b-n)/3,\ (1+b-n)/3,\ (2+b-n)/3,\ c,\ 3a-c+3/2\end{matrix}\right)=$$ $$=\frac{(1+3a-b)_n}{(1-b)_n}\,{}_4F_3\left(\begin{matrix}-n,\ 3a,\ c-1/2,\ 1+3a-c;\ 4\\ 1+3a-b,\ 2c-1,\ 2+6a-2c\end{matrix}\right).$$

8. $${}_6F_5\left(\begin{matrix}1,\ 3/2,\ (1+i)/2,\ (1+i)/2,\ (1-i)/2,\ (1-i)/2;\ 1\\ 1/2,\ (3+i)/2,\ (3+i)/2,\ (3-i)/2,\ (3-i)/2\end{matrix}\right)=1,175944\ldots$$

9. $${}_6F_5\left(\begin{matrix}1,\ 3/2,\ (2+i)/4,\ (2+i)/4,\ (2-i)/4,\ (2-i)/4;\ 1\\ 1/2,\ (6+i)/4,\ (6+i)/4,\ (6-i)/4,\ (6-i)/4\end{matrix}\right)=1,077536\ldots$$

10. $${}_6F_5\left(\frac{1}{4},\ \frac{1}{4},\ \frac{3}{4},\ \frac{3}{4},\ 1,\ \frac{3}{2};\ \frac{1}{2},\ \frac{5}{4},\ \frac{5}{4},\ \frac{7}{4},\ \frac{7}{4};\ 1\right)=1,030461\ldots$$

11. $${}_6F_5\left(\frac{1}{2},\ 1,\ \frac{3}{2},\ \frac{3}{2},\ \frac{3}{2},\ \frac{9}{4};\ \frac{5}{4},\ 2,\ 2,\ 2,\ 3;\ 1\right)=\frac{32}{5}\left(1-\frac{8}{\pi^2}\right).$$

7.7.3. Values of ${}_6F_5(a_1, \ldots, a_6; b_1, \ldots, b_5; -1)$.

1. $${}_6F_5\left(\begin{matrix}1/2,\ 1,\ a,\ a,\ 1-a,\ 1-a;\ -1\\ 3/2,\ 1+a,\ 1+a,\ 2-a,\ 2-a\end{matrix}\right)=$$ $$=\frac{\pi a^2(1-a)^2}{(1-2a)^4}\left[4-4\operatorname{cosec} a\pi+\frac{\pi(1-2a)}{\sin^2 a\pi}\cos a\pi\right]\quad [a\neq 1/2].$$

2. $$=\frac{5\pi^5}{1536}\quad [a=1/2].$$

3. $${}_6F_5\left(\begin{matrix}1,\ 3/2,\ a,\ a,\ 1-a,\ 1-a;\ -1\\ 1/2,\ 1+a,\ 1+a,\ 2-a,\ 2-a\end{matrix}\right)=\frac{\pi^2a^2(1-a)^2\cos a\pi}{(1-2a)\sin^2 a\pi}\quad [a\neq 1/2].$$

4. $${}_6F_5\left(\begin{matrix}2,\ 2,\ a,\ a,\ 2-a,\ 2-a;\ -1\\ 1,\ 1+a,\ 1+a,\ 3-a,\ 3-a\end{matrix}\right)=\frac{\pi a^2(2-a)^2}{4(1-a)}\left[1+\frac{\pi(1-a)}{\sin^2 a\pi}\cos a\pi\right]\quad [a\neq 1].$$

5. $${}_6F_5\left(\begin{matrix}a,\ a+1/3,\ a+2/3,\ 3a-1/3,\ 3a,\ 3a+1/3;\ -1\\ 1/3,\ 2/3,\ 2a,\ 2a+1/3,\ 2a+2/3\end{matrix}\right)=$$
$$=\frac{\sqrt{\pi}}{3}\Gamma(6a)\left[\frac{2^{1-9a}}{\Gamma((3a+1)/2)\,\Gamma(9a/2)}+\frac{4\cdot 3^{-9a/2}\sqrt{\pi}}{\Gamma[2/3,\ 3a,\ 3a+1/3]}\cos\frac{9a-1}{6}\pi\right].$$

6. $${}_6F_5\left(\begin{matrix}1/2,\ 1/2,\ 1/2,\ 1/2,\ 1/2,\ 5/4;\ -1\\ 1/4,\ 1,\ 1,\ 1,\ 1\end{matrix}\right)=2\Gamma^{-4}\left(\frac{3}{4}\right).$$

7.8. THE FUNCTION ${}_7F_6(a_1, \ldots, a_7;\ b_1, \ldots, b_6;\ z)$

7.8.1. Values of ${}_7F_6(a_1, \ldots, a_7;\ b_1, \ldots, b_6;\ 1)$.

1. $${}_7F_6\left(\begin{matrix}a,\ a/2+1,\ b,\ c,\ d,\ e,\ f;\ 1\\ a/2,\ 1+a-b,\ 1+a-c,\ 1+a-d,\ 1+a-e,\ 1+a-f\end{matrix}\right)=$$
$$=\Gamma\left[\begin{matrix}1+a-b,\ 1+a-c,\ 1+a-d,\ 1+a-f,\ 1+a-b-c-d,\\ 1+a,\ 1+a-b-c,\ 1+a-b-d,\ 1+a-c-d,\end{matrix}\right.$$
$$\left.\begin{matrix}1+a-b-c-f,\ 1+a-b-d-f,\ 1+a-c-d-f\\ 1+a-b-f,\ 1+a-c-f,\ 1+a-d-f,\ 1+a-b-c-d-f\end{matrix}\right]$$
$[e=2a-b-c-d-f+1]$.

2. $${}_7F_6\left(\begin{matrix}1,\ a,\ a,\ a,\ 1-a,\ 1-a,\ 1-a;\ 1\\ 1+a,\ 1+a,\ 1+a,\ 2-a,\ 2-a,\ 2-a\end{matrix}\right)=$$
$$=\frac{\pi a^3(1-a)^3}{(1-2a)^5}\left[6\operatorname{ctg} a\pi-3\pi(1-2a)\operatorname{cosec}^2 a\pi+\pi^2(1-2a)^2\frac{\cos a\pi}{\sin^3 a\pi}\right]$$ $[a\neq 1/2]$.

3. $$=\frac{\pi^6}{960}$$ $[a=1/2]$.

4. $${}_7F_6\left(\begin{matrix}1/2,\ 1/2,\ 1,\ a,\ a,\ 1-a,\ 1-a;\ 1\\ 3/2,\ 3/2,\ 1+a,\ 1+a,\ 2-a,\ 2-a\end{matrix}\right)=$$
$$=\frac{\pi a^2(1-a)^2}{(1-2a)^5}[2\pi(1-2a)-6\operatorname{ctg} a\pi+\pi(1-2a)\operatorname{cosec}^2 a\pi]$$ $[a\neq 1/2]$.

5. $$=\frac{\pi^6}{960}$$ $[a=1/2]$.

6. $${}_7F_6\left(\begin{matrix}1,\ 3/2,\ 3/2,\ a,\ a,\ 1-a,\ 1-a;\ 1\\ 1/2,\ 1/2,\ 1+a,\ 1+a,\ 2-a,\ 2-a\end{matrix}\right)=$$
$$=\frac{\pi a^2(1-a)^2}{1-2a}[2\operatorname{ctg} a\pi+\pi(1-2a)\operatorname{cosec}^2 a\pi]$$ $[a\neq 1/2]$.

7. $${}_7F_6\left(\begin{matrix}1,\ a,\ a,\ a,\ 2-a,\ 2-a,\ 2-a;\ 1\\ 1+a,\ 1+a,\ 1+a,\ 3-a,\ 3-a,\ 3-a\end{matrix}\right)=$$
$$=\frac{a^3(2-a)^3}{16(1-a)^6}\left[8-3\pi^2(1-a)^2\operatorname{cosec}^2 a\pi+3\pi(1-a)\operatorname{ctg} a\pi+2\pi^3(1-a)^3\frac{\cos a\pi}{\sin^3 a\pi}\right]$$
$[a\neq 1]$.

8. $$=\frac{\pi^6}{945}$$ $[a=1]$.

9. $${}_7F_6\left(\begin{matrix}1,\ 1,\ 1,\ a,\ a,\ 2-a,\ 2-a;\ 1\\ 2,\ 2,\ 1+a,\ 1+a,\ 3-a,\ 3-a\end{matrix}\right)=$$
$$=\frac{a^2(2-a)^2}{12(1-a)^6}[2\pi^2(1-a)^2-12-9\pi(1-a)\operatorname{ctg} a\pi+3\pi^2(1-a)^2\operatorname{cosec}^2 a\pi]$$ $[a\neq 1]$.

10. $$=\frac{\pi^6}{945}$$ $[a=1]$.

11. $${}_7F_6\left(\begin{matrix}-n,\ a,\ a/2+1,\ b,\ c,\ d,\ 2a-b-c-d+n+1;\ 1\\ a/2,\ 1+a-b,\ 1+a-c,\ 1+a-d,\ 1+a+n,\ b+c+d-a-n\end{matrix}\right)=$$
$$=\frac{(1+a)_n(1+a-b-c)_n(1+a-b-d)_n(1+a-c-d)_n}{(1+a-b)_n(1+a-c)_n(1+a-d)_n(1+a-b-c-d)_n}.$$

12. $${}_7F_6\left(\begin{matrix}-n,\ a,\ a/2+1,\ b,\ b+1/2,\ a-2b,\ 2a-2b+n+1;\ 1\\ a/2,\ 1+a-b,\ 1/2+a-b,\ 1+2b,\ 2b-a-n,\ 1+a+n\end{matrix}\right)=$$
$$=\frac{(1+a)_n\,(1+2a-4b)_n}{(1+a-2b)_n\,(1+2a-2b)_n}\qquad [a\neq 0;\ b\neq -1/2].$$

13. $${}_7F_6\left(\begin{matrix}-n,\ 1,\ 3/2,\ a,\ 1-a,\ b,\ 1-b;\ 1\\ 1/2,\ n+2,\ 1+a,\ 2-a,\ 1+b,\ 2-b\end{matrix}\right)=$$
$$=\frac{(-1)^n n!\,(n+1)!\,ab}{(a-b)\,(1-a-b)}(1-a)\,(1-b)\left\{\Gamma\left[\begin{matrix}-n-a\\ 2+n-a\end{matrix}\right]-\Gamma\left[\begin{matrix}-n-b\\ 2+n-b\end{matrix}\right]\right\}.$$

14. $${}_7F_6\left(\begin{matrix}1/4,\ 1/4,\ 1/4,\ 3/4,\ 3/4,\ 3/4,\ 1;\ 1\\ 5/4,\ 5/4,\ 5/4,\ 7/4,\ 7/4,\ 7/4\end{matrix}\right)=\frac{27\pi}{256}(12-6\pi-\pi^2).$$

7.8.2. Values of ${}_7F_6(a_1,\ \ldots,\ a_7;\ b_1,\ \ldots,\ b_6;\ -1)$.

1. $${}_7F_6\left(\begin{matrix}1,\ 3/2,\ 3/2,\ a,\ a,\ 1-a,\ 1-a;\ -1\\ 1/2,\ 1/2,\ 1+a,\ 1+a,\ 2-a,\ 2-a\end{matrix}\right)=$$
$$=\frac{a^2(1-a)^2}{1-2a}[4\beta(a)-(1-2a)(\beta'(a)+\beta'(1-a))-2\pi\operatorname{cosec}a\pi]\qquad [a\neq 1/2].$$

2. $${}_7F_6\left(\begin{matrix}1,\ 1,\ 1,\ a,\ a,\ 2-a,\ 2-a;\ -1\\ 2,\ 2,\ 1+a,\ 1+a,\ 3-a,\ 3-a\end{matrix}\right)=$$
$$=\frac{a^2(2-a)^2}{12(1-a)^6}\left[\pi^2(1-a)^2+12-9\pi(1-a)\operatorname{cosec}a\pi+3\pi^2(1-a)^2\frac{\cos a\pi}{\sin^2 a\pi}\right]\qquad [a\neq 1].$$

3. $$=\frac{31\pi^6}{32\cdot 945}\qquad [a=1].$$

4. $${}_7F_6\left(\begin{matrix}1,\ 3/2,\ 3/2,\ (1+i)/2,\ (1+i)/2,\ (1-i)/2,\ (1-i)/2;\ -1\\ 1/2,\ 1/2,\ (3+i)/2,\ (3+i)/2,\ (3-i)/2,\ (3-i)/2\end{matrix}\right)=0{,}738632\ldots$$

5. $${}_7F_6\left(\begin{matrix}1,\ 3/2,\ 3/2,\ (2+i)/4,\ (2+i)/4,\ (2-i)/4,\ (2-i)/4;\ -1\\ 1/2,\ 1/2,\ (6+i)/4,\ (6+i)/4,\ (6-i)/4,\ (6-i)/4\end{matrix}\right)=0{,}876963\ldots$$

6. $${}_7F_6\left(\begin{matrix}1/4,\ 1/4,\ 3/4,\ 3/4,\ 1,\ 3/2,\ 3/2;\ -1\\ 1/2,\ 1/2,\ 5/4,\ 5/4,\ 7/4,\ 7/4\end{matrix}\right)=0{,}961977\ldots$$

7.9. THE FUNCTIONS ${}_8F_7(a_1,\ \ldots,\ a_8;\ b_1,\ \ldots,\ b_7;\ z)$ AND ${}_9F_8(a_1,\ \ldots,\ a_9;\ b_1,\ \ldots,\ b_8;\ z)$

7.9.1. Values of ${}_8F_7(a_1,\ \ldots,\ a_8;\ b_1,\ \ldots,\ b_7;\ \pm 1)$.

1. $${}_8F_7\left(\begin{matrix}2,\ 2,\ a,\ a,\ a,\ 2-a,\ 2-a,\ 2-a;\ \pm 1\\ 1,\ 1+a,\ 1+a,\ 1+a,\ 3-a,\ 3-a,\ 3-a\end{matrix}\right)=$$
$$=\frac{\pi a^3(2-a)^3}{16(a-1)^3}\left[\left\{\begin{matrix}\operatorname{ctg}a\pi\\ \operatorname{cosec}a\pi\end{matrix}\right\}-\frac{\pi(1-a)}{\sin^2 a\pi}\left\{\begin{matrix}1\\ \cos a\pi\end{matrix}\right\}-\frac{\pi^2(1-a)^2}{\sin^3 a\pi}\left\{\begin{matrix}2\cos a\pi\\ 1+\cos^2 a\pi\end{matrix}\right\}\right]\qquad [a\neq 1].$$

2. $${}_8F_7\left(\begin{matrix}1,\ 3/2,\ 3/2,\ 3/2,\ a,\ a,\ 1-a,\ 1-a;\ -1\\ 1/2,\ 1/2,\ 1/2,\ 1+a,\ 1+a,\ 2-a,\ 2-a\end{matrix}\right)=\frac{\pi a^2(1-a)^2}{\sin a\pi}[4+\pi(1-2a)\operatorname{ctg}a\pi].$$

7.9.2. Values of ${}_9F_8(a_1,\ \ldots,\ a_9;\ b_1,\ \ldots,\ b_8;\ 1)$.

1. $${}_9F_8\left(\begin{matrix}1,\ 3/2,\ 3/2,\ a,\ a,\ a,\ 1-a,\ 1-a,\ 1-a;\ 1\\ 1/2,\ 1/2,\ 1+a,\ 1+a,\ 1+a,\ 2-a,\ 2-a,\ 2-a\end{matrix}\right)=$$
$$=\frac{\pi a^3(1-a)^3}{(1-2a)^3}[\pi(1-2a)\operatorname{cosec}^2 a\pi+\pi^2(1-2a)^2\operatorname{cosec}^2 a\pi\operatorname{ctg}a\pi-2\operatorname{ctg}a\pi]\qquad [a\neq 1/2].$$

2. $${}_9F_8\left(\begin{matrix}-n,\ 1,\ 3/2,\ 3/2,\ 3/2,\ a,\ 1-a,\ b,\ 1-b;\ 1\\ 1/2,\ 1/2,\ 1/2,\ n+2,\ 1+a,\ 2-a,\ 1+b,\ 2-b\end{matrix}\right)=$$
$$=\frac{n!\,(n+1)!}{(a+b-1)\,(a-b)}\left[\frac{(2b-1)^2 a(a-1)}{(b+1)_n(2-b)_n}-\frac{(2a-1)^2 b(b-1)}{(a+1)_n(2-a)_n}\right].$$

3. $${}_9F_8\left(\begin{matrix}1/4,\ 1/4,\ 1/4,\ 1/4,\ 3/4,\ 3/4,\ 3/4,\ 3/4,\ 1;\ 1\\ 5/4,\ 5/4,\ 5/4,\ 5/4,\ 7/4,\ 7/4,\ 7/4,\ 7/4\end{matrix}\right)=\frac{27\pi}{512}(\pi^3-6\pi^2+30\pi-60).$$

7.10 THE FUNCTION $_{q+1}F_q(a_1, \ldots, a_{q+1}; b_1, \ldots, b_q; z)$

7.10.1. Representations of $_{q+1}F_q(a_1, \ldots, a_{q+1}; b_1, \ldots, b_q; z)$.

1. $$_{q+1}F_q\left(\begin{matrix} a, b_1, \ldots, b_q; z \\ b_1+1, \ldots, b_{q+1} \end{matrix}\right) = \sum_{k=1}^{q} {}_2F_1(a, b_k; b_k+1; z) \prod_{\substack{l=1 \\ l\neq k}}^{q} \frac{b_l}{b_l-b_k}.$$

2. $$_{q+1}F_q\left(\begin{matrix} \Delta(q+1, a); z \\ \Delta(q, a+1) \end{matrix}\right) = (1-x)^a$$ $[q^q(1-x)^{q+1}z = -(q+1)^{q+1}x]$.

3. $$_{q+1}F_q\left(\begin{matrix} \Delta(q+1, a); z \\ \Delta(q, a) \end{matrix}\right) = \frac{(1-x)^a}{1+qx}$$ [z see 7.10.1.2].

4. $$_{q+1}F_q\left(\begin{matrix} 1, a, \ldots, a; z \\ a+1, \ldots, a+1 \end{matrix}\right) = a^q \Phi(z, q, a).$$

5. $$_{q+1}F_q\left(\begin{matrix} 1, m, \ldots, m; z \\ m+1, \ldots, m+1 \end{matrix}\right) = \frac{m^q}{z^m}\left[\mathrm{Li}_q(z) - \sum_{k=1}^{m-1} \frac{z^k}{k^q}\right].$$

6. $$_{q+1}F_q\left(\begin{matrix} 1, m, \ldots, m; z \\ m+2, \ldots, m+2 \end{matrix}\right) = \frac{m(m+1)}{z^{m+1}}\left[(1-z)\ln(1-z) + (1-z)\sum_{k=1}^{q-1} \frac{z^k}{k} + \frac{z^m}{m}\right].$$

7. $$_{q+1}F_q\left(\begin{matrix} 1, \Delta(q, 1); z \\ \Delta(q, m+2) \end{matrix}\right) = m(m+1) \sum_{k=0}^{m-1} (-1)^{k-1} \binom{m-1}{k} z^{-k-1} \times$$
$$\times \left[\sum_{l=1}^{[k/q]} \frac{z^{(k-lq+1)/q}}{k-lq+1} + \frac{2}{q} \sum_{l=1}^{[(q-1)/2]} \sin\frac{2l}{q}(k+1)\pi \operatorname{arctg} \frac{z^{1/q}\sin(2l\pi/q)}{1-z^{1/q}\cos(2l\pi/q)} + \right.$$
$$+ \frac{1}{q} \sum_{l=1}^{[(q-1)/2]} \cos\frac{2l}{q}(k+1)\pi \ln\left(z^{2/q} - 2z^{1/q}\cos\frac{2l\pi}{q} + 1\right) +$$
$$\left. + \frac{1}{q}\ln(1-z) + (-1)^{k-1}\frac{1+(-1)^q}{2q}\ln(1+z)\right].$$

8. $$_{q+1}F_q\left(\begin{matrix} 1, 1, \ldots, 1; z \\ 2, \ldots, 2 \end{matrix}\right) = z^{-1}\,\mathrm{Li}\,q(z) = \Phi(z, q, 1).$$

9. $$_{q+1}F_q\left(\begin{matrix} 1, 1, \ldots & \ldots, 1; z \\ \underbrace{2, \ldots, 2}_{q-n}, & \underbrace{3, \ldots, 3}_{n} \end{matrix}\right) = I(q, n),$$

$$I(q, n) = \frac{(-1)^q 2^n}{z}\left[\frac{1}{z}\sum_{k=0}^{n-1} \frac{(q)_k}{k!}\mathrm{Li}_{n-k}(z) + \sum_{k=0}^{q-1} (-1)^{q-k} \frac{(n)_k}{k!}\mathrm{Li}_{q-k}(z) - \binom{n+q-1}{q}\right].$$

10. $$_{q+1}F_q\left(\begin{matrix} 1, \overbrace{1, \ldots, 1}^{q-n}, \overbrace{2, \ldots, 2}^{n}; z \\ 3, \ldots \quad \ldots, 3 \end{matrix}\right) = I(q-n, q)$$ [$I(q, n)$ see 7.10.1.9].

11. $$_{q+1}F_q\left(\begin{matrix} 2, 2, \ldots, 2; z \\ 1, \ldots, 1 \end{matrix}\right) = \frac{1}{z}\left(z\frac{d}{dz}\right)^{q+1}\frac{1}{1-z} = \sum_{k=0}^{q} \frac{(k+1)(-z)^k}{(1-z)^{k+2}} \times$$
$$\times \left[\sum_{l=0}^{k} (-1)^l \binom{k}{l} (l+1)^q\right].$$

7.10.2. Values of ${}_{q+1}F_q(a_1, \ldots, a_{q+1}; b_1, \ldots, b_q; \pm 1)$.

1. $$\lim_{z\to 1}\left[(1-z)^{-c}\,{}_{q+1}F_q\left(\begin{matrix}a_1, a_2, \ldots, a_{q+1}; z\\ b_1, \ldots, b_q\end{matrix}\right)\right]=\Gamma\left[\begin{matrix}b_1, b_2, \ldots, b_q, -c\\ a_1, a_2, \ldots, a_{q+1}\end{matrix}\right]$$
$$\left[c=\sum_{k=1}^{q}(b_k-a_k)-a_{q+1};\ \operatorname{Re} c>0\right].$$

2. $${}_{q+1}F_q\left(\begin{matrix}a, b_1, b_2, \ldots, b_q; 1\\ b_1+n_1, \ldots, b_q+n_q\end{matrix}\right)=$$
$$=\Gamma(1-a)\sum_{k=1}^{q}\frac{1}{(n_k-1)!}\Gamma\left[\begin{matrix}b_k+n_k\\ b_k-a+1\end{matrix}\right]\sum_{j=0}^{n_k-1}\frac{(1-n_k)_j\,(b_k)_j}{j!\,(b_k-a+1)_j}\prod_{\substack{l=1\\ l\neq k}}^{q}\frac{(b_l)_{n_l}(1-b_l+b_k-n_l)_j}{(b_l-b_k)_{n_l}(1-b_l+b_k)_j}$$
$$\left[\operatorname{Re} a<\sum_{k=1}^{q}n_k,\ n_k=0, 1, 2, \ldots;\ b_k+n_k\neq 0, -1, -2, \ldots \text{ for } k=1, 2, \ldots, q;\right.$$
$$\left.b_k-b_l\neq 1-n_k, 2-n_k, \ldots, n_l-1 \text{ for } k, l=1, 2, \ldots, q;\ k\neq l\right].$$

3. $${}_{q+1}F_q\left(\begin{matrix}1, b_1, \ldots, b_q; \pm 1\\ b_1+1, \ldots, b_q+1\end{matrix}\right)=\mp\prod_{j=1}^{q}b_j\left[\sum_{k=1}^{q}\prod_{\substack{l=1\\ l\neq k}}^{q}(b_l-b_k)^{-1}\left\{\begin{matrix}\psi(b_k)\\ \beta(b_k)\end{matrix}\right\}\right]$$
$$\left[b_k\neq b_l;\ k, l=1, 2, \ldots, q;\ k\neq l,\ \left\{\begin{matrix}q=2, 3, 4, \ldots\\ q=1, 2, 3, \ldots\end{matrix}\right\}\right].$$

4. $${}_{q+1}F_q\left(\begin{matrix}1, a, a, b_1, \ldots, b_{q-2}; \pm 1\\ a+1, a+1, b_1+1, \ldots, b_{q-2}+1\end{matrix}\right)=$$
$$=\mp a^2\prod_{j=1}^{q-2}b_j\left\{\sum_{k=1}^{q-2}\left[\prod_{\substack{l=1\\ l\neq k}}^{q-2}(b_l-b_k)^{-1}(a-b_k)^{-2}\left\{\begin{matrix}\psi(b_k)\\ \beta(b_k)\end{matrix}\right\}-\right.\right.$$
$$\left.\left.-(b_k-a)^{-1}\prod_{l=1}^{q-2}(b_l-a)^{-1}\left\{\begin{matrix}\psi(a)\\ \beta(a)\end{matrix}\right\}\right]-\prod_{l=1}^{q-2}(b_l-a)^{-1}\left\{\begin{matrix}\psi'(a)\\ \beta'(a)\end{matrix}\right\}\right\}$$
$$[b_k\neq b_l;\ k, l=1, 2, \ldots, q-2;\ k\neq l;\ q\geqslant 2].$$

5. $${}_{q+1}F_q\left(\begin{matrix}1, a, a, a, b_1, \ldots, b_{q-3}; \pm 1\\ a+1, a+1, a+1, b_1+1, \ldots, b_{q-3}+1\end{matrix}\right)=$$
$$=\mp a^3\prod_{j=1}^{q-3}b_j\left\{\sum_{k=1}^{q-3}\left[\prod_{\substack{l=1\\ l\neq k}}^{q-3}(b_l-b_k)^{-1}(a-b_k)^{-3}\left\{\begin{matrix}\psi(b_k)\\ \beta(b_k)\end{matrix}\right\}+\right.\right.$$
$$\left.+(b_k-a)^{-2}\prod_{l=1}^{q-3}(b_l-a)^{-1}\left\{\begin{matrix}\psi(a)+(b_k-a)\psi'(a)\\ \beta(a)+(b_k-a)\beta'(a)\end{matrix}\right\}\right]+$$
$$\left.+\sum_{\substack{j, k=1\\ j\neq k}}^{q-3}(b_j-a)^{-1}(b_k-a)^{-1}\prod_{l=1}^{q-3}(b_l-a)^{-1}\left\{\begin{matrix}\psi(a)\\ \beta(a)\end{matrix}\right\}+\frac{1}{2}\prod_{l=1}^{q-3}(b_l-a)^{-1}\left\{\begin{matrix}\psi''(a)\\ \beta''(a)\end{matrix}\right\}\right\}$$
$$[b_k\neq b_l;\ k, l=1, 2, \ldots, q-3;\ k\neq l,\ q\geqslant 3].$$

6. $_{q+1}F_q\left(\begin{matrix}a,\ b,\ \ldots,\ b;\ 1\\ b+1,\ \ldots,\ b+1\end{matrix}\right)=\frac{(-1)^{q-1}b^q}{(q-1)!}\Gamma(1-a)\left(\frac{d}{db}\right)^{q-1}\Gamma\left[\begin{matrix}b\\ 1+b-a\end{matrix}\right]$

[Re $a<q$; $q=1,\ 2,\ 3,\ \ldots$].

7. $_{q+1}F_q\left(\begin{matrix}a,\ a,\ \ldots,\ a;\ \pm1\\ a+1,\ \ldots,\ a+1\end{matrix}\right)=\pm\frac{(-1)^q a^q}{(q-1)!}\left\{\begin{matrix}\psi^{(q-1)}(a)\\ \beta^{(q-1)}(a)\end{matrix}\right\}$ $\left[\left\{\begin{matrix}q=2,\ 3,\ 4,\ \ldots\\ q=1,\ 2,\ 3,\ \ldots\end{matrix}\right\}\right]$.

8. $_{q+1}F_q\left(\begin{matrix}a,\ 2,\ 2,\ \ldots,\ 2;\ 1\\ b,\ 1,\ \ldots,\ 1\end{matrix}\right)=\frac{(-1)^q(b-1)(b-2)}{(q-1)!\,(a-1)(a-b+1)}\sum_{k=1}^{q}(-1)^k\frac{k!\,(a-1)_k}{(a-b+2)_k}S_q^{(k)}$

[Re $(a-b)<-q-1$; $q\geqslant1$].

9. $_{q+1}F_q\left(\begin{matrix}-n,\ a_1,\ \ldots,\ a_q;\ 1\\ b_1,\ \ldots,\ b_{q-1},\ c\end{matrix}\right)=\frac{(\sigma_{q-1}-s_{q-1})_n(b_{q-1}-a_q)_n}{(\sigma_{q-1}-s_q)_n(b_{q-1})_n}\times$

$$\times\prod_{k=1}^{q-2}\sum_{n_k=0}^{n-m_{k-1}}\frac{(\sigma_k-s_k+m_{k-1})_{n_k}(b_k-a_{k+1})_{n_k}(m_{k-1}-n)_{n_k}}{n_k!\,(\sigma_{q-1}-s_{q-1}+m_{k-1})_{n_k}(1-b_{q-1}+a_q+m_{k-1}-n)_{n_k}}\times$$

$$\times\frac{\prod_{j=k+2}^{q}(a_j+m_{k-1})_{n_k}}{\prod_{l=k}^{q-2}(b_l+m_{k-1})_{n_k}}$$

$\left[c=s_q-\sigma_{q-1}-n+1,\ s_k=\sum_{j=1}^{k}a_j,\ \sigma_k=\sum_{j=1}^{k}b_j,\ m_l=\sum_{k=1}^{l}n_k,\ m_0=0,\ q=2,\ 3,\ 4,\ \ldots\right]$.

10. $_{q+1}F_q\left(\begin{matrix}-n,\ a_1,\ \ldots,\ a_q;\ 1\\ a_1-n_1,\ \ldots,\ a_q-n_q\end{matrix}\right)=R\delta_{s,\,n}\,[s=1,\ 2,\ 3,\ \ldots,\ n]$,

11. $=R\left[-s(a_1-n_1+n)+\sum_{k=2}^{q}n_k\sum_{j=1}^{k-1}(a_j-a_{j+1}+n_{j+1})\right]$ [$s=n+1$],

12. $=R\left\{\frac{1}{2}\sum_{k=1}^{q}n_k(n_k-1)\sum_{j=0}^{k-1}(a_j-a_{j+1}+n_{j+1})\left[-1+\sum_{j=0}^{k-1}(a_j-a_{j+1}+n_{j+1})\right]+\right.$

$\left.+\sum_{k=1}^{q-1}n_k\sum_{j=0}^{k-1}(a_j-a_{j+1}+n_{j+1})\sum_{l=k+1}^{q}n_l\left[-1+\sum_{i=0}^{l-1}(a_i-a_{i+1}+n_{i+1})\right]\right\}$ [$s=n+2$],

$\left[a_j\neq0,\ -1,\ -2,\ \ldots;\ j=1,\ 2,\ \ldots,\ q;\ s=\sum_{j=1}^{q}n_j,\ R=n!\prod_{j=1}^{q}\frac{1}{(1-a_j)_{n_j}};\ a_0=-n;\ n_0=0\right.$

$\left.n_j=1,\ 2,\ 3,\ \ldots\text{ for } j=1,\ 2,\ \ldots,\ q\right]$.

13. $_{q+1}F_q\left(\begin{matrix}-n,\ a,\ \ldots,\ a;\ 1\\ a-1,\ \ldots,\ a-1\end{matrix}\right)=S(q,\ n)$,

$S(q,\ n)=0$ [$q=0,\ 1,\ 2,\ \ldots,\ n-1$]

$S(n,\ n)=(1-a)^{-n}n!$,

$S(n+1,\ n)=(1-a)^{-n-1}(n+1)!\left(1-a-\frac{n}{2}\right)$,

$S(n+2,\ n)=\frac{(1-a)^{-n-2}}{24}(n+2)!\,[12(a+n-1)(a-1)+n(3n+1)]$,

$(a-1)S(q,\ n)=(a+n-1)S(q-1,\ n)-nS(q-1,\ n-1)$,

$S(q,\ 0)=1$.

14. $${}_{q+1}F_q\left(\begin{matrix}-qn-1,\ a,\ \ldots,\ a;\ 1\\ a-n,\ \ldots,\ a-n\end{matrix}\right)=\left[(-1)^n\frac{(-a-qn)_n}{(1-a)_n}\right]^q-\left[(-1)^n\frac{(a)_{qn+1}}{(a-n)_{qn+1}}\right]^q.$$

15. $${}_{q+1}F_q\left(\begin{matrix}-n,\ -n,\ \ldots,\ -n,\ 1;\ 1\\ -n-1,\ \ldots,\ -n-1,\ a\end{matrix}\right)=\frac{n!\,(a-1)}{(n+1)^{q-1}}\sum_{k=1}^{q}\frac{S_q^{(k)}}{(n-k+1)!\,(n-k+a)}.$$

16. $${}_{2q+2}F_{2q+1}\left(\begin{matrix}-n,\ \Delta(q+1,\ a),\ \Delta(q,\ b);\ 1\\ \Delta(q,\ a+1),\ \Delta(q+1,\ b-1)\end{matrix}\right)=\sum_{k=0}^{n}\frac{(-n)_k\,(a)_{qk+k}\,(b)_{qk}}{k!\,(a+1)_{qk}\,(b-1)_{qk+k}}.$$

17. $${}_{2q+2}F_{2q+1}\left(\begin{matrix}-n,\ \Delta(q+1,\ a),\ \Delta(q,\ b);\ 1\\ \Delta(q,\ a+1),\ \Delta(q+1,\ b-n)\end{matrix}\right)=\frac{(1+a-b)_n}{(1-b)_n}.$$

18. $${}_{2q+2}F_{2q+1}\left(\begin{matrix}-n,\ \Delta(q+1,\ a),\ \Delta(q,\ b);\ 1\\ \Delta(q,\ a),\ \Delta(q+1,\ 1+b-n)\end{matrix}\right)=-\frac{(a-b)_n}{(b+nq)\,(1-b)_{n-1}}.$$

19. $${}_{2q}F_{2q-1}\left(\begin{matrix}-n,\ 1,\ 2,\ 2,\ \ldots,\ 2;\ 1\\ n+3,\ 1,\ 1,\ \ldots,\ 1\end{matrix}\right)=I_q,$$

$$I_q=0 \qquad [q=2,\ 3,\ 4,\ \ldots,\ n+1],$$

$$I_{n+2}=\frac{(-1)^n\,n!\,(n+2)!}{2},$$

$$I_{n+3}=\frac{(-1)^n\,[(n+2)!]^2\,(2n+3)}{12}.$$

20. $${}_{2q+1}F_{2q}\left(\begin{matrix}-n,\ 1,\ 3/2,\ 3/2,\ \ldots,\ 3/2;\ 1\\ n+2,\ 1/2,\ 1/2,\ \ldots,\ 1/2\end{matrix}\right)=A_q,$$

$$A_{n+1}=(-1)^n\,2^{2n}\,n!\,(n+1)!,$$

$$A_{n+2}=\frac{(-1)^n}{3}\,2^{2n}\,[(n+1)!]^2\,(2n+1)\,(2n+3).$$

21. $${}_{q+1}F_q\left(\begin{matrix}-n,\ 1-n/2,\ 1-n/2,\ \ldots,\ 1-n/2;\ 1\\ -n/2,\ -n/2,\ \ldots,\ -n/2\end{matrix}\right)=2n^{-q}\sum_{k=0}^{[n/2]}\frac{(-n)_k}{k!}\,(n-2k)^q$$

$$[q=n,\ n+2,\ n+4,\ \ldots;\ n=1,\ 3,\ 5,\ \ldots].$$

22. $${}_{q+1}F_q\left(\begin{matrix}-n,\ 1/2-n,\ 1-n/2,\ 1-n/2,\ \ldots,\ 1-n/2;\ \pm 1\\ 1/2,\ -n/2,\ -n/2,\ \ldots,\ -n/2\end{matrix}\right)=0$$

$$\left[\left\{\begin{matrix}q=2,\ 4,\ 6,\ \ldots\\ q+n=2,\ 4,\ 6,\ \ldots\end{matrix}\right\}\right].$$

23. $${}_{q+1}F_q\left(\begin{matrix}-n,\ -1/2-n,\ 1-n/2,\ 1-n/2,\ \ldots,\ 1-n/2;\ \pm 1\\ 3/2,\ -n/2,\ -n/2,\ \ldots,\ -n/2\end{matrix}\right)=0$$

$$\left[\left\{\begin{matrix}q=2,\ 4,\ 6,\ \ldots\\ q+n=2,\ 4,\ 6,\ \ldots\end{matrix}\right\}\right].$$

24. $${}_{q+1}F_q\left(\begin{matrix}-n,\ 1-n/2,\ 1-n/2,\ \ldots,\ 1-n/2;\ \pm 1\\ -n/2,\ -n/2,\ \ldots,\ -n/2\end{matrix}\right)=0$$

$$\left[\left\{\begin{matrix}q=n+2l+1;\ l=0,\ \pm 1,\ \pm 2,\ \ldots\\ q=2l+1;\ l=0,\ 1,\ 2,\ \ldots\end{matrix}\right\}\right].$$

25. $_{q+1}F_q\left(\begin{matrix}\Delta(q,\ -n),\ 1;\ 1\\ \Delta(q,\ 1)\end{matrix}\right)=\frac{2^n}{q}\sum_{k=1}^{q}\cos^n\frac{2k-1}{2q}\pi$ $[n>q-1]$.

26. $_{q+1}F_q\left(\begin{matrix}-n,\ 2,\ 2,\ \ldots,\ 2;\ -1\\ 1,\ 1,\ \ldots,\ 1\end{matrix}\right)=\frac{1}{n+1}\left[\left(x\frac{d}{dx}\right)^{q+1}\{(1+x)^{n+1}-1\}\right]_{x=1}$.

27. $_{q+1}F_q\left(\begin{matrix}1,\ 1,\ \ldots,\ 1;\ 1\\ 2,\ \ldots,\ 2\end{matrix}\right)=S_q$ $[q=2,\ 3,\ 4,\ \ldots]$,

$$S_q=\frac{(-1)^q}{(q-1)!}\psi^{(q-1)}(1)=\zeta(q),$$

$$S_{2q}=\frac{2^{2q-1}\pi^{2q}}{(2q)!}|B_{2q}| \qquad [q=1,\ 2,\ 3,\ \ldots],$$

$S_2=\frac{\pi^2}{6}=1{,}64493407\ldots$, $S_3=1{,}20205690\ldots$,

$S_4=\frac{\pi^4}{90}=1{,}08232323\ldots$, $S_5=1{,}03692776\ldots$,

$S_6=\frac{\pi^6}{945}=1{,}01734306\ldots$, $S_7=1{,}00834928\ldots$,

$S_8=\frac{\pi^8}{9450}=1{,}00407736\ldots$, $S_9=1{,}00200839\ldots$,

$S_{10}=\frac{\pi^{10}}{93555}=1{,}00099458\ldots$

28. $_{q+1}F_q\left(\begin{matrix}1,\ 1,\ \ldots,\ 1;\ -1\\ 2,\ \ldots,\ 2\end{matrix}\right)=T_q$ $[q=1,\ 2,\ 3,\ \ldots]$,

$$T_{2q}=\frac{2^{2q-1}-1}{(2q)!}\pi^{2q}|B_{2q}| \qquad [q=1,2,3,\ldots],$$

$T_1=\ln 2=0{,}69314718\ldots$, $T_2=\frac{\pi^2}{12}=0{,}82246703\ldots$,

$T_3=0{,}90154268\ldots$, $T_4=\frac{7\pi^4}{720}=0{,}94703283\ldots$,

$T_5=0{,}97211977\ldots$, $T_6=\frac{31\pi^6}{30240}=0{,}98555109\ldots$,

$T_7=0{,}99259382\ldots$, $T_8=\frac{127\pi^8}{1209600}=0{,}99623300\ldots$,

$T_9=0{,}99809430\ldots$, $T_{10}=\frac{73\pi^{10}}{6842880}=0{,}99903951\ldots$

29. $_{q+1}F_q\left(\begin{matrix}1,\ 1,\ \ldots,\ 1;\ 1\\ 3,\ \ldots,\ 3\end{matrix}\right)=$

$$=\frac{(-1)^q 2^q}{(q-1)!}\sum_{k=1}^{[q/2]}\frac{(2q-2k-1)!\,(2\pi)^{2k}}{(q-2k)!\,(2k)!}|B_{2k}|+(-1)^{q-1}2^q\binom{2q-1}{q-1}.$$

30. $_{q+1}F_q\left(\begin{matrix}1,\ 1,\ \ldots,\ 1;\ -1\\ 3,\ \ldots,\ 3\end{matrix}\right)=$

$$=(-1)^{q-1}2^{q+1}\left[\sum_{k=1}^{[(q-1)/2]}\binom{2q-2k-2}{q-1}(1-2^{-2k})\zeta(2k+1)-\frac{1}{2}\binom{2q-1}{q-1}+\binom{2q-2}{q-1}\ln 2\right].$$

31. $$ {}_{q+1}F_q\left(\begin{matrix}1,\ 1,\ \ldots & & \ldots,\ 1;\ \pm 1\\ \underbrace{2,\ \ldots,\ 2}_{q-n}, & \underbrace{3,\ \ldots,\ 3}_{n} & \end{matrix}\right)=J(q,\ n), $$

$$ J(q,\ n)=\pm(-1)^q 2^n\left[\sum_{k=0}^{n-2}\binom{q+k-1}{k}\left\{\begin{matrix}1\\ 1-2^{1+k-n}\end{matrix}\right\}\zeta(n-k)\pm\right. $$

$$ \pm\sum_{k=0}^{q-2}(-1)^{q-k}\binom{n+k-1}{k}\left\{\begin{matrix}1\\ 1-2^{1+k-q}\end{matrix}\right\}\zeta(q-k)-(\pm 1)^{n-q}\times $$

$$ \left.\times\binom{q+n-1}{n-1}+\left\{\begin{matrix}0\\ 1\end{matrix}\right\}\binom{q+n-2}{q-1}2\ln 2\right]. $$

32. $$ {}_{q+1}F_q\left(\begin{matrix}1,\ \overbrace{1,\ \ldots,\ 1}^{q-n},\ \overbrace{2,\ \ldots,\ 2}^{n};\ \pm 1\\ 3,\ 3,\ \ldots\ \ \ldots,\ 3\end{matrix}\right)=J(q-n,\ q) $$ [$J(q,\ n)$ see 7.10.2.31].

33. $$ {}_{q+1}F_q\left(\begin{matrix}1,\ 1/2,\ \ldots,\ 1/2;\ \pm 1\\ 3/2,\ \ldots,\ 3/2\end{matrix}\right)=I^{\pm}(q), $$

$$ I^{\pm}(q)=\left\{\begin{matrix}(1-2^{-q})\,\zeta(q)\\ 2^{-2q}\,[\zeta(q,\ 1/4)-\zeta(q,\ 3/4)]\end{matrix}\right\}\qquad [q=2,\ 3,\ 4,\ \ldots], $$

$$ I^{+}(2n)=\frac{2^{2n}-1}{2\,(2n)!}\pi^{2n}\,|B_{2n}|,\quad I^{-}(2n+1)=\frac{1}{2\,(2n)!}\left(\frac{\pi}{2}\right)^{2n+1}|E_{2n}|, $$

$$ I^{-}(1)=\frac{\pi}{4}=0{,}78539816\ldots, $$

$I^{-}(2)=\mathbf{G}=0{,}91596559\ldots,$ $\qquad I^{+}(2)=\frac{\pi^2}{8}=1{,}23370055\ldots,$

$I^{-}(3)=\frac{\pi^3}{32}=0{,}96894615\ldots,$ $\qquad I^{+}(3)=\frac{7}{8}\zeta(3)=1{,}05179979\ldots,$

$I^{-}(4)=0{,}98894455\ldots,$ $\qquad I^{+}(4)=\frac{\pi^4}{96}=1{,}01467803\ldots,$

$I^{-}(5)=\frac{5\pi^5}{1536}=0{,}99615783\ldots,$ $\qquad I^{+}(5)=1{,}00452376\ldots,$

$I^{-}(6)=0{,}99868522\ldots,$ $\qquad I^{+}(6)=\frac{\pi^6}{960}=1{,}00144708\ldots,$

$I^{-}(7)=\frac{61\pi^7}{184320}=0{,}99955451\ldots,$ $\qquad I^{+}(7)=1{,}00047155\ldots,$

$I^{-}(8)=0{,}99984999\ldots,$ $\qquad I^{+}(8)=\frac{17\pi^8}{161280}=1{,}00015518\ldots,$

$I^{-}(9)=\frac{277\pi^9}{8257536}=0{,}99994968\ldots,$ $\qquad I^{+}(9)=1{,}00005135\ldots,$

$I^{-}(10)=0{,}99998316\ldots,$ $\qquad I^{+}(10)=\frac{31\pi^{10}}{2903040}=1{,}00001704\ldots$

34. $${}_{q+1}F_q\left(\begin{matrix}1,\ 1/2,\ \ldots,\ 1/2;\ 1\\ 5/2,\ \ldots,\ 5/2\end{matrix}\right)=$$
$$=\frac{(-1)^q 3^q}{2^{2q}(q-1)!}\sum_{k=1}^{[q/2]}\frac{(2q-2k-1)!\,(2^{2k}-1)}{(q-2k)!\,(2k)!}(2\pi)^{2k}B_{2k}+\frac{(-1)^{q-1}}{2}3^q.$$

35. $${}_{q+1}F_q\left(\begin{matrix}1,\ 1/2,\ \ldots,\ 1/2;-1\\ 5/2,\ \ldots,\ 5/2\end{matrix}\right)=$$
$$=\frac{(-1)^{q-1}3^q\pi}{2^{2q}(q-1)!}\sum_{k=0}^{[(q-1)/2]}\frac{(2q-2k-2)!\,\pi^{2k}}{(q-2k-1)!\,(2k)!}E_{2k}+\frac{(-1)^q}{2}3^q.$$

36. $${}_{q+1}F_q\left(\begin{matrix}1,\ 1/2,\ \ldots & \ldots,\ 1/2;\ 1\\ \underbrace{3/2,\ \ldots,\ 3/2}_{q-n}, & \underbrace{5/2,\ \ldots,\ 5/2}_{n}\end{matrix}\right)=A(q,\ n),$$

$$A(q,\ n)=\frac{(-1)^q 3^n}{2^{q+n}}\left[\sum_{k=0}^{n-2}\binom{q+k-1}{k}(2^{n-k}-1)\,\zeta(n-k)+\right.$$
$$\left.+\sum_{k=0}^{q-2}(-1)^{q-k}\binom{n+k-1}{k}(2^{q-k}-1)\,\zeta(q-k)-\sum_{k=0}^{q-1}\binom{q+k-1}{k}2^{n-k}\right].$$

37. $${}_{q+1}F_q\left(\begin{matrix}1,\ \overbrace{1/2,\ \ldots,\ 1/2}^{q-n},\ \overbrace{3/2,\ \ldots,\ 3/2}^{n};\ 1\\ 5/2,\ \ldots \qquad \ldots,\ 5/2\end{matrix}\right)=A(q-n,\ q)$$ [$A(q,\ n)$ see 7.10.2.44].

38. $${}_{q+1}F_q\left(\begin{matrix}1,\ 1/2,\ \ldots & \ldots,\ 1/2;\ -1\\ \underbrace{3/2,\ \ldots,\ 3/2}_{q-n}, & \underbrace{5/2,\ \ldots,\ 5/2}_{n}\end{matrix}\right)=C(q,\ n),$$

$$C(q,\ n)=\frac{(-1)^{q-1}3^n}{2^{q+n}}\left\{\sum_{k=0}^{n-2}\binom{q+k-1}{k}2^{k-n}\left[\zeta\left(n-k,\ \frac{1}{4}\right)-\zeta\left(n-k,\ \frac{3}{4}\right)\right]-\right.$$
$$-\sum_{k=0}^{q-2}(-1)^{q-k}\binom{n+k-1}{k}2^{k-q}\left[\zeta\left(q-k,\ \frac{1}{4}\right)-\zeta\left(q-k,\ \frac{3}{4}\right)\right]-$$
$$\left.-\sum_{k=0}^{n-1}\binom{q+k-1}{k}2^{n-k}+\binom{q+n-2}{q-1}\pi\right\}.$$

39. $${}_{q+1}F_q\left(\begin{matrix}1,\ \overbrace{1/2,\ \ldots,\ 1/2}^{q-n},\ \overbrace{3/2,\ \ldots,\ 3/2}^{n};\ -1\\ 5/2,\ \ldots \qquad \ldots,\ 5/2\end{matrix}\right)=C(q-n,\ q)$$

[$C(q,\ n)$ see 7.10.2.38].

7.11. THE FUNCTIONS OF KUMMER ${}_1F_1(a; b; z)$ AND TRICOMI $\Psi(a, b; z)$

7.11.1. Representations of ${}_0F_0(z)$ and ${}_1F_2(a; b; z)$.

1. ${}_0F_0(z) = e^z$.

	a	b	${}_1F_1(a;\ b;\ z)$
2	a	b	$e^z {}_1F_1(b-a;\ b;\ -z)$
3	a	b	$z^{-b/2} e^{z/2} M_{\rho,\,\sigma}(z)$ $\quad[\rho=b/2-a,\ 2\sigma=b-1]$
4	a	a	e^z
5	a	$2a$	$\Gamma\left(a+\frac{1}{2}\right)\left(\frac{z}{4}\right)^{1/2-a} e^{z/2} I_{a-1/2}\left(\frac{z}{2}\right)$
6	a	$2a-n$	$\Gamma\left(a-n-\frac{1}{2}\right)\left(\frac{z}{4}\right)^{n-a+1/2} e^{z/2} \sum_{k=0}^{n} (-1)^k \frac{(-n)_k (2a-2n-1)_k}{(2a-n)_k\, k!} \times \left(a+k-n-\frac{1}{2}\right) I_{a+k-n-1/2}\left(\frac{z}{2}\right)$
7	a	$2a+n$	$\Gamma\left(a-\frac{1}{2}\right)\left(\frac{z}{4}\right)^{1/2-a} e^{z/2} \sum_{k=0}^{n} \frac{(-n)_k (2a-1)_k}{(2a+n)_k\, k!} \times \left(a+k-\frac{1}{2}\right) I_{a+k-1/2}\left(\frac{z}{2}\right)$
8	a	$a-n$	$\frac{(-1)^n\, n!}{(1-a)_n} e^z L_n^{a-n-1}(-z)$
9	a	$\frac{1}{2}$	$\frac{2^{a-1}}{\sqrt{\pi}} \Gamma\left(a+\frac{1}{2}\right) e^{z/2} \left[D_{-2a}(-\sqrt{2z})+D_{-2a}(\sqrt{2z})\right]$
10	a	$\frac{3}{2}$	$\frac{2^{a-5/2}}{\sqrt{\pi z}} \Gamma\left(a-\frac{1}{2}\right) e^{z/2} \left[D_{1-2a}(-\sqrt{2z})-D_{1-2a}(\sqrt{2z})\right]$
11	n	b	$\frac{b-1}{(n-1)!}\left(\frac{\partial}{\partial z}\right)^{n-1} \left[z^{n-b} e^z \gamma(b-1,\ z)\right]$
12	n	m	$\frac{(m-2)!\,(1-m)_n}{(n-1)!} z^{1-m} \left[\sum_{k=0}^{m-n-1} \frac{(1+n-m)_k}{k!\,(2-m)_k} z^k - e^z \sum_{k=0}^{n-1} \frac{(1-n)_k}{k!\,(2-m)_k} (-z)^k\right]$ $\quad[n<m]$
13	n	$n+1$	$\frac{(-1)^n\, n!}{z^n}\left[1-e^z \sum_{k=0}^{n-1} \frac{(-1)^k z^k}{k!}\right]$
14	1	b	$(b-1)\, z^{1-b} e^z \gamma(b-1,\ z)$
15	1	m	$(m-1)!\, z^{1-m}\left[e^z - \sum_{k=0}^{m-2} \frac{z^k}{k!}\right]$ $\quad[m=1,\ 2,\ 3,\ldots]$
16	2	m	$\frac{(m-1)!}{z^{m-1}}(z-m+2)\left[e^z - \sum_{k=0}^{m-3} \frac{z^k}{k!}\right] + \frac{(m-2)(m-1)}{z}$ $\quad[m=2,\ 3,\ 4\ldots]$
17	$-n$	b	$\frac{n!}{(b)_n} L_n^{b-1}(z)$
18	$-n$	$-2n$	$\frac{n!\, z^{n+1/2}}{(2n)!\,\sqrt{\pi}} e^{z/2} K_{n+1/2}\left(\frac{z}{2}\right)$

	a	b	${}_1F_1(a;\ b;\ z)$
19	$-n$	$\frac{1}{2}$	$(-1)^n \frac{n!}{(2n)!} H_{2n}(\sqrt{z})$
20	$-n$	$\frac{3}{2}$	$\frac{(-1)^n n!}{2(2n+1)!\sqrt{z}} H_{2n+1}(\sqrt{z})$

21. $ {}_1F_1(a;\ b;\ z)+\Gamma\begin{bmatrix} a-b+1,\ b-1 \\ a,\ 1-b \end{bmatrix} z^{1-b}{}_1F_1(a-b+1;\ 2-b;\ z)=\Gamma\begin{bmatrix} a-b+1 \\ 1-b \end{bmatrix}\Psi(a, b; z).$

Note that for certain other values of the parameters a and b the function ${}_1F_1(a;b;z)$ can be expressed in terms of the functions given in section 7.11.1–2 by means of the recurrence formulae

22. $a\,{}_1F_1(a+1;\ b;\ z)=(z+2a-b)\,{}_1F_1(a;\ b;\ z)+(b-a)\,{}_1F_1(a-1;\ b;\ z),$
23. $z\,{}_1F_1(a+1;\ b+1;\ z)=b\,{}_1F_1(a+1;\ b;\ z)-b\,{}_1F_1(a;\ b;\ z),$
24. $a\,{}_1F_1(a+1;\ b+1;\ z)=(a-b)\,{}_1F_1(a;\ b+1;\ z)+b\,{}_1F_1(a;\ b;\ z).$

7.11.2. Particular values of ${}_1F_1(a;\ b;\ z)$.

	a	b	${}_1F_1(a;\ b;\ z)$
1	$-\frac{1}{2}$	$\frac{1}{2}$	$e^z-\sqrt{\pi z}\,\mathrm{erfi}(\sqrt{z})$
2	$-\frac{1}{2}$	1	$e^{z/2}\left[(1-z)\,I_0\left(\frac{z}{2}\right)+zI_1\left(\frac{z}{2}\right)\right]$
3	$-\frac{1}{2}$	$\frac{3}{2}$	$\frac{1}{2}\left[e^z+\frac{1-2z}{2}\sqrt{\frac{\pi}{z}}\,\mathrm{erfi}(\sqrt{z})\right]$
4	$-\frac{1}{2}$	2	$\frac{1}{3}e^{z/2}\left[(3-2z)\,I_0\left(\frac{z}{2}\right)-(1-2z)\,I_1\left(\frac{z}{2}\right)\right]$
5	$-\frac{1}{2}$	$\frac{5}{2}$	$\frac{3}{16z}\left[\frac{1}{2}(1+4z-4z^2)\sqrt{\frac{\pi}{z}}\,\mathrm{erfi}(\sqrt{z})-(1-2z)\,e^z\right]$
6	$-\frac{1}{2}$	3	$\frac{4}{15z}e^{z/2}\left[2z(2-z)\,I_0\left(\frac{z}{2}\right)-(1+2z-2z^2)\,I_1\left(\frac{z}{2}\right)\right]$
7	$-\frac{1}{2}$	$\frac{7}{2}$	$\frac{5}{64z^2}\left[\frac{1}{2}(3+6z+12z^2-8z^3)\sqrt{\frac{\pi}{z}}\,\mathrm{erfi}(\sqrt{z})-(3+4z-4z^2)\,e^z\right]$
8	$-\frac{1}{2}$	4	$\frac{4}{35z^2}e^{z/2}\left[z(1+10z-4z^2)\,I_0\left(\frac{z}{2}\right)-(4+5z+6z^2-4z^3)\,I_1\left(\frac{z}{2}\right)\right]$
9	$\frac{1}{2}$	$-\frac{1}{2}$	$(1-2z)\,e^z$
10	$\frac{1}{2}$	1	$e^{z/2}I_0\left(\frac{z}{2}\right)$
11	$\frac{1}{2}$	$\frac{3}{2}$	$\frac{1}{2}\sqrt{\frac{\pi}{z}}\,\mathrm{erfi}(\sqrt{z})$
12	$\frac{1}{2}$	2	$e^{z/2}\left[I_0\left(\frac{z}{2}\right)-I_1\left(\frac{z}{2}\right)\right]$
13	$\frac{1}{2}$	$\frac{5}{2}$	$\frac{3}{4z}\left[\frac{1+2z}{2}\sqrt{\frac{\pi}{z}}\,\mathrm{erfi}(\sqrt{z})-e^z\right]$
14	$\frac{1}{2}$	3	$\frac{4}{3z}e^{z/2}\left[zI_0\left(\frac{z}{2}\right)-(1+z)\,I_1\left(\frac{z}{2}\right)\right]$

	a	b	${}_1F_1(a;\,b;\,z)$
15	$\frac{1}{2}$	$\frac{7}{2}$	$\frac{15}{32z^2}\left[\frac{1}{2}(3+4z+4z^2)\sqrt{\frac{\pi}{z}}\,\mathrm{erfi}\left(\sqrt{z}\right)-(3+2z)\,e^z\right]$
16	$\frac{1}{2}$	4	$\frac{4}{5z^2}\,e^{z/2}\left[z\,(1+2z)\,I_0\left(\frac{z}{2}\right)-(4+3z+2z^2)\,I_1\left(\frac{z}{2}\right)\right]$
17	1	$-\frac{1}{2}$	$1-2z-2\sqrt{\pi}\,z^{3/2}\,e^z\,\mathrm{erf}\left(\sqrt{z}\right)$
18	1	$\frac{1}{2}$	$1+\sqrt{\pi z}\,e^z\,\mathrm{erf}\left(\sqrt{z}\right)$
19	1	$\frac{3}{2}$	$\frac{1}{2}\sqrt{\frac{\pi}{z}}\,e^z\,\mathrm{erf}\left(\sqrt{z}\right)$
20	1	2	$z^{-1}(e^z-1)$
21	1	$\frac{5}{2}$	$\frac{3}{2z}\left[\frac{1}{2}\sqrt{\frac{\pi}{z}}\,e^z\,\mathrm{erf}\left(\sqrt{z}\right)-1\right]$
22	1	3	$2z^{-2}(e^z-1-z)$
23	1	$\frac{7}{2}$	$\frac{5}{4z^2}\left[\frac{3}{2}\sqrt{\frac{\pi}{z}}\,e^z\,\mathrm{erf}\left(\sqrt{z}\right)-3-2z\right]$
24	1	4	$3z^{-3}(2e^z-2-2z-z^2)$
25	$\frac{3}{2}$	$-\frac{1}{2}$	$(1-4z-4z^2)\,e^z$
26	$\frac{3}{2}$	$\frac{1}{2}$	$(1+2z)\,e^z$
27	$\frac{3}{2}$	1	$e^{z/2}\left[(1+z)\,I_0\left(\frac{z}{2}\right)+zI_1\left(\frac{z}{2}\right)\right]$
28	$\frac{3}{2}$	2	$e^{z/2}\left[I_0\left(\frac{z}{2}\right)+I_1\left(\frac{z}{2}\right)\right]$
29	$\frac{3}{2}$	$\frac{5}{2}$	$\frac{3}{2z}\left[e^z-\frac{1}{2}\sqrt{\frac{\pi}{z}}\,\mathrm{erfi}\left(\sqrt{z}\right)\right]$
30	$\frac{3}{2}$	3	$\frac{4}{z}\,e^{z/2}\,I_1\left(\frac{z}{2}\right)$
31	$\frac{3}{2}$	$\frac{7}{2}$	$\frac{15}{8z^2}\left[3e^z-\frac{3+2z}{2}\sqrt{\frac{\pi}{z}}\,\mathrm{erfi}\left(\sqrt{z}\right)\right]$
32	$\frac{3}{2}$	4	$\frac{4}{z^2}\,e^{z/2}\left[(4+z)\,I_1\left(\frac{z}{2}\right)-zI_0\left(\frac{z}{2}\right)\right]$
33	2	$-\frac{1}{2}$	$1-4z-2z^2-z^{3/2}(5+2z)\sqrt{\pi}\,e^z\,\mathrm{erf}\left(\sqrt{z}\right)$
34	2	$\frac{1}{2}$	$1+z+\frac{\sqrt{\pi z}}{2}(3+2z)\,e^z\,\mathrm{erf}\left(\sqrt{z}\right)$
35	2	1	$(1+z)\,e^z$
36	2	$\frac{3}{2}$	$\frac{1}{2}\left[1+\frac{1+2z}{2}\sqrt{\frac{\pi}{z}}\,e^z\,\mathrm{erf}\left(\sqrt{z}\right)\right]$
37	2	$\frac{5}{2}$	$\frac{3}{4z}\left[1-\frac{1-2z}{2}\sqrt{\frac{\pi}{z}}\,e^z\,\mathrm{erf}\left(\sqrt{z}\right)\right]$
38	2	3	$2z^{-2}[1-(1-z)\,e^z]$
39	2	$\frac{7}{2}$	$\frac{15}{8z^2}\left[3-\frac{3-2z}{2}\sqrt{\frac{\pi}{z}}\,e^z\,\mathrm{erf}\left(\sqrt{z}\right)\right]$
40	2	4	$6z^{-3}[2+z-(2-z)\,e^z]$
41	$\frac{5}{2}$	$-\frac{1}{2}$	$\frac{1}{3}(3-18z-36z^2-8z^3)\,e^z$

	a	b	${}_1F_1\,(a;\,b;\,z)$
42	$\frac{5}{2}$	$\frac{1}{2}$	$\frac{1}{3}(3+12z+4z^2)\,e^z$
43	$\frac{5}{2}$	1	$\frac{1}{3}\,e^{z/2}\left[(3+6z+2z^2)\,I_0\left(\frac{z}{2}\right)+2z\,(2+z)\,I_1\left(\frac{z}{2}\right)\right]$
44	$\frac{5}{2}$	$\frac{3}{2}$	$\frac{1}{3}(3+2z)\,e^z$
45	$\frac{5}{2}$	2	$\frac{1}{3}\,e^{z/2}\left[(3+2z)\,I_0\left(\frac{z}{2}\right)+(1+2z)\,I_1\left(\frac{z}{2}\right)\right]$
46	$\frac{5}{2}$	3	$\frac{4}{3z}\,e^{z/2}\left[zI_0\left(\frac{z}{2}\right)-(1-z)\,I_1\left(\frac{z}{2}\right)\right]$
47	$\frac{5}{2}$	$\frac{7}{2}$	$\frac{5}{4z^2}\left[\frac{3}{2}\sqrt{\frac{\pi}{z}}\operatorname{erfi}(\sqrt{z})-(3-2z)\,e^z\right]$
48	$\frac{5}{2}$	4	$\frac{4}{z^2}\,e^{z/2}\left[zI_0\left(\frac{z}{2}\right)-(4-z)\,I_1\left(\frac{z}{2}\right)\right]$
49	3	$-\frac{1}{2}$	$\frac{1}{2}\left[2-12z-13z^2-2z^3-\frac{z^2}{2}(35+28z+4z^2)\sqrt{\frac{\pi}{z}}\,e^z\operatorname{erf}(\sqrt{z})\right]$
50	3	$\frac{1}{2}$	$\frac{1}{4}\left[4+9z+2z^2+\frac{1}{2}(15+20z+4z^2)\sqrt{\pi z}\,e^z\operatorname{erf}(\sqrt{z})\right]$
51	3	1	$\frac{1}{2}(2+4z+z^2)\,e^z$
52	3	$\frac{3}{2}$	$\frac{1}{8}\left[5+2z+\frac{1}{2}(3+12z+4z^2)\sqrt{\frac{\pi}{z}}\,e^z\operatorname{erf}(\sqrt{z})\right]$
53	3	2	$\frac{1}{2}(2+z)\,e^z$
54	3	$\frac{5}{2}$	$\frac{3}{16z}\left[1+2z-\frac{1}{2}(1-4z-4z^2)\sqrt{\frac{\pi}{z}}\,e^z\operatorname{erf}(\sqrt{z})\right]$
55	3	$\frac{7}{2}$	$\frac{15}{32z^2}\left[\frac{1}{2}(3-4z+4z^2)\sqrt{\frac{\pi}{z}}\,e^z\operatorname{erf}(\sqrt{z})-3+2z\right]$
56	3	4	$3z^{-3}\,[(2-2z+z^2)\,e^z-2]$
57	$\frac{7}{2}$	$-\frac{1}{2}$	$\frac{1}{15}(15-120z-360z^2-160z^3-16z^4)\,e^z$
58	$\frac{7}{2}$	$\frac{1}{2}$	$\frac{1}{15}(15+90z+60z^2+8z^3)\,e^z$
59	$\frac{7}{2}$	1	$\frac{1}{15}\,e^{z/2}\left[(15+45z+28z^2+4z^3)\,I_0\left(\frac{z}{2}\right)+z\,(23+24z+4z^2)\,I_1\left(\frac{z}{2}\right)\right]$
60	$\frac{7}{2}$	$\frac{3}{2}$	$\frac{1}{15}(15+20z+4z^2)\,e^z$
61	$\frac{7}{2}$	2	$\frac{1}{15}\,e^{z/2}\left[(15+18z+4z^2)\,I_0\left(\frac{z}{2}\right)+(3+14z+4z^2)\,I_1\left(\frac{z}{2}\right)\right]$
62	$\frac{7}{2}$	$\frac{5}{2}$	$\frac{1}{5}(5+2z)\,e^z$
63	$\frac{7}{2}$	3	$\frac{4}{15z}\,e^{z/2}\left[2z\,(2+z)\,I_0\left(\frac{z}{2}\right)-(1-2z-2z^2)\,I_1\left(\frac{z}{2}\right)\right]$
64	$\frac{7}{2}$	4	$\frac{4}{5z^2}\,e^{z/2}\left[(4-3z+2z^2)\,I_1\left(\frac{z}{2}\right)-z\,(1-2z)\,I_0\left(\frac{z}{2}\right)\right]$

	a	b	${}_1F_1(a;\ b;\ z)$
65	4	$-\frac{1}{2}$	$\frac{1}{12}\left[12-96z-165z^2-52z^3-4z^4-\right.$ $\left.-\frac{z^{3/2}}{2}(315+378z+108z^2+8z^3)\sqrt{\pi}\,e^z\operatorname{erf}(\sqrt{z})\right]$
66	4	$\frac{1}{2}$	$\frac{1}{24}\left[24+87z+40z^2+4z^3+\right.$ $\left.+\frac{1}{2}(105+210z+84z^2+8z^3)\sqrt{\pi z}\,e^z\operatorname{erf}(\sqrt{z})\right]$
67	4	1	$\frac{1}{6}(6+18z+9z^2+z^3)\,e^z$
68	4	$\frac{3}{2}$	$\frac{1}{48}\left[33+28z+4z^2+\frac{1}{2}(15+90z+60z^2+8z^3)\sqrt{\frac{\pi}{z}}\,e^z\operatorname{erf}(\sqrt{z})\right]$
69	4	2	$\frac{1}{6}(6+6z+z^2)\,e^z$
70	4	$\frac{5}{2}$	$\frac{1}{32z}\left[3+16z+4z^2-\frac{1}{2}(3-18z-36z^2-8z^3)\sqrt{\frac{\pi}{z}}\,e^z\operatorname{erf}(\sqrt{z})\right]$
71	4	3	$\frac{1}{3}(3+z)\,e^z$
72	4	$\frac{7}{2}$	$\frac{5}{64z^2}\left[\frac{1}{2}(3-6z+12z^2+8z^3)\sqrt{\frac{\pi}{z}}\,e^z\operatorname{erf}(\sqrt{z})-3+4z+4z^2\right]$

7.11.3. Representations of ${}_1F_1(a;\ b;\ -z)$.

	a	b	${}_1F_1(a;\ b;\ -z)$
1	a	$a+1$	$az^{-a}\gamma(a,\ z)$
2	a	$a+2$	$(a+1)\left[e^{-z}+(z-a)\,z^{-a-1}\gamma(a+1,\ z)\right]$
3	a	$\frac{1}{2}$	$\frac{2^{-a-1/2}}{\sqrt{\pi}}\Gamma(1-a)\,e^{-z/2}\left[D_{2a-1}(-\sqrt{2z})+D_{2a-1}(\sqrt{2z})\right]$
4	a	$\frac{3}{2}$	$\frac{2^{-a-1}}{\sqrt{\pi z}}\Gamma(1-a)\,e^{-z/2}\left[D_{2a-2}(-\sqrt{2z})-D_{2a-2}(\sqrt{2z})\right]$

7.11.4. Representations and particular values of $\Psi(a, b; z)$.

	a	b	$\Psi(a, b; z)$
1	a	b	$\Gamma\left[\begin{matrix}1-b\\1+a-b\end{matrix}\right] {}_1F_1(a; b; z)+\Gamma\left[\begin{matrix}b-1\\a\end{matrix}\right] z^{1-b}{}_1F_1(1+a-b; 2-b; z)$
2	a	b	$z^{1-b}\Psi(1+a-b, 2-b; z)$
3	a	b	$z^{-b/2}e^{z/2}W_{\varkappa, \mu}(z)$ $[\varkappa=b/2-a,\ 2\mu=b-1]$
4	a	a	$e^z\Gamma(1-a, z)$
5	a	$2a$	$\pi^{-1/2}z^{1/2-a}e^{z/2}K_{a-1/2}\left(\frac{z}{2}\right)=$
6			$=\frac{\sqrt{\pi}}{2}z^{1/2-a}\exp\left[\frac{z}{2}+\frac{\pi i}{2}\left(a+\frac{1}{2}\right)\right]H^{(1)}_{a-1/2}\left(\frac{iz}{2}\right)$
7	a	$\frac{1}{2}$	$2^a e^{z/2}D_{-2a}(\sqrt{2z})$
8	a	$\frac{3}{2}$	$2^{a-1/2}z^{-1/2}e^{z/2}D_{1-2a}(\sqrt{2z})$
9	a	2	$\frac{\Gamma(2-a)}{z}e^{z/2}k_{2-2a}\left(\frac{z}{2}\right)$
10	n	b	$\frac{1}{(n-1)(2-b)_{n-1}}\left(\frac{d}{dz}\right)^{n-1}[z^{n-b}e^z\Gamma(b-1, z)]$
11	1	b	$z^{1-b}e^z\Gamma(b-1, z)$
12	$-n$	b	$(-1)^n n! L_n^{b-1}(z)$
13	$\frac{1-n}{2}$	$\frac{3}{2}$	$2^{-n}z^{-1/2}H_n(\sqrt{z})$
14	$\frac{1}{2}$	$\frac{1}{2}$	$\sqrt{\pi}e^z\operatorname{erfc}(\sqrt{z})$
15	1	1	$-e^z\operatorname{Ei}(-z)$

7.12. THE FUNCTION ${}_2F_2(a_1,\ a_2;\ b_1,\ b_2;\ z)$ and ${}_qF_q(a_1,\ \ldots,\ a_q;\ b_1,\ \ldots,\ b_q;\ z)$

7.12.1. Representations of ${}_2F_2(a_1,\ a_2;\ b_1,\ b_2;\ z)$.

	a_1	a_2	b_1	b_2	${}_2F_2(a_1,\ a_2;\ b_1,\ b_2;\ z)$
1	a	b	$a-1$	c	${}_1F_1(b;\ c;\ z)+\dfrac{bz}{(a-1)c}\,{}_1F_1(b+1;\ c+1;\ z)$
2	a	b	$a+n$	$b+1$	$\dfrac{(a)_n}{(a-b)_n}\left[{}_1F_1(b;\ b+1;\ z)-\right.$ $\left.-b\sum\limits_{k=1}^{n}\dfrac{(a-b)_{k-1}}{(a)_k}\,{}_1F_1(a;\ a+k;\ z)\right]$
3	a	b	$a+1$	$b+1$	$\dfrac{a}{a-b}\,{}_1F_1(b;\ b+1;\ z)+\dfrac{b}{b-a}\,{}_1F_1(a;\ a+1;\ z)$
4	1	a	b	$1+2a-b$	$(b-1)(2a-b)\left(\dfrac{z}{2}\right)^{1/2-a}e^{z/2}h_{a-3/2,\ 1/2+a-b}\left(\dfrac{z}{2}\right)$
5	1	a	2	b	$\dfrac{b}{az}\,[{}_1F_1(a;\ b;\ z)-1]$ $[a\neq 1]$
6	$\dfrac{1}{2}$	a	$\dfrac{3}{2}$	$a+1$	$\dfrac{a}{2a-1}\left[\sqrt{\dfrac{\pi}{z}}\operatorname{erfi}(\sqrt{z})-(-z)^{-a}\gamma(a,\ -z)\right]$

7. $${}_2F_2(1,\ a;\ a+1,\ b;\ z)+\frac{b-1}{a-b+1}\,{}_2F_2(1,\ a;\ a+1,\ 2+a-b;\ -z)=$$
$$=\frac{a}{a-b+1}\,{}_1F_1(a-b+1;\ a-b+2;\ z)\,{}_1F_1(b-1;\ b;\ -z).$$

7.12.2. Particular values of ${}_2F_2(a_1,\ a_2;\ b_1,\ b_2;\ z)$.

	a_1	a_2	b_1	b_2	${}_2F_2(a_1,\ a_2;\ b_1,\ b_2;\ z)$
1	$-\dfrac{1}{2}$	1	$\dfrac{1}{2}$	2	$\dfrac{1}{3z}\left[(1+2z)e^z-2\sqrt{\pi}\,z^{3/2}\operatorname{erfi}(\sqrt{z})-1\right]$
2	$-\dfrac{1}{2}$	1	$\dfrac{1}{2}$	3	$\dfrac{2}{15z^2}\left[(3+2z+4z^2)e^z-4\sqrt{\pi}\,z^{5/2}\operatorname{erfi}(\sqrt{z})-3-5z\right]$
3	$-\dfrac{1}{2}$	1	$\dfrac{3}{2}$	2	$\dfrac{1}{3z}\left[\dfrac{3-2z}{2}\sqrt{\pi z}\operatorname{erfi}(\sqrt{z})+1-(1-z)e^z\right]$
4	$-\dfrac{1}{2}$	1	$\dfrac{3}{2}$	3	$\dfrac{2}{15z^2}\left[1+5z-(1+4z-2z^2)e^z+z^{3/2}(5-2z)\sqrt{\pi}\operatorname{erfi}(\sqrt{z})\right]$
5	$-\dfrac{1}{2}$	1	2	2	$\dfrac{2}{9z}\left\{3-e^{z/2}\left[(3-6z+2z^2)I_0\left(\dfrac{z}{2}\right)+2z(2-z)I_1\left(\dfrac{z}{2}\right)\right]\right\}$
6	$-\dfrac{1}{2}$	1	2	$\dfrac{5}{2}$	$\dfrac{1}{8z}\left[8-(5-2z)e^z-\dfrac{1}{2}(3-12z+4z^2)\sqrt{\dfrac{\pi}{z}}\operatorname{erfi}(\sqrt{z})\right]$
7	$-\dfrac{1}{2}$	1	2	3	$\dfrac{4}{45z}\left\{15+e^{z/2}\left[(3-14z+4z^2)I_1\left(\dfrac{z}{2}\right)-\right.\right.$ $\left.\left.-(15-18z+4z^2)I_0\left(\dfrac{z}{2}\right)\right]\right\}$
8	$-\dfrac{1}{2}$	1	$\dfrac{5}{2}$	3	$\dfrac{1}{10z^2}\left[(4-9z+2z^2)e^z-\dfrac{1}{2}(15-20z+4z^2)\sqrt{\pi z}\times\right.$ $\left.\times\operatorname{erfi}(\sqrt{z})-4+20z\right]$

	a_1	a_2	b_1	b_2	${}_2F_2(a_1, a_2;\ b_1, b_2;\ z)$
9	$-\frac{1}{2}$	1	3	3	$\frac{16}{225z^2}\left\{e^{z/2}\left[(15-45z+28z^2-4z^3)\,I_0\left(\frac{z}{2}\right)+z(23-24z+4z^2)\,I_1\left(\frac{z}{2}\right)\right]-15+\frac{75}{2}z\right\}$
10	$-\frac{1}{2}$	$\frac{3}{2}$	$\frac{1}{2}$	$\frac{1}{2}$	$e^z-2\sqrt{\pi z}\,\mathrm{erfi}\,(\sqrt{z})$
11	$-\frac{1}{2}$	$\frac{3}{2}$	$\frac{1}{2}$	1	$e^{z/2}\left[(1-2z)\,I_0\left(\frac{z}{2}\right)+2zI_1\left(\frac{z}{2}\right)\right]$
12	$-\frac{1}{2}$	$\frac{3}{2}$	$\frac{1}{2}$	2	$\frac{e^{z/2}}{3}\left[(3-4z)\,I_0\left(\frac{z}{2}\right)+(1+4z)\,I_1\left(\frac{z}{2}\right)\right]$
13	$-\frac{1}{2}$	$\frac{3}{2}$	$\frac{1}{2}$	$\frac{5}{2}$	$\frac{3}{8z}\left[(1+2z)\,e^z-\frac{1+4z^2}{2}\sqrt{\frac{\pi}{z}}\,\mathrm{erfi}(\sqrt{z})\right]$
14	$-\frac{1}{2}$	$\frac{3}{2}$	$\frac{1}{2}$	3	$\frac{4e^{z/2}}{15z}\left[z(3-4z)\,I_0\left(\frac{z}{2}\right)+(3+z+4z^2)\,I_1\left(\frac{z}{2}\right)\right]$
15	$-\frac{1}{2}$	2	$\frac{1}{2}$	1	$\frac{1}{2}\left[2e^z-3\sqrt{\pi z}\,\mathrm{erfi}\,(\sqrt{z})\right]$
16	$-\frac{1}{2}$	2	$\frac{1}{2}$	3	$\frac{2}{5z^2}\left[1-(1-z-2z^2)\,e^z-2\sqrt{\pi}\,z^{5/2}\,\mathrm{erfi}\,(\sqrt{z})\right]$
17	$-\frac{1}{2}$	2	1	1	$\frac{e^{z/2}}{2}\left[(2-3z)\,I_0\left(\frac{z}{2}\right)+3zI_1\left(\frac{z}{2}\right)\right]$
18	$-\frac{1}{2}$	2	1	$\frac{3}{2}$	$\frac{1}{4}\left[3e^z+\frac{1}{2}(1-6z)\sqrt{\frac{\pi}{z}}\,\mathrm{erfi}\,(\sqrt{z})\right]$
19	$-\frac{1}{2}$	2	1	$\frac{5}{2}$	$\frac{3}{32z}\left[(1+6z)\,e^z-\frac{1}{2}(1-4z+12z^2)\sqrt{\frac{\pi}{z}}\,\mathrm{erfi}\,(\sqrt{z})\right]$
20	$-\frac{1}{2}$	2	1	3	$\frac{2e^{z/2}}{15z}\left[z(7-6z)\,I_0\left(\frac{z}{2}\right)+(2-z+6z^2)\,I_1\left(\frac{z}{2}\right)\right]$
21	$-\frac{1}{2}$	2	$\frac{3}{2}$	3	$\frac{2}{15z^2}\left[(1-z+3z^2)\,e^z-1+\frac{z^{3/2}}{2}(5-6z)\sqrt{\pi}\,\mathrm{erfi}\,(\sqrt{z})\right]$
22	$-\frac{1}{2}$	2	$\frac{5}{2}$	3	$\frac{1}{20z^2}\left[8-(8+7z-6z^2)\,e^z+\frac{\sqrt{\pi z}}{2}(15+20z-12z^2)\,\mathrm{erfi}\,(\sqrt{z})\right]$
23	$-\frac{1}{2}$	2	3	3	$\frac{8}{225z^2}\left\{30-e^{z/2}\left[(30-15z-34z^2+12z^3)\,I_0\left(\frac{z}{2}\right)+z(31+22z-12z^2)\,I_1\left(\frac{z}{2}\right)\right]\right\}$
24	$-\frac{1}{2}$	$\frac{5}{2}$	$\frac{1}{2}$	$\frac{1}{2}$	$\frac{1}{3}\left[(3-2z)\,e^z-8\sqrt{\pi z}\,\mathrm{erfi}\,(\sqrt{z})\right]$
25	$-\frac{1}{2}$	$\frac{5}{2}$	$\frac{1}{2}$	1	$\frac{e^{z/2}}{3}\left[3(1-3z)\,I_0\left(\frac{z}{2}\right)+7zI_1\left(\frac{z}{2}\right)\right]$
26	$-\frac{1}{2}$	$\frac{5}{2}$	$\frac{1}{2}$	$\frac{3}{2}$	$\frac{1}{3}\left[3e^z-4\sqrt{\pi z}\,\mathrm{erfi}\,(\sqrt{z})\right]$
27	$-\frac{1}{2}$	$\frac{5}{2}$	$\frac{1}{2}$	2	$\frac{e^{z/2}}{9}\left[(9-16z)\,I_0\left(\frac{z}{2}\right)+(1+16z)\,I_1\left(\frac{z}{2}\right)\right]$
28	$-\frac{1}{2}$	$\frac{5}{2}$	$\frac{1}{2}$	3	$\frac{4e^{z/2}}{45z}\left[4(3-4z)\,zI_0\left(\frac{z}{2}\right)-(3-4z-16z^2)\,I_1\left(\frac{z}{2}\right)\right]$
29	$-\frac{1}{2}$	$\frac{5}{2}$	1	$\frac{3}{2}$	$\frac{e^{z/2}}{3}\left[(3-4z)\,I_0\left(\frac{z}{2}\right)+4zI_1\left(\frac{z}{2}\right)\right]$

	a_1	a_2	b_1	b_2	${}_2F_2(a_1, a_2; b_1, b_2; z)$
30	$-\frac{1}{2}$	$\frac{5}{2}$	$\frac{3}{2}$	$\frac{3}{2}$	$\frac{1}{3}\left[2e^z+\frac{1-4z}{2}\sqrt{\frac{\pi}{z}}\operatorname{erfi}(\sqrt{z})\right]$
31	$-\frac{1}{2}$	$\frac{5}{2}$	$\frac{3}{2}$	2	$\frac{e^{z/2}}{9}\left[(9-8z)I_0\left(\frac{z}{2}\right)-(1-8z)I_1\left(\frac{z}{2}\right)\right]$
32	$-\frac{1}{2}$	$\frac{5}{2}$	$\frac{3}{2}$	3	$\frac{4e^{z/2}}{45z}\left[(11-8z)zI_0\left(\frac{z}{2}\right)+(1-3z+8z^2)I_1\left(\frac{z}{2}\right)\right]$
33	$-\frac{1}{2}$	3	$\frac{1}{2}$	1	$\frac{1}{8}\left[2(4-z)e^z-15\sqrt{\pi z}\operatorname{erfi}(\sqrt{z})\right]$
34	$-\frac{1}{2}$	3	$\frac{1}{2}$	2	$\frac{1}{4}\left[4e^z-5\sqrt{\pi z}\operatorname{erfi}(\sqrt{z})\right]$
35	$-\frac{1}{2}$	3	1	1	$\frac{e^{z/2}}{4}\left[4(1-2z)I_0\left(\frac{z}{2}\right)+7zI_1\left(\frac{z}{2}\right)\right]$
36	$-\frac{1}{2}$	3	1	$\frac{3}{2}$	$\frac{1}{16}\left[13e^z+\frac{3}{2}(1-10z)\sqrt{\frac{\pi}{z}}\operatorname{erfi}(\sqrt{z})\right]$
37	$-\frac{1}{2}$	3	1	2	$\frac{e^{z/2}}{4}\left[(4-5z)I_0\left(\frac{z}{2}\right)+5zI_1\left(\frac{z}{2}\right)\right]$
38	$-\frac{1}{2}$	3	1	$\frac{5}{2}$	$\frac{3}{128z}\left[(1+30z)e^z-\frac{1}{2}(1-12z+60z^2)\sqrt{\frac{\pi}{z}}\operatorname{erfi}(\sqrt{z})\right]$
39	$-\frac{1}{2}$	3	$\frac{3}{2}$	2	$\frac{1}{8}\left[5e^z+\frac{1}{2}(3-10z)\sqrt{\frac{\pi}{z}}\operatorname{erfi}(\sqrt{z})\right]$
40	$-\frac{1}{2}$	3	2	2	$\frac{e^{z/2}}{6}\left[(6-5z)I_0\left(\frac{z}{2}\right)-(1-5z)I_1\left(\frac{z}{2}\right)\right]$
41	$-\frac{1}{2}$	3	2	$\frac{5}{2}$	$\frac{3}{64z}\left[\frac{1}{2}(1+12z-20z^2)\sqrt{\frac{\pi}{z}}\operatorname{erfi}(\sqrt{z})-(1-10z)e^z\right]$
42	$\frac{1}{2}$	1	$\frac{3}{2}$	2	$\frac{1}{z}\left[\sqrt{\pi z}\operatorname{erfi}(\sqrt{z})+1-e^z\right]$
43	$\frac{1}{2}$	1	$\frac{3}{2}$	3	$\frac{2}{3z^2}\left[1+3z-(1+2z)e^z+2\sqrt{\pi}z^{3/2}\operatorname{erfi}(\sqrt{z})\right]$
44	$\frac{1}{2}$	1	2	2	$\frac{2}{z}\left\{1-e^{z/2}\left[(1-z)I_0\left(\frac{z}{2}\right)+zI_1\left(\frac{z}{2}\right)\right]\right\}$
45	$\frac{1}{2}$	1	2	$\frac{5}{2}$	$\frac{3}{2z}\left[2-e^z-\frac{1-2z}{2}\sqrt{\frac{\pi}{z}}\operatorname{erfi}(\sqrt{z})\right]$
46	$\frac{1}{2}$	1	2	3	$\frac{4}{3z}\left\{3+e^{z/2}\left[(1-2z)I_1\left(\frac{z}{2}\right)-(3-2z)I_0\left(\frac{z}{2}\right)\right]\right\}$
47	$\frac{1}{2}$	1	$\frac{5}{2}$	3	$\frac{2}{z^2}\left[(1-z)e^z-\frac{z}{2}(3-2z)\sqrt{\frac{\pi}{z}}\operatorname{erfi}(\sqrt{z})-1+3z\right]$
48	$\frac{1}{2}$	1	3	3	$\frac{8}{9z^2}\left\{2e^{z/2}\left[(3-6z+2z^2)I_0\left(\frac{z}{2}\right)+2z(2-z)I_1\left(\frac{z}{2}\right)\right]-6+9z\right\}$
49	$\frac{1}{2}$	2	1	1	$\frac{e^{z/2}}{2}\left[(2+z)I_0\left(\frac{z}{2}\right)+zI_1\left(\frac{z}{2}\right)\right]$
50	$\frac{1}{2}$	2	1	$\frac{3}{2}$	$\frac{1}{2}\left[e^z+\frac{1}{2}\sqrt{\frac{\pi}{z}}\operatorname{erfi}(\sqrt{z})\right]$
51	$\frac{1}{2}$	2	1	$\frac{5}{2}$	$\frac{3}{8z}\left[e^z-\frac{1-2z}{2}\sqrt{\frac{\pi}{z}}\operatorname{erfi}(\sqrt{z})\right]$
52	$\frac{1}{2}$	2	1	3	$\frac{2e^{z/2}}{3z}\left[zI_0\left(\frac{z}{2}\right)+(2-z)I_1\left(\frac{z}{2}\right)\right]$

	a_1	a_2	b_1	b_2	${}_2F_2(a_1, a_2; b_1, b_2; z)$
53	$\frac{1}{2}$	2	$\frac{3}{2}$	3	$\frac{2}{3z^2}\left[(1-z)e^z-1+\sqrt{\pi}\,z^{3/2}\operatorname{erfi}(\sqrt{z})\right]$
54	$\frac{1}{2}$	2	$\frac{5}{2}$	3	$\frac{1}{z^2}\left[2-(2+z)e^z+\frac{\sqrt{\pi z}}{2}(3+2z)\operatorname{erfi}(\sqrt{z})\right]$
55	$\frac{1}{2}$	2	3	3	$\frac{8}{9z^2}\left\{6-e^{z/2}\left[(6-3z-2z^2)I_0\left(\frac{z}{2}\right)+z(5+2z)I_1\left(\frac{z}{2}\right)\right]\right\}$
56	$\frac{1}{2}$	$\frac{5}{2}$	1	$\frac{3}{2}$	$\frac{e^{z/2}}{3}\left[(3+z)I_0\left(\frac{z}{2}\right)+zI_1\left(\frac{z}{2}\right)\right]$
57	$\frac{1}{2}$	$\frac{5}{2}$	$\frac{3}{2}$	$\frac{3}{2}$	$\frac{1}{3}\left[\sqrt{\frac{\pi}{z}}\operatorname{erfi}(\sqrt{z})+e^z\right]$
58	$\frac{1}{2}$	$\frac{5}{2}$	$\frac{3}{2}$	2	$\frac{e^{z/2}}{3}\left[3I_0\left(\frac{z}{2}\right)-I_1\left(\frac{z}{2}\right)\right]$
59	$\frac{1}{2}$	$\frac{5}{2}$	$\frac{3}{2}$	3	$\frac{4e^{z/2}}{9z}\left[2zI_0\left(\frac{z}{2}\right)+(1-2z)I_1\left(\frac{z}{2}\right)\right]$
60	$\frac{1}{2}$	3	1	1	$\frac{e^{z/2}}{4}\left[(4+4z+z^2)I_0\left(\frac{z}{2}\right)+z(3+z)I_1\left(\frac{z}{2}\right)\right]$
61	$\frac{1}{2}$	3	1	$\frac{3}{2}$	$\frac{1}{8}\left[\frac{3}{2}\sqrt{\frac{\pi}{z}}\operatorname{erfi}(\sqrt{z})+(5+2z)e^z\right]$
62	$\frac{1}{2}$	3	1	2	$\frac{e^{z/2}}{4}\left[(4+z)I_0\left(\frac{z}{2}\right)+zI_1\left(\frac{z}{2}\right)\right]$
63	$\frac{1}{2}$	3	1	$\frac{5}{2}$	$\frac{3}{32z}\left[(1+4z)e^z-\frac{1-6z}{2}\sqrt{\frac{\pi}{z}}\operatorname{erfi}(\sqrt{z})\right]$
64	$\frac{1}{2}$	3	$\frac{3}{2}$	2	$\frac{1}{4}\left[\frac{3}{2}\sqrt{\frac{\pi}{z}}\operatorname{erfi}(\sqrt{z})+e^z\right]$
65	$\frac{1}{2}$	3	2	2	$\frac{e^{z/2}}{2}\left[2I_0\left(\frac{z}{2}\right)-I_1\left(\frac{z}{2}\right)\right]$
66	$\frac{1}{2}$	3	2	$\frac{5}{2}$	$\frac{3}{16z}\left[\frac{1+6z}{2}\sqrt{\frac{\pi}{z}}\operatorname{erfi}(\sqrt{z})-e^z\right]$
67	1	1	2	2	$\frac{1}{z}[\operatorname{Ei}(z)-\ln(-z)-\mathbf{C}]$
68	1	1	2	3	$\frac{2}{z^2}\{1+z-e^z-z[\mathbf{C}+\ln(-z)-\operatorname{Ei}(z)]\}$
69	1	1	3	3	$\frac{4}{z^2}\{1+2z-e^z+(1-z)[\mathbf{C}+\ln(-z)-\operatorname{Ei}(z)]\}$
70	1	$\frac{3}{2}$	$\frac{1}{2}$	$\frac{1}{2}$	$1+2z+2\sqrt{\pi z}(1+z)e^z\operatorname{erf}(\sqrt{z})$
71	1	$\frac{3}{2}$	$\frac{1}{2}$	2	$\frac{1}{z}[1-(1-2z)e^z]$
72	1	$\frac{3}{2}$	$\frac{1}{2}$	$\frac{5}{2}$	$\frac{3}{z}\left[1-\frac{1-z}{2}\sqrt{\frac{\pi}{z}}e^z\operatorname{erf}(\sqrt{z})\right]$
73	1	$\frac{3}{2}$	$\frac{1}{2}$	3	$\frac{2}{z^2}[3+z-(3-2z)e^z]$
74	1	$\frac{3}{2}$	2	2	$\frac{2}{z}\left[e^{z/2}I_0\left(\frac{z}{2}\right)-1\right]$
75	1	$\frac{3}{2}$	2	$\frac{5}{2}$	$\frac{3}{z}\left[\frac{1}{2}\sqrt{\frac{\pi}{z}}\operatorname{erfi}(\sqrt{z})-1\right]$

	a_1	a_2	b_1	b_2	${}_2F_2(a_1, a_2; b_1, b_2; z)$
76	1	$\frac{3}{2}$	2	3	$\frac{4}{z}\left\{e^{z/2}\left[I_0\left(\frac{z}{2}\right)-I_1\left(\frac{z}{2}\right)\right]-1\right\}$
77	1	$\frac{3}{2}$	$\frac{5}{2}$	3	$\frac{6}{z^2}\left[1-z-e^z+\sqrt{\pi z}\,\mathrm{erfi}\,(\sqrt{z})\right]$
78	1	$\frac{3}{2}$	3	3	$\frac{8}{z^2}\left\{2-z-2e^{z/2}\left[(1-z)\,I_0\left(\frac{z}{2}\right)+zI_1\left(\frac{z}{2}\right)\right]\right\}$
79	1	2	3	3	$\frac{4}{z^2}\left[\mathrm{Ei}\,(z)-z-\mathbf{C}-\ln(-z)\right]$
80	1	$\frac{5}{2}$	$\frac{1}{2}$	$\frac{1}{2}$	$\frac{1}{3}\left[3+14z+4z^2+4\,(2+4z+z^2)\,\sqrt{\pi z}\,e^z\,\mathrm{erf}\,(\sqrt{z})\right]$
81	1	$\frac{5}{2}$	$\frac{1}{2}$	$\frac{3}{2}$	$\frac{1}{3}\left[3+2z+2\sqrt{\pi z}\,(2+z)\,e^z\,\mathrm{erf}\,(\sqrt{z})\right]$
82	1	$\frac{5}{2}$	$\frac{1}{2}$	2	$\frac{1}{3z}\left[1-(1-4z-4z^2)\,e^z\right]$
83	1	$\frac{5}{2}$	$\frac{1}{2}$	3	$\frac{2}{3z^2}\left[(3-4z+4z^2)\,e^z-3+z\right]$
84	1	$\frac{5}{2}$	$\frac{3}{2}$	$\frac{3}{2}$	$\frac{1}{3}\left[1+(1+z)\sqrt{\frac{\pi}{z}}\,e^z\,\mathrm{erf}\,(\sqrt{z})\right]$
85	1	$\frac{5}{2}$	$\frac{3}{2}$	2	$\frac{1}{3z}\left[(1+2z)\,e^z-1\right]$
86	1	$\frac{5}{2}$	$\frac{3}{2}$	3	$\frac{2}{3z^2}\left[1-z-(1-2z)\,e^z\right]$
87	1	$\frac{5}{2}$	2	2	$\frac{2}{3z}\left\{e^{z/2}\left[(1+z)I_0\left(\frac{z}{2}\right)+zI_1\left(\frac{z}{2}\right)\right]-1\right\}$
88	1	$\frac{5}{2}$	2	3	$\frac{4}{3z}\left\{e^{z/2}\left[I_0\left(\frac{z}{2}\right)+I_1\left(\frac{z}{2}\right)\right]-1\right\}$
89	1	$\frac{5}{2}$	3	3	$\frac{8}{3z^2}\left[2e^{z/2}I_0\left(\frac{z}{2}\right)-2-z\right]$
90	1	3	$\frac{1}{2}$	2	$\frac{1}{2}\left[2+z+\frac{\sqrt{\pi z}}{2}\,(5+2z)\,e^z\,\mathrm{erf}\,(\sqrt{z})\right]$
91	1	3	$\frac{3}{2}$	2	$\frac{1}{4}\left[1+\frac{3+2z}{2}\sqrt{\frac{\pi}{z}}\,e^z\,\mathrm{erf}\,(\sqrt{z})\right]$
92	1	3	2	2	$\frac{1}{2z}\left[(1+z)\,e^z-1\right]$
93	1	3	2	$\frac{5}{2}$	$\frac{3}{8z}\left[\frac{1+2z}{2}\sqrt{\frac{\pi}{z}}\,e^z\,\mathrm{erf}\,(\sqrt{z})-1\right]$
94	$\frac{3}{2}$	$\frac{3}{2}$	$\frac{1}{2}$	$\frac{1}{2}$	$(1+8z+4z^2)\,e^z$
95	$\frac{3}{2}$	$\frac{3}{2}$	$\frac{1}{2}$	1	$e^{z/2}\left[(1+4z+2z^2)\,I_0\left(\frac{z}{2}\right)+2z\,(1+z)\,I_1\left(\frac{z}{2}\right)\right]$
96	$\frac{3}{2}$	$\frac{3}{2}$	$\frac{1}{2}$	2	$e^{z/2}\left[(1+2z)\,I_0\left(\frac{z}{2}\right)-(1-2z)\,I_1\left(\frac{z}{2}\right)\right]$
97	$\frac{3}{2}$	$\frac{3}{2}$	$\frac{1}{2}$	$\frac{5}{2}$	$\frac{3}{z}\left[\frac{1}{2}\sqrt{\frac{\pi}{z}}\,\mathrm{erfi}\,(\sqrt{z})-(1-z)\,e^z\right]$
98	$\frac{3}{2}$	$\frac{3}{2}$	$\frac{1}{2}$	3	$\frac{4e^{z/2}}{z}\left[zI_0\left(\frac{z}{2}\right)-(3-z)\,I_1\left(\frac{z}{2}\right)\right]$
99	$\frac{3}{2}$	2	$\frac{1}{2}$	$\frac{1}{2}$	$1+6z+2z^2+\sqrt{\pi z}\,(3+7z+2z^2)\,e^z\,\mathrm{erf}\,(\sqrt{z})$

	a_1	a_2	b_1	b_2	${}_2F_2(a_1, a_2; b_1, b_2; z)$
100	$\frac{3}{2}$	2	$\frac{1}{2}$	1	$(1+5z+2z^2)e^z$
101	$\frac{3}{2}$	2	$\frac{1}{2}$	$\frac{5}{2}$	$\frac{3}{2z}\left[\frac{1-z+2z^2}{2}\sqrt{\frac{\pi}{z}}e^z \operatorname{erf}(\sqrt{z})-1+z\right]$
102	$\frac{3}{2}$	2	$\frac{1}{2}$	3	$\frac{2}{z^2}[(3-3z+2z^2)e^z-3]$
103	$\frac{3}{2}$	2	1	1	$\frac{e^{z/2}}{2}\left[(2+5z+2z^2)I_0\left(\frac{z}{2}\right)+z(3+2z)I_1\left(\frac{z}{2}\right)\right]$
104	$\frac{3}{2}$	2	1	$\frac{5}{2}$	$\frac{3}{4z}\left[\frac{1}{2}\sqrt{\frac{\pi}{z}}\operatorname{erfi}(\sqrt{z})-(1-2z)e^z\right]$
105	$\frac{3}{2}$	2	1	3	$\frac{2e^{z/2}}{z}\left[zI_0\left(\frac{z}{2}\right)-(2-z)I_1\left(\frac{z}{2}\right)\right]$
106	$\frac{3}{2}$	2	$\frac{5}{2}$	3	$\frac{6}{z^2}\left[e^z-1-\frac{\sqrt{\pi z}}{2}\operatorname{erfi}(\sqrt{z})\right]$
107	$\frac{3}{2}$	2	3	3	$\frac{8}{z^2}\left\{e^{z/2}\left[(2-z)I_0\left(\frac{z}{2}\right)+zI_1\left(\frac{z}{2}\right)\right]-2\right\}$
108	$\frac{3}{2}$	$\frac{5}{2}$	$\frac{1}{2}$	$\frac{1}{2}$	$\frac{1}{3}(3+42z+44z^2+8z^3)e^z$
109	$\frac{3}{2}$	$\frac{5}{2}$	$\frac{1}{2}$	1	$\frac{e^{z/2}}{3}\left[(3+21z+20z^2+4z^3)I_0\left(\frac{z}{2}\right)+z(7+16z+4z^2)I_1\left(\frac{z}{2}\right)\right]$
110	$\frac{3}{2}$	$\frac{5}{2}$	$\frac{1}{2}$	2	$\frac{e^{z/2}}{3}\left[(3+10z+4z^2)I_0\left(\frac{z}{2}\right)-(1-6z-4z^2)I_1\left(\frac{z}{2}\right)\right]$
111	$\frac{3}{2}$	$\frac{5}{2}$	$\frac{1}{2}$	3	$\frac{4e^{z/2}}{3z}\left[(3-2z+2z^2)I_1\left(\frac{z}{2}\right)+2z^2I_0\left(\frac{z}{2}\right)\right]$
112	$\frac{3}{2}$	3	$\frac{1}{2}$	$\frac{1}{2}$	$\frac{1}{2}\left[2+21z+15z^2+2z^3+\frac{\sqrt{\pi z}}{2}(15+55z+32z^2+4z^3)e^z \operatorname{erf}(\sqrt{z})\right]$
113	$\frac{3}{2}$	3	$\frac{1}{2}$	1	$\frac{1}{2}(2+16z+13z^2+2z^3)e^z$
114	$\frac{3}{2}$	3	$\frac{1}{2}$	2	$\frac{1}{2}(2+7z+2z^2)e^z$
115	$\frac{3}{2}$	3	$\frac{1}{2}$	$\frac{5}{2}$	$\frac{3}{8z}\left[\frac{1-z+8z^2+4z^3}{2}\sqrt{\frac{\pi}{z}}e^z \operatorname{erf}(\sqrt{z})-1+3z+2z^2\right]$
116	$\frac{3}{2}$	3	1	1	$\frac{e^{z/2}}{4}\left[2(2+8z+6z^2+z^3)I_0\left(\frac{z}{2}\right)+z(7+10z+2z^2)I_1\left(\frac{z}{2}\right)\right]$
117	$\frac{3}{2}$	3	1	2	$\frac{e^{z/2}}{4}\left[(4+7z+2z^2)I_0\left(\frac{z}{2}\right)+z(5+2z)I_1\left(\frac{z}{2}\right)\right]$
118	$\frac{3}{2}$	3	1	$\frac{5}{2}$	$\frac{3}{16z}\left[\frac{1}{2}\sqrt{\frac{\pi}{z}}\operatorname{erfi}(\sqrt{z})-(1-6z-4z^2)e^z\right]$
119	$\frac{3}{2}$	3	2	2	$\frac{e^{z/2}}{2}\left[(2+z)I_0\left(\frac{z}{2}\right)+(1+z)I_1\left(\frac{z}{2}\right)\right]$
120	$\frac{3}{2}$	3	2	$\frac{5}{2}$	$\frac{3}{8z}\left[(1+2z)e^z-\frac{1}{2}\sqrt{\frac{\pi}{z}}\operatorname{erfi}(\sqrt{z})\right]$
121	2	2	$\frac{1}{2}$	1	$\frac{1}{2}\left[2+7z+2z^2+\frac{\sqrt{\pi z}}{2}(9+16z+4z^2)e^z \operatorname{erf}(\sqrt{z})\right]$
122	2	2	1	1	$(1+3z+z^2)e^z$

	a_1	a_2	b_1	b_2	${}_2F_2(a_1, a_2; b_1, b_2; z)$
123	2	2	1	$\frac{3}{2}$	$\frac{1}{4}\left[3+2z+\frac{1}{2}(1+8z+4z^2)\sqrt{\frac{\pi}{z}}\,e^z\operatorname{erf}(\sqrt{z})\right]$
124	2	2	1	$\frac{5}{2}$	$\frac{3}{8z}\left[\frac{1+4z^2}{2}\sqrt{\frac{\pi}{z}}\,e^z\operatorname{erf}(\sqrt{z})-1+2z\right]$
125	2	2	1	3	$\frac{2}{z^2}[(1-z+z^2)e^z-1]$
126	2	2	3	3	$\frac{4}{z^2}[e^z-1+\mathbf{C}+\ln(-z)-\operatorname{Ei}(z)]$
127	2	$\frac{5}{2}$	$\frac{1}{2}$	$\frac{1}{2}$	$\frac{1}{3}[3+36z+28z^2+4z^3+2(6+24z+15z^2+2z^3)\sqrt{\pi z}\,e^z\operatorname{erf}(\sqrt{z})]$
128	2	$\frac{5}{2}$	$\frac{1}{2}$	1	$\frac{1}{3}(3+27z+24z^2+4z^3)e^z$
129	2	$\frac{5}{2}$	$\frac{1}{2}$	$\frac{3}{2}$	$\frac{1}{3}[3+8z+2z^2+(6+9z+2z^2)\sqrt{\pi z}\,e^z\operatorname{erf}(\sqrt{z})]$
130	2	$\frac{5}{2}$	$\frac{1}{2}$	3	$\frac{2}{3z^2}[3-(3-3z+4z^2)e^z]$
131	2	$\frac{5}{2}$	1	1	$\frac{e^{z/2}}{6}\left[(6+27z+22z^2+4z^3)I_0\left(\frac{z}{2}\right)+z(11+18z+4z^2)I_1\left(\frac{z}{2}\right)\right]$
132	2	$\frac{5}{2}$	1	$\frac{3}{2}$	$\frac{1}{3}(3+7z+2z^2)e^z$
133	2	$\frac{5}{2}$	1	3	$\frac{2e^{z/2}}{3z}\left[z(1+2z)I_0\left(\frac{z}{2}\right)+(2-z+2z^2)I_1\left(\frac{z}{2}\right)\right]$
134	2	$\frac{5}{2}$	$\frac{3}{2}$	$\frac{3}{2}$	$\frac{1}{3}\left[2+z+\frac{1+5z+2z^2}{2}\sqrt{\frac{\pi}{z}}\,e^z\operatorname{erf}(\sqrt{z})\right]$
135	2	$\frac{5}{2}$	$\frac{3}{2}$	3	$\frac{2}{3z^2}[(1-z+2z^2)e^z-1]$
136	2	$\frac{5}{2}$	3	3	$\frac{8}{3z^2}\left\{2+e^{z/2}\left[zI_1\left(\frac{z}{2}\right)-(2-z)I_0\left(\frac{z}{2}\right)\right]\right\}$
137	2	3	$\frac{1}{2}$	1	$\frac{1}{8}\left[8+51z+32z^2+4z^3+\frac{\sqrt{\pi z}}{2}(45+130z+68z^2+8z^3)e^z\operatorname{erf}(\sqrt{z})\right]$
138	2	3	1	1	$\frac{1}{2}(2+10z+7z^2+z^3)e^z$
139	2	3	1	$\frac{3}{2}$	$\frac{1}{16}\left[13+20z+4z^2+\frac{1}{2}(3+42z+44z^2+8z^3)\sqrt{\frac{\pi}{z}}\,e^z\operatorname{erf}(\sqrt{z})\right]$
140	2	3	1	$\frac{5}{2}$	$\frac{3}{32z}\left[\frac{1}{2}(1+2z+20z^2+8z^3)\sqrt{\frac{\pi}{z}}\,e^z\operatorname{erf}(\sqrt{z})-1+8z+4z^2\right]$
141	$\frac{5}{2}$	$\frac{5}{2}$	$\frac{1}{2}$	$\frac{1}{2}$	$\frac{1}{9}(9+216z+392z^2+160z^3+16z^4)e^z$
142	$\frac{5}{2}$	$\frac{5}{2}$	$\frac{1}{2}$	1	$\frac{e^{z/2}}{9}\left[(9+108z+168z^2+72z^3+8z^4)I_0\left(\frac{z}{2}\right)+4z(6+27z+16z^2+2z^3)I_1\left(\frac{z}{2}\right)\right]$
143	$\frac{5}{2}$	$\frac{5}{2}$	$\frac{1}{2}$	$\frac{3}{2}$	$\frac{1}{9}(9+66z+52z^2+8z^3)e^z$

	a_1	a_2	b_1	b_2	${}_2F_2(a_1, a_2; b_1, b_2; z)$
144	$\frac{5}{2}$	$\frac{5}{2}$	$\frac{1}{2}$	2	$\frac{e^{z/2}}{9}\left[(9+52z+44z^2+8z^3)\, I_0\left(\frac{z}{2}\right)-(1-20z-36z^2-8z^3)\, I_1\left(\frac{z}{2}\right)\right]$
145	$\frac{5}{2}$	$\frac{5}{2}$	$\frac{1}{2}$	3	$\frac{4e^{z/2}}{9z}\left[z(3+8z+4z^2)\, I_0\left(\frac{z}{2}\right)-(3-z-4z^2-4z^3)\, I_1\left(\frac{z}{2}\right)\right]$
146	$\frac{5}{2}$	$\frac{5}{2}$	1	$\frac{3}{2}$	$\frac{e^{z/2}}{9}\left[(9+33z+24z^2+4z^3)\, I_0\left(\frac{z}{2}\right)+z(15+20z+4z^2)\, I_1\left(\frac{z}{2}\right)\right]$
147	$\frac{5}{2}$	$\frac{5}{2}$	$\frac{3}{2}$	2	$\frac{e^{z/2}}{9}\left[(9+14z+4z^2)\, I_0\left(\frac{z}{2}\right)+(1+10z+4z^2)\, I_1\left(\frac{z}{2}\right)\right]$
148	$\frac{5}{2}$	$\frac{5}{2}$	$\frac{3}{2}$	3	$\frac{4e^{z/2}}{9z}\left[2z(1+z)\, I_0\left(\frac{z}{2}\right)+(1+2z^2)\, I_1\left(\frac{z}{2}\right)\right]$
149	$\frac{5}{2}$	3	$\frac{1}{2}$	$\frac{1}{2}$	$\frac{1}{3}\left[3+60z+80z^2+25z^3+2z^4+\frac{1}{2}(30+180z+183z^2+52z^3+4z^4)\sqrt{\pi z}\, e^z \operatorname{erf}(\sqrt{z})\right]$
150	$\frac{5}{2}$	3	$\frac{1}{2}$	1	$\frac{1}{6}(6+84z+123z^2+44z^3+4z^4)\, e^z$
151	$\frac{5}{2}$	3	$\frac{1}{2}$	$\frac{3}{2}$	$\frac{1}{6}\left[6+30z+17z^2+2z^3+\frac{1}{2}(30+75z+36z^2+4z^3)\sqrt{\pi z}\, e^z \operatorname{erf}(\sqrt{z})\right]$
152	$\frac{5}{2}$	3	$\frac{1}{2}$	2	$\frac{1}{6}(6+39z+28z^2+4z^3)\, e^z$
153	$\frac{5}{2}$	3	1	1	$\frac{e^{z/2}}{12}\left[(12+84z+107z^2+40z^3+4z^4)\, I_0\left(\frac{z}{2}\right)+z(25+73z+36z^2+4z^3)\, I_1\left(\frac{z}{2}\right)\right]$
154	$\frac{5}{2}$	3	1	$\frac{3}{2}$	$\frac{1}{6}(6+24z+15z^2+2z^3)\, e^z$
155	$\frac{5}{2}$	3	1	2	$\frac{e^{z/2}}{12}\left[(12+39z+26z^2+4z^3)\, I_0\left(\frac{z}{2}\right)+z(19+22z+4z^2)\, I_1\left(\frac{z}{2}\right)\right]$
156	$\frac{5}{2}$	3	$\frac{3}{2}$	$\frac{3}{2}$	$\frac{1}{12}\left[9+11z+2z^2+\frac{1}{2}(3+27z+24z^2+4z^3)\sqrt{\frac{\pi}{z}}\, e^z \operatorname{erf}(\sqrt{z})\right]$
157	$\frac{5}{2}$	3	$\frac{3}{2}$	2	$\frac{1}{6}(6+9z+2z^2)\, e^z$
158	$\frac{5}{2}$	3	2	2	$\frac{e^{z/2}}{6}\left[2(3+4z+z^2)\, I_0\left(\frac{z}{2}\right)+(1+6z+2z^2)\, I_1\left(\frac{z}{2}\right)\right]$
159	3	3	$\frac{1}{2}$	1	$\frac{1}{32}\left[32+351z+386z^2+108z^3+8z^4+\frac{1}{2}(225+1000z+872z^2+224z^3+16z^4)\sqrt{\pi z}\, e^z \operatorname{erf}(\sqrt{z})\right]$
160	3	3	$\frac{1}{2}$	2	$\frac{1}{16}\left[16+69z+36z^2+4z^3+\frac{1}{2}(75+170z+76z^2+8z^3)\sqrt{\pi z}\, e^z \operatorname{erf}(\sqrt{z})\right]$

	a_1	a_2	b_1	b_2	${}_2F_2(a_1, a_2; b_1, b_2; z)$
161	3	3	1	1	$\frac{1}{4}(4+32z+38z^2+12z^3+z^4)e^z$
162	3	3	1	$\frac{3}{2}$	$\frac{1}{64}\left[55+162z+76z^2+8z^3+ +\frac{1}{2}(9+216z+392z^2+160z^3+16z^4)\sqrt{\frac{\pi}{z}}e^z\operatorname{erf}(\sqrt{z})\right]$
163	3	3	1	2	$\frac{1}{4}(4+14z+8z^2+z^3)e^z$
164	3	3	1	$\frac{5}{2}$	$\frac{3}{128z}\left[\frac{1}{2}(1+8z+104z^2+96z^3+16z^4)\times \times\sqrt{\frac{\pi}{z}}e^z\operatorname{erf}(\sqrt{z})-1+34z+44z^2+8z^3\right]$
165	3	3	$\frac{3}{2}$	2	$\frac{1}{32}\left[23+24z+4z^2+\frac{1}{2}(9+66z+52z^2+8z^3)\sqrt{\frac{\pi}{z}}e^z\operatorname{erf}(\sqrt{z})\right]$
166	3	3	2	2	$\frac{1}{4}(4+5z+z^2)e^z$
167	3	3	2	$\frac{5}{2}$	$\frac{3}{64z}\left[1+12z+4z^2-\frac{1}{2}(1-10z-28z^2-8z^3)\sqrt{\frac{\pi}{z}}e^z\operatorname{erf}(\sqrt{z})\right]$

7.12.3. Representations of ${}_3F_3(a_1, a_2, a_3; b_1, b_2, b_3; -z)$.

1. $${}_3F_3\left(\begin{matrix}1, 1, a; -z\\ 2, 2, a+1\end{matrix}\right)=\frac{a}{(a-1)z}\left[\mathbf{C}+\ln z-\frac{1}{a-1}-\operatorname{Ei}(-z)-z^{1-a}\gamma(a-1, z)\right].$$

7.12.4. Representations of ${}_qF_q((a_q); (b_q); z)$.

1. $${}_qF_q\left(\begin{matrix}a_1, \ldots, a_q; z\\ a_1+1, \ldots, a_q+1\end{matrix}\right)=$$
$$=\sum_{k=1}^{q}{}_1F_1(a_k; a_k+1; z)\prod_{\substack{l=1\\ l\neq k}}^{q}\frac{a_l}{a_l-a_k}=\sum_{k=1}^{q}a_k(-z)^{-a_k}\gamma(a_k, -z)\prod_{\substack{l=1\\ l\neq k}}^{q}\frac{a_l}{a_l-a_k}.$$

2. $${}_qF_q\left(\begin{matrix}a, \ldots, a; z\\ a-1, \ldots, a-1\end{matrix}\right)=(a-1)^{-q}\left[\frac{d^q}{dt^q}e^{at-t+ze^t}\right]_{t=0}=(a-1)^{-q}e^zR_q(z).$$

The polynomial $R_q(z)$ satisfies the recurrence relation
$$R_{q+1}(z)=(z+a-1)R_q(z)+z\frac{d}{dz}R_q(z), \quad R_0(z)=1,$$
$$R_1(z)=z+a-1, \quad R_2(z)=(z+a-1)^2+z, \quad R_3(z)=(z+a-1)^3+3z(z+a-1)+z.$$

3. $${}_qF_q\left(\begin{matrix}2, \ldots, 2; z\\ 1, \ldots, 1\end{matrix}\right)=I_q,$$
$$I_q=\frac{d}{dz}\left(z\frac{d}{dz}\right)^q e^z=\sum_{k=0}^{q}\frac{(-1)^k}{(k+1)!}\sum_{l=0}^{k}(-1)^l\binom{k+1}{l+1}(l+1)^{q+1}z^ke^z,$$
$$I_0=e^z, \quad I_1=(z+1)e^z, \quad I_2=(z^2+3z+1)e^z, \quad I_3=(z^3+6z^2+7z+1)e^z,$$
$$I_4=(z^4+10z^3+25z^2+15z+1)e^z, \quad I_5=(z^5+15z^4+65z^3+90z^2+31z+1)e^z,$$
$$I_6=(z^6+21z^5+140z^4+350z^3+301z^2+63z+1)e^z,$$
$$I_7=(z^7+28z^6+266z^5+1050z^4+1701z^3+966z^2+127z+1)e^z.$$

7.13. THE FUNCTION ${}_0F_1(b; z)$

7.13.1. Representations and particular values of ${}_0F_1(b; \pm z)$.

	b	${}_0F_1(b; -z)$	${}_0F_1(b; z)$
1	b	$\Gamma(b)\, z^{(1-b)/2} J_{b-1}(2\sqrt{z})$	$\Gamma(b)\, z^{(1-b)/2} I_{b-1}(2\sqrt{z})$
2	$-\frac{5}{2}$	$\left(1-\frac{8}{5}z\right)\cos 2\sqrt{z}+$ $+2\left(1-\frac{4}{15}z\right)\sqrt{z}\sin 2\sqrt{z}$	$\left(1+\frac{8}{5}z\right)\operatorname{ch} 2\sqrt{z}-$ $-2\left(1+\frac{4}{15}z\right)\sqrt{z}\operatorname{sh} 2\sqrt{z}$
3	$-\frac{3}{2}$	$\left(1-\frac{4}{3}z\right)\cos 2\sqrt{z}+$ $+2\sqrt{z}\sin 2\sqrt{z}$	$\left(1+\frac{4}{3}z\right)\operatorname{ch} 2\sqrt{z}-$ $-2\sqrt{z}\operatorname{sh} 2\sqrt{z}$
4	$-\frac{1}{2}$	$\cos 2\sqrt{z}+2\sqrt{z}\sin 2\sqrt{z}$	$\operatorname{ch} 2\sqrt{z}-2\sqrt{z}\operatorname{sh} 2\sqrt{z}$
5	$\frac{1}{2}$	$\cos 2\sqrt{z}$	$\operatorname{ch} 2\sqrt{z}$
6	$\frac{3}{2}$	$\dfrac{\sin 2\sqrt{z}}{2\sqrt{z}}$	$\dfrac{\operatorname{sh} 2\sqrt{z}}{2\sqrt{z}}$
7	$\frac{5}{2}$	$\frac{3}{8}z^{-3/2}(\sin 2\sqrt{z}-2\sqrt{z}\cos 2\sqrt{z})$	$\frac{3}{8}z^{-3/2}(2\sqrt{z}\operatorname{ch} 2\sqrt{z}-\operatorname{sh} 2\sqrt{z})$
8	$\frac{7}{2}$	$\frac{45}{32}z^{-5/2}\left(1-\frac{4}{3}z\right)\sin 2\sqrt{z}-$ $-\frac{45}{16z^2}\cos 2\sqrt{z}$	$\frac{45}{32}z^{-5/2}\left(1+\frac{4}{3}z\right)\operatorname{sh} 2\sqrt{z}-$ $-\frac{45}{16z^2}\operatorname{ch} 2\sqrt{z}$
9	$\frac{9}{2}$	$\frac{105}{64z^3}\left[\frac{3}{2\sqrt{z}}(5-8z)\sin 2\sqrt{z}-\right.$ $\left.-(15-4z)\cos 2\sqrt{z}\right]$	$\frac{105}{64z^3}\left[(15+4z)\operatorname{ch} 2\sqrt{z}-\right.$ $\left.-\frac{3}{2\sqrt{z}}(5+8z)\operatorname{sh} 2\sqrt{z}\right]$

Other functions of the form ${}_0F_1(\frac{1}{2}\pm n; z)$ are also elementary and are expressible in terms of the above functions by means of the recurrence formulae

10. ${}_0F_1(b+2; z)=\dfrac{b(b+1)}{z}[{}_0F_1(b; z)-{}_0F_1(b+1; z)]$,

11. ${}_0F_1(b-1; z)={}_0F_1(b; z)+\dfrac{z}{b(b-1)}\,{}_0F_1(b+1; z)$.

For $b\neq\frac{1}{2}\pm n$, the function ${}_0F_1(b;z)$ is not expressible in terms of elementary functions.

12. ${}_0F_1(b; z)+\Gamma\begin{bmatrix}b-1\\1-b\end{bmatrix} z^{1-b}\,{}_0F_1(2-b; z)=\dfrac{2}{\Gamma(1-b)}\,z^{(1-b)/2}K_{1-b}(2\sqrt{z})$,

13. ${}_0F_1(b;-z)-\cos b\pi\,\Gamma\begin{bmatrix}b-1\\1-b\end{bmatrix} z^{1-b}\,{}_0F_1(2-b;-z)=-\dfrac{\pi}{\Gamma(1-b)}\,z^{(1-b)/2}Y_{1-b}(2\sqrt{z})$,

14. ${}_0F_1(b; -z)-e^{\pm ib\pi}\,\Gamma\begin{bmatrix}b-1\\1-b\end{bmatrix} z^{1-b}\,{}_0F_1(2-b; -z)=$

$$=\pm\frac{\pi i}{\Gamma(1-b)}\,z^{(1-b)/2}H^{((3\mp 1)/2)}_{1-b}(2\sqrt{z}).$$

7.14. THE FUNCTION ${}_1F_2\,(a;\ b_1,\ b_2;\ z)$

7.14.1. Representations of ${}_1F_2\,(a;\ b_1,\ b_2;\ z)$.

	a	b_1	b_2	${}_1F_2\,(a;\ b_1,\ b_2;\ z^2)$
1	a	$a-1$	b	$\Gamma(b)\,z^{1-b}\left[I_{b-1}(2z)+\dfrac{z}{2(a-1)}I_b(2z)\right]$
2	a	$a-\frac{1}{2}$	$2a-2$	$\dfrac{2}{a-1}\Gamma^2\left(a-\dfrac{1}{2}\right)\left(\dfrac{z}{2}\right)^{4-2a}I_{a-3/2}(z)\,I_{a-5/2}(z)-\dfrac{a-2}{a-1}\Gamma^2\left(a-\dfrac{1}{2}\right)\left(\dfrac{z}{2}\right)^{3-2a}I^2_{a-3/2}(z)$
3	a	$a-\frac{1}{2}$	$2a-1$	$\Gamma^2\left(a-\dfrac{1}{2}\right)\left(\dfrac{z}{2}\right)^{3-2a}\left[I^2_{a-3/2}(z)+I^2_{a-1/2}(z)\right]$
4	a	$a-\frac{1}{2}$	$2a$	$2\Gamma\left(a-\dfrac{1}{2}\right)\Gamma\left(a+\dfrac{1}{2}\right)\left(\dfrac{z}{2}\right)^{2-2a}I_{a-1/2}(z)\,I_{a-3/2}(z)-\Gamma^2\left(a+\dfrac{1}{2}\right)\left(\dfrac{z}{2}\right)^{1-2a}I^2_{a-1/2}(z)$
5	a	$a-\frac{1}{2}$	$2a+1$	$\Gamma^2\left(a+\dfrac{1}{2}\right)\left[\left(\dfrac{z}{2}\right)^{1-2a}I^2_{a-1/2}(z)+\dfrac{z^2}{2(2a-1)}\left(\dfrac{z}{2}\right)^{-2a}I_{a+1/2}(z)\,I_{a-1/2}(z)\right]$
6	a	$a+\frac{1}{2}$	$2a-1$	$\Gamma\left(a-\dfrac{1}{2}\right)\Gamma\left(a+\dfrac{1}{2}\right)\left(\dfrac{z}{2}\right)^{2-2a}I_{a-3/2}(z)\,I_{a-1/2}(z)$
7	a	$a+\frac{1}{2}$	$2a$	$\Gamma^2\left(a+\dfrac{1}{2}\right)\left(\dfrac{z}{2}\right)^{1-2a}I^2_{a-1/2}(z)$
8	a	$a+\frac{1}{2}$	$2a+1$	$\Gamma^2\left(a+\dfrac{1}{2}\right)\left(\dfrac{z}{2}\right)^{1-2a}\left[I^2_{a-1/2}(z)-I^2_{a+1/2}(z)\right]$
9	$\frac{1}{2}$	b	$2-b$	$\dfrac{\pi(1-b)}{\sin b\pi}I_{1-b}(z)\,I_{b-1}(z)$
10	$\frac{3}{2}$	b	$3-b$	$\dfrac{\pi(b-1)(b-2)}{\sin b\pi}\left[I_{1-b}(z)\,I_{b-1}(z)+I_{2-b}(z)\,I_{b-2}(z)\right]$
11	1	$\frac{3}{2}$	b	$\dfrac{\sqrt{\pi}}{2}\Gamma(b)\,z^{1/2-b}\,\mathbf{L}_{b-3/2}(2z)$
12	1	2	b	$\dfrac{b-1}{z^2}\left[\Gamma(b-1)\,z^{2-b}I_{b-2}(2z)-1\right]$
13	$-n$	$\frac{1}{2}-n$	$-2n$	$2^{2n-1}\dfrac{\pi(n!)^2}{[(2n)!]^2}z^{2n+1}\left[I^2_{-n-1/2}(z)-I^2_{n+1/2}(z)\right]$

4. $${}_1F_2\left(a;\ a-\frac{1}{2},\ 2a-1;\ z\right)+\frac{1}{2a-1}\,{}_1F_2\left(a-1;\ a+\frac{1}{2},\ 2a-1;\ z\right)=$$
$$=\Gamma^2\left(a-\frac{1}{2}\right)\left(\frac{\sqrt{z}}{2}\right)^{2-2a}\left[\sqrt{z}\,I^2_{a-3/2}(\sqrt{z})+(a-1)\,I_{a-3/2}(\sqrt{z})\,I_{a-1/2}(\sqrt{z})\right].$$

5. $${}_1F_2\left(a;\ a-\frac{1}{2},\ 2a-1;\ z\right)+{}_1F_2\left(a-1;\ a-\frac{1}{2},\ 2a-1;\ z\right)=$$
$$=2^{2a-2}\,\Gamma^2\left(a-\frac{1}{2}\right)z^{3/2-a}\,I^2_{a-3/2}(\sqrt{z}).$$

6. $${}_1F_2\left(a;\ a-\frac{1}{2},\ 2a-1;\ z\right)-\frac{2a-3}{2a-1}\,{}_1F_2\left(a-1;\ a-\frac{3}{2},\ 2a-1;\ z\right)=$$
$$=\frac{2^{2a-2}}{2a-1}\Gamma^2\left(a-\frac{1}{2}\right)z^{3/2-a}\,I^2_{a-3/2}(\sqrt{z}).$$

17. $${}_1F_2\left(a;\, a-\frac{1}{2},\, 2a-3;\, z\right)+\frac{1}{2a-2}\,{}_1F_2\left(a-2;\, a-\frac{1}{2},\, 2a-3;\, z\right)=$$
$$=\frac{2a-3}{a-1}2^{2a-5}\Gamma^2\left(a-\frac{3}{2}\right)z^{2-a}\left[\sqrt{z}\,I_{a-5/2}^2(\sqrt{z})-\left(a-\frac{5}{2}\right)I_{a-5/2}(\sqrt{z})\times\right.$$
$$\left.\times I_{a-3/2}(\sqrt{z})\right].$$

18. $${}_1F_2\left(a;\, a-\frac{1}{2},\, 2a-3;\, z\right)+\frac{1}{a-1}\,{}_1F_2\left(a-2;\, a-\frac{1}{2},\, 2a-4;\, z\right)=$$
$$=\frac{\Gamma(a-3/2)\,\Gamma(a-1/2)}{a-1}\left(\frac{\sqrt{z}}{2}\right)^{4-2a}\left[\sqrt{z}\,I_{a-5/2}^2(\sqrt{z})+\right.$$
$$\left.+(3-a)\,I_{a-5/2}(\sqrt{z})\,I_{a-3/2}(\sqrt{z})\right].$$

19. $${}_1F_2\left(\frac{3}{2};\, b,\, 3-b;\, z\right)+(2b-3)\,{}_1F_2\left(\frac{1}{2};\, b,\, 3-b;\, z\right)=$$
$$=\frac{2\pi}{\sin b\pi}(1-b)(2-b)\,I_{1-b}(\sqrt{z})\,I_{b-1}(\sqrt{z}).$$

20. $${}_1F_2\left(\frac{5}{2};\, b,\, 4-b;\, z\right)-\frac{(2b-3)(2b-5)}{3}\,{}_1F_2\left(\frac{1}{2};\, b,\, 4-b;\, z\right)=$$
$$=\frac{4\pi(1-b)(2-b)(3-b)}{3\sin b\pi}\,I_{2-b}(\sqrt{z})\,I_{b-2}(\sqrt{z}).$$

7.14.2. Particular values of ${}_1F_2(a;\, b_1,\, b_2;\, z)$.

NOTATION: $U(z)=\frac{1}{z}\int_0^z I_0(2t)\,dt=I_0(2z)+\frac{\pi}{2}[I_0(2z)\mathbf{L}_1(2z)-I_1(2z)\mathbf{L}_0(2z)]$.

	a	b_1	b_2	${}_1F_2(a;\, b_1,\, b_2;\, z^2)$
1	$-\frac{1}{2}$	$\frac{1}{2}$	$\frac{1}{2}$	$\operatorname{ch} 2z-2z\operatorname{shi}(2z)$
2	$-\frac{1}{2}$	$\frac{1}{2}$	1	$I_0(2z)-4z^2U(z)+2zI_1(2z)$
3	$-\frac{1}{2}$	$\frac{1}{2}$	$\frac{3}{2}$	$\frac{1}{2}\left[\operatorname{ch} 2z+\frac{\operatorname{sh} 2z}{2z}-2z\operatorname{shi}(2z)\right]$
4	$-\frac{1}{2}$	$\frac{1}{2}$	2	$\frac{1}{3}\left[2I_0(2z)+(1+4z^2)\frac{I_1(2z)}{z}-8z^2U(z)\right]$
5	$-\frac{1}{2}$	$\frac{1}{2}$	$\frac{5}{2}$	$\frac{3}{16z^2}\left[(1+2z^2)\operatorname{ch} 2z-(1-2z^2)\frac{\operatorname{sh} 2z}{2z}-4z^3\operatorname{shi}(2z)\right]$
6	$-\frac{1}{2}$	$\frac{1}{2}$	3	$\frac{2}{15z^2}\left[(3+4z^2)I_0(2z)-(3-2z^2-8z^4)\frac{I_1(2z)}{z}-16z^4U(z)\right]$
7	$-\frac{1}{2}$	$\frac{1}{2}$	$\frac{7}{2}$	$\frac{5}{32z^4}\left[(3+3z^2+2z^4)\frac{\operatorname{sh} 2z}{2z}-(3-z^2-2z^4)\operatorname{ch} 2z-4z^5\operatorname{shi}(2z)\right]$
8	$-\frac{1}{2}$	$\frac{1}{2}$	4	$\frac{2}{35z^4}\left[(30+9z^2+4z^4+16z^6)\frac{I_1(2z)}{z}-2(15-3z^2-4z^4)I_0(2z)-32z^6U(z)\right]$
9	$-\frac{1}{2}$	1	1	$(1-2z^2)I_0^2(z)+2z^2I_1^2(z)+2zI_0(z)I_1(z)$
10	$-\frac{1}{2}$	1	$\frac{3}{2}$	$\frac{1}{2}[I_0(2z)+2zI_1(2z)+(1-4z^2)U(z)]$
11	$-\frac{1}{2}$	1	2	$\frac{1}{3}[(3-4z^2)I_0^2(z)-(1-4z^2)I_1^2(z)+4zI_0(z)I_1(z)]$
12	$-\frac{1}{2}$	1	$\frac{5}{2}$	$\frac{3}{32z^2}[(1+4z^2)I_0(2z)-2z(1-4z^2)I_1(2z)-(1-4z^2)^2U(z)]$

	a	b_1	b_2	${}_1F_2(a;\, b_1,\, b_2;\, z^2)$
13	$-\frac{1}{2}$	1	3	$\frac{4}{15}\left[4(1-z^2)\,I_0^2(z)+(1-2z^2+4z^4)\,\frac{I_1^2(z)}{z^2}-(1-4z^2)\,\frac{I_0(z)\,I_1(z)}{z}\right]$
14	$-\frac{1}{2}$	1	$\frac{7}{2}$	$\frac{5}{256z^4}\,[(9-12z^2+48z^4-64z^6)\,U(z)-(9-16z^4)\,I_0(2z)+2z\,(9-8z^2+16z^4)\,I_1(2z)]$
15	$-\frac{1}{2}$	1	4	$\frac{4}{35z^4}\,[4z\,(2-z^2+2z^4)\,I_0(z)\,I_1(z)-2z^2(1-5z^2+4z^4)\,I_0^2(z)-(8-3z^2+6z^4-8z^6)\,I_1^2(z)]$
16	$-\frac{1}{2}$	$\frac{3}{2}$	$\frac{3}{2}$	$\frac{1}{4}\left[\operatorname{ch} 2z+\frac{\operatorname{sh} 2z}{2z}+(1-2z^2)\,\frac{\operatorname{shi}(2z)}{z}\right]$
17	$-\frac{1}{2}$	$\frac{3}{2}$	2	$\frac{1}{3}\left[I_0(2z)-(1-2z^2)\,\frac{I_1(2z)}{z}+(3-4z^2)\,U(z)\right]$
18	$-\frac{1}{2}$	$\frac{3}{2}$	$\frac{5}{2}$	$\frac{3}{32z^2}\left[(1+2z^2)\,\frac{\operatorname{sh} 2z}{2z}-(1-2z^2)\operatorname{ch} 2z+4z\,(1-z^2)\operatorname{shi}(2z)\right]$
19	$-\frac{1}{2}$	$\frac{3}{2}$	3	$\frac{2}{15z^2}\left[(1-4z^2+4z^4)\,\frac{I_1(2z)}{z}-(1-2z^2)\,I_0(2z)+2z^2(5-4z^2)\,U(z)\right]$
20	$-\frac{1}{2}$	$\frac{3}{2}$	$\frac{7}{2}$	$\frac{5}{128z^4}\left[(3-4z^2+4z^4)\operatorname{ch} 2z-(3-4z^4)\,\frac{\operatorname{sh} 2z}{2z}+4z^3(3-2z^2)\operatorname{shi}(2z)\right]$
21	$-\frac{1}{2}$	$\frac{3}{2}$	4	$\frac{2}{35z^4}\left[2(3-2z^2+2z^4)\,I_0(2z)-(6-z^2+12z^4-8z^6)\,\frac{I_1(2z)}{z}+4z^4(7-4z^2)\,U(z)\right]$
22	$-\frac{1}{2}$	2	2	$\frac{2}{9}\left[2(3-2z^2)\,I_0^2(z)-4(1-z^2)\,I_1^2(z)-(3-4z^2)\,\frac{I_0(z)\,I_1(z)}{z}\right]$
23	$-\frac{1}{2}$	2	$\frac{5}{2}$	$\frac{1}{16z^2}\,[(3+24z^2-16z^4)\,U(z)-(3-4z^2)\,I_0(2z)-2z\,(5-4z^2)\,I_1(2z)]$
24	$-\frac{1}{2}$	2	3	$\frac{4}{45}\left[2(9-4z^2)\,I_0^2(z)-(3+14z^2-8z^4)\,\frac{I_1^2(z)}{z^2}-4(3-2z^2)\,\frac{I_0(z)\,I_1(z)}{z}\right]$
25	$-\frac{1}{2}$	2	$\frac{7}{2}$	$\frac{5}{384z^4}\,[(9-24z^2+16z^4)\,I_0(2z)-(9-36z^2-144z^4+64z^6)\,U(z)-2z\,(9+32z^2-16z^4)\,I_1(2z)]$
26	$-\frac{1}{2}$	2	4	$\frac{4}{105z^4}\,[z^2(3+48z^2-16z^4)\,I_0^2(z)+(12-15z^2-40z^4+16z^6)\,I_1^2(z)-4z\,(3+9z^2-4z^4)\,I_0(z)\,I_1(z)]$
27	$-\frac{1}{2}$	$\frac{5}{2}$	$\frac{5}{2}$	$\frac{9}{128z^2}\left[2(1+4z^2-2z^4)\,\frac{\operatorname{shi}(2z)}{z}-(3-2z^2)\operatorname{ch} 2z-(1-2z^2)\,\frac{\operatorname{sh} 2z}{2z}\right]$
28	$-\frac{1}{2}$	$\frac{5}{2}$	3	$\frac{1}{20z^2}\left[(15+40z^2-16z^4)\,U(z)-(7-4z^2)\,I_0(2z)-2(4+9z^2-4z^4)\,\frac{I_1(2z)}{z}\right]$
29	$-\frac{1}{2}$	$\frac{5}{2}$	$\frac{7}{2}$	$\frac{15}{256z^4}\left[(3-3z^2+2z^4)\,\frac{\operatorname{sh} 2z}{2z}-(3+5z^2-2z^4)\operatorname{ch} 2z+2z\,(3+6z^2-2z^4)\operatorname{shi}(2z)\right]$

	a	b_1 b_2	${}_1F_2\,(a;\ b_1,\ b_2;\ z^2)$
30	$-\frac{1}{2}$	$\frac{5}{2}$ 4	$\frac{3}{70z^4}\Big[2\,(4-10z^2-13z^4+4z^6)\,\frac{I_1\,(2z)}{z}-$ $-(8+11z^2-4z^4)\,I_0\,(2z)+z^2\,(35+56z^2-16z^4)\,U\,(z)\Big]$
31	$-\frac{1}{2}$	3 3	$\frac{8}{225z^2}\Big[(15+56z^2-16z^4)\,I_0^2\,(z)-(31+48z^2-16z^4)\,I_1^2(z)-$ $-2\,(15+22z^2-8z^4)\,\frac{I_0\,(z)\,I_1\,(z)}{z}\Big]$
32	$-\frac{1}{2}$	3 $\frac{7}{2}$	$\frac{1}{96z^4}\Big[(45+180z^2+240z^4-64z^6)\,U\,(z)-(45+48z^2-16z^4)\,I_0\,(2z)-$ $-2z\,(51+56z^2-16z^4)\,I_1\,(2z)\Big]$
33	$-\frac{1}{2}$	3 4	$\frac{16}{525z^4}\Big[z^2\,(45+76z^2-16z^4)\,I_0^2\,(z)-(30+71z^2+$ $+68z^4-16z^6)\,I_1^2\,(z)-z\,(75+64z^2-16z^4)\,I_0\,(z)\,I_1\,(z)\Big]$
34	$-\frac{1}{2}$	$\frac{7}{2}$ $\frac{7}{2}$	$\frac{25}{1024z^4}\Big[2\,(9+18z^2+18z^4-4z^6)\,\frac{\operatorname{shi}\,(2z)}{z}-$ $-(15+12z^2-4z^4)\,\frac{\operatorname{sh}2z}{2z}-(21+16z^2-4z^4)\operatorname{ch}2z\Big]$
35	$-\frac{1}{2}$	$\frac{7}{2}$ 4	$\frac{1}{112z^4}\Big[(315+420z^2+336z^4-64z^6)U(z)-(123+72z^2-16z^4)I_0(2z)-$ $-2\,(96+117z^2+80z^4-16z^6)\,\frac{I_1\,(2z)}{z}\Big]$
36	$-\frac{1}{2}$	4 4	$\frac{16}{1225z^4}\Big[2\,(105+122z^2+104z^4-16z^6)\,I_0^2\,(z)+$ $+(389+324z^2+192z^4-32z^6)\,I_1^2\,(z)-$ $-2\,(210+173z^2+92z^4-16z^6)\,\frac{I_0\,(z)\,I_1\,(z)}{z}\Big]$
37	$\frac{1}{4}$	$\frac{1}{2}$ $\frac{5}{4}$	$\frac{1}{4}\sqrt{\frac{\pi}{2z}}\,[\operatorname{erf}\,(\sqrt{2z})+\operatorname{erfi}\,(\sqrt{2z})]$
38	$\frac{1}{2}$	1 1	$I_0^2\,(z)$
39	$\frac{1}{2}$	1 $\frac{3}{2}$	$U\,(z)$
40	$\frac{1}{2}$	1 2	$I_0^2\,(z)-I_1^2\,(z)$
41	$\frac{1}{2}$	1 $\frac{5}{2}$	$\frac{3}{8z^2}\,[I_0\,(2z)-2zI_1\,(2z)-(1-4z^2)\,U\,(z)]$
42	$\frac{1}{2}$	1 3	$\frac{4}{3}\Big[I_0^2\,(z)-\frac{I_0\,(z)\,I_1\,(z)}{z}+(1-z^2)\,\frac{I_1^2(z)}{z^2}\Big]$
43	$\frac{1}{2}$	1 $\frac{7}{2}$	$\frac{15}{128z^4}\,[(9-8z^2+16z^4)\,U\,(z)-(9+4z^2)\,I_0\,(2z)+$ $+2z\,(9-4z^2)\,I_1\,(2z)]$
44	$\frac{1}{2}$	1 4	$\frac{4}{5z^4}\,[2z\,(4-z^2)\,I_0\,(z)\,I_1\,(z)-2z^2\,(1-z^2)\,I_0^2\,(z)-$ $-(8-z^2+2z^4)\,I_1^2\,(z)]$

	a	b_1	b_2	${}_1F_2(a;\, b_1,\, b_2;\, z^2)$
45	$\frac{1}{2}$	$\frac{3}{2}$	$\frac{3}{2}$	$\frac{\operatorname{shi}(2z)}{2z}$
46	$\frac{1}{2}$	$\frac{3}{2}$	2	$2U(z)-\frac{I_1(2z)}{z}$
47	$\frac{1}{2}$	$\frac{3}{2}$	$\frac{5}{2}$	$\frac{3}{8z^2}\left[2z\operatorname{shi}(2z)-\operatorname{ch}2z+\frac{\operatorname{sh}2z}{2z}\right]$
48	$\frac{1}{2}$	$\frac{3}{2}$	3	$\frac{2}{3z^2}\left[4z^2U(z)-I_0(2z)+(1-2z^2)\frac{I_1(2z)}{z}\right]$
49	$\frac{1}{2}$	$\frac{3}{2}$	$\frac{7}{2}$	$\frac{15}{64z^4}\left[(3-2z^2)\operatorname{ch}2z-(3+2z^2)\frac{\operatorname{sh}2z}{2z}+4z^3\operatorname{shi}(2z)\right]$
50	$\frac{1}{2}$	$\frac{3}{2}$	4	$\frac{2}{5z^4}\left[2(3-z^2)I_0(2z)-(6+z^2+4z^4)\frac{I_1(2z)}{z}+8z^4U(z)\right]$
51	$\frac{1}{2}$	2	2	$2I_0^2(z)-I_1^2(z)-\frac{I_0(z)I_1(z)}{z}$
52	$\frac{1}{2}$	2	$\frac{5}{2}$	$\frac{3}{4z^2}[(1+4z^2)U(z)-I_0(2z)-2zI_1(2z)]$
53	$\frac{1}{2}$	2	3	$\frac{4}{3}\left[2I_0^2(z)-\frac{2}{z}I_0(z)I_1(z)-(1+2z^2)\frac{I_1^2(z)}{z^2}\right]$
54	$\frac{1}{2}$	2	$\frac{7}{2}$	$\frac{15}{64z^4}[(3-4z^2)I_0(2z)-2z(3+4z^2)I_1(2z)-(3-8z^2-16z^4)U(z)]$
55	$\frac{1}{2}$	2	4	$\frac{4}{5z^4}[z^2(1+4z^2)I_0^2(z)+(4-3z^2-4z^4)I_1^2(z)-4z(1+z^2)I_0(z)I_1(z)]$
56	$\frac{1}{2}$	$\frac{5}{2}$	$\frac{5}{2}$	$\frac{9}{16z^2}\left[(1+2z^2)\frac{\operatorname{shi}(2z)}{z}-\frac{\operatorname{sh}2z}{2z}-\operatorname{ch}2z\right]$
57	$\frac{1}{2}$	$\frac{5}{2}$	3	$\frac{1}{z^2}\left[(3+4z^2)U(z)-I_0(2z)-2(1+z^2)\frac{I_1(2z)}{z}\right]$
58	$\frac{1}{2}$	$\frac{5}{2}$	$\frac{7}{2}$	$\frac{45}{128z^4}\left[(3-2z^2)\frac{\operatorname{sh}2z}{2z}-(3+2z^2)\operatorname{ch}2z+4z(1+z^2)\operatorname{shi}(2z)\right]$
59	$\frac{1}{2}$	$\frac{5}{2}$	4	$\frac{6}{5z^4}\left[(2-3z^2-2z^4)\frac{I_1(2z)}{z}-(2+z^2)I_0(2z)+z^2(5+4z^2)U(z)\right]$
60	$\frac{1}{2}$	3	3	$\frac{8}{9z^2}\left[(3+4z^2)I_0^2(z)-(5+4z^2)I_1^2(z)-2(3+2z^2)\frac{I_0(z)I_1(z)}{z}\right]$
61	$\frac{1}{2}$	3	$\frac{7}{2}$	$\frac{5}{16z^4}[(9+24z^2+16z^4)U(z)-(9+4z^2)I_0(2z)-2z(7+4z^2)I_1(2z)]$
62	$\frac{1}{2}$	3	4	$\frac{16}{15z^4}[2z^2(3+2z^2)I_0^2(z)-2(3+4z^2+2z^4)I_1^2(z)-$ $-z(9+4z^2)I_0(z)I_1(z)]$
63	$\frac{1}{2}$	$\frac{7}{2}$	$\frac{7}{2}$	$\frac{225}{512z^4}\left[2(3+4z^2+2z^4)\frac{\operatorname{shi}(2z)}{z}-(5+2z^2)\operatorname{ch}2z-(7+2z^2)\frac{\operatorname{sh}2z}{2z}\right]$
64	$\frac{1}{2}$	$\frac{7}{2}$	4	$\frac{3}{8z^4}\Big[(45+40z^2+16z^4)U(z)-(13+4z^2)I_0(2z)-$ $-2(16+11z^2+4z^4)\frac{I_1(2z)}{z}\Big]$
65	$\frac{1}{2}$	4	4	$\frac{16}{25z^4}\Big[2(15+11z^2+4z^4)I_0^2(z)-4(15+7z^2+2z^4)\frac{I_0(z)I_1(z)}{z}-$ $-(47+26z^2+8z^4)I_1^2(z)\Big]$

	a	b_1 b_2	${}_1F_2(a;\ b_1,\ b_2;\ z^2)$
66	$\frac{3}{4}$	$\frac{3}{2}$ $\frac{7}{4}$	$\frac{3}{8z}\sqrt{\frac{\pi}{2z}}\left[\operatorname{erfi}(\sqrt{2z})-\operatorname{erf}(\sqrt{2z})\right]$
67	1	$\frac{1}{4}$ $\frac{3}{4}$	$1+\sqrt{\frac{\pi z}{2}}\left[e^{2z}\operatorname{erf}(\sqrt{2z})-e^{-2z}\operatorname{erfi}(\sqrt{2z})\right]$
68	1	$\frac{1}{2}$ $\frac{1}{2}$	$1+\pi z\mathbf{L}_0(2z)$
69	1	$\frac{1}{2}$ $\frac{3}{2}$	$1+\frac{\pi}{2}\mathbf{L}_1(2z)$
70	1	$\frac{1}{2}$ 2	$\frac{1}{2z^2}[1-\operatorname{ch}2z+2z\operatorname{sh}2z]$
71	1	$\frac{1}{2}$ $\frac{5}{2}$	$\frac{3\pi}{4z^2}[z\mathbf{L}_0(2z)-\mathbf{L}_1(2z)]$
72	1	$\frac{1}{2}$ 3	$\frac{1}{2z^4}[(3+4z^2)\operatorname{ch}2z-6z\operatorname{sh}2z-3+2z^2]$
73	1	$\frac{1}{2}$ $\frac{7}{2}$	$\frac{5}{8z^4}[8z^2-6\pi z\mathbf{L}_0(2z)+3\pi(2+z^2)\mathbf{L}_1(2z)]$
74	1	$\frac{1}{2}$ 4	$\frac{3}{4z^6}[2z(15+4z^2)\operatorname{sh}2z-3(5+8z^2)\operatorname{ch}2z+15-6z^2+2z^4]$
75	1	$\frac{3}{4}$ $\frac{5}{4}$	$\frac{1}{4}\sqrt{\frac{\pi}{2z}}\left[e^{2z}\operatorname{erf}(\sqrt{2z})+e^{-2z}\operatorname{erfi}(\sqrt{2z})\right]$
76	1	$\frac{5}{4}$ $\frac{7}{4}$	$\frac{3}{16z}\sqrt{\frac{\pi}{2z}}\left[e^{2z}\operatorname{erf}(\sqrt{2z})-e^{-2z}\operatorname{erfi}(\sqrt{2z})\right]$
77	1	$\frac{3}{2}$ $\frac{3}{2}$	$\frac{\pi}{4z}\mathbf{L}_0(2z)$
78	1	$\frac{3}{2}$ 2	$\frac{1}{2z^2}[\operatorname{ch}2z-1]$
79	1	$\frac{3}{2}$ $\frac{5}{2}$	$\frac{3\pi}{8z^2}\mathbf{L}_1(2z)$
80	1	$\frac{3}{2}$ 3	$\frac{1}{2z^4}(1-2z^2-\operatorname{ch}2z+2z\operatorname{sh}2z)$
81	1	$\frac{3}{2}$ $\frac{7}{2}$	$\frac{5}{16z^4}[3\pi z\mathbf{L}_0(2z)-3\pi\mathbf{L}_1(2z)-4z^2]$
82	1	$\frac{3}{2}$ 4	$\frac{3}{4z^6}[(3+4z^2)\operatorname{ch}2z-6z\operatorname{sh}2z-3+2z^2-2z^4]$
83	1	$\frac{7}{4}$ $\frac{9}{4}$	$\frac{15}{64z^{5/2}}\sqrt{\frac{\pi}{2}}\left[e^{2z}\operatorname{erf}(\sqrt{2z})+e^{-2z}\operatorname{erfi}(\sqrt{2z})\right]-\frac{15}{16z^2}$
84	1	2 2	$\frac{1}{z^2}[I_0(2z)-1]$
85	1	2 $\frac{5}{2}$	$\frac{3}{2z^2}\left[\frac{\operatorname{sh}2z}{2z}-1\right]$
86	1	2 3	$\frac{2}{z^2}\left[\frac{I_1(2z)}{z}-1\right]$
87	1	2 $\frac{7}{2}$	$\frac{5}{8z^4}\left[3\operatorname{ch}2z-3\frac{\operatorname{sh}2z}{2z}-4z^2\right]$
88	1	2 4	$\frac{3}{z^4}\left[2I_0(2z)-\frac{2}{z}I_1(2z)-z^2\right]$
89	1	$\frac{5}{2}$ $\frac{5}{2}$	$\frac{9}{4z^2}\left[\frac{\pi}{4z}\mathbf{L}_0(2z)-1\right]$

	a	b_1 b_2	${}_1F_2(a;\ b_1,\ b_2;\ z^2)$
90	1	$\frac{5}{2}$ 3	$\frac{3}{2z^4}[\operatorname{ch} 2z-1-2z^2]$
91	1	$\frac{5}{2}$ $\frac{7}{2}$	$\frac{15}{32z^4}[3\pi\mathbf{L}_1(2z)-8z^2]$
92	1	$\frac{5}{2}$ 4	$\frac{9}{4z^6}[2z\operatorname{sh} 2z-\operatorname{ch} 2z+1-2z^2-2z^4]$
93	1	3 3	$\frac{4}{z^4}[I_0(2z)-1-z^2]$
94	1	3 $\frac{7}{2}$	$\frac{5}{2z^4}\left[3\frac{\operatorname{sh} 2z}{2z}-3-2z^2\right]$
95	1	3 4	$\frac{6}{z^4}\left[\frac{2}{z}I_1(2z)-2-z^2\right]$
96	1	$\frac{7}{2}$ $\frac{7}{2}$	$\frac{25}{16z^4}\left[\frac{9\pi}{4z}\mathbf{L}_0(2z)-9-4z^2\right]$
97	1	$\frac{7}{2}$ 4	$\frac{15}{4z^6}[3\operatorname{ch} 2z-3-6z^2-2z^4]$
98	1	4 4	$\frac{9}{z^6}[4I_0(2z)-4-4z^2-z^4]$
99	$\frac{3}{2}$	$\frac{1}{2}$ $\frac{1}{2}$	$\operatorname{ch} 2z+2z\operatorname{sh} 2z$
100	$\frac{3}{2}$	$\frac{1}{2}$ 1	$I_0(2z)+2zI_1(2z)$
101	$\frac{3}{2}$	$\frac{1}{2}$ 2	$2I_0(2z)-\frac{1}{z}I_1(2z)$
102	$\frac{3}{2}$	$\frac{1}{2}$ $\frac{5}{2}$	$\frac{3}{2z^2}\left[(1+2z^2)\frac{\operatorname{sh} 2z}{2z}-\operatorname{ch} 2z\right]$
103	$\frac{3}{2}$	$\frac{1}{2}$ 3	$\frac{2}{z^2}\left[(3+2z^2)\frac{I_1(2z)}{z}-3I_0(2z)\right]$
104	$\frac{3}{2}$	$\frac{1}{2}$ $\frac{7}{2}$	$\frac{15}{4z^4}\left[(3+z^2)\operatorname{ch} 2z-(3+5z^2)\frac{\operatorname{sh} 2z}{2z}\right]$
105	$\frac{3}{2}$	$\frac{1}{2}$ 4	$\frac{6}{z^4}\left[2(5+z^2)I_0(2z)-(10+7z^2)\frac{I_1(2z)}{z}\right]$
106	$\frac{3}{2}$	1 1	$I_0(z)[I_0(z)+2zI_1(z)]$
107	$\frac{3}{2}$	1 2	$I_0^2(z)+I_1^2(z)$
108	$\frac{3}{2}$	1 $\frac{5}{2}$	$\frac{3}{4z^2}[2zI_1(2z)-I_0(2z)+U(z)]$
109	$\frac{3}{2}$	1 3	$4\frac{I_1(z)}{z}\left[I_0(z)-\frac{I_1(z)}{z}\right]$
110	$\frac{3}{2}$	1 $\frac{7}{2}$	$\frac{15}{32z^4}[(9+8z^2)I_0(2z)-(9-4z^2)U(z)-18zI_1(2z)]$
111	$\frac{3}{2}$	1 4	$\frac{4}{z^4}[2z^2I_0^2(z)+(8+z^2)I_1^2(z)-8zI_0(z)I_1(z)]$
112	$\frac{3}{2}$	2 2	$\frac{2}{z}I_0(z)I_1(z)$
113	$\frac{3}{2}$	2 $\frac{5}{2}$	$\frac{3}{2z^2}[I_0(2z)-U(z)]$
114	$\frac{3}{2}$	2 3	$\frac{4}{z^2}I_1^2(z)$

	a	b_1	b_2	${}_1F_2(a;\ b_1,\ b_2;\ z^2)$
115	$\frac{3}{2}$	2	$\frac{7}{2}$	$\frac{15}{16z^4}[(3-4z^2)U(z)-3I_0(2z)+6zI_1(2z)]$
116	$\frac{3}{2}$	2	4	$\frac{4}{z^4}[4zI_0(z)I_1(z)-z^2I_0^2(z)-(4-z^2)I_1^2(z)]$
117	$\frac{3}{2}$	$\frac{5}{2}$	$\frac{5}{2}$	$\frac{9}{8z^3}[\mathrm{sh}\,2z-\mathrm{shi}\,(2z)]$
118	$\frac{3}{2}$	$\frac{5}{2}$	3	$\frac{6}{z^2}\left[\frac{I_1(2z)}{z}-U(z)\right]$
119	$\frac{3}{2}$	$\frac{5}{2}$	$\frac{7}{2}$	$\frac{45}{32z^4}\left[3\,\mathrm{ch}\,2z-3\frac{\mathrm{sh}\,2z}{2z}-2z\,\mathrm{shi}\,(2z)\right]$
120	$\frac{3}{2}$	$\frac{5}{2}$	4	$\frac{6}{z^4}\left[2I_0(2z)-(2-z^2)\frac{I_1(2z)}{z}-2z^2U(z)\right]$
121	$\frac{3}{2}$	3	3	$\frac{8}{z^2}\left[I_1^2(z)-I_0^2(z)+\frac{2}{z}I_0(z)I_1(z)\right]$
122	$\frac{3}{2}$	3	$\frac{7}{2}$	$\frac{15}{4z^4}[3I_0(2z)-(3+4z^2)U(z)+2zI_1(2z)]$
123	$\frac{3}{2}$	3	4	$\frac{16}{z^4}[zI_0(z)I_1(z)-z^2I_0^2(z)+(2+z^2)I_1^2(z)]$
124	$\frac{3}{2}$	$\frac{7}{2}$	$\frac{7}{2}$	$\frac{225}{64z^4}\left[\mathrm{ch}\,2z+5\frac{\mathrm{sh}\,2z}{2z}-(3+2z^2)\frac{\mathrm{shi}\,(2z)}{z}\right]$
125	$\frac{3}{2}$	$\frac{7}{2}$	4	$\frac{15}{2z^4}\left[I_0(2z)-(9+4z^2)U(z)+2(4+z^2)\frac{I_1(2z)}{z}\right]$
126	$\frac{3}{2}$	4	4	$\frac{16}{z^4}\left[2(6+z^2)\frac{I_0(z)I_1(z)}{z}-2(3+z^2)I_0^2(z)+(7+2z^2)I_1^2(z)\right]$
127	2	$\frac{1}{2}$	$\frac{1}{2}$	$1+2z^2+\pi z^2\mathbf{L}_1(2z)+\frac{3\pi}{2}z\mathbf{L}_0(2z)$
128	2	$\frac{1}{2}$	1	$\mathrm{ch}\,2z+z\,\mathrm{sh}\,2z$
129	2	$\frac{1}{2}$	$\frac{3}{2}$	$\frac{1}{2}\left[2+\frac{\pi}{2}\mathbf{L}_1(2z)+\pi z\mathbf{L}_0(2z)\right]$
130	2	$\frac{1}{2}$	$\frac{5}{2}$	$\frac{3}{8z^2}[\pi(1+2z^2)\mathbf{L}_1(2z)-\pi z\mathbf{L}_0(2z)+4z^2]$
131	2	$\frac{1}{2}$	3	$\frac{1}{2z^4}[3-3(1+2z^2)\,\mathrm{ch}\,2z+2z(3+2z^2)\,\mathrm{sh}\,2z]$
132	2	$\frac{1}{2}$	$\frac{7}{2}$	$\frac{15}{16z^4}[2\pi z(3+z^2)\mathbf{L}_0(2z)-\pi(6+5z^2)\mathbf{L}_1(2z)-8z^2]$
133	2	$\frac{1}{2}$	4	$\frac{3}{2z^6}[(15+27z^2+4z^4)\,\mathrm{ch}\,2z-2z(15+7z^2)\,\mathrm{sh}\,2z-15+3z^2]$
134	2	1	1	$I_0(2z)+zI_1(2z)$
135	2	1	$\frac{3}{2}$	$\frac{1}{2}\left[\frac{\mathrm{sh}\,2z}{2z}+\mathrm{ch}\,2z\right]$
136	2	1	$\frac{5}{2}$	$\frac{3}{8z^2}\left[(1+4z^2)\frac{\mathrm{sh}\,2z}{2z}-\mathrm{ch}\,2z\right]$
137	2	1	3	$\frac{2}{z^2}\left[(1+z^2)\frac{I_1(2z)}{z}-I_0(2z)\right]=\frac{2}{z}I_2'(2z)$
138	2	1	$\frac{7}{2}$	$\frac{15}{32z^4}\left[(9+4z^2)\,\mathrm{ch}\,2z-(9+16z^2)\frac{\mathrm{sh}\,2z}{2z}\right]$
139	2	1	4	$\frac{6}{z^4}\left[(4+z^2)I_0(2z)-(4+3z^2)\frac{I_1(2z)}{z}\right]$

	a	b_1 b_2	${}_1F_2(a;\ b_1,\ b_2;\ z^2)$
140	2	$\frac{3}{2}$ $\frac{3}{2}$	$\frac{1}{2}\left[1+\frac{\pi}{2}\mathbf{L}_1(2z)+\frac{\pi}{4z}\mathbf{L}_0(2z)\right]$
141	2	$\frac{3}{2}$ $\frac{5}{2}$	$\frac{3\pi}{8z^2}\left[z\mathbf{L}_0(2z)-\frac{1}{2}\mathbf{L}_1(2z)\right]$
142	2	$\frac{3}{2}$ 3	$\frac{1}{2z^4}[(1+2z^2)\operatorname{ch}2z-1-2z\operatorname{sh}2z]$
143	2	$\frac{3}{2}$ $\frac{7}{2}$	$\frac{15}{32z^4}[4z^2-3\pi z\mathbf{L}_0(2z)+\pi(3+2z^2)\mathbf{L}_1(2z)]$
144	2	$\frac{3}{2}$ 4	$\frac{3}{2z^6}[2z(3+z^2)\operatorname{sh}2z-(3+5z^2)\operatorname{ch}2z+3-z^2]$
145	2	$\frac{5}{2}$ $\frac{5}{2}$	$\frac{9}{8z^2}\left[1+\frac{\pi}{2}\mathbf{L}_1(2z)-\frac{\pi}{4z}\mathbf{L}_0(2z)\right]$
146	2	$\frac{5}{2}$ 3	$\frac{3}{2z^4}[1-\operatorname{ch}2z+z\operatorname{sh}2z]$
147	2	$\frac{5}{2}$ $\frac{7}{2}$	$\frac{45\pi}{64z^4}[2z\mathbf{L}_0(2z)-3\mathbf{L}_1(2z)]$
148	2	$\frac{5}{2}$ 4	$\frac{9}{2z^6}[(1+z^2)\operatorname{ch}2z-2z\operatorname{sh}2z-1+z^2]$
149	2	3 3	$\frac{4}{z^4}[1-I_0(2z)+zI_1(2z)]$
150	2	3 $\frac{7}{2}$	$\frac{15}{4z^4}\left[2-3\frac{\operatorname{sh}2z}{2z}+\operatorname{ch}2z\right]$
151	2	3 4	$\frac{12}{z^4}\left[1+I_0(2z)-\frac{2}{z}I_1(2z)\right]$
152	2	$\frac{7}{2}$ $\frac{7}{2}$	$\frac{225}{32z^4}\left[3+\frac{\pi}{2}\mathbf{L}_1(2z)-\frac{3\pi}{4z}\mathbf{L}_0(2z)\right]$
153	2	$\frac{7}{2}$ 4	$\frac{45}{4z^6}[z\operatorname{sh}2z-2\operatorname{ch}2z+2+2z^2]$
154	2	4 4	$\frac{36}{z^6}[zI_1(2z)-2I_0(2z)+2+z^2]$
155	$\frac{5}{2}$	$\frac{1}{2}$ $\frac{1}{2}$	$\frac{1}{3}[(3+4z^2)\operatorname{ch}2z+10z\operatorname{sh}2z]$
156	$\frac{5}{2}$	$\frac{1}{2}$ 1	$\frac{1}{3}[(3+4z^2)I_0(2z)+8zI_1(2z)]$
157	$\frac{5}{2}$	$\frac{1}{2}$ $\frac{3}{2}$	$\operatorname{ch}2z+\frac{2z}{3}\operatorname{sh}2z$
158	$\frac{5}{2}$	$\frac{1}{2}$ 2	$\frac{1}{3}\left[4I_0(2z)-(1-4z^2)\frac{I_1(2z)}{z}\right]$
159	$\frac{5}{2}$	$\frac{1}{2}$ 3	$\frac{2(3+4z^2)}{3z^2}\left[I_0(2z)-\frac{I_1(2z)}{z}\right]$
160	$\frac{5}{2}$	$\frac{1}{2}$ $\frac{7}{2}$	$\frac{5}{2z^4}\left[(3+6z^2+2z^4)\frac{\operatorname{sh}2z}{2z}-(3+2z^2)\operatorname{ch}2z\right]$
161	$\frac{5}{2}$	$\frac{1}{2}$ 4	$\frac{2}{z^4}\left[(30+27z^2+4z^4)\frac{I_1(2z)}{z}-6(5+2z^2)I_0(2z)\right]$
162	$\frac{5}{2}$	1 1	$\frac{1}{3}\left[(3+2z^2)I_0^2(z)+2z^2I_1^2(z)+8zI_0(z)I_1(z)\right]$
163	$\frac{5}{2}$	1 $\frac{3}{2}$	$I_0(2z)+\frac{2z}{3}I_1(2z)$
164	$\frac{5}{2}$	1 2	$\frac{1}{3}\left[3I_0^2(z)+I_1^2(z)+4zI_0(z)I_1(z)\right]$

	a	b_1	b_2	${}_1F_2(a;\ b_1,\ b_2;\ z^2)$
165	$\frac{5}{2}$	1	3	$\frac{4}{3}\left[I_0^2(z)-\frac{1}{z}I_0(z)I_1(z)+(1+z^2)\frac{I_1^2(z)}{z^2}\right]$
166	$\frac{5}{2}$	1	$\frac{7}{2}$	$\frac{5}{16z^4}[9U(z^2)-3(3+4z^2)I_0(2z)+2z(9+4z^2)I_1(2z)]$
167	$\frac{5}{2}$	1	4	$\frac{4}{z^4}[2z(4+z^2)I_0(z)I_1(z)-2z^2I_0^2(z)-(8+3z^2)I_1^2(z)]$
168	$\frac{5}{2}$	$\frac{3}{2}$	$\frac{3}{2}$	$\frac{1}{3z}[\operatorname{sh}2z+z\operatorname{ch}2z]$
169	$\frac{5}{2}$	$\frac{3}{2}$	2	$\frac{1}{3}\left[\frac{I_1(2z)}{z}+2I_0(2z)\right]$
170	$\frac{5}{2}$	$\frac{3}{2}$	3	$\frac{2}{3z^2}\left[(1+2z^2)\frac{I_1(2z)}{z}-I_0(2z)\right]$
171	$\frac{5}{2}$	$\frac{3}{2}$	$\frac{7}{2}$	$\frac{5}{8z^4}\left[(3+2z^2)\operatorname{ch}2z-3(1+2z^2)\frac{\operatorname{sh}2z}{2z}\right]$
172	$\frac{5}{2}$	$\frac{3}{2}$	4	$\frac{2}{z^4}\left[2(3+z^2)I_0(2z)-(6+5z^2)\frac{I_1(2z)}{z}\right]$
173	$\frac{5}{2}$	2	2	$\frac{2}{3}\left[I_0^2(z)+I_1^2(z)+\frac{1}{z}I_0(z)I_1(z)\right]$
174	$\frac{5}{2}$	2	3	$\frac{4}{3z^2}I_1(z)[2zI_0(z)-I_1(z)]$
175	$\frac{5}{2}$	2	$\frac{7}{2}$	$\frac{5}{8z^4}[(3+4z^2)I_0(2z)-3U(z)-6zI_1(2z)]$
176	$\frac{5}{2}$	2	4	$\frac{4}{z^4}[(4+z^2)I_1^2(z)+z^2I_0^2(z)-4zI_0(z)I_1(z)]$
177	$\frac{5}{2}$	3	3	$\frac{8}{3z^2}\left[I_1^2(z)+I_0^2(z)-\frac{2}{z}I_0(z)I_1(z)\right]$
178	$\frac{5}{2}$	3	$\frac{7}{2}$	$\frac{5}{2z^4}[3U(z)-3I_0(2z)+2zI_1(2z)]$
179	$\frac{5}{2}$	3	4	$\frac{16}{z^4}I_1(z)[zI_0(z)-2I_1(z)]$
180	$\frac{5}{2}$	$\frac{7}{2}$	$\frac{7}{2}$	$\frac{75}{16z^4}\left[3\frac{\operatorname{shi}(2z)}{2z}-2\frac{\operatorname{sh}2z}{z}+\operatorname{ch}2z\right]$
181	$\frac{5}{2}$	$\frac{7}{2}$	4	$\frac{15}{z^4}\left[3U(z)+I_0(2z)-4\frac{I_1(2z)}{z}\right]$
182	$\frac{5}{2}$	4	4	$\frac{48}{z^4}\left[2I_0^2(z)-\frac{4}{z}I_0(z)I_1(z)-I_1^2(z)\right]$
183	3	$\frac{1}{2}$	$\frac{1}{2}$	$\frac{1}{8}[8+36z^2+16\pi z^2\mathbf{L}_1(2z)+\pi z(15+4z^2)\mathbf{L}_0(2z)]$
184	3	$\frac{1}{2}$	1	$\frac{1}{4}[2(2+z^2)\operatorname{ch}2z+7z\operatorname{sh}2z]$
185	3	$\frac{1}{2}$	$\frac{3}{2}$	$\frac{1}{16}[16+8z^2+12\pi z\mathbf{L}_0(2z)+\pi(3+4z^2)\mathbf{L}_1(2z)]$
186	3	$\frac{1}{2}$	2	$\operatorname{ch}2z+\frac{z}{2}\operatorname{sh}2z$
187	3	$\frac{1}{2}$	$\frac{5}{2}$	$\frac{3}{32z^2}[\pi(1+4z^2)\mathbf{L}_1(2z)-\pi z(1-4z^2)\mathbf{L}_0(2z)+12z^2]$
188	3	$\frac{1}{2}$	$\frac{7}{2}$	$\frac{15}{64z^4}[8z^2+8z^4-2\pi z(3+2z^2)\mathbf{L}_0(2z)+\pi(6+7z^2+4z^4)\mathbf{L}_1(2z)]$

	a	b_1	b_2	${}_1F_2(a;\ b_1,\ b_2;\ z^2)$
189	3	$\frac{1}{2}$	4	$\frac{3}{4z^6}\left[2z(15+10z^2+2z^4)\operatorname{sh}2z-5(3+6z^2+2z^4)\operatorname{ch}2z+15\right]$
190	3	1	1	$\frac{1}{2}\left[(2+z^2)I_0(2z)+3zI_1(2z)\right]$
191	3	1	$\frac{3}{2}$	$\frac{1}{16}\left[10\operatorname{ch}2z+(3+4z^2)\frac{\operatorname{sh}2z}{z}\right]$
192	3	1	2	$I_0(2z)+\frac{z}{2}I_1(2z)$
193	3	1	$\frac{5}{2}$	$\frac{3}{32z^2}\left[(1+8z^2)\frac{\operatorname{sh}2z}{2z}-(1-4z^2)\operatorname{ch}2z\right]$
194	3	1	$\frac{7}{2}$	$\frac{15}{128z^4}\left[(9+20z^2+16z^4)\frac{\operatorname{sh}2z}{2z}-(9+8z^2)\operatorname{ch}2z\right]$
195	3	1	4	$\frac{3}{z^4}\left[(4+4z^2+z^4)\frac{I_1(2z)}{z}-2(2+z^2)I_0(2z)\right]$
196	3	$\frac{3}{2}$	$\frac{3}{2}$	$\frac{1}{8}\left[5+\frac{\pi}{4z}(3+4z^2)\mathbf{L}_0(2z)+2\pi\mathbf{L}_1(2z)\right]$
197	3	$\frac{3}{2}$	2	$\frac{1}{4}\left[3\frac{\operatorname{sh}2z}{2z}+\operatorname{ch}2z\right]$
198	3	$\frac{3}{2}$	$\frac{5}{2}$	$\frac{3}{64z^2}\left[4\pi z\mathbf{L}_0(2z)+8z^2-\pi(1-4z^2)\mathbf{L}_1(2z)\right]$
199	3	$\frac{3}{2}$	$\frac{7}{2}$	$\frac{15}{128z^4}\left[\pi z(3+4z^2)\mathbf{L}_0(2z)-\pi(3+4z^2)\mathbf{L}_1(2z)-4z^2\right]$
200	3	$\frac{3}{2}$	4	$\frac{3}{4z^6}\left[(3+6z^2+2z^4)\operatorname{ch}2z-2z(3+2z^2)\operatorname{sh}2z-3\right]$
201	3	2	2	$\frac{1}{2}\left[I_0(2z)+\frac{I_1(2z)}{z}\right]$
202	3	2	$\frac{5}{2}$	$\frac{3}{16z^2}\left[\operatorname{ch}2z-(1-4z^2)\frac{\operatorname{sh}2z}{2z}\right]$
203	3	2	$\frac{7}{2}$	$\frac{15}{64z^4}\left[(3+4z^2)\operatorname{ch}2z-(3+8z^2)\frac{\operatorname{sh}2z}{2z}\right]$
204	3	2	4	$\frac{3}{z^4}\left[(2+z^2)I_0(2z)-2(1+z^2)\frac{I_1(2z)}{z}\right]$
205	3	$\frac{5}{2}$	$\frac{5}{2}$	$\frac{9}{32z^2}\left[1-\frac{\pi}{4z}(1-4z^2)\mathbf{L}_0(2z)\right]$
206	3	$\frac{5}{2}$	$\frac{7}{2}$	$\frac{45}{256z^4}\left[8z^2-4\pi z\mathbf{L}_0(2z)+\pi(3+4z^2)\mathbf{L}_1(2z)\right]$
207	3	$\frac{5}{2}$	4	$\frac{9}{4z^6}\left[z(2+z^2)\operatorname{sh}2z-(1+2z^2)\operatorname{ch}2z+1\right]$
208	3	$\frac{7}{2}$	$\frac{7}{2}$	$\frac{225}{128z^4}\left[\frac{\pi}{4z}(3+4z^2)\mathbf{L}_0(2z)-2\pi\mathbf{L}_1(2z)-3\right]$
209	3	$\frac{7}{2}$	4	$\frac{45}{16z^6}\left[2(2+z^2)\operatorname{ch}2z-5z\operatorname{sh}2z-4\right]$
210	3	4	4	$\frac{18}{z^6}\left[(2+z^2)I_0(2z)-3zI_1(2z)-2\right]$
211	$\frac{7}{2}$	$\frac{1}{2}$	$\frac{1}{2}$	$\frac{1}{15}\left[3(5+16z^2)\operatorname{ch}2z+2z(33+4z^2)\operatorname{sh}2z\right]$
212	$\frac{7}{2}$	$\frac{1}{2}$	1	$\frac{1}{15}\left[(15+44z^2)I_0(2z)+2z(23+4z^2)I_1(2z)\right]$
213	$\frac{7}{2}$	$\frac{1}{2}$	$\frac{3}{2}$	$\frac{1}{15}\left[(15+4z^2)\operatorname{ch}2z+18z\operatorname{sh}2z\right]$

	a	b_1	b_2	${}_1F_2(a;\ b_1,\ b_2;\ z^2)$
214	$\frac{7}{2}$	$\frac{1}{2}$	2	$\frac{1}{15}\left[2(9+4z^2)I_0(2z)-(3-28z^2)\frac{I_1(2z)}{z}\right]$
215	$\frac{7}{2}$	$\frac{1}{2}$	$\frac{5}{2}$	$\operatorname{ch}2z+\frac{2z}{5}\operatorname{sh}2z$
216	$\frac{7}{2}$	$\frac{1}{2}$	3	$\frac{2}{15z^2}\left[3(1+4z^2)I_0(2z)-(3+6z^2-8z^4)\frac{I_1(2z)}{z}\right]$
217	$\frac{7}{2}$	$\frac{1}{2}$	4	$\frac{2}{5z^4}\left[2(15+9z^2+4z^4)I_0(2z)-3(10+11z^2+4z^4)\frac{I_1(2z)}{z}\right]$
218	$\frac{7}{2}$	1	1	$\frac{1}{15}[(15+22z^2)I_0^2(z)+18z^2I_1^2(z)+2z(23+4z^2)I_0(z)I_1(z)]$
219	$\frac{7}{2}$	1	$\frac{3}{2}$	$\frac{1}{15}[(15+4z^2)I_0(2z)+16zI_1(2z)]$
220	$\frac{7}{2}$	1	2	$\frac{1}{15}[(15+4z^2)I_0^2(z)+(3+4z^2)I_1^2(z)+28zI_0(z)I_1(z)]$
221	$\frac{7}{2}$	1	$\frac{5}{2}$	$I_0(2z)+\frac{2z}{5}I_1(2z)$
222	$\frac{7}{2}$	1	3	$\frac{4}{15}\left[4I_0^2(z)+(1+2z^2)\frac{I_1^2(z)}{z^2}-(1-4z^2)\frac{I_0(z)I_1(z)}{z}\right]$
223	$\frac{7}{2}$	1	4	$\frac{4}{5z^4}[(8+5z^2+2z^4)I_1^2(z)+2z^2(1+z^2)I_0^2(z)-4z(2+z^2)I_0(z)I_1(z)$
224	$\frac{7}{2}$	$\frac{3}{2}$	$\frac{3}{2}$	$\frac{1}{15}\left[7\operatorname{ch}2z+2(2+z^2)\frac{\operatorname{sh}2z}{z}\right]$
225	$\frac{7}{2}$	$\frac{3}{2}$	2	$\frac{1}{15}\left[12I_0(2z)+(3+4z^2)\frac{I_1(2z)}{z}\right]$
226	$\frac{7}{2}$	$\frac{3}{2}$	$\frac{5}{2}$	$\frac{1}{5}\left[2\frac{\operatorname{sh}2z}{z}+\operatorname{ch}2z\right]$
227	$\frac{7}{2}$	$\frac{3}{2}$	3	$\frac{2}{15z^2}\left[(1+4z^2)\frac{I_1(2z)}{z}-(1-4z^2)I_0(2z)\right]$
228	$\frac{7}{2}$	$\frac{3}{2}$	4	$\frac{2}{5z^4}\left[(6+7z^2+4z^4)\frac{I_1(2z)}{z}-2(3+2z^2)I_0(2z)\right]$
229	$\frac{7}{2}$	2	2	$\frac{2}{15}\left[6I_0^2(z)+4I_1^2(z)+(3+4z^2)\frac{I_0(z)I_1(z)}{z}\right]$
230	$\frac{7}{2}$	2	$\frac{5}{2}$	$\frac{1}{5}\left[2I_0(2z)+3\frac{I_1(2z)}{z}\right]$
231	$\frac{7}{2}$	2	3	$\frac{4}{15z^2}[4zI_0(z)I_1(z)+2z^2I_0^2(z)-(1-2z^2)I_1^2(z)]$
232	$\frac{7}{2}$	2	4	$\frac{4}{5z^4}[4z(1+z^2)I_0(z)I_1(z)-z^2I_0^2(z)-(4+3z^2)I_1^2(z)]$
233	$\frac{7}{2}$	$\frac{5}{2}$	$\frac{5}{2}$	$\frac{3}{10z^2}\left[\operatorname{ch}2z-(1-2z^2)\frac{\operatorname{sh}2z}{2z}\right]$
234	$\frac{7}{2}$	$\frac{5}{2}$	3	$\frac{2}{5z^2}\left[I_0(2z)-(1-2z^2)\frac{I_1(2z)}{z}\right]$
235	$\frac{7}{2}$	$\frac{5}{2}$	4	$\frac{6}{5z^4}\left[2(1+z^2)I_0(2z)-(2+3z^2)\frac{I_1(2z)}{z}\right]$
236	$\frac{7}{2}$	3	3	$\frac{8}{15z^2}\left[I_0^2(z)-I_1^2(z)-2(1-2z^2)\frac{I_0(z)I_1(z)}{z}\right]$
237	$\frac{7}{2}$	3	4	$\frac{16}{5z^4}[z^2I_0^2(z)+(2+z^2)I_1^2(z)-3zI_0(z)I_1(z)]$

	a	b_1	b_2	${}_1F_2(a;\ b_1,\ b_2;\ z^2)$
238	$\frac{7}{2}$	4	4	$\frac{48}{5z^4}\left[2(2+z^2)\frac{I_0(z)I_1(z)}{z}-2I_0^2(z)-3I_1^2(z)\right]$
239	4	$\frac{1}{2}$	$\frac{1}{2}$	$\frac{1}{24}\left[24+174z^2+8z^4+\pi z^2(71+4z^2)\mathbf{L}_1(2z)+\frac{\pi z}{2}(105+68z^2)\mathbf{L}_0(2z)\right]$
240	4	$\frac{1}{2}$	1	$\frac{1}{24}[6(4+5z^2)\operatorname{ch}2z+z(57+4z^2)\operatorname{sh}2z]$
241	4	$\frac{1}{2}$	$\frac{3}{2}$	$\frac{1}{96}[96+112z^2+\pi(15+52z^2)\mathbf{L}_1(2z)+2\pi z(45+4z^2)\mathbf{L}_0(2z)]$
242	4	$\frac{1}{2}$	2	$\frac{1}{12}[2(6+z^2)\operatorname{ch}2z+11z\operatorname{sh}2z]$
243	4	$\frac{1}{2}$	$\frac{5}{2}$	$\frac{1}{64z^2}[\pi(3+18z^2+8z^4)\mathbf{L}_1(2z)-3\pi z(1-12z^2)\mathbf{L}_0(2z)+68z^2+16z^4]$
244	4	$\frac{1}{2}$	3	$\operatorname{ch}2z+\frac{z}{3}\operatorname{sh}2z$
245	4	$\frac{1}{2}$	$\frac{7}{2}$	$\frac{5}{128z^4}[8z^2+32z^4-2\pi z(3+3z^2-4z^4)\mathbf{L}_0(2z)+$ $+3\pi(2+3z^2+4z^4)\mathbf{L}_1(2z)]$
246	4	1	1	$\frac{1}{6}[(6+7z^2)I_0(2z)+z(11+z^2)I_1(2z)]$
247	4	1	$\frac{3}{2}$	$\frac{1}{96}\left[2(33+4z^2)\operatorname{ch}2z+3(5+16z^2)\frac{\operatorname{sh}2z}{z}\right]$
248	4	1	2	$\frac{1}{6}[(6+z^2)I_0(2z)+5zI_1(2z)]$
249	4	1	$\frac{5}{2}$	$\frac{1}{64z^2}\left[(3+36z^2+16z^4)\frac{\operatorname{sh}2z}{2z}-(3-32z^2)\operatorname{ch}2z\right]$
250	4	1	3	$I_0(2z)+\frac{z}{3}I_1(2z)$
251	4	1	$\frac{7}{2}$	$\frac{5}{256z^4}\left[3(3+8z^2+16z^4)\frac{\operatorname{sh}2z}{2z}-(9+12z^2-16z^4)\operatorname{ch}2z\right]$
252	4	$\frac{3}{2}$	$\frac{3}{2}$	$\frac{1}{96}\left[66+8z^2+\pi(23+4z^2)\mathbf{L}_1(2z)+\frac{\pi}{2z}(15+44z^2)\mathbf{L}_0(2z)\right]$
253	4	$\frac{3}{2}$	2	$\frac{1}{24}\left[(15+4z^2)\frac{\operatorname{sh}2z}{2z}+9\operatorname{ch}2z\right]$
254	4	$\frac{3}{2}$	$\frac{5}{2}$	$\frac{1}{128z^2}[2\pi(9+4z^2)z\mathbf{L}_0(2z)-\pi(3-28z^2)\mathbf{L}_1(2z)+64z^2]$
255	4	$\frac{3}{2}$	3	$\frac{1}{6}\left[5\frac{\operatorname{sh}2z}{2z}+\operatorname{ch}2z\right]$
256	4	$\frac{3}{2}$	$\frac{7}{2}$	$\frac{5}{256z^4}[3\pi z(1+4z^2)\mathbf{L}_0(2z)-\pi(3+6z^2-8z^4)\mathbf{L}_1(2z)-4z^2+16z^4]$
257	4	2	2	$\frac{1}{6}\left[4I_0(2z)+(2+z^2)\frac{I_1(2z)}{z}\right]$
258	4	2	$\frac{5}{2}$	$\frac{1}{32z^2}\left[(3+4z^2)\operatorname{ch}2z-3(1-8z^2)\frac{\operatorname{sh}2z}{2z}\right]$
259	4	2	3	$\frac{1}{3}\left[\frac{2}{z}I_1(2z)+I_0(2z)\right]$
260	4	2	$\frac{7}{2}$	$\frac{5}{128z^4}\left[(3+8z^2)\operatorname{ch}2z-(3+12z^2-16z^4)\frac{\operatorname{sh}2z}{2z}\right]$
261	4	$\frac{5}{2}$	$\frac{5}{2}$	$\frac{3}{128z^2}\left[6+8z^2-\frac{\pi}{2z}(3-20z^2)\mathbf{L}_0(2z)-\pi(1-4z^2)\mathbf{L}_1(2z)\right]$

	a	b_1	b_2	${}_1F_2(a;\ b_1,\ b_2;\ z^2)$
262	4	$\frac{5}{2}$	3	$\frac{1}{8z^2}\left[3\operatorname{ch}2z-(3-4z^2)\frac{\operatorname{sh}2z}{2z}\right]$
263	4	$\frac{5}{2}$	$\frac{7}{2}$	$\frac{15}{512z^4}[16z^2-2\pi z(3-4z^2)\mathbf{L}_0(2z)+\pi(3+4z^2)\mathbf{L}_1(2z)]$
264	4	3	3	$\frac{2}{3z^2}\left[I_0(2z)-(1-z^2)\frac{I_1(2z)}{z}\right]$
265	4	3	$\frac{7}{2}$	$\frac{5}{32z^4}\left[3\frac{\operatorname{sh}2z}{2z}-(3-4z^2)\operatorname{ch}2z\right]$
266	4	$\frac{7}{2}$	$\frac{7}{2}$	$\frac{75}{512z^4}\left[\frac{\pi}{2z}(3-4z^2)\mathbf{L}_0(2z)-\pi(1-4z^2)\mathbf{L}_1(2z)-6+8z^2\right]$

7.14.3. Representations of ${}_1F_2(a;\ b_1,\ b_2;\ -z)$.

1. $${}_1F_2(1;\ b,\ 2-b;\ -z)-\frac{2(b-1)}{(1-2b)(3-2b)}\,{}_1F_2\left(1;\ \frac{1}{2}+b,\ \frac{5}{2}-b;\ -z\right)=\frac{2\pi(b-1)}{\sin 2b\pi}\mathbf{J}_{2b-2}(2\sqrt{z}).$$

2. $${}_1F_2(1;\ b,\ 2-b;\ -z)+\frac{4(b-1)\sqrt{z}\operatorname{ctg}^2 b\pi}{(1-2b)(3-2b)}\,{}_1F_2\left(1;\ \frac{1}{2}+b,\ \frac{5}{2}-b;\ -z\right)=\frac{\pi(b-1)}{\sin^2 b\pi}\mathbf{E}_{2b-2}(2\sqrt{z}).$$

	a	b_1	b_2	${}_1F_2(a;\ b_1,\ b_2;\ -z^2)$
3	a	$a+1$	b	$2^{1+b-2a}a\Gamma(b)z^{1-2a}[2aJ_{b-1}(2z)s_{2a-b-1,\ b-2}(2z)-$ $-J_{b-2}(2z)s_{2a-b,\ b-1}(2z)]$
4	a	$\frac{1}{2}$	$a+1$	$2^{1-2a}az^{-2a}[\Gamma(2a)\cos a\pi-C(2z,\ 2a)]$
5	a	$\frac{3}{2}$	$a+1$	$-2^{1-2a}az^{-2a}[\Gamma(2a-1)\cos a\pi+S(2z,\ 2a-1)]$
6	1	b	c	$2^{4-b-c}(b-1)(c-1)z^{2-b-c}s_{b+c-3,\ b-c}(2z)$
7	1	b	b	$2^{5-2b}z^{2-2b}(b-1)^2S_{2b-3,\ 0}(2z)-$ $-\Gamma^2(b)z^{2-2b}[\sin b\pi Y_0(2z)+\cos b\pi J_0(2z)]$
8	1	b	$b+\frac{1}{2}$	$\Gamma(2b)(2z)^{1-2b}U_{2b-1}(4z,\ 0)$
9	1	b	$2-b$	$\pi(b-1)[\mathbf{E}_{2b-2}(2z)+\operatorname{ctg}b\pi\mathbf{J}_{2b-2}(2z)]=$
10				$=\frac{\pi(b-1)}{\sin 2b\pi}[\mathbf{J}_{2b-2}(2z)+\mathbf{J}_{2-2b}(2z)]$
11	1	b	$3-b$	$\frac{\pi}{z}(1-b)(2-b)[\mathbf{E}_{3-2b}(2z)-\operatorname{ctg}b\pi\mathbf{J}_{3-2b}(2z)]=$
12				$=\frac{\pi(1-b)(2-b)}{z\sin 2b\pi}[\mathbf{J}_{2b-3}(2z)-\mathbf{J}_{3-2b}(2z)]$
13	1	$\frac{3}{2}$	b	$\frac{\sqrt{\pi}}{2}z^{1/2-b}\Gamma(b)\mathbf{H}_{b-3/2}(2z)$

7.14.4. Particular values of ${}_1F_2(a;\ b_1,\ b_2;\ -z)$.

	a	b_1	b_2	${}_1F_2\ (a;\ b_1,\ b_2;\ -z^2)$
1	$\frac{1}{4}$	$\frac{1}{2}$	$\frac{5}{4}$	$\frac{\sqrt{\pi}}{2\sqrt{z}}\, C\,(2z)$
2	$\frac{1}{2}$	$\frac{3}{2}$	$\frac{3}{2}$	$\frac{1}{2z}\,\mathrm{Si}\,(2z)$
3	$\frac{3}{4}$	$\frac{3}{2}$	$\frac{7}{4}$	$\frac{3\sqrt{\pi}}{4z^{3/2}}\, S\,(2z)$
4 5	1	$\frac{1}{4}$	$\frac{3}{4}$	$1-2\sqrt{\pi z}\,[\sin 2zC\,(2z)-\cos 2zS\,(2z)]=1-\sqrt{2\pi z}\,U_{3/2}\,(4z,\ 0)$
6 7	1	$\frac{3}{4}$	$\frac{5}{4}$	$\frac{1}{2}\sqrt{\frac{\pi}{z}}\,[\cos 2zC\,(2z)+\sin 2zS\,(2z)]=\frac{1}{2}\sqrt{\frac{\pi}{2z}}\,U_{1/2}\,(4z,\ 0)$
8	1	$\frac{5}{4}$	$\frac{7}{4}$	$\frac{3\sqrt{\pi}}{8z^{3/2}}\,[\sin 2zC\,(2z)-\cos 2zS\,(2z)]$
9	1	$\frac{7}{4}$	$\frac{9}{4}$	$\frac{15}{32z^{5/2}}\left\{2z^{1/2}-\sqrt{\pi}\,[\cos 2zC\,(2z)+\sin 2zS\,(2z)]\right\}$

7.15. THE FUNCTION ${}_2F_3\,(a_1,\ a_2;\ b_1,\ b_2,\ b_3;\ z)$

7.15.1. Representations of ${}_2F_3\,(a_1,\ a_2;\ b_1,\ b_2,\ b_3;\ z)$.

1. $${}_2F_3\left(a,\ 1-a;\ \frac{1}{2},\ b,\ 2-b;\ z\right)+\frac{4\,(2a-1)\,(b-1)\sqrt{z}}{(2b-1)\,(2b-3)}\,{}_2F_3\left(a+\frac{1}{2},\ \frac{3}{2}-a;\ \frac{3}{2},\ b+\frac{1}{2},\ \frac{5}{2}-b;\ z\right)={}_1F_1\,(a+b-1;\ 2b-1;\ 2\sqrt{z})\,{}_1F_1\,(a-b+1;\ 3-2b;\ 2\sqrt{z}).$$

	a_1	a_2	b_1	b_2	b_3	${}_2F_3\ (a_1,\ a_2;\ b_1,\ b_2,\ b_3;\ z)$
2	a	b	$a+b$	$\frac{a+b}{2}$	$\frac{a+b+1}{2}$	${}_1F_1\,(a;\,a+b;\,2\sqrt{z})\,{}_1F_1\,(a;\,a+b;\,-2\sqrt{z})$
3	a	$a+\frac{1}{2}$	$2a$	b	$2a-b+1$	${}_0F_1\left(b;\ \frac{z}{4}\right)\,{}_0F_1\left(1+2a-b;\ \frac{z}{4}\right)=$
4						$=\Gamma\,(b)\,\Gamma\,(2a-b+1)\left(\frac{\sqrt{z}}{2}\right)^{1-2a}\times$ $\times\, I_{b-1}\,(\sqrt{z})\,I_{2a-b}\,(\sqrt{z})$
5	a	$a+\frac{1}{2}$	$1\mp\frac{1}{2}$	b	$b+\frac{1}{2}$	$\frac{1}{2}\left[\frac{2b-1}{2\,(2a-1)\sqrt{z}}\right]^{(1\mp 1)/2}\times$ $\times\left[{}_1F_1\left(2a-\frac{1\mp 1}{2};\ 2b-\frac{1\mp 1}{2};\ 2\sqrt{z}\right)\pm\right.$ $\left.\pm\,{}_1F_1\left(2a-\frac{1\mp 1}{2};\ 2b-\frac{1\mp 1}{2};\ -2\sqrt{z}\right)\right]$
6	a	$a+\frac{1}{2}$	$\frac{1}{2}$	$2a$	$2a+\frac{1}{2}$	$\mathrm{ch}\,\sqrt{z}\,{}_0F_1\left(2a+\frac{1}{2};\ \frac{z}{4}\right)=\Gamma\left(2a+\frac{1}{2}\right)\times$
7						$\times\left(\frac{\sqrt{z}}{2}\right)^{1/2-2a}\mathrm{ch}\,\sqrt{z}\,I_{2a-1/2}\,(\sqrt{z})$

	a_1	a_2	b_1	b_2	b_3	${}_2F_3(a_1, a_2; b_1, b_2, b_3; z)$
8	a	$a+\frac{1}{2}$	$\frac{3}{2}$	$2a$	$2a-\frac{1}{2}$	$\frac{\operatorname{sh}\sqrt{z}}{\sqrt{z}}\,{}_0F_1\left(2a-\frac{1}{2};\ \frac{z}{4}\right)=$
9						$=\Gamma\left(2a-\frac{1}{2}\right)2^{2a-3/2}z^{1/4-a}\operatorname{sh}\sqrt{z}\times I_{2a-3/2}(\sqrt{z})$
10	$n+1$	$n-\frac{1}{2}$	1	$2n$	$2n+1$	$(2n-1)!\left(\frac{2}{\sqrt{z}}\right)^{2n-1}\left\{I_0(\sqrt{z})\times I_{2n-1}(\sqrt{z})-\frac{2\sqrt{z}}{4n^2-1}[I_1(\sqrt{z})\times I_{2n-1}(\sqrt{z})-I_0(\sqrt{z})I_{2n}(\sqrt{z})]-\frac{2n-1}{2n+1}I_1(\sqrt{z})I_{2n}(\sqrt{z})\right\}$
11	$-n$	$b+n$	b	$\frac{b}{2}$	$\frac{b+1}{2}$	$\left[\frac{n!}{(b)_n}\right]^2 L_n^{b-1}(2\sqrt{z})\,L_n^{b-1}(-2\sqrt{z})$
12	$-n$	$\frac{1}{2}-n$	$\frac{3}{2}$	$-2n$	$-2n-\frac{1}{2}$	$\frac{(2n)!}{(4n+1)!}2^{2n-1/2}\sqrt{\pi}\,z^{n+1/4}\times[\operatorname{ch}\sqrt{z}\,I_{2n+3/2}(\sqrt{z})-\operatorname{sh}\sqrt{z}\,I_{-2n-3/2}(\sqrt{z})]$

7.15.2. Particular values of ${}_2F_3(a_1, a_2; b_1, b_2, b_3; z)$.

	a_1	a_2	b_1	b_2	b_3	${}_2F_3(a_1, a_2; b_1, b_2, b_3; z)$
1	$\frac{1}{4}$	$\frac{3}{4}$	$\frac{1}{2}$	$\frac{1}{2}$	1	$\operatorname{ch}\sqrt{z}\,I_0(\sqrt{z})$
2	$\frac{1}{2}$	1	$\frac{3}{4}$	$\frac{5}{4}$	$\frac{3}{2}$	$\frac{\pi}{8\sqrt{z}}\operatorname{erf}(\sqrt[4]{4z})\operatorname{erfi}(\sqrt[4]{4z})$
3	$\frac{1}{2}$	1	$\frac{5}{4}$	$\frac{3}{2}$	$\frac{7}{4}$	$\frac{3}{2\sqrt{z}}\operatorname{erf}(\sqrt[4]{4z})\operatorname{erfi}(\sqrt[4]{4z})+\frac{3}{\sqrt{\pi}}(4z)^{-3/4}\left[e^{-2\sqrt{z}}\operatorname{erfi}(\sqrt[4]{4z})-e^{2\sqrt{z}}\operatorname{erf}(\sqrt[4]{4z})\right]$
4	$\frac{3}{4}$	$\frac{5}{4}$	$\frac{1}{2}$	$\frac{3}{2}$	2	$\frac{2\operatorname{ch}\sqrt{z}}{\sqrt{z}}I_1(\sqrt{z})$
5	$\frac{3}{4}$	$\frac{5}{4}$	1	$\frac{3}{2}$	$\frac{3}{2}$	$\frac{\operatorname{sh}\sqrt{z}}{\sqrt{z}}I_0(\sqrt{z})$
6	1	1	$\frac{3}{2}$	2	2	$\frac{1}{z}[\operatorname{chi}(2\sqrt{z})-\ln(2\sqrt{z})-\mathbf{C}]$
7	$\frac{5}{4}$	$\frac{7}{4}$	$\frac{3}{2}$	2	$\frac{5}{2}$	$\frac{2}{z}\operatorname{sh}\sqrt{z}\,I_1(\sqrt{z})$

7.15.3. Representations and particular values of $_2F_3(a_1,\ a_2;\ b_1,\ b_2,\ b_3;\ -z)$.

	a_1	a_2	b_1	b_2	b_3	$_2F_3(a_1, a_2; b_1, b_2, b_3; -z)$
1	$-n$	$-n-\frac{1}{2}$	$-2n-1$	$-2n-\frac{3}{2}$	$\frac{3}{2}$	$\frac{(2n+2)!}{(4n+4)!}\sqrt{\pi}\,(2\sqrt{z})^{2n+3/2}\times$ $\times\left[\sin\sqrt{z}\,Y_{2n+5/2}(\sqrt{z})-\right.$ $\left.-\cos\sqrt{z}\,J_{2n+5/2}(\sqrt{z})\right]$
2	$-\frac{n}{2}$	$\frac{1-n}{2}$	$\frac{1}{2}$	$-n$	$\frac{1}{2}-n$	$\frac{n!\,\sqrt{\pi}}{(2n)!}\,2^{n-1/2}z^{(2n+1)/4}\left[\sin\sqrt{z}\times\right.$ $\left.\times J_{n+1/2}(\sqrt{z})-\cos\sqrt{z}\,Y_{n+1/2}(\sqrt{z})\right]$
3	1	1	$\frac{3}{2}$	2	2	$\frac{1}{z}\left[\mathrm{C}+\ln(2\sqrt{z})-\mathrm{ci}(2\sqrt{z})\right]$

7.16. FUNCTIONS OF THE FORM $_0F_q((b_q);\ z),\ q=2,\ 3,\ \ldots$

7.16.1. Particular values of $_0F_2(b_1,\ b_2;\ z)$.

	b_1	b_2	$_0F_2(b_1, b_2; z)$ $\left[y=3\sqrt[3]{z}\right]$
1	$\frac{1}{3}$	$\frac{2}{3}$	$\frac{1}{3}\left[e^{y}+2e^{-y/2}\cos\left(\sqrt{3}\,\frac{y}{2}\right)\right]$
2	$\frac{2}{3}$	$\frac{4}{3}$	$\frac{1}{3y}\left[e^{y}-2e^{-y/2}\cos\left(\sqrt{3}\,\frac{y}{2}+\frac{\pi}{3}\right)\right]$
3	$\frac{4}{3}$	$\frac{5}{3}$	$\frac{2}{3y^2}\left[e^{y}-2e^{-y/2}\cos\left(\sqrt{3}\,\frac{y}{2}-\frac{\pi}{3}\right)\right]$

7.16.2. Representations and particular values of $_0F_3(b_1,\ b_2,\ b_3;\ z)$.

1. $$_0F_3\left(\frac{1}{2},\ a,\ a+\frac{1}{2};\ z\right)+\frac{2\sqrt{z}}{a}\,{}_0F_3\left(\frac{3}{2},\ a+\frac{1}{2},\ a+1;\ z\right)={}_0F_1(2a;\ 4\sqrt{z}).$$

	b_1	b_2	b_3	$_0F_3(b_1, b_2, b_3; z)$ $\left[y=2\sqrt[4]{4z},\ x=4\sqrt[4]{z}\right]$
2	b_1	b_2	b_3	$\prod_{k=1}^{3}(b_k-1)\,\frac{d}{dz}\,{}_0F_3(b_1-1,\ b_2-1,\ b_3-1;\ z)$
3	b_1	b_2	b_3	$\frac{z^{1-b_1}}{b_1}\frac{d}{dz}\left[z^{b_1}{}_0F_3(b_1+1,\ b_2,\ b_3;\ z)\right]$
4	a	$a+\frac{1}{2}$	$2a$	$\Gamma^2(2a)\,(2\sqrt{z})^{1-2a}\left[\mathrm{ber}^2_{2a-1}(y)+\mathrm{bei}^2_{2a-1}(y)\right]$
5	a	$a+\frac{1}{2}$	$2a-1$	$\Gamma(2a-1)\,\Gamma(2a)\left(\frac{y}{2}\right)^{3-4a}\times$ $\times\left[\mathrm{ber}_{2a-2}(y)\,\mathrm{bei}'_{2a-2}(y)-\mathrm{ber}'_{2a-2}(y)\,\mathrm{bei}_{2a-2}(y)\right]$
6	a	$a+\frac{1}{2}$	$2a+1$	$2\Gamma(2a)\,\Gamma(2a+1)\left(\frac{y}{2}\right)^{1-4a}\times$ $\times\left[\mathrm{ber}_{2a}(y)\,\mathrm{ber}'_{2a}(y)+\mathrm{bei}_{2a}(y)\,\mathrm{bei}'_{2a}(y)\right]$

	b_1	b_2	b_3	${}_0F_3(b_1, b_2, b_3; z)$ $\left[y=2\sqrt[4]{4z},\ x=4\sqrt[4]{z}\right]$
7	$\frac{1}{2}$	a	$a+\frac{1}{2}$	$2^{-2a}\Gamma(2a)\, z^{(1-2a)/4}\,[I_{2a-1}(x)+J_{2a-1}(x)]$
8	$\frac{3}{2}$	a	$a+\frac{1}{2}$	$2^{-2a-1}\Gamma(2a)\, z^{-a/2}\,[I_{2a-2}(x)-J_{2a-2}(x)]$
9	n	$n-m-\frac{1}{2}$	$n-m-l-\frac{1}{2}$	$\frac{(-1)^l (n-1)!\,[(1/2)_{n-1}]^2\, z^{l+m-n+3/2}}{2\,[(3/2-n)_m]^2\,(3/2+m-n)_l}\times$ $\times\left(\frac{d}{dz}\right)^l\left(\frac{d}{dz}z\frac{d}{dz}\right)^m z^{n-3/2}\left(\frac{d}{dz}\right)^{n-1}[J_0(x)+I_0(x)]$ $[n=1, 2, 3, \ldots]$
10	1	$n+\frac{1}{2}$	$n+\frac{1}{2}$	$\frac{[(1/2)_n]^2}{2}\sum_{k=0}^{n}\binom{n}{k}^2 (n-k)!\, z^k\left(\frac{d}{dz}\right)^{n+k}[J_0(x)+$ $+I_0(x)]$
11	$\frac{1}{4}$	$\frac{1}{2}$	$\frac{3}{4}$	$\frac{1}{2}(\operatorname{ch} x+\cos x)$
12	$\frac{1}{2}$	$\frac{1}{2}$	1	$\operatorname{ber}(t)=\frac{1}{2}[J_0(x)+I_0(x)]$ $\left[t=4\sqrt[4]{z}\, e^{\pi i/4}\right]$
13	$\frac{1}{2}$	$\frac{3}{4}$	$\frac{5}{4}$	$\frac{1}{8\sqrt[4]{z}}(\operatorname{sh} x+\sin x)$
14	$\frac{1}{2}$	1	1	$\operatorname{ber}^2(y)+\operatorname{bei}^2(y)$
15	$\frac{1}{2}$	1	$\frac{3}{2}$	$\frac{1}{4\sqrt[4]{z}}[I_1(x)+J_1(x)]$
16	$\frac{3}{4}$	$\frac{5}{4}$	$\frac{3}{2}$	$\frac{1}{16\sqrt{z}}(\operatorname{ch} x-\cos x)$
17	1	1	$\frac{3}{2}$	$\frac{1}{2\sqrt[4]{z}}\{\operatorname{bei}_1(y)\,[\operatorname{ber}(y)-\operatorname{bei}(y)]-$ $-\operatorname{ber}_1(y)\,[\operatorname{ber}(y)+\operatorname{bei}(y)]\}$
18	1	$\frac{3}{2}$	$\frac{3}{2}$	$\frac{1}{8\sqrt{z}}[I_0(x)-J_0(x)]$
19	1	$\frac{3}{2}$	2	$\frac{1}{2\sqrt{z}}[\operatorname{ber}_1^2(y)+\operatorname{bei}_1^2(y)]$
20	$\frac{5}{4}$	$\frac{3}{2}$	$\frac{7}{4}$	$\frac{3}{64z^{3/4}}(\operatorname{sh} x-\sin x)$
21	$\frac{3}{2}$	$\frac{3}{2}$	2	$\frac{1}{8}z^{-3/4}[I_1(x)-J_1(x)]$

7.16.3. Representations and particular values of ${}_0F_3(b_1, b_2, b_3; -z)$

1. $${}_0F_3\left(\frac{1}{2}, a, 2-a; -z\right) \pm \frac{8(a-1)\sqrt{z}}{(2a-1)(2a-3)}\,{}_0F_3\left(\frac{3}{2}, a+\frac{1}{2}, \frac{5}{2}-a; -z\right)= ={}_0F_1(2a-1; \pm 2\sqrt{z})\,{}_0F_1(3-2a; \mp 2\sqrt{z})=$$

2. $$=\frac{2\pi(a-1)}{\sin 2a\pi} I_{\pm(2a-2)}(y)\, J_{\pm(2-2a)}(y) \qquad \left[y=2\sqrt[4]{4z}\right]$$

	b_1	b_2	b_3	${}_0F_3(b_1, b_2, b_3;\ -z)$ $\left[y=2\sqrt[4]{4z},\ x=4\sqrt[4]{z}\right]$
3	a	$a+\frac{1}{2}$	$2a$	${}_0F_1(2a;\ 2\sqrt{z})\,{}_0F_1(2a;\ -2\sqrt{z})=$
4				$=(2\sqrt{z})^{1-2a}\,\Gamma^2(2a)\,J_{2a-1}(y)\,I_{2a-1}(y)$
5	$\frac{1}{2}$	a	$a+\frac{1}{2}$	$\Gamma(2a)\,(2\sqrt[4]{z})^{1-2a}\times$ $\times\left\{\cos\frac{3(2a-1)}{4}\pi\,\mathrm{ber}_{2a-1}(x)+\sin\frac{3(2a-1)}{4}\pi\,\mathrm{bei}_{2a-1}(x)\right\}$
6	$\frac{3}{2}$	a	$a+\frac{1}{2}$	$-\Gamma(2a)\,(2\sqrt[4]{z})^{-2a}\times$ $\times\left[\cos\frac{3a\pi}{2}\,\mathrm{ber}_{2a-2}(x)+\sin\frac{3a\pi}{2}\,\mathrm{bei}_{2a-2}(x)\right]$
7	$\frac{1}{2}$	a	$2-a$	$\frac{\pi(a-1)}{\sin 2a\pi}\left[J_{2-2a}(y)\,I_{2a-2}(y)+J_{2a-2}(y)\,I_{2-2a}(y)\right]$
8	$\frac{3}{2}$	a	$3-a$	$\frac{\pi(1-a)(2-a)}{\sin 2a\pi}\left[J_{2a-3}(y)\,I_{3-2a}(y)-J_{3-2a}(y)\,I_{2a-3}(y)\right]$
9	1	$\frac{1}{2}-n$	$\frac{1}{2}-n$	$\frac{z^{n+1/2}}{[(1/2)_n]^2}\left(\frac{d}{dz}\,z\,\frac{d}{dz}\right)^n z^{-1/2}\,\mathrm{ber}\,(4\sqrt[4]{z})$
10	$-\frac{1}{2}$	$-\frac{1}{2}$	1	$\mathrm{ber}(x)-2\sqrt[4]{4z}\,[\mathrm{ber}_1(x)+\mathrm{bei}_1(x)]-4\sqrt{z}\,\mathrm{bei}(x)=$
11				$=\mathrm{ber}(x)-x\,\mathrm{ber}'(x)-4\sqrt{z}\,\mathrm{bei}(x)$
12	$\frac{1}{4}$	$\frac{1}{2}$	$\frac{3}{4}$	$\mathrm{ch}\,y\cos y$
13	$\frac{1}{2}$	$\frac{1}{2}$	1	$\mathrm{ber}(x)$
14	$\frac{1}{2}$	$\frac{3}{4}$	$\frac{5}{4}$	$\frac{1}{4\sqrt[4]{4z}}\,[\mathrm{sh}\,y\cos y+\mathrm{ch}\,y\sin y]$
15	$\frac{1}{2}$	1	$\frac{3}{2}$	$\frac{2}{x}\,\mathrm{bei}'(x)$
16	$\frac{3}{4}$	$\frac{5}{4}$	$\frac{3}{2}$	$\frac{1}{8\sqrt{z}}\,\mathrm{sh}\,y\sin y$
17	1	$\frac{3}{2}$	$\frac{3}{2}$	$\frac{1}{4\sqrt{z}}\,\mathrm{bei}(x)$
18	1	$\frac{5}{2}$	$\frac{5}{2}$	$\frac{9}{64z^{3/2}}\,[\mathrm{bei}(x)-x\,\mathrm{bei}'(x)+4\sqrt{z}\,\mathrm{ber}(x)]$
19	$\frac{5}{4}$	$\frac{3}{2}$	$\frac{7}{4}$	$\frac{3}{32\sqrt{2}\,z^{3/4}}\,[\mathrm{ch}\,y\sin y-\mathrm{sh}\,y\cos y]$
20	$\frac{3}{2}$	$\frac{3}{2}$	2	$-\frac{1}{4z^{3/4}}\,\mathrm{ber}'(x)$

7.16.4. Representations of ${}_0F_4(b_1, b_2, b_3, b_4;\ z)$ и ${}_0F_{q-1}((b_{q-1});\ z)$.

1. $${}_0F_4\left(\frac{1}{5}, \frac{2}{5}, \frac{3}{5}, \frac{4}{5};\ z\right)=\frac{1}{5}e^y+\frac{2}{5}\left[\exp\left(y\cos\frac{2\pi}{5}\right)\cos\left(y\sin\frac{2\pi}{5}\right)+\right.$$
$$\left.+\exp\left(y\cos\frac{4\pi}{5}\right)\cos\left(y\sin\frac{4\pi}{5}\right)\right]\qquad\left[y=5\sqrt[5]{z}\right].$$

2. $${}_0F_{q-1}\left(\frac{1}{q}, \frac{2}{q}, \ldots, \frac{q-1}{q};\ z\right)=\frac{1}{q}\sum_{k=0}^{q-1}\exp\left[q\,(e^{2k\pi i}z)^{1/q}\right].$$

7.17. FUNCTIONS OF THE FORM $_pF_0(-n, (a_{p-1}); z)$, $p=2, 3\ldots$

7.17.1. Representations of $_2F_0(-n, a; z)$.

	a_1	a_2	$_2F_0(a_1, a_2; z)$
1	$-n$	a	$(a)_n(-z)^n\,{}_1F_1(-n;\ 1-a-n;\ -z^{-1})=$
2			$=n!\,z^nL_n^{-a-n}(-z^{-1})$
3	$-n$	$n+1$	$\frac{1}{2}\sqrt{\frac{\pi}{z}}e^{-1/(2z)}\left[I_{n+1/2}\left(\frac{1}{2z}\right)+I_{-n-1/2}\left(\frac{1}{2z}\right)\right]=$
4			$=\frac{e^{-1/(2z)}}{\sqrt{-\pi z}}K_{n+1/2}\left(-\frac{1}{2z}\right)$
5	$-n$	$n+1$	$\frac{1}{2}i^{\pm(n+3/2)}\sqrt{-\frac{\pi}{z}}e^{-1/(2z)}H_{n+1/2}^{\binom{1}{2}}\left(\mp\frac{i}{2z}\right)$ $\left[\left\{\begin{matrix}0<\arg z\leqslant 3\pi/2\\ -\pi/2<\arg z\leqslant 0\end{matrix}\right\}\right]$

7.17.2. Representations of $_2F_0(-n, a; -z)$.

	a_1	a_2	$_2F_0(a_1, a_2; -z)$
1	$-n$	n	$n!(-z)^nL_n^{-2n}(\frac{1}{z})$
2	$-n$	$n+1$	$\frac{(-1)^n}{2}\sqrt{\frac{\pi}{z}}e^{1/(2z)}\left[I_{-n-1/2}\left(\frac{1}{2z}\right)-I_{n+1/2}\left(\frac{1}{2z}\right)\right]$
3	$-\frac{n}{2}$	$\frac{1-n}{2}$	$2^{-n}z^{n/2}H_n(z^{-1/2})$
4	$-n$	2	$n+1$ $[z(n+1)=-1]$

7.17.3. Representations of $_3F_0(-n, a_1, a_2; z)$.

1. $_3F_0\left(-n,\ n+1,\ \frac{1}{2};\ z\right)=\frac{(-1)^n\pi}{2\sqrt{z}}\left[I_{-n-1/2}^2\left(\frac{1}{\sqrt{z}}\right)-I_{n+1/2}^2\left(\frac{1}{\sqrt{z}}\right)\right].$

2. $_3F_0\left(-n,\ n+1,\ \frac{1}{2};\ -z\right)=\frac{\pi}{2\sqrt{z}}\left[J_{n+1/2}^2\left(\frac{1}{\sqrt{z}}\right)+Y_{n+1/2}^2\left(\frac{1}{\sqrt{z}}\right)\right].$

7.18. VARIOUS HYPERGEOMETRIC FUNCTIONS

7.18.1. Representations of ${}_1F_q(a;\ (b_q);\ z)$.

1. $${}_1F_q\left(1;\ \frac{m+1}{q},\ \frac{m+2}{q},\ \ldots,\ \frac{m+q}{q};\ z\right)=$$
$$=\frac{m!}{q^{m+1}z^{m/q}}\left[\sum_{k=0}^{q-1}\frac{e^{q\theta_k z^{1/q}}}{\theta_k^m}-q^{m+1}\sum_{k=1}^{[m/q]}\frac{z^{m/q-k}}{q^{qk}(m-qk)!}\right]$$
$$\left[\theta_k=\sqrt[q]{1}=\cos\frac{2k\pi}{q}+i\sin\frac{2k\pi}{q}=e^{2k\pi i/q}\right].$$

2. $${}_1F_4(1;\ n,\ b_1,\ b_2,\ b_3;\ z)=$$
$$=(n-1)!\left[(-z)^{1-n}\prod_{k=1}^{3}(1-b_k)_{n-1}\,{}_0F_3(b_1-n+1,\ b_2-n+1,\ b_3-n+1;\ z)-\right.$$
$$\left.-\sum_{k=1}^{n-1}\prod_{l=1}^{3}(1-b_l)_k\frac{(-z)^{-k}}{(n-k-1)!}\right]\qquad [n=1,\ 2,\ 3,\ldots].$$

7.18.2. Representations of ${}_3F_8(a_1,\ a_2,\ a_3;\ b_1,\ \ldots,\ b_8;\ z)$.

1. $${}_3F_8\left(a,\ a+\frac{1}{3},\ a+\frac{2}{3};\ b,\ \frac{b}{2},\ \frac{b+1}{2},\ \frac{3a}{2},\ \frac{3a+1}{2},\ 1+3a-b,\right.$$
$$\left.\frac{1+3a-b}{2},\ 1+\frac{3a-b}{2};\ -z\right)=$$
$$={}_0F_2\left(b,\ 1+3a-b;\ 8\sqrt{\frac{z}{27}}\right){}_0F_2\left(b,\ 1+3a-b;\ -8\sqrt{\frac{z}{27}}\right).$$

7.18.3. Representations of ${}_4F_1(-n,\ a_1,\ a_2,\ a_3;\ b;\ z)$.

	a_1	a_2	a_3	a_4	b	${}_4F_1(a_1,\ a_2,\ a_3,\ a_4;\ b;\ z)$
1	$-n$	$2a+n$	a	$a+\frac{1}{2}$	$2a$	$(-1)^n(n!)^2 2^{-2n}z^n L_n^{-2a-2n}\left(\frac{2}{\sqrt{z}}\right)L_n^{-2a-2n}\left(-\frac{2}{\sqrt{z}}\right)$
2	$-n$	$1+n$	a	$1-a$	$\frac{1}{2}$	$\frac{\pi}{2\sqrt{z}\sin a\pi}[J_{a+n}(z^{-1/2})J_{-a+n+1}(z^{-1/2})+J_{-a-n}(z^{-1/2})J_{a-n-1}(z^{-1/2})]$
3	$-n$	$2+n$	a	$2-a$	$\frac{3}{2}$	$\frac{\pi}{4z(n+1)(1-a)\sin a\pi}[J_{a+n}(z^{-1/2})\times J_{-a+n+2}(z^{-1/2})-J_{-a-n}(z^{-1/2})J_{a-n-2}(z^{-1/2})]$
4	$-\frac{n}{2}$	$\frac{1-n}{2}$	$\frac{1+n}{2}$	$1+\frac{n}{2}$	$\frac{1}{2}$	$(-1)^n\sqrt{\frac{\pi}{2}}z^{-1/4}\left[\sin\left(\frac{1}{\sqrt{z}}+\frac{\pi n}{2}\right)J_{n+1/2}(z^{-1/2})+\cos\left(\frac{1}{\sqrt{z}}-\frac{\pi n}{2}\right)J_{-n-1/2}(z^{-1/2})\right]$
5	$-\frac{n}{2}$	$\frac{1-n}{2}$	$\frac{3+n}{2}$	$2+\frac{n}{2}$	$\frac{3}{2}$	$\frac{\sqrt{2\pi}}{z^{3/4}(n+1)(n+2)}\left[\sin\left(\frac{1}{\sqrt{z}}-\frac{\pi n}{2}\right)J_{n+3/2}(z^{-1/2})-(-1)^n\cos\left(\frac{1}{\sqrt{z}}-\frac{\pi n}{2}\right)J_{-n-3/2}(z^{-1/2})\right]$

Chapter 8. THE MEIJER *G*-FUNCTION AND THE FOX *H*-FUNCTION

8.1 INTRODUCTION

In this chapter we set out the properties of the Meijer *G*-function and the Fox *H*-function and give tables of Mellin transforms combined with tables of representations in terms of the *G*- and *H*-functions for a wide class of elementary and special functions. These functions $f(x)$ have the property that their Mellin transforms

$$f^*(s)=\int_0^\infty x^{s-1}f(x)\,dx \tag{1}$$

can be expressed as the ratio of products of gamma functions

$$f^*(s)=\prod_{j,k,l,m}\frac{\Gamma(a_j+\alpha_j s)\,\Gamma(b_k-\beta_k s)}{\Gamma(c_l+\gamma_l s)\,\Gamma(d_m-\delta_m s)}, \tag{2}$$

where a_j, b_k, c_l, d_m are complex constants, $\alpha_j, \beta_k, \gamma_l, \delta_m$ are positive constants. In some cases, there correspond to values of $f^*(s)$ in Table 8.4 functions $f(x)$ for which the integral (1) diverges, but one of the integrals

$$\frac{1}{2\pi i}\int_L f^*(s)\,x^{-s}\,ds=f(x),$$

converges, where $L=L_{-\infty}$ or $L_{+\infty}$, and L_- (or L_+) is the left (or right) loop (see 8.2); we denote this correspondence by the symbol $f(x)\overset{L_{\pm\infty}}{\longleftarrow}f^*(s)$.

The index of particular cases of the Meijer *G*-function and the Fox *H*-function enables one to find expressions for them in terms of elementary and special functions. Use of the Mellin transform makes it possible to calculate the integrals of products of the functions occurring in Table 8.4.

Suppose that the integral is given in the form

$$I(t)=\int_0^\infty f_1(x)\,f_2\left(\frac{t}{x}\right)\frac{dx}{x};$$

then by formula 8.4.1.2, its Mellin transform $I^*(s)$ is equal to the product of the Mellin transforms $f_1^*(s)$, $f_2^*(s)$ of the functions $f_1(x)$, $f_2(x)$ and hence is also of the form (2). By using the formula for the inverse Mellin transform

$$f(t)=\frac{1}{2\pi i}\int_{\gamma-i\infty}^{\gamma+i\infty}f^*(s)\,t^{-s}\,ds$$

and formula 2.2.1.7, the value of $I(t)$ can be found.

EXAMPLE: We shall calculate the integral

$$I(t)=\int_0^\infty x^{\alpha-1}e^{-x-x/t}\,dx.$$

It is a convolution (for the Mellin transform) of the functions

$$f_1(x)=x^\alpha e^{-x},\qquad f_2(x)=e^{-1/x}.$$

According to formulae 8.4.3.1, 8.4.1.5, 8.4.3.2,

$$f_1^*(s)=\Gamma(\alpha+s),\quad \operatorname{Re}(\alpha+s)>0,$$
$$f_2^*(s)=\Gamma(-s),\qquad \operatorname{Re}s<0.$$

The Mellin transform $I^*(s)$ of the integral $I(t)$ has the form

$$I^*(s)=\Gamma(\alpha+s)\,\Gamma(-s),\qquad -\operatorname{Re}\alpha<\operatorname{Re}s=\gamma<0.$$

By using formulae 8.4.2.5, 8.4.1.7 or 2.2.1.7, we now obtain

$$\int_0^\infty x^{\alpha-1}e^{-x-x/t}\,dx=\Gamma(\alpha)\left(1+\frac{1}{t}\right)^{-\alpha}\qquad\left[\operatorname{Re}\alpha,\ \operatorname{Re}\left(1+\frac{1}{t}\right)>0\right].$$

We also point out that Table 8.4 can be extended by means of the formulae of section 7.3–18, since formula 8.4.51.1 associates with each of the functions ${}_pF_q((a_p);(b_q);z)$ of Table 7.3–18 a function $f^*(s)$ of the form (2).

8.2. THE MEIJER G-FUNCTION $G_{pq}^{mn}\left(z\,\middle|\,\begin{matrix}(a_p)\\(b_q)\end{matrix}\right)$

8.2.1. Definition and notation.

The Meijer G-function of order (m,n,p,q), where $0\leqslant m\leqslant q$, $0\leqslant n\leqslant p$ is defined by the formula

1. $$G_{pq}^{mn}\left(z\,\middle|\,\begin{matrix}(a_p)\\(b_q)\end{matrix}\right)\equiv G_{pq}^{mn}\left(z\,\middle|\,\begin{matrix}a_1,\ \ldots,\ a_p\\b_1,\ \ldots,\ b_q\end{matrix}\right)=$$
$$=\frac{1}{2\pi i}\int_L\Gamma\begin{bmatrix}b_1+s,\ \ldots,\ b_m+s,\ 1-a_1-s,\ \ldots,\ 1-a_n-s\\a_{n+1}+s,\ \ldots,\ a_p+s,\ 1-b_{m+1}-s,\ \ldots,\ 1-b_q-s\end{bmatrix}z^{-s}\,ds;$$

the integral converges if one of the following conditions holds:

1) $L=L_{i\infty}$; $c^*>0$, $|\arg z|<c^*\pi$;

2) $L=L_{i\infty}$; $c^*\geqslant 0$, $|\arg z|=c^*\pi$, $(q-p)\gamma<-\operatorname{Re}\mu$;

3) $L=L_{-\infty}$; $p<q$, $0<|z|<\infty$ or $p=q$, $0<|z|<1$ or $p=q$, $c^*\geqslant 0$, $|z|=1$, $\operatorname{Re}\mu<0$;

4) $L=L_{+\infty}$; $p>q$, $0<|z|<\infty$ or $p=q$, $|z|>1$ or $p=q$, $c^*\geqslant 0$, $|z|=1$, $\operatorname{Re}\mu<0$;

Here

$$c^*=m+n-\frac{p+q}{2},\qquad \mu=\sum_{j=1}^q b_j-\sum_{j=1}^p a_j+\frac{p-q}{2}+1,\quad \gamma=\lim_{\substack{s\to\infty\\ s\in L_{i\infty}}}\operatorname{Re}s.$$

The contour $L_{-\infty}$ (or $L_{+\infty}$) is a left (or right) loop lying in a horizontal strip, starting at the point $-\infty+i\varphi_1$ (or $+\infty+i\varphi_1$), leaving all the poles of the integrand of the form $s=-b_j-k$, $j=1,2,\ldots,m$, on the left, and all the poles of the form $s=1-a_l+k$, $l=1,2,\ldots,n$, $k=0,1,2,\ldots$ on the right of the contour and finishing at the point $-\infty+i\varphi_2$ (or $+\infty+i\varphi_2$), where $\varphi_1<\varphi_2$. The contour $L_{i\infty}$ starts at the point $\gamma-i\infty$, finishes at the point $\gamma+i\infty$ and separates these poles in the same way as $L_{\pm\infty}$.

If in the definition of the G-function, the integrand only has factors with parameters b_k (or only with a_k), then the notation $G^{m0}_{0q}\left(z\,\middle|\,{}^{\cdot}_{(b_q)}\right)$ or $G^{0n}_{p0}\left(z\,\middle|\,{}^{(a_p)}_{\cdot}\right)$ is used.

8.2.2. Basic properties.

1. The G-function is symmetric with respect to the parameters $a_1,\ldots,a_n$ in (a_p), $a_{n+1},\ldots,a_p$ in (a_p), $b_1,\ldots,b_m$ in (b_q) and $b_{m+1},\ldots,b_q$ in (b_q) individually.

2. $G^{0n}_{pq}\left(z\,\middle|\,\begin{matrix}(a_p)\\(b_q)\end{matrix}\right)=0$ under conditions (1), (3) or (2), (3) of 8.2.1.

Representation as a combination of generalized hypergeometric functions:

3. $G^{mn}_{pq}\left(z\,\middle|\,\begin{matrix}(a_p)\\(b_q)\end{matrix}\right)=$

$$=\sum_{k=1}^{m}\Gamma\left[\begin{matrix}b_1-b_k,\ \overset{*}{\ldots},\ b_m-b_k,\ 1+b_k-a_1,\ \ldots,\ 1+b_k-a_n\\ a_{n+1}-b_k,\ \ldots,\ a_p-b_k,\ 1+b_k-b_{m+1},\ \ldots,\ 1+b_k-b_q\end{matrix}\right]z^{b_k}\times$$

$$\times{}_pF_{q-1}\left(\begin{matrix}1+b_k-(a_p);\\ 1+b_k-(b_q)'\end{matrix}\ (-1)^{p-m-n}z\right)$$

[$p\leqslant q$; conditions 1)–3) and $b_j-b_k\neq 0,\ \pm1,\ \pm2,\ \ldots;\ j\neq k;\ j,\ k=1,\ 2,\ \ldots,\ m$].

4. $$=\sum_{k=1}^{n}\Gamma\left[\begin{matrix}a_k-a_1,\ \overset{*}{\ldots},\ a_k-a_n,\ 1+b_1-a_k,\ \ldots,\ 1+b_m-a_k\\ a_k-b_{m+1},\ \ldots,\ a_k-b_q,\ 1+a_{n+1}-a_k,\ \ldots,\ 1+a_p-a_k\end{matrix}\right]z^{a_k-1}\times$$

$$\times{}_qF_{p-1}\left(\begin{matrix}1+(b_q)-a_k;\\ 1+(a_p)'-a_k\end{matrix}\ \frac{(-1)^{q-m-n}}{z}\right)$$

[$p\geqslant q$; conditions 1), 2), 4) and $a_j-a_k\neq 0,\ \pm1,\ \pm2,\ \ldots;\ j\neq k;\ j,\ k=1,\ 2,\ \ldots,\ n$].

Here $(a_p)'-a_k\equiv a_1-a_k,\ \ldots,\ a_{k-1}-a_k,\ a_{k+1}-a_k,\ \ldots,\ a_p-a_k$ and $\Gamma\,[a_1-a_k,\ \overset{*}{\ldots},\ a_m-a_k]=\Gamma\,(a_1-a_k)\ldots\Gamma\,(a_{k-1}-a_k)\,\Gamma\,(a_{k+1}-a_k)\ldots\Gamma\,(a_m-a_k)$ (the component a_k-a_k is absent). Cases when some of the differences b_j-b_k, a_j-a_k are integers lead to more complicated expressions.

Expansions into sums of G-functions:

5. $$G^{mn}_{pq}\left(z\,\middle|\,\begin{matrix}(a_p)\\(b_q)\end{matrix}\right)=\pi^{m+n-p}\sum_{k=1}^{m}\frac{\sin\,(a_{n+1}-b_k)\,\pi\ldots\sin\,(a_p-b_k)\,\pi}{\sin\,(b_1-b_k)\,\pi\overset{*}{\ldots}\sin\,(b_m-b_k)\,\pi}\times$$

$$\times e^{\pi i\,(m+n-p-1)\,b_k}\,G^{1p}_{pq}\left((-1)^{p-m-n+1}z\,\middle|\,\begin{matrix}(a_p)\\ b_k,\ (b_q)'_k\end{matrix}\right),$$

6. $$=\pi^{m+n-q}\sum_{k=1}^{n}\frac{\sin\,(a_k-b_{m+1})\,\pi\ldots\sin\,(a_k-b_q)\,\pi}{\sin\,(a_k-a_1)\pi\overset{*}{\ldots}\sin\,(a_k-a_n)\,\pi}\times$$

$$\times e^{\pi i\,(m+n-q-1)\,(a_k-1)}\,G^{q1}_{pq}\left((-1)^{q-m-n+1}z\,\middle|\,\begin{matrix}a_k,\ (a_p)'_k\\ (b_q)\end{matrix}\right)$$

[the factors $\sin(b_k-b_k)\pi$ and $\sin(a_k-a_k)\pi$ are absent], $(b_q)'_k=b_1,\ \ldots,\ b_{k-1},\ b_{k+1},\ \ldots,\ b_q$.

7. The function $G^{mn}_{pq}\left(z\,\middle|\,\begin{matrix}(a_p)\\(b_q)\end{matrix}\right)$ is analytic with respect to z in the sector $|\arg z|<c^*\pi$. For $c^*>0$, $p=q$, functions satisfying conditions (3) and (4) can be analytically continued into each other through the circle $|z|=1$ in this sector. In the case $c^*\leqslant 0$, the sector of analyticity is missing, but for $p=q$, the two above-mentioned analytic functions for $|z|<1$ and $|z|>1$ have the same limit as $z\to 1$ along the ray $\arg z=0$, if $\operatorname{Re}\mu<0$, $c^*=0$. For $|z|>1$, $p=q$, $c^*=1$, the principal branch of the function $G^{mn}_{pq}\left(z\,\middle|\,{}^{(a_p)}_{(b_q)}\right)$ is selected by drawing a cut along $(-\infty,-1)$. If $|z|>1$, $p=q$ and $c^*>1$, then for $0\leqslant\arg z\leqslant 2\pi$, the cut is not necessary.

Formulae for lowering the order and degeneracy:

8. $G_{pq}^{mn}\left(z\,\middle|\,\begin{matrix}(a_p)\\(b_{q-1}),\ a_1\end{matrix}\right)=G_{p-1,\,q-1}^{m,\ n-1}\left(z\,\middle|\,\begin{matrix}a_2,\ \ldots,\ a_p\\(b_{q-1})\end{matrix}\right).$

9. $G_{pq}^{mn}\left(z\,\middle|\,\begin{matrix}(a_{p-1}),\ b_1\\(b_q)\end{matrix}\right)=G_{p-1,\ q-1}^{m-1,\ n}\left(z\,\middle|\,\begin{matrix}(a_{p-1})\\b_2,\ \ldots,\ b_q\end{matrix}\right).$

10. $\lim\limits_{|a_1|\to\infty}\left[\frac{1}{\Gamma(1-a_1)}G_{pq}^{mn}\left(-\frac{z}{a_1}\,\middle|\,\begin{matrix}(a_p)\\(b_q)\end{matrix}\right)\right]=G_{p-1,\ q}^{m,\ n-1}\left(z\,\middle|\,\begin{matrix}a_2,\ \ldots,\ a_{p-1}\\(b_q)\end{matrix}\right).$

11. $\lim\limits_{|a_p|\to\infty}\left[\Gamma(a_p)\,G_{pq}^{mn}\left(\frac{z}{a_p}\,\middle|\,\begin{matrix}(a_p)\\(b_q)\end{matrix}\right)\right]=G_{p-1,\ q}^{mn}\left(z\,\middle|\,\begin{matrix}(a_{p-1})\\(b_q)\end{matrix}\right).$

12. $\lim\limits_{|b_1|\to\infty}\left[\frac{1}{\Gamma(b_1)}G_{pq}^{mn}\left(b_1z\,\middle|\,\begin{matrix}(a_p)\\(b_q)\end{matrix}\right)\right]=G_{p,\ q-1}^{m-1,\ n}\left(z\,\middle|\,\begin{matrix}(a_p)\\b_2,\ \ldots,\ b_q\end{matrix}\right).$

13. $\lim\limits_{|b_q|\to\infty}\left[\Gamma(1-b_q)\,G_{pq}^{mn}\left(-b_qz\,\middle|\,\begin{matrix}(a_p)\\(b_q)\end{matrix}\right)\right]=G_{p,\ q-1}^{mn}\left(z\,\middle|\,\begin{matrix}(a_p)\\(b_{q-1})\end{matrix}\right).$

Symmetry and translation formulae:

14. $G_{pq}^{mn}\left(z\,\middle|\,\begin{matrix}(a_p)\\(b_q)\end{matrix}\right)=G_{qp}^{nm}\left(\frac{1}{z}\,\middle|\,\begin{matrix}1-(b_q)\\1-(a_p)\end{matrix}\right)$ [$\arg(1/z)=-\arg z$].

15. $z^{\alpha}G_{pq}^{mn}\left(z\,\middle|\,\begin{matrix}(a_p)\\(b_q)\end{matrix}\right)=G_{pq}^{mn}\left(z\,\middle|\,\begin{matrix}(a_p)+\alpha\\(b_q)+\alpha\end{matrix}\right).$

Relations of special type:

16. $G_{pq}^{mn}\left(z\,\middle|\,\begin{matrix}a,\ (a_{p-1})\\(b_{q-1}),\ a\pm l\end{matrix}\right)=(-1)^l\,G_{p,\ q}^{m+1,\ n-1}\left(z\,\middle|\,\begin{matrix}(a_{p-1}),\ a\\a\pm l,\ (b_{q-1})\end{matrix}\right).$

17. $G_{pq}^{mn}\left(z\,\middle|\,\begin{matrix}(a_{p-1}),\ b\pm l\\b,\ (b_{q-1})\end{matrix}\right)=(-1)^l\,G_{p,\ q}^{m-1,\ n+1}\left(z\,\middle|\,\begin{matrix}b\pm l,\ (a_{p-1})\\(b_{q-1}),\ b\end{matrix}\right).$

18. $G_{pq}^{mn}\left(z\,\middle|\,\begin{matrix}a,\ (a_{p-2}),\ a\pm l\\a,\ (b_{q-1})\end{matrix}\right)=(-1)^l\,G_{p-1,\ q-1}^{m-1,\ n}\left(z\,\middle|\,\begin{matrix}a\pm l,\ (a_{p-2})\\(b_{q-1})\end{matrix}\right).$

19. $G_{pq}^{mn}\left(z\,\middle|\,\begin{matrix}(a_p)\\(b_q)\end{matrix}\right)=(2\pi)^{-(k-1)c^*}k^{\mu}G_{kp,\ kq}^{km,\ kn}\left(\frac{z^k}{k^{k(q-p)}}\,\middle|\,\begin{matrix}\Delta(k,a_1),\ \ldots,\ \Delta(k,a_p)\\\Delta(k,b_1),\ \ldots,\ \Delta(k,b_q)\end{matrix}\right).$

Relations between contiguous functions:

20. $(b_1-a_1+1)\,G_{pq}^{mn}\left(z\,\middle|\,\begin{matrix}(a_p)\\(b_q)\end{matrix}\right)=G_{pq}^{mn}\left(z\,\middle|\,\begin{matrix}a_1-1,\ a_2,\ \ldots,\ a_p\\(b_q)\end{matrix}\right)+$

$+G_{pq}^{mn}\left(z\,\middle|\,\begin{matrix}(a_p)\\b_1+1,\ b_2,\ \ldots,\ b_q\end{matrix}\right)$ [$m,\ n\geqslant1$].

21. $(a_p-a_1)\,G_{pq}^{mn}\left(z\,\middle|\,\begin{matrix}(a_p)\\(b_q)\end{matrix}\right)=G_{pq}^{mn}\left(z\,\middle|\,\begin{matrix}a_1-1,\ a_2,\ \ldots,\ a_p\\(b_q)\end{matrix}\right)+$

$+G_{pq}^{mn}\left(z\,\middle|\,\begin{matrix}(a_{p-1}),\ a_p-1\\(b_q)\end{matrix}\right)$ [$1\leqslant n\leqslant p-1$].

22. $G_{pq}^{mn}\left(z\,\middle|\,\begin{matrix}(a_{p-1}),\ a_1+\nu\\a_1+\nu,\ (b_{q-1})\end{matrix}\right)=(\nu+1)\,G_{p-1,\ q-1}^{m-1,\ n}\left(z\,\middle|\,\begin{matrix}(a_{p-1})\\(b_{q-1})\end{matrix}\right)-$

$-G_{p-1,\ q-1}^{m-1,\ n}\left(z\,\middle|\,\begin{matrix}a_1-1,\ a_2,\ \ldots,\ a_{p-1}\\(b_{q-1})\end{matrix}\right)$ [$1\leqslant n\leqslant p-1$].

23. $G_{pq}^{mn}\left(z\,\middle|\,\begin{matrix}(a_{p-1}),\ a_1+\nu-l\\a_1+\nu,\ (b_{q-1})\end{matrix}\right)=(-1)^l\sum_{k=0}^{l}\binom{l}{k}(-\nu)_{l-k}\times$

$\times G_{p-1,\ q-1}^{m-1,\ n}\left(z\,\middle|\,\begin{matrix}a_1-k,\ a_2,\ \ldots,\ a_{p-1}\\(b_{q-1})\end{matrix}\right).$

Formulae of the connection with functions of the argument $ze^{\pm i\pi}$.

24. $$G_{pq}^{mn}\left(z\,\middle|\,\begin{matrix}(a_p)\\(b_q)\end{matrix}\right)=\frac{1}{2\pi i}\left[e^{ib_{m+1}\pi}G_{p,q}^{m+1,n}\left(ze^{-i\pi}\,\middle|\,\begin{matrix}(a_p)\\(b_q)\end{matrix}\right)-e^{-ib_{m+1}\pi}G_{p,q}^{m+1,n}\left(ze^{i\pi}\,\middle|\,\begin{matrix}(a_p)\\(b_q)\end{matrix}\right)\right]\quad[m\leqslant q-1].$$

25. $$G_{pq}^{mn}\left(z\,\middle|\,\begin{matrix}(a_p)\\(b_q)\end{matrix}\right)=\frac{1}{2\pi i}\left[e^{ia_{n+1}\pi}G_{p,q}^{m,n+1}\left(ze^{-i\pi}\,\middle|\,\begin{matrix}(a_p)\\(b_q)\end{matrix}\right)-e^{-ia_{n+1}\pi}G_{p,q}^{m,n+1}\left(ze^{i\pi}\,\middle|\,\begin{matrix}(a_p)\\(b_q)\end{matrix}\right)\right]\quad[n\leqslant p-1].$$

26. $$G_{pq}^{mn}\left(z\,\middle|\,\begin{matrix}(a_p)\\0,\ 1/2,\ (b_{q-2})\end{matrix}\right)=\pi\left[G_{p,q}^{m-1,n}\left(ze^{i\pi}\,\middle|\,\begin{matrix}(a_p)\\0,\ (b_{q-2}),\ 1/2\end{matrix}\right)+iG_{p,q}^{m-1,n}\left(ze^{i\pi}\,\middle|\,\begin{matrix}(a_p)\\1/2,\ (b_{q-2}),\ 0\end{matrix}\right)\right].$$

27. $$G_{pq}^{mn}\left(z\,\middle|\,\begin{matrix}a+ib,\ a-ib,\ (a_{p-2})\\(b_q)\end{matrix}\right)=$$
$$=\frac{\pi i}{\sin 2b\pi}e^{-ia\pi}\left[e^{b\pi}G_{p,q}^{m,n-1}\left(ze^{i\pi}\,\middle|\,\begin{matrix}a+ib,\ (a_{p-2}),\ a-ib\\(b_q)\end{matrix}\right)-e^{-b\pi}G_{p,q}^{m,n-1}\left(ze^{i\pi}\,\middle|\,\begin{matrix}a-ib,\ (a_{p-2}),\ a+ib\\(b_q)\end{matrix}\right)\right].$$

28. $$G_{qq}^{mn}\left(z\,\middle|\,\begin{matrix}(a_q)\\(b_q)\end{matrix}\right)=\frac{\pi^{m+n-q}}{\sin\psi_q\pi}\sum_{h=1}^{m}e^{(m+n-q)b_h\pi i}\,\frac{\prod\limits_{k=n+1}^{q}\sin(a_k-b_h)\pi}{\sin((b_m)'-b_h)\pi}\times$$
$$\times G_{q+2,\,q+2}^{1,\,q+1}\left((-1)^{q-m-n}z\,\middle|\,\begin{matrix}b_h,\ (a_q),\ b_h+\psi_q\\b_h,\ (b_q),\ b_h+\psi_q\end{matrix}\right)\quad[|z|<1;\ \psi_q\neq 0,\ \pm1,\ \pm2,\ldots].$$

29. $$G_{qq}^{mn}\left(z\,\middle|\,\begin{matrix}(a_q)\\(b_q)\end{matrix}\right)=\frac{\pi^{m+n-q}}{\sin\psi_q\pi}\sum_{h=1}^{n}e^{(m+n-q)(1-a_h)\pi i}\,\frac{\prod\limits_{k=m+1}^{p}\sin(a_h-b_k)\pi}{\sin(a_h-(a_n)')\pi}\times$$
$$\times G_{q+2,\,q+2}^{q+1,\,1}\left((-1)^{m+n-q}z\,\middle|\,\begin{matrix}a_h,\ (a_q),\ a_h+\psi_q\\a_h,\ (b_q),\ a_h+\psi_q\end{matrix}\right)$$
$$[|z|>1;\ \psi_q\neq 0,\ \pm1,\ \pm2,\ \ldots;\ |\arg(1-(-1)^{q-m-n}z)|<\pi].$$

Here $\sin((b_m)'-b_h)\pi=\prod\limits_{\substack{k=1\\k\neq h}}^{m}\sin(b_k-b_h)\pi.$

Differentiation formulae:

30. $$\frac{d}{dz}\left[z^{\sigma}G_{pq}^{mn}\left(z\,\middle|\,\begin{matrix}(a_p)\\(b_q)\end{matrix}\right)\right]=z^{\sigma-1}G_{p+1,\,q+1}^{m,\,n+1}\left(z\,\middle|\,\begin{matrix}-\sigma,\ (a_p)\\(b_q),\ 1-\sigma\end{matrix}\right)=$$
$$=-z^{\sigma-1}G_{p+1,\,q+1}^{m+1,\,n}\left(z\,\middle|\,\begin{matrix}(a_p),\ -\sigma\\1-\sigma,\ (b_q)\end{matrix}\right)=(\sigma+b_1)\,z^{\sigma-1}G_{pq}^{mn}\left(z\,\middle|\,\begin{matrix}(a_p)\\(b_q)\end{matrix}\right)-z^{\sigma-1}G_{pq}^{mn}\left(z\,\middle|\,\begin{matrix}(a_p)\\b_1+1,\ b_2,\ \ldots,\ b_q\end{matrix}\right).$$

31. $$z\frac{d}{dz}G_{pq}^{mn}\left(z\,\middle|\,\begin{matrix}(a_p)\\(b_q)\end{matrix}\right)=G_{pq}^{mn}\left(z\,\middle|\,\begin{matrix}a_1-1,\ a_2,\ \ldots,\ a_p\\(b_q)\end{matrix}\right)+(a_1-1)\,G_{pq}^{mn}\left(z\,\middle|\,\begin{matrix}(a_p)\\(b_q)\end{matrix}\right)\quad[n\geqslant 1].$$

32. $$z^k\frac{d^k}{dz^k}G_{pq}^{mn}\left(z\,\middle|\,\begin{matrix}(a_p)\\(b_q)\end{matrix}\right)=G_{p+1,\,q+1}^{m,\,n+1}\left(z\,\middle|\,\begin{matrix}0,\ (a_p)\\(b_q),\ k\end{matrix}\right).$$

33. $z^k \frac{d^k}{dz^k} G_{pq}^{mn}\left(\frac{1}{z}\middle|\begin{matrix}(a_p)\\(b_q)\end{matrix}\right) = (-1)^k G_{p+1,\ q+1}^{m,\ n+1}\left(\frac{1}{z}\middle|\begin{matrix}1-k,\ (a_p)\\(b_q),\ 1\end{matrix}\right).$

34. $\frac{d^k}{dz^k}\left[z^{a_1-1} G_{pq}^{mn}\left(\frac{1}{z}\middle|\begin{matrix}(a_p)\\(b_q)\end{matrix}\right)\right] = (-1)^k z^{a_1-k-1} G_{pq}^{mn}\left(\frac{1}{z}\middle|\begin{matrix}a_1-k,\ a_2,\ \dots,\ a_p\\(b_q)\end{matrix}\right)$ $[n \geqslant 1]$.

35. $\frac{d^k}{dz^k}\left[z^{a_p-1} G_{pq}^{mn}\left(\frac{1}{z}\middle|\begin{matrix}(a_p)\\(b_q)\end{matrix}\right)\right] = z^{a_p-k-1} G_{pq}^{mn}\left(\frac{1}{z}\middle|\begin{matrix}(a_{p-1}),\ a_p-k\\(b_q)\end{matrix}\right)$ $[n \leqslant p-1]$.

36. $\frac{d}{dz}\left[z^{1-a_1} G_{pq}^{mn}\left(z\middle|\begin{matrix}(a_p)\\(b_q)\end{matrix}\right)\right] = z^{-a_1} G_{pq}^{mn}\left(z\middle|\begin{matrix}a_1-1,\ a_2,\ \dots,\ a_p\\(b_q)\end{matrix}\right)$ $[n \geqslant 1]$.

37. $\frac{d}{dz}\left[z^{1-a_p} G_{pq}^{mn}\left(z\middle|\begin{matrix}(a_p)\\(b_q)\end{matrix}\right)\right] = -z^{-a_p} G_{pq}^{mn}\left(z\middle|\begin{matrix}(a_{p-1}),\ a_p-1\\(b_q)\end{matrix}\right)$ $[n \leqslant p-1]$.

38. $\frac{d^k}{dz^k}\left[z^{-b_1} G_{pq}^{mn}\left(z\middle|\begin{matrix}(a_p)\\(b_q)\end{matrix}\right)\right] = (-1)^k z^{-b_1-k} G_{pq}^{mn}\left(z\middle|\begin{matrix}(a_p)\\b_1+k,\ b_2,\ \dots,\ b_q\end{matrix}\right)$ $[m \geqslant 1]$.

39. $\frac{d}{dz}\left[z^{-b_q} G_{pq}^{mn}\left(z\middle|\begin{matrix}(a_p)\\(b_q)\end{matrix}\right)\right] = z^{-b_q-1} G_{pq}^{mn}\left(z\middle|\begin{matrix}(a_p)\\(b_{q-1}),\ b_q+1\end{matrix}\right)$ $[m \leqslant q-1]$.

40. $z^k \frac{d^k}{dz^k} G_{pq}^{mn}\left(\omega z^{r/l}\middle|\begin{matrix}(a_p)\\(b_q)\end{matrix}\right) =$

$$= (2\pi)^{-(l-1)c^*} l^{\mu} r^k G_{lp+r,\ lq+r}^{lm,\ ln+r}\left(\omega^l l^{l(p-q)} z^r\middle|\begin{matrix}\Delta(r,\ 0),\ \Delta(l,\ a_1),\ \dots,\ \Delta(l,\ a_p)\\\Delta(l,\ b_1),\ \dots,\ \Delta(l,\ b_q),\ \Delta(r,\ k)\end{matrix}\right)$$

$[l,\ r=1,\ 2,\ \dots;\ c^*,\ \mu$ see 8.2.1].

41. $z^k \frac{d^k}{dz^k} G_{pq}^{mn}\left(\frac{\omega}{z^{r/l}}\middle|\begin{matrix}(a_p)\\(b_q)\end{matrix}\right) =$

$$= (2\pi)^{-(l-1)c^*} l^{\mu} r^k G_{lp+r,\ lq+r}^{lm+r,\ ln}\left(\frac{\omega^l l^{l(p-q)}}{z^r}\middle|\begin{matrix}\Delta(l,\ a_1),\ \dots,\ \Delta(l,\ a_p),\ \Delta(r,\ 1-k)\\\Delta(r,\ 1),\ \Delta(l,\ b_1),\ \dots,\ \Delta(l,\ b_q)\end{matrix}\right)$$

$[l,\ r=1,\ 2,\ \dots;\ c^*,\ \mu$ see 8.2.1].

Differentiating a G-function with respect to the parameters does not in general lead to G-functions. The differential equation satisfied by the function $G_{pq}^{mn}\left(z\middle|\begin{matrix}(a_p)\\(b_q)\end{matrix}\right)$ has the form

42. $$\left[(-1)^{m+n-p} z \prod_{j=1}^{p}\left(z\frac{d}{dz}-a_j+1\right) - \prod_{k=1}^{q}\left(z\frac{d}{dz}-b_k\right)\right] u = 0.$$

This equation of order $\max(p, q)$ has two ($z=0, \infty$ for $p \neq q$) or three ($z=0, (-1)^{m+n-q}, \infty$ for $p=q$) singular points. If $p<q$ (or $p>q$), then $z=0$ (or $z=\infty$) is a regular, and $z=\infty$ (or $z=0$) is a non-regular or essential singularity; if $p=q$, then all three singularities are regular.

43. $u = u_k^0(z) = e^{(m+n-p-1) b_k \pi i} G_{pq}^{1p}\left((-1)^{p-m-n+1} z\middle|\begin{matrix}(a_p)\\b_k,\ (b_q)'_k\end{matrix}\right)$

$[k=1,\ 2,\ \dots,\ q;\ b_j-b_k \neq 0,\ \pm 1,\ \pm 2,\ \dots;\ j \neq k;\ j,\ k=1,\ 2,\ \dots,\ q]$,

44. $u = u_j^1(z) = R_{jh}^{(q)}(z) = G_{q+2,\ q+2}^{2,\ q+2}\left(z_1\middle|\begin{matrix}b_h,\ b_j,\ (a_q)\\b_h,\ b_j,\ (b_q)\end{matrix}\right)$ $[j=1,\ 2,\ \dots,\ q;\ j \neq h]$,

45. $u = u_h^1(z) = \xi_{qh}(z,\ \psi_q)$ $[\psi_q \neq 0,\ \pm 1,\ \pm 2,\ \dots]$,

46. $u = u_h^1(z) = \varphi_q(z)$ $[\psi_q = 0,\ \pm 1,\ \pm 2,\ \dots]$,

$$\psi_q = \sum_{j=1}^{q}(a_j-b_j)-1, \quad z_1 = (-1)^{q-m-n} z = e^{(q-m-n)\pi i} z,$$

where the function $\xi_{qh}(z,\ \psi_q)$ can be represented by the formulae

47. $\xi_{qh}(z,\ \psi_q) = \left[G_{qq}^{q0}\left(z_1\middle|\begin{matrix}(a_q)\\(b_q)\end{matrix}\right) + e^{\mp\psi_q \pi i} G_{qq}^{0q}\left(z_1\middle|\begin{matrix}(a_q)\\(b_q)\end{matrix}\right)\right]$ $\left[\left\{\begin{matrix}-\pi < \arg(1-z_1) \leqslant 0\\0 \leqslant \arg(1-z_1) < \pi\end{matrix}\right\}\right]$.

48. $\xi_{qh}(z, \psi_q)=z_1^{b_h}(1-z_1)^{\psi_q}\sum_{j=0}^{\infty}\frac{c_{jqh}}{\Gamma(\psi_q+j+1)}(1-z_1)^j \quad [|1-z_1|<1; |\arg(1-z_1)|<\pi]$.

Here $c_{0qh}=1$, $c_{j11}=0$, $j=1,2,\ldots$, while the remaining coefficients c_{jqh} are defined by the relations

$$c_{12h}=R(\psi_2+b_h), \qquad 2c_{22h}=R(\psi_2+b_h+1)\,c_{12h}, \ldots,$$
$$jc_{j2h}=R(\psi_2+b_h+j-1)\,c_{j-1,2,h};$$
$$c_{13h}=\Delta R(\psi_3+b_h-1)-Q(\psi_3+b_h),$$
$$2c_{23h}=[\Delta R(\psi_3+b_h)-Q(\psi_3+b_h+1)]\,c_{13h}-R(\psi_3+b_h),$$

. . . .

$$jc_{j3h}=[\Delta R(\psi_3+b_h+j-2)-Q(\psi_3+b_h+j-1)]\,c_{j-1,3,h}-R(\psi_3+b_h+j-2)\,c_{j-2,3,h};$$
$$c_{1qh}=\frac{\Delta^{q-2}R(\psi_q+b_h-q+2)}{(q-2)!}-\frac{\Delta^{q-3}Q(\psi_q+b_h-q+3)}{(q-3)!},$$
$$2c_{2qh}=\left[\frac{\Delta^{q-2}R(\psi_q+b_h-q+3)}{(q-2)!}-\frac{\Delta^{q-3}Q(\psi_q+b_h-q+4)}{(q-3)!}\right]c_{1qh}-$$
$$-\left[\frac{\Delta^{q-3}R(\psi_q+b_h-q+3)}{(q-3)!}-\frac{\Delta^{q-4}Q(\psi_q+b_h-q+4)}{(q-4)!}\right],$$

. . . .

$$jc_{jqh}=\sum_{k=1}^{q-2}(-1)^{q-k}\left[\frac{\Delta^{k}R(\psi_q+b_h-q+j+1)}{k!}-\frac{\Delta^{k-1}Q(\psi_q+b_h-q+j+2)}{(k-1)!}\right]\times$$
$$\times c_{j+k-q+1,q,h}+(-1)^q R(\psi_q+b_h-q+j+1)\,c_{j-q+1,q,h},$$

where Δ is the difference operator of 7.2.3.74–75.

49. $R(x)=\prod_{j=1}^{q}(x+1-a_j)$, $Q(x)=\prod_{j=1}^{q}(x-b_j)$,

where the following relations hold

$$c_{jqh}=\xi_{qh}((-1)^{m+n-q}, -j), \quad \xi_{qh}(z, -l)=z_1^{b_h}\sum_{k=0}^{\infty}\frac{c_{k+l,q,h}}{k!}(1-z_1)^k$$
$$[|1-z_1|<1;\ l=0, 1, 2, \ldots].$$

The function $\varphi_q(z)$ is the same as the right hand sides of any of the corresponding formulae 8.2.2.53–56 given below.

In general, the functions $\xi_{qh}(z,\psi_q)$ and $\varphi_q(z)$ have, respectively, power (for $\operatorname{Re}\psi_q<0$; $\psi_q\neq-1,-2,\ldots$) and logarithmic-power singularities at the singular point $z=(-1)^{m+n-q}$ with a possible discontinuity on passing across the circle $|z|=1$ (when $m+n\leqslant q$), while all the functions $R_{jh}^{(q)}(z)$ are continuous at this point.

We denote by $u_k^{\infty}(z)$ the system obtained from $u_k^0(z)$ by simultaneous interchange of the positions of z and z^{-1}, $m, n, p, q, (a_p), (b_q)$ and $n, m, q, p, 1-(b_q), 1-(a_p)$ respectively.

The functions $u_k^0(z)$, $k=1, 2, \ldots, q$; $u_k^{\infty}(z)$, $k=1, 2, \ldots, p$; $R_{jh}^{(q)}(z)$, $j=1, 2, \ldots, q$, $j\neq h$, and $\xi_{qh}(z, \psi_q)$ (for $\psi_q\neq 0, \pm1, \pm2, \ldots$) or $\varphi_q(z)$ (for $\psi_q=0, \pm1, \pm2, \ldots$) then form three fundamental systems of solutions of the differential equation 8.2.2.42 respectively in neighbourhoods of the singular points $z=0$ for $p\leqslant q$, $z=\infty$ for $p\geqslant q$ and $z=(-1)^{m+n-q}$ for $p=q$.

In the case $p=q$, this equation is invariant under the interchange of the positions of the variables indicated above and $q+1$ of its solutions

$$u_0^0(z)=G_{pq}^{mn}\left(z\left|\begin{matrix}(a_p)\\(b_q)\end{matrix}\right.\right) \text{ and } u_k^0(z) \text{ or } R_{jh}^{(q)}(z),\ \xi_{qh}(z,\psi_q),\ \varphi_q(z)) \text{ or } u_k^{\infty}(z),$$

are interrelated by the formulae 8.2.2.5 (or 8.2.2.6 or 8.2.2.50, 53–57).

The representations of the functions $G_{qq}^{mn}\left(z\left|\begin{matrix}(a_q)\\(b_q)\end{matrix}\right.\right)$ in a neighbourhood of the singular point $z=(-1)^{m+n-q}$ have different forms depending on the position of z (inside or outside the circle $|z|=1$) and on the quantity ψ_q:

50. $G_{qq}^{mn}\left(z\left|\begin{matrix}(a_q)\\(b_q)\end{matrix}\right.\right)=$

$$=-\frac{\pi^{m+n-q}}{\sin\psi_q\pi}\sum_{h=1}^{m}e^{(m+n-q)\,b_h\pi i}\frac{\prod\limits_{k=n+1}^{q}\sin(a_k-b_h)\pi}{\sin((b_m)'-b_h)\pi}[\zeta_{qh}(z)+\xi_{qh}(z,\ \psi_q)]$$

[$|\arg(1-z_1)|<\pi$; $\psi_q\neq 0, \pm1, \pm2, \ldots$, $\zeta_{qh}(z)$ see p. 624],

here $\sin((b_m)'-b_h)\pi=\prod\limits_{\substack{k=1\\k\neq h}}^{m}\sin(b_k-b_h)\pi$; in particular, formula 8.2.2.50 reduces to the relations

51. $G_{qq}^{mn}\left(z\left|\begin{matrix}(a_q)\\(b_q)\end{matrix}\right.\right)=$

$$=-\frac{\pi^{m+n-q}}{\sin\psi_q\pi}\sum_{h=1}^{m}e^{(m+n-q)\,b_h\pi i}\frac{\prod\limits_{k=n+1}^{q}\sin(a_k-b_h)\pi}{\sin((b_m)'-b_h)\pi}\left[G_{qq}^{q0}\left((-1)^{q-m-n}z\left|\begin{matrix}(a_q)\\(b_q)\end{matrix}\right.\right)-\right.$$

$$\left.-\frac{1}{\pi^2}\sum_{j=1}^{q}\sin(b_h-b_j)\pi\,\frac{\sin((a_q)-b_j)\pi}{\sin((b_q)'-b_j)\pi}\,G_{q+2,\ q+2}^{2,\ q+2}\left((-1)^{q-m-n}z\left|\begin{matrix}b_h,\ b_j,\ (a_q)\\b_h,\ b_j,\ (b_q)\end{matrix}\right.\right)\right]$$

[$|z|<1$; $\psi_q\neq 0, \pm1, \pm2, \ldots$].

52. $G_{qq}^{mn}\left(z\left|\begin{matrix}(a_q)\\(b_q)\end{matrix}\right.\right)=$

$$=-\frac{\pi^{m+n-q}}{\sin\psi_q\pi}\sum_{h=1}^{n}e^{(m+n-q)(1-a_h)\pi i}\frac{\prod\limits_{k=m+1}^{q}\sin(a_h-b_k)\pi}{\sin(a_h-(a_n)')\pi}\left[G_{qq}^{0q}\left((-1)^{m+n-q}z\left|\begin{matrix}(a_q)\\(b_q)\end{matrix}\right.\right)-\right.$$

$$\left.-\frac{1}{\pi^2}\sum_{j=1}^{q}\sin(a_h-a_j)\pi\,\frac{\sin((b_q)-a_j)\pi}{\sin((a_q)'-a_j)\pi}\,G_{q+2,\ q+2}^{q+2,\ 2}\left((-1)^{m+n-q}z\left|\begin{matrix}a_h,\ a_j,\ (a_q)\\a_h,\ a_j,\ (b_q)\end{matrix}\right.\right)\right]$$

[$|z|>1$; $\psi_q\neq 0, \pm1, \pm2, \ldots$; $|\arg(1-(-1)^{q-m-n}z)|<\pi$],

while for integral ψ_q, it takes the form

53. $$G_{qq}^{mn}\left(z\left|\begin{matrix}(a_q)\\(b_q)\end{matrix}\right.\right)=\sum_{h=1}^{m}z^{b_h}\sum_{k=0}^{l-1}d_{kqh}(z_1-1)^k-(-1)^l\pi^{m+n-q-1}\times$$

$$\times\sum_{h=1}^{m}e^{(m+n-q)\,b_h\pi i}\frac{\prod\limits_{k=n+1}^{q}\sin(a_k-b_h)\pi}{\sin((b_m)'-b_h)\pi}[\xi_{qh}(z,\ l)\ln(1-z_1)-\theta_{qh}^{+}(z,l)]$$

[$\psi_q=l$; $l=0, 1, 2, \ldots$; $|z|<1$].

54. $$G_{qq}^{mn}\left(z\left|\begin{matrix}(a_q)\\(b_q)\end{matrix}\right.\right)=\sum_{h=1}^{n} z^{a_h-1}\sum_{k=0}^{l-1} d^{*}_{kqh}(z_2-1)^k-(-1)^l\pi^{m+n-q-1}\times$$
$$\times\sum_{h=1}^{n} e^{(m+n-q)(1-a_h)\pi i}\frac{\prod\limits_{k=m+1}^{q}\sin(a_h-b_k)\pi}{\sin(a_h-(a_n)')\pi}[\xi^{*}_{qh}(z,l)\ln(1-z_2)-\theta^{+*}_{qh}(z,l)]$$
$$[\psi_q=l;\ l=0,1,2,\ldots;\ |z|>1;\ |\arg(1-z_2)|<\pi;\ z_2=(-1)^{q-m-n}z^{-1}],$$

55. $$G_{qq}^{mn}\left(z\left|\begin{matrix}(a_q)\\(b_q)\end{matrix}\right.\right)=\sum_{h=1}^{m}\frac{e^{(m+n-q)b_h\pi i}}{\pi^{q-m-n+1}}\frac{\prod\limits_{k=n+1}^{q}\sin(a_k-b_h)\pi}{\sin((b_m)'-b_h)\pi}\left\{(l-1)!\,z_1^{b_h}(1-z_1)^{-l}\times\right.$$
$$\left.\times\sum_{j=0}^{l-1}\frac{c_{jqh}}{(1-l)_j}(1-z_1)^j+(-1)^{l-1}\xi_{qh}(z,-l)\ln(1-z_1)+(-1)^l\theta^{-}_{qh}(z,l)\right\}$$
$$[\psi_q=-l;\ l=1,2,\ldots;\ |z|<1].$$

56. $$G_{qq}^{mn}\left(z\left|\begin{matrix}(a_q)\\(b_q)\end{matrix}\right.\right)=$$
$$=\pi^{m+n-q-1}\sum_{h=1}^{n}e^{(m+n-q)(1-a_h)\pi i}\frac{\prod\limits_{k=m+1}^{q}\sin(a_h-b_k)\pi}{\sin(a_h-(a_n)')\pi}\times$$
$$\times\left\{(l-1)!\,z_2^{1-a_h}(1-z_2)^{-l}\sum_{j=0}^{l-1}\frac{c^{*}_{jqh}}{(1-l)_j}(1-z_2)^j+(-1)^{l-1}\xi^{*}_{qh}(z,-l)\ln(1-z_2)+\right.$$
$$\left.+(-1)^l\theta^{-*}_{qh}(z,l)\right\}\qquad[\psi_q=-l;\ l=1,2,\ldots;\ |z|>1;\ |\arg(1-z_2)|<\pi;\ z_2=(-1)^{q-m-n}z^{-1}].$$

The function $\zeta_{qh}(z)$ is representable in any of the following forms:

$$\zeta_{qh}(z)=z_1^{b_h}\sum_{j=0}^{\infty}b_{jqh}(z_1-1)^j\qquad[|1-z_1|<1],$$
$$b_{jqh}=\frac{1}{2\pi i}\int_{\gamma-i\infty}^{\gamma+i\infty}\frac{t^{-b_h}\xi_{qh}((-1)^{m+n-q}t,\psi_q)}{(t-1)^{j+1}}dt\qquad[0<\gamma<1],$$
$$\zeta_{qh}(z)=-G_{q+2,\,q+2}^{1,\,q+1}\left(z_1\left|\begin{matrix}b_h,\ (a_q),\ \psi_q+b_h\\b_h,\ (b_q),\ \psi_q+b_h\end{matrix}\right.\right)-G_{qq}^{q0}\left(z_1\left|\begin{matrix}(a_q)\\(b_q)\end{matrix}\right.\right)\qquad[|z|<1],$$
$$\zeta_{qh}(z)=\frac{z_1^{b_h}}{2\pi i}\int_{\gamma-i\infty}^{\gamma+i\infty}\frac{t^{-b_h}\xi_{qh}((-1)^{m+n-q}t,\psi_q)}{t-z_1}dt$$
$$[0<\gamma<1;\ z_1\text{ is to the right of }(\gamma-i\infty,\gamma+i\infty)].$$

57. $\zeta_{qh}(z)=-\dfrac{1}{\pi^2}\sum\limits_{j=1}^{q}\sin(b_j-b_h)\pi\dfrac{\sin(b_j-(a_q))\pi}{\sin(b_j-(b_q)')\pi}R_{jh}^{(q)}(z)$, where the functions $R_{jh}^{(q)}(z)$ are defined by equation 8.2.2.44. The remaining coefficients and functio[ns] have the representations

$$d_{kqh}=\Gamma\left[\begin{matrix}(b_m)'-b_h,\ 1+b_h-(a_n)+k\\a_{n+1}-b_h,\ \ldots,\ a_q-b_h,\ 1+b_h-b_{m+1}+k,\ \ldots,\ 1+b_h-b_q+k\end{matrix}\right]\times$$
$$\times\frac{(1+b_h-a_{n+1})_k\ldots(1+b_h-a_q)_k}{\Pi(1+b_h-(b_m))_k}\,{}_qF_{q-1}\left(\begin{matrix}1+b_h-(a_q)+k;\ 1\\1+b_h-(b_q)'_h+k\end{matrix}\right),$$

$$\theta_{qh}^{\pm}(z, l) = -\frac{z_1^{b_h}}{2\pi i}\int_{\gamma-i\infty}^{\gamma+i\infty} t^{-b_h}\left(\frac{1-z_1}{1-t}\right)^{(l\pm l)/2}\ln(1-t)\,\frac{\xi_{qh}((-1)^{m+n-q}t, l)}{t-z_1}\,dt$$

$0<\gamma<1$; z_1 is to the right of $(\gamma-i\infty, \gamma+i\infty)$; $\operatorname{Re}(b_h-a_j)>-(l\pm l)/2-1$; $j=1, 2, \dots, q$; $b_h-a_j\neq 0, -1, -2, \dots, -(l\pm l)/2]$,

$$\theta_{qh}^{\pm}(z, l) = z_1^{b_h}(1-z_1)^{(l\pm l)/2}\sum_{k=0}^{\infty} e_{kqh}^{\pm}(1-z_1)^k \qquad [|1-z_1|<1],$$

$$e_{kqh}^{\pm} = \frac{1}{2\pi i}\int_{\gamma-i\infty}^{\gamma+i\infty} t^{-b_h}\ln(1-t)\,\frac{\xi_{qh}((-1)^{m+n-q}t, l)}{(1-t)^{(l\pm l)/2+k+1}}\,dt$$

$0<\gamma<1$; $\operatorname{Re}(b_h-a_j)>-(l\pm l)/2-1$; $j=1, 2, \dots, q$; $b_h-a_j\neq 0, -1, -2, \dots, -(l\pm l)/2]$,

where $\xi_{qh}^*(z, \psi_q)$, $\theta_{qh}^{\pm *}(z, l)$, c_{jqh}^*, d_{kqh}^* are obtained from $\xi_{qh}(z, \psi_q)$, $\theta_{qh}^{\pm}(z, l)$, c_{jqh}, d_{kqh} by a simultaneous interchange of the positions of z and z^{-1}, m and n, a_h and $1-b_h$, b_h and $1-a_h$. The functions $\theta_{qh}^{\pm}(z, l)$ are continuous at the point $z=(-1)^{m+n-q}$.

The value of the G-function at this point is expressed by the equation

58. $$G_{qq}^{mn}\left((-1)^{m+n-q}\,\middle|\,\begin{matrix}(a_q)\\(b_q)\end{matrix}\right) = -\frac{\pi^{m+n-q}}{\sin\psi_q\pi}\sum_{h=1}^{m} b_{0qh}e^{(m+n-q)b_h\pi i}\times$$

$$\times\frac{\prod_{k=n+1}^{q}\sin(a_k-b_h)\pi}{\sin((b_m)'-b_h)\pi} \qquad [m+n>q;\ \operatorname{Re}\psi_q>0],$$

where the right hand side reduces to ratios of products of gamma functions only in particular cases. If the following condition holds:

$$\prod_{k=1}^{l-q+1} R(b_h+k)=0 \qquad \left(\prod_{k=1}^{l-q+1} R(1-a_h+k)=0\right)$$

$$[h=1, 2, \dots, q],$$

where $R(x)$ has the form 8.2.2.49, then the second term containing the logarithm in formula 8.2.2.53 (or 8.2.2.54) is absent. If the relation $\xi_{qh}(z, l)=0$, then $\theta_{qh}^{-}(z, l)=0$ (while if $\xi_{qh}^{*}(z, l)=0$, then $\theta_{qh}^{-*}(z, l)=0$) and formula 8.2.2.55 (or 8.2.2.56) does not contain the logarithmic term. For the latter requirements to hold, it is necessary and sufficient that the following conditions respectively hold:

$$\prod_{j=1}^{l} R(b_h-j)=0 \qquad \left(\prod_{j=1}^{l} R(1-a_h-j)=0\right) \qquad [h=1, 2, \dots, q].$$

If $m+n-q>0$, then the function $G_{qq}^{ml}\left(z\,\middle|\,\begin{matrix}(a_q)\\(b_q)\end{matrix}\right)$ is analytic in the sector $|\arg z|<(m+n-q)\pi$ and the right hand sides of formulae 8.2.2.51–52, 53–54, 55–56 can respectively be analytically continued into each other across the circle $|z|=1$. If, furthermore, $m+n-p\geqslant 2$, then for any ψ_q, the points $z=\pm 1$ are not singular if the arguments $\pi k, k=0, \pm 1, \pm 2, \dots$ (where $|k|<m+n-q$) are assigned to them. The coefficients of the G_{qq}^{q0}- and G_{qq}^{0q}-functions in formulae 8.2.2.51–52 then vanish, that is,

$$\sum_{h=1}^{m} e^{kb_h\pi i}\frac{\prod_{k=n+1}^{q}\sin(a_k-b_h)\pi}{\sin((b_m)'-b_h)\pi}=0 \qquad [|k|<m+n-q;\ k=0, \pm 1, \pm 2, \dots],$$

$$\sum_{h=1}^{n} e^{ka_h\pi i} \frac{\prod_{k=m+1}^{q} \sin(a_h-b_k)\pi}{\sin(a_h-(a_n)')\pi}=0 \quad [|k|<m+n-q;\ k=0,\ \pm 1,\ \pm 2,\ \ldots$$

If $m+n-q=1$ and $|\arg z|<\pi$, then at the singular point $z=-1$, th function $G_{qq}^{mn}\left(z\left|\begin{matrix}(a_q)\\(b_q)\end{matrix}\right.\right)$ is continuous for $\operatorname{Re}\psi_q=0$, bounded for $\operatorname{Re}\psi_q=0$ $\psi_q\neq 0$, has in general a logarithmic singularity for $\psi_q=0$ (see 8.2.2.53–56) while for $\operatorname{Re}\psi_q<0$, it has a power singularity of order $-\psi_q$ to which, fo integral ψ_q, a logarithmic singularity can also be added, the followin equality being valid:

$$59.\quad \lim_{z\to -1}\left[(1+z)^{-\psi_q}G_{qq}^{mn}\left(z\left|\begin{matrix}(a_q)\\(b_q)\end{matrix}\right.\right)\right]=\Gamma(-\psi_q)\sum_{h=1}^{m} e^{b_h\pi i}\frac{\prod_{k=n+1}^{q}\sin(a_k-b_h)\pi}{\sin((b_m)'-b_h)\pi}.$$

If $m+n-q\leqslant 0$, the function $G_{qq}^{mn}\left(z\left|\begin{matrix}(a_q)\\(b_q)\end{matrix}\right.\right)$ is a piecewise analytic functio of z with a discontinuity on the circle $|z|=1$. If $m+n-q=0$, then at th singular point $z=1$, this function is continuous for $\operatorname{Re}\psi_q>0$, bounded fo $\operatorname{Re}\psi_q=0$, $\psi_q\neq 0$, has, in general, a logarithmic singularity for $\psi_q=0$, $mn\neq 0$ while for $\operatorname{Re}\psi_q<0$, has a power singularity of order $-\psi_q$, to which a logarithmic singularity can also be added when ψ_q is an integer. Here th asymptotics of the function $G_{qq}^{mn}\left(z\left|\begin{matrix}(a_q)\\(b_q)\end{matrix}\right.\right)$ as $z\to 1$ depend on the way approaches 1 (from outside or inside the disc $|z|\leqslant 1$). In particular, th following equalities hold:

$$60.\quad \lim_{\substack{z\to 1\\|z|<1}}\left[(1-z)^{-\psi_q}G_{qq}^{mn}\left(z\left|\begin{matrix}(a_q)\\(b_q)\end{matrix}\right.\right)\right]=\frac{\Gamma(-\psi_q)}{\pi}\sum_{h=1}^{m}\frac{\prod_{k=n+1}^{q}\sin(a_k-b_h)\pi}{\sin((b_m)'-b_h)\pi},$$

$$61.\quad \lim_{\substack{z\to 1\\|z|>1\\z\notin[1,\infty)}}\left[(z-1)^{-\psi_q}G_{qq}^{mn}\left(z\left|\begin{matrix}(a_q)\\(b_q)\end{matrix}\right.\right)\right]=\frac{\Gamma(-\psi_q)}{\pi}\sum_{h=1}^{n}\frac{\prod_{k=m+1}^{q}\sin(a_h-b_k)\pi}{\sin(a_h-(a_n)')\pi}.$$

With regard to other properties and applications of the Meijer G function, see [1, 11, 13, 46].

8.3. THE FOX *H*-FUNCTION $H_{pq}^{mn}\left[z\left|\begin{matrix}[\mathbf{a}_p, \mathbf{A}_p]\\ [\mathbf{b}_q, \mathbf{B}_q]\end{matrix}\right.\right]$

8.3.1. Definition and notation.

The Fox H-function of order (m,n,p,q), where $0\leqslant m\leqslant q$, $0\leqslant n\leqslant p$, is defined by the formula

$$1.\quad H_{pq}^{mn}\left[z\left|\begin{matrix}[a_p, A_p]\\ [b_q, B_q]\end{matrix}\right.\right]\equiv H_{pq}^{mn}\left[z\left|\begin{matrix}(a_1, A_1), & \ldots, & (a_p, A_p)\\ (b_1, B_1), & \ldots, & (b_q, B_q)\end{matrix}\right.\right]=$$

$$=\frac{1}{2\pi i}\int_L \frac{\prod_{j=1}^{m}\Gamma(b_j+B_js)\prod_{j=1}^{n}\Gamma(1-a_j-A_js)}{\prod_{j=n+1}^{p}\Gamma(a_j+A_js)\prod_{j=m+1}^{q}\Gamma(1-b_j-B_js)}z^{-s}\,d$$

$$A_i,\quad B_j>0,\quad i=1,2,\ldots,p,\quad j=1,2,\ldots,q,$$

$$[a_p, A_p]=(a_1, A_1),\ \ldots,\ (a_p, A_p),\quad [b_q, B_q]=(b_1, B_1),\ \ldots,\ (b_q, B_q).$$

The integral converges if one of the following conditions holds:

1) $L=L_{i\infty}$; $a^*>0$, $|\arg z|<a^*\pi/2$;

2) $L=L_{i\infty}$; $a^*\geqslant 0$, $|\arg z|=a^*\pi/2$, $\gamma\Delta<-\operatorname{Re}\mu$;

3) $L=L_{-\infty}$; $\Delta>0$, $0<|z|<\infty$ or $\Delta=0$, $0<|z|<\beta$ or $\Delta=0$, $a^*\geqslant 0$, $|z|=\beta$, $\operatorname{Re}\mu<0$;

4) $L=L_{+\infty}$; $\Delta<0$, $0<|z|<\infty$ or $\Delta=0$, $|z|>\beta$ or $\Delta=0$, $a^*\geqslant 0$, $|z|=\beta$, $\operatorname{Re}\mu<0$;

here

$$a^*=\sum_{j=1}^{n}A_j-\sum_{j=n+1}^{p}A_j+\sum_{j=1}^{m}B_j-\sum_{j=m+1}^{q}B_j,$$

$$c^*=m+n-\frac{p+q}{2},$$

$$\beta=\prod_{j=1}^{p}A_j^{-A_j}\prod_{j=1}^{q}B_j^{B_j},$$

$$\gamma=\lim_{\substack{s\to\infty\\ s\in L_{i\infty}}}\operatorname{Re}s,$$

$$\Delta=\sum_{j=1}^{q}B_j-\sum_{j=1}^{p}A_j,$$

$$\mu=\sum_{j=1}^{q}b_j-\sum_{j=1}^{p}a_j+\frac{p-q}{2}+1.$$

The contour $L_{-\infty}$ (or $L_{+\infty}$) is a left (or right) loop situated in a horizontal strip, starting at the point $-\infty+i\varphi_1$ (or $+\infty+i\varphi_1$), leaves all the poles of the functions $\Gamma(b_j+B_js)$, $j=1,2,\ldots,m$, to the left, and of the functions $\Gamma(1-a_j-A_js)$, $j=1,2,\ldots,n$, to the right of the contour, and finishes at the point $-\infty+i\varphi_2$ (or $+\infty+i\varphi_2$), where $\varphi_1<\varphi_2$. The contour $L_{i\infty}$ starts at the point $\gamma-i\infty$, finishes at $\gamma+i\infty$ and separates the above poles in the same way as $L_{\pm\infty}$.

8.3.2. Basic properties.

The H-function is symmetric with respect to the pairs $(a_1, A_1),\ldots,(a_n, A_n)$ of $[a_p, A_p]$, $(a_{n+1}, A_{n+1}),\ldots,(a_p, A_p)$ of $[a_p, A_p]$, $(b_1\, B_1),\ldots,(b_m, B_m)$ of $[b_q, B_q]$ and $(b_{m+1}, B_{m+1}),\ldots,(b_q, B_q)$ of $[b_q, B_q]$ individually.

$H_{pq}^{0,n}\left[z\left|\begin{matrix}[a_p,\ A_p]\\ [b_q,\ B_q]\end{matrix}\right.\right]=0$ when the conditions (1), (3) or (2), (3) of 8.3.1. hold.

Representation in the form of a series:

$$H_{pq}^{mn}\left[z\left|\begin{matrix}[a_p,\ A_p]\\ [b_q,\ B_q]\end{matrix}\right.\right]=$$

$$=\sum_{i=1}^{m}\sum_{k=0}^{\infty}\frac{\prod\limits_{j=1,\,j\neq i}^{m}\Gamma(b_j-(b_i+k)B_j/B_i)}{\prod\limits_{j=m+1}^{q}\Gamma(1-b_j+(b_i+k)B_j/B_i)}\times$$

$$\times\frac{\prod\limits_{j=1}^{n}\Gamma(1-a_j+(b_i+k)A_j/B_i)}{\prod\limits_{j=n+1}^{p}\Gamma(a_j-(b_i+k)A_j/B_i)}\ \frac{(-1)^k z^{(b_i+k)/B_i}}{k!B_i}$$

$\geqslant 0$; conditions 1)–3) and $B_k(b_j+l)\neq B_j(b_k+s)$, $j\neq k$; $j, k=1, 2, \ldots, m$; $l, s=0, 1, 2, \ldots$].

4. $$H_{pq}^{mn}\left[z\left|\begin{matrix}[a_p, A_p]\\ [b_q, B_q]\end{matrix}\right.\right]=\sum_{i=1}^{n}\sum_{k=0}^{\infty}\frac{\prod_{j=1;\, j\neq i}^{n}\Gamma(1-a_j-(1-a_i+k)A_j/A_i)}{\prod_{j=n+1}^{p}\Gamma(a_j+(1-a_i+k)A_j/A_i)}\times$$
$$\times\frac{\prod_{j=1}^{m}\Gamma(b_j+(1-a_i+k)B_j/A_i)}{\prod_{j=m+1}^{q}\Gamma(1-b_j-(1-a_i+k)B_j/A_i)}\frac{(-1)^k z^{-(1-a_i+k)/A_i}}{k!\,A_i}$$
[$\Delta\leqslant 0$; conditions 1), 2), 4) and $A_k(1-a_j+l)\neq A_j(1-a_k+s)$, $j\neq k$; $j, k=1, 2, \ldots, l, s=0, 1, 2, \ldots$].

5. The function $H_{pq}^{mn}\left[z\left|\begin{matrix}[a_p, A_p]\\ [b_q, B_q]\end{matrix}\right.\right]$ is analytic with respect to z in the sector $|\arg z|<a^*\pi/2$. If $a^*>0$, $\Delta=0$, then functions satisfying conditions (3), (4) can be analytically continued into each other through the circle $|z|=\beta$ in this sector. In the case $a^*\leqslant 0$, the sector of analyticity is absent, but if $\Delta=0$, the above two functions which are analytic for $|z|<\beta$ and $|z|>\beta$ have the same limit as $z\to\beta$ along the ray $\arg z=0$ when $a^*=0$, $\operatorname{Re}\mu<0$.

Formulae for lowering the order. Symmetry and translation formulae:

6. $$H_{pq}^{mn}\left[z\left|\begin{matrix}[a_p, A_p]\\ [b_{q-1}, B_{q-1}], (a_1, A_1)\end{matrix}\right.\right]=H_{p-1,\, q-1}^{m,\, n-1}\left[z\left|\begin{matrix}(a_2, A_2), \ldots, (a_p, A_p)\\ [b_{q-1}, B_{q-1}]\end{matrix}\right.\right].$$

7. $$H_{pq}^{mn}\left[z\left|\begin{matrix}[a_p, A_p]\\ [b_q, B_q]\end{matrix}\right.\right]=H_{qp}^{nm}\left[\frac{1}{z}\left|\begin{matrix}[1-b_q, B_q]\\ [1-a_p, A_p]\end{matrix}\right.\right].$$

8. $$z^{\alpha}H_{pq}^{mn}\left[z\left|\begin{matrix}[a_p, A_p]\\ [b_q, B_q]\end{matrix}\right.\right]=H_{pq}^{mn}\left[z\left|\begin{matrix}[a_p+\alpha A_p, A_p]\\ [b_q+\alpha B_q, B_q]\end{matrix}\right.\right].$$

Relations of special form:

9. $$H_{pq}^{mn}\left[z\left|\begin{matrix}(a, A), [a_{p-1}, A_{p-1}]\\ [b_{q-1}, B_{q-1}], (a\pm l, A)\end{matrix}\right.\right]=(-1)^l H_{p,\, q}^{m+1,\, n-1}\left[z\left|\begin{matrix}[a_{p-1}, A_{p-1}], (a, A)\\ (a\pm l, A), [b_{q-1}, B_{q-1}]\end{matrix}\right.\right]$$

10. $$H_{pq}^{mn}\left[z\left|\begin{matrix}[a_{p-1}, A_{p-1}], (b\pm l, B)\\ (b, B), [b_{q-1}, B_{q-1}]\end{matrix}\right.\right]=(-1)^l H_{p,\, q}^{m-1,\, n+1}\left[z\left|\begin{matrix}(b\pm l, B), [a_{p-1}, A_{p-1}]\\ [b_{q-1}, B_{q-1}], (b, B)\end{matrix}\right.\right]$$

11. $$H_{pq}^{mn}\left[z\left|\begin{matrix}(a, A), [a_{p-2}, A_{p-2}], (a\pm l, A)\\ (a, A), [b_{q-1}, B_{q-1}]\end{matrix}\right.\right]=$$
$$=(-1)^l H_{p-1,\, q-1}^{m-1,\, n}\left[z\left|\begin{matrix}(a\pm l, A), [a_{p-2}, A_{p-2}]\\ [b_{q-1}, B_{q-1}]\end{matrix}\right.\right]$$

12. $$H_{pq}^{mn}\left[z\left|\begin{matrix}[a_p, A_p]\\ [b_q, B_q]\end{matrix}\right.\right]=(2\pi)^{-(k-1)c^*}k^{\mu}H_{kp,\, kq}^{km,\, kn}\left[(zk^{-\Delta})^k\left|\begin{matrix}(\Delta(k, a_p), A_p)\\ (\Delta(k, b_q), B_q)\end{matrix}\right.\right].$$

Relations between adjacent functions:

13. $$(b_1A_1-a_1B_1+B_1)H_{pq}^{mn}\left[z\left|\begin{matrix}[a_p, A_p]\\ [b_q, B_q]\end{matrix}\right.\right]=$$
$$=B_1H_{pq}^{mn}\left[z\left|\begin{matrix}(a_1-1, A_1), (a_2, A_2), \ldots, (a_p, A_p)\\ [b_q, B_q]\end{matrix}\right.\right]+$$
$$+A_1H_{pq}^{mn}\left[z\left|\begin{matrix}[a_p, A_p]\\ (b_1+1, B_1), (b_2, B_2), \ldots, (b_q, B_q)\end{matrix}\right.\right]$$ [$m, n\geqslant$

14. $$H_{pq}^{mn}\left[z\left|\begin{matrix}[a_{p-1}, A_{p-1}], (a_1+\alpha, A_1)\\ (a_1+\alpha+1, A_1), [b_{q-1}, B_{q-1}]\end{matrix}\right.\right]=$$
$$=(\alpha+1)H_{p-1,\, q-1}^{m-1,\, n}\left[z\left|\begin{matrix}[a_{p-1}, A_{p-1}]\\ [b_{q-1}, B_{q-1}]\end{matrix}\right.\right]-$$
$$-H_{p-1,\, q-1}^{m-1,\, n}\left[z\left|\begin{matrix}(a_1-1, A_1), (a_2, A_2), \ldots, (a_{p-1}, A_{p-1})\\ [b_{q-1}, B_{q-1}]\end{matrix}\right.\right]$$

Differentiation formulae:

15. $$\frac{d}{dz}\left\{z^{\sigma} H_{pq}^{mn}\left[z\left|\begin{matrix}[a_p, A_p]\\ [b_q, B_q]\end{matrix}\right.\right]\right\}=z^{\sigma-1}H_{p+1,\ q+1}^{m,\ n+1}\left[z\left|\begin{matrix}(-\sigma,\ 1),\ [a_p,\ A_p]\\ [b_q,\ B_q],\ (1-\sigma,\ 1)\end{matrix}\right.\right].$$

16. $$\frac{d^k}{dz^k}\left\{z^{-\sigma(1-a_1)/A_1}H_{pq}^{mn}\left[z^{-\sigma}\left|\begin{matrix}[a_p, A_p]\\ [b_q, B_q]\end{matrix}\right.\right]\right\}=$$

$$=\left(-\frac{\sigma}{A_1}\right)^k z^{-k-\sigma(1-a_1)/A_1}H_{pq}^{mn}\left[z^{-\sigma}\left|\begin{matrix}(a_1-k,\ A_1),\ (a_2,\ A_2),\ \dots,\ (a_p,\ A_p)\\ [b_q,\ B_q]\end{matrix}\right.\right]$$

$[n \geqslant 1;\ \sigma=A_1$ for $k>1]$.

17. $$\frac{d^k}{dz^k}\left\{z^{-\sigma b_1/B_1}H_{pq}^{mn}\left[z^{\sigma}\left|\begin{matrix}[a_p, A_p]\\ [b_q, B_q]\end{matrix}\right.\right]\right\}=$$

$$=\left(-\frac{\sigma}{B_1}\right)^k z^{-k-\sigma b_1/B_1}H_{pq}^{mn}\left[z^{\sigma}\left|\begin{matrix}[a_p,\ A_p]\\ (b_1+k,\ B_1),\ (b_2,\ B_2),\ \dots,\ (b_q,\ B_q)\end{matrix}\right.\right]$$

$[m \geqslant 1;\ \sigma=B_1$ for $k>1]$.

18. $$\frac{d^k}{dz^k}H_{pq}^{mn}\left[(cz+d)^{\sigma}\left|\begin{matrix}[a_p, A_p]\\ [b_q, B_q]\end{matrix}\right.\right]=\frac{c^k}{(cz+d)^k}H_{p+1,\ q+1}^{m,\ n+1}\left[(cz+d)^{\sigma}\left|\begin{matrix}(0,\sigma),[a_p,A_p]\\ [b_q,B_q],(k,\sigma)\end{matrix}\right.\right]$$

$[\sigma>0]$.

19. $$\frac{d^k}{dz^k}H_{pq}^{mn}\left[\frac{1}{(cz+d)^{\sigma}}\left|\begin{matrix}[a_p, A_p]\\ [b_q, B_q]\end{matrix}\right.\right]=$$

$$=\frac{c^k}{(cz+d)^k}H_{p+1,\ q+1}^{m,\ n+1}\left[\frac{1}{(cz+d)^{\sigma}}\left|\begin{matrix}[a_p,\ A_p],\ (1-k,\ \sigma)\\ (1,\ \sigma),\ [b_q,\ B_q]\end{matrix}\right.\right]\qquad [\sigma>0].$$

20. $$\prod_{j=1}^{r}\left(z\frac{d}{dz}-c_j\right)z^{\alpha}H_{pq}^{mn}\left[z^{\sigma}\omega\left|\begin{matrix}[a_p, A_p]\\ [b_q, B_q]\end{matrix}\right.\right]=$$

$$=z^{\alpha}H_{p+r,\ q+r}^{m,\ n+r}\left[z^{\sigma}\omega\left|\begin{matrix}[c_r-\alpha,\ \sigma],\ [a_p,\ A_p]\\ [b_q,\ B_q],\ [c_r-\alpha+1,\sigma]\end{matrix}\right.\right]\qquad [\sigma>0].$$

Representations in terms of the Meijer G-function:

21. $$H_{pq}^{mn}\left[z\left|\begin{matrix}[a_p, 1]\\ [b_q, 1]\end{matrix}\right.\right]=G_{pq}^{mn}\left(z\left|\begin{matrix}(a_p)\\ (b_q)\end{matrix}\right.\right).$$

22. $$H_{pq}^{mn}\left[z\left|\begin{matrix}[a_p, A_p]\\ [b_q, B_q]\end{matrix}\right.\right]=(2\pi)^{c^*-ka^*/2}k^{\mu}MG_{\tilde p\ \tilde q}^{\tilde m\ \tilde n}\left(\left(\frac{z}{\beta k^{\Delta}}\right)^k\left|\begin{matrix}\Delta(k_p,\ a_p)\\ \Delta(l_q,\ b_q)\end{matrix}\right.\right)$$

$[A_1,\ \dots,\ A_p,\ B_1,\ \dots,B_q>0$ are rational quantities].

Here k is the L.C.M. of all the denominators of the quantities $A_1,\ \dots,\ A_p$, $B_1,\ \dots,\ B_q$, $k_i=kA_i$, $i=1,\ 2,\ \dots,\ p$, $l_j=kB_j$, $j=1,\ 2,\ \dots,\ q$,

$$M=\prod_{i=1}^{p}A_i^{1/2-a_i}\prod_{j=1}^{q}B_j^{b_j-1/2},\quad \tilde m=\sum_{j=1}^{m}l_j,\quad \tilde n=\sum_{j=1}^{n}k_j$$

$$\tilde p=\sum_{j=1}^{p}k_j,\quad \tilde q=\sum_{j=1}^{q}l_j$$

[$a^*,\ c^*,\ \beta,\ \mu,\ \Delta$ see 8.3.1].

For other properties of the H-function and applications, see [50, 61].

8.4. TABLE OF MELLIN TRANSFORMS AND REPRESENTATIONS OF ELEMENTARY AND SPECIAL FUNCTIONS IN TERMS OF THE MEIJER G-FUNCTION AND THE FOX H-FUNCTION

In this section we give the Mellin transform formulae

$$f^*(s)=\int_0^\infty x^{s-1}f(x)\,dx$$

for various elementary and special functions $f(x)$ for which $f^*(s)$ has the form

$$\prod_{j,k,l,m}\frac{\Gamma(a_j+A_js)\,\Gamma(b_k-B_ks)}{\Gamma(c_l+C_ls)\,\Gamma(d_m-D_ms)},\ \text{where}\ A_j, B_k, C_l, D_m>0.$$

The functions $f(x)$ are given in the left hand column, and their Mellin transforms $f^*(s)$ in the right hand column. The notation $f(x)\overset{L_{\pm\infty}}{\longleftarrow}f^*(s)$ means that the integral $\frac{1}{2\pi i}\int_{L_{\pm\infty}}f^*(s)\,x^{-s}\,ds$ exists and is equal to $f(x)$, while the integral $\int_0^\infty x^{s-1}(fx)\,dx$ diverges in the classical sense; for the definition of the Mellin transform in the sense of the theory of generalized functions, see, for example, [36]. The contours $L_{\pm\infty}$ are loops (see 8.2.1) dividing the s-plane into regions D^+ and D^- (D^+ to the left and D^- to the right of $L_{\pm\infty}$). The conditions for the representability of functions $f(x)$ in terms of the Meijer G-function (or Fox H-function) indicated in the right hand column can be weakened; see 8.2.1, 8.2.4 (1)–(2) (or 8.3.1, 8.3.2).

8.4.1. Formulae of general form.

1	$\frac{1}{2\pi i}\int_{\gamma-i\infty}^{\gamma+i\infty}f^*(s)\,x^{-s}\,ds$	$f^*(s)$
2	$\int_0^\infty K_1\left(\frac{x}{t}\right)K_2(t)\frac{dt}{t}$	$K_1^*(s)\,K_2^*(s)$
3	$x^\tau\int_0^\infty t^{\sigma-1}K_1(t^r)\,K_2\left(\frac{t}{x^{1/p}}\right)dt$ $[r=p/q>0]$	$qK_1^*\left(qs+q\tau+\frac{\sigma q}{p}\right)K_2^*(-ps-p\tau)$
4	$K(ax)$ $[a>0]$	$a^{-s}K^*(s)$
5	$x^\sigma K(x)$	$K^*(s+\sigma)$
6	$K(x^p)$ $[p\neq 0]$	$\frac{1}{\lvert p\rvert}K^*\left(\frac{s}{p}\right)$
7	$K\left(\frac{1}{x}\right)$	$K^*(-s)$
8	$K^{(n)}(x)$ при $\lim_{x\to 0}x^{s-k-1}K^{(k)}(x)=0$ $[k=0,1,2,\ldots,n-1]$	$\Gamma\left[\begin{matrix}n+1-s\\1-s\end{matrix}\right]K^*(s-n)$
9	$\left(x\frac{d}{dx}\right)^nK(x)$	$(-s)^nK^*(s)$

10	$\left(\frac{d}{dx}x\right)^n K(x)$	$(1-s)^n K^*(s)$
11	$\left(x^{1-\sigma}\frac{d}{dx}\right)^n K(x)$ $[\sigma\neq 0]$	$(-\sigma)^n \Gamma\left[\begin{matrix} s/\sigma \\ s/\sigma-n\end{matrix}\right] K^*(s-\sigma n)$

8.4.2. Power functions and algebraic functions.

NOTATION: $A_1=\pi^{-1/2}(\mp\nu/2)^{1-2r}$, $A_2=2^{2r}\sqrt{\pi}\nu^{1-2r}$, $A_3=2^{4^r+\nu/2-1}\times\sqrt{\pi}\nu^{1-2r}$.

1	$x^\nu H(1-x)=G_{11}^{10}\left(x\,\middle\vert\,\begin{matrix}\nu+1\\ \nu\end{matrix}\right)$	$\Gamma\left[\begin{matrix}s+\nu\\ s+\nu+1\end{matrix}\right]=\frac{1}{s+\nu}$ $[\operatorname{Re}(s+\nu)>0]$
2	$x^\nu H(x-1)=G_{11}^{01}\left(x\,\middle\vert\,\begin{matrix}\nu+1\\ \nu\end{matrix}\right)$	$\Gamma\left[\begin{matrix}-\nu-s\\ 1-\nu-s\end{matrix}\right]=-\frac{1}{s+\nu}$ $[\operatorname{Re}(s+\nu)<0]$
3	$(1-x)_+^{\beta-1}=\Gamma(\beta)\,G_{11}^{10}\left(x\,\middle\vert\,\begin{matrix}\beta\\ 0\end{matrix}\right)$	$\Gamma(\beta)\,\Gamma\left[\begin{matrix}s\\ s+\beta\end{matrix}\right]$ $[\operatorname{Re} s,\ \operatorname{Re}\beta>0]$
4	$(x-1)_+^{\beta-1}=\Gamma(\beta)\,G_{11}^{01}\left(x\,\middle\vert\,\begin{matrix}\beta\\ 0\end{matrix}\right)$	$\Gamma(\beta)\,\Gamma\left[\begin{matrix}1-\beta-s\\ 1-s\end{matrix}\right]$ $[\operatorname{Re}\beta>0;\ \operatorname{Re} s<1-\operatorname{Re}\beta]$
5	$(1+x)^{-\rho}=\frac{1}{\Gamma(\rho)}G_{11}^{11}\left(x\,\middle\vert\,\begin{matrix}1-\rho\\ 0\end{matrix}\right)$	$\frac{1}{\Gamma(\rho)}\Gamma[s,\ \rho-s]$ $[0<\operatorname{Re} s<\operatorname{Re}\rho]$
6	$\frac{1}{1-x}=\pi G_{22}^{11}\left(x\,\middle\vert\,\begin{matrix}0,\ 1/2\\ 0,\ 1/2\end{matrix}\right)$	$\pi\Gamma\left[\begin{matrix}s, & 1-s\\ s+1/2, & 1/2-s\end{matrix}\right]$ $[0<\operatorname{Re} s<1]$
7	$\lvert 1-x\rvert^{-\rho}=\frac{\pi}{\Gamma(\rho)\cos(\rho\pi/2)}\times$ $\times G_{22}^{11}\left(x\,\middle\vert\,\begin{matrix}1-\rho, & (1-\rho)/2\\ 0, & (1-\rho)/2\end{matrix}\right)$	$\frac{\pi}{\Gamma(\rho)\cos(\rho\pi/2)}\times$ $\times\Gamma\left[\begin{matrix}s, & \rho-s\\ s+(1-\rho)/2, & (1+\rho)/2-s\end{matrix}\right]$ $[0<\operatorname{Re} s<\operatorname{Re}\rho<1]$
8	$(1-x)_+^{\beta-1}+\frac{\sin(\beta-\gamma)\pi}{\sin\gamma\pi}(x-1)_+^{\beta-1}=$ $=\frac{\pi}{\sin\gamma\pi\Gamma(1-\beta)}G_{22}^{11}\left(x\,\middle\vert\,\begin{matrix}\beta, & \gamma\\ 0, & \gamma\end{matrix}\right)$	$\frac{\pi}{\sin\gamma\pi\Gamma(1-\beta)}\Gamma\left[\begin{matrix}s, & 1-\beta-s\\ s+\gamma, & 1-\gamma-s\end{matrix}\right]$ $[0<\operatorname{Re} s<1-\operatorname{Re}\beta<1]$
9	$\frac{x^\sigma-1}{x-1}=\frac{\sin\sigma\pi}{\pi}G_{22}^{22}\left(x\,\middle\vert\,\begin{matrix}0, & \sigma\\ 0, & \sigma\end{matrix}\right)$	$\frac{\sin\sigma\pi}{\pi}\Gamma[s,\ s+\sigma,\ 1-s,\ 1-\sigma-s]$ $[\lvert\operatorname{Re}\sigma\rvert<1;\ (0,\ -\operatorname{Re}\sigma)<\operatorname{Re} s<(1,1-\operatorname{Re}\sigma)]$
10	$\left[1+\sum_{k=1}^{n}x^{k/(n+1)}\right]^{-1}=$ $=\pi G_{22}^{11}\left(x\,\middle\vert\,\begin{matrix}0,\ 1/2\\ 0,\ 1/2\end{matrix}\right)-$ $-\pi G_{22}^{11}\left(x\,\middle\vert\,\begin{matrix}(n+1)^{-1}, & (n+3)/(2n+2)\\ (n+1)^{-1}, & (n+3)/(2n+2)\end{matrix}\right)$	$\pi\Gamma\left[\begin{matrix}s, & 1-s\\ s+1/2, & 1/2-s\end{matrix}\right]-$ $-\pi\Gamma\left[\begin{matrix}s+(n+1)^{-1}, & n(n+1)^{-1}-s\\ s+(n+3)(2n+2)^{-1}, & (n-1)\times(2n+2)^{-1}-s\end{matrix}\right]$ $[0<\operatorname{Re} s<n(n+1)^{-1}]$
11	$\left[1-\sum_{k=1}^{2n}(-1)^{k-1}x^{k/(2n+1)}\right]^{-1}=$ $=G_{11}^{11}\left(x\,\middle\vert\,\begin{matrix}0\\ 0\end{matrix}\right)+G_{11}^{11}\left(x\,\middle\vert\,\begin{matrix}(2n+1)^{-1}\\ (2n+1)^{-1}\end{matrix}\right)$	$\Gamma[s,\ 1-s]+$ $+\Gamma[s+(2n+1)^{-1},\ 2n(2n+1)^{-1}-s]$ $[0<\operatorname{Re} s<2n(2n+1)^{-1}]$

12	$\dfrac{(\sqrt{1+x}\pm 1)}{(1+x)^r}=$ $=A_1 G_{22}^{12}\left(x\left|\begin{matrix}(1+v)/2,\ 1-2r+v/2\\(v\mp v)/2,\ (v\pm v)/2\end{matrix}\right.\right)$ $[r=0,\ 1/2]$	$A_1\Gamma\left[\begin{matrix}s+\dfrac{v\mp v}{2},\ \dfrac{1-v}{2}-s,\ 2r-\dfrac{v}{2}-s\\1-(v\pm v)/2-s\end{matrix}\right]$ $[-\operatorname{Re}(v\mp v)/2<\operatorname{Re}s<r-\operatorname{Re}v/2]$
13	$\dfrac{(\sqrt{1+x}\pm\sqrt{x})^v}{(1+x)^r}=$ $=A_1 G_{22}^{21}\left(x\left|\begin{matrix}1-r\pm v/2,\ 1-r\mp v/2\\0,\ 1/2\end{matrix}\right.\right)$	$A_1\Gamma\left[\begin{matrix}s,\ s+1/2,\ r\mp v/2-s\\s-r\mp v/2+1\end{matrix}\right]$ $[0<\operatorname{Re}s<r\mp\operatorname{Re}v/2]$
14	$(x^2+2x\cos\gamma+1)^{-1}=$ $=-\dfrac{\pi}{\sin\gamma}H_{22}^{11}\left(x\left|\begin{matrix}(0,1),\ (-\gamma/\pi,\gamma/\pi)\\(0,1),\ (-\gamma/\pi,\gamma/\pi)\end{matrix}\right.\right)$	$-\dfrac{\sin(\gamma s-\gamma)}{\sin\gamma}\,\Gamma[s,\ 1-s]$ $[0<\operatorname{Re}s<2;\ |\gamma|<\pi]$
15	$\dfrac{1+x\cos\gamma}{x^2+2x\cos\gamma+1}=$ $=\pi H_{22}^{11}\left(x\left|\begin{matrix}(0,1),\ (1/2,\gamma/\pi)\\(0,1),\ (1/2,\gamma/\pi)\end{matrix}\right.\right)$	$\cos(\gamma s)\,\Gamma[s,\ 1-s]$ $[0<\operatorname{Re}s<1;\ |\gamma|<\pi]$
16	$(x^2+2x\cos\gamma+1)^{-2}$	$\pi\dfrac{(s-1)\cos(\gamma s-2\gamma)\sin\gamma-\sin(\gamma s-\gamma)}{2\sin^3\gamma\sin s\pi}$ $[0<\operatorname{Re}s<4;\ |\gamma|<\pi]$
17	$(1-x)_+^{-r}\left[(1+\sqrt{1-x})^v-\right.$ $\left.-(-1)^{2r}(1-\sqrt{1-x})^v\right]=$ $=A_2 G_{22}^{20}\left(x\left|\begin{matrix}(v+1)/2,\ v/2+1-2r\\0,\ v\end{matrix}\right.\right)$ $[r=0,\ 1/2]$	$A_2\Gamma\left[\begin{matrix}s,\ s+v\\s+(v+1)/2,\ s+v/2+1-2r\end{matrix}\right]$ $[\operatorname{Re}s>0,\ -\operatorname{Re}v]$
18	$(x-1)_+^{-r}\left[(\sqrt{x}+\sqrt{x-1})^v-\right.$ $\left.-(-1)^{2r}(\sqrt{x}-\sqrt{x-1})^v\right]=$ $=A_2 G_{22}^{02}\left(x\left|\begin{matrix}1-r-v/2,\ 1-r+v/2\\0,\ 1/2\end{matrix}\right.\right)$ $[r=0,\ 1/2]$	$A_2\Gamma\left[\begin{matrix}r+v/2-s,\ r-v/2-s\\1-s,\ 1/2-s\end{matrix}\right]$ $[\operatorname{Re}s<r-|\operatorname{Re}v|/2]$
19	$(1-x)_+^{-r}\left[(\sqrt{1+\sqrt{x}}+\sqrt{1-\sqrt{x}})^v-\right.$ $\left.-(-1)^{2r}(\sqrt{1+\sqrt{x}}-\sqrt{1-\sqrt{x}})^v\right]=$ $=A_3 G_{22}^{20}\left(x\left|\begin{matrix}(v+2)/4,\ v/4+1-2r\\0,\ v/2\end{matrix}\right.\right)$ $[r=0,\ 1/2]$	$A_3\Gamma\left[\begin{matrix}s,\ s+v/2\\s+(v+2)/4,\ s+v/4+1-2r\end{matrix}\right]$ $[\operatorname{Re}s>0,\ -\operatorname{Re}v/2]$
20	$(x-1)_+^{-r}\left[(\sqrt{\sqrt{x}+1}+\sqrt{\sqrt{x}-1})^v-\right.$ $\left.-(-1)^{2r}(\sqrt{\sqrt{x}+1}-\sqrt{\sqrt{x}-1})^v\right]=$ $=A_3 G_{22}^{02}\left(x\left|\begin{matrix}1-r+v/4,\ 1-r-v/4\\0,\ 1/2\end{matrix}\right.\right)$ $[r=0,\ 1/2]$	$A_3\Gamma\left[\begin{matrix}r+v/4-s,\ r-v/4-s\\1-s,\ 1/2-s\end{matrix}\right]$ $[\operatorname{Re}s<r-|\operatorname{Re}v|/4]$
21	$(1+\sqrt{x})^{-\rho}+|1-\sqrt{x}|^{-\rho}=$ $=2^{\rho+1}\sin\dfrac{\rho\pi}{2}\,\Gamma(1-\rho)\times$ $\times G_{22}^{11}\left(x\left|\begin{matrix}1-\rho/2,\ (1-\rho)/2\\0,\ 1/2\end{matrix}\right.\right)$	$2^{\rho+1}\sin\dfrac{\rho\pi}{2}\,\Gamma(1-\rho)\times$ $\times\Gamma\left[\begin{matrix}s,\ \rho/2-s\\s+(1-\rho)/2,\ 1/2-s\end{matrix}\right]$ $[0<\operatorname{Re}s<\operatorname{Re}\rho/2<1/2]$

22	$\sqrt{1+(-1)^l\sqrt{1-x}}\,H(1-x)+$ $+\sqrt{\sqrt{x}-(-1)^l\sqrt{x-1}}\,H(x-1)=$ $=2^{-3/2}\sqrt{\pi}\times$ $\times G_{22}^{11}\left(x\,\middle\vert\,\begin{matrix}(3+2l)/4, & (5-2l)/4\\ l/2, & (1-l)/2\end{matrix}\right)$ [$l=0,\ 1$]	$2^{-3/2}\sqrt{\pi}\times$ $\times\Gamma\begin{bmatrix}s+l/2, & (1-2l)/4-s\\ s+(5-2l)/4, & (1+l)/2-s\end{bmatrix}$ [$-l/2<\operatorname{Re}s<(1-2l)/4$]

8.4.3. The exponential function.

1	$e^{-x}=G_{01}^{10}\left(x\,\middle\vert\,{}^{\cdot}_{0}\right)$	$\Gamma(s)$ [$\operatorname{Re}s>0$]
2	$e^{-1/x}=G_{10}^{01}\left(x\,\middle\vert\,{}^{1}_{\cdot}\right)$	$\Gamma(-s)$ [$\operatorname{Re}s<0$]
3	$e^{-x}-\sum_{k=0}^{n}\frac{(-x)^k}{k!}=$ $=(-1)^{n-1}G_{12}^{11}\left(x\,\middle\vert\,\begin{matrix}n+1, & \\ n+1, & 0\end{matrix}\right)$	$\Gamma(s)$ [$-n-1<\operatorname{Re}s<-n$]
4	$e^{-1/x}-\sum_{k=0}^{n}\frac{(-x)^{-k}}{k!}=$ $=(-1)^{n-1}G_{21}^{11}\left(x\,\middle\vert\,\begin{matrix}-n, & 1\\ -n, & \end{matrix}\right)$	$\Gamma(-s)$ [$n<\operatorname{Re}s<n+1$]
5	$e^{x}=\quad\overset{L_{-\infty}}{\longleftarrow}$ $=\frac{\pi}{\sin c\pi}G_{12}^{10}\left(x\,\middle\vert\,\begin{matrix}1-c & \\ 0, & 1-c\end{matrix}\right)$	$\frac{\pi}{\sin c\pi}\Gamma\begin{bmatrix}s & \\ s+1-c, & c-s\end{bmatrix}$ or $-\frac{\psi(s)+\mathbf{C}}{\Gamma(1-s)}$ [c is arbitrary; $s=-k\in D+$, $k=0, 1, 2, \ldots$].
6	$e^{1/x}=\quad\overset{L_{+\infty}}{\longleftarrow}$ $=\frac{\pi}{\sin c\pi}G_{21}^{01}\left(x\,\middle\vert\,\begin{matrix}1, & c\\ c & \end{matrix}\right)$	$\frac{\pi}{\sin c\pi}\Gamma\begin{bmatrix}-s & \\ s+c, & 1-c-s\end{bmatrix}$ or $-\frac{\psi(-s)+\mathbf{C}}{\Gamma(s+1)}$ [c is arbitrary; $s=k\in D^-$, $k=0, 1, 2, \ldots$].
7	$e^{x}-\sum_{k=0}^{n}\frac{x^k}{k!}=\quad\overset{L_{-\infty}}{\longleftarrow}$ $=\frac{(-1)^{n-1}\pi}{\sin c\pi}G_{23}^{11}\left(x\,\middle\vert\,\begin{matrix}n+1, & 1-c & \\ n+1, & 1-c, & 0\end{matrix}\right)$	$\frac{\pi}{\sin c\pi}\Gamma\begin{bmatrix}s & \\ s+1-c, & c-s\end{bmatrix}$ or $-\frac{\psi(s)+\mathbf{C}}{\Gamma(1-s)}$ [c is arbitrary; $s=-k\in D^-$, $k=0, 1, \ldots, n$; $s=-k\in D+$, $k=n+1, n+2, \ldots$].
8	$e^{1/x}-\sum_{k=0}^{n}\frac{x^{-k}}{k!}=\quad\overset{L_{+\infty}}{\longleftarrow}$ $=\frac{(-1)^{n-1}\pi}{\sin c\pi}G_{32}^{11}\left(x\,\middle\vert\,\begin{matrix}-n, & c, & 1\\ -n, & c & \end{matrix}\right)$	$\frac{\pi}{\sin c\pi}\Gamma\begin{bmatrix}-s & \\ s+c, & 1-c-s\end{bmatrix}$ or $-\frac{\psi(-s)+\mathbf{C}}{\Gamma(s+1)}$ [c is arbitrary; $s=k\in D+$, $k=1, 2, \ldots, n$; $s=k\in D^-$, $k=n+1, n+2, \ldots$].

8.4.4. The hyperbolic functions.

NOTATION $A_4=\dfrac{\pi^{3/2}}{\sin c\pi}$, $A_5=\dfrac{2^{1-n}\pi^{3/2}}{\sin b\pi}$, $\delta=\begin{Bmatrix}1\\0\end{Bmatrix}$.

1	$\begin{Bmatrix}\operatorname{sh} 2\sqrt{x}\\ \operatorname{ch} 2\sqrt{x}\end{Bmatrix}= A_4G_{13}^{10}\left(x\,\middle\vert\,\begin{matrix}1+\delta/2-c\\ \delta/2,\ 1+\delta/2-c,\ (1-\delta)/2\end{matrix}\right)$	$\overset{L_{-\infty}}{\longleftarrow}$	$A_4\Gamma\begin{bmatrix}s+\delta/2\\ s+1+\frac{\delta}{2}-c,\ c-\frac{\delta}{2}-s,\ \frac{1+\delta}{2}-s\end{bmatrix}$ [$c\neq 1$ is arbitrary; $s=-\delta/2-k\in D^+$, $k=0, 1, 2\ldots$]
2	$\begin{Bmatrix}\operatorname{sh}(2x^{-1/2})\\ \operatorname{ch}(2x^{-1/2})\end{Bmatrix}= A_4G_{31}^{01}\left(x\,\middle\vert\,\begin{matrix}1-\delta/2,\ c-\delta/2,\ (1+\delta)/2\\ c-\delta/2\end{matrix}\right)$	$\overset{L_{+\infty}}{\longleftarrow}$	$A_4\Gamma\begin{bmatrix}\delta/2-s\\ s+c-\frac{\delta}{2},\ s+\frac{1+\delta}{2},\ 1+\frac{\delta}{2}-c-s\end{bmatrix}$ [$c\neq 1$ is arbitrary; $s=\delta/2+k\in D^-$, $k=0, 1, 2,\ldots$]
3	$\operatorname{sh}^n\sqrt{x}$	$\overset{L_{-\infty}}{\longleftarrow}$	$A_5\sum\limits_{k=0}^{[(n-1)/2]}(-1)^k\binom{n}{k}\left(\frac{2}{n-2k}\right)^{2s}\times$ $\times\Gamma\begin{bmatrix}s+\gamma\\ s+\gamma-b+1,\ b-\gamma-s,\ 1/2+\gamma-s\end{bmatrix}$ [$b\neq 1$ is arbitrary; $\gamma=(1-(-1)^n)/4$; $s=-\gamma-k\in D^+$, $k=1, 2, 3, \ldots$; $s=0\in D^-$ for $n=2, 4, 6, \ldots$, $s=-1/2\in D^+$ for $n=1, 3, 5,\ldots$]
4	$\operatorname{ch}^n\sqrt{x}-c$ [$c=0$ for $n=1, 3, 5, \ldots$; $c=1$ or $c=2^{-n}\binom{n}{n/2}$ for $n=2, 4, 6,\ldots$]	$\overset{L_{-\infty}}{\longleftarrow}$	$A_5\sum\limits_{k=0}^{[(n-1)/2]}\binom{n}{k}\left(\frac{2}{n-2k}\right)^{2s}\times$ $\times\Gamma\begin{bmatrix}s\\ s-b+1,\ b-s,\ 1/2-s\end{bmatrix}$ [$b\neq 1$ is arbitrary; $s=-k\in D^+$, $k=1, 2, 3, \ldots$; $s=0\in D^-$ for $c=1$ and $n=2, 4, 6,\ldots$; $s=0\in D^+$ otherwise]
5	$\operatorname{sh}^n\dfrac{1}{\sqrt{x}}$	$\overset{L_{+\infty}}{\longleftarrow}$	$A_5\sum\limits_{k=0}^{[(n-1)/2]}(-1)^k\binom{n}{k}\left(\frac{2}{n-2k}\right)^{-2s}\times$ $\times\Gamma\begin{bmatrix}\gamma-s\\ s+\gamma+1/2,\ s+b-\gamma,\ 1+\gamma-b-s\end{bmatrix}$ [$b\neq 1$ is arbitrary; $\gamma=(1-(-1)^n)/4$; $s=\gamma+k\in D^-$, $k=1, 2, 3, \ldots$; $s=0\in D^+$ for $n=2, 4, 6, \ldots$, $s=1/2\in D^-$ for $n=1, 3, 5,\ldots$]
6	$\operatorname{ch}^n\dfrac{1}{\sqrt{x}}-c$ [$c=0$ for $n=1, 3, 5, \ldots$; $c=1$ or $c=2^{-n}\binom{n}{n/2}$ for $n=2, 4, 6,\ldots$]	$\overset{L_{+\infty}}{\longleftarrow}$	$A_5\sum\limits_{k=0}^{[(n-1)/2]}\binom{n}{k}\left(\frac{2}{n-2k}\right)^{-2s}\times$ $\times\Gamma\begin{bmatrix}-s\\ s+1/2,\ s+b,\ 1-b-s\end{bmatrix}$ [$b\neq 1$ is arbitrary; $s=k\in D^-$, $k=1, 2, 3, \ldots$; $s=0\in D^+$ for $c=1$ and $n=2, 4, 6, \ldots$; $s=0\in D^-$ otherwise]
7	$e^{-x/2}\operatorname{sh}\dfrac{x}{2}=\dfrac{1}{2}G_{12}^{11}\left(x\,\middle\vert\,\begin{matrix}1,\\ 1,\ 0\end{matrix}\right)$		$\frac{1}{2}\Gamma\begin{bmatrix}s+1,\ -s\\ 1-s\end{bmatrix}$ [$-1<\operatorname{Re} s<0$]
8	$e^{-1/(2x)}\operatorname{sh}\dfrac{1}{2x}=\dfrac{1}{2}G_{21}^{11}\left(x\,\middle\vert\,\begin{matrix}0,\ 1\\ 0\end{matrix}\right)$		$\frac{1}{2}\Gamma\begin{bmatrix}s,\ 1-s\\ s+1\end{bmatrix}$ [$0<\operatorname{Re} s<1$]

8.4.5. The trigonometric functions.

NOTATION: $\delta=\begin{Bmatrix}1\\0\end{Bmatrix}$.

1	$\sin 2\sqrt{x}=\sqrt{\pi}\,G_{02}^{10}\left(x\,\middle\vert\,\begin{matrix}\cdot\ \ \cdot\\ 1/2,\ 0\end{matrix}\right)$	$\sqrt{\pi}\,\Gamma\begin{bmatrix}s+1/2\\ 1-s\end{bmatrix}$	$[\lvert \operatorname{Re} s\rvert<1/2]$
2	$\cos 2\sqrt{x}=\sqrt{\pi}\,G_{02}^{10}\left(x\,\middle\vert\,\begin{matrix}\cdot\ \ \cdot\\ 0,\ 1/2\end{matrix}\right)$	$\sqrt{\pi}\,\Gamma\begin{bmatrix}s\\ 1/2-s\end{bmatrix}$	$[0<\operatorname{Re} s<1/2]$
3	$\begin{Bmatrix}\sin 2\sqrt{x}\\ \cos 2\sqrt{x}\end{Bmatrix}=\sqrt{\pi}\,G_{02}^{10}\left(x\,\middle\vert\,\begin{matrix}\cdot\ \ \cdot\\ \delta/2,\ (1-\delta)/2\end{matrix}\right)$	$\sqrt{\pi}\,\Gamma\begin{bmatrix}s+\delta/2\\ (\delta+1)/2-s\end{bmatrix}$	$[-\delta/2<\operatorname{Re} s<1/2]$
4	$\begin{Bmatrix}\sin 2x^{-1/2}\\ \cos 2x^{-1/2}\end{Bmatrix}=\sqrt{\pi}\,G_{20}^{01}\left(x\,\middle\vert\,\begin{matrix}1-\delta/2,\ (1+\delta)/2\\ \cdot\ \ \cdot\end{matrix}\right)$	$\sqrt{\pi}\,\Gamma\begin{bmatrix}\delta/2-s\\ s+(\delta+1)/2\end{bmatrix}$	$[-1/2<\operatorname{Re} s<\delta/2]$
5	$a\begin{Bmatrix}\sin 2\sqrt{x}\\ \cos 2\sqrt{x}\end{Bmatrix}\mp\sqrt{x}\begin{Bmatrix}\cos 2\sqrt{x}\\ \sin 2\sqrt{x}\end{Bmatrix}=\sqrt{\pi}\,G_{13}^{20}\left(x\,\middle\vert\,\begin{matrix}a\\ \delta/2,\ a+1,\ (1-\delta)/2\end{matrix}\right)$ $\overset{L_{-\infty}}{\longleftarrow}$	$\sqrt{\pi}\,\Gamma\begin{bmatrix}s+a+1,\ s+\delta/2\\ s+a,\ (\delta+1)/2-s\end{bmatrix}$	$[s=-\delta/2-k\in D^{+},\ k=0,1,2,\ldots]$
6	$a\begin{Bmatrix}\sin 2x^{-1/2}\\ \cos 2x^{-1/2}\end{Bmatrix}\mp x^{-1/2}\begin{Bmatrix}\cos 2x^{-1/2}\\ \sin 2x^{-1/2}\end{Bmatrix}=\sqrt{\pi}\,G_{31}^{02}\left(x\,\middle\vert\,\begin{matrix}1-\delta/2,\ -a,\ (\delta+1)/2\\ 1-a\end{matrix}\right)$ $\overset{L_{+\infty}}{\longleftarrow}$	$\sqrt{\pi}\,\Gamma\begin{bmatrix}1+a-s,\ \delta/2-s\\ s+(\delta+1)/2,\ a-s\end{bmatrix}$	$[s=\delta/2+k\in D^{-},\ k=0,1,2,\ldots]$
7	$\begin{Bmatrix}\sin(2\sqrt{x}+\gamma\pi)\\ \cos(2\sqrt{x}+\gamma\pi)\end{Bmatrix}=\sqrt{\pi}\,G_{13}^{20}\left(x\,\middle\vert\,\begin{matrix}\gamma+(1-\delta)/2\\ 0,\ 1/2,\ \gamma+(1-\delta)/2\end{matrix}\right)$	$\sqrt{\pi}\,\Gamma\begin{bmatrix}s,\ s+1/2\\ s+\gamma+\frac{1-\delta}{2},\ \frac{1+\delta}{2}-\gamma-s\end{bmatrix}$	$[0<\operatorname{Re} s<1/2]$
8	$\begin{Bmatrix}\sin(2x^{-1/2}+\gamma\pi)\\ \cos(2x^{-1/2}+\gamma\pi)\end{Bmatrix}=\sqrt{\pi}\,G_{31}^{02}\left(x\,\middle\vert\,\begin{matrix}1,\ 1/2,\ (1+\delta)/2-\gamma\\ (1+\delta)/2-\gamma\end{matrix}\right)$	$\sqrt{\pi}\,\Gamma\begin{bmatrix}-s,\ 1/2-s\\ s+\frac{1+\delta}{2}-\gamma,\ \gamma+\frac{1-\delta}{2}-s\end{bmatrix}$	$[-1/2<\operatorname{Re} s<0]$
9	$\gamma\sin(2\sqrt{x}+\gamma\pi)-\sqrt{x}\cos(2\sqrt{x}+\gamma\pi)=-\sqrt{\pi}\,G_{13}^{20}\left(x\,\middle\vert\,\begin{matrix}\gamma\\ 0,\ 1/2,\ \gamma+1\end{matrix}\right)$ $\overset{L_{-\infty}}{\longleftarrow}$	$-\sqrt{\pi}\,\Gamma\begin{bmatrix}s,\ s+1/2\\ s+\gamma,\ -\gamma-s\end{bmatrix}$	$[s=-k,\ -1/2-k\in D^{+},\ k=0,1,2,\ldots]$
10	$\gamma\sin\left(\frac{2}{\sqrt{x}}+\gamma\pi\right)-\frac{1}{\sqrt{x}}\cos\left(\frac{2}{\sqrt{x}}+\gamma\pi\right)=-\sqrt{\pi}\,G_{31}^{02}\left(x\,\middle\vert\,\begin{matrix}1/2,\ 1,\ -\gamma\\ 1-\gamma\end{matrix}\right)$ $\overset{L_{+\infty}}{\longleftarrow}$	$-\sqrt{\pi}\,\Gamma\begin{bmatrix}-s,\ 1/2-s\\ s-\gamma,\ \gamma-s\end{bmatrix}$	$[s=k,\ 1/2+k\in D^{-},\ k=0,1,2,\ldots]$

11	$\sin^n \sqrt{x}$	$2^{1-n}\sqrt{\pi}\times \sum_{k=0}^{[(n-1)/2]} (-1)^{[n/2]+k}\binom{n}{k}\left(\frac{2}{n-2k}\right)^{2s}\times \Gamma\left[\begin{matrix} s+\gamma \\ 1/2+\gamma-s\end{matrix}\right]$ $[-n/2 < \operatorname{Re} s < \gamma;\ \gamma=(1-(-1)^n)/4]$
12	$\cos^n \sqrt{x}-c$ $\left[c=0 \text{ for } n=1,\ 3,\ 5,\ \ldots;\ c=1 \text{ or } c=c_1=2^{-n}\binom{n}{n/2} \text{ for } n=2,\ 4,\ 6,\ \ldots\right]$	$2^{1-n}\sqrt{\pi}\times \sum_{k=0}^{[(n-1)/2]} \binom{n}{k}\left(\frac{2}{n-2k}\right)^{2s}\Gamma\left[\begin{matrix} s \\ 1/2-s\end{matrix}\right]$ $[0 < \operatorname{Re} s < 1/2 \text{ for } c=0 \text{ and } c=c_1;\ -1 < \operatorname{Re} s < 0 \text{ for } c=1]$
13	$\sin^n \frac{1}{\sqrt{x}}$	$2^{1-n}\sqrt{\pi}\times \sum_{k=0}^{[(n-1)/2]} (-1)^{[n/2]+k}\binom{n}{k}\times \left(\frac{n-2k}{2}\right)^{2s}\Gamma\left[\begin{matrix} \gamma-s \\ s+\gamma+1/2\end{matrix}\right]$ $[-\gamma < \operatorname{Re} s < n/2;\ \gamma=(1-(-1)^n)/4]$
14	$\cos^n \frac{1}{\sqrt{x}}-c$ $\left[c=0 \text{ for } n=1,\ 3,\ 5,\ldots;\ c=1 \text{ or } c=c_1=2^{-n}\binom{n}{n/2} \text{ for } n=2,\ 4,\ 6,\ \ldots\right]$	$2^{1-n}\sqrt{\pi}\times \sum_{k=0}^{[(n-1)/2]} \binom{n}{k}\left(\frac{n-2k}{2}\right)^{2s}\Gamma\left[\begin{matrix} -s \\ s+1/2\end{matrix}\right]$ $[-1/2 < \operatorname{Re} s < 0 \text{ for } c=0 \text{ and } c=c_1;\ 0 < \operatorname{Re} s < 1 \text{ and } c=1]$
15	$e^{-x\cos\gamma}\left\{\begin{matrix}\sin(x\sin\gamma)\\ \cos(x\sin\gamma)\end{matrix}\right\}= =\pi H_{12}^{10}\left(x\left\vert\begin{matrix}((1-\delta)/2,\ \gamma/\pi)\\ (0,\ 1),\ ((1-\delta)/2,\ \gamma/\pi)\end{matrix}\right.\right)$	$\Gamma(s)\left\{\begin{matrix}\sin\gamma s\\ \cos\gamma s\end{matrix}\right\}$ $[\lvert\gamma\rvert < \pi/2;\ \operatorname{Re} s > -(1\pm 1)/2]$
16	$e^{-x^{-1}\cos\gamma}\left\{\begin{matrix}\sin(x^{-1}\sin\gamma)\\ \cos(x^{-1}\sin\gamma)\end{matrix}\right\}= =\pi H_{21}^{01}\left(x\left\vert\begin{matrix}(1,\ 1),\ ((1+\delta)/2,\ \gamma/\pi)\\ ((1+\delta)/2,\ \gamma/\pi)\end{matrix}\right.\right)$	$\mp\Gamma(-s)\left\{\begin{matrix}\sin\gamma s\\ \cos\gamma s\end{matrix}\right\}$ $[\lvert\gamma\rvert < \pi/2;\ \operatorname{Re} s < (1\pm 1)/2]$
17	$e^{-3\sqrt[3]{x}/2}\left\{\begin{matrix}\sin\left(3\sqrt[6]{27x^2}/2\right)\\ \cos\left(3\sqrt[6]{27x^2}/2\right)\end{matrix}\right\}= =\frac{\sqrt{3}}{2}G_{14}^{30}\left(x\left\vert\begin{matrix}(1-\delta)/2\\ 0,\ 1/3,\ 2/3,\ (1-\delta)/2\end{matrix}\right.\right)$	$\frac{\sqrt{3}}{2}\Gamma\left[\begin{matrix}s,\ s+1/3,\ s+2/3\\ s+(1-\delta)/2,\ (1+\delta)/2-s\end{matrix}\right]$ $[\operatorname{Re} s > -\delta/3]$
18	$e^{-3\sqrt[6]{27x}}\left\{\begin{matrix}\sin 3\sqrt[6]{x}\\ \cos 3\sqrt[6]{x}\end{matrix}\right\}=\frac{\sqrt{3}}{4\pi^{3/2}}\times \times G_{06}^{50}\left(x\left\vert\begin{matrix}\cdots\\ \frac{\delta}{2},\ \frac{1}{6},\ \frac{1}{3},\ \frac{2}{3},\ \frac{5}{6},\ \frac{1-\delta}{2}\end{matrix}\right.\right)$	$\frac{\sqrt{3}}{4\pi^{3/2}}\times \Gamma\left[\begin{matrix}s+\frac{1}{6},\ s+\frac{1}{3},\ s+\frac{2}{3},\ s+\frac{5}{6},\ s+\frac{\delta}{2}\\ (1+\delta)/2-s\end{matrix}\right]$ $[\operatorname{Re} s > -\delta/6]$

19	$e^{-2\sqrt[4]{4x}}\left\{\begin{matrix}\sin 2\sqrt[4]{4x}\\ \cos 2\sqrt[4]{4x}\end{matrix}\right\}=$ $=\frac{1}{\sqrt{2\pi}}G_{04}^{30}\left(x\middle\vert\begin{matrix}\cdot\ \cdot\ \cdot\ \cdot\ \cdot\ \cdot\ \cdot\ \cdot\ \cdot\\ \delta/2,\ 1/4,\ 3/4,\ (1-\delta)/2\end{matrix}\right)$	$\frac{1}{\sqrt{2\pi}}\Gamma\begin{bmatrix}s+1/4,\ s+3/4,\ s+\delta/2\\ (1+\delta)/2-s\end{bmatrix}$ [Re $s>-\delta/4$]
20	$\left\{\begin{matrix}\mathrm{sh}\, 2\sqrt[4]{4x}\sin 2\sqrt[4]{4x}\\ \mathrm{ch}\, 2\sqrt[4]{4x}\cos 2\sqrt[4]{4x}\end{matrix}\right\}=\ \overset{L_{-\infty}}{\longleftarrow}$ $=\sqrt{2\pi^3}\,G_{04}^{10}\left(x\middle\vert\begin{matrix}\cdot\ \cdot\ \cdot\ \cdot\ \cdot\ \cdot\ \cdot\ \cdot\ \cdot\\ \delta/2,\ 1/4,\ 3/4,\ (1-\delta)/2\end{matrix}\right)$	$\sqrt{2\pi^3}\,\Gamma\begin{bmatrix}s+\delta/2\\ 1/4-s,\ 3/4-s,\ (1+\delta)/2-s\end{bmatrix}$ [$s=-\delta/2-k\in D+$, $k=0,\ 1,\ 2,\ \ldots$]
21	$\left\{\begin{matrix}\mathrm{sh}\, 2\sqrt[4]{4/x}\sin 2\sqrt[4]{4/x}\\ \mathrm{ch}\, 2\sqrt[4]{4/x}\cos 2\sqrt[4]{4/x}\end{matrix}\right\}=\ \overset{L_{+\infty}}{\longleftarrow}$ $=\sqrt{2\pi^3}\times$ $\times\, G_{40}^{01}\left(x\middle\vert\begin{matrix}1-\delta/2,\ 1/4,\ 3/4,\ (1+\delta)/2\\ \cdot\ \cdot\ \cdot\ \cdot\ \cdot\ \cdot\ \cdot\ \cdot\ \cdot\end{matrix}\right)$	$\sqrt{2\pi^3}\,\Gamma\begin{bmatrix}\delta/2-s\\ s+1/4,\ s+3/4,\ s+(\delta+1)/2\end{bmatrix}$ [$s=\delta/2+k\in D^-$, $k=0,\ 1,\ 2,\ \ldots$]
22	$\left\{\begin{matrix}\mathrm{ch}\, 2\sqrt[4]{4x}\sin 2\sqrt[4]{4x}\\ \mathrm{sh}\, 2\sqrt[4]{4x}\cos 2\sqrt[4]{4x}\end{matrix}\right\}=\ \overset{L_{-\infty}}{\longleftarrow}$ $=\pm\sqrt{2\pi^3}\,G_{15}^{20}\left(x\middle\vert\begin{matrix}\delta/2\\ 1/4,\ 3/4,\ 0,\ 1/2,\ \delta/2\end{matrix}\right)$	$\pm\sqrt{2\pi^3}\times$ $\times\Gamma\begin{bmatrix}s+\frac{1}{4},\ s+\frac{3}{4}\\ s+\frac{\delta}{2},\ \frac{1}{2}-s,\ 1-s,\ 1-\frac{\delta}{2}-s\end{bmatrix}$ [$s=-1/4-k,\ -3/4-k\in D+$, $k=0,\ 1,\ 2,\ \ldots$]
23	$\left\{\begin{matrix}\mathrm{ch}\, 2\sqrt[4]{4/x}\sin 2\sqrt[4]{4/x}\\ \mathrm{sh}\, 2\sqrt[4]{4/x}\cos 2\sqrt[4]{4/x}\end{matrix}\right\}=\ \overset{L_{+\infty}}{\longleftarrow}$ $=\pm\sqrt{2\pi^3}G_{51}^{02}\left(x\middle\vert\begin{matrix}1/4,\ 3/4,\ 1/2,\ 1,\ 1-\delta/2\\ 1-\delta/2\end{matrix}\right)$	$\pm\sqrt{2\pi^3}\times$ $\times\Gamma\begin{bmatrix}\frac{1}{4}-s,\ \frac{3}{4}-s\\ s+\frac{1}{2},\ s+1,\ s+1-\frac{\delta}{2},\ \frac{\delta}{2}-s\end{bmatrix}$ [$s=1/4+k,\ 3/4+k\in D^-$, $k=0,\ 1,\ 2,\ \ldots$]

8.4.6. The logarithmic function

1	$\ln xH(1-x)=-G_{22}^{20}\left(x\middle\vert\begin{matrix}1,\ 1\\ 0,\ 0\end{matrix}\right)$	$-\Gamma\begin{bmatrix}s,\ s\\ s+1,\ s+1\end{bmatrix}=-\frac{1}{s^2}$ [Re $s>0$]
2	$\ln xH(x-1)=G_{22}^{02}\left(x\middle\vert\begin{matrix}1,\ 1\\ 0,\ 0\end{matrix}\right)$	$\Gamma\begin{bmatrix}-s,\ -s\\ 1-s,\ 1-s\end{bmatrix}=\frac{1}{s^2}$ [Re $s<0$]
3	$\ln^n xH(1-x)=$ $=(-1)^n n!G_{n+1,\ n+1}^{n+1,\ 0}\left(x\middle\vert\begin{matrix}1,\ \ldots,\ 1\\ 0,\ \ldots,\ 0\end{matrix}\right)$	$(-1)^n n!\,\Gamma^{n+1}\begin{bmatrix}s\\ s+1\end{bmatrix}$ [Re $s>0$]
4	$\ln^n xH(x-1)=$ $=n!\,G_{n+1,\ n+1}^{0,\ n+1}\left(x\middle\vert\begin{matrix}1,\ \ldots,\ 1\\ 0,\ \ldots,\ 0\end{matrix}\right)$	$n!\Gamma^{n+1}\begin{bmatrix}-s\\ 1-s\end{bmatrix}$ [Re $s<0$]
5	$\ln(1+x)=G_{22}^{12}\left(x\middle\vert\begin{matrix}1,\ 1\\ 1,\ 0\end{matrix}\right)$	$\Gamma\begin{bmatrix}s+1,\ -s,\ -s\\ 1-s\end{bmatrix}=-\Gamma[s,\ -s]$ [$-1<$ Re $s<0$]
6	$\ln\left(1+\frac{1}{x}\right)=G_{22}^{21}\left(x\middle\vert\begin{matrix}0,\ 1\\ 0,\ 0\end{matrix}\right)$	$\Gamma\begin{bmatrix}s,\ s,\ 1-s\\ s+1\end{bmatrix}=-\Gamma[s,\ -s]$ [$0<$ Re $s<1$]
7	$\ln\vert 1-x\vert=\pi G_{33}^{12}\left(x\middle\vert\begin{matrix}1,\ 1,\ 1/2\\ 1,\ 0,\ 1/2\end{matrix}\right)$	$\pi\Gamma\begin{bmatrix}s+1,\ -s,\ -s\\ s+1/2,\ 1/2-s,\ 1-s\end{bmatrix}$ [$-1<$ Re $s<0$]

8	$\ln\left\lvert 1-\frac{1}{x}\right\rvert=\pi G_{33}^{21}\left(x\,\middle\vert\,{0,\ 1,\ 1/2 \atop 0,\ 0,\ 1/2}\right)$	$\pi\Gamma\begin{bmatrix}s,\ s,\ 1-s\\ s+1/2,\ s+1,\ 1/2-s\end{bmatrix}$ [$0<\operatorname{Re}s<1$]
9	$\ln\left(1+\sum_{k=1}^{n}x^{\pm k/(n+1)}\right)$	$\pi\Gamma\begin{bmatrix}\Delta(n+1,\ (n+1)s),\ \Delta(n+1,\ -(n+1)s)\\ \Delta\left(n+1,\frac{1}{2}+(n+1)s\right),\ \Delta\left(n+1,\frac{1}{2}-(n+1)s\right)\end{bmatrix}-\pi\Gamma\begin{bmatrix}s,\ -s\\ s+1/2,\ 1/2-s\end{bmatrix}$ $\left[\begin{Bmatrix}-(n+1)^{-1}<\operatorname{Re}s<0\\ 0<\operatorname{Re}s<(n+1)^{-1}\end{Bmatrix}\right]$
10	$\ln\left(1+\sum_{k=1}^{2n}(-1)^k x^{\pm k/(2n+1)}\right)$	$(2\pi)^{-2n}\Gamma[\Delta(2n+1,\ (2n+1)s),\ \Delta(2n+1,\ -(2n+1)s)]-\Gamma[s,\ -s]$ $\left[\begin{Bmatrix}-(2n+1)^{-1}<\operatorname{Re}s<0\\ 0<\operatorname{Re}s<(2n+1)^{-1}\end{Bmatrix}\right]$
11	$\frac{\ln x}{x-1}=G_{22}^{22}\left(x\,\middle\vert\,{0,\ 0 \atop 0,\ 0}\right)$	$\Gamma[s,\ s,\ 1-s,\ 1-s]$ [$0<\operatorname{Re}s<1$]
12	$\frac{\ln x}{x-y}=\frac{1}{y}G_{22}^{22}\left(\frac{x}{y}\,\middle\vert\,{0,\ 0 \atop 0,\ 0}\right)-\frac{\pi\ln y}{y}G_{22}^{11}\left(\frac{x}{y}\,\middle\vert\,{0,\ 1/2 \atop 0,\ 1/2}\right)$	$y^{s-1}\Gamma[s,\ 1-s]\times\left(\Gamma[s,\ 1-s]-\frac{\pi\ln y}{\Gamma[s+1/2,\ 1/2-s]}\right)$ [$y>0;\ 0<\operatorname{Re}s<1$]
13	$\frac{\ln x}{x+1}=-\pi G_{33}^{22}\left(x\,\middle\vert\,{0,\ 0,\ 1/2 \atop 0,\ 0,\ 1/2}\right)$	$-\pi\Gamma\begin{bmatrix}s,\ s,\ 1-s,\ 1-s\\ s+1/2,\ 1/2-s\end{bmatrix}$ [$0<\operatorname{Re}s<1$]
14	$\frac{\ln x}{x+y}$	$y^{s-1}\Gamma[s,\ 1-s]\times\left(\ln y-\pi\Gamma\begin{bmatrix}s,\ 1-s\\ s+1/2,\ 1/2-s\end{bmatrix}\right)$ [$y>0;\ 0<\operatorname{Re}s<1$]
15	$\frac{\ln x}{(x+y)^2}$	$y^{s-2}\Gamma[s,\ 2-s]\times\left(\ln y+\Gamma\begin{bmatrix}s-1\\ s\end{bmatrix}-\pi\Gamma\begin{bmatrix}s,\ 1-s\\ s+1/2,\ 1/2-s\end{bmatrix}\right)$ [$y>0;\ 0<\operatorname{Re}s<2$]
16	$\frac{\ln^2 x}{x+1}$	$2\Gamma[s,\ s,\ s,\ 1-s,\ 1-s,\ 1-s]-\pi^2\Gamma[s,\ 1-s]$ [$0<\operatorname{Re}s<1$]
17	$\ln(1+2x^{\pm1}\cos\gamma+x^{\pm2})$	$-2\cos(s\gamma)\Gamma[s,\ -s]$ $\left[\lvert\gamma\rvert<\pi,\ \begin{Bmatrix}-1<\operatorname{Re}s<0\\ 0<\operatorname{Re}s<1\end{Bmatrix}\right]$
18	$\ln\left\lvert\frac{1+\sqrt{x}}{1-\sqrt{x}}\right\rvert=\pi G_{22}^{11}\left(x\,\middle\vert\,{1/2,\ 1 \atop 1/2,\ 0}\right)$	$\pi\Gamma\begin{bmatrix}s+1/2,\ 1/2-s\\ s+1,\ 1-s\end{bmatrix}$ [$\lvert\operatorname{Re}s\rvert<1/2$]
19	$\frac{\ln(\sqrt{1+x}\pm\sqrt{x})}{(1+x)^r}$ [$r=0,\ 1/2$]	$\mp\frac{(-\pi)^{2r}}{2\sqrt{\pi}}\Gamma\begin{bmatrix}s+1/2,\ s+r,\ 1/2-s,\ r-s\\ s+1-r,\ 1/2+r-s\end{bmatrix}$ [$-1/2<\operatorname{Re}s<r$]

20	$(1+x)^{-r}\ln\frac{\sqrt{1+x}\pm 1}{\sqrt{x}}$ $[r=0,\ 1/2]$	$\mp\frac{(-\pi)^{2r}}{2\sqrt{\pi}}\times$ $\times\Gamma\left[\begin{matrix}s,\ s+1/2-r,\ 2r-s,\ 1/2+r-s\\ s+1/2,\ 1-s\end{matrix}\right]$ $[0<\operatorname{Re}s<r+1/2]$
21	$\ln(\sqrt{x}\pm\sqrt{x-1})\,H(x-1)=$ $=\pm\frac{\sqrt{\pi}}{2}G_{22}^{02}\left(x\left\vert\begin{matrix}1,\ 1\\ 0,\ 1/2\end{matrix}\right.\right)$	$\pm\frac{\sqrt{\pi}}{2}\Gamma\left[\begin{matrix}-s,\ -s\\ 1-s,\ 1/2-s\end{matrix}\right]$ $[\operatorname{Re}s<0]$
22	$\ln\frac{1\pm\sqrt{1-x}}{\sqrt{x}}H(1-x)=$ $=\pm\frac{\sqrt{\pi}}{2}G_{22}^{20}\left(x\left\vert\begin{matrix}1/2,\ 1\\ 0,\ 0\end{matrix}\right.\right)$	$\pm\frac{\sqrt{\pi}}{2}\Gamma\left[\begin{matrix}s,\ s\\ s+1/2,\ s+1\end{matrix}\right]$ $[\operatorname{Re}s>0]$
23	$\ln\frac{1+\sqrt{1+x}}{2}=$ $=\frac{1}{2\sqrt{\pi}}G_{33}^{13}\left(x\left\vert\begin{matrix}1,\ 1,\ 1/2\\ 1,\ 0,\ 0\end{matrix}\right.\right)$	$\frac{1}{2\sqrt{\pi}}\Gamma\left[\begin{matrix}s+1,\ -s,\ -s,\ 1/2-s\\ 1-s,\ 1-s\end{matrix}\right]$ $[-1<\operatorname{Re}s<0]$
24	$\ln\frac{\sqrt{x}+\sqrt{1+x}}{2\sqrt{x}}=$ $=\frac{1}{2\sqrt{\pi}}G_{33}^{31}\left(x\left\vert\begin{matrix}0,\ 1,\ 1\\ 0,\ 0,\ 1/2\end{matrix}\right.\right)$	$\frac{1}{2\sqrt{\pi}}\Gamma\left[\begin{matrix}s,\ s,\ s+1/2,\ 1-s\\ s+1,\ s+1\end{matrix}\right]$ $[0<\operatorname{Re}s<1]$
25	$\ln^2(\sqrt{x}+\sqrt{1+x})=$ $=\frac{\sqrt{\pi}}{2}G_{33}^{13}\left(x\left\vert\begin{matrix}1,\ 1,\ 1\\ 1,\ 0,\ 1/2\end{matrix}\right.\right)$	$\frac{\sqrt{\pi}}{2}\Gamma\left[\begin{matrix}s+1,\ -s,\ -s,\ -s\\ 1/2-s,\ 1-s\end{matrix}\right]$ $[-1<\operatorname{Re}s<0]$
26	$\ln^2\frac{1+\sqrt{1+x}}{\sqrt{x}}=$ $=\frac{\sqrt{\pi}}{2}G_{33}^{31}\left(x\left\vert\begin{matrix}0,\ 1,\ 1/2\\ 0,\ 0,\ 0\end{matrix}\right.\right)$	$\frac{\sqrt{\pi}}{2}\Gamma\left[\begin{matrix}s,\ s,\ s,\ 1-s\\ s+1/2,\ s+1\end{matrix}\right]$ $[0<\operatorname{Re}s<1]$

8.4.7. The inverse trigonometric functions.

NOTATION: $\delta=\begin{Bmatrix}1\\0\end{Bmatrix}$.

1	$\begin{Bmatrix}\arcsin\sqrt{x}\\ \arccos\sqrt{x}\end{Bmatrix}H(1-x)=$ $=\frac{\pi}{2}\begin{Bmatrix}1\\0\end{Bmatrix}G_{11}^{10}\left(x\left\vert\begin{matrix}1\\0\end{matrix}\right.\right)\mp$ $\mp\frac{\sqrt{\pi}}{2}G_{22}^{20}\left(x\left\vert\begin{matrix}1,\ 1\\ 0,\ 1/2\end{matrix}\right.\right)$	$\frac{\pi}{2}\begin{Bmatrix}1\\0\end{Bmatrix}\Gamma\left[\begin{matrix}s\\ s+1\end{matrix}\right]\mp$ $\mp\frac{\sqrt{\pi}}{2}\Gamma\left[\begin{matrix}s+1/2,\ s\\ s+1,\ s+1\end{matrix}\right]$ $[\operatorname{Re}s>-(1\pm 1)/4]$
2	$\begin{Bmatrix}\arcsin x^{-1/2}\\ \arccos x^{-1/2}\end{Bmatrix}H(x-1)=$ $=\frac{\pi}{2}\begin{Bmatrix}1\\0\end{Bmatrix}G_{11}^{01}\left(x\left\vert\begin{matrix}1\\0\end{matrix}\right.\right)\mp$ $\mp\frac{\sqrt{\pi}}{2}G_{22}^{02}\left(x\left\vert\begin{matrix}1/2,\ 1\\ 0,\ 0\end{matrix}\right.\right)$	$\frac{\pi}{2}\begin{Bmatrix}1\\0\end{Bmatrix}\Gamma\left[\begin{matrix}-s\\ 1-s\end{matrix}\right]\mp$ $\mp\frac{\sqrt{\pi}}{2}\Gamma\left[\begin{matrix}1/2-s,\ -s\\ 1-s,\ 1-s\end{matrix}\right]$ $[\operatorname{Re}s<(1\pm 1)/4]$

3	$\operatorname{arctg}\sqrt{x}= =\arccos\frac{1}{\sqrt{1+x}}=\arcsin\sqrt{\frac{x}{1+x}}= =\frac{1}{2}G_{22}^{12}\left(x\,\middle\vert\,\begin{matrix}1/2,\ 1\\1/2,\ 0\end{matrix}\right)$	$\frac{1}{2}\Gamma\left[\begin{matrix}s+1/2,\ 1/2-s,\ -s\\1-s\end{matrix}\right]$ $[-1/2<\operatorname{Re}s<0]$
4	$\operatorname{arctg}\frac{1}{\sqrt{x}}= =\arccos\sqrt{\frac{x}{1+x}}=\arcsin\frac{1}{\sqrt{1+x}}= =\frac{1}{2}G_{22}^{21}\left(x\,\middle\vert\,\begin{matrix}1/2,\ 1\\0,\ 1/2\end{matrix}\right)$	$\frac{1}{2}\Gamma\left[\begin{matrix}s,\ s+1/2,\ 1/2-s\\s+1\end{matrix}\right]$ $[0<\operatorname{Re}s<1/2]$
5	$(1-x)_{+}^{(\delta-1)/2}\left\{\begin{matrix}\sin(\nu\arccos\sqrt{x})\\\cos(\nu\arccos\sqrt{x})\end{matrix}\right\}= =\left(\frac{\nu}{2}\right)^{\delta}\sqrt{\pi}G_{22}^{20}\left(x\,\middle\vert\,\begin{matrix}\frac{1+\nu+\delta}{2},\ \frac{1-\nu+\delta}{2}\\0,\ \frac{1}{2}\end{matrix}\right)$	$\left(\frac{\nu}{2}\right)^{\delta}\sqrt{\pi}\,\Gamma\left[\begin{matrix}s,\ s+\frac{1}{2}\\s+\frac{1+\nu+\delta}{2},\ s+\frac{1-\nu+\delta}{2}\end{matrix}\right]$ $[\operatorname{Re}s>0]$
6	$(x-1)_{+}^{(\delta-1)/2}\left\{\begin{matrix}\sin(\nu\arccos x^{-1/2})\\\cos(\nu\arccos x^{-1/2})\end{matrix}\right\}= =\left(\frac{\nu}{2}\right)^{\delta}\sqrt{\pi}\,G_{22}^{02}\left(x\,\middle\vert\,\begin{matrix}1/2,\ \delta\\\nu/2,\ -\nu/2\end{matrix}\right)$	$\left(\frac{\nu}{2}\right)^{\delta}\sqrt{\pi}\,\Gamma\left[\begin{matrix}1/2-s,\ 1-\delta-s\\1+\nu/2-s,\ 1-\nu/2-s\end{matrix}\right]$ $[\operatorname{Re}s>0]$
7	$(1+x)^{\nu/2}\left\{\begin{matrix}\sin(\nu\operatorname{arctg}\sqrt{x})\\\cos(\nu\operatorname{arctg}\sqrt{x})\end{matrix}\right\}= =\mp\frac{2^{-\nu-1}}{\Gamma(-\nu)}G_{22}^{12}\left(x\,\middle\vert\,\begin{matrix}1+\nu/2,\ (1+\nu)/2\\\delta/2,\ (1-\delta)/2\end{matrix}\right)$	$\mp\frac{2^{-\nu-1}}{\Gamma(-\nu)}\times$ $\times\Gamma\left[\begin{matrix}s+\frac{\delta}{2},\ -\frac{\nu}{2}-s,\ \frac{1-\nu}{2}-s\\(1+\delta)/2-s\end{matrix}\right]$ $[-\delta/2<\operatorname{Re}s<-\operatorname{Re}\nu/2]$
8	$(1+x)^{\nu/2}\left\{\begin{matrix}\sin(\nu\operatorname{arctg}x^{-1/2})\\\cos(\nu\operatorname{arctg}x^{-1/2})\end{matrix}\right\}= =\mp\frac{2^{-\nu-1}}{\Gamma(-\nu)}G_{22}^{21}\left(x\,\middle\vert\,\begin{matrix}1+\frac{\nu-\delta}{2},\ \frac{1+\nu+\delta}{2}\\0,\ \frac{1}{2}\end{matrix}\right)$	$\mp\frac{2^{-\nu-1}}{\Gamma(-\nu)}\Gamma\left[\begin{matrix}s+1/2,\ s,\ (\delta-\nu)/2-s\\s+(1+\nu+\delta)/2\end{matrix}\right]$ $[0<\operatorname{Re}s<(\delta-\operatorname{Re}\nu)/2]$

8.4.8. The inverse hyperbolic functions.

NOTATION: $A_6=\frac{1}{\sqrt{\pi}}\left(\frac{\nu}{2}\right)^{1-2r}\left\{\begin{matrix}\cos(r-\nu/2)\pi\\\sin(r-\nu/2)\pi\end{matrix}\right\},\ \delta=\left\{\begin{matrix}1\\0\end{matrix}\right\}.$

1	$(1+x)^{-r}\left\{\begin{matrix}\operatorname{sh}(\nu\operatorname{Arsh}\sqrt{x})\\\operatorname{ch}(\nu\operatorname{Arsh}\sqrt{x})\end{matrix}\right\}= =A_6G_{22}^{12}\left(x\,\middle\vert\,\begin{matrix}1+\nu/2-r,\ 1-\nu/2-r\\\delta/2,\ (1-\delta)/2\end{matrix}\right)$ $[r=0,\ 1/2]$	$A_6\Gamma\left[\begin{matrix}s+\delta/2,\ \nu/2+r-s,\ r-\nu/2-s\\(1+\delta)/2-s\end{matrix}\right]$ $[-\delta/2<\operatorname{Re}s<r-\vert\operatorname{Re}\nu\vert/2]$

2	$(1+x)^{-r}\begin{Bmatrix}\operatorname{sh}(\nu\,\operatorname{Arsh}x^{-1/2})\\ \operatorname{ch}(\nu\,\operatorname{Arsh}x^{-1/2})\end{Bmatrix}=$ $=A_6G_{22}^{21}\left(x\,\middle\vert\,\begin{matrix}1-r-\delta/2,\ (1+\delta)/2-r\\ \nu/2,\ -\nu/2\end{matrix}\right)$ [$r=0,\ 1/2$]	$A_6\Gamma\left[\begin{matrix}s+\nu/2,\ s-\nu/2,\ r+\delta/2-s\\ s+(1+\delta)/2-r\end{matrix}\right]$ [$\vert\operatorname{Re}\nu\vert/2<\operatorname{Re}s<r+\delta/2$]
3	$(1-x)_+^{(\delta-1)/2}\begin{Bmatrix}\operatorname{sh}(\nu\,\operatorname{Arth}\sqrt{1-x})\\ \operatorname{ch}(\nu\,\operatorname{Arth}\sqrt{1-x})\end{Bmatrix}=$ $=\sqrt{\pi}\left(\frac{\nu}{2}\right)^{\delta}G_{22}^{20}\left(x\,\middle\vert\,\begin{matrix}1/2,\ \delta\\ \nu/2,\ -\nu/2\end{matrix}\right)$	$\sqrt{\pi}\left(\frac{\nu}{2}\right)^{\delta}\Gamma\left[\begin{matrix}s+\nu/2,\ s-\nu/2\\ s+1/2,\ s+\delta\end{matrix}\right]$ [$\operatorname{Re}s>\vert\operatorname{Re}\nu\vert/2$]
4	$(x-1)_+^{(\delta-1)/2}\begin{Bmatrix}\operatorname{sh}(\nu\,\operatorname{Arth}\sqrt{1-x^{-1}})\\ \operatorname{ch}(\nu\,\operatorname{Arth}\sqrt{1-x^{-1}})\end{Bmatrix}=$ $=\sqrt{\pi}\left(\frac{\nu}{2}\right)^{\delta}G_{22}^{02}\left(x\,\middle\vert\,\begin{matrix}\frac{1+\delta-\nu}{2},\ \frac{1+\delta+\nu}{2}\\ 0,\ \frac{1}{2}\end{matrix}\right)$	$\sqrt{\pi}\left(\frac{\nu}{2}\right)^{\delta}\Gamma\left[\begin{matrix}\frac{1+\nu-\delta}{2}-s,\ \frac{1-\nu-\delta}{2}-s\\ 1-s,\ \frac{1}{2}-s\end{matrix}\right]$ [$\operatorname{Re}s<(1-\delta-\vert\operatorname{Re}\nu\vert)/2$]
5	$\vert 1-x\vert^{\nu}\begin{Bmatrix}\operatorname{sh}(\nu\operatorname{Arth}[2\sqrt{x}(1+x)^{-1}])\\ \operatorname{ch}(\nu\operatorname{Arth}[2\sqrt{x}(1+x)^{-1}])\end{Bmatrix}=$ $=\mp\sqrt{\pi}\,\Gamma\left[\begin{matrix}\nu+\frac{1}{2}\\ -\nu\end{matrix}\right]\times$ $\times G_{22}^{11}\left(x\,\middle\vert\,\begin{matrix}1+\nu-\frac{\delta}{2},\ \frac{1+\delta}{2}+\nu\\ \frac{\delta}{2},\ \frac{1-\delta}{2}\end{matrix}\right)$	$\mp\sqrt{\pi}\,\Gamma\left[\begin{matrix}\nu+1/2\\ -\nu\end{matrix}\right]\times$ $\times\Gamma\left[\begin{matrix}s+\frac{\delta}{2},\ \frac{\delta}{2}-\nu-s\\ s+\nu+\frac{1+\delta}{2},\ \frac{1+\delta}{2}-s\end{matrix}\right]$ [$\operatorname{Re}\nu>-1/2,\ -\delta/2<\operatorname{Re}s<\delta/2-\operatorname{Re}\nu$]

8.4.9. The polylogarithm $\operatorname{Li}_n(x)$.

1	$\operatorname{Li}_n(-x)=-G_{n+1,\,n+1}^{1,\,n+1}\left(x\,\middle\vert\,\begin{matrix}1,\ 1,\ \ldots,\ 1\\ 1,\ 0,\ \ldots,\ 0\end{matrix}\right)$	$(-s)^{1-n}\,\Gamma[s,\ -s]$	[$-1<\operatorname{Re}s<0$]
2	$\operatorname{Li}_n\left(-\frac{1}{x}\right)=-G_{n+1,\,n+1}^{n+1,\,1}\left(x\,\middle\vert\,\begin{matrix}0,\ 1,\ \ldots,\ 1\\ 0,\ 0,\ \ldots,\ 0\end{matrix}\right)$	$s^{1-n}\Gamma[s,\ -s]$	[$0<\operatorname{Re}s<1$]

8.4.10. The function $\Phi(x,s,v)$

1	$\Phi(-x,\ n,\ v)=$ $=G_{n+1,\,n+1}^{1,\,n+1}\left(x\,\middle\vert\,\begin{matrix}0,\ 1-v,\ \ldots,\ 1-v\\ 0,\ -v,\ \ldots,\ -v\end{matrix}\right)$	$(-1)^n\dfrac{\Gamma[s,\ 1-s]}{(s-v)^n}$ [$0<\operatorname{Re}s<1;\ \operatorname{Re}s<\operatorname{Re}v$]
2	$\Phi\left(-\frac{1}{x},\ n,\ v\right)=$ $=G_{n+1,\,n+1}^{n+1,\,1}\left(x\,\middle\vert\,\begin{matrix}1,\ v+1,\ \ldots,\ v+1\\ 1,\ v,\ \ldots,\ v\end{matrix}\right)$	$\dfrac{\Gamma[s+1,\ -s]}{(s+v)^n}$ [$-1<\operatorname{Re}s<0;\ \operatorname{Re}s>-\operatorname{Re}v$]
3	$\Phi(1-x,\ 1,\ v)=\dfrac{1}{\Gamma(v)}G_{22}^{22}\left(x\,\middle\vert\,\begin{matrix}0,\ 1-v\\ 0,\ 0\end{matrix}\right)$	$\dfrac{1}{\Gamma(v)}\Gamma[s,\ s,\ 1-s,\ v-s]$ [$0<\operatorname{Re}s<1,\ \operatorname{Re}v$]
4	$\Phi\left(1-\frac{1}{x},\ 1,\ v\right)=\dfrac{1}{\Gamma(v)}G_{22}^{22}\left(x\,\middle\vert\,\begin{matrix}1,\ 1\\ 1,\ v\end{matrix}\right)$	$\dfrac{1}{\Gamma(v)}\Gamma[s+1,\ s+v,\ -s,\ -s]$ [$-1,\ -\operatorname{Re}v<\operatorname{Re}s<0$]

5	$(1-x)_+^{v}\,\Phi(1-x,\ 1,\ v)=$ $=\Gamma(v)\,G_{22}^{20}\left(x\left\vert\begin{matrix}1,\ v\\0,\ 0\end{matrix}\right.\right)$	$\Gamma(v)\,\Gamma\left[\begin{matrix}s, & s\\s+1, & s+v\end{matrix}\right]$ [Re $v>-1$; $v\neq 0$; Re $s>0$]
6	$(x-1)_+^{v}\,\Phi\left(1-\dfrac{1}{x},\ 1,\ v\right)=$ $=\Gamma(v)\,G_{22}^{02}\left(x\left\vert\begin{matrix}v+1,\ v+1\\1,\ v\end{matrix}\right.\right)$	$\Gamma(v)\,\Gamma\left[\begin{matrix}-v-s, & -v-s\\-s, & 1-v-s\end{matrix}\right]$ [Re $v>-1$; $v\neq 0$; Re $(s+v)<0$]
7	$(x-1)_+^{v}\,\Phi(1-x,\ 1,\ v)=$ $=\Gamma(v)\,G_{22}^{02}\left(x\left\vert\begin{matrix}1,\ v\\0,\ 0\end{matrix}\right.\right)$	$\Gamma(v)\,\Gamma\left[\begin{matrix}-s, & 1-v-s\\1-s, & 1-s\end{matrix}\right]$ [Re $v>-1$; $v\neq 0$; Re $s<0,\ 1-$Re v]
8	$(1-x)_+^{v}\,\Phi\left(1-\dfrac{1}{x},\ 1,\ v\right)=$ $=\Gamma(v)\,G_{22}^{20}\left(x\left\vert\begin{matrix}v+1,\ v+1\\1,\ v\end{matrix}\right.\right)$	$\Gamma(v)\,\Gamma\left[\begin{matrix}s+1, & s+v\\s+v+1, & s+v+1\end{matrix}\right]$ [Re $v>-1$; $v\neq 0$; Re $s>-1,\ -$Re v]

8.4.11. The integral exponential function Ei (x).

1	$\mathrm{Ei}(-x)=-G_{12}^{20}\left(x\left\vert\begin{matrix}1\\0,\ 0\end{matrix}\right.\right)$	$-\Gamma\left[\begin{matrix}s,\ s\\s+1\end{matrix}\right]=-\dfrac{\Gamma(s)}{s}$ [Re $s>0$]
2	$\mathrm{Ei}\left(-\dfrac{1}{x}\right)=-G_{21}^{02}\left(x\left\vert\begin{matrix}1,\ 1\\0\end{matrix}\right.\right)$	$-\Gamma\left[\begin{matrix}-s,\ -s\\1-s\end{matrix}\right]=\dfrac{\Gamma(-s)}{s}$ [Re $s<0$]
3	$e^{x}\,\mathrm{Ei}(-x)=-G_{12}^{21}\left(x\left\vert\begin{matrix}0\\0,\ 0\end{matrix}\right.\right)$	$-\Gamma[s,\ s,\ 1-s]$ [$0<$ Re $s<1$]
4	$e^{-x}\,\mathrm{Ei}(x)=-\pi G_{23}^{21}\left(x\left\vert\begin{matrix}0,\ 1/2\\0,\ 0,\ 1/2\end{matrix}\right.\right)$	$-\pi\Gamma\left[\begin{matrix}s,\ s,\ 1-s\\s+1/2,\ 1/2-s\end{matrix}\right]$ [$0<$ Re $s<1$]
5	$e^{1/x}\,\mathrm{Ei}\left(-\dfrac{1}{x}\right)=-G_{21}^{12}\left(x\left\vert\begin{matrix}1,\ 1\\1\end{matrix}\right.\right)$	$-\Gamma[s+1,\ -s,\ -s]$ [$-1<$ Re $s<0$]
6	$e^{-1/x}\,\mathrm{Ei}\left(\dfrac{1}{x}\right)=-\pi G_{32}^{12}\left(x\left\vert\begin{matrix}1,\ 1,\ 1/2\\1,\ 1/2\end{matrix}\right.\right)$	$-\pi\Gamma\left[\begin{matrix}s+1,\ -s,\ -s\\s+1/2,\ 1/2-s\end{matrix}\right]$ [$-1<$ Re $s<0$]
7	$\mathrm{Ei}(-2\sqrt{x})\,\mathrm{Ei}(2\sqrt{x})=$ $=\dfrac{\sqrt{\pi}}{2}\,G_{24}^{31}\left(x\left\vert\begin{matrix}0,\ 1\\0,\ 0,\ 0,\ 1/2\end{matrix}\right.\right)$	$\dfrac{\sqrt{\pi}}{2}\,\Gamma\left[\begin{matrix}s,\ s,\ s,\ 1-s\\s+1,\ 1/2-s\end{matrix}\right]$ [$0<$ Re $s<1$]
8	$\mathrm{Ei}\left(-\dfrac{2}{\sqrt{x}}\right)\mathrm{Ei}\left(\dfrac{2}{\sqrt{x}}\right)=$ $=\dfrac{\sqrt{\pi}}{2}\,G_{42}^{13}\left(x\left\vert\begin{matrix}1,\ 1,\ 1,\ 1/2\\1,\ 0\end{matrix}\right.\right)$	$\dfrac{\sqrt{\pi}}{2}\,\Gamma\left[\begin{matrix}s+1,\ -s,\ -s,\ -s\\s+1/2,\ 1-s\end{matrix}\right]$ [$-1<$ Re $s<0$]

8.4.12. The integral sines Si (x), si (x) and cosine ci (x).

NOTATION: $\delta=\left\{\begin{matrix}1\\0\end{matrix}\right\}$.

1	$\mathrm{Si}(2\sqrt{x})=$ $=\dfrac{\sqrt{\pi}}{2}\,G_{13}^{11}\left(x\left\vert\begin{matrix}1\\1/2,\ 0,\ 0\end{matrix}\right.\right)$	$\dfrac{\sqrt{\pi}}{2}\Gamma\left[\begin{matrix}s+1/2,\ -s\\1-s,\ 1-s\end{matrix}\right]=-\dfrac{\sqrt{\pi}}{2s}\Gamma\left[\begin{matrix}s+1/2\\1-s\end{matrix}\right]$ [$-1/2<$ Re $s<0$]

2	$\mathrm{Si}\left(\frac{2}{\sqrt{x}}\right)=\frac{\sqrt{\pi}}{2}G_{31}^{11}\left(x\left\|\begin{matrix}1/2,\ 1,\ 1\\ 0\end{matrix}\right.\right)$	$\frac{\sqrt{\pi}}{2}\Gamma\left[\begin{matrix}s,\ 1/2-s\\ s+1,\ s+1\end{matrix}\right]=\frac{\sqrt{\pi}}{2s}\Gamma\left[\begin{matrix}1/2-s\\ s+1\end{matrix}\right]$ $[0<\mathrm{Re}\, s<1/2]$
3	$\mathrm{si}\,(2\sqrt{x})=-\frac{\sqrt{\pi}}{2}G_{13}^{20}\left(x\left\|\begin{matrix}1\\ 0,\ 1/2,\ 0\end{matrix}\right.\right)$	$-\frac{\sqrt{\pi}}{2}\Gamma\left[\begin{matrix}s,\ s+1/2\\ s+1,\ 1-s\end{matrix}\right]=-\frac{\sqrt{\pi}}{2s}\Gamma\left[\begin{matrix}s+1/2\\ 1-s\end{matrix}\right]$ $[0<\mathrm{Re}\, s<1]$
4	$\mathrm{ci}\,(2\sqrt{x})=-\frac{\sqrt{\pi}}{2}G_{13}^{20}\left(x\left\|\begin{matrix}1\\ 0,\ 0,\ 1/2\end{matrix}\right.\right)$	$-\frac{\sqrt{\pi}}{2}\Gamma\left[\begin{matrix}s,\ s\\ s+1,\ 1/2-s\end{matrix}\right]=-\frac{\sqrt{\pi}}{2s}\Gamma\left[\begin{matrix}s\\ 1/2-s\end{matrix}\right]$ $[0<\mathrm{Re}\, s<1]$
5	$\left\{\begin{matrix}\mathrm{si}\,(2\sqrt{x})\\ \mathrm{ci}\,(2\sqrt{x})\end{matrix}\right\}=-\frac{\sqrt{\pi}}{2}G_{13}^{20}\left(x\left\|\begin{matrix}1\\ 0,\ \delta/2,\ (1-\delta)/2\end{matrix}\right.\right)$	$-\frac{\sqrt{\pi}}{2}\Gamma\left[\begin{matrix}s,\ s+\delta/2\\ s+1,\ (1+\delta)/2-s\end{matrix}\right]=-\frac{\sqrt{\pi}}{2s}\Gamma\left[\begin{matrix}s+\delta/2\\ (1+\delta)/2-s\end{matrix}\right]$ $[0<\mathrm{Re}\, s<1]$
6	$\left\{\begin{matrix}\mathrm{si}\,(2x^{-1/2})\\ \mathrm{ci}\,(2x^{-1/2})\end{matrix}\right\}=-\frac{\sqrt{\pi}}{2}G_{31}^{02}\left(x\left\|\begin{matrix}1-\delta/2,\ 1,\ (\delta+1)/2\\ 0\end{matrix}\right.\right)$	$-\frac{\sqrt{\pi}}{2}\Gamma\left[\begin{matrix}-s,\ \delta/2-s\\ s+(\delta+1)/2,\ 1-s\end{matrix}\right]=\frac{\sqrt{\pi}}{2s}\Gamma\left[\begin{matrix}\delta/2-s\\ s+(\delta+1)/2\end{matrix}\right]$ $[-1<\mathrm{Re}\, s<0]$
7	$\left\{\begin{matrix}\sin 2\sqrt{x}\\ \cos 2\sqrt{x}\end{matrix}\right\}\mathrm{Si}\,(2\sqrt{x})-\left\{\begin{matrix}\cos 2\sqrt{x}\\ \sin 2\sqrt{x}\end{matrix}\right\}\mathrm{ci}\,(2\sqrt{x})=\mp\frac{\pi^{3/2}}{2}\times G_{24}^{21}\left(x\left\|\begin{matrix}\frac{1-\delta}{2},\ 1-\frac{3\delta}{2}\\ \frac{1-\delta}{2},\ \frac{1-\delta}{2},\ 1-\frac{\delta}{2},\ -\frac{\delta}{2}\end{matrix}\right.\right)$	$\mp\frac{\pi^{3/2}}{2}\Gamma\left[\begin{matrix}s+\frac{1-\delta}{2},\ s+\frac{1-\delta}{2},\ \frac{1+\delta}{2}-s\\ s+1-\frac{3\delta}{2},\ \frac{\delta}{2}-s,\ \frac{\delta}{2}+1-s\end{matrix}\right]$ $[(\delta-1)/2<\mathrm{Re}\, s<1/2]$
8	$\left\{\begin{matrix}\sin(2x^{-1/2})\\ \cos(2x^{-1/2})\end{matrix}\right\}\mathrm{Si}\left(\frac{2}{\sqrt{x}}\right)-\left\{\begin{matrix}\cos(2x^{-1/2})\\ \sin(2x^{-1/2})\end{matrix}\right\}\mathrm{ci}\left(\frac{2}{\sqrt{x}}\right)=\mp\frac{\pi^{3/2}}{2}\times G_{42}^{12}\left(x\left\|\begin{matrix}\frac{1+\delta}{2},\ \frac{1+\delta}{2},\ \frac{\delta}{2},\ 1+\frac{\delta}{2}\\ \frac{1+\delta}{2},\ \frac{3\delta}{2}\end{matrix}\right.\right)$	$\mp\frac{\pi^{3/2}}{2}\Gamma\left[\begin{matrix}s+\frac{1+\delta}{2},\ \frac{1-\delta}{2}-s,\ \frac{1-\delta}{2}-s\\ s+\frac{\delta}{2},\ s+\frac{\delta}{2}+1,\ 1-\frac{3\delta}{2}-s\end{matrix}\right]$ $[-1/2<\mathrm{Re}\, s<(1-\delta)/2]$
9	$\left\{\begin{matrix}\sin 2\sqrt{x}\\ \cos 2\sqrt{x}\end{matrix}\right\}\mathrm{si}\,(2\sqrt{x})\pm\left\{\begin{matrix}\cos 2\sqrt{x}\\ \sin 2\sqrt{x}\end{matrix}\right\}\mathrm{ci}\,(2\sqrt{x})=-\frac{1}{2\sqrt{\pi}}G_{13}^{31}\left(x\left\|\begin{matrix}(1\mp 1)/4\\ 0,\ (1\mp 1)/4,\ 1/2\end{matrix}\right.\right)$	$-\frac{1}{2\sqrt{\pi}}\times\Gamma\left[s,\ s+\frac{1}{2},\ s+\frac{1-\delta}{2},\ \frac{1+\delta}{2}-s\right]$ $[0<\mathrm{Re}\, s<1/2]$

10	$\begin{Bmatrix}\sin(2x^{-1/2})\\ \cos(2x^{-1/2})\end{Bmatrix}\operatorname{si}\left(\frac{2}{\sqrt{x}}\right)\pm$ $\pm\begin{Bmatrix}\cos(2x^{-1/2})\\ \sin(2x^{-1/2})\end{Bmatrix}\operatorname{ci}\left(\frac{2}{\sqrt{x}}\right)=$ $=-\frac{1}{2\sqrt{\pi}}G_{31}^{13}\left(x\,\middle\vert\,\begin{matrix}1,\ (3\pm1)/4,\ 1/2\\ (3\pm1)/4,\end{matrix}\right)$	$-\frac{1}{2\sqrt{\pi}}\times$ $\times\Gamma\left[s+\frac{1+\delta}{2},\ -s,\ \frac{1}{2}-s,\ \frac{1-\delta}{2}-s\right]$ $[-1/2<\operatorname{Re}s<0]$
11	$\operatorname{si}^2(2\sqrt{x})+\operatorname{ci}^2(2\sqrt{x})=$ $=\frac{1}{2\sqrt{\pi}}G_{24}^{41}\left(x\,\middle\vert\,\begin{matrix}0,\ 1\\ 0,\ 0,\ 0,\ 1/2\end{matrix}\right)$	$-\frac{1}{2\sqrt{\pi}}\Gamma\left[s,\ s,\ s+\frac{1}{2},\ -s\right]$ $[0<\operatorname{Re}s<1]$
12	$\operatorname{si}^2\left(\frac{2}{\sqrt{x}}\right)+\operatorname{ci}^2\left(\frac{2}{\sqrt{x}}\right)=$ $=\frac{1}{2\sqrt{\pi}}G_{42}^{14}\left(x\,\middle\vert\,\begin{matrix}1,\ 1,\ 1,\ 1/2\\ 1,\ 0\end{matrix}\right)$	$-\frac{1}{2\sqrt{\pi}}\Gamma\left[s,\ -s,\ -s,\ \frac{1}{2}-s\right]$ $[-1<\operatorname{Re}s<0]$

8.4.13. The integral hyperbolic sine $\operatorname{shi}(x)$ and cosine $\operatorname{chi}(x)$.

1	$\operatorname{shi}(2\sqrt{x})=$ $\overleftarrow{L_{-\infty}}$ $=-\frac{\pi^{3/2}}{2\cos c\pi}G_{24}^{11}\left(x\,\middle\vert\,\begin{matrix}1,\ c\\ 1/2,\ 0,\ 0,\ c\end{matrix}\right)$	$\frac{\pi^{3/2}}{2\cos c\pi}\Gamma\left[\begin{matrix}s+1/2,\ -s\\ s+c,\ 1-c-s,\ 1-s,\ 1-s\end{matrix}\right]$ [$c\neq1/2$ is arbitrary; $s=-1/2-k\in D^+$, $s=0\in D^-$, $k=0,\ 1,\ 2,\ \ldots$]
2	$\operatorname{shi}\left(\frac{2}{\sqrt{x}}\right)=$ $\overleftarrow{L_{+\infty}}$ $=-\frac{\pi^{3/2}}{2\cos c\pi}G_{42}^{11}\left(x\,\middle\vert\,\begin{matrix}1/2,\ 1,\ 1,\ 1-c\\ 0,\ 1-c\end{matrix}\right)$	$-\frac{\pi^{3/2}}{2\cos c\pi}\times$ $\times\Gamma\left[\begin{matrix}s,\ 1/2-s\\ s+1,\ s+1,\ s-c+1,\ c-s\end{matrix}\right]$ [$c\neq1/2$ is arbitrary; $s=1/2+k\in D^-$, $s=0\in D^+$, $k=0,\ 1,\ 2,\ldots$]
3	$\operatorname{shi}(2i\sqrt{x})=\frac{i\sqrt{\pi}}{2}G_{13}^{11}\left(x\,\middle\vert\,\begin{matrix}1\\ 1/2,\ 0,\ 0\end{matrix}\right)$	$\frac{i\sqrt{\pi}}{2}\Gamma\left[\begin{matrix}s+1/2,\ -s\\ 1-s,\ 1-s\end{matrix}\right]=$ $=-\frac{i\sqrt{\pi}}{2s}\Gamma\left[\begin{matrix}s+1/2\\ 1-s\end{matrix}\right]$ $[-1/2<\operatorname{Re}s<0]$
4	$\operatorname{shi}\left(\frac{2i}{\sqrt{x}}\right)=\frac{i\sqrt{\pi}}{2}G_{31}^{11}\left(x\,\middle\vert\,\begin{matrix}1/2,\ 1,\ 1\\ 0\end{matrix}\right)$	$\frac{i\sqrt{\pi}}{2}\Gamma\left[\begin{matrix}s,\ 1/2-s\\ s+1,\ s+1\end{matrix}\right]=$ $=\frac{i\sqrt{\pi}}{2s}\Gamma\left[\begin{matrix}1/2-s\\ s+1\end{matrix}\right]$ $[0<\operatorname{Re}s<1/2]$
5	$\operatorname{chi}(2i\sqrt{x})-\frac{\pi i}{2}=$ $=-\frac{\sqrt{\pi}}{2}G_{13}^{20}\left(x\,\middle\vert\,\begin{matrix}1\\ 0,\ 0,\ 1/2\end{matrix}\right)$	$-\frac{\sqrt{\pi}}{2}\Gamma\left[\begin{matrix}s,\ s\\ s+1,\ 1/2-s\end{matrix}\right]=$ $=-\frac{\sqrt{\pi}}{2s}\Gamma\left[\begin{matrix}s\\ 1/2-s\end{matrix}\right]$ $[0<\operatorname{Re}s<1]$
6	$\operatorname{chi}\left(\frac{2i}{\sqrt{x}}\right)-\frac{\pi i}{2}=$ $=-\frac{\sqrt{\pi}}{2}G_{31}^{02}\left(x\,\middle\vert\,\begin{matrix}1,\ 1,\ 1/2\\ 0\end{matrix}\right)$	$-\frac{\sqrt{\pi}}{2}\Gamma\left[\begin{matrix}-s,\ -s\\ s+1/2,\ 1-s\end{matrix}\right]=$ $=\frac{\sqrt{\pi}}{2s}\Gamma\left[\begin{matrix}-s\\ s+1/2\end{matrix}\right]$ $[-1<\operatorname{Re}s<0]$

8.4.14. The error functions $\operatorname{erf}(x), \operatorname{erfc}(x)$ and $\operatorname{erfi}(x)$.

NOTATION: $A_7 = -\dfrac{\pi\sqrt{2}}{\cos c\pi}$.

1	$\operatorname{erf}(\sqrt{x}) = \dfrac{1}{\sqrt{\pi}} G_{12}^{11}\left(x \middle\vert \begin{matrix} 1 \\ 1/2, 0 \end{matrix}\right)$	$\dfrac{1}{\sqrt{\pi}} \Gamma\left[\begin{matrix} s+1/2, & -s \\ 1-s \end{matrix}\right] = -\dfrac{1}{\sqrt{\pi}\, s} \Gamma\left(s+\dfrac{1}{2}\right)$ $[-1/2 < \operatorname{Re} s < 0]$
2	$\operatorname{erfc}(\sqrt{x}) = \dfrac{1}{\sqrt{\pi}} G_{12}^{20}\left(x \middle\vert \begin{matrix} 1 \\ 0, 1/2 \end{matrix}\right)$	$\dfrac{1}{\sqrt{\pi}} \Gamma\left[\begin{matrix} s, & s+1/2 \\ s+1 \end{matrix}\right] = \dfrac{1}{\sqrt{\pi}\, s} \Gamma\left(s+\dfrac{1}{2}\right)$ $[\operatorname{Re} s > 0]$
3	$\left\{\begin{matrix} \operatorname{erf}(\sqrt{x}) \\ \operatorname{erfc}(\sqrt{x}) \end{matrix}\right\}$	$\mp \dfrac{1}{\sqrt{\pi}} \Gamma\left[\begin{matrix} s+1/2, & s \\ s+1 \end{matrix}\right] = \mp \dfrac{1}{\sqrt{\pi}\, s} \Gamma\left(s+\dfrac{1}{2}\right)$ $\left[\left\{\begin{matrix} -1/2 < \operatorname{Re} s < 0 \\ \operatorname{Re} s > 0 \end{matrix}\right\}\right]$
4	$\left\{\begin{matrix} \operatorname{erf}(x^{-1/2}) \\ \operatorname{erfc}(x^{-1/2}) \end{matrix}\right\}$	$\pm \dfrac{1}{\sqrt{\pi}} \Gamma\left[\begin{matrix} s, & 1/2-s \\ s+1 \end{matrix}\right] = \pm \dfrac{1}{\sqrt{\pi}\, s} \Gamma\left(\dfrac{1}{2}-s\right)$ $\left[\left\{\begin{matrix} 0 < \operatorname{Re} s < 1/2 \\ \operatorname{Re} s < 0 \end{matrix}\right\}\right]$
5	$e^{-x} \operatorname{erf}(i\sqrt{x}) = iG_{12}^{11}\left(x \middle\vert \begin{matrix} 1/2 \\ 1/2, 0 \end{matrix}\right)$	$i\Gamma\left[\begin{matrix} s+1/2, & 1/2-s \\ 1-s \end{matrix}\right]$ $[\lvert \operatorname{Re} s \rvert < 1/2]$
6	$e^{-1/x} \operatorname{erf}\left(\dfrac{i}{\sqrt{x}}\right) = iG_{21}^{11}\left(x \middle\vert \begin{matrix} 1/2, 1 \\ 1/2 \end{matrix}\right)$	$i\Gamma\left[\begin{matrix} s+1/2, & 1/2-s \\ s+1 \end{matrix}\right]$ $[\lvert \operatorname{Re} s \rvert < 1/2]$
7	$e^{x} \operatorname{erfc}(\sqrt{x}) = \dfrac{1}{\pi} G_{12}^{21}\left(x \middle\vert \begin{matrix} 1/2 \\ 0, 1/2 \end{matrix}\right)$	$\dfrac{1}{\pi} \Gamma\left[s, s+\dfrac{1}{2}, \dfrac{1}{2}-s\right]$ $[0 < \operatorname{Re} s < 1/2]$
8	$e^{1/x} \operatorname{erfc}\left(\dfrac{1}{\sqrt{x}}\right) = \dfrac{1}{\pi} G_{21}^{12}\left(x \middle\vert \begin{matrix} 1/2, 1 \\ 1/2 \end{matrix}\right)$	$\dfrac{1}{\pi} \Gamma\left[s+\dfrac{1}{2}, -s, \dfrac{1}{2}-s\right]$ $[-1/2 < \operatorname{Re} s < 0]$
9	$\operatorname{erf}(\sqrt{2}\, x^{1/4}) \operatorname{erfi}(\sqrt{2}\, x^{1/4}) \overset{L_{-\infty}}{\longleftarrow} = A_7 G_{35}^{12}\left(x \middle\vert \begin{matrix} 1/2, 1, c \\ 1/2, 1/4, 3/4, 0, c \end{matrix}\right)$	$A_7 \times \Gamma\left[\begin{matrix} s+1/2, & -s, & 1/2-s \\ s+c, & 1-c-s, & 1/4-s, 3/4-s, 1-s \end{matrix}\right]$ $[c \neq 1/2$ is arbitrary; $s = -1/2 - k \in D^+$, $s = 0, 1/2 + k \in D^-$, $k = 0, 1, 2, \ldots]$
10	$\operatorname{erf}\left(\sqrt[4]{\dfrac{4}{x}}\right) \operatorname{erfi}\left(\sqrt[4]{\dfrac{4}{x}}\right) \overset{L_{+\infty}}{\longleftarrow} = A_7 G_{53}^{21}\left(x \middle\vert \begin{matrix} 1/2, 1/4, 3/4, 1, 1-c \\ 0, 1/2, 1-c \end{matrix}\right)$	$A_7 \times \Gamma\left[\begin{matrix} s, & s+\dfrac{1}{2}, & \dfrac{1}{2}-s \\ s+1, s+\dfrac{1}{4}, s+\dfrac{3}{4}, s-c+1, c-s \end{matrix}\right]$ $[c \neq 1/2$ is arbitrary; $s = 1/2 + k \in D^-$, $s = 0, -1/2 - k \in D^+$, $k = 0, 1, 2, \ldots]$
11	$\operatorname{erf}(i\sqrt{2}\, x^{1/4}) \operatorname{erfc}(\sqrt{2}\, x^{1/4}) = \dfrac{i}{\pi\sqrt{2}} G_{24}^{31}\left(x \middle\vert \begin{matrix} 1/2, 1 \\ 1/4, 1/2, 3/4, 0 \end{matrix}\right)$	$\dfrac{i}{\pi\sqrt{2}} \times \Gamma\left[\begin{matrix} s+\dfrac{1}{4}, & s+\dfrac{1}{2}, & s+\dfrac{3}{4}, & \dfrac{1}{2}-s \\ s+1, & 1-s \end{matrix}\right]$ $[-1/4 < \operatorname{Re} s < 1/2]$

12	$\operatorname{erf}\left(i\sqrt[4]{\frac{4}{x}}\right)\operatorname{erfc}\left(\sqrt[4]{\frac{4}{x}}\right)=$ $=\frac{i}{\pi\sqrt{2}}G_{42}^{13}\left(x\,\middle\|\,\begin{matrix}1/4,\ 1/2,\ 3/4,\ 1\\ 1/2,\ 0\end{matrix}\right)$	$\frac{i}{\pi\sqrt{2}}\times$ $\times\Gamma\left[\begin{matrix}s+1/2,\ 1/4-s,\ 1/2-s,\ 3/4-s\\ s+1,\ 1-s\end{matrix}\right]$ $[-1/2<\operatorname{Re}s<1/4]$

8.4.15. The Fresnel integrals $S(x)$ and $C(x)$.

NOTATION: $\delta=\left\{\begin{matrix}1\\0\end{matrix}\right\}$.

1	$S(2\sqrt{x})=\frac{1}{2}G_{13}^{11}\left(x\,\middle\|\,\begin{matrix}1\\ 3/4,\ 0,\ 1/4\end{matrix}\right)$	$\frac{1}{2}\Gamma\left[\begin{matrix}s+3/4,\ -s\\ 1-s,\ 3/4-s\end{matrix}\right]=$ $=-\frac{1}{2s}\Gamma\left[\begin{matrix}s+3/4\\ 3/4-s\end{matrix}\right]$ $[-3/4<\operatorname{Re}s<0]$
2	$C(2\sqrt{x})=\frac{1}{2}G_{13}^{11}\left(x\,\middle\|\,\begin{matrix}1\\ 1/4,\ 0,\ 3/4\end{matrix}\right)$	$\frac{1}{2}\Gamma\left[\begin{matrix}s+1/4,\ -s\\ 1-s,\ 1/4-s\end{matrix}\right]=$ $=-\frac{1}{2s}\Gamma\left[\begin{matrix}s+1/4\\ 1/4-s\end{matrix}\right]$ $[-1/4<\operatorname{Re}s<0]$
3	$\left\{\begin{matrix}S(2\sqrt{x})\\ C(2\sqrt{x})\end{matrix}\right\}=$ $=\frac{1}{2}G_{13}^{11}\left(x\,\middle\|\,\begin{matrix}1\\ (1+2\delta)/4,\ 0,\ (3-2\delta)/4\end{matrix}\right)$	$\frac{1}{2}\Gamma\left[\begin{matrix}s+(1+2\delta)/4,\ -s\\ 1-s,\ (1+2\delta)/4-s\end{matrix}\right]=$ $=-\frac{1}{2s}\Gamma\left[\begin{matrix}s+(1+2\delta)/4\\ (1+2\delta)/4-s\end{matrix}\right]$ $[-(1+2\delta)/4<\operatorname{Re}s<0]$
4	$\left\{\begin{matrix}S(2x^{-1/2})\\ C(2x^{-1/2})\end{matrix}\right\}=$ $=\frac{1}{2}G_{31}^{11}\left(x\,\middle\|\,\begin{matrix}(2\mp 1)/4,\ (2\pm 1)/4,\ 1\\ 0\end{matrix}\right)$	$\frac{1}{2}\Gamma\left[\begin{matrix}s,\ (1+2\delta)/4-s\\ s+1,\ s+(1+2\delta)/4\end{matrix}\right]=$ $=\frac{1}{2s}\Gamma\left[\begin{matrix}(1+2\delta)/4-s\\ s+(1+2\delta)/4\end{matrix}\right]$ $[0<\operatorname{Re}s<(1+2\delta)/4]$
5	$\frac{1}{2}-\left\{\begin{matrix}S(2\sqrt{x})\\ C(2\sqrt{x})\end{matrix}\right\}=$ $=\frac{1}{2}G_{13}^{20}\left(x\,\middle\|\,\begin{matrix}1\\ 0,\ (2\pm 1)/4,\ (2\mp 1)/4\end{matrix}\right)$	$\frac{1}{2}\Gamma\left[\begin{matrix}s,\ s+(1+2\delta)/4\\ s+1,\ (1+2\delta)/4-s\end{matrix}\right]=$ $=\frac{1}{2s}\Gamma\left[\begin{matrix}s+(1+2\delta)/4\\ (1+2\delta)/4-s\end{matrix}\right]$ $[0<\operatorname{Re}s<3/4]$
6	$\frac{1}{2}-\left\{\begin{matrix}S(2x^{-1/2})\\ C(2x^{-1/2})\end{matrix}\right\}=$ $=\frac{1}{2}G_{31}^{02}\left(x\,\middle\|\,\begin{matrix}(2\mp 1)/4,\ 1,\ (2\pm 1)/4\\ 0\end{matrix}\right)$	$\frac{1}{2}\Gamma\left[\begin{matrix}(1+2\delta)/4-s,\ -s\\ s+(1+2\delta)/4,\ 1-s\end{matrix}\right]=$ $=-\frac{1}{2s}\Gamma\left[\begin{matrix}(1+2\delta)/4-s\\ s+(1+2\delta)/4\end{matrix}\right]$ $[-3/4<\operatorname{Re}s<0]$
7	$\left\{\begin{matrix}\sin 2\sqrt{x}\\ \cos 2\sqrt{x}\end{matrix}\right\}S(2\sqrt{x})\pm$ $\pm\left\{\begin{matrix}\cos 2\sqrt{x}\\ \sin 2\sqrt{x}\end{matrix}\right\}C(2\sqrt{x})=$ $=\pm\sqrt{\frac{\pi^3}{2}}G_{13}^{11}\left(x\,\middle\|\,\begin{matrix}(2\mp 1)/4\\ (2\mp 1)/4,\ 0,\ 1/2\end{matrix}\right)$	$\pm\frac{\pi^{3/2}}{\sqrt{2}}\Gamma\left[\begin{matrix}s+(3-2\delta)/4,\ (1+2\delta)/4-s\\ 1/2-s,\ 1-s\end{matrix}\right]$ $[(2\delta-3)/4<\operatorname{Re}s<(1+\delta)/4]$
8	$\left\{\begin{matrix}\sin(2x^{-1/2})\\ \cos(2x^{-1/2})\end{matrix}\right\}S\left(\frac{2}{\sqrt{x}}\right)\pm$	$\pm\frac{\pi^{3/2}}{\sqrt{2}}\Gamma\left[\begin{matrix}s+(1+2\delta)/4,\ (3-2\delta)/4-s\\ s+1/2,\ s+1\end{matrix}\right]$ $[-(1+\delta)/4<\operatorname{Re}s<(3-2\delta)/4]$

	$\pm\begin{Bmatrix}\cos(2x^{-1/2})\\ \sin(2x^{-1/2})\end{Bmatrix} C\left(\frac{2}{\sqrt{x}}\right)=$ $=\pm\sqrt{\frac{\pi^3}{2}}\, G_{31}^{11}\left(x\,\middle\vert\,\begin{matrix}(2\pm1)/4,\ 1/2,\ 1\\ (2\pm1)/4\end{matrix}\right)$	
9	$\begin{Bmatrix}\sin 2\sqrt{x}\\ \cos 2\sqrt{x}\end{Bmatrix}\left[\frac12 - S(2\sqrt{x})\right]\pm$ $\pm\begin{Bmatrix}\cos 2\sqrt{x}\\ \sin 2\sqrt{x}\end{Bmatrix}\left[\frac12 - C(2\sqrt{x})\right]=$ $=(2\pi)^{-3/2}\, G_{13}^{31}\left(x\,\middle\vert\,\begin{matrix}(2\mp1)/4\\ (2\mp1)/4,\ 0,\ 1/2\end{matrix}\right)$	$\frac{1}{(2\pi)^{3/2}}\times$ $\times\Gamma\left[s,\ s+\frac12,\ s+\frac{3-2\delta}{4},\ \frac{1+2\delta}{4}-s\right]$ [$0<\operatorname{Re} s<(1+2\delta)/4$]
10	$\begin{Bmatrix}\sin(2x^{-1/2})\\ \cos(2x^{-1/2})\end{Bmatrix}\left[\frac12 - S\left(\frac{2}{\sqrt{x}}\right)\right]\pm$ $\pm\begin{Bmatrix}\cos(2x^{-1/2})\\ \sin(2x^{-1/2})\end{Bmatrix}\left[\frac12 - C\left(\frac{2}{\sqrt{x}}\right)\right]=$ $=(2\pi)^{-3/2}\, G_{31}^{13}\left(x\,\middle\vert\,\begin{matrix}(2\pm1)/4,\ 1/2,\ 1\\ (2\pm1)/4\end{matrix}\right)$	$\frac{1}{(2\pi)^{3/2}}\times$ $\times\Gamma\left[s+\frac{1+2\delta}{4},\ -s,\ \frac12-s,\ \frac{3-2\delta}{4}-s\right]$ [$-(1+2\delta)/4<\operatorname{Re} s<0$]
11	$\left[\frac12 - S(2\sqrt{x})\right]^2+$ $+\left[\frac12 - C(2\sqrt{x})\right]^2=\frac{1}{2\pi^2\sqrt{2}}\times$ $\times\, G_{24}^{41}\left(x\,\middle\vert\,\begin{matrix}1/2,\ 1\\ 0,\ 1/4,\ 1/2,\ 3/4\end{matrix}\right)$	$\frac{1}{2\pi^2\sqrt{2}}\times$ $\times\Gamma\left[\begin{matrix}s,\ s+\frac14,\ s+\frac12,\ s+\frac34,\ \frac12-s\\ s+1\end{matrix}\right]$ [$0<\operatorname{Re} s<1/2$]
12	$\left[\frac12 - S\left(\frac{2}{\sqrt{x}}\right)\right]^2+$ $+\left[\frac12 - C\left(\frac{2}{\sqrt{x}}\right)\right]^2=$ $=\frac{1}{2\pi^2\sqrt{2}}\, G_{42}^{14}\left(x\,\middle\vert\,\begin{matrix}1/4,\ 1/2,\ 3/4,\ 1\\ 1/2,\ 0\end{matrix}\right)$	$\frac{1}{2\pi^2\sqrt{2}}\times$ $\times\Gamma\left[\begin{matrix}s+\frac12,\ -s,\ \frac14-s,\ \frac12-s,\ \frac34-s\\ 1-s\end{matrix}\right]$ [$-1/2<\operatorname{Re} s<0$]

8.4.16. The incomplete gamma-functions $\gamma(\nu, x)$ and $\Gamma(\nu, x)$.

NOTATION: $A_8=2^{\nu}\sqrt{\pi}\begin{Bmatrix}1\\ i\end{Bmatrix}$, $A_9=-\frac{\pi\Gamma(\nu)}{\sin(\nu-c)\pi}$,
$A_{10}=\frac{\pi^{-1/2}}{\Gamma(1-\nu)}\begin{Bmatrix}1\\ i\end{Bmatrix}$, $A_{11}=\frac{2^{\nu-1}}{\sqrt{\pi}\,\Gamma(1-\nu)}$, $\delta=\begin{Bmatrix}1\\ 0\end{Bmatrix}$.

1	$\gamma(\nu, x)=G_{12}^{11}\left(x\,\middle\vert\,\begin{matrix}1\\ \nu,\ 0\end{matrix}\right)$	$\Gamma\left[\begin{matrix}s+\nu,\ -s\\ 1-s\end{matrix}\right]=-\frac{1}{s}\Gamma(s+\nu)$ [$-\operatorname{Re}\nu<\operatorname{Re} s<0$]
2	$\Gamma(\nu, x)=G_{12}^{20}\left(x\,\middle\vert\,\begin{matrix}1\\ 0,\ \nu\end{matrix}\right)$	$\Gamma\left[\begin{matrix}s,\ s+\nu\\ s+1\end{matrix}\right]=\frac{1}{s}\Gamma(s+\nu)$ [$\operatorname{Re} s>0,\ -\operatorname{Re}\nu$]
3	$\begin{Bmatrix}\gamma(\nu, x)\\ \Gamma(\nu, x)\end{Bmatrix}$	$\mp\Gamma\left[\begin{matrix}s+\nu,\ s\\ s+1\end{matrix}\right]=\mp\frac{1}{s}\Gamma(s+\nu)$ $\left[\begin{Bmatrix}-\operatorname{Re}\nu<\operatorname{Re} s<0\\ \operatorname{Re} s>0,\ -\operatorname{Re}\nu\end{Bmatrix}\right]$

4	$\left\{\begin{matrix}\gamma(\nu,\, x^{-1})\\ \Gamma(\nu,\, x^{-1})\end{matrix}\right\}$	$\pm\Gamma\left[\begin{matrix}s,\ \nu-s\\ s+1\end{matrix}\right]=\pm\frac{1}{s}\Gamma(\nu-s)$ $\left[\left\{\begin{matrix}0<\operatorname{Re}s<\operatorname{Re}\nu\\ \operatorname{Re}s<0,\ \operatorname{Re}\nu\end{matrix}\right\}\right]$
5	$\gamma(\nu,\, 2i\sqrt{x})-\gamma(\nu,\, -2i\sqrt{x})=$ $=2^{\nu}i\sqrt{\pi}\,G_{13}^{20}\left(x\,\middle\vert\,\begin{matrix}1\\ (\nu+1)/2,\ \nu/2,\ 0\end{matrix}\right)$	$2^{\nu}i\sqrt{\pi}\,\Gamma\left[\begin{matrix}s+(\nu+1)/2,\ s+\nu/2\\ s+1,\ 1-s\end{matrix}\right]$ $[-\operatorname{Re}\nu/2<\operatorname{Re}s<1-\operatorname{Re}\nu/2]$
6	$\gamma\left(\nu,\, \frac{2i}{\sqrt{x}}\right)-\gamma\left(\nu,\, -\frac{2i}{\sqrt{x}}\right)=$ $=2^{\nu}i\sqrt{\pi}\,G_{31}^{02}\left(x\,\middle\vert\,\begin{matrix}(1-\nu)/2,\ 1-\nu/2,\ 1\\ 0\end{matrix}\right)$	$2^{\nu}i\sqrt{\pi}\,\Gamma\left[\begin{matrix}\nu/2-s,\ (1+\nu)/2-s\\ s+1,\ 1-s\end{matrix}\right]$ $[\operatorname{Re}\nu/2-1<\operatorname{Re}s<\operatorname{Re}\nu/2]$
7	$e^{i\nu\pi/2}\Gamma(\nu,\, -2i\sqrt{x})\pm$ $\pm e^{-i\nu\pi/2}\Gamma(\nu,\, 2i\sqrt{x})=$ $=A_8G_{13}^{20}\left(x\,\middle\vert\,\begin{matrix}1\\ 0,\ (1+\nu-\delta)/2,\ (\delta+\nu)/2\end{matrix}\right)$	$A_8\Gamma\left[\begin{matrix}s,\ s+(1+\nu-\delta)/2\\ s+1,\ 1-(\delta+\nu)/2-s\end{matrix}\right]$ $[0,\ (\delta-\operatorname{Re}\nu-1)/2<\operatorname{Re}s<1-\operatorname{Re}\nu/2]$
8	$e^{i\nu\pi/2}\Gamma\left(\nu,\, -\frac{2i}{\sqrt{x}}\right)\pm$ $\pm e^{-i\nu\pi/2}\Gamma\left(\nu,\, \frac{2i}{\sqrt{x}}\right)=$ $=A_8G_{31}^{02}\left(x\,\middle\vert\,\begin{matrix}1,\ \frac{1+\delta-\nu}{2},\ 1-\frac{\delta+\nu}{2}\\ 0\end{matrix}\right)$	$A_8\Gamma\left[\begin{matrix}-s,\ (1+\nu-\delta)/2-s\\ s+1-(\delta+\nu)/2,\ 1-s\end{matrix}\right]$ $[\operatorname{Re}\nu/2-1<\operatorname{Re}s<0,\ (1+\operatorname{Re}\nu-\delta)/2]$
9	$e^{x}\gamma(\nu,\, x)=$ $\overleftarrow{L_{-\infty}}$ $=A_9G_{23}^{11}\left(x\,\middle\vert\,\begin{matrix}\nu,\ c\\ \nu,\ 0,\ c\end{matrix}\right)$	$A_9\Gamma\left[\begin{matrix}s+\nu,\ 1-\nu-s\\ s+c,\ 1-c-s,\ 1-s\end{matrix}\right]$ [$c\neq\nu$ is arbitrary; $s=-\nu-k\in D^{+}$, $s=1-\nu+k\in D^{-}$, $k=0,\ 1,\ 2,\ \ldots$].
10	$e^{1/x}\gamma\left(\nu,\, \frac{1}{x}\right)=$ $\overleftarrow{L_{+\infty}}$ $=A_9G_{32}^{11}\left(x\,\middle\vert\,\begin{matrix}1-\nu,\ 1,\ 1-c\\ 1-\nu,\ 1-c\end{matrix}\right)$	$A_9\Gamma\left[\begin{matrix}s+1-\nu,\ \nu-s\\ s+1,\ s+1-c,\ c-s\end{matrix}\right]$ [$c\neq\nu$ is arbitrary $s=\nu+k\in D^{-}$, $s=\nu-1-k\in D^{+}$, $k=0,\ 1,\ 2,\ \ldots$]
11	$e^{-x}\gamma(\nu,\, e^{\pi i}x)=$ $=e^{\nu\pi i}\Gamma(\nu)\,G_{12}^{11}\left(x\,\middle\vert\,\begin{matrix}\nu\\ \nu,\ 0\end{matrix}\right)$	$e^{\nu\pi i}\Gamma(\nu)\,\Gamma\left[\begin{matrix}s+\nu,\ 1-\nu-s\\ 1-s\end{matrix}\right]$ $[-\operatorname{Re}\nu<\operatorname{Re}s<1-\operatorname{Re}\nu]$
12	$e^{-x}\gamma\left(\nu,\, \frac{e^{\pi i}}{x}\right)=$ $=e^{\nu\pi i}\Gamma(\nu)\,G_{21}^{11}\left(x\,\middle\vert\,\begin{matrix}1-\nu,\ 1\\ 1-\nu\end{matrix}\right)$	$e^{\nu\pi i}\Gamma(\nu)\,\Gamma\left[\begin{matrix}s+1-\nu,\ \nu-s\\ s+1\end{matrix}\right]$ $[\operatorname{Re}\nu-1<\operatorname{Re}s<\operatorname{Re}\nu]$
13	$e^{x}\Gamma(\nu,\, x)=\frac{1}{\Gamma(1-\nu)}\,G_{12}^{21}\left(x\,\middle\vert\,\begin{matrix}\nu\\ 0,\ \nu\end{matrix}\right)$	$\frac{1}{\Gamma(1-\nu)}\Gamma[s,\ s+\nu,\ 1-\nu-s]$ $[0,\ -\operatorname{Re}\nu<\operatorname{Re}s<1-\operatorname{Re}\nu]$
14	$e^{1/x}\Gamma\left(\nu,\, \frac{1}{x}\right)=$ $=\frac{1}{\Gamma(1-\nu)}\,G_{21}^{12}\left(x\,\middle\vert\,\begin{matrix}1,\ 1-\nu\\ 1-\nu\end{matrix}\right)$	$\frac{1}{\Gamma(1-\nu)}\Gamma[s+1-\nu,\ -s,\ \nu-s]$ $[\operatorname{Re}\nu-1<\operatorname{Re}s<0,\ \operatorname{Re}\nu]$

15	$e^{-2i\sqrt{x}}\Gamma(\nu,-2i\sqrt{x}) \pm e^{2i\sqrt{x}}\Gamma(\nu,2i\sqrt{x}) = A_{10}\times G_{24}^{32}\left(x\,\middle\vert\,\begin{matrix}\frac{1+\nu}{2}, \frac{\nu}{2}\\ \frac{1+\nu}{2}, \frac{\nu}{2}, \frac{1-\delta}{2}, \frac{\delta}{2}\end{matrix}\right)$	$A_{10}\times\Gamma\begin{bmatrix}s+\frac{\nu+1}{2}, s+\frac{\nu}{2}, s+\frac{1-\delta}{2}, \frac{1-\nu}{2}-s, 1-\frac{\nu}{2}-s\\ 1-\frac{\delta}{2}-s\end{bmatrix}$ [$-\mathrm{Re}\,\nu/2$, $(\delta-1)/2<\mathrm{Re}\,s<(1-\mathrm{Re}\,\nu)/2$]
16	$e^{-2i/\sqrt{x}}\Gamma\left(\nu,-\frac{2i}{\sqrt{x}}\right) \pm e^{2i/\sqrt{x}}\Gamma\left(\nu,\frac{2i}{\sqrt{x}}\right) = A_{10}G_{42}^{23}\left(x\,\middle\vert\,\begin{matrix}\frac{1-\nu}{2}, 1-\frac{\nu}{2}, \frac{1+\delta}{2}, 1-\frac{\delta}{2}\\ \frac{1-\nu}{2}, 1-\frac{\nu}{2}\end{matrix}\right)$	$A_{10}\Gamma\begin{bmatrix}s+\frac{1-\nu}{2}, s+1-\frac{\nu}{2}, \frac{\nu+1}{2}-s, \frac{\nu}{2}-s, \frac{1-\delta}{2}-s\\ s+1-\frac{\delta}{2}\end{bmatrix}$ [$(\mathrm{Re}\,\nu-1)/2<\mathrm{Re}\,s<(1-\delta)/2$, $\mathrm{Re}\,\nu/2$]
17	$e^{i\nu\pi/2-2i\sqrt{x}}\Gamma(\nu,-2i\sqrt{x}) \pm e^{-i\nu\pi/2+2i\sqrt{x}}\Gamma(\nu,2i\sqrt{x}) = A_{10}G_{13}^{31}\left(x\,\middle\vert\,\begin{matrix}(1+\nu-\delta)/2\\ 0, 1/2, (1+\nu-\delta)/2\end{matrix}\right)$	$A_{10}\times\Gamma\left[s, s+\frac{1}{2}, s+\frac{1+\nu-\delta}{2}, \frac{1+\delta-\nu}{2}-s\right]$ [0, $(\delta-\mathrm{Re}\,\nu-1)/2<\mathrm{Re}\,s<(1+\delta-\mathrm{Re}\,\nu)/2$]
18	$e^{i\nu\pi/2-2i/\sqrt{x}}\Gamma\left(\nu,-\frac{2i}{\sqrt{x}}\right) \pm e^{-i\nu\pi/2+2i/\sqrt{x}}\Gamma\left(\nu,\frac{2i}{\sqrt{x}}\right) = A_{10}G_{31}^{13}\left(x\,\middle\vert\,\begin{matrix}1/2, 1, (1+\delta-\nu)/2\\ (1+\delta-\nu)/2\end{matrix}\right)$	$A_{10}\times\Gamma\left[s+\frac{1+\delta-\nu}{2}, \frac{1+\nu-\delta}{2}-s, -s, \frac{1}{2}-s\right]$ [$(\mathrm{Re}\,\nu-\delta-1)/2<\mathrm{Re}\,s<0$, $(1+\mathrm{Re}\,\nu-\delta)/2$]
19	$\Gamma(\nu,2i\sqrt{x})\Gamma(\nu,-2i\sqrt{x}) = A_{11}G_{24}^{41}\left(x\,\middle\vert\,\begin{matrix}\nu, 1\\ 0, \nu/2, (\nu+1)/2, \nu\end{matrix}\right)$	$A_{11}\times\Gamma\begin{bmatrix}s, s+\nu, s+\frac{\nu}{2}, s+\frac{\nu+1}{2}, 1-\nu-s\\ s+1\end{bmatrix}$ [0, $-\mathrm{Re}\,\nu<\mathrm{Re}\,s<1-\mathrm{Re}\,\nu$]
20	$\Gamma\left(\nu,\frac{2i}{\sqrt{x}}\right)\Gamma\left(\nu,-\frac{2i}{\sqrt{x}}\right) = A_{11}G_{42}^{14}\left(x\,\middle\vert\,\begin{matrix}1, 1-\nu/2, (1-\nu)/2, 1-\nu\\ 1-\nu, 0\end{matrix}\right)$	$A_{11}\times\Gamma\begin{bmatrix}s+1-\nu, -s, \nu-s, \frac{\nu}{2}-s, \frac{\nu+1}{2}-s\\ 1-s\end{bmatrix}$ [$\mathrm{Re}\,\nu-1<\mathrm{Re}\,s<0$, $\mathrm{Re}\,\nu$]

8.4.17. The generalized Fresnel integrals $S(x,\nu)$ and $C(x,\nu)$.

NOTATION: $A_{12}=2^{\nu-1}\sqrt{\pi}$, $\delta=\begin{Bmatrix}1\\0\end{Bmatrix}$.

1	$\begin{Bmatrix}S(2\sqrt{x},\nu)\\ C(2\sqrt{x},\nu)\end{Bmatrix} = A_{12}G_{13}^{20}\left(x\,\middle\vert\,\begin{matrix}1\\ 0, \frac{\delta+\nu}{2}, \frac{1-\delta+\nu}{2}\end{matrix}\right)$	$A_{12}\Gamma\begin{bmatrix}s, s+\frac{\delta+\nu}{2}\\ s+1, \frac{1+\delta-\nu}{2}-s\end{bmatrix} = A_{12}\frac{1}{s}\Gamma\begin{bmatrix}s+\frac{\delta+\nu}{2}\\ \frac{1+\delta-\nu}{2}-s\end{bmatrix}$ [0, $-(\delta+\mathrm{Re}\,\nu)/2<\mathrm{Re}\,s<1-\mathrm{Re}\,\nu/2$]

2	$\begin{Bmatrix} S(2x^{-1/2}, \nu) \\ C(2x^{-1/2}, \nu) \end{Bmatrix} = A_{12} G_{31}^{02}\left(x \middle\vert \begin{matrix} 1, 1-(\delta+\nu)/2, (1+\delta-\nu)/2 \\ 0 \end{matrix}\right)$	$A_{12}\Gamma\left[\begin{matrix} -s, (\delta+\nu)/2-s \\ s+(1+\delta-\nu)/2, 1-s \end{matrix}\right] = -A_{12}\frac{1}{s}\Gamma\left[\begin{matrix} (\delta+\nu)/2-s \\ s+(1+\delta-\nu)/2 \end{matrix}\right]$ [Re $\nu/2-1<$ Re $s<0$, $(\delta+\text{Re }\nu)/2$]

8.4.18. The parabolic cylinder function $D_\nu(x)$.

NOTATION: $A_{13}=2^{\nu/2+1}\begin{Bmatrix} \cos(\nu\pi/2) \\ \sin(\nu\pi/2) \end{Bmatrix}$, $A_{14}=\frac{\pi^{3/2}}{2^{\nu/2}\Gamma(-\nu)}\begin{Bmatrix} \operatorname{cosec} c\pi \\ \sec c\pi \end{Bmatrix}$, $A_{15}=\frac{2^{-3/2}}{\pi\Gamma(-\nu)}$, $\delta=\begin{Bmatrix} 1 \\ 0 \end{Bmatrix}$.

1	$e^{-x/2}D_\nu(\sqrt{2x})=2^{\nu/2}G_{12}^{20}\left(x \middle\vert \begin{matrix} (1-\nu)/2 \\ 0, 1/2 \end{matrix}\right)$	$2^{\nu/2}\Gamma\left[\begin{matrix} s, s+1/2 \\ s+(1-\nu)/2 \end{matrix}\right]$ [Re $s>0$]
2	$e^{-1/(2x)}D_\nu\left(\sqrt{\frac{2}{x}}\right)=2^{\nu/2}G_{21}^{02}\left(x \middle\vert \begin{matrix} 1/2, 1 \\ (\nu+1)/2 \end{matrix}\right)$	$2^{\nu/2}\Gamma\left[\begin{matrix} -s, 1/2-s \\ (1-\nu)/2-s \end{matrix}\right]$ [Re $s<0$]
3	$e^{x/2}D_\nu(\sqrt{2x})=\frac{2^{-\nu/2-1}}{\sqrt{\pi}\Gamma(-\nu)}G_{12}^{21}\left(x \middle\vert \begin{matrix} 1+\nu/2 \\ 0, 1/2 \end{matrix}\right)$	$\frac{2^{-\nu/2-1}}{\sqrt{\pi}\Gamma(-\nu)}\Gamma\left[s, s+\frac{1}{2}, -\frac{\nu}{2}-s\right]$ [$0<$ Re $s<-$Re $\nu/2$]
4	$e^{1/(2x)}D_\nu\left(\sqrt{\frac{2}{x}}\right)=\frac{2^{-\nu/2-1}}{\sqrt{\pi}\Gamma(-\nu)}G_{21}^{12}\left(x \middle\vert \begin{matrix} 1/2, 1 \\ -\nu/2 \end{matrix}\right)$	$\frac{2^{-\nu/2-1}}{\sqrt{\pi}\Gamma(-\nu)}\Gamma\left[s-\frac{\nu}{2}, -s, \frac{1}{2}-s\right]$ [Re $\nu/2<$ Re $s<0$]
5	$e^{-x/2}D_\nu(-\sqrt{2x})=2^{\nu/2}G_{23}^{21}\left(x \middle\vert \begin{matrix} (1-\nu)/2, (1+\nu)/2 \\ 0, 1/2, (1+\nu)/2 \end{matrix}\right)$	$2^{\nu/2}\Gamma\left[\begin{matrix} s, s+1/2, (\nu+1)/2-s \\ s+(\nu+1)/2, (1-\nu)/2-s \end{matrix}\right]$ [$0<$ Re $s<$ (Re $\nu+1)/2$]
6	$e^{-(2x)^{-1}}D_\nu\left(-\sqrt{\frac{2}{x}}\right)=2^{\nu/2}G_{32}^{12}\left(x \middle\vert \begin{matrix} 1/2, 1, (1-\nu)/2 \\ (1+\nu)/2, (1-\nu)/2 \end{matrix}\right)$	$2^{\nu/2}\Gamma\left[\begin{matrix} s+(\nu+1)/2, -s, 1/2-s \\ s+(1-\nu)/2, (1+\nu)/2-s \end{matrix}\right]$ [$-$(Re $\nu+1)/2<$ Re $s<0$]
7	$e^{-x/2}\left[D_\nu(\sqrt{2x})\pm D_\nu(-\sqrt{2x})\right]=A_{13}G_{12}^{11}\left(x \middle\vert \begin{matrix} (1-\nu)/2 \\ (1-\delta)/2, \delta/2 \end{matrix}\right)$	$A_{13}\Gamma\left[\begin{matrix} s+(1-\delta)/2, (1+\nu)/2-s \\ 1-\delta/2-s \end{matrix}\right]$ [$(\delta-1)/2<$ Re $s<$ (Re $\nu+1)/2$]
8	$e^{-(2x)^{-1}}\left[D_\nu\left(\sqrt{\frac{2}{x}}\right)\pm D_\nu\left(-\sqrt{\frac{2}{x}}\right)\right]=A_{13}G_{21}^{11}\left(x \middle\vert \begin{matrix} (1+\delta)/2, 1-\delta/2 \\ (1+\nu)/2 \end{matrix}\right)$	$A_{13}\Gamma\left[\begin{matrix} s+(\nu+1)/2, (1-\delta)/2-s \\ s+1-\delta/2 \end{matrix}\right]$ [$-$(Re $\nu+1)/2<$ Re $s<(1-\delta)/2$]
9	$e^{x/2}\left[D_\nu(\sqrt{2x})\pm D_\nu(-\sqrt{2x})\right]\overset{L_{-\infty}}{\longleftarrow}=A_{14}G_{23}^{11}\left(x \middle\vert \begin{matrix} 1+\nu/2, c \\ (1-\delta)/2, \delta/2, c \end{matrix}\right)$	$A_{14}\Gamma\left[\begin{matrix} s+(1-\delta)/2, -\nu/2-s \\ s+c, 1-c-s, 1-\delta/2-s \end{matrix}\right]$ [$c\neq(1-\delta)/2$ is arbitrary; $s=(\delta-1)/2-k\in D^+$, $s=-\nu/2+k\in D^-$, $k=0, 1, 2\ldots$]

10	$e^{(2x)^{-1}}\left[D_\nu\left(\sqrt{\frac{2}{x}}\right)\pm \pm D_\nu\left(-\sqrt{\frac{2}{x}}\right)\right] \overset{L_{+\infty}}{\longleftarrow} = A_{14}G_{32}^{11}\left(x\,\middle\vert\,\begin{matrix}(1+\delta)/2,\ 1-\delta/2,\ 1-c\\ -\nu/2,\ 1-c\end{matrix}\right)$	$A_{14}\Gamma\left[\begin{matrix}s-\nu/2,\ (1-\delta)/2-s\\ s+1-\delta/2,\ s+1-c,\ c-s\end{matrix}\right]$ [$c\neq(1-\delta)/2$ is arbitrary; $s=(1-\delta)/2+k\in D^-$, $s=\nu/2-k\in D^+$, $k=0,\ 1,\ 2,\ \ldots$]
11	$D_\nu(2e^{\pi i/4}x^{1/4})\,D_\nu(2e^{-\pi i/4}x^{1/4})= =A_{15}G_{24}^{41}\left(x\,\middle\vert\,\begin{matrix}1+\nu/2,\ (1-\nu)/2\\ 0,\ 1/4,\ 1/2,\ 3/4\end{matrix}\right)$	$A_{15}\Gamma\left[\begin{matrix}s,\ s+\frac{1}{4},\ s+\frac{1}{2},\ s+\frac{3}{4},\ -\frac{\nu}{2}-s\\ s+(1-\nu)/2\end{matrix}\right]$ [$0<\operatorname{Re}s<-\operatorname{Re}\nu/2$]
12	$D_\nu\left(\frac{2e^{\pi i/4}}{x^{1/4}}\right)D_\nu\left(\frac{2e^{-\pi i/4}}{x^{1/4}}\right)= =A_{15}G_{42}^{14}\left(x\,\middle\vert\,\begin{matrix}1/4,\ 1/2,\ 3/4,\ 1\\ -\nu/2,\ (1+\nu)/2\end{matrix}\right)$	$A_{15}\Gamma\left[\begin{matrix}s-\frac{\nu}{2},\ -s,\ \frac{1}{4}-s,\ \frac{1}{2}-s,\ \frac{3}{4}-s\\ (1-\nu)/2-s\end{matrix}\right]$ [$\operatorname{Re}\nu/2<\operatorname{Re}s<0$]
13	$D_{-\nu-1}(2x^{1/4})\,D_\nu(2x^{1/4})= =\frac{1}{2\sqrt{\pi}}G_{24}^{40}\left(x\,\middle\vert\,\begin{matrix}1+\nu/2,\ (1-\nu)/2\\ 0,\ 1/4,\ 1/2,\ 3/4\end{matrix}\right)$	$\frac{1}{2\sqrt{\pi}}\Gamma\left[\begin{matrix}s,\ s+1/4,\ s+1/2,\ s+3/4\\ s+(1-\nu)/2,\ s+1+\nu/2\end{matrix}\right]$ [$\operatorname{Re}s>0$]
14	$D_{-\nu-1}\left(\frac{2}{x^{1/4}}\right)D_\nu\left(\frac{2}{x^{1/4}}\right)= =\frac{1}{2\sqrt{\pi}}G_{42}^{04}\left(x\,\middle\vert\,\begin{matrix}1/4,\ 1/2,\ 3/4,\ 1\\ -\nu/2,\ (1+\nu)/2\end{matrix}\right)$	$\frac{1}{2\sqrt{\pi}}\Gamma\left[\begin{matrix}-s,\ 1/4-s,\ 1/2-s,\ 3/4-s\\ 1+\nu/2-s,\ (1-\nu)/2-s\end{matrix}\right]$ [$\operatorname{Re}s<0$]
15	$D_{-n}^2(i\sqrt{2x})-D_{-n}^2(-i\sqrt{2x})= =\frac{(-1)^n\,2\pi i}{[(n-1)!]^2}G_{12}^{11}\left(x\,\middle\vert\,\begin{matrix}1/2\\ n-1/2,\ 0\end{matrix}\right)$	$\frac{(-1)^n\,2\pi i}{[(n-1)!]^2}\Gamma\left[\begin{matrix}s+n-1/2,\ 1/2-s\\ 1-s\end{matrix}\right]$ [$1/2-n<\operatorname{Re}s<1/2;\ n=1,\ 2,\ 3,\ \ldots$]
16	$D_{-n}^2\left(i\sqrt{\frac{2}{x}}\right)-D_{-n}^2\left(-i\sqrt{\frac{2}{x}}\right)= =\frac{(-1)^n\,2\pi i}{[(n-1)!]^2}G_{21}^{11}\left(x\,\middle\vert\,\begin{matrix}3/2-n,\ 1\\ 1/2\end{matrix}\right)$	$\frac{(-1)^n\,2\pi i}{[(n-1)!]^2}\Gamma\left[\begin{matrix}s+1/2,\ n-1/2-s\\ s+1\end{matrix}\right]$ [$-1/2<\operatorname{Re}s<n-1/2;\ n=1,\ 2,\ 3,\ \ldots$]

8.4.19. The Bessel function $J_\nu(x)$.

NOTATION: $A_{16}=\frac{2}{\sqrt{\pi}}\left\{\begin{matrix}\cos\,[(\mu+\nu)\,\pi/2]\\ \sin\,[(\mu+\nu)\,\pi/2]\end{matrix}\right\},\quad \delta=\left\{\begin{matrix}1\\ 0\end{matrix}\right\}.$

1	$J_\nu(2\sqrt{x})=G_{02}^{10}\left(x\,\middle\vert\,\begin{matrix}\cdot\quad\cdot\\ \nu/2,\ -\nu/2\end{matrix}\right)$	$\Gamma\left[\begin{matrix}s+\nu/2\\ 1+\nu/2-s\end{matrix}\right]$ [$-\operatorname{Re}\nu/2<\operatorname{Re}s<3/4$]
2	$J_\nu\left(\frac{2}{\sqrt{x}}\right)=G_{20}^{01}\left(x\,\middle\vert\,\begin{matrix}1-\nu/2,\ 1+\nu/2\\ \cdot\quad\cdot\end{matrix}\right)$	$\Gamma\left[\begin{matrix}\nu/2-s\\ s+\nu/2+1\end{matrix}\right]$ [$-3/4<\operatorname{Re}s<\operatorname{Re}\nu/2$]
3	$\left\{\begin{matrix}\sin\sqrt{x}\\ \cos\sqrt{x}\end{matrix}\right\}J_\nu(\sqrt{x})=\frac{1}{\sqrt{2}}\times \times G_{24}^{12}\left(x\,\middle\vert\,\begin{matrix}\frac{1}{4},\ \frac{3}{4}\\ \frac{\delta+\nu}{2},\ -\frac{\nu}{2},\ \frac{1-\delta+\nu}{2},\ \frac{1-\nu}{2}\end{matrix}\right)$	$\frac{1}{\sqrt{2}}\Gamma\left[\begin{matrix}s+\frac{\delta+\nu}{2},\ \frac{1}{4}-s,\ \frac{3}{4}-s\\ 1+\frac{\nu}{2}-s,\ \frac{1+\delta-\nu}{2}-s,\ \frac{1+\nu}{2}-s\end{matrix}\right]$ [$-(\delta+\operatorname{Re}\nu)/2<\operatorname{Re}s<1/4$]

4	$\begin{Bmatrix}\sin x^{-1/2}\\ \cos x^{-1/2}\end{Bmatrix} J_\nu\left(\frac{1}{\sqrt{x}}\right)=\frac{1}{\sqrt{2}}G_{42}^{21}\times$ $\times\left(x\,\middle\vert\,\begin{matrix}1-\frac{\delta+\nu}{2},\ 1+\frac{\nu}{2},\ \frac{1+\delta-\nu}{2},\ \frac{1+\nu}{2}\\ \frac{1}{4},\ \frac{3}{4}\end{matrix}\right)$	$\frac{1}{\sqrt{2}}\Gamma\left[\begin{matrix}s+\frac{1}{4},\ s+\frac{3}{4},\ \frac{\delta+\nu}{2}-s\\ s+1+\frac{\nu}{2},\ s+\frac{1+\delta-\nu}{2},\ s+\frac{1+\nu}{2}\end{matrix}\right]$ $[-1/4<\mathrm{Re}\, s<(\mathrm{Re}\,\delta+\nu)/2]$
5	$\begin{Bmatrix}\sin(\sqrt{x}+\mu\pi)\\ \cos(\sqrt{x}+\mu\pi)\end{Bmatrix} J_\nu(\sqrt{x})=$ $=\frac{1}{\sqrt{2}}G_{35}^{22}\left(x\,\middle\vert\,\begin{matrix}1/4,\ 3/4,\ (1-\delta+\nu)/2+\mu\\ \nu/2,\ (\nu+1)/2,\ (1-\delta+\nu)/2+\mu,\ -\nu/2,\ (1-\nu)/2\end{matrix}\right)$	$\frac{1}{\sqrt{2}}\Gamma\left[\begin{matrix}s+\frac{\nu}{2},\ s+\frac{\nu+1}{2},\ \frac{1}{4}-s,\ \frac{3}{4}-s\\ s+\mu+\frac{1-\delta+\nu}{2},\ \frac{1+\delta-\nu}{2}-\mu-s,\ 1+\frac{\nu}{2}-s,\ \frac{1+\nu}{2}-s\end{matrix}\right]$ $[-\mathrm{Re}\,\nu/2<\mathrm{Re}\, s<1/4]$
6	$\begin{Bmatrix}\sin(x^{-1/2}+\mu\pi)\\ \cos(x^{-1/2}+\mu\pi)\end{Bmatrix} J_\nu\left(\frac{1}{\sqrt{x}}\right)=$ $=\frac{1}{\sqrt{2}}G_{53}^{22}\left(x\,\middle\vert\,\begin{matrix}\frac{1-\nu}{2},\ 1-\frac{\nu}{2},\ \frac{1+\delta-\nu}{2}-\mu,\ 1+\frac{\nu}{2},\ \frac{1+\nu}{2}\\ \frac{1}{4},\ \frac{3}{4},\ \frac{1+\delta-\nu}{2}-\mu\end{matrix}\right)$	$\frac{1}{\sqrt{2}}\Gamma\left[\begin{matrix}s+\frac{1}{4},\ s+\frac{3}{4},\ \frac{\nu}{2}-s,\ \frac{1+\nu}{2}-s\\ s+\frac{\nu}{2}+1,\ s+\frac{\nu+1}{2},\ s+\frac{1+\delta-\nu}{2}-\mu,\ \frac{1-\delta+\nu}{2}+\mu-s\end{matrix}\right]$ $[-1/4<\mathrm{Re}\, s<\mathrm{Re}\,\nu/2]$
7	$\sin\sqrt{x}\,J_\nu(\sqrt{x})\pm\cos\sqrt{x}\,J_{-\nu}(\sqrt{x})=$ $=-\sqrt{2}\sin\frac{2\nu\mp1}{4}\pi\times$ $\times G_{24}^{21}\left(x\,\middle\vert\,\begin{matrix}(2\pm1)/4,\ (2\mp1)/4\\ (1+\nu)/2,\ -\nu/2,\ \nu/2,\ (1-\nu)/2\end{matrix}\right)$	$-\sqrt{2}\sin\frac{2\nu\mp1}{4}\pi\times$ $\times\Gamma\left[\begin{matrix}s+\frac{1+\nu}{2},\ s-\frac{\nu}{2},\ \frac{2\mp1}{4}-s\\ s+\frac{2\mp1}{4},\ 1-\frac{\nu}{2}-s,\ \frac{1+\nu}{2}-s\end{matrix}\right]$ $[\mathrm{Re}\,\nu/2,\ -(1+\mathrm{Re}\,\nu)/2<\mathrm{Re}\, s<(2\mp1)/4]$
8	$\sin\frac{1}{\sqrt{x}}J_\nu\left(\frac{1}{\sqrt{x}}\right)\pm\cos\frac{1}{\sqrt{x}}\times$ $\times J_{-\nu}\left(\frac{1}{\sqrt{x}}\right)=-\sqrt{2}\sin\frac{2\nu\mp1}{4}\pi\times$ $\times G_{42}^{12}\left(x\,\middle\vert\,\begin{matrix}\frac{1-\nu}{2},\ 1+\frac{\nu}{2},\ 1-\frac{\nu}{2},\ \frac{1+\nu}{2}\\ \frac{2\mp1}{4},\ \frac{2\pm1}{4}\end{matrix}\right)$	$-\sqrt{2}\sin\frac{2\nu\mp1}{4}\pi\times$ $\times\Gamma\left[\begin{matrix}s+\frac{2\mp1}{4},\ \frac{1+\nu}{2}-s,\ -\frac{\nu}{2}-s\\ s+1-\frac{\nu}{2},\ s+\frac{\nu+1}{2},\ \frac{2\mp1}{4}-s\end{matrix}\right]$ $[(\pm1-2)/4<\mathrm{Re}\, s<-\mathrm{Re}\,\nu/2,\ (\mathrm{Re}\,\nu+1)/2]$
9	$J_\nu^2(\sqrt{x})=\frac{1}{\sqrt{\pi}}G_{13}^{11}\left(x\,\middle\vert\,\begin{matrix}1/2\\ \nu,\ -\nu,\ 0\end{matrix}\right)$	$\frac{1}{\sqrt{\pi}}\Gamma\left[\begin{matrix}s+\nu,\ 1/2-s\\ 1+\nu-s,\ 1-s\end{matrix}\right]$ $[-\mathrm{Re}\,\nu<\mathrm{Re}\, s<1/2]$
10	$J_\nu^2\left(\frac{1}{\sqrt{x}}\right)=$ $=\frac{1}{\sqrt{\pi}}G_{31}^{11}\left(x\,\middle\vert\,\begin{matrix}1-\nu,\ 1+\nu,\ 1\\ 1/2\end{matrix}\right)$	$\frac{1}{\sqrt{\pi}}\Gamma\left[\begin{matrix}s+1/2,\ \nu-s\\ s+1,\ s+\nu+1\end{matrix}\right]$ $[-1/2<\mathrm{Re}\, s<\mathrm{Re}\,\nu]$
11	$J_{-\nu}(\sqrt{x})J_\nu(\sqrt{x})=$ $=\frac{1}{\sqrt{\pi}}G_{13}^{11}\left(x\,\middle\vert\,\begin{matrix}1/2\\ 0,\ -\nu,\ \nu\end{matrix}\right)$	$\frac{1}{\sqrt{\pi}}\Gamma\left[\begin{matrix}s,\ 1/2-s\\ 1+\nu-s,\ 1-\nu-s\end{matrix}\right]$ $[0<\mathrm{Re}\, s<1/2]$

12	$J_{-\nu}\left(\frac{1}{\sqrt{x}}\right)J_\nu\left(\frac{1}{\sqrt{x}}\right)= =\frac{1}{\sqrt{\pi}}G_{31}^{11}\left(x\middle\vert\begin{matrix}1,\ 1+\nu,\ 1-\nu\\ 1/2\end{matrix}\right)$	$\frac{1}{\sqrt{\pi}}\Gamma\left[\begin{matrix}s+1/2,\ -s\\ s+\nu+1,\ s-\nu+1\end{matrix}\right]$ $[-1/2<\operatorname{Re}s<0]$
13	$J_{\nu-1}(\sqrt{x})J_\nu(\sqrt{x})= =\frac{1}{\sqrt{\pi}}G_{13}^{11}\left(x\middle\vert\begin{matrix}0\\ \nu-1/2,\ -1/2,\ 1/2-\nu\end{matrix}\right)$	$\frac{1}{\sqrt{\pi}}\Gamma\left[\begin{matrix}s+\nu-1/2,\ 1-s\\ 3/2-s,\ 1/2+\nu-s\end{matrix}\right]$ $[1/2-\operatorname{Re}\nu<\operatorname{Re}s<1]$
14	$J_{\nu-1}\left(\frac{1}{\sqrt{x}}\right)J_\nu\left(\frac{1}{\sqrt{x}}\right)= =\frac{1}{\sqrt{\pi}}G_{31}^{11}\left(x\middle\vert\begin{matrix}3/2-\nu,\ 1/2+\nu,\ 3/2\\ 1\end{matrix}\right)$	$\frac{1}{\sqrt{\pi}}\Gamma\left[\begin{matrix}s+1,\ \nu-1/2-s\\ s+3/2,\ s+\nu+1/2\end{matrix}\right]$ $[-1<\operatorname{Re}s<\operatorname{Re}\nu-1/2]$
15	$J_\mu(\sqrt{x})J_\nu(\sqrt{x})= =\frac{1}{\sqrt{\pi}}G_{24}^{12}\left(x\middle\vert\begin{matrix}0,\ \frac{1}{2}\\ \frac{\mu+\nu}{2},\ -\frac{\mu+\nu}{2},\ \frac{\mu-\nu}{2},\ \frac{\nu-\mu}{2}\end{matrix}\right)$	$\frac{1}{\sqrt{\pi}}\Gamma\left[\begin{matrix}s+\frac{\mu+\nu}{2},\ \frac{1}{2}-s,\ 1-s\\ 1+\frac{\mu+\nu}{2}-s,\ 1+\frac{\nu-\mu}{2}-s,\ 1+\frac{\mu-\nu}{2}-s\end{matrix}\right]$ $[-\operatorname{Re}(\mu+\nu)/2<\operatorname{Re}s<1/2]$
16	$J_\mu\left(\frac{1}{\sqrt{x}}\right)J_\nu\left(\frac{1}{\sqrt{x}}\right)= =\frac{1}{\sqrt{\pi}}G_{42}^{21}\left(x\middle\vert\begin{matrix}1-\frac{\mu+\nu}{2},\ 1+\frac{\mu+\nu}{2},\ 1-\frac{\mu-\nu}{2},\ 1+\frac{\mu-\nu}{2}\\ \frac{1}{2},\ 1\end{matrix}\right)$	$\frac{1}{\sqrt{\pi}}\Gamma\left[\begin{matrix}s+\frac{1}{2},\ s+1,\ \frac{\mu+\nu}{2}-s\\ s+\frac{\mu+\nu}{2}+1,\ s-\frac{\mu-\nu}{2}+1,\ s+\frac{\mu-\nu}{2}+1\end{matrix}\right]$ $[-1/2<\operatorname{Re}s<\operatorname{Re}(\mu+\nu)/2]$
17	$J_{-\nu}^2(\sqrt{x})+J_\nu^2(\sqrt{x})= =\frac{2\cos\nu\pi}{\sqrt{\pi}}G_{24}^{21}\left(x\middle\vert\begin{matrix}1/2,\ 0\\ \nu,\ -\nu,\ 0,\ 0\end{matrix}\right)$	$\frac{2\cos\nu\pi}{\sqrt{\pi}}\Gamma\left[\begin{matrix}s+\nu,\ s-\nu,\ 1/2-s\\ s,\ 1-s,\ 1-s\end{matrix}\right]$ $[\vert\operatorname{Re}\nu\vert<\operatorname{Re}s<1/2]$
18	$J_{-\nu}^2\left(\frac{1}{\sqrt{x}}\right)+J_\nu^2\left(\frac{1}{\sqrt{x}}\right)= =\frac{2\cos\nu\pi}{\sqrt{\pi}}G_{42}^{12}\left(x\middle\vert\begin{matrix}1-\nu,\ 1+\nu,\ 1,\ 1\\ 1/2,\ 1\end{matrix}\right)$	$\frac{2\cos\nu\pi}{\sqrt{\pi}}\Gamma\left[\begin{matrix}s+1/2,\ \nu-s,\ -\nu-s\\ s+1,\ s+1,\ -s\end{matrix}\right]$ $[-1/2<\operatorname{Re}s<-\vert\operatorname{Re}\nu\vert]$
19	$J_{-\nu}^2(\sqrt{x})-J_\nu^2(\sqrt{x})= =\frac{2\sin\nu\pi}{\sqrt{\pi}}G_{13}^{20}\left(x\middle\vert\begin{matrix}1/2\\ \nu,\ -\nu,\ 0\end{matrix}\right)$	$\frac{2}{\sqrt{\pi}}\sin\nu\pi\Gamma\left[\begin{matrix}s+\nu,\ s-\nu\\ s+1/2,\ 1-s\end{matrix}\right]$ $[\vert\operatorname{Re}\nu\vert<\operatorname{Re}s<1]$
20	$J_{-\nu}^2\left(\frac{1}{\sqrt{x}}\right)-J_\nu^2\left(\frac{1}{\sqrt{x}}\right)= =\frac{2}{\sqrt{\pi}}\sin\nu\pi\, G_{31}^{02}\left(x\middle\vert\begin{matrix}1+\nu,\ 1-\nu,\ 1\\ 1/2\end{matrix}\right)$	$\frac{2}{\sqrt{\pi}}\sin\nu\pi\Gamma\left[\begin{matrix}\nu-s,\ -\nu-s\\ s+1,\ 1/2-s\end{matrix}\right]$ $[-1<\operatorname{Re}s<-\vert\operatorname{Re}\nu\vert]$

21	$J_{-\mu}(\sqrt{x})\,J_{-\nu}(\sqrt{x}) \pm \pm J_{\mu}(\sqrt{x})\,J_{\nu}(\sqrt{x}) = A_{16}G_{24}^{21}\left(x\,\middle\vert\,\begin{matrix}\frac{\delta}{2},\ \frac{1-\delta}{2}\\ \frac{\mu+\nu}{2},\ -\frac{\mu+\nu}{2},\ \frac{\mu-\nu}{2},\ \frac{\nu-\mu}{2}\end{matrix}\right)$	$A_{16}\Gamma\begin{bmatrix}s+\frac{\mu+\nu}{2},\ s-\frac{\mu+\nu}{2},\ 1-\frac{\delta}{2}-s\\ s+\frac{1-\delta}{2},\ 1+\frac{\mu-\nu}{2}-s,\ 1-\frac{\mu-\nu}{2}-s\end{bmatrix}$ [$\vert\mathrm{Re}\,(\mu+\nu)\vert/2<\mathrm{Re}\,s<1/2$]
22	$J_{-\mu}\left(\frac{1}{\sqrt{x}}\right)J_{-\nu}\left(\frac{1}{\sqrt{x}}\right) \pm \pm J_{\mu}\left(\frac{1}{\sqrt{x}}\right)J_{\nu}\left(\frac{1}{\sqrt{x}}\right) = A_{16}G_{42}^{12}\left(x\,\middle\vert\,\begin{matrix}1+\frac{\mu+\nu}{2},\ 1-\frac{\mu+\nu}{2},\ 1+\frac{\mu-\nu}{2},\ 1-\frac{\mu-\nu}{2}\\ 1-\frac{\delta}{2},\ \frac{1+\delta}{2}\end{matrix}\right)$	$A_{16}\Gamma\begin{bmatrix}s+1-\frac{\delta}{2},\ \frac{\mu+\nu}{2}-s,\ -\frac{\mu+\nu}{2}-s\\ s+\frac{\mu-\nu}{2}+1,\ s-\frac{\mu-\nu}{2}+1,\ \frac{1-\delta}{2}-s\end{bmatrix}$ [$-1/2<\mathrm{Re}\,s<-\vert\mathrm{Re}(\mu+\nu)\vert/2$]

8.4.20. The Neumann function $Y_{\nu}(x)$.

NOTATION: $A_{17}=\dfrac{\sin(\nu-\mu)\pi}{\pi^{5/2}}$, $\delta=\begin{Bmatrix}1\\0\end{Bmatrix}$.

1	$Y_{\nu}(2\sqrt{x}) = G_{13}^{20}\left(x\,\middle\vert\,\begin{matrix}-(\nu+1)/2\\ \nu/2,\ -\nu/2,\ -(\nu+1)/2\end{matrix}\right)$	$\Gamma\begin{bmatrix}s+\nu/2,\ s-\nu/2\\ s-(\nu+1)/2,\ (\nu+3)/2-s\end{bmatrix}$ [$\vert\mathrm{Re}\,\nu\vert/2<\mathrm{Re}\,s<3/4$]
2	$Y_{\nu}\left(\frac{2}{\sqrt{x}}\right) = G_{31}^{02}\left(x\,\middle\vert\,\begin{matrix}1-\nu/2,\ 1+\nu/2,\ (3+\nu)/2\\ (3+\nu)/2\end{matrix}\right)$	$\Gamma\begin{bmatrix}\nu/2-s,\ -\nu/2-s\\ s+(\nu+3)/2,\ -(\nu+1)/2-s\end{bmatrix}$ [$-3/4<\mathrm{Re}\,s<-\vert\mathrm{Re}\,\nu\vert/2$]
3	$\begin{Bmatrix}\sin\sqrt{x}\\ \cos\sqrt{x}\end{Bmatrix}Y_{\nu}(\sqrt{x}) = \frac{1}{\sqrt{2}}G_{35}^{22}\left(x\,\middle\vert\,\begin{matrix}1/4,\ 3/4,\ (\delta-\nu-1)/2\\ (\delta+\nu)/2,\ (\delta-\nu)/2,\ (\delta-\nu-1)/2,\ (1-\delta-\nu)/2,\ (1-\delta+\nu)/2\end{matrix}\right)$	$\frac{1}{\sqrt{2}}\Gamma\begin{bmatrix}s+\frac{\delta+\nu}{2},\ s+\frac{\delta-\nu}{2},\ \frac{1}{4}-s,\ \frac{3}{4}-s\\ s+\frac{\delta-\nu-1}{2},\ \frac{3-\delta+\nu}{2}-s,\ \frac{1+\delta+\nu}{2}-s,\ \frac{1+\delta-\nu}{2}-s\end{bmatrix}$ [$(\vert\mathrm{Re}\,\nu\vert-\delta)/2<\mathrm{Re}\,s<1/4$]
4	$\begin{Bmatrix}\sin x^{-1/2}\\ \cos x^{-1/2}\end{Bmatrix}Y_{\nu}\left(\frac{1}{\sqrt{x}}\right) = \frac{1}{\sqrt{2}}G_{53}^{22}\left(x\,\middle\vert\,\begin{matrix}1-(\nu+\delta)/2,\ 1+(\nu-\delta)/2,\ (3+\nu-\delta)/2,\ (1+\nu+\delta)/2,\ (1-\nu+\delta)/2\\ 1/4,\ 3/4,\ (3+\nu-\delta)/2\end{matrix}\right)$	$\frac{1}{\sqrt{2}}\Gamma\begin{bmatrix}s+\frac{1}{4},\ s+\frac{3}{4},\ \frac{\delta+\nu}{2}-s,\ \frac{\delta-\nu}{2}-s\\ s+\frac{1+\delta+\nu}{2},\ s+\frac{1+\delta-\nu}{2},\ s+\frac{3-\delta+\nu}{2},\ \frac{\delta-\nu-1}{2}-s\end{bmatrix}$ [$-1/4<\mathrm{Re}\,s<(\delta-\vert\mathrm{Re}\,\nu\vert)/2$]

5	$\begin{Bmatrix}\sin\sqrt{x}\\ \cos\sqrt{x}\end{Bmatrix} J_\nu(\sqrt{x}) \pm \begin{Bmatrix}\cos\sqrt{x}\\ \sin\sqrt{x}\end{Bmatrix} Y_\nu(\sqrt{x}) = \mp\sqrt{2}\, G_{24}^{30}\left(x\,\middle\vert\,\begin{matrix}1/4,\ 3/4\\ \nu/2,\ (\nu+1)/2,\ (1-\nu-\delta)/2,\ (\delta-\nu)/2\end{matrix}\right)$	$\mp\sqrt{2}\,\Gamma\begin{bmatrix}s+(1+\nu)/2,\ s+(1-\nu-\delta)/2,\ s+\nu/2\\ s+1/4,\ s+3/4,\ 1+(\nu-\delta)/2-s\end{bmatrix}$ $[(\operatorname{Re}\nu+\delta-1)/2,\ -\operatorname{Re}\nu/2<\operatorname{Re}s<3/4]$
6	$\begin{Bmatrix}\sin x^{-1/2}\\ \cos x^{-1/2}\end{Bmatrix} J_\nu\left(\frac{1}{\sqrt{x}}\right) \pm \begin{Bmatrix}\cos x^{-1/2}\\ \sin x^{-1/2}\end{Bmatrix} Y_\nu\left(\frac{1}{\sqrt{x}}\right) = \mp\sqrt{2}\, G_{42}^{03}\left(x\,\middle\vert\,\begin{matrix}1-\frac{\nu}{2},\ \frac{1-\nu}{2},\ \frac{1+\nu+\delta}{2},\ 1+\frac{\nu-\delta}{2}\\ \frac{1}{4},\ \frac{3}{4}\end{matrix}\right)$	$\mp\sqrt{2}\,\Gamma\begin{bmatrix}(1+\nu)/2-s,\ (1-\nu-\delta)/2-s,\ \nu/2-s\\ s+(\nu-\delta)/2+1,\ 1/4-s,\ 3/4-s\end{bmatrix}$ $[-3/4<\operatorname{Re}s<(1-\delta-\operatorname{Re}\nu)/2,\ \operatorname{Re}\nu/2]$
7	$\begin{Bmatrix}\sin\sqrt{x}\\ \cos\sqrt{x}\end{Bmatrix} J_\nu(\sqrt{x}) \mp \begin{Bmatrix}\cos\sqrt{x}\\ \sin\sqrt{x}\end{Bmatrix} Y_\nu(\sqrt{x}) = \frac{\cos\nu\pi}{\pi^2\sqrt{2}}\, G_{24}^{32}\left(x\,\middle\vert\,\begin{matrix}\frac{1}{4},\ \frac{3}{4}\\ \frac{\nu}{2},\ \frac{1+\nu}{2},\ \frac{1-\nu-\delta}{2},\ \frac{\delta-\nu}{2}\end{matrix}\right)$	$\frac{\cos\nu\pi}{\pi^2\sqrt{2}}\,\Gamma\begin{bmatrix}s+(1+\nu)/2,\ s+(1-\nu-\delta)/2,\ s+\nu/2,\ 1/4-s,\ 3/4-s\\ 1+(\nu-\delta)/2-s\end{bmatrix}$ $[(\operatorname{Re}\nu+\delta-1)/2,\ -\operatorname{Re}\nu/2<\operatorname{Re}s<1/4]$
8	$\begin{Bmatrix}\sin x^{-1/2}\\ \cos x^{-1/2}\end{Bmatrix} J_\nu\left(\frac{1}{\sqrt{x}}\right) \mp \begin{Bmatrix}\cos x^{-1/2}\\ \sin x^{-1/2}\end{Bmatrix} Y_\nu\left(\frac{1}{\sqrt{x}}\right) = \frac{\cos\nu\pi}{\pi^2\sqrt{2}}\, G_{42}^{23}\left(x\,\middle\vert\,\begin{matrix}1-\nu/2,\ (1-\nu)/2,\ (1+\nu-\delta)/2,\ 1+(\nu-\delta)/2\\ 1/4,\ 3/4\end{matrix}\right)$	$\frac{\cos\nu\pi}{\pi^2\sqrt{2}}\,\Gamma\begin{bmatrix}s+1/4,\ s+3/4,\ \nu/2-s,\ (1+\nu)/2-s,\ (1-\nu-\delta)/2-s\\ s+(\nu-\delta)/2+1\end{bmatrix}$ $[-1/4<\operatorname{Re}s<(1-\delta-\operatorname{Re}\nu)/2,\ \operatorname{Re}\nu/2]$
9	$J_\nu(\sqrt{x})\,Y_\nu(\sqrt{x}) = -\frac{1}{\sqrt{\pi}}\, G_{13}^{20}\left(x\,\middle\vert\,\begin{matrix}1/2\\ 0,\ \nu,\ -\nu\end{matrix}\right)$	$-\frac{1}{\sqrt{\pi}}\,\Gamma\begin{bmatrix}s,\ s+\nu\\ s+1/2,\ 1+\nu-s\end{bmatrix}$ $[0,\ -\operatorname{Re}\nu<\operatorname{Re}s<1]$

10	$J_\nu\left(\frac{1}{\sqrt{x}}\right)Y_\nu\left(\frac{1}{\sqrt{x}}\right)=-\frac{1}{\sqrt{\pi}}G_{31}^{02}\left(x\middle\vert\begin{matrix}1,\ 1-\nu,\ 1+\nu\\ 1/2\end{matrix}\right)$	$-\frac{1}{\sqrt{\pi}}\Gamma\left[\begin{matrix}-s,\ \nu-s\\ s+\nu+1,\ 1/2-s\end{matrix}\right]$ $[-1<\operatorname{Re}s<0,\ \operatorname{Re}\nu]$
11	$J_{-\nu}(\sqrt{x})Y_\nu(\sqrt{x})=\frac{1}{\sqrt{\pi}}G_{24}^{21}\left(x\middle\vert\begin{matrix}1/2,\ -\nu-1/2\\ 0,\ -\nu,\ \nu,\ -\nu-1/2\end{matrix}\right)$	$\frac{1}{\sqrt{\pi}}\Gamma\left[\begin{matrix}s,\ s-\nu,\ 1/2-s\\ s-\nu-1/2,\ 1-\nu-s,\ 3/2+\nu-s\end{matrix}\right]$ $[0,\ \operatorname{Re}\nu<\operatorname{Re}s<1/2]$
12	$J_{-\nu}\left(\frac{1}{\sqrt{x}}\right)Y_\nu\left(\frac{1}{\sqrt{x}}\right)=\frac{1}{\sqrt{\pi}}G_{42}^{12}\left(x\middle\vert\begin{matrix}1,\ 1+\nu,\ 1-\nu,\ 3/2+\nu\\ 1/2,\ 3/2+\nu\end{matrix}\right)$	$\frac{1}{\sqrt{\pi}}\Gamma\left[\begin{matrix}s+1/2,\ -\nu-s,\ -s\\ s+1-\nu,\ s+\nu+3/2,\ -1/2-\nu-s\end{matrix}\right]$ $[-1/2<\operatorname{Re}s<0,\ -\operatorname{Re}\nu]$
13	$J_{\nu+1}(\sqrt{x})Y_\nu(\sqrt{x})=\frac{1}{\sqrt{\pi}}G_{24}^{21}\left(x\middle\vert\begin{matrix}1/2,\ 0\\ \nu+1/2,\ 1/2,\ -\nu-1/2,\ -1/2\end{matrix}\right)$	$\frac{1}{\sqrt{\pi}}\Gamma\left[\begin{matrix}s+\nu+1/2,\ s+1/2,\ 1/2-s\\ s,\ \nu+3/2-s,\ 3/2-s\end{matrix}\right]$ $[-1/2,\ -1/2-\operatorname{Re}\nu<\operatorname{Re}s<1/2]$
14	$J_{\nu+1}\left(\frac{1}{\sqrt{x}}\right)Y_\nu\left(\frac{1}{\sqrt{x}}\right)=\frac{1}{\sqrt{\pi}}G_{42}^{12}\left(x\middle\vert\begin{matrix}1/2-\nu,\ 1/2,\ 3/2+\nu,\ 3/2\\ 1/2,\ 1\end{matrix}\right)$	$\frac{1}{\sqrt{\pi}}\Gamma\left[\begin{matrix}s+1/2,\ 1/2-s,\ 1/2+\nu-s\\ s+\nu+3/2,\ s+3/2,\ -s\end{matrix}\right]$ $[-1/2<\operatorname{Re}s<1/2+\operatorname{Re}\nu,\ 1/2]$
15	$J_{\nu+2}(\sqrt{x})Y_\nu(\sqrt{x})=\frac{1}{\sqrt{\pi}}G_{24}^{21}\left(x\middle\vert\begin{matrix}0,\ 1/2\\ \nu+1,\ 1,\ -\nu-1,\ -1\end{matrix}\right)$	$\frac{1}{\sqrt{\pi}}\Gamma\left[\begin{matrix}s+\nu+1,\ s+1,\ 1-s\\ s+1/2,\ 2+\nu-s,\ 2-s\end{matrix}\right]$ $[-1,\ -1-\operatorname{Re}\nu<\operatorname{Re}s<1]$
16	$J_{\nu+2}\left(\frac{1}{\sqrt{x}}\right)Y_\nu\left(\frac{1}{\sqrt{x}}\right)=\frac{1}{\sqrt{\pi}}G_{42}^{12}\left(x\middle\vert\begin{matrix}0,\ -\nu,\ 2,\ \nu+2\\ 1,\ 1/2\end{matrix}\right)$	$\frac{1}{\sqrt{\pi}}\Gamma\left[\begin{matrix}s+1,\ 1-s,\ 1+\nu-s\\ s+2,\ s+\nu+2,\ 1/2-s\end{matrix}\right]$ $[-1<\operatorname{Re}s<1,\ \operatorname{Re}\nu+1]$
17	$J_\mu(\sqrt{x})Y_\nu(\sqrt{x})=\frac{1}{\sqrt{\pi}}G_{35}^{22}\left(x\middle\vert\begin{matrix}0,\ 1/2,\ (\mu-\nu-1)/2\\ (\mu+\nu)/2,\ (\mu-\nu)/2,\ (\mu-\nu-1)/2,\ -(\mu+\nu)/2,\ (\nu-\mu)/2\end{matrix}\right)$	$\frac{1}{\sqrt{\pi}}\Gamma\left[\begin{matrix}s+\frac{\mu+\nu}{2},\ s+\frac{\mu-\nu}{2},\ \frac{1}{2}-s,\ 1-s\\ s+\frac{\mu-\nu-1}{2},\ \frac{3+\nu-\mu}{2}-s,\ 1+\frac{\mu+\nu}{2}-s,\ 1+\frac{\mu-\nu}{2}-s\end{matrix}\right]$ $[(\vert\operatorname{Re}\nu\vert-\operatorname{Re}\mu)/2<\operatorname{Re}s<1/2]$
18	$J_\mu\left(\frac{1}{\sqrt{x}}\right)Y_\nu\left(\frac{1}{\sqrt{x}}\right)=\frac{1}{\sqrt{\pi}}G_{53}^{22}\left(x\middle\vert\begin{matrix}1-\frac{\mu+\nu}{2},\ 1+\frac{\nu-\mu}{2},\ \frac{3+\nu-\mu}{2},\ 1+\frac{\mu+\nu}{2},\ 1+\frac{\mu-\nu}{2}\\ 1/2,\ 1,\ (3+\nu-\mu)/2\end{matrix}\right)$	$\frac{1}{\sqrt{\pi}}\Gamma\left[\begin{matrix}s+\frac{1}{2},\ s+1,\ \frac{\mu+\nu}{2}-s,\ \frac{\mu-\nu}{2}-s\\ s+\frac{3+\nu-\mu}{2},\ s+\frac{\mu+\nu}{2}+1,\ s+\frac{\mu-\nu}{2}+1,\ \frac{\mu-\nu-1}{2}-s\end{matrix}\right]$ $[-1/2<\operatorname{Re}s<(\operatorname{Re}\mu-\vert\operatorname{Re}\nu\vert)/2]$

19	$J_\nu(\sqrt{x})\,Y_{-\nu}(\sqrt{x})+J_{-\nu}(\sqrt{x})\times Y_\nu(\sqrt{x})=-\frac{2}{\sqrt{\pi}}\,G_{13}^{20}\left(x\,\middle\|\,\begin{matrix}1/2\\ \nu,\ -\nu,\ 0\end{matrix}\right)$	$-\frac{2}{\sqrt{\pi}}\,\Gamma\left[\begin{matrix}s-\nu,\ s+\nu\\ s+1/2,\ 1-s\end{matrix}\right]$ [$\|\mathrm{Re}\,\nu\|<\mathrm{Re}\,s<1$]
20	$J_\nu\left(\frac{1}{\sqrt{x}}\right)Y_{-\nu}\left(\frac{1}{\sqrt{x}}\right)+J_{-\nu}\left(\frac{1}{\sqrt{x}}\right)Y_\nu\left(\frac{1}{\sqrt{x}}\right)=-\frac{2}{\sqrt{\pi}}\,G_{31}^{02}\left(x\,\middle\|\,\begin{matrix}1-\nu,\ 1+\nu,\ 1\\ 1/2\end{matrix}\right)$	$-\frac{2}{\sqrt{\pi}}\,\Gamma\left[\begin{matrix}-\nu-s,\ \nu-s\\ s+1,\ 1/2-s\end{matrix}\right]$ [$-1<\mathrm{Re}\,s<-\|\mathrm{Re}\,\nu\|$]
21	$J_\nu(\sqrt{x})\,Y_\mu(\sqrt{x})+J_\mu(\sqrt{x})\,Y_\nu(\sqrt{x})=-\frac{2}{\sqrt{\pi}}\,G_{24}^{30}\left(x\,\middle\|\,\begin{matrix}0,\ \frac{1}{2}\\ \frac{\mu+\nu}{2},\ \frac{\mu-\nu}{2},\ \frac{\nu-\mu}{2},\ -\frac{\mu+\nu}{2}\end{matrix}\right)$	$-\frac{2}{\sqrt{\pi}}\,\Gamma\left[\begin{matrix}s+\frac{\mu+\nu}{2},\ s+\frac{\mu-\nu}{2},\ s+\frac{\nu-\mu}{2}\\ s,\ s+\frac{1}{2},\ 1+\frac{\mu+\nu}{2}-s\end{matrix}\right]$ [$(\|\mathrm{Re}\,\nu\|-\mathrm{Re}\,\mu)/2,\ \mathrm{Re}\,(\mu-\nu)/2<\mathrm{Re}\,s<1$]
22	$J_\nu\left(\frac{1}{\sqrt{x}}\right)Y_\mu\left(\frac{1}{\sqrt{x}}\right)+J_\mu\left(\frac{1}{\sqrt{x}}\right)Y_\nu\left(\frac{1}{\sqrt{x}}\right)=-\frac{2}{\sqrt{\pi}}\,G_{42}^{03}\left(x\,\middle\|\,\begin{matrix}1-\frac{\mu+\nu}{2},\ 1+\frac{\nu-\mu}{2},\ 1+\frac{\mu-\nu}{2},\ 1+\frac{\mu+\nu}{2}\\ \frac{1}{2},\ 1\end{matrix}\right)$	$-\frac{2}{\sqrt{\pi}}\,\Gamma\left[\begin{matrix}\frac{\mu+\nu}{2}-s,\ \frac{\mu-\nu}{2}-s,\ \frac{\nu-\mu}{2}-s\\ s+\frac{\mu+\nu}{2}+1,\ -s,\ \frac{1}{2}-s\end{matrix}\right]$ [$-1<\mathrm{Re}\,s<(\mathrm{Re}\,\mu-\|\mathrm{Re}\,\nu\|)/2,\ \mathrm{Re}\,(\nu-\mu)/2$]
23	$J_\nu(\sqrt{x})\,Y_{-\nu}(\sqrt{x})-J_{-\nu}(\sqrt{x})\,Y_\nu(\sqrt{x})=\frac{\sin 2\nu\pi}{\pi^{5/2}}\,G_{13}^{31}\left(x\,\middle\|\,\begin{matrix}1/2\\ 0,\ \nu,\ -\nu\end{matrix}\right)$	$\frac{\sin 2\nu\pi}{\pi^{5/2}}\,\Gamma\left[s,\ s-\nu,\ s+\nu,\ \frac{1}{2}-s\right]$ [$\|\mathrm{Re}\ \|<\mathrm{Re}\,s<1/2$]
24	$J_\nu\left(\frac{1}{\sqrt{x}}\right)Y_{-\nu}\left(\frac{1}{\sqrt{x}}\right)-J_{-\nu}\left(\frac{1}{\sqrt{x}}\right)Y_\nu\left(\frac{1}{\sqrt{x}}\right)=\frac{\sin 2\nu\pi}{\pi^{5/2}}\,G_{31}^{13}\left(x\,\middle\|\,\begin{matrix}1,\ 1-\nu,\ 1+\nu\\ 1/2\end{matrix}\right)$	$\frac{\sin 2\nu\pi}{\pi^{5/2}}\,\Gamma\left[s+\frac{1}{2},\ -s,\ \nu-s,\ -\nu-s\right]$ [$-1/2<\mathrm{Re}\,s<-\|\mathrm{Re}\,\nu\|$]

25	$J_\nu(\sqrt{x})\,Y_\mu(\sqrt{x}) - J_\mu(\sqrt{x})\,Y_\nu(\sqrt{x}) = A_{17}G_{24}^{32}\left(x \middle\vert \begin{matrix} 0,\ \frac{1}{2} \\ \frac{\mu+\nu}{2},\ \frac{\mu-\nu}{2},\ \frac{\nu-\mu}{2},\ -\frac{\mu+\nu}{2} \end{matrix}\right)$	$A_{17}\Gamma\left[\begin{matrix} s+\frac{\mu+\nu}{2},\ s+\frac{\mu-\nu}{2},\ 1+\frac{\mu+\nu}{2}-s \\ s+\frac{\nu-\mu}{2},\ \frac{1}{2}-s,\ 1-s \end{matrix}\right]$ [Re$(\mu-\nu)/2$, $(\lvert \mathrm{Re}\,\nu\rvert - \mathrm{Re}\,\mu)/2 < \mathrm{Re}\,s < 1/2$]
26	$J_\nu\left(\frac{1}{\sqrt{x}}\right)Y_\mu\left(\frac{1}{\sqrt{x}}\right) - J_\mu\left(\frac{1}{\sqrt{x}}\right)Y_\nu\left(\frac{1}{\sqrt{x}}\right) = A_{17}G_{42}^{23}\left(x \middle\vert \begin{matrix} 1-\frac{\mu+\nu}{2},\ 1-\frac{\mu-\nu}{2},\ 1+\frac{\mu-\nu}{2},\ 1+\frac{\mu+\nu}{2} \\ \frac{1}{2},\ 1 \end{matrix}\right)$	$A_{17}\Gamma\left[\begin{matrix} s+1/2,\ s+1,\ (\mu+\nu)/2-s \\ s+(\mu+\nu)/2+1,\ (\mu-\nu)/2-s,\ (\nu-\mu)/2-s \end{matrix}\right]$ [$-1/2 < \mathrm{Re}\,s < \mathrm{Re}(\nu-\mu)/2$, $(\mathrm{Re}\,\mu - \lvert \mathrm{Re}\,\nu\rvert)/2$]
27	$Y_\nu^2(\sqrt{x}) = \frac{2}{\sqrt{\pi}}G_{24}^{30}\left(x \middle\vert \begin{matrix} 1/2,\ 1/2-\nu \\ 0,\ \nu,\ -\nu,\ 1/2-\nu \end{matrix}\right) + \frac{1}{\sqrt{\pi}}G_{13}^{11}\left(x \middle\vert \begin{matrix} 1/2 \\ \nu,\ -\nu,\ 0 \end{matrix}\right)$	$\frac{2}{\sqrt{\pi}}\Gamma\left[\begin{matrix} s,\ s+\nu,\ s-\nu \\ s+1/2,\ s+1/2-\nu,\ 1/2+\nu-s \end{matrix}\right] + \frac{1}{\sqrt{\pi}}\Gamma\left[\begin{matrix} s+\nu,\ 1/2-s \\ 1+\nu-s,\ 1-s \end{matrix}\right]$ [$\lvert \mathrm{Re}\,\nu\rvert < \mathrm{Re}\,s < 1/2$]
28	$Y_\nu^2\left(\frac{1}{\sqrt{x}}\right) = \frac{2}{\sqrt{\pi}}G_{42}^{03}\left(x \middle\vert \begin{matrix} 1,\ 1-\nu,\ 1+\nu,\ 1/2+\nu \\ 1/2+\nu,\ 1/2 \end{matrix}\right) + \frac{1}{\sqrt{\pi}}G_{31}^{11}\left(x \middle\vert \begin{matrix} 1-\nu,\ 1+\nu,\ 1 \\ 1/2 \end{matrix}\right)$	$\frac{2}{\sqrt{\pi}}\Gamma\left[\begin{matrix} -s,\ \nu-s,\ -\nu-s \\ s+\nu+1/2,\ 1/2-s,\ 1/2-\nu-s \end{matrix}\right] + \frac{1}{\sqrt{\pi}}\Gamma\left[\begin{matrix} s+1/2,\ \nu-s \\ s+1,\ s+\nu+1 \end{matrix}\right]$ [$-1/2 < \mathrm{Re}\,s < -\lvert \mathrm{Re}\,\nu\rvert$]
29	$Y_{-\nu}(\sqrt{x})\,Y_\nu(\sqrt{x}) = \frac{2}{\sqrt{\pi}}G_{24}^{30}\left(x \middle\vert \begin{matrix} 1/2,\ 1/2 \\ 0,\ \nu,\ -\nu,\ 1/2 \end{matrix}\right) + \frac{1}{\sqrt{\pi}}G_{13}^{11}\left(x \middle\vert \begin{matrix} 1/2 \\ 0,\ \nu,\ -\nu \end{matrix}\right)$	$\frac{2}{\sqrt{\pi}}\Gamma\left[\begin{matrix} s,\ s+\nu,\ s-\nu \\ s+1/2,\ s+1/2,\ 1/2-s \end{matrix}\right] + \frac{1}{\sqrt{\pi}}\Gamma\left[\begin{matrix} s,\ 1/2-s \\ 1-\nu-s,\ 1+\nu-s \end{matrix}\right]$ [$\lvert \mathrm{Re}\,\nu\rvert < \mathrm{Re}\,s < 1/2$]
30	$Y_{-\nu}\left(\frac{1}{\sqrt{x}}\right)Y_\nu\left(\frac{1}{\sqrt{x}}\right) = \frac{2}{\sqrt{\pi}}G_{42}^{03}\left(x \middle\vert \begin{matrix} 1,\ 1-\nu,\ 1+\nu,\ 1/2 \\ 1/2,\ 1/2 \end{matrix}\right) + \frac{1}{\sqrt{\pi}}G_{31}^{11}\left(x \middle\vert \begin{matrix} 1,\ 1-\nu,\ 1+\nu \\ 1/2 \end{matrix}\right)$	$\frac{2}{\sqrt{\pi}}\Gamma\left[\begin{matrix} -s,\ \nu-s,\ -\nu-s \\ s+1/2,\ 1/2-s,\ 1/2-s \end{matrix}\right] + \frac{1}{\sqrt{\pi}}\Gamma\left[\begin{matrix} s+1/2,\ -s \\ s+1-\nu,\ s+1+\nu \end{matrix}\right]$ [$-1/2 < \mathrm{Re}\,s < -\lvert \mathrm{Re}\,\nu\rvert$]

31	$Y_{1/2-\nu}(\sqrt{x})\,Y_{\nu}(\sqrt{x})=$ $=\frac{2}{\sqrt{\pi}}G_{24}^{30}\left(x\,\middle\vert\,\begin{matrix}0,\ \frac{1}{2}\\ -\frac{1}{4},\ \frac{1}{4}-\nu,\ \nu-\frac{1}{4},\ \frac{1}{4}\end{matrix}\right)+$ $+\frac{1}{\sqrt{\pi}}G_{24}^{12}\left(x\,\middle\vert\,\begin{matrix}0,\ \frac{1}{2}\\ \frac{1}{4},\ -\frac{1}{4},\ \nu-\frac{1}{4},\ \frac{1}{4}-\nu\end{matrix}\right)$	$\frac{2}{\sqrt{\pi}}\Gamma\begin{bmatrix}s-1/4,\ s+1/4-\nu,\ s+\nu-1/4\\ s,\ s+1/2,\ 3/4-s\end{bmatrix}+$ $+\frac{1}{\sqrt{\pi}}\Gamma\begin{bmatrix}s+1/4,\ 1/2-s,\ 1-s\\ 5/4-s,\ 5/4-\nu-s,\ 3/4+\nu-s\end{bmatrix}$ [1/4, $\vert 1/4-\mathrm{Re}\,\nu\vert<\mathrm{Re}\,s<1/2$]
32	$Y_{1/2-\nu}\left(\frac{1}{\sqrt{x}}\right)Y_{\nu}\left(\frac{1}{\sqrt{x}}\right)=$ $=\frac{2}{\sqrt{\pi}}G_{42}^{03}\left(x\,\middle\vert\,\begin{matrix}\frac{5}{4},\ \frac{3}{4}+\nu,\ \frac{5}{4}-\nu,\ \frac{3}{4}\\ \frac{1}{2},\ 1\end{matrix}\right)+\frac{1}{\sqrt{\pi}}G_{42}^{21}\left(x\,\middle\vert\,\begin{matrix}\frac{3}{4},\ \frac{5}{4},\ \frac{5}{4}-\nu,\ \frac{3}{4}+\nu\\ \frac{1}{2},\ 1\end{matrix}\right)$	$\frac{2}{\sqrt{\pi}}\Gamma\begin{bmatrix}-1/4-s,\ 1/4-\nu-s,\ \nu-1/4-s\\ s+3/4,\ -s,\ 1/2-s\end{bmatrix}+\frac{1}{\sqrt{\pi}}\Gamma\begin{bmatrix}s+1,\ s+1/2,\ 1/4-s\\ s+5/4,\ s+5/4-\nu,\ s+3/4+\nu\end{bmatrix}$ [$-1/2<\mathrm{Re}\,s<-1/4,\ -\vert 1/4-\mathrm{Re}\,\nu\vert$]
33	$Y_{\mu}(\sqrt{x})\,Y_{\nu}(\sqrt{x})=$ $=\frac{2}{\sqrt{\pi}}G_{35}^{40}\left(x\,\middle\vert\,\begin{matrix}0,\ \frac{1}{2},\ \frac{1-\mu-\nu}{2}\\ \frac{\mu+\nu}{2},\ \frac{\mu-\nu}{2},\ \frac{\nu-\mu}{2},\ -\frac{\mu+\nu}{2},\ \frac{1-\mu-\nu}{2}\end{matrix}\right)+$ $+\frac{1}{\sqrt{\pi}}G_{24}^{12}\left(x\,\middle\vert\,\begin{matrix}0,\ \frac{1}{2}\\ \frac{\mu+\nu}{2},\ -\frac{\mu+\nu}{2},\ \frac{\nu-\mu}{2},\ \frac{\mu-\nu}{2}\end{matrix}\right)$	$\frac{2}{\sqrt{\pi}}\Gamma\begin{bmatrix}s+\frac{\mu+\nu}{2},\ s+\frac{\mu-\nu}{2},\ s+\frac{\nu-\mu}{2},\ s-\frac{\mu+\nu}{2}\\ s,\ s+\frac{1}{2},\ s+\frac{1-\mu-\nu}{2},\ \frac{1+\mu+\nu}{2}-s\end{bmatrix}+$ $+\frac{1}{\sqrt{\pi}}\Gamma\begin{bmatrix}s+\frac{\mu+\nu}{2},\ \frac{1}{2}-s,\ 1-s\\ 1+\frac{\mu+\nu}{2}-s,\ 1+\frac{\mu-\nu}{2}-s,\ 1+\frac{\nu-\mu}{2}-s\end{bmatrix}$ [$(\vert\mathrm{Re}\,\mu\vert+\vert\mathrm{Re}\,\nu\vert)/2<\mathrm{Re}\,s<1/2$]

34	$Y_\mu\left(\frac{1}{\sqrt{x}}\right)Y_\nu\left(\frac{1}{\sqrt{x}}\right)=$ $=\frac{2}{\sqrt{\pi}}G^{04}_{53}\left(x\,\middle\vert\,\begin{matrix}1-\frac{\mu+\nu}{2},\ 1+\frac{\nu-\mu}{2},\ 1+\frac{\mu-\nu}{2},\ 1+\frac{\mu+\nu}{2},\ \frac{1+\mu+\nu}{2}\\ \frac{1}{2},\ 1,\ \frac{1+\mu+\nu}{2}\end{matrix}\right)+$ $+\frac{1}{\sqrt{\pi}}G^{21}_{42}\left(x\,\middle\vert\,\begin{matrix}1-\frac{\mu+\nu}{2},\ 1+\frac{\mu+\nu}{2},\ 1+\frac{\mu-\nu}{2},\ 1+\frac{\nu-\mu}{2}\\ \frac{1}{2},\ 1\end{matrix}\right)$	$\frac{2}{\sqrt{\pi}}\Gamma\left[\begin{matrix}(\mu+\nu)/2-s,\ (\mu-\nu)/2-s,\ (\nu-\mu)/2-s,\ (\mu+\nu)/2-s\\ s+(\mu+\nu+1)/2,\ -s,\ 1/2-s,\ (1-\mu-\nu)/2-s\end{matrix}\right]+$ $+\frac{1}{\sqrt{\pi}}\Gamma\left[\begin{matrix}s+\frac{1}{2},\ s+1,\ \frac{\mu+\nu}{2}-s\\ s+\frac{\mu+\nu}{2}+1,\ s+\frac{\mu-\nu}{2}+1,\ s+\frac{\nu-\mu}{2}+1\end{matrix}\right]$ [$-1/2<\mathrm{Re}\,s<-(\vert\mathrm{Re}\,\mu\vert+\vert\mathrm{Re}\,\nu\vert)/2$]
35	$J_\nu^2(\sqrt{x})+Y_\nu^2(\sqrt{x})=$ $=\frac{2\cos\nu\pi}{\pi^{5/2}}G^{31}_{13}\left(x\,\middle\vert\,\begin{matrix}1/2\\ 0,\ \nu,\ -\nu\end{matrix}\right)$	$\frac{2\cos\nu\pi}{\pi^{5/2}}\Gamma\left[s,\ s+\nu,\ s-\nu,\ \frac{1}{2}-s\right]$ [$\vert\mathrm{Re}\,\nu\vert<\mathrm{Re}\,s<1/2$]
36	$J_\nu^2\left(\frac{1}{\sqrt{x}}\right)+Y_\nu^2\left(\frac{1}{\sqrt{x}}\right)=$ $=\frac{2\cos\nu\pi}{\pi^{5/2}}G^{13}_{31}\left(x\,\middle\vert\,\begin{matrix}1,\ 1-\nu,\ 1+\nu\\ 1/2\end{matrix}\right)$	$\frac{2\cos\nu\pi}{\pi^{5/2}}\Gamma\left[s+\frac{1}{2},\ -s,\ \nu-s,\ -\nu-s\right]$ [$-1/2<\mathrm{Re}\,s<-\vert\mathrm{Re}\,\nu\vert$]
37	$J_\nu^2(\sqrt{x})-Y_\nu^2(\sqrt{x})=$ $=-\frac{2}{\sqrt{\pi}}G^{30}_{24}\left(x\,\middle\vert\,\begin{matrix}1/2,\ 1/2-\nu\\ 0,\ \nu,\ -\nu,\ 1/2-\nu\end{matrix}\right)$	$-\frac{2}{\sqrt{\pi}}\Gamma\left[\begin{matrix}s,\ s+\nu,\ s-\nu\\ s+1/2,\ s+1/2-\nu,\ 1/2+\nu-s\end{matrix}\right]$ [$\vert\mathrm{Re}\,\nu\vert<\mathrm{Re}\,s<1$]
38	$J_\nu^2\left(\frac{1}{\sqrt{x}}\right)-Y_\nu^2\left(\frac{1}{\sqrt{x}}\right)=$ $=-\frac{2}{\sqrt{\pi}}G^{03}_{42}\left(x\,\middle\vert\,\begin{matrix}1,\ 1-\nu,\ 1+\nu,\ 1/2+\nu\\ 1/2,\ 1/2+\nu\end{matrix}\right)$	$-\frac{2}{\sqrt{\pi}}\Gamma\left[\begin{matrix}-s,\ \nu-s,\ -\nu-s\\ s+\nu+1/2,\ 1/2-\nu-s,\ 1/2-s\end{matrix}\right]$ [$-1<\mathrm{Re}\,s<-\vert\mathrm{Re}\,\nu\vert$]
39	$J_{-\nu}(x^{\pm1/2})J_\nu(x^{\pm1/2})+$ $+Y_{-\nu}(x^{\pm1/2})Y_\nu(x^{\pm1/2})=$ $=\cos\nu\pi\left[J_\nu^2(x^{\pm1/2})+Y_\nu^2(x^{\pm1/2})\right]$	[see 8.4.20. 35–36]
40	$J_{-\nu}(\sqrt{x})J_\nu(\sqrt{x})-$ $-Y_{-\nu}(\sqrt{x})Y_\nu(\sqrt{x})=$ $=-\frac{2}{\sqrt{\pi}}G^{30}_{24}\left(x\,\middle\vert\,\begin{matrix}1/2,\ 1/2\\ 0,\ \nu,\ -\nu,\ 1/2\end{matrix}\right)$	$-\frac{2}{\sqrt{\pi}}\Gamma\left[\begin{matrix}s,\ s+\nu,\ s-\nu\\ s+1/2,\ s+1/2,\ 1/2-s\end{matrix}\right]$ [$\vert\mathrm{Re}\,\nu\vert<\mathrm{Re}\,s<1$]
41	$J_{-\nu}\left(\frac{1}{\sqrt{x}}\right)J_\nu\left(\frac{1}{\sqrt{x}}\right)-$ $-Y_{-\nu}\left(\frac{1}{\sqrt{x}}\right)Y_\nu\left(\frac{1}{\sqrt{x}}\right)=$ $=-\frac{2}{\sqrt{\pi}}G^{03}_{42}\left(x\,\middle\vert\,\begin{matrix}1,\ 1-\nu,\ 1+\nu,\ 1/2\\ 1/2,\ 1/2\end{matrix}\right)$	$-\frac{2}{\sqrt{\pi}}\Gamma\left[\begin{matrix}-s,\ \nu-s,\ -\nu-s\\ s+1/2,\ 1/2-s,\ 1/2-s\end{matrix}\right]$ [$-1<\mathrm{Re}\,s<-\vert\mathrm{Re}\,\nu\vert$]

42	$J_{\mu}(\sqrt{x})\,J_{\nu}(\sqrt{x})+Y_{\mu}(\sqrt{x})\,Y_{\nu}(\sqrt{x})=\frac{\cos\mu\pi}{\pi^{5/2}}\,G_{24}^{32}\left(x\,\middle\vert\,\begin{matrix}0,\ \frac{1}{2}\\ \frac{\mu+\nu}{2},\ -\frac{\mu+\nu}{2},\ \frac{\mu-\nu}{2},\ \frac{\nu-\mu}{2}\end{matrix}\right)+\frac{\cos\nu\pi}{\pi^{5/2}}\,G_{24}^{32}\left(x\,\middle\vert\,\begin{matrix}0,\ \frac{1}{2}\\ \frac{\mu+\nu}{2},\ -\frac{\mu+\nu}{2},\ \frac{\nu-\mu}{2},\ \frac{\mu-\nu}{2}\end{matrix}\right)$	$\frac{1}{\pi^{5/2}}\,\Gamma\left[s+\frac{\mu+\nu}{2},\ s-\frac{\mu+\nu}{2},\ \frac{1}{2}-s,\ 1-s\right]\left\{\cos\mu\pi\,\Gamma\left[\begin{matrix}s+(\mu-\nu)/2\\ 1+(\mu-\nu)/2-s\end{matrix}\right]+\cos\nu\pi\,\Gamma\left[\begin{matrix}s+(\nu-\mu)/2\\ 1+(\nu-\mu)/2-s\end{matrix}\right]\right\}$ $[(\lvert\operatorname{Re}\mu\rvert+\lvert\operatorname{Re}\nu\rvert)/2<\operatorname{Re}s<1/2]$
43	$J_{\mu}\left(\frac{1}{\sqrt{x}}\right)J_{\nu}\left(\frac{1}{\sqrt{x}}\right)+Y_{\mu}\left(\frac{1}{\sqrt{x}}\right)Y_{\nu}\left(\frac{1}{\sqrt{x}}\right)=\frac{\cos\mu\pi}{\pi^{5/2}}\,G_{42}^{23}\left(x\,\middle\vert\,\begin{matrix}1-\frac{\mu+\nu}{2},\ 1+\frac{\mu+\nu}{2},\ 1+\frac{\nu-\mu}{2},\ 1+\frac{\mu-\nu}{2}\\ \frac{1}{2},\ 1\end{matrix}\right)+\frac{\cos\nu\pi}{\pi^{5/2}}\,G_{42}^{23}\left(x\,\middle\vert\,\begin{matrix}1-\frac{\mu+\nu}{2},\ 1+\frac{\mu+\nu}{2},\ 1+\frac{\mu-\nu}{2},\ 1+\frac{\nu-\mu}{2}\\ \frac{1}{2},\ 1\end{matrix}\right)$	$\frac{1}{\pi^{5/2}}\,\Gamma\left[s+\frac{1}{2},\ s+1,\ \frac{\mu+\nu}{2}-s,\ -\frac{\mu+\nu}{2}-s\right]\times\left\{\cos\mu\pi\,\Gamma\left[\begin{matrix}(\mu-\nu)/2-s\\ s+(\mu-\nu)/2+1\end{matrix}\right]+\cos\nu\pi\,\Gamma\left[\begin{matrix}(\nu-\mu)/2-s\\ s+(\nu-\mu)/2+1\end{matrix}\right]\right\}$ $[-1/2<\operatorname{Re}s<-(\lvert\operatorname{Re}\mu\rvert+\lvert\operatorname{Re}\nu\rvert)/2]$
44	$J_{\mu}(\sqrt{x})\,J_{\nu}(\sqrt{x})-Y_{\mu}(\sqrt{x})\,Y_{\nu}(\sqrt{x})=-\frac{2}{\sqrt{\pi}}\,G_{35}^{40}\left(x\,\middle\vert\,\begin{matrix}0,\ \frac{1}{2},\ \frac{1-\mu-\nu}{2}\\ \frac{\mu+\nu}{2},\ \frac{\mu-\nu}{2},\ \frac{\nu-\mu}{2},\ -\frac{\mu+\nu}{2},\ \frac{1-\mu-\nu}{2}\end{matrix}\right)$	$-\frac{2}{\sqrt{\pi}}\,\Gamma\left[\begin{matrix}s+\frac{\mu+\nu}{2},\ s+\frac{\mu-\nu}{2},\ s+\frac{\nu-\mu}{2},\ s-\frac{\mu+\nu}{2}\\ s,\ s+\frac{1}{2},\ s+\frac{1-\mu-\nu}{2},\ \frac{1+\mu+\nu}{2}-s\end{matrix}\right]$ $[(\lvert\operatorname{Re}\mu\rvert+\lvert\operatorname{Re}\nu\rvert)/2<\operatorname{Re}s<1]$

45	$J_\mu\left(\frac{1}{\sqrt{x}}\right)J_\nu\left(\frac{1}{\sqrt{x}}\right)-Y_\mu\left(\frac{1}{\sqrt{x}}\right)Y_\nu\left(\frac{1}{\sqrt{x}}\right)=-\frac{2}{\sqrt{\pi}}G^{04}_{53}\left(x\,\middle\vert\,\begin{matrix}1-(\mu+\nu)/2,\ 1+(\nu-\mu)/2,\ 1+(\mu-\nu)/2,\ 1+(\mu+\nu)/2,\ (1+\mu+\nu)/2\\ 1/2,\ 1,\ (1+\mu+\nu)/2\end{matrix}\right)$	$-\frac{2}{\sqrt{\pi}}\Gamma\left[\begin{matrix}\frac{\mu+\nu}{2}-s,\ \frac{\mu-\nu}{2}-s,\ \frac{\nu-\mu}{2}-s,\ -\frac{\mu+\nu}{2}-s\\ s+\frac{\mu+\nu+1}{2},\ -s,\ \frac{1}{2}-s,\ \frac{1-\mu-\nu}{2}-s\end{matrix}\right]$ $[-1<\operatorname{Re} s<-(\vert\operatorname{Re}\mu\vert+\vert\operatorname{Re}\nu\vert)/2]$

8.4.21. The Hankel functions $H^{(1)}_\nu(x)$ and $H^{(2)}_\nu(x)$.

1	$H^{(1)}_\nu(x^{\pm 1/2})\,H^{(2)}_\nu(x^{\pm 1/2})=J^2_\nu(x^{\pm 1/2})+Y^2_\nu(x^{\pm 1/2})$	[see 8.4.20. 35—36]

8.4.22. The modified Bessel function $I_\nu(x)$.

NOTATION: $A_{18}=\frac{\sqrt{\pi}}{\sin(c-\nu)\pi}$, $A_{19}=\frac{(-1)^\delta\pi}{\sqrt{2}}\left\{\begin{matrix}\sec(c-\nu/2)\pi\\ \operatorname{cosec}(c-\nu/2)\pi\end{matrix}\right\}$, $A_{20}=\frac{\sqrt{\pi}}{\sin c\pi}$, $A_{21}=\frac{\sqrt{\pi}}{\cos(c-\nu)\pi}$, $A_{22}=\sqrt{\pi}\operatorname{cosec}\left(c-\frac{\mu+\nu}{2}\pi\right)$, $A_{23}=\pi^{-3/2}\sin(\mu+\nu)\pi$, $\delta=\left\{\begin{matrix}1\\0\end{matrix}\right\}$.

1	$I_\nu(2\sqrt{x})=\frac{\pi}{\sin c\pi}G^{10}_{13}\left(x\,\middle\vert\,\begin{matrix}c+\nu/2\\ \nu/2,\ -\nu/2,\ c+\nu/2\end{matrix}\right)$ $\overleftarrow{L_{-\infty}}$	$\frac{\pi}{\sin c\pi}\Gamma\left[\begin{matrix}s+\frac{\nu}{2}\\ s+\frac{\nu}{2}+c,\ 1+\frac{\nu}{2}-s,\ 1-c-\frac{\nu}{2}-s\end{matrix}\right]$ [$c\neq 0$ is arbitrary; $s=-\nu/2-k\in D^+$, $k=0, 1, 2, \ldots$]
2	$I_\nu\left(\frac{2}{\sqrt{x}}\right)=\frac{\pi}{\sin c\pi}G^{01}_{31}\left(x\,\middle\vert\,\begin{matrix}1-\nu/2,\ 1+\nu/2,\ 1-c-\nu/2\\ 1-c-\nu/2\end{matrix}\right)$ $\overleftarrow{L_{+\infty}}$	$\frac{\pi}{\sin c\pi}\Gamma\left[\begin{matrix}\nu/2-s\\ s+\nu/2+1,\ s+1-c-\nu/2,\ \nu/2+c-s\end{matrix}\right]$ [$c\neq 0$ is arbitrary; $s=\nu/2+k\in D^-$, $k=0, 1, 2, \ldots$]
3	$e^{-x/2}I_\nu\left(\frac{x}{2}\right)=\frac{1}{\sqrt{\pi}}G^{11}_{12}\left(x\,\middle\vert\,\begin{matrix}1/2\\ \nu,\ -\nu\end{matrix}\right)$	$\frac{1}{\sqrt{\pi}}\Gamma\left[\begin{matrix}s+\nu,\ 1/2-s\\ 1+\nu-s\end{matrix}\right]$ $[-\operatorname{Re}\nu<\operatorname{Re} s<1/2]$
4	$e^{-1/(2x)}I_\nu\left(\frac{1}{2x}\right)=\frac{1}{\sqrt{\pi}}G^{11}_{21}\left(x\,\middle\vert\,\begin{matrix}1-\nu,\ 1+\nu\\ 1/2\end{matrix}\right)$	$\frac{1}{\sqrt{\pi}}\Gamma\left[\begin{matrix}s+1/2,\ \nu-s\\ s+\nu+1\end{matrix}\right]$ $[-1/2<\operatorname{Re} s<\operatorname{Re}\nu]$
5	$e^{x/2}I_\nu\left(\frac{x}{2}\right)=A_{18}G^{11}_{23}\left(x\,\middle\vert\,\begin{matrix}1/2,\ c\\ \nu,\ -\nu,\ c\end{matrix}\right)$ $\overleftarrow{L_{-\infty}}$	$A_{18}\Gamma\left[\begin{matrix}s+\nu,\ 1/2-s\\ s+c,\ 1-c-s,\ 1+\nu-s\end{matrix}\right]$ [$c\neq\nu$ is arbitrary $s=-\nu-k\in D^+$, $s=1/2+k\in D^-$, $k=0, 1, 2, \ldots$]

6	$e^{1/(2x)} I_\nu\left(\frac{1}{2x}\right) = \overset{L_{+\infty}}{\longleftarrow}$ $= A_{18} G_{32}^{11}\left(x \middle\vert \begin{matrix}1-\nu, 1+\nu, 1-c\\ 1/2, 1-c\end{matrix}\right)$	$A_{18}\Gamma\begin{bmatrix} s+1/2, \nu-s\\ s+\nu+1, s+1-c, c-s\end{bmatrix}$ [$c\neq\nu$ is arbitrary; $s=\nu+k\in D^-$, $s=-1/2-k\in D^+$, $k=0, 1, 2, \dots$]
7	$\begin{Bmatrix}\operatorname{sh}\sqrt{x}\\ \operatorname{ch}\sqrt{x}\end{Bmatrix} I_\nu(\sqrt{x}) = \overset{L_{-\infty}}{\longleftarrow}$ $= A_{19}G_{35}^{12}\left(x\middle\vert\begin{matrix}\frac14, \frac34, c\\ \frac{\nu+\delta}{2}, -\frac{\nu}{2}, \frac{1-\nu}{2}, \frac{1+\nu-\delta}{2}, c\end{matrix}\right)$	$A_{19}\Gamma\begin{bmatrix} s+\frac{\nu+\delta}{2}, \frac14-s, \frac34-s\\ s+c, 1-c-s, 1+\frac{\nu}{2}-s, \frac{1+\nu}{2}-s, \frac{1+\delta-\nu}{2}-s\end{bmatrix}$ [$c\neq(\nu+\delta)/2$ is arbitrary; $s=-(\nu+\delta)/2-k\in D^+$, $s=1/4+k$, $3/4+k\in D^-$, $k=0, 1, 2, \dots$]
8	$\begin{Bmatrix}\operatorname{sh}x^{-1/2}\\ \operatorname{ch}x^{-1/2}\end{Bmatrix} I_\nu\left(\frac{1}{\sqrt{x}}\right) = \overset{L_{+\infty}}{\longleftarrow}$ $= A_{19}G_{53}^{21}\left(x\middle\vert\begin{matrix}1-\frac{\nu+\delta}{2}, 1+\frac{\nu}{2}, (1+\nu)/2, (1+\delta-\nu)/2, 1-c\\ \frac14, \frac34, 1-c\end{matrix}\right)$	$A_{19}\Gamma\begin{bmatrix}s+\frac14, s+\frac34, \frac{\nu+\delta}{2}-s\\ s+\frac{\nu}{2}+1, s+\frac{\nu+1}{2}, s+\frac{1+\delta-\nu}{2}, s+1-c, c-s\end{bmatrix}$ [$c\neq(\nu+\delta)/2$ is arbitrary; $s=(\nu+\delta)/2+k\in D^-$, $s=-1/4-k$, $-3/4-k\in D^+$, $k=0, 1, 2, \dots$]
9	$\operatorname{sh}\sqrt{x} I_\nu(\sqrt{x}) - \operatorname{ch}\sqrt{x} I_{-\nu}(\sqrt{x}) =$ $= -\frac{\cos\nu\pi}{\pi\sqrt{2}}\times$ $\times G_{24}^{22}\left(x\middle\vert\begin{matrix}\frac14, \frac34\\ \frac{\nu+1}{2}, -\frac{\nu}{2}, \frac{\nu}{2}, \frac{1-\nu}{2}\end{matrix}\right)$	$-\frac{\cos\nu\pi}{\pi\sqrt{2}}\times$ $\times\Gamma\begin{bmatrix}s+\frac{\nu+1}{2}, s-\frac{\nu}{2}, \frac14-s, \frac34-s\\ 1-\frac{\nu}{2}-s, \frac{1+\nu}{2}-s\end{bmatrix}$ [$-(\operatorname{Re}\nu+1)/2$, $\operatorname{Re}\nu/2<\operatorname{Re}s<1/4$]
10	$\operatorname{sh}\frac{1}{\sqrt{x}} I_\nu\left(\frac{1}{\sqrt{x}}\right) - \operatorname{ch}\frac{1}{\sqrt{x}}\times$ $\times I_{-\nu}\left(\frac{1}{\sqrt{x}}\right) = -\frac{\cos\nu\pi}{\pi\sqrt{2}}\times$ $\times G_{42}^{22}\left(x\middle\vert\begin{matrix}\frac{1-\nu}{2}, 1+\frac{\nu}{2}, 1-\frac{\nu}{2}, \frac{1+\nu}{2}\\ \frac14, \frac34\end{matrix}\right)$	$-\frac{\cos\nu\pi}{\pi\sqrt{2}}\times$ $\times\Gamma\begin{bmatrix}s+\frac14, s+\frac34, -\frac{\nu}{2}-s, \frac{1+\nu}{2}-s\\ s+1-\frac{\nu}{2}, s+\frac{\nu+1}{2}\end{bmatrix}$ [$-1/4<\operatorname{Re}s<(\operatorname{Re}\nu+1)/2$, $-\operatorname{Re}\nu/2$]
11	$J_\nu(2\sqrt[4]{4x}) I_\nu(2\sqrt[4]{4x}) = \overset{L_{-\infty}}{\longleftarrow}$ $= \sqrt{\pi} G_{04}^{10}\left(x\middle\vert\begin{matrix}\cdot & \cdot & \cdot & \cdot\\ \nu/2, & -\nu/2, & 0, & 1/2\end{matrix}\right)$	$\sqrt{\pi}\Gamma\begin{bmatrix}s+\nu/2\\ 1+\nu/2-s, 1/2-s, 1-s\end{bmatrix}$ [$s=-\nu/2-k\in D^+$, $k=0, 1, 2, \dots$]
12	$J_\nu\left(2\sqrt[4]{\frac{4}{x}}\right) I_\nu\left(2\sqrt[4]{\frac{4}{x}}\right) = \overset{L_{+\infty}}{\longleftarrow}$ $= \sqrt{\pi} G_{40}^{01}\left(x\middle\vert\begin{matrix}1-\nu/2, & 1+\nu/2, & 1/2, & 1\\ \cdot & \cdot & \cdot & \cdot\end{matrix}\right)$	$\sqrt{\pi}\Gamma\begin{bmatrix}\nu/2-s\\ s+\nu/2+1, s+1/2, s+1\end{bmatrix}$ [$s=\nu/2+k\in D^-$, $k=0, 1, 2, \dots$]

13	$J_{-\nu}\left(2\sqrt[4]{4x}\right) I_\nu\left(2\sqrt[4]{4x}\right) = \sqrt{\pi}\, G^{20}_{15}\left(x\,\middle\vert\,\begin{matrix}(\nu+1)/2\\ 0, 1/2, -\nu/2, \nu/2, (\nu+1)/2\end{matrix}\right)$ $\xleftarrow{L_{-\infty}}$	$\sqrt{\pi}\,\Gamma\begin{bmatrix} s,\ s+\frac{1}{2} \\ s+\frac{\nu+1}{2},\ 1+\frac{\nu}{2}-s,\ 1-\frac{\nu}{2}-s,\ \frac{1-\nu}{2}-s\end{bmatrix}$ $[s=-k,\ -1/2-k\in D^+,\ k=0, 1, 2, \ldots]$
14	$J_{-\nu}\left(2\sqrt[4]{\frac{4}{x}}\right) I_\nu\left(2\sqrt[4]{\frac{4}{x}}\right) = \sqrt{\pi}\times G^{02}_{51}\left(x\,\middle\vert\,\begin{matrix}1/2, 1, 1+\nu/2, 1-\nu/2, (1-\nu)/2\\ (1-\nu)/2\end{matrix}\right)$ $\xleftarrow{L_{+\infty}}$	$\sqrt{\pi}\,\Gamma\begin{bmatrix} -s,\ \frac{1}{2}-s \\ s+1+\frac{\nu}{2},\ s+1-\frac{\nu}{2},\ s+\frac{1-\nu}{2},\ \frac{1+\nu}{2}-s\end{bmatrix}$ $[s=k,\ 1/2+k\in D^-,\ k=0, 1, 2, \ldots]$
15	$I_\nu^2(\sqrt{x}) = A_{19} G^{11}_{24}\left(x\,\middle\vert\,\begin{matrix}1/2, c\\ \nu, -\nu, 0, c\end{matrix}\right)$ $\xleftarrow{L_{-\infty}}$	$A_{19}\Gamma\begin{bmatrix} s+\nu,\ 1/2-s \\ s+c,\ 1-c-s,\ 1+\nu-s,\ 1-s\end{bmatrix}$ $[c\neq\nu$ is arbitrary; $s=-\nu-k\in D^+,\ s=1/2+k\in D^-,\ k=0, 1, 2, \ldots]$
16	$I_\nu^2\left(\frac{1}{\sqrt{x}}\right) = A_{19} G^{11}_{42}\left(x\,\middle\vert\,\begin{matrix}1-\nu, 1+\nu, 1, 1-c\\ 1/2, 1-c\end{matrix}\right)$ $\xleftarrow{L_{+\infty}}$	$A_{19}\Gamma\begin{bmatrix} s+1/2,\ \nu-s \\ s+1,\ s+\nu+1,\ s+1-c,\ c-s\end{bmatrix}$ $[c\neq\nu$ is arbitrary; $s=\nu+k\in D^-,\ s=-1/2-k\in D^+,\ k=0, 1, 2, \ldots]$
17	$I_{-\nu}(\sqrt{x})\, I_\nu(\sqrt{x}) = A_{20} G^{11}_{24}\left(x\,\middle\vert\,\begin{matrix}1/2, c\\ 0, c, -\nu, \nu\end{matrix}\right)$ $\xleftarrow{L_{-\infty}}$	$A_{20}\Gamma\begin{bmatrix} s,\ 1/2-s \\ s+c,\ 1-c-s,\ 1+\nu-s,\ 1-\nu-s\end{bmatrix}$ $[c\neq 0$ is arbitrary; $s=-k\in D^+,\ s=1/2+k\in D^-,\ k=0, 1, 2, \ldots]$
18	$I_{-\nu}\left(\frac{1}{\sqrt{x}}\right) I_\nu\left(\frac{1}{\sqrt{x}}\right) = A_{20} G^{11}_{42}\left(x\,\middle\vert\,\begin{matrix}1, 1-c, 1+\nu, 1-\nu\\ 1/2, 1-c\end{matrix}\right)$ $\xleftarrow{L_{+\infty}}$	$A_{20}\Gamma\begin{bmatrix} s+1/2,\ -s \\ s+1-c,\ s+\nu+1,\ s-\nu+1,\ c-s\end{bmatrix}$ $[c\neq 0$ is arbitrary; $s=k\in D^-,\ s=-1/2-k\in D^+,\ k=0, 1, 2, \ldots]$
19	$I_{\nu-1}(\sqrt{x})\, I_\nu(\sqrt{x}) = A_{21} G^{11}_{24}\left(x\,\middle\vert\,\begin{matrix}0, c\\ \nu-1/2, c, -1/2, 1/2-\nu\end{matrix}\right)$ $\xleftarrow{L_{-\infty}}$	$A_{21}\Gamma\begin{bmatrix} s+\nu-\frac{1}{2},\ 1-s \\ s+c,\ 1-c-s,\ \frac{3}{2}-s,\ \frac{1}{2}+\nu-s\end{bmatrix}$ $[c\neq\nu-1/2$ is arbitrary; $s=1/2-\nu-k\in D^+,\ s=1+k\in D^-,\ k=0, 1, 2, \ldots]$
20	$I_{\nu-1}\left(\frac{1}{\sqrt{x}}\right) I_\nu\left(\frac{1}{\sqrt{x}}\right) = A_{21} G^{11}_{42}\left(x\,\middle\vert\,\begin{matrix}3/2-\nu, 1-c, 3/2, 1/2+\nu\\ 1, 1-c\end{matrix}\right)$ $\xleftarrow{L_{+\infty}}$	$A_{21}\Gamma\begin{bmatrix} s+1,\ \nu-\frac{1}{2}-s \\ s+\frac{3}{2},\ s+\nu+\frac{1}{2},\ s+1-c,\ c-s\end{bmatrix}$ $[c\neq\nu-1/2$ is arbitrary; $s=\nu-1/2+k\in D^-,\ s=-1-k\in D^+,\ k=0, 1, 2, \ldots]$
21	$I_\mu(\sqrt{x})\, I_\nu(\sqrt{x}) = A_{22} G^{12}_{35}\left(x\,\middle\vert\,\begin{matrix}0, \frac{1}{2}, c\\ \frac{\mu+\nu}{2}, -\frac{\mu+\nu}{2}, \frac{\nu-\mu}{2}, \frac{\mu-\nu}{2}, c\end{matrix}\right)$ $\xleftarrow{L_{-\infty}}$	$A_{22}\Gamma\begin{bmatrix} s+\frac{\mu+\nu}{2},\ \frac{1}{2}-s,\ 1-s \\ s+c,\ 1-c-s,\ 1+\frac{\mu+\nu}{2}-s,\ 1+\frac{\mu-\nu}{2}-s,\ 1+\frac{\nu-\mu}{2}-s\end{bmatrix}$ $[c\neq(\mu+\nu)/2$ is arbitrary; $s=-(\mu+\nu)/2-k\in D^+;\ s=1/2+k,\ 1+k\in D^-,\ k=0, 1, 2, \ldots]$

22	$I_\mu\left(\frac{1}{\sqrt{x}}\right) I_\nu\left(\frac{1}{\sqrt{x}}\right) = \xleftarrow{L_{+\infty}}$ $= A_{22} G^{21}_{53}\left(x \middle\vert \begin{matrix} 1-\frac{\mu+\nu}{2}, 1+\frac{\mu+\nu}{2}, 1+\frac{\mu-\nu}{2}, 1+\frac{\nu-\mu}{2}, 1-c \\ \frac{1}{2}, 1, 1-c \end{matrix}\right)$	$A_{22}\Gamma\left[\begin{matrix} s+\frac{1}{2}, s+1, \frac{\mu+\nu}{2}-s \\ s+\frac{\mu+\nu}{2}+1, s+\frac{\mu-\nu}{2}+1, s+\frac{\nu-\mu}{2}+1, s+1-c, c-s \end{matrix}\right]$ [$c \neq (\mu+\nu)/2$ is arbitrary; $s=(\mu+\nu)/2+k \in D^-$; $s = -1/2-k, -1-k \in D^+$, $k=0, 1, 2, \ldots$]
23	$I^2_{-\nu}(\sqrt{x}) - I^2_{\nu}(\sqrt{x}) = \frac{\sin 2\nu\pi}{\pi^{3/2}} G^{21}_{13}\left(x \middle\vert \begin{matrix} 1/2 \\ \nu, -\nu, 0 \end{matrix}\right)$	$\frac{\sin 2\nu\pi}{\pi^{3/2}} \Gamma\left[\begin{matrix} s+\nu, s-\nu, 1/2-s \\ 1-s \end{matrix}\right]$ [$\vert \mathrm{Re}\,\nu \vert < \mathrm{Re}\,s < 1/2$]
24	$I^2_{-\nu}\left(\frac{1}{\sqrt{x}}\right) - I^2_{\nu}\left(\frac{1}{\sqrt{x}}\right) = \frac{\sin 2\nu\pi}{\pi^{3/2}} G^{12}_{31}\left(x \middle\vert \begin{matrix} 1-\nu, 1+\nu, 1 \\ 1/2 \end{matrix}\right)$	$\frac{\sin 2\nu\pi}{\pi^{3/2}} \Gamma\left[\begin{matrix} s+1/2, \nu-s, -\nu-s \\ s+1 \end{matrix}\right]$ [$-1/2 < \mathrm{Re}\,s < -\vert \mathrm{Re}\,\nu \vert$]
25	$I_{-\mu}(\sqrt{x})\, I_{-\nu}(\sqrt{x}) - I_\mu(\sqrt{x}) \times I_\nu(\sqrt{x}) = A_{23} \times G^{22}_{24}\left(x \middle\vert \begin{matrix} 0, \frac{1}{2} \\ \frac{\mu+\nu}{2}, -\frac{\mu+\nu}{2}, \frac{\nu-\mu}{2}, \frac{\mu-\nu}{2} \end{matrix}\right)$	$A_{23}\Gamma\left[\begin{matrix} s+\frac{\mu+\nu}{2}, s-\frac{\mu+\nu}{2}, \frac{1}{2}-s, 1-s \\ 1+\frac{\mu-\nu}{2}-s, 1+\frac{\nu-\mu}{2}-s \end{matrix}\right]$ [$\vert \mathrm{Re}(\mu+\nu)\vert/2 < \mathrm{Re}\,s < 1/2$]
26	$I_{-\mu}\left(\frac{1}{\sqrt{x}}\right) I_{-\nu}\left(\frac{1}{\sqrt{x}}\right) - I_\mu\left(\frac{1}{\sqrt{x}}\right) I_\nu\left(\frac{1}{\sqrt{x}}\right) = A_{23} \times G^{22}_{42}\left(x \middle\vert \begin{matrix} 1-\frac{\mu+\nu}{2}, 1+\frac{\mu+\nu}{2}, 1+\frac{\mu-\nu}{2}, 1+\frac{\nu-\mu}{2} \\ \frac{1}{2}, 1 \end{matrix}\right)$	$A_{23}\Gamma\left[\begin{matrix} s+\frac{1}{2}, s+1, \frac{\mu+\nu}{2}-s, -\frac{\mu+\nu}{2}-s \\ s+\frac{\mu-\nu}{2}+1, s+\frac{\nu-\mu}{2}+1 \end{matrix}\right]$ [$-1/2 < \mathrm{Re}\,s < -\vert \mathrm{Re}\,(\mu+\nu)\vert/2$]

8.4.23. The MacDonald function $K_\nu(x)$.

NOTATION: $A_{24} = \frac{1}{2\sqrt{\pi}}\left\{\begin{matrix}\cos(\nu\pi/2) \\ \sin(\nu\pi/2)\end{matrix}\right\}$, $A_{25} = \frac{1}{\sqrt{\pi}}\left\{\begin{matrix}\cos[(\mu-\nu)\pi/2] \\ \sin[(\mu-\nu)\pi/2]\end{matrix}\right\}$, $\delta = \left\{\begin{matrix}1 \\ 0\end{matrix}\right\}$.

1	$K_\nu(2\sqrt{x}) = \frac{1}{2} G^{20}_{02}\left(x \middle\vert \begin{matrix} \cdot \quad \cdot \\ \nu/2, -\nu/2 \end{matrix}\right)$	$\frac{1}{2}\Gamma\left[s+\frac{\nu}{2}, s-\frac{\nu}{2}\right]$ [$\mathrm{Re}\,s > \vert \mathrm{Re}\,\nu \vert/2$]
2	$K_\nu\left(\frac{2}{\sqrt{x}}\right) = \frac{1}{2} G^{02}_{20}\left(x \middle\vert \begin{matrix} 1-\nu/2, 1+\nu/2 \\ \cdot \quad \cdot \end{matrix}\right)$	$\frac{1}{2}\Gamma\left[\frac{\nu}{2}-s, -\frac{\nu}{2}-s\right]$ [$\mathrm{Re}\,s < -\vert \mathrm{Re}\,\nu \vert/2$]

3	$e^{-x/2}K_\nu\left(\frac{x}{2}\right)=\sqrt{\pi}\,G_{12}^{20}\left(x\,\middle\vert\,\begin{matrix}1/2\\ \nu,\ -\nu\end{matrix}\right)$	$\sqrt{\pi}\,\Gamma\begin{bmatrix}s+\nu,\ s-\nu\\ s+1/2\end{bmatrix}$ [Re $s>\vert\text{Re}\,\nu\vert$]
4	$e^{-1/(2x)}K_\nu\left(\frac{1}{2x}\right)=\sqrt{\pi}\,G_{21}^{02}\left(x\,\middle\vert\,\begin{matrix}1-\nu,\ 1+\nu\\ 1/2\end{matrix}\right)$	$\sqrt{\pi}\,\Gamma\begin{bmatrix}\nu-s,\ -\nu-s\\ 1/2-s\end{bmatrix}$ [Re $s<-\vert\text{Re}\,\nu\vert$]
5	$e^{x/2}K_\nu\left(\frac{x}{2}\right)=\frac{\cos\nu\pi}{\sqrt{\pi}}\,G_{12}^{21}\left(x\,\middle\vert\,\begin{matrix}1/2\\ \nu,\ -\nu\end{matrix}\right)$	$\frac{\cos\nu\pi}{\sqrt{\pi}}\,\Gamma\left[s+\nu,\ s-\nu,\ \frac{1}{2}-s\right]$ [$\vert\text{Re}\,\nu\vert<\text{Re}\,s<1/2$]
6	$e^{1/(2x)}K_\nu\left(\frac{1}{2x}\right)=\frac{\cos\nu\pi}{\sqrt{\pi}}\,G_{21}^{12}\left(x\,\middle\vert\,\begin{matrix}1-\nu,\ 1+\nu\\ 1/2\end{matrix}\right)$	$\frac{\cos\nu\pi}{\sqrt{\pi}}\,\Gamma\left[s+\frac{1}{2},\ \nu-s,\ -\nu-s\right]$ [$-1/2<\text{Re}\,s<-\vert\text{Re}\,\nu\vert$]
7	$\left\{\begin{matrix}\text{sh}\,\sqrt{x}\\ \text{ch}\,\sqrt{x}\end{matrix}\right\}K_\nu(\sqrt{x})=\frac{1}{2\sqrt{2}}\times G_{24}^{22}\left(x\,\middle\vert\,\begin{matrix}1/4,\ 3/4\\ \frac{\delta+\nu}{2},\ \frac{\delta-\nu}{2},\ \frac{1-\delta+\nu}{2},\ \frac{1-\delta-\nu}{2}\end{matrix}\right)$	$\frac{1}{2\sqrt{2}}\times\Gamma\begin{bmatrix}s+\frac{\delta+\nu}{2},\ s+\frac{\delta-\nu}{2},\ \frac{1}{4}-s,\ \frac{3}{4}-s\\ \frac{1+\delta-\nu}{2}-s,\ \frac{1+\delta+\nu}{2}-s\end{bmatrix}$ [$(\vert\text{Re}\,\nu\vert-\delta)/2<\text{Re}\,s<1/4$]
8	$\left\{\begin{matrix}\text{sh}\,x^{-1/2}\\ \text{ch}\,x^{-1/2}\end{matrix}\right\}K_\nu\left(\frac{1}{\sqrt{x}}\right)=\frac{1}{2\sqrt{2}}\times G_{42}^{22}\left(x\,\middle\vert\,\begin{matrix}1-\frac{\delta+\nu}{2},\ 1-\frac{\delta-\nu}{2},\ \frac{1+\delta-\nu}{2},\ \frac{1+\delta+\nu}{2}\\ \frac{1}{4},\ \frac{3}{4}\end{matrix}\right)$	$\frac{1}{2\sqrt{2}}\,\Gamma\begin{bmatrix}s+1/4,\ s+3/4,\ (\delta+\nu)/2-s,\ (\delta-\nu)/2-s\\ s+(1+\delta-\nu)/2,\ s+(1+\delta+\nu)/2\end{bmatrix}$ [$-1/4<\text{Re}\,s<(\delta-\vert\text{Re}\,\nu\vert)/2$]
9	$J_\nu(2\sqrt{2}x^{1/4})\,K_\nu(2\sqrt{2}x^{1/4})=\frac{1}{4\sqrt{\pi}}\times G_{04}^{30}\left(x\,\middle\vert\,\begin{matrix}\cdot\ \cdot\ \cdot\ \cdot\ \cdot\ \cdot\\ 0,\ 1/2,\ \nu/2,\ -\nu/2\end{matrix}\right)$	$\frac{1}{4\sqrt{\pi}}\,\Gamma\begin{bmatrix}s,\ s+1/2,\ s+\nu/2\\ \nu/2+1-s\end{bmatrix}$ [Re $s>0$, $-\text{Re}\,\nu/2$]
10	$J_\nu\left(2\sqrt[4]{\frac{4}{x}}\right)K_\nu\left(2\sqrt[4]{\frac{4}{x}}\right)=\frac{1}{4\sqrt{\pi}}\,G_{40}^{03}\left(x\,\middle\vert\,\begin{matrix}1/2,\ 1,\ 1-\nu/2,\ 1+\nu/2\\ \cdot\ \cdot\ \cdot\ \cdot\ \cdot\ \cdot\end{matrix}\right)$	$\frac{1}{4\sqrt{\pi}}\,\Gamma\begin{bmatrix}-s,\ 1/2-s,\ \nu/2-s\\ s+\nu/2+1\end{bmatrix}$ [Re $s<0$, Re $\nu/2$]
11	$[J_{-\nu}(2\sqrt{2}x^{1/4})\pm J_\nu(2\sqrt{2}x^{1/4})]\times K_\nu(2\sqrt{2}x^{1/4})=A_{24}G_{04}^{30}\left(x\,\middle\vert\,\begin{matrix}\cdot\ \cdot\ \cdot\ \cdot\ \cdot\ \cdot\\ \delta/2,\ \nu/2,\ -\nu/2,\ (1-\delta)/2\end{matrix}\right)$	$A_{24}\Gamma\begin{bmatrix}s+\delta/2,\ s+\nu/2,\ s-\nu/2\\ (1+\delta)/2-s\end{bmatrix}$ [Re $s>\vert\text{Re}\,\nu\vert/2$]
12	$\left[J_{-\nu}\left(2\sqrt[4]{\frac{4}{x}}\right)\pm J_\nu\left(2\sqrt[4]{\frac{4}{x}}\right)\right]\times K_\nu\left(2\sqrt[4]{\frac{4}{x}}\right)=A_{24}\times G_{40}^{03}\left(x\,\middle\vert\,\begin{matrix}1-\frac{\delta}{2},\ 1-\frac{\nu}{2},\ 1+\frac{\nu}{2},\ \frac{1+\delta}{2}\\ \cdot\ \cdot\ \cdot\ \cdot\ \cdot\ \cdot\end{matrix}\right)$	$A_{24}\Gamma\begin{bmatrix}\delta/2-s,\ \nu/2-s,\ -\nu/2-s\\ s+(1+\delta)/2\end{bmatrix}$ [Re $s<-\vert\text{Re}\,\nu\vert/2$]

13	$J_1\left(2\sqrt[4]{4x}\right)K_0\left(2\sqrt[4]{4x}\right)+J_0\left(2\sqrt[4]{4x}\right)K_1\left(2\sqrt[4]{4x}\right)=\frac{1}{2\sqrt{2\pi}}G^{30}_{04}\left(x\middle\vert\begin{matrix}\cdot\;\cdot\;\cdot\;\cdot\;\cdot\;\cdot\\-1/4,\,1/4,\,3/4,\,-1/4\end{matrix}\right)$	$\frac{1}{2\sqrt{2\pi}}\Gamma\left[\begin{matrix}s-1/4,\; s+1/4,\; s+3/4\\5/4-s\end{matrix}\right]$ [Re $s>1/4$]
14	$J_1\left(2\sqrt[4]{\frac{4}{x}}\right)K_0\left(2\sqrt[4]{\frac{4}{x}}\right)+J_0\left(2\sqrt[4]{\frac{4}{x}}\right)K_1\left(2\sqrt[4]{\frac{4}{x}}\right)=\frac{1}{2\sqrt{2\pi}}G^{03}_{40}\left(x\middle\vert\begin{matrix}1/4,\,3/4,\,5/4,\,5/4\\\cdot\;\cdot\;\cdot\;\cdot\;\cdot\;\cdot\end{matrix}\right)$	$\frac{1}{2\sqrt{2\pi}}\Gamma\left[\begin{matrix}-1/4-s,\; 1/4-s,\; 3/4-s\\s+5/4\end{matrix}\right]$ [Re $s<-1/4$]
15	$\frac{\pi}{2}Y_0\left(4\sqrt[4]{x}\right)-K_0\left(4\sqrt[4]{x}\right)=-\frac{\pi}{2}G^{20}_{04}\left(x\middle\vert\begin{matrix}\cdot\;\cdot\;\cdot\;\cdot\;\cdot\\0,\,0,\,1/2,\,1/2\end{matrix}\right)$	$-\frac{\pi}{2}\Gamma\left[\begin{matrix}s,\; s\\1/2-s,\; 1/2-s\end{matrix}\right]$ [$0<$ Re $s<3/8$]
16	$\frac{\pi}{2}Y_0\left(\frac{4}{\sqrt[4]{x}}\right)-K_0\left(\frac{4}{\sqrt[4]{x}}\right)=-\frac{\pi}{2}G^{02}_{40}\left(x\middle\vert\begin{matrix}1,\,1,\,1/2,\,1/2\\\cdot\;\cdot\;\cdot\;\cdot\;\cdot\end{matrix}\right)$	$-\frac{\pi}{2}\Gamma\left[\begin{matrix}-s,\; -s\\s+1/2,\; s+1/2\end{matrix}\right]$ [$-3/8<$ Re $s<0$]
17	$Y_\nu\left(2\sqrt{2}x^{1/4}\right)K_\nu\left(2\sqrt{2}x^{1/4}\right)=-\frac{1}{4\sqrt{\pi}}\times G^{40}_{15}\left(x\middle\vert\begin{matrix}(1-\nu)/2\\0,\,1/2,\,\nu/2,\,-\nu/2,\,(1-\nu)/2\end{matrix}\right)$	$-\frac{1}{4\sqrt{\pi}}\Gamma\left[\begin{matrix}s,\; s+1/2,\; s+\nu/2,\; s-\nu/2\\s+(1-\nu)/2,\; (1+\nu)/2-s\end{matrix}\right]$ [Re $s>\vert$Re $\nu\vert/2$]
18	$Y_\nu\left(2\sqrt[4]{\frac{4}{x}}\right)K_\nu\left(2\sqrt[4]{\frac{4}{x}}\right)=-\frac{1}{4\sqrt{\pi}}\times G^{04}_{51}\left(x\middle\vert\begin{matrix}\frac{1}{2},\,1,\,1-\frac{\nu}{2},\,1+\frac{\nu}{2},\,\frac{1+\nu}{2}\\\frac{1+\nu}{2}\end{matrix}\right)$	$-\frac{1}{4\sqrt{\pi}}\times\Gamma\left[\begin{matrix}-s,\; 1/2-s,\; \nu/2-s,\; -\nu/2-s\\s+(1+\nu)/2,\; (1-\nu)/2-s\end{matrix}\right]$ [Re $s<-\vert$Re $\nu\vert/2$]
19	$I_\nu(\sqrt{x})K_\nu(\sqrt{x})=\frac{1}{2\sqrt{\pi}}\times G^{21}_{13}\left(x\middle\vert\begin{matrix}1/2\\0,\,\nu,\,-\nu\end{matrix}\right)$	$\frac{1}{2\sqrt{\pi}}\Gamma\left[\begin{matrix}s,\; s+\nu,\; 1/2-s\\1+\nu-s\end{matrix}\right]$ [$0,\ -$Re $\nu<$ Re $s<1/2$]
20	$I_\nu\left(\frac{1}{\sqrt{x}}\right)K_\nu\left(\frac{1}{\sqrt{x}}\right)=\frac{1}{2\sqrt{\pi}}\times G^{12}_{31}\left(x\middle\vert\begin{matrix}1,\,1-\nu,\,1+\nu\\1/2\end{matrix}\right)$	$\frac{1}{2\sqrt{\pi}}\Gamma\left[\begin{matrix}s+1/2,\; -s,\; \nu-s\\s+\nu+1\end{matrix}\right]$ [$-1/2<$ Re $s<0$, Re ν]
21	$I_\mu(\sqrt{x})K_\nu(\sqrt{x})=\frac{1}{2\sqrt{\pi}}\times G^{22}_{24}\left(x\middle\vert\begin{matrix}0,\,\frac{1}{2}\\\frac{\mu+\nu}{2},\,\frac{\mu-\nu}{2},\,\frac{\nu-\mu}{2},\,-\frac{\mu+\nu}{2}\end{matrix}\right)$	$\frac{1}{2\sqrt{\pi}}\Gamma\left[\begin{matrix}s+(\mu+\nu)/2,\; 1+(\mu-\nu)/2-s,\; s+(\mu-\nu)/2,\; 1/2-s,\; 1-s\\1+(\mu+\nu)/2-s\end{matrix}\right]$ [$(\vert$Re $\nu\vert-$Re $\mu)/2<$ Re $s<1/2$]

22	$I_\mu\left(\frac{1}{\sqrt{x}}\right)K_\nu\left(\frac{1}{\sqrt{x}}\right)=\frac{1}{2\sqrt{\pi}}G_{42}^{22}\left(x\left\vert\begin{matrix}1-\frac{\mu+\nu}{2}, 1+\frac{\nu-\mu}{2}, 1+\frac{\mu-\nu}{2}, 1+\frac{\mu+\nu}{2}\\ \frac{1}{2}, 1\end{matrix}\right.\right)$	$\frac{1}{2\sqrt{\pi}}\Gamma\left[\begin{matrix}s+1/2, s+1, (\mu+\nu)/2-s, (\mu-\nu)/2-s\\ s+(\mu-\nu)/2+1, s+(\mu+\nu)/2+1\end{matrix}\right]$ [$-1/2<\operatorname{Re}s<(\operatorname{Re}\mu-\vert\operatorname{Re}\nu\vert)/2$]
23	$[I_{-\nu}(\sqrt{x})+I_\nu(\sqrt{x})]K_\nu(\sqrt{x})=\frac{\cos\nu\pi}{\sqrt{\pi}}G_{13}^{21}\left(x\left\vert\begin{matrix}1/2\\ \nu, -\nu, 0\end{matrix}\right.\right)$	$\frac{\cos\nu\pi}{\sqrt{\pi}}\Gamma\left[\begin{matrix}s+\nu, s-\nu, 1/2-s\\ 1-s\end{matrix}\right]$ [$\vert\operatorname{Re}\nu\vert<\operatorname{Re}s<1/2$]
24	$\left[I_{-\nu}\left(\frac{1}{\sqrt{x}}\right)+I_\nu\left(\frac{1}{\sqrt{x}}\right)\right]\times K_\nu\left(\frac{1}{\sqrt{x}}\right)=\frac{\cos\nu\pi}{\sqrt{\pi}}G_{31}^{12}\left(x\left\vert\begin{matrix}1+\nu, 1-\nu, 1\\ 1/2\end{matrix}\right.\right)$	$\frac{\cos\nu\pi}{\sqrt{\pi}}\Gamma\left[\begin{matrix}s+1/2, \nu-s, -\nu-s\\ s+1\end{matrix}\right]$ [$-1/2<\operatorname{Re}s<-\vert\operatorname{Re}\nu\vert$]
25	$I_\nu(\sqrt{x})K_\mu(\sqrt{x})\pm I_\mu(\sqrt{x})\times K_\nu(\sqrt{x})=A_{25}\times G_{24}^{31}\left(x\left\vert\begin{matrix}\frac{\delta}{2}, \frac{1-\delta}{2}\\ \frac{\mu+\nu}{2}, \frac{\nu-\mu}{2}, \frac{\mu-\nu}{2}, -\frac{\mu+\nu}{2}\end{matrix}\right.\right)$	$A_{25}\Gamma\left[\begin{matrix}s+\frac{\mu+\nu}{2}, s+\frac{\nu-\mu}{2}, s+\frac{\mu-\nu}{2}, 1-\frac{\delta}{2}-s\\ s+\frac{1-\delta}{2}, 1+\frac{\mu+\nu}{2}-s\end{matrix}\right]$ [$\operatorname{Re}(\mu-\nu)/2, (\vert\operatorname{Re}\nu\vert-\operatorname{Re}\mu)/2<\operatorname{Re}s<1-\delta/2$]
26	$I_\nu\left(\frac{1}{\sqrt{x}}\right)K_\mu\left(\frac{1}{\sqrt{x}}\right)\pm I_\mu\left(\frac{1}{\sqrt{x}}\right)\times K_\nu\left(\frac{1}{\sqrt{x}}\right)=A_{25}\times G_{42}^{13}\left(x\left\vert\begin{matrix}1-\frac{\mu+\nu}{2}, 1+\frac{\mu-\nu}{2}, 1+\frac{\nu-\mu}{2}, 1+\frac{\mu+\nu}{2}\\ 1-\frac{\delta}{2}, \frac{1+\delta}{2}\end{matrix}\right.\right)$	$A_{25}\Gamma\left[\begin{matrix}s+1-\frac{\delta}{2}, \frac{\mu+\nu}{2}-s, \frac{\nu-\mu}{2}-s, \frac{\mu-\nu}{2}-s\\ s+\frac{\mu+\nu}{2}+1, \frac{1-\delta}{2}-s\end{matrix}\right]$ [$\delta/2-1<\operatorname{Re}s<\operatorname{Re}(\nu-\mu)/2, (\operatorname{Re}\mu-\vert\operatorname{Re}\nu\vert)/2$]
27	$K_\nu^2(\sqrt{x})=\frac{\sqrt{\pi}}{2}G_{13}^{30}\left(x\left\vert\begin{matrix}1/2\\ 0, \nu, -\nu\end{matrix}\right.\right)$	$\frac{\sqrt{\pi}}{2}\Gamma\left[\begin{matrix}s, s+\nu, s-\nu\\ s+1/2\end{matrix}\right]$ [$\operatorname{Re}s>\vert\operatorname{Re}\nu\vert$]
28	$K_\nu^2\left(\frac{1}{\sqrt{x}}\right)=\frac{\sqrt{\pi}}{2}G_{31}^{03}\left(x\left\vert\begin{matrix}1, 1-\nu, 1+\nu\\ 1/2\end{matrix}\right.\right)$	$\frac{\sqrt{\pi}}{2}\Gamma\left[\begin{matrix}-s, \nu-s, -\nu-s\\ 1/2-s\end{matrix}\right]$ [$\operatorname{Re}s<-\vert\operatorname{Re}\nu\vert$]
29	$K_{\nu-1}(\sqrt{x})K_\nu(\sqrt{x})=\frac{\sqrt{\pi}}{2}G_{13}^{30}\left(x\left\vert\begin{matrix}0\\ -1/2, \nu-1/2, -\nu+1/2\end{matrix}\right.\right)$	$\frac{\sqrt{\pi}}{2}\Gamma\left[\begin{matrix}s+\nu-1/2, s-\nu+1/2, s-1/2\\ s\end{matrix}\right]$ [$\operatorname{Re}s>1/2, \vert\operatorname{Re}\nu-1/2\vert$]
30	$K_{\nu-1}\left(\frac{1}{\sqrt{x}}\right)K_\nu\left(\frac{1}{\sqrt{x}}\right)=$	$\frac{\sqrt{\pi}}{2}\Gamma\left[\begin{matrix}-1/2-s, \nu-1/2-s, 1/2-\nu-s\\ -s\end{matrix}\right]$ [$\operatorname{Re}s<-1/2, -\vert\operatorname{Re}\nu-1/2\vert$]

	$=\frac{\sqrt{\pi}}{2} G_{31}^{03}\left(x \,\middle|\, \begin{matrix} 3/2,\ 3/2-\nu,\ 1/2+\nu \\ 1 \end{matrix}\right)$	
31	$K_\mu(\sqrt{x})\,K_\nu(\sqrt{x})=\frac{\sqrt{\pi}}{2}\times G_{24}^{40}\left(x \,\middle|\, \begin{matrix} 0,\ \frac{1}{2} \\ \frac{\mu+\nu}{2}, \frac{\mu-\nu}{2}, \frac{\nu-\mu}{2}, -\frac{\mu+\nu}{2} \end{matrix}\right)$	$\frac{\sqrt{\pi}}{2}\Gamma\left[\begin{matrix} s+\frac{\mu+\nu}{2}, s+\frac{\mu-\nu}{2}, s+\frac{\nu-\mu}{2}, s-\frac{\mu+\nu}{2} \\ s, s+\frac{1}{2} \end{matrix}\right]$ [Re s > (| Re μ |+| Re ν |)/2]
32	$K_\mu\left(\frac{1}{\sqrt{x}}\right)K_\nu\left(\frac{1}{\sqrt{x}}\right)=\frac{\sqrt{\pi}}{2}\times G_{42}^{04}\left(x \,\middle|\, \begin{matrix} 1-\frac{\mu+\nu}{2}, 1+\frac{\nu-\mu}{2}, 1+\frac{\mu-\nu}{2}, 1+\frac{\mu+\nu}{2} \\ \frac{1}{2}, 1 \end{matrix}\right)$	$\frac{\sqrt{\pi}}{2}\Gamma\left[\begin{matrix} \frac{\mu+\nu}{2}-s, \frac{\mu-\nu}{2}-s, \frac{\nu-\mu}{2}-s, -\frac{\mu+\nu}{2}-s \\ -s, \frac{1}{2}-s \end{matrix}\right]$ [Re s<−(| Re μ |+| Re ν |)/2]
33	$K_\nu(\sqrt{i}\,2\sqrt[4]{4x})\,K_\nu(\sqrt{-i}\,2\sqrt[4]{4x})=\frac{1}{8\sqrt{\pi}}G_{04}^{40}\left(x \,\middle|\, \begin{matrix} \cdot\ \cdot\ \cdot\ \cdot\ \cdot\ \cdot\ \cdot \\ 0, 1/2, \nu/2, -\nu/2 \end{matrix}\right)$	$\frac{1}{8\sqrt{\pi}}\Gamma\left[s, s+\frac{1}{2}, s+\frac{\nu}{2}, s-\frac{\nu}{2}\right]$ [Re s > | Re ν |/2]
34	$K_\nu\left(\sqrt{i}\,2\sqrt[4]{\frac{4}{x}}\right)\times K_\nu\left(-i2\sqrt[4]{\frac{4}{x}}\right)=\frac{1}{8\sqrt{\pi}}\times G_{40}^{04}\left(x \,\middle|\, \begin{matrix} 1/2, 1, 1-\nu/2, 1+\nu/2 \\ \cdot\ \cdot\ \cdot\ \cdot\ \cdot\ \cdot\ \cdot\ \cdot\ \cdot \end{matrix}\right)$	$\frac{1}{8\sqrt{\pi}}\Gamma\left[-s, \frac{1}{2}-s, \frac{\nu}{2}-s, -\frac{\nu}{2}-s\right]$ [Re s <−| Re ν |/2]

8.4.24. The integral Bessel functions $Ji_\nu(x)$, $Yi_\nu(x)$ and $Ki_\nu(x)$.

1	$Ji_\nu(2\sqrt{x})=\frac{1}{2}G_{13}^{20}\left(x \,\middle|\, \begin{matrix} 1 \\ 0, \nu/2, -\nu/2 \end{matrix}\right)$	$\frac{1}{2}\Gamma\left[\begin{matrix} s, s+\nu/2 \\ s+1, 1+\nu/2-s \end{matrix}\right]=\frac{1}{2s}\Gamma\left[\begin{matrix} s+\nu/2 \\ 1+\nu/2-s \end{matrix}\right]$ [Re s > 0, −Re ν/2]
2	$Ji_\nu\left(\frac{2}{\sqrt{x}}\right)=\frac{1}{2}G_{31}^{02}\left(x \,\middle|\, \begin{matrix} 1, 1-\nu/2, 1+\nu/2 \\ 0 \end{matrix}\right)$	$\frac{1}{2}\Gamma\left[\begin{matrix} \nu/2-s, -s \\ s+\nu/2+1, 1-s \end{matrix}\right]=-\frac{1}{2s}\Gamma\left[\begin{matrix} \nu/2-s \\ s+\nu/2+1 \end{matrix}\right]$ [Re s < 0, Re ν/2]
3	$Yi_\nu(2\sqrt{x})=\frac{1}{2}G_{24}^{30}\left(x \,\middle|\, \begin{matrix} 1, -(\nu+1)/2 \\ 0, \nu/2, -\nu/2, -(1+\nu)/2 \end{matrix}\right)$	$\frac{1}{2}\Gamma\left[\begin{matrix} s, s+\nu/2, s-\nu/2 \\ s+1, s-(\nu+1)/2, (3+\nu)/2-s \end{matrix}\right]$ [Re s > 0]
4	$Yi_\nu\left(\frac{2}{\sqrt{x}}\right)=\frac{1}{2}\times G_{42}^{03}\left(x \,\middle|\, \begin{matrix} 1, 1-\nu/2, 1+\nu/2, (3+\nu)/2 \\ 0, (3+\nu)/2 \end{matrix}\right)$	$\frac{1}{2}\Gamma\left[\begin{matrix} \nu/2-s, -\nu/2-s, -s \\ s+(\nu+3)/2, -(1+\nu)/2-s, 1-s \end{matrix}\right]$ [Re s < 0]

5	$Ki_\nu(2\sqrt{x})=\frac{1}{4}G_{13}^{30}\left(x\,\middle\vert\,{1 \atop 0,\,\nu/2,\,-\nu/2}\right)$	$\frac{1}{4}\Gamma\left[{s,\ s+\nu/2,\ s-\nu/2 \atop s+1}\right]$ [Re $s>0$]
6	$Ki_\nu\left(\frac{2}{\sqrt{x}}\right)=\frac{1}{4}G_{31}^{03}\left(x\,\middle\vert\,{1,\ 1-\nu/2,\ 1+\nu/2 \atop 0}\right)$	$\frac{1}{4}\Gamma\left[{\nu/2-s,\ -\nu/2-s,\ -s \atop 1-s}\right]$ [Re $s<0$]

8.4.25. The Struve functions $\mathbf{H}_\nu(x)$ and $\mathbf{L}_\nu(x)$.

NOTATION: $A_{26}=-\frac{\cos\nu\pi}{\pi^2}$, $A_{27}=-\pi\sec\left(c-\frac{\nu}{2}\right)\pi$, $A_{28}=\frac{1}{\pi}\begin{Bmatrix}1\\ \cos\nu\pi\end{Bmatrix}$.

1	$\mathbf{H}_\nu(2\sqrt{x})=G_{13}^{11}\left(x\,\middle\vert\,{(\nu+1)/2 \atop (\nu+1)/2,\,-\nu/2,\,\nu/2}\right)$	$\Gamma\left[{s+(1+\nu)/2,\ (1-\nu)/2-s \atop 1+\nu/2-s,\ 1-\nu/2-s}\right]$ [$-(1+\mathrm{Re}\,\nu)/2<\mathrm{Re}\,s<(1-\mathrm{Re}\,\nu)/2,\ 3/4$]
2	$\mathbf{H}_\nu\left(\frac{2}{\sqrt{x}}\right)=G_{31}^{11}\left(x\,\middle\vert\,{(1-\nu)/2,\ 1+\nu/2,\ 1-\nu/2 \atop (1-\nu)/2}\right)$	$\Gamma\left[{s+(1-\nu)/2,\ (1+\nu)/2-s \atop s+1+\nu/2,\ s+1-\nu/2}\right]$ [$-3/4,\ (\mathrm{Re}\,\nu-1)/2<\mathrm{Re}\,s<(\mathrm{Re}\,\nu+1)/2$]
3	$Y_\nu(2\sqrt{x})-\mathbf{H}_\nu(2\sqrt{x})=A_{26}G_{13}^{31}\left(x\,\middle\vert\,{(\nu+1)/2 \atop (\nu+1)/2,\ \nu/2,\ -\nu/2}\right)$	$A_{26}\times\Gamma\left[s+\frac{\nu}{2},\ s+\frac{\nu+1}{2},\ s-\frac{\nu}{2},\ \frac{1-\nu}{2}-s\right]$ [$\lvert\mathrm{Re}\,\nu\rvert/2<\mathrm{Re}\,s<(1-\mathrm{Re}\,\nu)/2$]
4	$Y_\nu\left(\frac{2}{\sqrt{x}}\right)-\mathbf{H}_\nu\left(\frac{2}{\sqrt{x}}\right)=A_{26}\times G_{31}^{13}\left(x\,\middle\vert\,{(1-\nu)/2,\ 1+\nu/2,\ 1-\nu/2 \atop (1-\nu)/2}\right)$	$A_{26}\times\Gamma\left[s+\frac{1-\nu}{2},\ \frac{\nu}{2}-s,\ \frac{\nu+1}{2}-s,\ -\frac{\nu}{2}-s\right]$ [$(\mathrm{Re}\,\nu-1)/2<\mathrm{Re}\,s<-\lvert\mathrm{Re}\,\nu\rvert/2$]
5	$\mathbf{L}_\nu(2\sqrt{x})=A_{27}G_{24}^{11}\left(x\,\middle\vert\,{(\nu+1)/2,\ c \atop (\nu+1)/2,\ c,\ \nu/2,\ -\nu/2}\right)$ $\overleftarrow{L_{-\infty}}$	$A_{27}\times\Gamma\left[{s+\frac{\nu+1}{2},\ \frac{1-\nu}{2}-s \atop s+c,\ 1-c-s,\ 1-\frac{\nu}{2}-s,\ 1+\frac{\nu}{2}-s}\right]$ [$c\neq(\nu+1)/2$ is arbitrary; $s=-(\nu+1)/2-k\in D^+$, $s=(1-\nu)/2+k\in D^-$, $k=0,1,2,\ldots$]
6	$\mathbf{L}_\nu\left(\frac{2}{\sqrt{x}}\right)=A_{27}\times G_{42}^{11}\left(x\,\middle\vert\,{\frac{1-\nu}{2},\ 1-c,\ 1-\frac{\nu}{2},\ 1+\frac{\nu}{2} \atop \frac{1-\nu}{2},\ 1-c}\right)$ $\overleftarrow{L_{+\infty}}$	$A_{27}\times\Gamma\left[{s+\frac{1-\nu}{2},\ \frac{1+\nu}{2}-s \atop s+1-\frac{\nu}{2},\ s+1+\frac{\nu}{2},\ s+1-c,\ c-s}\right]$ [$c\neq(\nu+1)/2$ is arbitrary; $s=(\nu+1)/2+k\in D^-$, $s=(\nu-1)/2-k\in D^+$, $k=0,1,2,\ldots$]
7	$I_{\pm\nu}(2\sqrt{x})-\mathbf{L}_\nu(2\sqrt{x})=A_{28}G_{13}^{21}\left(x\,\middle\vert\,{(\nu+1)/2 \atop (\nu+1)/2,\ \pm\nu/2,\ \mp\nu/2}\right)$	$A_{28}\Gamma\left[{s+(\nu+1)/2,\ s\pm\nu/2,\ (1-\nu)/2-s \atop 1\pm\nu/2-s}\right]$ [$\begin{Bmatrix}-\mathrm{Re}\,\nu/2\\ -(\mathrm{Re}\,\nu+1)/2,\ \mathrm{Re}\,\nu/2\end{Bmatrix}<\mathrm{Re}\,s<(1-\mathrm{Re}\,\nu)/2$]
8	$I_{\pm\nu}\left(\frac{2}{\sqrt{x}}\right)-\mathbf{L}_\nu\left(\frac{2}{\sqrt{x}}\right)=A_{28}\times G_{31}^{12}\left(x\,\middle\vert\,{(1-\nu)/2,\ 1\mp\nu/2,\ 1\pm\nu/2 \atop (1-\nu)/2}\right)$	$A_{28}\times\Gamma\left[{s+(1-\nu)/2,\ (\nu+1)/2-s,\ \pm\nu/2-s \atop s+1\pm\nu/2}\right]$ [$(\mathrm{Re}\,\nu-1)/2<\mathrm{Re}\,s<\begin{Bmatrix}\mathrm{Re}\,\nu/2\\ (\mathrm{Re}\,\nu+1)/2,\ -\mathrm{Re}\,\nu/2\end{Bmatrix}$]

8.4.26. The functions of Weber $\mathbf{E}_\nu(x)$, $\mathbf{E}_\nu^\mu(x)$ and Anger $\mathbf{J}_\nu(x)$, $\mathbf{J}_\nu^\mu(x)$.

NOTATION: $A_{29}=-\dfrac{\sin\nu\pi}{2\pi^2}$, $A_{30}=2\begin{Bmatrix}\cos(\nu\pi/2)\\ \sin(\nu\pi/2)\end{Bmatrix}$, $\delta=\begin{Bmatrix}1\\0\end{Bmatrix}$.

1	$\begin{Bmatrix}\mathbf{E}_\nu(2\sqrt{x})\\ \mathbf{J}_\nu(2\sqrt{x})\end{Bmatrix}= G_{35}^{22}\left(x\,\middle\vert\,\begin{matrix}0,\ 1/2,\ (1+\delta-\nu)/2\\ 0,\ 1/2,-\nu/2,\ \nu/2,\ (1+\delta-\nu)/2\end{matrix}\right)$	$\Gamma\left[\begin{matrix}s,\ s+\frac{1}{2},\ \frac{1}{2}-s,\ 1-s\\ s+\frac{1+\delta-\nu}{2},\ 1+\frac{\nu}{2}-s,\ 1-\frac{\nu}{2}-s,\ \frac{1+\nu-\delta}{2}-s\end{matrix}\right]$ $[0<\operatorname{Re}s<1/2]$
2	$\begin{Bmatrix}\mathbf{E}_\nu(2x^{-1/2})\\ \mathbf{J}_\nu(2x^{-1/2})\end{Bmatrix}= G_{53}^{22}\left(x\,\middle\vert\,\begin{matrix}\frac{1}{2},1,\frac{\nu}{2}+1,1-\frac{\nu}{2},\frac{1+\nu-\delta}{2}\\ \frac{1}{2},\ 1,\ \frac{1+\nu-\delta}{2}\end{matrix}\right)$	$\Gamma\left[\begin{matrix}s+\frac{1}{2},\ s+1,\ -s,\ \frac{1}{2}-s\\ s+1+\frac{\nu}{2},\ s+1-\frac{\nu}{2},\ s+\frac{1+\nu-\delta}{2},\ \frac{1+\delta-\nu}{2}-s\end{matrix}\right]$ $[-1/2<\operatorname{Re}s<0]$
3	$\begin{Bmatrix}\mathbf{E}_\nu^\mu(2\sqrt{x})\\ \mathbf{J}_\nu^\mu(2\sqrt{x})\end{Bmatrix}=2^\mu\times G_{35}^{22}\left(x\,\middle\vert\,\begin{matrix}-\frac{\mu}{2},\ \frac{1-\mu}{2},\ \frac{1+\delta-\nu}{2}\\ 0,\frac{1}{2},-\frac{\mu+\nu}{2},\frac{\nu-\mu}{2},\frac{1+\delta-\nu}{2}\end{matrix}\right)$	$2^\mu\Gamma\left[\begin{matrix}s,\ s+\frac{1}{2},\ \frac{\mu}{2}+1-s,\ \frac{\mu+1}{2}-s\\ s+\frac{1+\delta-\nu}{2},\ 1+\frac{\mu+\nu}{2}-s,\ 1+\frac{\mu-\nu}{2}-s,\ \frac{1+\nu-\delta}{2}-s\end{matrix}\right]$ $[0<\operatorname{Re}s<3/4,\ (\operatorname{Re}\mu+1)/2]$
4	$\begin{Bmatrix}\mathbf{E}_\nu^\mu(2x^{-1/2})\\ \mathbf{J}_\nu^\mu(2x^{-1/2})\end{Bmatrix}=2^\mu\times G_{53}^{22}\left(x\,\middle\vert\,\begin{matrix}\frac{1}{2};\ 1,\ \frac{\mu+\nu}{2}+1,\ \frac{\mu-\nu}{2}+1,\ \frac{1+\nu-\delta}{2}\\ \frac{\mu}{2}+1,\ \frac{\mu+1}{2},\ \frac{1+\nu-\delta}{2}\end{matrix}\right)$	$2^\mu\Gamma\left[\begin{matrix}s+\frac{\mu}{2}+1,\ s+\frac{\mu+1}{2};\ -s,\ \frac{1}{2}-s\\ s+\frac{\mu+\nu}{2}+1,\ s+\frac{\mu-\nu}{2}+1;\ s+\frac{1+\nu-\delta}{2},\ \frac{1+\delta-\nu}{2}-s\end{matrix}\right]$ $[-3/4,\ -(\operatorname{Re}\mu+1)/2<\operatorname{Re}s<0]$
5	$J_\nu(2\sqrt{x})-\mathbf{J}_\nu(2\sqrt{x})= A_{29}G_{24}^{32}\left(x\,\middle\vert\,\begin{matrix}0,\ 1/2\\ 0,\ 1/2,\ \nu/2,\ -\nu/2\end{matrix}\right)$	$A_{29}\Gamma\left[\begin{matrix}s,\ s+1/2,\ s+\nu/2,\ 1-s,\ 1/2-s\\ 1+\nu/2-s\end{matrix}\right]$ $[-\operatorname{Re}\nu/2,\ 0<\operatorname{Re}s<1/2]$
6	$J_\nu\left(\frac{2}{\sqrt{x}}\right)-\mathbf{J}_\nu\left(\frac{2}{\sqrt{x}}\right)= A_{29}G_{42}^{23}\left(x\,\middle\vert\,\begin{matrix}1/2,\ 1,\ 1-\nu/2,\ 1+\nu/2\\ 1/2,\ 1\end{matrix}\right)$	$A_{29}\Gamma\left[\begin{matrix}s+1/2,\ s+1,\ -s,\ 1/2-s,\ \nu/2-s\\ s+\nu/2+1\end{matrix}\right]$ $[-1/2<\operatorname{Re}s<0,\ \operatorname{Re}\nu/2]$

7	$J_\nu(2\sqrt{x}) \pm J_{-\nu}(2\sqrt{x}) = A_{30} G_{13}^{11}\left(x \middle\vert \begin{matrix} (1-\delta)/2 \\ (1-\delta)/2,\ -\nu/2,\ \nu/2 \end{matrix}\right)$	$A_{30}\Gamma\begin{bmatrix} s+(1-\delta)/2,\ (1+\delta)/2-s \\ 1+\nu/2-s,\ 1-\nu/2-s \end{bmatrix}$ $[(\delta-1)/2<\operatorname{Re} s<(2+\delta)/4]$
8	$J_\nu\left(\frac{2}{\sqrt{x}}\right) \pm J_{-\nu}\left(\frac{2}{\sqrt{x}}\right) = A_{30} G_{31}^{11}\left(x \middle\vert \begin{matrix} (1+\delta)/2,\ 1+\nu/2,\ 1-\nu/2 \\ (1+\delta)/2 \end{matrix}\right)$	$A_{30}\Gamma\begin{bmatrix} s+(\delta+1)/2,\ (1-\delta)/2-s \\ s+1+\nu/2,\ s+1-\nu/2 \end{bmatrix}$ $[-(2+\delta)/4<\operatorname{Re} s<(1-\delta)/2]$
9	$2\cos\frac{\nu\pi}{2} I_\nu(2\sqrt{x}) - J_\nu(2i\sqrt{x}) - J_\nu(-2i\sqrt{x}) = -\frac{\sin\nu\pi}{\pi} G_{13}^{21}\left(x \middle\vert \begin{matrix} 0 \\ 0,\ \nu/2,\ -\nu/2 \end{matrix}\right)$	$-\frac{\sin\nu\pi}{\pi}\Gamma\begin{bmatrix} s,\ s+\nu/2,\ 1-s \\ 1+\nu/2-s \end{bmatrix}$ $[0,\ -\operatorname{Re}\nu/2<\operatorname{Re} s<1]$
10	$2\cos\frac{\nu\pi}{2} I_\nu\left(\frac{2}{\sqrt{x}}\right) - J_\nu\left(\frac{2i}{\sqrt{x}}\right) - J_\nu\left(-\frac{2i}{\sqrt{x}}\right) = -\frac{\sin\nu\pi}{\pi} G_{31}^{12}\left(x \middle\vert \begin{matrix} 1,\ 1-\nu/2,\ 1+\nu/2 \\ 1 \end{matrix}\right)$	$-\frac{\sin\nu\pi}{\pi}\Gamma\begin{bmatrix} s+1,\ -s,\ \nu/2-s \\ s+\nu/2+1 \end{bmatrix}$ $[-1<\operatorname{Re} s<0,\ \operatorname{Re}\nu/2]$

8.4.27. The Lommel functions $s_{\mu,\nu}(x)$ and $S_{\mu,\nu}(x)$.

NOTATION: $A_{31} = 2^{\mu-1}\Gamma\left(\frac{\mu-\nu+1}{2}\right)\Gamma\left(\frac{\mu+\nu+1}{2}\right)$, $A_{32} = \frac{2^{\mu-1}}{\Gamma((1-\mu-\nu)/2)\,\Gamma((1-\mu+\nu)/2)}$.

1	$s_{\mu,\nu}(2\sqrt{x}) = A_{31} G_{13}^{11}\left(x \middle\vert \begin{matrix} (\mu+1)/2 \\ (\mu+1)/2,\ \nu/2,\ -\nu/2 \end{matrix}\right)$	$A_{31}\Gamma\begin{bmatrix} s+(\mu+1)/2,\ (1-\mu)/2-s \\ 1+\nu/2-s,\ 1-\nu/2-s \end{bmatrix}$ $[-(1+\operatorname{Re}\mu)/2<\operatorname{Re} s<(1-\operatorname{Re}\mu)/2,\ 3/4]$
2	$s_{\mu,\nu}\left(\frac{2}{\sqrt{x}}\right) = A_{31} G_{31}^{11}\left(x \middle\vert \begin{matrix} (1-\mu)/2,\ 1-\nu/2,\ 1+\nu/2 \\ (1-\mu)/2 \end{matrix}\right)$	$A_{31}\Gamma\begin{bmatrix} s+(1-\mu)/2,\ (\mu+1)/2-s \\ s+1+\nu/2,\ s+1-\nu/2 \end{bmatrix}$ $[-3/4,\ (\operatorname{Re}\mu-1)/2<\operatorname{Re} s<(\operatorname{Re}\mu+1)/2]$
3	$S_{\mu,\nu}(2\sqrt{x}) = A_{32} G_{13}^{31}\left(x \middle\vert \begin{matrix} (\mu+1)/2 \\ (\mu+1)/2,\ \nu/2,\ -\nu/2 \end{matrix}\right)$	$A_{32}\Gamma\left[s+\frac{1+\mu}{2},\ s+\frac{\nu}{2},\ s-\frac{\nu}{2},\ \frac{1-\mu}{2}-s\right]$ $[-(1+\operatorname{Re}\mu)/2,\ \lvert\operatorname{Re}\nu\rvert/2<\operatorname{Re} s<(1-\operatorname{Re}\mu)/2]$
4	$S_{\mu,\nu}\left(\frac{2}{\sqrt{x}}\right) = A_{32} G_{31}^{13}\left(x \middle\vert \begin{matrix} \frac{1-\mu}{2},\ 1-\frac{\nu}{2},\ 1+\frac{\nu}{2} \\ \frac{1-\mu}{2} \end{matrix}\right)$	$A_{32}\Gamma\left[s+\frac{1-\mu}{2},\ \frac{1+\mu}{2}-s,\ \frac{\nu}{2}-s,\ -\frac{\nu}{2}-s\right]$ $[(\operatorname{Re}\mu-1)/2<\operatorname{Re} s<(\operatorname{Re}\mu+1)/2,\ \lvert\operatorname{Re}\nu\rvert/2]$

8.4.28. The Kelvin functions $\mathrm{ber}_\nu(x)$, $\mathrm{bei}_\nu(x)$, $\mathrm{ker}_\nu(x)$ and $\mathrm{kei}_\nu(x)$.

NOTATION: $\delta=\left\{\begin{matrix}1\\0\end{matrix}\right\}$.

1	$\left\{\begin{matrix}\mathrm{ber}_\nu(4\sqrt[4]{x})\\ \mathrm{bei}_\nu(4\sqrt[4]{x})\end{matrix}\right\}=\pi\times G_{15}^{20}\left(x\,\middle\vert\,\begin{matrix}\nu+\frac{\delta}{2}\\ \frac{\nu}{4},\frac{\nu+2}{4},\frac{2-\nu}{4},-\frac{\nu}{4},\frac{\delta}{2}+\nu\end{matrix}\right)$ $\xleftarrow{L_{-\infty}}$	$\pi\Gamma\left[\begin{matrix}s+\frac{\nu}{4},\ s+\frac{\nu+2}{4}\\ s+\frac{\delta}{2}+\nu,\ \frac{2+\nu}{4}-s,\ 1+\frac{\nu}{4}-s,\ 1-\frac{\delta}{2}-\nu-s\end{matrix}\right]$ $[s=-\nu/4-k,\ -(\nu+2)/4-k\in D^+,\ k=0,\,1,\,2,\ldots]$
2	$\left\{\begin{matrix}\mathrm{ber}_\nu(4x^{-1/4})\\ \mathrm{bei}_\nu(4x^{-1/4})\end{matrix}\right\}=\pi\times G_{51}^{02}\left(x\,\middle\vert\,\begin{matrix}1-\frac{\nu}{4},\frac{2-\nu}{4},\frac{2+\nu}{4},1+\frac{\nu}{4},1-\nu-\frac{\delta}{2}\\ 1-\nu-\frac{\delta}{2}\end{matrix}\right)$ $\xleftarrow{L_{+\infty}}$	$\pi\Gamma\left[\begin{matrix}\frac{\nu}{4}-s,\ \frac{\nu+2}{4}-s\\ s+\frac{2+\nu}{4},\ s+1+\frac{\nu}{4},\ s+1-\nu-\frac{\delta}{2},\ \frac{\delta}{2}+\nu-s\end{matrix}\right]$ $[s=\nu/4+k,\ (\nu+2)/4+k\in D^-,\ k=0,\,1,\,2,\ldots]$
3	$\left\{\begin{matrix}\mathrm{ber}(4\sqrt[4]{x})\\ \mathrm{bei}(4\sqrt[4]{x})\end{matrix}\right\}=\pi G_{04}^{10}\left(x\,\middle\vert\,\begin{matrix}\cdot\quad\cdot\quad\cdot\quad\cdot\\ (1-\delta)/2,\ 0,\ 1/2,\ \delta/2\end{matrix}\right)$ $\xleftarrow{L_{-\infty}}$	$\pi\Gamma\left[\begin{matrix}s+(1-\delta)/2\\ 1/2-s,\ 1-s,\ 1-\delta/2-s\end{matrix}\right]$ $[s=(\delta-1)/2-k\in D^+,\ k=0,\,1,\,2,\ldots]$
4	$\left\{\begin{matrix}\mathrm{ber}(4x^{-1/4})\\ \mathrm{bei}(4x^{-1/4})\end{matrix}\right\}=\pi G_{40}^{01}\left(x\,\middle\vert\,\begin{matrix}(1+\delta)/2,\ 1/2,\ 1,\ 1-\delta/2\\ \cdot\quad\cdot\quad\cdot\quad\cdot\end{matrix}\right)$ $\xleftarrow{L_{+\infty}}$	$\pi\Gamma\left[\begin{matrix}(1-\delta)/2-s\\ s+1/2,\ s+1,\ s+1-\delta/2\end{matrix}\right]$ $[s=(1-\delta)/2+k\in D^-,\ k=0,\,1,\,2,\ldots]$
5	$\left\{\begin{matrix}\mathrm{ber}'(4\sqrt[4]{x})\\ \mathrm{bei}'(4\sqrt[4]{x})\end{matrix}\right\}=\mp\pi\times G_{04}^{10}\left(x\,\middle\vert\,\begin{matrix}\cdot\quad\cdot\quad\cdot\quad\cdot\\ \frac{1+2\delta}{4},\ \frac{1}{4},\ -\frac{1}{4},\ \frac{3-2\delta}{4}\end{matrix}\right)$ $\xleftarrow{L_{-\infty}}$	$\mp\pi\Gamma\left[\begin{matrix}s+(1+2\delta)/4\\ 3/4-s,\ 5/4-s,\ (1+2\delta)/4-s\end{matrix}\right]$ $[s=-(1+2\delta)/4-k\in D^+,\ k=0,\,1,\,2,\ldots]$
6	$\left\{\begin{matrix}\mathrm{ber}'(4x^{-1/4})\\ \mathrm{bei}'(4x^{-1/4})\end{matrix}\right\}=\mp\pi G_{40}^{01}\left(x\,\middle\vert\,\begin{matrix}\frac{3-2\delta}{4},\ \frac{3}{4},\ \frac{5}{4},\ \frac{1+2\delta}{4}\\ \cdot\quad\cdot\quad\cdot\quad\cdot\end{matrix}\right)$ $\xleftarrow{L_{+\infty}}$	$\mp\pi\Gamma\left[\begin{matrix}(1+2\delta)/4-s\\ s+3/4,\ s+5/4,\ s+(1+2\delta)/4\end{matrix}\right]$ $\left[s=(1+2\delta)/4+k\in D^-,\ k=0,\,1,\,2,\ldots,\ \delta=\left\{\begin{matrix}1\\0\end{matrix}\right\}\right]$
7	$\mathrm{ber}_\nu^2(2\sqrt[4]{4x})+\mathrm{bei}_\nu^2(2\sqrt[4]{4x})=\pi^{3/2}G_{15}^{10}\left(x\,\middle\vert\,\begin{matrix}(\nu+1)/2\\ \nu/2,\ -\nu/2,\ 0,\ 1/2,\ (1+\nu)/2\end{matrix}\right)$ $\xleftarrow{L_{-\infty}}$	$\pi^{3/2}\Gamma\left[\begin{matrix}s+\frac{\nu}{2}\\ s+\frac{\nu+1}{2},\ \frac{\nu}{2}+1-s,\ 1-s,\ \frac{1}{2}-s,\ \frac{1-\nu}{2}-s\end{matrix}\right]$ $[s=-\nu/2-k\in D^+,\ k=0,\,1,\,2,\ldots]$

8 $\mathrm{ber}_\nu^2\left(2\sqrt[4]{\frac{4}{x}}\right)+\mathrm{bei}_\nu^2\left(2\sqrt[4]{\frac{4}{x}}\right)=\pi^{3/2}\times G_{51}^{01}\left(x\,\middle\vert\,\begin{matrix}1-\frac{\nu}{2},\ 1+\frac{\nu}{2},\ \frac{1}{2},\ 1,\ \frac{1-\nu}{2}\\ \frac{1-\nu}{2}\end{matrix}\right)$ $\xleftarrow{L_{+\infty}}$

$\pi^{3/2}\Gamma\left[\begin{matrix}\frac{\nu}{2}-s\\ s+\frac{\nu}{2}+1,\ s+1,\ s+\frac{1}{2},\ s+\frac{1-\nu}{2},\ \frac{\nu+1}{2}-s\end{matrix}\right]$

$[s=\nu/2+k\in D^-,\ k=0,\ 1,\ 2,\ldots]$

9 $\mathrm{ber}_\nu\left(2\sqrt[4]{4x}\right)\begin{Bmatrix}\mathrm{ber}'_\nu\left(2\sqrt[4]{4x}\right)\\ \mathrm{bei}'_\nu\left(2\sqrt[4]{4x}\right)\end{Bmatrix}\pm\mathrm{bei}_\nu\left(2\sqrt[4]{4x}\right)\times\begin{Bmatrix}\mathrm{bei}'_\nu\left(2\sqrt[4]{4x}\right)\\ \mathrm{ber}'_\nu\left(2\sqrt[4]{4x}\right)\end{Bmatrix}=\frac{\pi^{3/2}}{\sqrt{2}}G_{15}^{10}\left(x\,\middle\vert\,\begin{matrix}\frac{3+2\nu-2\delta}{4}\\ \frac{2\nu\mp 1}{4},\ \frac{3+2\nu-2\delta}{4},\ \delta-\frac{1}{4},\ \frac{1}{4},\ \frac{1-2\nu-2\delta}{4}\end{matrix}\right)$ $\xleftarrow{L_{-\infty}}$

$\frac{\pi^{3/2}}{\sqrt{2}}\Gamma\left[\begin{matrix}s+\frac{2\nu\mp 1}{4}\\ s+\frac{3+2\nu-2\delta}{4},\ \frac{3+2\nu+2\delta}{4}-s,\ \frac{1+2\delta-2\nu}{4}-s,\ \frac{3}{4}-s,\ \frac{5}{4}-\delta-s\end{matrix}\right]$

$[s=-(2\nu\mp 1)/4-k\in D^+,\ k=0,\ 1,\ 2,\ldots]$

10 $\mathrm{ber}_\nu\left(2\sqrt[4]{\frac{4}{x}}\right)\begin{Bmatrix}\mathrm{ber}'_\nu\left(2\sqrt[4]{4/x}\right)\\ \mathrm{bei}'_\nu\left(2\sqrt[4]{4/x}\right)\end{Bmatrix}\pm\mathrm{bei}_\nu\left(2\sqrt[4]{\frac{4}{x}}\right)\times\begin{Bmatrix}\mathrm{bei}'_\nu\left(2\sqrt[4]{4/x}\right)\\ \mathrm{ber}'_\nu\left(2\sqrt[4]{4/x}\right)\end{Bmatrix}=\frac{\pi^{3/2}}{\sqrt{2}}G_{51}^{01}\left(x\,\middle\vert\,\begin{matrix}1-\frac{2\nu\mp 1}{4},\ \frac{3+2\nu+2\delta}{4},\ \frac{1+2\delta-2\nu}{4},\ \frac{3}{4},\ \frac{5}{4}-\delta\\ \frac{1+2\delta-2\nu}{4}\end{matrix}\right)$ $\xleftarrow{L_{+\infty}}$

$\frac{\pi^{3/2}}{\sqrt{2}}\Gamma\left[\begin{matrix}\frac{2\nu\mp 1}{4}-s\\ s+\frac{3+2\nu+2\delta}{4},\ s+\frac{1+2\delta-2\nu}{4},\ s+\frac{3}{4},\ s+\frac{5}{4}-\delta,\ \frac{3+2\nu-2\delta}{4}-s\end{matrix}\right]$

$[s=(2\nu\mp 1)/4+k\in D^-,\ k=0,\ 1,\ 2,\ldots]$

11 $\begin{Bmatrix}\mathrm{ker}_\nu\left(4\sqrt[4]{x}\right)\\ \mathrm{kei}_\nu\left(4\sqrt[4]{x}\right)\end{Bmatrix}=\pm\frac{1}{4}\times G_{15}^{40}\left(x\,\middle\vert\,\begin{matrix}\frac{\delta+\nu}{2}\\ \frac{\nu}{4},\ \frac{\nu+2}{4},\ -\frac{\nu}{4},\ \frac{2-\nu}{4},\ \frac{\nu+\delta}{2}\end{matrix}\right)$

$\pm\frac{1}{4}\Gamma\left[\begin{matrix}s+\frac{\nu}{4},\ s+\frac{\nu+2}{4},\ s-\frac{\nu}{4},\ s+\frac{2-\nu}{4}\\ s+\frac{\delta+\nu}{2},\ 1-\frac{\nu+\delta}{2}-s\end{matrix}\right]$

$[\mathrm{Re}\, s>|\mathrm{Re}\,\nu|/4]$

12	$\left\{\begin{matrix}\operatorname{ker}_\nu(4x^{-1/4})\\ \operatorname{kei}_\nu(4x^{-1/4})\end{matrix}\right\}= \pm\frac{1}{4}G_{51}^{04}\left(x\left\vert\begin{matrix}1-\frac{\nu}{4}, \frac{2-\nu}{4}, 1+\frac{\nu}{4}, \frac{2+\nu}{4}, 1-\frac{\nu+\delta}{2}\\ 1-\frac{\delta+\nu}{2}\end{matrix}\right.\right)$	$\pm\frac{1}{4}\Gamma\left[\begin{matrix}\frac{\nu}{4}-s, \frac{\nu+2}{4}-s, -\frac{\nu}{4}-s, \frac{2-\nu}{4}-s\\ s+1-\frac{\nu+\delta}{2}, \frac{\delta+\nu}{2}-s\end{matrix}\right]$ [Re $s<-\vert\mathrm{Re}\,\nu\vert/4$]
13	$\left\{\begin{matrix}\operatorname{ker}(4\sqrt[4]{x})\\ \operatorname{kei}(4\sqrt[4]{x})\end{matrix}\right\}= \pm\frac{1}{4}G_{04}^{30}\left(x\left\vert\begin{matrix}\cdot\ \ \cdot\ \ \cdot\ \ \cdot\\ 0, 1/2, (1-\delta)/2, \delta/2\end{matrix}\right.\right)$	$\pm\frac{1}{4}\Gamma\left[\begin{matrix}s, s+1/2, s+(1-\delta)/2\\ 1-\delta/2-s\end{matrix}\right]$ [Re $s>0$]
14	$\left\{\begin{matrix}\operatorname{ker}(4x^{-1/4})\\ \operatorname{kei}(4x^{-1/4})\end{matrix}\right\}= \pm\frac{1}{4}G_{40}^{03}\left(x\left\vert\begin{matrix}1/2, 1, (1+\delta)/2, 1-\delta/2\\ \cdot\ \ \cdot\ \ \cdot\ \ \cdot\end{matrix}\right.\right)$	$\pm\frac{1}{4}\Gamma\left[\begin{matrix}-s, 1/2-s, (1-\delta)/2-s\\ s+1-\delta/2\end{matrix}\right]$ [Re $s<0$]
15	$\left\{\begin{matrix}\sin(3\nu\pi/4)\\ \cos(3\nu\pi/4)\end{matrix}\right\}\operatorname{ker}_\nu(4\sqrt[4]{x})\pm \left\{\begin{matrix}\cos(3\nu\pi/4)\\ \sin(3\nu\pi/4)\end{matrix}\right\}\operatorname{kei}_\nu(4\sqrt[4]{x})= \mp\frac{1}{4}G_{04}^{30}\left(x\left\vert\begin{matrix}\cdot\ \ \cdot\ \ \cdot\ \ \cdot\\ \frac{\nu}{4}, \frac{\nu+2}{4}, \frac{2\delta-\nu}{4}, \frac{2-2\delta-\nu}{4}\end{matrix}\right.\right)$	$\mp\frac{1}{4}\Gamma\left[\begin{matrix}s+\frac{\nu}{4}, s+\frac{\nu+2}{4}, s+\frac{2\delta-\nu}{4}\\ \frac{2+2\delta+\nu}{4}-s\end{matrix}\right]$ [Re $s>-\mathrm{Re}\,\nu/4, (\mathrm{Re}\,\nu-2\delta)/4$]
16	$\left\{\begin{matrix}\sin(3\nu\pi/4)\\ \cos(3\nu\pi/4)\end{matrix}\right\}\operatorname{ker}_\nu\left(\frac{4}{\sqrt[4]{x}}\right)\pm \left\{\begin{matrix}\cos(3\nu\pi/4)\\ \sin(3\nu\pi/4)\end{matrix}\right\}\operatorname{kei}_\nu\left(\frac{4}{\sqrt[4]{x}}\right)= \mp\frac{1}{4}G_{40}^{03}\left(x\left\vert\begin{matrix}1-\frac{\nu}{4}, \frac{2-\nu}{4}, 1+\frac{\nu-2\delta}{4}, \frac{2+2\delta+\nu}{4}\\ \cdot\ \ \cdot\ \ \cdot\ \ \cdot\end{matrix}\right.\right)$	$\mp\frac{1}{4}\Gamma\left[\begin{matrix}\frac{\nu}{4}-s, \frac{\nu+2}{4}-s, \frac{2\delta-\nu}{4}-s\\ s+\frac{2+2\delta+\nu}{4}\end{matrix}\right]$ [Re $s<\mathrm{Re}\,\nu/4, (2\delta-\mathrm{Re}\,\nu)/4$]
17	$\operatorname{ker}_\nu^2(2\sqrt[4]{4x})+\operatorname{kei}_\nu^2(2\sqrt[4]{4x})= \frac{1}{8\sqrt{\pi}}G_{04}^{40}\left(x\left\vert\begin{matrix}\cdot\ \ \cdot\ \ \cdot\ \ \cdot\\ 0, 1/2, \nu/2, -\nu/2\end{matrix}\right.\right)$	$\frac{1}{8\sqrt{\pi}}\Gamma\left[s, s+\frac{1}{2}, s+\frac{\nu}{2}, s-\frac{\nu}{2}\right]$ [Re $s>\vert\mathrm{Re}\,\nu\vert/2$]

18	$\ker_\nu^2\left(2\sqrt[4]{\frac{4}{x}}\right)+\mathrm{kei}_\nu^2\left(2\sqrt[4]{\frac{4}{x}}\right)=$ $=\frac{1}{8\sqrt{\pi}}G_{40}^{04}\left(x\,\middle\vert\,\begin{matrix}1/2,\ 1,\ 1-\nu/2,\ 1+\nu/2\\ \cdot\quad\cdot\quad\cdot\quad\cdot\end{matrix}\right)$	$\frac{1}{8\sqrt{\pi}}\Gamma\left[-s,\ \frac{1}{2}-s,\ \frac{\nu}{2}-s,\ -\frac{\nu}{2}-s\right]$ [Re $s<-\vert\mathrm{Re}\,\nu\vert/2$].

8.4.29. The Airy functions $\mathrm{Ai}\,(x)$ and $\mathrm{Bi}\,(x)$.

1	$\mathrm{Ai}\left(\sqrt[3]{9x}\right)=\frac{1}{2\pi\sqrt[6]{3}}G_{02}^{20}\left(x\,\middle\vert\,\begin{matrix}\cdot\quad\cdot\\ 0,\ 1/3\end{matrix}\right)$	$\frac{1}{2\pi\sqrt[6]{3}}\Gamma\left[s,\ s+\frac{1}{3}\right]$ [Re $s>0$]
2	$\mathrm{Ai}\left(\sqrt[3]{\frac{9}{x}}\right)=\frac{1}{2\pi\sqrt[6]{3}}G_{20}^{02}\left(x\,\middle\vert\,\begin{matrix}2/3,\ 1\\ \cdot\quad\cdot\end{matrix}\right)$	$\frac{1}{2\pi\sqrt[6]{3}}\Gamma\left[-s,\ \frac{1}{3}-s\right]$ [Re $s<0$]
3	$e^{-x/2}\,\mathrm{Ai}\left(\left(\frac{3x}{4}\right)^{2/3}\right)=$ $=\frac{1}{\sqrt{\pi}\sqrt[6]{48}}G_{12}^{20}\left(x\,\middle\vert\,\begin{matrix}5/6\\ 0,\ 2/3\end{matrix}\right)$	$\frac{1}{\sqrt{\pi}\sqrt[6]{48}}\Gamma\left[\begin{matrix}s,\ s+2/3\\ s+5/6\end{matrix}\right]$ [Re $s>0$]
4	$e^{-1/(2x)}\,\mathrm{Ai}\left(\left(\frac{3}{4x}\right)^{2/3}\right)=$ $=\frac{1}{\sqrt{\pi}\sqrt[6]{48}}G_{21}^{02}\left(x\,\middle\vert\,\begin{matrix}1/3,\ 1\\ 1/6\end{matrix}\right)$	$\frac{1}{\sqrt{\pi}\sqrt[6]{48}}\Gamma\left[\begin{matrix}-s,\ 2/3-s\\ 5/6-s\end{matrix}\right]$ [Re $s<0$]
5	$e^{x/2}\,\mathrm{Ai}\left(\left(\frac{3x}{4}\right)^{2/3}\right)=$ $=\frac{1}{2\pi^{3/2}\sqrt[6]{48}}G_{12}^{21}\left(x\,\middle\vert\,\begin{matrix}5/6\\ 0,\ 2/3\end{matrix}\right)$	$\frac{1}{2\pi^{3/2}\sqrt[6]{48}}\Gamma\left[s,\ s+\frac{2}{3},\ \frac{1}{6}-s\right]$ [$0<\mathrm{Re}\,s<1/6$]
6	$e^{1/(2x)}\,\mathrm{Ai}\left(\left(\frac{3}{4x}\right)^{2/3}\right)=$ $=\frac{1}{2\pi^{3/2}\sqrt[6]{48}}G_{21}^{12}\left(x\,\middle\vert\,\begin{matrix}1/3,\ 1\\ 1/6\end{matrix}\right)$	$\frac{1}{2\pi^{3/2}\sqrt[6]{48}}\Gamma\left[s+\frac{1}{6},\ -s,\ \frac{2}{3}-s\right]$ [$-1/6<\mathrm{Re}\,s<0$]
7	$\left\{\begin{matrix}\mathrm{sh}\sqrt{x}\\ \mathrm{ch}\sqrt{x}\end{matrix}\right\}\mathrm{Ai}\left(\sqrt[3]{\frac{9x}{4}}\right)=$ $=\frac{(2/3)^{1/6}}{4\pi}\times$ $\times G_{24}^{22}\left(x\,\middle\vert\,\begin{matrix}5/12,\ 11/12\\ \frac{3\delta+2}{6},\ \frac{\delta}{2},\ \frac{5-3\delta}{6},\ \frac{1-\delta}{2}\end{matrix}\right)$	$\frac{(2/3)^{1/6}}{4\pi}\times$ $\times\Gamma\left[\begin{matrix}s+\frac{3\delta+2}{6},\ s+\frac{\delta}{2},\ \frac{1}{12}-s,\ \frac{7}{12}-s\\ \frac{3\delta+1}{6}-s,\ \frac{\delta+1}{2}-s\end{matrix}\right]$ $\left[-\delta/2<\mathrm{Re}\,s<1/12,\ \delta=\left\{\begin{matrix}1\\ 0\end{matrix}\right\}\right]$
8	$\left\{\begin{matrix}\mathrm{sh}\,x^{-1/2}\\ \mathrm{ch}\,x^{-1/2}\end{matrix}\right\}\mathrm{Ai}\left(\sqrt[3]{\frac{9}{4x}}\right)=$ $=\frac{(2/3)^{1/6}}{4\pi}\times$ $\times G_{42}^{22}\left(x\,\middle\vert\,\begin{matrix}\frac{4-3\delta}{6},\ 1-\frac{\delta}{2},\ \frac{1+3\delta}{6},\ \frac{1+\delta}{2}\\ \frac{1}{12},\ \frac{7}{12},\end{matrix}\right)$	$\frac{(2/3)^{1/6}}{4\pi}\times$ $\times\Gamma\left[\begin{matrix}s+\frac{1}{12},\ s+\frac{7}{12},\ \frac{3\delta+2}{6}-s,\ \frac{\delta}{2}-s\\ s+\frac{3\delta+1}{6},\ s+\frac{\delta+1}{2}\end{matrix}\right]$ $\left[-1/12<\mathrm{Re}\,s<\delta/2,\ \delta=\left\{\begin{matrix}1\\ 0\end{matrix}\right\}\right]$

9	$J_{1/3}\left(2\sqrt[4]{4x}\right)\mathrm{Ai}\left(\sqrt[3]{18}\,x^{1/6}\right)=$ $=\dfrac{(2/3)^{1/6}}{4\pi^{3/2}}\times$ $\times G^{30}_{04}\left(x\left\|\begin{array}{c}\cdot\quad\cdot\quad\cdot\quad\cdot\\1/12,\ 7/12,\ 1/4,\ -1/12\end{array}\right.\right)$	$\dfrac{(2/3)^{1/6}}{4\pi^{3/2}}\,\Gamma\left[\begin{array}{c}s+1/12,\ s+7/12,\ s+1/4\\13/12-s\end{array}\right]$ $[\mathrm{Re}\,s>-1/12]$
10	$J_{1/3}\left(2\sqrt[4]{\dfrac{4}{x}}\right)\mathrm{Ai}\left(\dfrac{\sqrt[3]{18}}{x^{1/6}}\right)=$ $=\dfrac{(2/3)^{1/6}}{4\pi^{3/2}}\times$ $\times G^{03}_{40}\left(x\left\|\begin{array}{c}11/12,\ 5/12,\ 3/4,\ 13/12\\\cdot\quad\cdot\quad\cdot\quad\cdot\end{array}\right.\right)$	$\dfrac{(2/3)^{1/6}}{4\pi^{3/2}}\,\Gamma\left[\begin{array}{c}1/12-s,\ 7/12-s,\ 1/4-s\\s+13/12\end{array}\right]$ $[\mathrm{Re}\,s<1/12]$
11	$I_{1\,3}\left(\sqrt{x}\right)\mathrm{Ai}\left(\sqrt[3]{\dfrac{9x}{4}}\right)=$ $=\dfrac{1}{2\pi^{3/2}\sqrt[6]{12}}\,G^{21}_{13}\left(x\left\|\begin{array}{c}2/3\\1/6,\ 1/2,\ -1/6\end{array}\right.\right)$	$\dfrac{1}{2\pi^{3/2}\sqrt[6]{12}}\,\Gamma\left[\begin{array}{c}s+1/6,\ s+1/2,\ 1/3-s\\7/6-s\end{array}\right]$ $[-1/6<\mathrm{Re}\,s<1/3]$
12	$I_{1/3}\left(\dfrac{1}{\sqrt{x}}\right)\mathrm{Ai}\left(\sqrt[3]{\dfrac{9}{4x}}\right)=$ $=\dfrac{1}{2\pi^{3/2}\sqrt[6]{12}}\,G^{12}_{31}\left(x\left\|\begin{array}{c}5/6,\ 1/2,\ 7/6\\1/3\end{array}\right.\right)$	$\dfrac{1}{2\pi^{3/2}\sqrt[6]{12}}\,\Gamma\left[\begin{array}{c}s+1/3,\ 1/6-s,\ 1/2-s\\s+7/6\end{array}\right]$ $[-1/3<\mathrm{Re}\,s<1/6]$
13	$I_{\nu}\left(\sqrt{x}\right)\mathrm{Ai}\left(\sqrt[3]{\dfrac{9x}{4}}\right)=$ $=\dfrac{1}{2\pi^{3/2}\sqrt[6]{12}}\times$ $\times G^{22}_{24}\left(x\left\|\begin{array}{c}\dfrac{1}{6},\ \dfrac{2}{3}\\[4pt]\dfrac{3\nu+2}{6},\ \dfrac{\nu}{2},\ \dfrac{2-3\nu}{6},\ -\dfrac{\nu}{2}\end{array}\right.\right)$	$\dfrac{1}{2\pi^{3/2}\sqrt[6]{12}}\,\Gamma\left[\begin{array}{c}s+\dfrac{3\nu+2}{6},\ s+\dfrac{\nu}{2},\\[4pt]\dfrac{4+3\nu}{6}-s,\ 1+\dfrac{\nu}{2}-s\\[6pt]\dfrac{1}{3}-s,\ \dfrac{5}{6}-s\end{array}\right]$ $[-\mathrm{Re}\,\nu/2<\mathrm{Re}\,s<1/3]$
14	$I_{\nu}\left(\dfrac{1}{\sqrt{x}}\right)\mathrm{Ai}\left(\sqrt[3]{\dfrac{9}{4x}}\right)=$ $=\dfrac{1}{2\pi^{3/2}\sqrt[6]{12}}\times$ $G^{22}_{42}\left(x\left\|\begin{array}{c}\dfrac{4-3\nu}{6},\ 1-\dfrac{\nu}{2},\ \dfrac{4+3\nu}{6},\ 1+\dfrac{\nu}{2}\\[4pt]\dfrac{1}{3},\ \dfrac{5}{6}\end{array}\right.\right)$	$\dfrac{1}{2\pi^{3/2}\sqrt[6]{12}}\,\Gamma\left[\begin{array}{c}s+\dfrac{1}{3},\ s+\dfrac{5}{6},\\[4pt]s+\dfrac{3\nu+4}{6},\ s+\dfrac{\nu}{2}+1\\[6pt]\dfrac{3\nu+2}{6}-s,\ \dfrac{\nu}{2}-s\end{array}\right]$ $[-1/3<\mathrm{Re}\,s<\mathrm{Re}\,\nu/2]$
15	$\mathrm{Ai}^2\left(\sqrt[3]{\dfrac{9x}{4}}\right)=$ $=\dfrac{1}{2\sqrt{\pi}\sqrt[6]{12}}\,G^{30}_{13}\left(x\left\|\begin{array}{c}5/6\\0,\ 1/3,\ 2/3\end{array}\right.\right)$	$\dfrac{1}{2\sqrt{\pi}\sqrt[6]{12}}\,\Gamma\left[\begin{array}{c}s,\ s+1/3,\ s+2/3\\s+5/6\end{array}\right]$ $[\mathrm{Re}\,s>1/6]$

16	$\mathrm{Ai}^2\left(\sqrt[3]{\frac{9}{4x}}\right)=\frac{1}{2\sqrt{\pi}\sqrt[6]{12}}\times G_{31}^{03}\left(x\,\middle\|\,\begin{matrix}1,1/3,2/3\\1/6\end{matrix}\right)$	$\frac{1}{2\sqrt{\pi}\sqrt[6]{12}}\times\Gamma\left[\begin{matrix}-s,1/3-s,2/3-s\\5/6-s\end{matrix}\right]$ [Re $s<-1/6$]
17	$K_\nu(\sqrt{x})\,\mathrm{Ai}\left(\sqrt[3]{\frac{9x}{4}}\right)=\frac{1}{2\sqrt{\pi}\sqrt[6]{12}}\times G_{24}^{40}\left(x\,\middle\|\,\begin{matrix}\frac{1}{6},\frac{2}{3}\\\frac{3\nu+2}{6},\frac{\nu}{2},\frac{2-3\nu}{6},-\frac{\nu}{2}\end{matrix}\right)$	$\frac{1}{2\sqrt{\pi}\sqrt[6]{12}}\Gamma\left[\begin{matrix}s+\frac{3\nu+2}{6},s+\frac{\nu}{2},s+\frac{2-3\nu}{6},s-\frac{\nu}{2}\\s+\frac{1}{6},s+\frac{2}{3}\end{matrix}\right]$ [Re $s>\|\mathrm{Re}\,\nu\|/2$]
18	$K_\nu\left(\frac{1}{\sqrt{x}}\right)\mathrm{Ai}\left(\sqrt[3]{\frac{9}{4x}}\right)=\frac{1}{2\sqrt{\pi}\sqrt[6]{12}}\times G_{42}^{04}\left(x\,\middle\|\,\begin{matrix}\frac{4-3\nu}{6},1-\frac{\nu}{2},\frac{4+3\nu}{6},1+\frac{\nu}{2}\\\frac{1}{3},\frac{5}{6}\end{matrix}\right)$	$\frac{1}{2\sqrt{\pi}\sqrt[6]{12}}\Gamma\left[\begin{matrix}\frac{3\nu+2}{6}-s,\frac{\nu}{2}-s,\frac{2-3\nu}{6}-s,-\frac{\nu}{2}-s\\\frac{1}{6}-s,\frac{2}{3}-s\end{matrix}\right]$ [Re $s<-\|\mathrm{Re}\,\nu\|/2$]
19	$\mathrm{Bi}(\sqrt[3]{9x})=\frac{2\pi}{3^{1/6}}G_{24}^{20}\left(x\,\middle\|\,\begin{matrix}1/6,\ 2/3\\0,\ 1/3,\ 1/6,\ 2/3\end{matrix}\right)$ $\overset{L_{-\infty}}{\longleftarrow}$	$\frac{2\pi}{3^{1/6}}\Gamma\left[\begin{matrix}s,\ s+1/3\\s+1/6,s+2/3,5/6-s,1/3-s\end{matrix}\right]$ [$s=-k,\ -k-1/3\in D^+,\ k=0,1,2,\ldots$]
20	$\mathrm{Bi}\left(\sqrt[3]{\frac{9}{x}}\right)=\frac{2\pi}{3^{1/6}}G_{42}^{02}\left(x\,\middle\|\,\begin{matrix}2/3,\ 1,\ 1/3,\ 5/6\\1/3,\ 5/6\end{matrix}\right)$ $\overset{L_{+\infty}}{\longleftarrow}$	$\frac{2\pi}{3^{1/6}}\Gamma\left[\begin{matrix}-s,\ 1/3-s\\s+5/6,\ s+1/3,\ 1/6-s,\ 2/3-s\end{matrix}\right]$ [$s=k,\ 1/3+k\in D^-,\ k=0,1,2,\ldots$]
21	$e^{-x/2}\,\mathrm{Bi}\left(\left(\frac{3x}{4}\right)^{2/3}\right)=\frac{1}{\sqrt{\pi}\sqrt[6]{48}}G_{23}^{21}\left(x\,\middle\|\,\begin{matrix}5/6,\ 1/3\\0,\ 2/3,\ 1/3\end{matrix}\right)$	$\frac{1}{\sqrt{\pi}\sqrt[6]{48}}\Gamma\left[\begin{matrix}s,\ s+2/3,\ 1/6-s\\s+1/3,\ 2/3-s\end{matrix}\right]$ [$0<\mathrm{Re}\,s<1/6$]
22	$e^{-1/(2x)}\,\mathrm{Bi}\left(\left(\frac{3}{4x}\right)^{2/3}\right)=\frac{1}{\sqrt{\pi}\sqrt[6]{48}}G_{32}^{12}\left(x\,\middle\|\,\begin{matrix}1/3,\ 1,\ 2/3\\1/6,\ 2/3\end{matrix}\right)$	$\frac{1}{\sqrt{\pi}\sqrt[6]{48}}\Gamma\left[\begin{matrix}s+1/6,\ -s,\ 2/3-s\\s+2/3,\ 1/3-s\end{matrix}\right]$ [$-1/6<\mathrm{Re}\,s<0$]
23	$K_\nu(\sqrt{x})\,\mathrm{Bi}\left(\sqrt[3]{\frac{9x}{4}}\right)=\frac{1}{2\sqrt{\pi}\sqrt[6]{12}}\left[G_{24}^{22}\left(x\,\middle\|\,\begin{matrix}\frac{1}{6},\frac{2}{3}\\\frac{\nu}{2},-\frac{\nu}{2},\end{matrix}\right.\right.$	$\frac{1}{2\sqrt{\pi}\sqrt[6]{12}}\times\left\{\Gamma\left[\begin{matrix}s+\frac{\nu}{2},s-\frac{\nu}{2},\frac{1}{3}-s,\frac{5}{6}-s\\\frac{4-3\nu}{6}-s,\frac{4+3\nu}{6}-s\end{matrix}\right]+\right.$

	$\left.\left.\frac{2+3\nu}{6}, \frac{2-3\nu}{6}\right)+G_{24}^{22}\left(x\,\middle\vert\,\begin{matrix}\frac{1}{6}, \frac{2}{3}\\ \frac{3\nu+2}{6}, \frac{2-3\nu}{6}, -\frac{\nu}{2}, \frac{\nu}{2}\end{matrix}\right)\right]$	$\left.+\Gamma\left[\begin{matrix}s+\frac{3\nu+2}{6}, s+\frac{2-3\nu}{6}, 1+\frac{\nu}{2}-s, 1-\frac{\nu}{2}-s\\ \frac{1}{3}-s, \frac{5}{6}-s\end{matrix}\right]\right\}$ [$\vert\operatorname{Re}\nu\vert/2<\operatorname{Re}s<1/3$]
24	$K_\nu\left(\frac{1}{\sqrt{x}}\right)\operatorname{Bi}\left(\sqrt[3]{\frac{9}{4x}}\right)=\frac{1}{2\sqrt{\pi}\sqrt[6]{12}}\left[G_{42}^{22}\left(x\,\middle\vert\,\begin{matrix}1-\nu/2, 1+\nu/2, (4-3\nu)/6, (4+3\nu)/6\\ 1/3, 5/6\end{matrix}\right)+G_{42}^{22}\left(x\,\middle\vert\,\begin{matrix}\frac{4-3\nu}{6}, \frac{4+3\nu}{6}, 1+\frac{\nu}{2}, 1-\frac{\nu}{2}\\ \frac{1}{3}, \frac{5}{6}\end{matrix}\right)\right]$	$\frac{1}{2\sqrt{\pi}\sqrt[6]{12}}\times\left\{\Gamma\left[\begin{matrix}s+\frac{1}{3}, s+\frac{5}{6}, \frac{\nu}{2}-s, -\frac{\nu}{2}-s\\ s+\frac{4-3\nu}{6}, s+\frac{4+3\nu}{6}\end{matrix}\right]+\Gamma\left[\begin{matrix}s+\frac{1}{3}, s+\frac{5}{6}, \frac{3\nu+2}{6}-s, \frac{2-3\nu}{6}-s\\ s+1+\frac{\nu}{2}, s+1-\frac{\nu}{2}\end{matrix}\right]\right\}$ [$-1/3<\operatorname{Re}s<-\vert\operatorname{Re}\nu\vert/2$]
25	$K_{1/3}(\sqrt{x})\operatorname{Bi}\left(\sqrt[3]{\frac{9x}{4}}\right)=\frac{1}{2\sqrt{\pi}\sqrt[6]{12}}G_{13}^{21}\left(x\,\middle\vert\,\begin{matrix}2/3\\ 1/2, -1/6, 1/6\end{matrix}\right)$	$\frac{1}{2\sqrt{\pi}\sqrt[6]{12}}\Gamma\left[\begin{matrix}s+1/2, s-1/6, 1/3-s\\ 5/6-s\end{matrix}\right]$ [$1/6<\operatorname{Re}s<1/3$]
26	$K_{1/3}\left(\frac{1}{\sqrt{x}}\right)\operatorname{Bi}\left(\sqrt[3]{\frac{9}{4x}}\right)=\frac{1}{2\sqrt{\pi}\sqrt[6]{12}}G_{31}^{12}\left(x\,\middle\vert\,\begin{matrix}1/2, 7/6, 5/6\\ 1/3\end{matrix}\right)$	$\frac{1}{2\sqrt{\pi}\sqrt[6]{12}}\times\Gamma\left[\begin{matrix}s+1/3, 1/2-s, -1/6-s\\ s+5/6\end{matrix}\right]$ [$-1/3<\operatorname{Re}s<-1/6$]

8.4.30. The Legendre polynomials $P_n(x)$.

NOTATION: $A_{33}=\frac{\Gamma(1/2+n-l)}{l!}$, $A_{34}=\frac{(-1)^l}{l!}i^{2\varepsilon}\Gamma\left(\frac{1}{2}+n-l\right)$.

1	$P_n(2x-1)\,H(1-x)=G_{22}^{20}\left(x\,\middle\vert\,\begin{matrix}-n, n+1\\ 0, 0\end{matrix}\right)$	$\Gamma\left[\begin{matrix}s, s\\ s+n+1, s-n\end{matrix}\right]=(-1)^n\Gamma\left[\begin{matrix}s\\ s+n+1\end{matrix}\right](1-s)_n$ [$\operatorname{Re}s>0$]
2	$P_n\left(\frac{2}{x}-1\right)H(x-1)=G_{22}^{02}\left(x\,\middle\vert\,\begin{matrix}1, 1\\ -n, n+1\end{matrix}\right)$	$\Gamma\left[\begin{matrix}-s, -s\\ 1+n-s, -n-s\end{matrix}\right]=(-1)^n\Gamma\left[\begin{matrix}-s\\ 1+n-s\end{matrix}\right](s+1)_n$ [$\operatorname{Re}s<0$]

3	$P_n(2x-1)H(x-1)=$ $=G_{22}^{02}\left(x\left\vert\begin{matrix}-n,\ n+1\\ 0,\ 0\end{matrix}\right.\right)$	$\Gamma\left[\begin{matrix}n+1-s,\ -n-s\\ 1-s,\ 1-s\end{matrix}\right]=$ $=\Gamma\left[\begin{matrix}-n-s\\ 1-s\end{matrix}\right](1-s)_n$ [Re $s<-n$]
4	$P_n\left(\frac{2}{x}-1\right)H(1-x)=$ $=G_{22}^{20}\left(x\left\vert\begin{matrix}1,\ 1\\ -n,\ n+1\end{matrix}\right.\right)$	$\Gamma\left[\begin{matrix}s-n,\ s+n+1\\ s+1,\ s+1\end{matrix}\right]=\Gamma\left[\begin{matrix}s-n\\ s+1\end{matrix}\right](s+1)_n$ [Re $s>n$]
5	$(1+x)^{-n-1}P_n\left(\frac{1-x}{1+x}\right)=$ $=\frac{1}{(n!)^2}G_{22}^{12}\left(x\left\vert\begin{matrix}-n,\ -n\\ 0,\ 0\end{matrix}\right.\right)$	$\frac{1}{(n!)^2}\Gamma\left[\begin{matrix}s,\ n+1-s,\ n+1-s\\ 1-s\end{matrix}\right]=$ $=\frac{1}{(n!)^2}\Gamma[s,\ n+1-s](1-s)_n$ [$0<$ Re $s<n+1$]
6	$P_n(\sqrt{x})H(1-x)=$ $=G_{22}^{20}\left(x\left\vert\begin{matrix}(1-n)/2,\ 1+n/2\\ 0,\ 1/2\end{matrix}\right.\right)$	$\Gamma\left[\begin{matrix}s,\ s+1/2\\ s+1+n/2,\ s+(1-n)/2\end{matrix}\right]=$ $=(-1)^l\Gamma\left[\begin{matrix}s+\varepsilon\\ s+1+n/2\end{matrix}\right]\left(\frac{1}{2}+\varepsilon-s\right)_l$ [Re $s>-\varepsilon$; $l=[n/2]$; $2\varepsilon=n-2l$]
7	$P_n\left(\frac{1}{\sqrt{x}}\right)H(x-1)=$ $=G_{22}^{02}\left(x\left\vert\begin{matrix}1/2,\ 1\\ -n/2,\ (n+1)/2\end{matrix}\right.\right)$	$\Gamma\left[\begin{matrix}-s,\ 1/2-s\\ 1+n/2-s,\ (1-n)/2-s\end{matrix}\right]=$ $=(-1)^l\Gamma\left[\begin{matrix}\varepsilon-s\\ 1+n/2-s\end{matrix}\right]\left(s+\frac{1}{2}+\varepsilon\right)_l$ [Re $s<\varepsilon$; $l=[n/2]$; $2\varepsilon=n-2l$]
8	$P_n(\sqrt{x})H(x-1)=$ $=G_{22}^{02}\left(x\left\vert\begin{matrix}(1-n)/2,\ 1+n/2\\ 0,\ 1/2\end{matrix}\right.\right)$	$\Gamma\left[\begin{matrix}-n/2-s,\ (1+n)/2-s\\ 1/2-s,\ 1-s\end{matrix}\right]=$ $=\Gamma\left[\begin{matrix}-n/2-s\\ 1-\varepsilon-s\end{matrix}\right]\left(\frac{1}{2}+\varepsilon-s\right)_l$ [Re $s<-n/2$; $l=[n/2]$; $2\varepsilon=n-2l$]
9	$P_n\left(\frac{1}{\sqrt{x}}\right)H(1-x)=$ $=G_{22}^{20}\left(x\left\vert\begin{matrix}1/2,\ 1\\ -n/2,\ (n+1)/2\end{matrix}\right.\right)$	$\Gamma\left[\begin{matrix}s-n/2,\ s+(n+1)/2\\ s+1/2,\ s+1\end{matrix}\right]=$ $=\Gamma\left[\begin{matrix}s-n/2\\ s+1-\varepsilon\end{matrix}\right]\left(s+\varepsilon+\frac{1}{2}\right)_l$ [Re $s>n/2$; $l=[n/2]$; $2\varepsilon=n-2l$]
10	$(1-x)_+^{\varepsilon-1/2}P_n(\sqrt{1-x})=$ $=(-1)^lA_{33}G_{22}^{20}\left(x\left\vert\begin{matrix}-l,\ \varepsilon+(n+1)/2\\ 0,\ 0\end{matrix}\right.\right)$	$A_{33}\Gamma\left[\begin{matrix}s,\ 1+l-s\\ s+1/2+n-l,\ 1-s\end{matrix}\right]=$ $=A_{33}\Gamma\left[\begin{matrix}s\\ s+1/2+n-l\end{matrix}\right](1-s)_l$ [Re $s>0$; $l=[n/2]$; $2\varepsilon=n-2l$]
11	$(x-1)_+^{\varepsilon-1/2}P_n\left(\sqrt{1-\frac{1}{x}}\right)=$ $=(-1)^lA_{33}G_{22}^{02}\left(x\left\vert\begin{matrix}1/2+\varepsilon,\ 1/2+\varepsilon\\ -n/2,\ (1+n)/2\end{matrix}\right.\right)$	$A_{33}\Gamma\left[\begin{matrix}s+(n+1)/2,\ 1/2-\varepsilon-s\\ s+\varepsilon+1/2,\ 1+n/2-s\end{matrix}\right]=$ $=A_{33}\Gamma\left[\begin{matrix}1/2-\varepsilon-s\\ 1+n/2-s\end{matrix}\right]\left(\frac{1}{2}+\varepsilon+s\right)_l$ [Re $s<1/2-\varepsilon$; $l=[n/2]$; $2\varepsilon=n-2l$]
12	$(x-1)_+^{\varepsilon-1/2}P_n(i\sqrt{x-1})=$ $=A_{34}G_{22}^{02}\left(x\left\vert\begin{matrix}-l,\ 1/2+n-l\\ 0,\ 0\end{matrix}\right.\right)$	$A_{34}\Gamma\left[\begin{matrix}1/2+l-n-s,\ 1+l-s\\ 1-s,\ 1-s\end{matrix}\right]=$ $=A_{34}\Gamma\left[\begin{matrix}1/2+l-n-s\\ 1-s\end{matrix}\right](1-s)_l$ [Re $s<1/2+l-n$; $l=[n/2]$; $2\varepsilon=n-2l$]
13	$(1-x)_+^{\varepsilon-1/2}P_n\left(i\sqrt{\frac{1}{x}-1}\right)=$ $=A_{34}G_{22}^{20}\left(x\left\vert\begin{matrix}\varepsilon+1/2,\ \varepsilon+1/2\\ (n+1)/2,\ -n/2\end{matrix}\right.\right)$	$A_{34}\Gamma\left[\begin{matrix}s+(n+1)/2,\ s-n/2\\ s+\varepsilon+1/2,\ s+\varepsilon+1/2\end{matrix}\right]=$ $=A_{34}\Gamma\left[\begin{matrix}s-n/2\\ s+\varepsilon+1/2\end{matrix}\right]\left(\frac{1}{2}+\varepsilon+s\right)_l$ [Re $s>n/2$; $l=[n/2]$; $2\varepsilon=n-2l$]

14	$(1+x)^{-(n+1)/2}P_n\left(\frac{1}{\sqrt{1+x}}\right)=$ $=\frac{2^n}{n!\sqrt{\pi}}G_{22}^{12}\left(x\,\middle\vert\,\begin{matrix}(1-n)/2, & -n/2\\ 0, & 0\end{matrix}\right)$	$\frac{2^n}{n!\sqrt{\pi}}\Gamma\left[\begin{matrix}s,\ (n+1)/2-s,\ 1+n/2-s\\ 1-s\end{matrix}\right]$ $=\frac{2^n}{n!\sqrt{\pi}}\Gamma\left[s,\ \frac{1}{2}+n-l-s\right](1-s)_l$ $[0<\operatorname{Re} s<1/2+n-l;\ l=[n/2]]$
15	$(1+x)^{-(n+1)/2}P_n\left(\sqrt{\frac{x}{x+1}}\right)=$ $=\frac{2^n}{n!\sqrt{\pi}}G_{22}^{21}\left(x\,\middle\vert\,\begin{matrix}(1-n)/2, & (1-n)/2\\ 0, & 1/2\end{matrix}\right)$	$\frac{2^n}{n!\sqrt{\pi}}\Gamma\left[\begin{matrix}s,\ s+1/2,\ (1+n)/2-s\\ s+(1-n)/2\end{matrix}\right]=$ $=\frac{2^n}{n!\sqrt{\pi}}\Gamma\left[s+\varepsilon,\ \frac{n+1}{2}-s\right]\left(\frac{1-n}{2}+s\right)_l$ $[-\varepsilon<\operatorname{Re} s<(n+1)/2;\ l=[n/2];\ 2\varepsilon=n-2l]$
16	$P_n\left(\frac{x+1}{2\sqrt{x}}\right)H(1-x)=$ $=G_{22}^{20}\left(x\,\middle\vert\,\begin{matrix}(1-n)/2, & 1+n/2\\ -n/2, & (n+1)/2\end{matrix}\right)$	$\Gamma\left[\begin{matrix}s-n/2,\ s+(n+1)/2\\ s+1+n/2,\ s+(1-n)/2\end{matrix}\right]=$ $=\Gamma\left[\begin{matrix}s-n/2\\ s+1+n/2\end{matrix}\right]\left(s+\frac{1-n}{2}\right)_n$ $[\operatorname{Re} s>n/2]$
17	$P_n\left(\frac{x+1}{2\sqrt{x}}\right)H(x-1)=$ $=G_{22}^{02}\left(x\,\middle\vert\,\begin{matrix}1+n/2, & (1-n)/2\\ (1+n)/2, & -n/2\end{matrix}\right)$	$\Gamma\left[\begin{matrix}-n/2-s,\ (n+1)/2-s\\ 1+n/2-s,\ (1-n)/2-s\end{matrix}\right]=$ $=(-1)^n\Gamma\left[\begin{matrix}-n/2-s\\ 1+n/2-s\end{matrix}\right]\left(\frac{1-n}{2}+s\right)_n$ $[\operatorname{Re} s<-n/2]$
18	$(1+x)^{-(n+1)/2}P_n\left(\frac{2+x}{2\sqrt{1+x}}\right)=$ $=\frac{1}{n!\sqrt{\pi}}G_{22}^{12}\left(x\,\middle\vert\,\begin{matrix}1/2, & -n\\ 0, & 0\end{matrix}\right)$	$\frac{1}{n!\sqrt{\pi}}\Gamma\left[\begin{matrix}s,\ 1/2-s,\ n+1-s\\ 1-s\end{matrix}\right]=$ $=\frac{1}{n!\sqrt{\pi}}\Gamma\left[s,\ \frac{1}{2}-s\right](1-s)_n$ $[0<\operatorname{Re} s<1/2]$
19	$(1+x)^{-(n+1)/2}P_n\left(\frac{2x+1}{2\sqrt{x^2+x}}\right)=$ $=\frac{1}{n!\sqrt{\pi}}G_{22}^{21}\left(x\,\middle\vert\,\begin{matrix}(1-n)/2, & (1-n)/2\\ -n/2, & (n+1)/2\end{matrix}\right)$	$\frac{1}{n!\sqrt{\pi}}\Gamma\left[\begin{matrix}s+\frac{n+1}{2},\ s-\frac{n}{2},\ \frac{1+n}{2}-s\\ s+\frac{1-n}{2}\end{matrix}\right]=$ $=\frac{1}{n!\sqrt{\pi}}\Gamma\left[s-\frac{n}{2},\ \frac{1+n}{2}-s\right]\times$ $\times\left(s+\frac{1-n}{2}\right)_n$ $[n/2<\operatorname{Re} s<(n+1)/2]$

8.4.31. The Chebyshev polynomials of the first kind $T_n(x)$.

NOTATION: $A_{35}=\left(\frac{n}{2}\right)^{2\varepsilon}\sqrt{\pi},\quad A_{36}=(-1)^l\left(\frac{ni}{2}\right)^{2\varepsilon}\sqrt{\pi}.$

1	$(1-x)_+^{-1/2}T_n(2x-1)=$ $=\sqrt{\pi}G_{22}^{20}\left(x\,\middle\vert\,\begin{matrix}1/2-n, & 1/2+n\\ 0, & 1/2\end{matrix}\right)$	$\sqrt{\pi}\Gamma\left[\begin{matrix}s,\ s+1/2\\ s+n+1/2,\ s+1/2-n\end{matrix}\right]=$ $=(-1)^n\sqrt{\pi}\Gamma\left[\begin{matrix}s\\ s+n+1/2\end{matrix}\right]\left(\frac{1}{2}-s\right)_n$ $[\operatorname{Re} s>0]$
2	$(x-1)_+^{-1/2}T_n\left(\frac{2}{x}-1\right)=$ $=\sqrt{\pi}G_{22}^{02}\left(x\,\middle\vert\,\begin{matrix}0, & 1/2\\ n, & -n\end{matrix}\right)$	$\sqrt{\pi}\Gamma\left[\begin{matrix}1/2-s,\ 1-s\\ 1+n-s,\ 1-n-s\end{matrix}\right]=$ $=(-1)^n\sqrt{\pi}\Gamma\left[\begin{matrix}1/2-s\\ 1+n-s\end{matrix}\right](s)_n$ $[\operatorname{Re} s<1/2]$
3	$(x-1)_+^{-1/2}T_n(2x-1)=$	$\sqrt{\pi}\Gamma\left[\begin{matrix}1/2+n-s,\ 1/2-n-s\\ 1-s,\ 1/2-s\end{matrix}\right]=$

	$=\sqrt{\pi}\,G_{22}^{02}\left(x\,\middle\vert\,\begin{matrix}1/2+n,\ 1/2-n\\ 0,\ 1/2\end{matrix}\right)$	$=\Gamma\begin{bmatrix}1/2-n-s\\ 1-s\end{bmatrix}\left(\frac{1}{2}-s\right)_n$ [Re s < 1/2 − n]
4	$(1-x)_+^{-1/2}T_n\left(\frac{2}{x}-1\right)=$ $=\sqrt{\pi}\,G_{22}^{20}\left(x\,\middle\vert\,\begin{matrix}0,\ 1/2\\ n,\ -n\end{matrix}\right)$	$\sqrt{\pi}\,\Gamma\begin{bmatrix}s-n,\ s+n\\ s+1/2,\ s\end{bmatrix}=$ $=\sqrt{\pi}\,\Gamma\begin{bmatrix}s-n\\ s+1/2\end{bmatrix}(s)_n$ [Re s > n]
5	$(1+x)^{-n}T_n\left(\frac{1-x}{1+x}\right)=$ $=\frac{2^{2n-1}}{(2n-1)!}\,G_{22}^{12}\left(x\,\middle\vert\,\begin{matrix}1-n,\ 1/2-n\\ 0,\ 1/2\end{matrix}\right)$	$\frac{2^{2n-1}}{(2n-1)!}\,\Gamma\begin{bmatrix}s,\ n-s,\ n+1/2-s\\ 1/2-s\end{bmatrix}=$ $=\frac{2^{2n-1}}{(2n-1)!}\,\Gamma[s,\ n-s]\left(\frac{1}{2}-s\right)_n$ [0 < Re s < n; n = 1, 2, 3, . . .]
6	$(1-x)_+^{-1/2}T_n(\sqrt{x})=$ $=\sqrt{\pi}\,G_{22}^{20}\left(x\,\middle\vert\,\begin{matrix}(1+n)/2,\ (1-n)/2\\ 0,\ 1/2\end{matrix}\right)$	$\sqrt{\pi}\,\Gamma\begin{bmatrix}s,\ s+1/2\\ s+(n+1)/2,\ s+(1-n)/2\end{bmatrix}=$ $=(-1)^l\sqrt{\pi}\,\Gamma\begin{bmatrix}s+\varepsilon\\ s+(n+1)/2\end{bmatrix}\times$ $\times\left(\frac{1}{2}+\varepsilon-s\right)_l$ [Re s > −ε; l = [n/2]; 2ε = n − 2l]
7	$(x-1)_+^{-1/2}T_n\left(\frac{1}{\sqrt{x}}\right)=$ $=\sqrt{\pi}\,G_{22}^{02}\left(x\,\middle\vert\,\begin{matrix}0,\ 1/2\\ n/2,\ -n/2\end{matrix}\right)$	$\sqrt{\pi}\,\Gamma\begin{bmatrix}1/2-s,\ 1-s\\ 1+n/2-s,\ 1-n/2-s\end{bmatrix}=$ $=(-1)^l\sqrt{\pi}\,\Gamma\begin{bmatrix}1/2+\varepsilon-s\\ 1+n/2-s\end{bmatrix}(s+\varepsilon)_l$ [Re s < 1/2 + ε; l = [n/2]; 2ε = n − 2l]
8	$(x-1)_+^{-1/2}T_n(\sqrt{x})=$ $=\sqrt{\pi}\,G_{22}^{02}\left(x\,\middle\vert\,\begin{matrix}(1+n)/2,\ (1-n)/2\\ 0,\ 1/2\end{matrix}\right)$	$\sqrt{\pi}\,\Gamma\begin{bmatrix}(1-n)/2-s,\ (1+n)/2-s\\ 1/2-s,\ 1-s\end{bmatrix}=$ $=\sqrt{\pi}\,\Gamma\begin{bmatrix}(1-n)/2-s\\ 1-\varepsilon-s\end{bmatrix}\left(\frac{1}{2}+\varepsilon-s\right)_l$ [Re s < (1 − n)/2; l = [n/2]; 2ε = n − 2l]
9	$(1-x)_+^{-1/2}T_n\left(\frac{1}{\sqrt{x}}\right)=$ $=\sqrt{\pi}\,G_{22}^{20}\left(x\,\middle\vert\,\begin{matrix}0,\ 1/2\\ n/2,\ -n/2\end{matrix}\right)$	$\sqrt{\pi}\,\Gamma\begin{bmatrix}s-n/2,\ s+n/2\\ s,\ s+1/2\end{bmatrix}=$ $=\sqrt{\pi}\,\Gamma\begin{bmatrix}s-n/2\\ s+1/2-\varepsilon\end{bmatrix}(s+\varepsilon)_l$ [Re s > n/2; l = [n/2]; 2ε = n − 2l]
10	$(1-x)_+^{\varepsilon-1/2}T_n(\sqrt{1-x})=$ $=(-1)^lA_{35}G_{22}^{20}\left(x\,\middle\vert\,\begin{matrix}1/2-l,\ 1/2+l+2\varepsilon\\ 0,\ 1/2\end{matrix}\right)$	$A_{35}\Gamma\begin{bmatrix}s,\ 1/2+l-s\\ s+1/2+n-l,\ 1/2-s\end{bmatrix}=$ $=A_{35}\Gamma\begin{bmatrix}s\\ s+1/2+n-l\end{bmatrix}\left(\frac{1}{2}-s\right)_l$ [Re s > 0; l = [n/2]; 2ε = n − 2l]
11	$(x-1)_+^{\varepsilon-1/2}T_n\left(\sqrt{1-\frac{1}{x}}\right)=$ $=(-1)^lA_{35}G_{22}^{02}\left(x\,\middle\vert\,\begin{matrix}\varepsilon,\ \varepsilon+1/2\\ n/2,\ -n/2\end{matrix}\right)$	$A_{35}\Gamma\begin{bmatrix}s+n/2,\ 1/2-\varepsilon-s\\ s+\varepsilon,\ 1+n/2-s\end{bmatrix}=$ $=A_{35}\Gamma\begin{bmatrix}1/2-\varepsilon-s\\ 1+n/2-s\end{bmatrix}(\varepsilon+s)_l$ [Re s < 1/2 − ε; l = [n/2]; 2ε = n − 2l]
12	$(x-1)_+^{\varepsilon-1/2}T_n(i\sqrt{x-1})=$ $=A_{36}G_{22}^{02}\left(x\,\middle\vert\,\begin{matrix}1/2+2\varepsilon+l,\ 1/2-l\\ 1/2,\ 0\end{matrix}\right)$	$A_{36}\Gamma\begin{bmatrix}1/2+l-n-s,\ 1/2+l-s\\ 1-s,\ 1/2-s\end{bmatrix}=$ $=A_{36}\Gamma\begin{bmatrix}1/2+l-n-s\\ 1-s\end{bmatrix}\left(\frac{1}{2}-s\right)_l$ [Re s < 1/2 + l − n; l = [n/2]; 2ε = n − 2l]
13	$(1-x)_+^{\varepsilon-1/2}T_n\left(i\sqrt{\frac{1}{x}-1}\right)=$ $=A_{36}G_{22}^{20}\left(x\,\middle\vert\,\begin{matrix}1/2+\varepsilon,\ \varepsilon\\ -n/2,\ n/2\end{matrix}\right)$	$A_{36}\Gamma\begin{bmatrix}s+n/2,\ s-n/2\\ s+\varepsilon,\ s+\varepsilon+1/2\end{bmatrix}=$ $=A_{36}\Gamma\begin{bmatrix}s-n/2\\ s+\varepsilon+1/2\end{bmatrix}(\varepsilon+s)_l$ [Re s > n/2; l = [n/2]; 2ε = n − 2l]

14	$(1+x)^{-n/2}T_n\left(\frac{1}{\sqrt{1+x}}\right)=$ $=\frac{2^{n-1}}{(n-1)!}G_{22}^{12}\left(x\middle\vert\begin{matrix}1-n/2, & (1-n)/2\\ 0, & 1/2\end{matrix}\right)$	$\frac{2^{n-1}}{(n-1)!}\Gamma\left[\begin{matrix}s, n/2-s, (n+1)/2-s\\ 1/2-s\end{matrix}\right]=$ $=\frac{2^{n-1}}{(n-1)!}\Gamma[s, n-l-s]\left(\frac{1}{2}-s\right)_l$ $[0<\operatorname{Re}s<n-l;\ l=[n/2];\ 2\varepsilon=n-2l;\ n=1, 2, 3\ldots]$
15	$(1+x)^{-n/2}T_n\left(\sqrt{\frac{x}{x+1}}\right)=$ $=\frac{2^{n-1}}{(n-1)!}G_{22}^{21}\left(x\middle\vert\begin{matrix}1-n/2, & (1-n)/2\\ 0, & 1/2\end{matrix}\right)$	$\frac{2^{n-1}}{(n-1)!}\Gamma\left[\begin{matrix}s, s+1/2, n/2-s\\ s+(1-n)/2\end{matrix}\right]=$ $=\frac{2^{n-1}}{(n-1)!}\Gamma\left[s+\varepsilon, \frac{n}{2}-s\right]\left(\frac{1-n}{2}+s\right)_l$ $[-\varepsilon<\operatorname{Re}s<n/2;\ l=[n/2];\ 2\varepsilon=n-2l]$
16	$(x-1)_+^{-1/2}T_n(8x^2-8x+1)=$ $=\sqrt{\pi}G_{22}^{02}\left(x\middle\vert\begin{matrix}1/2+2n, & 1/2-2n\\ 0, & 1/2\end{matrix}\right)$	$\sqrt{\pi}\Gamma\left[\begin{matrix}1/2-2n-s, 1/2+2n-s\\ 1-s, 1/2-s\end{matrix}\right]=$ $=\sqrt{\pi}\Gamma\left[\begin{matrix}1/2-2n-s\\ 1-s\end{matrix}\right]\left(\frac{1}{2}-s\right)_{2n}$ $[\operatorname{Re}s<1/2-2n]$
17	$(1-x)_+^{-1/2}T_n\left(\frac{x^2-8x+8}{x^2}\right)=$ $=\sqrt{\pi}G_{22}^{20}\left(x\middle\vert\begin{matrix}0, & 1/2\\ -2n, & 2n\end{matrix}\right)$	$\sqrt{\pi}\Gamma\left[\begin{matrix}s-2n, s+2n\\ s, s+1/2\end{matrix}\right]=$ $=\sqrt{\pi}\Gamma\left[\begin{matrix}s-2n\\ s+1/2\end{matrix}\right](s)_{2n}$ $[\operatorname{Re}s>2n]$.

8.4.32. The Chebyshev polynomials of the second kind $U_n(x)$.

NOTATION: $A_{37}=\frac{n+1}{2}\sqrt{\pi}$, $A_{38}=\left(\frac{n+1}{2}\right)^{2\varepsilon}\sqrt{\pi}$, $A_{39}=(-1)^l\times$ $\times\left(\frac{n+1}{2}i\right)^{2\varepsilon}\sqrt{\pi}$.

1	$(1-x)_+^{1/2}U_n(2x-1)=$ $=A_{37}G_{22}^{20}\left(x\middle\vert\begin{matrix}3/2+n, & -1/2-n\\ 0, & -1/2\end{matrix}\right)$	$A_{37}\Gamma\left[\begin{matrix}s, s-1/2\\ s+n+3/2, s-n-1/2\end{matrix}\right]=$ $=(-1)^nA_{37}\Gamma\left[\begin{matrix}s\\ s+n+3/2\end{matrix}\right]\left(\frac{3}{2}-s\right)_n$ $[\operatorname{Re}s>0]$
2	$(x-1)_+^{1/2}U_n\left(\frac{2}{x}-1\right)=$ $=A_{37}G_{22}^{02}\left(x\middle\vert\begin{matrix}3/2, & 2\\ -n, & n+2\end{matrix}\right)$	$A_{37}\Gamma\left[\begin{matrix}-1/2-s, -1-s\\ 1+n-s, -1-n-s\end{matrix}\right]=$ $=(-1)^nA_{37}\Gamma\left[\begin{matrix}-1/2-s\\ 1+n-s\end{matrix}\right](2+s)_n$ $[\operatorname{Re}s<-1/2]$
3	$(x-1)_+^{1/2}U_n(2x-1)=$ $=A_{37}G_{22}^{02}\left(x\middle\vert\begin{matrix}-1/2-n, & 3/2+n\\ 0, & -1/2\end{matrix}\right)$	$A_{37}\Gamma\left[\begin{matrix}3/2+n-s, -1/2-n-s\\ 1-s, 3/2-s\end{matrix}\right]=$ $=A_{37}\Gamma\left[\begin{matrix}-1/2-n-s\\ 1-s\end{matrix}\right]\left(\frac{3}{2}-s\right)_n$ $[\operatorname{Re}s<-1/2-n]$
4	$(1-x)_+^{1/2}U_n\left(\frac{2}{x}-1\right)=$ $=A_{37}G_{22}^{20}\left(x\middle\vert\begin{matrix}3/2, & 2\\ n+2, & -n\end{matrix}\right)$	$A_{37}\Gamma\left[\begin{matrix}s-n, s+n+2\\ s+3/2, s+2\end{matrix}\right]=$ $=A_{37}\Gamma\left[\begin{matrix}s-n\\ s+3/2\end{matrix}\right](s+2)_n$ $[\operatorname{Re}s>n]$
5	$(1+x)^{-n-2}U_n\left(\frac{1-x}{1+x}\right)=$ $=\frac{2^{2n}}{(2n+1)!}G_{22}^{12}\left(x\middle\vert\begin{matrix}-n-1, & -n-1/2\\ 0, & -1/2\end{matrix}\right)$	$\frac{1}{n!\,(3/2)_n}\Gamma\left[\begin{matrix}s, n+2-s, n+3/2-s\\ 3/2-s\end{matrix}\right]=$ $=\frac{1}{n!\,(3/2)_n}\Gamma[s, n+2-s]\left(\frac{3}{2}-s\right)_n$ $[0<\operatorname{Re}s<n+2]$

6	$(1-x)_+^{1/2}U_n(\sqrt{x})=$ $=A_{37}G_{22}^{20}\left(x\,\middle\vert\,\begin{matrix}(n+3)/2,\ (1-n)/2\\ 0,\ 1/2\end{matrix}\right)$	$A_{37}\Gamma\left[\begin{matrix}s,\ s+1/2\\ s+(n+3)/2,\ s+(1-n)/2\end{matrix}\right]=$ $=(-1)^lA_{37}\Gamma\left[\begin{matrix}s+\varepsilon\\ s+(n+3)/2\end{matrix}\right]\times$ $\times\left(\frac{1}{2}+\varepsilon-s\right)_l$ [Re $s>-\varepsilon$; $l=[n/2]$; $2\varepsilon=n-2l$]
7	$(x-1)_+^{1/2}U_n\left(\frac{1}{\sqrt{x}}\right)=$ $=A_{37}G_{22}^{02}\left(x\,\middle\vert\,\begin{matrix}1,\ 3/2\\ -n/2,\ 1+n/2\end{matrix}\right)$	$A_{37}\Gamma\left[\begin{matrix}-1/2-s,\ -s\\ 1+n/2-s,\ -n/2-s\end{matrix}\right]=$ $=(-1)^lA_{37}\Gamma\left[\begin{matrix}\varepsilon-1/2-s\\ 1+n/2-s\end{matrix}\right](s+\varepsilon+1)_l$ [Re $s<\varepsilon-1/2$; $l=[n/2]$; $2\varepsilon=n-2l$]
8	$(x-1)_+^{1/2}U_n(\sqrt{x})=$ $=A_{37}G_{22}^{02}\left(x\,\middle\vert\,\begin{matrix}(3+n)/2,\ (1-n)/2\\ 0,\ 1/2\end{matrix}\right)$	$A_{37}\Gamma\left[\begin{matrix}-(n+1)/2-s,\ (n+1)/2-s\\ 1/2-s,\ 1-s\end{matrix}\right]=$ $=A_{37}\Gamma\left[\begin{matrix}-(n+1)/2-s\\ 1-\varepsilon-s\end{matrix}\right]\left(\frac{1}{2}+\varepsilon-s\right)_l$ [Re $s<-(n+1)/2$; $l=[n/2]$; $2\varepsilon=n-2l$]
9	$(1-x)_+^{1/2}U_n\left(\frac{1}{\sqrt{x}}\right)=$ $=A_{37}G_{22}^{20}\left(x\,\middle\vert\,\begin{matrix}1,\ 3/2\\ -n/2,\ 1+n/2\end{matrix}\right)$	$A_{37}\Gamma\left[\begin{matrix}s-n/2,\ s+1+n/2\\ s+1,\ s+3/2\end{matrix}\right]=$ $=A_{37}\Gamma\left[\begin{matrix}s-n/2\\ s+3/2-\varepsilon\end{matrix}\right](s+\varepsilon+1)_l$ [Re $s>n/2$; $l=[n/2]$; $2\varepsilon=n-2l$]
10	$(1-x)_+^{\varepsilon-1/2}U_n(\sqrt{1-x})=$ $=(-1)^lA_{38}\times$ $\times G_{22}^{20}\left(x\,\middle\vert\,\begin{matrix}-l-1/2,\ 1/2+l+2\varepsilon\\ 0,\ -1/2\end{matrix}\right)$	$A_{38}\Gamma\left[\begin{matrix}s,\ 3/2+l-s\\ s+1/2+n-l,\ 3/2-s\end{matrix}\right]=$ $=A_{38}\Gamma\left[\begin{matrix}s\\ s+1/2+n-l\end{matrix}\right]\left(\frac{3}{2}-s\right)_l$ [Re $s>0$; $l=[n/2]$; $2\varepsilon=n-2l$]
11	$(x-1)_+^{\varepsilon-1/2}U_n\left(\sqrt{1-\frac{1}{x}}\right)=$ $=(-1)^lA_{38}G_{22}^{02}\left(x\,\middle\vert\,\begin{matrix}1/2+\varepsilon,\ 1+\varepsilon\\ n/2+1,\ -n/2\end{matrix}\right)$	$A_{38}\Gamma\left[\begin{matrix}s+1+n/2,\ 1/2-\varepsilon-s\\ s+\varepsilon+1,\ 1+n/2-s\end{matrix}\right]=$ $=A_{38}\Gamma\left[\begin{matrix}1/2-\varepsilon-s\\ 1+n/2-s\end{matrix}\right](1+\varepsilon+s)_l$ [Re $s<1/2-\varepsilon$; $l=[n/2]$; $2\varepsilon=n-2l$]
12	$(x-1)_+^{\varepsilon-1/2}U_n(i\sqrt{x-1})=$ $=A_{39}G_{22}^{02}\left(x\,\middle\vert\,\begin{matrix}1/2+l+2\varepsilon,\ -1/2-l\\ -1/2,\ 0\end{matrix}\right)$	$A_{39}\Gamma\left[\begin{matrix}1/2+l-n-s,\ 3/2+l-s\\ 1-s,\ 3/2-s\end{matrix}\right]=$ $=A_{39}\Gamma\left[\begin{matrix}1/2+l-n-s\\ 1-s\end{matrix}\right]\left(\frac{3}{2}-s\right)_l$ [Re $s<1/2+l-n$; $l=[n/2]$; $2\varepsilon=n-2l$]
13	$(1-x)_+^{\varepsilon-1/2}U_n\left(i\sqrt{\frac{1}{x}-1}\right)=$ $=A_{39}G_{22}^{20}\left(x\,\middle\vert\,\begin{matrix}1+\varepsilon,\ 1/2+\varepsilon\\ -n/2,\ 1+n/2\end{matrix}\right)$	$A_{39}\Gamma\left[\begin{matrix}s+1+n/2,\ s-n/2\\ s+\varepsilon+1,\ s+\varepsilon+1/2\end{matrix}\right]=$ $=A_{39}\Gamma\left[\begin{matrix}s-n/2\\ s+\varepsilon+1/2\end{matrix}\right](1+\varepsilon+s)_l$ [Re $s>n/2$; $l=[n/2]$; $2\varepsilon=n-2l$]
14	$(1+x)^{-n/2-1}U_n\left(\frac{1}{\sqrt{1+x}}\right)=$ $=\frac{2^n}{n!}G_{22}^{12}\left(x\,\middle\vert\,\begin{matrix}-n/2,\ -(n+1)/2\\ 0,\ -1/2\end{matrix}\right)$	$\frac{2^n}{n!}\Gamma\left[\begin{matrix}s,\ n/2+1-s,\ (n+3)/2-s\\ 3/2-s\end{matrix}\right]=$ $=\frac{2^n}{n!}\Gamma[s,\ 1+n-l-s]\left(\frac{3}{2}-s\right)_l$ [$0<$ Re $s<1+n-l$; $l=[n/2]$; $2\varepsilon=n-2l$]
15	$(1+x)^{-n/2-1}U_n\left(\sqrt{\frac{x}{x+1}}\right)=$ $=\frac{2^n}{n!}G_{22}^{21}\left(x\,\middle\vert\,\begin{matrix}-n/2,\ (1-n)/2\\ 0,\ 1/2\end{matrix}\right)$	$\frac{2^n}{n!}\Gamma\left[\begin{matrix}s,\ s+1/2,\ 1+n/2-s\\ s+(1-n)/2\end{matrix}\right]=$ $=\frac{2^n}{n!}\Gamma\left[s+\varepsilon,\ \frac{n}{2}+1-s\right]\left(\frac{1-n}{2}+s\right)_l$ [$-\varepsilon<$ Re $s<n/2+1$; $l=[n/2]$; $2\varepsilon=n-2l$]

16	$(1+x)^{-n/2-1}U_n\left(\frac{2+x}{2\sqrt{1+x}}\right)=$ $=\frac{1}{n!}G_{22}^{12}\left(x\,\middle\vert\,\begin{matrix}0,\ -n-1\\0,\ -1\end{matrix}\right)$	$\frac{1}{n!}\Gamma\left[\begin{matrix}s,\ 1-s,\ 2+n-s\\2-s\end{matrix}\right]=$ $=\frac{1}{n!}\Gamma[s,\ 1-s]\,(2-s)_n$ [$0<\mathrm{Re}\,s<1$]
17	$(1+x)^{-n/2-1}U_n\left(\frac{2x+1}{2\sqrt{x^2+x}}\right)=$ $=\frac{1}{n!}G_{22}^{21}\left(x\,\middle\vert\,\begin{matrix}-n/2,\ 1-n/2\\-n/2,\ 1+n/2\end{matrix}\right)$	$\frac{1}{n!}\Gamma\left[\begin{matrix}s+n/2+1,\ s-n/2,\ 1+n/2-s\\s+1-n/2\end{matrix}\right]=$ $=\frac{1}{n!}\Gamma\left[s-\frac{n}{2},\ 1+\frac{n}{2}-s\right]\times$ $\times\left(s+1-\frac{n}{2}\right)_n$ [$n/2<\mathrm{Re}\,s<n/2+1$]
18	$(2x-1)(x-1)_+^{1/2}\,U_n(8x^2-8x+1)=$ $=A_{37}G_{22}^{02}\left(x\,\middle\vert\,\begin{matrix}-2n-3/2,\ 2n+5/2\\0,\ -1/2\end{matrix}\right)$	$A_{37}\Gamma\left[\begin{matrix}-2n-3/2-s,\ 2n+5/2-s\\1-s,\ 3/2-s\end{matrix}\right]=$ $=A_{37}\Gamma\left[\begin{matrix}-2n-3/2-s\\1-s\end{matrix}\right]\left(\frac{3}{2}-s\right)_{2n+1}$ [$\mathrm{Re}\,s<-2n-3/2$]
19	$(2-x)(1-x)_+^{1/2}\,U_n\left(\frac{x^2-8x+8}{x^2}\right)=$ $=A_{37}G_{22}^{20}\left(x\,\middle\vert\,\begin{matrix}5/2,\ 3\\2n+4,\ -2n\end{matrix}\right)$	$A_{37}\Gamma\left[\begin{matrix}s-2n,\ s+2n+4\\s+5/2,\ s+3\end{matrix}\right]=$ $=A_{37}\Gamma\left[\begin{matrix}s-2n\\s+5/2\end{matrix}\right](s+3)_{2n+1}$ [$\mathrm{Re}\,s>2n$]
20	$(1-x^2)_+\,U_n\left(\frac{x^2+1}{2x}\right)=$ $=2(n+1)G_{22}^{20}\left(x\,\middle\vert\,\begin{matrix}n+3,\ 1-n\\n+2,\ -n\end{matrix}\right)$	$2(n+1)\Gamma\left[\begin{matrix}s+n+2,\ s-n\\s+n+3,\ s+1-n\end{matrix}\right]=$ $=2(n+1)\frac{1}{(s+n+2)(s-n)}$ [$\mathrm{Re}\,s>n$]
21	$(x^2-1)_+\,U_n\left(\frac{x^2+1}{2x}\right)=$ $=2(n+1)G_{22}^{02}\left(x\,\middle\vert\,\begin{matrix}1-n,\ 3+n\\-n,\ 2+n\end{matrix}\right)$	$2(n+1)\Gamma\left[\begin{matrix}n-s,\ -2-n-s\\n+1-s,\ -1-n-s\end{matrix}\right]=$ $=2(n+1)\frac{1}{(s+n+2)(s-n)}$ [$\mathrm{Re}\,s<-2-n$]
22	$\frac{x+2}{(x+1)^{n+2}}U_n\left(\frac{x^2+2x+2}{2(x+1)}\right)=$ $=\frac{1}{(2n+1)!}G_{22}^{12}\left(x\,\middle\vert\,\begin{matrix}0,\ -2n-2\\0,\ -1\end{matrix}\right)$	$\frac{1}{(2n+1)!}\Gamma\left[\begin{matrix}s,\ 1-s,\ 3+2n-s\\2-s\end{matrix}\right]=$ $=\frac{1}{(2n+1)!}\Gamma[s,\ 1-s]\,(2-s)_{2n+1}$ [$0<\mathrm{Re}\,s<1$]
23	$\frac{2x+1}{(x+1)^{n+2}}U_n\left(\frac{2x^2+2x+1}{2x(x+1)}\right)=$ $=\frac{1}{(2n+1)!}G_{22}^{21}\left(x\,\middle\vert\,\begin{matrix}-n,\ 1-n\\-n,\ 2+n\end{matrix}\right)$	$\frac{1}{(2n+1)!}\Gamma\left[\begin{matrix}s-n,\ s+n+2,\ n+1-s\\s+1-n\end{matrix}\right]=$ $=\frac{1}{(2n+1)!}\Gamma[s-n,\ n+1-s]\times$ $\times(s+1-n)_{2n+1}$ [$n<\mathrm{Re}\,s<n+1$]

8.4.33. The Laguerre polynomials $L_n^{\lambda}(x)$ and $L_n(x)$.

1	$L_n^{\lambda}(x)=$ $\overset{L_{-\infty}}{\longleftarrow}$ $=\Gamma(\lambda+n+1)\,G_{12}^{10}\left(x\,\middle\vert\,\begin{matrix}n+1\\0,\ -\lambda\end{matrix}\right)$	$\Gamma(\lambda+n+1)\,\Gamma\left[\begin{matrix}s\\s+n+1,\ 1+\lambda-s\end{matrix}\right]$ [$s=-k\in D^+,\ k=0,\ 1,\ 2,\ \ldots$]
2	$L_n^{\lambda}\left(\frac{1}{x}\right)=$ $\overset{L_{+\infty}}{\longleftarrow}$ $=\Gamma(\lambda+n+1)\,G_{21}^{01}\left(x\,\middle\vert\,\begin{matrix}1,\ \lambda+1\\-n\end{matrix}\right)$	$\Gamma(\lambda+n+1)\,\Gamma\left[\begin{matrix}-s\\s+\lambda+1,\ 1+n-s\end{matrix}\right]$ [$s=k\in D^-,\ k=0,\ 1,\ 2,\ \ldots$]

3	$e^{-x}L_n^\lambda(x)=\frac{1}{n!}G_{12}^{11}\left(x\left\vert\begin{matrix}-n-\lambda\\0,\ -\lambda\end{matrix}\right.\right)$	$\frac{1}{n!}\Gamma\left[\begin{matrix}s,\ 1+n+\lambda-s\\1+\lambda-s\end{matrix}\right]=$ $=\frac{1}{n!}\Gamma(s)(1+\lambda-s)_n$ [Re $s>0$]
4	$e^{-1/x}L_n^\lambda\left(\frac{1}{x}\right)=\frac{1}{n!}G_{21}^{11}\left(x\left\vert\begin{matrix}1,\ \lambda+1\\\lambda+n+1\end{matrix}\right.\right)$	$\frac{1}{n!}\Gamma\left[\begin{matrix}s+\lambda+n+1,\ -s\\s+\lambda+1\end{matrix}\right]=$ $=\frac{1}{n!}\Gamma(-s)(s+\lambda+1)_n$ [Re $s<0$]
5	$e^{-x}L_n(x)=\frac{(-1)^n}{n!}G_{12}^{20}\left(x\left\vert\begin{matrix}-n\\0,\ 0\end{matrix}\right.\right)$	$\frac{1}{n!}\Gamma\left[\begin{matrix}s,\ 1+n-s\\1-s\end{matrix}\right]=\frac{1}{n!}\Gamma(s)(1-s)_n$ [Re $s>0$]
6	$e^{-1/x}L_n\left(\frac{1}{x}\right)=\frac{(-1)^n}{n!}G_{21}^{02}\left(x\left\vert\begin{matrix}1,\ 1\\n+1\end{matrix}\right.\right)$	$\frac{1}{n!}\Gamma\left[\begin{matrix}s+n+1,\ -s\\s+1\end{matrix}\right]=$ $=\frac{1}{n!}\Gamma(-s)(s+1)_n$ [Re $s<0$]

8.4.34. The Hermite polynomials $H_n(x)$.

1	$H_n(\sqrt{x})=$ $\xleftarrow{L_{-\infty}}$ $=(-1)^{[n/2]}\sqrt{\pi}\,n!\,G_{12}^{10}\left(x\left\vert\begin{matrix}n/2+1\\\varepsilon,\ 1/2-\varepsilon\end{matrix}\right.\right)$	$(-1)^{[n/2]}\sqrt{\pi}\,n!\,\Gamma\left[\begin{matrix}s+\varepsilon\\s+\frac{n}{2}+1,\ \frac{1}{2}+\varepsilon-s\end{matrix}\right]$ [$s=-\varepsilon-k\in D^+$, $k=0, 1, 2, \ldots$; $\varepsilon=n/2-[n/2]$]
2	$H_n\left(\frac{1}{\sqrt{x}}\right)=$ $\xleftarrow{L_{+\infty}}$ $=(-1)^{[n/2]}\sqrt{\pi}\,n!\,G_{21}^{01}\left(x\left\vert\begin{matrix}1-\varepsilon,\\-n/2\\1/2+\varepsilon\end{matrix}\right.\right)$	$(-1)^{[n/2]}\sqrt{\pi}\,n!\,\Gamma\left[\begin{matrix}\varepsilon-s\\s+\varepsilon+1/2,\ 1+n/2-s\end{matrix}\right]$ [$s=\varepsilon+k\in D^-$, $k=0, 1, 2, \ldots$; $\varepsilon=n/2-[n/2]$]
3	$e^{-x}H_n(\sqrt{x})=$ $=2^nG_{12}^{20}\left(x\left\vert\begin{matrix}(1-n)/2\\0,\ 1/2\end{matrix}\right.\right)$	$2^n\Gamma\left[\begin{matrix}s,\ s+1/2\\s+(1-n)/2\end{matrix}\right]=$ $=(-1)^l2^n\Gamma(s+\varepsilon)\left(\frac{1}{2}+\varepsilon-s\right)_l$ [Re $s>-\varepsilon$; $l=[n/2]$; $2\varepsilon=n-2l$]
4	$e^{-1/x}H_n\left(\frac{1}{\sqrt{x}}\right)=$ $=2^nG_{21}^{02}\left(x\left\vert\begin{matrix}1,\ 1/2\\(n+1)/2\end{matrix}\right.\right)$	$2^n\Gamma\left[\begin{matrix}-s,\ 1/2-s\\(1-n)/2-s\end{matrix}\right]=$ $=(-1)^l2^n\Gamma(\varepsilon-s)\left(\frac{1}{2}+\varepsilon+s\right)_l$ [Re $s<\varepsilon$; $l=[n/2]$; $2\varepsilon=n-2l$]

8.4.35. The Gegenbauer polynomicals $C_n^\lambda(x)$.

NOTATION: $A_{40}=\frac{(2\lambda)_n\Gamma(\lambda+1/2)}{n!}$, $A_{41}=\frac{(-1)^n(2\lambda)_n}{n!\,\Gamma(1/2-\lambda)}$, $A_{42}=\frac{1}{n!\,(\lambda+1/2)_n\Gamma(2\lambda)}$, $A_{43}=\frac{\Gamma(1-2\lambda)}{n!\,(\lambda+1/2)_n}$, $A_{44}=\frac{(\lambda)_{n-l}}{l!}\sqrt{\pi}$, $A_{45}=\frac{(-1)^l(\lambda)_{n-l}}{l!}\times i^{2\varepsilon}\sqrt{\pi}$, $A_{46}=\frac{(-2)^n}{n!}\Gamma(1-\lambda)$, $A_{47}=\frac{2^n}{n!\,\Gamma(\lambda)}$, $A_{48}=\frac{\Gamma(2\lambda+n)}{n!}$, $A_{49}=\frac{1}{n!\,\Gamma(\lambda)}$, $A_{50}=\frac{\Gamma(1-\lambda)}{n!}$.

1	$(1-x)_+^{\lambda-1/2}C_n^{\lambda}(2x-1)=$ $=A_{40}G_{22}^{20}\left(x\,\middle\vert\,\begin{matrix}\lambda+n+1/2,\ 1/2-\lambda-n\\ 0,\ 1/2-\lambda\end{matrix}\right)$	$A_{40}\Gamma\begin{bmatrix}s,\ s-\lambda+1/2\\ s+\lambda+n+1/2,\ s+1/2-n-\lambda\end{bmatrix}=$ $=(-1)^n A_{40}\Gamma\begin{bmatrix}s\\ s+\lambda+n+\frac{1}{2}\end{bmatrix}\times$ $\times\left(\frac{1}{2}+\lambda-s\right)_n$ [Re $s>0$; Re $\lambda>-1/2$]
2	$(x-1)_+^{\lambda-1/2}C_n^{\lambda}\left(\frac{2}{x}-1\right)=$ $=A_{40}G_{22}^{02}\left(x\,\middle\vert\,\begin{matrix}\lambda+1/2,\ 2\lambda\\ -n,\ 2\lambda+n\end{matrix}\right)$	$A_{40}\Gamma\begin{bmatrix}1/2-\lambda-s,\ 1-2\lambda-s\\ 1+n-s,\ 1-2\lambda-n-s\end{bmatrix}=$ $=(-1)^n A_{40}\Gamma\begin{bmatrix}1/2-\lambda-s\\ 1+n-s\end{bmatrix}(2\lambda+s)_n$ [Re $s<1/2-$Re λ; Re $\lambda>-1/2$]
3	$(x-1)_+^{\lambda-1/2}C_n^{\lambda}(2x-1)=$ $=A_{40}G_{22}^{02}\left(x\,\middle\vert\,\begin{matrix}1/2+\lambda+n,\ 1/2-\lambda-n\\ 0,\ 1/2-\lambda\end{matrix}\right)$	$A_{40}\Gamma\begin{bmatrix}1/2+n+\lambda-s,\ 1/2-n-\lambda-s\\ 1-s,\ 1/2+\lambda-s\end{bmatrix}=$ $=A_{40}\Gamma\begin{bmatrix}1/2-n-\lambda-s\\ 1-s\end{bmatrix}\left(\frac{1}{2}+\lambda-s\right)_n$ [Re $s<1/2-n-$Re λ; Re $\lambda>-1/2$]
4	$(1-x)_+^{\lambda-1/2}C_n^{\lambda}\left(\frac{2}{x}-1\right)=$ $=A_{40}G_{22}^{20}\left(x\,\middle\vert\,\begin{matrix}\lambda+1/2,\ 2\lambda\\ -n,\ n+2\lambda\end{matrix}\right)$	$A_{40}\Gamma\begin{bmatrix}s-n,\ s+n+2\lambda\\ s+\lambda+1/2,\ s+2\lambda\end{bmatrix}=$ $=A_{40}\Gamma\begin{bmatrix}s-n\\ s+\lambda+1/2\end{bmatrix}(s+2\lambda)_n$ [Re $s>n$; Re $\lambda>-1/2$]
5	$(1+x)^{\lambda-1/2}C_n^{\lambda}(2x+1)=$ $=A_{41}G_{22}^{12}\left(x\,\middle\vert\,\begin{matrix}1/2-\lambda-n,\ 1/2+\lambda+n\\ 0,\ 1/2-\lambda\end{matrix}\right)$	$A_{41}\Gamma\begin{bmatrix}s,\ 1/2+\lambda+n-s,\ 1/2-\lambda-n-s\\ 1/2+\lambda-s\end{bmatrix}=$ $=A_{41}\Gamma\left[s,\ \frac{1}{2}-\lambda-n-s\right]\left(\frac{1}{2}+\lambda-s\right)_n$ [$0<$ Re $s<1/2-n-$Re λ]
6	$(1+x)^{\lambda-1/2}C_n^{\lambda}\left(\frac{2}{x}+1\right)=$ $=A_{41}G_{22}^{21}\left(x\,\middle\vert\,\begin{matrix}1/2+\lambda,\ 2\lambda\\ 2\lambda+n,\ -n\end{matrix}\right)$	$A_{41}\Gamma\begin{bmatrix}s+2\lambda+n,\ s-n,\ 1/2-\lambda-s\\ s+2\lambda\end{bmatrix}=$ $=A_{41}\Gamma\left[s-n,\ \frac{1}{2}-\lambda-s\right](s+2\lambda)_n$ [$n<$ Re $s<1/2-$Re λ]
7	$(1+x)^{-n-2\lambda}C_n^{\lambda}\left(\frac{1-x}{1+x}\right)=$ $=A_{42}G_{22}^{12}\left(x\,\middle\vert\,\begin{matrix}1-2\lambda-n,\ 1/2-\lambda-n\\ 0,\ 1/2-\lambda\end{matrix}\right)$	$A_{42}\Gamma\begin{bmatrix}s,\ n+2\lambda-s,\ n+\lambda+1/2-s\\ \lambda+1/2-s\end{bmatrix}=$ $=A_{42}\Gamma[s,\ n+2\lambda-s]\left(\frac{1}{2}+\lambda-s\right)_n$ [$0<$ Re $s<n+2$ Re λ]
8	$(1-x)_+^{-n-2\lambda}C_n^{\lambda}\left(\frac{1+x}{1-x}\right)=$ $=A_{43}G_{22}^{20}\left(x\,\middle\vert\,\begin{matrix}1-2\lambda-n,\ 1/2-\lambda-n\\ 0,\ 1/2-\lambda\end{matrix}\right)$	$A_{43}\Gamma\begin{bmatrix}s,\ s+1/2-\lambda\\ s+1-n-2\lambda,\ s+1/2-\lambda-n\end{bmatrix}=$ $=(-1)^n A_{43}\Gamma\begin{bmatrix}s\\ s+1-n-2\lambda\end{bmatrix}\times$ $\times\left(\frac{1}{2}+\lambda-s\right)_n$ [Re $s>0$; Re $\lambda<1/2-n$]
9	$(x-1)_+^{-n-2\lambda}C_n^{\lambda}\left(\frac{x+1}{x-1}\right)=$ $=A_{43}G_{22}^{02}\left(x\,\middle\vert\,\begin{matrix}1-2\lambda-n,\ 1/2-\lambda-n\\ 0,\ 1/2-\lambda\end{matrix}\right)$	$A_{43}\Gamma\begin{bmatrix}n+2\lambda-s,\ n+\lambda+1/2-s\\ 1-s,\ \lambda+1/2-s\end{bmatrix}=$ $=A_{43}\Gamma\begin{bmatrix}n+2\lambda-s\\ 1-s\end{bmatrix}\left(\frac{1}{2}+\lambda-s\right)_n$ [Re $s<n+2$ Re λ; Re $\lambda<1/2-n$]

10	$\lvert 1-x\rvert^{-n-2\lambda} C_n^{\lambda}\left(\frac{1+x}{\lvert 1-x\rvert}\right)= \frac{(-1)^n\sqrt{\pi}\,\Gamma(1/2-\lambda)}{2^{2\lambda-1}n!\,(\lambda+1/2)_n\,\Gamma(\lambda)}\times G_{22}^{11}\left(x\,\middle\vert\begin{matrix}1-2\lambda-n,\ 1/2-\lambda-n\\ 0,\ 1/2-\lambda\end{matrix}\right)$	$\frac{(-1)^n\sqrt{\pi}\,\Gamma(1/2-\lambda)}{2^{2\lambda-1}n!\,(\lambda+1/2)_n\,\Gamma(\lambda)}\times \Gamma\left[\begin{matrix}s,\ 2\lambda+n-s\\ s+1/2-\lambda-n,\ \lambda+1/2-s\end{matrix}\right]$ [$0<\operatorname{Re}s<n+2\operatorname{Re}\lambda$; $\operatorname{Re}\lambda<1/2-n$]
11	$(1-x)_+^{\lambda-1/2}C_n^{\lambda}(\sqrt{x})= A_{40}G_{22}^{20}\left(x\,\middle\vert\begin{matrix}\lambda+(n+1)/2,\ (1-n)/2\\ 0,\ 1/2\end{matrix}\right)$	$A_{40}\Gamma\left[\begin{matrix}s,\ s+1/2\\ s+\lambda+(n+1)/2,\ s+(1-n)/2\end{matrix}\right]= (-1)^l A_{40}\Gamma\left[\begin{matrix}s+\varepsilon\\ s+\lambda+(n+1)/2\end{matrix}\right]\times\left(\frac{1}{2}+\varepsilon-s\right)_l$ [$\operatorname{Re}s>-\varepsilon$; $\operatorname{Re}\lambda>-1/2$; $l=[n/2]$; $2\varepsilon=n-2l$]
12	$(x-1)_+^{\lambda-1/2}C_n^{\lambda}\left(\frac{1}{\sqrt{x}}\right)= A_{40}G_{22}^{12}\left(x\,\middle\vert\begin{matrix}\lambda,\ \lambda+1/2\\ -n/2,\ \lambda+n/2\end{matrix}\right)$	$A_{40}\Gamma\left[\begin{matrix}1/2-\lambda-s,\ 1-\lambda-s\\ 1+n/2-s,\ 1-\lambda-n/2-s\end{matrix}\right]= (-1)^l A_{40}\Gamma\left[\begin{matrix}1/2+\varepsilon-\lambda-s\\ 1+n/2-s\end{matrix}\right](s+\lambda+\varepsilon)_l$ [$\operatorname{Re}s<1/2+\varepsilon-\operatorname{Re}\lambda$; $\operatorname{Re}\lambda>-1/2$; $l=[n/2]$; $2\varepsilon=n-2l$]
13	$(x-1)_+^{\lambda-1/2}C_n^{\lambda}(\sqrt{x})= A_{40}G_{22}^{02}\left(x\,\middle\vert\begin{matrix}(1+n)/2+\lambda,\ (1-n)/2\\ 0,\ 1/2\end{matrix}\right)$	$A_{40}\Gamma\left[\begin{matrix}(1-n)/2-\lambda-s,\ (1+n)/2-s\\ 1/2-s,\ 1-s\end{matrix}\right]= A_{40}\Gamma\left[\begin{matrix}(1-n)/2-\lambda-s\\ 1-\varepsilon-s\end{matrix}\right]\left(\frac{1}{2}+\varepsilon-s\right)_l$ [$\operatorname{Re}s<(1-n)/2-\operatorname{Re}\lambda$; $\operatorname{Re}\lambda>-1/2$; $l=[n/2]$; $2\varepsilon=n-2l$]
14	$(1-x)_+^{\lambda-1/2}C_n^{\lambda}\left(\frac{1}{\sqrt{x}}\right)= A_{40}G_{22}^{20}\left(x\,\middle\vert\begin{matrix}\lambda,\ \lambda+1/2\\ -n/2,\ \lambda+n/2\end{matrix}\right)$	$A_{40}\Gamma\left[\begin{matrix}s-n/2,\ s+\lambda+n/2\\ s+\lambda,\ s+\lambda+1/2\end{matrix}\right]= A_{40}\Gamma\left[\begin{matrix}s-n/2\\ s+1/2+\lambda-\varepsilon\end{matrix}\right](s+\varepsilon+\lambda)_l$ [$\operatorname{Re}s>n/2$; $\operatorname{Re}\lambda>-1/2$; $l=[n/2]$; $2\varepsilon=n-2l$]
15	$(1-x)_+^{\varepsilon-1/2}C_n^{\lambda}(\sqrt{1-x})= (-1)^l A_{44}G_{22}^{20}\left(x\,\middle\vert\begin{matrix}\frac{1}{2}+n-l,\ \frac{1}{2}-\lambda-l\\ 0,\ \frac{1}{2}-\lambda\end{matrix}\right)$	$A_{44}\Gamma\left[\begin{matrix}s,\ 1/2+\lambda+l-s\\ s+1/2+n-l,\ 1/2+\lambda-s\end{matrix}\right]= A_{44}\Gamma\left[\begin{matrix}s\\ s+1/2+n-l\end{matrix}\right]\left(\frac{1}{2}+\lambda-s\right)_l$ [$\operatorname{Re}s>0$; $l=[n/2]$; $2\varepsilon=n-2l$]
16	$(x-1)_+^{\varepsilon-1/2}C_n^{\lambda}\left(\sqrt{1-\frac{1}{x}}\right)= (-1)^l A_{44}G_{22}^{02}\left(x\,\middle\vert\begin{matrix}1/2+\varepsilon,\ \lambda+\varepsilon\\ -n/2,\ \lambda+n/2\end{matrix}\right)$	$A_{44}\Gamma\left[\begin{matrix}s+n/2+\lambda,\ 1/2-\varepsilon-s\\ s+\lambda+\varepsilon,\ 1+n/2-s\end{matrix}\right]= A_{44}\Gamma\left[\begin{matrix}1/2-\varepsilon-s\\ 1+n/2-s\end{matrix}\right](\lambda+\varepsilon+s)_l$ [$\operatorname{Re}s<1/2-\varepsilon$; $l=[n/2]$; $2\varepsilon=n-2l$]
17	$(x-1)_+^{\varepsilon-1/2}C_n^{\lambda}(i\sqrt{x-1})= A_{45}G_{22}^{02}\left(x\,\middle\vert\begin{matrix}\frac{1}{2}+l+2\varepsilon,\ \frac{1}{2}-\lambda-l\\ \frac{1}{2}-\lambda,\ 0\end{matrix}\right)$	$A_{45}\Gamma\left[\begin{matrix}1/2+l-n-s,\ 1/2+\lambda+l-s\\ 1-s,\ 1/2+\lambda-s\end{matrix}\right]= A_{45}\Gamma\left[\begin{matrix}1/2+l-n-s\\ 1-s\end{matrix}\right]\left(\frac{1}{2}+\lambda-s\right)_l$ [$\operatorname{Re}s<1/2+l-n$; $l=[n/2]$; $2\varepsilon=n-2l$]

18	$(1-x)_+^{\varepsilon-1/2}C_n^{\lambda}\left(i\sqrt{\frac{1}{x}-1}\right)= =A_{45}G_{22}^{20}\left(x\left\vert\begin{matrix}\lambda+\varepsilon,\ 1/2+\varepsilon\\ -n/2,\ \lambda+n/2\end{matrix}\right.\right)$	$A_{45}\Gamma\begin{bmatrix}s+\lambda+n/2,\ s-n/2\\ s+\lambda+\varepsilon,\ s+\varepsilon+1/2\end{bmatrix}= A_{45}\Gamma\begin{bmatrix}s-n/2\\ s+\varepsilon+1/2\end{bmatrix}(\lambda+\varepsilon+s)_l$ [Re $s>n/2$; $l=[n/2]$; $2\varepsilon=n-2l$]
19	$(1-x)_+^{-\lambda-n/2}C_n^{\lambda}\left(\frac{1}{\sqrt{1-x}}\right)= =A_{46}G_{22}^{20}\left(x\left\vert\begin{matrix}(1-n)/2-\lambda,\ 1-\lambda-n/2\\ 0,\ 1/2-\lambda\end{matrix}\right.\right)$	$A_{46}\Gamma\begin{bmatrix}s,\ s+1/2-\lambda\\ s+(1-n)/2-\lambda,\ s+1-\lambda-n/2\end{bmatrix}= =(-1)^l A_{46}\Gamma\begin{bmatrix}s\\ s+1+l-n-\lambda\end{bmatrix}\times \times\left(\frac{1}{2}+\lambda-s\right)_l$ [Re $s>0$; Re $\lambda<1-n$; $l=[n/2]$; $2\varepsilon=n-2l$]
20	$(x-1)_+^{-\lambda-n/2}C_n^{\lambda}\left(\sqrt{\frac{x}{x-1}}\right)= =A_{46}G_{22}^{02}\left(x\left\vert\begin{matrix}1-\lambda-n/2,\ (1-n)/2\\ 0,\ 1/2\end{matrix}\right.\right)$	$A_{46}\Gamma\begin{bmatrix}\lambda+n/2-s,\ (n+1)/2-s\\ 1/2-s,\ 1-s\end{bmatrix}= =(-1)^l A_{46}\Gamma\begin{bmatrix}\lambda+n/2-s\\ 1-\varepsilon-s\end{bmatrix}\left(\frac{1-n}{2}+s\right)_l$ [Re $s<n/2+$Re λ; Re $\lambda<1-n$; $l=[n/2]$; $2\varepsilon=n-2l$]
21	$(1+x)^{-\lambda-n/2}C_n^{\lambda}\left(\frac{1}{\sqrt{1+x}}\right)= =A_{47}G_{22}^{12}\left(x\left\vert\begin{matrix}1-\lambda-n/2,\ (1-n)/2-\lambda\\ 0,\ 1/2-\lambda\end{matrix}\right.\right)$	$A_{47}\Gamma\begin{bmatrix}s,\ \lambda+n/2-s,\ \lambda+(n+1)/2-s\\ \lambda+1/2-s\end{bmatrix}= =A_{47}\Gamma[s,\ \lambda+n-l-s]\left(\frac{1}{2}+\lambda-s\right)_l$ [$0<$ Re $s<$ Re $\lambda+n-l$; $l=[n/2]$; $2\varepsilon=n-2l$]
22	$(1+x)^{-\lambda-n/2}C_n^{\lambda}\left(\sqrt{\frac{x}{x+1}}\right)= =A_{47}G_{22}^{21}\left(x\left\vert\begin{matrix}1-\lambda-n/2,\ (1-n)/2\\ 0,\ 1/2\end{matrix}\right.\right)$	$A_{47}\Gamma\begin{bmatrix}s,\ s+1/2,\ \lambda+n/2-s\\ s+(1-n)/2\end{bmatrix}= =A_{47}\Gamma\left[s+\varepsilon,\ \frac{n}{2}+\lambda-s\right]\left(\frac{1-n}{2}+s\right)_l$ [$-\varepsilon<$ Re $s<n/2+$Re λ; $l=[n/2]$; $2\varepsilon=n-2l$]
23	$(1-x)_+^{2\lambda-1}C_n^{\lambda}\left(\frac{x+1}{2\sqrt{x}}\right)= =A_{48}G_{22}^{20}\left(x\left\vert\begin{matrix}2\lambda+n/2,\ \lambda-n/2\\ -n/2,\ \lambda+n/2\end{matrix}\right.\right)$	$A_{48}\Gamma\begin{bmatrix}s-n/2,\ s+\lambda+n/2\\ s+2\lambda+n/2,\ s+\lambda-n/2\end{bmatrix}= =A_{48}\Gamma\begin{bmatrix}s-n/2\\ s+2\lambda+n/2\end{bmatrix}\left(s+\lambda-\frac{n}{2}\right)_n$ [Re $s>n/2$; Re $\lambda>0$]
24	$(x-1)_+^{2\lambda-1}C_n^{\lambda}\left(\frac{x+1}{2\sqrt{x}}\right)= =A_{48}G_{22}^{02}\left(x\left\vert\begin{matrix}2\lambda+n/2,\ \lambda-n/2\\ -n/2,\ \lambda+n/2\end{matrix}\right.\right)$	$A_{48}\Gamma\begin{bmatrix}1-2\lambda-n/2-s,\ 1+n/2-\lambda-s\\ 1+n/2-s,\ 1-\lambda-n/2-s\end{bmatrix}= =(-1)^n A_{48}\Gamma\begin{bmatrix}1-2\lambda-n/2-s\\ 1+n/2-s\end{bmatrix}\times \times\left(\lambda-\frac{n}{2}+s\right)_n$ [Re $s<1-n/2-2$ Re λ; Re $\lambda>0$]
25	$(1+x)^{-\lambda-n/2}C_n^{\lambda}\left(\frac{2+x}{2\sqrt{1+x}}\right)= =A_{49}G_{22}^{12}\left(x\left\vert\begin{matrix}1-\lambda,\ 1-2\lambda-n\\ 0,\ 1-2\lambda\end{matrix}\right.\right)$	$A_{49}\Gamma\begin{bmatrix}s,\ \lambda-s,\ 2\lambda+n-s\\ 2\lambda-s\end{bmatrix}= =A_{49}\Gamma[s,\ \lambda-s]\,(2\lambda-s)_n$ [$0<$ Re $s<$ Re λ]
26	$(1+x)^{-\lambda-n/2}C_n^{\lambda}\left(\frac{2x+1}{2\sqrt{x^2+x}}\right)= =A_{49}G_{22}^{21}\left(x\left\vert\begin{matrix}1-\lambda-n/2,\ \lambda-n/2\\ \lambda+n/2,\ -n/2\end{matrix}\right.\right)$	$A_{49}\Gamma\begin{bmatrix}s+\lambda+n/2,\ s-n/2,\ \lambda+n/2-s\\ s+\lambda-n/2\end{bmatrix}= =A_{49}\Gamma\left[s-\frac{n}{2},\ \lambda+\frac{n}{2}-s\right]\times \times\left(s+\lambda-\frac{n}{2}\right)_n$ [$n/2<$ Re $s<n/2+$Re λ]

27	$(1-x)_+^{-\lambda-n/2}C_n^{\lambda}\left(\frac{2-x}{2\sqrt{1-x}}\right)=$ $=(-1)^n A_{50}G_{22}^{20}\left(x\,\middle\|\,\begin{matrix}1-2\lambda-n,\ 1-\lambda\\ 0,\ 1-2\lambda\end{matrix}\right)$	$A_{50}\Gamma\left[\begin{matrix}s,\ 2\lambda+n-s\\ s+1-\lambda,\ 2\lambda-s\end{matrix}\right]=$ $=A_{50}\Gamma\left[\begin{matrix}s\\ s+1-\lambda\end{matrix}\right](2\lambda-s)_n$ [Re $s>0$; Re $\lambda<1-n$]
28	$(x-1)_+^{-\lambda-n/2}C_n^{\lambda}\left(\frac{2x-1}{2\sqrt{x^2-x}}\right)=$ $=(-1)^n A_{50}G_{22}^{02}\left(x\,\middle\|\,\begin{matrix}1-\lambda-n/2,\ \lambda-n/2\\ \lambda+n/2,\ -n/2\end{matrix}\right)$	$A_{50}\Gamma\left[\begin{matrix}s+\lambda+n/2,\ \lambda+n/2-s\\ s+\lambda-n/2,\ 1+n/2-s\end{matrix}\right]=$ $=A_{50}\Gamma\left[\begin{matrix}\lambda+n/2-s\\ 1+n/2-s\end{matrix}\right]\left(s+\lambda-\frac{n}{2}\right)_n$ [Re $s<n/2+$Re λ; Re $\lambda<1-n$]

8.4.36. The Jacobi polynomials $P_n^{(\rho,\sigma)}(x)$.

NOTATION: $A_{51}=\frac{\Gamma(\rho+n+1)}{n!}$, $A_{52}=\frac{1}{n!\,\Gamma(-\sigma-n)}$, $A_{53}=\frac{1}{n!\,\Gamma(\rho+\sigma+n+1)}$, $A_{54}=\frac{\Gamma(-\rho-\sigma-n)}{n!}$.

1	$(1-x)_+^{\rho}P_n^{(\rho,\sigma)}(2x-1)=$ $=A_{51}G_{22}^{20}\left(x\,\middle\|\,\begin{matrix}\rho+n+1,\ -\sigma-n\\ 0,\ -\sigma\end{matrix}\right)$	$A_{51}\Gamma\left[\begin{matrix}s,\ s-\sigma\\ s+\rho+n+1,\ s-\sigma-n\end{matrix}\right]=$ $=(-1)^n A_{51}\Gamma\left[\begin{matrix}s\\ s+\rho+n+1\end{matrix}\right](1+\sigma-s)_n$ [Re $s>0$; Re $\rho>-1$]
2	$(x-1)_+^{\rho}P_n^{(\rho,\sigma)}\left(\frac{2}{x}-1\right)=$ $=A_{51}G_{22}^{02}\left(x\,\middle\|\,\begin{matrix}\rho+1,\ \rho+\sigma+1\\ -n,\ \rho+\sigma+n+1\end{matrix}\right)$	$A_{51}\Gamma\left[\begin{matrix}-\rho-s,\ -\rho-\sigma-s\\ 1+n-s,\ -\rho-\sigma-n-s\end{matrix}\right]=$ $=(-1)^n A_{51}\Gamma\left[\begin{matrix}-\rho-s\\ 1+n-s\end{matrix}\right]\times$ $\times(1+\rho+\sigma+s)_n$ [Re $s<-$Re ρ; Re $\rho>-1$]
3	$(x-1)_+^{\rho}P_n^{(\rho,\sigma)}(2x-1)=$ $=A_{51}G_{22}^{02}\left(x\,\middle\|\,\begin{matrix}-\sigma-n,\ \rho+n+1\\ 0,\ -\sigma\end{matrix}\right)$	$A_{51}\Gamma\left[\begin{matrix}1+n+\sigma-s,\ -\rho-n-s\\ 1+\sigma-s,\ 1-s\end{matrix}\right]=$ $=A_{51}\Gamma\left[\begin{matrix}-\rho-n-s\\ 1-s\end{matrix}\right](1+\sigma-s)_n$ [Re $s<-n-$Re ρ; Re $\rho>-1$]
4	$(1-x)_+^{\rho}P_n^{(\rho,\sigma)}\left(\frac{2}{x}-1\right)=$ $=A_{51}G_{22}^{20}\left(x\,\middle\|\,\begin{matrix}\rho+\sigma+1,\ \rho+1\\ \rho+\sigma+n+1,\ -n\end{matrix}\right)$	$A_{51}\Gamma\left[\begin{matrix}s+\rho+\sigma+n+1,\ s-n\\ s+\rho+\sigma+1,\ s+\rho+1\end{matrix}\right]=$ $=A_{51}\Gamma\left[\begin{matrix}s-n\\ s+\rho+1\end{matrix}\right](s+\rho+\sigma+1)_n$ [Re $s>n$; Re $\rho>-1$]
5	$(1+x)^{\sigma}P_n^{(\rho,\sigma)}(2x+1)=$ $=A_{52}G_{22}^{12}\left(x\,\middle\|\,\begin{matrix}\sigma+n+1,\ -\rho-n\\ 0,\ -\rho\end{matrix}\right)$	$A_{52}\Gamma\left[\begin{matrix}s,\ -\sigma-n-s,\ 1+\rho+n-s\\ 1+\rho-s\end{matrix}\right]=$ $=A_{52}\Gamma[s,\ -\sigma-n-s]\,(1+\rho-s)_n$ [$0<$ Re $s<-n-$Re σ]
6	$(1+x)^{\sigma}P_n^{(\rho,\sigma)}\left(\frac{2}{x}+1\right)=$ $=A_{52}G_{22}^{21}\left(x\,\middle\|\,\begin{matrix}\sigma+1,\ \rho+\sigma+1\\ -n,\ \rho+\sigma+n+1\end{matrix}\right)$	$A_{52}\Gamma\left[\begin{matrix}s-n,\ s+\rho+\sigma+n+1,\ -\sigma-s\\ s+\rho+\sigma+1\end{matrix}\right]=$ $=A_{52}\Gamma[s-n,\ -\sigma-s]\,(s+\rho+\sigma+1)_n$ [$n<$ Re $s<-$Re σ]
7	$(1+x)^{-n-\rho-\sigma-1}P_n^{(\rho,\sigma)}\left(\frac{1-x}{1+x}\right)=$ $=A_{53}G_{22}^{12}\left(x\,\middle\|\,\begin{matrix}-\rho-\sigma-n,\ -\rho-n\\ 0,\ -\rho\end{matrix}\right)$	$A_{53}\Gamma\left[\begin{matrix}s,\ n+\rho+\sigma+1-s,\ \rho+n+1-s\\ \rho+1-s\end{matrix}\right]=$ $=A_{53}\Gamma[s,\ n+\rho+\sigma+1-s]\,(\rho+1-s)_n$ [$0<$ Re $s<n+1+$Re $(\rho+\sigma)$]

8	$(1-x)_+^{-n-\rho-\sigma-1} P_n^{(\rho,\sigma)}\left(\frac{1+x}{1-x}\right)=$ $=(-1)^n A_{54} G_{22}^{20}\left(x \middle\vert \begin{matrix} -\rho-n, -\rho-\sigma-n \\ 0, -\rho \end{matrix}\right)$	$A_{54}\Gamma\begin{bmatrix} s, \rho+n+1-s \\ s-n-\rho-\sigma, 1+\rho-s \end{bmatrix}=$ $=A_{54}\Gamma\begin{bmatrix} s \\ s-n-\rho-\sigma \end{bmatrix}(1+\rho-s)_n$ [Re $s>0$; Re $(\rho+\sigma)<-2n$]
9	$(x-1)_+^{-n-\rho-\sigma-1} P_n^{(\rho,\sigma)}\left(\frac{x+1}{x-1}\right)=$ $=(-1)^n A_{54} G_{22}^{02}\left(x \middle\vert \begin{matrix} -\rho-\sigma-n, -\sigma-n \\ -\sigma, 0 \end{matrix}\right)$	$A_{54}\Gamma\begin{bmatrix} s-\sigma, n+1+\rho+\sigma-s \\ s-\sigma-n, 1-s \end{bmatrix}=$ $(-1)^n A_{54}\Gamma\begin{bmatrix} n+1+\rho+\sigma-s \\ 1-s \end{bmatrix}(1+\sigma-s)_n$ [Re $s<n+1+$Re $(\rho+\sigma)$; Re $(\rho+\sigma)<-2n$]

8.4.37. The Laguerre function $L_\nu(x)$.

1	$e^{-x}L_\nu(x)=$ $=\frac{1}{\Gamma(\nu+1)} G_{12}^{11}\left(x \middle\vert \begin{matrix} -\nu \\ 0, 0 \end{matrix}\right)$	$\frac{1}{\Gamma(\nu+1)}\Gamma\begin{bmatrix} s, 1+\nu-s \\ 1-s \end{bmatrix}$ [$0<$ Re $s<1+$Re ν, $\nu\neq 0, 1, 2, \ldots$] or [Re $s>0$, $\nu=0, 1, 2, \ldots$].
2	$e^{-1/x}L_\nu\left(\frac{1}{x}\right)=$ $=\frac{1}{\Gamma(\nu+1)} G_{21}^{11}\left(x \middle\vert \begin{matrix} 1, 1 \\ 1+\nu \end{matrix}\right)$	$\frac{1}{\Gamma(\nu+1)}\Gamma\begin{bmatrix} s+\nu+1, -s \\ s+1 \end{bmatrix}$ [$-1-$Re $\nu<$ Re $s<0$, $\nu\neq 0, 1, 2, \ldots$] or [Re $s<0$, $\nu=0, 1, 2, \ldots$].

8.4.38. The Bateman function $k_\nu(x)$.

NOTATION: $A_{55}=\frac{1}{\Gamma(1+\nu/2)}$, $A_{56}=-\frac{\sin(\nu\pi/2)}{\pi\Gamma(1-\nu/2)}$, $A_{57}=\frac{2\sin(\nu\pi/2)}{\pi^{3/2}\nu}$.

1	$e^{-x/2}k_\nu\left(\frac{x}{2}\right)=A_{55}G_{12}^{20}\left(x \middle\vert \begin{matrix} 1-\nu/2 \\ 0, 1 \end{matrix}\right)$	$A_{55}\Gamma\begin{bmatrix} s, s+1 \\ s+1-\nu/2 \end{bmatrix}$ [Re $s>0$]
2	$e^{-1/(2x)}k_\nu\left(\frac{1}{2x}\right)=A_{55}G_{21}^{02}\left(x \middle\vert \begin{matrix} 0, 1 \\ \nu/2 \end{matrix}\right)$	$A_{55}\Gamma\begin{bmatrix} -s, 1-s \\ 1-\nu/2-s \end{bmatrix}$ [Re $s<0$]
3	$e^{x/2}k_\nu\left(\frac{x}{2}\right)=A_{56}G_{12}^{21}\left(x \middle\vert \begin{matrix} 1+\nu/2 \\ 0, 1 \end{matrix}\right)$	$A_{56}\Gamma\left[s, s+1, -\frac{\nu}{2}-s\right]$ [$0<$ Re $s<-$Re $\nu/2$]
4	$e^{1/(2x)}k_\nu\left(\frac{1}{2x}\right)=A_{56}G_{21}^{12}\left(x \middle\vert \begin{matrix} 0, 1 \\ -\nu/2 \end{matrix}\right)$	$A_{56}\Gamma\left[s-\frac{\nu}{2}, -s, 1-s\right]$ [Re $\nu/2<$ Re $s<0$]
5	$k_{-\nu}(\sqrt{x})\,k_\nu(\sqrt{x})=$ $=A_{57}G_{24}^{40}\left(x \middle\vert \begin{matrix} 1+\nu/2, 1-\nu/2 \\ 0, 1/2, 1, 1 \end{matrix}\right)$	$A_{57}\Gamma\begin{bmatrix} s, s+1/2, s+1, s+1 \\ s+1+\nu/2, s+1-\nu/2 \end{bmatrix}$ [Re $s>0$]
6	$k_{-\nu}\left(\frac{1}{\sqrt{x}}\right)k_\nu\left(\frac{1}{\sqrt{x}}\right)=$ $=A_{57}G_{42}^{04}\left(x \middle\vert \begin{matrix} 0, 0, 1/2, 1 \\ -\nu/2, \nu/2 \end{matrix}\right)$	$A_{57}\Gamma\begin{bmatrix} -s, 1-s, 1-s, 1/2-s \\ 1+\nu/2-s, 1-\nu/2-s \end{bmatrix}$ [Re $s<0$]

8.4.39. The Lommel function $U_\nu(x, z)$.

1	$U_\nu(4\sqrt{x}, 0)=\sqrt{\pi}\, G_{13}^{11}\left(x\left\vert\begin{matrix}\nu/2,\\ \nu/2, 0, 1/2\end{matrix}\right.\right)$	$\sqrt{\pi}\,\Gamma\left[\begin{matrix}s+\nu/2,\ 1-\nu/2-s\\ 1/2-s,\ 1-s\end{matrix}\right]$ $[-\mathrm{Re}\,\nu/2<\mathrm{Re}\,s<1/2,\ 1-\mathrm{Re}\,\nu/2$
2	$U_\nu\left(\frac{4}{\sqrt{x}}, 0\right)=$ $=\sqrt{\pi}\, G_{31}^{11}\left(x\left\vert\begin{matrix}1-\nu/2,\ 1/2,\ 1\\ 1-\nu/2\end{matrix}\right.\right)$	$\sqrt{\pi}\,\Gamma\left[\begin{matrix}s+1-\nu/2,\ \nu/2-s\\ s+1/2,\ s+1\end{matrix}\right]$ $[-1/2,\ \mathrm{Re}\,\nu/2-1<\mathrm{Re}\,s<\mathrm{Re}\,\nu/2$

8.4.40. The complete elliptic integrals $\mathbf{K}(x)$, $\mathbf{E}(x)$, $\mathbf{D}(x)$.

NOTATION: $A_{58}=\frac{\pi}{2\sqrt{2}\,\Gamma^2(1/4)}$, $A_{59}=\frac{\pi}{2\sqrt{2}\,\Gamma^2(3/4)}$, $A_{60}=\frac{\pi}{2\Gamma^2(1/4)}$

$A_{61}=\frac{\pi}{2\Gamma^2(3/4)}$.

1	$\mathbf{K}(i\sqrt{x})=\frac{1}{2}G_{22}^{12}\left(x\left\vert\begin{matrix}1/2,\ 1/2\\ 0,\ 0\end{matrix}\right.\right)$	$\frac{1}{2}\Gamma\left[\begin{matrix}s,\ 1/2-s,\ 1/2-s\\ 1-s\end{matrix}\right]$	$[0<\mathrm{Re}\,s<1/2$
2	$\mathbf{K}\left(\frac{i}{\sqrt{x}}\right)=\frac{1}{2}G_{22}^{21}\left(x\left\vert\begin{matrix}1,\ 1\\ 1/2,\ 1/2\end{matrix}\right.\right)$	$\frac{1}{2}\Gamma\left[\begin{matrix}s+1/2,\ s+1/2,\ -s\\ s+1\end{matrix}\right]$	$[-1/2<\mathrm{Re}\,s<0$
3	$\mathbf{K}(\sqrt{1-x})\,H(1-x)=$ $=\frac{\pi}{2}G_{22}^{20}\left(x\left\vert\begin{matrix}1/2,\ 1/2\\ 0,\ 0\end{matrix}\right.\right)$	$\frac{\pi}{2}\Gamma\left[\begin{matrix}s,\ s\\ s+1/2,\ s+1/2\end{matrix}\right]$	$[\mathrm{Re}\,s>0$
4	$\mathbf{K}\left(\sqrt{1-\frac{1}{x}}\right)H(x-1)=$ $=\frac{\pi}{2}G_{22}^{02}\left(x\left\vert\begin{matrix}1,\ 1\\ 1/2,\ 1/2\end{matrix}\right.\right)$	$\frac{\pi}{2}\Gamma\left[\begin{matrix}-s,\ -s\\ 1/2-s,\ 1/2-s\end{matrix}\right]$	$[\mathrm{Re}\,s<0$
5	$(1+x)^{-1/2}\mathbf{K}\left(\frac{1}{\sqrt{1+x}}\right)=$ $=\frac{1}{2}G_{22}^{21}\left(x\left\vert\begin{matrix}1/2,\ 1/2\\ 0,\ 0\end{matrix}\right.\right)$	$\frac{1}{2}\Gamma\left[\begin{matrix}s,\ s,\ 1/2-s\\ s+1/2\end{matrix}\right]$	$[0<\mathrm{Re}\,s<1/2$
6	$(1+x)^{1/2}\mathbf{K}\left(\sqrt{\frac{x}{1+x}}\right)=$ $=\frac{1}{2}G_{22}^{12}\left(x\left\vert\begin{matrix}1/2,\ 1/2\\ 0,\ 0\end{matrix}\right.\right)$	$\frac{1}{2}\Gamma\left[\begin{matrix}s,\ 1/2-s,\ 1/2-s\\ 1-s\end{matrix}\right]$	$[0<\mathrm{Re}\,s<1/2$
7	$\frac{H(1-x)}{1+\sqrt{x}}\mathbf{K}\left(\frac{1-\sqrt{x}}{1+\sqrt{x}}\right)=$ $=\frac{\pi}{4}G_{22}^{20}\left(x\left\vert\begin{matrix}1/2,\ 1/2\\ 0,\ 0\end{matrix}\right.\right)$	$\frac{\pi}{4}\Gamma\left[\begin{matrix}s,\ s\\ s+1/2,\ s+1/2\end{matrix}\right]$	$[\mathrm{Re}\,s>0$
8	$\frac{H(x-1)}{\sqrt{x}+1}\mathbf{K}\left(\frac{\sqrt{x}-1}{\sqrt{x}+1}\right)=$ $=\frac{\pi}{4}G_{22}^{02}\left(x\left\vert\begin{matrix}1/2,\ 1/2\\ 0,\ 0\end{matrix}\right.\right)$	$\frac{\pi}{4}\Gamma\left[\begin{matrix}1/2-s,\ 1/2-s\\ 1-s,\ 1-s\end{matrix}\right]$	$[\mathrm{Re}\,s<1/2$

9	$\frac{1}{1+\sqrt{x}}\,\mathbf{K}\left(\left\lvert\frac{1-\sqrt{x}}{1+\sqrt{x}}\right\rvert\right)= \frac{1}{4\pi}\,G_{22}^{22}\left(x\,\middle\vert\,\begin{matrix}1/2,\ 1/2\\ 0,\ 0\end{matrix}\right)$	$\frac{1}{4\pi}\,\Gamma\left[s,\ s,\ \frac{1}{2}-s,\ \frac{1}{2}-s\right]$ [0 < Re s < 1/2]
10	$(1-\sqrt{1+x})\,\mathbf{K}\left(\frac{2+x-2\sqrt{1+x}}{x}\right)= -\frac{1}{4}\,G_{22}^{12}\left(x\,\middle\vert\,\begin{matrix}3/2,\ 3/2\\ 1,\ 1\end{matrix}\right)$	$-\frac{1}{4}\,\Gamma\begin{bmatrix}s+1,\ -1/2-s,\ -1/2-s\\ -s\end{bmatrix}$ [−1 < Re s < −1/2]
11	$(\sqrt{x}-\sqrt{1+x})\,\mathbf{K}\left(1+2x-2\sqrt{x^2+x}\right)= -\frac{1}{4}\,G_{22}^{21}\left(x\,\middle\vert\,\begin{matrix}1/2,\ 1/2\\ 0,\ 0\end{matrix}\right)$	$-\frac{1}{4}\,\Gamma\begin{bmatrix}s,\ s,\ 1/2-s\\ 1/2+s\end{bmatrix}$ [0 < Re s < 1/2]
12	$H(1-x)\,\mathbf{K}\left(\sqrt{\frac{1-\sqrt{x}}{2}}\right)= \frac{\pi}{2}\,G_{22}^{20}\left(x\,\middle\vert\,\begin{matrix}3/4,\ 3/4\\ 0,\ 1/2\end{matrix}\right)$	$\frac{\pi}{2}\,\Gamma\begin{bmatrix}s,\ s+1/2\\ s+3/4,\ s+3/4\end{bmatrix}$ [Re s > 0]
13	$H(x-1)\,\mathbf{K}\left(\frac{\sqrt{\sqrt{x}-1}}{\sqrt{2}\,x^{1/4}}\right)= \frac{\pi}{2}\,G_{22}^{02}\left(x\,\middle\vert\,\begin{matrix}1/2,\ 1\\ 1/4,\ 1/4\end{matrix}\right)$	$\frac{\pi}{2}\,\Gamma\begin{bmatrix}-s,\ 1/2-s\\ 3/4-s,\ 3/4-s\end{bmatrix}$ [Re s < 0]
14	$\frac{H(1-x)}{\sqrt{1+\sqrt{x}}}\,\mathbf{K}\left(\sqrt{\frac{1-\sqrt{x}}{1+\sqrt{x}}}\right)= \frac{\pi}{2\sqrt{2}}\,G_{22}^{20}\left(x\,\middle\vert\,\begin{matrix}1/4,\ 3/4\\ 0,\ 0\end{matrix}\right)$	$\frac{\pi}{2\sqrt{2}}\,\Gamma\begin{bmatrix}s,\ s\\ s+3/4,\ s+1/4\end{bmatrix}$ [Re s > 0]
15	$\frac{H(x-1)}{\sqrt{1+\sqrt{x}}}\,\mathbf{K}\left(\sqrt{\frac{\sqrt{x}-1}{\sqrt{x}+1}}\right)= \frac{\pi}{2\sqrt{2}}\,G_{22}^{02}\left(x\,\middle\vert\,\begin{matrix}3/4,\ 3/4\\ 1/2,\ 0\end{matrix}\right)$	$\frac{\pi}{2\sqrt{2}}\,\Gamma\begin{bmatrix}1/4-s,\ 1/4-s\\ 1-s,\ 1/2-s\end{bmatrix}$ [Re s < 1/4]
16	$\sqrt{1-\sqrt{1-x}}\,H(1-x)\times \mathbf{K}\left(\sqrt{\frac{2}{x}}\sqrt{x-1+\sqrt{1-x}}\right)= \frac{\pi}{2}\,G_{22}^{20}\left(x\,\middle\vert\,\begin{matrix}3/4,\ 5/4\\ 1/2,\ 1/2\end{matrix}\right)$	$\frac{\pi}{2}\,\Gamma\begin{bmatrix}s+1/2,\ s+1/2\\ s+5/4,\ s+3/4\end{bmatrix}$ [Re s > −1/2]
17	$\sqrt{\sqrt{x}-\sqrt{x-1}}\,H(x-1)\times \mathbf{K}\left(\sqrt{2(1-x+\sqrt{x^2-x})}\right)= \frac{\pi}{2}\,G_{22}^{02}\left(x\,\middle\vert\,\begin{matrix}3/4,\ 3/4\\ 0,\ 1/2\end{matrix}\right)$	$\frac{\pi}{2}\,\Gamma\begin{bmatrix}1/4-s,\ 1/4-s\\ 1-s,\ 1/2-s\end{bmatrix}$ [Re s < 1/4]
18	$(\sqrt{1+x}+1)^{-1/2}\,\mathbf{K}\left(\sqrt{\frac{\sqrt{1+x}-1}{\sqrt{1+x}+1}}\right)= A_{58}\,G_{22}^{12}\left(x\,\middle\vert\,\begin{matrix}3/4,\ 3/4\\ 0,\ 0\end{matrix}\right)$	$A_{58}\,\Gamma\begin{bmatrix}s,\ 1/4-s,\ 1/4-s\\ 1-s\end{bmatrix}$ [0 < Re s < 1/4]

19	$(\sqrt{1+x}+\sqrt{x})^{-1/2}\times$ $\times K\left(\sqrt{\frac{\sqrt{1+x}-\sqrt{x}}{\sqrt{1+x}+\sqrt{x}}}\right)=$ $=A_{58}G_{22}^{21}\left(x\left\vert\begin{matrix}3/4, & 3/4\\0, & 0\end{matrix}\right.\right)$	$A_{58}\Gamma\left[\begin{matrix}s, s, 1/4-s\\s+3/4\end{matrix}\right]$ [$0<\operatorname{Re}s<1/4$]
20	$\frac{(1+x)^{-1/2}}{\sqrt{\sqrt{1+x}+1}}K\left(\sqrt{\frac{\sqrt{1+x}-1}{\sqrt{1+x}+1}}\right)=$ $=A_{59}G_{22}^{12}\left(x\left\vert\begin{matrix}1/4, & 1/4\\0, & 0\end{matrix}\right.\right)$	$A_{59}\Gamma\left[\begin{matrix}s, 3/4-s, 3/4-s\\1-s\end{matrix}\right]$ [$0<\operatorname{Re}s<3/4$]
21	$\frac{(1+x)^{-1/2}}{\sqrt{\sqrt{1+x}+\sqrt{x}}}\times$ $\times K\left(\sqrt{\frac{\sqrt{1+x}-\sqrt{x}}{\sqrt{1+x}+\sqrt{x}}}\right)=$ $=A_{59}G_{22}^{21}\left(x\left\vert\begin{matrix}1/4, & 1/4\\0, & 0\end{matrix}\right.\right)$	$A_{59}\Gamma\left[\begin{matrix}s, s, 3/4-s\\s+1/4\end{matrix}\right]$ [$0<\operatorname{Re}s<3/4$]
22	$(\sqrt{1+x}-\sqrt{x})^{1/2}\times$ $\times K\left(x^{1/4}\sqrt{2(\sqrt{1+x}-\sqrt{x})}\right)=$ $=A_{60}G_{22}^{12}\left(x\left\vert\begin{matrix}3/4, & 3/4\\0, & 0\end{matrix}\right.\right)$	$A_{60}\Gamma\left[\begin{matrix}s, 1/4-s, 1/4-s\\1-s\end{matrix}\right]$ [$0<\operatorname{Re}s<1/4$]
23	$(\sqrt{1+x}-1)^{1/2}\times$ $\times K\left(\sqrt{\frac{2}{x}}\sqrt{\sqrt{1+x}-1}\right)=$ $=A_{60}G_{22}^{21}\left(x\left\vert\begin{matrix}5/4, & 5/4\\1/2, & 1/2\end{matrix}\right.\right)$	$A_{60}\Gamma\left[\begin{matrix}s+1/2, s+1/2, -1/4-s\\s+5/4\end{matrix}\right]$ [$-1/2<\operatorname{Re}s<-1/4$]
24	$\frac{\sqrt{\sqrt{1+x}-\sqrt{x}}}{\sqrt{1+x}}\times$ $\times K\left(x^{1/4}\sqrt{2(\sqrt{1+x}-\sqrt{x})}\right)=$ $=A_{61}G_{22}^{12}\left(x\left\vert\begin{matrix}1/4, & 1/4\\0, & 0\end{matrix}\right.\right)$	$A_{61}\Gamma\left[\begin{matrix}s, 3/4-s, 3/4-s\\1-s\end{matrix}\right]$ [$0<\operatorname{Re}s<3/4$]
25	$\frac{\sqrt{\sqrt{1+x}-1}}{\sqrt{1+x}}\times$ $\times K\left(\sqrt{\frac{2}{x}}\sqrt{\sqrt{1+x}-1}\right)=$ $=A_{61}G_{22}^{21}\left(x\left\vert\begin{matrix}3/4, & 3/4\\1/2, & 1/2\end{matrix}\right.\right)$	$A_{61}\Gamma\left[\begin{matrix}s+1/2, s+1/2, 1/4-s\\s+3/4\end{matrix}\right]$ [$-1/2<\operatorname{Re}s<1/4$]
26	$\frac{1}{1+\sqrt{x}}K\left(\frac{2\sqrt[4]{x}}{1+\sqrt{x}}\right)=$ $=\frac{\pi}{2}G_{22}^{11}\left(x\left\vert\begin{matrix}1/2, & 1/2\\0, & 0\end{matrix}\right.\right)$	$\frac{\pi}{2}\Gamma\left[\begin{matrix}s, 1/2-s\\s+1/2, 1-s\end{matrix}\right]$ [$0<\operatorname{Re}s<1/2$]

27	$\frac{H(1-x)}{\sqrt{1-x}+1}\mathbf{K}\left(\frac{2\sqrt[4]{1-x}}{\sqrt{1-x}+1}\right)=\frac{\pi}{2}G_{22}^{20}\left(x\,\middle\vert\,\begin{matrix}1/2, & 1/2\\ 0, & 0\end{matrix}\right)$	$\frac{\pi}{2}\Gamma\left[\begin{matrix}s, s\\ s+1/2, s+1/2\end{matrix}\right]$	[Re s > 0]
28	$\frac{H(x-1)}{\sqrt{x}+\sqrt{x-1}}\mathbf{K}\left(\frac{2\sqrt[4]{x^2-x}}{\sqrt{x}+\sqrt{x-1}}\right)=\frac{\pi}{2}G_{22}^{02}\left(x\,\middle\vert\,\begin{matrix}1/2, & 1/2\\ 0 & 0\end{matrix}\right)$	$\frac{\pi}{2}\Gamma\left[\begin{matrix}1/2-s, 1/2-s\\ 1-s, 1-s\end{matrix}\right]$	[Re s < 1/2]
29	$(\sqrt{1+x}-\sqrt{x})\mathbf{K}\left(\frac{2\sqrt[4]{x^2+x}}{\sqrt{1+x}+\sqrt{x}}\right)=\frac{1}{2}G_{22}^{12}\left(x\,\middle\vert\,\begin{matrix}1/2, & 1/2\\ 0, & 0\end{matrix}\right)$	$\frac{1}{2}\Gamma\left[\begin{matrix}s, 1/2-s, 1/2-s\\ 1-s\end{matrix}\right]$	[0 < Re s < 1/2]
30	$(\sqrt{1+x}-1)\mathbf{K}\left(\frac{2\sqrt[4]{1+x}}{\sqrt{1+x}+1}\right)=\frac{1}{2}G_{22}^{21}\left(x\,\middle\vert\,\begin{matrix}3/2, & 3/2\\ 1, & 1\end{matrix}\right)$	$\frac{1}{2}\Gamma\left[\begin{matrix}s+1, s+1, -1/2-s\\ s+3/2\end{matrix}\right]$	[−1 < Re s < −1/2]
31	$(1+x)^{-1/4}\mathbf{K}\left(\sqrt{\frac{\sqrt{1+x}-\sqrt{x}}{2\sqrt{1+x}}}\right)=\frac{1}{2\sqrt{2}}G_{22}^{21}\left(x\,\middle\vert\,\begin{matrix}3/4, & 3/4\\ 0, & 1/2\end{matrix}\right)$	$\frac{1}{2\sqrt{2}}\Gamma\left[\begin{matrix}s, s+1/2, 1/4-s\\ s+3/4\end{matrix}\right]$	[0 < Re s < 1/4]
32	$(1+x)^{-1/4}\mathbf{K}\left(\sqrt{\frac{\sqrt{1+x}-1}{2\sqrt{1+x}}}\right)=\frac{1}{2\sqrt{2}}G_{22}^{12}\left(x\,\middle\vert\,\begin{matrix}1/4, & 3/4\\ 0, & 0\end{matrix}\right)$	$\frac{1}{2\sqrt{2}}\Gamma\left[\begin{matrix}s, 1/4-s, 3/4-s\\ 1-s\end{matrix}\right]$	[0 < Re s < 1/4]
33	$\mathbf{K}^2\left(i\sqrt{\frac{\sqrt{1+x}-1}{2}}\right)=\frac{\sqrt{\pi}}{4}G_{33}^{13}\left(x\,\middle\vert\,\begin{matrix}1/2, & 1/2, & 1/2\\ 0, & 0, & 0\end{matrix}\right)$	$\frac{\sqrt{\pi}}{4}\Gamma\left[\begin{matrix}s, 1/2-s, 1/2-s, 1/2-s\\ 1-s, 1-s\end{matrix}\right]$	[0 < Re s < 1/2]
34	$\mathbf{K}^2\left(i\sqrt{\frac{\sqrt{1+x}-\sqrt{x}}{2\sqrt{x}}}\right)=\frac{\sqrt{\pi}}{4}G_{33}^{31}\left(x\,\middle\vert\,\begin{matrix}1, & 1, & 1\\ 1/2, & 1/2, & 1/2\end{matrix}\right)$	$\frac{\sqrt{\pi}}{4}\Gamma\left[\begin{matrix}s+1/2, s+1/2, s+1/2, -s\\ s+1, s+1\end{matrix}\right]$	[−1/2 < Re s < 0]
35	$\frac{1}{1+x}\mathbf{E}(i\sqrt{x})=G_{22}^{12}\left(x\,\middle\vert\,\begin{matrix}-1/2, & 1/2\\ 0, & 0\end{matrix}\right)$	$\Gamma\left[\begin{matrix}s, 1/2-s, 3/2-s\\ 1-s\end{matrix}\right]$	[0 < Re s < 1/2]
36	$\frac{1}{1+x}\mathbf{E}\left(\frac{i}{\sqrt{x}}\right)=G_{22}^{21}\left(x\,\middle\vert\,\begin{matrix}0, & 0\\ -1/2, & 1/2\end{matrix}\right)$	$\Gamma\left[\begin{matrix}s-1/2, s+1/2, 1-s\\ s\end{matrix}\right]$	[1/2 < Re s < 1]
37	$\left(\frac{\pi}{2}\right)^{(1\pm 1)/2}x^{(1\mp 1)/4}-\mathbf{E}(i\sqrt{x})$	$\frac{1}{4}\Gamma\left[\begin{matrix}s, -1/2-s, 1/2-s\\ 1-s\end{matrix}\right]$	[(−1∓1)/2 < Re s < ∓1/2]
38	$\left(\frac{\pi}{2}\right)^{(1\pm 1)/2}x^{-(1\mp 1)/4}-\mathbf{E}\left(\frac{i}{\sqrt{x}}\right)$	$\frac{1}{4}\Gamma\left[\begin{matrix}s-1/2, s+1/2, -s\\ s+1\end{matrix}\right]$	[±1/2 < Re s < (1 ± 1)/2]

39	$H(1-x)\,\mathbf{E}(\sqrt{1-x})= =\frac{\pi}{2}\,G_{22}^{20}\left(x\,\middle|\,\begin{matrix}1/2,\ 3/2\\ 0,\ 1\end{matrix}\right)$	$\frac{\pi}{2}\,\Gamma\left[\begin{matrix}s,\ s+1\\ s+1/2,\ s+3/2\end{matrix}\right]$	[Re $s>0$]
40	$H(x-1)\,\mathbf{E}\left(\sqrt{1-\frac{1}{x}}\right)= =\frac{\pi}{2}\,G_{22}^{02}\left(x\,\middle|\,\begin{matrix}0,\ 1\\ -1/2,\ 1/2\end{matrix}\right)$	$\frac{\pi}{2}\,\Gamma\left[\begin{matrix}-s,\ 1-s\\ 1/2-s,\ 3/2-s\end{matrix}\right]$	[Re $s<0$]
41	$\frac{1}{\sqrt{1+x}}\,\mathbf{E}\left(\frac{1}{\sqrt{1+x}}\right)= =G_{22}^{21}\left(x\,\middle|\,\begin{matrix}1/2,\ 1/2\\ 0,\ 1\end{matrix}\right)$	$\Gamma\left[\begin{matrix}s,\ s+1,\ 1/2-s\\ s+1/2\end{matrix}\right]$	[$0<$ Re $s<1/2$]
42	$\frac{1}{\sqrt{1+x}}\,\mathbf{E}\left(\sqrt{\frac{x}{1+x}}\right)= =G_{22}^{12}\left(x\,\middle|\,\begin{matrix}-1/2,\ 1/2\\ 0\ \ 0\end{matrix}\right)$	$\Gamma\left[\begin{matrix}s,\ 1/2-s,\ 3/2-s\\ 1-s\end{matrix}\right]$	[$0<$ Re $s<1/2$]
43	$\frac{H(1-x)}{\sqrt{1-\sqrt{1-x}}}\times \mathbf{E}\left((1-x)^{1/4}\sqrt{2(1-\sqrt{1-x})}\,x^{-1/2}\right)= =\frac{\pi}{2}\,G_{22}^{20}\left(x\,\middle|\,\begin{matrix}1/4,\ 3/4\\ -1/2,\ 1/2\end{matrix}\right)$	$\frac{\pi}{2}\,\Gamma\left[\begin{matrix}s-1/2,\ s+1/2\\ s+1/4,\ s+3/4\end{matrix}\right]$	[Re $s>1/2$]
44	$\frac{H(x-1)}{\sqrt{\sqrt{x}-\sqrt{x-1}}}\times \mathbf{E}\left(\sqrt{2\sqrt{x-1}\,(\sqrt{x}-\sqrt{x-1})}\right)= =\frac{\pi}{2}\,G_{22}^{02}\left(x\,\middle|\,\begin{matrix}1/4,\ 5/4\\ 0,\ 1/2\end{matrix}\right)$	$\frac{\pi}{2}\,\Gamma\left[\begin{matrix}3/4-s,\ -1/4-s\\ 1-s,\ 1/2-s\end{matrix}\right]$	[Re $s<-1/4$]
45	$\frac{(1+x)^{-1/2}}{\sqrt{\sqrt{1+x}-\sqrt{x}}}\times \mathbf{E}\left(x^{1/4}\sqrt{2(\sqrt{1+x}-\sqrt{x})}\right)= =\frac{2\pi}{\Gamma^2(1/4)}\,G_{22}^{12}\left(x\,\middle|\,\begin{matrix}-1/4,\ 3/4\\ 0,\ \ 0\end{matrix}\right)$	$\frac{2\pi}{\Gamma^2(1/4)}\,\Gamma\left[\begin{matrix}s,\ 5/4-s,\ 1/4-s\\ 1-s\end{matrix}\right]$	[$0<$ Re $s<1/4$]
46	$\frac{(1+x)^{-1/2}}{\sqrt{\sqrt{1+x}-1}}\times \mathbf{E}\left(\sqrt{\frac{2}{x}}\sqrt{(\sqrt{1+x}-1)}\right)= =\frac{2\pi}{\Gamma^2(1/4)}\,G_{22}^{21}\left(x\,\middle|\,\begin{matrix}1/4,\ 1/4\\ -1/2,\ 1/2\end{matrix}\right)$	$\frac{2\pi}{\Gamma^2(1/4)}\,\Gamma\left[\begin{matrix}s+1/2,\ s-1/2,\ 3/4-s\\ s+1/4\end{matrix}\right]$	[$1/2<$ Re $s<3/4$]
47	$\frac{H(1-x)}{1-\sqrt{1-x}}\,\mathbf{E}\left(\frac{2\sqrt[4]{1-x}}{1+\sqrt{1-x}}\right)= =\frac{\pi}{2}\,G_{22}^{20}\left(x\,\middle|\,\begin{matrix}1/2,\ 1/2\\ -1,\ 1\end{matrix}\right)$	$\frac{\pi}{2}\,\Gamma\left[\begin{matrix}s+1,\ s-1\\ s+1/2,\ s+1/2\end{matrix}\right]$	[Re $s>1$]
48	$\frac{H(x-1)}{\sqrt{x}-\sqrt{x-1}}\,\mathbf{E}\left(\frac{2\sqrt[4]{x^2-x}}{\sqrt{x}+\sqrt{x-1}}\right)= =\frac{\pi}{2}\,G_{22}^{02}\left(x\,\middle|\,\begin{matrix}-1/2,\ 3/2\\ 0,\ \ 0\end{matrix}\right)$	$\frac{\pi}{2}\,\Gamma\left[\begin{matrix}3/2-s,\ -1/2-s\\ 1-s,\ 1-s\end{matrix}\right]$	[Re $s<-1/2$]

49	$\mathbf{D}(i\sqrt{x}) = \frac{1}{2} G_{22}^{12}\left(x \,\middle|\, \begin{matrix} -1/2, & 1/2 \\ 0, & -1 \end{matrix}\right)$	$\frac{1}{2}\Gamma\left[\begin{matrix} s, & 1/2-s, & 3/2-s \\ 2-s \end{matrix}\right]$ $[0 < \operatorname{Re} s < 1/2]$
50	$\mathbf{D}\left(\frac{i}{\sqrt{x}}\right) = \frac{1}{2} G_{22}^{21}\left(x \,\middle|\, \begin{matrix} 1, & 2 \\ 1/2, & 3/2 \end{matrix}\right)$	$\frac{1}{2}\Gamma\left[\begin{matrix} s+1/2, & s+3/2, & -s \\ s+2 \end{matrix}\right]$ $[-1/2 < \operatorname{Re} s < 0]$

8.4.41. The Legendre functions of the first kind $P_\nu^\mu(x)$ and $P_\nu(x)$.

NOTATION:

$$A_{62} = -\frac{\sin \nu\pi}{\pi\Gamma(1-\mu+\nu)\Gamma(-\mu-\nu)}, \quad A_{63} = \frac{1}{\Gamma(1-\mu+\nu)\Gamma(-\mu-\nu)},$$

$$A_{64} = -\frac{\sin \nu\pi}{\pi}, \quad A_{65} = \frac{1}{\Gamma(-\nu)\Gamma(-\mu-\nu)}, \quad A_{66} = \frac{2^{-\mu-1}\pi^{-1}}{\Gamma(1-\mu+\nu)\Gamma(-\mu-\nu)},$$

$$A_{67} = \frac{2^{\mu}}{\Gamma(r+(1-\mu+\nu)/2)\Gamma(r-(\mu+\nu)/2)}, \quad A_{68} = \frac{2^{-\nu-1}}{\sqrt{\pi}\,\Gamma(-\mu-\nu)},$$

$$A_{69} = \frac{\Gamma(1/2-\mu)}{\sqrt{\pi}}, \quad A_{70} = \frac{2^{\mu+1}\pi}{\Gamma(1-(\mu-\nu)/2)\Gamma((1-\mu-\nu)/2)}, \quad A_{71} = -\frac{\sin \nu\pi}{\pi^{3/2}};$$

$$A_{72} = \frac{1}{\sqrt{\pi}\,\Gamma(1-\mu+\nu)\Gamma(1-\mu-\nu)}, \quad A_{73} = \frac{\pi^{-1/2}}{\Gamma(1-\mu\pm\nu)\Gamma(1\mp 1-\mu\mp\nu)}.$$

1	$(1-x)_+^{-\mu/2} P_\nu^\mu(2x-1) = G_{22}^{20}\left(x \,\middle|\, \begin{matrix} 1-\mu/2+\nu, & -\mu/2-\nu \\ \mu/2, & -\mu/2 \end{matrix}\right)$	$\Gamma\left[\begin{matrix} s+\mu/2, & s-\mu/2 \\ s+\nu+1-\mu/2, & s-\nu-\mu/2 \end{matrix}\right]$ $[\operatorname{Re} s > \lvert\operatorname{Re}\mu\rvert/2;\ \operatorname{Re}\mu < 1]$
2	$(x-1)_+^{-\mu/2} P_\nu^\mu\left(\frac{2}{x}-1\right) = G_{22}^{02}\left(x \,\middle|\, \begin{matrix} 1, & \mu+1 \\ \mu-\nu, & 1+\mu+\nu \end{matrix}\right)$	$\Gamma\left[\begin{matrix} -s, & -\mu-s \\ 1+\nu-\mu-s, & -\nu-\mu-s \end{matrix}\right]$ $[\operatorname{Re}\mu < 1;\ \operatorname{Re} s < 0,\ -\operatorname{Re}\mu]$
3	$(x-1)_+^{-\mu/2} P_\nu^\mu(2x-1) = G_{22}^{02}\left(x \,\middle|\, \begin{matrix} 1-\mu/2+\nu, & -\mu/2-\nu \\ \mu/2, & -\mu/2 \end{matrix}\right)$	$\Gamma\left[\begin{matrix} \mu/2-\nu-s, & 1+\mu/2+\nu-s \\ 1-\mu/2-s, & 1+\mu/2-s \end{matrix}\right]$ $[\operatorname{Re} s < 1+\operatorname{Re}(\mu/2+\nu),\ \operatorname{Re}(\mu/2-\nu);\ \operatorname{Re}\mu<1]$
4	$(1-x)_+^{-\mu/2} P_\nu^\mu\left(\frac{2}{x}-1\right) = G_{22}^{20}\left(x \,\middle|\, \begin{matrix} 1, & 1-\mu \\ -\nu, & 1+\nu \end{matrix}\right)$	$\Gamma\left[\begin{matrix} s-\nu, & s+\nu+1 \\ s+1-\mu, & s+1 \end{matrix}\right]$ $[\operatorname{Re}\mu < 1;\ \operatorname{Re} s > \operatorname{Re}\nu,\ -1-\operatorname{Re}\nu]$
5	$\lvert 1-x\rvert^{\mu/2} P_\nu^\mu(2x-1) = A_{62} G_{22}^{22}\left(x \,\middle|\, \begin{matrix} \mu/2-\nu, & 1+\mu/2+\nu \\ \mu/2, & -\mu/2 \end{matrix}\right)$	$A_{62}\Gamma\left[s+\frac{\mu}{2},\ s-\frac{\mu}{2},\ 1+\nu-\frac{\mu}{2}-s,\ -\nu-\frac{\mu}{2}-s\right]$ $[\lvert\operatorname{Re}\mu\rvert/2 < \operatorname{Re} s < 1+\operatorname{Re}(\nu-\mu/2),\ -\operatorname{Re}(\nu+\mu/2)]$
6	$\lvert 1-x\rvert^{\mu/2} P_\nu^\mu\left(\frac{2}{x}-1\right) = A_{62} G_{22}^{22}\left(x \,\middle|\, \begin{matrix} 1, & \mu+1 \\ \nu+1, & -\nu \end{matrix}\right)$	$A_{62}\Gamma[s+\nu+1,\ s-\nu,\ -s,\ -\mu-s]$ $[\operatorname{Re}\nu,\ -1-\operatorname{Re}\nu < \operatorname{Re} s < 0,\ -\operatorname{Re}\mu]$

7	$\|1-x\|^{-\mu/2}P_{\nu}^{\mu}(2x-1)=$ $=G_{22}^{20}\left(x\,\middle\|\,\begin{matrix}1-\mu/2+\nu, & -\mu/2-\nu\\ \mu/2, & -\mu/2\end{matrix}\right)+$ $+G_{22}^{02}\left(x\,\middle\|\,\begin{matrix}1-\mu/2+\nu, & -\mu/2-\nu\\ \mu/2, & -\mu/2\end{matrix}\right)$	$\Gamma\left[\begin{matrix}s+\mu/2, & s-\mu/2\\ s+1+\nu-\mu/2, & s-\nu-\mu/2\end{matrix}\right]+$ $+\Gamma\left[\begin{matrix}\mu/2-\nu-s, & 1+\nu+\mu/2-s\\ 1-\mu/2-s, & 1+\mu/2-s\end{matrix}\right]$ [$\operatorname{Re}\mu<1$; $\|\operatorname{Re}\mu\|/2<\operatorname{Re}s<1+\operatorname{Re}(\mu/2+\nu)$, $\operatorname{Re}(\mu/2-\nu)$]
8	$\|1-x\|^{-\mu/2}P_{\nu}^{\mu}\left(\frac{2}{x}-1\right)=$ $=G_{22}^{20}\left(x\,\middle\|\,\begin{matrix}1, & 1-\mu\\ \nu+1, & -\nu\end{matrix}\right)+$ $+G_{22}^{02}\left(x\,\middle\|\,\begin{matrix}1, & 1-\mu\\ \nu+1, & -\nu\end{matrix}\right)$	$\Gamma\left[\begin{matrix}s+\nu+1, & s-\nu\\ s+1, & s+1-\mu\end{matrix}\right]+$ $+\Gamma\left[\begin{matrix}\mu-s, & -s\\ 1+\nu-s, & -\nu-s\end{matrix}\right]$ [$\operatorname{Re}\mu<1$; $\operatorname{Re}\nu$, $-1-\operatorname{Re}\nu<\operatorname{Re}s<0$, $\operatorname{Re}\mu$]
9	$(1+x)^{\mu/2}P_{\nu}^{\mu}(2x+1)=$ $=A_{63}G_{22}^{12}\left(x\,\middle\|\,\begin{matrix}\mu/2-\nu, & 1+\nu+\mu/2\\ -\mu/2, & \mu/2\end{matrix}\right)$	$A_{63}\Gamma\left[\begin{matrix}s-\frac{\mu}{2}, & 1+\nu-\frac{\mu}{2}-s, & -\nu-\frac{\mu}{2}-s\\ 1-\mu/2-s\end{matrix}\right]$ [$\operatorname{Re}\mu/2<\operatorname{Re}s<1+\operatorname{Re}(\nu-\mu/2)$, $-\operatorname{Re}(\nu+\mu/2)$]
10	$(1+x)^{\mu/2}P_{\nu}^{\mu}\left(\frac{2}{x}+1\right)=$ $=A_{63}G_{22}^{21}\left(x\,\middle\|\,\begin{matrix}\mu+1, & 1\\ \nu+1, & -\nu\end{matrix}\right)$	$A_{63}\Gamma\left[\begin{matrix}s+\nu+1, & s-\nu, & -\mu-s\\ s+1\end{matrix}\right]$ [$\operatorname{Re}\nu$, $-1-\operatorname{Re}\nu<\operatorname{Re}s<-\operatorname{Re}\mu$]
11	$(1+x)^{-\mu/2}P_{\nu}^{\mu}(2x+1)=$ $=A_{64}G_{22}^{12}\left(x\,\middle\|\,\begin{matrix}1+\nu-\mu/2, & -\nu-\mu/2\\ -\mu/2, & \mu/2\end{matrix}\right)$	$A_{64}\Gamma\left[\begin{matrix}s-\mu/2, & \mu/2-\nu-s, & 1+\nu+\mu/2-s\\ 1-\mu/2-s\end{matrix}\right]$ [$\operatorname{Re}\mu/2<\operatorname{Re}s<\operatorname{Re}(\mu/2-\nu)$, $1+\operatorname{Re}(\mu/2+\nu)$]
12	$(1+x)^{-\mu/2}P_{\nu}^{\mu}\left(\frac{2}{x}+1\right)=$ $=A_{64}G_{22}^{21}\left(x\,\middle\|\,\begin{matrix}1, & 1-\mu\\ \nu+1, & -\nu\end{matrix}\right)$	$A_{64}\Gamma\left[\begin{matrix}s-\nu, & s+\nu+1, & -s\\ s+1-\mu\end{matrix}\right]$ [$\operatorname{Re}\nu$, $-1-\operatorname{Re}\nu<\operatorname{Re}s<0$]
13	$(1+x)^{\nu}P_{\nu}^{\mu}\left(\frac{1-x}{1+x}\right)=$ $=A_{65}G_{22}^{12}\left(x\,\middle\|\,\begin{matrix}1+\nu-\mu/2, & 1+\nu+\mu/2\\ -\mu/2, & \mu/2\end{matrix}\right)$	$A_{65}\Gamma\left[\begin{matrix}s-\mu/2, & \mu/2-\nu-s, & -\mu/2-\nu-s\\ 1-\mu/2-s\end{matrix}\right]$ [$\operatorname{Re}\mu/2<\operatorname{Re}s<-\operatorname{Re}\nu-\|\operatorname{Re}\mu\|/2$]
14	$(1+x)^{\nu}P_{\nu}^{\mu}\left(\frac{x-1}{x+1}\right)=$ $=A_{65}G_{22}^{21}\left(x\,\middle\|\,\begin{matrix}1+\nu+\mu/2, & 1+\nu-\mu/2\\ \mu/2, & -\mu/2\end{matrix}\right)$	$A_{65}\Gamma\left[\begin{matrix}s-\mu/2, & s+\mu/2, & -\mu/2-\nu-s\\ s+1+\nu-\mu/2\end{matrix}\right]$ [$\|\operatorname{Re}\mu\|/2<\operatorname{Re}s<-\operatorname{Re}(\mu/2+\nu)$]
15	$\|1-x\|^{\nu}P_{\nu}^{\mu}\left(\frac{1+x}{\|1-x\|}\right)=$ $=\Gamma\left[\begin{matrix}\nu+1\\ -\nu-\mu\end{matrix}\right]G_{22}^{11}\left(x\,\middle\|\,\begin{matrix}1+\nu+\frac{\mu}{2}, & 1+\nu-\frac{\mu}{2}\\ -\frac{\mu}{2}, & \frac{\mu}{2}\end{matrix}\right)$	$\Gamma\left[\begin{matrix}\nu+1\\ -\nu-\mu\end{matrix}\right]\times$ $\times\Gamma\left[\begin{matrix}s-\mu/2, & -\nu-\mu/2-s\\ s+\nu+1-\mu/2, & 1-\mu/2-s\end{matrix}\right]$ [$\operatorname{Re}\mu/2<\operatorname{Re}s<-\operatorname{Re}(\nu+\mu/2)$; $\operatorname{Re}\nu>-1$]

16	$\frac{\lvert 1-x\rvert^{\mu}}{\sqrt{1+x}}P^{\mu}_{-1/4}\left(\frac{1-6x+x^2}{(1+x)^2}\right)=$ $=2^{-\mu-1}\Gamma\left[\begin{matrix}1/2+\mu\\1/2-2\mu\end{matrix}\right]\times$ $\times G^{11}_{22}\left(x\left\lvert\begin{matrix}(1+3\mu)/2,\ (1+\mu)/2\\-\mu/2,\ \mu/2\end{matrix}\right.\right)$	$2^{-\mu-1}\Gamma\left[\begin{matrix}1/2+\mu\\1/2-2\mu\end{matrix}\right]\times$ $\times\Gamma\left[\begin{matrix}s-\mu/2,\ (1-3\mu)/2-s\\s+(1+\mu)/2,\ 1-\mu/2-s\end{matrix}\right]$ [Re $\mu/2<$ Re $s<(1-3$ Re $\mu)/2$; Re $\mu>-1/2$]
17	$\frac{\lvert 1-x\rvert^{\mu}}{\sqrt{1+x}}P^{\mu}_{-1/4}\left(\frac{-1+6x-x^2}{(1+x)^2}\right)=$ $=\frac{\cos\mu\pi}{2^{\mu+1/2}\pi^2}\Gamma\left[\begin{matrix}1/2+\mu\\1/2-2\mu\end{matrix}\right]\times$ $\times G^{22}_{22}\left(x\left\lvert\begin{matrix}(1+3\mu)/2,\ (1+\mu)/2\\\mu/2,\ -\mu/2\end{matrix}\right.\right)$	$\frac{\cos\mu\pi}{2^{\mu+1/2}\pi^2}\Gamma\left[\begin{matrix}1/2+\mu\\1/2-2\mu\end{matrix}\right]\times$ $\times\Gamma\left[s+\frac{\mu}{2},\ s-\frac{\mu}{2},\ \frac{1-\mu}{2}-s,\ \frac{1-3\mu}{2}-s\right]$ [$\lvert$Re $\mu\rvert/2<$ Re $s<(1-$Re $\mu)/2$, $(1-3$ Re $\mu)/2$]
18	$\frac{\lvert 1-x\rvert^{-\mu}}{\sqrt{1+x}}P^{\mu}_{-1/4}\left(\frac{1-6x+x^2}{(1+x)^2}\right)=$ $=\frac{\Gamma(1/2-\mu)}{2^{\mu}\sqrt{\pi}}\times$ $\times G^{11}_{22}\left(x\left\lvert\begin{matrix}(1-\mu)/2,\ (1-3\mu)/2\\-\mu/2,\ \mu/2\end{matrix}\right.\right)$	$\frac{\Gamma(1/2-\mu)}{2^{\mu}\sqrt{\pi}}\Gamma\left[\begin{matrix}s-\mu/2,\ (1+\mu)/2-s\\s+(1-3\mu)/2,\ 1-\mu/2-s\end{matrix}\right]$ [Re $\mu/2<$ Re $s<(1+$Re $\mu)/2$; Re $\mu<1/2$]
19	$\frac{\lvert 1-x\rvert^{-\mu}}{\sqrt{1+x}}P^{\mu}_{-1/4}\left(\frac{-1+6x-x^2}{(1+x)^2}\right)=$ $=\frac{2^{-\mu-1/2}\Gamma(1/2-\mu)}{\sqrt{\pi}\sin\mu\pi}\times$ $\times\left[G^{11}_{22}\left(x\left\lvert\begin{matrix}(1-\mu)/2,\ (1-3\mu)/2\\-\mu/2,\ \mu/2\end{matrix}\right.\right)-\right.$ $\left.-\cos 2\mu\pi G^{11}_{22}\left(x\left\lvert\begin{matrix}(1-3\mu)/2,\ (1-\mu)/2\\\mu/2,\ -\mu/2\end{matrix}\right.\right)\right]$	$\frac{2^{-\mu-1/2}\Gamma(1/2-\mu)}{\sqrt{\pi}\sin\mu\pi}\times$ $\times\left\{\Gamma\left[\begin{matrix}s-\mu/2,\ (1+\mu)/2-s\\s+(1-3\mu)/2,\ 1-\mu/2-s\end{matrix}\right]-\right.$ $\left.-\cos 2\mu\pi\,\Gamma\left[\begin{matrix}s+\mu/2,\ (1+3\mu)/2-s\\s+(1-\mu)/2,\ 1+\mu/2-s\end{matrix}\right]\right\}$ [Re $\mu<1/2$; $\lvert$Re $\mu\rvert/2<$ Re $s<(1+$Re $\mu)/2$, $(1+3$ Re $\mu)/2$]
20	$(1-x)_{+}^{-\mu/2}P^{\mu}_{\nu}(\sqrt{x})=$ $=2^{\mu}G^{20}_{22}\left(x\left\lvert\begin{matrix}(1-\mu-\nu)/2,\ 1+(\nu-\mu)/2\\0,\ 1/2\end{matrix}\right.\right)$	$2^{\mu}\Gamma\left[\begin{matrix}s,\ s+1/2\\s+(1-\mu-\nu)/2,\ s+1+(\nu-\mu)/2\end{matrix}\right]$ [Re $s>0$; Re $\mu<1$]
21	$(x-1)_{+}^{-\mu/2}P^{\mu}_{\nu}\left(\frac{1}{\sqrt{x}}\right)=$ $=2^{\mu}G^{02}_{22}\left(x\left\lvert\begin{matrix}1-\mu/2,\ (1-\mu)/2\\(1+\nu)/2,\ -\nu/2\end{matrix}\right.\right)$	$2^{\mu}\Gamma\left[\begin{matrix}\mu/2-s,\ (\mu+1)/2-s\\(1-\nu)/2-s,\ 1+\nu/2-s\end{matrix}\right]$ [Re $s<$ Re $\mu/2$; Re $\mu<1$]
22	$(x-1)_{+}^{-\mu/2}P^{\mu}_{\nu}(\sqrt{x})=$ $=2^{\mu}G^{02}_{22}\left(x\left\lvert\begin{matrix}(1-\mu-\nu)/2,\ 1+(\nu-\mu)/2\\0,\ 1/2\end{matrix}\right.\right)$	$2^{\mu}\Gamma\left[\begin{matrix}(1+\mu+\nu)/2-s,\ (\mu-\nu)/2-s\\1-s,\ 1/2-s\end{matrix}\right]$ [Re $\mu<1$; Re $s<(1+$Re $(\mu+\nu))/2$, Re $(\mu-\nu)/2$]
23	$(1-x)_{+}^{-\mu/2}P^{\mu}_{\nu}\left(\frac{1}{\sqrt{x}}\right)=$ $=2^{\mu}G^{20}_{22}\left(x\left\lvert\begin{matrix}1-\mu/2,\ (1-\mu)/2\\(1+\nu)/2,\ -\nu/2\end{matrix}\right.\right)$	$2^{\mu}\Gamma\left[\begin{matrix}s+(1+\nu)/2,\ s-\nu/2\\s+1-\mu/2,\ s+(1-\mu)/2\end{matrix}\right]$ [Re $\mu<1$; Re $s>$ Re $\nu/2$, $-(1+$Re $\nu)/2$]
24	$\lvert 1-x\rvert^{\mu/2}P^{\mu}_{\nu}(\sqrt{x})=$ $=A_{66}G^{22}_{22}\left(x\left\lvert\begin{matrix}(1+\mu-\nu)/2,\ 1+(\mu+\nu)/2\\0,\ 1/2\end{matrix}\right.\right)$	$A_{66}\Gamma\left[s,\ s+\frac{1}{2},\ \frac{1+\nu-\mu}{2}-s,\ -\frac{\mu+\nu}{2}-s\right]$ [$0<$ Re $s<[1+$Re $(\nu-\mu)]/2$, $-$Re $(\mu+\nu)/2$]

25	$\lvert 1-x\rvert^{\mu/2}P_\nu^\mu\left(\frac{1}{\sqrt{x}}\right)= A_{66}G_{22}^{22}\left(x\left\lvert\begin{matrix}1+\mu/2,\ (1+\mu)/2\\ (1+\nu)/2,\ -\nu/2\end{matrix}\right.\right)$	$A_{66}\Gamma\left[s+\frac{1+\nu}{2},\ s-\frac{\nu}{2},\ -\frac{\mu}{2}-s,\ \frac{1-\mu}{2}-s\right]$ [$\operatorname{Re}\nu/2,\ -(1+\operatorname{Re}\nu)/2<\operatorname{Re}s<-\operatorname{Re}\mu/2$]
26	$\lvert 1-x\rvert^{-\mu/2}P_\nu^\mu(\sqrt{x})= 2^\mu G_{22}^{20}\left(x\left\lvert\begin{matrix}(1-\mu-\nu)/2,\ 1+(\nu-\mu)/2\\ 0,\ 1/2\end{matrix}\right.\right)+2^\mu G_{22}^{02}\left(x\left\lvert\begin{matrix}(1-\mu-\nu)/2,\ 1+(\nu-\mu)/2\\ 0,\ 1/2\end{matrix}\right.\right)$	$2^\mu\Gamma\left[\begin{matrix}s,\ s+1/2\\ s+(1-\mu-\nu)/2,\ s+1+(\nu-\mu)/2\end{matrix}\right]+2^\mu\Gamma\left[\begin{matrix}(1+\mu+\nu)/2-s,\ (\mu-\nu)/2-s\\ 1-s,\ 1/2-s\end{matrix}\right]$ [$\operatorname{Re}\mu<1;\ 0<\operatorname{Re}s<(1+\operatorname{Re}(\mu+\nu))/2,\ \operatorname{Re}(\mu-\nu)/2$]
27	$\lvert 1-x\rvert^{-\mu/2}P_\nu^\mu\left(\frac{1}{\sqrt{x}}\right)= 2^\mu G_{22}^{02}\left(x\left\lvert\begin{matrix}1-\mu/2\ (1-\mu)/2\\ (1+\nu)/2,\ -\nu/2\end{matrix}\right.\right)+2^\mu G_{22}^{20}\left(x\left\lvert\begin{matrix}1-\mu/2,\ (1-\mu)/2\\ (1+\nu)/2,\ -\nu/2\end{matrix}\right.\right)$	$2^\mu\Gamma\left[\begin{matrix}\mu/2-s,\ (\mu+1)/2-s\\ (1-\nu)/2-s,\ 1+\nu/2-s\end{matrix}\right]+2^\mu\Gamma\left[\begin{matrix}s+(1+\nu)/2,\ s-\nu/2\\ s+1-\mu/2,\ s+(1-\mu)/2\end{matrix}\right]$ [$\operatorname{Re}\nu/2,\ -(1+\operatorname{Re}\nu)/2<\operatorname{Re}s<\operatorname{Re}\mu/2;\ \operatorname{Re}\mu<1$]
28	$(1+x)^{-r}P_\nu^\mu(\sqrt{1+x})= A_{67}G_{22}^{12}\left(x\left\lvert\begin{matrix}(1-\nu)/2-r,\ 1+\nu/2-r\\ -\mu/2,\ \mu/2\end{matrix}\right.\right)$ [$r=0,\ 1/2$]	$A_{67}\Gamma\left[\begin{matrix}s-\frac{\mu}{2},\ r+\frac{1+\nu}{2}-s,\ r-\frac{\nu}{2}-s\\ 1-\frac{\mu}{2}-s\end{matrix}\right]$ [$\operatorname{Re}\mu/2<\operatorname{Re}s<r+(1+\operatorname{Re}\nu)/2,\ r-\operatorname{Re}\nu/2$]
29	$(1+x)^{-r}P_\nu^\mu\left(\sqrt{1+\frac{1}{x}}\right)= A_{67}G_{22}^{21}\left(x\left\lvert\begin{matrix}1+\mu/2-r,\ 1-\mu/2-r\\ -\nu/2,\ (1+\nu)/2\end{matrix}\right.\right)$ [$r=0,\ 1/2$]	$A_{67}\Gamma\left[\begin{matrix}s-\nu/2,\ s+(\nu+1)/2,\ r-\mu/2-s\\ s+1-\mu/2-r\end{matrix}\right]$ [$\operatorname{Re}\nu/2,\ -(\operatorname{Re}\nu+1)/2<\operatorname{Re}s<r-\operatorname{Re}\mu/2$]
30	$(1+x)^{\nu/2}P_\nu^\mu\left(\frac{1}{\sqrt{1+x}}\right)= A_{68}G_{22}^{12}\left(x\left\lvert\begin{matrix}1+\nu/2,\ (1+\nu)/2\\ -\mu/2,\ \mu/2\end{matrix}\right.\right)$	$A_{68}\Gamma\left[\begin{matrix}s-\mu/2,\ -\nu/2-s,\ (1-\nu)/2-s\\ 1-\mu/2-s\end{matrix}\right]$ [$\operatorname{Re}\mu/2<\operatorname{Re}s<-\operatorname{Re}\nu/2$]
31	$(1+x)^{\nu/2}P_\nu^\mu\left(\sqrt{\frac{x}{1+x}}\right)= A_{68}G_{22}^{21}\left(x\left\lvert\begin{matrix}1+(\nu+\mu)/2,\ 1+(\nu-\mu)/2\\ 0,\ 1/2\end{matrix}\right.\right)$	$A_{68}\Gamma\left[\begin{matrix}s,\ s+1/2,\ -(\nu+\mu)/2-s\\ s+1+(\nu-\mu)/2\end{matrix}\right]$ [$0<\operatorname{Re}s<-\operatorname{Re}(\nu+\mu)/2$]
32	$(1-x)_+^{-\mu}P_\nu^\mu\left(\frac{1+x}{2\sqrt{x}}\right)= A_{69}G_{22}^{20}\left(x\left\lvert\begin{matrix}1+\nu/2-\mu,\ (1-\nu)/2-\mu\\ -\nu/2,\ (1+\nu)/2\end{matrix}\right.\right)$	$A_{69}\Gamma\left[\begin{matrix}s-\nu/2,\ s+(1+\nu)/2\\ s+\nu/2+1-\mu,\ s+(1-\nu)/2-\mu\end{matrix}\right]$ [$\operatorname{Re}\mu<1/2;\ \operatorname{Re}s>\operatorname{Re}\nu/2,\ -(1+\operatorname{Re}\nu)/2$]
33	$(x-1)_+^{-\mu}P_\nu^\mu\left(\frac{1+x}{2\sqrt{x}}\right)= A_{69}G_{22}^{02}\left(x\left\lvert\begin{matrix}1+\nu/2-\mu,\ (1-\nu)/2-\mu\\ -\nu/2,\ (1+\nu)/2\end{matrix}\right.\right)$	$A_{69}\Gamma\left[\begin{matrix}\mu-\nu/2-s,\ \mu+(1+\nu)/2-s\\ 1+\nu/2-s,\ (1-\nu)/2-s\end{matrix}\right]$ [$\operatorname{Re}\mu<1/2;\ \operatorname{Re}s<\operatorname{Re}(\mu-\nu/2),\ \operatorname{Re}(\mu+\nu/2)+1/2$]
34	$\lvert 1-x\rvert^{\mu}P_\nu^\mu\left(\frac{1+x}{2\sqrt{x}}\right)=$	$\frac{\cos\mu\pi}{\pi^{3/2}}\Gamma\left[\begin{matrix}\mu+1/2\\ -\nu-\mu,\ 1+\nu-\mu\end{matrix}\right]\times$

No.	$f(x)$	$F(s)$
	$=\frac{\cos\mu\pi}{\pi^{3/2}}\Gamma\left[\begin{matrix}\mu+1/2\\ -\nu-\mu,\ 1+\nu-\mu\end{matrix}\right]\times$ $\times G_{22}^{22}\left(x\left\vert\begin{matrix}1+\mu+\nu/2,\ (1-\nu)/2+\mu\\ (1+\nu)/2,\ -\nu/2\end{matrix}\right.\right)$	$\times\Gamma\left[s+\frac{\nu+1}{2},\ s-\frac{\nu}{2},\ -\mu-\frac{\nu}{2}-s,\ \frac{1+\nu}{2}-\mu-s\right]$ [Re $\nu/2$, $-(1+\text{Re}\,\nu)/2<\text{Re}\,s<-\text{Re}\,(\mu+\nu/2)$, $1/2+\text{Re}\,(\nu/2-\mu)$]
35	$\vert 1-x\vert^{-\mu}P_\nu^\mu\left(\frac{1+x}{2\sqrt{x}}\right)=$ $=\frac{\Gamma(1/2-\mu)}{\sqrt{\pi}\cos\nu\pi}\left\{\sin(\mu-\nu)\pi\times\right.$ $\times G_{22}^{11}\left(x\left\vert\begin{matrix}1+\frac{\nu}{2}-\mu,\ \frac{1-\nu}{2}-\mu\\ -\frac{\nu}{2},\ \frac{1+\nu}{2}\end{matrix}\right.\right)+$ $\left.+\sin(\mu+\nu)\pi G_{22}^{11}\left(x\left\vert\begin{matrix}\frac{1-\nu}{2}-\mu,\ 1+\frac{\nu}{2}-\mu\\ \frac{1+\nu}{2},\ -\frac{\nu}{2}\end{matrix}\right.\right)\right\}$	$\frac{\Gamma(1/2-\mu)}{\sqrt{\pi}\cos\nu\pi}\left\{\sin(\mu-\nu)\pi\times\right.$ $\times\Gamma\left[\begin{matrix}s-\nu/2,\ \mu-\nu/2-s\\ s+(1-\nu)/2-\mu,\ (1-\nu)/2-s\end{matrix}\right]+$ $\left.+\sin(\mu+\nu)\pi\Gamma\left[\begin{matrix}s+\frac{1+\nu}{2},\ \frac{1+\nu}{2}+\mu-s\\ s+1+\frac{\nu}{2}-\mu,\ 1+\frac{\nu}{2}-s\end{matrix}\right]\right\}$ [Re $\mu<1/2$; Re $\nu/2$, $-(1+\text{Re}\,\nu)/2<\text{Re}\,s<1/2+\text{Re}\,(\mu+\nu/2)$, Re $(\mu-\nu/2)$]
36	$(1+x)^{\nu/2}P_\nu^\mu\left(\frac{2+x}{2\sqrt{1+x}}\right)=$ $=\frac{\pi^{-1/2}}{\Gamma(-\mu-\nu)}G_{22}^{12}\left(x\left\vert\begin{matrix}1+\nu,\ 1/2\\ -\mu,\ \mu\end{matrix}\right.\right)$	$\frac{1}{\sqrt{\pi}\,\Gamma(-\mu-\nu)}\times$ $\times\Gamma\left[\begin{matrix}s-\mu,\ -\nu-s,\ 1/2-s\\ 1-\mu-s\end{matrix}\right]$ [Re $\mu<\text{Re}\,s<1/2$, $-\text{Re}\,\nu$]
37	$(1+x)^{\nu/2}P_\nu^\mu\left(\frac{1+2x}{2\sqrt{x^2+x}}\right)=$ $=\frac{\pi^{-1/2}}{\Gamma(-\mu-\nu)}G_{22}^{21}\left(x\left\vert\begin{matrix}1+\nu/2+\mu,\ 1+\nu/2-\mu\\ (1+\nu)/2,\ -\nu/2\end{matrix}\right.\right)$	$\frac{1}{\sqrt{\pi}\,\Gamma(-\mu-\nu)}\times$ $\times\Gamma\left[\begin{matrix}s+(1+\nu)/2,\ s-\nu/2,\ -\nu/2-\mu-s\\ s+1+\nu/2-\mu\end{matrix}\right]$ [Re $\nu/2$, $-(1+\text{Re}\,\nu)/2<\text{Re}\,s<-\text{Re}\,(\mu+\nu/2)$]
38	$\frac{\vert 1\pm\sqrt{x}\vert}{\vert 1-x\vert^{-2\nu}}P_\nu^{2\nu+1}\left(\frac{1\pm 6\sqrt{x}+x}{(1\mp\sqrt{x})^2}\right)=$ $=\sqrt{\pi}\Gamma\left[\begin{matrix}3/2+2\nu\\ -1-3\nu,\ -\nu\end{matrix}\right]\times$ $\times G_{22}^{11}\left(x\left\vert\begin{matrix}(10\nu+1)/4,\ (6\nu+5)/4\\ -(2\nu+1)/4,\ (2\nu+1)/4\end{matrix}\right.\right)$	$\sqrt{\pi}\Gamma\left[\begin{matrix}3/2+2\nu\\ -1-3\nu,\ -\nu\end{matrix}\right]\times$ $\times\Gamma\left[\begin{matrix}s-(1+2\nu)/4,\ -(10\nu-3)/4-s\\ s+(5+6\nu)/4,\ (3-2\nu)/4-s\end{matrix}\right]$ [$(1+2\,\text{Re}\,\nu)/4<\text{Re}\,s<-(3+10\,\text{Re}\,\nu)/4$; Re $\nu>-3/4$]
39	$\frac{\vert 1\pm\sqrt{x}\vert}{\vert 1-x\vert^{-2\nu}}P_\nu^{2\nu+1}\left(\frac{-1\mp 6\sqrt{x}-x}{(1\mp\sqrt{x})^2}\right)=$ $=-\frac{\cos 2\nu\pi}{2\pi^{3/2}}\Gamma\left[\begin{matrix}2\nu+3/2\\ -1-3\nu,\ -\nu\end{matrix}\right]\times$ $\times G_{22}^{22}\left(x\left\vert\begin{matrix}(6\nu+5)/4,\ (10\nu+7)/4\\ (2\nu+1)/4,\ -(2\nu+1)/4\end{matrix}\right.\right)$	$-\frac{\cos 2\nu\pi}{2\pi^{3/2}}\Gamma\left[\begin{matrix}2\nu+3/2\\ -\nu,\ -1-3\nu\end{matrix}\right]\times$ $\times\Gamma\left[s+\frac{2\nu+1}{4},\ s-\frac{2\nu+1}{4},\ -\frac{1+6\nu}{4}-s,\ -\frac{3+10\nu}{4}-s\right]$ [Re $\nu<-1/3$; $-(1+2\,\text{Re}\,\nu)/4$, $(1+2\,\text{Re}\,\nu)/4<\text{Re}\,s<-(1+6\,\text{Re}\,\nu)/4$, $-(3+10\,\text{Re}\,\nu)/4$]

40	$\frac{\lvert 1\pm\sqrt{x}\rvert}{\lvert 1-x\rvert^{-2\nu}} P_\nu^{-2\nu-1}\left(\frac{1\pm 6\sqrt{x}+x}{(1\mp\sqrt{x})^2}\right) = -\frac{\sin\nu\pi}{\sqrt{\pi}}\Gamma\left(2\nu+\frac{3}{2}\right)\times G_{22}^{11}\left(x\,\middle\vert\,\begin{matrix}(6\nu+5)/4, & (10\nu+7)/4\\ (2\nu+1)/4, & -(2\nu+1)/4\end{matrix}\right)$	$-\frac{\sin\nu\pi}{\sqrt{\pi}}\Gamma\left(2\nu+\frac{3}{2}\right)\times\Gamma\begin{bmatrix} s+(2\nu+1)/4, & -(1+6\nu)/4-s\\ s+(10\nu+7)/4, & (5+2\nu)/4-s\end{bmatrix}$ $[\operatorname{Re}\nu>-3/4;\ -(1+2\operatorname{Re}\nu)/4<\operatorname{Re}s<-(1+6\operatorname{Re}\nu)/4]$
41	$\frac{\lvert 1\pm\sqrt{x}\rvert}{\lvert 1-x\rvert^{-2\nu}}P_\nu^{-2\nu-1}\left(\frac{-1\mp 6\sqrt{x}-x}{(1\mp\sqrt{x})^2}\right)=\frac{\Gamma(2\nu+3/2)}{2\pi^{5/2}\cos\nu\pi}\left\{\sin 3\nu\pi\times G_{22}^{11}\left(x\,\middle\vert\,\begin{matrix}(10\nu+7)/4, & (6\nu+5)/4\\ -(2\nu+1)/4, & (2\nu+1)/4\end{matrix}\right)+\sin\nu\pi\, G_{22}^{11}\left(x\,\middle\vert\,\begin{matrix}(6\nu+5)/4, & (10\nu+7)/4\\ (2\nu+1)/4, & -(2\nu+1)/4\end{matrix}\right)\right\}$	$\frac{\Gamma(2\nu+3/2)}{2\pi^{5/2}\cos\nu\pi}\left\{\sin 3\nu\pi\times\Gamma\begin{bmatrix}s-(2\nu+1)/4, & -(3+10\nu)/4-s\\ s+(6\nu+5)/4, & (3-2\nu)/4-s\end{bmatrix}+\sin\nu\pi\,\Gamma\begin{bmatrix}s+\frac{2\nu+1}{4}, & -\frac{6\nu+1}{4}-s\\ s+\frac{10\nu+7}{4}, & \frac{2\nu+5}{4}-s\end{bmatrix}\right\}$ $[-3/4<\operatorname{Re}\nu<-1/3;\ \lvert\operatorname{Re}(2\nu+1)\rvert/4<\operatorname{Re}s<-(1+6\operatorname{Re}\nu)/4,\ -(3+10\operatorname{Re}\nu)/4]$
42	$\frac{\lvert 1-x\rvert^{\nu+1/2}}{\sqrt{1+\sqrt{x}}}P_\nu^{\nu+1/2}\left(\frac{2\sqrt[4]{x}}{1+\sqrt{x}}\right)=-\frac{\sin\nu\pi}{2^{\nu+3/2}\pi^2}\Gamma\begin{bmatrix}\nu+1\\ -2\nu-1/2\end{bmatrix}\times G_{22}^{22}\left(x\,\middle\vert\,\begin{matrix}\nu+5/4, & \nu+1\\ 0, & 1/4\end{matrix}\right)$	$-\frac{\sin\nu\pi}{2^{\nu+3/2}\pi^2}\Gamma\begin{bmatrix}\nu+1\\ -2\nu-1/2\end{bmatrix}\times\Gamma\left[s,\ s+\frac{1}{4},\ -\nu-\frac{1}{4}-s,\ -\nu-s\right]$ $[0<\operatorname{Re}s<-1/4-\operatorname{Re}\nu]$
43	$\frac{\lvert 1-x\rvert^{\nu+1/2}}{\sqrt{1+\sqrt{x}}}P_\nu^{-\nu-1/2}\left(\frac{2\sqrt[4]{x}}{1+\sqrt{x}}\right)=-\frac{\Gamma(\nu+1)}{2^{\nu+1/2}\sqrt{\pi}}\left\{\sin\left(\nu+\frac{1}{4}\right)\pi\times G_{22}^{11}\left(x\,\middle\vert\,\begin{matrix}\nu+5/4, & \nu+1\\ 0, & 1/4\end{matrix}\right)+\cos\left(\nu+\frac{1}{4}\right)\pi\, G_{22}^{11}\left(x\,\middle\vert\,\begin{matrix}\nu+1, & \nu+5/4\\ 1/4, & 0\end{matrix}\right)\right\}$	$-\frac{\Gamma(\nu+1)}{2^{\nu+1/2}\sqrt{\pi}}\left\{\sin\left(\nu+\frac{1}{4}\right)\pi\times\Gamma\begin{bmatrix}s, & -1/4-\nu-s\\ s+\nu+1, & 3/4-s\end{bmatrix}+\cos\left(\nu+\frac{1}{4}\right)\pi\times\Gamma\begin{bmatrix}s+1/4, & -\nu-s\\ s+\nu+5/4, & 1-s\end{bmatrix}\right\}$ $[-1<\operatorname{Re}\nu<-1/4;\ 0<\operatorname{Re}s<-1/4-\operatorname{Re}\nu]$
44	$(1-x)_+^{-1/2}\left[P_\nu^\mu(\sqrt{1-x})+P_\nu^\mu(-\sqrt{1-x})\right]=A_{70}G_{22}^{20}\left(x\,\middle\vert\,\begin{matrix}(1+\nu)/2, & -\nu/2\\ \mu/2, & -\mu/2\end{matrix}\right)$	$A_{70}\Gamma\begin{bmatrix}s+\mu/2, & s-\mu/2\\ s+(1+\nu)/2, & s-\nu/2\end{bmatrix}$ $[\operatorname{Re}s>\lvert\operatorname{Re}\mu\rvert/2]$
45	$(x-1)_+^{-1/2}\left[P_\nu^\mu\left(\sqrt{1-\frac{1}{x}}\right)+P_\nu^\mu\left(-\sqrt{1-\frac{1}{x}}\right)\right]=A_{70}G_{22}^{02}\left(x\,\middle\vert\,\begin{matrix}(1-\mu)/2, & (1+\mu)/2\\ -\nu/2, & (1+\nu)/2\end{matrix}\right)$	$A_{70}\Gamma\begin{bmatrix}(1+\mu)/2-s, & (1-\mu)/2-s\\ 1+\nu/2-s, & (1-\nu)/2-s\end{bmatrix}$ $[\operatorname{Re}s<(1-\lvert\operatorname{Re}\mu\rvert)/2]$
46	$\left[P_\nu^\mu(\sqrt{1+x})\right]^2=\frac{A_{63}}{\sqrt{\pi}}G_{33}^{13}\left(x\,\middle\vert\,\begin{matrix}-\nu, & 1+\nu, & 1/2\\ -\mu, & 0, & \mu\end{matrix}\right)$	$\frac{A_{63}}{\sqrt{\pi}}\Gamma\begin{bmatrix}s-\mu, & 1+\nu-s, & 1/2-s, & -\nu-s\\ 1-s, & 1-\mu-s\end{bmatrix}$ $[\operatorname{Re}\mu<\operatorname{Re}s<-\operatorname{Re}\nu,\ 1+\operatorname{Re}\nu]$

47	$\left[P_\nu^\mu\left(\sqrt{1+\frac{1}{x}}\right)\right]^2 = \frac{A_{63}}{\sqrt{\pi}} G_{33}^{31}\left(x\,\middle\vert\,\begin{matrix}1+\mu, & 1, & 1-\mu\\ 1+\nu, & -\nu, & 1/2\end{matrix}\right)$	$\frac{A_{63}}{\sqrt{\pi}}\Gamma\begin{bmatrix}s+\nu+1, s+1/2, s-\nu, -\mu-s\\ s+1, s+1-\mu\end{bmatrix}$ [Re ν, $-1-\mathrm{Re}\,\nu<\mathrm{Re}\,s<-\mathrm{Re}\,\mu$]
48	$P_\nu^{-\mu}(\sqrt{1+x})\,P_\nu^\mu(\sqrt{1+x}) = A_{71}G_{33}^{13}\left(x\,\middle\vert\,\begin{matrix}-\nu, & \nu+1, & 1/2\\ 0, & -\mu, & \mu\end{matrix}\right)$	$A_{71}\Gamma\begin{bmatrix}s, -\nu-s, 1+\nu-s, 1/2-s\\ 1+\mu-s, 1-\mu-s\end{bmatrix}$ [$0<\mathrm{Re}\,s<-\mathrm{Re}\,\nu$, $1+\mathrm{Re}\,\nu$]
49	$P_\nu^{-\mu}\left(\sqrt{\frac{1+x}{x}}\right)P_\nu^\mu\left(\sqrt{\frac{1+x}{x}}\right) = A_{71}G_{33}^{31}\left(x\,\middle\vert\,\begin{matrix}1, & 1+\mu, & 1-\mu\\ 1/2, & 1+\nu, & -\nu\end{matrix}\right)$	$A_{71}\Gamma\begin{bmatrix}s-\nu, s+\nu+1, s+1/2, -s\\ s+\mu+1, s-\mu+1\end{bmatrix}$ [Re ν, $-1-\mathrm{Re}\,\nu<\mathrm{Re}\,s<0$].
50	$\frac{P_\nu^\mu(\sqrt{1+x})\,P_{-\nu}^{\mu}(\sqrt{1+x})}{\sqrt{1+x}} = A_{72}G_{33}^{13}\left(x\,\middle\vert\,\begin{matrix}\nu, & -\nu, & 1/2\\ -\mu, & 0, & \mu\end{matrix}\right)$	$A_{72}\Gamma\begin{bmatrix}s-\mu, 1/2-s, 1+\nu-s, 1-\nu-s\\ 1-s, 1-\mu-s\end{bmatrix}$ [$\mathrm{Re}\,\mu<\mathrm{Re}\,s<1-\lvert\mathrm{Re}\,\nu\rvert$, $1/2$]
51	$(1+x)^{-1/2}P_\nu^\mu\left(\sqrt{1+\frac{1}{x}}\right)\times P_{-\nu}^\mu\left(\sqrt{1+\frac{1}{x}}\right) = A_{72}G_{33}^{31}\left(x\,\middle\vert\,\begin{matrix}1/2+\mu, & 1/2, & 1/2-\mu\\ 0, & 1/2+\nu, & 1/2-\nu\end{matrix}\right)$	$A_{72}\Gamma\begin{bmatrix}s, s+\frac{1}{2}+\nu, s+\frac{1}{2}-\nu, \frac{1}{2}-\mu-s\\ s+\frac{1}{2}, s+\frac{1}{2}-\mu\end{bmatrix}$ [0, $\lvert\mathrm{Re}\,\nu\rvert-1/2<\mathrm{Re}\,s<1/2-\mathrm{Re}\,\mu$]
52	$\frac{P_\nu^{\mu\pm1}(\sqrt{1+x})\,P_\nu^\mu(\sqrt{1+x})}{\sqrt{1+x}} = A_{73}G_{33}^{13}\left(x\,\middle\vert\,\begin{matrix}0, & \frac{1}{2}+\nu, & -\frac{1}{2}-\nu\\ -\mu\mp\frac{1}{2}, & -\frac{1}{2}, & \pm\frac{1}{2}+\mu\end{matrix}\right)$	$A_{73}\Gamma\begin{bmatrix}s-\mu\mp\frac{1}{2}, 1-s, \frac{1}{2}-\nu-s, \frac{3}{2}+\nu-s\\ \frac{3}{2}-s, 1\mp\frac{1}{2}-\mu-s\end{bmatrix}$ [$\mathrm{Re}\,\mu\pm1/2<\mathrm{Re}\,s<1/2-\mathrm{Re}\,\nu$, $3/2+\mathrm{Re}\,\nu$]
53	$(1+x)^{-1/2}P_\nu^{\mu\pm1}\left(\sqrt{\frac{1+x}{x}}\right)\times P_\nu^\mu\left(\sqrt{\frac{1+x}{x}}\right) = A_{73}G_{33}^{31}\left(x\,\middle\vert\,\begin{matrix}\frac{1\pm1}{2}+\mu, & 1, & \frac{1\mp1}{2}-\mu\\ \frac{1}{2}, & -\nu, & \nu+1\end{matrix}\right)$	$A_{73}\Gamma\begin{bmatrix}s+\frac{1}{2}, s-\nu, s+\nu+1, \frac{1\mp1}{2}-\mu-s\\ s+1, s+\frac{1\mp1}{2}-\mu\end{bmatrix}$ [Re ν, $-1-\mathrm{Re}\,\nu<\mathrm{Re}\,s<(1\mp1)/2-\mathrm{Re}\,\mu$]

54	$P_\nu^\mu(\sqrt{1+x})\,P^{\pm\nu\pm 1/2}_{\pm\mu-1/2}\left(\sqrt{1+\frac{1}{x}}\right)= = \frac{1}{\sqrt{2\pi}\,\Gamma((1\mp 1)/2-\mu\mp\nu)}\times \times G^{22}_{33}\left(x\,\middle\vert\,\begin{matrix}\frac{3}{4}, \frac{3\pm 2}{4}\pm\nu, \frac{3\mp 2}{4}\mp\nu\\ \frac{1}{4}, \frac{1}{4}-\mu, \frac{1}{4}+\mu\end{matrix}\right)$	$\frac{1}{\sqrt{2}\,\pi\Gamma((1\mp 1)/2-\mu\mp\nu)}\times \times\Gamma\left[\begin{matrix}s+\frac{1}{4}, s+\frac{1}{4}-\mu, \frac{1}{4}-s, \frac{1\mp 2}{4}\mp\nu-s\\ s+\frac{3\mp 2}{4}\mp\nu, \frac{3}{4}-\mu-s\end{matrix}\right]$ [$-1/4$, $\mathrm{Re}\,\mu-1/4<\mathrm{Re}\,s<1/4$, $(1\mp 2)/4-\mathrm{Re}\,\nu$]
55	$P_\nu(\sqrt{1+x}-\sqrt{x})\,P_\nu(\sqrt{1+x}+\sqrt{x})= =\frac{A_{71}}{2}G^{14}_{44}\left(x\,\middle\vert\,\begin{matrix}(1-\nu)/2, -\nu/2, 1+\nu/2, (1+\nu)/2\\ 0, 0, 0, 1/2\end{matrix}\right)$	$\frac{A_{71}}{2}\Gamma\left[\begin{matrix}s, \frac{1+\nu}{2}-s, 1+\frac{\nu}{2}-s, -\frac{\nu}{2}-s, \frac{1-\nu}{2}-s\\ \frac{1}{2}-s, 1-s, 1-s\end{matrix}\right]$ [$0<\mathrm{Re}\,s<(1+\mathrm{Re}\,\nu)/2$, $-\mathrm{Re}\,\nu/2$]
56	$P_\nu\left(\frac{\sqrt{1+x}-1}{\sqrt{x}}\right)P_\nu\left(\frac{\sqrt{1+x}+1}{\sqrt{x}}\right)= =\frac{A_{71}}{2}G^{41}_{44}\left(x\,\middle\vert\,\begin{matrix}1/2, 1, 1, 1\\ (\nu+1)/2, \nu/2+1, -\nu/2, (1-\nu)/2\end{matrix}\right)$	$\frac{A_{71}}{2}\Gamma\left[\begin{matrix}s+\frac{\nu+1}{2}, s+\frac{\nu}{2}+1, s-\frac{\nu}{2}, s+\frac{1-\nu}{2}, -s\\ s+\frac{1}{2}, s+1, s+1\end{matrix}\right]$ [$\mathrm{Re}\,\nu/2$, $-(1+\mathrm{Re}\,\nu)/2<\mathrm{Re}\,s<0$]

8.4.42. The Legendre functions of the second kind $Q_\nu^\mu(x)$ and $Q_\nu(x)$.

NOTATION:

$$A_{74}=\frac{e^{i\mu\pi}}{2}\Gamma^{(1\pm 1)/2}\left[\begin{matrix}1+\mu+\nu\\1-\mu+\nu\end{matrix}\right],\quad A_{75}=\frac{e^{i\mu\pi}}{2}\Gamma(\nu+1)\,\Gamma(\mu+\nu+1),$$

$$A_{76}=\frac{\pi\operatorname{cosec}\mu\pi}{2\Gamma(-\nu)\,\Gamma(-\mu-\nu)},\quad A_{77}=\frac{\pi\operatorname{cosec}\mu\pi}{2\Gamma(\nu+1)\,\Gamma(\nu-\mu+1)},$$

$$A_{78}=\frac{2^{-\mu-3/2}\pi}{\sin\mu\pi}\Gamma\left[\begin{matrix}1/2+\mu\\1/2-2\mu\end{matrix}\right],\quad A_{79}=2^{\mu-1}e^{i\mu\pi}\Gamma\left[\begin{matrix}1+(\mu+\nu)/2-r\\(1-\mu+\nu)/2\end{matrix}\right],$$

$$A_{80}=-\frac{\Gamma(\mu+\nu+1)}{2^{\nu+2}\sqrt{\pi}},\quad A_{81}=\frac{2^{\nu-1}\sqrt{\pi}}{\Gamma(1-\mu+\nu)},\quad A_{82}=\frac{e^{i\mu\pi}\sqrt{\pi}}{2^{\nu+1}}\Gamma(\mu+\nu+1),$$

$$A_{83}=\frac{e^{i\mu\pi}}{\sqrt{\pi}}\cos\mu\pi\Gamma(\mu+\nu+1),\quad A_{84}=\frac{e^{i\mu\pi}\sqrt{\pi}}{\Gamma(1-\mu+\nu)},\quad A_{85}=e^{i\mu\pi}\sqrt{\pi}\,\Gamma(\mu+\nu+1),$$

$$A_{86}=\frac{e^{i\mu\pi}}{\sqrt{2}\,\Gamma(1-\mu+\nu)},\quad A_{87}=\frac{e^{i\mu\pi}}{\sqrt{2}\,(\mu+\nu+1)^{(1\pm 1)/2}\,\Gamma(\nu-\mu+(5\mp 1)/2)},$$

$$A_{88}=\frac{e^{i\mu\pi}}{\sqrt{2}\,(1+\mu+\nu)\,\Gamma(1-\mu+\nu)},\quad A_{89}=\frac{\sqrt{\pi}}{2}e^{2i\mu\pi}\Gamma\left[\begin{matrix}1+\mu+\nu\\1-\mu+\nu\end{matrix}\right],$$

$$A_{90}=\frac{e^{2i\mu\pi}}{2\sqrt{\pi}}\cos\mu\pi\Gamma(\mu+\nu+1)\,\Gamma(\mu-\nu),$$

$$A_{91}=\frac{\sqrt{\pi}}{2}e^{2i\mu\pi}\,\Gamma\begin{bmatrix}\mu+\nu+(1\pm1)/2\\ \nu-\mu+(3\pm1)/2\end{bmatrix}.$$

1	$(1+x)^{\pm\mu/2}\,Q_\nu^\mu(2x+1)=$ $=A_{74}G_{22}^{21}\left(x\,\middle\vert\,\begin{matrix}\pm\mu/2-\nu,\pm\mu/2+\nu+1\\ \mu/2,\ -\mu/2\end{matrix}\right)$	$A_{74}\Gamma\begin{bmatrix}s+\mu/2,\ s-\mu/2,\ \nu+1\mp\mu/2-s\\ s+\nu+1\pm\mu/2\end{bmatrix}$ $[\vert\operatorname{Re}\mu\vert/2<\operatorname{Re}s<1+\operatorname{Re}(\nu\mp\mu/2)]$
2	$(1+x)^{\pm\mu/2}\,Q_\nu^\mu\left(\frac{2}{x}+1\right)=$ $=A_{74}G_{22}^{12}\left(x\,\middle\vert\,\begin{matrix}1,\ 1\pm\mu\\ \nu+1,\ -\nu\end{matrix}\right)$	$A_{74}\Gamma\begin{bmatrix}s+\nu+1,\ \mp\mu-s,\ -s\\ \nu+1-s\end{bmatrix}$ $[-1-\operatorname{Re}\nu<\operatorname{Re}s<0,\ \mp\operatorname{Re}\mu]$
3	$e^{-i\mu\pi}(x-1)_+^{-\mu/2}Q_\nu^\mu(2x-1)+$ $+(1-x)_+^{-\mu/2}Q_\nu^\mu(2x-1)=$ $=\frac{\sin\nu\pi\sin(\mu-\nu)\pi}{2\pi\sin\mu\pi}\times$ $\times G_{22}^{22}\left(x\,\middle\vert\,\begin{matrix}1+\nu-\mu/2,\ -\nu-\mu/2\\ \mu/2,\ -\mu/2\end{matrix}\right)-$ $-\frac{\pi}{2\sin\mu\pi}G_{22}^{02}\left(x\,\middle\vert\,\begin{matrix}1+\nu-\mu/2,-\nu-\mu/2\\ \mu/2,\ -\mu/2\end{matrix}\right)-$ $-\frac{\pi}{2}\operatorname{ctg}\mu\pi G_{22}^{20}\left(x\,\middle\vert\,\begin{matrix}1+\nu-\mu/2,\ -\nu-\mu/2\\ \mu/2,\ -\mu/2\end{matrix}\right)$	$\frac{\sin\nu\pi\sin(\mu-\nu)\pi}{2\pi\sin\mu\pi}\Gamma\left[s-\frac{\mu}{2},\ s+\frac{\mu}{2},\right.$ $\left.\frac{\mu}{2}-\nu-s,\ \frac{\mu}{2}+\nu+1-s\right]-\frac{\pi}{2\sin\mu\pi}\times$ $\times\Gamma\begin{bmatrix}\mu/2-\nu-s,\ 1+\nu+\mu/2-s\\ 1+\mu/2-s,\ 1-\mu/2-s\end{bmatrix}-$ $-\frac{\pi}{2}\operatorname{ctg}\mu\pi\Gamma\begin{bmatrix}s-\mu/2,\ s+\mu/2\\ s-\mu/2-\nu,\ s-\mu/2+\nu+1\end{bmatrix}$ $[\operatorname{Re}\mu<1;\ \mu\neq-1,\ -2,\ \ldots,\ \vert\operatorname{Re}\mu\vert/2<$ $<\operatorname{Re}s<\operatorname{Re}(\mu/2-\nu),\ 1+\operatorname{Re}(\mu/2+\nu)]$
4	$(1-x)_+^\nu\,Q_\nu^\mu\left(\frac{1+x}{1-x}\right)=$ $=A_{75}G_{22}^{20}\left(x\,\middle\vert\,\begin{matrix}1+\nu+\mu/2,\ 1+\nu-\mu/2\\ \mu/2,\ -\mu/2\end{matrix}\right)$	$A_{75}\Gamma\begin{bmatrix}s+\mu/2,\ s-\mu/2\\ s+\nu+1+\mu/2,\ s+\nu+1-\mu/2\end{bmatrix}$ $[\operatorname{Re}\nu>-1;\ \operatorname{Re}s>\vert\operatorname{Re}\mu\vert/2]$
5	$(x-1)_+^\nu\,Q_\nu^\mu\left(\frac{x+1}{x-1}\right)=$ $=A_{75}G_{22}^{02}\left(x\,\middle\vert\,\begin{matrix}1+\nu+\mu/2,\ 1+\nu-\mu/2\\ \mu/2,\ -\mu/2\end{matrix}\right)$	$A_{75}\Gamma\begin{bmatrix}\mu/2-\nu-s,\ -\mu/2-\nu-s\\ 1+\mu/2-s,\ 1-\mu/2-s\end{bmatrix}$ $[\operatorname{Re}\nu>-1;\ \operatorname{Re}s<-\operatorname{Re}\nu-\vert\operatorname{Re}\mu\vert/2]$
6	$\vert1-x\vert^\nu\,Q_\nu^\mu\left(\frac{1+x}{\vert1-x\vert}\right)=$ $=A_{75}\left\{G_{22}^{20}\left(x\,\middle\vert\,\begin{matrix}1+\nu+\mu/2,1+\nu-\mu/2\\ \mu/2,\ -\mu/2\end{matrix}\right)+\right.$ $\left.+G_{22}^{02}\left(x\,\middle\vert\,\begin{matrix}1+\nu+\mu/2,\ 1+\nu-\mu/2\\ \mu/2,\ -\mu/2\end{matrix}\right)\right\}$	$A_{75}\left\{\Gamma\begin{bmatrix}s+\mu/2,\ s-\mu/2\\ s+\nu+1+\mu/2,\ s+\nu+1-\mu/2\end{bmatrix}+\right.$ $\left.+\Gamma\begin{bmatrix}\mu/2-\nu-s,\ -\mu/2-\nu-s\\ 1+\mu/2-s,\ 1-\mu/2-s\end{bmatrix}\right\}$ $[\operatorname{Re}\nu>-1;\ \vert\operatorname{Re}\mu\vert/2<\operatorname{Re}s<$ $<-\operatorname{Re}\nu-\vert\operatorname{Re}\mu\vert/2]$
7	$\vert1-x\vert^{-\nu-1}\,Q_\nu^\mu\left(\frac{1+x}{\vert1-x\vert}\right)=$ $=\frac{e^{i\mu\pi}}{2\Gamma(1+\nu)\,\Gamma(1-\mu+\nu)}\times$ $\times G_{22}^{22}\left(x\,\middle\vert\,\begin{matrix}\mu/2-\nu,\ -\mu/2-\nu\\ \mu/2,\ -\mu/2\end{matrix}\right)$	$\frac{e^{i\mu\pi}}{2\Gamma(1+\nu)\,\Gamma(1-\mu+\nu)}\Gamma\left[s+\frac{\mu}{2},\right.$ $\left.s-\frac{\mu}{2},\ 1+\nu+\frac{\mu}{2}-s,\ 1+\nu-\frac{\mu}{2}-s\right]$ $[2\nu\neq-2,\ -3,\ -4,\ \ldots,\ \vert\operatorname{Re}\mu\vert/2<$ $<\operatorname{Re}s<1+\operatorname{Re}\nu-\vert\operatorname{Re}\mu\vert/2]$

8	$(1+x)^{\nu} Q_{\nu}^{\mu}\left(\frac{1-x}{1+x}\right)=A_{76}\left\{\frac{\sin(\mu-\nu)\pi}{\sin(\mu+\nu)\pi}\times G_{22}^{12}\left(x\middle\vert\begin{matrix}1+\nu+\mu/2,\ 1+\nu-\mu/2\\ \mu/2,\ -\mu/2\end{matrix}\right)+\cos\mu\pi G_{22}^{12}\left(x\middle\vert\begin{matrix}1+\nu+\mu/2, 1+\nu-\mu/2\\ -\mu/2,\ \mu/2\end{matrix}\right)\right\}$	$A_{76}\Gamma\left[\frac{\mu}{2}-\nu-s,\ -\frac{\mu}{2}-\nu-s\right]\times\left\{\frac{\sin(\mu-\nu)\pi}{\sin(\mu+\nu)\pi}\Gamma\left[\begin{matrix}s+\mu/2\\ 1+\mu/2-s\end{matrix}\right]+\cos\mu\pi\Gamma\left[\begin{matrix}s-\mu/2\\ 1-\mu/2-s\end{matrix}\right]\right\}$ $[\mu\neq 0, \pm 1, \pm 2, \ldots, \vert\operatorname{Re}\mu\vert/2<\operatorname{Re}s<-\operatorname{Re}\nu-\vert\operatorname{Re}\mu\vert/2]$
9	$(1+x)^{\nu} Q_{\nu}^{\mu}\left(\frac{x-1}{x+1}\right)=A_{76}\left\{\frac{\sin(\mu-\nu)\pi}{\sin(\mu+\nu)\pi}\times G_{22}^{21}\left(x\middle\vert\begin{matrix}1+\nu-\mu/2,\ 1+\nu+\mu/2\\ \mu/2,\ -\mu/2\end{matrix}\right)+\cos\mu\pi G_{22}^{21}\left(x\middle\vert\begin{matrix}1+\nu+\mu/2, 1+\nu-\mu/2\\ -\mu/2,\ \mu/2\end{matrix}\right)\right\}$	$A_{76}\Gamma\left[s+\frac{\mu}{2},\ s-\frac{\mu}{2}\right]\times\left\{\frac{\sin(\mu-\nu)\pi}{\sin(\mu+\nu)\pi}\Gamma\left[\begin{matrix}\mu/2-\nu-s\\ s+\nu+\mu/2+1\end{matrix}\right]+\cos\mu\pi\Gamma\left[\begin{matrix}-\mu/2-\nu-s\\ s+\nu-\mu/2+1\end{matrix}\right]\right\}$ $[\mu\neq 0, \pm 1, \pm 2, \ldots, \vert\operatorname{Re}\mu\vert/2<\operatorname{Re}s<-\operatorname{Re}\nu-\vert\operatorname{Re}\mu\vert/2]$
10	$(1+x)^{-\nu-1} Q_{\nu}^{\mu}\left(\frac{1-x}{1+x}\right)=A_{77}\left\{\cos\mu\pi\times G_{22}^{12}\left(x\middle\vert\begin{matrix}\mu/2-\nu,\ -\mu/2-\nu\\ -\mu/2,\ \mu/2\end{matrix}\right)-G_{22}^{12}\left(x\middle\vert\begin{matrix}\mu/2-\nu,\ -\mu/2-\nu\\ \mu/2,\ -\mu/2\end{matrix}\right)\right\}$	$A_{77}\Gamma\left[1+\nu-\frac{\mu}{2}-s,\ 1+\nu+\frac{\mu}{2}-s\right]\times\left\{\cos\mu\pi\Gamma\left[\begin{matrix}s-\mu/2\\ 1-\mu/2-s\end{matrix}\right]-\Gamma\left[\begin{matrix}s+\mu/2\\ 1+\mu/2-s\end{matrix}\right]\right\}$ $[\mu\neq 0, \pm 1, \pm 2, \ldots; \vert\operatorname{Re}\mu\vert/2<\operatorname{Re}s<1+\operatorname{Re}\nu-\vert\operatorname{Re}\mu\vert/2]$
11	$(1+x)^{-\nu-1} Q_{\nu}^{\mu}\left(\frac{x-1}{x+1}\right)=A_{77}\left\{\cos\mu\pi\times G_{22}^{21}\left(x\middle\vert\begin{matrix}\mu/2-\nu,\ -\mu/2-\nu\\ \mu/2,\ -\mu/2\end{matrix}\right)-G_{22}^{21}\left(x\middle\vert\begin{matrix}-\mu/2-\nu,\ \mu/2-\nu\\ \mu/2,\ -\mu/2\end{matrix}\right)\right\}$	$A_{77}\Gamma\left[s+\frac{\mu}{2},\ s-\frac{\mu}{2}\right]\times\left\{\cos\mu\pi\Gamma\left[\begin{matrix}1+\nu-\mu/2-s\\ s-\mu/2-\nu\end{matrix}\right]-\Gamma\left[\begin{matrix}1+\nu+\mu/2-s\\ s+\mu/2-\nu\end{matrix}\right]\right\}$ $[\mu\neq 0, \pm 1, \pm 2, \ldots, \vert\operatorname{Re}\mu\vert/2<\operatorname{Re}s<1+\operatorname{Re}\nu-\vert\operatorname{Re}\mu\vert/2]$
12	$\frac{\vert 1-x\vert^{\mu}}{\sqrt{1+x}} Q_{(-2\pm 1)/4}^{\mu}\left(\frac{1-6x+x^2}{(1+x)^2}\right)=A_{78}\left\{\operatorname{cosec}(\pm\mu-1/4)\pi\times G_{22}^{11}\left(x\middle\vert\begin{matrix}(1+\mu)/2,\ (1+3\mu)/2\\ \mu/2,\ -\mu/2\end{matrix}\right)+\sqrt{2}\cos\mu\pi G_{22}^{11}\left(x\middle\vert\begin{matrix}(1+3\mu)/2, (1+\mu)/2\\ -\mu/2,\ \mu/2\end{matrix}\right)\right\}$	$A_{78}\left\{\frac{1}{\sin(\pm\mu-1/4)\pi}\times\Gamma\left[\begin{matrix}s+\mu/2,\ (1-\mu)/2-s\\ s+(1+3\mu)/2,\ 1+\mu/2-s\end{matrix}\right]+\sqrt{2}\cos\mu\pi\times\Gamma\left[\begin{matrix}s-\mu/2,\ (1-3\mu)/2-s\\ s+(\mu+1)/2,\ 1-\mu/2-s\end{matrix}\right]\right\}$ $[-1/2<\operatorname{Re}\mu<1/4;\ \vert\operatorname{Re}\mu\vert/2<\operatorname{Re}s<(1-\operatorname{Re}\mu)/2,\ (1-3\operatorname{Re}\mu)/2]$

13	$\frac{\lvert 1-x\rvert^{\mu}}{\sqrt{1+x}}\, Q^{\mu}_{(-2\pm 1)/4}\left(\frac{-1+6x-x^2}{(1+x)^2}\right)= \mp A_{78}\left\{\operatorname{ctg}(\pm\mu-1/4)\,\pi\times G^{11}_{22}\left(x\,\middle\vert\,\begin{matrix}(1+\mu)/2,\ (1+3\mu)/2\\ \mu/2,\ -\mu/2\end{matrix}\right)+G^{11}_{22}\left(x\,\middle\vert\,\begin{matrix}(1+3\mu)/2,\ (1+\mu)/2\\ -\mu/2,\ \mu/2\end{matrix}\right)\right\}$	$\mp A_{78}\left\{\operatorname{ctg}\left(\pm\mu-\frac{1}{4}\right)\pi\times\Gamma\left[\begin{matrix}s+\mu/2,\ (1-\mu)/2-s\\ s+(1+3\mu)/2,\ \ 1+\mu/2-s\end{matrix}\right]+\Gamma\left[\begin{matrix}s-\mu/2,\ (1-3\mu)/2-s\\ s+(1+\mu)/2,\ 1-\mu/2-s\end{matrix}\right]\right\}$ [$-1/2<\mathrm{Re}\,\mu<1/4$; $\lvert\mathrm{Re}\,\mu\rvert/2<\mathrm{Re}\,s<(1-\mathrm{Re}\,\mu)/2,\ (1-3\,\mathrm{Re}\,\mu)/2$]
14	$e^{-i\mu\pi}(x-1)_+^{-\mu/2}\,Q^{\mu}_{\nu}(\sqrt{x})+(1-x)_+^{-\mu/2}\,Q^{\mu}_{\nu}(\sqrt{x})=\frac{2^{\mu-2}\sin(\nu-\mu)\pi}{\pi\sin\mu\pi}\times G^{22}_{22}\left(x\,\middle\vert\,\begin{matrix}(1-\mu-\nu)/2,\ 1+(\nu-\mu)/2\\ 0,\ 1/2\end{matrix}\right)+\frac{2^{\mu-1}\pi}{\sin\mu\pi}\left\{G^{02}_{22}\left(x\,\middle\vert\,\begin{matrix}\frac{1-\mu-\nu}{2},\ 1+\frac{\nu-\mu}{2}\\ 0,\ \frac{1}{2}\end{matrix}\right)+\cos\mu\pi\, G^{20}_{22}\left(x\,\middle\vert\,\begin{matrix}\frac{1-\mu-\nu}{2},\ 1+\frac{\nu-\mu}{2}\\ 0,\ \frac{1}{2}\end{matrix}\right)\right\}$	$\frac{2^{\mu-2}\sin(\nu-\mu)\pi}{\pi\sin\mu\pi}\times\Gamma\left[s,\ s+\frac{1}{2},\ \frac{\mu+\nu+1}{2}-s,\ \frac{\mu-\nu}{2}-s\right]+\frac{2^{\mu-1}\pi}{\sin\mu\pi}\left\{\Gamma\left[\begin{matrix}\frac{\mu+\nu+1}{2}-s,\ \frac{\mu-\nu}{2}-s\\ 1-s,\ \frac{1}{2}-s\end{matrix}\right]+\cos\mu\pi\,\Gamma\left[\begin{matrix}s,\ s+\frac{1}{2}\\ s+\frac{1-\nu-\mu}{2},\ s+1+\frac{\nu-\mu}{2}\end{matrix}\right]\right\}$ [$\mathrm{Re}\,\mu<1$; $\mu\neq 0,\ -1,\ -2,\ \ldots$; $0<\mathrm{Re}\,s<(1+\mathrm{Re}\,(\mu+\nu))/2,\ \mathrm{Re}\,(\mu-\nu)/2$]
15	$(1+x)^{-r}\,Q^{\mu}_{\nu}(\sqrt{1+x})=A_{79}G^{21}_{22}\left(x\,\middle\vert\,\begin{matrix}(1-\nu)/2-r,\ 1+\nu/2-r\\ \mu/2,\ -\mu/2\end{matrix}\right)$ [$r=0,\ 1/2$]	$A_{79}\,\Gamma\left[\begin{matrix}s+\mu/2,\ s-\mu/2,\ r+(\nu+1)/2-s\\ s+1+\nu/2-r\end{matrix}\right]$ [$\lvert\mathrm{Re}\,\mu\rvert/2<\mathrm{Re}\,s<(\mathrm{Re}\,\nu+1)/2+r$]
16	$(1+x)^{-r}\,Q^{\mu}_{\nu}\left(\sqrt{1+\frac{1}{x}}\right)=A_{79}G^{12}_{22}\left(x\,\middle\vert\,\begin{matrix}1-r-\mu/2,\ 1-r+\mu/2\\ (\nu+1)/2,\ -\nu/2\end{matrix}\right)$ [$r=0,\ 1/2$]	$A_{79}\,\Gamma\left[\begin{matrix}s+\frac{\nu+1}{2},\ r+\frac{\mu}{2}-s,\ r-\frac{\mu}{2}-s\\ 1+\frac{\nu}{2}-s\end{matrix}\right]$ [$-(1+\mathrm{Re}\,\nu)/2<\mathrm{Re}\,s<r-\lvert\mathrm{Re}\,\mu\rvert/2$
17	$(1+x)^{\nu/2}\,Q^{\mu}_{\nu}\left(\frac{1}{\sqrt{1+x}}\right)=A_{80}\left\{G^{22}_{33}\left(x\,\middle\vert\,\begin{matrix}1+\frac{\nu}{2},\ \frac{1+\nu}{2},\ \nu-\frac{\mu}{2}\\ \frac{\mu}{2},\ -\frac{\mu}{2},\ \nu-\frac{\mu}{2}\end{matrix}\right)+\right.$	$A_{80}\left\{\Gamma\left[\begin{matrix}s+\frac{\mu}{2},\ s-\frac{\mu}{2},\ -\frac{\nu}{2}-s,\ \frac{1-\nu}{2}-s\\ s+\nu-\frac{\mu}{2},\ 1-\nu+\frac{\mu}{2}-s\end{matrix}\right]+\right.$

	$+\cos(\mu+\nu)\,\pi G_{22}^{12}\left(x\left\vert\begin{matrix}1+\frac{\nu}{2},\ \frac{1+\nu}{2}\\ -\frac{\mu}{2},\ \frac{\mu}{2}\end{matrix}\right.\right)\Bigg\}$	$+\cos(\mu+\nu)\,\pi\Gamma\left[\begin{matrix}s-\frac{\mu}{2},\ -\frac{\nu}{2}-s,\\ 1-\frac{\mu}{2}-s,\ \frac{1-\nu}{2}-s\end{matrix}\right]\Bigg\}$ $[\vert\operatorname{Re}\mu\vert/2<\operatorname{Re}s<-\operatorname{Re}\nu/2]$
18	$(1+x)^{\nu/2}\,Q_\nu^\mu\left(\sqrt{\frac{x}{1+x}}\right)=$ $=A_{80}\left\{G_{33}^{22}\left(x\left\vert\begin{matrix}1+\frac{\nu-\mu}{2},\ 1+\frac{\nu+\mu}{2},\ 1+\frac{\mu-\nu}{2}\\ 0,\ \frac{1}{2},\ 1+\frac{\mu-\nu}{2}\end{matrix}\right.\right)+\right.$ $\left.+\cos(\mu+\nu)\,\pi G_{22}^{21}\left(x\left\vert\begin{matrix}1+\frac{\mu+\nu}{2},\ 1+\frac{\nu-\mu}{2}\\ 0,\ \frac{1}{2}\end{matrix}\right.\right)\right\}$	$A_{80}\left\{\Gamma\left[\begin{matrix}s,\ s+\frac{1}{2},\ \frac{\mu-\nu}{2}-s,\ -\frac{\mu+\nu}{2}-s\\ s+1+\frac{\mu-\nu}{2},\ \frac{\nu-\mu}{2}-s\end{matrix}\right]+\right.$ $\left.+\cos(\mu+\nu)\,\pi\Gamma\left[\begin{matrix}s,\ s+\frac{1}{2},\ -\frac{\mu+\nu}{2}-s\\ s+1+\frac{\nu-\mu}{2}\end{matrix}\right]\right\}$ $[0<\operatorname{Re}s<-(\operatorname{Re}\nu+\vert\operatorname{Re}\mu\vert)/2]$
19	$(1+x)^{-(\nu+1)/2}\,Q_\nu^\mu\left(\frac{1}{\sqrt{1+x}}\right)=$ $=A_{81}G_{33}^{22}\left(x\left\vert\begin{matrix}(1-\nu)/2,\ -\nu/2,\ (1+\mu)/2\\ \mu/2,\ -\mu/2,\ (1+\mu)/2\end{matrix}\right.\right)$	$A_{81}\,\Gamma\left[\begin{matrix}s+\frac{\mu}{2},\ s-\frac{\mu}{2},\ \frac{1+\nu}{2}-s,\ 1+\frac{\nu}{2}-s\\ s+\frac{1+\mu}{2},\ \frac{1-\mu}{2}-s\end{matrix}\right]$ $[\vert\operatorname{Re}\mu\vert<\operatorname{Re}s<(1+\operatorname{Re}\nu)/2]$
20	$(1+x)^{-(\nu+1)/2}\,Q_\nu^\mu\left(\sqrt{\frac{x}{1+x}}\right)=$ $=A_{81}G_{33}^{22}\left(x\left\vert\begin{matrix}\frac{1-\mu-\nu}{2},\ \frac{1+\mu-\nu}{2},\ -\frac{\mu+\nu}{2}\\ 0,\ \frac{1}{2},\ -\frac{\mu+\nu}{2}\end{matrix}\right.\right)$	$A_{81}\,\Gamma\left[\begin{matrix}s,\ s+\frac{1}{2},\ \frac{1+\nu+\mu}{2}-s,\ \frac{1+\nu-\mu}{2}-s\\ s-\frac{\mu+\nu}{2},\ 1+\frac{\mu+\nu}{2}-s\end{matrix}\right]$ $[0<\operatorname{Re}s<(1+\operatorname{Re}\nu-\vert\operatorname{Re}\mu\vert)/2]$

21	$(1-x)_+^{\nu/2} Q_\nu^\mu\left(\frac{1}{\sqrt{1-x}}\right)= = A_{82} G_{22}^{20}\left(x \middle\| \begin{matrix}(1+\nu)/2, & 1+\nu/2 \\ \mu/2, & -\mu/2\end{matrix}\right)$	$A_{82}\,\Gamma\left[\begin{matrix}s+\mu/2, & s-\mu/2 \\ s+(1+\nu)/2, & s+1+\nu/2\end{matrix}\right]$ [$\operatorname{Re}\nu>-3/2$; $\operatorname{Re}s>\|\operatorname{Re}\mu\|/2$]
22	$(x-1)_+^{\nu/2} Q_\nu^\mu\left(\sqrt{\frac{x}{x-1}}\right)= = A_{82} G_{22}^{02}\left(x \middle\| \begin{matrix}1+(\nu-\mu)/2, & 1+(\nu+\mu)/2 \\ 0, & 1/2\end{matrix}\right)$	$A_{82}\Gamma\left[\begin{matrix}(\mu-\nu)/2-s, & -(\mu+\nu)/2-s \\ 1/2-s, & 1-s\end{matrix}\right]$ [$\operatorname{Re}\nu>-3/2$; $\operatorname{Re}s<-(\operatorname{Re}\nu+\|\operatorname{Re}\mu\|)/2$]
23	$\|1-x\|^{-\mu} Q_\nu^\mu\left(\frac{1+x}{2\sqrt{x}}\right)= =e^{i\mu\pi}\sqrt{\pi}\,\Gamma\left(\frac{1}{2}-\mu\right)\times \times G_{22}^{11}\left(x \middle\| \begin{matrix}(1-\nu)/2-\mu, & 1+\nu/2-\mu \\ (\nu+1)/2, & -\nu/2\end{matrix}\right)$	$e^{i\mu\pi}\sqrt{\pi}\,\Gamma\left(\frac{1}{2}-\mu\right)\times \times\Gamma\left[\begin{matrix}s+(\nu+1)/2, & \mu+(\nu+1)/2-s \\ s+1+\nu/2-\mu, & 1+\nu/2-s\end{matrix}\right]$ [$\operatorname{Re}\mu<1/2$; $-(1+\operatorname{Re}\nu)/2<\operatorname{Re}s<$ $<1/2+\operatorname{Re}(\mu+\nu/2)$]
24	$(1+x)^{\nu/2} Q_\nu^\mu\left(\frac{2+x}{2\sqrt{1+x}}\right)= =A_{83} G_{22}^{21}\left(x \middle\| \begin{matrix}1/2, & 1+\nu \\ \mu, & -\mu\end{matrix}\right)$	$A_{83}\,\Gamma\left[\begin{matrix}s+\mu, & s-\mu, & 1/2-s \\ s+\nu+1\end{matrix}\right]$ [$\|\operatorname{Re}\mu\|<\operatorname{Re}s<1/2$]
25	$(1+x)^{\nu/2} Q_\nu^\mu\left(\frac{2x+1}{2\sqrt{x^2+x}}\right)= = A_{83} G_{22}^{12}\left(x \middle\| \begin{matrix}1+\nu/2-\mu, & 1+\nu/2+\mu \\ (1+\nu)/2, & -\nu/2\end{matrix}\right)$	$A_{83}\Gamma\left[\begin{matrix}s+\frac{\nu+1}{2}, & \mu-\frac{\nu}{2}-s, & -\mu-\frac{\nu}{2}-s \\ \frac{\nu}{2}+1-s\end{matrix}\right]$ [$-(1+\operatorname{Re}\nu)/2<\operatorname{Re}s<-\operatorname{Re}\nu/2-\|\operatorname{Re}\mu\|$]
26	$(1+x)^{-(\nu+1)/2} Q_\nu^\mu\left(\frac{2+x}{2\sqrt{1+x}}\right)= = A_{84} G_{22}^{21}\left(x \middle\| \begin{matrix}-\nu, & 1/2 \\ \mu, & -\mu\end{matrix}\right)$	$A_{84}\Gamma\left[\begin{matrix}s+\mu, & s-\mu, & \nu+1-s \\ s+1/2\end{matrix}\right]$ [$\|\operatorname{Re}\mu\|<\operatorname{Re}s<1+\operatorname{Re}\nu$]
27	$(1+x)^{-(\nu+1)/2} Q_\nu^\mu\left(\frac{2x+1}{2\sqrt{x^2+x}}\right)= = A_{84} G_{22}^{12}\left(x \middle\| \begin{matrix}\frac{1-\nu}{2}-\mu, & \frac{1-\nu}{2}+\mu \\ \frac{1+\nu}{2}, & -\frac{\nu}{2}\end{matrix}\right)$	$A_{84}\Gamma\left[\begin{matrix}s+\frac{\nu+1}{2}, & \frac{\nu+1}{2}+\mu-s, & \frac{\nu+1}{2}-\mu-s \\ \frac{\nu}{2}+1-s\end{matrix}\right]$ [$-(1+\operatorname{Re}\nu)/2<\operatorname{Re}s<1/2+\operatorname{Re}\nu/2-\|\operatorname{Re}\mu\|$]
28	$(1-x)_+^{\nu/2} Q_\nu^\mu\left(\frac{2-x}{2\sqrt{1-x}}\right)= = A_{85} G_{22}^{20}\left(x \middle\| \begin{matrix}1/2, & 1+\nu \\ \mu, & -\mu\end{matrix}\right)$	$A_{85}\Gamma\left[\begin{matrix}s+\mu, & s-\mu \\ s+1/2, & s+\nu+1\end{matrix}\right]$ [$\operatorname{Re}\nu>-3/2$; $\operatorname{Re}s>\|\operatorname{Re}\mu\|$]

29	$(x-1)_+^{\nu/2} Q_\nu^\mu\left(\frac{2x-1}{2\sqrt{x^2-x}}\right)=$ $= A_{85} G_{22}^{02}\left(x\,\middle\vert\,\begin{matrix}1+\nu/2-\mu, & 1+\nu/2+\mu\\ (1+\nu)/2, & -\nu/2\end{matrix}\right)$	$A_{85}\,\Gamma\begin{bmatrix}\mu-\nu/2-s, & -\mu-\nu/2-s\\ (1-\nu)/2-s, & 1+\nu/2-s\end{bmatrix}$ [Re $\nu > -3/2$; Re $s <$ −Re $\nu/2-\lvert$Re $\mu\rvert$]
30	$\frac{\lvert 1-x\rvert^{-2\nu-1}}{1+\sqrt{x}}\, Q_\nu^{2\nu+1}\left(\frac{\pm 1\mp 6\sqrt{x}\pm x}{(1+\sqrt{x})^2}\right)$	$\pm\frac{\sqrt{\pi}\,\Gamma(-2\nu-1/2)}{4\cos\nu\pi}\left\{\begin{Bmatrix}1\\ \cos 3\nu\pi\end{Bmatrix}\times\Gamma\begin{bmatrix}s+(2\nu+1)/4, & (10\nu+7)/4-s\\ s-(6\nu+1)/4, & (2\nu+5)/4-s\end{bmatrix}+\begin{Bmatrix}\cos 2\nu\pi\\ \cos\nu\pi\end{Bmatrix}\Gamma\begin{bmatrix}s-\frac{2\nu+1}{4}, & \frac{6\nu+5}{4}-s\\ s-\frac{10\nu+3}{4}, & \frac{3-2\nu}{4}-s\end{bmatrix}\right\}$ [$-2/3 <$ Re $\nu < -1/4$; $\lvert 1+2$ Re $\nu\rvert/4 <$ Re $s < (7+10$ Re $\nu)/4$, $(5+6$ Re $\nu)/4$]
31	$\frac{\lvert 1-x\rvert^{-2\nu-1}}{\lvert 1-\sqrt{x}\rvert}\, Q_\nu^{2\nu+1}\left(\frac{1+6\sqrt{x}+x}{(1-\sqrt{x})^2}\right)=$ $=\frac{e^{2i\nu\pi}}{4\pi^{3/2}}\cos 2\nu\pi\,\Gamma\left(-2\nu-\frac{1}{2}\right)\times$ $\times G_{22}^{22}\left(x\,\middle\vert\,\begin{matrix}-(10\nu+3)/4, & -(6\nu+1)/4\\ (2\nu+1)/4, & -(2\nu+1)/4\end{matrix}\right)$	$\frac{e^{2i\nu\pi}\cos 2\nu\pi}{4\pi^{3/2}}\,\Gamma\left(-2\nu-\frac{1}{2}\right)\times$ $\times\Gamma\left[s+\frac{2\nu+1}{4},\ s-\frac{2\nu+1}{4},\ \frac{10\nu+7}{4}-s,\ \frac{6\nu+5}{4}-s\right]$ [$-2/3 <$ Re $\nu < -1/4$; $\lvert 1+2$ Re $\nu\rvert/4 <$ Re $s < (7+10$ Re $\nu)/4$, $(5+6$ Re $\nu)/4$]
32	$\frac{\lvert 1-x\rvert^{\pm 2\mu}}{\sqrt{1+\sqrt{x}}}\times$ $\times Q_{2\mu-1/2}^{\mu}\left(\frac{1+6\sqrt{x}+x}{4x^{1/4}(1+\sqrt{x})}\right)=$ $=e^{i\mu\pi}\sqrt{2}\,\Gamma\left(\frac{1}{2}\pm 2\mu\right)\times$ $\times\begin{Bmatrix}\cos\mu\pi\\ \pi\Gamma^{-1}(\mu+1/2)\end{Bmatrix}\times$ $G_{22}^{11}\left(x\,\middle\vert\,\begin{matrix}(5-4\mu)/8\pm 2\mu, & (5+4\mu)/8\pm 2\mu\\ (4\mu+1)/8, & (1-4\mu)/8\end{matrix}\right)$	$e^{i\mu\pi}\sqrt{2}\,\Gamma\left(\frac{1}{2}\pm 2\mu\right)\begin{Bmatrix}\cos\mu\pi\\ \pi\Gamma^{-1}(\mu+1/2)\end{Bmatrix}\times$ $\times\Gamma\begin{bmatrix}s+(4\mu+1)/8, & (3+4\mu)/8\mp 2\mu-s\\ s+(5+4\mu)/8\pm 2\mu, & (4\mu+7)/8-s\end{bmatrix}$ [$-(5\mp 1)^{-1} <$ Re $\mu < (3\mp 1)^{-1}$; $-(1+4$ Re $\mu)/8 <$ Re $s < 3/8+(1/2\mp 2)$ Re μ]
33	$P_\nu^{\pm\mu}(\sqrt{1+x})\, Q_\nu^\mu(\sqrt{1+x})=$ $=\frac{A_{74}}{\sqrt{\pi}}\, G_{33}^{22}\left(x\,\middle\vert\,\begin{matrix}1/2, & -\nu, & 1+\nu\\ 0, & \mp\mu, & \pm\mu\end{matrix}\right)$	$\frac{A_{74}}{\sqrt{\pi}}\,\Gamma\begin{bmatrix}s, & s\mp\mu, & 1/2-s, & 1+\nu-s\\ s+\nu+1, & 1\mp\mu-s & & \end{bmatrix}$ [0, $\pm$ Re $\mu <$ Re $s < 1/2$, $1+$Re ν]
34	$P_\nu^{\pm\mu}\left(\sqrt{1+\frac{1}{x}}\right) Q_\nu^\mu\left(\sqrt{1+\frac{1}{x}}\right)=$ $=\frac{A_{74}}{\sqrt{\pi}}\, G_{33}^{22}\left(x\,\middle\vert\,\begin{matrix}1, & 1\pm\mu, & 1\mp\mu\\ 1/2, & 1+\nu, & -\nu\end{matrix}\right)$	$\frac{A_{74}}{\sqrt{\pi}}\,\Gamma\begin{bmatrix}s+1/2, & s+\nu+1, & -s, & \mp\mu-s\\ s+\mu\mp 1, & 1+\nu-s & & \end{bmatrix}$ [$-1/2$, $-1-$Re $\nu <$ Re $s < 0$, $\mp$ Re μ]

35	$P_{-\mu-1/2}^{-\nu-1/2}\left(\sqrt{1+\frac{1}{x}}\right) Q_{\nu}^{\mu}\left(\sqrt{1+x}\right) = A_{86} G_{33}^{31}\left(x \middle\| \begin{matrix} 1/4-\nu, 5/4+\nu, 3/4 \\ 1/4, 1/4+\mu, 1/4-\mu \end{matrix}\right)$	$A_{86}\Gamma\begin{bmatrix} s+\frac{1}{4}, s+\frac{1}{4}+\mu, s+\frac{1}{4}-\mu, \nu+\frac{3}{4}-s \\ s+\frac{5}{4}+\nu, s+\frac{3}{4} \end{bmatrix}$ [$\|\mathrm{Re}\,\mu\|-1/4 < \mathrm{Re}\,s < \mathrm{Re}\,\nu+3/4$]
36	$P_{-\mu-1/2}^{-\nu-1/2}\left(\sqrt{1+x}\right) Q_{\nu}^{\mu}\left(\sqrt{1+\frac{1}{x}}\right) = A_{86} G_{33}^{13}\left(x \middle\| \begin{matrix} 3/4, 3/4-\mu, 3/4+\mu \\ 3/4+\nu, 1/4, -1/4-\nu \end{matrix}\right)$	$A_{86}\Gamma\begin{bmatrix} s+\nu+3/4, 1/4-s, 1/4+\mu-s, 1/4-\mu-s \\ 3/4-s, 5/4+\nu-s \end{bmatrix}$ [$-3/4-\mathrm{Re}\,\nu < \mathrm{Re}\,s < 1/4-\|\mathrm{Re}\,\mu\|$]
37	$P_{\mu-1/2}^{\nu+1/2}\left(\sqrt{1+\frac{1}{x}}\right) Q_{\nu}^{\mu}\left(\sqrt{1+x}\right) = \frac{A_{83}}{\sqrt{2\pi}} G_{33}^{31}\left(x \middle\| \begin{matrix} 3/4, 5/4+\nu, 1/4-\nu \\ 1/4, 1/4+\mu, 1/4-\mu \end{matrix}\right)$	$\frac{A_{83}}{\sqrt{2\pi}}\Gamma\begin{bmatrix} s+1/4, s+1/4+\mu, s+1/4+\mu, 1/4-s \\ s+\nu+5/4, s+1/4-\nu \end{bmatrix}$ [$\|\mathrm{Re}\,\mu\|-1/4 < \mathrm{Re}\,s < 1/4$]
38	$P_{\mu-1/2}^{\nu+1/2}\left(\sqrt{1+x}\right) Q_{\nu}^{\mu}\left(\sqrt{1+\frac{1}{x}}\right) = \frac{A_{83}}{\sqrt{2\pi}} G_{33}^{13}\left(x \middle\| \begin{matrix} 3/4, 3/4-\mu, 3/4+\mu \\ 1/4, -1/4-\nu, 3/4+\nu \end{matrix}\right)$	$\frac{A_{83}}{\sqrt{2\pi}}\Gamma\begin{bmatrix} s+1/4, 1/4-s, 1/4+\mu-s, 1/4-\mu-s \\ 5/4+\nu-s, 1/4-\nu-s \end{bmatrix}$ [$-1/4 < \mathrm{Re}\,s < 1/4-\|\mathrm{Re}\,\mu\|$]
39	$(1+x)^{-1/2} P_{\mu-1/2}^{-\nu-1/2\mp1}\left(\sqrt{1+\frac{1}{x}}\right) \times Q_{\nu}^{\mu}\left(\sqrt{1+x}\right) = A_{87} G_{33}^{31}\left(x \middle\| \begin{matrix} \frac{-1\mp2}{4}-\nu, \frac{3}{4}, \nu+\frac{3\pm2}{4} \\ \frac{1}{4}, \frac{1}{4}+\mu, \frac{1}{4}-\mu \end{matrix}\right)$	$A_{87}\Gamma\begin{bmatrix} s+\frac{1}{4}, s+\frac{1}{4}+\mu, s+\frac{1}{4}-\mu, \frac{5\pm2}{4}+\nu-s \\ s+\frac{3}{4}, s+\nu+\frac{3\pm2}{4} \end{bmatrix}$ [$\|\mathrm{Re}\,\mu\|-1/4 < \mathrm{Re}\,s < (5\pm2)/4+\mathrm{Re}\,\nu$]
40	$(1+x)^{-1/2} P_{\mu-1/2}^{-\nu-1/2\mp1}\left(\sqrt{1+x}\right) \times Q_{\nu}^{\mu}\left(\sqrt{1+\frac{1}{x}}\right) = A_{87} G_{33}^{13}\left(x \middle\| \begin{matrix} 1/4, 1/4-\mu, 1/4+\mu \\ \nu+(3\pm2)/4, -1/4, (-1\mp2)/4-\nu \end{matrix}\right)$	$A_{87}\Gamma\begin{bmatrix} s+\nu+\frac{3\pm2}{4}, \frac{3}{4}-s, \frac{3}{4}+\mu-s, \frac{3}{4}-\mu-s \\ \frac{5}{4}-s, \frac{5\pm2}{4}+\nu-s \end{bmatrix}$ [$-(3\pm2)/4-\mathrm{Re}\,\nu < \mathrm{Re}\,s < 3/4-\|\mathrm{Re}\,\mu\|$]
41	$(1+x)^{-1/2} P_{\mu+1/2}^{-\nu-1/2}\left(\sqrt{1+\frac{1}{x}}\right) \times Q_{\nu}^{\mu}\left(\sqrt{1+x}\right) = A_{88} G_{33}^{31}\left(x \middle\| \begin{matrix} -1/4-\nu, 1/4, 3/4+\nu \\ -1/4, 3/4+\mu, -1/4-\mu \end{matrix}\right)$	$A_{88}\Gamma\begin{bmatrix} s-1/4, s+\mu+3/4, s-\mu-1/4, 5/4+\nu-s \\ s+1/4, s+\nu+3/4 \end{bmatrix}$ [$1/4, 1/4+\mathrm{Re}\,\mu, -3/4-\mathrm{Re}\,\mu < \mathrm{Re}\,s < 5/4+\mathrm{Re}\,\nu$]

42	$(1+x)^{-1/2} P_{\mu+1/2}^{-\nu-1/2}(\sqrt{1+x}) \times Q_{\nu}^{\mu}\left(\sqrt{1+\frac{1}{x}}\right) = A_{88} G_{33}^{13}\left(x \middle\| \begin{matrix} 3/4, & -1/4-\mu, & 3/4+\mu \\ 3/4+\nu, & 1/4, & -1/4-\nu \end{matrix}\right)$	$A_{88}\Gamma\begin{bmatrix} s+\nu+3/4, & 1/4-s, & 5/4+\mu-s, & 1/4-\mu-s \\ 3/4-s, & & & 5/4+\nu-s \end{bmatrix}$ [$-3/4-\mathrm{Re}\,\nu<\mathrm{Re}\,s<1/4, 5/4+\mathrm{Re}\,\mu, 1/4-\mathrm{Re}\,\mu$]
43	$[Q_{\nu}^{\mu}(\sqrt{1+x})]^2 = A_{89} G_{33}^{31}\left(x \middle\| \begin{matrix} -\nu, & 1/2, & 1+\nu \\ 0, & \mu, & -\mu \end{matrix}\right)$	$A_{89}\Gamma\begin{bmatrix} s, & s+\mu, & s-\mu, & 1+\nu-s \\ s+1/2, & s+\nu+1 & & \end{bmatrix}$ [$\|\mathrm{Re}\,\mu\| < \mathrm{Re}\,s < 1+\mathrm{Re}\,\nu$]
44	$\left[Q_{\nu}^{\mu}\left(\sqrt{1+\frac{1}{x}}\right)\right]^2 = A_{89} G_{33}^{13}\left(x \middle\| \begin{matrix} 1, & 1-\mu, & 1+\mu \\ 1+\nu, & 1/2, & -\nu \end{matrix}\right)$	$A_{89}\Gamma\begin{bmatrix} s+\nu+1, & -s, & \mu-s, & -\mu-s \\ 1/2-s, & 1+\nu-s & & \end{bmatrix}$ [$-1-\mathrm{Re}\,\nu < \mathrm{Re}\,s < -\|\mathrm{Re}\,\mu\|$]
45	$Q_{-\nu-1}^{\mu}(\sqrt{1+x})\, Q_{\nu}^{\mu}(\sqrt{1+x}) = A_{90} G_{33}^{31}\left(x \middle\| \begin{matrix} 1/2, & \nu+1, & -\nu \\ 0, & -\mu, & \mu \end{matrix}\right)$	$A_{90}\Gamma\begin{bmatrix} s, & s-\mu, & s+\mu, & 1/2-s \\ s+\nu+1, & s-\nu & & \end{bmatrix}$ [$\|\mathrm{Re}\,\mu\| < \mathrm{Re}\,s < 1/2$]
46	$Q_{-\nu-1}^{\mu}\left(\sqrt{\frac{1+x}{x}}\right) Q_{\nu}^{\mu}\left(\sqrt{\frac{1+x}{x}}\right) = A_{90} G_{33}^{13}\left(x \middle\| \begin{matrix} 1, & 1+\mu, & 1-\mu \\ 1/2, & -\nu, & \nu+1 \end{matrix}\right)$	$A_{90}\Gamma\begin{bmatrix} s+1/2, & -s, & -\mu-s, & \mu-s \\ 1+\nu-s, & -\nu-s & & \end{bmatrix}$ [$-1/2 < \mathrm{Re}\,s < -\|\mathrm{Re}\,\mu\|$]
47	$\frac{Q_{\nu\pm1}^{\mu}(\sqrt{1+x})\, Q_{\nu}^{\mu}(\sqrt{1+x})}{\sqrt{1+x}} = A_{91} G_{33}^{31}\left(x \middle\| \begin{matrix} \frac{\mp1-1}{2}-\nu, & \frac{1}{2}, & \frac{1\pm1}{2}+\nu \\ 0, & \mu, & -\mu \end{matrix}\right)$	$A_{91}\Gamma\begin{bmatrix} s, & s+\mu, & s-\mu, & (3\pm1)/2+\nu-s \\ s+1/2, & s+(1\pm1)/2+\nu & & \end{bmatrix}$ [$\|\mathrm{Re}\,\mu\| < \mathrm{Re}\,s < (3\pm1)/2+\mathrm{Re}\,\nu$]
48	$(1+x)^{-1/2} Q_{\nu\pm1}^{\mu}\left(\sqrt{\frac{1+x}{x}}\right) \times Q_{\nu}^{\mu}\left(\sqrt{\frac{1+x}{x}}\right) = A_{91} G_{33}^{13}\left(x \middle\| \begin{matrix} 1/2, & 1/2-\mu, & 1/2+\mu \\ \nu+1\pm1/2, & 0, & \mp1/2-\nu \end{matrix}\right)$	$A_{91}\Gamma\begin{bmatrix} s+\nu\pm1/2+1, & 1/2-s, & 1/2+\mu-s, & 1/2-\mu-s \\ 1-s, & & & 1\pm1/2+\nu-s \end{bmatrix}$ [$-\mathrm{Re}\,\nu\mp1/2-1 < \mathrm{Re}\,s < 1/2-\|\mathrm{Re}\,\mu\|$]
49	$(1+x)^{-1/2} Q_{\nu}^{\mu}(\sqrt{1+x})\, Q_{\nu}^{\mu+1}(\sqrt{1+x}) = -A_{89} G_{33}^{31}\left(x \middle\| \begin{matrix} -1/2-\nu, & 0, & 1/2+\nu \\ -1/2, & -1/2-\mu, & 1/2+\mu \end{matrix}\right)$	$-A_{89}\Gamma\begin{bmatrix} s-1/2, & s-1/2-\mu, & s+1/2+\mu, & 3/2+\nu-s \\ s, & & & s+1/2+\nu \end{bmatrix}$ [$\|1/2+\mathrm{Re}\,\mu\| < \mathrm{Re}\,s < 3/2+\mathrm{Re}\,\nu$]
50	$(1+x)^{-1/2} Q_{\nu}^{\mu}\left(\sqrt{1+\frac{1}{x}}\right) \times Q_{\nu}^{\mu+1}\left(\sqrt{1+\frac{1}{x}}\right) = -A_{89} G_{33}^{13}\left(x \middle\| \begin{matrix} 1, & 1+\mu, & -\mu \\ 1+\nu, & 1/2, & -\nu \end{matrix}\right)$	$-A_{89}\Gamma\begin{bmatrix} s+\nu+1, & -s, & -\mu-s, & 1+\mu-s \\ 1/2-s, & & & 1+\nu-s \end{bmatrix}$ [$-1 < \mathrm{Re}\,\nu < \mathrm{Re}\,s < 0, 1+\mathrm{Re}\,\mu, -\mathrm{Re}\,\mu$]

51	$Q_\nu^\mu(\sqrt{1+x})\,Q_{\pm\mu-1/2}^{\pm\nu\pm1/2}\left(\sqrt{1+\frac{1}{x}}\right)=$ $=\frac{e^{(\mu\pm\nu\pm1/2)\pi i}}{2\sqrt{2}}\Gamma(\mu+\nu+1)\times$ $\times\Gamma^{(1\mp1)/2}\left[\begin{matrix}-\mu-\nu\\1-\mu+\nu\end{matrix}\right]\times$ $\times G_{33}^{22}\left(x\left\vert\begin{matrix}3/4,\ 1/4-\nu,\ 5/4+\nu\\1/4,\ 1/4\pm\mu,\ 1/4\mp\mu\end{matrix}\right.\right)$	$\frac{e^{(\mu\pm\nu\pm1/2)\pi i}}{2\sqrt{2}}\Gamma(\mu+\nu+1)\times$ $\times\Gamma^{(1\mp1)/2}\left[\begin{matrix}-\mu-\nu\\1-\mu+\nu\end{matrix}\right]\times$ $\times\Gamma\left[\begin{matrix}s+1/4,\ s\pm\mu+1/4,\ 1/4-s,\ 3/4+\nu-s\\s+\nu+5/4,\ 3/4\pm\mu-s\end{matrix}\right]$ $[-1/4,\ -1/4\mp\mathrm{Re}\,\mu<\mathrm{Re}\,s<1/4,\ 3/4+\mathrm{Re}\,\nu]$

8.4.43. The Whittaker function $M_{\rho,\sigma}(x)$.
(See also 8.4.45)

NOTATION:

$$A_{92}=\Gamma\left[\begin{matrix}2\sigma+1\\1/2\mp\rho+\sigma\end{matrix}\right],\qquad A_{93}=2\sqrt{\pi}\,\Gamma\left[\begin{matrix}2\sigma+1,\ 2\sigma+1\\1/2-\rho+\sigma,\ 1/2+\rho+\sigma\end{matrix}\right].$$

1	$e^{\pm x/2}M_{\rho,\sigma}(\mp x)=$ $=A_{92}G_{12}^{11}\left(x\left\vert\begin{matrix}1\pm\rho\\\sigma+1/2,\ 1/2-\sigma\end{matrix}\right.\right)$	$A_{92}\Gamma\left[\begin{matrix}s+\sigma+1/2,\ \mp\rho-s\\\sigma+1/2-s\end{matrix}\right]$ $[-1/2-\mathrm{Re}\,\sigma<\mathrm{Re}\,s<\mp\mathrm{Re}\,\rho;\ 2\sigma\neq-1,\ -2,\ -3,\ \ldots]$
2	$e^{\pm1/(2x)}M_{\rho,\sigma}\left(\mp\frac{1}{x}\right)=$ $=A_{92}G_{21}^{11}\left(x\left\vert\begin{matrix}1/2-\sigma,\ 1/2+\sigma\\\mp\rho\end{matrix}\right.\right)$	$A_{92}\Gamma\left[\begin{matrix}s\mp\rho,\ 1/2+\sigma-s\\s+\sigma+1/2\end{matrix}\right]$ $[\pm\mathrm{Re}\,\rho<\mathrm{Re}\,s<\mathrm{Re}\,\sigma+1/2;\ 2\sigma\neq-1,\ -2,\ -3,\ \ldots]$
3	$M_{\rho,\sigma}(2i\sqrt{x})M_{\rho,\sigma}(-2i\sqrt{x})=$ $=A_{93}G_{24}^{12}\left(x\left\vert\begin{matrix}1-\rho,\ 1+\rho\\1/2+\sigma,\ 1/2-\sigma,\ 1/2,\ 1\end{matrix}\right.\right)$	$A_{93}\Gamma\left[\begin{matrix}s+\sigma+1/2,\ \rho-s,\ -\rho-s\\1/2+\sigma-s,\ -s,\ 1/2-s\end{matrix}\right]$ $[-1/2-\mathrm{Re}\,\sigma<\mathrm{Re}\,s<-\vert\mathrm{Re}\,\rho\vert]$
4	$M_{\rho,\sigma}\left(\frac{2i}{\sqrt{x}}\right)M_{\rho,\sigma}\left(-\frac{2i}{\sqrt{x}}\right)=$ $=A_{93}G_{42}^{21}\left(x\left\vert\begin{matrix}1/2-\sigma,\ 1/2+\sigma,\ 0,\ 1/2\\\rho,\ -\rho\end{matrix}\right.\right)$	$A_{93}\Gamma\left[\begin{matrix}s+\rho,\ s-\rho,\ 1/2+\sigma-s\\s,\ s+1/2,\ s+\sigma+1/2\end{matrix}\right]$ $[\vert\mathrm{Re}\,\rho\vert<\mathrm{Re}\,s<\mathrm{Re}\,\sigma+1/2]$

8.4.44. The Whittaker function $W_{\rho,\sigma}(x)$.
(See also 8.4.46)

NOTATION:

$$A_{94}=\frac{1}{\Gamma(1/2-\rho-\sigma)\,\Gamma(1/2-\rho+\sigma)},\qquad A_{95}=\frac{1}{\sqrt{\pi}}\Gamma\left[\begin{matrix}2\sigma+1\\1/2-\rho+\sigma\end{matrix}\right],$$

$$A_{96}=\frac{\pi^{-1/2}}{\Gamma(1/2-\rho-\sigma)\,\Gamma(1/2-\rho+\sigma)}.$$

1	$e^{-x/2}W_{\rho,\sigma}(x)=$ $=G_{12}^{20}\left(x\left\vert\begin{matrix}1-\rho\\1/2+\sigma,\ 1/2-\sigma\end{matrix}\right.\right)$	$\Gamma\left[\begin{matrix}s+\sigma+1/2,\ s+1/2-\sigma\\s+1-\rho\end{matrix}\right]$ $[\mathrm{Re}\,s>\vert\mathrm{Re}\,\sigma\vert-1/2]$
2	$e^{-1/(2x)}W_{\rho,\sigma}\left(\frac{1}{x}\right)=$ $=G_{21}^{02}\left(x\left\vert\begin{matrix}1/2-\sigma,\ 1/2+\sigma\\\rho\end{matrix}\right.\right)$	$\Gamma\left[\begin{matrix}1/2+\sigma-s,\ 1/2-\sigma-s\\1-\rho-s\end{matrix}\right]$ $[\mathrm{Re}\,s<1/2-\vert\mathrm{Re}\,\sigma\vert]$

3	$e^{x/2}W_{\rho,\sigma}(x)= A_{94}G_{12}^{21}\left(x\,\middle\vert\,\begin{matrix}1+\rho\\ 1/2-\sigma,\ 1/2+\sigma\end{matrix}\right)$	$A_{94}\Gamma\left[s+\frac{1}{2}-\sigma,\ s+\frac{1}{2}+\sigma,\ -\rho-s\right]$ $[\vert\mathrm{Re}\,\sigma\vert-1/2<\mathrm{Re}\,s<-\mathrm{Re}\,\rho]$
4	$e^{1/(2x)}W_{\rho,\sigma}\left(\frac{1}{x}\right)= A_{94}G_{21}^{12}\left(x\,\middle\vert\,\begin{matrix}1/2+\sigma,\ 1/2-\sigma\\ -\rho\end{matrix}\right)$	$A_{94}\Gamma\left[s-\rho,\ \frac{1}{2}-\sigma-s,\ \frac{1}{2}+\sigma-s\right]$ $[\mathrm{Re}\,\rho<\mathrm{Re}\,s<1/2-\vert\mathrm{Re}\,\sigma\vert]$
5	$M_{-\rho,\sigma}(2\sqrt{x})\,W_{\rho,\sigma}(2\sqrt{x})= A_{95}G_{24}^{31}\left(x\,\middle\vert\,\begin{matrix}1+\rho,\ 1-\rho\\ 1/2,\ 1,\ 1/2+\sigma,\ 1/2-\sigma\end{matrix}\right)$	$A_{95}\Gamma\begin{bmatrix}s+1/2,\ s+1,\ s+1/2+\sigma,\ -\rho-s\\ s+1-\rho,\ 1/2+\sigma-s\end{bmatrix}$ $[-1/2,\ -1/2-\mathrm{Re}\,\sigma<\mathrm{Re}\,s<-\mathrm{Re}\,\rho]$
6	$M_{-\rho,\sigma}\left(\frac{2}{\sqrt{x}}\right)W_{\rho,\sigma}\left(\frac{2}{\sqrt{x}}\right)= A_{95}G_{42}^{13}\left(x\,\middle\vert\,\begin{matrix}0,\ 1/2,\ 1/2-\sigma,\ 1/2+\sigma\\ -\rho,\ \rho\end{matrix}\right)$	$A_{95}\Gamma\begin{bmatrix}s-\rho,\ 1/2-s,\ 1-s,\ 1/2+\sigma-s\\ s+\sigma+1/2,\ 1-\rho-s\end{bmatrix}$ $[\mathrm{Re}\,\rho<\mathrm{Re}\,s<1/2,\ \mathrm{Re}\,\sigma+1/2]$
7	$W_{-\rho,\sigma}(2\sqrt{x})\,W_{\rho,\sigma}(2\sqrt{x})= \frac{1}{\sqrt{\pi}}G_{24}^{40}\left(x\,\middle\vert\,\begin{matrix}1+\rho,\ 1-\rho\\ 1/2,\ 1,\ 1/2+\sigma,\ 1/2-\sigma\end{matrix}\right)$	$\frac{1}{\sqrt{\pi}}\Gamma\begin{bmatrix}s+\frac{1}{2},\ s+1,\ s+\frac{1}{2}+\sigma,\ s+\frac{1}{2}-\sigma\\ s+1+\rho,\ s+1-\rho\end{bmatrix}$ $[\mathrm{Re}\,s>\vert\mathrm{Re}\,\sigma\vert-1/2]$
8	$W_{-\rho,\sigma}\left(\frac{2}{\sqrt{x}}\right)W_{\rho,\sigma}\left(\frac{2}{\sqrt{x}}\right)= \frac{1}{\sqrt{\pi}}G_{42}^{04}\left(x\,\middle\vert\,\begin{matrix}0,\ 1/2,\ 1/2-\sigma,\ 1/2+\sigma\\ -\rho,\ \rho\end{matrix}\right)$	$\frac{1}{\sqrt{\pi}}\Gamma\begin{bmatrix}\frac{1}{2}-s,\ 1-s,\ \frac{1}{2}+\sigma-s,\ \frac{1}{2}-\sigma-s\\ 1+\rho-s,\ 1-\rho-s\end{bmatrix}$ $[\mathrm{Re}\,s<1/2-\vert\mathrm{Re}\,\sigma\vert]$
9	$W_{\rho,\sigma}(2i\sqrt{x})\,W_{\rho,\sigma}(-2i\sqrt{x})= A_{96}G_{24}^{41}\left(x\,\middle\vert\,\begin{matrix}1+\rho,\ 1-\rho\\ 1/2,\ 1,\ 1/2+\sigma,\ 1/2-\sigma\end{matrix}\right)$	$A_{96}\Gamma\begin{bmatrix}s+\frac{1}{2},\ s+1,\ s+\frac{1}{2}+\sigma,\ s+\frac{1}{2}-\sigma,\ \rho-s\\ s+1-\rho\end{bmatrix}$ $[\vert\mathrm{Re}\,\sigma\vert-1/2<\mathrm{Re}\,s<-\mathrm{Re}\,\rho]$
10	$W_{\rho,\sigma}\left(\frac{2i}{\sqrt{x}}\right)W_{\rho,\sigma}\left(-\frac{2i}{\sqrt{x}}\right)= A_{96}G_{42}^{14}\left(x\,\middle\vert\,\begin{matrix}0,\ 1/2,\ 1/2+\sigma,\ 1/2-\sigma\\ -\rho,\ \rho\end{matrix}\right)$	$A_{96}\Gamma\begin{bmatrix}s-\rho,\ \frac{1}{2}-s,\ 1-s,\ \frac{1}{2}+\sigma-s,\ \frac{1}{2}-\sigma-s\\ 1-\rho-s\end{bmatrix}$ $[\mathrm{Re}\,\rho<\mathrm{Re}\,s<1/2-\vert\mathrm{Re}\,\sigma\vert]$

8.4.45. The confluent hypergeometric function of Kummer ${}_1F_1(a;\ b;\ x)$.

(See also 8.4.43)

NOTATION: $A_{97}=2^{a-b+1}\sqrt{\pi}\begin{Bmatrix}1\\i\end{Bmatrix}\Gamma\begin{bmatrix}b\\a\end{bmatrix},\ \delta=\begin{Bmatrix}1\\0\end{Bmatrix}.$

1	${}_1F_1(a;\ b;\ -x)=\Gamma\begin{bmatrix}b\\a\end{bmatrix}\times G_{12}^{11}\left(x\,\middle\vert\,\begin{matrix}1-a\\0,\ 1-b\end{matrix}\right)$	$\Gamma\begin{bmatrix}b\\a\end{bmatrix}\Gamma\begin{bmatrix}s,\ a-s\\b-s\end{bmatrix}$ $[0<\operatorname{Re}s<\operatorname{Re}a;\ b\neq 0,\ -1,\ -2,\ \ldots]$
2	${}_1F_1\left(a;\ b;\ -\frac{1}{x}\right)=\Gamma\begin{bmatrix}b\\a\end{bmatrix}G_{21}^{11}\left(x\,\middle\vert\,\begin{matrix}1,\ b\\a\end{matrix}\right)$	$\Gamma\begin{bmatrix}b\\a\end{bmatrix}\Gamma\begin{bmatrix}s+a,\ -s\\s+b\end{bmatrix}$ $[-\operatorname{Re}a<\operatorname{Re}s<0;\ b\neq 0,\ -1,\ -2,\ \ldots]$
3	$e^{ia\pi/2}{}_1F_1(a;\ b;\ -2i\sqrt{x})\pm e^{-ia\pi/2}{}_1F_1(a;\ b;\ 2i\sqrt{x})=A_{97}G_{24}^{21}\left(x\,\middle\vert\,\begin{matrix}\frac{1+\delta-a}{2},\ 1-\frac{\delta+a}{2}\\0,\ \frac{1}{2},\ 1-\frac{b}{2},\ \frac{1-b}{2}\end{matrix}\right)$	$A_{97}\Gamma\begin{bmatrix}s,\ s+\frac{1}{2},\ \frac{1+a-\delta}{2}-s\\s-\frac{a+\delta}{2}+1,\ \frac{b}{2}-s,\ \frac{1+b}{2}-s\end{bmatrix}$ $[0<\operatorname{Re}s<(1-\delta+\operatorname{Re}a)/2,\ (1+\operatorname{Re}(b-a))/2]$
4	$e^{ia\pi/2}{}_1F_1\left(a;\ b;\ -\frac{2i}{\sqrt{x}}\right)\pm e^{-ia\pi/2}{}_1F_1\left(a;\ b;\ \frac{2i}{\sqrt{x}}\right)=A_{97}G_{42}^{12}\left(x\,\middle\vert\,\begin{matrix}1/2,\ 1,\ b/2,\ (1+b)/2\\(1+a-\delta)/2,\ (a+\delta)/2\end{matrix}\right)$	$A_{97}\Gamma\begin{bmatrix}s+\frac{1+a-\delta}{2},\ -s,\ \frac{1}{2}-s\\s+\frac{b}{2},\ s+\frac{b+1}{2},\ 1-\frac{a+\delta}{2}-s\end{bmatrix}$ $[(\delta-\operatorname{Re}a-1)/2,\ (\operatorname{Re}(a-b)-1)/2<\operatorname{Re}s<0]$
5	$e^{-x}{}_1F_1(a;\ b;\ x)=\Gamma\begin{bmatrix}b\\b-a\end{bmatrix}G_{12}^{11}\left(x\,\middle\vert\,\begin{matrix}1+a-b\\0,\ 1-b\end{matrix}\right)$	$\Gamma\begin{bmatrix}b\\b-a\end{bmatrix}\Gamma\begin{bmatrix}s,\ b-a-s\\b-s\end{bmatrix}$ $[0<\operatorname{Re}s<\operatorname{Re}(b-a);\ b\neq 0,\ -1,\ -2,\ \ldots]$
6	$e^{-1/x}{}_1F_1\left(a;\ b;\ \frac{1}{x}\right)=\Gamma\begin{bmatrix}b\\b-a\end{bmatrix}G_{21}^{11}\left(x\,\middle\vert\,\begin{matrix}1,\ b\\b-a\end{matrix}\right)$	$\Gamma\begin{bmatrix}b\\b-a\end{bmatrix}\Gamma\begin{bmatrix}s+b-a,\ -s\\s+b\end{bmatrix}$ $[\operatorname{Re}(a-b)<\operatorname{Re}s<0;\ \ b\neq 0,\ -1,\ -2,\ \ldots]$

8.4.46. The confluent hypergeometric function of Tricomi $\Psi(a,\ b;\ x)$.

(See also 8.4.44)

NOTATION:

$$A_{98}=\frac{1}{\Gamma(a)\,\Gamma(a-b+1)},\quad A_{99}=\frac{2^{a-b}}{\sqrt{\pi}\,\Gamma(a)\,\Gamma(a-b+1)}\begin{Bmatrix}1\\i\end{Bmatrix},$$

$$B_1=2^{1-a}\sqrt{\pi}\begin{Bmatrix}1\\i\end{Bmatrix},\quad B_2=\frac{2^{-b}}{\sqrt{\pi}}\Gamma\begin{bmatrix}b\\a\end{bmatrix},\quad \delta=\begin{Bmatrix}1\\0\end{Bmatrix}.$$

1	$\Psi(a,\ b;\ x)=A_{98}G_{12}^{21}\left(x\,\middle\vert\,\begin{matrix}1-a\\0,\ 1-b\end{matrix}\right)$	$A_{98}\Gamma[s,\ s+1-b,\ a-s]$ $[0,\ \operatorname{Re}b-1<\operatorname{Re}s<\operatorname{Re}a]$
2	$\Psi\left(a,\ b;\ \frac{1}{x}\right)=A_{98}G_{21}^{12}\left(x\,\middle\vert\,\begin{matrix}1,\ b\\a\end{matrix}\right)$	$A_{98}\Gamma[s+a,\ -s,\ 1-b-s]$ $[-\operatorname{Re}a<\operatorname{Re}s<0,\ 1-\operatorname{Re}b]$

3	$\Psi\left(a,\ b;\ -2i\sqrt{x}\right)\pm\Psi\left(a,\ b;\ 2i\sqrt{x}\right)=$ $=A_{99}G_{24}^{32}\left(x\,\middle\vert\,\begin{matrix}(1-a)/2,\ 1-a/2\\(1-\delta)/2,\ (1-b)/2,\ 1-b/2,\ \delta/2\end{matrix}\right)$	$A_{99}\Gamma\left[\begin{matrix}s+\frac{1-\delta}{2},\ s+\frac{1-b}{2},\ s+1-\frac{b}{2},\ 1-\frac{\delta}{2}-s\\ \frac{1+a}{2}-s,\ \frac{a}{2}-s\end{matrix}\right]$ $[(\operatorname{Re} b-1)/2,\ (\delta-1)/2<\operatorname{Re} s<\operatorname{Re} a/2]$
4	$\Psi\left(a,\ b;\ -\frac{2i}{\sqrt{x}}\right)\pm\Psi\left(a,\ b;\ \frac{2i}{\sqrt{x}}\right)=$ $=A_{99}G_{42}^{23}\left(x\,\middle\vert\,\begin{matrix}\frac{\delta+1}{2},\frac{b+1}{2},\frac{b}{2},\ 1-\frac{\delta}{2}\\ \frac{a+1}{2},\ \frac{a}{2}\end{matrix}\right)$	$A_{99}\Gamma\left[\begin{matrix}s+\frac{a+1}{2},\ s+\frac{a}{2},\ \frac{1-\delta}{2}-s,\ s+1-\frac{\delta}{2}\\ \frac{1-b}{2}-s,\ \frac{2-b}{2}-s\end{matrix}\right]$ $[-\operatorname{Re} a/2<\operatorname{Re} s<(1-\operatorname{Re} b)/2,\ (1-\delta)/2]$
5	$e^{ia\pi/2}\,\Psi\left(a,\ b;\ 2i\sqrt{x}\right)\pm$ $\pm e^{-ia\pi/2}\,\Psi\left(a,\ b;\ -2i\sqrt{x}\right)=$ $=A_{99}G_{24}^{41}\left(x\,\middle\vert\,\begin{matrix}(1+\delta-a)/2,\ 1-(\delta+a)/2\\0,\ 1/2,\ (1-b)/2,\ 1-b/2\end{matrix}\right)$	$A_{99}\Gamma\left[\begin{matrix}s,\ s+1/2,\ s+(1-b)/2,\ s+1-b/2,\ s+1-(\delta+a)/2\\(1-\delta+a)/2-s\end{matrix}\right]$ $[0,\ (\operatorname{Re} b-1)/2<\operatorname{Re} s<(1-\delta+\operatorname{Re} a)/2]$
6	$e^{ia\pi/2}\,\Psi\left(a,\ b;\ \frac{2i}{\sqrt{x}}\right)\pm$ $\pm e^{-ia\pi/2}\,\Psi\left(a,\ b;\ -\frac{2i}{\sqrt{x}}\right)=$ $=A_{99}G_{42}^{14}\left(x\,\middle\vert\,\begin{matrix}1/2,\ 1,\ b/2,\ (1+b)/2\\(1-\delta+a)/2,\ (\delta+a)/2\end{matrix}\right)$	$A_{99}\Gamma\left[\begin{matrix}s+\frac{1-\delta+a}{2},\ -s,\ \frac{1}{2}-s,\ 1-\frac{\delta+a}{2}-s\\ \frac{1-b}{2}-s,\ 1-\frac{b}{2}-s\end{matrix}\right]$ $[(\delta-\operatorname{Re} a-1)/2<\operatorname{Re} s<0,\ (1-\operatorname{Re} b)/2]$
7	$e^{-x}\Psi\left(a,\ b;\ x\right)=G_{12}^{20}\left(x\,\middle\vert\,\begin{matrix}1+a-b\\0,\ 1-b\end{matrix}\right)$	$\Gamma\left[\begin{matrix}s,\ s+1-b\\s+a+1-b\end{matrix}\right]$ $[\operatorname{Re} s>0,\ \operatorname{Re} b-1]$
8	$e^{-1/x}\Psi\left(a,\ b;\ \frac{1}{x}\right)=G_{21}^{02}\left(x\,\middle\vert\,\begin{matrix}1,\ b\\b-a\end{matrix}\right)$	$\Gamma\left[\begin{matrix}-s,\ 1-b-s\\1+a-b-s\end{matrix}\right]$ $[\operatorname{Re} s<0,\ 1-\operatorname{Re} b]$
9	$e^{2i\sqrt{x}}\,\Psi\left(a,\ b;\ -2i\sqrt{x}\right)\pm$ $\pm e^{-2i\sqrt{x}}\,\Psi\left(a,\ b;\ 2i\sqrt{x}\right)=$ $=B_1G_{24}^{30}\left(x\,\middle\vert\,\begin{matrix}\frac{1-a-b}{2},\ \frac{a-b}{2}+1\\1-\frac{b}{2},\ \frac{1-b}{2},\ \frac{1-\delta}{2},\ \frac{\delta}{2}\end{matrix}\right)$	$B_1\Gamma\left[\begin{matrix}s+1-\frac{b}{2},\ s+\frac{1-b}{2},\ s+\frac{1-\delta}{2}\\ s+\frac{1-a-b}{2},\ s+\frac{a-b}{2}+1,\ 1-\frac{\delta}{2}-s\end{matrix}\right]$ $[(\operatorname{Re} b-1)/2,\ (\delta-1)/2<\operatorname{Re} s<1/2]$
10	$e^{2i/\sqrt{x}}\,\Psi\left(a,\ b;\ -\frac{2i}{\sqrt{x}}\right)\pm$ $\pm e^{-2i/\sqrt{x}}\,\Psi\left(a,\ b;\ \frac{2i}{\sqrt{x}}\right)=$ $=B_1G_{42}^{03}\left(x\,\middle\vert\,\begin{matrix}b/2,\ (b+1)/2,\ (\delta+1)/2,\ 1-\delta/2\\(1+a+b)/2,\ (b-a)/2\end{matrix}\right)$	$B_1\Gamma\left[\begin{matrix}1-b/2-s,\ (1-b)/2-s,\ (1-\delta)/2-s\\ s+1-\delta/2,\ (1-a-b)/2-s,\ 1+(a-b)/2-s\end{matrix}\right]$ $[-1/2<\operatorname{Re} s<(1-\operatorname{Re} b)/2,\ (1-\delta)/2]$

11	${}_1F_1(a;\ b;\ -2\sqrt{x})\,\Psi(a,\ b;\ 2\sqrt{x}) = B_2 G^{31}_{24}\left(x\,\middle\vert\,\begin{matrix}1-a,\ 1+a-b\\ 0,\ (1-b)/2,\ 1-b/2,\ 1-b\end{matrix}\right)$	$B_2\Gamma\begin{bmatrix}s,\ s+(1-b)/2,\ s+1-b/2,\ a-s\\ s+a-b+1,\ b-s\end{bmatrix}$ [0, (Re $b-1)/2 <$ Re $s <$ Re a]
12	${}_1F_1\left(a;\ b;\ -\frac{2}{\sqrt{x}}\right)\Psi\left(a,\ b;\ \frac{2}{\sqrt{x}}\right) = B_2 G^{13}_{42}\left(x\,\middle\vert\,\begin{matrix}1,\ (1+b)/2,\ b/2,\ b\\ a,\ b-a\end{matrix}\right)$	$B_2\Gamma\begin{bmatrix}s+a,\ -s,\ (1-b)/2-s,\ 1-b/2-s\\ s+b,\ 1+a-b-s\end{bmatrix}$ [−Re $a <$ Re $s < 0$, (1 − Re b)/2]
13	$e^{-2\sqrt{x}}\,\Psi(b-a,\ b;\ 2\sqrt{x})\times\Psi(a,\ b;\ 2\sqrt{x}) = \frac{2^{-b}}{\sqrt{\pi}}\times G^{40}_{24}\left(x\,\middle\vert\,\begin{matrix}a-b+1,\ 1-a\\ 0,\ (1-b)/2,\ 1-b/2,\ 1-b\end{matrix}\right)$	$\frac{2^{-b}}{\sqrt{\pi}}\Gamma\begin{bmatrix}s,\ s+\frac{1-b}{2},\ s+1-\frac{b}{2},\ s-b+1\\ s+a-b+1,\ s+1-a\end{bmatrix}$ [Re $s > 0$, (Re $b - 1)/2$, Re $b - 1$]
14	$e^{-2/\sqrt{x}}\,\Psi\left(b-a,\ b;\ \frac{2}{\sqrt{x}}\right)\times\Psi\left(a,\ b;\ \frac{2}{\sqrt{x}}\right) = \frac{2^{-b}}{\sqrt{\pi}}\times G^{04}_{42}\left(x\,\middle\vert\,\begin{matrix}1,\ b,\ b/2,\ (1+b)/2\\ a,\ b-a\end{matrix}\right)$	$\frac{2^{-b}}{\sqrt{\pi}}\Gamma\begin{bmatrix}-s,\ \frac{1-b}{2}-s,\ 1-\frac{b}{2}-s,\ 1-b-s\\ 1+a-b-s,\ 1-a-s\end{bmatrix}$ [Re $s < 0$, 1 − Re b, (1 − Re b)/2].

8.4.47. The function ${}_0F_1(b;\ x)$.
(See also 8.4.19, 8.4.22))

NOTATION: $B_3 = \frac{2^{b+c-2}}{\sqrt{\pi}}\,\Gamma(b)\,\Gamma(c)$.

1	${}_0F_1(b;\ -x) = \Gamma(b)\,G^{10}_{02}\left(x\,\middle\vert\,\begin{matrix}\cdot\ \ \cdot\\ 0,\ 1-b\end{matrix}\right)$	$\Gamma(b)\,\Gamma\begin{bmatrix}s\\ b-s\end{bmatrix}$ [$0 <$ Re $s < (1+2$ Re $b)/4$].
2	${}_0F_1(b;\ -1/x) = \Gamma(b)\,G^{01}_{20}\left(x\,\middle\vert\,\begin{matrix}1,\ b\\ \cdot\ \ \cdot\end{matrix}\right)$	$\Gamma(b)\,\Gamma\begin{bmatrix}-s\\ s+b\end{bmatrix}$ [$-(1+2$ Re $b)/4 <$ Re $s < 0$].
3	${}_0F_1\left(b;\ -\frac{x}{4}\right){}_0F_1\left(c;\ -\frac{x}{4}\right) = B_3 G^{12}_{24}\left(x\,\middle\vert\,\begin{matrix}(3-b-c)/2,\ 1-(b+c)/2\\ 0,\ 2-b-c,\ 1-b,\ 1-c\end{matrix}\right)$	$B_3\Gamma\begin{bmatrix}s,\ (b+c-1)/2-s,\ (b+c)/2-s\\ b+c-1-s,\ b-s,\ c-s\end{bmatrix}$ [$0 <$ Re $s <$ (Re $(b+c)-1)/2$]
4	${}_0F_1\left(b;\ -\frac{1}{4x}\right){}_0F_1\left(c;\ -\frac{1}{4x}\right) = B_3 G^{21}_{42}\left(x\,\middle\vert\,\begin{matrix}1,\ b+c-1,\ b,\ c\\ (b+c-1)/2,\ (b+c)/2\end{matrix}\right)$	$B_3\Gamma\begin{bmatrix}s+(b+c-1)/2,\ s+(b+c)/2,\ -s\\ s+b+c-1,\ s+b,\ s+c\end{bmatrix}$ [$(1-$ Re $(b+c))/2 <$ Re $s < 0$]

8.4.48. The function ${}_1F_2(a;\ b_1,\ b_2;\ x)$.

NOTATION: $B_4 = \Gamma\begin{bmatrix}b_1,\ b_2\\ a\end{bmatrix}$.

1	${}_1F_2(a;\ b_1,\ b_2;\ -x) = B_4 G^{11}_{13}\left(x\,\middle\vert\,\begin{matrix}1-a\\ 0,\ 1-b_1,\ 1-b_2\end{matrix}\right)$	$B_4\Gamma\begin{bmatrix}s,\ a-s\\ b_1-s,\ b_2-s\end{bmatrix}$ [$0 <$ Re $s <$ Re a, $1/4 +$ Re $(b_1+b_2-a)/2$; $b_1, b_2 \neq 0, -1, -2, \ldots$]

2	${}_1F_2\left(a;\ b_1,\ b_2;\ -\frac{1}{x}\right)=$ $= B_4 G_{31}^{11}\left(x\,\middle\vert\,\begin{matrix}1,\ b_1,\ b_2\\ a\end{matrix}\right)$	$B_4\Gamma\left[\begin{matrix}s+a,\ -s\\ s+b_1,\ s+b_2\end{matrix}\right]$ $[-\operatorname{Re} a,\ \operatorname{Re}(a-b_1-b_2)/2-1/4<\operatorname{Re} s<0;$ $b_1, b_2\neq 0, -1, -2, \ldots]$

8.4.49. The Gauss hypergeometric function ${}_2F_1(a,\ b;\ c;\ x)$.

NOTATION:

$$B_5=\frac{\Gamma(b-c+1)}{(c)_n},\quad B_6=\frac{\Gamma(1-b)}{(c)_n},\quad B_7=\frac{\Gamma(1-a)}{(c)_n},$$

$$B_8=\frac{\Gamma(a-c+1)}{(c)_n},\quad B_9=\Gamma\left[\begin{matrix}c\\ c-a,\ c-b\end{matrix}\right],\quad B_{10}=\frac{1}{2}\Gamma\left[\begin{matrix}2a+b+1\\ 2a,\ a+b+1\end{matrix}\right],$$

$$B_{11}=\Gamma\left[\begin{matrix}c\\ a,\ b,\ c-a,\ c-b\end{matrix}\right],\quad B_{12}=\Gamma\left[\begin{matrix}c\\ a,\ c-b\end{matrix}\right],\quad B_{13}=\Gamma\left[\begin{matrix}c,\ c-2a\\ 2a\end{matrix}\right],$$

$$B_{14}=\Gamma\left[\begin{matrix}c,\ 2a-c+1\\ 2c-2a-1\end{matrix}\right],\quad B_{15}=\Gamma\left[\begin{matrix}a+b\mp 1/2,\ 1/2-a+b\\ 2a-(1\pm 1)/2\end{matrix}\right],$$

$$B_{16}=\Gamma\left[\begin{matrix}b+1/2,\ 1/2-a+b\\ a\end{matrix}\right],\quad B_{17}=\Gamma\left[\begin{matrix}b+1/2,\ 1/2+a-b\\ 2b-a\end{matrix}\right],$$

$$B_{18}=\frac{3^{2-3a}\sqrt{\pi}}{2\Gamma(a+1/6)\,\Gamma(3a-1)},\quad B_{19}=\frac{3^{2-3a}\sqrt{\pi}}{2\Gamma(4/3-a)\,\Gamma(3a-1)},$$

$$B_{20}=2^{-6a}\Gamma\left[\begin{matrix}4a+2/3\\ a+1/6,\ 3a\end{matrix}\right],\quad B_{21}=\frac{1}{2\sqrt{\pi}}\Gamma\left[\begin{matrix}4a+2/3\\ 6a\end{matrix}\right],\quad B_{22}=\frac{\sqrt{\pi}}{\Gamma(1/3-a)\,\Gamma(3a)},$$

$$B_{23}=\frac{2^{2a+2b-1}}{\sqrt{\pi}}\Gamma\left[\begin{matrix}a+b+1/2,\ a+b+1/2\\ 2a,\ 2b\end{matrix}\right],$$

$$B_{24}=\frac{2^{2a+2b+\delta-1}}{\sqrt{\pi}}\Gamma\left[\begin{matrix}a+b+1/2,\ a+b+\delta+1/2\\ 2a+\delta,\ 2b+1\end{matrix}\right],$$

$$B_{25}=\frac{(1-2a-2b)\cos(a-b)\pi}{2\sqrt{\pi}\cos(a+b)\pi},\quad B_{26}=\frac{\sqrt{\pi}}{2^{c-1}}\Gamma\left[\begin{matrix}c,\ c\\ a,\ b,\ c-a,\ c-b\end{matrix}\right],$$

$$B_{27}=\frac{2^{a+b-1}}{\sqrt{\pi}}\Gamma\left[\begin{matrix}c,\ a+b-c+1\\ a,\ b\end{matrix}\right],$$

$$B_{28}=\frac{2^{a+b-1}}{\sqrt{\pi}}\Gamma\left[\begin{matrix}(a+b+1)/2,\ (a+b+1)/2\\ a,\ b\end{matrix}\right],\quad \delta=\left\{\begin{matrix}1\\ 0\end{matrix}\right\}.$$

1	$(1-x)_+^{b-c-n}\,{}_2F_1(-n,\ b;\ c;\ x)=$ $=(-1)^n B_5 G_{22}^{20}\left(x\,\middle\vert\,\begin{matrix}1-c-n,\ 1+b-c\\ 0,\ 1-c\end{matrix}\right)$	$B_5\Gamma\left[\begin{matrix}s,\ c+n-s\\ s+1+b-c,\ c-s\end{matrix}\right]=$ $=B_5\Gamma\left[\begin{matrix}s\\ s+1+b-c\end{matrix}\right](c-s)_n$ $[\operatorname{Re}(b-c)>n-1;\ \operatorname{Re} s>0]$
2	$(x-1)_+^{b-c-n}\,{}_2F_1\left(-n,\ b;\ c;\ \frac{1}{x}\right)=$ $=(-1)^n B_5 G_{22}^{02}\left(x\,\middle\vert\,\begin{matrix}1+c-b-n,\ 2c-b-n\\ 2c-b,\ 2c-2b-n\end{matrix}\right)$	$B_5\Gamma\left[\begin{matrix}s+2c-b,\ b+n-c-s\\ s+2c-b-n,\ 1+n+2b-2c-s\end{matrix}\right]=$ $=(-1)^n B_5\Gamma\left[\begin{matrix}b+n-c-s\\ 1+n+2b-2c-s\end{matrix}\right]\times$ $\times(1+b-2c-s)_n$ $[\operatorname{Re}(b-c)>n-1,\ \operatorname{Re} s<n+\operatorname{Re}(b-c)]$

3	$(x-1)_+^{b-c-n}\,{}_2F_1(-n,\ b;\ c;\ x)= =(-1)^n B_5 G_{22}^{02}\left(x\,\middle\vert\,\begin{matrix}1+b-c,\ 1-c-n\\ 1-c,\ 0\end{matrix}\right)$	$B_5\,\Gamma\begin{bmatrix}s+1-c,\ c-b-s\\ s+1-c-n,\ 1-s\end{bmatrix}= =(-1)^n B_5\Gamma\begin{bmatrix}c-b-s\\ 1-s\end{bmatrix}(c-s)_n$ [Re $(b-c)>n-1$; Re $s<$ Re $(c-b)$]
4	$(1-x)_+^{b-c-n}\,{}_2F_1\left(-n,\ b;\ c;\ \frac{1}{x}\right)= =(-1)^n B_5 G_{22}^{20}\left(x\,\middle\vert\,\begin{matrix}b-n,\ 1+b-c-n\\ -n,\ b\end{matrix}\right)$	$B_5\Gamma\begin{bmatrix}s-n,\ 1+n-b-s\\ s+1+b-c-n,\ 1-b-s\end{bmatrix}= =B_5\Gamma\begin{bmatrix}s-n\\ s+1+b-c-n\end{bmatrix}(1-b-s)_n$ [Re $(b-c)>n-1$; Re $s>n$]
5	$(1-x)_+^{-b}\,{}_2F_1\left(-n,\ b;\ c;\ \frac{1}{1-x}\right)= =B_6 G_{22}^{20}\left(x\,\middle\vert\,\begin{matrix}1-b,\ c-b\\ 0,\ n+c-b\end{matrix}\right)$	$(-1)^n B_6\Gamma\begin{bmatrix}s,\ 1+b-c-s\\ s+1-b,\ 1+b-c-n-s\end{bmatrix}= =B_6\Gamma\begin{bmatrix}s\\ s+1-b\end{bmatrix}(s+c-b)_n$ [Re $s>0$; Re $b<1-n$]
6	$(x-1)_+^{-b}\,{}_2F_1\left(-n,\ b;\ c;\ \frac{x}{x-1}\right)= =B_6 G_{22}^{02}\left(x\,\middle\vert\,\begin{matrix}1-c-n,\ 1-b\\ 0,\ 1-c\end{matrix}\right)$	$(-1)^n B_6\Gamma\begin{bmatrix}s+1-c,\ b-s\\ s+1-c-n,\ 1-s\end{bmatrix}= =B_6\Gamma\begin{bmatrix}b-s\\ 1-s\end{bmatrix}(c-s)_n$ [Re $s<$ Re $b<1-n$]
7	$(1-x)_+^{-b}\,{}_2F_1\left(-n,\ b;\ c;\ \frac{x}{x-1}\right)= =(-1)^n B_6 G_{22}^{20}\left(x\,\middle\vert\,\begin{matrix}1-c-n,\ 1-b\\ 0,\ 1-c\end{matrix}\right)$	$B_6\Gamma\begin{bmatrix}s,\ c+n-s\\ s+1-b,\ c-s\end{bmatrix}= =B_6\Gamma\begin{bmatrix}s\\ s+1-b\end{bmatrix}(c-s)_n$ [Re $b<1-n$; Re $s>0$]
8	$(x-1)_+^{-b}\,{}_2F_1\left(-n,\ b;\ c;\ \frac{1}{1-x}\right)= =(-1)^n B_6 G_{22}^{02}\left(x\,\middle\vert\,\begin{matrix}1+b,\ b+c\\ b+c+n,\ 2b\end{matrix}\right)$	$B_6\Gamma\begin{bmatrix}s+b+c+n,\ -b-s\\ s+b+c,\ 1-2b-s\end{bmatrix}= =B_6\Gamma\begin{bmatrix}-b-s\\ 1-2b-s\end{bmatrix}(s+b+c)_n$ [Re $s<-$Re b; Re $b<1-n$]
9	$H(1-x)\,{}_2F_1(a,\ c+n;\ c;\ x)= =(-1)^n B_7 G_{22}^{20}\left(x\,\middle\vert\,\begin{matrix}1-c-n,\ 1-a\\ 0,\ 1-c\end{matrix}\right)$	$B_7\Gamma\begin{bmatrix}s,\ c+n-s\\ s+1-a,\ c-s\end{bmatrix}= =B_7\Gamma\begin{bmatrix}s\\ s+1-a\end{bmatrix}(c-s)_n$ [Re $a<1-n$; Re $s>0$]
10	$H(x-1)\,{}_2F_1\left(a,\ c+n;\ c;\ \frac{1}{x}\right)= =(-1)^n B_7 G_{22}^{02}\left(x\,\middle\vert\,\begin{matrix}1,\ c\\ c+n,\ a\end{matrix}\right)$	$B_7\Gamma\begin{bmatrix}s+c+n,\ -s\\ s+c,\ 1-a-s\end{bmatrix}= =B_7\Gamma\begin{bmatrix}-s\\ 1-a-s\end{bmatrix}(s+c)_n$ [Re $a<1-n$; Re $s<0$]
11	$(1-x)_+^{-c-n}\,{}_2F_1\left(a,\ c+n;\ c;\ \frac{x}{x-1}\right)= =(-1)^n B_8 G_{22}^{20}\left(x\,\middle\vert\,\begin{matrix}1-c-n,\ 1+a-c\\ 0,\ 1-c\end{matrix}\right)$	$B_8\Gamma\begin{bmatrix}s,\ c+n-s\\ s+1+a-c,\ c-s\end{bmatrix}= =B_8\Gamma\begin{bmatrix}s\\ s+1+a-c\end{bmatrix}(c-s)_n$ [Re $(a-c)>n-1$; Re $s>0$]
12	$(x-1)_+^{-c-n}\,{}_2F_1\left(a,\ c+n;\ c;\ \frac{1}{1-x}\right)= =(-1)^n B_8 G_{22}^{02}\left(x\,\middle\vert\,\begin{matrix}1-c-n,\ -n\\ 0,\ -a-n\end{matrix}\right)$	$B_8\Gamma\begin{bmatrix}s,\ c+n-s\\ s-n,\ 1+a+n-s\end{bmatrix}= =(-1)^n B_8\Gamma\begin{bmatrix}c+n-s\\ 1+a+n-s\end{bmatrix}(1-s)_n$ [Re $s<n+$ Re c; Re $(a-c)>n-1$]
13	${}_2F_1(a,\ b;\ c;\ -x)= =\Gamma\begin{bmatrix}c\\ a,\ b\end{bmatrix}G_{22}^{12}\left(x\,\middle\vert\,\begin{matrix}1-a,\ 1-b\\ 0,\ 1-c\end{matrix}\right)$	$\Gamma\begin{bmatrix}c\\ a,\ b\end{bmatrix}\Gamma\begin{bmatrix}s,\ a-s,\ b-s\\ c-s\end{bmatrix}$ [$0<$ Re $s<$ Re a, Re b; $c\neq 0,\ -1,\ -2,\ \ldots$]

14	${}_2F_1\left(a,\ b;\ c;\ -\frac{1}{x}\right)= =\Gamma\begin{bmatrix}c\\ a,\ b\end{bmatrix} G_{22}^{21}\left(x\,\middle\vert\,\begin{matrix}1,\ c\\ a,\ b\end{matrix}\right)$	$\Gamma\begin{bmatrix}c\\ a,\ b\end{bmatrix}\Gamma\begin{bmatrix}s+a,\ s+b,\ -s\\ s+c\end{bmatrix}$ [$-\mathrm{Re}\,a,\ -\mathrm{Re}\,b<\mathrm{Re}\,s<0;\ c\neq 0,\ -1,\ -2,\ \ldots$]
15	$(1+x)^{a+b-c}\,{}_2F_1(a,\ b;\ c;\ -x)= =B_9G_{22}^{12}\left(x\,\middle\vert\,\begin{matrix}1+a-c,\ 1+b-c\\ 0,\ 1-c\end{matrix}\right)$	$B_9\Gamma\begin{bmatrix}s,\ c-a-s,\ c-b-s\\ c-s\end{bmatrix}$ [$0<\mathrm{Re}\,s<\mathrm{Re}\,(c-a),\ \mathrm{Re}\,(c-b);\ c\neq 0,\ -1,\ -2,\ldots$]
16	$(1+x)^{a+b-c}\,{}_2F_1\left(a,\ b;\ c;-\frac{1}{x}\right)= =B_9G_{22}^{21}\left(x\,\middle\vert\,\begin{matrix}1+a+b-c,\ a+b\\ a,\ b\end{matrix}\right)$	$B_9\Gamma\begin{bmatrix}s+a,\ s+b,\ c-a-b-s\\ s+a+b\end{bmatrix}$ [$-\mathrm{Re}\,a,\ -\mathrm{Re}\,b<\mathrm{Re}\,s<\mathrm{Re}\,(c-a-b);\ c\neq 0,\ -1,\ -2,\ \ldots$]
17	$(1+x)^{-a}\,{}_2F_1(a,\ b;\ 2a+b+1;\ -x)= =B_{10}G_{33}^{13}\left(x\,\middle\vert\,\begin{matrix}-2a-2b,\ 1-a-b,\ 1-2a\\ 0,\ -2a-b,\ 1-2a-2b\end{matrix}\right)$	$B_{10}\Gamma\begin{bmatrix}s,\ 1+2a+2b-s,\ a+b-s,\ 2a-s\\ 2a+b+1-s,\ 2a+2b-s\end{bmatrix}$ [$0<\mathrm{Re}\,s<2\,\mathrm{Re}\,a,\ \mathrm{Re}\,(a+b);\ 2a+b\neq -1,\ -2,\ -3,\ \ldots$]
18	$(1+x)^{-a}\,{}_2F_1\left(a,\ b;\ 2a+b+1;\ -\frac{1}{x}\right)= =B_{10}G_{33}^{31}\left(x\,\middle\vert\,\begin{matrix}1-a,\ a+b+1,\ a+2b\\ a,\ b,\ a+2b+1\end{matrix}\right)$	$B_{10}\Gamma\begin{bmatrix}s+a,\ s+b,\ s+a+2b+1,\ a-s\\ s+a+b+1,\ s+a+2b\end{bmatrix}$ [$-\mathrm{Re}\,a,\ -\mathrm{Re}\,b<\mathrm{Re}\,s<\mathrm{Re}\,a;\ 2a+b\neq -1,\ -2,\ -3,\ \ldots$]
19	${}_2F_1(a,\ b;\ c;\ x)\,H(1-x)+ +x^{-a}\Gamma\begin{bmatrix}c,\ 1-b\\ c-a,\ 1+a-b\end{bmatrix}\times \times{}_2F_1\left(a,\ 1+a-c;\ 1+a-b;\ \frac{1}{x}\right)\times \times H(x-1)=\Gamma\begin{bmatrix}c,\ 1-b\\ a\end{bmatrix}\times \times G_{22}^{11}\left(x\,\middle\vert\,\begin{matrix}1-a,\ 1-b\\ 0,\ 1-c\end{matrix}\right)$	$\Gamma\begin{bmatrix}c,\ 1-b\\ a\end{bmatrix}\Gamma\begin{bmatrix}s,\ a-s\\ s+1-b,\ c-s\end{bmatrix}$ [$\mathrm{Re}\,(c-a-b)>-1;\ 0<\mathrm{Re}\,s<\mathrm{Re}\,a;\ 1-b,\ c\neq 0,\ -1,\ -2,\ \ldots$]
20	${}_2F_1(a,\ b;\ c;\ 1-x)= =B_{11}G_{22}^{22}\left(x\,\middle\vert\,\begin{matrix}1-a,\ 1-b\\ 0,\ c-a-b\end{matrix}\right)$	$B_{11}\Gamma\,[s,\ s+c-a-b,\ a-s,\ b-s]$ [$\mathrm{Re}\,(a+b-c),\ 0<\mathrm{Re}\,s<\mathrm{Re}\,a,\ \mathrm{Re}\,b;\ c\neq 0,\ -1,\ -2,\ \ldots$]
21	${}_2F_1\left(a,\ b;\ c;\ 1-\frac{1}{x}\right)= =B_{11}G_{22}^{22}\left(x\,\middle\vert\,\begin{matrix}1,\ 1+a+b-c\\ a,\ b\end{matrix}\right)$	$B_{11}\Gamma\,[s+a,\ s+b,\ -s,\ c-a-b-s]$ [$-\mathrm{Re}\,a,\ -\mathrm{Re}\,b<\mathrm{Re}\,s<0,\ \mathrm{Re}\,(c-a-b);\ c\neq 0,\ -1,\ -2,\ \ldots$]
22	$(1-x)_+^{c-1}\,{}_2F_1(a,\ b;\ c;\ 1-x)= =\Gamma(c)\,G_{22}^{20}\left(x\,\middle\vert\,\begin{matrix}c-a,\ c-b\\ 0,\ c-a-b\end{matrix}\right)$	$\Gamma(c)\,\Gamma\begin{bmatrix}s,\ s+c-a-b\\ s+c-a,\ s+c-b\end{bmatrix}$ [$\mathrm{Re}\,s>0,\ \mathrm{Re}\,(a+b-c);\ \mathrm{Re}\,c>0$]
23	$(x-1)_+^{c-1}\,{}_2F_1\left(a,\ b;\ c;\ 1-\frac{1}{x}\right)= =\Gamma(c)\,G_{22}^{02}\left(x\,\middle\vert\,\begin{matrix}c,\ a+b\\ a,\ b\end{matrix}\right)$	$\Gamma(c)\,\Gamma\begin{bmatrix}1-c-s,\ 1-a-b-s\\ 1-a-s,\ 1-b-s\end{bmatrix}$ [$\mathrm{Re}\,c>0;\ \mathrm{Re}\,s<1-\mathrm{Re}\,c,\ 1-\mathrm{Re}\,(a+b)$]
24	$(x-1)_+^{c-1}\,{}_2F_1(a,\ b;\ c;\ 1-x)= =\Gamma(c)\,G_{22}^{02}\left(x\,\middle\vert\,\begin{matrix}c-a,\ c-b\\ 0,\ c-a-b\end{matrix}\right)$	$\Gamma(c)\,\Gamma\begin{bmatrix}1+a-c-s,\ 1+b-c-s\\ 1-s,\ 1+a+b-c-s\end{bmatrix}$ [$\mathrm{Re}\,c>0;\ \mathrm{Re}\,s<1+\mathrm{Re}\,(a-c),\ 1+\mathrm{Re}\,(b-c)$]

25	$(1-x)_+^{c-1}\,{}_2F_1\left(a,\ b;\ c;\ 1-\frac{1}{x}\right)= \Gamma(c)\,G_{22}^{20}\left(x\,\middle\vert\,\begin{matrix}c,\ a+b\\ a,\ b\end{matrix}\right)$	$\Gamma(c)\,\Gamma\left[\begin{matrix}s+a,\ s+b\\ s+c,\ s+a+b\end{matrix}\right]$ [$\mathrm{Re}\,c>0$; $\mathrm{Re}\,s>-\mathrm{Re}\,a,\ -\mathrm{Re}\,b$]
26	$(1+x)^{-a}\,{}_2F_1\left(a,\ b;\ c;\ \frac{x}{x+1}\right)= B_{12}G_{22}^{12}\left(x\,\middle\vert\,\begin{matrix}1-a,\ 1+b-c\\ 0,\ 1-c\end{matrix}\right)$	$B_{12}\Gamma\left[\begin{matrix}s,\ a-s,\ c-b-s\\ c-s\end{matrix}\right]$ [$0<\mathrm{Re}\,s<\mathrm{Re}\,a,\ \mathrm{Re}\,(c-b)$; $c\neq 0,\ -1,\ -2,\ \ldots$]
27	$(1+x)^{-a}\,{}_2F_1\left(a,\ b;\ c;\ \frac{1}{1+x}\right)= B_{12}G_{22}^{21}\left(x\,\middle\vert\,\begin{matrix}1-a,\ c-a\\ 0,\ c-a-b\end{matrix}\right)$	$B_{12}\Gamma\left[\begin{matrix}s,\ s+c-a-b,\ a-s\\ s+c-a\end{matrix}\right]$ [$0,\ \mathrm{Re}\,(a+b-c)<\mathrm{Re}\,s<\mathrm{Re}\,a$; $c\neq 0,\ -1,\ -2,\ \ldots$]
28	$(1+x)^{-2a}\,{}_2F_1\left(a,a+\frac{1}{2};c;\frac{4x}{(1+x)^2}\right)= B_{13}G_{22}^{11}\left(x\,\middle\vert\,\begin{matrix}1-2a,\ c-2a\\ 0,\ 1-c\end{matrix}\right)$	$B_{13}\Gamma\left[\begin{matrix}s,\ 2a-s\\ s+c-2a,\ c-s\end{matrix}\right]$ [$\mathrm{Re}\,(c-2a)>0$; $0<\mathrm{Re}\,s<2\,\mathrm{Re}\,a$; $c\neq 0,\ -1,\ -2,\ \ldots$]
29	$\frac{\vert 1-x\vert^{4a-2c+1}}{(1+x)^{2a}}\times {}_2F_1\left(a,a+\frac{1}{2};c;\frac{4x}{(1+x)^2}\right)= B_{14}G_{22}^{11}\left(x\,\middle\vert\,\begin{matrix}2+2a-2c,\ 1+2a-c\\ 0,\ 1-c\end{matrix}\right)$	$B_{14}\Gamma\left[\begin{matrix}s,\ 2c-2a-1-s\\ s+1+2a-c,\ c-s\end{matrix}\right]$ [$\mathrm{Re}\,(2a-c)>-1$; $0<\mathrm{Re}\,s<2\,\mathrm{Re}\,(c-a)-1$; $c\neq 0,\ -1,\ -2,\ \ldots$]
30	$\frac{(1+x)^{(1\pm 1)/2}}{\vert 1-x\vert^{2a}}\times {}_2F_1\left(a,\ b;\ a+b\mp\frac{1}{2};-\frac{4x}{(1-x)^2}\right)= B_{15}G_{22}^{11}\left(x\,\middle\vert\,\begin{matrix}(3\pm 1)/2-2a,\ 1/2+b-a\\ 0,\ 1\pm 1/2-a-b\end{matrix}\right)$	$B_{15}\Gamma\left[\begin{matrix}s,\ 2a-(1\pm 1)/2-s\\ s+b-a+1/2,\ a+b\mp 1/2-s\end{matrix}\right]$ [$\mathrm{Re}\,(b-a)>-1/2$; $0<\mathrm{Re}\,s<2\,\mathrm{Re}\,a-(1\pm 1)/2$; $a+b\mp 1/2\neq 0,\ -1,\ -2,\ \ldots$]
31	$\vert 1\pm\sqrt{x}\vert^{-2a}\times {}_2F_1\left(a,\ b;\ 2b;\ \pm\frac{4\sqrt{x}}{(1\pm\sqrt{x})^2}\right)= B_{16}G_{22}^{11}\left(x\,\middle\vert\,\begin{matrix}1-a,\ 1/2+b-a\\ 0,\ 1/2-b\end{matrix}\right)$	$B_{16}\Gamma\left[\begin{matrix}s,\ a-s\\ s+b-a+1/2,\ b+1/2-s\end{matrix}\right]$ [$\mathrm{Re}\,(b-a)>-1/2$; $0<\mathrm{Re}\,s<\mathrm{Re}\,a$; $b+1/2\neq 0,\ -1,\ -2,\ \ldots$]
32	$\frac{\vert 1\mp\sqrt{x}\vert^{2a}}{\vert 1-x\vert^{2b}}\times {}_2F_1\left(a,\ b;\ 2b;\ \pm\frac{4\sqrt{x}}{(1\pm\sqrt{x})^2}\right)= B_{17}G_{22}^{11}\left(x\,\middle\vert\,\begin{matrix}1+a-2b,\ 1/2+a-b\\ 0,\ 1/2-b\end{matrix}\right)$	$B_{17}\Gamma\left[\begin{matrix}s,\ 2b-a-s\\ s+a-b+1/2,\ b+1/2-s\end{matrix}\right]$ [$\mathrm{Re}\,(a-b)>-1/2$; $0<\mathrm{Re}\,s<\mathrm{Re}\,(2b-a)$; $b+1/2\neq 0,\ -1,\ -2,\ \ldots$]
33	$(1+\sqrt{1+x})^{-a}\times {}_2F_1\left(a,\ b;\ a+1;\ \frac{1-\sqrt{1+x}}{2}\right)= \frac{2^{b-1}a}{\sqrt{\pi}}\,G_{33}^{13}\left(x\,\middle\vert\,\begin{matrix}1-a,\ \frac{1-a-b}{2},\ 1-\frac{a+b}{2}\\ 0,\ 1-a-b,\ -a\end{matrix}\right)$	$\frac{2^{b-1}a}{\sqrt{\pi}}\Gamma\left[\begin{matrix}s,\ a-s,\ \frac{a+b}{2}-s,\ \frac{a+b+1}{2}-s\\ a+b-s,\ a+1-s\end{matrix}\right]$ [$0<\mathrm{Re}\,s<\mathrm{Re}\,a,\ \mathrm{Re}\,(a+b)/2$]

34	$(\sqrt{x}+\sqrt{1+x})^{-a}\times$ $\times{}_2F_1\left(a, b; a+1; \frac{\sqrt{x}-\sqrt{1+x}}{2\sqrt{x}}\right)=$ $=\frac{2^{b-1}a}{\sqrt{\pi}}G_{33}^{31}\left(x\,\middle\vert\,\begin{matrix}1-\frac{a}{2}, & 1+\frac{a}{2}, & \frac{a}{2}+b\\ \frac{a}{2}, & \frac{b}{2}, & \frac{b+1}{2}\end{matrix}\right)$	$\frac{2^{b-1}a}{\sqrt{\pi}}\Gamma\left[\begin{matrix}s+\frac{a}{2}, s+\frac{b}{2}, s+\frac{b+1}{2}, \frac{a}{2}-s\\ s+\frac{a}{2}+1, s+\frac{a}{2}+b\end{matrix}\right]$ [$-\mathrm{Re}\,a/2$, $-\mathrm{Re}\,b/2<\mathrm{Re}\,s<\mathrm{Re}\,a/2$]
35	$(9+x)(3-x)^{-3a}\times$ $\times{}_2F_1\left(a, a+\frac{1}{3}; \frac{3}{2}; \frac{x(x+9)^2}{(x-3)^3}\right)=$ $=B_{18}G_{22}^{12}\left(x\,\middle\vert\,\begin{matrix}5/6-a, 2-3a\\ 0, -1/2\end{matrix}\right)$	$B_{18}\Gamma\left[\begin{matrix}s, a+1/6-s, 3a-1-s\\ 3/2-s\end{matrix}\right]$ [$0<\mathrm{Re}\,s<\mathrm{Re}\,a+1/6$, $3\,\mathrm{Re}\,a-1$]
36	$(1+9x)(3x-1)^{-3a}\times$ $\times{}_2F_1\left(a, a+\frac{1}{3}; \frac{3}{2}; \frac{(1+9x)^2}{(1-3x)^3}\right)=$ $=B_{18}G_{22}^{21}\left(x\,\middle\vert\,\begin{matrix}2-3a, 5/2-3a\\ 0, 7/6-2a\end{matrix}\right)$	$B_{18}\Gamma\left[\begin{matrix}s, s+7/6-2a, 3a-1-s\\ s+5/2-3a\end{matrix}\right]$ [0, $2\,\mathrm{Re}\,a-7/6<\mathrm{Re}\,s<3\,\mathrm{Re}\,a-1$]
37	$(9+8x)(3+4x)^{-3a}\times$ $\times{}_2F_1\left(a, a+\frac{1}{3}; \frac{3}{2}; \frac{x(9+8x)^2}{(3+4x)^3}\right)=$ $=B_{19}G_{22}^{12}\left(x\,\middle\vert\,\begin{matrix}a-1/3, 2-3a\\ 0, -1/2\end{matrix}\right)$	$B_{19}\Gamma\left[\begin{matrix}s, 4/3-a-s, 3a-1-s\\ 3/2-s\end{matrix}\right]$ [$0<\mathrm{Re}\,s<4/3-\mathrm{Re}\,a$, $3\,\mathrm{Re}\,a-1$]
38	$(8+9x)(4+3x)^{-3a}\times$ $\times{}_2F_1\left(a, a+\frac{1}{3}; \frac{3}{2}; \frac{(8+9x)^2}{(4+3x)^3}\right)=$ $=B_{19}G_{22}^{21}\left(x\,\middle\vert\,\begin{matrix}2-3a, 5/2-3a\\ 0, 7/3-4a\end{matrix}\right)$	$B_{19}\Gamma\left[\begin{matrix}s, s+7/3-4a, 3a-1-s\\ s+5/2-3a\end{matrix}\right]$ [0, $4\,\mathrm{Re}\,a-7/3<\mathrm{Re}\,s<3\,\mathrm{Re}\,a-1$]
39	$(4+x)^{-3a}\times$ $\times{}_2F_1\left(a, a+\frac{1}{3}; 2a+\frac{5}{6}; \frac{27x^2}{(4+x)^3}\right)=$ $=B_{20}G_{22}^{12}\left(x\,\middle\vert\,\begin{matrix}5/6-a, 1-3a\\ 0, 1/3-4a\end{matrix}\right)$	$B_{20}\Gamma\left[\begin{matrix}s, a+1/6-s, 3a-s\\ 4a+2/3-s\end{matrix}\right]$ [$0<\mathrm{Re}\,s<\mathrm{Re}\,a+1/6$, $3\,\mathrm{Re}\,a$]
40	$(1+4x)^{-3a}\times$ $\times{}_2F_1\left(a, a+\frac{1}{3}; 2a+\frac{5}{6}; \frac{27x}{(1+4x)^3}\right)=$ $=B_{20}G_{22}^{21}\left(x\,\middle\vert\,\begin{matrix}1-3a, 2/3+a\\ 0, 1/6-2a\end{matrix}\right)$	$B_{20}\Gamma\left[\begin{matrix}s, s+1/6-2a, 3a-s\\ s+a+2/3\end{matrix}\right]$ [0, $2\,\mathrm{Re}\,a-1/6<\mathrm{Re}\,s<3\,\mathrm{Re}\,a$]

1	$(8+9x)^{-2a}\times {}_2F_1\left(a, a+\frac{1}{2}, a+\frac{5}{6}; -\frac{27x^2(1+x)}{(8+9x)^2}\right)= B_{21}G_{22}^{12}\left(x\,\middle\vert\,\begin{matrix}1/2-3a, & 1-3a\\ 0, & 1/3-4a\end{matrix}\right)$	$B_{21}\Gamma\begin{bmatrix}s, \ 3a+1/2-s, \ 3a-s\\ 4a+2/3-s\end{bmatrix}$ [$0<\operatorname{Re} s<3\operatorname{Re} a$]
2	$(9+8x)^{-2a}\times {}_2F_1\left(a, a+\frac{1}{2}; a+\frac{5}{6}; -\frac{27(1+x)}{x(9+8x)^2}\right)= B_{21}G_{22}^{21}\left(x\,\middle\vert\,\begin{matrix}1-2a, & 2/3+2a\\ a, & a+1/2\end{matrix}\right)$	$B_{21}\Gamma\begin{bmatrix}s+a, \ s+a+1/2, \ 2a-s\\ s+2a+2/3\end{bmatrix}$ [$-\operatorname{Re} a<\operatorname{Re} s<2\operatorname{Re} a$]
3	$(1+x)^{-a}\times {}_2F_1\left(a, \frac{1}{6}-a; \frac{1}{2}; -\frac{x(9+8x)^2}{27(1+x)}\right)= B_{22}G_{22}^{12}\left(x\,\middle\vert\,\begin{matrix}2/3+a, & 1-3a\\ 0, & 1/2\end{matrix}\right)$	$B_{22}\Gamma\begin{bmatrix}s, \ 1/3-a-s, \ 3a-s\\ 1/2-s\end{bmatrix}$ [$0<\operatorname{Re} s<1/3-\operatorname{Re} a, 3\operatorname{Re} a$]
4	$(1+x)^{-a}\times {}_2F_1\left(a, \frac{1}{6}-a; \frac{1}{2}; -\frac{(8+9x)^2}{27x^2(1+x)}\right)= B_{22}G_{22}^{21}\left(x\,\middle\vert\,\begin{matrix}1-a, & 1/2-a\\ 2a, & 1/3\end{matrix}\right)$	$B_{22}\Gamma\begin{bmatrix}s+2a, \ s+1/3, \ a-s\\ s+1/2-a\end{bmatrix}$ [$-1/3, -2\operatorname{Re} a<\operatorname{Re} s<\operatorname{Re} a$]
5	$\left[{}_2F_1\left(a, b; a+b+\frac{1}{2}; -x\right)\right]^2= B_{23}G_{33}^{13}\left(x\,\middle\vert\,\begin{matrix}1-2a, & 1-2b, & 1-a-b\\ 0, & 1/2-a-b, & 1-2a-2b\end{matrix}\right)$	$B_{23}\Gamma\begin{bmatrix}s, \ 2a-s, \ 2b-s, \ a+b-s\\ a+b+1/2-s, \ 2a+2b-s\end{bmatrix}$ [$0<\operatorname{Re} s<2\operatorname{Re} a, 2\operatorname{Re} b, \operatorname{Re}(a+b)$]
6	$\left[{}_2F_1\left(a, b; a+b+\frac{1}{2}; -\frac{1}{x}\right)\right]^2= B_{23}G_{33}^{31}\left(x\,\middle\vert\,\begin{matrix}1, & 1/2+a+b, & 2a+2b\\ 2a, & 2b, & a+b\end{matrix}\right)$	$B_{23}\Gamma\begin{bmatrix}s+2a, \ s+2b, \ s+a+b, \ -s\\ s+a+b+1/2, \ s+2a+2b\end{bmatrix}$ [$-2\operatorname{Re} a, -2\operatorname{Re} b, -\operatorname{Re}(a+b)<\operatorname{Re} s<0$]
7	${}_2F_1\left(a, b; a+b+\frac{1}{2}; -x\right)\times {}_2F_1\left(a+\frac{1\pm 1}{2}, b+1; a+b+1\pm\frac{1}{2}; -x\right)= B_{24}G_{33}^{13}\left(x\,\middle\vert\,\begin{matrix}1-2a-\delta, & -2b, & 1-a-b-\delta\\ 0, & \frac{1}{2}-a-b-\delta, & 1-2a-2b-\delta\end{matrix}\right)$	$B_{24}\times\Gamma\begin{bmatrix}s, \ 2a+\delta-s, \ 2b+1-s, \ a+b+\delta-s\\ 2a+2b+\delta-s, \ a+b+\delta+1/2-s\end{bmatrix}$ [$0<\operatorname{Re} s<2\operatorname{Re} a+\delta, 2\operatorname{Re} b+1, \operatorname{Re}(a+b)+\delta$]

48	${}_2F_1\left(a, b; a+b+\frac{1}{2}; -\frac{1}{x}\right){}_2F_1\left(a+\frac{1\pm 1}{2}, b+1; a+b+1\pm\frac{1}{2}; -\frac{1}{x}\right)=$ $=B_{24}G_{33}^{31}\left(x\middle\vert\begin{matrix}1, \frac{1}{2}+a+b+\delta, 2a+2b+\delta\\ 2a+\delta, 2b+1, a+b+\delta\end{matrix}\right)$	$B_{24}\times\Gamma\begin{bmatrix}s+2a+\delta, s+2b+1, s+a+b+\delta, -s\\ s+2a+2b+\delta, s+a+b+\delta+\frac{1}{2}\end{bmatrix}$ [$-2\operatorname{Re}a-\delta$, $-2\operatorname{Re}b-1$, $-\operatorname{Re}(a+b)-\delta < \operatorname{Re}s < 0$
49	${}_2F_1\left(a, b; a+b+\frac{1}{2}; -x\right)\times$ $\times{}_2F_1\left(\frac{1}{2}-a, \frac{1}{2}-b; \frac{3}{2}-a-b; -x\right)=$ $=B_{25}G_{33}^{13}\left(x\middle\vert\begin{matrix}\frac{1}{2}, \frac{1}{2}+a-b, \frac{1}{2}+b-a\\ 0, \frac{1}{2}-a-b, a+b-\frac{1}{2}\end{matrix}\right)$	$B_{25}\times\Gamma\begin{bmatrix}s, \frac{1}{2}-s, \frac{1}{2}+a-b-s, \frac{1}{2}+b-a-s\\ \frac{1}{2}+a+b-s, \frac{3}{2}-a-b-s\end{bmatrix}$ [$0 < \operatorname{Re}s < 1/2-\vert\operatorname{Re}(a-b)\vert$]
50	${}_2F_1\left(a, b; a+b+\frac{1}{2}; -\frac{1}{x}\right)\times$ $\times{}_2F_1\left(\frac{1}{2}-a, \frac{1}{2}-b; \frac{3}{2}-a-b; -\frac{1}{x}\right)=$ $=B_{25}G_{33}^{31}\left(x\middle\vert\begin{matrix}1, \frac{1}{2}+a+b, \frac{3}{2}-a-b\\ \frac{1}{2}, \frac{1}{2}+a-b, \frac{1}{2}+b-a\end{matrix}\right)$	$B_{25}\Gamma\begin{bmatrix}s+\frac{1}{2}, s+\frac{1}{2}+a-b, s+\frac{1}{2}+b-a, -s\\ s+\frac{1}{2}+a+b, s+\frac{3}{2}-a-b\end{bmatrix}$ [$\vert\operatorname{Re}(a-b)\vert-1/2 < \operatorname{Re}s <$
51	${}_2F_1\left(a, b; c; 1-(\sqrt{1+x}-\sqrt{x})^2\right)\times$ $\times{}_2F_1\left(a, b; c; 1-(\sqrt{1+x}+\sqrt{x})^2\right)=$ $=B_{26}G_{44}^{14}\left(x\middle\vert\begin{matrix}1-a, 1-b, 1+a-c, 1+b-c\\ 0, 1-c, 1-\frac{c}{2}, \frac{1-c}{2}\end{matrix}\right)$	$B_{26}\Gamma\begin{bmatrix}s, a-s, b-s, c-a-s, c-b-s\\ c-s, c/2-s, (c+1)/2-s\end{bmatrix}$ [$0 < \operatorname{Re}s < \operatorname{Re}a$, $\operatorname{Re}b$, $\operatorname{Re}(c-a)$, $\operatorname{Re}(c-b$

52	${}_2F_1\left(a, b; c; 1-\frac{(\sqrt{1+x}-1)^2}{x}\right)\times {}_2F_1\left(a, b; c; 1-\frac{(\sqrt{1+x}+1)^2}{x}\right)= B_{26}\, G^{41}_{44}\left(x\,\middle\vert\,\begin{matrix}1, c, c/2, (c+1)/2\\ a, b, c-a, c-b\end{matrix}\right)$	$B_{26}\Gamma\begin{bmatrix}s+a, s+b, s+c-a, s+c-b, -s\\ s+c, s+c/2, s+(c+1)/2\end{bmatrix}$ [$-\operatorname{Re} a, -\operatorname{Re} b, \operatorname{Re}(a-c), \operatorname{Re}(b-c)<\operatorname{Re} s<0$]
53	${}_2F_1\left(a, b; 1+a+b-c; \frac{1-\sqrt{1+x}}{2}\right)\times {}_2F_1\left(a, b; c; \frac{1-\sqrt{1+x}}{2}\right)= B_{27}G^{14}_{44}\left(x\,\middle\vert\,\begin{matrix}1-a, 1-b, \frac{1-a-b}{2}, 1-\frac{a+b}{2}\\ 0, 1-c, 1-a-b, c-a-b\end{matrix}\right)$	$B_{27}\Gamma\begin{bmatrix}s, a-s, b-s, \frac{a+b}{2}-s, \frac{a+b+1}{2}-s\\ c-s, a+b-s, 1+a+b-c-s\end{bmatrix}$ [$0<\operatorname{Re} s<\operatorname{Re} a, \operatorname{Re} b, \operatorname{Re}(a+b)/2$]
54	${}_2F_1\left(a, b; 1+a+b-c; \frac{\sqrt{x}-\sqrt{1+x}}{2\sqrt{x}}\right)\times {}_2F_1\left(a, b; c; \frac{\sqrt{x}-\sqrt{1+x}}{2\sqrt{x}}\right)= B_{27}G^{41}_{44}\left(x\,\middle\vert\,\begin{matrix}1, c, a+b, 1+a+b-c\\ a, b, \frac{a+b}{2}, \frac{a+b+1}{2}\end{matrix}\right)$	$B_{27}\Gamma\begin{bmatrix}s+a, s+b, s+(a+b)/2, s+(a+b+1)/2, -s\\ s+c, s+a+b, s+1+a+b-c\end{bmatrix}$ [$-\operatorname{Re} a, -\operatorname{Re} b, -\operatorname{Re}(a+b)/2<\operatorname{Re} s<0$]
55	$\left[{}_2F_1\left(a, b; \frac{a+b+1}{2}; \frac{1-\sqrt{1+x}}{2}\right)\right]^2= B_{28}G^{13}_{33}\left(x\,\middle\vert\,\begin{matrix}1-a, 1-b, 1-(a+b)/2\\ 0, 1-a-b, (1-a-b)/2\end{matrix}\right)$	$B_{28}\Gamma\begin{bmatrix}s, a-s, b-s, (a+b)/2-s\\ a+b-s, (a+b+1)/2-s\end{bmatrix}$ [$0<\operatorname{Re} s<\operatorname{Re} a, \operatorname{Re} b, \operatorname{Re}(a+b)/2$]
56	$\left[{}_2F_1\left(a, b; \frac{a+b+1}{2}; \frac{\sqrt{x}-\sqrt{1+x}}{2\sqrt{x}}\right)\right]^2= B_{28}G^{31}_{33}\left(x\,\middle\vert\,\begin{matrix}1, a+b, (a+b+1)/2\\ a, b, (a+b)/2\end{matrix}\right)$	$B_{28}\Gamma\begin{bmatrix}s+a, s+b, s+(a+b)/2, -s\\ s+a+b, s+(a+b+1)/2\end{bmatrix}$ [$-\operatorname{Re} a, -\operatorname{Re} b, -\operatorname{Re}(a+b)/2<\operatorname{Re} s<0$]

8.4.50. The function ${}_3F_2(a_1, a_2, a_3; b_1, b_2; x)$.

NOTATION:

$$B_{29}=\Gamma\begin{bmatrix}b_1, b_2\\ a_1, a_2, a_3\end{bmatrix},\quad B_{30}=2\Gamma\begin{bmatrix}2a-b-1\\ 2a-2b-2, -b-1\end{bmatrix},$$

$$B_{31}=\frac{1}{2\sqrt{\pi}}\Gamma\begin{bmatrix}a-b+1, a-c+1\\ a, a-b-c+1\end{bmatrix},\quad B_{32}=2^{\delta}\Gamma\begin{bmatrix}c, 2a+b-c-\delta+1\\ 2a, 2a-c-\delta+1, c-b\end{bmatrix},$$

$$B_{33}=\frac{3^{\delta}}{2\pi}\Gamma\begin{bmatrix}b, 3a-b-\delta+3/2\\ 3a\end{bmatrix},\quad \delta=\begin{Bmatrix}1\\0\end{Bmatrix}.$$

1	${}_3F_2(a_1, a_2, a_3; b_1, b_2; -x)= B_{29}G^{13}_{33}\left(x\,\middle\vert\,\begin{matrix}1-a_1, 1-a_2, 1-a_3\\ 0, 1-b_1, 1-b_2\end{matrix}\right)$	$B_{29}\Gamma\begin{bmatrix}s, a_1-s, a_2-s, a_3-s\\ b_1-s, b_2-s\end{bmatrix}$ [$0<\operatorname{Re} s<\operatorname{Re} a_j$, $j=1, 2, 3$].

2	${}_3F_2\left(a_1, a_2, a_3; b_1, b_2; -\frac{1}{x}\right) = B_{29}G_{33}^{31}\left(x\,\middle\|\,\begin{matrix}1, b_1, b_2\\ a_1, a_2, a_3\end{matrix}\right)$	$B_{29}\Gamma\begin{bmatrix}s+a_1, s+a_2, s+a_3, -s\\ s+b_1, s+b_2\end{bmatrix}$ [$-\mathrm{Re}\,a_j<\mathrm{Re}\,s<0,\ j=1, 2, 3$]
3	$(1+x)^{2b+1}\,{}_3F_2(a, 2a-2, b; a-1, 2a-b-1; -x) = B_{30}G_{33}^{13}\left(x\,\middle\|\,\begin{matrix}b+2, 4+2b-2a, \frac{3}{2}+b-a\\ 0, \frac{5}{2}+b-a, 2+b-2a\end{matrix}\right)$	$B_{30}\Gamma\begin{bmatrix}s, -1-b-s, 2a-2b-3-s, a-b-\frac{1}{2}-s\\ a-b-\frac{3}{2}-s, 2a-b-1-s\end{bmatrix}$ [$0<\mathrm{Re}\,s<-\mathrm{Re}\,b-1,\ 2\mathrm{Re}\,(a-b)-3$]
4	$(1+x)^{2b+1}\,{}_3F_2\left(a, 2a-2, b; a-1, 2a-b-1; -\frac{1}{x}\right) = B_{30}G_{33}^{31}\left(x\,\middle\|\,\begin{matrix}2+2b, a+b-\frac{1}{2}, 2a+b\\ b, 2a-2, a+b+\frac{1}{2}\end{matrix}\right)$	$B_{30}\Gamma\begin{bmatrix}s+b, s+2a-2, s+a+b+\frac{1}{2}, -2b-1-s\\ s+a+b-\frac{1}{2}, s+2a+b\end{bmatrix}$ [$-\mathrm{Re}\,b,\ 2-2\,\mathrm{Re}\,a<\mathrm{Re}\,s<-2\mathrm{R}b-1$]
5	$(\sqrt{1+x}-1)^a\,{}_3F_2\left(a, b, c; 1+a-b, 1+a-c; 1+\frac{2}{x}(1-\sqrt{1+x})\right) = B_{31}G_{33}^{13}\left(x\,\middle\|\,\begin{matrix}a/2+1, (a+1)/2, b+c\\ a, b, c\end{matrix}\right)$	$B_{31}\Gamma\begin{bmatrix}s+a, -\frac{a}{2}-s, \frac{1-a}{2}-s, 1-b-c-s\\ 1-b-s, 1-c-s\end{bmatrix}$ [$-\mathrm{Re}\,a<\mathrm{Re}\,s<-\mathrm{Re}\,a/2,\ 1-\mathrm{Re}\,(b+c)$]
6	$(\sqrt{1+x}-\sqrt{x})^a\,{}_3F_2(a, b, c; 1+a-b, 1+a-c; 1+2(x-\sqrt{x^2+x})) = B_{31}G_{33}^{31}\left(x\,\middle\|\,\begin{matrix}1-\frac{a}{2}, 1+\frac{a}{2}-b, 1+\frac{a}{2}-c\\ 0, \frac{1}{2}, 1+\frac{a}{2}-b-c\end{matrix}\right)$	$B_{31}\Gamma\begin{bmatrix}s, s+\frac{1}{2}, s+1+\frac{a}{2}-b-c, \frac{a}{2}-s\\ s+1+\frac{a}{2}-b, s+1+\frac{a}{2}-c\end{bmatrix}$ [$0,\ \mathrm{Re}\,(b+c-a/2)-1<\mathrm{Re}\,s<\mathrm{Re}\,a/2$]
7	$(1-x)^{(1\pm 1)/2}(1+x)^{-2a}\,{}_3F_2\left(a, a+\frac{1}{2}, b; c, \frac{1\mp 1}{2}+2a+b-c; \frac{4x}{(1+x)^2}\right) = B_{32}G_{44}^{14}\left(x\,\middle\|\,\begin{matrix}1+\delta-2a, \frac{3}{2}-a-\delta, \delta+c-2a, 1+b-c\\ 0, \frac{3}{2}-a, 1-c, \delta+c-b-2a\end{matrix}\right)$	$B_{32}\Gamma\begin{bmatrix}s, 2a-\delta-s, a+\delta-\frac{1}{2}-s, 1+2a-c-\delta-s, c-b-s\\ a-\frac{1}{2}-s, c-s, 1+2a+b-c-\delta-s\end{bmatrix}$ [$0<\mathrm{Re}\,s<2\,\mathrm{Re}\,a-\delta,\ \mathrm{Re}\,(2a-c)+1-\delta,\ \mathrm{Re}\,(c-b)$]

8	$(8-x)^{(1\pm1)/2}(4+x)^{-3a}{}_3F_2\left(a, a+\frac{1}{3}, a+\frac{2}{3}; b, 3a-b+1\mp\frac{1}{2}; \frac{27x^2}{(4+x)^3}\right)= B_{33}G_{44}^{14}\left(x\,\middle\vert\,\begin{matrix}1+\delta-3a, \frac{3}{2}-b, \frac{5}{3}-2a-\delta, b+\delta-3a\\ 0, \frac{5}{3}-2a, 2-2b, 2\delta+2b-6a-1\end{matrix}\right)$	$B_{33}\Gamma\left[\begin{matrix}s, 3a-\delta-s, 2a+\delta-\frac{2}{3}-s, b-\frac{1}{2}-s, 3a-b-\delta+1-s\\ 2a-\frac{2}{3}-s, 2b-1-s, 6a-2b+2-2\delta-s\end{matrix}\right]$ [$0<\operatorname{Re}s<3\operatorname{Re}a-\delta, \operatorname{Re}b-1/2, \operatorname{Re}(3a-b)+1-\delta$]
9	$(1-8x)^{(1\pm1)/2}(1+4x)^{-3a}{}_3F_2\left(a, a+\frac{1}{3}, a+\frac{2}{3}; b, 3a-b+1\mp\frac{1}{2}; \frac{27x}{(1+4x)^3}\right)= \mp B_{33}G_{44}^{41}\left(x\,\middle\vert\,\begin{matrix}1+\delta-3a, \delta-a-\frac{2}{3}, 2b+\delta-3a-1, 3a-2b+2-\delta\\ 0, 2\delta-a-\frac{2}{3}, 1-b, b+\delta-3a-\frac{1}{2}\end{matrix}\right)$	$\mp B_{33}\Gamma\left[\begin{matrix}s, s+2\delta-a-\frac{2}{3}, s+1-b, s+b+\delta-3a-\frac{1}{2}, 3a-\delta-s\\ s+\delta-a-\frac{2}{3}, s+2b+\delta-3a-1, s+3a-2b+2-\delta\end{matrix}\right]$ [$0, \operatorname{Re}(3a-b)+1/2-\delta, \operatorname{Re}b-1<\operatorname{Re}s<3\operatorname{Re}a-\delta$]

8.4.51. Various functions of hypergeometric type

1	${}_pF_q(a_1, a_2, \ldots, a_p; b_1, \ldots, b_q; -x)= \Gamma\left[\begin{matrix}b_1, b_2, \ldots, b_q\\ a_1, a_2, \ldots, a_p\end{matrix}\right]\times G_{p, q+1}^{1, p}\left(x\,\middle\vert\,\begin{matrix}1-a_1, \ldots, 1-a_p\\ 0, 1-b_1, \ldots, 1-b_q\end{matrix}\right)$	$\Gamma\left[\begin{matrix}b_1, b_2, \ldots, b_q\\ a_1, \ldots, a_p\end{matrix}\right]\times\Gamma\left[\begin{matrix}s, a_1-s, \ldots, a_p-s\\ b_1-s, b_2-s, \ldots, b_q-s\end{matrix}\right]$ [[$0<\operatorname{Re}s<\min_{1\leqslant j\leqslant p}\operatorname{Re}a_j$, $b_k\neq 0, -1, -2, \ldots, k=1, 2, \ldots, q$ and either 1) $q=p-1$, $q=p$ or 2) $q=p+1$, $\operatorname{Re}s<\frac{1}{4}-\frac{1}{2}\operatorname{Re}\left(\sum_{j=1}^{p}a_j-\sum_{k=1}^{q}b_k\right)$] or [$q>p-1$; $s=-k\in D^{+}$, $s=a_j+k\in D^{-}$, $k=0, 1, 2, \ldots$, $j=1,2,\ldots,p$; for the correspondence symbol $\xleftarrow{L-\infty}$]
2	$(1-x)_{+}^{c-1}F_3\left(a, a', b, b', c; 1-x, 1-\frac{1}{x}\right)= \Gamma(c)\,G_{33}^{30}\left(x\,\middle\vert\,\begin{matrix}a'+b', c-a, c-b\\ a', b', c-a-b\end{matrix}\right)$	$\Gamma(c)\,\Gamma\left[\begin{matrix}s+a', s+b', s+c-a-b\\ s+a'+b', s+c-a, s+c-b\end{matrix}\right]$ [$\operatorname{Re}s>-\operatorname{Re}a', -\operatorname{Re}b', \operatorname{Re}(a+b-c)$; $\operatorname{Re}c>0$]

3	$(x-1)_+^{c-1}\times F_3\left(a, a', b, b', c; 1-x, 1-\frac{1}{x}\right)=\Gamma(c)\,G_{33}^{03}\left(x\middle\vert\begin{matrix}a'+b', c-a, c-b\\ a', b', c-a-b\end{matrix}\right)$	$\Gamma(c)\,\Gamma\left[\begin{matrix}1-a'-b'-s, 1+a-c-s, 1+b-c-s\\ 1-a'-s, 1-b'-s, 1+a+b-c-s\end{matrix}\right]$ $[\operatorname{Re} c>0;\ \operatorname{Re} s<1-\operatorname{Re}(a'+b'), 1+\operatorname{Re}(a-c), 1+\operatorname{Re}(b-c)]$
4	$J_\nu^\mu(x)=H_{02}^{10}\left[x\middle\vert\begin{matrix}\cdot\quad\cdot\\ (0,1),(-\nu,\mu)\end{matrix}\right]$	$\Gamma\left[\begin{matrix}s\\ 1+\nu-\mu s\end{matrix}\right]$ $[\vert\mu\vert<1;\ \operatorname{Re} s>0$ or $\mu=1;\ 0<\operatorname{Re} s<3/4+\operatorname{Re}\nu/2]$
5	$\mathbf{H}_\nu^\mu(2\sqrt{x})=H_{13}^{11}\left[x\middle\vert\begin{matrix}\left(\frac{\nu+1}{2},1\right)\\ \left(\frac{\nu+1}{2},1\right),\left(\frac{\nu}{2},1\right),(\mu(1+\nu)/2-\nu-1/2,\mu)\end{matrix}\right]$	$\Gamma\left[\begin{matrix}s+(\nu+1)/2, (1-\nu)/2-s\\ 1-\nu/2-s, 3/2+\nu-\mu(1+\nu)/2-\mu s\end{matrix}\right]$ $[\vert\mu\vert<1;\ -(\operatorname{Re}\nu+1)/2<\operatorname{Re} s<(1-\operatorname{Re}\nu)/2$ or $\mu=1;\ -(\operatorname{Re}\nu+1)/2<\operatorname{Re} s<3/4, (1-\operatorname{Re}\nu)/2]$
6	$J_{\nu,\lambda}^\mu(2\sqrt{x})=H_{13}^{11}\left[x\middle\vert\begin{matrix}\left(\frac{\nu}{2}+\lambda,1\right)\\ \left(\frac{\nu}{2}+\lambda,1\right),\left(\frac{\nu}{2},1\right),\left(\mu\left(\lambda+\frac{\nu}{2}\right)-\lambda-\nu,\mu\right)\end{matrix}\right]$	$\Gamma\left[\begin{matrix}s+\nu/2+\lambda, 1-\nu/2-\lambda-s\\ 1-\nu/2-s, 1+\lambda+\nu-\mu(\lambda+\nu/2)-\mu s\end{matrix}\right]$ $[\vert\mu\vert<1;\ -\operatorname{Re}(\nu/2+\lambda)<\operatorname{Re} s<1-\operatorname{Re}(\nu/2+\lambda)$ or $\mu=1;\ -\operatorname{Re}(\nu/2+\lambda)<\operatorname{Re} s<3/4, 1-\operatorname{Re}(\nu/2+\lambda)]$
7	$E_\rho(-x;\mu)=H_{12}^{11}\left[x\middle\vert\begin{matrix}(0,1)\\ (0,1),(1-\mu,\rho^{-1})\end{matrix}\right]$	$\Gamma\left[\begin{matrix}s, 1-s\\ \mu-\rho^{-1}s\end{matrix}\right]$ $[\rho>1/2;\ 0<\operatorname{Re} s<1$ or $\rho=1/2;\ 0<\operatorname{Re} s<1, \operatorname{Re}\mu/2]$.
8	$E(p; a_r:q; b_k:x)=G_{q+1,p}^{p,1}\left(x\middle\vert\begin{matrix}1, b_1, \ldots, b_q\\ a_1, \ldots, a_p\end{matrix}\right)$	$\Gamma\left[\begin{matrix}s+a_1, s+a_2, \ldots, s+a_p, -s\\ s+b_1, \ldots, s+b_q\end{matrix}\right]$ $\left[-\min \operatorname{Re} a_r<\operatorname{Re} s<0\right.$ and either 1) $0\leqslant q\leqslant p,\ 1\leqslant r\leqslant p$ or 2) $q=p+1$, $\left.\operatorname{Re} s>-\frac{1}{4}+\frac{1}{2}\operatorname{Re}\left(\sum_{r=1}^{p}a_r-\sum_{k=1}^{q}b_k\right)\right]$
9	$G_{pq}^{mn}\left(x\middle\vert\begin{matrix}a_1, \ldots, a_p\\ b_1, \ldots, b_q\end{matrix}\right)=H_{pq}^{mn}\left[x\middle\vert\begin{matrix}(a_1,1), \ldots, (a_p,1)\\ (b_1,1), \ldots, (b_q,1)\end{matrix}\right]$ $[0\leqslant m\leqslant q,\ 0\leqslant n\leqslant p]$	$\Gamma\left[\begin{matrix}s+b_1, \ldots, s+b_m, 1-a_1-s, \ldots, 1-a_n-s\\ s+a_{n+1}, \ldots, s+a_p, 1-b_{m+1}-s, \ldots, 1-b_q-s\end{matrix}\right]$ $\left[-\min_{1\leqslant k\leqslant m}\operatorname{Re} b_k<\operatorname{Re} s<1-\max_{1\leqslant j\leqslant n}\operatorname{Re} a_j\right.$ and either 1) $2(m+n)>p+q$, or 2) $2(m+n)=p+q,\ \vert q-p\vert\geqslant 2$, $(q-p)\operatorname{Re} s<\frac{q-p+1}{2}+\operatorname{Re}\left(\sum_{j=1}^{p}a_j-\sum_{k=1}^{q}b_k\right)$, or 3) $m+n=p,\ q=p\geqslant 1,\ \operatorname{Re}\sum_{j=1}^{p}(a_j-b_j)>$

		$>0 \Big]$ or $[2(m+n) < p+q; \ s=-b_k-l\in D+,$ $k=1, 2, ..., m; \ s=1-a_j+l\in D-, \ j=1, 2, ...$ $..., n; \ l=0, 1, 2, ...;$ for the correspondence symbols $\overset{L-\infty}{\longleftarrow}$ when $q>p, \ x>0$ or $q=p,$ $0<x<1)$ and $\overset{L+\infty}{\longleftarrow}$ (when $q<p, \ x>0$ or $q=p, \ x>1)]$
10	$\Sigma_A(x)$ $\qquad\qquad [\Delta>0],$ $\Sigma_A(x) H(1-x) + \Sigma_B(1/x) H(x-1)$ $\qquad\qquad\qquad\qquad [\Delta=0],$ $\Sigma_B(1/x) \qquad\qquad [\Delta<0]$ $\qquad\qquad\qquad [\Delta=A+D-B-C]$ $=G^{A,\,B}_{B+C,\,A+D}\left(x \left\vert \begin{array}{l} 1-b_1, ..., 1-b_B, \\ a_1, ..., a_A, c_1, ..., c_C \\ \quad 1-d_1, ... \\ \qquad ..., 1-d_D \end{array}\right.\right)$ $\qquad\qquad\qquad [A, B, C, D = 0, 1, 2, ...]$	$\Gamma\left[\begin{array}{l} s+(a), (b)-s \\ s+(c), (d)-s \end{array}\right] =$ $=\Gamma\left[\begin{array}{l} s+a_1, ..., s+a_A, b_1-s, ... \\ s+c_1, ..., s+c_C, d_1-s, ... \\ \qquad\qquad\qquad ..., b_B-s \\ \qquad\qquad\qquad ..., d_D-s \end{array}\right]$ $[\underset{1\leqslant j\leqslant A}{-\min}\ \mathrm{Re}\,a_j < \mathrm{Re}\,s < \underset{1\leqslant k\leqslant B}{\min}\ \mathrm{Re}\,b_k$ and either 1) $A+B>C+D,$ or 2) $A+B=C+$ $+D, \vert\Delta\vert\geqslant 2, \Delta\,\mathrm{Re}\,s < 1/2-\mathrm{Re}\,\nu,$ $\nu=\sum_{j=1}^A a_j+\sum_{k=1}^B b_k-\sum_{j=1}^C c_j-\sum_{k=1}^D d_k.$ or 3) $A=C, B=D, \mathrm{Re}\,\nu<0]$ or $[A+B<$ $<C+D; \ s=-a_j-l\in D+, \ j=1, 2, ..., A; \ s=$ $=b_k+l\in D-, \ k=1, 2, ..., B; \ l=0, 1, 2, ...; -$ for the correspondence symbols $\overset{L-\infty}{\longleftarrow}$ (when $\Delta>0,$ $x>0$ or $\Delta=0, 0<x<1)$ and $\overset{L+\infty}{\longleftarrow}$ (when $\Delta<0, x>0$ or $\Delta=0, x>1)]$
11	$H^{mn}_{pq}\left[x \left\vert \begin{array}{l} (a_1, A_1), ..., (a_p, A_p) \\ (b_1, B_1), ..., (b_q, B_q) \end{array}\right.\right]$ $\qquad\qquad [0\leqslant m\leqslant p, \ 0\leqslant n\leqslant q]$	$\Gamma\left[\begin{array}{l} B_1 s+b_1, ..., B_m s+b_m, \\ A_{n+1}s+a_{n+1}, ..., A_p s+a_p, \\ 1-a_1-A_1 s, ..., 1-a_n-A_n s, \\ 1-b_{m+1}-B_{m+1}s, ..., 1-b_q-B_q s \end{array}\right]$ $\left[\begin{array}{l} A_j>0, j=1, 2, ..., p; \ B_k>0, \ k=1, 2, ... \\ ..., q; \ \underset{1\leqslant k\leqslant m}{-\min}\ \dfrac{\mathrm{Re}\,b_k}{B_k} < \mathrm{Re}\,s < \underset{1\leqslant j\leqslant n}{\min}\ \dfrac{1-\mathrm{Re}\,a_j}{A_j} \\ \text{and either 1) } a^*>0, \text{ or 2) } a^*=0, \Delta\,\mathrm{Re}\,s< \\ <\dfrac{q-p}{2}-1+\mathrm{Re}\left(\sum_{j=1}^p a_j-\sum_{k=1}^q b_k\right) \end{array}\right]$ or $\left[\begin{array}{l} a^*<0; \ s=-\dfrac{b_k+l}{B_k}\in D+, \ k=1, 2, ..., m; \ s= \\ =\dfrac{1-a_j+l}{A_j}\in D-, \ j=1, 2, ..., n; \ l=0, 1, 2...; - \\ \text{for the correspondence symbols } \overset{L-\infty}{\longleftarrow} \ (\text{when} \\ \Delta>0, x>0 \text{ or } \Delta=0, 0<x<\beta) \text{ and } \overset{L+\infty}{\longleftarrow} \\ (\text{ when } \Delta<0, x>0 \text{ or } \Delta=0, x>\beta) \end{array}\right].$ Here $a^*=\sum_{j=1}^n A_j-\sum_{j=n+1}^p A_j+\sum_{k=1}^m B_k-\sum_{k=m+1}^q B_k,$ $\Delta=\sum_{k=1}^q B_k-\sum_{j=1}^p A_j, \ \beta=\prod_{j=1}^p A_j^{-A_j}\prod_{k=1}^q B_k^{B_k}.$

8.4.52. Index of particular cases of the Meijer *G*-function and the Fox *H*-function.

Below we give the numbers of the subsections and formulae of section 8.4 containing particular cases of the Meijer *G*-function $G_{pq}^{mn}(z)$ and the Fox *H*-function $H_{pq}^{mn}(z)$ corresponding to certain values of m, n, p, q. Since $G_{pq}^{mn}(z)$ is transformed into $G_{qp}^{nm}(z)$ on replacing z by $1/z$, only numbers of formulae will be indicated for which $p \leqslant q$ $(m \geqslant n$ for $p=q)$. The corresponding formulae with $p>q\,(m<n, p=q)$ can be found in the adjacent rows. In the "type" column we note the position and signs of s in the gamma functions occurring on the right hand sides of the formulae. The values of m, n, p, q and the type are interrelated: m and n denote the number of plus and minus signs for s in the numerator, while $q-m$ and $p-n$ denote the number of minus and plus signs for s in the denominator. The most general relations of a given type are indicated by boldface numerals.

Particular cases of the Fox *H*-function are listed at the end.

Type	*m n p q*	Formula numbers
$(+)$	1 0 0 1	3.**1**, 3
$\begin{pmatrix} + \\ + \end{pmatrix}$	1 0 1 1	2.1—2, **3—4**
$(+-)$	1 1 1 1	2.**5**, 11
$\begin{pmatrix} + \\ - \end{pmatrix}$	1 0 0 2	5.1—3, 11—12; 19.**1**; **47.1**
$(++)$	2 0 0 2	23.**1**; 29.1
$\begin{pmatrix} + & \\ + & - \end{pmatrix}$	1 0 1 2	3.5
$\begin{pmatrix} + & - \\ & - \end{pmatrix}$	1 1 1 2	3.3; 4.7; 14.1, 5; 16.1, 11; 18.7; 43.**1**; 45.1, 5
$\begin{pmatrix} + & + \\ + & \end{pmatrix}$	2 0 1 2	11.1; 14.2; 16.2; 18.1; 23.3; 29.3; 44.**1**; 46.**7**
$(++-)$	2 1 1 2	11.3; 14.7; 16.13; 18.3; 23.5; 29.5; 44.**3**; 46.**1**
$\begin{pmatrix} + & - \\ + & - \end{pmatrix}$	1 1 2 2	2.6—8, 10, 21—23; 6.18; 8.5; 40.26; 41.15—16, 18—19, 35, 38, 40—41, 43; 42.12—13, 23, 30, 32; 49.**19**, 28—32
$\begin{pmatrix} + & + \\ + & + \end{pmatrix}$	2 0 2 2	2.17—20; 6.1—2, 22; 7.1—2, 5—6; 8.3—4; 10.5—8; 30.1—4, 6—13, 16—17; 31.1—4, 6—13, 16—17; 32.1—4, 6—13, 18—21; 35.1—4, 8—9, 11—20, 23—24, 27—28; 36.1—4, 8—9; 40.3—4, 7—8, 12—17, 27—28, 39—40, 43—44, 47—48; 41.1—4, 7—8, 20—23, 26—27, 32—33, 44—45; 42.4—6, 21—22, 28—29; 49.1—12, **22—25**
$\begin{pmatrix} + & + & - \\ + & & \end{pmatrix}$	2 1 2 2	2.12—13; 6.5; 7.3—4, 7—8; 8.1—2; 30.5, 14—15, 18—19; 31.5, 14—15; 32.5, 14—15, 16—17, 22—23; 35. 5—7, 21—22, 25—26; 36.5—7; 40.1—2, 5—6, 10—11, 18—25, 29—32, 35—38, 41—42, 45—46, 49—50; 41.9—14, 28—31, 36—37; 42.1—2, 8—11, 15—16, 24—27; 49.**13**—16, 26—27, 35—44
$(++--)$	2 2 2 2	2.9; 6.11; 10.3—4; 40.9; 41.5—6, 17, 24—25, 34, 39, 42; 42.7, 31; 49.**20—21**
$\begin{pmatrix} + & & \\ + & - & - \end{pmatrix}$	1 0 1 3	4.1, 3—4; 22.1
$\begin{pmatrix} + & - \\ - & - \end{pmatrix}$	1 1 1 3	12.1; 13.3; 15.1—3, 7; 19.9, 11, 13; 25.1; 26.7; 27.1; 48.**1**
$\begin{pmatrix} + & + \\ + & - \end{pmatrix}$	2 0 1 3	5.5, 7, 9; 12.3—5; 13.5; 15.5; 16.5, 7; 19.19; 20.1, 9, 19; 24.1
$\begin{pmatrix} + & + & - \\ & & - \end{pmatrix}$	2 1 1 3	22.23; 23.19, 23; 25.7; 26.9; 29.11, 25

Type	m n p q	Formula numbers
(+++ / +)	3 0 1 3	23.27, 29; 24.5; 29.15
(+++−)	3 1 1 3	12.9; 15.9; 16.17; 20.23, 35, 39; 21.1; 25.3; 27.3
(+− / +−−)	1 1 2 3	3.7; 16.9; 18.9; 22.5
(++− / +−)	2 1 2 3	11.4; 18.5; 29.21
(++− / ++−)	2 1 3 3	6.8
(++−− / +−)	2 2 3 3	6.13; 41.54; 42.19—20, 33—34, 51
(+++ / +++)	3 0 3 3	51.2—3
(+++− / ++)	3 1 3 3	6.23—26; 40.33—34; 41.46—53; 42.35—50; 49.17—18, 33—34, 45—50, 55—56; 50.1—2, 3—6
(+ / −−−)	1 0 0 4	5.20; 22.11; 28.3, 5
(++ / −−)	2 0 0 4	23.15
(+− / +−−−)	1 1 2 4	13.1; 22.15, 17, 19; 25.5
(++ / ++−−)	2 0 2 4	29.19
(++− / +−−)	2 1 2 4	12.7; 19.7, 17, 21; 20.11, 13, 15; 45.3
(+−− / −−−)	1 2 2 4	19.3, 15; 43.3; 47.3
(++−− / −−)	2 2 2 4	22.9, 25; 23.7, 21; 29.7, 13, 23
(+++ / −)	3 0 0 4	5.19; 23.9, 11, 13; 28.13, 15; 29.9
(+++ / +−)	3 0 1 4	5.17
(+++ / ++−)	3 0 2 4	20.5, 21, 37, 40; 24.3; 46.9
(+++− / +−)	3 1 2 4	11.7; 14.11; 23.25; 44.5; 46.11
(+++−− / −)	3 2 2 4	16.15; 20.7, 25, 42; 26.5; 46.3
(++++)	4 0 0 4	23.33; 28.17
(++++ / ++)	4 0 2 4	18.13; 23.31; 29.17; 44.7; 46.13
(++++− / +)	4 1 2 4	12.11; 15.11; 16.19; 18.11; 44.9; 46.5
(++++− / +++)	4 1 4 4	41.55—56; 49.51—54; 50.7—9
(+ / +−−−−)	1 0 1 5	28.7,9
(++ / +−−−)	2 0 1 5	5.22; 22.13; 28.1
(+−− / +−−−−)	1 2 3 5	14.9; 22.7, 21
(++−− / +−−−)	2 2 3 5	19.5; 20.3, 17; 26.1, 3

Type	$m\ n\ p\ q$	Formula numbers
$\begin{pmatrix} ++++ \\ +- \end{pmatrix}$	4 0 1 5	23.17; 28.11
$\begin{pmatrix} ++++ \\ +++- \end{pmatrix}$	4 0 3 5	20.44
$\begin{pmatrix} +++++\pm \\ \quad\quad\quad\quad - \end{pmatrix}$	5 0 0 6	5.18
	$m\ n\ q\ q$	6.3, 4, 9—10; 9.1; 10.1
	$m\ n\ p\ q$	51.1, 8—10
H-function		2.14—16; 5.15—16; 6.17; 51.4—7, 9—11

Since the Meijer *G*-function is a particular case of the Fox *H*-function, each of the formulae of Table 8.4 gives corresponding particular values of the *H*-function.

Appendix I. **SOME PROPERTIES OF INTEGRALS, SERIES, PRODUCTS AND OPERATIONS WITH THEM**

I.1 INTRODUCTION

This appendix contains information on the convergence of improper integrals, series and products, as well as formulae of various operations with integrals and series which may prove useful for calculations with them. As is well known, there are no universal convergence tests providing exhaustive information in their application to any improper integral or series. All convergence tests have well-defined spheres of application.

To describe the behaviour of a function $\varphi(x)$ in terms of a known function $\psi(x)$ we shall use the following symbols defining the relation of order of magnitude.

Let the functions φ and ψ of the variable $x \in B$ be defined on some set B and let a be a (finite or infinite) limit point of B.

The formulae

$$\varphi(x) \sim \psi(x) \qquad (x \longrightarrow a,\ x \in B),$$
$$\varphi(x) = o(\psi(x)) \qquad (x \longrightarrow a,\ x \in B),$$
$$\varphi(x) = O(\psi(x)) \qquad (x \longrightarrow a,\ x \in B)$$

denote the following respectively:

1) there exists $\lim\limits_{x \to a} \dfrac{\varphi(x)}{\psi(x)} = 1,\ x \in B;$

2) there exists $\lim\limits_{x \to a} \dfrac{\varphi(x)}{\psi(x)} = 0,\ x \in B;$

3) there exists a constant $M > 0$, not depending on $x \in B$, such that $|\varphi(x)| < M|\psi(x)|$ as $x \to a, x \in B$. For example, $\sin^2 x \sim x^2 (x \to 0)$; $\dfrac{1}{x^2} = o\left(\dfrac{1}{x}\right) (x \longrightarrow \infty)$; $\cos x = O(1)\ (x \longrightarrow \infty)$.

Given a sequence of real numbers $x_k, k = 1, 2, \ldots$, the set of all its (finite or infinite) limit points has a greatest and least element. The greatest element is called the upper limit of the sequence and is denoted by $\overline{\lim\limits_{k \to \infty}}\, x_k$, ; the least element is called the lower limit and is denoted by $\underline{\lim\limits_{k \to \infty}}\, x_k$. For example, if $x_k = (-1)^k$, then $\underline{\lim\limits_{k \to \infty}}\, x_k = -1$, $\overline{\lim\limits_{k \to \infty}}\, x_k = 1$.

See, for example, [9, 16, 19, 24, 26] for proofs and examples.

I.2. CONVERGENCE OF INTEGRALS AND OPERATIONS WITH THEM

I.2.1. Integrals along unbounded curves.

A number of mathematical and applied problems lead to integrals along unbounded curves when the contour of integration has unbounded length.

Suppose that a complex-valued function $f(z)$, $z \in C$, is given where C is a simple piecewise-smooth oriented unbounded curve; by a curve is meant a set of points in the extended complex plane that can be represented as the image of an interval $a \leqslant t \leqslant b$ of the real axis under an appropriate map $z = z(t)$ [9].

We consider to begin with the case when curve C has one of its end points at the point a and the other end point goes off to infinity. We denote by C_l the segment of the curve C of length l measured from the point a. We suppose that $f(z)$ has no singular points on C, that is, points z where $f(z)$ becomes infinite, and that it is integrable over every C_l. If the limit

$$\lim_{l \to \infty} \int_{C_l} f(z)\, dz = \int_C f(z)\, dz$$

exists and is finite, then it is called the value of the improper integral of the function $f(z)$ along the unbounded contour C.

In this case one says that the integral converges; if the integral does not exist or is infinite, the integral is said to diverge.

Note that if the length l of each finite segment C_l of the curve C lying inside a disc of radius R with centre at a has the estimate $l = O(R)$ as $R \to \infty$, and the function $f(z)$ is integrable along C_l and $|f(z)| \leqslant B|z|^{-1-\delta}, \delta > 0$, for large $|z|$ (where B is a constant), then the integral $\int_C f(z)\, dz$ converges.

If both ends of the curve C go off to infinity, then the improper integral along C is defined by the formula

$$\int_C f(z)\, dz = \int_{C_a^+} f(z)\, dz - \int_{C_a^-} f(z)\, dz,$$

where C_a^+ and C_a^- are the two portions into which the curve C is divided by an arbitrary point a, the end point of the curve C^+ goes off to infinity in the direction of the orientation of the curve C, while the end point of the curve C^- goes off to infinity in the direction opposite to the orientation of the curve C.

In particular, if the curve C is the Ox axis and the function $f(x)$ is integrable over any interval (a, b), $-\infty < a < b < \infty$, then

$$\int_{-\infty}^{\infty} f(x)\, dx = \lim_{\substack{a \to -\infty \\ b \to +\infty}} \int_a^b f(x)\, dx.$$

If this limit does not exist or is infinite, but the limit

$$\lim_{a \to \infty} \int_{-a}^{a} f(x)\, dx = \int_{-\infty}^{\infty} f(x)\, dx$$

exists and is finite, then it is called the principal value of the improper integral $\int_{-\infty}^{\infty} f(x)\, dx$ and is denoted by the same symbol.

I.2.2. Criteria for the convergence of integrals with infinite limits of non-negative functions.

We consider the integral

$$\int_a^{\infty} f(x)\, dx, \quad f(x) \geqslant 0, \quad a > -\infty, \tag{1}$$

where $f(x)$ is integrable over any interval $(a, b), a < b < \infty$. If $a = -\infty$ in the integral (1), then this integral can be split up into the sum of two integrals between the limits $-\infty$ to 0 and 0 to ∞, and the convergence of each of them is considered separately.

We give some criteria for the convergence of the integral (1).

1. If

$$f(x)=O(x^{-1-\delta}),\quad \delta>0,\quad x\longrightarrow\infty,$$

then the integral (1) converges. If

$$f(x)\sim\frac{c}{x},\quad c>0,\quad x\longrightarrow\infty,$$

then the integral (1) diverges. If

$$f(x)=O(x^{-1}\ln^{-1-\lambda}x),\quad \lambda>0,\quad x\longrightarrow\infty,$$

then the integral (1) converges.

2. The Cauchy criterion. Let

$$f(x)\sim\frac{\varphi(x)}{x^{\lambda}},\quad x\longrightarrow\infty.$$

If $\lambda>1$ and $\varphi(x)\leqslant c<\infty$, then the integral (1) converges. If $\lambda\leqslant 1$ and $\varphi(x)\geqslant c>0$, then the integral (1) diverges.

3. If for any b, $a<b<\infty$,

$$\int_a^b f(x)\,dx\leqslant M<\infty,$$

where M is a constant, then the integral (1) converges.

4. If $f(x)\leqslant g(x)$, then the convergence of $\int_a^\infty g(x)\,dx$ implies that of the integral (1), while the divergence of the integral (1) implies that of the integral $\int_a^\infty g(x)\,dx$.

5. Suppose that the following limit exists:

$$\lim_{x\to\infty}\frac{f(x)}{g(x)}=c,\qquad 0\leqslant c\leqslant\infty.$$

If $c<\infty$ and the integral $\int_a^\infty g(x)\,dx$ converges, then so does the integral (1). If $c>0$ and the integral $\int_a^\infty g(x)\,dx$ diverges, then so does the integral (1).

6. Let $f(x)$ be a differentiable function. If $\overline{\lim_{x\to\infty}}\frac{f'(x)}{f(x)}<0$, then the integral (1) converges. If $\underline{\lim}_{x\to\infty}\frac{f'(x)}{f(x)}>0$, then the integral (1) diverges.

7. Let $\alpha>0$ and let $\varphi(x)$ be a differentiable function such that $\lim\limits_{x\to\infty}\varphi(x)=\infty$. If $0<\alpha<1$ and

$$\varliminf_{x\to\infty}\frac{\varphi'(x)\,\alpha^{\varphi(x)}}{f(x)}>0,$$

then the integral (1) converges. If $\alpha>1$ and

$$\varlimsup_{x\to\infty}\frac{\varphi'(x)\,\alpha^{\varphi(x)}}{f(x)}<\infty,$$

then the integral (1) diverges.

8. Let $\varphi(x)$ be a differentiable increasing function on $[a,\infty)$ such that $a\leqslant x<\varphi(x)$. Then the integral (1) converges if

$$\varlimsup_{x\to\infty}\frac{\varphi'(x)\,f[\varphi(x)]}{f(x)}<1,$$

and diverges if

$$\varliminf_{x\to\infty}\frac{\varphi'(x)\,f[\varphi(x)]}{f(x)}>1.$$

9. If $f(x)$ is differentiable and there exists a differentiable function $g(x)<0$ such that

$$\varliminf_{x\to\infty}\frac{(f(x)\,g(x))'}{f(x)}>0,$$

then the integral (1) converges. If

$$\varlimsup_{x\to\infty}\frac{(f(x)\,g(x))'}{f(x)}<0$$

and the integral $\int\limits_a^\infty\frac{1}{g(x)}\,dx$ diverges then so does the integral (1).

10. Suppose that $f(x)$ and $\varphi(x)$ are differentiable functions and

$$\frac{f'(x)}{f(x)}=\varphi(x)\,(1+o(1)),\quad\left(\frac{1}{\varphi(x)}\right)'=\beta+o(1),\quad x\longrightarrow\infty;$$

where β is a constant, $\varphi(x)<0$ and $x>a$. Then the integral (1) converges for $\beta>-1$ and diverges for $\beta<-1$.

11. Suppose that $f(x)$ is a differentiable function and

$$\frac{f'(x)}{f(x)}=cx^{\alpha}\,(1+o(1)),\qquad x\longrightarrow\infty.$$

Then the integral (1) converges if $c<0$, $\alpha>-1$ or $c<-1, \alpha=-1$. The integral (1) diverges if $c>0$ and α is arbitrary, or $\alpha<-1$ and c is arbitrary, or $c>-1$ and $\alpha=-1$.

12. Suppose that $f(x)$ and $\varphi(x)$ are differentiable functions and

$$\frac{f'(x)}{f(x)}=-\frac{1}{x}+\frac{\varphi(x)}{x}\,(1+o(1)),\quad\left(\frac{1}{\varphi(x)}\right)'=\frac{1}{x}\,(\alpha+o(1)),\quad x\longrightarrow\infty,$$

where α is a constant, $\varphi(x)<0$ and $x>a$. The integral (1) converges for $\alpha>-1$ and diverges for $\alpha<-1$.

13. Let $f(x)$ and $\varphi(x)$ be differentiable functions and

$$\frac{f'(x)}{f(x)}=-\sum_{k=0}^{n}\frac{1}{\lambda_k(x)}+\frac{\varphi(x)}{\lambda_n(x)}(1+o(1)), \qquad x\longrightarrow\infty,$$
$$\left(\frac{1}{\varphi(x)}\right)'=\frac{\alpha+o(1)}{\lambda_n(x)}, \qquad x\longrightarrow\infty,$$
$$\lambda_0(x)=x,\quad \lambda_k(x)=x\underbrace{\ln\ln\ldots\ln}_{k\,\text{times}} x,\quad \varphi(x)<0,\ x>a.$$

Then the integral (1) converges for $\alpha>-1$ and diverges for $\alpha<-1$.

14. Let $f(x)>0$ be a differentiable function and suppose that for some $\alpha>0$,

$$\lim_{x\to\infty}[f^{\alpha-2}(x)f'(x)]=-\beta\neq 0.$$

Then the integral (1) converges for $\beta>0$ and diverges for $\beta<0$.

I.2.3. Convergence tests for integrals with infinite limits of arbitrary functions.

1. If the integral $\int_a^\infty |f(x)|\,dx$, converges, then so does the integral

$$\int_a^\infty f(x)\,dx, \tag{2}$$

which in this case is said to be absolutely convergent.

If the integral (2) converges but the integral $\int_a^\infty |f(x)|\,dx$ diverges, then the integral (2) is said to be conditionally convergent.

See I.2.2. with regard to convergence of the integrals $\int_a^\infty |f(x)|\,dx$.

2. Let

$$f(x)=\frac{\varphi(x)}{x^\lambda}[1+O(x^{-\varepsilon})], \qquad x\longrightarrow\infty, \qquad \varepsilon>0.$$

If $\operatorname{Re}\lambda>1$ and $|\varphi(x)|\leqslant c<\infty$, then the integral (2) converges absolutely.

3. The Abel test. If the integral (2) converges and the function $g(x)$ is monotone and bounded on $[a,\infty)$, then the integral

$$\int_a^\infty f(x)\,g(x)\,dx \tag{3}$$

converges.

4. The Dirichlet test. If on each finite integral (a,b), $a<b<\infty$,

$$\left|\int_a^b f(x)\,dx\right|\leqslant M,$$

where M does not depend on b, and $g(x)$ converges monotonically to zero as $x\to\infty$, then the integral (3) converges.

5. Suppose that $\gamma,\lambda>0$ and

$$\begin{aligned} f(x)&=x^\alpha\cos bx^\gamma[1+O(x^{-\delta_1})],\\ g(x)&=\cos cx^\lambda[1+O(x^{-\delta_2})],\end{aligned} \qquad x\to\infty,$$

where $\delta_1,\delta_2>\operatorname{Re}\alpha+1$. Then the integral (3) converges for $\operatorname{Re}\alpha<\max(\gamma,\lambda)-1$ (if $\gamma\neq\lambda$ or $b\neq c$) or for $\operatorname{Re}\alpha<-1$ (if $\gamma=\lambda$ and $b=c$).

I.2.4. Integrals of unbounded functions along bounded curves.

The definition of improper integral for the case when the integrand becomes infinite at several points of the contour need only be given for the case of single singularity, since an integral with several singularities can then be split up into a sum of finitely many integrals with a single singularity.

Let $z=a$ be the only singular point of the function $f(z)$, $z\in C$, where C is a bounded simple smooth curve. If the limit

$$\int_C f(z)\,dz=\lim_{\varepsilon_1,\,\varepsilon_2\to 0}\left[\int_{C'} f(z)\,dz+\int_{C''} f(z)\,dz\right]$$

exists and is finite, where C', C'' are the portions of the curve C remaining after removing a small neighbourhood of the point a, and ε_1, ε_2 are the lengths of the removed portions of C measured from the point a, then it is called the improper integral of $f(z)$ along the contour C (with singularity at the point a).

If the limit does not exist or is infinite as ε_1, ε_2 tend to zero independently of each other (that is, the integral diverges), but a finite limit exists as ε_1, $\varepsilon_2\to 0$ subject to the extra condition that the removed neighbourhood of the point a is inside the disc of radius ε with centre at the point a, then this limit is called the principal value of the improper integral and is denoted by the same symbol.

Suppose, in particular, that C is a segment $[a,b]$ of the Ox axis and that the function $f(x)$ defined on the interval $[a,b)$ is integrable over any interval $[a,b-\varepsilon]$, $\varepsilon>0$, but is unbounded as $x\to b-0$. If the limit

$$\lim_{\varepsilon\to 0}\int_a^{b-\varepsilon} f(x)\,dx=\int_a^b f(x)\,dx$$

exists and is finite, then it is called the improper integral of the function $f(x)$ over $[a,b]$. In this case the integral is said to converge. If the limit does not exist, the integral is called divergent.

The convergence of an improper integral for the case when the function $f(x)$ is unbounded as $x\to a+0$ is defined similarly.

Suppose that the function $f(x)$ has one singular point c in the interval $[a,b]$, $a<c<b$, at which it is unbounded. Then a convergent improper integral from a to b is defined by the formula

$$\int_a^b f(x)\,dx=\lim_{\substack{\varepsilon_1\to 0\\ \varepsilon_2\to 0}}\left[\int_0^{c-\varepsilon_1} f(x)\,dx+\int_{c+\varepsilon_2}^{b} f(x)\,dx\right],$$

where $\varepsilon_1,\varepsilon_2>0$. If this limit does not exist or is infinite but the limit exists as $\varepsilon_1=\varepsilon_2\to 0$, then it is called the principal value of the improper integral $\int_a^b f(x)\,dx$ and is denoted by the same symbol.

I.2.5. Convergence tests for integrals of non-negative unbounded functions.

Suppose that the non-negative function $f(x)$ is integrable over any interval $[a,c]$, $a<c<b$ and $f(x)\to\infty$ as $x\to b-0$. We consider the improper integral

$$\int_a^b f(x)\,dx,\qquad f(x)\geqslant 0. \tag{1}$$

1. If for some $\delta>0$

$$f(x)=O\left((b-x)^{-1+\delta}\right), \qquad x\to b-0,$$

then the integral (4) converges. If

$$f(x)\sim\frac{M}{b-x}, \qquad x\to b-0,$$

where $M>0$ is a constant, then the integral (4) diverges.

If

$$f(x)=O\left[(b-x)^{-1}\ln^{-1-\lambda}(b-x)\right], \qquad x\to b-0,\quad \lambda>0,$$

then the integral (4) diverges.

2. Let $\varphi(x)>0$. If

$$\varliminf_{x\to b-0}\frac{\varphi(x)}{f(x)}>0$$

and the integral $\int\limits_a^b \varphi(x)\,dx$ converges, then so does the integral (4). If

$$\varlimsup_{x\to b-0}\frac{\varphi(x)}{f(x)}<\infty$$

and the integral $\int\limits_a^b \varphi(x)\,dx$ diverges, then so does (4).

3. If $f(x)$ is differentiable and there exists a differentiable function $g(x)<0$ such that

$$\varliminf_{x\to b-0}\frac{(f(x)\,g(x))'}{f(x)}>0,$$

then the integral (4) converges. If

$$\varlimsup_{x\to b-0}\frac{(f(x)\,g(x))'}{f(x)}<0,$$

and the integral $\int_a^b \frac{dx}{g(x)}$ diverges, then so does the integral (4).

4. If $f(x)$ is differentiable and

$$\varliminf_{x\to b-0}\frac{(x-b)\,f'(x)}{f(x)}>-1,$$

then the integral (4) converges. If

$$\varlimsup_{x\to b-0}\frac{(x-b)\,f'(x)}{f(x)}<-1,$$

then the integral (4) diverges.

5. If $f(x)$ and $\varphi(x)$ are differentiable and

$$\frac{f'(x)}{f(x)}=\varphi(x)\,(1+o(1)), \qquad \left(\frac{1}{\varphi(x)}\right)'=\beta+o(1), \quad x\to b-0,$$

where β is a constant, $\varphi(x)>0$, $a\leqslant x<b$, then the integral (4) converges for $\beta<-1$ and diverges for $\beta>-1$.

6. If $f(x)$ is differentiable and

$$\frac{f'(x)}{f(x)}=c\,(b-x)^{\alpha}\,(1+o(1)), \qquad x\to b-0,$$

where c is a constant, then the integral (4) converges for $\alpha=-1$, $c<1$ or for $-1<\alpha<0$, $c>0$, and diverges for $\alpha=-1$, $c>1$ or for $\alpha<-1$, $c>0$.

7. If $f(x)$ and $\varphi(x)$ are differentiable and

$$\frac{f'(x)}{f(x)}=\frac{1}{b-x}+\frac{\varphi(x)}{b-x}(1+o(1)), \qquad x\to b-0,$$

$$\left(\frac{1}{\varphi(x)}\right)'=\frac{\beta+o(1)}{b-x}, \qquad x\to b-0,$$

where β is a constant and $\varphi(x)$ does not change sign over $[a,b)$, then the integral (4) converges for $\beta<-1$, $\varphi(x)>0$ or for $\beta>-1$, $\varphi(x)<0$ and diverges for $\beta>-1$, $\varphi(x)>0$ or for $\beta<-1$, $\varphi(x)<0$.

I.2.6. Uniform convergence of functions and integrals depending on a parameter.

1. Suppose that the function $f(t,x)$ is defined for $t\in D$, $x\in\Delta$, and t_0 is a limit point of D. If for any $x\in\Delta$ the limit $\lim_{t\to t_0} f(t,x)=\varphi(x)$ exists and for any $\varepsilon>0$ there exists a number $\delta_\varepsilon>0$ not depending on x such that whenever $|t-t_0|<\delta_\varepsilon$, the inequality $|f(t,x)-\varphi(x)|<\varepsilon$ holds for all $x\in\Delta$, then $f(t,x)$ is said to converge to $\varphi(x)$ uniformly with respect to x in Δ as $t\to t_0$.

2. Suppose that the function $f(t,x)$ is defined for all $x\geqslant a$, and all t in some region D and that for every $t\in D$ the following integral exists:

$$\int_a^\infty f(t,x)\,dx. \tag{5}$$

If for any $\varepsilon>0$ there exists a number $a_\varepsilon\geqslant a$ not dependent on t such that for all $c>a_\varepsilon$, the inequality $\left|\int_c^\infty f(t,x)\,dx\right|<\varepsilon$ holds for all values of $t\in D$, then the integral is said to be uniformly convergent with respect to t in D.

3. The Cauchy criterion. The integral (5) converges uniformly in D if and only if for any $\varepsilon>0$ there exists a number $a_\varepsilon\geqslant a$ not depending on t, such that the inequality $\left|\int_{a'}^{a''} f(t,x)\,dx\right|<\varepsilon$ holds for all $t\in D$ provided only that $a''>a'\geqslant a_\varepsilon$.

4. Suppose that the function $f(t,x)$ is defined for all $x\in[a,b)$, $-\infty<a<b<\infty$ and for all t in some region D; suppose further that the following (proper or improper) integral exists for each $t\in D$:

$$\int_a^b f(t,x)\,dx. \tag{6}$$

If for any $\varepsilon>0$ there exists a number $\delta_\varepsilon>0$ not depending on t such that for all $0<\eta<\delta_\varepsilon$, the inequality $\left|\int_{b-\eta}^b f(t,x)\,dx\right|<\varepsilon$ holds for all values of $t\in D$, then the integral (6) is said to be uniformly convergent with respect to t in the region D.

We give some tests for the uniform convergence of the integral (6).

Suppose that the function $f(t,x)$ is integrable with respect to x over any interval $[a,c]$, $c>a$, for all t in D.

5. The Weierstrass test. If there exists a non-negative function $\varphi(x)$ of x only which is integrable over $[a,\infty)$, and is such that for all $t\in D$

$$|f(t,x)|\leqslant\varphi(x)$$

for $x\geqslant a$, then the integral (6), $b=\infty$, converges uniformly with respect to t in D.

6. The Abel test. If the integral (6), with $b=\infty$, converges uniformly with respect to t in the region D, and the function $g(t,x)$ is uniformly bounded, that is, $|g(t,x)|\leqslant M$, $x\geqslant a$, $t\in D$, where M is a constant not depending on t or x, then the integral

$$\int_a^{\infty} f(t, x)\, g(t, x)\, dx \tag{7}$$

converges uniformly with respect to t in D.

7. The Dirichlet test. If the integral $\int_a^c f(t, x)\, dx$ is uniformly bounded as a function of t and c, that is, $\left|\int_a^c f(t\ x)\, dx\right|\leqslant M, c\geqslant a,\ t\in D$, where M is a constant not depending on t or c, and $g(t,x)\to 0$ as $x\to\infty$ uniformly with respect to $t\in D$ and monotonically in x, then the integral (7) converges uniformly with respect to t in D.

I.2.7. Operations with integrals depending on a parameter.

We give some properties of improper integrals depending on a parameter.

1. Suppose that the function $f(t,x)$ is integrable with respect to x (as a proper integral) over the interval $[a,b]$ for any $b>a$ and for all t in some set D having t_0 as a limit point, and $f(t,x)\to\varphi(x)$ as $t\to t_0$ uniformly with respect to x in $[a,b]$.

Then

$$\lim_{t\to t_0}\int_a^b f(t, x)\, dx=\int_a^b \lim_{t\to t_0} f(t, x)\, dx = \int_a^b \varphi(x)\, dx.$$

If, in addition, the integral (5) converges uniformly with respect to $t\in D$, then

$$\lim_{t\to t_0}\int_a^{\infty} f(t, x)\, dx=\int_a^{\infty} \lim_{t\to t_0} f(t, x)\, dx=\int_a^{\infty} \varphi(x)\, dx.$$

This last formula is also valid when the function $f(t,x)\geqslant 0$ is continuous in x over $[a,\infty)$ and is monotone increasing and converges to a continuous integrable function as $t\to t_0$.

If the function $f(t,x)$ satisfies the conditions given at the beginning of this subsection and the function $g(x)$ is absolutely integrable (possibly as an improper integral) over $[a,b]$, then

$$\lim_{t\to t_0}\int_a^b f(t, x)\, g(x)\, dx=\int_a^b \lim_{t\to t_0} f(t, x)\, g(x)\, dx.$$

2. Suppose that we are given a sequence of functions $f_k(x)$, $k=1,2,\dots$, which are integrable (as proper integrals) over the interval $[a, b]$. If for all $x\in[a,b]$, $\lim_{k\to\infty} f_k(x)=\varphi(x)$ uniformly with respect to x in $[a,b]$, then the function $\varphi(x)$ is integrable over $[a,b]$ and

$$\lim_{k\to\infty}\int_a^b f_k(x)\, dx=\int_a^b \lim_{k\to\infty} f_k(x)\, dx = \int_a^b \varphi(x)\, dx.$$

3. Suppose that for $x\geqslant a$ and $t\in[\alpha,\beta]$, the function $f(t,x)$ is continuous with respect to x and has a partial derivative $\frac{\partial}{\partial t} f(t, x)$, which is continuous or simply integrable with respect to x and uniformly bounded in its domain of definition.

Then for any $t\in[\alpha,\beta]$

$$\left(\int_a^b f(t, x)\, dx\right)'=\int_a^b \frac{\partial}{\partial t} f(t, x)\, dx, \quad a<b<\infty.$$

If, moreover, the integral (5) converges for all $t\in[\alpha,\beta]$ and the integral $\int_a^\infty \frac{\partial}{\partial t} f(t,x)\,dx$ converges uniformly with respect to t in $[\alpha,\beta]$, then for any $t\in[\alpha,\beta]$

$$\left(\int_a^\infty f(t,x)\,dx\right)'=\int_a^\infty \frac{\partial}{\partial t} f(t,x)\,dx.$$

If the function $f(t,x)$ satisfies the above conditions and the function $g(x)$ is absolutely integrable (possibly as an improper integral) over $[a,b]$, then

$$\left(\int_a^b f(t,x)\,g(x)\,dx\right)'=\int_a^b \frac{\partial}{\partial t} f(t,x)\,g(x)\,dx.$$

4. Suppose that the function $f(t,x)$ is continuous for $x\geqslant a$ and $t\in[\alpha,\beta]$. Then

$$\int_\alpha^\beta\left(\int_a^b f(t,x)\,dx\right)dt=\int_a^b\left(\int_\alpha^\beta f(t,x)\,dt\right)dx,\quad a<b<\infty.$$

If, moreover, the integral (5) converges uniformly with respect to t in $[\alpha,\beta]$, then this integral is continuous with respect to t in $[\alpha,\beta]$ and

$$\int_\alpha^\beta\left(\int_a^\infty f(t,x)\,dx\right)dt=\int_a^\infty\left(\int_\alpha^\beta f(t,x)\,dt\right)dx.$$

This last formula is also valid when the function $f(t,x)\geqslant 0$ is continuous and the integral (5) is continuous for $t\in[\alpha,\beta]$.

If the function $f(t,x)$ satisfies the above conditions and $g(x)$ is absolutely integrable (possibly as an improper integral) over $[a,b]$, then

$$\int_\alpha^\beta\left(\int_a^b g(x)\,f(t,x)\,dx\right)dt=\int_a^b g(x)\left(\int_\alpha^\beta f(t,x)\,dt\right)dx.$$

5. Suppose that $f(t,x)$ is continuous for $x\geqslant a$, $t\geqslant\alpha$ and the integrals

$$\int_a^\infty f(t,x)\,dx,\qquad \int_\alpha^\infty f(t,x)\,dt$$

converge uniformly with respect to t and x respectively, in any finite interval.

If one of the iterated integrals

$$\int_\alpha^\infty\left(\int_a^\infty |f(t,x)|\,dx\right)dt,\quad \int_a^\infty\left(\int_\alpha^\infty |f(t,x)|\,dt\right)dx$$

exists, then

$$\int_\alpha^\infty\left(\int_a^\infty f(t,x)\,dx\right)dt=\int_a^\infty\left(\int_\alpha^\infty f(t,x)\,dt\right)dx.$$

If $f(t,x)\geqslant 0$, then this formula holds if the uniform convergence hypothesis is replaced by the hypothesis that the integrals given in the condition are continuous.

6. If $f(t,x)$ is analytic in t and integrable with respect to x over $[a,\infty)$ for all t in some region D, and the integral (5) converges uniformly in D, then it is an analytic function in D.

I.3. CONVERGENCE OF SERIES AND PRODUCTS AND OPERATIONS WITH THEM

I.3.1. Basic notions.

1. Suppose that we are given a numerical or functional sequence

$$\{a_k\}_{k=1}^\infty=a_1,\,a_2,\,a_3,\,\ldots$$

If the limit

$$\lim_{n\to\infty}\sum_{k=1}^{n} a_k=\sum_{k=1}^{\infty} a_k$$

exists and is finite, then it is called the sum of the series. In this case we say that the series

$$\sum_{k=1} a_k$$

converges. If the limit does not exist or is infinite, then the series is said to be divergent; the a_k are called the terms of the series, and $\sum_{k=n+1}^{\infty} a_k$ the n-th remainder of the series.

2. The Cauchy criterion. The series $\sum_{k=1}^{\infty} a_k$ converges if and only if for any $\varepsilon>0$ there exists an integer N_ε such that

$$\left|\sum_{k=n}^{m} a_k\right|\leqslant\varepsilon,$$

if $m\geqslant n\geqslant N_\varepsilon$.

3. A necessary condition for the convergence of a series. If the series $\sum_{k=1}^{\infty} a_k$ converges, then $a_k\to 0$ as $k\to\infty$.

4. A numerical series with non-negative terms

$$\sum_{k=1}^{\infty} a_k, \qquad a_k\geqslant 0, \tag{1}$$

is called a positive series.

Suppose that we are also given the series

$$\sum_{k=1}^{\infty} b_k, \qquad b_k\geqslant 0, \tag{2}$$

and

$$\sum_{k=1}^{\infty} f(k), \qquad f(x)>0, \tag{3}$$

where $f(x)$ is monotone decreasing for $x>1$.

The connection between the convergence of improper integrals of non-negative functions $f(x)$ and that of the corresponding positive series $f(k)$ is reflected by the following result (integral comparison test).

5. The positive series (3) and the integral $\int_1^{\infty} f(x)\,dx$ converge or diverge simultaneously.

I.3.2. Convergence tests for positive series.

1. If $a_k=O(k^{-1-\delta})$ for $\delta>0$, $k\to\infty$, then the series (1) converges.

If $a_k\sim ck^{-1+\delta}$ for $c>0$, $\delta\geqslant 0$, $k\to\infty$, then the series (1) diverges.

2. The comparison test. If $a_k\leqslant b_k$ for all sufficiently large k and the series (2) converges, then so does the series (1).

If $a_k\geqslant b_k$ for all sufficiently large k and the series (2) diverges, then so does the series (1).

3. Comparison test. Suppose the following limit exists:

$$\lim_{k\to\infty}\frac{a_k}{b_k}=c, \qquad 0\leqslant c\leqslant\infty.$$

If $c<\infty$ and the series (2) converges, then so does (1). If $c>0$ and the series (2) diverges, then so does (1).

4. If the following inequality holds for all sufficiently large k:

$$\frac{a_{k+1}}{a_k} \leqslant \frac{b_{k+1}}{b_k}, \qquad b_k \neq 0,$$

then the convergence of the series (2) implies that of the series (1), while the divergence of (1) implies that of (2).

5. The Kummer test. If there exists as sequence of positive numbers $c_k, k=1,2,\ldots$, such that for all sufficiently large k

$$c_k \frac{a_k}{a_{k+1}} - c_{k+1} \geqslant q > 0,$$

then the series (1) converges. If for all sufficiently large k

$$c_k \frac{a_k}{a_{k+1}} - c_{k+1} \leqslant 0$$

and the series $\sum_{k=1}^{\infty} c_k^{-1}$ diverges, then so does the series (1).

If there exists a sequence of positive numbers c_k, $k=1,2,\ldots$, such that

$$\varliminf_{k\to\infty} \left(c_k \frac{a_k}{a_{k+1}} - c_{k+1} \right) > 0,$$

then the series (1) converges.

If

$$\varlimsup_{k\to\infty} \left(c_k \frac{a_k}{a_{k+1}} - c_{k+1} \right) < 0$$

and the series $\sum_{k=1}^{\infty} c_k^{-1}$ diverges, then so does the series (1).

If

$$\varliminf_{k\to\infty} \frac{1}{a_k} (c_k a_k - c_{k+1} a_{k+1}) > 0,$$

then the series (1) converges.

If

$$\varlimsup_{k\to\infty} \frac{1}{a_k} (c_k a_k - c_{k+1} a_{k+1}) < 0$$

and the series $\sum_{k=1}^{\infty} c_k^{-1}$ diverges, then so does the series (1).

6. The generalized Kummer test. If there exists a sequence of positive numbers $c_k, k=1,2,\ldots$, such that for some fixed l

$$\varliminf_{k\to\infty} \frac{1}{a_k} (c_{k+l} a_{k+l} - c_{k+l+1} a_{k+l+1}) > 0,$$

then the series (1) converges. If

$$\varlimsup_{k\to\infty} \frac{1}{a_k} (c_{k+l} a_{k+l} - c_{k+l+1} a_{k+l+1}) < 0$$

and the series $\sum_{k=1}^{\infty} c_k^{-1}$ diverges, then so does the series (1).

7. The Bertrand test. If

$$\underline{\lim}_{k\to\infty}\left\{\ln k\left[k\left(\frac{a_k}{a_{k+1}}-1\right)-1\right]\right\}>1,$$

then the series (1) converges.

If

$$\overline{\lim}_{k\to\infty}\left\{\ln k\left[k\left(\frac{a_k}{a_{k+1}}-1\right)-1\right]\right\}<1,$$

then the series (1) diverges.

8. The Cauchy test. If for all sufficiently large k

$$\sqrt[k]{a_k}\leqslant q<1,$$

then the series (1) converges. If for all sufficiently large k

$$\sqrt[k]{a_k}>1,$$

then the series (1) diverges.

If

$$\overline{\lim}_{k\to\infty}\sqrt[k]{a_k}<1,$$

then the series (1) converges.

If

$$\overline{\lim}_{k\to\infty}\sqrt[k]{a_k}>1,$$

then the series (1) diverges.

If for some $q<1$

$$\underline{\lim}_{k\to\infty}\frac{q^k}{a_k}>0,$$

then the series (1) converges.

If for $q\geqslant 1$

$$0\leqslant\overline{\lim}_{k\to\infty}\frac{q^k}{a_k}<\infty,$$

then the series (1) diverges.

9. The Jamé test. If for all sufficiently large k

$$\left(1-\sqrt[k]{a_k}\right)\frac{k}{\ln k}\geqslant q>1,$$

then the series (1) converges.

If

$$\left(1-\sqrt[k]{a_k}\right)\frac{k}{\ln k}\leqslant 1,$$

then the series (1) diverges.

10. The d'Alembert test. If for all sufficiently large k

$$\frac{a_{k+1}}{a_k}\leqslant q<1,$$

then the series (1) converges.

If for all sufficiently large k

$$\frac{a_{k+1}}{a_k}\geqslant 1,$$

then the series (1) diverges.

If

$$\overline{\lim_{k\to\infty}} \frac{a_{k+1}}{a_k} < 1,$$

then the series (1) converges.

If

$$\underline{\lim_{k\to\infty}} \frac{a_{k+1}}{a_k} > 1,$$

then the series (1) diverges.

11. The generalized d'Alembert test. If for some $\alpha > 0$

$$\underline{\lim_{k\to\infty}} \frac{a_k^{\alpha} - a_{k+1}^{\alpha}}{a_k} > 0,$$

then the series (1) converges.

If

$$\overline{\lim_{k\to\infty}} \frac{a_k^{\alpha} - a_{k+1}^{\alpha}}{a_k} < 0,$$

then the series (1) diverges.

12. The Raabe test. If for all sufficiently large k

$$k\left(\frac{a_k}{a_{k+1}} - 1\right) \geqslant q > 1,$$

then the series (1) converges.

If for all sufficiently large k

$$k\left(\frac{a_k}{a_{k+1}} - 1\right) \leqslant 1,$$

then the series (1) diverges.

If

$$\underline{\lim_{k\to\infty}} k\left(1 - \frac{a_{k+1}}{a_k}\right) > 1,$$

then the series (1) converges.

If

$$\overline{\lim_{k\to\infty}} k\left(1 - \frac{a_{k+1}}{a_k}\right) < 1,$$

then the series (1) diverges.

13. Let

$$\frac{a_{k+1}}{a_k} = 1 + \varphi(k)(1 + o(1)), \qquad k \to \infty,$$

where

$$\varphi(k) = o(1), \qquad \frac{1}{\varphi(k)} - \frac{1}{\varphi(k-1)} = a + o(1), \qquad k \to \infty.$$

If $\varphi(k) < 0$ for all sufficiently large k and $a > -1$, then the series (1) converges.

If $a < -1$ and the series $\sum_{k=1}^{\infty} \varphi(k)$ diverges, then so does the series (1).

14. The Gauss test. If

$$\frac{a_{k+1}}{a_k} = \frac{k^{\alpha} + pk^{\alpha-1}(1 + O(1))}{k^{\alpha} + qk^{\alpha-1}(1 + O(1))}, \qquad k \to \infty,$$

then the series (1) converges for $q - p > 1$ and diverges for $q - p \leqslant 1$.

Let

$$\frac{a_k}{a_{k+1}}=\lambda+\frac{\mu}{k}+O\left(\frac{1}{k^2}\right), \qquad k\to\infty.$$

Then the series (1) converges if $\lambda>1$ or $\lambda=1$, $\mu>1$ and diverges if $\lambda<1$ or $\lambda=1$, $\mu<1$.

15. Suppose that

$$\frac{a_{k+1}}{a_k}=1+\frac{c+o(1)}{k^{\alpha}}, \qquad k\to\infty.$$

If $\alpha<1$, $c<0$ or $\alpha=1$, $c<-1$, then the series (1) converges.

If $c>0$, α is arbitrary, or $c>-1$, $\alpha=1$, or c is arbitrary, $\alpha>1$, then the series (1) diverges.

16. Suppose that

$$\frac{a_{k+1}}{a_k}=1-\frac{1}{k}+\frac{\varphi(k)}{k}(1+o(1)), \qquad k\to\infty,$$

$$\frac{1}{\varphi(k)}-\frac{1}{\varphi(k-1)}=\frac{a+o(1)}{k}, \qquad k\to\infty.$$

If $\varphi(k)<0$ for all sufficiently large k and $a>-1$, then the series (1) converges.

If $a<-1$ and the series $\sum_{k=1}^{\infty} k^{-1}\varphi(k)$ diverges, then so does the series (1).

17. Suppose that $\dfrac{a_{k+1}}{a_k}=1+\dfrac{c+o(1)}{k\ln^{\alpha}k}$, $\qquad k\to\infty.$

If $c<0$, $\alpha<0$, then the series (1) converges.

If $c>0$ and α is arbitrary or $\alpha>0$ and c is arbitrary, then the series (1) diverges.

18. Suppose that

$$\frac{a_{k+1}}{a_k}=1-\sum_{j=0}^{n}\frac{1}{\lambda_j(k)}+\frac{\varphi(k)}{\lambda_n(k)}(1+o(1)), \qquad k\to\infty,$$

$$\frac{1}{\varphi(k)}-\frac{1}{\varphi(k-1)}=\frac{\alpha+o(1)}{\lambda_n(k)}, \qquad k\to\infty,$$

where $\lambda_0(k)=k$, $\lambda_j(k)=k\ln k\ln\ln k\ldots\underbrace{\ln\ln\ldots\ln}_{j\,\text{times}} k$.

If $\varphi(k)<0$ for all sufficiently large k and $\alpha>-1$, then the series (1) converges. If $\alpha<-1$ and the series $\sum_{k=m}^{\infty}\lambda_n^{-1}(k)\varphi(k)$ $(\underbrace{\ln\ln\ldots\ln}_{n\,\text{times}} m>0)$ diverges, then so does the series (1).

19. The logarithmic test. If for all sufficiently large k

$$\frac{\ln a_k}{\ln k}\leqslant -q<-1,$$

then the series (1) converges.

If

$$\frac{\ln a_k}{\ln k}\geqslant -1,$$

then the series (1) diverges.

20. If the terms of the series (1) are monotone decreasing, then the series (1) converges or diverges simultaneously with the series

$$\sum_{k=1}^{\infty} m^k a_{m^k},$$

where m is any natural number.

21. The Lobachevskii test. If the terms of the series (1) tend monotonically to zero, then the series (1) converges or diverges simultaneously with the series

$$\sum_{m=0}^{\infty} 2^{-m} p_m,$$

where p_m is the largest index of the terms a_k satisfying the inequality $a_k \geqslant 2^{-m}$, $k=1,2,\ldots,p_m$.

22. Let $F(x)>0$ be an integrable function over $(1,\infty)$. If the integral $\int_1^{\infty} F(x)\,dx$ converges and

$$\underline{\lim}_{k\to\infty} \frac{1}{f(k)} \int_k^{k+1} F(x)\,dx > 0,$$

where $f(x)>0$ is monotone decreasing for $x \geqslant 1$, then the series (3) converges.

If

$$\int_1^{\infty} F(x)\,dx$$

diverges and

$$\overline{\lim}_{k\to\infty} \frac{1}{f(k)} \int_k^{k+1} F(x)\,dx < \infty,$$

then the series (3) diverges.

23. If there exists a function $F(x)<0$ that is differentiable for $x>1$ and is such that

$$\inf_{0\leqslant\theta\leqslant 1}\left[\underline{\lim}_{k\to\infty} \frac{F'(k+\theta)}{f(k)}\right] > 0,$$

where $f(x)>0$ is monotone decreasing for $x>1$, then the series (1) converges.

24. Let $f(x)>0$ be monotone decreasing for $x \geqslant 1$ and suppose that there exists a function $g(x)<0$ such that for $x \geqslant 1$ the function $F(x)=f(x)g(x)$ is twice differentiable and $F''(x)=o(F'(x))$ as $x\to\infty$.

If

$$\underline{\lim}_{x\to\infty} \frac{F'(x)}{f(x)} > 0,$$

then the series (3) converges.

If

$$\overline{\lim}_{x\to\infty} \frac{F'(x)}{f(x)} < 0$$

and the series $\sum_{k=1}^{\infty} 1/g(k)$ diverges, then so does the series (3).

25. Let $f(x)$ be a monotone decreasing twice differentiable function for $x \geqslant 1$ and suppose that

$$f''(x)=o(f'(x)), \quad \frac{f'(x)}{f(x)}=\varphi(x)(1+o(1)), \qquad x \to \infty,$$

$$\varphi(x)=o(1), \quad \left(\frac{1}{\varphi(x)}\right)'=a+o(1), \quad \left(\frac{1}{\varphi(x)}\right)''=o\left(\left(\frac{1}{\varphi(x)}\right)'\right), \quad x \to \infty.$$

If $\varphi(x)<0$ for $x>1$ and $a>-1$, then the series (3) converges.

If $a<-1$ and the series $\sum_{k=1}^{\infty} \varphi(k)$ diverges, then so does the series (3).

26. The Ermakov test. Let $f(x)>0$ be a monotone decreasing continuous function for $x>1$, and $\gamma(x)$ a differentiable increasing function such that $\gamma(x)>x$.

If

$$\overline{\lim_{k\to\infty}} \frac{\gamma'(k) f(\gamma(k))}{f(k)} < 1,$$

then the series (3) converges.

If

$$\underline{\lim_{k\to\infty}} \frac{\gamma'(k) f(\gamma(k))}{f(k)} > 1,$$

then the series (3) diverges.

27. If $f(x)>0$ is a continuous monotone decreasing function, $\lim_{x\to\infty} f(x)=0$ and $\delta(x)>0$ is a continuous increasing function for $x>1$, where $\delta(x)\to\infty$ as $x\to\infty$, $\delta(x+1)>\delta(x)$ and

$$\delta(x+2)-\delta(x+1) \leqslant A[\delta(x+1)-\delta(x)], \qquad A>0,$$

then the positive series

$$\sum_{k=1}^{\infty} f(k), \quad \sum_{k=1}^{\infty} f(\delta(k))[\delta(k+1)-\delta(k)]$$

converge or diverge simultaneously.

28. Let $f(x)>0$ be a continuous monotone decreasing function for $x>1$, and $\delta(x)>0$ a differentiable unboundedly increasing function such that $\lim_{x\to\infty} \delta(x)=\infty$ and $\delta'(x)f(\delta(x))$ is monotone decreasing for sufficiently large x.

Then the positive series $\sum_{k=1}^{\infty} f(k)$, $\sum_{k=1}^{\infty} \delta'(k) f(\delta(k))$ converge or diverge simultaneously.

I.3.3. Convergence tests for arbitrary series.

Suppose that we are given an arbitrary numerical series

$$\sum_{k=1}^{\infty} a_k. \tag{4}$$

1. If the series $\sum_{k=1}^{\infty} |a_k|$, converges, then so does the series (4), in which case the latter series is said to be absolutely convergent.

If the series (4) converges but the series $\sum_{k=1}^{\infty} |a_k|$ diverges, then the series (4) is said to be conditionally convergent. Convergence tests for the series $\sum_{k=1}^{\infty} |a_k|$ are listed in **I.3.2.**

2. Let

$$a_k = \frac{b}{k^\lambda}[1+o(k^{-\varepsilon})], \quad k\to\infty, \quad b\neq 0, \ \varepsilon>0.$$

If $\operatorname{Re}\lambda>1$, then the series (4) is absolutely convergent. If $\operatorname{Re}\lambda\leqslant 1$, then (4) is absolutely divergent (but may be conditionally convergent).

3. The Abel test. If the series (4) converges and the numbers b_k form a bounded monotone sequence,

$$|b_k|\leqslant B, \qquad k=1, 2, \ldots,$$

then the series

$$\sum_{k=1}^{\infty} a_k b_k \tag{5}$$

converges.

4. The Dirichlet test. If the partial sums of the series (4) form a bounded sequence, that is,

$$\left|\sum_{k=1}^{n} a_k\right|\leqslant B, \qquad n=1, 2, \ldots,$$

where B does not depend on n and the numbers b_k form a monotone decreasing sequence with $b_k\to 0$, then the series (5) converges.

5. The Leibnitz test. If the absolute values of the terms of the alternating series

$$\sum_{k=1}^{\infty}(-1)^k a_k, \qquad a_k>0,$$

are monotone decreasing, that is, $a_k\geqslant a_{k+1}$, and

$$\lim_{k\to\infty} a_k=0,$$

then this series converges. In this case we have the following estimate of the remainder term:

$$\left|\sum_{k=n+1}^{\infty}(-1)^k a_k\right|<a_{n+1}.$$

I.3.4. Tests for the uniform convergence of series depending on a parameter.

1. Suppose that the functions $u_k(z)$, $k=1,2,\ldots$, are defined on some set D. Then a series of functions

$$\sum_{k=1}^{\infty} u_k(z), \qquad z\in D, \tag{6}$$

that is convergent at each point $z\in D$ is said to be uniformly convergent over the set D if for any $\varepsilon>0$ there exists a number N_ε not depending on z such that for all $N\geqslant N_\varepsilon$ the inequality

$$\left|\sum_{k=N+1}^{\infty} u_k(z)\right|<\varepsilon$$

holds for all $z\in D$.

2. The Cauchy criterion. The series (6) converges uniformly on the set D if and only if for any $\varepsilon>0$ there exists as integer N_ε not depending on z such that the inequality

$$\left|\sum_{k=n}^{m} u_k(z)\right| \leqslant \varepsilon$$

holds for all $z \in D$ when $m \geqslant n \geqslant N_\varepsilon$.

We give below some tests for the uniform convergence of series depending on a parameter.

3. The Weierstrass test. If for all $z \in D$ the functions $u_k(z)$, $k=1,2,\ldots$, satisfy the inequalities

$$|u_k(z)| \leqslant c_k, \qquad k=1, 2, \ldots,$$

where the series $\sum_{k=1}^{\infty} c_k$ is convergent, then the series (6) converges uniformly on D.

4. The Abel test. If the series (6) is uniformly convergent on the set D and the functions $v_k(z)$ form a monotone sequence for each z and are uniformly bounded over D, that is,

$$|v_k(z)| \leqslant M, \qquad k=1, 2, \ldots,$$

where M does not depend on k and on $z \in D$, then the series

$$\sum_{k=1}^{\infty} u_k(z)\, v_k(z) \tag{7}$$

converges uniformly on D.

5. The Dirichlet test. If the partial sums of the series (6) are uniformly bounded over D, that is, $\left|\sum_{k=1}^{N} u_k(z)\right| \leqslant M, \qquad N=1, 2, \ldots,$

where M does not depend on $z \in D$ or on N, and the functions $v_k(z)$ form a monotone sequence which converges uniformly to zero on D, then the series (7) converges uniformly on D.

6. The Dini test. Suppose that the terms of the series (6) are continuous and positive for $z \in [a, b]$. If the sum of the series is also continuous in $[a, b]$, then the series (6) converges uniformly on $[a, b]$.

I.3.5. Operations with series.

1. The convergent series (4) has the associativity property:

$$\sum_{k=1}^{\infty} (a_{n_{k-1}+1} + \ldots + a_{n_k}) = \sum_{k=1}^{\infty} a_k, \qquad n_0 = 0,$$

$n_1, n_2, \ldots$ being a subsequence of the natural number series.

2. The Cauchy theorem. An absolutely convergent series (4) possesses the rearrangement property, that is, each series obtained from (4) by a rearrangement of the terms is also convergent and has the same sum.

3. The Riemann theorem. A conditionally convergent series (4) does not have the rearrangement property; there exists for any A a rearrangement of the terms of the series (4) such that the rearranged series has the sum A.

4. Convergent series can be added (or subtracted) and multiplied by a constant factor: $\quad c\sum_{k=1}^{\infty} a_k \pm d\sum_{k=1}^{\infty} b_k = \sum_{k=1}^{\infty} (ca_k + db_k).$

5. If two series are absolutely convergent, they can be multiplied together and the product series is absolutely convergent:

$$\left(\sum_{k=1}^{\infty} a_k\right)\left(\sum_{l=1}^{\infty} b_l\right) = \sum_{k,l=1}^{\infty} a_k b_l.$$

6. If the series (6) converges uniformly on a set D which has a limit point a, and each of the functions $u_k(z)$ has a finite limit as $z \to a$, then it is possible to pass to the limit in the series (6):

$$\lim_{z \to a} \sum_{k=1}^{\infty} u_k(z) = \sum_{k=1}^{\infty} \lim_{z \to a} u_k(z).$$

7. If the functions $u_k(x)$, $k=1, 2, \ldots$, are integrable over $[a, b]$, $a<b\leqslant\infty$, and the series (6) is uniformly convergent on $[a, b]$, then its sum is integrable over $[a, b]$ and the series (6) can be integrated term by term:

$$\int_a^b \left(\sum_{k=1}^{\infty} u_k(x) \right) dx = \sum_{k=1}^{\infty} \int_a^b u_k(x)\, dx.$$

This formula also holds when the functions $u_k(x)$, $k=1, 2, \ldots$, are continuous and positive on $[a, b]$, $b\leqslant\infty$, and the sum of the series (6) is continuous and integrable on this interval.

8. Suppose that the functions $u_k(x)$, $k=1, 2, \ldots$, are defined and have continuous derivatives on $[a, b,]$. If the series (6) is convergent and the series $\sum_{k=1}^{\infty} u_k'(x)$ is uniformly convergent on $[a, b]$, then the series (6) can be differentiated term by term on $[a, b]$: $\left(\sum_{k=1}^{\infty} u_k(x) \right)' = \sum_{k=1}^{\infty} u_k'(x)$.

9. Suppose that the functions $u_k(x)$, $k=1, 2, \ldots$, are defined and have finite derivatives on $[a, b]$. If the series (6) converges at some point and the series $\sum_{k=1}^{\infty} u_k'(x)$ is uniformly convergent on $[a, b]$, then the series (6) is uniformly convergent on $[a, b]$ and $\left(\sum_{k=1}^{\infty} u_k(x) \right)' = \sum_{k=1}^{\infty} u_k'(x)$.

10. If the functions $u_k(z)$, $k=1, 2, \ldots$, are analytic in the domain D and the series (6) converges uniformly in D, then its sum is analytic in D.

I.3.6. Power series.

Power series are particular cases of series of functions of the form (6) where $u_k(z)=a_{k-1}(z-z_0)^{k-1}$:

$$f(z) = \sum_{k=0}^{\infty} a_k (z-z_0)^k = a_0 + a_1 (z-z_0) + a_2 (z-z_0)^2 + \ldots \tag{8}$$

For each power series (8) that is not everywhere divergent (apart from the point $z=z_0$), the region of convergence D consists of points of some disc $|z-z_0|<R$ with centre at $z=z_0$ and radius $R=\left(\overline{\lim_{k \to \infty}} \sqrt[k]{|a_k|} \right)^{-1}$ together perhaps with a subset of the circle $|z-z_0|=R$. The power series (8) is absolutely convergent at each point inside this disc and is uniformly convergent on any region strictly inside the disc (that is, for $|z-z_0|\leqslant r<R$). The series is divergent outside the disc. If the series converges at all points of the complex plane, then $R=\infty$ and its sum $f(z)$ is an entire function. On the boundary of the disc of convergence, there is at least one singular point of the analytic function $f(z)$ defined by the sum of the series (8).

If the function $f(x)$ has derivatives of all orders in a neighbourhood U of the point x_0 and is such that $r_n(x) \to 0$ as $n \to \infty$ throughout U, where

$$r_n(x) = f(x) - \sum_{k=0}^{n} \frac{f^{(k)}(x_0)}{k!} (x-x_0)^k,$$

then it is representable in this neighbourhood as a Taylor series in powers of $x-x_0$:

$$f(x)=\sum_{k=0}^{\infty} a_k (x-x_0)^k, \qquad a_k=\frac{f^{(k)}(x_0)}{k!}.$$

If a function can be expanded in a power series, then this expansion is unique and is its Taylor series. Its remainder term $r_n(x)$ can be expressed in the Lagrange form

$$r_n(x)=\frac{(x-x_0)^{n+1}}{(n+1)!} f^{(n+1)}(x_0+\theta(x-x_0)), \qquad 0<\theta<1,$$

in the Cauchy form

$$r_n(x)=\frac{(x-x_0)^{n+1}}{n!}(1-\theta)^n f^{(n+1)}(x_0+\theta(x-x_0)), \qquad 0<\theta<1,$$

in the Roche form

$$r_n(x)=\frac{(x-x_0)^{n+1}}{n!\,(p+1)}(1-\theta)^{n-p-1} f^{(n+1)}(x_0+\theta(x-x_0)), \qquad 0<p\leqslant n, \quad 0<\theta<1;$$

in the integral form

$$r_n(x)=\frac{1}{n!}\int_{x_0}^{x} f^{(n+1)}(t)(x-t)^n\,dt,$$

or in the Schlömilch form

$$r_n(x)=\frac{\psi(x-x_0)-\psi(0)}{\psi'[(x-x_0)(1-\theta)]}\,\frac{(x-x_0)^n(1-\theta)^n}{n!} f^{(n+1)}(x_0+\theta(x-x_0)), \qquad 0<\theta<1,$$

where $\psi(x)$ is any function satisfying the conditions:

1) $\psi(t)$ and $\psi'(t)$ are continuous in $(0, x-x_0)$; $\psi'(t)$ does not change sign in $(0, x-x_0)$.

Let

$$\sum_{k=0}^{\infty} a_k z^k = S(z), \qquad |z|<R_1, \tag{9}$$

$$\sum_{l=0}^{\infty} b_l z^l = T(z), \qquad |z|<R_2,$$

$r=\min(R_1, R_2)$.

The following operations can be performed with power series.

1. Power series can be added inside their common circle of convergence;

$$\sum_{l=0}^{m}\left(\sum_{k=0}^{\infty} a_k^l z^k\right)=\sum_{k=0}^{\infty}\left(\sum_{l=0}^{m} a_k^l\right) z^k, \quad |z|<\inf_{l=1,2,\ldots,m} r_l, \qquad 1\leqslant m\leqslant\infty,$$

r_l is the radius of convergence of the series $\sum_{k=0}^{\infty} a_k^l z^k$.

2. Power series can be multiplied together:

$$\left(\sum_{k=0}^{\infty} a_k z^k\right)\left(\sum_{l=0}^{\infty} b_l z^l\right)=\sum_{k=0}^{\infty}\left(\sum_{j=0}^{k} a_j b_{k-j}\right) z^k, \qquad |z|<r.$$

3. A power series can be raised to a power:

$$\left(\sum_{k=0}^{\infty} a_k z^k\right)^n = \sum_{k=0}^{\infty} c_k z^k, \qquad |z| < R_1$$

$$c_0 = a_0^n,\ a_0 \neq 0,\ c_k = \frac{1}{a_0 k} \sum_{l=1}^{k} (nl - k + l)\, a_l c_{k-l},\ k = 1, 2, \ldots$$

4. Power series can be divided:

$$\frac{\sum_{k=0}^{\infty} a_k z^k}{\sum_{k=0}^{\infty} b_k z^k} = \frac{1}{b_0} \sum_{k=0}^{\infty} c_k z^k, \qquad b_0 \neq 0,$$

where

$$c_k + \frac{1}{b_0} \sum_{l=1}^{k} c_{k-l} b_l - a_k = 0$$

or

$$c_k = \frac{(-1)^k}{b_0^k} \begin{vmatrix} a_0 b_1 - a_1 b_0 & b_0 & 0 & \ldots & 0 \\ a_0 b_2 - a_2 b_0 & b_1 & b_0 & \ldots & 0 \\ a_0 b_3 - a_3 b_0 & b_2 & b_1 & \ldots & 0 \\ \ldots & \ldots & \ldots & \ldots & \ldots \\ a_0 b_{k-1} - a_{k-1} b_0 & b_{k-2} & b_{k-3} & \ldots & b_0 \\ a_0 b_k - a_k b_0 & b_{k-1} & b_{k-2} & \ldots & b_1 \end{vmatrix}.$$

5. Inside its circle of convergence $|z - z_0| \leqslant R$ a power series (8) can be integrated and differentiated term by term any number of times; furthermore, its radius of convergence R does not change. In particular, for the series (9) with $|z| < R_1$, we have the equalities:

$$\sum_{k=0} \frac{a_k}{nk+m} z^{nk+m} = \int_0 x^{m-1} S(x^n)\, dx,$$

$$\sum_{k=0}^{\infty} (nk + m + 1)\, a_k z^{nk+m} = \frac{d}{dz} [z^{m+1} S(z^n)],$$

$$\sum_{k=0}^{\infty} k^m a_k z^k = \left(z \frac{d}{dz}\right)^m S(z).$$

6. Substitution of a series into a series:

$$\sum_{l=0}^{\infty} b_l \left(\sum_{k=1}^{\infty} a_k z^k\right)^l = \sum_{k=0}^{\infty} c_k z^k, \quad |z| < R_1, \qquad \sum_{k=0}^{\infty} |a_k z^k| < R_2,$$

$$c_k = \sum_{l_1 + 2l_2 + \ldots + kl_k = k} \frac{|l|!}{l_1! \ldots l_k!} b_{|l|} a_1^{l_1} \ldots a_k^{l_k}, \quad |l| = l_1 + \ldots + l_k;$$

$$c_0 = b_0, \quad c_1 = a_1 b_1, \quad c_2 = a_1 b_2 + a_2 b_1,$$
$$c_3 = a_3 b_1 + 2a_1 a_2 b_2 + a_1^3 b_3,$$
$$c_4 = a_4 b_1 + 2a_1 a_3 b_2 + a_2^2 b_2 + 3a_1^2 a_2 b_3 + a_1^4 b_4.$$

7. Let $n \neq 0$ and $l = 1, 2, \ldots, n-1$.
Then

$$\sum_{k=0}^{\infty} a_{nk+l} z^{nk+l} = \frac{1}{n} \sum_{j=0}^{n-1} e^{-2jl\pi i/n} \sum_{k=0}^{\infty} a_k e^{2jk\pi i/n} z^k \qquad [|z| < R_1].$$

8. The inverse of a power series. Let

$$w = f(z) = \sum_{k=1}^{\infty} a_k z^k, \qquad a_1 \neq 0.$$

Then

$$z = f^{-1}(w) = \sum_{m=1}^{\infty} c_m w^m,$$

$$c_m = \sum_{\alpha_2 + 2\alpha_3 + \ldots + (m-1)\alpha_m = m-1} \frac{(m + \alpha_1 + \alpha_2 + \ldots + \alpha_m - 1)!}{m! \alpha_2! \ldots \alpha_m! a_1^m} \times$$

$$\times \left(-\frac{a_2}{a_1}\right)^{\alpha_2} \ldots \left(-\frac{a_m}{a_1}\right)^{\alpha_m};$$

where the summation is carried out over all sets of integers $\alpha_2, \alpha_3, \ldots, \alpha_m$ satisfying the equation $\alpha_2 + 2\alpha_3 + \ldots + (m-1)\alpha_m = m-1$.

9. The Lagrange series. If $z = a + \zeta\psi(z)$, then

$$F(z(\zeta)) = F(a) + \sum_{k=1}^{\infty} \frac{\zeta^k}{k!} \frac{\partial^{k-1}}{\partial z^{k-1}} [F'(z) \psi^k(z)] |_{z=a}.$$

10. The Bürman–Lagrange series. If $w = f(z) = z\psi(z)$, $\psi'(a) \neq 0$, $\psi(a) = 0$, then

$$F(z) = F(f^{-1}(w)) = F(a) + \sum_{k=1}^{\infty} \frac{w^k}{k!} \frac{\partial^{k-1}}{\partial z^{k-1}} \left[\frac{F'(z)}{\psi^k(z)}\right] \Bigg|_{z=a}.$$

I.3.7. Trigonometric series.

A series of the form

$$\alpha_0 + \sum_{k=1}^{\infty} (\alpha_k \cos kx + \beta_k \sin kx), \tag{10}$$

where α_k, β_k are constants (coefficients of the series), is called a trigonometric series.

By the Fourier series of a function $f(x)$ defined and absolutely integrable over the interval $(-l, l)$, is meant the series of the form

$$\frac{a_0}{2} + \sum_{k=1}^{\infty} \left(a_k \cos \frac{k\pi x}{l} + b_k \sin \frac{k\pi x}{l}\right), \tag{11}$$

the coefficients (Fourier coefficients) of which are defined by the formulae

$$a_k = \frac{1}{l} \int_{-l}^{l} f(t) \cos \frac{k\pi t}{l}\, dt, \qquad b_k = \frac{1}{l} \int_{-l}^{l} f(t) \sin \frac{k\pi t}{l}\, dt. \tag{12}$$

1. If the function $f(x)$ can be expanded in a trigonometric series (11) over $(-l, l)$, then this series is its Fourier series, that is, its coefficients a_k, b_k are given by formulae (3).

2. If the function $f(x)$ is piecewise-differentiable in the interval $(-l, l)$ then its Fourier series converges at each point x and has the sum $S(x) = \frac{f(x+0) + f(x-0)}{2}$, $x \in (-l, l)$. If furthermore, $f(x)$ is continued at x, then $S(x) = f(x)$.

3. The Fourier coefficients a_k and b_k of an absolutely integrable function f converge to zero as $k\to\infty$:

$$a_k,\quad b_k\longrightarrow 0.$$

4. A function $f(x)$ defined on an interval can be expanded in a series of the form

$$S_C(x)=\frac{a_0}{2}+\sum_{k=1}^{\infty} a_k\cos\frac{k\pi x}{l},\qquad a_k=\frac{2}{l}\int_0^l f(t)\cos\frac{k\pi t}{l}\,dt,$$

or in a series of the form

$$S_S(x)=\sum_{k=1}^{\infty} b_k\sin\frac{k\pi x}{l},\qquad b_k=\frac{2}{l}\int_0^l f(t)\sin\frac{k\pi t}{l}\,dt.$$

In the first case, $f(x)$ is defined on $(-l,0)$ so as to be an even function: $f(-x)=f(x)$, while in the second case it is defined to be an odd function: $f(-x)=-f(x)$. If $f(x)$ is a piecewise-differentiable function, then at the point of continuity $S_C(x)=f(x)$ or $S_S=f(x)$, $-l<x<l$.

I.3.8. Asymptotic series.

Let $\{\psi_k(x)\}_{k=1}^{\infty}$, $x\in D$,— be an asymptotic sequence of functions as $x\to x_0$, that is, $\psi_{k+1}(x)=o\ (\psi_k(x))$, $x\to x_0$, $x\in D$, and let $\varphi_k(x)$ be functions defined on D. A formal series $\sum_{k=1}^{\infty}\varphi_k(x)$ is called an asymptotic series (or expansion) of the function $f(x)$ with respect to the asymptotic sequence $\{\psi_k(x)\}$ if for any integer $N\geqslant 1$

$$f(x)-\sum_{k=1}^{N}\varphi_k(x)=o\,(\psi_N(x)),\qquad x\longrightarrow x_0.$$

In this case one writes

$$f(x)\sim\sum_{k=1}^{\infty}\varphi_k(x),\quad \{\psi_k(x)\},\qquad x\longrightarrow x_0.$$

If $\varphi_k(x)=a_k\psi_k(x)$, where the a_k are constants, then the relation

$$f(x)\sim\sum_{k=1}^{\infty}a_k\psi_k(x),\qquad x\longrightarrow x_0,$$

is called an asymptotic series (or expansion) of the function $f(x)$.

An important role is played by asymptotic power series for which $\psi_k=x^{-k}$ $(x\to\infty)$ or $\psi_k=(x-x_0)^k$ $(x\to x_0)$. Asymptotic series admit certain operations of algebra and analysis. Corresponding statements also hold for functions of a complex variable.

Examples of asymptotic series:

$$J_\nu(x)\sim\sqrt{\frac{2}{\pi x}}\left[\cos\left(x-\frac{\pi\nu}{2}-\frac{\pi}{4}\right)\sum_{k=0}^{\infty}(-1)^k a_{2k}x^{-2k}-\right.$$
$$\left.-\sin\left(x-\frac{\pi\nu}{2}-\frac{\pi}{4}\right)\sum_{k=0}^{\infty}(-1)^k a_{2k+1}x^{-2k-1}\right],\ \{x^{-k}\},\ x\longrightarrow+\infty$$
$$a_k=\frac{\Gamma(\nu+k+1/2)}{2^k\,k!\,\Gamma(\nu-k+1/2)};$$

$$e^{-x}I_0(x)=\frac{1}{\sqrt{2\pi}}\sum_{k=0}^{\infty}\frac{(-1)^k\,\Gamma\left(\frac{1}{2}+k\right)}{k!\,\Gamma\left(\frac{1}{2}-k\right)(2x)^{k+1/2}},\quad x\to+\infty.$$

I.3.9. Infinite products.

Let $a_1, a_2, a_3, \ldots$ be a given sequence of numbers or functions.

If the following limit exists and is finite and non-zero:

$$\lim_{n\to\infty}\prod_{k=1}^{n}a_k=\prod_{k=1}^{\infty}a_k, \tag{13}$$

then it is called the value of the infinite product. In this case, the product $\prod_{k=1}^{\infty}a_k$ is said to converge; otherwise it is said to diverge.

1. A necessary condition for the convergence of a product.

If the product (13) converges, then $a_k\to 1$ as $k\to\infty$.

2. The connection between products and series. The product (13) converges if and only if the series $\sum_{k=1}^{\infty}\ln a_k$ converges.

In this case, we have the equation

$$\sum_{k=1}^{\infty}\ln a_k=\ln\prod_{k=1}^{\infty}a_k.$$

3. If for sufficiently large k, $a_k>1$ (or $a_k<1$), then the product (13) converges if and only if the series $\sum_{k=1}^{\infty}(a_k-1)$ converges.

4. If the series $\sum_{k=1}^{\infty}(a_k-1)$, $\sum_{k=1}^{\infty}(a_k-1)^2$ converge, then the product (13) also converges.

Appendix II. CERTAIN SPECIAL FUNCTIONS AND THEIR PROPERTIES

II.1 THE BINOMIAL COEFFICIENTS $\binom{a}{b}$

$$\binom{a}{k}=(-1)^k\frac{(-a)_k}{k!}=\frac{a(a-1)\ldots(a-k+1)}{k!}\qquad [k=1, 2, 3, \ldots].$$

$$\binom{a}{0}=1,\quad \binom{a}{-k}=0\qquad [k=1, 2, 3, \ldots; a\neq-1, -2, -3, \ldots].$$

$$\binom{n}{k}=\frac{n!}{k!\,(n-k)!}\qquad [k=1, 2, \ldots, n].$$

$$\binom{n}{k}=0\qquad [k=-1, -2, -3, \ldots], \text{ or } [k>n], \text{ or } [n=0;\ k=1, 2, \ldots].$$

$$\binom{n}{k}=\binom{n}{n-k}\qquad [0\leqslant k\leqslant n].$$

$$\binom{a}{k}=(-1)^k\binom{-a+k-1}{k},\quad \binom{a}{b}=\frac{\Gamma(a+1)}{\Gamma(b+1)\,\Gamma(a-b+1)},$$

$$\binom{a}{b}=\binom{a}{a-b} \qquad [b\neq -k,\ k+a, \text{ where } k=-1,-2,-3,\ldots]$$

$$\binom{a}{b+1}=\frac{a}{b+1}\binom{a-1}{b}=\frac{a-b}{b+1}\binom{a}{b}, \quad \binom{a}{b}+\binom{a}{b+1}=\binom{a+1}{b+1}.$$

$$\binom{n+a}{n}\binom{b}{n+a}=\binom{b}{n}\binom{b-n}{a}=\binom{b}{a}\binom{b-a}{n}.$$

$$\binom{-1/2}{n}=\frac{(-1)^n}{2^{2n}}\binom{2n}{n}=(-1)^n\frac{(2n-1)!!}{(2n)!!}.$$

$$\binom{1/2}{n}=\frac{(-1)^{n-1}}{2^{2n-1}n}\binom{2n-2}{n-1}=(-1)^{n-1}\frac{(2n-3)!!}{(2n-2)!!\,n}.$$

$$\binom{n+1/2}{2n+1}=(-1)^n 2^{-4n-1}\binom{2n}{n}, \quad \binom{2n+1/2}{n}=2^{2n-1}\binom{4n+1}{2n}.$$

$$\binom{n}{1/2}=\frac{2^{2n+1}}{\pi}\binom{2n}{n}^{-1}, \quad \binom{n}{n/2}=\frac{2^{2n}}{\pi}\binom{n}{(n-1)/2}^{-1}.$$

II.2. THE POCHHAMMER SYMBOL $(a)_k$

$$(a)_k=a(a+1)\ldots(a+k-1)=\frac{\Gamma(a+k)}{\Gamma(a)}=(-1)^k\frac{\Gamma(1-a)}{\Gamma(1-a-k)}.$$

$$(a)_0=1, \quad (a)_{-k}=\frac{\Gamma(a-k)}{\Gamma(a)}=\frac{(-1)^k}{(1-a)_k} \qquad [a\neq 1,2,\ldots,k;\ k=1,2,3,\ldots]$$

$$(a)_k=(a)_n(a+n)_{k-n}=(-1)^k(1-a-k)_k.$$

$$(n)_k=\frac{(n+k-1)!}{(n-1)!}, \quad (1)_k=k!.$$

$$(1/2)_k=2^{-2k}\frac{(2k)!}{k!}, \qquad (3/2)_k=2^{-2k}\frac{(2k+1)!}{k!}.$$

$$(a)_{n+k}=(a)_n(a+n)_k, \qquad (a)_{n-k}=\frac{(-1)^k(a)_n}{(1-a-n)_k}.$$

$$(a)_{nk}=\left(\frac{a}{n}\right)_k\left(\frac{a+1}{n}\right)_k\ldots\left(\frac{a+n-1}{n}\right)_k n^{nk}, \qquad (a)_{2k}=\left(\frac{a}{2}\right)_k\left(\frac{a+1}{2}\right)_k 2^{2k}.$$

$$(nk)!=\left(\frac{1}{n}\right)_k\left(\frac{2}{n}\right)_k\ldots\left(\frac{n-1}{n}\right)_k k!\,n^{nk}, \quad (2k)!=\left(\frac{1}{2}\right)_k k!\,2^{2k},$$

$$(nk+m)!=m!\left(\frac{m+1}{n}\right)_k\left(\frac{m+2}{n}\right)_k\ldots\left(\frac{m+n}{n}\right)_k n^{nk}, \quad (2k+1)!=\left(\frac{3}{2}\right)_k k!\,2^{2k}.$$

$$(a+b)_k=\sum_{j=0}^{k}(-1)^j\binom{k}{j}(a+j)_{k-j}(-b)_j=\sum_{j=0}^{k}\binom{k}{j}(a+j)_{k-j}(1+b-j)_j.$$

$$(a+mk)_{nk}=\frac{(a)_{mk+nk}}{(a)_{mk}}, \quad (a+k)_k=\frac{(a)_{2k}}{(a)_k}, \quad (a+m)_k=\frac{(a)_k(a+k)_m}{(a)_m}.$$

$$(a-mk)_{nk}=(-1)^{mk}(a)_{nk-mk}(1-a)_{mk} \qquad [n\geqslant m],$$

$$=(-1)^{nk}\frac{(1-a)_{mk}}{(1-a)_{mk-nk}} \qquad [n<m].$$

$$(a-k)_k=(-1)^k(1-a)_k, \quad (a-m)_k=\frac{(1-a)_m(a)_k}{(1-a-k)_m}.$$

$$(a+mk)_{n\pm lk}=\frac{(a)_n(a+n)_{mk\pm lk}}{(a)_{mk}}, \qquad (a+k)_{n-k}=\frac{(a)_n}{(a)_k}.$$

$$(a-mk)_{n\pm lk}=\frac{(-1)^{lk}(a)_n(1-a)_{mk}}{(1-a-n)_{mk\mp lk}}, \quad (a-k)_{n-k}=\frac{(-1)^k(a)_n(1-a)_k}{(1-a-n)_{2n}}.$$

II.3. THE GAMMA FUNCTION $\Gamma(z)$

Definition:

$$\Gamma(z)=\int_0^\infty t^{z-1}e^{-t}\,dt \qquad [\operatorname{Re} z>0]$$

or

$$\Gamma(z)=\lim_{n\to\infty}\frac{n!\,n^z}{(z)_{n+1}} \qquad [z\neq 0,\ -1,\ -2,\ \ldots].$$

For $\operatorname{Re} z\leqslant 0$, the function $\Gamma(z)$ can also be defined as an analytic continuation of the above integral or by the equation

$$\Gamma(z)=\int_0^\infty t^{z-1}\left(e^{-t}-\sum_{k=0}^{n}\frac{(-t)^k}{k!}\right)dt \qquad [-n-1<\operatorname{Re} z<-n].$$

$\Gamma(z)$ is an analytic function over the entire z-plane except for the points $z=0, -1, -2, \ldots$, at which it has simple poles with residues

$$\operatorname*{res}_{z=-k}\Gamma(z)=\frac{(-1)^k}{k!} \qquad [k=0,\ 1,\ 2,\ \ldots].$$

$$\Gamma(z+1)=z\Gamma(z),\quad \Gamma(n+1)=n!,\quad \Gamma(1)=\Gamma(2)=1.$$

$$\Gamma(z+k)=(z)_k\,\Gamma(z),\qquad \Gamma(z-k)=\frac{(-1)^k}{(1-z)_k}\Gamma(z).$$

$$\Gamma(z+nk)=\Gamma(z)\left(\frac{z}{n}\right)_k\left(\frac{z+1}{n}\right)_k\cdots\left(\frac{z+n-1}{n}\right)_k n^{nk},$$

$$\Gamma(z-nk)=\frac{(-1)^{nk}\,\Gamma(z)\,n^{-nk}}{\left(\frac{1-z}{n}\right)_k\left(\frac{2-z}{n}\right)_k\cdots\left(\frac{n-z}{n}\right)_k}.$$

$$\Gamma(z+2k)=\Gamma(z)\left(\frac{z}{2}\right)_k\left(\frac{z+1}{2}\right)_k 2^{2k},\quad \Gamma(z-2k)=\frac{\Gamma(z)\,2^{-2k}}{\left(\frac{1-z}{2}\right)_k\left(\frac{2-z}{2}\right)_k}.$$

$$\Gamma\left(n+\frac{1}{2}\right)=\frac{\sqrt{\pi}}{2^n}(2n-1)!!,\quad \Gamma\left(\frac{1}{2}\right)=\sqrt{\pi},\quad \Gamma\left(\frac{3}{2}\right)=\frac{\sqrt{\pi}}{2},$$

$$\Gamma\left(\frac{1}{2}-n\right)=(-1)^n\frac{2^n\sqrt{\pi}}{(2n-1)!!},\quad \Gamma\left(-\frac{1}{2}\right)=-2\sqrt{\pi},\quad \Gamma\left(-\frac{3}{2}\right)=\frac{4\sqrt{\pi}}{3},$$

$$\Gamma\left(n+\frac{1}{3}\right)=\frac{1\cdot 4\cdot 7\ldots(3n-2)}{3^n}\Gamma\left(\frac{1}{3}\right),\quad \Gamma\left(\frac{1}{3}\right)=2{,}678938\ldots,$$

$$\Gamma\left(n+\frac{2}{3}\right)=\frac{2\cdot 5\cdot 8\ldots(3n-1)}{3^n}\Gamma\left(\frac{2}{3}\right),\quad \Gamma\left(\frac{2}{3}\right)=1{,}354118\ldots,$$

$$\Gamma\left(n+\frac{1}{4}\right)=\frac{1\cdot 5\cdot 9\ldots(4n-3)}{4^n}\Gamma\left(\frac{1}{4}\right),\quad \Gamma\left(\frac{1}{4}\right)=3{,}625600\ldots,$$

$$\Gamma\left(n+\frac{3}{4}\right)=\frac{3\cdot 7\cdot 11\ldots(4n-1)}{4^n}\Gamma\left(\frac{3}{4}\right),\quad \Gamma\left(\frac{3}{4}\right)=1{,}225417\ldots,$$

$$\Gamma(z)\,\Gamma(-z)=-\frac{\pi}{z\sin\pi z},\quad \Gamma(z)\,\Gamma(1-z)=\frac{\pi}{\sin\pi z},$$

$$\Gamma\left(\frac{1}{2}+z\right)\Gamma\left(\frac{1}{2}-z\right)=\frac{\pi}{\cos\pi z},\quad \Gamma(1+z)\,\Gamma(1-z)=\frac{\pi z}{\sin\pi z}.$$

$$\frac{\Gamma(z+1)}{\Gamma(w+1)\,\Gamma(z-w+1)}=\binom{z}{w}.$$

$$\Gamma(nz)=(2\pi)^{(1-n)/2}n^{nz-1/2}\prod_{k=0}^{n-1}\Gamma\left(z+\frac{k}{n}\right).$$

$$\Gamma(2z)=\frac{2^{2z-1}}{\sqrt{\pi}}\Gamma(z)\Gamma\left(z+\frac{1}{2}\right).$$

$$\Gamma(3z)=\frac{3^{3z-1/2}}{2\pi}\Gamma(z)\Gamma\left(z+\frac{1}{3}\right)\Gamma\left(z+\frac{2}{3}\right).$$

$$|\Gamma(ix)|^2=\frac{\pi}{x\operatorname{sh}\pi x},\quad \left|\Gamma\left(\frac{1}{2}+ix\right)\right|^2=\frac{\pi}{\operatorname{ch}\pi x}.$$

$$|\Gamma(1+ix)|^2=\frac{\pi x}{\operatorname{sh}\pi x},\quad \Gamma\left(\frac{1}{4}+ix\right)\Gamma\left(\frac{3}{4}-ix\right)=\frac{\pi\sqrt{2}}{\operatorname{ch}\pi x+i\operatorname{sh}\pi x}.$$

$$\Gamma(z_0+\varepsilon)=\Gamma(z_0)\left\{1+\psi(z_0)\varepsilon+[\psi'(z_0)+\psi^2(z_0)]\frac{\varepsilon^2}{2}+\right.$$
$$\left.+[\psi''(z_0)+3\psi(z_0)\psi'(z_0)+\psi^3(z_0)]\frac{\varepsilon^3}{6}+O(\varepsilon^4)\right\}\qquad [\varepsilon\to 0;\ z_0\neq 0,-1,-2,\ldots].$$

$$\Gamma(-k+\varepsilon)=\frac{(-1)^k}{k!\,\varepsilon}\left\{1+\psi(k+1)\varepsilon+[\pi^2+3\psi^2(k+1)-3\psi'(k+1)]\frac{\varepsilon^2}{6}+\right.$$
$$\left.+[\pi^2\psi(k+1)+\psi^3(k+1)-3\psi(k+1)\psi'(k+1)+\psi''(k+1)]\frac{\varepsilon^3}{6}+O(\varepsilon^4)\right\}\qquad [\varepsilon\to 0].$$

$$\Gamma^{\pm 1}(z+1)=1\pm\sum_{k=1}^{\infty}a_kz^k,\quad a_k=\frac{1}{k}\sum_{l=1}^{k}(-1)^l b_l a_{k-l},$$
$$b_1=\mathbf{C},\quad b_l=\zeta(l),\quad l=2,3,4,\ldots\qquad [|z|<1].$$

$$\ln\Gamma(z+1)=-\mathbf{C}z+\sum_{k=2}^{\infty}(-1)^k\zeta(k)\frac{z^k}{k}\qquad [|z|<1],$$

$$=\frac{1}{2}\ln\frac{\pi z}{\sin\pi z}-\mathbf{C}z-\sum_{k=1}^{\infty}\zeta(2k+1)\frac{z^{2k+1}}{2k+1}\qquad [|z|<1].$$

$$\Gamma(z)=\sqrt{2\pi}\,z^{z-1/2}e^{-z}\left[1+O\left(\frac{1}{z}\right)\right]\qquad [z\to\infty;\ |\arg z|<\pi].$$

$$n!=\sqrt{2\pi}\,n^{n+1/2}e^{-n}\left[1+O\left(\frac{1}{n}\right)\right]\qquad [n\to\infty].$$

$$|\Gamma(x+iy)|=\sqrt{2\pi}\,|y|^{x-1/2}e^{-\pi|y|/2}\left[1+O\left(\frac{1}{y}\right)\right]\qquad [|y|\to\infty].$$

$$\Gamma\left[\begin{matrix}z+a\\ z+b\end{matrix}\right]=z^{a-b}\left[1+O\left(\frac{1}{z}\right)\right]\qquad [z\to\infty].$$

II.4 THE PSI-FUNCTION $\psi(z)$

Definition:

$$\psi(z)=\frac{\Gamma'(z)}{\Gamma(z)}=\frac{a}{dz}\ln\Gamma(z).$$

$\psi(z)$ is an analytic function over the entire z-plane except at the points $z=0,-1,-2,\ldots$, at which it has simple poles with residues

$$\operatorname*{res}_{s=-k}\psi(z)=-1\qquad [k=0,1,2,\ldots].$$

$$\psi(z+n)=\psi(z)+\sum_{k=0}^{n-1}\frac{1}{k+z},\ \psi(z-n)=\psi(z)-\sum_{k=1}^{n}\frac{1}{z-k},$$

$$\psi(z+1)=\psi(z)+\frac{1}{z}, \quad \psi(n+1)=-\mathbf{C}+\sum_{k=1}^{n}\frac{1}{k}, \quad \psi(1)=-\mathbf{C}.$$

$$\mathbf{C}=\lim_{n\to\infty}\left(\sum_{k=1}^{n}\frac{1}{k}-\ln n\right)=0{,}5772156649\ldots$$

$$\psi\left(\frac{1}{2}\pm n\right)=-\mathbf{C}-2\ln 2+2\sum_{k=0}^{n-1}\frac{1}{2k+1}, \quad \psi\left(\frac{1}{2}\right)=-\mathbf{C}-2\ln 2.$$

$$\psi\left(\frac{1}{3}\right)=-\mathbf{C}-\frac{\pi}{2\sqrt{3}}-\frac{3}{2}\ln 3, \quad \psi\left(\frac{2}{3}\right)=-\mathbf{C}+\frac{\pi}{2\sqrt{3}}-\frac{3}{2}\ln 3.$$

$$\psi\left(\frac{1}{4}\right)=-\mathbf{C}-\frac{\pi}{2}-3\ln 2, \quad \psi\left(\frac{3}{4}\right)=-\mathbf{C}+\frac{\pi}{2}-3\ln 2.$$

$$\psi\left(\frac{p}{q}\right)=-\mathbf{C}-\ln(2q)-\frac{\pi}{2}\operatorname{ctg}\frac{p\pi}{q}+2\sum_{k=1}^{[(q-1)/2]}\cos\frac{2pk\pi}{q}\ln\sin\frac{k\pi}{q}$$

$$[p=1,\,2,\,\ldots,\,q-1;\; q=2,\,3,\,4,\,\ldots].$$

$$\psi(z)-\psi(-z)=-\pi\operatorname{ctg}\pi z-\frac{1}{z}, \qquad \psi(z)-\psi(1-z)=-\pi\operatorname{ctg}\pi z.$$

$$\psi(1+z)-\psi(1-z)=\frac{1}{z}-\pi\operatorname{ctg}\pi z, \qquad \psi\left(\frac{1}{2}+z\right)-\psi\left(\frac{1}{2}-z\right)=\pi\operatorname{tg}\pi z.$$

$$\psi\left(\frac{z}{2}\right)-\psi\left(\frac{z+1}{2}\right)=-2\beta(z), \quad \psi(n)-\psi\left(\frac{n+1}{2}\right)=((-1)^n+1)\ln 2+\sum_{k=1}^{n-1}\frac{(-1)^{n+k}}{k}.$$

$$\psi(nz)=\frac{1}{n}\sum_{k=0}^{n-1}\psi\left(z+\frac{k}{n}\right)+\ln n, \qquad \psi(2z)=\frac{1}{2}\psi(z)+\frac{1}{2}\psi\left(z+\frac{1}{2}\right)+\ln 2.$$

$$\operatorname{Re}\psi(ix)=\operatorname{Re}\psi(-ix)=\operatorname{Re}\psi(1+ix)=\operatorname{Re}\psi(1-ix)=-\mathbf{C}+x^2\sum_{k=1}^{\infty}\frac{1}{k(k^2+x^2)}.$$

$$\operatorname{Im}\psi(ix)=\frac{1}{2x}+\frac{\pi}{2}\operatorname{cth}\pi x, \qquad \operatorname{Im}\psi(1+ix)=-\frac{1}{2x}+\frac{\pi}{2}\operatorname{cth}\pi x=x\sum_{k=1}^{\infty}\frac{1}{k^2+x^2}.$$

$$\operatorname{Im}\psi\left(\frac{1}{2}+ix\right)=\frac{\pi}{2}\operatorname{th}\pi x. \qquad \psi(z)=-\mathbf{C}+(z-1)\sum_{k=0}^{\infty}\frac{1}{(k+1)(k+z)}.$$

$$\psi(-k+\varepsilon)=-\frac{1}{\varepsilon}\left\{1-\psi(k+1)\,\varepsilon+\left[\psi'(k+1)-\frac{\pi^2}{3}\right]\varepsilon^2+O(\varepsilon^3)\right\} \qquad [\varepsilon\to 0].$$

$$\psi^{(n)}(z)=(-1)^{n+1}\,n!\sum_{k=0}^{\infty}\frac{1}{(k+z)^{n+1}}.$$

$$\psi^{(n)}(z+1)=\psi^{(n)}(z)+(-1)^n\,n!\,z^{-n-1}$$

$$=(-1)^{n+1}\sum_{k=0}^{\infty}\frac{(-1)^k(n+k)!}{k!}\zeta(n+k+1)\,z^k \qquad [\,|z|<1].$$

$$\psi^{(n)}(z+m)=\psi^{(n)}(z)+(-1)^n\,n!\sum_{k=0}^{m-1}\frac{1}{(k+z)^{n+1}}.$$

$$\psi^{(n)}(1-z)=(-1)^n\psi^{(n)}(z)+(-1)^n\pi\frac{d^n}{dz^n}\operatorname{ctg}\pi z.$$

$$\psi^{(n)}(mz)=\ln m+\frac{1}{m^{n+1}}\sum_{k=0}^{m-1}\psi^{(n)}\left(z+\frac{k}{m}\right) \qquad [n=1, 2, \ldots].$$

$$\psi^{(n)}(m+1)=(-1)^n n!\left[-\zeta(n+1)+\sum_{k=1}^{m}\frac{1}{k^{n+1}}\right].$$

$$\psi^{(n)}(1)=(-1)^{n+1}n!\,\zeta(n+1) \qquad [n=1, 2, \ldots].$$

$$\psi^{(n)}\left(\frac{1}{2}\right)=(-1)^{n+1}n!\,(2^{n+1}-1)\,\zeta(n+1) \qquad [n=1, 2, \ldots].$$

$$\psi'(1)=\frac{\pi^2}{6}, \qquad \psi'(m)=\frac{\pi^2}{6}-\sum_{k=1}^{m-1}\frac{1}{k^2}.$$

$$\psi'\left(\frac{1}{2}\right)=\frac{\pi^2}{2}, \qquad \psi'\left(\frac{1}{2}\pm m\right)=\frac{\pi^2}{2}\mp 4\sum_{k=1}^{m}\frac{1}{(2k-1)^2}.$$

II.5. THE POLYLOGARITHM $\mathrm{Li}_\nu(z)$

Definition:

$$\mathrm{Li}_\nu(z)=\sum_{k=1}^{\infty}\frac{z^k}{k^\nu} \qquad [|z|<1 \text{ or } |z|=1, \operatorname{Re}\nu>1].$$

For $|z|\geqslant 1$, the function $\mathrm{Li}_\nu(z)$ is defined as the analytic continuation of this series. Its principal branch has the representations

$$\mathrm{Li}_\nu(z)=\frac{z}{\Gamma(\nu)}\int_0^\infty\frac{t^{\nu-1}}{e^t-z}\,dt \qquad [\operatorname{Re}\nu>0;\ |\arg(1-z)|<\pi],$$

$$=\frac{iz}{2}\int_{\gamma-i\infty}^{\gamma+i\infty}\frac{(-z)^s}{(1+s)^\nu\sin s\pi}\,ds=$$

$$=\frac{1}{2\pi i}\int_{\gamma-i\infty}^{\gamma+i\infty}\frac{\Gamma(s)\,\Gamma(-s)}{(-s)^{\nu-1}}(-z)^{-s}\,ds$$

$$[-1<\gamma<0;\ |\arg(-z)|<\pi].$$

$\mathrm{Li}_2(z)$ is the Euler dilogarithm.

$$\mathrm{Li}_\nu(z)=z\Phi(z, \nu, 1), \qquad \mathrm{Li}_\nu(1)=\zeta(\nu).$$

$$\mathrm{Li}_\nu(z^m)=m^{\nu-1}\sum_{k=1}^{m}\mathrm{Li}_\nu(e^{2\pi ik/m}z). \qquad \frac{d^m}{dz^m}\mathrm{Li}_\nu(z)=z^{-m}\sum_{k=1}^{m}S_m^{(k)}\mathrm{Li}_{\nu-k}(z)$$

[$S_m^{(k)}$ is the Stirling number of the first kind].

$$\mathrm{Li}_\nu(z)-\mathrm{Li}_\nu(e^{2\pi i}z)=\frac{2\pi i}{\Gamma(\nu)}\ln^{\nu-1}z \qquad [|\arg(1-z)|<\pi].$$

$$\mathrm{Li}_\nu(z)+e^{\nu\pi i}\mathrm{Li}_\nu\left(\frac{1}{z}\right)=\frac{(2\pi)^\nu}{\Gamma(\nu)}e^{\nu\pi i/2}\zeta\left(1-\nu, \frac{\ln z}{2\pi i}\right).$$

$$\mathrm{Li}_{-n}(z)+(-1)^n\mathrm{Li}_{-n}\left(\frac{1}{z}\right)=0 \qquad [n=1, 2, \ldots].$$

$$\operatorname{Li}_n(z)+(-1)^n \operatorname{Li}_n\left(\frac{1}{z}\right)=-\frac{(2\pi i)^n}{n!} B_n\left(\frac{\ln z}{2\pi i}\right)=-\sum_{k=0}^{[n/2]} c_{k,n} \ln^{n-2k}(-z);$$

$$c_{0,n}=\frac{1}{n!}, \qquad c_{1,n}=\frac{\pi^2}{6(n-2)!},$$

$$c_{k,n}=-\frac{2\operatorname{Li}_{2k}(-1)}{(n-2k)!}=(-1)^{k-1}\frac{2(2^{2k-1}-1)\pi^{2k}B_{2k}}{(2k)!(n-2k)!} \qquad [k=0, 1, \ldots, [n/2]].$$

$$\operatorname{Li}_n(z)+(-1)^n \operatorname{Li}_n\left(\frac{1}{z}\right)=$$

$$=\sum_{k=0}^{n-1}\frac{\ln^k w}{k!}\left[(-1)^k \operatorname{Li}_{n-k}(wz)+(-1)^n \operatorname{Li}_{n-k}\left(\frac{1}{wz}\right)\right]+(-1)^{n+1}\frac{\ln^n w}{n!} \qquad [n=1, 2, \ldots].$$

$$\operatorname{Li}_n(z)+\operatorname{Li}_n(-z)=2^{1-n}\operatorname{Li}_n(z^2).$$

$$\operatorname{Li}_n(iz)+\operatorname{Li}_n(-iz)=4^{1-n}\operatorname{Li}_n(z^4)-2^{1-n}\operatorname{Li}_n(z^2).$$

$$\operatorname{Li}_n(iz)-\operatorname{Li}_n(-iz)=2i\sum_{k=0}^{\infty}\frac{(-1)^k z^{2k+1}}{(2k+1)^n} \qquad [|z|<1].$$

$$\operatorname{Li}_n(\rho e^{i\varphi}) \pm \operatorname{Li}_n(\rho e^{-i\varphi})=2\sum_{k=1}^{\infty}\frac{\rho^k}{k^n}\begin{Bmatrix}\cos k\varphi\\ i\sin k\varphi\end{Bmatrix} \qquad [0\leqslant\rho<1;\ -\pi<\varphi\leqslant\pi].$$

$$\operatorname{Li}_n(z)=\int_0^z t^{-1}\operatorname{Li}_{n-1}(t)\,dt \qquad [n=1, 2, \ldots].$$

$$\operatorname{Li}_0(z)=\frac{z}{1-z}, \qquad \operatorname{Li}_1(z)=-\ln(1-z).$$

$$\operatorname{Li}_{2n}(1)=(-1)^{n-1}\frac{(2\pi)^{2n}}{2(2n)!}B_{2n} \qquad [n=1, 2, \ldots].$$

$$\operatorname{Li}_{2n}(-1)=(-1)^n\frac{2^{2n-1}-1}{(2n)!}\pi^{2n}B_{2n}.$$

$$\operatorname{Li}_2(z)=-\int_0^z\frac{\ln(1-t)}{t}\,dt=\int_0^1\frac{\ln t}{t-z^{-1}}\,dt=2\sqrt{z}\int_0^1\frac{\ln\left[1+\sqrt{z}(1-t)\right]}{1-t\sqrt{z}}\,dt$$

$$[|\arg(1-z)|<\pi].$$

$$\operatorname{Li}_2(z)=-\operatorname{Li}_2(1-z)+\frac{\pi^2}{6}-\ln z\ln(1-z) \qquad [|\arg z|, |\arg(1-z)|<\pi].$$

$$\operatorname{Li}_2(z)=-\operatorname{Li}_2\left(\frac{1}{z}\right)-\frac{1}{2}\ln^2 z+\pi i\ln z+\frac{\pi^2}{3} \qquad [|\arg(-z)|<\pi].$$

$$\operatorname{Li}_2(z)=\operatorname{Li}_2\left(\frac{1}{1-z}\right)+\frac{1}{2}\ln^2(1-z)-\ln(-z)\ln(1-z)-\frac{\pi^2}{6} \qquad [|\arg(-z)|<\pi].$$

$$\operatorname{Li}_2(wz)=\operatorname{Li}_2(w)+\operatorname{Li}_2(z)+\operatorname{Li}_2\left(\frac{wz-w}{1-w}\right)+\operatorname{Li}_2\left(\frac{wz-z}{1-z}\right)+\frac{1}{2}\ln^2\frac{1-w}{1-z}.$$

$$\operatorname{Li}_2(1)=\frac{\pi^2}{6}, \qquad \operatorname{Li}_2(-1)=-\frac{\pi^2}{12},$$

$$\operatorname{Li}_2\left(\frac{1}{2}\right)=\frac{\pi^2}{12}-\frac{1}{2}\ln^2 2, \qquad \operatorname{Li}_2(\pm i)=-\frac{\pi^2}{48}\pm i\mathbf{G}.$$

II.6 THE GENERALIZED FRESNEL INTEGRALS $S(z,\nu)$ and $C(z,\nu)$

Definitions:

$$S(x,\ \nu)=\int_x^\infty t^{\nu-1}\sin t\,dt \qquad [\operatorname{Re}\nu<1],$$

$$C(x,\ \nu)=\int_x^\infty t^{\nu-1}\cos t\,dt \qquad [\operatorname{Re}\nu<1].$$

These functions are defined on the z-plane as analytic continuations of the above integrals.

$$\begin{Bmatrix} S(z,\nu) \\ C(z,\nu) \end{Bmatrix}=\begin{Bmatrix} \sin(\nu\pi/2) \\ \cos(\nu\pi/2) \end{Bmatrix}\Gamma(\nu)-\frac{z^{\delta+\nu}}{\delta+\nu}\,{}_1F_2\left(\frac{\delta+\nu}{2};\ \frac{\delta+\nu}{2}+1,\ \delta+\frac{1}{2};\ -\frac{z^2}{4}\right)$$
$$\left[\delta=\begin{Bmatrix} 1 \\ 0 \end{Bmatrix}\right].$$

$$\begin{Bmatrix} S(z,\ \nu) \\ C(z,\ \nu) \end{Bmatrix}=\frac{(-i)^{(1\pm1)/2}}{2}\left[e^{i\nu\pi/2}\Gamma(\nu,\ -iz)\mp e^{-i\nu\pi/2}\Gamma(\nu,\ iz)\right].$$

$$\frac{d}{dz}\begin{Bmatrix} S(z,\ \nu) \\ C(z,\ \nu) \end{Bmatrix}=-z^{\nu-1}\begin{Bmatrix} \sin z \\ \cos z \end{Bmatrix}.$$

$$\begin{Bmatrix} S(z,0) \\ C(z,0) \end{Bmatrix}=-\begin{Bmatrix} \operatorname{si}(z) \\ \operatorname{ci}(z) \end{Bmatrix}, \qquad \begin{Bmatrix} S(z,\ 1/2) \\ C(z,\ 1/2) \end{Bmatrix}=\sqrt{2\pi}\left[\frac{1}{2}-\begin{Bmatrix} S(z) \\ C(z) \end{Bmatrix}\right].$$

II.7. THE GENERALIZED ZETA-FUNCTION $\zeta(z,\nu)$

Definition:

$$\zeta(z,\ v)=\sum_{k=0}^{\infty}\frac{1}{(k+v)^z} \qquad [\operatorname{Re}z>1;\ v\neq0,\ -1,\ -2,\ldots],$$

$$=\frac{1}{\Gamma(z)}\int_0^\infty\frac{t^{z-1}e^{-vt}}{1-e^{-t}}\,dt \qquad [\operatorname{Re}z>1;\ \operatorname{Re}v>0].$$

For $\operatorname{Re}z\leqslant1, z\neq1$, the function $\zeta(z,v)$ is defined as the analytic continuation of the above representations. At the point $z=1$. the function $\zeta(z,v)$ has a simple pole with residue 1.

$$\zeta(z,\ v+n)=\zeta(z,\ v)-\sum_{k=0}^{n-1}\frac{1}{(k+v)^z}. \qquad \zeta(z,\ 1)=\zeta(z),$$

$$\zeta\left(z,\ \frac{1}{2}\right)=(2^z-1)\,\zeta(z). \qquad \zeta(0,\ v)=\frac{1}{2}-v,$$

$$\zeta(-n,\ v)=-\frac{1}{n+1}B_{n+1}(v). \qquad \frac{\partial}{\partial z}\zeta(0,\ v)=\ln\Gamma(v)-\frac{1}{2}\ln(2\pi).$$

$$\lim_{z\to1}\left[\zeta(z,\ v)-\frac{1}{z-1}\right]=-\psi(v) \qquad [\operatorname{Re}v>0].$$

For particular values of $\zeta(z,\ v)$ see [17], §§ 5.1.3—5.

II.8. THE BERNOULLI POLYNOMIALS $B_n(z)$ AND THE BERNOULLI NUMBERS B_n

Definitions:

$$\frac{te^{zt}}{e^t-1}=\sum_{k=0}^{\infty} B_k(z)\frac{t^k}{k!} \qquad [\,|t|<2\pi],$$

$$\frac{t}{e^t-1}=\sum_{k=0}^{\infty} B_k\frac{t^k}{k!} \qquad [\,|t|<2\pi].$$

$$B_n(z)=\sum_{k=0}^{n}\binom{n}{k}B_k z^{n-k}, \qquad B_n=\sum_{k=0}^{n}\frac{1}{k+1}\sum_{j=0}^{k}(-1)^j\binom{k}{j}j^n.$$

$$B_n(z+1)=B_n(z)+nz^{n-1}, \qquad B_n(z+a)=\sum_{k=0}^{n}\binom{n}{k}a^{n-k}B_k(z).$$

$$B_n(1-z)=(-1)^nB_n(z), \qquad B_n(-z)=(-1)^n[B_n(z)+nz^{n-1}].$$

$$B_n(mz)=m^{n-1}\sum_{k=0}^{m-1}B_n\left(z+\frac{k}{m}\right) \qquad [m=1,2,\ldots].$$

$$B_n'(z)=nB_{n-1}(z) \qquad [n=1,2,\ldots].$$

$$B_n(z)=2^{-n}\sum_{k=0}^{n}\binom{n}{k}B_{n-k}E_k(2z).$$

$$B_n(m)=B_n+n\sum_{k=1}^{m-1}k^{n-1} \qquad [m,n=1,2,\ldots].$$

$$B_n(0)=(-1)^nB_n(1)=B_n.$$

$$B_{2n}\left(\frac{1}{6}\right)=B_{2n}\left(\frac{5}{6}\right)=\frac{1}{2}(1-2^{1-2n})(1-3^{1-2n})B_{2n}.$$

$$B_n\left(\frac{1}{4}\right)=(-1)^nB_n\left(\frac{3}{4}\right)=-2^{-n}(1-2^{1-n})B_n-4^{-n}nE_{n-1} \qquad [n=1,2,\ldots].$$

$$B_{2n}\left(\frac{1}{3}\right)=B_{2n}\left(\frac{2}{3}\right)=-\frac{1}{2}(1-3^{1-2n})B_{2n}. \qquad B_n\left(\frac{1}{2}\right)=-(1-2^{1-n})B_n.$$

$$B_0(z)=1, \qquad B_1(z)=z-\frac{1}{2}, \qquad B_2(z)=z^2-z+\frac{1}{6},$$

$$B_3(z)=z^3-\frac{3}{2}z^2+\frac{1}{2}z, \qquad B_4(z)=z^4-2z^3+z^2-\frac{1}{30},$$

$$B_5(z)=z^5-\frac{5}{2}z^4+\frac{5}{3}z^3-\frac{1}{6}z, \qquad B_6(z)=z^6-3z^5+\frac{5}{2}z^4-\frac{1}{2}z^2+\frac{1}{42}.$$

$$B_n=\sum_{k=0}^{n}\binom{n}{k}B_k, \qquad B_n=B_n(0).$$

$$B_{2n+1}=0 \quad [n=1,2,\ldots], \qquad B_{2n}=\frac{(-1)^{n-1}(2n)!}{2^{2n-1}\pi^{2n}}\zeta(2n).$$

$$B_0=1, \qquad B_1=-\frac{1}{2}, \qquad B_2=\frac{1}{6}, \qquad B_4=-\frac{1}{30}, \qquad B_6=\frac{1}{42},$$

$$B_8=-\frac{1}{30}, \qquad B_{10}=\frac{5}{66}, \qquad B_{12}=-\frac{691}{2730}.$$

II.9. THE EULER POLYNOMIALS $E_n(z)$ AND THE EULER NUMBERS E_n

Definitions:

$$\frac{2e^{zt}}{e^t+1}=\sum_{k=0}^{\infty} E_k(z)\frac{t^k}{k!} \qquad [|t|<\pi].$$

$$\frac{1}{\operatorname{ch} t}=\sum_{k=0}^{\infty} E_k\frac{t^k}{k!} \qquad [|t|<\pi/2].$$

$$E_n(z)=\sum_{k=0}^{n}\binom{n}{k}\frac{E_k}{2^k}\left(z-\frac{1}{2}\right)^{n-k}.$$

$$E_n(z+1)=2z^n-E_n(z), \qquad E_n(z+a)=\sum_{k=0}^{n}\binom{n}{k}a^{n-k}E_k(z).$$

$$E_n(1-z)=(-1)^n E_n(z), \qquad E_n(-z)=(-1)^{n+1}[E_n(z)-2z^n].$$

$$E_n(mz)=m^n\sum_{k=0}^{m-1}(-1)^k E_n\left(z+\frac{k}{m}\right) \qquad [m=1, 3, \ldots].$$

$$=-\frac{2}{n+1}m^n\sum_{k=0}^{m-1}(-1)^k B_{n+1}\left(z+\frac{k}{m}\right) \qquad [m=2, 4, \ldots].$$

$$E_n'(z)=nE_{n-1}(z) \qquad [n=1, 2, \ldots].$$

$$E_n(z)=\frac{2^{n+1}}{n+1}\left[B_{n+1}\left(\frac{z+1}{2}\right)-B_{n+1}\left(\frac{z}{2}\right)\right].$$

$$=\frac{2}{n+1}\left[B_{n+1}(z)-2^{n+1}B_{n+1}\left(\frac{z}{2}\right)\right].$$

$$=\frac{4}{(n+1)(n+2)}\sum_{k=0}^{n}\binom{n+2}{k}(2^{n-k+2}-1)B_{n-k+2}B_k(z).$$

$$E_n(0)=-E_n(1),$$

$$=-\frac{2}{n+1}(2^{n+1}-1)B_{n+1} \qquad [n=1, 2, \ldots].$$

$$E_{2n-1}\left(\frac{1}{3}\right)=-E_{2n-1}\left(\frac{2}{3}\right)$$

$$=-\frac{1}{2n}(1-3^{1-2n})(2^{2n}-1)B_{2n} \qquad [n=1, 2, \ldots].$$

$$E_n\left(\frac{1}{2}\right)=2^{-n}E_n.$$

$$E_0(z)=1, \quad E_1(z)=z-\frac{1}{2}, \quad E_2(z)=z^2-z,$$

$$E_3(z)=z^3-\frac{3}{2}z^2+\frac{1}{4}, \quad E_4(z)=z^4-2z^3+z,$$

$$E_5(z)=z^5-\frac{5}{2}z^4+\frac{5}{2}z^2-\frac{1}{2}, \quad E_6(z)=z^6-3z^5+5z^3-3z.$$

$$E_n=2^nE_n\left(\frac{1}{2}\right), \quad E_{2n+1}=0 \qquad [n=1, 2, \ldots].$$

$$E_0=1, \quad E_2=-1, \quad E_4=5, \quad E_6=-61, \quad E_8=1385,$$

$$E_{10}=-50\,521, \quad E_{12}=2\,702\,765.$$

II.10. THE STRUVE FUNCTIONS $\mathbf{H}_\nu(z)$ AND $\mathbf{L}_\nu(z)$

Definitions:

$$\left\{\begin{matrix}\mathbf{H}_\nu(z)\\ \mathbf{L}_\nu(z)\end{matrix}\right\}=\frac{2^{-\nu}z^{\nu+1}}{\sqrt{\pi}\,\Gamma(\nu+3/2)}\,{}_1F_2\left(1;\ \frac{3}{2}+\nu,\ \frac{3}{2};\ \mp\frac{z^2}{4}\right).$$

$$2\nu\left\{\begin{matrix}\mathbf{H}_\nu(z)\\ \mathbf{L}_\nu(z)\end{matrix}\right\}=\left\{\begin{matrix}z\mathbf{H}_{\nu-1}(z)+z\mathbf{H}_{\nu+1}(z)\\ z\mathbf{L}_{\nu-1}(z)-z\mathbf{L}_{\nu+1}(z)\end{matrix}\right\}-\frac{z^{\nu+1}}{2^\nu\sqrt{\pi}\,\Gamma(\nu+3/2)}.$$

$$2\frac{d}{dz}\left\{\begin{matrix}\mathbf{H}_\nu(z)\\ \mathbf{L}_\nu(z)\end{matrix}\right\}=\left\{\begin{matrix}\mathbf{H}_{\nu-1}(z)-\mathbf{H}_{\nu+1}(z)\\ \mathbf{L}_{\nu-1}(z)+\mathbf{L}_{\nu+1}(z)\end{matrix}\right\}+\frac{(z/2)^\nu}{\sqrt{\pi}\,\Gamma(\nu+3/2)}.$$

$$\frac{d}{dz}\left[z^{\pm\nu}\mathbf{H}_\nu(z)\right]=\pm z^{\pm\nu}\mathbf{H}_{\nu\mp1}(z)+\left\{\begin{matrix}0\\1\end{matrix}\right\}\frac{2^{-\nu}}{\sqrt{\pi}\,\Gamma(\nu+3/2)}.$$

$$\frac{d}{dz}\left[z^{\pm\nu}\mathbf{L}_\nu(z)\right]=z^{\pm\nu}\mathbf{L}_{\nu\mp1}(z)+\left\{\begin{matrix}0\\1\end{matrix}\right\}\frac{2^{-\nu}}{\sqrt{\pi}\,\Gamma(\nu+3/2)}.$$

$$\frac{d}{dz}\mathbf{H}_0(z)=\frac{2}{\pi}-\mathbf{H}_1(z),\qquad \frac{d}{dz}\mathbf{L}_0(z)=\frac{2}{\pi}+\mathbf{L}_1(z).$$

$$\mathbf{H}_\nu\left(ze^{m\pi i}\right)=e^{m(\nu+1)\pi i}\mathbf{H}_\nu(z)\qquad [m=0,\ \pm1,\ \pm2,\ \dots].$$

$$\mathbf{L}_\nu(z)=-ie^{-\nu\pi i/2}\mathbf{H}_\nu\left(ze^{\pi i/2}\right).$$

$$\mathbf{H}_{n+1/2}(z)=Y_{n+1/2}(z)+\frac{(z/2)^{n-1/2}}{n!\sqrt{\pi}}\sum_{k=0}^{n}\left(\frac{1}{2}\right)_k(-n)_k\left(-\frac{z^2}{4}\right)^{-k},$$

$$\mathbf{H}_{-n-1/2}(z)=(-1)^nJ_{n+1/2}(z),\qquad \mathbf{L}_{-n-1/2}(z)=I_{n+1/2}(z),$$

$$\mathbf{H}_{\pm1/2}(z)=\sqrt{\frac{2}{\pi z}}\left\{\begin{matrix}1-\cos z\\ \sin z\end{matrix}\right\},\qquad \mathbf{H}_{3/2}(z)=\sqrt{\frac{z}{2\pi}}\left(1+\frac{2}{z^2}\right)-$$

$$-\sqrt{\frac{2}{\pi z}}\left(\sin z+\frac{\cos z}{z}\right).$$

II.11. THE FUNCTIONS OF WEBER $\mathbf{E}_\nu(z)$, $\mathbf{E}_\nu^\mu(z)$ AND ANGER $\mathbf{J}_\nu(z)$, $\mathbf{J}_\nu^\mu(z)$

Definitions:

$$\left\{\begin{matrix}\mathbf{E}_\nu(z)\\ \mathbf{J}_\nu(z)\end{matrix}\right\}=\frac{1}{\pi}\int_0^\pi\left\{\begin{matrix}\sin(\nu t-z\sin t)\\ \cos(\nu t-z\sin t)\end{matrix}\right\}dt=$$

$$=\frac{1}{\nu\pi}\left\{\begin{matrix}1-\cos\nu\pi\\ \sin\nu\pi\end{matrix}\right\}{}_1F_2\left(1;\ 1+\frac{\nu}{2},\ 1-\frac{\nu}{2};\ -\frac{z^2}{4}\right)\mp$$

$$\mp\frac{z}{\pi(1-\nu^2)}\left\{\begin{matrix}1+\cos\nu\pi\\ \sin\nu\pi\end{matrix}\right\}{}_1F_2\left(1;\ \frac{3+\nu}{2},\ \frac{3-\nu}{2};\ -\frac{z^2}{4}\right)$$

$$\left\{\begin{matrix}\mathbf{E}_\nu^\mu(z)\\ \mathbf{J}_\nu^\mu(z)\end{matrix}\right\}=\frac{1}{\pi}\int_0^\pi(2\sin t)^\mu\left\{\begin{matrix}\sin(\nu t-z\sin t)\\ \cos(\nu t-z\sin t)\end{matrix}\right\}dt=$$

$$=\left\{\begin{matrix}\sin\left(\frac{\nu\pi}{2}\right)\\ \cos\left(\frac{\nu\pi}{2}\right)\end{matrix}\right\}\Gamma\left[\begin{matrix}\mu+1\\ \frac{\mu+\nu}{2}+1,\ \frac{\mu-\nu}{2}+1\end{matrix}\right]{}_2F_3\left(\begin{matrix}\frac{\mu+1}{2},\ \frac{\mu}{2}+1;\ -\frac{z^2}{4}\\ \frac{1}{2},\ \frac{\mu+\nu}{2}+1,\ \frac{\mu-\nu}{2}+1\end{matrix}\right)\mp$$

$$\mp\frac{z}{2}\left\{\begin{matrix}\cos(\nu\pi/2)\\ \sin(\nu\pi/2)\end{matrix}\right\}\Gamma\left[\begin{matrix}\mu+2\\ (\mu+\nu+3)/2,\ (\mu-\nu+3)/2\end{matrix}\right]{}_2F_3\left(\begin{matrix}\mu/2+1,\ (\mu+3)/2;\ -z^2/4\\ 3/2,\ (\mu+\nu+3)/2,\ (\mu-\nu+3)/2\end{matrix}\right).$$

$$2\nu\begin{Bmatrix}\mathbf{E}_\nu(z)\\ \mathbf{J}_\nu(z)\end{Bmatrix}=z\begin{Bmatrix}\mathbf{E}_{\nu-1}(z)+\mathbf{E}_{\nu+1}(z)\\ \mathbf{J}_{\nu-1}(z)+\mathbf{J}_{\nu+1}(z)\end{Bmatrix}+\frac{2}{\pi}\begin{Bmatrix}1-\cos\nu\pi\\ \sin\nu\pi\end{Bmatrix}.$$

$$2\frac{d}{dz}\begin{Bmatrix}\mathbf{E}_\nu(z)\\ \mathbf{J}_\nu(z)\end{Bmatrix}=\begin{Bmatrix}\mathbf{E}_{\nu-1}(z)-\mathbf{E}_{\nu+1}(z)\\ \mathbf{J}_{\nu-1}(z)-\mathbf{J}_{\nu+1}(z)\end{Bmatrix}.$$

$$\begin{Bmatrix}\mathbf{E}_\nu(z)\\ \mathbf{J}_\nu(z)\end{Bmatrix}=\mp\operatorname{ctg}\nu\pi\begin{Bmatrix}\mathbf{J}_\nu(z)\\ \mathbf{E}_\nu(z)\end{Bmatrix}\pm\operatorname{cosec}\nu\pi\begin{Bmatrix}\mathbf{J}_{-\nu}(z)\\ \mathbf{E}_{-\nu}(z)\end{Bmatrix}.$$

$$\mathbf{E}_n(z)=\frac{2z^{n-1}}{(2n-1)!!\,\pi}\sum_{k=0}^{[(n-1)/2]}\left(\frac{1}{2}\right)_k\left(\frac{1}{2}-n\right)_k\left(-\frac{4}{z^2}\right)^k-\mathbf{H}_n(z).$$

$$\mathbf{E}_{-n}(z)=\frac{(-1)^{n+1}\,2\,(2n-3)!!}{\pi z^{n-1}}\sum_{k=0}^{[(n-1)/2]}\frac{(-z^2/4)^k}{(3/2)_k\,(3/2-n)_k}-\mathbf{H}_{-n}(z).$$

$$\mathbf{E}_0(z)=-\mathbf{H}_0(z),\qquad \mathbf{E}_1(z)=\frac{2}{\pi}-\mathbf{H}_1(z).$$

$$\mathbf{J}_n(z)=J_n(z)\qquad [n=0,\ \pm1,\ \pm2,\ \ldots]$$

$$\mathbf{J}_{\pm1/2}(z)=\mathbf{E}_{\mp1/2}(z)=\sqrt{\frac{2}{\pi z}}\{\cos z\,[C(z)\mp S(z)]\pm\sin z\,[C(z)\pm S(z)]\}.$$

II.12. THE LOMMEL FUNCTIONS $s_{\mu,\nu}(z)$ AND $S_{\mu,\nu}(z)$

Definitions:

$$s_{\mu,\nu}(z)=\frac{z^{\mu+1}}{(\mu+1)^2-\nu^2}\,{}_1F_2\left(1;\ \frac{\mu-\nu+3}{2},\ \frac{\mu+\nu+3}{2};\ -\frac{z^2}{4}\right)\qquad [\mu\pm\nu\neq-1,\ -3,\ -5,\ \ldots].$$

$$S_{\mu,\nu}(z)=s_{\mu,\nu}(z)+2^{\mu-1}\Gamma\left(\frac{\mu-\nu+1}{2}\right)\Gamma\left(\frac{\mu+\nu+1}{2}\right)\left[\sin\frac{\mu-\nu}{2}\pi J_\nu(z)-\right.$$
$$\left.-\cos\frac{\mu-\nu}{2}\pi Y_\nu(z)\right].$$

$$s_{\mu,-\nu}(z)=s_{\mu,\nu}(z),\qquad S_{\mu,-\nu}(z)=S_{\mu,\nu}(z).$$

$$s_{\mu+2,\nu}(z)=z^{\mu+1}-[(\mu+1)^2-\nu^2]\,s_{\mu,\nu}(z),$$

$$2\nu s_{\mu,\nu}(z)=(\mu+\nu-1)\,zs_{\mu-1,\nu-1}(z)-(\mu-\nu-1)\,zs_{\mu-1,\nu+1}(z),$$

$$\frac{d}{dz}s_{\mu,\nu}(z)=\mp\frac{\nu}{z}s_{\mu,\nu}(z)+(\mu\pm\nu-1)\,s_{\mu-1,\nu\mp1}(z),$$

$$2\frac{d}{dz}s_{\mu,\nu}(z)=(\mu+\nu-1)\,s_{\mu-1,\nu-1}(z)+(\mu-\nu-1)\,s_{\mu-1,\nu+1}(z)$$

(the last 4 formulae are valid with $s_{\mu,\nu}(z)$ replaced everywhere by $S_{\mu,\nu}(z)$).

$$s_{\nu,\nu}(z)=2^{\nu-1}\sqrt{\pi}\,\Gamma\left(\nu+\frac{1}{2}\right)\mathbf{H}_\nu(z).$$

$$S_{\nu,\nu}(z)=2^{\nu-1}\sqrt{\pi}\,\Gamma\left(\nu+\frac{1}{2}\right)[\mathbf{H}_\nu(z)-Y_\nu(z)].$$

$$s_{\nu+1,\nu}(z)=z^\nu-2^\nu\Gamma(\nu+1)\,J_\nu(z),\qquad S_{\nu+1,\nu}(z)=z^\nu.$$

$$\lim_{\mu\to\nu}\frac{s_{\mu-1,\nu}(z)}{\Gamma(\mu-\nu)}=2^{\nu-1}\Gamma(\nu)\,J_\nu(z).$$

$$s_{(\pm1-1)/2,\nu}(z)=\frac{\nu^{(\pm1-1)/2}\pi}{2\sin\nu\pi}[\pm\mathbf{J}_\nu(z)-\mathbf{J}_{-\nu}(z)].$$

$$S_{(\pm1-1)/2,\nu}(z)=\frac{\nu^{(\pm1-1)/2}\pi}{2\sin\nu\pi}[\pm\mathbf{J}_\nu(z)-\mathbf{J}_{-\nu}(z)\mp J_\nu(z)+J_{-\nu}(z)].$$

$$S_{0,\,2n+1}(z)=\frac{z}{2n+1}O_{2n+1}(z),\qquad S_{1,\,2n}(z)=zO_{2n}(z).$$

$$s_{1,\,\nu}(z)=1+\nu^2 s_{-1,\,\nu}(z),\quad S_{1,\,\nu}(z)=1+\nu^2 S_{-1,\,\nu}(z).\quad S_{1/2,\,1/2}(z)=\frac{1}{\sqrt{z}},$$

$$S_{3/2,\,1/2}(z)=\sqrt{z}.\quad S_{-1\pm 1/2,\,1/2}(z)=-\frac{1}{\sqrt{z}}\left[\begin{Bmatrix}\cos z\\ \sin z\end{Bmatrix}\operatorname{si}(z)\mp\begin{Bmatrix}\sin z\\ \cos z\end{Bmatrix}\operatorname{ci}(z)\right].$$

$$\mathbf{E}_\nu(z)=-\frac{1}{\pi}[(1+\cos\nu\pi)\,s_{0,\,\nu}(z)+\nu(1-\cos\nu\pi)\,s_{-1,\,\nu}(z)].$$

II.13. THE KELVIN FUNCTIONS $\operatorname{ber}_\nu(z)$, $\operatorname{bei}_\nu(z)$, $\operatorname{ker}_\nu(z)$ and $\operatorname{kei}_\nu(z)$

Definitions:

$$\operatorname{ber}_\nu(x)\pm i\operatorname{bei}_\nu(x)=J_\nu(xe^{\pm 3\pi i/4})=e^{\pm\nu\pi i}J_\nu(xe^{\mp\pi i/4})=$$
$$=e^{\pm\nu\pi i/2}I_\nu(xe^{\pm\pi i/4})=e^{\pm 3\nu\pi i/2}I_\nu(xe^{\mp 3\pi i/4});$$

$$\operatorname{ker}_\nu(x)\pm i\operatorname{kei}_\nu(x)=e^{\mp\nu\pi i/2}K_\nu(xe^{\pm\pi i/4})=$$
$$=\pm\frac{\pi i}{2}H_\nu^{(1)}(xe^{\pm 3\pi i/4})=\mp\frac{\pi i}{2}e^{\mp\nu\pi i}H_\nu^{(2)}(xe^{\mp\pi i/4}).$$

$$\begin{Bmatrix}\operatorname{ber}_\nu(z)\\ \operatorname{bei}_\nu(z)\end{Bmatrix}=\left(\frac{z}{2}\right)^\nu\sum_{k=0}^{\infty}\begin{Bmatrix}\cos[(3\nu+2k)\pi/4]\\ \sin[(3\nu+2k)\pi/4]\end{Bmatrix}\frac{(z/2)^{2k}}{k!\,\Gamma(\nu+k+1)}.$$

$$\begin{Bmatrix}\operatorname{ker}_\nu(z)\\ \operatorname{kei}_\nu(z)\end{Bmatrix}=\pm\frac{z^\nu\Gamma(-\nu)}{2^{\nu+1}}\sum_{k=0}^{\infty}\begin{Bmatrix}\cos[(\nu-2k)\pi/4]\\ \sin[(\nu-2k)\pi/4]\end{Bmatrix}\frac{(z/2)^{2k}}{(\nu+1)_k\,k!}\pm$$
$$\pm 2^{\nu-1}z^{-\nu}\Gamma(\nu)\sum_{k=0}^{\infty}\begin{Bmatrix}\cos[(3\nu-2k)\pi/4]\\ \sin[(3\nu-2k)\pi/4]\end{Bmatrix}\frac{(z/2)^{2k}}{(1-\nu)_k k!}\qquad[\nu\neq\pm n].$$

$$\begin{Bmatrix}\operatorname{ker}_n(z)\\ \operatorname{kei}_n(z)\end{Bmatrix}=\pm\frac{z^{-n}}{2^{1-n}}\sum_{k=0}^{n-1}\begin{Bmatrix}\cos[(3n+2k)\pi/4]\\ \sin[(3n+2k)\pi/4]\end{Bmatrix}\frac{(n-k-1)!}{k!}\left(\frac{z}{2}\right)^{2k}-\ln\frac{z}{2}\begin{Bmatrix}\operatorname{ber}_n(z)\\ \operatorname{bei}_n(z)\end{Bmatrix}\pm$$
$$\pm\frac{\pi}{4}\begin{Bmatrix}\operatorname{bei}_n(z)\\ \operatorname{ber}_n(z)\end{Bmatrix}+\frac{z^n}{2^{n+1}}\sum_{k=0}^{\infty}\begin{Bmatrix}\cos[(3n+2k)\pi/4]\\ \sin[(3n+2k)\pi/4]\end{Bmatrix}\frac{\psi(n+k+1)+\psi(k+1)}{(n+k)!\,k!}\left(\frac{z}{2}\right)^{2k},$$

$$\begin{Bmatrix}\operatorname{ber}(z)\\ \operatorname{bei}(z)\end{Bmatrix}=\begin{Bmatrix}\operatorname{ber}_0(z)\\ \operatorname{bei}_0(z)\end{Bmatrix}=\sum_{k=0}^{\infty}\frac{(-1)^k}{[(2k+1-\delta)!]^2}\left(\frac{z^2}{4}\right)^{2k+1-\delta}\qquad\left[\delta=\begin{Bmatrix}1\\ 0\end{Bmatrix}\right],$$

$$\begin{Bmatrix}\operatorname{ker}(z)\\ \operatorname{kei}(z)\end{Bmatrix}=\begin{Bmatrix}\operatorname{ker}_0(z)\\ \operatorname{kei}_0(z)\end{Bmatrix}=$$
$$=-\ln\frac{z}{2}\begin{Bmatrix}\operatorname{ber}(z)\\ \operatorname{bei}(z)\end{Bmatrix}\pm\frac{\pi}{4}\begin{Bmatrix}\operatorname{bei}(z)\\ \operatorname{ber}(z)\end{Bmatrix}+\sum_{k=0}^{\infty}(-1)^k\frac{\psi(2k+2-\delta)}{[(2k+1-\delta)!]^2}\left(\frac{z^2}{4}\right)^{2k+1-\delta}\quad\left[\delta=\begin{Bmatrix}1\\ 0\end{Bmatrix}\right].$$

$$\begin{Bmatrix}\operatorname{ber}_{-n}(z)\\ \operatorname{bei}_{-n}(z)\end{Bmatrix}=\begin{Bmatrix}\operatorname{ber}_n(-z)\\ \operatorname{bei}_n(-z)\end{Bmatrix}=(-1)^n\begin{Bmatrix}\operatorname{ber}_n(z)\\ \operatorname{bei}_n(z)\end{Bmatrix},$$

$$\begin{Bmatrix}\operatorname{ker}_{-n}(z)\\ \operatorname{kei}_{-n}(z)\end{Bmatrix}=(-1)^n\begin{Bmatrix}\operatorname{ker}_n(z)\\ \operatorname{kei}_n(z)\end{Bmatrix}\qquad[n=0,1,2,\ldots].$$

$$\nu\sqrt{2}(f_\nu-g_\nu)=-z[f_{\nu+1}+f_{\nu-1}],$$

$$2\sqrt{2}\frac{d}{dz}f_\nu=f_{\nu+1}+g_{\nu+1}-f_{\nu-1}-g_{\nu-1},\quad \frac{d}{dz}f_\nu=\pm\frac{\nu}{z}f_\nu\pm\frac{1}{\sqrt{2}}(f_{\nu\pm1}+g_{\nu\pm1}),$$

for any pair of functions

$$\left.\begin{matrix} f_\nu=\operatorname{ber}_\nu(z) \\ g_\nu=\operatorname{bei}_\nu(z)\end{matrix}\right\},\quad \left.\begin{matrix} f_\nu=\operatorname{bei}_\nu(z) \\ g_\nu=-\operatorname{ber}_\nu(z)\end{matrix}\right\},\quad \left.\begin{matrix} f_\nu=\operatorname{ker}_\nu(z) \\ g_\nu=\operatorname{kei}_\nu(z)\end{matrix}\right\},\quad \left.\begin{matrix} f_\nu=\operatorname{kei}_\nu(z) \\ g_\nu=-\operatorname{ker}_\nu(z)\end{matrix}\right\}.$$

$$\sqrt{2}\begin{Bmatrix}\operatorname{ber}(z)\\ \operatorname{bei}(z)\end{Bmatrix}'=\pm\operatorname{ber}_1(z)+\operatorname{bei}_1(z),\quad \sqrt{2}\begin{Bmatrix}\operatorname{ker}(z)\\ \operatorname{kei}(z)\end{Bmatrix}'=\pm\operatorname{ker}_1(z)+\operatorname{kei}_1(z).$$

$$\begin{Bmatrix}\operatorname{ber}_{-\nu}(z)\\ \operatorname{bei}_{-\nu}(z)\end{Bmatrix}=\pm\begin{Bmatrix}\cos\nu\pi\\ \sin\nu\pi\end{Bmatrix}\operatorname{ber}_\nu(z)+\begin{Bmatrix}\sin\nu\pi\\ \cos\nu\pi\end{Bmatrix}\operatorname{bei}_\nu(z)+\frac{2}{\pi}\sin\nu\pi\begin{Bmatrix}\operatorname{ker}_\nu(z)\\ \operatorname{kei}_\nu(z)\end{Bmatrix}.$$

$$\begin{Bmatrix}\operatorname{ker}_{-\nu}(z)\\ \operatorname{kei}_{-\nu}(z)\end{Bmatrix}=\begin{Bmatrix}\cos\nu\pi\\ \sin\nu\pi\end{Bmatrix}\operatorname{ker}_\nu(z)\mp\begin{Bmatrix}\sin\nu\pi\\ \cos\nu\pi\end{Bmatrix}\operatorname{kei}_\nu(z).$$

II.14. THE AIRY FUNCTIONS $\mathbf{Ai}(z)$, $\mathbf{Bi}(z)$

Definitions:

$$\operatorname{Ai}(z)=\frac{1}{\pi}\sqrt{\frac{z}{3}}K_{1/3}\left(\frac{2}{3}z^{3/2}\right),\ \operatorname{Bi}(z)=\sqrt{\frac{z}{3}}\left[I_{-1/3}\left(\frac{2}{3}z^{3/2}\right)+I_{1/3}\left(\frac{2}{3}z^{3/2}\right)\right].$$

$$\begin{Bmatrix}\operatorname{Ai}(z)\\ \operatorname{Bi}(z)\end{Bmatrix}=\frac{3^{(-5\mp3)/12}}{\Gamma(2/3)}\sum_{k=0}^{\infty}\left(\frac{1}{3}\right)_k\frac{(3z^3)^k}{(3k)!}\mp\frac{3^{(-1\mp3)/12}}{\Gamma(1/3)}z\sum_{k=0}^{\infty}\left(\frac{2}{3}\right)_k\frac{(3z^3)^k}{(3k+1)!}.$$

$$\operatorname{Ai}\left(e^{\pm2\pi i/3}z\right)=\frac{1}{2}e^{\pm\pi i/3}\left[\operatorname{Ai}(z)\mp i\operatorname{Bi}(z)\right],$$

$$\begin{Bmatrix}\operatorname{Ai}(z)\\ \operatorname{Bi}(z)\end{Bmatrix}+e^{2\pi i/3}\begin{Bmatrix}\operatorname{Ai}(e^{2\pi i/3}z)\\ \operatorname{Bi}(e^{2\pi i/3}z)\end{Bmatrix}+e^{-2\pi i/3}\begin{Bmatrix}\operatorname{Ai}(e^{-2\pi i/3}z)\\ \operatorname{Bi}(e^{-2\pi i/3}z)\end{Bmatrix}=0,$$

$$\operatorname{Bi}(z)=e^{\pi i/6}\operatorname{Ai}(e^{2\pi i/3}z)+e^{-\pi i/6}\operatorname{Ai}(e^{-2\pi i/3}z).$$

II.15. THE INTEGRAL BESSEL FUNCTIONS $Ji_\nu(z)$, $Yi_\nu(z)$, $Ki_\nu(z)$

Definitions:

$$Ji_\nu(x)=\int_x^\infty J_\nu(t)\frac{dt}{t},\quad \begin{Bmatrix}Ki_\nu(x)\\ Yi_\nu(x)\end{Bmatrix}=\int_x^\infty\begin{Bmatrix}K_\nu(t)\\ Y_\nu(t)\end{Bmatrix}\frac{dt}{t}.$$

These functions are defined on the z-plane as analytic continuations of the above integrals by means of the series

$$Ji_\nu(z)=-\frac{z^\nu}{2^\nu\nu^2\Gamma(\nu)}\,{}_1F_2\left(\frac{\nu}{2};\ \frac{\nu}{2}+1,\ \nu+1;\ -\frac{z^2}{4}\right)+\frac{1}{\nu},$$

$$\begin{Bmatrix}Yi_\nu(z)\\ Ki_\nu(z)\end{Bmatrix}=\pm\frac{1}{2\nu}\begin{Bmatrix}\operatorname{ctg}(\nu\pi/2)\\ \pi\operatorname{cosec}(\nu\pi/2)\end{Bmatrix}\pm\frac{\Gamma(-\nu)}{2\pi\nu}\begin{Bmatrix}\cos\nu\pi\\ \pi\end{Bmatrix}\left(\frac{z}{2}\right)^\nu\times$$

$$\times{}_1F_2\left(\frac{\nu}{2};\ 1+\frac{\nu}{2},\ 1+\nu;\ \mp\frac{z^2}{4}\right)\mp\frac{\Gamma(\nu)}{2\nu\pi}\begin{Bmatrix}1\\ \pi\end{Bmatrix}\left(\frac{z}{2}\right)^{-\nu}\times$$

$$\times{}_1F_2\left(-\frac{\nu}{2};\ 1-\frac{\nu}{2},\ 1-\nu;\ \mp\frac{z^2}{4}\right)\qquad[\nu\neq\pm n],$$

$$\begin{Bmatrix}Yi_n(z)\\ Ki_n(z)\end{Bmatrix}=\begin{Bmatrix}0\\ 1\end{Bmatrix}(-1)^{(n+1)/2}\frac{1-(-1)^n}{4n}\pi\pm\frac{1}{2\pi}\begin{Bmatrix}1\\ \pi\end{Bmatrix}\left(\frac{z}{2}\right)^{-n}\times$$

$$\times\sum_{\substack{k=0\\ k\neq n/2}}^{n-1}\frac{(n-k-1)!}{k!\,(2k-n)}\left(\pm\frac{z^2}{4}\right)^k+\frac{(\pm1)^{n-1}}{2\pi}\begin{Bmatrix}1\\ \pi\end{Bmatrix}\left(\frac{z}{2}\right)^n\sum_{k=0}^{\infty}\frac{(\mp z^2/4)^k}{k!\,(k+n)!\,(2k+n)}\times$$

$$\times\left[\psi(k+1)+\psi(k+n+1)+\frac{2}{2k+n}-2\ln\frac{z}{2}\right]-$$

$$-(\pm1)^{n/2+1}\frac{1+(-1)^n}{4n\pi}\begin{Bmatrix}1\\ \pi\end{Bmatrix}\left[2\psi\left(\frac{n}{2}\right)+\frac{2}{n}-2\ln\frac{z}{2}\right]\qquad[n=1,2,3,\ldots].$$

$$\left\{\begin{matrix} Yi_0(z) \\ Ki_0(z) \end{matrix}\right\} = \pm \frac{1}{2\pi} \left\{\begin{matrix} 1 \\ \pi \end{matrix}\right\} \sum_{k=1}^{\infty} \frac{(\mp z^2/4)^k}{(k!)^2\, k} \left[\psi(k+1) + \frac{1}{2k} - \ln \frac{z}{2} \right] \mp$$

$$\mp \frac{1}{2\pi} \left\{\begin{matrix} 1 \\ \pi \end{matrix}\right\} \left[\mathbf{C}^2 - \frac{\pi^2}{6} + 2\mathbf{C} \ln \frac{z}{2} + \ln^2 \frac{z}{2} \right].$$

$$Ji_{-n}(z) = (-1)^n\, Ji_n(z), \quad Yi_{-n}(z) = (-1)^n\, Yi_n(z), \quad Ki_{-\nu}(z) = Ki_\nu(z).$$

$$z \frac{d}{dz} Ji_\nu(z) = -J_\nu(z), \quad z \frac{d}{dz} \left\{\begin{matrix} Yi_\nu(z) \\ Ki_\nu(z) \end{matrix}\right\} = -\left\{\begin{matrix} Y_\nu(z) \\ K_\nu(z) \end{matrix}\right\},$$

$$Ji_{\pm 1/2}(z) = \pm Yi_{\mp 1/2}(z) = \sqrt{\frac{2}{\pi}} \left\{\begin{matrix} S(z, -1/2) \\ C(z, -1/2) \end{matrix}\right\}. \quad Ki_{\pm 1/2}(z) = \sqrt{\frac{\pi}{2}}\, \Gamma\left(-\frac{1}{2}, z\right).$$

II.16. THE INCOMPLETE ELLIPTIC INTEGRALS $F(\varphi, k)$, $E(\varphi, k)$, $D(\varphi, k)$, $\Pi(\varphi, \nu, k)$, $\Lambda_0(\varphi, \beta, k)$ AND THE COMPLETE ELLIPTIC INTEGRALS $\mathbf{K}(k)$, $\mathbf{E}(k)$, $\mathbf{D}(k)$

Definitions:

$$0 < k < 1, \quad k' = \sqrt{1-k^2}, \quad \Delta t = \sqrt{1-k^2 \sin^2 t},$$

$$\beta = \arcsin \sqrt{\frac{\nu - k^2}{\nu k'^2}}, \quad k^2 < \nu < 1, \quad q = \sqrt{(1-\nu)\left(1-\frac{k^2}{\nu}\right)},$$

$$q_1 = \sqrt{(1-\nu_1)\left(1-\frac{k^2}{\nu_1}\right)}, \quad \nu = \frac{\nu_1 - k^2}{\nu_1 - 1}, \quad \nu_1 < 0,$$

$$\operatorname{tg}(\psi - \varphi) = k' \operatorname{tg} \varphi, \quad \sin \chi = \frac{(1+k)\sin\varphi}{1+k\sin^2\varphi}.$$

Definitions:

$$F(\varphi, k) = \int_0^{\varphi} \frac{dt}{\Delta t} = \int_0^{\sin\varphi} \frac{dt}{\sqrt{(1-t^2)(1-k^2t^2)}},$$

$$E(\varphi, k) = \int_0^{\varphi} \Delta t\, dt = \int_0^{\sin\varphi} \frac{\sqrt{1-k^2t^2}}{\sqrt{1-t^2}}\, dt,$$

$$D(\varphi, k) = \int_0^{\varphi} \frac{\sin^2 t\, dt}{\Delta t} = \int_0^{\sin\varphi} \frac{t^2\, dt}{\sqrt{(1-t^2)(1-k^2t^2)}},$$

$$\Pi(\varphi, \nu, k) = \int_0^{\varphi} \frac{dt}{(1-\nu\sin^2 t)\,\Delta t} = \int_0^{\sin\varphi} \frac{dt}{(1-\nu t^2)\sqrt{(1-t^2)(1-k^2t^2)}}.$$

$$\Lambda_0(\varphi, \beta, k) = \frac{2q}{\pi}\, \Pi(\varphi, \nu, k).$$

$$\mathbf{K}(k) = F\left(\frac{\pi}{2}, k\right), \quad \mathbf{E}(k) = E\left(\frac{\pi}{2}, k\right), \quad \mathbf{D}(k) = D\left(\frac{\pi}{2}, k\right).$$

$$\left\{\begin{matrix} F(-\varphi, k) \\ E(-\varphi, k) \end{matrix}\right\} = -\left\{\begin{matrix} F(\varphi, k) \\ E(\varphi, k) \end{matrix}\right\}, \quad \left\{\begin{matrix} F(n\pi \pm \varphi, k) \\ E(n\pi \pm \varphi, k) \end{matrix}\right\} = \pm \left\{\begin{matrix} F(\varphi, k) \\ E(\varphi, k) \end{matrix}\right\} + 2n \left\{\begin{matrix} \mathbf{K}(k) \\ \mathbf{E}(k) \end{matrix}\right\}.$$

$$D(\varphi, k) = \frac{1}{k^2}\, [F(\varphi, k) - E(\varphi, k)].$$

$$F\left(\psi, \frac{1-k'}{1+k'}\right) = (1+k')\, F(\varphi, k).$$

$$E\left(\psi, \frac{1-k'}{1+k'}\right)=\frac{2}{1+k'}\,[E(\varphi, k)+k'F(\varphi, k)]-\frac{1-k'}{1+k'}\sin\psi.$$

$$F\left(\chi, \frac{2\sqrt{k}}{1+k}\right)=(1+k)\,F(\varphi, k).$$

$$E\left(\chi, \frac{2\sqrt{k}}{1+k}\right)=\frac{1}{1+k}\left[2E(\varphi, k)-k'^2F(\varphi, k)+k\frac{\sin 2\varphi}{1+k\sin^2\varphi}\sqrt{1-k^2\sin^2\varphi}\right].$$

$$\frac{\partial F(\varphi, k)}{\partial k}=\frac{1}{k'^2}\left[\frac{1}{k}E(\varphi, k)-\frac{k'^2}{k}F(\varphi, k)-\frac{k\sin 2\varphi}{2\sqrt{1-k^2\sin^2\varphi}}\right],$$

$$\frac{\partial E(\varphi, k)}{\partial k}=\frac{1}{k}\,[E(\varphi, k)-F(\varphi, k)]=-kD(\varphi, k).$$

$$\mathbf{E}(k)\,\mathbf{K}(k')+\mathbf{E}(k')\,\mathbf{K}(k)-\mathbf{K}(k)\,\mathbf{K}(k')=\frac{\pi}{2}.$$

$$\mathbf{D}(k)=\frac{1}{k^2}\,[\mathbf{K}(k)-\mathbf{E}(k)],\qquad \mathbf{K}\left(\frac{1-k'}{1+k'}\right)=\frac{1+k'}{2}\,\mathbf{K}(k),$$

$$\mathbf{E}\left(\frac{1-k'}{1+k'}\right)=\frac{1}{1+k'}\,[\mathbf{E}(k)+k'\mathbf{K}(k)],\qquad \mathbf{K}\left(\frac{2\sqrt{k}}{1+k}\right)=(1+k)\,\mathbf{K}(k),$$

$$\mathbf{E}\left(\frac{2\sqrt{k}}{1+k}\right)=\frac{1}{1+k}\,[2\mathbf{E}(k)-k'^2\mathbf{K}(k)],\qquad \mathbf{K}\left(i\frac{k}{k'}\right)=k'\mathbf{K}(k),$$

$$\mathbf{K}\left(\frac{1}{k}\right)=k\mathbf{K}(k)+i\mathbf{K}(k'),\quad \mathbf{K}\left(\sin\frac{\pi}{4}\right)=\mathbf{K}\left(\frac{\sqrt{2}}{2}\right)=\frac{1}{4\sqrt{\pi}}\,\Gamma^2\left(\frac{1}{4}\right),$$

$$\frac{d\mathbf{K}(k)}{dk}=\frac{1}{k}\left[\frac{1}{k'^2}\mathbf{E}(k)-\mathbf{K}(k)\right],\qquad \frac{d\mathbf{E}(k)}{dk}=\frac{1}{k}\,[\mathbf{E}(k)-\mathbf{K}(k)],$$

$$\mathbf{K}(k)=\frac{\pi}{2}\,{}_2F_1\left(\frac{1}{2}, \frac{1}{2};\ 1;\ k^2\right),\quad \mathbf{E}(k)=\frac{\pi}{2}\,{}_2F_1\left(-\frac{1}{2}, \frac{1}{2};\ 1;\ k^2\right),$$

$$\mathbf{D}(k)=\frac{\pi}{4}\,{}_2F_1\left(\frac{1}{2}, \frac{3}{2};\ 2;\ k^2\right).$$

For values of $\mathbf{K}(k)$, $\mathbf{E}(k)$ and $\mathbf{D}(k)$ for certain k See Ch. 7.

$$q_1\Pi(\varphi, \nu_1, k)=q\Pi(\varphi, \nu, k)+\frac{k^2}{\sqrt{-\nu\nu_1}}F(\varphi, k)+\operatorname{arctg}\frac{\sqrt{-\nu\nu_1}\sin 2\varphi}{2\sqrt{1-k^2\sin^2\varphi}}$$

$[\nu_1 < 0;\ k^2 < \nu < 1]$.

$$\Pi\left(\frac{\pi}{2}, \nu, k\right)=\sin^2\psi\mathbf{K}(k)+\frac{\sin 2\psi}{2\sqrt{1-k'^2\sin^2\psi}}\Big[F(\psi, k')\,\mathbf{K}(k)-$$

$$-E(\psi, k')\,\mathbf{K}(k)-F(\psi, k')\,\mathbf{E}(k)+\frac{\pi}{2}\Big]$$

$[\nu=\operatorname{ctg}^2\psi;\ 0 < \psi \leqslant \pi/2]$.

$$=\mathbf{K}(k)+\frac{2\sqrt{1-k'^2\sin^2\psi}}{k'^2\sin 2\psi}\Big[F(\psi, k')\,\mathbf{K}(k)-E(\psi, k')\,\mathbf{K}(k)-$$

$$-F(\psi, k')\,\mathbf{E}(k)+\frac{\pi}{2}\Big]$$

$[\nu=k'^2\sin\psi-1;\ -1 \leqslant \nu \leqslant -k^2;\ 0 \leqslant \psi \leqslant \pi/2]$.

$$=\mathbf{K}(k)+\frac{\operatorname{tg}\psi}{\sqrt{1-k^2\sin^2\psi}}\,[E(\psi, k)\,\mathbf{K}(k)-F(\psi, k)\,\mathbf{E}(k)]$$

$[\nu=-k^2\sin^2\psi;\ -k^2 \leqslant \nu \leqslant 0;\ 0 \leqslant \psi \leqslant \pi/2]$.

$$\Pi(\varphi, 0, k)=F(\varphi, k).$$

$$\Pi(\varphi, -1, k)=F(\varphi, k)-\frac{1}{k'^2}E(\varphi, k)+\frac{\sin\varphi\sqrt{1-k^2\sin^2\varphi}}{k'^2\cos\varphi}.$$

$$\Pi(\varphi, -k^2, k) = \frac{1}{k'^2} E(\varphi, k) - \frac{k^2 \sin 2\varphi}{2k'^2 \sqrt{1-k^2\sin^2\varphi}}.$$

$$\Pi(\varphi, \nu, 0) = \frac{1}{\sqrt{1-\nu}} \operatorname{arctg}\left(\sqrt{1-\nu}\,\operatorname{tg}\varphi\right) \qquad [\nu < 1].$$

$$\Pi(\varphi, 0, 0) = \varphi.$$

$$\frac{\pi}{2}\Lambda_0\left(\frac{\pi}{2}, \beta, k\right) = \mathbf{K}(k) E(\beta, k') - [\mathbf{K}(k) - \mathbf{E}(k)] F(\beta, k').$$

$$\Lambda_0\left(\frac{\pi}{2}, \beta, 0\right) = \sin\beta, \quad \Lambda_0\left(\frac{\pi}{2}, \beta, 1\right) = \frac{2\beta}{\pi}.$$

$$\Lambda_0\left(\frac{\pi}{2}, 0, k\right) = 0, \quad \Lambda_0\left(\frac{\pi}{2}, \frac{\pi}{2}, k\right) = 1.$$

$$\Lambda_0\left(\frac{\pi}{2}, n\pi \pm \beta, k\right) = 2n \pm \Lambda_0\left(\frac{\pi}{2}, \beta, k\right). \quad \Lambda_0\left(\frac{\pi}{2}, n\frac{\pi}{2}, k\right) = n.$$

II.17. THE BATEMAN FUNCTION $k_\nu(z)$

Definition:

$$k_\nu(z) = \frac{1}{\Gamma(1+\nu/2)} W_{\nu/2,\,1/2}(2z) = \frac{e^{-z}}{\Gamma(1+\nu/2)} \Psi\left(-\frac{\nu}{2}, 0; 2z\right) =$$

$$= \frac{ze^{-z}}{\Gamma(1+\nu/2)} \Psi\left(1-\frac{\nu}{2}, 2; 2z\right) = \frac{2}{\pi}\sin\frac{\nu\pi}{2} e^{-z} \times$$

$$\times \left\{\frac{1}{\nu} - z\sum_{k=0}^{\infty} \frac{(1-\nu/2)_k (2z)^k}{(k!)^2 (k+1)} \left[2\psi(k+1) + \frac{1}{k+1} - \psi\left(k-\frac{\nu}{2}+1\right) - \ln(2z)\right]\right. \qquad [\nu \neq 0, 2, 4, \ldots].$$

$$\frac{d^n}{dz^n}\left[z^{n\pm\nu/2-1} e^{\mp z} k_\nu(z)\right] = (-1)^n \left(1 \pm \frac{\nu}{2}\right)_n z^{\pm\nu/2-1} e^{\mp z} k_{\nu\pm 2n}(z).$$

II.18. THE LEGENDRE FUNCTIONS $P_\nu(z)$, $P_\nu^\mu(z)$, $Q_\nu(z)$, $Q_\nu^\mu(z)$

Definitions:

$$P_\nu^\mu(z) = \frac{1}{\Gamma(1-\mu)} \left(\frac{z+1}{z-1}\right)^{\mu/2} {}_2F_1\left(-\nu, \nu+1; 1-\mu; \frac{1-z}{2}\right)$$
$$[|\arg(z-1)| < \pi;\ \mu \neq m;\ m = 1, 2, \ldots].$$

$$P_\nu^m(z) = (z^2-1)^{m/2} \left(\frac{d}{dz}\right)^m P_\nu(z) \qquad [|\arg(z-1)| < \pi;\ m = 1, 2, \ldots].$$

$$P_\nu^\mu(x) = \frac{1}{\Gamma(1-\mu)} \left(\frac{1+x}{1-x}\right)^{\mu/2} {}_2F_1\left(-\nu, \nu+1; 1-\mu; \frac{1-x}{2}\right)$$
$$[-1 < x < 1;\ \mu \neq m;\ m = 1, 2, \ldots].$$

$$P_\nu^m(x) = (-1)^m (1-x^2)^{m/2} \left(\frac{d}{dx}\right)^m P_\nu(x) \qquad [-1 < x < 1;\ m = 1, 2, \ldots].$$

$$P_\nu(z) = P_\nu^0(z) = {}_2F_1\left(-\nu, 1+\nu; 1; \frac{1-z}{2}\right) \qquad [|\arg(z+1)| < \pi].$$

$$Q_\nu^\mu(z) = \frac{e^{i\mu\pi}\sqrt{\pi}}{2^{\nu+1}} \Gamma\begin{bmatrix}\mu+\nu+1\\ \nu+3/2\end{bmatrix} z^{-\mu-\nu-1} (z^2-1)^{\mu/2} \times$$

$$\times {}_2F_1\left(\frac{\mu+\nu+1}{2}, \frac{\mu+\nu}{2}+1; \nu+\frac{3}{2}; \frac{1}{z^2}\right)$$
$$[|\arg z|,\ |\arg(z\pm 1)| < \pi;\ \nu + 1/2,\ \mu+\nu \neq -1, -2, -3, \ldots].$$

$$Q^{\mu}_{-n-3/2}(z)=\frac{e^{i\mu\pi}\sqrt{\pi}\,\Gamma(\mu+n+3/2)}{2^{n+3/2}(n+1)!}z^{-\mu-n-3/2}(z^2-1)^{\mu/2}\times$$
$$\times{}_2F_1\left(\frac{2\mu+2n+3}{4},\ \frac{2\mu+2n+5}{4};\ n+2;\ \frac{1}{z^2}\right)$$
$$[|\arg z|,\ |\arg(z\pm 1)|<\pi;\ \mu+\nu\neq -1,\ -2,\ -3,\dots].$$

$$Q^{\mu}_{\nu}(x)=\frac{e^{-i\mu\pi}}{2}\left[e^{-i\mu\pi/2}Q^{\mu}_{\nu}(x+i0)+e^{i\mu\pi/2}Q^{\mu}_{\nu}(x-i0)\right]=$$
$$=\frac{\pi}{2\sin\mu\pi}\left[P^{\mu}_{\nu}(x)\cos\mu\pi-\Gamma\begin{bmatrix}\nu+\mu+1\\ \nu-\mu+1\end{bmatrix}P^{-\mu}_{\nu}(x)\right]$$
$$[-1<x<1;\ \mu\neq\pm m;\ \mu+\nu\neq -1,\ -2,\ -3,\dots],$$
$$=(-1)^m(1-x^2)^{m/2}\left(\frac{d}{dx}\right)^m Q_{\nu}(x)\qquad[\mu=m;\ \nu\neq -m-1,\ -m-2,\dots],$$
$$=(-1)^m\,\Gamma\begin{bmatrix}\nu-m+1\\ \nu+m+1\end{bmatrix}Q^{m}_{\nu}(x)\qquad[\mu=-m;\ \nu\neq m-1,\ m-2,\dots].$$
$$Q_{\nu}(z)=Q^{0}_{\nu}(z)\qquad[|\arg(z-1)|<\pi].$$

Transformation formulae:

$$P^{\mu}_{-\nu-1}(z)=P^{\mu}_{\nu}(z),\quad P^{\mu}_{-\nu-1}(x)=P^{\mu}_{\nu}(x)\qquad[-1<x<1].$$
$$P^{-m}_{\nu}(z)=\Gamma\begin{bmatrix}\nu-m+1\\ \nu+m+1\end{bmatrix}P^{m}_{\nu}(z).$$
$$P^{-m}_{\nu}(x)=(-1)^m\,\Gamma\begin{bmatrix}\nu-m+1\\ \nu+m+1\end{bmatrix}P^{m}_{\nu}(x)\qquad[-1<x<1].$$
$$P^{m}_{n}(z)=P^{m}_{n}(x)=0\qquad[m>n].$$
$$P^{m}_{n}(-x)=(-1)^{m+n}P^{m}_{n}(x)\qquad[-1<x<1].$$
$$P^{\mu}_{\nu}(x)=\frac{1}{2}\left[e^{i\mu\pi/2}P^{\mu}_{\nu}(x+i0)+e^{i\mu\pi/2}P^{\mu}_{\nu}(x-i0)\right]\qquad[-1<x<1].$$
$$P^{\mu}_{\nu}(x)=e^{i\mu\pi/2}P^{\mu}_{\nu}(x+i0)=e^{-i\mu\pi/2}P^{\mu}_{\nu}(x-i0)\qquad[-1<x<1].$$
$$Q^{-\mu}_{\nu}(z)=e^{-2i\mu\pi}\,\Gamma\begin{bmatrix}\nu-\mu+1\\ \nu+\mu+1\end{bmatrix}Q^{\mu}_{\nu}(z).\quad Q^{m}_{\nu}(z)=(z^2-1)^{m/2}\left(\frac{d}{dz}\right)^m Q_{\nu}(z).$$
$$Q^{\mu}_{\nu}(-z)=-e^{\pm i\nu\pi}Q^{\mu}_{\nu}(z)\qquad[\operatorname{Im} z\gtrless 0].$$
$$Q^{m}_{n}(-x)=(-1)^{m+n+1}Q^{m}_{n}(x)\qquad[-1<x<1].$$
$$P^{\mu}_{\nu}(z)=\frac{ie^{i\nu\pi}}{\Gamma(-\mu-\nu)}\sqrt{\frac{2}{\pi}}(z^2-1)^{-1/4}Q^{-\nu-1/2}_{-\mu-1/2}\left(\frac{z}{\sqrt{z^2-1}}\right)\qquad[\operatorname{Re} z>0].$$
$$P^{\mu}_{\nu}(i\operatorname{ctg}\varphi)=\sqrt{\frac{2}{\pi}}e^{i(\nu+1/2)\pi}\frac{\sqrt{\sin\varphi}}{\Gamma(-\mu-\nu)}Q^{-\nu-1/2}_{-\mu-1/2}(\cos\varphi-i0)$$
$$[0<\varphi<\pi/2].$$
$$P^{-\mu}_{\nu}(z)=\frac{e^{-i\mu\pi}\sin(\nu-\mu)\pi}{\pi\cos\nu\pi}\Gamma\begin{bmatrix}\nu-\mu+1\\ \nu+\mu+1\end{bmatrix}\left[Q^{\mu}_{\nu}(z)-Q^{\mu}_{-\nu-1}(z)\right].$$
$$P^{-\mu}_{\nu}(z)=\Gamma\begin{bmatrix}\nu-\mu+1\\ \nu+\mu+1\end{bmatrix}\left\{P^{\mu}_{\nu}(z)-\frac{2}{\pi}e^{-i\mu\pi}\sin\mu\pi\,Q^{\mu}_{\nu}(z)\right\}.$$
$$P^{-\mu}_{\nu}(x)=\Gamma\begin{bmatrix}\nu-\mu+1\\ \nu+\mu+1\end{bmatrix}\left\{\cos\mu\pi\,P^{\mu}_{\nu}(x)-\frac{2}{\pi}\sin\mu\pi\,Q^{\mu}_{\nu}(x)\right\}$$
$$[-1<x<1].$$
$$P^{\mu}_{\nu}(-z)=e^{\mp i\nu\pi}P^{\mu}_{\nu}(z)-\frac{2}{\pi}\sin(\mu+\nu)\pi e^{-i\mu\pi}Q^{\mu}_{\nu}(z)\qquad[\operatorname{Im} z\gtrless 0].$$
$$P^{\mu}_{\nu}(-x)=\cos(\mu+\nu)\pi P^{\mu}_{\nu}(x)-\frac{2}{\pi}\sin(\mu+\nu)\pi Q^{\mu}_{\nu}(x)\qquad[0<x<1].$$

$$Q_\nu^\mu(z)=\sqrt{\frac{\pi}{2}}\,e^{i\mu\pi}\Gamma(\mu+\nu+1)(z^2-1)^{-1/4}P_{-\mu-1/2}^{-\nu-1/2}\left(\frac{z}{\sqrt{z^2-1}}\right)$$

[Re $z>0$].

$$Q_\nu^\mu(i\operatorname{ctg}\varphi)=\exp\left[i\pi\left(\mu-\frac{\nu+1}{2}\right)\right]\sqrt{\frac{\pi\sin\varphi}{2}}\,\Gamma(\mu+\nu+1)\,P_{-\nu-1/2}^{-\mu-1/2}(\cos\varphi)$$

[$0<\varphi<\pi/2$].

$$Q_\nu^\mu(z)=\frac{\pi e^{i\mu\pi}}{2\sin\mu\pi}\left[P_\nu^\mu(z)-\Gamma\begin{bmatrix}\nu+\mu+1\\ \nu-\mu+1\end{bmatrix}P_\nu^{-\mu}(z)\right].$$

$$Q_\nu^\mu(x)=\frac{\pi}{2\sin\mu\pi}\left\{\cos\mu\pi P_\nu^\mu(x)-\Gamma\begin{bmatrix}\nu+\mu+1\\ \nu-\mu+1\end{bmatrix}P_\nu^{-\mu}(x)\right\}\qquad[-1<x<1].$$

$$Q_\nu^\mu(z)=\frac{\pi e^{i\mu\pi}}{2\sin(\mu+\nu)\pi}\left[e^{\mp i\nu\pi}P_\nu^\mu(z)-P_\nu^\mu(-z)\right]\qquad[\operatorname{Im} z\gtrless 0].$$

$$Q_{-\nu-1}^\mu(z)-Q_\nu^\mu(z)=e^{i\mu\pi}\cos\nu\pi\,\Gamma(\mu+\nu+1)\,\Gamma(\mu-\nu)\,P_\nu^{-\mu}(z).$$

$$\sin(\mu+\nu)\pi Q_\nu^\mu(z)+\sin(\mu-\nu)\pi Q_{-\nu-1}^\mu(z)=\pi e^{i\mu\pi}\cos\mu\pi\,P_\nu^\mu(z).$$

$$\sin(\mu+\nu)\pi Q_\nu^\mu(x)+\sin(\mu-\nu)\pi Q_{-\nu-1}^\mu(x)=\pi\cos\mu\pi\cos\nu\pi P_\nu^\mu(x)$$

[$-1<x<1$].

$$Q_\nu^{-\mu}(x)=\Gamma\begin{bmatrix}\nu-\mu+1\\ \nu+\mu+1\end{bmatrix}\left\{\cos\mu\pi Q_\nu^\mu(x)+\frac{\pi}{2}\sin\mu\pi P_\nu^\mu(x)\right\}\qquad[-1<x<1].$$

$$Q_\nu^\mu(-x)=-\cos(\mu+\nu)\pi Q_\nu^\mu(x)-\frac{\pi}{2}\sin(\mu+\nu)\pi P_\nu^\mu(x)\qquad[0<x<1].$$

$$e^{-i\mu\pi/2}Q_\nu^\mu(x+i0)-e^{i\mu\pi/2}Q_\nu^\mu(x-i0)=i\pi e^{i\mu\pi}P_\nu^\mu(x)\qquad[-1<x<1].$$

$$Q_\nu^\mu(x\pm i0)=e^{(1\pm 1/2)i\mu\pi}\left[Q_\nu^\mu(x)\mp\frac{i\pi}{2}P_\nu^\mu(x)\right]\qquad[-1<x<1].$$

Recurrence formulae:
(if $z\notin[-1,1]$, then $A=B=C=D=1$; if $z=x\in[-1,1]$, then one must set $A=-D=-B^{-1}=-\frac{\sqrt{1-x^2}}{\sqrt{z^2-1}}$, $C=-1$ and replace z by x after cancelling fractions):

$$(2\nu+1)z\begin{Bmatrix}P_\nu^\mu(z)\\ Q_\nu^\mu(z)\end{Bmatrix}=(\nu-\mu+1)\begin{Bmatrix}P_{\nu+1}^\mu(z)\\ Q_{\nu+1}^\mu(z)\end{Bmatrix}+(\nu+\mu)\begin{Bmatrix}P_{\nu-1}^\mu(z)\\ Q_{\nu-1}^\mu(z)\end{Bmatrix}.$$

$$\begin{Bmatrix}P_{\nu+1}^\mu(z)-P_{\nu-1}^\mu(z)\\ Q_{\nu+1}^\mu(z)-Q_{\nu-1}^\mu(z)\end{Bmatrix}=A(2\nu+1)\sqrt{z^2-1}\begin{Bmatrix}P_\nu^{\mu-1}(z)\\ Q_\nu^{\mu-1}(z)\end{Bmatrix}.$$

$$\begin{Bmatrix}P_\nu^{\mu+2}(z)\\ Q_\nu^{\mu+2}(z)\end{Bmatrix}+2B(\mu+1)\frac{z}{\sqrt{z^2-1}}\begin{Bmatrix}P_\nu^{\mu+1}(z)\\ Q_\nu^{\mu+1}(z)\end{Bmatrix}=C(\nu-\mu)(\nu+\mu+1)\begin{Bmatrix}P_\nu^\mu(z)\\ Q_\nu^\mu(z)\end{Bmatrix}.$$

$$\begin{Bmatrix}P_{\nu-1}^\mu(z)-zP_\nu^\mu(z)\\ Q_{\nu-1}^\mu(z)-zQ_\nu^\mu(z)\end{Bmatrix}=A(\mu-\nu-1)\sqrt{z^2-1}\begin{Bmatrix}P_\nu^{\mu-1}(z)\\ Q_\nu^{\mu-1}(z)\end{Bmatrix}.$$

$$\begin{Bmatrix}zP_\nu^\mu(z)-P_{\nu+1}^\mu(z)\\ zQ_\nu^\mu(z)-Q_{\nu+1}^\mu(z)\end{Bmatrix}=-A(\mu+\nu)\sqrt{z^2-1}\begin{Bmatrix}P_\nu^{\mu-1}(z)\\ Q_\nu^{\mu-1}(z)\end{Bmatrix}.$$

$$(\nu-\mu)z\begin{Bmatrix}P_\nu^\mu(z)\\ Q_\nu^\mu(z)\end{Bmatrix}-(\nu+\mu)\begin{Bmatrix}P_{\nu-1}^\mu(z)\\ Q_{\nu-1}^\mu(z)\end{Bmatrix}=D\sqrt{z^2-1}\begin{Bmatrix}P_\nu^{\mu+1}(z)\\ Q_\nu^{\mu+1}(z)\end{Bmatrix}.$$

$$(\nu-\mu+1)\begin{Bmatrix}P_{\nu+1}^\mu(z)\\ Q_{\nu+1}^\mu(z)\end{Bmatrix}-(\nu+\mu+1)z\begin{Bmatrix}P_\nu^\mu(z)\\ Q_\nu^\mu(z)\end{Bmatrix}=D\sqrt{z^2-1}\begin{Bmatrix}P_\nu^{\mu+1}(z)\\ Q_\nu^{\mu+1}(z)\end{Bmatrix}.$$

$$(\nu-\mu)(\nu-\mu+1)\begin{Bmatrix}P_{\nu+1}^\mu(z)\\ Q_{\nu+1}^\mu(z)\end{Bmatrix}-(\nu+\mu)(\nu+\mu+1)\begin{Bmatrix}P_{\nu-1}^\mu(z)\\ Q_{\nu-1}^\mu(z)\end{Bmatrix}=$$

$$=D(2\nu+1)\sqrt{z^2-1}\begin{Bmatrix}P_\nu^{\mu+1}(z)\\ Q_\nu^{\mu+1}(z)\end{Bmatrix}.$$

Representations in terms of the Gauss hypergeometric function:

$$P_\nu^\mu(z)=\frac{\sin\mu\pi}{\pi}\Gamma\left[\begin{matrix}\mu+\nu+1,\ \mu-\nu\\ \mu+1\end{matrix}\right]\left(\frac{z+1}{z-1}\right)^{\mu/2}\left[{}_2F_1\left(-\nu,\ \nu+1;\ \mu+1;\ \frac{1+z}{2}\right)-\right.$$
$$\left.-\frac{\sin\nu\pi}{\sin\mu\pi}e^{\mp i\mu\pi}\left(\frac{z-1}{z+1}\right)^{\mu}{}_2F_1\left(-\nu,\ \nu+1;\ \mu+1;\ \frac{1-z}{2}\right)\right]\qquad [\operatorname{Im} z \gtrless 0].$$

$$P_\nu^m(z)=\frac{2^{-m}}{m!}\Gamma\left[\begin{matrix}\nu+m+1\\ \nu-m+1\end{matrix}\right](z^2-1)^{m/2}{}_2F_1\left(m+\nu+1,\ m-\nu;\ m+1;\ \frac{1-z}{2}\right).$$

$$P_\nu^m(x)=\frac{(-2)^{-m}}{m!}\Gamma\left[\begin{matrix}\nu+m+1\\ \nu-m+1\end{matrix}\right](1-x^2)^{m/2}{}_2F_1\left(m+\nu+1,\ m-\nu;\ m+1;\ \frac{1-x}{2}\right)$$
$$[-1<x<1].$$

$$\left\{\begin{matrix}P_\nu^\mu(x)\\ Q_\nu^\mu(x)\end{matrix}\right\}=\pm\frac{2^{\mu-(1\mp 1)/2}(1-x^2)^{-\mu/2}}{\pi^{\pm 1/2}}\left\{\begin{matrix}\cos[(\mu+\nu)\pi/2]\\ \sin[(\mu+\nu)\pi/2]\end{matrix}\right\}\times$$
$$\times\Gamma\left[\begin{matrix}(1+\nu+\mu)/2\\ 1+(\nu-\mu)/2\end{matrix}\right]{}_2F_1\left(-\frac{\nu+\mu}{2},\ \frac{1+\nu-\mu}{2};\ \frac{1}{2};\ x^2\right)+$$
$$+\frac{2^{\mu+(1\pm 1)/2}x(1-x^2)^{-\mu/2}}{\pi^{\pm 1/2}}\left\{\begin{matrix}\sin[(\mu+\nu)\pi/2]\\ \cos[(\mu+\nu)\pi/2]\end{matrix}\right\}\times$$
$$\times\Gamma\left[\begin{matrix}1+(\nu+\mu)/2\\ (1+\nu-\mu)/2\end{matrix}\right]{}_2F_1\left(\frac{1-\mu-\nu}{2},\ 1+\frac{\nu-\mu}{2};\ \frac{3}{2};\ x^2\right)\qquad [-1<x<1].$$

$$Q_\nu^\mu(z)=\frac{\pi e^{i\mu\pi}}{2\sin(\mu+\nu)\pi\Gamma(1-\mu)}\left[e^{\mp i\nu\pi}\left(\frac{z+1}{z-1}\right)^{\mu/2}{}_2F_1\left(-\nu,\ \nu+1;\ 1-\mu;\ \frac{1-z}{2}\right)-\right.$$
$$\left.-\left(\frac{z-1}{z+1}\right)^{\mu/2}{}_2F_1\left(-\nu,\ \nu+1;\ 1-\mu;\ \frac{1+z}{2}\right)\right]\qquad [\operatorname{Im} z\gtrless 0],$$
$$=\frac{e^{i\mu\pi}}{2}\Gamma\left[\begin{matrix}1+\mu+\nu,\ \mu-\nu\\ 1+\mu\end{matrix}\right]\left(\frac{z+1}{z-1}\right)^{\mu/2}\left[{}_2F_1\left(-\nu,\ \nu+1;\ \mu+1;\ \frac{1+z}{2}\right)-\right.$$
$$\left.-e^{\mp i\nu\pi}\left(\frac{z-1}{z+1}\right)^{\mu}{}_2F_1\left(-\nu,\ \nu+1;\ \mu+1;\ \frac{1-z}{2}\right)\right]\qquad [\operatorname{Im} z\gtrless 0].$$

$$Q_\nu^\mu(x)=\frac{1}{2}\Gamma\left[\begin{matrix}1+\mu+\nu,\ -\mu\\ 1-\mu+\nu\end{matrix}\right]\left(\frac{1-x}{1+x}\right)^{\mu/2}{}_2F_1\left(-\nu,\ \nu+1;\ \mu+1;\ \frac{1-x}{2}\right)+$$
$$+\frac{\cos\mu\pi}{2}\Gamma(\mu)\left(\frac{1+x}{1-x}\right)^{\mu/2}{}_2F_1\left(-\nu,\ \nu+1;\ 1-\mu;\ \frac{1-x}{2}\right)\qquad [-1<x<1].$$

In Ch. 3.2 of [1], 72 representations are given of the form

$$P_\nu^\mu(z)=A_1\,{}_2F_1(a_1,\ b_1;\ c_1;\ \zeta)+A_2\,{}_2F_1(a_2,\ b_2;\ c_2;\ \zeta).$$
$$e^{-i\mu\pi}Q_\nu^\mu(z)=A_3\,{}_2F_1(a_3,\ b_3;\ c_3;\ \zeta)+A_4\,{}_2F_1(a_4,\ b_4;\ c_4;\ \zeta).$$

Particular cases:

$$P_\nu^{-\nu}(z)=\frac{2^{-\nu}}{\Gamma(\nu+1)}(z^2-1)^{\nu/2},\quad P_\nu^{-\nu}(x)=\frac{2^{-\nu}}{\Gamma(\nu+1)}(1-x^2)^{\nu/2}\qquad [-1<x<1].$$

$$P_\nu^{\pm 1/2}(z)=\frac{2^{\mp 1/2}(z^2-1)^{-1/4}}{\sqrt{\pi}(2\nu+1)^{(1\mp 1)/2}}\left[(z+\sqrt{z^2-1})^{\nu+1/2}\pm(z+\sqrt{z^2-1})^{-\nu-1/2}\right].$$

$$P_\nu^{\pm 1/2}(x)=\sqrt{\frac{2}{\pi\sqrt{1-x^2}}}\left\{\begin{matrix}\cos[(\nu+1/2)\arccos x]\\ (\nu+1/2)^{-1}\sin[(\nu+1/2)\arccos x]\end{matrix}\right\}\qquad [-1<x<1].$$

$$P_0^\mu(z)=\frac{1}{\Gamma(1-\mu)}\left(\frac{z+1}{z-1}\right)^{\mu/2},\quad P_0^\mu(x)=\frac{1}{\Gamma(1-\mu)}\left(\frac{1+x}{1-x}\right)^{\mu/2}\qquad [-1<x<1].$$

$$P_\nu^{-1}(x)=\frac{(1-x^2)^{-1/2}}{\nu(\nu+1)}\frac{d}{dx}P_\nu(x)\qquad [\nu\neq 0;\ -1<x<1].$$

$$P_0^{-1}(z)=P_{-1}^{-1}(z)=\sqrt{\frac{z-1}{z+1}}.$$

$$P_{1/2}^{-1}(z)=\frac{4}{3\pi}\sqrt{\frac{z-1}{2}}\left[\mathbf{E}\left(\sqrt{\frac{z-1}{z+1}}\right)+\mathbf{K}\left(\sqrt{\frac{z-1}{z+1}}\right)-\mathbf{D}\left(\sqrt{\frac{z-1}{z+1}}\right)\right].$$

$$P_{-1/2}^{-1}(z)=\frac{4\sqrt{2(z-1)}}{\pi(z+1)}\mathbf{D}\left(\sqrt{\frac{z-1}{z+1}}\right).$$

$$P_1^1(x)=-\sqrt{1-x^2},\quad P_2^1(x)=-3x\sqrt{1-x^2},\quad P_2^2(x)=3(1-x^2).$$

$$P_3^1(x)=-\frac{3}{2}\sqrt{1-x^2}(5x^2-1),\quad P_3^2(x)=15x(1-x^2),\quad P_3^3(x)=-15(1-x^2)^{3/2}.$$

$$P_\nu(z)\big|_{\nu=n}=P_n(z)\qquad [P_n(z)\text{ is the Legendre polynomial}].$$

$$P_{1/2}(z)=\frac{2\sqrt{2(z+1)}}{\pi}\left[\mathbf{E}\left(\sqrt{\frac{z-1}{z+1}}\right)-\frac{1}{z+1}\mathbf{K}\left(\sqrt{\frac{z-1}{z+1}}\right)\right].$$

$$P_{-1/2}(z)=\frac{2}{\pi}\sqrt{\frac{2}{z+1}}\mathbf{K}\left(\sqrt{\frac{z-1}{z+1}}\right).$$

$$P_{n+1/2}(x)P_{n-1/2}(-x)+P_{n+1/2}(-x)P_{n-1/2}(x)=\frac{(-1)^n 4}{(2n+1)\pi}.$$

$$Q_\nu^{\pm 1/2}(z)=\frac{2^{\mp 1/2}i\sqrt{\pi}(z^2-1)^{-1/4}}{(2\nu+1)^{(1\mp 1)/2}}\left(z+\sqrt{z^2-1}\right)^{-\nu-1/2}.$$

$$Q_{-1/2}^{-1}(z)=2\sqrt{\frac{2}{z-1}}\mathbf{E}\left(\sqrt{\frac{2}{z+1}}\right).$$

$$Q_{1/2}^{-1}(z)=\frac{2}{3}\sqrt{\frac{2}{z-1}}\left[2\mathbf{D}\left(\sqrt{\frac{2}{z+1}}\right)+\mathbf{E}\left(\sqrt{\frac{2}{z+1}}\right)-2\mathbf{K}\left(\sqrt{\frac{2}{z+1}}\right)\right].$$

$$Q_\nu(z)=\frac{1}{2}P_\nu(z)\left[\ln\frac{z+1}{z-1}-2\mathbf{C}-2\psi(\nu+1)\right]-$$

$$-\frac{\sin\nu\pi}{\pi}\sum_{k=1}^{\infty}\Gamma(k-\nu)\Gamma(k+\nu+1)\frac{\psi(k+1)+\mathbf{C}}{(k!)^2}\left(\frac{1-z}{2}\right)^k.$$

$$Q_{1/2}(z)=\sqrt{\frac{2}{z+1}}\left[2\mathbf{D}\left(\sqrt{\frac{2}{z+1}}\right)-\mathbf{K}\left(\sqrt{\frac{2}{z+1}}\right)\right].$$

$$Q_{-1/2}(z)=\sqrt{\frac{2}{z+1}}\mathbf{K}\left(\sqrt{\frac{2}{z+1}}\right).$$

$$Q_n(z)=\frac{1}{2}P_n(z)\left[\ln\frac{z+1}{z-1}-2\psi(n+1)-2\mathbf{C}\right]+$$

$$+\sum_{k=0}^{n}\frac{(-1)^k(n+k)!}{(k!)^2(n-k)!}[\psi(k+1)-\mathbf{C}]\left(\frac{1-z}{2}\right)^k$$

$$Q_n(z)=\frac{1}{2}P_n(z)\ln\frac{z+1}{z-1}-W_{n-1}(z),\qquad Q_n(x)=\frac{1}{2}P_n(x)\ln\frac{1+x}{1-x}-W_{n-1}(x),$$

where

$$W_{n-1}(z)=\sum_{k=0}^{[(n-1)/2]}\frac{2n-4k-1}{(n-k)(2k+1)}P_{n-2k-1}(z)=\sum_{k=1}^{n}\frac{1}{k}P_{k-1}(z)P_{n-k}(z),$$

$$W_0(z)=1,\quad W_1(z)=\frac{3z}{2},\quad W_2(z)=-\frac{5z^2}{2}+\frac{2}{3},\quad Q_n(-z)=(-1)^{n+1}Q_n(z).$$

$$Q_0(z)=\frac{1}{2}\ln\frac{z+1}{z-1},\quad Q_0(x)=\frac{1}{2}\ln\frac{1+x}{1-x}=\operatorname{Arth}x\qquad [-1<x<1].$$

$$Q_{n+1/2}(x)\,Q_{n-1/2}(-x)+Q_{n+1/2}(-x)\,Q_{n-1/2}(x)=\frac{(-1)^n\pi}{2n+1}.$$

$$\begin{Bmatrix}P_\nu^\mu(0)\\ Q_\nu^\mu(0)\end{Bmatrix}=\pm 2^{\mu-(1\mp1)/2}\pi^{\mp1/2}\begin{Bmatrix}\cos[(\mu+\nu)\pi/2]\\ \sin[(\mu+\nu)\pi/2]\end{Bmatrix}\Gamma\begin{bmatrix}(1+\nu+\mu)/2\\ 1+(\nu-\mu)/2\end{bmatrix}.$$

$$\frac{d}{dx}\begin{Bmatrix}P_\nu^\mu(x)\\ Q_\nu^\mu(x)\end{Bmatrix}\Bigg|_{x=0}=(\mu+\nu)\begin{Bmatrix}P_{\nu-1}^\mu(0)\\ Q_{\nu-1}^\mu(0)\end{Bmatrix}=$$

$$=2^{\mu+(1\pm1)/2}\pi^{\mp1/2}\begin{Bmatrix}\sin[(\mu+\nu)\pi/2]\\ \cos[(\mu+\nu)\pi/2]\end{Bmatrix}\Gamma\begin{bmatrix}(\nu+\mu)/2+1\\ (\nu-\mu+1)/2\end{bmatrix}.$$

$$P_\nu^m(0)\frac{d}{dx}P_\nu^m(x)\Big|_{x=0}=(-1)^m\frac{\sin\nu\pi}{\pi}\Gamma\begin{bmatrix}\nu+m+1\\ \nu-m+1\end{bmatrix}.$$

Differentation formulae:

$$\frac{d}{dz}\begin{Bmatrix}P_\nu^\mu(z)\\ Q_\nu^\mu(z)\end{Bmatrix}=(z^2-1)^{-1}\left[(\nu-\mu+1)\begin{Bmatrix}P_{\nu+1}^\mu(z)\\ Q_{\nu+1}^\mu(z)\end{Bmatrix}-(\nu+1)z\begin{Bmatrix}P_\nu^\mu(z)\\ Q_\nu^\mu(z)\end{Bmatrix}\right]=$$

$$=(z^2-1)^{-1}\left[\nu z\begin{Bmatrix}P_\nu^\mu(z)\\ Q_\nu^\mu(z)\end{Bmatrix}-(\mu+\nu)\begin{Bmatrix}P_{\nu-1}^\mu(z)\\ Q_{\nu-1}^\mu(z)\end{Bmatrix}\right]=$$

$$=B\frac{(\mu+\nu)(\nu-\mu+1)}{\sqrt{z^2-1}}\begin{Bmatrix}P_\nu^{\mu-1}(z)\\ Q_\nu^{\mu-1}(z)\end{Bmatrix}-\frac{\mu z}{z^2-1}\begin{Bmatrix}P_\nu^\mu(z)\\ Q_\nu^\mu(z)\end{Bmatrix}\quad [B \text{ see p. } 775].$$

$$P_\nu^{-\mu}(x)\frac{dP_\nu^\mu(x)}{dx}-P_\nu^\mu(x)\frac{dP_\nu^{-\mu}(x)}{dx}=\frac{2\sin\mu\pi}{\pi(1-x^2)}.$$

$$P_\nu^\mu(z)\frac{d}{dz}Q_\nu^\mu(z)-Q_\nu^\mu(z)\frac{d}{dz}P_\nu^\mu(z)=\frac{1}{1-z^2}\begin{Bmatrix}e^{i\mu\pi}\\ 1\end{Bmatrix}\Gamma\begin{bmatrix}\nu+\mu+1\\ \nu-\mu+1\end{bmatrix}$$

$$\left[\begin{Bmatrix}z\notin[-1,1]\\ z=x\in(-1,1)\end{Bmatrix}\right].$$

Derivatives with respect to the index:

$$\frac{\partial P_\nu^\mu(x)}{\partial\nu}=\frac{1}{\Gamma(1-\mu)}\left(\frac{1+x}{1-x}\right)^{\mu/2}\sum_{k=1}^{\infty}\frac{(-\nu)_k(\nu+1)_k}{(1-\mu)_k\,k!}\times$$

$$\times[\psi(\nu+k+1)-\psi(\nu-k+1)]\left(\frac{1-x}{2}\right)^k$$

$$[\nu\neq 0,\ \pm1,\ \pm2,\ \dots;\ \operatorname{Re}\mu<1].$$

$$\left[\frac{\partial}{\partial\nu}P_\nu^\mu(x)\right]_{\nu=-1/2}=0.\quad \left[\frac{\partial}{\partial\nu}P_\nu^{-1}(x)\right]_{\nu=0}=-\sqrt{\frac{1-x}{1+x}}-\sqrt{\frac{1+x}{1-x}}\ln\frac{1+x}{2}.$$

$$\left[\frac{\partial}{\partial\nu}P_\nu^{-1}(x)\right]_{\nu=1}=-\frac{1}{4}\frac{(1-x)^{3/2}}{(1+x)^{1/2}}+\frac{\sqrt{1-x^2}}{2}\ln\frac{1+x}{2}.\quad \left[\frac{\partial}{\partial\nu}P_\nu(x)\right]_{\nu=0}=\ln\frac{1+x}{2}.$$

$$\left[\frac{\partial}{\partial\nu}P_\nu(x)\right]_{\nu=1/2}=-\left[\frac{\partial}{\partial\nu}P_\nu(x)\right]_{\nu=-3/2}=-4\sqrt{1-x^2}\,P_{-1/2}(x).$$

The differential equation

$$(1-z^2)\frac{d^2u}{dz^2}-2z\frac{du}{dz}+\left[\nu(\nu+1)-\frac{\mu^2}{1-z^2}\right]u=0$$

has the following fundamental systems of equations for $z\notin(-\infty,1]$ $(z=x\in[-1,1])$:

$P_\nu^\mu(z),\ Q_\nu^\mu(z)\quad (P_\nu^\mu(x),\ Q_\nu^\mu(x))\qquad [\nu+\mu,\ \nu-\mu\neq 0,\ \pm1,\ \pm2,\ \dots]$.

$P_\nu^\mu(z),\ P_\nu^{-\mu}(z)\quad (P_\nu^\mu(x),\ P_\nu^{-\mu}(x))\qquad [\nu+\mu,\ \nu-\mu=0,\ \pm1,\ \pm2,\ \dots;\ \mu\neq 0,\ \pm1,\ \pm2,\ \dots]$.

$P_n^m(z),\ Q_n^m(z)\quad (P_n^m(x),\ Q_n^m(x))\qquad [\mu=\pm m,\ \nu=n$ or $\nu=-n-1,\ n\geqslant m]$.

$P_n^{-m}(z),\ Q_n^m(z)\quad (P_n^{-m}(x),\ Q_n^m(x))\qquad [\mu=\pm m,\ \nu=n$ or $\nu=-n-1,\ n<m]$.

II.19. The MacROBERT *E*-FUNCTIONS $E(p; a_r : q; b_s : z)$

Definition:

$$E(p;\ a_r:q;\ b_s:z)=E((a_p);\ (b_q);\ z)=$$

$$=E\left(\begin{matrix}(a_p);\ z\\(b_q)\end{matrix}\right)=E\left(\begin{matrix}(a_p)\\(b_q);\ z\end{matrix}\right)=G^{p,\,1}_{q+1,\,p}\left(z\,\middle|\,\begin{matrix}1,\ (b_q)\\(a_p)\end{matrix}\right).$$

$$E(p;\ a_r:q;\ b_s:z)=\Gamma\left[\begin{matrix}a_1,\ \ldots,\ a_p\\b_1,\ \ldots,\ b_q\end{matrix}\right]{}_pF_q((a_p);\ (b_q);\ -z^{-1})$$

$[p\leqslant q;\ |z|>0]$ or $[p=q+1;\ |z|>1]$,

$$=\sum_{k=1}^{p}\Gamma\left[\begin{matrix}(a_p)'-a_k,\ a_k\\b_1-a_k,\ \ldots,\ b_q-a_k\end{matrix}\right]z^{a_k}\,{}_{q+1}F_{p-1}\left(\begin{matrix}a_k,\ 1+a_k-(b_q);\ (-1)^{p-q}z\\1+a_k-(a_p)'\end{matrix}\right)$$

$[p>q+1;\ |z|<\infty]$ or $[p=q+1;\ |z|<1]$;

here the prime $'$ means that the component a_k-a_k is omitted from the vector $(a_p)-a_k$.

$$E(p;\ a_r:q;\ b_1,\ \ldots,\ b_{q-1},\ a_1:z)=E(p-1;\ a_2,\ \ldots,\ a_p:q-1;\ b_1,\ \ldots,\ b_{q-1}:z).$$

$$E(p;\ a_1,\ \ldots,\ a_{p-1},\ b_1:q;\ b_s:z)=E(p-1;\ a_1,\ \ldots,\ a_{p-1}:q-1;\ b_2,\ \ldots,\ b_q:z).$$

$$a_1zE(p;\ a_r:q;\ b_s:z)=zE(p;\ a_1+1,\ a_2,\ \ldots,\ a_p:q;\ b_s:z)+E(p;\ a_r+1:q;\ b_s+1:z).$$

$$(b_1-1)\,zE(p;\ a_r:q;\ b_s:z)=$$

$$=zE(p;\ a_r:q;\ b_1-1,\ b_2,\ \ldots,\ b_q:z)+E(p;\ a_r+1:q;\ b_s+1:z).$$

$$\left(z^2\frac{d}{dz}\right)^n E(p;\ a_r:q;\ b_s:z)=E(p;\ a_r+n:q;\ b_s+n:z).$$

$$\frac{d^n}{dz^n}[z^{-a_1}E(p;\ a_r:q;\ b_s:z)]=(-1)^n z^{-a_1-n}E(p;\ a_1+n,\ a_2,\ \ldots,\ a_p:q;\ b_s:z).$$

$$\frac{d^n}{dz^n}E(p;\ a_r:q;\ b_s:cz^{-m})=m^nz^{-n}E(p+m;\ a_r,\ \Delta(m,\ 1):q+m;\ b_s,\ \Delta(m,\ 1-n):cz^{-m})$$

$[m=1,\ 2,\ \ldots]$.

Other relations can be obtained from formulae for the Meijer *G*-function (see Ch. 8).

II.20. THE JACOBI ELLIPTIC FUNCTIONS cn *u*, dn *u*, sn *u*

Definition: let $u=\int\limits_0^{\varphi}\frac{dt}{\sqrt{1-k^2\sin^2 t}}$; then

$$\operatorname{am} u\equiv\operatorname{am}(u,\ k)=\varphi,\quad \operatorname{sn} u=\sin\varphi,\quad \operatorname{cn} u=\cos\varphi,$$

$$\operatorname{dn} u=\sqrt{1-k^2\sin^2\varphi},\quad \operatorname{sn}(-u)=-\operatorname{sn} u,\quad \operatorname{cn}(-u)=\operatorname{cn} u,\ \operatorname{dn}(-u)=\operatorname{dn} u.$$

$$\operatorname{sn}(u\pm v)=\frac{\operatorname{sn} u\operatorname{cn} v\operatorname{dn} v\pm\operatorname{sn} v\operatorname{cn} u\operatorname{dn} u}{1-k^2\operatorname{sn}^2 u\operatorname{sn}^2 v},\quad \operatorname{cn}(u\pm v)=\frac{\operatorname{cn} u\operatorname{cn} v\mp\operatorname{sn} u\operatorname{sn} v\operatorname{dn} u\operatorname{dn} v}{1-k^2\operatorname{sn}^2 u\operatorname{sn}^2 v},$$

$$\operatorname{dn}(u\pm v)=\frac{\operatorname{dn} u\operatorname{dn} v\mp k^2\operatorname{sn} u\operatorname{sn} v\operatorname{cn} u\operatorname{cn} v}{1-k^2\operatorname{sn}^2 u\operatorname{sn}^2 v}.$$

$$\operatorname{sn}\frac{u}{2}=\pm\frac{1}{k}\sqrt{\frac{1-\operatorname{dn} u}{1+\operatorname{cn} u}}=\pm\sqrt{\frac{1-\operatorname{cn} u}{1+\operatorname{dn} u}},$$

$$\operatorname{cn}\frac{u}{2}=\pm\sqrt{\frac{\operatorname{cn} u+\operatorname{dn} u}{1+\operatorname{dn} u}}=\pm\frac{k'}{k}\sqrt{\frac{1-\operatorname{dn} u}{\operatorname{dn} u-\operatorname{cn} u}},$$

$$\operatorname{dn}\frac{u}{2}=\pm\sqrt{\frac{\operatorname{cn} u+\operatorname{dn} u}{1+\operatorname{cn} u}}=\pm k'\sqrt{\frac{1-\operatorname{cn} u}{\operatorname{dn} u-\operatorname{cn} u}},\quad k'=\sqrt{1-k^2}.$$

$$\operatorname{dn}^2 u=\frac{\operatorname{dn} 2u+k^2\operatorname{cn} 2u+1-k^2}{1+\operatorname{dn} 2u},\quad \operatorname{sn}^2 u+\operatorname{cn}^2 u=1,\quad \operatorname{dn}^2 u+k^2\operatorname{sn}^2 u=1.$$

$$\frac{d}{du}\operatorname{sn} u=\operatorname{cn} u\operatorname{dn} u,\quad \frac{d}{du}\operatorname{cn} u=-\operatorname{sn} u\operatorname{dn} u,\quad \frac{d}{du}\operatorname{dn} u=-k^2\operatorname{sn} u\operatorname{cn} u.$$

II.21. THE WEIERSTRASS ELLIPTIC FUNCTIONS

$\wp(u)$, $\zeta(u)$, $\sigma(u)$

$$\wp(u)=\frac{1}{u^2}\sum_{\substack{m,\,n=0\\ m+n\neq 0}}^{\infty}\left[\frac{1}{(u-2m\omega_1-2n\omega_2)^2}-\frac{1}{(2m\omega_1+2n\omega_2)^2}\right],$$

$$\zeta(u)=\frac{1}{u}-\int_0^u\left[\wp(u)-\frac{1}{u^2}\right]du,\qquad \sigma(u)=u\exp\left\{\int_0^u\left[\zeta(u)-\frac{1}{u}\right]du\right\}.$$

$$\wp(-u)=\wp(u),\quad \zeta(-u)=-\zeta(u),\quad \sigma(-u)=-\sigma(u).$$

$$g_2=60\sum_{\substack{m,\,n=0\\ m+n\neq 0}}^{\infty}(m\omega_1+n\omega_2)^{-4},\quad g_3=140\sum_{\substack{m,\,n=0\\ m+n\neq 0}}^{\infty}(m\omega_1+n\omega_2)^{-6}.$$

$$g_2=-4(e_1e_2+e_2e_3+e_3e_1),\ g_3=4e_1e_2e_3,\ e_1+e_2+e_3=0,$$

where e_1, e_2, e_3 are the roots of the equation $4z^3-g_2z-g_3=0$.

II.22. THE THETA-FUNCTION $\theta_j(z, q)$, $j=0, 1, 2, 3, 4$

Definitions (throughout $|q|<1$):

$$\theta_0(z, q)=1+2\sum_{k=1}^{\infty}(-1)^k q^{k^2}\cos 2kz.\quad \theta_1(z, q)=2\sum_{k=0}^{\infty}(-1)^k q^{(k+1/2)^2}\sin(2k+1)z.$$

$$\theta_2(z, q)=2\sum_{k=0}^{\infty}q^{(k+1/2)^2}\cos(2k+1)z.\quad \theta_3(z, q)=1+2\sum_{k=1}^{\infty}q^{k^2}\cos 2kz.$$

$$\theta_4(z, q)\equiv\theta_0(z, q).\quad \theta_i(-z, q)=\theta_i(z, q),\quad i=0,\ 2,\ 3,\quad \theta_1(-z, q)=-\theta_1(z, q).$$

$$\begin{Bmatrix}\theta_0(z+\pi, q)\\ \theta_3(z+\pi, q)\end{Bmatrix}=\begin{Bmatrix}\theta_0(z, q)\\ \theta_3(z, q)\end{Bmatrix},\quad \begin{Bmatrix}\theta_1(z+\pi, q)\\ \theta_2(z+\pi, q)\end{Bmatrix}=-\begin{Bmatrix}\theta_1(z, q)\\ \theta_2(z, q)\end{Bmatrix}.$$

$$\begin{Bmatrix}\theta_0(z+\tau\pi, q)\\ \theta_3(z+\tau\pi, q)\end{Bmatrix}=\mp\frac{1}{q}e^{-2iz}\begin{Bmatrix}\theta_0(z, q)\\ \theta_3(z, q)\end{Bmatrix},$$

$$\begin{Bmatrix}\theta_1(z+\tau\pi, q)\\ \theta_2(z+\tau\pi, q)\end{Bmatrix}=\mp\frac{1}{q}e^{-2iz}\begin{Bmatrix}\theta_1(z, q)\\ \theta_2(z, q)\end{Bmatrix}\qquad [q=e^{i\tau\pi};\ \operatorname{Im}\tau>0].$$

$$\begin{Bmatrix}\theta_0(z+\pi/2, q)\\ \theta_3(z+\pi/2, q)\end{Bmatrix}=\begin{Bmatrix}\theta_3(z, q)\\ \theta_0(z, q)\end{Bmatrix},\quad \begin{Bmatrix}\theta_1(z+\pi/2, q)\\ \theta_2(z+\pi/2, q)\end{Bmatrix}=\pm\begin{Bmatrix}\theta_2(z, q)\\ \theta_1(z, q)\end{Bmatrix},$$

$$\begin{Bmatrix}\theta_0(z+\tau\pi/2, q)\\ \theta_1(z+\tau\pi/2, q)\end{Bmatrix}=iq^{-1/4}e^{-iz}\begin{Bmatrix}\theta_1(z, q)\\ \theta_0(z, q)\end{Bmatrix},$$

$$\begin{Bmatrix}\theta_2(z+\tau\pi/2, q)\\ \theta_3(z+\tau\pi/2, q)\end{Bmatrix}=q^{-1/4}e^{-iz}\begin{Bmatrix}\theta_3(z, q)\\ \theta_2(z, q)\end{Bmatrix}\qquad [q=e^{i\tau\pi};\ \operatorname{Im}\tau>0].$$

$$\theta_{\left\{\substack{2\\3}\right\}}(2z, q)=\frac{1}{2}\left[\theta_3(z, q^{1/4})\mp\theta_0(z, q^{1/4})\right].$$

$$\theta_3^2(0, q)=\frac{2}{\pi}\mathbf{K}(k),\qquad q=\exp\left[-\pi\frac{\mathbf{K}(\sqrt{1-k^2})}{\mathbf{K}(k)}\right].$$

II.23. THE MATHIEU FUNCTIONS

By the Mathieu functions $\mathrm{ce}_n(z,q)$, $\mathrm{se}_n(z,q)$ (periodic or of the first kind) we mean the periodic solutions of the Mathieu equation

$$\frac{d^2y}{dz^2}+(a-2q\cos 2z)\,y=0,\qquad k=2\sqrt{q}.$$

If $\sqrt{q}$ is a real number, then there exist an infinite number of eigenvalues a and periodic solutions $y(z)=y(z+2\pi)$ corresponding to them. They can be written in the form

$$\mathrm{ce}_{2n}(z, q)=\sum_{r=0}^{\infty} A_{2r}^{(2n)} \cos 2rz,$$

$$=\frac{\mathrm{ce}_{2n}(\pi/2, q)}{A_0^{(2n)}} \sum_{r=0}^{\infty} (-1)^r A_{2r}^{(2n)} J_{2r}(2\sqrt{q}\cos z),$$

$$=\frac{\mathrm{ce}_{2n}(0, q)}{A_0^{(2n)}} \sum_{r=0}^{\infty} (-1)^r A_{2r}^{(2n)} I_{2r}(2\sqrt{q}\sin z).$$

$$\mathrm{ce}_{2n+1}(z, q)=\sum_{r=0}^{\infty} A_{2r+1}^{(2n+1)} \cos (2r+1) z,$$

$$=-\frac{\mathrm{ce}_{2n+1}'(\pi/2)}{\sqrt{q} A_1^{(2n+1)}} \sum_{r=0}^{\infty} (-1)^r A_{2r+1}^{(2n+1)} J_{2r+1}(2\sqrt{q}\cos z),$$

$$=\frac{\mathrm{ce}_{2n+1}(0, q)}{\sqrt{q}\, A_1^{(2n+1)}} \operatorname{ctg} z \sum_{r=0}^{\infty} (-1)^r (2r+1) A_{2r+1}^{(2n+1)} I_{2r+1}(2\sqrt{q}\sin z).$$

$$\mathrm{se}_{2n+1}(z, q)=\sum_{r=0}^{\infty} B_{2r+1}^{(2n+1)} \sin (2k+1) z,$$

$$=\frac{\mathrm{se}_{2n+1}(\pi/2, q)}{\sqrt{q}\, B_1^{(2n+1)}} \operatorname{tg} z \sum_{r=0}^{\infty} (-1)^r (2r+1) B_{2r+1}^{(2n+1)} J_{2r+1}(2\sqrt{q}\cos z),$$

$$=\frac{\mathrm{se}_{2n+1}'(0, q)}{\sqrt{q}\, B_1^{(2n+1)}} \sum_{r=0}^{\infty} (-1)^r B_{2r}^{(2n+1)} I_{2r+1}(2\sqrt{q}\sin z).$$

$$\mathrm{se}_{2n+2}(z, q)=\sum_{r=0}^{\infty} B_{2r+2}^{(2n+2)} \sin (2r+2) z,$$

$$=-\frac{\mathrm{se}_{2n+2}'(\pi/2, q)}{qB_2^{(2n+2)}} \operatorname{tg} z \sum_{r=0}^{\infty} (-1)^r (2r+2) B_{2r+2}^{(2n+2)} J_{2r+2}(2\sqrt{q}\cos z),$$

$$=\frac{\mathrm{se}_{2n+2}'(0, q)}{qB_2^{(2n+2)}} \operatorname{ctg} z \sum_{r=0}^{\infty} (-1)^r (2r+2) B_{2r+2}^{(2n+2)} I_{2r+2}(2\sqrt{q}\sin z).$$

The eigenvalues and coefficients of the series are found from recurrence relations [12]. The prime denotes differentiation with respect to z.

The normalization of the Mathieu functions:

$$\int_0^{2\pi} y^2(x)\, dx=\pi.$$

There exists along with each periodic solution of the Mathieu equation a second aperiodic solution linearly independent of it (called the Mathieu function of the second kind). The aperiodic solution corresponding to $\mathrm{ce}_n(z,q)$ that is an odd function of z is denoted by $\mathrm{fe}_n(z,q)$. The even aperiodic solution corresponding to $se_n(z,q)$ is denoted by $ge_n(z,q)$. They are connected by the relations

$$\mathrm{fe}_n(z, q) = C_n(q)\,[z\,\mathrm{ce}_n(z, q) + f_n(z, q)],$$
$$\mathrm{ge}_n(z, q) = S_n(q)\,[z\,\mathrm{se}_n(z, q) + q_n(z, q)],$$

where $f_n(z,q)$, $q_n(z,q)$ are certain functions, and $C_n(q)$, $S_n(q)$ do not depend on z.

We have the following expansions:

$$\mathrm{fe}_{2n}(z, q) = -\frac{\pi\,\mathrm{fe}'_{2n}(0, q)}{2\mathrm{ce}_{2n}(\pi/2, q)} \sum_{r=0}^{\infty} (-1)^r A_{2r}^{(2n)}\,\mathrm{Im}\left[J_r(\sqrt{q}\,e^{iz})\,Y_r(\sqrt{q}\,e^{-iz})\right],$$

$$\mathrm{fe}_{2n+1}(z, q) = \frac{\pi\sqrt{q}\,\mathrm{fe}'_{2n+1}(0, q)}{2\,\mathrm{ce}'_{2n+1}(\pi/2, q)} \sum_{r=0}^{\infty} (-1)^r A_{2r+1}^{(2n+1)}\,\mathrm{Im}\left[J_r(\sqrt{q}\,e^{iz})Y_{r+1}(\sqrt{q}\,e^{-iz}) + J_{r+1}(\sqrt{q}\,e^{iz})\,Y_r(\sqrt{q}\,e^{-iz})\right],$$

$$\mathrm{ge}_{2n+1}(z, q) = -\frac{\pi\sqrt{q}\,\mathrm{ge}_{2n+1}(0, q)}{2\,\mathrm{se}_{2n+1}(\pi/2, q)} \sum_{r=0}^{\infty} B_{2r+1}^{(2n+1)}\,\mathrm{Re}\left[J_r(\sqrt{q}\,e^{iz})\,Y_{r+1}(\sqrt{q}\,e^{-iz}) - J_{r+1}(\sqrt{q}\,e^{iz})\,Y_r(\sqrt{q}\,e^{-iz})\right],$$

$$\mathrm{ge}_{2n+2}(z, q) = -\frac{\pi q\,\mathrm{ge}_{2n+2}(0, q)}{2\,\mathrm{se}'_{2n+2}(\pi/2, q)} \times$$
$$\times \sum_{r=0}^{\infty} (-1)^r\,\mathrm{Re}\left[J_r(\sqrt{q}\,e^{iz})\,Y_{r+2}(\sqrt{q}\,e^{-iz}) - J_{r+2}(\sqrt{q}\,e^{iz})\,Y_r(\sqrt{q}\,e^{-iz})\right].$$

The modified Mathieu functions of the first kind

$$\mathrm{Ce}_n(z, q) = \mathrm{ce}_n(iz, q), \qquad \mathrm{Se}_n(z, q) = -i\,\mathrm{se}_n(iz, q)$$

and of the second kind

$$\mathrm{Fe}_n(z, q) = -i\,\mathrm{fe}_n(iz, q), \qquad \mathrm{Ge}_n(z, q) = \mathrm{ge}_n(iz, q)$$

satisfy the equation

$$\frac{d^2y}{dz^2} - (a - 2q\,\mathrm{ch}\,2z)\,y = 0, \qquad k = 2\sqrt{q}.$$

This equation has, along with the solution $\mathrm{Ce}_{2n}(z,q)$, which admits the representations

$$\mathrm{Ce}_{2n}(z, q) = \frac{\mathrm{ce}_{2n}(0, q)}{A_0^{(2n)}} \sum_{r=0}^{\infty} A_{2r}^{(2n)}\,J_{2r}(2\sqrt{q}\,\mathrm{sh}\,z)$$
$$= \frac{\mathrm{ce}_{2n}(\pi/2, q)}{A_0^{(2n)}} \sum_{r=0}^{\infty} (-1)^r A_{2r}^{(2n)}\,J_{2r}(2\sqrt{q}\,\mathrm{ch}\,z),$$

the solution

$$\mathrm{Fe}\,y_{2n}(z, q) = \frac{\mathrm{ce}_{2n}(0, q)}{A_0^{(2n)}} \sum_{r=0}^{\infty} A_{2r}^{(2n)}\,Y_{2r}(2\sqrt{q}\,\mathrm{sh}\,z) \qquad [|\mathrm{sh}\,z| > 1;\ \mathrm{Re}\,z > 0],$$
$$= \frac{\mathrm{ce}_{2n}(\pi/2, q)}{A_0^{(2n)}} \sum_{r=0}^{\infty} (-1)^r A_{2r}^{(2n)}\,Y_{2r}(2\sqrt{q}\,\mathrm{ch}\,z) \qquad [|\mathrm{ch}\,z| > 1],$$

which is linearly independent of it and is obtained by replacing $J_{2r}(z)$ by $Y_{2r}(z)$.

The other pairs of solutions have analogous representations:

$$\mathrm{Ce}_{2n+1}(z, q) = -\frac{\mathrm{ce}'_{2n+1}(\pi/2, q)}{\sqrt{q}\,A_1^{(2n+1)}} \sum_{r=0}^{\infty} (-1)^r A_{2r+1}^{(2n+1)} J_{2r+1}(2\sqrt{q}\,\mathrm{ch}\,z),$$

$$= \frac{\mathrm{ce}_{2n+1}(0, q)}{\sqrt{q}\,A_1^{(2n+1)}} \mathrm{cth}\,z \sum_{r=0}^{\infty} (2r+1) A_{2r+1}^{(2n+1)} J_{2r+1}(2\sqrt{q}\,\mathrm{sh}\,z)$$

and $\mathrm{Fey}_{2n+1}(z, q)$, obtained by replacing $J_{2r+1}(z)$ by $Y_{2r+1}(z)$;

$$\mathrm{Se}_{2n+1}(z, q) = \frac{\mathrm{se}'_{2n+1}(0, q)}{\sqrt{q}\,B_1^{(2n+1)}} \sum_{r=0}^{\infty} B_{2r+1}^{(2n+1)} J_{2r+1}(2\sqrt{q}\,\mathrm{sh}\,z),$$

$$= \frac{\mathrm{se}_{2n+1}(\pi/2, q)}{\sqrt{q}\,B_1^{(2n+1)}} \mathrm{th}\,z \sum_{r=0}^{\infty} (-1)^r (2r+1) B_{2r+1}^{(2n+1)} J_{2r+1}(2\sqrt{q}\,\mathrm{ch}\,z)$$

and $\mathrm{Gey}_{2n+1}(z, q)$, obtained by replacing $J_{2r+1}(z)$ by $Y_{2r+1}(z)$;

$$\mathrm{Se}_{2n+2}(z, q) = \frac{\mathrm{se}'_{2n+2}(0, q)}{qB_2^{(2n+2)}} \mathrm{cth}\,z \sum_{r=0}^{\infty} (2r+2) B_{2r+2}^{(2n+2)} J_{2r+2}(2\sqrt{q}\,\mathrm{sh}\,z),$$

$$= -\frac{\mathrm{se}'_{2n+2}(\pi/2, q)}{qB_2^{(2n+2)}} \mathrm{th}\,z \sum_{r=0}^{\infty} (-1)^r (2r+2) B_{2r+2}^{(2n+2)} J_{2r+2}(2\sqrt{q}\,\mathrm{ch}\,z)$$

and $\mathrm{Gey}_{2n+2}(z, q)$, obtained by replacing $J_{2r+2}(z)$ by $Y_{2r+2}(z)$.

The modified Mathieu functions of the third kind $\mathrm{Me}_n^{(j)}(z, q)$, $\mathrm{Ne}_n^{(j)}(z, q)$, $j=1,2$, are obtained by replacing $J_n(z)$ by $H_n^{(j)}(z)$ in the above series.

Other modified Mathieu functions of the third kind are:

$$\mathrm{Fek}_{2n}(z, q) = \frac{\mathrm{ce}_{2n}(0, q)}{\pi A_0^{(2n)}} \sum_{r=0}^{\infty} (-1)^r A_{2r}^{(2n)} K_{2r}(-2i\sqrt{q}\,\mathrm{sh}\,z)$$

$$[\,|\mathrm{sh}\,z| > 1;\quad \mathrm{Re}\,z > 0],$$

$$\mathrm{Fek}_{2n+1}(z, q) = \frac{\mathrm{ce}_{2n+1}(0, q)}{\pi\sqrt{q}\,A_1^{(2n+1)}} \mathrm{cth}\,z \sum_{r=0}^{\infty} (-1)^r (2r+1) A_{2r+1}^{(2n+1)} K_{2r+1}(-2i\sqrt{q}\,\mathrm{sh}\,z)$$

$$[\,|\mathrm{sh}\,z| > 1;\quad \mathrm{Re}\,z > 0],$$

$$\mathrm{Gek}_{2n+1}(z, q) = \frac{\mathrm{se}_{2n+1}(\pi/2, q)}{\pi\sqrt{q}\,B_1^{(2n+1)}} \mathrm{th}\,z \sum_{r=1}^{\infty} (2r+1) B_{2r+1}^{(2n+1)} K_{2r+1}(-2i\sqrt{q}\,\mathrm{ch}\,z),$$

$$\mathrm{Gek}_{2n+2}(z, q) = \frac{\mathrm{se}'_{2n+2}(\pi/2, q)}{\pi q B_2^{(2n+2)}} \mathrm{th}\,z \sum_{r=0}^{\infty} (2r+2) B_{2r+2}^{(2n+2)} K_{2r+2}(-2i\sqrt{q}\,\mathrm{ch}\,z).$$

Formulae connecting these functions:

$$\mathrm{Ce}_{2n}(z, q) + i\,\mathrm{Fey}_{2n}(z, q) = \mathrm{Me}_{2n}^{(1)}(z, q) = -2i\,\mathrm{Fek}_{2n}(z, q),$$

$$\mathrm{Ce}_{2n+1}(z, q) + i\mathrm{Fey}_{2n+1}(z, q) = \mathrm{Me}_{2n+1}^{(1)}(z, q) = -2\,\mathrm{Fek}_{2n+1}(z, q),$$

$$\mathrm{Se}_{2n+1}(z, q) + i\mathrm{Gey}_{2n+1}(z, q) = \mathrm{Ne}_{2n+1}^{(1)}(z, q) = -2\,\mathrm{Gek}_{2n+1}(z, q),$$

$$\mathrm{Se}_{2n+2}(z, q) + i\,\mathrm{Gey}_{2n+2}(z, q) = \mathrm{Ne}_{2n+2}^{(1)}(z, q) = -2i\,\mathrm{Gek}_{2n+2}(z, q),$$

$$\mathrm{Ce}_n(z, q) - i\,\mathrm{Fey}_n(z, q) = \mathrm{Me}_n^{(2)}(z, q),\quad \mathrm{Se}_n(z, q) - i\,\mathrm{Gey}_n(z, q) = \mathrm{Ne}_n^{(2)}(z, q),$$

$$\mathrm{Fey}_n(z, q) = \frac{\mathrm{Fey}_n(0, q)}{\mathrm{Ce}_n(0, q)} \mathrm{Ce}_n(z, q) + \frac{\mathrm{Fey}'_n(0, q)}{\mathrm{Fe}'_n(0, q)} \mathrm{Fe}_n(z, q),$$

$$\mathrm{Gey}_n(z, q)=\frac{\mathrm{Gey}_n'(0, q)}{\mathrm{Se}_n'(0, q)}\mathrm{Se}_n(z, q)+\frac{\mathrm{Gey}_n(0, q)}{\mathrm{Ge}_n(0, q)}\mathrm{Ge}_n(z, q),$$

$$\mathrm{Fey}_{2n}(-iz, q)=\frac{\mathrm{Fey}_{2n}(0, q)}{\mathrm{ce}_{2n}(0, q)}\mathrm{ce}_{2n}(z, q)-i\frac{\mathrm{Fey}_{2n}'(0, q)}{\mathrm{fe}_{2n}'(0, q)}\mathrm{fe}_{2n}(z, q),$$

$$\mathrm{Gey}_{2n+1}(-iz, q)=\frac{\mathrm{Gey}_{2n+1}'(0, q)}{\mathrm{se}_{2n+1}'(0, q)}\mathrm{se}_{2n+1}(z, q)+\frac{\mathrm{Gey}_{2n+1}(0, q)}{\mathrm{ge}_{2n+1}(0, q)}\mathrm{ge}_{2n+1}(z, q).$$

$$\mathrm{Me}_{2n}^{(1)}(z, \pm q)=-2i\,\mathrm{Fek}_{2n}(z, \pm q),\quad \mathrm{Me}_{2n+1}^{(1)}(z, q)=-2\,\mathrm{Fek}_{2n+1}(z, q),$$

$$\mathrm{Me}_{2n+1}^{(1)}(z, -q)=2i\,\mathrm{Fek}_{2n}(z, -q),\quad \mathrm{Ne}_{2n+1}^{(1)}(z, q)=-2\,\mathrm{Gek}_{2n+1}(z, q),$$

$$\mathrm{Ne}_{2n+1}^{(1)}(z, -q)=2i\,\mathrm{Gek}_{2n+1}(z, -q),\quad \mathrm{Ne}_{2n+2}^{(1)}(z, \pm q)=-2i\,\mathrm{Gek}_{2n+2}(z, \pm q).$$

Some properties:

$$\mathrm{ce}_0(x, 0)=\frac{1}{\sqrt{2}},\quad \mathrm{ce}_n(x, 0)=\cos nx\ (n\neq 0),\quad \mathrm{se}_n(x, 0)=\sin x.$$

$$\left\{\begin{matrix}\mathrm{ce}_n(m\pi+z, q)\\ \mathrm{se}_n(m\pi+z, q)\end{matrix}\right\}=\pm\left\{\begin{matrix}\mathrm{ce}_n(m\pi-z, q)\\ \mathrm{se}_n(m\pi-z, q)\end{matrix}\right\},$$

$$\left\{\begin{matrix}\mathrm{ce}_n(\pi/2+z, q)\\ \mathrm{se}_n(\pi/2+z, q)\end{matrix}\right\}=\pm(-1)^n\left\{\begin{matrix}\mathrm{ce}_n(\pi/2-z, q)\\ \mathrm{se}_n(\pi/2-z, q)\end{matrix}\right\},$$

$$\left\{\begin{matrix}\mathrm{ce}_n(z+\pi, q)\\ \mathrm{se}_n(z+\pi, q)\end{matrix}\right\}=(-1)^n\left\{\begin{matrix}\mathrm{ce}_n(z, q)\\ \mathrm{se}_n(z, q)\end{matrix}\right\},$$

$$\left\{\begin{matrix}\mathrm{ce}_{2n}(z, -q)\\ \mathrm{ce}_{2n+1}(z, -q)\end{matrix}\right\}=(-1)^n\left\{\begin{matrix}\mathrm{ce}_{2n}(\pi/2-z, q)\\ \mathrm{se}_{2n+1}(\pi/2-z, q)\end{matrix}\right\},$$

$$\left\{\begin{matrix}\mathrm{se}_{2n+1}(z, -q)\\ \mathrm{se}_{2n+2}(z, -q)\end{matrix}\right\}=(-1)^n\left\{\begin{matrix}\mathrm{ce}_{2n+1}(\pi/2-z, q)\\ \mathrm{se}_{2n+2}(\pi/2-z, q)\end{matrix}\right\},$$

$$\left\{\begin{matrix}\mathrm{fe}_{2n}(z, -q)\\ \mathrm{fe}_{2n+1}(z, -q)\end{matrix}\right\}=\mp(-1)^n\left\{\begin{matrix}\mathrm{fe}(\pi/2-z, q)\\ \mathrm{ge}_{2n+1}(\pi/2-z, q)\end{matrix}\right\},$$

$$\left\{\begin{matrix}\mathrm{ge}_{2n+1}(z, -q)\\ \mathrm{ge}_{2n+2}(z, -q)\end{matrix}\right\}=(-1)^n\left\{\begin{matrix}\mathrm{fe}_{2n+1}(\pi/2-z, q)\\ \mathrm{ge}_{2n+2}(\pi/2-z, q)\end{matrix}\right\},$$

$$\left\{\begin{matrix}\mathrm{Ce}_{2n}(z, -q)\\ \mathrm{Ce}_{2n+1}(z, -q)\end{matrix}\right\}=(-1)^n\left\{\begin{matrix}\mathrm{Ce}_{2n}(z+\pi i/2, q)\\ -i\,\mathrm{Se}_{2n+1}(z+\pi i/2, q)\end{matrix}\right\},$$

$$\left\{\begin{matrix}\mathrm{Se}_{2n+1}(z, -q)\\ \mathrm{Se}_{2n+2}(z, -q)\end{matrix}\right\}=(-1)^{n+1}\left\{\begin{matrix}i\mathrm{Ce}_{2n+1}(z+\pi i/2, q)\\ \mathrm{Se}_{2n+2}(z+\pi i/2, q)\end{matrix}\right\},$$

$$\left\{\begin{matrix}\mathrm{Fe}_{2n}(z, -q)\\ \mathrm{Fe}_{2n+1}(z, -q)\end{matrix}\right\}=(-1)^n\left\{\begin{matrix}\mathrm{Fe}_{2n}(z+\pi i/2, q)\\ -i\mathrm{Ge}_{2n+1}(z+\pi i/2, q)\end{matrix}\right\},$$

$$\left\{\begin{matrix}\mathrm{Ge}_{2n+1}(z, -q)\\ \mathrm{Ge}_{2n+2}(z, -q)\end{matrix}\right\}=(-1)^{n+1}\left\{\begin{matrix}i\mathrm{Fe}_{2n+1}(z+\pi i/2, q)\\ \mathrm{Ge}_{2n+2}(z+\pi i/2, q)\end{matrix}\right\}.$$

In the formulae given below we use the following notation:

$$p_{2n}=\frac{\mathrm{ce}_{2n}(0, q)\,\mathrm{ce}_{2n}(\pi/2, q)}{A_0^{(2n)}},\quad p_{2n+1}=-\frac{\mathrm{ce}_{2n+1}(0, q)\,\mathrm{ce}_{2n+1}'(\pi/2, q)}{\sqrt{q}\,A_1^{(2n+1)}},$$

$$s_{2n+1}=\frac{\mathrm{se}_{2n+1}(0, q)\,\mathrm{se}_{2n+1}(\pi/2, q)}{\sqrt{q}\,B_1^{(2n+1)}},\quad s_{2n+2}=\frac{\mathrm{se}_{2n+2}'(0, q)\,\mathrm{se}_{2n+2}'(\pi/2, q)}{qB_2^{(2n+2)}}.$$

$$\sum_{n=0}^{\infty}A_0^{(2n)}A_{2r}^{(2n)}=\frac{1}{2}\delta_{0,r}.$$

$$\sum_{n=0}^{\infty} A_{2r+2}^{(2n)} A_{2s+2}^{(2n)} = \sum_{n=0}^{\infty} A_{2r+1}^{(2n+1)} A_{2s+1}^{(2n+1)} =$$

$$= \sum_{n=0}^{\infty} B_{2r+1}^{(2n+1)} B_{2s+1}^{(2n+1)} = \sum_{n=0}^{\infty} B_{2r+2}^{(2n+2)} B_{2s+2}^{(2n+2)} = \delta_{r,s}.$$

$$\sum_{r=0}^{\infty} (-1)^r A_{2r}^{(2n)} \cos 2rw J_{2r}(z) = \frac{1}{p_{2n}} \operatorname{ce}_{2n}(u) \operatorname{Ce}_{2n}(v).$$

$$\sum_{r=0}^{\infty} (-1)^r A_{2r+1}^{(2n+1)} \cos(2r+1) w J_{2r+1}(z) = -\frac{1}{p_{2n+1}} \operatorname{ce}_{2n+1}(u, q) \operatorname{Ce}_{2n+1}(v, q).$$

$$\sum_{r=0}^{\infty} (-1)^r B_{2r+1}^{(2n+1)} \sin(2r+1) w J_{2r+1}(z) = \frac{1}{s_{2n+1}} \operatorname{se}_{2n+1}(u, q) \operatorname{Se}_{2n+1}(v, q),$$

$$\sum_{r=0}^{\infty} (-1)^r B_{2r+2}^{(2n+2)} \sin(2r+2) w J_{2r+2}(z) = -\frac{1}{s_{2n+2}} \operatorname{se}_{2n+2}(u, q) \operatorname{Se}_{2n+2}(v, q),$$

where $= 2\sqrt{q(\operatorname{ch}^2 v - \sin^2 u)}$, $\operatorname{tg} w = \operatorname{tg} u \operatorname{th} v$.

$$\sum_{r=0}^{\infty} (-1)^r A_{2r}^{(2n)} J_r^2(\sqrt{q}) = \frac{(A_0^{(2n)})^2}{\operatorname{ce}_{2n}(\pi/2, q)},$$

$$\sum_{r=0}^{\infty} (-1)^r A_{2r}^{(2n)} J_r(\sqrt{q} e^{iz}) J_r(\sqrt{q} e^{-iz}) = \frac{A_0^{(2n)}}{p_{2n}} \operatorname{ce}_{2n}(z, q),$$

$$\sum_{r=0}^{\infty} (-1)^r A_{2r+1}^{(2n+1)} [J_r(\sqrt{q} e^{iz}) J_{r+1}(\sqrt{q} e^{-iz}) + J_{r+1}(\sqrt{q} e^{iz}) J_r(\sqrt{q} e^{-iz})] =$$

$$= \frac{A_1^{(2n+1)}}{p_{2n+1}} \operatorname{ce}_{2n+1}(z, q),$$

$$\sum_{r=0}^{\infty} (-1)^r B_{2r+1}^{(2n+1)} [J_r(\sqrt{q} e^{iz}) J_{r+1}(\sqrt{q} e^{-iz}) - J_{r+1}(\sqrt{q} e^{iz}) J_r(\sqrt{q} e^{-iz})] =$$

$$= \frac{iB_1^{(2n+1)}}{s_{2n+1}} \operatorname{se}_{2n+1}(z, q),$$

$$\sum_{r=0}^{\infty} (-1)^r B_{2r+2}^{(2n+2)} [J_r(\sqrt{q} e^{iz}) J_{r+2}(\sqrt{q} e^{-iz}) - J_{r+2}(\sqrt{q} e^{iz}) J_r(\sqrt{q} e^{-iz})] =$$

$$= -\frac{iB_2^{(2n+2)}}{s_{2n+2}} \operatorname{se}_{2n+2}(z, q),$$

$$\sum_{r=0}^{\infty} (-1)^r A_{2r}^{(2n)} I_r(\sqrt{q} e^{-iz}) I_r(\sqrt{q} e^{iz}) = (-1)^n \frac{A_0^{(2n)}}{p_{2n}} \operatorname{ce}_{2n}(z, -q),$$

$$\sum_{r=0}^{\infty} (-1)^r A_{2r}^{(2n)} J_r(\sqrt{q} e^{-z}) J_r(\sqrt{q} e^{z}) = \frac{A_0^{(2n)}}{p_{2n}} \operatorname{Ce}_{2n}(z, q),$$

$$\sum_{r=0}^{\infty} (-1)^r B_{2r+1}^{(2n+1)} [J_r(\sqrt{q} e^{-z}) J_{r+1}(\sqrt{q} e^{z}) - J_{r+1}(\sqrt{q} e^{-z}) J_r(\sqrt{q} e^{z})] =$$

$$= \frac{B_1^{(2n+1)}}{s_{2n+1}} \operatorname{Se}_{2n+1}(z, q),$$

$$\sum_{r=0}^{\infty}(-1)^r A_{2r+1}^{(2n+1)}\left[J_r(\sqrt{q}e^{-z})J_{r+1}(\sqrt{q}e^{z})+J_{r+1}(\sqrt{q}e^{-z})J_r(\sqrt{q}e^{z})\right]=$$
$$=\frac{A_1^{(2n+1)}}{p_{2n+1}}\operatorname{Ce}_{2n+1}(z,\ q),$$

$$\sum_{r=0}^{\infty}(-1)^r B_{2r+2}^{(2n+2)}\left[J_r(\sqrt{q}e^{-z})J_{r+2}(\sqrt{q}e^{-z})-J_{r+2}(\sqrt{q}e^{-z})J_r(\sqrt{q}e^{z})\right]=$$
$$=-\frac{B_2^{(2n+2)}}{s_{2n+2}}\operatorname{Se}_{2n+2}(z,\ q).$$

$$\sum_{n=0}^{\infty}A_{2r}^{(2n)}\operatorname{ce}_{2n}(z,\ q)=\cos 2rz-\frac{1}{2}\delta_{0,\ r},\quad \sum_{n=0}^{\infty}A_{2r+1}^{(2n+1)}\operatorname{ce}_{2n+1}(z,\ q)=\cos(2r+1)z,$$

$$\sum_{n=0}^{\infty}B_{2r+1}^{(2n+1)}\operatorname{se}_{2n+1}(z,\ q)=\sin(2r+1)z,\quad \sum_{n=0}^{\infty}B_{2r+2}^{(2n+2)}\operatorname{se}_{2n+2}(z,\ q)=\sin(2r+2)z.$$

$$\sum_{n=0}^{\infty}\frac{A_0^{(2n)}A_{2m}^{(2n)}}{\operatorname{ce}_{2n}(\pi/2,\ q)}\operatorname{ce}_{2n}(z,\ q)=\frac{1}{2}\left(1-\frac{1}{2}\delta_{m,0}\right)J_m(2\sqrt{q}\cos z).$$

$$\sum_{n=0}^{\infty}\frac{A_0^{(2n)}}{\operatorname{ce}_{2n}(\pi/2,\ q)}\operatorname{ce}_{2n}(u,\ q)\operatorname{ce}_{2n}(z,\ q)=\frac{1}{2}\cos(2\sqrt{q}\cos u\cos z),$$

$$\sum_{n=0}^{\infty}\frac{A_0^{(2n)}}{\operatorname{ce}_{2n}(0,\ q)}\operatorname{ce}_{2n}(u,\ q)\operatorname{ce}_{2n}(z,\ q)=\frac{1}{2}\operatorname{ch}(2\sqrt{q}\sin u\sin z),$$

$$\sum_{n=0}^{\infty}\frac{A_1^{(2n+1)}}{\operatorname{ce}'_{2n+1}(\pi/2,\ q)}\operatorname{ce}_{2n+1}(u,\ q)\operatorname{ce}_{2n+1}(z,\ q)=-\frac{1}{2\sqrt{q}}\sin(2\sqrt{q}\cos u\cos z),$$

$$\sum_{n=0}^{\infty}\frac{A_1^{(2n+1)}}{\operatorname{ce}_{2n+1}(0,\ q)}\operatorname{ce}_{2n+1}(u,\ q)\operatorname{ce}_{2n+1}(z,\ q)=\cos u\cos z\operatorname{ch}(2\sqrt{q}\sin u\sin z),$$

$$\sum_{n=0}^{\infty}\frac{B_1^{(2n+1)}}{\operatorname{se}'_{2n+1}(0,\ q)}\operatorname{se}_{2n+1}(u,\ q)\operatorname{se}_{2n+1}(z,\ q)=\frac{1}{2\sqrt{q}}\operatorname{sh}(2\sqrt{q}\sin u\sin z),$$

$$\sum_{n=0}^{\infty}\frac{B_1^{(2n+1)}}{\operatorname{se}_{2n+1}(\pi/2,\ q)}\operatorname{se}_{2n+1}(u,\ q)\operatorname{se}_{2n+1}(z,\ q)=\sin u\sin z\cos(2\sqrt{q}\cos u\cos z),$$

$$\sum_{n=0}^{\infty}\frac{B_2^{(2n+2)}}{\operatorname{se}'_{2n+2}(\pi/2,\ q)}\operatorname{se}_{2n+2}(u,q)\operatorname{se}_{2n+2}(z,\ q)=-\frac{1}{\sqrt{q}}\sin u\sin z\cos(2\sqrt{q}\cos u\cos z),$$

$$\sum_{n=0}^{\infty}\frac{B_2^{(2n+2)}}{\operatorname{se}'_{2n+2}(0,\ q)}\operatorname{se}_{2n+2}(u,\ q)\operatorname{se}_{2n+2}(z,\ q)=\frac{1}{\sqrt{q}}\cos u\cos z\operatorname{sh}(2\sqrt{q}\sin u\sin z).$$

$$\sum_{n=0}^{\infty}\frac{1}{p_{2n}}\operatorname{Ce}_{2n}(u,\ q)\operatorname{ce}_{2n}(v,\ q)\operatorname{ce}_{2n}(z,\ q)=\frac{1}{2}\cos(2\sqrt{q}\,x\cos z)\cos(\sqrt{2q}\,y\sin z),$$

$$\sum_{n=0}^{\infty}\frac{\operatorname{Ce}_{2n+1}(u,q)}{p_{2n+1}}\operatorname{ce}_{2n+1}(v,\ q)\operatorname{ce}_{2n+1}(z,\ q)=\frac{1}{2}\sin(2\sqrt{q}\,x\cos z)\cos(\sqrt{2q}\,y\sin z),$$

$$\sum_{n=0}^{\infty}\frac{\operatorname{Se}_{2n+1}(u,\ q)}{s_{2n+1}}\operatorname{se}_{2n+1}(v,\ q)\operatorname{se}_{2n+1}(z,\ q)=\frac{1}{2}\cos(2\sqrt{q}\,x\cos z)\sin(\sqrt{2q}\,y\sin z),$$

$$\sum_{n=0}^{\infty} \frac{\mathrm{Se}_{2n+2}(u, q)}{s_{2n+2}} \mathrm{se}_{2n+2}(v, q)\, \mathrm{se}_{2n+2}(z, q) = \frac{1}{2} \cos\left(2\sqrt{q}\, x \cos z\right) \sin\left(\sqrt{2q}\, y \sin z\right),$$

where $=\mathrm{ch}\, u \sin v \cos \varphi$, $y = \mathrm{ch}\, u \sin v \sin \varphi$, $e^{2i\varphi} = \frac{\mathrm{ch}\,(u+iv)-z}{\mathrm{ch}\,(u-iv)-z}$.

II.24. THE POLYNOMIALS OF NEUMANN $O_n(z)$ AND SCHLÄFLI $S_n(z)$

$$O_n(z) = \frac{n}{4} \sum_{k=0}^{[n/2]} \frac{(n-k-1)!}{k!} \left(\frac{z}{2}\right)^{2k-n-1} \qquad [n=1, 2, \ldots].$$

$$O_{-n}(z) = (-1)^n O_n(z) \qquad [n=1, 2, \ldots].$$

$$O_0(z) = z^{-1}, \quad O_1(z) = z^{-2}, \quad O_2(z) = z^{-1} + 4z^{-3}.$$

$$O_0'(z) = -O_1(z), \quad 2O_n'(z) = O_{n-1}(z) - O_{n+1}(z) \qquad [n=1, 2, \ldots].$$

$$(n-1)\, O_{n+1}(z) + (n+1)\, O_{n-1}(z) - \frac{2(n^2-1)}{z} O_n(z) = \frac{2n}{z} \sin^2 \frac{n\pi}{2} \qquad [n=1, 2, \ldots].$$

$$nzO_{n\pm1}(z) - (n^2-1)\, O_n(z) = (\mp n - 1)\, zO_n'(z) + n \sin^2(n\pi/2).$$

$$O_{2n}(z) = z^{-1} S_{1,\,2n}(z), \quad O_{2n+1}(z) = (2n+1)\, z^{-1} S_{0,\,2n}(z).$$

$$S_n(z) = \sum_{k=0}^{[n/2]} \frac{(n-k-1)!}{k!} \left(\frac{z}{2}\right)^{2k-n} \qquad [n=1, 2, \ldots].$$

$$S_0(z) = 1, \quad nS_n(z) = 2z\, O_n(z) - 2\cos^2(n\pi/2).$$

II.25. THE FUNCTIONS $\nu(z)$, $\nu(z, \rho)$, $\mu(z, \lambda)$, $\mu(z, \lambda, \rho)$

Definitions:

$$\nu(z) = \int_0^\infty \frac{z^t}{\Gamma(t+1)}\, dt, \quad \nu(z, \rho) = \int_0^\infty \frac{z^{t+\rho}}{\Gamma(t+\rho+1)}\, dt, \quad \mu(z, \lambda) = \int_0^\infty \frac{t^\lambda z^t}{\Gamma(t+1)}\, dt,$$

$$\mu(z, \lambda, \rho) = \int_0^\infty \frac{t^\lambda z^{t+\rho}}{\Gamma(t+\rho+1)}\, dt = \qquad [\mathrm{Re}\, \lambda > -1].$$

$$= \frac{(-1)^m}{\Gamma(\lambda+m+1)} \int_0^\infty t^{\lambda+m} \frac{d^m}{dt^m} \left[\frac{z^{t+\rho}}{\Gamma(t+\rho+1)}\right] dt \qquad [\mathrm{Re}\, \lambda > -m-1].$$

$$\mu(z, -m, \rho) = (-1)^{m-1} \frac{d^{m-1}}{d\rho^{m-1}} \left[\frac{z^\rho}{\Gamma(\rho+1)}\right] \qquad [m=1, 2, \ldots].$$

$$\nu(z) = \nu(z, 0) = \mu(z, 0) = \mu(z, 0, 0), \qquad \nu(z, \rho) = \mu(z, 0, \rho),$$

$$\mu(z, \lambda) = \mu(z, \lambda, 0) = z\mu(z, \lambda-1, -1), \qquad z\nu(z, \rho-1) - \rho\nu(z, \rho) = \mu(z, 1, \rho),$$

$$\mu(z, \lambda+1, \rho) = z\mu(z, \lambda, \rho-1) - \rho\mu(z, \lambda, \rho).$$

$$\frac{d^n}{dz^n} \nu(z) = \nu(z, -n), \qquad \frac{d^n}{dz^n} \nu(z, \rho) = \nu(z, \rho-n),$$

$$\frac{d^n}{dz^n} \mu(z, \lambda) = \mu(z, \lambda, -n), \qquad \frac{d^n}{dz^n} \mu(z, \lambda, \rho) = \mu(z, \lambda, \rho-n).$$

BIBLIOGRAPHY

1. A. Erdélyi (ed.), *Higher Transcendental Functions,* Vols. 1–3 [Bateman Manuscript Project], McGraw-Hill, New York, 1953–1955.
2. A. Erdélyi (ed.), *Tables of Integral Transforms,* Vols. 1 and 2 [Bateman Manuscript Project], McGraw-Hill, New York, 1954.
3. Yu. A. Brychkov, H.-J. Glaeske and O. I. Marichev, Factorization of integral transforms of convolution type, *Itogi Nauk i Tekhn. VINITI. Mat. Analiz* **21** (1983), 3–41.
4. G. N. Watson, *Treatise on the Theory of Bessel Functions,* 2nd Ed., Cambridge University Press, Cambridge, 1952.
5. N. Ya. Vilenkin, *Special Functions and the Theory of Group Representations,* Amer. Math. Soc., Providence, R.I. (1968).
6. E. W. Hobson, *The Theory of Spherical and Ellipsoidal Harmonics,* Cambridge, 1931.
7. I. S. Gradshtein and I. M. Ryzhik, *Nauka,* Moscow, 1971; *Table of Integrals, Series and Products,* Academic Press, New York, 1980.
8. V. A. Ditkin and A. P. Prudnikov, Spravochnik po operatsionnomu ichisleniyu (Handbook of operational calculus) Vysshaya Shkola, Moscow (1965). French Translation: Formulaire pour la calcul opérationel, Masson & Cie, Paris (1967).
9. L. D. Kudryatsev, *A Course of Mathematical Analysis,* Vols. 1 and 2, Vysshaya Schkola, 1981.
10. N. N. Lebedev, *Special Functions and their Applications,* Prentice-Hall Inc., Englewood Cliffs, N.J., 1965.
11. Y. L. Luke, *The Special Mathematical Functions and Their Approximations,* Vols. 1 and 2, Academic Press, 1969.
12. N. W. McLachlan, *Theory and Applications of Mathieu Functions,* Oxford, 1947.
13. O. I. Marichev, *Handbook of Integral Transforms of Higher Transcendental Functions: Theory and Algorithmic Tables,* Ellis Horwood, Chichester, 1983.
14. V. V. Karpenko, E. T. Kolesov and Yu. S. Yakovlev (Eds.), *Mathematical Tables,* 3rd Ed., Leningrad, 1978.
15. A. F. Nikiforov and V. B. Uvarov, *Special Functions of Mathematical Physics,* Nauka, Moscow, 1978; French translation: *Eléments de la Théorie des Fonctions Spéciales,* Mir, Moscow, 1976.
16. S. M. Nikol'skii, *A Course of Mathematical Analysis,* Vols. 1 and 2, Nauka, Moscow, 1975.
17. A. P. Prudnikov, Yu. A. Brychkov and O. I. Marichev, *Integrals and Series. Elementary Functions* [Vol. 1 of the present work], Nauka, Moscow, 1983. English translation: Gordon and Breach, London, 1986.
18. A. P. Prudnikov, Yu. A. Brychkov and O. I. Marichev, *Integrals and Series. Elementary Functions* [Vol. 2 of the present work], Nauka, Moscow, 1983. English translation: Gordon and Breach, London, 1986.
19. G. S. Salekhov, L. M. Muratov and V. E. Pospeev, *Calculation of Series and Improper Integrals,* Izdat. Kazansk. Univ., Kazan', 1973.
20. L. J. Slater, *Confluent Hypergeometric Functions,* Cambridge University Press, New York, 1960.

21. M. Abramowitz and I. A. Stegun (Eds.), *Handbook of Mathematical Functions,* Dover, New York, 1965.
22. E. C. Titchmarsh, *Introduction to the Theory of Fourier Integrals,* Oxford University Press, Oxford, 1937.
23. E. C. Titchmarsh, *The Zeta-Function of Riemann,* Cambridge University Press, Cambridge, 1930.
24. E. T. Whittaker and G. N. Watson, *A Course of Modern Analysis,* Cambridge University Press, Cambridge, 1952.
25. Yu. F. Filippov, *Tables of Indefinite Integrals and Higher Transcendental Functions,* Idzat. Khar'kovsk. Univ., Khar'kov, 1983.
26. G. M. Fikhtengol'ts, *A Course of Differential and Integral Calculus,* Vols. 1–3, Nauka, Moscow, 1966. German translation: *Differential- und Integralrechnung,* VEB Deutsche Verlag der Wissenschaften, Berlin, 1966–1968.
27. P. I. Khadzhi, *The Error Function,* Inst. Priklad. Fiz. Akad. Nauk Mold. SSR, Kishinev, 1971.
28. E. Jahnke, F. Emde and F. Lösch, *Tables of Higher Functions,* 3rd ed., McGraw-Hill, New York, 1960.
29. R. P. Agarwal, *Generalised Hypergeometric Series,* Asia Publ. House, New York, 1963.
30. P. Appell, *Sur les Fonctions Hypergéometriques de Plusieurs Variables,* Gauthier-Villars, Paris, 1925.
31. P. Appell and J. Kampé de Fériet, *Fonctions Hypergéometriques et Hypersphériques,* Gauthier-Villars, Paris, 1926.
32. W. N. Bailey, *Generalized Hypergeometric Series,* Cambridge University Press, London and New York, 1935.
33. H. Buchholz, *The Confluent Hypergeometric Function with Special Emphasis on its Applications,* Springer-Verlag, Berlin, 1969.
34. P. F. Byrd and M. D. Friedman, *Handbook of Elliptic Integrals for Engineers and Scientists,* Springer-Verlag, Berlin, 1971.
35. S. Colombo, *Les Transformations de Mellin et de Hankel, Applications à la Physique Mathématique,* CNRS, Paris, 1959.
36. S. Colombo et J. Lavoine, *Transformation de Laplace et de Mellin. Formulaires. Mode d'Utilisation,* Mem. Sci. Math., Paris, 1962.
37. G. Doetsch, H. Kniess and D. Voelker, *Tabellen zur Laplace—Transformation,* Springer-Verlag, Berlin, Göttingen, 1947.
38. G. Doetsch, *Handbuch der Laplace—Transformation,* Vols. 1–4, Birkhäuser Verlag, Basel, 1950–1956.
39. U. M. Edwards, *Riemann's Zeta Function,* Academic Press, New York, 1974.
40. H. Exton, *Multiple Hypergeometric Functions and Applications,* Horwood, Chichester, 1976.
41. H. Exton, *Handbook of Hypergeometric Integrals: Theory, Applications, Tables, Computer Programs,* Horwood, Chichester, 1978.
42. E. R. Hansen, *A Table of Series and Products,* Prentice-Hall, Englewood Cliffs and London, 1978.
43. J. Igusa, *Theta Functions,* Springer-Verlag, Berlin, 1972.
44. S. Lang, *Elliptic Functions,* Addison Wesley, New York, 1973.
45. L. Lewin, *Dilogarithms and Associated Functions,* MacDonald and Co., London, 1958.

46. Y. L. Luke, *The Special Functions and Their Approximations,* Vols. 1 and 2, Academic Press, New York, 1969.
47. T. M. MacRobert, *Functions of a Complex Variable,* Macmillan, London and New York, 1962.
48. W. Magnus, F. Oberhettinger and R. P. Soni, *Formulas and Theorems for the Special Functions of Mathematical Physics,* 3rd ed., Springer-Verlag, New York and Berlin, 1966.
49. A. M. Mathai and R. K. Saxena, Generalized hypergeometric functions with applications in statistics and physical sciences, Lecture Notes in Math., no. 348, Springer-Verlag, Heidelberg and New York, 1973.
50. A. M. Mathai and R. K. Saxena, *The H-Function with Applications in Statistics and Other Disciplines,* Halsted Press Book, New York, London, Sydney and Toronto, 1978.
51. J. Meixner, F. W. Schötke and G. Wolf, *Mathieu Functions and Spheroidal Functions and Their Mathematical Foundations,* Springer-Verlag, Berlin, 1957.
52. F. Oberhettinger, *Tabellen zur Fourier Transformation,* Springer-Verlag, Berlin, 1957.
53. F. Oberhettinger and T. P. Higgins, Tables of Lebedev, Mehler and generalized Mehler transforms, Math. Note no. 246, Boeing Scientific Research Laboratories, Seattle, Washington, 1961.
54. F. Oberhettinger, *Tables of Bessel Transforms,* Springer-Verlag, New York, 1972.
55. F. Oberhettinger, *Fourier Expansions. A Collection of Formulas,* Academic Press, New York, 1973.
56. F. Oberhettinger and L. Badii, *Tables of Laplace Transforms,* Springer-Verlag, Berlin, Heidelberg and New York, 1973.
57. F. Oberhettinger, *Tables of Mellin Transforms,* Springer-Verlag, New York, Heidelberg and Berlin, 1974.
58. G. E. Roberts and H. Kaufman, *Table of Laplace Transforms,* McAinsh and Co., Philadelphia; W. B. Saunders and Co., London, 1966.
59. L. J. Slater, *Generalized Hypergeometric Functions,* Cambridge University Press, London and New York, 1966.
60. H. M. Srivastava and R. G. Buschman, *Convolution Integral Equations with Special Function Kernels,* Wiley Eastern Ltd., New Delhi, Bangalore, 1977.
61. H. M. Srivastava, K. C. Gupta and S. P. Goyal, *The H-Functions of One or Two Variables with Applications,* South Asian Publishers, New Delhi, 1982.
62. F. Tricomi, *Elliptische Funktionen,* Academische Verlagsgesellschaft, Leipzig, 1948.
63. F. Tricomi, Fonctions hypergéometriques confluentes, 140, Mem. Sci. Math. Fasc., Paris; Gauthier-Villars, 1960.
64. D. Voelker and G. Doetsch, *Die Zweidimensionale Laplace Transformation,* Birkhäuser, Basel, 1950.
65. A. D. Wheelon, *Tables of Summable Series and Integrals Involving Bessel Functions,* Holden-Day Inc., San Francisco, 1968.

INDEX OF NOTATIONS FOR FUNCTIONS AND CONSTANTS

$A(x)$ is an algebraic function or power function with arbitrary exponent.

$\operatorname{am} u \equiv \operatorname{am}(u, k)$ is the Jacobi elliptic function, $u = \int\limits_0^{\operatorname{am} u} \frac{dt}{\sqrt{1-k^2 \sin^2 t}}$

$\operatorname{Ai}(z) = \frac{1}{\pi} \sqrt{\frac{z}{3}} K_{1/3}\left(\frac{2}{3} z^{3/2}\right)$ is the Airy function.

$\arccos z$, $\operatorname{arcctg} z$, $\arcsin z$, $\operatorname{arctg} z$ are the inverse trigonometric functions.

$\operatorname{Arch} z$, $\operatorname{Arcth} z$, $\operatorname{Arsh} z$, $\operatorname{Arth} z$ are the inverse hyperbolic functions.

$\arg z$ is the argument of the complex number z $\left(z = |z| e^{i \arg z}\right)$

B_n are the Bernoulli numbers.

$B_n(z)$ are the Bernoulli polynomials.

$\operatorname{bei}_\nu(z)$, $\operatorname{ber}_\nu(z)$, $\operatorname{bei}(z) \equiv \operatorname{bei}_0(z)$, $\operatorname{ber}(z) \equiv \operatorname{ber}_0(z)$ are the Kelvin functions

$\operatorname{Bi}(z) = \sqrt{\frac{z}{3}} \left[I_{-1/3}\left(\frac{2}{3} z^{3/2}\right) + I_{1/3}\left(\frac{2}{3} z^{3/2}\right)\right]$ is the Airy function.

$\mathbf{C} = -\psi(1) = 0{,}5772156649\ldots$ is the Euler–Masceroni constant.

$C(x) = \frac{1}{\sqrt{2\pi}} \int\limits_0^x \frac{\cos t}{\sqrt{t}}\, dt$ is the Fresnel cosine integral.

$C(x, \nu) = \int\limits_x^\infty t^{\nu-1} \cos t\, dt$ $[\operatorname{Re} \nu < 1]$ is the generalized Fresnel cosine integral.

$C_n^\lambda(z) = \frac{(2\lambda)_n}{n!}\, {}_2F_1\left(-n, n+2\lambda; \lambda+\frac{1}{2}; \frac{1-z}{2}\right)$ are the Gegenbauer polynomials.

$C_\nu^\lambda(z) = \frac{\Gamma(2\lambda+\nu)}{\Gamma(2\lambda)\Gamma(\nu+1)}\, {}_2F_1\left(-\nu, 2\lambda+\nu; \lambda+\frac{1}{2}; \frac{1-z}{2}\right)$ is the Gegenbauer function.

$\operatorname{Ce}_n(z, q)$ is the modified Mathieu function of the 1st kind.

$\operatorname{ce}_n(z, q)$ is the periodic Mathieu function of the 1st kind.

$\operatorname{ch} z = \frac{e^z + e^{-z}}{2}$ is the hyperbolic function.

$\operatorname{chi}(x) = \mathbf{C} + \ln x + \int\limits_0^x \frac{\operatorname{ch} t - 1}{t}\, dt$ is the integral hyperbolic cosine.

$\operatorname{ci}(x) = -\int\limits_x^\infty \frac{\cos t}{t}\, dt$ is the integral cosine.

$\operatorname{Cl}_2(z) = -\int\limits_0^z \ln\left(2 \sin \frac{t}{2}\right) dt$ is the Clausen integral.

$\operatorname{cn} u = \cos(\operatorname{am} u)$ is the Jacobi elliptic function.

$\cos z = \operatorname{ch}(iz)$, $\operatorname{cosec} z = \dfrac{1}{\sin z}$ are the trigonometric functions.

$\operatorname{cosech} z = \dfrac{1}{\operatorname{sh} z}$ is the hyperbolic function.

$\operatorname{ctg} z = \dfrac{\cos z}{\sin z} = i \operatorname{cth}(iz)$ is the trigonometric function.

$\operatorname{cth} z = \dfrac{\operatorname{ch} z}{\operatorname{sh} z}$ — is the hyperbolic function.

$D+$ $(D-)$ are the parts of the s-plane lying to the left (or to the right) of the contours $L_{\pm\infty}$

$$\mathbf{D}(k) = \int_0^{\pi/2} \frac{\sin^2 t\, dt}{\sqrt{1-k^2\sin^2 t}} = D\left(\frac{\pi}{2},\ k\right)$$ is the complete elliptic integral.

$$D(\varphi,\ k) = \int_0^{\varphi} \frac{\sin^2 t\, dt}{\sqrt{1-k^2\sin^2 t}}$$ is the elliptic integral.

$$D_\nu(z) = 2^{\nu/2} e^{-z^2/4} \Psi\left(-\frac{\nu}{2},\ \frac{1}{2};\ \frac{z^2}{2}\right)$$ is the parabolic cylinder function.

$\operatorname{dn} u = \sqrt{1-k^2\sin^2(\operatorname{am} u)}$ is the elliptic Jacobi function.

$$\mathbf{E}(k) = \int_0^{\pi/2} \sqrt{1-k^2\sin^2 t}\, dt$$ is the complete elliptic integral of the 2nd kind.

$$E(\varphi,\ k) = \int_0^{\varphi} \sqrt{1-k^2\sin^2 t}\, dt$$ is the elliptic integral of the 2nd kind.

E_n is the Euler number.

$E_n(z)$ is the Euler polynomial.

$$\mathbf{E}_\nu(z) = \frac{1}{\pi}\int_0^{\pi} \sin(\nu t - z\sin t)\, dt$$ is the Weber function.

$$E_\rho(z;\ \mu) = \sum_{k=0}^{\infty} \frac{z^k}{\Gamma(\mu+\rho^{-1}k)}$$ $[\rho > 0]$ is the function of Mittag–Leffler type.

$$\operatorname{Ei}(x) = \int_{-\infty}^{x} \frac{e^t}{t}\, dt$$ is the integral exponential function.

$$E(p;\ a_r:\ q;\ b_s:\ z) \equiv E((a_p);\ (b_q);\ z) \equiv E\left(\begin{matrix}(a_p); & z\\ (b_q) & \end{matrix}\right) \equiv E\left(\begin{matrix}(a_p)\\ (b_q);\ z\end{matrix}\right) =$$
$$= G^{p,\,1}_{q+1,\,p}\left(z \,\middle|\, \begin{matrix}1,\ b_1,\ \ldots,\ b_q\\ a_1,\ \ldots,\ a_p\end{matrix}\right)$$ is the MacRobert E-function.

$e = 2{,}718281828459\ldots$ is the number e.

$e^z = \exp z$ is the exponential function.

$$\operatorname{erf}(x) = \frac{2}{\sqrt{\pi}}\int_0^{x} e^{-t^2}\, dt$$ is the error function.

$$\operatorname{erfc}(x) = 1 - \operatorname{erf}(x) = \frac{2}{\sqrt{\pi}}\int_x^{\infty} e^{-t^2}\, dt$$ is the complementary error function.

$$\operatorname{erfi}(x) = \frac{2}{\sqrt{\pi}}\int_0^{x} e^{t^2}\, dt$$ is the error function of an imaginary argument.

$$F(\varphi,\ k) = \int_0^{\varphi} \frac{dt}{\sqrt{1-k^2\sin^2 t}}$$ is the elliptic integral of the 1st kind.

$\mathrm{Se}_n(z, q)$, $\mathrm{Fe}_n(z, q)$, $\mathrm{Fey}_n(z, q)$, $\mathrm{Fek}_n(z, q)$ are the second aperiodic solutions of the Mathieu equation.

$$ {}_2F_1(a, b; c; z) \equiv F(a, b; c; z) \equiv {}_2F_1\left(\begin{matrix} a, & b; & z \\ c & & \end{matrix}\right) \equiv {}_2F_1\left(\begin{matrix} a, & b \\ c; & z \end{matrix}\right) = \sum_{k=0}^{\infty} \frac{(a)_k (b)_k}{(c)_k} \frac{z^k}{k!} \quad [|z| < 1], $$

$$ = \Gamma\left[\begin{matrix} c \\ b, \ c-b \end{matrix}\right] \int_0^1 t^{b-1} (1-t)^{c-b-1} (1-tz)^{-a} \, dt \quad [\mathrm{Re}\, c > \mathrm{Re}\, b > 0; \ |\arg(1-z)| < \pi] $$

is the Gauss hypergeometric function.

$$ {}_pF_q(a_1, \ldots, a_p; b_1, \ldots, b_q; z) \equiv {}_pF_q((a_p); (b_q); z) \equiv {}_pF_q\left(\begin{matrix} a_1, & \ldots, & a_p; & z \\ b_1, & \ldots, & b_q & \end{matrix}\right) \equiv $$

$$ \equiv {}_pF_q\left(\begin{matrix} (a_p); & z \\ (b_q) & \end{matrix}\right) \equiv {}_pF_q\left(\begin{matrix} (a_p) & \\ (b_q); & z \end{matrix}\right) = \sum_{k=0}^{\infty} \frac{(a_1)_k (a_2)_k \ldots (a_p)_k}{(b_1)_k (b_2)_k \ldots (b_q)_k} \frac{z^k}{k!} $$ is the generalized hypergeometric function

$$ {}_1F_1(a; b; z) = \sum_{k=0}^{\infty} \frac{(a)_k z^k}{(b)_k k!} $$ is the Kummer confluent hypergeometric function.

$$ F_A^{(n)}(a, b_1, \ldots, b_n; c_1, \ldots, c_n; z_1, \ldots, z_n) = $$

$$ = \sum_{k_1, \ldots, k_n = 0}^{\infty} \frac{(a)_{k_1 + \ldots + k_n} (b_1)_{k_1} \ldots (b_n)_{k_n}}{(c_1)_{k_1} \ldots (c_n)_{k_n}} \frac{z_1^{k_1} \ldots z_n^{k_n}}{k_1! \ldots k_n!} \quad \left[\sum_{j=1}^{n} |z_j| < 1\right] $$ is the Lauricella function.

$$ F_C^{(n)}(a, b; c_1, \ldots, c_n; z_1, \ldots, z_n) = \sum_{k_1, \ldots, k_n = 0}^{\infty} \frac{(a)_{k_1 + \ldots + k_n} (b)_{k_1 + \ldots + k_n}}{(c_1)_{k_1} \ldots (c_n)_{k_n}} \frac{z_1^{k_1} \ldots z_n^{k_n}}{k_1! \ldots k_n!} $$

$$ \left[\sum_{j=1}^{n} \sqrt{|z_j|} < 1\right] $$ is the Lauricella function.

$F_j(\ldots; w, z)$ $[j = 1, 2, 3, 4]$ are the Appell functions:

$$ F_1(a, b, b'; c; w, z) = \sum_{k, l=0}^{\infty} \frac{(a)_{k+l} (b)_k (b')_l}{(c)_{k+l}} \frac{w^k z^l}{k! \, l!} \quad [|w|, |z| < 1], $$

$$ F_2(a, b, b'; c, c'; w, z) = \sum_{k, l=0}^{\infty} \frac{(a)_{k+l} (b)_k (b')_l}{(c)_k (c')_l} \frac{w^k z^l}{k! \, l!} \quad [|w| + |z| < 1], $$

$$ F_3(a, a', b, b'; c; w, z) = \sum_{k, l=0}^{\infty} \frac{(a)_k (a')_l (b)_k (b')_l}{(c)_{k+l}} \frac{w^k z^l}{k! \, l!} \quad [|w|, |z| < 1], $$

$$ F_4(a, b; c, c'; w, z) = \sum_{k, l=0}^{\infty} \frac{(a)_{k+l} (b)_{k+l}}{(c)_k (c')_l} \frac{w^k z^l}{k! \, l!} \quad [\sqrt{|w|} + \sqrt{|z|} < 1] $$

$$ \mathrm{G} = \sum_{k=0}^{\infty} \frac{(-1)^k}{(2k+1)^2} = 0{,}9159655942\ldots $$ is the Catalan constant.

$$ G_{p,q}^{m,n}\left(z \,\middle|\, \begin{matrix} a_1, & \ldots, & a_p \\ b_1, & \ldots, & b_q \end{matrix}\right) \equiv G_{p,q}^{m,n}\left(z \,\middle|\, \begin{matrix} (a_p) \\ (b_q) \end{matrix}\right) = $$

$$ = \frac{1}{2\pi i} \int_L \Gamma\left[\begin{matrix} b_1 + s, & \ldots, & b_m + s, & 1 - a_1 - s, & \ldots, & 1 - a_n - s \\ a_{n+1} + s, & \ldots, & a_p + s, & 1 - b_{m+1} - s, & \ldots, & 1 - b_q - s \end{matrix}\right] z^{-s} \, ds, $$

$L = L_{\pm\infty}$, $L_{i\infty}$, is the Meijer G-function.

$\mathrm{ge}_n(z, q)$, $\mathrm{Ge}_n(z, q)$, $\mathrm{Gey}_n(z, q)$, $\mathrm{Gek}_n(z, q)$ are the second aperiodic solutions of the Mathieu equati

$H(x)=\begin{cases}1, & x \geqslant 0,\\ 0, & x<0\end{cases}$ is the Heaviside function.

$\mathbf{H}_\nu(z)=\frac{2}{\sqrt{\pi}}\left(\frac{z}{2}\right)^{\nu+1}\frac{1}{\Gamma(\nu+3/2)}\,{}_1F_2\left(1;\ \frac{3}{2},\ \nu+\frac{3}{2};\ -\frac{z^2}{4}\right)$ is the Struve function.

$H_\nu^{(1)}(z)=J_\nu(z)+iY_\nu(z)$, $H_\nu^{(2)}(z)=J_\nu(z)-iY_\nu(z)$ are the Bessel functions of the 3rd kind (Hankel fun of the 1st and 2nd kinds).

$H_n(z)=(-1)^n e^{z^2}\frac{d^n}{dz^n}e^{-z^2}$ are the Hermite polynomials.

$$h_{\mu,\nu}(z)=\frac{z^{\mu+1}e^{-z}}{(\mu+1)^2-\nu^2}\,{}_2F_2\left(1,\ \mu+\frac{3}{2};\ \mu-\nu+2,\ \mu+\nu+2;\ 2z\right)$$

$$H_{\mu,\nu}(z)=h_{\mu,\nu}(z)-\frac{\sqrt{\pi}\,\Gamma(\mu-\nu+1)\,\Gamma(\mu+\nu+1)}{2^{\mu+1}\Gamma(\mu+3/2)}\left[I_\nu(z)+\frac{\sin(\nu-\mu)\pi}{\pi\cos\mu\pi}K_\nu(z)\right]$$

$\mathbf{H}_\nu^\mu(z)=J_{\nu,1/2}^\mu(z)$ $(\mathbf{H}_\nu^2(z)=\mathbf{H}_\nu(z))$ is the generalized Struve function.

$H_{pq}^{mn}\left[z\left|\begin{matrix}(a_1, A_1), & \dots, & (a_p, A_p)\\ (b_1, B_1), & \dots, & (b_q, B_q)\end{matrix}\right.\right]\equiv H_{pq}^{mn}\left[z\left|\begin{matrix}[a_p, A_p]\\ [b_q, B_q]\end{matrix}\right.\right]$ is the Fox H-function.

$I_\nu(z)=\frac{1}{\Gamma(\nu+1)}\left(\frac{z}{2}\right)^\nu {}_0F_1\left(\nu+1;\ \frac{z^2}{4}\right)=e^{-\nu\pi i/2}J_\nu\left(e^{\pi i/2}z\right)$ is the modified Bessel function of the 1st kind (Bessel function of an imaginary argument).

$\mathrm{Im}\,z$ is the imaginary part of the complex number $z=x+iy$ $(\mathrm{Im}\,z=y)$

$J_\nu(z)=\frac{1}{\Gamma(\nu+1)}\left(\frac{z}{2}\right)^\nu {}_0F_1\left(\nu+1;\ -\frac{z^2}{4}\right)$ is the Bessel function of the 1st kind.

$\mathbf{J}_\nu(z)=\frac{1}{\pi}\int\limits_0^\pi \cos(\nu t-z\sin t)\,dt$ is the Anger function.

$J_\nu^\mu(z)=\sum\limits_{k=0}^\infty \frac{(-z)^k}{k!\,\Gamma(k\mu+\nu+1)}$ $[\mu>-1]$ is the Bessel–Maitland function.

$Ji_\nu(x)=\int\limits_x^\infty J_\nu(t)\frac{dt}{t}$ is the integral Bessel function of the 1st kind.

$$J(x, y)=1-e^{-y}\int\limits_0^x e^{-t}I_0(2\sqrt{yt})\,dt$$

$\mathbf{K}(k)=\int\limits_0^{\pi/2}\frac{dt}{\sqrt{1-k^2\sin^2 t}}=F\left(\frac{\pi}{2},\ k\right)$ is the complete elliptic integral of the 1st kind.

$K_\nu(z)=\frac{\pi[I_{-\nu}(z)-I_\nu(z)]}{2\sin\nu\pi}$ $[\nu\neq n]$, $K_n(z)=\lim\limits_{\nu\to n}K_\nu(z)$ $[n=0,\ \pm1,\ \pm2,\ \dots]$ is the MacDonald function (modified Bessel function of the 3rd kind).

$k_\nu(z)=\frac{1}{\Gamma(\nu/2+1)}W_{\nu/2,\,1/2}(2z)$ is the Bateman function.

$\mathrm{kei}_\nu(z)$, $\mathrm{ker}_\nu(z)$, $\mathrm{kei}(z)=\mathrm{kei}_0(z)$, $\mathrm{ker}(z)=\mathrm{ker}_0(z)$ is the Kelvin function.

$Ki_\nu(x)=\int\limits_x^\infty K_\nu(t)\frac{dt}{t}$ is the modified integral Bessel function.

$L_{\pm\infty}$, $L_{i\infty}$ are the contours in the Mellin–Barnes integrals for the G- and H-functions.

$\mathbf{L}_\nu(z)=e^{-(\nu+1)\pi i/2}\mathbf{H}_\nu\left(e^{\pi i/2}z\right)$ is the modified Struve function.

$L_\nu(z)={}_1F_1(-\nu;\ 1;\ z)$ is the Laguerre function.

$L_n(z)=L_n^0(z)$ are the Laguerre polynomials.

$L_n^{\lambda}(z)=\frac{z^{-\lambda}e^{z}}{n!}\frac{d^n}{dz^n}\left(z^{n+\lambda}e^{-z}\right)$ are the generalized Laguerre polynomials.

$$\mathrm{Li}_{\nu}(z)=\sum_{k=1}^{\infty}\frac{z^k}{k^{\nu}}\qquad [|z|<1],$$

$$=\frac{z}{\Gamma(\nu)}\int_0^{\infty}\frac{t^{\nu-1}\,dt}{e^t-z}\qquad [\mathrm{Re}\,\nu>0;\quad |\arg(1-z)|<\pi]$$ is the polylogarithm of order ν.

$\mathrm{Li}_2(z)$ is the Euler dilogarithm.

$\mathrm{li}(z)=\mathrm{Ei}(\ln z),\quad \mathrm{li}(x)=\int_0^x\frac{dt}{\ln t}$ is the integral logarithm.

$\ln z=\ln|z|+i\arg z$ is the natural logarithm $[z=|z|\,e^{i(\arg z+2\pi k)},\quad k=0,\quad \pm1,\quad \pm2,\quad \ldots]$

$M_{\varkappa,\,\mu}(z)=z^{\mu+1/2}e^{-z/2}\,{}_1F_1\left(\mu-\varkappa+\frac{1}{2};\ 2\mu+1;\ z\right)$ is the Whittaker confluent hypergeometric

nction.

$O_n(z)=\frac{n}{4}\sum_{k=0}^{[n/2]}\frac{(n-k-1)!}{k!}\left(\frac{z}{2}\right)^{2k-n-1}$ are the Neumann polynomials.

$P_n(z)=\frac{2^{-n}}{n!}\frac{d^n}{dz^n}(z^2-1)^n$ are the Legendre polynomials.

$P_{\nu}(z)\equiv P_{\nu}^{0}(z)={}_2F_1\left(-\nu,\ 1+\nu;\ 1;\ \frac{1-z}{2}\right)$ $[\,|\arg(1+z)|<\pi]$ are the Legendre functions of the

kind.

$$P_{\nu}^{\mu}(z)=\frac{1}{\Gamma(1-\mu)}\left(\frac{z+1}{z-1}\right)^{\mu/2}{}_2F_1\left(-\nu,\ \nu+1;\ 1-\mu;\ \frac{1-z}{2}\right)$$
$$[\,|\arg(z\pm1)|<\pi;\ \mu\neq m;\quad m=1,\ 2,\ \ldots]$$

$$P_{\nu}^{m}(z)=(z^2-1)^{m/2}\left(\frac{d}{dz}\right)^m P_{\nu}(z)\qquad [\,|\arg(z-1)|<\pi;\quad m=1,\ 2,\ \ldots]$$

$$P_{\nu}^{\mu}(x)=\frac{1}{\Gamma(1-\mu)}\left(\frac{1+x}{1-x}\right)^{\mu/2}{}_2F_1\left(-\nu,\ \nu+1;\ 1-\mu;\ \frac{1-x}{2}\right)$$
$$[-1<x<1;\ \mu\neq m;\quad m=1,\ 2,\ \ldots]$$

$$P_{\nu}^{m}(x)=(-1)^m(1-x^2)^{m/2}\left(\frac{d}{dx}\right)^m P_{\nu}(x)\qquad [-1<x<1;\ m=1,\ 2,\ \ldots]$$

the associated Legendre function of the 1st kind.

$$P_n^{(\rho,\,\sigma)}(z)=\frac{(-1)^n}{2^n n!}(1-z)^{-\rho}(1+z)^{-\sigma}\frac{d^n}{dz^n}\left[(1-z)^{\rho+n}(1+z)^{\sigma+n}\right]=$$
$$=\frac{(\rho+1)_n}{n!}\,{}_2F_1\left(-n,\ \rho+\sigma+n+1;\ \rho+1;\ \frac{1-z}{2}\right)$$ are the Jacobi polynomials.

$Q_{\nu}(z)\equiv Q_{\nu}^{0}(z)$ is the Legendre function of the 2nd kind.

$$Q_{\nu}^{\mu}(z)=\frac{e^{i\mu\pi}\sqrt{\pi}}{2^{\nu+1}}\Gamma\begin{bmatrix}\mu+\nu+1\\ \nu+3/2\end{bmatrix}z^{-\mu-\nu-1}(z^2-1)^{\mu/2}\times$$
$$\times{}_2F_1\left(\frac{\mu+\nu+1}{2},\ \frac{\mu+\nu}{2}+1;\ \nu+\frac{3}{2};\ \frac{1}{z^2}\right)$$
$$[\,|\arg z|,\ |\arg(z\pm1)|<\pi;\ \nu+1/2,\ \mu+\nu\neq-1,\ -2,\ -3,\ \ldots]$$

$$Q^{\mu}_{-n-3/2}(z)=\frac{e^{i\mu\pi}\sqrt{\pi}\,\Gamma(\mu+n+3/2)}{2^{n+3/2}(n+1)!}z^{-\mu-n-3/2}(z^2-1)^{\mu/2}\times.$$
$$\times{}_2F_1\left(\frac{2\mu+2n+3}{4},\ \frac{2\mu+2n+5}{4};\ n+2;\ \frac{1}{z^2}\right)$$
$$[\,|\arg(z\pm1)|,\ |\arg z|<\pi;\ \mu+\nu\neq-1,\ -2,\ -3,\ \ldots]$$

$$Q_\nu^\mu(x)=\frac{e^{-i\mu\pi}}{2}\left[e^{-i\mu\pi/2}Q_\nu^\mu(x+i0)+e^{-\mu\pi/2}Q_\nu^\mu(x-i0)\right]=$$

$$=\frac{\pi}{2\sin\mu\pi}\left[P_\nu^\mu(x)\cos\mu\pi-\Gamma\begin{bmatrix}\nu+\mu+1\\ \nu-\mu+1\end{bmatrix}P_\nu^{-\mu}(x)\right]$$

$[-1<x<1;\ \mu\neq\pm m;\ \mu+\nu\neq-1,\ -2,\ -3,\ \ldots]$,

$$=(-1)^m(1-x^2)^{m/2}\left(\frac{d}{dx}\right)^m Q_\nu(x)\quad [\mu=m;\ \nu\neq-m-1,\ -m-2,\ \ldots],$$

$$=(-1)^m\Gamma\begin{bmatrix}\nu-m+1\\ \nu+m+1\end{bmatrix}Q_\nu^m(x)\quad [\mu=-m;\ \nu\neq-m-1,\ -m-2,\ \ldots]$$ is the associated Legendre function of the 2nd kind.

$$Q_n^{(\rho,\sigma)}(z)=2^{\rho+\sigma+n}\,\Gamma\begin{bmatrix}\rho+n+1,\ \sigma+n+1\\ \rho+\sigma+2n+2\end{bmatrix}\frac{(z+1)^\sigma}{(z-1)^{\rho+n+1}}\times$$

$$\times{}_2F_1\left(n+1,\ \rho+n+1;\ \rho+\sigma+2n+2;\ \frac{2}{1-z}\right)$$

$[|\arg(z\pm1)|<\pi]$ is the Jacobi function of the 2nd kir

$\operatorname{Re} z$ is the real part of the complex number $z=x+iy$ $(\operatorname{Re} z=x)$

$\operatorname*{res}_{z=a}\varphi(z)$ is the residue of the function $\varphi(z)$ at the point a.

$$S(x)=\frac{1}{\sqrt{2\pi}}\int_0^x\frac{\sin t}{\sqrt{t}}\,dt$$ is the Fresnel sine integral.

$$S(x,\ \nu)=\int_x^\infty t^{\nu-1}\sin t\,dt\ [\operatorname{Re}\nu<1]$$ is the generalized Fresnel sine integral.

$$S_n(z)=\sum_{k=0}^{[n/2]}\frac{(n-k-1)!}{k!}\left(\frac{z}{2}\right)^{2k-n},\quad S_0(z)=1$$ are the Schläffli polynomials.

$$S_n^{(m)}=\sum_{k=0}^{n-m}(-1)^k\binom{n+k-1}{n-m+k}\binom{2n-m}{n-m-k}\sigma_{n-m+k}^k$$ are the Stirling numbers of the 1st kind.

$$S_{\mu,\nu}(z)=s_{\mu,\nu}(z)+2^{\mu-1}\Gamma\begin{bmatrix}\nu,\ (1+\mu+\nu)/2\\ (1-\mu+\nu)/2\end{bmatrix}\left(\frac{z}{2}\right)^{-\nu}{}_0F_1\left(1-\nu;\ -\frac{z^2}{4}\right)+$$

$$+2^{\mu-1}\Gamma\begin{bmatrix}-\nu,\ (1+\mu-\nu)/2\\ (1-\mu-\nu)/2\end{bmatrix}\left(\frac{z}{2}\right)^{\nu}{}_0F_1\left(1+\nu;\ -\frac{z^2}{4}\right)$$ is the Lommel function.

$\operatorname{se}_n(z,\ q)$ — is the periodic Mathieu function.

$\operatorname{Se}_n(z,\ q)$ — is the Mathieu function of an imaginary argument.

$$s_{\mu,\nu}(z)=\frac{z^{\mu+1}}{(\mu-\nu+1)(\mu+\nu+1)}\,{}_1F_2\left(1;\ \frac{\mu-\nu+3}{2},\ \frac{\mu+\nu+3}{2};\ -\frac{z^2}{4}\right)$$ is the Lommel function.

$$\sec z=\frac{1}{\cos z}$$ is the trigonometric function.

$$\operatorname{sech} z=\frac{1}{\operatorname{cn} z}$$ is the hyperbolic function.

$$\operatorname{sgn} x=\begin{cases}1, & x>0,\\ 0, & x=0,\\ -1, & x<0\end{cases}$$

$$\operatorname{sh} z=\frac{e^z-e^{-z}}{2}$$ is the hyperbolic function.

$$\operatorname{shi}(x)=\int_0^x\frac{\operatorname{sh} t}{t}\,dt=-i\operatorname{Si}(ix)$$ is the integral hyperbolic sine.

$$\operatorname{Si}(x)=\int_0^x\frac{\sin t}{t}\,dt$$ is the integral sine.

$$\operatorname{si}(x)=\operatorname{Si}(x)-\frac{\pi}{2}=-\int_x^\infty\frac{\sin t}{t}\,dt$$ is the integral sine.

$\sin z = -i \operatorname{sh}(iz)$ is the trigonometric function.

$\operatorname{sn} u = \sin(\operatorname{am} u)$ is the Jacobi elliptic function.

$T_n(z) = \cos(n \arccos z) = F\left(-n, n; \frac{1}{2}; \frac{1-z}{2}\right)$ are the Chebyshev polynomials of the 1st kind.

$\operatorname{tg} z = \frac{\sin z}{\cos z} = -i \operatorname{th}(iz)$ is the trigonometric function.

$\operatorname{th} z = \frac{\operatorname{sh} z}{\operatorname{ch} z}$ is the hyperbolic function.

$U_n(z) = \frac{\sin[(n+1)\arccos z]}{\sqrt{1-z^2}} = (n+1)\,{}_2F_1\left(-n, n+2; \frac{3}{2}; \frac{1-z}{2}\right)$ are the Chebyshev polynomials of the 2nd kind.

$U_\nu(w, z) = \sum_{k=0}^{\infty} (-1)^k \left(\frac{w}{z}\right)^{2k+\nu} J_{2k+\nu}(z)$ is the Lommel function of two variables.

$W_{\varkappa,\mu}(z) = z^{\mu+1/2} e^{-z/2} \Psi\left(\mu - \varkappa + \frac{1}{2}, 2\mu+1; z\right)$ is the Whittaker confluent hypergeometric function.

$Y_\nu(z) = \frac{\cos \nu\pi J_\nu(z) - J_{-\nu}(z)}{\sin \nu\pi}$ $[\nu \neq n]$, $Y_n(z) = \lim_{\nu \to n} Y_\nu(z)$ $[n = 0, \pm 1, \pm 2, \ldots]$ is the Neumann function (Bessel function of the 2nd kind).

$y_n(z, a, b) = {}_2F_0\left(-n, a-1+n; -\frac{z}{b}\right)$ are the generalized Bessel polynomials.

$y_n(z) = {}_2F_0\left(-n, n+1; -\frac{z}{2}\right)$ are the Bessel polynomials.

$Yi_\nu(x) = \int_x^\infty Y_\nu(t) \frac{dt}{t}$ is the integral Bessel function of the 2nd kind.

$\mathrm{B}(\alpha, \beta) = \frac{\Gamma(\alpha)\Gamma(\beta)}{\Gamma(\alpha+\beta)}$ is the beta-function.

$$\mathrm{B}_x(\alpha, \beta) = \int_0^x t^{\alpha-1}(1-t)^{\beta-1}\,dt \qquad [\operatorname{Re}\alpha > 0;\ x < 1],$$

$= \frac{x^\alpha}{\alpha}\,{}_2F_1(\alpha, 1-\beta; \alpha+1; x)$ $[|\arg x|, |\arg(1-x)| < \pi]$ is the incomplete beta-function.

$$\beta(z) = \frac{1}{2}\left[\psi\left(\frac{z+1}{2}\right) - \psi\left(\frac{z}{2}\right)\right]$$

$\Gamma(z) = \int_0^\infty t^{z-1} e^{-t}\,dt$ $[\operatorname{Re} z > 0]$ is the gamma-function.

$\Gamma(\nu, x) = \int_x^\infty t^{\nu-1} e^{-t}\,dt = e^{-x}\Psi(1-\nu, 1-\nu; x)$ is the complementary incomplete gamma-function.

$\gamma(\nu, x) = \Gamma(\nu) - \Gamma(\nu, x) = \int_0^x t^{\nu-1} e^{-t}\,dt = \frac{x^\nu}{\nu}\,{}_1F_1(\nu; \nu+1; -x)$ $[\operatorname{Re}\nu > 0]$ is the incomplete gamma-function.

$$\Gamma\begin{bmatrix} a_1, \ldots, a_p \\ b_1, \ldots, b_q \end{bmatrix} \equiv \frac{\prod_{k=1}^{p} \Gamma(a_k)}{\prod_{l=1}^{q} \Gamma(b_l)}, \quad \Gamma[a_1, \ldots, a_p] \equiv \prod_{k=1}^{p} \Gamma(a_k)$$

$$\Gamma\begin{bmatrix} (a_p) \\ (b_q) \end{bmatrix} \equiv \Gamma\begin{bmatrix} a_1, \ldots, a_p \\ b_1, \ldots, b_q \end{bmatrix}, \qquad \Gamma[(a_p)] \equiv \Gamma[a_1, \ldots, a_p]$$

$$\Gamma\left[\begin{matrix}(a)+s, & (b)-s\\ (c)+s, & (d)-s\end{matrix}\right] \equiv \frac{\prod\limits_{j=1}^{A}\Gamma(a_j+s)\prod\limits_{j=1}^{B}\Gamma(b_j-s)}{\prod\limits_{j=1}^{C}\Gamma(c_j+s)\prod\limits_{j=1}^{D}\Gamma(d_j-s)}$$

$$\Delta(k, a)=\frac{a}{k}, \quad \frac{a+1}{k}, \ \ldots, \ \frac{a+k-1}{k}$$

$$\Delta(k, (a_p))=\frac{(a_p)}{k}, \quad \frac{(a_p)+1}{k}, \ \ldots, \ \frac{(a_p)+k-1}{k}$$

$\delta_{m,n}=\begin{cases}0, & m\neq n,\\ 1, & m=n\end{cases}$ is the Kronecker symbol.

$\zeta(z)=\sum\limits_{k=1}^{\infty}\frac{1}{k^z}$ [$\operatorname{Re} z>1$] is the Riemann zeta-function.

$\zeta(z, v)=\sum\limits_{k=0}^{\infty}\frac{1}{(v+k)^z}$ [$\operatorname{Re} z>1;\ v\neq 0, -1, -2, \ldots$] is the generalized Hurwitz zeta-function.

$\theta_j(z, q)$ [$j=0, 1, 2, 3, 4$] are the theta-functions:

$$\theta_0(z, q)=1+2\sum_{k=1}^{\infty}(-1)^k q^{k^2}\cos 2kz,$$

$$\theta_1(z, q)=2\sum_{k=0}^{\infty}(-1)^k q^{(k+1/2)^2}\sin(2k+1)z,$$

$$\theta_2(z, q)=2\sum_{k=0}^{\infty}q^{(k+1/2)^2}\cos(2k+1)z,$$

$$\theta_3(z, q)=1+2\sum_{k=1}^{\infty}q^{k^2}\cos 2kz,$$

$$\theta_4(z, q)\equiv\theta_0(z, q)$$

$$\hat{\theta}_{(1)\atop(4)}(x,q)=\frac{1}{(\pi q)^{1/2}}\left\{\sum_{k=0}^{\infty}(\mp 1)^k\exp\left[-\frac{(x+k\mp 1/2)^2}{q}\right]-\sum_{k=-1}^{-\infty}(\mp 1)^k\exp\left[-\frac{(x+k\mp 1/2)^2}{q}\right]\right\}$$

$$\hat{\theta}_{(2)\atop(3)}(x,q)=\frac{1}{(\pi q)^{1/2}}\left\{\sum_{k=0}^{\infty}(\mp 1)^k\exp\left[-\frac{(x+k)^2}{q}\right]-\sum_{k=-1}^{-\infty}(\mp 1)^k\exp\left[-\frac{(x+k)^2}{q}\right]\right\}$$

$\Lambda_0(\varphi, \beta, k)$ is the elliptic function.

$$\lambda(z, a)=\int_0^a z^{-t}\Gamma(t+1)\,dt; \qquad \mu(z, \lambda)=\int_0^\infty\frac{t^\lambda z^t\,dt}{\Gamma(t+1)}; \qquad \mu(z, \lambda, \rho)=\int_0^\infty\frac{t^\lambda z^{t+\rho}\,dt}{\Gamma(t+\rho+1)}$$

[$\operatorname{Re}\lambda>-$

$\nu(z)=\int_0^\infty \frac{z^t\,dt}{\Gamma(t+1)}$; $\nu(z,\rho)=\int_0^\infty \frac{z^{t+\rho}}{\Gamma(t+\rho+1)}\,dt$

$\Pi(\varphi,\nu,k)=\int_0^\varphi \frac{dt}{(1-\nu\sin^2 t)\sqrt{1-k^2\sin^2 t}}$ is the elliptic integral of the 3rd kind.

$\wp(u)$ is the Weierstrass function.

$\sigma(u)$ is the Weierstrass sigma-function.

$$\Xi_1(a,a',b;c;w,z)=\sum_{k,l=0}^{\infty}\frac{(a)_k(a')_l(b)_k}{(c)_{k+l}}\frac{w^k z^l}{k!\,l!} \qquad [|w|<1]$$

$$\Xi_2(a,b;c;w,z)=\sum_{k,l=0}^{\infty}\frac{(a)_k(b)_k}{(c)_{k+l}}\frac{w^k z^l}{k!\,l!} \qquad [|w|<1]$$

$\sigma_n^m=\frac{1}{m!}\sum_{k=0}^{m}(-1)^{m-k}\binom{m}{k}k^n$ are the Stirling numbers of the 2nd kind.

$\Sigma_A(z)$, $\Sigma_B(1/z)$ $(\Sigma_0(z)\equiv 0)$ are the functions of hypergeometric type:

$$\Sigma_A(z)=\sum_{j=1}^{A} z^{a_j}\Gamma\left[\begin{matrix}(a)'-a_j, & (b)+a_j\\ (c)-a_j, & (d)+a_j\end{matrix}\right]{}_{B+C}F_{A+D-1}\left(\begin{matrix}(b)+a_j;\ 1+a_j-(c);\ (-1)^{C-A}z\\ 1+a_j-(a)',\ (d)+a_j\end{matrix}\right)$$

$[a_j-a_k\neq 0,\ \pm1,\ \pm2,\ \ldots;\ j\neq k;\ j,\ k=1,\ 2,\ \ldots,\ A]$,

$$\Sigma_B(1/z)=\sum_{k=1}^{B} z^{-b_k}\Gamma\left[\begin{matrix}(b)'-b_k, & (a)+b_k\\ (d)-b_k, & (c)+b_k\end{matrix}\right]{}_{A+D}F_{B+C-1}\left(\begin{matrix}(a)+b_k,\ 1+b_k-(d);\ \frac{(-1)^{D-B}}{z}\\ 1+b_k-(b)',\ (c)+b_k\end{matrix}\right)$$

$[b_j-b_k\neq 0,\ \pm1,\ \pm2,\ \ldots;\ j\neq k;\ j,\ k=1,\ 2,\ \ldots,\ B]$.

$$\Phi(z,s,v)=\sum_{k=0}^{\infty}\frac{z^k}{(v+k)^s} \qquad [|z|<1;\ v\neq 0,\ -1,\ -2,\ \ldots]$$

$$\Phi_1(a,b;c;w,z)=\sum_{k,l=0}^{\infty}\frac{(a)_{k+l}(b)_k}{(c)_{k+l}}\frac{w^k z^l}{k!\,l!} \qquad [|w|<1]$$

$$\Phi_2(b,b';c;w,z)=\sum_{k,l=0}^{\infty}\frac{(b)_k(b')_l}{(c)_{k+l}}\frac{w^k z^l}{k!\,l!},\qquad \Phi_3(b;c;w,z)=\sum_{k,l=0}^{\infty}\frac{(b)_k}{(c)_{k+l}}\frac{w^k z^l}{k!\,l!}$$

$\Psi(a,c,z)=\Gamma\left[\begin{matrix}1-c\\ 1+a-c\end{matrix}\right]{}_1F_1(a;c;z)+\Gamma\left[\begin{matrix}c-1\\ a\end{matrix}\right]z^{1-c}{}_1F_1(1+a-c;\ 2-c;\ z)$ is the Tricomi confluent hypergeometric function.

$$\Psi_1(a,b;c,c';w,z)=\sum_{k,l=0}^{\infty}\frac{(a)_{k+l}(b)_k}{(c)_k(c')_l}\frac{w^k z^l}{k!\,l!} \qquad [|w|<1]$$

$$\Psi_2(a;c,c';w,z)=\sum_{k,l=0}^{\infty}\frac{(a)_{k+l}}{(c)_k(c')_l}\frac{w^k z^l}{k!\,l!}$$

$\psi(z)=[\ln\Gamma(z)]'=\frac{\Gamma'(z)}{\Gamma(z)}$ is the psi-function.

$\psi^{(n)}(z)=\frac{d^n}{dz^n}\psi(z)$.

INDEX OF MATHEMATICAL SYMBOLS

$(a)=a_1, a_2, \ldots, a_A$; $(a_p)=a_1, a_2, \ldots, a_p$ are special vectors.
$(a_p-b_p)=a_1-b_1, a_2-b_2, \ldots, a_p-b_p$
$(a)+s=a_1+s, a_2+s, \ldots, a_A+s$; $(a_p)+s=a_1+s, \ldots, a_p+s$;
$(a)'-a_j=a_1-a_j, \ldots, a_{j-1}-a_j\, a_{j+1}-a_j, \ldots, a_A-a_j$ $[1 \leqslant j \leqslant A]$
$(a_p)'-a_j=a_1-a_j, \ldots, a_{j-1}-a_j, a_{j+1}-a_j, \ldots, a_p-a_j$ $[1 \leqslant j \leqslant p]$
$(a)_k=a(a+1)\ldots(a+k-1)$ $[k=1, 2, 3, \ldots]$, $(a)_0=1$ is the Pochhammer symbol.
$[a_p, A_p]=(a_1, A_1), (a_2, A_2), \ldots, (a_p, A_p)$
$n!=1\cdot 2\cdot 3 \ldots (n-1)\, n=(1)_n$, $0!=1!=(-1)!=1$
$(2n)!!=2\cdot 4\cdot 6 \ldots (2n-2)\, 2n=2^n n!$
$(2n+1)!!=1\cdot 3\cdot 5 \ldots (2n+1)=\frac{2^{n+1}}{\sqrt{\pi}}\Gamma\left(n+\frac{3}{2}\right)=\left(\frac{3}{2}\right)_n 2^n$

$$n!!=\begin{cases}(2k)!!, & n=2k,\\ (2k+1)!!, & n=2k+1,\end{cases} \qquad 0!!=(-1)!!=1$$

$\binom{n}{k}=\frac{n(n-1)\ldots(n-k+1)}{k!}=\frac{n!}{k!\,(n-k)!}=\frac{(-1)^k(-n)_k}{k!}$, $\binom{n}{0}=1$ are the binomial coefficients.

$\operatorname{Re} a, \operatorname{Re} b>c$ means that $\operatorname{Re} a>c$ and $\operatorname{Re} b>c$.

$[x]=n$ $[n \leqslant x<n+1,\ n=0, \pm 1, \pm 2, \ldots]$ is the integer part of the number x.

$x_+^\lambda=\begin{cases}x^\lambda, & x>0,\\ 0, & x<0\end{cases}$ is the truncated power function.

$|x|=\begin{cases}x, & x \geqslant 0,\\ -x, & x<0\end{cases}$ is the modulus of the number x.

$\bar{z}=x-iy$ $[z=x+iy]$

$|z|=\sqrt{x^2+y^2}$

$\varphi(x)=O(\chi(x))$, $x \to x_0$ ($\varphi(x)=o(\chi(x))$, $x \to x_0$) means that the ration $\frac{\varphi(x)}{\chi(x)}$ is bounded (or tends to zero) as $x \to x_0$.

$\varphi^*(s)=\int_0^\infty x^{s-1}\varphi(x)\,dx$ is the Mellin transform of the function $\varphi(x)$.

$\varphi(x)=\frac{1}{2\pi i}\int_{\gamma-i\infty}^{\gamma+i\infty}\varphi^*(s)\,x^{-s}\,ds$ $[x>0]$ is the inverse Mellin transform.

$\varphi(x) \xleftarrow{L\pm\infty} \varphi^*(s)$ is the correspondence symbol (the integral $\frac{1}{2\pi i}\int_L \varphi^*(s)\,z^{-s}\,ds=\varphi(x)$ exists only for $L=L_{\pm\infty}$; for $L=L_{i\infty}$ it diverges)

$\frac{1}{2\pi i}\int_L \Gamma\left[\begin{matrix}(a)+s, & (b)-s\\ (c)+s, & (d)-s\end{matrix}\right] z^{-s}\,ds$ $[L=L_{\pm\infty}, L_{i\infty}]$ is the Mellin–Barnes integral.

$$\prod (a_p)_k=\prod_{j=1}^{p}(a_j)_k$$

$$\prod ((a_p)+b)_k=\prod_{j=1}^{p}(a_j+b)_k$$

$$\prod_{k=m}^{n} a_k=a_m a_{m+1}\ldots a_n \quad [n \geqslant m],$$
$$=1 \quad [n<m]$$

$$\prod_{k=1}^{\infty} a_k(z)=\lim_{n\to\infty}\prod_{k=1}^{n} a_k(z)$$

$$\sum_{k=m}^{n} a_k=a_m+a_{m+1}+\ldots+a_n \quad [n \geqslant m],$$
$$=0 \quad [n<m]$$

$$\sum_{k=1}^{\infty} a_k(z)=\lim_{n\to\infty}\sum_{k=1}^{n} a_k(z),$$